W9-BTG-083

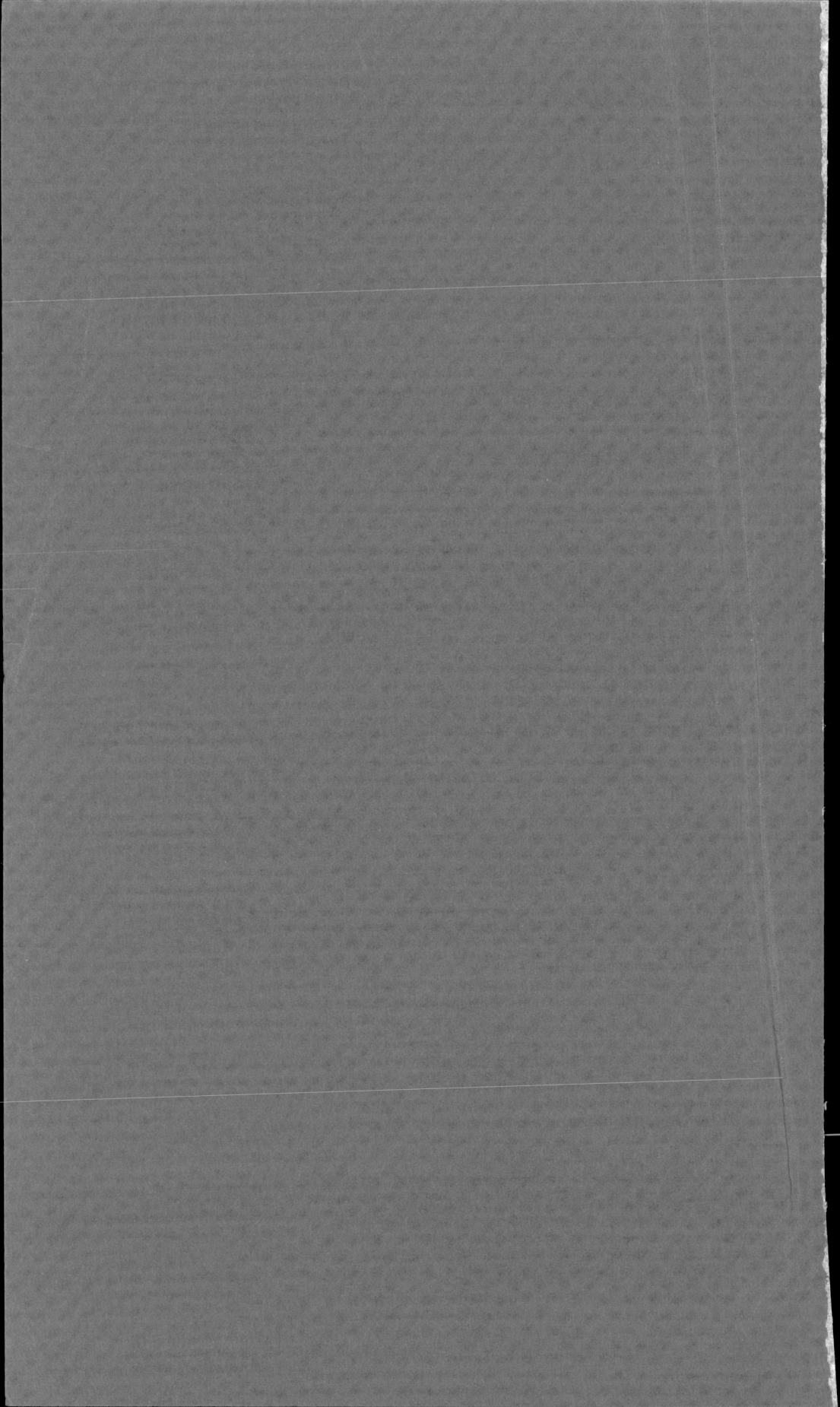

Smithells Metals Reference Book

Smithells Metals Reference Book

Sixth Edition

Editor
Eric A Brandes *BSc, ARCS, CEng, FIM*
In association with Fulmer Research Institute Ltd

Butterworths
London Boston Singapore Sydney Wellington Durban Toronto

First published 1949
Second edition 1955
Third edition 1962
Fourth edition 1967
Fifth edition 1976
Reprinted 1978
Sixth edition 1983

British Library Cataloguing in Publication Data

Smithells, Colin James
Smithells metals reference book.—6th ed.
1. Metals—Handbooks, manuals, etc. II. Brandes, Eric A.
669'. 00212 TA459

ISBN 0-408-71053-5

Typeset by Interprint Limited Malta.
Printed in England by Robert Hartnoll Ltd., Bodmin, Cornwall

Preface to the Sixth Edition

This edition is the first to have been prepared subsequent to the death of Dr Colin Smithells. The Metals Reference Book is a continuing memorial to him, and his name has been included in the title. In the preparation, the principles established by Colin Smithells have been retained. A convenient summary of data relating to metallurgy is presented with values mainly in the form of tables and with descriptive matter reduced to a minimum.

Although SI units are used throughout, many tables also give traditional units where these are still current. Full unit conversion tables are included.

The values or formulations given are selected by the contributors as the most reliable but for a particular critical review the reader should consult the references. In the case of mechanical properties data, the values are for guidance only; for design purposes it is essential to consult the relevant specifications.

All chapters have been reviewed and several have been extensively recast — in particular those on: Metallography, Hard Metals, Friction and Wear, Guide to Corrosion, and Solders and Brazing.

A new chapter on Vapour Deposited Coatings, extensively referenced, will provide a useful summary of data in this field. Tables on Superplasticity have also been added.

I wish to thank all the contributors and the organizations to which they belong for their cooperation and help, and in particular the Fulmer Research Institute for assistance in the preparation of this edition.

Chalfont St. Peter,
Bucks

E.A.B.

Acknowledgements

Assistance given by the following organizations is gratefully acknowledged:

British Aluminium Co. Ltd
British Ceramic Research Association
Bureau International des Poids & Mésures
Copper Development Association
Cosworth Research & Development Ltd
Culham Labs., UKAEA
European Space Agency
Fulmer Research Institute
Henry Wiggin Ltd
Imperial College of Science & Technology
International Carbide Data
Lead & Zinc Development Association
Magnesium Elektron Ltd
Manganese Centre
Radiochemical Centre (Amersham International)
University of Birmingham
University of Dundee
University of Manchester Institute of Science & Technology
University of Nottingham
University of Sheffield
Warren Springs Laboratory
Wild Barfield Ltd

The Editor and the Publishers desire to thank all those who have kindly authorized the reproduction of diagrams and tables and in particular the following:

American Ceramic Society, Columbus, Ohio
American Institute of Mining and Metallurgical Engineers, New York
American Society for Metals, Cleveland, Ohio
American Society for Testing Materials, Philadelphia, Pa.
British Standards Institute, London
Cavendish Laboratory, Cambridge
Chapman & Hall Ltd, London
Edward Arnold & Co. Ltd, London
Institute of Physics, London
Institution of Gas Engineers, London
Izvest. Akad. Nauk SSSR
Johnson, Matthey & Co. Ltd, London
John Wiley & Sons, Inc., New York
Journal of Scientific Instruments, London
Longmans, Green & Co. Ltd, London
McGraw-Hill Book Co. Inc., New York
Melbourne University Press, Victoria, Australia
Metals Society, London
Oxford University Press, Oxford
Pergamon Press Ltd
Physical Review, New York
Reinhold Publishing Corporation, New York
Royal Society, London
Society of Chemical Industry, London
Taylor & Francis Ltd, London
Zeitschrift für Physikalische Chemie, Leipzig

Contributors

Editor

E. A. Brandes, CEng, BSc(Lond.), ARCS, FIM

Contributors to this edition	*Chapter*
N. J. Archer, BA (Cantab.), PhD	35
L. C. Archibald, BSc (Nott.), PhD	32
B. J. Boden, CEng, BSc, PhD, MIM, FRIC, FICorrT	31
E. A. Brandes, CEng, BSc(Lond.), ARCS, FIM	2, 3, 11, 14.1
G. B. Brook, CEng, BMet(Sheff.), FIM, M(Surrey)	10, 15.3
J. A. Brookes, CEng, BSc(Eng) (Met), MIM	23
V. A. Calcutt, MIQA	14.4, 22.2
G. R. Campbell	22.12
J. Campbell, CEng, MA(Cantab.), MMet(Sheff.), PhD	14.2, 26
W. C. Campbell-Heselwood, CEng, BSc, FIMC, FIM, FInstP, FINDT	14.10
A. R. Chivers, MA	14.8, 22.7
Dr Coppin	1
G. J. Davies, CEng, BE(NZ), PhD(NZ), MA	22.9, 22.10
M. Deighton, CEng, BSc(Dunelm), PhD, MIM	21
B. Dunn, CEng, BTech, MIMM	34
M. Finlan, BSc, MInstP	3, 4
I. Fitzpatrick, CEng, BSc, PhD(Manc.)	15.1
R. F. Flint, ALA	36
P. J. Foster, CEng, BSc Tech, PhD, MIChE, MInstF	28
T. I. Fowle, BSc(Eng), FIMechE	24
R. Freeman, BSc(Lond.), MInstP	19
T. G. Gooch, CEng, BSc(Birm.), MSc(Eng)	33
R. Grimes, BSc(Lond.), PhD, ARSM	14.3, 22.1
B. H. Hanson, BSc(Lond.)	14.7, 14.9, 22.6, 22.9
D. Inman, CEng, BSc(Lond.), PhD, DSc, DHonCausa, MIMM, FRSC	9
R. O. Jenkins, ARCS, DIC, PhD, FInstP	18
A. D. LeClaire, BA(Cantab.), FInstP	13
F. T. Louis, CEng, MIM	14.6, 22.5
J. H. Megaw, BSc(QUB), PhD, MInstP	30
M. A. Moore, CEng, BSc(Met.) (Wales), PhD	25
L. G. Palethorpe, FRIC	29
F. T. Palin, MPhil, Dip.Ceram, AICeram, MIRefEng	27
T. J. Quinn, BSc, DPhil, MInstP	16
E. E. Riches, CEng, MSc, MIEE, MInstP	20
R. A. Shelton, CEng, BSc(Lond.), PhD, AMIMM	8
C. C. Smith, BSc(Manc.), PhD	15.2
R. Smith, BSc(Birm.), MINucE	22.3
D. E. J. Talbot, CEng, MSc, AIM	12
W. Unsworth, FRIC	14.5, 22.4
T. J. Veasey, CEng, BSc(Birm.), PhD, MIMM	7
M. B. Waldron, PhD, FIM	22.11
A. J. Wall, BSc, PhD	14.8, 22.7
N. Waterman, CEng, BSc(Wales), PhD, MIM, MInstP	4, 5, 6
M. J. Wheeler, BTech, PhD, FInstP	17

Contents

10 Metallography 10–1

11 Equilibrium diagrams 11–1

12 Gas–metal systems 12–1

13 Diffusion in metals 13–1

1 First aid

1.1 Laboratory accidents

These notes aim to provide guidance for dealing with laboratory accidents, in particular those where action on the spot could affect the outcome. Prevention, by strict adherence to a code of safe practices, is preferable.

1.1.1 First aid box

Unless there are statutory requirements, the following contents are suggested (quantities represent minimum for 10 people):

Sterilized lint dressings (in packets): Large—3
Medium—3
Small (finger)—6
Unbleached triangular bandages (36 in × 51 in)—2
Crepe bandages (3 in)—2
Sterile cotton wool ($\frac{1}{2}$-oz packets)—6
Sterile eye pads—2
Adhesive wound dressings (assorted)—12
Cetrimide lotion $\frac{1}{2}\%$ (cleansing/antiseptic)—8 oz

Items other than dressings may be required according to circumstances. Some examples will be found in the text to follow.

1.1.2 Unconsciousness and 'shock'

A state of faintness, sometimes leading to unconsciousness, may occur even after trivial injuries. Premonitory signs are pallor and delayed response to questions. If these are noted, or the patient feels unwell, act at once.

1. Remove from source of danger.
2. Lie patient down.
3. Loosen clothing at neck.
4. Raise legs, if condition does not improve.

After serious injuries, especially if appreciable loss of blood has occurred or continues, the condition may worsen. The patient will be restless, pale, and sweating; with sighing respiration. He may vomit.

5. Look for, and arrest, external haemorrhage if necessary.
6. Arrange prompt transfer to hospital.
7. Cover lightly to prevent chilling, but do not apply external heat.

If unconscious the above principles apply, but in addition:

8. Place (and transport) in the semi-prone position.

Position half-way between lying on side and lying fact downwards, maintained by keeping the upper knee drawn up. This will reduce the risk of asphyxia from inhalation of vomit.

Give nothing by mouth to a patient who:

(a) is unconscious,
(b) is too shocked to sit up,
(c) has suspected internal injuries, or undiagnosed abdominal pain,
(d) is likely to require surgical treatment shortly.

1.1.3 Chemical eye splashes

The victim can seldom help himself, due to involuntary spasm causing the eyelids to be held tightly closed. Act immediately and firmly.

1. Lie patient down.
2. Hold eyelids apart with forefinger and thumb of one hand.
3. Irrigate eye with fresh tap water or distilled water.
   Direct a stream of water at the eyeball and inner and outer surfaces of the lids for at least 15 minutes, continuing for an hour in the case of *caustic* and *hydrofluoric acid* splashes.
4. Obtain medical attention.

The need for prompt hospital or medical treatment does not affect the absolute priority to be given to immediate and adequate initial irrigation, which may have to be continued until patient is under medical care.

Preventive measures cannot be stressed too strongly, and should include the use of protective eyewear in situations of risk.

1.1.4 Chemical skin burns and splashes

Burns may be caused by local irritants such as strong acids, caustic alkalis, phenol, bromine, and organic solvents; but other chemicals can cause harm by absorption through the skin (e.g. aniline, carbon disulphide, dimethyl sulphate, hydrocyanic acid, nitrobenzene, and organic compounds of arsenic, lead and mercury).

TREATMENT

1. Remove affected clothing and irrigate copiously with water.
   Continue irrigation for at least 20 minutes, holding part under running tap if feasible. At sites where there is a risk of large body areas being involved it is best to install a shower, or a bath kept filled with water into which a badly splashed person can at once immerse himself.
   Sulphuric acid. First wash with anhydrous ethyl alcohol, provided delay is not caused.
   Phenol. Rub skin with swabs soaked in glycerol, polyethylene glycol (PEG), or a 70/30 PEG/methylated spirit mixture, for at least 10 minutes. If these chemicals are not immediately available, rub site with water-soaked swab.
   Phosphorus burns should at once be wetted and the phosphorus particles then removed under water with forceps or by gentle scraping. Neutralize remainder by washing with copper sulphate 2% for 5 minutes, followed by sodium bicarbonate 2% to remove the precipitated copper.
   Hydrofluoric acid burns should be thoroughly washed with abundant water. Calcium gluconate gel* should then be massaged into and around the affected area with clean fingers. Continue to do this until 15 minutes after the pain has subsided, or medical treatment is available.
2. Apply sterile dressings and refer for medical attention.
   If it is impracticable or would cause delay to apply individual dressings to large or multiple burnt areas, clothe patient in clean towels or white coat, then rugs or coats.

* Obtainable from Industrial Pharmaceutical Service Ltd., Hampden Road, Sale, Cheshire.

1.1.5 Non-chemical burns

Immediately run cold water on to the burned area until no longer painful.

Small superficial areas with unbroken skin

1. Gently cleanse with cetrimide ½% using sterile cotton wool.
2. Apply sterile dressing.
   Arrange medical attention for all but the most trivial burns, as their full extent and depth is seldom apparent initially.

Small areas with blistering, rawness, or charring

1. Apply sterile dressing.
2. Arrange medical treatment.

Extensive burns

Send urgently to hospital with minimum disturbance.
Sterile dry dressings may be applied to exposed burns provided delay is not caused.
Do not attempt to remove affected clothing.

All burns (including chemical *q.v.*)

Do not prick blisters.
Do not apply oil or grease.
Do not apply adhesive plaster.
Do not apply dressings to facial burns.

1.1.6 Haemorrhage

Clean-cut wounds, as by broken glass, are apt to bleed freely.

1. Apply pressure to bleeding point.
2. Raise affected limb above heart level.
   Patients with lower limb haemorrhage must lie down. Patients with head or arm injuries should be propped in sitting position, but watch for shock, (*see* 'Shock and unconsciousness').
3. Look for and remove any superficial glass fragments projecting from the wound.
4. Apply pad of sterile wound dressing, and bandage firmly.
   Choose pad larger than wound. On unwrapping pad from packet avoid touching side to be applied to wound.
   In larger wounds, or if appreciable bleeding continues, apply pad of cotton wool or a further wound dressing over the first. Firm pressure can then be applied by a crepe bandage.
5. Arrange medical treatment.
   Except in trivial wounds it is best to continue to keep the affected limb above heart level until medical care is available. A sling may be used for the upper limb, but patients with leg injuries which have bled freely are best transported as stretcher cases with the limb supported on cushions or rolled up coats.

Trivial wounds (which do not go through the full thickness of the skin and do not gape), or *superficial abrasions*

1. Clean with cetrimide ½%, using sterile cotton wool.
2. Apply sterile dressing.

1.1.7 Poisons by mouth

CHEMICALS ENTERING THE MOUTH

Rinse out vigorously and repeatedly with water. Also gargle and spit out.

SWALLOWED CHEMICALS

1. Give water to drink.
2. Arrange urgent medical attention.

Detail nature and quantity of poison ingested, and the treatment given. Save specimen (source or vomit) for analysis if necessary.

1.1.8 Gassing accidents

TREATMENT *and see* CYANIDE POISONING

1. Remove patient from contaminated atmosphere.
   Rescuers must protect themselves by use of suitable equipment or breath-holding. Specific masks are available for moderate concentrations of most gases but *carbon monoxide*, *carbon dioxide*, and *nickel carbonyl* require self-contained breathing apparatus or mask with airline.
2. Remove contaminated clothing and cleanse skin with water, if necessary.
3. Keep completely at rest while awaiting medical help or transport.
   To exercise a gassed patient may jeopardize his chances of recovery. Certain gases or vapours (e.g. *phosgene*, *nickel carbonyl*, *nitrous fumes*, *phosphorus oxychloride*, *methyl bromide*) can cause delayed effects, the victim at first apparently little affected becoming ill several hours later.
4. Administer oxygen, if available, to those in need.
   Blue lips will be an indication, but patients poisoned by carbon monoxide or cyanide will have a deceptive pink colour.
5. Attempt artificial respiration if breathing ceases.

1.1.9 Cyanide poisoning

It is safest not to take risks with cyanides (hydrogen cyanide and soluble inorganic cyanide salts) unless a suitably briefed doctor is close by. He will require, on the spot, antidote ampoules and facilities for intravenous administration.

FIRST AID MEASURES (HAVING SENT FOR MEDICAL HELP)

1. Keep patient at rest.
2. Cleanse skin and remove affected clothing, if necessary.
3. Administer amyl nitrite by inhalation.
   Break ampoule of amyl nitrite (3 minims) in handkerchief, holding to patient's nose and mouth for 15–30 seconds. Repeat after a few seconds, using fresh ampoules, for 3–5 minutes.
4. If cyanide has been taken by mouth, give antidote:

Mixture *A*	Ferrous sulphate	158 g
	Citric acid	3 g
	Water, cold	to 1000 ml
Mixture *B*	Exsiccated sodium carbonate	60 g
	Water	to 1000 ml

   Mix equal parts of *A* and *B* and give a tumblerful. Mixture *A* should not be kept for more than a month.
5. If breathing stops, give artificial respiration.
   Not mouth-to-mouth method (use resuscitator or manual method). Continue use of amyl nitrite while giving artificial respiration, and give oxygen if available.

1.1.10 Artificial respiration

INDICATION

Artificial respiration is indicated if the patient has stopped breathing. The presence of breathing can usually be detected by holding the back of the hand in front of the patient's nose and mouth.

METHODS

Lacking special apparatus, the 'Mouth-to-Mouth' method is the best. If it cannot be used (e.g. cyanide poisoning or injuries to patient's face), intermittent manual chest compression as by the Holger Nielsen method will be necessary.

THE 'MOUTH-TO-MOUTH' METHOD

1. Lie patient on back.
2. Kneel to one side of patient's head, facing it.
3. Clear mouth of any obstruction (handkerchief-covered finger to remove mucus).
4. Pinch patient's nostrils closed with finger and thumb.
5. Hold up patient's jaw with other hand, extending his head back.
6. Encircle patient's lips with your lips.
7. Blow into patient's mouth until chest is seen to rise.
8. Withdraw mouth and take breaths. Patient will exhale spontaneously.
9. Repeat items 6 to 8 at rate of about one cycle every 4 seconds.

Continue until spontaneous respiration returns and persists, or until more experienced advice is obtainable. As an alternative to 4 and 6 it is sometimes more effective to encircle the patient's nostrils with the lips, keeping his mouth closed.

2 Introductory tables

2.1 Conversion factors

Conversion factors into and from SI units are given in Table 2.5. The table can also be used to convert from one traditional unit to another. Convenient multiples or sub-multiples of SI units can be derived by the application of the prefix multipliers given in Table 2.4. Table 2.6 gives commonly required conversions.

The majority of the conversion factors are based upon equivalents given in BS 350: Part 1:1959 'Conversion Factors and Tables'.

Throughout the conversions the acceleration due to gravity (g) has been taken as the standard acceleration 9.806 65 $m\,s^{-1}$. Units containing the word force like 'pounds force' are converted to SI units using this value of g.

The B.t.u. conversions are based on the definition accepted by the 5th International Conference on Properties of Steam, London, 1956, that 1 B.t.u. $lb^{-1} = 2.326\ J\,g^{-1}$ exactly. Conversions to joules are given for three calories; calories (IC) is the 'international table calorie' redefined by the 1956 conference referred to above as 4.1868 J. Calories (15 °C) refers to the calorie defined by raising the temperature of water at 15 °C by 1 °C and calories (US thermochemical) is the 'defined' calorie used in some USA work and is defined at 4.184 J exactly.

The conversions are grouped in alphabetical order of the physical property to which they relate but are not alphabetical within the groups.

2.1.1 SI units

In this edition quantities are expressed in SI (Système International) units. Where c.g.s. units have been used previously only SI units are given. However, familiar units in general technical use have been retained where they bear a simple power of ten relation to the strict SI unit. For instance density is given as $g\,cm^{-3}$ and not as $kg\,m^{-3}$. Where Imperial units have been used (e.g. in Mechanical Properties, etc.) data are given in both SI units and Imperial units.

The basic units of the SI system are given in Table 2.1, derived units with special names and symbols in Table 2.2 and derived units without special names in Table 2.3.

Multiples and sub-multiples of SI units are formed by prefixes to the name of the unit. The prefixes are shown in Table 2.4. The prefixed unit is written without a hyphen — for instance a thousand million newtons is written giganewton — symbol GN. The name of the unit is written with a small letter even when the symbol has a capital letter, e.g. ampere, symbol A. In the case of the kilogram, the multiple or sub-multiple is applied to the gram — for instance a thousand kilograms is written Mg.

In this edition stress is expressed in Pascals (Pa). A pascal (Pa) is identical to a newton per square metre ($N\,m^{-2}$) and a megapascal (MPa) is identical to a newton per square millimetre ($N\,mm^{-2}$).

PRINTED FORM OF UNITS AND NUMBERS

The symbol for a unit is in upright type and unaltered by the plural. It is not followed by a full stop unless it is at the end of a sentence. Only symbols of units derived from proper names are in the upper case.

When units are multiplied they will be printed with a space between them. Negative indices are used for units expressed as a quotient. Thus newtons per square metre will be $\mathrm{N\,m^{-2}}$ and metres per second will be $\mathrm{m\,s^{-1}}$.

The prefix to a unit symbol is written before the unit symbol without a space between and a power index applies to both the symbols. Thus square centimetres is $\mathrm{cm^2}$ and not $\mathrm{(cm)^2}$.

Numbers are printed with the decimal point as a full stop. For long numbers, a space and not a comma is given between every three digits. For example $\pi = 3.141\,592\,653$. When a number is entirely decimal it will begin with a zero, e.g. 0.5461. If two numbers are multiplied, a × sign is used as the operator.

HEADING OF COLUMNS IN TABLES AND LABELLING OF GROUPS

The rule adopted in this edition is that the quantity is obtained by multiplying the unit and its multiple given at the column head by the number in the table.

For example when tabulating a stress of 2×10^5 Pa the heading is stress, below which appears 10^5 Pa, with 2.0 appearing in the table. If no units are given in the column heading, the values given are numbers only. In graphs the power of ten and units by which the point on the graph must be multiplied are given on the axis label.

TEMPERATURES

In 1968 the International Committee of Weights and Measures adopted a new temperature scale (IPTS–68) to replace the then existing scale IPTS–48. By the change numerical values of temperature approximated much more closely to modern determinations of thermodynamic temperature and the scale was extended to lower temperatures. Unless otherwise stated the temperatures given are to IPTS–48. Conversion from IPTS–48 to IPTS–68 can be made by adding the appropriate values from Table 2.7. For further information about IPTS–68 *see* C. R. Barber 'International Practical Temperature Scale of 1968', *Nature*, 1969, *222*, 929.

Table 2.1 BASIC SI UNITS

Quantity	*Name of unit*	*Unit symbol*
Length	metre	m
Mass	kilogram	kg
Time	second	s
Electric current	ampere	A
Temperature	kelvin	K
Luminous intensity	candela	cd
Amount of substance	mole	mol
Plane angle	radian	rad
Solid angle	steradian	sr

From 'Quantities, Units and Symbols', Royal Society, 1971.

Table 2.2 DERIVED SI UNITS WITH SPECIAL NAMES

Quantity	*Name of unit*	*Symbol*	*Derivation*
Activity (radioactivity)	becquerel	Bq	$\mathrm{s^{-1}}$
Absorbed dose (of radiation)	gray	Gy	$\mathrm{J\,kg^{-1}}$
Dose equivalent (of radiation)	sievert	Sv	$\mathrm{J\,kg^{-1}}$
Energy	joule	J	$\mathrm{N\,m}$
Force	newton	N	$\mathrm{J\,m^{-1}}$
Stress or pressure	pascal	Pa	$\mathrm{N\,m^{-2}}$
Power	watt	W	$\mathrm{J\,s^{-1}}$
Electric charge	coulomb	C	$\mathrm{A\,s}$
Electric potential	volt	V	$\mathrm{W\,A^{-1}}$
Electric resistance	ohm	Ω	$\mathrm{V\,A^{-1}}$
Electric capacitance	farad	F	$\mathrm{C\,V^{-1}}$
Electric conductance	siemens	S	$\mathrm{A\,V^{-1}}$
Magnetic flux	weber	Wb	$\mathrm{V\,s}$

Table 2.2 DERIVED SI UNITS WITH SPECIAL NAMES—*continued*

Quantity	*Name of unit*	*Symbol*	*Derivation*
Inductance	henry	H	$V s A^{-1}$
Magnetic flux density	tesla	T	$Wb m^{-2}$
Luminous flux	lumen	lm	cd sr
Illumination	lux	lx	$cd sr m^{-2}$
Frequency	hertz	Hz	s^{-1}

From 'Quantities, Units and Symbols', Royal Society, 1971.
Note: Symbols derived from proper names begin with a capital letter.

Table 2.3 SOME DERIVED SI UNITS WITHOUT SPECIAL NAMES

Quantity	*SI unit*	*Symbol*
Area	square metre	m^2
Acceleration	metre/second squared	$m s^{-2}$
Angular velocity	radian/second	$rad s^{-1}$
Calorific value	joule/kilogram	$J kg^{-1}$
Concentration	mole/cubic metre	$mol m^{-3}$
Current density	ampere/square metre	$A m^{-2}$
Density	kilogram/cubic metre	$kg m^{-3}$
Diffusion coefficient	square metre/second	$m^2 s^{-1}$
Electrical conductivity	siemens/metre	$S m^{-1}$
Electric field strength	volt/metre	$V m^{-1}$
Electrical resistivity	ohm metre	Ωm
Entropy	joule/kelvin	$J K^{-1}$
Exposure (to radiation)	coulomb/kilogram	$C kg^{-1}$
Heat capacity	joule/kelvin	$J K^{-1}$
Heat flux density	watt/square metre	$W m^{-2}$
Latent heat	joule/kilogram	$J kg^{-1}$
Magnetic field strength	ampere/metre	$A m^{-1}$
Molar volume	cubic metre/mole	$m^3 mol^{-1}$
Moment of inertia	kilogram/square metre	$kg m^{-2}$
Moment of force	newton metre	N m
Molar heat capacity	joule/kelvin mole	$J K^{-1} mol$
Permittivity	farad/metre	$F m^{-1}$
Permeability	henry/metre	$H m^{-1}$
Radioactivity	l/second	s^{-1}
Speed (velocity)	metre/second	ms^{-1}
Specific volume	cubic metre/kilogram	$m^3 kg^{-1}$
Specific heat-mass	joule/kilogram kelvin	$J kg^{-1} K^{-1}$
Specific heat-volume	joule/cubic metre kelvin	$J m^{-3} K^{-1}$
Surface tension	newton/metre	$N m^{-1}$
Thermal conductivity	watt/metre kelvin	$W m^{-1} K^{-1}$
Thermoelectric power	volt/kelvin	$V K^{-1}$
Viscosity-kinematic	square metre/second	$m^2 s^{-1}$
Viscosity–dynamic	pascal second	Pa s
Volume	cubic metre	m^3
Wave number	l/metre	m^{-1}

Table 2.4 PREFIXES FOR MULTIPLES AND SUB-MULTIPLES USED IN THE SI SYSTEM OF UNITS

Sub-multiple	*Prefix*	*Symbol*	*Multiple*	*Prefix*	*Symbol*
10^{-1}	deci	d	10	deca	da
10^{-2}	centi	c	10^2	hecto	h
10^{-3}	milli	m	10^3	kilo	k
10^{-6}	micro	μ	10^6	mega	M
10^{-9}	nano	n	10^9	giga	G
10^{-12}	pico	p	10^{12}	tera	T
10^{-15}	femto	f	10^{15}	peta	P
10^{-18}	atto	a	10^{18}	exa	E

From 'Quantities, Units and Symbols', Royal Society, 1971.

Table 2.5 CONVERSION FACTORS

To convert B to A multiply by	A	B	To convert A to B multiply by
	Acceleration		
10^2	centimetres/second squared	metres/second squared	10^{-2}
$3.937\,008 \times 10$	inches/second squared	metres/second squared	2.54×10^{-2}
3.280 84	feet/second squared	metres/second squared	3.048×10^{-1}
$1.019\,716 \times 10^{-1}$	standard acceleration due to gravity	metres/second squared	9.806 65
	Angle—plane		
$2.062\,65 \times 10^5$	seconds	radians	$4.848\,14 \times 10^{-6}$
$3.437\,75 \times 10^3$	minutes	radians	$2.908\,88 \times 10^{-4}$
$5.729\,58 \times 10$	degrees	radians	$1.745\,33 \times 10^{-2}$
$1.591\,55 \times 10^{-1}$	revolutions	radians	6.283 20
	Angular velocity		
$5.729\,58 \times 10$	degrees/second	radians/second	$1.745\,33 \times 10^{-2}$
$1.591\,55 \times 10^{-1}$	revolutions/second	radians/second	6.283 20
$3.437\,75 \times 10^3$	degrees/minute	radians/second	$2.908\,88 \times 10^{-4}$
9.549 27	revolutions/minute	radians/second	$1.047\,20 \times 10^{-1}$
	Area		
10^{28}	barn	square metres	10^{-28}
$1.550\,003 \times 10^3$	square inches	square metres	$6.451\,6 \times 10^{-4}$
$1.076\,391 \times 10$	square feet	square metres	$9.290\,3 \times 10^{-2}$
1.195 990	square yards	square metres	$8.361\,27 \times 10^{-1}$
$3.861\,02 \times 10^{-7}$	square miles	square metres	$2.589\,99 \times 10^6$
$2.471\,052 \times 10^{-4}$	acres	square metres	$4.046\,86 \times 10^3$
10^{-4}	hectares	square metres	10^4
2.471	acres	hectares	4.047×10^{-1}
2.5×10^{-1}	acres	roods	4
$1.562\,5 \times 10^{-3}$	square miles	acres	6.40×10^2
	Calorific value—volume basis		
$2.683\,92 \times 10^{-2}$	British thermal units/cubic foot	joules/cubic metre	$3.725\,89 \times 10$
$4.308\,86 \times 10^{-11}$	therms/UK gallon	joules/cubic metre	$2.320\,80 \times 10^{10}$
$2.388\,46 \times 10^{-4}$	kilocalories/cubic metre	joules/cubic metre	$4.186\,8 \times 10^3$
	Calorific value—mass basis		
4.299×10^{-4}	British thermal units/pound	joules/kilogram	2.326×10^3
$2.388\,46 \times 10^{-4}$	kilocalories/kilogram	joules/kilogram	$4.186\,8 \times 10^3$
	Capacity—see *Volume*		
	Compressibility		
10^{-1}	square centimetres/dyne	metres/newton	10
	Density		
10^{-3}	grams/cubic centimetre	kilograms/cubic metre	10^3
$1.603\,59 \times 10^{-1}$	ounces/gallon (UK)	kilograms/cubic metre	6.236 03
$6.242\,80 \times 10^{-2}$	pounds/cubic foot	kilograms/cubic metre	$1.601\,85 \times 10$
$3.612\,73 \times 10^{-5}$	pounds/cubic inch	kilograms/cubic metre	$2.767\,99 \times 10^4$
$1.002\,241 \times 10^{-2}$	pounds/gallon (UK)	kilograms/cubic metre	$9.977\,64 \times 10$
$8.345\,434 \times 10^{-3}$	pounds/gallon (US)	kilograms/cubic metre	$1.198\,26 \times 10^2$
$7.015\,673 \times 10$	grains/gallon (UK)	kilograms/cubic metre	$1.425\,38 \times 10^{-2}$
	Diffusion coefficient		
10^4	square centimetres/second	square metres/second	10^{-4}
	Electric charg		
$2.997\,93 \times 10^9$	electrostatic units	coulombs	$3.335\,64 \times 10^{-10}$
10^{-1}	electromagnetic units	coulombs	10

Table 2.5 CONVERSION FACTORS—*continued*

To convert B to A multiply by	*A*	*B*	*To convert A to B multiply by*
	Electric current		
2.99793×10^{9}	electrostatic units	amperes	3.33564×10^{-10}
10^{-1}	electromagnetic units	amperes	10
	Electric current density		
2.99793×10^{5}	electrostatic units	amperes/square metre	3.33564×10^{-6}
10^{-5}	electromagnetic units	amperes/square metre	10^{5}
10^{-4}	amperes/square centimetre	amperes/square metre	10^{4}
6.452×10^{-4}	amperes/square inch	amperes/square metre	1.55×10^{3}
9.2902×10^{-2}	amperes/square foot	amperes/square metre	1.0764×10
	Energy—work—heat		
2.777778×10^{-7}	kilowatt hours	joules	3.6×10^{6}
1.01972×10^{-1}	kilogram force metres	joules	9.80665
2.37304×10	foot poundals	joules	4.21401×10^{-2}
7.37562×10^{-1}	foot pounds force	joules	1.35582
3.72506×10^{-7}	horsepower hours	joules	2.68452×10^{6}
9.86894×10^{-3}	litre atmospheres	joules	1.01328×10^{2}
2.38846×10^{-4}	kilocalories (IC)	joules	4.1868×10^{3}
8.85034	inch pounds force	joules	1.1299×10^{-1}
9.47813×10^{-4}	British thermal units	joules	1.05506×10^{3}
10^{7}	ergs	joules	10^{-7}
6.241808×10^{18}	electron volts	joules	1.6021×10^{-19}
9.47813×10^{-9}	therms	joules	1.05506×10^{8}
2.38846×10^{-1}	calories (IC)	joules	4.1868
2.389201×10^{-1}	calories (15 °C)	joules	4.1855
2.390057×10^{-1}	calories (US thermochemical)	joules	4.184
	Entropy		
2.38846×10^{-1}	calories (IC)/degree centigrade	joules/kelvin	4.1868
5.26562×10^{-4}	British thermal unit/degree Fahrenheit	joules/kelvin	1.89911×10^{3}
	Force		
10^{5}	dynes	newtons	10^{-5}
3.59694	ounces force	newtons	0.278014
1.01972×10^{2}	grams force	newtons	9.80665×10^{-3}
2.24809×10^{-1}	pounds force	newtons	4.44822
7.23301	poundals	newtons	0.138255
1.00361×10^{-4}	UK tons force	newtons	9.96402×10^{3}
1.124047×10^{-4}	US tons force	newtons	8.896422×10^{3}
1.01972×10^{-1}	kilograms force	newtons	9.80665
	Fracture toughness		
1.01972×10^{-4}	(kilograms force/square centimetre)$\sqrt{(\text{centimetre})}$	newtons/$\sqrt{(\text{metre}^3)}$	9.80655×10^{3}
9.10042×10^{-7}	(kilopounds force/square inch) $\sqrt{(\text{inch})}$	newtons/$\sqrt{(\text{metre}^3)}$	1.09885×10^{6}
4.06273×10^{-7}	(tons force/square inch)$\sqrt{(\text{inch})}$	newtons/$\sqrt{(\text{metre}^3)}$	2.4614×10^{6}
3.16226×10^{-6}	hectobars$\sqrt{(\text{millimetre})}$	newtons/$\sqrt{(\text{metre}^3)}$	3.1623×10^{5}
	Heat—see *Energy*		
	Heat flow rate—see *Power*		
	Latent heat		
4.29923×10^{-4}	British thermal units/pound	joules/kilogram	2.326×10^{3}
2.38846×10^{-4}	calories/gram	joules/kilogram	4.1868×10^{3}
3.34552×10^{-1}	foot pounds force/pound	joules/kilogram	2.98907
1.01972×10^{-1}	kilogram force metres/kilogram	joules/kilogram	9.80665

Table 2.5 CONVERSION FACTORS—*continued*

To convert B to A multiply by	*A*	*B*	*To convert A to B multiply by*
		Leak rate	
$7.500\,64 \times 10^{3}$	lusec (micron Hg litre/second)	joules/second	$1.333\,22 \times 10^{-4}$
		Length	
10^{10}	angstroms (Å)	metres	10^{-10}
10^{6}	microns (μ)	metres	10^{-6}
$9.979\,84 \times 10^{9}$	kx units	metres	$1.002\,02 \times 10^{-10}$
$3.937\,01 \times 10$	inches	metres	2.54×10^{-2}
$3.280\,84$	feet	metres	3.048×10^{-1}
$1.093\,61$	yards	metres	9.144×10^{-1}
$6.213\,71 \times 10^{-4}$	miles	metres	$1.609\,344 \times 10^{3}$
$5.396\,118 \times 10^{-4}$	miles (naut UK)	metres	$1.853\,184 \times 10^{3}$
$5.399\,568 \times 10^{-4}$	miles (naut Int)	metres	1.852×10^{3}
$1.\dot{8}\dot{1} \times 10^{-1}$	rods, poles or perches	yards	5.5
2.5×10^{-1}	chains	rods, poles, etc.	4.0
10^{-1}	furlongs	chains	10.0
1.25×10^{-1}	miles (UK)	furlongs	8.0
$1.\dot{6}\dot{6} \times 10^{-1}$	fathoms	feet	6.0
$8.\dot{3}\dot{3} \times 10^{-3}$	cable lengths	fathoms	1.2×10^{2}
$1.644\,7 \times 10^{-4}$	nautical miles	feet	6.080×10^{3}
$1.89\dot{3}\dot{9} \times 10^{-4}$	miles (UK)	feet	5.280×10^{3}
	Magnetic conversions—see Magnetic units and conversion factors, chapter 20		
		Moment of force—see *Energy*	
		Moment of inertia	
10^{7}	grams centimetre squared	kilograms metre squared	10^{-7}
$3.417\,17 \times 10^{3}$	pounds inch squared	kilograms metre squared	$2.926\,40 \times 10^{-4}$
$2.373\,04 \times 10$	pounds foot squared	kilograms metre squared	$4.214\,01 \times 10^{-2}$
$5.467\,47 \times 10^{4}$	ounces inch squared	kilograms metre squared	$1.829\,00 \times 10^{-5}$
		Momentum	
$7.233\,01$	foot pounds/second	kilogram metres/second	$1.382\,55 \times 10^{-1}$
10^{5}	gram centimetres/second	kilogram metres/second	10^{-5}
		Mass	
$5.643\,819 \times 10^{2}$	drams (Av)	kilograms	$1.771\,85 \times 10^{-3}$
$3.527\,399 \times 10$	ounces (Av)	kilograms	$2.834\,95 \times 10^{-2}$
$2.204\,624$	pounds (Av)	kilograms	$4.535\,92 \times 10^{-1}$
$1.574\,731 \times 10^{-1}$	stones (Av)	kilograms	$6.350\,293$
$7.873\,650 \times 10^{-2}$	quarters (Av)	kilograms	$1.270\,059 \times 10$
$1.968\,415 \times 10^{-2}$	hundredweights (Av)	kilograms	$5.080\,23 \times 10$
$9.842\,035 \times 10^{-4}$	tons (Av)	kilograms	$1.016\,05 \times 10^{3}$
$1.543\,236 \times 10^{4}$	grains or minims (Apoth)	kilograms	$6.479\,89 \times 10^{-5}$
$7.716\,180 \times 10^{2}$	scruples (Apoth)	kilograms	$1.295\,978 \times 10^{-3}$
$2.572\,063 \times 10^{2}$	drams (Apoth)	kilograms	$3.887\,93 \times 10^{-3}$
$3.215\,072 \times 10$	ounces (Apoth or Troy)	kilograms	$3.110\,35 \times 10^{-2}$
$1.543\,237 \times 10^{4}$	grains (Troy)	kilograms	$6.479\,885 \times 10^{-5}$
10^{-3}	tonnes (metric)	kilograms	10^{3}
$1.102\,311 \times 10^{-3}$	tons (short 2000 lb)	kilograms	$9.071\,85 \times 10^{2}$
5.0×10^{3}	metric carats	kilograms	2.0×10^{-4}
		Mass per unit area	
$8.921\,80 \times 10^{3}$	pounds/acre	kilograms/square metre	$1.120\,85 \times 10^{-4}$
$1.843\,348$	pounds/square yard	kilograms/square metre	$5.424\,912 \times 10^{-1}$
$2.048\,16 \times 10^{-1}$	pounds/square foot	kilograms/square metre	$4.882\,432$
$2.949\,357 \times 10$	ounces/square yard	kilograms/square metre	$3.390\,57 \times 10^{-2}$
$3.227\,055$	ounces/square foot	kilograms/square metre	$3.051\,52 \times 10^{-1}$

Table 2.5 CONVERSION FACTORS—*continued*

To convert B to A multiply by	*A*	*B*	*To convert A to B multiply by*
	Mass per unit length		
$5.599\,73 \times 10^{-2}$	pounds/inch	kilograms/metre	$1.785\,80 \times 10$
$6.719\,71 \times 10^{-1}$	pounds/foot	kilograms/metre	$1.488\,16$
$2.015\,91$	pounds/yard	kilograms/metre	$4.960\,55 \times 10^{-1}$
	Power—Heat flow rate		
10^{7}	ergs/second	watts	10^{-7}
$3.412\,14$	British thermal units/hour	watts	$2.930\,71 \times 10^{-1}$
$8.598\,45 \times 10^{-1}$	kilocalories (IC)/hour	watts	1.163
$7.375\,61 \times 10^{-1}$	foot pounds force/second	watts	$1.355\,82$
$2.388\,46 \times 10^{-1}$	calories (IC)/second	watts	$4.186\,8$
$1.341\,022 \times 10^{-3}$	horsepower	watts	7.457×10^{2}
	Pressure—see *Stress*		
	Radioactivity		
2.7×10^{-11}	curie	becquerel	3.7×10^{10}
	Radiation-absorbed dose		
10^{2}	rem	sievert	10^{-2}
	Radiation exposure		
3.876×10^{3}	roentgen	coulomb/kilogram	2.58×10^{-4}
	Specific heat capacity—mass basis		
10^{-3}	joules/gram degree centigrade	joules/kilogram kelvin	10^{3}
$2.388\,46 \times 10^{-4}$	calories/gram degree centigrade	joules/kilogram kelvin	$4.186\,8 \times 10^{3}$
$2.388\,46 \times 10^{-4}$	British thermal units/ pound degree Fahrenheit	joules/kilogram kelvin	$4.186\,8 \times 10^{3}$
$1.858\,63 \times 10^{-1}$	foot pounds force/ pound degree Fahrenheit	joules/kilogram kelvin	$5.380\,32$
$1.019\,72 \times 10^{-1}$	kilogram force metres/ kilogram degree centigrade	joules/kilogram kelvin	$9.806\,65$
	Specific heat—volume basis		
10^{-6}	joules/cubic centimetre degree centigrade	joules/cubic metre kelvin	10^{6}
$2.388\,459 \times 10^{-4}$	kilocalories/cubic metre degree centigrade	joules/cubic metre kelvin	$4.186\,8 \times 10^{3}$
$1.491\,066 \times 10^{-5}$	British thermal units/cubic foot degree Fahrenheit	joules/cubic metre kelvin	$6.706\,61 \times 10^{4}$
	Stress		
$1.450\,377 \times 10^{-4}$	pounds force/square inch	newtons/square metre	$6.894\,76 \times 10^{3}$
$6.474\,881 \times 10^{-8}$	UK tons force/square inch	newtons/square metre	$1.544\,43 \times 10^{7}$
10	dynes/square centimetre	newtons/square metre	10^{-1}
10^{-5}	bars	newtons/square metre	10^{5}
10^{-7}	hectobars	newtons/square metre	10^{7}
$1.019\,716 \times 10^{-7}$	kilograms force/square millimetre	newtons/square metre	$9.806\,65 \times 10^{6}$
$7.500\,638 \times 10^{-3}$	torrs	newtons/square metre	$1.333\,22 \times 10^{2}$
$7.500\,638 \times 10^{-3}$	millimetres of mercury	newtons/square metre	$1.333\,22 \times 10^{2}$
$7.500\,638$	micron of mercury	newtons/square metre	$1.333\,22 \times 10^{-1}$
$9.869\,233 \times 10^{-6}$	atmospheres	newtons/square metre	$1.013\,250 \times 10^{5}$
1	pascals	newtons/square metre	1
$6.474\,8807 \times 10^{-2}$	UK tons force/square inch	megapascals	$1.544\,43 \times 10$
	Surface tension		
10^{3}	dynes/centimetre	newtons/metre	10^{-3}
$6.852\,178 \times 10^{-2}$	pounds force/foot	newtons/metre	$1.459\,39 \times 10$
$5.710\,148 \times 10^{-3}$	pounds force/inch	newtons/metre	$1.751\,268 \times 10^{2}$
	Temperature interval		
1	degrees Celsius (centigrade)	kelvins	1

Table 2.5 CONVERSION FACTORS—*continued*

To convert B to A multiply by	*A*	*B*	*To convert A to B multiply by*
1.8	degrees Fahrenheit	kelvins	$5.\dot{5}\dot{5} \times 10^{-1}$
1.8	degrees Rankine	kelvins	$5.\dot{5}\dot{5} \times 10^{-1}$
	Thermal conductivity		
10^{-2}	watts/centimetre degree centigrade	watts/metre kelvin	10^{2}
$2.388\,46 \times 10^{-3}$	calories/centimetre second degree centigrade	watts/metre kelvin	$4.186\,8 \times 10^{2}$
$8.598\,45 \times 10^{-1}$	kilocalories/metre hour degree centigrade	watts/metre kelvin	1.163
$5.777\,91 \times 10^{-1}$	British thermal unit/foot hour degree Fahrenheit	watts/metre kelvin	1.730 73
6.933 47	British thermal unit inch/square foot hour degree Fahrenheit	watts/metre kelvin	$1.442\,28 \times 10^{-1}$
	Time		
$1.\dot{6}\dot{6} \times 10^{-2}$	minutes	seconds	6.0×10
$2.\dot{7}\dot{7} \times 10^{-4}$	hours	seconds	3.600×10^{3}
$1.157\,41 \times 10^{-5}$	days	seconds	8.64×10^{4}
$1.653\,44 \times 10^{-6}$	weeks	seconds	6.048×10^{5}
$3.170\,98 \times 10^{-8}$	years	seconds	$3.153\,6 \times 10^{7}$
$1.141\,552\,5 \times 10^{-4}$	years	hours	8.760×10^{3}
	Torque—see *Energy*		
	Velocity		
3.280 84	feet/second	metres/second	0.304 8
$1.958\,504 \times 10^{2}$	feet/minute	metres/second	5.08×10^{-3}
3.6	kilometres/hour	metres/second	0.277
$3.728\,227 \times 10^{-2}$	miles/minute	metres/second	$2.682\,240 \times 10$
2.236 94	miles/hour	metres/second	0.447 04
1.942 60	UK knots	metres/second	0.514 773
1.943 85	International knots	metres/second	0.514 444
$1.13\dot{6} \times 10^{-2}$	UK miles/minute	feet/second	8.8×10
	Viscosity—dynamic		
10	poise	newton seconds/square metre	10^{-1}
$1.019\,72 \times 10^{-1}$	kilogram force seconds/square metre	newton seconds/square metre	9.806 65
$6.719\,71 \times 10^{-1}$	poundal seconds/square foot	newton seconds/square metre	1.488 16
$2.088\,542 \times 10^{-2}$	pound force seconds/square foot	newton seconds/square metre	$4.788\,03 \times 10$
	Viscosity—kinematic		
10^{4}	stokes	square metres/second	10^{-4}
$1.550\,03 \times 10^{3}$	square inches/second	square metres/second	$6.451\,6 \times 10^{-4}$
$1.076\,392 \times 10$	square feet/second	square metres/second	$9.290\,3 \times 10^{-2}$
$5.580\,011 \times 10^{6}$	square inches/hour	square metres/second	$1.792\,111 \times 10^{-7}$
$3.875\,009 \times 10^{4}$	square feet/hour	square metres/second	$2.580\,639 \times 10^{-5}$
3.6×10^{3}	square metres/hour	square metres/second	$2.\dot{7}\dot{7} \times 10^{-4}$
	Volume		
$6.102\,37 \times 10^{4}$	cubic inches	cubic metres	$1.638\,71 \times 10^{-5}$
$3.531\,473 \times 10$	cubic feet	cubic metres	$2.831\,68 \times 10^{-2}$
1.307 95	cubic yards	cubic metres	$7.645\,55 \times 10^{-1}$
10^{3}	litres	cubic metres	10^{-3}
$2.199\,69 \times 10^{2}$	gallons (UK)	cubic metres	$4.546\,09 \times 10^{-3}$
$2.641\,72 \times 10^{2}$	gallons (US)	cubic metres	$3.785\,41 \times 10^{-3}$
$1.759\,755 \times 10^{-3}$	pints (UK)	cubic centimetres	$5.682\,61 \times 10^{2}$
$2.199\,69 \times 10^{-1}$	gallons (UK)	litres	4.546 09
$3.519\,508 \times 10$	fluid ounces (UK)	litres	$2.841\,306 \times 10^{-2}$
	Work—see *Energy*		

Table 2.6 COMMONLY REQUIRED CONVERSIONS

Acceleration	$g = 32$ feet/second squared	=	9.806 65 metres/second squared	$m\,s^{-2}$
Angle	1 radian	=	57.295 8 degrees	°
Area	1 acre	=	4 046.86 square metres	m^2
	1 hectare	=	10 000 square metres	m^2
Density	1 gram/cubic centimetre	=	1 000 kilograms/cubic metre	$kg\,m^{-3}$
Energy	1 calorie (IC)	=	4.186 8 joules	J
	1 kilowatt hour	=	3.6 megajoules	MJ
	1 British thermal unit	=	1055.06 joules	J
	1 erg	=	10^{-7} joules	J
	1 therm	=	105.506 megajoules	MJ
	1 horsepower hour	=	2.684 52 megajoules	MJ
Force	1 dyne	=	10^{-5} newtons	N
	1 pound force	=	4.448 22 newtons	N
	1 UK ton force	=	9964.02 newtons	N
	1 kilogram force	=	9.806 65 newtons	N
Length	1 angstrom unit (Å)	=	10^{-10} metres	m
	1 micron (μm)	=	10^{-6} metres	m
	1 micron (μm)	=	$0.039\,37 \times 10^{-3}$ inches	in
	1 thousandth of an inch	=	25.4 micrometres	μm
	1 inch	=	2.54 centimetres	cm
	1 foot	=	30.48 centimetres	cm
	1 yard	=	91.44 centimetres	cm
	1 mile	=	1.609 344 kilometres	km
Mass	1 ounce (Av)	=	28.349 5 grams	g
	1 ounce (Troy)	=	31.103 5 grams	g
	1 pound (Av)	=	453.592 grams	g
	1 hundredweight	=	50.802 3 kilograms	kg
	1 UK ton (Av)	=	1 016.05 kilograms	kg
	1 short ton (2 000 lbs)	=	907.185 kilograms	kg
	1 carat (metric)	=	0.2 grams	g
Power	1 horsepower	=	745.7 watts	W
Stress	1 pound force/square inch (p.s.i.)	=	6.894 76 kilopascals	kPa
	1 UK ton force/square inch	=	15.444 3 megapascals	MPa
	1 bar	=	100 kilopascals	kPa
	1 hectobar	=	10 megapascals	MPa
	1 kilogram force/square centimetre	=	98.006 5 kilopascals	kPa
	1 kilogram force/square millimetre	=	9.806 65 megapascals	MPa
	1 torr = 1 millimetre of mercury	=	133.322 pascals	Pa
	1 atmosphere	=	101.325 kilopascals	kPa
	1 pascal	=	1 newton/square metre	$N\,m^{-2}$
Surface tension	1 dyne/centimetre	=	1 millinewton/metre	$mN\,m^{-1}$
Velocity	1 foot/second	=	1.097 28 kilometres/hour	$km\,h^{-1}$
	1 mile/hour	=	1.609 344 kilometres/hour	$km\,h^{-1}$
Volume	1 cubic inch	=	16.387 1 cubic centimetres	cm^3
	1 cubic foot	=	28.316 8 cubic decimetres	dm^3
	1 cubic yard	=	0.764 555 cubic metres	m^3
	1 litre	=	1 cubic decimetre	dm^3
	1 litre	=	1.759 75 UK pints	pint
	1 UK gallon	=	4.546 09 cubic decimetres	dm^3
	1 UK gallon	=	0.160 544 cubic feet	ft^3

Table 2.7 CORRECTIONS TO TEMPERATURE VALUES IPTS–48 TO IMPLEMENT IPTS–68 (IPTS–68) – (IPTS–48) IN °C

t_{68}	**0**	**−10**	**−20**	**−30**	**−40**	**−50**	**−60**	**−70**	**−80**	**−90**	**−100**	t_{68} °C
−100	0.022	0.013	0.003	−0.006	−0.013	0.013	−0.005	0.007	0.012	(0.008 at O_2 point)		**−100**
−0	0.000	0.006	0.012	0.018	0.024	0.029	0.032	0.034	0.033	0.029	0.022	**−0**

t_{68} °C	**0**	**10**	**20**	**30**	**40**	**50**	**60**	**70**	**80**	**90**	**100**	t_{68} °C
0	0.000	−0.004	−0.007	−0.009	−0.010	−0.010	−0.010	−0.008	−0.006	−0.003	0.000	**0**
100	0.000	0.004	0.007	0.012	0.016	0.020	0.025	0.029	0.034	0.038	0.043	**100**
200	0.043	0.047	0.051	0.054	0.058	0.061	0.064	0.067	0.069	0.071	0.073	**200**
300	0.073	0.074	0.075	0.076	0.077	0.077	0.077	0.077	0.077	0.076	0.076	**300**
400	0.076	0.075	0.075	0.075	0.074	0.074	0.074	0.075	0.076	0.077	0.079	**400**
500	0.079	0.082	0.085	0.089	0.094	0.100	0.108	0.116	0.126	0.137	0.150	**500**
600	0.150	0.165	0.182	0.200	0.23	0.25	0.28	0.31	0.34	0.36	0.39	**600**
700	0.39	0.42	0.45	0.47	0.50	0.53	0.56	0.58	0.61	0.64	0.67	**700**
800	0.67	0.70	0.72	0.75	0.78	0.81	0.84	0.87	0.89	0.92	0.95	**800**
900	0.95	0.98	1.01	1.04	1.07	1.10	1.12	1.15	1.18	1.21	1.24	**900**
1 000	1.24	1.27	1.30	1.33	1.36	1.39	1.42	1.44	—	—	—	**1 000**

t_{68} °C	**0**	**100**	**200**	**300**	**400**	**500**	**600**	**700**	**800**	**900**	**1 000**	t_{68} °C
1 000	—	1.5	1.7	1.8	2.0	2.2	2.4	2.6	2.8	3.0	3.2	**1 000**
2 000	3.2	3.5	3.7	4.0	4.2	4.5	4.8	5.0	5.3	5.6	5.9	**2 000**
3 000	5.9	6.2	6.5	6.9	7.2	7.5	7.9	8.2	8.6	9.0	9.3	**3 000**

From BS 1826:1952 Amendment No. 1, 2 February 1970. Example: 1000 °C according to IPTS–48 would be corrected to 1001.24 °C to conform to IPTS–68.

Table 2.8 CORROSION CONVERSION FACTORS

The following conversion factors relating loss in weight and depth of penetration are useful in the assessment of corrosion.

density of metal in grams/cubic centimetre $= d$
density of metal in kilograms/cubic metre $= 10^3 d$

To convert B to A multiply by	*A*	*B*	*To convert A to B multiply by*
$10^{-3}d^{-1}$	millimetres	grams/square metre	$10^3 d$
$3.65 \times 10^{-1}d^{-1}$	millimetres/year	grams/square metre per day	$2.74d$
$8.76d^{-1}$	millimetres/year	grams/square metre per hour	$1.14 \times 10^{-1}d$
$3.937 \times 10^{-2}d^{-1}$	thousandths of an inch (mils)	grams/square metre	$2.54 \times 10d$
$1.44 \times 10d^{-1}$	mils/year	grams/square metre per day	$6.96 \times 10^{-2}d$
$3.45 \times 10^2 d^{-1}$	mils/year	grams/square metre per hour	$2.90 \times 10^{-3}d$
$1.201 \times 10d^{-1}$	mils	ounces/square foot	$8.326 \times 10^{-2}d$

Table 2.9 TEST SIEVE MESH NUMBERS CONVERTED TO NOMINAL APERTURE SIZE FROM BS 410:1969

Wire cloth test sieves were designated by the mesh count or number. This method, widely used until 1962, was laid down in previous British Standards—BS 410. Sieves are now designated by aperture size: *see* BS 410:1969, for full details.

The table gives the previously used mesh numbers with the corresponding nominal aperture sizes, the preferred average wire diameters in the test sieves and the tolerances.

			Aperture tolerances		
Mesh No.	*Nominal aperture size* mm	*Preferred average wire diameter in test sieve* mm	*Max. tolerance for size of an individual aperture* mm +	*Tolerance for average aperture size* mm ±	*Intermediate tolerance* mm +
3	5.60	1.60	0.50	0.17	0.34
3½	4.75	1.60	0.43	0.14	0.29
4	4.00	1.40	0.40	0.12	0.28
5	3.35	1.25	0.34	0.10	0.23
6	2.80	1.12	0.31	0.084	0.20
7	2.36	1.00	0.26	0.071	0.17
8	2.00	0.90	0.24	0.060	0.16
10	1.70	0.80	0.20	0.051	0.14
12	1.40	0.71	0.18	0.042	0.11
14	1.18	0.63	0.17	0.035	0.11
16	1.00	0.56	0.15	0.030	0.09
	μm	μm	μm +	μm ±	μm +
18	850	500	128	30	79
22	710	450	114	28	71
25	600	400	102	24	66
30	500	315	90	20	55
36	425	280	81	17	51
44	355	224	71	14	43
52	300	200	64	15	40
60	250	160	58	13	36
72	212	140	53	12	33

Table 2.9 TEST SIEVE MESH NUMBERS—*continued*

			Aperture tolerances		
Mesh No.	*Nominal aperture size* μm	*Preferred average wire diameter in test sieve* μm	*Max. tolerance for size of an individual aperture* μm +	*Tolerance for average aperture size* μm ±	*Intermediate tolerance* μm +
85	180	125	51	11	31
100	150	100	48	9.4	29
120	125	90	46	8.1	27
150	106	71	43	7.4	25
170	90	63	43	6.6	25
200	75	50	41	6.1	24
240	63	45	41	5.3	23
300	53	36	38	4.8	21
350	45	32	38	4.8	21
400	38	30	36	4.0	20

Notes:

(1) No aperture size shall exceed the nominal by more than the maximum tolerance.
(2) The average aperture size shall not be greater or smaller than the nominal by more than the average tolerance size.
(3) Not more than 6% of the apertures shall be above the nominal size by more than the intermediate tolerance.

For perforated plate sieve sizes with square or round holes–*see* BS 410:1969. Other national standards for test sieves may be found for France in NF X11–501, for Germany in DIN 4188 and for USA in ASTM E11–61.

2.1.2 Temperature scale conversions

The absolute unit of temperature, symbol K, is the kelvin which is 1/273.16 of the thermodynamic temperature of the triple point of water—*see* Section 16. Practical temperature scales are Celsius (previously Centigrade) symbol °C, and Fahrenheit, symbol °F. An absolute scale based on Fahrenheit is the Rankine, symbol °R.

Where K, C, F and R, represent the same temperature on the Kelvin, Celsius, Fahrenheit and Rankine scales, conversion formulae are:

$$K = C + 273.15 \qquad C = \tfrac{5}{9}(F - 32)$$
$$F = \tfrac{9}{5}C + 32 \qquad R = F + 459.67$$

Rapid approximate conversions between Celsius and Fahrenheit scales can be obtained from Figure 2.1.

2.2 Mathematical formulae

2.2.1 Algebra

IDENTITIES

$$a^2 - b^2 = (a-b)(a+b)$$
$$a^2 + b^2 = (a-\mathrm{i}b)(a+\mathrm{i}b) \text{ where } \mathrm{i} = \sqrt{(-1)}$$
$$a^3 - b^3 = (a-b)(a^2+ab+b^2)$$
$$a^3 + b^3 = (a+b)(a^2-ab+b^2)$$
$$a^n - b^n = (a-b)(a^{n-1}+a^{n-2}b+\ldots+b^{n-1})$$
$$a^n + b^n = (a+b)(a^{n-1}-a^{n-2}b+\ldots+b^{n-1}) \text{ when } n \text{ is odd}$$
$$(a \pm b)^2 = a^2 \pm 2ab + b^2$$
$$(a \pm b)^3 = a^3 \pm 3a^2b + 3ab^2 \pm b^3$$
$$(a+b)^n = a^n + na^{n-1}b + \frac{n(n-1)}{2!}a^{n-2}b^2 + \ldots + \frac{n(n-1)(n-2)\ldots(n-r+1)}{r!}a^{n-r}b^r + \ldots + b^n$$

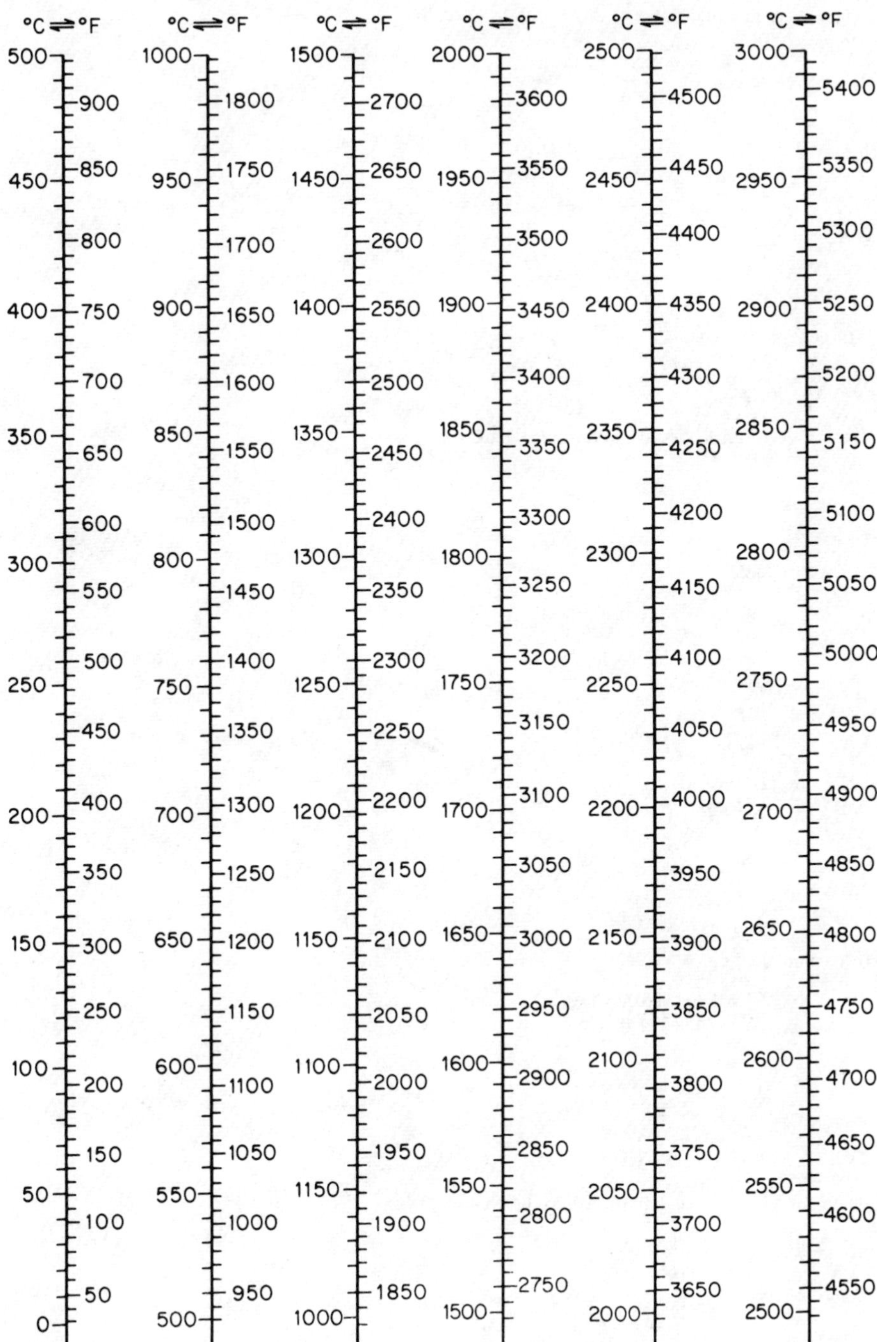

Figure 2.1 *Nomogram for approximate interconversion between Celsius and Fahrenheit temperature scales*

RATIO AND PROPORTION

If

$$\frac{a}{b}=\frac{c}{d}$$

then

$$\frac{a+b}{b}=\frac{c+d}{d}$$

and

$$\frac{a-b}{b}=\frac{c-d}{d}$$

In general,

$$\frac{a}{b}=\frac{c}{d}=\frac{e}{f}=\ldots=\left(\frac{pa^n+qc^n+re^n+\ldots}{pb^n+qd^n+rf^n+\ldots}\right)^{1/n}$$

where p, q, r and n are any quantities whatever.

LOGARITHMS

If

$$a^x=N, \text{ then } x=\log_a N$$

$$\log_a MN=\log_a M+\log_a N$$

$$\log_a\frac{M}{N}=\log_a M-\log_a N$$

$$\log_a(M^p)=p\log_a M$$

$$\log_a(M^{1/r})=\frac{1}{r}\log_a M$$

$$\log_b N=\frac{1}{\log_a b}\times\log_a N$$

In particular,

$$\log_e N=2.302\,585\,09\times\log_{10} N$$
$$\log_{10} N=0.434\,294\,48\times\log_e N$$

THE QUADRATIC EQUATION

The general quadratic equation may be written

$$ax^2+bx+c=0$$

Solution $x=\dfrac{-b\pm\sqrt{(b^2-4ac)}}{2a}$

If

$$\Delta=(b^2-4ac)$$

the roots are real and equal if $\Delta=0$
the roots are imaginary and unequal if $\Delta<0$
the roots are real and unequal if $\Delta>0$

Also, if the roots are α and β,

$$\alpha+\beta=-\frac{b}{a}$$

$$\alpha\beta=\frac{c}{a}$$

THE CUBIC EQUATION

The general cubic equation may be written

$$y^3+a_1y^2+a_2y+a_3=0$$

If we put $y=x-\frac{1}{3}a_1$ the equation reduces to $x^3+ax+b=0$, where

$$a=a_2-\tfrac{1}{3}a_1^2 \quad\text{and}\quad b=\tfrac{2}{27}a_1^3-\tfrac{1}{3}a_1a_2+a_3$$

Solution $x=z+v$ or $-\dfrac{z+v}{2}\pm i\sqrt{3}\left(\dfrac{z-v}{2}\right)$,

where

$$z=\sqrt[3]{\left[-\tfrac{1}{2}b+\sqrt{\left(\frac{b^2}{4}+\frac{a^3}{27}\right)}\right]}$$

and

$$v=\sqrt[3]{\left[-\tfrac{1}{2}b-\sqrt{\left(\frac{b^2}{4}+\frac{a^3}{27}\right)}\right]}$$

Alternatively,*

$$x=2\sqrt{\left(\frac{-a}{3}\right)}\cos\frac{\theta+2k\pi}{3}\,(k=0,\ 1,\ 2)$$

where

$$\cos\theta=-\frac{b}{2}\left(\frac{-a^3}{27}\right)^{-1/2}$$

If

$$\Delta=\frac{b^2}{4}+\frac{a^3}{27}$$

there are two equal and one unequal root if $\Delta=0$
three real roots if $\Delta<0$
one real and two complex roots if $\Delta>0$

2.2.2 Series and progressions

NUMERICAL SERIES

$$1+2+3+\ldots+n=\frac{n}{2}(n+1)$$

$$1^2+2^2+3^2+\ldots+n^2=\frac{n}{6}(n+1)(2n+1)$$

*The second form of the solution is particularly useful when $\Delta<0$ (i.e. in the case of three real roots).

$$1^3+2^3+3^3+\ldots+n^3=\frac{n^2}{4}(n+1)^2$$

ARITHMETIC PROGRESSION

$$a,\ a+d,\ a+2d,\ \ldots\ a+(n-1)d$$

$$S_n=\frac{n}{2}[2a+(n-1)d]$$

where S_n denotes the sum to n terms.

GEOMETRIC PROGRESSION

$$a,\ ar,\ ar^2\ \ldots\ ar^{n-1}$$

$$S_n=\frac{a(r^n-1)}{(r-1)}=\frac{a(1-r^n)}{(1-r)}$$

where S_n denotes the sum to n terms.
If

$$r^2<1 \text{ and } n\to\infty,\ S_\infty=\frac{a}{(1-r)}$$

TAYLOR'S SERIES

$$f(x)=f(a)+(x-a)f'(a)+\frac{(x-a)^2}{2!}f''(a)+\frac{(x-a)^3}{3!}f'''(a)+\ldots$$

MACLAURIN'S SERIES

$$f(x)=f(0)+xf'(0)+\frac{x^2}{2!}f''(0)+\frac{x^3}{3!}f'''(0)+\ldots$$

In the following series, the region of convergence is indicated in parentheses. If no region is shown, the series is convergent for all values of x.

BINOMIAL SERIES

$$(1\pm x)^n=1\pm nx+\frac{n(n-1)}{2!}x^2\pm\frac{n(n-1)(n-2)}{3!}x^3+\ldots\qquad(x^2<1)$$

$$(1\pm x)^{-n}=1\mp nx+\frac{n(n+1)}{2!}x^2\mp\frac{n(n+1)(n+2)}{3!}x^3+\ldots\qquad(x^2<1)$$

LOGARITHMIC SERIES

$$\log_e(1\pm x)=\pm x-\frac{x^2}{2}\pm\frac{x^3}{3}-\frac{x^4}{4}\pm\ldots\qquad(x^2<1)$$

$$\log_e\left(\frac{1+x}{1-x}\right)=2\left(x+\frac{x^3}{3}+\frac{x^5}{5}+\ldots\right)\qquad(x^2<1)$$

$$\log_e x=2\left[\frac{x-1}{x+1}+\frac{1}{3}\left(\frac{x-1}{x+1}\right)^3+\frac{1}{5}\left(\frac{x-1}{x+1}\right)^5+\ldots\right]\qquad(x>0)$$

$$\log_e(a+x)=\log_e a+2\left[\frac{x}{2a+x}+\frac{1}{3}\left(\frac{x}{2a+x}\right)^3+\frac{1}{5}\left(\frac{x}{2a+x}\right)^5+\ldots\right]\qquad (a>0,\ x+a>0)$$

EXPONENTIAL SERIES

$$e=1+1+\frac{1}{2!}+\frac{1}{3!}+\frac{1}{4!}+\ldots$$

$$e^x=1+x+\frac{x^2}{2!}+\frac{x^3}{3!}+\frac{x^4}{4!}+\ldots$$

$$a^x=1+x\log_e a+\frac{(x\log_e a)^2}{2!}+\frac{(x\log_e a)^3}{3!}+\ldots$$

TRIGONOMETRIC SERIES

$$\sin x=x-\frac{x^3}{3!}+\frac{x^5}{5!}-\frac{x^7}{7!}+\ldots$$

$$\cos x=1-\frac{x^2}{2!}+\frac{x^4}{4!}-\frac{x^6}{6!}+\ldots$$

$$\tan x=x+\frac{x^3}{3}+\frac{2x^5}{15}+\frac{17x^7}{315}+\frac{62x^9}{2835}+\ldots\qquad\left(-\frac{\pi}{2}<x<\frac{\pi}{2}\right)$$

$$\sin^{-1} x=x+\frac{1}{2}\cdot\frac{x^3}{3}+\frac{1}{2}\cdot\frac{3}{4}\cdot\frac{x^5}{5}+\frac{1}{2}\cdot\frac{3}{4}\cdot\frac{5}{6}\cdot\frac{x^7}{7}+\ldots\qquad(-1\le x\le 1)$$

$$\cos^{-1} x=\frac{\pi}{2}-\sin^{-1} x$$

$$\tan^{-1} x=x-\frac{x^3}{3}+\frac{x^5}{5}-\frac{x^7}{7}+\ldots\qquad(-1\le x\le 1)$$

If

$$x=1,\ \frac{\pi}{4}=1-\frac{1}{3}+\frac{1}{5}-\frac{1}{7}+\ldots$$

$$\cot^{-1} x=\frac{\pi}{2}-\tan^{-1} x$$

SERIES FOR HYPERBOLIC FUNCTIONS

$$\sinh x=x+\frac{x^3}{3!}+\frac{x^5}{5!}+\frac{x^7}{7!}+\ldots$$

$$\cosh x=1+\frac{x^2}{2!}+\frac{x^4}{4!}+\frac{x^6}{6!}+\ldots$$

$$\tanh x=x-\frac{x^3}{3}+\frac{2x^5}{15}-\frac{17x^7}{315}+\frac{62x^9}{2835}\ldots\qquad\left(-\frac{\pi}{2}<x<\frac{\pi}{2}\right)$$

$$\sinh^{-1} x=x-\frac{1}{2}\cdot\frac{x^3}{3}+\frac{1}{2}\cdot\frac{3}{4}\cdot\frac{x^5}{5}-\frac{1}{2}\cdot\frac{3}{4}\cdot\frac{5}{6}\cdot\frac{x^7}{7}+\ldots\qquad(-1<x<1)$$

$$\tanh^{-1} x=x+\frac{x^3}{3}+\frac{x^5}{5}+\frac{x^7}{7}+\ldots\qquad(-1<x<1)$$

$$\coth^{-1} x = \frac{1}{x} + \frac{1}{3x^3} + \frac{1}{5x^5} + \frac{1}{7x^7} + \ldots \qquad (-1 < x < 1)$$

2.2.3 Trigonometry

DEFINITIONS AND SIMPLE RELATIONSHIPS

$$\sin A = \frac{a}{b}$$

$$\cos A = \frac{c}{b}$$

$$\tan A = \frac{a}{c}$$

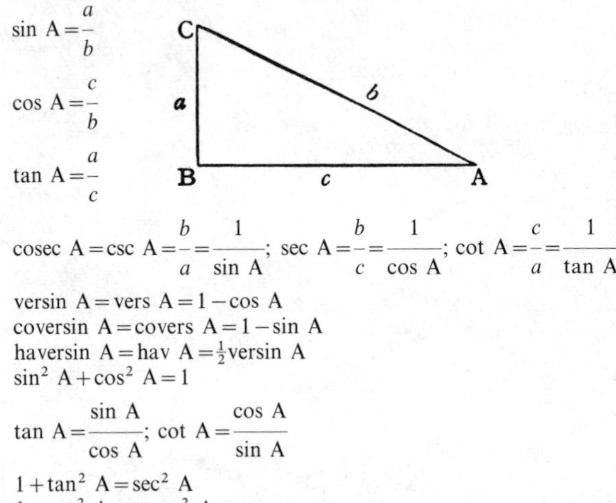

$$\text{cosec } A = \csc A = \frac{b}{a} = \frac{1}{\sin A}; \ \sec A = \frac{b}{c} = \frac{1}{\cos A}; \ \cot A = \frac{c}{a} = \frac{1}{\tan A}$$

$$\text{versin } A = \text{vers } A = 1 - \cos A$$
$$\text{coversin } A = \text{covers } A = 1 - \sin A$$
$$\text{haversin } A = \text{hav } A = \tfrac{1}{2}\text{versin } A$$
$$\sin^2 A + \cos^2 A = 1$$

$$\tan A = \frac{\sin A}{\cos A}; \ \cot A = \frac{\cos A}{\sin A}$$

$$1 + \tan^2 A = \sec^2 A$$
$$1 + \cot^2 A = \text{cosec}^2 A$$

RADIAN MEASURE

π radians $= 180°$, $\pi = 3.141\,59 \ldots$

COMPOUND ANGLES

$$\sin (A+B) = \sin A \cos B + \cos A \sin B$$
$$\cos (A+B) = \cos A \cos B - \sin A \sin B$$
$$\sin (A-B) = \sin A \cos B - \cos A \sin B$$
$$\cos (A-B) = \cos A \cos B + \sin A \sin B$$

$$\tan (A+B) = \frac{\tan A + \tan B}{1 - \tan A \tan B}$$

$$\tan (A-B) = \frac{\tan A - \tan B}{1 + \tan A \tan B}$$

$$\sin 2A = 2 \sin A \cos A$$

$$\cos 2A = \cos^2 A - \sin^2 A = 1 - 2\sin^2 A = \cos^2 A - 1$$

$$\tan 2A = \frac{2 \tan A}{1 - \tan^2 A}$$

$$\sin 3A = 3 \sin A - 4 \sin^3 A$$
$$\cos 3A = 4 \cos^3 A - 3 \cos A$$

$$\left.\begin{aligned} \sin A &= \frac{2t}{1+t^2} \\ \cos A &= \frac{1-t^2}{1+t^2} \end{aligned}\right\} \text{where } t = \tan\frac{A}{2}$$

$$\sin A+\sin B=2\sin\frac{A+B}{2}\cos\frac{A-B}{2}$$

$$\sin A-\sin B=2\cos\frac{A+B}{2}\sin\frac{A-B}{2}$$

$$\cos A+\cos B=2\cos\frac{A+B}{2}\cos\frac{A-B}{2}$$

$$\cos B-\cos A=2\sin\frac{A+B}{2}\sin\frac{A-B}{2}$$

Table 2.10 SIGN AND VALUE OF THE FUNCTIONS BETWEEN 0° AND 360°

Degrees	0°		90°		180°		270°		360°	30°	45°	60°
Quadrant		1		2		3		4				
sin	0	+	1	+	0	−	−1	−	0	1/2	$1/\sqrt{2}$	$\sqrt{3}/2$
cos	1	+	0	−	−1	−	0	+	1	$\sqrt{3}/2$	$1/\sqrt{2}$	1/2
tan	0	+	∞	−	0	+	∞	−	0	$1/\sqrt{3}$	1	$\sqrt{3}$
cot	∞	+	0	−	∞	+	0	−	∞	$\sqrt{3}$	1	$1/\sqrt{3}$

Table 2.11 SUPPLEMENTARY AND COMPLEMENTARY ANGLES

$x=$	$90\pm\alpha$	$180\pm\alpha$	$270\pm\alpha$	$n360\pm\alpha$(or $\pm\alpha$)
$\sin x$	$\pm\cos\alpha$	$\mp\sin\alpha$	$-\cos\alpha$	$\pm\sin\alpha$
$\cos x$	$\mp\sin\alpha$	$-\cos\alpha$	$\pm\sin\alpha$	$+\cos\alpha$
$\tan x$	$\mp\cot\alpha$	$\pm\tan\alpha$	$\mp\cot\alpha$	$\pm\tan\alpha$
$\cot x$	$\mp\tan\alpha$	$\pm\cot\alpha$	$\mp\tan\alpha$	$\pm\cot\alpha$

PROPERTIES OF TRIANGLES

$$\frac{a}{\sin A}=\frac{b}{\sin B}=\frac{c}{\sin C}=2R,\ \text{where } R=\text{radius of circumcircle}$$

$$a^2=b^2+c^2-2bc\cos A$$

$$\tan\frac{B-C}{2}=\frac{b-c}{b+c}\cot\frac{A}{2}$$

$$a=b\cos C+c\cos B$$

Area of triangle $=\triangle=\frac{1}{2}ab\sin C=\sqrt{[s(s-a)(s-b)(s-c)]}$, where $s=\frac{1}{2}(a+b+c)$

$$\sin\frac{A}{2}=\sqrt{\left[\frac{(s-b)(s-c)}{bc}\right]}$$

$$\cos\frac{A}{2}=\sqrt{\left[\frac{s(s-a)}{bc}\right]}$$

$$\tan\frac{A}{2}=\sqrt{\left[\frac{(s-b)(s-c)}{s(s-a)}\right]}$$

$$A+B+C=180°$$

$$\sin A+\sin B+\sin C=4\cos\frac{A}{2}\cos\frac{B}{2}\cos\frac{C}{2}$$

$$\cos A+\cos B+\cos C=4\sin\frac{A}{2}\sin\frac{B}{2}\sin\frac{C}{2}+1$$

$$\tan A+\tan B+\tan C=\tan A\tan B\tan C$$

$$\cot\frac{A}{2}+\cot\frac{B}{2}+\cot\frac{C}{2}=\cot\frac{A}{2}\cot\frac{B}{2}\cot\frac{C}{2}$$

HYPERBOLIC FUNCTIONS

$$\sinh x=\tfrac{1}{2}(e^{x}-e^{-x})$$

$$\cosh x=\tfrac{1}{2}(e^{x}+e^{-x})$$

$$\tanh x=\frac{\sinh x}{\cosh x}=\frac{e^{x}-e^{-x}}{e^{x}+e^{-x}}$$

$$\operatorname{sech} x=\frac{1}{\cosh x}=\frac{2}{e^{x}+e^{-x}}$$

$$\operatorname{cosech} x=\frac{1}{\sinh x}=\frac{2}{e^{x}-e^{-x}}$$

$$\cosh^2 x-\sinh^2 x=1$$

$$\sinh(-x)=-\sinh x$$

$$\cosh(-x)=\cosh x$$

$$\tanh(-x)=-\tanh x$$

$$\sinh(x+y)=\sinh x\cosh y+\cosh x\sinh y$$

$$\cosh(x+y)=\cosh x\cosh y+\sinh x\sinh y$$

(*See also* 'Series and progressions'.)

2.2.4 Mensuration

PLANE FIGURES

Parallelogram. Area = base × altitude
Triangle. Area = $\frac{1}{2}$ × base × altitude
Trapezium. Area = $\frac{1}{2}$(sum of parallel sides) × altitude
Circle. Radius = r

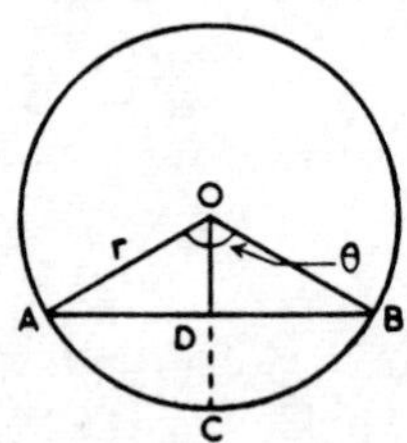

θ = AÔB (*measured in radians*) OD = x

Area = πr^2; circumference = $2\pi r$
Length of arc ACB = $r\theta$
Area of sector OACB = $\frac{1}{2}r^2\theta$

Length of chord AB = $2r\sin\frac{\theta}{2}$

Area of segment ACB = $\frac{1}{2}r^2(\theta-\sin\theta)=\frac{\pi r^2}{2}-\left[x\sqrt{(r^2-x^2)}+r^2\sin^{-1}\left(\frac{x}{r}\right)\right]$

Ellipse. Area $=\pi ab$, where a and b are the semi-axes

$$\text{Perimeter}=2\pi\sqrt{\left(\frac{a^2+b^2}{2}\right)} \text{ (approximate)}$$

SOLID FIGURES

Rectangular prism. Sides a, b and c
Surface area $=2(ab+bc+ca)$
Volume $=abc$
Diagonal $=\sqrt{(a^2+b^2+c^2)}$

Sphere. Radius $=r$
Surface area $=4\pi r^2$; volume $=\frac{4}{3}\pi r^3$
Curved area of spherical segment, height $h=2\pi rh$
Volume of spherical segment, height $h=\frac{1}{3}\pi h^2(3r-h)$
Curved area of spherical zone between two parallel planes, distance l apart $=2\pi rl$

Spherical shell. Internal radius $=r$, external radius $=R$
Volume of shell $=\frac{4}{3}\pi(R^3-r^3)$

$$=\tfrac{4}{3}\pi t^3+4\pi trR$$

$$=4\pi r'^2 t+\frac{\pi}{3}t^3$$

where $t=\text{thickness}=R-r$

and $r'=\text{mean radius}=\frac{1}{2}(R+r)$

*Right circular cylinder.** Height $=h$, radius $=r$
Area of curved surface $=2\pi rh$
Total surface area $=2\pi r(h+r)$
Volume $=\pi r^2 h$

Right circular cone.† Radius of base $=r$, height $=h$
Area of curved surface $=\pi r\sqrt{(r^2+h^2)}$

$$\text{Volume}=\frac{\pi}{3}r^2 h$$

Area of curved surface of frustrum (radii R and r) $=\pi(R+r)\sqrt{[h^2+(R-r)^2]}$

Volume of frustrum (radii R and r) $=\dfrac{\pi h}{3}(R^2+Rr+r^2)$

2.2.5 Co-ordinate geometry (two dimensions, rectangular axes)

The length d of the straight line joining the points A(x_1, y_1) and B(x_2, y_2) is given by

$$d=\pm\sqrt{[(x_2-x_1)^2+(y_2-y_1)^2]}$$

If P(x, y) is a point on AB such that $\dfrac{\text{AP}}{\text{PB}}=\dfrac{m}{n}$,

$$x=\frac{mx_2+nx_1}{m+n};\ y=\frac{my_2+ny_1}{m+n}$$

STRAIGHT LINE

The general equation to a straight line is,

$$ax+by+c=0$$

* Volume of *any* cylinder = area of base × vertical height.
† Volume of *any* cone or pyramid = $\frac{1}{3}$[area of base × perpendicular distance from vertex to base].

Slope $= -\dfrac{a}{b}$. Intercept on y-axis $= -\dfrac{c}{b}$

Alternative forms are

1. $\dfrac{y_2-y_1}{x_2-x_1}=\dfrac{y-y_1}{x-x_1}$ for the line joining the points (x_1, y_1) and (x_2, y_2)
2. $y=mx+c$ for the line of slope, m, cutting y-axis at point $(0, c)$.
3. $y-y_1=m(x-x_1)$ for the line through the point (x_1, y_1), slope m.
4. $\dfrac{x}{a}+\dfrac{y}{b}=1$ for the line making intercepts of a and b on the axes of x and y, respectively.
5. $x\cos\alpha+y\sin\alpha=p$, where p is the length of the perpendicular from the origin to the line and α is the angle between the perpendicular and the positive direction of the x-axis.

The perpendicular distance d from the point (x_1, y_1) to the line $ax+by+c=0$ is

$$d=\pm\frac{ax_1+by_1+c}{\sqrt{(a^2+b^2)}}$$

The angle θ between two lines of slopes m_1 and m_2 is given by:

$$\tan\theta=\frac{m_1-m_2}{1+m_1m_2}$$

If the two lines are parallel, $m_1=m_2$
If the two lines are perpendicular, $m_1m_2=-1$

TRIANGLE

The area of a triangle, vertices (x_1, y_1), (x_2, y_2) and (x_3, y_3) is

$$\triangle=\tfrac{1}{2}(x_1y_2-x_2y_1+x_2y_3-x_3y_2+x_3y_1-x_1y_3)$$

CIRCLE

The general equation to a circle is

$$x^2+y^2+2gx+2fy+c=0$$

centre $(-g, -f)$; radius $=\sqrt{(g^2+f^2-c)}$

Equation to tangent at the point (x_1, y_1) on the circle is

$$xx_1+yy_1+g(x+x_1)+f(y+y_1)+c=0$$

An alternative form of the equation to the circle is

$(x-a)^2+(y-b)^2=R^2$, centre (a, b); radius $=R$

ELLIPSE

$\dfrac{x^2}{a^2}+\dfrac{y^2}{b^2}=1$. Centre at origin and semi-axes a and b

Equation to tangent at point (x_1, y_1) is $\dfrac{xx_1}{a^2}+\dfrac{yy_1}{b^2}=1$

PARABOLA

$y^2=4ax$. Vertex at origin, *latus rectum* $=2a$

Equation to tangent at point (x_1, y_1) is $yy_1=2a(x+x_1)$

HYPERBOLA

$\frac{x^2}{a^2}-\frac{y^2}{b^2}=1$. Centre at origin and semi-axes a and b

Equation to tangent at point (x_1, y_1) is $\frac{xx_1}{a^2}-\frac{yy_1}{b^2}=1$

The asymptotes are $\frac{x}{a}\pm\frac{y}{b}=0$

2.2.6 Calculus*

DIFFERENTIALS†

$dax=a\ dx$
$d(u+v)=du+dv$
$duv=v\ dv+v\ du$
$d\frac{u}{v}=\frac{v\ du-u\ dv}{v^2}$
$dx^n=nx^{n-1}\ dx$
$dx^y=yx^{y-1}\ dx+x^y \log_e x\ dy$

$de^x=e^x\ dx$
$de^{ax}=ae^{ax}\ dx$
$da^x=a^x \log_e a\ dx$
$d \log_e x=x^{-1}\ dx$
$d \log_a x=x^{-1} \log_a e\ dx$
$dx^x=x^x(1+\log_e x)dx$

$d \sin x=\cos x\ dx$
$d \cos x=-\sin x\ dx$
$d \tan x=\sec^2 x\ dx$
$d \cot x=-\operatorname{cosec}^2 x\ dx$
$d \sec x=\tan x \sec x\ dx$
$d \operatorname{cosec} x=-\cot x \cdot \operatorname{cosec} x\ dx$
$d \operatorname{vers} x=\sin x\ dx$
$d \sin^{-1} x=(1-x^2)^{-1/2}\ dx$
$d \cos^{-1} x=-(1-x^2)^{-1/2}\ dx$
$d \tan^{-1} x=(1+x^2)^{-1}\ dx$
$d \cot^{-1} x=-(1+x^2)^{-1}\ dx$
$d \sec^{-1} x=x^{-1}(x^2-1)^{-1/2}\ dx$
$d \operatorname{cosec}^{-1} x=-x^{-1}(x^2-1)^{-1/2}\ dx$

$d \operatorname{vers}^{-1} x=(2x-x^2)^{-1/2}\ dx$
$d \sinh x=\cosh x\ dx$
$d \cosh x=\sinh x\ dx$
$d \tanh x=\operatorname{sech}^2 x\ dx$
$d \coth x=-\operatorname{cosech}^2 x\ dx$
$d \operatorname{sech} x=-\operatorname{sech} x \tanh x\ dx$
$d \operatorname{cosech} x=-\operatorname{cosech} x \coth x\ dx$
$d \sinh^{-1} x=(x^2+1)^{-1/2}\ dx$
$d \cosh^{-1} x=(x^2-1)^{-1/2}\ dx$
$d \tanh^{-1} x=(1-x^2)^{-1}\ dx$
$d \coth^{-1} x=-(x^2-1)^{-1}\ dx$
$d \operatorname{sech}^{-1} x=-x^{-1}(1-x^2)^{-1/2}\ dx$
$d \operatorname{cosech}^{-1} x=-x^{-1}(x^2+1)^{-1/2}\ dx$

INTEGRALS

Elementary forms

1. $\int a\ dx=ax$
2. $\int a \cdot f(x)\ dx=a \int f(x)\ dx$
3. $\int \phi(y)\ dx=\int \frac{\phi(y)}{y'} dy,\quad$ where $y'=\frac{dy}{dx}$
4. $\int (u+v)\ dx=\int u\ dx+\int v\ dx,\quad$ where u and v are any functions of x
5. $\int u\ dv=uv-\int v\ du$

---

* From *Handbook of Chemistry and Physics*, Cleveland, Ohio, 1945.
† Differentials have been written as above for ease in use, e.g. $dax=adx$ instead of $(d/dx)ax=a$, which is the mathematically correct form.

6. $\int u\frac{dv}{dx}dx = uv - \int v\frac{dx}{dx}dx$

7. $\int x^n\,dx = \frac{x^{n+1}}{n+1}, \qquad (n \neq -1)$

8. $\int \frac{f'(x)\,dx}{f(x)} = \log f(x), \qquad [df(x) = f'(x)\,dx]$

9. $\int \frac{dx}{x} = \log x$, or $\log(-x)$

10. $\int \frac{f'(x)\,dx}{2\sqrt{[f(x)]}} = \sqrt{[f(x)]} \qquad [df(x) = f'(x)\,dx]$

11. $\int e^x\,dx = e^x$

12. $\int e^{ax}\,dx = \frac{e^{ax}}{a}$

13. $\int b^{ax}\,dx = \frac{b^{ax}}{a \log b}$

14. $\int \log x\,dx = x \log x - x$

15. $\int a^x \log a\,dx = a^x$

16. $\int \frac{dx}{a^2+x^2} = \frac{1}{a}\tan^{-1}\left(\frac{x}{a}\right)$, or $-\frac{1}{a}\cot^{-1}\left(\frac{x}{a}\right)$

17. $\int \frac{dx}{a^2-x^2} = \frac{1}{a}\tanh^{-1}\left(\frac{x}{a}\right)$, or $\frac{1}{2a}\log\frac{a+x}{a-x}$

18. $\int \frac{dx}{x^2-a^2} = -\frac{1}{a}\coth^{-1}\left(\frac{x}{a}\right)$, or $\frac{1}{2a}\log\frac{x-a}{x+a}$

19. $\int \frac{dx}{\sqrt{(a^2-x^2)}} = \sin^{-1}\left(\frac{x}{a}\right)$, or $-\cos^{-1}\left(\frac{x}{a}\right)$

20. $\int \frac{dx}{\sqrt{(x^2 \pm a^2)}} = \log[x + \sqrt{(x^2 \pm a^2)}]$

21. $\int \frac{dx}{x\sqrt{(x^2-a^2)}} = \frac{1}{a}\cos^{-1}\left(\frac{a}{x}\right)$

22. $\int \frac{dx}{x\sqrt{(a^2 \pm x^2)}} = -\frac{1}{a}\log\left[\frac{a+\sqrt{(a^2 \pm x^2)}}{x}\right]$

23. $\int \frac{dx}{x\sqrt{(a+bx)}} = \frac{2}{\sqrt{(-a)}}\tan^{-1}\sqrt{\left[\frac{(a+bx)}{-a}\right]}$, or $\frac{-2}{\sqrt{(a)}}\tanh^{-1}\sqrt{\left[\frac{(a+bx)}{a}\right]}$

More complex integrals are to be found in *Handbook of Chemistry and Physics*, Cleveland, Ohio, 1945.

DIFFERENTIAL EQUATIONS

Equations of the first order

1. $\frac{dy}{dx} = f(x)\phi(y)$ (Variable separable type)

   Solution $\int \frac{dy}{\phi(y)} = \int f(x)\,dx + A$

2a. $\dfrac{dy}{dx}=f\left(\dfrac{y}{x}\right)$ (Homogeneous equation)

Let $v=\dfrac{y}{x}$

Solution $\displaystyle\int\frac{dv}{f(v)-v}=\log x+A$

2b. $\dfrac{dy}{dx}=\dfrac{ax+by+c}{a'x+b'y+c'}$ (Reducible to homogeneous form if $a'b-ab'\neq 0$)

Let $x=X+h$; $y=Y+k$, where h and k are the values of x and y which satisfy equations

$$\left.\begin{aligned}ax+by+c=0\\ a'x+b'y+c'=0\end{aligned}\right\}$$

Solution obtained as in (2a) by putting $Y=vX$

3. $M\,dx+N\,dy=0$ where M and N are functions of x and y and

$$\frac{\partial M}{\partial y}=\frac{\partial N}{\partial x} \quad \text{(Exact equation)}$$

Here, a function $u(x, y)$ exists such that

$$du=M\,dx+N\,dy$$

Solution $u=\displaystyle\int M\,dx\,(y\text{ constant})+\int(\text{terms in }N\text{ independent of }x)\,dy=\text{constant}$

4. $\dfrac{dy}{dx}+Py=Q$ (Linear equation)

where P and Q are functions of x only (or constants).

Solution $y\,e^{\int P\,dx}=\displaystyle\int Q\,e^{\int P\,dx}dx+A$ (Integrating factor $=e^{\int P\,dx}$)

5. $\dfrac{dy}{dx}+Py=Qy^n$ (Bernouilli's equation)

where P and Q are functions of x only (or constants).

Multiplying both sides of the equation by $(1-n)y^{-n}$, and putting $z=y^{1-n}$ the equation becomes

$$\frac{dz}{dx}+(1-n)Pz=(1-n)Q$$

This equation is linear.

Solution obtained as in 4. [Integrating factor $=e^{\int(1-n)P\,dx}$]

6. $y=x\dfrac{dy}{dx}+f\left(\dfrac{dy}{dx}\right)$ (Clairaut's form)

Let $\dfrac{dy}{dx}=p$ and differentiate the equation with respect to x,

$$\frac{dp}{dx}[x+f'(p)]=0$$

Solution $\dfrac{dp}{dx}=0$ or $x+f'(p)=0$

i.e. $p=\text{constant}=A$
or $y=Ax+B$

This solution is obtained by eliminating p from this equation and the original equation. The solution contains no arbitrary constant and is known as the *singular solution.*

Equations of the second order

7. $\dfrac{d^2y}{dx^2}=f(x)$

Solution $y=\int F(x)\,dx+Ax+B$, where $F(x)=\int f(x)\,dx$

8. $\dfrac{d^2y}{dx^2}=f(y)$

Let $\dfrac{dy}{dx}=p;\ \tfrac{1}{2}p^2=\int f(y)\,dy+A$

Integration of this equation of the first order may then lead to the required solution.

9.* $a\dfrac{d^2y}{dx^2}+b\dfrac{dy}{dx}+cy=0$ (Linear equation, constant coefficients, RHS$=0$)

Auxiliary equation. $am^2+bm+c=0$
Solution
(a) If $b^2>4ac$ (roots m_1 and m_2), $Ae^{m_1x}+Be^{m_2x}$
(b) If $b^2=4ac$ (two equal roots, m_1), $e^{m_1x}\,(Ax+B)$
(c) If $b^2<4ac$ (roots $m_1\pm im_2$), $e^{m_1x}\,(A\sin m_2x+B\cos m_2x)$
or $Ce^{m_1x}\sin(m_2x+\alpha)$
or $Ce^{m_1x}\cos(m_2x-\alpha)$

10.* $a\dfrac{d^2y}{dx^2}+b\dfrac{dy}{dx}+cy=P$ (Linear equation, constant coefficients, RHS a function of x only)

Solution $y=$complementary function$+$particular integral.
The *complementary function* is the solution of the equation when $P=0$ [*see* (9)].
The *particular integral* is *any* solution of the equation which involves no arbitrary constants.
The following examples of values of the particular integral may be useful:
(a) $P=\text{constant}=m$

Particular solution $y=\dfrac{m}{c}$

(b) $P=p+qx+rx^2+\ldots$ (p, q, r … are constants)
Particular solution $y=A+Bx+Cx^2+\ldots$
If $C\neq 0$, P and y are of the same degree.

If $C=0$, P and $\dfrac{dy}{dx}$ are of the same degree.

(c) $P=k\,e^{mx}$
Particular solution $y=A\,e^{mx}$
If e^{mx} is a term of the complementary function, try $y=Ax\,e^{mx}$
If $x\,e^{mx}$ is a term of the complementary function, try $y=Ax^2\,e^{mx}$.
(d) $P=l\sin nx+m\cos nx$ (l and m constants, or zero)
Particular solution $y=A\sin nx+B\cos nx$
If $C\sin nx+D\cos nx$ is a term of the complementary function try
$Ax\sin nx+Bx\cos$

Note: The constants A. B. C. D, etc. in the particular solutions are evaluated by substitution in the original equation.

* Reproduced with permission from G. W. Caunt, *Introduction to the Infinitesimal Calculus*, Oxford University Press, 1928.

3 General physical and chemical constants

Table 3.1 ATOMIC WEIGHTS AND ATOMIC NUMBERS OF THE ELEMENTS

Name	*Symbol*	*Atomic number* %	*International atomic weights* 1971* $^{12}C = 12$	*Name*	*Symbol*	*Atomic number* %	*International atomic weights* 1971* $^{12}C = 12$
Actinium	Ac	89	(227)	Lanthanum	La	57	138.905_5
Aluminium	Al	13	26.981 54	Lawrencium	Lr	103	(256)
Americium	Am	95	(243)	Lead	Pb	82	207.2
Antimony	Sb	51	121.7_5	Lithium	Li	3	6.94_1
Argon	Ar	18	39.94_8	Lutetium	Lu	71	174.97
Arsenic	As	33	74.921 6	Magnesium	Mg	12	24.305
Astatine	At	85	(210)	Manganese	Mn	25	54.938 0
Barium	Ba	56	137.3_4	Mendelevium	Md	101	(258)
Berkelium	Bk	97	(247)	Mercury	Hg	80	200.5_9
Beryllium	Be	4	9.012 18	Molybdenum	Mo	42	95.9_4
Bismuth	Bi	83	208.980 4	Neodymium	Nd	60	144.2_4
Boron	B	5	10.81	Neon	Ne	10	20.17_9
Bromine	Br	35	79.904	Neptunium	Np	93	237.048 2
Cadmium	Cd	48	112.40	Nickel	Ni	28	58.7_1
Caesium	Cs	55	132.905 4	Niobium			
Calcium	Ca	20	40.08	(Columbium)	Nb	41	92.906 4
Californium	Cf	98	(251)	Nitrogen	N	7	14.006 7
Carbon	C	6	12.011	Nobelium	No	102	(255)
Cerium	Ce	58	140.12	Osmium	Os	76	190.2
Chlorine	Cl	17	35.453	Oxygen	O	8	15.999_4
Chromium	Cr	24	51.996	Palladium	Pd	46	106.4
Cobalt	Co	27	58.933 2	Phosphorus	P	15	30.973 76
Copper	Cu	29	63.54_6	Platinum	Pt	78	195.0_9
Curium	Cm	96	(247)	Plutonium	Pu	94	(244)
Dysprosium	Dy	66	162.5_0	Polonium	Po	84	(209)
Einstenium	Es	99	(254)	Potassium	K	19	39.09_8
Erbium	Er	68	167.2_6	Praseodymium	Pr	59	140.907 7
Europium	Eu	63	151.96	Promethium	Pm	61	(145)
Fermium	Fm	100	(257)	Protactinium	Pa	91	231.035 9
Fluorine	F	9	18.998 40	Radium	Ra	88	226.025 4
Francium	Fr	87	(223)	Radon	Rn	86	(222)
Gadolinium	Gd	64	157.2_5	Rhenium	Re	75	186.2
Gallium	Ga	31	69.72	Rhodium	Rh	45	102.905 5
Germanium	Ge	32	72.5_9	Rubidium	Rb	37	85.467_8
Gold	Au	79	196.966 5	Ruthenium	Ru	44	101.0_7
Hafnium	Hf	72	178.4_9	Samarium	Sm	62	150.4
Helium	He	2	4.002 60	Scandium	Sc	21	44.955 9
Holmium	Ho	67	164.930 4	Selenium	Se	34	78.9_6
Hydrogen	H	1	1.007 9	Silicon	Si	14	28.08_6
Indium	In	49	114.82	Silver	Ag	47	107.868
Iodine	I	53	126.904 5	Sodium	Na	11	22.989 77
Iridium	Ir	77	192.2_2	Strontium	Sr	38	87.62
Iron	Fe	26	55.84_7	Sulphur	S	16	32.06
Krypton	Kr	36	83.80	Tantalum	Ta	73	180.947_9

Table 3.1 ATOMIC WEIGHTS AND ATOMIC NUMBERS OF THE ELEMENTS—*continued*

Name	*Symbol*	*Atomic number* %	*International atomic weights* 1971* $^{12}C=12$	*Name*	*Symbol*	*Atomic number* %	*International atomic weights* 1971* $^{12}C=12$
Technetium	Tc	43	(97)	Tungsten	W	74	183.8_5
Tellurium	Te	52	127.6_0	Uranium	U	92	238.029
Terbium	Tb	65	158.925 4	Vanadium	V	23	50.941_4
Thallium	Tl	81	204.3_7	Xenon	Xe	54	131.30
Thorium	Th	90	232.038 1	Ytterbium	Yb	70	173.0_4
Thulium	Tm	69	168.934 2	Yttrium	Y	39	88.905 9
Tin	Sn	50	118.6_9	Zinc	Zn	30	65.38
Titanium	Ti	22	47.9_0	Zirconium	Zr	40	91.22

* Atomic Weights of the Elements 1971, *Pure and Applied Chemistry* 1972, **30** (3/4), 637–649. A value given in brackets denotes the mass number of the isotope of longest known half-life. Because of natural variation in the relative abundance of the isotopes of some elements, their atomic weights may vary. Apart from this they are considered reliable to ±1 in the last digit, or ±3 if that digit is subscript.

Table 3.2 GENERAL PHYSICAL CONSTANTS

The probable errors of the various quantities may be obtained from the reference given at the end of the table.

Quantity	*Symbol*	*Value*	*Units*
Acceleration due to gravity (standard)	g_0	9.806 65	$m\ s^{-2}$
Atmospheric pressure (standard)	A_0	$1.013\,25 \times 10^5$	Pa
Atomic mass unit	amu	$1.660\,53 \times 10^{-27}$	kg
Atomic weight of electron	$N_A m_c$	$5.485\,97 \times 10^{-4}$	u*
Avogadro number	N_A	$6.022\,17 \times 10^{23}$	mol^{-1}
Boltzmann's constant	k	$1.380\,62 \times 10^{-23}$	$J\,K^{-1}$
Bohr radius	a_0	$5.291\,77 \times 10^{-11}$	m
Bohr magneton	μ_B	$9.274\,09 \times 10^{-24}$	$J\,T^{-1}$
Charge in electrolysis of 1 g hydrogen	F/H	$9.572\,378 \times 10^4$	C
Compton wavelength of electron	λ_C	$2.426\,31 \times 10^{-12}$	m
Classical electron radius	r_0	$2.817\,94 \times 10^{-15}$	m
Compton wavelength of proton	λ_{C_p}	$1.321\,44 \times 10^{-15}$	m
Compton wavelength of neutron	λ_{C_n}	$1.319\,62 \times 10^{-15}$	m
Density of the earth (average)	δ	5.518×10^3	$kg\,m^{-3}$
(core)		1.072×10^4	$kg\,m^{-3}$
Density of mercury (0°C, A_0)	D_0	$1.359\,508 \times 10^4$	$kg\ m^{-3}$
Density of water (max)	$\delta_m(H_2O)$	$9.999\,72 \times 10^2$	$kg\ m^{-3}$
Electronic charge	e	$1.602\,192 \times 10^{-19}$	C
Electron rest mass	m_e	$9.109\,558 \times 10^{-31}$	kg
Electron volt energy	E_0	$1.602\,192 \times 10^{-19}$	J
Electron magnetic moment	μ_e	$9.284\,851 \times 10^{-24}$	$J\ T^{-1}$
Faraday constant	F	$9.648\,670 \times 10^4$	$C\ mol^{-1}$
Fine structure constant	α	$7.297\,351 \times 10^{-3}$	—
Gas constant	R_0	8.314 34	$J\,K^{-1}\,mol^{-1}$
Gravitational constant	G	$6.673\,2 \times 10^{-11}$	$Nm^2\ kg^{-2}$
Ice point (absolute value)	T_0	$2.731\,5 \times 10^2$	K
Litre (12th CGPM 1964)	l	1.0 exactly $\times 10^{-3}$	m^3
(1963 weights and measures)		$1.000\,028 \times 10^{-3}$	m^3
Neutron rest mass	M_n	$1.674\,920 \times 10^{-27}$	kg
Planck's constant	h	$6.626\,196 \times 10^{-34}$	Js
Proton rest mass	M_p	$1.672\,614 \times 10^{-27}$	kg
Radiation constant–first	C_1	$4.992\,579 \times 10^{-24}$	Jm
Radiation constant–second	C_2	$1.438\,833 \times 10^{-2}$	mK
Rydberg's constant	$R\infty$	$1.097\,373 \times 10^7$	m^{-1}
Stefan–Boltzmann constant	σ	$5.669\,61 \times 10^{-8}$	$Wm^{-2}K^4$
Velocity of light	c	$2.997\,925 \times 10^8$	$m\ s^{-1}$
Volume of ideal gas (0°C, A_0)	V_0	$2.241\,36 \times 10^{-2}$	$m^3\ mol^{-1}$
Wien's constant	$\lambda_m T$	$2.897\,8 \times 10^{-3}$	mK
Zeeman displacement		$4.668\,58 \times 10$	$m\ Wb^{-1}$

* u = unified atomic mass unit (amu) based on $u = \frac{1}{12}$ of the mass of ^{12}C.

REFERENCES

B. N. Taylor, W. H. Parker and D. N. Langenberg (1969), *Rev. Mod. Phys.*, **41**, 375.
G. W. C. Kaye and T. H. Laby (1966), *Tables of Physical and Chemical Constants*, 13th edn, Longmans.

Table 3.3 MOMENTS OF INERTIA

Moment of inertia $= Mk^2$ where M is mass and k radius of gyration

Body	*Dimensions*	*Axis**	k^2
Uniform thin rod	length l	Through centre, perpendicular to rod	$l^2/12$
		Through end, perpendicular to rod	$l^2/3$
Rectangular lamina	sides a and b	Through centre, perpendicular to plane of lamina	$(a^2+b^2)/12$
		Through centre, parallel to side b	$a^2/12$
Circular lamina	radius r	Through centre, perpendicular to plane of lamina	$r^2/2$
		Through any diameter	$r^2/4$
Annular lamina	radii r_1 and r_2	Through centre perpendicular to plane of lamina	$(r_1^2+r_2^2)/2$
		Through any diameter	$(r_1^2+r_2^2)/4$
Rectangular solid	sides a, b and c	Through centre, perpendicular to face ab	$(a^2+b^2)/12$
Sphere	radius r	Through any diameter	$2r^2/5$
Spherical shell	radii r_1 and r_2	Through any diameter	$2(r_1^5-r_2^5)/5(r_1^3-r_2^3)$
Thin spherical shell	radius r	Through any diameter	$2r^2/3$
Right circular cylinder	radius r length l	Longitudinal axis through centre	$r^2/2$
		Through centre perpendicular to longitudinal axis	$r^2/4+l^2/12$
Hollow circular cylinder	radii r_1 and r_2 length l	Longitudinal axis through centre	$(r_1^2+r_2^2)/2$
		Through centre perpendicular to longitudinal axis	$(r_1^2+r_2^2)/4+l^2/12$
Thin hollow circular cylinder	radius r length l	Longitudinal axis through centre	r^2
		Through centre perpendicular to longitudinal axis	$r^2/2+l^2/12$
Right circular cone	height h base radius r	Longitudinal axis through apex	$3r^2/10$
		Through centre of gravity perpendicular to longitudinal axis	$3(r^2+h^2/4)/20$
Ellipsoid	semi-axis a, n and c	Through centre along axis a	$(b^2+c^2)/5$

* If the moment of inertia I_q about an axis through the centre of gravity is known, then the moment, I, about any other parallel axis may be obtained from

$$I = I_q + Mh^2$$

where h is the distance from the centre of gravity to the parallel axis.

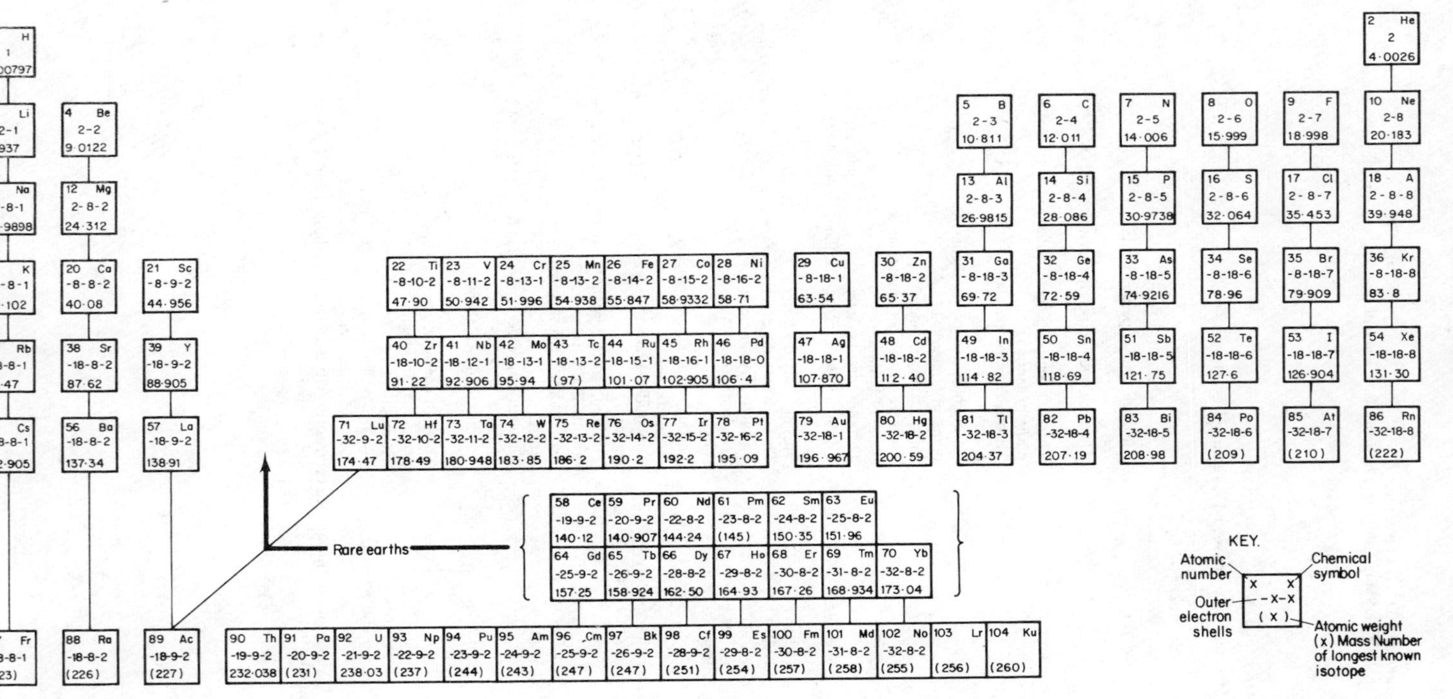

Figure 3.1 *The Periodic Table*

3.1 Radioactive isotopes and radiation sources

Tables 3.4, 3.5 and 3.6 are so arranged as to assist in selecting an isotope with a given half-life and decay radiation energy, for gamma, beta and positron emitters.

Tables 3.7, 3.8, 3.9, 3.10 and 3.11 list the most commonly used commercially available alpha, beta and neutron sources.

Information on the selection of radiochemical isotope can be obtained from *The American Institute of Physics Handbook*, 2nd edn, Section 8, McGraw-Hill, New York, 1963.

Table 3.4 POSITRON EMITTERS

Isotope	*Half-life*	β^+ *energies* MeV %
Oxygen-15	2 min	1.7–100
Nitrogen-13	10 min	1.2–100
Bromine-80	18 min	0.87–3
Carbon-11	20 min	0.97–100
Manganese-52m	21 min	2.63–100
Gallium-68	68 min	0.82–1
		1.89–86
Fluorine-18	110 min	0.649–97
Scandium-44	3.9 h	1.467–91.5
Iron-52 (Daughter: 52mMn)	8.3 h	0.80–~57
Gallium-66	9.5 h	0.92–4
		4.15–44
		others–3
Copper-64	12.84 h	0.66–19
Arsenic-72	26 h	1.84–3
		2.50–56
		3.34–17
		others–2
Bromine-77	58 h	0.336–0.7
Yttrium-87	80 h	0.7–0.3
Iodine-124	4.0	0.9–0.5
		1.6–14
		2.2–11
Manganese-52	5.7 d	0.58–29
Caesium-132	6.48 d	0.6–1.2
Iodine-126	13 d	0.46–0.3
		1.11–1.0
Vanadium-48	16 d	0.70–56
Arsenic-74	18 d	0.91–26.1
		1.51–3.6
Rubidium-84	33 d	0.8–10.9
		1.63–9.7
Cobalt-58	71 d	0.485–14.8
Cobalt-56	77.3 d	1.50–18
Yttrium-88	106.5 d	0.6–0.2
Zinc-65	245 d	0.325–1.7
Sodium-22	2.6 yr	0.54–90.5
		1.83–~0.06
Aluminium-26	7.4 × 10yr	1.16–85

Table 3.5 BETA ENERGIES

Half-life *β-energy* MeV	*<1 hour*	*1–10 hours*	*10 hours–1 day*
0–0.3		^{105}Ru, ^{117}Yb	
0.3–0.5	^{70}Ga, ^{116m}In	^{56}Mn, ^{165}Dy	**^{28}Mg,** ^{43}K, ^{77}Ge, ^{97}Zr, ^{194}Ir, ^{197}Pt
0.5–0.7	^{70}Ga, ^{101}Mo, ^{116m}In	**^{65}Ni** ^{105}Ru, **^{127}Te,** ^{152m}Eu, ^{171}Er	**^{64}Cu,** **^{72}Ga,** **^{130}I,** ^{142}Pr, **^{159}Gd** ^{187}W, **^{197}Pt**
0.7–1.0	^{49}Ca, ^{66}Cu, **^{69}Zn** ^{101}Mo, **^{116m}In,** **^{117}In,** ^{144}Pr, ^{199}Pt	**^{56}Mn,** **^{75}Ge,** **^{85m}Kr,** ^{105}Ru, **^{117}Cd,** ^{132}I, ^{139}Ba, ^{149}Nd, ^{165}Dy	**^{43}K** **^{72}Ga** ~ ^{77}Ge **^{159}Gd,** ^{194}Ir
1.0–1.5	**^{38}Cl,** ^{51}Ti, ^{80}Br, **^{81}Se,** ^{101}Mo, **^{101}Tc,** ^{108}Ag, **^{123}Sn,** ^{128}I, ^{155}Sm, ^{199}Pt, **^{233}Th**	**^{31}Si,** **^{41}A,** **^{56}Mn,** ^{65}Ni, **^{75}Ge,** **^{97}Nb,** **^{105}Ru,** **^{132}I,** **^{149}Nd,** **^{165}Dy,** **^{171}Er,** **^{176m}Lu,** **^{177}Yb**	**^{24}Na,** ^{43}K, ^{77}Ge, **^{109}Pd,** **^{130}I,** **^{187}W**
1.5–2.0	**^{27}Mg,** **^{49}Ca,** **^{49}Sc,** ^{66}Cu, **^{70}Ga,** **^{80}Br,** ^{101}Mo, ^{104}Rh, **^{108}Ag,** ^{128}I, **^{155}Sm,** ^{199}Pt	^{105}Ru, ^{132}I, **^{152m}Eu**	**^{42}K,** ^{43}K, ^{72}Ga, ^{77}Ge, **^{97}Zr,** **^{188}Re,** ^{194}Ir
2.0–3.0	**^{28}Al,** ^{38}Cl, **^{51}Ti** **^{52}V** **^{66}Cu,** **^{88}Rb,** ^{101}Mo, **^{104}Rh,** ^{128}I, **^{144}Pr**	^{41}A, **^{56}Mn,** **^{65}Ni,** ^{132}I, **^{139}Ba**	^{28}Mg/^{28}Al, ^{72}Ga, **^{77}Ge,** **^{142}Pr,** **^{188}Re,** **^{194}Ir**
>3.0	**^{38}Cl,** **^{88}Rb**		**^{42}K,** ^{72}Ga

AND HALF-LIVES

1–10 days	*10–100 days*	*100 days–1 year*	*>1 year*
^{76}As, ^{77}As, **^{105}Rh,** ^{31}I, ^{132}Te, ^{133}Xe, ^{43}Ce, ^{166}Ho, ^{175}Yb, ^{77}Lu, ^{186}Re, ^{198}Au, **^{99}Au**	**^{33}P** **^{35}S** **^{59}Fe** **^{95}Nb,** **^{103}Ru,** ^{115m}Cd, ^{124}Sb, $^{129m/129}$Te, ^{147}Nd, **^{191}Os,** ^{192}Ir, **^{203}Hg** **^{233}Pa**	**^{45}Ca,** ^{106}Ru, ^{110m}Ag, ^{144}Ce, ^{182}Ta	**^{3}H,** **^{14}C** **^{63}Ni,** ^{85}Kr, ^{87}Rb, **^{99}Tc,** **^{125}Sb,** **^{129}I,** **^{134}Cs** **^{147}Pm,** ^{152}Eu, ^{154}Eu, **^{155}Eu,** **^{210}Pb,** **^{227}Ac** ^{228}Ra, ^{235}U + D.P., ^{238}U/^{234}Th
^{47}Sc, **^{67}Cu,** ^{77}As, **^{82}Br,** ^{99}Mo, **^{121}Sn,** ^{31}I, **^{133}Xe,** ^{151}Pm, ^{61}Tb, ^{166}Ho, **^{169}Er,** **^{75}Yb,** **^{177}Lu,** ^{199}Au, ^{22}Rn + D.P.	**^{46}Sc,** **^{59}Fe,** ^{91}Y, **^{95}Zr,** ^{103}Ru, ^{115m}Cd, ^{126}I ^{140}Ba, **^{141}Ce,** ^{147}Nd, ^{160}Tb, **^{181}Hf,** **^{185}W**	^{123m}Sn, **^{144}Ce,** ^{182}Ta,	**^{60}Co,** ^{125}Sb, ^{134}Cs, ^{152}Eu, ^{226}Ra + D.P., ^{227}Ac + D.P., ^{228}Th + D.P.
^{47}Ca, **^{47}Sc,** ^{67}Cu, ^{76}As, **^{77}As,** **^{105}Rh,** ^{11}Ag, ^{115}Cd, **^{131}I,** ^{43}Ce, **^{153}Sm,** ^{161}Tb, ^{93}Os, ^{22}Rn + D.P.	^{86}Rb, ^{103}Ru, $^{114m/114}$In, ^{115m}Cd, **^{124}Sb,** $^{129m/129}$Te ^{140}Ba, ^{141}Ce, **^{160}Tb,** **^{192}Ir,** ^{233}Pa	^{110m}Ag, ^{182}Ta	**^{85}Kr,** **^{90}Sr,** ^{125}Sb, **^{134}Cs,** **^{137}Cs,** ^{152}Eu, **^{154}Eu,** ^{226}Ra + D.P., ^{228}Th + D.P., ^{238}U/^{234m}Pa
^{99}Mo, ^{111}Ag, ^{115}Cd, ^{22}Sb, ^{131}I, ^{140}La, ^{43}Ce, ^{149}Pm, ^{151}Pm, ^{53}Sm, ^{166}Ho, ^{186}Re, ^{93}Os, **^{198}Au,** ^{22}Rn + D.P.	^{74}As, ^{84}Rb, ^{95}Zr, ^{124}Sb, ^{126}I, $^{129m/129}$Te ^{140}Ba, **^{143}Pr,** **^{147}Nd,** ^{160}Tb	^{127m}Te, ^{144}Ce/^{144}Pr, **^{170}Tm**	**^{36}Cl,** ^{134}Cs, ^{154}Eu, **^{204}Tl,** ^{226}Ra + D.P.
^{76}As, **^{99}Mo,** **^{111}Ag,** **^{15}Cd,** **^{122}Sb,** **^{140}La,** **^{43}Ce,** **^{149}Pm,** **^{151}Pm,** **^{86}Re,** **^{193}Os,** ^{198}Au, **^{10}Bi**	^{46}Sc, ^{74}As, **^{89}Sr,** ^{124}Sb, ^{126}I, $^{129m/129}$Te, **^{140}Ba/^{140}La**	^{106}Ru/^{106}Rh, **^{123m}Sn**	^{40}K, ^{60}Co, ^{134}Cs, ^{137}Cs, ^{152}Eu, ^{227}Ac + D.P., ^{228}Ra/^{228}Ac, ^{228}Th + D.P., ^{238}U/^{234m}Pa
^{47}Ca, ^{76}As, **^{122}Sb,** ^{40}La, **^{166}Ho,** ^{22}Rn + D.P.	**^{32}P,** ^{59}Fe, **^{86}Rb,** **^{91}Y,** **$^{114m/114}$In,** **^{115m}Cd,** ^{124}Sb, ^{160}Tb	^{106}Ru/^{106}Rh	^{154}Eu, ^{226}Ra + D.P., ^{228}Ra/^{228}Ac, ^{228}Th + D.P.
^{76}As, **^{90}Y** ^{140}La	^{124}Sb	^{106}Ru/^{106}Rh, $^{110m/110}$Ag, **^{144}Ce/^{144}Pr**	**^{90}Sr/^{90}Y,** ^{228}Th + D.P., ^{228}Ra/^{228}Ac, ^{238}U/^{234m}Pa
^{22}Rn + D.P.		**^{106}Ru/^{106}Rh**	^{226}Ra + D.P.

Table 3.6 GAMMA ENERGIE

Half-life *γ-energy* MeV	*<1 hour*	*1–10 hours*	*10 hours–1 day*
0–0.3	^{70}Ga, ^{81m}Se, ^{94m}Nb, ^{101}Mo, ^{101}Tc, ^{104m}Rh, ^{116m}In, ^{117}In, **^{123}Sn,** **^{155}Sm,** ^{199}Pt	^{52}Fe, **^{75}Ge,** ^{80m}Br, **^{85m}Kr,** **^{99m}Tc,** ^{105}Ru/^{105m}Rh, ^{117}Cd, ^{127}Te, ^{134m}Cs, **^{139}Ba,** ^{149}Nd, ^{152m}Eu, ^{165}Dy, **^{171}Er,** ^{176}Lu, ^{177}Yb, **^{180m}Hf**	^{28}Mg, ^{43}K, **^{77}Ge,** ^{109}Pd/^{109}Ag, **^{123}I,** ^{159}Gd, ^{187}W, ^{185}Re, ^{194}Ir, **^{197m}Hg,** ^{197}Pt
0.3–0.5	**^{51}Ti,** **^{101}Tc,** **^{116m}In,** **^{128}I,** ^{199}Pt	^{65}Ni, **^{85m}Kr,** **^{87m}Sr,** ^{105}Ru, ^{132}I, ^{149}Nd, ^{152m}Eu, ^{165}Dy, **^{171}Er,** **^{180m}Hf**	^{28}Mg, **^{42}K,** **^{43}K,** **^{69m}Zn,** ^{77}Ge, **^{87m}Y,** ^{130}I, **^{159}Gd,** ^{187}W, ^{194}Ir, ^{197m}Hg/^{197m}Au
0.5–0.7	**^{11}C,** **^{13}N,** **^{15}O,** ^{51}Ti, ^{80}Br, ^{101}Mo, ^{101}Tc, ^{104}Rh, **^{108}Ag,** **^{117}In,** ^{128}I, ^{144}Pr, ^{199}Pt	**^{18}F,** **^{44}Sc,** **^{52}Fe,** **^{66}Ga,** **^{68}Ga,** **^{97}Nb,** ^{105}Ru, **^{132}I,** ^{149}Nd, ^{165}Dy	**^{43}K,** **^{64}Cu** ^{72}Ga. ^{77}Ge, **^{130}I,** ^{187}W, ^{194}Ir
	(Including isotopes giving 0.51 MeV gamma rays from annihilation of positrons)		
0.7–1.0	**^{27}Mg,** ^{51}Ti, ^{66}Cu, **^{88}Rb,** ^{101}Mo, ^{116m}In, ^{128}I, ^{199}Pt	**^{56}Mn,** ^{66}Ga, **^{105}Ru,** **^{132}I,** ^{152m}Eu, ^{165}Dy, ^{177}Yb	^{28}Mg, **^{72}Ga,** ^{77}Ge, **^{97}Zr/^{97m}Nb,** **^{130}I,** ^{187}W, ^{194}Ir
1.0–1.5	**^{27}Mg,** **^{52}V,** **^{66}Cu,** ^{70}Ga, ^{81}Se, ^{88}Rb, ^{101}Mo, **^{116m}In,** ^{144}Pr	^{31}Si, **^{41}A,** **^{44}Sc** **^{52}Fe/^{52m}Mn,** **^{65}Ni,** **^{66}Ga,** ^{68}Ga, ^{105}Ru, ^{117}Cd, ^{132}I, ^{139}Ba, ^{152m}Eu, ^{177}Yb	**^{24}Na,** **^{28}Mg,** ^{43}K, ^{64}Cu, ^{72}Ga, ^{77}Ge, ^{97}Zr, ^{130}I, ^{194}Ir
1.5–2.0	**^{28}Al,** **^{38}Cl,** **^{88}Rb,** ^{101}Mo, ^{116m}In	^{44}Sc, **^{56}Mn**	^{28}Mg/^{28}Al ^{42}K, ^{72}Ga, ^{97}Zr, **^{142}Pr,** ^{194}Ir
2.0–3.0	**^{38}Cl,** ^{88}Rb, ^{101}Mo, **^{116m}In** ^{144}Pr	^{44}Sc, **^{56}Mn,** **^{66}Ga**	**^{24}Na,** **^{72}Ga,** ^{77}Ge, ^{97}Zr, ^{194}Ir
>3.0	**^{49}Ca** ^{88}Rb	^{56}Mn, ^{66}Ga	

ND HALF-LIVES

–10 days	*10–100 days*	*100 days–1 year*	*>1 year*
^{47}**Sc**, ^{67}**Cu**, ^{67}**Ga**, ^{72}**Se**, ^{77}As, ^{77}**Br**, ^{97}Ru, ^{99}Mo, ^{111}Ag, ^{11}**In** ^{131}I, ^{132}Te, ^{3m}**Ba**, ^{133m}**Xe**, ^{133}**Xe**, ^{43}**Ce**, ^{149}Pm, ^{151}Pm, ^{53}**Sm**, ^{161}Tb, ^{166}Ho, ^{75}Yb, ^{177}Lu, ^{186}Re, ^{93}Os, ^{197}**Hg**, ^{199}**Au**, ^{06}Bi, ^{222}Rn+DP., ^{24}Ra ^{225}Ac	^{59}Fe, ^{73}As ^{83}Rb ^{103}Pd/^{103m}Rh, ^{103}Ru/^{103m}Rh, ^{105}Ag, ^{114m}In, ^{125}I, ^{129m}/^{129}Te, ^{131}**Ba**, ^{131m}**Xe**, ^{140}Ba, ^{141}**Ce**, ^{147}Nd, ^{160}Tb, ^{169}Yb, ^{175}Hf, ^{181}**Hf**, ^{183}**Re**, ^{185}Os, ^{191}Os/^{191m}Ir, ^{192}Ir, ^{203}**Hg**, ^{223}Ra, ^{233}Pa	^{49}V ^{57}**Co**, ^{75}**Se** ^{110m}Ag, ^{113}Sn, ^{119m}Sn, ^{127m}Te, ^{139}**Ce**, ^{144}Ce, ^{153}Gd, ^{170}**Tm**, ^{181}W, ^{182}Ta, ^{195}**Au**	^{44}**Ti**, ^{109}**Cd**, ^{125}Sb, ^{129}I, ^{133}Ba, ^{152}**Eu**, ^{154}Eu, ^{155}**Eu**, ^{208}Po, ^{210}Pb, ^{226}Ra+D.P., ^{227}Ac+D.P., ^{228}Ra/^{228}Ac, ^{228}Th+D.P., ^{230}Th, ^{231}Pa, ^{232}U, ^{233}U, ^{235}U+D.P., ^{237}Np, ^{238}U/^{234}Th, ^{239}Pu, ^{241}**Am**
^{47}Ca, ^{67}Ga, ^{77}Br, ^{87}**Y**, ^{97}Ru, ^{99}Mo, ^{05}**Rh**, ^{111}Ag, ^{115}Cd, ^{31}**I**, ^{140}**La**, ^{143}Ce, ^{51}Pm, ^{175}Yb, ^{177}Lu, ^{93}Os, ^{198}**Au**, ^{206}Bi, ^{22}Rn+D.P.	^{7}**Be**, ^{51}**Cr**, ^{59}Fe, ^{103}**Ru**, ^{105} **Ag**, ^{115m}Cd, ^{126}I, ^{129m}/^{129}Te, ^{131}**Ba**, ^{140}**Ba**/^{140}**La**, ^{147}Nd, ^{160}Tb, ^{169}Yb, ^{175}**Hf**, ^{181}**Hf**, ^{192}**Ir** ^{223}Ra, ^{233}**Pa**	^{75}Se, ^{110m}Ag, ^{113}**Sn**/^{113m}**In**	^{125}**Sb**, ^{133}**Ba**, ^{134}Cs, ^{152}Eu, ^{226}Ra+DP., ^{227}Ac+D. P., ^{231}Pa
^{52}**Mn**, ^{72}**As**, ^{76}**As** ^{77}As, ^{77}Br, ^{82}**Br**, ^{87}Y, ^{97}Ru, ^{115}Cd, ^{22}**Sb**, ^{124}**I**, ^{131}I, ^{32}**Cs**, ^{143}Ce, ^{153}Sm, ^{86}Re, ^{198}Au, ^{206}**Bi** ^{22}Rn+D P.	^{48}**V**, ^{56}Co, ^{58}**Co**, ^{74}**As**, ^{83}**Rb**, ^{84}Rb, ^{85}**Sr**/^{85m}**Rb**, ^{103}Ru, ^{105}Ag, ^{114m}In, ^{124}**Sb**, ^{126}**I**, ^{140}Ba, ^{147}Nd, ^{185}**Os**, ^{192}**Ir** ^{223}Ra	^{57}Co, ^{65}Zn, ^{88}Y, ^{106}**Ru**/^{106} **Rh**, ^{110m}**Ag**, ^{144}Ce/^{144} Pr	^{22}**Na**, ^{26}**Al**, ^{85}Kr, ^{125}Sb, ^{134}**Cs**, ^{137}**Cs**/^{137m}**Ba**, ^{154}Eu, ^{207}**Bi**, ^{208}Po, ^{226}Ra+D. P., ^{227}Ac+D. P., ^{228}Th+D. P.,
^{47}Ca, ^{52}**Mn**, ^{72}**As**, ^{77}Br, ^{82}**Br** ^{99}Mo, ^{24}I, ^{131}I, ^{140}La, ^{43}Ce, ^{151}Pm, ^{186}Re, ^{6}**Bi**	^{46}**Sc**, ^{48}**V**, ^{56}**Co**, ^{58}**Co**, ^{83}Rb, ^{84}**Rb**, ^{89}Sr, ^{95}**Nb**, ^{95}**Zr**, ^{114m}In, ^{115m}Cd, ^{124}Sb, ^{126}I, ^{129m}/^{129} Te, ^{140}Ba/^{140}La, ^{160}Tb, ^{185}Os	^{54}**Mn**, ^{88}**Y**, ^{110m}**Ag**, ^{182}Ta, ^{210}Po	^{134}**Cs**, ^{152}Eu, ^{154}Eu, ^{207}Bi, ^{227}Ac+D. P., ^{228}Ra/^{228}Ac, ^{228}Th+D. P., ^{238}U/^{234m}Pa
^{47}**Ca**, ^{52}**Mn** ^{76}**As**, ^{82}**Br**, ^{122}Sb, ^{124}I, ^{32}Cs, ^{143}Ce, ^{166}Ho, ^{98}Au, ^{206}Bi, ^{2}Rn+D.P.	^{46}**Sc**, ^{48}**V**, ^{56}**Co**, ^{59}**Fe**, ^{84}Rb, ^{86}**Rb**, ^{91}Y, ^{114m}/^{114}In, ^{115m}Cd, ^{124}Sb, ^{126}I, ^{129m}/^{129}Te, ^{131}Ba, ^{160}Tb	^{65}**Zn**, ^{110m}Ag, ^{123m}Sn, ^{144}Ce/^{144}Pr, ^{182}Ta	^{22}**Na**, ^{26}Al, ^{40}K ^{60}**Co**, ^{134}Cs, ^{152}Eu, ^{154}**Eu**, ^{207}Bi/^{207}Pb, ^{226}Ra+D. P., ^{228}Th+D. P.
^{6}As, ^{124}I, ^{140}**La** ^{6}Ho, ^{206}**Bi**, ^{2}Rn+D.P.	^{56}Co, ^{58}Co, ^{84}Rb, ^{124}**Sb**, ^{140}**Ba**/^{140}**La**	^{88}**Y**, ^{110m}Ag	^{26}**Al**, ^{154}Eu, ^{207}Bi, ^{226}Ra+D. P., ^{228}Ra/^{228}Ac, ^{228}Th+D. P.
^{6}As, ^{124}I, ^{140}La, ^{2}Rn+D.P.	^{48}V, ^{56}Co, ^{124}Sb, ^{140}Ba/^{140}La	^{88}Y, ^{106}Ru/^{106}Rh, ^{144}Ce/^{144}Pr	^{207}Bi, ^{226}Ra+D.P., ^{228}Th+D. P.
	^{56}Co		

Table 3.7 NUCLIDES FOR ALPHA SOURCES

Nuclide	*Half-life*	*α-energies* MeV	*Associated β and γ radiation* MeV
Americium-241	458 yr	5.44, 5.48	γ_{max} 0.060
Lead-210 (+daughters)	22 yr	5.305	β_{max} 1.17
			γ_{max} 0.8 (very weak)
Plutonium-238	86 yr	5.352, 5.452, 5.495	γ 0.043 5 (very weak)
Plutonium-239	2.44×10^4 yr	5.096, 5.134, 5.147	γ_{max} 0.051 (weak)
Polonium-210	138.4 d	5.305	γ 0.8 (very weak)
Radium-226 (+daughters)	1 620 yr	4.589–7.68	β_{max} 3.26 γ_{max} 2.43

Table 3.8 NUCLIDES FOR BETA SOURCES

Nuclide	*Half-life*	β_{max} MeV	*Associated α and γ radiation* MeV
Carbon-14	5 730 yr	0.159	—
Cerium-144 + praseodymium-144	285 d	2.98	γ 0.034–2.18
Iron-55	2.7 yr	0.0052	γ 0.0059–0.0065
Krypton-85	10.6 yr	0.67	γ 0.51
Lead-210 + bismuth-210	22 yr	1·17	α 5.305
			γ_{max} 0.8 (very weak)
Nickel-63	100 yr	0.066	—
Promethium-147	2.6 yr	0.225	—
Ruthenium-106 + rhodium-106	1.0 yr	3.6	γ 0.51–2.9
Strontium-90 + yttrium-90	28 yr	2.27	—
Thallium-204	3.76 yr	0.77	—
Tritium	12.26 yr	0.018	—
Yttrium-90	64.2 h	2.27	—

Table 3.9 NUCLIDES FOR NEUTRON SOURCES–POLONIUM 210 (ALPHA, N) SOURCES WITH VARIOUS TARGETS

Target	Neutrons s^{-1} $curie^{-1}$	*Neutron energy* (MeV) *Mean*	*Maximum*
Aluminium	0.02×10^6	—	2.7
Beryllium	2.5×10^6	4.3	10.8
Boron	0.2×10^6	—	5.0
Fluorine-19	0.1×10^6	1.4	2.8
Lithium	0.05×10^6	0.48	1.32
Magnesium	0.03×10^6	—	—
Oxygen-18	1.0×10^6	—	4.3
Sodium	0.04×10^6	—	—
Mock fission	0.04×10^6	1.6	10.8

Table 3.10 NUCLIDES FOR NEUTRON SOURCES–(GAMMA, N) SOURCES

Nuclide	*Half-life*	*Target*	*Observed* neutrons s^{-1} $curie^{-1}$	*Observed neutron energy* keV
Antimony-124	60 d	Beryllium	1.6×10^6	24.8
Radium-226 (+daughters)	1 620 yr	Beryllium	1.3×10^6	700 (max)
Radium-226 (+ daughters)	1 620 yr	Deuterium (heavy water)	—	120
Thorium-228 (+daughters)	1.91 yr	Beryllium	—	827
Thorium-228 (+daughters)	1.91 yr	Deuterium (heavy water)	1.2×10^6	197

Table 3.11 NUCLIDES FOR NEUTRON SOURCES–(ALPHA, N) SOURCES WITH BERYLLIUM TARGETS

Nuclide	*Half-life*	Neutrons s^{-1} curie^{-1}	*Gamma emission* mrad h^{-1} *at* 1 m *from* 10^6 neutrons s^{-1}
Actinium-227 (+daughters)	22 yr	1.8×10^7	8
Americium-241	458 yr	2.7×10^6	<0.1
Lead-210 (+daughters)	22 yr	2.3×10^6	8.8
Plutonium-239	2.44×10^4 yr	1.4×10^6	1.7
Polonium-210	138.4 d	2.5×10^6	0.04
Radium-226 (+daughters)	1 620 yr	1.3×10^7	60
Thorium-228 (+daughters)	1.91 yr	2.5×10^7	30

Table 3.12 SPONTANEOUS FISSION NEUTRON SOURCE

Nuclide	*Half-life*	Neutrons s^{-1} mg^{-1}	*Gamma emission* mrad h^{-1} *at* 1 m *from* 10^6 neutrons s^{-1}
Californium-252	2.65 yr (effective)	2.3×10^9	0.07

4 X-ray analysis of metallic materials

4.1 Introduction

X-rays are very short wavelength electromagnetic waves. The wavelengths' range (typically 0.5–2.5 Å, or 5.25 Nm, but much longer wavelength X-rays occur) is similar to the typical interatomic spacing of metallic materials, and this permits the identification and detailed structural analysis of crystalline materials by X-ray diffraction. The interaction of X-rays with the electronic structure of atoms permits elemental analysis and investigation of the bonding state of chemical compounds, the relevant techniques being known as X-ray fluorescence and X-ray photoelectron spectroscopy (XPS).

This chapter reviews the application of X-rays to the investigation of metallic materials, and gives reference data for those involved with the use of X-rays in metals industries. Reference to more complete information is given where appropriate.

4.2 Excitation of X-rays

X-rays are produced (excited) when an electron experiences a sudden acceleration or (more commonly in practice) deceleration as, for example, when a beam of fast-moving electrons strikes an atom. If an electron is completely stopped by the impact, and its energy completely converted into radiation, an X-ray quantum will be emitted having frequency ν and wavelength λ given by the quantum relationship

$$E = Ve = h\nu = \frac{hc}{\lambda}$$

where E = energy of electron, V = accelerating voltage, e = charge on electron, h = Planck's constant, c = velocity of light: which, by substitution of numerical values, reduces to

$$\lambda = \frac{12.34}{V}\,\text{Å}$$

where V is expressed in kilovolts.

In an X-ray tube, the energy of the electrons is converted immediately into radiation in only a few cases: most electrons undergo repeated collisions and dissipate the greater part of their energy as heat. In such cases the energy E available for radiation is reduced, and appears as X-rays of longer wavelength. The radiation is therefore not monochromatic, and the above equation gives the minimum wavelength occurring in the emergent beam, the so-called 'short wave cut-off'.

If the intensity of the radiation is plotted against the wavelength, for relatively low applied voltages a curve is obtained rising from zero to a more or less pronounced maximum, falling again to a low value with increasing wavelength. It has been found that for this curve of so-called 'white radiation' both the total radiation and the intensity of the peak are closely proportional to the atomic number of the target material. Thus if white radiation is required a target of one of the heavy metals should be selected, and since the total intensity (the area under the curve) is proportional to the square of the applied voltage the latter should be as high as possible.

On increasing the voltage beyond a certain limit, which is different for different targets, a line spectrum begins to appear, superimposed upon the continuous or white spectrum, the wavelengths of the lines being constant and characteristic of the target. The lowest critical voltage at which these lines appear is that which endows the bombarding electrons with just sufficient energy to eject electrons from one of the inner shells in the target atoms. Each vacancy as it occurs is filled by an electron jumping in from one of the outer shells, the jump being accompanied by an emission of energy of frequency ν given by the Bohr relationship

$$h\nu = W_1 - W_2$$

where $W_1 - W_2$ represents the change in potential energy of the system and h is Planck's constant.

The K series of lines is excited when electrons are ejected from the K shell and their places taken by electrons from the L, M or N shells. It consists of four lines $K\alpha_1$, $K\alpha_2$, $K\beta_1$ and $K\beta_2$, the α lines being associated with jumps from the L shell, β_1 with those from the M shell, and β_2 with those for the N shell. Other series of lines are emitted when the bombarding electrons eject electrons from the L, M or N shells. Of these series the K is the one used in determinations of crystal structure. K, L, M and N series lines are used in X-ray fluorescence methods (*see* below) to identify the atomic elements present in a sample. If V_0 is the critical voltage at which the K lines first appear and e the electronic mass, the amount of energy required to dislodge an electron from the K shell is $V_0 e$, and this will be associated with the emission of radiation of frequency $h\nu_0$ and of wavelength h/λ_0. This wavelength is termed the K absorption edge of the element.
The empirical expression

$$I = k(V - V_0)^{1.7}$$

where k is a constant, enables the intensity I of the characteristic spectrum to be calculated from a knowledge of the tube voltage V and the minimum excitation voltage V_0. The relative intensity of the line spectrum to that of the continuous spectrum is greatest if the tube is operated at about four times the critical voltage V_0.

One method of monochromatizing the line spectrum is to filter the radiation by means of a foil of suitable thickness made of an element having its K absorption edge between the $K\alpha$ and $K\beta$ lines emitted by the target. In this way the intensity of the $K\alpha$ radiation is increased relative to that of the $K\beta$ and shorter wavelengths, advantage being taken of the very pronounced difference in absorption occurring on either side of the absorption edge. Generally speaking, the element preceding the target material in the periodic system will act as a suitable filter.

Table 4.1 lists wavelengths, excitation potentials and β-filters. Table 4.2 gives excitation potentials for characteristic X-ray spectra. Table 4.3 gives K emission spectra and K absorption edges, while Table 4.4 gives L emission spectra and L_{111} absorption edges.

The two α lines cannot be separated by filtering. Where they are not resolved on a diffraction pattern an average wavelength should be used, giving α_1 double the weight of α_2, corresponding to its greater intensity.

Where the degree of monochromatism given by filtering is inadequate a method may be used involving the use of X-rays diffracted from a set of planes in a single crystal of some suitable compound, such as quartz, calcite, rock salt or pentaerythritol. By setting this crystal at the correct angle to the incident beam the $K\alpha$ wavelengths alone can be reflected, and a spectrum obtained free from any white radiation and giving extreme clarity of diffraction pattern. Focusing of the beam, resulting in greater intensity, can be obtained by hollow-grinding the crystal used as the reflector. Suitable crystals, with notes on their merits, are listed in Table 4.5.

4.2.1 X-ray wavelengths

The X-unit defined by Siegbahn and very nearly equal to 10^{-3} Å was based on the atomic spacing in calcite as determined from the expression:

$$2d = \left(\frac{4M}{\rho NV}\right)^{1/3}$$

where M = molecular weight of calcite, ρ = density of calcite, N = Avogadro's number, V = volume of calcite rhombohedron with unit distance between the faces considered.

kX units were widely used up to about 1945 but were then gradually replaced by absolute Ångströms as more precise methods for determining X-ray wavelengths were introduced. The nanometer (10^{-9} m) is now declared to be the standard unit for X-ray studies. However, in view of the much greater familiarity with Å, and the very simple conversion factor, in this edition the Ångström unit is used unless the contrary is stated.

1 kX unit = 1.002 02 Ångström units
1 Ångström unit, Å = 10^{-10} m or 10^{-1} nanometers

Table 4.1 WAVELENGTHS, EXCITATION POTENTIALS, AND β FILTERS FREQUENTLY USED IN CRYSTALLOGRAPHIC WORK

Target					*β-filter*				
							Mass absorption coefficient	*Material content*	
Element	*Atomic No.*	*Line*	*Wavelength** Å	*Excitation potential*† kV	*Element*	*Absorption edge*	μ/ρ	$\mathrm{g\,cm^{-2}}$	*Thickness* mm
Cr	24	$K\alpha_1$	2.289 62	5.98	V	2.267 5	77.3	0.009	0.016
		$K\alpha_2$	2.293 51						
		$K\alpha$	2.290 9						
		$K\beta_1$	2.084 80						
Mn	25	$K\alpha_1$	2.101 75	6.54	Cr	2.070 1	71	0.011	0.016
		$K\alpha_2$	2.105 69						
		$K\alpha$	2.103 1						
		$K\beta_1$	1.910 15						
Fe	26	$K\alpha_1$	1.935 97	7.1	Mn	1.895 4	61.9	0.012	0.016
		$K\alpha_2$	1.939 91						
		$K\alpha$	1.937 3						
		$K\beta_1$	1.756 53						
Co	27	$K\alpha_1$	1.788 92	7.71	Fe	1.742 9	58.6	0.014	0.018
		$K\alpha_2$	1.792 78						
		$K\alpha$	1.790 2						
		$K\beta_1$	1.620 75						
Ni	28	$K\alpha_1$	1.657 84	8.29	Co	1.607 2	51.6	0.015	0.018
		$K\alpha_2$	1.661 69						
		$K\alpha$	1.659 1						
		$K\beta_1$	1.500 10						
Cu	29	$K\alpha_1$	1.540 51	8.86	Ni	1.486 9	48.0	0.019	0.021
		$K\alpha_2$	1.544 33						
		$K\alpha$	1.541 8						
		$K\beta_1$	1.392 17						
W	74	$L\alpha_1$	1.476 35	12.1	Cu	1.380 2	42.0	0.019	0.021
		$L\alpha_2$	1.487 42						
Zn	30	$K\alpha_1$	1.435 11	9.65	Cu	1.380 2	42.0	0.019	0.021
		$K\alpha_2$	1.438 94						
		$K\alpha$	1.436 4						
		$K\beta_1$	1.295 22						
Au	79	$L\alpha_1$	1.276 39	14.4	Ga	1.195 7	37.0	0.028	0.047
		$L\alpha_2$	1.287 77						
		$K\alpha_2$	0.785 88						
		$K\alpha_2$	0.790 10						
		$K\alpha$	0.787 29	80.5	Sr	0.769 69	18.1	0.053	0.210
Zr	40	$K\beta$	0.701 70						
Mo	42	$K\alpha_1$	0.709 26	20	Zr	0.688 8	17.2	0.069	0.108
		$K\alpha_2$	0.713 54						
		$K\alpha$	0.710 7						
		$K\beta_1$	0.632 25						
Rh	45	$K\alpha_1$	0.613 25	23.2	Ru	0.559 5	15.4	0.077	0.064
		$K\alpha_2$	0.617 61						
		$K\alpha$	0.614 7						
		$K\beta_1$	0.545 59						
Pd	46	$K\alpha_1$	0.585 42	24.4	Rh	0.534 1	14.6	0.091	0.073
		$K\alpha_2$	0.589 80		or				
		$K\alpha$	0.586 9		Ru	0.559 5			
		$K\beta_1$	0.520 52						
Ag	47	$K\alpha_1$	0.559 36	25.5	Pd	0.509 0	13.1	0.096	0.079
		$K\alpha_2$	0.563 78		or				
		$K\alpha$	0.560 9		Rh	0.534 1			
		$K\beta_1$	0.497 01						

* $\lambda K\alpha$ is here defined as $(2\lambda K\alpha_1 + \lambda K\alpha_2)/3$.

† The optimum voltage for operating a tube with raw alternating currents is approximately 5 times the excitation potential, and 4 times the excitation potential with fully smoothed direct current, but normally 80 kV cannot be exceeded owing to the danger of electrical breakdown.

Table 4.2 EXCITATION POTENTIALS IN kV FOR CHARACTERISTIC X-RAY SPECTRA

Atomic No. and element	K	L	M	N	*Atomic No. and element*	K	L	M	N
11 Na	1.07	—	—	—	49 In	27.80	4.21	0.83	0.12
12 Mg	1.30	—	—	—	50 Sn	29.06	4.42	0.88	0.13
13 Al	1.55	—	—	—	51 Sb	30.35	4.69	0.94	0.15
14 Si	1.83	—	—	—	52 Te	31.66	4.93	1.01	0.17
15 P	2.13	—	—	—	53 I	33.01	5.16	1.08	0.19
16 S	2.46	—	—	—	55 Cs	35.80	5.68	1.21	0.23
17 Cl	2.81	0.24	—	—	56 Ba	37.24	5.92	1.29	0.25
19 K	3.69	0.34	—	—	57 La	38.75	6.24	1.36	0.27
20 Ca	4.02	0.40	—	—	58 Ce	40.27	6.53	1.43	0.29
21 Sc	4.48	0.46	—	—	59 Pr	41.81	6.81	1.51	0.30
22 Ti	4.94	0.53	—	—	60 Nd	43.37	7.10	1.58	0.32
23 V	5.44	0.60	—	—	62 Sm	46.63	7.71	1.72	0.35
24 Cr	5.96	0.68	—	—	63 Eu	48.29	8.02	1.80	0.36
25 Mn	6.51	0.76	—	—	64 Gd	50.00	8.35	1.88	0.38
26 Fe	7.08	0.85	—	—	65 Tb	51.76	8.67	1.96	0.40
27 Co	7.67	0.92	—	—	66 Dy	53.55	9.01	2.04	0.42
28 Ni	8.29	1.01	—	—	67 Ho	55.36	9.35	2.13	0.43
29 Cu	8.94	1.10	—	—	68 Er	57.22	9.71	2.22	0.45
30 Zn	9.62	1.19	—	—	69 Tm	59.07	10.06	2.31	0.47
31 Ga	10.32	1.29	—	—	70 Yb	61.02	10.45	2.41	0.50
32 Ge	11.05	1.41	—	—	71 Lu	63.01	10.82	2.50	0.51
33 As	11.81	1.52	—	—	72 Hf	65.01	11.23	2.60	0.54
34 Se	12.59	1.65	—	—	73 Ta	67.09	11.63	2.69	0.57
35 Br	13.41	1.78	—	—	74 W	69.18	12.04	2.80	0.59
37 Rb	15.13	2.05	—	—	75 Re	71.28	12.46	2.91	—
38 Sr	16.03	2.20	—	—	76 Os	73.54	12.91	3.03	0.64
39 Y	16.96	2.38	—	—	77 Ir	75.77	13.35	3.15	0.67
40 Zr	17.92	2.52	0.43	0.05	78 Pt	78.02	13.80	3.28	0.71
41 Nb	18.90	2.69	0.48	0.05	79 Au	80.42	14.29	3.41	0.79
42 Mo	19.91	2.86	0.51	0.06	80 Hg	82.69	14.77	3.55	0.82
44 Ru	22.02	3.21	0.59	0.06	81 Tl	85.28	15.27	3.69	0.86
45 Rh	23.12	3.39	0.62	0.07	82 Pb	87.66	15.79	3.84	0.89
46 Pd	24.23	3.60	0.67	0.08	83 Bi	90.03	16.31	3.99	0.96
47 Ag	25.40	3.81	0.72	0.10	90 Th	109.27	20.36	5.17	1.33
48 Cd	26.59	4.00	0.77	0.11	92 U	115.54	21.66	5.54	1.44

Table 4.3 K EMISSION LINES AND K ABSORPTION EDGES IN Å

Line transition intensity rel. to Kα_1	α_2 KL_{II} 50	α_1 KL_{III} 100	β_3 KM_{II} 15	β_1 KM_{III} 15	β_2 $KN_{II,III}$ 5	K *Absorption edge*
3 Li	228.0		—	—	—	226.5
4 Be	114.0		—	—	—	—
5 B	67.6		—	—	—	—
6 C	44.7		—	—	—	43.7
7 N	31.6		—	—	—	31.0
8 O	23.62		—	—	—	23.3
9 F	18.32		—	—	—	—
10 Ne	14.61		—	14.452	—	—
11 Na	11.910 1		—	11.575	—	—
12 Mg	9.890 0		—	9.521	—	9.511 7
13 Al	8.341 73	8.339 34	—	7.960	—	7.951 1
14 Si	7.127 91	7.125 42	—	6.753	—	6.744 6
15 P	6.160	6.157	—	5.796	—	5.786 6
16 S	5.374 96	5.372 16	—	5.032	—	5.018 2
17 Cl	4.730 7	4.727 8	—	4.403 4	—	4.396 9
18 A	4.194 74	4.191 80	—	3.886 0	—	3.870 7
19 K	3.744 5	3.741 4	—	3.453 9	—	3.436 45

Table 4.3 K EMISSION LINES AND K ABSORPTION EDGES IN Å—*continued*

Line transition intensity rel. to Kα_1	α_2 KL$_{11}$ 50	α_1 KL$_{111}$ 100	β_3 KM$_{11}$ 15	β_1 KM$_{111}$ 15	β_2 KN$_{11,111}$ 5	K *Absorption edge*
20 Ca	3.361 66	3.358 39	—	3.089 7	—	3.070 16
21 Sc	3.034 2	3.030 9	—	2.779 6	—	2.757 3
22 Ti	2.752 16	2.748 51	—	2.513 91	—	2.497 3
23 V	2.507 38	2.503 56	—	2.284 40	—	2.269 0
24 Cr	2.293 61	2.289 70	—	2.084 87	—	2.070 1
25 Mn	2.105 78	2.101 82	—	1.910 21	—	1.896 4
26 Fe	1.939 98	1.936 04	—	1.756 61	1.744 2	1.743 3
27 Co	1,792 85	1.788 97	—	1.620 79	1.608 9	1.608 1
28 Ni	1.661 75	1.657 91	—	1.500 14	1.488 6	1.488 0
29 Cu	1.544 39	1.540 56	1.392 6	1.392 22	1.381 09	1.380 4
30 Zn	1.439 00	1.435 16	—	1.295 25	1.283 72	1.283 3
31 Ga	1.343 99	1.340 08	1.208 35	1.207 89	1.196 00	1.195 7
32 Ge	1.258 01	1.254 05	1.129 36	1.128 94	1.116 86	1.116 5
33 As	1.179 87	1.175 88	1.057 83	1.057 30	1.045 00	1.045 0
34 Se	1.108 82	1.104 77	0.992 68	0.992 18	0.979 92	0.979 78
35 Br	1.043 82	1.039 74	0.933 27	0.932 79	0.920 46	0.919 95
36 Kr	0.984 1	0.980 1	0.879 0	0.878 5	0.866 1	0.865 47
37 Rb	0.929 69	0.925 553	0.829 21	0.828 68	0.816 45	0.815 49
38 Sr	0.879 43	0.875 26	0.783 45	0.782 92	0.770 81	0.769 69
39 Y	0.833 05	0.828 84	0.741 26	0.740 72	0.728 64	0.727 62
40 Zr	0.790 15	0.785 93	0.702 28	0.701 73	0.689 93	0.688 77
41 Nb	0.750 44	0.746 20	0.666 34	0.665 76	0.654 16	0.652 91
42 Mo	0.713 59	0.709 30	0.632 87	0.632 29	0.620 99	0.619 77
43 Tc	0.679 32^{+}	0.675 02^{+}	0.601 88^{+}	0.601 30^{+}	0.590 24^{+}	(0.589 1)
44 Ru	0.647 41	0.643 08	0.573 07	0.572 48	0.561 66	0.560 05
45 Rh	0.617 63	0.613 28	0.546 20	0.545 61	0.535 03	0.533 8
46 Pd	0.589 82	0.585 45	0.521 12	0.520 52	0.510 23	0.509 2
47 Ag	0.563 80	0.559 41	0.497 69	0.487 07	0.487 03	0.485 8
48 Cd	0.539 42	0.535 01	0.475 73	0.475 10	0.465 33	0.464 09
49 In	0.516 54	0.512 11	0.455 18	0.454 54	0.445 00	0.443 88
50 Sn	0.435 24	0.425 92	0.424 67	0.431 84	0.431 75	0.424 68
51 Sb	0.474 83	0.470 35	0.417 74	0.417 09	0.407 97	0.406 63
52 Te	0.455 78	0.451 30	0.400 66	0.399 99	0.391 10	0.389 72
53 I	0.437 83	0.433 32	0.384 56	0.383 91	0.375 23^{+}	0.373 79
54 Xe	0.420 87^{+}	0.416 34^{+}	0.369 41^{+}	0.368 72^{+}	0.360 26^{+}	0.358 49
55 Cs	0.404 84	0.400 29	0.355 05	0.354 36	0.346 11	0.344 74
56 Ba	0.389 67	0.385 11	0.341 51	0.340 81	0.332 77	0.331 37
57 La	0.375 31	0.370 74	0.328 69	0.327 98	0.320 12	0.318 42
58 Ce	0.361 68	0.357 09	0.316 52	0,315 82	0.308 16	0.306 47
59 Pr	0.348 75	0.344 14	0.304 98	0.304 26	0.296 79	0.295 16
60 Nd	0.336 47	0.331 85	0.294 03	0.293 30	0.286 1^{+}	0.284 51
61 Pm	0.324 80	0.320 16	0.283 63^{+}	0.282 90^{+}	0.275 9^{+}	0.274 3
62 Sm	0.313 70	0.309 04	0.273 76	0.273 01	0.266 2	0.264 62
63 Eu	0.303 12	0.298 45	0.264 33	0.263 58	0.256 92	0.255 52
64 Gd	0.293 04	0.288 35	0.255 34	0.254 60	0.248 16	0.246 80
65 Tb	0.283 42	0.278 72	0.246 83	0.246 08	0.239 7^{+}	0.238 40
66 Dy	0.274 25	0.269 53	0.238 62	0.237 88	0.231 7^{+}	0.230 46
67 Ho	0.265 486	0.260 76	0.230 83	0.230 12	0.224 1^{+}	0.222 90
68 Er	0.257 11	0.252 37	0.223 41	0.222 66	0.216 7^{+}	0.215 66
69 Tm	0.249 10	0.244 34	0.216 36	0.215 56	0.209 8^{+}	0.208 9
70 Yb	0.241 42	0.236 66	0.209 6^{+}	0.208 84	0.203 3^{+}	0.202 23
71 Lu	0.234 08	0.229 30	0.203 09^{+}	0.202 31^{+}	0.196 9^{+}	0.195 84
72 Hf	0.227 02	0.222 22	0.196 86^{+}	0.196 07	0.190 8^{+}	0.189 81
73 Ta	0.220 31	0.215 50	0.190 89	0.190 09	0.185 10	0.183 93
74 W	0.213 83	0.209 01	0.185 18	0.184 37	0.179 51	0.178 37
75 Re	0.207 61	0.202 78	0.179 70	0.178 88	0.174 15	0.173 11
76 Os	0.201 64	0.196 79	0.174 43	0.173 61	0.169 90	0.167 80
77 Ir	0.195 90	0.191 05	0.169 37	0.168 54	0.164 05	0.162 86
78 Pt	0.190 38	0.185 51	0.164 50	0.163 68	0.159 29	0.158 16
79 Au	0.185 08	0.180 20	0.159 81	0.158 98	0.154 72	0.153 44
80 Hg	0.179 96	0.175 07	0.155 32	0.154 49	0.150 30	0.149 23
81 Tl	0.175 04	0.170 14	0.150 98	0.150 14	0.146 04	0.144 70

Table 4.3 K EMISSION LINES AND K ABSORPTION EDGES IN Å—*continued*

Line transition intensity rel. to Kα_1	α_2 KL_{11} 50	α_1 KL_{111} 100	β_3 KM_{11} 15	β_1 KM_{111} 15	β_2 $KN_{11,111}$ 5	K *Absorption edge*
82 Pb	0.170 29	0.165 38	0.146 81	0.145 97	0.142 01	0.140 77
83 Bi	0.165 72	0.160 79	0.142 78	0.141 95	0.138 07	0.137 06
84 Po	0.161 30⁺	0.156 36⁺	0.138 92⁺	0.138 07⁺	0.134 28⁺	—
85 At	0.157 05⁺	0.152 10⁺	0.135 17⁺	0.134 32⁺	0.130 67⁺	—
86 Rn	0.152 94⁺	0.147 98⁺	0.131 55⁺	0.130 69⁺	0.127 08⁺	—
87 Fr	0.148 96⁺	0.143 99⁺	0.128 07⁺	0.127 19⁺	0.123 68	—
88 Ra	0.145 12⁺	0.140 14⁺	0.124 69⁺	0.123 82⁺	0.120 39	—
89 Ac	0.141 41⁺	0.136 42⁺	0.121 43⁺	0.120 55⁺	0.117 21⁺	—
90 Th	0.137 83	0.132 81	0.118 27	0.117 40	0.114 15⁺	0.112 93
91 Pa	0.134 34⁺	0.129 33⁺	0.115 23⁺	0.114 35⁺	0.111 18⁺	—
92 U	0.130 97	0.125 95	0.112 30	0.111 39	0.108 27⁺	0.106 80

⁺ Interpolated values.

Table 4.4 L EMISSION SPECTRA AND ABSORPTION EDGES IN Å

Line transition intensity rel. to α_1	l Liii Mi 30	η Lii Mi 10	α_2 Liii Miv 10	α_1 Liii Mv 100	β_1 Lii Miv 80	β_4 Li Mii 20	β_3 Li Miii 30	β_2 Liii Nv 60	β_5 Lii Oiv, v 60	γ_1 Lii Niv 40	*Absorption edges —* Li	Lii	Liii
17 Cl	67.90	67.33	—	—	—	—	—	—	—	—	52.084	61.366	61.672
18 A	56.30⁺	55.9⁺	—	—	—	—	—	—	—	—	43.192	50.390	50.803
19 K	47.74	47.24	—	—	—	—	—	—	—	—	36.352	42.020	42.452
20 Ca	40.96	40.46		36.33	—	—	—	—	—		31.068	35.417	35.827
21 Sc	35.59	35.13		31.35	31.02	—	—	—	—	—	26.831	30.161	30.457
22 Ti	31.36	30.89		27.42	27.05	—	—	—	—	—	23.389	26.831	27.184
23 V	27.77	27.34		24.25	23.88	—	—	—	—	—	20.523	23,702	24,070
24 Cr	24.78	24.30		21.64	21.27	—	—	—	—	—	18.256	21.226	21.596
25 Mn	22.29	21.85		19.45	19.11	—	—	—	—		16.268	18.896	19.248
26 Fe	20.15	19.75		17.59	17.26	—	—	—	—	—	14.601	17.169	17.484
27 Co	18.292	17.87		15.972	15,666	—	—	—	—	—	13.343	15.534	15.831
28 Ni	16.693	16.27		14.561	14,271	—	—	—	—	—	12,267	14,135	14,448
29 Cu	15.286	14.90		13.336	13.053	—	—	—	—	—	11.269	12.994	13.258
30 Zn	14.02	13.68		12.254	11,983	—	—	—	—		10.330	11.840	12.106
31 Ga	12.953	12.597		11.292	11.023	—	—	—	—	—	9.535	10.613	10.855
32 Ge	11.965	11.609		10.436 1	10.175	—	—	—	—	—	8.729	9.965	10.228
33 As	11.072	10.734		9.670 9	9.414 1	—	—	—	—	—	8.107	9.128 1	9.376 7
34 Se	10.294	9.962		8.990 0	8.735 8	—	—	—	—	—	7.467	8.421 2	8.662 4
35 Br	9.585	9.255		8.374 6	8.125 1	—	—	—	—		6,925	7.752 3	7.987 1
36 Kr	—	—		7.817⁺	7.576⁺	—	—	—	—	—	6.456	7.165 3	7.422 7
37 Rb	8.363 6	8.041 5	7.325 1	7.318 3	7.075 9	6.820 7	6.787 6	—	—	—	6.006	6.653 8	6.875 2
38 Sr	7.836 2	7.517 1	6.869 7	6.862 8	6.623 9	6.402 6	6.367 2	—	—	—	5.604	6.185 6	6.399 6
39 Y	7.356 3	7.040 6	6.455 8	6.448 8	6.212 0	6.018 6	5.983 2	—	—	—	5.193 1	5.709 8	5.914 1
40 Zr	6.918 5	6.606 9	6.077 8	6.070 5	5.836 0	5.668 1	5.633 0	5.586 3	—	5.384 3	4.893 8	5.370 9	5.573 7
41 Nb	6.517 6	6.210 9	5.731 9	5.724 3	5.492 3	5.345 5	5.310 2	5.237 9	—	5.036 1	4,591 1	5.024 7	5.226 0
42 Mo	6.150 8	5.847 5	5.414 4	5.406 6	5.177 1	5.048 8	5.013 3	4.923 2	—	4.725 8	4.320 7	4.713 3	4.909 3
43 Tc	—	—	—	5.114 8⁺	4.887 3⁺	—	—	—	—	—	4.064 3	4.427 1	4.625 4
44 Ru	5.503 5	5.205 0	4.853 8	4.845 8	4.620 6	4.523 0	4.486 6	4.371 8	—	4.182 2	3.841 3	4.176 5	4.366 3
45 Rh	5.216 9	4.921 7	4.605 5	4.597 4	4.374 1	4.288 8	4.252 2	4.131 0	—	3.943 7	3.641 6	3.949 0	4.138 9
46 Pd	4.952 5	4.660 5	4.375 9	4.367 7	3.990 2	4.071 1	4.034 6	3.908 9	—	3.724 6	3.430 0	3.713 6	3.896 9

Table 4.4 L EMISSION SPECTRA AND ABSORPTION EDGES IN Å—*continued*

Line transition intensity rel. to α_1	l Liii Mi 30	η Lii Mi 10	α_2 Liii Miv 10	α_1 Liii Mv 100	β_1 Lii Miv 80	β_4 Li Mii 20	β_3 Li Miii 30	β_2 Liii Nv 60	β_5 Liii Oiv, v 60	γ_1 Lii Niv 40	L *Absorption edges* Li	Lii	Liii
47 Ag	4.7076	4.4183	4.1629	4.1544	3.9347	3.8702	3.8331	3.7034	—	3.5226	3.2382	3.4948	3.6729
48 Cd	4.4801	4,1932	3.9650	3.9564	3.7382	3.6820	3.6450	3.5141	—	3.3356	3.0843	3.3224	3.5007
49 In	4.2687	3.9833	3.7807	3.7719	3.5553	3.5070	3.4698	3.3384	—	3.1621	2.9333	3.1550	3.3215
50 Sn	4.0717	3.7888	3.6089	3.5999	3.3849	3.3434	3.3059	3.1751	—	3.0012	2.7888	2.9949	3.1695
51 Sb	3.8883	3.6077	3.4484	3.4394	3.2257	3.1901	3.1526	3.0234	—	2.8516	2.6330	2.8230	2.9964
52 Te	3.7170	3.4383	3.2985	3.2892	3.0768	3.0466	3.0089	2.8822	—	2.7124	2.5027	2.6825	2.8516
53 I	3.5575	3.2798	3.1579	3.1486	2.9374	2.9121	2.8743	2.7505	—	2.5824	2.3898	2.5532	2.7190
54 Xe	—	—	—	3.0166$^+$	—	—	—	—	—	—	2.2745	2.4306	2.5933
55 Cs	3.2670	2.9932	2.9020	2.8924	2.6837	2.6666	2.6285	2.5118	—	2.3480	2.1725	2.3127	2.4723
56 Ba	3.1355	2.8627	2.7855	2.7760	2.5682	2.5553	2.5164	2.4044	—	2.2415	2.0834	2.2022	2.3611
57 La	3.006	2.740	2.6753	2.6657	2.4589	2.4493	2.4105	2.3030	—	2.1418	1.9789	2.1003	2.2579
58 Ce	2.8917	2.6203	2.5706	2.5615	2.3561	2.3497	2.3109	2.2087	—	2.0487	1.8908	2.0094	2.1641
59 Pr	2.7841	2.512	2.4729	2.4630	2.2588	2.2550	2.2172	2.1194	—	1.9611	1.8131	1.9231	2.0771
60 Nd	2.6760	2.4094	2.3807	2.3704	2.1669	2.1669	2.1268	2.0360	—	1.8779	1.7376	1.8424	1.9945
61 Pm	—	—	2.2926	2.2822	2.0797	—	2.0421	1.9559	—	1.7989	1.6684	1.7658	1.9189
62 Sm	2.4823	2.2182	2.2106	2.1998	1.9981	2.0010	1.9624	1.8822	1.8470	1.7272	1.6011	1.6944	1.8446
63 Eu	2.3948	2.1315	2.1315	2.1209	1.9203	1.9255	1.8867	1.8118	1.7772	1.6574	1.5382	1.6259	1.7749
64 Gd	2.3122	2.0494	2.0578	2.0468	1.8468	1.8540	1.8150	1.7455	1.7130	1.5924	1.4787	1.5608	1.7096
65 Tb	2.2352	1.9730	1.9875	1.9765	1.7768	1.7864	1.7472	1.6830	1.6510	1.5303	1.4227	1.5011	1.6484
66 Dy	2.1589	1.8974	1.9199	1.9088	1.7106	1.7210	1.6822	1.6237	1.5884	1.4727	1.3693	1.4436	1.5903
67 Ho	2.0860	1.8264	1.8561	1.8450	1.6475	1.6595	1.6203	1.5671	1.5378	1.4174	1.3194	1.3900	1.5353
68 Er	2.015	1.7566	1.7955	1.7843	1.5873	1.6007	1.5616	1.5140	1.4848	1.3641	1.2709	1.3372	1.4824
69 Tm	1.9550	1.6963	1.7381	1.7268$^+$	1.5304	1.5448	1.5063	1.4640	1.4349	1.3153	1.2271	1.2886	1.4314
70 Yb	1.8942	1.6356	1.6829	1.6719	1.4757	1.4914	1.4523	1.4155	1.3870	1.2677	1.1814	1.2415	1.3852
71 Lu	1.8360	1.5779	1.6303	1.6195	1.4236	1.4406	1.4014	1.3701	1.3418	1.2223	1.1401	1.1975	1.3404
72 Hf	1.7815	1.5233	1.5805	1.5696	1.3741	1.3922	1.3530	1.3264	1.2976	1.1790	1.0986	1.1531	1.2957
73 Ta	1.7284	1.4711	1.5329	1.5220	1.3270	1.3458	1.3068	1.2845	1.2555	1.1379	1.0608	1.1126	1.2543
74 W	1.6782	1.4211	1.4874	1.4764	1.2818	1.3016	1.2627	1.2446	1.2155	1.0986	1.0250	1.0744	1.2153
75 Re	1.6306	1.3734	1.4440	1.4329	1.2386	1.2592	1.2203	1.2066	1.1772	1.0610	0.9901	1.0365	1.1772
76 Os	1.5850	1.3279	1.4023	1.3912	1.1973	1.2184	1.1796	1.1698	1.1405	1.0250	0.9557	1.0013	1.1414

77 Ir	1.540 9	1.284 5	1.362 5	1.351 3	1.157 8	1.179 6	1.140 9	1.135 3	1.105 9	0.990 9	0.924 25	0.967 00	1.106 0
78 Pt	1.499 5	1.242 9	1.324 3	1.313 0	1.119 9	1.142 2	1.103 9	1.102 0	1.072 4	0.958 0	0.894 05	0.934 84	1.073 1
79 Au	1.459 6	1.202 7	1.287 7	1.276 4	1.083 5	1.106 5	1.067 9	1.070 2	1.040 4	0.926 5	0.863 78	0.902 77	1.040 3
80 Hg	1.421 6	1.164	1.252 6	1.241 2	1.048 7	1.072 2	1.033 6	1.039 8	1.009 9	0.896 5	0.835 31	0.877 90	1.009 4
81 Tl	1.384 8	1.127 7	1.218 8	1.207 4	1.015 1	1.039 2	1.000 6	1.010 3	0.980 6	0.867 5	0.807 87	0.843 55	0.979 68
82 Pb	1.349 9	1.092 4	1.186 5	1.175 0	0.982 9	1.007 5	0.969 1	0.982 2	0.952 6	0.839 7	0.781 53	0.815 52	0.951 12
83 Bi	1.316 1	1.058 6	1.155 4	1.143 9	0.952 0	0.976 9	0.938 6	0.955 2	0.925 6	0.813 1	0.756 49	0.789 10	0.924 59
84 Po	1.282 9	—	1.125 5^{+}	1.113 9	0.922 0	0.947 5	0.909 1	0.929 4	0.899 6	0.787 5	0.732 19	0.763 77	0.897 61
85 At	—	—	1.096 7^{+}	1.085 0^{+}	0.893 5^{+}	—	0.881 4^{+}	—	—	0.762 9	0.709 15	0.738 73	0.872 34
86 Rn	—	—	1.069 0^{+}	1.057 2^{+}	0.866 1^{+}	—	0.854 4^{+}	—	—	0.739 3^{+}	0.686 75	0.715 29	0.848 45
87 Fr	—	—	1.042 3	1.030 5	0.839 4	—	0.827 9^{+}	0.858	—	0.716 5^{+}	0.665 37	0.692 90	0.825 29
88 Ra	1.167 2	0.907 4	1.016 6	1.004 7	0.813 8	0.840 7	0.802 7	0.835 4	0.806 3	0.694 6	0.644 61	0.671 14	0.802 84
89 Ac	—	—	0.991 8^{+}	0.979 9^{+}	0.789 0^{+}	—	0.778 2^{+}	—	—	0.673 5^{+}	0.624 8	0.650 0	0.781 6
90 Th	1.115 1	0.854 5	0.967 9	0.956 0	0.765 2	0.792 6	0.754 8	0.793 5	0.764 7	0.653 1	0.606 1	0.630 1	0.761 5
91 Pa	1.090 8	0.829 5	0.944 8^{+}	0.932 8	0.742 3	0.769 9	0.732 3	0.773 7	0.745 2	0.633 6^{+}	0.587 5	0.610 6	0.741 4
92 U	1.067 1	0.805 1	0.922 6	0.910 6	0.720 0	0.748 0	0.710 3	0.754 7	0.726 3	0.614 8	0.569 7	0.591 9	0.722 3
93 Np	1.042 8	0.780 9	0.901 1	0.889 1	0.698 5	0.726 7	0.689 2	0.736 2	0.708 1	0.596 5	0.553 1	0.574 2	0.703 9
94 Pu	1.022 6	0.759 1	0.880 3	0.868 3	0.677 7	0.706 2	0.668 7	0.718 5	0.690 7	0.578 9	0.536 6	0.557 1	0.686 4

$^{+}$Interpolated values.

Table 4.5 CRYSTALS FOR PRODUCING MONOCHROMATIC X-RAYS

			Properties of reflection		Properties of crystal			
Crystal	*Reflection*	*Spacing* Å	*Peak intensity*	*Breadth*	*Crystal imperfection*	*Stability*	*Mechanical properties*	*Special uses*
β alumina	0002 0004	11.24 5.62	Weak Weak–medium	Moderate	Great	Perfect	Hard, brittle	For long wavelengths, but usable crystals hard to obtain
Mica	001 004	10.1 2.53	Weak	Small	Negligible for selected specimens	Fair	Flexible, easily cleaved	For point-focusing devices; exhibits irradiation effects
Gypsum	020	7.60	Medium–strong	Very small	Good specimens hard to find	Poor	Soft, flexible	For small-angle scattering; focusing long wavelengths
Pentaerythritol	002	4.40	Very strong	Moderate	Great	Poor	Soft, easily deformed	General purposes; exhibits irradiation effects
Quartz	1011	3.35	Weak–medium	Very small	Negligible	Perfect	Can be elastically bent	For small-angle scattering; focusing
Potassium bromide	200	3.29	Medium–strong	Moderate	Negligible	Slightly deliquescent	—	—
Fluorite	111 220	3.16 1,94	Medium–strong Very strong	Moderate	Small	Perfect	Moderately hard	For eliminating harmonics; general purposes; short wavelengths
Urea nitrate	002	3.14	Strong	Very large	Very great	Very poor	Very easily deformed	For large specimens; soon decays
Calcite	200	3.04	Medium	Small	Negligible	Perfect	Moderately soft	For small-angle scattering; isolation of α_1 or α_2
Rock salt	200	2.82	Medium–strong	Large	Great	Slightly deliquescent	Can be plastically bent in warm supersaturated saline	For focusing
Aluminium	111	2.33	Very strong	Moderate to large	—	Good	Soft, can be seeded and grown to shape, then plastically shaped at room temperature	For focusing; diffuse scattering
Diamond	111	2.05	Weak	Very small	Negligible	Perfect	Very hard	For eliminating harmonics
Lithium fluoride	200	2.01	Very strong	Small-moderate	Negligible	Perfect	Hard, can be plastically bent at high temperature	For focusing; diffuse scattering; general purposes
Graphite	002	3.35	Very strong	Small	Negligible	Perfect	Easily shaped	—

4.3 X-ray techniques

The techniques, involving X-rays, which are commonly employed in the investigation of metallic materials are X-radiography, X-ray diffraction analysis, X-ray small angle scattering, X-ray fluorescence, and X-ray photoelectron spectroscopy (XPS). The basis of these techniques and their metallurgical applications are given in Table 4.6.

4.3.1 X-ray diffraction

Diffraction effects are produced when a beam of X-rays passes through the three-dimensional array of atoms which constitutes a crystal. Each atom scatters a fraction of the incident beams, and if certain conditions are fulfilled then the scattered waves reinforce to give a diffracted beam. These conditions are expressed by the Bragg equation:

$$n\lambda = 2d_{hkl}\sin\theta$$

where n is an integer, λ the wavelength of the incident beam, d_{hkl} the interplanar spacing of the hkl planes, and θ the angle of incidence on the hkl planes.

In the case of a single crystal randomly oriented with respect to a monochromatic X-ray beam it is unlikely that any planes will be in the correct orientation to satisfy the Bragg equation. Hence when single crystals (or more usually individual grains) of metallic substances are investigated by X-rays, either the crystal is rotated (e.g. as in the single crystal camera—*see* Table 4.7), or continuous (white) radiation is used as in the Laue techniques (*see* Table 4.7).

In the case of powdered or fine grained samples of crystalline materials, enough crystals are in the correct position to satisfy the Bragg condition, and the diffraction pattern may be recorded directly either with a photographic film by the Debye–Scherrer method, or by scanning an electronic counter through the diffraction spectra—the familiar powder diffractometry method.

The experimental methods of X-ray diffraction and their most common metallurgical applications are reviewed in Table 4.7. More information on some of the more important applications to metallurgy is given below.

4.3.2 Phase identification and quantitative measurement

Phase identification and quantification depend on the accurate measurement of the position and intensity of diffracted spectra. These can be measured by either film or diffractometer techniques. However, the speed and ease with which patterns can be obtained with a diffractometer, together with the possibility of automation, make this the preferred technique in most laboratories.

For routine phase identification the observed d spacings and relative intensities are compared with standard X-ray data listed in the 'X-ray Powder Dates File' published by the Joint Committee on Powder Diffraction Standards (JCPDS).[1] The JCPDS file contains data for over 35 000 materials and is regularly updated. Standard manual and computer search procedures are available.

The formulae for calculating the interplanar d_{hkl} spacings from lattice parameters with other useful information on crystal symmetry and geometry are listed in Tables 4.8a and 4.8b.

Quantitative determinations are carried out by comparing the integrated intensities of selected diffractions. Powders are usually analysed by adding a known fraction of a standard calibrating powder and comparing the intensity of a diffraction from the internal standard with the intensity of a diffraction from the component to be analysed. Most metallurgical samples of interest, however, are solid and in this case either diffractions from different phases are measured to give their volume ratio or a calibration curve intensity phase volume is constructed for precisely defined diffraction settings.

In all measurements the operator should be aware of factors which can lead to wrong volumes for accurate preferred orientations, which can be identified by increasing several diffraction peaks; also, uneven grain size can cause errors. All surfaces should be carefully prepared to ensure that they are truly representative of the bulk material, for example retained austenite can be transformed to martensite by rough polishing.

The accuracy of spectra position and intensity measurement obtainable by camera and diffractometer techniques can be comparable, as has been demonstrated by Hanawault,[2] and the results also shows the superiority of the Guinier camera for photographic methods.

The theoretical values of the intensities X-ray diffraction aspect depend on a number of factors, which are illustrated below in the description of retained austenite determination.

Table 4.6 METALLURGICAL INVESTIGATION TECHNIQUES INVOLVING X-RAYS

Technique	*Principle of technique*	*Metallurgical applications*
X-radiography	Sample placed between divergent X-ray beam and detector (usually a photographic film). Dark areas on negative show regions of low X-ray absorption (i.e. low density in specimen such as cracks) and vice versa	Crack and other defect detection in castings, welded fabrications, etc. Very widely used inspection and quality assurance technique
X-ray diffraction	Beam of X-rays of specific wavelength (λ) 'reflected' at certain angles (θ) by crystal planes of appropriate spacing (d) which satisfies the Bragg equation, i.e. $n\lambda = 2d \sin\theta$ Angular position of diffraction peaks, shape of peaks, and intensity peaks give information on crystal structure and physical state	Identification and quantitative analysis of crystalline chemical compounds (phase analysis and quantification, e.g. retained austenite determination) Crystal structure determination Residual macro- and micro-stress analysis Texture (preferred grain orientation) measurement Detection of cold work and imperfections (by stacking faults, etc.) N.B. All results obtained from thin (typically 50 μm) surface layer
Small angle X-ray scattering	Large regions (e.g. 10 – > 10 000 Å) of inhomogeneous electron density distributions (e.g. vacancy clusters of Guinier–Preston zones) cause scattered radiation pattern near main beam	Determination of size, shape and composition (in terms of electron density) of Guinier–Preston zones, vacancy clusters, etc. in metallic transmission technique therefore limited to thin samples—typically 10–50 μm
X-ray fluorescence	Electrons ejected from inner shells of atoms cause emission of X-rays (fluorescence) characteristic of atomic species—exciting radiation may be photons (γ and X-rays), positive ions or electrons. Fluorescent X-rays analysed by wavelength or energy dispersive spectrometers.	Elemental analyses Na to U routine. C, N, O, F require specifically designed equipment. Hence used for routine analysis of composition of ores, semi-finished and finished metals and their alloys Quantitative analysis normally employs calibrated standards. N.B. Results are obtained from surface layer of approx. 2um thickness
X-ray photo-electron spectroscopy (XPS)	Electrons from inner shells of atoms are ejected by X-radiation of specific wavelength. Energy of ejected electron measured by spectrometer gives information on binding energy of electron shells	Qualitative and quantitative analysis of surface layers, XPS signals typically come from depths of less than 50 Å. Elemental analysis He to U including, C, N, O, F. Information on the state of bonding of analysed elements

Table 4.7 X-RAY DIFFRACTION TECHNIQUES

	Experimental Technique	*Description*	*Metallurgical applications*
SINGLE CRYSTAL	Rotating crystal, e.g. Weissenberg camera or single crystal diffractometer	Small single crystal (usually just visible to the naked eye) is mounted on a two-axis goniometer which is rotated and irradiated with monochromatic X-rays to produce diffraction spectra. Spectra position and intensity are detected and recorded by photographic film (Weissenberg camera) or electronic counter (diffractometer)	Determination of unit cell dimensions Crystal structure analysis (i.e. determination of atom positions and thermal vibrations) etc.
SINGLE CRYSTAL	Back reflection or transmission Laue camera	Stationary sample, usually large (i.e. > 1 mm max. dimension) grained metallic material is irradiated with 'white' X-rays (i.e. range of wavelengths). X-rays pass through hole in centre of first film which records back-reflected X-ray spectra. Position and symmetry relationships of diffraction spectra give orientation of crystal (grain)	Texture studies on deformed (worked) metallic materials Confirmation of crystal symmetry system Determination of crystal orientation
POWDER OR FINE-GRAINED SAMPLE	Debye-Scherrer camera	Cylindrical compact of powder or wire sample (approx. 5 mm long and 0.5–1 mm in diameter) rotated (to avoid texture effects) in monochromatic X-ray beam. Diffraction cones intersect narrow cylindrical film (coaxial with sample) to give line spectrum	Phase identification Quantitative analysis of mixtures of chemical compounds usually with aid of calibrated standards Composition of alloy phases by correlation of lattice parameters with varying constituent elements
POWDER OR FINE-GRAINED SAMPLE	Glancing angle camera	Diffraction spectra obtained by irradiating surface or edge sample at glancing angle and recording one half of total diffraction spectra (other half absorbed by sample) on photographic film	Examination of bulk samples. Simultaneous detection and analysis of surface film (e.g. oxides) and parent metal substrate
POWDER OR FINE-GRAINED SAMPLE	Gunier camera	Flat powdered specimens positioned on circumference of camera are irradiated by a diverging beam of X-rays from a curved crystal for focusing monochromator Four specimens can be simultaneously exposed	Improved resolution and inherently low background aids identification and comparison of specimens with small differences of structure. Diffractions at low Bragg angles (high *d* values) can be studied
POWDER OR FINE-GRAINED SAMPLE	Diffractometry	Sample of approximate area 1 cm^2 is irradiated by monochromatic or filtered X-rays and diffraction spectra recorded by scanning with electronic counter. Can be automated	Phase identification and quantification (usually with calibrated standards) Studies of crystal imperfections, e.g. stacking faults, microstrain, etc. by detailed measurement of spectra profiles With appropriate attachments the following are possible: Residual stress measurement Texture (preferred orientation) determination Thermal expansion parameter measurement Phase transition monitoring

Table 4.8a CRYSTAL GEOMETRY

System	d_{hkl} = *Interplanar spacing*	V = *Vol. of unit cell*
Cubic	$\frac{1}{d^2}=\frac{h^2+k^2+l^2}{a^2}$	$V=a^3$
Tetragonal	$\frac{1}{d^2}=\frac{h^2}{a^2}+\frac{k^2}{a^2}+\frac{l^2}{c^2}$	$V=a^2c$
Orthorhombic	$\frac{1}{d^2}=\frac{h^2}{a^2}+\frac{k^2}{b^2}+\frac{l^2}{c^2}$	$V=abc$
Rhombohedral*	$\frac{1}{d^2}=\frac{(h^2+k^2+l^2)\sin^2\alpha+2(hk+kl+hl)(\cos^2\alpha-\cos\alpha)}{a^2(1-3\cos^2\alpha+2\cos^3\alpha)}$	$V=a^3\sqrt{(1-3\cos^2\alpha+2\cos^3\alpha)}$
Hexagonal†	$\frac{1}{d^2}=\frac{4}{3}\left(\frac{h^2+hk+k^2}{a^2}\right)+\frac{l^2}{c^2}$	$V=\frac{\sqrt{3}}{2}a^2c=0.866\,a^2c$
Monoclinic	$\frac{1}{d^2}=\frac{h^2}{a^2\sin^2\beta}+\frac{k^2}{b^2}+\frac{l^2}{c^2\sin^2\beta}-\frac{2hl\cos\beta}{ac\sin^2\beta}$	$V=abc\sin\beta$
Triclinic	$\frac{1}{d^2}=\frac{1}{V^2}(s_{11}h^2+s_{22}k^2+s_{33}l^2+2s_{12}hk+2s_{23}kl+2s_{13}hl)$	$V=abc\sqrt{(1-\cos^2\alpha-\cos^2\beta-\cos^2\gamma+2\cos\alpha\cos\beta\cos\gamma)}$

System	ϕ = *angle between planes* $h_1k_1l_1$ *and* $h_2k_2l_2$
Cubic	$\cos\phi = \dfrac{h_1h_2+k_1k_2+l_1l_2}{\sqrt{[(h_1^2+k_1^2+l_1^2)(h_2^2+k_2^2+l_2^2)]}}$
Tetragonal	$\cos\phi = \dfrac{(h_1h_2/a^2)+(k_1k_2/a^2)+(l_1l_2/c^2)}{\sqrt{\{[(h_1^2/a^2)+(k_1^2/a^2)+(l_1^2/c^2)][(h_2^2/a^2)+(k_2^2/b^2)+(l_2^2/c^2)]\}}}$
Orthorhombic	$\cos\phi = \dfrac{(h_1h_2/a^2)+(k_1k_2/b^2)+(l_1l_2/c^2)}{\sqrt{\{[(h_1^2/a^2)+(k_1^2/b^2)+(l_1^2/c^2)][(h_2^2/a^2)+(k_2^2/b^2)+(l_2^2/c^2)]\}}}$
Rhombohedral*	$\cos\phi = \dfrac{(h_1h_2+k_1k_2+l_1l_2)\sin^2\alpha+(k_1l_2+k_2l_1+l_1h_2+l_2h_1+h_1k_2+h_2k_1)(\cos^2\alpha-\cos\alpha)}{\sqrt{\{[(h_1^2+k_1^2+l_1^2)\sin^2\alpha+2(h_1k_1+k_1l_1+h_1l_1)(\cos^2\alpha-\cos\alpha)][(h_2^2+k_2^2+l_2^2)\sin^2\alpha+2(h_2k_2+k_2l_2+h_2l_2)(\cos^2\alpha-\cos\alpha)]\}}}$
Hexagonal†	$\cos\phi = \dfrac{h_1h_2+k_1k_2+\frac{1}{2}(h_1k_2+h_2k_1)+\frac{3}{4}\cdot(a^2/c^2)\cdot l_1l_2}{\sqrt{\{[h_1^2+k_1^2+h_1k_1+\frac{3}{4}(a^2/c^2)l_1^2][h_2^2+k_2^2+h_2k_2+\frac{3}{4}(a^2/c^2)\cdot l_2^2]\}}}$
Monoclinic	$\cos\phi = \dfrac{(h_1h_2/a^2)+(k_1k_2\sin^2\beta/b^2)+(l_1l_2/c^2)-[(l_1h_2+l_2h_1)\cos\beta/ac]}{\sqrt{\{[(h_1^2/a^2)+(k_1^2\sin^2\beta/b^2)+(l_1^2/c^2)-(2h_1l_1\cos\beta/ac)][(h_2^2/a^2)+(k_2^2\sin^2\beta/b^2)+(l_2^2/c^2)-(2h_2l_2\cos\beta/ac)]\}}}$
Triclinic	$\cos\phi = \dfrac{d_{h_1k_1l_1}\cdot d_{h_2k_2l_2}}{V^2}[s_{11}h_1h_2+s_{22}k_1k_2+s_{33}l_1l_2+s_{23}(k_1l_2+k_2l_1)+s_{13}(l_1h_2+l_2h_1)+s_{12}(h_1k_2+h_2k_1)]$ where $s_{11}=b^2c^2\sin^2\alpha$ $s_{12}=abc^2(\cos\alpha\cos\beta-\cos\gamma)$ $s_{22}=a^2c^2\sin^2\beta$ $s_{23}=a^2bc(\cos\beta\cos\gamma-\cos\alpha)$ $s_{33}=a^2b^2\sin^2\gamma$ $s_{13}=ab^2c(\cos\gamma\cos\alpha-\cos\beta)$

* Rhombohedral axes. † Hexagonal axes, co-ordinates *hkil* where $i=-(h+k)$.

Table 4.8b ANGLES BETWEEN CRYSTALLOGRAPHIC PLANES IN CRYSTALS OF THE CUBIC SYSTEM

(*HKL*)	(*hkl*)	*Values of* α, *the angle between* (*HKL*) *and* (*hkl*)						
100	100	0°	90°					
	110	45°	90°					
	111	54° 44′						
	210	26° 34′	63° 26′	90°				
	211	35° 16′	65° 54′					
	221	48° 11′	70° 32′					
	310	18° 26′	71° 34′	90°				
	311	25° 14′	72° 27′					
	320	33° 41′	56° 19′	90°				
	321	36° 43′	57° 42′	74° 30′				
110	110	0°	60°	90°				
	111	35° 16′	90°					
	210	18° 26′	50° 46′	71° 34′				
	211	30°	54° 44′	73° 13′	90°			
	221	19° 28′	45°	76° 22′	90°			
	310	26° 34′	47° 52′	63° 26′	77° 5′			
	311	31° 29′	64° 46′	90°				
	320	11° 19′	53° 58′	66° 54′	78° 41′			
	321	19° 6′	40° 54′	55° 28′	67° 48′	79° 6′		
111	111	0°	70° 32′					
	210	39° 14′	75° 2′					
	211	19° 28′	61° 52′	90°				
	221	15° 48′	54° 44′	78° 54′				
	310	43° 5′	68° 35′					
	311	29° 30′	58° 31′	79° 58′				
	320	61° 17′	71° 19′					
	321	22° 12′	51° 53′	72° 1′	90°			
210	210	0°	36° 52′	53° 8′	66° 25′	78° 28′	90°	
	211	24° 6′	43° 5′	56° 47′	79° 29′	90°		
	221	26° 34′	41° 49′	53° 24′	63° 26′	72° 39′	90°	
	310	8° 8′	58° 3′	45°	64° 54′	73° 34′		
	311	19° 17′	47° 36′	66° 8′	82° 15′			
	320	7° 7′	29° 45′	41° 55′	60° 15′	68° 9′	75° 38′	82° 53′
	321	17° 1′	33° 13′	53° 18′	61° 26′	70° 13′	83° 8′	90°
211	211	0°	33° 33′	48° 11′	60°	70° 32′	80° 24′	
	221	17° 43′	35° 16′	47° 7′	65° 54′	74° 12′	82° 12′	
	310	25° 21′	49° 48′	58° 55′	75° 2′	82° 35′		
	311	19° 8′	42° 24′	60° 30′	75° 45′	90°		
	320	25° 9′	37° 37′	55° 33′	63° 5′	83° 30′		
	321	10° 54′	29° 12′	40° 12′	49° 6′	56° 56′		
		70° 54′	77° 24′	83° 44′	90°			
221	221	0°	27° 16′	38° 57′	63° 37′	83° 37′	90°	
	310	32° 31′	42° 27′	58° 12′	65° 4′	83° 57′		
	311	25° 14′	45° 17′	59° 50′	72° 27′	84° 14′		
	320	22° 24′	42° 18′	49° 40′	68° 18′	79° 21′	84° 42′	
	321	11° 29′	27° 1′	36° 42′	57° 41′	63° 33′	74° 30′	
		79° 44′	84° 53′					
310	310	0°	25° 51′	36° 52′	53° 8′	72° 33′	84° 16′	
	311	17° 33′	40° 17′	55° 6′	67° 35′	79° 1′	90°	
	320	15° 15′	37° 52′	52° 8′	74° 45′	84° 58′		
	321	21° 37′	32° 19′	40° 29′	47° 28′	53° 44′	59° 32′	
		65°	75° 19′	85° 9′	90°			
311	311	0°	35° 6′	50° 29′	62° 58′	84° 47′		
	320	23° 6′	41° 11′	54° 10′	65° 17′	75° 28′	85° 12′	
	321	14° 46′	36° 19′	49° 52′	61° 5′	71° 12′	82° 44′	
320	320	0°	22° 37′	46° 11′	62° 31′	67° 23′	72° 5′	90°
	321	15° 30′	27° 11′	35° 23′	48° 9′	53° 37′	58° 45′	63° 36′
		72° 45′	77° 9′	85° 45′	90°			
321	321	0°	21° 47′	31°	38° 13′	44° 25′	50°	60°
		64° 37′	69° 4′	73° 24′	81° 47′	85° 54′		

DETERMINATION OF RETAINED AUSTENITE IN STEEL

An important example of quantitative phase analysis by X-ray diffraction is the determination of retained austenite in steels. The method is based on the comparison of the integrated diffracted X-ray intensities of selected (*hkl*) reflections of the martensite and austenite phases. The necessary formulae and reference data are given below; for more details of the experimental methods the definitive paper by Durnin and Ridal[3] should be consulted.

The integrated intensity of a diffraction line is given by the equation:

$$I_{(hkl)} = n^2\ Vm\ (LP)\ e^{-2m}(Ff)^2 \tag{4.1}$$

in which $I_{(hkl)}$ = integrated intensity for a special (*hkl*) reflection; n = number of cells in cm^3; V = volume exposed to the X-ray beam; (LP) = Lorentz–Polarization factor; m = multiplicity of (*hkl*); e^{-2m} = Debye–Waller temperature factor; F = structure factor; and f = atomic factor.

For n^2V we may substitute V/v^2, in which v is the volume of the unit cell.

If the ratio between the integrated intensities of martensite and austenite is denoted by P:

$$P = \frac{I\ \text{martensite}\ (\alpha)}{I\ \text{austenite}\ (\gamma)} = \frac{V_\alpha\ v_\gamma^{\,2} m_\alpha (LP)_\alpha e_\alpha^{\,-2m}(F_\alpha f_\alpha)^2}{V_\gamma\ v_\alpha^{\,2} m_\gamma (LP)_\gamma e_\gamma^{\,-2m}(F_\gamma f_\gamma)^2} \tag{4.2}$$

Each factor is determined from the International Tables[4] and depends on the reflection used. A factor G is then determined for each combination of α and γ peaks used; hence:

$$P = \frac{V_\alpha}{V_\gamma} \times \frac{1}{G} \tag{4.3}$$

If α and γ are the only phases present:

$$V_\gamma = \frac{1}{1+GP} \tag{4.4}$$

Hence, measurement of the ratio (P) of two diffraction peaks and calculation of the factor G will give the volume fraction of austenite V_γ.

The factors involved in the calculation of G for two steels—16.8%Ni–Fe(0.35%C) and N C MV (a Ni–Cr–Mo–V steel with composition wt % 0.43C; 0.31Si; 0.57Mn; 0.009S; 0.005P; 1.69Ni; 1.36Cr; 1.08Mo; 0.24V; 0.11Cu), Mo, Co and Cr radiation and a selection of *hkl* peaks have been extracted from the 'International Tables for X-Ray Crystallography'[4] by Durnin and Ridal[3] and are presented in Table 4.9.

These factors may be used to calculate G for different radiations and peaks. The results are presented in Table 4.10.

When the alloy compositions are being investigated the factors which make up G must be determined from the International Tables.[4]

Accuracy obtainable using a diffractometer is in the region of 0.5% for the range 1.5–38 volume percentage of austenite. X-ray diffraction determination accuracy thus compares favourably with other techniques such as metallography, dilatometry and saturation magnetization intensity methods which are all inaccurate below 10% austenite content.

The main source of error in X-ray determination of retained austenite comes from overlapping carbide peaks. The carbides and their diffraction peaks most likely to cause problems are summarized in Table 4.11.

RESIDUAL STRESS MEASUREMENT[5, 6]

X-ray diffraction is the only non-destructive method of measuring residual stresses in metallic specimens. However, for successful measurement and application of the results the following factors must be remembered:

1. X-rays detect the change in lattice spacing of certain surface grains. Hence strain is measured directly and not stress. To convert accurately the strain measurements to stress, the X-ray elastic constants for the materials under investigation must be known. These may be determined experimentally. Use of macroscopic, i.e. average elastic constants, for the material can lead to substantial errors.

Table 4.9 INTENSITY FACTORS FOR DIFFERENT RADIATIONS AND PEAKS[3]

Material	*Radiation*	*Factor*	*Peak* α200	α211	γ200	γ220	γ311
16.8%Ni–Fe	Mo	Bragg angle, θ	14.41	17.73	11.49	16.32	19.21
NCMV	Mo		14.40	17.70	11.49	16.32	19.21
16.8%Ni–Fe	Co		38.67	49.89	29.99	44.92	55.85
16.8%Ni–Fe	Cr		53.15	78.05	39.74	64.65	—
Both compositions	All	Multiplicity, m	6	24	6	12	24
Both	Mo	Lorentz and	29.46	18.84	47.56	22.55	15.78
	Co	polarization	3.44	2.73	5.79	2.83	2.96
	Cr	(LP)	2.81	9.26	3.29	4.01	—
16.8%Ni–Fe	Mo	Debye–Waller	0.910	0.869	0.943	0.889	0.847
NCMV	Mo	temp. e^{-2m}	0.910	0.869	0.943	0.889	0.848
16.8%Ni–Fe	Co		0.908	0.869	0.941	0.889	0.852
16.8%Ni–Fe	Cr		0.912	0.869	0.943	0.889	—
16.8%Ni–Fe	Mo	Atomic scattering,	15.1	13.4	17.0	14.0	12.8
NCMV	Mo	*fo*	14.7	13.1	16.6	13.7	12.5
16.8%Ni–Fe	Co		10.78	9.14	12.69	9.84	8.60
16.8%Ni–Fe	Cr		13.19	11.54	15.09	12.24	—
Both	All	Structure factor, F	2	2	4	4	4
Both	All	$1/V^2$ (V is volume of unit cell)	1.79×10^{-3} *kx* units		4.68×10^{-4} *kx* units		

Table 4.10 AUSTENITE DETERMINATION FACTOR *G* FOR DIFFERENT RADIATIONS AND PEAK COMBINATIONS[3]

Material	*Radiation*	*Peak combination* α200–γ200	α200–γ220	α200–γ311	α211–γ200	α211–γ220	α211–γ311
16.8%Ni–Fe	Mo	2.22	1.36	1.50	1.16	0.71	0.78
NCMV	Mo	2.23	1.38	1.51	1.15	0.72	0.78
16.8%Ni–Fe	Co	2.52	1.40	2.15	1.09	0.61	0.93
16.8%Ni–Fe	Cr	1.66	2.50	—	0.17	0.26	—

Table 4.11 INTERFERENCE OF ALLOY CARBIDE LINES WITH AUSTENITE AND MARTENSITE LINES[3]

Austenite and martensite 'd' spacings	Fe_3C	M_6C	V_4C_3	Mo_2C or W_2C	WC	$Cr_{23}C_6$	Cr_7C_3
(200)γ 1.80 Å	Clear	Clear	Clear	Clear	Clear	Strong overlap	Strong overlap
(200)α 1.43 Å	Clear	Strong overlap	Clear	Clear	Weak overlap	Weak overlap	Weak overlap
(220)γ 1.27 Å	Weak overlap	Strong overlap	Weak overlap	Strong overlap	Weak overlap	Medium overlap	Clear
(211)α 1.17 Å	Strong overlap	Clear	Clear	Weak overlap	Weak overlap	Medium overlap	Medium overlap
(311)γ 1.08 Å	Weak overlap	Medium overlap	Clear	Weak overlap	Clear	Strong overlap	Clear

2. As with all X-ray techniques measurements are obtained from the surface layers of a specimen. As the residual stress perpendicular to a free surface must be zero, only two-dimensional stress systems within the surface can be measured directly. Stripping of surface layers with appropriate mathematical calculations or direct experimental observation, e.g. with strain gauges, can permit the determination of sub-surface stresses. However, these procedures are complex. It follows that specimen surfaces should be free of scale or other layers opaque to X-rays. Surface roughness should not exceed 20 μm Ra.

3. It is important to know the plastic deformation history of the specimens under examination. Specimens which have been plastically deformed in tension, compression, bending, rolling, or drawing, exhibit a pseudo-macrostress[7] which causes inaccurate measurement of true residual macrostress by X-ray diffraction. True residual macrostress is defined[7] as being reasonably constant (in magnitude and direction) over fairly large distances (several grain diameters). It is measurable by dissection, because removal of part of the specimen causes strain in the remainder. It produces an X-ray line shift when the specimen is rotated to a new orientation with respect to the incident X-ray beam. Pseudo-macrostress does not cause strain on dissection, and is believed to be a special kind of microstress. Microstress varies over a small distance (one grain diameter or less). It causes X-ray line broadening.

Specimens which have been plastically deformed by peening, grinding or machining do not exhibit pseudo-macrostress, and hence accurate non-destructive determination of macrostress can be made on these specimens by X-ray diffraction methods. The same is true of specimens where the source of plastic or elastic deformation is remote from the surface being analysed. This is particularly important as this category includes stresses due to casting, welding, quenching and phase transformations.

4. The accuracy of residual stress measurements for specimens which can be mounted on an X-ray diffractometer is typically of the order of $\pm 1.5\text{–}3.0 \times 10^7\ \mathrm{Nm^{-2}}$ ($\pm 1-2$ ton in^{-2}). Better accuracy has been achieved under ideal conditions. For large specimens where portable X-ray techniques are used (usually with a reflection camera) accuracy is typically around $\pm 7.5-10^7\,\mathrm{Nm^{-2}}$ ($\pm 5-7$ ton in^{-2}).

4.4 X-ray results

Table 4.12 GLIDE ELEMENTS AND FRACTURE PLANES OF METAL CRYSTALS

		Low temperatures		*Elevated temperatures*		*Most closely packed*		
Structure	*Metal*	*Glide plane*	*Glide direction*	*Glide plane*	*Glide direction*	*Lattice plane*	*Lattice direction*	*Fracture plane*
Cubic, face centred	Al							
	Cu			Approximately 450 °C		*1* (111)	*1* $[10\bar{1}]$	—
	Ag			(100)	[101]	*2* (100)	*2* [100]	—
	Au					*3* (110)	*3* [112]	—
	Pb							
	Ni	(111)	$[10\bar{1}]$	—	—	—	—	—
	Cu-Au							
	α-Cu-Zn							
	α-Cu-Al							
	Al-Cu							
	Al-Zn							
	Au-Ag							
Cubic, body centred	α-Fe	(100)	$[11\bar{1}]$	—	—			
		(112)	$[11\bar{1}]$	—	—	*1* (101)	*1* [111]	
		(123)	$[11\bar{1}]$	—	—	*2* (100)	*2* [100]	(001)
	W	(112)	$[11\bar{1}]$	—	—	*3* (111)	*3* [110]	
	Mo	(112)	$[11\bar{1}]$	(110)	$[1\bar{1}1]$	—	—	—
	K	(123)	$[11\bar{1}]$	(123)	$[11\bar{1}]$	—	—	—
	Na	(112)	$[11\bar{1}]$	(110)	$[11\bar{1}]$	—	—	—
		—	—	(123)	$[11\bar{1}]$	—	—	—
	β-Cu-Zn	(110)	$[11\bar{1}]$	—	—	—	—	—
		(112)(?)	—	—	—	—	—	—
	α-Fe-Si, 5% Si	(110)	—	—	—	—	—	

Table 4.12 GLIDE ELEMENTS AND FRACTURE PLANES OF METAL CRYSTALS—*continued*

		Low temperatures		*Elevated temperatures*		*Most closely packed*		
Structure	*Metal*	*Glide plane*	*Glide direction*	*Glide plane*	*Glide direction*	*Lattice plane*	*Lattice direction*	*Fracture plane*
Hexagonal, close packed	Mg, Zn, Cd, Be, Zn-Cd, Zn-Sn	(0001)	$[11\bar{2}0]$	Approximately 225 °C $(10\bar{1}1)$ *or* $(10\bar{1}2)$	$[11\bar{2}0]$	(0001)	$[11\bar{2}0]$	(0001), $(10\bar{1}1)$, $(10\bar{1}2)$, $(10\bar{1}0)$
Tetragonal	β-Sn	(110)	[001]	Approximately 150 °C		*1*(100)	*1*[001]	—
	(white)	(100)	[001]	(110)	$[\bar{1}11]$	*2*(110)	*2*[111]	—
		(101)	—	—	—	*3*(101)	*3*[100]	—
		(121)	$[10\bar{1}]$	—	—	—	*4*[101]	—
Rhombo-hedral	As	—	—	—	—	*1*$(\bar{1}10)$	*1*$[\bar{1}01]$	(111), (110)
	Sb	(111)	$[10\bar{1}]$ and	—	—	*2*$(11\bar{1})$	—	(111), (110), $(11\bar{1})$
	Bi	$(11\bar{1})$	[101]	—	—	—	—	(111), $(11\bar{1})$, (110)
	Hg	(100) and complex	—	—	—	—	—	—
Hexagonal	Te	$(10\bar{1}0)$	$[11\bar{2}0]$	—	—	$(10\bar{1}1)$	—	$(10\bar{1}0)$

Table 4.13 TWINNING IN METALS*

Lattice	*Metal*	*Twinning plane*	*Twinning direction*
Body centred cubic	α-Fe	(112)	[111]
	β-Cu-Zn	(112)	[111]
	W	(112)	[111]
Hexagonal close packed	Cd	$(10\bar{1}2)$	—
	Zn	$(10\bar{1}2)$	—
	Zn-Cd	$(10\bar{1}2)$	—
	Zn-Sn	$(10\bar{1}2)$	—
	Be	$(10\bar{1}2)$	—
	Mg	$(10\bar{1}2)$	—
Face centred cubic	Al	(111)	$[11\bar{2}]$
	Cu	(111)	$[11\bar{2}]$
	Ag	(111)	$[11\bar{2}]$
	Au	(111)	$[11\bar{2}]$
	Cu-Zn	(111)	$[11\bar{2}]$
	Cu-Al	(111)	$[11\bar{2}]$
	Al-Cu	(111)	$[11\bar{2}]$
	Al-Zn	(111)	$[11\bar{2}]$
	Au-Ag	(111)	$[11\bar{2}]$
	Cu-Au	(111)	$[11\bar{2}]$
Rhombohedral	Bi	(011)	—
	As	(011)	—
	Sb	(011)	—
Tetragonal	β-Sn (white)	(331)	—

* From C. S. Barrett, 'Structure of Metals', McGraw-Hill, New York, 1943.

Table 4.14 PREFERRED ORIENTATIONS IN METALS*

	Nature of process							
	Casting	*Drawing*	*Compression*	*Torsion*	*Rolling*		*Annealing after cold rolling*	
	Parallel to axis of				*Parallel to rolling*			
Metal	*Columnar crystals*	*Wire*	*Compression*	*Torsion*	*Plane*	*Direction*	*Plane*	*Direction*
Al	[100]	[111] }	[110] }	[111] }				
Cu	[100] }				(110)	[1$\bar{1}$2] }	(001)	[100]
Ni		*1*[111] }						
Au	[100]	*2*[100]						
Pd	}							
Pb	[100] }	[111]		}				
		1[100]			(110)	[1$\bar{1}$2]		
Ag	[100] }	*2*[111]		}			(311)	[$\bar{1}$12]
				1[110]	*1*(001)	[110]	*1*(001)	[110]
α-Fe	[100] }		[111], ([100]?)	*2*[112]	*2*(112)	[1$\bar{1}$0]	*2*(111)	[11$\bar{2}$]
					3(111)	[11$\bar{2}$]	*3*(112)	[1$\bar{1}$0]
Mo	}	[110]		}				
Ta					(001)	[110]		
W	}			}				
Mg	[11$\bar{2}$0]	(0001)	[0001]		(0001)		(0001)	
Zn }		at 18°			at 20°	at 20°	at 20°	at 20°
	[0001]	(0001)			(0001)	[11$\bar{2}$0]	(0001)	[11$\bar{2}$0]
	is ⊥							
Cd }					at 30°			
					(0001)			
Zr		(0001)			(0001)			

* From W. Boas, 'Introduction to the Physics of Metals and Alloys', University Press, Melbourne, 1947.

Table 4.15 FIBRE TEXTURES OF DRAWN WIRE[7]

	Percentage of crystals	
Metal	*With* [100] *parallel to axis of wire*	*With* [111] *parallel to axis of wire*
Al	0	100
Cu	40	60
Au	50	50
Ag	75	25

Table 4.16 TEXTURES IN ELECTRODEPOSITS*

Metal	*Fibre textures*
Ni	[100]; [100] + [110]; [112]
Cu	[110]; [100]
Ag	[111] + [100]; [111]; [110]
Pb	[112]
Au	[110]
Fe	[111]; [112]
Co	[110]
Cr	[100] + [111] (f.c.c.); [0001] (hexagonal)
Sn	[111]; [001]
Cd	[11$\bar{2}$2]
Bi	[211]; [100]

* From C. S. Barrett, 'Structure of Metals', McGraw-Hill, New York, 1943.

Table 4.17 TEXTURES IN EVAPORATED AND SPUTTERED FILMS*

	Metal deposited	*Texture*	*Technique*
Face centred cubic	Ag	[111]; [100]; [110]	Evaporated
	Al	[111]; [100]; [110]	Evaporated
	Au	[110]; [111]	Evaporated
	Pt	[100]; [111]	Sputtered
	Pd, Cu, Ni	[111]	Evaporated
Body centred cubic	Fe	[111]	Evaporated
	Mo	[110]	Evaporated
Hexagonal	Cd, Zn	[0001]	Evaporated
Rhombohedral	Bi	[111]; [110]	Evaporated

* From C. S. Barrett, 'Structure of Metals', McGraw-Hill, New York, 1943.

Table 4.18 TEXTURES OF CAST METALS*

Structure	*Metal*	*Normal to cold surface*
Body centred cubic	Fe–Si (4.3% Si)	[100]
	β-Brass	[100]
Face centred cubic	Al, Cu, Ag, Au, Pb, α-Brass	[100]
Hexagonal close packed†	Cd ($c/a=1.885$)	Columnar grains, [001]; (100) ‖ to surface Chilled surface, [001]
	Zn ($c/a=1.856$)	Columnar grains, [001]; (100) ‖ to surface Chilled surface, [001]
	Mg ($c/a=1.624$)	Columnar grains, [100]; (205) ‖ to surface and (001) 37° from it
Rhombohedral	Bi	[111]
Tetragonal	β-Sn	[110]

* From C. S. Barrett, 'Structure of Metals', McGraw-Hill, New York, 1943.
† Three indices system; equivalent indices in four indices systems are as follows: (001) = (0001) = basal plane; [100] = [$2\bar{1}\bar{1}0$] = diagonal axis of type I = close packed row of atoms in basal plane; [100] normal to surface = (120) parallel to surface.

4.5 The determination of crystal structure[8]

The number of atoms or molecules associated with the unit cell may be calculated by means of the formula

$$n=6.0227\cdot\frac{\rho V}{M}\cdot 10^{23}$$

where n is no. of atoms or molecules, ρ the density, V the volume of the cell as given in the appropriate formula of Table 4.8a and M the atomic or molecular weight.

Table 4.19 gives the correlation between missing X-ray reflections and the translation and symmetry elements responsible for the extinctions.

Pauling's rules for ionic structures may be stated thus:[9]

1. A co-ordinated polyhedron of anions is formed about each cation, the cation/anion distance being determined by the radius sum and the co-ordination number of the cation by the radius ratio.

2. In a stable co-ordination structure the total strength of the valency bonds which reach an anion from all the neighbouring cations is equal to the charge of the anion.
3. The existence of edges, and particularly of faces, common to two anion polyhedra decreases the stability of the structure.
4. Cations with high valency and small co-ordination number tend not to share polyhedron elements with each other.
5. The environment of all chemically similar anions in a structure tends to be similar.

4.5.1 Atomic and ionic radii

It has been found that the distance between the centres of ions in ionic crystals, and between atoms in metals and alloys, approximately obeys an additive law, each ion or atom being packed into the structure as if it were a sphere of a definite size. The radius of this sphere, the ionic or atomic radius, is not, however, a constant for any given ion or atom but depends on its environment. V. M. Goldschmidt[10] has shown that in passing from 12-fold to 8-fold co-ordination there is a 3% contraction in the radius, from 12-fold to 6-fold one of 4%, and from 12-fold to 4-fold one of 12%. For purposes of comparison, therefore, atomic radii are best expressed as those appropriate to one particular degree of co-ordination, and that for close packing, 12-fold has been adopted. In the case of ionic crystals, the radii chosen are those for 6-fold co-ordination.

Table 4.19 SYMMETRY INTERPRETATIONS OF EXTINCTIONS*

Class of reflection	*Condition for non-extinction (n = an integer)*	*Interpretation of extinction*	*Symbol of symmetry element*
hkl	$h+k+l=2n$	Body centred lattice	*I*
	$h+k=2n$	*C*-centred lattice	*C*
	$h+l=2n$	*B*-centred lattice	*B*
	$k+l=2n$	*A*-centred lattice	*A*
	$\left\{\begin{array}{l} h+k=2n \\ h+l=2n \\ k+l=2n \end{array}\right\}$ ⇔ *h*, *k*, *l*, all even or all odd	Face centred lattice	*F*
	$-h+k+l=3n$	Rhombohedral lattice indexed on hexagonal reference system	*R*
	$h+k+l=3n$	Hexagonal lattice indexed on rhombohedral reference system	*H*
0kl	$k=2n$	(100) glide plane, component $b/2$	$b(P, B, C)$
	$l=2n$	(100) glide plane, component $c/2$	$c(P, C, I)$
	$k+l=2n$	(100) glide plane, component $(b/2)+(c/2)$	$n(P)$
	$k+l=4n$	(100) glide plane, component $(b/4)+(c/4)$	$d(F)$
h0l	$h=2n$	(010) glide plane, component $a/2$	$a(P, A, I)$
	$l=2n$	(010) glide plane, component $c/2$	$c(P, A, C)$
	$h+l=2n$	(010) glide plane, component $(a/2)+(c/2)$	$n(P)$
	$h+l=4n$	(010) glide plane, component $(a/4)+(c/4)$	$d(F)$, (B)
hk0	$h=2n$	(001) glide plane, component $a/2$	$a(P, B, I)$
	$k=2n$	(001) glide plane, component $b/2$	$b(P, A, B)$
	$h+k=2n$	(001) glide plane, component $(a/2)+(b/2)$	$n(P)$
	$h+k=4n$	(001) glide plane, component $(a/4)+(b/4)$	$d(F)$
hhl	$l=2n$	$(1\bar{1}0)$ glide plane, component $c/2$	$c(P, C, F)$
	$h=2n$	$(1\bar{1}0)$ glide plane, component $(a/2)+(b/2)$	$b(C)$
	$h+l=2n$	$(1\bar{1}0)$ glide plane, component $(a/4)+(b/4)+(c/4)$	$n(C)$
	$2h+l=4n$	$(1\bar{1}0)$ glide plane, component $(a/2)+(b/4)+(c/4)$	$d(I)$
h00	$h=2n$	[100] screw axis, component $a/2$	2_1, 4_2
	$h=4n$	[100] screw axis, component $a/4$	4_1, 4_3

* From M. J. Buerger, 'X-ray Crystallography', John Wiley & Sons, New York, 1942.

Table 4.19 SYMMETRY INTERPRETATIONS OF EXTINCTIONS—*continued*

Class of reflection	*Condition for non-extinction (n = an integer)*	*Interpretation of extinction*	*Symbol of symmetry element*
$0k0$	$k=2n$	[010] screw axis, component $b/2$	2_1, 4_2
	$k=4n$	[010] screw axis, component $b/4$	4_1, 4_3
$00l$	$l=2n$	[001] screw axis, component $c/2$	2_1, 4_2, 6_3
	$l=3n$	[001] screw axis, component $c/3$	3_1, 3_2, 6_2, 6_4
	$l=4n$	[001] screw axis, component $c/4$	4_1, 4_2
	$l=6n$	[001] screw axis, component $c/6$	6_1, 6_2
$hh0$	$h=2n$	[110] screw axis, component $(a/2)+(b/2)$	2_1

* From M. J. Buerger, 'X-ray Crystallography', John Wiley & Sons, New York, 1942.

Table 4.20 CORRELATION BETWEEN RADIUS RATIO AND CO-ORDINATION

Co-ordination round A	*Arrangement of X ions*	*Radius ratio RA:RX*
8	Corners of cube	1–0.732
6	Corners of regular octahedron	0.732 – 0.414 (6 and 4, bracketed)
4	Corners of square	
4	Corners of regular tetrahedron	0.414 – 0.225
3	Corners of equilateral triangle	0.225–0.155
2	Linear	0.155 – 0

Table 4.21 ATOMIC AND IONIC RADII

1	*2*	*3*	*4*	*5*	*6*	*7*	*8*	*9*	*10*
			As element				*In ionic crystals*		
Atomic number	*Symbol*	*Type of structure*	*Co-ordination No.*	*Inter-atomic distances*	*Gold-schmidt at. radii*	*State of ionization*	*Gold-schmidt ionic radii*	*Polarizing power*	*Polariza-bility*
1	H	c.p.h	6, 6	—	0.46	H	1.54	0.62	—
2	He	—	—	—	—	—	—	—	—
3	Li	b.c.c.	8	3.03	1.57	Li^+	0.78	1.64	0.075
4	Be	c.p.h.	6, 6	2.22; 2.28	1.13	Be^{2+}	0.34	17.30	0.028
5	B	—	—	—	0.97	B^{3+}	0.2	—	0.014
6	C	d.	4	1.54	0.77				
		hex.	3	1.42	—	C^{4+}	<0.2	—	—
7	N	cub.	—	—	0.71	N^{5+}	0.1–0.2	—	—
8	O	orthorh.	—	—	0.60	O^{2-}	1.32	1.15	3.1
9	F	—	—	—	—	F^-	1.33	0.57	0.99
10	Ne	f.c.c.	12	3.20	1.60	—	—	—	—
11	Na	b.c.c.	8	3.71	1.92	Na^+	0.98	1.04	0.21
12	Mg	c.p.h.	6, 6	3.19; 3.20	1.60	Mg^{2+}	0.78	3.29	0.12
13	Al	f.c.c.	12	2.86	1.43	Al^{3+}	0.57	9.23	0.065
14	Si	d.	4	2.35	[1.17]	Si^{4-}	1.98	—	—
						Si^{4+}	0.39	26.30	0.043
15	P	orthorh.	3	2.18	[1.09]	P^{5+}	0.3–0.4	—	—
16	S	f.c. orthorh.	—	2.12	[1.04]	S^{2-}	1.74	0.66	7.25
						S^{6+}	0.34	51.90	—
17	Cl	orthorh.	1	2.14	[1.07]	Cl^-	1.81	0.30	3.05
18	A	f.c.c.	12	3.84	1.92	—	—	—	—
19	K	b.c.c.	8	4.62	2.38	K^+	1.33	0.57	0.85
20	Ca	f.c.c.	12	3.93	1.97	Ca^{2+}	1.06	1.78	0.57
		c.p.h.	6,6	3.98; 3.99	2.00				

Table 4.21 ATOMIC AND IONIC RADII—*continued*

1	*2*	*3*	*4*	*5*	*6*	*7*	*8*	*9*	*10*
			As element				*In ionic crystals*		
Atomic number	*Symbol*	*Type of structure*	*Co-ordination No.*	*Inter-atomic distances*	*Gold-schmidt at. radii*	*State of ionization*	*Gold-schmidt ionic radii*	*polarizing power*	*polariza-bility*
21	Sc	f.c.c.	12	3.20	1.60	Sc^{3+}	0.83	4.35	0.38
		c.p.h.	6, 6	3.23; 3.30	1.64				
22	Ti	c.p.h.	6, 6	2.91; 2.95	1.47	Ti^{2+}	0.76	—	—
						Ti^{3+}	0.69	—	—
						Ti^{4+}	0.64	9.76	0.27
23	V	b.c.c.	8	2.63	1.36	V^{3+}	0.65	—	—
						V^{4+}	0.61	—	—
						V^{5+}	~0.4	—	—
24	Cr	b.c.c. (α)	8	2.49	1.28	Cr^{3+}	0.64	—	—
		c.p.h. (β)	6, 6	2.71; 2.72	1.36	Cr^{6+}	0.3–0.4	—	—
25	Mn	cub. (α)	—	2.24–2.96	[1.12]	Mn^{2+}	0.91	—	—
		cub. (β)	—	2.36–2.68	[1.18]	Mn^{3+}	0.70	—	—
		f.c.t. (γ)	8, 4	2.58; 2.67	~1.37	Mn^{4+}	0.52	—	—
26	Fe	b.c.c. (α)	8	2.48	1.28	Fe^{2+}	0.87	—	—
		f.c.c. (γ)	12	2.52	1.26	Fe^{3+}	0.67	—	—
27	Co	c.p.h. (α)	6, 6	2.49; 2.51	1.25	Co^{2+}	0.82	—	—
		f.c.c. (β)	12	2.51	1.26	Co^{3+}	0.65	—	—
28	Ni	c.p.h. (α)	6, 6	2.49; 2.49	1.25	Ni^{2+}	0.78	—	—
		f.c.c. (β)	12	2.49	1.25				
29	Cu	f.c.c.	12	2.55	1.28	Cu^{+}	0.96	—	—
30	Zn	c.p.h.	6, 6	2.66; 2.91	1.37	Zn^{2+}	0.83	2.90	—
31	Ga	orthorh.	—	2.43–2.79	1.35	Ga^{2+}	0.62	7.80	—
32	Ge	d.	4	2.44	1.39	Ge^{4+}	0.44	20.66	—
33	As	r.	3, 3	2.51; 3.15	[1.25]	As^{3+}	0.69	—	—
						As^{5+}	~0.4	—	—
34	Se	hex.	2, 4	2.32; 3.46	[1.16]	Se^{2-}	1.91	0.55	6.4
						Se^{6+}	0.3–0.4	—	—
35	Br	orthorh.	1	2.38	[1.19]	Br^{-}	1.96	0.26	4.17
36	Kr	f.c.c.	12	3.94	1.97	—	—	—	—
37	Rb	b.c.c.	8	4.87	2.51	Rb^{+}	1.49	0.45	1.81
38	Sr	f.c.c.	12	4.30	2.15	Sr^{2+}	1.27	1.24	1.42
39	Y	c.p.h.	6, 6	3.59; 3.66	1.81	Y^{3+}	1.06	2.67	1.04
40	Zr	c.p.h.	6, 6	3.16; 3.22	1.60	Zr^{4+}	0.87	5.28	—
		b.c.c.	8	3.12	1.61				
41	Nb	b.c.c.	8	2.85	1.47	Nb^{4+}	0.69	—	—
						Nb^{5+}	0.69	10.50	—
42	Mo	b.c.c.	8	2.72	1.40	Mo^{4+}	0.68	—	—
						Mo^{6+}	0.65	—	—
43	Tc	—	—	—	—	—	—	—	—
44	Ru	c.p.h.	6, 6	2.64; 2.70	1.34	Ru^{4+}	0.65	—	—
45	Rh	f.c.c.	12	2.68	1.34	Rh^{3+}	0.68	—	—
						Rh^{4+}	0.65	—	—
46	Pd	f.c.c.	12	2.75	1.37	Pd^{2+}	0.50	—	—
47	Ag	f.c.c.	12	2.88	1.44	Ag^{+}	1.13	0.78	—
48	Cd	c.p.h.	6, 6	2.97; 3.29	1.52	Cd^{2+}	1.03	1.88	—
49	In	f.c.t.	4, 8	3.24; 3.37	1.57	In^{3+}	0.92	3.54	—
50	Sn	d.	4	2.80	1.58	Sn^{4-}	2.15	—	—
		tetra.	4, 2	3.02; 3.18	—	Sn^{4+}	0.74	7.30	—
51	Sb	r.	3, 3	2.90; 3.36	1.61	Sb^{3+}	0.90	—	—
52	Te	hex.	2, 4	2.86; 3.46	[1.43]	Te^{2-}	2.11	0.45	9.6
						Te^{4+}	0.89	—	—
53	I	orthorh	—	2.70	[1.36]	I^{-}	2.20	0.21	6.28
						I^{5+}	0.94	—	—
54	Xe	f.c.c.	12	4.36	2.18	—	—	—	—
55	Cs	b.c.c.	8	5.24	2.70	Cs^{+}	1.65	0.37	2.79

Table 4.21 ATOMIC AND IONIC RADII—*continued*

1	2	3	4	5	6	7	8	9	10
				As element				*In ionic crystals*	
Atomic number	*Symbol*	*Type of structure*	*Co-ordination No.*	*Interatomic distances*	*Goldschmidt at. radii*	*Stage of ionization*	*Goldschmidt ionic radii*	*Polarizing power*	*polarizability*
56	Ba	b.c.c.	8	4.34	2.24	Ba^{2+}	1.43	0.98	2.08
57	La	c.p.h.	6, 6	3.72; 3.75	1.87	La^{3+}	1.22	2.01	1.56
		f.c.c.	12	3.75	1.87				
58	Ce	c.p.h.	6, 6	3.63; 3.65	1.82	Ce^{3+}	1.18	—	—
		f.c.c.	12	3.63	1.82	Ce^{4+}	1.02	3.84	1.20
59	Pr	hex.	6, 6	3.63; 3.66	1.83	Pr^{3+}	1.16	—	—
		f.c.c.	12	3.64	1.82	Pr^{4+}	1.00	—	—
60	Nd	hex.	6, 6	3.62; 3.65	1.82	Nd^{3+}	1.15	—	—
61	—	—	—	—	—	—	—	—	—
62	Sm	—	—	—	—	Sm^{3+}	1.13	—	—
63	Eu	b.c.c.	8	3.96	2.04	Eu^{3+}	1.13	—	—
64	Gd	c.p.h.	6, 6	3.55; 3.62	1.80	Gd^{3+}	1.11	—	—
65	Tb	c.p.h.	6, 6	3.51; 3.59	1.77	Tb^{3+}	1.09	—	—
						Tb^{4+}	0.89	—	—
66	Dy	c.p.h.	6, 6	3.50; 3.58	1.77	Dy^{3+}	1.07	—	—
67	Ho	c.p.h.	6, 6	3.48; 3.56	1.76	Ho^{3+}	1.05	—	—
68	Er	c.p.h.	6, 6	3.46; 3.53	1.75	Er^{3+}	1.04	—	—
69	Tm	c.p.h.	6, 6	3.45; 3.52	1.74	Tm^{3+}	1.04	—	—
70	Yb	f.c.c.	12	3.87	1.93	Yb^{3+}	1.00	—	—
71	Lu	c.p.h.	6, 6	3.44; 3.51	1.73	Lu^{3+}	0.99	—	—
72	Hf	c.p.h.	6, 6	3.13; 3.20	1.59	Hf^{4+}	0.84	—	—
73	Ta	b.c.c.	8	2.85	1.47	Ta^{5+}	0.68	—	—
74	W	b.c.c. (α)	8	2.74	1.41	W^{4+}	0.68	—	—
		cub. (β)	12; 2, 4	2.82; 2.52	1.41	W^{6+}	0.65	—	—
				2.82					
75	Re	c.p.h.	6, 6	2.73; 2.76	1.38	—	—	—	—
76	Os	c.p.h.	6, 6	2.67; 2.73	1.35	Os^{4+}	0.67	—	—
77	Ir	f.c.c.	12	2.71	1.35	Ir^{4+}	0.66	—	—
78	Pt	f.c.c.	12	2.77	1.38	Pt^{2+}	0.52	—	—
						Pt^{4+}	0.55	—	—
79	Au	f.c.c.	12	2.88	1.44	Au^{+}	1.37	—	—
80	Hg	r.	6	3.00	1.55	Hg^{2+}	1.12	1.59	—
81	Tl	c.p.h.	6, 6	3.40; 3.45	1.71	Tl^{+}	1.49	—	—
		b.c.c.	8	3.36	1.73	Tl^{3+}	1.06	2.72	—
82	Pb	f.c.c.	12	3.49	1.75	Pb^{4-}	2.15	—	—
						Pb^{2+}	1.32	—	—
						Pb^{4+}	0.84	5.67	—
83	Bi	r.	3, 3	3.11; 3.47	1.82	Bi^{3+}	1.20	—	—
84	Po	monocl.	—	2.81	[1.40]	—	—	—	—
85	At	—	—	—	—	—	—	—	—
86	Rn	—	—	—	—	—	—	—	—
87	Fr	—	—	—	—	—	—	—	—
88	Ra	—	—	—	—	Ra^{+}	1.52	—	—
89	Ac	—	—	—	—	—	—	—	—
90	Th	f.c.c.	12	3.60	1.80	Th^{4+}	1.10	3.31	—
91	Pa	—	—	—	—	—	—	—	—
92	U	orthorh.	—	2.76	[1.38]	U^{4+}	1.05	—	—
93	Np	—	—	—	—	—	—	—	—
94	Pu	—	—	—	—	—	—	—	—
95	Am	—	—	—	—	—	—	—	—
96	Cm	—	—	—	—	—	—	—	—

Values for atomic and ionic radii, with structural data for the free elements, are given in Table 4.21. The symbols used in column 3 to indicate the lattice types of the free elements are as follows:

cub. = cubic
f.c.c.; b.c.c. = face or body centred cubic
d. = diamond structure, two interpenetrating f.c.c. lattices
t.; f.c.t. = tetragonal, or face centred tetragonal
orthorh.; f.c. orthorh. = orthorhombic, or face centred orthorhombic
monocl. = monoclinic
r. = rhombohedral
hex. = hexagonal
c.p.h. = close-packed hexagonal

A co-ordination symbol *x* in column 4 indicates that each atom has *x* equidistant nearest neighbours, at a distance from it (in kX-units) specified in column 5. The symbol *x*, *y* indicates that a given atom has *x* equidistant nearest neighbours, and *y* equidistant neighbours lying a small distance further away. These distances are given in column 5. In complex structures, such as α-Mn, where the co-ordination is not exact, no symbol is used, and the range of distances between near neighbours is given in column 5.

The Goldschmidt atomic radii given in column 6 are the radii appropriate to 12-fold co-ordination In the case of the f.c.c. and c.p.h. metals the radius given is one-half of the measured interatomic distance, or of the mean of the two distances for the hexagonal packing. In the case of the b.c.c. metals, where the measured interatomic distances are for 8-fold co-ordination, a numerical correction has been applied. In some cases, where the pure element crystallizes in a structure having a low degree of co-ordination, or where the co-ordination is not exact, it is possible to find some compound or solid solution in which the element exists in 12-fold co-ordination, and hence to calculate its appropriate radius. In a few cases no correction for co-ordination has been attempted, and here the figures, given in parentheses, are one-half of the smallest interatomic distances. It should be emphasized that the Goldschmidt radii must not be regarded as constants subject only to correction for co-ordination and applicable to all alloy systems: they may vary with the solvent or with the degree of ionization, and they depend to some extent on the filling of the Brillouin zones.

Ionic radii vary largely with the valency, and to a smaller extent with co-ordination. The values given in column 8 are appropriate to 6-fold co-ordination, and have been derived either by direct measurement or by methods similar to those outlined for the atomic radii. All are based, ultimately, on the value of 1.32 Å obtained for O^{2+} ions by Wasastjerna,[11] using refractivity measurements. Ionic radii are also affected by the charge on neighbouring ions: thus in CaF_2 the fluorine ion is 3% smaller than in KF, where the metal ion carries a smaller charge. It is not possible to give a simple correction factor, applicable to all ions: the effect is specific and is especially marked in structures of low co-ordination. Figures in arbitrary units indicating the power of one ion to bring about distortion in a neighbour (its 'polarizing power'), and indicating the susceptibility of an ion to such distortion (its 'polarizability') are given in columns 9 and 10, respectively.

4.6 X-ray fluorescence

X-ray fluorescence occurs after an electron has been ejected from a shell surrounding the nucleus of an atom. The X-radiation is characteristic of the atom from which the electron has been ejected, and hence provides a means of identifying the atomic species. The ejection of an electron may be induced by irradiating the sample with photons (X or γ-rays) electrons, protons, charged particles or, indeed, any radiation capable of creating vacancies in the inner shells of the atoms of interest in the sample. The relative merits of each technique are given in Table 4.22. A further comparison of X-ray or radio-isotope sources for X-ray fluorescent spectroscopy is given in Table 4.23. Details of suitable available isotope sources are given in Table 4.24.

Analysis of fluorescent X-rays is achieved by wavelength dispersion using crystal analyser (or several in a multichannel instrument), or by energy dispersion with solid-state detectors. Wavelength dispersion offers more accurate quantitative analysis, especially for the detection of small concentrations of elements where X-ray spectra from several elements overlap. Energy dispersion is preferred when rapid or quantitative analysis is required of an unknown sample.

Example of the detection limits for X-ray excited samples are given in Tables 4.25 and 4.26, and for ion excited samples in Table 4.27.

Accuracy levels for elemental analysis are typically:

for X-ray excitation	better than 1%.
for electron and ion excitation	1–2%.

These values can be improved with very carefully calibrated standards, but are frequently much worse, especially when the specimen surface is rough. Unlike X-ray diffraction, powdered samples are the most difficult sample form to analyse.

Table 4.22 COMPARISON OF X-RAY FLUORESCENCE TECHNIQUES

Exciting radiation	*Advantages*	*Limitations*
Electrons	High-intensity energy-regulated sources easily produced Can be focused into submicron spot size Low cost Good light element detection	Specimen must be in vacuum with source Signal-to-background ratio relatively poor
Positive ions	Better signal-to-background ratio than electrons Can be focused Very sensitive to low concentrations	Specimens must be in vacuum with source Expensive equipment
Photons X-rays or γ-rays	Convenient — specimen need not be in vacuum Widely used Wavelength can be chosen for maximum sensitivity for element of interest	Cannot be focused Not as sensitive to small samples as positive ion excited methods Light elements ($<$ Mg difficult)

Table 4.23 COMPARISON OF X-RAY AND RADIOACTIVE SOURCES FOR X-RAY FLUORESCENT SPECTROSCOPY

Radiation source	*Advantages*	*Limitations*
X-rays	Controllable high-intensity source which can be switched off when not required Intensity can be 10^2–10^4 times that of radio-isotope source	Bulky, expensive equipment — requires high voltage generator
Radio isotopes	Cheaper, portable, and smaller than X-ray systems Can be built into process plant for local on-stream analysis	Permanent radioactive hazard Low intensity means long exposure times and/or larger samples Relatively small number of available isotopes (*see* Table 4.24)

Table 4.24 RADIO-ISOTOPE SOURCES FOR X-RAY FLUORESCENCE

Nuclide	*Half-life*	*Emission energies* keV
^{55}Fe	2.7 years	5.9
^{109}Cd	453 days	22.1, 87.7
^{125}I	60 days	27
^{241}Am	458 years	12.17, 60
^{57}Co	270 days	6.4, 1.22, 144
^{238}Pu	86.4 years	12–17

Source: Jaklevic and Goulding, in ref. 12, p. 33.

Table 4.25 3σ DETECTION LIMITS FOR X-RAY EXCITED SAMPLES

	Experimental conditions		
	Wavelength dispersion Cr tube 2 500 W 100 s *analysis time*	*Energy dispersive Ag tube* 10 min *analysis time*	
		Ag *filter* 1.8 W	Ag + *W filter* 22 W
Element	3σ *Detection limits in* 10^{-9} gcm^{-2}		
Mg	2	80	
Al	3	40	
Si	3	20	
S	9	12	
Cl	9	16	
Ca	2	28	
Fe	18		140 Fe impurity in Be window
Zn	7		30
Br	28		20
Pb	30		50

Source: J. V. Gilfrich, in ref. 12, p. 405.

Table 4.26 3σ DETECTION LIMITS FOR BULK SAMPLES

	Experimental conditions: Wavelength dispersion; All measurements in p.p.m.	
Element	*Iron and steel sample W target,* 2 240 W 10 min *analysis time*	Fe *and* Ni *base alloys W target* 2 025 W 100 s
Si	170	4
P		35
S		8
Ti	1.0	
V	1.9	
Cr	4.0	1
Mn	1.4	5
Ni	5.4	
Ca	8.5	12
As	6.8	
Zr	4.6	
Mo	4.5	22
Sn	3.9	

Source: J. V. Gilfrich, in ref. 12, p. 408.

Table 4.27 EXAMPLES OF DETECTION LIMITS FOR ION EXCITATION

Sample mount	*Ions*	*Energy* MeV	*Current or charge*	*Time*	*Detection limit*	*Detection limit criterion*
1 mg cm^{-2} VYNS	α	50	1 nA	400 s	Cu 1.9×10^{-12} g Sn 3.2×10^{-12} g Pb 5.5×10^{-12} g	P/B = 0.1
10–20 μg cm^{-2} carbon or nitrocellulose	p^+	5	5 μC	100–200 s	K 1×10^{-9} g cm^{-2} Cu 2×10^{-9} g cm^{-2} Br 1×10^{-9} g cm^{-2} Au 5×10^{-9} g cm^{-2}	3σ Bgd
	α	5	5 μC	100–200 s	K 1×10^{-9} g cm^{-2} Br 10×10^{-9} g cm^{-2} Au 20×10^{-9} g cm^{-2}	3σ Bgd
40 μg cm^{-2} carbon	p^+	1.5	5 μA	30 min	Ca 0.3×10^{-12} g Cu 1×10^{-12} g Ba 20×10^{-12} g Pb 10×10^{-12} g	100 counts above Bgd

Table 4.27 EXAMPLES OF DETECTION LIMITS FOR ION EXCITATION—*continued*

Sample mount	*Ions*	*Energy* MeV	*Current or charge*	*Time*	*Detection limit*	*Detection limit criterion*
4 μm Mylar	p^+	1	10 μC	500 s	Ca 7×10^{-9} g cm^{-2} Zn 18×10^{-9} g cm^{-2} Zr 300×10^{-9} g cm^{-2} Pb 90×10^{-9} g cm^{-2}	3σ Bgd
		0.3	10 μC	500 s	Ca 3×10^{-9} g cm^{-2} Zn 5×10^{-9} g cm^{-2} Zr 30×10^{-9} g cm^{-2} Pb 23×10^{-9} g cm^{-2}	3σ Bgd

Source: J. V. Gilfrich, in ref. 2, p. 406.

4.7 Radiation screening

In using X-rays, radio-isotope or accelerator-based sources of radiation, exposure to individuals must be controlled to be as low as is possible but in any event to be less than the values indicated in Table 4.28. In most countries, persons processing, using, selling or transporting radioactive materials must be licensed by a national authority. In England the authority is the Department of the Environment, in Wales it is the Welsh Office, in Scotland it is the Scottish Development Department.

4.7.1 Definitions

EXPOSURE

Exposure is a measure of the intensity of ionizing radiation multiplied by time. It is measured in roentgens (R) or coulombs per kilogramme (C kg^{-1}) in SI system units. An exposure of one roentgen implies the production of a stated number of ion pairs per unit mass of air.

ABSORBED DOSE

Absorbed dose is a measure of the energy absorbed in a stated material when exposed to ionizing radiation under stated conditions. It is measured in rads, or in grays (Gy) in SI system units.

DOSE EQUIVALENT

An absorbed dose has a biological effect which depends upon the type of ionizing radiation and on the end-effect under consideration. The relationship between the absorbed dose (A) and dose equivalent (D) is determined by the quality factor (Q) (originally described as relative biological effectiveness) in the equation

$$D = A \times Q$$

Dose equivalent is measured in rems, or in sieverts (Sv) in SI system units. Q is approximately 1 for X-rays, γ-rays and electrons; it is in the range of approximately 1–11 for neutrons, and can be as high as 20 for α particles.

BODY DOSE (EQUIVALENT)

This is the dose received at a specified position in the body, usually the gonads or the blood-forming organs. It is normally taken as the penetrating component of absorbed dose measured by a film badge at or near the chest or trunk of the body, plus any neutron absorbed dose multiplied by a quality factor of 10. It is also normally assumed that this is the dose received by the whole body.

The SI unit of activity is the becquerel (Bq), equal to one nuclear transformation per second. 3.7×10^{10} Bq is exactly equivalent to 1 Ci (curie). The SI unit of absorbed dose is the gray (Gy) equal to one joule per kilogram (J kg^{-1}). One gray equals 100 rads exactly.

The SI unit of dose equivalent is the sievert (Sv) equal to one joule per kilogram (J kg^{-1}). One sievert equals 100 rems exactly. The SI unit of exposure is the coulomb per kilogram (C kg^{-1}); 1 coulomb per kilogram is 3 876 roentgens.

Table 4.28 CURRENT DOSE LIMITS

Organ, tissue, part of body	*Adult male occupationally exposed* (Sv × 100 = rem)
Gonads, red bone marrow or whole body	0.05 Sv per year 0.03 Sv per quarter
Hands, forearms, feet and ankles	0.75 Sv per year 0.40 Sv per quarter
Other bones and skin, thyroid	0.30 Sv per year 0.15 Sv per quarter
Any other organ	0.15 Sv per year 0.08 Sv per quarter

ADULT FEMALE OCCUPATIONALLY EXPOSED PERSONS

As in Table 4.28, but abdomen quarterly dose not to exceed 0.008 Sv; after declaration of pregnancy, abdomen dose for remainder of pregnancy not to exceed 0.01 Sv.

YOUNG PERSONS

No person under 16 years of age should be employed in work which may involve exposure to ionizing radiation. Persons who are over 16 but under 18 years of age are limited to 0.3 times the maximum doses set out in Table 4.28.

OTHER PERSONS

All other persons are limited to one tenth of the annual maximum permissible doses set out in Table 4.28.

Table 4.29 THICKNESS OF CONCRETE REQUIRED TO REDUCE BROAD BEAM PULSATING OR CONSTANT POTENTIAL X-RAYS BY THE TRANSMISSION FACTOR GIVEN

Transmission factor		*1/10*	*1/100*	*1/1 000*	*1/10 000*	*1/100 000*	*1/1 000 000*
Tube potential kV	*Total filtration* mm Al			*Concrete thickness* cm			
50	1.0	1	2	3	4.5	6	7
70	1.5	2	4.5	8	12	15.5	19.5
100	2.0	3.5	8.5	14	19.5	25	30.5
125	3.0	4.0	10.5	17	23.5	30	36.5
150	3.0	5.5	12.5	19.5	26.5	33.5	41
200	3.0	6.5	15	23.5	32	40.5	49
250	3.0	7.5	17	26	35	44.5	54
300	3.0	8.0	18	28	38	48	58
300	3.0 Cu	13	22	31.5	41	50.5	60
400	3.0 Cu	14	24	34	44	54	64

Table 4.30 THICKNESS OF LEAD REQUIRED TO REDUCE BROAD BEAM PULSATING POTENTIAL X-RAYS BY THE TRANSMISSION FACTOR GIVEN

Transmission factor		*1/10*	*1/100*	*1/1 000*	*1/10 000*	*1/100 000*	*1/1 000 000*
Tube potential kV	*Total filtration* mm Al	*Lead thickness* cm					
50	1.8	0.01	0.02	0.04	0.065	0.085	0.11
75	3.8	0.015	0.05	0.10	0.155	0.215	0.27
100	4.0	0.023	0.075	0.15	0.24	0.330	0.417
125	3.2	0.03	0.085	0.165	0.26	0.36	0.46
150	3.2	0.035	0.10	0.185	0.28	0.38	0.48
200	3.3	0.055	0.13	0.25	0.39	0.53	0.68
250	3.0	0.08	0.24	0.46	0.72	1.02	1.32
300	4.0 Cu	0.17	0.46	0.88	1.32	1.80	2.30

Table 4.31 THICKNESS OF LEAD REQUIRED TO REDUCE BROAD BEAM CONSTANT POTENTIAL X-RAYS BY THE TRANSMISSION FACTOR GIVEN

Transmission factor		*1/10*	*1/100*	*1/1 000*	*1/10 000*	*1/100 000*	*1/1 000 000*
Tube potential kV	*Total filtration* Al Cu mm	*Lead thickness* cm					
50	2 or 0.07	0.005	0.015	0.03	0.05	0.07	0.09
75	2 or 0.07	0.015	0.04	0.08	0.125	0.19	0.25
100	2 or 0.07	0.02	0.07	0.15	0.23	0.32	0.415
150	2 or 0.07	0.03	0.095	0.175	0.27	0.37	0.47
200	2 or 0.07	0.05	0.14	0.255	0.385	0.52	0.66
250	— 0.5	0.08	0.24	0.44	0.70	1.0	1.32
300	— 0.5	0.16	0.40	0.70	1.08	1.5	1.94
400	— 3.0	0.34	0.84	1.52	2.32	—	—

NOTES TO TABLES 4.29–4.31

1. Dose in this context is the attributable dose equivalent, i.e. including contributions from internal and external irradiations where appropriate, and contributions from ionizing radiations with different quality factors.
2. The whole body dose is normally assumed to be the penetrating component of the dose recorded on a film badge worn on or near the chest.

FURTHER READING—RADIATION SCREENING

Recommendations of the International Commission on Radiological Protection, ICRP Publication 9, Pergamon, Oxford, 1966.

Protection against Ionising Radiation from External Sources, ICRP Publication 15, Pergamon, Oxford, 1970.

Data for Protection against Ionising Radiation from External Sources, ICRP Publication 21, Pergamon, Oxford, 1973.

Recommendations of the International Commission on Radiological Protection, ICRP Publication 26, Pergamon, Oxford, 1977.

Basic Radiation Protection Criteria, NCRP Report No. 39, 1971.

Radiation Protection Design Guidelines for 0.1 to 100 MeV Particle Accelerator Facilities. NCRP Report No. 51, 1977.

Concrete and Lead Screening for X-rays, 'Handbook of Radiological Protection', National Radiation Protection Board, 1971.

REFERENCES

1. Joint Committee on Powder Diffraction Standards (JC & DS).
2. J. D. Hanawalt, 'Advances in X-Ray Analysis', Vol. 20, p. 63, Plenum Press, New York, 1976.
3. J. Durnin and K. A. Ridal, *Journal of the Iron and Steel Institute*, Jan. 1968, p. 60.
4. 'International Tables for X-Ray Crystallography', Kynoch Press, 1962.
5. 'Residual stress measurement by X-Ray Diffraction', J.786a, Society of Automotive Engineers, New York, 1971.
6. D. Kirk, *Strain*, Jan. 1971, p. 7.
7. B. D. Cullity, 'Advances in X-Ray Analysis', Vol. 20, p. 259, Plenum Press, New York, 1976.
8. W. G. Wyckoff, 'The Structure of Crystals', New York, 1931.
   W. H. and W. L. Bragg, 'The Crystalline State', London, 1939.
   R. C. Evans, 'An Introduction to Crystal Chemistry', Cambridge, 1939.
   W. Hume-Rothery, 'The Structure of Metals and Alloys', London, 1944.
9. L. Pauling, *J. Am. Chem. Soc.*, 1929, **51**, 1010.
10. V. M. Goldschmidt, 'Geochemische Verteilungsgesetze der Elemente', Oslo, 1923–1038. 'Kristallchemie in Handwörterbuch d. Naturwissensch', Jena, 1934.
11. J. A. Wasastjerna, *Comment. Phys.-Math., Helsinaf*, 1923, **1**, 38.
12. H. K. Herglotz and L. S. Birks (Eds), 'X-Ray Spectrometry', Marcel Dekker, New York, 1978.

5 Crystallography

5.1 The structure of crystals

5.1.1 Translation groups

Metals and alloys, like most solid matter, are aggregates of crystals; they are built up of units, consisting of small groups of atoms regularly and indefinitely repeated throughout the body by parallel translations. If the co-ordinates of the atoms within such a group are given, then three independent translations represented by vectors $\boldsymbol{a}, \boldsymbol{b}, \boldsymbol{c}$, which are not all parallel to the same plane, suffice to specify the position of any other atom in the crystal. Let the vector from an arbitrary origin to an atom be

$$r = x\boldsymbol{a} + y\boldsymbol{b} + z\boldsymbol{c}$$

then atoms of the same kind will be found at all points

$$r_n = (n_1 + x)\boldsymbol{a} + (n_2 + y)\boldsymbol{b} + (n_3 + z)\boldsymbol{c}$$

where n_1, n_2, n_3 may be any positive or negative integers. Such a succession of regularly arranged points in space constitutes a space lattice.

The lattice may be regarded either as a system of translations relating identical points in a structure, or as a system of points arranged in parallel and equidistant nets, each net consisting of series of parallel and equidistant rows in which the points are spaced at equal distances. The points of such an array can be arbitrarily arranged in an infinite number of ways in parallel equidistant linear rows or planar nets; they can, in other words, be referred to an infinite number of systems of three primitive vectors, but investigation has shown that any structure possessing the symmetry observed in crystals can be referred to one of 14 lattices, defined by its primitive vectors and by the character of its unit cell, the latter being the parallelepiped formed by the three translations selected as units. In general, unit translations are selected so as to give the simplest cell having edges as short as possible, but there are several cases in which a more complex cell is chosen so as to display the symmetry of the lattice, or its relation to other lattices, to greater advantage.

The system of three vectors $\boldsymbol{a}, \boldsymbol{b}, \boldsymbol{c}$, is described by their lengths a, b, c and by the angles between them: $(bc) = \alpha$, $(ca) = \beta$, $(ab) = \gamma$. The face of the unit cell which is parallel to the plane of the (a) and (b) axes, and which therefore intersects the (c) axis at distance c from the origin is termed the c face. Similarly, the face parallel to the b–c axial plane is the a face, and that parallel to the a–c axial plane the b face.

The simplest cell, having points only at its corners, is termed 'primitive' and is given the symbol Γ (Schoenflies) or P (Hermann). Other cells, termed 'face centred', have points at the corners and at the centres of two or more of their faces, and are given symbols indicating the faces carrying these additional points. Thus A, B, C, F represent centring on the a, b, c and all faces respectively. Finally there is the 'body centred' cell, having points at its corners and one additional point at the intersection of the body diagonals. This is given the symbol I (Hermann). Centred cells are indicated by Schoenflies by dashes. The 14 lattices are listed in Table 5.1.

5.1.2 Symmetry elements

Symmetry elements may be classified as axes, planes, and centres. A body has an axis of symmetry when rotation through a definite angle about some line through it (the axis of rotation) causes it to

Table 5.1 THE FOURTEEN LATTICES

System	Axes	Angles	Unit cells	Symbols	
				Schoenflies	Hermann
Triclinic	$a \neq b \neq c$	$\alpha \neq \beta \neq \gamma$	Primitive	Γtr	P
Monoclinic	$a \neq b \neq c$	$\alpha = \gamma = 90°$ β obtuse	*1* Primitive *2* c face centred	Γm $\Gamma m'$	P C*
Orthorhombic	$a \neq b \neq c$	$\alpha = \beta = \gamma = 90°$	*1* Primitive *2* Face centred *3* All face centred *4* Body centred	Γo $\Gamma o'(c)$ $\Gamma o''$ F'''	P C† F† I
Tetragonal	$a = b \neq c$	$\alpha = \beta = \gamma = 90°$	*1* Primitive *2* Body centred	Γt $\Gamma t'$	P I
Cubic	$a = b = c$	$\alpha = \beta = \gamma = 90°$	*1* Primitive *2* All face centred *3* Body centred	Γc $\Gamma c'$ $\Gamma c''$	P F I
Rhombohedral	$a = b = c$	$\alpha = \beta = \gamma \neq 90°$	Primitive	Γrh	R
Hexagonal	$a = b \neq c$	$\alpha = \beta = 90°$ $\gamma = 120°$	Primitive	Γh	C‡

* By choosing different a and b axes the c-centred monoclinic cell will be seen to be equivalent to a primitive cell in which $a = b \neq c$, $\alpha = \beta = 90°$, $\gamma \neq \alpha$. Even if these axes are selected, the symbol C is retained for this cell.
† By suitable change of axes it is possible to convert the orthorhombic C and F cells into primitive and body centred cells respectively having $a = b \neq c$, $\alpha = \beta = 90°$, γ obtuse. The symbols for these alternative settings remain C and F.
‡ This is a special form of the alternative setting described in the preceding note for the orthorhombic C cell. It is therefore given the symbol C.

assume its original aspect. Crystals have been observed to have axes of 2-, 3-, 4- and 6-fold reflection, involving coincidence after rotation through 180°, 120°, 90° and 60° respectively. If a plane can be passed through a body such that every point on one side of the plane stands in mirror-image relationship to a corresponding point on the other side, the plane is said to be a reflecting plane or plane of symmetry. A point within a body is a centre of symmetry or centre of inversion if a line drawn from any point of the body to the centre and extended to an equal distance beyond it encounters a corresponding point. Other symmetry operations are:

Rotary reflection, involving rotation through a definite angle, combined with reflection in a plane normal to the axis.
Screw axes of rotation, combining rotation about an n-fold axis with a translation of a specified length in the direction of the axis.
Glide planes, combining reflection with a translation parallel to the plane of the mirror.

In the case of screw axes, the amount of the shift must be a rational fraction of the translation along the same axis, the denominator of the fraction being the multiplicity of the rotation. Thus for a 6-fold axis, the shift may be 1/6, 2/6, 3/6... of the translation. In the case of the glide plane, the shift must be one half of some translation in that plane. Thus it may be $a/2$ or $b/2$ parallel to the a and b axes, or of half the face diagonal in the direction parallel to that diagonal. If the cell is centred on that particular face, the shift may be one half of the distance to the centre, i.e. of the face diagonal.

5.1.3 The point group

The point group may be defined as a group of symmetry elements distributed about a point in space, and may be conveniently visualized as an assembly of points generated by the operation of the symmetry elements in question upon a single point having co-ordinates xyz referred to specified axes, the symmetry elements passing through the origin. Thus a symmetry plane, passing through the origin and containing the x and y axes, will generate, from the point xyz, an equivalent point of which the co-ordinates are $xy\bar{z}$. These two points serve to characterize the point group.

The 32-point groups define all the ways in which axes, planes and centres of symmetry can be distributed so as to intersect in a point in space, and correspond to the 32 classes of morphological crystallography.

5.1.4 The space group

The space group may be defined as an extended network of symmetry elements distributed about the points of a space lattice, and may be visualized as an assembly of points generated by the operation of symmetry elements on a series of points situated identically in each cell of the lattice. Whereas in the point group the repeated operation of any symmetry element must ultimately bring each point back to its original position, in the space group an operation need only bring the point to an analogous position in the same or in another cell of the lattice. Thus in the space group the more complex symmetry operation of screw axes and glide planes are possible, combining translation with reflection and rotation.

Point groups, placed at the points of space lattices belonging to the same system of symmetry, give rise to the simplest of the 230 space groups. The remainder are generated by replacing the simple planes and axes of the point group by glide planes and screw axes.

5.2 The Schoenflies system of point- and space-group notation

The symmetry elements chosen by Schoenflies are axes of n-fold rotation, reflection planes and centres of inversion. The symbols assigned to the various point groups are as follows:

C_n = groups having a single n-fold axis.
C_n^h = groups having a single vertical n-fold axis, together with a horizontal reflection plane
C_n^v = groups having a single vertical n-fold axis, together with n vertical reflection planes
D_n = groups having an n-fold axis and n two-fold axes at right angles to it
V = a symbol frequently used as an alternative to D_2
O = the two cubic groups which possess the maximum possible number of rotation axes, namely four 4-fold axes parallel to the cube edges, four 3-fold axes parallel to the cube diagonals and six 2-fold axes parallel to the face diagonals
T = the three remaining groups of the cubic system
S_n = groups having an n-fold axis of rotary reflection

The suffix i signifies a centre of inversion.
The suffix s signifies a single plane of symmetry.
The suffix d signifies a diagonal reflection plane, bisecting the angle between two horizontal axes.

The symbols for the space groups are simple modifications of those for the point groups: the index and subscript of the point group are combined to give the subscript of the space group symbol, and an index is added representing the order in which Schoenflies deduced the symmetry of the group. Thus C_{nh}^m represents the mth group derived from the point group C_n^h.

Table 5.2 gives the point groups, their symbols and elements of symmetry, the crystal classes with which they correspond, and the co-ordinates of equivalent points.

5.3 The Hermann–Mauguin system of point- and space-group notation

The symbols used by Hermann and Mauguin to indicate the various symmetry operations are as follows:

Rotation axes: the number 2, 3, 4 or 6 denoting the multiplicity.
Screw axes: the symbol denoting the multiplicity of rotation, with a subscript indicating the magnitude of the shift. The complete set of screw axes is 2_1; 3_1; 3_2; 4_1, 4_2, 4_3; 6_1, 6_2, 6_3, 6_4, 6_5.
Axes of rotary reflection: $\bar{2}$, $\bar{3}$, $\bar{4}$, $\bar{6}$, the numeral indicating the multiplicity.
Centre of inversion: $\bar{1}$.
Reflection plane: m.
Glide plane: with shift in the a direction: a
with shift in the b direction: b
with shift of $\frac{1}{2}$ the face diagonal: n
with shift of $\frac{1}{2}$ the centring translation: d.

The full space group symbol consists of the translation (or lattice) symbol followed by symbols of the symmetry elements associated with specified crystallographic directions in a specified order. The direction associated with a reflection or glide plane is that of its normal: no direction can be specified for a centre of inversion.

Table 5.2 POINT-GROUP

System	Class no.	Schoenflies symbol	Crystal class: Schoenflies	Crystal class: Dana	Crystal class: Miers
Triclinic	1	C_1	Hemihedry	Asymmetric	Asymmetric
	2	C_i	Holohedry	Normal	Central
Monoclinic	3	$C_s = C_1^h$	Hemihedry	Clinohedral	Planar
	4	C_2	Hemimorphic hemihedry	Hemimorphic	Digonal polar
	5	C_2^h	Holohedry	Normal	Digonal equatorial
Orthorhombic	6	C_2^v	Hemimorphic hemihedry	Hemimorphic	Didigonal polar
	7	$V = D_2$	Enantiomorphic hemihedry	Sphenoidal	Digonal holoaxial
	8	$V^h = D_2^i$	Holohedry	Normal	Didigonal equatorial
Tetragonal	9	S_4	Tetartohedry of 2nd sort	Tetartohedral	Tetragonal alternating
	10	$V^4 = D_2^d$	Hemihedry of 2nd sort	Sphenoidal	Ditetragonal alternating
	11	C_4	Tetartohedry	Pyramidal hemimorphic	Tetragonal polar
	12	C_4^h	Paramorphic hemihedry	Pyramidal	Tetragonal equatorial
	13	C_4^v	Hemimorphic hemihedry	Hemimorphic	Ditetragonal polar
	14	D_4	Enantiomorphic hemihedry	Trapezohedral	Tetragonal holoaxial
	15	D_4^h	Holohedry	Normal	Ditetragonal equatorial
Cubic	16	T	Tetartohedry	Tetartohedral	Tesseral polar
	17	T^h	Paramorphic hemihedry	Pyritohedral	Tesseral central
	18	T^d	Hemimorphic hemihedry	Tetrahedral	Ditesseral polar
	19	O	Enantiomorphic hemihedry	Plagihedral	Tesseral holoaxial
	20	O^h	Holohedry	Normal	Ditesseral central
Rhombohedral	21	C_3	Tetartohedry	Not named	Trigonal polar
	22	$C_3^i = S_5$	Hexagonal tetartohedry of 2nd sort	Trirhombohedral	Hexagonal alternating
	23	C_3^v	Hemimorphic hemihedry	Ditrigonal pyramidal	Ditrigonal polar
	24	D_3	Enantiomorphic hemihedry	Trapezohedral	Trigonal holoaxial
	25	D_3^d	Holohedry	Rhombohedral	Dihexagonal alternating
Hexagonal	26	C_3^n	Trigonal paramorphic hemihedry	Not named	Trigonal equatorial
	27	D_3^h	Trigonal holohedry	Trigonotype	Ditrigonal equatorial
	28	C_6	Tetartohedry	Pyramidal hemimorphic	Hexagonal polar
	29	C_6^h	Paramorphic hemihedry	Pyramidal	Hexagonal equatorial
	30	C_6^v	Hemimorphic hemihedry	Hemimorphic	Dihexagonal polar
	31	D_6	Enantiomorphic hemihedry	Trapezohedral	Hexagonal holoaxial
	32	D_6^h	Holohedry	Normal	Dihexagonal equatorial

NOTATION (SCHOENFLIES)

Typical example	*Symmetry elements*	*Co-ordinates of equivalent points*
—	None	xyz
Copper sulphate	Centre of inversion	$xyz,\ \bar{x}\bar{y}\bar{z}$
—	Horizontal reflecting plane	$xyz,\ xy\bar{z}$
Tartaric acid	2-fold axis	$xyz,\ \bar{x}\bar{y}z$
Gypsum	2-fold axis and horizontal plane	$xyz,\ \bar{x}\bar{y}z,\ xy\bar{z},\ \bar{x}\bar{y}\bar{z}$
Topaz	2-fold axis and vertical plane	$xyz,\ \bar{x}\bar{y}z,\ \bar{x}yz,\ x\bar{y}z$
Sulphur	Three 2-fold axes	$xyz,\ x\bar{y}\bar{z},\ \bar{x}y\bar{z},\ \bar{x}\bar{y}z$
Barytes	Three 2-fold axes and horizontal plane	$xyz,\ x\bar{y}\bar{z},\ \bar{x}y\bar{z},\ \bar{x}\bar{y}z,\ xy\bar{z},\ x\bar{y}z,\ \bar{x}yz,\ \bar{x}\bar{y}\bar{z}$
—	4-fold rotary reflection	$xyz,\ \bar{y}x\bar{z},\ \bar{x}\bar{y}z,\ y\bar{x}\bar{z}$
Chalcopyrite	4-fold axis, two 2-fold axes and diagonal vertical plane	$xyz,\ x\bar{y}\bar{z},\ \bar{x}y\bar{z},\ \bar{x}\bar{y}z,\ yxz,\ \bar{y}x\bar{z},\ y\bar{x}\bar{z},\ y\bar{x}\bar{z},\ \bar{y}\bar{x}z$
Wulfenite	4-fold axis	$xyz,\ \bar{y}xz,\ \bar{x}\bar{y}z,\ y\bar{x}z$
Scheelite	4-fold axis, horizontal plane	$xyz,\ \bar{y}xz,\ \bar{x}\bar{y}z,\ y\bar{x}z,\ xy\bar{z},\ yx\bar{z},\ \bar{x}\bar{y}\bar{z},\ y\bar{x}\bar{z}$
—	4-fold axis, vertical plane	$xyz,\ \bar{y}xz,\ \bar{x}\bar{y}z,\ y\bar{x}z,\ \bar{x}yz,\ yxz,\ x\bar{y}z,\ \bar{y}\bar{x}z$
—	4-fold axis and four 2-fold axes	$xyz,\ \bar{y}xz,\ \bar{x}\bar{y}z,\ y\bar{x}z,\ x\bar{y}\bar{z},\ \bar{y}\bar{x}\bar{z},\ \bar{x}y\bar{z},\ yxz$
Zircon	4-fold axis, four 2-fold axes, horizontal plane	$xyz,\ \bar{y}xz,\ \bar{x}\bar{y}z,\ y\bar{x}z,\ x\bar{y}\bar{z},\ \bar{y}\bar{x}\bar{z},\ \bar{x}y\bar{z},\ \bar{y}x\bar{z}$ $xy\bar{z},\ \bar{y}x\bar{z},\ \bar{x}\bar{y}\bar{z},\ y\bar{x}\bar{z},\ x\bar{y}z,\ \bar{y}\bar{x}z,\ \bar{x}yz,\ yxz$
Ullmanite	Three 2-fold axes coincident with cube axes, *and* Four 3-fold axes coincident with cube diagonals	$xyz,\ x\bar{y}\bar{z},\ \bar{x}y\bar{z},\ \bar{x}\bar{y}z$ $zxy,\ \bar{z}x\bar{y},\ \bar{z}\bar{x}y,\ z\bar{x}\bar{y}$ $yzx,\ \bar{y}\bar{z}x,\ y\bar{z}\bar{x},\ \bar{y}z\bar{x}$
Pyrites	As for T plus a horizontal plane	Those of T plus $xy\bar{z},\ x\bar{y}z,\ \bar{x}yz,\ \bar{x}\bar{y}\bar{z}$ $zx\bar{y},\ zxy,\ \bar{z}\bar{x}\bar{y},\ z\bar{x}y$ $yz\bar{x},\ \bar{y}\bar{z}\bar{x},\ y\bar{z}x,\ \bar{y}zx$
Blende	As for T plus a diagonal vertical plane	Those of T plus $yxz,\ \bar{y}x\bar{z},\ y\bar{x}\bar{z},\ \bar{y}\bar{x}z$ $xzy,\ x\bar{z}\bar{y},\ \bar{x}\bar{z}y,\ \bar{x}z\bar{y}$ $zyx,\ \bar{z}\bar{y}x,\ \bar{z}y\bar{x},\ z\bar{y}\bar{x}$
Cuprite	As for T plus six 2-fold axes coincident with face diagonals thereby converting the three original 2-fold axes into 4-folds	Those of T plus $\bar{y}\bar{x}\bar{z},\ y\bar{x}z,\ \bar{y}xz,\ yx\bar{z}$ $\bar{x}\bar{z}\bar{y},\ \bar{x}zy,\ xz\bar{y},\ x\bar{z}y$ $\bar{z}\bar{y}\bar{x},\ zy\bar{x},\ z\bar{y}x,\ \bar{z}yx$
Galena	As for O plus a horizontal plane	Those of all the four preceding classes
—	3-fold axis	$xyz,\ zxy,\ yzx$ referred to rhombohedral axes
Dioptase	3-fold axis and centre of inversion	$xyz,\ zxy,\ yzx,\ \bar{x}\bar{y}\bar{z},\ \bar{z}\bar{x}\bar{y},\ \bar{y}\bar{z}\bar{x}$ (rhombohedral axes)
Tourmaline	3-fold axis and vertical plane	$xyz,\ zxy,\ yzx,\ yxz,\ xzy,\ zyx$ (rhombohedral axes)
Quartz	3-fold axis and three 2-fold axes	$xyz,\ zxy,\ yzx,\ \bar{y}\bar{x}\bar{z},\ \bar{x}\bar{z}\bar{y},\ \bar{z}\bar{y}\bar{x}$ (rhombohedral axes)
Calcite	3-fold axis, three 2-fold axes and diagonal vertical planes	$xyz,\ zxy,\ yzx,\ \bar{y}\bar{x}\bar{z},\ \bar{x}\bar{z}\bar{y},\ \bar{z}\bar{y}\bar{x}$ $\bar{x}\bar{y}\bar{z},\ \bar{z}\bar{x}\bar{y},\ \bar{y}\bar{z}\bar{x},\ yxz,\ xzy,\ zyx$ (rhombohedral axes)
—	3-fold axis and horizontal plane	$xyz,\ (y-x)\bar{x}z,\ \bar{y}(x-y)z,\ xy\bar{z},\ (y-x)\bar{x}\bar{z},\ \bar{y}(x-y)\bar{z}$ (hexagonal axes)
—	3-fold axis, three 2-fold axes and horizontal plane	$xyz,\ (y-x)\bar{x}z,\ \bar{y}(x-y)z,\ (x-y)\bar{y}\bar{z},\ yx\bar{z},\ \bar{x}(y-x)\bar{z}$ $xy\bar{z},\ (y-x)\bar{x}\bar{z},\ \bar{y}(x-y)\bar{z},\ (x-y)\bar{y}z,\ yxz,\ \bar{x}(y-x)z$ (hexagonal axes)
Nepheline	6-fold axes	$xyz,\ y(y-x)z,\ (y-x)\bar{x}z,\ \bar{x}\bar{y}z,\ \bar{y}(x-y)z,\ (x-y)xz$ (hexagonal axes)
Apatite	6-fold axis and horizontal plane	Those of C_6 plus $xy\bar{z},\ y(y-x)\bar{z},\ (y-x)\bar{x}\bar{z},\ \bar{x}\bar{y}\bar{z},\ \bar{y}(x-y)\bar{z},\ (x-y)x\bar{z}$ (hexagonal axes)
Greenockite	6-fold axis and vertical plane	Those of C_6 plus $\bar{x}(y-x)z,\ (y-x)yz,\ yxz,\ x(x-y)z,\ (x-y)\bar{y}z,\ \bar{y}\bar{x}z$ (hexagonal axes)
—	6-fold axis and six 2-fold axes	Those of C_6 plus $\bar{x}(y-x)\bar{z},\ (y-x)y\bar{z},\ xy\bar{z},\ x(x-y)\bar{z}\ (x-y)\bar{y}\bar{z},\ \bar{y}\bar{x}\bar{z}$ (hexagonal axes)
Beryl	6-fold axis, six 2-fold axes and horizontal plane	Those of all the four preceding classes

The specified directions are:

Triclinic system: none
Monoclinic system: the ***b*** direction, i.e. the *b* axis.
Orthorhombic system: the ***a***, ***b***, ***c*** directions, in that order.
Tetragonal, *Hexagonal*, *Rhombohedral* } *systems:* the ***a***, ***b***, and (***a***–***b***) directions, in that order. The direction represented by the vector difference is one of the diagonals of the *c*-face.
Cubic system: the directions ***c***, (***a*** + ***b*** + ***c***) and (***a*** − ***b***), in that order, i.e. the *c*-axis the cube diagonal and a diagonal of the *c*-face.

If a symmetry axis has a symmetry plane normal to it, the two symbols are combined in the form of a fraction, thus $\frac{2}{m}$, $\frac{4_1}{d}$, alternatively written $2/m$, $4_1/d$.

If one of the specified crystallographic directions has no symmetry element associated with it, this is indicated by inserting the symbol 1 in the appropriate position in the space group symbol. The 1 may be omitted without risk of misunderstanding if it occurs at the end of the space group symbol. Symmetry symbols may also be omitted if they can be derived from those already indicated. These abbreviated symbols are termed 'short'.

As already explained, the symmetry of a space group can be derived from that of a point group by placing the latter at the points of the various lattices appropriate to the crystal system, and by using glide planes and screw axes as well as reflection planes and rotation axes. Thus the point group symbol will contain no symbol for the lattice; its symmetry planes will be indicated by *m* and its axes by numbers specifying the multiplicity and without subscripts. Thus the point group $2/m$ will be associated with the space groups $P(2/m)$; $P(2_1/m)$; $P(2/c)$; $P(2_1/c)$; $C(2/m)$; and $C(2/c)$.

5.3.1 Notes on the space-group tables

For a full description of the space groups, reference should be made to the *Internationale Tabellen zur Bestimmung von Kristallstrukturen.*

If there are *n* symmetry elements associated with any space group, their operation upon any single point having co-ordinates *xyz* will give rise to a total of *n* points which may be termed geometrically equivalent. If, however, the co-ordinates *xyz* are such that the point lies, say, on an axis or plane of symmetry, then the number of equivalent positions will be reduced, while if it lies at the intersection of two elements the number will be reduced still further. A knowledge of these so-called special positions is of importance, because experience has shown that they are the positions which are frequently occupied by the atoms or ions in an actual crystal. In sodium chloride, for example, the four sodium ions are situated in one set of four equivalent positions, those having co-ordinates 000, $\frac{1}{2}\frac{1}{2}0$, $\frac{1}{2}0\frac{1}{2}$, $0\frac{1}{2}\frac{1}{2}$, whilst the four chlorine ions are situated in another set of four points, having co-ordinates $\frac{1}{2}\frac{1}{2}\frac{1}{2}$, $00\frac{1}{2}$, $0\frac{1}{2}0$, $\frac{1}{2}00$. The co-ordinates of all the special positions for each space group were given by R. W. G. Wyckoff in *The Analytical Expression of the Results of the Theory of Space Groups* (Washington Carnegie Institution, 1930) and are also listed in the *Internationale Tabellen.*

The last column of Table 5.3 gives the missing x-ray reflections characteristic of each space group. If the unit cell is centred on one or more faces, or is body centred, certain reflections will be absent, because in directions corresponding to the missing reflections the waves scattered by the atoms at the face or body centres will be exactly out of phase with those scattered by the atoms at the cell corners. In other words, the spacings of certain planes are halved, and odd-order reflections from these planes are destroyed. Thus with the body-centred lattice all reflections are absent for which $(h + k + l)$ is odd. Again, a glide plane halves the spacings in the direction of glide, and a 2-fold screw axis halves those along the axis. Consequently, odd order reflections are missing in these directions. Similarly with a 3-fold axis; the only reflections occurring in the direction of the axis are those for which (l) is a multiple of three.

In Table 5.3, x-ray reflections of the type indicated do not occur unless indices which are underlined are even, or unless the sum of indices joined together by brackets is even. Thus:

$00\underline{l}$ means that reflections will not occur unless l is even.
$0\underbrace{kl}$ means that reflections will not occur unless $(k + l)$ is even.
$\underbrace{hk}0$ means that reflections will not occur unless both h and k are even.
$\underbrace{hkl}$ means that reflections will not occur unless $(h + k + l)$ is even.

$\underline{\underbrace{hkl}}$ means that reflections will not occur unless the sums of any two indices are even.

$hh\underline{l}$ means that reflections will not occur if the first two indices are equal, unless the third index, l, is even.

A subscript 3, 4 or 6 means that the marked index, or the sum of the marked indices, must be a multiple of that number for reflections to occur. Thus

$0\underbrace{kl}_4$ means that reflections will not occur unless $(k + l)$ is a multiple of 4.

$\underbrace{hkl}_3$ means that reflections will not occur unless $h + 2k + l$ is a multiple of 3.

Table 5.3 THE HERMANN–MAUGUIN SYSTEM OF POINT- AND SPACE-GROUP NOTATION

Space group Hermann–Mauguin Full	Short	*Schoenflies*	*Missing spectra*
Triclinic system			
Class 1—C_1			
	$P1$	C_1^1	—
Class $\bar{1}$—C_i			
	$P\bar{1}$	C_i^1	—
Monoclinic system			
Class m—C_s			
	Pm	C_s^1	—
	Pc	C_s^2	$h0\underline{l}$
	Cm	C_s^3	$\underbrace{hk}l$
	Cc	C_s^4	$\underbrace{hk}l$, $h0\underline{l}$
Class 2—C_2			
	$P2$	C_2^1	—
	$P2_1$	C_2^2	$0\underline{k}0$
	$C2$	C_2^3	$\underbrace{hk}l$
Class $\frac{2}{m}$—C_{2h}			
	$P\frac{2}{m}$	C_{2h}^1	—
	$P\frac{2_1}{m}$	C_{2h}^2	$0\underline{k}0$
	$P\frac{2}{c}$	C_{2h}^4	$h0\underline{l}$
	$P\frac{2_1}{c}$	C_{2h}^5	$h0\underline{l}$, $0\underline{k}0$
	$C\frac{2}{m}$	C_{2h}^3	$\underbrace{hk}l$
	$C\frac{2}{c}$	C_{2h}^6	$\underbrace{hk}l$, $h0\underline{l}$
Orthorhombic system			
Class $mm2$ (short mm)—C_{2v}			
$Pmm2$	Pmm	C_{2v}^1	—
$Pmc2_1$	Pmc	C_{2v}^2	$h0\underline{l}$
Class $mm2$ (short mm)—C_{2v} continued			
$Pma2$	Pma	C_{2v}^4	$\underline{h}0l$
$Pmn2_1$	Pmn	C_{2v}^7	$\underbrace{h0}l$
$Pcc2$	Pcc	C_{2v}^3	$0k\underline{l}$, $h0\underline{l}$
$Pca2_1$	Pca	C_{2v}^5	$0k\underline{l}$, $\underline{h}0l$
$Pcn2$	Pcn	C_{2v}^6	$0k\underline{l}$, $\underbrace{h0l}$
$Pba2$	Pba	C_{2v}^8	$0\underline{k}l$, $\underline{h}0l$
$Pbn2_1$	Pbn	C_{2v}^9	$0\underline{k}l$, $\underbrace{h0l}$
$Pnn2$	Pnn	C_{2v}^{10}	$0\underbrace{kl}$, $\underbrace{h0l}$
$Cmm2$	Cmm	C_{2v}^{11}	$\underbrace{hk}l$
$Cmc2_1$	Cmc	C_{2v}^{12}	$\underbrace{hk}l$, $\underline{h}0\underline{l}$
$Ccc2$	Ccc	C_{2v}^{13}	$\underbrace{hk}l$, $0\underline{kl}$, $\underline{h}0\underline{l}$
$Amm2$	Amm	C_{2v}^{14}	$h\underbrace{kl}$
$Ama2$	Ama	C_{2v}^{16}	$h\underbrace{kl}$, $\underline{h}0\underline{l}$
$Abm2$	Abm	C_{2v}^{15}	$h\underbrace{kl}$, $0\underline{kl}$
$Aba2$	Aba	C_{2v}^{17}	$h\underbrace{kl}$, $0\underline{kl}$, $\underline{h}0\underline{l}$
$Fmm2$	Fmm	C_{2v}^{18}	$\underline{\underbrace{hkl}}$
$Fdd2$	Fdd	C_{2v}^{19}	$\underline{\underbrace{hkl}}$, $0\underbrace{kl}_4$, $\underbrace{h0l}_4$
$Imm2$	Imm	C_{2v}^{20}	$\underbrace{hkl}$
$Ima2$	Ima	C_{2v}^{22}	$\underbrace{hkl}$, $\underline{h}0\underline{l}$
$Iba2$	Iba	C_{2v}^{21}	$\underbrace{hkl}$, $0\underline{kl}$, $\underline{h}0\underline{l}$
Class 222 (short 22)—$D_2(V)$			
$P222$		$D_2^1(V^1)$	—
$P222_1$		$D_2^2(V^2)$	$00\underline{l}$
$P2_12_12$		$D_2^3(V^3)$	$\underline{h}00$, $0\underline{k}0$
$P2_12_12_1$		$D_2^4(V^4)$	$\underline{h}00$, $0\underline{k}0$, $00\underline{l}$
$C222$		$D_2^6(V^5)$	$\underbrace{hk}l$
$C222_1$		$D_2^5(V^6)$	$\underbrace{hk}l$, $00l$
$F222$		$D_2^7(V^7)$	$\underline{\underbrace{hkl}}$
$I222$		$D_2^8(V^8)$	$\underbrace{hkl}$
$I2_12_12_1$		$D_2^9(V^9)$	$\underbrace{hkl}$

Table 5.3 THE HERMANN–MAUGUIN SYSTEM OF POINT- AND SPACE-GROUP NOTATION—*continued*

Space group Hermann–Mauguin *Full*	*Short*	*Schoenflies*	*Missing spectra*
Class $\frac{2}{m}\frac{2}{m}\frac{2}{m}$ (short *mmm*)—$D_{2h}(V_h)$			
$P\frac{2}{m}\frac{2}{m}\frac{2}{m}$	*Pmmm*	$D^1_{2h}(V_h^1)$	—
$P\frac{2_1}{m}\frac{2_1}{m}\frac{2}{n}$	*Pmmn*	$D^{13}_{2h}(V_h^{13})$	*hk*0
$P\frac{2_1}{n}\frac{2_1}{n}\frac{2}{m}$	*Pnnm*	$D^{12}_{2h}(V_h^{12})$	0*kl*, *h*0*l*
$P\frac{2}{n}\frac{2}{n}\frac{2}{n}$	*Pnnn*	$D^2_{2h}(V_h^2)$	0*kl*, *h*0*l*, *hk*0
$P\frac{2_1}{m}\frac{2}{m}\frac{2}{a}$	*Pmma*	$D^5_{2h}(V_h^5)$	*hk*0
$P\frac{2}{n}\frac{2_1}{n}\frac{2}{a}$	*Pnna*	$D^6_{2h}(V_h^6)$	0*kl*, *h*0*l*, *hk*0
$P\frac{2}{m}\frac{2}{n}\frac{2_1}{a}$	*Pmna*	$D^7_{2h}(V_h^7)$	*h*0*l*, *hk*0
$P\frac{2_1}{n}\frac{2_1}{m}\frac{2_1}{a}$	*Pnma*	$D^{16}_{2h}(V_h^{16})$	0*kl*, *hk*0
$P\frac{2_1}{b}\frac{2_1}{a}\frac{2}{m}$	*Pbam*	$D^9_{2h}(V_h^9)$	0*kl*, *h*0*l*
$P\frac{2}{b}\frac{2}{a}\frac{2}{n}$	*Pban*	$D^4_{2h}(V_h^4)$	0*kl*, *h*0*l*, *hk*0
$P\frac{2}{c}\frac{2}{c}\frac{2}{m}$	*Pccm*	$D^3_{2h}(V_h^3)$	0*kl*, *h*0*l*
$P\frac{2_1}{c}\frac{2_1}{c}\frac{2}{n}$	*Pccn*	$D^{10}_{2h}(V_h^{10})$	0*kl*, *h*0*l*, *hk*0
$P\frac{2}{b}\frac{2_1}{c}\frac{2_1}{m}$	*Pbcm*	$D^{11}_{2h}(V_h^{11})$	0*kl*, *h*0*l*
$P\frac{2_1}{b}\frac{2}{c}\frac{2_1}{n}$	*Pbcn*	$D^{14}_{2h}(V_h^{14})$	0*kl*, *h*0*l*, *hk*0
$P\frac{2_1}{b}\frac{2_1}{c}\frac{2_1}{a}$	*Pbca*	$D^{15}_{2h}(V_h^{15})$	0*kl*, *h*0*l*, *hk*0
$P\frac{2_1}{c}\frac{2}{c}\frac{2}{a}$	*Pcca*	$D^8_{2h}(V_h^8)$	0*kl*, *h*0*l*, *hk*0
$C\frac{2}{m}\frac{2}{m}\frac{2}{m}$	*Cmmm*	$D^{19}_{2h}(V_h^{19})$	*hkl*
$C\frac{2}{m}\frac{2}{m}\frac{2}{a}$	*Cmma*	$D^{21}_{2h}(V_h^{21})$	*hkl*, *hk*0
$C\frac{2}{c}\frac{2}{c}\frac{2}{m}$	*Cccm*	$D^{20}_{2h}(V_h^{20})$	*hkl*, 0*kl*, *h*0*l*
$C\frac{2}{c}\frac{2}{c}\frac{2}{a}$	*Ccca*	$D^{22}_{2h}(V_h^{22})$	*hkl*, 0*kl*, *h*0*l*, *hk*0
$C\frac{2}{m}\frac{2}{c}\frac{2_1}{m}$	*Cmcm*	$D^{17}_{2h}(V_h^{17})$	*hkl*, *h*0*l*

Space group Hermann–Mauguin *Full*	*Short*	*Schoenflies*	*Missing spectra*
Class $\frac{2}{m}\frac{2}{m}\frac{2}{m}$ (short *mmm*)—$D_{2h}(V_h)$—*continued*			
$C\frac{2}{m}\frac{2}{c}\frac{2_1}{a}$	*Cmca*	$D^{18}_{2h}(V_h^{18})$	*hkl*, *h*0*l*, *hk*0
$F\frac{2}{m}\frac{2}{m}\frac{2}{m}$	*Fmmm*	$D^{23}_{2h}(V_h^{23})$	*hkl*
$F\frac{2}{d}\frac{2}{d}\frac{2}{d}$	*Fddd*	$D^{24}_{2h}(V_h^{24})$	*hkl*, $0kl_4$, $h0l_4$, $hk0_4$
$I\frac{2}{m}\frac{2}{m}\frac{2}{m}$	*Immm*	$D^{25}_{2h}(V_h^{25})$	*hkl*
$I\frac{2}{b}\frac{2}{a}\frac{2}{m}$	*Ibam*	$D^{26}_{2h}(V_h^{26})$	*hkl*, 0*kl*, *h*0*l*
$I\frac{2_1}{m}\frac{2_1}{m}\frac{2_1}{a}$	*Imma*	$D^{28}_{2h}(V_h^{28})$	*hkl*, *hk*0
$I\frac{2_1}{b}\frac{2_1}{c}\frac{2_1}{a}$	*Ibca*	$D^{27}_{2h}(V_h^{27})$	*hkl*, 0*kl*, *h*0*l*, *hk*0
Tetragonal system			
Class $\bar{4}$—S_4			
	$P\bar{4}$	S_4^1	—
	$I\bar{4}$	S_4^2	*hkl*
Class 4—C_4			
	*P*4	C_4^1	—
	$P4_2$	C_4^3	00*l*
	$P4_1$, $P4_3$	C_4^2, C_4^4	$00l_4$
	*I*4	C_4^5	*hkl*
	$I4_1$	C_4^6	*hkl*, $00l_4$
Class $\frac{4}{m}$—C_{4h}			
	$P\frac{4}{m}$	C^1_{4h}	—
	$P\frac{4}{n}$	C^3_{4h}	*hk*0
	$P\frac{4_2}{m}$	C^2_{4h}	00*l*
	$P\frac{4_2}{n}$	C^4_{4h}	*hk*0, 00*l*
	$I\frac{4}{m}$	C^5_{4h}	*hkl*
	$I\frac{4_1}{a}$	C^6_{4h}	*hkl*, *hk*0 $00l_4$
Class $\bar{4}2m$ (or, in other orientation, $\bar{4}m2$)—$D_{2d}(V_d)$			
	$P\bar{4}2m$	$D^1_{2d}(V^1_d)$	—

Table 5.3 THE HERMANN–MAUGUIN SYSTEM OF POINT- AND SPACE-GROUP NOTATION—*continued*

Space group Hermann–Mauguin *Full*	*Short*	*Schoen-flies*	*Missing spectra*
Class $\bar{4}2m$ (or, in other orientation, $\bar{4}m2$)—$D_{2d}(V_d)$—*continued*:			
	$P\bar{4}2c$	$D_{2d}^{2}(V_d^{2})$	hhl
	$P\bar{4}2_1m$	$D_{2d}^{3}(V_d^{3})$	$h00$
	$P\bar{4}2_1c$	$D_{2d}^{4}(V_d^{4})$	hhl, $h00$
	$P\bar{4}m2$	$D_{2d}^{5}(V_d^{5})$	—
	$P\bar{4}c2$	$D_{2d}^{6}(V_d^{6})$	$0kl$
	$P\bar{4}b2$	$D_{2d}^{7}(V_d^{7})$	$0kl$
	$P\bar{4}n2$	$D_{2d}^{8}(V_d^{8})$	$0kl$
	$I\bar{4}2m$	$D_{2d}^{11}(V_d^{11})$	hkl
	$I\bar{4}2d$	$D_{2d}^{12}(V_d^{12})$	hkl, hhl_4
	$I\bar{4}m2$	$D_{2d}^{9}(V_d^{9})$	hkl
	$I\bar{4}c2$	$D_{2d}^{10}(V_d^{10})$	hkl, $0kl$
Class $4mm$—C_{4v}			
$P4mm$	$P4mm$	C_{4v}^{1}	—
$P4_2mc$	$P4mc$	C_{4v}^{7}	hhl
$P4_2cm$	$P4cm$	C_{4v}^{3}	$0kl$
$P4cc$	$P4cc$	C_{4v}^{5}	$0kl$, hhl
$P4bm$	$P4bm$	C_{4v}^{2}	$0kl$
$P4_2bc$	$P4bc$	C_{4v}^{8}	$0kl$, hhl
$P4_2nm$	$P4nm$	C_{4v}^{4}	$0kl$
$P4nc$	$P4nc$	C_{4v}^{6}	$0kl$, hhl
$I4mm$	$I4mm$	C_{4v}^{9}	hkl
$I4cm$	$I4cm$	C_{4v}^{10}	hkl, $0kl$
$I4_1md$	$I4md$	C_{4v}^{11}	hkl, hhl_4
$I4_1cd$	$I4cd$	C_{4v}^{12}	hkl, $0kl$ hhl_4
Class 422 (short 42)—D_4			
$P422$	$P42$	D_4^{1}	—
$P4_222$	$P4_22$	D_4^{5}	$00l$
$P4_122$	$P4_12$	D_4^{3}	$00l_4$
$P4_322$	$P4_32$	D_4^{7}	
$P42_12$	$P42_1$	D_4^{2}	$h00$
$P4_22_12$	$P4_22_1$	D_4^{6}	$00l$, $h00$
$P4_12_12$	$P4_12_1$	D_4^{4}	$00l_4$, $h00$
$P4_32_12$	$P4_32_1$	D_4^{8}	
$I422$	$I422$	D_4^{9}	hkl
$I4_122$	$I4_12$	D_4^{10}	hkl, $00l_4$
Class $\frac{4}{m}\frac{2}{m}\frac{2}{m}$ (short $4/mmm$)—D_{4h}			
$P\frac{4}{m}\frac{2}{m}\frac{2}{m}$	$P4/mmm$	D_{4h}^{1}	—
Class $\frac{4}{m}\frac{2}{m}\frac{2}{m}$ (short $4/mmm$)—D_{4h}—*continued*			
$P\frac{4_2}{m}\frac{2}{m}\frac{2}{c}$	$P4/mmc$	D_{4h}^{9}	hhl
$P\frac{4_2}{m}\frac{2}{c}\frac{2}{m}$	$P4/mcm$	D_{4h}^{10}	$0kl$
$P\frac{4}{m}\frac{2}{c}\frac{2}{c}$	$P4/mcc$	D_{4h}^{2}	$0kl$, hhl
$P\frac{4}{m}\frac{2_1}{b}\frac{2}{m}$	$P4/mbm$	D_{4h}^{5}	$0kl$
$P\frac{4_2}{m}\frac{2_1}{b}\frac{2}{c}$	$P4/mbc$	D_{4h}^{13}	$0kl$, hhl
$P\frac{4_2}{m}\frac{2_1}{n}\frac{2}{m}$	$P4/mnm$	D_{4h}^{14}	$0kl$
$P\frac{4}{m}\frac{2_1}{n}\frac{2}{c}$	$P4/mnc$	D_{4h}^{6}	$0kl$, hhl
$P\frac{4}{n}\frac{2_1}{m}\frac{2}{m}$	$P4/nmm$	D_{4h}^{7}	$hk0$
$P\frac{4_2}{n}\frac{2_1}{m}\frac{2}{c}$	$P4/nmc$	D_{4h}^{15}	$hk0$, hhl
$P\frac{4_2}{n}\frac{2_1}{c}\frac{2}{m}$	$P4/ncm$	D_{4h}^{16}	$hk0$, $0kl$
$P\frac{4}{n}\frac{2_1}{c}\frac{2}{c}$	$P4/ncc$	D_{4h}^{8}	$hk0$, $0kl$, hhl
$P\frac{4}{n}\frac{2}{b}\frac{2}{m}$	$P4/nbm$	D_{4h}^{3}	$hk0$, $0kl$
$P\frac{4_2}{n}\frac{2}{b}\frac{2}{c}$	$P4/nbc$	D_{4h}^{11}	$hk0$, $0kl$, hhl
$P\frac{4_2}{n}\frac{2}{n}\frac{2}{m}$	$P4/nnm$	D_{4h}^{12}	$hk0$, $0kl$
$P\frac{4}{n}\frac{2}{n}\frac{2}{c}$	$P4/nnc$	D_{4h}^{4}	$hk0$, $0kl$, hhl
$I\frac{4}{m}\frac{2}{m}\frac{2}{m}$	$I4/mmm$	D_{4h}^{17}	hkl
$I\frac{4}{m}\frac{2}{c}\frac{2}{m}$	$I4/mcm$	D_{4h}^{18}	hkl, $0kl$
$I\frac{4_1}{a}\frac{2}{m}\frac{2}{d}$	$I4/amd$	D_{4h}^{19}	hkl, $hk0$, hhl_4
$I\frac{4_1}{a}\frac{2}{c}\frac{2}{d}$	$I4/acd$	D_{4h}^{20}	hkl, $hk0$, $0kl$, hhl_4
Cubic system			
Class 23—T			
	$P23$	T^{1}	—
	$P2_13$	T^{4}	$h00$

Table 5.3 THE HERMANN–MAUGUIN SYSTEM OF POINT- AND SPACE-GROUP NOTATION—*continued*

Space group Hermann–Mauguin Full	Short	Schoenflies	Missing spectra
Class 23—*T*—*continued*			
	$F23$	T^2	hkl
	$I23$	T^3	hkl
	$I2_13$	T^5	hkl
Class $\frac{2}{m}\bar{3}$ (short $m3$)—T_h			
$P\frac{2}{m}\bar{3}$	$Pm3$	T_h^1	—
$P\frac{2}{n}\bar{3}$	$Pn3$	T_h^2	$hk0$
$P\frac{2_1}{a}\bar{3}$	$Pa3$	T_h^6	$hk0$
$F\frac{2}{m}\bar{3}$	$Fm3$	T_h^3	hkl
$F\frac{2}{d}\bar{3}$	$Fd3$	T_h^4	hkl, $hk0_4$
$I\frac{2}{m}\bar{3}$	$Im3$	T_h^5	hkl
$I\frac{2_1}{a}\bar{3}$	$Ia3$	T_h^7	hkl, $hk0$
Class $\bar{4}3m$—T_d			
	$P\bar{4}3m$	T_d^1	—
	$P\bar{4}3n$	T_d^4	hh
	$F\bar{4}3m$	T_d^2	hkl
	$F\bar{4}3c$	T_d^5	hkl, khl
	$I\bar{4}3m$	T_d^3	hkl
	$I\bar{4}3d$	T_d^6	hkl, hhl_4
Class 432—O			
$P432$	$P43$	O^1	—
$P4_232$	$P4_23$	O^2	$h00$
$P4_132$	$P4_13$	O^7	h_400 (for O^7 and O^6)
$P4_332$	$P4_33$	O^6	
$F432$	$F43$	O^3	hkl
$F4_132$	$F4_13$	O^4	hkl, h_400
$I432$	$I43$	O^5	hkl
$I4_132$	$I4_13$	O^8	hkl, h_400
Class $\frac{4}{m}\bar{3}\frac{2}{m}$ short $m3m$)—O_h			
$P\frac{4}{m}\bar{3}\frac{2}{m}$	$Pm3m$	O_h^1	—
Class $\frac{4}{m}\bar{3}\frac{2}{m}$(short $m3m$)—O_h—*continued*			
$P\frac{4_2}{m}\bar{3}\frac{2}{n}$	$Pm3n$	O_h^3	hhl
$P\frac{4_2}{n}\bar{3}\frac{2}{m}$	$Pn3m$	O_h^4	$hk0$
$P\frac{4}{n}\bar{3}\frac{2}{n}$	$Pn3n$	O_h^2	$hk0$, hhl
$F\frac{4}{m}\bar{3}\frac{2}{m}$	$Fm3m$	O_h^5	hkl
$F\frac{4}{m}\bar{3}\frac{2}{c}$	$Fm3c$	O_h^6	hkl, hhl
$F\frac{4_1}{d}\bar{3}\frac{2}{m}$	$Fd3m$	O_h^7	hkl, $hk0_4$
$F\frac{4_1}{d}\bar{3}\frac{2}{c}$	$Fd3c$	O_h^8	hkl, $hk0_4$, hhl
$I\frac{4}{m}\bar{3}\frac{2}{m}$	$Im3m$	O_h^9	hkl
$I\frac{4_1}{a}\bar{3}\frac{2}{d}$	$Ia3d$	O_h^{10}	hkl, $hk0$, hhl_4

Rhombohedral system (all indices and multiplicities referred to hexagonal axes)

Full	Short	Schoenflies	Missing spectra
Class 3—C_3			
	$C3$	C_3^1	—
	$C3_1, C3_2$	C_3^2, C_3^3	$00l$
	$R3$	C_3^4	hkl_3
Class $\bar{3}$—C_{3i}			
	$C\bar{3}$	C_{3i}^1	—
	$R\bar{3}$	C_{3i}^2	hkl_3
Class $3m$(to indicate orientation distinguish $3m1$ and $31m$)—C_{3}^{v}			
	$C3m1$	C_{3v}^1	—
	$C3c1$	C_{3v}^3	$h0l$
	$C31m$	C_{3v}^2	—
	$C31c$	C_{3v}^4	hhl
	$R3m$	C_{3v}^5	hkl_3
	$R3c$	C_{3v}^6	hkl_3, $h0l$

Table 5.3 THE HERMANN–MAUGUIN SYSTEM OF POINT- AND SPACE-GROUP NOTATION—*continued*

Space group Hermann–Mauguin Full	*Short*	*Schoenflies*	*Missing spectra*
Class 32 (to indicate orientation distinguish 321 and 312)—$D3$			
$C321$		D_3^2	—
$C3_121$		D_3^4	$00\underline{l}$ (with D_3^6)
$C3_221$		D_3^6	
$C312$		D_3^1	—
$C3_112$		D_3^3	$00\underline{l}$ (with D_3^5)
$C3_212$		D_3^5	
$R32$		D_3^7	$\underbrace{hkl}_{3}$
Class $\bar{3}\frac{2}{m}$ (short $\bar{3}m$) (to indicate orientation distinguish $\bar{3}ml$ and $\bar{3}lm$)—D_{3d}			
$C\bar{3}\frac{2}{m}1$	$C\bar{3}m1$	D_{3d}^3	—
$C\bar{3}\frac{2}{c}1$	$C\bar{3}c1$	D_{3d}^4	$h0\underline{l}$
$C\bar{3}1\frac{2}{m}$	$C\bar{3}1m$	D_{3d}^1	—
$C\bar{3}1\frac{2}{c}$	$C\bar{3}1c$	D_{3d}^2	$hh\underline{l}$
$R\bar{3}\frac{2}{m}$	$R\bar{3}m$	D_{3d}^5	$\underbrace{hkl}_{3}$
$R\bar{3}\frac{2}{c}$	$R\bar{3}c$	D_{3d}^6	$\underbrace{hkl}_{3}$, $h0\underline{l}$

Hexagonal system

Space group Hermann–Mauguin Full	*Short*	*Schoenflies*	*Missing spectra*
Class $\bar{6}$—C_{3h}			
$C\bar{6}$		C_{3h}^1	—
Class $\bar{6}$—C_6			
$C6$		C_6^1	—
$C6_3$		C_6^6	$00\underline{l}$
$C6_2$, $C6_4$		C_6^4, C_6^5	$00\underline{l}$
$C6_1$, $C6_5$		C_6^2, C_6^3	$00\underline{l}$
Class $\frac{6}{m}$—C_{6h}			
$C\frac{6}{m}$		C_{6h}^1	—
$C\frac{6_3}{m}$		C_{6h}^2	$00\underline{l}$
Class $\bar{6}2m$ (in other orientation $\bar{6}m2$)—D_{3h}			
$C\bar{6}2m$		D_{3h}^3	—
$C\bar{6}2c$		D_{3h}^4	$hh\underline{l}$
$C\bar{6}m2$		D_{3h}^1	—
$C\bar{6}c2$		D_{3h}^2	$h0\underline{l}$
Class $6mm$—C_{6v}			
$C6mm$	$C6mm$	C_{6v}^1	—
$C6_3mc$	$C6mc$	C_{6v}^4	$hh\underline{l}$
$C6_3cm$	$C6cm$	C_{6v}^3	$h0\underline{l}$
$C6cc$	$C6cc$	C_{6v}^2	$h0\underline{l}$, $hh\underline{l}$
Class 622—D_6			
$C622$	$C62$	D_6^1	—
$C6_322$	$C6_32$	D_6^6	$00\underline{l}$
$C6_222$	$C6_22$	D_6^4	$00\underline{l}$ (with D_6^5)
$C6_422$	$C6_42$	D_6^5	
$C6_122$	$C6_12$	D_6^2	$00\underline{l}$ (with D_6^3)
$C6_522$	$C6_52$	D_6^3	
Class $\frac{6}{m}\frac{2}{m}\frac{2}{m}$ (short $6/mmm$)—D_{6h}			
$C\frac{6}{m}\frac{2}{m}\frac{2}{m}$	$C6/mmm$	D_{6h}^1	—
$C\frac{6_3}{m}\frac{2}{m}\frac{2}{c}$	$C6/mmc$	D_{6h}^4	$hh\underline{l}$
$C\frac{6_3}{m}\frac{2}{c}\frac{2}{m}$	$C6/mcm$	D_{6h}^3	$h0\underline{l}$
$C\frac{6}{m}\frac{2}{c}\frac{2}{c}$	$C6/mcc$	D_{6h}^2	$h0\underline{l}$, $hh\underline{l}$

6 Crystal chemistry

6.1 Structures of metals, metalloids and their compounds

The elements have been arranged in the following order:

1. Group Ia—Li, Na, K, Rb, Cs
2. Group IIa—Be, Mg, Ca, Sr, Ba
3. Group IIIa—Sc, Y, La, Ce, Pr, Nd, Eu, Gd, Tb, Dy, Ho, Er, Tm, Yb, Lu, Ac
4. Group IVa—Ti, Zr, Hf, Th, Pa, U, Np, Pu, Am, Cm
5. Group Va—V, Nb, Ta
6. Group VIa—Cr, Mo, W
7. Group VIIa—Mn, Tc, Re
8. Group VIII—Fe, Co, Ni; Ru, Rh, Pd, Os, Ir, Pt
9. Group Ib—Cu, Ag, Au
10. Group IIb—Zn, Cd, Hg
11. Group IIIb—Al, Ga, In, Tl
12. Group IVb—Si, Ge, Sn, Pb
13. Group Vb—As, Sb, Bi
14. Metalloids etc.—B, C, P, N, Te, Se, S

A compound or solid solution composed of the elements ABCD ... is placed under that element, A, B, C, D,... which occurs *last* in the above list. If there are several compounds containing this particular element, they are arranged in the following order:

(a) Compounds containing the same elements—in the order of increasing content of the second element.

(b) Compounds of different elements—in the order in which the other elements occur in the above list.

For example, Na_2K is described under K, because K comes after Na in the list; MgSr and Mg_2Sr are both entered under Sr, and are described in that order, in accordance with (a); Li_2Ca and Mg_2Ca are entered under Ca, and described in that order, in accordance with (b).

The second column of Table 6.1 gives the symbol for the structural type to which the element or compound is assigned, the notation used being that of *Strukturbericht*.* Detailed descriptions of the structural types are given in Table 6.2.

The third column of Table 6.1 gives the lattice constants of the various elements and compounds. For cubic crystals the single parameter a is given, in kX units; for tetragonal and hexagonal crystals a and c, in that order; for orthorhombic, a, b and c; for rhombohedral a, α; for monoclinic a, b, c, β.

Temperatures given in parentheses are those at which allotropic or polymorphic modifications are stable; figures such as $A = 28$, $M = 4$, also given in parentheses, refer to the number of atoms (A) or molecules (M) included in the unit cell; alternative structures are given where the available evidence is insufficient to permit of a decision being reached between them, and in such cases the authorities are quoted. Finally, it should be noted that in the case of the sulphides, only those having relatively simple structures have been included.

* Strukturbericht of Z. *Krystallographie*, Leipzig.

Table 6.1 STRUCTURES OF METALS, METALLOIDS AND THEIR COMPOUNDS

Element or compound	*Structure type*	*Lattice constants, remarks*	*Refs.*
Group Ia: Li, Na, K, Rb, Cs			
Li	*A*2	3.50	1
	*A*3	3.09; 4.83 (spontaneous transformation at −195 °C)	
	*A*1	4.40 (produced by deformation at low temp.)	
Na	*A*2	4.28	
	*A*1	5.34 (incomplete transformation by working at −253 °C)	
K	*A*2	5.31	
Na_2K	*C*14	7.48; 12.27	
Rb	*A*1	5.70	
Cs	*A*1	6.13	
Group IIa: Be, Mg, Ca, Sr, Ba			
Be	*A*3	2.28; 3.57	
Mg	*A*3	3.21; 5.21	
$Be_{13}Mg$	$D2_3$	10.166	2
Ca (α)	*A*1	5.57 (< ~300 °C)	
(γ)	*A*3	3.94; 6.46 (> ~450 °C; a third modification of unknown structure exists in the range 300–450 °C)	
Li_2Ca	*C*14	6.25; 10.23	
$Be_{13}Ca$	$D2_3$	10.312	2
Mg_2Ca	*C*14	6.22; 10.10	
Sr	*A*1	6.05	
MgSr	*B*2	3.90	
Mg_2Sr	*C*14	6.43; 10.47	
Ba	*A*2	5.01 (5.000 at 5 K)	3
Mg_2Ba	*C*14	6.64; 10.66	
Group IIIa: Sc, Y and Rare Earths			
Sc (α)	*A*1	4.53	
(β)	*A*3	3.308; 5.267	207
Y	*A*3	3.63; 5.75	
La (α)	*A*3	3.770; 12.159 (room temp.)	4
(β)	*A*1	5.31 (above 340 °C)	4
MgLa	*B*2	3.97	
Mg_2La	*C*15	8.71	
Mg_3La	$D0_3$	7.47	
Ce (α)	*A*3	3.65; 5.96	
(β)	*A*1	5.161 2	4
(β′)	*A*1	4.84 (under 15 000 atm. pressure; also formed at 90 K under atm. pressure)	
$Be_{13}Ce$	$D2_3$	10.38	
MgCe	*B*2	3.90	
Mg_2Ce	*C*15	8.70 (range 615 − 750 °C)	
Mg_3Ce	$D0_3$	7.42	
Pr	*A*1	5.15	
	*A*3	3.672; 11.835	4
MgPr	*B*2	3.88	
Mg_3Pr	$D0_2$	7.37	
Nd	*A*3	3.66; 11.80	207
Eu	*A*2	4.582 0	208
Gd	*A*3	3.63; 5.79	209
Tb	*A*3	3.59; 5.66	
Dy	*A*3	3.58; 5.65	
Ho	*A*3	3.56; 5.62	
Er	*A*3	3.55; 5.58	210
Tm	*A*3	3.52; 5.56	
Yb	*A*1	5.47	
Lu	*A*3	3.51; 5.56	
Ac	cubic f.c.	5.311	6
Group IVa: Ti, Zr, Hf, Th and Pa, U, Np, Pu, Am, Cm			
Ti (α)	*A*3	2.95; 4.68 (< ~900 °C)	
(β)	*A*2	3.29 (> ~900 °C; *a*-spacing by extrapolation to 100% purity and room temp. from Ta–Ti solid solutions)	

Table 6.1 STRUCTURES OF METALS, METALLOIDS AND THEIR COMPOUNDS—*continued*

Element or compound	*Structure type*	*Lattice constants, remarks*	*Refs.*
Group IVa: Ti, Zr, Hf, Th and Pa, U, Np, Pu, Am, Cm—*continued*			
$BeLi_4$	hex.	10.92; 8.94 ($M=6$)	211
Be_2Ti	$C15$	6.44	
$Be_{12}Ti$	$D2a$	29.44; 7.33	
Zr (α)	$A3$	3.23; 5.14 (< ~840 °C)	7
(β)	$A2$	3.61 (> ~840 °C)	7
Be_5Zr	$D^1_{6h}\ P_6/mmm$	5.564; 3.485	212
$Be_{17}Zr_2$	$R\bar{3}m.$	4.694; 83.02	212
Be_2Zr	$C32$	3.82; 3.24	8
$Be_{13}Zr$	$D2_3$	10.05	
Hf (α)	$A3$	3.19; < 1310 °C	9, 10
(β)	$A2$	~ 3.50 (*a*-spacing by extrapolation from Nb–Hf solid solutions)	
Hf_2Be_{17}	hex.	7.50; 10.94	213
Th	$A1$	5.08	
Th (ζ)	orthorh.	9.820; 8.164; 6.681	11
$Be_{13}Th$	$D2_3$	10.40	
Mg_2Th	$C36$	6.086; 19.64 (at and below 700 °C)	12
	$C15$	8.570 (at and above 800 °C)	12
Pa	Aa	3.925; 3.238	13
U (α)	$A20$	2.854; 5.869; 4.955	214
(β)	tetr.	10.759; 5.656 (range 640–760 °C)	14
("β")	Ab	10.763; 5.652 at 720 °C	14
		10.590; 5.634 at room temp.	15
(γ)	$A2$	3.47 (> 760 °C)	
UH_3	$A15$	6.644 } the $A15$ type holds only for the U positions	
UD_3	$A15$	6.633 } the $A15$ type holds only for the U positions	
$Be_{13}U$	$D2_3$	1 026	
TiU_2	$C32$	4.828; 2.847	16
Zr–U	cubic f.c.	10.68	17
Np (α)	Ac	(< 278 °C)	
(β)	Ad	4.90; 3.39 (range 278–540 °C)	
(γ)	$A2$ (?)	3.52	18
$Be_{13}Np$	cubic f.c.	10.266 to 10.256	19
Pu (α)	monocl.	6.183 5; 4.824 4; 10.973; β = 101.80	20
(β)	monocl.	9.284; 10.463; 7.859 B = 92.13° A = 34	215
(δ)	cubic f.c.	4.637 9	21
(δ')	tetr.	3.220; 4.496 9	21
(ε)	cubic b.c.	3.634 8	21
Am	$P6_3/mma$	3.468 1; 11.240	216
Group Va: V, Nb, Ta			
V	$A2$	3.024	22
Be_2V	$C14$	4.39; 7.13	
Be_2Nb_3	P_4/mbm	6.49; 3.35 (M = 2)	217
$Be_{17}Nb_2$	$R\bar{3}m$	5.599; 82.84° (M = 1)	218
Nb	$A2$	3.294	
Ta	$A2$	3.30	
ZrV_3	$C14$	5.28; 8.65	
TiV (ω)	orthorh.	6.205; 6.597; 13.63	219
Group VIa: Cr, Mo, W			
Cr (α)	$A2$	2.89 (< 1 840 °C)	
(β)	$A1$	~3.8 (> 1 840 °C)	
(γ)	$A12$	8.72 (electrolytic)	
~ CrH	$B4$	2.72; 4.43	
~ CrH_2	$C1$	3.86	
Be_2Cr	$C14$	4.24; 6.92	
TiCr (ω)	orthorh.	6.203; 6.498; 13.63	219
$TiCr_2$	$C15$	6.93 (Ti positions partly replaced by Cr)	
$ZrCr_2$ (1)	$C14$	5.079; 8.262	24
(2)	$C15$	7.195	24
$NbCr_2$	$C15$	6.98	

Table 6.1 STRUCTURES OF METALS, METALLOIDS AND THEIR COMPOUNDS—*continued*

Element or compound	*Structure type*	*Lattice constants, remarks*	*Refs.*
Group VIa: Cr, Mo, W—*continued*			
$TaCr_2$ (1)	$C15$	6.95	
(2)	$C14$	4.92; 8.05	
Mo	$A2$	3.140	
Be_2Mo	$C14$	4.43; 7.34	
$Be_{12}Mo$	tetr.	7.271; 4.234	25
$Be_{13}Mo$	tetr.	10.27; 4.29 (M = 4)	
$Be_{20}Mo$	O_h^7	11.64	220
$BeMo_3$	O_h^3	4.89	
MoTc (ω)	orthorh.	6.231; 6.500; 13.52	219
$ZrMo_3$	$A15$	4.94 (doubted by Duwez[26])	
$ZrMo_2$	$C15$	7.58	26
U_2Mo (γ)	tetr.	3.427; 3.279	27
W (α)	$A2$	3.16	
(β)	$A15$	5.04 (unstable)	
Be_2W	$C14$	44.4; 7.24	
WBe_{20}	O_h^7	11.64	220
ZrW_2	$C15$	7.61	
Group VIIa: Mn, Tc, Re			
Mn (α)	$A12$	8.90 (<742 °C)	
(β)	$A13$	6.30 (724–1 191 °C) (A = 20)	28
	$\sim A13$	12.58 (A = 160)	29
(γ)	$A6$	3.77; 3.53 (metastable below 250 °C)	
(δ)	$A1$	3.72 (>1 191 °C)	30
YMn_2	O^7Fd3m	7.680 (M = 8)	221
YMn_4	b.c.tetr.	8.808; 12.521 (M = 12)	222
YMn_{12}	b.c.tetr.	8.541; 4.785	
Be_2Mn	$C14$	4.23; 6.91	
$GdMn_2$	$C15$	7.74	
$TiMn_2$	$C14$	4.825; 7.917	31
Ti_2Mn	$E9_3$	11.29 (actually Ti_4Mn_2 σ)	
$ZrMn_2$	$C14$	5.03; 8.22	
$ThMn_{12}$	$D2b$	8.74; 4.95	
Th_6Mn_{23}	$D8a$	12.52	
$ThMn_2$	$C14$	5.48; 8.95	
UMn_2	$C15$	7.16	
U_6Mn	$D2c$	10.29; 524	
VMn_3	$D8b$	—	
$NbMn_2$	$C14$	4.87; 7.96	
$TaMn_2$	$C14$	4.86; 7.94	
$CrMn_3$	$D8b$	—	
Tc	$A3$	2.74; 4.39	
Re	$A3$	2.76; 4.45	
Be_2Re	$C14$	4.35; 7.09	
$ZrRe_2$	$C14$	5.270 1; 8.634 9	223
$HfRe_2$	$C14$	5.247 8; 8.593 4	
URe_2	orthorh.	< 180°. 5.600; 9.180; 8.460	224
	hex.	> 180°. 5.433; 8.561	
Group VIII: Fe, Co, Ni; Ru, Rh, Pd; Os, Ir, Pt			
Fe (α)	$A2$	2.86 (<900° and > 1 400°)	
(γ)	$A1$	3.56 (900–1 400°; constant extrapolated for room temp.)	
Be_2Fe	$C14$	4.21; 6.83	
Be_5Fe	$C15$	5.88 (stable only > ~ 1 000°)	
$Be_{11}Fe$	hex	4.14; 10.73 (A = 18)	
$CeFe_2$	$C15$	—	
Gd_2Fe_3	cubic	8.25 (M = 6)	225
$GdFe_3$	rhomb.	4.72; 39° 46 (M = 6)	
Gd_2Fe_7	orthorh.	5.71; 6.78; 7.15 (M=2)	
$GdFe_4$	hex.	5.15; 6.64 (M = 2)	
$GdFe_5$	hex.	4.92; 4.11 (M = 1)	
Gd_2Fe_{17}	hex.	8.39; 8.53 (M = 2)	

Table 6.1 STRUCTURES OF METALS, METALLOIDS AND THEIR COMPOUNDS—*continued*

Element or compound	*Structure type*	*Lattice constants, remarks*	*Refs.*
Group VIII: Fe, Co, Ni; Ru, Rh, Pd; Os Ir, Pt—*continued*			
$GdFe_2$	*C*15	7.43	
YFe_2	O^7Fd3m	7.357 (M = 8)	221
$TiFe_2$	*C*14	4.77; 7.75	
TiFe	*A*2	2.97 (probably ordered as *B*2-type)	
Ti_3Fe_3O	$E9_2$	11.15 }	32
Ti_4Fe_2O	$E9_3$	11.28 }	32
Ti_2Fe	$E9_2$	11.31 (Duwez, Taylor)	33
$ZrFe_2$	*C*15	7.04	
$Zr_{0.8}Fe_{2.3}$	*C*36	4.95; 16.12 (doubted by Hayes[34])	
$ThFe_5$	*D*2*d*	—	
U_6Fe	*D*2*c*	10.31; 5.24	
UFe_2	*Fd*3*m*	7.766	226
VFe	*D*8*b*	—	
$NbFe_2$	*C*14	4.82; 7.87	
$TaFe_2$	*C*14	4.80; 7.84	
CrFe (σ)	*D*8*b*	8.800; 4.544	35
MoFe (σ)	*D*8*b*	9.188; 4.812	35
Mo_6Fe_7	$D8_5$	8.97; 30° 39′	
WFe_2	*C*14	4.73; 7.70	
W_6Fe_7	$D8_5$	9.02; 30° 31′	
$MnFe_4$	*A*3	2.53; 4.08 (not an equilibrium phase)	
Co (α)	A_3	2.505 9; 4.065 9	227
(β)	A1	3.54 (>390 °C)	
BeCO	*B*2	2.61	
$Be_{21}Co_5$	*A*12	7.66(?)	
YCo_2	O^7Fd3m	7.216 (M=8)	221
La_3Co	*Pnmc.*	7.279; 10.088; 6.578 (M = 4)	228
$CeCO_5$	*D*2*d*	4.96; 4.06	
Co_5Sm	*P*6/*mmm*	5.004; 3.971	232
Co_5Tb	*P*6/*mmm*	4.947; 3.982	
Co_5Ho	*P*6/*mmm*	4.910; 3.996	
Gd_3Co	orthorh.	5.17; 6.72; 5.94 (M = 2)	235
GdCo	orthorh.	3.90; 4.87; 4.22 (M = 2)	
Gd_2Co_3	cubic	7.98 (M = 6)	
$GdCo_2$	cubic	7.3 (M = 8)	
$GdCo_3$	rhomb.	4.80; 41° 32′ (M=6)	
$GdCo_4$	hex.	5.47; 6.02 (M = 2)	
$GdCo_5$	hex.	4.97; 3.97 (M = 1)	
$CeCO_2$	*C*15	7.15	
$TiCO_2$	*C*36	4.72; 15.39 (Co-rich)	
	*C*15	6.73	
TiCo	*A*2	2.99 (probably ordered as *B*2-type)	
Ti_4Co_2O	$E9_3$	11.30 (Rostoker[32])	
Ti_2Co	$E9_3$	11.28 (Dewez, Taylor[33])	
$ZrCo_2$	*C*15	6.89	
$ThCo_5$	*D*2*d*	4.95; 4.04	
UCo_2	*C*15	6.99	
UCo	*Ba*	6.36	
U_6Co	*D*2*c*	10.36; 5.21	
VCo_3	$P\bar{6}m2$	5.032; 12.27 (M = 6)	229
VCo	*D*8*b*	—	
V_3Co	*A*15	4.68	
$Nb_{0.8}Co_{2.2}$	*C*36	4.73; 14.43	
$NbCo_2$	*C*15	6.75	
$Ta_{0.8}Co_{2.2}$	*C*36	4.72; 15.39 }	230
Co_3Ta (α)	f.c.c.	3.647 (M = 1) }	230
(β)	hex.	9.411; 15.50 (M = 32) }	230
Co_2Ta (α)	hex.	4.797; 7.827 }	230
(β)	f.c.c.	6.729 }	230
(γ)	hex.	4.700; 15.42 }	230
$TaCo_2$	*C*15	6.72	
~ CrCo	*D*8*b*	8.77; 4.54	36

Table 6.1 STRUCTURES OF METALS, METALLOIDS AND THEIR COMPOUNDS— *continued*

Element or compound	*Structure type*	*Lattice constants, remarks*	*Refs.*
Group VIII: Fe, Co, Ni; Ru, Rh, Pd; Os, Ir, Pt— *continued*			
$Cr_{17}Co_{13}$ (σ)	tetr.	8.81; 4.56	37
$MoCo_3$	$D0_{19}$	5.12; 4.11	
Mo_6Co_7	$D8_5$	8.98; 30° 48′	
Mo_3Co_2	$D8b$	—	
WCo_3	$D0_{19}$	5.12; 4.12	
W_6Co_7	$D8_5$	8.95; 30° 41′	
FeCo	$B2$	2.850 4 } transf. temp. ~730 °C; constants at room temp.	
	$A2$	2.848 8	
Ni	$A1$	3.52	
	$A3$	2.56–2.65; 4.17–4.32 } probably exists only in presence of H or N	
	$A6$	3.99; 3.76	
BeNi	$B2$	2.60	
$Be_{21}Ni_5$	$D8_{1,3}$	7.56	
$MgNi_2$	C_{36}	4.81; 15.77	
Mg_2Ni	*Ca*	5.18; 13.19	
$CaNi_5$	*D2d*	4.95; 3.94	
YMi_2	O^7Fd3m	7.181 (M=8)	221
YNi_3	hex.		
YNi	orthorh.	4.10; 5.51; 7.12 (M=4)	231
YNi_5	hex.	4.883; 3.967 (M=1)	
$LaNi_5$	*D2d*	4.95; 4.00	
$LaNi_2$	C15	7.25	
$CeNi_5$	*D2d*	4.86; 4.00	
$CeNi_2$	C15	7.19	
$CeNi_3$	$P6_3/mmc$	4.98; 16.54	233
$PrNi_5$	*D2d*	4.94; 3.97	
$PrNi_2$	*C*15	7.19	
Gd_3Ni	orthorh.	5.15; 6.70; 6.23 (M=2)	234
Gd_3Ni_2	tetr.	7.28; 8.61 (M=4)	
GdNi	orthorh.	3.8; 5.2; 4.2 (M=2)	
$GdNi_2$	rhomb.	4.25; 31° 53′ (M=6)	
Gd_2Ni_7	orthorh.	6.05; 6.22; 7.03 (M=2)	
$GdNi_4$	hex.	5.35; 5.83 (M=2)	
$GdNi_5$	hex.	4.90; 3.97 (M=1)	
Gd_2Ni_{17}	hex.	8.18; 8.47 (M=2)	
$GdNi_5$	*D2d*	4.90; 3.98	
Ni_5Sm	*P6/mmm*	4.924; 3.974	
Ni_5Tb	*P6/mmm*	4.984; 3.966	232
Ni_5Ho	*P6/mmm*	4.871; 3.966	
Ni_5Yb	*P6/mmm*	4.841; 3.965	
Ti_2Ni	O_h^7–*Fd3m*	11.278 (A=96)	236
$TiNi_3$	$D0_{24}$	5.10; 8.31	
TiNi	*A*2	3.01 (probably ordered as *B*2-type)	
Ti_2Ni	$E9_3$	11.29	
ζZrNi	b.c.c.	6.702	237
Ti_4Ni_2O	$E9_3$	11.30	
$ZrNi_5$	*C*15	6.71	38
Zr_2Ni	D_h^{18}l$_4$ *mcm*	6.477; 5.241	238
ZrNi	D_{2h}^{17} *cmcm*	3.268; 9.937; 4.101	
Hf_2Ni	analogous to Zr_2Ni		
HfNi	analogous to *ZrNi*		
$ThNi_5$	*D2d*	4.92; 3.99	
$ThNi_2$	*C*32	—	
UNi_5	*C*15	6.78	
UNi_2	*C*14	4.97; 8.25	
U_6Ni	*D2c*	10.37; 5.21	
$PuNi_4$	monocl. 62/*m*	4.87; 8.46; 10.27; β=100° (M=6)	239
$PuNi_3$	rhomb	6.22; 30° 44′ (M=3)	240
PuNi	orthorh. *cmcm*	3.59; 10.21; 4.22	241
VNi_3	$D0_{22}$	3.54; 7.22	
VNi_2	orh.	2.61; 3.54; 2.57 (A=2; deformed *A*1-type, perhaps ordered)	
~VNi	*D8b*	8.97; 4.64	
$NbNi_3$	*A*3 (deformed)	—	

Table 6.1 STRUCTURES OF METALS, METALLOIDS AND THEIR COMPOUNDS—*continued*

Element or compound	*Structure type*	*Lattice constants, remarks*	*Refs.*
Group VIII: Fe, Co, Ni; Ru, Rh, Pd; Os, Ir, Pt—*continued*			
$TaNi_3$	*D0a*	5.11; 4.25; 4.54	
Cr_2Ni	*A*2(tetr. deformed)	($c/a = 1.09$; metastable intermediate by quenching)	
~(Cr, Mo)$_2$Ni	*D8b*	—	
$MoNi_4$	*D*1*a*	5.72; 3.56	
$MoNi_3$	type *TiCu*$_3$		
	orthorh.	5.064; 4.224; 4.448	242
MoNi	*A*3	2.54; 4.18	
WNi_4	*D*1*a*	5.73; 3.55	
	cubic b.c.	12.79	
$MnNi_3$	$L1_2$	3.59 (<510°C)	
MnNi	$L1_0$	3.74; 3.52 (< ~700°C)	
	*A*2	2.97 (range ~ 700–900°C; constant at 745°)	
$FeNi_3$	Li_2	3.54 (<586°C)	
Fe_5Cr_5Ni	cubic b.c.	8.88 (powder diagram similar *A*12-type)	
Ru	*A*3	2.70; 4.27	
	hex.	2.706 0; 4.253 7	243
Ru_2Be_3	b.c.c.	11.42	252
$RuBe_2$	hex.	5.90; 9.10	
Ru_3Be_{10}	b.c.c.	11.03	
Ru_2Ce	*C*15	7.79	244
TiRu	*B*2	3.06	40
$ZrRu_2$	*C*14	5.13; 8.49	
URu_3	*C*12	3.980	39
RuU_2	monocl.	13.106; 3.343; 5.202; 96°9.2′ ($M = 4$)	245
Rh	*A*1	3.80	
	cubic	9.21 ($A = 48$; electrolytic)	
WRh_2	*A*3	2.73; 4.38	
RhMg	*CsCl* type	3.099	246
RhCa	*C*15–O_h^7	7.525	247
RhSr	C_5Cl type	3.206	246
RhBe	*C*15 type O_h^7	7.852 ($M = 8$)	247
RhSr	*C*15 type O_h^7	7.706 ($M = 8$)	247
Rh_3Ti	Cu_3Au type	3.822 }	248
Rh_3Hf	Cu_3Au type	3.911 }	248
Rh_3Nb	Cu_3Au type	3.865	
Rh_3Zr	Cu_3Au type	3.927	
Rh_3V	Cu_3Au type	3.795	
Rh_3Ta	Cu_3Au type	3.86	
Pd	*A*1	3.88	
BePd	*B*2	2.81	
Be_5Pd	*C*15	5.98 (ordered as Be_4Be, Pd)	
	*C*15	5.98 [disordered as Be_4(Be, Pd)$_2$]	
PdCa	*C*15O_h^7*Fd*3*m*	7.665 }	247
PdSr	*C*15O_h^7*Fd*3*m*	7.826 }	247
PdBa	*C*15O_h^7*Fd*3*m*	7.983 }	247
$TiPd_3$	$D0_{24}$	5.48; 8.96	
Pd_3Zr	$TiNi_3$ type	5.612; 9.235 }	248
Pd_3Hf	$TiNi_3$ type	5.595; 9.192 }	248
Pd_4Th	Cu_3Au type	4.110	249
$PdTi_2$	*C*11*B*	3.090; 10.084	342
$PdZr_2$	*C*11*B*	3.306; 10.894	342
$PdHf_2$	*C*11*B*	3.251; 11.061	342
PdMg	*B*2	3.12	341
UPd'_3	$D0_{24}$	5.757; 9.621	39
Pd_2V	f.c. tetr.	3.88; 3.72	250
Pd_3V	f.c. tetr.	3.84; 3.85	
$FePd_3$	$L1_2$	3.84	
FePd	$L1_0$	3.85; 3.72	
Os	*A*3	2.73; 4.31	
TiOs	*B*2	3.07	40
$ZrOs_2$	*C*14	5.18; 8.51	
UOs	*C*15	7.497 4	39
TaOs(σ)		9.934; 5.189	41

Table 6.1 STRUCTURES OF METALS, METALLOIDS AND THEIR COMPOUNDS—*continued*

Element or compound	*Structure type*	*Lattice constants, remarks*	*Refs.*
Group VIII: Fe, Co, Ni; Ru, Rh, Pd; Os, Ir, Pt—*continued*			
Cr_3Os	*A*15	4.677	42
WOs (σ)		9.686; 5.012	41
Ir	*A*1	3.83	
IrCa	$C15O_h^7$	7.545	
IrSr	$C15O_h^7$	7.700	247
Ir_3Ti	Cu_3Au type	3.822	250
Ir_3Hf	Cu_3Au type	3.911	
Ir_3Nb	Cu_3Au type	3.865	
Ir_3Zr	Cu_3Au type	3.943	
Ir_3V	Cu_3Au type	3.812	
Ir_3Tc	Cu_3Au type	3.889	
TaIr (σ)		9.938; 5.172	41
$ZrIr_2$	*C*14	—	
UIr_2	*C*15	7.4939	39
$\sim W_2Ir_3$	*A*3	2.75; 4.42	
Pt	*A*1	3.92	
$Be_{21}Pt_5$	$D8_{1-3}$ (deformed)	—	
PtCa	$C15O_h^7$	7.629	
PtSr	$C15O_h^7$	7.777	
PtBa	$C15O_h^7$	7.920	
PtMg	*B*20	4.86	341
Pt_3Mg	$L6_0$	3.88; 3.72	341
$CePt_2$	*C*15	7.73	
$TiPt_3$	$L1_2$	3.89	
Ti_3Pt	*A*15	5.033	43
$ZrPt_3$	$D0_{24}$	5.63; 9.21	
Pt_3Hf	$TiNi_3$ type	5.636; 9.208	248
Pt_3U	$D0_{19}$	5.752; 4.889	39
Pt_2U	orthorh.	5.60; 9.68; 4.12	251
$FePt_3$	$L1_2$	3.88 (<700 °C)	
FePt	$L1_0$	3.86; 3.76 (<1 300 °C)	
Fe_3Pt	$L1_2$	3.75 (<850 °C)	
$CoPt_3$	$L1_2$	3.831 (constant at 700 °C; transformation to disordered state ~750 °C; constant at 800 °C = 3.829)	
CoPt	$L1_0$	3.80; 3.70 (constant at 700 °C; transformation temp. to disordered *A*1-type ≈825 °C)	
NiPt	$L1_0$	3.82; 3.58 (<600 °C)	
Group Ib: Cu, Ag, Au			
Cu	*A*1	3.61	
BeCu (γ)	*B*2	2.70	
(γ′)	tetr.	2.79; 2.54 } intermediate phases during precipitation	
(γ″)	monocl.	2.54; 2.54; 3.24; 85° 25′ } intermediate phases during precipitation	
$Be_{2.4}Cu$	*C*15	5.94 (report *A*2-type with *a* = 2.79) (Misch[44], Kossolapow[45])	
$MgCu_2$	*C*15	7.03	
Mg_2Cu	*Cb*	5.27; 9.05; 18.21	
$CaCu_5$	*D*2*d*	5.10; 4.08	
$LaCu_5$	*D*2*d*	5.17; 4.12	
CeCu	orthorh. *Pnma*	7.30; 4.30; 6.36	253
$CeCu_2$	orthorh. 1*mma*	4.43; 7.05; 7.45	254
$CeCu_6$	orthorh. *Pnma*	8.12; 5.102; 10.162 (*M* = 4)	255
$CeCu_4$	*D*2*d*	5.510; 4.102	46
Cn_5Tb	$P6/mmm.D_6^1h$	4.96; 4.15	256
Cu_5Sm	$P6/mmm.D_6^1h$	5.074; 4.099	256
Ti_3Cu	*L*1*a*	4.16; 3.59	
$CuTi_2$	f.c.tetr.	4.164; 3.611	257
Ti_2Cu	$E9_3$	11.24	
Ti_4Cu_2O	$E9_3$	11.47	
TiCu (δ)	*L*2*a*	4.44; 2.86	
(γ)	*B*11	3.11; 5.89	
$TiCu_3$ (β)	*Ae*	2.59; 4.53; 4.35 (> ~600 °C)	
(β)	*D*0*a*	5.15; 4.34; 4.52 (< ~600 °C)	

Table 6.1 STRUCTURES OF METALS, METALLOIDS AND THEIR COMPOUNDS—*continued*

Element or compound	*Structure type*	*Lattice constants, remarks*	*Refs.*
Group Ib: Cu, Ag, Au—*continued*			
$CuZr_2$	*C*11*B*	3.220 4; 11.183 2	342
Zr_3Cu	*L*1*a*	4.54; 3.72	
$CuHf_2$	*C*11*B*	3.169 5; 11.133 3	342
$ThCu_2$	*C*32	4.35; 3.47	47
Th_2Cu	*C*16	7.28; 5.74	47
UCu_5	*C*15	7.03	
$FeNi_3Cu_6$	tetr.	3.584; 3.548 (a phase of said composition decomposes into the Cu-poor tetr. intermediate phase + Cu-rich cubic phase with $a=3.592$. In the final state 2 cubic phases coexist with $a=3.592$ and $a=3.568$)	
$FeNi_3Cu_3$	tetr.	3.577; 3.598 (a phase of said composition decomposes into the Cu-rich tetr. intermediate phase + a Cu-poor cubic phase with 3.567. In the final state two cubic phases coexist with $a=3.590+3.567$)	
$PdCu_3$	$L1_2$	3.65	
	tetr.	9.74; 7.31 (quenched 460 °C)	
PdCu	*B*2	2.96	
$PtCu_3$	$L1_2$	3.68	
PtCu	$L1_1$	7.58; 90° 54′	
Pt_3Cu	*L*1*b*	—	
Ag	*A*1	4.08	
LiAg	*B*2	3.17	
Li_3Ag	$D8_{1-3}$	9.94	
Be_2Ag	*C*15	6.29	
MgAg	*B*2	3.32	
Mg_3Ag	hex.	4.93; 7.81	
$CuAg_2$	*C*14	5.72; 9.35 (trace Mg or Ag)	
Ca_5Ag_3	tetr.$D\text{–}8_1$	8.039; 15.011	258
LaAg	*B*2	3.77	
CeAg	*B*2	3.74	
PrAg	*B*2	3.73	
TiAg	$L1_0$	4.10; 4.07	
ZrAg	$L1_0$	—	
$AgZr_2$	*C*11*B*	3.246 4; 12.003 7	342
Th_2Ag	C16	7.56; 5.84	48
$PtAg_3$	$L1_2$	3.89 (<800 °C)	
Pt_3Ag	$L1_2$	3.89	
Au	*A*1	4.07	
$NaAu_2$	*C*15	7.79	
Na_2Au	*C*16	7.40; 5.51	
BeAu	*B*20	4.67	
Be_5Au	*C*15	6.69	
MgAu	*B*2	3.26	
Mg_3Au	$D0_{18}$	4.63; 8.44	
$\sim TiAu_6$	tetr.	4.07; 3.98 ($A=4$)	
$\sim TiAu_2$	*A*3	2.79; 4.77	
Ti_3Au	$L1_2$	—	
	*A*15	5.096	43
Th_2Au	*C*16	7.42; 5.95	48
V_3Au	*A*15	4.88	49
Nb_3Au	*A*15	5.21	49
MnAu	*B*2 (tetr. deformed)	3.28; 3.14	
Au_2Mn	Tet.$I_4\text{–}mmm$	3.363, c/a = 2.555	259
Au_4Mn	b.c.tet. (cf. Ni_4Mo)	C^5_{4h}	260
$PtAu_3$	$L1_2$	(produced by mechanical deformation)	
CuAu	$L1_0$	3.98; 3.72	
	$L1_0$ (orthorh. deformed)	($c/a=1.003$)	
Cu_3Au	Ll_2	3.747 4 a (disordered) = 3.752 8	
Group IIb: Zn, Cd, Hg			
Zn	*A*3	2.66; 4.94	
$LiZn_9$	*A*3	2.78; 4.39	

Table 6.1 STRUCTURES OF METALS, METALLOIDS AND THEIR COMPOUNDS—*continued*

Element or compound	*Structure type*	*Lattice constants, remarks*	*Refs.*
Group IIb: Zn, Cd, Hg—*continued*			
$Li_{0.8}Zn_2$	*Ck*	4.36; 2.51	
LiZn	*B*32	6.21	
$NaZn_{13}$	$D2_3$	12.28	
KZn_{13}	$D2_3$	12.36	
Mg_2Zn_{11}	*D*8*c*	8.55	
$MgZn_2$	*C*14	5.21; 8.54	
$CaZn_{13}$	$D2_2$	12.13	
$CaZn_5$	*D*2*d*	5.40; 4.22	
$SrZn_5$	orthorh. *Pmcn*	5.32; 6.72; 13.15 ($M=4$)	50
$SrZn_{13}$	$D2_3$	12.22	
$BaZn_5$	orthorh. *Amam*	5.32; 8.44; 10.78 ($M=4$)	50
$BaZn_{13}$	$D2_3$	12.33	
$LaZn_5$	*D*2*d*	5.43; 4.23	
LaZn	*B*2	3.75	
CeZn	*B*2	3.70	
PrZn	*B*2	3.67	
$TiZn_3$	$L1_2$	3.932 2	51
$TiZn_2$	*C*14	5.064; 8.210	51
TiZn	*B*2	—	
Th_2Zn	*C*16	7.62; 5.64	47, 52
$ThZn_2$	hex.	0.93; 7.39	261
$ThZn_4$	b.c.tetr.	4.273; 10.309	261
$ThZn_9$	*D*2*d*	5.24; 4.45	52
Th_4Zn_7	$D1_3$	9.03; 13.20	52
Th–Zn (*x* phase)	cubic f.c.	5.68	52
$ZrZn_4$	f.c.c.*Fd*/3*m*	14.11	262
$ZrZn_6$	orthorh.	17.8; 12.5; 8.68	262
Zr_3Zn_2	tetr.$P4_2mm$	7.633; 6.965	263
$NbZn_2$	hex.*P*63/*mmc*	5.05; 16.32	264
$CrZn_{11}$	hex.	12.9; 30.5	
$MnZn_{13}$ (ξ)	monocl.	13.7; 7.6; 5.1; 128° 44′	
$MnZn_9$ (δ_1)	hex.	12.8; 57.6	
Mn_5Zn_{21} (Γ)	$D8_{1-3}$	9.14	
$MnZn_3$(α)	$L1_2$	3.85 (<320 °C)	
$\sim MnZn_2$(ε)	*A*3	2.75; 4.44 (>350 °C)	
$\sim$MnZn (β)	*A*2	3.05(high temp.)	
(β_1)	*A*13	(low temp.)	
$FeZn_{13}$(ξ)	monocl.	13.65; 7.61; 5.1; 128° 44′ ($A\approx 28$)	
$FeZn_9$(δ_1)	hex.	12.8; 57.6 ($A\approx 555$)	
Fe_5Zn_{21}	$D8_{1-3}$	8.98	
$CoZn_{1-3}$ (ξ)	monocl.	13.46; 7.49; 5.06; 127° 05′	
Co_5Zn_{21}	$D8_{1-3}$	8.92	
CoZn (β_1)	*A*13	6.33	
(β)	*A*2(?)	(>920 °C)	
Ni_5Zn_{21}	$D8_{1-3}$	8.90	
NiZn (β_1)	$L1_0$	2.73; 3.19 (high temp.)	
(β)	*B*2	2.91	
NiMgZn	*C*15	6.96	
Rh_5Zn_{21}	$D8_{1-3}$	—	
Pd_2Zn	$C37^1$ type	5.35; 7.65; 4.14	341
$PdZn_{11}$ (η)	similar to *A*3	$c/a\approx 1.55$	
$Pd_{18}Zn_{82}$ (γ)	similar to $D8_{1-3}$	9.09 } γ is more similar to γ-brass-str. than γ^1	
$Pd_{22}Zn_{78}$ (γ^1)	similar to $D8_{1-3}$	—	
Pd_2Zn_3 (β_1)	*B*2	3.04 (> ~600 °C)	
PdZn (β)	$L1_0$	4.14; 3.50	53
Pd_2Zn	*B*2	3.05 (> ~600 °C)	
Pt_5Zn_{22} (γ)	$D8_{1-3}$	18.08	
$\sim PtZn_3$ (γ_1)	similar to $D8_{1-3}$		
$PtZn_2$ (ξ)	*C*32	4.10; 2.74	
PtZn (ξ)	$L1_0$	4.03; 3.47	
Pt_3Zn	$L1_2$	3.89 (<≈800 °C)	

Table 6.1 STRUCTURES OF METALS, METALLOIDS AND THEIR COMPOUNDS—*continued*

Element or compound	*Structure type*	*Lattice constants, remarks*	*Refs.*
Group IIb: Zn, Cd, Hg—*continued*			
$CuZn_5$ (ε)	*A*3	2.75; 4.29	
$CuZn_3$ (δ)	*A*2	3.01 (stable between 550–700 °C)	
Cu_5Zn_8 (γ)	*D*8_2	8.84	
CuZn (β)	*A*2	(>450 °C) 2.95	
(β')	*B*2	(< ~450 °C) 2.94	
~Cu_6Zn_4	tetr. f.c.	3.678; 3.602 (intermediate state during precipitation α from β at 225 °C)	
~Cu_6Zn_4	tetr. f.c.	3.690; 3.650 (intermediate state during precipitation α from β at 250 °C)	
Cu_2TiZn	*CsCl* (B2) type	2.95	265
	tetr. b.c.	2.945; 3.007 (intermediate state during precipitation α from β at 250 °C)	
$AgZn_3$	*A*3	2.81; 4.42	
Ag_5Zn_8	*D*8_2	9.340 7	54
AgZn (β)	*A*2	3.16 (high temp.)	
(β')	*B*2	3.16 (medium temp.)	
(ξ)	*Bb*	7.64; 2.82 (<260 °C)	
$(Ag_{0.07}Zn_{0.93})_2Mg$	~*C*36	—	
$(Ag_{0.10}Zn_{0.90})_2Mg$	~*C*36	—	
$(Ag_{0.25}Zn_{0.75})_2Mg$	*C*36	5.24; 17.27	
$(Ag_{0.4}Zn_{0.6})_2Mg$	*C*15	7.46	
$AuZr_2$	*C*11*B*	3.28; 11.6	342
$AuHf_2$	*C*11*B*	3.230 9; 11.605 7	342
$AuZn_6$ (ε)	*A*3	2.81; 4.37	
$AuZn_3$	*cub* O_h^3 *Pm*3*n*	7.887	266
$AuZn_2$	cubic (?)	11.2 (A ≈ 90)	
Au_5Zn_3	*D*8_{1-3}	9.25	
AuZn	*B*2	3.12	
Au_3Zn (α)	*A*1	4.031 (>420 °C; quenchable)	
(α')	tetr.	4.026; 4.107 (between 270–415 °C; not quenchable)	
(α'')	*L*1*c*	3.948; 8.306 (<260 °C)	
Cd	*A*3	2.973; 5.605	
$LiCd_3$	*A*3	3.08; 4.89	
LiCd	*B*32	6.69	
Li_3Cd	*A*1	4.25	
KCd_{13}	*D*2_3	13.78	
$RbCd_{13}$	*D*2_3	13.88	
$CsCd_{13}$	*D*2_3	13.89	
$MgCd_3$	*D*0_{19}	6.2; 5.1	
MgCd	*B*19	5.00; 3.22; 5.27	
Mg_3Cd	*D*0_{19}	6.26; 5.07 (<160 °C)	
$CaCd_2$	*C*14	5.98; 9.64	
SrCd	cubic	4.003	55
BaCd	cubic	4.207	55
$BaCd_{11}$	tetr.	12.02; 7.74 ($M=4$)	56
LaCd	*B*2	3.90	
CeCd	*B*2	3.86	
PrCd	*B*2	3.82	
TiCd	tetr. *B*11 type	2.904; 8.954 ($M=4$)	267
Ti_2Cd	tetr. *C*11 type	2.865; 13.42 ($M=6$)	267
$Zr_{(1-x)}Cd_{(x)}$	cubic f.c.	4.376 8	51
	tetr.	3.124 3; 4.300 8	51
$NaCd_2$	O_h^7–*Fd*3*m*		268
Ni_5Cd_{21}	*D*8_{1-3}	9.76	
$Pd_{17.5}Cd_{82.5}$ (γ)	*D*8_{1-3}	9.94	
$Pd_{22}Cd_{78}$ (γ_2)	~*D*8_{1-3}	9.92	
Pd_2Cd_2	A_2	3.25 (ordered?)	

Table 6.1 STRUCTURES OF METALS, METALLOIDS AND THEIR COMPOUNDS—*continued*

Element or compound	*Structure type*	*Lattice constants, remarks*	*Refs.*
Group IIb: Zn, Cd, Hg—*continued*			
PdCd	$L1_0$	4.31; 3.65	53
~$PtCd_5$ (γ)	$D8_{1.3}$	9.88	
$PtCd_2$ (ξ)	~$C32$	—	
PtCd	$L1_0$	4.24; 3.91	53
Cu_5Cd_3	$D8_2$	9.62	
$AgCd_3$	$A3$	3.07; 4.81	
Ag_5Cd_3	$D8_{1-3}$	9.96	
AgCd	h.c.p.	2.9710; 4.8279 at 25 °C	269
AgCd (β)	$A2$	(>443 °C) 3.32	
(β₁)	$A3$	(186–475 °C) 2.98; 4.81	
(β′)	$B2$	(<210 °C) 3.33	
$AuCd_3$	~$L1_2$	4.11	
$AuCd_2$	$A3$ (?)	—	
AuCd	orthorh.D_{2h}^{17}-*cmcm*	at −196 °C 3.116; 4.890; 4.779	
AuCd ($β_1$)	$B2$	3.33 (>60–80 °C)	
($β^1$)	$B19$	3.15; 4.85; 4.75 (< ~60 °C)	
$Au_{55}Cd_{45}$ ($α_3$)	$A1$ (*rhbdr* deformed)	5.48; 12.62 (Owen[57], denied by Byström[58])	
Au_2Cd ($α_2$)	~$A3$	2.91; 4.79	
Au_3Cd ($α_1$)	tetr. $L1_2$	4.116; 4.131 (< ~400 °C)	
ZnCuCd	$C15$	7.15	
Hg	$A10$	2.9925; 70° 44.6′ (78 K): 2.9863; 70° 44.6′ (5 K)	59
$LiHg_3$	$D0_{19}$	6.24; 4.79	
LiHg	$B2$	3.23	
Li_3Hg	$D0_3$	6.55	
NaHg	orthorh.	7.19; 10.79; 5.21 ($M=8$)	60
Na_3Hg_2	tetr.	8.52; 7.80 ($M=4$)	60
$NaHg_2$	$C32$	5.029; 3.230 ($M=1$)	60
KHg	triclinic	6.59; 6.76; 7.06 (α=106° 5′, β=101° 52′; γ=92° 47′)	61, 62
K_5Hg_7	orthorh. *Pbcm*	10.06; 19.45; 8.34 ($M=4$)	270
KHg_2	orthorh.	8.10; 5.16; 8.77 ($M=4$)	61, 62
$MgHg_2$	$C11$	3.83; 8.78	
MgHg	$B2$	3.44	
Mg_5Hg_3	$D8_8$	8.24; 5.92	
Mg_3Hg	$D0_{18}$	4.86; 8.64	
SrHg	cubic	3.922	55
BaHg	cubic	4.125	55
$BaHg_{11}$	$D2e$	9.60	
$LaHg_4$	cubic	10.97 ($A=54$; similar to $D8_{1-3}$-type)	
$LaHg_3$	$D0_{19}$ (?)	3.40; 4.95 (cf. UHg_2)	
$LaHg_2$	$C32$	4.95; 3.63	
LaHg	$A2$	3.84 (ordered as $B2$-type?)	
CeHg	$A2$	3.81 (ordered as $B2$-type?)	
PrHg	$A2$	3.79 (ordered as $B2$-type?)	
TiHg	$L1_0$	3.009; 4.041	51
Ti_3Hg	$A15$	5.1888	51
	$L1_2$	4.1654 ($A=4$)	51
Hg_3Th	hex.c.p.	3.361; 4.905	271
HgTh	f.c.c.	4.80	
ZrHg	$L1_0$	3.15	51
Zr_3Hg	cubic, *Pm3L*	5.5583	51
$ZrHg_3$	cubic, *Pm3L*	4.3652	51
$ThHg_3$	hex.	3.38; 4.72	47
$ThAg_3$	hex.c.p.	3.361; 4.905	272
ThAg	f.c.c.	4.80	272
NdHg	$A2$	3.77 (ordered as $B2$-type)	
UHg_4	cubic b.c.	3.63 (with superstructure lines)	
UHg_3	$A3$	3.33; 4.89	
UHg_2	$C32$	4.99; 3.23	
$NiHg_3$	$B2$ (?)	3.00	
PdHg	$L1_0$	4.28; 3.69	

Table 6.1 STRUCTURES OF METALS, METALLOIDS AND THEIR COMPOUNDS—*continued*

Element or compound	*Structure type*	*Lattice constants, remarks*	*Refs.*
Pt_3Hg	*A*1	3.92	341
~CuHg	$D8_{1-3}$	9.41	
Ag_3Hg_4	$D8_{1-3}$	10.02	
$Ag_{5.5}Hg_{4.5}$	*A*3	2.99; 4.85	
Au_3Hg	*A*3	2.91; 4.78	
~Zn_3Hg_3	hex.	2.71; 5.48	
CdHg	tetr.	3.885; $c/a=0.747$ at 31% Hg 3.98; $c/a=0.721$ at 71.8% Hg	63
CdHg (β)	hex.	3.206 2; 2.985 6	64
Group IIIb: Al, Ga, In, Tl			
Al	*A*1	4.04	
LiAl	*B*32	6.36	
Mg_2Al_3 (β)	*Fd*3*m*	28.13 ($A=1\,172$)	
~MgAl (δ′)	*A*12 (deformed)		
~$Mg_{17}Al_{12}$ (δ)	*A*12	10.54	
γMgAl	*A*12	10.57 (38.5 at.% Al): 10.448 (48.4 at.% Al)	65
$Al_{18}Mg_3Cr_2$	$O_h^7 Fd3m$	14.53; a/cell 184	273
Y_3Al	tetr.	8.239; 7.648 ($M=4$)	274
YAl	orthorh.	3.884; 11.522; 4.385 ($M=4$)	
$CaAl_4$	$D1_3$	4.35; 11.07	
$CaAl_2$	*C*15	8.02	
SrAl	cubic b.c.	15.8 ($A=116$)	
$SrAl_4$	$D1_3$	4.45; 11.05	
$BaAl_4$	$D1_3$	4.566; 11.250	66
$LaAl_4$	$D1_3$	4.42; 10.21	
$LaAl_2$	*C*15	8.14	
CeAl	orthorh. *Cmc*2	9.27; 7.68; 5.76	67
$CeAl_4$	$D1_3$	4.37; 10.10	
$CeAl_2$	*C*15	8.06	
Ce_3Al (α)	*A*3	7.04; 5.451	67
Ce_3Al (β)	$L1_2$	4.985	67
NdAl	*B*2	3.73	
$TiAl_3$	$D0_{22}$	5.43; 8.58	
Ti_2Al		5.775; 4.638	68
Ti–Al (α_2)	hex.	5.76; 4.65	60
~TiAl	$L1_0$	3.99; 4.07	
Al_6Ti	isom. with $MnAl_{16}$	6.58; 7.63; 9.00	275
Ti_3Al	DO_{19} struct.	5.77; 4.62	276
$ZrAl_3$	$D0_{23}$	4.01; 17.29	
Zr_3Al	$L1_2$	4.372	82
Zr_2Al	hex. Pb_3mmc	4.893 9; 5.928 3	277
Zr_2Al_3	orthorh. *Fdd*2	9.601; 13.906; 5.57	278
Zr_4Al_3	hex.C_{3h}^1–P_6^1	5.433; 5.390 ($M=1$)	279
Zr_3Al_2	tetr.D_{4h}^{14}–P_2^4/mnm	7.630; 6.998	280
HfAl	orthorh.*Cmcm*	3.25; 10.83; 4.28 ($M=4$)	281
Hf_2Al_3	orthorh.	9.52; 13.76; 5.52 ($M=8$)	282
$ThAl_3$	DO_{19}	6.500; 4.626	48
$ThAl_2$	*C*32	4.388; 4.162	48
Th_3Al_2	tetr. *P*4/*mbm*	8.125; 4.217	48
Th_2Al	*C*16	7.614; 5.857	48
UAl_4	*D*1*b*	4.41; 6.27; 13.71	
UAl_3	$L1_2$	4.28	
UAl_2	*C*15	7.80	
$NpAl_2$	*C*15	7.785	70
$NpAl_3$	$L1_1$	4.262	70
$NpAl_4$	*D*1*b*	4.42; 6.26; 13.71	70
$PuAl_3$	hex.	6.10; 14.47 ($M=6$)	81
VAl_3	$D0_{22}$	5.33; 8.31	
V_4Al_{23}	hex. $P6_3/mmc$	7.693; 17.04	71
$NbAl_3$	$D0_{22}$	5.43; 8.58	
Nb_2Al	tetr.	9.943; 5.586 30 a/cell	283

Table 6.1 STRUCTURES OF METALS, METALLOIDS AND THEIR COMPOUNDS—*continued*

Element or compound	*Structure type*	*Lattice constants, remarks*	*Refs.*
Group IIIb; Al, Ga, In, Tl—*continued*			
Nb_3Al	O_h^3–$Pm3n$	5.187 ($M=2$)	337
$TaAl_3$	$D0_{22}$	5.42; 8.54	
$CrAl_7$	orthorh.	19.99; 34.51; 12.47	72
	monocl.	20.43; 7.62; 25.31; 155° 10	73
	orthorh.	24.8; 24.7; 30.2 ($M=145$)	74
Cr_2Al_{11}	orthorh.	24.8; 24.7; 30.2 ($M=93$)	74 (*cf.* 72)
Cr_5Al_6	$D8_{10}$	7.79; 109° 8′	
CrAl (?)	$\sim A2$	3 × 3.02	
Cr_2Al	$C11$	3.00; 8.63	
$CrMg_{1.5}Al_{12}$	cubic f.c.	14.65	75
$CrMg_2Al_{12}$	cubic	14.68 ($M=16$)	
$Cr_2Mg_3Al_{18}$ (E phase)	$Fd3m$	14.55	76
$MoAl_{12}$	b.c.c.	7.575	284
$MoAl_7$	monocl.	5.12; 13.0; 13.5 ($\beta=95°$)	284
WAl_4	monocl.	5.272; 17.771; 5.218; 100° 12′ (a/cell = 30)	285
Mo_3Al	monocl.	4.950	
WAl_5	hex. $P6_3$	4.902 0; 8.857 0	78
WAl_{12}	cubic b.c., $Im3$	7.580	77
$MnAl_6$	orthorh.	6.497 8; 7.551 8; 8.870 3	79
$MnAl_4$	hex.	28.35; 12.36	
MnAl (y)	orthorh.	14.79; 12.60; 12.43	
(x)	monocl. or tricl.	$a \approx b \approx 5.1$	
Mn_3Al_{10}	hex.$6/mmm$	7.543; 7.898	287
Mn_4Al_{11}	tricl.	5.092; 8.862; 5.047; α 85° 19′; β 100° 24′; γ 105° 20′	286
$MnAl_3$	orthorh.$Pnma$	14.79; 12.42; 12.59 ($M=36$)	288
(Mn, Cr)Al_{12}	cubic b.c., $Im3$	7.507	77
(Mn, Si)$_2Al_6$	similar to $D8_{11}$	—	
Tc_2Al_3	trigonal	4.16; 5.13	289
$TcAl_{12}$	b.c.c.	7.527 0	289
$TcAl_4$	monocl.	5.1; 17.0; 5.1 ($\beta=100°$)	289
$RcAl_6$	isomorph with *$MnAl_6$*	6.59; 7.61; 9.02	275
Fe_2Al_7 (λ_1)	rhomb. (?)	—	
Fe_2Al_5	$D8_{11}$	7.660; 6.390; 4.195	80
$FeAl_2$ (ν)	monocl. (?)	—	
$FeAl_3$	D_{2h}^{23}	47.43; 14.46; 8.08 ($A=400$)	
FeAl (β_2)	$B2$	2.90	
Fe_2Al (B_1)	$D0_3$	5.78	
$Fe_2Al+C_{0.5}$	$L'1_2$	3.75	
Co_2Al_9	$D8d$	8.56; 6.29; 6.21; 94° 45′	
Co_2Al_5	$D8_{11}$	7.59; 7.66	
CoAl	$B2$	2.85	
$NiAl_3$	$D0_{20}$	6.60; 7.35; 4.80	
Ni_2Al_3	$D5_{13}$	4.03; 4.89	
NiAl	$B2$	2.88	
Ni_3Al	$L1_3$	3.56	
Ni_2TiAl	$L2_1$	5.86	
$NiFe_2Al$ (β')	$B2$	≈ (Ni, Fe) (Al, Fe)	
$NiFeAl_9$	$D8d$		87
$\sim NiFe_2Al_9$	$D8d$		88
RuAl	*cub.*	3.03	290
$RuAl_2$	CaC_2 type	4.40; c/a 1.45	
$RuAl_3$	hex.	4.81; c/a 1.63	
Pd_2Al_3	$D15_{13}$	4.21; 5.15	
PdAl	monoc.	9.68; 15.14; 5.24	83
	$B2$	3.03 (> 700 °C)	
OsAl	$B2$	2.999	83
IrAl	$B2$	2.977	83
PtAl	$B20$	4.855	83
$PtAl_2$	$C1$	5.91	
Pt_5Al_3	D_{2h}^9	5.41; 10.70; 3.95 ($M=2$)	338

Table 6.1 STRUCTURES OF METALS, METALLOIDS AND THEIR COMPOUNDS—*continued*

Element or compound	*Structure type*	*Lattice constants, remarks*	*Refs.*
Group IIIb: Al, Ga, In, Tl—*continued*			
Pt_3Al	Au_3Cu type	3.876	
$CuAl_2$	*C*16	6.05; 4.87 (stable form)	
	tetr.	8.2; 11.6 (metastable form, *e.g.* precipitated from supersaturated solid solution)	
	*C*1 (deformed into tetr.)	$a=5.71$; $c/a=1.017$ (metastable)	
CuAl (η)	orthorh.	6.88; 4.08; 9.87 (49–50 at. % *A*1; similar to $-D5_{13}$)	
Cu_4Al_3 (ζ)	monocl.	7.06; 4.04; 10.00; 90° 38′ (42–43 at. % *A*1; similar to $-D5_{13}$)	
Cu_3Al_2 (γ_2)	$D8_{1-3}$	8.69 ($A=50.5$ to 49.5; 38–41 at. % *A*1)	
$Cu_{32}Al_{19}$ (γ_1)	$\sim D8_{1-3}$	~8.71 ($A=51$ for pseudo-cubic cell: monocl.)	
Cu_9Al_4 (γ)	$D8_3$	8.685 to 8.704 ($A=52.2$ to 51.8; 31–35 at. % *A*1)	
Cu_2Al	*A*3	2.60; 4.22 (Kurdjumow's martensitic γ'-phase)	
	~*A*3, D_{2h}^{13}	4.51; 5.20; 4.22	
Cu_3Al (β)	*A*2	(>570 °C) 2.94	
(β_1)	$D0_2$	(~300 °C) 5.84 (unstable)	
(β')	~*A*1 (?)	(<300 °C)	
$\sim Mg_4CuAl_6$	*D*8*e*	14.25	
$\sim Mg_2Cu_6Al_5$	*D*8*c*	8.31	
$\sim MgCuAl_2$	*E*1*a*	4.00; 9.23; 7.14	
$Mg_6Cu_3Al_7$	cubic b.c.	12.09 ($A=94$)	
$Mg_2Cu_2Al_5$	tetr.	5.71; 7.94	
$MgCu_{0.9}Al_{1.1}$	*C*36	5.09; 16.57	
MgCuAl	*C*14	5.09; 8.40	
Mn_2CuAl_{12}	orthorh.	7.96; 24.06; 12.48 (Phragmén[89]; identical work with $Cu_3Mn_3Al_{20}$ Raynor; *r*-AlCuMn Petri; and $MnCuAl_6$ Hanemann's '*T*'; references *see* Phragmén[89])	
$\sim Mn_3CuAl_{12}$	orthorh.	14.79; 12.60; 12.43 (Hanemann's '*Y*')	89
(Mn, Cu)Al (δ)	*B*2	—	
$(Mn, Cu)_3Al$ (β)	$D0_3$	5.9	
$MnCu_2Al$ (β')	$L2_1$	5.95	
$FeCu_{10}Al_{10}$ (ζ), $FeCu_{10}Al_{18}$ (χ)	monocl. (?)	(similar to $D15_{13}$)	
$FeCu_2Al_7$	*E*9*a*	6.32; 14.78	
(Fe, Cu) $(Cu, Al)_6$	orthorh.	6.43; 7.45; 8.77 (isom. $MnAl_6$)	
$NiCu_3Al_6$	*B*2 (deformed) isom. with $ZnCu_3Al_3$	—	
$\sim NiCu_4Al_7$	rhomb.	4.105; 39.97	84
(T-phases)	cubic	14.6	84
Ag_2Al (γ)	*A*3	2.90; 4.71	
Ag_3Al (β')	*A*13	6.92	
(β)	*A*2	3.24	
$Ca(Ag, Al)_2$	*C*14	—	
$AuAl_2$	*C*1	5.99	
$\sim Au_3Al$	$D8_{1-3}$ (deformed)	(hex.?)	
$\sim Au_4Al$	*A*13	6.91	
Zn_3Al_2	$L1_2$ (?)	4.04	
Mg_{32} $(Zn, Al)_{49}$	*D*8*e*	14.2	
$\sim ZnCu_2Al$ (T)	*A*2	2.92	
$\sim ZnCu_3Al_3$ (T′)	*B*2 (deformed) isom. with $NiCu_3Al_6$	—	
Ga	*A*11	4.52; 4.51; 7.65	
LiGa	*B*32	6.20	
Mg_5Ga_2	*D*8*g*	13.72; 7.02; 6.02	
Mg_2Ga	~*C*22	7.85; 6.94	

Table 6.1 STRUCTURES OF METALS, METALLOIDS AND THEIR COMPOUNDS—*continued*

Element or compound	*Structure type*	*Lattice constants, remarks*	*Refs.*
Group IIIb; Al, Ga, In, Tl—*continued*			
$CaGa_2$	*C*32	4.31; 4.31	
$LaGa_2$	*C*32	4.32; 4.40	
$CeGa_2$	*C*32	4.30; 4.31	
$PrGa_2$	*C*32	4.28; 4.28	
α_2GaTi_3	hex.D_0^{19}		291
$GaTi_2$	hex.Ni_2In type		
TiGa (α_2)	$D0_{19}$	5.76; 4.64	85
Ti_2Ga	hex.	4.51; 5.50	86
$TiGa_3$	$D0_{22}$	5.55; 8.09	
$ZrGa_3$	$D0_{22}$	5.61; 8.71	
Zr_5Ga_3	hex.D_8^8	8.04; 5.71	291
Nb_3Ga	O_h^3Pm3m	5.171	337
V_3Ga	O_h^3Pm3m	4.816	337
Mo_3Ga	O_h^3Pm3m	4.932	
CoGa	*B*2	2.86	83
$\sim NiGa_4$	*D*8*f*	8.41	
Ni_2Ga_3	$D5_{13}$	4.05; 4.89	
NiGa	*B*2	2.87	
Ni_3Ga_2	~*B*8	—	
Ni_2Ga	*B*8	3.99; 4.97	
	*A*6	3.75; 3.38	
RhGa	*B*2	—	
PdGa	*B*20	4.88	
$PtGa_2$	*C*1	5.91	
Pt_2Ga_3	$D5_{13}$	4.22; 5.17	
PtGa	*B*20	4.90	
Pt_4Ga	monocl.	7.74; 7.74; 7.785 ~ 90°	293
Pt_3Ga	$L1_2$	3.892	
Pt_5Ga_3	orthorh.	8.031; 3.94; 7.44	
$PtGa_6$	orthorh.	15.946; 12.034; 8.87	
$CuGa_2$ (θ)	~*C*38(tetr.)	2.83; 5.84 (A = 3)	
$\sim Cu_9Ga_4$ (γ)	$D8_{1-3}$	8.71 (between 500 and 836 °C; 29–35% Ga; A = 52; electrons/cell = 85; disordered)	
(γ_1)	$D8_{1-3}$	8.71 (< 645 °C; 30–37% Ga; A = 52; electrons/cell = 86; ordered)	
(γ_2)	$D8_{1-2}$	8.70 (< 485 °C; 33–37% Ga; A = 49; electrons/cell = 86)	
(γ_3)	$D8_{1-3}$	8.65 (< 468 °C; 37–43% Ga; A = 47; electrons/cell = 86)	
Cu_4Ga	*A*3	2.59; 4.23	
Ag_3Ga (γ)	*D*0*b*	7.80; 288 (< ~440 °C)	
(β)	*A*3	2.93; 4.75 (> ~440 °C)	
$AuGa_2$	*C*1	6.06	
AuGa	*B*31	6.38; 6.25; 3.41	
In	*A*6	4.58; 4.94	
LiIn	*B*32	6.79	
NaIn	*B*32	7.30	
$MgIn_2$	LI_2	4.60	
~MgIn (β'')	$L1_0$	3.24; 4.38 (< ~300 °C)	
$\sim Mg_{70}In_{30}$ (β')	$L1_2$	(< ~300 °C)	
β-Mg–In	*A*1	(> ~300 °C; ranging from ~20–50 at. % In)	
~MgIn (β''')	orthorh.	(at ~300 °C, between B″ and β; $b/a \approx 1$)	
Mg_2In	~*C*22	8.40; 6.96 (< 300 °C)	
α_2Ti_3In	hex. D_0^{19}	5.89; 4.76	291
Mn_3In	$D8_{1-3}$	9.44	
ZrIn	L_2^1 type	4.461	291
$ThIn_3$	cubic *L*12	4.695	292
$\sim NiIn_3$ (η)	*D*8*f*	9.18	
Ni_2In_3 (δ')	$D5_{13}$	4.39; 5.30	
NiIn (δ)	*B*2	3.09 (> 760 °C)	
(ε)	*B*35	4.54; 4.34 (< 840 °C; constant after Hellner[90]; Makarow[91] has a = 5.20; c = 4.34)	
Ni_2In (β)	*B*8	4.19; 5.15	
Ni_3In (γ)	$D0_{19}$	5.32; 4.24	
$\sim PdIn_3$	*D*8*f*	9.42	

Table 6.1 STRUCTURES OF METALS, METALLOIDS AND THEIR COMPOUNDS—*continued*

Element or compound	*Structure type*	*Lattice constants, remarks*	*Refs.*
Group IIIb: Al, Ga, In, Tl—*continued*			
Pd_2In_3	$D5_{13}$	4.52; 5.49	
PdIn	$B2$	3.25	
Pd_3In	$A6$	4.06; 3.79	
Pt_3In_7	$D8f$	9.42	
$PtIn_2$	$C1$	6.35	
Pt_3In	$L6_0$ type	3.93; 3.87	342
Pt_3In_2	$D5_{13}$	4.52; 5.50	
Pt_2In_3	$B8$	—	
Cu_3In	$B8$	4.28; 5.25	
$\sim Cu_7In_3$ (γ)	$D8_{1-3}$	9.24 (>610 °C)	
(δ)	tetr. (?)	8.97; 9.14 (<630 °C; constants and symmetry derived by powder method Reynolds[92]; structure is said to be similar to $D8_{1-3}$-type. Similarity to $B8$-type was proposed by Hellner[93])	
Cu_4In	$\sim A2$	$n \times 3.01$	
Cu_2MnIn	$L2_1$	6.18	
AuIn	c.c. pseudo orthorh.	4.29; 10.57; 3.55 (α 90.54°; β 90.00°; γ 90.17°)	80
Au_3In	A_2	72.914; 4.775	293
CdIn (β)	hex.	3.211 2; 2.992 8	64
Tl	$A3$	3.45; 5.51 (<230 °C)	
	$A2$	3.87 (>230 °C)	
LiTl	$B2$	3.42	
NaTl	$B32$	7.47	
MgTl	$B2$	3.63	
Mg_2Tl	$\sim C22$	8.11; 7.34	
Mg_5Tl_2	$D8g$	15.17; 7.30; 6.16	
$CaTl_3$	$L1_2$	4.79	
CaTl	$B2$	3.85	
SrTl	$B2$	4.02	
$LaTl_3$	$A3$	3.45; 5.52	
LaTl	$B2$	3.91	
CeTl	$B2$	3.89	
$ThTl_3$	cubic L_{12}	4.748	292
PrTl	$B2$	3.86	
Pd_2Tl	$B8$	—	
Pd_3Tl	$L60$ type	4.12; 3.84	342
PtTl	$B35$	5.61; 4.64	
$\sim HgTl_8$	$A2$	3.82	
$\sim Hg_5Tl_2$	$A1$	4.66 (perhaps with superstructure)	
$HgLi_2Tl$	$B2$	3.35	
In_3Tl	$A2$	3.81	
Group IVb: Si, Ge, Sn, Pb			
Si	$A4$	5.42	
$NaSi_2$	tetr. (?)	4.98; 16.7	
Mg_2Si	$C1$	6.34	
$CaSi_2$	$C12$	10.4; 20° 30′	
CaSi	Bc	3.91; 4.59; 10.80	102
Ca_2Si	$C23$		
$BaSi_1$	AlB_2 type	4.38; 4.82	294
$LaSi_2$	Cc	4.27; 13.72	
$CeSi_2$	Cc	4.14; 13.81	
$PrSi_2$	Cc	4.14; 13.64	
$NdSi_2$	Cc	4.10; 13.53	
$SmSi_2$	Cc (orthorh. deformed)	4.04; 13.33	
YSi	orthorh. $D_{2h}^{17}Cmcm$	4.25; 10.52; 3.82 ($M = 4$)	295
YSi_2	Cc (orthorh. deformed)	—	
TiSi	αhex.		
	βb.c.c.		296
$TiSi_2$	$C49$	3.6; 13.76; 3.60	94
Ti_5Si_3	$D8_8$	7.47; 5.16	
Zr_2Si	$D8_8$ or $C16$	6.56; 5.36	95, 96

Table 6.1 STRUCTURES OF METALS, METALLOIDS AND THEIR COMPOUNDS—*continued*

Element or compound	*Structure type*	*Lattice constants, remarks*	*Refs.*
Group IVb: Si, Ge, Sn, Pb—*continued*			
Zr_5Si_3	*D*8$_8$ or *C*16	7.87; 5.54	95
$ZrSi_2$	*C*49	3.72; 14.69; 3.66	94
$HfSi_2$	*C*49	3.69; 14.46; 3.64	94
$ThSi_2$	*Cc*	4.13; 14.35	
USi_2 (α)	*Cc*	3.97; 13.71 } Zachariasen[97]	
(β)	*C*32	3.85; 4.06 }	
(?)	*L*1$_2$	4.05 (Bramer[98]) (this phase might be due to *Al*-contamination)	
USi	*B*27	5.65; 7.65; 3.90	
U_3Si_2	*D*5*a*	7.33; 3.90	
U_3Si	*D*0*c*	6.02; 8.68	
$NpSi_2$	*Cc*	3.96; 13.67	
$PuSi_2$	*C*32	3.884; 4.082	99
VSi_2	*C*40	4.56; 6.36	
V_3Si	*A*15	4.71	
V_5Si_3	*D*8$_8$	7.12; 4.832	100
$NbSi_2$	*C*40	4.79; 6.58	
Nb_5Si_3	*D*8$_8$	7.52; 5.238	100
$TaSi_2$	*C*40	4.77; 6.55	
$CrSi_2$	*C*40	4.42; 6.35	
CrSi	*B*20	4.62	
Cr_3Si_2	isom. with Cr_3Ge_2	—	
Cr_3Si	*A*15	4.56	
$MoSi_2$	*C*11	3.19; 7.83	101
Mo_3Si	*A*15	4.89	
Mo_5Si_3	*D*8$_8$	7.27; 4.992	100
WSi_2	*C*11	3.20; 7.81	101
$MnSi_2$	tetr.	5.51; 17.42	
MnSi	*B*20	4.55	
Mn_5Si_3	*D*8$_8$	6.90; 4.80	
Mn_3Si	*A*2	2.85	
$ReSi_2$	*C*11	3.12; 7.66	
$FeSi_2$	tetr.	2.69; 5.13	
FeSi	*B*20	4.49	
Fe_5Si_3	*D*8$_8$	6.73; 4.70	
Fe_3Si	*D*0$_3$	5.64	
$CoSi_2$	*C*1	5.36	
CoSi	*B*20	4.44	
Co_2Si	*C*37	7.10; 4.91; 3.73	
$NiSi_2$	*C*1	5.40	
(ζ)	hex.	12.60; 15.28 (Osawa[104]; the existence of this phase is questioned by Schubert[103])	
NiSi	*B*20	4.44 (questioned by Schubert[103])	105
(η)	tetr.	7.65; 8.45 (Osawa[104], questioned by Toman[106])	
	Bd	5.62; 5.18; 3.34	106
Ni_3Si_2 (ε)	—	—	104
Ni_2Si (θ)	*B*8	3.81; 4.89 (>1 200 °C)	106
(δ)	*Cd*	7.03; 4.99; 3.72 (<1 200 °C)	106
Ni_5Si_2 (γ)	hex.	7.67; 9.75 (?)	104
Ni_3Si (β_1)	*L*1$_2$	3.50 (<1 040 °C)	
Ru_2Si_3	tetr.	5.52; 4.46	
RuSi (1)	*B*2	2.90	
(2)	cubic	4.70	
$RuSi_{1.0-0.1}$	*FeSi* type	4.703	297
Ru_2Si			
RhSi	*B*20	4.672	107
$RhSi_{0.5}$	orthorh.		297
Rh_2Si			
Rh_5Si_3	orthorh.	10.07; 5.30; 3.88 }	
Rh_3Si_2	*B*8	3.94; 5.047 }	293
RhSi	*B*31	6.36; 5.53; 3.06 }	
PdSi	*B*31	6.12; 5.59; 3.37	

Table 6.1 STRUCTURES OF METALS, METALLOIDS AND THEIR COMPOUNDS—*continued*

Element or compound	*Structure type*	*Lattice constants, remarks*	*Refs.*
Group IVb: Si, Ge, Sn, Pb—*continued*			
Pd_2Si	*C*22	6.48; 3.42	108
Os_2Si_3	tetr.	5.57; 4.47	
OsSi	C_8Cl type	2.960	297
Ir_3Si_2	hex.	3.96; 5.12 (*B*8-type?)	
$IrSi_{0.3}$	*BC* tetr.		297
$IrSi_3$	Na_3As type	4.35; 6.30	
Ir_3Si	U_3Si type	5.22; 7.95	293
Ir_2Si	Ni_2Si type	7.615; 5.28; 3.98	293
Ir_3Si_2	B_8	3.96; 5.12	293
$IrSi_3$	hex.	4.31; 6.61	293
PtSi	*B*31	5.92; 5.58; 3.60	
$\sim Pt_2Si$	tetr.	2.77; 2.95	109
Pt_2Si	*C*22	6.76; 3.45	110
BeZrSi	hex.	3.71; 7.19	8
Pd_4Al_3Si	*B*20	4.830	83
Cu_3Si (η) and $Cu_{15}Si_4$ (δ)	$D8_{1-3}$ (?) deformed	—	
$Cu_{15}Si_4$ (ε)	$D8_6$	9.69	
Cu_5Si	*A*13	6.21	
$\sim Cu_7Si$	*A*3	2.57; 4.18	
$Cu_{16}Mg_6Si_7$	cubic f.c.	11.67 ($A=116$)	111
Cu_3SiMg_2	*C*14	5.00; 7.87	
	*C*36	5.0; 16.0 (>870 °C)	
$AlNaSi_4$	$\sim C38$	4.13; 7.40	
$\sim Al_3CrSi$	cubic f.c.	—	
$Al_{13}Cr_4Si_4$	cubic	10.917	112
$Al_5(Mn, Si)_2$	similar to $D8_{11}$	—	
$Al_{21}Mn_3Si_5$	cubic	12.63	
$\sim Al_9Mn_3Si$	*E*9*c*	7.51; 7.74	
AlFeSi (α)	hex.	12.3; 26.2	113
$Al_{21}Fe_3Si_5$	cubic	12.52	
$Al_9Fe_2Si_2$	monocl.	6.11; 6.11; 41.4; 91° (pseudo-tetr.)	
Al_4FeSi_2	tetr.	6.13; 9.46	
$AlNi_2Si$	*B*20	4.55	83
Al_8FeMg_3Si	*E*9*b*	6.62; 7.92	
$Al_5Cu_2Mg_8Si_6$	tetr.	10.30; 4.04	
Ge	*A*4	5.65	
Mg_2Ge	*C*1	6.37	
GaGe	*Bc*	4.001; 4.575; 10.845	114
$CaGe_2$	*C*12	10.49; 21° 42′	
Ca_2Ge	*C*23		102
Y_5Ge_3	D_8^8	8.47; 6.35	298
$CeGe_2$	$\alpha ThSi_2$ type	4.202; 14.153	299
$PrGe_2$	*Cc*	4.25; 13.94	
TiGe	orthorh. C_{2v}^1	3.80; 5.22; 6.82	300
$TiGe_2$	*C*54	8.58; 5.02; 8.85	
Ti_5Ge_3	$D8_8$	7.54; 5.22	
$ZrGe_2$	*C*49	3.80; 15.01; 3.76	
Zr_5Ge_3	$D8_8$	7.99; 5.59	115
Hf_5Ge_3	hex. D_{6h}^3–$P6_3mcm$	7.88; 5.53	298
$\alpha ThGe_2$	tetr.	4.106; 14.193	301
ThGe	cubic	6.033	301
Th_3Ge_2	tetr.	7.971; 4.170	301
$ThGe_3$	cubic	11.72	301
V_3Ge	*A*15	4.76	
$NbGe_2$	*C*40	4.96; 6.77	
$TaGe_2$	*C*40	4.95; 6.74	
Ta_5Ge_3	$D8_8$	7.58; 5.23	115
Cr_3Ge	*A*15	4.645	337
CrGe	*B*20	4.78	
Cr_5Ge_3	tetr.	9.41; 4.78	115
Mo_3Ge	*A*15	4.93	
Mn_5Ge_3	$\sim B8$	(high temp. mod.)	

Table 6.1 STRUCTURES OF METALS, METALLOIDS AND THEIR COMPOUNDS—*continued*

Element or compound	*Structure type*	*Lattice constants, remarks*	*Refs.*
Group IVb: Si, Ge, Sn, Pb–*continued*			
$Mn_{3.25}Ge$	$D0_{19}$	5.35; 4.37 (at low temp.)	
	$A3$	(at high temp.)	
$FeGe_2$	$C16$	5.90; 4.94	
Fe_2Ge	$B8$	4.03; 5.02 ($Fe_{1.8}Ge$)	
CoGe	Ni_3Sn_4 type	11.64; 3.80; 4.94 (β=101.10°; M=8)	293
Co_5Ge_8	b.c.tetr.	7.64; 5.814 (A=26)	
$CoGe_2$	Ce	5.65; 5.65; 10.8	
NiGe	$B31$	5.80; 5.37; 3.42	
Ni_2Ge	$B8$	3.95; 5.04 ($Ni_{1.8}Ge$)	
Ni_3Ge	$L1_2$	3.56	
GeRu	B_{20}–$FeSi$ type	4.546	302
Ge_3Ru_2	tetr.	5.709; 4.650	
PdGe	$B31$	6.25; 5.77; 3.47	
Pd_2Ge	$C22$	6.67; 3.52	108
$OsGe_2$	monocl. $C2/m$	8.995; 3.094; 7.685 (β=119° 10′; M=4)	
Ir_4Ge_5	tetr.	5.64; 4×4.56	293
$IrGe_4$	hex.	6.211; 7.77	
Ir_8Ge_7	$D8$f	8.74	
IrGe	$B31$	6.27; 5.60; 3.48	
PtGe	$B31$	6.08; 5.72; 3.69	
Pt_2Ge	$C22$	6.67; 3.52	108
Pt_3Ge	$L1$, monocl.	7.931; 7.767; 7.767 (β=90.06)	293
Pt_3Ge_2	$B31$	7.544; 3.423; 12.236	
Pt_2Ge_3	$B31$	3×5.48; 3.37; 6.22	
$PtGe_2$	$C35$	6.18; 5.76; 2.908	
PtGe	$A1$	3.91	
~Cu_3Ge (1)	$A2$ (rhomb. deformed)	4.16; 4.91 (>~600 °C)	
(2)	$A3$ (orthorh. deformed)	2.64; 4.54; 4.19 (<~600 °C)	
(3)	~$A2$	(27–28 at. % Ge; between 620–700 °C; superstructure with holes and $a'=3a$)	
~Cu_5Ge	$A3$	2.65; 4.28	
Co_4GaGe_3	$B20$	4.63	83
$Ni_{50}Ga_{42}Ge_8$	$B20$	4.64	83
Pd_5Al_4Ge	$B20$	4.86	83
Rh_5GaGe_4	$B20$	4.822	83
Sn grey	$A4$	6.49 (<13.2 °C)	
metallic	$A5$	5.83; 3.18 (>13.2 °C)	
$Na_{15}Sn_4$	~DS_6, orthorh.	9.79; 22.78; 5.65 (A=38 to 40)	
Mg_2Sn	$C1$	6.75	
CaSn	Bc	4.349; 4.821; 11.52	114
$CaSn_3$	$L1_2$	4.73	
$LaSn_2$	$L1_2$	4.77	
$CeSn_3$	$L1_2$	4.71	
$PrSn_3$	$L1_2$	4.70	
Ti_5Sn_3	$D8_8$	8.05; 5.45	
Ti_3Sn	$D0_{19}$	5.92; 4.76	
Hf_5Sn_3	hex. D_8^8	8.39; 5.82	304
$ThSn_3$	cubic $L12$	4.718	292
USn_3	$L1_2$	4.62	
$MnSn_2$	$C16$	6.65; 5.43	
Mn_3Sn_2	$B8$	4.37; 5.48 (at low temp. superstructure)	
Mn_2Sn	$B8$	4.39; 5.46	
~$Mn_{11}Sn_3$	$D0_{19}$	5.67; 4.53 (<~1 000 °C)	
$FeSn_2$	$C16$	6.52; 5.31	
FeSn	$B35$	5.29; 4.44	
Fe_3Sn_2	monocl.	13.53; 5.34; 9.20; 103° (M=8)	
~Fe_3Sn_2	$B8$	4.23; 5.21	
Fe_3Sn	$D0_{19}$	5.46; 4.36 (760–900 °C)	
$CoSn_2$	$C16$	6.35; 5.44	
CoSn	$B35$	5.27; 4.25	

Table 6.1 STRUCTURES OF METALS, METALLOIDS AND THEIR COMPOUNDS—*continued*

Element or compound	*Structure type*	*Lattice constants, remarks*	*Refs.*
Group IVb: Si, Ge, Sn, Pb–*continued*			
Co_2Sn_2	*B*8	(high temp.) 4.12; 5.18	
Co_3Sn_2	~*B*8	(<550 °C), $a \doteq 4 \times 4.089$; $c = 5.198$	
	or orthorh.	8.18; 7.08; 5.20	
Co_2MnSn	*D*0$_3$	5.991	116
Ni_3Sn_4	*D*7*a*	12.20; 4.06; 5.22; 105° 3′	
~Ni_3Sn_2	*B*8	4.08; 5.18 (at low temp. superstructure)	
	tetr.	9.20; 8.58 (*A* = 50) (>900 °C)	
Ni_3Sn	*A*3	(>900 °C)	
	*D*0$_{19}$	(<900 °C), 5.28; 4.23	
Ni_4Sn	tetr. (?)	5.11; 4.88 (*A* = 10)	
Ni_2MgSn	L 2$_1$	6.10	
Ni_2MnSn	*D*0$_3$	6.045	116
Ru_3Sn_7	*D*8*f*	9.35	
RhSn	*B*20	5.12	
~Rh_3Sn_2	*B*8	4.33; 5.54	
$RhSn_2$ (1)	*C*16	6.40; 5.64 (> 500 °C)	
(2)	*Ce*	6.32; 11.97 (<500 °C)	117
	Cf	6.38; 17.88	118
$PdSn_4$	*D*1*c*	6.38; 6.41; 11.47	
$PdSn_2$ (1)	*Ce*	6.48; 12.15	117
(2)	*Cf*	6.55; 24.57 (low temp. mod.)	118
	~*B*31 (monocl. deformed)	6.18; 3.93; 6.38; 88.5° (quenched from high temp.)	
PdSn	*B*31	6.13; 3.87; 6.32	
Pd_3Sn_2	*B*8	4.39; 5.70	
Pd_3Sn	*L*1$_2$	3.97	
Ir_3Sn_7	*D*8*f*	9.36	
$IrSn_2$	*C*1	6.34	
IrSn	*B*8	3.99; 5.57	
$PtSn_4$	*D*1*c*	6.38; 6.41; 11.33	
$PtSn_2$	*C*1	6.43	
Pt_2Sn_3	*D*5*b*	4.33; 12.96	
PtSn	*B*8	4.10; 5.43	
Pt_3Sn	*D*0$_{19}$	4.00	
Cu_6Sn_5 (η)	*B*8	4.19; 5.09	119
	~*B*8	20.95; 25.43	120
~Cu_3Sn (ε)	*A*3	2.76; 4.32	119
	~*A*3 (orthorh. deformed)	5.51; 38.18; 4.32	120
~Cu_3Sn (ε')	~*A*3 (orthorh. deformed)	9.93; 5.50; 8.46	121
γCu_3Sn	orthorh.	4.772; 5.514; 4.335 (700 °C)	122
~$Cu_{20}Sn_6$	~*D*8$_{1-3}$	7.32; 7.85 D'_3d; (*A* = 26)	
~$Cu_{31}Sn_8$	~*D*8$_{1-3}$	17.92 = 2 × 8.96	
~Cu_5Sn (β)	*A*2	2.97 (stable > 600 °C)	
(β'')	orthorh.	4.55; 5.36; 4.31 (*A* = 8; at low temp.)	
Cu_2MnSn	*L*2$_1$	6.15	
$(Cu, Ni)_3Sn$	*D*0$_3$	5.59	
Ag_3Sn	*Ae*	2.99; 5.14; 4.77	
~Ag_5Sn	*A*3	2.94; 4.77	
Ag_6Mg_3Sn	*L*2$_1$	6.60 [structural formula: $Ag_2Mg(Mg, Ag, Sn)$]	
$AuSn_4$	*D*1*c*	6.43; 6.47; 11.58	
$AuSn_2$	orthorh.	6.845; 6.990; 11.760	80
AuSn	*B*8	4.31; 5.51	
Au_6Sn	*A*3	2.92; 4.77	
$ZnSn_2$	*C*54	9.55; 5.63; 9.90	123
CdSn (β)	hex.	3.226 3; 2.996 3	64
$CdSn_9$	hex.	3.21; 2.99	63
~$HgSn_{15}$ (1)	*Af*	3.20; 2.98 } orthorh. deformed on the Sn-poor end	
(2)	orthorh.	3.20; 5.55; 2.98 } orthorh. deformed on the Sn-poor end	
~$InSn_7$	tetr. (?)	5.61; 3.54 (similar to *A*5-type?)	
$InSn_4$	hex.	3.21; 2.99	63

Table 6.1 STRUCTURES OF METALS, METALLOIDS AND THEIR COMPOUNDS—*continued*

Element or compound	*Structure type*	*Lattice constants, remarks*	*Refs.*
Group IVb: Si, Ge, Sn, Pb–*continued*			
In_3Sn	$L1_2$ (deformed into tetr.)	4.94; 4.40	
Pb	$A1$	4.95	
LiPb	$B2$	3.52	
Li_3Pb	cubic f.c.	6.687	124
Li_7Pb_2	hex.	4.751; 8.589	124
Li_8Pb_3	monocl. $C2/m$	8.240; 4.757; 11.03; $\beta=104° 25'$ ($M=2$)	125
$Li_{10}Pb_3$	$D8_{1-3}$	10.08	
$Li_{22}Pb_5$	f.c.c.	20.8 ($M=16$)	305
Na_2Pb_5	$L1_2$	4.87	
NaPb	tetr.	10.580; 17.746	126
$Na_{15}Pb_4$	$D8_6$	13.29	
KPb_2	$C14$	6.66; 10.76	127
Mg_2Pb	$C1$	6.80	
$CaPb_3$	$L1_2$	4.89	
$SrPb_3$	$L1_2$ (deformed into tetr.)	4.96; 5.03	
$LaPb_3$	$L1_2$	4.89	
$CePb_3$	$L1_2$	4.86	
$PrPb_3$	$L1_2$	4.86	
Ti_4Pb	$D0_{19}$	5.962; 4.814	128
Zr_3Pb_3	$D8_8$	8.51; 5.85	123
UPb_3	$L1_2$	4.79 (because U and Pb scatter similarly no decision between $L1_2$ and $A1$-type was possible)	
$ThPb_3$	cubic $L12$	4.856	292
$RhPb_2$	$C16$	6.65; 5.85	
$PdPb_2$	$C16$	6.84; 5.82	
PdPb	monocl.	7.08; 8.43; 5.56; 71°	
Pd_3Pb_2	$B8$	4.47; 5.71	
Pd_3Pb	$L1_2$	4.01	
IrPb	$B8$	3.99; 5.56	
$PtPb_4$	$D1d$	6.65; 5.97	
PtPb	$B8$	4.25; 5.46	
Pt_3Pb	$A1$ (or $L1_2$)	4.05 (because Pt and Pb scatter similarly no decision between $L1_2$ and $A1$-type was possible)	
Ag_4Pb	hex.	2.92; 4.76	129
$AuPb_2$	$C16$	7.31; 5.64	
Au_2Pb	$C15$	7.91	
$\sim In_3Pb$	$A6$	4.85; 4.50	
Tl_3Pb	cubic	4.88 } strong indications that ordered structures $PbTl_3$ and	
Tl_7Pb	cubic	4.86 } $PbTl_7$ exist	
Group Vb: As, Sb, Bi			
As	$A7$	5.59; 84° 36′	
Li_3As	$D0_{18}$	4.39; 7.81	
LiAs	monocl.P^2_{1h}	5.79; 5.24; 10.70 ($\beta=117.4°$; $M=8$)	306
Na_3As	$D0_{18}$	5.09; 8.98	
K_3As	$D0_{18}$	5.78; 10.22	
Mg_3As_2	$D5_3$	12.33	
MgLiAs	$C1$	6.21	
LaAs	$B1$	6.13	
CeAs	$B1$	6.06	
PrAs	$B1$	6.00	
NdAs	$B1$	5.96	
ZrAs	hex.D^4_6–$P6_3/mmc$	3.80; 12.87	307
$ZrAs_2$	C_{23} type orthorh.	6.80; 9.02; 3.68	
TiAs	$B8$	3.63; 6.14	130
	isom. TiP	3.64; 12.3	130
CrAs	$B31$	6.21; 5.73; 3.48	
Cr_2As	$C38$	3.61; 6.33	
$\sim$MnAs (1)	$B8$	3.72; 5.70	131
(2)	$B31$		132

Table 6.1 STRUCTURES OF METALS, METALLOIDS AND THEIR COMPOUNDS—*continued*

Element or compound	*Structure type*	*Lattice constants, remarks*	*Refs.*
Group Vb: As, Sb, Bi—*continued*			
Mn_2As	*C*38	3.76; 6.27	
Mn_3As	*D*0*d*	3.78; 3.78; 16.26	
$FeAs_2$	*C*18	2.86; 5.20; 5.92	
FeAs	*B*31	6.02; 5.43; 3.37	
Fe_2As	*C*38	3.63; 5.97	
$CoAs_3$	$D0_2$	8.18 (Skutterudite)	
$CoAs_{3-x}$	$\sim D0_2$	8.27 (Speiskobalt, Smaltite)	
$CoAs_2$	*C*18	—	
CoAs	*B*31	5.96; 5.15; 3.51	
$NiAs_{3-x}$	$\sim D0_2$	8.26	
$NiAs_2$	*C*18	3.53; 4.78; 5.78 (Rammelsbergite)	
	D_{2h}^{11}	5.74; 5.81; 11.41 ($M \approx 8$) (para-Rammelsbergite)	
NiAs	*B*8	3.61; 5.03	
Ni_4As_3	tetr.	6.84; 21.83 ($M \approx 11$)	
Ni_5As_2	hex.	6.80; 12.48	
$PdAs_2$	*C*2	5.97	
$PtAs_2$	*C*2	5.96	
Cu_3As	T_d^6	9.59 ($M = 16$) (Domeykite)	
	$D0_{21}$	7.09; 7.23 (synthetic)	
$\sim Cu_3As$	*A*3	2.58; 4.22 (part of Algodonite and Whitneyite)	
CuMgAs	*C*38	3.95; 6.23	
$\sim Ag_8As$	*A*3	2.89; 4.72	
AgMgAs	*C*1	6.24	
$ZnAs_2$	orthorh.	7.72; 7.99; 36.28 ($M = 32$)	133
Zn_3As_2	$D5_9$	11.78; 23.65	
ZnLiAs	*C*1	5.91	
ZnNaAs	*C*1	5.90	134
ZnCuAs	*C*1	5.87	
ZnAgAs	*C*1	5.90	134
Cd_3As_2	$D5_9$	8.95; 12.68	
AlAs	*B*3	5.62	
$AiLi_3As_2$	$\sim C1$	11.87; 11.98; 12.11 [orthorh, deformed (D_{2h}^{27}) superstructure]	
$GeAs_2$	orthorh. *Pbam*	14.76; 10.16; 3.728 ($M = 8$)	308
GaAs	B_3	5.653 4	309
InAs	B_3	6.050 84	309
SnAs	*B*1	5.72	
$\sim Sn_3As_2$	cubic, deformed	2.91; 88° 54′ ($A = 1$)	
Sb	*A*7	6.22; 87° 24′	
Li_3Sb (α)	$D0_{18}$	4.70; 8.31	
(β)	$D0_3$	6.56	
NaSb	isostruct. with LiAs	6.80; 6.34; 12.48 ($\beta = 117.6°$)	306
Na_3Sb	$D0_{18}$	5.36; 9.50	
K_3Sb	$D0_{18}$	6.03; 10.69	
Cs_3Sb	*B*32	9.147–9.188	135
Mg_3Sb_2	$D5_2$	4.57; 7.23	
MgLiSb	*C*1	6.61	
LsSb	*B*1	6.48	
CeSb	*B*1	6.40	
PrSb	*B*1	6.35	
NdSb	*B*1	6.31	
$TiSb_2$	*C*16	6.65; 5.80	
TiSb	*B*8	4.06; 6.29	
Ti_3Sb	cubic *βw* type	5.218 6	310
	tetr. b.c. 1–4*mcm*	10.465; 5.263 9 ($M = 32$)	
Ti_4Sb	$D0_{19}$	5.95; 4.80 [structural formula: $Ti_3(Ti_{0.2}Sb_{0.8})$]	131
ThSb	*B*1	6.305	136
Th_3Sb_4	cubic b.c.	9.353	136
$ThSb_2$	tetr. P4/*nmn*	4.344; 9.154	136
VSb_2	*C*16	6.54; 5.62	131
V_3Sb	O_h^3–*Pm*3*m*	4.932 ($M = 2$)	337
Nb_3Sb	O_h^3–*Pm*3*m*	5.262 ($M = 2$)	337
$CrSb_2$	*C*18	3.27; 6.02; 6.86	

Table 6.1 STRUCTURES OF METALS, METALLOIDS AND THEIR COMPOUNDS—*continued*

Element or compound	*Structure type*	*Lattice constants, remarks*	*Refs.*
Group Vb: As, Sb, Bi—*continued*			
CrSb	*B*8	4.11; 5.47	
MnSb	*B*8	4.12; 5.78	
Mn_2Sb	*C*38	4.078; 6.557	137
$FeSb_2$	*C*18	3.19; 5.82; 6.52	
FeSb to Fe_3Sb_2	*B*8	4.06–4.12; 5.13–5.17	
$CoSb_2$	*C*18	3.21; 5.78	
CoSb	*B*8	3.87; 5.18	
CoMnSb	*C*1	5.89	
$NiSb_2$	*C*18	3.21; 5.63; 6.23	
NiSb	*B*8	3.91; 5.13	
Ni_2Sb	tetr.	5.79; 6.00 ($A \approx 15$) (Ni_5Sb_2?)	
Ni_3Sb	cubic	6.05	
NiMgSb	*C*1	6.04	
Ni_2MgSb	$L2_1$	6.05	
NiMnSb	*C*1	5.90	
Ni_2MnSb	$L2_1$	6.00	
RhSb	*B*31	6.32; 5.94; 3.87	
$PdSb_2$	*C*2	6.44	
PdSb	*B*8	4.07; 5.58	
Pd_5Sb_3	*B*8	4.44; 5.77	80
$IrSb_2$	monocl.	6.6; 6.5; 6.7; $B = 115°$	13
$PtSb_2$	*C*2	6.43	
PtSb	*B*8	4.13; 5.47	
Cu_2Sb	orthorh.	2.78; 4.77; 4.38	80
Cu_2Sb (γ)	*C*38	3.97; 6.07	
Cu_5Sb_2 (β)	tetr. (?)	(>440 °C), 9.01; 8.57	
(β')	DO_3 (tetr. deformed) incompl. filled (*cf.* Cu_3Sb)	(<440 °C) (unstable)	
$Cu_{11}Sb_4$ to Cu_3Sb (θ)	~*A*3	(<400 °C), peritectic from β and γ (2 ×)2.45; (2 ×)4.34	
~Cu_3Sb	DO_3	(>400 °C) 6.00	
Cu_9Sb_2 (δ)	~*A*3	10.84; 8.61 ($A = 54$)	
$Cu_{11}Sb_2$ (η)	~*A*3 (orthorh.)	9.29; 8.18; 8.63	139
	*A*3	2.68; 4.32	140
CuMgSb	*C*1	6.15	
CuMnSb	*C*1	6.05	
(Cu, Ni)$_3$Sb	DO_3	5.86	
Ag_3Sb	*Ae*	—	
Ag_3Sb to Ag_7Sb	~*A*3	2.99–2.92; 4.84–4.80 (deformed similar phases)	
AgMgSb	~*C*38	—	
$AuSb_2$	*C*2	6.64	
ZnSb	*Be*	6.22; 7.74; 8.12	
$ZnMg_2Sb_2$	$D5_2$	4.37; 7.15	
CdSb	*Be*	6.47; 8.25; 8.53	
Cd_3Sb_2	monocl.	72.0; 13.51; 6.16; 100° 14′ ($M = 4$)	
CdCuSb	*C*1	6.26	
AlSb	*B*3	6.135 5	309
GaSb	*B*3	6.095 4	309
InSb	*B*3	6.478 77	309
Tl_7Sb_2	$L2_2$	11.59	
SnSb	*B*1 (rhomb. deformed)	6.13; 89° 44′	
~$AsSb_2$	cubic f.c.	11.06–11.12	
Bi	*A*7	6.54; 87° 34′	
LiBi	$L1_0$	4.75; 4.25	
Li_3Bi	DO_3	6.71	
NaBi	$L1_0$	4.90; 4.80	
Na_3Bi	DO_{18}	5.45; 9.66	
KBi_2	*C*15	9.50	

Table 6.1 STRUCTURES OF METALS, METALLOIDS AND THEIR COMPOUNDS—*continued*

Element or compound	*Structure type*	*Lattice constants, remarks*	*Refs.*
Group Vb: As, Sb, Bi—*continued*			
K_3Bi	$D0_{18}$	6.18; 10.93	
Mg_3Bi_2	$D5_2$	4.67; 7.40	
MgLiBi	$C1$	6.75	
LaBi	$B1$	6.57	
CeBi	$B1$	6.49	
PrBi	$B1$	6.45	
Ti_2Bi_{17}	hex.	7.34; 10.73	220
$\sim Ti_4Bi$	$\sim D0_{19}$		
Ti_3Bi	tetr.	6.020; 8.204	312
$ZrBr_2$	orthorh.	10.2; 15.5; 4.0 ($n=8$)	311
Th_3Bi_4	$D7_3$	9.559 ($M=4$)	141
$ThBi_2$	$C38$	4.492; 4.298; ($M=2$)	141
UBi	$B1$	6.36	
Ta_2Bi_{17}	hex.	7.39; 10.76	220
$MnBi_2$	tetr. (?)	5.83; 5.35 ($M=2$)	
Mn_2Bi_3	orthorh. (?)	4.31; 5.25; 6.31	
MnBi	$B8$	4.30; 6.12	
NiBi	$B8$	4.07; 5.35	
NiMgBi	$C1$	6.15	
$ReBi_{20}$	O_h^7	11.54	220
RhBi	$B8$	4.08; 5.66	
Pd_5Bi_3	$B8$	4.50; 5.80	80
$PtBi_2$	$C2$	6.68	
CuMgBi	$C1$	6.26	
Au_2Bi	$C15$	7.94	
InBi	$B10$	5.000; 4.773	142
In_2Bi	$C32$	5.50; 3.29 (disordered)	
$TiBi_2$	$C32$	5.65; 3.37 (disordered)	
Ti–Bi (γ)	$C32$	5.64; 3.37	65
$\sim Tl_6Bi$	$A1$	4.85	
Pb_3Bi	$A3$	3.48; 5.78	
Metalloids, *etc.*: B, C, P, N, Po, Te, Se, S			
B (needles)	tetr.		
(plates)	orthorh. (?)	17.86; 8.93; 10.13	144
(graphitic)	tetr.	8.57; 8.13 }	145
("more crystal-line")	hex.	11.98; 9.54 }	
B	tetr.	8.75; 5.06	313
$\sim Be_2B$	$C1$	4.3 (or Be_5B_2?)	
Be_2B	CaF_2 type	4.663	314
BeB_2	hex.$P6/mmm$	9.79; 9.55	
BeB_6	tetr.		
MgB_2	hex.	3.084; 3.522; ($M=1$)	146
CaB_6	$D2_1$	4.15	
SrB_6	$D2_1$	4.19	
BaB_6	$D2_1$	4.28	
YB_6	$D2_1$	4.07	
LaB_6	$D2_1$	4.15	
LaB_{13}	f.c.c. O_h^6	10.44	220
EuB_6	cubic	4.163	315
TbB_6	cubic O_h^1	4.11	316
TbB_4	tetr.	7.13; 4.07	
CeB_6	$D2_1$	4.13	
CeB_4	$D1e$	7.21; 4.09	
PrB_6	$D2_1$	4.12	
NdB_6	$D2_1$	4.12	
GdB_6	$D2_1$	4.12	
ErB_6	$D2_1$	4.10	
YbB_6	$D2_1$	4.13	
Ti_2B_5	$D8h$	2.98; 13.98	147
TiB_2	$C32$	3.03; 3.22	

Table 6.1 STRUCTURES OF METALS, METALLOIDS AND THEIR COMPOUNDS—*continued*

Element or compound	*Structure type*	*Lattice constants, remarks*	*Refs.*
Metalloids, *etc.*: B, C, P, N, Po, Te, Se, S—*continued*			
TiB	orthorh.	6.12; 3.06; 4.56	148 (149, 150)
Ti_2B	tetr.	6.11; 4.56 ($M=4$; lattice primitive)	147
ZrB_{12}	*D2f*	7.41	151
ZrB_2	*C32*	3.17; 3.53	151
ZrB	cubic f.c.	4.65	148 (149, 150)
ThB_6	$D2_1$	4.2	
ThB_4	*D1e*	7.26; 4.11	
UB_{12}	*D2f*	7.47	
UB_4	*D1e*	7.08; 3.98	
PuB	*NaCl* type	4.92	317
PuB_2	AlB_2 type	3.18; 3.94	
PuB_4	P_4/mbm	7.10; 4.014	
PuB_6	CaB_6 type	4.115–4.140	
VB_2	*C32*	3.00; 3.06	
V_3B_4	orthorh.	3.030; 13.18; 2.986	152
VB	*Bf*	3.10; 8.17; 2.98	
NbB_2	*C32*	3.09; 3.30	
Nb_3B_4	*D7b*	3.31; 14.08; 3.14	
NbB	*Bf*	3.30; 8.72; 3.17	
Nb_9B	cubic	4.21 ($> \sim 800\,°C$)	
TaB_2	*C32*	3.08; 3.27	
Ta_3B_4	*D7b*	3.29; 14.0; 3.13	
TaB	*Bf*	3.28; 0.867; 3.16	
Ta_2B	*C16*	5.78; 4.86	
CrB_2	*C32*	2.97; 3.07	
Cr_3B_4	*D7b*	2.98; 13.02; 2.95	
Cr_3B_4	1*mma*	2.986; 13.02; 2.952	318
CrB	*Bf*	2.96; 7.81; 2.94	
Cr_5B_3	tetr.	5.46; 10.64	153
Cr_2B	*C16*	5.18; 4.31	153
Cr_4B	isom. *MnB*	4.26; 7.38; 14.71	153
Mo_2B_5	*D8i*	3.01; 20.93	
MoB_2	*C32*	3.05; 3.11	
MoB	*Bg*	3.11; 16.97	
Mo_2B	*C16*	5.54; 4.74	
W_2B_5	*D8h*	2.98; 13.87	
WB_2	hex. (?)	6.35; 16.4 ($M=8$) or 8.24; 15.60 ($M=12$)	
WB (1)	*Bf*	3.19; 8.40; 3.07 ($>1\,850\,°C$)	
(2)	*Bg*	3.12; 16.93	
W_2B	*C16*	5.56; 4.74	
Mn_3B_4	*D7b*	3.03; 12.86; 2.96	
MnB	*B27*	4.15; 5.56; 2.98	
Mn_2B	*C16*	5.15; 4.21	
Mn_4B	*D1f*	14.53; 7.29; 4.21	
ReB_2	hex.	2.900; 7.478 ($M=2$)	319
FeB	*B27*	4.05; 5.50; 2.95	
Fe_2B	*C16*	5.10; 4.24	
CoB	*B27*	3.95; 5.34; 3.04	
Co_2B	*C16*	5.01; 4.21	
Co_3B	orthorh.	4.41; 5.23; 6.63	154
Ni_2B	*C16*	4.98; 4.24	
Ni_3B	orthorh.	4.389; 5.211; 6619	154
RuB	b.c.c.	7.02	320
RuB_2	h.c.p.		
Rh_2B	orthorh.	5.42; 3.98; 7.44: ($M=4$)	155
$\sim Pd_3B_2$	hex.	6.48; 3.42	
Pd_5B_2	monocl. *C2/a*	12.786; 4.955; 5.472 ($\beta=97°\ 2'$; $M=4$)	321

Table 6.1 STRUCTURES OF METALS, METALLOIDS AND THEIR COMPOUNDS—*continued*

Element or compound	*Structure type*	*Lattice constants, remarks*	*Refs.*
Metalloids, *etc.*: B, C, P, N, Po, Te, Se, S—*continued*			
Pd_3B	orthorh. Fe_3C type	5.463; 7.567; 4.852	
OsB	cubic	7.03	
~PtB	tetr.	2.77; 2.95	
AlB_2	*C*32	3.00; 3.25	
AlB_{10}	orthorh. *B*-2*b*-21	8.881; 9.100; 5.690 ($M=2$)	339
AlB_{12} (Mod. I)	monocl. f.c.	8.51; 10.98; 9.40; 110° 54′ ($M=8$)	
(Mod. II)	tetr.	10.28; 14.30 ($A=196$)	
$Al_6Mo_7B_7$	—	—	
B_3Si	hex. *R*3*m*	6.319; 12.713	322
C	*A*4	3.57 (diamond)	
	*A*9	2.46; 6.71	
KC_8	~*A*9	4.97; 21.35	
KC_{16}	~*A*9	4.94; 17.45	
RbC_8, RbC_{16}, CsC_8, CsC_{16}	similar	—	
Be_2C	*C*1	4.33	
MgC_2	tetr.	5.55; 5.03 ($M=4$; similar to ThC_2?)	
Mg_2C_3	hex.	7.45; 10.61 ($M=8$)	
CaC_2	*C*11	5.48; 6.37	
SrC_2	*C*11	5.81; 6.68	
BaC_2	*C*11	6.22; 7.06	
LaC_2	b.c.tetr.	3.94; 6.572	323
La_2C_3	b.c.c.	8.803–8.819	
CeC_2	*C*11	5.48; 6.48	
PrC_2	*C*11	5.44; 6.38	
NdC_2	*C*11	5.41; 6.23	
TiC	*B*1	4.32 ($TiC_{0.3}$ has $\alpha=4.26$)	
ZrC	*B*1	4.67	
HfC	*B*1	4.46	
ThC_2	~*C*11	5.85; 5.28	156
	Cg	6.53; 4.24; 6.56; 104°	157
ThC	*B*1	5.34 (complete solid solution with ThC_2 at ~2 300 °C Wilhelm[158])	
$\sim UC_{2.3}$	*C*11	3.51; 5.97	159
$UC_{0.4}$	*C*11	3.54; 5.97	160
UC_2	*C*11	3.55; 6.00	161
U_2C_3	*D*5*c*	8.09	162
UC	*B*1	4.95	
Pu_2C_3	*D*5*c*	7.13	
PuC	*B*1	4.91	
VC	*B*1	4.17	
$\sim V_6C$	*L*′3	2.86; 4.54	
NbC	*B*1	4.424–4.457	163
NbC (α)	b.c.c. *A*2	3.301 2	324
(β)	hex.	3.120–3.128; 4.957–4.974	
(δ)	cubic *B*1	4.430 9–4.469 0	
(ζ)	isom. with TaC_3	—	
Nb_2C	*hch*	3.12; 4.95	163
Nb_3C to Nb_4C	*L*′1	~4.40	
TaC	*B*1	4.45	
Ta_2C	*L*′3	3.09; 4.93	
Ta to $TaC_{0.03}$	cubic b.c.	3.306	164
$\beta TaC_{0.38}$ to $TaC_{0.5}$	hex	3.101; 4.937	164
$\gamma TaC_{0.38}$ to $TaC_{0.5}$	*B*1	4.420	164
$\delta TaC_{0.38}$ to $TaC_{0.5}$	hex.	2.46; 6.69	164

Table 6.1 STRUCTURES OF METALS, METALLOIDS AND THEIR COMPOUNDS—*continued*

Element or compound	*Structure type*	*Lattice constants, remarks*	*Refs*
Metalloids, *etc.*: B, C, P, N, Po, Te, Se, S—*continued*			
Cr_7C_3	C^4_{3v}	13.98; 4.52($A\approx 80$) (isom. with Mn_7C_3)	
Cr_3C_2	$D5_{10}$	11.46; 5.52; 2.82	
$Cr_{23}C_6$	$D8_4$	10.64	
MoC (γ)	*Bh*	2.90; 2.81	
(γ')	*Bi*	2.93; 10.97	
MoC_{1-x}	*B*1	4.27 (perhaps only as substructure)	
Mo_2C	*L'*3	3.01; 4.74	
WC	*Bh*	2.90; 2.83	
W_2C	*L'*3	2.99; 4.71	
$W_2Cr_{21}C_6$	$D8_4$ ordered	—	
Mn_7C_2	C^4_{3v}	13.87; 4.53 ($A\approx 80$) (isom. with Cr_7C_3)	
$Mn_{23}C_6$	$D8_4$	—	
$Fe_{20}C_9$	orthorh.	9.06; 15.69; 7.93 ($M=4$); or hex. with $a'=2a$, $c'=c$, $M'=8$	165
Fe_2C	hex	2.76; 4.35	166
Fe_3C	$D0_{11}$	4.52; 5.09; 6.75 (Cementite)	
('ε')	hex.		167
εFeC	hex. P^6_322	4.767; 4.354	325
Fe (+0.25% C)	*L'*2 (deformed tetr.)	2.842; 3.008 } Martensite at low temp.	
(+0.75% C)	*L'*2 (deformed tetr.)	2.850; 2.939 } Martensite at low temp.	
Fe (+C)	*A*1	3.6 (Austenite at high temp.)	
$\sim Fe_6Al_2C$	$L'1_2$	3.72–3.78	
Fe_3Mo_3C	$E9_3$	11.1	
$Fe_{21}Mo_2C_6$	$D8_4$ordered	—	
Fe_3W_3C	$E9_3$	11.04	
$Fe_{21}W_2C_6$	$D8_4$ordered	—	
$Fe_3W_{10}C_4$	hex	~7.85; ~7.85 (isom, Co, Ni-compounds)	
$Fe_3W_6C_2$	cubic	~11.25 (isom. Co, Ni-compounds)	
$\sim(Cr_1Fe)_2C$	cubic f.c.	3.62 > 1 000 °C (below 1 000 °C Cr_7C_3-type)	
Co_2C	orthorh.	2.910; 4.469; 4.426	326
Co_3C	orthorh.	4.483; 5.033; 6.731	
Co_3W_3C	$E9_3$	11.01	
$Co_3W_9C_4$	hex.	7.286; 7.286	168
$Co_3W_{10}C_4$	hex.	7.85; 7.85 (isom. Fe, Ni-compounds)	
$Co_2W_6C_2$	cubic f.c.	11.25	
Ni_3C	hex	2.628; 4.306	169
Ni_3W_3C (θ)	$E9_3$	11.217;	170
Ni_5W_6C (η)	cubic	10.873	170
$Ni_3W_6C_2$	cubic f.c.	~11.25	
$Ni_3W_{10}C_4$	hex.	~7.85; ~7.85	
$Ni_3W_{16}C_6$	hex.	7.818 3; 7.818 0	170
Al_4C_3	$D7_1$	3.33; 24.94	
$AlMn_3C$	$L'1_2$	3.83	
$\sim AlFe_3C$	$L'1_2$	3.72–3.78	
SiC	*B*3	4.35 } there are several other modifications described with the *B*5-type	
	*B*5	3.08; 10.08 } there are several other modifications described with the *B*5-type	
B_4C	*D*1*g*	5.60; 12.12	
$B_{44}Al_3C_2$	D^{12}_{4h}	12.55; 10.18 ($M=4.31$) (diamond-type boron)	
	monocl.	17.64; 25.0; 10.26; ≈90° (?) (graphite-type boron)	
P (red)	cubic	11.31 ($A=66$)	
(white)	*I*432	18.51 ($A=224$; other possible space groups: 1m3m and I43m)	
Li_3P	$D0_{18}$	4.26; 7.58	
Na_3P	$D0_{18}$	4.98; 8.80	
Be_3P_2	$D5_3$	10.15	
Mg_3P_2	$D5_3$	12.01	
MgLiP	~*C*1	6.01	
AlP	B_3	5.42	
GaP	B_2	5.4505	309
InP	B_3	5.868 75	309
LaP	*B*1	6.01	
CeP	*B*1	5.90	
PrP	*B*1	5.86	
NdP	*B*1	5.83	

Table 6.1 STRUCTURES OF METALS, METALLOIDS AND THEIR COMPOUNDS—*continued*

Element or compound	*Structure type*	*Lattice constants, remarks*	*Refs.*
Metalloids, *etc.*: B, C, P, N, Po, Te, Se, S—*continued*			
Th_3P_4	$D7_3$	8.60	
Th_4P_3	$B1$	5.82	
U_3P_4	$D7_3$	8.20	
UP	$B1$	5.59	
Np_3P_4	$D7_3$	—	
V_3P	S_4^2	(similar to Cr_3P)	
CrP	$B31$	5.93; 5.36; 3.12	
Cr_3P	S_4^2	9.13; 4.56 [isom. with (V, Mn, Ni, Fe)$_3$P]	
WP	$B8$ (orthorh. deformed)	(?)	
MnP	$B31$	5.91; 5.25; 3.17	
Mn_2P	$C22$	6.07; 3.45	
Mn_3P	S_4^2	9.16; 4.59 (isom. Cr_3P)	
FeP_2	$C18$	2.73; 4.98; 5.66	
FeP	$B31$	5.78; 5.18; 3.09	
Fe_2P	$C22$	5.93; 3.45	
Fe_3P	S_4^2	9.10; 4.45 (isom. Cr_3P)	
CoP	$B31$	5.59; 5.07; 3.27	
Co_2P	$C23$	6.66; 5.71; 3.53	
Ni_2P	$C22$	5.85; 3.37	
Ni_7P_3	cubic b.c. (?)	8.63 ($M=6$)	
Ni_3P	S_4^2	8.92; 4.39 (isom. Cr_3P)	
Rh_2P	$C1$	5.51	
Ir_2P	$C1$	5.54	
PtP_2	$C2$	5.68	
Cu_3P	$D0_{21}$	6.94; 7.14	
ZnP_2	tetr.	5.07; 18.65 (isom. with CdP_2)	
Zn_3P_2	$D5_9$	8.10; 11.45	
ZnLiP	$\sim C1$	5.77	
CdP_2	tetr	5.28; 19.70 (isom. with ZnP_2)	
Cd_3P_2	$D5_9$	8.75; 12.28	
AlP	B_3	5.42	
$AlLi_3P_2$	$\sim C1$	11.47; 11.61; 11.73 (orthorh. deformed, with superstructure)	
GaP	$B3$	5.44	
InP	$B3$	5.86	
Li_3N	$\sim C32$	3.66; 3.88	
Be_3N_2	$D5_3$	8.13	
Mg_3N_2	$D5_3$	9.95	
MgLiN	$C1$	4.97	
Ca_3N_2 (α)	$D5_3$	11.40 (high temp. mod.)	
(β)	pseudo-hex.	3.55; 4.11 at 300 °C (low temp. mod.)	
ScN	$B1$	4.44	
LaN	$B1$	5.28	
CeN	$B1$	5.01	
PrN	$B1$	5.16	
NdN	$B1$	5.14	
EuN	$B1$	5.007	172
GdN	$B1$	4.99	
SmN	$B1$	5.048 1	172
YbN	$B1$	4.785 2	172
TiN	$B1$	4.24	
$TiLi_5N_3$	T_h^7 fluorspar	9.73	173
ZrN	B	4.63	
Th_2N_3	$D5_2$	3.88; 6.18	
UN_2	$C1$	5.32	
U_2N_3	$D5_3$	10.70	
UN	$B1$	4.89	
NpN	$B1$	4.89	
PuN	$B1$	4.90	
~VN	$B1$	4.13	
$VN_{0.71}$	$B1$	4.07	
$VN_{0.4}$	$L'3$	2.84; 4.55 [superstructure $a' = a\sqrt{3}$; similar to (Fe, Ni, Co)$_2$N] Co}$_2$N	

Table 6.1 STRUCTURES OF METALS, METALLOIDS AND THEIR COMPOUNDS—*continued*

Element or compound	*Structure type*	*Lattice constants, remarks*	*Refs.*
Metalloids, *etc.*: B, C, P, N, Po, Te, Se, S—*continued*			
NbN (1)	*B*1	4.38	
(2)	hex.	2.93; 5.45	
(3)	*Bi*	2.95; 11.25	
$\sim NbN_{0.75}$	tetr.	4.38; 4.31	
$\sim Nb_2N$	*L*′3	3.05; 4.96	
TaN	hex.	5.181; 2.902	174
Ta_2N	*L*′3	3.06; 4.96	
CrN	*B*1	4.14	
Cr_2N	*L*′3	2.76; 4.46	
MoN	*Bh*	2.86; 2.80	
Mo_2N (β)	*L*′6	4.18; 4.02	
(γ)	*L*′1	4.17	
W_2N (β)	*L*′1	4.12	
(γ)	cubic	4.12 (4 W in simple cubic cell; perhaps $a' = 4 \times 4.12$)	
Mn_3N_2	*L*′6	4.19; 4.03	
Mn_4N	*L*′1	3.85	
$ReN_{0.4}$	*L*′1	3.92	
Fe_2N (ξ)	orthorh.	5.52; 4.83; 4.43 (deformed *A*3-type with superstructure; detailed description *see* Jack[167, 171])	
Fe_2N (ε)	hex.	2.76; 4.43 } complex superstructures; *see* Jack[171]	
Fe_3N (ε)	hex.	2.72; 4.39 }	
Fe_4N (ε)	hex.	2.66; 4.34 }	
Fe_4N (γ')	*L*′1	3.80	
Fe_8N (α'')	*D*2*g*	5.72; 6.29	
Co_2N	*L*′3 (orthorh. deformed)	2.84; 4.63; 4.33	
Co_3N	~*L*′3	2.66; 4.35 ($a^* = 2a$)	
$CoLi_2N$	~*C*32	3.74; 3.62	
Ni_3N	~*L*′3	2.67; 4.31 (isom. ε-Fe_3N)	
$\sim NiLi_2N$	~*C*32	3.77; 3.52	
Cu_3N	$D0_9$	3.81	
$\sim CuLi_2N$	~*C*32	3.68; 3.77	
Zn_3N_2	$D5_3$	9.74	
ZnLiN	*C*1	4.89 (ordered)	
Cd_3N_2	$D5_3$	10.79	
AlN	*B*4	3.10; 4.97	
$AiLi_3N_2$	*E*9*d*	9.48	
GaN	*B*4	3.18; 5.17	
$GaLi_3N_2$	*E*9*d*	9.61	
InN	*B*4	3.53; 5.69	
α Si_3N_4	hex. D_d^2-P31c	7.758; 5.623	327
β Si_3N_4	hex. C_{6h}^2-P6/3*m*	7.603; 2.909	
Si_2ON	orthorh.	5.498; 8.877; 4.853	
Si_xN	hex.	4.534; 4.556	
Si_3N_4	orthorh.	13.38; 8.60; 7.74	
$SiLi_5N_3$	T_h^7 fluorspar	9.43	173
Ge_3N_4	orthorh.	13.84; 9.06; 8.18	
$GeLi_5N_3$	T_h^7 fluorspar	9.66	173
BN	*Bk*	2.50; 6.66	
Al_5C_3N	$E9_4$	3.28; 21.55	
Po (α)	*Ah*	3.35	
(β)	*Ai*	3.37; 98° 13′	
PbPo	*B*1	—	
Te	*A*8	4.45; 5.88	175
Li_2Te	*C*1	6.50	
Na_2Te	*C*1	7.32	
K_2Te	*C*1	8.15	
$K_{2.67}$(Sb, Te)	~*C*1	8.40 } a continuous series of unit crystals between K_2Te and K_3Sb appears probable at high temp.	
$K_{2.4}$(Te, Sb)	~*C*1	8.32 }	
$K_{2.25}$(Te, Sb)	~*C*1	8.2 }	
BeTe	*B*3	5.61	

Table 6.1 STRUCTURES OF METALS, METALLOIDS AND THEIR COMPOUNDS—*continued*

Element or compound	*Structure type*	*Lattice constants, remarks*	*Refs.*
Metalloids, *etc.*: B, C, P, N, Po, Te, Se, S—*continued*			
MgTe	*B*4	4.53; 7.38	
CaTe	*B*1	6.34	
SrTe	*B*1	6.65	
BaTe	*B*1	6.99	
EuTe	*B*1	6.57	
YbTe	*B*1	6.34	
$TiTe_2$	*C*6	3.77; 6.54	
TiTe	*B*8	3.83; 6.39	
Ti_5Te_4	tetr. $14/mc_45$	10.164; 3.7720	328
UTe	*B*1	6.151	176
U_3Te_4	cubic	9.378	176
UTe_2	tetr.	3.998; $c/a = 1.865$	176
VTe	*B*8	3.81; 6.12	
~CrTe	*B*8	3.89–3.98; 5.91–6.21	
$MoTe_2$	hex. *P*63–*mmc*	3.519; 13.964	329
WTe_2	orthorh.	3.490; 6.277; 14.07	177
$MnTe_2$	*C*2	6.94	
MnTe	*B*8	4.12; 6.70	
$FeTe_2$	*C*18	3.85; 5.34; 6.26	
FeTe	*B*8	3.80; 5.65	
$CoTe_2$	*C*6	3.78; 5.40	
	*C*18	3.88; 5.30; 6.30	
CoTe	*B*8	3.88; 5.37	
$NiTe_2$	*C*6	3.86; 5.30	
NiTe	*B*8	3.96; 5.35	
$RuTe_2$	*C*2	6.36	
$PdTe_2$	*C*6	4.03; 5.12	
PdTe	*B*8	4.13; 5.66	
$OsTe_2$	*C*2	6.37	
$PtTe_2$	*C*6	4.01; 5.20	
Cu_2Te	*Ch*	4.24; 7.27	
Cu_2Te	cubic f.c.	6.10 ($A = 12$)	80
CuTe	orthorh.	—	80
$Cu_{2-x}Te$	~*C*38	3.98; 6.12	
Ag_2Te	~*C*1	(>150 °C) 6.57 (high temp. mod.)	
	orthorh.	13.0; 12.7; 12.2 (low temp. mod.)	
	monocl.	5.98; 6.31; 5.56; 75° 24′ (Hessite)	
$(Au, Ag)Te_2$	*C*34	7.18; 4.40; 5.07; ~90° (Calaverite)	
	*C*46	16.51; 8.80; 4.45 (Krennerite)	
$AuAgTe_2$	*E*1*b*	8.94; 4.48; 14.59; 145° 24′	
ZnTe	*B*3	6.09	
CdTe	*B*3	6.46	
HgTe	*B*3	6.429 (Coloradoite)	175
Ga_2Te_3	~*B*3	5.89	
Te_3Al_2	hex. Wurtzite type	4.07; 6.93	330
InTe	*B*37	8.42; 7.12	63,80
In_2Te_3	~*B*3	6.16	
TeTl	b.c.tetr.	12.950; 6.175	331
Te_3Tl_2	monocl.	13.5; 6.5; 7.9 ($\beta = 73°$)	
γ-Te_3Tl_2	b.c. tetr.	8.92; 12.63	
SnTe	*B*1	6.28	
PbTe	*B*1	6.44 (Altaite)	
As_2Te_3	monocl.	14.4; 4.05; 9.92: $B = 97°$	178
Sb_2Te_3	hex	4.24; 29.90	179
Te_2SbTl	$R\bar{3}$-n–D^5_{3d}	8.177 ($\alpha = 31°$ 24′)	332
Te_2BiTl	$R\bar{3}$-n–D^5_{3d}	8.137 ($\alpha = 32°$ 18′)	
Bi_2Te_3	*C*33	10.45; 24° 8′ (Tellurobismuthite)	
Se (γ)	*A*8	4.36; 4.95	
(α)	*Ak*	9.05; 9.07; 11.61; 90° 46′	
(β)	*A*1	12.85; 8.07; 9.31; 93° 8′	
Li_2Se	*C*1	6.01	
Na_2Se	*C*1	6.80	

Table 6.1 STRUCTURES OF METALS, METALLOIDS AND THEIR COMPOUNDS—*continued*

Element or compound	*Structure type*	*Lattice constants, remarks*	*Refs.*
Metalloids, *etc.*: B, C, P, N, Po, Te, Sc, S—*continued*			
K_2Se	*C*1	7.68	
BeSe	*B*3	5.13	
MgSe	*B*1	5.45	
CaSe	*B*1	5.91	
SrSe	*B*1	6.23	
BaSe	*B*1	6.59	
Y_5Se_3	D^8_8	8.40; 6.30	333
EuSe	*B*1	6.17	
YbSe	*B*1	5.87	
$TiSe_2$	*C*6	3.54; 5.99	
TiSe	*B*8	3.56; 6.22	
$ZrSe_3$	monocl.	5.41; 3.77; 9.45 (β=97.5°)	340
$ZrSe_2$	*C*6	3.79; 6.18	
U_3Se_4	cubic Th_3P_4 type	8.804	334
U_2Se_3	orthorh. Sb_2S_3 type	11.33; 10.941; 4.06	334
$ThSe_2$	*C*23	4.98; 7.50, 9.38	
Th_7Se_{12}	hex.	11.56; 4.35	
Th_2Se_3	$D5_8$	11.32; 11.55; 4.26	
ThSe	*B*1	5.86	
USe_3	monocl.	5.68; 4.06; 19.26: (d=7.25)	180
$VSe_{1.6}$ to VSe_2	*C*6	3.35; 6.12	
VSe	*B*8	3.58; 5.98	
Cr_2Se_3	*B*8	3.60; 5.77	
CrSe	~*B*8 (monocl.)	55–58 at. % Se: 6.30; 3.60; 5.85; 90° 30′	
	*B*8	50–54 at. % Se: 3.68; 6.02	
$CrNaSe_2$	$F5_1$	3.71; 20.29	
$CrK_{0.5}Se_2$	~$F5_1$	3.44; 24.2	
$CrRbSe_2$	$F5_1$	3.43; 26.9	
WSe_2	*C*7	3.29; 12.97	
$MnSe_2$	*C*2	6.42	
MnSe (α)	*B*1	5.45	
(β)	*B*3	5.82	
(γ)	*B*4	4.12; 6.72	
$FeSe_2$	*C*18	3.58; 4.79; 5.72	
FeSe	*B*8 to *B*8 monocl.	3.64; 5.96 } 50–56 at. % Se	
	deformed	6.25; 3.58; 5.81; 91° } 50–56 at. % Se	
	*B*10	44 at. % Se: 3.77; 5.52	
$CoSe_2$	*C*2	5.85	
CoSe	*B*8	3.61; 5.28	
$NiSe_2$	*C*2	6.02	
NiSe	*B*8	3.66; 5.33	
	*B*13	9.84; 3.18	
$RuSe_2$	*C*2	5.92	
$RhSe_2$	*C*2	6.015 (63.6%): 5.985 (71.4%)	181
$OsSe_2$	*C*2	5.93	
$PtSe_2$	*C*6	3.72; 5.06	
CuSe	~*B*18	3.94; 17.25 (superstructure $a'=12a$) (Klockmannite)	
Cu_2Se	~*C*1	5.84 (Berzelianite)	
Ag_2Se	~*C*1	(>133 °C), 4.98 (at 170 °C) (Naumannite)	
ZnSe	*B*3	5.66	
$ZnCr_2Se$	$H1_1$	10.44 ('normal' $H1_1$-type)	182
CdSe	*B*3	6.04	
	*B*4	4.30; 7.01	
$CdCr_2Se_4$	$H1_1$	10.72 ('normal' $H1_1$-type)	182
HgSe	*B*3	6.074 (Tiemannite)	175
GaSe	hex.	3.73; 15.89	63, 184, 185
	hex.	3.74; 23.86	63,
Ga_2Se_3	~*B*3	5.43	184
In_2Se_3	h.c.p. (room temp.)	4.01; 19.24	186

Table 6.1 STRUCTURES OF METALS, METALLOIDS AND THEIR COMPOUNDS—*continued*

Element or compound	*Structure type*	*Lattice constants, remarks*	*Refs.*
Metalloids, *etc.*: B, C, P, N, Po, Te, Se, S—*continued*			
TlSe	*B*37	8.02; 7.00	
$GeSe_2$	orthorh.	12.96; 6.93; 22.09 ($M=24$)	
GeSe	orthorh.		
	$D_{2h}^{16}-Pcmn$	4.38; 3.82; 10.79 ($M=4$)	336
SnSe	orthorh.	4.46; 4.19; 11.57	187
PbSe	*B*1	6.14 (Clausthalite)	
Sb_2Se_3	$D5_8$	11.58; 11.68; 3.98	
Bi_2Se_3	? type	4.14; 28.59 ($A=15$)	80
	rhombic	9.8; $\alpha=24.4°$	188
S (orthorh.)	*A*16	10.4; 12.9; 24.5	
(monocl.)	C_2^5h	10.9; 10.9; 11.0; 83° 16′ ($A=48$)	
(rhomb.)	*C*3*i*	10.9; 4.26 ($A=18$)	
(fibre)	monocl.	26.4; 9.3; 12.3; 79° 15′ ($C=9.3=$fibre axis; $A=112$)	
Li_2S	*C*1	5.71	
Na_2S	*C*1	6.53	
K_2S	*C*1	7.39	
Rb_2S	*C*1	7.65	
BeS	*B*3	4.86	
MgS	*B*1	5.19	
CaS	*B*1	5.69 (Oldhamite)	
SrS	*B*1	6.01	
BaS	*B*1	6.37	
La_2S_3	$D7_3$	8.71	
Ce_2S_3	$D7_3$	8.617	
Ce_3S_4	$D7_3$	8.608	
CeS	*B*1	5.77	
DyS	*B*1	5.96	
Ac_2S_3	$D7_3$	8.97	
TiS_2	*C*6	3.40; 5.69	
ThS_2	C_{23}	4.26; 7.25; 8.60	
Th_7S_{12}	*D*8*k*	11.04; 3.98	
Th_2S_3	$D5_8$	10.97; 10.83; 3.95	
ThS	*B*1	5.67	
ZrS_2	*C*6	3.68; 5.85	
U_2S_3	$D5_8$	10.39; 10.63; 3.88	
US	*B*1	5.47	
Np_2S_3	$D5_8$	10.3; 10.6; 3.85	
Pu_2S_3	$D7_3$	8.44	
PuS	*B*1	5.53	
Am_2S_3	$D7_3$	8.43	
Vs to $V_{0.84}S$	*B*8	3.36; 5.81	
TaS_2	*C*6 (or *C*7)	3.40; 5.90	
CrS	~*B*8	50.0–52.4 at. % S: 12.00; 11.52	
CrS	*B*8	52.4–54.2 at. % S; 3.45; 5.75	
	*B*8 with monocl. deformed	55–59 at. % S: 5.95; 3.42; 5.63; 91° 44′	
	*B*8 with ordered vacant sites	~60 at. % S	
$CrNaS_2$	$F5_1$	3.51; 19.57	
$CrKS_2$	$F5_1$	3.62; 21.16	
$CrRbS_2$	~$F5_1$	3.59; 16.20	
MoS_2	*C*7	3.15; 12.30 (Molybdenite)	
WS_2	*C*7	3.15; 12.3 (Tungstenite)	
MnS_2	*C*2	6.10	
MnS	*B*1	5.21 (Alabandite)	
	*B*3	5.60	
	*B*4	3.98; 6.43	
$MnCr_2S_4$	$H1_1$	10.06	
FeS_2	*C*2	5.40 (Pyrites)	
	*C*18	4.44; 5.41; 3.83 (Marcasite)	
~FeS	~*B*8	~3.4; ~5.7	183, 189

Table 6.1 STRUCTURES OF METALS, METALLOIDS AND THEIR COMPOUNDS—*continued*

Element or compound	*Structure type*	*Lattice constants, remarks*	*Refs.*
Metalloids, *etc.*: B, C, P, N, Po, Te, Se, S—*continued*			
Room temp. modifications:			
(α′)	~*B*8	50.0–51.0 at. % S: 5.96–5.97; 11.74–11.58	
(α″)	*B*8	51.0–52.3 at. % S: 3.44; 5.79–5.74	
(β)	*B*8	52.3–53.5 at. % S: 3.44–3.43; 5.74–5.69	
High temp. modifications:			
(β)	*B*8	~100 °C to ~320 °C, 50.0–~54.0 at. % S	
(δ)	*B*8	>~320 °C, lattice parameters different from β-modification (~0.5%) with different temp. coeff.	
$Fe_{1.78}S_2$	*B*8 (monocl. deformed)	5.94; 3.43; 5.69; 89° 38′ (natural material)	190
$Fe_{2-x}S_2$	~*B*8	3.44; 5.82 (superstructure with $a' = 3 \times 3.44$ and $c' = 2 \times 5.82$; Graham[191])	
Fe_7S_8	~*B*8	6.86; 11.9; 22.7; 89° 33′ (detailed superstructure proposed by Bertant[192])	
$FeKS_2$	*F*5*a*	7.05; 11.28; 5.40; 112° 30′	
$FeCr_2S_4$	$H1_1$	9.97	193
	$D7_2$	9.97	194
CoS_2	*C*2	5.52	
Co_3S_4	$D7_2$	9.38 (Linnaeite)	
CoS	*B*8	51–53 at. % S: 3.37–3.36; 5.18–5.16 (Jaipurite)	
Co_9S_8	$D8_9$	9.91	
$CoCr_2S_4$	$H1_1$	9.91	
NiS_2	*C*2	5.68	
Ni_2S_3	monocl.	$b = 3.2$ (Parkerite)	
Ni_3S_4	$D7_2$	9.46	
NiS	*B*8	3.43; 5.33 (high temp. mod.)	
NiS	*B*13	9.59; 3.15 (low temp. mod.)(Millerite)	
Ni_6S_5	D_{2h}^{17}	11.22; 16.56; 3.27 ($M = 4$)	
Ni_3S_2	*D*5*e*	4.04; 90° 18′	
(Ni, Fe)$_9S_8$	$D8_9$	10.1 (Pentlandite)	
Ni_2FeS_4	$D7_2$	9.45	
RuS_2	*C*2	5.59	
RhS_2	*C*2	5.57	
PdS	*B*34	6.43; 6.63	
OsS_2	*C*2	5.64	
PtS_2	*C*6	3.54; 5.02	
PtS	*B*17	4.91; 6.10 (Cooperite)	
(Pt, Pd, Ni)S	*B*34	6.37; 6.58 (Braggite)	
CuS	*B*18	3.75; 16.2 (Covellite)	
$Cu_{1.8}S$	~*C*1	5.56	
Cu_2S (γ)	orthorh.	11.90; 27.28; 13.41 (C_{2v}^{15} AbZm; S ≈ hex. cl. packed; 105 °C)	
(β)	$D6_h^4$	3.89; 6.68 ($M = 2$; structures proposed by Belov[195] and Ueda[196] differ *cf.*[197])	
Cu_3VS_4	$H2_4$	5.38 (Sulvanite)	
$CuFeS_2$	$E1_1$	5.24; 10.30 (Chalcopyrite)	
$CuFe_2S_3$	*E*9*e*	6.23; 11.12; 6.46 (Cubanite)	
Cu_5FeS_4	~*B*3	2 × 5.49 > 220 °C	
	orthorh.	21.94; 21.94; 10.97 ($M = 32$; superstructure of orthorh. deformed *B*3-type)(Bornite)	
$CuCo_2S_4$	$H1_1$	9.46 (Carrollite)	
Ag_2S	~*C*1	4.88 (> 180 °C)(Acanthite)	
	monocl.	(< 180 °C)	
$AgFeS_2$	$E1_1$	5.66; 10.30	
ZnS	*B*3	5.40 (Zincblende)(low temp. mod.)	
	*B*4	3.811; 6.23 (Würtzite)(high temp. mod.)	
	*B*5	3.806; 12.44	
	cf. SiC	3.813; 18.69	
	cf. SiC	3.822; 46.79	
$ZnAl_2S_4$	$H1_1$	9.97 (< ~1 000 °C)	
	*B*4	3.76; 6.13 > ~1 000 °C	
$ZnCr_2S_4$	$H1_1$	9.9	
CdS	*B*4	4.13; 6.69 (Greenockite)	
	*B*3	5.81	

Table 6.1 STRUCTURES OF METALS, METALLOIDS AND THEIR COMPOUNDS—*continued*

Element or compound	*Structure type*	*Lattice constants, remarks*	*Refs.*
Metalloids, *etc*.: B, C, P, N, Po, Te, Se, S—*continued*			
$CdMg_2S_4$	$H1_1$	10.80	
$CdCr_2S_4$	$H1_1$	10.19	
HgS	*B*3	5.84 (Metacinnabarite)	
	*B*9	4.14; 9.49 (Cinnabar)	
$HgCr_2S_4$	$H1_1$	10.21	
Al_2S_3	~*B*4	3.70; 5.94 (with superstructure)	
Ga_2S_3 (α)	*B*3	5.18 < 550 °C	
(β)	*B*4	3.69; 6.03 > 550 °C	
InS	*Pmnm*	3.93; 4.43; 10.62: (*A* = 8)	198
In_2S_3 (α)	~*B*1	5.37 < 300 °C	
(β)	$H1_1$	10.74 > 300 °C	
In_2BeS_4	$H1_1$	10.77	
In_2CaS_4	$H1_1$	10.77	
In_2MgS_4	$H1_1$	10.69	
In_2MnS_4	$H1_1$	10.69	
In_2FeS_4	$H1_1$	10.60	
In_2CoS_4	$H1_1$	10.56	
In_2NiS_4	$H1_1$	10.46	
In_2HgS_4	$H1_1$	10.81	
TlS	*B*37	7.79; 6.80	
Tl_2S	~*C*6	12.20; 18.17	
SiS_2	*C*42	5.60; 5.53; 9.55	
GeS_2	*C*44	11.66; 22.34; 6.86	
GeS	*B*16	4.29; 10.42; 3.64	
(Fe, Ge)Cu_3S_4	~*B*3	5.29 (Germanite)	
SnS_2	*C*6	3.64; 5.87	
SnS	*B*29	3.98; 4.33; 11.18 (Herzenbergite)	
$SnCu_2FeS_4$	$H2_6$	5.46; 10.28 (Stannite)	
PbS	*B*1	5.92 (Galena)	
$PbSnS_2$	*B*29	4.04; 4.28; 11.33 (Teallite)	
AsS	*B*1	9.27; 13.50; 6.56; 106° 37′ (Realgar)	
As_2S_3	*D*5*f*	11.46; 9.56; 4.21; 90° ± 1° (Orpiment)	
FeAsS	$E0_7$	9.6; 5.7; 6.4; ~90° (Arsenopyrite)	
CoAsS	*C*2	5.61 (Cobaltite)	
NiAsS	*C*2	5.66 (Gersdorffite)	
CuAsS	~*B*3 (orthorh.)	3.78; 5.47; 11.47 (Hautite)	
(Cu, Fe)$_3AsS_3$	cubic; ~*B*3	10.2 (Tetrahedrite, Fahlerz)	
Sb_2S_3	$D5_8$	11.3; 11.5; 3.9 (Stibnite)	
FeSbS	$E0_7$	10.00; 5.93; 6.73; ~90° (Gudmundite)	
NiSbS	$F0_1$	5.90 (Ullmannite)	
$CuSbS_2$	$F5_6$	6.01; 3.78; 14.46 (Wolfsbergite)	
(Cu, Fe)$_3$, (Sb, As)S_3	cubic; ~*B*3	10.2–10.3 (Tetrahedrite, Fahlerz)	
Bi_2S_3	$D5_8$	11.13; 11.27; 3.97 (Bismuthite)	
$NaBiS_2$	*B*1	5.76	
$KBiS_2$	*B*1	6.01	
$CuBiS_2$	$F5_6$	6.13; 3.89; 14.51 (Emplectite)	
$AgBiS_2$	*B*1	(> 210 °C), 5.64	
	*B*1 (orthorh. deformed)	(< 210 °C), 8.08; 7.82; 5.65	
Bi_2Te_2S	*C*33	10.31; 24° 10′ (Tetradymite)	

Table 6.2 STRUCTURAL DETAILS

*A*1 (Cu type)

Cubic: $O_h^5 - Fm3m$; $a = 3.61$; $A = 4$
Co-ordinates: $4Cu(O_h)$: $000, \frac{1}{2}\frac{1}{2}0$ ↷

*A*2 (W type)

Cubic: $O_h^9 - Im3m$; $a = 3.16$; $A = 2$
Co-ordinates: $2W(O_h)$: $000; \frac{1}{2}\frac{1}{2}\frac{1}{2}$

*A*3 (Mg type)

Hexagonal: $D_{6h}^4 - P6_3/mmc$; $a = 3.20$, $c = 5.20$; $A = 2$
Co-ordinates: $2Mg(D_{3h})$: $\frac{2}{3}\frac{1}{3}0; \frac{1}{3}\frac{2}{3}\frac{1}{2}$

*A*4 (Diamond type)

Cubic: $O_h^7 - Fd3m$; $a = 3.56$; $A = 8$
Co-ordinates: $8C(T_d)$: $000; \frac{1}{2}\frac{1}{2}0$; ↷ ; $\frac{1}{4}\frac{1}{4}\frac{1}{4}; \frac{3}{4}\frac{3}{4}\frac{1}{4}$; ↷

*A*5 (Tin type)

Tetragonal: $D_{4h}^{19} - I4/amd$; $a = 5.82$, $c = 3.17$; $A = 4$
Co-ordinates: $4Sn(D_{2d})$: $000; \frac{1}{2}\frac{1}{2}\frac{1}{2}; \frac{1}{2}0\frac{1}{4}; 0\frac{1}{2}\frac{3}{4}$

*A*6 (In type)

Tetragonal: $D_{4h}^{17} - F4/mmm$; $a = 4.58$, $c = 4.94$; $A = 4$
Co-ordinates: $4In(D_{4h})$: $000; \frac{1}{2}\frac{1}{2}0$; ↷

*A*7 (As type)

Rhombohedral: $D_{3d}^5 - R\bar{3}m$
Co-ordinates: Rhombohedral (I), $2As(C_{3v})$: $\pm(xxx)$
Rhombohedral (II), $8As(C_{3v})$: $(000; \frac{1}{2}\frac{1}{2}0$; ↷ $) \pm (xxx)$
Hexagonal (III), $6As(C_{3v})$: $(000; \frac{2}{3}\frac{1}{3}\frac{1}{3}; \frac{1}{3}\frac{2}{3}\frac{2}{3}) \pm (00x)$

	Rhomb. I $A=2$		*Rhomb. II* $A=8$		*Hexagonal III* $A=6$			c/a
	a	α	a	α	a	c	x	
As	4.12	54° 10′	5.57	84° 38′	3.75	10.50	0.226	2.80
Sb	4.50	57° 06′	6.20	87° 24′	4.30	11.24	0.233	2.62
Bi	4.74	57° 14′	6.57	87° 32′	4.54	11.84	0.237	2.61
Simple cub.	—	60°	—	90°	—	—	—	2.45

*A*8 (Se type)

Hexagonal: $D_3^4–P3_121$ (and $D_3^6–P3_221$)
Co-ordinates: $3Se(C_2)$: $x00; \bar{x}\bar{x}\frac{1}{3}; 0x\frac{2}{3}$ (and $x00; \bar{x}\bar{x}\frac{2}{3}; 0x\frac{1}{3}$)

	a	c	c/a	x
Se	4.36	4.95	1.14	0.22
Te	4.45	5.92	1.33	0.27
Simple cub.	—	—	1.23	0.33

Table 6.2 STRUCTURAL DETAILS—*continued*

*A*9 (Graphite type)

(α) Hexagonal: $D^4_{6h}-P6_3/mmc$ (if $z=0$); or
$C^4_{6v}-P6_3/mc$ (if $z\neq 0$); $a=2.46$, $c=6.7$; $A=4$
Co-ordinates: $2C(D_{3h}$ or $C_{3v})$: $000; 00\frac{1}{2}$
$2C(D_{3h}$ or $C_{3v})$: $\frac{1}{3}\frac{2}{3}z; \frac{2}{3}\frac{1}{3}(\frac{1}{2}+z)$; $z\approx 0$ (or very probably $z=0$)
(β) Rhombohedral: $D^5_{3d}-R\bar{3}m$; $a=2.46$, $c=10.1$; $A=6$
Co-ordinates: $6C(C_{3v})$: $(000; \frac{2}{3}\frac{1}{3}\frac{1}{3}; \frac{1}{3}\frac{2}{3}\frac{2}{3})\pm(00x)$, with $x\approx\frac{1}{6}$ (or very probably $x=\frac{1}{6}$)

*A*10 (Hg type)

Rhombohedral: $D^5_{3d}-R\bar{3}m$
Co-ordinates: Rhombohedral (I): $1\text{Hg}(D_{3d})$: 000
Rhombohedral (II): $2\text{Hg}(D_{3d})$: $000; \frac{1}{2}\frac{1}{2}\frac{1}{2}$
Rhombohedral (III): $4\text{Hg}(D_{3d})$: $000; \frac{1}{2}\frac{1}{2}0; \circlearrowright$
Hexagonal (IV): $3\text{Hg}(D_{3d})$: $000; \frac{1}{3}\frac{2}{3}\frac{1}{3}; \frac{2}{3}\frac{1}{3}\frac{2}{3}$

	Rhombohedral			*Hexagonal*	*Ideal cubic*			
	I	*II*	*III*	*IV*	*I*	*II*	*III*	*Hex.*
a	3.00	4.90	4.38	3.46	$\frac{a}{2}\sqrt{2}$	$\frac{a}{2}\sqrt{6}$	*a*	$\frac{a}{2}\sqrt{2}$
α	70° 32′	41° 25′	98° 15′	—	60°	33° 33′	90°	—
c	—	—	—	6.71	—	—	—	$a\sqrt{3}$
A	1	2	4	3	1	2	4	3
c/a	—	—	—	1.94	—	—	—	2.45

*A*11 (Ga type)

Orthorhombic: $D^{18}_{2h}-Abma$; $a=4.52$, $b=4.51$, $c=7.64$; $A=8$
Co-ordinates: $8\text{Ga}(C_s)$: $(000; 0\frac{1}{2}\frac{1}{2})\pm(x0z; \frac{1}{2}+x, \frac{1}{2}, \bar{z})$ with $x=0.079$; $z=0.153$

*A*12 (α-Mn type)

Cubic: $T^3_d-I\bar{4}3m$; $a=8.89$; $A=58$
Co-ordinates: $(000; \frac{1}{2}\frac{1}{2}\frac{1}{2})+2\text{Mn}(T_d)$: 000
$+8\text{Mn}(C_{3v})$: $xxx; \bar{x}\bar{x}x; \circlearrowright$; with $x=0.32$
$+24\text{Mn}(C_s)$: $xxz; \circlearrowright; \bar{x}\bar{x}z; \circlearrowright; \bar{x}x\bar{z}; \circlearrowright : x\bar{x}\bar{z}; \circlearrowright$; with $x=0.36$; $z=0.04$
$+24\text{Mn}(C_s)$ with similar co-ordinates but with $x=0.09$; $z=0.28$

*A*13 (β-Mn type)

Cubic: O^6-P4_33 and O^7-P4_13; $a=6.30$; $A=20$
Co-ordinates: $8\text{Mn}(C_3)$: $xxx; (\frac{1}{2}+x)(\frac{1}{2}-x)\bar{x}; \circlearrowright; (\frac{3}{4}-x)(\frac{3}{4}-x)(\frac{3}{4}-x); (\frac{1}{4}-x)(\frac{3}{4}+x)(\frac{1}{4}+x); \circlearrowright$;
with $x=0.061$
$12\text{Mn}(C_2)$: $\frac{3}{8}\bar{x}(\frac{3}{4}+x); \circlearrowright; \frac{7}{8}(\frac{1}{2}+x)(\frac{1}{4}-x); \circlearrowright; \frac{1}{8}x(\frac{1}{4}+x); \circlearrowright; \frac{5}{8}(\frac{1}{2}-x)(\frac{3}{4}-x); \circlearrowright$;
with $x=0.206$

An alternative structure has been proposed (Wilson[9]) with space group O^6_h-Fm3c; $a=12.58$; $A=160$

*A*15 (β-W or Cr_3Si type)

Cubic: O^3_h-Pm3n; $a=5.04$ or 4.56; $A=8$
Co-ordinates: 2W or $2\text{Si}(T_h)$: $000; \frac{1}{2}\frac{1}{2}\frac{1}{2}$
6W or $6\text{Cr}(D_{2d})$: $\frac{1}{2}0\frac{1}{4}; \circlearrowright; \frac{1}{2}0\frac{3}{4}; \circlearrowright$

*A*16 (orh. S type)

Orthorhombic: $D^{24}_{2h}-Fddd$; $a=10.48$, $b=12.92$, $c=24.55$; $A=128$
Co-ordinates: 4 times $32\text{S}(C_1)$ in $(000; \frac{1}{2}\frac{1}{2}0; \circlearrowright)+xyz; \bar{x}y\bar{z}; (\frac{1}{4}-x)(\frac{1}{4}-y)(\frac{1}{4}-z); (\frac{1}{4}+x)(\frac{1}{4}-y)(\frac{1}{4}+z)$;
$x\bar{y}\bar{z}; \bar{x}\bar{y}z; (\frac{1}{4}-x)(\frac{1}{4}+y)(\frac{1}{4}+z); (\frac{1}{4}+x)(\frac{1}{4}+y)(\frac{1}{4}-z)$

Table 6.2 STRUCTURAL DETAILS—*continued*

	x	y	z
S I	−0.017	0.083	0.072
S II	−0.094	0.161	0.200
S III	−0.167	0.105	0.125
S IV	−0.094	0.028	0.250

*A*20 (α-U type)

Orthorhombic: $D_{2h}^{17}-Cmcm$; $a=2.85$, $b=5.87$, $c=4.95$; $A=4$
Co-ordinates: $4\text{U}(C_{2v})$: $(000; \frac{1}{2}\frac{1}{2}0)+0y\frac{1}{4}; 0\bar{y}\frac{3}{4}$; $y=0$, 105

A_a (Pa type)

Tetragonal: $D_{4h}^{17}-I^4/mmm$; $a=3.93$, $c=3.24$; $A=2$
Co-ordinates: $2\text{Pa}(D_{4h})$: $000; \frac{1}{2}\frac{1}{2}\frac{1}{2}$

A_b (β-U type)

The structure of pure β-U is not yet known.
It is very similar to β-U with 1.4 at. % Cr contamination.
The structure of such a product has been determined by Tucker[15].
Tetragonal: $C_{4v}^{4}-P4nm$; $a=10.52$, $c=5.57$; $A=30$
Co-ordinates: $2\text{U}(C_{2v})$: $00z; \frac{1}{2}\frac{1}{2}(\frac{1}{2}+z)$; $z=0.66$
$4\text{U}(C_s)$: $xxz; \bar{x}\bar{x}z; (\frac{1}{2}+x)(\frac{1}{2}-x)(\frac{1}{2}+z); (\frac{1}{2}-x)(\frac{1}{2}+x)(\frac{1}{2}+z)$; $x=0.11$; $z=0.23$
$4\text{U}(C_s)$ in similar position with $x=0.32$; $z=0.00$
$4\text{U}(C_s)$ in similar position with $x=0.68$; $z=0.50$
$8\text{U}(C_1)$: $xyz; \bar{x}\bar{y}z; (\frac{1}{2}+x)(\frac{1}{2}-y)(\frac{1}{2}+z); (\frac{1}{2}-x)(\frac{1}{2}+y)(\frac{1}{2}+z); xyz; \bar{y}\bar{x}z; (\frac{1}{2}+y)(\frac{1}{2}-x)(\frac{1}{2}+z); (\frac{1}{2}-y)(\frac{1}{2}+x)(\frac{1}{2}+z)$; with $x=0.56$; $y=0.24$; $z=0.25$
$8\text{U}(C_1)$ in similar position with $x=0.38$; $y=0.04$; $z=0.20$
Thewlis[14] compared the lattice constants at 720 °C of pure β-U and Cr containing β-U.

	a	c
Pure β-U at 720 °C	10.759	5.656
1.4 at % Cr-U alloy at 720 °C	10.763	5.652
1.4 at % Cr-U alloy at 20 °C	10.590	5.634

A_c (α-Np)

Orthorhombic: $D_{2h}^{16}-Pmcn$; $a=4.72$, $b=4.89$, $c=3.66$; $A=8$
Co-ordinates: $4\text{Np}(C_s)$: $\pm(\frac{1}{4}yz); \pm(\frac{1}{4}, \frac{1}{2}-y, \frac{1}{2}+z)$; $y=0.208$; $z=0.036$
$4\text{Np}(C_s)$ in similar positions with $y=0.842$; $z=0.319$

A_d (β-Np)

Tetragonal: $D_4^2-P42_1$; $a=4.90$, $c=3.39$; $A=4$
Co-ordinates: $2\text{Np}(D_2)$: $000, \frac{1}{2}\frac{1}{2}0$
$2\text{Np}(C_4)$: $\frac{1}{2}0z; 0\frac{1}{2}\bar{z}$; $z=0.38$

A_e (β′-$TiCu_3$ type)

Orthorhombic: $D_{2h}^{17}-Cmcm$; $a=2.59$, $b=4.53$, $c=4.35$; $A=4$
Co-ordinates: 4Ti or Cu(C_{2v}): $(000; \frac{1}{2}\frac{1}{2}0)+0y\frac{1}{4}; 0\bar{y}\frac{3}{4}$; $y=0.345$

A_f ($HgSn_{10}$ type)

Hexagonal: $D_{6h}^{1}-P6/mmm$; $a=3.20$, $c=2.98$; $A=1$
Co-ordinates: 1Hg or Sn(D_{6h}): 000

Table 6.2 STRUCTURAL DETAILS—*continued*

A_g (B type)

Tetragonal: $D_{2d}^{8}-P\bar{4}n2$; $a=8.73$, $c=5.03$; $A=50$
Co-ordinates: $2\mathrm{B}(S_4)$: $00\frac{1}{2}$; $\frac{1}{2}\frac{1}{2}0$
6 times $8\mathrm{B}(C_1)$: xyz; $(\frac{1}{2}-x)(\frac{1}{2}+y)(\frac{1}{2}+z)$
$\bar{x}\bar{y}z$; $(\frac{1}{2}+x)(\frac{1}{2}-y)(\frac{1}{2}+z)$
$\bar{y}x\bar{z}$; $(\frac{1}{2}+y)(\frac{1}{2}+x)(\frac{1}{2}-z)$
$y\bar{x}\bar{z}$; $(\frac{1}{2}-y)(\frac{1}{2}-x)(\frac{1}{2}-z)$

	B I	*B II*	*B III*	*B IV*	*B V*	*B VI*
x	0.328	0.095	0.223	0.078	0.127	0.250
y	0.095	0.328	0.078	0.223	0.127	0.250
x	0.395	0.395	0.105	0.105	0.395	−0.078

A_h (α-Po type)

Cubic: O_h^1-Pm3m; $a=3.35$; $A=1$
Co-ordinates: $1\mathrm{Po}(O_h)$: 000

A_i (β-Po)

Rhombohedral: $D_{3d}^{5}-R\bar{3}m$; $a=3.37$; $\alpha=98°\ 13'$; $A=1$
Co-ordinates: $1\mathrm{Po}(D_{3d}^{5})$: 000

A_k (α-Se)

Monoclinic: $C_{2h}^{5}-P2_1/n$; $a=9.05$, $b=9.07$, $c=11.61$; $\beta=90°\ 46'$; $A=32$
Co-ordinates: 8 times $4\mathrm{Se}(C_1)$: $\pm(xyz)\pm(\frac{1}{2}+x,\ \frac{1}{2}-y,\ \frac{1}{2}+z)$

	Se *I*	Se *II*	Se *III*	Se *IV*	Se *V*	Se *VI*	Se *VII*	Se *VIII*
x	0.321	0.427	0.317	0.134	−0.081	−0.156	−0.084	0.131
y	0.486	0.664	0.637	0.820	0.686	0.733	0.520	0.597
z	0.237	0.357	0.535	0.556	0.521	0.328	0.229	0.134

A_l (β-Se)

Monoclinic: $C_{2h}^{5}-P2_1/a$; $a=12.85$, $b=8.07$, $c=9.31$; $\beta=93°\ 08'$; $A=32$
Co-ordinates: 8 times $4\mathrm{Se}(C_1)$: $\pm(xyz)\pm(\frac{1}{2}+x,\ \frac{1}{2}-y,\ z)$

	Se *I*	Se *II*	Se *III*	Se *IV*	Se *V*	Se *VI*	Se *VII*	Se *VIII*
x	0.334	0.227	0.080	0.102	0.159	0.340	0.409	0.459
y	0.182	0.221	0.397	0.578	0.832	0.832	0.763	0.476
z	0.436	0.245	0.238	0.050	0.157	0.141	0.366	0.336

*B*1 (NaCl type)

Cubic: O_h^5-Fm3m; $a=5.63$; $A=8$
Co-ordinates: $(000;\ \frac{1}{2}\frac{1}{2}0;\ \circlearrowleft)+4\mathrm{Na}(O_h)$: 000
$+4\mathrm{Cl}(O_h)$: $\frac{1}{2}\frac{1}{2}\frac{1}{2}$

*B*2 (CsCl type)

Cubic: O_h^1-Pm3m; $a=4.11$; $A=2$
Co-ordinates: $\mathrm{Cs}(O_h)$: 000; $\mathrm{Cl}(O_h)$: $\frac{1}{2}\frac{1}{2}\frac{1}{2}$

Table 6.2 STRUCTURAL DETAILS—*continued*

*B*3 [Sphalerite (ZnS) type]

Cubic: $T_d^2 - F\bar{4}3m$; $a = 5.42$; $A = 8$
Co-ordinates: (000; $\frac{1}{2}\frac{1}{2}0$; ↻)+4Zn(T_d): 000
+4S(T_d): $\frac{1}{4}\frac{1}{4}\frac{1}{4}$

*B*4 [Wurtzite (ZnS) type]

Hexagonal: $C_{6v}^4 - P6_3mc$; $a = 3.81$, $c = 6.23$; $A = 4$
Co-ordinates: 2Zn(C_{3v}): $\frac{1}{3}\frac{2}{3}0$; $\frac{2}{3}\frac{1}{3}\frac{1}{2}$
2S(C_{3v}): $\frac{1}{3}\frac{2}{3}z$; $\frac{2}{3}\frac{1}{3}(\frac{1}{2}+z)$; $z \approx \frac{3}{8}$

*B*8 (α-NiAs type; β-Ni_2In type)

Between the main types (α) and (β) there exist a number of intermediate arrangements due to the variation of the stoichiometric formulae. The axial ratio *c*/*a* may change from the value 1.75 (in type α) to 1.22 (in type β). Similarly, there is virtually a continuous change from the *B*8 type to the *C*6 type.

α-NiAs type

Hexagonal: $D_{6h}^4 - P6_3/mmc$; $a = 3.61$, $c = 5.03$, $c/a = 1.39$; $A = 4$
Co-ordinates: 2Ni(D_{3d}): 000; 00$\frac{1}{2}$
2As(D_{3h}): $\frac{1}{3}\frac{2}{3}\frac{1}{4}$; $\frac{2}{3}\frac{1}{3}\frac{3}{4}$

β-Ni_2In type

Hexagonal: $D_{6h}^4 - P6_3/mmc$; $a = 4.19$, $c = 5.15$, $c/a = 1.23$; $A = 6$
2Ni(D_{3d}): 000; 00$\frac{1}{2}$; 2Ni(D_{3h}): $\frac{1}{3}\frac{2}{3}\frac{3}{4}$; $\frac{2}{3}\frac{1}{3}\frac{1}{4}$
2In(D_{3h}): $\frac{1}{3}\frac{2}{3}\frac{1}{4}$; $\frac{2}{3}\frac{1}{3}\frac{3}{4}$

*B*9 [Cinnabar (HgS) type]

Hexagonal: $D_3^4 - P3_121$ and $D_3^6 - P3_221$; $a = 4.14$, $c = 9.49$; $A = 6$
Co-ordinates: 3Hg(C_2): $x00$; $\bar{x}\bar{x}\frac{1}{3}$; $0x\frac{2}{3}$; $x = 0.33$
3S(C_2): $x0\frac{1}{2}$; $\bar{x}\bar{x}\frac{5}{6}$; $0x\frac{1}{6}$; $x = 0.21$

*B*10 (LiOH type)

Tetragonal: $D_{4h}^7 - P4/nmm$; $a = 3.55$, $c = 4.33$; $A = 4$
Co-ordinates: 2Li(D_{2d}): 000; $\frac{1}{2}\frac{1}{2}0$
2OH(C_{4v}): $0\frac{1}{2}z$; $\frac{1}{2}0\bar{z}$; $z = 0.20$
For FeSe: $z = 0.26$

*B*11 (PbO type)

Tetragonal: $D_{4h}^7 - P4/nmm$; $a = 3.98$, $c = 5.01$; $A = 4$
Co-ordinates: 2Pb(C_{4v}): $0\frac{1}{2}z$; $\frac{1}{2}0\bar{z}$; $z = 0.24$
2O(C_{4v}): the same with $z = 0.74$
For γ-TiCu: z(Ti) = 0.65; z(Cu) = 0.10

*B*13 [Millerite (NiS) type]

Rhombohedral: $C_{3v}^5 - R3m$; $a = 5.64$; $\alpha = 116° 35'$; $A = 6$
Co-ordinates: 3Ni(C_s): xxz; ↻ ; $x = 0$; $z = 0.264$
3S(C_s): the same with $x = 0.714$; $z = 0.361$

Table 6.2 STRUCTURAL DETAILS—*continued*

*B*16 (GeS type)

Orthorhombic: $D_{2h}^{16}-Pbnm$; $a=4.29$, $b=10.42$, $c=3.64$; $A=8$
Co-ordinates: 4Ge(C_s): $\pm(xy\tfrac{1}{4})$; $\pm[(\tfrac{1}{2}-x)(\tfrac{1}{2}+x)\tfrac{1}{4}]$; $x=0.167$; $y=0.375$
4S(C_s): the same with $x=0.111$; $y=0.139$

*B*17 [Cooperite (PtS) type]

Tetragonal: $D_{4h}^{9}-P4/mmc$; $a=3.47$, $c=6.12$; $A=4$
Co-ordinates: 2Pt(D_{2h}): $0\tfrac{1}{2}0$; $\tfrac{1}{2}0\tfrac{1}{2}$; 2S($D_{2d}$): $00\tfrac{1}{4}$; $00\tfrac{3}{4}$

*B*18 [Covellite (CuS) type]

Hexagonal: $D_{6h}^{4}-P6_3/mmc$; $a=3.80$, $c=16.4$; $A=12$
Co-ordinates: 2Cu(D_{3h}): $\pm(\tfrac{2}{3}\tfrac{1}{3}\tfrac{1}{4})$
4Cu(C_{3v}): $\pm(\tfrac{1}{3}\tfrac{2}{3}z)$; $\pm(\tfrac{1}{3}, \tfrac{2}{3}, \tfrac{1}{2}-z)$; $z=0.107$
2S(D_{3h}): $\pm(\tfrac{1}{3}\tfrac{2}{3}\tfrac{1}{4})$
4S(C_{3v}): $\pm(00z)$; $\pm(0, 0, \tfrac{1}{2}-z)$; $z=0.063$

*B*19 (AuCd type)

Orthorhombic: $D_{2h}^{5}-Pmcm$; $a=3.14$, $b=4.85$, $c=4.75$; $A=4$
Co-ordinates: 2Au(C_{2v}): $\pm(0y\tfrac{1}{4})$; $y=0.805$
2Cd(C_{2v}): $\pm(\tfrac{1}{2}y\tfrac{1}{4})$; $y=0.315$
For MgCd: y(Mg)$=0.818$; y(Cd)$=0.323$

*B*20 (FeSi type)

Cubic: T^4-P2_13; $a=4.48$; $A=8$
Co-ordinates: 4Fe(C_3): xxx; $(\tfrac{1}{2}+x)(\tfrac{1}{2}-x)\bar{x}$; $\circlearrowleft$; $x=0.137$
4Si(C_3): the same with $x=-0.158$
For BeAu: x(Be)$=-0.156$; x(Au)$=0.150$
For RhSn: x(Rh)$=0.142$; x(Sn)$=0.159$

*B*27 (FeB type)

Orthorhombic: $D_{2h}^{16}-Pbnm$; $a=4.05$, $b=5.50$, $c=2.95$; $A=8$
Co-ordinates: 4Fe(C_s): $\pm(xy\tfrac{1}{4})$; $\pm(\tfrac{1}{2}-x, \tfrac{1}{2}+y, \tfrac{1}{4})$; $x=0.125$; $y=0.180$
4B(C_s): the same with $x=0.61$; $y=0.036$
For MnB: x(Mn)$=0.125$; y(Mn)$=0.180$; x(B)$=0.614$; y(B)$=0.031$
For USi: x(U)$=0.125$; y(U)$=0.180$; x(Si)$=0.611$; y(Si)$=0.028$
For TiB: x(Ti) $= 0.123$; y(Ti) $= 0.177$; x(B) $= 0.603$; y(B) $= 0.029$

*B*29 (SnS type)

Orthorhombic: $D_{2h}^{16}-Pmcn$; $a=3.98$, $b=4.33$, $c=11.18$; $A=8$
Co-ordinates: 4Sn(C_s): $\pm(\tfrac{1}{4}yz)$; $\pm(\tfrac{1}{4}, \tfrac{1}{2}-y, \tfrac{1}{2}+z)$; $y=0.115$; $z=0.118$
4S(C_s): the same with $y=0.478$; $z=0.850$
If this description, given in Strukturbericht vol. **3**, p. 14, is transformed to the following, it is virtually identical with the *B*16 (GeS type).
Orthorhombic: $D_{2h}^{16}-Pbnm$; $a=4.33$, $b=11.18$, $c=3.98$; $A=8$
Co-ordinates: 4Sn(C_s): $\pm(xy\tfrac{1}{4})$; $\pm(\tfrac{1}{2}-x, \tfrac{1}{2}+y, \tfrac{1}{4})$; $x=0.115$; $y=0.382$
4S(C_s): the same with $x=0.022$; $y=0.150$

*B*31 (MnP type)

Orthorhombic: $D_{2h}^{16}-Pcmn$; $a=5.91$, $b=3.17$, $c=5.25$; $A=8$
Co-ordinates: 4Mn(C_s): $\pm(x\tfrac{1}{4}z)$; $\pm(\tfrac{1}{2}-x, \tfrac{1}{4}, \tfrac{1}{2}+z)$; $x=0.20$; $z=0.005$
4P(C_s): the same with $x=0.57$; $z=0.19$
For AuGa: x(Au)$=0.184$; z(Au)$=0.010$; x(Ga)$=0.590$; z(Ga)$=0.195$

Table 6.2 STRUCTURAL DETAILS—*continued*

For PdSi: $x(\mathrm{Pd})=0.190$; $z(\mathrm{Pd})=0.070$; $x(\mathrm{Si})=0.570$; $z(\mathrm{Si})=0.190$
For PtSi: $x(\mathrm{Pt})=0.195$; $z(\mathrm{Pt})=0.010$; $x(\mathrm{Si})=0.590$; $z(\mathrm{Si})=0.195$
For NiGe: $x(\mathrm{Ni})=0.190$; $z(\mathrm{Ni})=0.005$; $x(\mathrm{Ge})=0.583$; $z(\mathrm{Ge})=0.188$
For PdGe: $x(\mathrm{Pd})=0.188$; $z(\mathrm{Pd})=0.005$; $x(\mathrm{Ge})=0.595$; $z(\mathrm{Ge})=0.190$
For IrGe: $x(\mathrm{Ir})=0.192$; $z(\mathrm{Ir})=0.010$; $x(\mathrm{Ge})=0.590$; $z(\mathrm{Ge})=0.185$
For PtGe: $x(\mathrm{Pt})=0.195$; $z(\mathrm{Pt})=0.010$; $x(\mathrm{Ge})=0.590$; $z(\mathrm{Ge})=0.195$
For PdSn: $x(\mathrm{Pd})=0.182$; $z(\mathrm{Pd})=0.007$; $x(\mathrm{Sn})=0.590$; $z(\mathrm{Sn})=0.182$
For RhSb: $x(\mathrm{Rh})=0.192$; $z(\mathrm{Rh})=0.010$; $x(\mathrm{Sb})=0.590$; $z(\mathrm{Sb})=0.195$
For NiSi: $x(\mathrm{Ni})=0.184$; $z(\mathrm{Ni})=0.006$; $x(\mathrm{Si})=0.580$; $z(\mathrm{Si})=0.170$

*B*32 (NaTl type)

Cubic: O_h^7-Fd3m; $a=7.47$; $A=16$
Co-ordinates: $(000;\ \frac{1}{2}\frac{1}{2}0;\ \circlearrowright)+8\mathrm{Na}(T_d)$: $000;\ \frac{1}{4}\frac{1}{4}\frac{1}{4}$
$+8\mathrm{Tl}(T_d)$: $\frac{1}{2}\frac{1}{2}\frac{1}{2};\ \frac{3}{4}\frac{3}{4}\frac{3}{4}$

*B*34 (PdS type)

Tetragonal: $C_{4h}^2-P4_2/m$; $a=6.43$, $c=6.63$; $A=16$
Co-ordinates: $2\mathrm{Pd}(S_4)$: $00\frac{1}{4};\ 00\frac{3}{4}$; $2\mathrm{Pd}(C_{2h})$: $0\frac{1}{2}0;\ \frac{1}{2}0\frac{1}{2}$
$4\mathrm{Pd}(C_s)$: $\pm(xy0);\ \pm(\bar{x}y\frac{1}{2})$; $x=0.475$; $y=0.250$
$8\mathrm{S}(C_1)$: $\pm(xyz);\ \pm(xy\bar{z});\ \pm(y,\ \bar{x},\ \frac{1}{2}+z);\ \pm(y,\ \bar{x},\ \frac{1}{2}-z)$ with $x=0.20$; $y=0.32$; $z=0.22$

*B*35 (CoSn type)

Hexagonal: D_{6h}^1-P6/mmm; $a=5.27$, $c=4.25$; $A=6$
Co-ordinates: $1\mathrm{Sn}(D_{6h})$: 000; $2\mathrm{Sn}(D_{3h})$: $\frac{1}{3}\frac{2}{3}\frac{1}{2};\ \frac{2}{3}\frac{1}{3}\frac{1}{2}$
$3\mathrm{Co}(D_{2h})$: $\frac{1}{2}00;\ 0\frac{1}{2}0;\ \frac{1}{2}\frac{1}{2}0$

*B*37 (TlSe type)

Tetragonal: $D_{4h}^{18}-I4/mcm$; $a=8.02$, $c=7.00$; $A=16$
Co-ordinates: $(000;\ \frac{1}{2}\frac{1}{2}\frac{1}{2})+4\mathrm{Tl}(D_4)$: $00\frac{1}{4};\ 00\frac{3}{4}$; $+4\mathrm{Tl}(D_{2d})$: $\frac{1}{2}0\frac{1}{4};\ \frac{1}{2}0\frac{3}{4}$
$+8\mathrm{Se}(C_{2v})$: $\pm(x,\ \frac{1}{2}+x,\ 0);\ \pm(\frac{1}{2}+x,\ \bar{x},\ 0)$; $x=0.179$

B_a (UCo type)

Cubic: T^5-I2_13; $a=6.36$; $A=16$
Co-ordinates: $(000;\ \frac{1}{2}\frac{1}{2}\frac{1}{2})+8\mathrm{U}(C_3)$: $xxx;\ (\frac{1}{2}+x)(\frac{1}{2}-x)\bar{x};\ \circlearrowright$; $x=0.035$
$+8\mathrm{Co}(C_3)$: the same with $x=0.294$

B_b(ζ-AgZn type)

Hexagonal: $C_{3i}^1-P\bar{3}$; $a=7.64$; $c=2.82$; $A=9$
Co-ordinates: $1\mathrm{Zn}(C_{3i})$: 000; $+2\mathrm{Zn}(C_3)$: $\frac{1}{3}\frac{2}{3}z;\ \frac{2}{3}\frac{1}{3}z$; $z\approx\frac{3}{4}$
$(1.5\mathrm{Zn}+4.5\mathrm{Ag})\ (C_1)$: $\pm(xyz);\ \pm(\bar{y},\ x-y,\ z);\ \pm(y-x,\ \bar{x},\ z)$ with $x=0.350$; $y=0.032$; $z=0.750$

B_c (CaSi type)

Orthorhombic: $D_{2h}^{17}-Cmmc$; $a=3.91$, 4.59, 10.80; $A=8$
Co-ordinates: $(000;\ 0\frac{1}{2}\frac{1}{2})+4\mathrm{Ca}(C_{2v})$: $\pm(\frac{1}{4}0z)$; $z=0.36$
$+4\mathrm{Si}(C_{2v})$: the same with $z=0.07$

By choosing different axes and origin from those given in the original paper, this type becomes virtually identical with the *Bf*(CrB type):

$D_{2h}^{17}-Cmcm$; $a=4.59$, $b=10.80$, $c=3.91$; $y(\mathrm{Ca})=0.14$; $y(\mathrm{Si})=0.43$

Table 6.2 STRUCTURAL DETAILS—*continued*

B_d (η-NiSi)

Orthorhombic: $D_{2h}^{16}-Pbnm$; $a=5.62$, $b=5.18$, $c=3.34$; $A=8$
Co-ordinates: 4Ni(C_s): $xy0$; $\bar{x}\bar{y}\frac{1}{2}$; $(\frac{1}{2}-x)\ (\frac{1}{2}+y)0$; $(\frac{1}{2}+x)\ (\frac{1}{2}-y)\frac{1}{2}$; $x=0.184$; $y=0.006$
4Si(C_s): the same with $x=0.080$; $y=0.330$

By choosing different axes and origin from those given in the original paper, this type becomes identical with the $B31$(MnP type):
$D_{2h}^{16}-Pcmn$; $a=5.62$, $b=3.34$, $c=5.18$; x(Ni)$=0.184$; z(Ni)$=0.006$; x(Si)$=0.580$; z(Si)$=0.170$

B_e (CdSb type)

Orthorhombic: $D_{2h}^{15}-Pbca$; $a=6.47$, $b=8.25$, $c=8.53$; $A=16$
Co-ordinates: 8Sb(C_1): $\pm(xyz)$; $\pm(\frac{1}{2}+x, \frac{1}{2}-y, \bar{z})$; $\pm(\bar{x}, \frac{1}{2}+y, \frac{1}{2}-z)$; $\pm(\frac{1}{2}-x, \bar{y}, \frac{1}{2}+z)$
8Cd(C_1): the same

	Sb			Cd *or* Zn		
	x	y	z	x	y	z
CdSb	0.136	0.072	0.108	0.456	0.119	−0.128
ZnSb	0.142	0.081	0.111	0.461	0.103	−0.122

B_f (CrB type)

Orthorhombic: $D_{2h}^{17}-Cmcm$; $a=2.97$, $b=7.86$, $c=2.93$; $A=8$
Co-ordinates: $(000; \frac{1}{2}\frac{1}{2}0)+$4Cr($C_{2v}$): $\pm(0y\frac{1}{4})$; $y=0.146$
+4B(C_{2v}): the same with $y=0.440$
For NbB: y(Nb)$=0.146$; y(B)$=0.444$
For CaSi: y(Ca)$=0.14$; y(Si)$=0.43$ (cf. Bc type)

B_g (MoB type)

Tetragonal: $D_{4h}^{19}-I4/amd$; $a=3.11$, $c=16.97$; $A=16$
Co-ordinates: $(000, \frac{1}{2}\frac{1}{2}\frac{1}{2})+$8Mo($C_{2v}$): $\pm(00z)$; $\pm(0, \frac{1}{2}, \frac{1}{4}+z)$; $z=0.197$
+8B(C_{2v}): the same with $z=0.35$

B_h (WC type)

Hexagonal: $D_{6h}^{1}-P6/mmm$; $a=2.91$, $c=2.84$; $A=2$
Co-ordinates: 1W(D_{6h}): 000; 1C(D_{3h}): $\frac{1}{3}\frac{2}{3}\frac{1}{2}$, $\frac{2}{3}\frac{1}{3}\frac{1}{2}$

B_i (γ'-MoC type)

Hexagonal: $D_{6h}^{4}-P6_3/mmc$; $a=2.93$, $c=10.97$; $A=8$
Co-ordinates: 4Mo(C_{3h}): $\frac{1}{3}\frac{2}{3}z$; $\frac{1}{3}\frac{2}{3}(\frac{1}{2}-z)$; $\frac{2}{3}\frac{1}{3}(\frac{1}{2}+z)$; $\frac{2}{3}\frac{1}{3}\bar{z}$; $z\approx\frac{1}{8}$
4C: in holes

B_k (BN type)

Hexagonal: $D_{6h}^{4}-P6_3/mmc$; $a=2.50$, $c=6.66$; $A=4$
Co-ordinates: 2B(D_{3h}): $\frac{1}{3}\frac{2}{3}\frac{1}{4}$; $\frac{2}{3}\frac{1}{3}\frac{3}{4}$; 2N($D_{3h}$): $\frac{1}{3}\frac{2}{3}\frac{3}{4}$, $\frac{2}{3}\frac{1}{3}\frac{1}{4}$

Table 6.2 STRUCTURAL DETAILS—*continued*

B_l [Realgar (AsS) type]

Monoclinic: $C_{2h}^5 - P2_1n$; $a=9.27$, $b=13.50$, $c=6.56$; $\beta=106°\ 37'$; $A=32$
Co-ordinates: 4 times 4As(C_1) and 4 times 4S(C_1) in:
$\pm(xyz)$; $\pm(\frac{1}{2}+x, \frac{1}{2}-x, \frac{1}{2}+z)$

	As *I*	As *II*	As *III*	As *IV*	S *I*	S *II*	S *III*	S *IV*
x	0.118	0.425	0.318	0.038	0.346	0.213	0.245	0.115
y	0.024	−0.140	−0.127	−0.161	0.008	0.024	−0.225	−0.215
z	−0.241	−0.142	0.181	−0.290	−0.295	0.120	−0.363	0.048

B_m (TiB type)

Orthorhombic: $D_{2h}^{16} - Pnma$; $a=6.12$, $b=3.06$, $c=4.56$; $A=8$
Co-ordinates: 4Ti(C_s): $\pm(x\frac{1}{4}z)$; $\pm(\frac{1}{2}-x, \frac{3}{4}, \frac{1}{2}+z)$; $x=0.177$; $z=0.123$
4B(C_s): the same with $x=0.029$; $z=0.603$
If the axes are changed from those of the original paper, this type becomes identical with the *B*27 (FeB) type:

$D_{2h}^{16} - Pbnm$; $a=4.56$, $b=6.12$, $c=3.06$; x(Ti)$=0.123$; y(Ti)$=0.177$;
x(B)$=0.603$; y(B)$=0.029$

*C*1 (CaF_2 type—MgAgAs type)

(α) Cubic: $O_h^5 - Fm3m$; $a=5.45$; $A=12$
Co-ordinates: (000; $\frac{1}{2}\frac{1}{2}0$; ↻)+4Ca(O_h): 000
+8F(T_d): $\pm(\frac{1}{4}\frac{1}{4}\frac{1}{4})$
In those cases in which the *F*-position is occupied by two components in an ordered fashion—for example in As(MgAg)—the space group is changed to
(β) Cubic: $T_d^2 - F\bar{4}3m$; $a=6.24$; $A=12$
Co-ordinates: (000; $\frac{1}{2}\frac{1}{2}0$; ↻)+4As(T_d): 000
+4Ag(T_d): $\frac{1}{4}\frac{1}{4}\frac{1}{4}$
+4Mg(T_d): $\frac{3}{4}\frac{3}{4}\frac{3}{4}$

*C*2 [Pyrites (FeS_2) type]

Cubic: $T_h^6 - Pa3$; $a=5.40$; $A=12$
Co-ordinates: 4Fe(S_{3i}): 000; $\frac{1}{2}\frac{1}{2}0$; ↻
8S(C_3): $\pm(xxx)$; $\pm(\frac{1}{2}+x, \frac{1}{2}-x, \bar{x}$; ↻); $x=0.386$
For MnS_2, $x=0.401$; for CoAsS and NiAsS (random distribution of As and S) $x=0.385$; for $PtBi_2$: $x=0.38$.

*C*6 (CdI_2 type)

Hexagonal: $D_{3d}^3 - P\bar{3}m1$; $a=4.24$; $c=6.84$; $A=3$
Co-ordinates: 1Cd(D_{3d}): 000; 2I(C_{3v}): $\frac{1}{3}\frac{2}{3}z$; $\frac{2}{3}\frac{1}{3}\bar{z}$; $z\approx\frac{1}{4}$
There is virtually a continuous change from this type to the *B*8 type.

*C*7 (MoS_2 type)

Hexagonal: $D_{6h}^4 - P6_3/mmc$; $a=3.15$, $c=12.30$; $A=6$
Co-ordinates: 2Mo(D_{3h}): $\frac{1}{3}\frac{2}{3}\frac{1}{4}$; $\frac{2}{3}\frac{1}{3}\frac{3}{4}$
4S(C_{3v}): $\frac{1}{3}\frac{2}{3}z$; $\frac{2}{3}\frac{1}{3}\bar{z}$; $\frac{1}{3}\frac{2}{3}(\frac{1}{2}-z)$; $\frac{2}{3}\frac{1}{3}(\frac{1}{2}+z)$; $z=0.62$

Table 6.2 STRUCTURAL DETAILS—*continued*

*C*11*a* (CaC_2 type)

Tetragonal: $D_{4h}^{17}-I4/mmm$; $a=3.87$, $c=6.37$, $c/a=1.65$; $A=6$
Co-ordinates $(000;\ \frac{111}{222})$ $+2Ca(D_{4h})$: 000
$+4C(C_{4v})$: $\pm(00z)$; $z=0.38$

*C*11*b* ($MoSi_2$ type)

Tetragonal: $D_{4h}^{17}-I/mmm$; $a=3.20$, $c=7.86$, $c/a=2.46$; $A=6$
Co-ordinates: $(000;\ \frac{111}{222})$ $+2Mo(D_{4h})$: 000
$+4Si(C_{4v})$: $\pm(00z)$; $z\approx\frac{1}{3}$
This type is a superstructure of the $A\alpha$(Pa) type.

*C*12 ($CaSi_2$ type)

Rhombohedral: $D_{3d}^{5}-R\bar{3}m$; $a=10.4$; $\alpha=21°30'$; $A=6$
Co-ordinates: $2Ca(C_{3v})$: $\pm(xxx)$; $x=0.083$
$2Si(C_{3v})$: the same with $x=0.185$
$2Si(C_{3v})$: the same with $x=0.352$
Hexagonal axes: $a=3.88$, $c=30.4$; $A=18$
Co-ordinates: $(000;\ \frac{211}{333};\ \frac{122}{333})$ $+6Ca(C_{3v})$: $\pm(00x)$; $x=0.083$
$+6Si(C_{3v})$: the same with $x=0.185$
$+6Si(C_{3v})$: the same with $x=0.352$

*C*14 ($MgZn_2$ type)

Hexagonal: $D_{6h}^{4}-P6_3/mmc$; $c=5.15$, $c=8.48$; $A=12$
Co-ordinates: $4Mg(C_{3v})$: $\pm(\frac{1}{3}\frac{2}{3}z;\ \frac{1}{3},\ \frac{2}{3},\ \frac{1}{2}-z)$; $z\approx\frac{1}{16}=0.062$
$2Zn(D_{3d})$: 000; $00\frac{1}{2}$
$6Zn(C_{2v})$: $\pm(x,\ 2x,\ \frac{1}{4};\ \bar{2}\bar{x},\ \bar{x},\ \frac{1}{4};\ x\bar{x}\ \frac{1}{4})$; $x\approx-\frac{1}{6}=-0.170$

*C*15 ($MgCu_2$ type)

Cubic: O_h^7-Fd3m; $a=7.01$; $A=24$
Co-ordinates: $(000;\ \frac{11}{22}0;\ \circlearrowright)$ $+8Mg(T_d)$: 000; $\frac{111}{444}$
$+16Cu(D_{3d})$: $\frac{555}{888};\ \frac{775}{888};$

*C*16 ($CuAl_2$ type)

Tetragonal: $D_{4h}^{18}-I4/mcm$; $a=6.05$, $c=4.88$; $A=12$
Co-ordinates: $(000;\ \frac{111}{222})$ $+4Cu(D_4)$: $\pm(00\frac{1}{4})$
$+8Al(C_{2v})$: $\pm(x,\ \frac{1}{2}+x,\ 0;\ \frac{1}{2}+x,\ \bar{x},\ 0)$; $x=0.158$

	$AuNa_2$	$MnSn_2$	$FeSn_2$	$CoSn_2$	$RhSn_2$	$TiSb_2$	VSb_2	Ta_2B	Mo_2B	W_2B	Mn_2B
x	0.160	0.159	0.160	0.116	0.161	0.158	0.158	0.167	0.170	0.170	0.163

For the compounds $FeGe_2$, (Rh,Pd,Au)Pb_2 no deviation from $x=0.158$ has been reported.

*C*18 [Marcasite (FeS_2) type]

Orthorhombic: $D_{2h}^{12}-Pnnm$; $a=4.44$, $b=5.41$, $c=3.38$; $A=6$
Co-ordinates: $2Fe(C_{2h})$: 000; $\frac{111}{222}$
$4S(C_s)$: $\pm(xy0)$; $\pm(\frac{1}{2}+x,\ \frac{1}{2}-y,\ \frac{1}{2})$; $x=0.20$; $y=0.38$
For $FeAs_2$: $x=0.18$; $y=0.36$
For $NiAs_2$: $x=0.22$; $y=0.37$
For $SeSb_2$: $x=0.18$; $y=0.36$
For FeP_2: $x=0.16$; $y=0.37$
For $CoTe_2$: $x=0.22$; $y=0.36$

Table 6.2 STRUCTURAL DETAILS—*continued*

For $FeTe_2$: $x=0.22$; $y=0.36$
For $FeSe_2$: $x=0.21$; $y=0.37$

*C*22 (Fe_2Pe_2P type)

Hexagonal: D_3^2-P321; $a=5.85$, $c=3.45$; $A=9$
Co-ordinates: $3Fe(C_2)$: $\bar{x}00$; $0\bar{x}0$; $xx0$; $x=0.26$
$3Fe(C_2)$: $x0\frac{1}{2}$; $0x\frac{1}{2}$; $\bar{x}x\frac{1}{2}$; $x=0.40$
$1P(D_3)$: $00\frac{1}{2}$
$2P(C_3)$: $\pm(\frac{1}{3}\frac{2}{3}z)$; $z\approx\frac{1}{8}=0.125$

*C*23 ($PbCl_2$ type)

Orthorhombic: $D_{2h}^{16}-Pmnb$; $a=4.53$, $b=7.61$, $c=9.03$; $A=12$
Co-ordinates: $4Pb(C_s)$: $\pm(\frac{1}{4}yz)$; $\pm(\frac{1}{4})\frac{1}{2}+y, \frac{1}{2}-z)$; $y=0.246$; $z=0.905$
$4Cl(C_s)$: the same with $y=0.85$; $z=0.93$
$4Cl(C_s)$: the same with $y=0.95$; $z=0.33$

For Co_2P, Ni_2Si* and ThS_2 the parameters are:

	P	Co	Co	Th	S	S	Si	Ni	Ni
y	0.250	0.862	0.970	0.250	0.850	0.965	0.236	0.825	0.958
z	0.900	0.930	0.333	0.875	0.942	0.320	0.886	0.937	0.297

*C*32 (AlB_2 type)

Hexagonal: D_{6h}^1-P6/mmm; $a=3.00$, $c=3.25$; $A=3$
Co-ordinates: $1Al(D_{6h})$: 000
$2B(D_{3h})$: $\frac{1}{3}\frac{2}{3}\frac{1}{3}$; $\frac{2}{3}\frac{1}{3}\frac{1}{2}$

*C*33 (Bi_2Te_2S type)

Rhombohedral: $D_{3d}^5-R\bar{3}m$; $a=10.31$; $\alpha=24°\,10'$; $A=5$
Co-ordinates: $2Bi(C_{3v})$: $\pm(xxx)$; $x=0.392$
$2Te(C_{3v})$: the same with $x=0.788$
$1S(D_{3d})$: 000

*C*34 [Calaverite ($AuTe_2$) type]

Monoclinic: C_{2h}^3-C2/m; $a=7.18$, $b=4.40$, $c=5.07$; $\beta=90°\,13'$; $A=6$
Co-ordinates: (000; $\frac{1}{2}\frac{1}{2}0$) $+2Au(C_{2h})$: 000
$+4Te(C_s)$: $\pm(x0z)$; $x=0.689$; $z=0.280$

*C*36 ($MgNi_2$ type)

Hexagonal: $D_{6h}^4-P6_3/mmc$; $a=4.81$, $c=15.77$; $A=24$
Co-ordinates: $4Mg(C_{3v})$: $+(\frac{1}{3}\frac{2}{3}z)$; $\pm(\frac{2}{3}, \frac{1}{3}, \frac{1}{2}+z)$; $z\approx\frac{27}{32}$
$4Mg(C_{3v})$: $\pm(00z)$; $\pm(0, 0, \frac{1}{2}+z)$; $z\approx\frac{3}{32}$
$6Ni(C_{2h})$: $\frac{1}{2}00$; $0\frac{1}{2}0$; $\frac{1}{2}\frac{1}{2}0$; $\frac{1}{2}0\frac{1}{2}$; $0\frac{1}{2}\frac{1}{2}$; $\frac{1}{2}\frac{1}{2}\frac{1}{2}$
$6Ni(C_{2v})$: $\pm(x, 2x, \frac{1}{4}; 2\bar{x}, \bar{x}, \frac{1}{4}; x, \bar{x}, \frac{1}{4})$; $x\approx\frac{1}{6}$
$4Ni(C_{3v})$: $\pm(\frac{1}{3}\frac{2}{3}z)$; $\pm(\frac{2}{3}, \frac{1}{3}, \frac{1}{2}+z)$; $z\approx\frac{1}{8}$

* If Toman's[199] description of δ-Ni_2Si is charged from *Pbnm* to *Pmnb*, and if the origin is chosen differently, this compound belongs to the C23 type.

Table 6.2 STRUCTURAL DETAILS—*continued*

*C*37 (Co_2Si type)

Orthorhombic: $D_{2h}^{16}-Pbnm$; $a=7.10$, $b=4.91$, $c=3.73$; $A=12$
Co-ordinates: $4Si(C_s)$: $\pm(xy\frac{1}{4})$; $\pm(\frac{1}{2}-x, \frac{1}{2}+y, \frac{1}{4})$; $x=0.440$; $y=0.070$
$4Co(C_s)$: the same with $x=0.103$; $y=0.090$
$4Co(C_s)$: the same with $x=0.772$; $y=0.193$
With a different choice of axes and origin, a similarity to the *C*23 ($PbCl_2$ type) becomes apparent:
Orthorhombic: $D_{2h}^{16}-Pmnb$; $a=3.73$, $b=4.91$, $c=7.10$; $A=12$

Co-ordinates: $4Si(C_s)$: $\pm(\frac{1}{4}yz)$; $\pm(\frac{1}{4}, \frac{1}{2}+y, \frac{1}{2}-z)$; $y=0.07$; $z=0.94$
$4Co(C_s)$: the same with $y=0.59$; $z=0.897$
$4Co(C_s)$: the same with $y=0.693$; $z=0.228$
With this orientation the lattice constants are virtually identical with those of δ-Ni_2Si ($a=3.73$, $b=4.99$, $c=7.03$) which has the *C*23 type. A redetermination of the Co_2Si type might lead to a still closer similarity than that appearing in this description.

*C*38 (Cu_2Sb type)

Tetragonal: $D_{4h}^{7}-P4/nmm$; $a=3.99$, $c=6.09$; $A=6$
Co-ordinates: $2Cu(D_{2d})$: $000, \frac{1}{2}\frac{1}{2}0$
$2Cu(C_{4v})$: $0\frac{1}{2}z; \frac{1}{2}0\bar{z}; z=0.27$
$2Sb(C_{4v})$: the same with $z=0.70$

For $Cu_{2-x}Te$; $z(Cu)=0.27$; $z(Te)=0.715$, with vacant sites in the $(0\frac{1}{2}z)$-position.
For Fe_2As and Cr_2As: $z(As)=0.735$; $z(Fe \text{ or } Cr)=0.33$. For CuMgAs: $z(As)=0.75$; $z(Cu, Mg)=0.335$. For $AlNaSi_4$: $z(Si)=0.79$; $z(Na, Al)=0.37$.
A very similar arrangement is shown by $CuGa_2$ with $a=2.83$, $c=5.84$. The 3 atoms per cell take the positions 000; $\frac{1}{2}\frac{1}{2}x$; $\frac{1}{2}\frac{1}{2}y$ with $x=0.70$; $y=0.27$. For comparison a larger cell with $a'=a\sqrt{2}$ can be chosen:

Tetragonal: $C_{4v}^{1}-P4mm$; $a'=4.03$, $c=5.84$; $A=6$
Co-ordinates: $2Ga(C_{4v})$: $000; \frac{1}{2}\frac{1}{2}0$
$2Ga(C_{4v})$: $0\frac{1}{2}z; \frac{1}{2}0z; z=0.70$
$2Cu(C_{4v})$: $0\frac{1}{2}z; \frac{1}{2}0z; z=0.27$

*C*40 ($CrSi_2$ type)

Hexagonal: D_6^4 (or D_6^5)$-P6_222$ (or $P6_422$); $a=4.42$, 6.35; $A=9$
Co-ordinates: $3Cr(D_2)$: $\frac{1}{2}00; \frac{1}{2}\frac{1}{2}\frac{1}{3}; 0\frac{1}{2}\frac{2}{3}$
$6Si(C_2)$: $\pm(x, 2x, 0)$; $x\bar{x}\frac{1}{3}$; $\bar{x}x\frac{1}{3}$; $2x, x, \frac{2}{3}$; $2\bar{x}, \bar{x}\frac{2}{3}$; $x\approx\frac{1}{6}$

*C*42 (SiS_2 type)

Orthorhombic: $D_{2h}^{26}-Icma$; $a=5.60$, $b=5.53$, $c=9.55$; $A=12$
Co-ordinates: $(000, \frac{1}{2}\frac{1}{2}\frac{1}{2})$ $+4Si(D_2)$: $\pm(0\frac{1}{4}0)$
$+8S(C_s)$: $\pm(x0z)$; $\pm(x\frac{1}{2}\bar{z})$; $x=0.208$; $z=0.119$

*C*44 (GeS_2 type)

Orthorhombic: $C_{2v}^{19}-Fdd$; $a=11.66$, $b=22.34$, $c=6.86$; $A=72$
Co-ordinates: $(000; \frac{1}{2}\frac{1}{2}0; \circlearrowleft)$ $+8Ge(C_2)$: $000; \frac{1}{4}\frac{1}{4}\frac{1}{4}$
$+16Ge(C_1)$: xyz; $\bar{x}\bar{y}z$; $(\frac{1}{4}-x)(\frac{1}{4}+y)(\frac{1}{4}+z)$; $(\frac{1}{4}+x)(\frac{1}{4}-y)(\frac{1}{4}+z)$ with $x=0.125$; $y=0.139$; $z=0.00$
$+16S(C_1)$: the same with $x=0.022$; $y=0.081$; $z=0.183$
$+16S(C_1)$: the same with $x=0.153$; $y=-0.014$; $z=-0.183$
$+16S(C_1)$: the same with $x=0.063$; $y=0.125$; $z=-0.278$

*C*46 [Krennerite ($AuTe_2$) type]

Orthorhombic: $C_{2v}^{4}-Pma$; $a=16.54$, $b=8.82$, $c=4.46$; $A=24$
Co-ordinates: $2Au(C_2)$: $00z; \frac{1}{2}0z; z=0$

Table 6.2 STRUCTURAL DETAILS—*continued*

$2Au(C_s)$: $\frac{1}{4}yz$; $\frac{3}{4}\bar{y}z$; $y=0.319$; $z=0.014$
$4Au(C_1)$: xyz; $\bar{x}\bar{y}z$; $(\frac{1}{2}-x)yz$; $(\frac{1}{2}+x)\bar{y}z$; $x=0.124$; $y=0.666$; $z=0.500$
$2Te(C_s)$: $\frac{1}{4}yz$; $\frac{3}{4}\bar{y}z$; $y=0.018$; $z=0.042$
$2Te(C_s)$: the same with $y=0.617$; $z=0.042$
$4Te(C_1)$: as 4Au (C_1) with $x=0.003$; $y=0.699$; $z=0.042$
$4Te(C_1)$: the same with $x=0.132$; $y=0.364$; $z=0.500$
$4Te(C_1)$: the same with $x=0.119$; $y=0.964$; $z=0.500$

$C49$ ($ZrSi_2$ type)

Orthorhombic: $D_{2h}^{17}-Cmcm$; $a=3.72$, $b=14.61$, $c=3.67$; $A=12$
Co-ordinates: $(000; \frac{1}{2}\frac{1}{2}0)$ $+4Zr(C_{2v})$: $\pm(0y\frac{1}{4})$; $y=0.106$
$+4Si(C_{2v})$: the same with $y=0.750$
$+4Si(C_{2v})$: the same with $y=0.355$

$C54$ ($TiSi_2$ type)

Orthorhombic: $D_{2h}^{24}-Fddd$; $a=8.24$, $b=4.77$, $c=8.52$; $A=24$
Co-ordinates: $(000, \frac{1}{2}\frac{1}{2}0; ⟲)$ $+8Ti(D_2)$: 000; $\frac{1}{4}\frac{1}{4}\frac{1}{4}$
$+16Si(C_2)$: $\pm(x00)$; $(\frac{1}{4}+x)\frac{1}{4}\frac{1}{4}$; $(\frac{1}{4}-x)\frac{1}{4}\frac{1}{4}$; $x\approx\frac{1}{3}$

C_a (Mg_2Ni type)

Hexagonal: $D_6^4(D_6^5)-P6_222(P6_422)$; $a=5.18$, $c=13.19$; $A=18$
Co-ordinates: $3Ni(D_2)$: $00\frac{1}{6}$; $00\frac{3}{6}$; $00\frac{5}{6}$
$3Ni(D_2)$: $0\frac{1}{2}\frac{1}{6}$; $\frac{1}{2}0\frac{3}{6}$; $\frac{1}{2}\frac{1}{2}\frac{5}{6}$
$6Mg(C_2)$: $\pm(\frac{1}{2}0z)$, $\frac{1}{2}\frac{1}{2}(\frac{1}{3}+z)$; $\frac{1}{2}\frac{1}{2}(\frac{1}{3}-z)$; $0\frac{1}{2}(\frac{2}{3}+z)$; $0\frac{1}{2}(\frac{2}{3}-z)$; $z=\frac{1}{9}$
$6Mg(C_2)$: $\pm(x, 2x, 0)$; $x\bar{x}\frac{1}{3}$; $\bar{x}x\frac{1}{3}$; $2x, x, \frac{2}{3}$; $2\bar{x}, \bar{x}\frac{2}{3}$; $x=\frac{1}{6}$

C_b (Mg_2Cu type)

Orthorhombic: $D_{2h}^{24}-Fddd$; $a=5.27$, $b=9.05$, $c=18.21$; $A=48$
Co-ordinates: $(000; \frac{1}{2}\frac{1}{2}0; ⟲)$ $+16Cu(C_2)$: $\pm(00z)$; $\frac{1}{4}\frac{1}{4}(\frac{1}{4}+z)$; $\frac{1}{4}\frac{1}{4}(\frac{1}{4}-z)$; $z=0.128$
$+16Mg(C_2)$: the same with $z=0.411$
$+16Mg(C_2)$: $\pm(0y0$; $\frac{1}{4}(\frac{1}{4}+y)\frac{1}{4}$; $\frac{1}{4}(\frac{1}{4}-y)\frac{1}{4}$; $y=0.161$

C_c ($ThSi_2$ type)

Tetragonal: $D_{4h}^{19}-I4/amd$; $a=4.13$, $c=14.35$; $A=12$
Co-ordinates: $(000; \frac{1}{2}\frac{1}{2}\frac{1}{2})$ $+4Th(D_{2d})$: 000; $0\frac{1}{2}\frac{1}{4}$
$+8Si(C_{2v})$: $\pm(00z)$; $0\frac{1}{2}(\frac{1}{4}+z)$; $0\frac{1}{2}(\frac{1}{4}-z)$; $z=0.417$

C_e ($CoGe_2$ type)

Orthorhombic: $C_{2v}^{17}-Aba$; $a\approx b=5.67$, $c=10.80$; $A=24$
Co-ordinates: $(000, 0\frac{1}{2}\frac{1}{2})$ $+4Co(C_2)$: $00z$; $\frac{1}{2}\frac{1}{2}z$; $z=-0.012$
$+4Co(C_2)$: the same with $z=-0.238$
$+8Ge(C_1)$: xyz; $\bar{x}\bar{y}z$; $(\frac{1}{2}-x)(\frac{1}{2}+y)z$; $(\frac{1}{2}+x)(\frac{1}{2}-y)z$; with $x=0.342$; $y=0.158$; $z=-\frac{1}{8}$
$+8Ge(C_1)$: the same with $x=y=\frac{1}{4}$; $z=\frac{1}{8}$

The real composition is $Co_{0.9}Ge_2$.

C_f

Tetragonal: a series of types composed of alternating sheets of the Fluorite—and the $C16$(Al_2Cu)—type. Whereas the length of the *a*-axis is approximately constant, the length of the *c*-axis varies due to the different possibilities of the sheet sequences.

Table 6.2 STRUCTURAL DETAILS—*continued*

	$RhSn_2$	$PdSn_2$
a	6.38	6.55
c	17.88	24.57

C_g(ThC_2 type)

Monoclinic: $C^6_{2h}-C2/c$; $a=6.53$, $b=4.24$, $c=6.56$; $\beta=104°$; $A=12$
Co-ordinates: (000; $\frac{1}{2}\frac{1}{2}0$) $+4Th(C_2)$: $\pm(0y\frac{1}{4})$; $y=0.202$
$+8C(C_1)$: $\pm(xyz)$; $\pm(x, \bar{y}, \frac{1}{2}+z)$; with $x=0.29$; $y=0.13$; $z=0.08$

C_h(Cu_2Te type)

Hexagonal: $D^1_{6h}-P6/mmm$; $a=4.24$, $c=7.27$; $A=6$
Co-ordinates: $2Te(C_{6v})$: $\pm(00z)$; $z=0.306$
$4Cu(C_{3v})$: $\pm(\frac{1}{3}\frac{2}{3}z)$; $\pm(\frac{1}{3}\frac{2}{3}\bar{z})$; $z=0.160$

C_k($\sim$ $LiZn_2$ type)

Hexagonal: $D^4_{6h}-P6_3/mmc$; $a=4.36$, $c=2.51$; $A\approx3$
Co-ordinates: $2Zn(D_{3h})$: $\pm(\frac{1}{3}\frac{2}{3}\frac{1}{4})$
$0.8Li(D_{3d})$: 000; $00\frac{1}{2}$; (?)

$D0_2$ ($CoAs_3$ type)

Cubic: T^5_h-Im3; $a=8.18$; $A=32$
Co-ordinates: (000; $\frac{1}{2}\frac{1}{2}\frac{1}{2}$) $+8Co(C_{3i})$: $\frac{1}{4}\frac{1}{4}\frac{1}{4}$; $\frac{3}{4}\frac{3}{4}\frac{1}{4}$; ↻
$+24As(C_s)$: $\pm(xy0$; ↻) $\pm(x\bar{y}0$; ↻); $x=0.35$; $y=0.15$

$D0_3$ (BiF_3 or $BiLi_3$ type)

Cubic: O^5_h-Fm3m; for $BiLi_3$: $a=6.71$; $A=16$
Co-ordinates: (000; $\frac{1}{2}\frac{1}{2}0$; ↻) $+4Bi(O_h)$: 000
$+4Li(O_h)$: $\frac{1}{2}\frac{1}{2}\frac{1}{2}$
$+8Li(T_d)$: $\pm(\frac{1}{4}\frac{1}{4}\frac{1}{4})$

$D0_9$ (ReO_3 or Cu_3N type)

Cubic: O^1_h-Pm3m; for Cu_3N; $a=3.81$; $A=4$
Co-ordinates: $IN(O_h)$: 000; $3Cu(D_{4h})$: $\frac{1}{2}00$; ↻

$D0_{11}$ (Fe_3C type)

Orthorhombic: $D^{16}_{2h}-Pbnm$; $a=4.51$, $b=5.08$, $c=6.73$; $A=16$
Co-ordinates: $4Fe(C_s)$: $\pm(xy\frac{1}{4})$; $\pm(\frac{1}{2}-x, \frac{1}{2}+y, \frac{1}{4})$; $x=0.833$; $y=0.040$
$8Fe(C_1)$: $\pm(xyz)$; x, y, $\frac{1}{2}-z$; $\pm(\frac{1}{2}-x, \frac{1}{2}+y, z; \frac{1}{2}-x, \frac{1}{2}+y, \frac{1}{2}-z)$;
$x=0.333$; $y=0.183$; $z=0.065$
$4C(C_s)$: $\pm(xy\frac{1}{4})$; $\pm(\frac{1}{2}-x, \frac{1}{2}+y, \frac{1}{4})$; $x=0.47$; $y=0.86$

$D0_{18}$ (Na_3As type)

Hexagonal: $D^4_{6h}-P6_3/mmc$; $a=5.09$, $c=8.98$; $A=8$
Co-ordinates: $2As(D_{3h})$: $\pm(\frac{1}{3}\frac{2}{3}\frac{1}{4})$
$2Na(D_{3h})$: $\pm(00\frac{1}{4})$
$4Na(C_{3v})$: $\pm(\frac{1}{3}\frac{2}{3}z$; $\frac{2}{3}$, $\frac{1}{3}$, $\frac{1}{2}+z)$; $z=0.583$

$D0_{19}$ (Mg_3Cd type)

Hexagonal: $D^4_{6h}-P6_3/mmc$; $a=6.26$, $c=5.07$; $A=8$
Co-ordinates: $2Cd(D_{3h})$: $\pm(\frac{1}{3}\frac{2}{3}\frac{1}{4})$
$6Mg(C_{2v})$: $\pm(2x, x, \frac{1}{4}$; $\bar{x}x\frac{1}{4}$; $\bar{x}, 2x, \frac{1}{4})$; $x\approx\frac{1}{6}$

Table 6.2 STRUCTURAL DETAILS—*continued*

DO_{20}(NiAl$_3$ type)

Orthorhombic: $D_{2h}^{16}-Pnma$; $a=6.60$, $b=7.35$, $c=4.80$; $A=16$

Co-ordinates: 4Ni(C_s): $\pm(x\frac{1}{4}z)$; $\pm(\frac{1}{2}+x, \frac{1}{4}, \frac{1}{2}-z)$; $x=-0.731$; $z=-0.055$

4Al(C_s): the same with $x=0.011$; $z=0.415$

8Al(C_1): $\pm(xyz)$; $\pm(\frac{1}{2}+x, \frac{1}{2}-y, \frac{1}{2}-z)$; $\pm(x, \frac{1}{2}-y, z)$; $\pm(\frac{1}{2}+x, y, \frac{1}{2}-z)$ with $x=0.174$; $y=0.053$; $z=0.856$

Following the choice of co-ordinates and of the origin in the original paper, NiAl$_3$ was described in the Strukturbericht as a separate type. However, another choice of co-ordinates and of origin shows that the structure is very similar to the $D0_{11}$(Fe$_3$C) type:

Orthorhombic: $D_{2h}^{16}-Pbnm$; $a=4.80$, $b=6.60$, $c=7.35$; $A=16$

Co-ordinates: 4Al(C_s): $\pm(xy\frac{1}{4})$; $\pm(\frac{1}{2}-x, \frac{1}{2}+y, \frac{1}{4})$; $x=0.915$; $y=0.011$

8Al(C_1): $\pm(xyz, xy\frac{1}{2}-z)$; $\pm(\frac{1}{2}-x, \frac{1}{2}+y, z; \frac{1}{2}-x, \frac{1}{2}+y, \frac{1}{2}-z)$; $x=0.356$; $y=0.174$; $z=0.053$

4Ni(C_s): $\pm(xy\frac{1}{4})$; $\pm(\frac{1}{2}-x, \frac{1}{2}+y, \frac{1}{4})$; $x=0.445$; $y=0.869$

$D0_{21}$(Cu$_3$P type)

Hexagonal: $D_{3d}^{4}-P\bar{3}c1$; $a=7.07$, $c=7.14$; $A=24$

Co-ordinates: 6P(C_2): $\pm(x0\frac{1}{4})$: $\pm(0x\frac{1}{4})$; $\pm(\bar{x}\bar{x}\frac{1}{4})$; $x=0.38$

2Cu(C_{3i}): 000, $00\frac{1}{2}$

4Cu(C_3): $\pm(\frac{1}{3}\frac{2}{3}z)$; $\pm(\frac{1}{3}, \frac{2}{3}, \frac{1}{2}+z)$; $z=0.17$

12Cu(C_1): $\pm[xyz;\ \bar{y}(x-y)z;\ (y-x)\bar{x}z;\ \bar{y}\bar{x}(\frac{1}{2}+z);\ x(x-y)(\frac{1}{2}+z);\ (x-y)y(\frac{1}{2}+x)]$; $x=0.69$; $y=0.07$; $z=0.08$

See remarks by Haraldsen[200]

$D0_{22}$(TiAl$_3$ type)

Tetragonal: $D_{4h}^{17}-I4/mmm$; $a=3.84$, $c=8.58$; $A=8$

Co-ordinates: $(000; \frac{111}{222})$ + 2Ti(D_{4h}): 000

+ 2Al(D_{4h}): $00\frac{1}{2}$

+ Al(D_{2d}): $0\frac{1}{2}\frac{1}{4}$; $\frac{1}{2}0\frac{1}{4}$

$D0_{23}$(ZrAl$_3$ type)

Tetragonal: $D_{4h}^{17}-I4/mmm$; $a=4.01$, $c=17.29$; $A=16$

Co-ordinates: $(000; \frac{111}{222})$ + 4Zr(C_{4v}): $\pm(00z)$; $z=0.122\approx\frac{1}{8}$

+ 4Al(D_{2h}): $0\frac{1}{2}0$; $\frac{1}{2}00$

+ 4Al(D_{2d}): $0\frac{1}{2}\frac{1}{4}$; $\frac{1}{2}0\frac{1}{4}$

+ 4Al(C_{4v}): $\pm(00z)$; $z=0.361\approx\frac{3}{8}$

$D0_{24}$(TiNi$_3$ type)

Hexagonal: $D_{6h}^{4}-P6_3/mmc$; $a=5.10$, $c=8.30$; $A=16$

Co-ordinates: 2Ti(D_{3d}): 000; $00\frac{1}{2}$; 2Ti(D_{3h}): $\frac{121}{334}$; $\frac{213}{334}$

6Ni(C_{2h}): $\frac{1}{2}00$; $0\frac{1}{2}0$; $\frac{1}{2}\frac{1}{2}0$; $\frac{1}{2}0\frac{1}{2}$; $0\frac{1}{2}\frac{1}{2}$; $\frac{111}{222}$

6Ni(C_{2v}): $\pm(x, 2x\frac{1}{4}; 2\bar{x}, \bar{x}, \frac{1}{4}; x\bar{x}\frac{1}{4})$; $x=\approx-\frac{1}{6}$

$D0_a$(β-TiCu$_3$ type)

Orthorhombic: $D_{2h}^{13}-Pmmn$; $a=5.15$, $b=4.34$, $c=4.52$; $A=8$

Co-ordinates: 2Ti(C_{2v}): $00z$; $\frac{1}{2}\frac{1}{2}\bar{z}$; $z=0.655$

2Cu(C_{2v}): $0\frac{1}{2}z$; $\frac{1}{2}0\bar{z}$; $z=0.345$

4Cu(C_s): $x0z$; $x0z$; $(\frac{1}{2}+x)\frac{1}{2}\bar{z}$; $(\frac{1}{2}-x)\frac{1}{2}\bar{z}$; $x=\frac{1}{4}$; $z=0.155$

$D0_b$(γ-Ag$_3$Ga type)

Hexagonal: $C_{3i}^{1}-P\bar{3}$; $a=7.80$, $c=2.88$; $A=9$

Co-ordinates: 2Ga(C_3): $\frac{1}{2}\frac{2}{3}z$; $\frac{2}{3}\frac{1}{3}\bar{z}$; $z\approx\frac{1}{4}$

6Ag(C_1): $\pm(xyz)$; $\pm(\bar{y}, x-y, z)$; $\pm(y-x, \bar{x}, z)$; $x\approx\frac{1}{3}$; $y\approx\frac{1}{3}$; $z\approx\frac{1}{4}$

1(Ag, Ga)(C_{3i}): 000

Table 6.2 STRUCTURAL DETAILS—*continued*

$D0_c$ (U_3Si type)

Tetragonal: $D_{4h}^{18}-I4/mcm$; $a=6.02, 8.70$; $A=16$
Co-ordinates: $(000; \frac{1}{2}\frac{1}{2}\frac{1}{2}$ $+4U(D_4)$: $\pm(00\frac{1}{4})$
$+8U(C_{2v})$: $\pm(x, \frac{1}{2}+x, 0)$; $\pm(\frac{1}{2}+x, \bar{x}, 0)$; $x=0.231$
$+4Si(D_{2d})$: $0\frac{1}{2}\frac{1}{4}$; $\frac{1}{2}0\frac{1}{4}$

$D0_d$ (Mn_3As type)

Orthorhombic: $D_{2h}^{13}-Pmmn$; $a\approx b=3.78$, $c=16.26$; $A=16$
Co-ordinates: 3 times $2Mn(C_{2v})$: $00z$; $\frac{1}{2}\frac{1}{2}z$;
$2As(C_{2v})$: the same
3 times $2Mn(C_{2v})$: $0\frac{1}{2}z$; $\frac{1}{2}0\bar{z}$;
$2As\ (C_{2v})$: the same

	(00z)-position				*(0½z)-position*			
	Mn *I*	Mn *II*	Mn *III*	As *I*	Mn *IV*	Mn *V*	Mn *VI*	As *II*
z	0.194	−0.194	−0.435	0.409	0.307	−0.307	−0.066	0.091

$D1_3$ ($BaAl_4$ type)

Tetragonal: $D_{4h}^{17}-I4/mmm$; $a=4.53$, $c=11.14$; $A=10$
Co-ordinates: $(000; \frac{1}{2}\frac{1}{2}\frac{1}{2})$ $+2Ba(D_{4h})$: 000
$+4Al(C_{4v})$: $\pm(00z)$; $z=0.380$
$+Al(D_{2d})$; $0\frac{1}{2}\frac{1}{4}$; $\frac{1}{2}0\frac{1}{4}$

$D1_a$ ($MoNi_4$ type)

Tetragonal: $C_{4h}^{5}-I_4/m$; $a=5.73$, $c=3.55$; $A=10$
Co-ordinates: $(000, \frac{1}{2}\frac{1}{2}\frac{1}{2})$ $+2Mo(C_{4h})$: 000
$+8Ni(C_s)$: $\pm(xy0)$; $\pm(y\bar{x}0)$; $x=0.400$; $y=0.200$

$D1_b$ (UAl_4 type)

Orthorhombic: C_{2v}^{20} (or D_{2h}^{28})$-I2ma$ (*or Imma*); $a=4.41$, $b=6.27$, $c=13.71$; $A=20$
Co-ordinates: $(000; \frac{1}{2}\frac{1}{2}\frac{1}{2})+4U(C_{2v})$: $\pm(0\frac{1}{4}z)$; $z=0.111$
$+4Al(C_{2v})$: the same with $z=-0.111$
$+4Al(C_{2h})$: $000, 0\frac{1}{2}\frac{1}{2}$
$+8Al(C_s)$: $\pm(0yz)$; $\pm(0, \frac{1}{2}-y, z)$; $y=-0.033$; $z=0.314$

$D1_c$ ($PtSn_4$ type)

Orthorhombic: $C_{2v}^{17}-Aba2$; $a=6.38$, $b=6.41$, $c=11.33$; $A=20$
Co-ordinates: $(000; 0\frac{1}{2}\frac{1}{2})$ $+4Pt(C_2)$: $00z$; $\frac{1}{2}\frac{1}{2}z$; $z=0$
$+2$ times $8Sn(C_1)$: xyz; $\bar{x}\bar{y}z$; $(\frac{1}{2}+x)(\frac{1}{2}-y)z$; $(\frac{1}{2}-x)(\frac{1}{2}+y)z$

	x	*y*	*z*
Sn 1	0.173	0.327	0.125
Sn II	0.33	0.17	−0.13

$D1_d$ ($PtPb_4$ type)

Tetragonal: $D_{4h}^{3}-P4/nbm$; $a=6.65$, $c=5.97$; $A=10$
Co-ordinates: $2Pt(D_4)$: 000; $\frac{1}{2}\frac{1}{2}0$
$8Pb(C_s)$: $x(\frac{1}{2}+x)z$; $\bar{x}(\frac{1}{2}-x)z$; $(\frac{1}{2}+x)\bar{x}z$; $(\frac{1}{2}-x)xz$; $x(\frac{1}{2}-x)\bar{z}$; $\bar{x}(\frac{1}{2}+x)\bar{z}$; $(\frac{1}{2}+x)x\bar{z}$;
$(\frac{1}{2}-x)\bar{x}\bar{z}$; $x=0.175$; $z=0.255$

Table 6.2 STRUCTURAL DETAILS—*continued*

$D1_e$ (UB_4 type)

Tetragonal: D_{4h}^5-P4/mbm; $a=7.08$, $c=3.98$; $A=20$
Co-ordinates: $4U(C_{2v})$: $\pm(x, \tfrac{1}{2}+x, 0; \tfrac{1}{2}+x, \bar{x}, 0)$; $x=0.31$
$4B(C_4)$: $\pm(00z; \tfrac{1}{2}\tfrac{1}{2}z)$; $z=0.2$
$4B(C_{2v})$: $\pm(x, \tfrac{1}{2}+x, \tfrac{1}{2}; \tfrac{1}{2}+x, \bar{x}, \tfrac{1}{2})$; $x=0.1$
$4B(C_s)$: $\pm[xy\tfrac{1}{2}; y\bar{x}\tfrac{1}{2}; (\tfrac{1}{2}+x)(\tfrac{1}{2}-y)\tfrac{1}{2}; (\tfrac{1}{2}+y)(\tfrac{1}{2}+x)\tfrac{1}{2}]x=0.2$; $y=0.04$

$D1_f$ (Mn_4B type)

Orthorhombic: $D_{2h}^{24}-Fddd$; $a=14.53$, $b=7.29$, $c=4.21$; $A=40$
Co-ordinates: $(000; \tfrac{1}{2}\tfrac{1}{2}0; \circlearrowleft)+16Mn(C_2)$: $\pm(0y0)$; $\tfrac{1}{4}(\tfrac{1}{4}+y)\tfrac{1}{4}$; $\tfrac{1}{4}(\tfrac{1}{4}-y)\tfrac{1}{4}$; $y=0.333$
$+16Mn(C_2)$: $\pm(x00)$; $(\tfrac{1}{4}+x)\tfrac{1}{4}\tfrac{1}{4}$; $(\tfrac{1}{4}-x)\tfrac{1}{4}\tfrac{1}{4}$; $x=0.083$
$+8B(C_2)$: at random in the same position with $x=0.375$

$D1_g$ (B_4C type)

Rhombohedral: $D_{3d}^5-R\bar{3}m$; $a=5.19$; $\alpha=66°\ 18'$; $A=15$
(Hexagonal setting: $a=5.60$, $c=12.12$; $A=45$)
Co-ordinates: $(000; \tfrac{1}{3}\tfrac{2}{3}\tfrac{1}{3}; \tfrac{2}{3}\tfrac{1}{3}\tfrac{2}{3})+3C(D_{3d})$: $00\tfrac{1}{2}+6C(C_{3v})$: $\pm(00z)$; $z=0.385$
$+18B(C_1)$: $\pm(x\bar{x}z)$; $\pm(x, 2x, z)$; $\pm(2\bar{x}, \bar{x}, z)$; $x=0.106$; $z=0.113$
$+18B(C_1)$: the same with $x=\tfrac{1}{6}$; $z=0.360$

$D2_1$ (CaB_6 type)

Cubic: O_h^1-Pm3m; $a=4.15$; $A=7$
Co-ordinates: $1Ca(O_h)$: 000
$6B(C_{4v})$: $\pm(\tfrac{1}{2}\tfrac{1}{2}x; \circlearrowleft)$; $x=0.20$

$D2_3$ ($NaZn_{13}$ type)

Cubic: O_h^6-Fm3c; $a=12.27$; $A=112$
Co-ordinates: $(000; \tfrac{1}{2}\tfrac{1}{2}0; \circlearrowleft)+8Na(O)$: $\pm(\tfrac{1}{4}\tfrac{1}{4}\tfrac{1}{4})$
$+8Zn(T_h)$: $000; \tfrac{1}{2}\tfrac{1}{2}\tfrac{1}{2}$
$+96Zn(C_s)$: $\pm(0yz; \circlearrowleft; \tfrac{1}{2}zy; \circlearrowleft; 0y\bar{z}; \circlearrowleft; \tfrac{1}{2}\bar{z}y; \circlearrowleft$; $y=0.1806$;
$z=0.1192$

For $Be_{13}U$, neutron diffraction gave: $y=0.178$; $z=0.112$

$D2_a$ ($TiBe_{12}$ type)

Hexagonal: $a=29.44$, $c=7.33$; $A=624$ (637?)

The structure shows a particular type of disorder that cannot yet be explained in detail. The approximate structure of a pseudocell was determined (Raeuchle and Rundle[201]):

Hexagonal: D_{6h}^1-P6/mmm; $a=4.23$, $c=7.33$; $A=13$
Co-ordinates: $1Ti(D_{6h})$: in (000) or in $(00\tfrac{1}{2})$
$2Be(C_{6v})$: $\pm(00z)$; $z\approx\tfrac{1}{4}$
$2Be(D_{3d})$: $\tfrac{1}{3}\tfrac{2}{3}0; \tfrac{2}{3}\tfrac{1}{3}0$; $2Be(D_{3d})$: $\tfrac{1}{3}\tfrac{2}{3}\tfrac{1}{2}; \tfrac{2}{3}\tfrac{1}{3}\tfrac{1}{2}$
$6Be(C_{2v})$: $\pm(\tfrac{1}{2}0z)$; $\pm(0\tfrac{1}{2}z)$; $\pm(\tfrac{1}{2}\tfrac{1}{2}z)$

$D2_b$ ($ThMn_{12}$ type)

Tetragonal: $D_{4h}^{17}-I4/mmm$; $a=8.74$, 4.95; $A=26$
Co-ordinates: $(000; \tfrac{1}{2}\tfrac{1}{2}\tfrac{1}{2})+2Th(D_{4h})$: 000
$+8Mn(C_{2h})$: $\tfrac{1}{4}\tfrac{1}{4}\tfrac{1}{4}; \tfrac{3}{4}\tfrac{3}{4}\tfrac{1}{4}; \circlearrowleft$
$+8Mn(C_{2v})$: $\pm(x00)$; $\pm(0x0)$; $x=0.361$
$+8Mn(C_{2v})$: $\pm(x\tfrac{1}{2}0)$; $\pm(\tfrac{1}{2}x0)$; $x=0.277$

Table 6.2 STRUCTURAL DETAILS—*continued*

*D*2$_c$ (U_6Mn type)

Tetragonal: $D_{4h}^{18}-I4/mcm$ (or subgroup); $a=10.29$, $b=5.24$; $A=28$
Co-ordinates: $(000;\ \frac{1}{2}\frac{1}{2}\frac{1}{2})+4\mathrm{Mn}(D_4)$: $\pm(00\frac{1}{4})$
$+8\mathrm{U}(C_{2v})$: $\pm(x,\ \frac{1}{2}+x,\ 0);\ \pm(\frac{1}{2}+x,\ \bar{x},\ 0);\ x=0.405$
$+16\mathrm{U}(C_s)$: $\pm(xy0;\ y\bar{x}0;\ x\bar{y}\frac{1}{2};\ yx\frac{1}{2});\ x=0.213;\ y=0.103$

*D*2$_d$ ($CaCu_5$ type)

Hexagonal: $D_{6h}^{1}-C6/mmm$; $a=5.10$, $c=4.08$; $A=6$
Co-ordinates: $1\mathrm{Ca}(D_{6h})$: 000
$2\mathrm{Cu}(D_{3h})$: $\frac{1}{3}\frac{2}{3}0;\ \frac{2}{3}\frac{1}{3}0$; $3\mathrm{Cu}(D_{2h})$: $\frac{1}{2}0\frac{1}{2};\ 0\frac{1}{2}\frac{1}{2};\ \frac{1}{2}\frac{1}{2}\frac{1}{2}$

*D*2$_e$ ($BaHg_{11}$ type)

Cubic: O_h^1-Pm3m; $a=9.60$; $A=36$
Co-ordinates: $3\mathrm{Ba}(D_{4h})$: $\frac{1}{2}00$; ⤸
$1\mathrm{Hg}(O_h)$: 000; $8\mathrm{Hg}\ (C_{3v})$: $\pm(xxx;\ xx\bar{x};$ ⤸$);\ x=0.155$
$12\mathrm{Hg}(C_{2v})$: $\pm(xx0;$ ⤸ $);\ \pm(x\bar{x}0;$ ⤸ $);\ x=0.345$
$12\mathrm{Hg}(C_{2v})$: $\pm(xx\frac{1}{2};$ ⤸ $);\ \pm(x\bar{x}\frac{1}{2};$ ⤸ $);\ x=0.275$

*D*2$_f$ (UB_{12} type)

Cubic: O_h^5-Fm3m; $a=7.47$; $A=52$
Co-ordinates: $(000;\ \frac{1}{2}\frac{1}{2}0;)$ ⤸ $)+4\mathrm{U}(O_h)$: 000
$+48\mathrm{B}(C_{2v})$: $\pm(\frac{1}{2}xx;$ ⤸ $);\ \pm(\frac{1}{2}x\bar{x};$ ⤸ $);\ x=\frac{1}{6}$

*D*2$_g$ (Fe_8N type)

Tetragonal: $D_{4h}^{17}-I4/mmm$; $a=5.72$, $c=6.29$; $A=18$
Co-ordinates: $(000;\ \frac{1}{2}\frac{1}{2}\frac{1}{2})+2\mathrm{N}(D_{4h})$: 000
$+4\mathrm{Fe}(D_{2d})$: $\frac{1}{2}0\frac{1}{4};\ 0\frac{1}{2}\frac{1}{4}$
$+4\mathrm{Fe}(C_{2v})$: $\pm(00z);\ z=0.56$
$+8\mathrm{Fe}(C_{2v})$: $\pm(xx0);\ \pm(x\bar{x}0);\ x=\frac{1}{4}$

*D*5$_2$ (La_2O_3 type)

Hexagonal: $D_{3d}^{3}-P\bar{3}m1$; $a=3.93$, $c=6.12$, $c/a=1.56$; $A=5$
Co-ordinates: $2\mathrm{La}(C_{3v})$: $\frac{1}{3}\frac{2}{3}z;\ \frac{2}{3}\frac{1}{3}\bar{z};\ z\approx0.23$
$1\mathrm{O}(D_{3d})$: 000; $2\mathrm{O}(C_{3v})$: $\frac{1}{3}\frac{2}{3}z;\ \frac{2}{3}\frac{1}{3}\bar{z};\ z\approx0.63$

Apart from the different c/a-values and small differences in the z-values this type is similar to the *D*5$_{13}$ (Ni_2Al_3) type.

*D*5$_3$ (Mn_2O_3 type)

Cubic: T_h^7-Ia3; $a=9.41$; $A=80$
Co-ordinates: $(000;\ \frac{1}{2}\frac{1}{2}\frac{1}{2})+8\mathrm{Mn}(C_{3i})$: $\frac{1}{4}\frac{1}{4}\frac{1}{4};\ \frac{1}{4}\frac{3}{4}\frac{3}{4};$ ⤸;
$+24\mathrm{Mn}(C_2)$: $\pm(x0\frac{1}{4};$ ⤸ $);\ \pm(x\frac{1}{2}\frac{3}{4};$ ⤸ $);\ x=-0.030$
$+48\mathrm{O}(C_1)$: $\pm(xyz;$ ⤸ $);\ \pm(x,\ \bar{y};\ \frac{1}{2}-z;$ ⤸ $);\ \pm(\frac{1}{2}+x,\ \bar{y},\ z;$ ⤸ $);\ \pm(x,\ \frac{1}{2}+y,$
$\bar{z};$ ⤸ $);\ x=0.39;\ y=0.15;\ z=0.38$

*D*5$_8$ (Sb_2S_3 type)

Orthorhombic: $D_{2h}^{16}-Pbnm$; $a=11.20$, $b=11.28$, $c=3.83$; $A=20$
Co-ordinates: $4\mathrm{Sb}(C_s)$: $\pm(xy\frac{1}{4});\ \pm(\frac{1}{2}-x,\ \frac{1}{2}+y,\ \frac{1}{4});\ x=0.33;\ y=0.03$
$4\mathrm{Sb}(C_s)$: the same with $x=-0.04$; $y=-0.15$
$4\mathrm{S}(C_s)$: the same with $x=0.88$; $y=0.05$
$4\mathrm{S}(C_s)$: the same with $x=-0.44$; $y=-0.13$
$4\mathrm{S}(C_s)$: the same with $x=0.19$; $y=0.21$

Table 6.2 STRUCTURAL DETAILS—*continued*

For other substances:

	M *I*	M *II*	S *I*	S *II*	S *III*
U_2S_3, *x*	0.311	−0.008	0.878	−0.439	0.206
y	−0.014	−0.195	0.053	−0.129	0.230
Th_2S_3, *x*	0.314	−0.019	0.878	−0.439	0.206
y	−0.022	−0.200	0.053	−0.129	0.230

$D5_9$ (Zn_3P_2 type)

Tetragonal: $D_{4h}^{15}-P4/nmc$; $a=8.10$, $c=11.45$; $A=40$
Co-ordinates: $4P(C_{2v})$: $\pm(00z)$; $\pm(\frac{1}{2}, \frac{1}{2}, \frac{1}{2}+z)$; $z=0.25$
$4P(C_{2v})$: $0\frac{1}{2}z$; $\frac{1}{2}0\bar{z}$; $0\frac{1}{2}(\frac{1}{2}+z)$; $\frac{1}{2}0(\frac{1}{2}-z)$; $z=0.24$
$8P(C_s)$: $\pm(xx0$; $\bar{x}x0$; $\frac{1}{2}+x, \frac{1}{2}+x, \frac{1}{2}$; $\frac{1}{2}-x, \frac{1}{2}+x, \frac{1}{2})$; $x=0.26$
$8Zn(C_s)$: $0xz$; $0\bar{x}z$; $x0\bar{z}$; $\bar{x}0\bar{z}$; $\frac{1}{2}(\frac{1}{2}+x)\ (\frac{1}{2}-z)$; $\frac{1}{2}(\frac{1}{2}-x)\ (\frac{1}{2}-z)$; $(\frac{1}{2}+x)\frac{1}{2}(\frac{1}{2}+z)$; $(\frac{1}{2}-x)$
$\frac{1}{2}(\frac{1}{2}+z)$; $x=0.22$; $z=0.10$
$8Zn(Cs)$: the same with $x=0.28$; $z=0.39$
$8Zn(C_s)$: the same with $x=0.26$; $z=0.65$

$D5_{10}$(Cr_3C_2 type)

Orthorhombic: $D_{2h}^{16}-Pbnm$; $a=11.46$ $b=5.52$, $c=2.85$; $A=20$
Co-ordinates: $4C(C_s)$: $\pm(xy\frac{1}{4})$; $\pm(\frac{1}{2}-x, \frac{1}{2}+y, \frac{1}{4})$; $x=0.11$; $y=-0.10$
$4C(C_s)$: the same with $x=-0.06$; $y=0.22$
$4Cr(C_s)$: the same with $x=4.406$; $y=0.03$
$4Cr(C_s)$: the same with $x=-0.230$; $y=0.175$
$4Cr(C_s)$: the same with $x=-0.070$; $y=-0.150$

$D5_{13}$ (Ni_2Al_3 type)

Hexagonal: $D_{3d}^{3}-C\bar{3}m$; $a=4.03$, $c=4.89$, $c/a=1.21$; $A=5$
Co-ordinates: $2Ni(C_{3v})$: $\frac{1}{3}\frac{2}{3}z$; $\frac{2}{3}\frac{1}{3}z$; $z=0.149$
$1Al(D_{3d})$: 000; $2A1(C_{3v})$: $\frac{1}{3}\frac{2}{3}z$; $\frac{2}{3}\frac{1}{3}z$; $z=0.648$

Apart from the different c/a-values and small differences in the z-values, this type is similar to the $D5_2$ (La_2O_3) type.

For Ni_2Ga_3: $z(Ni)=0.138$; $z(Ga)=0.625$
For Ni_2In_3: $z(Ni)=0.135$; $z(In)=0.641$

$D5_a$ (U_3Si_2 type)

Tetragonal: $D_{4h}^{5}-P4/mbm$; $a=7.33$, $c=3.90$; $A=10$
Co-ordinates: $2U(C_{4h})$: 000; $\frac{1}{2}\frac{1}{2}0$
$4U(C_{2v})$: $\pm(x, \frac{1}{2}+x, \frac{1}{2})$; $\pm(\frac{1}{2}+x, \bar{x}, \frac{1}{2})$; $x=0.181$
$4Si(C_{2v})$: $\pm(x, \frac{1}{2}+x, 0)$; $\pm(\frac{1}{2}+x, \bar{x}, \frac{1}{2})$; $x = 0.389$

$D5_b$ (Pt_2Sn_3 type)

Hexagonal: $D_{6h}^{4}-P6/mmc$; $a=4.34$, $c=12.96$; $A=10$
Co-ordinates: $4Pt(C_{3v})$: $\pm(\frac{1}{3}\frac{2}{3}z)$; $\pm(\frac{1}{3}, \frac{2}{3}, \frac{1}{2}-z)$; $z=0.14$
$4Sn(C_{3v})$: the same with $z=-0.07$
$2Sn(D_{3h})$: $\pm(00\frac{1}{4})$

$D5_c$ (Pu_2C_3 type)

Cubic: $T_d^{6}-I\bar{4}3d$; $a=8.13$; $A=40$
Co-ordinates: $(000; \frac{1}{2}\frac{1}{2}\frac{1}{2})+16Pu(C_3)$: xxx; $(x+\frac{1}{4})\ (x+\frac{1}{4})\ (x+\frac{1}{4})$; $(\frac{1}{2}+x, \frac{1}{2}-x, \bar{x})$; ↻; $(\frac{3}{4}+x)$
$(\frac{1}{4}-x)\ (\frac{3}{4}-x)$; ↻; $x=0.050$
$+24C(C_2)$: $y0\frac{1}{4}$; ↻; $(\frac{1}{2}-y)0\frac{3}{4}$; ↻; $(\frac{3}{4}+y)0\frac{3}{4}$; ↻; $(\frac{3}{4}-y)0\frac{1}{4}$; ↻; $y=0.28$

Table 6.2 STRUCTURAL DETAILS—*continued*

*D*5$_e$ (Ni_3S_2 type)

Rhombohedral: D_3^7-R32; $a=4.08$; $\alpha=89°\ 25'$; $A=5$
Co-ordinates: $3Ni(C_2)$: $\frac{1}{2}x\bar{x}$; ⤸ ; $x=\frac{1}{4}$
$2S(C_3)$: $\pm(xxx)$; $x=\frac{1}{4}$

*D*5$_f$ (As_2S_3 type)

Monoclinic: $C_{5h}^2-P2_1/n$; $a=11.48$, $b=9.58$, $c=4.23$; $\beta=90°\ 27'$; $A=20$
Co-ordinates: $4As(C_1)$: $\pm(xyz)$; $\pm(\frac{1}{2}-x, \frac{1}{2}+y, \frac{1}{2}-z)$
All the other atoms in the same position.

	As *I*	As *II*	S *I*	S *II*	S *III*
x	0.268	0.482	0.410	0.340	0.125
y	0.187	0.313	0.120	0.380	0.305
z	0.161	−0.339	0.454	−0.046	0.455

*D*7$_1$ (Al_4C_3 type)

Rhombohedral: $D_{3d}^5-R\bar{3}m$; $a=8.53$; $\alpha=22°\ 28'$; $A=7$
(Hexagonal setting: $a=3.32$, $c=24.95$; $A=21$)
Co-ordinates: $1C(D_{3d})$: 000
$2C(C_{3v})$: $\pm(xxx)$; $x=0.217$
$2Al(C_{3v})$: the same with $x=0.293$
$2Al(C_{3v})$: the same with $x=0.128$

*D*7$_2$ (Co_3S_4 type)

Cubic: O_h^7-Fd3m; $a=9.38$; $A=56$
Co-ordinates: (000; $\frac{1}{2}\frac{1}{2}0$; ⤸)+$8Co(T_d)$: 000; $\frac{1}{4}\frac{1}{4}\frac{1}{4}$
+$16Co(D_{3d})$: $\frac{5}{8}\frac{5}{8}\frac{5}{8}$; $\frac{7}{8}\frac{7}{8}\frac{7}{8}$; ⤸
+$32S(C_{3v})$: xxx; $x\bar{x}\bar{x}$; ⤸ ; $(\frac{1}{4}-x)(\frac{1}{4}-x)(\frac{1}{4}-x)$; $(\frac{1}{4}-x)(\frac{1}{4}-x)(\frac{1}{4}+x)$;
⤸; $x=-0.135$

An ordered variety of this type is described as *H*1 (Spinel) type.

*D*7$_3$ (Th_3P_4 type)

Cubic: $T_d^6-I\bar{4}3d$; $a=8.60$; $A=28$ (or less: ≈ 26)
Co-ordinates: (000; $\frac{1}{2}\frac{1}{2}\frac{1}{2}$)+$12Th(S_4)$: $\frac{1}{4}\frac{3}{8}0$; ⤸ ; $\frac{3}{4}\frac{1}{8}0$; ⤸
+$16P(C_3)$: xxx; $(\frac{1}{4}+x)(\frac{1}{4}+x)(\frac{1}{4}+x)$; $(\frac{1}{2}+x)(\frac{1}{2}+x)\bar{x}$; ⤸; $(\frac{3}{4}+x)(\frac{1}{4}-x)$
$(\frac{3}{4}-x)$; ⤸ ; $x=\approx\frac{1}{12}=0.083$

In the compounds (La, Ce, Ac, Pu, Am)$_2$S$_3$, $10\frac{2}{3}$ metal ions occupy the Th positions at random.

*D*7$_a$ (Ni_3Sn_4 type)

Monoclinic: C_{2h}^3-C2/m; $a=12.2$, $b=4.08$, $c=5.22$; $\beta=105°$; $A=14$
Co-ordinates: (000; $\frac{1}{2}\frac{1}{2}0$)+$2Ni(C_{2h})$: 000
+$4Ni(C_s)$: $\pm(x0z)$; $x=0.220$; $z=0.350$
+$4Sn(C_s)$: the same with $x=0.428$; $z=0.675$
+$4Sn(C_s)$: the same with $x=0.180$; $z=0.800$

*D*7$_b$ (Ta_3B_4 type)

Orthorhombic: $D_{2h}^{25}-Immm$; $a=3.29$, $b=14.00$, $c=3.13$; $A=14$
Co-ordinates: (000; $\frac{1}{2}\frac{1}{2}\frac{1}{2}$)+$2Ta(D_{2h})$; $00\frac{1}{2}$
+$4Ta(C_{2v})$: $\pm(0y0)$; $y=0.180$
+$4B(C_{2v})$: the same with $y=0.375$
+$4B(C_{2v})$: $\pm(0y\frac{1}{2})$; $y=0.444$

Table 6.2 STRUCTURAL DETAILS—*continued*

$D8_1$ (Fe_3Zn_{10} type)

Cubic: O_h^9-Im3m; $a=8.98$; $A=52$
Co-ordinates: $(000;\ \tfrac{1}{2}\tfrac{1}{2}\tfrac{1}{2})+12\mathrm{Fe}(C_{4v})$: $\pm(x00;\ \circlearrowright)$; $x\approx\tfrac{1}{3}$
$+16\mathrm{Zn}(C_{3v})$: $\pm(xxx;\ x\bar{x}\bar{x};\ \circlearrowright)$; $x\approx\tfrac{1}{6}$
$+24\mathrm{Zn}(C_{2v})$: $\pm(xx0;\ \circlearrowright;\ x\bar{x}0;\ \circlearrowright)$; $x\approx\tfrac{1}{3}$

$D8_2$ (Cu_5Zn_8 type)

Cubic: $T_d^3-I\bar{4}3m$; $a=8.84$; $A=52$
Co-ordinates: $(000,\ \tfrac{1}{2}\tfrac{1}{2}\tfrac{1}{2})+12\mathrm{Cu}(C_{2v})$: $\pm(x00;\ \circlearrowright)$; $x=0.355\approx\tfrac{1}{3}$
$+8\mathrm{Cu}(C_{3v})$: $xxx;\ x\bar{x}\bar{x};\ \circlearrowright$; $x=-0.172\approx-\tfrac{1}{6}$
$+8\mathrm{Zn}(C_{3v})$: the same with $x=0.110\approx+\tfrac{1}{6}$
$+24\mathrm{Zn}(C_s)$: $xxz;\ \circlearrowright;\ x\bar{x}\bar{z};\ \circlearrowright;\ \bar{x}\bar{x}z;\ \circlearrowright;\ \bar{x}x\bar{z};\ \circlearrowright$; $x=0.313\approx\tfrac{1}{3}$;
$z=0.036\approx0$

$D8_3$ (Cu_9Al_4 type)

Cubic: $T_d^1-P\bar{4}3m$; $a=8.69$; $A=52$
Co-ordinates: $6\mathrm{Cu}(C_{2v})$: $\pm(x00;\ \circlearrowright)$; $x=0.356\approx\tfrac{1}{3}$
$6\mathrm{Cu}(C_{2v})$: $+(x\tfrac{1}{2}\tfrac{1}{2};\ \circlearrowright)$; $x=0.856\approx\tfrac{1}{2}+\tfrac{1}{3}$
$4\mathrm{Cu}(C_{3v})$: $xxx;\ x\bar{x}\bar{x};\ \circlearrowright$; $x=-0.172\approx-\tfrac{1}{6}$
$4\mathrm{Cu}(C_{3v})$: the same with $x=0.331\approx\tfrac{1}{2}-\tfrac{1}{6}$
$4\mathrm{Cu}(C_{3v})$: the same with $x=0.601\approx\tfrac{1}{2}+\tfrac{1}{6}$
$4\mathrm{Al}(C_{3v})$: the same with $x=0.112\approx\tfrac{1}{6}$
$12\mathrm{Al}(C_s)$: $xxz;\ \circlearrowright;\ x\bar{x}\bar{z};\ \circlearrowright;\ \bar{x}\bar{x}z;\ \circlearrowright;\ \bar{x}x\bar{z}$; $x=0.812\approx\tfrac{1}{2}+\tfrac{1}{3}$; $z=0.536\approx\tfrac{1}{2}$
$12\mathrm{Cu}(C_s)$: the same with $x=0.312\approx\tfrac{1}{3}$; $z=0.036\approx0$

$D8_4$ ($Cr_{23}C_6$ type)

Cubic: O_h^5-Fm3m; $a=10.64$; $A=116$
Co-ordinates: $(000;\ \tfrac{1}{2}\tfrac{1}{2}0;\ \circlearrowright)+4\mathrm{Cr}(O_h)$: 000
$+8\mathrm{Cr}(T_d)$: $\pm(\tfrac{1}{4}\tfrac{1}{4}\tfrac{1}{4})$
$+32\mathrm{Cr}(C_{3v})$: $\pm(xxx;\ x\bar{x}\bar{x};\ \circlearrowright)$; $x=0.385$
$+48\mathrm{Cr}(C_{2v})$: $\pm(xx0;\ \circlearrowright;\ x\bar{x}0;\ \circlearrowright)$; $x=0.165$
$+24\mathrm{C}(C_{4v})$: $\pm(x00;\ \circlearrowright)$; $x=0.275$

$D8_5$ (Fe_7W_6 type)

Rhombohedral: $D_{3d}^5-R\bar{3}m$; $a=9.02$; $\alpha=30°\ 31'$; $A=13$
(Hexagonal setting: $a=4.74$, $c=25.75$; $A=39$)
Co-ordinates: $1\mathrm{Fe}(D_{3d})$: 000
$6\mathrm{Fe}(C_s)$: $\pm(xxz;\ \circlearrowright)$; $x=0.09$; $z=0.59$
$2\mathrm{W}(C_{3v})$: $\pm(xxx)$; $x=\tfrac{1}{6}=0.167$
$2\mathrm{W}(C_{3v})$: the same with $x=0.346$
$2\mathrm{W}(C_{3v})$: the same with $x=0.448$

$D8_6$ ($Cu_{15}Si_4$ type)

Cubic: $T_d^6-I\bar{4}3d$; $a=9.69$; $A=76$
Co-ordinates: $(000;\ \tfrac{1}{2}\tfrac{1}{2}\tfrac{1}{2})+12\mathrm{Cu}(S_4)$: $0\tfrac{1}{4}\tfrac{3}{8};\ \circlearrowright;\ 0\tfrac{3}{4}\tfrac{1}{8};\ \circlearrowright$;
$+48\mathrm{Cu}(C_1)$: $xyz;\ \circlearrowright;\ x\bar{y}(\tfrac{1}{2}-z);\ \circlearrowright;\ (\tfrac{1}{2}-x)y\bar{z};\ \circlearrowright;\ \bar{x}(\tfrac{1}{2}-y)z;\ \circlearrowright;\ (\tfrac{1}{4}+y)(\tfrac{1}{4}+x)$
$(\tfrac{1}{4}+z);\ \circlearrowright;\ (\tfrac{1}{4}-y)(\tfrac{1}{4}+x)(\tfrac{3}{4}-z);\ \circlearrowright;\ (\tfrac{1}{4}+y)(\tfrac{3}{4}-x)(\tfrac{1}{4}-z);\ \circlearrowright;\ (\tfrac{3}{4}-y)(\tfrac{1}{4}-x)(\tfrac{1}{4}+z);\ \circlearrowright$;
$x=0.12$; $y=0.16$; $z=0.04$
$+16\mathrm{Si}(C_3)$: $xxx;\ x\bar{x}(\tfrac{1}{2}-x);\ \circlearrowright;\ (\tfrac{1}{4}+x)(\tfrac{1}{4}+x)(\tfrac{1}{4}+x);\ (\tfrac{1}{4}-x)(\tfrac{1}{4}+x)(\tfrac{3}{4}-x);\ \circlearrowright$; $x=0.208$

Table 6.2 STRUCTURAL DETAILS—*continued*

D8_8 (Mn_5Si_3 type)

Hexagonal: $D^3_{6h}-P6_3/mcm$; $a=6.90$, $c=4.80$; $A=16$
Co-ordinates: $4\text{Mn}(D_3)$: $\frac{1}{3}\frac{2}{3}0$; $\frac{2}{3}\frac{1}{3}0$; $\frac{1}{3}\frac{2}{3}\frac{1}{2}$; $\frac{2}{3}\frac{1}{3}\frac{1}{2}$
$6\text{Mn}(C_{2v})$: $\pm(x0\frac{1}{4};\ 0x\frac{1}{4};\ \bar{x}\bar{x}\frac{1}{4})$; $x=0.23$
$6\text{Si}(C_{2v})$: the same with $x=0.60$
For Mg_5Hg_3: $x(\text{Mg})=0.25$; $x(\text{Hg})=0.615$

D8_9 (Co_9S_8 type)

Cubic: O^5_h-Fm3m; $a=9.91$; $A=68$
Co-ordinates: $(000;\ \frac{1}{2}\frac{1}{2}0;\ \circlearrowright)+4\text{Co}(O_h)$: $\frac{1}{2}\frac{1}{2}\frac{1}{2}$
$+32\text{Co}(C_{3v})$: $\pm(xxx;\ x\bar{x}\bar{x};\ \circlearrowright)$; $x\approx\frac{1}{8}$
$+8\text{S}(T_d)$: $\pm(\frac{1}{4}\frac{1}{4}\frac{1}{4})$
$+24\text{S}(C_{4v})$: $\pm(x00;\ \circlearrowright)$; $x\approx\frac{1}{4}$

D8_{10} (Cr_5Al_8 type)

Rhombohedral: $C^5_{3v}-R3m$; $a=7.79$; $\alpha=109°\ 8'$; $A=26$
(Hexagonal setting: $a=12.70$, $c=7.90$; $A=78$)
Co-ordinates: $1\text{Cr}(C_{3v})$: xxx; $x=0.097$
$3\text{Cr}(C_s)$: xxz; $\circlearrowright$; $x=-0.103$; $z=0.106$
$3\text{Cr}(C_s)$: the same with $x=0.170$; $z=-0.172$
$3\text{Cr}(C_s)$: the same with $x=0.003$; $z=0.352$
$1\text{Al}(C_{3v})$: xxx; $x=-0.164$
$3\text{Al}(C_s)$: xxz; $\circlearrowright$; $x=0.006$; $z=-0.352$
$3\text{Al}(C_s)$: the same with $x=0.291$; $z=0.058$
$3\text{Al}(C_3)$: the same with $x=-0.322$; $z=0.044$
$6\text{Al}(C_1)$: xyz; $\circlearrowright$; xzy; $\circlearrowright$; $x=0.330$; $y=-0.297$; $z=-0.042$

D8_{11} (Co_2Al_5 type)

Hexagonal: $D^4_{6h}-P6_3/mmc$; $a=7.66$, $c=7.59$; $A=28$
Co-ordinates: $2\text{Co}(D_{3h})$: $\frac{2}{3}\frac{1}{3}\frac{1}{4}$; $\frac{1}{3}\frac{2}{3}\frac{3}{4}$
$6\text{Co}(C_{2v})$: $\pm(x,\ 2x,\ \frac{1}{4};\ 2\bar{x},\ \bar{x},\ \frac{1}{4};\ x\bar{x}\frac{1}{4})$; $x=0.128$
$2\text{Al}(D_{3d})$: 000; $00\frac{1}{2}$
$6\text{Al}(C_{2v})$: $\pm(x,\ 2x,\ \frac{1}{4};\ 2\bar{x},\ \bar{x},\ \frac{1}{4};\ x\bar{x}\frac{1}{4})$; $x=0.467$
$12\text{Al}(C_s)$: $\pm(2x,\ x,\ z;\ \bar{x},\ 2\bar{x},\ z;\ \bar{x}xz;\ 2\bar{x},\ \bar{x},\ \frac{1}{2}+z;\ x,\ 2x,\ \frac{1}{2}+z;\ x,\ \bar{x},\ \frac{1}{2}+z)$; $x=0.196$; $z=0.061$

D8_a (Th_6Mn_{23} type)

Cubic: O^5_h-Fm3m; $a=12.52$; $A=116$
Co-ordinates: $(000;\ \frac{1}{2}\frac{1}{2}0;\ \circlearrowright)+24\text{Th}(C_{4v})$: $\pm(x00;\ \circlearrowright)$; $x=0.207$
$+4\text{Mn}(O_h)$: $\frac{1}{2}\frac{1}{2}\frac{1}{2}$
$+24\text{Mn}(D_{2h})$: $\frac{1}{4}\frac{1}{4}0$; $\circlearrowright$; $\frac{1}{4}\frac{3}{4}0$; $\circlearrowright$
$+32\text{Mn}(C_{3v})$: $\pm(xxx;\ x\bar{x}\bar{x};\ \circlearrowright)$; $x=0.378$
$+32\text{Mn}(C_{3v})$: the same with $x=0.178$

D8_b (σ type, as for example: V_3Ni_2)

(a) According to Bergman and Shoemaker[202]:
Tetragonal: $D^{14}_{4h}-P4/mnm$; for σ-FeCr: $a=8.80$, $c=4.55$; $A=30$
Co-ordinates: 2 atoms $A(D_{2h})$: 000; $\frac{1}{2}\frac{1}{2}\frac{1}{2}$
4 atoms $B(C_{2v})$: $\pm(xx0)$; $\pm(\frac{1}{2}+x,\ \frac{1}{2}-x,\ \frac{1}{2})$; $x=\frac{2}{5}=0.400$
8 atoms $C(C_s)$: $\pm(xy0)$; $\pm(yx0)$; $\pm(\frac{1}{2}+x,\ \frac{1}{2}-y,\ \frac{1}{2})$; $\pm(\frac{1}{2}+y,\ \frac{1}{2}-x,\ \frac{1}{2})$; $x=\frac{7}{15}=0.468$; $y=\frac{2}{15}=0.134$
8 atoms $D(C_s)$: the same with $x=\frac{11}{15}=0.735$; $y=\frac{1}{15}=0.067$
8 atoms $E(C_s)$: $\pm(xxz)$; $\pm(xx\bar{z})$; $\pm(\frac{1}{2}+x,\ \frac{1}{2}-x,\ \frac{1}{2}+z)$; $\pm(\frac{1}{2}+x,\ \frac{1}{2}-x,\ \frac{1}{2}-z)$; $x=\frac{11}{16}=0.183$; $z=\frac{1}{4}=0.250$

Table 6.2 STRUCTURAL DETAILS—*continued*

The space groups $C_v^4 - P4nm$ and $D_{2d}^8 - P\bar{4}n2$ could not be ruled out.

(b) According to Kasper, Decker and Belanger,[203] who investigated σ-CoCr: the same space group; $a=8.75$, $c=4.54$; $A=30$.
Co-ordinates: 2 atoms $A(D_{2h})$: $00\frac{1}{2}$; $\frac{1}{2}\frac{1}{2}0$
4 atoms B as under (a) but $x=0.100$
8 atoms C as under (a) but $x=0.373$; $y=0.027$
8 atoms D as under (a) but $x=0.573$; $y=0.227$
8 atoms E as under (a) but $x=0.300$; $z=\frac{1}{4}$

(c) Pearson and Christian,[204] who investigated σ-NiV, found best agreement between observed and calculated intensities when those atomic positions were used which were proposed by Tucker[15] for β-uranium [description as A_b (β-U type) on page **6**–38]; Tetragonal: C_{4v}^4–$P4nm$; for σ-NiV: $a=8.97$, $c=4.64$; $A=30$. An ordered arrangement is proposed with Ni taking the positions in which $z\approx 0$ or $\frac{1}{4}$ and with V taking the positions in which $z\approx\frac{1}{2}$ or $\frac{3}{4}$.

*D*8$_c$ ($Mg_2Cu_6Al_5$ type)

Cubic: $T_h^1 - Pm3$; $a=8.31$; $A=39$
Co-ordinates: 6Mg(C_{2v}): $\pm(x0\frac{1}{2}$; ↻); $x=0.32$
6Cu(or Zn I)(C_{2v}): ↻ $\pm(x00$; ↻); $x=0.225$
12Cu(or Zn II)(C_s): $\pm(\frac{1}{2}yz$; ↻); $\pm(\frac{1}{2}y\bar{z}$; ↻); $y=0.243$; $z=0.336$
1Al(or Zn III)(T_h): $\frac{1}{2}\frac{1}{2}\frac{1}{2}$
6Al(or Zn IV)(C_{2v}): $\pm(x\frac{1}{2}0$; ↻); $x=0.16$
8Al(or Zn V)(C_3): $\pm(xxx$; $x\bar{x}\bar{x}$; ↻); $x=0.215$
For Mg_2Zn_{11}: x(Zn I)=0.235; y(Zn II)=0.243; z(Zn II)=0.343; x(Zn IV)=0.160; x(Zn V)=0.222.

*D*8$_d$ (Co_2Al_9)

Monoclinic: $C_{2h}^5 - P2_1/a = 8.56$, $b=6.29$, $c=6.21$; $\beta=94°\ 46'$; $A=22$
Co-ordinates: 2Al I(C_i)=000; $0\frac{1}{2}\frac{1}{2}$
4 times Al, II, Al III, Al IV, Al V and Co in: $\pm(yyz)$; $\pm(\frac{1}{2}+x, \frac{1}{2}-y, z)$

	Co	Al *II*	Al *III*	Al *IV*	Al *V*
x	0.3335	0.2682	0.2309	0.9986	0.0417
y	0.6149	0.9619	0.2899	0.1931	0.6148
z	0.2646	0.4044	0.0889	0.3891	0.2159

*D*8$_e$ ($Mg_{32}X_{49}$ type)

Cubic: $T_h^5 - Im3$; $a=14.16$; $A=162$
Co-ordinates: $(000; \frac{1}{2}\frac{1}{2}\frac{1}{2})$ + 12MG(C_{2v}): $\pm(x0\frac{1}{2}$; ↻); $x=0.605$
+12Mg(C_{2v}): ±the same with $x=0.185$
+16Mg(C_3): $\pm(xxx$; $x\bar{x}\bar{x}$; ↻); $x=0.185$
+24Mg(C_s): $\pm(0yz$; ↻ ; $0y\bar{z}$; ↻); $y=0.300$; $z=0.115$
+2Al(T_h): 000
+24 atoms (83% Zn; 17% Al)(C_s): $\pm(0yz$; ↻ ; $0y\bar{z}$; ↻); $y=0.097$; $z=0.157$
+24 atoms (44% Zn; 56% Al)(C_s): the same with $y=0.195$; $z=0.310$
+48 atoms (48% Zn; 52% Al) (C_s): $\pm(xyz$; ↻ ; $x\bar{y}\bar{z}$; ↻ ; $\bar{x}y\bar{z}$; ↻ ; $\bar{x}\bar{y}z$; ↻); $x=0.160$; $y=0.190$; $z=0.400$

*D*8$_f$ (Ir_3Sn_7 type)

Cubic: $O_h^9 - Im3m$; $a=9.36$; $A=40$
Co-ordinates: $(000; \frac{1}{2}\frac{1}{2}\frac{1}{2})$ + 12Tr(C_{4v}): $\pm(x00$; ↻); $x=0.342$
+12Sn(D_{2d}): $\pm(\frac{1}{4}0\frac{1}{2})$, ↻)
+16Sn(C_{3v}): $(xxx$; $x\bar{x}\bar{x}$; ↻); $x=0.156$

Table 6.2 STRUCTURAL DETAILS—*continued*

D8_g (Mg_5Ga_2 type)

Orthorhombic: $D_{2h}^{26}-Ibam$; $a=13.72$, $b=7.00$, $c=6.02$; $A=28$
Co-ordinates: $(000; \frac{1}{2}\frac{1}{2}\frac{1}{2})+8Ga(C_s)$: $\pm(xy0)$; $\pm(x\bar{y}\frac{1}{2})$; $x=0.122$; $y=0.262$
$+8Mg(C_s)$: the same with $x=0.080$; $y=0.660$
$+8Mg(C_2)$: $\pm(x0\frac{1}{4})$; $\pm(\bar{x}0\frac{1}{4})$; $x=0.242$
$+4Mg(D_2)$: $\pm(00\frac{1}{4})$

[Unpublished work by E, Hellner; for Mg_5Tl_2: $x(Tl)\approx\frac{1}{8}$; $y(Tl)\approx\frac{1}{4}$]

D8_h (W_2B_5 type)

Hexagonal: $D_{6h}^{4}-P6_3/mmc$; $a=2.98$. $c=13.87$; $A=14$
Co-ordinates: $4W(C_{3v})$: $\pm(\frac{1}{3}\frac{2}{3}z)$; $\pm(\frac{1}{3},\frac{2}{3},\frac{1}{2}-z)$; $z=0.139$
$2B(D_{3h})$: $\pm(00\frac{1}{4})$; $+2B(D_{3h})$: $\pm(\frac{1}{3}\frac{2}{3}\frac{3}{4})$
$2B(D_{3d})$: 000; $00\frac{1}{2}$
$4B(C_{3v})$: $\pm(\frac{1}{3}\frac{2}{3}z)$; $\pm(\frac{1}{3},\frac{2}{3},\frac{1}{2}-z)$; $z=-0.028$

D8_i (Mo_2B_5 type)

Rhombohedral: $D_{3d}^{5}-R\bar{3}m$; $a=7.19$; $\alpha=24°\ 10'$; $A=7$
(Hexagonal setting: $a=3.01$, $c=20.93$; $A=21$)
Co-ordinates (for hex. setting): $(000; \frac{1}{3}\frac{2}{3}\frac{1}{3}; \frac{2}{3}\frac{1}{3}\frac{2}{3})+6Mo(C_{3v})$: $\pm(00z)$; $z=0.075$
$+6B(C_{3v})$: the same with $z=\frac{1}{3}$
$+6B(C_{3v})$: the same with $z=0.186$
$+3B(D_{3d})$: $00\frac{1}{2}$

D8_k (Th_7S_{12} type)

Hexagonal: $C_{6h}^{2}-P6_3/m$; $a=11.04$, $c=3.98$; $A=19$
Co-ordinates: $1Th(S_6)$: $\pm(00\frac{1}{4})$
$6Th(C_s)$: $\pm(xy\frac{1}{4})$; $\pm(\bar{y},x-y,\frac{1}{4})$; $\pm(y-x,\bar{x},\frac{1}{4})$; $x=0.153$; $y=-0.283$
$6S(C_s)$: the same with $x=0.514$; $y=0.375$
$6S(C_s)$: the same with $x=0.235$; $y=0\pm0.010$

E0_7 (FeAsS type)

Monoclinic: $C_{2h}^{5}-B2_1/d$; $a=9.51$, $b=5.65$, $c=6.42$; $\beta\approx90°$; $A=24$
Co-ordinates: $(000; \frac{1}{2}0\frac{1}{2})+8Fe(C_1)$: $\pm(xyz)$; $\pm(\frac{3}{4}+x,\frac{1}{2}-y,\frac{3}{4}+z)$; $x=0$; $y=0$; $z=0.275$
$+8As(C_1)$: the same with $x=0.147$; $y=0.128$; $z=0$
$+8S(C_1)$: the same with $x=0.167$; $y=0.132$; $z=0.500$

E1_1 ($CuFeS_2$ type)

Tetragonal: $D_{2d}^{12}-I\bar{4}2d$; $a=5.24$, $c=10.30$; $A=16$
Co-ordinates: $(000; \frac{1}{2}\frac{1}{2}\frac{1}{2})+4Cu(S_4)$: 000; $\frac{1}{2}0\frac{1}{4}$
$+4Fe(S_4)$: $00\frac{1}{2}$; $\frac{1}{2}0\frac{3}{4}$
$+8S(C_2)$: $\frac{1}{4}y\frac{1}{8}$; $\frac{3}{4}\bar{y}\frac{1}{8}$; $y\frac{3}{4}\frac{7}{8}$; $\bar{y}\frac{1}{4}\frac{7}{8}$; $y=0.27$

E1_a ($MgCuAl_2$ type)

Orthorhombic: $D_{2h}^{17}-Cmcm$; $a=4.00$, $b=9.28$, $c=7.14$; $A=16$
Co-ordinates: $(000; \frac{1}{2}\frac{1}{2}0)+4Mg(C_{2v})$: $\pm(0y\frac{1}{4})$; $y=0.072$
$+4Cu(C_{2v})$: $\pm(0y\frac{1}{4})$; $y=-0.222$
$+8Al(C_s)$: $\pm(0yz)$; $\pm(0,y,\frac{1}{2}-z)$; $y=0.356$; $z=0.056$

Table 6.2 STRUCTURAL DETAILS—*continued*

$E1_b$ [$AuAgTe_4$ (Sylvanite) type]

Monoclinic: C_{2h}^4-P2/c; $a=8.96$, $b=4.49$, $c=14.62$; $\beta=145°\ 26'$; $A=12$
Co-ordinates: $2Au(C_i)$: 000; $00\frac{1}{2}$
$2Ag(C_2)$: $\pm(0y\frac{1}{4})$; $y=0.433$
$4Te(C_1)$: $\pm(xyz)$; $\pm(x, \bar{y}, \frac{1}{2}+z)$; $x=0.298$; $y=0.031$; $z=0.999$
$4Te(C_1)$: the same with $x=0.277$; $y=0.425$; $z=0.235$

$E9_3$ (Fe_3W_3C type)

Cubic: O_h^7-Fd3m; $a=11.04$; $A=112$
Co-ordinates: $(000; \frac{1}{2}\frac{1}{2}0; \circlearrowright) + 16Fe(D_{3d})$: $\frac{5}{8}\frac{5}{8}\frac{5}{8}$; $\frac{5}{8}\frac{7}{8}\frac{7}{8}$; $\circlearrowright$
$+32Fe(C_{3v})$: xxx; $x\bar{x}\bar{x}$; $\circlearrowright$; $(\frac{1}{4}-x)(\frac{1}{4}-x)(\frac{1}{4}-x)$; $(\frac{1}{4}-x)(\frac{1}{4}+x)(\frac{1}{4}+x)$; $\circlearrowright$; $x=0.175$
$+48W(C_{2v})$: $\pm(x00; \circlearrowright)$; $(\frac{1}{4}+x)\frac{1}{4}\frac{1}{4}$; $\circlearrowright$; $(\frac{1}{4}-x)\frac{1}{4}\frac{1}{4}$; $x=0.195$
$+16C(C_{3d})$: $\frac{1}{8}\frac{1}{8}\frac{1}{8}$; $\frac{1}{8}\frac{3}{8}\frac{3}{8}$; $\circlearrowright$

$E9_4$ (Al_5C_3N type)

Hexagonal: $C_{6v}^4-P6_3mc_3$; $a=3.28$, $c=21.55$; $A=18$
Co-ordinates: $2Al(C_{3v})$: $00z$; $00(\frac{1}{2}+z)$; $z=0.150$
$2Al(C_{3v})$: the same with $z=0.345$
$2Al(C_{3v})$: $\frac{1}{3}\frac{2}{3}z$; $\frac{2}{3}\frac{1}{3}(\frac{1}{2}+z)$; $z=0.045$
$2Al(C_{3v})$: the same with $z=0.456$
$2Al(C_{3v})$: the same with $z=0.240$
$2C(C_{3v})$ the same with $z=0.133$
$2C(C_{3v})$: the same with $z=0.369$
$2C(C_{3v})$: $00z$; $00(\frac{1}{2}+z)$; $z=0.001$
$2N(C_{3v})$: the same with $z=0.250$

$E9_a$ ($FeCu_2Al_7$ type)

Tetragonal: $D_{4h}^6-P4mnc$; $a=6.32$, $c=14.78$; $A=40$
Co-ordinates: $4Fe(C_4)$: $\pm(00z)$; $\pm(\frac{1}{2}, \frac{1}{2}, \frac{1}{2}+z)$; $z=0.300$
$8Cu(C_s)$: $\pm(xy0)$; $\pm(\frac{1}{2}+x, \frac{1}{2}-y, \frac{1}{2})$; $\pm(\bar{y}x0)$; $\pm(\frac{1}{2}+y, \frac{1}{2}+x, \frac{1}{2})$; $x=0.278$; $y=0.092$
$4Al(C_4)$: $\pm(00z)$; $\pm(\frac{1}{2}, \frac{1}{2}, \frac{1}{2}+z)$; $z=0.122$
$8Al(C_2)$: $\pm(x, \frac{1}{2}+x, \frac{1}{4}; \bar{x}, \frac{1}{2}-x, \frac{1}{4}; \frac{1}{2}-x, x, \frac{1}{4})$; $x=0.167$
$16Al(C_1)$: $\pm(xyz; \bar{x}\bar{y}z; \bar{y}xz; y\bar{x}z; \frac{1}{2}+x, \frac{1}{2}-y, \frac{1}{2}+z; \frac{1}{2}-x, \frac{1}{2}+y, \frac{1}{2}+z; \frac{1}{2}+y, \frac{1}{2}+x, \frac{1}{2}+z; \frac{1}{2}-y, \frac{1}{2}-x, \frac{1}{2}+z)$; $x=0.203$; $y=0.414$; $z=0.100$

$E9_b$ ($FeMg_3Al_8Si_6$ type)

Hexagonal $D_{3h}^3-P\bar{6}2m$; $a=6.62$, $c=7.92$; $A=18$
Co-ordinates: $1Fe(D_{3h})$: 000
$3Mg(C_{2v})$: $x0\frac{1}{2}$; $0x\frac{1}{2}$; $\bar{x}\bar{x}\frac{1}{2}$; $x=0.445$
$1Al(D_{3h})$: $00\frac{1}{2}$
$3Al(C_{2v})$: $x00$; $0x0$; $\bar{x}\bar{x}0$; $x=0.403$
$4Al(C_3)$: $\pm(\frac{1}{3}\frac{2}{3}z)$; $\pm(\frac{1}{3}\frac{2}{3}\bar{z})$; $z=0.231$
$6Si(C_s)$: $x0z$; $x0\bar{z}$; $0xz$; $0x\bar{z}$; $\bar{x}\bar{x}z$; $\bar{x}\bar{x}\bar{z}$; $x=0.750$; $z=0.223$

$E9_c$ (Mn_3Al_9Si type)

Hexagonal: $D_{4h}^6-P6_3/mmc$; $a=7.51$, $c=7.74$; $A=26$
Co-ordinates: $6Mn(C_{2v})$: $\pm(x, 2x, \frac{1}{4})$; $\pm(2x, x, \frac{3}{4})$; $\pm(x\bar{x}\frac{1}{4})$; $x=0.120$
$6Al(C_{2v})$: the same with $x=0.458$
$12Al(C_s)$: $\pm(x, 2x, z; 2x, x, \bar{z}; x\bar{x}z; x, 2x, \frac{1}{2}-z; 2x, x, \frac{1}{2}+z; x, \bar{x}, \frac{1}{2}-z)$; $x=0.201$; $z=-0.067$
$2Si(D_{3d})$: 000; $00\frac{1}{2}$

Table 6.2 STRUCTURAL DETAILS—*continued*

*E*9$_d$ ($AlLi_3N_2$ type)

Cubic: $T_h^7 - Ia3$; $a = 9.48$; $A = 96$
Co-ordinates: $(000; \frac{1}{2}\frac{1}{2}\frac{1}{2}) + 16Al(C_3)$: $\pm(xxx)$; $\pm(\frac{1}{2}+x, \frac{1}{2}-x, \bar{x}; \circlearrowleft)$; $x = 0.115$
$+48Li(C_1)$: $\pm(xyz; \circlearrowleft)$; $\pm(x, \bar{y}, \frac{1}{2}-z; \circlearrowleft)$; $\pm(\frac{1}{2}-x, y, \bar{z}; \circlearrowleft)$; $\pm(\bar{x}, \frac{1}{2}-y, z; \circlearrowleft)$; $x = 0.160$; $y = 0.382$; $z = 0.110$
$+8N(C_{3i})$: $000; \frac{1}{2}\frac{1}{2}0; \circlearrowleft$
$+24N(C_2)$: $\pm(x0\frac{1}{4}; \circlearrowleft)$; $\pm(\bar{x}\frac{1}{2}\frac{1}{4}; \circlearrowleft)$; $x = 0.205$
For $GaLi_3N_2$: $x(Ga) = 0.117$; $x(Li) = 0.152$; $y(Li) = 0.381$; $z(Li) = 0.114$; $x(N) = 0.215$

*E*9$_e$[$CuFe_2S_3$ (Cubanite) type]

Orthorhombic: $D_{2h}^{16} - Pnma$; $a = 6.23$, $b = 11.12$, $c = 6.46$; $A = 24$
Co-ordinates: $4Cu(C_s)$: $\pm(x\frac{1}{4}z)$; $\pm(\frac{1}{2}-x, \frac{3}{4}, \frac{1}{2}+z)$; $x = \frac{1}{8}$; $z = \frac{7}{12}$
$8Fe(C_1)$: $\pm(xyz; \frac{1}{2}+x, \frac{1}{2}-y, \frac{1}{2}-z; \bar{x}, \frac{1}{2}+y, \bar{z}; \frac{1}{2}-x, \bar{y}, \frac{1}{2}+z)$; $x = \frac{1}{8}$; $y = \frac{1}{12}$; $z = \frac{1}{12}$
$8S(C_1)$: the same with $x = \frac{1}{4}$; $y = \frac{1}{12}$; $z = \frac{5}{12}$
$4S(C_s)$: $\pm(x\frac{1}{4}z)$; $\pm(\frac{1}{2}-x, \frac{3}{4}, \frac{1}{2}+z)$; $x = \frac{1}{4}$; $z = \frac{11}{12}$

*F*0$_1$ [NiSbS (Ullmannite) type]

Cubic: $T^4 - P2_13$; $a = 5.60$; $A = 12$
Co-ordinates: $4Ni(C_3)$: xxx; $(\frac{1}{2}+x)(\frac{1}{2}-x)\bar{x}$; $\circlearrowleft$; $x \approx 0$
$4Sb(C_3)$: the same with $x \approx 0.385$
$4S(C_3)$: the same with $x \approx 0.615$

*F*5$_1$ ($NaHF_2$ type)

Rhombohedral: $D_{3d}^5 - R\bar{3}m$; $a = 5.05$; $\alpha = 40° \, 2'$; $A = 4$
(Hexagonal setting: $a = 3.45$, $c = 13.90$, $c/a = 4.03$; $A = 12$)
Co-ordinates: $1Na(D_{3d})$: 000; $1H(D_{3d}) = \frac{1}{2}\frac{1}{2}\frac{1}{2}$
$2F(C_{3v})$: $\pm(xxx)$; $x = 0.410$
For $CaCN_2$: $\alpha = 43° \, 50'$; $c/a = 3.63$; $x = 0.37$
For $NaCrS_2$: $\alpha = 29° \, 48'$; $c/a = 5.59$; $x = 0.236$
For $NaCrSe_2$: $\alpha = 30° \, 18'$; $c/a = 5.49$; $x = 0.235$
For $RbCrSe_2$: $\alpha = 21° \, 33'$; $c/a = 7.85$

*F*5$_6$ ($CuSbS_2$ type)

Orthorhombic: $D_{2h}^{16} - Pnma$; $a = 6.01$, $b = 3.78$, $c = 14.46$; $A = 16$
Co-ordinates: $4Cu(C_s)$: $\pm(x\frac{1}{4}z)$; $\pm(\frac{1}{2}+x, \frac{1}{4}, \frac{1}{2}-z)$; $x = 0.25$; $z = 0.83$
$4Sb(C_s)$: the same with $x = 0.23$; $z = 0.06$
$4S(C_s)$: the same with $x = 0.63$; $z = 0.10$
$4S(C_s)$: the same with $x = 0.88$; $z = 0.83$

*F*5$_a$ ($KFeS_2$ type)

Monoclinic: $C_{2h}^6 - C2/c$; $a = 7.05$, $b = 11.28$, $c = 5.40$; $\beta = 112° \, 30'$; $A = 16$
Co-ordinates: $(000; \frac{1}{2}\frac{1}{2}0) + 4K(C_2)$: $\pm(0y\frac{1}{4})$; $y = 0.355$
$+Fe(C_2)$: the same with $y = -0.008$
$+8S(C_1)$: $\pm(xyz)$; $\pm(x, \bar{y}, \frac{1}{2}+z)$; $x = 0.195$; $y = 0.111$; $z = 0.10$

*H*1$_1$ [Spinel (Al_2MgO_4) type]

Cubic: $O_h^7 - Fd3m$; $a = 8.06$; $A = 56$
Co-ordinates: $(000; \frac{1}{2}\frac{1}{2}0; \circlearrowleft) + 8Mg(I_d)$: 000; $\frac{1}{4}\frac{1}{4}\frac{1}{4}$
$+16Al(D_{3d})$: $\frac{5}{8}\frac{5}{8}\frac{5}{8}$; $\frac{5}{8}\frac{7}{8}\frac{7}{8}$; $\circlearrowleft$
$+32O(C_{3v})$: xxx; $x\bar{x}\bar{x}$; $\circlearrowleft$; $(\frac{1}{4}-x)(\frac{1}{4}-x)(\frac{1}{4}-x)$; $(\frac{1}{4}-x, \frac{1}{4}+x, \frac{1}{4}+x)$; $\circlearrowleft$; $x \approx -\frac{1}{8}$

In some compounds, better agreement with observed intensities is obtained by assuming that the metal atoms are distributed at random among the 24 available sites, or that the trivalent element occupies all the 8-equivalent sites and half of the 16-equivalent ones. In some cases lattice sites may be vacant, e.g., γ-Al_2O_3 or In_2S_3.

Table 6.2 STRUCTURAL DETAILS—*continued*

*H*2$_4$[Cu_3VS_4 (Sulvanite) type]

Cubic: $T_d^1 - P\bar{4}3m$; $a = 5.37$; $A = 8$
Co-ordinates: 3Cu(D_{2d}): $\frac{1}{2}00$; $0\frac{1}{2}0$; $00\frac{1}{2}$
1V(T_d): 000
4S(C_{3v}): xxx; $x\bar{x}\bar{x}$; ⟲ ; $x = 0.235$

*H*2$_6$ [Stannite ($FeCu_2SnS_4$) type]

Tetragonal: $D_{2d}^{11} - I\bar{4}2m$; $a = 5.46$; $c = 10.72$; $A = 16$
Co-ordinates: (000; $\frac{1}{2}\frac{1}{2}\frac{1}{2}$) + 2Fe($D_{2d}$): 000
+ 2Sn(D_{2d}): $00\frac{1}{2}$
+ 4Cu(S_4): $0\frac{1}{2}\frac{1}{4}$; $\frac{1}{2}0\frac{1}{4}$
+ 8S(C_s): xxz; $\bar{x}\bar{x}z$; $x\bar{x}\bar{z}$; $\bar{x}x\bar{z}$; $x = 0.245$; $z = 0.132$

*L*1$_0$ (CuAu type)

Tetragonal: $D_{4h}^1 - C4/mmm$; $a = 3.98$m $c = 3.72$; $A = 4$
Co-ordinates: (000; $\frac{1}{2}\frac{1}{2}0$) + 2Cu(D_{4h}): 000
+ 2Au(D_{4h}): $\frac{1}{2}0\frac{1}{2}$
Superstructure of the *A*1 (Cu) type

*L*1 (CuPt type)

Rhombohedral: $D_{3d}^5 - R\bar{3}m$; $a = 7.56$; $\alpha = 90°\ 54'$; $A = 32$
Co-ordinates: (000; $\frac{1}{2}\frac{1}{2}0$; ⟲) + 16Cu(D_{3d}): 000; $\frac{1}{4}\frac{1}{4}\frac{1}{2}$; ⟲
+ 16Pt(D_{3d}): $\frac{1}{2}\frac{1}{2}\frac{1}{2}$; $\frac{3}{4}\frac{3}{4}0$ ⟲
Superstructure of *A*1 (Cu) type

*L*1$_2$ (Cu_3Au type)

Cubic: $O_h^1 - Pm3m$; $a = 3.75$; $A = 4$
Co-ordinates: 3Cu(D_{4h}): $\frac{1}{2}\frac{1}{2}0$; ⟲
1Au(O_h): 000
Superstructure of *A*1 (Cu) type

*L*1$_a$ (Pt_3Cu type) ·

Cubic: $O^3 - Fm3c$; $a \approx 5.6$; $A = 32$
Co-ordinates: (000; $\frac{1}{2}\frac{1}{2}0$; ⟲) + 4Pt(O): 000
+ 4Cu(O): $\frac{1}{2}\frac{1}{2}\frac{1}{2}$
+ 24(Pt, Cu)(D_2): $0\frac{1}{4}\frac{1}{4}$; ⟲ ; $\frac{1}{2}\frac{1}{4}\frac{1}{4}$; ⟲

This structure was suggested by Tang.[205] An alternative has been proposed by Schneider and Esch.[206]

*L*2$_1$ (Cu_2MnAl type)

Cubic: $O_h^5 - Fm3m$; $a = 5.90$; $A = 16$
Co-ordinates: (000; $\frac{1}{2}\frac{1}{2}0$; ⟲ + 4Al(O_h): 000
+ 8Cu(T_d): $\frac{1}{4}\frac{1}{4}\frac{1}{4}$; $\frac{3}{4}\frac{3}{4}\frac{3}{4}$
+ 4Mn(O_h): $\frac{1}{2}\frac{1}{2}\frac{1}{2}$

Superstructure of the *A*2 (W) type; this type is virtually identical with the D_{o3} (BiF_3 or $BiLi_3$) type

*L*2$_2$ (Tl_7Sb_2 type)

Cubic: $O_h^9 - Im3m$; $a = 11.59$; $A = 54$
Co-ordinates: (000; $\frac{1}{2}\frac{1}{2}\frac{1}{2}$) + 2Tl($O_h$): 000
+ 16Tl(C_{3v}): $\pm(xxx$, $x\bar{x}\bar{x}$; ⟲); $x = 0.17 \approx \frac{1}{6}$
+ 24Tl(C_{2v}): $\pm(xx0$; ⟲ ; $x\bar{x}0$; ⟲ ; $x = 0.35 \approx \frac{1}{3}$
+ 12Sb(C_{4v}): $\pm(x00$; ⟲); $x = 0.29 \approx \frac{1}{3}$

Table 6.2 STRUCTURAL DETAILS—*continued*

$L2_a$ (δ-TiCu type)

Tetragonal: D_{4h}^1-P4/mmm; $a=3.14$, $c=2.86$; $A=2$
Co-ordinates: 1Ti(D_{4h}): 000
1Cu(D_{4h}): $\frac{1}{2}\frac{1}{2}\frac{1}{2}$

L'_1 (Fe_4N type)

Cubic: O_h^1-Pm3m or O_h^5-Fm3m (depending on the distribution of the N atoms); $a=3.79$; $A=5$
Co-ordinates: 4Fe at 000; $\frac{1}{2}\frac{1}{2}0$; ↻
1N at $\frac{1}{4}\frac{1}{4}\frac{1}{4}$; or at $\frac{1}{2}\frac{1}{2}\frac{1}{2}$, probably the latter; or at random at $\frac{1}{4}\frac{1}{4}\frac{1}{4}$; $\frac{3}{4}\frac{3}{4}\frac{1}{4}$; ↻ ; and (or) at $\frac{3}{4}\frac{3}{4}\frac{3}{4}$; $\frac{1}{4}\frac{1}{4}\frac{3}{4}$; ↻ ; and (or) at $\frac{1}{2}\frac{1}{2}\frac{1}{2}$; $\frac{1}{2}00$; ↻

$L'1_2$ ($\sim AlFe_3C$ type)

Cubic: O_h^1-Pm3m; $a=3.76$; $A\approx 5$
Co-ordinates: 1Al(O_h): 000
3Fe(D_{4h}): $\frac{1}{2}00$; ↻
0.6 to 0.9C(O_h) at $\frac{1}{2}\frac{1}{2}\frac{1}{2}$

$L'2$ (Martensite type)

Tetragonal: $D_{4h}^{17}-I4/mmm$; $a=2.84$, $c=2.97$; $A=2Fe+$(up to) 0.12 C
Co-ordinates: 2Fe(D_{4h}) at 000; $\frac{1}{2}\frac{1}{2}\frac{1}{2}$
The C atoms at random: $\frac{1}{2}\frac{1}{2}0$ and (or) $00\frac{1}{2}$

$L'3$ (Interstitial $A3$ type)

Hexagonal: $D_{6h}^4-P6_3/mmc$ or D_{6h}^1-P6/mmm
Co-ordinates: 2 metal atoms (D_{3h}): $\frac{2}{3}\frac{1}{3}0$; $\frac{1}{3}\frac{2}{3}\frac{1}{2}$
C or N(C_{3v}): $\frac{1}{3}\frac{2}{3}z$; $\frac{2}{3}\frac{1}{3}(\frac{1}{2}+z)$; $\frac{1}{3}\frac{2}{3}\bar{z}$; $\frac{2}{3}\frac{1}{3}(\frac{1}{2}-z)$; $z\approx\frac{3}{8}$; or $00\frac{1}{4}$; $00\frac{3}{4}$

$L6_0$ (Ti_3Cu type)

Tetragonal: D_{4h}^1-P4/mmm; $a=4.16$, $c=3.59$; $A=4$
Co-ordinates: 1Cu(D_{4h}): 000
1Ti(D_{4h}): $\frac{1}{2}\frac{1}{2}0$
2Ti(D_{2h}): $0\frac{1}{2}\frac{1}{2}$; $\frac{1}{2}0\frac{1}{2}$

Tetragonal deformed $L1_2$ (Cu_3Au) type; superstructure of $A6$ (In) type

$L'6$ (Interstitial $A6$ type)

Tetragonal: $D_{4h}^{17}-F4/mmm$ (or D_{4h}^1-P4/mmm, depending on the distribution of the N atoms)
Co-ordinates: 4 metal atoms (D_{4h}): 000; $\frac{1}{2}\frac{1}{2}0$; ↻
N atoms in the holes: $\frac{1}{4}\frac{1}{4}\frac{1}{4}$; $\frac{1}{4}\frac{3}{4}\frac{3}{4}$; ↻ ; $\frac{3}{4}\frac{3}{4}\frac{3}{4}$; $\frac{3}{4}\frac{1}{4}\frac{1}{4}$; ↻ ; or $\frac{1}{2}\frac{1}{2}\frac{1}{2}$; $\frac{1}{2}00$; ↻

REFERENCES

1 E. A. Owen and G. I. Williams, *Proc. phys. Soc.*, 1954 (A), **67**, 895.
2 T. W. Baker and J. Williams, *Acta Cryst.*, 1955, **8**, 519.
3. C. S. Barrett, *J. chem. Physics*, 1956, **25**, 1123.
4. F. H. Spedding, A. H. Daane and K. W. Herrmann, *Trans. Amer. Inst. min. (metall.) Engrs.*, 1957 **209**, 895.
5. F. H. Ellinger, *US Atomic Energy Comm. Publn*, LADC–1460.
6. J. D. Farr, A. L. Giorgi, and M. G. Bowman, *US Atomic Energy Comm. Publn*, LA–1545, 1953.
7. K. Gordon, B. Skinner and H. L. Johnston, *US Atomic Energy Comm. Publn*, NP–4737, 1953.
8. J. W. Nielsen and N. C. Baenziger, *Acta Cryst.*, 1954, **7**, 132.
9. P. Duwez, *J. appl. Physics*, 1951, **22**, 1174.
10. J. D. Fast, *J. appl. Physics*, 1952, **23**, 350.
11. D. M. Poole, G. K. Williamson and J. A. C. Marples, *J. Inst. Metals*, 1957–58, **80**, 172.
12. D. T. Peterson, P. F. Djilak and C. L. Vold, *Acta Cryst.*, 1956, **9**, 1936.
13. P. A. Sellers, S. Fried, R. E. Elson and W. H. Zachariasen, *J. Amer. chem Soc.*, 1954, **76**, 5935.

14. J. Thewlis, *Acta Cryst.*, 1952, **5**, 790.
15. C. W. Tucker, *Acta Cryst.*, 1951, **4**, 425; 1952, **5**, 395.
16. A. G. Knapton, *Acta Cryst.*, 1954, **7**, 457.
17. A. M. Holden and W. E. Seymour, *Trans. Amer. Inst. min. (metall.) Engrs*, 1956, **206**, 1312.
18. W. H. Zachariasen, *Acta Cryst.*, 1952, **5**, 664.
19. O. J. C. Runnalls, *Acta Cryst.*, 1954, **7**, 222.
20. W. H. Zachariasen and F. Ellinger, *J. Chem. Physics*, 1957, **27**, 811.
21. F. H. Ellinger, *Trans. Amer. Inst. min. (metall.) Engrs*, 1956, **206**, 1256.
22. M. A. Gurevich and B. F. Ormont, *Fizika Metall.*, 1957, **4**, 112.
23. A. E. Austin and J. R. Doig, *Trans. Amer. Inst. min. (metall.) Engrs*, 1957, **209**, 27.
24. R. F. Domogala and D. J. McPherson, *US Atomic Energy Comm. Publn*, COO–100, 1952.
25. R. F. Raeuchle and F. W. von Batchelor, *Acta Cryst.*, 1955, **8**, 691.
26. P. Duwez and C. B. Jordan, *J. Am. chem. Soc.*, 1951, **73**, 5509.
27. E. K. Halteman, *Acta Cryst.*, 1957, **10**, 166.
28. E. Oehman, *Metallwirtschaft*, 1930, **9**, 825.
29. A. J. C. Wilson, *Bull. Am. phys. Soc.*, 1934, **9**, No. 16.
30. Z. S. Basinski and J. W. Christian, *J. Inst. Metals*, 1952, **80**, 659.
31. H. Margolin and Elmars Ence, *Trans. Amer. Inst. min. (metall.) Engrs*, 1954, **200**, 1267.
32. W. Rostoker, *J. Metals*, 1952, **4**, 209.
33. P. Duwez and J. L. Taylor, *J. Metals*, 1950, **2**, 1173.
34. E. T. Hayes, A. H. Robertson and W. L. O'Brien, *Trans. Am. Soc. Metals*, 1951, **43**, 888.
35. D. Bergman and D. P. Shoemaker, *Acta Cryst.*, 1954, **7**, 857.
36. G. J. Dickins, Abs. Dissert. Univ. Cambridge, 1954–55, 1957, 244.
37. G. J. Dickins, Audrey M. B. Douglas and W. H. Taylor, *Acta Cryst.*, 1956, **9**, 297.
38. Emma Smith and R. W. Guard, *Trans. Amer. Inst. min. (metall.) Engrs.*, 1957, **209**, 1189.
39. T. J. Heal and G. I. Williams, *Acta Cryst.*, 1955, **8**, 494.
40. C. B. Jordan, *Trans. Amer. Inst. min. (metall.) Engrs*, 1955, **203**, 832.
41. M. V. Nevitt and J. W. Downey, *Trans. Amer. Inst. min. (metall.), Engrs*, 1957, **209**, 1072.
42. R. M. Waterstrat and J. S. Kasper, *Trans. Amer. Inst. min. (metall.) Engrs*, 1957, **209**, 872.
43. P. Duwez and C. B. Jordan, *Acta Cryst.*, 1952, **5**, 213.
44. L. Misch, *Z. physikal Chem.*, 1935, (B), **29**, 42.
45. G. F. Kossolapow and A. K. Trapesnikow, *Metallwirtschaft*, 1935, **14**, 45.
46. A. Byström, P. Kierkegaard and O. Knop, *Acta Chem. Scand.*, 1956, **6**, 709.
47. N. C. Baenziger, R. E. Rundle and A. I. Snow, *Acta Cryst.*, 1956, **9**, 93.
48. J. R. Murray, *J. Inst. Metals*, 1955–56, **84**, 91.
49. E. A. Wood and B. T. Mattheas, *Acta Cryst.*, 1956, **9**, 534.
50. N. C. Baenziger and J. W. Conant, *Acta Cryst.*, 1956, **9**, 361.
51. P. Pietrokowsky, *Trans. Amer. Inst. min. (metall.) Engrs*, 1954, **200**, 219.
52. E. S. Makarov and L. S. Gudhov, *Krystallografiya*, 1956, **1**, 650.
53. H. Nowotny, E. Bauer and A. Stempfl, *Monatsh.*, 1950, **81**, 1164.
54. R. E. Marsh, *Acta Cryst.*, 1954, **7**, 379.
55. R. Ferro, *Acta Cryst.*, 1954, **7**, 781.
56. M. J. Sanderson and N. C. Baenziger, *Acta Cryst.*, 1953, **6**, 627.
57. E. A. Owen and E. A. O'D. Roberts, *J. Inst. Metals*, 1940, **66**, 389.
58. A. Byström and K. E. Almin, *Acta Chem. Scand.*, 1948, **1**, 76.
59. C. S. Barrett, *Acta Cryst.*, 1957, **10**, 58.
60. J. W. Nielsen and N. C. Baenziger, *Acta Cryst.*, 1954, **7**, 277.
61. E. J. Duwell and N. C. Baenziger, *Acta Cryst.*, 1955, **8**, 705.
62. N. C. Baenziger, E. J. Duwell and J. W. Conant, *US Atomic Energy Comm. Publn*, COO–127, 1954.
63. K. Schubert, U. Rösler, W. Mahler, E. Dorre and W. Schütt, *Z. Metallk.*, 1954, **45**, 643.
64. G. V. Raynor and J. A. Lee, *Acta Met.*, 1954, **2**, 616.
65. E. S. Makarov, *Dokl. Akad. Nauk. USSR*, 1950, **74**, 935.
66. D. K. Das and D. T. Pitman, *Trans. Amer. Inst. min. (metall.) Engrs*, 1957, **209**, 1175.
67. J. H. N. van Vucht, *Z. Metallk.*, 1957, **48**, 253.
68. E. Ence and H. Margoli, *Trans. Amer. Inst. min. (metall.) Engrs*, 1957, **209**, 484.
69. W. Köster and A. Sampaio, *Z. Metallk.*, 1957, **48**, 331.
70. O. J. C. Runnals, *J. Metals*, 1953, **5**, 1460.
71. J. F. Smith and E. A. Ray, *Acta Cryst.*, 1957, **10**, 169.
72. A. J. Bradley and S. S. Lu, *Z. Krist.*, 1937, **96**, 20.
73. W. Hofmann and H. Wiehr, *Z. Metallk.*, 1941, **33**, 369.
74. K. Little, J. N. Pratt and G. V. Raynor, *J. Inst. Metals*, 1951, **80**, 456.
75. K. Little, *J. Inst, Metals*, 1953–54, **82**, 463.
76. S. Samson, *Nature*, 1954, **173**, 1185.
77. J. Adam and J. B. Rich, *Acta Cryst.*, 1954, **7**, 813.
78. *idem. Acta Cryst.*, 1955, **8**, 349.
79. A. D. I. Nicol, *Acta Cryst.*, 1953, **6**, 285.
80. K. Schubert, U. Rösler, M. Kluge, K. Anderko and L. Härle, *Naturwiss.*, 1953, **40**, 269, 437.
81. A. C. Larsen, D. T. Cromer and C. N. Stambaugh, *Acta Cryst.*, 1957, **10**, 443.
82. J. H. Keeler, *US Atomic Energy Comm. Publn*, SO–2515, 1954.
83. P. Esslinger and K. Schubert, *Z. Metallk.*, 1957, **48**, 126.

84. M. G. Bown, *Acta Cryst.*, 1956, **9**, 70.
85. K. Anderko and U. Zwick, *Naturwiss.*, 1957, **44**, 510.
86. K. Anderko, *Naturwiss.*, 1957, **44**, 88.
87. M. B. Waldron, *J. Inst. Metals*, 1951, **79**, 103.
88. K. Robinson, *Acta Cryst.*, 1952, **5**, 401.
89. G. Phragmen, *J. Inst. Metals*, 1950, **77**, 489.
90. E. Hellner, *Z. Metallk.*, 1950, **41**, 401.
91. E. S. Makarov, *Izvest. Akad. Nauk S.S.S.R.*, 1943, (Khim), 264.
92. J. Reynolds, W. A. Wiseman and W. Hume-Rothery, *J. Inst. Metals*, 1952, **80**, 637.
93. E. Hellner and F. Laves, *Z. Naturforsch.*, 1947 (A), **2**, 180.
94. P. G. Cotter, J. A. Kohn and R. A. Potter, *J. Amer. Ceram. Soc.*, 1956, **39**, 11.
95. H. Schachner, H. Nowotny and R. Machenschalk, *Monatsh.*, 1953, **84**, 677.
96. P. Pietrokowsky, *Acta Crysta.*, 1954, **7**, 435.
97. W. H. Zachariasen, *Acta Cryst.*, 1949, **2**, 94.
98. G. Brauer and H. Haag, *Z. anorg. Chem.*, 1949, **259**, 197.
99. O. J. C. Runnals and R. R. Boucher, *Acta Cryst.*, 1955, **8**, 592.
100. H. Schachner, E. Cerwenka and H. Nowotny, *Monatsh.*, 1954, **85**, 245.
101. H. Nowotny, R. Kieffer and H. Schachner, *Monatsh.*, 1952, **83**, 1243.
102. P. Eckerlin and E. Wölfel, *Z. Anorg. Chem.*, 1955, **280**, 321.
103. K. Schubert and H. Pfisterer, *Z. Metallk.*, 1950, **41**, 438.
104. A. Osawa and M. Okamoto, *Sci. Rep. Tohoku Imp. Univ.*, 1939, (i), **27**, 326.
105. B. Boren, *Arkiv. Kemi Mineral. Geol.*, 1933, **11A**, (10), 1.
106. K. Toman, *Acta Cryst.*, 1951, **4**, 462.
107. S. Geller and E. A. Wood, *Acta Cryst.*, 1954, **7**, 441.
108. K. Anderko and K. Schubert, *Z. Metallk.*, 1953, **44**, 307.
109. J. H. Buddery and A. J. E. Welch, *Nature*, 1951, **167**, 362.
110. K. Schubert, *Naturwiss.*, 1952, **39**, 351.
111. G. Bergman and J. L. T. Waugh, *Acta Cryst.*, 1953, **6**, 93.
112. K. Robinson, *Acta Cryst.*, 1953, **6**, 854.
113. K. Robinson and P. J. Black, *Phil. Mag.*, 1953, **44**, 1392.
114. P. Eckerlin, H. J. Meyer and E. Wölfel, *Z. anorg. Chem.*, 1955, **281**, 322.
115. E. Parthé and J. T. Norton, *Acta Cryst.*, 1958, **11**, 14.
116. P. I. Kripyakevich, E. I. Gladyshevskv and O. S. Zarechnyuk, *Dokl. Akad. Nauk., S.S.S.R.*, 1954, **95**, 525.
117. K. Schubert and H. Pfisterer, *Z. Metallk.*, 1951, **41**, 433.
118. E. Hellner, *Fortschr. Mineral, Krist. Petrogr.*, 1951, **29–30**, 59.
119. A. Westgren and G. Phragmen, *Z. anorg. Chem.*, 1928, **175**, 80.
120. O. Carlssohn and G. Hägg, *Z. Krist.*, 1932, **83**, 308.
121. S. T. Knobejewski and W. P. Tarassova, *Zh. fiz. Khim.* (*J. phys. Chem.*), 1937, **9**, 681.
122. H. Knödler, *Acta Cryst.*, 1957, **10**, 86.
123. H. Nowotny and H. Schachner, *Monatsh.*, 1953, **84**, 169.
124. A. Zalkin and W. J. Ramsey, *J. phys. Chem.*, 1956, **60**, 234.
125. *idem*, *J. phys. Chem.*, 1956, **60**, 1275.
126. R. E. Marsh and D. P. Shoemaker, *Acta Cryst.*, 1953, **6**, 197.
127. D. Gilde, *Z. anorg. Chem.*, 1956, **284**, 142.
128. P. Farrar and H. Margolin, *Trans. Amer. Inst. min.* (*metall.*) *Engrs*, 1955, **203**, 101.
129. R. D. Heidenreich, *Acta Met.*, 1955, **3**, 79.
130. K. Bachmayer, H. Nowotny and A. Kohl, *Monatsh.*, 1955, **86**, 39.
131. H. Nowotny, R. Funk and J. Pesl, *Monatsh.*, 1951, **82**, 513.
132. K. E. Fylking, *Arkiv. Kem. Min. Geol.*, 1935, (B), **11**, (48), 1.
133. H. Cole, F. W. Chambers and H. M. Dunn, *Acta Cryst.*, 1956, **9**, 685.
134. H. Nowotny and B. Glatzl, *Monatsh.*, 1951, **82**, 720.
135. K. H. Jack and M. M. Wachtel, *Proc. Roy. Soc.*, 1957, *A*, **239**, 46.
136. R. Ferro, *Acta Cryst.*, 1956, **9**, 817.
137. Le Roy Heaton and N. S. Gingrich, *Acta Cryst.*, 1955, **8**, 207.
138. R. N. Kuzmin, G. S. Khdanov and N. N. Zhuravlev, *Kristallografiya*, 1957, **2**, 48.
139. A. Osawa and N. Shibata, *Sci. Rep. Tohoku Imp. Univ.*, 1939, (i), **28**, 1, 197.
140. W. Hofmann, *Z. Metallk.*, 1941, **33**, 61.
141. R. Ferro, *Acta Cryst.*, 1957, **10**, 476.
142. W. P. Binnie, *Acta Cryst.*, 1956, **9**, 686.
143. J. L. Hoard, S. Geller and R. E. Hughes, *J. Am. chem. Soc.*, 1951, **73**, 1892.
144. A. W. Laubengayer, D. T. Hurd, A. E. Newkirk and J. L. Hoard, *J. Am chem. Soc.*, 1943, **65**, 1924.
145. St. v. N. Szabo and C. W. Tobias, *J. Am. chem. Soc.*, 1949, **71**, 1882.
146. V. Russell, R. Hirst, F. A. Kanda and A. J. King, *Acta Cryst.*, 1953, **6**, 870.
147. B. Post and F. W. Glaser, *J. chem. Phys.*, 1952, **20**, 1050.
148. B. F. Decker and J. S. Kosher, *Acta Cryst.*, 1954, **7**, 77.
149. P. Ehrlich, *Z. anorg. Chem.*, 1949, **259**, 1.
150. R. Kiessling, *Acta Chem. Scand.*, 1950, **4**, 164.
151. F. W. Glaser and B. Post, *J. Metals*, 1953, **5**, 1117.
152. D. Moscowitz, *Trans. Amer. Inst. min.* (*metall.*) *Engrs*, 1956, **206**, 1325.
153. F. Bertant and P. Blum, *Compt. Rend.*, 1953, **236**, 1055.

154. Stig Rundqviet, *Nature*, 1958, **181**, 259.
155. R. W. Mooney and A. J. E. Welch, *Acta Cryst.*, 1954, **7**, 49.
156. M. von Stackelberg, *Z. physical Chem.*, 1930, **B**, **9**, 437.
157. E. B. Hunt and R. E. Rundle, *J. Am. chem. Soc.*, 1951, **73**, 4777.
158. H. A. Wilhelm and P. Chiotti, *Trans. Amer. Soc. Metals*, 1950, **42**, 1295.
159. R. E. Rundle, N. C. Baenziger, A. S. Wilson and R. A. McDonald, *J. Am. chem. Soc.*, 1948, **70**, 99.
160. U. Esch and A. Schnieder, *Z. anorg. Chem.*, 1948, **257**, 254.
161. L. M. Litz, A. B. Garrett and F. C. Croxton, *J. Am. chem. Soc.*, 1948, **70**, 1718.
162. M. W. Mallett, A. F. Gerds and D. A. Vaughan, *J. electrochem. Soc.*, 1951, **98**, 505.
163. G. Brauer, H. Renner and J. Wernet, *Z. anorg. Chem.*, 1954, **277**, 249.
164. V. I. Smirnova and B. F. Ormont, *Dokl. Akad. Nauk. SSSR*, 1954, **96**, 557.
165. K. H. Jack, *Proc. Roy. Soc.*, 1948, **195**, 56.
166. L. J. E. Hofer, E. M. Cohen and W. C. Peebles, *J. Am chem. Soc.*, 1949, **71**, 189.
167. K. H. Jack, *Acta Cryst.*, 1950, **3**, 392.
168. N. Schönberg, *Acta Met.*, 1954, **2**, 837.
169. S. Nagakura, *J. phys. Soc. Japan*, 1957, **12**, 482.
170. K. Whitehead and L. D. Brownlee, *Planseebar, Pulvermet*, 1956, **4**, 62.
171. K. H. Jack, *Acta Cryst.*, 1952, **5**, 404.
172. H. A. Eick, N. C. Baenziger and L. Eyring, *J. Am chem. Soc.*, 1956, **78**, 5987.
173. R. Jusa, H. H. Weber and C. Meyer-Simon, *Z. anorg. Chem.*, 1953, **273**, 48.
174. G. Brauer and K. H. Zapp, *Naturwiss*, 1953, **40**, 604.
175. U. Zorll, *Z. Physik*, 1954, **138**, 167.
176. R. Ferro, *Z. anorg. Chem.*, 1954, **275**, 320.
177. O. Knop and H. Haraldsen, *Canad. J. Chem.*, 1956, **34**, 1142.
178. J. Singer and C. W. Spencer, *Trans. Amer. Inst. min. (metall.) Engrs*, 1955, **203**, 144.
179. S. A. Semiletov, *Kristallografiya*, 1956, **1**, 403.
180. P. Khodadad and J. Flahaut, *Compte rendu.*, 1957, **244**, 462.
181. S. Geller and B. B. Cetlin, *Acta Cryst.*, 1955, **8**, 272.
182. H. Hahn and K. F. Schröder, *Z. anorg. Chem.*, 1952, **269**, 135.
183. H. Haraldsen, *Z. anorg. Chem.*, 1941, **246**, 169.
184. K. Schubert, E. Dörre and M. Kluge, *Z. Metallk.*, 1955, **46**, 216.
185. L. I. Tatarinova, Yu K. Auleitner and Z. G. Pinsker, *Kristallografiya*, 1956, **1**, 537.
186. H. Miyazawa and S. Sugaike, *J. phys. Soc. Japan*, 1957, **12**, 312.
187. A. Okazaki and I. Ueda, *J. phys. Soc. Japan*, 1956, **11**, 470.
188. K. Schubert and H. Frieke, *Z. Metallk.*, 1953, **44**, 457.
189. Structure Reports, 1951, **11**, 246.
190. A. Byström, *Arkiv Kemi. Mineral. Geol.*, 1945, **19B**, No. 8.
191. A. R. Graham, *Amer. Mineralogist*, 1949, **34**, 462.
192. F. Bertaut, *Compt. rend.*, 1952, **234**, 1295.
193. H. Hahn, *Z. anorg. Chem.*, 1951, **264**, 184.
194. D. Lundqvist, *Arkiv Kemi. Mineral. Geol.*, 1943, **17B**, No. 12.
195. N. V. Belov and V. P. Butuzov, *Dokl. Akad. Nauk. SSSR*, 1946, **54**, 717.
196. R. Ueda, *J. Phys. Soc. Japan*, 1949, **4**, 287.
197. Structure Reports, 1952, **12**, 156.
198. K. Schubert, E. Dörre and E. Günzel, *Naturwiss.*, 1954, **41**, 448.
199. K. Toman, *Acta Cryst.*, 1952, **5**, 329.
200. H. Haraldsen, *Z. anorg. Chem.*, 1939, **240**, 337.
201. R. F. Raeuchle and R. E. Rundle, *Acta Cryst.*, 1952, **5**, 85.
202. B. G. Bergman and D. P. Shoemaker, *J. chem. Physics*, 1951, **19**, 515.
203. J. S. Kasper, B. F. Decker and J. R. Belanger, *J. appl. Physcs*, 1951, **22**, 361.
204. W. B. Pearson and J. W. Christian, *Acta Cryst.*, 1952, **5**, 157.
205. Y.-C. Tang, *Acta Cryst.*, 1951, **4**, 377.
206. A. Schneider and U. Esch. *Z. Elektrochem.*, 1944, **50**, 290.
207. F. H. Spedding, A. H. Deane, G. Wakefield and B. H. Dennison, *Trans. metall. Soc. A.I.M.E.*, 1960, **218**, 608.
208. F. H. Spedding, A. H. Deane and J. J. Hanah, *Trans. metall. Soc., A.I.M.E.*, 1959, **212**, 179.
209. E. M. Savitsky, V. F. Terekhova and I. V. Burov, *Tsvet. Metally*, 1960, **33**, 59.
210. E. M. Savitsky, V. F. Terekhova and O. P. Naumking, *Tsvet. Metally*, 1960, **33**, 43.
211. D. V. Keller, F. A. Kanda and A. J. King, *J. phys. Chem.*, 1958, **62**, 732.
212 A. Zalkin, R. G. Bedford and D. E. Sands, *Acta Cryst.*, 1959, **12**, 701.
213. R. M. Paine and J. A. Carrabine, *Acta Cryst.*, 1960, **13**, 680.
214. E. F. Sturken and B. Post, *Acta Cryst.*, 1960, **13**, 852.
215. W. H. Zadareisen and F. H. Ellinger, *Acta Cryst.*, 1959, **12**, 175.
216. D. B. McWhan, *US Atomic Energy Comm. Publn*, UCRL 9695, 1964.
217. A. Zalkin, D. E. Sands and A. H. Krikorian, *Acta Cryst.*, 1960, **13**, 160.
218. *idem*, *Acta Cryst.*, 1960, **13**, 713.
219. S. A. Spakner, *Trans. metall. Soc. A.I.M.E.*, 1958, **212**, 57.
220. R. M. Paine and J. A. Carrabine, *Acta Cryst.*, 1960, **13**, 680.
221. B. J. Baudry, J. F. Haufling and A. H. Daane, *Acta Cryst.*, 1960, **13**, 743.
222. R. I. Myklebist and A. H. Deane, *Trans. metall. Soc. A.I.M.E.*, 1962, **224**, 354.

223. N. H. Krikorian, W. G. Witteman and M. S. Cowme, *J. phys. Chem.*, 1960, **64**, 1517.
224. B. A. Hatt, *Acta Cryst.*, 1961, **14**, 119.
225. V. F. Novy, R. C. Vickery and E. V. Kleben, *Trans. metall. Soc. A.I.M.E.*, 1961, **221**, 585.
226. G. Katz and A. J. Jacobs, *J. nucl. Mater.*, 1962, **5**, 338.
227. F. R. Morral, *J. Metals., N.Y.*, 1958, **10**, 662.
228. D. J. Cromer and A. C. Larsen, *Acta Cryst.*, 1961, **14**, 1226.
229. S. Saito, *Acta Cryst.*, 1959, **12**, 500.
230. M. Korchynsky and R. W. Fountain, *Trans. metall. Soc. A.I.M.E.*, 1959, **215**, 1053.
231. B. J. Baudry and A. H. Daane, *Trans. metall. Soc. A.I.M.E.*, 1960, **218**, 854.
232. S. E. Haszko, *Trans. metall. Soc. A.I.M.E.*, 1960, **218**, 763.
233. D. T. Cromer and G. E. Olsen, *Acta Cryst.*, 1959, **12**, 689.
234. V. F. Novy, R. C. Vickery and E. V. Kleben, *Trans. metall. Soc. A.I.M.E.*, 1961, **221**, 585.
235. *idem. Trans. metall. Soc. A.I.M.E.*, 1961, **221**, 588.
236. G. A. Yurks, J. W. Barton and J. S. Parr, *Acta Cryst.*, 1959, **12**, 909.
237. D. Kramer, *Trans. metall. Soc. A.I.M.E.*, 1959, **215**, 256.
238. J. F. Smith and W. L. Laram, *Acta. Cryst.*, 1962, **15**, 252.
239. D. T. Cromer and A. C. Larsen, *Acta Cryst.*, 1950, **13**, 909.
240. D. T. Cromer and C. E. Olsen, *Acta Cryst.*, 1959, **12**, 689.
241. D. T. Cromer and R. B. Roof, *Acta Cryst.*, 1959, **12**, 942.
242. S. Saito and P. A. Beck, *Trans. metall. Soc. A.I.M.E.*, 1959, **215**, 938.
243. E. Rudy, B. Kieffer and H. Fröhlich, *Z. Metallk.*, 1962, **53**, 90.
244. W. Obrowski, *Z. Metallk.*, 1962, **53,** 715.
245. A. F. Berndt, *Acta Cryst.*, 1961, **14**, 1301.
246. V. B. Crompton, *Acta Cryst.*, 1958, **11**, 446.
247. E. A. Wood and V. B. Compton, *Acta Cryst.*, 1958, **11**, 429.
248. A. E. Dwight and P. A. Beck, *Trans. metall. Soc. A.I.M.E.*, 1959, **215**, 976.
249. J. R. Thomson, *Proc. 11th Amer. Conf. on X-Ray*, 1963.
250. W. Köster and W. D. Hackl, *Z. Metallk.*, 1958–59, **12**, 647.
251. B. A. Hatt and G. I. Williams, *Acta Cryst.*, 1959, **12**, 685.
252. W. Obrowski, *Metall.*, 1963, **17**, 108.
253. A. C. Larsen and D. T. Cromer, *Acta Cryst.*, 1961, **14**, 514.
254. *idem, Acta Cryst.*, 1961, **14**, 73.
255. D. T. Cromer, A. C. Larsen and R. B. Roof, *Acta Cryst.*, 1960, **13**, 913.
256. S. C. Haszko, *Trans. metall. Soc. A.I.M.E.*, 1960, **218**, 4, 763.
257. E. Ence and M. Mayoln, *Trans. metall. Soc. A.I.M.E.*, 1961, **221**, 370.
258. R. P. Rand and L. D. Calvert, *Canad. J. Chem.*, 1962, **40**, 705.
259. E. O. Hall and J. Royan, *Acta Cryst.*, 1959, **12**, 607.
260. D. Watanabe, *J. Phys. Soc. Japan*, 1960, **15**, 1251.
261. P. Chiotti and K. J. Gill, *Trans. metall. Soc. A.I.M.E.*, 1961, **221**, 573.
262. *idem. Trans. metall. Soc. A.I.M.E.*, 1959, **215**, 892.
263. D. R. Petersen and H. W. Rinn, *Acta Cryst.*, 1961, **14**, 328.
264. C. L. Vold, *Acta Cryst.*, 1961, **14**, 1289.
265. W. Heine and U. Zwicker, *Z. Metallk.*, 1962, **53**, 386.
266. E. Günzel and K. Schubert, *Z. Metallik.*, 1958, **49**, 234.
267. R. V. Schablaski, B. S. Tani and M. G. Chesanov, *Trans. metall. Soc. A.I.M.E.*, 1962, **224**, 867.
268. S. Samson, *Nature, Lond.*, 1962, **195**, 259.
269. D. V. Masson and C. S. Barrett, *Trans. metall. Soc. A.I.M.E.*, 1958, **212**, 260.
270. E. I. Duwell and N. C. Baenziger, *Acta Cryst.*, 1960, **13**, 476.
271. W. Rostowker, *Trans. metall. Soc. A.I.M.E.*, 1958, **212**, 393.
272. R. F. Domagala, R. P. Elliott and W. Rostotier, *Trans. metall. Soc. A.I.M.E.*, 1958, **212**, 393.
273. S. Sansom, *Acta Cryst.*, 1958, **11**, 857.
274. T. Dagenham, *Acta Chem. Scand.*, 1963, **17**, 267.
275. L. M. d'Alte da Veiges, *Phil. Mag.*, 1962, **7**, 1247.
276. A. J. Goldat and J. G. Pair, *Trans. metall. Soc. A.I.M.E.*, 1961, **221**, 639.
277. C. J. Wilson and D. Sand, *Acta Cryst.*, 1961, **14**, 72.
278. T. J. Rensuf and C. A. Beevers, *Acta Cryst.*, 1961, **14**, 469.
279. C. G. Wilson, D. K. Thomas and F. J. Spooner, *Acta Cryst.*, 1960, **13**, 56.
280. C. G. Wilson and F. J. Spooner, *Acta Cryst.*, 1960, **13**, 4, 358.
281. L. E. Edshammer, *Acta Chem. Scand.*, 1961, **15**, 403.
282. *idem, Acta Chem. Scand.*, 1960, **14**, 2248.
283. C. R. McKinsey and G. M. Faubring, *Acta Cryst.*, 1959, **12**, 701.
284. J. W. H. Clare, *J. Inst. Mab.*, 1960–61, **89**, 232.
285. J. A. Bland and D. Clarke, *Acta Cryst.*, 1958, **11**, 231.
286. J. H. Bland, *Acta Cryst.*, 1958, **11**, 236.
287. M. A. Taylor, *Acta Cryst.*, 1959, **12**, 393.
288. *idem, Acta Cryst.*, 1961, **14**, 84.
289. L. M. d'Alte de Veiges and L. K. Walford, *Phil. Mag.*, 1963, **8**, 349.
290. W. Obrowski, *Metall.*, 1963, **17**, 108.
291. K. Andutro, *Z. Metallk.*, 1958, **49**, 165.
292. R. Ferro, *Acta Cryst.*, 1958, **11**, 737.

293. S. Bhan and K. Schubert, *Z. Metallk.*, 1960, **51**, 327.
294. E. I. Gladishevsky, *Dopov., Akad. Nauk ukr. R.S.R.*, 1959, **3**, 294.
295. E. Parthé, *Acta Cryst.*, 1959, **12**, 559.
296. B. J. Baudry and A. H. Daane, *M.A.*, 1962, 443.
297. L. M. Finme, *J. less common Metals*, 1962, **4**, 24.
298. E. Parthé, *Acta Cryst.*, 1960, **13**, 968.
299. E. I. Gladishevsky, *Dopov. Akad. Nauk ukr. R.S.R.*, 1959, **3**, 294.
300. N. Ageev and V. Samsonov, *Doklady Akad. Nauk S.S.S.R.*, 1957, **112**, 853.
301. A. G. Tharp, A. W. Searcy and H. Novotkny, *J. Electrochem. Soc.*, 1958, **105**, 473.
302. E. Raub and W. Fuzsche, *Z. Metallk.*, 1962, 1962, **53**, 779.
303. E. Weitz, L. Born and E. Hellness, *Z. Metallk.*, 1960, **51**, 228.
304. D. M. Bailey and J. F. Smith, *Acta Cryst.*, 1961, **14**, 57.
305. A. Zalkin and W. J. Ramsey, *J. Phys. Chem.*, 1958, **62**, 689.
306. D. T. Cromer, *Acta Cryst.*, 1959, **12**, 36.
307. W. Trzebratowski, S. Weglowski and K. Lukasgewag, *Roczniki Chem.*, 1958, **32**, 189.
308. J. H. Bryden, *Acta Cryst.*, 1962, **15**, 167.
309. C. Giesecki and H. Pfister, *Acta Cryst.*, 1958, **11**, 369.
310. A. Kockus, F. Gronvolde and J. Thorbioin, *Acta Chem. Scand.*, 1962, **16**, 1493.
311. V. N. Bykoff and V. V. Kazarnikov, *Kristallografiya*, 1959, **4**, 924.
312. I. Obinata, Y. Takechi and S. Saikewa, *Trans. Amer. Soc. Metals*, 1959, **52**, 156.
313. J. L. Hoard, R. E. Hughes and D. E. Sands, *J. Am. chem. Soc.*, 1958, **80**, 4507.
314. D. E. Sands, C. F. Cline, A. Zalkin and C. L. Hoenig, *Acta Cryst.*, 1961, **14**, 309.
315. G. V. Samonov, V. P. Dzeganovsky and I. A. Simashko, *Kristallografiya*, 1959, **4**, 119.
316. Y. B. Paderno, T. I. Serebaykova and G. V. Samsonov, *Dokl. Akad. Nauk S.S.S.R.*, 1959, **125**, 317.
317. B. J. MacDonald and W. I. Stuart, *Acta Cryst.*, 1960, **13**, 447.
318. M. Elfstrom, *Acta Chem. Scand.*, 1961, **15**, 1178.
319. S. Laplace and B. Poste, *Acta Cryst.*, 1962, **15**, 97.
320. W. Obrowski, *Metall.*, 1963, **17**, 108.
321. E. Steinberg, *Acta Chem. Scand.*, 1961, **15**, 861.
322. B. Magnussen and C. Brossit, *Acta Chem. Scand.*, 1962, **16**, 449.
323. M. Atoji, K. G. Schneider, A. H. Waane, R. E. Rundle and F. H. Spedding, *J. Am. chem. Soc.*, 1958, **80**, 1804.
324. G. Brauer and K. Lesser, *Z. Metallk.*, 1959, **50**, 8.
325. S. Nagakura, *J. phys. Soc. Japan*, 1959, **14**, 186.
326. *idem*, *J. phys. Soc. Japan*, 1961, **16**, 1213.
327. W. D. Forgang and B. F. Decker, *Trans. metall. Soc. A.I.M.E.*, 1958, **212**, 343.
328. F. Gronwold, A. Kjetshus and F. Raun, *Acta Cryst.*, 1961, **14**, 93.
329. D. Pustiner and R. E. Newnham, *Acta Cryst.*, 1961, **14**, 691.
330. M. S. Mirgalowskaya and E. V. Skudnova, *Izv. Akad. Nauk S.S.S.R.*, 1959, **4**, 148.
331. A. Stechen and P. Eckerlin, *Z. Metallk.*, 1960, **51**, 295.
332 E. F. Hockiup and J. C. White, *Acta Cryst.*, 1961, **14**, 328.
333. E. Parthé, *Acta Cryst.*, 1960, **13**, 865.
334. P. Khodad, *C. R. Akad. Sci., Paris*, 1960, **250**, 3998.
335. *idem*, *C. R. Akad. Sci., Paris*, 1959, **249,** 694.
336. A. Okasaki, *J. phys. Soc. Japan*, 1958, **13**, 1151.
337. E. A. Wood, V. B. Compton, B. T. Matthias and E. Carengurt, *Acta Cryst.*, 1958, **11**, 604.
338. W. Klemm, F. Darn and R. Huck, *Naturwiss*, 1958, **45**, 490.
339. J. A. John, G. Katz and A. A. Giardini, *Z. Krist.*, 1958, **111**, 52.
340. W. Krönert and K. Plieth, *Naturwiss.*, 1958, **45**, 416.
341. H. H. Stadelmeier and W. K. Hardy, *Z. Metallk.*, 1961, **52**, 391.
342. M. V. Nevitt and J. W. Downey, *Trans. metall. Soc. A.I.M.E.*, 1962, **224**, 195.

7 Metallurgically important minerals

Table 7.1 gives data on the minerals from which the more important metals are extracted. Those minerals of major importance are shown in **bold type** in column 2. Parentheses indicate that the element is recovered as a by-product in the extraction of another metal.

The chemical formulae assigned in column 3 are given only to indicate the nature of the minerals since they are not stoichiometric chemical compounds.

The mineral-producing countries are listed in order of decreasing production in column 6, and the major metal producers in column 8.

The figure for abundance given in column 1 is the amount of the metal in parts per million of the igneous rocks of the lithosphere.

BIBLIOGRAPHY

Mineralogical data

C. Palache, H. Berman and C. Frondel, 'Dana's System of Mineralogy, 7th edn, Chapman and Hall, London: Vol. I, 1944, Vol. II, 1951.
W. E. Ford, 'Textbook of Mineralogy', 4th edn, Wiley, New York, 1957.
H. H. Read, 'Rutley's Elements of Mineralogy', 25th edn, George Allen and Unwin, London, 1962.
M. P. Jones and M. G. Fleming, 'Identification of Mineral Grains', Elsevier, London, 1965.
'Mineral Facts and Problems Bulletin 630', Bureau of Mines, US Department of the Interior, Washington, 1965.
S. J. and M. G. Johnstone, 'Minerals for the Chemical and Allied Industries', 2nd edn, Chapman and Hall, London, 1961.
C. J. Smithells, 'Metals Reference Book', Vol. 1, Butterworths, London, 1962.
V. M. Goldschmidt, 'Geochemistry' (Ed. A. Muir), Oxford University Press, 1954.
D'Arcy George, 'Mineralogy of Uranium and Thorium Bearing Minerals', US Atomic Energy Commission, RMO 563, 1949.
G. Frondel, 'Systematic Mineralogy of Uranium and Thorium', Geological Survey Bulletin 1064, US Department of the Interior, 1958.
K. A. Vlasov, 'Geochemistry and Mineralogy of Rare Elements and Genetic Types of their Deposits', Israel Programme for Scientific Translations, Jerusalem, 1968.
W. Uytenbogaart and E. A. J. Buske, 'Tables for Microscopic Identification of Ore Minerals', 2nd edn, Elsevier, Amsterdam, 1971.

Economic factors

W. Ryan, 'Non-ferrous Metallurgy in the UK', Institution of Mining and Metallurgy, London, 1968.
J. D. Gilchrist, 'Extraction Metallurgy', Pergamon Press, Oxford, 1967.
Kirk-Othmer, 'Encyclopedia of Chemical Technology', 2nd edn, Interscience, New York, 1970.
'Mining Annual Review, 1979', Mining Journal, London, 1979.
'Mining Journal', London.
'Metal Bulletin', London.
World Mineral Statistics, 1972–76, Institute of Geological Sciences, HMSO, London, 1979.
World Metal Statistics, World Bureau of Metal Statistics, London.

Table 7.1 ORE GRADES AND SOURCES

Element abundance p.p.m.	*Minerals*	*Formulae*	*Metal content* %	*Specific gravity* g cm^{-3}	*Major mineral sources*	*Normal ore grade*	*Major metal producers*	*1979 world production of metal* tonnes
1	*2*	*3*	*4*	*5*	*6*	*7*	*8*	*9*
Aluminium 81 300	**Bauxite**	Hydrous aluminium and iron oxides	25–39	2.55	Australia, Guinea, Jamaica, Surinam, USSR, Guyana	25–39% Al	USA, USSR, Japan, Canada, W. Germany	15×10^6
Antimony	Senarmonite	Sb_2O_3	84	5.3	Bolivia, China, S. Africa, USSR, Canada, Thailand	2–25% Sb	—	64.5×10^3
	Valentinite	Sb_2O_3	84	5.60				
	Kermesite	Sb_2S_2O	75.3	4.60				
1	**Stibnite**	Sb_2S_3	71.7	4.5–4.6				
Arsenic	(**Arsenopyrite**)	FeAsS	40–49.9	5.9–6.3	Sweden, France, Namibia, Mexico	2–15% As	Sweden, France	~1300
	(Orpiment)	As_2S_3	60.9	3.49				
2	(Realgar)	AsS	70	3.48–3.56				
	(**Sulpharsenides** and **Arsenides** of Cu, Pb, Au, Sn)		15–23 (in flue dusts)					
Beryllium	**Bertrandite**	$Be_4Si_2O_7(OH)_2$	15.1	2.59–2.66	Brazil, USSR, Argentina, Zimbabwe USA	0.1–0.6% Be	USA, USSR	200–300
2	**Beryl**	$Be_3Al_2Si_6O_{18}$	5	2.7				
Bismuth (0.2)	(Native bismuth)	Bi	100	9.7–9.8	Peru, Bolivia, Australia, Mexico, Canada	0.05–0.8% Bi	Peru, Bolivia, Australia, USA	4.1×10^3 (1978)
	Bismuthinite	Bi_2S_3	81.2	6.4–6.5				
Cadmium (0.15)	(in **Sphalerite**)	—	Trace to 1.66	—	—	0.1–0.3% Cd (in Sphalerite)	USA, Japan, USSR, Germany, Belgium, Canada	18.4×10^3
Calcium	**Limestone**	Chiefly $CaCO_3$	40	2.71	Worldwide	Pure	USA, France, Germany	—
36 300	Brines		3.6 (Variable)					—
Cerium	Bastnaesite	(Ce, La, Di)(CO_3)F	50	4.9–5.2	Australia, India, Brazil, S. Africa, USA, Malaysia, China	0.1–3% Ce	—	37×10^3 (rare earth oxides)
(25)	**Monazite**	(Ce, La, Y, Th) PO_4	46	4.6–5.4				

Chromium 200	**Chromite**	$FeCr_2O_4$	38.2–46.5	4.5–4.8	USSR, S. Africa, Turkey, Zimbabwe, Phillipines, Albania, Finland	20–30% Cr	USSR, USA, UK, Japan	3×10^6 (Metal and alloys)
Cobalt	**(Safflorite)**	$(Co, Fe)As_2$	13–28	6.9–7.3	Zaire, Canada, Zambia, USSR, Cuba, Morocco	0.1–1% Co	—	24.25×10^3 (exl. USSR, China, Cuba)
	(Skutterudite)	$(Co, Ni)As_3$	11–21.7	6.5–6.9				
25	**(Smaltite)**	$(Co, Ni)As_2$	13.8–24	5.7–6.8				
	(Sphaerocobaltite)	$CoCO_3$	54	4.1				
	(Erythrite)	$(Co, Ni)_3(AsO_4)_2{\cdot}8H_2O$	18.8–26.6	3.06				
	(Asbolite)	Hydrous Mn oxide	<13	2.8–4.4				
	(Heterogenite)	Hydrous cobaltic/cobaltous oxide	Variable	3.44				
	(Cobaltite)	$(Co, Fe)AsS$	28.5–35.5	6.0–6.3				
	(Carrollite)	Co_2CuS_4	35–36	4.8				
	(Linnaeite)	Co_3S_4	58	4.8–5.0				
Copper	Native copper	Cu	100	8.95	*See* metal producers	0.4–3% Cu	USA, USSR, Japan, Zambia, Chile, Canada, Germany, Zaire, South Africa	7.9×10^6 (Free World mine production)
	Azurite	$Cu_3(OH)_2(CO_3)_2$	55	3.77				
70	**Malachite**	$Cu_2(OH)_2(CO_3)$	58	4.05				
	Cuprite	Cu_2O	88.8	5.9–6.2				
	Chrysocolla	$CuSiO_3{\cdot}2H_2O$	36.2	2.0–2.2				
	Brochantite	$Cu_4(SO_4)(OH)_6$	56	3.97				
	Chalcanthite	$CuSO_4{\cdot}5H_2O$	25	2.1–2.3				
	Atacamite	$Cu_2(OH)_3Cl$	59.4	3.76				
	Chalcocite	Cu_2S	79.9	5.5–5.8				
	Bornite	Cu_5FeS_4	63.5	4.9–5.4				
	Chalcopyrite	$CuFeS_2$	34.6	4.1–4.3				
	Covellite	CuS	66.7	4.6–4.76				
	Enargite	Cu_3AsS_4	45.7–49.0	4.45				
	Tennantite	$Cu_8As_2S_7$ variable	57.5 var.	4.4–4.5				
	Tetrahedrite	$(CuFe)_{12}Sb_4S_{13}$	25.0–45.7	4.6–5.1				
Gallium	(Gallite)	$CuGaS_2$	35.5	4.2	S. Africa, Jamaica, Australia	0.003–0.01% 0.001–0.05% Up to 1% Ga 0.005–0.02% (in associated minerals)	USA, Canada Switzerland	20 (estimated)
	(in **Bauxite**)		<0.01					
15	(in Coal ash)		<0.05					
	(in Germanite)		<1.85					
	(in **Sphalerite**)		<0.02					
Germanium	Renierite	$(Cu, Fe)_3(Fe, Ge, Zn, Sn)(S, As)_4$	6.4–7.8	4.46–4.59	Namibia, Zaire, USA	0.005–0.25% 0.03–0.3% Ge (in associated minerals)		~100
	Germanite	$(Cu, Ge^3)(S, As)$	6.0–10.2	4.5–4.6				
1–2	(in **Coal ash**)		<0.25					
	(in **Sphalerite**)		<0.3					

Table 7.1 ORE GRADES AND SOURCES—*continued*

Element abundance p.p.m.	*Minerals*	*Formulae*	*Metal content* %	*Specific gravity* g cm^{-3}	*Major mineral sources*	*Normal ore grade*	*Major metal producers*	*1979 world production of metal* tonnes
1	*2*	*3*	*4*	*5*	*6*	*7*	*8*	*9*
Gold	**Native gold**	Au	74–99.9	19.3	S. Africa, USSR, Canada USA, Australia, America	0.0007–0.003% Au 5–20 dwts-tonne^{-1} Au	(*See* mineral producers)	~1700
	Alloys with Hg							
(0.005)	Ag, Cu, Fe, Pd, Rh							
	Calaverite	$AuTe_2$	39–43.6	9.0–9.3				
	Nagyagite	$Pb_5Au(Te, Sb)_4S_{5-8}$	7.4–10.2	7.4				
	Krennerite	$AuTe_2$	30.7–43.9	8.6				
	Petzite	Ag_3AuTe_2	19.0–25.4	8.7–9.02				
	Sylvanite	$(Ag, Au)Te_2$	24.2–29.9	8.16				
	(Anode slime from Cu, Pb, Ag, Ni extraction)							
Hafnium	(in **Zircon**)		0.5–2.0		Australia, USA	0.5–2% Hf	USA	100 (estimated)
	(in Cyrtolite)		<31			Up to 31% Hf (in associated minerals)		
(4.5)								
Indium	(in **Sphalerite**)		<0.01		*See* metal producers	Up to 0.1% In (in associated minerals)	Canada, Japan, USA, Germany	40
	(in **Smithsonite**)		<0.01					
0.11								
Iron	Siderite	$FeCO_3$	48.3	3.48–3.96	USSR, USA, France, Australia, Canada, China, Sweden, Brazil, India	20–70% Fe	USSR, USA, Japan, Germany, France, UK	745 × 10^6 (steel)
	Goethite	$HFeO_2$	63.0	3.3–4.3				
50 000	**Hematite**	Fe_2O_3	69.9	4.9–5.3				
	Limonite	Hydrous iron oxides	60.0	2.7–4.3				
	Magnetite	$FeFe_2O_4$	72.4	5.2				
	Pyrite	FeS_2	46.5	4.8–5.02				
	Pyrrhotite	$Fe_{1-x}S(x=0–0.2)$	58.0–63.5	4.4–4.6				
	(Ilmenite)	$FeTiO_3$	1–13.5	4.7				
Lead	Cerussite	$PbCO_3$	77.0	6.55	USSR, Australia, USA, Mexico, Canada, Peru, Yugoslavia, Morocco, S. Africa, Sweden	2–15% Pb	USA, USSR, Germany, UK, Japan Australia	3.6 × 10^6 (World mine production)
	Anglesite	$PbSO_4$	68.0	6.38				
15	**Galena**	PbS	86.6	7.4–7.6				

Lithium	**Amblygonite**	$(Li, Na)(Al)PO_4(F, OH)$	4.8	3.0–3.1	Canada, USA, Brazil, Zimbabwe, Namibia, Argentina, USSR, Spain, Zaire	0.2–1.0% Li	USSR, USA	
	Eucryptite	$LiAlSiO_4$	5.5	2.67				
30	**Lepidolite**	$K(Li, Al)_3(Si, Al)_4O_{10}(F, OH)$	2.0	2.8–3.3				
	Petalite	$Li(Al, Si_4)O_{10}$	2.3	2.4				
	Spodumene	$LiAlSi_2O_6$	3.7	3.1–3.2				
	Brines		~0.05					
Magnesium	Dolomite	$CaMg(CO_3)_2$	22	2.8–2.9	Worldwide	0.1–0.2% Mg (in solution)	USA, USSR, Norway, Canada	260×10^3
	Brines							
20 900								
Manganese	Rhodochrosite	$MnCO_3$	47.8	3.7	USSR, S. Africa, Brazil, India, Australia	40–60% Mn	—	22×10^6 (ore)
	Hausmannite	$MnMn_2O_4$	71.6	4.8				
1 000	**Psilomelane**	$BaMn^2Mn_8O_{16}(OH)_4$	<51	4.7				
	Pyrolusite	βMnO_2	63	4.4–5.0				
	Manganite	$MnO(OH)$	62	4.3				
	Braunite	Oxide of Mn and Si	51–63	4.7–4.8				
Mercury	**Cinnabar**	HgS	86.2	8.1	USSR, Italy, Spain, Mexico, Canada, USA, Turkey, Yugoslavia	0.2–1% Hg	See mineral producers	7.5×10^3
(0.03)								
Molybdenum	(Wulfenite)	$PbMoO_4$	24.6–33.3	6.5–7.0	USA, Chile, USSR, Canada, Japan	0.002–0.2% Mo	See mineral producers	157×10^3 (Ore)
	Molybdenite	MoS_2	59.9	4.6–4.7				
1	Molybdite	MoO_3	66.7	4.5				
Nickel	**Garnierite**	$(Ni, Mg)_3Si_2O_5(OH)_4$	<46	2.2–2.8	Canada, USSR, New Caledonia, Cuba, Indonesia	0.1–2% Ni	Canada, New Caledonia, USSR, Japan, UK	698×10^3
	Pentlandite	$(Fe, Ni)_9S_8$	34–35	4.6–5.0				
80								
Niobium	**Columbite**	$(Fe, Mn)(Nb, Ta)_2O_6$	24–55	5.2–6.4	Brazil, Canada, Norway, USSR, USA, Nigeria, Zaire, Australia	0.4–3% Nb	USA, Canada	26×10^3
	Pyrochlore	$NaCaNb_2O_6F$	16–51	3.77–4.95				
(25)								
Platinum	**Native platinum**	Pt	60–90	14–19	USSR, S. Africa, Canada	—	See mineral producers	~100
	Sperrylite	$PtAs_2$	52.5–56.6	10.58				
(0.005)	Braggite	PtS	58–60	10.0				
	Cooperite	PtS	80–86	9.5				
	Alloyed with other PtGp, CuNiAu							

Table 7.1 ORE GRADES AND SOURCES—*continued*

Element abundance p.p.m.	*Minerals*	*Formulae*	*Metal content* %	*Specific gravity* g cm^{-3}	*Major mineral sources*	*Normal ore grade*	*Major metal producers*	*1979 world production of metal* tonnes
1	*2*	*3*	*4*	*5*	*6*	*7*	*8*	*9*
Potassium 27 000	Carnalite **Sylvite** Brines	$KMgCl_3.6H_2O$ KCl	14 52.4 0.5–16.6	1.60 1.99	USSR, Canada, USA, W. Germany, E. Germany, France	10–30% K 0.5–16% K	USA, Germany ' Japan, USSR	26×10^6 (Potash K_2O)
Rhenium (0.001)	(in **Molybdenite**)		<1.0		*See* associated mineral.	0.001–1% Re in molybdenite	USA, USSR, Germany, France, Sweden, Belgium, UK	7.2
Selenium 0.09	(Anode slime from Cu extraction)	—	3–28	—	—	3–28% Se in slime	USA, Canada, Belgium, Germany, Japan, Sweden	1500 (excl. USSR. China)
Sodium 28 300	**Halite** **Brines** Sea-water	NaCl	39.4 ~10.0 ~1.0	2.17	Worldwide	Pure ~10% Na ~1% Na	USA, USSR, Germany, UK, Japan, Australia	—
Silicon 277 000	**Quartz**	SiO_2	46.7	2.65	Brazil	Pure	USA, UK, Germany, France, Japan, Denmark, Italy	—
Silver (0.04–0.1)	Native silver **Argentite** **Polybasite** **Proustite** **Pyrargarite** **Stephanite** (By-product of Pb, Cu, Zn, Au, Ni, Sn)	Ag Ag_2S $(Ag, Cu)_{16}Sb_2S_{11}$ Ag_3AsS_3 Ag_3SbS_3 Ag_5SbS_4	95–98.5 87 58–74.3 64.5–65.4 59.5–60.8 68.3	10.1–11.1 7.2–7.4 6.1 5.5–5.64 5.77–5.87 6.25	Canada, Mexico, USA, USSR, Peru	0.01–0.1% Ag	See mineral producers	8.3×10^3 8.2×10^3 (New Mine Production)
Tantalum (2.1)	**Tantalite** (in Tin slags)	$(Fe, Mn)(Ta, Nb)_2O_6$	36–56 ~7	5.3–7.3	Canada, Brazil, Spain	0.1–0.5% Ta	USA	1.15×10^3 (excl. USSR)

Tellurium 0.01	(Anode slime from Cu extraction) (Flue dusts Cu, Ag, Au, Pb, Bi)	—	<8	—	*See* metal producers	Up to 8% in slimes	USA, Canada, USSR, Japan, Peru	290 (excl. USSR)
Thallium 1.3	(By-product from Zn, Cd refining) (From flue dusts smelting Pb, Cu, Zn ores)	—	0.1–0.5	—	*See* metal producers	0.05–0.1% Tl (in flue dusts)	USA, Belgium, Germany, USSR	—
Thorium 11.5	Thorianite	ThO_2	35–83	9.7	India, Brazil, Australia, Sri Lanka, Indonesia, USA	0.1–1% Th	Australia, Brazil	37×10^3 (rare earth oxides)
	Thorite	$ThSiO_4$	71	4.5–5.4				
	Monazite	(Ce, La, Y, Th)(PO_4)	5.1–10.8	4.6–5.4				
Tin 2	**Cassiterite**	SnO_2	78.6	6.8–7.1	Malaysia, Bolivia, Thailand, Indonesia	0.5–3% Sn (hard rock) >0.02% Sn (alluvial)	Malaysia, Thailand, UK	242×10^3
Titanium 4 400	**Ilmenite**	(Fe, Mg, Mn)TiO_3	31.2–33.6	4.5–5.0	USSR, Australia, USA, Sri Lanka, Brazil India, Canada	2–40% Ti	USA, USSR, Japan, UK	80.6×10^3
	Rutile	TiO_2	60	4.23				
Tungsten (3)	Ferberite	$FeWO_4$	60	7.2–7.5	China, USSR, USA, N. Korea, S. Korea, Bolivia	0.4–3% W	China, USSR	95×10^3 (in concentrates)
	Hubnerite	$MnWO_4$	60	7.2–7.5				
	Scheelite	$CaWO_4$	59–64	5.9–6.1				
	Wolframite	(Fe, Mn)WO_4	60	7.1–7.5				
Uranium 4	Brannerite	Oxide of Ti, U, Ca, minor Y, Th, Fe	37	4.5–5.4	USA, S. Africa, Canada, France	0.1–0.25% U	USA, Japan, UK, Germany	33×10^3 (in concentrates)
	Carnotite	$K_2(UO_2)_2(VO_4)_2.3H_2O$	52–54	<4.7				
	Uraninite **Pitchblende**	UO_2 (normally $U_{3-4}O_8$)	54–80.7	6.5–10.8				
	Autunite	$Ca(UO_2)_2(PO_4)_2 10\text{–}12H_2O$	48–51	3.1–3.2				
	Torbernite	$Cu(UO_2)_2(PO_4)_2.8\text{–}12H_2O$	47–50.6	3.2				
	Coffinite	Silicate of U	41–60	2.2–5.1				

Table 7.1 ORE GRADES AND SOURCES—*continued*

Element abundance p.p.m.	*Minerals*	*Formulae*	*Metal content* %	*Specific gravity* g cm^{-3}	*Major mineral sources*	*Normal ore grade*	*Major metal producers*	*1979 world production of metal* tonnes
1	*2*	*3*	*4*	*5*	*6*	*7*	*8*	*9*
Vanadium	**(Carnotite)**	$K_2(UO_2)_2(VO_4)_2.3H_2O$	11.5	<4.7	USSR, S. Africa, USA, Namibia, Finland, Norway	0.5–2% V	USSR, USA, S. Africa	62×10^3
	Descloizite	$(ZnCu)Pb(VO_4)(OH)$	9.0–13.0	~6.2				
150	**Roscoelite**	$2K_2O.2Al_2O_3(Mg, Fe)O$ $3V_2O_5.10SiO_2.4H_2O$	11.8	2.97				
	Vanadinite	$Pb_5(VO_4)_3Cl$	9.5–10.6	6.5–7.1				
	Patronite	VS_4	28.5	—				
	(Associated with iron ores)		0.8–0.9					
Zinc	Smithsonite	$ZnCO_3$	29.0–52.0	4.0–4.45	Canada, USA, Australia, Peru, Japan, Italy	5–15% Zn	USA, Japan, Canada, Australia, Germany	6.36×10^6 (World Mine Production)
	Hemimorphite	$Zn_4Si_2O_7(OH)_2.H_2O$	54	3.45				
50	**Sphalerite**	(Zn, Fe)S	51–67	3.9–4.1				
Zirconium	Baddeleyite	ZrO_2	71.4–73.4	5.4–6.0	Australia, Brazil, USA	Up to 10% Zr		728×10^3 (zircon sand)
220	**Zircon**	$ZrSiO_4$	49.7	4.2–4.7				

8 Thermochemical data

Except where otherwise indicated, the data given in these tables have been selected from three main sources: 'Selected Values of the Thermodynamic Properties of the Elements', by R. Hultgren, P. O. Desai, D. T. Hawkins, M. Gleiser and K. K. Kelley, and by the same authors, 'Selected Values of the Thermodynamic Properties of Binary Alloys'; also 'Metallurgical Thermochemistry', 5th edn., by O. Kubaschewski and C. B. Alcock. These works represent the most authoritative compilations of critically assessed data presently available, to which reference should be made for original sources or details of assessment.

8.1 Symbols

θ_m = melting point in °C
θ_e = boiling or sublimation point in °C at 760 mmHg
θ_t = transition temperature in °C
θ_s = sublimation point in °C
T = absolute temperature in K
L_m = latent heat of fusion
L_e = latent heat of vaporization
L_t = latent heat of transition
L_s = latent heat of sublimation
(L_m, L_e, L_t, L_s) in kJ mol^{-1} (or kJ g-atom^{-1}) or in Jg^{-1}
ΔV_m = volume change during melting $[(V_{liq} - V_{solid})/V_{solid}]\%$
ΔH_{298} = heat of formation at 298 K (25°C) in kJ mol^{-1} (or kJ g-atom^{-1}) or in Jg^{-1}. (The value of a heat evolved during a reaction is taken to be negative)
ΔG = maximum work (change of free energy) in kJ mol^{-}1 (or kJ g-atom^{-1})
S_{298} = standard entropy at 298 K (25°C) in Jk^{-1} mol^{-1} (or J g-atom^{-1})

p (mm Hg) = vapour or dissociation pressure in mmHg
N_1 = mol fraction of the first component
N_2 = mol fraction of the second component
C_p = specific heat in J k^{-1} mol^{-1}
c_p = specific heat in J g^{-1} k^{-1}

8.2 Changes of phase

Table 8.1 ELEMENTS
Latent heats and temperatures of fusion, vaporization and transition, and change in volume on melting

Element	*Melting point* θ_m °C	*Boiling point* θ_e °C	θ_t °C	L_m *at m.p.* kJ g-atom^{-1} *or* mol^{-1}	L, kJ g-atom^{-1} *or* mol^{-1} L_s *at* 25°C	L_e *at b.p.*	L_t kJ g-atom^{-1} *or* mol^{-1}	ΔV_m %
Ag	960.8	2 200	— —	11.09	284.2	257.8	—	(3.8)
Al	660.1	2 520	—	10.47	321.9	290.9	—	6.5
Am	—	2 600	—	—	—	238.6	—	—
As	817	603	—	$As_4 = 4As + 118.1$ kJ			—	10
Au	1 063	2 860	—	12.78	378.9	342.4	—	5.1

Table 8.1 ELEMENTS—*continued*

Element	*Melting point* θ_m °C	*Boiling point* θ_e °C	θ_t °C	L_m *at m.p.* kJ g-atom^{-1} *or* mol^{-1}	*L*, kJ g-atom^{-1} *or* mol^{-1}		L_t kJ g-atom^{-1} *or* mol^{-1}	ΔV_m %
					L_s *at* 25 °C	L_e *at b.p.*		
B	2 180	3 800	—	22.6	577.8	—	—	—
Ba	729	1 700	370	7.66	(192)	177.1	0.59	—
Be	1 287	2 470	1 254	12.22	324.4	292.6	2.55	—
Bi	271	1 564	—	10.89	207.2	179.2	—	−3.35
Br_2	−7.3	58	—	10.55	—	30.56	—	—
C (graph)	(3 800)	(5 000)	—	—	712	—	1.90 diam→graph	—
Ca	843	1 484	464	8.36	176.2	150.7	0.25	—
Cd	320.9	767	—	6.41	112.2	99.6	—	4.0
Ce	798	3 430	726	5.23	(407)	376.0	(2.9)	—
Cl_2	−101.0	−34.1	—	6.41	—	20.423	—	—
Co	1 495	2 930	{440 1 120}	(15.5)	425	—	0.25, 0.92	3.5
Cr	1 857	2 672	—	(20.9)	397	342.1	—	—
Cs	29.8	700	—	2.09	78.7	66.6	—	2.6
Cu	1 083.4	2 560	—	13.02	341.2	304.8	—	4.2
Dy	1 409	2 560	1 384	—	—	—	—	—
Er	1 522	2 860	1 470	—	—	—	—	—
Eu	826	1 490	—	—	—	—	—	4.8
F_2	−219.6	−188.0	—	1.595	—	6.531	—	—
Fe	1 536	2 860	{914 1 391}	15.2	398.6	340.4	{β5.11* γ0.563* $\gamma\rightarrow\delta$0.84}	3.5
Ga	29.7	2 420	—	5.594	285.0	270.5	—	−3.2
Gd	1 312	3 290	1 260	—	—	—	—	—
Ge	937	2 830	—	36.8	383.8	327.8	—	−5.1
H_2	−259.2	−252.5	—	0.117	—	0.909	—	(12.3)
Hf	2 227	4 600	1 940	24.07	611.3	571.1	(6.91)	—
Hg	−38.87	357	—	2.324	—	61.1	—	3.7
I_2	113.6	183	—	15.78	62.4	41.9	—	21.6
In	156.4	2 070	—	3.27	242.8	232.4	—	2.0
Ir	2 443	4 430	—	(26)	669.9	612.5	—	—
K	63.2	779	—	2.39	90.0	79.5	—	2.55
La	920	(3.420)	868	(8.37)	422.9	402.4	(2.9)	—
Li	181	1 324	—	2.93	161.6	147.8	—	(1.65)
Mg	649	1 090	—	8.79	146.5	127.7	—	4.12
Mn	1 244	2 060	{710 1 090 1 136}	(14.7)	291.0	231.1	{2.22 2.22 1.80}	(1.7)
Mo	2 620	4 610	—	35.6	664.5	590.3	—	—
N_2	−210.0	−195.8	−237.5	0.720	—	5.581	0.229 0	7.3
Na	97.8	883	—	2.64	108.9	98.0	—	2.5
Nb	2 467	4 740	—	29.3	722.2	683.7	—	—
Nd	1 016	3 070	862	7.14	323.6	—	2.98	—
Ni	1 453	2 910	358	17.16	429.6	374.3	0.58	4.5
Np	637	—	280, 577	—	—	—	—	—
O_2	−218.8	−183.0	{−249.5 −229.4}	0.445	—	6.8	{0.093 8 0.743 6}	7.4
Os	3 030	5 030	—	—	791	—	—	—
P (yellow)	44.1	280	—	2.64	{140.7(P_2) 58.8(P_4)}	{51.9(P_4)}	17.6 yell.→red	3.5
Pb	327.4	1 750	—	4.81	196.4	178.8	—	3.5
Pd	1 552	2 940	—	(16.7)	377.2	361.7	—	—
Po	246	965	(100)	—	—	100.9	—	—
Pr	932	3 510	798	(11.3)	—	—	—	—

* Only together with the tabulated heat capacities of iron: Table 8.10.

Table 8.1 ELEMENTS—*continued*

Element	*Melting point* θ_m °C	*Boiling point* θ_e °C	θ_t °C	L_m *at m.p.* kJ g-atom^{-1} *or* mol^{-1}	*L*, kJ g-atom^{-1} *or* mol^{-1} L_s *at* 25 °C	L_e *at b.p.*	L_t kJ g-atom^{-1} *or* mol^{-1}	ΔV_m %
Pt	1 769	4 100	—	(19.7)	545.0	469.2	—	—
Pu	640	3 420	122, 205, 318, 452, 476	2.9	343.7	352.0	3.39, 0.59, 0.54, 0.08, 1.84	−2.5
Ra	700	1 500	—	—	—	—	—	—
Rb	38.8	688	—	2.198	87.5	75.8	—	2.5
Re	3 180	5 690	—	33.5	779.2	(712)	—	—
Rh	1 966	3 700	(1 030)	(22.6)	556.0	(494)	—	—
Ru	2 250	4 250	—	—	—	—	—	—
S (rhomb.)	112.8	444.5	95.5	1.235	—	*	0.38	(5.1)
S (mon.)	119.0	—	—	—	—	—	—	—
Sb	630.5	1 590	—	19.89	—	(167)(Sb_2)	—	0.8
Sc	1 538	(2 870	1 334	—	376.0	—	—	—
Se (met.)	220.5	685	—	6.28	—	95.5(Se_2)	—	15.8
Si	1 412	3 270	—	50.66	450.1	384.8	—	(−10)
Sm	1 072	1 803	917	8.92	207.2	165.0	3.10	—
Sn	231.9	2 625	13	7.08	302.3	296.4	2.22	2.3
Sr	770	1 375	235, 540	(8.4)	177.1	154.5	—	—
Ta	3 015	5 370	—	(24.7)	782.5	—	—	—
Tb	1 360	3 220	1 290	—	—	—	—	—
Te	450	988	—	17.6	171.6(Te_2)	104.7(Te_2)	—	4.9
Th	1 750	4 790	1 325	—	576.1	(511)	—	—
Ti	1 667	3 285	882	(17.5)	469.3	425.8	3.34	—
Tl	304	1 473	234	4.3	180.9	166.2	0.38	2.2
U	1 132	4 400	662, 770	12.5	482.2	417.4	2.85, 4.878	—
V	1 902	3 410	—	16.74	510.2	457.2	—	—
W	3 400	5 555	—	35.2	847.8	(737)	—	—
Y	1 530	3 300	1 485	11.43	424.9	367.6	5.0	—
Yb	824	1 194	760	—	—	—	—	—
Zn	419.5	907	—	7.28	129.3	114.3	—	4.7
Zr	1 852	4 400	852	(19.3)	612.1	579.9	3.85	—

* L_e at b.p.: S_2, 106.4 (625 °C); S_4, 96.0 (625 °C); S_6, 66.2 (527 °C); S_8, 63.1 (490 °C).
ΔV_m, ref. 4.

Table 8.2a INTERMETALLIC COMPOUNDS

Latent heats and temperatures of fusion

If an intermetallic phase is completely disordered, the entropy of fusion (L_m/T_m) can generally be calculated additively from the entropies of fusion of the components. If it is completely ordered, $-19.146\,(N_1 \log N_1 + N_2 \log N_2)$ is as a rule to be added to the calculated entropy of fusion.

Phase	N_2 10^{-2}	θ_m °C	L_m kJ g-atom^{-1}	J g^{-1}
δ-Ag-Cd	67.5	592	8.46 ± 0.42	76.2
γ-Ag-Zn	61.8	664	7.79 ± 0.33	95.5
δ-Ag-Zn	72.1	632	8.75 ± 0.42	131.5
Al_2Cu	33.3	590–605	12.6 ± 0.8	320.3
γ-Al-Mg	57.2	455	8.8 ± 0.8	346.7
AuCd	50.0	627	8.96 ± 0.50	57.8
$AuPb_2$	66.7	254–300	8.00 ± 0.75	39.4
AuSn	50.0	418	12.81 ± 0.33	81.2
β-Au-Zn	50.0	760	12.31 ± 0.54	93.8
ε-Au-Zn	88.9	490	7.45 ± 0.38	93.4

Table 8.2a INTERMETALLIC COMPOUNDS—*continued*

Phase	N_2 10^{-2}	θ_m °C	L_m kJ g-atom^{-1}	J g^{-1}
Bi-In	50.0	110	—	7.33
Bi–In_2	66.7	—	—	4.81
δ-Bi-Tl	40.0	214	7.24±0.25	35.2
δ-Cd-Cu	38.4	555	9.71±0.21	103.8
δ-Cd-Cu	40.0	562	9.55±0.21	103.0
Cd–Mg	25.0	349–368	5.86	56.5
	50.0	415–430	6.05	87.9
	75.0	489–515	6.91	152.4
Cd_2Na	33.3	385	7.87±0.63	95.5
CdSb	50.0	456	16.0±0.33	136.9
GaLi	50.0	700–760	16.7±0.8	435.4
GaSb	50.0	703	25.1±1.7	131.0
Hg_2Na	33.3	350	8.8±0.8	63
γ-Hg–Tl	28.6	14.5	2.01±0.13	10.0
InLi	50.0	630	13.4±0.8	219.8
InSb	50.0	425	24.7±0.8	205.2
KNa_2	66.7	7	2.9±0.1	103.4
Mg_2Pb	33.3	550	13.4±1.3	155
$MgZn_2$	66.7	590	14.3±0.8	260
Na_5Pb_2	28.5	400	7.18±0.5	95
NaPb	50.0	368	8.4±0.4	71
β Na–Pb	71.5	320	7.1±0.4	46
NaTl	50.0	250–305	8.4±0.8	73.7
γ-Pb–Tl	87.5	329–339	5.23±0.17	25.5
γ-Pb–Tl	63.0	379·	5.65±0.17	27.6

Ref. 3.

Table 8.2b INTERMETALLIC COMPOUNDS
Latent heats and temperatures of transition
The method of measurement is subject to error. Most of the reported values are probably too low.

Phase	N_2 10^{-2}	*Transition*	θ_t °C	L_t *J g-atom*$^{-1}$
β-Ag–Cd	50.0	β′–β	211	712
AgZn	50.0	order–disorder	258	2 449
AuCu	50.0	order–disorder	408	1 779
$AuCu_3$	75.0	order–disorder	390	1 214
$AuSb_2$	66.7	β–γ	355	335
Cd_3Mg	27.0	order–disorder	95	963
CdMg	50.0	order–disorder	50–260	2 638
$CdMg_3$	75.3	order–disorder	80–165	1 256
CrFe	52.0	σ–α	805	1 633
CuPt	50.0	order–disorder	800	3 810
Cu_3Pt	20.0	order–disorder	610	1 968
β-Cu–Zn	50.0	β′–β	470	2 219
$FeNi_3$	74.3	order–disorder	506	2 721
$MnNi_3$	75	order–disorder	(500)	3 119
Pd_3Sb	25	β–β′	950	10 300
Zn_3Sb_2	40.0	—	409–455	6.071

Ref. 3.

Table 8.3 OTHER METALLURGICALLY IMPORTANT COMPOUNDS
Latent heats and temperatures of fusion, vaporization and transition
(If not stated otherwise, the values of the latent heats are for the temperatures of transition, fusion, evaporation or sublimation, respectively. Boiling and sublimation points are for 1 atm pressure of the undissociated molecules)

Compound	$\theta_m, \theta_e, \theta_s$ *or* θ_t °C	L_m, L_e, L_s *or* L_t kJ mol^{-1}
	Carbides	
CH_4	θ_m, −182.5	L_m, 0.938
	θ_e, −161.4	L_e, 8.323
Fe_3C	θ_t, 190	L_t, 0.75
	θ_m, 1 227	L_m, 51.62
Mn_3C	θ_t, 1 037	L_t, 15.1
	Halides	
AgCl	θ_m, 455	L_m, 13.0
	θ_e, 1 564	L_e, 177.9
AgBr	θ_m, 430	L_m, 9.2
	θ_e, 1 560	L_e, 192.2
AgI	θ_m, 557	L_m, 9.42
	θ_e, 1 506	L_e, 144.24
	Hex.→cub.	
	θ_t, 147	L_t, 6.07
AlF_3	θ_s, 1 280	L_s, 280.5
Al_2Cl_6	θ_m, 193	L_m, 71.2
	θ_s, 160	L_s, 111.8
Al_2Br_6	θ_m, 97	L_m, 22.6
	θ_e, 255	L_e, 45.6
AlI_3	θ_m, 191	L_m, 16.3
	θ_e, 385	L_e, 64.5
AsF_3	θ_m, −6	L_m, 10.38
	θ_e, 58	L_e, 29.7
AsF_5	θ_m −80	L_m, 11.47
	θ_e, −53	L_e, 20.9
$AsCl_3$	θ_m, −16	L_m, 10.13
	θ_e, 130	L_e, 31.4
$AsBr_3$	θ_m, 31	L_m, 11.7
	θ_e, 221	L_e, 47.7
AsI_3	θ_m, 142	L_m, 9.2
	θ_e, 424	L_e, 59.5
BaF_2	θ_m, 1 290	L_m, 28.5
	θ_e, 2 382	L_e, 270
$BaCl_2$	θ_m 962	L_m, 16.7
	θ_t, 922	L_t, 17.2
$BaBr_2$	θ_m, 854	L_m, 31.4
BaI_2	θ_m, 711	L_m, 26.4
$BeCl_2$	θ_m, 415	L_m, 8.7
	θ_t, 403	L_t, 16.8
	θ_e, 532	L_e, 104.7
$BeBr_2$	θ_m, 488	L_m, 18.9
	θ_e, 511	L_e, 100.0
	θ_s, 473	L_s, 125.6
BeI_2	θ_m, 480	L_m, 20.9
	θ_e, 482	L_e, 96.3
	θ_s, 488	L_s, (79.5)
$BiCl_3$	θ_m, 230	L_m, 23.9
	θ_e, 441	L_e, 72.4
$BiBr_3$	θ_m, 218	L_m, 21.8
	θ_e, 461	L_e, 75.4
CF_4	θ_m, −183.6	L_m, 0.701
	θ_e, −151	L_e, 13.063
CBr_4	θ_m, 90	L_m, 3.957
	θ_t, 47	L_t, 5.95
	θ_e, 190	L_e, 44.4
CCl_4	θ_m, −23	L_m, 2.51
	θ_e, 77	L_e, 30.6
CaF_2	θ_m, 1 418	L_m, 29.7
	θ_e, 2 510	L_e, 312.2
	θ_t, 1 151	L_t, 4.77
	Halides–continued	
$CaCl_2$	θ_m, 772	L_m, 28.5
$CaBr_2$	θ_m, 741	L_m, 28.9
CaI_2	θ_m, 779	L_m, (42)
CdF_2	θ_m, 1 072	L_m, 22.6
	θ_e, 1 750	L_e, 225.2
$CdCl_2$	θ_m, 568	L_m, 30.1
	θ_e, 961	L_e, 125.2
$CdBr_2$	θ_m, 565	L_m, 33.5
	θ_e, 863	L_e, 113.0
CdI_2	θ_m, 390	L_m, 20.9
	θ_e, 796	L_e, 106.3
$CeCl_3$	θ_m, 817	L_m, 53.6
$CeBr_3$	θ_m, 732	L_m, 51.9
CeI_3	θ_m, 760	L_m, 51.1
$CoCl_2$	θ_m, 740	L_m, 59.0
	θ_e, 1 025	L_e, 157.4
$CrCl_2$	θ_m, 815	L_m, 31.4
	θ_e, 1 304	L_e, 198.9
$CrCl_3$	θ_s, 945	L_s, 237.8
CsF	θ_m, 703	L_m, 21.8
	θ_e, 1 210	L_e, 155.7
CsCl	θ_m, 646	L_m, 20.5
	θ_e, 1 300	L_e, 159.9
	θ_t, 469	L_t, 2.5
CsBr	θ_m, 636	L_m, 23.66
	θ_e, 1 300	L_e, 150.7
CsI	θ_m, 621	L_m, 23.9
	θ_e, 1 280	L_e, 150.3
CuCl	θ_m, 430	L_m, 7.5
CuBr	θ_m, 488	L_m, (9.6)
CuI	θ_m, 588	L_m, (10.9)
$FeCl_2$	θ_m, 677	L_m, 43.1
	θ_e, 1 026	L_e, 126.4
$FeCl_3$	θ_m, 304	L_m, (43.1)
	θ_e, 315	L_e, 60.7 (Fe_2Cl_6)
FeI_2	θ_t, 375	L_t, 0.83
	θ_m, 590	L_m, 44.8
	θ_e, 935	L_e, 111.8
$GaCl_3$	θ_m, 78	L_m, 11.5
	θ_e, 302	L_e, 62.8
Ga_2Cl_6	θ_e, 201	L_e, 44.0
$GaBr_3$	θ_m, 122	L_m, 11.7
	θ_e, 314	L_e, 58.6
Ga_2Br_6	θ_e, 292	L_e, 50.2
GaI_3	θ_m, 212	L_m, 16.3
	θ_e, 349	L_e, 67.8
Ga_2I_6	θ_e, 462	L_e, 82.5
$GeCl_4$	θ_e, 84	L_e, 29.4
$GeBr_4$	θ_e, 189	L_e, 36.0
HF	θ_m, −83	L_m, 3.94
HCl	θ_m, −114.2	L_m, 1.99
	θ_e, −85.1	L_e, 16.161
HBr	θ_m, −86.9	L_m, 2.407
	θ_e, −67	L_e, 17.626
HI	θ_m, −50.8	L_m, 2.872
	θ_e, −35.4	L_e, 19.778
$HfCl_4$	θ_s, 316	L_s, 99.65
$HgCl_2$	θ_m, 278	L_m, 19.47
	θ_e, 304	L_e, 59.0
$HgBr_2$	θ_m, 238	L_m, 18.0
	θ_e, 319	L_e, 59.0

Table 8.3 OTHER METALLURGICALLY IMPORTANT COMPOUNDS—*continued*

Compound	θ_m, θ_e, θ_s *or* θ_t °C	L_m, L_e, L_s *or* L_t kJ mol^{-1}
	Halides—continued	
HgI_2 (yellow)	θ_m, 250	L_m, 18.8
	yellow→red	
	θ_t, 127	L_t, 2.72
HgI_2 (red)	θ_m, 256	L_m, 19.3
	θ_e, 354	L_m, 59.9
InCl	θ_m, 225	L_m, 9.2
	θ_e, 608	L_e, 88.4
$InCl_3$	θ_s, 498	L_s, 158.3
InBr	θ_m, 275	L_m, 24.3
	θ_e, 660	L_e, 95.0
$InBr_2$	θ_e, 630	L_e, 82.5
$InBr_3$	θ_s, 371	L_s, 108.4
InI	θ_m, 365	L_m, 22.4
	θ_e, 770	L_e, 96.7
IrF_6	θ_m, 44	L_m, 5.0
	θ_e, 53	L_e, 27.2
KF	θ_m, 857	L_m, 28.26
	θ_e, 1 510	L_e, 186.7
KCl	θ_m, 772	L_m, 26.6
	θ_e, 1 407	L_e, 162.4
KBr	θ_m, 740	L_m, 25.6
	θ_e, 1 383	L_e, 155.3
KI	θ_m, 685	L_m, 24.07
	θ_e, 1 330	L_e, 145.3
LiF	θ_m, 848	L_m, 26.8
	θ_e, 1 681	L_e, 214
LiCl	θ_m, 610	L_m, 19.89
	θ_e, 1 382	L_e, 150.7
LiBr	θ_m, 550	L_m, 17.6
	θ_e, 1 310	L_e, 148.2
LiI	θ_m, 469	L_m, 14.7
	θ_e, 1 171	L_e, 170.8
MgF_2	θ_m, 1 263	L_m, 58.2
	θ_e, 2 332	L_e, 292.2
$MgCl_2$	θ_m, 714	L_m, 43.1
	θ_e, 1 418	L_e, 136.9
$MnCl_2$	θ_m, 650	L_m, 37.7
	θ_e, 1 190	L_e, 123.9
MoF_5	θ_e, 214	L_e, 51.9
MoF_6	θ_m, 17	L_m, 4.2
	θ_e, 34	L_e, 28.1
$MoCl_5$	θ_m, 194	L_m, 34
	θ_e, 268	L_e, 68.8
NaF	θ_m, 992	L_m, 32.7
	θ_e, 1 710	L_e, 216.9
NaCl	θ_m, 801	L_m, 28.1
	θ_e, 1 465	L_e, 170.4
NaBr	θ_m, 750	L_m, 26.2
	θ_e, 1 392	L_e, 38.0
NaI	θ_m, 660	L_m, 23.7
	θ_e, 1 304	L_e, 159.5
NbF_5	θ_m, 77	L_m, 12.1
	θ_e, 233	L_e, 52.3
$NbCl_5$	θ_m, 205	L_m, 28.9
	θ_e, 250	L_e, 54.8
$NdCl_3$	θ_m, 760	L_m, 50.2
$NdBr_3$	θ_m, 684	L_m, 45.2
NH_4Cl	θ_t, 184	L_t, 4.3
NH_4Br	θ_t, 138	L_t, 3.2
$NiCl_2$	θ_m, 1 030	L_m, 77.5
$NiCl_2$	θ_s, 970	L_s, 225.3
$NiBr_2$	θ_s, 919	L_s, 224.8
NpF_6	θ_m, 54	L_m, 17.6
	θ_e, 55	L_e, 30.1
OsF_5	θ_e, 226	L_e, 65.7
	Halides—continued	
OsF_6	θ_m, 33	L_m, 7.5
	θ_e, 47.5	L_e, 28.1
PCl_3	θ_m, −92	L_m, 4.5
	θ_e, 74	L_e, 30.6
PCl_5	θ_s, 163	L_s, 64.9
PBr_3	θ_e, 174	L_e, 39.8
PbF_2	θ_m, 818	L_m, 17.4
	θ_e, 1 293	L_e, 160.4
$PbCl_2$	θ_m, 498	L_m, 24.3
	θ_e, 954	L_e, 123.9
$PbBr_2$	θ_m, 370	L_m, 18.0
	θ_e, 914	L_e, 116.0
PbI_2	θ_m, 410	L_m, 16.2
	θ_e, 872	L_e, 103.8
$PdCl_2$	θ_m, 678	L_m, 18.4
$PrCl_3$	θ_m, 786	L_m, 50.7
$PrBr_3$	θ_m, 693	L_m, 47.3
PrI_3	θ_m, 735	L_m, 53.2
PuF_3	θ_m, 1 426	L_m, (59.9)
	θ_e, (2120)	L_m, (321)
PuF_6	θ_m, 52	L_m, 17.6
	θ_e, 62	L_e, 30.1
$PuCl_3$	θ_m, 760	L_m, 63.6
	θ_e, 1 790	L_e, 186
$PuBr_3$	θ_m, 681	L_m, 51.5
	θ_e, 1 460	L_e, 193
RbF	θ_m, 775	L_m, 23.0
	θ_e, 1 390	L_e, 177.9
RbCl	θ_m, 717	L_m, (21)
	θ_e, 1 381	L_e, 165.8
RbBr	θ_m, 682	L_m, (18.4)
	θ_e, 1 352	L_e, 154.9
RbI	θ_m, 641	—
	θ_e, 1 304	L_e, 150.7
ReF_5	θ_e, 221	L_e, 58.2
ReF_6	θ_m, 19	L_m, 4.2
	θ_e, 34	L_e, 28.5
ReF_7	θ_e, 48	L_m, 7.5
	θ_e, 74	L_e, 36.0
SF_6	θ_m, −51	L_m, 5.9
	θ_s, −64	—
S_2Cl_2	θ_m, −80	—
	θ_e, 138	L_e, 36.4
$SbCl_3$	θ_m, 73	L_m, 12.6
	θ_e, 219	L_e, 43.5
$SbCl_5$	θ_m, 2	L_m, (10.5)
$SbBr_3$	θ_m, 97	L_m, 14.7
	θ_e, 280	L_e, (59.0)
SbI_3	θ_m, 170	L_m, 17.6
	θ_e, 401	L_e, 68.7
$ScCl_3$	θ_m, 966	L_m, 67.4
	θ_e, 967	L_s, 272.1
SeF_6	θ_m, −34	L_m, (8.4)
	θ_s, −46	L_s, 26.8
SiF_4	θ_s, −95	L_s, 25.8
$SiCl_4$	θ_m, −70	L_m, 7.75
	θ_e, 57	L_e, 28.9
$SiBr_4$	θ_e, 153	L_e, 38.1
SiI_4	θ_e, 301	L_e, 50.2
$SnCl_2$	θ_m, 247	L_m, 12.77
	θ_e, 652	L_e, 82.9
$SnCl_4$	θ_m, −33	L_m, 9.2
	θ_e, 115	L_e, 33.9
$SnBr_2$	θ_m, 232	L_e, 98.4
	θ_e, 639	L_m, (7.5)

Table 8.3 OTHER METALLURGICALLY IMPORTANT COMPOUNDS—*continued*

Compound	θ_m, θ_e, θ_s or θ_t °C	L_m, L_e, L_s or L_t kJ mol^{-1}
Halides—continued		
$SnBr_4$	θ_m, 30	L_m, 12.6
	θ_e, 205	L_e, 41.0
SnI_2	θ_m, 320	—
	θ_m, 714	L_e, 99.6
SnI_4	θ_e, 145	L_m, 19.3
	θ_e, 348	L_e, 50.2
SrF_2	θ_m, 1 477	L_m, 18.4
	θ_e, 2 480	L_e, 297.2
$SrCl_2$	θ_m, 874	L_m, 15.9
$SrCl_2$	θ_t, 730	L_t, 5.0
$SrBr_2$	θ_m, 657	L_m, 10.5
	θ_t, 645	L_t, 12.1
SrI_2	θ_m, 538	L_m, 19.7
$TaCl_5$	θ_m, 217	L_m, 36.8
	θ_e, 234	L_e, 50.2
$TaBr_5$	θ_m, 269	L_m, 45.6
	θ_e, 347	L_e, 62.4
TaI_5	θ_m, 497	—
	θ_e, 543	L_e, 75.8
TeF_4	θ_m, 130	L_m, 26.8
$TeCl_2$	θ_e, 322	L_e, 64.1
$TeCl_4$	θ_m, 224	L_m, 18.8
	θ_e, 392	L_e, 70.3
$ThCl_4$	θ_m, 770	—
	θ_e, 921	L_e, 152.8
$ThBr_4$	θ_m, 679	L_m, 54.4
	θ_e, 857	L_e, 144.4
ThI_4	θ_m, 566	L_m, 48.1
	θ_e, 837	L_e, 131.9
TiF_4	θ_s, 283	L_s, 90.0
$TiCl_4$	θ_m, −25	L_m, 9.37
	θ_e, 136	L_e, 36.22
$TiBr_4$	θ_m, 39	L_m, 13.0
	θ_e, 233	L_e, 44.4
TiI_2	θ_s, 1 170	L_s, 222.7
TiI_4	θ_m, 154	L_m, 19.8
	θ_e, 377	L_e, 56.1
TlF	θ_e, 700	L_e, 116.0
TlCl	θ_m, 430	L_m, 15.9
	θ_e, 816	L_e, 103.6
TlBr	θ_m, 460	L_m, 16.3
	θ_e, 825	L_e, 103.4
TlI	θ_e, 845	L_e, 103.8
	θ_m, 440	L_m, 14.8
UF_4	θ_m, 1 036	L_m, 42.7
	θ_e, 1 457	L_e, 221.9
UF_6	θ_m, 64	L_m, 19.3
	θ_s, 57	L_s, 48.1
UCl_4	θ_m, 590	L_m, 44.8
	θ_e, 789	L_e, 141.5
UBr_4	θ_m, 519	L_m, 55.3
	θ_e, 777	L_e, 113.5
VF_5	θ_e, 48.3	L_e, 46.5
VCl_4	θ_e, 160	L_e, 33.1
WF_6	θ_m, 0	L_m, 1.76
	θ_e, 17	L_e, 26.6
WCl_5	θ_e, 298	L_e, 52.8
	θ_m, 240	L_m, (17.6)
WCl_6	θ_m, 282	L_m, 6.7
	θ_e, 337	L_e, 58.2
ZnF_2	θ_m, 875	L_m, 41.8
	θ_e, 1 500	L_e, 184.2
$ZnCl_2$	θ_m, 318	L_m, 10.5
	θ_e, 732	L_e, 119.3
$ZnBr_2$	θ_m, 402	L_m, 15.7
	θ_e, 650	L_e, 98.4
ZnI_2	θ_m, 446	L_m, (18.8)
	θ_e, 727	L_e, 96
ZrF_4	θ_s, 908	L_s, 232.4
$ZrCl_4$	θ_s, 334	L_s, 103.8
$ZrBr_4$	θ_s, 356	L_s, 108.0
ZrI_4	θ_s, 431	L_s, 121.4
Nitrides		
Mg_3N_2	θ_t, 550	L_t, 0.46
	θ_t, 788	L_t, 0.92
Oxides		
As_4O_6	θ_t, −33	L_t, 5.7
	θ_m, 309	L_m, 36.8
	θ_e, 459	L_e, 59.9
BeO	θ_m, 2 580	80.8
	θ_e, 4 120	L_e, 471.0
CdO	θ_s, 1 559	L_s, 251
$Fe_{0.95}O$	θ_m, 1 378	L_m, 31.0
Fe_3O_4	θ_m, 1 597	L_m, 138.2
H_2O	θ_m, 0	L_m, 6.016
	θ_e, 100	L_e, 41.11
MgO	θ_m, 2 825	L_m, 77.0
MnO	θ_m, 1 875	L_m, 54
Mn_3O_4	θ_t, 1 172	L_t, 20.3
MoO_3	θ_m, 795	L_m, 48.4
	θ_e, 1 100	L_e, 192.6
NbO_2	θ_t, 817	L_e, 2.9
OsO_4 (white)	θ_m, 42	L_m, 9.6
	θ_e, 130	L_e, 39.57
P_4O_6	θ_e, 175	L_e, 43.5
PbO (yellow)	θ_m, 886	L_m, 29.3
	θ_e, 1 470	L_e, (214.8)
Re_2O_7	θ_m, 298	L_m, 64.1
	θ_e, 363	L_e, 75.4
SO_2	θ_m, −75.5	L_m, 7.5
	θ_e, −10	L_e, 24.95
SO_3	θ_m, 17 (α)	L_m, 2.1
	θ_s, 52 (γ)	L_s, 66.6
Sb_4O_6	θ_t, 570	L_t, (14.2)
	θ_m, 656	L_m, 109.9
	θ_e, 1 425	L_s, 214.4 (m.p.)
SeO_2	θ_s, 316	L_s, (102.6)
SiO_2	θ_t, 250	L_t, 1.3
(cristobalite)	θ_m, 1 713	L_m, 10.9
SnO_2	θ_t, 410	L_t, 1.88
	θ_t, 540	L_t, 1.26
SrO	θ_m, 2 460	L_s, 523 (25 °C)
Tc_2O_7	θ_m, 119	L_m, 48.1
	θ_e, 312	L_e, 58.6
TeO_2	θ_m, 733	L_s, 237.8 (m.p.)
TiO	θ_t, 991	L_t, 3.43
Ti_3O_5	θ_t, 177	L_t, 9.38
VO_2	θ_m, 1 550	L_m, 56.9
	θ_t, 67	L_t, 4.29
V_2O_5	θ_m, 670	L_m, 65.3
WO_3	θ_m, 1 473	L_s, 456 (m.p.)
ZrO_2	θ_t, 1 175	L_t, 5.9
Sulphides		
Ag_2S	θ_t, 176; 586	L_t, 5.9; —
	θ_m, 830	L_m, 11.3

Table 8.3 OTHER METALLURGICALLY IMPORTANT COMPOUNDS—*continued*

Compound	θ_m, θ_e, θ_s or θ_t °C	L_m, L_e, L_s or L_t kJ mol^{-1}	*Compound*	θ_m, θ_e, θ_s or θ_t °C	L_m, L_e, L_s or L_t kJ mol^{-1}
	Sulphides—continued			*Sulphides*—continued	
CS_2	θ_e, 45.2	L_e, 27.2	HgS	θ_t, 345	L_t, 4.2
Cu_2S	θ_t, 103	L_t, 3.85	H_2S	θ_m, −85.3	L_m, 2.43
	θ_t, 350	L_t, 0.8		θ_e, −60.2	L_e, 18.686
	θ_m, 1 130	L_m, 10.9	MnS	θ_m, 1 530	L_m, 26.8
FeS	θ_t, 138	L_t, 2.39			
	θ_t, 325	L_t, 0.50	Na_2S	θ_m, 978	L_m, (6.91)
	θ_m, 1 195	L_m, 32.36	Sb_4S_6	θ_m, 550	L_m, 126.3 (25 °C)
GeS	θ_m, 615	L_m, 21.4		—	L_s, 214.4 (m.p.)
	θ_s, 760	L_s, 145.3	ZnS	θ_t, 1 020	L_t, (13.4)

8.3 Heat, entropy and free energy of formation

Table 8.4 ELEMENTS
Standard entropies

Element	S_{298} J K^{-1}	*Element*	S_{298} J K^{-1}	*Element*	S_{298} J K^{-1}
Ag	42.7	Hf	44.0	Rh	31.8
Al	28.34	Hg	76.20	Ru	28.5
As	35.2	I	58.6	S (rhomb.)	31.90
Au	47.39	In	58.2		
B (cryst.)	5.9			S (monocl.)	32.57
		Ir	35.6	S_2 (gas)	227.8
Ba	(67.8)	K	63.6	S_4 (gas)	306.1
Be	9.50	La	56.9	S_6 (gas)	376.0
Bi	56.9	Li	29.1	S_8 (gas)	471.4
Br_2 (liq.)	152.4	Mg	32.5		
C (graph.)	5.69			Sb	45.6
		Mn	31.82	Sc	34.3
Ca	41.66	Mo	28.6	Se (met.)	42.3
Cd	51.5	N_2	191.63	Si	18.8
Ce	64.1	Na	51.29	Sm	69.5
Cl_2	223.24	Nb	36.55		
Co	30.06			Sn	51.5
		Nd	73.3	Ta	41.4
Cr	23.9	Ni	29.81	Te	49.61
Cs	82.9	O_2	205.24	Th	53.42
Cu	33.37	Os	32.7	Ti	30.31
Dy	74.9	P (yellow)	44.17		
Er	73.3			Tl	64.27
		Pb	64.9	U	50.2
F_2	203.1	Pd	37.89	V	29.3
Fe	27.2	Pr	73.3	W	33.5
Ga	41.0	Pt	41.9	Y	44.4
Gd	66.2	Pu	51.5		
Ge	31.19			Zn	41.7
		Rb	76.6	Zr	38.9
H_2	130.75	Re	37.3		

Table 8.5a INTERMETALLIC COMPOUNDS
Heats of formation in kJ and standard entropies

Phase or compound	$-\triangle H$ kJ g-atom^{-1}	kJ mol^{-1}	S_{298} J K^{-1} mol^{-1}
Ag–Au	*See* Table 8.5c		—
Ag–Cd	*See* Table 8.5c		—
Ag–Zn	*See* Table 8.5c		—
Al_2Au (α)	42.1	126.4	—
AlAu (β)	38.7	77.4	—
$AlAu_2$ (γ)	34.9	104.7	—
Al_4Ca	—	218.5	138.1
Al_2Ca	—	216.8	85.4
Al_4Co (γ)	32.2	—	—
Al_5Co_2 (ε)	42.0	—	—
AlCo (ξ)	55.2	—	—
Al_7Cr (θ)	13.4	—	—
$Al_{11}Cr_2$ (η)	15.1	—	—
Al_4Cr (ε)	17.2	—	—
Al_9Cr_4 (γ_4)	15.9	—	—
Al_8Cr_5 (γ_2)	15.1	—	—
$AlCr_2$ (β)	10.9	—	—
Al_2Cu (θ)	13.4	—	—
AlCu (η_2)	20.0	—	—
$AlCu_2$ (γ)	23.0	—	—
Al-Fe	*See* Table 8.5c		—
Al_4La (β′)	35.2	—	—
Al_2La	—	150.7	98.8
AlAs	—	122.6	60.3
AlSb	—	50.2	65.0
Al_3Ni (ε)	37.7	150.7	110.7
Al_3Ni_2 (δ)	56.5	282.6	136.5
AlNi (β)	59.2	118.5	54.1
$AlNi_3$ (α′)	37.7	150.8	113.8
Al_4Pu	36.2	181.0	—
Al_3Pu	45.2	180.8	—
Al_2Pu	47.3	141.9	—
Al_3Ti (γ)	35.6	142.3	94.6
AlTi (δ)	37.0	—	52.3
$AlTi_3$ (ε)	27.6	—	—
Al_4U	26.0	129.7	163.3
Al_3U	28.6	114.3	136.0
Al_2U	32.9	98.8	106.7
Al_3V (ξ)	27.6	—	—
AsIn	—	57.8	74.7
An–Cd	*See* Table 8.5c		—
Au–Cu	*See* Table 8.5c		—
Au–Ni	*See* Table 8.5c		—
Au–Pb	*See* Table 8.5c		—
$AuSb_2$	—	19.5	119.3
AuSn	*See* Table 8.5c		—
Au_3Zn (α)	(18.0)	—	—
AuZn (β)	(26.0)	—	—
Ba_3Bl_2	—	670	—
$BaMg_2$ (β)	2.1	6.3	—
Ba_2Pb	97.6	293.0	—
BaPb	75.0	150.0	—

Table 8.5a INTERMETALLIC COMPOUNDS—*continued*

Phase or compound	$-\triangle H$ kJ g-atom^{-1}	kJ mol^{-1}	S_{298} J K^{-1} mol^{-1}
$BaPb_3$	44.0	176.0	—
Ba_3Sb_2	146.5	733	—
Ba_2Sn	—	377.0	126.8
$BaSn_3$	—	194.6	188.0
Bi_2Ca_3	—	528.0	117.9
BiK_3	—	226.5	198.0
BiMn	9.8	19.7	—
BiNa	32.7	65.3	109.6
$BiNa_3$	47.6	190.5	160.0
BiNi	3.9	7.8	88.3
Bi–Tl	*See* Table 8.5c		—
$CaCd_3$	31.4	126.0	—
$CaMg_2$	—	39.3	—
Ca_2Pb	—	215.6	105.5
CaPb	—	119.7	80.8
Ca_3Sb_2	—	728.4	157.4
Ca_2Su	—	314	100.5
CaSn	—	159.0	70.7
$CaSn_3$	—	180.0	—
CaTl	81.6	163.2	—
CaZn (δ)	36.6	73.2	66.6
$CaZn_2$	31.4	94.2	101.7
$CaZn_5$ (ξ)	23.0	138.1	—
Cd–Hg	*See* Table 8.5c		—
Cd–Mg	*See* Table 8.5c		—
Cd–Sb	*See* Table 8.5c		—
Cd_3As_2	—	38.1	207.2
Cd_6Na (β)	7.5	—	—
Cd_2Na (γ)	11.7	—	—
CeMg (β)	8.0	—	—
Co_5As_2	—	111.3	223.1
Co_2As	18.8	56.5	96.3
Co_3As_2	—	113.8	160.3
CoAs	—	56.9	64.5
Co_2As_3	—	144.0	164.1
$CoAs_2$	—	92.1	92.9
CoSb (γ)	20.9	41.8	70.7
$CoSb_x(x=0.74\rightarrow0.96)$			
CoSn	14.7	29.3	71.6
Cr–Ni	*See* Table 8.5c		—
Cr_2Ta	9.0	27.0	88.1
Cu_7In_3 (δ)	8.5	—	—
Cu_2In (η)	8.0	—	—
Cu–Mg	*See* Table 8.5c		—
Cu–Pt	*See* Table 8.5c		—
$Cu_{11}Sb_2$ (η)	0.26	—	—
Cu_9Sb_2 (ϵ)	0.54	—	—
Cu_2Sb (γ)	4.2	—	—
$Cu_{31}Sn_8$(δ)	5.36	—	—
Cu_3Sn(ε)	7.5	—	—
Fe–Ni	*See* Table 8.5c		—
Fe_2Pu(ζ)	9.1	27.3	—
$FeSb_2$(ζ)	(9.6)	—	—
FeTi(ε)	20.3	40.6	—

Table 8.5a INTERMETALLIC COMPOUNDS—*continued*

Phase or compound	$-\Delta H$ kJ g-atom^{-1}	kJ mol^{-1}	S_{298} J K^{-1} mol^{-1}
$Fe_2U(\varepsilon)$	10.9	32.2	—
GaAs	—	81.6	64.2
GaSb	—	41.9	77.4
GeU	*See* Table 8.5c		—
Hg_8K	11.6	104.6	—
Hg_4K	18.0	90.0	—
Hg_3K	20.9	83.6	—
Hg_2K	38.7	77.4	—
HgK	28.1	56.1	—
Hg_3Li	28.2	113.0	—
Hg_2Li	34.8	104.4	—
HgLi	44.0	88.0	—
Hg_4Na	16.7	83.5	—
HgNa	21.4	42.7	—
Hg_2Na_3	18.8	94.4	—
Hg_2Na_5	13.4	93.8	—
$HgNa_3$	11.7	46.9	—
Hg_5Tl_2	*See* Table 8.5c		—
InSb	—	31.1	87.7
KNa_2	*See* Table 8.5c		—
K_3Bi	56.6	226.4	—
K_3Sb	47.1	188.4	—
KTl	28.3	56.5	—
$LaMg(\zeta)$	9.0	18.0	—
$Li_4Pb(\gamma)$	35.2	—	—
$LiPb(\zeta)$	30.6	—	—
Li_3Bi	—	232.3	—
Li_3Sb	—	180.0	—
LiTl	—	153.6	—
βLi–Sn	39.3	—	—
$Li_5Sn_2(\delta)$	40.2	—	—
$LiSn(\zeta)$	35.1	—	—
Mg_2Ge	—	115.2	73.0
Mg_2Sn	—	80.6	101.5
Mg_2Pb	—	48.1	119.3
Mg_3Sb_2	—	300.1	136.7
Mg_3Bi_2	25.3	126.8	191.9
Mg_2Ni	17.3	51.9	95.0
$MgNi_2$	18.8	56.5	88.7
$MgZn(\gamma)$	10.5	—	—
$MgZn_2(\varepsilon)$	10.9	—	—
$Mg_2Zn_{11}(\zeta)$	10.0	—	—
MnAs	—	57.3	—
Mn_2Sb	—	32.6	136.9
Na_3As	—	217.7	—
Na_3Sb	—	197.6	176.2
NaSb	—	66.2	99.8
Na_3Bi	—	190.5	(160)
NaBi	—	65.3	(110)
Na_4Sn	11.7	58.6	—
Na_2Sn	20.1	60.3	—
NaSn	25.1	50.2	—
Na–Pb	*See* Table 8.5c		—

Table 8.5a INTERMETALLIC COMPOUNDS—*continued*

Phase or compound	$-\Delta H$ kJ g-atom^{-1}	$-\Delta H$ kJ mol^{-1}	S_{298} J K^{-1} mol^{-1}
NaTl	*See* Table 8.5c		—
$NbCr_2$	—	20.9	83.7
$NbFe_2$	—	61.5	75.3
NbNi(γ)	22.6	—	—
$NbNi_3(\delta)$	31.8	—	—
Ni_2Ge	—	110.1	90.8
NiAs	—	72.0	51.9
$Ni_3Sb(\delta)$	18.8	—	—
$Ni_5Sb_2(\beta')$	21.8	—	—
NiSb(γ)	33.1	66.2	78.3
$NiSb_2(\varepsilon)$	24.7	—	—
$Ni_3Sn\ (\beta')$	25.2	103.0	131.4
$Ni_3Sn_2(\gamma)$	38.5	192.3	173.7
$Ni_3Sn_4\ (\delta)$	33.7	235.7	257.9
Ni_3Ti	35.1	140.3	104.6
NiTi	33.3	66.6	53.2
$NiTi_2$	27.9	83.7	83.6
Pb–Tl	*See* Table 8.5c		—
Pb–U	*See* Table 8.5c		—
Sb–Zn	*See* Table 8.5c		—
Ta Fe_2	—	57.8	106.7
$ThRe_2$	—	174.1	123.7
$ThMg_2$	—	31.4	92.5
UCd_{11}	—	45.6	583.5
UFe_2	—	32.2	104.7
URh_3	—	259.5	148.2
URu_3	—	217.6	108.4

Table 8.5b SELENIDES AND TELLURIDES

Heats of formation in kJ and standard entropies

Compound	$-\Delta H$ kJ mol^{-1}	S_{298} J K^{-1}	*Accuracy* kJ
Ag_2Se	43.5	150.3	1
Ag_2Te	36.0	153.6	0.5
Al_2Se_3	540.0	157.0	15
Al_2Te_3	319.0	188.4	4
As_2Se_3	102.6	194.6	18
As_2Te_3	37.7	226.5	7
$AuTe_2$	18.6	141.8	3
BaSe	393.5	89.6	40

Table 8.5b SELENIDES AND TELLURIDES—*continued*

Compound	$-\Delta H$ kJ mol^{-1}	S_{298} J K^{-1}	*Accuracy* kJ
BaTe	269.6	99.6	30
Bi_2Se_3	140.2	240.0	4
Bi_2Te	78.3	261.2	4
CaSe	368.4	69.1	25
CaTe	272.1	80.8	33
CdSe	144.8	83.3	14
CdTe	101.8	93.1	1
Cu_2Se	65.3	129.8	7
CuSe	41.9	78.3	5
Cu_2Te	41.9	134.8	11
$FeSe_{0.96}$	67.0	69.2	7
$FeSe_{1.14}$	66.1	87.7	4
Fe_3Se_4	212.2	280.0	3
$FeTe_{0.9}$	23.0	80.2	4
$FeTe_2$	72.0	100.3	5
GaSe	161.2	70.3	11
Ga_2Se_3	406.0	192.1	17
GaTe	123.5	85.4	11
Ga_2Te_3	272.1	222.7	13
GeSe	69.1	87.5	13
$GeSe_2$	113.0	112.6	21
GeTe	32.7	89.2	13
$H_2Se_{(g)}$	−29.3	218.9	1.3
$H_2Te_{(g)}$	−99.6	229.0	0.9
HgSe	43.3	100.9	15
HgTe	31.8	113.0	4.5
InSe	118.0	81.6	13
In_2Se_2	326.5	201.3	17
In_2Te	79.5	157.0	2
InTe	72.0	105.7	2
In_2Te_3	191.7	238.6	3
In_2Te_5	191.7	—	2.5
K_2Se	372.6	—	2.5
La_2Se_3	933.5	202.3	21
La_2Te_3	784.9	223.4	25
Li_2Se	401.8	71.2	40
Li_2Te	355.8	77.4	—
MgSe	272.9	62.8	17
MgTe	209.3	74.5	21
MnSe	154.9	90.8	11
MnTe	111.3	93.8	9
$MnTe_2$	125.6	145.0	40
Na_2Se	343.2	—	13
NaSe	194.2	62.8	21
Na_2Te	343.2	94.6	25
NaTe	173.3	83.7	11
$NaTe_3$	210.6	145.3	21
$NiSe_{1.05}$	74.9	75.2	4
$NiSe_{1.143}$	79.7	77.2	2.5
NiTe	35.7	80.1	2
PbSe	99.6	102.6	4
PbTe	69.1	110.1	1

Table 8.5b SELENIDES AND TELLURIDES—*continued*

Compound	$-\Delta H$ kJ mol^{-1}	S_{298} J K^{-1}	*Accuracy* kJ
Pt_5Se_4	265.4	327.3	25
Re_2Te_5	122.6	253.3	25
Sb_2Se_3	127.7	211.4	5
Sb_2Te_3	56.5	246.1	2
$SiSe_2$	146.5	94.2	40
SnSe	88.7	86.2	7
$SnSe_2$	124.7	118.0	5
SnTe	60.7	98.8	1.5
SrTe	397.7	80.8	38
SrTe	259.5	—	25
Tl_2Se	94.2	173.7	2.5
TlSe	61.3	102.6	1.5
Tl_2Te	80.4	174.1	12
USe	275.9	96.6	21
ZnSe	159.1	70.3	9
ZnTe	119.3	78.2	1

The above values of the heat of formation ΔH were measured calorimetrically at room temperature or at about 600 °C, or have been calculated from measurements of vapour pressure or electromotive force at different temperatures. They are probably correct to within $\pm 10\%$. As the molar heats of alloys are obtained nearly additively from the atomic heats of the components (Neumann and Kopp's rule) all the heats of formation (even if measured at higher temperatures) are probably valid also at room temperature within the limits of error mentioned above. The formulae of the compounds are given only to indicate composition, independent of whether the phases form a broad or narrow homogeneous field, or are ordered or disordered.

Table 8.5c INTERMETALLIC PHASES
Heats, entropies and free energies of formation

Phase	N_2	*Temp.* °C	$-\Delta H$ kJ g-atom^{-1}	$-\Delta G$ kJ g-atom^{-1}	ΔS J K^{-1} g-atom^{-1}	*Remarks*
Ag–Au, s.s.	0.55	600	4.02	8.08	5.02	
α Ag–Cd	0.40	400	7.16	—	4.15	
ζ Ag–Cd	0.564	400	6.78	—	4.52	Cd solid
β Ag–Cd	0.494	450	6.62	—	—	Cd solid
γ' Ag–Cd	0.60	400	8.37	—	1.88	Cd solid
ε Ag–Cd	0.70	400	6.45	—	2.85	Cd solid
AgMg	0.50	500	19.47	17.08	1.38	
Ag–Pd, s.s.	0.40	727	5.65	—	−2.09	
β Ag–Zn	0.50	600	3.14	10.34	8.25	Zn solid
ζ Ag–Zn	0.50	51	6.45	—	—	
γ Ag–Zn	0.612	600	4.48	10.09	6.41	Zn solid
θ Al–Fe	0.26	900	28.1	—	−3.94	Al solid
η Al–Fe	0.30	900	28.26	—	−3.56	Al solid
ζ Al–Fe	0.33	900	27.2	—	−2.34	Al solid
AlFe	0.50	900	27.2	—	0.13	Al solid
α Al–Zn	0.20	380	−4.02	—	7.45	
α Au–Cd	0.35	427	17.38	15.78	−2.26	Cd liquid
β Au–Cd	0.50	427	21.35	18.42	−4.19	Cd liquid
δ' Au–Cd	0.62	427	19.05	16.20	−3.48	Cd liquid
ε Au–Cd	0.70	427	19.26	15.83	−4.90	Cd liquid

ΔS = entropy of formation. N_2 = mol fraction of second component. s.s. = solid solution.

Table 8.5c INTERMETALLIC PHASES—*continued*

Phase	N_2	*Temp.* °C	$-\Delta H$ kJ g-atom^{-1}	$-\Delta G$ kJ g-atom^{-1}	ΔS J K^{-1} g-atom^{-1}	*Remarks*
Au–Cu, s.s.	0.58	500	5.32	9.67	5.65	
Au_3Cu	0.26	25	4.02	4.90	2.97	
AuCu I	0.50	25	8.96	8.96	0.00	
Au Cu II	0.50	400	6.03	8.79	4.10	
$AuCu_3$	0.75	25	6.87	7.24	1.26	
AuNi, s.s.	0.53	877	−7.5	—	8.71	
Au_2Pb	0.33	227	−0.96	1.34	4.61	
$AuPb_2$	0.67	227	2.09	1.80	−0.72	
AuSn	0.50	25	15.24	—	−0.23	
Bi–Tl, δ	0.47	150	2.81	5.28	5.86	
Bi–Tl, γ	0.80	150	3.18	5.15	4.69	
Bi_2U	0.333	25	34.2	—	−3.93	
Bi_4U_3	0.43	25	38.1	—	−5.0	
BiU	0.50	25	36.8	—	−4.96	
β Cd–Hg	0.50	25	4.35	4.06	−0.92	Hg liquid
Cd–Mg, s.s.	0.50	270	5.61	7.96	4.35	
Cd_3Mg	0.25	25	5.19	5.19	−0.04	
CdMg	0.50	25	8.37	8.08	−1.05	
$CdMg_3$	0.75	25	5.65	5.61	−0.17	
CdSb	0.50	25	6.67	6.42	−0.86	
CdTe	0.50	20	50.9	49.7	−4.12	
$CeHg_4$	0.80	342	66.6	7.54	−96.13	Hg gas
α Co–Fe, s.s.	0.50	870	6.6	—	—	
α Cr–Fe, s.s.	0.50	1 280	−5.9	−7.63	8.42	
σ Cr–Fe	0.55	750	−4.73	—	6.91	
Cr–Mo, s.s.	0.47	1 400	−7.24	—	7.70	
Cr–Ni, f.c.c.	0.50	1 265	−6.70	—	9.21	
Cu_2Mg	0.333	25	11.18	—	−8.4	
$CuMg_2$	0.667	25	9.6	—	—	
Cu–Ni, s.s.	0.65	700	−1.84	—	4.40	
Cu–Pd, s.s.	0.40	640	12.6	—	−5.32	
Cu–Pt, s.s.	0.50	640	15.5	—	−4.15	
Cu_2Sb	0.33	25	4.2	—	−3.56	
α Cu–Zn	0.39	25	9.38	10.22	2.81	
β Cu–Zn	0.50	—	9.17	—	4.19	
β′ Cu–Zn	0.50	25	11.72	12.14	1.47	
γ Cu–Zn	0.615	—	12.35	—	0.54	
ε Cu–Zn	0.79	25	7.20	7.49	1.00	
γ Fe–Mn	0.5	1 127	5.0	—	—	
γ Fe–Ni, s.s.	0.65	1 000	4.35	—	3.60	
$FeSb_2$	0.67	560	3.4	—	−0.8	
Ge_3U	0.25	1 100	26.8	—	−1.47	
Ge_2U	0.33	1 100	29.3	—	−1.3	
Ge_5U_3	0.375	1 100	30.2	—	−1.3	
GeU	0.50	1 100	30.77	—	−1.09	
Ge_3U_5	0.625	1 050	29.48	—	−0.38	
γ Hg–Tl	0.286	−59	−0.29	1.05	6.45	
β In–Sn	0.42	100	−2.1	—	—	
KNa_2	0.667	25	−0.75	0.63	4.61	Na_2K, Na, K; liquid
	0.667	25	−0.04	0.29	1.13	Na_2K, Na, K: solid
Na_5Pb_2	0.286	25	20.5	—	−6.7	

ΔS = entropy of formation. N_2 = mol fraction of second component. s.s. = solid solution.

Table 8.5c INTERMETALLIC PHASES—*continued*

Phase	N_2	*Temp.* °C	$-\Delta H$ kJ g-atom^{-1}	$-\Delta G$ kJ g-atom^{-1}	ΔS J K^{-1} g-atom^{-1}	*Remarks*
Na_9Pb_4	0.31	25	20.9	—	−6.3	
NaPb	0.50	25	23.0	—	−8.8	
$NaPb_3$	0.714	25	13.8	—	−2.1	
NaTl	0.50	25	16.54	—	−3.98	
α Pb–Tl	0.50	250	1.80	4.19	4.56	
β-Pb–Tl	0.85	270	2.81	2.81	0.0	
Pb_3U	0.25	700	19.7	—	−0.4	Pb solid
PbU	0.50	700	18.8	—	−1.3	Pb solid
SbZn	0.50	25	9.52	8.85	−2.26	
Sn_3U	0.25	950	23.9	—	−0.63	Sn solid
Sn_5U_3	0.375	950	27.2	—	−1.17	Sn solid
Sn_2U_3	0.60	1 000	26.8	—	−1.05	Sn solid
Th_2Zn	0.333	700	22.2	15.5	−7.75	
$ThZn_2$	0.667	700	40.2	25.5	−15.1	
$ThZn_4$	0.80	700	38.5	23.0	−15.1	
Th_2Zn_{17}	0.895	700	32.7	13.0	−16.96	
U_2Zn_{17}	0.895	700	24.3	7.5	−17.2	
ZrZn	0.50	500	63.6	32.2	−40.6	
$ZrZn_2$	0.667	500	56.1	30.1	−33.5	
$ZrZn_3$	0.75	500	54.22	28.5	−33.29	
$ZrZn_6$	0.86	500	40.6	18.8	−28.26	
$ZrZn_{14}$	0.933	500	28.1	9.42	−24.07	

ΔS = entropy of formation. N_2 = mol fraction of second component. s.s. = solid solution.

8.4 Metallic systems of unlimited mutual solubility

While mutual solubility in the solid state is usually limited, that in the liquid state is frequently unlimited. Thus, the curves of concentration against integral heat, free energy and entropy of formation are convex with a maximum or a minimum. The thermochemical values at the concentrations corresponding to these maxima and minima are given in Table 8.6.

According to van Laar the form of the heat of mixing curve may be represented by

$$\Delta H = \frac{aN_2(1-N_2)}{1+b\cdot N_2}$$

where N_2 is the mol fraction of the second component and a and b are constants for each binary system.

For a system, completely disordered, the entropy of formation is often represented by this equation:

$$\Delta S = -19.155\,(N_1 \log N_1 + N_2 \log N_2) \qquad \text{J K}^{-1}\ \text{g-atom}^{-1}$$

which has a maximum value of 5.78 J K^{-1} g-atom^{-1} at $N_1 = N_2 = 0.5$. Some deviations from this value may be due to experimental errors, but others are real.

For more detailed discussions *see* refs 1–3.

Table 8.6 LIQUID BINARY METALLIC SYSTEMS
Heats and entropies of formation

		Heat of formation		Entropy of formation	
System	°C	N_2	ΔH_{max} J g-atom^{-1}	N_2	ΔS_{max} J K^{-1} g-atom^{-1}
Ag–Al	1 000	0.30	−6 410	0.60	7.58
Ag–Au	1 077	0.50	−5 150	0.50	3.81
Ag–Bi	727*	0.75	1 700	0.50	6.17
Ag–Cu	1 150	0.45	4 260	0.50	6.27
Ag–In	827*	0.32	−5 680	0.70	3.82
Ag–Mg	1 050	0.50	1 215	—	—
Ag–Pb	1 000	0.53	3 720	0.47	7.62
Ag–Sb	977	0.20	−3 120	0.52	8.37
Ag–Sn	977	0.19	−3 220	0.56	6.68
Ag–Tl	702*	0.56	2 510	0.50	6.32
Al–Bi	900	0.40	7 340	—	—
Al–Cu	1 100	0.60	−9 435	0.60	9.25
Al–Ga	750	0.40	670	0.50	6.10
Al–Ge	927	0.53	−3 915	0.50	6.20
Al–In	900	0.45	5 735	0.50	6.36
Al–Mg	800	0.45	−3 390	0.50	4.94
Al–Sn	700	0.45	4 100	0.50	6.99
Al–Zn	727	0.51	2 575	0.50	6.71
Au–Bi	700*	0.50	625	0.50	6.97
Au–Cu	1 277	0.53	−4 400	0.53	6.90
Au–Pb	927*	{0.40 0.90	{−795 260	0.50	7.54
Au–Sn	550*	0.45	−11 595	0.42	7.91
Au–Tl	700*	0.50	140	0.50	7.86
Au–Zn	807*	0.52	−22 810	0.25	4.50
Bi–Cd	500	0.65	880	0.52	7.25
Bi–Cu	927†	0.60	5 420	0.50	7.59
Bi–Hg	321	0.70	555	0.50	5.55
Bi–In	627	0.55	−185	0.50	5.87
Bi–Pb	427	0.50	−1 100	0.50	5.94
Bi–Sb	927	0.50	560	0.50	8.15
Bi–Sn	327	0.50	105	0.50	5.48
Bi–Tl	477	0.62	−4 580	0.50	6.67
Bi–Zn	600	0.58	4 685	0.50	7.85
Cd–Ga	427	0.50	2 665	0.50	5.73
Cd–Hg	327	0.50	−2 625	0.43	5.28
Cd–In	527	0.46	1 435	0.50	6.32
Cd–Mg	650	0.50	−5 615	0.50	4.72
Cd–Pb	500	0.45	2 665	0.45	6.50
Cd–Sb	500†	0.47	−2 035	0.50	6.92
Cd–Sn	500	0.43	1 815	0.50	6.91
Cd–Tl	477	0.40	2 300	0.50	6.78
Cd–Zn	527	0.50	2 090	0.50	5.86
Co–Fe	1 590	0.45	−2 610	0.58	3.80
Cs–K	111	0.59	126	—	—
Cs–Na	111	0.65	1 020	—	—
Cu–In	800*	{0.23 0.82	{−3 390 305	0.55	4.83
Cu–Mg	827*	0.44	10 445	0.50	2.53
Cu–Pb	1 200	0.45	6 800	0.49	6.85
Cu–Sb	917*	{0.25 0.90	{−5 660 345	0.55	8.20

N_2 = mol fraction of second component.
* or † indicates that the standard state for the first or second component, respectively, is the hypothetical supercooled liquid.

Table 8.6 LIQUID BINARY METALLIC SYSTEMS—*continued*

System	°C	Heat of formation N_2	ΔH_{max} J g-atom^{-1}	Entropy of formation N_2	ΔS_{max} J K^{-1} g-atom^{-1}
Cu–Sn	1 127	{0.25, 0.90}	{−4 150, 218}	0.54	8.16
Cu–Tl	1 300	0.50	8 580	0.52	7.75
Fe–Si	1 600	0.48	−37 950	{0.05, 0.54, 0.95}	{1.05, −1.66, 0.65}
Ga–Zn	477	0.58	1 610	0.50	6.80
Hg–In	25†	0.48	−2 260	0.45	5.25
Hg–Na	400	0.40	−20 570	{(0.04), 0.39, 0.12}	{(0.4), −4.57, 0.71}
Hg–Pb	327	{0.20, 0.83}	{505, −160}	0.50	4.60
Hg–Sn	177†	0.40	905	0.50	5.17
Hg–Tl	310	0.28	−1 130	—	—
Hg–Zn	300†	0.33	440	0.42	4.89
In–Mg	700	0.60	−7 300	—	—
In–Pb	400	0.48	963	0.50	6.29
In–Sb	627	0.45	3 300	0.50	6.51
In–Sn	427	0.45	−200	0.48	6.76
In–Tl	500	0.42	575	0.50	5.48
In–Zn	427	0.58	3 235	0.53	6.99
K–Na	111	0.40	737	0.50	5.66
Mg–Pb	700	0.40	−10 050	0.53	5.21
Mg–Sn	500	0.40	−14 650	—	—
Mg–Tl	650	0.40	−7 040	0.57	6.66
Mg–Zn	700	0.60	−6 490	—	—
Na–Rb	111	0.42	1 260	—	—
Na–Pb	427	0.40	−17 100	{0.05, 0.20, 0.80}	{0.34, −2.85, 2.26}
Ni–Pd	1 600	0.50	1 200	0.50	5.41
Pb–Sb	632	0.55	−75	0.50	6.21
Pb–Sn	777	0.50	1 370	0.55	4.81
Pb–Tl	500	0.55	−1 090	0.50	5.22
Sb–Sn	632	0.50	−1 390	0.50	6.08
Sn–Tl	500	0.45	724	0.47	5.17
Sn–Zn	477	0.57	3 220	0.54	7.93

N_2 = mol fraction of second component.
* or † indicates that the standard state for the first or second component, respectively, is the hypothetical super-cooled liquid.

Table 8.7a LIQUID BINARY METALLIC SYSTEMS AND SOME SOLID SOLUTIONS
Partial molar free energies ($-\overline{\Delta G}$) in kJ
For the solution of 1 g-atom of metal C in a theoretically infinite amount of alloy of concentration N_c (mol fraction of the dissolved metal), $-\overline{\Delta G}$ is given in kJ.

System	*Metal C*	*Temp.*°C	N_c: 0.1	0.2	0.3	0.5	0.8
Ag–Al (sol.)	Al	449	37.36	s+s	20.33	s+s	s+s
Ag–Al (liq.)	Al	1 000	51.81	36.57	23.40	8.73	2.26
Ag–Au (sol.)	Ag	527	25.87	19.46	15.03	8.54	2.20
Ag–Au (liq.)	Ag	1 077	32.99	24.16	18.51	10.63	3.01
Ag–Cd (sol.)	Cd*	400*	34.87α	26.41α	19.01α	10.47ζ	l+s

*Note standard state for metal C.

Table 8.7a LIQUID BINARY METALLIC SYSTEMS AND SOME SOLID SOLUTIONS—*continued*

System	*Metal C*	*Temp°C*	N_c: 0.1	0.2	0.3	0.5	0.8
Ag–Cd (liq.)	Cd	950	42.41	29.53	20.91	10.03	2.40
Ag–Cu (liq.)	Cu	1 150	15.94	10.20	7.41	4.58	2.06
Ag–Hg (sol.)	Hg*	227*	19.28α	12.32α	6.24α	l+s	l+s
Ag–Mg (sol.)	Mg	500	63.56α	51.33α	s+s	32.57β′	l
Ag–Pb (liq.)	Pb	1 000	20.99	13.56	9.51	5.50	1.96
Ag–Pd (sol.)	Ag	927	25.85	21.00	18.16	13.52	3.49
Ag–Sn (liq.)	Sn	977	25.34	21.02	17.68	11.50	3.50
Ag–Tl (liq.)	Tl	702	s	l+s	(5.31)	2.77	1.20
Ag–Zn (sol.)	Zn*	600*	30.96α	26.04α	19.69α	11.57β	s+l
Al–Mg (liq.)	Mg	800	31.26	21.21	14.97	7.42	2.01
Al–Zn (sol.)	Zn	380	3.34	1.99	1.64	1.41	s+s
Al–Zn (liq.)	Zn	727	13.52	8.88	6.78	4.42	1.45
Au–Bi (liq.)	Bi	700	l+s	l+s	10.45	5.11	1.94
Au–Cd (sol.)	Cd*	427*	51.92α	44.12α	35.16α_2	18.84β	l
Au–Cd (liq.)	Cd	727	s	s	(38.0)	19.05	3.33
Au–Cu (sol.)	Cu	527	28.90	23.41	19.24	11.88	2.17
Au–Cu (liq.)	Cu	1 277	49.19	36.15	27.32	14.95	3.84
Au–Fe (sol.)	Fe	850	17.27	8.95	4.44	0.44	s+s
Au–Hg (sol.)	Hg*	227*	9.41α	3.20β	l+s	l+s	l+s
Au–Pb (liq.)	Pb	927	l+s	23.71	17.34	9.15	2.83
Au–Sn (liq.)	Sn	550	l+s	44.03	30.87	13.51	2 28
Au–Tl (liq.)	Tl	700	l+s	l+s	(13)	8.00	2.19
Au–Zn (liq.)	Zn	807	s	68.04	51.84	26.32	3.86
Bi–Cd (liq.)	Cd	500	15.28	11.10	8.61	5.07	1.34
Bi–Hg (liq.)	Hg	321	9.77	6.52	4.69	2.52	0.74
Bi–K (liq.)	K	575	87.44	78.57	70.33	48.10	l+s
Bi–Mg (liq.)	Mg	702	60.04	54.44	50.17	42.48	3.73
Bi–Pb (liq.)	Pb	427	17.24	12.52	9.48	5.27	1.43
Bi–Sn (liq.)	Sn	327	10.76	7.36	5.43	3.11	2.01
Bi–Tl (liq.)	Tl	477	25.15	21.10	17.73	10.90	2.50
Bi–Zn (liq.)	Zn	600	9.96	5.61	3.35	1.01	0.21
Cd–Cu (liq.)	Cu*	600*	16.01	12.34	10.22	6.60	l+s
Cd–Ga (liq.)	Ga	427	4.71	2.92	1.96	1.28	0.77
Cd–In (liq.)	In	527	11.37	8.15	6.35	3.84	1.34
Cd–Mg (sol.)	Mg	270	24.45	19.38	15.57	7.87	1.42
Cd–Na (liq.)	Na	400	21.30	11.42	6.21	2.02	0.76
Cd–Pb (liq.)	Pb	500	7.30	4.91	3.83	2.62	1.15
Cd–Sb (liq.)	Sb*	500*	24.62	18.97	14.19	6.29	l+s
Cd–Sn (liq.)	Sn	500	11.62	7.96	5.97	3.62	1.33
Cd–Zn (liq.)	Zn	527	8.88	5.62	4.18	2.53	1.04
Co–Fe (liq.)	Fe	1 590	30.29	21.09	16.05	10.21	3.65
Co–Fe (sol.)	Co, Fe	both solutes approx. ideal in γ phase					
Co–Pt (sol.)	Pt	1 000	64.59	48.82	37.08	19.75	4.35
Cr–Fe (solα)	Cr	1 327	23.75	16.60	12.82	8.10	2.70
Cr–Mo (sol.)	Cr	1 198	15.15	9.14	6.25	3.64	1.59
Cr–Ni (sol.)	Cr	1 277	35.76β	21.39β	12.24β	5.79β	α
Cr–V (sol.)	Cr	1 277	37.96	28.36	22.66	14.17	3.88
Cu–Fe (liq.)	Cu	1 600	8.86	4.81	3.78	2.85	1.66
Cu–Ni (sol.)	Ni	700	10.34	6.28	4.12	2.21	1.08
Cu–Pb (liq.)	Cu	1 200	12.18	6.74	4.53	2.88	1.96
Cu–Pt (sol.)	Cu	1 077	54.27	41.52	32.98	20.38	5.44
Cu–Sn (liq.)	Sn	1 127	57.37	30.57	18.93	8.87	2.84
Cu–Zn (sol.)	Zn*	500*	38.91α	29.67α	22.05α	(14.4β)	2.28ε
Cu–Zn (liq.)	Zn	1 060	42.58	31.28	23.66	12.94	—
Fe–Mn (liq.)	Mn	1 590	32.24	22.19	16.58	9.71	3.26

*Note standard state for metal *C*.

Table 8.7a LIQUID BINARY METALLIC SYSTEMS AND SOME SOLID SOLUTIONS—*continued*

System	*Metal C*	*Temp.°C*	N_c: 0.1	0.2	0.3	0.5	0.8
Fe–Ni (sol.)	Fe	927	—	—	13.02γ	7.45γ	2.30γ
Fe–Ni (liq.)	Ni	1 600	41.87	30.82	24.39	15.38	4.36
Fe–Ni (liq.)	Fe	1 600	48.15	33.54	23.75	11.74	3.60
Fe–Si (liq.)	Si	1 600	126.53	97.62	68.68	23.35	4.64
Fe–V (sol.)	V	1 327	59.24α	41.78α	30.21α	15.43α	3.91α
Ga–Mg (liq.)	Mg	650	43.89	35.68	28.86	17.50	3.04
Ga–Zn (liq.)	Zn	477	12.03	8.11	5.93	3.38	1.15
Hg–K (liq.)	K	327	69.98	44.39	25.56	6.33	1.09
Hg–Na (liq.)	Na	400	64.02	46.64	32.22	12.25	1.62
Hg–Pb (liq.)	Pb	327	68.61	56.51	47.76	33.45	1.22
Hg–Tl (liq.)	Tl*	25*	8.46	5.43	3.64	l+s	l+s
Hq–Zn (liq.)	Zn*	300*	8.12	5.55	4.08	2.26	l+s
In–Sb (liq.)	Sb	627	31.33	23.07	16.56	8.08	1.88
In–Zn (liq.)	Zn	427	6.41	3.53	2.05	1.21	0.51
K–Pb (liq.)	Pb	575	49.77	42.69	35.17	17.78	2.24
K–Tl (liq.)	Tl	525	27.17	24.20	21.65	14.58	3.22
Mg–Pb (liq.)	Mg	700	38.47	33.45	28.97	19.10	4.84
Mg–Sn (liq.)	Mg	800	55.98	47.77	40.53	25.86	6.26
Mg–Zn (liq.)	Mg	650	32.26	22.51	16.22	8.37	2.24
Mn–Ni (sol.)	Mn	777	61.76γ	43.10γ	33.35γ	17.22η	(2.1γ)
Na–Pb (liq.)	Na	427	45.92	38.70	30.82	19.05	3.14
Na–Sn (liq.)	Na	500	48.76	41.44	34.95	s	2.68
Na–Tl (liq.)	Na	400	41.06	32.05	24.07	10.47	1.47
Pb–Sb (liq.)	Pb	632	18.86	13.32	9.99	5.69	1.75
Pb–Sn (liq.)	Sn	777	9.27	7.37	6.57	4.79	1.80
Pb–Tl (liq.)	Tl	500	16.32	11.90	9.16	5.43	1.41
Sb–Sn (liq.)	Sn	632	22.73	16.40	12.34	6.89	1.97
Sb–Zn (liq.)	Zn	677	l+s	18.82	15.12	8.10	0.81
Sn–Tl (liq.)	Tl	450	9.93	6.71	5.06	3.15	1.20
Sn–Zn (liq.)	Zn	477	10.49	6.54	4.46	2.22	0.84

*Note standard state for metal *C*.

Table 8.7b LIQUID BINARY METALLIC SYSTEMS AND SOME SOLID SOLUTIONS
Partial molar heats of solution ($\overline{\Delta H}$) in kJ

System	*Metal* C	*Temp.*°C	N_c: 0.1	0.2	0.3	0.5	0.8
Ag–Al (liq.)	Al	1 000	−27.43	−18.71	−8.98	+5.53	+1.31
Ag–Au (sol.)	Ag	527	−14.28	−11.72	−9.29	−5.07	−0.88
Ag–Cd (sol.)	Cd*	400*	−23.42	−18.96	−15.42	—	—
Ag–Cu (liq.)	Cu	1 150	15.70	10.76	7.48	3.77	0.63
Ag–Mg (sol.)	Mg	500	−38.45	−31.18	(26.3)	—	—
Ag–Pb (liq.)	Ag	1 000	11.05	9.56	7.76	3.73	0.17
Ag–Zn (sol.)	Zn*	600*	−14.22	−11.08	− 9.41	−7.89β	l+s
Al–Mg (liq.)	Mg	800	−11.86	−9.14	−6.61	−2.68	−0.21
Al–Zn (sol.)	Zn	380	11.93	8.99	6.30	3.79	s+s
Al–Zn (liq.)	Zn	727	8.35	6.51	4.94	2.64	0.46
Au–Bi (liq.)	Bi	700	l+s	l+s	1.77	0.70	0.04
Au–Cd (sol.)	Cd*	427*	−55.26	−49.42	$-39.31\alpha_2$	−20.32β	ε+l
Au–Cu (sol.)	Cu	527	−12.37	−12.00	−10.92	−7.39	−0.37
Au–Fe (sol.)	Fe	850	18.48	14.14	12.02	9.67	s+s
Au–Pb (liq.)	Pb	927	−3.05	−2.18	−1.14	−0.09	+0.60
Au–Sn (liq.)	Sn	550	l+s	−32.83	−22.70	−9.31	−0.48
Au–Zn (liq.)	Zn	807	s	−61.04	−48.12	−25.20	−1.93
Bi–Cd (liq.)	Cd	500	2.29	1.81	1.41	1.13	0.63
Bi–In (liq.)	In	627	−5.02	−4.80	−3.64	−2.35	−0.19
Bi–Na (liq.)	Na	500	−54.93	−66.36	−68.34	(−54)	l+s

*Note standard state for metal *C*.

Table 8.7b LIQUID BINARY METALLIC SYSTEMS AND SOME SOLID SOLUTIONS—*continued*

System	*Metal C*	*Temp* °C	N_c: 0.1	0.2	0.3	0.5	0.8
Bi–Pb (liq.)	Pb	427	−3.35	−2.90	−2.29	−1.07	−0.09
Bi–Sn (liq.)	Sn	327	0.34	0.27	0.20	0.10	0.02
Bi–Tl (liq.)	Tl	477	−10.96	−9.33	−7.89	−7.18	−1.12
Bi–Zn (liq.)	Zn	600	12.22	10.81	9.11	6.20	1.59
Cd–Ga (liq.)	Cd	427	9.19	6.60	4.89	2.67	0.84
Cd–Hg (liq.)	Cd	327	−7.85	−7.09	−5.66	−2.47	0.31
Cd–Mg (sol.)	Mg	270	−14.88	−14.39	−12.29	−5.92	−0.34
Cd–In (liq.)	Cd	527	14.15	3.54	2.89	1.70	0.39
Cd–Pb (liq.)	Cd	500	7.89	6.54	5.28	3.14	0.84
Cd–Sb (liq.)	Cd	500	l+s	l+s	(−4.5)	−3.24	+0.45
Cd–Sn (liq.)	Cd	500	5.45	4.40	3.52	2.16	0.52
Cd–Tl (liq.)	Cd	477	6.49	5.56	4.65	2.80	0.65
Cd–Zn (liq.)	Zn	527	6.82	5.37	4.04	2.12	0.40
Cr–V (sol.)	Cr	1 277	−7.48	−2.62	+4.43	−7.70	−1.49
Cu–Ni (sol.)	Ni	700	3.52	3.96	3.94	2.93	0.69
Cu–Pb (liq.)	Cu	1 200	21.70	16.80	13.06	6.99	1.96
Cu–Pt (sol.)	Cu	1 077	−28.65	−27.41	−25.30	−13.10	−3.33
Cu–Zn (sol.)	Zn*	500*	−28.69	−24.65	−17.36	(−15.7β)	−2.40ε
Fe–Mn (sol.)	Mn	1 177	−12.36	−10.95	−9.59	−6.61	−1.65
Fe–Ni (sol.)	Ni	927	−4.31	—	−8.83	−5.99	—
Fe–Ni (liq.)	Ni	1 600	−9.77	—	−9.72	−7.22	—
Fe–Si (liq.)	Si	1 600	−125.15	−109.73	−81.20	−27.80	−1.82
Fe–V (sol.)	Fe	1 327	−9.29	−5.98	−0.57	+14.17	+10.71
Ga–Zn (liq.)	Zn	477	4.52	3.73	3.08	2.01	0.51
Hg–Na (liq.)	Na	400	−72.24	−61.52	−40.18	−10.88	−0.35
Hg–Pb (liq.)	Hg	327	−0.73	+0.16	+0.65	+0.96	+0.45
Hg–Sn (liq.)	Sn*	177*	3.61	2.23	1.41	0.62	l+s
Hg–Tl (liq.)	Tl*	25*	−4.37	−2.78	−2.06	l+s	l+s
In–Pb (liq.)	In	400	2.98	2.40	1.88	1.00	0.17
In–Sn (liq.)	In	427	−0.59	−0.46	−0.38	−0.21	−0.04
In–Tl (liq.)	Tl	350	2.12	1.51	1.05	0.42	0.04
In–Zn (liq.)	Zn	427	9.50	7.98	6.50	3.78	0.91
Na–Pb (liq.)	Na	427	−39.05	−37.55	−34.32	−23.86	−3.43
Na–Tl (liq.)	Na	400	−33.33	−30.22	−24.70	−10.98	−0.67
Pb–Sb (liq.)	Pb	732	−0.14	−0.20	−0.19	−0.04	+0.03
Pb–Sn (liq.)	Sn	777	4.69	3.48	2.59	1.34	0.25
Pb–Tl (sol.)	Tl	250	−4.50	−4.20	−3.98	−2.93	−1.08
Pb–Tl (liq.)	Tl	500	−3.11	−2.67	−2.18	−1.15	−0.15
Sb–Sn (liq.)	Sn	632	−4.02	−3.56	−2.91	−1.50	−0.15
Sb–Zn (liq.)	Zn	577	l+s	−4.73	−5.39	−2.70	+0.96
Sn–Tl (liq.)	Tl	450	2.70	1.94	1.28	0.41	0.08
Sn–Zn (liq.)	Zn	477	8.22	7.47	6.43	4.13	1.35

*Note standard state for metal *C*.

8.5 Metallurgically important compounds

In Tables 8.8a to 8.8j values of the heats and free energies of formation from the constituent elements are given throughout in kJ mol^{-1} of compound and standard entropies in J K^{-1} mol^{-1} of compound. It is to be noted that the minus signs appear in the captions, and therefore positive numerical values denote exothermic formation.

The standard state of a reactant element is the most stable form at the temperature indicated unless otherwise stated in the headings; the ideal diatomic gases were used as the standard states for sulphur, bromine and iodine in computing the free energy values at room temperature. The standard state of the compounds is the ideal stoichiometric proportion, and the state of aggregation where necessary is indicated by s=solid, m=liquid and g=gaseous. Many compounds are stable with slight deviations from stoichiometric proportion. Compounds which may exist over a range of composition are indicated by an asterisk (*). For such compounds thermochemical values must be used with caution; dissociation pressures calculated from the free energy values may show considerable deviations from measured dissociation pressures.

Free energy values are given at five temperatures; further values can be obtained by linear interpolation provided that no change in the state of aggregation occurs in the interval. Even then, a linear interpolation will generally give a value within the accuracy of the data. The limits of accuracy are given in the last column of the tables; they apply to the values of the heat and free energy of formation below, say, 1 500 K. The errors may be greater at 2 000 K.

Dissociation pressures of sulphides and phosphides have only been determined for parts of many systems, the data for the lowest oxidation steps often being absent. No total thermochemical values for the formation of the compounds from the elements can therefore be given in Tables 8.8f and 8.8j but dissociation pressures of the compounds rich in sulphur or phosphorus appear in Tables 8.9a and 8.9b, respectively.

The thermochemical values for the metal borides and carbides suggested below can only be used as a guide. They are not sufficiently reliable for accurate calculations—mainly for two reasons. (1) transition metals do not form stoichiometric compounds with boron and carbon but rather extensively homogeneity ranges of which the investigators have taken too little heed. Thus the data below are oversimplifications. (2) it is very doubtful whether the experimental zero-point entropies of borides and carbides are generally zero. More likely, 'frozen-in disorder' of the substances studied accounted for substantial zero-point entropies, so that the values in column 3 when obtained from low-temperature heat capacities are not truly 'standard entropies'.

Table 8.8a BORIDES
Heats and free energies of formation in kJ and standard entropies

Compound	$-\Delta H_{298}$ 25 °C	S_{298} 25 °C	$-\Delta G_{300}$ 27 °C	$-\Delta G_{500}$ 227 °C	$-\Delta G_{1000}$ 727 °C	$-\Delta G_{1500}$ 1227 °C	*Accuracy* ±kJ
AlB_2	67.0	—	—	—	—	—	12
AlB_{12}	201.0	—	—	—	—	—	15
CeB_6	351.6	74.1	344.0	338.9	326.2	313.5	100
Co_2B	125.6	59.9	123.8	122.5	119.5	116.4	30
CoB	94.2	30.6	92.6	91.1	89.8	86.2	25
CrB	75.3	24.1	73.6	72.5	69.6	66.8	25
CrB_2	94.2	26.2	91.4	89.5	84.7	80.0	25
Fe_2B	71.2	56.7	70.1	69.3	67.4	65.5	20
FeB	71.2	27.7	69.6	68.5	65.8	63.1	20
HfB_2	336.1	42.7	332.2	329.5	323.0	316.5	10
MgB_2	92.1	36.0	89.6	88.0	83.0	60.0	25
MgB_4	105.1	51.9	103.8	103.0	100.4	77.5	25
MnB	75.3	32.4	73.7	72.6	70.0	67.3	30
MnB_2	94.2	34.5	91.4	89.6	85.0	80.4	30
NbB_2	251.2	37.7	248.0	245.9	240.6	235.2	25
Ni_4B_3	311.9	114.7	305.2	300.7	289.7	278.6	60
NiB	100.5	30.1	98.8	97.7	94.9	92.1	25
TaB_2	209.3	44.4	206.7	204.9	200.5	196.1	40
TiB_2	342.0	28.5	319.9	317.2	310.4	303.6	5
UB_2	164.5	55.1	162.1	160.6	156.6	152.7	25
UB_4	245.7	71.2	244.9	244.4	243.1	241.8	40
UB_{12}	433.2	139.8	438.8	442.6	452.0	461.4	40
ZrB_2	324.0	36.0	319.6	316.7	309.3	302.0	10

Table 8.8b CARBIDES
Heats and free energies of formation in kJ and standard entropies

Compound	$-\Delta H_{298}$ 25 °C	S_{298} 25 °C	$-\Delta G_{300}$ 27 °C	$-\Delta G_{500}$ 227 °C	$-\Delta G_{1000}$ 727 °C	$-\Delta G_{1500}$ 1 227 °C	$-\Delta G_{2000}$ 1 727 °C	*Accuracy* ± kJ
Al_4C_3	215.8	88.7	203.6	—	174.1	153.3	122.2	8
B_4C	71.6	27.09	71.2	70.8	69.8	68.7	67.4	19
Be_2C	117.2	16.3	114.8	—	—	100.9	78.8	5
CaC_2	59.0	70.3	64.5(α)	69.5(α)	85.0(β)	103.0(β)	121.0(β)	13
Ce_2C_3	176.6	173.7	—	—	—	—	—	7
CeC_2	97.1	90.4	—	—	—	—	—	7
Co_2C	−16.7	74.5	−19.3	−12.1	−8.0	—	—	21
Cr_4C	98.4	135.6	100.0	72.0	104.2	118.3	—	9
Cr_7C_3	228.1	200.9	223.4	237.1	245.9	254.3	—	13
Cr_3C_2	109.7	85.4	108.6	109.8	115.3	122.4	—	13
Fe_3C	−25.1	104.7	−19.0	−14.2	−1.93	+4.90	—	4
CH_4	74.90(g)	186.3(g)	50.7(g)	32.7(g)	−19.3(g)	—	—	4
Mg_2C_3	−79.5	−62.8	—	—	—	—	—	33
MgC_2	−87.9	—	−84	—	—	—	—	21
Mn_7C_3	112.2	239.0	111.4	112.6	118.9	113.9	—	13
Nb_2C	186.3	64.0	—	—	—	—	—	13
NbC	138.1	35.1	—	—	—	—	—	—
SiC	67.0	16.54	—	58.2	55.3	53.2	41.9	9
Ta_2C	203.0	83.6	—	—	—	—	—	17
TaC	143.6	42.2	142.2	—	141.0	137.8	—	8
ThC	125.6	59	—	—	—	—	—	13
ThC_2	117.2	70.3	—	124.3	117.2	—	—	21
TiC	183.8	24.3	180.0	177.9	172.9	167.1	160.4	13
UC	90.9	59.5	92.1	93.4	96.7	95.5	92.5	13
U_2C_3	205.1	138.4	205.2	205.2	206.4	199.3	187.1	17
UC_2	96.3	67.8	98.4	100.0	101.7	99.6	93.8	13
VC	100.9	27.6	—	—	94.6	91.3	80.4	17
WC	37.7	41.8	37.3	36.4	35.6	34.8	33.5	13
W_2C	26.4	—	—	—	—	—	—	7
ZrC	202.0	33.1	198.3	196.2	190.7	184.4	—	21

Table 8.8c NITRIDES
Heats and free energies of formation in kJ and standard entropies

Compound	$-\Delta H_{298}$ 25 °C	S_{298} 25 °C	$-\Delta G_{300}$ 27 °C	$-\Delta G_{500}$ 227 °C	$-\Delta G_{1000}$ 727 °C	$-\Delta G_{1500}$ 1 227 °C	$-\Delta G_{2000}$ 1 727 °C	*Accuracy* ± kJ
AlN	318.6	20.2	287.1	266.1	—	—	—	5
BN	254.1	14.8	—	—	—	48.1	27.6	5
Ba_3N_2	341.1	(152.4)	292.2	244.1	123.9	—	—	38
Be_3N_2	589.9	34.3	534.1	—	—	—	—	5
Ca_3N_2	439.6	108.0	377	335	230	—	—	23
Cd_3N_2	−161.6	—	—	—	—	—	—	13
CeN	326.6	—	295.2	274.2	222	—	—	38
Co_3N	−8.4	98.8	—	—	—	—	—	21
CrN	123.1	—	96.7	78.7	40.6	—	—	5
Cr_2N	114.7	—	93.8	79.5	51.1	28.0	—	5
Cu_3N	−74.5	—	—	—	—	—	—	13
Fe_4N	10.9	(156.2)	−4.2	−12.77	−41.0	—	—	17
GaN	109.7	29.7	—	—	—	—	—	10
Ge_3N_4	65.3	(167)	—	—	—	—	—	9

Table 8.8c NITRIDES—*continued*

Compound	$-\Delta H_{298}$ 25 °C	S_{298} 25 °C	$-\Delta G_{300}$ 27 °C	$-\Delta G_{500}$ 227 °C	$-\Delta G_{1000}$ 727 °C	$-\Delta G_{1500}$ 1 227 °C	$-\Delta G_{2000}$ 1 727 °C	*Accuracy* +kJ
HfN	369.3	50.6	—	—	—	—	—	2.5
NH_3	46.1	192.47	16.7	−4.6	−61.5	—	—	2.1
InN	138.1	43.5	—	—	—	—	—	13
LaN	299.4	44.4	270.5	249.5	197.2	—	—	38
Li_3N	196.8	—	154.1	125.6	—	—	—	2.1
Mg_3N_2	461.8	93.7	399.4	359.6	258.3	—	—	13
Mn_4N	126.9	—	—	—	—	—	—	4
Mn_5N_2	201.8	—	156.2	125.6	49.4	—	—	4
Mo_2N	69.5	(87.9)	70.8	37.3	−8.8	—	—	17
NbN	234	—	—	—	—	—	—	11
Nb_2N	248.6	67.0	—	—	—	—	—	4
Ni_3N	−0.8	—	−27.2	—	—	—	—	4
Si_3N_4	745.1	113.0	622.2	571.9	410.3	251.2	64.5	17
Sr_3N_2	391.0	123.5	323.6	278.4	—	—	—	23
Ta_2N	270.9	—	—	—	—	—	—	13
TaN	252.4	42.7	214.8	202.2	166.6	134.0	103.8	13
Th_3N_4	1 298.0	—	1 185	1 114	925	737	548	84
TiN	336.6	30.1	308.1	289.7	243.3	195.9	148.2	8
UN	294.7	62.7	260.0	242.0	196.8	152.0	106.7	42
U_2N_2	708.5	128.9	—	—	—	—	—	17
VN	217.3	37.3	—	—	—	—	—	17
V_2N	264.5	53.4	—	—	—	—	—	17
Zn_3N_2	22.2	140.2	—	—	—	—	—	8
ZrN	365.5	38.9	336.2	317.8	270.9	223.2	—	7

Table 8.8d SILICIDES
Heats and free energies of formation in kJ and standard entropies

Compound	$-\Delta H_{298}$ 25 °C	S_{298} 25 °C	$-\Delta G_{300}$ 27 °C	$-\Delta G_{500}$ 227 °C	$-\Delta G_{1000}$ 727 °C	$-\Delta G_{1500}$ 1227 °C	*Accuracy* ±kJ
Ca_2Si	209.3	—	—	—	—	—	13
CaSi	150.7	—	—	—	—	—	9
$CaSi_2$	150.7	—	—	—	—	—	13
Co_2Si	117.2	—	—	—	—	—	10
CoSi	95.0	42.7	93.2	91.9	88.9	85.8	9
$CoSi_2$	98.8	64.0	97.7	97.0	95.1	95.5	12
Cr_3Si	92.1	86.3	90.8	90.0	87.9	85.8	18
Cr_5Si_3	211.4	16.91	209.4	280.0	204.6	201.2	35
CrSi	53.2	43.9	53.6	53.8	54.4	55.0	10
$CrSi_2$	80.0	58.6	78.5	77.5	75.1	72.7	13
FeSi	76.9	41.8	75.6	74.8	72.7	70.9	5
Mg_2Si	79.1	63.8	73.1	69.1	55.0	—	5
Mn_2Si	79.5	103.8	76.4	74.3	69.0	63.8	15
Mn_5Si_3	200.9	238.6	207.8	212.5	224.0	235.6	40
MnSi	60.7	46.5	59.5	58.6	56.6	54.5	13
Mn_4Si_7	308.5	228.1	299.3	293.1	277.7	262.3	65
Mo_3Si	116.4	106.5	117.0	117.4	118.3	119.3	7.5
Mo_5Si_3	310.2	208.0	312.8	314.5	318.8	323.1	25
$MoSi_2$	131.9	65.1	131.6	131.4	130.8	130.3	10
$NbSi_2$	138.1	69.9	136.8	136.0	133.9	131.7	45
Ni_2Si	142.7	—	—	—	—	—	11
Ni_3Si_2	232.3	—	—	—	—	—	20
NiSi	89.6	44.4	88.4	87.5	85.4	83.3	7.5
$NiSi_2$	94.2	65.3	93.6	93.1	92.1	91.0	13
Re_5Si_3	157.4	256.0	161.3	164.0	170.5	177.1	70

Table 8.8d SILICIDES—*continued*

Compound	$-\Delta H_{298}$ 25°C	S_{298} 25°C	$-\Delta G_{300}$ 27°C	$-\Delta G_{500}$ 227°C	$-\Delta G_{1000}$ 727°C	$-\Delta G_{1500}$ 1 227°C	*Accuracy* ±kJ
ReSi	52.7	55.4	52.5	52.4	52.0	51.7	21
$ReSi_2$	90.4	74.1	90.2	90.0	89.6	89.2	30
Ta_2Si	125.6	105.5	126.7	127.4	129.2	131.0	20
Ta_5Si_3	334.9	280.9	340.2	343.7	352.4	361.2	35
$TaSi_2$	119.3	75.3	118.2	117.5	115.6	113.8	13
Th_3Si_2	279.6	166.4	270.2	263.9	248.1	232.4	50
ThSi	126.0	62.8	123.2	121.3	116.6	111.9	17
Th_3Si_5	477.6	231.9	470.9	466.4	455.2	444.1	85
$ThSi_2$	170.8	89.2	170.3	169.9	169.0	168.1	30
Ti_5Si_3	579.8	—	—	—	—	—	65
TiSi	129.8	49.0	49.0	48.95	48.9	48.8	15
$TiSi_2$	134.4	61.1	132.0	131.0	127.6	124.2	21
U_3Si	92.1	167.4	91.5	91.1	90.1	89.1	13
U_3Si_2	170.8	197.6	173.6	175.5	180.2	184.9	11
USi	84.6	66.6	83.9	83.4	82.2	81.0	5
U_3Si_5	354.6	231.5	353.7	353.1	351.5	350.0	17
USi_2	129.8	82.0	127.1	126.9	124.0	121.1	3
USi_3	130.6	106.3	130.5	130.45	130.3	130.1	3

Table 8.8e OXIDES
Heats and free energies of formation in kJ and standard entropies

Compound	$-\Delta H_{298}$ 25°C	S_{298} 25°C	$-\Delta G_{300}$ 27°C	$-\Delta G_{500}$ 227°C	$-\Delta G_{1000}$ 727°C	$-\Delta G_{1500}$ 1 227°C	$-\Delta G_{2000}$ 1 727°C	*Accuracy* ±kJ
Ag_2O	30.6	121.8	10.68	—	—	—	—	2.1
Al_2O_3	1 678.2	51.1	1 584.0	1 520.8	1 362.4	1 146.9	—	17
As_2O_3	653.77	122.80	577.4	526.3	—	—	—	3.3
As_2O_5	914.8	105.5	771.6	675.7	—	—	—	6.3
B_2O_3	1 272.5	54.0	1 192.9	1 139.8	—	—	—	8
BaO	553.8	70.3	524.5	504.4	455.0	—	—	22
BaO_2	634.2	93.1	—	—	—	—	—	13
BeO	608.4	14.1	580.0	560.7	513.4	466.9	412.9	13
Bi_2O_3	570.7	151.6	496.6	—	—	—	—	17
CO	110.5	198.0	138.2	155.7	199.50	243.3	287.2	4
CO_2	393.77	213.9	394.4	394.8	395.2	395.7	396.1	4
CaO	634.3	39.8	603.7	584.1	534.7	481.9	405.7	4
CaO_2	659.4	—	—	—	—	—	—	4
CdO	259.4	54.8	229.7	212.1	204.1	106.6	—	4
Ce_2O_3	1 821.7	150.7	1 733.7	1 676.9	—	—	—	8
CeO_2	1 089.4	62.4	1 025.8	983.9	882.2	—	—	13
CoO	239.1	52.96	212.7	198.9	163.3	127.7	—	4
Co_3O_4	905.6	102.6	777.1	705.9	525.0	—	—	13
Cr_2O_3	1 130.4	81.2	1 051.3	991.0	861.6	731.0	601.2	17
CrO_2	582.8	51.0	—	—	—	—	—	8
CrO_3	579.9	71.9	—	—	—	—	—	10.5
Cs_2O	317.8	127.6	—	—	—	—	—	13
Cu_2O	167.5	93.8	144.9	129.8	95.5	59.5	—	4
CuO	155.3	42.7	127.3	108.9	66.6	—	—	4
Dy_2O_3	1 866.5	149.9	1 773.9	1 712.4	—	—	—	13
Er_2O_3	1 899.1	153.2	1 808.7	1 748.4	—	—	—	8
$Fe_{0.95}O$	264.6	58.82	241.2	228.6	197.2	165.8	—	13
Fe_3O_4	1 117.5	151.6	1 015.3	950.0	780.8	—	—	25
Fe_2O_3	821.9	87.5	—	695.0	556.8	—	—	21
Ga_2O	347.5	—	—	—	—	—	—	13

Table 8.8e OXIDES—*continued*

Compound	$-\Delta H_{298}$ 25 °C	S_{298} 25 °C	$-\Delta G_{300}$ 27 °C	$-\Delta G_{500}$ 227 °C	$-\Delta G_{1000}$ 727 °C	$-\Delta G_{1500}$ 1 227 °C	$-\Delta G_{2000}$ 1 727 °C	*Accuracy* ±kJ
Ga_2O_3	1 083.5	84.78	992.3	—	—	—	—	8
Gd_2O_3	1 817.1	150.7	1 730.4	1 651.7	—	—	—	13
GeO	30.7(g)	224.0(g)	54.2(g)	69.2(g)	105.7(g)	—	—	17
GeO_2	580.2	39.7	525.3	488.7	397.1	—	—	8
H_2O	286.0(m)	70.13(m)	237.0(m)	—	—	—	—	0.8
H_2O	242.0(g)	188.8(g)	228.6(g)	219.0(g)	192.6(g)	164.1(g)	134.4(g)	1.3
H_2O_2	187.1(m)	109.5(m)	118.1(m)	—	—	—	—	2.1
H_2O_2	135.2(g)	226.9(g)	—	—	—	—	—	1.3
HfO_2	1 113.7	59.5	1 053.4	1 014.0	919.4	828.1	739.0	4
HgO (red)	90.9	70.3	58.32	36.59	—	—	—	4
Ho_2O_3	1 882.4	158.3	1 792.4	1 732.1	—	—	—	13
In_2O_3	927.4	—	834.0	771.6	—	—	—	13
IrO_2	241.5	56.5	186.1	149.2	57.0	—	—	17
IrO_3	−13.4(g)	290.5(g)	—	—	—	85.0	—	76
K_2O	363.3	—	320.7	295.2	229.4	163.3	—	3
KO_2	284.6	122.6	—	—	—	—	—	2.1
La_2O_3	1 794.1	128.1	1 706.2	1 648.4	1 509.0	—	—	8
Li_2O	596.6	37.93	560.2	534.2	467.7	—	—	4
Li_2O_2	635.6	—	—	—	—	—	—	8
MgO	601.6	26.97	571.1	550.1	498.6	440.5	329.9	6.3
MnO	385.2	59.9	363.0	348.8	312.3	275.9	234.5	13
Mn_3O_4	1 387.5	148.6	1 280.3	1 209.6	1 035.8	—	—	13
Mn_2O_3	960.5	110.5	882.2	821.9	703.8	—	—	13
MnO_2	520.4	53.2	462.2	424.5	334.9	—	—	13
Mn_2O_7	728.9	—	—	—	—	—	—	10.5
MoO_2	588.7	50.0	533.4	496.5	404.3	—	—	4
MoO_3	746.1	77.9	668.6	617.1	486.9	—	—	4
Na_2O	415.2	75.1	376.1	350.1	285.0	—	—	8
Na_2O_2	515.0	94.6	—	—	—	—	—	6.3
NaO_2	261.7	116.0	—	—	—	—	—	4
NbO	419.8	46.0	391.2	373.1	326.5	—	—	17
NbO_2	799.3	54.55	743.2	705.5	612.1	518.3	425.0	8
Nb_2O_5	1 900.8	137.3	1 766.0	1 676.4	—	—	—	13
Nd_2O_3	1 809.1	158.6	—	—	—	—	—	17
NiO	240.7	38.1	213.1	194.7	150.7	108.4	57.4	13
NpO_2	1 030.0	80.4	—	—	—	—	—	42
OsO_4	393.9	136.9	—	—	—	—	—	13
P_2O_5	1 493.0	229.0	1 381.6	—	—	—	—	25
PbO	219.4	66.3	188.8	168.3	117.2	—	—	13
Pb_3O_4	719.0	211.4	600.9	519.7	325.4	—	—	25
PbO_2	274.6	71.9	215.3	175.8	76.9	—	—	8
PdO	112.6	39.3	82.2	62.0	11.3	—	—	17
Pr_2O_3	1 828.8	158.6	—	—	—	—	—	6
$PrO_{1.72}$	937.8	—	—	—	—	—	—	6.3
PrO_2	974.9	80.0	—	—	—	—	—	17
PtO_2	−168.7(g)	256.0(g)	—	—	—	157.8	—	8
PuO_2	1 058.4	82.5	1 003.6	968.8	882.2	795	—	21
Rb_2O	330.3	—	—	—	—	—	—	13
Rb_2O_3	527.5	—	—	—	—	—	—	42
RbO_2	284.7	—	—	—	—	—	—	25
ReO_2	432.9	62.8	379.0	343.0	253.2	—	—	13
ReO_3	611.3	80.8	—	—	—	—	—	21
Re_2O_7	1 249.1	207.5	1 073.7	959.0	—	—	—	8
Rh_2O_3	383.0	92.1	—	—	—	—	—	17

Table 8.8e OXIDES—*continued*

Compound	$-\Delta H_{298}$ 25 °C	S_{298} 25 °C	$-\Delta G_{300}$ 27 °C	$-\Delta G_{500}$ 227 °C	$-\Delta G_{1000}$ 727 °C	$-\Delta G_{1500}$ 1 227 °C	$-\Delta G_{2000}$ 1 727 °C	*Accuracy* ±kJ
RhO_2	195.9(g)	263.8(g)	—	—	—	157.8	—	4
RuO_2	304.4	60.7	252.3	217.6	130.9	44.2	—	8
RuO_3	78.3(g)	276.2(g)	—	—	—	—	—	—
RuO_4	180.9(g)	290.1(g)	—	—	—	—	—	—
SO	−5.0(g)	222.3(g)	—	73.2(g)	70.3(g)	79.1(g)	82.0(g)	8
SO_2	297.05(g)	248.07(g)	340.8(g)	326.6(g)	290.1(g)	254.1(g)	217.7(g)	2.1
SO_3	395.2(g)	256.2(g)	—	376.4(g)	295.6(g)	214.4(g)	—	6.3
Sb_2O_3	699.2	123.1	—	—	—	—	—	13
SbO_2	454.0	63.6	—	—	—	—	—	25
Sb_2O_5	1 008.0	125.2	—	—	—	—	—	63
Sc_2O_3	1 906.7	77.0	1 818.7	1 756.8	—	—	—	—
SeO_2	236.1	66.7	—	—	—	—	—	4
SiO	98.4(g)	211.64(g)	126.9(g)	144.5(g)	185.1(g)	223.7(g)	—	13
SiO_2	910.9(q)	41.5(q)	—	—	726.9	640.0	538.5	13
Sm_2O_3	1 833.0	151.1	1 744.2	1 685.2	—	—	—	13
SnO	286.4	56.5	—	—	—	—	—	4
SnO_2	580.7	52.3	519.4	478.3	—	—	—	4
SrO	592.3	55.5	562.4	542.3	492.3	—	—	13
SrO_2	633.7	59.0	—	—	—	—	—	17
Ta_2O_5	2 047.3	143.2	1 910.9	1 822.1	1 612.8	1 413.0	1 220.0	21
Tb_2O_3	1 828.8	—	—	—	—	—	—	8
$TbO_{1.71}$	934.9	—	—	—	—	—	—	4
$TbO_{1.8}$	947.9	—	—	—	—	—	—	4
TcO_2	433.3	58.6	—	—	—	—	—	13
Tc_2O_7	1 113.7	184.2	996.0	917.7	722.2	—	—	17
TeO_2	322.4	74.1	268.3	232.2	138.6	—	—	8
ThO_2	1 227.6	65.3	1 173.6	1 135.0	1 048.8	960.9	—	21
TiO	542.9	34.8	509.5	491.9	447.1	402.7	352.9	17
Ti_2O_3	1521.6	77.3	1 427.1	375.8	245.7	1 116.5	986.4	17
Ti_3O_5	2 457.2	129.4	2 309.4	2 224.4	2 010.1	1 797.0	1 581.8	13
TiO_2	944.1	50.2	862.1	827.3	739.5	652.7	564.8	16
Tl_2O	167.4	134.3	—	—	—	—	—	4
Tl_2O_3	390.6	137.3	—	—	—	—	—	25
Tm_2O_3	1 889.9	—	—	—	—	—	—	8
UO_2	1 085.2	77.9	1 032.0	996.9	913.6	827.7	740.6	13
U_4O_9	4 513	336.2	—	—	—	—	—	42
U_3O_8	3 575.9	282.6	—	—	—	—	—	21
UO_3	1 230.7	98.8	—	—	—	—	—	21
VO	432.0	38.9	413.5	398.5	360.8	322.9	—	25
V_2O_3	1 219.4	98.4	1 147.0	1 099.7	982.0	864.4	—	33
VO_2	713.7	51.5	678.4	643.8	567.2	—	—	21
V_2O_5	1 551.3	131.0	1 454.1	1 373.7	1 172.8	—	—	50
WO_3	838.6	75.9						
WO_2	589.9	50.6	—	480.6	405.7	334.5	—	13
Y_2O_3	1 906.7	99.2	1 817.9	1 759.3	1 618.6	1 482.1	1 348.1	21
Yb_2O_3	1 815.8	133.1	—	—	—	—	—	8
ZnO	350.8	43.5	320.7	301.0	256.6	174.6	82.0	8
ZrO_2	1 101.3	50.7	1 042.2	1 003.7	909.9	818.2	729.5	17

Table 8.8f SULPHIDES

Heats and free energies of formation in kJ and standard entropies

In calculating the heats of formation, rhombic sulphur is taken as the standard state. For the free energies of formation the perfect diatomic gas S_2 is taken as the standard state at all temperatures.

Compound	$-\Delta H_{298}$ 25 °C	S_{298} 25 °C	$-\Delta G_{300}$ 27 °C	ΔG_{500} 227 °C	$-\Delta G_{1000}$ 727 °C	$-\Delta G_{1500}$ 1 227 °C	$-\Delta G_{2000}$ 1 727 °C	*Accuracy* ±kJ
Ag_2S	31.8(α)	144.4(α)	180.4(α)	70.8(β)	53.2(β)	32.2(m)	—	2.1
Al_2S_3	723.9	123.5	—	—	—	—	—	17
As_2S_3	167.4	169.6	—	—	—	—	—	23
B_2S_3	252.4	92.1	—	—	—	—	—	9
BaS	443.8	78.3	481.9	463.9	419.1	—	—	21
BeS	234.0	35.2	272	247	201	—	—	21
Bi_2S_3	201.8	200.5	—	—	—	—	—	—
CS_2	−87.9(m)	151.6(m)	15.1(g)	16.3(g)	19.7(g)	23.4(g)	—	4
COS	138.5	231.5	207.2	208.9	212.4	218.5	223.1	13
CaS	476.4	56.5	—	491.9	444.2	393.1	319.9	8
CdS	149.4	69.1	185.5	—	—	—	—	4
CeS	456.7	78.3	496.5	479.8	437.9	396.0	354.2	21
Ce_3S_4	1 653.5	255.3	—	—	—	—	—	17
Ce_2S_3	1 188.8	180.4	—	—	—	—	—	13
$CoS_{0.89}$	94.6	52.3	125.2	110.5	73.3	—	—	8
Co_3S_4	359.2	184.6	—	427.9	—	—	—	13
CoS_2	153.2	69.1	—	—	—	—	—	17
CrS	155.7	64.0	—	—	—	—	—	13
Cs_2S	339.5	—	—	—	—	—	—	33
Cu_2S	79.6	120.9	128.5	118.1	100.9	—	—	8
CuS	52.3	66.6	—	—	—	—	—	4
FeS	100.5	60.3	140.2(α)	126.4(β)	97.3(β)	—	—	4
FeS_2	171.6	53.2	—	299.8	90.9	—	—	8
GaS	209.3	57.7	—	—	—	—	—	13
Ga_2S_3	514.0	139.8	—	—	—	—	—	13
GeS	76.2	66.2	46.5	36.7	—	—	—	33
H_2S	20.1(g)	205.6(g)	73.7(g)	65.10(g)	41.24(g)	16.3(g)	−8.0(g)	2.1
HgS	53.4	82.5	—	—	—	—	—	6.3
InS	133.9	69.1	—	—	—	—	—	13
In_5S_6	774.4	174.6	—	—	—	—	—	84
In_2S_3	355.8	163.7	—	—	—	—	—	17
Ir_2S_3	244.5	97.1	—	—	124.8	10.5	—	33
IrS_2	144.8	61.6	—	—	—	—	—	25
K_2S	428.7	—	—	—	—	—	—	14.7
La_2S_3	1 222.3	165.1	1 235.5	—	—	—	—	42
Li_2S	446.6	60.7	—	—	—	—	—	8
MgS	351.6	50.4	—	—	296.8	—	—	8
MnS	213.5	80.4	254.3	241.7	211.0	128.5	143.2(m)	6.3
MnS_2	223.9	99.9	—	—	—	—	—	10.5
Mo_2S_3	410.2	113.0	—	—	309.4	195.1	—	29
MoS_2	275.4	62.6	—	—	203.1	125.2	—	21
Na_2S	374.6	79.5	—	388.1	322.4	206.8	—	17
NaS	201.3	44.8	—	—	—	—	—	14.7
NaS_2	206.0	83.7	—	—	—	—	—	14.7
Ni_3S_2	216.0	134.0	—	252.0	—	—	—	13
NiS	94.2	53.0	—	—	—	—	—	6.3
OsS_2	100.5	—	181.7	144.9	66.2	−8.8	—	17
PbS	98.4	91.3	134.6	120.6	81.7	—	—	4
PtS	82.5	55.06	116.4	96.3	47.7	0.8	—	13
PtS_2	110.1	74.73	179.6	141.1	49.4	—	—	13
Rb_2S	348.3	133.1	—	—	—	—	—	25
ReS_2	139.0	60.7	211.4	177.1	97.1	21.4	—	25
RuS_2	206.0	—	242.8	207.7	125.2	46.9	—	21

Table 8.8f SULPHIDES—*continued*

Compound	$-\Delta H_{298}$ 25°C	S_{298} 25°C	$-\Delta G_{300}$ 27°C	$-\Delta G_{500}$ 227°C	$-\Delta G_{1000}$ 727°C	$-\Delta G_{1500}$ 1 227°C	$-\Delta G_{2000}$ 1 727°C	*Accuracy* ±kJ
S	−277.1(g)	167.9(g)	—	−146.5(g)	−115.1(g)	−83.7(g)	−53.2(g)	0.5
S_2	−129.8(g)	228.2(g)	0(g)	0(g)	0(g)	0(g)	0(g)	4
S_6	−103.0(g)	354.0(g)	179.6	121.8	−23.4	—	—	13
S_8	− 99.6(g)	423.3(g)	258.7	174.2	−37.7	—	—	33
Sb_2S_3	205.1	182.1	285	230	—	—	—	29
SiS_2	213.5	80.4	278.4	—	—	—	—	25
SnS	108.4	77.0	143.6	—	—	—	—	6.3
SnS_2	153.6	87.5	—	—	—	—	—	17
SrS	452.6	69.1	488.6	468.5	419	—	—	42
Th_2S_3	1 082.7	—	—	—	—	—	—	10.5
TiS	272.1	56.5	—	—	—	—	—	33
Tl_2S	95.0	159.0	—	—	—	—	—	4
WS_2	202.6	83.7	278.4	243.7	162.4	85.4	—	17
ZnS	205.3	57.8	235.4	219.1	174.3	90.4	−4.9	17

Table 8.8g HALIDES

Heats and free energies of formation in kJ and standard entropies

For the free energies of formation the standard states for the halogens at all temperatures are the perfect diatomic gases at a pressure of one atmosphere.

Compound	$-\Delta H_{298}$ 25°C	S_{298} 25°C	$-\Delta G_{300}$ 27°C	$-\Delta G_{500}$ 227°C	$-\Delta G_{1000}$ 727°C	$-\Delta G_{1500}$ 1 227°C	$-\Delta G_{2000}$ 1 727°C	*Accuracy* ±kJ
AgF	206.0	83.7	185.5	173.8	161.2(m)	146.5(m)	—	8
AgF_2	359.6	—	304.8	276.3	213.5(m)	—	—	13
AgCl	126.9	96.3	109.3	98.8	80.4(m)	69.9(m)	—	0.8
AgBr	99.2	107.2	97.1	86.7	69.5(m)	65.7(m)	—	0.8
AgI	62.4	115.6	77.0	66.6	50.2	46.1	—	0.8
AlF	265.4(g)	215.2(g)	—	—	—	—	—	4
AlF_3	1 511.1	66.6	1 418.5	1 377.9	—	—	—	0.8
AlCl	51.5(g)	227.8(g)	75.3(g)	94.6(g)	133.5(g)	164.9(g)	196.1(g)	2
$AlCl_3$	705.9	110.1	572.3(g)	566.5(g)	334.7(g)	500.7(g)	—	4
Al_2Cl_6	1 275.3(g)	475.8	—	1 238.0(g)	1 194.9(g)	—	—	2
$AlBr_3$	511.3	180.3	515.0	—	—	—	—	2
AlI_3	310.2	189.6	—	—	—	—	—	6
AsF_3	958.0(m)	180.5(m)	—	—	—	—	—	4
AsF_3	921.7(g)	289.3(g)	—	—	—	—	—	4
$AsCl_3$	335.8(m)	233.6(m)	294.8(m)	278.0(g)	254.1(g)	217.7(g)	—	25
$AsBr_3$	132.1(g)	363.9(g)	—	—	—	—	—	6
AsI_3	64.9	213.1	94.6	61.5	—	—	—	10.5
AuF_3	365.4	114.3	—	—	—	—	—	36
AuCl	34.8	92.9	17.6	5.4	—	—	—	4
$AuCl_3$	115.1	148.2	57.4	18.8	—	—	—	8
AuBr	18.4	119.0	16.3	4.6	—	—	—	4
$AuBr_3$	54.4	—	44.0	4.2	—	—	—	21
AuI	−0.8	—	12.6	0.0	—	—	—	4
BF_3	1 136.1(g)	254.6(g)	—	—	—	—	—	2.1
BCl_3	427.0(m)	206.0(m)	—	—	—	—	—	13
BCl_3	403.1(g)	290.1(g)	—	—	—	—	—	13
BBr_3	238.6(m)	228.9(m)	—	—	—	—	—	2
BBr_3	204.2(g)	324.5(g)	—	—	—	—	—	2
BaF_2	1 207.6	96.3	1 146.3	1 112.4	1 035.8	952.5	—	13
$BaCl_2$	860.8	123.7	—	—	—	—	—	3

Table 8.8g HALIDES—*continued*

Compound	$-\Delta H_{298}$ 25 °C	S_{298} 25 °C	$-\Delta G_{300}$ 27 °C	$-\Delta G_{500}$ 227 °C	$-\Delta G_{1000}$ 727 °C	$-\Delta G_{1500}$ 1 227 °C	$-\Delta G_{2000}$ 1 727 °C	*Accuracy* ±kJ
$BaBr_2$	755.3	148.6	736.0	702.5	622.2	553.5	—	13
BaI_2	602.9	165.2	617.6	584.5	507.9(m)	447.2	—	21
BeF	175.0(g)	205.78(g)	236.6(g)	255.0(g)	299.8(g)	341.6(g)	378.5(g)	17
BeF_2	1 027.4	53.4	963.4	—	—	—	—	13
BeCl	−8.3(g)	217.7(g)	16.3(g)	35.6(g)	81.6(g)	125.2(g)	163.3(g)	—
$BeCl_2$	494.0	75.8	452.2	427.1	385.2(m)	—	—	8
$BeBr_2$	353.7	106.3	322	297	—	—	—	25
BeI_2	192.5	120.6	186.7	159.9	—	—	—	25
BiF_3	910.8	122.6	821	775	—	—	—	25
BiCl	−25.1(g)	255.1(g)	—	—	52.3(g)	—	—	17
$BiCl_3$	379.3	171.6	328.5	294.7	250.7	174.5	—	8
BiI_3	150.7	224.8	131.5	88	—	—	—	13
CF_4	933.5(g)	262.5(g)	888(g)	858(g)	782(g)	707(g)	—	21
CCl_4	135.3(m)	214.8(m)	68.2	—	—	—	—	2.1
CCl_4	103(g)	309.4(g)	—	36.0	—	—	—	4
CBr_4	−50.2(g)	358.4(g)	—	—	—	—	—	17
CaF_2	1 220.2	68.87	1 167.1	1 132.1	1 050.7	970.3	—	9
$CaCl_2$	796.2	104.6	752.4	724	653	599(m)	—	5
$CaBr_2$	680.4	134.0	663.6	633.0	562.3	502	—	10.5
CaI_2	535.1	142	551.4	519.6	498.4	385.6(m)		21
CdF_2	697.1	83.6	648.1	613.4	538.0	—	—	5
$CdCl_2$	391.0	115.3	—	—	—	—	—	2.1
$CdBr_2$	314.8	139.0	301.4	272.1	202.2	—	—	8
CdI_2	204.2	158.4	—	—	—	—	—	4
$CeCl_3$	1 058.4	147.8	981.8	932.0	818.5	—	—	8
CeI_3	650.4	214.7	679.1	631.4	520.4	424.1(m)	—	21
CoF_2	692.9	82.1	646.4	618.4	547.6	—	—	13
CoF_3	811.2	95.4	744.3	706.3	602.3	—	—	33
$CoCl_2$	310.2	109.3	425.5	239.5	174.2	150.3	—	5
$CoBr_2$	221.0	134.0	203.9	177.1	122.7(m)	(87.1)(m)	—	13
CoI_2	87.9	153.2	118.9	85.0	30.6(m)	−5.0(m)	—	21
CrF_2	779.9	89.7	—	—	—	—	—	13
CrF_3	1 113.9	93.95	1 043	996	879	762(m)	—	21
CrF_4	1 199.5	—	1 118	1 065.5	—	—	—	25
$CrCl_2$	395.6	115.3	351.3	327.0	272.1	233.6(m)	252.9(g)	13
$CrCl_3$	556.7	123.01	491.9	445.0	327.4	—	—	21
CrI_2	158.3	—	179.6	151.1	88.3	46.5(m)	—	21
CrI_3	205.1	199.6	—	—	—	—	—	21
CsF	555.1	88.3	527.5	507.0	458.8	416.9	—	8
CsCl	433.3	100.1	404.4	384.3	336.6	298.9	—	5
CsBr	394.8	113.5	382.3	361.7	318.2	284.7	—	13
CsI	337.0	—	340.4	318.6	273.0	250.0	—	10.5
CuF_2	549.2	—	502.2	475.9	411.3	—	—	17
CuCl	137.3	87.1	117.2	105.9	93.8	81.6	—	8
$CuCl_2$	217.6	108.4	163.3	132.7	—	—	—	13
CuBr	104.3	96.3	101.3	89.6	70.8	55.7	—	4
$CuBr_2$	141.9	133.9	126.9	98.4	—	—	—	18.8
CuI	67.8	96.7	79.5	65.7	39.4	23.9	—	4
CuI_2	7.1	—	—	—	—	—	—	10.5
$ErCl_3$	959.2	146.9	—	—	—	—	—	4
FeF_2	711.6	87.1	667.6	640.8	574.3		—	25
FeF_3	1 042.3	98.4	975	929	824	728(m)	—	54
$FeCl_2$	341.2	120.2	304.4	279.7	222.7(m)	209.88	—	1.7
$FeCl_3$	400.7	142.3	—	—	—	—	—	4
$FeBr_2$	247.4	140.0	239.1	213.1	160.8	127.3(m)	—	4

Table 8.8g HALIDES—*continued*

Compound	$-\Delta H_{298}$ 25 °C	S_{298} 25 °C	$-\Delta G_{300}$ 27 °C	$-\Delta G_{500}$ 227 °C	$-\Delta G_{1000}$ 727 °C	$-\Delta G_{1500}$ 1 227 °C	$-\Delta G_{2000}$ 1 727 °C	*Accuracy* ±kJ
$FeBr_3$	265.9	173.7	250.4	213.5	—	—	—	17
FeI_2	116.4	170.0	149.5	122.7	68.2(m)	32.7(m)	—	17
$GaCl_3$	523.4	135.2	466.8	431(m)	—	—	—	6.3
$GaBr_3$	386.9	180.0	375.1	339(m)	—	—	—	4
GaI_3	239.4	203.8	248.3	209(m)	—	—	—	13
$GdCl_3$	1 005.3	146.1	—	—	—	—	—	4
GeF_4	1 192.5(g)	302.9(g)	—	—	—	—	—	2
$GeCl_4$	504.8(g)	347.5(g)	463	—	—	—	—	—
$GeBr_4$	330.8	396.9	309.0	266.3(g)	—	—	—	10.5
GeI_4	37.7(g)	429.1(g)	—	—	—	—	—	13
HF	272.7(g)	173.8(g)	273.4(g)	274.7(g)	277.2(g)	278.8(g)	279.7(g)	8
HCl	92.4(g)	186.94(g)	95.0(g)	96.7(g)	100.5(g)	103.4(g)	107.2(g)	1.3
HBr	36.4(g)	198.9(g)	57.65(g)	59.58(g)	63.35(g)	66.65(g)	69.75(g)	1.3
HI	−26.4(g)	206.8(g)	7.70(g)	9.67(g)	13.57(g)	17.2(g)	21.52(g)	1.3
HfF_4	1 931.8	136.0	—	—	—	—	—	8
$HfCl_4$	991.9	190.9	898.9	—	—	—	—	17
HgF	246.2	—	219.0	—	—	—	—	21
HgF_2	397.7	—	—	—	—	—	—	42
HgCl	131.9	98.4	104.7	85.4	—	—	—	1.3
$HgCl_2$	230.3	144.4	183.4	153.2	—	—	—	4
HgBr	103.4	111.4	92.1	72.9	—	—	—	1.3
$HgBr_2$	169.6	162.9	152.0	121.0	—	—	—	1.3
HgI	60.3	121.4	66.2	48.6	—	—	—	0.8
HgI_2	105.5	170.8	119.3	87.5	—	—	—	1.7
InCl	186.3	95.0	180	163(m)	—	—	—	10.5
$InCl_2$	362.9	122.2	315.1	—	—	—	—	18
$InCl_3$	537.5	141.0	465	425.0	327(m)	—	—	13
InBr	173.8	113.9	—	—	—	—	—	8
$InBr_3$	411.1	178.7	352	306	—	—	—	13
InI	116.4	131.5	—	—	—	—	—	8
InI_3	249.1	—	285	247(m)	—	—	—	13
IrCl	67.0	—	50	42	8	—	—	13
$IrCl_2$	138.2	—	99.2	71	8	—	—	13
$IrCl_3$	254.5	114.9	—	—	—	—	—	13
KCl	436.9	82.5	409.1	389.0	338.7	267.1(m)	223.6(g)	2
KBr	394.0	96.7	379.7	360.9	313.6	280.1(m)	—	3
KF	568.9	66.6	532.6	510.4	455.9	416.2(m)	—	3
KI	327.8	104.3	332.0	312.8	264.6(m)	228.2(m)	—	1.3
$LaCl_3$	1 071.4	(144.4)	999.8	954.2	841.1	757.4(m)	—	8
LaI_3	657.3	214.7	680.8	632.2	519.2	423(m)	—	18.8
LiF	617.5	35.6	589.1	565.6	521.2	518.1(m)	—	8
LiCl	405.7	59.0	381.4	364.3	323.6(m)	288.5(m)	—	8
LiBr	349.2	—	340.8	324.1	283.9	255.8(m)	—	8
LiI	271.3	—	279.7	265.0	227.8	201.8(m)	—	8
MgF_2	1 113.7	57.4	1 068.1	1 034.6	949.6	—	—	8
$MgCl_2$	642.3	89.6	591.6	559.8	484.8(m)	421.6(m)	349.6(g)	6.3
$MgBr_2$	524.5	117.2	509.4	473.1	404.4(m)	360(m)	—	13
MgI_2	360.1	129.8	377.6	347.5	289.7(m)	242.0(m)	—	13
MnF_2	795.5	93.16	754	724	657	611(m)	—	25
MnF_3	996.5	—	—	—	—	—	—	29
$MnCl_2$	482.3	118.28	442.5	417.0	360.9(m)	324.9(m)	—	6.3
$MnBr_2$	385.1	138.1	373.0	336.2	273.4(m)	231.5(m)	—	8
MnI_2	242.8	—	262.9	233.6	172.9(m)	131.0(m)	—	13
MoF_6	1 559.6(g)	350.6(g)	1 473.4	—	—	—	—	8
$MoCl_5$	527.4	238.6	—	—	—	—	—	10.5

Table 8.8g HALIDES—*continued*

Compound	$-\Delta H_{298}$ 25°C	S_{298} 25°C	$-\Delta G_{300}$ 27°C	$-\Delta G_{500}$ 227°C	$-\Delta G_{1000}$ 727°C	$-\Delta G_{1500}$ 1 227°C	$-\Delta G_{2000}$ 1 727°C	*Accuracy* ±kJ
NaF	575.6	51.5	542.1	521.6	469.3	—	—	4
NaCl	412.8	72.9	384.8	365.1	315.7	253.7	—	2
NaBr	361.95	86.9	351.4	333.6	288.3	259.8(m)	—	4
NaI	288.0	98.4	295.2	276.3	232.8(m)	202.2(m)	—	7
NbF_5	1 814.7	160.4	1 702	—	—	—	—	50
$NbCl_2$	407.4	117	—	—	—	—	—	8
$NbCl_{2.67}$	538.4	(137.3)	—	—	—	—	—	8
$NbCl_{3.13}$	601.2	(151.6)	—	—	—	—	—	8
$NbCl_4$	695.0	(184.2)	—	—	—	—	—	—
$NbCl_5$	797.6	(226)	—	—	—	—	—	—
$NbBr_5$	556.4	(306)	—	—	—	—	—	8
$NdCl_3$	1 028.3	—	957.9	913	808	729(m)	—	8
NdI_3	628.9	—	652.3	603	490	394(m)	—	13
NiF_2	657.2	73.7	612.3	582	515	452	—	6.3
$NiCl_2$	305.6	97.76	259.2	232.4	163.7	109.7	—	2.1
$NiBr_2$	212.2	136.0	202.2	174.2	100.9	71.6	—	8
NiI_2	78.3	154.0	116.4	87.1	9.6	−15.5(m)	—	8
NpF_3	1 507.2	118.5	—	—	—	—	—	33
$NpCl_3$	904.3	160.3	—	—	—	—	—	21
$NpCl_4$	199.7	—	—	—	—	—	—	21
PF_3	920.9	272.9(g)	—	—	—	—	—	33
PCl_3	321(m)	218.1(m)	—	—	—	—	—	3.3
PCl_3	288(g)	312.3(g)	268.7(g)	254.4(g)	—	—	—	4
PCl_5	367.1(g)	364.6(g)	294.5(g)	—	—	—	—	13
PBr_3	132.3(g)	348.8(g)	—	—	—	—	—	13
PI_3	18.0(g)	374.6(g)	—	—	—	—	—	21
PbF_2	677.3	161.5	630.8	597.8	530.8	463.8	—	5
PbF_4	930.7	—	845.3	791	—	—	—	18.8
$PbCl_2$	359.2	136.5	312.8	284.5	220.6	196.8	—	2
$PbBr_2$	277.6	161.6	263.8	234.9	171.2(m)	—	—	4
PbI_2	175.4	176.7	195.1	166.2	111.4	—	—	8.3
$PdCl_2$	180.0	103.8	138	105	25	—	—	13
$PdBr_2$	103.4	—	93.8	64.9	—	—	—	21
$PrCl_3$	1 057.4	(144.4)	985.2	940.4	833.6	754.0(m)	—	2
PrI_3	654.7	235.3	679.7	631.2	520.7	424.2	—	18.8
$PtCl_2$	110.9	(129.8)	69.9	41.0	−29	—	—	13
$PtCl_4$	236.9	(205.1)	152	100.0	—	—	—	17
$PtBr_2$	80.4	(154.9)	—	—	—	—	—	13
$PtBr_4$	140.7	(251.2)	117	63	—	—	—	17
PtI_2	16.7	—	—	—	—	—	—	13
PtI_4	72.9	(281.3)	97.1	15.5	—	—	—	33
PuF_3	1 553.0	113.0	—	—	—	—	—	17
$PuCl_3$	949.4	164.1	—	—	—	—	—	4
$PuBr_3$	831.9	191.3	—	—	—	—	—	4
RbF	553.4	73.7	527.7	508.4	466.3(m)	427.4(m)	—	8
RbCl	430.8	91.7	403.2	383.9	335.8(m)	298.5(m)	—	8
RbBr	389.4	108.9	378.5	360.9	316.9(m)	279.3(m)	—	8
RbI	328.7	118.1	334.5	317.4	273.0(m)	235.3(m)	—	8
ReF_6	1 163.9(m)	—	1 068(m)	—	—	—	—	50
$ReCl_3$	263.8	—	200.1	157.4	—	—	—	13
$ReBr_3$	164.5	—	145.7	101.7	—	—	—	2
RhCl	83.7	83.7	66.2	56.5	21	—	—	13
$RhCl_2$	163.3	121.4	123.1	94.2	29	—	—	13
$RhCl_3$	275.4	123.9	—	—	—	—	—	25
$RhBr_3$	209.3	188.4	—	—	—	—	—	25

Table 8.8g HALIDES—*continued*

Compound	$-\Delta H_{298}$ 25 °C	S_{298} 25 °C	$-\Delta G_{300}$ 27 °C	$-\Delta G_{500}$ 227 °C	$-\Delta G_{1000}$ 727 °C	$-\Delta G_{1500}$ 1 227 °C	$-\Delta G_{2000}$ 1 727 °C	*Accuracy* ±kJ
$RuCl_3$	253.3	127.7	182.1	136.1	26.4	—	—	13
$RuCl_4$	93.4(g)	374.7(g)	64.9(g)	45.6(g)	−5.0(g)	—	—	17
SbF_3	915.9	127.2	860.5	818.5	—	—	—	21
$SbCl_3$	382.7	187.1	—	297.7	—	—	—	2.1
$SbCl_5$	438.8(m)	—	357.1(m)	306.9(m)	—	—	—	13
$SbBr_3$	260.0	210.1	249.5	228.6(m)	—	—	—	38
SbI_3	100.5	215.5	133.6	111.0	—	—	—	33
ScF_3	1 658	90	—	—	—	—	—	50
$ScCl_3$	900.2	—	828.6	782.9	670	586(m)	—	50
$ScBr_3$	711.8	—	686.2	638.5	531.7	450.1(m)	—	50
SeF_6	1 117.4(g)	314.4(g)	1 014(g)	—	—	—	—	5
$SeCl_4$	188.4	194.6	—	—	—	—	—	6.3
SiF_4	1 607.7(g)	282.2(g)	—	—	—	—	—	25
$SiCl_2$	162.4(g)	282.0(g)	—	—	—	—	—	5
$SiCl_4$	695.8(m)	239.9(m)	598.3(m)	571.9(m)	—	—	—	8
$SiBr_4$	461.7(m)	279.6(m)	435(m)	389(m)	—	—	—	29
SiI_4	199.2	265.6	—	—	—	—	—	6
$SmCl_2$	818.9	—	—	—	—	—	—	4.6
$SnCl_2$	350.0	—	309.8	283.0	230.7(g)	—	—	6.3
$SnCl_4$	529.1(m)	258.7(m)	473.9(m)	436.3(g)	360.9(g)	—	—	6.3
$SnBr_2$	266.3	—	256.7	229.9	—	—	—	17
$SnBr_4$	348.2(g)	412.8(g)	322.6(g)	—	—	—	—	11
SnI_2	144.0	167.8	165.4	136.9	—	—	—	21
SrF_2	1 217.7	82.2	—	—	—	—	—	5
$SrCl_2$	829.0	114.9	780.8	755.7	686.6	628	—	5
$SrBr_2$	716.4	143.6	699.2	667.8	597.0	539.7	—	5
SrI_2	563.1	159.1	580.3	548.9	486.9	428.3	—	5
$TaCl_3$	553.5	(155)	—	—	—	—	—	—
$TaCl_4$	708.8	(193)	—	—	—	—	—	—
$TaCl_5$	860.4	(235)	—	—	—	—	—	—
$TaBr_5$	602.9	(306)	—	—	—	—	—	13
TeF_6	1 318.8(g)	337.9(g)	1 223	1 164	—	—	—	5
$TeCl_4$	324.0	200.9	—	—	—	—	—	10.5
$TeBr_4$	195.1	243.6	2 017.2	1 955.1	—	—	—	13
ThF_4	2 112.2	142.14	2 017.2	1 955.1	—	—	—	13
$ThCl_4$	1 187.1	190.5	1 095.7	1 032.0	879	—	—	4
$ThBr_4$	966.1	228.1	933	870.2	721.6	—	—	7
ThI_4	665	—	—	—	—	—	—	50
TiF_3	1 436.2	87.9	—	—	—	—	—	—
TiF_4	1 650.1	131.1	—	—	—	—	—	—
$TiCl_2$	515.7	87.4	—	—	—	—	—	—
$TiCl_3$	722.1	139.8	—	—	—	—	—	—
$TiCl_4$	804(m)	252(m)	—	—	—	—	—	—
$TiBr_2$	407.4	119.7	—	—	—	—	—	—
$TiBr_3$	550.2	171.7	—	—	—	—	—	—
$TiBr_4$	619.6	239.5	—	—	—	—	—	—
TiI_2	241.1	138.1	—	—	—	—	—	—
TiI_4	375.9	246.1	—	—	—	—	—	—
TlF	325.6	95.7	286.6	272.1	251.2	228.2	—	5
TlF_3	573.2	—	—	—	—	—	—	13
TlCl	205.2	113.0	185.1	170.4	150.7(m)	—	—	1.7
$TlCl_3$	315.2	152.4	—	—	—	—	—	13
TlBr	172.5	125.2	169.2	156.2	136.1(m)	—	—	2.9
TlI	124.3	127.7	137.3	123.5	102.6(m)	—	—	2.5
UF_4	1 884	151.9	1 793.2	1 736.7	1 595.6	1 464.1	—	25
UF_5	2 057.8	188.4	—	—	—	—	—	25

Table 8.8g HALIDES—*continued*

Compound	$-\Delta H_{298}$ 25 °C	S_{298} 25 °C	$-\Delta G_{300}$ 27 °C	$-\Delta G_{500}$ 227 °C	$-\Delta G_{1000}$ 727 °C	$-\Delta G_{1500}$ 1 227 °C	$-\Delta G_{2000}$ 1 727 °C	Accuracy ±kJ
UF_6	2 189.7	227.8	—	1 999.2(g)	—	—	—	25
UCl_3	893.9	159.1	827.7	785.0	678.3	—	—	13
UCl_4	1 051.7	198.5	962.5	904.3	773.3(m)	—	—	8
UCl_5	1 094.8	(243)	—	—	—	—	—	13
UCl_6	1 133.4	285.9	—	—	—	—	—	21
UBr_3	721.4	(188)	695.4	651.0	541.4	—	—	17
UBr_4	826.9	—	794.2	734.8	606.2(m)	—	—	13
UI_3	478.1	(239)	—	—	—	—	—	8
UI_4	529.6	—	565.2	512.9	410(m)	—	—	17
VF_4	1 404.0	121.4	—	—	—	—	—	13
VF_5	1 481.0(m)	209.3(m)	—	—	—	—	—	25
VCl_3	561.0	—	—	—	—	—	—	13
VCl_4	570.2(m)	235.3(m)	498.2(m)	—	—	—	—	13
VBr_2	347.5	125.6	—	—	—	—	—	25
VBr_3	447.9	142.4	—	—	—	—	—	25
VBr_4	393.6(g)	334.9(g)	—	—	—	—	—	25
VI_2	263.8	(147)	—	—	—	—	—	25
VI_3	280.5	(203.1)	—	—	—	—	—	25
WF_6	1 748.3(m)	(251.2)	1 643.7	—	—	—	—	29
WCl_6	414	(217)	310	211.9	100(g)	8(g)	−88(g)	29
YCl_3	974.3	(136.9)	907.7	863.3	769.1	685.4	—	13
YI_3	618.2	(207)	640.4	588.9	490.0	383.8(m)	—	42
$YbCl_2$	800.0	130.6	—	—	—	—	—	13
ZnF_2	764.9	73.7	721.0	700.0	622.6	—	—	13
$ZnCl_2$	416.6	108.4	369.7	339.1	278.8(m)	338.7(g)	—	1.3
$ZnBr_2$	327.8	136.9	313.6	284.7	237.4	—	—	2.1
ZnI_2	209.3	161.2	231.1	201.4	153.7	—	—	1.7
ZrF_4	1 912.5	104.7	1 820.0	—	—	—	—	13
$ZrCl_4$	982.6	186.3	893.0	835.3	—	—	—	4
$ZrBr_4$	760.3	—	—	—	—	—	—	8
ZrI_4	485.3	—	—	—	—	—	—	8

Table 8.8h SILICATES AND CARBONATES

Heats and free energies of formation from the constituent oxides in kJ and standard entropies. The standard state of carbon dioxide is the perfect gas under 1 atm pressure; that of silica is quartz.

Compound	$-\Delta H_{298}$ 25 °C	S_{298} 25 °C	$-\Delta G_{300}$ 27 °C	$-\Delta G_{500}$ 227 °C	$-\Delta G_{1000}$ 727 °C	$-\Delta G_{1500}$ 1 227 °C	Accuracy ±kJ
Ag_2CO_3	81.2	167.5	30.6	−2.9	—	—	13
Al_2SiO_5 (kyanite)	8.4	83.7	—	—	—	—	5
(andalusite)	5.9	99.4	—	—	—	—	5
(sillimanite)	3.0	96.3	—	—	—	—	5
$Al_6Si_2O_{13}$ (mullite)	17.2	275.0	—	—	—	—	3
$BaCO_3$ (witherite)	269.4	112.2	214.9	183.1	97.3	—	17
$BaSiO_3$	159.1	—	112.2	112.2	111.8	111.8	13
Ba_2SiO_4	270.0	—	270.0	270.5	—	—	13
Be_2SiO_4	19.7	64.5	—	—	—	—	3
$CaCO_3$ (calcite)	179.2	92.9	131.0	98.8	18.4	—	8
$CaSiO_3$ (wollastonite)	90.0	82.1	89.2	89.2	88.8	88.3	2.1
Ca_2SiO_4	136.9	120.6	128.1	128.9	131.5	134	6.3
Ca_3SiO_5	113.0	168.7	117.2	118.9	122.7	—	8
$CdCO_3$	98.8	—	48.1	14.6	—	—	13
$CoCO_3$	78.9	88.7	—	—	—	—	21
$CuCO_3$	47.5	87.9	—	—	—	—	4

Table 8.8h SILICATES AND CARBONATES—*continued*

Compound	$-\Delta H_{298}$ 25 °C	S_{298} 25 °C	$-\Delta G_{300}$ 27 °C	$-\Delta G_{500}$ 227 °C	$-\Delta G_{1000}$ 727 °C	$-\Delta G_{1500}$ 1227 °C	*Accuracy* ±kJ
$FeCO_3$ (siderite)	82.7	92.9	35.6	−0.4	—	—	13
Fe_2SiO_4 (fayalite)	34.3	145.3	—	—	25.1	18.0(m)	13
K_2CO_3	393.7	155.6	342.9	308.6	224.8	—	10
K_2SiO_3	274.2	169.6	260.8	260.4	259.6	—	13
$K_2Si_4O_9$	307.4	265.8	—	—	—	—	13
$K_2Si_2O_5$	299.3	182.1	—	—	—	—	13
Li_2CO_3	226.7	90.2	—	—	37.7	—	15
Li_2SiO_3	139.8	80.4	—	—	—	—	13
$MgCO_3$	117.0	65.7	66.6	32.7	−52.3	—	17
Mg_2SiO_4 (forsterite)	63.2	95.2	63.2	63.2	63.2	63.2	6.3
$MgSiO_3$ (clinoenstatite)	36.4	67.8	36.0	34.8	32.7	30.6	4
$MnCO_3$	116.8	85.8	60.3	22.6	—	—	8
Mn_2SiO_4 (tephroite)	47.3	—	—	—	—	—	8
$MnSiO_3$ (rhodonite)	24.7	89.2	20.9	18.4	12.1	5.9	4
Na_2CO_3	322.0	136.1	280.9	253.7	185.9	—	17
Na_4SiO_4	360.0	195.9	316.9	319.5	325.3	—	25
Na_2SiO_3	232.4	113.9	232.4	232.8	233.2	233.6	33
$Na_2Si_2O_5$	230.6	165.0	233.5	235.2	—	—	13
$NiCO_3$	46.9	85.4	—	—	—	—	13
$PbCO_3$ (cerussite)	87.29	131.0	42.3	12.1	—	—	13
Pb_2SiO_4	16.7	186.7	—	23.4	30.1	39.7	9
$PbSiO_3$	17.6	109.7	—	18.0	19.2	26.4	4
Rb_2CO_3	404.9	—	—	—	—	—	29
$SrCO_3$	234.9	97.1	183.4	149.1	63.6	−22.2	17
$SrSiO_3$	130.6	—	—	—	—	—	4
Sr_2SiO_4	209.3	—	—	—	—	—	8
$ZnCO_3$	70.8	82.5	18.4	−16.7	—	—	4
Zn_2SiO_4	32.6	131.5	31.8	31.4	31.1	28.8	8
$ZrSiO_4$	24.0	84.6	—	—	—	—	13

Table 8.8i COMPOUND (DOUBLE) OXIDES
Heats and free energies of formation from the constituent oxides in kJ and standard entropies

Compound	$-\Delta H_{298}$ 25 °C	S_{298} 25 °C	$-\Delta G_{300}$ 27 °C	$-\Delta G_{500}$ 227 °C	$-\Delta G_{1000}$ 727 °C	$-\Delta G_{1500}$ 1227 °C	*Accuracy* ±kJ
Al_2TiO_5	8.4	109.7	10.9	12.6	16.7	20.9	8
$BaAl_2O_4$	100.5	148.6	108.7	114.2	127.8	141.6	7
$Ba_3Al_2O_6$	188.4	—	—	—	—	—	9
BaH_7O_3	128.5	122.4	126.3	124.7	121.0	115.2	17
$BaMoO_3$	208.5	144.5	207.4	206.6	204.6	—	11
$BaUO_4$	205.1	168.7	204.9	204.8	204.6	204.3	18
$BeAl_2O_4$	13.0	66.3	13.5	13.8	14.5	15.3	13
CaV_2O_6	143.6	179.2	146.2	148.0	152.4	(156.8)	3
$Ca_2V_2O_7$	262.5	220.6	265.6	267.8	273.0	(240.3)	3
$Ca_3V_2O_8$	322.3	275.0	329.8	334.8	347.4	(314.7)	3
$Ca_3Al_2O_6$	8.4	205.5	18.9	25.9	43.6	61.1	5
$CaAl_2O_4$	15.0	114.1	22.0	26.6	38.2	49.8	3
$CaAl_4O_7$	12.6	177.9	23.4	30.6	48.6	66.6	3
$CaTiO_3$	80.8	93.8	81.9	82.6	84.4	86.2	5
$Ca_3Ti_2O_7$	209.3	234.8	213.7	216.7	224.1	231.5	50
$Ca_4Ti_3O_{10}$	297.2	328.6	362.7	306.4	315.7	324.9	50
$CaZrO_3$	30.6	93.8	31.6	32.3	33.9	35.5	10
$CaHfO_3$	31.4	100.5	31.8	32.0	32.6	33.2	20
$CaMoO_4$	166.6	122.6	168.1	169.1	171.6	(150.0)	5
$CaWO_4$	146.1	126.4	149.3	151.5	156.8	162.1	11

Table 8.8i COMPOUND (DOUBLE) OXIDES—*continued*

Compound	$-\Delta H_{298}$ 25°C	S_{298} 25°C	$-\Delta G_{300}$ 27°C	$-\Delta G_{500}$ 227°C	$-\Delta G_{1000}$ 727°C	$-\Delta G_{1500}$ 1227°C	*Accuracy* ±kJ
$CaUO_4$	135.6	143.2	137.0	137.9	140.2	142.5	20
$CaFe_2O_5$	40.2	188.8	48.3	53.8	67.4	81.0	5
$Ca_2Fe_2O_4$	20.9	145.2	36.5	46.8	72.8	98.8	11
$CaSnO_3$	72.8	83.7	70.3	68.6	64.4	60.2	3
Ca_2SnO_4	71.6	128.3	70.5	69.8	68.0	66.2	3
$CaTiO_3$	27.6	105.1	27.6	27.6	27.5	27.4	2
$CdAl_2O_4$	−15.1	125.2	−9.3	−5.5	4.1	13.7	5
$CdWO_4$	79.1	154.9	86.3	92.1	103.2	115.3	8
$CeAlO_3$	(13.6)	106.7	15.4	16.5	19.6	22.5	15
$CoTiO_3$	24.7	96.9	22.8	21.5	18.3	15.1	5
$CoWO_4$	62.0	126.4	61.3	60.7	59.5	58.3	5
$CoAl_2O_4$	33.5	99.6	32.3	31.3	29.1	26.9	5
$CoCr_2O_4$	59.9	126.8	57.7	56.2	52.6	48.9	3
$CoFe_2O_4$	27.6	142.7	28.3	28.8	29.9	30.1	3
$CuAlO_2$	11.9	67.0	10.4	9.4	6.8	4.3	3
$CuAl_2O_4$	−18.4	111.3	−13.1	−9.6	−0.8	8.0	3
$CuGaO_2$	7.1	83.3	5.4	4.3	1.5	—	3
$CuGa_2O_4$	−17.2	146.5	−11.5	−7.7	1.8	10.3	3
$CuFeO_2$	36.4	88.9	36.2	35.7	35.0	34.3	10
$CuFe_2O_4$	−10.0	146.4	(−5.0)	(−1.7)	6.6	14.9	10
$CuCr_2O_4$	−3.1	129.8	(−1.3)	(−0.2)	2.8	5.7	3
$FeTiO_3$	29.7	105.9	28.7	28.1	26.4	24.8	5
$FeAl_2O_4$	27.8	106.7	26.9	26.2	24.5	22.9	2
Fe_2ZnO_4	5.0	134.9	6.1	6.8	8.6	10.4	5
$FeMoO_4$	54.4	129.3	52.2	50.8	47.1	(20.0)	20
$FeWO_4$	75.3	131.9	74.4	73.9	72.4	71.0	13
Fe_2NiO_4	22.6	126.0	22.7	22.8	23.0	23.2	10
Fe_2CoO_4	27.6	142.7	28.3	28.7	29.9	31.0	5
$FeCr_2O_4$	52.1	146.9	54.1	55.6	58.0	61.5	3
FeV_2O_4	24.3	157.0	24.3	24.3	24.3	—	3
Li_2TiO_3	129.3	91.7	130.3	132.0	132.7	—	12
$LiAlO_2$	54.0	53.2	56.8	58.3	62.6	66.9	7
$LiFeO_2$	17.8	75.3	21.6	24.1	30.4	36.7	3
Mg_2TiO_4	—	103.8	—	—	—	—	—
$MgTiO_3$	23.0	74.5	22.2	21.6	20.2	18.8	3
$MgTi_2O_5$	—	127.3	—	—	—	—	—
$MgAl_2O_4$	35.6	80.6	36.4	36.9	38.2	39.5	3
$MgFe_2O_4$	15.5	123.9	18.3	20.8	25.5	30.2	3
$MgMn_2O_4$	11.3	—	—	—	—	—	—
$MgCr_2O_4$	41.9	105.9	41.2	40.7	39.6	38.5	5
$MgMoO_4$	54.0	118.9	58.2	61.0	68.1	(61.0)	5
$MgWO_4$	73.3	100.9	72.7	72.1	71.3	70.3	3
$Mg_2V_2O_7$	81.8	200.1	86.5	89.6	97.4	105.2	3
MgV_2O_6	49.2	160.7	50.3	51.0	52.8	—	3
$MnAl_2O_4$	37.7	103.8	35.5	34.2	30.6	27.1	5
$MnCr_2O_4$	69.1	134.0	67.0	65.6	62.0	58.5	5
$MnMoO_4$	61.1	136.0	60.6	60.3	59.5	(52.0)	12
Na_2TiO_3	213.5	121.8	212.4	211.7	209.9	(206.0)	20
$Na_2Ti_2O_5$	232.3	173.7	231.7	232.3	230.2	(224.0)	25
$Na_2Ti_3O_7$	237.3	234.0	239.7	241.2	246.1	(247.0)	25
$NaAlO_2$	87.5	70.7	89.8	91.3	95.2	99.0	5
Na_2CrO_4	334.5	185.9	346.1	353.9	373.3	392.6	14
$NaCrO_2$	101.5	83.3	103.0	103.1	106.6	109.4	2.5
$NaFeO_2$	44.0	88.3	46.1	47.5	51.0	54.5	10
Na_2MoO_4	306.0	159.5	308.0	309.3	312.6	(315.9)	50

Table 8.8i COMPOUND (DOUBLE) OXIDES—*continued*

Compound	$-\Delta H_{298}$ 25 °C	S_{298} 25 °C	$-\Delta G_{300}$ 27 °C	$-\Delta G_{500}$ 227 °C	$-\Delta G_{1000}$ 727 °C	$-\Delta G_{1500}$ 1 227 °C	*Accuracy* ±kJ
$NaVO_3$	162.0	113.9	165.3	167.5	173.0	—	8
$Na_4V_2O_7$	534.6	318.6	545.9	553.5	572.4	—	18
Na_3VO_4	357.5	189.6	361.0	363.3	369.2	—	13
$NiAl_2O_4$	6.1	98.4	8.9	10.7	15.3	19.9	2
$NiTiO_3$	18.0	80.2	15.5	13.9	9.7	5.6	9
$NiWO_4$	45.2	118.0	46.5	47.3	49.4	51.6	9
$NiCr_2O_4$	13.6	124.3	15.1	16.1	18.6	21.1	5
$NiGa_2O_4$	4.4	130.2	6.6	8.1	11.8	15.7	2
$PbTiO_3$	34.6	112.0	33.2	32.4	29.9	27.5	20
$Pb_2V_2O_7$	146.9	—	—	—	—	—	7
$Pb_3V_2O_8$	170.8	—	—	—	—	—	9
$PbMoO_4$	41.9	166.2	48.5	52.9	63.9	—	15
$PbWO_4$	58.6	167.4	66.1	71.2	83.8	96.4	9
$SrTiO_3$	134.0	108.4	134.8	135.3	136.5	138.8	4
Sr_2TiO_4	159.1	159.1	158.4	157.9	156.7	155.5	9
$SrHfO_3$	78.3	113.0	77.7	77.3	76.3	75.3	20
$SrAl_2O_4$	16.6	108.8	17.3	17.7	18.8	19.9	25
$SrMoO_4$	212.2	128.9	210.9	210.0	207.8	—	9
$SrWO_4$	186.3	134.0	187.0	187.5	188.8	190.0	20
Zn_2TiO_4	3.3	148.2	3.6	8.8	13.8	—	3
$ZnTiO_3$	6.7	95.0	7.0	7.2	7.7	—	3
$ZnFe_2O_4$	5.0	134.9	6.1	6.8	8.6	10.4	5
$ZnWO_4$	41.9	144.4	49.3	54.3	66.7	79.1	3
$ZnAl_2O_4$	45.0	87.1	42.7	41.2	37.3	33.5	2
$ZnCr_2O_4$	62.8	116.4	60.3	58.6	54.3	50.1	2

Table 8.8j PHOSPHIDES
Heats of formation in kJ and standard entropies

Compound	$-\Delta H_{298}$ 25 °C	S_{298} 25 °C	*Accuracy* ±kJ	*Compound*	$-\Delta H_{298}$ 25 °C	S_{298} 25 °C	*Accuracy* ±kJ
AgP_2	44.8	87.9	9	InP	75.3	59.8	9
AgP_3	69.1	105.5	11	Mg_3P_2	464.6	77.4	80
AlP	164.5	47.3	3	MnP	96.3	52.3	17
Au_2P_3	97.5	150.7	15	MnP_3	174.1	96.7	31
Ca_3P_2	506.5	123.9	25	Ni_3P	200.2	106.3	11
Co_2P	157.5	77.4	15	Ni_5P_2	391.4	185.0	19
CoP	125.6	50.2	21	Ni_2P	164.5	77.4	18
CoP_3	204.7	98.4	21	Ni_6P_5	556.3	276.3	60
Cu_3P	129.0	119.3	18	NiP_2	129.3	73.3	15
CuP_2	90.1	81.6	11	NiP_3	157.8	98.4	17
Fe_3P	164.1	101.7	9	SiP	62.0	32.7	15
Fe_2P	160.3	72.4	9	ThP	361.3	70.3	40
Fe_1P	138.1	—	13	Th_3P_4	1 195.1	247.0	110
FeP_2	221.0	—	17	Zn_3P_2	159.0	150.7	18
GaP	122.2	52.3	9	ZnP_2	101.7	60.3	13
GeP	27.2	61.1	11				

The standard state is white phosphorus and for P white→P red, $\Delta H = -17.4$ kJ g-atom^{-1}. With respect to red phosphorus, heats of formation are thus less negative per g-atom P by this amount.

Table 8.9a PHOSPHIDES
Dissociation pressures in mmHg*

Dissociating phase	Other condensed phase	Temp. °C	Phosphorus pressure mmHg		
			p_{P_2}	p_{P_4}	Σp
AgP_3(s)	AgP_2(s)	411–501		$\log p = -\frac{7\,766}{T} + 12.74$	
AgP_2(s)	Ag(s)	420–506		$\log p = -\frac{7\,186}{T} + 11.60$	
Au_2P_3(s)	Au(s)	501–700		$\log p = -\frac{9\,124}{T} + 12.21$	
CoP_3(s)	CoP(s)	934	15	27	42
		994	54	113	167
		1 037	120	280	400
CoP(s)	Co_2P(s)	1 215	—	—	24
CuP_2(s)	Cu_3P(s)	632	—	—	18
		672	—	—	65
		712	—	—	200
		762	—	—	698
FeP_2(s)	FeP(s)	892	16.5	78.5	95
		922	34	166	200
		953	63	329	392
		973	92	516	608
FeP(s)	Fe_2P(s)	1 175	12.5	—	13
		1 215	25.5	—	27
GeP(s)	Ge(s)	500	—	—	64
		524	—	—	156
		559	—	—	394
MnP_3(s)	MnP(s)	580–680		$\log p = -\frac{11\,856}{T} + 14.948$	
NiP_3(s)	NiP_2(s)	597	—	135	135
		644	—	460	460
		667	—	850	850
NiP_2(s)	Ni_6P_5(s)	740	4	109	113
		761	7	193	200
		780	11	323	334
		804	19	585	604
Ni_6P_5(s)	Ni_2P(s)	742	2	27	29
		761	3	43	46
		780	5	75	80
		806	11	172	183
		823	16	262	278
OsP_2(s)	Os(s)	1 190	—	—	8
ReP_3(s)	ReP_2(s)	863	11	57	68
		904	31	186	217
		935	63	430	493
		956	101	762	863
ReP_2(s)	ReP(s)	935	19	39	58
		986	52	120	172
		1 028	111	278	389
		1 059	187	494	681
ReP(s)	Re(s)	1110	8.5	—	9
		1 141	14	—	15
		1 172	23	—	24.5
		1 214	43	—	46.5

*To convert the pressures given in mmHg to pascals multiply by 133.322.

Table 8.9a PHOSPHIDES—*continued*

Dissociating phase	*Other condensed phase*	*Temp.* °C	*Phosphorus pressure* mmHg p_{P_2}	p_{P_4}	Σp
RhP_3(s)	RhP_2(s)	968	19	22	41
		1 012	39	45	84
		1 038	59	71	130
		1 067	98	125	223
RuP_2(s)	RuP(s)	1 190	34	—	34
SiP(s)	$SiP_{0.2}$(s)	1 010	33	34	67
		1 055	76	86	162
		1 068	96	115	211
		1 101	169	220	389
TaP_2(s)	TaP(s)	787	7	111	118
		816	14	221	235
		836	23	397	420
		846	29	494	523
VP_2(s)	VP(s)	657	—	148	148
		670	—	195	195
		680	—	298	298
		689	—	402	402
$WP_{0.96}$	$WP_{0.65}$	750	1.65×10^{-5}	—	1.65×10^{-5}
		850	4.6×10^{-4}	—	4.6×10^{-4}
		950	7.2×10^{-4}	—	7.2×10^{-4}
ZrP_2(s)	ZrP(s)	729	2	26	28
		774	5	74	79
		814	12	165	177
		848	25	357	382

Ref. 5.
*To convert the pressures given in mmHg to pascals multiply by 133.322.

Table 8.9b SULPHIDES
Dissociation pressures in mmHg‡
A = Dissociating phase. *B* = Other condensed phase.

Phases	*Temp.* °C	*Vapour pressure* mmHg p_{S_2}	Σp	*Phases*	*Temp.* °C	*Vapour pressure* mmHg p_{S_2}	Σp
A As_2S_2(g)	755–1 075	$\log K_p = \log \dfrac{p_{S_2^4}\cdot p_{As_4}\cdot p_{As_2}\cdot p_{As^2}}{p_{As_2S_2^4}} = -\dfrac{312\,900}{T} + 56.7$		*A* $CoS_{1.3}$(s)	700	—	31
					730	—	79
					760	—	162
				A CuS(s)	460	19	90
				B Cu_2S(s)	474	34	210
A $CoS_{1.92}$(s)	700	—	38	*A* $FeS_{1.94}$(s)	629	—	66
	730	—	97	*B* $FeS_{1.12}$(s)	649	—	170
	760	—	226		659	—	258
					669	—	445
A $CoS_{1.5}$(s)	700	—	40				
	730	—	86	*A* Ir_3S_8(s)	880	—	400
	760	—	200	*B* IrS_2(s)	—	—	—

Table 8.9b SULPHIDES—*continued*

Phases	*Temp.* °C	*Vapour pressure* mmHg p_{S_2}	Σp
A IrS_2(s)	880	—	52
B Ir_2S_3(s)	904	—	75
	944	—	153
A Ir_2S_3(s)	1 020	—	55
B Ir(s)	1 056	—	109
	1 073	—	148
A MnS_2(s)	408	—	424
B MnS(s)	—	—	—
A NiS_2(s)	650	—	42
	700	—	153
	730	—	325
	760	—	650
A OsS_2(s)	944	—	22
	994	—	85
B Os(s)	1 044	—	214
	1 094	—	490
A PdS_2(s)	451	16	88*
B PdS(s)	476	35	210*
	501	72	425*
A PtS_2(s)	616	—	38
B PtS(s)	651	—	98
	691	—	240
A PtS(s)	1 060	30	30
B Pt(s)	1 110	75	75
	1 186	249	249
A ReS_2(s)	1 110	13	13
B Re(s)	1 189	55	55
	1 225	96	96
A Rh_2S_5(s)	715	—	44
	757	—	130
B Rh_2S_3(s)	790	—	300
	830	—	610

Phases	*Temp.* °C	*Vapour pressure* mmHg p_{S_2}	Σp
A Rh_2S_3(s)	953	—	19
	1 003	—	56
B Rh_3S_4(s)	1 043	—	137
	1 083	—	300
A Rh_3S_4(s)	953	—	8
	1 003	—	24
B Rh_9S_8(s)	1 043	—	54
	1 083	—	120
A RuS_2(s)	1 123	3.0	3.0
	1 153	5.3	5.3
B Ru(s)	1 184	9.7	9.7
	1 208	15.3	15.3
A Th_3S_7(s)	651	—	43
	676	—	87
B ThS_2(s)	713	—	163
	754	—	381
A TiS_3(s)	500	38.5	91†
	525	83	223†
B TiS_2(s)	538	122	349†
	551	173	429
A US_3(s)	500	—	~2
	608	—	34
B US_2(s)	650	—	155
	700	—	470
A VS_4(s)	390	—	48
B V_2S_3(s)	412	—	131
	440	—	350
A $ZrS_{2.6}$(s)	783	52	52
	814	98	98
B ZrS_2(s)	847	187	187
	874	311	311

**t*°C	p_{S_6}	p_{S_8}
451	52	20
476	125	50
501	254	99

†*t*°C	p_{S_6}	p_{S_8}
500	43	9.5
525	112.5	27.5
538	181	46
551	280	76

Ref. 6.

‡To convert the pressures given in mmHg to pascals multiply by 133.322.

8.6 Molar heat capacities and specific heats

Table 8.10 ELEMENTS
Molar heat capacities

The molar heat capacity C_p may be given empirically by an equation of the form:
$C_p = 4.186\,8(a + 10^{-3}bT + 10^5cT^{-2})$, $\mathrm{J\,K^{-1}\,mol^{-1}}$
where a, b and c are constants and T=temperature in K.

The values of the constants are given below.

Element	*a*	*b*	*c*	*Remarks*	*Temp. range* K
Ag(s)	5.09	2.04	0.36		298–m.p.
Ag(l)	7.3	—	—		m.p.– 1 600
Al(s)	4.94	2.96	—		298–m.p.
Al(l)	7.0	—	—		m.p.– 1 273
As(s)	5.54	1.32	—		273–1 090
Au(s)	5.66	1.24	—		298–m.p.
Au(l)	7.0	—	—		m.p.– 1 575
B (amorph.)	3.835	2.39	−1.50		298–1 240
B (cryst.)	4.735	1.38	−2.20		298–1 100
Ba(β)	−1.36	19.2	—		673–m.p.
Ba(l)	10.5		—		m.p.–1 105
Be(s)	4.54	2.12	−0.82		298–1 173
Be(l)	6.08	0.515	—		1 560–2 200
Bi(s)	4.49	5.40	—		298–m.p.
Bi(l)	4.78	1.47	5.05		m.p.–820
Br_2(l)	17.2	—	—		300
Br_2(g)	8.93	0.11	−0.31		298–1 600
C (graph)	0.026	9.307	−0.354	$-4.155 \times 10^{-6}T^2$	298–1 100
	5.841	0.104			1 100–4 000
C (diam)	2.18	3.16	—		298–1 200
Ca(α)	6.064	−1.736	—	$+5.67 \times 10^{-6}T^2$	273–720
Ca(β)	−0.086	9.86	—		720–m.p.
Ca(l)	7.00	—	—		m.p.–1 250
Cd(s)	5.31	2.94	—		298–m.p.
Cd(l)	7.1	—	—		m.p.–1 100
Ce(α)	5.613	2.485		$+0.97 \times 10^{-6}T^2$	298–1 003
Ce(β)	9.05	—	—		1 003–m.p.
Ce(l)	9.35	—	—		m.p.–1 373
Cl_2(g)	8.82	0.06	−0.68		298–3 000
Co(α)	5.11	3.42	−0.21		298–650
Co(β)	3.30	5.86	—		718–1 400
Co(γ)	9.60	—	—		1 400–m.p.
Co(l)	9.65	—	—		m.p.–1 900
Cr(s)	5.84	2.36	−0.88		298–1 823
Cr(l)	9.4	—	—		m.p.
Cs(s)	7.42	—	—		298–m.p.
Cs(l)	7.62	—	—		m.p.–400
Cu(s)	5.41	1.40	—		298–m.p.
Cu(l)	7.5	—	—		m.p.–1 600
F_2(g)	8.29	0.44	−0.80		—
Fe(α, β, δ)	8.873	1.474	—	$-56.92\,T^{-1/2}$	298–m.p.
Fe(γ)	5.85	2.02	—		1 187–1 664
Fe(l)	10.0	—	—		m.p.–
Ga(s)	6.25	—	—		298–m.p.
Ga(l)	6.65	—	—		295–373
Ge(s)	5.16	1.40	—		298–m.p.
Ge(l)	6.60	—	—		m.p.–1 573
H_2(g)	6.52	0.78	0.12		298–3 000
Hf(s)	5.61	1.82	—		298–1 346
Hg(l)	6.61	—	—		298–b.p.
I_2(s)	9.59	11.90	—		298–387

Table 8.10 ELEMENTS—*continued*

Element	*a*	*b*	*c*	*Remarks*	*Temp. range* K
I_2(l)	19.2	—	—		387–456
I_2(g)	8.89	—	—		456–1 500
In(s)	5.81	2.50	—		298–m.p.
In(l)	7.24	−0.33	—		m.p.–800
Ir(s)	5.56	1.42	—		298–1 800
K(s)	6.04	3.12	—		298–m.p.
K(l)	8.886	−4.57	—	$+2.94\times10^{-6}T^2$	m.p.–1 037
La(s)	6.17	1.60	—		298–800
Li(s)	3.33	8.21	—		273–m.p.
Li(l)	5.85	1.31	2.07	$-467\times10^{-6}T^2$	m.p.–580
Mg(s)	5.33	2.45	−0.103		298–m.p.
Mg(l)	7.80	—	—		m.p.–1 100
Mn(α)	5.70	3.38	−0.375		298–1 000
Mn(β)	8.33	0.66	—		1 108–1 317
Mn(γ)	6.03	3.56	−0.443		1 374–1 410
Mn(δ)	11.10	—	—		1 410–1 450
Mn(l)	11.0	—	—		m.p.–b.p.
Mo(s)	5.77	0.28	—	$+2.26\times10^{-6}T^2$	298–2 500
N_2(g)	6.67	1.0	—		298–2 000
Na(s)	19.71	−88.27	—	$+150\times10^{-6}T^2$	273–m.p.
Na(l)	8.965	−4.594	—	$+2.542\times10^{-6}T^2$	m.p.–451
Nb(s)	5.66	0.96	—		298–1 900
Nd(α)	3.503	6.434	+1.07		298–1 128
Nd(β)	10.65	—	—		1 128–m.p.
Ni(α)	4.06	+7.04	—		298–630
Ni(β)	6.00	+1.80	—		630–m.p.
Ni(l)	9.20	—	—		m.p.–2 200
O_2(g)	7.16	1.00	−0.40		298–3 000
Os(s)	5.69	0.88	—		298–1 900
P (yell.)	4.57	3.78	—		273–m.p.
P (red)	4.05	3.56	—		298–800
P(l)	6.29	—	—		m.p.–370
Pb(s)	5.63	2.33	—		298–m.p.
Pb(l)	7.75	−0.74	—		m.p.–1 300
Pd(s)	5.80	1.38	—		298–1 828
Pr(α)	6.21	0.3	—		298–1 068
Pr(β)	9.19	—	—	$+3.11\times10^{-6}T^2$	1 068–m.p.
Pt(s)	5.80	1.28	—		298–2043
Pu(α)	5.91	5.8			298–395
Pu(β)	5.21	7.05			395–480
Pu(γ)	2.98	11.1			480–588
Pu(δ)	9.0	—	—		588–753
Pu(ε)	8.4	—	—		753–913
Pu(l)	10.0	—	—		913–2 000
Rb(s)	7.27	—	—		298–m.p.
Rb(l)	7.85	—	—		m.p.–373
Re(s)	5.80	0.95	—		298–2 300
Rh(s)	5.49	2.06	—		298–1 900
Ru(α)	5.28	1.1	—		298–2 000
S (rhomb.)	3.58	6.24	—		273–369
S (monocl.)	3.56	6.95	—		273–392
S(l)	5.4	5.5	—		m.p.–b.p.
S_2(g)	8.54	0.28	−0.79		298–2 000
Sb(s)	5.51	1.78	—		273–m.p.
Sb(l)	7.5	—	—		m.p.–1 000
Se(s)	4.53	5.5	—		273–490
Se(l)	7.00	—	—		490–600
Si(s)	5.72	0.59	−0.99		298–1 200
Si(l)	6.12	—	—		m.p.–1 873
Sm(α)	6.00	5.84	−0.61		298–1 190

Table 8.10 ELEMENTS—*continued*

Element	*a*	*b*	*c*	*Remarks*	*Temp. range* K
Sm(β)	11.22	—	—		1 190–1 345
Sm(l)	12.57	—	—		m.p.–1 398
Sn(s)	5.16	4.34	—		298–m.p.
Sn(l)	8.29	−2.2	—		m.p.–800
Ta(s)	6.65	−0.52	−0.45	$+0.47\times10^{-6}T^2$	298–2 300
Te(s)	4.58	5.25	—		273–m.p.
Te(l)	9.0	—	—		m.p.–873
Th(s)	5.63	3.04	—		298–1 300
Ti(α)	5.28	2.4	—		298–1 155
Ti(β)	4.74	1.90	—		1 155–1 350
Tl(α)	3.74	6.04	+0.67		298–506
Tl(β)	5.00	5.00	—		506–m.p.
Tl(1)	7.2	—	—		m.p.–1 760
U(α)	2.61	8.95	1.17		298–941
U(β)	10.0	—	—		941–1 048
U(γ)	9.1	—	—		1 048–m.p.
V(s)	4.90	2.58	+0.2		298–1 900
W(s)	5.74	0.76	—		298–2 000
Y(α)	5.72	1.805	+0.08		298–1 758
Y(β)	8.37	—	—		1 758–m.p.
Y(l)	9.51	—	—		m.p.–1 950
Zn(s)	5.35	2.40	—		298–m.p.
Zn(l)	7.5	—	—		m.p.–b.p.
Zr(α)	5.25	2.78	−0.91		298–1 135
Zr(β)	5.55	1.11	—		1 135–m.p.

Table 8.11 ALLOY PHASES AND INTERMETALLIC COMPOUNDS
Molar heat capacities and specific heats

Generally, Neumann and Kopp's rule applies better to intermetallic compounds than to inorganic salts. Thus, the molar heat capacity of an alloy may be calculated additively from the atomic heat capacities of the components. This relationship is useful as only a few specific heats of alloys have been measured.

In the following cases Neumann–Kopp's rule is obeyed to within ±3% in the temperature range 0–500 °C: Ag–Au; Ag_3Al; γ-Ag–Al; β-Ag–Mg; AlCu; Al_2Cu; $AlCu_3$; Al_3Mg_4; Bi–Cd; Co_2Sn; Cu_2Mg; $MgZn_2$; Mg_2Si; Pd–Sb

The heat capacities of heterogeneous alloys must always be calculated additively from those of the components.

In other cases the calculated values differ from the determined values by more than 3%. This result may be due in part to experimental errors.

C_p may be represented empirically by an equation of the form:

$$C_p = 4.1868\,(a + 10^{-3}bT + 10^5 cT^{-2})\,\mathrm{J\,K^{-1}\,mol^{-1}}$$

where *a*, *b* and *c* are constants and *T* is the temperature in K. The values of the constants are given below.

Phase	*a*	*b*	*c*	*Temp. range* K
AuCd	7.66	10.36	—	298–900
AuCd (liq.)	14.4	—	—	900–1 100
AuCu		*See* ref. 3		320–900
$AuCu_3$		*See* ref. 3		298–1 200
$AuPb_2$ (liq.)	23.1	—	—	527–800
$AuSb_2$(α)	17.12	4.64	—	298–628
AuSn	11.13	3.8	—	298–691
AuSn (liq.)	14.6	—	—	691–900
AuZn(β)	11.87	2.85	—	723–m.p.
AuZn (liq.)	13.6	—	—	m.p.–1 200

Table 8.11 ALLOY PHASES AND INTERMETALLIC COMPOUNDS—*continued*

Phase	*a*	*b*	*c*	*Temp. range* K
Bi_3Tl_2	31.5	—	—	298–487
Bi_3Tl_2 (liq.)	35.8	—	—	487–700
Cd–Cu		*See* ref. 2		292–373
Cd_3Cu_2 (liq.)	41.5	—	—	835–1 000
Cd_3Mg		*See* ref. 2		298–543
CdMg		*See* ref. 2		298–543
$CdMg_3$		*See* ref. 2		298–543
Co_7W_6	68.86	29.7	—	293–1 145
$Cr_{0.48}Fe_{0.52}(\alpha)$	5.02	2.78	—	1 100–1 400
$Cr_{0.48}Fe_{0.52}$	8.06	—	—	870–1 100
Cu–Ni		79–94% Ni *see Figure 8.1*		
CuPd	12.02	1.96	−1.16	298–1 200
Cu_3Pd	20.98	8.80	−1.16	298–1 200
Cu_2Sb	16.38	6.60	—	298–600
$Cu_3Sb(\beta)$	21.79	9.00	—	273–573
Cu–Zn		*See Figure 8.2*		
Fe_7W_6	72.08	24.0	—	293–1 145
FeW_2	28.29	9.90	−1.60	298–1 250
$MgNi_2$	15.67	7.30	—	298–900
$Ni_3Sn(\delta)$	20.78	10.2	—	273–943
Ni_4W	25.7	11.14	—	293–1 100
$PtSb_2$	15.27	5.06	—	298–900
PtSn	11.26	2.20	—	298–1 400

Cr–Ni. Atomic heats of alloys with 1.8–11 at. % Cr obey Kopp's law above the Curie temperature.

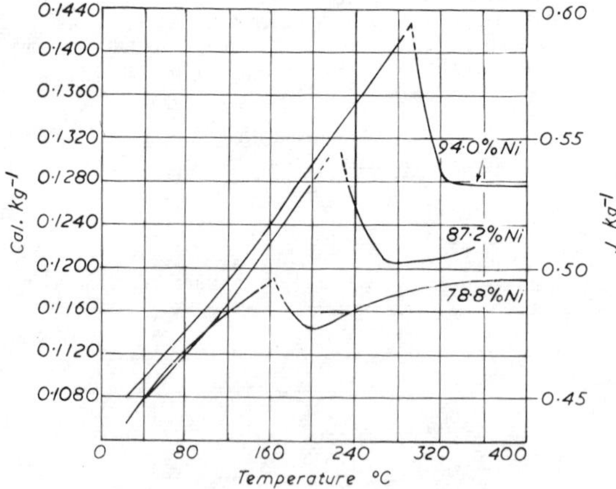

Figure 8.1 *Specific heat of* Cu–Ni *alloys*

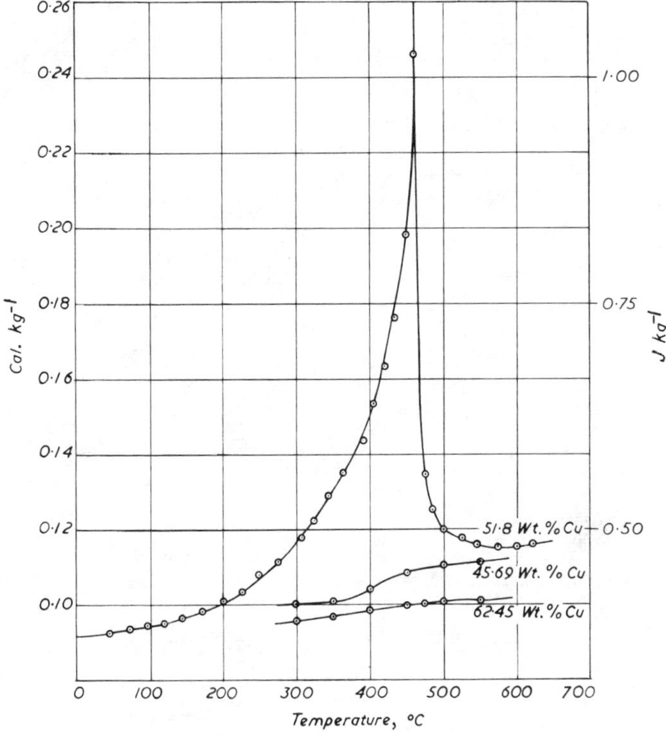

Figure 8.2 *Specific heat of* Cu–Zn *alloys*

Table 8.12a BORIDES
Molar heat capacities

C_p may be represented empirically by an equation of the form:

$C_p = 4.1868\ (a + 10^{-3}bT + 10^5 cT^{-2})\,\mathrm{J\ K^{-1}\ mol^{-1}}$

where a, b and c are constants and T = temperature in K. The values of the constants are given below.

Boride	*a*	*b*	*c*	*Temp. range* K
CrB	10.12	3.83	−2.40	298–1 200
CrB_2	9.63	10.7	—	298–1 200
Mo_2B	18.42	1.3	—	298–800
MoB	9.77	3.07	−1.13	298–1 200
MoB_2	7.92	(13)	—	600–1 200
NbB_2	11.01	9.38	−1.78	298–1 200
NiB	10.26	3.5	−2.69	298–1 300
Ni_3B_4	37.28	11.74	−9.03	298–1 300
TaB	7.82	5.8	—	500–1 200
TaB_2	14.21	4.49	−3.60	298–3 370
TiB_2	13.48	6.48	−4.17	298–3 190
UB_2	25.58	−9.62	−8.48	298–2 300
W_2B	18.46	1.44	−3.14	298–1 200
WB	13.89	−0.78	−5.02	298–1 200
W_2B_5	39.27	2.08	−16.72	298–1 200
ZrB_2	15.79	4.22	−3.51	298–1 200

Table 8.12b CARBIDES
Molar heat capacities

C_p may be represented empirically by an equation of the form:

$C_p = 4.186\,8(a + 10^{-3}bT + 10^{5}cT^{-2})\ \mathrm{J\ K^{-1}\ mol^{-1}}$

where a, b and c are constants and T = temperature in K. The values of the constants are given below.

Carbide	*a*	*b*	*c*		*Temp. range* K
Al_4C_3	36.97	6.866	−10.02		298–1 800
B_4C	22.99	5.40	−10.72		298–1 100
Be_2C	27.37	3.60	−14.23		430–1 200
$CaC_2(\alpha)$	16.40	2.84	−2.07		298–720
$CaC_2(\beta)$	15.40	2.00	—		720–1 275
Cr_3C_2	30.03	5.58	−7.4		298–1 500
Cr_7C_3	57.00	14.38	−10.1		298–1 500
Cr_4C	29.35	7.40	−5.02		298–1 700
$Fe_3C(\alpha)$	19.64	20.0	—		298–463
$Fe_3C(\beta)$	25.62	3.0	—		463–1 500
CH_4(g)	5.65	11.44	−0.46		298–1 500
$Mn_3C(\alpha)$	25.26	5.60	−4.07		298–1 310
$Mn_3C(\beta)$	38.00	—	—		1 310–1 500
$NbC_{0.5}$	7.94	1.50	−1.025		298–1 703
$NbC_{0.75}$	8.95	2.25	−1.26		298–1 763
NbC	10.79	1.73	−2.15		298–1 790
SiC	12.14	0.47	−11.76	$1.96 \times 10^{8}\,T^{-3}$	298–3 260
Ta_2C	15.88	3.33	−2.05		298–3 775
TaC	10.35	2.71	−2.10		298–4 270
$ThC_{1.94}$	15.17	2.89	−2.21		298–1 700
UC	13.40	1.02	−1.46		298–2 073
U_2C_3	29.9	3.06	−3.71		298–2 050
UC_2	16.5	2.04	−2.25		298–2 050
$VC_{0.88}$	8.69	3.18	−1.7		298–2 000
WC	10.37	2.06	−2.23	$0.24 \times 10^{-6}T^{2}$	298–2 500

Table 8.12c NITRIDES
Molar heat capacities

C_p may be represented empirically by an equation of the form:

$C_p = 4.186\,8\ (a + 10^{-3}bT + 10^{5}cT^{-2})\ \mathrm{J\ K^{-1}\ mol^{-1}}$

where a, b and c are constants and T = temperature in K. The values of the constants are given below.

Compound	*a*	*b*	*c*	*Remarks*	*Temp. range* K
AlN	8.22	4.05	−2.00		298–1 500
BN (hexag.)	16.29	0.675	—	$-854T^{-1/2}$	298–1 400
BN (cubic)	8.10	3.52	−5.51		298–1 200
Be_3N_2	12.89	24.75	−4.22		298–430
Ca_3N_2	33.17	3.70	−6.29		298–1 468
CeN	11.1	1.65	−1.73		298–2 000
Cr_2N	15.24	6.8	—		298–800
CrN	9.84	3.9	—		298–800
Cu_3N	21.7	—	—		298–373
Fe_4N	26.8	8.16	—		298–373

Table 8.12c NITRIDES—*continued*

Compound	*a*	*b*	*c*	*Remarks*	*Temp. range* K
Fe_2N	14.9	6.1	—		298–373
NH_3	7.11	6.00	−0.37		298–1 800
Li_3N	11.73	23.00	—		298–800
$Mg_3N_2(\alpha)$	22.81	7.30	—		298–823
$Mg_3N_2(\beta)$	29.60	—	—		823–1 061
$Mg_3N_2(\gamma)$	29.54	—	—		1 061–1 300
Mn_4N	21.16	30.5	—		298–800
Mn_5N_2	30.55	38.40	—		298–800
Mn_3N_2	22.32	22.40	—		298–800
Mo_2N	11.19	13.8	—		298–800
NbN	8.69	5.40	—		298–600
Si_3N_4	16.83	23.6	—		298–900
Ta_2N	16.85	4.22	−1.69		298–3000
TaN	13.21	0.65	−3.02		298–3360
ThN	11.34	2.28	−1.14		298–2000
Th_3N_4	27.78	31.8	—		298–800
TiN	11.91	0.94	−2.96		298–1 800
VN	10.94	2.10	−2.21		298–1 611
Zn_3N_2	19.00	22.5	—		298–700
ZrN	11.10	1.68	−1.72		298–1 700

Table 8.12d SILICIDES
Molar heat capacities

C_p may be given empirically by an equation of the form:

$C_p = 4.186\ (a + 10^{-3}bT + 10^{3}cT^{-2})\ \text{J K}^{-1}\ \text{mol}^{-1}$

The values of the constants *a*, *b* and *c* are given below.

Compound	*a*	*b*	*c*	*Temp. range* K
CoSi	11.75	2.89	−1.80	298–1 733
CoSi (liq.)	20.88	—	—	m.p.–1 900
$CoSi_2$	16.935	4.46	−2.37	298–m.p.
$CoSi_2$ (liq.)	27.75	—	—	m.p.–1 900
Cr_5Si_3	47.46	11.78	−6.12	298–1 300
CrSi	12.43	2.09	−2.01	298–1 700
$CrSi_2$	15.68	5.38	−1.855	298–1 730
$CrSi_2$ (liq.)	21.5	—	—	1 730–1 900
FeSi	10.72	4.30	—	298–900
Mg_2Si	17.52	3.58	−2.11	298–873
Mn_3Si	24.11	12.45	−3.52	298–950
Mn_5Si_3	48.13	12.94	−4.68	298–m.p.
MnSi	11.79	3.05	−1.53	298–m.p.
$Mn_{0.37}Si_{0.63}$	6.37	4.085	−1.16	298–1 425
Mo_3Si	20.52	5.42	+0.076	298–2 200
Mo_5Si_3	43.82	8.37	−2.87	298–2 200
$MoSi_2$	16.22	2.86	−1.57	298–2 200
Nb_5Si_3	45.21	7.36	−3.60	298–2 000
$NbSi_2$	15.10	3.67	−0.67	298–2 000
Ni_2Si	15.8	3.29	—	298–1 582

Table 8.12d SILICIDES—*continued*

Compound	*a*	*b*	*c*	*Temp. range* K
NiSi	11.65	1.47	−1.56	298–1 265
$Ni_{0.35}Si_{0.65}$	5.98	0.88	−0.86	298–1 200
Re_5Si_3	45.60	10.8	−3.36	298–1 500
Ta_5Si_3	42.95	9.35	−2.13	298–2 000
$TaSi_2$	17.51	1.84	−2.17	298–2 200
V_3Si	22.41	4.37	−1.66	298–1 400
VSi_2	17.08	2.78	−2.25	298–1 950
W_5Si_3	42.94	9.36	−2.13	298–2 200
WSi_2	16.21	2.64	−1.46	298–2 200

Table 8.12e OXIDES
Molar heat capacities

C_p may be given empirically by an equation of the form:

$C_p = 4.1868\ (a + 10^{-3}bT + 10^5cT^{-2})\ \mathrm{J\ K^{-1}\ mol^{-1}}$

The values of the constants *a*, *b* and *c* are given below.

Compound	*a*	*b*	*c*	*Remarks*	*Temp. range* K
Ag_2O	14.18	9.75	−1.0		298–500
Al_2O_3	25.48	4.25	−6.82		298–m.p.
As_2O_3	14.30	42.0	—		298–m.p.
B_2O_3 (cryst.)	13.63	17.45	−3.36		298–723
B_2O_3 (glass)	2.28	42.10	—		298–723
B_2O_3 (liq.)	30.45	—	—		900–1 800
BaO	12.74	1.04	−1.984		298–1 270
BeO	9.94	2.44	−4.15	$-32\times10^{-6}T^2$	298–2 835
Bi_2O_3 (α)	24.74	8.0	—		298–800
Bi_2O_3 (β)	35.0	—	—		978–1 097
CO	6.79	0.98	−0.11		298–2 500
CO_2	10.55	2.16	−2.04		298–2 500
CaO	11.86	1.08	−1.66		298–1 177
CdO	11.53	1.525	−1.17		298–1 500
CeO_2	15.49	4.23	−1.815		298–1 250
CoO	11.54	2.04	0.4		298–1 800
Co_3O_4	30.84	17.08	−5.72		298–1 000
Cr_2O_3	28.53	2.20	−3.74		350–1 800
Cu_2O	14.90	5.70	—		298–1 200
CuO	9.27	4.80	—		298–1 250
Eu_2O_3(α)	29.60	6.48	−2.08		298–895
Eu_2O_3(β)	31.06	4.16	—		895–1 802
Eu_2O_3 (cubic)	31.90	4.38	−3.04		298–1 350
$Fe_{0.947}O$	11.66	2.00	−0.67		298–m.p.
$Fe_{0.947}O$ (liq.)	16.30	—	—		m.p.–1 800
Fe_3O_4(α)	21.88	48.2	—		298–900
Fe_3O_4(β)	48.00	—	—		900–1 800
Fe_2O_3(α)	23.49	18.6	−3.55		298–950
Fe_2O_3(β)	36.0	—	—		950–1 050
Fe_2O_3(γ)	31.71	1.76	—		1 050–1 750
Gd_2O_3 (monocl.)	27.28	3.54	−2.54		298–1 802
Gd_2O_2 (cubic)	28.72	2.84	−3.88		298–1 550
H_2O (liq.)	18.03	—	—		298–373
H_2O(g)	7.17	2.56	0.08		298–2 500
H_2O_2(g)	12.50	2.84	−2.84		298–1 300
HgO	9.00	6.0	—	estim.	298–?

Table 8.12e OXIDES—*continued*

Compound	*a*	*b*	*c*	*Remarks*	*Temp. range* K
HfO_2	17.39	2.08	−3.48		298–1 800
IrO_2	17.33	1.93	−3.68		298–1 400
La_2O_3	28.86	3.08	−3.28		298–1 771
Li_2O(s)	14.94	6.08	−3.38		298–1 045
MgO	11.71	0.75	−2.80		298–3 098
MnO	11.11	1.94	−0.88		298–1 800
$Mn_3O_4(\alpha)$	34.64	10.82	−2.20		298–1 445
$Mn_3O_4(\beta)$	50.20	—	—		1 445–1 800
Mn_2O_3(s)	24.73	8.38	−3.23		298–1 350
MnO_2	16.60	2.44	−3.88		298–780
MoO_3	20.07	5.90	−3.68		298–1 068
$Na_2O(\alpha)$	13.26	16.78	0.99	$-7.3 \times 10^{-6}T^2$	298–1 023
$Na_2O(\beta)$	19.67	3.05	—		1 023–1 243
$Na_2O(\gamma)$	20.28	2.56	—		1 243–m.p.
Na_2O (liq.)	25.0	—	—		m.p. – 2 200
NbO	10.04	2.35	−0.78		298–1 700
$NbO_2(\alpha)$	14.68	6.16	−2.42		298–1 040
$NbO_2(\beta)$	22.20	—	—		1 040–1 200
Nb_2O_5	38.76	3.54	−7.32		298–1 780
$Nd_2O_3(\alpha)$	27.67	7.12	−2.84		298–1 395
$Nd_2O_3(\beta)$	37.20	—	—		1 395–1 795
$NiO(\alpha)$	−4.99	37.58	3.89		298–525
$NiO(\beta)$	13.88	—	—		525–565
$NiO(\gamma)$	11.18	2.02	—		565–1 800
P_2O_5	17.90	38.8	−3.73		298–700
P_4O_{10}(g)	73.6	—	—		631–1 400
PbO (red)	10.95	3.75	−1.0		298–1 900
PbO (yell.)	9.05	6.40	—		298–1 000
PbO_2	12.7	7.8	—	estim.	298–?
PdO	10.83	1.68	−0.80	$+0.09 \times 10^{-6}T^2$	298–1 200
Pr_6O_{11}	95.29	26.16	−9.31		298–1 172
Rh_2O	15.59	6.47	—		298–973
RhO	9.84	5.33	—		298–1 023
Rh_2O_3	20.74	13.8	—		298–973
SO(g)	7.70	0.84	−0.65		298–2 000
SO_2(g)	10.38	2.54	−1.42		298–1 800
SO_3(g)	13.70	6.42	−3.12		298–1 200
Sb_2O_3	19.10	17.1	—	estim.	298–930
SbO_2	11.30	8.1	—	estim.	298–1 198
Sb_2O_5	10.95	57.57	—		298–500
Sc_2O_3	23.17	5.64	—	estim.	298–2 500
α-quartz	10.49	0.24	−1.44		298–848
β-quartz	14.08	2.4	—		848–2 000
α-cristobalite	4.28	21.06	—		298–523
β-cristobalite	17.39	0.31	−9.90		523–2 000
α-tridymite	3.27	24.80	—		298–390
β-tridymite	13.64	2.64	—		390–2 000
SiO_2 (glass)	13.38	3.68	−3.45		298–2 000
$Sm_2O_3(\alpha)$	30.75	4.64	−4.30		298–1 195
$Sm_2O_3(\beta)$	36.90	—	—		1 195–1 798
Sm_2O_3 (cubic)	30.64	5.08	−3.96		298–1 150
SnO_2	17.66	2.40	−5.16		298–1 500
SrO	12.34	1.12	−1.806		298–1 270
Ta_2O_5	37.0	6.56	−5.92		298–1 800
TeO_2	15.58	3.48	−1.20		298–m.p.

Table 8.12e OXIDES—*continued*

Compound	*a*	*b*	*c*	*Remarks*	*Temp. range* K
TeO_2 (liq.)	26.95	0.52	—		m.p.–1 146
ThO_2	16.65	2.13	−2.24		298–2 500
$TiO(\alpha)$	10.57	3.6	−1.86		298–1 264
$TiO(\beta)$	11.85	3.0	—		1 264–1 800
$Ti_2O_3(\alpha)$	7.31	53.52	—		298–473
$Ti_2O_3(\beta)$	34.68	1.3	−10.2		473–1 800
$Ti_3O_5(\alpha)$	35.47	29.50	—		298–450
$Ti_3O_5(\beta)$	41.60	8.0	—		450–1 400
rutile	17.97	0.28	−4.35		298–1 800
anatase	17.83	0.50	−4.23		298–1 300
UO_2	19.20	1.62	−3.96		298–1 500
U_3O_8	67.5	8.83	−11.94		298–900
UO_3	22.1	2.64	−2.65		298–900
VO	11.32	3.22	−1.26		298–1 700
V_2O_3	29.35	4.76	−5.42		298–1 800
$VO_2(\alpha)$	14.96	—	—		298–345
$VO_2(\beta)$	17.85	1.70	−3.95		345–1 818
VO_2 (liq.)	25.5	—	—		1 818–1 900
V_2O_5	46.54	−3.90	−13.22		298–943
V_2O_5 (liq.)	45.60	—	—		943–1 500
WO_3	17.48	6.79	—	estim.	298–1 550
$Y_2O_3(\alpha)$	29.60	1.20	−4.78		298–1 330
$Y_2O_3(\beta)$	31.50	—	—		1 330–1 800
ZnO	11.71	1.22	−2.18		298–1 600
$ZrO_2(\alpha)$	16.64	1.80	−3.36		298–1 478
$ZrO_2(\beta)$	17.80	—	—		1 478–1 850

Table 8.12f SULPHIDES, SELENIDES AND TELLURIDES
Molar heat capacities
C_p may be given empirically by an equation of the form:
$C_p = 4.1868\,(a + 10^{-3}bT + 10^5 cT^{-2})\ \mathrm{J\,K^{-1}\,mol^{-1}}$
The values of the constants are given below.

Compound	*a*	*b*	*c*	*Remarks*	*Temp. range* K
$Ag_2S(\alpha)$	10.13	26.40	—		298–452
$Ag_2S(\beta)$	21.64	—	—		452–850
$Ag_2Se(\alpha)$	23.29	0.39	−4.0		298–406
$Ag_2Se(\beta)$	20.4	—	—		406–460
Bi_2S_3	26.25	9.8	—	estim.	—
Bi_2Te_3	36.0	13.05	−3.12		373–m.p.
CS_2(l)	18.4	—	—		298–b.p.
CS_2(g)	12.45	1.60	−1.80		298–1 800
CaS	10.80	1.85	—	estim.	298–2 000
CdS	10.65	3.30	—	estim.	273–1 273
CdTe	12.55	4.54	−1.76		298–m.p.
CoS	10.6	2.51	—	estim.	273–1 373
$Cu_2S(\alpha)$	19.50	—	—		298–376
$Cu_2S(\beta)$	23.25	—	—		376–623
$Cu_2S(\gamma)$	20.32	—	—		623–1 400
CuS	10.6	2.64	—	estim.	273–1 273
$Cu_2Se(\alpha)$	14.0	18.5	—		298–395
$Cu_2Se(\beta)$	20.1	—	—		383–488
$Cu_2Te(\zeta)$	20.9	—	—		841–950
$Cu_2Te(\epsilon)$	26.12	—	—		633–841

Table 8.12f SULPHIDES, SELENIDES AND TELLURIDES—*continued*

Compound	*a*	*b*	*c*	*Remarks*	*Temp. range* K
$Cu_2Te(\delta)$	32.0	—	—		590–633
$Cu_2Te(\gamma)$	27.0	—	—		531–590
$Cu_2Te(\beta)$	14.45	12.8	—		433–531
$Cu_2Te(\alpha)$	14.3	12.8	—		298–433
$FeS(\alpha)$	5.19	26.40	—		298–411
$FeS(\beta)$	17.40	—	—		411–598
$FeS(\gamma)$	12.20	2.38	—		598–1 468
FeS(l)	17.0	—	—		1 468–1 500
FeS_2	17.88	1.32	−3.05		298–1 000
$H_2S(g)$	7.81	2.96	−0.46		298–2 000
$H_2Se(g)$	7.59	3.50	−0.31		298–2 000
MnS	11.40	1.80	—		298–1 803
MnS(liq.)	16.00	—	—		1 803–2 000
$MoS_2(s)$	11.2	13.5	—	estim.	298–729
NiS	9.3	6.4	—		273–670
NiTe(s)	11.57	3.30	—		298–700
PbS	10.63	4.01	—		273–873
PtS	11.95	2.6	−2.12	estim.	298–1 100
PtS_2	14.07	7.08	−0.4	estim.	298–1 000
Sb_2S_3	24.2	13.2	—	estim.	273–821
$SnS(\alpha)$	8.53	7.48	0.9		298–875
$SnS(\beta)$	9.78	3.74	—		875–m.p.
SnS(l)	17.90	—	—		1 153–1 250
SnS_2	15.51	4.20	—		298–1 000
SnTe	16.88	3.67	—		273–603
$TiS_2(\alpha)$	8.08	27.34	—		298–420
$TiS_2(\beta)$	14.99	5.14	—		420–1 010
ZnS(s)	12.16	1.24	−1.36		298–1 200
ZnTe	11.11	2.61	—	—	298–m.p.

Table 8.12g HALIDES
Molar heat capacities

C_p may be given empirically by an equation of the form:

$C_p = 4.1868\ (a + 10^{-3}bT + 10^5cT^{-2})\ \text{J K}^{-1}\ \text{mol}^{-1}$

The values of the constants are given in the following table.

Compound	*a*	*b*	*c*	*Remarks*	*Temp. range* K
AgCl(s)	14.88	1.0	−2.70		298–728
AgCl(l)	16.0	—	—		728–900
AgBr(s)	7.93	15.40	—		298–703
AgBr(l)	14.9	—	—		m.p.–836
$AgI(\alpha)$	5.82	24.10	—		298–423
$AgI(\beta)$	13.5	—	—		423–600
$AlCl_3(s)$	13.25	28.00	—		273–m.p.
$AlCl_3(l)$	31.2	—	—		m.p.–504
AlCl(g)	9.0	—	−0.68		298–2 000
$AlBr_3(s)$	18.74	18.66	—		273–m.p.
$AlBr_3(l)$	29.5	—	—		m.p.–407
$AlI_3(s)$	16.88	22.66	—		273–m.p.
$AlI_3(l)$	28.8	—	—		m.p.–480
$AlF_3(\alpha)$	17.27	10.96	−2.30		298–727
$AlF_3(\beta)$	20.93	3.0	—		727–1 400

Table 8.12g HALIDES—*continued*

Compound	*a*	*b*	*c*	*Remarks*	*Temp. range* K
AlF(g)	8.9	—	−1.45		298–2 000
$AsCl_3$(l)	31.9	—	—	estim.	286–371
$AsCl_3$(g)	19.62	0.24	−1.42		298–1 000
AsF_3(l)	30.2	—	—		298
AsF_3(g)	18.18	1.66	−2.66		298–1 000
$BaCl_2$(α)	22.20	0.76	−4.0		273–1 198
$BaCl_2$(β)	26.61	—	—		1 198–m.p.
$BaCl_2$(l)	24.96	—	—		m.p.–1 339
$BaBr_2$(s)	15.96	6.22	—		487–1 126
CCl_4(g)	24.90	0.48	−4.74		298–1 000
CBr_4(α)	31.7	—	—		295–320
CBr_4(β)	33.0	—	—		320–m.p.
CBr_4(g)	25.03	0.60	−3.03		298–1 000
CF_4(g)	16.64	7.84	−4.00		298–1 200
$CaCl_2$(s)	17.18	3.04	−0.6		600–1 055
$CaCl_2$(l)	24.70	—	—		1 055–1 700
$CaBr_2$(s)	13.96	7.86	—		434–m.p.
$CaBr_2$(l)	27.38	—	—		m.p.–1 132
CaF_2(α)	14.30	7.28	0.47		298–1 424
CaF_2(β)	25.81	2.5	—		1 424–m.p.
CaF_2(l)	23.88	—	—		m.p.–1 800
$CoCl_2$(s)	14.41	14.60	—		298–1 000
$CrCl_2$(s)	15.23	5.96	—		298–m.p.
$CrCl_3$(s)	19.44	7.03	—		298–m.p.
CsBr(s)	11.6	2.59	—	estim.	273–909
CsCl(α)	12.78	1.23	−0.46		293–743
CsCl(β)	0.81	17.64	−0.89		743–m.p.
CsCl(l)	13.86	4.28	—		m.p.–1 170
CsF(s)	11.3	2.71	—	estim.	273–957
CsI(s)	7.06	10.36	1.93		298–m.p.
CsI(l)	−4.29	20.5	—		m.p.–1 170
CuCl(s)	5.87	19.2	—		298–703
CuCl(l)	15.8	—	—		703–1 200
CuCl(g)	8.92	—	−0.47		298–2 000
$CuCl_2$(s)	15.42	12.0	—		298–800
CuI(s)	12.1	2.86	—	estim.	273–675
CuI_2(s)	20.1	—	—	estim.	274–328
$FeCl_2$(s)	18.94	2.08	−1.17		600–950
$FeCl_2$(l)	24.42	—	—		950–1 100
$FeCl_3$(s)	29.56	—	−6.11		298–m.p.
HF(g)	6.43	0.82	0.26		298–2 000
HCl(g)	6.34	1.10	0.26		298–2 000
HBr(g)	6.25	1.40	0.26		298–1 600
HI(g)	6.29	1.42	0.22		298–2 000
HfF_4(s)	31.9	74.8	9.0		273–1 103
$HfCl_4$(s)	31.47	—	−2.38		298–485
HgCl(s)	11.05	3.70	—	estim.	273–798
$HgCl_2$(s)	15.3	10.3	—	estim.	273–553
HgI(s)	11.4	4.61	—	estim.	273–563
HgI_2(α)	18.50	—	—		298–403
HgI_2(β)	20.2	—	—		403–m.p.
HgI_2(l)	25.0	—	—		m.p.–603
KCl(s)	9.89	5.20	0.77		298–1 043
KCl(l)	16.0	—	—		1 043–1 200
KCl(g)	8.94	0.18	−0.24		298–2 000

Table 8.12g HALIDES—*continued*

Compound	*a*	*b*	*c*	*Remarks*	*Temp. range* K
KBr(s)	12.84	2.50	−2.84		600–1 000
KBr(g)	8.94	—	−0.17		298–2 000
KI(s)	21.10	−8.38	−10.38		600–1 000
KF(s)	11.02	3.12	—		298–1 130
LiCl(s)	11.0	3.39	—	estim.	273–887
LiBr(s)	11.5	3.02	—	estim.	273–825
LiF(s)	9.14	5.19	—		298–m.p.
LiF(l)	15.50	—	—		m.p.–1 170
LiI(s)	12.5	2.08	—	estim.	273–723
$MgCl_2$(s)	18.90	1.42	−2.06		600–987
$MgCl_2$(l)	22.10	—	—		987–1 500
MgF_2(s)	16.93	2.52	−2.20		298–m.p.
MgF_2(l)	22.57	—	—		1 538–1 800
$MnCl_2$(s)	18.04	3.16	−1.37		600–923
$MnCl_2$(l)	22.60	—	—		923–1 200
MnF_2(s)	14.79	5.70	−0.47		298–m.p.
NaCl(s)	10.98	3.90	—		298–1 073
NaCl(l)	16.0	—	—		m.p.–1 300
NaCl(g)	8.93	—	−0.41		298–2 000
NaBr(s)	11.87	2.10	—		298–550
NaBr(g)	8.93	—	−0.29		298–2 000
NaF(s)	10.40	3.88	−0.33		298–m.p.
NaI(s)	12.5	1.62	—	estim.	273–936
$NH_4Cl(\alpha)$	11.80	32.0	—		298–457.7
$NH_4Cl(\beta)$	5.00	34.0	—		457.7–500
$NiCl_2$(s)	17.50	3.16	−1.19		298–1 303
$NiCl_2$(l)	24.00	—	—		1 303–1 336
PF_3(g)	17.18	1.92	−3.88		298–2 000
PCl_3(l)	28.7	—	—	estim.	284–371
PCl_3(g)	19.15	0.74	−1.91		298–1 000
PCl_5(g)	31.42	0.20	−4.27		298–2 000
PBr_3(g)	19.81	—	−1.43		298–1 000
$PbCl_2$(s)	16.1	4.0	—		298–771
$PbCl_2$(l)	28.2	—	—		m.p.–851
$PbBr_2$(s)	18.59	2.20	—		298–643
$PbBr_2$(l)	27.6	—	—		m.p.–860
PbI_2(s)	18.00	4.70	—		298–685
PbI_2(l)	32.3	—	—		m.p.–776
PbF_2(s)	16.5	4.12	—	estim.	273–1 091
RbF(s)	7.97	9.2	1.21		298–m.p.
RbF(l)	−11.30	0.88	350.7		m.p.–1 200
RbCl(s)	11.5	2.49	—	estim.	273–987
RbBr(s)	11.6	2.55	—	estim.	273–954
RbI(s)	11.6	2.63	—	estim.	273–913
$SbCl_3$(s)	10.3	51.1	—	estim.	273–346
$SbBr_3$(s)	17.2	29.3	—	estim.	273–370
$SiCl_4$(g)	24.25	1.64	−2.75		298–1 000
$SiBr_4$(g)	25.19	0.64	−1.94		298–1 000
SiF_4(g)	21.86	3.15	−4.70		298–1 000
$SnCl_2$(s)	16.2	9.26	—	estim.	273–520
$SnCl_4$(l)	39.5	—	—	estim.	286–371
$SnCl_4$(g)	25.57	0.20	−1.87		298–1 000
$SnBr_4$(g)	25.80	—	−0.97		298–1 000
SnI_4(s)	19.4	36.0	—		298–b.p.
SnI_4(l)	40.1	—	—		b.p.–443

Table 8.12g HALIDES—*continued*

Compound	*a*	*b*	*c*	*Remarks*	*Temp. range* K
$TeCl_4$(s)	33.1	—	—		298–m.p.
$TeCl_4$(l)	53.2	—	—		m.p.–538
TeF_6(g)	35.33	1.62	−7.00		298–2 000
$TiCl_4$(l)	35.7	—	—	estim.	273–372
$TiCl_4$(g)	25.45	0.24	−2.36		298–2 000
$TiBr_4$(s)	28.0	—	—		298–m.p.
$TiBr_4$(l)	32.75	15.66	—		m.p.–423
TlCl(s)	12.00	2.00	—		298–700
TlCl(l)	14.2	—	—		m.p.–803
TlCl(g)	8.94	—	−0.25		298–2 000
TlBr(s)	11.07	4.95	—		298–733
TlBr(l)	25.25	−9.04	—		m.p.–800
UF_4(s)	25.7	7.00	−0.06		298–1 309
UF_6(s)	12.6	92.0	—		273–337
UF_6(g)	22.3	28.5	—		273–400
UCl_3(s)	20.8	7.75	1.05		298–900
UCl_4(s)	27.2	8.57	−0.79		298–800
UCl_4(l)	25.8	14.4	—		890–920
UBr_4(s)	31.4	4.92	−3.15		350–750
UI_4(s)	34.8	2.38	−4.72		380–720
UI_4(l)	39.6	—	—		820–870
VCl_2(s)	17.25	2.72	−0.71		298–1 200
VCl_3(s)	22.99	3.92	−1.68		298–900
$ZnCl_2$(s)	14.5	5.5	—		298–m.p.
$ZnCl_2$(l)	24.1	—	—		m.p.–1 000
$ZnBr_2$(s)	12.6	10.4	—		298–m.p.
$ZnBr_2$(l)	27.2	—	—		m.p.–1 000
$ZrCl_4$(s)	31.92	—	−2.91		298–604
$ZrBr_4$(s)	25.5	15.1	—	estim.	298–630

8.7 Vapour pressures

Table 8.13 ELEMENTS
Vapour pressures

The vapour pressure p (mmHg) of an element may be represented by an equation of the type:

$$\log p = -\frac{A}{T} + B + C \log T + 10^{-3} DT$$

The values of the constants A, B, C and D in this equation are given below.

Element	*A*	*B*	*C*	*D*	*Temp. range* K
Ag	14 710	11.66	−0.755	—	298–1 234
	14 260	12.23	−1.055	—	1 234–2 400
Al	16 450	12.36	−1.023	—	1 200–2 800
Am	13 700	13.97	−1.0	—	1 103–1 453
As_4	6 160	9.82	—	—	600–900
Au	19 820	10.81	−0.306	−0.16	298–1 336
	19 280	12.38	−1.01	—	1 336–3 240
B	29 900	13.88	−1.0	—	1 000–m.p.
Ba	9 730	7.83	—	—	750–983
	9 340	7.42	—	—	983–1 200
Be	10 734	9.067	—	−0.145	900–1 557
Bi	10 400	12.35	−1.26	—	m.p.–b.p.
Bi_2	10 730	18.1	−3.02	—	m.p.–b.p.

Table 8.13 ELEMENTS—*continued*

Element	*A*	*B*	*C*	*D*	*Temp. range* K
Ca	10 300	14.97	−1.76	—	713–m.p.
	9 600	12.55	−1.21	—	m.p.–b.p.
Ce	20 305	8.305	—	—	1 611–2 038
Cd	5 908	9.717	−0.232	−0.284	450–594
	5 819	12.287	−1.257	—	594–1 050
Co	22 210	10.817	—	−0.223	1 000–1 772
Cr	20 680	14.56	−1.31	—	298–m.p.
Cs	4 075	11.38	−1.45	—	280–1 000
Cu	17 870	10.63	−0.236	−0.16	298–1 356
	17 650	13.39	−1.273	—	1 356–2 870
Eu	8 980	8.16	—	—	696–900
Fe	21 080	16.89	−2.14	—	900–1 812
	19 710	13.27	−1.27	—	1 812–3 000
Ga	14 700	10.07	−0.5	—	m.p.–b.p.
Ge	20 150	13.28	−0.91	—	298–m.p.
	18 700	12.87	−1.16	—	m.p.–b.p.
Hf(α)	32 000	11.81	−0.5	—	298–2 023
Hf(β)	31 630	11.63	−0.5	—	2 023–m.p.
Hf	29 830	9.20	—	—	m.p.–b.p.
Hg	3 308	10.373	−0.8	—	298–630
I_2	3 578	17.72	−2.51	—	298–m.p.
	3 205	23.65	−5.18	—	m.p.–b.p.
In	12 580	9.79	−0.45	—	m.p.–b.p.
Ir	35 070	13.18	−0.7	—	298–m.p.
K	4 770	11.58	−1.370	—	350–1 050
La	22 120	10.39	−0.33	—	298–m.p.
	21 530	9.89	−0.33	—	m.p.–b.p.
Li	8 415	11.34	−1.0	—	m.p.–b.p.
Mg	7 780	11.41	−0.855	—	298–m.p.
	7 550	12.79	−1.41	—	m.p.–b.p.
Mn	14 850	17.88	−2.52	—	993–1 373
	13 900	17.27	−2.52	—	m.p.–b.p.
Mo	34 700	11.66	−0.236	−0.145	298–m.p.
Na	5 700	11.33	−1.718	—	400–1 200
Na_2	6 540	10.7	—	—	—
Nb	37 650	8.94	+0.715	−0.166	298–m.p.
Ni	22 500	13.60	−0.96	—	298–m.p.
	22 400	16.95	−2.01	—	m.p.–b.p.
P_4 (yell.)	3 530	19.09	−3.5	—	298–317
P_4	2 740	7.84	—	—	317–553
Pb	10 130	11.16	−0.985	—	600–2 030
Pd	19 800	11.82	−0.755	—	298–m.p.
	17 500	4.81	+1.0	—	m.p.–b.p.
Pr	17 190	8.10	—	—	1 425–1 692
Pt	29 200	13.24	−0.855	—	298–m.p.
	28 500	14.30	−1.26	—	m.p.–b.p.
Pu	17 590	7.90	—	—	1 392–1 793
Rb	4 560	12.00	−1.45	—	312–952
Re	40 800	14.20	−1.16	—	298–3 000
Rh	29 360	13.50	−0.88	—	298–m.p.
Ru	33 550	10.76	—	—	2 000–2 500
S_2	6 975	16.22	−1.53	−1.0	m.p.–b.p.
S_x	4 830	23.88	−5.0	—	m.p.–b.p.
Sb_x	11 560	22.40	−3.52	—	298–m.p.
Sb_2	11 170	18.54	−3.02	—	m.p.–b.p.
Sc(β)	19 700	13.07	−1.0	—	1 607–m.p.
Se_x	4 990	8.09	—	—	493–958

Table 8.13 ELEMENTS—*continued*

Element	*A*	*B*	*C*	*D*	*Temp. range* K
Si	20 900	10.84	−0.565	—	m.p.–b.p.
Sm	11 170	13.76	−1.56	~	298–m.p.
Sn	15 500	8.23	—	—	505–b.p.
Sr	9 450	13.08	−1.31	—	813–m.p.
	9 000	12.63	−1.31	—	m.p.–b.p.
Ta	40 800	10.29	—	—	298–m.p.
Te_2	9 175	19.68	−2.71	—	298–m.p.
	7 830	22.29	−4.27	—	m.p.–
Th	30 200	12.95	−1.0	—	298–m.p.
Ti(β)	24 400	13.18	−0.91	—	1 155–m.p.
Ti	23 200	11.74	−0.66	—	m.p.–b.p.
Tl	9 300	11.10	−0.892	—	700–1 800
Tm	12 550	9.18	—	—	807–1 219
U	25 580	18.58	−2.62	—	298–1 405
	24 090	13.20	−1.26	—	1 405–4 200
V	26 900	10.12	+0.33	−0.265	298–m.p.
W	44 000	8.76	+0.50	—	298–m.p.
Y	22 230	11.835	−0.66	—	298–m.p.
	22 280	16.13	−1.97	—	m.p.–b.p.
Zn	6 883	9.418	−0.050 3	−0.33	473–692.5
	6 670	12.00	−1.126	—	692.5–1 000
Zr	31 820	11.78	−0.50	—	1 125–m.p.
	30 300	9.38	—	—	m.p.–b.p.

Table 8.14 HALIDES AND OXIDES

Vapour pressures

$$\log p = \frac{A}{T} + B + C \log T + 10^{-3} D T \text{ (mmHg)}$$

Substance	*A*	*B*	*C*	*D*	*Temp. range* K
AgCl	−11 830	12.39	−0.30	−1.02	298–m.p.
AgCl	−11 320	17.34	−2.55	—	m.p.–b.p.
AgBr	−12 400	19.33	−2.97	—	m.p.–b.p.
AgI	−10 250	20.09	−3.52	—	m.p.–b.p.
AlF_3	−16 700	23.27	−3.02	—	298–s.p.
Al_2Cl_6	−6 360	9.66	3.77	−6.12	298–s.p.
Al_2Br_6	−5 280	20.81	−1.75	−4.08	298–m.p.
Al_2Br_6	−5 280	46.70	−12.59	—	m.p.–b.p.
Al_2I_6	−7 150	17.76	0.12	−4.96	298–m.p.
Al_2I_6	−6 760	46.67	−11.89	—	m.p.–b.p.
AmF_3	−24 600	36.87	−7.05	—	1 100–1 300
AsF_3	−4 150	61.38	−18.26	—	265–292
AsF_5	−1 088	7.72	—	—	m.p.–b.p.
$AsCl_3$	−2 660	24.76	−5.83	—	m.p.–b.p.
As_4O_6*	−5 282	10.91	—	—	373–573
As_4O_6†	−5 452	11.468	—	—	488–573
As_4O_6	−3 130	7.16	—	—	m.p.–b.p.
BCl_3	−2 115	27.56	−7.04	—	m.p.–b.p.
BBr_3	−2 710	28.36	−7.04	—	m.p.–b.p.
BI_3	−3 342	24.31	−5.4	—	m.p.–b.p.

* claudetite † arsenolite

Table 8.14 HALIDES AND OXIDES—*continued*

Substance	*A*	*B*	*C*	*D*	*Temp. range* K
B_2O_3	−16 960	6.64	—	—	1 300–1 650
BaF_2	−20 330	28.04	−5.03	—	m.p.–b.p.
BaO	−21 900	9.99	—	—	1 200–1 700
BeF_2	−13 000	24.56	−3.79	—	m.p.–b.p.
Be_2Cl_4	−8 970	37.0	−7.65	—	298–m.p.
$BeCl_2$	−7 870	27.15	−5.03	—	298–m.p.
$BeCl_2$	−7 220	26.28	−5.03	—	m.p.–b.p.
Be_2Br_4	−8 320	35.9	−7.65	—	298–s.p.
$BeBr_2$	−7 650	27.15	−5.03	—	298–m.p.
$BeBr_2$	−6 570	25.63	−5.03	—	m.p.–b.p.
Be_2I_4	−8 520	35.9	−7.65	—	298–m.p.
BeI_2	−7 000	26.5	−5.03	—	298–m.p.
BeI_2	−5 800	24.96	−5.03	—	m.p.–b.p.
$BiCl_3$	−6 200	12.83	—	—	298–m.p.
$BiCl_3$	−5 980	31.38	−7.04	—	m.p.–b.p.
$BiBr_3$	−6 190	31.40	−7.04	—	m.p.–b.p.
CCl_4	−2 400	23.60	−5.30	—	m.p.–b.p.
CBr_4	−2 650	8.78	—	—	298–m.p.
CBr_4	−2 330	7.89	—	—	m.p.–b.p.
$CaF_2(\alpha)$	−23 600	27.41	−4.525	—	298–1 424
$CaF_2(\beta)$	−23 350	27.23	−4.525	—	1 424–m.p.
CaF_2	−21 800	26.31	−4.525	—	m.p.–b.p.
$CaCl_2$	−13 570	9.22	—	—	1 110–1 281
CdF_2	−16 170	27.50	−5.03	—	m.p.–b.p.
$CdCl_2$	−9 270	17.46	−2.11	—	298–m.p.
$CdCl_2$	−9 183	25.907	−5.04	—	m.p.–b.p.
$CdBr_2$	−8 250	18.15	−2.5	—	298–m.p.
$CdBr_2$	−7 150	16.85	−2.5	—	m.p.–b.p.
CdI_2	−7 530	18.01	−2.5	—	298–m.p.
CdI_2	−6 720	16.79	−2.5	—	m.p.–b.p.
$CeCl_3$	−18 750	36.38	−7.05	—	298–m.p.
$CeBr_3$	−18 000	36.49	−7.05	—	298–m.p.
$CoCl_2$	−14 150	30.10	−5.03	—	298–m.p.
$CoCl_2$	−11 050	27.06	−5.03	—	m.p.–b.p.
$CrCl_2$	−14 000	15.14	−0.62	−0.58	298–m.p.
$CrCl_2$	−13 800	27.70	−5.03	—	m.p.–b.p.
$CrCl_3$	−13 950	17.49	−0.73	−0.77	298–s.p.
CrI_2	−16 080	25.92	−3.53	—	298–m.p.
CrO_2Cl_2	−3 340	34.94	−9.08	—	m.p.–b.p.
CsF	−10 930	17.51	−2.12	—	298–m.p.
CsF	−9 950	18.62	−2.84	—	m.p.–b.p.
CsCl	−10 800	19.99	−3.02	—	700–m.p.
CsCl	−9 815	20.38	−3.52	—	m.p.–b.p.
CsBr	−10 950	20.02	−3.02	—	700–m.p.
CsBr	−10 080	20.56	−3.52	—	m.p.–b.p.
CsI	−10 420	19.70	−3.02	—	600–m.p.
CsI	−9 678	20.35	−3.52	—	m.p.–b.p.
CuCl	−10 170	8.04	—	—	1 000–1 900
Cu_3Cl_3	−3 750	4.90	—	—	900–1 800
CuBr	−7 700	7.69	—	—	1 000–1 480
Cu_3Br_3	−4 010	4.88	—	—	1 000–2 000
$Cu_3I_3(\alpha)$	−9 463	11.14	—	—	629
$Cu_3I_3(\beta)$	−8 351	9.41	—	—	643–670
$Cu_3I_3(\gamma)$	−7 853	8.68	—	—	684–770
$FeCl_2$	−9 890	11.10	—	—	670–740

Table 8.14 HALIDES AND OXIDES—*continued*

Substance	*A*	*B*	*C*	*D*	*Temp. range* K
$FeBr_2$	−10 220	11.95	—	—	670–740
FeI_2	−12 180	29.59	−5.03	—	298–m.p.
FeI_2	−8 750	27.185	−5.535	—	m.p.–b.p.
Fe_2Cl_6	−9 540	45.53	−9.5	—	298–m.p.
$Fe(CO)_5$	−2 075	8.42	—	—	298–b.p.
$GaCl_2$	−4 886	29.14	−6.44	—	m.p.–b.p.
$GaBr_3$	−4 700	28.69	−6.44	—	m.p.–b.p.
$GeCl_4$	−2 940	34.27	−9.08	—	m.p.–b.p.
$GeBr_4$	−3 690	35.00	−9.05	—	m.p.–b.p.
GeI_4	−4 920	22.73	−4.02	—	298–m.p.
H_2O	−2 900	22.613	−4.65	—	m.p.–b.p.
H_2O_2	−3 560	29.68	−7.04	—	m.p.–
$HfCl_4$	−5 200	11.71	—	—	476–681
$HfI_4(\alpha)$	−10 700	19.56	—	—	575–597
$HfI_4(\beta)$	−7 360	13.97	—	—	598–645
$HfI_4(\gamma)$	−6 173	12.13	—	—	648–678
$HgCl_2$	−4 580	16.39	−2.0	—	298–m.p.
$HgBr_2$	−4 500	11.47	0.05	−1.51	298–m.p.
$HgBr_2$	−4 370	24.18	−5.03	—	m.p.–b.p.
HgI_2	−5 690	30.27	−6.47	—	298–m.p.
HgI_2	−4 620	25.72	−5.33	—	m.p.–b.p.
InCl	−4 640	8.03	—	—	m.p.–b.p.
$InCl_3$	−8 270	13.62	—	—	500–s.p.
InBr	−6 470	16.31	−2.01	—	298–m.p.
$InBr_2$	−4 480	7.48	—	—	m.p.–b.p.
$InBr_3$	−5 670	11.67	—	—	500–s.p.
InI	−6 730	15.74	−1.97	—	298–m.p.
IrF_6	−1 657	7.952	—	—	m.p.–b.p.
KF	−12 930	17.30	−2.06	—	298–m.p.
KF	−11 570	16.90	−2.32	—	m.p.–b.p.
KCl	−12 230	20.34	−3.0	—	298—m.p.
KCl	−10 710	18.91	−3.0	—	m.p.–b.p.
KBr	−11 110	16.60	−2.0	—	298–m.p.
KBr	−10 180	18.67	−3.0	—	m.p.–b.p.
KI	−11 000	16.99	−2.0	—	298–m.p.
KI	−10 050	20.41	−3.52	—	m.p.–b.p.
$LaCl_3$	−19 040	36.20	−7.05	—	298–m.p.
$LaBr_3$	−18 780	36.83	−7.05	—	298–m.p.
LaI_3	−18 390	37.00	−7.05	—	298–m.p.
LiF	−14 560	23.56	−4.02	—	m.p.–b.p.
LiCl	−10 760	22.30	−4.02	—	m.p.–b.p.
LiBr	−10 170	20.55	−3.52	—	m.p.–b.p.
LiI	−11 110	21.70	−3.52	—	m.p.–b.p.
MgF_2	−19 700	27.80	−5.03	—	m.p.–b.p.
$MgCl_2$	−10 840	25.53	−5.03	—	m.p.–b.p.
$MgBr_2$	−10 930	26.07	−5.03	—	m.p.–b.p.
MgI_2	−8 090	25.18	−5.03	—	m.p.–b.p.
MnF_2	−17 400	22.06	−3.02	—	m.p.–b.p.
$MnCl_2$	−10 606	23.68	−4.33	—	m.p.–b.p.
MoF_5	−2 772	8.58	—	—	m.p.–b.p.
MoF_6	−1 500	7.77	—	—	m.p.–b.p.
$MoOF_4$	−2 854	9.21	—	—	313–m.p.
$MoOF_4$	−2 671	8.716	—	—	m.p.–b.p.
$MoCl_5$	−5 210	13.1	—	—	298–m.p.
MoO_3	−15 230	27.16	−4.02	—	298–m.p.

Table 8.14 HALIDES AND OXIDES—*continued*

Substance	*A*	*B*	*C*	*D*	*Temp. range* K
MoO_3	−12 480	24.60	−4.02	—	m.p.–b.p.
NaF	−14 960	17.53	−2.01	—	298–m.p.
NaF	−13 500	17.93	−2.52	—	m.p.–b.p.
NaCl	−12 440	14.31	−0.90	−0.46	298–m.p.
NaCl	−11 530	20.77	−3.48	—	m.p.–b.p.
NaBr	−12 100	20.39	−3.0	—	298–m.p.
NaBr	−10 500	18.81	−3.0	—	m.p.–b.p.
NaI	−10 740	20.96	−3.52	—	m.p.–b.p.
NbF_5	−4 900	14.397	—	—	298–m.p.
NbF_5	−2 780	8.37	—	—	m.p.–b.p.
$NbCl_4$	−6 870	12.30	—	—	577–651
$NbCl_5$	−4 370	11.51	—	—	403–m.p.
$NbCl_5$	−2 870	8.37	—	—	m.p.–b.p.
$NbBr_5$	−4 085	9.33	—	—	m.p.–b.p.
$NbOCl_3$	−5 333	8.79	—	—	298–s.p.
$NdCl_3$	−18 220	36.27	−7.05	—	298–m.p.
$NdBr_3$	−17 650	36.51	−7.05	—	298–m.p.
NdI_3	−17 490	36.61	−7.05	—	298–m.p.
NiF_2	−14 650	20.28	−3.02	—	298–m.p.
$NiCl_2$	−13 300	21.88	−2.68	—	298–s.p.
$NiBr_2$	−13 110	16.68	−1.71	−0.35	298–s.p.
$Ni(CO)_4$	−1 530	7.73	—	—	298–b.p.
NpF_6	−2 892	18.48	−2.7	—	273–m.p.
	−1 913	14.61	−2.35	—	m.p.–350
OsF_5	−3 429	9.75	—	—	m.p.–b.p.
OsF_6	−1 858	8.726	—	—	273–m.p.
OsF_6	−1 473	7.47	—	—	m.p.–b.p.
OsO_4*	−2 955	9.64	—	—	273–329
OsO_4†	−2 580	10.70	—	—	273–315
OsO_4	−2 065	8.01	—	—	m.p.–b.p.
PCl_3	−2 370	22.74	−5.14	—	273–b.p.
PCl_5	−3 520	11.035	—	—	373–432
$POCl_3$	−1 830	7.72	—	—	273–b.p.
P_4O_{10}‡	−4 350	9.81	—	—	298–s.p.
PbF_2	−11 800	26.48	−5.03	—	m.p.–b.p.
$PbCl_2$	−9 890	15.36	−0.95	−0.91	298–m.p.
$PbCl_2$	−10 000	31.60	−6.65	—	m.p.–b.p.
$PbBr_2$	−9 320	18.44	−2.08	−0.34	298–m.p.
$PbBr_2$	−9 540	31.67	−6.76	—	m.p.–b.p.
PbI_2	−9 340	19.68	−2.35	−0.32	298–m.p.
PbI_2	−10 000	39.80	−9.21	—	m.p.–b.p.
PbO	−13 480	14.36	−0.92	−0.35	298–m.p.
PbO	−13 310	19.47	−2.77	—	m.p.–b.p.
$PrCl_3$	−18 490	36.31	−7.05	—	298–m.p.
$PrBr_3$	−17 800	36.53	−7.05	—	298–m.p.
PrI_3	−17 470	36.66	−7.05	—	298–m.p.
PuF_3	−24 950	36.91	−7.05	—	298–m.p.
PuF_3	−23 500	34.47	−6.45	—	m.p.–200
$PuCl_3$	−18 270	32.60	−5.34	—	298–m.p.
$PuCl_3$	−15 490	31.76	−6.45	—	m.p.–b.p.
$PuBr_3$	−17 460	31.32	−5.34	—	298–m.p.
$PuBr_3$	−15 030	32.34	−6.45	—	m.p.–b.p.
RbF	−11 230	18.26	−2.66	—	m.p.–b.p.
RbCl	−11 670	20.157	−3.0	—	298–m.p.

* yellow † white ‡ hexagonal

Table 8.14 HALIDES AND OXIDES—*continued*

Substance	*A*	*B*	*C*	*D*	*Temp. range* K
RbCl	−10 300	18.77	−3.0	—	m.p.–b.p.
RbBr	−11 510	20.155	−3.0	—	298–m.p.
RbBr	−10 220	18.805	−3.0	—	m.p.–b.p.
RbI	−10 280	20.64	−3.52	—	m.p.–b.p.
ReF_5	−3 037	9.024	—	—	m.p.–b.p.
ReF_6	−1 489	7.732	—	—	m.p.–b.p.
ReF_7	−2 206	13.045	−1.47	—	259–m.p.
ReF_7	−244	−21.585	+9.91	—	m.p.–b.p.
$ReOF_4$	−3 888	11.88	—	—	323–m.p.
$ReOF_4$	−3 206	10.09	—	—	m.p.–b.p.
$ReOF_5$	−1 959	8.62	—	—	303–m.p.
$ReOF_5$	−1 679	7.727	—	—	m.p.–b.p.
Re_2O_2F	−3 437	10.36	—	—	m.p.–b.p.
Re_2O_7	−7 300	15.000	—	—	273–m.p.
Re_2O_7	−3 950	9.10	—	—	m.p.–b.p.
$RuCl_3$	−16 750	30.53	−4.63	—	298–1 000
$SO_3\alpha$	−2 680	11.44	—	—	273–m.p.
$SO_3\beta$	−2 860	11.97	—	—	273–m.p.
$SO_3\gamma$	−3 610	14.00	—	—	273–m.p.
SO_3	−2 230	9.90	—	—	m.p.–b.p.
SO_2Cl_2	−1 660	7.65	—	—	m.p.–b.p.
SbF_5	−2 364	8.567	—	—	282–416
$SbCl_3$	−3 460	2.81	3.88	−5.6	298–m.p.
$SbCl_3$	−3 770	29.48	−7.04	—	m.p.–b.p.
$SbCl_5$	−2 530	8.56	—	—	m.p.–350
$SbBr_3$	−2 860	7.97	—	—	435–561
SbI_3	−3 450	7.99	—	—	510–629
Sb_4O_6*	−10 360	12.195	—	—	742–839
Sb_4O_6†	−9 625	11.312	—	—	742–914
Sb_4O_6	−3 900	5.137	—	—	929–1 073
$ScCl_3$	−14 200	14.37	—	—	1 065–1 233
$ScBr_3$	−13 780	14.35	—	—	1 042–1 200
ScI_3	−13 340	14.17	—	—	1 010–1 180
SeF_4	−2 457	9.44	—	—	m.p.–b.p.
SeO_2	−6 170	21.40	−3.02	—	298–s.p.
$SiCl_4$	−1 572	7.64	—	—	273–333
SiI_4	−3 863	23.38	−5.0	—	m.p.–b.p.
$SnCl_4$	−1 925	7.865	—	—	298–b.p.
$SnBr_4$	−3 510	27.63	−6.5	—	303–b.p.
SnI_4	−3 990	10.08	—	—	298–m.p.
SnI_4	−2 975	7.666	—	—	m.p.–b.p.
SrF_2	−21 660	28.04	−5.03	—	m.p.–b.p.
$TaCl_5$	−6 275	34.305	−7.04	—	298–m.p.
$TaCl_5$	−2 975	8.68	—	—	m.p.–b.p.
$TaBr_5$	−7 320	34.85	−7.04	—	298–m.p.
$TaBr_5$	−3 260	8.14	—	—	m.p.–b.p.
TaI_5	−6 660	31.61	−7.04	—	298–m.p.
TaI_5	−3 955	7.72	—	—	m.p.–b.p.
Tc_2O_7	−7 205	18.28	—	—	298–m.p.
Tc_2O_7	−3 570	9.00	—	—	m.p.–b.p.
TeF_4	−3 174	9.093	—	—	298–m.p.
TeF_4	−1 787	5.640	—	—	m.p.–467

* cubic †orthorhombic

Table 8.14 HALIDES AND OXIDES—*continued*

Substance	*A*	*B*	*C*	*D*	*Temp. range* K
TeF_6	−1 460	9.13	—	—	194–241
$TeCl_2$	−3 350	8.51	—	—	m.p.–b.p.
TeO_2	−13 940	23.51	−3.52	—	298–m.p.
$ThCl_4$	−12 900	14.30	—	—	974–1 043
$ThCl_4$	−7 980	9.57	—	—	1 043–1 186
$ThBr_4$	−9 630	11.73	—	—	903–951
$ThBr_4$	−7 550	9.56	—	—	955–1 126
ThI_3	−6 890	9.09	—	—	856–1 107
ThO_2	−34 890	10.87	—	—	2 500–2 900
TiF_4	−5 332	19.51	−2.57	—	298–s.p.
$TiCl_2$	−15 230	19.36	−2.51	—	298–m.p.
$TiCl_2$	−13 110	17.93	−2.51	—	m.p.–b.p.
$TiCl_3$	−9 620	21.47	−3.27	—	298–m.p.
$TiCl_4$	−2 919	25.129	−5.788	—	298–b.p.
$TiBr_4$	−3 706	27.08	−6.24	—	m.p.–b.p.
TiI_2	−12 500	16.90	−1.51	—	298–1 000
TiI_4	−3 054	7.576	—	—	430–643
TlF	−7 710	17.66	−2.18	—	298–m.p.
TlCl	−7 370	16.49	−2.11	—	298–m.p.
TlCl	−6 650	16.92	−2.62	—	m.p.–b.p.
TlBr	−7 420	16.18	−2.0	—	298–m.p.
TlBr	−6 840	18.26	−3.02	—	m.p.–b.p.
TlI	−7 270	15.85	−2.01	—	298–m.p.
TlI	−6 890	18.20	−3.02	—	m.p.–b.p.
UF_4	−16 400	22.60	−3.02	—	298–m.p.
UF_4	−15 300	29.05	−5.03	—	m.p.–b.p.
UF_6	−2 858	16.36	−1.91	—	273–s.p.
UCl_4	−11 350	23.21	−3.02	—	298–m.p.
UCl_4	−9 950	28.96	−5.53	—	m.p.–b.p.
UCl_6	−4 000	10.20	—	—	298–450
UBr_3	−16 420	22.95	−3.02	—	298–m.p.
UBr_3	−15 000	27.54	−5.03	—	m.p.–(b.p.)
UBr_4	−10 800	23.15	−3.02	—	298–m.p.
UBr_4	−8 770	27.93	−5.53	—	m.p.–b.p.
UI_4	−12 330	26.62	−3.52	—	298–m.p.
UI_4	−9 310	28.57	−5.53	—	m.p.–b.p.
UO_2	−33 120	25.69	−4.03	—	1 500–2 800
VF_5	−2 423	10.43	—	—	m.p.–b.p.
VCl_4	−2 875	25.56	−6.07	—	298–m.p.
$VOCl_3$	−1 921	7.70	—	—	298–b.p.
V_2O_5*	−7 100	5.05	—	—	m.p.–1 500
WF_6	−1 380	7.635	—	—	m.p.–b.p.
$WCl_4(\alpha)$	−3 996	9.615	—	—	458–503
$WCl_4(\beta)$	−3 588	8.795	—	—	503–554
WCl_4	−3 253	8.195	—	—	555–598
WCl_5	−3 670	9.50	—	—	413–m.p.
WCl_5	−2 760	7.72	—	—	m.p.–b.p.
$WCl_6\alpha$	−4 580	10.73	—	—	425–t.p.
$WCl_6\beta$	−4 080	9.73	—	—	t.p.–m.p.
WCl_6	−3.050	7.87	—	—	m.p.–b.p.
WOF_4	−3 605	10.96	—	—	298–m.p.
WOF_4	−3 125	9.69	—	—	m.p.–b.p.

* Apparent vapour pressures. V_2O_5 loses oxygen with increasing temperature.

Table 8.14 HALIDES AND OXIDES—*continued*

Substance	*A*	*B*	*C*	*D*	*Temp. range* K
WO_3	−24 600	15.63	—	—	1 000–m.p.
ZnF_2	−13 650	26.90	−5.03	—	m.p.–b.p.
$ZnCl_2$	−8 500	16.61	−1.50	—	298–m.p.
$ZnCl_2$	−8 440	26.37	−5.03	—	m.p.–b.p.
$ZnBr_2$	−7 120	16.21	−2.01	—	298–m.p.
ZnI_2	−6 450	14.70	−1.76	—	298–m.p.
ZnS	−13 980	8.98	—	—	970–1 280
ZrF_4	−14 700	30.80	−5.03	—	298–s.p.
$ZrCl_4$	−5 400	11.765	—	—	480–689
$ZrBr_4$	−6 780	19.60	−1.76	−1.65	298–s.p.
ZrI_4	−7 680	20.87	−2.164	−1.344	298–s.p.

REFERENCES—THERMOCHEMICAL DATA

1a. R. Hultgren, P. O. Desai, D. T. Hawkins, M. Gleiser, K. K. Kelley and D. D. Wagman, 'Selected Values of the Thermodynamic Properties of the Elements', American Society for Metals, Metals Park, Ohio 44073, 1973.
1b. R. Hultgren, P. O. Desai, D. T. Hawkins, M. Gleiser and K. K. Kelley, 'Selected Values of the Thermodynamic Properties of Binary Alloys', American Society for Metals, Metals Park, Ohio 44073, 1973.
2. O. Kubaschewski and C. B. Alcock, 'Metallurgical Thermochemistry', 5th edn, Pergamon, Oxford, 1979.
3. O. Kubaschewski and J. A. Catterall, 'Thermochemical Data of Alloys', Pergamon, Oxford, 1956.
4. A. Schneider and G. Heymer, NPL Symposium No. 9, 'Metallurgical Chemistry', HMSO, London, 1958.
5. E. F. Strotzer and M. Zumbusch, *Z. anorg. Chem.*, 1941, **247**, 415 (concluding paper of a series).
6. M. Zumbusch and W. Biltz, *Z. anorg. Chem.*, 1942, **249**, 1 (concluding paper of a series).

9 Physical properties of molten salts

In the following tables are given densities, electrical conductivities, surface tensions and coefficients of viscosity of pure molten salts, molten binary salt systems and other ionic melts. For comprehensive data and treatment, reference should be made to G. J. Janz, 'Molten Salts Handbook', 1967, Academic Press, New York/London.

Table 9.1 DENSITY OF PURE MOLTEN SALTS

The density of most pure molten salts varies almost linearly with temperature, and may be represented by the equation:

$$d_t = a - 10^{-3} \cdot bt$$

where d_t is the density in g cm^{-3}, a and b are constants, and t is the temperature in °C over appreciable ranges of temperature. Values of the constants and the appropriate temperature ranges are given below. Principal references are in **bold type**.

Substance	*a*	*b*	*Range of observations** °C	*References*
AgBr	6.025	1.04	m.p. to 820	**91, 60, 33,** 10
AgCl	5.257	0.849	467 to 637	**140, 97, 91,** 33
$AgClO_3$	4.2626	1.742	m.p. to 250	**18**
AgI	6.139	1.01	600 to 800	**33**
$AgNO_3$	4.167	1.00	m.p. to 410	**108, 100, 97, 18,** 112, 105, 93, 34, 10
$AlBr_3$	2.875	2.314	m.p. to 225	**89, 88, 38**
$AlCl_3$	1.805	2.5	194 to 250	**38**
AlI_3	3.70	2.5	m.p. to 250	**38**
Al_2O_3	5.259	1.127	2102 to 2352	**121**
$AsBr_3$	3.455	2.6	50 to 100	**96, 34,** 14
$AsCl_3$	2.205	2.18	m.p. to 130	**87, 8,** 1
AsF_3	$d_1 = 2.6659 = 3.839 \cdot 10^{-3}t + 4.35 \cdot 10^{-8}t^2$		0 to 60	**8**
		$d_{25} = 3.01$		**78**
AsF_5	2.047	5.34	m.p. to −53	**68**
BBr_3		$d_o = 2.650$		**15**
BCl_3		$d_{11} = 1.349$		**52**
BI_3		$d_{50} = 3.3$		**12**
B_2O_3	1.609	0.0867	1030 to 1310	**101, 114,** 19
$BaBr_2$		$d_{850} = 4.00$		**102**
$BaCl_2$	3.8292	0.6813	966 to 1081	**122,** 104, 103, 85, 19
BaF_2	5.502	0.999	1327 to 1727	**123**
$Ba(NO_2)_2$	3.448	0.70	—	**124**
$BeCl_2$	1.976	1.1	430 to 475	**47**
$BiBr_3$	5.248	2.6	270 to 330	**34**
$BiCl_3$	4.42	2.20	240 to 350	**120, 40, 34, 23**
$Bi_2(MoO_4)_3$		$d_{723} = 5.170$		**125**
$CaCl_2$	2.4108	0.4225	787 to 950	**122,** 99, 86, 22, 19, 92
CaF_2	3.072	0.391	1367 to 2027	**123,** 111

Table 9.1 DENSITY OF PURE MOLTEN SALTS—*continued*

Substance	*a*	*b*	*Range of observations** °C	*References*
$CdBr_2$	4.688	1.08	m.p. to 720	**91**
$CdCl_2$	3.858	0.825	m.p. to 800	**102, 91**
CdI_2	4.828	1.12	m.p. to 700	**102**
CeF_3	5.997	0.936	1 427 to 1 927	**123**
CsBr	3.911	1.22	m.p. to 860	**113, 34**
CsCl	3.478 5	1.065	667 to 907	**113, 34, 140**
CsF	4.548 9	1.280 6	712 to 912	**122**, 34
CsI	3.918	1.18	645 to 855	**113, 34**
$CsNO_3$	3.302 3	1.160 0_5	415 to 491	**126**, 34
Cs_2SO_4	2.956	0.586	1027 to 1477	**34**
Cu_2Cl_2	4.010	0.79	m.p. to 585	**47**
Cu_2S	6.76	0.75	1 150 to 1 400	**119**
FeS	3.85	0	1 250 to 1 450	**119**
$GaBr_2$	3.753	1.69	160 to 175	**127**
$GaBr_3$	3.507	2.95	m.p. to 230	**70**
$GaCl_2$	2.652	1.36	166 to 177	**127**
$GaCl_3$	2.223	2.05	m.p. to 195	**116, 70**, 9
Ga_2I_4	4.380	1.688	181 to 265	**128**
Ga_2I_6	4.128	2.377	185 to 255	**128**, 70
$HfCl_4$		$d_{435}=1.71$		**129**
$HgBr_2$	5.888 9	3.233 1	238 to 319	**130**, 26
$HgCl_2$	5.157 7	2.862 4	277 to 304	**130**, 26
Hg_2Cl_2	8.00	4.0	m.p. to 580	**47**
HgI_2	6.060 3	3.235 1	259 to 354	**130**, 26, 107
ICl	3.186	3.0	m.p. to 100	**8**, 5
$InBr_3$	3.674	1.5	450 to 530	**48**
InCl	4.055	1.4	269 to 365	**48**
$InCl_2$	3.43	1.6	268 to 437	**48**
$InCl_3$	3.37	2.1	597 to 666	**48**
InI_3	4.135	1.5	230 to 360	**48**
KBr	2.733	0.825 3	747 to 927	**113, 102, 140**
KCl	1.976 7	0.583 1	777 to 947	**146, 140, 102, 110**, 104, 98
K_2CO_3	2.293 4	0.442	907 to 1007	**141**
$K_2Cr_2O_7$	2.563 3	0.695	420 to 535	**142**
KF	2.468 5	0.651 5	881 to 1 037	**122**, 34
$KHSO_4$	2.232	0.767	207 to 230	**94**
KI	3.098 5	0.955 7	682 to 904	**110, 102**
K_2MoO_4	2.992 2	0.549 1	935 to 988	**134**, 34
KNO_2	2.005	0.700	440 to 500	**106**, 124
KNO_3	2.116	0.729	m.p. to 600	**106, 102, 77, 55, 34, 21, 18**, 124, 126
KOH	1.893	0.44	400 to 600	**45**, 37
KPO_3	2.455	0.43	990 to 1 200	**34**
K_2SO_4	2.472	0.545	1 100 to 1 300	**34**
K_2WO_4	3.844 7	0.727 3	944 to 1 053	**135**, 34
$LaBr_3$	5.008 9	0.096 0	796 to 912	**122**
$LaCl_3$	3.877 3	0.777 4	873 to 973	**122**, 47
LaF_3	5.607	0.682	1 477 to 2 177	**123**
LiBr	2.888	0.652	m.p. to 740	**113, 16**, 60
LiCl	1.766 0	0.432 8	627 to 777	**110, 102, 98, 34, 16**
$LiClO_4$	2.171 2	0.622 3	—	**136**
Li_2CO_3	2.10	0.373	737 to 847	**141**

Table 9.1 DENSITY OF PURE MOLTEN SALTS—*continued*

Substance	*a*	*b*	*Range of observations** °C	*References*
LiF	2.224 3	0.490 2	876 to 1 047	**122**, 34
LiI	3.540	0.918	m.p. to 670	**113**
Li_2MoO_4	3.200 8	0.415 2	781 to 963	**134**
$LiNO_3$	1.924	0.548	m.p. to 550	**34, 18**
Li_2SO_4	2.352 9	0.407	860 to 1 214	**143**, 34
Li_2WO_4	4.905 5	0.805 3	764 to 901	**135**
$MgCl_2$	1.894	0.302	727 to 967	**144**, 47, 83, 133
MgF_2	3.092	0.524	1 377 to 1 827	**123**
$MnCl_2$	2.639	0.44	650 to 850	**137**
MoF_6	2.637	4.91	m.p. to 34	**64**
MoO_3	4.443 6	1.498 3	821 to 918	**134**, 109
Na_3AlF_6	3.036	0.94	m.p. to 1 100	*See* Binary System Na_3AlF_6–NaF
NaBr	2.952	0.817	m.p. to 945	**113, 34, 16**
NaCl	1.991 1	0.543	803 to 1 030	**102, 147, 140, 146, 110,** 98
Na_2CO_3	2.357	0.449	867 to 1 007	**141**
NaF	2.502	0.560	997 to 1 057	**34**, 45
NaI	3.368	0.949	m.p. to 915	**113, 102, 34**
Na_2MoO_4	3.235	0.629	700 to 1 400	**35**, 109
$NaNO_2$	2.004	0.690	280 to 500	**106, 102**
$NaNO_3$	2.124 8	0.715	317 to 427	**106, 100, 102, 77, 34, 21, 18**
NaOH	1.937 4	0.478 4	320 to 450	**145**
$NaPO_3$	2.545	0.44	905 to 1 010	**34, 19**
Na_2SO_4	2.495	0.48	900 to 1 050	**34, 16**
Na_2WO_4	4.519 9	0.906 7	714 to 880	**135**, 35, 16
$NdBr_3$	4.762 6	0.777 9	695 to 860	**122**
NH_4NO_3	1.536	0.60	170 to 200	**118**
$Ni(CO)_4$	$d_t = 1.356\,1 - 2.213 \cdot 10^{-3}t - 4.10^{-6}t^2$		0 to 36	**13**
$OsSO_4$	4.504	4.17	43 to 150	**62**
PBr_3	2.924	2.48	0 to 200	**8, 1**
PCl_3	1.612	1.86	−80 to 75	**87, 29, 8**, 39, 1
$PbBr_2$	5.036	1.45	377 to 497	**91, 21**
$PbCl_2$	5.702	1.5	502 to 710	**91**, 85, 61, 60, 21
$PbCl_4$		$d_0 = 3.18$		**50**
PbI_2		$d_{383} = 5.625$		**10**
$PbMoO_4$		$d_{1107} = 5.213$		**125**
RbBr	3.446 4	1.072	700 to 910	**113, 34**
RbCl	2.880	0.883	m.p. to 925	**113, 47**, 34
RbF	3.707	1.011	820 to 1 005	**34**
RbI	3.638	1.14	655 to 905	**113, 34**
$RbNO_3$	2.782	0.97	350 to 550	**34**
Rb_2SO_4	3.260	0.665	1 100 to 1 310	**34**
ReF_6	3.776	8.51	m.p. to 47.6	**79, 76**
Re_2O_7		$d_{331} = 4.30$		**67**
ReO_3Cl	3.94	4.0	16 to 37	**67**
$ReOF_4$	3.921	5.1	40 to 60	**79**
S_2Cl_2	1.710	1.57	0 to 136	**87, 8**, 2
$SOCl_2$	1.677	1.97	0 to 69	**8**
SO_2Cl_2	1.708	2.11	0 to 70	**8**
$SbBr_3$		$d = 3.691$ at m.p.		**81, 31**, 27, 6, 3
$SbCl_3$†	2.849	2.268	m.p. to 165	**84**, 54, 41, 20, 3
$SbCl_5$	2.392	2.04	m.p. to 80	**65, 54**, 32, 17

Table 9.1 DENSITY OF PURE MOLTEN SALTS—*continued*

Substance	*a*	*b*	*Ranges of observations** °C	*References*
SbF_5		$d_{22.7}=2.993$		**17**
$ScCl_3$		$d_{1000}=1.63$		**47**
Se_2Cl_2		$d_{25}=2.774$		**51**
SeF_4		$d=2.8$ at room temp.		**56**
SeF_6		$d=2.3$ at m.p.		**69**
$SeOBr_2$		$d_{50}=3.38$		**36**
$SeOCl_2$	2.478	2.08	m.p. to 80	**63, 57,** 43
$SeOF_2$		$d=2.7$ at room temp.		**56**
$SiBr_4$	2.812	2.63	—	**8**
$SiCl_4$	1.523	2.08	−30 to 60	**87, 8,** 1
$SnCl_2$	3.6739	1.253	247 to 407	**47, 34**
$SnCl_4$	2.273	2.62	−19 to 113	**8, 1**
SnI_4	4.145	2.45	145 to 275	**53**
$SrCl_2$	3.2318	0.5781	893 to 1037	**122,** 19
SrF_2	4.579	0.751	1477 to 1927	**123**
$TeCl_4$	2.965	1.64	230 to 430	**59**
TeF_6	2.442	6.02	m.p. to −10	**69**
$ThCl_4$		$d=3.32$ at about 830		**47**
$TiBr_4$	3.043	2.25	40 to 120	**115**
$TiCl_4$	1.761	1.72	m.p. to 135	**73, 71, 8, 1,** 90, 58
TiI_4	3.755	2.19	166 to 270	**70**
TlBr	6.908_4	1.922_0	493 to 750	**131**
TlCl	6.402	1.80	m.p. to 640	**47**
$TlNO_3$	5.2672	1.75	210 to 430	**34**
UCl_4	2.50	1.7	—	**138**
V_2O_5		$d_{1000}=2.5$		**139**
$VOCl_3$	1.865	1.83	0 to 125	**25, 11, 8,** 4
WF_6	3.529	5.84	m.p. to 19	**64**
YCl_2	2.87	0.5	m.p. to 845	**47**
$ZnBr_2$	3.8512	0.959	397 to 627	**144, 72**
$ZnCl_2$	2.693	0.515	m.p. to 630	**117, 47,** 61
$ZrCl_4$		$d_{448}=1.54$		**129**

* Melting points will be found in Table 9.3.

† Between m.p. and 375 °C density of $SbCl_3$ given by:
$d_t=2.622-2.268\cdot10^{-8}\ (t-100)-0.32\cdot10^{-8}\ (t-100)^3$.
Between 375 °C and 505 °C:
$d_t=2.622-2.268\cdot10^{-3}\ (t-100)-0.32\cdot10^{-8}\ (t-100)^3+8.8\cdot10^{-12}(t-100)^4-3.4\cdot10^{-14}\ (t-100)^5$.

REFERENCES TO TABLE 9.1

1. I. Pierre, *Ann. Chim.* (*Phys.*), 1845, **15**, 325.
2. H. Kopp, *Ann. Chem.*, 1855, **95**, 307.
3. H. Kopp, *ibid.*, 1855, **95**, 350.
4. Roscoe, *Phil. Trans. R. Soc.*, 1868, **158**, 1.
5. Hannay, *J. chem. Soc.*, 1873, **26**, 815.
6. R. W. E. MacIvor, *Chem. News*, 1874, **29**, 179.
7. Ditte, *Compt. rend.*, 1877, **85**, 1069.
8. Nat. Res. Council, USA, 'International Critical Tables', vol. 3, p. 23, 1928.
9. Lecoq de Boisbaudran, *Chem. News*, 1881, **44**, 166; *Compt. rend.*, 1881, **93**, 294, 329, 815.
10. Rodwell, *Phil. Trans. R. Soc.*, 1882, **173**, 1125.
11. L'Hôte, *Compt. rend.*, 1885, **101**, 1151.
12. H. Moissan, *ibid.*, 1891, **112**, 718.
13. Mond and Nasini, *Z. phys. Chem.*, 1891, **8**, 150.
14. J. W. Retgers, *ibid.*, 1893, **11**, 328.
15. Ghira, *ibid.*, 1893, **12**, 765.
16. E. Brunner, *Z. anorg. Chem.*, 1904, **38**, 350.

17. O. Ruff and W. Plato, *Ber. dt. chem. Ges.*, 1904, **37**, 679.
18. H. M. Goodwin and R. D. Mailey, *Phys. Rev.*, 1907, **25**, 469.
19. K. Arndt and A. Gessler, *Z. Elektrochem.*, 1908 **14**, 665.
20. Z. Klemensiewicz, *Bull. int. Acad., Cracovie*, 1908, 487.
21. R. Lorenz, H. Frei and A. Jabs, *Z. phys., Chem.*, 1908, **61**, 468.
22. K. Arndt and W. Löwenstein, *Z. Elektrochem.*, 1909, **15**, 789.
23. A. H. W. Aten, *Z. phys. Chem.*, 1909, **66**, 641.
24. A. H. W. Aten, *ibid.*, 1910, **73**, 578.
25. Prandtl and Bleyer, *Z. anorg. Chem.*, 1910, **66**, 152.
26. E. B. R. Prideaux, *J. chem. Soc.*, 1910, **97**, 2032.
27. Izbekov and Plotnikov, *Z. anorg. Chem.*, 1911, **71**, 328.
28. K. Arndt and Kunze, *Z. Elektrochem.*, 1912, **18**, 994.
29. Körber, *Ann. Physik.*, 1912, **37**, 1014.
30. Sackur, *Z. phys. Chem.*, 1913., 1913, **83**, 297.
31. N. S. Kurnakov, Krotkov and Oksmann, *J. Russ. Phys.-Chem. Soc.*, 1915, **47**, 558.
32. E. Moles, *Z. phys. Chem.*, 1915, **90**, 74.
33. R. Lorenz and A. Höchberg, *Z. anorg. Chem.*, 1916, **94**, 288.
34. F. M. Jaeger, *ibid.*, 1917, **101**, 16.
35. F. M. Jaeger and B. Kapma, *ibid.*, 1920, **113**, 27.
36. V. Lenher, *J. Am. chem. Soc.*, 1922, **44**, 1668.
37. Meyer and Heck, *Z. phys. Chem.*, 1922, **100**, 316.
38. W. Blitz and A. Voigt, *Z. anorg. Chem.*, 1923, **126**, 39.
39. Timmermans, *Bull. Soc. Chim. Belg.*, 1923, **32**, 299.
40. A. Voigt and W. Blitz, *Z. anorg. Chem.*, 1924, **133**, 277.
41. N. S. Kurnakov, *ibid.*, 1924, **135**, 86.
42. Pascal and Allendorff, *ibid.*, 1924, **135**, 327.
43. C. W. Muehlberger and V. Lenher, *J. Am. chem. Soc.*, 1925, **47**, 1843.
44. Samsoen, *Compt. rend.*, 1925, **181**, 354.
45. K. Arndt, *Z. phys. Chem.*, 1926, **121**, 448.
46. H. V. A. Briscoe, P. L. Robinson and Stephenson, *J. chem. Soc.*, 1926, 39.
47. W. Klemm, *Z. anorg. Chem.*, 1926, **152**, 235.
48. W. Klemm, *ibid.*, 1926, **152**, 252.
49. W. Blitz and W. Klemm, *ibid.*, 1926, **152**, 267.
50. Friedrich, quoted by W. Blitz and W. Klemm, *ibid.*, 1926, **152**, 267.
51. V. Lenher and C. H. Kao, *J. Am. chem. Soc.*, 1926, **48**, 1550.
52. H. V. A. Briscoe, P. L. Robinson and H. C. Smith, *J. chem. Soc.*, 1927, 282.
53. Dortmann and Hildebrand, *J. Am. chem. Soc.*, 1927, **49**, 737.
54. S. Sugden and A. Freiman, *J. chem. soc.* 1927, 1185.
55. Dantuma, *Z. anorg. Chem.*, 1928, **175**, 33.
56. E. B. R. Prideaux and C. B. Cox, *J. chem. Soc.*, 1928, 740, 1606.
57. W. J. R. Henley and S. Sugden, *ibid.*, 1929, 1064.
58. F. B. Garner and S. Sugden, *ibid.*, 1929, 1298.
59. J. H. Simons, *J. Am. chem. Soc.*, 1930, **52**, 3491.
60. E. Salstrom, *ibid.*, 1930, **52**, 4647.
61. A. Wachter and J. Hildebrand, *ibid.*, 1930, **52**, 4656.
62. E. Ogawa, *Bull. Chem. Soc. Japan*, 1931, **6**, 315.
63. T. W. Parker and P. L. Robinson, *J. chem. Soc.*, 1931, 1316.
64. O. Ruff and A. Ascher, *Z. anorg. Chem.*, 1931, **196**, 417.
65. J. H. Simons and G. Jessop, *J. Am. chem. Soc.*, 1931, **53**, 1265.
66. E. van Aubel, *Bull. Acad. Belg.*, 1932 (5), **18**, 692.
67. H. V. A. Briscoe, P. L. Robinson and A. J. Rudge, *J. chem. Soc.*, 1932, 2675.
68. O. Ruff, A. Brader, O. Bretschneider, W. Menzel and H. Plaut, *Z. anorg. Chem.*, 1932, **206**, 59.
69. W. Klemm and P. Henkel, *ibid.*, 1932, **207**, 73.
70. W. Klemm and W. Tilk, *ibid.*, 1932, **207**, 161.
71. H. Ulich, E. Hertel and W. Nespital, *Z. phys. Chem.*, 1932, **B17**, 369.
72. E. Salstrom, *J. Am. chem. Soc.*, 1933, **55**, 1031.
73. T. Sugawa, *Sci. Rep. Tôhoku Univ.*, 1933, **22**, 959.
74. British Aluminium Co. Ltd., 1934, private commun.
75. S. Karpachev, A. Stromberg and O. Poltoratzkaya, *J. phys. Chem.* (*USSR*), 1934, **5**, 793.
76. W. Kwasnik, Diss, Breslau, T. H. 1934, p. 13.
77. K. Laybourne and W. Madgin, *J. chem. Soc.*, 1934, 1.
78. M. G. Malone and A. L. Ferguson, *J. chem. Physics*, 1934, **2**, 99.
79. O. Ruff and W. Kwasnik, *Z. anorg. Chem.*, 1934, **219**, 65.
80. P. Drossbach, 'Electrochemistry of Fused Salts', Berlin, 1938.
81. N. S. Kurnakov, N. K. Voskresenskaja and G. D. Gurovic, *Bull Acad. Sci., U.R.S.S., ser chim.*, 1938, 396.
82. Lundina, quoted by V. P. Mashovets, 'The Electro-Metallurgy of Aluminium', Russia, 1938.
83. W. Treadwell *et al.*, *Helv. Chim. Acta*, 1939, **22**, 445.
84. D. I. Zuravlev, *J. Phys. Chem.* (*USSR*), 1939, **13**, 684.
85. V. P. Barzakovskii, *Bull. Acad. Sci., U.R.S.S.* (Cl. Sci. chim.), 1940, 825.
86. V. P. Barzakovskii, *J. appl. Chem.* (*USSR*), 1940, **13**, 1117.

87. S. T. Bowden and A. R. Morgan, *Phil. Mag.*, 1940, **29**, 367.
88. E. Ya Gorenbein, *J. Gen. Chem. (USSR)*, 1947, **17**, 873.
89. E. Ya. Gorenbein, *ibid.*, 1948, **18**, 1427.
90. R. de Malleman and F. Suhner, *Compt. rend.*, 1948, **227**, 546.
91. N. K. Boardman, F. H. Dorman and E. Heymann, *J. phys. Chem.*, 1949, **53**, 375.
92. G. Fuseya and K. Ouchi, *J. Electrochem. Soc. Japan*, 1949, **17**, 254.
93. I. M. Bokhovkin, *J. Gen. Chem. (USSR)*, 1950, **20**, 397.
94. S. E. Rogers and A. R. Ubbelohde, *Trans. Faraday Soc.*, 1950, **46**, 1051.
95. A. Vayna, *Alluminio*, 1950, **19**, 541.
96. E. Ya. Gorenbein and E. E. Kriss, *J. phys. Chem. (USSR)*, 1951, **25**, 791.
97. R. C. Spooner and F. E. W. Wetmore, *Canad. J. Chem.*, 1951, **29**, 777
98. J. D. Edwards, C. S. Taylor, A. S. Russell and L. F. Maranville, *J. Electrochem. Soc.*, 1952, **99**, 527.
99. R. W. Huber, E. V. Potter and H. W. St. Clair, *Rep. Invest. US Bur. Mines No. 4858*, 1952.
100. J. Byrne, H. Fleming and F. E. W. Wetmore, *Canad. J. Chem.*, 1952, **30**, 922.
101. E. F. Riebling, *J. Am. Ceram. Soc.*, 1964, **47**, 478.
102. H. Bloom. I. W. Knaggs, J. J. Molloy and D. Welch, *Trans. Faraday Soc.*, 1953, **49**, 1458.
103. I. P. Vereshchetina and N. P. Luzhnaya, *Izvest. Sekt. Fiziko-Khim. Anal.*, 1954, **25**, 188.
104. J. S. Peake and M. R. Bothwell, *J. Am. chem. Soc.*, 1954, **76**, 2653.
105. V. D. Polyakov, *Izvest, Sekt. Fiziko-Khim. Anal.*, 1955, **26**, 147.
106. V. D. Polyakov and S. I. Berul, *ibid.*, 1955, **26**, 164.
107. V. D. Polyakov, *ibid.*, 1955, **26**, 191.
108. N. P. Popovskaya and P. I. Protsenko, *Zh. fiz. Khim.*, 1955, **29**, 225.
109. K. B. Morris, M. I. Cook, C. Z. Sykes and M. B. Templeman, *J. Am. chem. Soc.*, 1955, **77**, 851.
110. E. R. van Artsdalen and I. S. Yaffe, *J. phys. Chem.*, 1955, **59**, 118.
111. T. Baak, *Acta. Chem. Scand.*, 1955, **9**, 1406.
112. N. P. Luzhnaya, N. N. Evseeva and I. P. Vereshchetina; *Zh. neorg. Khim.*, 1956, **1**, 1490.
113. I. S. Yaffe and E. R. van Artsdalen, *J. phys. Chem.*, 1956, **60**, 1125.
114. J. D. Mackenzie, *Trans. Faraday Soc.*, 1956, **52**, 1564.
115. J. M. Blocher, R. F. Rolsten and I. E. Campbell, *J. electrochem. Soc.*, 1957, **104**, 553.
116. N. N. Greenwood and K. Wade, *J. Inorg. nuclear Chem.*, 1957, **3**, 349.
117. F. R. Duke and R. A. Fleming, *J. electrochem. Soc.*, 1957, **104**, 251.
118. S. Toshiaki and T. Ishibashi, *Sci. Papers Coll. Gen. Educ., Univ. Tokyo*, 1957, **7**, 53.
119. M. Bourgon, G. Derge and C. M. Pound, *Trans. Amer. Inst. Min. Met. Eng.*, 1958, **212**, 338.
120. F. J. Keneshea and D. Cubicciotti, *J. phys. Chem.*, 1958, **62**, 843.
121. A. D. Kirshenbaum and J. A. Cahill, *J. inorg. nucl. Chem.*, 1960, **14**, 283.
122. I. S. Yaffe and E. R. van Artsdalen, *Chem. Semi-Ann. Progr. Rep. No. 2159, Oak Ridge Natl Lab.*, 1956 p. 77.
123. A. D. Kirshenbaum, J. A. Cahill and C. S. Stokes, *J. inorg. nucl. Chem.*, 1960, **15**, 297.
124. P. I. Protsenko and A. Ya. Malakhova, *Zh. neorg. Khim.*, 1961, **6**, 1662.
125. K. B. Morris, M. McNair and G. Koops, *J. chem. Engng Data*, 1962, **7**, 224.
126 N. V. Smith and E. R. van Artsdalen, *Chem. Semi-ann. Progr. Rep. No. 2159 , Oak Ridge Natl Lab.*, 1956, p. 80.
127. N. N. Greenwood and I. J. Worrall, *J. chem. Soc.*, 1958, 1680.
128. E. F. Riebling and C. E. Erickson, *J. phys. Chem.*, 1963, **67**, 307.
129. L. A. Nisel'son, *Zh. neorg. Khim.*, 1961, **6**, 1242.
130. G. J. Janz and J. D. E. McIntyre, *J. electrochem. Soc.*, 1962, **109**, 842.
131. E. R. Buckle, P. E. Tsaoussoglou and A. R. Ubbelohde, *Trans. Faraday Soc.*, 1964, **60**, 684.
132. A. D. Kirshenbaum, J. A. Cahill, P. J. McGonigal and A. V. Grosse, *J. inorg. nucl. Chem.*, 1962, **24**, 1287.
133. J. N. Reding, *J. chem. Engng Data*, 1965, **10**, 1.
134. K. B. Morris and P. L. Robinson, *J. phys. Chem.*, 1964, **68**, 1194.
135. K. B. Morris and P. L. Robinson, *J. chem. Engng Data*, 1964, **9**, 444.
136. J. Padova and J. Soriano, *ibid.*, 1964, **9**, 510
137. I. G. Murgulescu and S. Zuca, *Acad. rep. pop. Romaine, Studii Cerc. Chim.*, 1959, **7**, 325.
138. T. Kuroda and T. Suzuki, *J. electrochem. Soc. Japan*, 1961, **29**, E215.
139. B. M. Lepinskikh, O. A. Esin and G. A. Teterin, *Zh. neorg. Khim.*, 1960, **5**, 642.
140. H. Schinke and F. Saverwald, *Z. anorg. allgem. Chem.*, 1956, **287**, 313.
141. G. J. Janz and M. R. Lorenz, *J. electrochem. Soc.*, 1961, **108**, 1052.
142. J. P. Frame, E. Rhodes and A. R. Ubbelohde, *Trans, Faraday Soc.*, 1959, **55**, 2039.
143. A. Kvist and A. Lunden, *Z. Naturforsch.*, 1965, **20a**, 235.
144. J. O'M. Bockris, A. Pilla and J. L. Barton, *Rev. Chim. Acad. Rep. Populaire. Roumaine*, 1962, **7**, 59.
145. V. D. Polyakov, *Izv. Sektova. Fiz. Khim. Analiza Inst. Obshch. Neorgan, Khim. Acad. Nauk SSSR*, 1955, **26**, 173, 191.
146. E. Vogel, H. Schinke and F. Saverwald, *Z. anorg. allgem. Chem.*, 1956, **284**, 131.
147. J. O'M. Bockris, A. Pilla and J. L. Barton, *J. phys. Chem.*, 1960, **64**, 507.

Table 9.2. DENSITIES OF MOLTEN BINARY SALT SYSTEMS AND OTHER MIXED IONIC MELTS
The density ($g\,cm^{-3}$) at temperature t(°C) and composition p(wt.%) of the first-named constituent is given as d_t, or the constants a and b in the equation $d_t = a - 10^{-3}\,bt$ or A, B and C in the equation $d_t = 10^6 At^{-2} - 10^3 Bt^{-1} + C$ are given together with the temperature range r(°C). Principal references are in **bold type.**

System								
AgBr–AgCl	*p*	0	27.8	46.9	71.6	100		
Ref. **25**	*a*	5.262	5.523	5.678	5.832	6.025		
	b	0.94	1.08	1.12	1.07	1.04		
	r	480–630	440–580	420–590	420–580	440–600		
AgBr–KBr	*p*	0	50.8	70.4	85.7	100	64.6	
Ref. **64, 25, 7**	*a*	2.706	3.686	4.376	5.156	6.023	4.077	
	b	0.80	0.98	1.03	1.12	1.05	1.03	
	r	750–800	593–700	380–600	380–600	440–600	546–629	
AgBr–LiBr	*p*	68.5						
Ref. **65**	*a*	4.504						
	b	0.877						
	r	517–555						
AgBr–NaBr	*p*	64.6						
Ref. **66**	*a*	4.311						
	b	0.9						
	r	607–619						
AgBr–RbBr	*p*	53.2						
Ref. **67**	*a*	4.470						
	b	1.23						
	r	514–624						
AgCl–$AgNO_3$	*p*	0	9.00	12.5	15.5	20.0	30.0	40.0
Ref. **33**	*a*	4.167	4.242	4.279	4.297	4.338	4.431	4.497
	b	1.00	1.00	1.00	1.00	1.00	1.00	0.90
	r				310–330			
AgCl–KCl	*p*	0	63.8	80.3	88.9	100		
Ref. **25, 7**	*a*	1.988	3.286	4.001	4.463	5.263		
	b	0.60	0.88	0.96	0.95	0.94		
	r	785–880	560–745	385–640	433–670	480–630		
AgCl–$PbCl_2$	*p*	0	11.0	19.4	24.7	31.0	41.0	66.9
Ref. **25**	*a*	5.702	5.660	5.634	5.582	5.543	5.520	5.387
	b	1.50	1.45	1.42	1.34	1.28	1.26	1.08
	r	516–710	520–700	470–680	445–670	444–680	380–700	478–660
AgI–$AgNO_3$	*p*	20	30	40	50	60	70	
Ref. **54, 30**	*a*	4.53	4.97	5.12	5.31	5.44	5.53	
	b	1.2	2.2	2.1	2.1	1.8	1.5	
	r			about 150–300				
$AgNO_3$–$Cd(NO_3)_2$	*p*	32.4	37.0	46.8	57.2	68.4	74.2	
Ref. **50**	*a*	1.721	1.762	1.843	1.930	2.003	2.032	
	b	1.01	1.05	1.10	1.15	1.15	1.14	
	r	210–290	160–290	160–290	190–290	190–290	210–290	
$AgNO_3$–HgI_2	*p*	13.8	19.9	27.2	35.9	46.6	52.8	67.9
Ref. **49**	*a*	5.826	5.732	5.584	5.364	5.188	5.074	4.930
	b	1.58	1.46	1.46	1.46	1.40	1.34	1.28
	r	160–240	160–240	100–240	120–240	100–240	100–240	120–240
$AgNO_3$–KNO_3	*p*	15.7	29.6	41.9	52.8	62.7	79.7	93.8
Ref. **57, 46**	*a*	2.284	2.534	2.654	2.856	2.986	3.516	3.765
	b	0.76	0.97	0.84	0.89	0.76	1.08	0.85
	r	350–400	300–400	250–400	250–400	250–400	170–350	200–400
$AgNO_3$–$NaNO_3$	*p*	0	10	20	30	40	60	80
Ref. **37**	*a*	2.124	2.226	2.357	2.486	2.616	2.974	3.439
	b	0.70	0.71	0.77	0.79	0.79	0.70	0.90
	r				about 290–370			

Table 9.2 DENSITIES OF MOLTEN BINARY SALT SYSTEMS AND OTHER MIXED IONIC MELTS—*continued*

System								
$AgNO_3$–NH_4NO_3	p	0	5	10				
Ref. **58**	d_{170}	1.432	1.479	1.529				
	d_{180}	1.426	1.474	1.523				
$AgNO_3$–$TlNO_3$	p	25	45	50	60	75	90	100
Ref. **26**	d_{100}	—	4.671	4.630	4.554	—	—	—
	d_{150}	—	4.575	4.526	4.435	—	—	—
	d_{200}	4.638	4.452	4.410	4.319	4.183	4.074	
	d_{225}	4.579	4.406	4.351	4.261	4.132	4.024	3.922
$AlBr_3$–$HgBr_2$	p	59.7	65.4	71.2	74.9	80.5	89.6	100
Ref. **21, 4**	d_{110}	3.577	3.415	3.276	3.173	3.030	2.827	2.624
	d_{140}	3.504	3.359	3.202	3.097	2.961	2.755	2.555
$AlBr_3$–KBr	p	81.8	83.2	84.7	86.2	87.5	88.0	
Ref. **28, 4**	d_{110}	2.818	2.815	2.810	2.804	2.798	2.790	
	d_{140}	2.775	2.771	2.764	2.755	2.749	2.741	
$2AlBr_3$–KCl	a	2.846						
Ref. **23**	b	1.445						
	r	80–170						
$AlBr_3$–$NaBr$	p	83.8	84.7	85.6	86.8	87.9	88.9	
Ref. **28**	d_{120}	2.827	2.820	2.817	2.809	2.798	2.786	
	d_{140}	2.797	2.792	2.787	(2.777)	2.767	2.756	
$2AlBr_3.NaBr$	a	3.005						
Ref. **23**	b	1.5						
	r	110–170						
$AlBr_3$–NH_4Br	p	84.5	87.9	89.0	89.4	90.8	100	
Ref. **21**	a	2.842	2.865	2.877	2.881	2.889	2.877	
	b	1.30	1.40	1.45	1.50	1.60	2.30	
	r				110–150			
$2AlBr_3.NH_4Br$	a	2.848						
Ref. **23**	b	1.36						
	r	110–160						
$AlBr_3$–$SbBr_3$	p	0	34.1	42.5	67.4	71.9	91.3	100
Ref. **20**, 4	d_{100}	3.697	3.402	3.318	3.034	2.971	2.737	2.644
	d_{140}	3.594	3.311	3.224	2.941	2.881	2.641	2.556
$AlBr_3.SbBr_3$	a	3.541						
Ref. **32, 23**	b	2.3						
	r	80–170						
$AlBr_3.SbBr_3$–	p	0	16.3	59.9	67.9	77.3	86.8	100
$AsBr_3$	d_{85}	3.232	3.231	3.286	3.304	3.310	3.330	3.346
Ref. **32**	d_{100}	3.193	3.191	3.247	3.263	3.274	3.295	3.313
$AlBr_3$–$ZnBr_2$	p	70.5	74.5	77.7	82.5	85.0	88.4	100
Ref. **20**	d_{100}	3.014	(2.957)	2.914	2.849	2.811	2.768	2.644
	d_{150}	2.915	2.850	2.811	2.740	2.704	2.657	2.534
$2AlBr_3.ZnBr_2$	t	100	140	180				
Ref. **23**	d	3.01	2.93	2.81				
$AlCl_3$–KCl**	p	50.4	53.7	62.5	64.2	71.2	76.1	84.2
Refs. **63, 45, 29,**	a	1.787	1.785	1.755	1.937	1.859	1.819	1.820
19	b	0.590	0.605	0.600	0.935	0.902	0.789	0.850
	r	600–800	600–800	600–800	260–350	240–280	190–270	175–225
$AlCl_3$–$LiCl$**	p	75.9	79.3	82.5	85.4	87.9	90.4	
Ref. **63, 19**	a	1.735	1.737	1.757	1.759	1.768	1.741	
	b	0.77	0.68	0.80	0.84	0.90	0.80	
	r	180–330	175–225	175–225	175–225	175–225	175–225	

Table 9.2 DENSITIES OF MOLTEN BINARY SALT SYSTEMS AND OTHER MIXED IONIC MELTS—*continued*

System								
$AlCl_3$–NaBr	*p*	58.6	61.0	65.7	70.3	74.8	79.3	
Ref. **63**	*a*	2.158	2.119	2.064	2.017	1.979	1.944	
	b	0.92	0.86	0.92	0.96	1.02	1.12	
	r		175–225				200–275	
$AlCl_3$–NaCl**	*p*	69.5	71.9	73.5	75.8	79.5	84.1	90.1
Ref. **63, 45, 16,** 19	*a*	1.848	1.858	1.839	1.829	1.792	1.787	1.810
	b	0.812	0.910	0.849	0.844	0.715	0.840	1.04
	r	220–280	160–210	150–210	150–210	150–210	175–225	175–225
$AlCl_3.NH_4Cl$	*t*	284	293	311	315	324	354	
Ref. **19**	*d*	1.475	1.470	1.445	1.440	1.425	1.420	
$AlCl_3$–RbCl	*p*	71.9	76.7					
Ref. **63**	*a*	1.992	1.975					
	b	0.96	1.00					
	r	175–225						
AlF_3–Na_3AlF_6*	*p*	0	5	10	15	20	25	30
Ref. **62, 38, 31, 24,**	d_{1000}	2.096	2.078	2.048	2.015	1.977	1.930	1.873
22, 15, 14, 13,	d_{1100}	2.002	1.987	1.965	1.935	1.894	1.839	1.775
10, 9, 6								
Al_2O_3–Na_3AlF_6	*p*	0	2.5	5	7.5	10	12.5	15
Ref. **62, 38, 14,**	d_{1000}	2.096	2.076	2.060	2.048	2.039	2.033	2.028
31, 6	d_{1100}	2.002	1.985	1.974	1.966	1.960	1.957	1.954
B_2O_3–BaO	*p*	20.7	30.0	39.8	49.2	59.7	67.9	
Ref. **44**	*a*	4.038	5.006	4.780	4.422	3.704	3.392	
	b	—	1.32	1.40	1.34	0.96	0.96	
	r	1 120	1 000–1 200	1 000–1 100	850–1 100	850–1 000	850–1 100	
B_2O_3–CaO	*p*	55.3	59.8	62.8	67.9	70.2	73.3	
Ref. **44**	*a*	2.926	2.901	2.876	2.871	2.932	2.844	
	b	0.44	0.45	0.46	0.51	0.59	0.56	
	r	1 160–1 200	1 110–1 210	1 140–1 190	1 100–1 200	1 060–1 100	900–1 200	
B_2O_3–K_2O	*p*	51.5	55.7	65.6	75.3	84.9	94.8	98.5
Ref. **40**	*a*	2.690	2.534	2.412	2.185	2.057	1.855	1.737
	b	0.905	0.687	0.500	0.310	0.295	0.255	0.217
	r	900–1 000	800–1 000	700–1 000	800–1 000	700–1 000	600–1 000	500–1 000
B_2O_3–Li_2O	*p*	71.3	85.2	89.4	93.5	97.2	100.0	
Ref. **40**	*a*	2.340	2.400	2.281	2.103	1.901	1.662	
	b	0.467	0.467	0.402	0.335	0.260	0.153	
	r	800–1 000	800–1 000	700–1 000	600–1 000	600–1 000	600–1 200	
B_2O_3–Na_2O	*p*	64.0	69.2	77.6	85.8	94.4	99.1	100.0
Ref. **40**	*a*	2.540	2.778	2.413	2.137	1.898	1.759	1.662
	b	0.57	0.86	0.435	0.278	0.235	0.222	0.153
	r	900–1 000	700–800	700–1 000	700–1 000	600–1 100	600–1 000	600–1 200
B_2O_3–SrO	*p*	44.9	49.6	56.8	61.6	65.2	70.0	
Ref. **44**	*a*	3.472	3.513	3.402	3.230	3.124	2.948	
	b	0.43	0.56	0.61	0.57	0.57	0.51	
	r	1 150–1 200	1 120–1 220	960–1 100	950–1 100	950–1 100	920–1 120	
$BaBr_2$–KBr	*p*	39.2	45.1	55.5	62.5	73.0	82.2	88.2
Ref. **41**	*a*	3.318	3.324	3.558	3.720	3.932	4.117	4.425
	b	0.937	0.849	0.896	0.930	0.906	0.875	1.02
	r	660–850	640–850	630–850	630–850	630–850	690–850	750–850
$BaCl_2$–$CdCl_2$	*p*	0	18.99	39.00	57.35	100		
Ref. **25**, 1	*a*	3.870	3.996	4.018	4.069	3.672		
	b	0.84	0.93	0.93	0.96	0.52		
	r	582–725	597–700	580–700	600–690	above 1 000		

Table 9.2 DENSITIES OF MOLTEN BINARY SALT SYSTEMS AND OTHER MIXED IONIC MELTS—*continued*

$BaCl_2$–KCl	*p*	20.4	29.6	47.4	58.2	70.1	82.8	91.4
Ref. **43**	*a*	2.199	2.237	2.554	2.744	3.035	3.268	3.456
	b	0.64	0.59	0.63	0.70	0.77	0.65	0.68
	r	790–900	790–890	790–890	800–890	790–880	820–910	880–940
$BaCl_2$–$MgCl_2$	*p*	9.95	25.0	50.2	75.1	90.0		
Ref. **69**, 35, 70	*a*	2.073	2.333	2.835	3.400	(3.642)		
	b	0.36	0.47	0.64	0.75	0.71		
	r			800–900				
$BaCl_2$–NaCl	*p*	44.0	54.4	62.7	69.7	75.3	80.1	84.4
Ref. **71**, 17	d_{725}	—	2.204	2.350	2.491	2.620	—	—
	d_{750}	—	2.192	2.334	2.471	2.590	—	—
	d_{775}	1.992	2.180	2.318	2.452	2.560	2.650	2.760
	d_{800}	1.986	2.169	2.302	2.432	2.531	2.616	2.730
$BaCl_2$–NH_4NO_3	*p*	0	1.5					
Ref. **58**	d_{180}	1.426	1.436					
$BaCl_2$–$PbCl_2$	*p*	0	10.71	15.57	24.83	100		
Ref. **25, 1**	*a*	5.702	5.430	5.356	5.156	3.662		
	b	1.50	1.35	1.36	1.27	0.52		
	r	516–710	565–700	575–690	660–710	above 1 000		
BaF_2–Na_3AlF_6	*p*	0	33.1	55.4	71.5			
Ref. **15**, 10	*a*	3.069	3.429	3.978	3.804			
	b	0.96	0.838	0.996	0.447			
	r	1 020–1 130	920–1 175	960–1 155	910–1 120			
BaF_2–NaF	*p*	0	45.2	67.5	80.7			
Ref. **15**	*a*	2.590	3.291	3.739	4.195			
	b	0.628	0.626	0.600	0.624			
	r	1 010–1 120	955–1 160	970–1 165	955–1 180			
$Ba(NO_2)_2$–	*p*	5.7	13.4					
$Ba(NO_3)_2$	d_{300}	3.240	3.243					
Ref. **72**	d_{320}	3.226	3.212					
	d_{340}	3.212	3.216					
$Ba(NO_2)_2$–KNO_2	*p*	59.0	73.7	80.0	86.8	92.5	97.6	
Ref. **72**	d_{280}	2.274	2.492	—	—	—	3.154	
	d_{320}	2.245	2.458	2.615	2.780	2.950	3.126	
	d_{340}	2.132	2.440	2.577	2.750	2.921	3.099	
	d_{360}	2.218	2.422	2.577	2.750	2.921	3.099	
$Ba(NO_3)_2$–KNO_3	*p*	9.54	25.8	33.1	39.7	45.9	51.5	57.0
Ref. **72**, 36, 73	d_{340}	1.914	2.014	2.068	2.118	—	—	—
	d_{380}	1.884	1.984	2.036	2.088	2.143	2.201	—
	d_{420}	1.885	1.953	2.004	2.056	2.111	2.168	2.228
	d_{460}	—	—	1.974	2.026	2.080	2.136	2.196
$Ba(NO_3)_2$–	*p*	0	2.5	5.0				
NH_4NO_3	d_{170}	1.432	1.453	1.474				
Ref. **58**	d_{180}	1.426	1.447	1.466				
BaO–SiO_2	*p*	22.1	39.0	52.3	63.0	72.0		
Ref. **74**	*a*	2.580	3.016	3.401	3.748	4.044		
	b	0.000	0.048	0.068	0.075	0.080		
	r			1 600–1 950				
BeF_2–LiF	*p*	51.5						
Ref. **75**	*a*	2.09						
	b	0.27						
	r	500–800						
Bi–$BiBr_3$	*p*	52.1						
Ref. **68**	d_{543}	6.69						

Table 9.2 DENSITIES OF MOLTEN BINARY SALT SYSTEMS AND OTHER MIXED IONIC MELTS—*continued*

System								
Bi–$BiCl_3$	*p*	0	7.5	14.7	19.7	23.8	97.9	100
Ref. **61**	*a*	4.42	4.61	4.87	5.07	5.24	10.39	10.39
	b	2.20	2.06	2.08	2.13	2.15	1.29	1.29
	r	240–330	290–400	270–420	290–450	310–440	330–440	310–440
$CaCl_2$–KCl	*p*	0	20	40	50	60	80	100
Ref. **8**	d_{800}	1.495	1.573	1.671	1.725	1.780	1.896	2.057
	d_{900}	1.434	1.517	1.613	1.667	1.734	1.850	2.009
$CaCl_2$–$MgCl_2$	*p*	22.6	27.7	42.6	59.4	76.6		
Ref. **35**	*a*	2.16	2.115	2.217	2.319	2.365		
	b	0.47	0.40	0.44	0.48	0.45		
	r	723–895	753–907	728–886	730–902	746–898		
$CaCl_2$–NaCl**	*p*	17.5	32.3	44.9	65.5	74.0	88.5	94.5
Ref. **71**, 27, 18, 8	d_{625}	—	—	—	1.899	1.944	—	—
	d_{700}	—	—	1.767	1.855	1.912	2.005	—
	d_{775}	1.612	1.679	1.730	1.830	1.879	1.974	2.013
	d_{850}	1.575	1.646	1.693	1.798	1.846	1.944	1.985
CaF_2–CaO	*p*	92.9	94.6	95.3	97.1	98.6	100	
Ref. **53**	d_{1545}	2.63	2.50	2.53	2.59	2.68	2.75	
CaF_2–Na_3AlF_6	*p*	0	10	20	30	40	50	
Ref. **38, 31, 15**	d_{1000}	2.096	2.162	2.223	2.283	2.334	2.366	
	d_{1100}	2.002	2.070	2.135	2.200	2.256	2.294	
CaF_2–NaF	*p*	0	26.9	48.0	65.0			
Ref. **15**	*a*	2.602	2.692	2.804	2.929			
	b	0.64	0.58	0.57	0.54			
	r	1 010–1 120	930–1 185	870–1 165	1 050–1 170			
$Ca(NO_3)_2$–NH_4HO_3	*p*	0	5	10				
Ref. **58**	d_{170}	1.432	1.460	—				
	d_{180}	1.426	1.453	1.480				
CaO–SiO_2	*p*	28.6	38.4	48.4	58.5			
Ref. **74**	*a*	2.578	2.651	2.758	2.840			
	b	0.066	0.064	0.084	0.103			
	r		1 600–1 950					
$CdBr_2CdCl_2$	*p*	0	38.55	55.46	73.64	100		
Ref. **25**	*a*	3.870	4.138	4.255	4.390	4.687		
	b	0.84	0.90	0.91	0.93	1.08		
	r	582–725	580–680	590–710	606–705	580–720		
$CdCl_2$–CdI_2	*p*	0	14.2	33.2	59.9	100		
Ref. **41**	*a*	4.828	4.638	4.367	4.187	3.839		
	b	1.12	1.06	0.88	0.87	0.80		
	r	380–700	360–700	440–700	520–700	560–700		
$CdCl_2KCl$	*p*	0	44.77	62.10	78.10	92.36	100	
Ref. **25, 7**	*a*	1.988	2.495	2.791	3.176	3.625	3.870	
	b	0.60	0.72	0.82	0.95	0.96	0.84	
	r	above 750	604–750	460–680	464–680	534–700	582–725	
$CdCl_2$–LiCl	*p*	0	59.1	81.2	92.9	100		
Ref. **41**	*a*	1.731	2.572	3.245	3.593	3.839		
	b	0.382	0.577	0.845	0.825	0.800		
	r	600–750	560–750	510–750	520–750	560–750		
$CdCl_2$–NaCl	*p*	0	62.08	71.38	79.64	85.23	91.66	100
Ref. **25, 7**	*a*	2.053	2.896	3.090	3.315	3.543	3.678	3.870
	b	0.63	0.83	0.86	0.92	1.04	0.95	0.84
	r	above 800	580–690	500–690	570–680	540–680	580–700	582–725

Table 9.2 DENSITIES OF MOLTEN BINARY SALT SYSTEMS AND OTHER MIXED IONIC MELTS—*continued*

System								
$CdCl_2$–$PbCl_2$	*p*	0	27.62	52.14	75.66	85.40	100	
Ref. **25**	*a*	3.870	4.305	4.726	5.222	5.402	5.702	
	b	0.84	1.02	1.18	1.39	1.43	1.50	
	r	582–725	540–680	515–700	480–680	545–680	510–710	
CdI_2–KI	*p*	0	28.0	52.5	68.8	81.6	92.6	100
Ref. **41**	*a*	3.108	3.469	3.769	4.120	4.431	4.645	4.828
	b	0.96	1.09	1.12	1.22	1.29	1.17	1.12
	r	680–800	540–800	400–800	190–700	300–700	360–700	380–700
$Cd(NO_3)_2$–KNO_3	*p*	50.0	55.7	65.7	74.1	81.2		
Ref. **50**	*a*	2.436	2.475	2.583	2.689	2.820		
	b	0.92	0.89	0.93	0.93	0.95		
	r	260–300	220–300	200–300	180–300	200–300		
Ce–$CeCl_3$	*p*	0.28	0.57	1.14	2.31	3.50	4.70	
Ref. **76**	d_{850}	3.165 5	3.169 5	3.180 0	3.205 5	3.239 0	3.287 5	
	d_{900}	3.127 5	3.132 7	3.144 0	3.170 0	3.204 1	3.249 0	
	d_{950}	3.087 7	3.093 2	3.104 0	3.133 3	3.174 0	3.228 8	
Cu_2Cl_2–KCl	*p*	0	11.49					
Ref. **5**	d_{800}	1.51	1.62					
Cu_2S–FeS	*p*	0	50	100				
Ref. **60**	*a*	3.85	5.10	6.76				
	b	—	0.62	0.75				
	r	1 250–1 450	1 100–1 500	1 150–1 400				
FeO–SiO_2	*p*	73.7	78.3	82.8	88.1	91.5	95.7	100
Ref. **55**	d_{1300}	3.67	3.81	4.00	4.16	4.32	4.61	4.90
KBr–KNO_3	*p*	0	17.2	37.0	54.1	70.2	86.9	100
Ref. **41**	*a*	2.116	2.186	2.285	2.415	2.476	2.635	2.725
	b	0.729	0.714	0.727	0.779	0.730	0.801	0.794
	r	330–600	330–600	340–600	440–600	530–800	650–800	740–800
KBr–NaCl	*p*	67.2	100					
Ref. **41**	*a*	2.447	2.725					
	b	0.723	0.794					
	r	750–800	740–800					
KBr–TlBr	*p*	5.7	15.2	37.0	41.4	57.6	60.5	74.6
Ref. **77**	*a*	6.493_5	6.908_4	—	5.732_8	—	—	3.978_7
	b	1.934_2	3.421_4	—	3.084_1	—	—	1.893_6
	A	—	—	1.781_6	—	3.218_3	6.425_3	—
	B	—	—	3.878_9	—	6.631_9	14.160_5	—
	C	—	—	4.192_8	—	6.087_0	10.347_8	—
	r	490–628	584–732	707–759	750–799	721–798	747–857	758–797
KBr–$ZnSO_4$	*p*	34.7	41.0	42.6	55.2	59.6	63.9	68.9
Ref. **42, 34**	*a*	3.090	3.003	2.985	3.105	2.939	2.940	2.965
	b	0.46	0.40	0.42	0.80	0.52	0.60	0.74
	r				500–550			
KCl–KI	*p*	0	2.8	7.8	13.3	26.8	41.2	64.4
Ref. **52**	*a*	3.098	3.030	2.914	2.873	2.684	2.498	2.274
	b	0.956	0.930	0.864	0.890	0.826	0.755	0.690
	r	680–900	680–900	710–900	640–910	620–900	680–920	710–900
KCl–KNO_3	*p*	6.4	15.3	23.5				
Ref. **78**	*a*	2.101	2.090	2.069				
	b	0.728	0.721	0.680				
	r	349–540	429–580	485–633				

Table 9.2 DENSITIES OF MOLTEN BINARY SALT SYSTEMS AND OTHER MIXED IONIC MELTS—*continued*

System								
KCl–LiCl**	p	0	28.2	42.6	55.8	72.2	87.6	100
Ref. **52, 12**	a	1.766	1.835	1.856	1.885	1.923	1.968	1.977
	b	0.433	0.489	0.507	0.528	0.561	0.509	0.583
	r	620–780	530–750	460–600	390–590	590–750	690–850	780–940
$KCl–MgCl_2$	p	10.7	20.4	31.2	41.5	49.7	59.0	80
Ref. **35**, 69, 11	a	1.896	1.993	1.990	1.924	1.946	1.944	1.946
	b	0.31	0.41	0.48	0.44	0.50	0.52	0.54
	r	706–886	695–890	756–901	734–894	707–880	707–881	704–876
$KCl–MnCl_2$	p	12.6	23.2	32.1	36.6	41.3	51.0	63.5
Ref. **79**	d_{500}	—	2.210	2.111	2.063	2.009	1.943	—
	d_{600}	2.337	2.157	2.051	2.007	1.955	1.881	1.795
	d_{700}	2.274	2.104	1.998	1.958	1.900	1.818	1.727
	d_{800}	2.212	2.051	1.941	1.898	1.845	1.755	1.668
KCl–NaBr	p	41.8	100					
Ref. **41**	a	2.449	1.986					
	b	0.728	0.582					
	r	750–800	770–800					
KCl–NaCl**	p	0	18.7	31.1	40.6	54.8	64.8	82.9
Ref. **52, 17**, 8	a	1.991	1.989	1.985	1.982	1.976	1.977	1.979
	b	0.543	0.554	0.560	0.557	0.568	0.575	0.581
	r	800–1 030	780–920	710–920	710–930	670–910	680–910	720–930
KCl–NaI	p	0	8.0	19.8	33.0	49.7	73.6	100
Ref. **41**	a	3.412	3.118	2.971	2.724	2.482	2.195	1.986
	b	1.00	0.824	0.900	0.815	0.727	0.625	0.582
	r	670–800	600–800	520–800	540–800	570–800	690–800	770–800
$KCl–PbCl_2$	p	0	5.51	13.21	22.94	100		
Ref. **25, 7**	a	5.702	5.145	4.513	3.960	1.988		
	b	1.50	1.42	1.28	1.13	0.60		
	r	516–700	565–700	580–680	490–680	above 750		
$KCl–ZnCl_2$	p	5.4	9.3	20.0	31.1	49.2	55.9	70.1
Ref. **59**	a	2.653	2.588	2.542	2.448	2.272	2.217	2.080
	b	0.55	0.50	0.62	0.66	0.60	0.60	0.57
	r	460–670	450–660	450–660	450–640	440–660	450–650	690–720
$KCl–ZnSO_4$	p	23.2	29.0	31.4	38.2	47.0	54.3	56.1
Ref. **42, 34**	a	3.024	2.884	2.821	2.681	2.619	2.511	2.491
	b	0.68	0.64	0.58	0.56	0.72	0.70	0.70
	r				475–550			
KF–LiF–NaF	p_{KF}	59.0						
eutectic	p_{LiF}	29.2						
Ref. **80**	a	2.47						
	b	0.68						
	r	600–800						
KI–NaCl	p	33.6	55.0	74.1	86.9	94.2	100	
Ref. **41**	a	2.253	2.484	2.727	2.893	3.034	3.108	
	b	0.632	0.706	0.819	0.858	0.965	0.960	
	r	720–800	680–800	560–800	580–800	640–800	680–800	
$K_2MoO_4–MoO_3$	p	17.5	27.5	43.0	52.2	60.0	79.4	91.0
Ref. **81**	a	3.997 4	3.937 1	3.781 3	3.653 4	3.428 2	3.163 7	3.036 0
	b	1.133 7	1.099 7	1.118 9	1.052 9	0.995 1	0.714 6	0.634 0
	r	761–869	614–750	637–792	632–779	574–766	783–933	882–978

Table 9.2 DENSITIES OF MOLTEN BINARY SALT SYSTEMS AND OTHER MIXED IONIC MELTS—*continued*

System								
KNO_2–KNO_3	p	0	26.5	35.9	50.7	100		
Ref. **48**, 72	a	2.116	2.068	2.062	2.052	2.005		
	b	0.73	0.69	0.70	0.70	0.70		
	r	340–500	340–500	340–500	380–500	440–500		
KNO_2–$NaNO_2$	p	0	17.8	29.1	45.1	55.2	74.2	100
Ref. **48**	a	2.004	1.959	1.951	1.961	1.951	1.963	2.005
	b	0.69	0.59	0.58	0.59	0.58	0.61	0.70
	r	380–500	260–500	260–500	260–500	350–500	350–500	440–500
KNO_2–$NaNO_3$	p	0	15	35	50	65	85	100
Ref. **48**	a	2.121	2.117	2.073	2.057	2.028	1.993	2.005
	b	0.68	0.70	0.66	0.68	0.66	0.63	0.70
	r	350–500	300–500	200–500	200–500	240–500	350–500	440–500
KNO_3–$LiNO_3$	p	16.7	26.7	40.9	59.5	71.0	81.9	92.2
Ref. **78**, 82	a	1.954	1.974	1.997	2.033	2.055	2.083	2.102
	b	0.599	0.623	0.638	0.683	0.696	0.729	0.735
	r	257–403	264–350	219–445	328–476	306–454	294–425	318–452
KNO_3–$NaNO_2$	p	0	20.5	44.0	59.4	73.1	89.3	100
Ref. **48**	a	2.004	1.991	2.028	2.048	2.061	2.100	2.116
	b	0.69	0.61	0.66	0.67	0.67	0.72	0.73
	r	380–500	300–500	200–500	200–500	200–500	240–500	340–500
KNO_3–$NaNO_3$**	p	0	20	40	60	80	100	
Ref. **48, 47, 36,**	a	2.121	2.127	2.127	2.126	2.126	2.116	
82, 8, 3	b	0.68	0.71	0.72	0.73	0.72	0.73	
	r	350–500	290–500	250–500	240–500	290–500	350–500	
KNO_3–NH_4NO_3	p	0	2.5	5.0	7.5	10.0		
Ref. **58**	d_{170}	1.432	1.441	1.451	1.462	1.473		
	d_{190}	1.420	1.429	1.439	1.450	1.460		
KNO_3–$Pb(NO_3)_2$	p	40	50	60	70	80	90	100
Ref. **36**	a	3.080	2.881	2.679	2.483	2.362	2.218	2.113
	b	1.00	1.00	0.908	0.750	0.787	0.709	0.730
	r	275–345	230–365	275–380	300–390	320–405	330–420	350–460
KNO_3–$Sr(NO_3)_2$	p	52.9	60	70	80	90	100	
Ref. **73, 36**	a	2.403	2.360	2.285	2.226	2.167	2.113	
	b	0.721	0.760	0.748	0.752	0.753	0.730	
	r	421–452	365–445	300–420	305–465	325–475	350–460	
K_2O–SiO_2	p	23.9	26.7	29.9	32.9	38.7	43.6	
Ref. **56, 39**	a	2.353	2.320	2.377	2.380	2.464	2.504	
	b	0.14	0.11	0.16	0.16	0.23	0.27	
	r			1 000–1 400				
K_2SO_4–$ZnSO_4$	p	26.59	32.66	35.15	36.77	41.41	49.92	58.86
Ref. **34**	d_{500}	(2.841)	2.751	2.731	2.728	2.680	2.592	2.509
	d_{550}	2.812	2.727	2.708	2.692	2.641	2.556	2.485
K_2WO_4–WO_3	p	37.4	50.5	58.3	67.4	76.6	85.0	92.2
Ref. **83**	a	5.889 8	5.486 5	5.332 6	4.955 3	4.532 2	4.244 4	4.077 1
	b	1.553 8	1.519 6	1.562 0	1.334 1	1.015 7	0.907 4	0.845 2
	r	897–999	772–943	655–783	682–851	803–910	860–987	904–1 029
$LiCl$–$LiNO_3$	p	6.4	13.3	20.8				
Ref. **78**	a	1.911	1.903	1.899				
	b	0.537	0.538	0.524				
	r	278–446	340–497	378–497				
$LiClO_4$ in	p_{LiClO_4}	5	10	15				
KNO_3–$NaNO_3$	a	2.120 1	2.127 5	2.139 4				
eutectic	b	0.711 0	0.723 2	0.744 1				
(57.3% KNO_3)	r	—	230–400	—				
Ref. **84**								

Table 9.2 DENSITIES OF MOLTEN BINARY SALT SYSTEMS AND OTHER MIXED IONIC MELTS—*continued*

System								
$LiClO_4$–$LiNO_3$	*p*	34.0	57.3	82.3				
Ref. **73**	*a*	2.014	2.088	2.134				
	b	0.610	0.629	0.629				
	r	240–357	198–347	225–336				
Li_2MoO_4–MoO_3	*p*	19.5	33.9	51.5	60.5	71.3	85.2	91.0
Ref. **81**	*a*	3.993 2	3.895 1	3.735 9	3.564 1	2.985 6	3.455 9	3.352 0
	b	0.950 2	0.845 1	0.772 3	0.644 8	0.031 7	0.593 0	0.531 7
	r	766–921	755–924	760–934	802–962	799–950	781–905	825–924
$LiNO_3$–NH_4NO_3	*p*	0	5					
Ref. **58**	d_{180}	1.426	1.430					
Li_2O–SiO_2	*p*	16.1	21.8	25.9	31.9	42.8	48.1	
Ref. **56, 39**	*a*	2.311	2.355	2.359	2.344	1.980	1.955	
	b	0.13	0.19	0.20	0.22	—	—	
	r		1 100–1 400		1 250–1 400		1 400	
Li_2WO_4–WO_3	*p*	54.6	59.7	69.0	77.5	85.2	92.3	
Ref. **83**	*a*	5.971 8	6.077 6	5.769 6	5.527 5	5.281 9	5.016 0	
	b	1.162 8	1.341 1	1.190 3	1.112 4	0.991 3	0.851 1	
	r	809–939	764–968	755–978	736–930	714–923	733–959	
$MgCl_2$–$NaCl$	*p*	20.0	45.7	60.6	75.8			
Ref. **35**	*a*	1.961	1.963	2.002	2.000			
	b	0.49	0.47	0.48	0.43			
	r	750–888	732–890	722–896	712–893			
MgF_2–Na_3AlF_6	*p*	3	6	9				
Ref. **62**	d_{950}	—	—	2.14				
	d_{980}	2.12	2.13	—				
MgO–SiO_2	*p*	32.8	35.5	38.3	41.1	44.0		
Ref. **74**	*a*	2.547	2.579	2.597	2.655	2.670		
	b	0.049	0.056	0.052	0.071	0.070		
	r			1 600–1 950				
MoO_3–Na_2MoO_4	*p*	21.4	26.4	51.2	62.2	71.2	83.7	86.2
Ref. **51**	*a*	2.79	3.63	3.92	4.02	3.88	3.74	3.81
	b	—	1.0	1.2	1.2	1.0	0.8	0.9
	r	700–830	690–790	660–760	650–810	650–750	730–840	780–880
NH_4Cl–NH_4NO_3	*p*	0	2.5	5	7.5	10		
Ref. **58**	d_{170}	1.432	1.426	1.419	1.412	1.403		
	d_{180}	1.420	1.414	1.407	1.403	—		
NH_4NO_3–$(NH_4)_2SO_4$	*p*	92.5	95.0	97.5	100			
Ref. **58**	p_{180}	—	1.435	1.430	1.427			
	d_{200}	1.425	1.421	1.417	1.416			
NH_4NO_3–$Pb(NO_3)_2$	*p*	89.9	95.0	100				
Ref. **58**	d_{170}	—	1.481	1.432				
	d_{180}	1.527	1.475	1.426				
NH_4NO_3–$Sr(NO_3)_2$	*p*	95	100					
Ref. **58**	d_{170}	1.466	1.432					
	d_{190}	1.453	1.420					
$NaCl$–NH_4NO_3	*p*	0	1.5					
Ref. **58**	d_{180}	1.426	1.432					
$NaCl$–$NaNO_3$	*p*	1.4	2.8	4.2	5.7	7.1	8.6	10.1
Ref. **71**	d_{350}	1.879	1.878	1.879	1.875	1.875	—	—
	d_{400}	1.842	1.841	1.843	1.839	1.839	1.840	1.836
	d_{450}	1.806	1.804	1.806	1.802	1.803	1.803	1.800

Table 9.2 DENSITIES OF MOLTEN BINARY SALT SYSTEMS AND OTHER MIXED IONIC MELTS—*continued*

System								
$NaCl–PbCl_2$	p	0	2.69	5.85	9.42	13.32	17.5	
Ref. **17**	d_{500}	4.96	4.72	4.34	4.21	3.96	3.66	
	d_{600}	4.82	4.57	4.22	4.08	3.77	3.50	
$NaF–Na_3AlF_6$†	p	0	10	20	40	60	80	100
Ref. **38, 31, 24, 22,**	d_{1000}	2.096	2.110	2.108	2.085	2.042	1.998	1.957
15, 14, 13, 10, 9, 6	d_{1100}	2.002	2.017	2.019	2.002	1.970	1.933	1.895
$NaF–UF_4–ZrF_4$	p_{NaF}	19.0						
Ref. **85**	p_{UF_4}	11.4						
	a	3.93						
	b	0.93						
	r	600–800						
$NaF–ZrF_4$	p	17.05	22.0	27.3	31.8	51.1		
Ref. **86**	a	3.83	3.71	3.61	3.52	3.23		
	b	0.91	0.89	0.87	0.86	0.81		
	r			300–800				
$NaNO_2 NaNO_3$	p	0	21.3	44.8	65.4	82.1	100	
Ref. **48, 41**	a	2.121	2.094	2.066	2.043	2.028	2.004	
	b	0.68	0.68	0.68	0.68	0.70	0.69	
	r	310–500	250–500	220–500	270–500	270–500	280–500	
$NaNO_3–NH_4NO_3$	p	0	2.5	5.0				
Ref. **58**	d_{170}	1.432	1.443	1.452				
	d_{180}	1.426	1.436	1.446				
$NaNO_3–Pb(NO_3)_2$	p	40	50	60	70	80	90	100
Ref. **36**	a	3.171	2.867	2.703	2.498	2.373	2.238	2.117
	b	1.12	0.806	0.856	0.726	0.746	0.695	0.670
	r	340–365	305–375	285–390	295–400	305–420	310–430	320–460
$Na_2O–P_2O_5$	$d_{p,t}=2.372+0.204p/(100-p)-0.338\times10^{-3}t$							
Ref. **87**	r_p	30.4–48.5						
	r_t	liquidus–1 070						
$Na_2O–SiO_2$	p	20.0	30.8	33.6	36.9	50.0		
Ref. **56, 39**	a	2.312	2.380	2.436	2.456	2.516		
	b	0.10	0.14	0.18	0.20	0.26		
	r	1 100–1 400		900–1 400		1 050–1 400		
$Na_2WO_4–WO_3$	p	45.9	50.9	55.9	65.1	74.2	83.0	91.6
Ref. **83**	a	5.963 6	5.711 3	5.882 5	4.415 9	5.232 5	4.997 6	3.934 4
	b	1.424 4	1.228 9	1.557 0	–0.031 2	1.226 2	1.116 7	–0.051 0
	r	782–926	775–898	758–899	737–882	688–880	654–815	685–880
$Nd(NO_3)_2$ in $KNO_3–NaNO_3$ eutectic (57.3% KNO_3)	$p_{Nd(NO_3)_2}$	1.3	6.4	12.0	21.4	29.1	35.3	
	a	2.128 5	2.158 7	2.218 6	2.274 1	2.337 1	2.418 3	
	b	0.731 7	0.725 4	0.757 1	0.686 7	0.708 7	0.753 3	
Ref. **84**	r				230–400			
$Nd(NO_3)_2$ in $KNO_3–LiClO_4–NaNO_3$ (45%–10%–45%)	$p_{Nd(NO_3)_2}$	35.3						
	a	2.412 8						
	b	0.707 1						
Ref. **84**	r	230–400						
$PbBr_2–PbCl_2$	p	0	24.46	57.19	87.88	100		
Ref. **25**, 2	a	5.702	5.840	6.025	5.264	6.338		
	b	1.50	1.52	1.55	1.71	1.65		
	r	516–570	492–620	465–640	410–600	505–600		
$PbCl_2–ZnCl_2$	p	67.1						
Ref. **89**	d_{510}	3.733						
	d_{553}	3.703						

Table 9.2 DENSITIES OF MOLTEN BINARY SALT SYSTEMS AND OTHER MIXED IONIC MELTS—*continued*

PbO–B_2O_3	p	57.9	68.1	76.3	82.9	88.2	92.9	96.6
Ref. **88**	d_{1050}	2.8	3.5	4.0	4.6	5.3	6.0	6.6
PbO–SiO_2	p	61.4	71.2	78.8	84.8	89.7	93.7	97.1
Ref. **88**	d_{1050}	4.2	4.9	5.7	6.0	6.7	7.0	7.4
PbO–V_2O_5	p	15.3	39.6	52.8	65.4	79.0	85.1	92.1
Ref. **90**	d_{1000}	3.1	4.0	5.0	5.0	6.2	6.8	7.8
PbO–SiO_2–V_2O_5	p_{PbO}	28.5	38.6	58.4	60.0	68.5	80.0	
Ref. **90**	p_{SiO_2}	10.3	10.4	10.4	10.1	10.6	10.0	
	d_{1000}	3.8	3.8	4.0	4.0	4.0	5.2	
SrO–SiO_2	p	48.1	53.4	58.5	63.3	67.8		
Ref. **74**	a	3.129	3.247	3.384	3.493	3.612		
	b	0.058	0.059	0.074	0.079	0.072		
	r			1 600–1 950				
$TlCl$–$ZnSO_4$	p	55.23	57.99	59.86	61.04	63.55	69.24	74.82
Ref. **34**	d_{450}	(4.116)	4.146	4.196	4.223	4.277	4.421	(4.573)
	d_{500}	4.076	4.094	4.155	4.177	4.227	4.376	4.512

* See also NaF–Na_3AlF_6.
† See also AlF_3–Na_3AlF_6
** See also 'Physical Properties Data Compilations Relevant to Energy Storage II Molten Salts—Data on Single and Multi-Component Salt Systems', Janz *et al.*, NSRDS–NBS 61.

REFERENCES TO TABLE 9.2

1. K. Arndt and A. Gessler, *Z. Electrochem.*, 1908, **14**, 665.
2. R. Lorenz, H. Frei, and A. Jabs, *Z. phys. Chem.*, 1908, **61**, 468.
3. Smith and Menzies, *Proc. R. Soc.*, Edinburgh, 1910, **30**, 432.
4. Izbekov and Plotnikov, *J. Russ. Phys.-Chem. Soc.*, 1911, **43**, 18.
5. Sackur, *Z. phys. Chem.*, 1913, **83**, 297.
6. Pascal and Jouniaux, *Bull. Soc. chim.*, France, 1914, **15**, 312; *Z. Elektrochem.*, 1916, **22**, 71.
7. F. M. Jaeger, *Z. anorg. Chem.*, 1917, **101**, 175.
8. C. Sandonnini, *Gazz. Chim. Ital.*, 1920, **51**, 289.
9. J. D. Edwards, F. C. Frary and Z. Jeffries, 'The Aluminium Industry', New York, 1930, p. 308.
10. N. Kameyama and A. Naka, *J. Soc. Chem. Ind., Japan*, 1931, **34**, 140.
11. S. V. Karpachev, A. G. Stromberg and O. Poltoratzkaya, *J. Phys. Chem. (USSR)*, 1934, **5**, 793.
12. S. V. Karpachev, A. G. Stromberg and V. N. Podchainova, *J. Gen. Chem. (USSR)*, 1935, **5**, 1517.
13. G. A. Abramov, *Legkie Metally*, 1936, **11**, 27.
14. Z. F. Lundina, *Trans. All Union Aluminium and Magnesium Inst.*, 1936, **13**, 5.
15. G. A. Abramov and P. A. Kozunov, *Trans. Leningrad Indust. Inst.*, 1939, *No. 1*, 60.
16. A. I. Kryagova, *J. Gen. Chem. (USSR)*, 1939, **9**, 2061.
17. V. P. Barzakovskii, *Bull. Acad. Sci., U.R.S.S.* (Class sci chim), 1940, 825.
18. V. P. Barzakovskii, *J. appl. Chem. (USSR)*, 1940, **13**, 1117.
19. Y. Yamaguti and S. Sisido, *J. chem. Soc. Japan*, 1941, **62**, 304.
20. E. Ya. Gorenbein, *J. Gen. Chem. (USSR)*, 1945, **15**, 729.
21. E. Ya. Gorenbein, *ibid.*, 1947, **17**, 873.
22. T. G. Pearson and J. Waddington, *Disc. Faraday Soc.*, 1947, No. 1, 307.
23. E. Ya. Gorenbein, *J. Gen. Chem. (USSR)*, 1948, **18**, 1427.
24. V. P. Mashovets, 'The Electrometallurgy of Aluminium,' 1948.
25. N. K. Boardman, F. H. Dorman and E. Heymann, *J. phys. Chem.*, 1949, **53**, 375.
26. I. M. Bokhovkin, *J. Gen. Chem. (USSR)*, 1949, **19**, 805.
27. G. Fuseya and K. Ouchi, *J. electrochem. Soc. Japan*, 1949, **17**, 254.
28. E. Ya. Gorenbein and E. E. Kriss, *J. Gen. Chem. (USSR)*, 1949, **19**, 1978.
29. H. Grothe, *Z. Elektrochem.*, 1949, **53**, 362.
30. I. M. Bokhovkin, *J. Gen. Chem. (USSR)*, 1950, **20**, 397.
31. A. Vayna, *Alluminio*, 1950, **19**, 541.
32. E. Ya. Gorenbein and E. E. Kriss, *J. phys. Chem. (USSR)*, 1951, **25**, 791.
33. R. C. Spooner and F. E. W. Wetmore, *Canad. J. Chem.*, 1951, **29**, 777.
34. I. P. Vereshchetina and N. P. Luzhnaya, *J. appl. Chem. (USSR)*, 1951, **24**, 148.
35. R. W. Huber, E. V. Potter and H. W. St. Clair, *U.S. Bur. Mines, Rep. Invest.* 4858, 1952.
36. K. Laybourne and W. M. Madgin, *J. chem. Soc.*, **1934**, 1.
37. J. Byrne, H. Fleming and F. E. W. Wetmore, *Canad. J. Chem.*, 1952, **30**, 922.

38. J. D. Edwards, C. S. Taylor, L. A. Cosgrove and A. S. Russell, *Trans. electrochem. Soc.*, 1953, **100**, 508.
39. L. Shartsis, S. Spinner and W. Capps, *J. Am. ceram. Soc.*, 1952, **35**, 155.
40. L. Shartsis, W. Capps and S. Spinner, *J. Am. ceram. Soc.*, 1953, **36**, 35.
41. H. Bloom, I. W. Knaggs, J. J. Molloy and D. Welch, *Trans. Faraday Soc.*, 1953, **49**, 1458.
42. N. P. Luzhnaya and I. P. Vereshchetina, *Izvest. Sekt. Fiziko-Khim. Anal.*, 1954, **24**, 192.
43. J. S. Peake and M. R. Bothwell, *J. Am. chem. Soc.*, 1954, **76**, 2653.
44. L. Shartsis and H. F. Shermer, *J. Am. ceram. Soc.*, 1954, **37**, 544.
45. R. Midorikawa, *J. electrochem. Soc. Japan*, 1954, **23**, 310.
46. V. D. Polyakov, *Izvest. Sekt. Fiziko-Khim. Anal.*, 1955, **26**, 147.
47. A. G. Bergman, I. S. Rassonskaya and N. E. Schmidt, *ibid.*, 1955, **26**, 156.
48. V. D. Polyakov and S. I. Berul, *ibid.*, 1935, **26**, 164.
49. V. P. Polyakov, *ibid.*, 1955, **26**, 191.
50. N. P. Popovskaya and P. I. Protsenko, *Zh. fiz. Khim.*, 1955, **29**, 225.
51. K. B. Morris, M. I. Cook, C. Z. Sykes and M. B. Templeman, *J. Am. chem. Soc.*, 1955, **77**, 851.
52. E. R. van Artsdalen and I. S. Yaffe. *J. phys. Chem.*, 1955, **59**, 118.
53. T. Baak, *Acta. Chem. Scand.*, 1955, **9**, 1406.
54. N. P. Luzhnaya, N. N. Evseeva and I. P. Vereshchetina, *Zh. neorg. Khim.*, 1956, **1**, 1490.
55. S. I. Popel and O. A. Esin, *Zhur. Priklad. Khim.*, 1956, **29**, 651.
56. J. O'M. Bockris, J. W. Tomlinson and J. L. White, *Trans. Faraday Soc.*, 1956, **52**, 299.
57. H. Bloom and D. C. Rhodes, *J. phys. Chem.*, 1956, **60**, 791.
58. S. Toshiaki and T. Ishibashi, *Sci. Papers Coll. Gen. Educ., Univ. Tokyo*, 1957, **7**, 53.
59. F. R. Duke and R. A. Fleming, *J. electrochem. Soc.*, 1957, **104**, 251.
60. M. Bourgon, G. Derge and C. M. Pound, *Trans. Amer, Inst. Min. Met. Eng.*, 1958, **212**, 338.
61. F. J. Keneshea and D. Cubicciotti, *J. phys. Chem.*, 1958, **62**, 843.
62. E. Vatslavik and A. I. Belyaev, *Zh. neorg. Khim.*, 1958, **3**, 1044.
63. R. H. Moss, *Univ. Microfilms* (*Ann Arbor. Mich.*), 1955, No. 12, 730.
64. E. J. Salstrom, *J. Am. chem. Soc.*, 1931, **53**, 3385.
65. E. J. Salstrom and J. H. Hildebrand, *ibid.*, 1930, **52**, 4650.
66. E. J. Salstrom and J. H. Hildebrand, *ibid.*, 1931, **53**, 1794.
67. E. J. Salstrom and J. H. Hildebrand, *ibid.*, 1932, **54**, 4252.
68. L. E. Topol and F. Y. Lieu, *J. phys. Chem.*, 1964, **68**, 851.
69. J. N. Reding, *J. chem. Engng Data*, 1965, **10**, 1.
70. N. V. Bondarenko and K. L. Strelets, *Zh. prikl. Khim.*, 1962, **35**, 1271.
71. I. P. Vereshchetina and N. P. Luzhnaya, *Inv. Sekt. fiz.-khim. Analiza*, 1954, **25**, 188.
72. P. I. Protsenko and A. Ya. Malakhova, *Zh. neorg. Khim.*, 1961, **6**, 1662.
73. G. F. Petersen, W. M. Ewing and G. P. Smith, *J. chem. Engng Data*, 1961, **6**, 540.
74. J. W. Tomlinson, M. S. R. Heynes and J. O'M. Bockris, *Trans. Faraday Soc.*, 1958, **54**, 1822.
75. B. C. Blanke, E. N. Bousquet, M. L. Curtis and E. L. Murphy, *Mound Lab., Miamisburg, Ohio. Memo, 1086* (1956)
76. G. W. Mellors and S. Senderoff, *J. phys. Chem.*, 1960, **64**, 294.
77. E. R. Buckle, P. E. Tsaoussoglou and A. R. Ubbelohde, *Trans Faraday Soc.*, 1964, **60**, 684.
78. G. P. Smith and G. F. Petersen, *J. chem. Engng Data*, 1961, **6**, 493.
79. I. G. Murgulescu and S. Zuca, *Studii, Cerc. Chim.*, 1959, **7**, 325.
80. M. Blander, W. R. Grimes, N. V. Smith and G. M. Watson, *J. phys. Chem.*, 1959, **63**, 1164.
81. K. B. Morris and P. L. Robinson, *ibid.*, 1964, **68**, 1194.
82. P. C. Papaioannou and G. W. Harrington, *ibid.*, 1964, **68**, 2424.
83. K. B. Morris and P. L. Robinson, *J. chem. Engng Data*, 1964, **9**, 444.
84. J. Padova and J. Soriano, *ibid.*, 1964, **9**, 510.
85. W. R. Grimes, N. V. Smith and G. M. Watson, *J. phys. Chem.*, 1958, **62**, 862.
86. J. H. Shaffer, W. R. Grimes and G. M. Watson, *ibid.*, 1959, **63**, 1999.
87. C. F. Callis, J. R. Van Wazer and J. S. Metcalf, *J. Am. chem. Soc.*, 1955, **77**, 1468.
88. J. O'M Bockris and G. W. Mellors, *J. phys. Chem.*, 1956, **60**, 1321.
89. A. Wachter and J. H. Hildebrand, *J. Am. chem. Soc.*, 1930, **52**, 4655.
90. B. M. Lepinskikh, O. A. Esin and G. A. Teterin, *Zh. neorg. Khim.*, 1960, **5**, 642.

Table 9.3 DENSITY OF SOME SOLID INORGANIC COMPOUNDS AT ROOM TEMPERATURE

Compound	*Density* g cm^{-3}
$AgBr$	6.47
$AgCl$	5.56
$AgClO_3$	4.43
$AgI(\alpha)$	5.683
$AgNO_2$	4.35
$AlBr_3$	2.64
$AlCl_3$	2.44
AlI_3	3.98
$AsBr_3$	3.54
BI_3	3.35
B_2O_3	1.84
$BaBr_2$	4.78
$BaCl_2$	3.856
$BaCO_3$	4.43
BaF_2	4.89
BaI_2	5.15
BaO	5.68
BaO_2	4.96
$BaSO_4$	4.50
$BeBr_2$	3.47
$BeCl_2$	1.90
BeF_2	1.99
BeO	3.00
$BiBr_3$	4.72
$BiCl_3$	4.75
C_2Cl_6	2.09
$CaBr_2$	3.353
$CaCl_2$	2.15
$CaCO_3$	2.71*
CaF_2	3.18
CaO	3.25/3.38
$CaSO_4$	2.96
$CdBr_2$	5.19
$CdCl_2$	4.05
$CsBr$	4.43
$CsCl$	3.99
CsF	4.12

Compound	*Density* g cm^{-3}
CsI	4.51
$CsNO_3$	3.69
Cs_2SO_4	4.24
Cu_2Cl_2	4.14
Cu_2O	6.0
$GaBr_3$	3.69
$GaCl_3$	2.47
GaI_3	4.15
H_3BO_3	1.44
$HgBr_2$	6.11
Hg_2Br_2	7.31
$HgCl_2$	5.44
Hg_2Cl_2	7.15
HgF_2	8.95
HgI_2	6.36
Hg_2I_2	7.7
$InBr_3$	4.74
$InCl$	4.19
$InCl_2$	3.66
$InCl_3$	3.46
InI_3	4.69
KBF_4	2.55
KBr	2.75
KCl	1.98
KCN	1.52
K_2CO_3	2.43
$K_2Cr_2O_7$	2.68
KF	2.48
$KHSO_4$	2.31
KI	3.13
$KMnO_4$	2.91
KNO_3	2.11
KOH	2.04
K_3PO_4	2.56
K_2SiF_6	3.08 (hex) 2.67 (cub)
K_2SO_4	2.66

Compound	*Density* g cm^{-3}
$LaCl_3$	3.84
$LiBr$	3.46
$LiCl$	2.07
Li_2CO_3	2.11
LiF	2.64
$LiNO_3$	2.38
Li_2SO_4	2.22
$MgCl_2$	2.32
MgF_2	3.0
MgO	3.58
MnO_2	4.9
Na_3AlF_6	2.90
$Na_2B_4O_7$	2.37
$NaBr$	3.20
$NaCl$	2.17
Na_2CO_3	2.53
NaF	2.56
NaI	3.67
$NaNO_3$	2.26
$NaOH$	2.13
$Na_4P_2O_7$	2.53
Na_2SO_4	2.70
Na_2WO_4	4.18
NH_4Cl	1.53
$Ni(CO)_4$†	1.32
NiO	6.67
OsO_4	4.91
PBr_3	2.85
$PbBr_2$	6.66
$PbCl_2$	5.85
$PbCl_4$†	3.18
PbI_2	6.16
$RbBr$	3.35
$RbCl$	2.80
RbF	3.56
RbI	3.55
$RbNO_3$	3.11

Compound	*Density* g cm^{-3}
Rb_2SO_4	3.61
ReF_6	4.25
Re_2O_7	6.10
$ReOF_4$	4.03
$SbBr_3$	4.15
$SbCl_3$	3.14
$SnCl_2$	3.95
SnI_4	4.47
$SrBr_2$	4.22
$SrCl_2$	3.05
SrF_2	4.24
SrI_2	4.55
SrO	4.7
$TeCl_4$	3.26
$ThCl_4$	4.59
$TiBr_4$	2.6
TiF_4	2.80
TiI_4	4.3
TiO_2	4.26 rutile 3.84 anatase
$TlBr$	7.56
$TlCl$	7.00
TlI	7.1
$TlNO_3$	5.8
WCl_5	3.88
WCl_6	3.52
WO_3	7.2
UCl_4	4.87
YCl_3	2.8
$ZnBr_2$	4.20
$ZnCl_2$	2.91
ZnI_2	4.74

* Calcite. † Liquid.

REFERENCES TO TABLE 9.3

'Handbook of Chemistry and Physics', 58th edn, Cleveland, Ohio, 1977–78

Gmelin, 'Handb. anorg. Chem.', 8th edn, Berlin.

J. W. Mellor, 'A Comprehensive Treatise on Inorganic Chemistry', London, 1922–1937.

Table 9.4 ELECTRICAL CONDUCTIVITY OF PURE MOLTEN SALTS
The conductivity in Ω^{-1} cm^{-1} is given at the melting point θ_m, and at 50° intervals beginning at θ, or as an equation in the temperature t, or at the temperatures given in brackets. All temperatures are in °C. Extrapolated conductivities are also bracketed. Principal references are in **bold type**.

Substance	θ_m °C	θ °C	Conductivity Ω^{-1} cm^{-1}: θ_m	θ	$\theta+50$	$\theta+100$	$\theta+150$	$\theta+200$	$\theta+250$	References
AgBr	430	450	2.8561	2.903	3.016	3.119	3.214	3.3	3.378	**133, 96, 19, 18,** 22
AgCl	455	500	3.817	3.972	4.131	4.276	4.407	4.523	4.626	**133, 96, 76, 19, 18,** 107, 22
$AgClO_3$	231	250	0.417	0.474	—	—	—	—	—	**13**
AgI	557	600	2.3	2.35	2.40	2.45	2.5	2.55	—	**19, 12**
$AgNO_3$	210	250	0.67	0.84	1.05	1.24	—	—	—	**106, 100, 88, 87, 86, 85, 79, 76, 55,** 71, 22, 13, 4
$AlBr_3$	97	150	($<10^{-8}$)	(0.02×10^{-6})	0.10×10^{-6}	0.19×10^{-6}	(0.26×10^{-6})	—	—	**27,** 70, 68, 66, 62
$AlCl_3$	193	200	4.6×10^{-7}	5.5×10^{-7}	11.1×10^{-7}	—	—	—	—	**37,** 60, 53, 52, 40
AlI_3	191	200	(1.205×10^{-6})	1.852×10^{-6}	5.645×10^{-6}	7.1×10^{-6}	—	—	—	**36, 37**
$AsBr_3$	31	35	—	2.6×10^{-7}	—	—	—	—	—	**59,** 7
$AsCl_3$	−16	—	1.0×10^{-7}	—	—	—	—	—	—	**39, 6,** 38
AsF_3	−13	0	1.56×10^{-5}	1.84×10^{-5}	2.92×10^{-5}	—	—	—	—	**74**
AsI_3	142					Poor				**8**
BCl_3	−107					Non-conductor				**6**
BF_3	−128.7	−120	—	$<5.0\times10^{-10}$	—	—	—	—	—	**74**
B_2O_3	570	600	—	1.2×10^{-6}	3.0×10^{-6}	5.2×10^{-6}	1.5×10^{-5}	2.3×10^{-5}	3.6×10^{-5}	**101, 99, 81,** 12, 10
		900	—	6.6×10^{-5}	9.7×10^{-5}	1.5×10^{-4}	2.1×10^{-4}	2.9×10^{-4}	4.0×10^{-4}	
		1 200	—	5.3×10^{-4}	—	—	—	—	—	
$BaBr_2$	847	850	—	1.178	1.307	1.440	1.572	1.705	—	**108**
$BaCl_2$	960	975	—	2.085		2.176 (1 000)	2.264 (1 025)	2.354 (1 050)		**108,** 109, 107, 105, 78, 64, 56, 9
						2.472 (1 075)	2.532 (1 100)			
BaI_2	711	750	—	0.784	0.910	1.024	1.136	1.248	1.361	**108**
$Ba(NO_2)_2$	—	280	—	0.182		0.235(300)	0.260(310)			**110**
$BeCl_2$	405	450	0.87×10^{-3}	2.986×10^{-3}	0.016	—	—	—	—	**111**

BeF_2	*c.* 800					Poor				**34**
BeI_2	510					Non-conductor				**5**
$BiBr_3$	—	250	—	0.3214	0.3813	0.4237	0.4487	0.4562	0.4462	**111**
		550	—	0.320	0.298	0.260	0.224	—	—	
$BiCl_3$	230	250	—	0.4122	0.4910	0.546	0.5772	0.5845	0.5680	**134**, 32
	550	—	—	0.515	0.468	—	—	—	—	
BiI_3	—	400	—	0.285	0.300	0.310	0.309	0.305	0.295	**111**
		700	—	0.285	0.265	0.250	0.227	0.205	—	
$Bi_2(MoO_4)_3$	—	708	—	0.274		0.333(747) 0.433(794) 0.480(817) 0.550(847)				**112**
$CaBr_2$	—	750	—	1.420	1.573	1.727	1.882	2.037	2.194	**108**
$CaCl_2$	—	800	—	2.116	2.342	2.515	2.778	3.006	—	**108**, 105, 104, 78, 64, 63, 57, 56, 46, 42, 36, 22, 14, 12, 9, 107, 109
CaF_2	1 420	1 500	—	3.9	4.1	—	—	—	—	**90**
CaI_2		800	—	1.172	1.292	1.390	1.486	1.566	—	**108**
$CdBr_2$	565	600	—	1.125	1.229	1.324	1.419	1.517	—	**108**, 65
$CdCl_2$	568	600	—	1.950	2.085	2.206	2.317	2.425	—	**108**, 91, 86, 82, 65, 31, 22
CdI_2	390	400	—	0.210	0.316	0.425	0.533	0.640	0.750	**108**, 82
$CeCl_3$	—	828	—	1.12		1.17(844) 1.21(858) 1.23(868) 1.29(886) 1.30(895) 1.38(931)				**113**
CeI_3	761	796	—	0.448		0.470(814) 0.499(836) 0.523(860)				**114**
CsBr	635	650	—	0.84	0.97	1.09	1.21	1.31	—	**98**
CsCl	645	650	—	1.12	1.27	1.42	1.57	1.71	1.83	**98**, **31**, 115
CsF	—	—		$\kappa_t=-4.511+1.642\times10^{-2}t-7.632\times10^{-6}t^2$ at 725–921						**109**
CsI	630	650	—	0.69	0.80	0.91	1.00	1.09	—	**98**
$CsNO_3$	417	—		$\kappa_t=-0.2452+1.8878_5\times10^{-3}t$ at 415–491						**116**, 21
CuCl	430	450	3.26	3.32	3.46	3.56	3.66	—	—	**86**, **31**
Cu_2S	1 130	1 150	60	70	80	90	100	120	130	**97**, **16**
		1 450	—	140	160	—	—	—	—	
$ErCl_3$	—	801	—	0.468		0.496(818) 0.531(839)				**117**
FeS	1 200	1 250	1 560	1 540	1 530	1 520	1 510	1 500	1 490	**97**
$GaBr_2$	—	—		$\log\kappa_t=1.142-865/(t+273)$ at 167–189						**118**
$GaBr_3$	125	—	5.0×10^{-8}	—	—	— —	—		—	41
$GaCl_2$	—	—		$\log\kappa_t=1.180-784/(t+273)$ at 167–176						**118**, 2
$GaCl_3$	77	—	10^{-8} (approx.)	—	—	— —	—		—	**41**

Table 9.4 ELECTRICAL CONDUCTIVITY OF PURE MOLTEN SALTS—*continued*

Substance	θ_m °C	θ °C	Conductivity Ω^{-1} cm^{-1} θ_m	θ	$\theta+50$	$\theta+100$	$\theta+150$	$\theta+200$	$\theta+250$	*References*
Ga_2I_4	—	—			log $\kappa_t = 2.887–1\,401/(t+273)$ at 150–211 log $\kappa_t = 1.292–1\,114/(t+273)$ at 211–279 log $\kappa_t = 0.460–653.4(t+273)$ at 279–350					**119**
Ga_2I_6	—	—			log $\kappa_t = -0.068–1\,041/(t+273)$ at 185–222 log $\kappa_t = -1.807–624.4/(t+273)$ at 222–284 log $\kappa_t = -2.660–147.6/(t+273)$ at 284–352 log $\kappa_t = -4.211\,2+822.6/(t+273)$ at 352–400					**119**, 41
$GdBr_3$	—	800	—	0.444			0.476(823) 0.500 (842)			**117**
$GdCl_3$	—	629	—	0.382			0.408(649) 0.465(675) 0.506(698)			**117**
$HgBr_2$	235	240	—	1.38×10^{-4}			2.08×10^{-4}(280) 2.85×10^{-4}(320)			**108**, 120, 75, 23
$HgCl_2$	277	280	—	3.20×10^{-5}			3.57×10^{-5}(290) 3.96×10^{-5}(300)			**108**, 120, 55, 23
Hg_2Cl_2	525	550	1.00	1.03	—	—	—	—	—	**36**
HgI_2	250	260	—	0.029 9	0.024 0			0.018 8(350)		**108**, 120, 83, 55, 23
$HoCl_3$	—	747	—	0.431			0.488(776) 0.526(800) 0.562(819)			**117**
$InBr_3$	436	450	0.168	0.167	0.162	0.156	—	—	—	**37**
InCl	225	250	0.88	1.04	1.33	1.62	(1.89)	—	—	**37**
$InCl_2$	235	250	0.23	0.26	0.36	0.46	0.56	0.64	0.72	**37**
$InCl_3$	498	500	(0.50)	(0.50)	(0.46)	0.41	0.37	0.32	(0.28)	**37**
InI_3	210	250	0.052	0.066	0.080	0.092	(0.100)	—	—	**37**
K_3AlF_6	1 035	1 050	2.33	2.38	—	—	—	—	—	**102**
KBF_4	—	550	—	1.075	1.180	1.243	—	—	—	**121**
KBr	742	—			$\kappa_t = -1.327_1 + 5.710_0\times10^{-3}t - 2.313_9\times10^{-6}t^2$ at 760–980					**122**, 98, 82, 31, 21, 70, 12, 4
KCl	770	800	2.12	2.22	2.35	2.46	2.56	2.65	2.73	**92, 82, 78, 77, 63, 44,** 107, 68, 56, 54, 49, 37, 22
		1 100	—	2.79	—	—	—	—	—	
$KClO_3$	368	—	4.19	—	—	—	—	—	—	**1**
K_2CO_3	895	900	2.020	2.035	2.178 3	2.322 1	—	—	—	**135**
$K_2Cr_2O_7$	398	400	0.200	0.204	0.298	0.389	(0.474)	—	—	**83, 11**
KF	857	900	—	4.4728	4.601 7	4.697 5	4.760 3	—	—	**109**, 121, 102, 47, 42, 29, 21
					$\kappa_t = -3.493 + 1.480\times10^{-2}t - 6.608\times10^{-6}t^2$ at 869–1 040					

$KHSO_4$	212	301	0.049	0.141	—	—	—	—	—	**72**
KI	682	700	1.82	1.32	1.42	1.52	1.60	1.66	—	**92, 82, 31,** 21, 12, 4
K_2MoO_4	—	—			$\kappa_t = -21.0210 + 4.4823 \times 10^{-2}t - 22.5066 \times 10^{-6}t^2$ at 931–988					**123**
KNH_2	330	340	—	0.389	—	—	—	—	—	**20**
KNO_3	337	350	0.62	0.66	0.82	0.96	1.10	1.24	1.38	**106, 93, 87, 86, 84, 83, 82, 61, 55, 22, 21,** 116, 15, 11, 1
KOH	410				*See* ref.					35
K_2SO_4	1 069	1 100	(1.78)	1.84	1.94	—	—	—	—	**9**
K_2TaF_7	—	750	—	0.750	0.919	0.981	(1.052)	—	—	**121**
K_2TiF_6	—	850	—	1.359	1.462	1.559	(1.650)	—	—	**121**, 124
K_2WO_4	—	—			$\kappa_t = -8.2953 + 1.7633 \times 10^{-2}t - 8.0295 \times 10^{-6}t^2$ at 946–1 024					**125**
$LaCl_3$	885	900	1.550	1.602	1.748	1.853 1	1.917 0	—	—	**36, 109**
Li_3AlF_6	—	800	—	3.45	3.65	3.80	3.95	—	—	**102**
LiBr	550	600	4.69	4.97	5.23	5.49	5.73	—	—	**98**, 65
LiCl	614	650	5.730 6	5.921 9	6.171 4	6.402 1	6.613 9	6.807	6.981 3	**92, 82, 77, 49, 31, 107**, 60, 42
$LiClO_3$	—	131.8	—	0.115 1		0.123 1(135.7) 0.125 8 (136.5) 0.137 0(140.8) 0.142 0(143.0)				**126**
LiF	844	875	—	8.663		8.889(915) 9.058(958) 9.216 (1 008) 9.306(1 037)				**121**, 109, 102, 42
LiI	465	500	—	3.800	4.067	4.250	4.417	4.570	4.703	**127**, 98
		800	—	4.830	4.947	5.045	—	—	—	
Li_2MoO_4	—	—			$\kappa t = 43.2596 - 10.71 \times 10^{-2}t + 72.668 \times 10^{-6}t^2$ at 790–939					**123**
$LiNO_3$	254	300	0.803 8	1.054 5	1.335 7	1.625 9	1.925 1	2.233 3	—	**138**, 13, 21
Li_2WO_4	—	—			$\kappa_t = -7.5067 + 1.8733 \times 10^{-2}t - 8.8034 \times 10^{-6}t^2$ at 762–903					**125**
$MgBr_2$	—	750	—	0.766	0.857	0.950	1.045	1.138	—	**108**
$MgCl_2$	714	750	1.014	1.085	1.183	1.283	1.383	1.483	—	**108**, 91, 78, 63, 54, 48, 44, 31
MgI_2	—	650	—	0.408	0.496	0.585	0.675	0.765	0.856	**108**
$MnCl_2$	—	850	—	(1.436)	1.578	1.690	—	—	—	**107**, 128
$MoCl_5$	194	200	—	0.3×10^{-6}	6.3×10^{-6}	—	—	—	—	**32**
MoO_3	795	—			$\kappa_t = -5.7537 + 1.4563 \times 10^{-2}t - 8.1949 \times 10^{-6}t^2$ at 823–891					**123**, 89
Na_3AlF_6	100 0	104 0	2.799	2.901	2.997	—	—	—	—	**102, 80, 77,** 73, 69, 51, 50, 29
NaBr	745	800	2.87	3.06	3.22	3.37	3.49	—	—	**98, 12,** 4

Table 9.4 ELECTRICAL CONDUCTIVITY OF PURE MOLTEN SALTS—*continued*

Substance	θ_m °C	θ °C	Conductivity Ω^{-1} cm^{-1} θ_m	θ	$\theta+50$	$\theta+100$	$\theta+150$	$\theta+200$	$\theta+250$	*References*
NaCl	800	850	3.58	3.74	3.87	4.02	4.16	4.29	4.39	**104, 92, 78, 77, 63,** 107‡
NaCN	562	700	—	1.15	1.27	1.39	—	—	—	**42**
Na_2CO_3	850	900	2.834 7	3.029	3.222	—	—	—	—	**135**
NaF	992	1 003	—	4.960		4.985(1 018) 5.082(1 047) 5.111(1 059) 5.179(1 086) 5.209(1 099) 5.271(1 122) 5.335(1 138)				**121**, 102, 80, 50, 29
$NaHSO_4$	182	200	0.067	0.094	0.168	0.241	—	—	—	**72**
NaI	655	700	2.30	2.41	2.55	2.68	2.80	2.92	—	**98, 82, 12,** 4
Na_2MoO_4	687	750	(1.15)	1.26	1.37	1.49	1.57	1.66	1.74	**89, 21**
		1 050	—	1.82	1.90	1.98	2.07	2.14	2.22	
		1 350	—	2.30	2.37	—	—	—	—	
$NaNH_2$	208	210	—	0.593	—	—	—	—	—	**20**
$NaNO_2$	270	300	—	1.34	1.62	1.89	2.16	—	—	**91, 82**
$NaNO_3$	310	350	0.94	1.16	1.36	1.55	1.73	—	—	**95, 93, 86, 82, 79, 55, 21,** 22, 11
NaOH	322	350	2.125	2.377	2.827	3.277	—	—	—	**35**
$NaPO_3$	625	650	0.36	0.425	0.55	0.675	0.80	0.925	1.05	**12**
		950		1.175	1.30	1.42	1.54			
Na_2SO_4	885	900	(2.19)	2.23	2.37	2.50	2.64	2.77	(2.91)	**9**
Na_2TaF_7	—	750	—	0.750	0.919	0.981	(1.052)	—	—	**121**
Na_2WO_4	696	—		$\kappa_t = -1.4108 + 0.3451 \times 10^{-2}\,t - 0.2058 \times 10^{-6}\,t^2$ at 706–871						**125**, 21
$NbCl_5$	210	228	—	0.22×10^{-6}	—	—	—	—	—	**26**
$NdBr_3$	—	713	—	0.466			0.498(731) 0.519(743) 0.541(759) 0.563(772) 0.603(797)			**117**
$NdCl_3$	761	800	(0.587)	0.692	0.821.	0.945	(1.058)	—	—	**32**
NdI_3	787	799	—	0.396			0.416(818) 0.440(842)			**114**
NH_4HSO_4	145	150	0.079	0.086	0.154	—	—	—	—	**72**
NH_4NO_3	169.6	200	0.31	0.41	(0.53)	—	—	—	—	**4,** 1
$PbBr_2$	370	400	(0.57)	0.68	0.85	1.03	1.19	—	—	**96, 11,** 22
$PbCl_2$	498	500	5.829 5	5.839 8	6.088	6.316 6				**136, 137, 96, 107,** 11
$PbCl_4$	−15	0	—	8.0×10^{-7}	—	—	—	—	—	**32**
PbI_2	—	450	—	0.432 6	0.462 8	0.493 9	0.530 6	0.390 6(402.7)		**127**
$PbMoO_4$	—	1 098	—	0.938		0.950(1 107)	0.958(1 116)	0.963(1 118)		**112**
$PrBr_3$	—	727	—	0.52			0.56(750) 0.60(770)			**117**

$PrCl_3$	—	—	$\kappa_t = -1.189+2.75\times10^{-3}t$ at 800–860							**129**, 32
PrI_3	738	763	—	0.399	0.426(786) 0.452(809)					**114**
RbBr	692	700	—	1.132	1.263	1.360	1.454	1.526	1.586	**127, 98**
		1 000	—	1.637	1.679	1.716	1.749	1.778	1.806	
RbCl	722	750	1.50	1.58	1.72	1.85	1.97	—	—	**98, 31**
RbI	647	650	—	0.86	0.96	1.04	1.12	1.19	1.25	**98**
$RbNO_3$	316	350	0.431	0.509	0.618	0.72	0.816 2	—	—	**21**
$SbBr_3$	97	100	228×10^{-6}	236×10^{-6}	355×10^{-6}	450×10^{-6}	—	—	—	**23**, 45
$SbCl_3$	73	—	0.5×10^{-6}	—	—	—	—	—	—	**43, 33**, 25
SbF_3	290	300	—	0.065	—	—	—	—	—	**94**
SbF_5	6	50	0.4×10^{-8}	5.85×10^{-8}	41.0×10^{-8}(80)		—	—	—	**74**
SbI_3	170	200	3.91×10^{-8}	3.91×10^{-8}	4.83×10^{-8}	5.68×10^{-8}	6.44×10^{-8}	(7.11×10^{-8})	—	**23**
$ScCl_3$	960	980	0.56	0.63	0.66 (1 000)		—	—	—	**37**, 32, 28
SiO_2	—	—	*ca.* 10^{-5}–*ca.* 10^{-4} at 1 800–2 600							**130**
$SnCl_2$	247	250	0.89	0.90	1.11	1.38	1.67	(1.93)	—	**86, 36**, 3
$SrBr_2$	—	650	—	0.728	0.980	1.142	1.295	1.446	1.600	**108**
$SrCl_2$	872	900	1.989	2.096	2.285	2.477	2.670	—	—	**108**, 9
SrI_2	—	550	—	0.516	0.637	0.758	0.875	0.989	1.100	**108**
		850	—	1.212	1.320	1.422	1.525	—	—	
$TaCl_5$	221	235	—	0.3×10^{-6}	—	—	—	—	—	**26**
$TeCl_2$	175	200	(0.005)	0.034	0.089	0.145	—	—	—	**32**
$TeCl_4$	224	250	0.100	0.131	0.186	—	—	—	—	**32**
$ThCl_4$	770	800	(0.54)	(0.61)	0.71		(0.86)		—	**37**
TiI_4	150	—	$1\times10^{-7}-6\times10^{-7}$ (180–240)							**41**
TlBr	460	—	$\kappa_t = -0.461_1+3.105_0\times10^{-3}t-0.732_6\times10^{-6}t^2$ at 480–705							**131**, 18
TlCl	430	450	1.088	1.164	1.350	1.532	1.700	(1.865)	—	**83, 17**, 22
$TlCl_3$	*c.* 60	—	$< 5.0\times10^{-5}$	—	—	—	—	—	—	**37**
TlI	440	450	(0.530)	0.551	0.651	0.747	0.840	—	—	**17**
$TlNO_3$	206	250	(0.35)	0.47	0.60	—	—	—	—	**84, 55**, 22
Tl_2S	448		*See* ref.							24
UCl_4	567	—	$\kappa_t = -0.318+1.22\times10^{-3}t$							**132**, 32
WCl_5	248	250	0.61×10^{-6}	0.67×10^{-6}	1.84×10^{-6}	—	—	—	—	**32**, 25
WCl_6	275	300	1.86×10^{-6}	2.60×10^{-6}	4.05×10^{-6} (320)		6.94×10^{-6} (330)		—	**32**, 25
YCl_3	721	750	0.42	0.48	0.58	0.69	0.79	—	—	**37**
$ZnBr_2$	—	400	—	0.020	0.043	0.085	0.145	0.215	0.310	**108**
$ZnCl_2$	331	350	—	0.007	0.020	0.050	0.090	0.156	0.243	**108**, 103, 31, 1
		650	—	0.350	0.462	—	—	—	—	
ZnI_2	—	450	—	0.060	0.118	0.190	0.267	—	—	**108**

‡ *See also* 64, 60, 58, 57, 56, 54, 53, 52, 48, 46, 42, 31, 22, 4.

REFERENCES TO TABLE 9.4

1. Foussereau, *J. Phys. Radium*, 1885, **4**, 189.
2. W. Hampe, *Chem. Zeit.*, 1888, **11**, 1109.
3. L. Graetz, *Wied Ann.*, 1890, **40**, 28.
4. M. Poincaré, *Ann. Chim.* (*Phys.*), 1890, **21**, 289.
5. P. Lebeau, *Compt. rend.*, 1898, **126**, 1272; *Ann. Chim.* (*Phys.*), 1899, (7) **16**, 491.
6. P. Walden, *Z. anorg. Chem.*, 1900, **25**, 209, 227.
7. P. Walden, *ibid.*, 1902, **29**, 371.
8. J, H. Mathews, *J. phys. Chem.*, 1905, **9**, 641.
9. K. Arndt, *Z. Elektrochem.*, 1906, **12**, 337.
10. K. Arndt, *Ber. deut. chem. Ges.*, 1907, **40**, 2938.
11. R. Lorenz and H. T. Kalmus, *Z. physikal. Chem.*, 1907, **59**, 17.
12. K. Arndt and A. Gessler, *Z. Elektrochem.*, 1908, **14**, 662.
13. H. M. Goodwin and R. D. Mailey, *Phys. Rev.*, 1908, **26**, 28.
14. K. Arndt and W. Löwenstein, *Z. Elektrochem.*, 1909, **15**, 789.
15. A. H. W. Aten, *Z. physikal. Chem.*, 1911, **78**, 1.
16. Gornemann and von Rauschenplat, *Metallurgie*, 1912, **9**, 473, 505.
17. C. Tubandt, 'Nernst Festschrift', Halle: 1912, p. 446.
18. C. Tubandt and E. Lorenz, *Z. physikal. Chem.*, 1914, **87**, 513.
19. R. Lorenz and A. Höchberg, *Z. anorg. Chem.*, 1916, **94**, 305.
20. Wöhler and Stang-Lund, *Z. Elektrochem.*, 1918, **24**, 261.
21. F. M. Jaeger and B. Kapma, *Z. anorg. Chem.*, 1920, **113**, 27.
22. C. Sandonnini, *Gazz. Chim. Ital.*, 1920, **50**, 289.
23. G. von Hevesy, *Kgl. Danske Vid. Selsk. Medd.*, 1921, *III*, 13.
24. Pélabon, *Compt. rend.*, 1921, **173**, 142.
25. W. Biltz, *Z. phys. Chem.*, 1922, **100**, 52.
26. W. Biltz and A. Voigt, *Z. anorg. Chem.*, 1922, **120**, 71.
27. W. Biltz and A. Voigt, *ibid.*, 1923, **126**, 39.
28. W. Biltz and W. Klemm, *ibid.*, 1923, **131**, 22.
29. K. Arndt and W. Kalass, *Z. Elektrochem.*, 1924, **30**, 12.
30. K. Arndt and G. Ploetz, *Z. phys. Chem.*, 1924, **110**, 237.
31. W. Biltz and W. Klemm, *ibid.*, 1924, **110**, 318.
32. A. Voigt and W. Biltz, *Z. anorg. Chem.*, 1924, **133**, 277.
33. Klemensiewicz, *Z. physikal. Chem.*, 1924, **113**, 28.
34. B. Neumann and H. Richter, *Z. Elektrochem.*, 1925, **21**, 484.
35. K. Arndt and G. Ploetz, *Z. phys. Chem.*, 1926, **121**, 439.
36. W. Klemm and W. Biltz, *Z. anorg. Chem.*, 1926, **152**, 225.
37. W. Biltz and W. Klemm, *ibid.*, 1926, **152**, 267.
38. M. Ussanowitsch. *Z. phys. Chem.*, 1929, *A* **140**, 429.
39. V. S. Finkelstein, *J. Russ. Phys.-Chem. Soc.*, 1930, **62**, 161.
40. V. A. Plotnikov and P. T. Kalita, *ibid.*, 1930, **62**, 2195.
41. W. Klemm and W. Tilk, *Z. anorg. Chem.*, 1932, **207**, 161.
42. E. Ryschkewitsch, *Z. Elektrochem*, 1933, **39**, 531.
43. M. Ussanowitsch and F. Terpugov, *Z. phys. Chem.*, 1933, *A* **165**, 39.
44. S. Karpachev. A. Stromberg and O. Poltoratzkaya, *J. phys. Chem.* (*USSR*), 1934, **5**, 793.
45. M. Ussanowitsch and V. Serebrennikov, *J. Gen. Chem.* (*USSR*), 1934, **4**, 230.
46. V. P. Barzakovskii, *Sbornik Trudov Pervoi Vsesoyuznoi Konferentzii Nevodnuim Rastvoram Ukrain. Akad. Nauk. Inst. Khim.* (Proc. 1st All-Union Conf. Nonaqueous Solutions), 1935, 143.
47. K. P. Batashev and A. Zhurin, *Metallurg.* (*USSR*), 1935, **10**, 67.
48. K. P. Batashev, *ibid.*, 1935, **10**, 100.
49. S. Karpachev, A. Stromberg and V. N. Podchainova, *J. Gen. Chem., Moscow*, 1935, **5**, 1517.
50. K. P. Batashev, *Legkie Metally*, 1936, **10**, 48.
51. J. W. Cuthbertson and J. Waddington, *Trans. Faraday Soc.*, 1936, **32**, 745.
52. Y. Yamaguti and S. Sisido, *J. Chem. Soc., Japan*, 1938, **59**, 1311.
53. A. I. Kryagova, *J. Gen. Chem.* (*USSR*), 1939, **9**, 2061.
54. A. A. Scherbakov and B. F. Markov, *J. phys. Chem.* (*USSR*), 1939, **13**, 621.
55. A. G. Bergman and I. M. Chagin, *Bull. Acad. Sci. URSS*, Cl. Sci. chim., 1940, 727.
56. V. P. Barzakovskii, *ibid.*, 1940, 825.
57. V. P. Barzakovskii, *J. appl. Chem.* (*USSR*), 1940, **13**, 1117.
58. N. A. Belozerskii and B. A. Freidlina, *ibid.*, 1941, **14**, 466.
59. A. Bernshtein, *J. Gen. Chem.* (*USSR*), 1941, **11**, 901.
60. Y. Yamaguti and S. Sisido, *J. Chem. Soc. Japan*, 1941, **62**, 304.
61. E. R. Natsvilischvili and A. G. Bergman, *Bull. Acad. Sci., URSS*, Cl. Sci. chim., 1943, 23.
62. E. Y. Gorenbein, *J. Gen. Chem.* (*USSR*), 1945, **15**, 720.
63. E. K. Lee and E. P. Pearson, *Trans. Electrochem. Soc.*, 1945, **88**, 171.
64. A. F. Alabyshev and N. Ya. Kulakovskaya, *Trudy Leningrad. Tekhnol. Inst. im. Leningrad. Soveta*, 1946, No. 12, 152.
65. H. Bloom and E. Heymann, *Proc. Roy. Soc.*, 1946–7, *A* **188**, 392.
66. E. Ya. Gorenbein, *J. Gen. Chem.* (*USSR*), 1947, **17**, 873.
67. N. M. Tarasova, *J. phys. Chem.* (*USSR*), 1947, **21**, 825.

68. E. Ya. Gorenbein, *J. Gen. Chem.* (*USSR*), 1948, **18**, 1427.
69. M. Frejaques, *Bull. Soc. Franc. Elec.*, 1949, **9**, 684.
70. E. Ya. Gorenbein and E. E. Kriss, *J. Gen. Chem.* (*USSR*), 1949, **19**, 1978.
71. I. M. Bokhovkin, *ibid.*, 1950, **20**, 397.
72. S. E. Rogers and A. R. Ubbelohde, *Trans. Faraday Soc.*, 1950, **46**, 1051.
73. A. Vayna, *Alluminio*, 1950, **19**, 215.
74. A. A. Woolf and N. N. Greenwood, *J. chem. Soc.*, 1950, 2200.
75. G. Jander and K. Brodersen, *Z. anorg. Chem.*, 1951, **264**, 57.
76. R. C. Spooner and F. E. W. Wetmore, *Canad. J. Chem.*, 1951, **29**, 777.
77. J. D. Edwards, C. S. Taylor, A. S. Russell and L. F. Maranville, *J. electrochem. Soc.*, 1952, **99**, 527.
78. R. W. Huber, E. V. Potter and H. W. St. Clair, *Rep. Invest. US Bur. Mines*, No. 4858, 1952.
79. J. Byrne, H. Fleming and F. E. W. Wetmore, *Canad. J. Chem.*, 1952, **30**, 922.
80. J. D. Edwards, C. S. Taylor, L. A. Cosgrove and A. S. Russell, *Trans. Electrochem. Soc.*, 1953, **100**, 508.
81. L. Shartsis, W. Capps and S. Spinner, *J. Am. ceram. Soc.*, 1953, **36**, 319.
82. H. Bloom, I. W. Knaggs, J. J. Molloy and D. Welch, *Trans. Faraday Soc.*, 1953, **49**, 1458.
83. I. N. Beylaev, *Izvest. Sekt. Fiziko-Khim. Anal.*, 1953, **23**, 176.
84. P. I. Protsenko and N. P. Popovskaya, *Zhur. obshch. Khim.*, 1954, **24**, 2119.
85. P. I. Protsenko and N. P. Popovskaya, *Zh. fiz. Khim.*, 1954, **28**, 299.
86. K. Sakai, *J. chem. Soc. Japan, Pure Chem. Sect.*, 1954, **75**, 182.
87. V. D. Polyakov, *Izvest. Sekt. Fiziko-Khim. Anal.*, 1955, **26**, 147.
88. P. I. Protsenko, *ibid.*, 1955, **26**, 173.
89. K. B. Morris, M. I. Cook, C. Z. Sykes and M. B. Templeman, *J. Am. chem. Soc.*, 1955, **77**, 851.
90. T. Baak, *Acta. Chem. Scand.*, 1955, **9**, 1406.
91. K. Sakai and S. Hayashi, *J. chem. Soc. Japan, Pure Chem. Sect.*, 1955, **76**, 101.
92. E. R. van Artsdalen and I. S. Yaffe, *J. phys. Chem.*, 1955, **59**, 118.
93. C. Kroger and P. Weisgerber, *Z. phys. Chem.* (*Frankfurt*), 1955, **5**, 192.
94. A. A. Woolf, *J. chem. Soc.*, 1955, 279.
95. Yu. K. Delimarskii, I. N. Sheiko and V. G. Fenchenko, *Zh. fiz. Khim.*, 1955, **29**, 1499.
96. B. S. Harrap and E. Heymann, *Trans. Faraday Soc.*, 1955, **51**, 259.
97. G. M. Pound, G. Derge and G. Osuch, *Trans. Amer. Inst. Min. Met. Eng.*, 1955, **203**, 481.
98. I. S. Yaffe and E. R. van Artsdalen, *J. phys. Chem.*, 1956, **60**, 1125.
99. J. D. Mackenzie, *Trans. Faraday Soc.*, 1956, **52**, 1564.
100. H. C. Cowen and H. J. Axon, *ibid.*, 1956, **52**, 242.
101. W. C. Phelps and R. E. Grace, *Trans. Amer. Inst. Min. Met. Eng.*, 1957, **209**, 1447.
102. E. W. Yim and M. Feinleib, *J. electrochem. Soc.*, 1957, **104**, 626.
103. F. R. Duke and R. A. Fleming, *ibid.*, 1957, **104**, 251.
104. J. B. Story and J. T. Clarke, *Trans. Amer. Inst. Min. Met. Eng.*, 1957, **209**, 1449.
105. V. A. Kochinashvili and V. P. Barzakovskii, *Zhur. priklad. Khim.*, 1957, **30**, 1755.
106. F. R. Duke and R. A. Fleming, *J. electrochem. Soc.*, 1958, **105**, 412.
107. H. Winterhager and L. Werner, *ForschBer. Wirtn.-u. VerkMinist. NRhein.-Westf.*, 1956, **341**.
108. J. O'M. Bockris, E. Crook, H. Bloom and N. E. Richards, *Proc. R. Soc.*, 1960, *A* **255**, 558.
109. I. S. Yaffe and E. R. van Artsdalen, *Chem. Semi-Ann. Progr. Rep.* No. 2159, *Oak Ridge Natl. Lab.*, 1956, p. 77.
110. P. I. Protsenko and O. N. Shokina, *Zh. neorg. Khim.*, 1960, **5**, 437.
111. L. F. Grantham and S. J. Yosim, *J. Chem. Phys.*, 1963, **38**, 1671.
112. K. B. Morris, M. McNair and G. Koops, *J. chem. Engng Data*, 1962, **7**, 224.
113. H. R. Bronstein, A. S. Dworkin and M. A. Bredig, *J. phys. Chem.*, 1962, **66**, 44.
114. A. S. Dworkin, R. A. Sallach, H. R. Bronstein, M. A. Bredig and J. D. Corbett, *ibid.*, 1963, **67**, 1145.
115. B. F. Markov and V. D. Prusyazhnyi, *Ukr. khim. Zh.*, 1962, **28**, 268.
116. N. V. Smith and E. R. van Artsdalen, *Chem. Semi-Ann. Progr. Rep.* No. 2159, *Oak Ridge Natl. Lab.*, 1956, p. 80.
117. A. S. Dworkin, H. R. Bronstein and M. A. Bredig, *J. phys. Chem.*, 1963, **67**, 2715.
118. N. N. Greenwood and I. J. Worrall, *J. chem. Soc.*, 1958, 1680.
119. E. F. Riebling and C. E. Erickson, *J. phys. Chem.*, 1963, **67**, 307.
120. G. J. Janz and J. D. E. McIntyre, *J. electrochem. Soc.*, 1962, **109**, 842.
121. H. Winterhager and L. Werner, *ForschBer. Wirt.-u. VerkMinist. NRhein.-Westf.*, 1957, **438**.
122. E. R. Buckle and P. E. Tsaoussoglou, *J. chem. Soc.*, 1964, 667.
123. K. B. Morris and P. L. Robinson, *J. phys. Chem.*, 1964, **68**, 1194.
124. F. M. Kolomitskii and V. D. Ponomarev, *Izv. Akad. Nauk, kazakh, SSR, Ser. metall. obog. ogneu.*, 1959, **1**, 21.
125. K. B. Morris and P. L. Robinson, *J. chem. Engang Data*, 1964, **9**, 444.
126. A. N. Campbell, E. M. Kartzmark and D. F. Williams, *Can. J. Chem.*, 1962, **40**, 890.
127. W. Karl and A. Klemm, *Z. Naturf.*, 1964, **19***A*, 1619.
128. I. G. Murgulescu and S. Zuca, *Studii Cerc. Chim.*, 1959, **7**, 325.
129. A. S. Dworkin, H. R. Bronstein and M. A. Bredig, *J. phys. Chem.*, 1962, **66**, 1201.
130. M. Panish, *ibid.*, 1959, **63**, 1337.
131. E. R. Buckle and P. E. Tsaoussoglou, *Trans. Faraday Soc.*, 1964, **60**, 2144.
132. T. Kuroda and T. Suzuki, *J. electrochem. Soc. Japan*, 1961, **29**, E215.
133. Y. Doucet and M. Bizouard, *Compt. Rend.*, 1960, **250**, 73.
134. L. F. Grantham, *J. Chem. Phys.*, 1965, **43**, 1415.
135. G. J. Janz and M. R. Lorenz, *J. electrochem. Soc.*, 1961, **108**, 1052.
136. H. Bloom and E. Heymann, *Proc. Roy. Soc.* (*London*), 1947, *A* **188**, 392.
137. M. F. Lantratov and O. F. Moiseeva, *Zh. Fiz. Khim.*, 1960, **34**, 367.
138. L. A. King and F. R. Duke, *J. electrochem Soc.*, 1964, **111**, 712.

Table 9.5 ELECTRICAL CONDUCTIVITY OF MOLTEN BINARY SALT SYSTEMS AND OTHER MIXED IONIC MELTS

The electrical conductivity in Ω^{-1} cm^{-1} at temperature t°C and composition p(wt.%) of the first-named constituent is given as κ_t, or empirical equations for κ as a function of temperature within the stated range are given at each composition. Principal references are in **bold type**.

System								
AgBr–AgCl	p	0	24.7	46.7	66.9	84.0	100	
Ref. **59**, **3**	κ_{450}	—	3.52	3.33	3.16	3.00	2.89	
	κ_{500}	3.90	3.70	3.50	3.31	3.16	3.02	
	κ_{550}	4.07	3.86	3.64	3.44	3.28	3.13	
	κ_{600}	4.21	3.99	3.77	3.55	3.38	3.21	
AgBr–KBr	p	52.2	61.9	78.0	90.1	100		
Ref. **59**	κ_{400}	—	—	1.42	1.98	—		
	κ_{500}	—	1.29	1.70	2.23	3.03		
	κ_{600}	1.43	1.55	1.91	2.43	3.21		
AgBr–KCl	p	38.6	62.6	79.0	91.0	100.0		
Ref. **77**, 78	κ_{500}	—	—	1.678	2.210	3.020		
	κ_{600}	—	1.680	1.942	2.432	3.215		
	κ_{700}	1.872	1.933	2.172	2.602	3.374		
	κ_{800}	2.127	2.150	2.340	2.742	3.505		
AgCl–AgNO$_3$	p	0	9.00	12.5	15.5	20.0	30.0	40.0
Ref. **36**	κ_{220}	0.706	0.705	0.710	0.712	—	—	—
	κ_{270}	0.926	0.926	0.931	0.932	0.950	1.005	1.092
	κ_{320}	1.132	1.132	1.138	1.138	1.161	1.223	1.312
AgCl–AlBr$_3$	p				13.29			
Ref. **13**	t	100	110	125	130	140	155	170
	κ	1.07	1.21	1.61	1.68	1.90	2.67	3.22
AgCl–KBr	p	23.2	44.6	64.4	83.0	100.0		
Ref. **77**, 78	κ_{500}	—	—	1.742	2.600	3.965		
	κ_{600}	1.400	1.570	2.008	2.871	4.277		
	κ_{700}	1.637	1.805	2.215	3.065	4.526		
	κ_{800}	1.815	1.980	2.390	3.220	4.716		
AgCl–KCl	p	50.2	63.0	72.5	87.9	100		
Ref. **59**	κ_{400}	—	—	—	2.30	—		
	κ_{500}	—	—	1.93	2.70	3.90		
	κ_{600}	—	2.01	2.24	3.00	4.21		
	κ_{700}	2.16	2.31	2.51	3.22	4.46		
AgCl–PbCl$_2$	p	0	11.2	25.6	43.5	76.3	100	
Ref. **59**	κ_{400}	—	—	1.27	1.67	2.62	—	
	κ_{500}	1.43	1.59	1.82	2.20	3.11	3.90	
	κ_{600}	1.92	2.07	2.27	2.61	3.46	4.21	
	κ_{700}	2.33	2.51	2.68	2.96	3.73	4.46	
AgCl–TlCl	p	0	20.6	37.0	58.0	77.2	100	
Ref. **3**	κ_{500}	1.215	1.470	1.711	2.260	2.925	3.653	
AgI–AgNO$_3$	p	0	10	20	30	40	50	60
Ref. **62**, **33**	κ_{150}	—	—	—	0.38	0.44	0.54	0.65
	κ_{200}	—	—	0.58	0.60	0.66	0.75	0.86
	κ_{250}	0.80	0.78	0.78	0.80	0.85	0.93	1.03
	κ_{300}	1.00	0.98	0.98	1.00	1.05	1.12	1.21
AgNO$_3$–Cd(NO$_3$)$_2$	p	23.5	32.3	41.7	51.8	62.6	74.1	100
Ref. **46**	κ_{150}	—	—	0.07	0.15	—	—	—
	κ_{200}	—	0.11	0.19	0.28	0.38	0.54	—
	κ_{250}	0.15	0.23	0.34	0.44	0.55	0.65	0.83
	κ_{300}	0.27	0.38	0.49	0.61	0.73	0.86	—
AgNO$_3$–HgI$_2$	p	4.0	13.8	27.2	35.9	46.6	59.9	77.0
Ref. **53**, **19**	κ_{100}	—	—	—	0.08	0.07	0.08	0.08
	κ_{150}	0.02	0.11	0.12	0.20	0.15	0.22	0.31
	κ_{200}	0.09	0.24	0.22	0.30	0.27	0.36	0.51

Table 9.5 ELECTRICAL CONDUCTIVITY OF MOLTEN BINARY SALT SYSTEMS AND OTHER MIXED IONIC MELTS—*continued*

System								
$AgNO_3$–KNO_3	p	0	20	40	60	80	90	100
Ref. **71**, **61**, **51**, 79	κ_{250}	—	—	0.47	0.54	0.66	0.74	0.84
	κ_{300}	—	0.56	0.64	0.73	0.85	0.93	1.05
	κ_{350}	0.66	0.73	0.81	0.91	1.04	1.13	1.24
$AgNO_3$–$LiNO_3$	p	0	38.0	62.0	78.6	90.7	100	
Ref. **61**	κ_{250}	0.78	0.79	0.80	0.81	0.82	0.83	
	κ_{300}	1.05	1.04	1.04	1.04	1.04	1.04	
$AgNO_3$–$NaNO_3$	p	0	10	20	30	40	60	100
Ref. **38**	κ_{290}	—	0.900	0.907	0.910	0.920	0.941	1.010
	κ_{330}	1.076	1.088	1.094	1.104	1.103	1.118	1.173
	κ_{370}	1.251	1.267	1.264	1.277	1.272	1.289	—
$AgNO_3$–$RbNO_3$	p	22.3	33.0	43.4	63.3	72.9	82.2	91.2
Ref. **52**	κ_{200}	—	0.24	0.28	0.37	0.43	0.49	0.57
	κ_{250}	0.33	0.38	0.42	0.54	0.61	0.69	0.77
	κ_{300}	0.47	0.52	0.58	0.71	0.78	0.87	0.96
$AgNO_3$–$TlNO_3$	p	10	35	45	50	60	65	90
Ref. **29**, 3	κ_{100}	—	—	0.137	0.123	0.145	—	—
	κ_{200}	0.336	0.405	0.424	0.427	0.468	0.488	0.560
	κ_{300}	0.607	0.712	0.735	0.753	0.808	0.824	0.943
$AlBr_3$–$HgBr_2$	p	59.7	60.5	63.9	68.0	79.0	89.6	92.2
Ref. **25**, 3	$10^2\cdot\kappa_{110}$	1.38	1.34	1.22	1.00	0.419	0.064	0.013
	$10^2\cdot\kappa_{1140}$	2.45	2.34	2.00	1.61	0.573	0.064	0.013
$AlBr_3$–KBr	p	81.8	83.2	84.7	86.2	87.5	88.0	
Ref. **30**, 23, 2	κ_{110}	0.036 0	0.033 1	0.030 1	0.026 9	(0.024 2)	0.023 3	
	κ_{140}	0.060 1	0.054 6	0.049 3	0.043 9	(0.039 6)	0.037 9	
$2AlBr_3\cdot KCl$	t	90	100	140	160	170		
Ref. **28**, 13	κ	0.023 2	0.029 5	0.058 7	0.074 2	0.082 9		
$4AlBr_3\cdot LiCl$	t	80	100	120	140	170	180	
Ref. **13**	κ	0.011 0	0.013 9	0.017 0	0.020 4	0.024 2	0.025 1	
$AlBr_3$–$NaBr$	p	83.8	84.7	85.6	86.8	87.9	88.9	
Ref. **30**	κ_{120}	0.057 8	0.053 3	0.049 2	0.044 1	0.038 2	0.034 6	
	κ_{130}	0.067 4	0.061 3	0.056 5	0.050 6	0.044 4	0.040 2	
	κ_{140}	0.077 8	0.068 7	0.064 1	—	0.050 0	0.045 7	
$2AlBr_3\cdot NaBr$	t	110	130	150	170			
Ref. **28**	κ	0.051 5	0.069 2	0.088 7	0.108 7			
$AlBr_3$–$NaCl$	p	90.73	93.69	94.83		at $p = 80.1$, $\kappa_{170} = 0.036\,5$		
Ref. **13**	κ_{110}	0.025 2	0.026 2	0.022 0				
	κ_{130}	0.041 5	0.031 0	0.025 2				
	κ_{150}	—	0.042 8	0.031 3				
$AlBr_3$–NH_4Br	p	84.50	86.90	88.98	89.42	90.81		
Ref. **25**	κ_{110}	0.022	0.019	0.017	0.016	0.014		
	κ_{130}	0.033	0.027	—	0.022	0.020		
	κ_{150}	0.043	0.036	0.031	0.029	0.025		
$2AlBr_3\cdot NH_4Br$	t	110	160					
Ref. **28**	κ	0.022 0	0.048 4					
$AlBr_3$–$SbBr_3$	p	0	7.58	19.74	26.65	42.45	74.69	86.91
Ref. **23**, 2	$10^3\cdot\kappa_{100}$	0	12	18	16	11	4	1
	$10^3\cdot\kappa_{140}$	0	16	25	26	20	7	1
$AlBr_3\cdot SbBr_3$	t	100	150	170				
Ref. **28**	$10^3\cdot\kappa$	11.6	26.3	32.2				
$AlBr_3\cdot SbBr_3$–$AsBr_3$	p	16.3	32.9	41.4	67.9	77.3	86.8	100
Ref. **35**	$10^2\cdot\kappa_{85}$	0.089	0.121	0.252	0.766	0.855	0.892	0.803
	$10^2\cdot\kappa_{100}$	0.082	0.116	0.264	0.938	1.125	1.228	1.155

Table 9.5 ELECTRICAL CONDUCTIVITY OF MOLTEN BINARY SALT SYSTEMS AND OTHER MIXED IONIC MELTS—*continued*

System								
$AlBr_3$–$ZnBr_2$	p	73.43	78.03	82.57	87.03	91.42		
Ref. **23**	κ_{100}	5.5	5	3	1	0		
	κ_{130}	12.5	10.5	8	—	0		
	κ_{150}	20	17	11	2	0		
$2AlBr_3 \cdot ZnBr_2$	t	100	140	180				
Ref. **28**	$10^3 \cdot \kappa$	5.15	15.12	30.62				
$AlCl_3$–KCl	p	64.73	73.30	80.11	81.55	89.14		
Ref. **21**	κ_{200}	—	0.200	0.165	0.157	0.105		
	κ_{250}	0.346	0.280	0.228	0.214	0.143		
	κ_{300}	0.458	0.357	0.290	(0.270)	(0.180)		
$AlCl_3$–KCl**	p	50.4	53.7	62.5				
Ref. **31**	κ_{600}	0.790	0.811	0.858				
	κ_{700}	0.900	0.935	1.005				
	κ_{800}	1.010	1.060	1.155				
$AlCl_3 \cdot KCl$	t	250	300	500	600	700	800	
Ref. **31, 21**	κ	0.353	0.479	0.706	0.862	1.018	1.174	
$AlCl_3 \cdot LiCl$**	t	174	184	217	259	303.5	327	351
Ref. **21**	κ	0.354	0.380	0.468	0.553	0.647	0.687	0.731
$AlCl_3$–NaCl**	p	69.78	71.03	72.01	76.06	76.95	81.25	82.51
Ref. **15**, 14, 5	κ_{200}	0.436	0.262	0.280	0.160	0.170	0.090	0.070
	κ_{250}	0.532	0.378	0.380	—	0.300	0.178	—
$AlCl_3$–NH_4Cl	p	71.37	78.89	85.46	90.89			
Ref. **21**	κ_{200}	—	—	0.177	0.114			
	κ_{300}	0.479	0.387	0.302	0.194			
AlF_3–Na_3AlF_6*	p	0	2	4	6	8	10	20
Ref. **72, 65, 39,**	κ_{1000}	2.8	2.7	2.7	2.6	2.6	2.5	(2.3)
34, 12	κ_{1050}	2.9	2.9	2.8	2.8	2.7	2.6	(2.4)
Al_2O_3–Na_3AlF_6	p	0	3	6	9	12	15	20
Ref. **72, 65, 39, 34,**	κ_{1000}	2.8	2.6	2.5	2.3	2.2	2.1	(1.8)
80, 12, 4	κ_{1100}	3.0	2.9	2.7	2.6	2.4	2.3	(2.0)
Al_2O_3–Li_3AlF_6—	$p_{Al_2O_3}$	6.4	3.2	8.8	3.6	9.8	4.2	
Na_3AlF_6	$p_{Na_3AlF_6}$	46.8	58.2	64.0	67.6	72.2	76.6	
Ref. **80**	κ_{1000}	3.44	—	3.10	3.24	2.43	2.67	
	$10^3\alpha$	5.2	—	4.7	4.8	5.5	5.4	
	$\kappa_t = \kappa_{1000} + \alpha(t - 1\,000)$							
Al_2O_3–SiO_2	p	3	5.5	8	10	12		
Ref. **81**	a	–0.30	0.08	–0.40	–0.30	–1.00		
	$10^{-3}b$	5.9	5.7	4.1	4.8	3.0		
	$\log \kappa_t = a - b(t+273)$ at 1 600–1 800							
B_2O_3–BaO	p	30.0	35.0	44.5	49.2	59.7	67.9	
Ref. **50**	κ_{900}	—	—	0.007	0.005	0.002	0.002	
	κ_{1000}	0.10	0.07	0.036	0.024	0.013	0.008	
	κ_{1100}	0.19	0.15	0.089	0.062	0.037	0.026	
	κ_{1200}	0.28	0.23	—	—	—	0.056	
B_2O_3–CaO	p	55.3	59.8	62.8	67.9	73.3		
Ref. **50**	κ_{1000}	—	—	—	0.025	0.013		
	κ_{1100}	—	0.12	0.10	0.069	0.047		
	κ_{1200}	0.23	0.22	0.20	0.151	0.100		
B_2O_3–K_2O	p	61.6	70.6	79.6	89.0	94.8	98.5	
Ref. **75, 42**	κ_{600}	0.016	0.003	—	1.8×10^{-4}	8.3×10^{-5}	1.1×10^{-5}	
	κ_{800}	0.186	0.056	0.022	6.0×10^{-3}	2.2×10^{-3}	2.2×10^{-4}	
	κ_{1000}	0.513	0.309	0.112	3.0×10^{-3}	9.1×10^{-3}	1.1×10^{-3}	

Table 9.5 ELECTRICAL CONDUCTIVITY OF MOLTEN BINARY SALT SYSTEMS AND OTHER MIXED IONIC MELTS—*continued*

System								
B_2O_3–Li_2O	p	79.9	87.1	92.0	95.5	97.2	98.5	
Ref. **75, 42**	κ_{700}	0.08	—	—	0.003	0.001	—	
	κ_{800}	0.87	0.10	0.03	0.011	0.004	0.001 2	
	κ_{900}	3.02	0.26	0.10	0.026	0.011	0.002 8	
	κ_{1000}	5.25	0.40	0.17	0.051	0.019	0.005 2	
B_2O_3–Na_2O	p	64.0	68.2	77.6	85.8	94.4	99.1	
Ref. **75, 42**	κ_{600}	—	—	—	6.5×10^{-4}	1.5×10^{-4}	1.6×10^{-5}	
	κ_{800}	—	0.33	0.08	0.022	4.8×10^{-3}	3.5×10^{-4}	
	κ_{1000}	1.74	1.32	0.40	0.107	2.1×10^{-2}	1.7×10^{-3}	
B_2O_3–PbO	p	88.2	92.7					
Ref. **69**	κ_{650}	5×10^{-7}	4×10^{-7}					
	κ_{750}	2×10^{-5}	0.6×10^{-5}					
	κ_{850}	8.5×10^{-5}	2.0×10^{-5}					
B_2O_3–SrO	p	44.9	49.6	56.8	61.6	65.2	70.0	
Ref. **50**	κ_{1000}	—	—	0.019	0.012	0.010	0.005	
	κ_{1100}	0.10	0.08	0.062	0.047	0.036	0.021	
	κ_{1200}	0.21	0.17	—	—	—	—	
$BaBr_2$–KBr	p	0	39.2	55.5	62.5	73.0	88.2	93.4
Ref. **43**	κ_{700}	—	1.17	1.03	0.85	0.89	—	—
	κ_{750}	1.63	1.29	1.16	0.96	1.00	0.93	—
	κ_{800}	1.76	1.41	1.27	1.06	1.13	1.06	0.98
	κ_{850}	1.87	1.52	1.39	1.17	1.25	1.19	1.12
$BaCl_2$–$CaCl_2$	p	17.3	32.0	44.6	55.6	65.3	73.8	88.3
Ref. **68**	κ_{800}	1.84	1.80	1.76	1.71	1.65	1.62	—
	κ_{900}	2.23	2.15	2.08	2.05	1.98	1.93	1.82
	κ_{1000}	2.57	2.51	2.44	2.38	2.32	2.27	2.14
	κ_{1100}	2.86	2.81	2.74	2.66	2.60	2.56	2.44
$BaCl_2$–$MgCl_2$†	p	13.4	31.0	49.7	73.2	75.7		
Ref. **37**	κ_{800}	1.405	1.620	1.70	1.61	1.57		
	κ_{900}	1.590	1.810	1.91	1.83	1.785		
	κ_{1000}	(1.775)	2.000	2.12	2.05	2.000		
$BaCl_2$–NaCl	p	28.4	47.1	59.8	70.4	78.1	84.2	93.4
Ref. **49, 17**	κ_{800}	3.00	2.55	2.26	2.00	1.54	1.20	—
	κ_{900}	3.26	2.90	2.68	2.50	2.32	2.25	2.00
	κ_{1000}	3.48	3.14	2.92	2.76	2.64	2.58	2.32
$BaCl_2$–$NaNO_3$	p	1.17	1.98	6.14	9.12			
Ref. **40**	κ_{350}	1.124	1.107	1.073	1.024			
	κ_{355}	1.143	1.128	1.093	1.052			
	κ_{360}	1.163	—	—	—			
BaF_2–NaF	p			66.6				
Ref. **11**	t	900	950	1 000	1 050	1 100		
	κ	4.027	4.335	4.602	5.051	5.319		
$Ba(NO_2)_2$–KNO_2	p	47.8	59.0	68.3	76.3	83.5	89.6	95.0
Ref. **82**	κ_{300}	—	0.494	0.438	0.393	—	—	—
	κ_{310}	—	0.528	0.472	0.428	0.381	—	0.311
	κ_{320}	—	0.560	0.505	0.464	0.415	0.386	0.345
	κ_{330}	—	0.594	0.542	0.498	0.450	0.418	0.377
	κ_{340}	0.670	0.628	0.572	0.534	0.486	0.452	0.411
	κ_{350}	0.702	0.660	0.610	0.570	0.520	0.484	0.445
$Ba(NO_2)_2$–KNO_3	p	26.8	31.3	43.8	54.8	64.5	73.3	81.0
Ref. **83**	κ_{320}	—	0.50	—	0.42	—	0.39	—
	κ_{340}	—	0.55	—	0.48	—	0.42	0.40
	κ_{360}	0.65	0.60	0.58	0.55	0.50	0.49	—
	κ_{380}	0.73	0.68	0.62	0.60	—	—	—

Table 9.5 ELECTRICAL CONDUCTIVITY OF MOLTEN BINARY SALT SYSTEMS AND OTHER MIXED IONIC MELTS—*continued*

$Ba(NO_2)_2$–$NaNO_2$	p	19.4	39.2	53.9	64.0	73.9	84.5	94.5
Ref. **82**	κ_{240}	—	—	0.598	0.488	0.372	0.258	—
	κ_{260}	—	0.852	0.700	0.588	0.464	0.334	0.200
	κ_{280}	1.120	0.957	0.802	0.688	0.554	0.410	0.260
	κ_{300}	1.232	1.064	0.910	0.788	0.646	0.486	0.320
	κ_{320}	1.343	1.170	1.016	0.888	0.738	0.564	0.380
	κ_{340}	1.454	1.275	1.120	0.985	0.828	—	—
$Ba(NO_3)_2$–KNO_2	p	50.1	61.0	70.2				
Ref. **83**	κ_{260}	(0.38)	0.30	(0.28)				
	κ_{280}	0.41	0.36	(0.31)				
	κ_{300}	0.49	0.41	0.35				
	κ_{320}	0.55	0.46	0.40				
$BeCl_2$–$NaCl$	p	39.1	45.2	49.2	58.3	63.6	70.8	83.2
Ref. **58**	κ_{300}	—	—	0.69	0.58	0.49	—	—
	κ_{400}	—	1.00	1.10	0.98	0.85	0.56	0.19
	κ_{500}	1.35	1.28	1.38	—	—	—	—
Bi–BiI_3	p	3.8	8.1	13.1	34.7	45.1	58.8	76.4
Ref. **84**	κ_{500}	0.5	0.8	2.1	200	600	1 400	3 100
Ca–$CaCl_2$	p	0.22	0.29	0.36	0.47	0.62	0.88	1.10
Ref. **85**	κ_{855}	2.50	2.52	2.55	2.58	2.61	2.68	2.73
$CaCl_2$–KCl	p	24.6	40.7	66.0	81.5			
Ref. **37**, 3	κ_{800}	1.770	1.500	1.415	1.560			
	κ_{900}	2.035	1.790	1.685	1.890			
$CaCl_2$–$MgCl_2$†	p	10.8	19.0	28.0	41.2	62.1	77.1	
Ref. **37**	κ_{700}	(1.18)	1.34	1.47	1.65	(1.79)	(1.87)	
	κ_{800}	1.37	1.53	1.68	1.90	2.07	2.20	
	κ_{900}	1.56	1.72	1.89	2.15	2.35	2.53	
$CaCl_2$–$NaCl$**	p	0	20	40	60	75	90	100
Ref. **67**, **49**, **17**, **18**, 8, 6, 3	κ_{100}	—	2.78	2.31	1.92	1.71	—	—
	κ_{800}	3.58	3.10	2.64	2.27	2.09	2.01	2.02
	κ_{900}	3.79	3.37	2.96	2.60	2.40	2.32	2.32
	κ_{1000}	4.00	3.58	3.22	2.92	2.71	2.67	2.68
CaF_2–CaO	p	90.7	92.6	94.5	96.4	98.2	100	
Ref. **55**	κ_{1500}	4.3	4.6	5.3	4.2	4.1	3.9	
	κ_{1545}	4.6	4.7	7.2	5.2	4.4	4.1	
CaF_2–Na_3AlF_6	p	0	5	10	20	30	40	
Ref. **65**, **39**, **26**, 34, 12	κ_{1000}	2.8	2.8	2.7	2.6	2.5	2.4	
	κ_{1100}	3.0	3.0	3.0	2.9	2.8	2.7	
CaF_2–NaF	p	0	10					
Ref. **4**, 11	κ_{1000}	3.15	3.0					
	κ_{1020}	3.3	3.1					
$Ca(NO_3)_2$–KNO_3	p	0	16.9	24.0	34.2	44.8	54.9	65.5
Ref. **22**	κ_{300}	—	—	0.375	0.330	0.270	0.195	—
	κ_{350}	0.666	0.570	0.525	0.460	0.390	0.315	0.255
	κ_{400}	0.818	0.705	0.660	0.605	0.522	0.430	0.355
CaO–SiO_2	p	20	30	35	40	45	50	55
Ref. **81**	a	2.17	2.19	2.22	2.51	3.19	2.39	2.06
	$10^{-3}b$	5.7	5.7	5.2	5.7	6.7	5.0	4.1
		$\log \kappa_t = a - b/(t+273)$ at 1 600–1 800						
$CdBr_2$–$CdCl_2$	p	0	27.1	49.7	69.0	85.6	100	
Ref. **24**	κ_{600}	1.95	1.7	1.5	1.35	1.2	1.05	
	κ_{640}	2.1	1.85	1.65	1.5	1.35	1.25	

Table 9.5 ELECTRICAL CONDUCTIVITY OF MOLTEN BINARY SALT SYSTEMS AND OTHER MIXED IONIC MELTS—*continued*

System								
$CdCl_2$–CdI_2	p	0	14.2	33.2	59.9	100		
Ref. **43**	κ_{400}	0.22	0.35	—	—	—		
	κ_{500}	0.41	0.56	0.78	—	—		
	κ_{600}	0.66	0.79	1.03	1.32	1.96		
	κ_{700}	0.95	1.05	1.30	1.57	2.09		
$CdCl_2$–KCl	p	40	50	60	70	80	90	100
Ref. **47, 24, 3**, 27	κ_{500}	—	1.10	1.11	1.19	1.33	1.52	—
	κ_{600}	—	1.24	1.27	1.36	1.51	1.71	1.96
	κ_{700}	1.55	1.48	1.48	1.60	1.77	1.97	2.18
	κ_{800}	1.81	1.73	1.66	1.78	1.94	2.09	2.25
$CdCl_2$–LiCl	p	0	59.1	81.2	92.9	100		
Ref. **43**	κ_{550}	—	—	2.8	2.3	—		
	κ_{650}	6.0	4.1	3.2	2.6	2.1		
	κ_{750}	6.4	4.5	3.5	3.0	2.3		
$CdCl_2$–NaCl	p	43.9	67.6	82.5	92.6	100		
Ref. **24**	κ_{550}	(2.25)	1.95	1.9	1.85	1.8		
	κ_{600}	(2.4)	2.15	2.1	2.0	1.9		
	κ_{700}	2.7	2.45	2.4	2.3	2.2		
$CdCl_2$–$PbCl_2$	p	0	20	40	60	80	100	
Ref. **56, 27, 24**	κ_{500}	1.44	1.68	—	—	—	—	
	κ_{600}	1.88	1.94	1.97	1.97	1.94	1.88	
	κ_{700}	2.28	2.35	2.36	2.31	2.26	2.20	
$CdCl_2$–TlCl	p	0	10	25	40	60	80	100
Ref. **44, 3**	κ_{500}	1.35	1.30	1.12	1.24	1.41	—	—
	κ_{600}	1.70	1.66	1.56	1.52	1.66	1.81	1.97
	κ_{700}	1.95	1.86	1.76	1.72	1.86	1.99	2.10
CdI_2–KI	p	0	19.7	35.5	64.3	75.3	89.8	100
Ref. **43**	κ_{300}	—	—	—	0.30	0.31	—	—
	κ_{500}	—	—	—	0.66	0.64	0.57	0.41
	κ_{700}	1.32	1.21	1.09	1.06	0.99	0.89	0.95
	κ_{800}	1.52	1.45	1.29	1.26	—	—	—
$Cd(NO_3)_2$–$CsNO_3$	p	39.4	44.6	54.8	64.5	73.8	82.9	
Ref. **45**	κ_{200}	—	—	0.09	0.07	—	—	
	κ_{250}	—	0.17	0.16	0.15	0.12	—	
	κ_{300}	0.26	0.25	0.24	0.23	0.20	0.15	
$Cd(NO_3)_2$–KNO_3	p	36.9	50.0	60.9	70.0	77.8	84.4	
Ref. **46**	κ_{200}	—	—	0.14	0.12	0.08	0.04	
	κ_{250}	—	—	0.26	0.23	0.18	0.12	
	κ_{300}	0.46	0.44	0.39	0.35	0.31	0.23	
$Cd(NO_3)_2$–$RbNO_3$	p	15.1	28.6	40.8	51.7	61.6	70.6	78.9
Ref. **46**	κ_{200}	—	—	0.12	0.10	0.10	0.08	—
	κ_{250}	—	0.22	0.21	0.20	0.19	0.16	—
	κ_{300}	0.31	0.34	0.32	0.31	0.29	0.26	0.20
$Cd(NO_3)_2$–$TlNO_3$	p	9.0	18.1	27.5	37.1	47.0	57.1	67.4
Ref. **45**	κ_{150}	—	—	0.14	0.11	0.08	—	—
	κ_{200}	—	0.29	0.25	0.22	0.17	—	—
	κ_{250}	0.47	0.31	0.38	0.34	0.28	0.23	—
	κ_{300}	0.60	0.54	0.50	0.46	0.40	0.34	0.26
Ce–$CeCl_3$	p	1.05	1.8	2.8	3.4	3.9	4.7	5.3
Ref. **95,** 96	κ_{855}	1.56	2.02	2.59	3.26	3.81	4.45	5.35
Ce–CeI_3	p	0.55	1.03	1.43	2.07	2.87	3.7	6.1
Ref. **86**	κ_{820}	1.00	1.92	2.74	4.75	8.88	15.5	*c.* 60
CsBr–NaCl	p	8.3	19.5	52.1	59.2			
Ref. **87**	κ_{800}	2.40	1.69	1.40	1.21			
	κ_{850}	2.55	1.79	1.50	1.30			

Table 9.5 ELECTRICAL CONDUCTIVITY OF MOLTEN BINARY SALT SYSTEMS AND OTHER MIXED IONIC MELTS—*continued*

System								
$CsBr–ZnSO_4$ Ref. **62**	p	39.8	63.8	79.9				
	κ_{505}	0.03	0.09	0.20				
	κ_{550}	0.05	0.13	0.26				
CsCl–NaBr Ref. **87**	p	29.4	52.2	71.0	87.0			
	κ_{800}	2.05	1.61	1.50	1.45			
	κ_{850}	2.15	1.70	1.58	1.55			
$CsCl–TiCl_3$ Ref. **88**	p	52.1	62.3	72.0	81.5	90.8		
	κ_{800}	0.7	0.7	0.8	0.9	1.2		
CuCl–KCl Ref. **47**	p	25.8	42.0	47.6	57.5	76.0	85.9	100
	κ_{450}	—	2.00	2.08	2.23	2.71	2.93	3.32
	κ_{500}	—	2.07	2.15	2.32	2.82	3.06	3.43
	κ_{550}	—	2.18	2.26	2.44	2.94	3.20	3.54
	κ_{600}	2.22	2.32	2.40	2.59	3.08	3.34	3.66
$CuO–V_2O_5$ Ref. **74**	p	10	30	40	50			
	κ_{800}	0.6	3.2	3.8	5.5			
	κ_{900}	1.4	5.0	6.0	10.0			
	κ_{1000}	2.2	—	—	—			
$Cu_2S–FeS$ Ref. **60**	p	0	25	35	50	65	75	100
	κ_{1100}	—	9.30	8.20	4.60	3.30	2.20	0.50
	κ_{1300}	1.530	9.40	8.30	5.20	4.60	3.80	1.30
	κ_{1500}	1.490	9.50	8.30	—	—	—	1.70
$Dy–DyCl_3$ Ref. **89**	p	0.0	0.9	2.7	4.9	7.8	12.0	16.7
	κ_{700}	0.42	0.46	0.54	0.65	0.76	0.90	0.98
$Er–ErCl_3$ Ref. **89**	p	0.0	1.0	3.0	3.8			
	κ_{820}	0.50	0.55	0.67	0.69			
$FeO–SiO_2$ Ref. **41**	p	55.9	63.3	68.6	72.4	75.4	78.7	84.9
	κ_{1300}	0.5	0.9	1.4	2.1	3.2	4.2	9.0
	κ_{1400}	0.8	1.6	2.3	4.0	5.0	7.8	—
$FeO–TiO_2$ Ref. **63**	p	32.6	35.7	56.5	62.1	91.9	97.2	
	κ_{1300}	48	50	40	38	145	—	
	κ_{1400}	48	47	34	33	155	225	
$Fe_2O_3–V_2O_5$ Ref. **74**	p	15.0	19.4	30.5				
	κ_{900}	1.4	3.8	6.8				
	κ_{1000}	2.0	5.8	11.6				
	κ_{1100}	3.6	10.6	—				
$Gd–GdBr_3$ Ref. **89**	p	0.0	0.4	1.3	2.3	2.7		
	κ_{820}	0.47	0.58	0.87	1.27	1.46		
$Gd–GdCl_3$ Ref. **89**	p	0.0	0.3	0.6	1.0			
	κ_{650}	0.41	0.43	0.45	0.49			
$HgBr_2–NH_4Br$ Ref. **32**	p	61.19	78.63	84.66	89.57	93.64	100	
	κ_{250}	0.4	0.5	0.6(2)	0.6	0.45	0	
	κ_{300}	0.55	0.7	0.8(2)	0.75	0.6	0	
$HgCl_2–HgI_2$ Ref. **19**	p	0	20	40	50	60	80	100
	κ_{225}	—	0.029 5	0.016 0	0.009 4	0.005 3	—	—
	κ_{250}	—	0.028 1	0.016 0	0.009 7	0.005 6	0.001 1	—
	κ_{300}	0.028 0	0.025 2	0.016 0	0.010 4	0.006 3	0.002 0	0.000 8
$HgCl_2–NH_4Cl$ Ref. **32**	p	71.42	81.82	83.54	86.12	92.21	93.84	95.31
	κ_{250}	0.7	0.9	0.7	0.8	0.55	0.78	0.4
	κ_{300}	0.95	1.15	1.1	1.15	0.75	1.0	0.6
$HgCl_2–TINO_3$ Ref. **19**	p	0	10	20	40	50	60	75
	κ_{225}	0.416	0.344	0.274	0.184	0.150	0.126	—
	κ_{250}	0.476	0.397	0.320	0.226	0.186	0.156	0.100
	κ_{275}	0.536	0.450	0.368	0.269	0.222	0.188	0.126

Table 9.5 ELECTRICAL CONDUCTIVITY OF MOLTEN BINARY SALT SYSTEMS AND OTHER MIXED IONIC MELTS—*continued*

System								
HgI_2–KI	p	73.3	80.4	86.5	91.6	96.1	100	
Ref. **44**	κ_{270}	—	0.48	0.64	0.64	0.43	0.03	
	κ_{290}	0.50	0.56	0.69	0.69	0.45	0.03	
	κ_{310}	0.56	0.64	0.75	0.75	—	—	
HgI_2–NH_4I	p	67.64	75.82	82.47	87.98	92.62	96.58	100
Ref. **32**	κ_{250}	0.3	0.4	0.5	0.6	0.55	0.37	0.03
	κ_{350}	0.6	0.5	0.73	0.77	0.7	0.4	0.03
HgI_2–$TlNO_3$	p	0	20	40	50	60	80	100
Ref. **19**	κ_{250}	0.476	0.302	0.174	0.154	0.136	0.142	—
	κ_{275}	0.536	0.348	0.210	0.182	0.161	0.156	0.030 5
	κ_{300}	0.596	0.395	0.248	0.214	0.187	0.171	0.028 0
Ho–$HoCl_3$	p	0.0	0.7	3.2	6.3	10.6		
Ref. **89**	κ_{800}	0.53	0.56	0.69	0.84	1.05		
K–KBr	p	1	2	3	4	6	8	
Ref. **73**	κ_{763}	5.6	11.7	22.4	40.8	104.7	229.2	
	κ_{870}	6.8	16.3	29.5	51.3	129.0	263.0	
K–KCl	p	1	2	4	6	8	10	
Ref. **73**	κ_{816}	3.8	6.2	14.1	31.6	64.6	117.5	
K–KF	p	1.35	2.72	4.10	4.80			
Ref. **90**	κ_{900}	4.5	6.7	12.0	18.0			
K–KI	p	0.5	1.0	1.5	2.0	—	—	—
Ref. **90, 91**	κ_{900}	4.5	8.5	13.5	19.0	—	—	—
	p	0.5	1.2	2.6	13.5	26.0	48.7	64.9
	κ_{700}	3.7	7.9	19.7	500	1 500	4 200	7 300
KBF_4–KF	p	30	40	50	60	70	80	90
Ref. **92**	κ_{600}	—	—	—	—	1.0	2.0	1.5
	κ_{700}	—	—	3.0	5.3	6.8	7.2	6.7
	κ_{800}	5.0	8.0	10.0	11.5	12.5	12.8	12.6
KBr–KNO_3	p	0	17.2	37.0	54.1	70.2	86.9	100
Ref. **43**	κ_{400}	0.82	0.79	0.78	—	—	—	—
	κ_{600}	1.38	1.34	1.29	1.30	1.31	—	—
	κ_{700}	—	—	—	—	1.53	1.53	—
	κ_{800}	—	—	—	—	1.76	1.76	1.76
KBr–NaCl	p	33.7	57.6	85.9	89.0			
Ref. **93**	κ_{800}	2.97	2.51	2.20	1.95			
	κ_{850}	3.05	2.60	2.29	2.05			
KBr–RbCl	p	19.7	39.5	59.5	81.3			
Ref. **94**	κ_{770}	1.64	1.64	1.64	1.65			
	κ_{800}	1.70	1.70	1.70	1.72			
	κ_{850}	1.80	1.80	1.80	1.82			
KBr–TlBr	p	2.2	5.3	9.7	18.4	30.3	53.9	72.0
Ref. **97**	a	0.489_1	0.441_9	2.008_7	2.390_4	0.475_4	0.808_8	0.426_5
	10^3b	3.239_7	2.957_2	7.546_5	8.864_5	3.284_7	4.267_7	5.332_0
	10^6c	0.835_7	0.499_1	3.845_7	4.904_9	0.845_0	1.478_0	0.865_2
	t range	537–701	473–640	650–762	619–717	626–791	687–796	718–783
				$\kappa_t = a + bt - ct^2$				
KBr–$ZnSO_4$	p	35.8	40.0	44.4	50.2	60.0	64.7	73.0
Ref. **48**	κ_{475}	0.10	—	—	0.13	0.21	0.24	0.34
	κ_{500}	0.12	0.15	0.14	0.16	0.25	0.29	0.41
	κ_{550}	0.16	0.21	0.19	0.23	0.33	0.39	0.55
KCl–KI	p	2.8	7.8	13.3	26.8	41.2	64.4	
Ref. **57**	κ_{700}	1.31	1.38	1.41	1.48	1.56	1.66	
	κ_{800}	1.52	1.56	1.61	1.70	1.80	1.96	
	κ_{900}	1.66	1.72	1.76	1.88	1.97	2.14	

Table 9.5 ELECTRICAL CONDUCTIVITY OF MOLTEN BINARY SALT SYSTEMS AND OTHER MIXED IONIC MELTS—*continued*

System								
KCl–LiCl**	p	0	28.2	42.6	55.8	72.2	87.6	100
Ref. **57, 10**	κ_{500}	—	—	2.31	1.87	—	—	—
	κ_{600}	—	3.51	2.87	2.40	1.90	—	—
	κ_{700}	6.17	4.10	3.42	2.88	2.35	2.04	—
	κ_{800}	6.62	4.56	3.82	3.24	2.72	2.36	2.24
$KCl–MgCl_2$	p	10	20	30	50	60	70	80
Ref. **56, 37, 16,**	κ_{600}	—	—	0.97	0.98	0.97	0.98	—
9, 7	κ_{700}	1.10	1.17	1.21	1.24	1.27	1.34	1.48
	κ_{800}	1.33	1.37	1.39	1.44	1.48	1.56	1.76
	κ_{900}	1.50	1.54	1.56	1.62	1.67	1.75	1.87
KCl–NaBr	p	22.0	32.0	51.4	74.0			
Ref. **93**	κ_{800}	2.64	2.40	2.29	2.21			
	κ_{850}	2.76	2.53	2.39	2.32			
KCl–NaCl**	p	18.7	32.1	40.6	54.8	64.8	82.9	
Ref. **57, 16**, 9, 6, 3	κ_{700}	—	—	2.52	2.35	2.21	—	
	κ_{800}	3.23	2.98	2.85	2.64	2.53	2.34	
	κ_{900}	3.50	3.24	3.11	2.88	2.76	2.56	
KCl–NaI	p	0	8.0	19.8	33.0	49.7	73.6	100
Ref. **43**	κ_{600}	—	1.88	1.68	1.54	1.51	—	—
	κ_{700}	2.41	2.20	2.04	1.86	1.84	1.87	—
	κ_{800}	2.68	2.51	2.37	2.16	2.44	2.17	2.22
$KCl–PbCl_2$	p	2.8	4.5	6.1	14.8	20.7	37.9	51.2
Ref. **98**, 47, 27, 24	κ_{450}	—	1.176	1.142	1.034	0.937	—	—
	κ_{500}	1.414	1.401	1.345	1.205	1.106	1.038	(1.110)
	κ_{550}	1.645	1.609	1.558	1.364	1.273	1.174	1.230
	κ_{600}	1.868	1.807	1.723	1.515	1.417	1.306	1.350
	κ_{650}	2.086	2.018	1.913	1.684	1.572	1.467	1.519
	κ_{700}	—	—	—	—	1.716	1.624	1.670
KCl–RbBr	p	9.9	22.7	39.8	63.9			
Ref. **94**	κ_{770}	1.42	1.58	1.70	1.88			
	κ_{800}	1.50	1.65	1.77	1.98			
	κ_{850}	1.59	1.73	1.89	2.09			
$KCl–SnCl_2$	p	0.0	4.1	8.7	14.1	20.3	27.8	31.9
Ref. **99**	κ_{300}	1.115	1.079	0.995	0.865	0.688	0.472	—
	κ_{350}	1.388	1.331	1.215	1.060	0.871	0.645	0.702
$KCl–TiCl_3$	p	32.6	42.0	53.0	65.9	81.4		
Ref. **88**	κ_{800}	1.0	1.05	1.3	1.5	1.8		
$KCl–ZnCl_2$	p	0	10.5	17.8	26.3	35.9	52.6	58.7
Ref. **66**	κ_{500}	0.09	0.40	0.64	0.70	0.76	0.78	0.79
	κ_{550}	0.16	0.51	0.77	0.82	0.89	0.92	0.95
	κ_{600}	0.26	0.63	0.90	0.95	1.01	1.05	1.09
	κ_{650}	0.37	0.76	1.02	1.06	1.14	1.19	1.24
$KCl–ZnSO_4$	p	19.0	24.0	29.3	33.7	39.8	50.0	56.8
Ref. **48**	κ_{475}	—	0.09	0.13	0.16	0.21	0.30	0.36
	κ_{500}	0.09	0.12	0.17	0.21	0.25	0.35	0.41
	κ_{550}	0.14	0.16	0.24	0.27	0.32	0.44	0.51
$K_2Cr_2O_7–KNO_3$	p	0	24.4	42.1	55.5	74.4	87.1	100
Ref. **44**	κ_{350}	0.56	0.52	0.42	0.35	0.22	0.17	—
	κ_{400}	0.72	0.66	0.64	0.48	0.34	0.26	0.20
	κ_{475}	0.95	0.88	0.74	0.67	0.52	0.41	0.35
KF–NaCl	p	33.20	49.85	66.54	100			
Ref. **6**	κ_{750}	2.6	2.9	3.2	—			
	κ_{850}	3.4	3.6	3.7	4.1			
	κ_{950}	4.2	4.3	(4.4)	4.5			

Table 9.5 ELECTRICAL CONDUCTIVITY OF MOLTEN BINARY SALT SYSTEMS AND OTHER MIXED IONIC MELTS—*continued*

System								
KI–NaCl	p	33.6	55.0	74.1	86.9	94.2	100	
Ref. **43**	κ_{600}	—	—	1.56	1.35	—	—	
	κ_{700}	—	2.24	1.87	1.64	1.47	1.32	
	κ_{800}	3.06	2.58	2.20	1.91	1.73	1.52	
$KI–ZnSO_4$	p	47.4	50.9	54.8	60.4	62.9	67.3	
Ref. **48**	κ_{440}	—	—	0.11	0.13	0.15	0.21	
	κ_{460}	0.09	0.08	0.13	0.16	0.18	0.24	
	κ_{480}	0.12	0.10	0.15	0.18	0.20	0.28	
$K_2MoO_4–MoO_3$	p	17.5	27.5	43.0	52.2	60.0	79.4	91.0
Ref. **100**	a	−3.660 6	−1.312 0	−1.982 0	−1.208 3	−2.728 0	6.064 4	4.250 6
	10^2b	0.942 8	0.336 5	0.534 5	0.333 3	0.832 9	−1.299 9	−0.853 7
	10^6c	−5.010 1	−1.058 9	−2.442 6	−1.206 1	−5.307 3	8.173 1	5.507 5
	t range	760–869	614–804	654–783	642–794	583–705	882–934	877–995
				$\kappa_t = a + bt + ct^2$				
$KNO_2–NaNO_2$	p	12.0	23.6	32.7	45.1	55.3	74.2	83.3
Ref. **82**	κ_{300}	1.262	1.176	1.108	1.052	0.998	—	—
	κ_{340}	1.472	1.380	1.308	1.244	1.185	1.040	—
	κ_{380}	—	1.586	1.510	1.434	1.366	1.204	1.130
	κ_{420}	—	—	—	1.625	1.548	1.365	1.287
	κ_{460}	—	—	—	—	—	1.525	1.442
KNO_3–**$LiNO_3$**	p	66.0						
Ref. **101**	$\kappa_t = -0.3788 + 3.40 \times 10^{-3}t$ at 160–240							
	$\kappa_t = -0.4037 + 3.47 \times 10^{-3}t$ at 240–400							
$KNO_3–NaNO_3$**	p	0	20	40	50	60	80	100
Ref. **19**, 101, 79, 1	κ_{300}	—	0.820	0.706	0.652	0.625	0.572	—
	κ_{400}	1.364	1.160	1.024	0.966	0.931	0.874	0.818
	κ_{500}	1.720	1.496	1.351	1.284	1.234	1.178	1.107
$KNO_3–RbNO_3$	p	7.1	18.6	31.4	40.7	56.0	73.3	86.0
Ref. **45**	κ_{300}	—	0.40	0.42	—	—	—	—
	κ_{350}	0.51	0.52	0.55	0.57	0.59	0.63	0.64
	κ_{400}	0.63	0.65	0.68	0.70	0.73	0.77	0.79
$K_2SO_4–ZnSO_4$	p	31.6	41.8	52.0				
Ref. **62**	κ_{450}	0.04	0.06	0.05				
	κ_{500}	0.06	0.08	0.08				
	κ_{550}	0.07	0.12	0.12				
$K_2TaF_7–Ta_2O_5$	p	78.0	83.4	89.0	94.5	100.0		
Ref. **102**	κ_{800}	0.71	0.80	0.87	0.91	0.92		
	κ_{900}	0.85	0.94	1.00	1.04	1.05		
$K_2TiF_6–TiO_2$	p	81.9	87.5	92.3	94.5	96.6	98.3	100
Ref. **102**	κ_{870}	1.18	1.22	1.28	1.30	1.31	1.39	1.40
	κ_{920}	1.29	1.33	1.37	1.39	1.40	1.48	1.49
$K_2WO_4–WO_3$	p	37.4	50.5	58.3	67.4	76.6	85.0	92.2
Ref. **103**	a	2.815 6	2.880 9	2.311 1	2.356 2	5.493 6	7.543 6	38.80
	10^2b	0.560 1	0.615 0	0.549 2	0.565 0	1.250 1	1.640 2	7.964 6
	10^6c	2.087 5	2.501 1	2.432 4	2.392 3	5.882 2	7.524 8	39.636
	t range	891–1 010	759–912	754–770	674–819	773–955	842–1 016	885–1 029
				$\kappa_t = -a + bt - ct^2$				
$La–LaBr_3$	p	0.0	0.4	1.2	2.5	4.1	5.5	6.5
Ref. **89**	κ_{820}	0.70	0.94	1.55	3.15	7.35	15.8	25.9
$La–LaCl_3$	p	0.3	1.1	2.0	3.0	4.6	6.2	
Ref. **85**	κ_{910}	1.51	1.84	2.46	3.21	5.02	7.21	
$La–LaI_3$	p	0.5	1.4	2.4	3.5	4.4		
Ref. **86**	κ_{840}	0.87	2.39	5.54	12.2	20.9		

Table 9.5 ELECTRICAL CONDUCTIVITY OF MOLTEN BINARY SALT SYSTEMS AND OTHER MIXED IONIC MELTS—*continued*

System								
Li_3AlF_6–Na_3AlF_6	p	20	30	50	80			
Ref. **80**, 65	κ_{1000}	3.05	3.38	3.80	4.33			
	$10^3\alpha$	4.8	4.8	4.8	4.8			
		$\kappa_t=\kappa_{1000}+\alpha(t-1\,000)$ at 762–936						
Li_2MoO_4–MoO_3	p	19.5	33.9	51.5	60.5	71.3	85.2	91.0
Ref. **100**	a	10.470 8	7.490 5	4.031 1	4.612 7	2.795 8	21.335 9	13.230 5
	10^2b	2.498 1	1.815 3	1.028 9	1.183 2	0.753 9	5.368 6	3.225 9
	10^6c	13.536	9.358 9	4.489 7	5.211 7	2.887	30.734	16.322
	t range	778–917	800–907	768–917	799–945	845–956	764–872	809–908
				$\kappa_t=-a+bt-ct^2$				
$LiNO_3$–$RbNO_3$	p	10.5	16.7	23.8	31.9	52.3	65.2	80.8
Ref. **52**	κ_{200}	—	0.20	0.24	—	—	—	—
	κ_{250}	—	0.33	0.36	0.41	0.53	0.61	—
	κ_{300}	0.43	0.47	0.51	0.56	0.72	0.82	0.96
Li_2WO_4–WO_3	p	54.6	59.7	69.0	77.5	85.2	92.3	
Ref. **103**	a	6.015 2	2.149 1	7.464 8	0.606 3	5.259 7	5.723 7	
	10^2b	1.333 6	0.455 8	1.764 2	0.108 6	1.277 3	1.404 8	
	10^6c	5.594 8	0.392 3	7.669 4	−1.989 8	4.823 7	5.797 8	
	t range	805–952	769–924	757–938	735–890	716–897	734–907	
				$\kappa_t=-a+bt-ct^2$				
$MgCl_2$–NaCl	p	23.2	39.1	57.2	69.0	81.5	90.3	
Ref. **37**, 56, 16, 9	κ_{700}	2.80	2.31	1.97	1.78	1.63	1.39	
	κ_{800}	2.96	2.49	2.16	1.98	1.82	1.57	
	κ_{900}	3.12	2.67	2.35	2.18	2.01	1.75	
MgF_2–Na_3AlF_6	p	5	10					
Ref. **72**	k_{950}	—	2.2					
	κ_{975}	2.5	—					
MnO–SiO_2	p	40	50	55	60	65	70	80
Ref. **81**	a	2.32	2.26	1.85	2.25	2.94	2.21	2.00
	$10^{-3}b$	4.8	4.1	3.1	3.4	4.6	2.5	1.7
			$\log\,\kappa_t=a-b/(t+273)$ at 1 600 – 1 800					
MoO_3–Na_2MoO_4	p	21.4	26.4	51.2	62.2	71.2	83.7	86.2
Ref. **54**	κ_{700}	1.06	1.07	0.86	0.80	0.75	0.63	—
	κ_{800}	1.45	1.41	1.17	1.11	1.08	0.75	0.91
	κ_{900}	—	—	—	—	—	—	1.16
Na–NaBr	p	0.25	0.5	0.75	1.0	1.5	2.0	
Ref. **73**	κ_{803}	4.0	6.3	7.4	8.2	—	—	
	κ_{894}	4.8	7.7	9.7	11.5	14.7	17.5	
Na–NaCl	p	0.25	0.5	0.75	1.0	1.5		
Ref. **73**	κ_{848}	4.3	4.9	5.3	5.7	6.4		
	κ_{893}	4.4	5.0	5.5	5.9	6.7		
Na–NaI	p	0.4	0.6	1.0	1.3	1.6	2.0	
Ref. **90**	κ_{700}	3.7	—	—	—	—	—	
	κ_{800}	4.5	3.5	6.5	—	—	—	
	κ_{900}	5.0	6.5	8.0	10.0	12.2	15.0	
Na_3AlF_6–NaF‡	p	0	20	40	60	80	90	100
Ref. **65, 39, 34**, 12	κ_{1000}	5.4	4.9	4.4	3.9	3.3	3.1	2.8
	κ_{1050}	5.7	5.1	4.6	4.0	3.5	3.2	2.9
	κ_{1100}	5.9	5.4	4.8	4.2	3.6	3.3	3.1
$NaBF_4$–NaF	p	40	50	60	70	80	90	
Ref. **92**	κ_{500}	—	—	1.5	3.0	3.5	4.0	
	κ_{600}	—	3.0	5.0	6.0	6.5	7.0	
	κ_{700}	—	5.0	8.0	10.0	10.5	10.5	
	κ_{800}	6.5	12.0	14.0	15.0	15.5	15.0	

Table 9.5 ELECTRICAL CONDUCTIVITY OF MOLTEN BINARY SALT SYSTEMS AND OTHER MIXED IONIC MELTS—*continued*

NaBr–NaOH	p	17.7	25.0	30.4	39.8	48.4	52.3	
Ref. **70**	κ_{300}	1.18	1.14	1.03	0.95	0.83	0.79	
	κ_{400}	2.00	1.96	1.83	1.75	1.63	1.59	
$NaCl–Na_2CO_3$	p	0	52.45	68.81				
Ref. **6**	κ_{700}	—	2.2	2.3				
	κ_{850}	2.4	3.1	3.0				
	κ_{1000}	2.8	3.1	3.4				
$NaCl–NbCl_5$	p	43.8	60.4	60.7	71.8	73.6	100	
Ref. **20**	κ_{800}	1.32	2.20	2.19	—	—	3.34	
	κ_{850}	1.83	2.60	2.59	2.98	2.94	3.60	
$NaCl–PbCl_2$	p	0	2.55	9.01	20.05			
Ref. **17**	κ_{500}	1.445	1.51	1.58	1.645			
	κ_{600}	1.91	1.95	2.005	2.07			
$NaCl–TiCl_3$	p	27.5	36.2	47.0	60.3	77.4		
Ref. **88**	κ_{800}	1.75	1.75	1.9	2.3	2.7		
$NaCl–ZrCl_4$	p	11.5	34.9	59.7	75.1	88.1	88.7	
Ref. **20, 104**	κ_{350}	0.46	—	—	—	—	—	
	κ_{800}	—	1.45	1.74	2.41	2.88	2.68	
	κ_{825}	—	—	1.83	2.83	3.08	—	
	κ_{850}	—	—	1.97	3.01	3.34	3.36	
$NaF–SrF_2$	p	—	—	33.3	—	—		
Ref. **11**	t	900	950	1 000	1 050	1 100		
	κ	4.441	4.798	4.961	5.407	5.642		
NaI–NaOH	p	21.5	35.0	44.8	52.8			
Ref. **70**	κ_{300}	1.14	0.87	0.72	0.63			
	κ_{400}	1.90	1.53	1.32	1.23			
$NaNO_2–NaNO_3$	p	0	21.3	44.8	65.4	82.1	100	
Ref. **43**	κ_{300}	—	1.02	1.11	1.19	1.25	1.34	
	κ_{350}	1.16	1.27	1.34	1.44	1.51	1.62	
	κ_{400}	1.36	1.51	1.58	1.69	1.78	1.89	
	κ_{450}	1.55	1.73	1.79	1.92	2.02	2.10	
$Na_2O–TiO_2$	p	28.3	41.7	44.6	51.7	60.0	65.5	71.6
Ref. **64**	κ_{1000}	—	0.32	0.38	0.47	0.56	0.70	0.86
	κ_{1100}	0.32	0.45	0.56	0.80	0.90	0.94	1.15
	κ_{1200}	0.74	1.03	1.16	1.55	1.76	2.10	2.54
$Na_2WO_4–WO_3$	p	45.9	50.9	55.7	65.1	74.2	83.0	91.6
Ref. **103**	a	4.450 9	7.001 3	4.563 4	7.275 4	3.136 8	6.074 3	2.748 1
	10^2b	0.962 5	1.547 6	1.032 6	1.720 1	0.739 9	1.550 6	0.671 8
	10^6c	4.036 4	7.032 0	4.455 5	8.560 0	2.422 0	7.194 3	2.495 8
	t range	780–923	750–888	767–905	741–881	689–850	653–803	701–869
				$\kappa_t = -a + bt - ct^2$				
$Nd–NdBr_3$	p	0.4	1.5	3.6	7.5	11.4	13.9	15.4
Ref. **89**	κ_{730}	0.53	0.61	0.79	0.94	1.01	0.96	0.92
$Nd–NdCl_3$	p	1.7	2.4	3.8	7.3	11.0	14.6	18.6
Ref. **85**	κ_{855}	1.20	1.24	1.32	1.54	1.67	1.77	1.84
$Nd–NdI_3$	p	0.3	0.9	2.1	5.2	8.1	11.1	11.5
Ref. **86**	κ_{820}	0.476	0.585	0.759	1.033	1.102	0.980	0.930
$PbBr_2–PbCl_2$	p	0	23.7	45.3	66.6	83.8	100	
Ref. **59, 3**	κ_{450}	—	—	1.02	0.97	0.91	0.84	
	κ_{500}	1.43	1.34	1.25	1.17	1.10	1.03	
	κ_{550}	1.69	1.56	1.45	1.36	1.26	1.19	

Table 9.5 ELECTRICAL CONDUCTIVITY OF MOLTEN BINARY SALT SYSTEMS AND OTHER MIXED IONIC MELTS—*continued*

$PbO–SiO_2$ Ref. **105**	p	61.4	71.2	78.8	84.8	89.7	93.7	97.1
	$\log \kappa_{900}$	−2.9	−2.0	−1.2	−0.7	−0.3	0.0	0.2
	$\log \kappa_{1150}$	−1.9	−1.2	−0.6	−0.2	0.0	0.2	0.6
$Pr–PrBr_3$ Ref. **89**	p	0.5	1.1	2.5	4.1	5.8	8.6	8.7
	κ_{740}	0.66	0.80	1.10	1.47	1.90	2.38	2.67
$Pr–PrI_3$ Ref. **86**	p	0.3	1.3	2.6	4.2	6.3	8.1	8.9
	κ_{780}	0.61	1.28	2.48	4.24	8.25	14.8	21.0
$RbCl–TiCl_3$ Ref. **88**	p	43.9	54.0	64.6	75.8	87.6		
	κ_{800}	0.95	0.95	1.0	1.2	1.4		
VCl_2 in KCl–NaCl (equi-molar: 56.1% KCl) Ref. **106**	p	9.6	12.8	17.9	22.4	29.7	39.7	52.0
	κ_{600}	0.22	0.48	2.34 (610)	1.30	1.18 (605)	1.46	0.050
	κ_{800}	3.68 (810)	3.77 (790)	3.94 (790)	3.94	3.85	3.68 (795)	2.93 (805)

* *See also* Na_3AlF_6–NaF.
** *See also* 'Physical Properties Data Compilations Relevant to Energy Storage II Molten Salts: Data on Single and Multi-Component Salt Systems', J. Janz *et al.*, NSRDS–NBS 61, Part II.
† Melts may contain up to 1% MgO.
‡ *See also* AlF_3–Na_3AlF_6.

REFERENCES TO TABLE 9.5

1. H. M. Goodwin and R. D. Mailey, *Phys. Rev.*, 1908, **26**, 28.
2. V. A. Izbekov and V. A. Plotnikov, *J. Russ. Phys.-Chem. Soc.*, 1911, **43**, 18.
3. C. Sandonnini, *Gazz. Chim. Ital.*, 1920, **50**, 289.
4. K. Arndt and W. Kalass, *Z. Elektrochem.*, 1924, **30**, 12.
5. V. A. Plotnikov and P. T. Kalita, *J. Russ. Phys.-Chem. Soc.*, 1930, **62**, 2195.
6. E. Ryschkewitsch, *Z. Elektrochem.*, 1933, **39**, 531.
7. S. V. Karpachev, A. G. Stromberg and O. Poltoratzkaya, *J. phys. Chem.* (*USSR*), 1934, **5**, 793.
8. V. P. Barzakovskii, Proc. 1st All-Union Conf. on Non-Aqueous Solutions, 1935, 153.
9. K. P. Batashev, *Metallurg.* (*USSR*), 1935, **10**, 100.
10. S. V. Karpachev, A. G. Stromberg and V. N. Podchainova, *J. Gen. Chem.* (*USSR*), 1935, **5**, 1517.
11. M. de Kay Thompson and A. L. Kaye, *Trans. Electrochem. Soc.*, 1935, **67**, 169.
12. K. P. Batashev, *Legkie Metal.*, 1936, **10**, 48, quoted by V. P. Mashovets, The Electrometallurgy of Aluminium, 1938.
13. Ya. P. Mezhennii, *Mém. Inst. Chem., Acad. Sci. Ukrain S.S.R.*, 1938, **4**, 413.
14. Y. Yamaguti and S. Sisido, *J. chem. Soc. Japan*, 1938, **59**, 1311.
15. A. I. Kryagova, *J. Gen. Chem.* (*USSR*), 1939, **9**, 2061.
16. A. A. Sherbakov and B. F. Markov, *J. phys. Chem.* (*USSR*), 1939, **13**, 621.
17. V. P. Barzakovskii. *Bull. Acad. Sci. U.R.S.S.*, Class Sci. chim., 1940, 825.
18. V. P. Barzakovskii, *J. appl. Chem.* (*USSR*), 1940, **13**, 1117.
19. A. G. Bergman and I. M. Chagin, *Bull. Acad. Sci. U.R.S.S.*, Class Sci. chim., 1940, 727.
20. N. A. Belozerskii and B. A. Freidlina, *J. appl. Chem.* (*USSR*), 1941, **14**, 466.
21. Y. Yamaguti and S. Sisido, *J. chem. Soc. Japan*, 1941, **62**, 304.
22. E. R. Natsvilishvili and A. G. Bergman, *Bull Acad. Sci., U.R.S.S.*, Class Sci. chim., 1943, 23.
23. E. Ya. Gorenbein, *J. Gen. Chem.* (*USSR*), 1945, **15**, 729.
24. H. Bloom and E. Heymann, *Proc. R. Soc.*, 1947, **188***A*, 392.
25. E. Ya. Gorenbein, *J. Gen. Chem.* (*USSR*), 1947, **17**, 873.
26. T. G. Pearson and J. Waddington, *Faraday Soc. Discussion*, 1947, No. 1, 307.
27. N. M. Tarasova, *J. phys. Chem.* (*USSR*), 1947, **21**, 825.
28. E. Ya. Gorenbein, *J. Gen. Chem.* (*USSR*), 1948, **18**, 1427.
29. I. I. Bokhovkin, *ibid.*, 1949, **19**, 805.
30. E. Ya. Gorenbein and E. E. Kriss, *ibid.*, 1949, **19**, 1978.
31. H. Grothe, *Z. Elektrochem.*, 1949, **53**, 362.
32. I. N. Belyaev and K. E. Mironov, *Dokl. Akad. Nauk SSSR*, 1950, **73**, 1217.
33. I. I. Bokhovkin, *J. Gen. Chem.* (*USSR*), 1950, **20**, 397.
34. A. Vayna, *Alluminio*, 1950, **19**, 215.
35. E. Ya. Gorenbein and E. E. Kriss, *J. phys. Chem.* (*USSR*), 1951, **25**, 791.
36. R. C. Spooner and F. E. W. Wetmore, *Canad. J. Chem.*, 1951, **29**, 777.
37. R. W. Huber, E. V. Potter and H. W. St. Clair, *Rep. Invest. US Bur. Mines, No. 4858*, 1952.
38. J. Byrne, H. Fleming and F. E. W. Wetmore, *Canad. J. Chem.*, 1952, **30**, 922.
39. J. D. Edwards, C. S. Taylor, L. A. Cosgrove and A. S. Russell, *Trans. electrochem. Soc.*, 1953, **100**, 508.

40. A. Bogorodski, *J. Soc. phys.-chim. russe*, 1905, **37**, 796.
41. K. Mori and Y. Matsushita, *Tetsu to Hagane*, 1952, **38**, 365.
42. L. Shartsis, W. Capps and S. Spinner, *J. Am. ceram. Soc.*, 1953, **36**, 319.
43. H. Bloom, I. W. Knaggs, J. J. Molloy and D. Welch, *Trans. Faraday Soc.*, 1953, **49**, 1458.
44. I. N. Belyaev, *Invest. Sekt. Fiziko-Khim. Anal.*, 1953, **23**, 176.
45. P. I. Protsenko and N. P. Popovskaya, *Zh. obshch. Khim.*, 1954, **24**, 2119.
46. P. I. Protsenko and N. P. Popovskaya, *Zh. fiz. Khim.*, 1954, **28**, 299.
47. K. Sakai, *J. Chem. Soc. Japan, Pure Chem. Sect.*, 1954, **75**, 186.
48. N. P. Luzhnaya and I. P. Vereshchetina, *Izvest. Sekt. Fiziko-Khim. Anal.*, 1954, **24**, 192.
49. I. P. Vereshchetina and N. P. Luzhnava, *ibid.*, 1954, **25**, 188.
50. L. Shartsis and H. F. Shermer, *J. Am. ceram. Soc.*, 1954, **37**, 544.
51. V. D. Polyakov, *Izvest. Sekt. Fiziko-Khim. Anal.*, 1955, **26**, 147.
52. P. I. Protsenko, *ibid.*, 1955, **26**, 173.
53. V. D. Polyakov, *ibid.*, 1955, **26**, 191.
54. K. B. Morris, M. I. Cook, C. Z. Sykes and M. B. Templeman, *J. Am. chem. Soc.*, 1955, **77**, 851.
55. T. Baak, *Acta. Chem. Scand.*, 1955, **9**, 1406.
56. K. Sakai and S. Hayashi, *J. chem. Soc. Japan, Pure Chem. Sect.*, 1955, **76**, 101.
57. E. R. van Artsdalen and I. S. Yaffe, *J. phys. Chem.*, 1955, **59**, 118.
58. Yu. K. Delimarskii, I. N. Sheiko and V. G. Fenchenko, *Zh. fiz. Khim.*, 1955, **29**, 1499.
59. B. S. Harrap and E. Heymann, *Trans. Faraday Soc.*, 1955, **51**, 259.
60. G. M. Pound, G. Derge and G. Osuch, *Trans. Amer. Inst. Min. Met. Eng.*, 1955, **203**, 481.
61. H. C. Cowen and H. J. Axon, *ibid.*, 1956, **52**, 242.
62. N. P. Luzhnaya, N. N. Evseeva and I. P. Vereshchetina, *Zh. neorg. Khim.*, 1956, **1**, 1490.
63. K. Mori, *Tetsu to Hagane*, 1956, **42**, 1024.
64. B. M. Lepinskikh, O. A. Esin and S. V. Sharrin, *Zh. priklad, Khin.*, 1956, **29**, 1813.
65. E. W. Yim and M. Feinleib, *J. electrochem. Soc.*, 1957, **104**, 626.
66. F. R. Duke and R. A. Fleming, *ibid.*, 1957, **104**, 251.
67. J. B. Story and J. T. Clarke, *Trans. Amer. Inst. Min. Met. Eng.*, 1957, **209**, 1449.
68. V. A. Kochinashvili and V. P. Barzakovskii, *Zh. priklad. Khim.*, 1957, **30**, 1755.
69. W. C. Phelps and R. E. Grace, *Trans. Amer. Inst. Min. Met. Eng.*, 1957, **209**, 1447.
70. S. Okado, S. Yashizawa, N. Watanabe and Y. Omota, *J. chem. Soc. Japan, Ind. Chem. Sect.*, 1957, **60**, 670.
71. F. R. Duke and R. A. Fleming, *J. electrochem. Soc.*, 1958, **105**, 412.
72. E. Vatslavik and A. I. Belyaev, *Zh. neorg. Khim.*, 1958, **3**, 1044.
73. H. R. Bronstein and M. A. Bredig, *J. Am. chem. Soc.*, 1958, **80**, 2077.
74. O. A. Esin and V. L. Zyazcv, *Izvest. Akad. Nauk, S.S.S.R.*, 1958, No. 6, 7.
75. K. A. Kostanyan, *Izvest. Akad. Nauk. Armyan. S.S.S.R.*, 1958, **11**, 65.
76. R. H. Moss, *Univ. Microfilms* (*Ann. Arbor, Mich.*), 1955, No. 12, 730.
77. Y. Doucet and M. Bizouard, *Compt. rend.*, 1960, **250**, 73.
78. B. F. Markov and V. D. Prusyazhnyi, *Ukr. khim. Zh.*, 1962, **28**, 653.
79. Y. Doucet and M. Bizouard, *Compt. rend.*, 1959, **248**, 1328.
80. V. P. Mashovets and V. I. Petrov, *Zh. prikl. Khim.*, 1959, **32**, 1528.
81. J. O'M. Bockris, J. A. Kitchener, S. Ignatowitz and J. W. Tomlinson, *Discuss. Faraday Soc.*, 1948, **4**, 265.
82. P. I. Protsenko and O. N. Shokina, *Zh. neorg. Khim.*, 1960, **5**, 437.
83. P. I. Protensko and A. Ya Malakhova, *ibid.*, 1960, **5**, 2307.
84. L. F. Grantham and S. J. Yosim, *J. chem. Phys.*, 1963, **38**, 1671.
85. A. S. Dworkin, H. R. Bronstein and M. A. Bredig, *Discuss. Faraday Soc.*, 1961, **32**, 188.
86. A. S. Dworkin, R. A. Sallach, H. R. Bronstein, M. A. Bredig and J. D. Corbett, *J. phys. Chem.*, 1963, **67**, 1145.
87. B. F. Markov and V. D. Prusyazhnyi, *Ukr. khim. Zh.*, 1962, **28**, 268.
88. R. V. Chernov and Yu. K. Delimarskii, *Zh. neorg. Khim.*, 1961, **6**, 2749.
89. A. S. Dworkin, H. R. Bronstein and M. A. Bredig, *J. phys. Chem.*, 1963, **67**, 2715.
90. H. R. Bronstein and M. A. Bredig, *ibid.*, 1961, **65**, 1220.
91. H. R. Bronstein, A. S. Dworkin and M. A. Bredig, *J. chem. Phys.*, 1961, **34**, 1843.
92. V. G. Selivanov and V. V. Stender, *Zh. neorg. Khim.*, 1959, **4**, 2058.
93. B. F. Markov and V. D. Prusyazhnyi, *Ukr. khim. Zh.*, 1962, **28**, 130.
94. B. F. Markov and V. D. Prusyazhnyi, *ibid.*, 1962, **28**, 419.
95. H. R. Bronstein, A. S. Dworkin and M. A. Bredig, *J. phys. Chem.*, 1962, **66**, 44.
96. G. W. Mellors and S. Senderoff, *ibid.*, 1960, **64**, 294.
97. E. R. Buckle and P. E. Tsaoussoglou, *Trans. Faraday Soc.*, 1964, **60**, 2144.
98. M. F. Lantratov and O. F. Moiseeva, *Zh. fiz. Khim.*, 1960, **34**, 367.
99. V. V. Rafal'skii, *Ukr. khim. Zh.*, 1960, **26**, 585.
100. K. B. Morris and P. L. Robinson, *J. phys. Chem.*, 1964, **68**, 1194.
101. P. C. Papaioannou and G. W. Harrington, *ibid.*, 1964, **68**, 2424.
102. H. Winterhager and L. Werner, *ForschBer. Wirt.-u. Verk Minist. NRhein.-Westf.*, 1957, **438.**
103. K. B. Morris and P. L. Robinson, *J. chem. Engng Data*, 1964, **9**, **9**, 444.
104. L. J. Howell and H. H. Kellogg, *Trans. Am. Inst. Min. Engrs*, 1959, **215**, 143.
105. J. O'M. Bockris and G. W. Mellors, *J. phys. Chem.*, 1956, **60**, 1321.
106. Yu. U. Samson, L. P. Ruzinov, N. S. Rezhemnukova and V. E. Baru, *Zh. fiz. Khim.*, 1964, **38**, 481.

Table 9.6 SURFACE TENSION OF PURE MOLTEN SALTS

The surface tension (mNm^{-1}) at temperature t (°C) is given as γ_t, or the constants a, b and t_0 in the equation $\gamma_t = a - b(t - t_0)$ are given for the temperature range r. Principal references are in **bold type.**

Salt		
AgBr	γ_{500}	152
Ref. **16, 1**	γ_{600}	148
	γ_{700}	146
AgCl	γ_{500}	176
Ref. **16, 5**	γ_{600}	171
	γ_{700}	166
AgI	γ_{600}	115
Ref. **16**		
$AgNO_3$	a	152.1
Ref. **22**, 27, 19	b	0.082
	t_0	212
	r	244–400
Al_2O_3	γ_{2050}	700
Ref. **28**		
B_2O_3	γ_{700}	67
Ref. **21**	γ_{1000}	83
	γ_{1300}	99
$BaBr_2$	γ_{900}	150
Ref. **20**	γ_{950}	147
	γ_{1000}	143
$BaCl_2$	γ_{966}	169.2
Ref. **23**, 15, 12, 9, 2	γ_{979}	168.3
	γ_{984}	168.6
	γ_{996}	168.1
	γ_{1000}	167.8
	γ_{1050}	164
$Ba(NO_3)_2$	a	134.8
Ref. **22**	b	0.015
	t_0	595
	r	600–660
$BiBr_3$	γ_{250}	66
Ref. **6**	γ_{350}	56
	γ_{450}	45
$BiCl_3$	γ_{275}	67
Ref. **6**	γ_{325}	59
	γ_{375}	53
BrF_3	γ_{12}	37
Ref. **18**	γ_{27}	36
	γ_{45}	34
BrF_5	γ_{9}	24
Ref. **18**	γ_{33}	22
$CaBr_2$	γ_{780}	117.31
Ref. **20**	γ_{800}	116.39
$CaCl_2$	γ_{800}	150.57
Ref. **28, 29**	γ_{1000}	140.07
CaI_2	γ_{800}	85
Ref. **20**	γ_{900}	83
	γ_{1000}	81
$Ca(NO_3)_2$	γ_{560}	101.5
Ref. **22**		
$CdBr_2$	γ_{600}	67
Ref. **16**	γ_{700}	65
$CdCl_2$	γ_{600}	83
Ref. **16**	γ_{700}	81
	γ_{800}	78
CsBr	γ_{650}	82.5
Ref. **28, 29**	γ_{800}	73.35
	γ_{1000}	61.147
CsCl	γ_{700}	87.50
Ref. **29, 6**, 12	γ_{900}	71.18
	γ_{1100}	54.86
CsF	γ_{700}	106.1
Ref. **28, 29**	γ_{900}	92.26
	γ_{1100}	78.46
CsI	γ_{650}	73.47
Ref. **28**, 29	γ_{800}	64.77
	γ_{1000}	53.17
$CsNO_3$	a	92.5
Ref. **22**, 6	b	0.069
	t_0	414
	r	421–597
Cs_2SO_4	γ_{1000}	114.25
Ref. **30**	γ_{1250}	98.84
	γ_{1500}	83.75
FeO	γ_{1400}	584.0
Ref. **28**		
$GaCl_2$	a	56.6
Ref. **24**	b	0.18
	t_0	170
	r	166–170
$GaCl_3$	γ_{80}	27
Ref. **17**	γ_{110}	24
	γ_{140}	21
GeO_2	γ_{1000}	241
Ref. **21**	γ_{1200}	253
	γ_{1400}	265
$HgBr_2$	γ_{241}	64
Ref. **8**	γ_{276}	60
$HgCl_2$	γ_{293}	56
Ref. **8**		
IF_5	γ_{18}	31
Ref. **18**	γ_{28}	28
KBr	γ_{750}	90
Ref. **16, 11, 6**, 27	γ_{800}	86
	γ_{850}	82
	γ_{900}	78
KCl	γ_{800}	98
Ref. **19, 16, 15, 12, 9, 6,**	γ_{900}	90
27	γ_{1000}	82
	γ_{1100}	75

Table 9.6 SURFACE TENSION OF PURE MOLTEN SALTS—*continued*

Salt		
$K_2Cr_2O_7$ Ref. **30**	γ_{420}	140.28
	γ_{470}	138.28
	γ_{520}	136.28
KF Ref. **28, 29,** 30	γ_{900}	138.12
	γ_{1100}	123.12
	γ_{1300}	108.12
KI Ref. **27**	a	138.7
	b	0.087
	t_0	0
	r	(m.p. + 10)–(m.p. + 210)
K_2MoO_4 Ref. **30**	γ_{900}	151.47
	γ_{1200}	132.87
	γ_{1500}	114.27
KNO_2 Ref. **22**	a	107.6
	b	0.080
	t_0	435
	r	445–501
KNO_3 Ref. **27**	γ_{350}	108.61
	γ_{450}	102.21
	γ_{600}	92.61
	γ_{800}	79.81
KPO_3 Ref. **30**	γ_{900}	193.54
	γ_{1200}	171.94
	γ_{1500}	150.34
K_2SO_4 Ref. **13**	γ_{1100}	143
	γ_{1300}	130
	γ_{1500}	116
K_2WO_4 Ref. 30	γ_{900}	162.41
	γ_{1200}	134.51
	γ_{1500}	106.61
La_2O_3 Ref. **7**	γ_{2320}	560
LiCl Ref. **28, 29**	γ_{600}	137.14
	γ_{800}	123.22
	γ_{1000}	109.30
LiF Ref. **28, 29**	γ_{850}	250.46
	γ_{1050}	228.60
	γ_{1250}	206.74
$LiNO_2$ Ref. **22**, 6	a	115.4
	b	0.053
	t_0	255
	r	276–425
Li_2SO_4 Ref. **13, 6**	γ_{900}	223
	γ_{1000}	216
	γ_{1100}	209
	γ_{1200}	202
NH_4NO_3 Ref. **22**	a	101.9
	b	0.105
	t_0	170
	r	170–220.
Na_3AlF_6 Ref. 31	γ_{1000}	134.06
	γ_{1010}	132.78
	γ_{1020}	131.50
NaBr Ref. **16, 6,** 27	γ_{750}	106
	γ_{800}	103
	γ_{950}	92
	γ_{1150}	79
NaCl Ref. **6,** 27	γ_{800}	116.42
	γ_{900}	107.12
	γ_{1000}	97.82
	γ_{1100}	88.52
NaF Ref. **6,** 31	γ_{1000}	185.21
	γ_{1200}	168.81
	γ_{1400}	152.41
	γ_{1500}	144.21
NaI Ref. **27**	a	147.4
	b	0.090
	t_0	0
	r	(m.p. + 10)–(m.p. + 210)
Na_2MoO_4 Ref. 30	γ_{700}	211.68
	γ_{900}	196.28
	γ_{1100}	180.88
	γ_{1200}	173.18
$NaNO_2$ Ref. **22,** 27	a	121.2
	b	0.041
	t_0	277
	r	291–384
$NaNO_3$ Ref. 27	γ_{310}	116.21
	γ_{400}	112.70
	γ_{500}	108.80
	γ_{600}	104.90
$NaPO_3$ Ref. **6,** 30	γ_{800}	200.74
	γ_{1000}	186.34
	γ_{1200}	171.94
	γ_{1500}	150.34
Na_2SO_4 Ref. **13, 6**	γ_{900}	193
	γ_{950}	190
	γ_{1000}	186
	γ_{1050}	183
Na_2WO_4 Ref. **6,** 30	γ_{700}	201.46
	γ_{1000}	182.0
	γ_{1300}	162.46
	γ_{1600}	143.0
$PbCl_2$ Ref. 27	γ_{500}	137.12
	γ_{600}	126.12
	γ_{700}	115.12
P_2O_5 Ref. **21**	γ_{100}	60
	γ_{200}	58
	γ_{400}	54
RbBr Ref. **28, 29**	γ_{700}	99.77
	γ_{900}	88.17
	γ_{1100}	76.57

Table 9.6 SURFACE TENSION OF PURE MOLTEN SALTS—*continued*

Salt		
RbCl	γ_{750}	94.90
Ref. **28, 29**	γ_{900}	82.49
	γ_{1000}	74.22
	γ_{1150}	61.82
RbF	γ_{800}	163.83
Ref. 28, 29	γ_{900}	153.60
	γ_{1000}	143.37
RbI	γ_{700}	76.86
Ref. **28, 29**	γ_{900}	63.18
	γ_{1000}	56.34
$RbNO_3$	γ_{300}	109.53
Ref. **32**	γ_{500}	95.53
	γ_{700}	81.53
Rb_2SO_4	γ_{1100}	130.8
Ref. **30**	γ_{1300}	120.4
	γ_{1500}	110
SiO_2	γ_{1000}	278
Ref. **21**	γ_{1500}	295
	γ_{2000}	313
$SnCl_2$	γ_{300}	96.74
Ref. **28, 29**	γ_{400}	88.44
	γ_{500}	80.14
$SrBr_2$	γ_{700}	147
Ref. **20**	γ_{800}	143
	γ_{900}	138
	γ_{1000}	134
$SrCl_2$	γ_{900}	168
Ref. **20**	γ_{950}	165
	γ_{1000}	162
SrI_2	γ_{600}	112
Ref. **20**	γ_{750}	106
	γ_{850}	102
	γ_{950}	98
$Sr(NO_3)_2$	γ_{615}	128.4
Ref. **22**		
$TlNO_3$	a	94.8
Ref. **22**, 6	b	0.078
	t_0	206
	r	226–458
Tl_2S	a	215.6
Ref. **25**	b	0.0356
	t_0	445
	r	500–700
V_2O_5	γ_{1000}	86
Ref. **26**		

REFERENCES TO TABLE 9.6

1. A. Gradenwitz, *Ann. Physik.*, 1899, **67**, 467.
2. Z. Motylewski, *Z. anorg. Chem.*, 1904, **38**, 410.
3. R. Lorenz and F. Kaufler, *Ber. dt. chem. Ges.*, 1908, **41**, 3727.
4. R. Lorenz and A. Liebmann, *Z. phys. Chem.*, 1913, **83**, 459.
5. R. Lorenz, A. Liebmann and A. Hochberg, *Z. anorg. Chem.*, 1916, **94**, 301.
6. F. M. Jaeger, *Z. anorg. Chem.*, 1917, **101**, 1.
7. H. V. Wartenberg, G. Wehner and E. Suran, *Nachr. Ges. Wiss. Göttingen*, 1936, **2**, 65.
8. E. B. R. Prideaux and J. R. Jarrett, *J. chem. Soc.*, 1938, 1203.
9. V. P. Barzakovskii, *Bull. acad. sci. URSS. Classe sci. chim.*, 1940, 825.
10. P. P. Kozakevich and A. F. Kononenko, *J. phys. Chem* (*USSR*), 1940, **14**, 1118.
11. K. Semenchenko and L. P. Shikhobolova, *ibid.*, 1947, **21**, 613
12. K. Semenchenko, *ibid.*, 1947, **21**, 707.
13. K. Semenchenko, *ibid.*, 1947, **21**, 1387.
14. A. Vajna, *Alluminio*, 1951, **20**, 29.
15. J. S. Peake and M. R. Bothwell, *J. Am. chem. Soc.*, 1954, **76**, 2625.
16. N. K. Boardman, A. R. Palmer and E. Heymann, *Trans. Faraday Soc.*, 1955, **51**, 277.
17. N. N. Greenwood and K. Wade, *J. inorg. nuclear Chem.*, 1957, **3**, 349.
18. M. T. Rogers and E. E. Carver. *J. Phys. Chem.*, 1958, **62**, 952.
19. J. L. Dahl and F. R. Duke, *US Atomic Energy Comm.*, 1958, ISC-923.
20. R. B. Ellis, J. E. Smith and E. B. Baker, *J. phys. Chem.*, 1958, **62**, 766.
21. W. D. Kingery, *J. Am. ceram. Soc.*, 1959, **42**, 6.
22. C. C. Addison and J. M. Coldrey, *J. chem. Soc.*, 1961, 468.
23. I. D. Sokolova and N. K. Voskresenskaya, *Zh. prikl. Khim.*, 1962, **36**, 955.
24. N. N. Greenwood and I. J. Worrall, *J. chem. Soc.*, 1958, 1680.
25. V. B. Lazarev and M. N. Abdusalyamova, *Izv. Akad. Nauk SSSR, Ser. Khim.*, 1964, 1104.
26. B. M. Lepinskikh, O. A. Esin and G. A. Teterin, *Zh. neorg. Khim.*, 1960, **5**, 642.
27. H. Bloom, F. G. Davis and D. W. James, *Trans. Faraday Soc.*, 1960, **56**, 1179.
28. O. K. Sokolov, *Izv. Akad. Nauk SSSR Met. Gorn. Delv*, 1963, **4**, 59.
29. R. B. Ellis and W. S. Wilcox, Work performed under U. S. At. Energy Comm; T–10–7622, 1962, pp. 128–36.

30. 'International Critical Tables', McGraw-Hill, New York, 1933.
31. H. Bloom and B. W. Burrows, 'Proc 1st Australian Conf. Electrochem' (J. A. Friend and F. Gutman eds), p. 882. Pergamon Press, Oxford, 1964.
32. S. D. Gromakov and A. I. Kostromin, *Univ. in V. I UP Yanova-Lenina Khim.*, 1955, **115**, 93.

Table 9.7 SURFACE TENSION OF MOLTEN BINARY SALT SYSTEMS AND OTHER MIXED IONIC MELTS
The surface tension (mN m^{-1}) at temperature t(°C) and composition p(wt.%) of the first-named constituent is given as γ_t (or σ_t), or the constants a, b and t_0 in the equation $\gamma_t = a - b(t - t_0)$ are given for the temperature range r. Principal references are in **bold type.**

System							
$AgBr–AgCl$	p	0	24.7	46.7	66.4	84.0	100
Ref. **16**	σ_{500}	176	169	164	158	155	152
	σ_{600}	171	164	160	155	152	148
$AgBr–AgI$	p	0	16.7	34.8	54.5	76.2	100
Ref. **16**	σ_{500}	—	122	127	133	141	152
	σ_{600}	115	118	123	130	138	148
$AgBr–KBr$	p	0	28.3	51.3	70.3	86.3	100
Ref. **16**	σ_{700}	—	96	99	104	113	146
	σ_{750}	90	92	96	101	110	146
$AgCl–KCl$	p		32.6	56.3	72.6	88.6	100
Ref. **16**	σ_{600}		112	117	123	136	171
	σ_{700}		105	110	117	129	166
$AgCl–PbCl_2$	p	0	11.4	25.6	43.7	67.4	100
Ref. **16**	σ_{500}	137	139	144	150	160	176
	σ_{600}	127	129	134	142	152	171
$AgNO_3–CsNO_3$	p	22.5	46.5	72.4	88.8		
Ref. **19**	a	126.1	130.3	140.1	148.4		
	b	0.075	0.072	0.073	0.072		
	t_0	0	0	0	0		
	r		liquidus–*c.* 400				
$AgNO_3–KNO_3$	p	20	40	60	80	95	100
Ref. **18**, 24. 19	σ_{250}	—	—	124	129	139	147
	σ_{300}	—	118	121	126	136	144
	σ_{350}	113	114	117	122	133	140
$AgNO_3–LiNO_3$	p	45.0	71.2	88.1	100		
Ref. **19**	a	137.8	145.7	153.5	163.7		
	b	0.064	0.068	0.067	0.066		
	t_0	0	0	0	0		
	r		liquidus–*c.* 400				
$AgNO_3–NaNO_3$	p	24.1	33.3	60.0	75.0	90.0	95.0
Ref. **18**, 24, 19	σ_{300}	—	—	126	130	136	137
	σ_{350}	119	120	123	127	133	135
	σ_{400}	115	116	—	—	—	—
$AgNO_3–RbNO_3$	p	27.8	53.5	77.7	91.3		
Ref. **19**	a	134.0	136.9	143.4	151.7		
	b	0.077	0.073	0.070	0.069		
	t_0	0	0	0	0		
	r		liquidus–*c.* 400				
$AlF_3–Na_3AlF_6$	p	2.5	5	10	15		
Ref. **11, 1**	σ_{1000}	142	137	127	118.		
$Al_2O_3–FeO$	p	5.6	6.9				
Ref. **4**	σ_{1410}	598	604				
$Al_2O_3–Na_3AlF_6$	p	2.5	5	10	15		
Ref. **11, 1**	σ_{1000}	143	140	136	134		

Table 9.7 SURFACE TENSION OF MOLTEN BINARY SALT SYSTEMS AND OTHER MIXED IONIC MELTS—*continued*

System								
B_2O_3–BaO	p	20.5	34.6	44.4	54.7	64.2	75.1	84.5
Ref. **14**	σ_{900}	—	—	239	197	146	88	79
	σ_{1100}	315	257	229	187	138	92	87
	σ_{1300}	294	243	221	181	135	106	96
B_2O_3–CaO	p	51.2	62.8	73.3	83.1	93.7	98.1	
Ref. **14**	σ_{900}	—	—	176	—	81	81	
	σ_{1100}	302	235	150	87	87	88	
	σ_{1300}	297	221	140	96	95	96	
B_2O_3–K_2O	p	51.5	55.7	65.6	75.3	84.9	94.8	98.5
Ref. **13**	σ_{700}	—	—	—	140	107	80	75
	σ_{900}	139	153	160	143	112	85	80
	σ_{1100}	—	—	149	142	118	93	88
B_2O_3–Li_2O	p	71.3	79.9	87.1	92.0	95.5	100	
Ref. **13**	σ_{700}	—	221	—	120	84	74	
	σ_{900}	231	207	158	121	92	80	
	σ_{1100}	—	—	152	125	101	88	
B_2O_3–Na_2O	p	64.0	73.6	82.2	85.8	94.4	99.1	
Ref. **13**	σ_{700}	—	196	146	—	83	77	
	σ_{900}	197	185	147	130	91	81	
	σ_{1100}	—	—	149	134	99	89	
B_2O_3–PbO	p	0	15.9	33.6	50.9	65.6	78.4	100
Ref. **10, 8**	σ_{700}	—	162	160	—	—	—	—
	σ_{900}	132	163	144	92	78	79	80
	σ_{1000}	135	163	—	—	—	—	83
B_2O_3–SrO	p	35.4	49.6	61.6	70.0	89.0	96.0	
Ref. **14**	σ_{900}	—	—	194	137	80	—	
	σ_{1100}	303	247	178	123	87	86	
	σ_{1300}	292	237	166	122	95	93	
B_2O_3–ZnO	p	20.5	29.4	35.4	41.4	50.0	90.0	100
Ref. **9**	σ_{900}	—	—	235	184	80	78	80
	σ_{1100}	—	283	232	179	88	86	86
	σ_{1300}	355	287	238	185	—	95	94
$BaCl_2$–KCl	p	0	20.4	47.4	59.0	67.2	82.2	92.8
Ref. **15**	σ_{850}	93	98	106	110	114	129	—
	σ_{900}	89	94	103	106	112	128	144
	σ_{950}	—	—	—	—	—	121	141
$BaCl_2$–Li_2SO_4	p	0	9.0	17.3	44.6	65.3	84.9	100
Ref. **6**	σ_{1000}	220	198	192	167	163	161	175
	σ_{1050}	216	190	180	164	158	159	172
$BaCl_2$–$NaCl$	p	0	28.4	47.1	65.7	78.1	93.4	100
Ref. **3**, 20	σ_{900}	107	109	117	126	133	160	—
	σ_{1000}	100	103	111	120	126	154	171
$CaCl_2$–$NaCl$*	p	0	17.4	44.8	65.5	88.3	100	
Ref. **3**	σ_{800}	114	115	121	126	137	148	
	σ_{1000}	100	100	108	116	125	140	
CaF_2–Na_3AlF_6	p	2.5	5	10	15			
Ref. **11**	σ_{1000}	149	150	152	155			
CaO–FeO	p	0	5.5	15.4	22.3			
Ref. **4**	σ_{1410}	585	555	543	573			
$CdBr_2$–$CdCl_2$	p	0	20.9	41.3	61.7	80.9	100	
Ref. **16**	σ_{600}	83	78	74	71	68	67	
	σ_{700}	81	76	72	68	66	65	

Table 9.7 SURFACE TENSION OF MOLTEN BINARY SALT SYSTEMS AND OTHER MIXED IONIC MELTS—*continued*

System								
$CdCl_2$–KCl	p		38.1	62.1	78.7	90.7	100	
Ref. **16**	σ_{600}		100	93	90	88	82	
	σ_{700}		94	87	84	84	81	
$CdCl_2$–NaCl	p	0	44.0	67.6	82.5	92.6	100	
Ref. **16**	σ_{700}	—	108	98	90	85	81	
	σ_{800}	114	101	92	85	81	78	
$CdCl_2$–$PbCl_2$	p	0	14.1	31.8	49.7	72.4	100	
Ref. **16**	σ_{600}	126	116	107	98	90	83	
	σ_{700}	116	108	101	93	87	81	
Cr_2O_3–FeO	p	0	3.1					
Ref. **4**	σ_{1420}	585	588					
CsCl–Li_2SO_4	p	0	1.6	3.2	4.8	15.2	61.8	100
Ref. **7, 6**	σ_{900}	224	205	193	189	163	102	72
	σ_{1000}	220	201	188	185	160	92	64
	σ_{1100}	211	194	181	177	154	—	—
CsCl–$PbCl_2$	p	12.4	27.4	38.6	49.0	64.7		
Ref. **18**	σ_{500}	116	—	—	—	103		
	σ_{600}	—	97	—	93	94		
	σ_{625}	—	94	91	90	92		
$CsNO_3$–KNO_3	p	39.1	66.0	85.3				
Ref. **19**	a	133.7	129.4	125.1				
	b	0.079	0.077	0.074				
	t_0	0	0	0				
	r		liquidus–*c*. 400					
$CsNO_3$–$LiNO_3$	p	48.5	73.9	89.4				
Ref. **19**	a	124.3	124.3	122.9				
	b	0.070	0.076	0.075				
	t_0	0	0	0				
	r		liquidus–*c*. 400					
$CsNO_3$–$NaNO_3$	p	43.1	67.6	87.3				
Ref. **19**	a	130.5	127.3	123.4				
	b	0.068	0.074	0.072				
	t_0	0	0	0				
	r		liquidus–*c*. 400					
Cs_2O–SiO_2	p	47.2	54.0	69.0	78.9			
Ref. **21**	γ_{1300}	166.1	165.1	144.3	120.5			
	γ_{1400}	163.8	162.5	—	—			
FeO–MnO	p	90.5	94.0	100				
Ref. **4**	σ_{1410}	555	567	585				
FeO–SiO_2	p	68.1	76.9	84.1	95.8	100		
Ref. **4**	σ_{1410}	409	468	503	563	585		
FeO–TiO_2	p	81.9	85.1	100				
Ref. **4**	σ_{1410}	510	522	585				
KBr–KCl	p	0	28.6	51.7	70.7	86.5	100	
Ref. **16**	σ_{750}	—	97	95	93	92	90	
	σ_{800}	99	95	93	89	87	86	
KBr–NaBr	p	0	22.4	43.5	63.4	82.2	100	
Ref. **16**	σ_{700}	—	103	98	96	95	—	
	σ_{800}	103	96	92	88	87	86	
KBr–Na_2SO_4	p	0	4.2	21.8	45.6	100		
Ref. **5**	σ_{900}	193	181	140	116	81		

Table 9.7 SURFACE TENSION OF MOLTEN BINARY SALT SYSTEMS AND OTHER MIXED IONIC MELTS—*continued*

System								
$KCl–K_2SO_4$	p	0	0.9	2.2	4.5			
Ref. **7**	σ_{1075}	144	140	135	133			
$KCl–Li_2SO_4$	p	0	7.0	26.6	40.2	66.8	85.8	100
Ref. **7, 6**	σ_{900}	224	182	145	128	109	97	91
	σ_{1000}	220	178	142	123	101	90	85
	σ_{1100}	211	173	135	116	92	81	75
$KCl–MgCl_2$	p	8.0	16.4	34.4	54.1	75.9	87.6	
Ref. **17**	σ_{700}	76	79	83	85	93	97	
	σ_{800}	69	72	76	78	86	90	
	σ_{900}	62	65	69	71	79	83	
KCl–NaCl*	p	0	24.2	46.0	65.6	83.6	100	
Ref. **18, 17, 3**	σ_{700}	—	—	114	111	—	—	
	σ_{750}	—	115	111	107	104	—	
	σ_{800}	114	111	107	104	101	96	
KCl–NaI	p	7.9	19.5	32.6	49.0	73.2	84.9	
Ref. **24**	a	135.2	128.0	121.7	124.4	127.2	128.5	
	b	0.066	0.063	0.057	0.063	0.069	0.070	
	t_0	0	0	0	0	0	0	
	r		(m.p. + 10) – (m.p. + 210)					
$KCl–Na_2SO_4$	p	0	2.7	14.8	34.2	100		
Ref. **5**	σ_{900}	193	183	153	129	92		
$KCl–PbCl_2$	p	0	5.5	15.2	28.6	51.7		
Ref. **18, 16**	σ_{500}	137	125	118	117	118		
	σ_{600}	117	114	109	107	107		
$KI–Na_2SO_4$	p	0	5.8	11.5	28.0	53.9	100	
Ref. **5**	σ_{900}	193	177	157	127	100	73	
$KNO_2–KNO_3$	p	17.4	35.8	55.9	77.0			
Ref. **24**	a	129.6	131.3	130.6	131.8			
	b	0.059	0.061	0.055	0.052			
	t_0	0	0	0	0			
	r		(m.p. + 10)–(m.p. + 210)					
$KNO_3–LiNO_3$	p	0	32.9	59.5	81.5	100		
Ref. **19**	a	129.9	127.9	129.6	133.6	139.8		
	b	0.055	0.056	0.062	0.070	0.081		
	t_0	0	0	0	0	0		
	r		liquidus–*c.* 400					
$KNO_3–NaNO_3$*	p	0	40	60	100			
Ref. **18, 16**, 19	σ_{350}	117	115	114	112			
	σ_{400}	114	112	111	109			
	σ_{450}	111	109	107	105			
$KNO_3–RbNO_3$	p	0	18.6	40.7	67.4			
Ref. **19**	a	134.3	135.0	136.9	138.5			
	b	0.083	0.082	0.083	0.083			
	t_0	0	0	0	0			
	r		liquidus–*c.* 400					
$K_2O–SiO_2$	p	23.9	26.7	29.9	32.9	38.7	43.6	
Ref. **12**, 21	σ_{1000}	—	227	227	227	225	—	
	σ_{1200}	222	223	220	219	218	215	
	σ_{1400}	219	219	216	214	210	206	
$K_2SO_4–RbCl$	p	92.9	96.5	98.6	100			
Ref. **7**	σ_{1075}	134	139	141	143			
$LiCl–PbCl_2$	p	6.1	7.8	12.8	13.4	24.9	27.1	
Ref. **18**	σ_{500}	133	134	134	133	—	—	
	σ_{550}	130	129	129	129	128	—	
	σ_{600}	124	—	—	—	—	124	

Table 9.7 SURFACE TENSION OF MOLTEN BINARY SALT SYSTEMS AND OTHER MIXED IONIC MELTS—*continued*

LiCl–RbCl	p	0	75.9	86.9	91.9	94.5	97.2	100
Ref. **7**	σ_{750}	96	113	118	119	121	122	127
Li_2O–SiO_2	p	12.9	17.8	23.9	27.8	33.5	43.1	49.1
Ref. **12**, 21	σ_{1100}	—	315	328	338	—	—	—
	σ_{1300}	311	317	328	334	352	369	381
	σ_{1400}	316	317	328	332	349	364	374
$LiNO_3$–$RbNO_3$	p	13.5	31.8	58.3				
Ref. **19**	a	130.0	129.6	125.0				
	b	0.075	0.074	0.059				
	t_0	0	0	0				
	r		liquidus–*c.* 400					
Li_2SO_4–NaCl*	p	0	38.8	65.5	81.6	94.5	98.7	100
Ref. **7, 6**	σ_{900}	109	131	148	168	198	208	224
	σ_{1000}	104	125	143	164	194	204	220
	σ_{1100}	95	116	134	157	187	196	211
Li_2SO_4–RbCl	p	0	38.0	57.9	68.1	78.6	89.2	100
Ref. **7, 6**	σ_{900}	83	105	126	140	155	179	224
	σ_{1000}	74	97	120	134	149	174	220
	σ_{1100}	66	92	114	—	145	168	211
$MgCl_2$–NaCl	p	17.9	33.0	56.7	74.7	88.7	94.7	
Ref. **17**	σ_{700}	110	103	95	87	79	75	
	σ_{800}	104	97	89	81	73	69	
	σ_{900}	98	91	83	75	67	63	
Na_3AlF_6–NaF	p	85	90	95	97.5			
Ref. **11, 1**	σ_{1000}	161	159	155	152			
NaCl–$NaNO_3$	p	0	9.2					
Ref. **5**	σ_{470}	110	110					
NaCl–Na_2SO_4	p	0	0.8	1.7	4.3			
Ref. **7**	σ_{900}	192	189	184	179			
NaCl–$PbCl_2$	p	0	4.9	11.7	15.1	18.2		
Ref. **18, 3**, 24	σ_{500}	—	133	131	—	—		
	σ_{550}	131	127	124	—	—		
	σ_{575}	128	124	122	122	122		
NaF–$NaNO_3$	p	0	9.8					
Ref. **5**	σ_{560}	108	108					
NaI–$NaNO_3$	p	0	10.1					
Ref. **5**	σ_{312}	119	119					
NaI–Na_2SO_4	p	0.5	1.1	2.1	3.2	5.3	10.5	
Ref. **2**	σ_{900}	168	167	160	158	154	141	
$NaNO_2$–$NaNO_3$	p	16.9	35.4	55.0	76.3			
Ref. **24**	a	127.9	127.1	128.9	126.8			
	b	0.039	0.035	0.038	0.030			
	t_0	0	0	0	0			
	r		(m.p. + 10) – (m.p. + 210)					
$NaNO_3$–$RbNO_3$	p	16.1	36.5	63.4				
Ref. **19**	a	132.5	133.7	135.4				
	b	0.076	0.073	0.068				
	t_0	0	0	0				
	r	liquidus–*c.* 400						
Na_2O–P_2O_5	$\gamma_t = 150.6 + 155p/(100-p) - 0.0379t$							
Ref. **22**	r_p	30.4–39.5						
	r_t	liquidus–1 050						

Table 9.7 SURFACE TENSION OF MOLTEN BINARY SALTS SYSTEMS AND OTHER MIXED IONIC MELTS—*continued*

Na_2O–SiO_2	p	20.0	30.8	33.6	36.9	50.0		
Ref. **12**, **2**, 21	σ_{1000}	277	284	286	288	—		
	σ_{1200}	276	280	281	283	295		
	σ_{1400}	273	274	274	276	284		
Na_2SO_4RbCl	p	0	1.7	3.4	4.3	8.6		
Ref. **7**	σ_{1050}	183	173	171	169	162		
$PbCl_2$–RbCl	p	46.7	62.8	69.0	83.1	90.3	98.2	
Ref. **7**	σ_{475}	—	112	112	116	124	—	
	σ_{525}	—	107	107	111	118	130	
	σ_{575}	104	104	104	107	113	124	
PbO–SiO_2	p	65.1	69.9	75.7	82.5	84.7	90.6	96.8
Ref. **8**, 23	σ_{900}	—	—	217	199	192	174	134
	σ_{1100}	234	232	221	204	196	179	145
	σ_{1300}	235	230	223	209	—	183	158
PbO–V_2O_5	p	15.3	39.6	52.8	65.4	79.0	85.1	92.1
Ref. **23**	γ_{1000}	92	135	168	174	205	202	192
PbO–SiO_2–V_2O_5	p_{PbO}	28.5	38.6	58.4	60.0	68.5	80.0	
Ref. **23**	p_{SiO_2}	10.3	10.4	10.4	10.1	10.6	10.0	
	γ_{1000}	128	142	150	190	210	202	
Rb_2O–SiO_2	p	39.1	43.9	49.5	59.5	67.3		
Ref. **21**	γ_{1200}	—	—	—	175.1	155.0		
	γ_{1300}	200.1	192.7	188.0	173.4	146.3		
	γ_{1400}	197.1	188.8	183.5	170.9	—		

* *See also:* 'Physical Properties Data Compilations Relevant to Energy Storage II Molten Salts: Data on Single and Multi-Component Salt Systems', J. Janz *et al.*, NSRDS–NBS 61, Part II.

REFERENCES TO TABLE 9.7

1. E. Elchardus, *Compt. rend.*, 1938, **206**, 1460.
2. C. W. Parmelee and C. G. Harman, *Univ. Illinois Eng. Exptl. Sta. Bull.*, 1939, No. 311, 29.
3. V. P. Barzakovskii, *Bull acad. sci. U.R.S.S., Classe sci. chim.*, 1940, 825.
4. P. P. Kozakevich and A. F. Kononenko, *J. phys. Chem.* (*USSR*), 1940, **14**, 1118.
5. K. Semenchenko and L. P. Shikhobolova, *ibid.*, 1947, **21**, 613.
6. K, Semenchenko, *ibid.*, 1947, **21**, 707.
7. K. Semenchenko, *ibid.*, 1947, **21**, 1387.
8. L. Shartsis, S. Spinner and A. W. Smock, *J. Am. ceram. Soc.*, 1948, **31**, 23.
9. L. Shartsis and R. Canga, *J. Res. Nat. Bur. Stand.*, 1949, **43**, 221.
10. S. Carlen, *Trans. Chalmers Univ. Tech. Gottenburg*, 1949, No. 85.
11. A. Vajna, *Alluminio*, 1951, **20**, 29.
12. L. Shartsis and S. Spinner, *J. Res. Nat. Bur. Stand.*, 1951, **46**, 385.
13. L. Shartsis and W. Capps, *J. Am. ceram. Soc.*, 1952, **35**, 169.
14. L. Shartsis and H. F. Shermer, *ibid.*, 1954, **37**, 544.
15. J. S. Peake and M. R. Bothwell, *J. Am. chem. Soc.*, 1954, **76**, 2656.
16. N. K. Boardman, A. R. Palmer and E. Heymann, *Trans. Faraday Soc.*, 1955, **51**, 277.
17. O. G. Desyatnikov, *Zh. priklad. Khim.*, 1956, **29**, 870.
18. J. H. Dahl and F. R. Duke, *J. phys. Chem.*, 1958, **62**, 1498.
19. G. Bertozzi and G. Sternheim, *ibid.*, 1964, **68**, 2908.
20. I. D. Sokolova and N. K. Voskresenskaya, *Zh. prkl. Khim.*, 1962, **36**, 955.
21. A. A. Appen and S. S. Kayalova, *Dokl. Akad. Nauk., SSSR, Ser. fiz. Khim.*, 1962, **145**, 592.
22. C. F. Callis, J. R. Van Wazer and J. S. Metcalf, *J. Am. chem. Soc.*, 1955, **77**, 1468.
23. B. M. Lepinskikh, O. A. Esin and G. A. Teterin, *Zh. neorg. Khim.*, 1960, **5**, 642.
24. H. Bloom, F. G. Davis and D. W. James, *Trans. Faraday Soc.*, 1960, **56**, 1179.

Table 9.8 VISCOSITY OF PURE MOLTEN SALTS

The viscosity (centipoise) at temperature t(°C) is given as η_t, or the constants a and b in the equation $\log \eta_t = a + b/(t+273)$ are given for the temperature range r. Principal references are in **bold type**.

Salt		
AgBr	η_{450}	3.3
Ref. **22, 11, 9**	η_{550}	2.4
	η_{650}	1.7
	η_{800}	1.2
AgCl	η_{500}	2.05
Ref. **22, 11, 9**	η_{600}	1.60
	η_{700}	1.25
	η_{800}	1.05
AgI	η_{600}	3.0
Ref. **11**, 9	η_{700}	2.3
	η_{800}	1.7
$AgNO_3$	η_{230}	4.10
Ref. **25, 21, 17, 11, 9**	η_{280}	3.05
	η_{330}	2.40
	η_{350}	2.20
B_2O_3	η_{600}	158 000
Ref. **24, 20, 4**	η_{800}	25 100
	η_{1000}	6 300
	η_{1200}	2 000
$BaCl_2$	η_{1000}	4.506
Ref. 2, **6**		
$BiCl_3$	η_{260}	32
Ref. **5**	η_{300}	23
	η_{340}	18
$CaCl_2$	η_{800}	3.021
Ref. **6**, 7	η_{900}	1.870
	η_{1000}	1.248
$CdBr_2$	η_{600}	2.60
Ref. **18**	η_{640}	2.35
	η_{680}	2.10
$CdCl_2$	η_{600}	2.35
Ref. **23, 18, 14**	η_{700}	1.85
	η_{800}	1.55
CuCl	η_{500}	2.80
Ref. **14**	η_{600}	1.95
	η_{700}	1.40
$HgBr_2$	η_{255}	2.196
Ref. 8, **13**	η_{265}	2.008
	η_{275}	1.843
$HgCl_2$	η_{281}	1.768
Ref. **32**	η_{287}	1.738
	η_{292}	1.694
	η_{299}	1.600
	η_{306}	1.543
HgI_2	η_{268}	2.669
Ref. **32**	η_{292}	2.244
	η_{314}	1.995
	η_{334}	1.715
	η_{358}	1.458
KBr	η_{750}	1.150
Ref. 7, **14**, 16	η_{800}	1.022
	η_{900}	0.831
KCl	η_{800}	1.094
Ref. **14**, 16, 26	η_{900}	0.841
	η_{1000}	0.673
$K_2Cr_2O_7$	η_{400}	13.79
Ref. **27**, 29	η_{450}	9.665
	η_{500}	7.091
KI	η_{750}	1.362
Ref. 30	η_{800}	1.205
	η_{900}	0.973
KNO_2	a	−0.87
Ref. **33**	b	960
	r	418–450
KNO_3	η_{350}	2.705
Ref. 27, 16, 31, **34**, 35	η_{400}	2.090
	η_{450}	1.673
	η_{550}	1.163
KOH	η_{400}	2.3
Ref. **10**	η_{500}	1.3
	η_{600}	0.8
LiBr	η_{550}	1.815
Ref. 7, **14**	η_{700}	1.096
	η_{850}	0.757
LiI	η_{450}	2.50
Ref. **14**	η_{550}	1.70
	η_{650}	1.30
$LiNO_3$	η_{260}	5.5
Ref. **17, 12, 11, 3**	η_{300}	4.0
	η_{350}	2.9
Na_3AlF_6	η_{1007}	6.7
Ref. **19**	η_{1017}	6.5
	η_{1050}	6.0
NaBr	η_{762}	1.345
Ref. **14**, 16	η_{766}	1.332
	η_{780}	1.288
NaCl	η_{800}	1.463
Ref. 16, **34**, 35	η_{900}	1.009
	η_{1000}	0.737
NaI	η_{650}	1.581
Ref. **30**	η_{750}	1.168
	η_{900}	0.818
$NaNO_2$	a	−1.07
Ref. **33**	b	868
	r	282–310
$NaNO_3$	η_{300}	3.156
Ref. 31, **34**, 36, 37	η_{400}	1.901
	η_{500}	1.305
NaOH	η_{350}	4.0
Ref. **10**	η_{450}	2.2
	η_{550}	1.5
$NaPO_3$	η_{650}	1 250
Ref. **28, 4**	η_{750}	440
	η_{850}	210
	η_{900}	150
$PbBr_2$	η_{360}	10.5
Ref. **22, 17, 1**	η_{400}	7.3
	η_{450}	5.0
	η_{550}	3.0
$PbCl_2$	η_{500}	4.6
Ref. **22, 1**	η_{600}	2.8
	η_{700}	1.9
$TlNO_3$	a	−1.04
Ref. **33**	b	565
	r	207–250

REFERENCES TO TABLE 9.8

1. R. Lorenz and H. T. Kalmus, *Z. phys. Chem.*, 1907, **59**, 244.
2. V. T. Slavyanskii, *Dokl. Akad. Nauk. SSSR*, 1947, **58**, 1077.
3. H. M. Goodwin and R. D. Mailey, *Phys. Rev.*, 1908, **26**, 28.
4. K. Arndt, *Z. chem. Apparat.*, 1908, **3**, 549.
5. A. H. W. Aten, *Z. phys. Chem.*, 1909, **66**, 641.
6. G. J. Janz and R. D. Reeves, *Advan. electrochem. Eng.*, 1967, **5**.
7. S. Karpachev and A. Stromberg, *Zh. Fiz. Khim.*, 1938, **11**, 852.
8. R. S. Dantuma, *Z. anorg. allgem. Chem.*, 1938, **175**, 1.
9. R. Lorenz and A. Hoechberg, *Z. anorg. Chem.*, 1916, **94**, 317.
10. K. Arndt and G. Ploetz, *Z. phys. Chem.*, 1926, **121**, 439.
11. E. van Aubel, *Bull. sci. acad. roy. Belg.*, 1926, **12**, 374.
12. R. S. Dantuma, *Z. anorg. allg. Chem.*, 1928, **175**, 1.
13. G. Jander and K. Broderson, *Z. anorg. allgem. Chem.*, 1951, **264**, 57.
14. I. G. Murgulescu and S. Zuca, *Z. physik. Chem.* (*Leipzig*), 1961, **218**, 379.
15. A. G. Stromberg, *Zh. Fiz. Khim.*, 1939, **13**, 436.
16. C. E. Fawsitt, *Proc. roy. Soc.* (*London*), 1908, 93.
17. K. S. Evstropev, *Akad. Nauk. SSSR., Otdel. Tekh. Nauk. Inst. Mash. Sov.*, 1945, **3**, 61.
18. H. Bloom, B. S. Harrap and E. Heymann, *Proc. R. Soc.*, 1948, **A194**, 237.
19. A. Vajna, *Alluminio*, 1950, **19**, 133.
20. L. Shartsis, W. Capps and S. Spinner, *J. Am. ceram. Soc.*, 1953, **36**, 319.
21. F. A. Pugsley and F. E. W. Wetmore, *Canad. J. Chem.*, 1954, **32**, 839.
22. B. S. Harrap and E. Heymann, *Trans. Faraday Soc.*, 1955, **51**, 259.
23. B. S. Harrap and E. Heymann, *ibid.*, 1955, **51**, 268.
24. J. D. Mackenzie, *ibid.*, 1956, **52**, 1564.
25. N. P. Luzhnaya, N. N. Evseeva and I. P. Vereshchetina, *Zh. neorg. Khim.*, 1956, **1**, 1490.
26. S. Karpachev, *Zh. Obshch. Khim.*, 1935, **5**, 625.
27. R. Lorenz and T. Kalmus, *Z. Physik. Chem.*, 1907, **59**, 244.
28. G. G. Nozadze, *Soobshch. Akad. Nauk. Gruzin. S.S.S.R.*, 1957, **19**, 567.
29. J. P. Frame, E. Rhodes and A. R. Ubbelohde, *Trans. Faraday Soc.*, 1959, **55**, 2039.
30. I. G. Murgulescu and S. Zuca, *Rev. Roumaine. Chim.*, 1965, **10**, 123.
31. H. M. Goodwin and R. D. Mailey, *Phys. Rev.*, 1906, **23**, 22; *ibid.*, 1907, **25**, 469; *ibid.*, 1908, **26**, 28.
32. G. J. Janz and J. D. E. McIntyre, *J. electrochem. Soc.*, 1962, **109**, 842.
33. J. P. Frame, E. Rhodes and A. R. Ubbelohde, *Trans. Faraday Soc.*, 1959, **55**, 2039.
34. R. S. Dantuma, *Z. Anorg. allgem. Chem.*, 1938, **175**, 1.
35. K. Ogawa, *Nippon Kinzoku Gakkaishi*, 1950, **14B**, 49.
36. R. Lorenz and H. T. Kalmus, *Z. physik. Chem.*, 1907, **59**, 17.
37. C. E. Fawsitt, *Proc. roy. Soc.* (*London*), 1908, **A80**, 290.

Table 9.9 VISCOSITY OF MOLTEN BINARY SALT SYSTEMS AND OTHER MIXED IONIC MELTS

The viscosity (centipoise) at temperature t(°C) and composition p(wt.%) of the first-named constituent is given as η_t, or the constants a and b in the equation $\log \eta_t = a + b/(t + 273)$ are given for the temperature range r. Principal references are in **bold type**.

AgBr–AgCl	p	0	32.5	46.7	66.9	84.0	100
Ref. **12**	η_{440}	—	2.58	2.75	2.97	3.12	3.38
	η_{520}	1.98	2.10	2.18	2.35	2.46	2.69
	η_{600}	1.68	1.74	1.74	1.98	2.09	2.28
AgBr–KBr	p	57.1	66.0	78.0	86.0	100	
Ref. **12**	η_{400}	—	—	3.55	3.55	—	
	η_{500}	—	2.55	2.40	2.46	2.83	
	η_{600}	1.73	1.74	1.78	1.86	2.27	
AgCl–KCl	p	54.3	61.1	80.4	88.9	100	
Ref. **12**	η_{500}	—	—	2.10	2.12	2.08	
	η_{600}	1.63	1.61	1.56	1.62	1.66	
	η_{700}	1.28	1.24	1.24	1.28	1.40	
AgCl–$PbCl_2$	p	0	9.8	24.1	44.5	67.6	100
Ref. **12**	η_{500}	4.56	4.13	3.47	2.95	2.47	2.08
	η_{600}	2.75	2.59	2.30	2.12	1.84	1.66
	η_{700}	1.87	1.99	1.84	1.76	1.65	1.40

Table 9.9 VISCOSITY OF MOLTEN BINARY SALT SYSTEMS AND OTHER MIXED IONIC MELTS—*continued*

$AgI–AgNO_3$	p	0	25.7	47.9	67.4			
Ref. **17**	η_{150}	—	9.6	13.6	19.6			
	η_{200}	—	6.4	8.6	12.6			
	η_{250}	4.2	5.2	6.5	8.7			
	η_{300}	3.8	4.8	5.5	7.0			
$AgNO_3–HgI_2$	p	9.8	13.8	19.9	27.2	35.9	46.6	56.2
Ref. **15, 21**	η_{110}	—	2.3	—	—	19.5	13.8	9.6
	η_{150}	3.2	—	7.4	7.0	5.6	4.2	3.2
	η_{200}	1.5	2.3	2.4	2.2	1.9	1.8	1.4
$AgNO_3–KNO_3$	p	29.6	41.9	52.8	62.7	71.6	79.7	83.5
Ref. **14, 21**	η_{150}	—	—	—	18.2	19.1	18.1	—
	η_{200}	—	—	—	8.9	9.6	9.2	8.9
	η_{250}	—	7.4	6.4	5.6	6.0	6.4	6.0
	η_{300}	5.8	5.1	4.8	4.1	4.3	4.6	4.5
$AlF_3–Na_3AlF_6$	p	2.5	5	10	15			
Ref. **7**	η_{1000}	6.5	6.3	5.7	4.8			
$Al_2O_3–Na_3AlF_6$	p	2.5	5	10	15			
Ref. **7**	η_{1000}	6.9	6.9	7.1	10.9			
$B_2O_3–BaO$	p	44.5	49.2	54.9	59.7	64.4	67.9	
Ref. **10**	η_{850}	—	30 900	35 500	35 500	33 900	25 100	
	η_{900}	2 750	6 170	8 710	9 550	9 550	7 080	
	η_{950}	980	1 900	3 160	3 710	3 240	2 950	
	η_{1000}	—	—	—	1 260	1 260	1 180	
$B_2O_3–K_2O$	p	61.6	70.6	84.9	89.0	94.8	98.5	100
Ref. **9**	η_{600}	6.8×10^5	5.1×10^6	—	1.5×10^5	0.9×10^5	1.3×10^5	1.6×10^5
	η_{800}	780	4 680	4 360	3 900	5 130	13 800	21 400
	η_{900}	—	960	1 350	1 290	2 400	6 310	11 500
	η_{1000}	—	—	—	450	1 120	3 710	6 460
$B_2O_3–Li_2O$	p	85.2	87.1	92.0	95.5	97.2	100	
Ref. **9**	η_{600}	—	—	—	355 000	141 000	158 000	
	η_{800}	3 630	3 020	5 370	3 630	7 240	21 400	
	η_{900}	460	520	790	980	3 090	11 500	
	η_{1000}	—	—	250	320	1 440	6 460	
$B_2O_3–Na_2O$	p	68.2	69.2	77.6	85.8	94.4	99.1	100
Ref. **9**	η_{600}	—	3.1×10^6	—	1.9×10^6	1.3×10^5	1.3×10^5	1.6×10^5
	η_{800}	930	1 290	6 920	5 890	3 800	12 000	21 400
	η_{900}	—	—	870	1 380	1 580	5 750	11 500
	η_{1000}	—	—	—	—	810	2 630	6 460
$B_2O_3–NaPO_3$	p	0	50	95	99.5			
Ref. **18**	η_{900}	150	450	4 730	11 000			
$BaCl_2–NaCl$	p	45.5	55.6	63.8	70.5	75.3	80.1	84.4
Ref. **19**, 5, **21**	η_{725}	—	3.2	3.42	4.05	4.6	—	—
	η_{775}	2.40	2.58	2.82	3.22	3.48	3.70	3.85
	η_{825}	2.00	2.23	2.50	2.80	3.00	3.15	3.28
	η_{875}	1.75	2.02	2.36	2.55	2.72	2.84	3.05
$BaO–SiO_2$	p	31.7	46.1	53.1	63.2	71.7		
Ref. **16**	η_{1500}	—	11 000	3 200	1 700	—		
	η_{1700}	19 500	2 140	850	540	190		
	η_{1800}	10 000	1 250	530	400	150		
$CaCl_2–NaCl$*	p	0	17.4	44.8	65.5	88.3	100	
Ref. **5**, 19	η_{800}	1.59	1.65	2.59	3.49	4.36	4.92	
	η_{900}	1.00	1.16	2.11	2.95	3.69	4.22	
	η_{1000}	0.70	1.00	1.79	2.60	3.31	3.74	
$CaF_2–Na_3AlF_6$	p	2.5	5	10	15			
Ref. **7**	η_{1000}	6.9	7.0	7.3	8.0			

Table 9.9 VISCOSITY OF MOLTEN BINARY SALT SYSTEMS AND OTHER MIXED IONIC MELTS—*continued*

$CaO–SiO_2$	p	29.1	37.1	41.9	47.9	51.9	55.9	
Ref. **11, 3**	η_{1500}	—	1 440	765	—	288	—	
	η_{1600}	—	730	405	218	157	113	
	η_{1700}	1 360	392	235	133	96	74	
	η_{1800}	850	250	150	88	66	54	
$CdBr_2–CdCl_2$	p	0	27.2	49.9	69.1	85.6	100	
Ref. **6**	η_{600}	2.3	2.4	2.4	2.5	2.5	2.6	
	η_{660}	2.0	2.0	2.1	2.1	2.1	2.2	
$CdCl_2–KCl$	p	40.3	59.5	67.3	71.9	82.4	88.8	100
Ref. **13**	η_{500}	—	2.53	2.54	2.62	2.97	3.17	—
	η_{600}	1.76	1.67	1.67	1.73	1.92	2.15	2.31
	η_{700}	1.23	1.21	1.20	1.24	1.40	1.57	1.83
$CdCl_2–NaCl$	p	53.9	75.8	79.0	82.5	90.2	100	
Ref. **13**	η_{500}	—	2.68	2.71	2.84	2.99	—	
	η_{600}	—	1.78	1.80	1.88	2.02	2.31	
	η_{700}	1.37	1.30	1.27	1.39	1.50	1.83	
$CdCl_2–PbCl_2$	p	0	18.0	26.8	37.9	65.9	100	
Ref. **13**	η_{520}	4.22	3.75	3.50	3.36	—	—	
	η_{600}	2.75	2.60	2.47	2.38	2.26	2.31	
	η_{680}	2.00	2.02	1.88	1.85	1.85	1.90	
$KCl–MgCl_2$	p	37.6	48.5	58.9	59.9	69.1	78.8	
Ref. **2, 21**	η_{500}	3.26	—	2.78	2.69	—	—	
	η_{600}	2.19	1.55	1.51	1.70	1.44	—	
	η_{700}	—	1.22	—	—	0.99	1.25	
KCl–NaCl*	p	0	56.0	79.3	100			
Ref. **5**	η_{800}	1.59	1.17	1.07	1.13			
	η_{900}	1.00	0.90	0.81	0.89			
$KCl–PbCl_2$	p	0	5.2	8.8	18.3	21.8	34.7	43.3
Ref. **13**	η_{500}	4.56	4.02	3.66	3.34	3.37	—	—
	η_{600}	2.75	2.42	2.27	2.06	2.10	2.15	—
	η_{700}	1.87	1.65	1.56	1.42	1.47	1.48	1.50
$K_2O–SiO_2$	p	3.9	9.5	15.9	23.9	29.9	38.7	43.6
Ref. **16, 8**	η_{1000}	—	—	—	32 800	15 100	7 940	4 570
	η_{1200}	—	—	—	2 240	1 350	661	355
	η_{1400}	—	—	—	372	200	81	47
	η_{1600}	468 000	107 000	24 000	—	—	—	—
$Li_2O–SiO_2$	p	12.0	16.1	21.8	25.9	28.9	33.2	37.9
Ref. **16, 8**	η_{1100}	—	50 100	12 000	4 470	—	—	—
	η_{1300}	42 700	8 320	2 190	1 000	510	220	72
	η_{1400}	17 800	3 890	1 120	580	300	140	52
	η_{1500}	8 320	2 040	710	400	200	100	41
$MgO–SiO_2$	p	34.8	35.5	36.2	40.1	41.4		
Ref. **16**	η_{1650}	610	460	350	250	190		
	η_{1750}	360	270	200	150	120		
	η_{1800}	280	210	180	120	110		
$Na_3AlF_6–NaF$	p	85	90	95	97.5			
Ref. **7**	η_{1000}	4.8	5.9	6.4	6.7			
$NaCl–NaNO_3$	p	1.4	2.8	4.2	5.7	7.1	8.6	10.1
Ref. **19**	η_{310}	3.20	3.27	3.68	—	—	—	—
	η_{350}	2.57	2.67	2.79	2.83	3.00	—	—
	η_{400}	2.06	2.21	2.32	2.18	2.11	2.20	2.39
	η_{450}	1.72	1.91	1.98	1.96	1.94	1.73	2.18
$Na_2O–P_2O_5$	p	30.4	32.0	36.0	39.6	43.3	46.5	48.4
Ref. **20**	a	−0.84	−0.63	−0.26	0.11	0.38	0.58	0.59
	$10^{-3}b$	3.57	3.19	2.45	1.87	1.45	1.10	1.00
	r	741–1 013	707–1 001	657–1 017	822–1 002	942–1 064	1 030–1 077	995–1 070

Table 9.9 VISCOSITY OF MOLTEN BINARY SALT SYSTEMS AND OTHER MIXED IONIC MELTS—*continued*

Na_2O–SiO_2	p	15.2	20.0	30.8	36.9	40.9	45.2	50.0
Ref. **16, 8, 4, 1**	η_{900}	—	4.6×10^6	2.4×10^5	3.2×10^5	1.8×10^5	—	—
	η_{1100}	—	263 000	50 100	26 900	15 100	4 070	890
	η_{1300}	118 000	33 100	7 940	4 270	2 340	690	160
	η_{1400}	41 700	16 600	4 900	2 570	1 350	390	100
$PbBr_2$–$PbCl_2$	p	0	28.4	63.1	80.2	100		
Ref. **12**	η_{450}	—	—	5.53	5.24	4.83		
	η_{500}	4.56	4.45	4.03	3.94	3.73		
	η_{550}	3.54	3.38	3.14	3.05	2.97		
SiO_2–SrO	p	36.4	42.0	46.0	57.8	62.9	69.7	
Ref. **16**	η_{1650}	47	360	620	1 380	4 220	—	
	η_{1750}	160	210	370	780	2 240	5 820	
	η_{1800}	130	180	300	580	1 660	4 300	

* *See also* 'Physical Properties Data Compilations Relevant to Energy Storage II, Molten Salts, Data on Single and Multi-Component Salt Systems', Janz *et al.*, NSRDS–NBS, 61.

REFERENCES TO TABLE 9.9

1. C. L. Babcock, *J. Am. ceram. Soc.*, 1934, **17**, 329.
2. S. Karpachev and A. Stromberg, *Z. anorg. allg. Chem.*, 1935, **222**, 78.
3. J. R. Rait and R. Hay, *J. R. Tech. Coll. (Glasgow)*, 1938, **4**, 252.
4. E. Preston, *J. Soc. Glass Tech.*, 1938, **22**, 45.
5. V. P. Barzakovskii, *Bull. Acad. Sci. URSS, Classe, sci. chim.*, 1940, 825.
6. H. Bloom, B. S. Harrap and E. Heymann, *Proc. R. Soc.*, 1948, **A194**, 237.
7. A. Vajna, *Alluminio*, 1950, **19**, 133.
8. L. Shartsis, S. Spinner and W. Capps, *J. Am. ceram. Soc.*, 1952, **35**, 155.
9. L. Shartsis, W. Capps and S. Spinner, *ibid.*, 1953, **36**, 319.
10. L. Shartsis, S. Spinner, and H. F. Shermer, *ibid.*, 1954, **37**, 544.
11. J. O'M. Bockris and D. C. Lowe, *Proc. R. Soc.*, 1954, **A226**, 1167.
12. B. S. Harrap and E. Heymann, *Trans. Faraday Soc.*, 1955, **51**, 259.
13. B. S. Harrap and E. Heymann, *ibid.*, 1955, **51**, 268.
14. V. D. Polyakov, *Izvest. Sekt. Fiziko-Khim. Anal.*, 1955, **26**, 147.
15. V. D. Polyakov, *ibid.*, 1955, **26**, 191.
16. J. O'M. Bockris, J. D. Mackenzie and J. A. Kitchener, *Trans. Faraday Soc.*, 1955, **51**, 1734.
17. N. P. Luzhnaya, N. N. Evseeva and I. P. Vereshchetina, *Zh. neorg. Khim.*, 1956, **1**, 1490.
18. G. G. Nozadze, *Soobshch. Akad. Nauk. Gruzin. SSSR*, 1957, **19**, 567.
19. I. P. Vereshchetina and N. P. Luzhnaya, *Izv. Sekt. fiz.-khim. Analiza*, 1954, **25**, 188.
20. C. F. Callis, J. R. Van Wazer and J. S. Metcalf, *J. Am. chem. Soc.*, 1955, **77**, 1471.
21. G. J. Janz, 'Molten Salts Handbook', Academic Press, London, 1967.

10 Metallography

Metallography can be defined as the study of the structure of materials and alloys by the examination of specially prepared surfaces. Its original scope was limited by the resolution and depth of field in focus by the imaging of light reflected from the metallic surface. These limitations have been overcome by both transmission and scanning electron microscopy (TEM, STEM and SEM). The analysis of X-rays generated by the interaction of electron beams with atoms at or near the surface, by wavelength or energy dispersive detectors (WDX, EDX), has added quantitative determination of local composition, e.g. of intermetallic compounds, to the deductions from the well-developed etching techniques. X-ray photoelectron microscopy (XPS or ESCA) now enables the metallographer to analyse the atoms in the outermost surface layer to a depth of a few atoms (0.3–5.0 nm) and provides information about the chemical environment of the atom. Auger spectroscopy uses a low-energy electron beam instead of X-rays to excite atoms, and analysis of the Auger electrons produced provides similar information about the atoms from which the Auger electron is ejected.

Nevertheless, the conventional optical techniques still have a significant role to play and their interpretation is extended and reinforced by the results of the electronic techniques.

10.1 Macroscopic examination

For examination of large-scale features—grain structure of castings, coarse grain in wrought products, porosity in castings, major defects, or distribution of alloying elements and impurities on a large scale (e.g. to study 'segregation') it is necessary to prepare large surfaces free from major distortion, but extreme smoothness and structural perfection are not required.

The required section is cut by sawing, abrasive slitting wheels or machining with adequate cooling and lubrication, and is normally finished by fine machining, followed by etching if necessary. Grinding on abrasive cloth or paper, which may be followed by polishing with proprietary metal polish, is sometimes beneficial, but vigorous polishing, especially with power-driven machines, may cause the metal to flow over defects such as porosity. Examination for porosity is usually best carried out on a fine machined surface.

The final machining operation should be done with a single sharp tool, for instance by planing, turning or milling with a fly-cutter, rather than by the use of a milling cutter. For soft metals (e.g. copper, lead, pure aluminium) the shape of the tool is important; it should have a rounded nose and adequate front clearance to prevent rubbing, and it should have a large top rake (the softer the metal, the larger the rake required) so that it presents almost a chisel edge to the specimen. For harder metals more orthodox tools may be used.

Illumination of unetched specimens for photomacrographs to show porosity requires a broad source of illumination. The sky (without direct sunlight) is sometimes the most suitable source.

Etching reagents for macroscopic work are listed in Table 10.1. Directions for 'sulphur-printing', to show the distribution of sulphide in steel, are included.

10.2 Microscopic examination

Metallographic specimens are normally prepared for examination under the microscope by cutting out the piece to be examined (preferably not more than 3 cm dia.), carefully removing the disturbed surface layer (by turning or filing with a sharp tool) and then rubbing the surface with successively finer abrasives until a smooth polished surface is obtained, sensibly free from disturbing effects from the cutting and grinding; the clean, smooth, undistorted surface is then attacked chemically, or otherwise, by etching reagents which reveal the structure of the metal. Any mechanical method of cutting or smoothing the surface produces distortion of the metal near the surface, and it may produce local heating; the objective is to make the disturbed layer successively thinner at each stage until it is negligible or can be removed by etching. The thickness of the disturbed layer is in the range 10–100 µm for emery or silicon carbide papers with hand grinding. Some or all of the mechanical grinding and polishing can often he replaced by chemical of electrochemical polishing methods, by which the metal is attacked in such a way that protuberances are preferentially dissolved and the flat undisturbed metal surface is laid bare, usually with a saving of time and frequently with an improvement in result.

For some purposes, e.g. study of slip processes involving individual dislocations, electron microscopical studies of fine structure, and quantitative microhardness testing under light loads, electro-polishing is almost indispensable. In general, the type of finish required varies somewhat with the magnification to be used in examination. High-power examination demands great perfection of small areas, but relatively large-scale undulations, such as may sometimes occur on electropolished specimens, are unimportant. At lower powers detail may be less important, but widely spaced imperfections and undulations are liable to become obtrusive.

Table 10.1 ETCHING REAGENTS FOR MACROSCOPIC EXAMINATION

Material		*Reagent**		*Remarks*
A. *Aluminium base*				
1. Aluminium and its alloys	(a)	Concentrated Keller's Reagent		Can be diluted with up to 50ml water
		Nitric acid (1.40)	100 ml	
		Hydrochloric acid (1.19)	50 ml	
		Hydrofluoric acid (40%)	1½ ml	
	(b)	Nitric acid (1.40)	30 ml	Widely applicable, but very vigorous
		Hydrochloric acid (1.19)	30 ml	
		2% conc. hydrofluoric acid	30 ml	
	(c)	Tucker's Reagent		Use fresh
		Nitric acid (1.40)	15 ml	
		Hydrochloric acid (1.19)	45 ml	
		Hydrofluoric acid (40%)	15 ml	
		Water	25 ml	
	(d)	10% sodium hydroxide in water		Use at 60–70°C
2. Unalloyed aluminium and Al–Cn alloys	(e)	Flick's Reagent		
		Hydrochloric acid	15 ml	Wash in warm water after etching and clear by dipping in concentrated nitric acid
		Hydrofluoric acid	10 ml	
		Water	90 ml	
3. Aluminium–silicon	(f)	Hume-Rothery's Reagent		For high-silicon alloys. Fine polish undesirable. Immerse specimen 5–10 s, remove, and brush away deposited copper or remove it with 50% nitric acid in water
		Cupric chloride	15 g	
		Water	100 ml	
4. Aluminium–copper	(g)	Keller's Reagent 2½% nitric acid (1.40) 1½% hydrochloric acid (1.19) ½% hydrofluoric acid (40%) Rem. water		More frequently used as micro-etch
5. Aluminium–magnesium	(h)	5% cupric chloride 3% nitric acid (1.40) Rem. water		Clear surface with strong nitric acid
6. Aluminium–copper–silicon	(g)	Keller's Reagent (as above)		
	(i)	Nitric acid (1.40)	15 ml	
		Hydrochloric acid (1.19)	10 ml	
		Hydrofluoric acid (40%)	5 ml	
		Water	70 ml	
7. Aluminium–copper–magnesium–nickel	(j)	Zeerleder's Reagent		
		Hydrochloric acid (1.19)	20 ml	
		Nitric acid (1.40)	15 ml	
		Hydrofluoric acid (40%)	5 ml	
		Water	60 ml	

* Acids are concentrated, unless otherwise indicated, e.g. with specific gravity.

Table 10.1 ETCHING REAGENTS FOR MACROSCOPIC EXAMINATION—*continued*

Material	*Reagent**		*Remarks*
B. *Copper base*			
1. Copper and copper alloys generally	(a) Alcoholic ferric chloride		
	Ethyl alcohol	96 ml	Avoid use of water, for washing or staining may result. Use alcohol or acetone instead. Grain contrast
	Ferric chloride (anhydrous)	59 g	
	Hydrochloric acid (1.19)	2 ml	
	(b) Acid aqueous ferric chloride		
	Ferric chloride	25 g	(a) and (b) require moderately high standard of surface finish
	Hydrochloric acid (1.40)	25 ml	
	Water	100 ml	
	(c) Concentrated nitric acid (140)		A rapid etch suitable for roughly prepared surfaces. Addition of a trace of silver nitrate (5%) enhances contrast
	Nitric acid (1.40)	50 ml	
	Water	50 ml	
	(d) 10% ferric chloride in water	10 ml	To reveal strains in brasses
	5% chromium trioxide in saturated brine	10 ml	
	20% acetic acid in water	20 ml	
	(e) A. 1% mercuric nitrate in distilled water B. 1% nitric acid (1.40) in water Mix A and B in equal proportions		Time required to induce cracks is indication of residual stress
	(f) Chromium trioxide	40 g	Good for alloys with silicon and silicon bronzes
	Ammonium chloride	7.5 g	
	Nitric acid (1.40)	50 ml	
	Sulphuric acid (1.84)	8 ml	
	Distilled water	100 ml	
C. *Iron and steel*			
	(a) 50% hydrochloric acid in water		Use hot (70–80°C) for up to 1 h. Shows segregation, porosity, cracks useful for examination of welds for soundness
	(b) 20% sulphuric acid in water		Use hot (80°C) for 10–20 min. Scrub lightly to remove carbonaceous deposit. Purpose as (a). Mixtures of (a) and (b) are also used similarly
	(c) 25% nitric acid in water		Purposes as (a) and (b). May be used cold if more convenient
	(d) 10% ammonium persulphate in water		Grain contrast etch. Apply with swab. Reveals grain growth and recrystallization at welds.
	(e) Stead's Reagent		For revealing phosphorus segregation and primary dendritic structure of cast steels. Dissolve the salts in the acid with addition of a minimum of water. Phosphorus segregate unattacked, also eutectic cells in cast iron
	Cupric chloride	10 g	
	Magnesium chloride	40 g	
	Hydrochloric acid (1.19)	20 ml	
	Alcohol to 1 litre		
	(f) Fry's Reagent		
	Cupric chloride	90 g	To reveal strain lines in mild steel. Heat specimen to 150–250 °C for 15–30 min before etching. Etch for 1–3 min while rubbing with a soft cloth. Rinse with alcohol.
	Hydrochloric acid	120 ml	
	Water	100 ml	
	(g) Humphrey's Reagent		Reveals dendritic structure of cast steels. First treat surface with 8% copper ammonium chloride solution and then with (g) for $\frac{1}{2}$–$1\frac{1}{2}$ h. Remove copper deposit (loosely adherent), dry and rub surface lightly with abrasive
	Copper ammonium chloride	120 g	
	Hydrochloric acid (1.19)	50 ml	
	Water	1 litre	
	(h) 5–10% nitric acid in alcohol		Etch for up to $\frac{1}{2}$ h. Reveals cracks and carbon segregation. More controlled than aqueous acids
	(j) Sulphur-printing 3% sulphuric acid in water		Soak photographic printing paper in the acid and remove surplus acid with blotting paper. Lay paper face down on the clean steel surface and

* Acids are concentrated, unless otherwise indicated, e.g. with specific gravity.

Table 10.1 ETCHING REAGENTS FOR MACROSCOPIC EXAMINATION—*continued*

Material	*Reagent**		*Remarks*
C. *Iron and steel—continued*			'squeegee' into close contact. After 2 min remove paper, wash it and fix in 6% sodium thiosulphate in water. Brown coloration on the paper indicates local segregation of sulphides
	(k) Dithizone process for lead distribution		*See* p. 10.38, *Lead in steels*. Analogous to sulphur-printing
	(l) Marble's Reagent Hydrochloric acid (1.19) Saturated aqueous solution of cupric sulphate	 50 ml 25 ml	Austenitic steels. High temperature steels. Fe–Cr–Ni casting alloys. Also shows depth of nitriding
	(m) Oberhoffer's Reagent Hydrochloric acid (1.19) Ferric chloride Stannous chloride Water Ethanol (acid added last) Rinse in 20% hydrochloric acid in ethanol	 42 ml 30 g 0.5 g 500 ml 500 ml	Good surface preparation needed. Steel castings. Darkens Fe-rich areas, reveals segregation and primary cast structure
	(n) Klemm's Reagent Saturated aqueous solution of sodium thiosulphate Sodium metabisulphite (can be increased for contrast)	 50 ml 1 g	Phosphorus distribution in cast steel and cast iron. Grain contrast
D. *Lead base*			
Lead and lead alloys generally	(a) Russell's Reagent A. 80 ml nitric acid (1.40) in 220 ml water B. 45 g ammonium molybdate in 300 ml water		Grain contrast etch; removes deformed layer Mix equal parts of A and B immediately before use. Swab for 10–30 s Rinse in water
	(b) Ammonium molybdate Citric acid Water	10 g 25 g 100 ml	Bright etch revealing grain structure, defects, etc.
	(c) Worner and Worner's Reagent Acetic acid, glacial '100 vol.' hydrogen peroxide	75 ml 25 ml	Chemical polish revealing defects, etc. Specimen must be dry and water content of solution as low as possible *N.B.*—Avoid all heating, as lead alloys recrystallize very readily
	(d) Nitric acid (1.40) Water (distilled)	20 ml 80 ml	Immerse 5–10 min. Grain contrast, laminations, welds. Up to 50% nitric acid can be used
	(e) Glacial acetic acid Nitric acid (1.40) Glycerol	20 ml 20 ml 80 ml	Macrostructure of alloy with Ca, Sb and Sn. Use fresh only. Several minutes needed
	(f) Glacial acetic acid Nitric acid (1.40) Hydrogen peroxide (30%) Water (distilled)	20 ml 20 ml 20 ml 50 ml	2–10 s by swabbing. Good for alloys with Bi, Te or Ni
E. *Magnesium base*	(a) Picric acid (64%) saturated in ethanol (96%) Glacial acetic acid Distilled water	50 ml 20 ml 20 ml	Grain size. Flow lines in forging (wash precipitate in hot water). Etch for up to 3 min
	(b) Ammonium persulphate Distilled water	2 ml 98 ml	Flow lines in forgings
	(c) Nitric acid (1.40) Water	20 ml 80 ml	Internal defects in casts. Useful for Mg–Mn and Mg–Zr. Etch for up to 3 min
	(d) Glacial acetic acid Water	10 ml 90 ml	General defects; flow lines, segregation. Etch for up to 3 min
F. *Nickel base*			
	(a) Nitric acid (1.40) Acetic acid	50 ml 50 ml	Welds, Ni–Cr–Fe alloys

* Acids are concentrated, unless otherwise indicated, e.g. with specific gravity.

Table 10.1 ETCHING REAGENTS FOR MACROSCOPIC EXAMINATION—*continued*

Material	*Reagent**		*Remarks*
F. *Nickel base—continued*			
	(b) Aqua regia		As (a) *See also* ref. 1. p. 10.69
	Nitric acid (1.19)	25 ml	
	Hydrochloric acid (1.19)	75 ml	
G. *Tin base*			
	(a) Sat. soln of ammonium polysulphide in water (wipe off surface film)		Grain structure; suitable most tin alloys (etching time 20–30 min)
	(b) FeCl	10 g	Sn–Sb alloys (up to 3 min)
	Hydrochloric acid (1.18)	2 ml	
	Water	100 ml	
H. *Zinc base*			
Zinc and zinc alloys	(a) Concentrated hydrochloric acid (1.19)		Good grain contrast
Zinc-rich alloys	(b) 5% hydrochloric acid in alcohol		HCl can be increased to 50%. Wash under running water to remove reaction products
	(c) Sodium sulphate (3.5 g if hydrated)	1.5 g	Better than above for Zn–Cu alloys
	Chromium trioxide	20 g	
	Water	100 ml	
I. *Other metals*			
Many of these require etching in agressive solutions comprising various mixtures of HCl, HNO_3 and HF	(a) Hydrochloric acid (1.18)	50 ml	Platinum metals group, especially Ru, Os, Rh
	Nitric acid (1.40)	20 ml	
	Hydrofluoric acid (40%)	30 ml	
	(b) Hydrochloric acid (1.19)	30 ml	Cr, Mo, W, V, Nb, Ta
	Nitric acid (1.40)	15 ml	
	Hydrofluoric acid (40%)	30 ml	
Nitric acid/HF etches: These do not appear to be very sensitive to composition. HF should be 5–10%. Heating to 60–80 °C will accelerate etching, e.g. for Ti	(c) Nitric acid (1.40)	30–45 ml	Highly alloyed Ti, Hf, Zr; also Cr, W, Mo, V
	Hydrofluoric acid (40%)	10 ml	
	Water	60–45 ml	
Aqueous HCl HNO_3 etches. The reactivity can be reduced by adding water	(d) Hydrochloric acid (1.19)	66 ml	Gold, platinum, palladium. Used for cobalt alloy if added to 34 ml water
	Nitric acid (1.40)	34 ml	
Acidified hydrogen peroxide etch	(e) Hydrofluoric acid	10 ml	Dilute Ti, Hf and Zr alloys
	Hydrogen peroxide (30%)	45–60 ml	
	Water	45–30 ml	
Nitric acid in alcohol	(f) Nitric acid (1.40)	10 ml	Silver. (*Note*: for safety methanol must be used. It is dangerous to add more than 5% nitric acid to ethanol)
	Methanol	90 ml	
Hydrochloric acid etches	(g) Hydrochloric acid (1.19)	10 ml	Be and its alloys, especially for large grain sizes
	Water	90 ml	
	Ammonium chloride	4 g or 2 g	
	Picric acid	2 g	
	(h) Hydrochloric acid (1.19)	50 ml	Cobalt alloys
	Water	50 ml	
	60–80 °C for 30–60 min		

* Acids are concentrated, unless otherwise indicated, e.g. with specific gravity.

The most frequent novices' errors are to fail completely to remove the distorted metal beneath the original cut surface, to change the structure by overheating the specimen, to carry abrasives over (by lack of cleanliness) from a coarse stage of grinding or polishing to a finer one, and to develop false structures by staining through faulty drying after etching. Preparation of an unfamiliar material must be checked by repeated etching and repolishing to see that the structure remains constant as more metal is removed.

The early stages of preparation are common to most metals and types of specimen. Fine polishing may have to be varied to suit the metal. Etching is peculiar to the metal under examination and the feature of the structure to be investigated.

MOUNTING

Specimens of irregular shape, great fragility or very small size are best mounted in plastic. Several specimens, if of similar materials, may be prepared in the same mount, with a saving of time. For critical work a first-class finish is easiest to obtain on a rather small specimen, and this is best mounted for ease of handling except when electrolytic or chemical polishing is used. Edge-sections (e.g. sections through plated coatings) must almost inevitably be mounted.

The basic method is to place the specimen face-down in a die, cover it with plastic and apply the treatment needed to make the plastic set. The mount is conveniently 2–3 cm in dia. × approx. 1 cm high. Thermosetting, thermoplastic and cold-setting plastics are used. Very hard materials (especially tungsten wires) are sometimes mounted in low-melting-point glass. In many laboratories the majority of specimens are mounted.

It is essential to verify that the structure of the metal will not be materially affected by any heat and pressure applied in forming the mount. Some 'cold-setting' plastics become hot while setting.

Some plastics used, with their characteristics, are listed in Table 10.2.

Table 10.2 PLASTIC USED FOR MOUNTING

Plastics	*Type*	*Remarks*
Phenolic (e.g. 'Bakelite')	Thermosetting	Needs controlled heat and pressure. Sufficiently inert to most solvents. Normal grades good for general work but have high shrinkage; mineral-filled type (Bakelite × 262/2) preferable for edge-sections. If curing insufficient, e.g. too low a temperature, the mount is soft and is attacked by acetone
Polymethyl methacrylate (Perspex, Diakon)	Thermoplastic*	Needs controlled heat and pressure. Gives clear mount. Attacked by acetone. Rather soft
(NHP)	Warm-setting	Two-ingredient version. Polymer + catalyst + monomer. Can be used as casting resin, cold-setting resin with some pressure or warm-setting resin with pressure
Polyester (e.g. 'Marco' grade 26C)	Cold-setting	Several ingredients to be mixed for each batch, but gives good mounts without heat or pressure. Inert to usual solvents
Epoxy resins (e.g. Araldite)	Various	'Araldite' Grade D, a liquid casting resin, gives good mounts without heat or pressure. Inert to usual solvents
Phenolic varnish	Thermosetting liquid	For vacuum impregnation of oxide films, etc. (*see* text)
Polyvinyl chloride	Thermoplastic*	Low shrinkage. Inert to usual solvents but attacked by glacial acetic acid[2]
Diallyl phthalate (prepolymer)	Thermosetting	Needs controlled heat (130–140 °C) and pressure. Low shrinkage good polishing characteristics[3]

* Must be cooled under pressure to low temperature to solidify before ejection.

Thermoplastics, such as polymethyl methacrylate, and thermosetting resins, such as 'Bakelite', are convenient for routine work because they are available as powders immediately ready for use, but they require a press, and normally only one size of cylindrical mount would be available. Cold-setting resins may be formed simply in a container consisting of a short piece of tube standing on a glass plate, and are therefore suitable for occasional use and odd shapes and sizes.

To examine a surface critically in section, support it if possible by plating (e.g. with copper or nickel) by applying an evaporated coating, or by wrapping with aluminium foil and mounting under pressure (this method is useful for measuring the thickness of anodic or similar transparent films). Fragile oxide or other films may be held together by vacuum impregnation: use a vacuum desiccator and tap funnel to run resin varnish round the specimen in a rough vacuum (e.g. at about 10 torr residual pressure), remove the specimen and container to an oven (at 80 °C for Bakelite grade NPA) and heat until the resin is polymerized. A similar technique may be used with casting resins (if sufficiently fluid) which set without heat, although impregnation is liable to be less effective than with the very fluid hot varnish.

GRINDING

Emery or silicon-carbide cloths and papers are normally used. Silicon carbide is preferred because it is harder, has sharper particles and cuts at a faster rate. Use strips 20–30 cm × about 8 cm laid flat on plate glass, and rub the specimen to and fro on the strip. Start with not finer than 80 grit, and rub until all traces of saw cuts are removed. Turn the specimen through 90° and rub until the first set of emery scratches are removed. Repeat at least once, because the depth of the deformed layer is several times the depth of the residual scratches. Then progress to the next finer paper or cloth, turning the specimen through 90°, and again rub until the previous scratches are removed, then to the next finer paper similarly, until grade 600 silicon carbide paper is reached or, for softer alloys such as aluminium, the finest emery paper is reached (usually grade 4/0, but grade 6/0 is sometimes useful). A fine paraffin oil (e.g. 'white spirit') should be flooded over the papers to act as a lubricant, or they should be continuously washed with water or white spirit. For soft metals, a more viscous liquid paraffin is preferred to avoid pick-up of silicon carbide or emery in the surface of the specimen. Slowly rotating silicon carbide discs continuously washed with water are frequently used.

For very hard metals diamond hones[4] and laps[106] have been used for grinding.

Metals containing constituents of widely differing hardness may develop undesirable relief when ground on fine papers. An alternative is to use a lead lap. Lead foil is stretched over a glass plate and is flooded with white spirit. Fine abrasive (e.g. alumina) is worked into the surface by placing some on the wet surface and working it in with a steel disc. Any loose abrasive remaining is washed off, and, in use, the plate carrying the lap is mounted at a slight tilt in a dish and the surface is washed continuously with a slow stream of white spirit to remove loose particles.

MECHANICAL POLISHING

Mechanical polishing is often done in two stages, with a coarse and a fine abrasive or polishing agent, respectively. The coarse polishing stage is carried out at 300 r.p.m., uses a low nap or napless cloth such as Selvyt or synthetic cloth. (The nap is intended to retain the abrasive without causing relief effects.) It is fed with a suspension of a relatively coarse abrasive. The final polishing stage is carried out with finer abrasive at a lower rotational speed (100 r.p.m.) using medium nap cloth, preferably dense electroflocked terylene fibres bonded to a chemically resistant backing. Polishing agents include α-alumina, γ-alumina, magnesium oxide, chromium oxide, proprietary metal polishes and diamond dust. The polishing agent may have a cutting action or it may produce a 'flowed' layer on the surface or both. The modern tendency is to use cutting, rather than flowing, polishing agents, and diamond dust is now preferred. α-Alumina (a fast-cutting hard material) may be made by roasting aluminium sulphate to 1400 °C (a high proportion of α is obtained at 1200 °C) and can be used without further treatment. γ-Alumina suitable for fine polishing may be made by heating to 950 °C. Suitable magnesium oxide is obtainable cheaply from medical suppliers. Magnesium oxide is slowly converted to carbonate when damp, so polishing cloths, if kept overnight, are cleaned with dilute acid and thoroughly washed. Diamond powder of up to 12 μm diameter is used for rough polishing and 0–1 or 0–$\frac{1}{2}$ μm diameter for fine polishing (usually to be followed briefly with γ-alumina, as it leaves very fine scratches). The powder may simply be rubbed into cloth which is kept lubricated with white spirit (a plastic rim pressed on to the polishing wheel conserves the powder), or may be made into a cream. The recipe below[5] is recommended:

The ingredients are:

Stearic acid	12.5 g
Triethanolamine	6 ml
Water	25 ml
Diamond powder	0.5 g

The stearic acid is melted and heated to 80–90 °C. The triethanolamine and most of the water are mixed and heated to the same temperature range, a small amount of wetting agent and the diamond powder are added, and the abrasive is shaken into uniform suspension. The molten stearic acid is stirred vigorously with a mechanical stirrer and the abrasive suspension is introduced rapidly. The water not used in the original suspension can be used to wash in any abrasive remaining in the container. Continue stirring until the emulsion cools and thickens.

Where it is particularly required to avoid relief effects in specimens containing constituents of widely differing hardness, diamond dust may be used on a pile-free nylon or terylene cloth.

Some metals are readily stained or corroded in the presence of water, and for these a non-aqueous polishing mixture, normally diamond with white spirit, is preferred. In borderline cases the use of distilled water, rather than tap water, helps to avoid staining.

After polishing by any method, the specimen must be thoroughly washed and dried as described under *Etching* (p. **10**–16), or washed and etched immediately. The specimen should be flooded with water, then with alcohol or acetone to remove all water and finally dried with a blast of hot air. If the polish or etch is non-aqueous, wash with alcohol or acetone.

Polish attack is a method of hastening polishing by the simultaneous use of an etching agent. For instance, ammonia is used with advantage on the pad in polishing copper alloys. The action is thought to depend on the enhanced chemical activity of the 'flowed' layer.

Attack polishing in a deep layer of liquid is done by mounting a polythene pot on the spindle of the polishing machine, with the polishing pad in it and submerged in the liquid.[6] Table 10.3 gives reagents for use with various metals by this method. Several solutions have also been proposed for magnesium alloys.[7]

Table 10.3 ATTACK POLISHING CONDITIONS FOR VARIOUS METALS AND ALLOYS USING TERYLENE-COVERED LAPS[6]

Material	*Solution**		*Time* min	*Remarks*
Uranium	CrO_3 H_2O HNO_3 (1.40)	50 g 100 ml 10 ml	20–30	Medium contrast under polarized light, no pitting, good resistance to oxidation
Zirconium	HNO_3 (1.40) Glycerol	50 ml 150 ml	1–10	Good contrast under polarized light. Slight grain relief
Bismuth	HNO_3 (1.40) Glycerol	50 ml 150 ml	3–5	Good contrast under polarized light. Requires less pressure than usual
Chromium	$(COOH)_2$ H_2O	15 g 150 cm	5–10	Bright polish revealing oxides, etc.
Molybdenum and tungsten	Pot. ferricyanide Sodium hydroxide Water	3.5 g 1 g 300 ml		

* Acids are concentrated, unless otherwise indicated, e.g. with specific gravity.

ELECTROLYTIC POLISHING

Very full reviews have been given by Jacquet,[8] Tegart,[9] and Petzow[1] which may be consulted for individual references. A comparison with mechanical methods has been made by Samuels.[10]

The specimen is made with the anode in a suitable solution, and conditions are adjusted so that the hills on the surface are dissolved much more rapidly than the valleys, and when enough metal has been removed a smooth surface is obtained. The condition for polishing often corresponds to a nearly flat (i.e. constant current) region in the curve for cell current versus voltage. As the voltage is increased (*see* Figure 10.1), etching (AB) is replaced by film formation (BC). The voltage then increases and the current falls slightly as the film disappears and polishing conditions are established (CD). At higher voltages, gas evolution occurs with pitting. Near E gas evolution is rapid and polishing continues but the region just below D is preferred. By reducing the voltage to

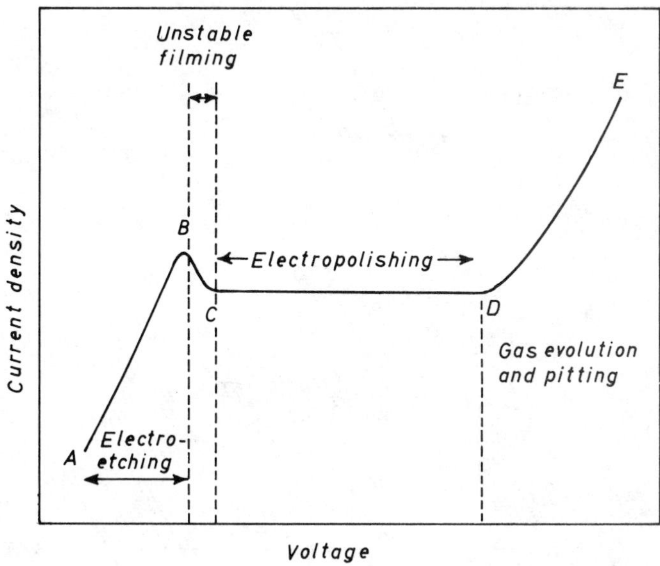

Figure 10.1 *Idealized relationship between current density and voltage in electropolishing cell.*

below B, the specimen can be etched in the same operation. For many specimens electropolishing leads to a great saving in time, and it reliably produces surfaces free from strain provided sufficient metal is removed in the process. It tends to exaggerate porosity and is unsuitable for highly porous specimens. Inclusions are often removed, though not invariably, and their place taken by severe pits. Many two-phase and complex alloys, however, can be successfully polished.

Apparatus. To cover the widest range of applications a d.c. supply of 4–5 A at voltages variable up to at least 60 V is required, but some solutions require only 2 V. Accurate voltage regulation is essential, and a rectifier set fed from a variac, a tapped battery or a potentiometer circuit across a constant d.c. source is recommended. Published recommendations for particular solutions sometimes state the voltage, and sometimes the current density, required. It is preferable to work on voltage, as the current density for a given electrode condition is much affected by temperature and other variables. If both are stated, but cannot be simultaneously obtained, the solution is probably wrong; if it is not, the current density should be disregarded. Two general cell arrangments are used: with electrodes in a beaker of still or gently stirred solution, and with flowing or pumped electrolyte.[11–13] The first arrangement is easily set up and often suffices; the second is more powerful but requires more complicated apparatus (obtainable commercially, however). The characteristics are quite different: with flowing electrolyte a good polish may be obtained with more strongly conducting solutions, and hence with higher current densities, and it is therefore frequently possible to remove more metal in polishing and to start with a more roughly prepared surface. A small area of an article may be electropolished by the use of electrolyte flowing from a vertical jet above the article, the jet itself containing a projecting wire to act as cathode.[13] In suitable conditions, polishing of an area already rubbed with emery may be completed in 3–10 s. Apparatus for this method is also available commercially.

Jacquet has described a device (the 'Ellapol') in which an electrolyte is applied to the surface by a small swab surrounding the cathode. The device can conveniently be used to polish a small area of a large component *in situ* (*see* e.g., refs. 14–16).

Solutions for electropolishing particular metals are listed in Table 10.4. Table 10.4 is not a complete list, but should cover most requirements. More detailed solutions are given in refs. 1, 2, 8 and 9. Minor differences between solutions are often a consequence of the cell used. The most widely useful solutions are methyl alcohol–nitric acid mixtures, strong solutions of phosphoric acid and mixtures of perchloric acid with alcohol, acetic acid or acetic anhydride. Mixtures of perchloric acid with acetic anhydride, although frequently the best polishing agents, are often explosive and deserve respect. They must be kept cold in use; plastics (especially cellulose) and bismuth must be kept away from them, and they must not be stored in the laboratory as they are liable to explode without apparent reason. The explosion of a few hundred millilitres is not likely

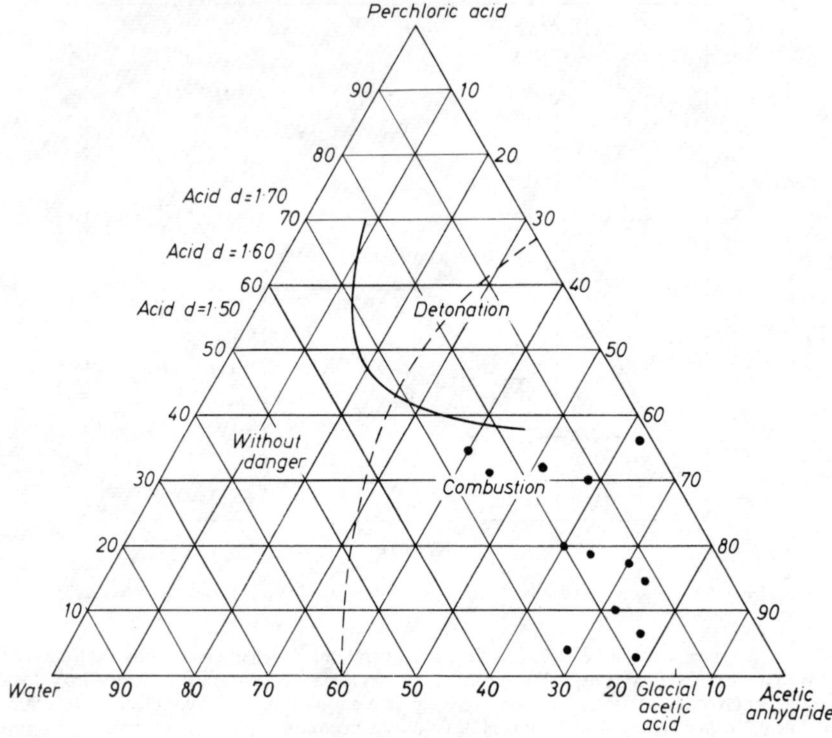

Figure 10.2 *Characteristics of perchloric acid/acetic anhydride/water solutions (after Jacquet[8, 17] and Petzow[1]).* ● = *Typical electrolytes*

to do great physical damage, but larger quantities should not be used. The limits of the dangerous mixtures, according to Jacquet, the originator,[8, 17] are indicated in Figure 10.2. Perchloric acid must always be added to the acetic anhydride–water mixture to avoid compositions in the detonation zone.

Table 10.4a ELECTROLYTIC POLISHING SOLUTIONS FOR VARIOUS METALS AND ALLOYS

Because of the considerable number of solutions published in the literature, a selection has been made on the basis of (a) wide usage, (b) simplicity of composition, and (c) least danger. References 1 and 2 provide a wider range of compositions. Temperatures should be in the range 15–35°C. Cooling should be used to avoid temperatures above 35°C unless stated otherwise.

Composition of solution		Usage	Cell voltage	Time	Cathode
1 Ethanol	800 ml	Al alloys (not Al–Si)	30–80	15–60 s	Stainless steel
Distilled water	14 ml	Most steels	35–65	15–60 s	Stainless steel
Perchloric acid (1.61)	60 ml	Lead alloys	10–35	15–60 s	Stainless steel
		Zinc alloys	20–60	15–60 s	Stainless steel
		Magnesium alloys	20–40	up to 2 min	Nickel
2 Ethanol	800 ml	Al alloys	35–80	15–60 s	Stainless steel
Perchloric acid (1.61)	200 ml	Stainless steels	35–80	15–60 s	Stainless steel
		Pb alloys	15–35	15–60 s	Stainless steel
		Zn alloys	20–60	15–60 s	Stainless steel
		and many other metals			
3 Ethanol	940 ml	Stainless steel	30–45	15–60 s	Stainless steel
Distilled water	6 ml	Thorium	30–40	15–60 s	Stainless steel
Perchloric acid (1.61)	54 ml				

Table 10.4(a) ELECTROLYTIC POLISHING SOLUTIONS FOR VARIOUS METALS AND ALLOYS—*continued*

Composition of solution		*Usage*	*Cell voltage*	*Time*	*Cathode*
4 Ethanol Water 2-Butoxyethanol Perchloric acid (1.61) Glycerol (100 ml) can replace butoxyethanol	700 ml 120 ml 100 ml 80 ml	Al alloy Steel, cast iron Ni, Sn, Ag, Be Ti, Zr, U, Pb Complex steels and nickel alloy general use	30–65	15–60 s	Stainless steel
5 Ethanol Distilled water Ether Perchloric acid (1.61)	760 ml 30 ml 190 ml 20 ml	Al alloys including Al–Si alloys Fe–Si alloys Sb Preferred solution for Al alloys	30–60	15–60 s	Stainless steel
6 Methanol Water (distilled) 2-Butoxyethanol Perchloric acid	590 ml 6 ml 350 ml 54 ml	Germanium and silicon Titanium Vanadium Zirconium	25–35 ~60 ~30 ~70	30–60 s 45 s 3–5 s* 15 s	Stainless steel Stainless steel Stainless steel Stainless steel
7 Glacial acetic acid Perchloric acid (1.61)	940 ml 60 ml	Cr, Ti, U, Zr, Fe Cast iron, all steels, V Re and many other metals	20–60	up to 5 min	Stainless steel
8 Glacial acetic acid Perchloric acid (1.61)	900 ml 100 ml	Ti, Zr, U steels Superalloys	10–60	up to 2 min	Stainless steel
9 Glacial acetic acid Perchloric acid (1.61)	800 ml 200 ml	U, Ti, Zr, Al steels Superalloys	40–100	up to 15 min	Stainless steel
10 Glacial acetic acid Perchloric acid (1.61)	700 ml 300 ml	Nickel, Pb, especially Pb–Sb alloys	40–100	up to 5 min	Stainless steel
11 Phosphoric acid (1.75)		Cobalt, Fe–Si alloys	1–2	up to 5 min	Stainless steel
12 Distilled water Phosphoric acid (1.75)	300 ml 700 ml	Cu, Cu alloys (not Cu–Sn) Stainless steels—rinse in 20% H_3PO_4	1–1.6	10–40 min	Copper
13 Distilled water Phosphoric acid	600 ml 400 ml	Brasses, Cu–Fe, Cu–Co Co, Cd	1–2	up to 15 min	Copper or stainless steel
14 Distilled water Ethanol Phosphoric acid (1.75)	200 ml 400 ml 400 ml	Al, Mg, Ag	25–30 at 40 °C	4–6 min	Aluminium
15 Ethanol Glycerol Phosphoric acid (1.75)	300 ml 300 ml 300 ml	U (preferred solution)	20–30	4–6 min	Aluminium
16 Ethanol Glycerol Phosphoric acid (1.75)	500 ml 250 ml 250 ml	Mn, Mn–Cu	18	up to 10 min	Stainless steel
17 Ethanol Phosphoric acid (1.75)	625 ml 375 ml	Mg, Zn alloy	1.5–2.5	up to 30 min	Stainless steel
18 Ethanol Ethylene glycol Phosphoric acid (1.75)	445 ml 275 ml 275 ml	U alloys	18–20	up to 15 min	Stainless steel
19 Distilled water Sulphuric acid (1.75)	750 ml 250 ml	Stainless steel iron, nickel Molybdenum	1.5–6.0 1.5–6.0	up to 10 min 1 min	Stainless steel Stainless steel
20 Methanol Sulphuric acid (1.84)	875 ml 125 ml	Molybdenum	6–18 Keep below 27 °C	1 min	Stainless steel

*With vanadium, give several 3–5 s bursts and avoid heating.

Table 10.4(a) ELECTROLYTIC POLISHING SOLUTIONS FOR VARIOUS METALS AND ALLOYS—*continued*

Composition of solution		*Usage*	*Cell voltage*	*Time*	*Cathode*
21 Distilled water Chromium trioxide	830 ml 170 g	Zn, Al bronze Brass	1.5–12	up to 1 min	Stainless steel
22 Distilled water Phosphoric acid (1.75) Sulphuric acid (1.84)	450 ml 390 ml 160 ml	Tin Tin bronzes (high tin) (rinse in 20% H_3PO_4)	2	up to 15 min	Copper
23 Distilled water Phosphoric acid (1.75) Sulphuric acid (1.84)	330 ml 580 ml 90 ml	Tin Tin bronzes (low tin <6%) (rinse in 20% H_3PO_4)	2	up to 15 min	Copper
24 Distilled water Chromium trioxide Phosphoric acid (1.75) Sulphuric acid (1.84)	170 ml 105 g 460 ml 390 ml	Stainless steel (use at 35–40 °C)	2	up to 60 min	Stainless steel
25 Distilled water Chromium trioxide Phosphoric acid (1.75) Sulphuric acid (1.84)	240 ml 80 g 650 ml 130 ml	Stainless steel Alloy steels (use at 40–50 °C)	2	up to 60 min	Stainless steel
26 Hydrofluoric acid (40%) Sulphuric acid (1.84)	100 ml 900 ml	Tantalum Niobium (use at ~40 °C)		5–15 min	Graphite
27 Glycerol Glacial acetic acid Nitric acid (1.40) (*Warning:* This solution will decompose vigorous if kept, especially if cathode left in it. Throw away solution as soon at finished with)	750 ml 125 ml 125 ml	Bismuth	12	1–5 min	Stainless steel
28 Methanol Hydrochloric acid (1.19) Sulphuric acid (1.84) Keep cool below 2 °C. Avoid water contamination.	685 ml 225 ml 90 ml	Molybdenum	19–35	30 s	Stainless steel
29 Ethanol n-Butyl alcohol Hydrated aluminium trichloride Anhydrous Zn chloride	885 ml 100 ml 109 g 250 g	Ti and most other alloys	25–50	5 min	Stainless steel
30 The above diluted with 120 ml distilled water		Zinc	20–40	up to 3 min	Stainless steel
31 Glycerol Hydrofluoric acid (40%) Nitric acid (1.40) As 27—will decompose on standing and must be thrown away as soon as possible	870 ml 43 ml 87 ml	Zirconium	9–12	up to 10 min	Stainless steel
32 Potassium cyanide Potassium carbonate Gold chloride Distilled water to 1 000 ml	80 g 40 g 50 g	Gold, silver	7–5	2–4 min	Graphite
33 Sodium cyanide Potassium ferrocyanide Distilled water to 1 000 ml	100 g 100 g	Silver	2–5	up to 1 min	Graphite

Table 10.4(a) ELECTROLYTIC POLISHING SOLUTIONS FOR VARIOUS METALS AND ALLOYS—*continued*

Composition of solution		*Usage*	*Cell voltage*	*Time*	*Cathode*
34 Sodium hydroxide Distilled water to 1 000 ml	100 g	Tungsten lead	6	10 min	Graphite
35 Methanol Nitric acid (1.40) *Warning:* Do not keep longer than necessary. May become explosive. On no account substitute ethanol for methanol	600 ml 330 ml	Ni, Cu, Zn, Ni–Cu Cu–Zn, Ni–Cr Stainless steel, In, Co Very versatile	40–70	10–60 s	Stainless steel

Table 10.4(b) RECOMMENDED ELECTROPOLISHING SOLUTION FROM TABLE 10.4(a) FOR SPECIFIC METALS AND ALLOYS

Alloy	*Electrolyte* (No. in Table 10.4 (a))
Aluminium	1, 2, 4, 9, 14
Aluminium–silicon	5
Antimony	5
Beryllium	4
Bismuth	27
Cadmium	13
Cast iron	4, 7
Chromium	7
Cobalt	11, 13, 35
Copper and alloys	12, 13, 35
Copper–tin alloys	22, 23
Copper–zinc alloys	13, 21
Germanium	6
Gold	32
Hafnium	4
Indium	35
Iron-base alloys	4, 7, 8, 9, 19
Lead	1, 2, 4, 10, 34
Magnesium	1, 14, 17
Manganese	16
Molybdenum	19, 20, 28
Nickel and superalloys	4, 8, 9, 10, 19, 35
Niobium	26
Rhenium	7
Silver	4, 14, 32, 33
Stainless steels	1, 2, 3, 4, 7, 8, 9, 12, 19, 24, 25, 35
Steels: carbon and alloy	1, 4, 7, 8, 9, 19, 25
Tantalum	26
Thorium	3
Tin	4, 22, 23
Titanium	4, 6, 7, 8, 9, 29
Tungsten	34
Uranium	4, 7, 8, 9, 15, 17
Vanadium	6, 7
Zinc	1, 2, 17, 21, 30, 35
Zirconium	4, 6, 7, 8, 9, 31

CHEMICAL POLISHING

Chemical polishing is usually adopted as a quick method of obtaining a passable result, rather than as a method of preparing a perfect surface. However, where it is difficult to prepare a work-free surface by other means, as with some very soft metals or where other difficulties are encountered, it may provide the best method of preliminary or final preparation.

In general, a ground or turned specimen is held in the polishing agent until a polish is obtained, and it is then etched or washed and dried, as appropriate. Reagents are listed in Table 10.5.

Table 10.5 REAGENTS FOR CHEMICAL POLISHING[1,2,9]

Metal	*Reagent**		*Time*	*Temperature* °C	*Remarks*
Aluminium and alloys	Sulphuric acid (1.84) Orthophosphoric acid Nitric acid	25 ml 70 ml 5 ml	30 s–2 min	85	Very useful for studying alloys containing intermetallic compounds, e.g. Al–Cu, Al–Fe and Al–Si alloys
Beryllium	Sulphuric acid (1.84) Orthophosphoric acid (1.75) Chromic acid Water	1 ml 14 ml 20 g 100 ml	Several min	49–50	Rate of metal removal is approx. 1 μm min^{-1}. Passive film formed may be removed by immersion for 15–30 s in 10% sulphuric acid
Cadmium	Nitric acid (1.4) Water	75 ml 25 ml	5–10 s	20	Cycles of dipping for a few seconds, followed immediately by washing in a rapid stream of water are used until a bright surface is obtained
Copper	Nitric acid Orthophosphoric acid Glacial acetic acid	33 ml 33 ml 33 ml	1–2 min	60–70	Finish is better when copper oxide is absent
Copper alloys	Nitric acid Hydrochloric acid Orthophosphoric acid Glacial acetic acid	30 ml 10 ml 10 ml 50 ml	1–2 min	70–80	Specimen should be agitated
Copper–zinc alloys	Nitric acid (1.40) Water	80 ml 20 ml	5 s	40	Use periods of 5 s immersion followed immediately by washing in a rapid stream of water. Slight variations in composition are needed for α–β and β–γ brasses to prevent differential attack. With β–γ alloys, a dull film forms and this can be removed by immersion in a saturated solution of chromic acid in fuming nitric acid for a few seconds followed by washing
Germanium	Hydrofluoric acid Nitric acid Glacial acetic acid	15 ml 25 ml 15 ml 3–4 drops	5–10 s	20	—
Hafnium	Nitric acid Water Hydrofluoric acid	45 ml 45 ml 8–10 ml	5–10 s	20	As for zirconium
Iron	Nitric acid Hydrofluoric acid (40%) Water	3 ml 7 ml 30 ml	2–3 min	60–70	Dense brown viscous layer forms on surface; layer is soluble in solution. Low carbon steels can also be polished, but the cementite is attacked preferentially
Irons and steels	Distilled water Oxalic acid (100 g l^{-1}) Hydrogen peroxide (30%)	80 ml 28 ml 4 ml	15 min	35	The solution must be prepared freshly before use. Careful washing is necessary before treatment. A microstructure is obtained similar to that produced by mechanical polishing, followed by etching with Nital

* Acids are concentrated, unless otherwise indicated.

Table 10.5 REAGENTS FOR CHEMICAL POLISHING—*continued*

Metal	*Reagent**		*Time*	*Temperature °C*	*Remarks*
Lead	Hydrogen peroxide (30%) Glacial acetic acid	80 ml 80 ml	Periods of 5–10 s	20	Use Russell's reagent (Table 10.1) to check that any flowed layer has been removed before final polishing in this reagent
Magnesium	Fuming nitric acid Water	75 ml 25 ml	Periods of 3 s	20	The reaction reaches almost explosive violence after about a minute, but if allowed to continue it ceases after several minutes, leaving a polished surface ready for examination. Specimen should be washed immediately after removal from solution
Nickel	Nitric acid (1.40) Sulphuric acid (1.84) Orthophosphoric acid (1.70) Glacial acetic acid	30 ml 10 ml 10 ml 50 ml	$\frac{1}{2}$–1 min	85–95	This solution gives a very good polish
Silicon	Nitric acid (1.40) Hydrofluoric acid (40%)	20 ml 5 ml	5–10 s	20	1:1 mixture also used
Tantalum	Sulphuric acid (1.84) Nitric acid (1.40) Hydrofluoric acid (40%)	50 ml 20 ml 20 ml	5–10 s	20	Solution is useful for preparing surfaces prior to anodizing
Titanium	Hydrofluoric acid (40%) Hydrogen peroxide (30%) Water	10 ml 60 ml 30 ml	30–60 s	—	Swab till satisfactory
	Hydrofluoric acid (40%) Nitric acid (1.40) Lactic acid (90%)	10 ml 10 ml 30 ml			Few seconds to several minutes according to alloy
Zinc	Fuming nitric acid Water	75 ml 25 ml	5–10 s	20	As for cadmium
	Chromium trioxide Sodium sulphate Nitric acid (1.40) Water to 100 ml	20 g 1.5 g 5 ml	3 min – 30 min	20	Solution must be replaced frequently
Zirconium (also Hafnium)	Acid ammonium fluoride Nitric acid (1.40) Fluosilicic acid Water	10 g 40 ml 20 ml 100 ml	$\frac{1}{2}$–1 min	30–40	Rate of dissolution varies markedly with temperature and is about 20–60 μm min^{-1} in the given range
	Nitric acid (1.40) Water Hydrofluoric acid (40%)	40–45 ml 40–45 ml 10–15 ml	5–10 s repeated	—	Reaction is vigorous at air/solution interface, and specimen is therefore held near surface of liquid. Hydrogen peroxide (30%) can be used in place of water.

* Acids are concentrated, unless otherwise indicated.

ETCHING

Specimens should first be examined without etching. This reveals features which have a significant difference in reflectivity from the main structure, such as differences in colour and relief due to phases of large difference in hardness. Non-metallic inclusions, cracks, porosity and various kinds of pit can be recognized clearly and should be recorded because subsequent etching treatments can change both shapes and colours.

In order to obtain the maximum resolution from the optical microscope with minimum reflections from stray light, the microscope must be set up using the 'Köhler' principle of illumination.[25] Most modern microscopes are constructed to achieve this principle and it is only necessary to adjust the two iris diaphragms. The first of these, usually called the field diaphragm, should project an image sharply in focus on the specimen and should be adjusted so that the image falls just outside the field of view. The aperture diaphragm should be sharply in focus on the rear of the objective lens. It can be viewed by removing the eyepiece and should be adjusted (through an auxiliary lens on some microscopes) so that the image is centrally located on the rear of the objective and it illuminates 90% of the lens area. If after these adjustments the image is too bright, it should be dimmed by either reducing the light intensity or interposing a filter. Reduction of the aperture reduces the resolution achieved by the lens, emphasizes differences in level and can introduce artefacts.

To emphasize small differences in surface topography, several techniques are available which are usually provided on good optical microscopes. These include:

1. Dark field illumination[2,26]

By this technique a specimen is illuminated by an annulus of light which passes up the outside of the objective and is focussed as a cone by a concave reflector. Thus the normal beam of light is not used to form the image. Instead, the light scattered by angled surfaces is focused to make the image and thus the contrast is reversed. Cracks, inclusions and defects are seen as bright features on a black background.

2. Phase contrast

This technique converts differences in surface height (which produce differences in phase) into differences in intensity which the eye can detect more readily. To achieve this, an annular disc is placed at the front focal plane of the condenser lens of the microscope. This is then imaged at the back focal plane of the objective where is placed a transparent phase retarding ring of the same shape and size as the image of the annulus. The light passing through the ring is retarded (or it can be advanced) by one quarter of the wavelength of the monochromatic light used. The diffracted light from the specimen passes outside this ring and is not retarded (or advanced) and the image formed as light and dark contrast. The low areas appear dark and the high areas light for phase retardation (or the opposite for phase advancement). This technique is sensitive to surface changes of approx. 5 nm.

3. Interference microscopy[27–30]

Interference microscopes have been described by several authors but the most sensitive and useful techniques have been developed by Tolansky[27]. His multiple beam interferometry can be used with conventional microscopes at magnifications of up to about 250. Monochromatic light is essential and a parallel beam normal to the surface is used. An optical flat, silvered or aluminized to give about 95% reflectivity, is placed in contact with the specimen and is slightly tilted to produce a thin wedge between the two. The light is repeatedly reflected between the specimen and the optical flat, and interference takes place to produce very thin, sharp, dark fringes. The spacing can be varied by the tilt of the plate. Where a change in the surface of the specimen occurs, e.g. a step or depression, the light path occurs and the interference fringe is displaced one way or the other. A total fringe displacement corresponds to a change in height of half the wavelength of light used and so it is possible to measure surface displacements of about 5–50 nm. To obtain sharp fringes, the reflectivity of the metallic surface should be the same as the reference plate, i.e. approximately 95%, and it may be necessary to aluminize the specimen surface for the best results.

Normarski has designed a very sensitive microscope for detecting steps in the surface of a specimen.[26,30] He uses a conventional polarizing microscope, into which a double quartz prism is inserted between the objective and the analyser. If a step is present, this produces two images slightly displaced to one another and interference between these produces light and dark fringes, the spacing of which can be varied by adjusting the prism. A modification of the technique produces interference contrast in images similar to those produced by phase contrast.

4. Polarized light

The use of polarized light is an extremely powerful method of studying inclusions and structures of unetched and electropolished surfaces.

The equipment needed includes a monochromatic source of illumination, a polarizer (Nicol prism or polaroid filter) which can be rotated to change the plane of polarization, and an analyser of comparable material which can also be rotated. When the analyser is oriented so that its plane of polarization is at 90° to that of the polarizer, an isotropic specimen will appear black if the objectives and the microscope are well adjusted and no depolarization occurs. The objectives should be strain free for the highest sensitivity.

Bausch and Lomb use a Foster prism,[25] which produces the effect of 'crossed polars' with isotropic materials. A rotatable quarter wave plate is added so that the effect of rotating the analyser towards the parallel position can be achieved.

With isotropic metals and the polarizer and analyser set in the 'crossed' position, no light reaches the eyepiece. However, with an optically anisotropic material, e.g. a hexagonal metal like beryllium, magnesium or zinc, the reflected beam becomes elliptically polarized and the intensity of the component normal to the plane of polarization of the incident light depends on the orientation of the anisotropic structure. Thus the intensity of light which passes through the analyser will be dependent on the orientation of the structure to the surface of the specimen and to the incident beam. The image will vary from dark to bright, according to orientation and the grain structure of an anisotropic material will be revealed. On rotating the specimen, the intensity of light passing through the analyser from any one grain will pass from minimum to maximum intensity every 45° to give four maxima per revolution at 90° to each other, with a minimum intensity at 45° to each maximum. If the analyser and polarizer are not quite crossed and differ by a few degrees, only two maxima and minima occur. The difference in contrast between the maxima and minima is much greater under these conditions and improves the sensitivity of the method, especially for weakly anisotropic or pleochroic materials.

The method can be used for studying grain structures, twins and martensites. In uranium alloys, for instance, the isotropic gamma phase can be distinguished from weakly anisotropic retained beta phase and the strongly anisotropic variants of the alpha phase. Martensites can be distinguished from retained beta phase in copper base alloys. In other systems, such as aluminium, grain structure can be revealed by anodizing to produce an anisotropic oxide film which has orientation related to the underlying lattice.

The other useful application is in the identification of inclusions. Glassy inclusions such as silicates display the so-called 'optical cross'. Others are optically anisotropic, e.g. MnO and MnS can be distinguished in steels. Other inclusions are pleochroic and display characteristic colours under crossed polars. Examples are the grey Cu_2O phase in copper alloys which is blood red under polarized light, and Cr_2O_3 which changes from blue grey to a beautiful emerald green. References 41–44 and 26 should be consulted for more detail.

The anisotropic metals include:

Antimony	Tin
Beryllium	Titanium
Cadmium	Uranium
Hafnium	Zinc
Magnesium	Zirconium

If correctly prepared, these metals will reveal their structure under crossed polars.

The following metals can be made to respond to polarized light by anodizing to produce an anisotropic film, or by deep etching to produce an uneven surface with etch pits.[44]

Aluminium	Molybdenum
Chromium	Nickel
Copper	Tungsten
Iron	Vanadium
Manganese	

5. Colour (based mainly on ref. 44)

Only two metals demonstrate natural colour: gold and copper.

Others can be rendered coloured by:

(*a*) Optical methods such as polarized light, especially of etched structures with a sensitive tint plate inserted between polarizer and analyser. This is used on Bausch and Lomb microscopes and converts shades of grey into shades of blue and magenta. Brighter colours are obtained in etched structures. All these methods change shades of grey to colour contrast.

(*b*) By producing interference films by heat tinting or by chemical processes. This can colour phases according to their reactivity or grains according to their orientation. Most frequently, the reagents used leave thin films of oxide, sulphide, chromate, phosphate or molybdate which cause colouration by interference of light. The colour depends on the thickness of films. This method is most appropriate for distinguishing inclusions or intermetallic compounds. The more useful applications are given in Tables 10.6 to 10.8. More details and illustrations are given in ref. 44.

6. Physical methods

(*a*) *Cathodic vacuum etching* (*ion etching*). The specimen is made a cathode in a high voltage gas discharge. High energy ions such as argon are accelerated at voltages of 1–10 kV and gas pressures of 10 μm. This bombardment removes atoms at a rate dependent on orientation, presence of grain boundaries, intermetallic compounds.

(*b*) *Thermal etching*. On heating specimens, e.g. in vacuum or inert atmosphere, atoms are lost from regions of low binding energy, e.g. grain boundaries. Surface tension forces lead to changes in surface topography at grain boundaries, leaving a structure characteristic of the high temperature.

7. Chemical and electrochemical etching

Etching is usually an oxidation process. In general, elements with electrode potentials more negative than hydrogen will pass into solution in many solutions, the rate depending on the local environment, so resulting in grain boundary attack or outlining of phases or other structures. For elements with electrode potentials more positive than hydrogen, or for elements which polarize, solutions containing oxidizing agents are needed. Making the specimen the anode in a low voltage cell has the same effect. Indeed, most electropolishing solutions will cause etching if the voltage is reduced, usually by a factor of 10.

Detailed etching procedures for individual metals are given below but, as a general principle, iron alloys are usually etched in dilute oxidizing agents, e.g. nitric acid or picric acid in alcohol. Stainless steels are usually etched in weak oxidizing reagents in alkaline solution, e.g. alkaline ferricyanides. Virtually the same result is achieved electrolytically in a 1% potassium hydroxide solution. While claims are made by many authors that subtle selective etches can be produced by minor changes in composition of the etching solution, these are often difficult to achieve and reproduce and need considerable skill and experience by the metallographer to achieve.

In the case of copper alloys, most etchants require an ammoniacal atmosphere plus an oxidizing reagent such as air (by swabbing), hydrogen peroxide or dichromates, permanganates, persulphates, etc. Electrolytic etching in an oxidizing acid, e.g. 1% chromium trioxide often suffices, the control of etching being achieved by varying the voltage.

Many pure metals are notoriously difficult to etch, e.g. aluminium. If there are no natural impurities present to induce grain boundary attack, etchants should be used which can chemically deposit elements which can be reduced and diffuse into the boundaries, e.g. gallium from a gallium salt solution.

Reduction reactions can also be used and these cause staining or colouration of phases, especially intermetallic compounds.

A wide variety of etching reagents has been devised by empirical methods. Some of these are reproduced but the perceptive reader will see that these can be modified easily to cope with new compositions. Small amounts of some components are needed to control pH or potential. Some solutions are not stable and their effectiveness will change with time. Others containing mixtures of oxidizing agents and organic chemicals can become very dangerous with time and should only be retained for short times and be discarded immediately after use.

Reference is given to simple electrolytic etches and it is recommended that these be used with a potentiostat[18,19] to control the potential at a known and reproducible value. This will aid reliable and consistent etching.[20,21,22]

8. Washing and drying

On completion of etching (with most etches) the specimen should at once be flooded with water, washed free from water with acetone or alcohol (avoid ether) and dried in a blast of hot air from a hair dryer or equivalent. With non aqueous etches it is often preferable to do the initial washing with alcohol or acetone rather than water.

Porous specimens are often easily stained by residual etching solution seeping out of the pores. This trouble can usually be avoided by prolonging the washing and drying stages, and in severe cases by prolonged soaking of the specimen in alcohol or acetone before final drying. If this fails the specimen may be treated with a clear lacquer immediately after drying.

Table 10.6 COLOUR ETCHES FOR STEEL

Reagent		*Procedure*	*Comments*
1 Saturated sodium thiosulphate	50 ml	Immersion for times dependent on stainless steel	ϵ martensite–white α martensite–black Austenite–grey
Potassium metabisulphate	5 g		
2 20% HCl in water	100 ml	Immersion for 5–10 s	Carbon and tool steels Martensite–blue Bainite–red Carbides–white
Ammonium bifluoride	2 g		
Potassium metabisulphite	1 g		
3 As above		As above	Stainless steels Martensite–blue

Table 10.7 COLOUR ETCHES FOR CAST IRON

Reagent		*Procedure*	*Comments*
1 HCl (35%)	2 ml	5–6 min immersion	Fe_3C–red/violet
Selenic acid	0.5 ml	2–3 min if pre-etched	Ferrite–white
Ethanol	100 ml	in 2% Nital	Phosphides–blue/green
2 Sodium thiosulphate	24 g	Pre-etch in 2% Nital	Phosphides–yellow/brown
Lead acetate	2.4 g	Immerse until surface	Sulphides–white
Citric acid	3 g	blue/violet	Ferrite and carbides } blue/violet
Sodium nitrite	0.2 g		
Water	100 ml		
3 Sodium thiosulphate	24 g	Pre-etch in 2% Nital	
Citric acid	3 g	(a) Immerse 20–40 s	(a) Ferrite–red/violet
Cadmium chloride	2 g	(b) Immerse 50–90 s	(b) Phosphides–orange/brown
Water	100 ml		Carbides–blue/violet Ferrite–yellow

Table 10.8 COLOUR ETCHES FOR ALUMINIUM ALLOYS

Reagent		*Procedure*	*Comments*
1 Sodium molybdate	3 g	Immersion by trial	AlCuFeMn script–blue
Ammonium bifluoride	2 g		FeSi Al–brown/blue
HCl (35%)	5 ml		Ni_3Al, FeNiAlg–brown
Water	100 ml		$CuAl_2$–pale blue
2 Chromic trioxide	20 g	Immerse 5–30 s	If alloy susceptible to intergranular corrosion, grain boundaries coloured black
Sulphuric acid (1.84)	20 g		
Ammonium bifluoride	5 g		
Water	100 ml		
3 20% sulphuric acid (1.84)		Immerse 30 s at 70 °C	AlCuFeMn–brown
Water	80 ml		Mg_2Si–brown to black Outlines other phases
4 Sodium hydroxide	10%	Immerse 5 s at 70 °C	$CuAl_2$ } brown
Water	90 ml		Mg_2Si } brown

Table 10.9 ETCHING REAGENTS FOR DISLOCATIONS
(Taken largely from Lovell, Vogel and Wernick[61])

Metal or alloy	*Reagent**		*Remarks*
Aluminium (99.99%) (*see* ref. 62)	Hydrochloric acid Nitric acid Hydrofluoric acid	50% 4% 3%	—
	Hydrochloric acid Nitric acid Hydrofluoric acid	50 ml 47 ml 3 ml	Lacombe and Beaujard's reagent[69, 70]
	Hydrofluoric acid Hydrochloric acid Nitric acid Hydrogen peroxide (29% w/v)	37 ml 18 ml 9 ml 36 ml	Make: *A*, 49: 51 HF: H_2O_2 *B*, 65: 35 HCl: HNO_3 Mix in ratio A:B = 5:2, care required Keep at 0–15 °C in use (ref. 62)
Antimony	Hydrofluoric acid Nitric acid Acetic acid Bromine	3 pts 5 pts 3 pts 3 drops	Electrolytic etch on cleaved surface. 2–3 s
	Hydrofluoric acid Superoxol†	1 pt 1 pt	Electrolytic etch. 1 s
Bismuth	1% Iodine in methyl alcohol		Cleaved surface. 15 s
Brass (65% Cu–35% Zn)	0.2% Sodium thiosulphate		Electrolytic etch. 10 A dm^{-2} 18–20°C. Remove film with hydrochloric acid
Brass (Alpha)[63]	Saturated aqueous ferric chloride Hydrochloric acid	50 ml 2 drops	—
Brass (Beta)[63]	Saturated aqueous ammonium molybdate Hydrochloric acid	30 ml 6–7 drops	Immerse the electrolytically polished surface for 30 min
Columbium (Niobium)	Sulphuric acid Hydrofluoric acid Water Superoxol†	10 ml 10 ml 10 ml a few drops	Agitate specimen in solution
Copper (pure)	Saturated ferric chloride solution Hydrochloric acid Acetic acid Bromine	4 pts 4 pts 1 pt a few drops	Rinse in ammonia solution *See* Ruff[64] for further solutions and references
Germanium (also 0.2 at. % boron, 6.0 at. % silicon, 0.2 at. % tin)	Hydrofluoric acid Nitric acid Acetic acid Bromine	3 pts 5 pts 3 pts 3 drops	3–5 s. Polish etch. 600-grit carborundum ground surface
Germanium	Potassium ferricyanide Potassium hydroxide Water	8 g 12 g 100 ml	600-grit carborundum ground surface. 2–5 min. Boiling solution
Iron (99.96%)	4% metanitrobenzosulphonic acid in ethyl alcohol		Long etch. Rinse in alcohol. Result questionable
Iron	(a) 1% nitric acid in ethyl alcohol (b) 0.5% picric acid in methyl alcohol		1 min in (a) followed by rinse in methyl alcohol and 5 min in (b). Pits appear only in specimen cooled slowly from 750° to 800°C‡
	Fry's reagent Table 10.1, C(f)		10 s etch of chemically polished surface

* Acids are concentrated, unless otherwise indicated.
† Superoxol contains hydrogen peroxide (30%) 1 pt, hydrofluoric acid (40%) 1 pt, water 4 pts.

Table 10.9 ETCHING REAGENTS FOR DISLOCATIONS—*continued*

Metal or alloy	*Reagent**		*Remarks*
Iron (99.99%)	Disa Electropol solution A–Z		Electrolytic etch. Observation by electron microscopy
Iron	2% Nital containing 2% of saturated picral		15 min
	Saturated picral		4 min. Anneal to decorate dislocations‡
Iron-silicon (3.25 Si)	Acetic acid Chromium trioxide Water	133 ml 25 g 7 ml	Electrolytic etch. 3 A dm^{-2}. 17–19°C. Decorate with 0.004% carbon at 770°C or above in low-pressure acetylene atmosphere‡
Nickel–manganese	Orthophosphoric acid Ethyl alcohol	100 ml 100 ml	Electrolytic etch. 2 min. 200 A dm^{-2} Copper cathode 40°C
Silicon	Hydrofluoric acid Nitric acid Acetic acid	1 pt 3 pts 12 pts	15 min or longer. Chemically polished surface
	Hydrofluoric acid Nitric acid 3% aqueous mercuric nitrate	4 pts 2 pts 4 pts	600-grit carborundum ground surface
	Hydrofluoric acid Nitric acid Acetic acid 3% aqueous mercuric nitrate	3 pts 5 pts 3 pts 1.5–2 pts	600-grit carborundum ground surface. Use de-ionized water. Utensils and specimen must be dry. 2 min
	Hydrofluoric acid Nitric acid Water Silver nitrate	160 ml 80 ml 160 ml 8 g	600-grit carborundum ground surface
	Hydrofluoric acid Chromium trioxide water	 50 g per 100 g	Use in ratio: 2:1 by vol. for large etch pits 1:1 for medium etch pits 2:3 for small etch pits Time 15 s[65]
Tellurium	Hydrofluoric acid Nitric acid Acetic acid	3 pts 5 pts 6 pts	1 min etch. Cleaved surface
Zinc	Chromium trioxide Hydrated sodium sulphate Water	160 g 50 g 500 ml	Immerse with mild agitation for 1 min. Chemically polish before etching. Dip in solution of 320 g chromium trioxide per litre after etching, to remove stain. Decorate with 0.1 atomic %cadmium. Anneal at 300–400°C†. Age 1 week at room temperature
Zinc with 0.002% tin[66]	Saturated aqueous ammonium tungsgtate Saturated aqueous ammonium molybdate Hydrochloric acid	35 ml 5 drops 5 drops	Etch by immersing for about 5 min. Agitate to remove adherent layer. Quench from 400°C and anneal 100–400°C‡

* Acids are concentrated, unless otherwise indicated.
‡ Note that this heat treatment must alter the dislocation structure.

Etching to Reveal Magnetic Domains. See under IRON AND STEEL, *Microferrographic Technique*, p. **10**–39.

TAPER SECTIONING

Taper-sectioning[25,26] is used to give an apparent magnification of 10 times by using a plane of section to cut the surface at a shallow angle, typically 6° (i.e. $\sin^{-1} \frac{1}{10}$) so that the vertical features of the surface are given an apparent relative magnification of about 10 times.

Taper sectioning requires plating of the surface to be examined, followed by mounting in plastic at a suitable angle. The plating may conveniently be done with nickel or copper from conventional plating baths. Assuming that the surface is ready to be examined, it must not be altered during the plating procedure; cleaning before plating must be confined to methods such as solvent washing and cathodic alkaline degreasing, which have no appreciable effect on the particular surface under examination. The mounting procedure is evident from Figure 10.3.

This method may also be used for examining thin intermediate layers, for example in electroplating or in the study of diffusion couples.

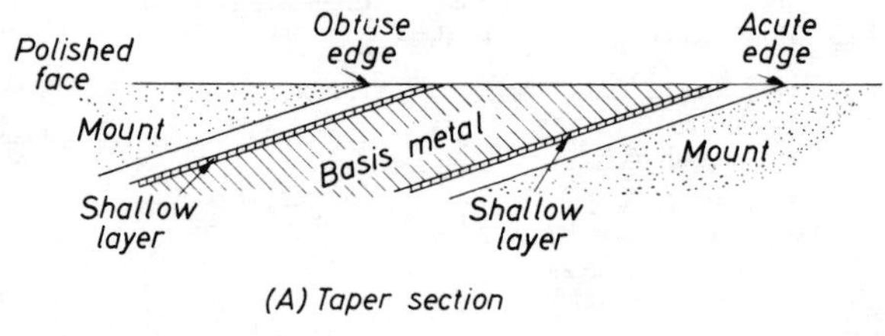

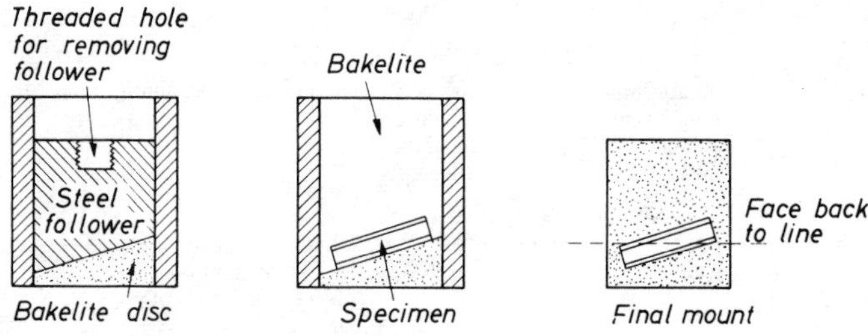

Figure 10.3 *Taper sectioning (from Vickers Projection Microscope Handbook)*

10.3 Metallographic methods for specific metals

10.3.1 Aluminium

PREPARATION

Aluminium and its alloys are soft and easily scratched or distorted during preparation. For cutting specimens, sharp saw-blades should be used with light pressure to avoid local overheating. Specimens may be ground on emery papers by the usual methods, but the papers should preferably have been already well used, and lubricated or coated with a paraffin oil ('white spirit' is suitable), paraffin wax or a solution of paraffin wax in paraffin oil. Silicon carbide papers (down to 600-grit) which can be well washed with water are preferred for harder alloys, the essential point being to avoid the embedding of abrasive particles in the metal. For pure soft aluminium, a high viscosity paraffin is needed to avoid this. Polishing is carried out in two stages: initial polishing with fine α-alumina, proprietary metal polish, or diamond, and final polishing with γ-alumina or fine magnesia, using a slowly rotating wheel (not above 150 rev. min^{-1}). It is essential to use properly graded or levigated abrasives and it is preferable to use distilled water only; it is an advantage to boil new polishing cloths in water for some hours in order to soften the fibres. Many aluminium alloys contain hard particles of various intermetallic compounds, and polishing times should in general be as short as possible owing to the danger of producing excessive relief. Relief may be minimized by experience and skill in polishing; blanket felt may with advantage be substituted for velveteen or selvyt cloth as a polishing pad, while the use of parachute silk on a cork pad is also useful for avoiding relief in the initial stages of the process, but a better general alternative is to use diamond polishing, followed by a very brief final polishing with magnesia.

Many aluminium alloys contain the reactive compound Mg_2Si. If this constituent is suspected, white spirit should be substituted for water during all but the initial stages of wet polishing, to avoid loss of the reactive particles by corrosion.

The microtome technique (*see* 10.3.11, *Lead*, p. **10**–41) has been successfully applied to the softer aluminium alloys.

It should be noted that some aluminium alloys are liable to undergo precipitation reactions at the temperatures used to cure thermosetting mounting resins; this applies particularly to aluminium–magnesium alloys, in which grain boundary precipitates may be induced.

Electropolishing is often rapid and convenient (*see* Tables 10.4(a) and 10.4(b).

ETCHING

The range of aluminium alloys now in use contains many complex alloy systems. A relatively large number of etching reagents have therefore been developed, and only those whose use has become more or less standard practice are given in Table 10.11. Many etches are designed to render the distinction between the many possible microconstituents easier, and the type of etching often depends on the magnification to be used. The identification of constituents, which is best accomplished by using cast specimens where possible, depends to a large extent on distinguishing between the colours of particles, so that the illumination should be as near as possible to daylight quality. It is recommended that a set of specially prepared standard specimens, containing various known metallographic constituents, be used for comparison.

It is very easy to obtain anomalous etching effects, such as ranges of colour in certain types of particles, and carefully standardized procedure is necessary. It should be remembered that the form and colour of the microconstituents may vary according to the degree of dispersion brought about by mechanical treatments, and also that the etching characteristics of a constituent may vary according to the nature of the other constituents present in the same section.

Some etching reagents for aluminium require the use of a high temperature; in such cases the specimen should be preheated to this temperature by immersion in hot water before etching. For washing purposes, a liberal stream of running water is advisable.

Table 10.10 MICROCONSTITUENTS WHICH MAY BE ENCOUNTERED IN ALUMINIUM ALLOYS

Microconstituent	*Appearance in unetched polished sections*
Al_3Mg_2	Faint, white. Difficult to distinguish from the matrix.
Mg_2Si	Slate grey to blue. Readily tarnishes on exposure to air and may show irridescent colour effects. Often brown if poorly prepared. Forms Chinese script eutectic.
$CaSi_2$	Grey. Easily tarnished
$CuAl_2$	Whitish, with pink tinge. A little in relief; usually rounded
$NiAl_3$	Light grey, with a purplish pink tinge
Co_2Al_9	Light grey
$FeAl_3$[1]	Lavender to purplish grey; parallel-sided blades with longitudinal markings
$MnAl_6$	Flat grey. The other constituents of binary aluminium–manganese alloys ($MnAl_4$, $MnAl_3$ and 'δ') are also grey and appear progressively darker. May form hollow parallelograms
$CrAl_7$	Whitish grey; polygonal. Rarely attacked by etches
Silicon	Slate grey. Hard, and in relief. Often primary with polygonal shape—use etch to outline
α(AlMnSi)[2]	Light grey, darker and more buff than $MnAl_6$
β(AlMnSi)[2]	Darker than α(AlMnSi), with a more bluish grey tint. Usually occurs in long needles
Al_2CuMg	Like $CuAl_2$ but with bluish tinge
Al_6Mg_4Cu	Flat, faint and similar to matrix

Table 10.10 MICROCONSTITUENTS WHICH MAY BE ENCOUNTERED IN ALUMINIUM ALLOYS —*continued*

Microconstituent	*Appearance in unetched polished sections*
(AlCuMn)[3]	Grey
α(AlFeSi)[4]	Purplish grey. Often occurs in Chinese-script formation. Isomorphous with α(AlMnSi)
β(AlFeSi)[4]	Light grey. Usually has a needle-like formation
(AlCuFe)[5]	Grey α phase lighter than β phase (*see* Note 5)
(AlFeMn)[6]	Flat grey, like $MnAl_6$
(AlCuNi)	Purplish grey
(AlFeSiMg)[7]	Pearly grey
$FeNiAl_9$	Very similar to and difficult to distinguish from $NiAl_3$
(AlCuFeMn)	Light grey
$Ni_4Mn_{11}Al_{60}$	Purplish grey
$MgZn_2$	Faint white; no relief

In Table 10.10, constituents are designated by symbols denoting the compositions upon which they appear to be based, or by the elements, in parentheses, of which they are composed. The latter nomenclature is adopted where the composition is unknown, not fully established, or markedly variable. The superscript numbers in column 1 refer to the following notes:

(1) On very slow cooling under some conditions, $FeAl_3$ decomposes into Fe_2Al_7 and Fe_2Al_5. The former is micrographically indistinguishable from $FeAl_3$. The simpler formula is retained for consistency with most of the original literature.

(2) α(AlMnSi) is present in all slowly solidified aluminium–manganese–silicon alloys containing more than 0.3% of manganese and 0.2% of silicon, while β(AlMnSi), a different ternary compound, occurs above approximately 3% of manganese for alloys containing more than approximately 1.5% of silicon. α(AlMnSi) has a variable composition in the region of 30% of manganese and 10–15% of silicon. The composition of β(AlMnSi) is around 35% of manganese and 5–10% of silicon.

(3) (AlCuMn) is a ternary compound with a relatively large range of homogeneity based on the composition $Cu_2Mn_3Al_{20}$.

(4) α(AlFeSi) may contain approximately 30% of iron and 8% of silicon, while β(AlFeSi) may contain approximately 27% of iron and 15% of silicon. Both constituents may occur at low percentages of iron and silicon.

(5) The composition of this phase is uncertain. Two ternary phases exist. α(AlCuFe) resembles $FeAl_3$; β(AlCuFe) forms long needles.

(6) The phase denoted as (AlFeMn) is a solid solution of iron in $MnAl_6$.

(7) This constituent is only likely to be observed at high silicon contents.

Table 10.11 ETCHING REAGENTS FOR ALUMINIUM AND ITS ALLOYS

No.	*Reagent*		*Remarks*	
1	Hydrofluoric acid (40%)	0.5 ml	15 s immersion is recommended. Particles of all common micro-constituents are outlined. Colour indications:	
	Hydrochloric acid (1.19)	1.5 ml		
	Nitric acid (1.40)	2.5 ml	Mg_2Si and $CaSi_2$:	blue to brown
	Water	95.5 ml	α(AlFeSi) and (AlFeMn):	darkened
			β(AlCuFe):	light brown
	(Keller's etch)†		$MgZn_2$, $NiAl_3$, (AlCuFeMn), Al_2CuMg and Al_6CuMg:	brown to black
			α(AlCuFe) and (AlCuMn):	blackened
			Al_3Mg_2:	heavily outlined and pitted
			The colours of other constituents are little altered. Not good for high Si alloys	
2	Hydrofluoric acid (40%)	0.5 ml	15 s swabbing is recommended. This reagent removes surface flowed layers, and reveals small particles of constituents, which are usually fairly heavily outlined. There is little grain contrast in the matrix.	
	Water	99.5 ml		

* These are isomorphous and the colour depends on the proportion of Mn and Fe.
† Sodium fluoride can be used in place of HF in mixed acid etches.

Table 10.11 ETCHING REAGENTS FOR ALUMINIUM AND ITS ALLOYS—*continued*

No.	*Reagent*		*Remarks*	
2 (*cont.*)			Colour indications:	
			Mg_2Si and $CaSi_2$:	blue
			$FeAl_3$ and $MnAl_6$:	slightly darkened
			$NiAl_3$:	brown (irregular)
			α(AlFeSi):	dull brown
			(AlCrFe):	light brown
			Co_2Al_9:	dark brown
			(AlFeMn):	brownish tinge
			α(AlCuFe), (AlCuMg) and (AlCuMn):	blackened
			α(AlMnSi), β(AlMnSi) and (AlCuFeMn) may appear light brown to black	
			β(AlFeSi) is coloured red brown to black	
			The remaining possible constituents are little affected	
3	Sulphuric acid (1.84)	20 ml	30 s immersion at 70 °C; the specimen is quenched in cold water.	
	Water	80 ml	Colour indications:	
			Mg_2Si, Al_3Mg_2 and $FeAl_3$:	violently attacked, blackened and may be dissolved out
			$CaSi_2$:	blue
			α(AlMnSi) and β(AlMnSi):	rough and attacked
			$NiAl_3$ and (AlCuNi):	slightly darkened
			β(AlFeSi):	slightly darkened and pitted
			α(AlFeSi), (AlCuMg) and (AlCuFeMn): outlined and blackened	
			Other constituents are not markedly affected	
4	Nitric acid (1.40)	25 ml	Specimens are immersed for 40 s at 70 °C and quenched in cold water.	
	Water	75 ml	Most constituents (not $MnAl_6$) are outlined. Colour indications:	
			β(AlCuFe) is slightly darkened	
			Al_3Mg_2 and AlMnSi: attacked and darkened slightly	
			Mg_2Si, $CuAl_2$, (AlCuNi) and (AlCuMg) are coloured brown to black	
5	Sodium hydroxide	1 g	Specimens are etched by swabbing for 10 s. All usual constituents are heavily outlined, except for Al_3Mg_2 (which may be lightly outlined) and (AlCrFe) which is both unattacked and uncoloured. Colour indications:	
	Water	99 ml	$FeAl_3$ and $NiAl_3$:	slightly darkened
			(AlCuMg):	light brown
			α(AlFeSi):	dull brown*
			α(AlMnSi):	rough and attacked; slightly darkened*
			$MnAl_6$ and (AlFeMn):	coloured brown to blue (uneven attack)
			$MnAl_4$:	tends to be darkened
			The colours of other constituents are only slightly altered	
6	Sodium hydroxide	10 g	Specimens immersed for 5 s at 70 °C, and quenched in cold water. Colour indications:	
	Water	90 ml	β(AlFeSi):	slightly darkened
			$Mn_{11}Ni_4Al_{60}$:	light brown
			β(AlCuFe):	light brown and pitted
			$CuAl_2$:	light to dark brown
			$FeAl_3$:	dark brown
			($FeAl_3$ is more rapidly attacked in the presence of $CuAl_2$ than when alone)	
			$MnAl_6$, $NiAl_3$, (AlFeMn), $CrAl_7$ and AlCrFe:	blue to brown
			α(AlFeSi), α(AlCuFe), $CaSi_2$ and (AlCuMn):	blackened

* These are isomorphous and the colour depends on the proportion of Mn and Fe.
† Sodium fluoride can be used in place of HF in mixed acid etches.

Table 10.11 ETCHING REAGENTS FOR ALUMINIUM AND ITS ALLOYS—*continued*

No.	*Reagent*		*Remarks*
7	Sodium hydroxide Sodium carbonate (in water)	3%–5% 3%–5%	Useful for sensitive etching where reproducibility is essential. In general, the effects are similar to those of Reagent 5, but the tendency towards colour variations for a given constituent is diminished. Particularly useful for distinguishing $FeNiAl_9$ (dark blue) from $NiAl_3$ (brown). Potassium salts can be used.
8	Nitric acid Hydrofluoric acid Glycerol	20 ml 20 ml 60 ml	A reliable reagent for grain boundary etching, especially if the alternate polish and etch technique is adopted. The colours of particles are somewhat accentuated
9	Nitric acid, 1% to 10% by vol. in alcohol		Recommended for aluminium–magnesium alloys. Al_3Mg_2 is coloured brown. 5–20% chromium trioxide can be used
10	Picric acid Water	4 g 96 ml	Etching for 10 min darkens $CuAl_2$, leaving other constituents unaffected. Like reagent 4
11	Orthophosphoric acid Water	9 ml 91 ml	The reagent is used cold. Recommended for aluminium–magnesium alloys in which it darkens any grain boundaries containing thin β-precipitates. Specimen is immersed for a long period (up to 30 min). Mg_2Si is coloured black, Al_3Mg_2 a light grey, and the ternary (AlMnFe) phase a dark grey
12	Nitric acid		10 s immersion colours Al_6CuMg_4 greenish brown and distinguishes it from Al_2CuMg, which is slightly outlined but not otherwise affected
13	Nitric acid (density 1.2) Water Ammonium molybdate, $(NH_4)_6Mo_7O_{24}$, $4H_2O$	20 ml 20 ml 3 g	20 ml of reagent are mixed with 80 ml alcohol. Specimens are immersed, and well washed with alcohol after etching. Brilliant and characteristic colours are developed on particles of intermetallic compounds. The effects depend on the duration of etching, and for differentiation purposes standardisation against known specimens is advised
14	Sodium hydroxide (various strengths, with 1 ml of zinc chloride per 100 ml of solution)		Generally useful for revealing the grain structure of commercial aluminium alloy sheet[67]
15	Hydrochloric acid (37%) Hydrofluoric acid (38%) Water	15.3 ml 7.7 ml 77.0 ml	Recommended (30 s immersion at room temperature) for testing the diffusion of copper through claddings of aluminium, aluminium–manganese–silicon, or aluminium–manganese on aluminium–copper–magnesium sheet. Zinc contents up to 2% in the clad material do not influence the result[68]
16	Ammonium oxalate Ammonium hydroxide, 15% in water	1 g 100 ml	Develops grain boundaries in aluminium–magnesium–silicon alloys. Specimens are etched for 5 min at 80 °C in a solution freshly prepared for each experiment

* These are isomorphous and the colour depends on the proportion of Mn and Fe.
† Sodium fluoride can be used in place of HF in mixed acid etches.

Electrolytic etching for aluminium alloys. In addition to the reagents given for aluminium in Table 10.11, the following solutions have been found useful for a restricted range of aluminium-rich alloys:

1. The following solution has been used for grain orientation studies:

Orthophosphoric acid (density 1.65)	53 ml
Distilled water	26 ml
Diethylene glycol monoethyl ether	20 ml
Hydrofluoric acid (48%)	1 ml

The specimen should be at room temperature and electrolysis is carried out at 40 V and less than 0.1 A dm^{-2}. An etching time of 1.5–2 min is sufficient for producing grain contrast in polarized light after electropolishing.

2. The solution below is also used for the same purpose and is more reliable for some alloys:

Ethyl alcohol	49 ml
Water	49 ml
Hydrofluoric acid	2 ml (quantity not critical)

The specimen is anodized in this solution at 30 V for 2 min at room temperature. A glass dish must be used. Not suitable for high-copper alloys.

3. For aluminium alloys containing up to 7% of magnesium:

Nitric acid (density 1.42)	2 ml
40% hydrofluoric acid	0.1 ml
Water	98 ml

Electrolysis is carried out at a current density of $0.3\,A\,dm^{-2}$ and a potential of 2 V. The specimen is placed 7.6 cm from a carbon cathode.

4. For cast duralumin:

Citric acid	100 g
Hydrochloric acid	3 ml
Ethyl alcohol	20 ml
Water	977 ml

Electrolysis is carried out at $0.2\,A\,dm^{-2}$ and a potential of 12 V.

5. For commercial aluminium:

Hydrofluoric acid (40%)	10 ml
Glycerol	55 ml
Water	35 ml

This reagent, used for 5 min at room temperature, with a current density of $1.5\,A\,dm^{-2}$ and a voltage of 7–8 V, is suitable for revealing the grain structure after electropolishing.[72]

6. For distinguishing between the phases present in aluminium-rich aluminium–copper–magnesium alloys, electrolytic etching in either ammonium molybdate solution or 0.880 ammonia has been recommended. In both cases, Al_2CuMg is hardly affected, $CuAl_2$ is blackened, Al_6Mg_4Cu is coloured brown, while Mg_2Al_3 is thrown into relief without change of colour.[73]

GRAIN-COLOURING ETCH[74]

For many aluminium alloys containing copper, and especially for binary aluminium-copper alloys, it is found that Reagent No. 1 of Table 10.11 gives copper films on cubic faces which are subject to preferential attack and greater roughening of the surface. Subsequent etching with 1% caustic soda solution converts the copper into bronze-coloured cuprous oxide, and a brilliant and contrasting representation of the underlying surfaces is obtained. The technique is of use in orientation studies in so far as the films are dark and unbroken on (100) surfaces, but shrink on drying on other surfaces. In particular (111) faces have a bright yellow colour with a fine network on drying, which has no preferred orientation, while (110) faces develop lines (cracks in film) which are parallel to a cube edge.

ETCHING TO PRODUCE ETCH PITS[69]

The orientation of crystal grains may be determined by developed etch pits on specimens previously electropolished (*see* Table 10.4). The following reagents develop pits with facets parallel to the (100) cube planes:

	a	*b*	*c*
Fuming nitric acid	15 ml	15 ml	47 ml
Pure hydrochloric acid	45 ml	46 ml	50 ml
Hydrofluoric acid	15 ml	10 ml	3 ml
Distilled water	25 ml	29 ml	—

Mixture *c* is recommended where only a few large well-formed etch figures are preferred.
Etching with dry gaseous hydrochloric acid gives pits with facets parallel to (111) planes.

10.3.2 Antimony and bismuth

PREPARATION

Antimony as usually obtained is hard and brittle, and easily breaks up into fragments. Mounting in a suitably hard mounting medium is thus generally necessary, after which the usual methods are applied.

ETCHING

1. (a)	Water HNO_3 (1.40)	22 ml 8 ml	Mix equal quantities of (a) and (b) immediately before use ~1 min	Sb and Sb–Bi alloys
(b)	Water Ammonium molybdate	30 ml 4.5 g		
2.	Water Citric acid Ammonium molybdate	100 ml 25 g 10 g	~1 min	Grain structure of Sb and Bi alloys

10.3.3 Beryllium

PREPARATION

Conventional methods may be employed, with due regard to the toxicity of beryllium. All operations which produce Be dust must be done in a glove box. Do not inhale any beryllium particles. The metal easily twins mechanically, and light hand grinding on lubricated abrasive papers is preferred. A suspension of alumina in 5% aqueous oxalic acid may be used as polishing medium on worsted serge or other short-nap cloth (cloths with longer nap may lead to pitting). The abrasive and oxalic acid are placed on the cloth before polishing begins, and the oxalic acid solution only is added as polishing proceeds. If there are soft constituents present, iron oxide (Fe_2O_3) may with advantage be substituted for alumina. Alternatively, magnesia in 30% hydrogen peroxide may be used as the polishing medium.

Other methods of preparation have been discussed,[43,76] including *electropolishing* (*see* Table 10.4).

ETCHING

1.	Hydrofluoric acid (40%) Ethyl alcohol	10 ml 90 ml

On immersion for 10–30 s microconstituents are in general outlined, and some colour differentiations are observed. The identifications of constituents commonly met with in commercial beryllium, both in the etched and unetched condition, are as shown in Table 10.12.

2.	Water Sulphuric acid (1.84)	95 ml 5 ml	1–15 s	Be alloys
3.	Water Oxalic acid	100 ml 10 g	2–16 min	Outlines precipitates first then grain boundaries. Better used electrolytically

Table 10.12 MICROCONSTITUENTS IN COMMERCIAL BERYLLIUM

Constituent	*Appearance unetched*	*Effect of etching*
Carbide	Hard angular grey particles, staining in air to many colours, finally brown	No effect on colour
Silicon-rich phase	Light blue grey	Not stained, but outlined
Nitride	Needle-like particles, darker grey than carbide	Unaffected
$TiBe_{12}$	Angular particles, slightly pink	Unaffected
$CaBe_{13}$	Light yellow, usually angular	Unaffected
$FeBe_{12}$	Almost invisible unetched	Coloured reddish brown
$MnBe_{12}$	Similar to iron, but slightly pink	Outlined
Boride	Reddish particles in grain boundaries	Coloured blue or purple
$MoBe_{22}$	Almost invisible unetched	Coloured chocolate brown
UBe_{13}	Invisible unetched	Phase revealed as a yellow or green dendritic structure
Aluminium	Bright yellow, often appearing speckled. Soft and difficult to polish	Outlined
Magnesium	Very soft, and white in appearance	Removed
$ZrBe_{13}$	Barely distinguishable dendritic phase	Outlined and coloured light blue

Beryllium can also be examined by polarized light.

10.3.4 Cadmium

PREPARATION

The preparation of cadmium for metallographic examination should be carried out in the same manner as for zinc (*see Zinc*, p. **10**–57). The metal and its alloys are soft, and very liable to mechanical twinning on deformation. *Electrolytic polishing* is satisfactory[25,43] (*see* Table 10.4).

ETCHING

	Reagent		*Conditions*	*Remarks*
1.	Ethanol	2 ml	Few seconds	Most alloys
	Nitric acid (1.40)	98 ml	to a minute	Thallium also
2.	Water	100 ml	Up to a	Eutectics of
	Hydrochloric acid (1.19)	25 ml	minute	Cd
	Ferric chloride	8 g		
3.	Water	40 ml		
	Hydrofluoric acid (40%)	10 ml	5–10 s	Most Cd alloys also thallium
	Hydrogen peroxide (30%)	10 ml		and indium

10.3.5 Chromium

PREPARATION

Normal methods may be applied successfully to chromium and its alloys, and a final electrolytic polish or attack polish is recommended (*see* Tables 10.3 and 10.4). Hard or decorative chromium plate may be mounted in hard plastic and prepared in section by grinding in the usual way, finishing with lead laps and finally with diamond polishing.

ETCHING

Table 10.13 ETCHING REAGENTS FOR CHROMIUM

No.	*Reagent**		*Remarks*
1	Dilute hydrofluoric acid		After electropolishing with the reagent listed in Table 10.4, the specimen is agitated for a few seconds in dilute hydrofluoric acid
2	Hydrochloric acid (concentrated)		Shows striations in electrodeposits
3	Nitric acid Hydrochloric acid Glycerol	10 ml 20 ml 30 ml	Suitable for alloys. Also used electrolytically; specimen is made anode at 4 V, 45 s. Proportions 1:3:2 also used. This decomposes on standing, especially in contact with stainless steel. *Beware!*
4	Sulphuric acid (10%)		Used hot with swabbing

* Acids are concentrated, unless otherwise indicated.

10.3.6 Cobalt

PREPARATION

Normal methods, carefully applied, will suffice.

ETCHING

Cobalt may be etched with some of the reagents used for nickel and iron alloys, but some alloys are strongly resistant to attack. *See* Table 10.14.

Table 10.14 ETCHING REAGENTS FOR COBALT AND ITS ALLOYS

No.	*Reagent**		*Remarks*
1	Hydrochloric acid Nitric acid Acetic acid Water	60 ml 15 ml 15 ml 15 ml	The solution should be aged for 1 hour. Grain boundaries and the general structure of alloys are revealed. May also be used electrolytically. Of wide application, including hard metals
2	Nitric acid Hydrochloric acid Glycerol	10 ml 20 ml 30 ml	Reveals general structure. May be used electrolytically (*see* Table 10.13, Reagent 3. Same precautions
3	Potassium ferricyanide Potassium hydroxide Water	10 g 10 g 100 ml	For hard cobalt–chromium alloys containing carbon. Used at approx. 70 °C, 10–20 s. Can be replaced by electrolytic 3% KOH. Suitable for carbides
4	Chromium trioxide Sulphuric acid Water	2–10 g 10 ml 90 ml	Used electrolytically, the specimen being made the anode at 6 V

* Acids are concentrated, unless otherwise indicated.

10.3.7 Copper

PREPARATION

In general, preparation of copper and copper alloys by the usual techniques calls for no extra or special precautions. Copper itself is relatively soft, so that care is necessary to avoid excessive straining. Abrasive papers must be lubricated. Final polishing is normally carried out with slightly ammoniacal preparations, giving polish attack. Examine initially *unetched* (Table 10.15 No. 32).

ETCHING

Most homogeneous copper alloys have much the same etching characteristics as copper, and the same etching reagents as are used for copper may be tried. Alloys containing more than one phase usually etch easily, but careful attention must be paid to the time of etching. When using chromic acid solutions, sulphate ions should preferably be absent, as their presence leads to uneven action. In general, etches comprise (a) ammonia plus oxidizing agent, (b) acidified chromic acid or dichromates; the latter is good as an electrolytic etch.

Table 10.15 ETCHING REAGENTS FOR COPPER AND ITS ALLOYS

Principal use	*No.*	*Reagent*		*Remarks*
Copper, copper alloys in general. Brass, bronze and nickel–silver	*1*	Ammonium hydroxide Water Hydrogen peroxide (30 vol.)	50 ml 50 ml 20 ml	Used for copper, and many copper-rich alloys. Gives a grain boundary etch, and also tends to darken the α solid solution, leaving the β solid solution lighter. The hydrogen peroxide content may be varied. Less is required the lower the copper content (*see* Reagent 2)
	2	Ammonium hydroxide Water Hydrogen peroxide (30 vol.)	50 ml 50 ml 10 ml	Used for bronze, 70:30 and 60:40 brasses; this etch may with advantage be followed by a ferric chloride etch (Reagent 6) to darken the β areas in duplex alloys
	3	Ammonium hydroxide Water Ammonium persulphate, $2\frac{1}{2}\%$ solution	10 ml 10 ml 10 ml	Recommended for polish attack on copper and copper alloys
	4	Ammonium persulphate Water	10 g 100 ml	Used cold or boiling for copper, brass, bronze, nickel-silver and aluminium-bronze. Tends to produce relief effects. Good for aluminium bronzes followed by Reagent 23 to darken β martensite (retained β is pink) and then Reagent 24 to etch $\gamma_2(\delta)$ brown
	5	Dilute ammonium hydroxide solution (10–50%)		May be used for polish or swab attack on brass and bronze. The oxidising action of atmospheric oxygen is necessary for the process
	6	Ferric chloride, various strengths and compositions. To 100 parts of water are added: Hydrochloric acid (1.19): 20, 10, 50, 25, 6, 25, 1, 10	Ferric chloride (g): 1, 5, 5, 8, 19, 25, 10†, 3‡	Used as a general reagent for copper, brass, bronze, nickel–silver, aluminium–bronze and other copper-rich alloys. It darkens the β constituent in brasses and gives grain contrast following ammoniacal or chromic acid etches. The most suitable composition should be found by trial and error in specific cases. This reagent generally emphasizes scratches in imperfectly prepared specimens, and tends to roughen the surface. For sensitive work it is frequently a great advantage to replace the water in the reagent by a 50:50 water–alcohol mixture or by pure alcohol

† Usually used with 1 part of chromic acid CrO_3.
‡ Used with 1 part of cupric chloride and 0.05 parts of stannous chloride.

Table 10.15 ETCHING REAGENTS FOR COPPER AND ITS ALLOYS—*continued*

Principal use	*No.*	*Reagent*	*Remarks*
Copper, copper alloys in general. Brass, bronze and nickel–silver (continued)	*7*	Ethyl alcohol (commercial) 96 ml Ferric chloride (anhydrous) 59 g Hydrochloric acid 2 ml	Dilute 5:1 with alcohol. Wash with alcohol or acetone. More delicate and controllable than aqueous solutions
	8	Chromic acid CrO_3 saturated in water	Used for copper, brass, bronze and nickel-silver. The etching time is 1 to 1½ min. Grain boundaries are attacked and β constituents are coloured pale yellow, while the primary solid solution is coloured dark yellow
	9	Chromic acid, 100–150 g l^{-1} of water, 1–2 drops of hydrochloric acid are added immediately before use to a 50 ml portion	Used for copper, brass, bronze, nickel-silver, and recommended for revealing the silicides of Ni, Co, Cr and Fe in certain alloys. Alternate polishing and etching is recommended, when a grain contrast is obtained. The primary solid solution is not attacked. The β copper-zinc phase is coloured light yellow while the δ copper-tin phase is coloured brown to black. The reagent may be followed by a ferric chloride etch
	10	Chromic acid 25 g Nitric acid (1.40) 40 ml Water 35 ml	This reagent is useful for distinguishing the general constituents of many copper alloys. The γ and δ phases are particularly well shown up as shining blue crystals
	11	Sulphuric acid (1.84) 5 ml Potassium dichromate saturated in water 100 ml	Etching time ½ to 1 min. Suitable for copper, but attacks oxide inclusions strongly
	12	Nitric acid, various strengths	Used for any purpose requiring deep etching and the removal of a thick layer of surface material. The times of etching are short and difficult to control
	13	Sulphuric acid (1.84) 5 ml Hydrogen peroxide (10 vol.) 100 ml	Etching time 1 to 1½ min. Used for pure copper, but attacks oxide
	14	Silver nitrate, 100 g l^{-1} of water, following exposure to hydrogen sulphide	This technique has been used for 70:30 and 60:40 brasses, and produces satisfactory contrast
	15	Silver nitrate, 20 g l^{-1} of water	Immerse for 30 s, and wash off silver stain under water. Useful in certain cases for pure copper
	16	Ammonium hydroxide 10 ml Ammonium oxalate saturated in water 30 ml	Used for high-zinc brasses
	17	Ammonium hydroxide 10 ml Potassium arsenate saturated in water 30 ml	Used for high-zinc brasses
	18	Copper ammonium chloride. 100 g l^{-1} of water, plus ammonium hydroxide to slight alkalinity	Copper, brass and nickel-silver. Specially recommended for darkening large β areas in duplex brasses
	19	Ammonia 20 ml Potassium permanganate (0.4%) in water 30 ml	Used for pure copper. Etching time 2 to 3 min. Liable to produce staining
	20	Bromine water, saturated	Etching time 30 to 60 s. May be used satisfactorily with pure copper if the coating which forms is removed by washing in strong ammonia

Table 10.15 ETCHING REAGENTS FOR COPPER AND ITS ALLOYS—*continued*

Principal use	*No.*	*Reagent*		*Remarks*
Aluminium–bronze	*4*	See above		As for Reagent 4
	6	See above		As for Reagent 6. Darkens β
	10	See above		As for Reagent 10. Also generally useful for aluminium-bronzes
	21	Chromic acid Nitric acid (1.40) Water	20 g 50 ml 30 ml	Recommended for aluminium-bronze after pretreatment with 10% hydrofluoric acid solution in order to remove surface oxide films
	22	Chromic acid Nitric acid (1.40) Water	20 g 5 ml 75 ml	As for Reagent 21
	23	Nitric acid (1.40) Hydrogen peroxide (100 vol.)	0.5 ml 99 ml	Darkens martensite
	24	Sod. hydroxide Water	10 g 100 ml	γ_2 (δ) etched brown
Aluminium–bronze	*25*	Ferric nitrate Ammonium nitrate Nitric acid Water	20 g 20 g 2 ml 500 ml	Good for complex aluminium bronzes
	26	Ammonium hydroxide Water Hydrogen peroxide 3% solution	25 ml 25 ml 20 ml	Age 1 week in loosely stoppered bottle before use. To distinguish eutectoid ($\alpha+\gamma_2$) from acicular $\beta.\gamma_2$ attacked
Copper alloys with beryllium, silicon, manganese and chromium	*8*	—		As for Reagent 8. Is also useful for manganese bronze
	27	Potassium dichromate Water Sodium chloride (saturated) Sulphuric acid (s.g. 1.84)	2 g 100 ml 4 ml 8 ml	Used for copper, and copper alloys with beryllium, manganese and silicon. Also suitable for nickel–silver, bronzes and chromium–copper alloys. This reagent should be followed by a ferric chloride etch to give added contrast
High-nickel alloys	*28*	Nitric acid (s.g. 1.42) Glacial acetic acid Water	50 ml 25 ml 25 ml	Recommended for high-nickel alloys which might prove resistant to attack by the more usual reagents, and for bright copper electrodeposits. 1:1 nitric acid : acetic acid is also used
Copper–nickel–aluminium	*29*	Acetic acid, 75% Nitric acid Acetone	30 ml 20 ml 30 ml	Recommended for copper-rich copper–nickel–aluminium alloys. NiAl shows as dove-grey rectangular needles; Ni_3Al is globular and darker grey. The γ_2 phase of the aluminium–copper system is a pale grey; *see also* Reagent 24
Copper–silicon alloys	*30*	Hydrogen peroxide (30 vol.) Water Potassium hydroxide (20%) Ammonium hydroxide (s.g. 0.90)	20 ml 25 ml 5 ml 50 ml	Used specifically to distinguish the κ phase in copper–silicon alloys, and in ternary and more complex alloys based on this system.
Leaded copper and bearing metals of this material	*31*	Nitric acid Glacial acetic acid Glycerol	20 ml 20 ml 80 ml	General reagent for leaded copper, bronzes, etc. *See* below for further notes on leaded coppers. Darkens lead
	32	Trichloroacetic acid Water Ammonium hydroxide to make 100 ml	20 g 20 ml	As for Reagent 31. Etches and outlines lead constituent, bringing it into prominence. Monochloroacetic acid may be substituted

Table 10.15 ETCHING REAGENTS FOR COPPER AND ITS ALLOYS—*continued*

Principal use	*No.*	*Reagent*	*Remarks*
Macro-etching	*33*	Heat tinting	This technique occasionally gives useful results. Phosphor bronzes react the most favourably
	34	No etching	Several constituents which may be present may be observed in unetched sections. Thus the selenide and telluride appear blue grey, while the oxide appears blue. Metallic bismuth appears a very pale blue grey. Cuprous oxide, unless the particles are very small, appears ruby-red in polarised light or with dark-field illumination; selenide and telluride remain dark. Zinc oxide (in brass) appears transparent and anisotropic in polarised light, blue-grey with ordinary illumination

LEADED COPPER ALLOYS

Difficulties can arise in the preparation and examination of leaded alloys, owing to the different hardnesses of the constituents, and also because specimens are particularly subject to smearing of the lead over the copper matrix. The standard methods of preparation properly applied, i.e. hand grinding on lubricated papers, followed preferably by diamond polishing, are quite satisfactory. Lead particles may be observed on properly polished unetched surfaces; it is essential that any flowed polished layer should be removed before observation by etching (Reagent 9 is suitable) and the surface again lightly polished before examination. For the best results it may be advisable to repeat the alternate etching and polishing two or three times. If it is necessary to etch, Reagents 31 and 32 of Table 10.15 may be used to etch the lead, or ferric chloride or chromic acid reagents to attack the copper-rich matrix.

The dithizone process described under *Iron and steel* (p. **10**–38, **10**–39) is also applicable to brasses. (*See also Electrolytic etching*, below.)

Table 10.16 ELECTROLYTIC ETCHING OF COPPER ALLOYS

Principal use	*No.*	*Reagent*		*Remarks*
General	*1*	Ferrous sulphate Sodium hydroxide Sulphuric acid (1.84) Water	30 g 4 g 100 ml 1900 ml	May be used to develop contrast after the use of ammoniacal hydrogen peroxide reagents. β in brasses is darkened. A current density of 0.1 A dm^{-2} at 8–10 V is suitable
Cupro–nickel and nickel–silvers	*2*	Citric acid	100 g/l	Also useful for brasses
	3	Ammonium molybdate in excess of ammonia		Also useful for brasses
	4	Glacial acetic acid Nitric acid (1.40) Water	5 ml 10 ml 85 ml	This reagent tends to minimize the effect on the microstructure of the coring which usually occurs with these alloys
Brasses	*5*	Ammonium acetate 100 g l^{-1} of water		Current density 0.3 A dm^{-2}
	6	Ammonium sulphate 10 g l^{-1} of water		Current density 0.3 A dm^{-2}
	7	0.10 M ammonium acetate 0.50 M sodium thiosulphate 14 M ammonium hydroxide Distilled water	10 ml 30 ml 30 ml 30 ml	Carried out at 31 A dm^{-2}. The etching time is approximately inversely proportional to the copper content of the material

Table 10.16 ELECTROLYTIC ETCHING OF COPPER ALLOYS —*continued*

Principal use	*No.*	*Reagent*		*Remarks*
	8	Chromic acid CrO_3 170 g l^{-1} of water		Useful for distinguishing the γ and ε phases of the copper–zinc system. At current densities above 23 A dm^{-2} the γ phase is attacked, but not the ε phase. At low current densities the order of the attack is reversed. The zinc-rich solid solution is attacked under both conditions.
Aluminium–bronze and copper–beryllium alloys	*9*	Chromic acid Water	10 g 900 ml	This reagent is satisfactory for all stages of heat treatment, and is useful for following the stages of precipitation in the age-hardening of copper–beryllium alloys. Distilled water must be used as tap water leads to staining. A potential of 6 V, with an aluminium cathode, is satisfactory
Leaded copper and brass	*10*	Dilute sulphuric acid (up to 10% by vol.)		Electrolytic etching at 6 V, with a carbon cathode, has been recommended for lead-bearing copper and brass, in order to avoid misinterpretation of the structure due to surface flow

ELECTROLYTIC ETCHING OF COPPER ALLOYS

Copper alloys in general are particularly suitable for electrolytic etching, and this technique frequently gives good results with alloys (e.g. high-nickel alloys) which are otherwise difficult to etch. Solutions which have been found effective are given in Table 10.16; potential differences and current densities must be adjusted to suit specific materials. Reagents listed for electrolytic polishing of copper (Table 10.4) may also be tried.

A very sensitive etching of homogeneous copper alloys may be obtained by electrolytically polishing in phosphoric acid solution (1000 g l^{-1}), and short-circuiting the electrodes when polishing is complete. The polarisation current set up gives anodic action at the crystal boundaries only.

10.3.8 Gold

PREPARATION

Gold and its alloys are soft, and care in the application of the usual techniques is necessary.

ETCHING

The five solutions given in Table 10.17 are of general suitability.

Table 10.17 ETCHING REAGENTS FOR GOLD AND ITS ALLOYS

No.	*Reagent*		*Remarks*
1	Potassium cyanide, 10% in water Ammonium persulphate, 10% in water	10 ml 10 ml	Used for gold and its alloys. A fresh solution, warmed if necessary, must be used for each experiment. The etching time varies from $\frac{1}{2}$ to 3 min. The attack may be speeded up by the addition of 2% of potassium iodide, but this is liable to give staining effects
2	Tincture of iodine, 50% solution in aqueous potassium iodide		Used for gold alloys. With silver–gold alloys a silver iodide film may form. This may be removed by immersion in potassium cyanide solution

Table 10.17 ETCHING REAGENTS FOR GOLD AND ITS ALLOYS—*continued*

No.	*Reagent*		*Remarks*
3	Aqua regia (20 ml conc. nitric acid + 80 ml conc. hydrochloric acid)		The hot solution is used. If much silver is present a silver chloride film may form; this may be removed by ammonium hydroxide or potassium cyanide solutions. Use fresh only; liable to decompose with evolution of chlorine and NO_2
4	Potassium sulphide solution		Solution used hot; particularly useful for gold–nickel alloys
5	Chromic acid (chromium trioxide) Hydrochloric acid (1.19)	3 g	Gold-rich alloys (up to 1 min)

ELECTROLYTIC ETCHING OF GOLD

The microstructure of gold may be developed by anodic treatment in concentrated hydrochloric acid to which a little ferric chloride has been added.

Dilute solutions of hydrochloric acid, potassium cyanide or potassium cyanide plus potassium iodide may also be used.

10.3.9 Indium

PREPARATION

The pure metal and many of its alloys are very soft, and should in general be prepared by the methods recommended for lead or cadmium. It is frequently sufficient to cut suitable specimens with a sharp razor blade. After very light grinding, which may be carried out with carborundum in soap solution on broadcloth, specimens may be polished with alumina suspended in soap solution on silk nap cloth.[77,78]

ETCHING

Etching reagents containing hydrofluoric acid, nitric acid, or mixtures of the two acids, are in general satisfactory, and the following are recommended for indium-rich alloys:[77,78]

Note that an acid solution plus an oxidizing agent is the basis of etches for indium.

1. Potassium dichromate	1.3 g	
Sulphuric acid conc. (1.84)	4.5 ml	
Saturated sodium chloride solution	2.7 ml	
Hydrofluoric acid (40%)	17.7 ml	
Nitric acid conc. (1.40)	8.8 ml	
Water	66.3 ml	Use up to 1 min
2. Hydrochloric acid (1.19)	20 ml	
Picric acid	4 g	
Ethyl alcohol	400 ml	

For alloys containing bismuth it is necessary to increase the proportions of alcohol and hydrochloric acid.

10.3.10 Iron and steel

PREPARATION

Iron and low carbon steels are particularly susceptible to surface deformation during preparation, so that grinding must be carefully carried through, without undue straining or pressure. The development of heat in the surface must also be avoided; the heating of a thin skin, as a result of defective grinding, followed by rapid loss of heat by condition inwards, may easily produce martensitic patches on the surface of some unhardened alloy steels, so that in these cases special care should be taken. In the same way, retained austenite can be produced in tool steels if uncorrectly ground. Specimens may be adequately cooled in cold water. In general, it is an advantage to remove the cold-worked surface layer progressively by alternate polishing and etching. Reagents 1 and 2 in Table 10.18 are suitable. Polishing with diamond powder, using a coarse grade (8 μm) for preliminary polishing and a fine grade ($\frac{1}{2}$–1 μm) for a second stage, gives very good results with good retention of inclusions. This may be followed by a brief alumina polish.

Galvanized steel is best prepared by diamond polishing, using white spirit as a lubricant. This avoids attack of the zinc due to potential differences set up between the zinc and the basis metal with aqueous polishing media. The preparation is very flat, thus facilitating examination of diffusion layers.

A similar technique may be employed for tin-plate.

MICROGRAPHIC CONSTITUENTS OBSERVED

The metallography of iron and steel is complex, and the various constituents likely to be observed may be very briefly summarized as follows:

1. *Ingot iron and wrought iron.* These consist mainly of *ferrite* (α-iron; body centred cubic crystal structure), the etching characteristics of which may be affected by phosphorus, manganese or silicon in solid solution, or by the presence of slag inclusions. Carbon is usually present to the extent of 0.15–0.25%, and is revealed as *cementite* (iron carbide Fe_3C). Sulphur may be present as *iron sulphide.*

2. *Normalized and annealed carbon steels.** The microconstituents present vary according to the carbon content. Below 0.9% of carbon, specimens consist of *ferrite* and a closely intermingled eutectoid mixture of *ferrite* and *cementite* which is known as *pearlite* and occurs as a characteristically finely laminated structure. At 0.9% of carbon, specimens are entirely *pearlite.* Above 0.9% of carbon the constituents are a network of massive *cementite* surrounding *pearlite* areas. Normalized and annealed steels differ in the character and extent of the pearlitic areas, which are usually coarser in the latter case. Very rapid cooling gives very fine, almost unresolvable pearlite. Heating at 700 °C leads to the formation of *globular cementite,* resulting from the coalescence of eutectic cementite. Prolonged heating of high carbon steels at 800 °C decomposes cementite into *ferrite* and free *graphite,* which must be considered as a possible constituent.

3. *Hardened and tempered carbon steels.* At a high temperature (above that at which the transformation of α- into γ-iron takes place) carbon steels consist of *austenite,* a solid solution of carbon in γ-iron. This is rarely retained by quenching (some austenite is retained in highly alloyed steels such as the tool steels). In most steels a variety of decomposition products may arise depending on the severity of quenching locally. The structures obtained are:

(a) Martensite, which requires most rapid quenching. This is recognized as an intersecting system of parallel or lenticular needles (acicular structure). In highly alloyed steels some residual austenite may be retained in between the needles. Electron microscopy is needed for a detailed study of martensite. In low and medium carbon steels and maraging steels, *lath martensite* is formed and has a parallel-banded structure containing high densities of dislocation. This is characteristic of martensite formed at relatively high temperatures (i.e. 200 °C and above).
In high carbon steels and iron–nickel steels, *twinned acicular martensite* is formed. This is often lenticular with internal twinning. It forms below about 150 °C. Mixtures of lath and acicular martensite are found in medium carbon steels.

(b) Bainite is formed at lower rates of quenching. It is a non-lamellar aggregate of ferrite and cementite.

Upper bainite comprises bundles of parallel laths of ferrite between which carbide precipitates. While it can form by continuous cooling, it is more usually a consequence of isothermal decomposition of austensite at 400–500 °C. It etches up more darkly the lower the temperature at which it forms, but the structure is only seen in clear detail by electron microscopy. The transformation does not usually go to completion and residual austenite transforms to martensite at lower temperatures.

---

* By 'normalizing' is meant the reheating of a steel to a temperature at which it consists of a solid solution of carbon in γ-iron, followed by free cooling in air. By 'annealing' is meant the reheating of a steel to a similar temperature for an appreciable time, followed by slow cooling, usually in a furnace.

Lower bainite is more usually found in steels quenched too slowly to form fully martensitic structures. Like upper bainite, it is diffusion controlled but forms at lower temperatures, usually below 350 °C. It is difficult to distinguish from a tempered martensite by optical examination and needs electron microscopy for full identification. It is acicular or plate-like with subsidiary plates or needles nucleated from existing plates (unlike upper bainite, which is nucleated from austenitic boundaries). Lower bainite comprises ferrite plates or needles with carbides precipitated internally on one orientation. (In tempered martensite, two or more orientations of carbides are found.) Lower bainite is also free from twins.

(c) Pearlite. This is a lamellar eutectoid of iron and cementite, the spacing of which is dependent on the temperature at which it forms. It nucleates as nodules which grow from prior austenitic grain boundaries or in low carbon steels from ferrite/austenite interfaces.

(d) In alloy steels, manganese sulphides replace iron sulphides.

(e) In cast irons, similar structures are found but the higher carbon content results in either primary cementite or flakes of graphite depending on alloy content and/or cooling rate. Graphite can also be present as a eutectic or as nodules. By careful preparation, the structure of graphite can be revealed by polarized light. Phosphorus introduces a characteristic eutectic of ferrite and Fe_3P or a ternary eutectic with Fe_3C.

ETCHING OF IRON, STEEL AND CAST IRONS

Etching reagents are used for two purposes:

1. To reveal the general structure of the steel for which a few reagents can be used for most steels.
2. To differentiate various carbides or to differentiate carbides from nitrides and other constituents. For these an enormous variety of quite complex reagents has been developed. It is often difficult to reproduce the effects claimed by those skilled in the use of these reagents. In more complex steels, these reagents are used hot. In general, most reagents are based on an alkaline solution containing a mild oxidizing reagent and a chemical likely to cause selective staining of the minor phases. It is often simpler and equally effective to use a 1–10% caustic potash solution electrolytically with the specimen as anode. Many solutions are recommended to be used hot for the more highly alloyed and etch-resistant steels, but they can equally well be used cold as electrolytic etches. In Table 10.18 a few tried and reliable etches are given which should be used before considering the more complex etches in the literature.

LEAD IN STEELS

Lead particles in free-cutting steels can be revealed by three reagents:

1. 10% ammonium acetate in water	Brown stains on lead particles after 30 s immersion.
2. 30 g potass. dichromate 225 ml water (hot to aid solution) 30 ml acetic acid added to cold solution	10–20 s etch reveals lead as yellow to gold particles under cross-polars with polarized light. (Steel not etched)
3. 1 g potass. cyanide 100 ml water, mixed with 0.25 g diphenyl diacarbazone in 10 ml chloroform	Etch first in picrate. Rinse and dry. Then swab with this solution. Lead coloured red, especially under polarized light

The general distribution of lead can be displayed by the following:

1. Etched ground surface in nitric acid, wash and dry.
2. Cover for 5 min with absorbent paper soaked in mixture of 50% glacial acetic acid and 50% of 10% chromium trioxide in water.
3. Strip paper and wash with 10% acetic acid leaving yellow lead chromatic colouration. Wash with water.

4. Develop paper in either the third etchant above or in 1% potassium cyanide in 100 ml water plus 10 ml of 0.1% solution of dithizone in chloroform. After washing and drying, the distribution of lead is indicated by red spots. (This is generally appliable to distribution of lead in metals, e.g. in brasses.)

INCLUSIONS IN STEEL

Many of these can be recognized by shape and colour without etching and can be identified positively by EDAX analysis on scanning electron microscopes. The following notes are a useful guide to a quick optical assessment.

Iron phosphide	Brilliant white, especially in cast irons. Distinguished from cementite (Fe_3C) by alkaline potass. ferricyanide which darkens phosphide before cementite or by alkaline sod, picrate which darkens cementite
Iron nitride	Bluish grey. Not attacked by alkaline potass. ferricyanide. Coloured yellow in 4% picric acid in ethanol
Iron sulphide	Brownish yellow
Manganese sulphide	Dove grey. Ferrous sulphide is attacked by 1% oxalic acid in water; manganese sulphide is attacked by 10% chromic acid in water. Both deformed by hot rolling
Chromium oxide	Dark bluish grey; brilliant green under polarized light (crossed polars)
Silicates	Iron silicate dark grey; manganese silicate somewhat lighter, greenish tint. Often glassy as spheres which show 'optical cross' under polarized light
Titanium nitride	Sections of golden yellow cubes. If large amounts, dendritic growth from corners of cube but this is rare
Zirconium nitride	As TiN but more lemon yellow
Alumina	Angular particles in groups not elongated by rolling
Silica	Dark angular particles, often not elongated by rolling

FERROMAGNETIC ANALYSIS

To distinguish between magnetic and non-magnetic phases, e.g. ferrite from austenite in iron–chromium–nickel alloys, a thin film of a colloidal suspension of magnetic particles is applied to the surface of a specimen subject to a magnetic field. The particles concentrate on the magnetic constituent, usually in a characteristic banded or mosaic pattern. The following suspension has been recommended.[80]

A coarse flaky precipitate of magnetite is prepared by dissolving 2 g of $FeCl_2 \cdot 4H_2O$ and 5.4 g of $FeCl_3 \cdot 6H_2O$ in 300 ml of hot water, and adding, with constant stirring, 5 g of caustic soda in 50 ml water. The precipitate is filtered, washed with water and with 0.1 N hydrochloric acid. On transferring the precipitate to 1 l of 0.5% soap solution and boiling, a colloidal suspension is obtained, which should be filtered from the unsuspended residue. The suspension may now be applied to the surface of the specimen, which is placed in the field of an electric magnet. The magnetic properties of the particles gradually deteriorate owing to oxidation. The addition of small amounts of photographic reducing agents has been found an advantage in this connection.[80]

Table 10.18 ETCHING OF IRON STEEL AND CAST IRONS
(*See also* refs. 1 and 2 for extended list of etchants)

No.	*Etchant*		*Conditions*	*Remarks*
1	Nitric acid (1.40) Ethanol (Nital)	1.5–5 ml to 100 ml	5–30 s depending on steel	Ferrite g.b.'s in low carbon steels. Darkens pearlite and gives contrast with ferrite or cementite network. Etches martensite and its decomposition products in many steels. Better than Picral for low alloy steels and for ferritic grain boundaries

Table 10.18 ETCHING OF IRON STEEL AND CAST IRONS—*continued*
(*See also* refs. 1 and 2 for extended list of etchants)

No.	*Etchant*		*Conditions*	*Remarks*
2	Picric acid Ethanol (Picral)	4 g 100 ml	5–30 s depending on steel	Similar to Reagent 1 but gives more uniform etch of pearlite. Better for detail of pearlite and martensite. Reveals undissolved carbides in martensite and gives better distinction of carbides in spheroidized steels. Differentiates pearlite and bainite
3	50/50 mixture of Reagents 1 and 2		5–30 s depending on steel	Used for low alloy steels. Gives lower contrast but more even etching than Reagent 1
4	HCl (1.19) Picric acid Ethanol (5 ml HCl and 1 g picric acid usually referred to as Vilella's reagent)	1–5 ml 1–4 g to 100 ml	5–10 s	Attacks prior austenite boundaries; increased contrast between grains. (Attack of boundaries in martensite alloys increased if tempered for 30 min at 310 °C)
5	Chromium trioxide Water Sodium hydroxide (add carefully)	16 g 145 ml 80 g	10–30 min in boiling solution	Reveals intergranular oxidation of medium carbon alloy (nickel) steels
6	Sod. metabisulphite Water	8–20 g to 100 ml	5–60 s	Increases contrast in martensitic steels; distinguishes pearlite, bainite and martensite in high carbon alloy steels
7	Picric acid Sod. hydroxide Water	2 g 25 g to 100 ml	Either 10 s to 2 m boiling soln. or electrolytically cold at 0.5–2 Aft^{-2} for 2 min (stainless steel cathode)	Blackens cementite (in alloy with 10% Cr). No effect on M_7C_3, $M_{23}C_6$, M_6C or MC
8	Potass ferricyanide Potass. or sod. hydroxide (Murakami's reagent)	10 g 10 g to 100 ml	2–20 min at 20–50 °C (or electrolytically as Reagent 7	Cementite in alloys >10%Cr stained black. M_7C_3, $M_{23}C_6$ and iron phosphide stained. In cast irons, immerse 10 s at 80 °C to darken iron phosphide, then 30 s at 80 °C to darken cementite unaffected by 10 s immersion. In stainless steels, carbides darken, sigma phase blue, ferrite yellow
9	Nitric acid (1.40) Hydrochloric acid (1.19) Glycerol (also known as Vilella's reagent but do not confuse with Reagent 4; also called glyceregia)	10 ml 20 ml 30 ml	Immerse up to 30 s or use electrolytically with stainless steel cathode *Do not keep, discard when yellow; gives off chlorine and* NO_x. *Do not leave in contact with stainless steel*	Etches high chromium cast irons, stainless steel and high chromium steels
10	Sod. hydroxide Water	45 g to 100 ml	5–60 s Electrolytic with 1–3 V d.c. stainless steel or Pt cathode	In stainless steels, sigma and chi phases yellow to reddish brown (chi etches before sigma), ferrite blue grey, carbides outlined after longer etch
11	Potass. permanganate Sod. hydroxide Water	10 g 4 g to 100 ml	1–10 min in boiling soln. Electrolytic for 5–30 s as Reagent 10	In tool steels, blackens M_7C_3 and $M_{23}C_6$, outlines MC (VC)
12	Ferric chloride Hydrochloric acid (1.19) Water (Kalling's reagent)	2 g 5 ml 30 ml	Immerse 1–5 min	In tool steels attacks ferrite and martensite, outlines carbides, leaves austenite unattacked

Table 10.19 ELECTROLYTE ETCHING OF STEELS
Some etches in Table 10.18 can be used electrolytically.
The following are used exclusively electrolytically

No.	*Etchant*		*Conditions*	*Remarks*	
1	Lead acetate Water	10 g to 100 ml	2 V d.c. Stainless steel cathode 5–20 s	In Fe–Cr–Ni cast alloys Sigma phase Ferrite Austenite Carbides	 blue red Dark blue Pale blue Yellow
2	Ammonium hydroxide (0.88)		2–6V d.c. Pt cathode 30–60s	Stainless and high alloy steels. Etches carbides; sigma phase unattacked	
3	Chromium trioxide Water	10 g to 100 ml	3–6 V d.c. Pt cathode 5–60s	Stainless steels. Carbides outlined and attacked. Austenite, ferrite, phosphide attacked in that order	
4	Oxalic acid Water	10 g to 100 ml	6 V d.c. Stainless steel cathode 5–60 s	Outlines carbides first then grains in stainless steels	
5	Potass, or sod. hydroxide Water	1 g to 100 ml	1–6 V d.c. Stainless steel cathode 5–30 s	By control of voltage, can be made to etch various phases sequentially. High voltages outline carbides and grains. Low voltages selectively stain phases in sequence. Establish behaviour on a given composition (which applies to most electrolytic etches)	

10.3.11 Lead

PREPARATION

The preparation of lead and its alloys for micrographical examination is difficult, owing to the softness of the metal. Samples should be cut very gently with a sharp, oiled hacksaw. Grinding may be carried out by hand on well paraffined emery paper, and a very gentle pressure used (otherwise the surface will be badly smeared). It is advisable to finish the grinding on a well-worn paper which has been preserved from contamination by atmospheric dust. The use of a sharp, chisel-edged tool on a lathe, or of a microtome is, however, greatly to be preferred to grinding. A number of progressively finer cuts is taken, the finest possible cut being used at the last stage. The flowed layers remaining after this operation are removed by alternate etching (Reagent 15 in Table 10.20 is suitable) and polishing (e.g. with 'Bluebell' metal polish on Selvyt cloth). This is continued until all evidence of distorted or recrystallized metal is removed. A final polish is then given using a slurry of fine magnesia and water on a Selvyt cloth. Alternatively, the pad may be dressed with soap and polishing done with fine γ-alumina. Chemical polishing in 75 ml acetic acid and 25 ml hydrogen peroxide is also recommended (*see* Table 10.1).

ETCHING

Satisfactory light etches are obtained only if the degree of polish is high and the flowed layer thin. In many cases it is possible to improve results by alternate light etching and polishing as stated above. In all etching experiments fresh solutions should be used, and the specimen well washed and dried with hot air after rinsing with acetone. Rapid tarnishing (oxidation) of polished and etched samples may be avoided by attention to the drying procedure (p. **10**–18). Surfaces properly finished may be preserved unchanged for some days. Certain alloys, particularly antimonial, may be preserved by washing in strong sodium hydroxide solution, rinsing in dilute soap solution, and drying. A transparent preservative film is formed on the surface.

Suitable reagents are given in Table 10.20.

ELECTROLYTIC ETCHING OF LEAD ALLOYS

Lead and its alloys may be etched electrolytically, in many cases with a saving of time and with better control, in a solution containing 40 ml of perchloric acid and 60 ml of water. Good results are obtained in 10 s at a potential of 2 V.

Table 10.20 ETCHING REAGENTS FOR LEAD AND ITS ALLOYS

No.	*Etchant**		*Remarks*
1	Nitric acid		Recommended for pure lead. Specimen is alternately etched and washed until the desired result is obtained. Macro etch mainly
2	Acetic acid 5% in alcohol		Slow-acting grain contrast etch for lead
3	Perchloric acid 600 g l^{-1} of water		Of general application
4	Acetic acid Hydrogen peroxide (30 vol.)	30 ml 10 ml	This is recommended as a general-purpose reagent for lead, and its alloys with tin, antimony, calcium, sulphur, selenium and tellurium, as well as many other metals. The proportions may have to be adjusted slightly to suit individual cases. The reagent may be made somewhat less vigorous by dissolving a little lead in it before use. The usual etching time is 3–5 s. The action is exothermic and may give rise to recrystallization and grain growth if continued for too long a time. The reagent should not be kept for more than 1 h. It is recommended that the surface of the specimen be cleaned after etching with nitric acid. (*See also* Table 10.5)
5	Acetic acid Nitric acid Water	30 ml 40 ml 160 ml	Used at 40–42 °C for lead, and lead-tin alloys rich in lead
6	Acetic acid Nitric acid Glycerol	10 ml 10 ml 40 ml	This reagent is recommended when the alternate etching and polishing technique is employed. Used for lead, lead–calcium alloys and lead–antimony and lead–cadmium alloys. Also used for ternary lead–cadmium–antimony. The lead-rich matrix is grey, the compound SbCd blue, the antimony-rich solid solution whitish yellow, and the cadmium-rich solid solution brownish yellow. This reagent may be used hot (up to 80 °C). (Beware of recrystallization!)
7	Hydrochloric acid Nitric acid Alcohol	10 ml 5 ml 85 ml	Used for eutectic lead–tin alloys
8	Hydrochloric acid Ferric chloride Water	30 ml 10 g to 150 ml	Used mostly for lead–antimony alloys
9	Hydrochloric acid, (1.19)		General reagent for grain boundary etching. Usually good for alloys containing antimony
10	Hydrochloric acid, 1–5% in alcohol		Used for lead–tin alloys. For higher tin contents the 5% reagent is preferable
11	Nitric acid, 5% in alcohol		Produces grain contrast in lead-rich alloys
12	Nitric acid, 10% in water		Used for lead-rich alloys. May be improved by the addition of a little chromic acid
13	Sodium hydroxide saturated in water		Used for lead–tin alloys
14	Silver nitrate, 5–10% in water		Swab etch used for lead–tin and lead–antimony alloys. Useful for anti-friction alloys in general
15	Molybdic acid Ammonium hydroxide Water Filter solution and add to nitric acid	100 g 140 ml 240 ml 60 ml	This reagent, applied by the alternate swab and wash technique, gives a rapid etching which is very effective for removing thick layers of cold-worked metal. It may be followed by Reagent 5 An alternative is Russell's reagent (*see* Table 10.1)
16	Citric acid Ammonium molybdate Water	25 g 10 g 100 ml	A useful grain boundary etch for lead and dilute lead alloys. Often used after chemical polish

* Acids are concentrated, unless otherwise indicated.

10.3.12 Magnesium

PREPARATION

1. Magnesium is soft and readily forms mechanical twins and so deformed layers should be avoided.
2. Abrasives and polishing media tend to become embedded. Therefore use papers well-covered with paraffin making sure the deformed layer is removed.
3. Some phases in magnesium alloys are attacked by water. If these are present use paraffin or ethanol as lubricant.
4. Some very hard intermetallics can be present. Therefore keep polishing times short to avoid relief.

The recommended procedure is to grind carefully to 600-grit silicon carbide papers. Then polish with fine α-alumina slurry or 4–6 μm diamond paste. This is followed by polishing on a fine cloth using light magnesia paste made with distilled water or a chemical attack polish of 1 g MgO, 20 ml ammon. tartrate soln. (10%) in 120 ml of distilled water. In reactive alloys, white spirit replaces distilled water and chemical attack methods avoided.

ETCHING

The general grain structure is revealed by examination under cross-polars. This will also detect mechanical twins formed during preparation. A selection of etching reagents suitable for magnesium and its alloys is given in Table 10.21. Of these, 4 and 1 are the most generally useful reagents for cast alloys, while 16 is a useful macro-etchant and, followed by 4, is invaluable for showing up the grain structure in wrought alloys.

The appearance of constituents after etching. The micrographic appearances of the commonly occurring microconstituents in cast alloys are as given in Table 10.22.

Table 10.21 ETCHING REAGENTS FOR MAGNESIUM AND ITS ALLOYS

No.	*Etchant*		*Remarks*
1	Nitric acid Diethylene glycol Distilled water	1 ml 75 ml 24 ml	This reagent is recommended for general use, particularly with cast, die-cast and aged alloys. Specimens are immersed for 10–15 s, and washed with hot distilled water. The appearance of common constituents following this treatment is outlined in Table 10.22. Mg–RE and Mg–Ih alloys also
2	Nitric acid Glacial acetic acid Water Diethylene glycol	1 ml 20 ml 19 ml 60 ml	Recommended for solution–heat-treated castings, and wrought alloys. Grain boundaries are revealed. The proportions are somewhat critical. Use 1–10 s
3	Citric acid Water	5 g 95 ml	This reagent reveals grain boundaries, and should be applied by swabbing. Polarized light is an alternative
4	Nitric acid, 2% in alcohol		A generally useful reagent
5	Nitric acid, 8% in alcohol		Etching time 4–6 s. Recommended for cast, extruded and rolled magnesium–manganese alloys
6	Nitric acid, 4% in alcohol		Used for magnesium-rich alloys containing other phases, which are coloured light to dark brown
7	Nitric acid, 5% in water		Etching time 1–3 s. Recommended for cast and forged alloys containing approximately 9% of aluminium
8	Oxalic acid 20 g l^{-1} in water		Etching time 6–10 s. Used also for extruded magnesium–manganese alloys
9	Acetic acid, 10% in water		Etching time 3–4 s. Used for magnesium–aluminium alloys with 3% of aluminium

Table 10.21 ETCHING REAGENTS FOR MAGNESIUM AND ITS ALLOYS—*continued*

No.	*Etchant*		*Remarks*
10	Tartaric acid 20 g l^{-1} of water		Etching time 6 s } These reagents are recommended for magnesium–aluminium alloys with 3 to 6% of aluminium
11	Orthophosphoric acid, 13% in glycerol		Etching time 12 s }
12	Tartaric acid 100 g l^{-1} of water		Used for wrought alloys. Mg_2Si is roughened and pitted. 10 s to 2 min for Mg–Mn–Al–Zn alloys. Grain ??? in cast alloys
13	Citric acid and nitric acid in glycerol		Used for magnesium–cerium and magnesium–zirconium alloys. The magnesium-rich matrix is darkened and the other phases left white
14	Orthophosphoric acid Picric acid Ethyl alcohol	0.7 ml 4 g 100 ml	Recommended for solution–heat-treated castings. The specimen is lightly swabbed, or immersed with agitation for 10–20 s. The magnesium-rich matrix is darkened, and other phases (except Mg_2Sn) are little affected. The maximum contrast between the matrix and $Mg_{17}Al_{12}$ is developed. The darkening of the matrix is due to the development of a film, which must not be harmed by careless drying
15	Picric acid saturated in 95% alcohol Glacial acetic acid	10 ml 1 ml	A grain boundary etching reagent; especially for Dow metal (Al 3% Zn 1%, Mn 0.3%). Reveals cold work and twins
16	Picric acid, 5% in ethyl alcohol Glacial acetic acid Distilled water	50 ml 20 ml 20 ml	Useful for magnesium–aluminium–zinc alloys. On etching for 15 s an amorphous film is produced on the polished surface. When dry, the film cracks parallel to the trace of the basal plane in each grain. The reagent may be used to reveal changes of composition within grains, and other special purposes
17	Picric acid, 5% in ethyl alcohol Glacial acetic acid Distilled water	50 ml 16 ml 20 ml	As for Reagent 16, but suitable for a more restricted range of alloy composition[86]
18	Picric acid, 5% in ethyl alcohol Glacial acetic acid Nitric acid (1.40)	100 ml 5 ml 3 ml	General reagent[86]
19	Picric acid, 5% in ethyl alcohol Distilled water	10 ml 10 ml	Mg_2Si is coloured dark blue and manganese-bearing constituents are left unaffected[86]
20	Hydrofluoric acid (40%) Distilled water	10 ml 90 ml	Useful for magnesium–aluminium–zinc alloys. $Mg_{17}Al_{12}$ is darkened, and $Mg_3Al_2Zn_3$ is left unetched. If the specimen is now immersed in dilute picric acid solution (1 vol. of 5% picric acid in alcohol and 9 vol. of water) the matrix turns yellow, and the ternary compound remains white[86]
21	Picric acid, 5% in ethyl alcohol Distilled water Glacial acetic acid	100 ml 10 ml 5 ml	Reveals grain-boundaries in both cast and wrought alloys. This reagent is useful for differentiating between grains of different orientations, and for revealing internally stressed regions[86]
22	Nitric acid conc.		Recommended for pure metal only. Specimen is immersed in the cold acid. After 1 min a copious evolution of NO_2 occurs, and then almost ceases. At the end of the violent stage, the specimen is removed, washed and dried. Surfaces of very high reflectivity result, and grain boundaries are revealed

ELECTROLYTIC ETCHING OF MAGNESIUM ALLOYS

This has been recommended for forged alloys. The specimen is anodically treated in 10% aqueous sodium hydroxide containing 0.06 g l^{-1} of copper. A copper cathode is used, and a current density of 0.53 A dm^{-2} is applied at 4 V. After etching, the specimen is successively washed with 5% sodium hydroxide, distilled water and alcohol, and is finally dried.

NON-METALLIC INCLUSIONS IN MAGNESIUM-BASE ALLOYS[87,88]

The detection and identification of accidental flux and other inclusions in magnesium alloys involves the exposure of a prepared surface to controlled conditions of humidity, when corrosion occurs at the site of certain inclusions, others being comparatively unaffected. The corrosion product or the inclusion may then be examined by microchemical techniques.

The surface to be examined should be carefully machined and polished by standard procedures. The polishing time should be short, and alcohol or other solvent capable of dissolving flux must be avoided. As soon as possible the prepared specimens are placed in a humidity chamber, having been protected in transit by wrapping in paper. A suitable degree of humidity is provided by the

Table 10.22 THE MICROGRAPHIC APPEARANCE OF CONSTITUENTS OF MAGNESIUM ALLOYS

Microconstituent	*Appearance in polished sections, etched with Reagent 1 (zirconium-free alloys)*
$Mg_{17}Al_{12}$[(1)]	White, sharply outlined and brought into definite relief
$MgZn_2$[(2)]	Appearance very similar to that of $Mg_{17}Al_{12}$
$Mg_3Al_2Zn_3$[(3)]	Appearance similar to those of $Mg_{17}Al_{12}$ and $MgZn_2$
Mg_2Si[(4)]	Watery blue green; the phase usually has a characteristic Chinese-script formation, but may appear in massive particles. Relief less than for manganese
Mg_2Sn[(4)]	Tan to brown or dark blue, depending on duration of etching. Individual particles may differ in colour
Manganese[(5)]	Grey particles, usually rounded and in relief. Little affected by etching
(MgMnAl)[(5)]	Grey particles, angular in shape and in relief. Little affected by etching
Microconstituent	*Appearance in polished sections etched with Reagent 4 (zirconium-bearing alloys)*
Primary Zr (undissolved in molten alloy)	Hard, coarse, pinkish grey rounded particles, readily visible before etching
Zinc-rich particles[(6)]	Fine, dark particles, loosely clustered and comparatively inconspicuous before etching
Mg_9Ce	Compound or divorced eutectic in grain boundaries. Appearance hardly changed by few per cent of zinc or silver
Mg_5Th	Compound or divorced eutectic in grain boundaries (bluish). Appearance hardly changed by few per cent of zinc if Zn exceeds Th
(?) Mg–Th–Zn	Brown acicular phase. Appears in Mg–Th–Zn–Zr alloys when Th $\geqslant$ Zn
$MgZn_2$	Compound or divorced eutectic in grain boundaries. Absent from alloys containing RE or Th

The superscript numbers in column 1 refer to the following notes:

(1) This is the γ-phase of the magnesium–aluminium system; it is also frequently called Mg_4Al_3 or Mg_3Al_2.
(2) Although the phase MgZn may be observed in equilibrium conditions, $MgZn_2$ is frequently encountered in cast alloys.
(3) This ternary compound occurs in alloys based on the ternary system magnesium–aluminium–zinc, and may be associated with $Mg_{17}Al_{12}$.
(4) Blue unetched.
(5) These constituents are best observed in the unetched condition.
(6) Alloys of zirconium with interfering elements such as Fe, Al, Si, N and H, separating as a Zr-rich precipitate in the liquid alloy. Co-precipitation of various impurities makes the particles of indefinite composition.

Note: The microstructure of all zirconium-bearing cast alloys with satisfactory dissolved zirconium content is characterised by Zr-rich coring in the centre of most grains. In the wrought alloys zirconium is precipitated from the cored areas during preheating or working, resulting in longitudinal striations of fine precipitate which become visible on etching.

For descriptions of the metallography of magnesium alloys refs. 83–85 should be consulted.

air above a saturated solution of sodium thiosulphate. The presence of corrosive inclusions is indicated by the development of corrosion spots. At this stage the corroded area may be lightly ground away to expose the underlying structure for microexamination so that the micrographic features which are holding the flux become visible. With other specimens, or with the same specimens re-exposed to the humid conditions, identification of the inclusions may be proceeded with, as follows:

1. *Detection of chloride*

The corrosion product is scraped off, and dissolved on a microscope slip in 5% aqueous nitric acid. A 1% silver nitrate solution is then added, and a turbidity of silver chloride indicates the presence of the chloride ion. The solution of the corrosion product should preferably be heated before adding the silver nitrate to remove any sulphide ion, which also gives rise to turbidity. Alternatively, a 10% solution of chromium trioxide may be added directly to the corrosion spot, when chloride is indicated by an evolution of gas bubbles from the metal surface, and the development of a brown stain. This method is less specific than the silver nitrate method, and may give positive reactions in the presence of relatively large amounts of sulphates and nitrates.

2. *Detection of calcium*

Scrapings of corrosion product are dissolved in a small watch glass on a hot plate in 2 ml water and one drop of glacial acetic acid. To the hot solution a few drops of saturated ammonium oxalate solution are added. The presence of calcium is indicated by turbidity or precipitation. Spectroscopic identification of calcium in the solution is also possible.

3. *Detection of boric acid in inclusions*

Scrapings of corrosion product and metal are placed in a test tube with 1 ml of water. The inclusion dissolves, and complete solution of the sample is effected by adding a small portion of sulphuric acid (density 1.84) from 9 ml carefully measured and contained in a graduated cylinder. When solution is complete, the remainder of the acid is added and the mixture is well shaken; 0.5 ml of a 0.1% solution of quinalizarin in 93% (by wt.) of sulphuric acid is now added, mixed in, and allowed to stand for 5 min. A blue colour indicates the presence of boric acid. The colour in the absence of boric acid varies from bluish violet to red according to the dilution of the acid, which must thus be carefully controlled as described.

4. *Detection of nitride*

A drop of Nessler's solution applied directly to the metal surface in the presence of nitride, gives an orange brown precipitate, which may take about 1 min to develop. This test should be made on freshly prepared surfaces on which no water has been used, since decomposition of nitride to oxide occurs in damp air.

5. *Detection of sulphide*

The corrosion product is added to a few drops of water slightly acidified with nitric acid. A drop of the solution placed on a silver surface gives rise to a dark stain if sulphide was present in the corrosion product. Sulphur printing may also be applied.

6. *Detection of iron*

The corrosion product is dissolved in hydrochloric acid. A drop of nitric acid is added with several drops of distilled water. In the presence of iron, the addition of a crystal of ammonium thiocyanate develops a blood-red colouration.

In all the above tests, a simultaneous *blank* test should be carried out.

Iron-printing, analogous to sulphur-printing, can be applied using cleaned photographic paper impregnated with a freshly prepared solution of potassium ferricyanide and potassium ferrocyanide acidified with nitric or hydrochloric acid.

10.3.13 Molybdenum

PREPARATION

Normal methods are used but it is difficult to obtain a scratch-free underformed surface and it is preferred to use attack polishing. This requires a fine α or γ alumina slurry with 1 g potass. ferricyanide, 0.5 g sod. hydroxide, 100 ml water.

ETCHING

Etchant		*Conditions*	*Remarks*
1. (a) Potass. hydroxide	10 g	Mix equal amounts of (a) and (b) as needed	Grain boundary etch
Water	to 100 ml		
(b) Pot. ferricyanide	10 g		
Water	to 100 ml		
2. Ammonia (0.88)	50 ml	Boil for up to 10 min	General etch
Hydrogen peroxide (3%)	50 ml		
Water	50 ml		

10.3.14 Nickel

PREPARATION

Great care must be taken to produce polishes of a high quality, with as little deformation as possible at the surface, since it is difficult to remove much surface material from nickel during etching.

ETCHING

Since nickel is generally resistant to corrosive media, etching involves the use of vigorous reagents, which tend to form etch pits and to dissolve out inclusions. The difficulty of etching increases with the purity of the metal, and alternate polishing and etching is frequently necessary. In general, etching reagents (Table 10.23) produce grain boundary effects without much grain contrast. Low nickel copper–nickel alloys may be etched by the reagents recommended for the purpose under *Copper*. Heat-resisting nickel-base alloys are etched with the solutions used for stainless steels (Table 10.18).

Table 10.23 ETCHING REAGENTS FOR NICKEL AND ITS ALLOYS

No.	*Etchant*		*Remarks*
1	Nitric acid (1.40)	10 ml	Pure nickel, and nickel-chromium alloys. Grain boundaries etched
	Hydrochloric acid (1.19)	20 ml	
	Glycerol	30 ml	
2	Nitric acid (1.40)	10 ml	Used for pure nickel, cupro-nickel, Monel metal and nickel-silver
	Acetic acid	10 ml	
	Acetone	10 ml	
3	Nitric acid (1.40)	50 ml	Useful for nickel and most nickel-rich alloys
	Glacial acetic acid	25 ml	
	Water	25 ml	
4	Nitric acid, 2% in alcohol, mixed in various proportions with hydrochloric acid, 2% in alcohol		Used for pure nickel
5	Hydrochloric acid	30 ml	Pure nickel
	Ferric chloride	10 g	
	Water	120 ml	
6	Hydrochloric acid (1.19)	20 ml	Pure nickel
	Ferric chloride	10 g	
	Water	30 ml	

Table 10.23 ETCHING REAGENTS FOR NICKEL AND ITS ALLOYS —*continued*

No.	*Etchant*		*Remarks*
7	Sulphuric acid (1.84) Hydrogen peroxide (10%)	1 ml 10 ml	Pure nickel
8	Sulphuric acid (1.84) Potassium dichromate (saturated in water)	10 ml 50 ml	Pure nickel
9	Ammonium persulphate 100 $g l^{-1}$ of water		Cast nickel
10	Ammonium per sulphate, 10% in water Potassium cyanide, 10% in water	10 ml 10 ml	Pure nickel. Also recommended for cross-sections of nickel plated steels. The steel is not affected but may be etched by Reagent 2 for *Iron and steel*
11	Ammonium hydroxide		Used as a polish attack for nickel plate
12	Ammonium hydroxide Hydrogen peroxide 30%	85 ml 15 ml	Used for nickel-silvers

ELECTROLYTIC ETCHING OF NICKEL AND ITS ALLOYS

This technique frequently gives better results than ordinary immersion or swab etching. Suitable solutions are summarized in Table 10.24.

Table 10.24 ELECTROLYTIC ETCHING OF NICKEL ALLOYS

No.	*Etchant*		*Conditions*	*Remarks*
1	Sulphuric acid (1.84) Hydrogen peroxide (30%) Water	22 ml 12 ml to 100 ml	6 V d.c. Pt cathode 5–30 s	General etch
2	Sulphuric acid (1.84) Water	5 ml to 100 ml	6 V d.c. Pt cathode 5–15 s	All Ni–base alloys, especially Ni–Cu and Ni–Cr. Delineates carbides
3	Ammon. persulphate Water	10 g to 100 ml	6 V d.c. Nickel cathode	Most alloys, especially Ni–Fe Ni–Cr cast alloys
4	Chromium trioxide Water	6 g to 100 ml	As etchant 3	As etchant 3
5	Nitric acid (1.40) Glacial acetic acid Water	10 ml 5 ml 85 ml	1.5 V d.c. Pt cathode 20–60 s (*do not keep*)	Grain contrast for all nickel alloys
6	Oxalic acid Water	10 g 100 ml	6 V d.c. Stainless steel cathode 10–15 s	Superalloys, Ni–Au, Ni–Cr, Ni–Mo alloys

Table 10.24 ELECTROLYTIC ETCHING OF NICKEL ALLOYS—*continued*

No.	*Etchant*		*Conditions*	*Remarks*
7	Phosphoric acid (1.71) Water (if necessary, sulphuric acid (1.84))	70 ml 30 ml 15 ml	2–10 V d.c. Ni cathode 5–60 s	Superalloys (Nimonics) Ni–Cr, Ni–Fe
8	Phosphoric acid (1.71) Sulphuric acid (1.84) Chromium trioxide	85 ml 5 ml 8 g	10 V d.c. Pt cathode 5–30 s	Ni–base superalloys. Gamma prime paper. Ti and Nb segregation

10.3.15 Niobium

PREPARATION

Niobium is soft and tough but can be prepared by methods used for stainless steel. It is best polished using a polish attack method on an acid-resistant cloth using gamma alumina and a solution of:

Hydrofluoric acid	2 ml
Nitric acid	5 ml
Lactic acid	30 ml

This is slow (approx. 1 hour) and needs an automatic polisher as the solution is very corrosive.

Table 10.25 ETCHING REAGENTS FOR NIOBIUM ALLOYS

No.	*Etchant*		*Conditions*	*Remarks*
1	Hydrochloric acid (1.19) Sulphuric acid (1.84) Nitric acid (1.40) Water	15 ml 15 ml 8 ml 62 ml	Immerse 10–60 s	The only general-purpose etch free from HF
2	Hydrofluoric acid (40%) Nitric acid (1.40) Lactic acid	10 ml 10 ml 30 ml	Immerse 15–20 s	General etch
3	Hydrofluoric acid (40%) Nitric acid (1.40) Water	10 ml 20 ml 70 ml	Immerse up to 3 min *or* Use electrolytically 12–30 V d.c. Pt cathode up to 1 min	Grain boundary etch, especially at low voltages, electrolytically

10.3.16 Platinum group metals

PREPARATION

On polishing platinum and many of its analogues, considerable surface flow may occur, and this makes the subsequent etching operations difficult to control, and gives a poor appearance to the microstructure.

ETCHING

The reagents given in Table 10.25 will produce satisfactory microstructures if the preparation and polishing have been carefully done. However, owing to the almost invariable presence of a flowed layer after mechanical polishing, it is frequently recommended that electrolytic etching be used.

Table 10.26 ETCHING REAGENTS FOR PLATINUM GROUP

No.	*Etchant*		*Conditions*	*Remarks*
1	Aqua Regia			
	Nitric acid (1.4)	34 ml	Up to 1 min. May need to be warmed	Pt, Pd, Rh alloys
	Hydrochloric acid (1.19)	66 ml		
2	Potass. ferricyanide	3.5 g	Several minutes	Most alloys, including Os alloys.
	Sod. hydroxide	1 g		
	Water	150 ml		
3	Hydrochloric acid (1.19)	20 ml	Several minutes	Ru alloys.
	Hydrogen peroxide (3%)	1 ml		
	Water	80 ml		
4	Potass. cyanide	5 g	Electrolytic 1–5 V a.c. Pt cathode 1–2 min	Pt. alloys.
	Water	100 ml		
5	Hydrochloric acid (1.19)	20 ml	Electrolytic with Pt or graphite electrode.	
	Sod. chloride	25 g		
	Water	65 ml	25 s 10 V a.c.	Rh alloys
			1 min 1.5 V a.c.	Pt–Rh alloys
			1–2 min 20 V a.c.	Ir alloys
			1 min 6 V a.c.	Pt alloys
			1 min 5–20 V a.c.	Ru-base alloys
6	Hydrochloric acid (1.19)	10 ml	Electrolytic 30 s, 10 V a.c. Graphite cathode	Os, Pd, Pt, Ir.
	Ethanol	90 ml		
7	Hydrochloric acid (1.19)		1–2 min electrolytic 5 V a.c. Pt or graphite electrode	Rh, Pt. Grain contrast

10.3.17 Silicon

PREPARATION

Lump silicon, which is very hard and brittle, may be prepared for metallographic examination by methods similar to those recommended for hard alloys. No standard procedure appears to have been developed. Both electrolytic polishing and chemical polishing have met with some success (*see* Tables 10.4b and 10.4). Coarse abrasives leave rows of pits and so it is preferred to use fine silicon carbide papers where possible.

ETCHING

Polished surfaces of silicon may be etched with 5% aqueous hydrofluoric acid, to which various amounts of concentrated nitric acid may be added. Microstructures developed in this way show angular grains, with twin markings, together with the particles of other constituents due to the presence of impurities (e.g. iron, aluminium and calcium). The precise nature of the impurity particles is not clear.

Commercial silicon may contain inclusions of slag and unreduced quartz. Silicate inclusions may be detected by etching a polished section for 3 h in a stream of chlorine. The specimen is then immersed in concentrated hydrofluoric acid for 10 min in order to etch and attack the quartz inclusions (*see also* Table 10.9, p. **10**–21).

10.3.18 Silver

PREPARATION

Silver is comparatively soft, so that extreme care and cleanliness must be maintained. Otherwise normal methods of preparation are suitable. In polishing, proprietary metal polishing media containing ammonia or ammoniacal compounds should be avoided.

Table 10.27 ETCHING REAGENTS FOR SILVER AND ITS ALLOYS

No.	*Etchant*	*Remarks*
1	Ammonium hydroxide and hydrogen peroxide—various proportions	The silver-rich matrix is in general unaffected. Other phases (e.g. the β silver–antimony phase) are often coloured blue to brown
2	Ammonium hydroxide 50 ml Hydrogen peroxide (3%) 10–30 ml	Recommended for silver, silver–nickel and silver–palladium alloys. Also useful for the examination of silver-soldered joints
3	Sulphuric acid (10% in water) to which a few crystals of chromic acid CrO_3 have been added (2 g)	This reagent reveals the grain structure of silver and silver-rich alloys
4	Solution containing 7.6 g l^{-1} of chromic acid CrO_3 and 8 g l^{-1} of sulphuric acid	Useful general etching reagent. Used as a sensitive etching reagent for silver–copper alloys
5	Ferric chloride 20 g l^{-1} of water	Recommended for silver solders
6	Potassium cyanide, 10% in water 10 ml Ammonium persulphate, 10% in water 10 ml	Etch for pure silver and dilute alloys. Duration 1–2 min
7	Solution A: 50 : 50 nitric acid in water 100 ml Potassium dichromate 2 g Solution B: Chromic acid CrO_3 20 g Sodium sulphate 1.5 g Water 100 ml	Used for silver alloys in general. Solution A is diluted to 20 vol. and an equal amount of solution B added. The reagent is applied by gentle swabbing or with a camel hair brush. A loose film of silver chromate should form if the reagent is working correctly. If the film is adherent, more of solution A should be added. If no film forms, more of solution B is required
8	Chromic acid, CrO_3, 0.2% and sulphuric acid, 0.2% in water	Used (1 min immersion) for silver and silver-rich alloys
9	Potassium dichromate saturated in water 100 ml Sodium chloride saturated in water 2 ml Sulphuric acid 10 ml	Silver and silver-rich alloys Silver solders

Table 10.27 ETCHING REAGENTS FOR SILVER AND ITS ALLOYS—*continued*

No.	*Etchant*		*Remarks*	
10	Chromic acid and hydrogen peroxide in water. Various proportions		General reagent. Composition adjustments must be made to suit specific cases	
11	Sodium hydroxide (10%) Potass. ferricyanide (30%)	10 ml 10 ml	5–15 s (dilute with equal vol. of water if too fast)	Ag alloys with W, Mo and WC

Pure silver and dilute silver alloys are difficult to etch, but several solutions will give good results on duplex or more complex alloys. The strengths of the reagents, unless otherwise noted in Table 10.27, should be adjusted to the specific alloys to be examined.

ELECTROLYTIC ETCHING OF SILVER

In many cases excellent grain boundary and grain contrast etching is obtained by electrolysis (specimen as anode) in one of the following solutions:

Table 10.28 ELECTROLYTIC ETCHING OF SILVER ALLOYS

No.	*Etchant*	*Remarks*
1	Citric acid 100 g l^{-1} of water	15 s to 1 min. 6 V d.c. Ag cathode. Most alloys
2	Ammoniacal ammonium molybdate	Molybdic acid is dissolved in an excess of strong ammonia. The composition is not critical provided ammonia is in excess. The optimum potential and current density should be experimentally established for each case
3	Potassium cyanide 50 g l^{-1} of water	This reagent is used particularly for silver when it is in contact with other metals, as in plated articles. Optimum conditions should again be established experimentally
4	Hydrofluoric acid + a little stannous chloride	Used for silver–tin alloys containing more than 73% of silver

See also reagents listed for electrolytic polishing of silver, Table 10.4b. The silver cyanide–potassium cyanide–potassium carbonate reagent in Table 10.4b is particularly convenient, since, on reduction of the polishing potential from 1.5 to 0.5 V for 90 s, etching occurs.

10.3.19 Tantalum

PREPARATION

For the preparation of tantalum (pure and commercial) it is recommended that polishing should be carried out on felt (5–10 min), using the following mixture as polishing medium:

Alumina, carefully levigated	35 g
Hydrofluoric acid (60%)	20 ml
Ammonium fluoride	20 g
Distilled water	1 000 ml

See also Niobium.

ETCHING

The reagents given in Table 10.29 have been found useful.

Table 10.29 ETCHING REAGENTS FOR TANTALUM

No.	*Etchant*		*Remarks*
1	Ammonium fluoride, 20% in water	10 ml	Used at 50–60 °C (ineffective when cold). Immersion for 1 min etches up the structure of tantalum without colouring inclusions of tantalum sulphide, Ta_2S_5
	Hydrofluoric acid, (40%) 60% in water	10 ml	
2	Ammonium fluoride, 20% in water	10 ml	Used at 60 °C for 1–2 min. Etching effects are similar to those for Reagent 1
	Sulphuric acid (1.84)	20 ml	
3	Ammonium fluoride, 20% in water	10 ml	Used at 60 °C; the matrix is usually not adequately etched but Ta_2S_5 is blackened. This reagent may be used after Reagent 1 or 2 to identify Ta_2S_5
	Nitric acid	10 ml	
4	Ammonium fluoride, 20% in water		Used for 5–6 min at 80 °C. The grain structure of the matrix is developed, and Ta_2S_5 is not affected
5	Ammonium fluoride, 20% in water	20 ml	This reagent is used boiling, and colours Ta_2S_5 brown. The matrix is not affected
	Hydrogen peroxide	10 ml	
6	Sulphuric acid	25 ml	For general structure[89]
	Nitric acid	10 ml	
	Hydrofluoric acid	10 ml	

10.3.20 Tin

PREPARATION

Because of its relative softness, the same precautions should be observed as for lead and its alloys. Very careful hand-grinding is necessary in order to avoid surface recrystallization, and again the microtome technique (*see Lead*) may be used with advantage. The flowed layer on the surface may be removed before the final etching treatment by immersion in 10–20% aqueous hydrochloric acid. The technique involving alternate polishing and etching very frequently proves advantageous. If specimens are mounted for metallographic preparation, care should be used, since tin is liable to flow under the pressures necessary for the usual mounting processes.

For further details of methods for tin and tin alloys, refs. 1, 2, 91 and 92 may be consulted.

ETCHING

Tin is anisotropic and the grain structure can be revealed under crossed-polars. The reagents of Table 10.30 have been recommended for tin-rich alloys. In general, etching times are not critical, and the progress of the treatment should be judged visually.

Table 10.30 ETCHING REAGENTS FOR TIN AND ITS ALLOYS

No.	*Etchant**		*Remarks*
1	Nitric acid, 2% in alcohol		This reagent is a general one for tin-rich alloys, and particularly for tin–cadmium, tin–antimony and tin–iron alloys. The tin-rich matrix is darkened, and the intermetallic compounds usually little affected
2	Nitric acid, 5% in alcohol		Tin, tin–cadmium and tin–iron alloys
3	Nitric acid	10 ml	Mainly used as a reagent for tin. Used at 38–40 °C
	Acetic acid	30 ml	
	Glycerol	50 ml	

* Acids are concentrated, unless otherwise indicated.

Table 10.30 ETCHING REAGENTS FOR TIN AND ITS ALLOYS—*continued*

No.	*Etchant**		*Remarks*
4	Nitric acid Acetic acid Glycerol	10 ml 10 ml 80 ml	Used, at 38–40 °C, for tin–lead alloys, especially tin-rich alloys. Pb is blackened.
5	Picric acid and nitric acid in alcohol—proportions variable		Useful for tin–iron alloys in contact with steel.
6	Hydrochloric acid, 1–5% in alcohol		Useful for tin–lead, tin–cadmium, tin–iron, tin–antimony–copper alloys.
7	Hydrochloric acid Ferric chloride Water	2 ml 10 g 95 ml	Useful for tin-rich alloys in general and for Babbitt metal. It is often an advantage to add alcohol to this reagent. Any cadmium-rich solid solution present is stained black, while the β-phase of the tin–cadmium system turns brown
8	Hydrochloric acid Ferric chloride Water Alcohol	5 ml 2 g 30 ml 60 ml	Reveals general structure of tin and alloys without lead. Tin–iron and tin–copper compounds unattacked
9	Stannous chloride in acid solution		Tin-rich alloys in general. The composition may be varied to suit individual alloy systems
10	Potassium dichromate, in dilute acidified aqueous solution (composition variable)		Recommended for tin–cadmium alloys
11	Potassium ferricyanide in caustic soda (composition variable)		Used for tin–cadmium–antimony alloys. It distinguishes SbSn (with tin and cadmium in solid solution), which is not affected, from CdSb in tin–cadmium–antimony alloys
12	Ammonium persulphate Water	5–10 g 100 ml	Used particularly for tinplate. The tin is heavily darkened, leaving the basis metal unattacked. Most intermetallic compounds of tin are also unattacked. More dilute solution gives grain boundary etch with tin and alloys
13	Nitric acid and hydrofluoric acid in glycerol		Various strengths are recommended for tinplate. The contrast between plating and steel may be improved by the use of Reagent 2 (Table 10.17)
14	Sodium sulphide 20% in water, with a few drops of hydrochloric acid		Useful for tin-rich tin–antimony–copper alloys. The phase SbSn is not affected, but the phase Cu_6Sn_5 is coloured brown[93]
15	Silver nitrate Water	5 g 100 ml	Darkens primary and eutectic lead in lead-rich lead–tin alloys[91]

* Acids are concentrated, unless otherwise indicated.

ELECTROLYTIC ETCHING OF TIN ALLOYS

A 10% aqueous solution of hydrochloric acid or 20% sulphuric acid may be used at a very low current density; this is especially useful for tin–iron alloys. Satisfactory results are also obtained from:

Glacial acetic acid	130 ml
Perchloric acid	50 ml

used at a current density of 3–6 A dm^{-2}. (Take care; *see* Section 10.2, p. **10**–8, *Electrolytic polishing*).

10.3.21 Titanium

PREPARATION

The preparation of titanium samples by ordinary methods of grinding is straightforward but needs care; final polishing is difficult. Specimens are easily scratched, and mechanical working of the surface during polishing causes twin-formation which may obscure other metallographic features.

Other 'false' structures may be caused by the presence of local, randomly dispersed areas of cold work, which give a duplex appearance to homogeneous specimens. Electrolytic polishing of surfaces ground wet by ordinary methods to the 000 grade of emery paper is therefore recommended[95] (*see* Table 10.4b).

Mechanical polishing, if preferred, may be carried out with diamond preparation, with a final fine polish (if required) with alumina, both with a trace of hydrofluoric acid.

Examination for hydride is carried out in polarised light between crossed polaroids; the hydride then appears bright and anisotropic. This also reveals the grain structure of α-titanium.

ETCHING

The presence of surface oxide films on titanium and its alloys necessitates the use of strongly acid etchants. Those given in Table 10.31 are useful.

Table 10.31 ETCHING REAGENTS FOR TITANIUM AND ITS ALLOYS

No.	*Etchant*		*Conditions*	*Remarks*
1	Hydrofluoric acid (40%) Nitric acid (1.40) Lactic acid	1–3 ml 10 ml 30 ml	5–30 s	Mainly unalloyed titanium; reveals hydrides
2	Hydrofluoric acid (40%) Nitric acid (1.40) Lactic acid	1 ml 30 ml 30 ml	5–30 s	As Etchant 1
3	Hydrofluoric acid (40%) Nitric acid (1.40) Water (Kroll's reagent)	1–3 ml 2–6 ml to 100 ml	3–10 s	Most useful general etch
4	Hydrofluoric acid (40%) Nitric acid (1.40) Lactic acid	10 ml 10 ml 30 ml	5–30 s	Chemical polish and g.b. etch
5	Potassium hydroxide (40%) Hydrogen peroxide (30%) Water (can be varied to suit alloy)	10 ml 5 ml 20 ml	3–20 s	Useful for α/β alloys. α is attacked or stained. β unattacked
6	Hydrofluoric acid (40%) Nitric acid Glycerol	20 ml 20 ml 40 ml	5–15 s	General purpose, TiAlSn alloys
7	Hydrofluoric acid Nitric acid (1.40) Glycerol Water	1 ml 25 ml 45 ml 20 ml	3–20 s	TiAlSn alloys

10.3.22 Tungsten

PREPARATION

Specimens, which are extremely hard, must be ground flat on an emery or carborundum wheel. Fine grinding may be carried out in the usual manner, and polishing, which is relatively easy, is done with graded alumina suspensions. In view of the difficulty of etching (*see* below) it has been recommended that tungsten should be prepared by a polish attack.[97] The specimen may be taken directly from the No. 0 emery paper to a polishing wheel which is moistened with a cold suspension of the polishing medium in an etching reagent (e.g. 3.5 g of potassium ferrocyanide and 1 g of caustic soda in 150 ml distilled water) and polished for approximately 15 s. After rinsing, the specimen is dipped in the etching reagent alone for 5 s, and is again rinsed and dried. This process results in considerable saving of time. Many commercial tungsten products are in the form of small complex shapes, fine wires and filaments, and this raises special problems in mounting for metallographic examination. It is recommended that small complicated specimens should be embedded in a low melting point glass which has been melted into a groove or cavity in a metal holder. The composite sample may then be ground and polished in the usual manner to expose the section required. Ordinary mounting agents are too soft for satisfactory use with tungsten. A suitable glass may be prepared by fusing 85 g of lead peroxide with 15 g of boric acid. This may be darkened by a little manganese dioxide, and melts at approximately 450 °C.

ETCHING

The etching of tungsten is difficult, but Murakami's etch is most often used. Inclusions (e.g. thorium, uranium or calcium oxides) may be seen in the polished, unetched surface, while the general structure of filaments which have been heated *in vacuo* or in a reducing atmosphere may often be observed owing to evaporation effects. A selection of suitable etching reagents is given in Table 10.32.

ELECTROLYTIC ETCHING OF TUNGSTEN

This may be carried out in a mixture of 25 ml of normal aqueous sodium hydroxide and 20 ml of hydrogen peroxide. The current density and potential are somewhat critical, and should be carefully controlled after investigation to find the optimum conditions for specific cases.

Sodium hydroxide solution of 0.025 normal strength has also been used, at a current density of 5 A dm^{-2}.

Table 10.32 ETCHING REAGENTS FOR TUNGSTEN

No.	*Etchant*		*Remarks*
1	Sodium hydroxide, 10% in water	10 ml	This reagent is used cold and, on immersion of the specimen for approximately 10 s, develops grain boundaries (Murakami's reagent)
	Potassium ferricyanide, 10% in water	10 ml	
2	Hydrogen peroxide, 3% in water		This reagent develops grain boundaries, but only after some 30–90 s in the boiling reagent
3	Potassium ferricyanide	305 g	Recommended for deep etching of single crystal bars and wires in order to produce etch-pits for the investigation of orientation
	Caustic soda	44.5 g	
	Water	1 000 ml	
4	Hydrofluoric acid	5 ml	Swab 10–20 s, rinse and dry. Follow with Murakami's reagent (Etchant 1)
	Nitric acid (1.40)	10 ml	
	Lactic acid	30 ml	

10.3.23 Uranium

PREPARATION

Conventional methods are used but as the metal is toxic, pyrophoric when finely divided, and an α-particle emitter, it is essential to keep the metal wet during cutting and to carry out the operation behind a screen so that particles cannot be ingested.

Final polishing may be carried out with magnesia or alumina on Selvyt or terylene cloth.

Chromium oxide (green) is also used but tends to smear the structure and pull out inclusions. Polish attack methods are preferred and give structures which can be examined under polarized light with crossed polars, which is the most useful method of examination.[43]

Polish attack solutions are:

1. Dilute hydrofluoric acid and nitric acid (1 ml + 5 ml respectively in 100 ml water).
2. 50 g chromium trioxide, 10 ml nitric acid, 100 ml water is less aggressive but, like the other, needs careful handling to avoid contact with the skin. This gives the best results for polarized light examination.

ETCHING

Polarized light is the most useful technique.

α uranium— high grain contrast under crossed polars about 1° off the extinction position. The various transformation products, granular α, martensites etc., can be distinguished.

β uranium— weakly anisotropic when retained but clearly distinguished from α and γ.

γ uranium— optically isotropic and therefore distinguished from α and β. In uranium alloys with molybdenum, niobium and rhenium, γ_0, tetragonal with $c/a = 0.5$ is optically active and the grain structure is visible under crossed polars.

On ageing, these alloys develop γ_0 with $c/a < 0.5$ and this is readily detected by polarized light metallography.

Note that chemical etching may give quite different structures, e.g. in U–Ti alloys. Chemical etching is responsive to the manner of decomposition of γ uranium and will reveal the sequence of decomposition, e.g. distribution of U_2Ti, UZr_2 etc., whereas polarized light reveals the final grain structure which may have little relevance to the etched structure.[45]

Table 10.33 ETCHING REAGENTS FOR URANIUM ALLOYS

No.	*Etchant*		*Conditions*	*Remarks*
1	Nitric acid (1.40) Glacial acetic acid Glycerol	30 ml 30 ml 30 ml	5–60 s	Most U alloys; distinguishes precipitated phases, e.g. U_2Ti, U_2Mo, UZr_2, eutectoid nucleation at grain boundaries (like pearlite in steel)
2	Orthophosphoric acid Diethylene glycol Ethanol	30 ml 30 ml 40 ml	Electrolytic 20 V stainless steel cathode 5–30 s	Will electropolish at high voltages (30–40 V), stains phases differentially at 10–20 V
3	Hydrofluoric acid (40%) Nitric acid (1.40) Glycerol	10 ml 40 ml 40 ml	5–10 s	U–Mo, U–Nb, U–Zr alloys (gamma phase)
4	Hydrofluoric acid (40%) Nitric acid (1.40) Lactic acid (or distilled water)	1 ml 30 ml 30 ml	5–30 s	U–Be, U–Nb, U–Zr, U–Mo. Also U–Al alloys. UAl_2, UAl_3, UAl_4 stained differentially according to time

10.3.24 Zinc

PREPARATION

Though harder than lead and tin, zinc and its alloys must be carefully prepared, since careless grinding, involving distortion of the surface, produces mechanical twinning in a comparatively thick surface layer. Owing to these deep surface effects, at least 2 mm must be removed from a sawn or filed surface, overheating being avoided, as it is likely to cause recrystallization of the surface layers. Also, the successive grindings should be carried on past the stage at which scratches due to the previous treatment have disappeared. The depth of the deformed layer may be much greater than the deepest scratch. Coarse grained specimens of the pure metal need the most care in this respect; find grained metal, and most of the common alloys, are less liable to such deep surface distortion.

Polishing may be carried out on a slowly rotating wheel, and it is often an advantage to etch and polish alternatively, in order to avoid the production of a distorted layer during the polishing process. Since zinc alloys often contain hard particles of intermetallic compounds, this method tends to produce excessive relief in some cases, and must be used with discretion. A suitable etching reagent for this purpose is a solution containing 200 g of chromic acid (CrO_3), 15 g of anhydrous sodium sulphate and 1 000 ml of water.

Table 10.34 ETCHING REAGENTS FOR ZINC AND ITS ALLOYS

No.	*Etchant*	*Remarks*
1	Hydrochloric acid, 1% in alcohol	These reagents are of general applicability to zinc and many zinc-rich alloys. Etchant 3 is suitable for examination of microconstituents at high magnifications
2	Sodium hydroxide 100 g l^{-1} of water	
3	Nitric acid, 1–2% in alcohol	Rinse in 20% chromic acid in water to avoid stains. Good for Zn–Fe layers in galvanized samples
4	Nitric acid 94 ml, Chromic acid 6 ml } Stock solution	A few drops of this stock solution are added to 100 ml of water immediately before use. The resulting solution is generally useful, particularly for the recognition of small amounts of other microconstituents in zinc
5	Chromic acid, CrO_3 200 g Sodium sulphate, Na_2SO_4 15 g Water 1 000 ml	General reagent for commercial zinc and zinc alloys. If a film of stain results, this may be removed by immersion in a 20% solution of chromic acid in water
6	Chromic acid, CrO_3 50 g Sodium sulphate, Na_2SO_4 4 g Water 1 000 ml	As for Etchant 5. This composition is recommended for die-castings
7	Chromic acid, CrO_3 200 g Sodium sulphate, Na_2SO_4 7.5 ml Water 1 000 ml	Recommended for zinc-rich alloys containing copper. Subsequent immersion in 20% chromic acid solution is again helpful
8	Chromic acid, CrO_3 200 g Sodium sulphate, Na_2SO_4 7 g Sodium fluoride 2 g Water 1 000 ml	Recommended for die-casting alloys containing aluminium, as the aluminium-rich microconstituent is satisfactorily etched with this reagent. In the presence of copper, staining may result; to prevent this, immerse after etching, without washing, in a solution of 50 g chromic acid and 4 g sodium sulphate in 1 000 ml water (*see* Etchant 5).
9	Solution prepared thus: mix 51 ml of concentrated potassium hydroxide solution with 50 ml of water and 20 ml of concentrated copper nitrate solution. Stir in 25 g of powdered potassium cyanide. Filter, and add 2.5 ml of concentrated citric acid solution before use	Etch by immersion for 10–20 s. The zinc-rich phase is coloured dark brown to black. The iron-phase in commercial alloys usually appears as white rods, and the lead-phase as round white spots

ETCHING

The reagents listed in Table 10.34 may be used for zinc and its alloys.

ELECTROLYTIC ETCHING OF ZINC ALLOYS

Comparatively little work has been done on electrolytic etching of zinc alloys, since the microstructures are relatively simple and adequately brought out by the above reagents. For alloys containing copper, however, a 20% aqueous chromic acid solution has been recommended, the specimen being the anode.

Ternary alloys containing aluminium and copper may be conveniently etched electrolytically in a solution made by adding 20 drops of hydrochloric acid to 50 ml alcohol.

DEVELOPMENT OF MICROSTRUCTURES WITHOUT ETCHING

Zinc and many zinc-rich alloys develop satisfactory microstructures if the polished surface is allowed to remain exposed to the air for 1–3 days. Similar results may be obtained by heating at 100 °C for a shorter time, but care must be taken to ensure that the use of the high temperature does not lead to any structural modification.

10.3.25 Zirconium

PREPARATION

In general, standard methods of preparation may be used, but difficulties may be encountered in connection with the production of twin markings by surface deformation.[104] Further hand or machine grinding on successive grades of silicon carbide papers gives good results, particularly when a considerable proportion of hard intermetallic compound is present. Alundum papers may also be used, but tend to fragment compound particles. The final polish is given on two wheels used successively. On the first wheel, diamond dust (5–8 μm) is used with kerosene on aeroplane silk, or billiard cloth if the specimen is subject to scratching. The second wheel employs synthetic sapphire with water as lubricant.

The use of plastic laps and attack polishing is also recommended (Table 10.3, p. **10**–8), while chemical polishing (Table 10.5, p. **10**–14) is often effective.

ETCHING

The etching reagents for zirconium and its alloys are summarized in Table 10.35.

Table 10.35 ETCHING REAGENTS FOR ZIRCONIUM

No.	*Etchant*		*Remarks*
1	Nitric acid (1.40) Hydrofluoric acid (1.19) Glycerol	20 ml 20 ml 60 ml	The reaction rate may be increased by heating the sample in a stream of hot water before immersion in the reagent. Conversely the reaction rate is decreased by chilling the specimen
2	Glycerol Hydrofluoric acid (1.19) Nitric acid	60 ml 20 ml 10 ml	On etching for 3–5 s, microconstituents are outlined and differentiated, and carbides unattacked. In the presence of moisture the reagent tends to stain the specimen
3	Glycerol Hydrofluoric acid (1.19) Nitric acid (1.40) Water	16 ml 2 ml 1 ml 2–4 ml	Etching times of 1–2 s are used. The reagent is useful for alloys which are not satisfactorily etched by Etchant 2
4	Hydrofluoric acid (1.19) Water Nitric acid (1.40)	20 ml 80 ml 1 ml	Short etching times are necessary (1–2 s). The reagent is similar to Etchant 3 but is more drastic

ELECTROLYTIC ETCHING

Solutions for electrolytic etching are given in Table 10.36.

Table 10.36 ELECTROLYTIC ETCHING OF ZIRCONIUM

No.	*Etchant*		*Remarks*
1	Ethyl alcohol Perchloric acid (density 1.2) 2-Butoxy-ethanol	70 ml 20 ml 10 ml	Suitable for cast zirconium
2	Acetic acid Perchloric acid (density 1.59)	1 000 ml 50 ml	Suitable for worked and annealed material
3	Ethyl alcohol Hydrochloric acid conc.	30 ml 10 ml	Etching time 10–20 s at 1 A dm^{-2}
4	Ethyl alcohol Distilled water Perchloric acid	450 ml 70 ml 25 ml	As for Etchant 3

10.3.26 Bearing metals (lead–tin–antimony), low melting solders, and type metals

PREPARATION

Bearing metals consist of hard particles of intermetallic compounds set in a matrix of soft lead-rich or tin-rich material. Much the same applies to printing metals and solders. The precautions to be observed are in general those to be observed for lead alloys and tin alloys, and a microtome may be used with advantage. Very light pressure during grinding is essential, and care must be taken, during polishing, to ensure that the soft matrix is not smeared over the surfaces of hard particles, and that the latter are not obtained in excessive relief.

ETCHING

All the reagents listed for lead and tin alloys may be tried for revealing the structure of bearing metals, the choice depending on whether the alloy is lead-rich or tin-rich. Of the reagents used for lead, the most useful for the present purpose are:

Nos. 4, 5, 6, 8, 11 and 12;

of the reagents for tin,

Nos. 1, 6 and 7

are suitable.

Further reagents are summarized in Table 10.37.

Table 10.37 ETCHING REAGENTS FOR BEARING METALS

No.	*Etchant*	*Remarks*
1	Iodine in potassium iodide solution	Type metals containing small amounts of zinc
2	Iodine, 10% in alcohol	Type metals containing large amounts of zinc
3	Silver nitrate (2–5%) in water	Useful for bearing metals in general

ELECTROLYTIC ETCHING

Bearing metals may be polished electrolytically in a solution containing 60% perchloric acid and acetic anhydride in the ratio of 1 part to 4 parts by volume. If the voltage is reversed after polishing, sensitive etches of most bearing metal materials may obtained.

10.3.27 Cemented carbides and other hard alloys

PREPARATION

Owing to the hardness of these materials, normal methods of preparation cannot be used. In one method, specimens are ground flat on an abrasive wheel, and polished with successively finer grades of diamond powder suspended in olive oil. Polishing should be carried out on a horizontal wheel, the surface of which may be hard-wood, felt or leather. The specimen should be kept cool with running water.

Alternatively, a diamond hone[4] or diamond-impregnated plastic laps may be preferred.[106] The surface of the lap is moulded from 6 g plastic: 1.25 g (6 carats) diamond, giving 0.8 mm thickness over a $7\frac{1}{2}$ cm diameter disc. Two laps are used, with 300 mesh diamond for roughing and with 4–8 μm diamond for intermediate grinding. In use they are lubricated with light paraffin oil. Finishing is by normal diamond polishing on cloth, in two stages.

ETCHING

A number of etching reagents have been developed for these materials, and a selection is given in Table 10.38.

ELECTROLYTIC ETCHING OF TUNGSTEN CARBIDE–COBALT AND TUNGSTEN CARBIDE–TITANIUM CARBIDE–COBALT ALLOYS

Two solutions have been recommended:

1. For tungsten carbide in a matrix of cobalt:

Sodium hydroxide	10%	in water
Potassium ferricyanide	10%	

On electrolysis at 2 V, the carbide particles are attacked, and the cobalt is almost unaffected.

2. For tungsten and titanium carbides in a matrix of cobalt:

Nitric acid conc.	10 ml
Hydrofluoric acid	10 ml

Titanium carbide is attacked, while tungsten carbide is not attacked.
The cobalt-rich matrix is attacked and dissolved.

Table 10.38 ETCHING REAGENTS FOR CEMENTED CARBIDES, ETC.

No.	*Etchant*	*Remarks*
1	Potassium hydroxide, 20% in water Potassium ferricyanide, 20% in water Equal volumes are mixed	The cold solution is used as a general reagent
2	Potassium hydroxide 10 g Potassium ferricyanide 10 g Water 10 ml	The boiling solution is used as a general reagent

Table 10.38 ETCHING REAGENTS FOR CEMENTED CARBIDES, ETC.—*continued*

No.	*Etchant*		*Remarks*
3	Aqua regia (20 ml nitric acid + 80 ml hydrochloric acid)		General reagent
4	Hydrofluoric acid Nitric acid conc.	70 ml 30 ml	Rapidly attacks eutectic material in iron–tungsten carbide alloys and iron–molybdenum carbide alloys
5	Picric acid, 2% in alcohol		Develops the eutectic structure in iron–tungsten carbide alloys and iron–molybdenum carbide alloys. The carbides are differentiated
6	Nitric acid, 3% in alcohol		The crystal boundaries in iron–tungsten carbide and iron–molybdenum carbide alloys are developed
7	Mixtures of 5 and 6		Generally effective for this class of material
8	Phosphoric acid Hydrogen peroxide	10 ml 10 ml	Used after successive treatments in Etchants 4 and 1 for molybdenum carbide–titanium carbide–cobalt materials. The titanium carbide with tungsten in solution and the cobalt are darkened; the tungsten carbide with titanium in solution is relatively unaffected[105]
9	Dilute ammonium sulphide		Used after successive treatments in Etchants 4, 1 and 8 for similar materials as recommended for Etchant 8. The carbides are unaffected, but the cobalt-rich matrix is darkened, and differentiated from titanium carbide[105]
10	Potassium permanganate Potassium hydroxide Water	10 g 5 g to 100 ml	WC grey, TiC, pink, TaC gold β phase rapidly attacked

ELECTROLYTIC ETCHING OF IRON–TUNGSTEN–CARBON AND IRON–MOLYBDENUM–CARBON ALLOYS

The use of a 5% solution of potassium ferrocyanide in 5% sodium hydroxide has been suggested. The metallographic effects are very similar to those of Etchant 5 in Table 10.38.

10.3.28 Powdered and sintered metals

PREPARATION

The preparation of powders, pressed compacts, and sintered metals is a specialized process, and requires special methods.

For the examination of powders, a suitable mounting may be made in plastic. The medium chosen will depend on many factors such as the melting point, reactivity and hardness of the metal of which the powder is composed. The plastic mounting may be made simply by mixing the metal powder with an appropriate amount of the mounting medium, and heating and pressing the composite sample in the usual way. Grinding, polishing and etching are then carried out as for the metal involved.

In pressed compacts and sintered materials, voids are present; these tend to pick up grinding and polishing materials, and to absorb etching reagents, so that difficulty may be experienced at all stages of the preparation. Comparatively little work has been carried out to determine precisely the best technique for the metallographic preparation of such specimens but impregnation with plastic is sometimes helpful (*see* section 10.2—*Microscopic examination, mounting*, p. **10**–6).

10.4 Electron metallography

The application of electron microscopy to the study of the microstructure of metals and alloys is termed 'electron metallography'. There are basically two types of electron microscope: one operating primarily as a transmission instrument and the other known as a scanning electron microscope which operates in the reflection or emission modes. Combined scanning/transmission electron microscopes (STEM) are increasing in popularity.

10.4.1 Transmission electron microscopy

As electrons are readily scattered by metals, the materials and components used industrially cannot in general be examined directly in the transmission microscope, so that special preparation techniques are required.[107–113] In *the replica technique* the surface topography of the metal is reproduced in a thin film of a substance which is not decomposed by the action of the electron beam, usually carbon. Generally, the technique involves polishing and etching a section through the metal, as in optical metallography, when the structure so developed can be reproduced either using a two-stage plastic/carbon or a single-stage direct carbon replica. In the former case the plastic most often used is cellulose triacetate sheets of which 20–250 µm thick are softened in acetone, laid on to the prepared metal surface and as the acetone evaporates the plastic hardens and contracts into the surface features. It is then carefully removed from the metal, and a layer of carbon 200–300 Å thick, is evaporated *in vacuo* on to the replica face, after which the plastic is dissolved in acetone leaving a secondary carbon replica. The more direct technique is to evaporate the carbon layer directly on to the prepared metal surface. Then because the carbon layer is somewhat porous it may be removed by re-etching the surface to dissolve the metal away from the carbon film. A particularly useful variation of this technique is the carbon extraction replica, which is used for the examination of multiphase alloys. The etching of the specimen, both prior to and after deposition of the carbon film, is controlled so that one phase, usually the matrix, is dissolved at a much faster rate than the other phase(s). In this way the second phase precipitate particles present in most alloys, except those of aluminium, can be extracted from the matrix on to the carbon replica film and because these particles are attached to the carbon, their distribution in the replica is normally identical to that in the original bulk metal sample. Prior to the examination in the microscope the pieces of carbon replica are collected on to fine mesh support grids, usually made of copper. The resolution attainable in carbon replicas is typically 20 Å and they may be usefully examined at all magnifications up to about 40 000 times. The carbon extraction replica technique is the most popular replication procedure used in the examination of all types of steels, copper alloys, nickel and cobalt-based superalloys, titanium alloys and uranium alloys. Details of replica techniques suitable for each of these types of alloy are given in Table 10.39. In addition to the replication of metallographically prepared surfaces, replica transmission microscopy is also used to study the fracture surfaces of failed metallic components, a technique known as electron fractography. The procedure is basically as described above, except that the specimen is not prepared in any way except possibly careful cleaning to remove corrosion products, the plastic or carbon being deposited directly on to the fracture surface. The contrast in such replicas may be enhanced by shadowing the surface of the carbon with a heavy metal such as gold or palladium.

Table 10.39 EXTRACTION REPLICA TECHNIQUES FOR INDUSTRIAL ALLOYS

Alloy	*Technique*	*Reference*
Copper alloys such as aluminium bronze, aluminium silicon bronze and cupro-nickels	(a) Etch in ferric chloride and hydrochloric acid in water for approx. 30 s (b) Deposit carbon (c) Re-etch in the acid ferric chloride solution for 2–5 min, wash in alcohol, float off replicas in distilled water	109
	(a) Etch in saturated aqueous solution of sodium bisulphate for approx. 2 min (b) Deposit carbon (c) Re-etch for several minutes in sodium bisulphate solution, back with layer of Bex† film to dry strip. Dissolve Bex in acetone	109
	(a) Etch in alcoholic ferric chloride (b) Deposit carbon (c) Re-etch in either Disapol* D2 electrolyte at 5 V or E5 electrolyte at 30 V	128

* Disapol is a trade name of H. Struers Chemiske Laborotorium, Skindergade 38, Copenhagen.
† Bex is a proprietary brand of cellulose triacetate obtainable from Polaron Equipment Ltd., Watford, England.

Table 10.39 EXTRACTION REPLICA TECHNIQUES FOR INDUSTRIAL ALLOYS—*continued*

Alloy	*Technique*	*Reference*
Mild and low alloy steels	(a) Etch in ethyl alcohol + 2/5% nitric acid for 10–40 s (b) Deposit carbon (c) Strip by etching as in (a)	129
	(a) Etch in 5 g cupric chloride, 40 ml hydrochloric acid, 30 ml water, 25 ml ethyl alcohol (b) Deposit carbon (c) Strip by etching as in (a) also electropolish in ethyl alcohol + 10% nitric acid at 10 V Also use above techniques, without step (a), for fracture surfaces	129
High chromium and alloy steels	(a) Etch in methyl alcohol + 10% hydrochloric acid (b) Deposit carbon (c) Strip by etching as in (a)	130
High speed steels	(a) Electropolish in solution containing 50 ml perchloric acid (s.g. 1.54), 950 ml glacial acetic acid, stainless steel cathode, 1 $A\,cm^{-2}$ with applied voltage of 64 V (b) Etch in Vilella's reagent (c) Deposit carbon (d) Strip by etching in solution of 5% nitric acid in alcohol containing few drops of hydrofluoric acid (etching time ~ 1.5 min)	131
Stainless steels	(a) Electropolish in 5% perchloric acid, 95% glacial acetic acid + 10 $g\,l^{-1}$ nickel chloride + 20 $g\,l^{-1}$ chromic anhydride at 45 to 60 V (b) Etch in 20% hydrochloric acid + 80% of 10% aqueous solution of chromic anhydride at 6.5 V for approx. 5 s (c) Deposit carbon (d) Strip by etching in aqueous solution containing 50% hydrochloric acid for several hours	109
Nickel alloys	For extraction of carbides (a) Mechanical polish to $\frac{1}{4}$ μm diamond, electropolish in 10% perchloric acid + 90% acetic acid at 40 V (b) Etch 15 s in 25 g ferric chloride, 25 g cuprous chloride, 100 ml nitric acid, 300 ml hydrochloric acid (c) Deposit carbon (d) Re-etch for 30 s as in (b) (e) Strip carbon by electropolishing at 40 V in 20% perchloric acid in ethyl alcohol	109
	For extraction of γ' particles (a) Polish as in (a) above (b) Etch in 10% phosphoric acid in water at 30 V for 1 s (c) Deposit carbon (d) Strip carbon by electropolishing at 40 V in 20% perchloric acid in ethyl alcohol	109
Titanium alloys	For α alloys (a) Electropolish in methyl alcohol, butyl cellosolve, perchloric acid solution (b) Etch in 2% hydrofluoric acid in saturated aqueous solution of oxalic acid (c) Strip replicas in 1 part hydrofluoric acid, 1 part nitric acid, 30 parts water	132
	For $\alpha + \beta$ alloys (a) Electropolish as (a) above (b) Etch in aqueous solution 1% hydrofluoric acid, 2% nitric acid (c) Strip replicas as (c) above	133

Table 10.39 EXTRACTION REPLICA TECHNIQUES FOR INDUSTRIAL ALLOYS—*continued*

Alloy	*Technique*	*Reference*
Titanium alloys (*cont.*)	(a) Etch for 5 s in aqueous solution containing 1% hydrofluoric acid, 2% nitric acid (b) Deposit carbon (c) Strip by scoring carbon into squares and etching for 20 s in 1 part hydrofluoric acid, 1 part nitric acid, 30 parts water	109
Uranium alloys	(a) Mechanical polish down to $\frac{1}{4}$ μm diamond (b) Polish-attack for 20 min on terylene cloth using 50 g l^{-1} of chromic anhydride, 100 ml glacial acetic acid, 100 ml water + γ alumina (c) Electrolytic etch, 2% citric acid, 0.5% nitric acid, 97.5% water at 6 V for approx. 40 s (d) Deposit carbon (e) Score carbon film into squares, electropolish in bath given in (c) at 15 V for 10–20 min. Float off in water	109
	(a) Mechanical grind to 6 (b) Electropolish in 50 ml phosphoric acid, 50 ml water. Open circuit of 30 V, c.d. of 4.65 kAm^{-2} for 40 s (c) De-oxidize in 75% sulphuric acid, 18% glycerol, 7% water at 10 V, initial current density is 2.32 kAm^{-2} falls to 0.7 A on completion of deoxidation (d) Etch in solution of 2% chromic anhydride in 25% acetic acid at 2 V (c.d. 75 Am^{-2}) for approx. 20 s with platinum cathode (e) Deposit carbon (f) Strip by etching as in (d) for 10 to 20 s. If unsuccessful, electropolish for few seconds as in (b)	109

The second type of specimen preparation procedure, *the thin foil technique* involves reducing the thickness of the metal until it can be penetrated by the electron beam of the microscope. The limiting thickness depends on the atomic number of the metal and the electron accelerating voltage at which the microscope is operated. Thus for conventional instruments operating at 100 kV the thickness decreases from about 2000 Å for aluminium to about 300 Å for uranium, whereas using a high voltage microscope at 1000 kV an aluminium specimen of up to about 9 μm can be penetrated. The usual method for reducing the thickness of the specimen is by electropolishing[114–119] and a widely used procedure requires that the starting specimen is in the form of a disc up to 3 mm diameter and 0.5 mm thick, which can be readily prepared from the bulk material by processes such as grinding or machining.[120–125] These discs are then electropolished until a small hole(s) is formed near the centre, when it is usually found that the metal adjacent to the edge of the hole is sufficiently thin to transmit electrons. Using these disc techniques thin foils can easily be produced from selected areas or from a required distance with respect to a reference surface. Electron microscopes and associated techniques are now capable of resolutions of 3–5 Å. Line imaging using phase contrast enables lattice planes to be resolved. High resolution can only be attained in thin foils of ferromagnetic materials such as ferritic steels, by ensuring that the bulk of specimen is as small as possible with the perforation occurring in the centre of the sample, to minimize the effects of the foil on the magnetic field of the objective lens. Modification of the structure of a metal may occur during the preparation of thin foils or during subsequent examination in the microscope. Thus defect structures may be altered because of the relaxation of long range stresses during thinning or due to mechanical damage of the thin foil by careless handling. Also metastable alloys may undergo a phase transformation when in thin foil form, either because the surface acts as a preferential nucleating site for nucleation and growth transformations, or because in materials which undergo a phase transformation by shear, the large stresses produced by the transformation may be relaxed at the surface of the foil. The thin foil technique has been used for studying in detail the microstructure of all the common industrial alloys, and some suitable preparation procedures are given in Table 10.40.

Table 10.40 THIN FOIL TECHNIQUES FOR INDUSTRIAL ALLOYS[107]

Material	*Technique*	*Solution*	*Conditions*
Ag–Zn	Jet	9% KCN, 91% H_2O	
Al	Window	1 part nitric acid, 3 parts methanol	−70 °C, 40 V
Al–Cu	Chemical	94 parts phosphoric acid, 6 parts nitric acid	
Al–Au alloys	Window	20% perchloric acid, 80% methanol	16–18 V, 100 cm^{-2}, −55 to −65 °C
Au	Window (3 mm discs)	100 ml CH_3COOH, 20 ml HCl, 3 ml H_2O	Stainless steel cathode, 32 V, 15 °C
	Electrolytic	(a) 15% glycerol, 35% 50% HCl (b) 133 ml glacial CH_3COOH, 25 g CrO_3, 7 ml H_2O	30 V, −30 °C, 1.5 A cm^{-2}—fast polish Finish after (a) at 22 V, 0.8 A cm^{-2}, 0 °C
Be, Be–Cu	Window	20% perchloric acid, 80% ethanol	−30 °C, 40 V
Co–Fe	(a) Chemical (b) Jet	50% phosphoric acid, 50% hydrogen peroxide 20% perchloric acid, 80% methanol	 −20 °C
Cu–Ni–Fe alloys	Chemical	20 ml acetic acid, 10 ml nitric acid, 4 ml HCl	
Cu–Ti	Jet	750 ml acetic acid, 300 ml phosphoric acid, 150 g chromic acid, 30 ml distilled H_2O	30–40 V, 45–55 A
Fe–Al–C	Chemical	HF, H_2O, H_2O_2 (1 : 3 : 16)	
Fe_3Si	Window	1% perchloric acid, 2.5% hydrofluoric acid in methanol	−77 °C
Gd–Ce alloys	Window	1% perchloric acid, 99% methyl alcohol	−77 °C
Hf	(a) Jet (b) Jet	45 parts nitric acid, 45 parts H_2O, 8 parts HF 2 parts perchloric acid, 98 parts ethanol	 40–45 V, 16–20 mA, −77 °C
In	Window: from 0.1 mm thick	33% HNO_3, 67% methanol	−40 °C
Mg	 Chemical	(a) 20% $HClO_4$, 80% ethanol (b) 375 ml H_3PO_4, 625 ml ethanol 6% nitric acid, 94% H_2O	Fast polish, 0 °C Finish at 15 V, 0 °C, lowest current 3 °C
Mo	Jet	1 part sulphuric acid, 7 parts methanol	10 V
Nb	Chemical	60% HNO_3, 40% HF	0 °C
Nb alloys	Window	100 ml lactic acid, 100 ml sulphuric acid, 20 ml HF	40 °C, ~0.4 A cm^{-2}
Ni steels Stainless steels	Chemical	50 ml 60% H_2O_2, 50 ml H_2O, 7 ml HF	Place specimen in H_2O_2 solution, then add HF until reaction starts; produces 1000 Å foils which can then be finished by electro-polishing by standard chromic-acetic electropolishing (courtesy Republic Steel Corp.)
Ni alloys Cu–Ni–Fe	Jet	1 part HNO_3, 3 parts CH_3OH	6–10 V, 18–30 mA, −30 °C

Table 10.40 THIN FOIL TECHNIQUES FOR INDUSTRIAL ALLOYS[107]—*continued*

Material	*Technique*	*Solution*	*Conditions*
Ni	Jet	20% perchloric acid in ethanol	
Ni–Al alloys	Jet	20% sulphuric acid, 80% methanol	Dished at 150 V, final polish in same solution at 10–12 V
Ni and Ni–Fe	Jet	8 parts 50% sulphuric acid, 3 parts glycerine	Less than 10 °C, 1.3 A cm^{-2}
Rene 95	Jet	250 ml methanol, 12 ml perchloric acid, 150 ml butylcellusolve	−35 °C, 65 V
Th	Window	5% perchloric acid, 95% methanol	−77 °C, 3–10 V
Pt	Electrolytic	H_2SO_4 (96%), HNO_3 (65%), H_3PO_4 (80%)	0.2–0.5 A cm^{-2}, 20–30 °C
Re	Window	Ethyl alcohol, perchloric acid, butoxyethanol (6 : 3 : 1)	−40 °C, 35 V
Ta	Chemical	3 parts HNO_3, 1 part HF	−10 °C, immerse in Petri dish after lacquering
Ta	Jet	1 part sulphuric acid, 5 parts methanol	−5 °C, current density of 0.5 A cm^{-2}
Ti	Jet	1 part HF, 9 parts sulphuric acid	
Ti–Nb	Chemical	4 parts nitric acid, 1 part HF	
TiC	Jet	6 parts nitric acid, 2 parts HF, 3 parts acetic acid	100 V
Ti and Ti alloys	Jet	30 ml $HClO_4$, 175 ml *n*-butanol, 300 ml CH_3OH	15 V, 0.1 A cm^{-2}, may need ion thinning at end to remove contamination layer
U	Jet	1 part perchloric acid, 10 parts methanol, 6 parts butylcellusolve	−20 °C, 35–45 V
UO_2 (sintered)	Chemical	10 ml HOAc, 20 ml sat. CrO_3, 5 ml 40% HF, 7 ml HNO_3 (d=1.42)	
	or Chemical	20 ml H_3PO_4 (d=1.75), 10 ml HOAc, 2 ml HNO_3 (d=1.42)	
V	Chemical	2 parts HF, 1 part nitric acid	Less than 10 °C
V	Jet	20% sulphuric acid in methanol	
Zn–Al	Chemical	50–70% nitric acid in water	
Zn–Al	Window	90% methanol, 10% perchloric acid	
Zr–25% Ti	Chemical	50 ml nitric acid, 40 ml H_2O, 10 ml HF	
Zr alloys	Jet	5% perchloric acid, 90% ethanol	70 V, below −50 °C

Selected area electron diffraction analysis is extensively used for the identification of second phase particles in carbon extraction replicas and thin foils, and individual particles down to ~1000 Å diameter can be analysed. Electron diffraction effects can also be obtained from other microstructural features such as GP zones in age-hardening alloys, order–disorder transformations and stacking faults, so that the interpretation of diffraction patterns can be difficult and requires extreme care.[107, 108, 112, 126, 127] Diffraction patterns from extraction replicas are normally calibrated (to determine the camera constant) by evaporating a substance whose interplanar spacings are known accurately, such as gold or thallium chloride, on to the carbon film. In the case of second phase particles in thin foils, the diffraction patterns are usually calibrated using the matrix phase. Spot patterns are obtained from single crystal particles and the patterns are effectively an enlarged image of a plane of the reciprocal lattice of the crystal. If a large number of small particles are contributing to the diffraction pattern, a ring pattern is obtained. In either case the interplanar spacings can be calculated from the equation: $d=(L\lambda)/r$, where L is the camera length, λ the wave-length of the electrons, and r is half the distance separating corresponding spots in the single crystal spot pattern, or the radius of the diffraction ring in a polycrystalline ring pattern. L is usually obtained from the calibration pattern. By calculating the d values for 3 or 4 sets of diffraction spots or rings and determining the symmetry of the diffraction pattern, it is possible to index the spots or rings in terms of their Miller indices and therefore to determine the crystal structures and parameters of the diffracting phase. The best accuracy possible with selected area electron diffraction is about 0.1% but is more commonly in the range 0.5–1.0%.

However, to identify unambiguously the structure of a single particle it is often necessary to tilt the diffracting crystal into two or three prominent diffracting zones. Also, as the contrast in images of thin foils arises mainly from Bragg diffraction of the electrons, the contrast from fine second phase particles and defect structures depends critically upon the orientation of the foil with respect to the electron beam. Therefore modern instruments incorporate special stages which allow the specimen to be tilted and rotated. It is also useful to have a facility for cooling the specimen, to reduce the amount of contamination which can occur if one area of the foil is examined for a prolonged period. Stages designed to enable the specimen to be heated or strained are of only limited value in electron metallography, because due to the high surface area: volume ratio, the rate and morphology of structural changes which occur during heating or straining of thin foils are usually different from those occurring in bulk material.

Modern microscopes now have fitted to them detectors to collect the X-rays generated by the interaction between the electron beam and the metallic foil specimen. The wavelengths of X-rays are then separated by either dispersive or non-dispersive systems and from this a semiquantitative analysis of the elements present can be obtained, usually from a volume of about 1–2 µm. The electron microscope for microanalysis (EMMA) has been developed to make more accurate analyses from areas of <0.5 µm.

10.4.2 Scanning electron microscopy

The use of scanning electron microscopy to study the surface structure of metals has increased tremendously over the last few years (for example, *see* refs. 143–145). This has been due to a number of advantages which this form of microscopy enjoys compared to optical or transmission electron microscopy, namely:

1. The depth of focus in the image is very large, at least 300 times that of a conventional optical microscope, and the resolution is typically better than 100 Å.
2. Compared to transmission electron microscopy the preparation required to produce a sample for examination by scanning microscopy is extremely simple. Normally all that is required is to mount a specimen, typically 3 cm dia. and 0.5 cm high, on to an aluminium stub using a conducting adhesive. If the sample contains non-conducting material on the surface, which will tend to charge-up during examination, it is also necessary to deposit a thin conducting film, usually of carbon or a gold/palladium alloy, by evaporation in a vacuum.
3. Images obtained by a scanning microscope are generally much easier to interpret than those from a transmission microscope, so that it is often possible to obtain the desired result from an examination of the image as it is formed on the viewing screen.

In the scanning microscope a beam of electrons with energies in the range 1–30 kV is focused to a spot $\sim$100 Å in diameter and made to scan the surface to be examined in a rectangular raster. Primary electrons are reflected and secondary electrons emitted from the surface and are focused with an electrostatic electrode on to a biased scintillator. The light produced is transmitted via a perspex light pipe to a photomultiplier and the signal generated is used to modulate the brightness of an oscilloscope spot which traverses a raster in exact synchronism with the electron beam at the specimen surface. The magnification of the projected image is easily adjusted, because it is simply the ratio of the rasters of the electron beam and oscilloscope. In modern scanning microscopes the X-rays emitted from the specimen surface under the action of the primary electron beam can also be analysed to yield information on the chemical composition. Each element in the specimen surface emits X-rays of characteristic wavelengths and energies. The wavelengths may be separated by a dispersive technique using a spectrometer and different energy levels may be distinguished with high resolution X-ray detectors, such as the lithium drifted silicon semiconductor radiation detector. This latter technique is particularly suitable for use in conjunction with scanning microscopes, in which the requirements of image resolution necessitate the use of lower electron beam intensities than in purpose-built X-ray microprobes.

10.4.3 Electron spectroscopy

This is a technique for studying the energy distribution of electrons ejected from a material on irradiation by a source of ionizing radiation such as X-rays, ultra-violet light or electrons.

When X-ray radiation is used as the ionizing source, the technique is called X-ray photoelectron spectroscopy (XPS) or electron spectroscopy for chemical analysis (ESCA). If ultra-violet light is used, the technique is called ultra-violet photoelectron spectroscopy (UPS).

Both electron and high-energy X-ray beams can also cause a secondary electron to be emitted from the atoms and the analysis of the energy of these electrons (Auger electrons) is known as Auger spectroscopy (AES).

The analysis of XPS and Auger spectroscopy has now been developed to an extent that not only can the atomic species be identified, but also information can be obtained about its binding energy, charge and valency. As the photoelectrons and Auger electrons can only escape from a region of 3–50 Å below the surface, the region analysed is truly only the surface and not more than a few atomic spacings below the surface. Not only can the individual elements be clearly resolved, but it is also possible to detect accurately changes in binding energy when the element exists in forms with different valencies.

The latest instruments combine scanning electron microscopy and energy dispersive analysis of X-rays with scanning Auger spectroscopy, so that it is possible to compare electron images with Auger images for specific elements. An analysis of X-rays generated from a depth of 0.5 μm can be contrasted with a simultaneous analysis of Auger electrons from the same area, but from a depth of 3–50 Å.

10.5 Quantitative image analysis

Image analysis may be defined as the science of making geometric and densitometric measurements and classifications of selected features in images formed by optical, transmission electron and scanning electron microscopes and from photographs, X-rays and radiographs. A number of image analysing computers are available which allow fast, accurate measurements and classifications, so that tasks in R & D, quality control and routine assessment which previously were impossible are now both possible and economic. The term ‘quantitative image analysis’ is virtually synonymous with Quantimet, a trade name of Image Analysis Computers Ltd., Cambridge, England. In modern image analysing computers the image is scanned by a vidicon or plumbicon scanner, whose output is passed to a detector. This either selects features for measurement in terms of their common grey level characteristics or alternatively the scanner output can be passed to a densitometer, which will assess the optical density at each point in a feature and compute its integrated density. The signal from the detector can be amended to allow for imperfect feature definition or can be passed directly to one or more of several alternative modules which make the required measurements. Alternatively, or additionally, results can be passed to data output systems, such as a desk-top calculator, or to a supervisor module, which screen large numbers of fields of view and select only those with interesting features. The entire operation can be automatically controlled by a simple programmer or it can be manually controlled by switches.

A large number of types of measurements can be made using an image analysing computer including (1) number, (2) area, (3) length, (4) perimeter, (5) mean linear intercept, (6) form factor, (7) optical density, (8) size distribution, (9) pattern or shape recognition. Typical uses include quality control of inclusions in steels, fibre classification, size distribution of powders, phase concentrations and grain morphology and pore sizing in composites.

REFERENCES

1. G. Petzow, 'Metallographic Etching', Amer. Soc. Metals, 1978.
2. 'Metals Handbook', 8th edn, Vol. 8, Amer. Soc. Metals, 1973.
3. C. A. Godden, *Met. Progress*, 1961, **79**, 121.
4. L. P. Tarasov and C. O. Lundberg, *Met. Progress*, 1949, **55**, 183.
5. L. E. Samuels, *J. Inst. Metals*, 1952–53, **81**, 471.
6. V. J. Haddrell, E. C. Sykes and B. W. Mott, *J. Inst. Metals*, 1955–56, **84**, 112.
7. V. J. Haddrell, *J. Inst. Metals*, 1963–64, **92**, 121.
8. P. A. Jacquet, 'Electrolytic and Chemical Polishing', *Met. Reviews*, 1956, **1** (2), 157.
9. W. J. McG. Tegart, 'The Electrolytic and Chemical Polishing of Metals in Research and Industry', 2nd edn, Pergamon, Oxford, 1959.
10. L. E. Samuels, *Metallurgia*, 1962, **66**, 187.
11. E. Knuth-Winterfeldt, *Trans. Instruments and Measurements Conf., Stockholm*, 1949, p. 223. (*See also* P. Jacquet, *ONERA Publication No. 51*, 1952. In French.)
12. E. C. Sykes, V. J. Haddrell, H. R. Haines and B. W. Mott, *J. Inst Metals*, 1954–55, **83**, 166.
13. E. Knuth-Winterfeldt, *Mét. et Corrosion*, 1948, **23**, 5.
14. P. A. Jacquet, *Compt. Rend.*, 1956, **243**, 1066.
15. P. A. Jacquet, *Proc. Am. Soc. Test. Mater.*, 1957, **57**, 1290.
16. P. A. Jacquet, *Rech. Aéronautique*, 1962, **90**, 15.
17. M. Médard, P. Jacquet and R. Sartorius, *Rev. Mét.*, 1949, **46**, 549.
18. A. Hickling, *Trans. Farad. Soc.*, 1941, **38**, 27.
19. M. H. Roberts, *Brit. J. appl. Phys.*, 1954, **5**, 351.
20. C. Edeleanu, *Metallurgia*, 1954, **50**, 113.
21. C. Edeleanu, *J. Iron Steel Inst.*, 1957, **185**, 482; *J. Iron Steel Inst.*, 1958, **188**, 122.
22. V. Čihel and M. Pražák, *J. Iron Steel Inst.*, 1959, **193**, 360.
23. C. S. Smith, *J. Inst. Metals*, 1927, **38**, 133.
24. L. Holland, 'Vacuum Deposition of Thin Films', Chapman & Hall, London, 1956, p. 432.
25. C. P. Shillaber, 'Photomicrography in Theory and Practice', Wiley, London, 1944.
26. K. W. Andrews, 'Physical Metallurgy: Techniques and Applications', Allen & Unwin, London, 1973.
27. S. Tolansky, 'Multiple Beam Interferometry of Surfaces and Films', Clarendon Press, 1948.
28. F. W. Cuckow, *J. Iron Steel Inst.*, Jan. 1949, **161**, 1.
29. D. McLean, *Met. Treatment*, Feb. 1951, 51.
30. G. Nomarski and A. R. Weill, *Rev. Mét.*, 1955, **52**, 121.
31. E. Knuth-Winterfeldt, *Mikroskopie*, 1950, **5**, 184.
32. P. A. Jacquet, *Compt. Rend.*, 1948, **277**, 556; *Rev. Mét.*, 1949, **46**, 214.
33. P. A. Jacquet, *Rev. Mét.*, 1957, **54**, 663.
34. H. J. Merchant, *J. Iron Steel Inst.*, 1947, **155**, 179.
35. G. Bassi, *Z. Metallkunde*, 1961, **52**, 141.
36. *idem*, *Z. Metallkunde*, 1960, **51**, 219.
37. A. de Sy and H. Haemers, *Metal Progress*, 1948, **53**, 368.
38. A. Hone amd E. C. Pearson, *Metal Progress*, 1948, **53**, 363.
39. A. de Sy and H. Haemers, *Aluminium*, 1942, **24**, 96.
40. J. A. Verö, *Mitt. berg. u. Hüttenmänn. Abt. Univ. Tech. Wiss., Sopron*, 1948–49, **17**, 23.
41. G. K. T. Conn and F. J. Bradshaw (Ed.), 'Polarised Light in Metallography', Butterworth, London, 1952.
42. L. B. Larke and E. H. Wicks, *Metallurgia*, Jan. 1950, **41**, 172.
43. B. W. Mott and H. R. Haines, *J. Inst. Metals*, 1951–52, **80**, 629.
44. F. Beraha and B. Shpigler, 'Colour Metallography', Amer. Soc. Metals, 1977.
45. G. B. Brook and R. I. Saunderson, 2nd Charlottesville Conference, 1981, *AMMRC*.
46. I. Epelboin and M. Froment, *Métaux, Corrosion, Ind.*, Jan. 1958, **33** (389), 1.
47. Z. S. Basinski and J. W. Christian, *J. Inst. Metals*, 1951–52, **80**, 659.
48. N. K. Chen and R. Maddin, *Trans. AIMME*, 1951, **191**, 937.
49. G. L. Hopkin, J. E. Jones, A. R. Moss and D. O. Pickman, *J. Inst. Metals*, 1953–54, **82**, 361.
50. F. R. Cortes, *Met. Progress*, 1961, **80**, 97.
51. T. P. Hoar and J. A. S. Mowat, *Electrodep. Tech. Soc. Reprint No. 1*, 1950.
52. R. Shuttleworth, R. King and B. Chalmers, *Metal Treatment*, 1947, **14**, 161.
53. O. J. Krudtaa and K. Stokland, *Met. Progress*, 1960, **77**, 101.
54. M. D. Smith and R. W. K. Honeycombe, *J. Inst. Metals*, 1954–55, **83**, 421.
55. H. R. Haines and B. W. Mott, *J. Sci. Instruments*, 1953, **30**, 459.
56. R. Osadchuk, W. P. Koster and J. F. Kahles, *Met. Progress*, Oct. 1953, **64**(4), p. 129.
57. B. W. Mott and H. R. Haines, *J. Inst. Metals*, 1951–52, **80**, 621.
58. W. Evans, *Trans. Canad. Inst. Min. Metall.*, 1960, **63**, 617.
59. H. J. Engell, *Arch. Eisenhüttenw.*, 1958, **29**, 73.
60. D. W. Bloor, *Metallurgia*, 1962, **66**, 139.
61. L. C. Lovell, F. L. Vogel and J. H. Wernick, *Met. Progress*, May 1959, **75**, 96.
62. D. J. Barber, *Phil. Mag.*, 1962, **7**, 1925.
63. G. Bassi and J. P. Hugo, *J. Inst. Metals*, 1958–59, **87**, 155.
64. A. W. Ruff, *J. appl. Phys.*, 1962, **33**, 3392.
65. E. Sirtl and A. Alder, *Z. Metallkunde*, 1961, **52**, 529.
66. G. Bassi and J. P. Hugo, *J. Inst. Metals*, 1958–59, **87**, 376.

67. L. J. Barker, *Iron Age*, 1949, **163**, 74.
68. H. J. Seemann and M. Dudek, *Metalloberfläche*, 1948, **2**, 84.
69. P. Lacombe and L. Beaujard, *Rev. Mét.*, 1947, **44**, 71; *J. Inst. Metals*, 1948, **74**, 1.
70. R. W. Cahn, *J. Inst. Metals*, 1949–50, **76**, 121.
71. F. Erdmann-Jesnitzer and W. Bernhardt, *Metall.*, 1957, **11**, 1032.
72. J. Hérenguel and F. Santini, *Métaux et Corrosion*, 1946, **21**, 131.
73. N. S. Bromelle and H. W. L. Phillips, *J. Inst. Metals*, 1949, **75**, 529.
74. H. Kostron, *Z. Metallkunde*, 1948, **39**, 333.
75. M. C. Udy, G. K. Manning and L. W. Eastwood, *Trans. AIMME*, 1949, **185**, 779.
76. A. R. Kaufmann, P. Gordon and D. W. Lillie, *Met. Progress*, 1949, **56**(5), 664.
77. F. N. Rhines, W. M. Urquhart and H. R. Hoge, *Trans. Am. Soc. Metals*, 1947, **39**, 694.
78. S. C. Carapella and E. A. Peretti, *Met. Progress*, 1949, **56**, 666.
79. I. R. Lambourn, *Australasian Engineer*, Nov. 1948, 39.
80. P. F. Weinrich, *Australasian Engineer*, Nov. 1948, 42.
81. R. W. Turner, Report No. 538, *BCIRA Journal*, March 1960, **8** (2), 238.
82. D. H. Rowland, *Trans. Am. Soc. Metals*, 1948, **40**, 983.
83. E. F. Emley, 'Principles of Magnesium Technology', Pergamon, Oxford, 1966.
84. G. E. Holdeman, *Trans. AIME*, 1956, **64**, 698.
85. E. F. Emley, *British Welding J.*, 1957, **4**, 307.
86. P. F. George, *Trans. Am. Soc. Metals*, 1947, **38**, 686.
87. E. F. Emley, *J. Inst. Metals*, 1949, **75**, 431.
88. E. F. Emley, *J. Inst. Metals*, 1949, **75**, 481.
89. R. Bakish, *J. Electrochem. Soc.*, 1958, **105**, 574.
90. G. L. Miller, 'Tantalum and Niobium', Butterworth, London, 1948.
91. B. L. Eyre, *Metallurgia*, August 1958, **58**, 95.
92. L. T. Greenfield and J. E. Davies, 'The Preparation of Tin and Tin Alloys for Microscopical Examination', Tin Research Inst., Middlesex, 1951.
93. J. V. Harding and W. I. Pell-Walpole, *J. Inst. Metals*, 1948, **75**, 115.
94. R. F. Smart, R. M. Angles and D. A. Robins, *J. Inst. Metals*, 1960–61, **89**, 349.
95. D. A. Sutcliff, J. I. M. Forsyth and J. A. Reynolds, *Metallurgia*, 1950, **41**, 283.
96. H. R. Odgen and F. C. Holden, 'Metallography of Titanium Alloys', Titanium Met. Lab. Report 103, Battelle Memorial Inst., Columbus, Ohio, May 1958.
97. H. Woods, *Met. Progress*, 1947, **51**, 261.
98. J. D. Grogan and C. J. Birkett Clews, *J. Inst. Metals*, 1950, **77**, 571.
99. D. Boyd Metz and H. W. Woods, *US Atomic Energy Com. Publ.*, 1950 (Sep. 42).
100. J. F. R. Ambler and G. F. Slattery, *J. nucl. Mater.*, 1961, **4**, 90.
101. W. M. Justusson, *J. nucl. Mater.*, 1961, **4**, 37.
102. C. H. Schramm, P. Gordon and A. R. Kaufmann, *Trans. AIME* (*J. of Metals*), 1950, **188**, 195.
103. P. Gordon and A. R. Kaufmann, *Trans. AIME* (*J. of Metals*), 1950, **188**, 182.
104. A. H. Roberson, *Met. Progress*, 1949, **56**, 667.
105. R. Kieffer, *Metallforschung*, 1947, **2**, 236.
106. L. E. Samuels, *J. Aust. Inst. Metals*, 1960, **5**, 63.
107. G. Thomas and M. J. Garrinje, 'Transmission Electron Microscopy of Metals', Wiley, New York, 1979.
108. D. Kay, Ed., 'Techniques for Electron Microscopy', Blackwell, Oxford: 1965.
109. I. S. Brammar and M. A. P. Dewey, 'Specimen Preparation for Electron Metallography', Blackwell, Oxford, 1966.
110. P. J. Goodhew, 'Specimen Preparation in Materials Science', Part I, Vol. 1 of *Practical Methods in Electron Microscopy*, North-Holland, Amsterdam, 1972.
111. 'Advances in Electron Metallography', ASTM Spec. Tech. Report 396, 1966.
112. P. B. Hirsch, A. Howie, R. B. Nicholson, D. W. Pashley and M. J. Whelan, 'Electron Microscopy of Thin Crystals', Butterworth, London, 1965.
113. 'Electron Fractography', ASTM Spec. Tech. Report 436, 1967, 453, 1969 and 493, 1971.
114. W. Bollmann, *Phys. Rev.*, 1956, **103**, 1588.
115. D. Brandon and J. Nutting, *Acta. Met.*, 1959, **7**, 101.
116. R. B. Nicholson, G. Thomas and J. Nutting, *Br. J. appl. Phys.*, 1958, **9**, 25.
117. R. M. Fisher, ASTM Spec. Tech. Publication 262, 1959, p. 104.
118. P. R. Strutt, *Rev. Sci. Instrum.*, 1961, **32**, 411.
119. R. C. Glenn and J. C. Raley, ASTM Spec. Tech. Publication 339, 1963, p. 60.
120. G. Blankenburgs and M. J. Wheeler, *J. Inst. Metals*, 1964, **92**, 337.
121. M. A. P. Dewey and T. G. Lewis, *J. Sci. Instrum*, 1963, **40**, 385.
122. G. W. Briers, D. W. Dawe, M. A. P. Dewey and I. S. Brammar, *J. Inst. Metals*, 1964, **93**, 77.
123. A. R. Davis, *J. Inst. Metals*, 1968, **96**, 61.
124. J. E. Bainbridge and L. Thorne, *J. nucl. Mater.*, 1970, **34**, 202.
125. J. A. F. Gidley and R. A. Davies, *J. Sci. Instrum.*, 1967, **44**, 297.
126. K. W. Andrews, D. J. Dyson and S. R. Keown, 'Interpretation of Electron Diffraction Patterns', Hilger and Watts Ltd., London, 1971.
127. B. E. P. Beeston, R. W. Horne and R. Markham, 'Electron Diffraction and Optical Diffraction Techniques', Part II, Vol. 1 of *Practical Methods in Electron Microscopy*, North-Holland, Amsterdam, 1972.
128. M. Vowles and J. Billingham, Fulmer Research Institute private communication, 1971.
129. T. J. Baker and M. A. P. Dewey, Fulmer Research Institute private communication, 1970.

130. P. W. Teare and N. T. Williams, *J. Iron Steel Inst.*, 1963, **201**, 125.
131. C. H. White and R. W. K. Honeycombe, *J. Iron Steel Inst.*, 1961, **197**, 21.
132. A. L. Dalton, D. Webster and H. C. Child, *Iron and Steel Inst.*, Spec. Report 70, 1961.
133. N. A. Neilsen, ASTM Spec. Publication 245, 1958.
134. B. Lux and W. Bollmann, *Cobalt*, 1961, **12**, 32.
135. P. R. Swann and J. Nutting, *J. Inst. Metals*, 1960, **88**, 478.
136. M. A. P. Dewey, Fulmer Research Institute private communication, 1967.
137. M. Vowles and J. Billingham, Fulmer Research Institute private communication, 1972.
138. R. C. Glenn and R. D. Schoone, *Rev. Sci. Instrum.*, 1964, **35**, 1223.
139. R. D. Schoone and E. A. Fischione, *Rev. Sci. Instrum.*, 1966, **37**, 1351.
140. A. F. Rowcliffe, *J. Inst. Metals*, 1966, **94**, 263.
141. R. W. Gardiner and P. G. Partridge, *J. Sci. Instrum.*, 1967, **44**, 63.
142. J. M. Capenos and F. H. Froes, *J. Sci. Instrum.*, 1969, **2**, 735.
143. P. R. Thornton, 'Scanning Electron Microscopy. Application to Materials and Device Science', Chapman & Hall, London, 1968.
144. 'Scanning Electron Microscopy', *Proceedings of Annual Scanning Electron Microscope Symposium*, Illinois Institute of Technology, Chicago, 1968–1972.
145. F. E. Astbury and C. Baker, *Metals and Materials*, 1967, **1**, 10, 323.

11 Equilibrium diagrams

11.1 Index of binary diagrams

11.2 Equilibrium diagrams

References to individual diagrams are given at the end of this section. When no reference is given the reader should consult constitution of Binary Alloys by M. Hansen and K. Anderko, McGraw-Hill (1958) and the supplements by R. P. Elliott (1965) and F. A. Shunk (1969).

Ag–Al

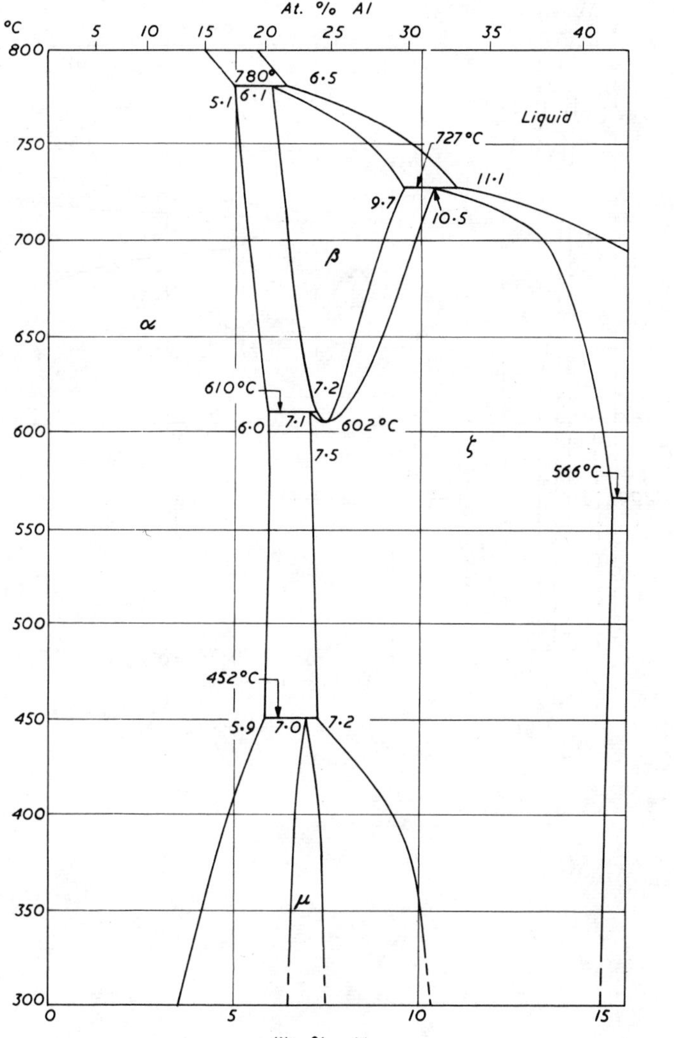

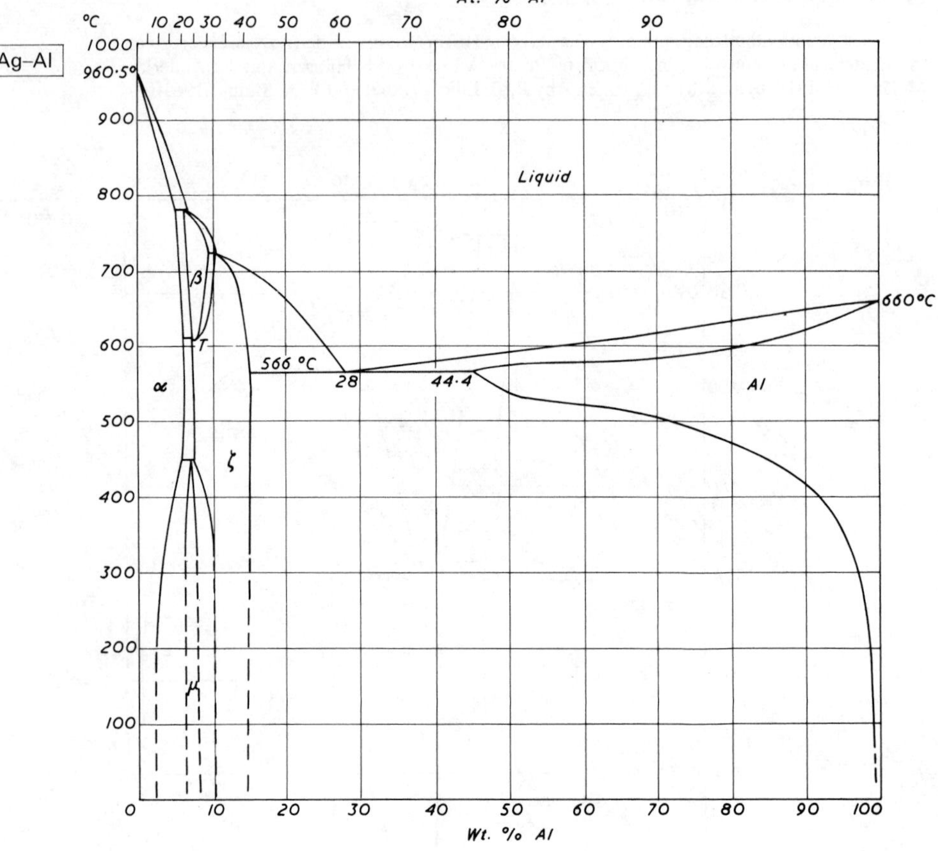
Ag–Al
At. % Al
10 20 30 40 50 60 70 80 90
°C
1000
960·5°
900
800
700
600
500
400
300
200
100
0
Liquid
660°C
β
T
566 °C
28
44·4
α
Al
ζ
μ
0 10 20 30 40 50 60 70 80 90 100
Wt. % Al

Ag–As

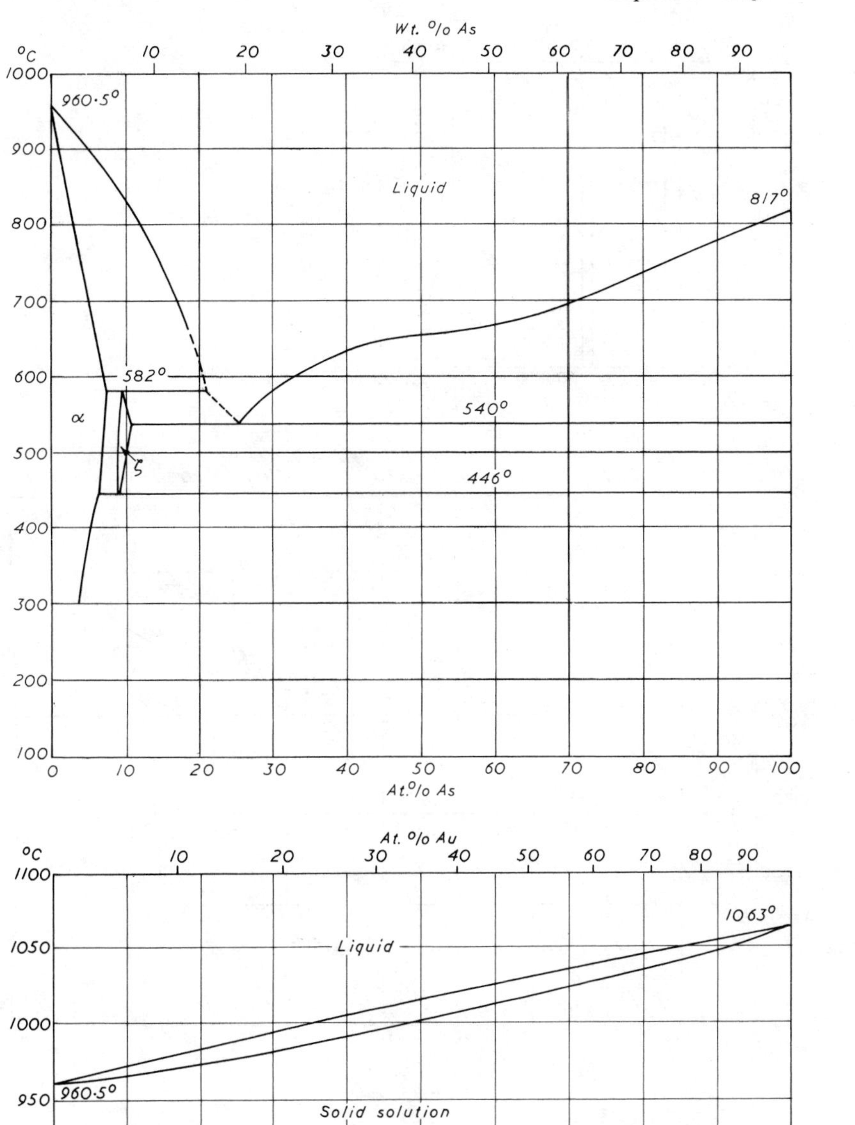
Wt. % As
°C
960·5°
Liquid
817°
582°
540°
α
ζ
446°
At.% As
Ag–Au
At. % Au
°C
1063°
Liquid
960·5°
Solid solution
Wt. % Au

Ag–Ba

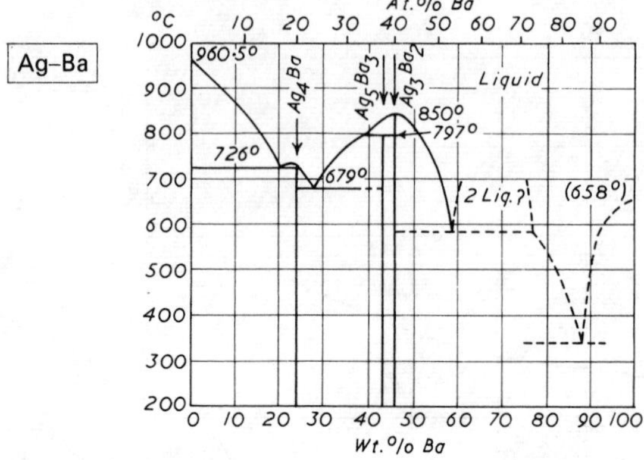
At.% Ba
°C
960·5°
Ag4Ba
Ag5Ba3
Ag3Ba2
Liquid
850°
797°
726°
679°
2 Liq.?
(658°)
Wt.% Ba

Ag–Be

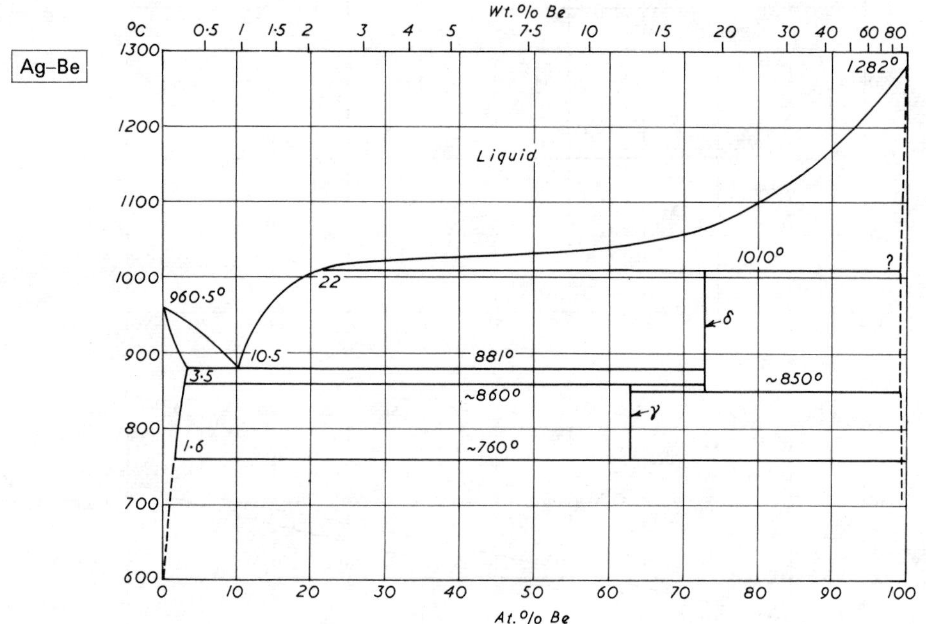
Wt.% Be
°C
1282°
Liquid
1010°
?
22
960·5°
δ
10·5
881°
3·5
~850°
~860°
γ
1·6
~760°
At.% Be

Ag–Bi

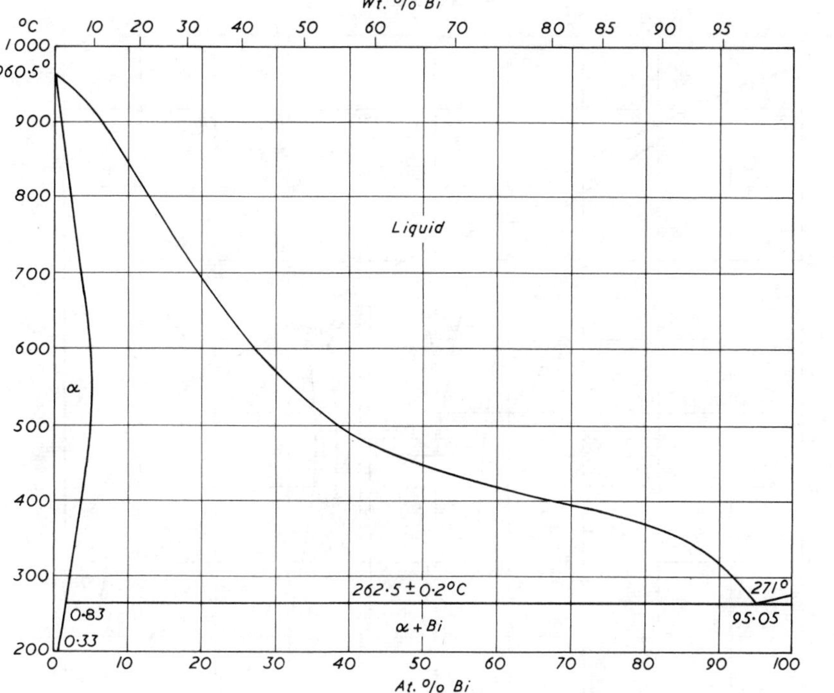
Wt. % Bi
°C
960·5°
Liquid
α
262·5 ± 0·2°C
271°
0·83
0·33
95·05
α + Bi
At. % Bi

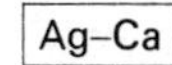

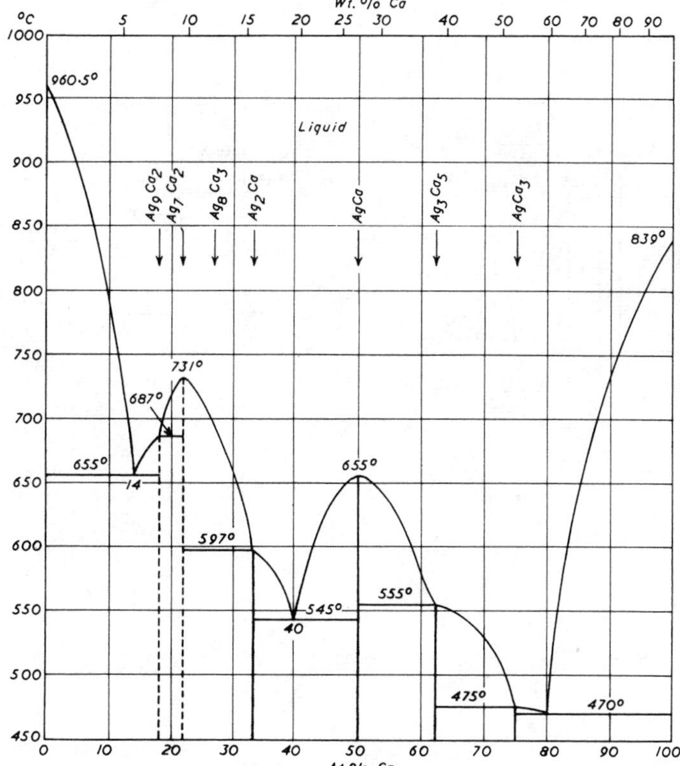
Wt. % Ca
°C
960·5°
Liquid
Ag9Ca2
Ag7Ca2
Ag8Ca3
Ag2Ca
AgCa
Ag3Ca5
AgCa3
839°
731°
687°
655°
14
655°
597°
545°
40
555°
475°
470°
At. % Ca

Ag–Cd

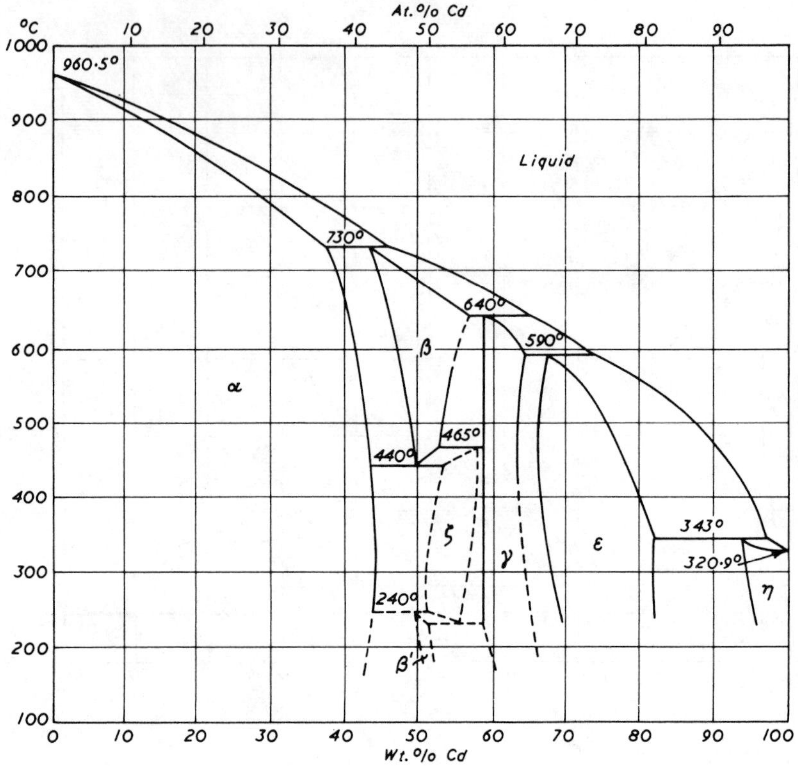

At. % Cd
Wt. % Cd
Liquid
960·5°
730°
640°
590°
465°
440°
240°
343°
320·9°
α
β
ζ
γ
ε
η
β′

Ag–Ce

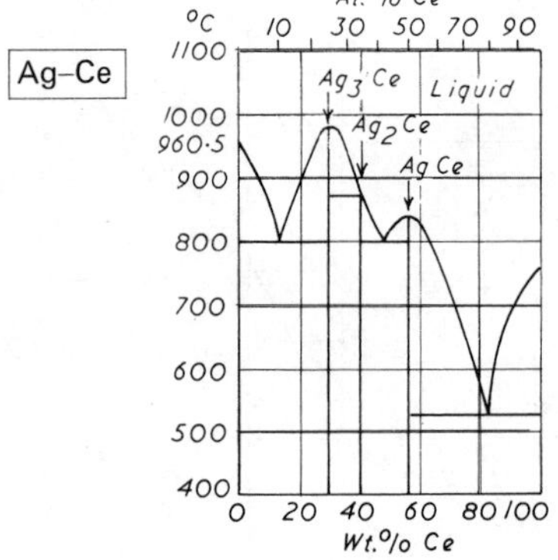

Ag–Cr

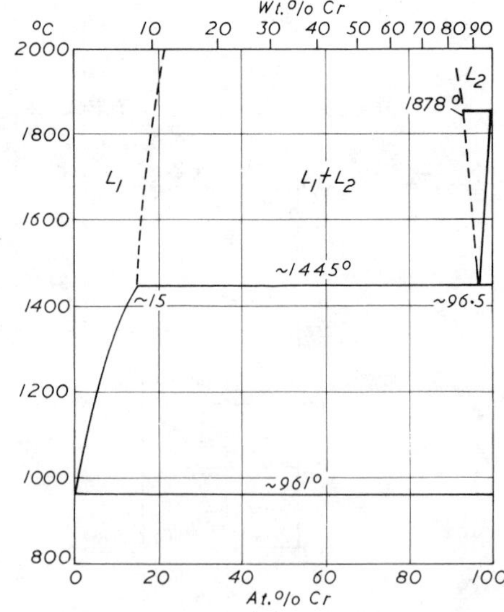

Ag–Cu

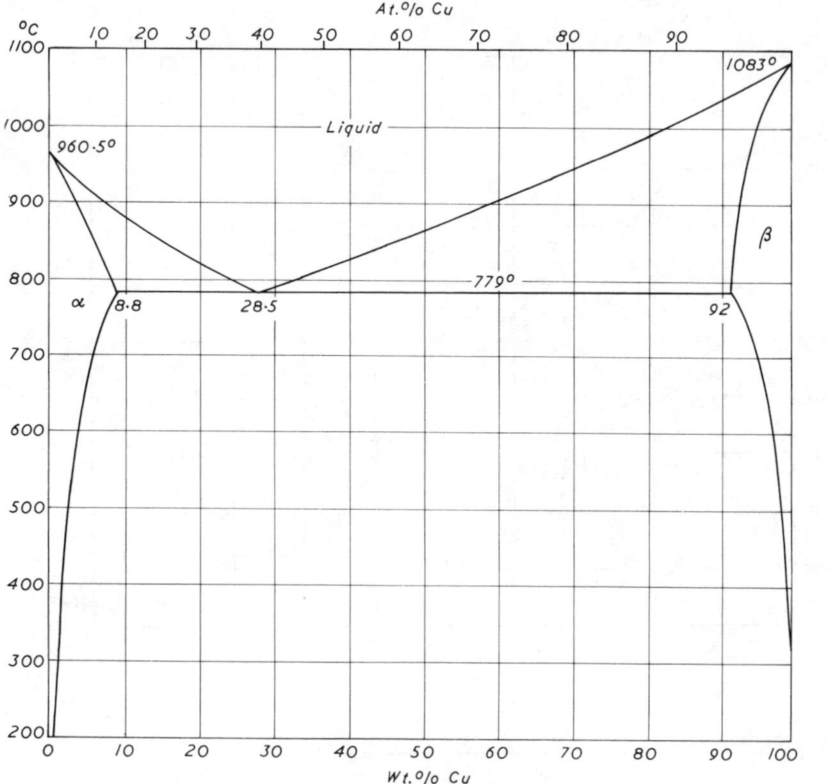

At.% Cu
°C
Liquid
1083°
960·5°
β
α
8·8
28·5
779°
92
Wt.% Cu

Ag–Dy

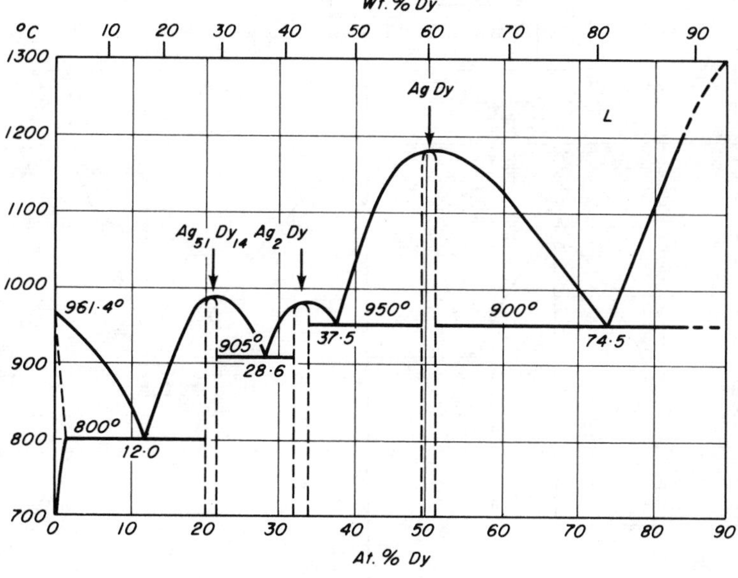

Wt.% Dy
°C
Ag Dy
L
Ag51 Dy14
Ag2 Dy
961·4°
905°
28·6
950°
900°
37·5
74·5
800°
12·0
At.% Dy

Ag–Ga

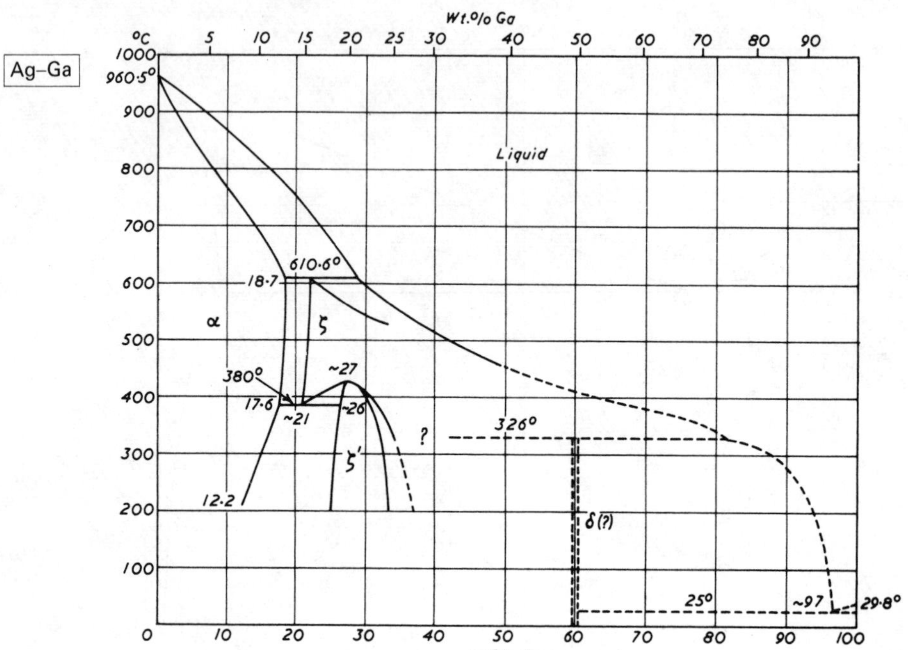

Wt.% Ga
°C
5
10
15
20
25
30
40
50
60
70
80
90
1000
960·5°
900
800
700
600
500
400
300
200
100
0
Liquid
610·6°
18·7
α
ζ
~27
380°
17·6
~21
~26
ζ'
?
326°
12·2
δ(?)
25°
~97
29·8°
10
20
30
40
50
60
70
80
90
100
At.% Ga

Ag–Gd

Wt. % Gd
°C
10
20
30
40
50
60
70
80
90
1400
1300
1200
1100
1000
900
800
700
600
500
35 At.% Gd
830°
825°
37 At.% Gd
1315°
AgGd
Ag3Gd
Ag2Gd
1100°
Liquid
975°
960·5°
830°
825°
780°
790°
11
77·5
0
10
20
30
40
50
60
70
80
90
100
At.% Gd

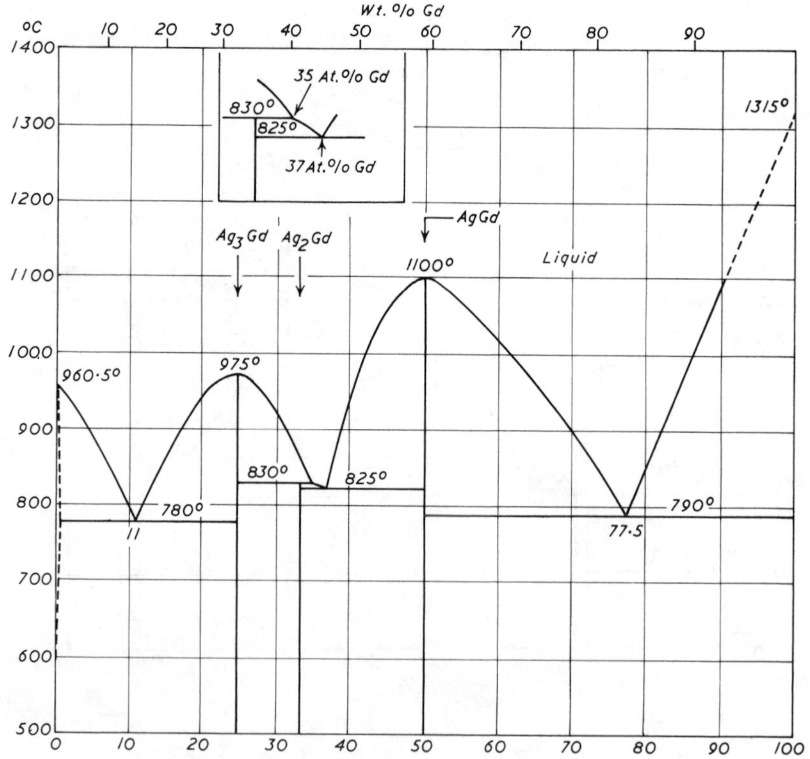

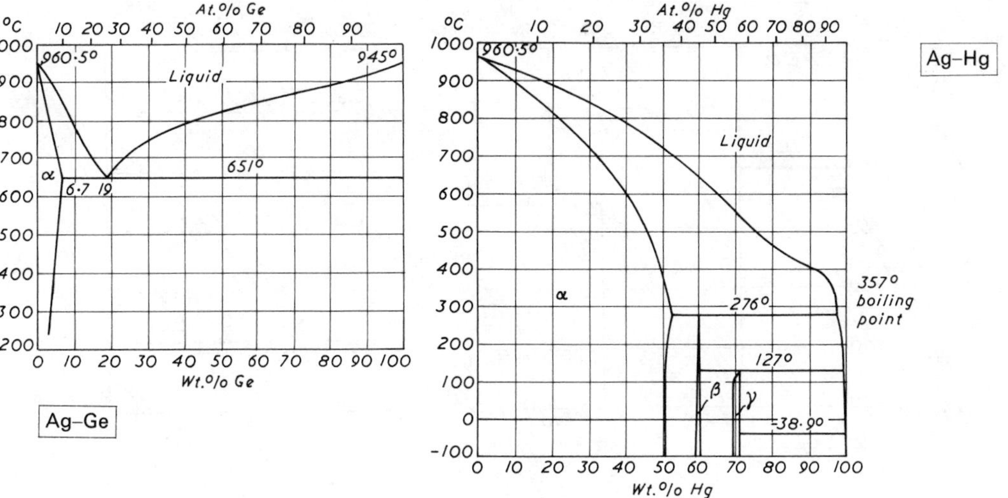

At.% Ge
°C
960·5°
945°
Liquid
651°
α
6·7 19
Wt.% Ge

Ag–Ge

At.% Hg
°C
960·5°
Liquid
α
276°
357°
boiling
point
127°
β
γ
-38·9°
Wt.% Hg

Ag–Hg

At.% In
°C
960·5°
Liquid
695°
670°
660°
β
α
γ
205°
156·4°
187°
166°
144°
α′
ε
ϕ
Wt.% In
°C
695°
Liquid
670°
β
α
γ
660°
Wt.% In

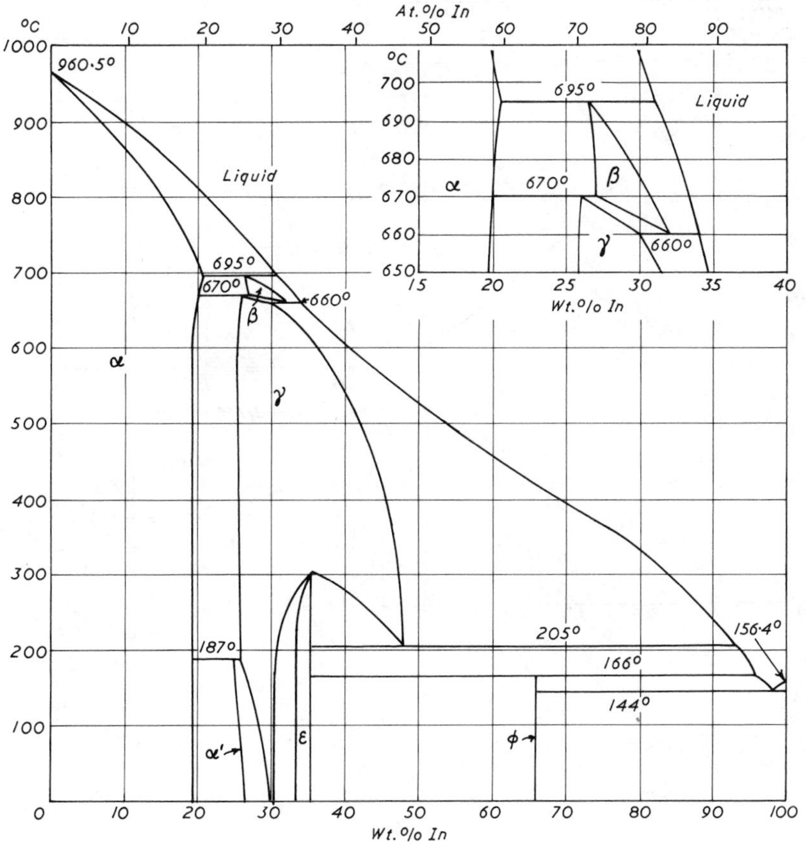

Ag–In

Ag–La

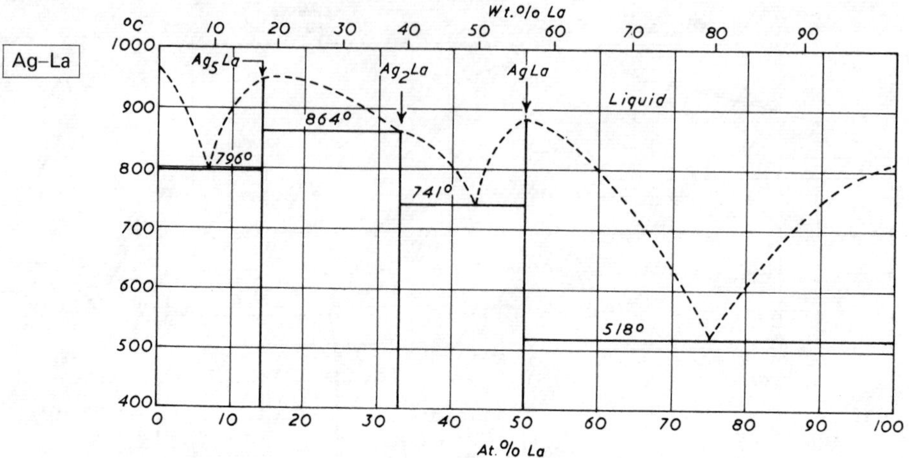
Wt.% La
°C
Ag5La
Ag2La
AgLa
Liquid
864°
796°
741°
518°
At.% La

Ag–Li

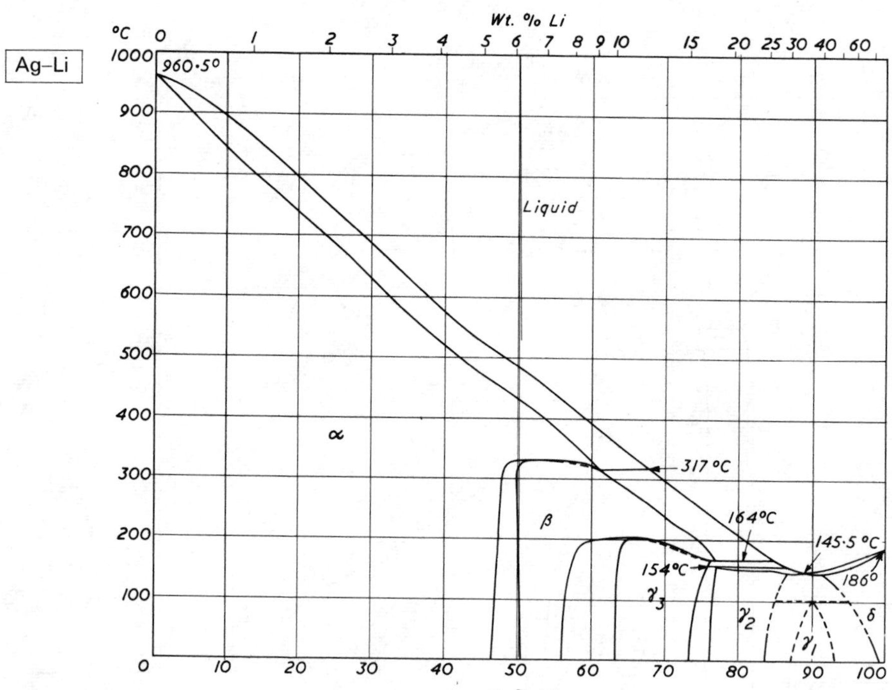
Wt. % Li
°C
960·5°
Liquid
α
β
317 °C
164°C
145·5 °C
186°
154°C
γ3
γ2
γ1
δ
At. % Li

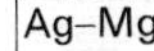
Ag–Mg

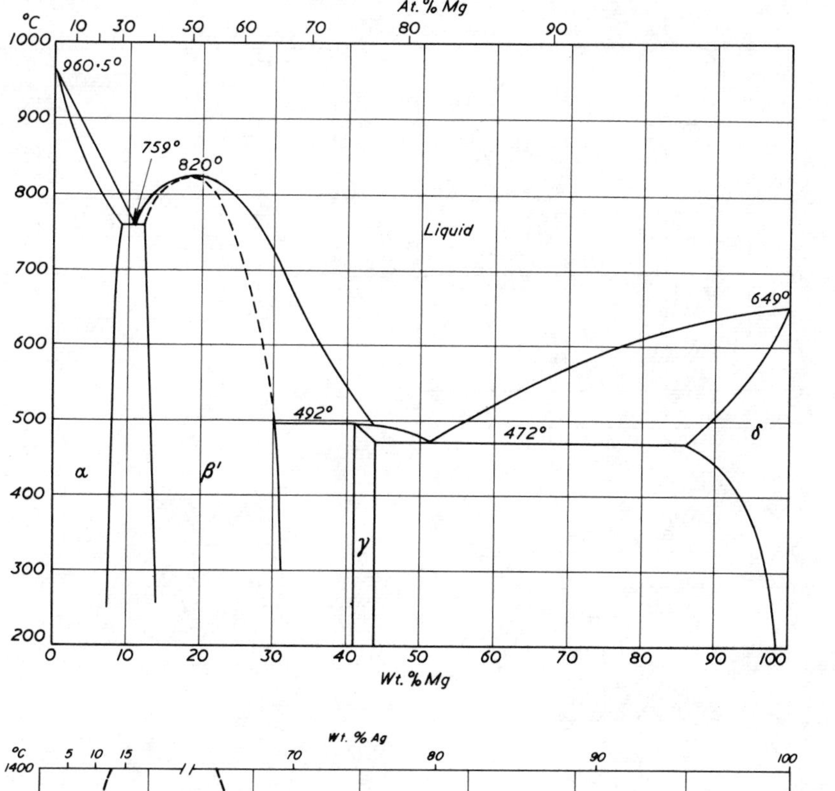
At. % Mg
°C
10
30
50
60
70
80
90
1000
960·5°
900
759°
820°
800
Liquid
700
649°
600
500
492°
472°
δ
α
β'
400
γ
300
200
0
10
20
30
40
50
60
70
80
90
100
Wt. % Mg

Ag–Mn

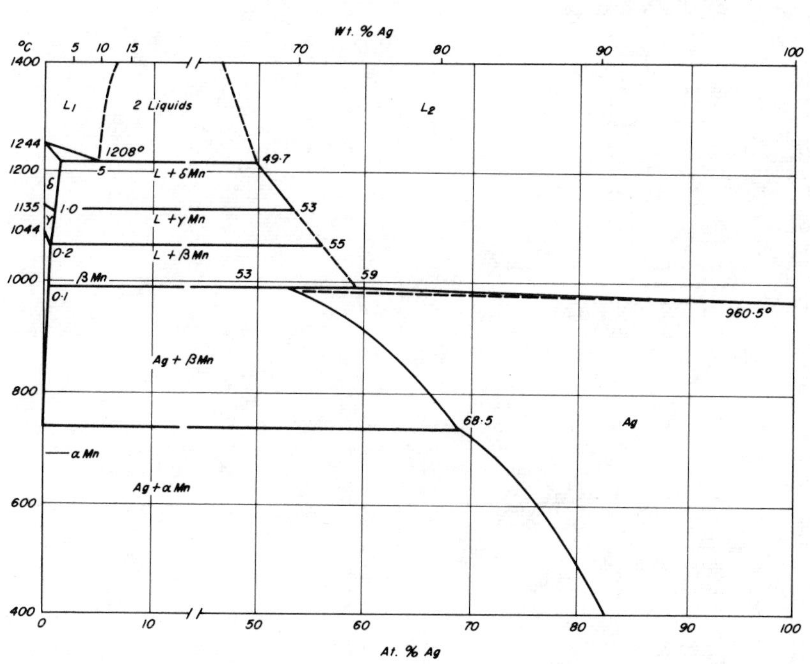
Wt. % Ag
°C
5
10
15
70
80
90
100
1400
L1
2 Liquids
L2
1244
1208°
49·7
1200
5
L + δMn
δ
1135
1·0
53
L + γ Mn
1044
γ
55
0·2
L + βMn
1000
βMn
53
59
0·1
960·5°
Ag + βMn
800
68·5
Ag
αMn
Ag + αMn
600
400
0
10
50
60
70
80
90
100
At. % Ag

Ag–Na

Ag–Ni

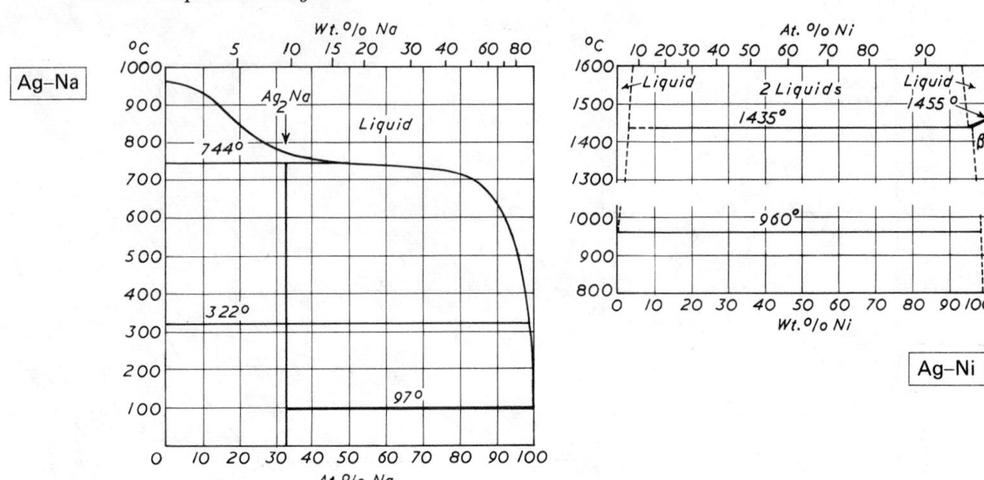

Ag–O

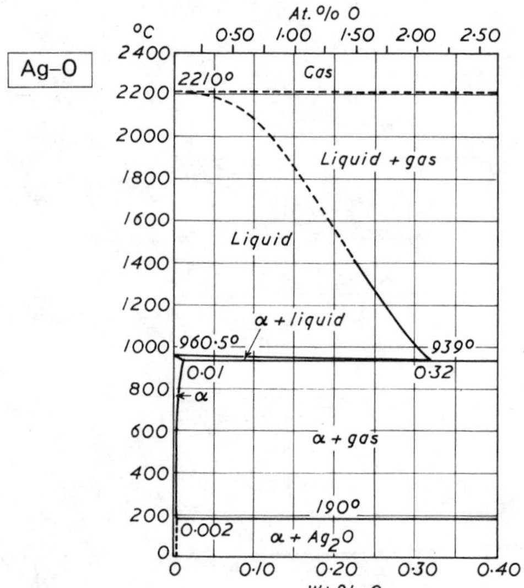

Ag–O

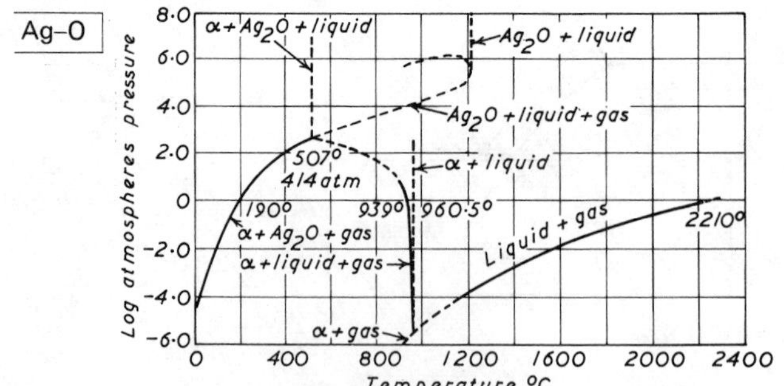

Pressure-temperature relationships in the Ag-O system

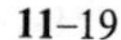

Ag–P

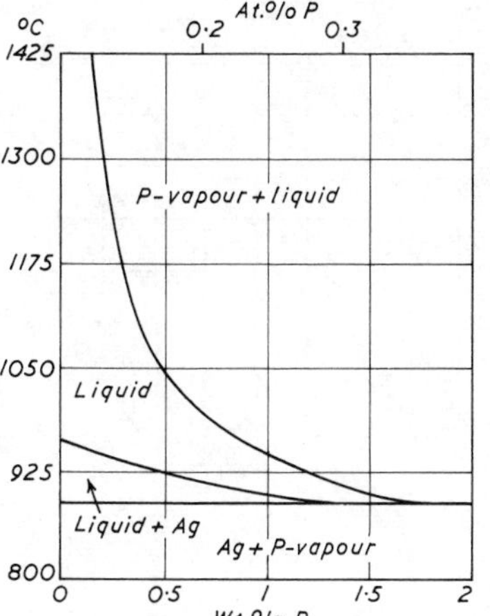
°C
At.% P
0·2
0·3
1425
1300
1175
1050
925
800
P-vapour + liquid
Liquid
Liquid + Ag
Ag + P-vapour
0
0·5
1
1·5
2
Wt.% P

Ag–Pb

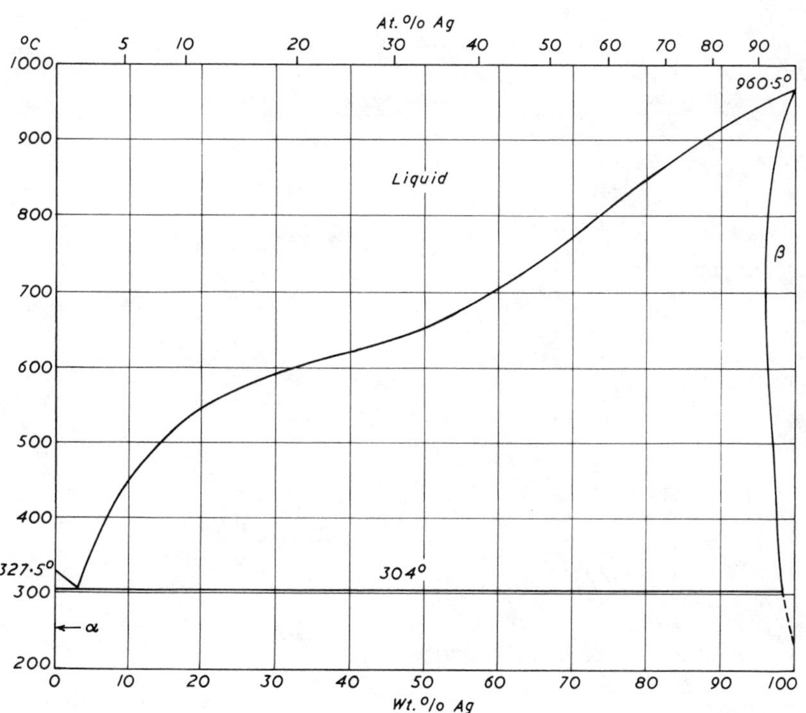
At.% Ag
°C
5
10
20
30
40
50
60
70
80
90
1000
900
800
700
600
500
400
300
200
960·5°
Liquid
β
327·5°
304°
α
0
10
20
30
40
50
60
70
80
90
100
Wt.% Ag

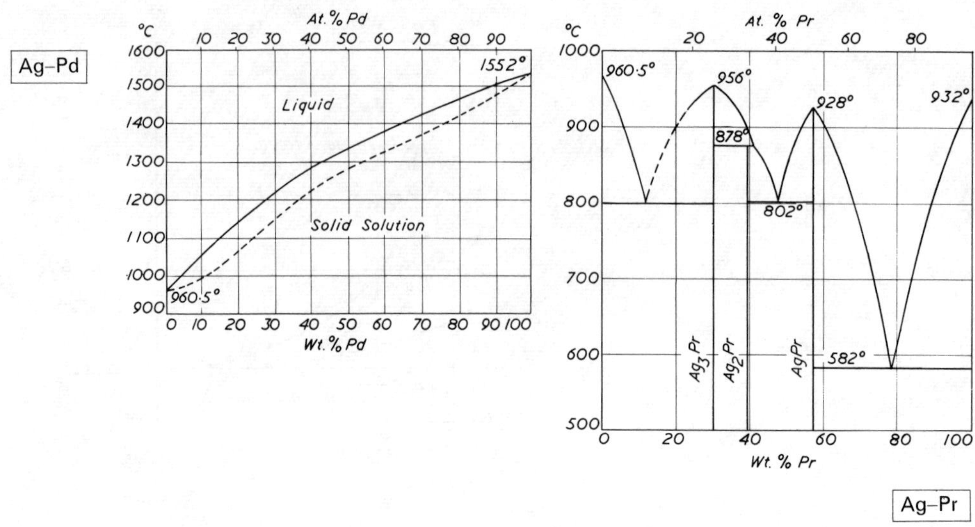
Ag–Pd
At.% Pd
°C
Liquid
Solid Solution
1552°
960·5°
Wt.% Pd
Ag–Pr
At. % Pr
960·5°
956°
878°
928°
932°
802°
582°
Ag_3Pr
Ag_2Pr
AgPr
Wt. % Pr

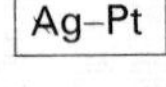
Ag–Pt

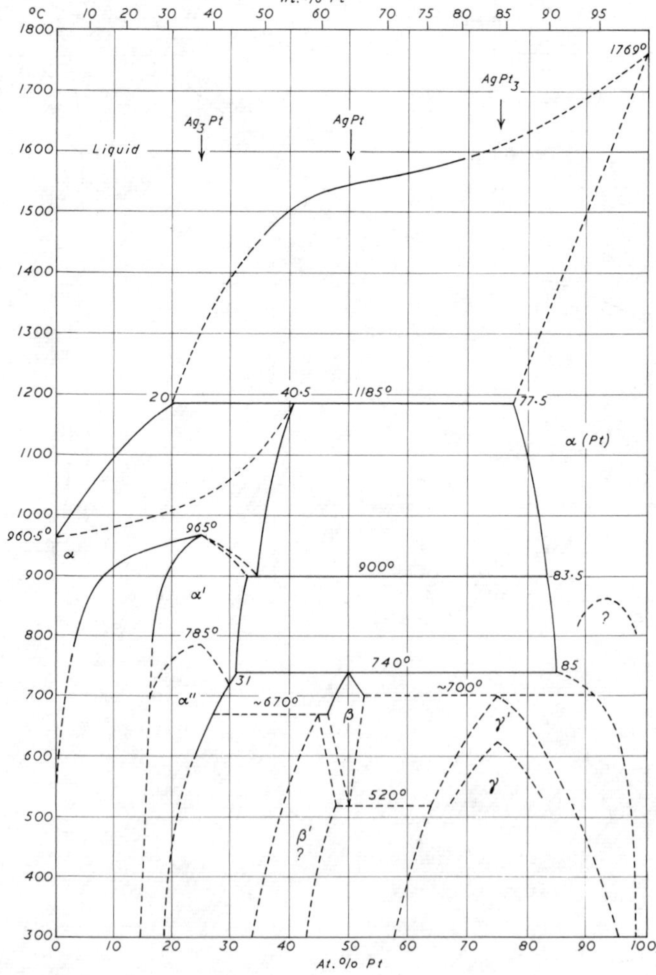
Wt. % Pt
°C
1769°
$AgPt_3$
Ag_3Pt
AgPt
Liquid
20
40·5
1185°
77·5
α (Pt)
960·5°
965°
α
α′
900°
83·5
785°
?
740°
31
85
α″
~670°
~700°
β
γ′
γ
520°
β′
?
At. % Pt

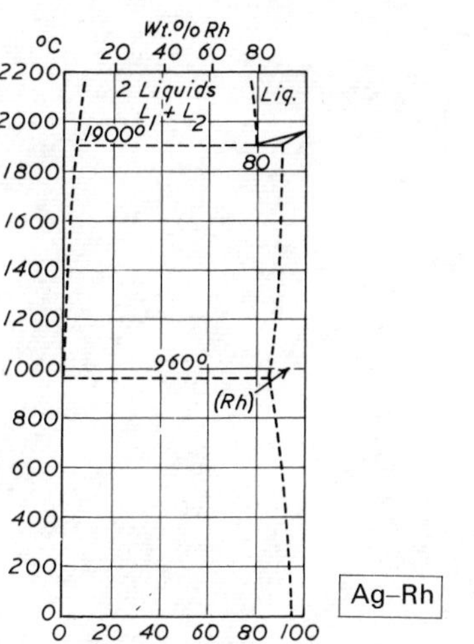

°C
Wt.% Rh
20 40 60 80
2 Liquids
$L_1 + L_2$
Liq.
1900°
80
960°
(Rh)
2200 2000 1800 1600 1400 1200 1000 800 600 400 200 0
0 20 40 60 80 100
At.% Rh
Ag–Rh

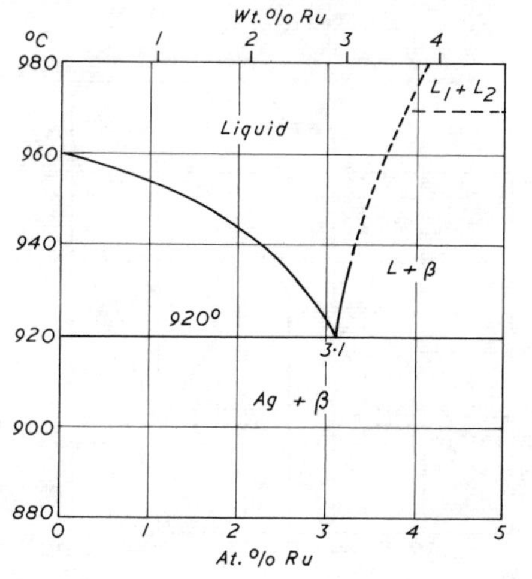

°C
Wt.% Ru
1 2 3 4
$L_1 + L_2$
Liquid
$L + \beta$
920°
3·1
$Ag + \beta$
980 960 940 920 900 880
0 1 2 3 4 5
At.% Ru

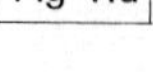

Ag–Ru

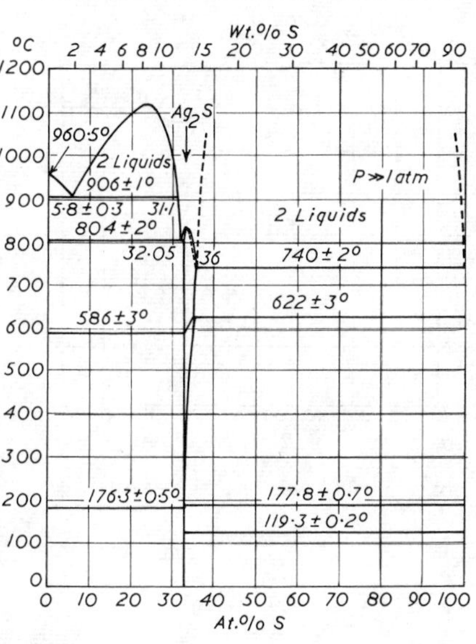

°C
Wt.% S
2 4 6 8 10 15 20 30 40 50 60 70 90
960·5°
Ag_2S
2 Liquids
906 ± 1°
P ≫ 1 atm
5·8 ± 0·3
31·1
2 Liquids
804 ± 2°
32·05
36
740 ± 2°
622 ± 3°
586 ± 3°
176·3 ± 0·5°
177·8 ± 0·7°
119·3 ± 0·2°
1200 1100 1000 900 800 700 600 500 400 300 200 100 0
0 10 20 30 40 50 60 70 80 90 100
At.% S

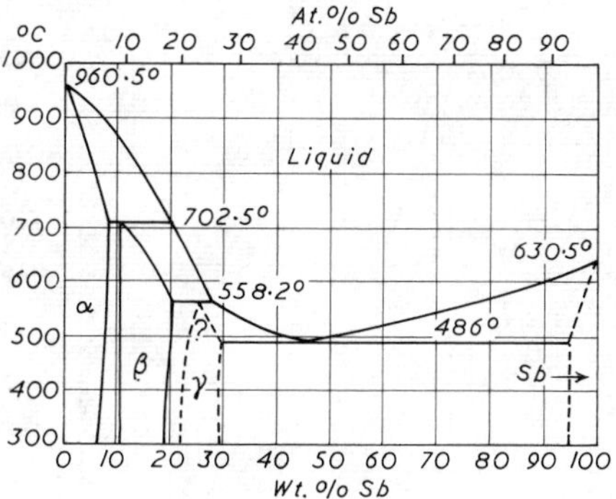

°C
At.% Sb
10 20 30 40 50 60 70 80 90
960·5°
Liquid
702·5°
630·5°
558·2°
α
486°
β
?
γ
Sb →
1000 900 800 700 600 500 400 300
0 10 20 30 40 50 60 70 80 90 100
Wt.% Sb

Ag–Sb

Ag–S

Ag–Sc

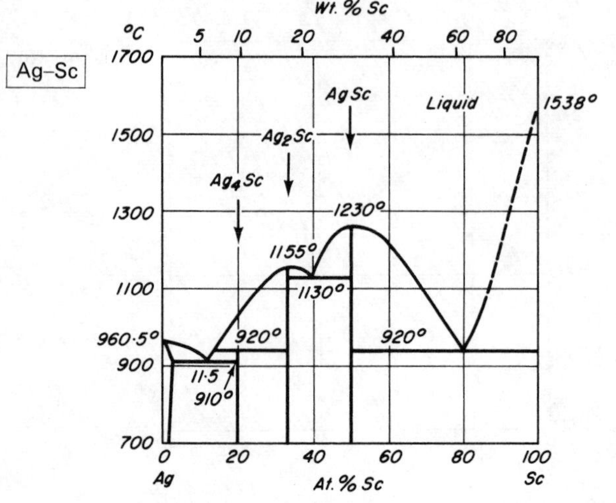

Ag–Se

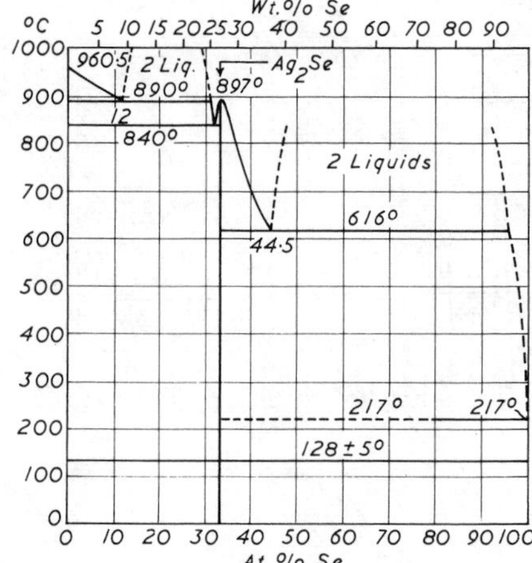

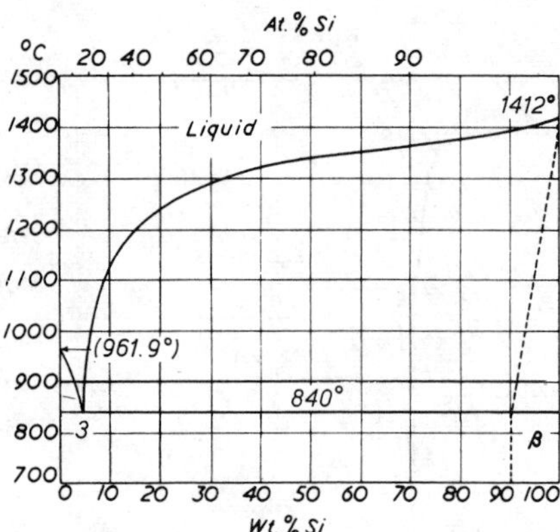

Ag–Si

Ag–Sm

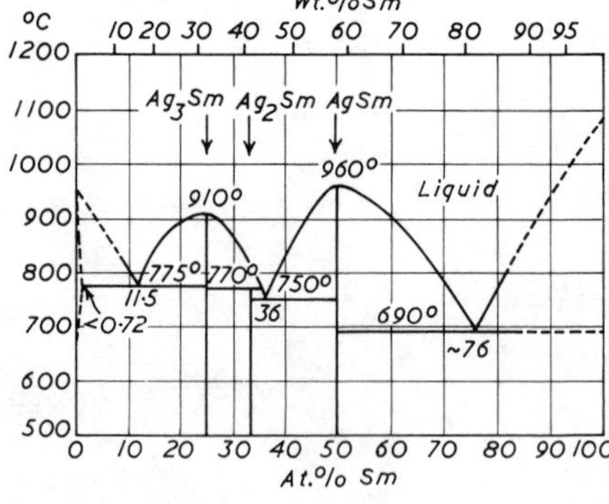

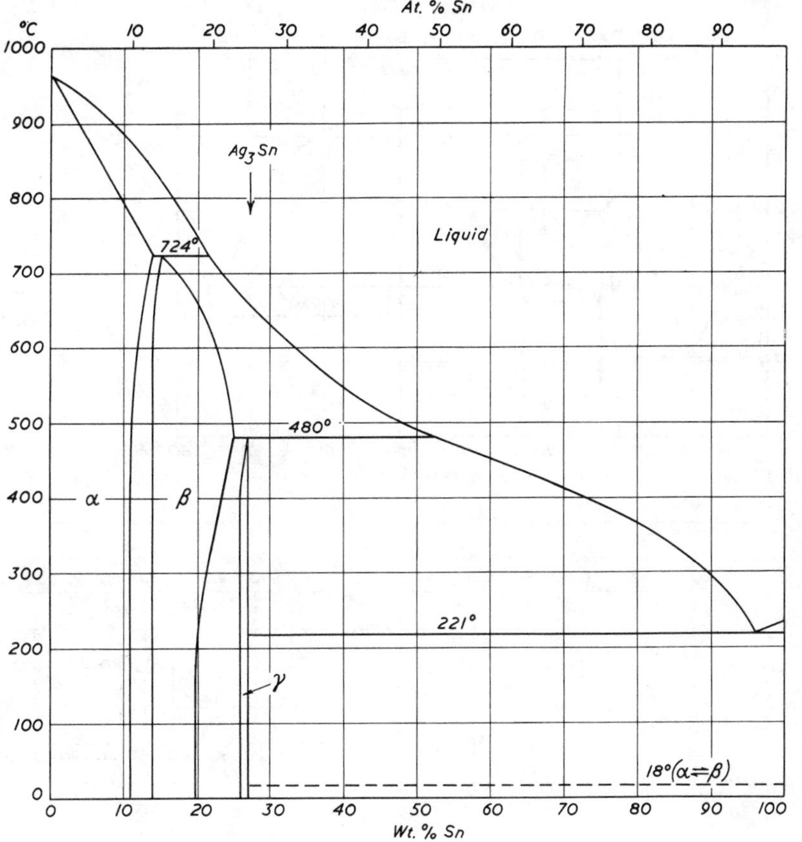
At. % Sn
°C
Ag₃Sn
Liquid
724°
480°
221°
α
β
γ
18°(α⇌β)
Wt. % Sn

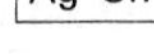
Ag–Sn

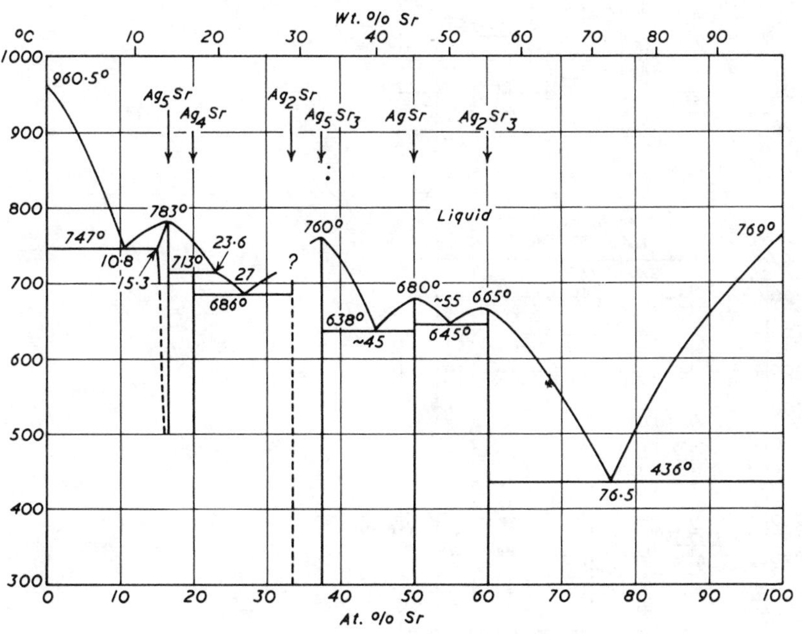
Wt. % Sr
°C
960·5°
Ag5Sr
Ag4Sr
Ag2Sr
Ag5Sr3
AgSr
Ag2Sr3
Liquid
783°
747°
10·8
15·3
713°
23·6
27
?
686°
760°
638°
~45
680°
~55
665°
645°
769°
436°
76·5
At. % Sr
Ag–Sr

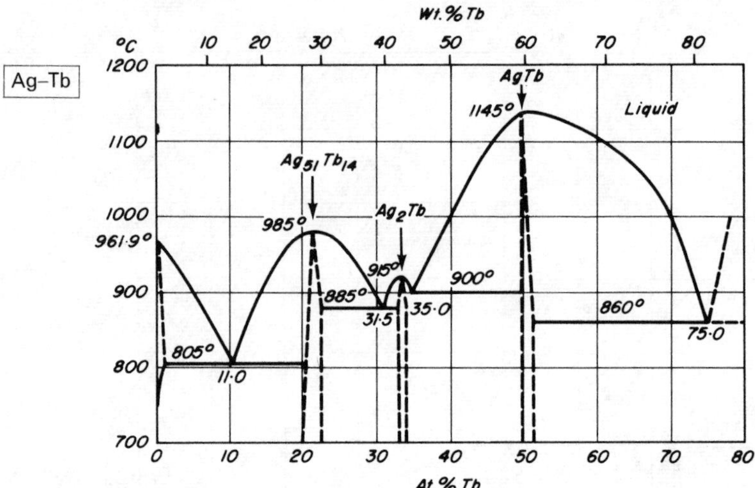
Ag–Tb
Wt.% Tb
At % Tb
°C
AgTb
1145°
Liquid
$Ag_{51}Tb_{14}$
985°
Ag_2Tb
961·9°
915°
900°
885°
31·5
35·0
860°
75·0
805°
11·0

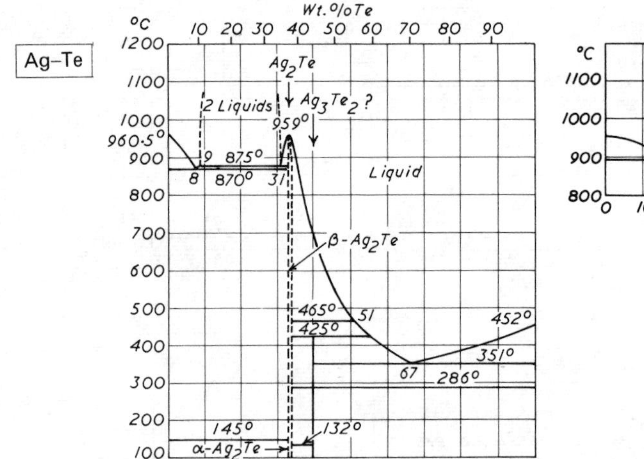
Ag–Te
Wt.%Te
At.%Te
°C
Ag_2Te
2 Liquids
Ag_3Te_2 ?
959°
960·5°
9
875°
8
870°
31
Liquid
β-Ag_2Te
465°
51
452°
425°
351°
67
286°
145°
132°
α-Ag_2Te

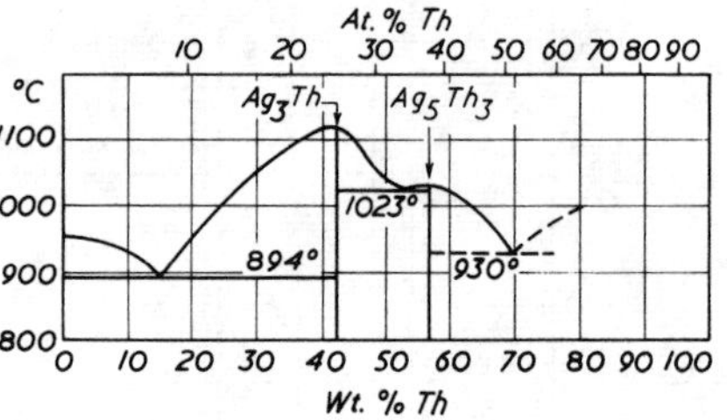
At.% Th
Wt. % Th
°C
Ag_3Th
Ag_5Th_3
1023°
894°
930°
Ag–Th

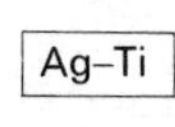
Ag–Ti

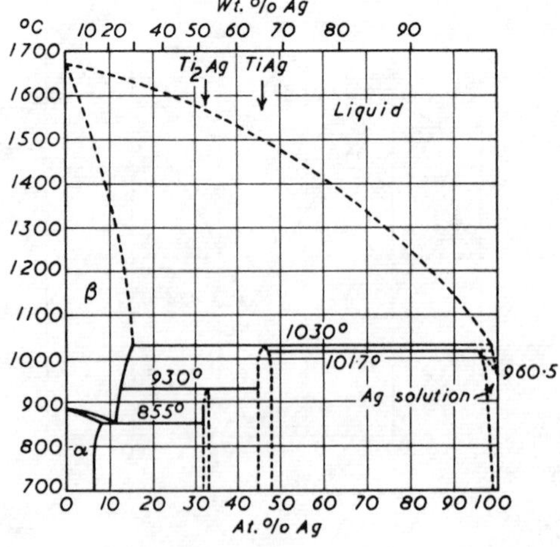
Wt. % Ag
At. % Ag
°C
Ti_2Ag
TiAg
Liquid
β
1030°
1017°
960·5
930°
Ag solution
855°
α

Ag–Tl

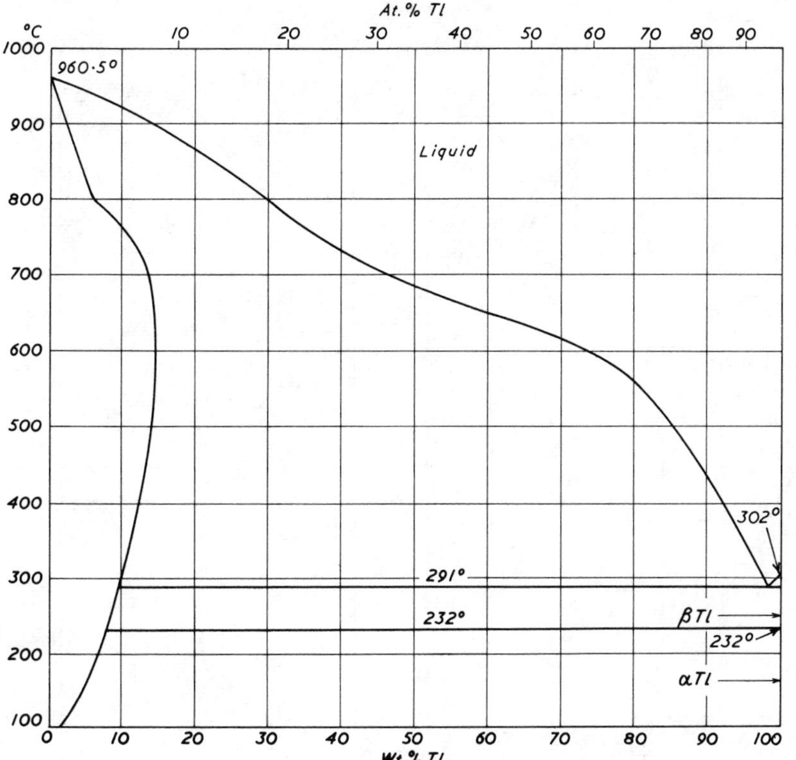
At.% Tl
°C
960·5°
Liquid
302°
291°
232°
βTl
232°
αTl
Wt.% Tl

Ag–U

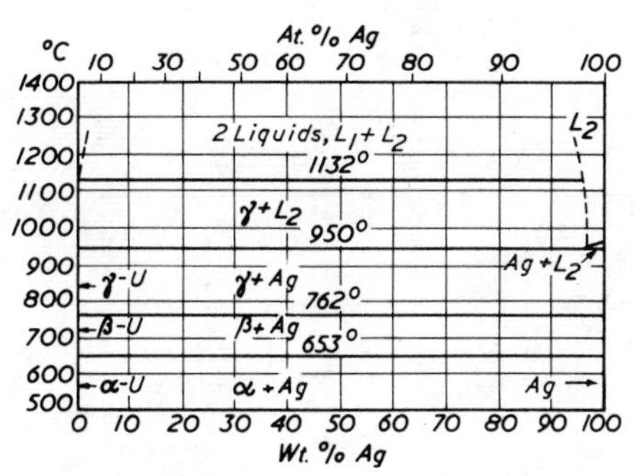
At.% Ag
°C
2 Liquids, L_1 + L_2
1132°
L_2
γ+L_2
950°
Ag+L_2
γ-U
γ+Ag
762°
β-U
β+Ag
653°
α-U
α+Ag
Ag
Wt.% Ag

Ag–Y

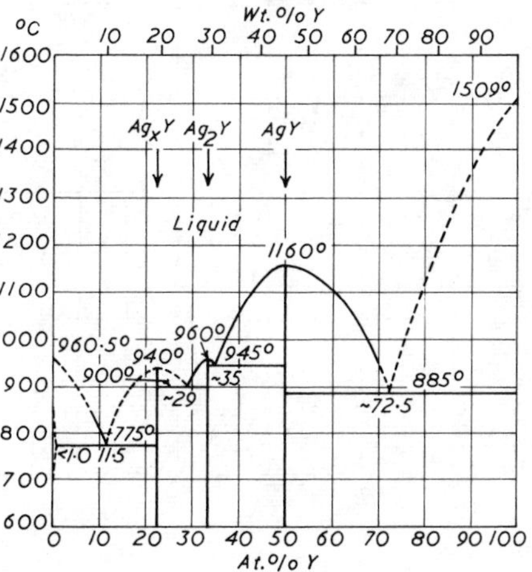
Wt.% Y
°C
1509°
Ag_xY
Ag_2Y
AgY
Liquid
1160°
960°
960·5°
940°
945°
900°
~29
~35
885°
~72·5
775°
<1·0
11·5
At.% Y

Ag–Yb

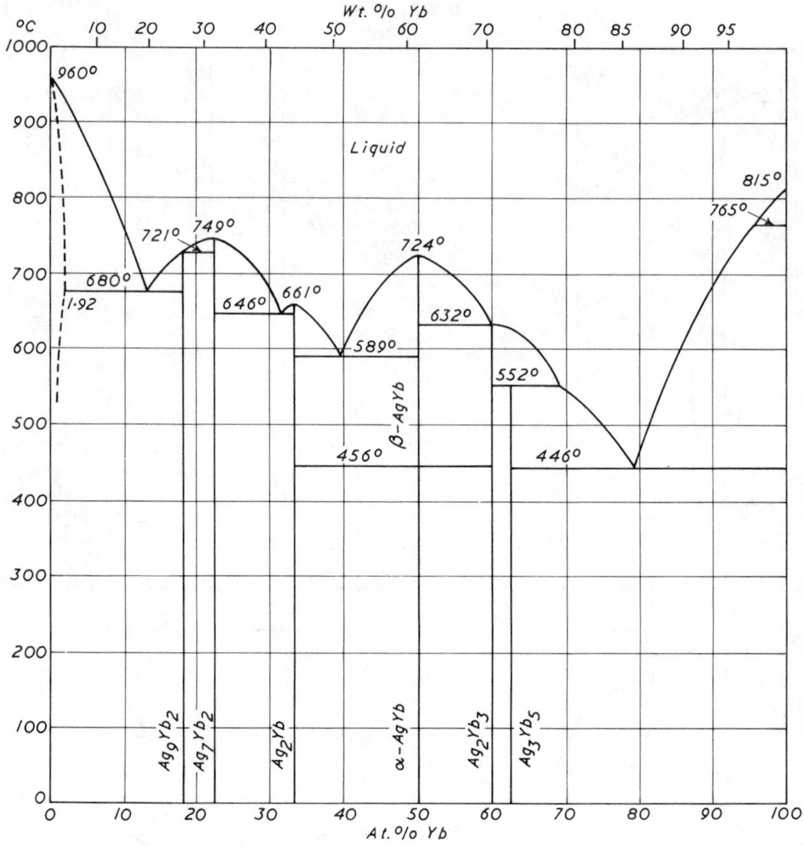

Wt. % Yb
°C
Liquid
960°
815°
765°
721°
749°
724°
680°
1·92
646°
661°
632°
589°
552°
456°
446°
β-AgYb
Ag_9Yb_2
Ag_7Yb_2
Ag_2Yb
α-AgYb
Ag_2Yb_3
Ag_3Yb_5
At. % Yb

Ag–Zn

At. % Zn
°C
960·5°
Liquid
710°
661°
631°
α
β
γ
ε
419°
431°
η
258°
274°
ξ
Wt. % Zn

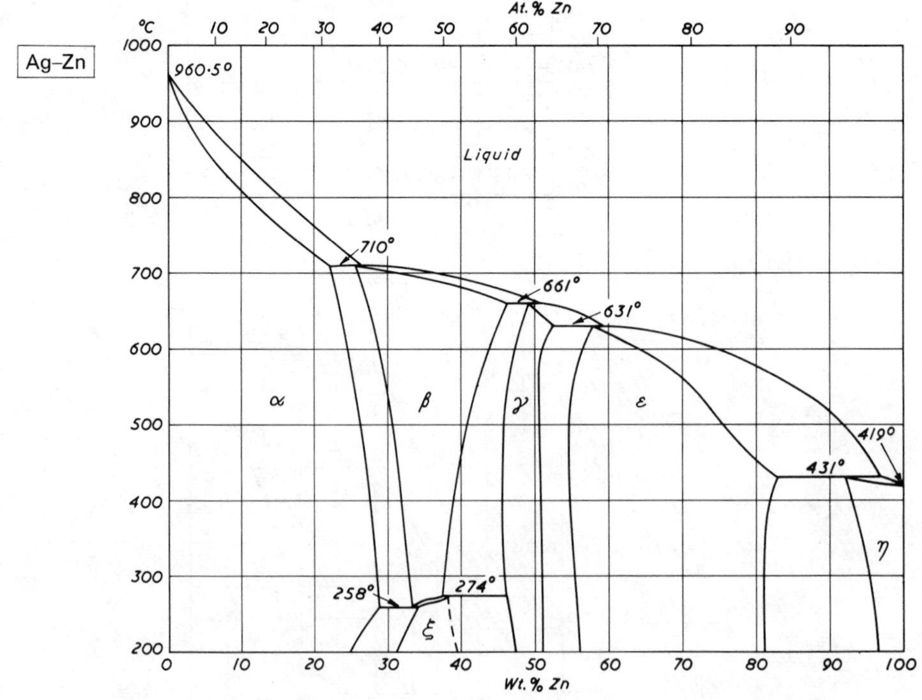

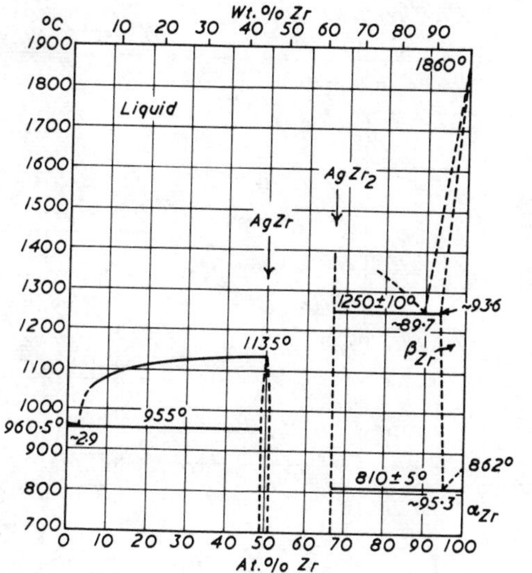

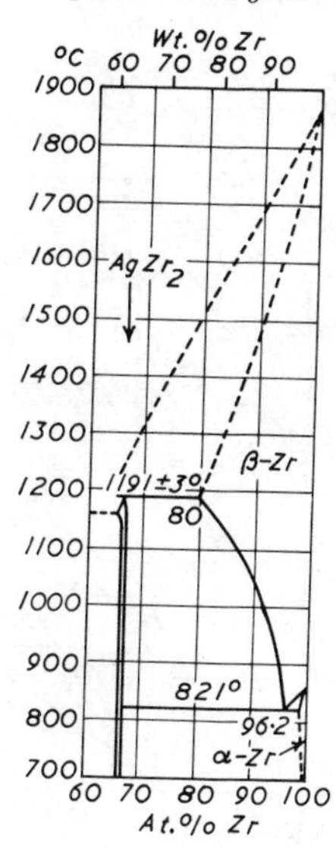

Ag–Zr

Ag–Zr

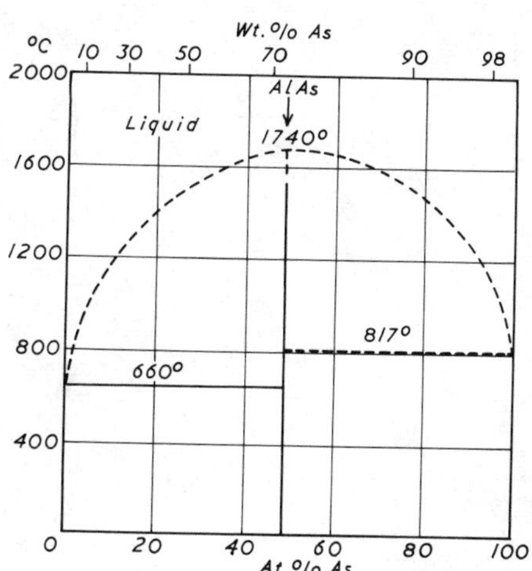

Al–As

Al–Au

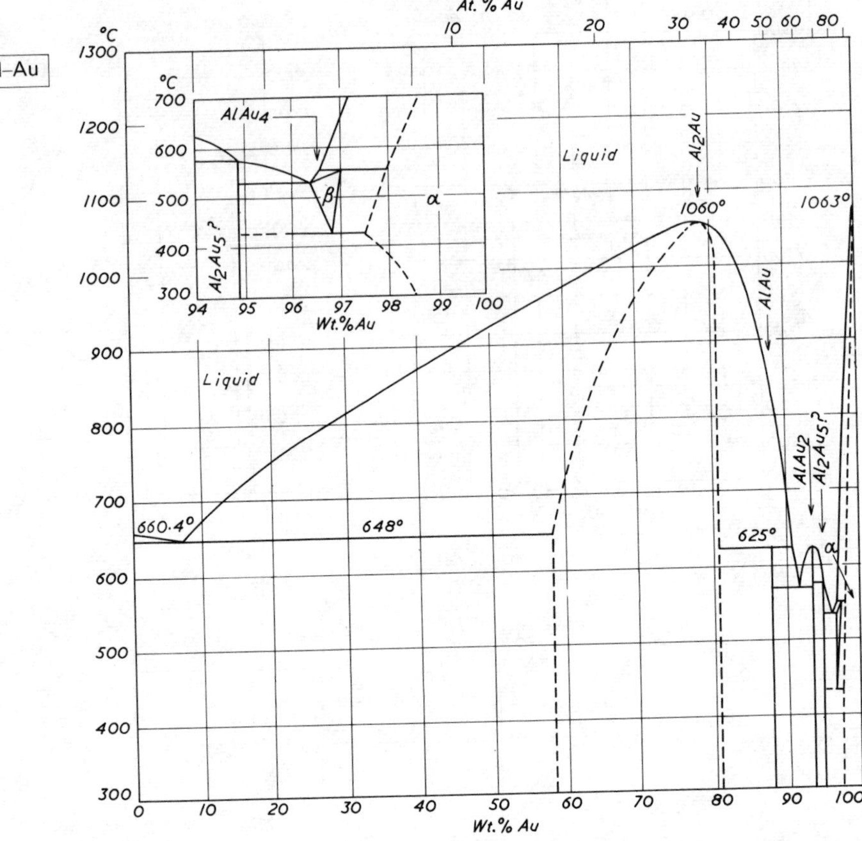
At. % Au
10
20
30
40
50
60
80
°C
1300
1200
1100
1000
900
800
700
600
500
400
300
700
600
500
400
300
94
95
96
97
98
99
100
Wt.% Au
AlAu4
β
α
Al2Au5?
Liquid
Al2Au
1060°
1063°
AlAu
Liquid
660·4°
648°
625°
AlAu2
Al2Au5?
α
0
10
20
30
40
50
60
70
80
90
100
Wt.% Au

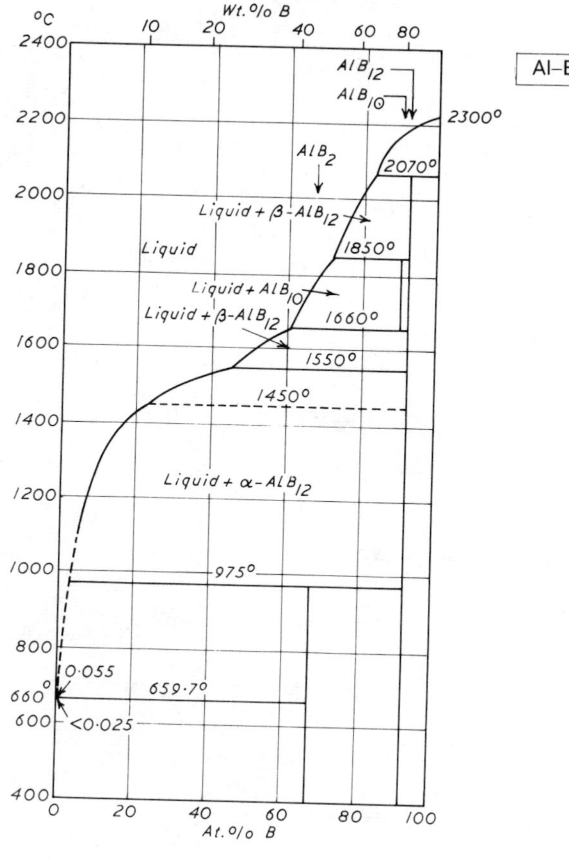

°C
Wt. % B
10
20
40
60
80
2400
2200
2000
1800
1600
1400
1200
1000
800
660°
600
400
AlB12
AlB10
AlB2
2300°
2070°
Liquid + β-AlB12
1850°
Liquid
Liquid + AlB10
Liquid + β-AlB12
1660°
1550°
1450°
Liquid + α-AlB12
975°
0·055
659·7°
<0·025
0
20
40
60
80
100
At. % B

Al–B

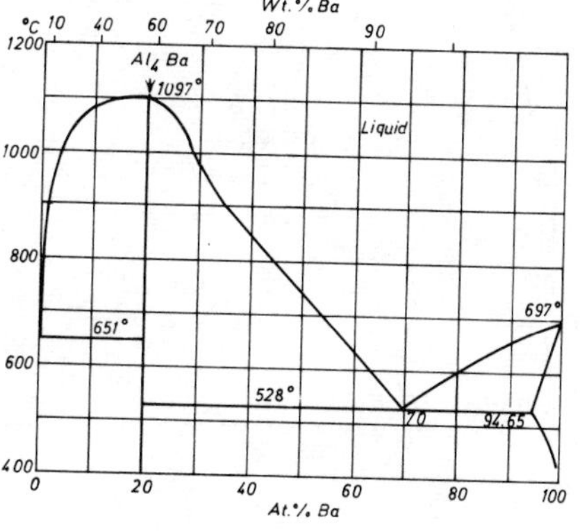

Wt.% Ba
°C
10
40
60
70
80
90
1200
Al4 Ba
1097°
Liquid
1000
800
600
400
651°
697°
528°
70
94.65
0
20
40
60
80
100
At.% Ba

Al–Ba

Al–Be

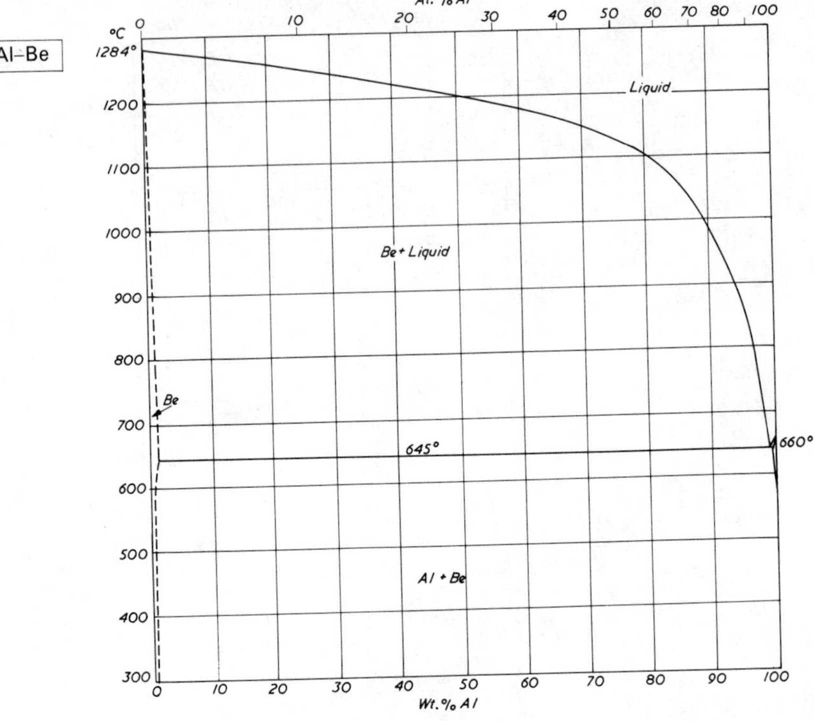

At.% Al
0
10
20
30
40
50
60
70
80
100
°C
1284°
1200
1100
1000
900
800
700
600
500
400
300
Liquid
Be+Liquid
Be
645°
660°
Al + Be
0
10
20
30
40
50
60
70
80
90
100
Wt.% Al

Al–Bi

Wt.% Bi
10
40
60
70
80
85
90
95
98
°C
1100
1000
900
800
700
660°
600
500
400
300
200
Liquid
2 Liquids
657°
0·45 at.% Bi
Al + Liquid
271°
Al + Bi
0
10
20
30
40
50
60
70
80
90
100
At.% Bi

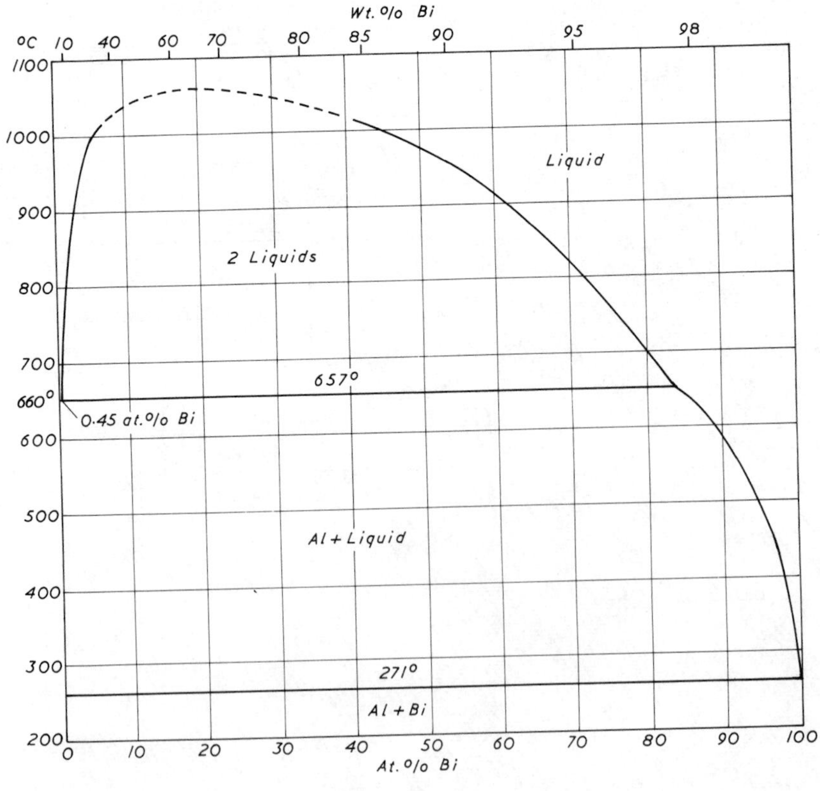

Al–Ca

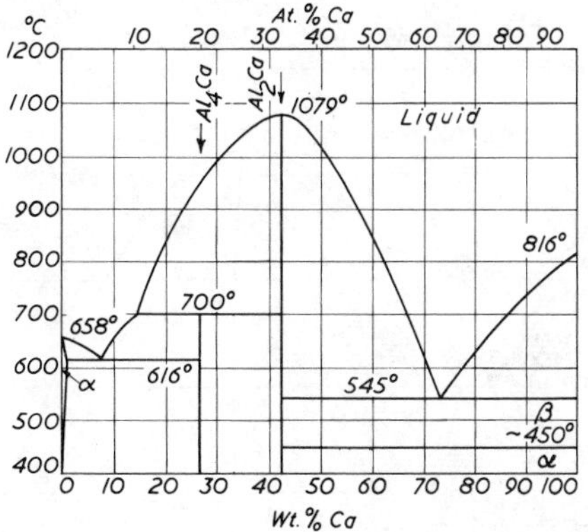

Al–Cd

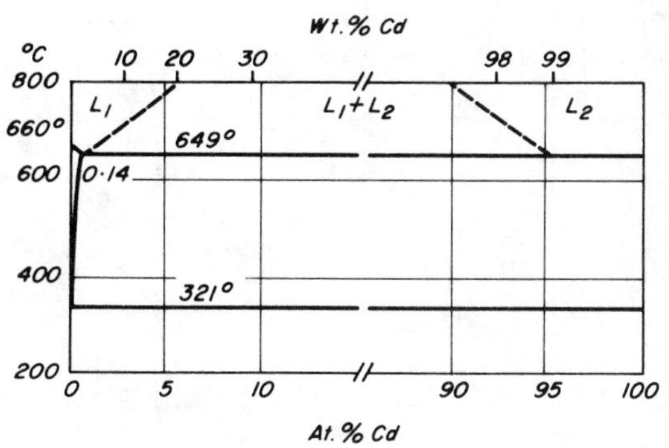

Al–Ce

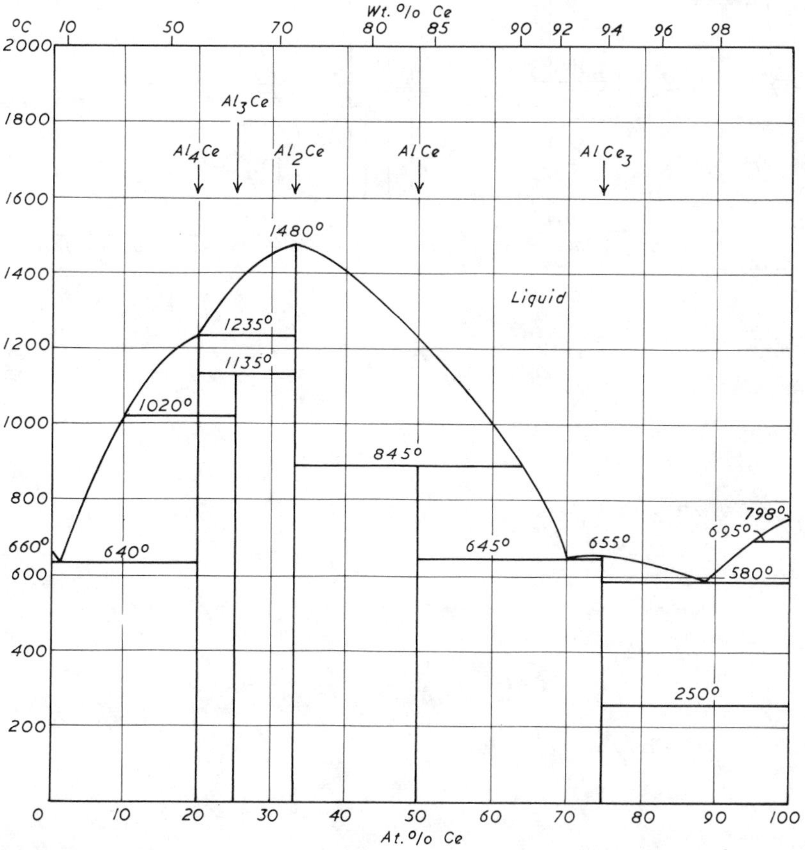

Wt. % Ce

°C

Al_3Ce

Al_4Ce

Al_2Ce

$AlCe$

$AlCe_3$

1480°

Liquid

1235°

1135°

1020°

845°

798°

695°

660°

640°

645°

655°

580°

250°

At. % Ce

Al–Co

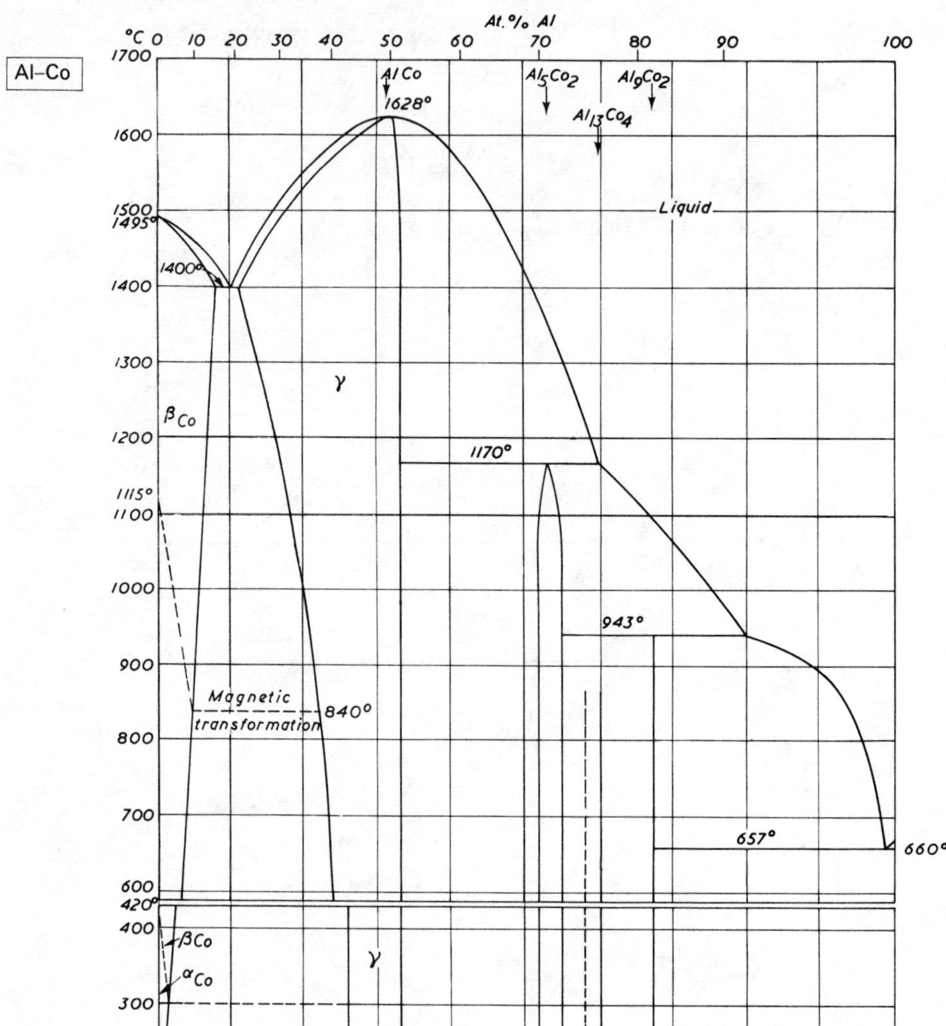

At.% Al
°C
Wt.% Al
AlCo
1628°
Al_5Co_2
$Al_{13}Co_4$
Al_9Co_2
Liquid
1495°
1400°
γ
β_{Co}
1170°
1115°
943°
Magnetic
transformation
840°
657°
660°
420°
βCo
α_{Co}
γ

Al–Cr

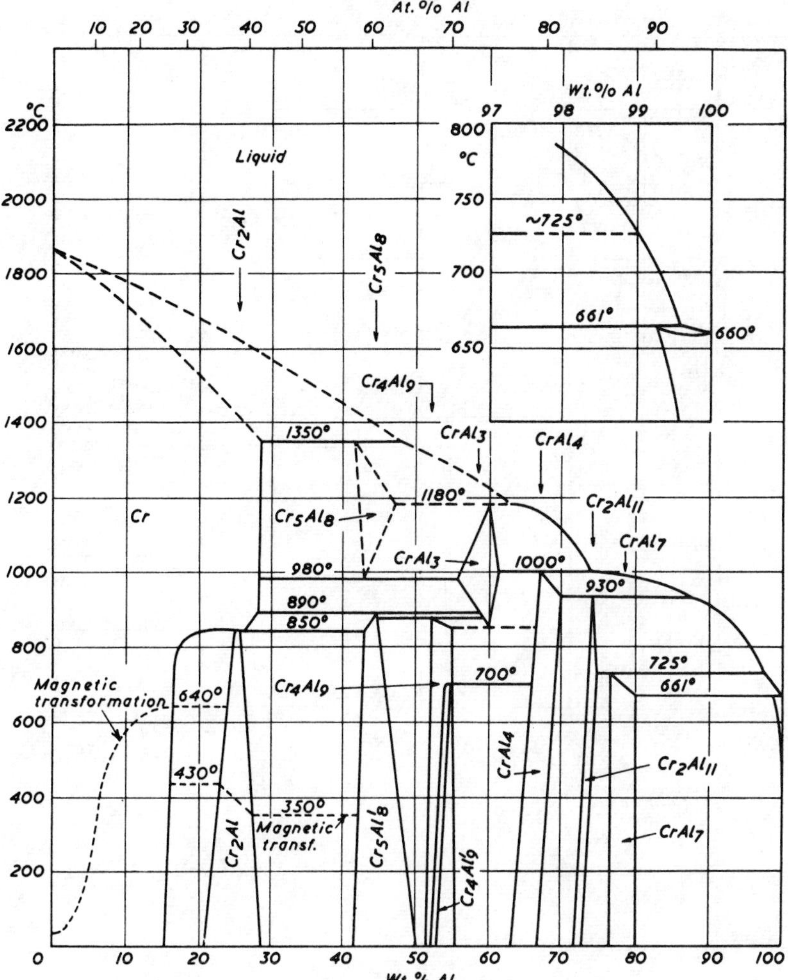
At. % Al
10
20
30
40
50
60
70
80
90
Wt. % Al
97
98
99
100
800
°C
750
700
650
~725°
661°
660°
°C
2200
2000
1800
1600
1400
1200
1000
800
600
400
200
0
Liquid
Cr2Al
Cr5Al8
Cr4Al9
CrAl3
CrAl4
1350°
1180°
Cr2Al11
CrAl7
Cr
Cr5Al8
CrAl3
980°
1000°
930°
890°
850°
700°
725°
661°
Magnetic transformation
640°
Cr4Al9
430°
350°
Magnetic transf.
Cr2Al
Cr5Al8
Cr4Al9
CrAl4
Cr2Al11
CrAl7
10
20
30
40
50
60
70
80
90
100
Wt. % Al

Al–Cu

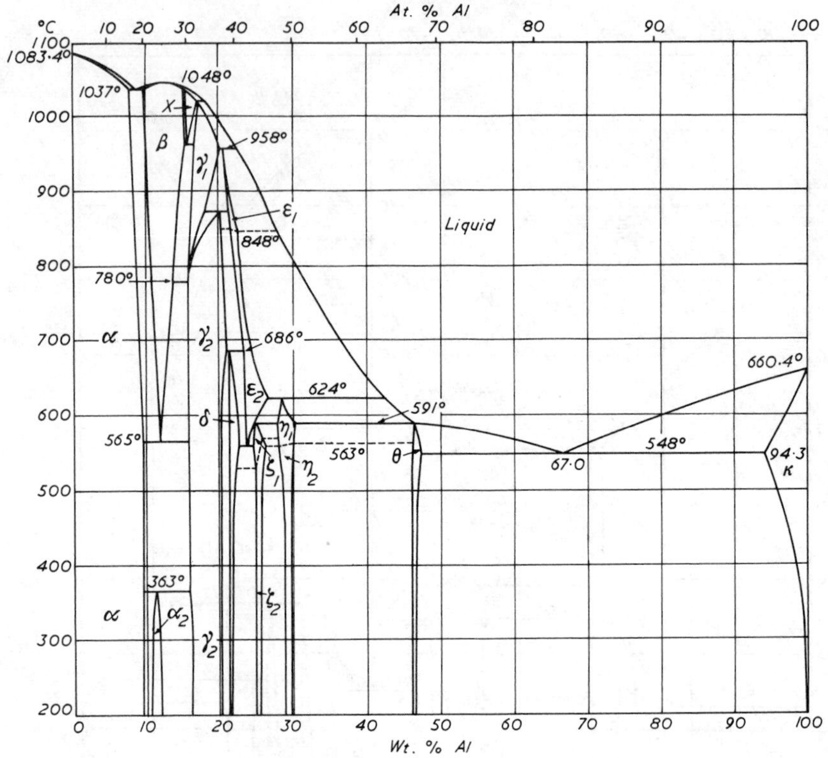

At. % Al
°C
1083·4°
1037°
1048°
X
β
γ_1
958°
ε_1
848°
Liquid
780°
α
γ_2
686°
ε_2
624°
591°
660·4°
δ
η_1
565°
563°
θ
548°
67·0
94·3
κ
ζ_1
η_2
363°
ζ_2
α
α_2
γ_2
Wt. % Al

Al–Dy

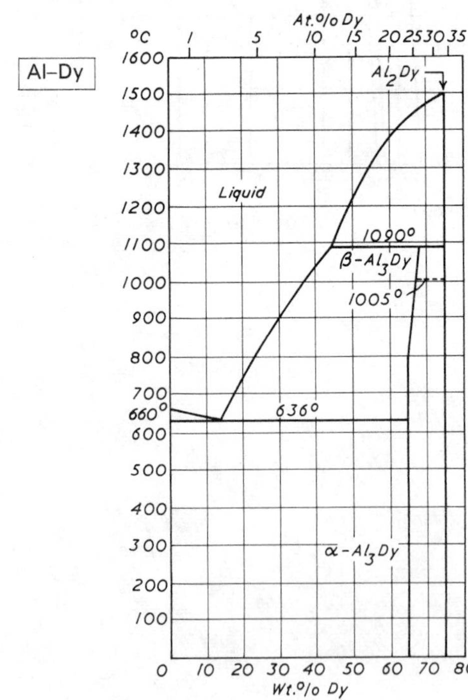

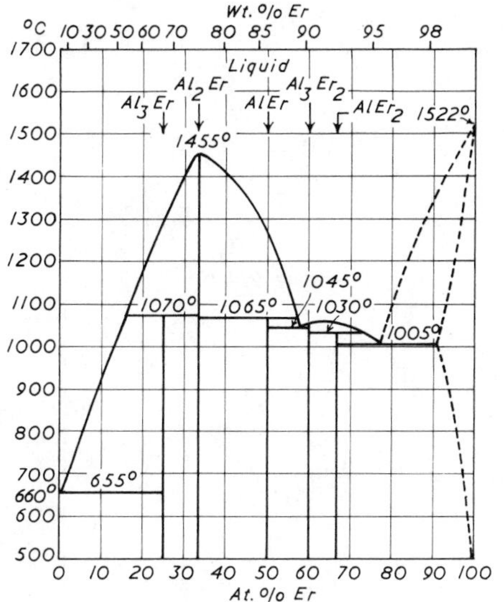

Al–Er

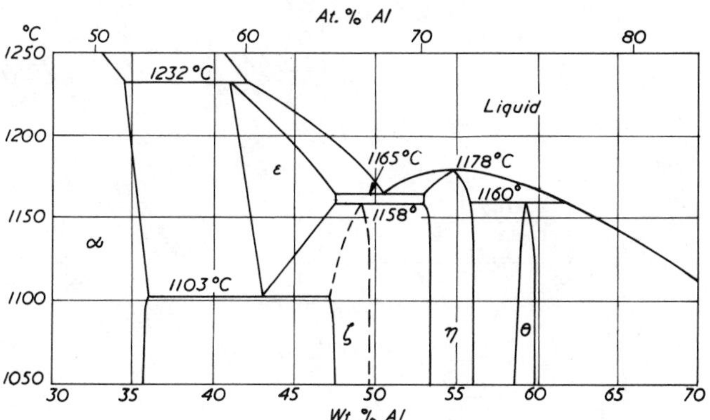

Al–Fe

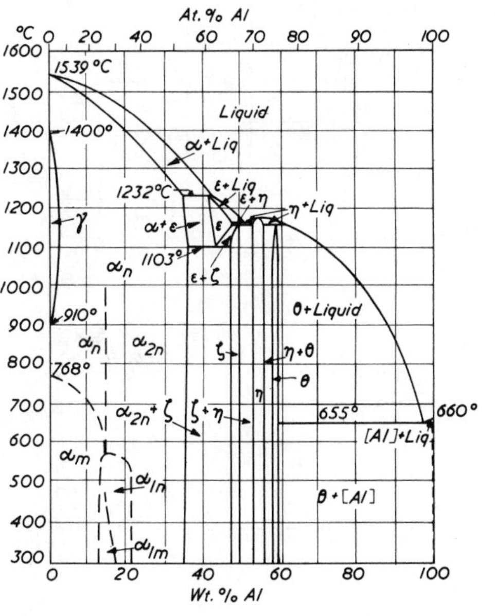

Al–Fe

Al–Ga

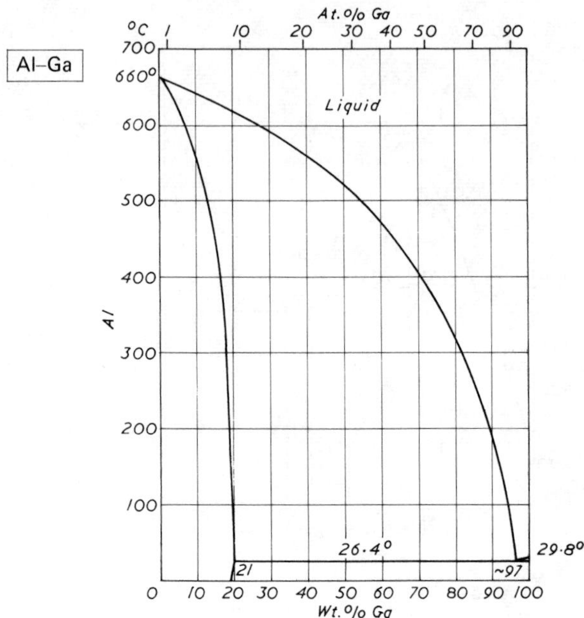

Al–Gd

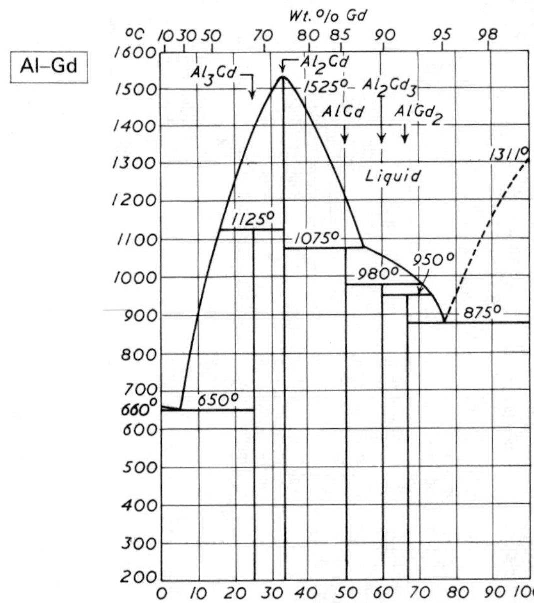

Al–Ge

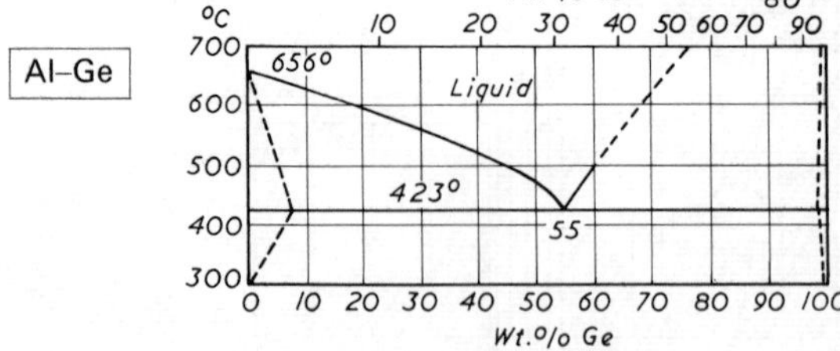

Al–Hf

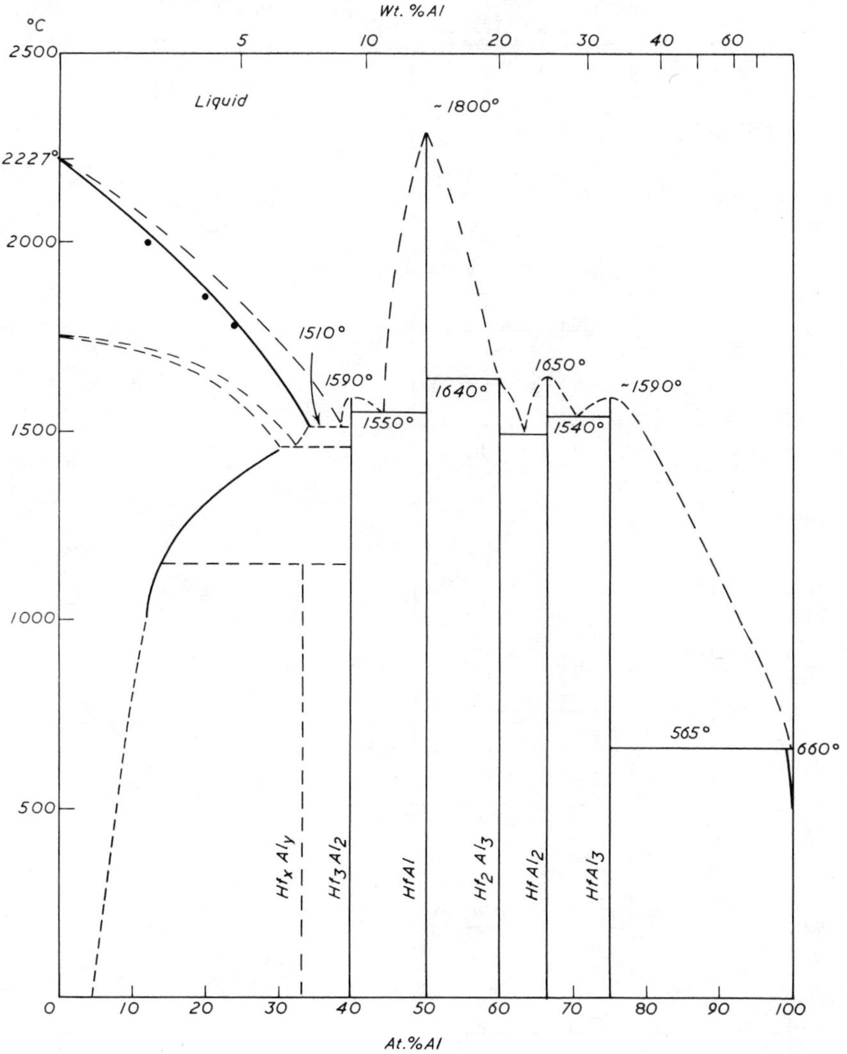

Wt. %Al
°C
2500
2227°
2000
1500
1000
500
Liquid
~1800°
1510°
1590°
1640°
1650°
~1590°
1550°
1540°
565°
660°
$Hf_x Al_y$
Hf_3Al_2
HfAl
Hf_2Al_3
$HfAl_2$
$HfAl_3$
At.%Al

Al–Hg

At.% Hg
°C
660°
Liquid
Liquid + α
−38·9°
α
α+Hg
Wt.% Hg

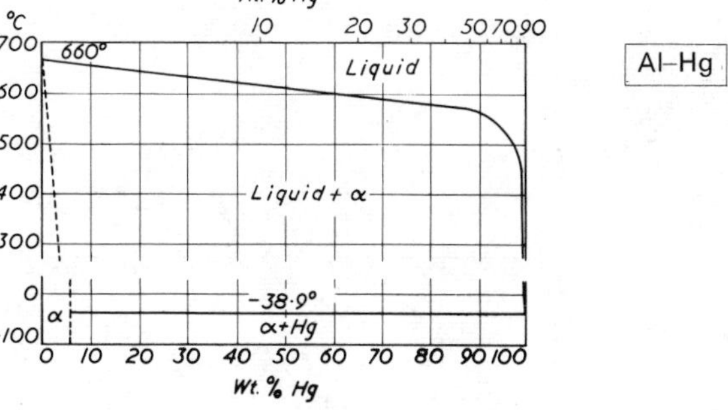

Al–In

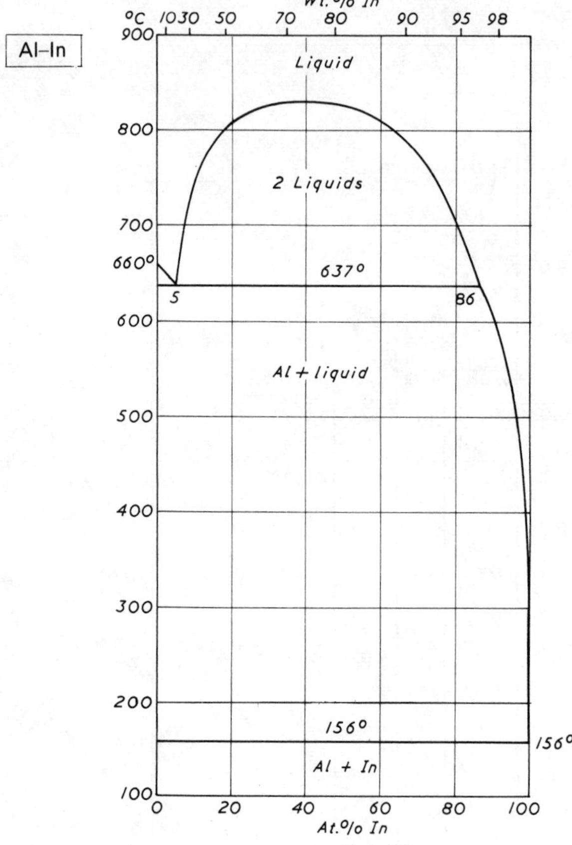
Wt.% In
°C 10 30 50 70 80 90 95 98
900
800
700
660°
600
500
400
300
200
100
Liquid
2 Liquids
637°
5
86
Al + liquid
156°
156°
Al + In
0 20 40 60 80 100
At.% In

Al–K

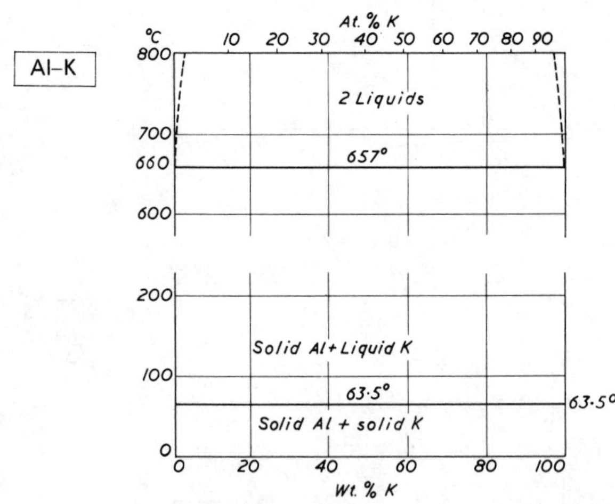
At.% K
°C 10 20 30 40 50 60 70 80 90
800
700
660
600
200
100
0
2 Liquids
657°
Solid Al + Liquid K
63·5°
63·5°
Solid Al + solid K
0 20 40 60 80 100
Wt.% K

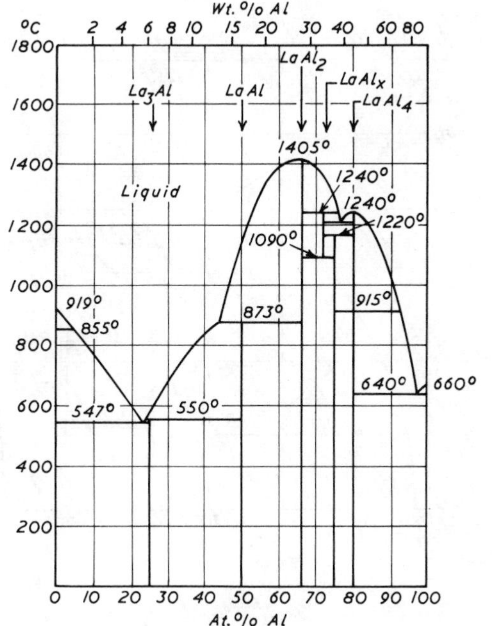

Wt. % Al
°C
2 4 6 8 10 15 20 30 40 60 80
1800
1600
1400
1200
1000
800
600
400
200
La3Al
LaAl
LaAl2
LaAlx
LaAl4
1405°
1240°
1240°
1220°
1090°
Liquid
919°
855°
873°
915°
640°
660°
547°
550°
0 10 20 30 40 50 60 70 80 90 100
At. % Al

Al–La

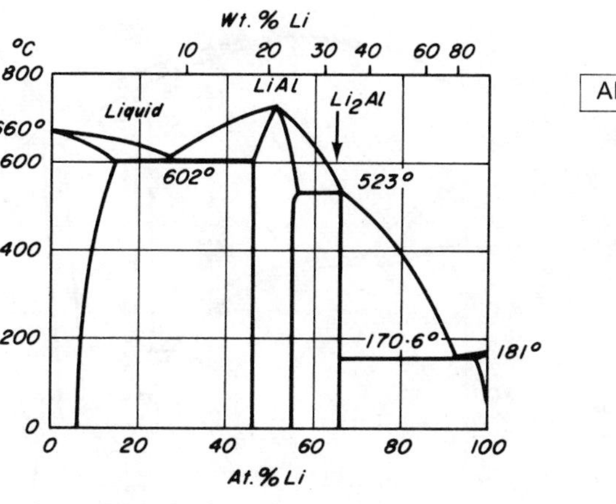

Wt. % Li
°C
10 20 30 40 60 80
800
660°
600
400
200
0
LiAl
Li2Al
Liquid
602°
523°
170·6°
181°
0 20 40 60 80 100
At. % Li

Al–Li

Al–Mg

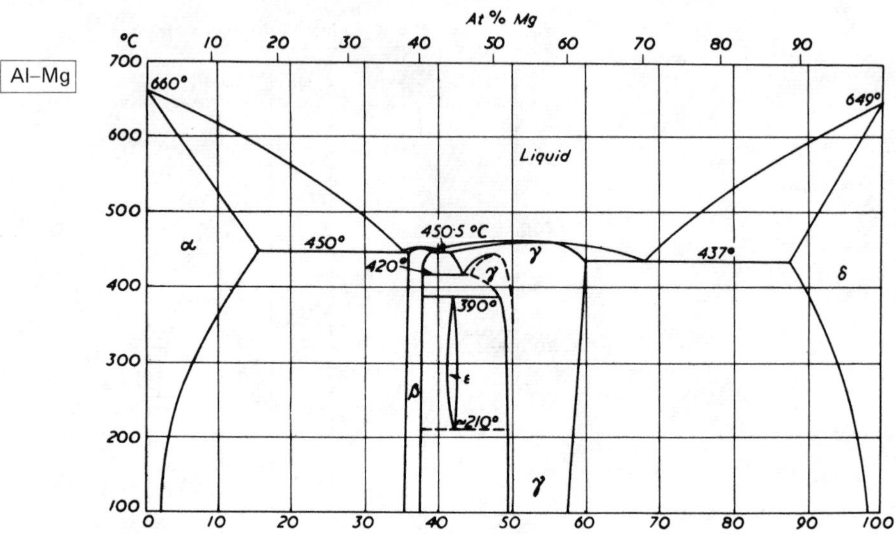
At % Mg
°C
660°
649°
Liquid
450·5 °C
450°
437°
420°
390°
~210°
α
β
γ
δ
ε
Wt. % Mg

Al–Mn

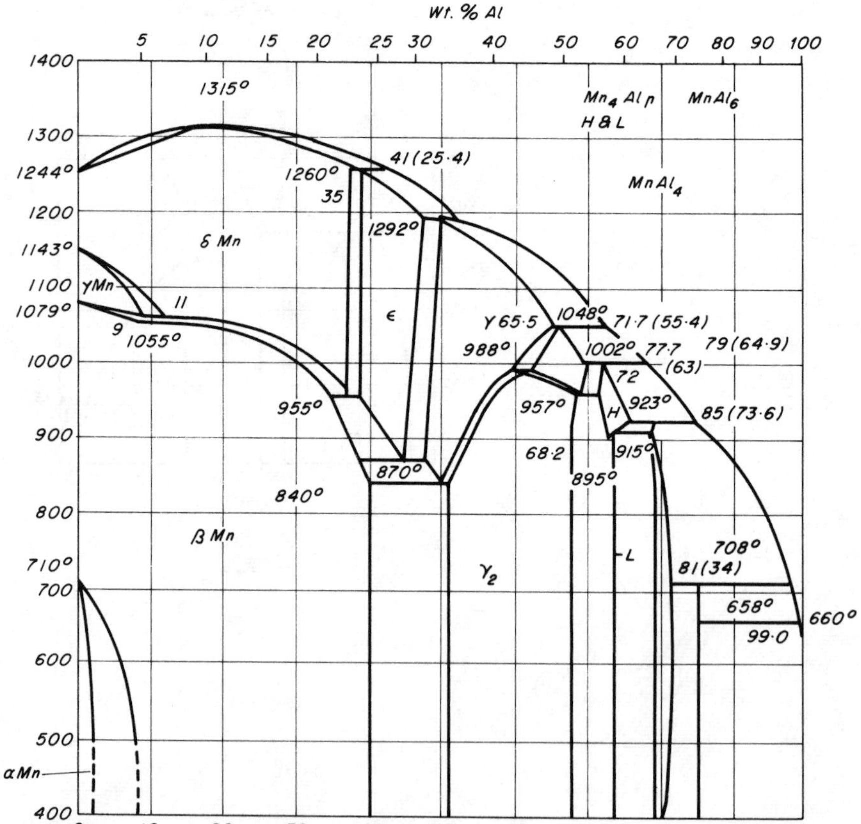

Wt. % Al
1315°
Mn4 Al p
H & L
MnAl6
41 (25·4)
1260°
35
1244°
MnAl4
1292°
δ Mn
1143°
γMn
11
1079°
9
1055°
ϵ
Y 65·5
1048°
71·7 (55·4)
988°
1002°
77·7
79 (64·9)
(63)
72
955°
957°
923°
H
85 (73·6)
68·2
915°
870°
895°
840°
β Mn
L
708°
710°
γ2
81 (34)
658°
660°
99·0
αMn
At. % Al

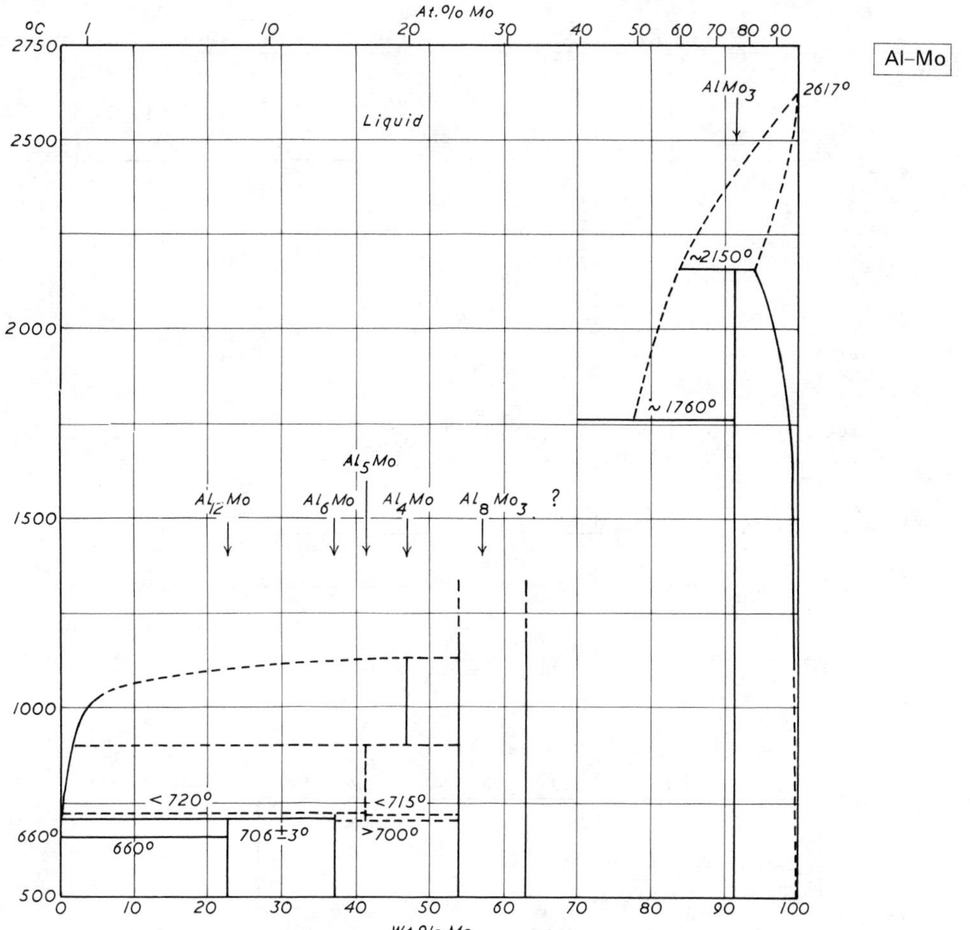
Al–Mo
At.% Mo
°C
2750
2500
2000
1500
1000
660°
500
Liquid
AlMo3
2617°
~2150°
~1760°
Al5Mo
Al12Mo
Al6Mo
Al4Mo
Al8Mo3
?
< 720°
< 715°
> 700°
706±3°
660°
Wt.% Mo

Al–Na

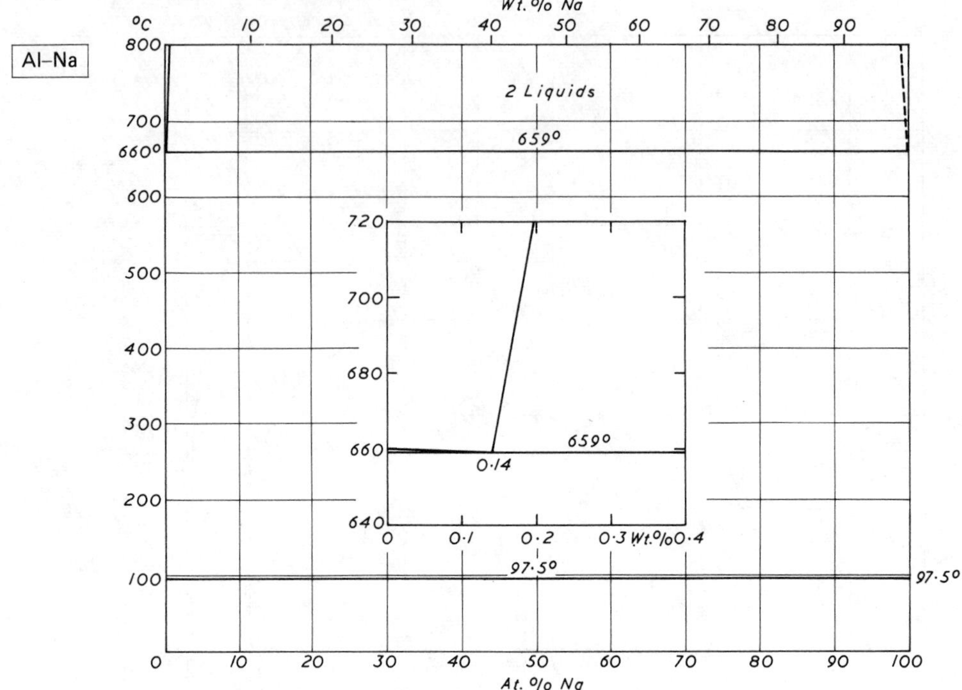
Wt. % Na
°C
2 Liquids
659°
660°
0·14
Wt.%
97·5°
At. % Na

Al–Nb

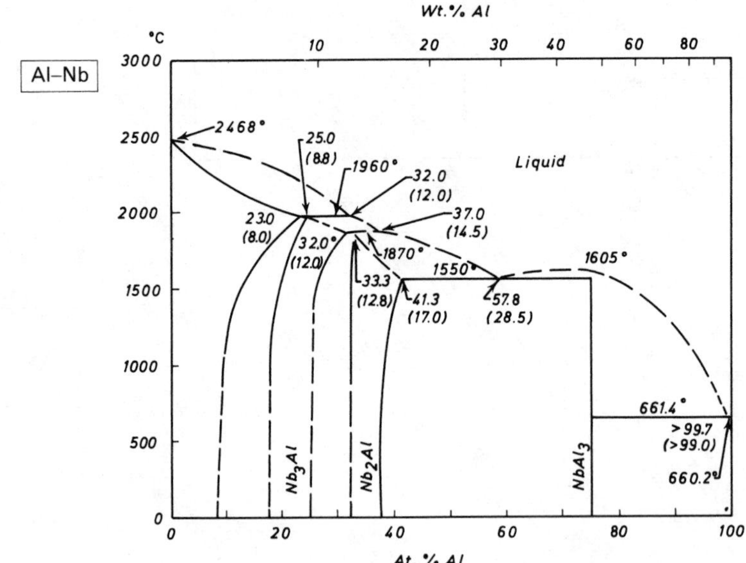
Wt.% Al
°C
2468°
25.0
(8.8)
1960°
32.0
(12.0)
Liquid
37.0
(14.5)
23.0
(8.0)
32.0°
(12.0)
1870°
1550°
1605°
33.3
(12.8)
41.3
(17.0)
57.8
(28.5)
661.4°
>99.7
(>99.0)
660.2°
Nb3Al
Nb2Al
NbAl3
At. % Al

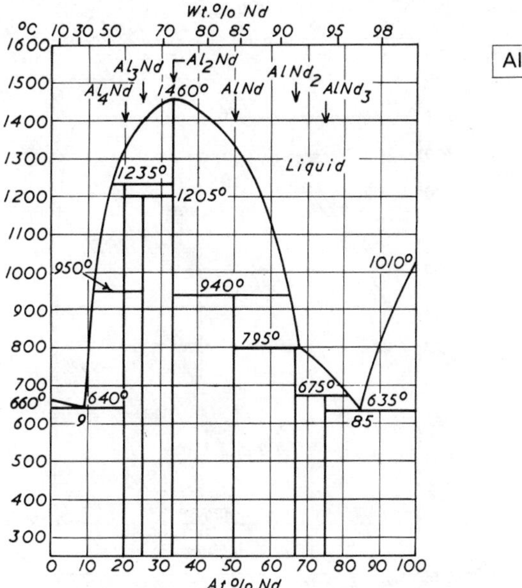
Wt.% Nd
°C
10 30 50 70 80 85 90 95 98
Al$_3$Nd
Al$_2$Nd
Al$_4$Nd
1460°
AlNd
AlNd$_2$
AlNd$_3$
Liquid
1235°
1205°
950°
940°
1010°
795°
675°
640°
635°
9
85
660°
1600
1500
1400
1300
1200
1100
1000
900
800
700
600
500
400
300
0 10 20 30 40 50 60 70 80 90 100
At.% Nd

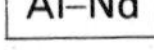
Al–Nd

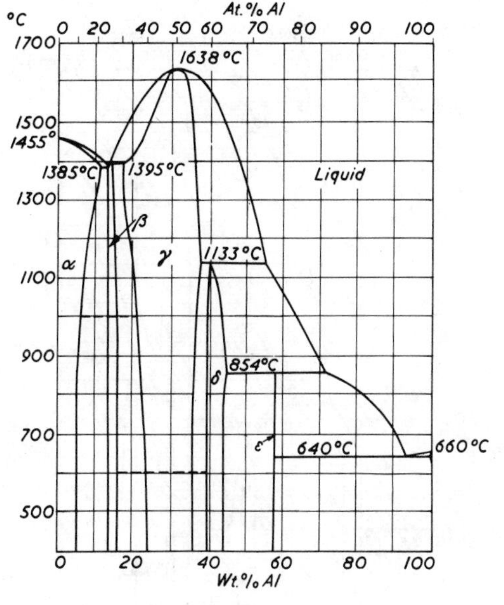
At.% Al
°C
0 20 40 50 60 70 80 90 100
1700
1638 °C
1500
1455°
1385°C
1395°C
Liquid
1300
β
α
γ
1133°C
1100
900
δ
854°C
700
ε
640°C
660°C
500
0 20 40 60 80 100
Wt.% Al

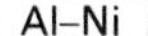
Al–Ni

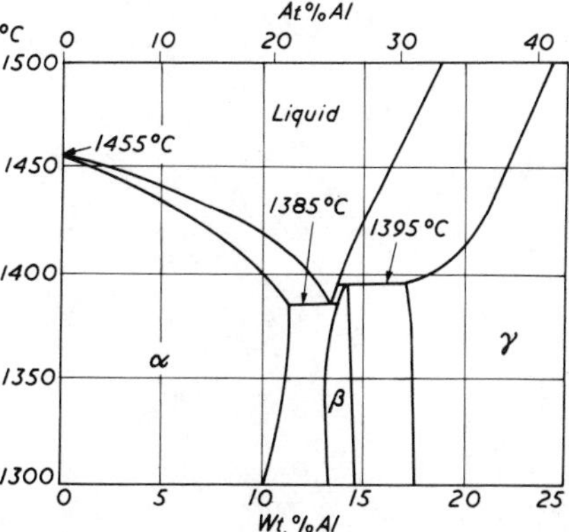
At% Al
°C
0 10 20 30 40
1500
Liquid
1455°C
1450
1385°C
1395°C
1400
α
γ
1350
β
1300
0 5 10 15 20 25
Wt.% Al

Al–Ni

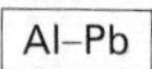
Al–Pb

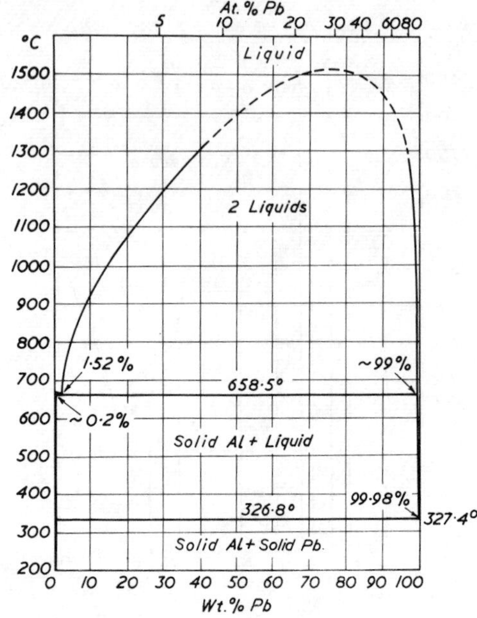
At. % Pb
5
10
20
30 40
60 80
°C
Liquid
2 Liquids
1·52 %
~99 %
658·5°
~0·2%
Solid Al + Liquid
99·98 %
326·8°
327·4°
Solid Al + Solid Pb
Wt. % Pb

Al–Pd

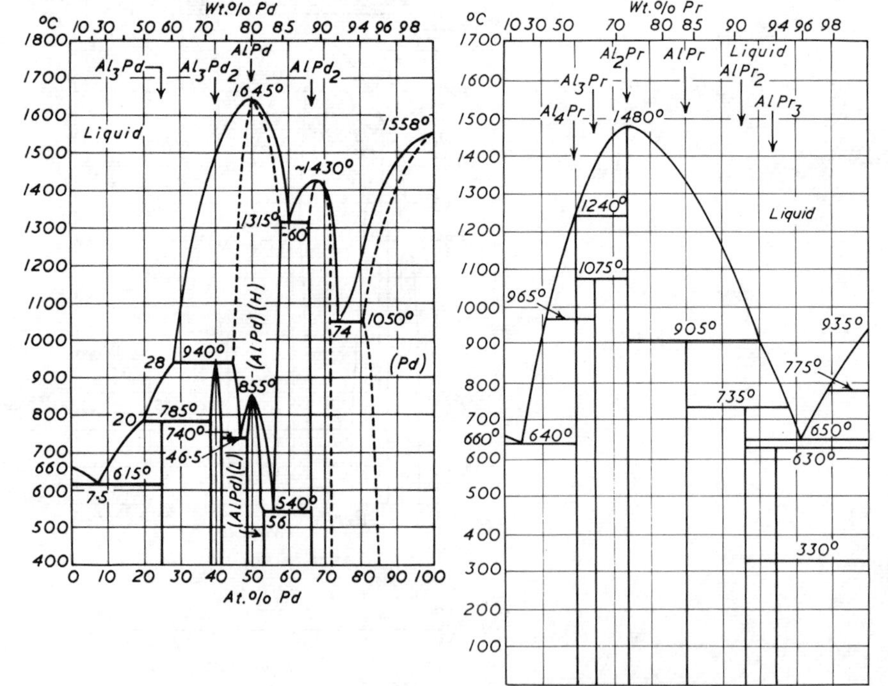
Wt. % Pd
°C
Liquid
Al3Pd
Al3Pd2
AlPd
AlPd2
1645°
1558°
~1430°
1315°
60
74
1050°
(AlPd)(H)
(Pd)
28
940°
855°
20
785°
740°
46·5
(AlPd)(L)
615°
7·5
540°
56
At. % Pd
Wt. % Pr
Al2Pr
AlPr
Liquid
Al3Pr
AlPr2
Al4Pr
1480°
AlPr3
1240°
1075°
965°
905°
935°
775°
735°
640°
650°
630°
330°
At. % Pr

Al–Pr

Al–Pt

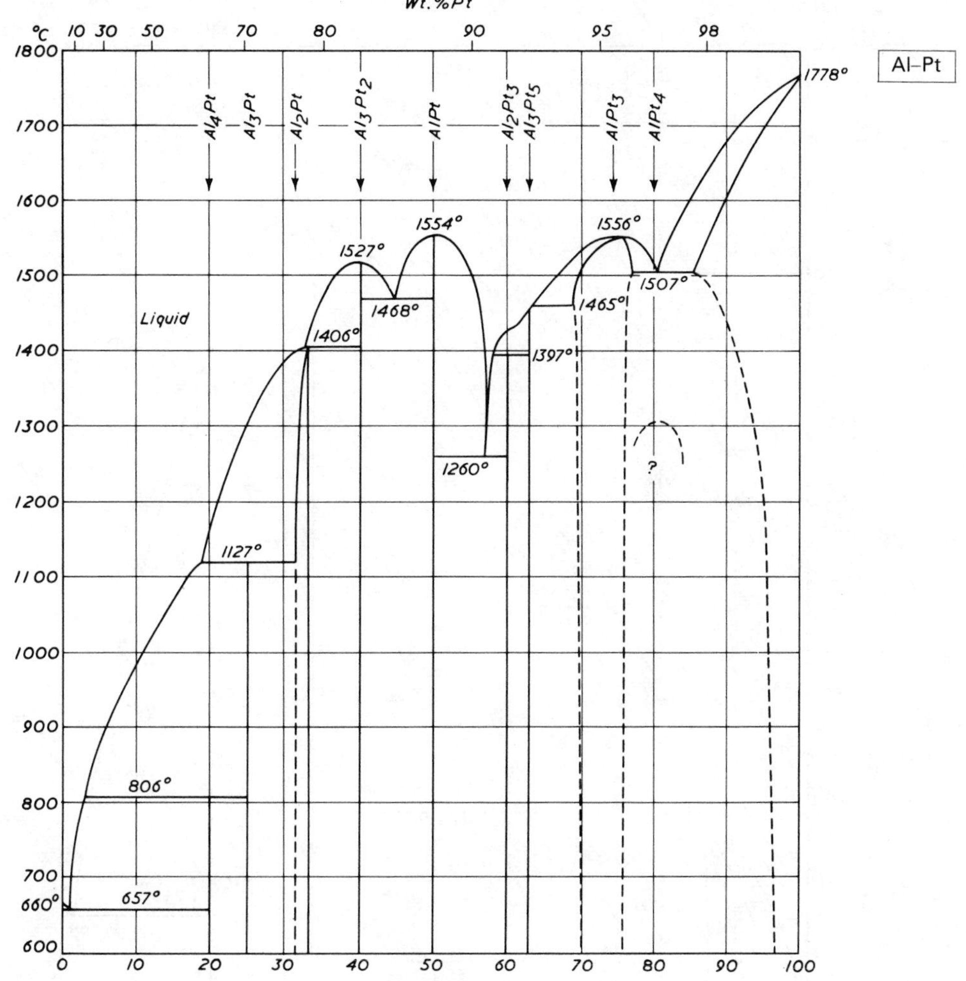

Wt. %Pt
°C
10 30 50 70 80 90 95 98
1800
1700
1600
1500
1400
1300
1200
1100
1000
900
800
700
660°
600
Al4Pt
Al3Pt
Al2Pt
Al3Pt2
AlPt
Al2Pt3
Al3Pt5
AlPt3
AlPt4
1778°
1554°
1556°
1527°
1468°
1507°
1465°
Liquid
1406°
1397°
1260°
?
1127°
806°
657°
0 10 20 30 40 50 60 70 80 90 100
At. %Pt

Al–Pu

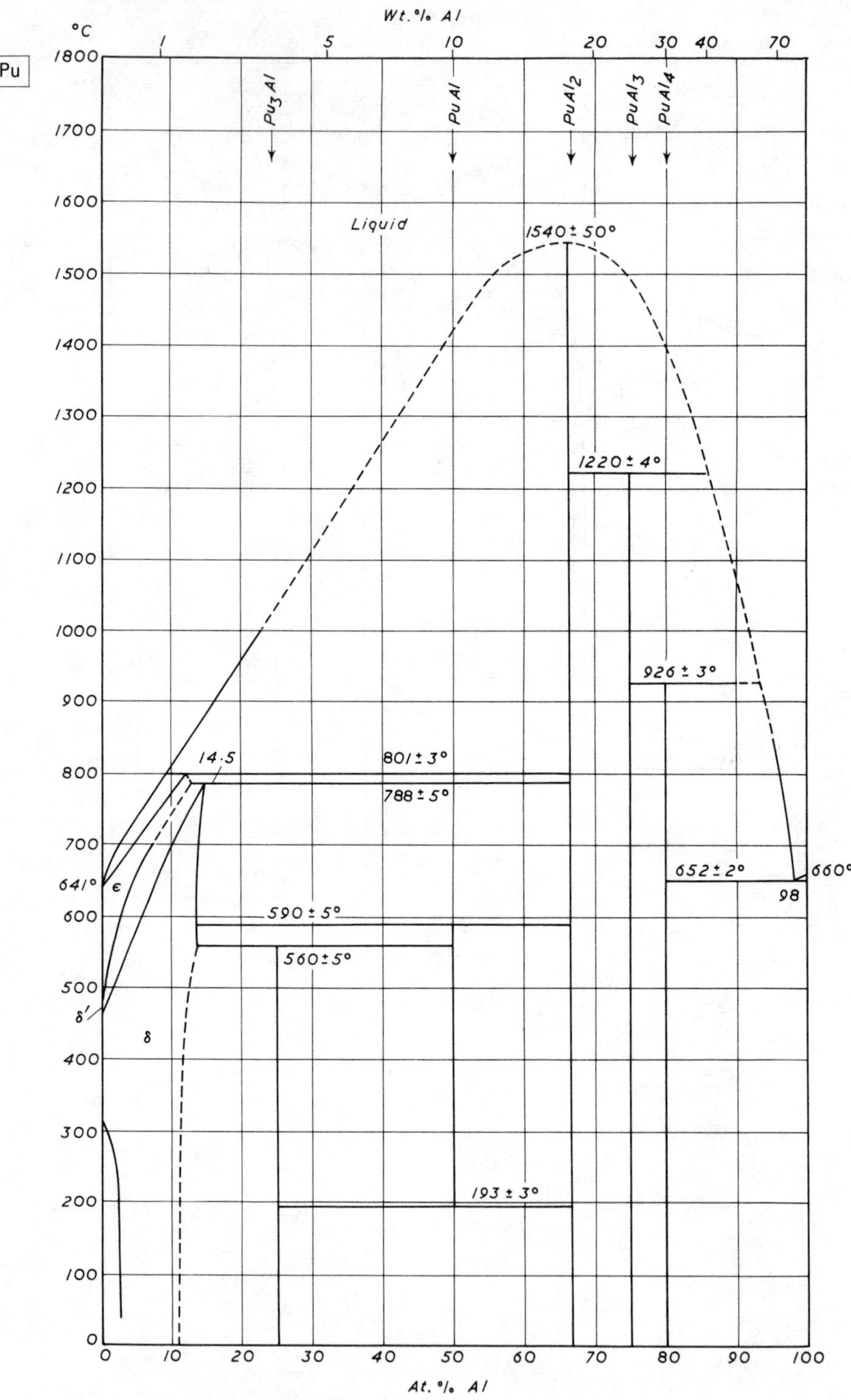
Wt.% Al
°C
1
5
10
20
30
40
70
1800
1700
1600
1500
1400
1300
1200
1100
1000
900
800
700
641°
600
500
400
300
200
100
0
Pu3Al
PuAl
PuAl2
PuAl3
PuAl4
Liquid
1540 ± 50°
1220 ± 4°
926 ± 3°
14.5
801 ± 3°
788 ± 5°
652 ± 2°
660°
98
590 ± 5°
560 ± 5°
193 ± 3°
ε
δ'
δ
0
10
20
30
40
50
60
70
80
90
100
At.% Al

Al–Re

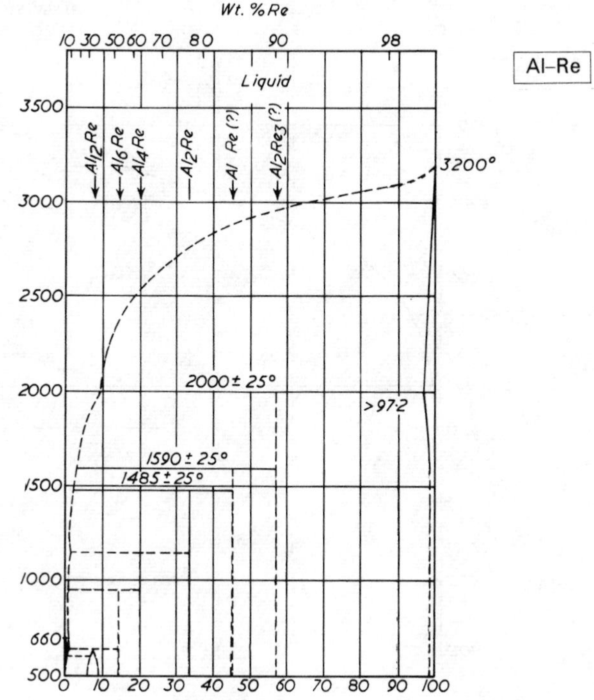

Wt. % Re
10 30 50 60 70 80 90 98
Liquid
$Al_{12}Re$
Al_6Re
Al_4Re
Al_2Re
Al Re (?)
Al_2Re_3 (?)
3200°
2000 ± 25°
>97·2
1590 ± 25°
1485 ± 25°
3500
3000
2500
2000
1500
1000
660
500
0 10 20 30 40 50 60 70 80 90 100
At. % Re

Al–Ru

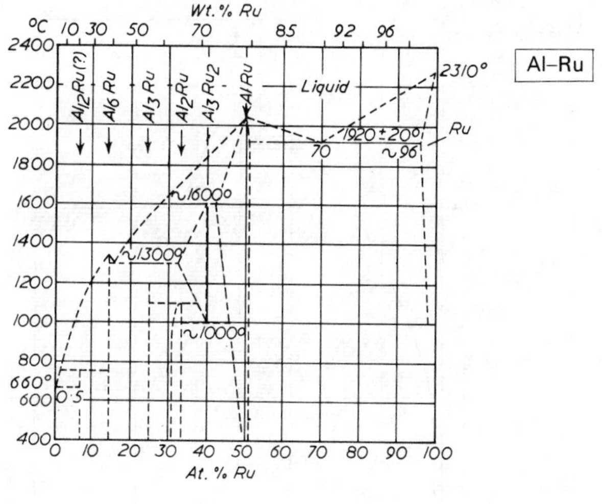

Wt. % Ru
°C 10 30 50 70 85 92 96
$Al_{12}Ru$ (?)
Al_6Ru
Al_3Ru
Al_2Ru
Al_3Ru_2
AlRu
Liquid
2310°
1920 ± 20°
Ru
70
~96
~1600°
~1300°
~1000°
660°
0·5
2400
2200
2000
1800
1600
1400
1200
1000
800
600
400
0 10 20 30 40 50 60 70 80 90 100
At. % Ru

Al–Sb

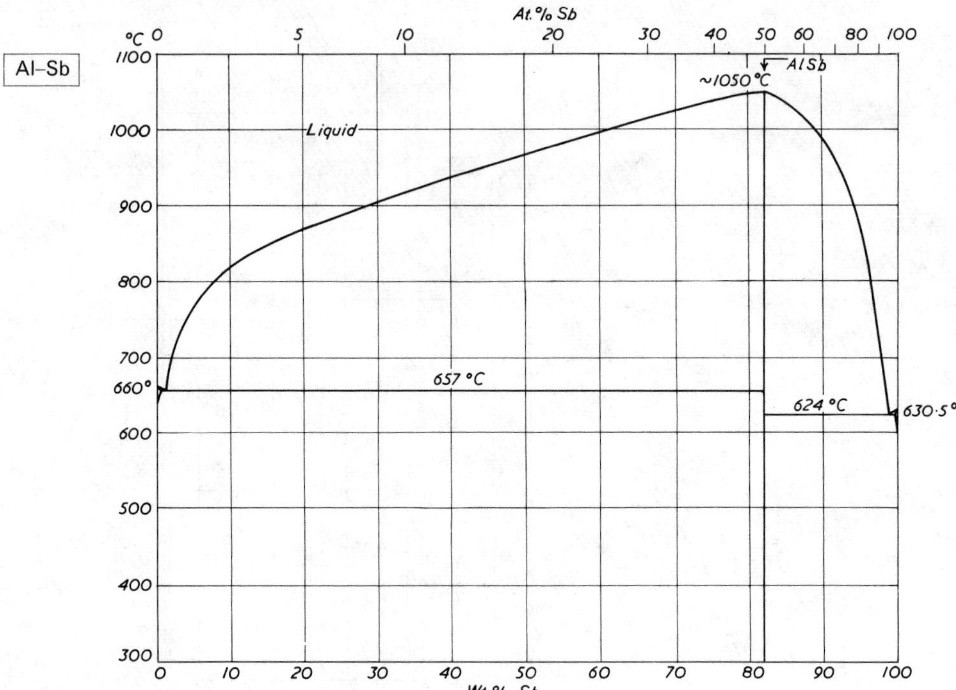

At.% Sb
°C
1100
1000
900
800
700
660°
600
500
400
300
AlSb
~1050 °C
Liquid
657 °C
624 °C
630·5°
Wt.% Sb

Al–Sc

Wt.% Sc
°C
1200
800
660°
400
Al3Sc
Al2Sc
AlSc
Al2Sc
1420°
Liquid
1320 ± 7°
~1300°
1195 ± 7°
1150 ± 7°
~43
945 ± 5°
~87
665 ± 2°
1539°
β-Sc
1350°
α-Sc
At.% Sc

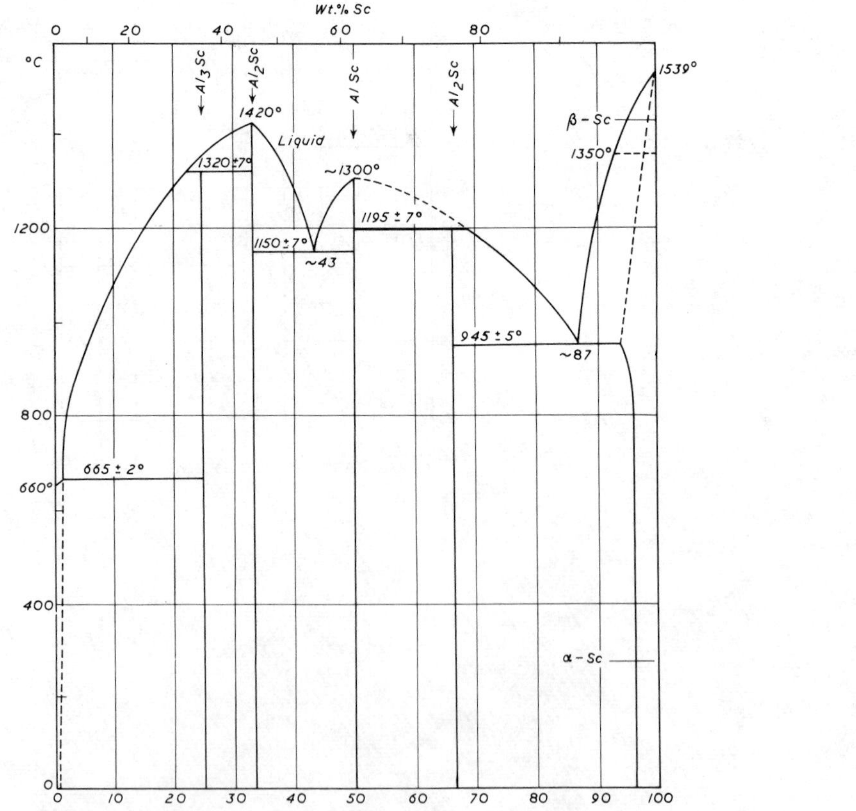

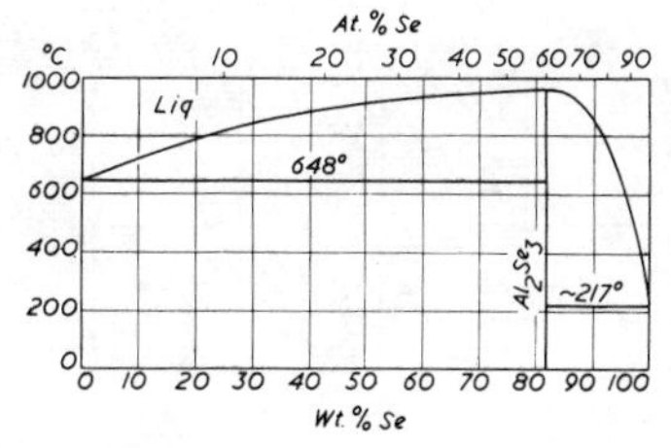

Al–Se

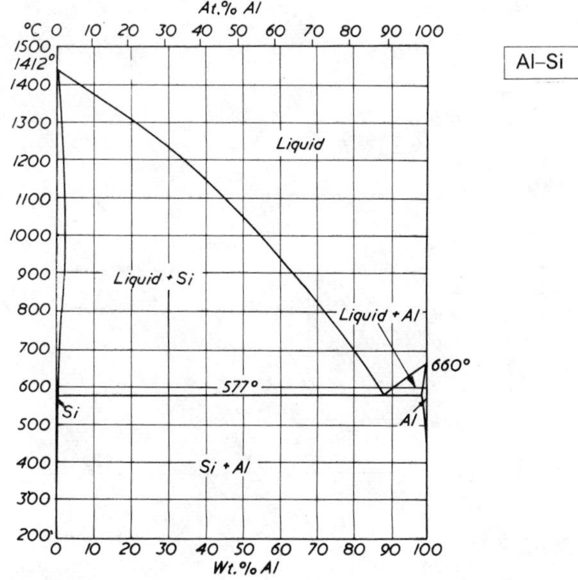

Al–Si

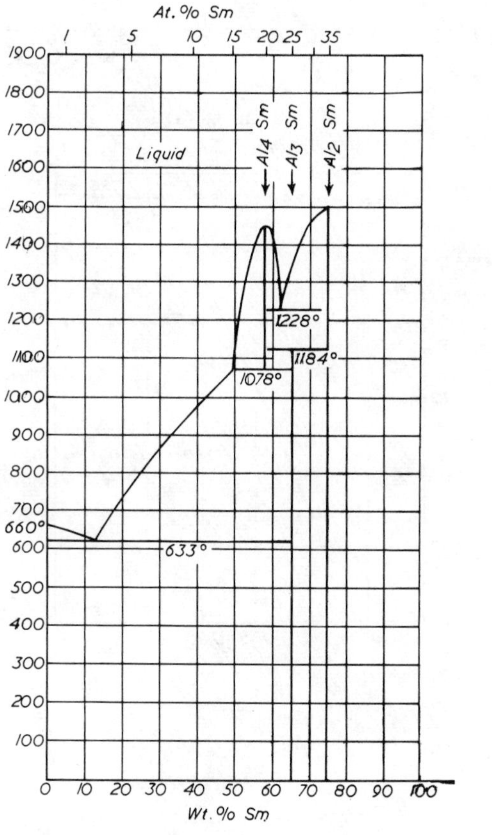

Al–Sm

Al–Sn

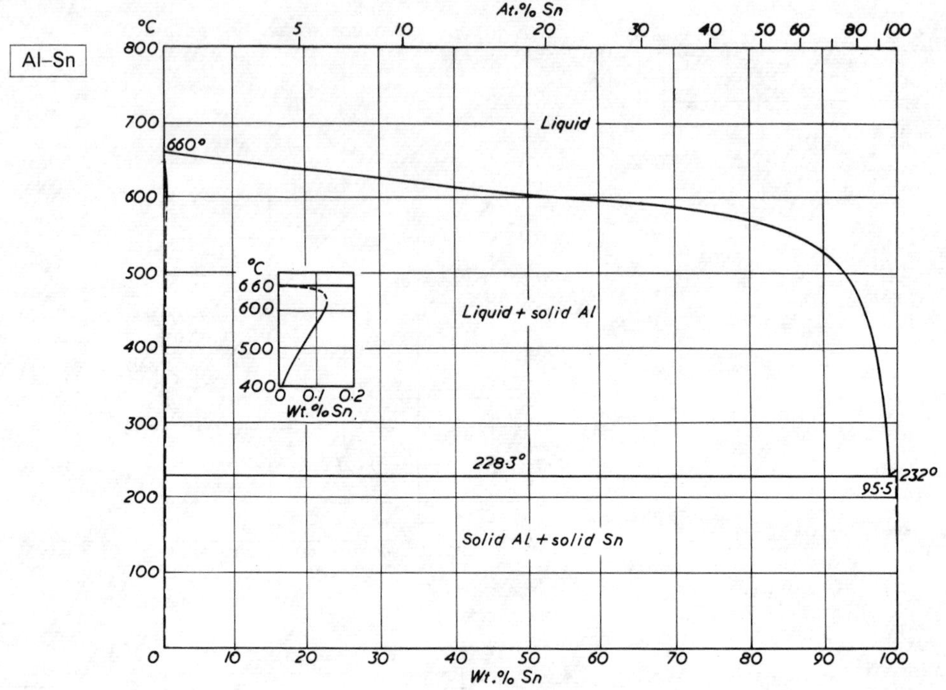
At.% Sn
°C
5
10
20
30
40
50 60
80 100
800
700
Liquid
660°
600
500
°C
660
600
500
400
0 0·1 0·2
Wt.% Sn.
Liquid + solid Al
400
300
228·3°
232°
95·5
200
Solid Al + solid Sn
100
0
10
20
30
40
50
60
70
80
90
100
Wt.% Sn

Al–Ta

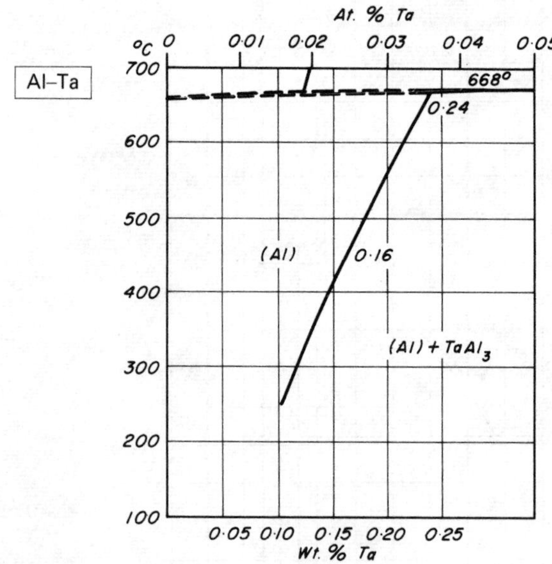
At. % Ta
°C
0
0·01
0·02
0·03
0·04
0·05
700
668°
0·24
600
500
(Al)
0·16
400
(Al) + TaAl$_3$
300
200
100
0·05 0·10 0·15 0·20 0·25
Wt. % Ta

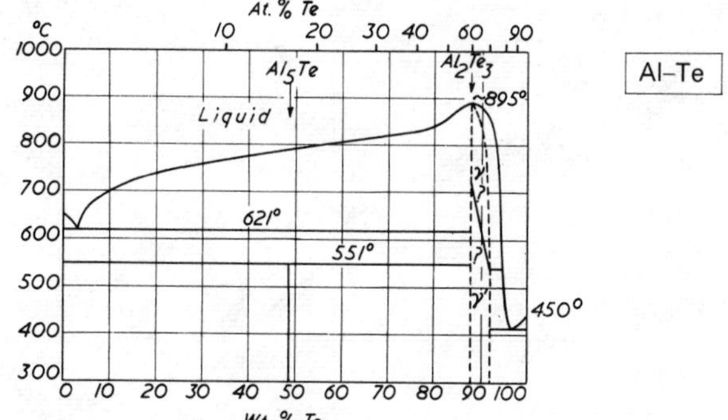

Al–Te

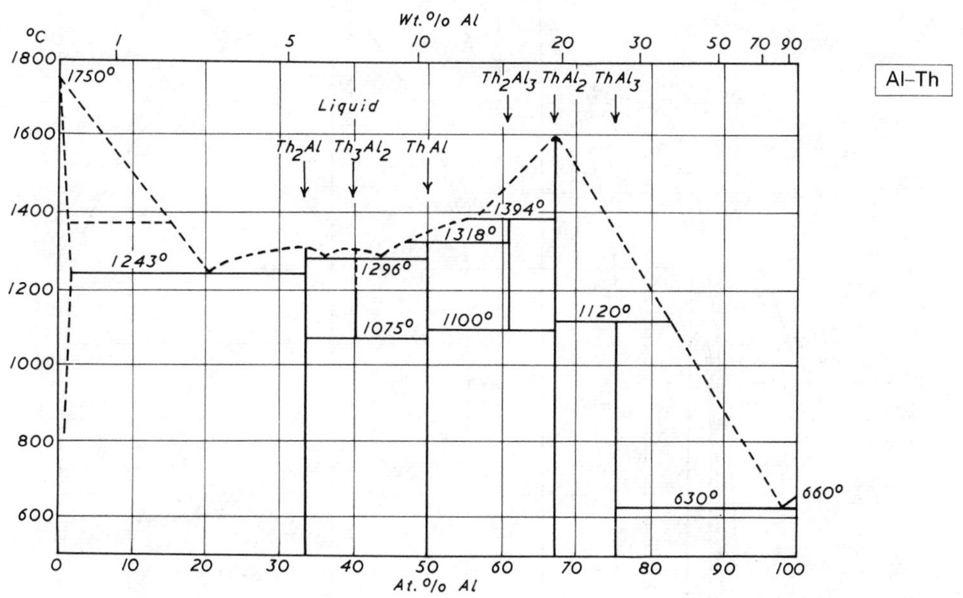

Al–Th

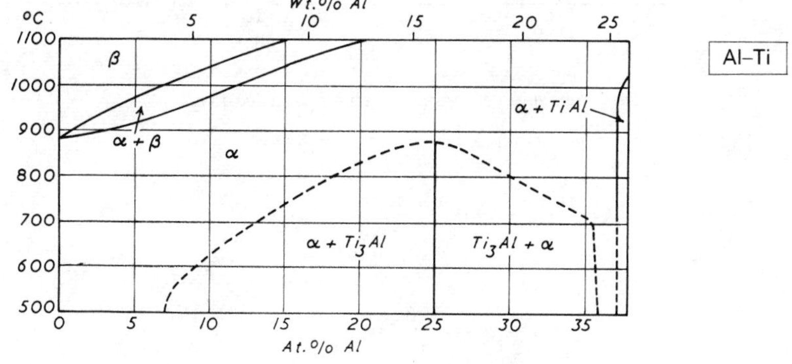

Al–Ti

Al–Ti

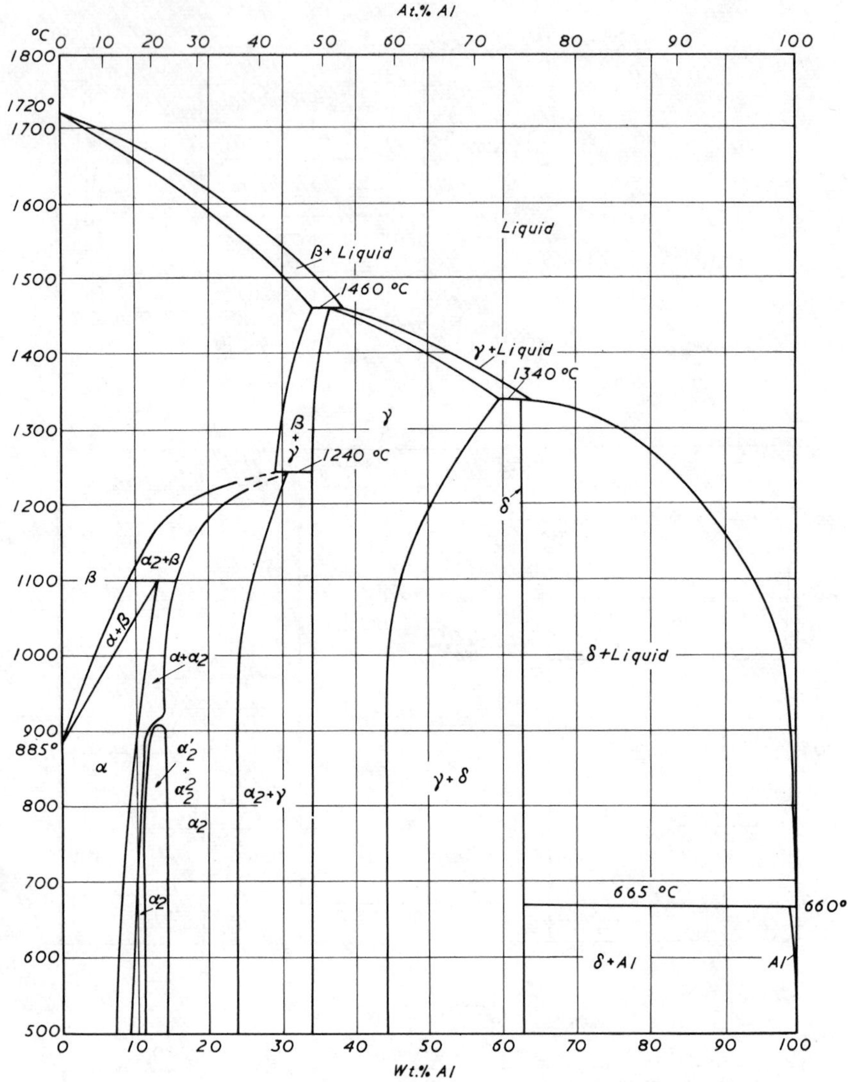

See alternative versions of Ti-rich end on page **11**-53

Al–Ti

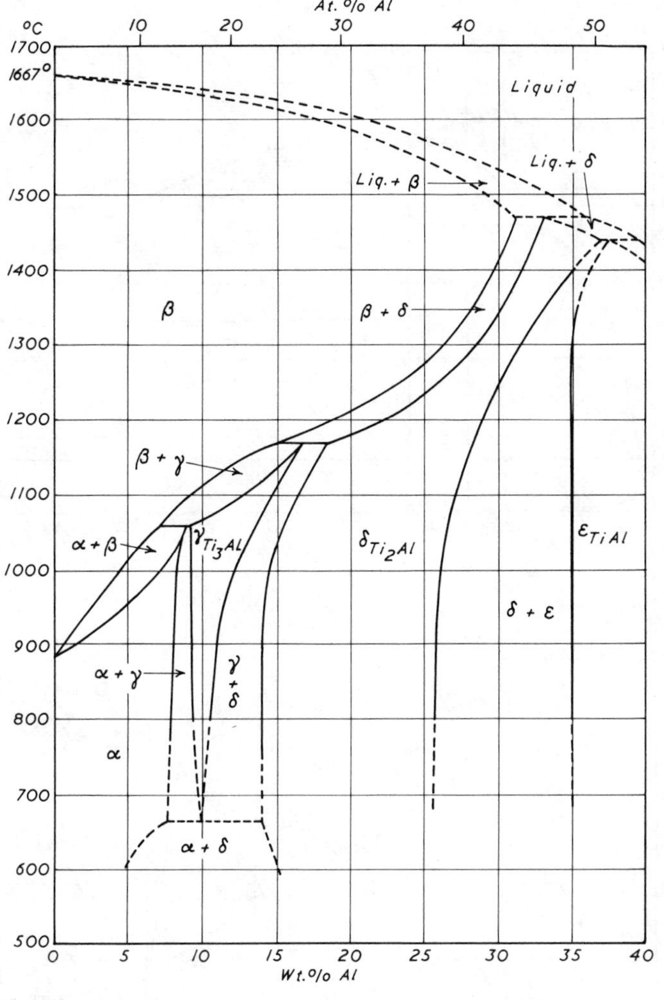
At. % Al
°C
1667°
Liquid
Liq.+ δ
Liq.+ β
β
β + δ
β + γ
α + β
γ_{Ti_3Al}
δ_{Ti_2Al}
ε_{TiAl}
δ + ε
α + γ
γ + δ
α
α + δ
Wt.% Al

Al–Tl

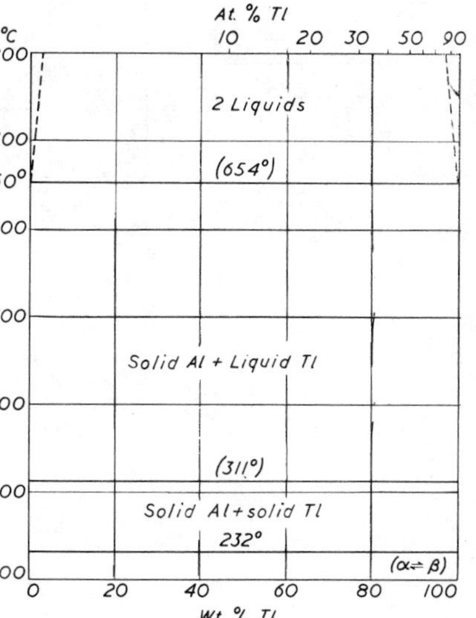
At. % Tl
°C
2 Liquids
(654°)
660°
Solid Al + Liquid Tl
(311°)
Solid Al + solid Tl
232°
(α⇌β)
Wt. % Tl

Al–U

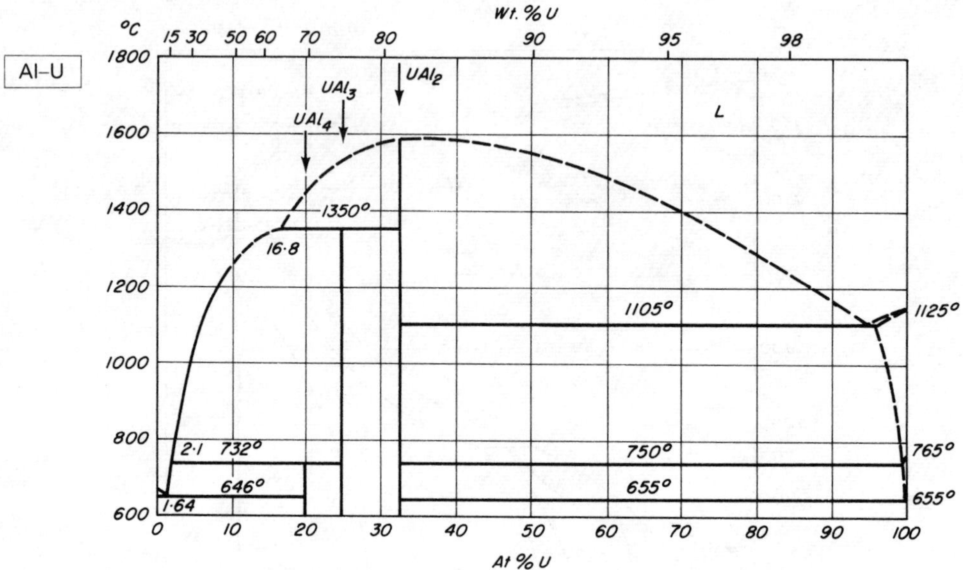
Wt.% U
°C
15 30 50 60 70 80 90 95 98
1800
1600
1400
1200
1000
800
600
UAl2
UAl3
UAl4
L
1350°
16·8
1105°
1125°
2·1 732°
750°
765°
646°
655°
655°
1·64
0 10 20 30 40 50 60 70 80 90 100
At % U

Al–V

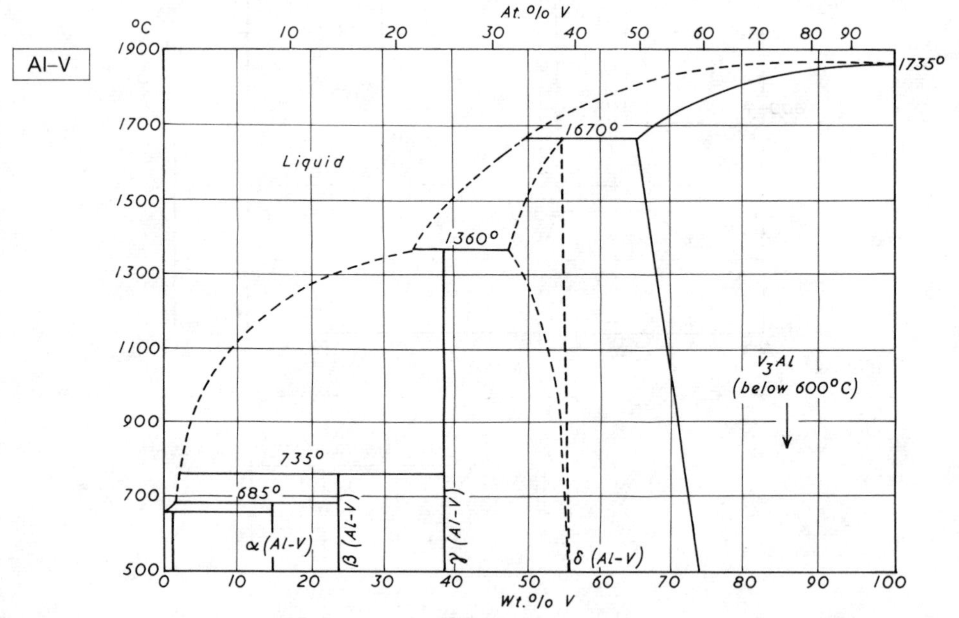
At. % V
°C
10 20 30 40 50 60 70 80 90
1900
1735°
1700
1670°
Liquid
1500
1360°
1300
1100
V3Al
(below 600°C)
900
735°
700
685°
α (Al-V)
β (Al-V)
γ (Al-V)
δ (Al-V)
500
0 10 20 30 40 50 60 70 80 90 100
Wt. % V

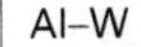

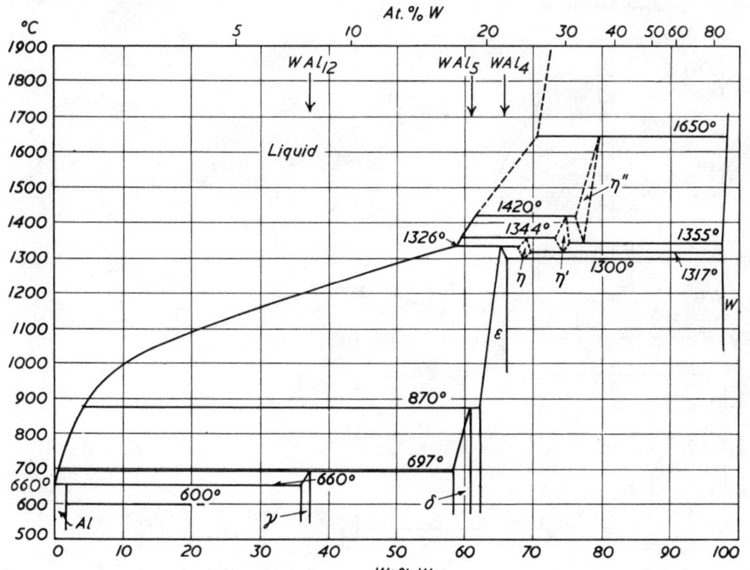
At.% W
5
10
20
30
40
50
60
80
°C
1900
1800
1700
1600
1500
1400
1300
1200
1100
1000
900
800
700
660°
600
500
WAl12
WAl5
WAl4
Liquid
1650°
η″
1420°
1344°
1326°
1355°
η
η′
1300°
1317°
W
ε
870°
697°
660°
600°
δ
γ
Al
0
10
20
30
40
50
60
70
80
90
100
Wt.% W

Al–Y

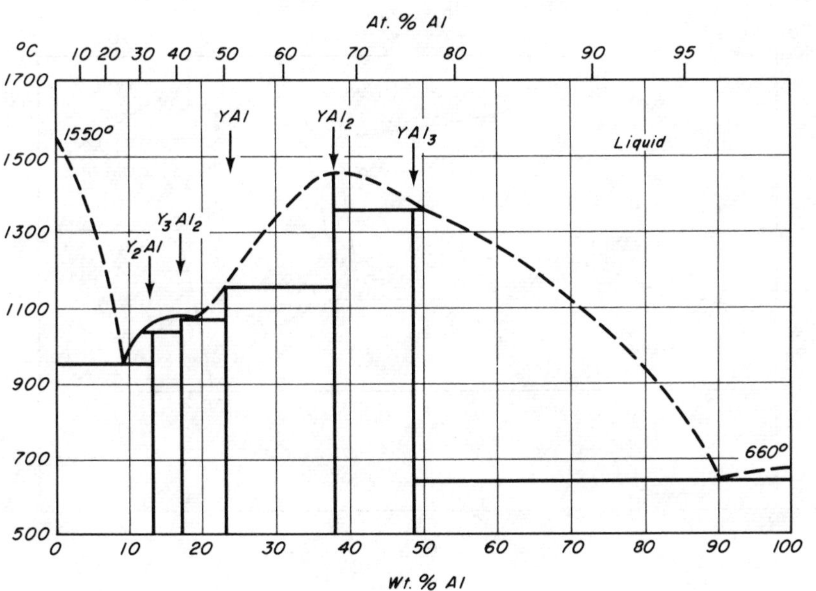
At. % Al
10
20
30
40
50
60
70
80
90
95
°C
1700
1500
1300
1100
900
700
500
1550°
YAl
YAl2
YAl3
Liquid
Y3Al2
Y2Al
660°
0
10
20
30
40
50
60
70
80
90
100
Wt. % Al

Al–Yb

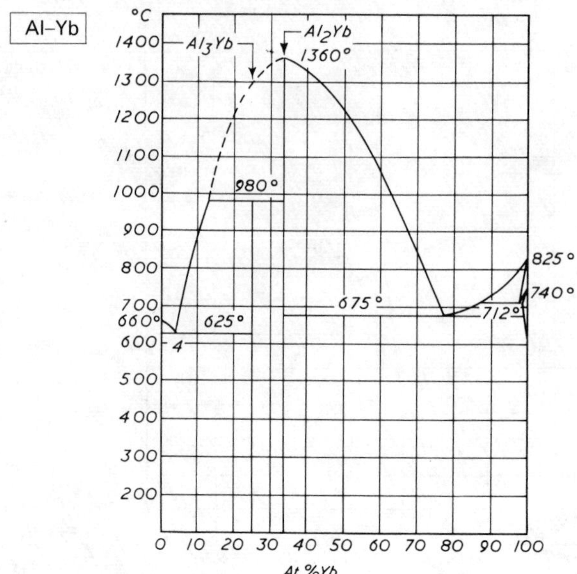
°C
1400
1300
1200
1100
1000
900
800
700
660°
600
500
400
300
200
0 10 20 30 40 50 60 70 80 90 100
At.%Yb
Al3Yb
Al2Yb
1360°
980°
825°
740°
675°
712°
625°
4

Al–Zn

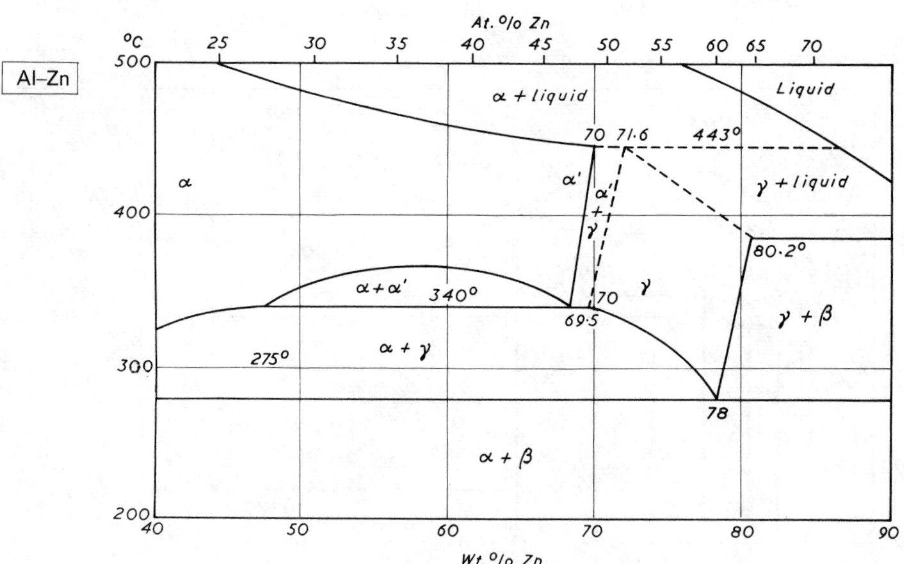
At.% Zn
25 30 35 40 45 50 55 60 65 70
°C
500
400
300
200
40 50 60 70 80 90
Wt.% Zn
α + liquid
Liquid
70 71·6 443°
α
α'
α' + γ
γ + liquid
80·2°
α+α' 340°
70
γ
69·5
γ + β
275°
α + γ
78
α + β

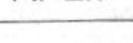
Al–Zn

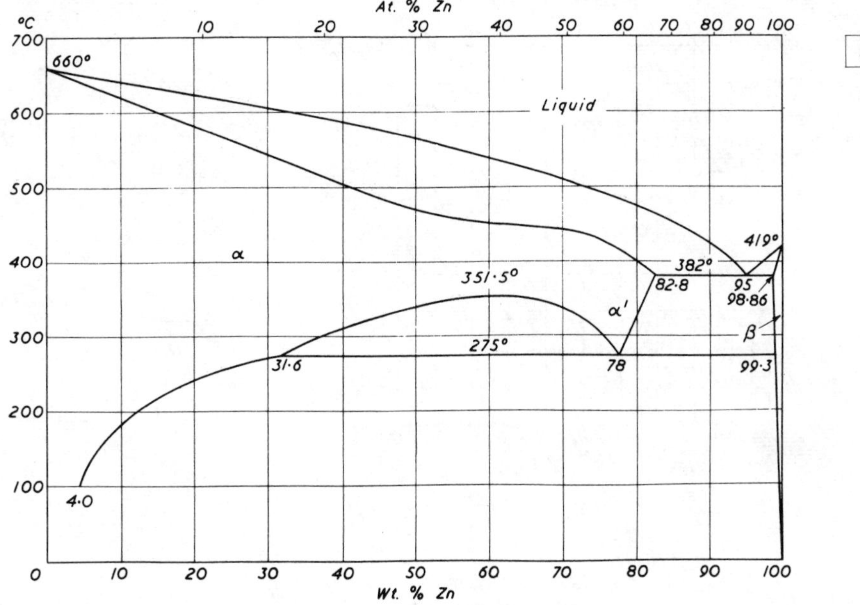
At. % Zn
°C
700
600
500
400
300
200
100
0
10
20
30
40
50
60
70
80
90
100
660°
Liquid
α
419°
382°
351·5°
82·8
95
98·86
α'
β
275°
31·6
78
99·3
4·0
Wt. % Zn

Al–Zr

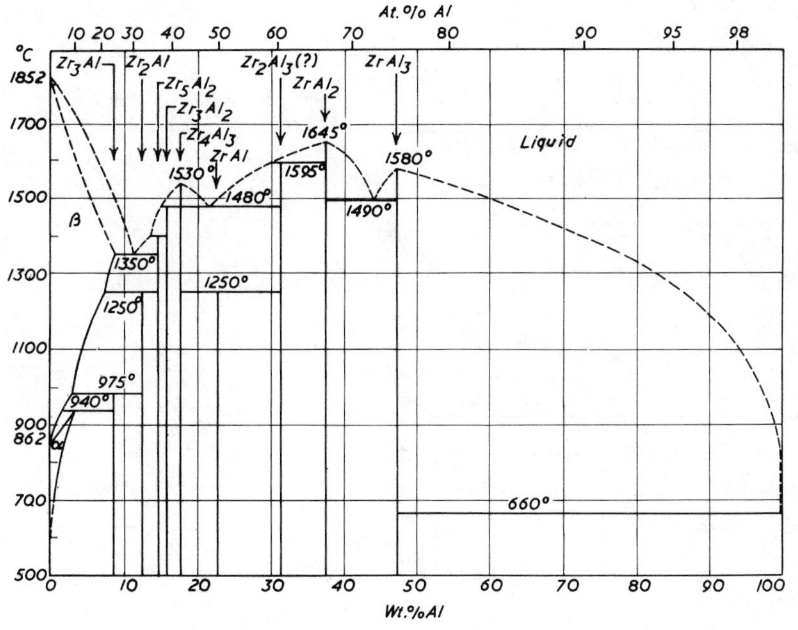
At.% Al
10 20 30 40 50 60 70 80 90 95 98
°C
1852
1700
1500
1300
1100
900
862
700
500
Zr_3Al
Zr_2Al
Zr_5Al_2
Zr_3Al_2
Zr_4Al_3
ZrAl
Zr_2Al_3(?)
$ZrAl_2$
$ZrAl_3$
1645°
1580°
Liquid
1530°
1480°
1595°
1490°
β
1350°
1250°
1250°
975°
940°
α
660°
0 10 20 30 40 50 60 70 80 90 100
Wt.% Al

Am–Pu

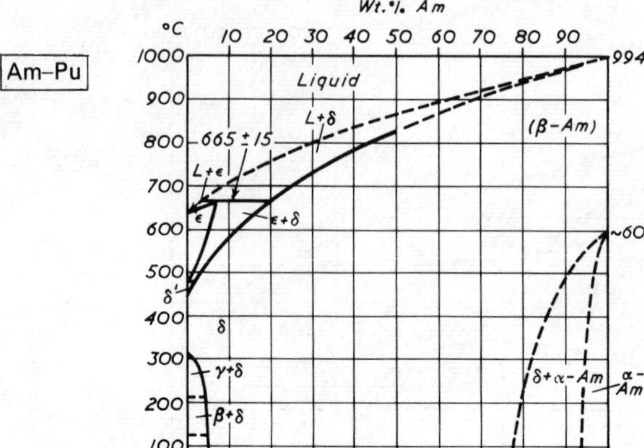

As–Au

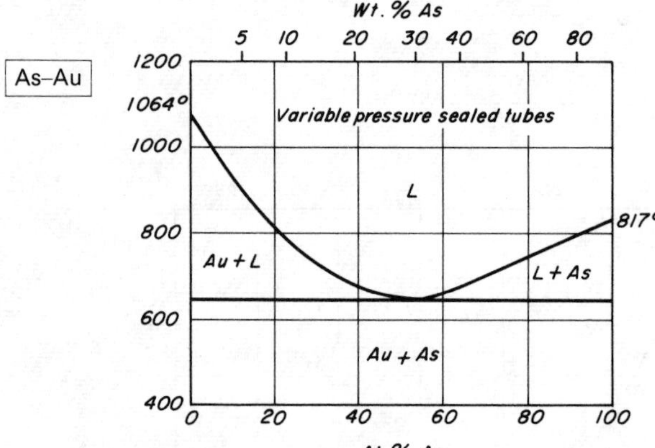

As–Bi

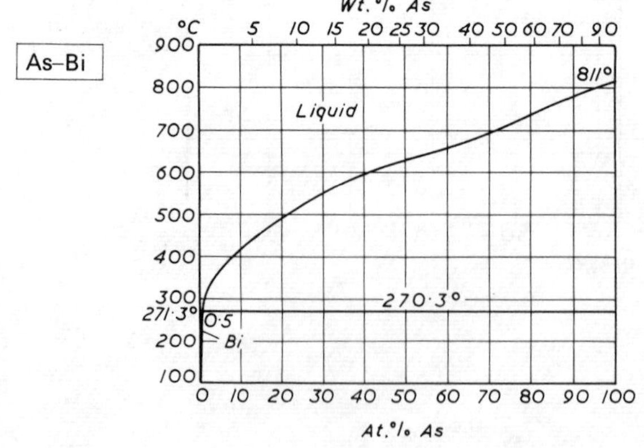

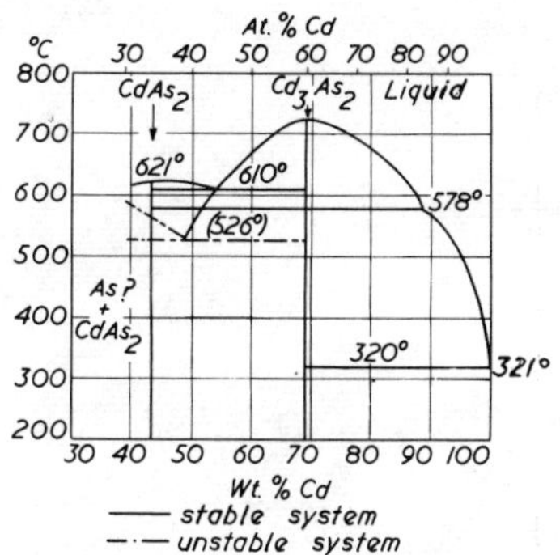

As–Cd

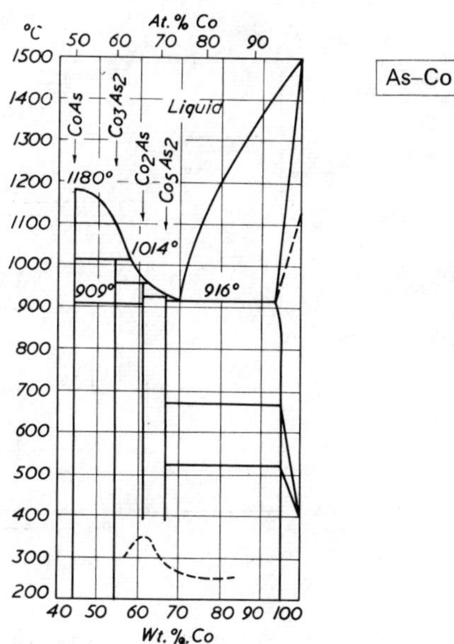

As–Co

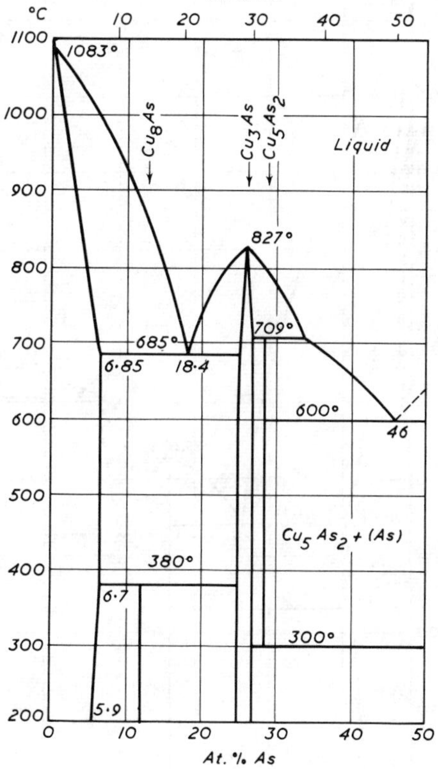

As–Cu

As–Eu

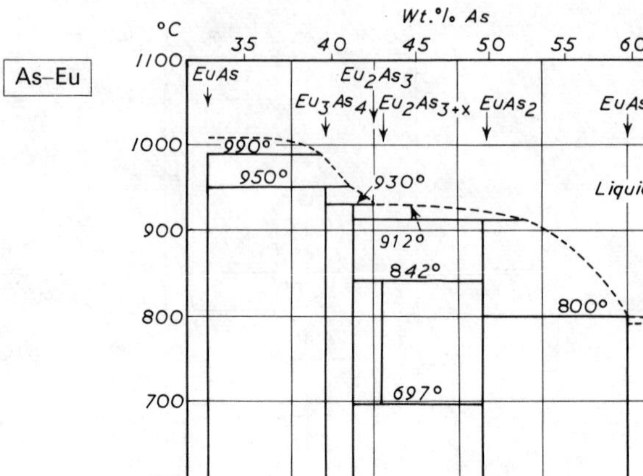

As–Fe

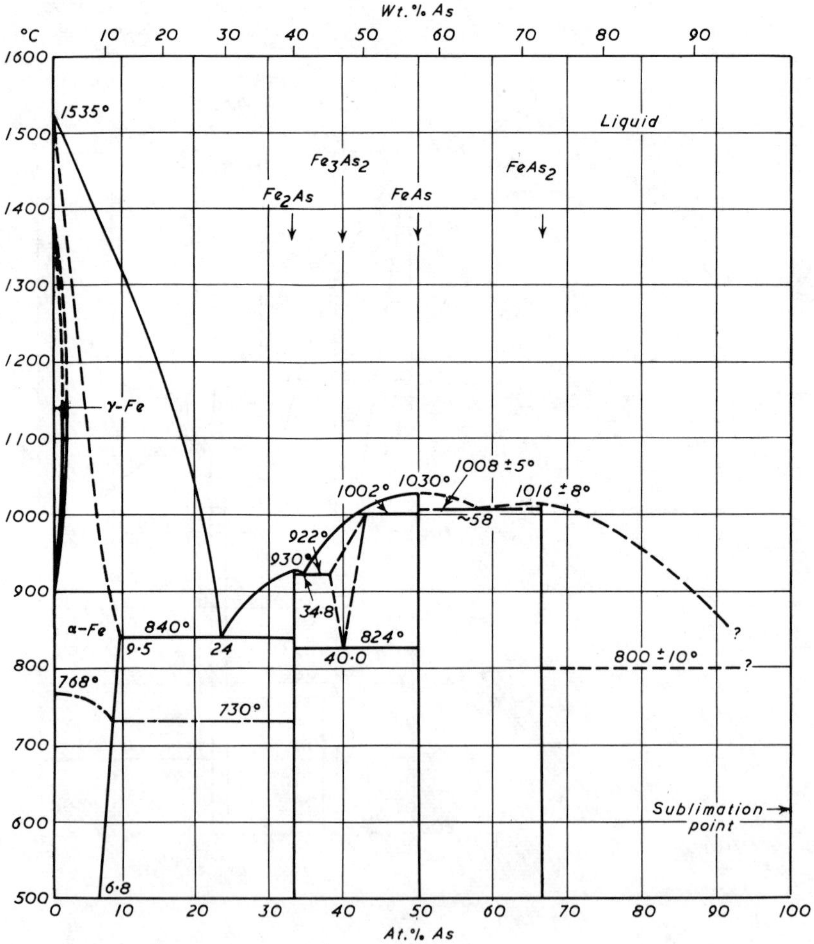

Wt.% As
°C
10 20 30 40 50 60 70 80 90
1600 1500 1400 1300 1200 1100 1000 900 800 700 600 500
1535°
Liquid
Fe_3As_2
$FeAs_2$
Fe_2As
FeAs
γ-Fe
1008 ± 5°
1002°
1030°
1016 ± 8°
~58
922°
930°
34·8
α-Fe
840°
824°
9·5
24
40·0
800 ± 10°
?
768°
730°
Sublimation point
6·8
0 10 20 30 40 50 60 70 80 90 100
At.% As

As–Ga

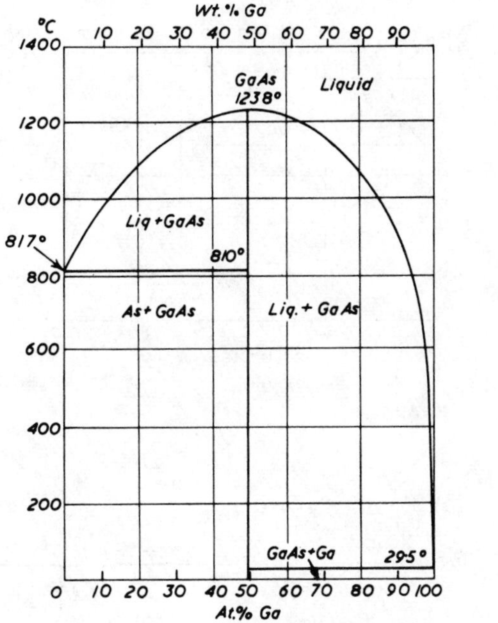
Wt.% Ga
°C
10 20 30 40 50 60 70 80 90
1400
1200
1000
800
600
400
200
0
GaAs
1238°
Liquid
Liq + GaAs
817°
810°
As + GaAs
Liq. + GaAs
GaAs + Ga
29.5°
10 20 30 40 50 60 70 80 90 100
At.% Ga

As–Ge

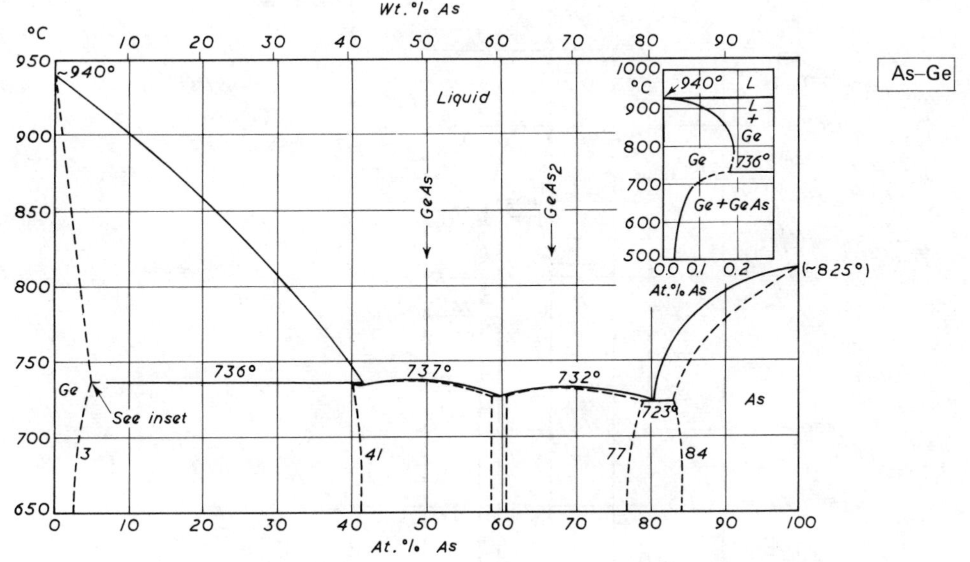
Wt.% As
°C
10 20 30 40 50 60 70 80 90
950
900
850
800
750
700
650
~940°
Liquid
GeAs
GeAs2
736°
737°
732°
723°
Ge
See inset
As
13
41
77
84
(~825°)
0 10 20 30 40 50 60 70 80 90 100
At.% As
1000
°C
940°
L
L + Ge
Ge
736°
Ge + GeAs
900
800
700
600
500
0.0 0.1 0.2
At.% As

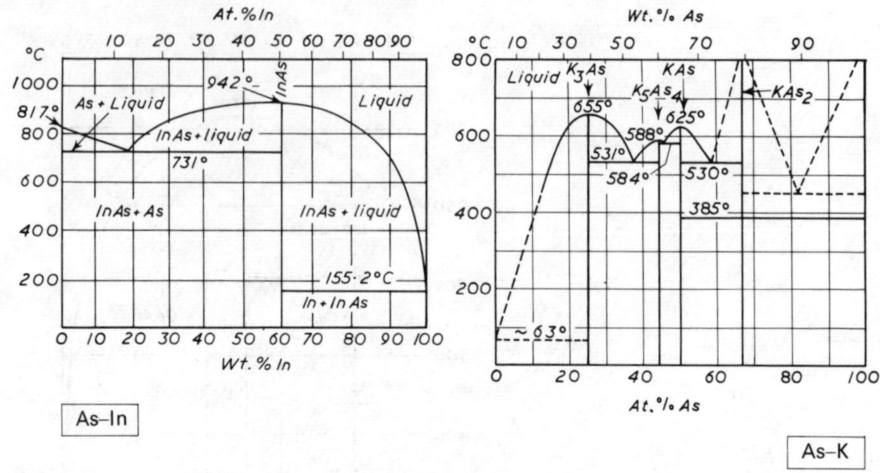
At.% In
°C
1000
817°
800
600
400
200
As + Liquid
942°
InAs
Liquid
InAs + liquid
731°
InAs + As
InAs + liquid
155·2°C
In + In As
Wt.% In
As–In
Wt.°/. As
°C
800
600
400
200
Liquid
K_3As
KAs
K_5As_4
KAs_2
655°
625°
588°
531°
584°
530°
385°
~63°
At.°/. As
As–K

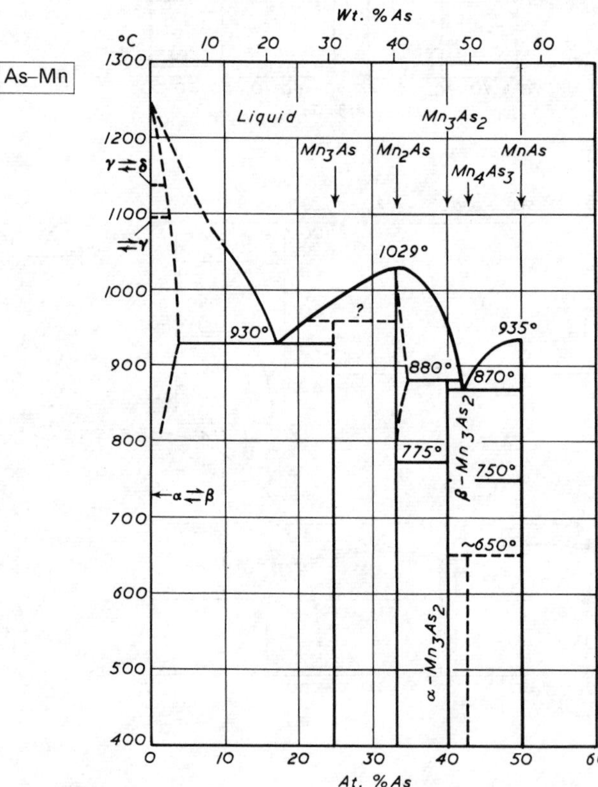
Wt. % As
°C
As–Mn
Liquid
Mn_3As_2
Mn_3As
Mn_2As
MnAs
Mn_4As_3
γ ⇄ δ
⇄ γ
1029°
?
935°
930°
880°
870°
775°
750°
~650°
β-Mn_3As_2
α-Mn_3As_2
← α ⇄ β
At. % As

As–Ni

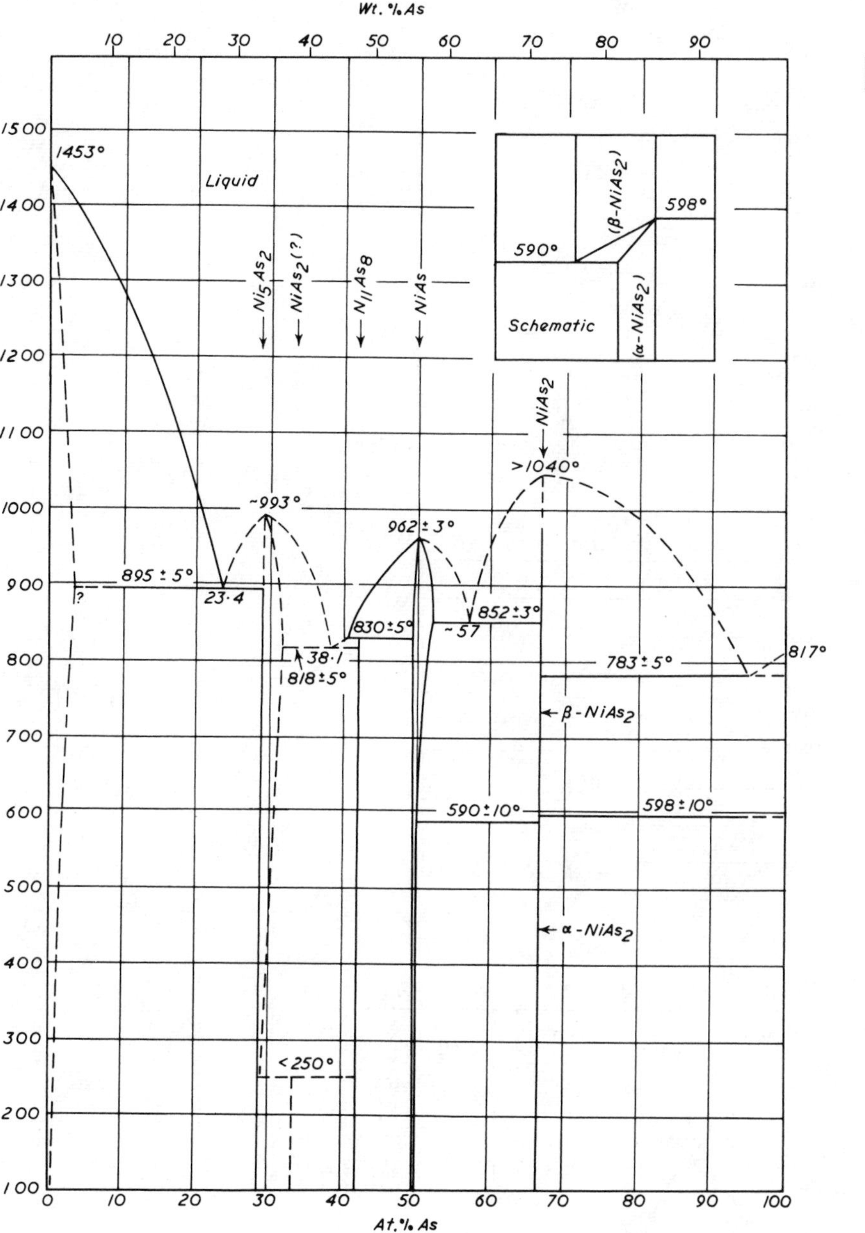

Wt. % As
At. % As
Liquid
1453°
Ni5As2
NiAs2(?)
Ni11As8
NiAs
NiAs2
~993°
962 ± 3°
>1040°
895 ± 5°
23·4
?
830 ± 5°
~57
852 ± 3°
38·1
818 ± 5°
783 ± 5°
817°
β-NiAs2
590 ± 10°
598 ± 10°
α-NiAs2
<250°
Schematic
590°
598°
(β-NiAs2)
(α-NiAs2)

As–P

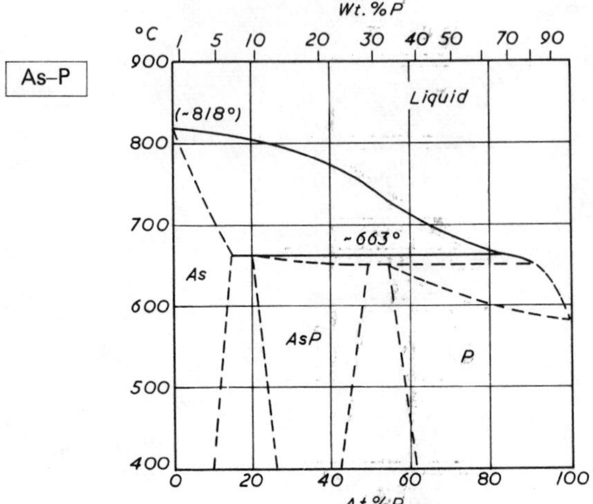

As–Pb

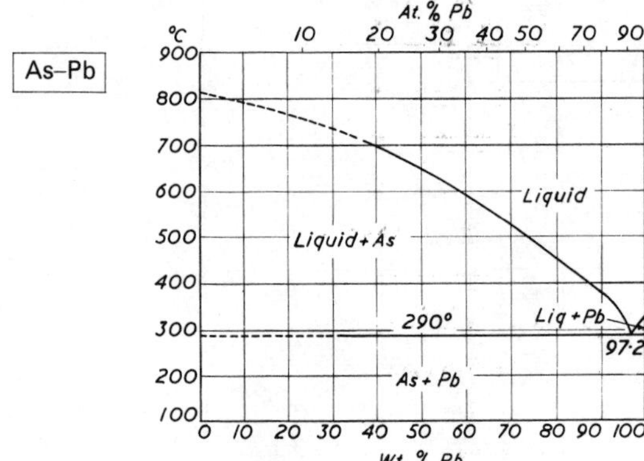

As–Pd

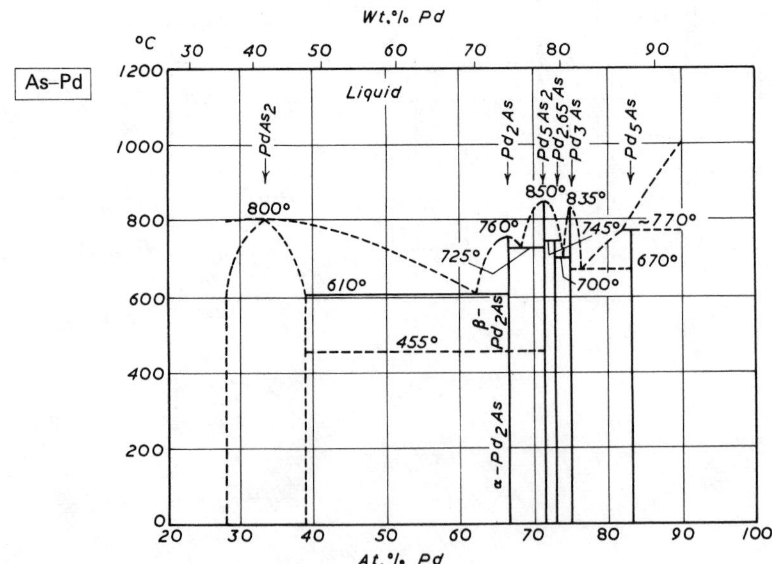

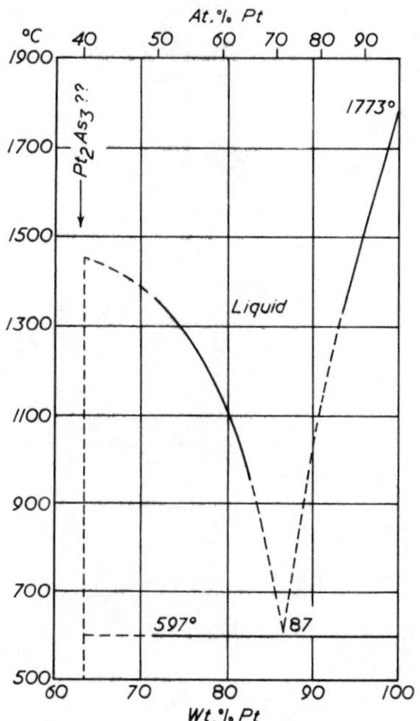

As–Pt

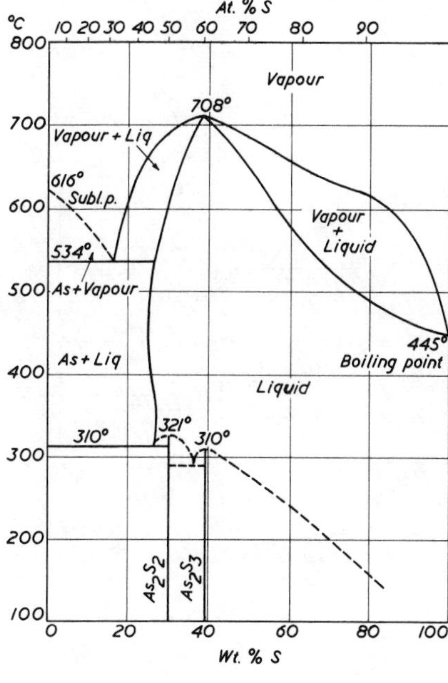

As–S

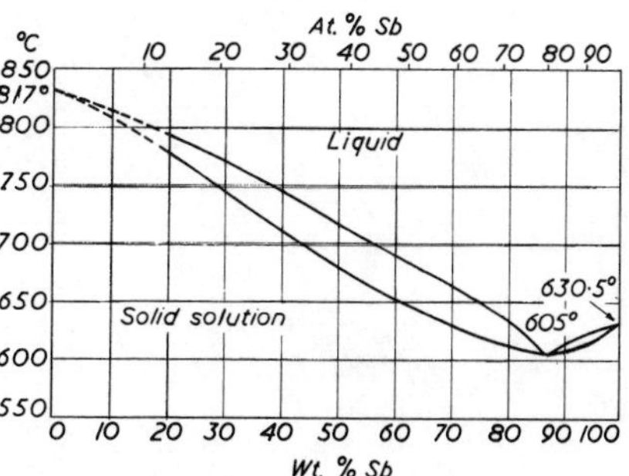

As–Sb

As–Si

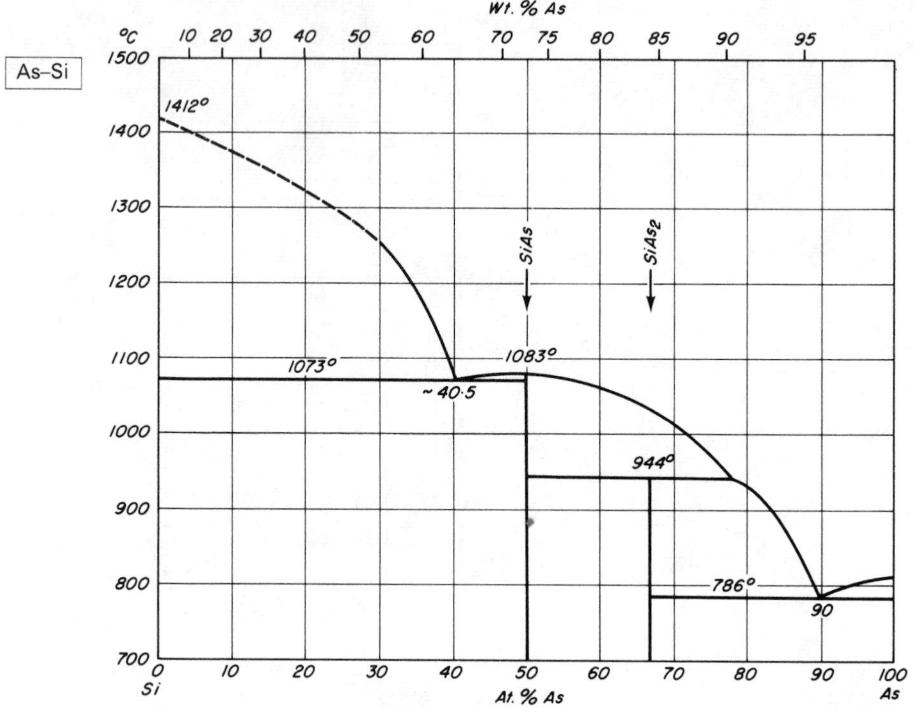

As–Sn

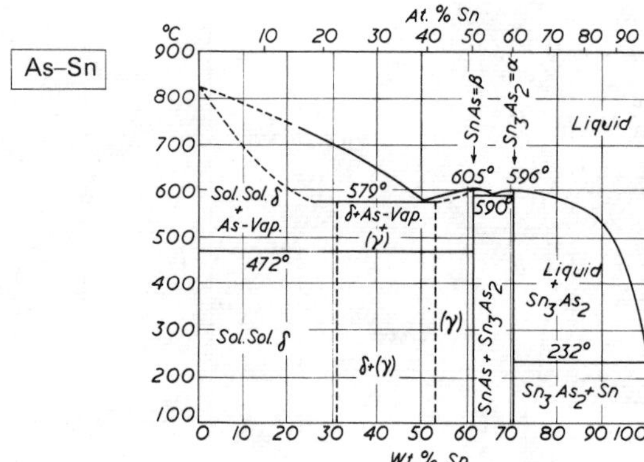

See alternative version of Sn-rich end

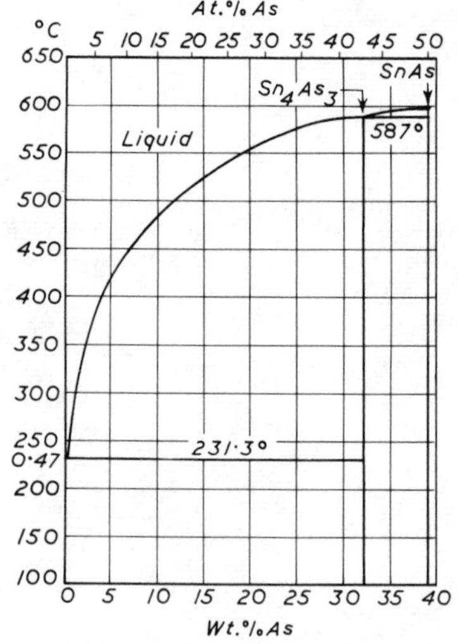

As–Sn

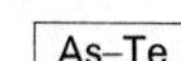
As–Te

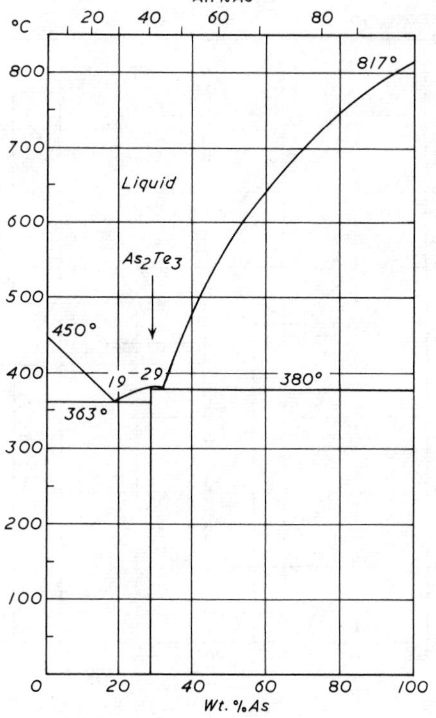
At.% As
°C
20
40
60
80
817°
800
700
Liquid
600
As_2Te_3
500
450°
400
19
29
380°
363°
300
200
100
0
20
40
60
80
100
Wt.% As

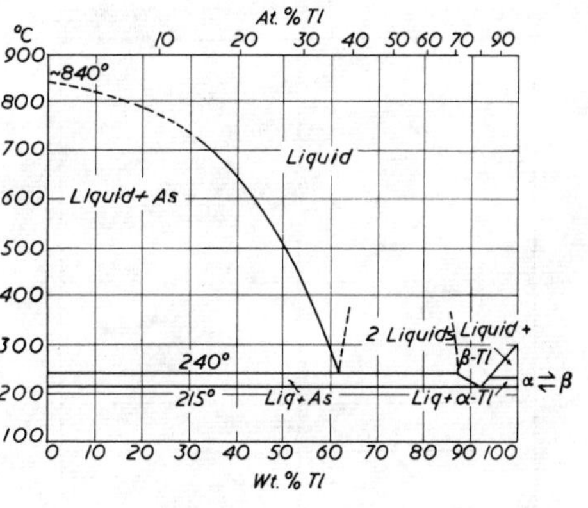
At. % Tl
°C
900
~840°
800
700
Liquid
600
Liquid+As
500
400
300
2 Liquids
Liquid + β-Tl
240°
α ⇄ β
200
215°
Liq+As
Liq+α-Tl
100
0 10 20 30 40 50 60 70 80 90 100
Wt. % Tl
As–Tl

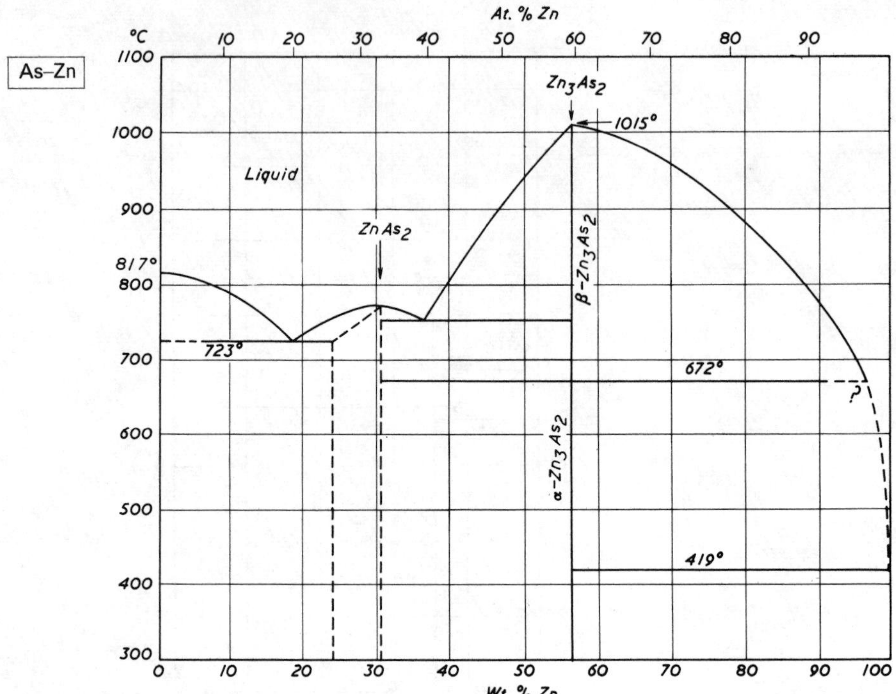
As–Zn
At. % Zn
°C
Liquid
Zn_3As_2
1015°
$ZnAs_2$
817°
723°
672°
?
β-Zn_3As_2
α-Zn_3As_2
419°
Wt. % Zn

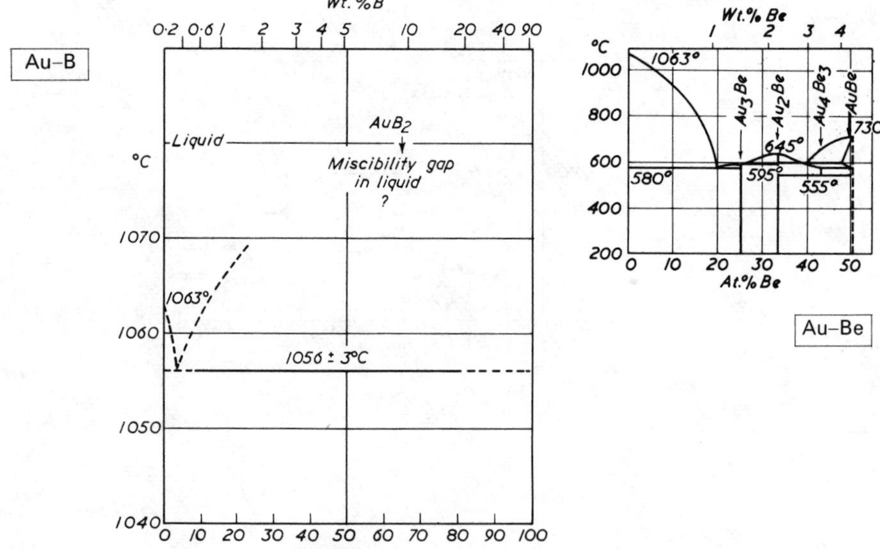
Au–B
Wt. % B
°C
Liquid
AuB_2
Miscibility gap in liquid
?
1063°
1056 ± 3°C
At. % B
Au–Be
Wt.% Be
°C
1063°
Au_3Be
Au_2Be
Au_4Be_3
AuBe
730°
645°
580°
595°
555°
At.% Be

Au–Bi

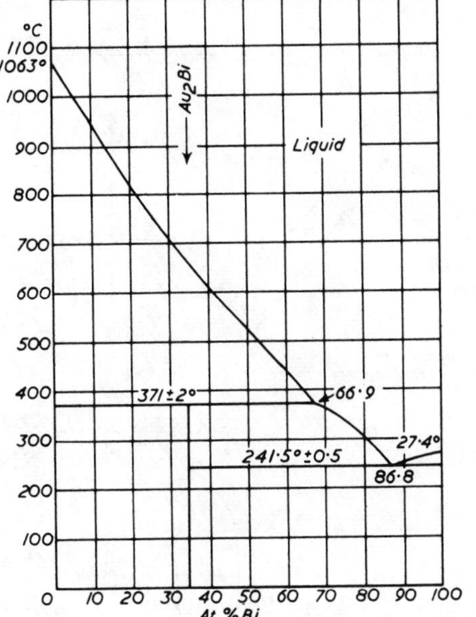
°C
1100
1063°
1000
900
800
700
600
500
400
300
200
100
0
Au2Bi
Liquid
371±2°
66·9
27·4°
241·5°±0·5
86·8
10 20 30 40 50 60 70 80 90 100
At. % Bi

Au–Ca

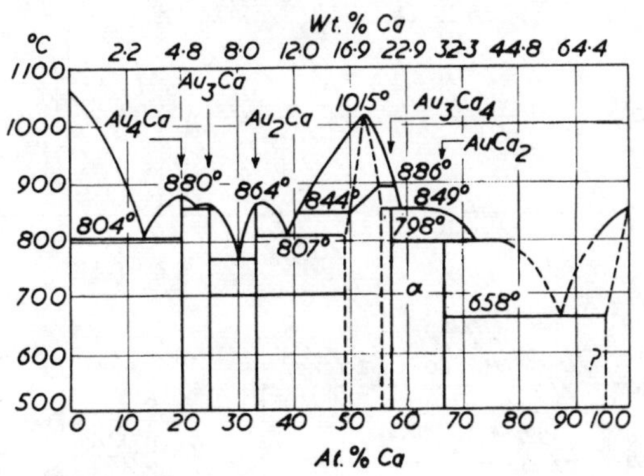
Wt. % Ca
2·2 4·8 8·0 12·0 16·9 22·9 32·3 44·8 64·4
°C
1100
1000
900
800
700
600
500
Au3Ca
Au4Ca
Au2Ca
1015°
Au3Ca4
AuCa2
880°
864°
844°
886°
849°
804°
807°
798°
α
658°
?
0 10 20 30 40 50 60 70 80 90 100
At. % Ca

Au–Cd

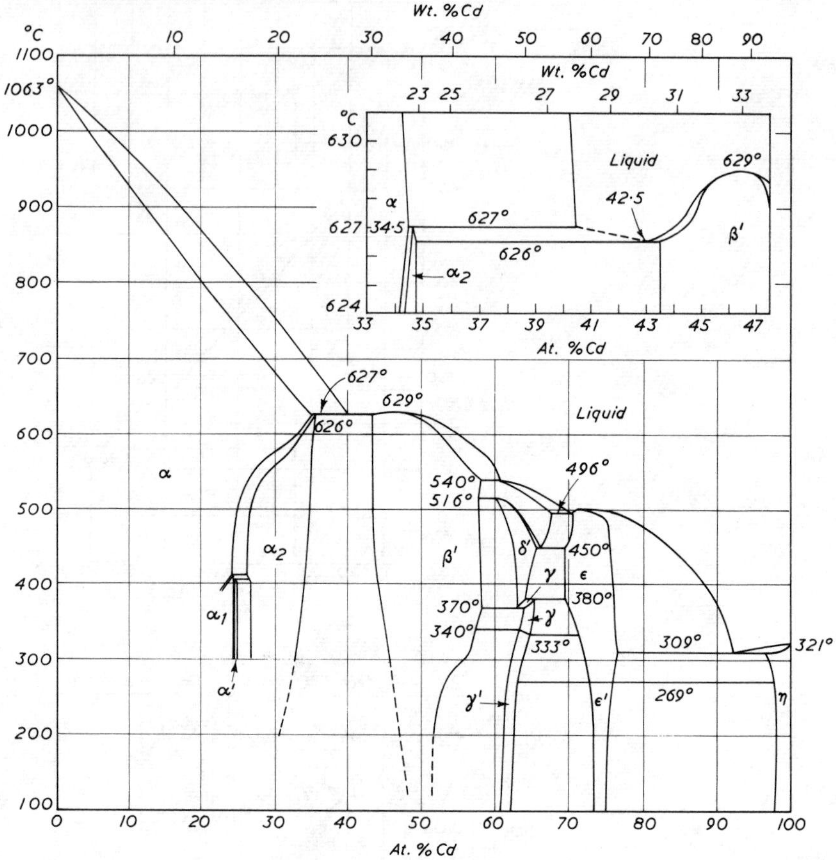

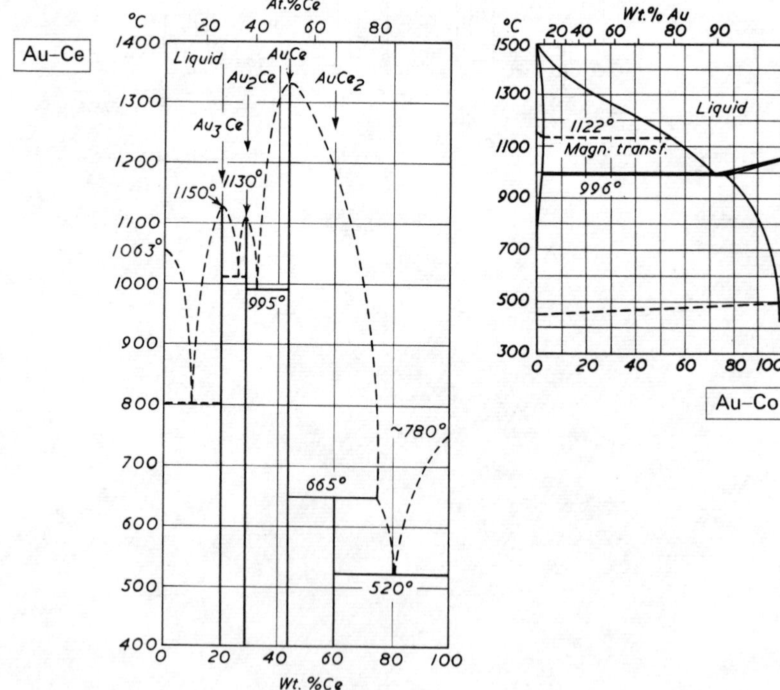
Au–Ce
At.%Ce
°C
20
40
60
80
1400
1300
1200
1100
1063°
1000
900
800
700
600
500
400
Liquid
Au2Ce
AuCe
AuCe2
Au3Ce
1150°
1130°
995°
~780°
665°
520°
0
20
40
60
80
100
Wt. %Ce
Wt.% Au
°C
20
40
60
80
90
1500
1300
Liquid
1122°
1100
Magn. transf.
900
996°
700
500
300
0
20
40
60
80
100
Au–Co

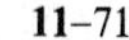

Au–Cr

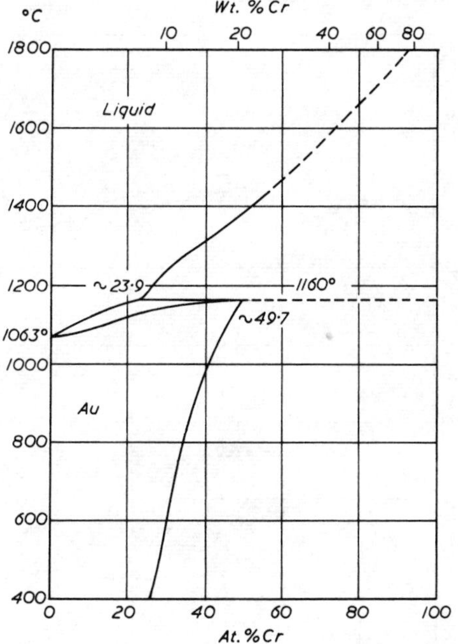
°C
Wt. % Cr
10
20
40
60
80
1800
1600
1400
1200
1063°
1000
800
600
400
Liquid
~23·9
1160°
~49·7
Au
0
20
40
60
80
100
At. % Cr

Au–Cs

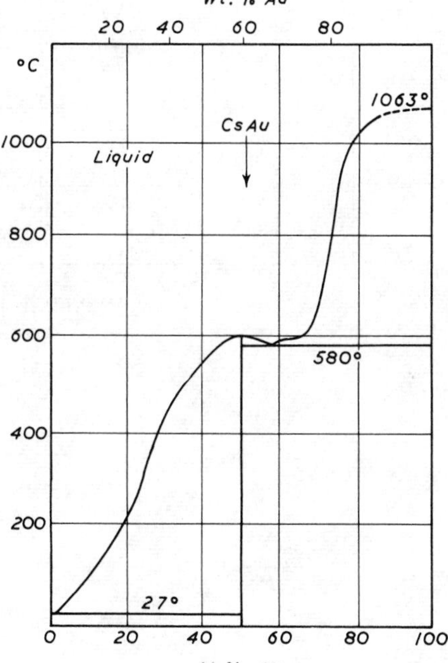
Wt. % Au
20
40
60
80
°C
1000
800
600
400
200
1063°
CsAu
Liquid
580°
27°
0
20
40
60
80
100
At. % Au

Au–Cu

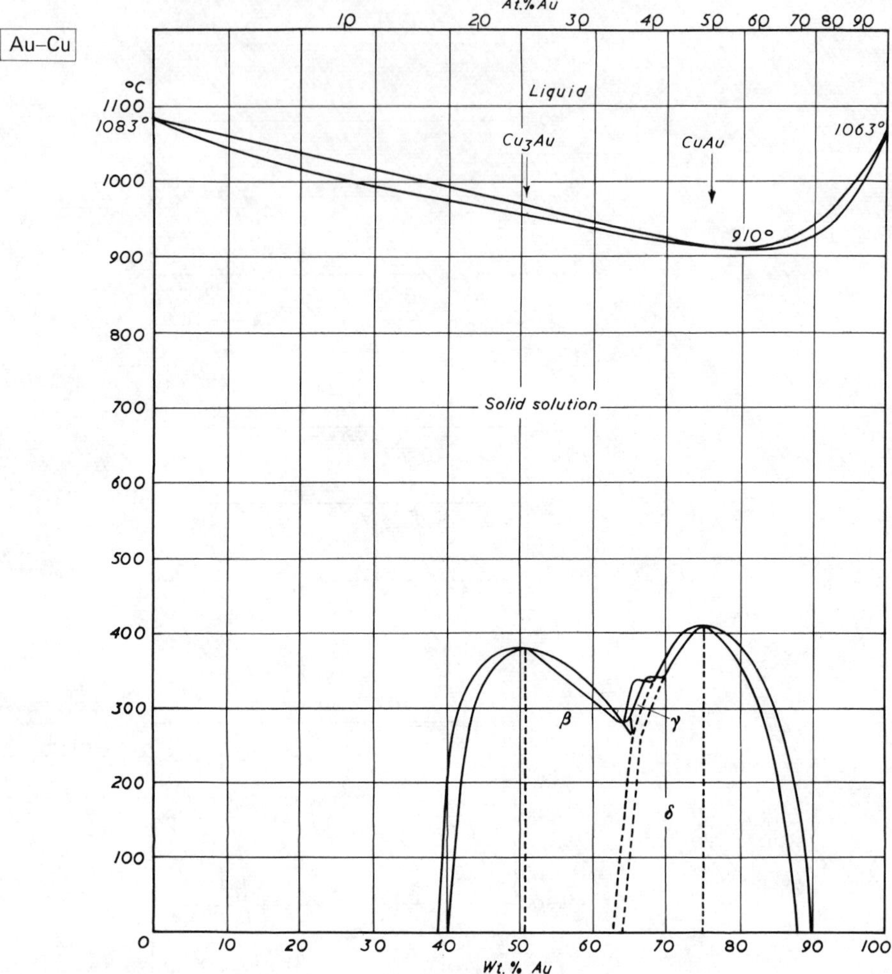
At.% Au
10
20
30
40
50
60
70
80
90
°C
1100
1083°
1000
900
800
700
600
500
400
300
200
100
0
Liquid
Cu_3Au
CuAu
1063°
910°
Solid solution
β
γ
δ
10
20
30
40
50
60
70
80
90
100
Wt.% Au

Au–Dy

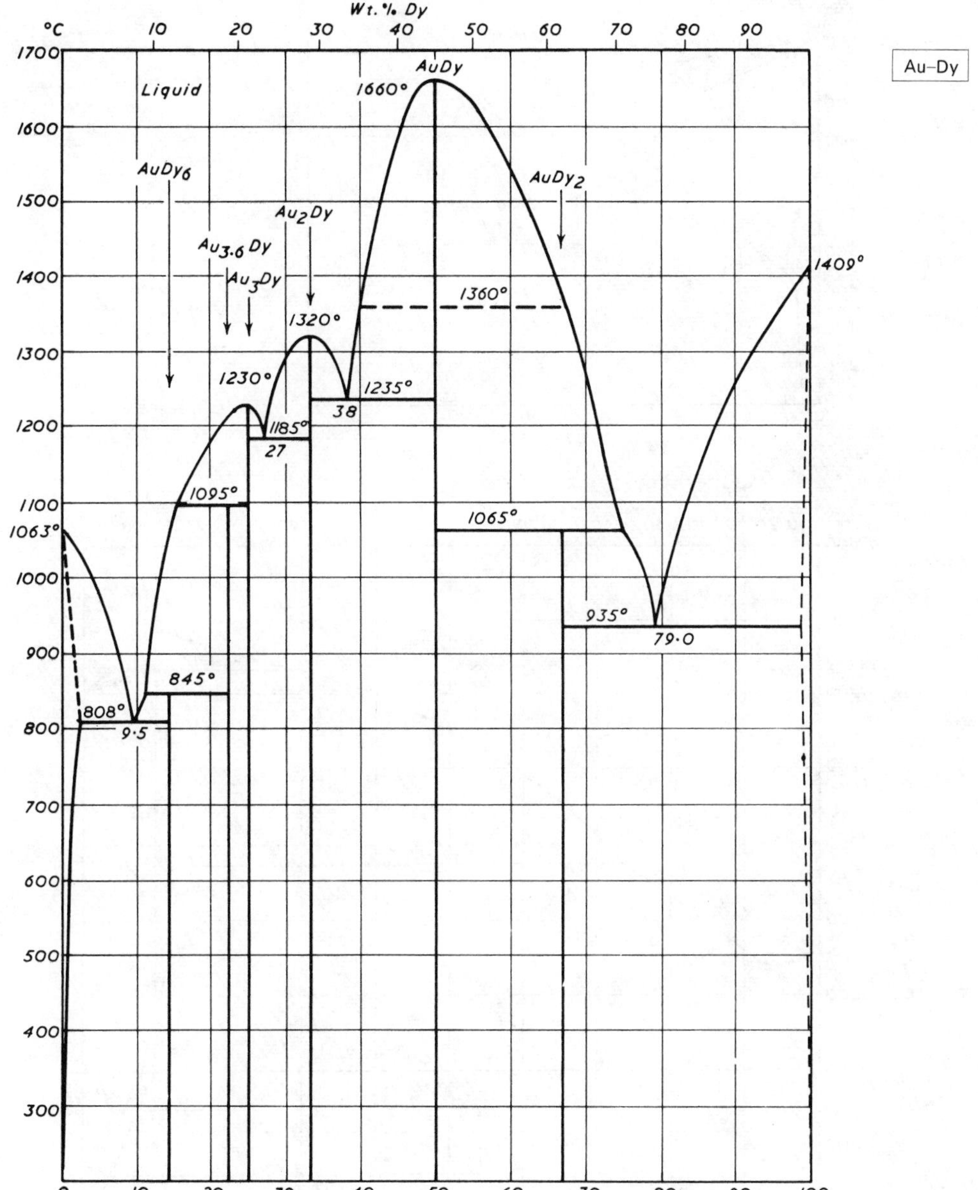
Wt.% Dy
°C
Liquid
AuDy
1660°
AuDy6
AuDy2
Au2Dy
Au3.6Dy
Au3Dy
1360°
1409°
1320°
1230°
1235°
38
1185°
27
1095°
1065°
935°
79·0
845°
808°
9·5
1063°
At.% Dy

Au–Er

°C
1100
1063°
1000
900
800
700
600
500
400
300
200
100
Liquid
812°
(5·7)
Solid solution
+
compound
0 1 2 3 4 5 6 7 8 9 10 11 12 13
At.% Er

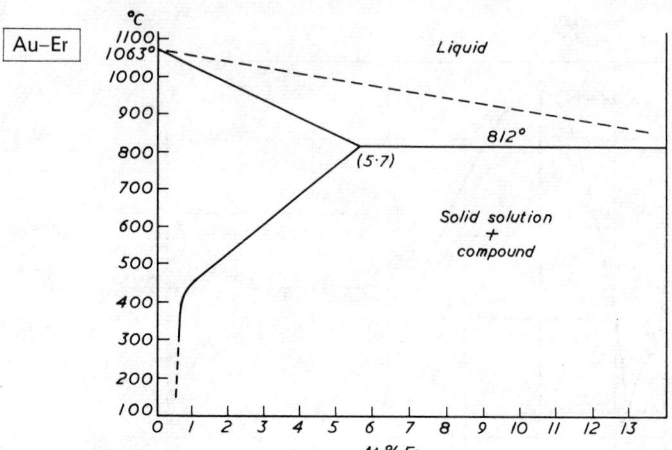

Au–Fe

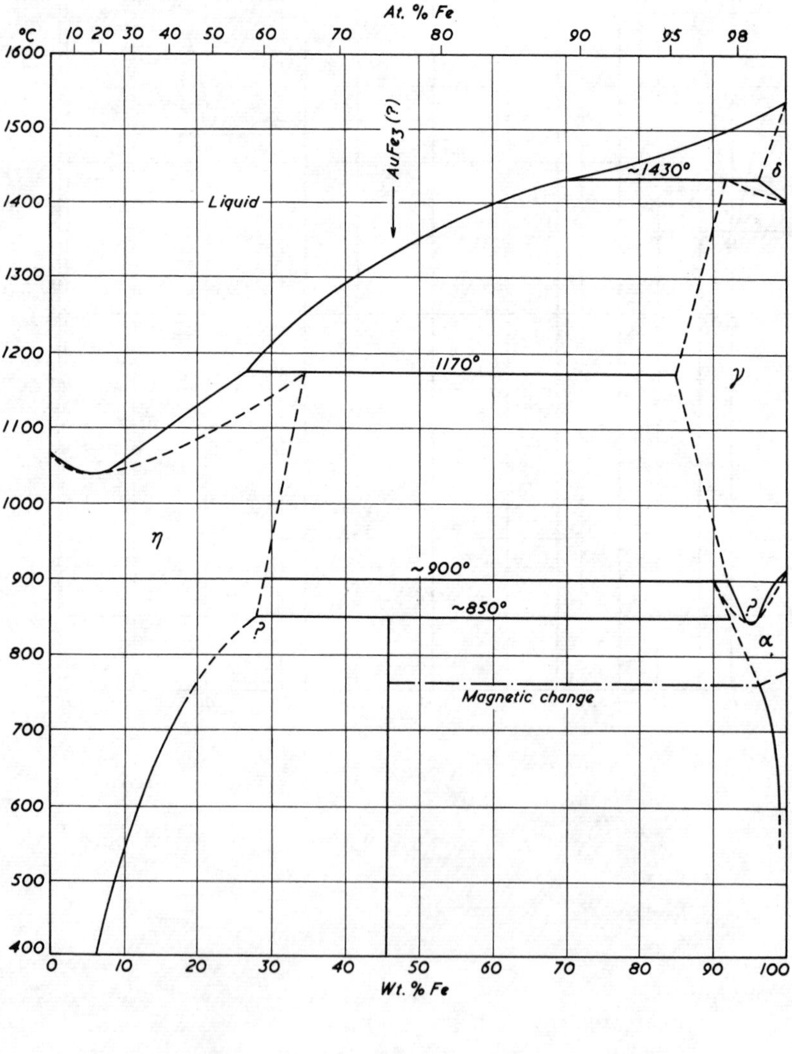

At. % Fe
°C
10 20 30 40 50 60 70 80 90 95 98
1600
1500
1400
1300
1200
1100
1000
900
800
700
600
500
400
AuFe3 (?)
Liquid
~1430°
δ
1170°
γ
η
~900°
~850°
?
?
α
Magnetic change
0 10 20 30 40 50 60 70 80 90 100
Wt. % Fe

Au–Ga

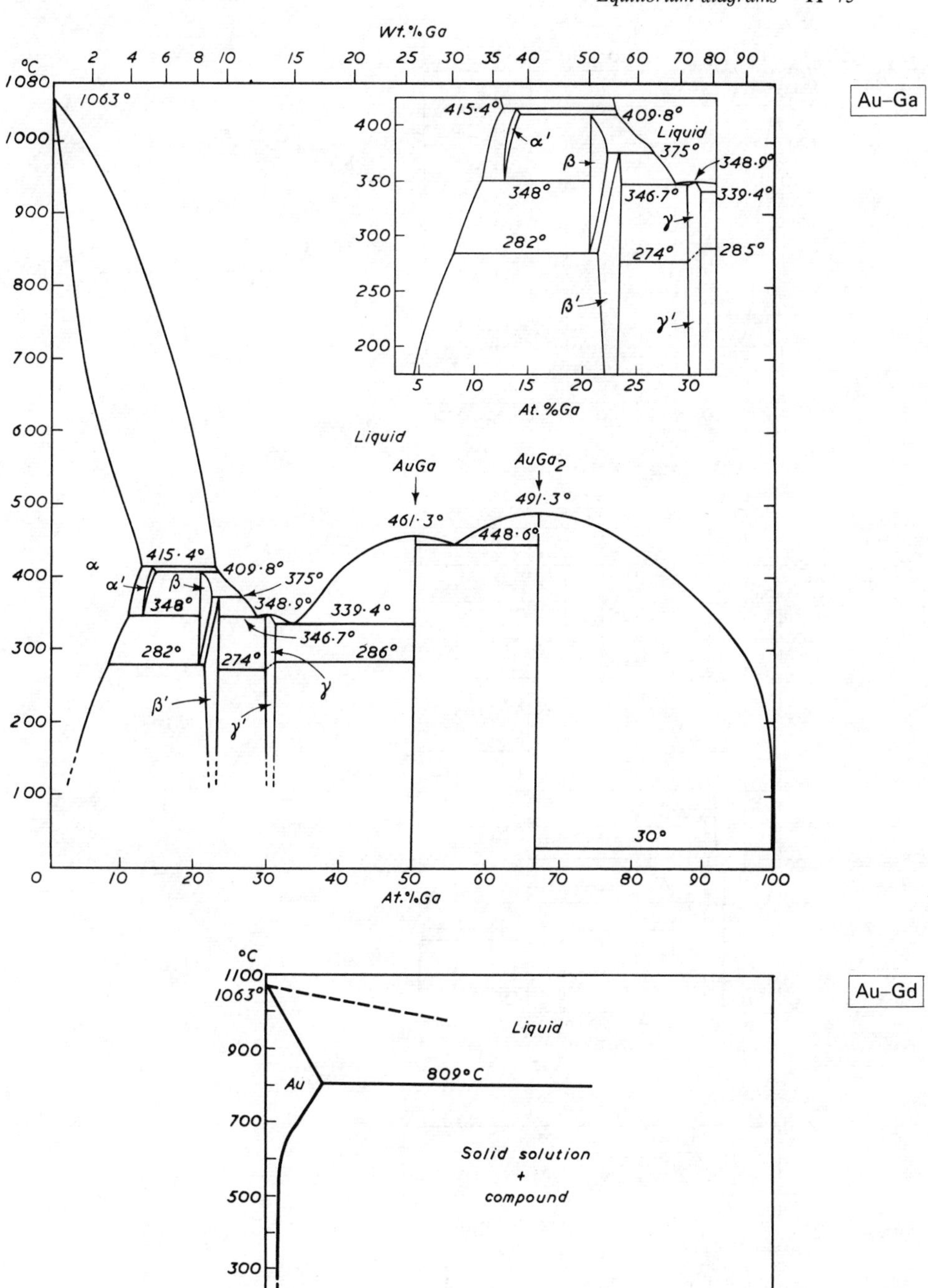
Wt.% Ga
2 4 6 8 10 15 20 25 30 35 40 50 60 70 80 90
°C
1080
1000
900
800
700
600
500
400
300
200
100
1063°
415·4°
409·8°
Liquid
375°
348·9°
α'
β
350
348°
346·7°
339·4°
γ
300
282°
274°
285°
250
β'
γ'
200
5 10 15 20 25 30
At.%Ga
Liquid
AuGa
AuGa$_2$
491·3°
461·3°
448·6°
415·4°
α
α'
β
409·8°
375°
348°
348·9°
339·4°
346·7°
282°
274°
286°
γ
β'
γ'
30°
0 10 20 30 40 50 60 70 80 90 100
At.% Ga

Au–Gd

°C
1100
1063°
Liquid
900
809°C
Au
700
Solid solution
+
compound
500
300
100
0 1 2 3 4 5 6 7
At. % Gd

Au–Ge

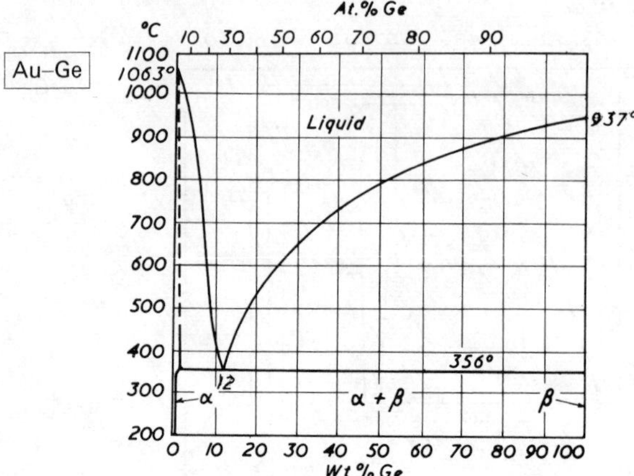

Au–Hg

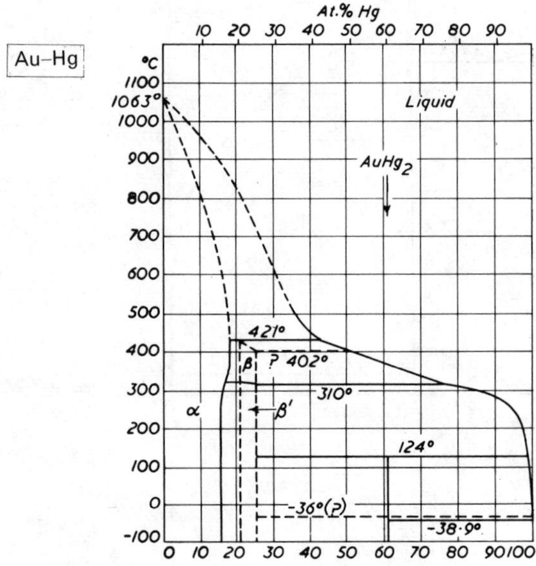

Au–In

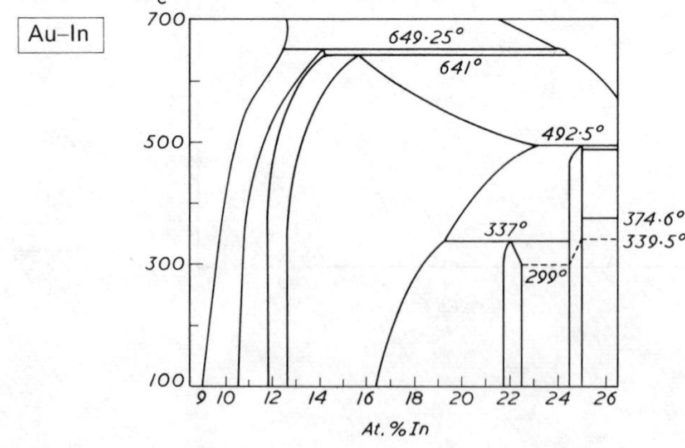

Au–In

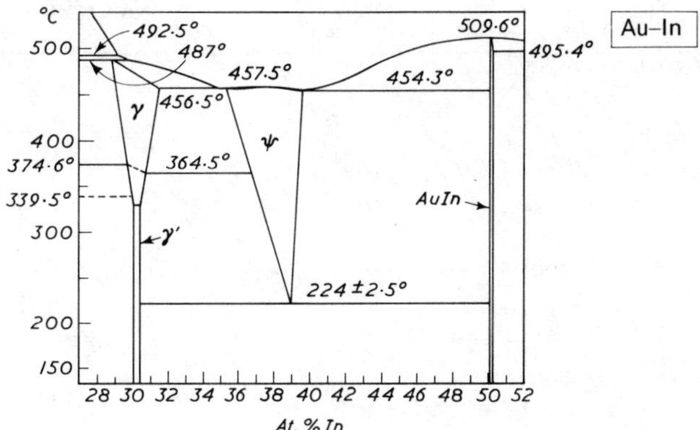

Au–K

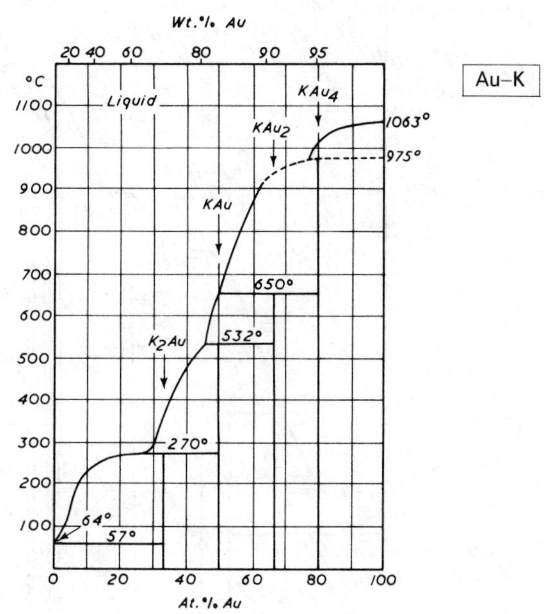

Au–La

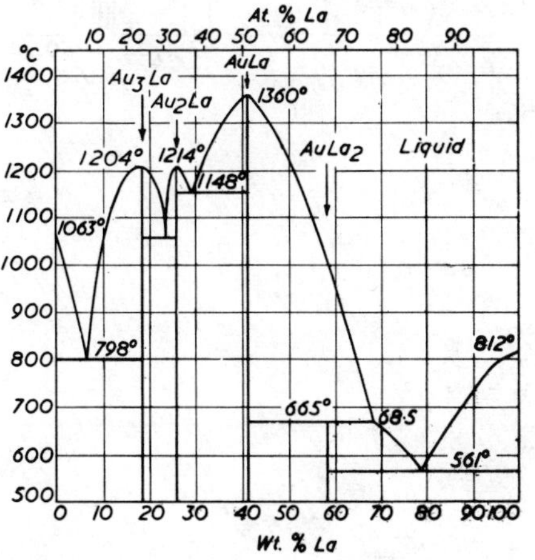

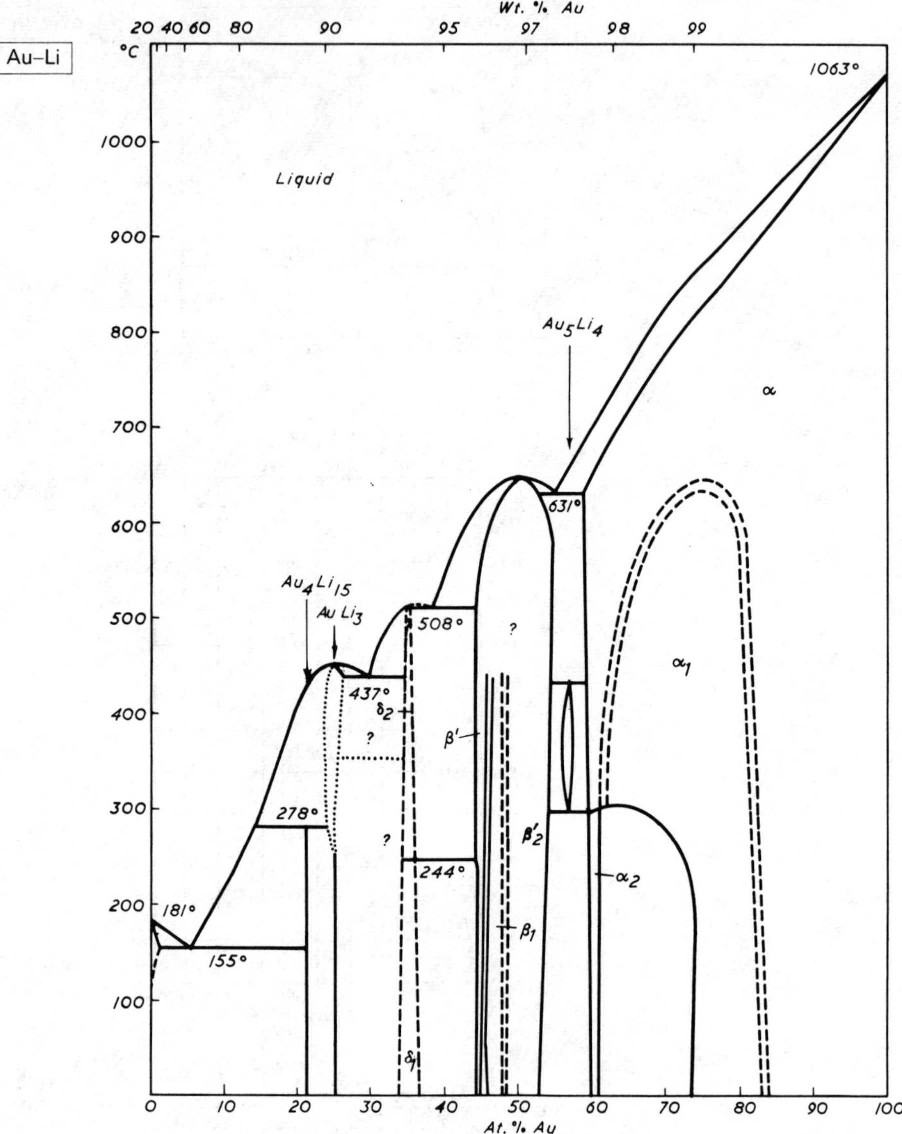
Au–Li
Wt. % Au
20 40 60 80 90 95 97 98 99
°C
1000
900
800
700
600
500
400
300
200
100
Liquid
1063°
Au_5Li_4
α
631°
Au_4Li_{15}
AuLi_3
508°
?
437°
δ_2
?
β′
α_1
278°
?
244°
β'_2
α_2
181°
155°
β_1
δ_1
0 10 20 30 40 50 60 70 80 90 100
At. % Au

Au–Lu

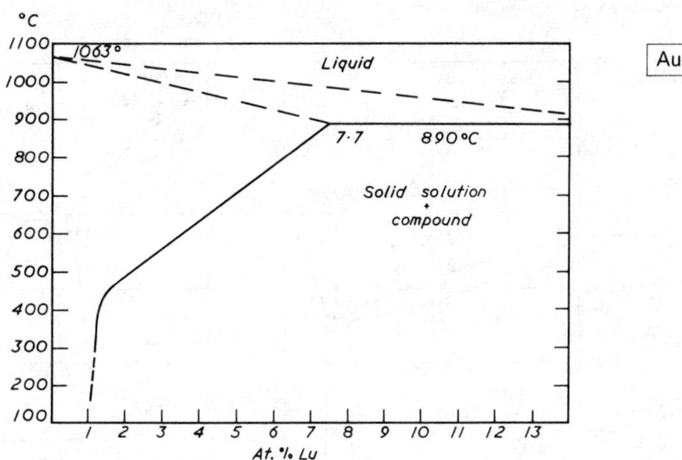
°C
1063°
Liquid
7·7
890°C
Solid solution
+
compound
At.% Lu

Au–Mg

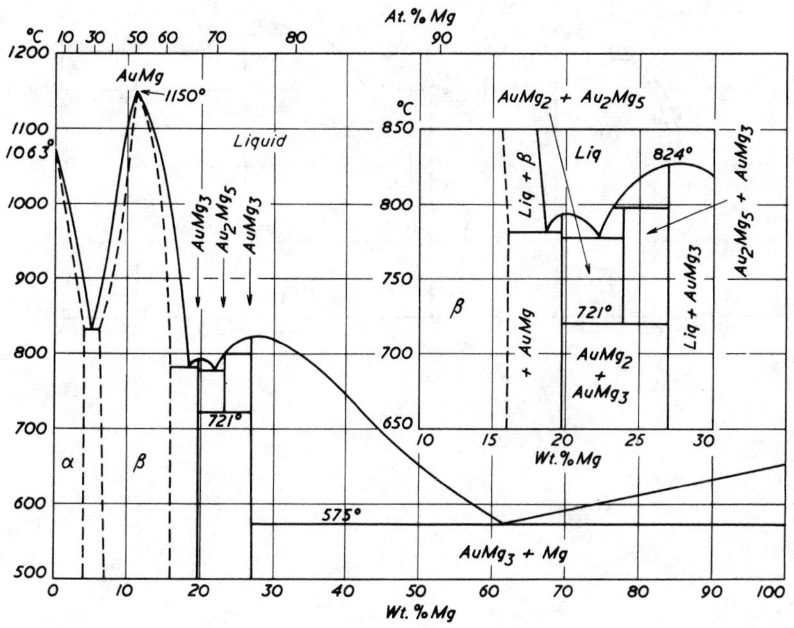
At.% Mg
°C
AuMg
1150°
Liquid
1063°
AuMg3
Au2Mg5
AuMg3
721°
α
β
575°
AuMg3 + Mg
Wt.% Mg
AuMg2 + Au2Mg5
Liq + β
Liq
824°
Au2Mg5 + AuMg3
Liq + AuMg3
721°
β
+ AuMg
AuMg2
+
AuMg3
Wt.% Mg

Au–Mn

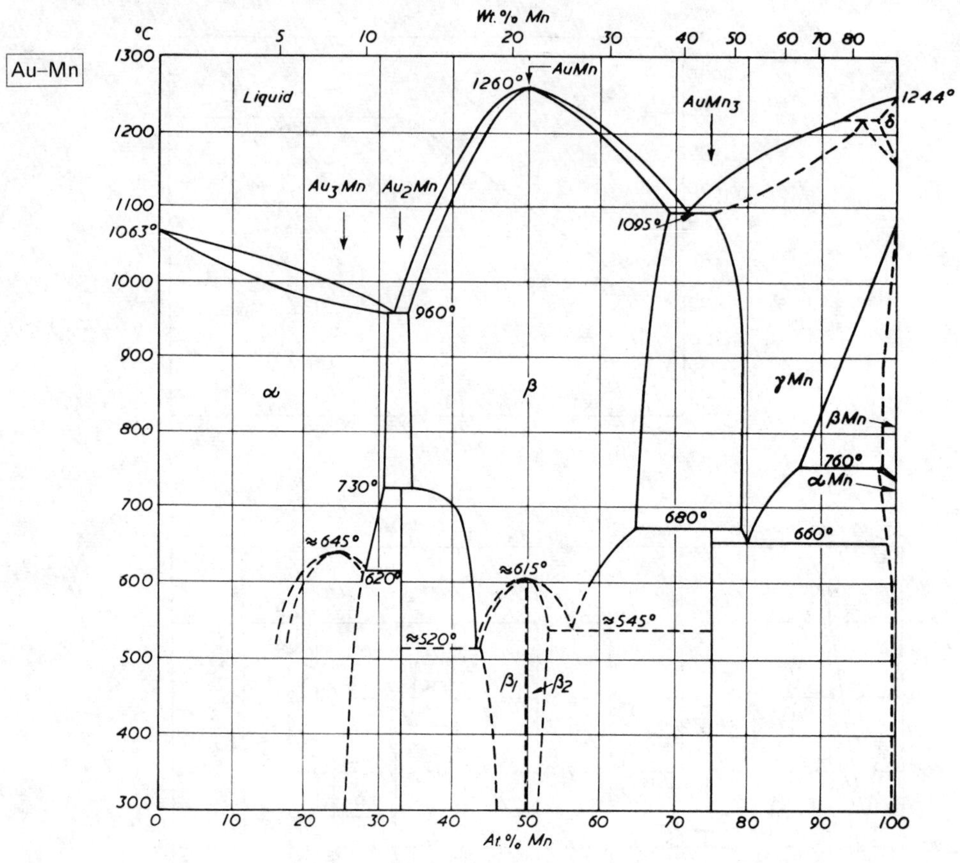

Wt.% Mn
°C
5
10
20
30
40
50
60
70
80
Liquid
AuMn
1260°
AuMn3
1244°
δ
Au3Mn
Au2Mn
1063°
1095°
960°
α
β
γMn
βMn
760°
αMn
730°
680°
660°
≈645°
620°
≈615°
≈545°
≈520°
β1
β2
At.% Mn

Au–Na

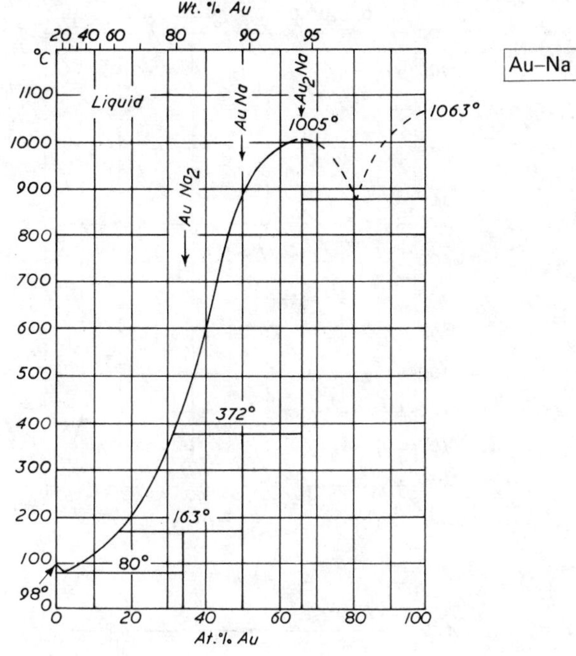
Wt. % Au
20 40 60 80 90 95
°C
1100
1000
900
800
700
600
500
400
300
200
100
Liquid
Au Na
Au_2Na
1005°
1063°
$AuNa_2$
372°
163°
80°
98°
0 20 40 60 80 100
At. % Au

Au–Nb

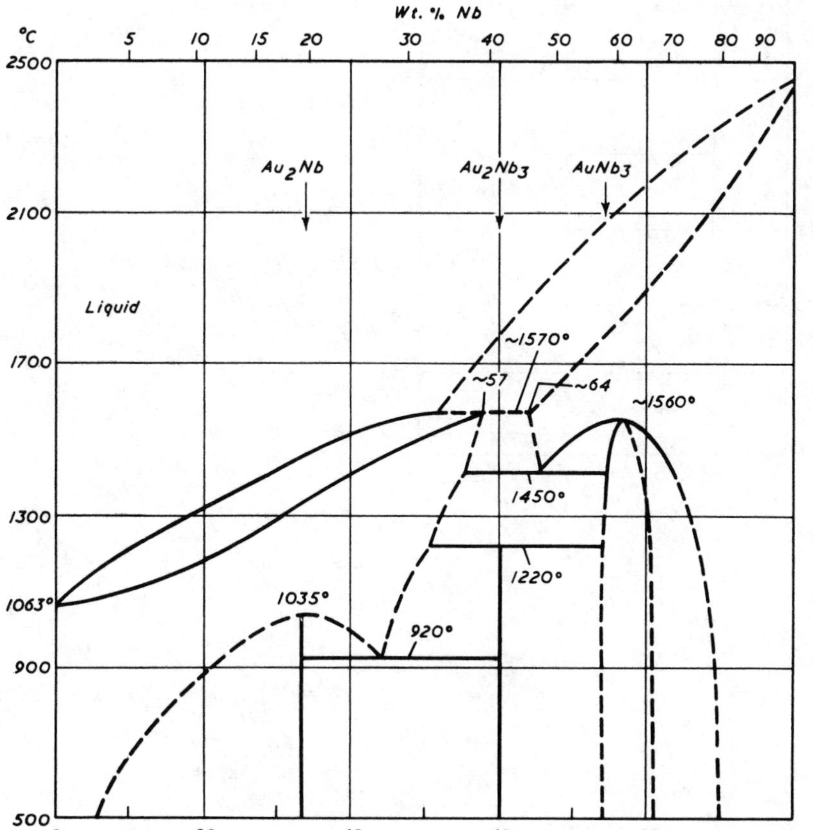

Au–Ni

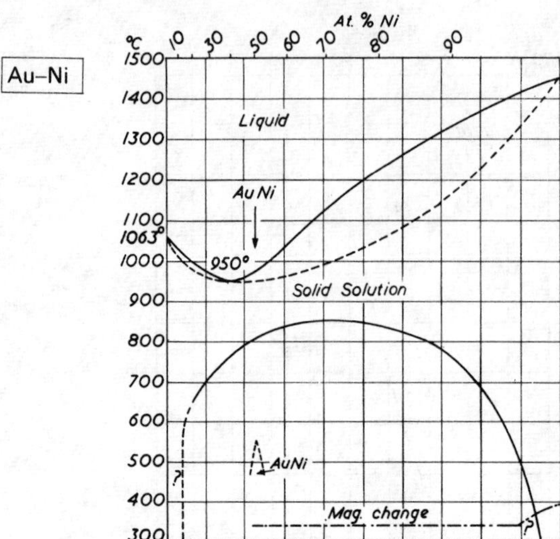
At. % Ni
°C
Liquid
AuNi
1063°
950°
Solid Solution
AuNi
Mag. change
Wt. % Ni

Au–P

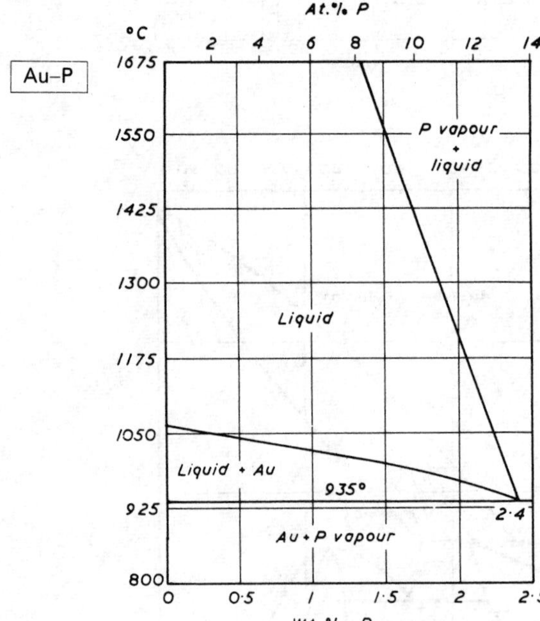
At.% P
°C
P vapour + liquid
Liquid
Liquid + Au
935°
2·4
Au + P vapour
Wt.% P

Au–Pb

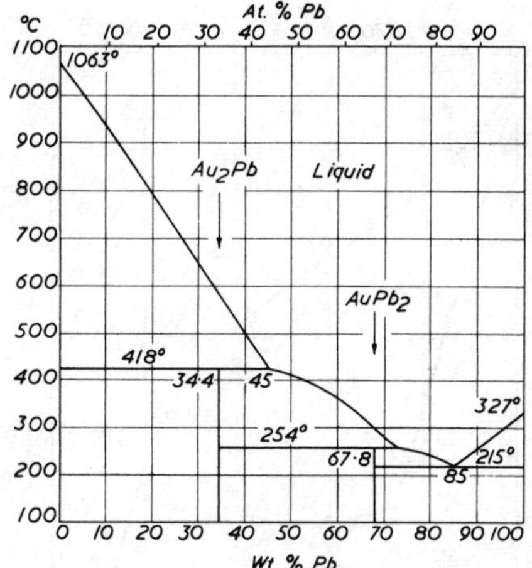

Au–Pd

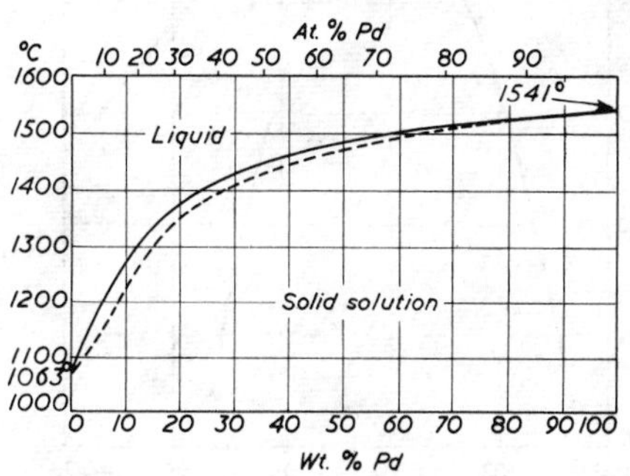

Au–Pr

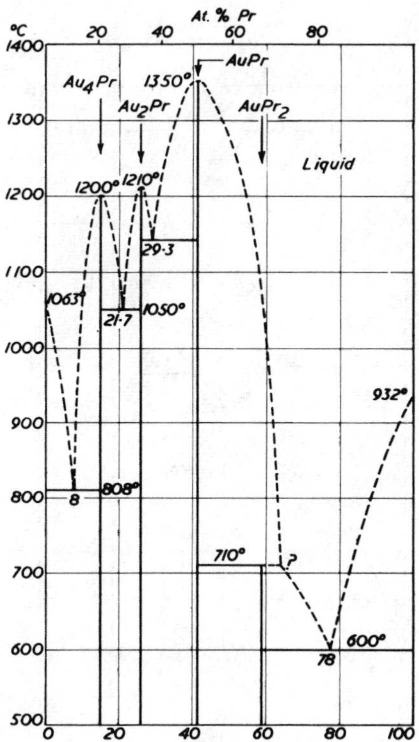

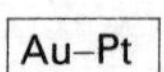
Au–Pt

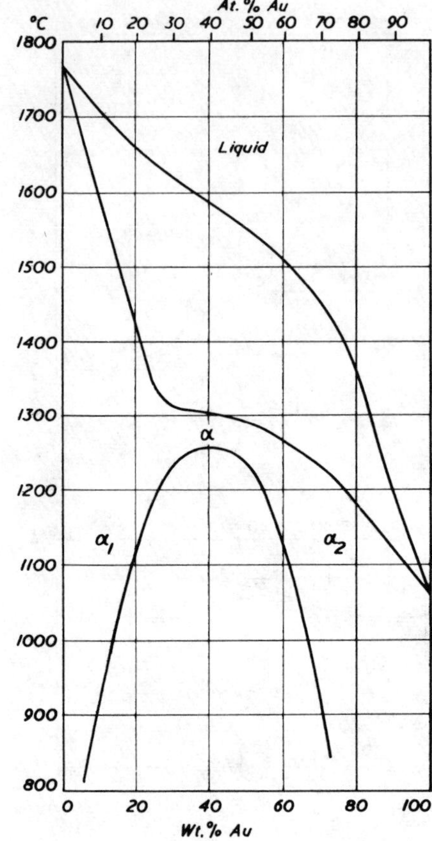
At.% Au
°C
10
20
30
40
50
60
70
80
90
1800
1700
1600
1500
1400
1300
1200
1100
1000
900
800
Liquid
α
α1
α2
0
20
40
60
80
100
Wt.% Au

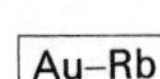
Au–Rb

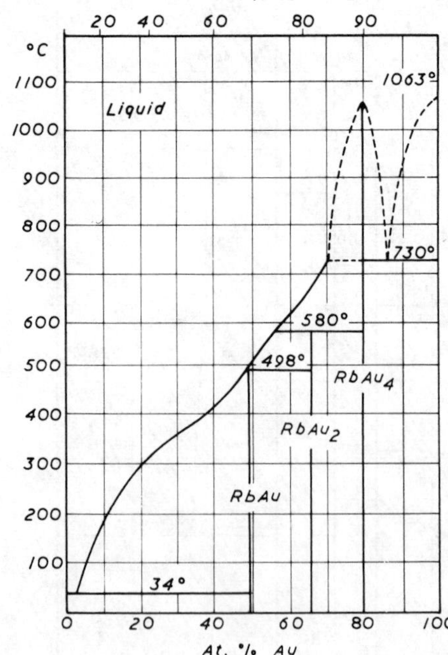
Wt.% Rb
20
40
60
80
90
°C
1100
1000
900
800
700
600
500
400
300
200
100
Liquid
1063°
730°
580°
498°
RbAu4
RbAu2
RbAu
34°
0
20
40
60
80
100
At.% Au

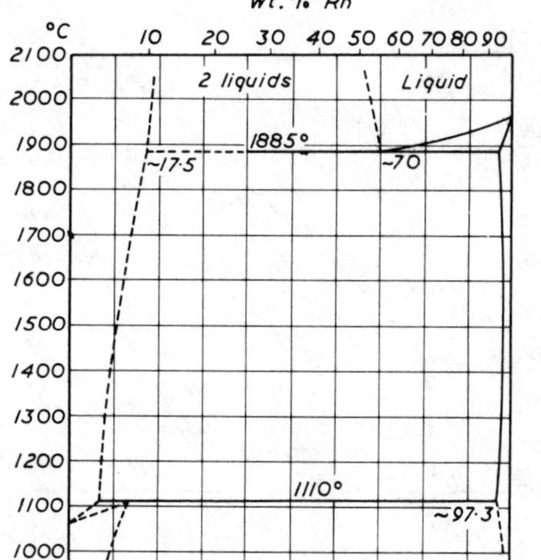

Au–Rh

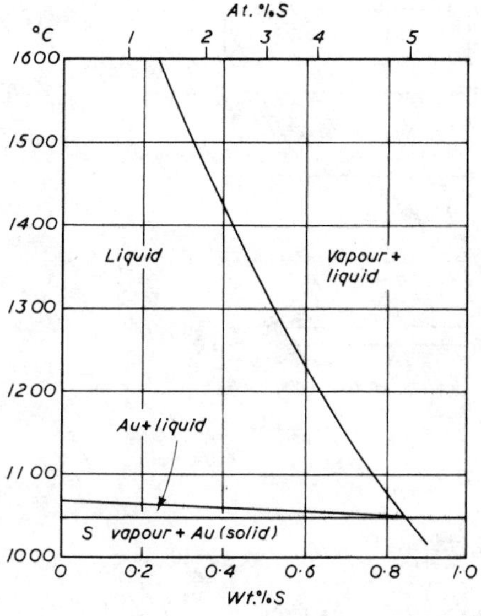

Au–S

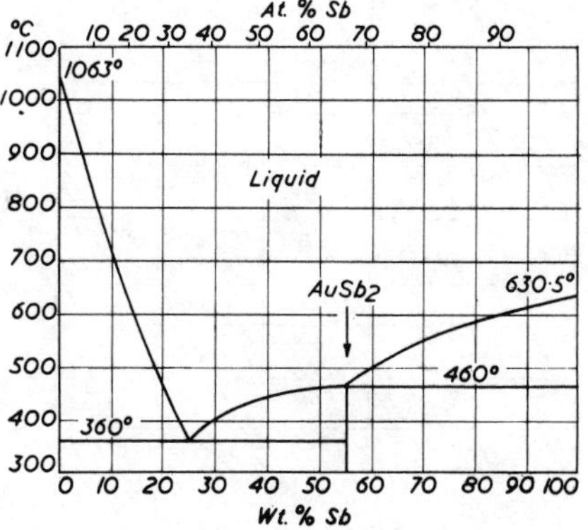

Au–Sb

Au–Sc

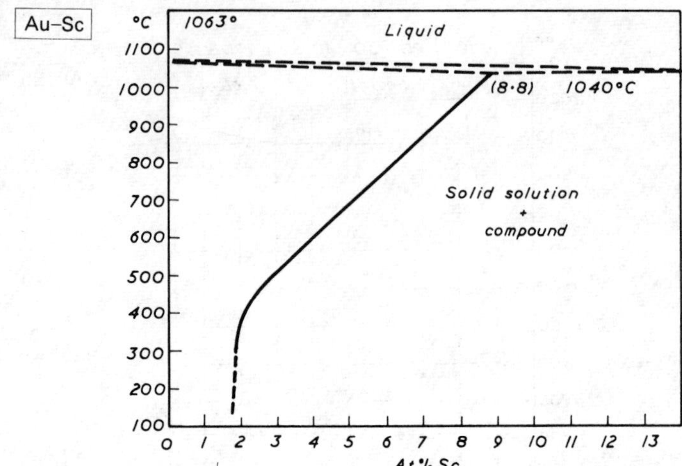

Au–Se

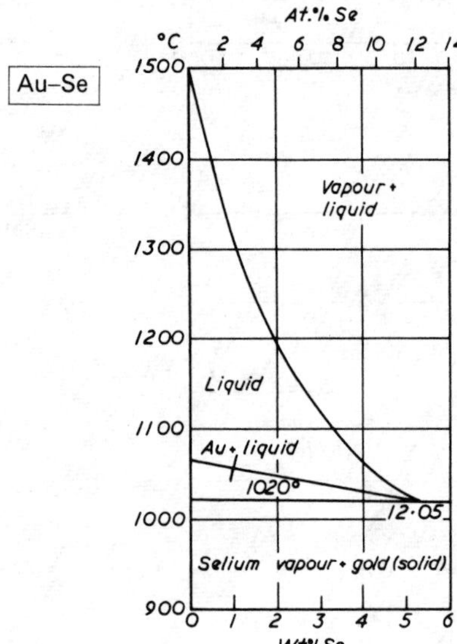

Au–Se

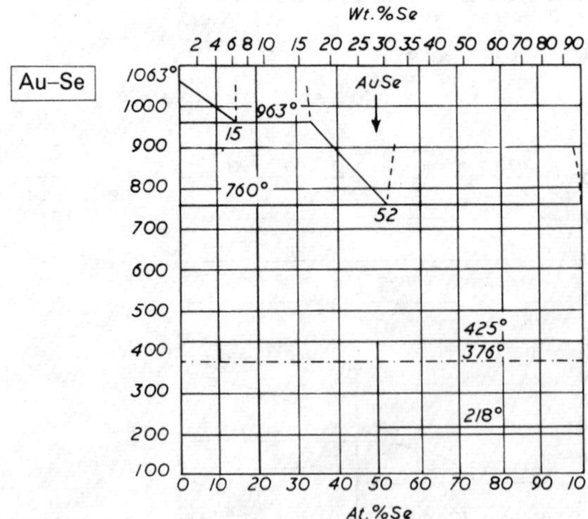

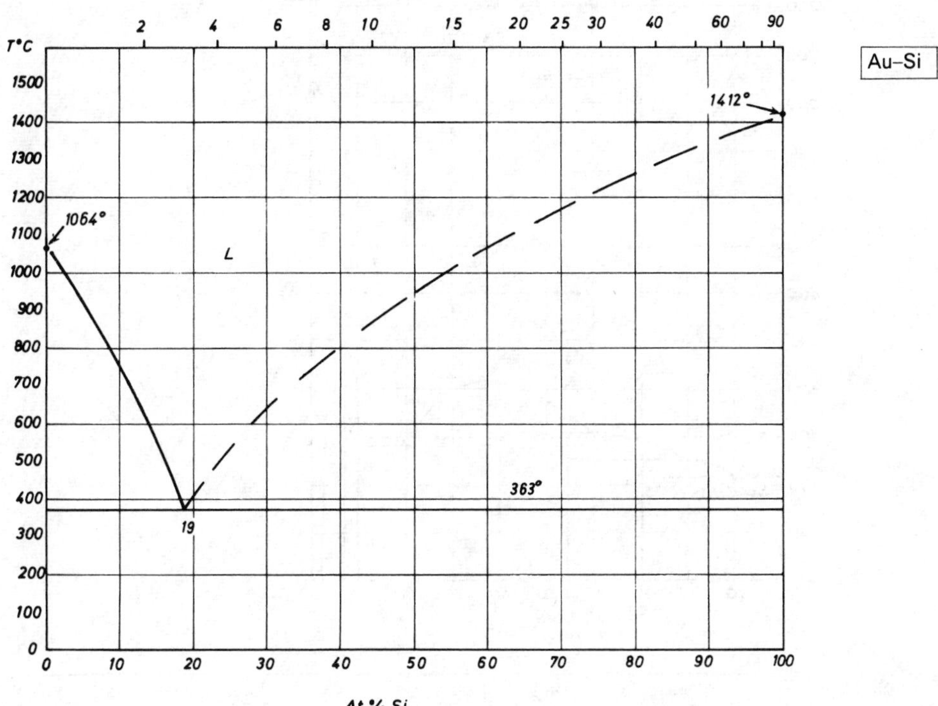
Wt.% Si
T°C
Au–Si
1412°
1064°
L
363°
19
At.% Si

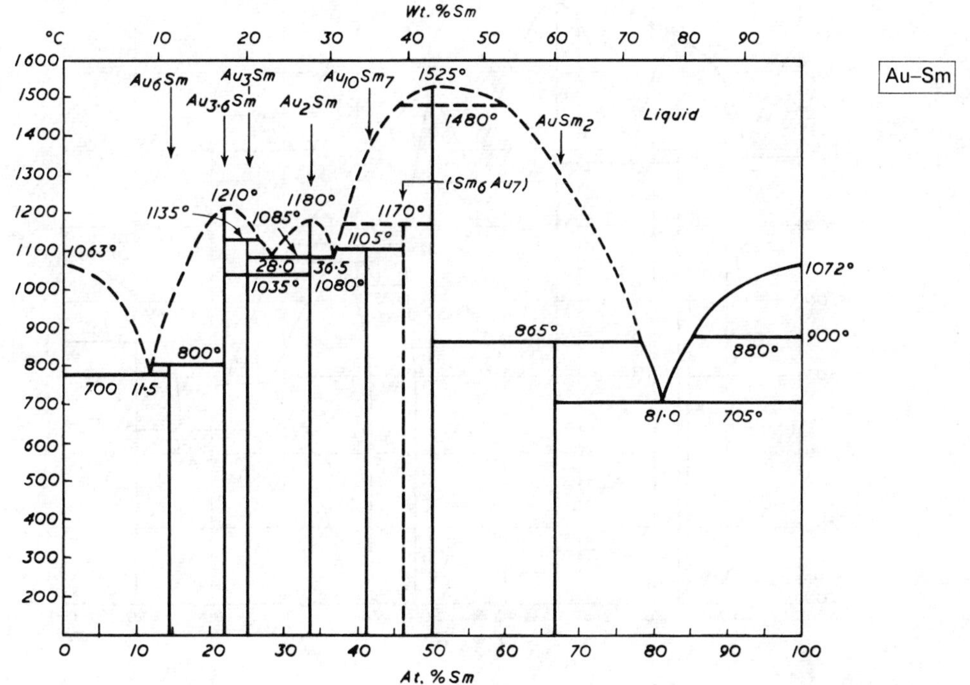
Wt. % Sm
°C
Au–Sm
Au6Sm
Au3Sm
Au3·6Sm
Au10Sm7
Au2Sm
1525°
1480°
AuSm2
Liquid
(Sm6Au7)
1210°
1180°
1135°
1085°
1170°
1105°
1063°
28·0
36·5
1035°
1080°
1072°
865°
900°
880°
800°
700
11·5
81·0
705°
At. % Sm

Au–Sn

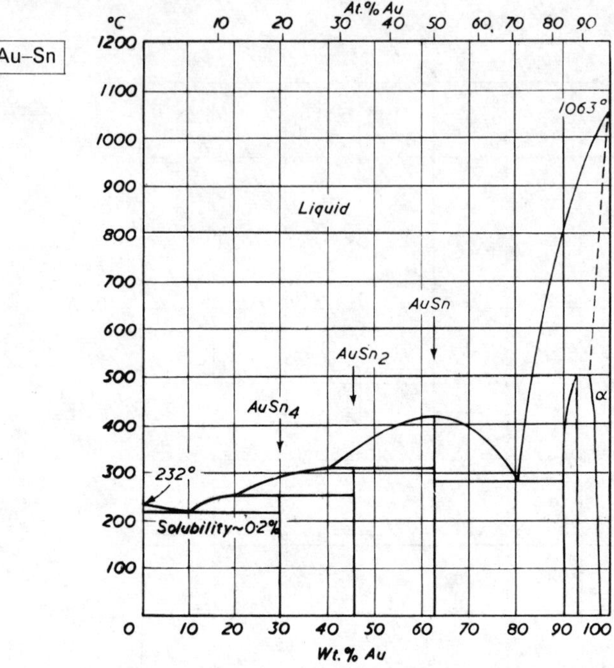
At.% Au
°C
10
20
30
40
50
60
70
80
90
1200
1100
1000
900
800
700
600
500
400
300
200
100
0
1063°
Liquid
AuSn
AuSn2
AuSn4
α
232°
Solubility~0.2%
10
20
30
40
50
60
70
80
90
100
Wt.% Au

Au–Sr

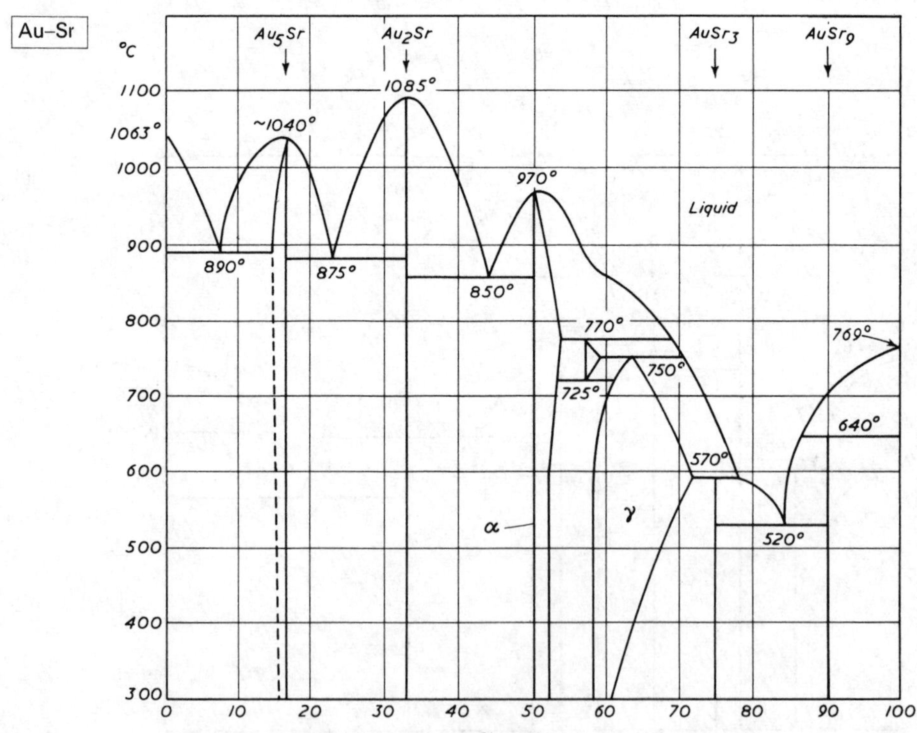
°C
Au5Sr
Au2Sr
AuSr3
AuSr9
1100
1063°
1000
900
800
700
600
500
400
300
~1040°
1085°
970°
Liquid
890°
875°
850°
770°
769°
750°
725°
640°
570°
520°
α
γ
0
10
20
30
40
50
60
70
80
90
100
At.%Sr

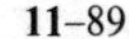

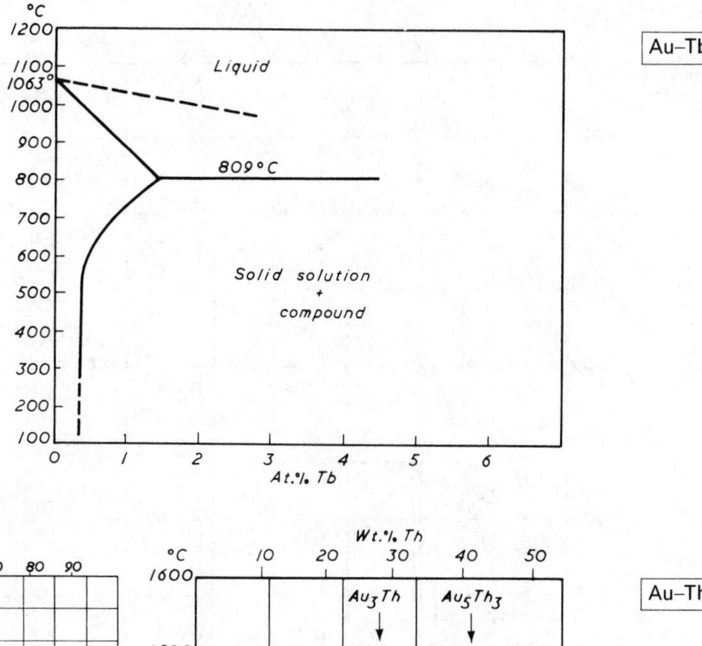

Au–Tb

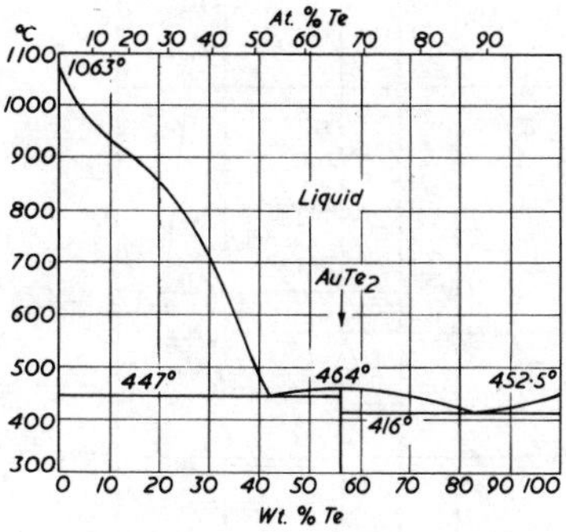

Au–Te

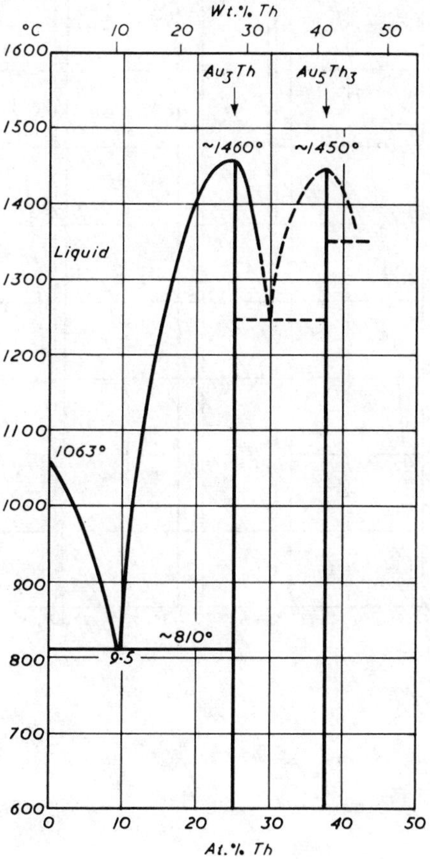

Au–Th

Au–Ti

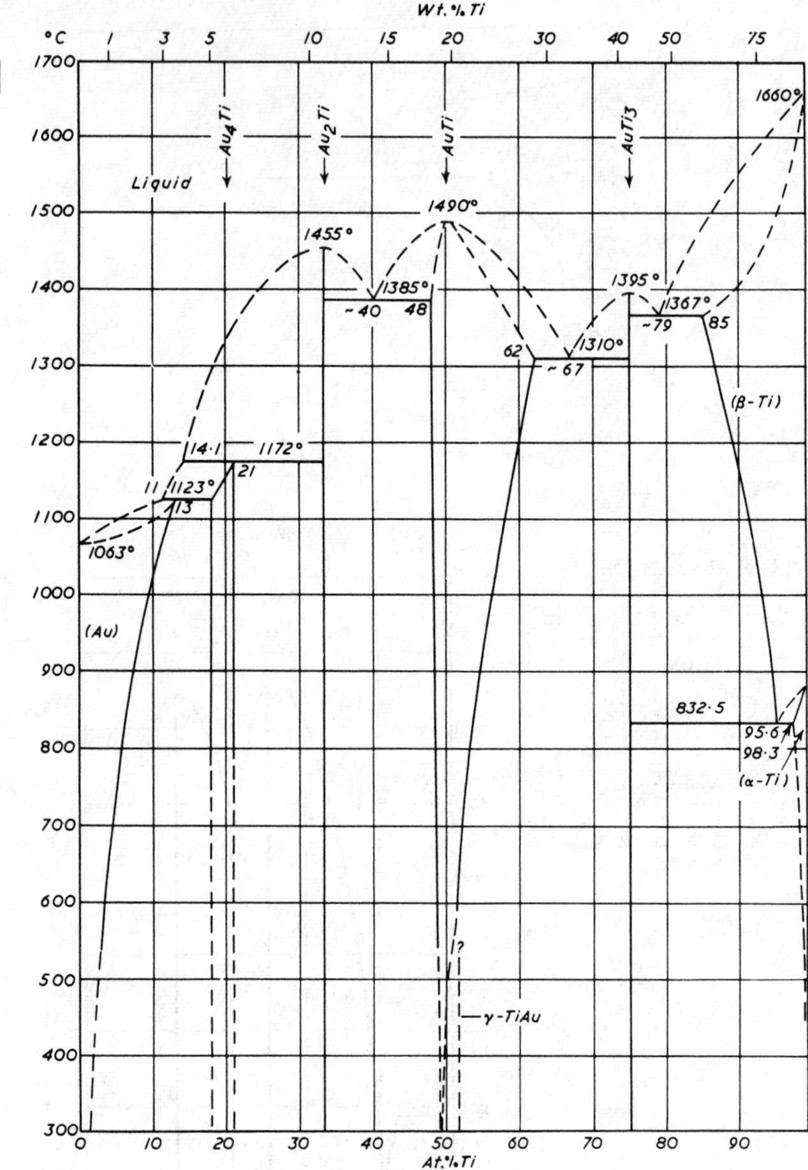
Wt.% Ti
°C
1700
1600
1500
1400
1300
1200
1100
1000
900
800
700
600
500
400
300
1 3 5 10 15 20 30 40 50 75
Au4Ti
Au2Ti
AuTi
AuTi3
Liquid
1660°
1490°
1455°
1395°
1385°
1367°
~40
48
~79
85
62
1310°
~67
(β-Ti)
14·1
1172°
21
11
1123°
13
1063°
(Au)
832·5
95·6
98·3
(α-Ti)
1?
γ-TiAu
0 10 20 30 40 50 60 70 80 90
At.% Ti

Au–Tl

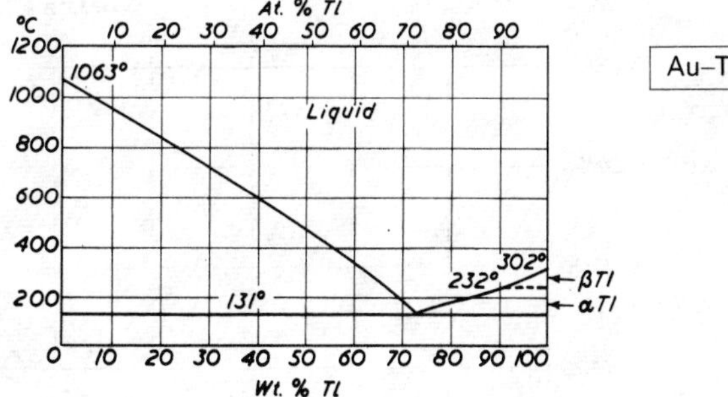

Au–Tm

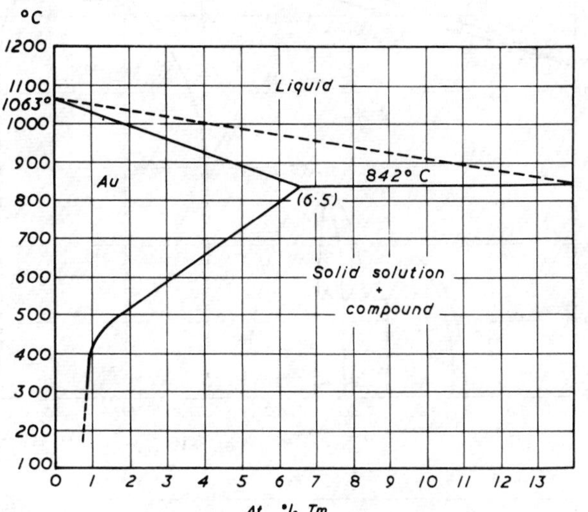

Au–U

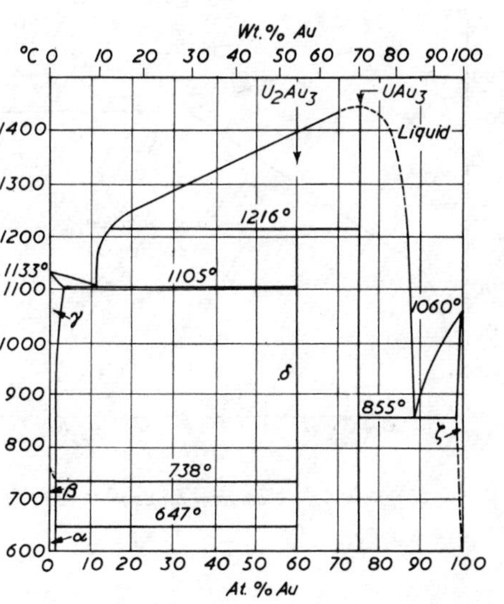

Au–V

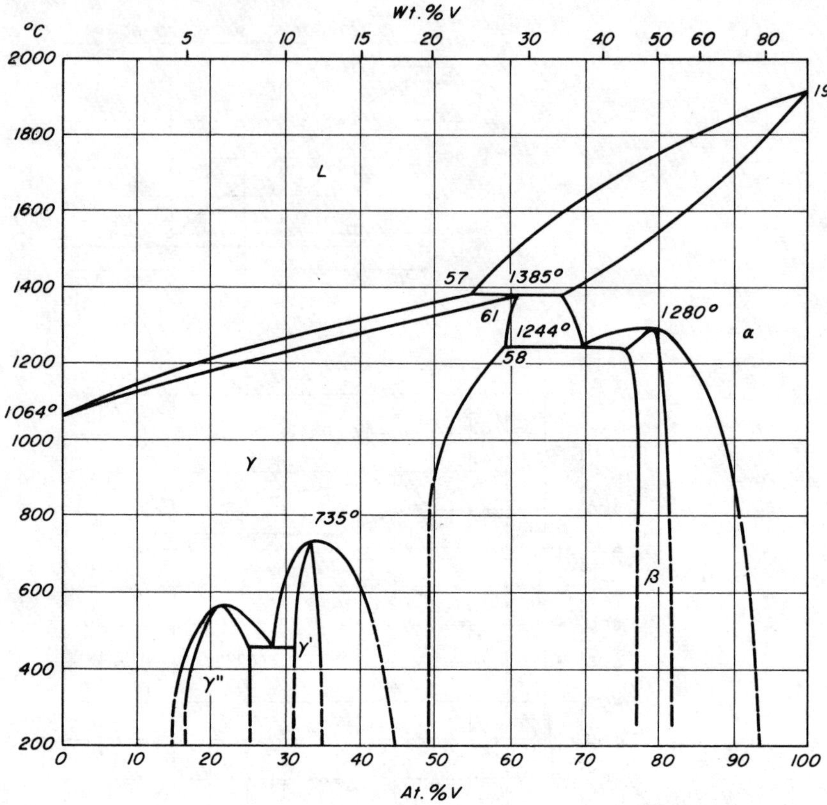

Au–Yb

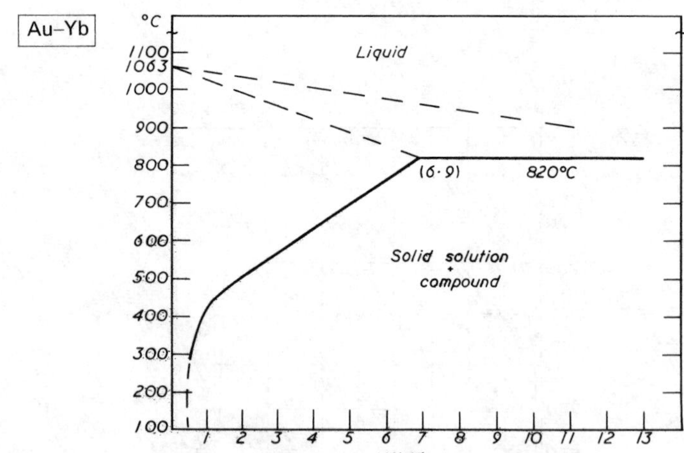

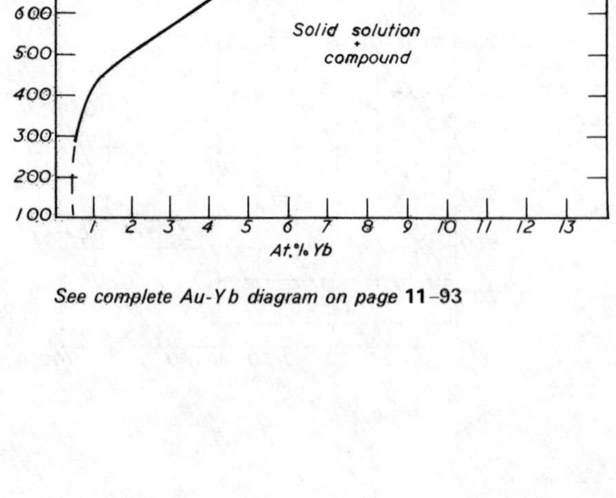

See complete Au-Yb diagram on page 11–93

Au–Yb

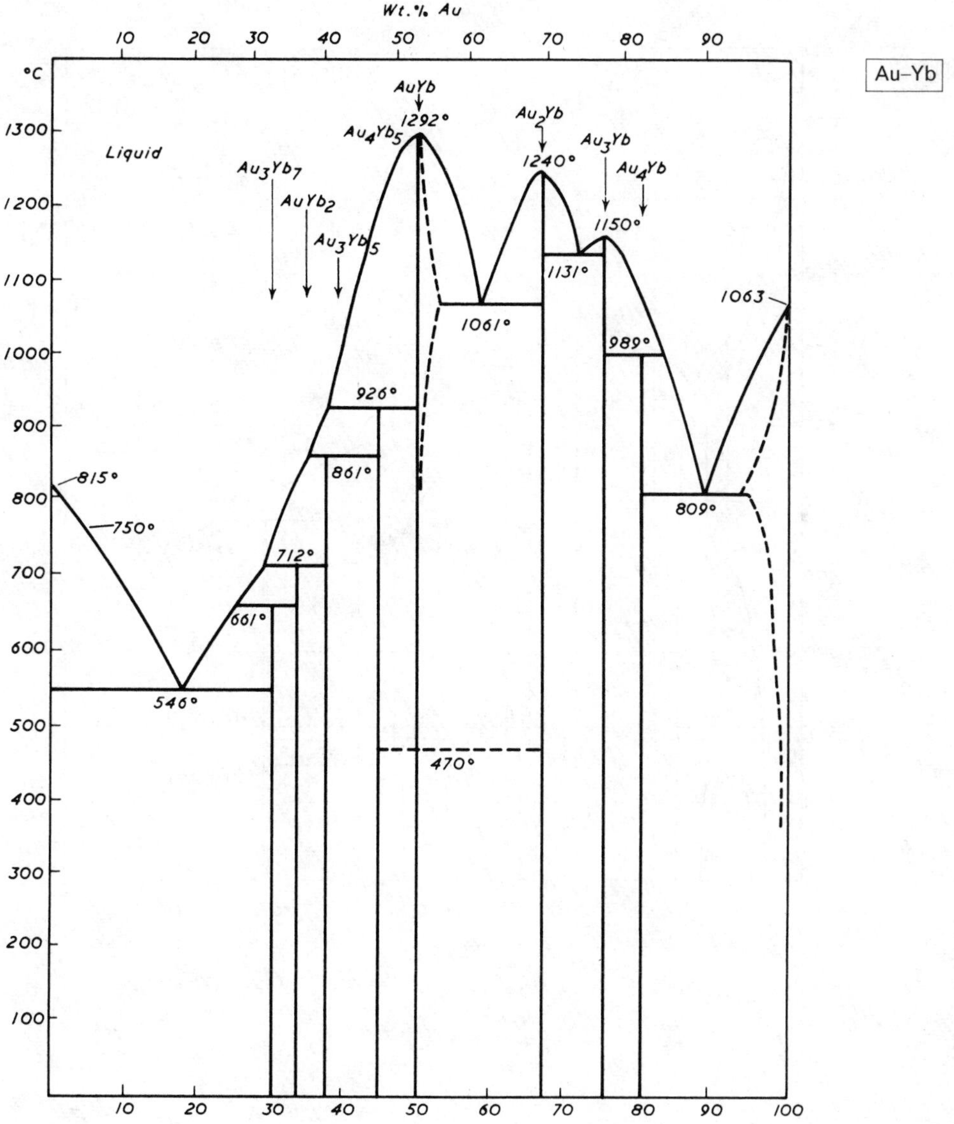
Wt.% Au
10
20
30
40
50
60
70
80
90
°C
1300
1200
1100
1000
900
800
700
600
500
400
300
200
100
Liquid
AuYb
1292°
Au_4Yb_5
Au_3Yb_7
$AuYb_2$
Au_3Yb_5
Au_2Yb
1240°
Au_3Yb
Au_4Yb
1150°
1131°
1061°
1063
989°
926°
861°
815°
750°
712°
661°
546°
470°
809°
10
20
30
40
50
60
70
80
90
100
At.% Au

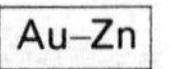
Au–Zn
At.% Zn
°C
1063°
Au_3Zn
AuZn
$AuZn_3$
725°
Liquid
642°
644°
626°
α
γ
515°
475°
$\gamma+\gamma_1$
416°
423°
419·4°
425°
404°
α_1
β
γ_1
ε
270°
η
225°
?
150°
α_2
150°
γ_2
150°
Wt. % Zn

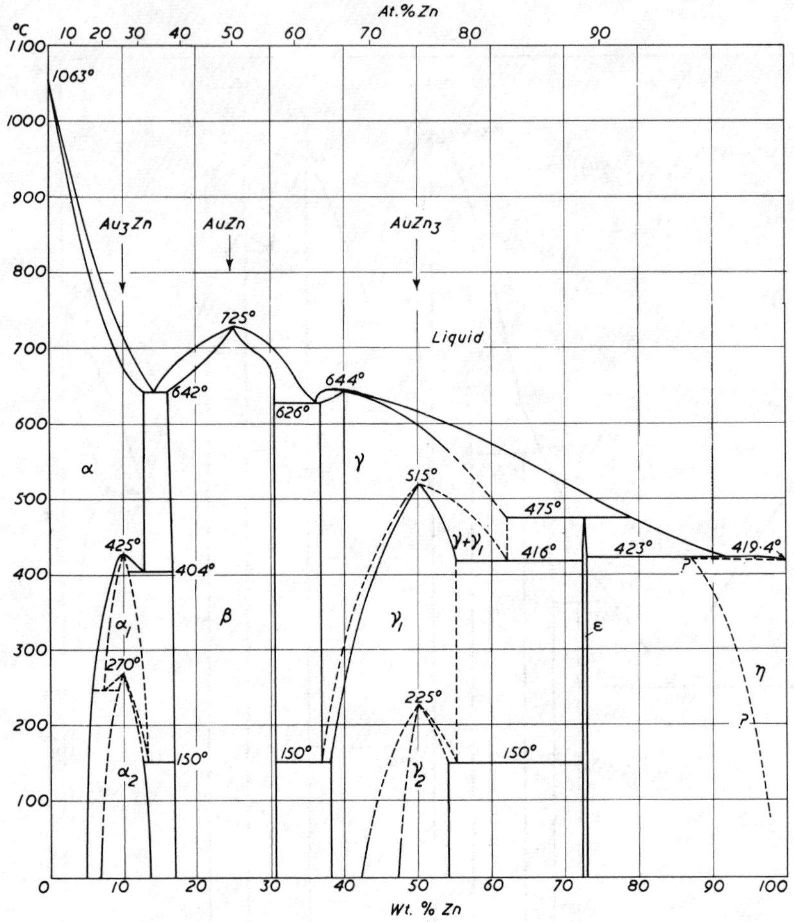

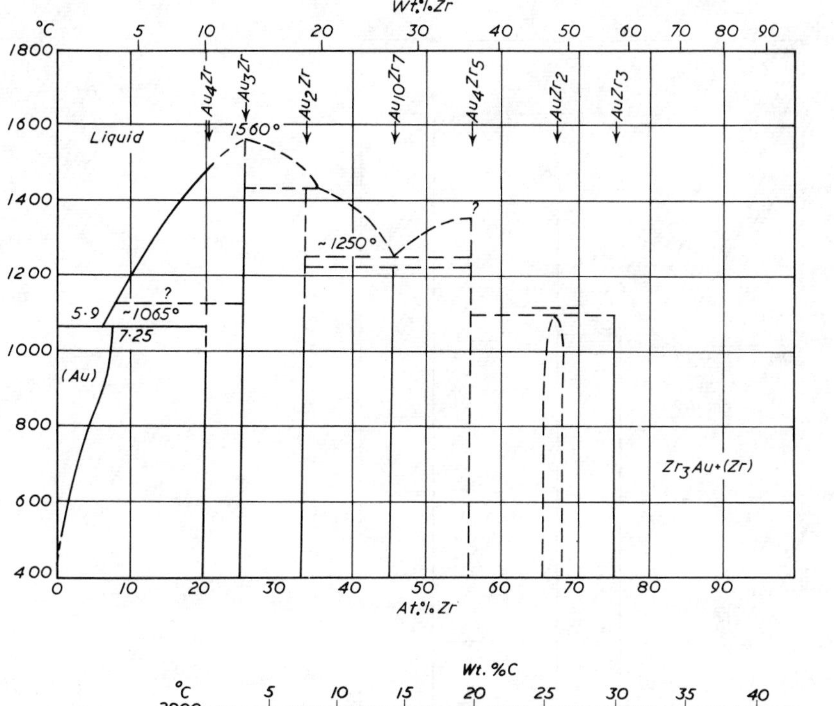
Wt.% Zr
°C
Au4Zr
Au3Zr
Au2Zr
Au10Zr7
Au4Zr5
AuZr2
AuZr3
Liquid
1560°
~1250°
?
5·9
~1065°
7·25
(Au)
Zr3Au+(Zr)
At.% Zr

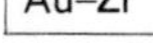
Au–Zr

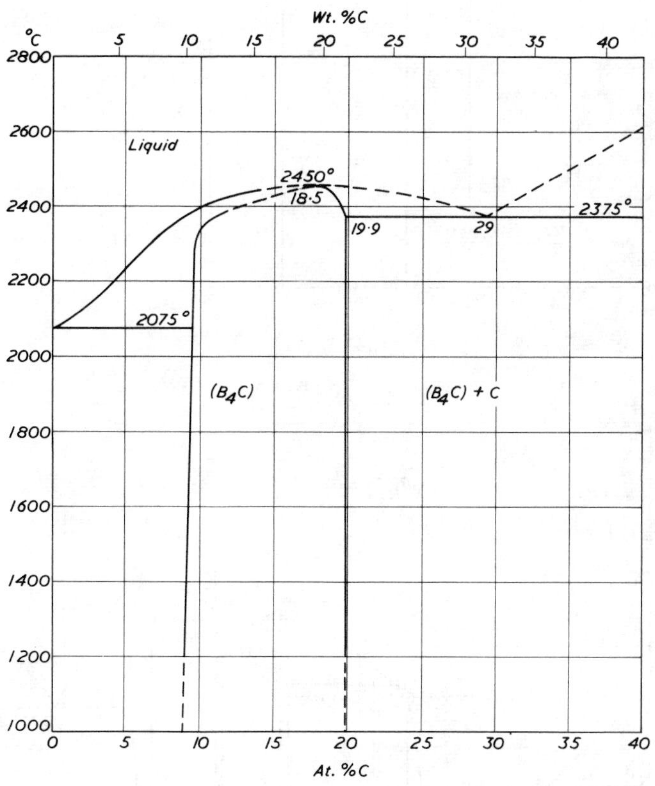
Wt. %C
°C
Liquid
2450°
18·5
19·9
29
2375°
2075°
(B_4C)
(B_4C) + C
At. %C

B–C

B–Co

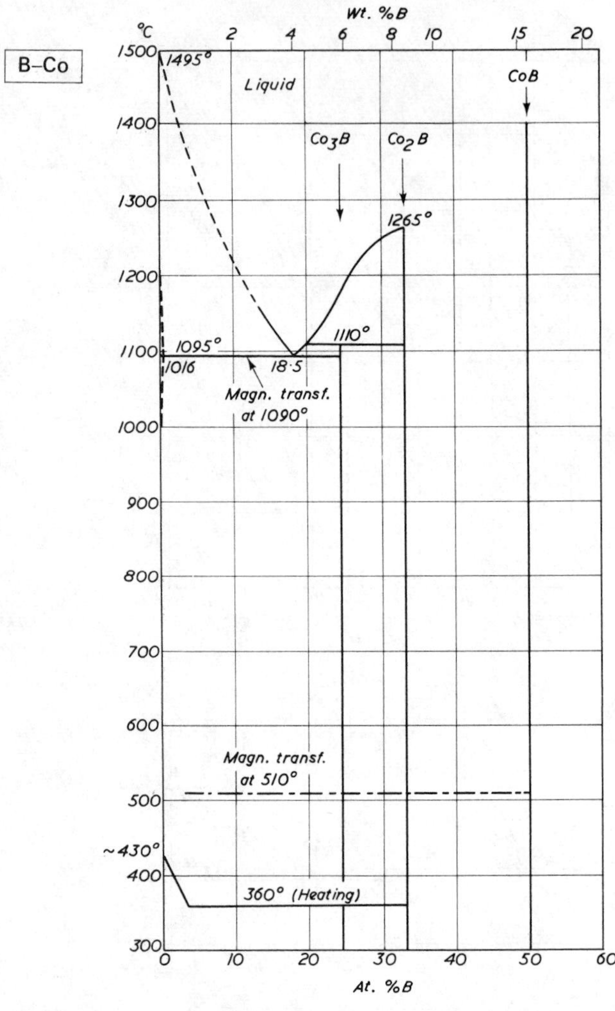
Wt. %B
2
4
6
8
10
15
20
°C
1500
1400
1300
1200
1100
1000
900
800
700
600
500
400
300
1495°
Liquid
CoB
Co_3B
Co_2B
1265°
1110°
1095°
1016
18·5
Magn. transf.
at 1090°
Magn. transf.
at 510°
~430°
360° (Heating)
0
10
20
30
40
50
60
At. %B

B–Cr

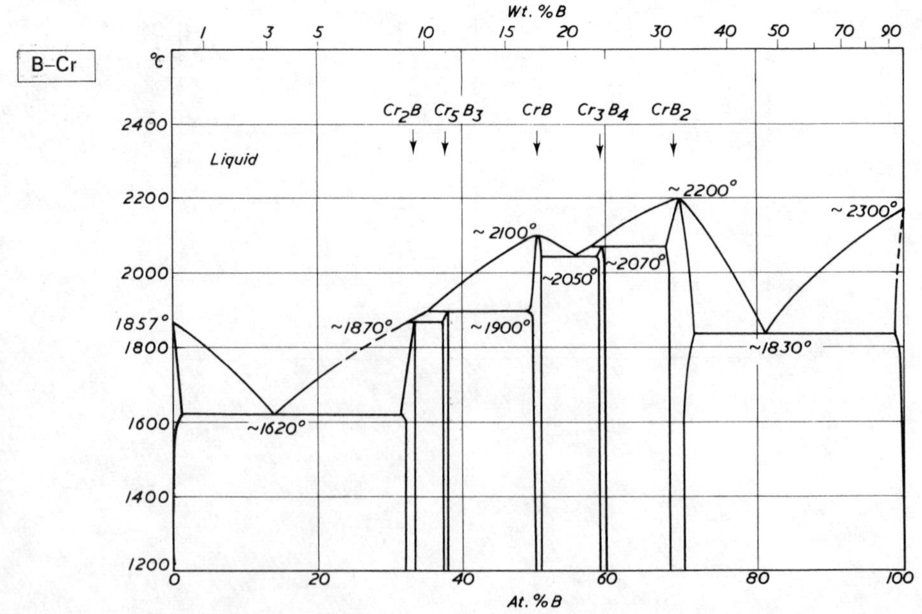
Wt. %B
1
3
5
10
15
20
30
40
50
70
90
°C
2400
2200
2000
1857°
1800
1600
1400
1200
Liquid
Cr_2B
Cr_5B_3
CrB
Cr_3B_4
CrB_2
~2200°
~2300°
~2100°
~2050°
~2070°
~1870°
~1900°
~1830°
~1620°
0
20
40
60
80
100
At. %B

B–Cu

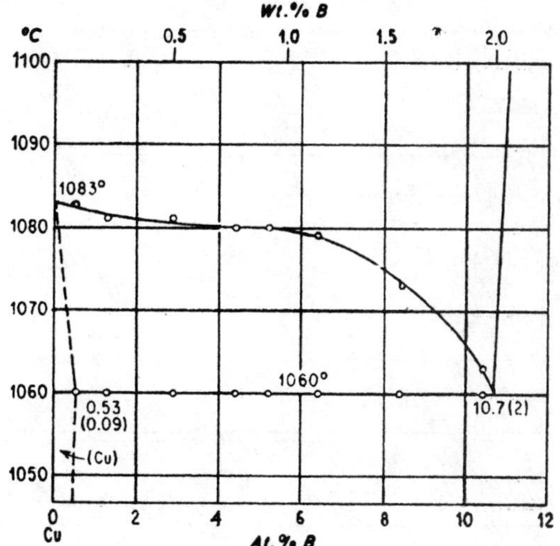
Wt.% B
°C
0.5
1.0
1.5
2.0
1100
1090
1080
1070
1060
1050
1083°
1060°
0.53
(0.09)
(Cu)
10.7(2)
0
Cu
2
4
6
8
10
12
At.% B

B–Eu

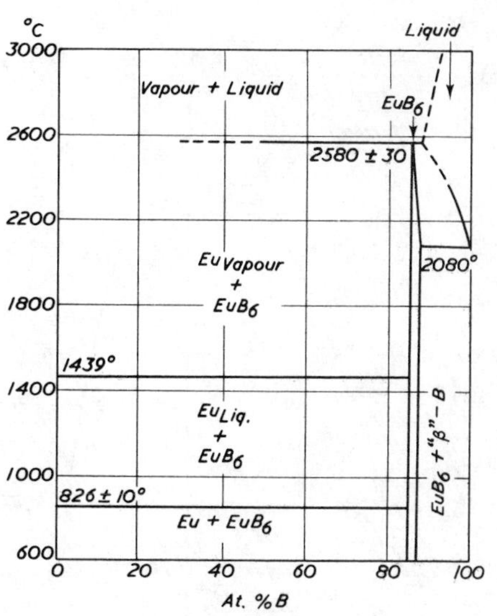
°C
3000
2600
2200
1800
1400
1000
600
Liquid
Vapour + Liquid
EuB6
2580 ± 30
2080°
Eu Vapour
+
EuB6
1439°
Eu Liq.
+
EuB6
826 ± 10°
Eu + EuB6
EuB6 + "β"– B
0
20
40
60
80
100
At. % B

B–Fe

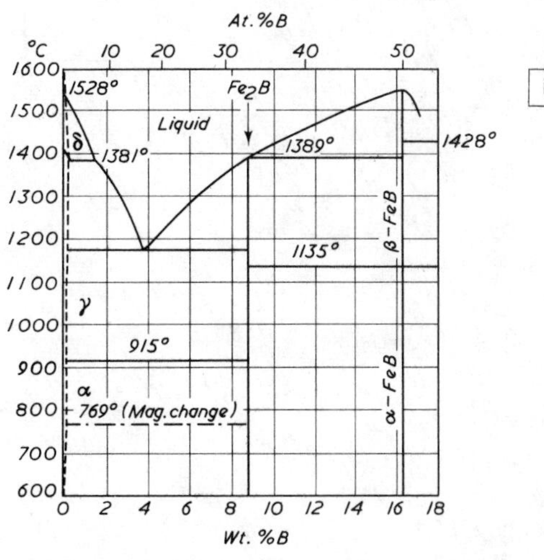
At.% B
°C
10
20
30
40
50
1600
1500
1400
1300
1200
1100
1000
900
800
700
600
1528°
Fe2B
Liquid
δ
1381°
1389°
1428°
1135°
β – FeB
γ
915°
α
769° (Mag. change)
α – FeB
0
2
4
6
8
10
12
14
16
18
Wt. % B

B–Ge

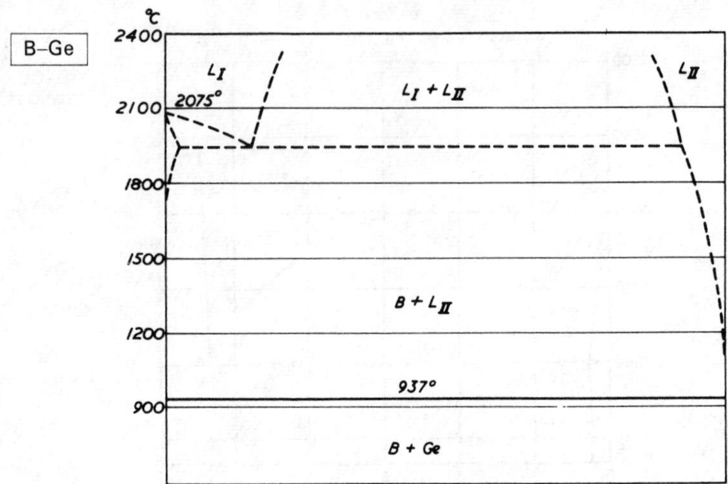

B–Hf

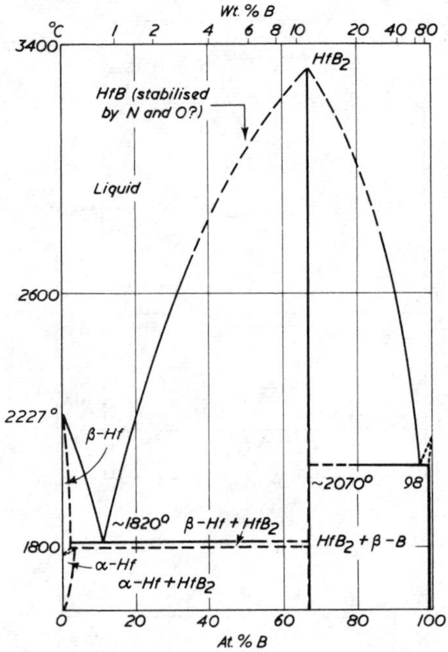

Wt. % B
°C
1 2 4 6 8 10 20 40 80
3400
2600
2227°
1800
HfB_2
HfB (stabilised by N and O?)
Liquid
β-Hf
~1820°
β-Hf + HfB_2
~2070°
98
HfB_2 + β-B
α-Hf
α-Hf + HfB_2
0 20 40 60 80 100
At. % B

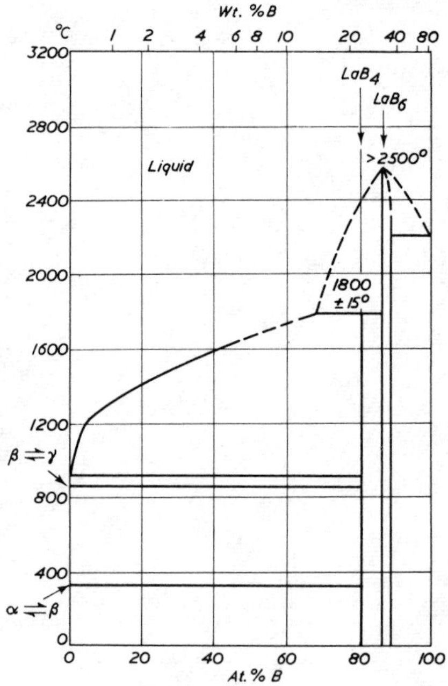

B–La

Wt. %B
°C
1 2 4 6 8 10 20 40 80
3200
2800
2400
2000
1600
1200
800
400
0
LaB_4
LaB_6
> 2500°
Liquid
1800 ± 15°
β ⇌ γ
α ⇌ β
0 20 40 60 80 100
At. % B

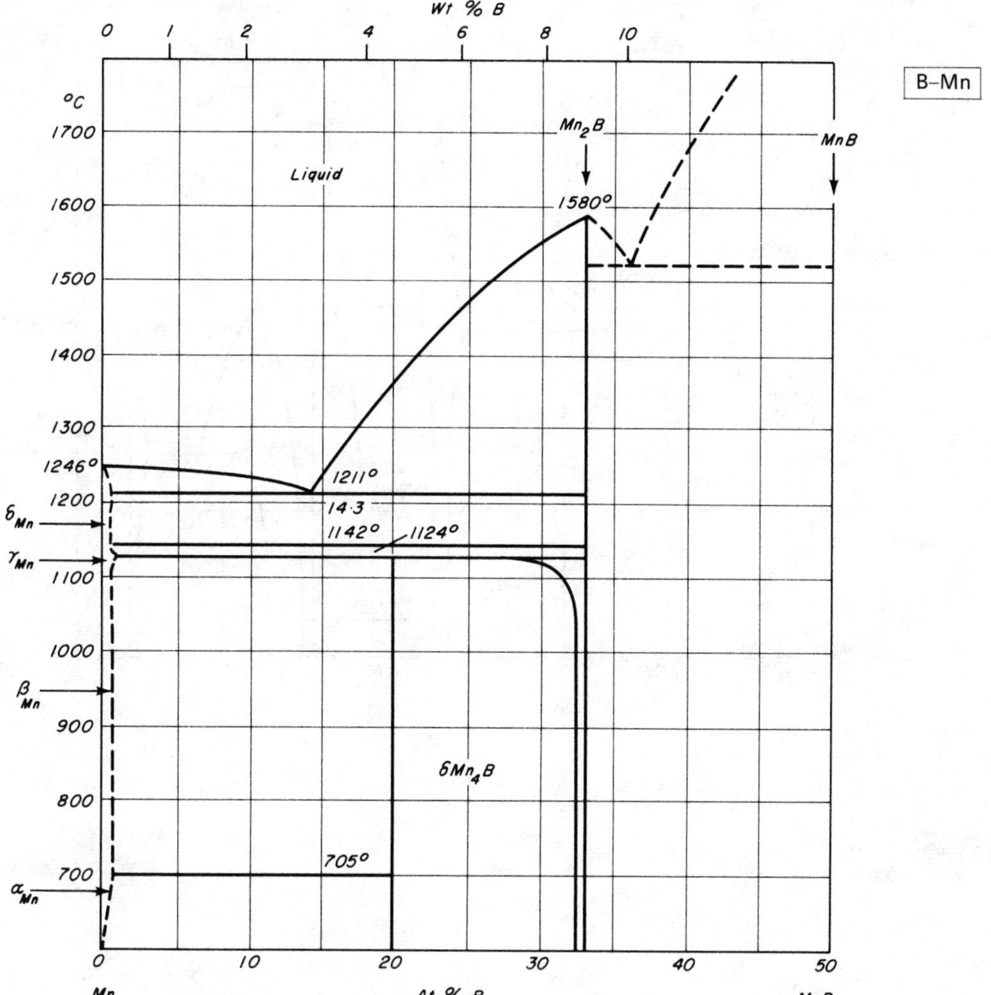
B–Mn
Wt % B
0
1
2
4
6
8
10
°C
1700
1600
1500
1400
1300
1200
1100
1000
900
800
700
Liquid
Mn_2B
MnB
1580°
1246°
1211°
14·3
1142°
1124°
δ_{Mn}
γ_{Mn}
β_{Mn}
α_{Mn}
δMn_4B
705°
0
10
20
30
40
50
Mn
At. % B
MnB

B–Mo

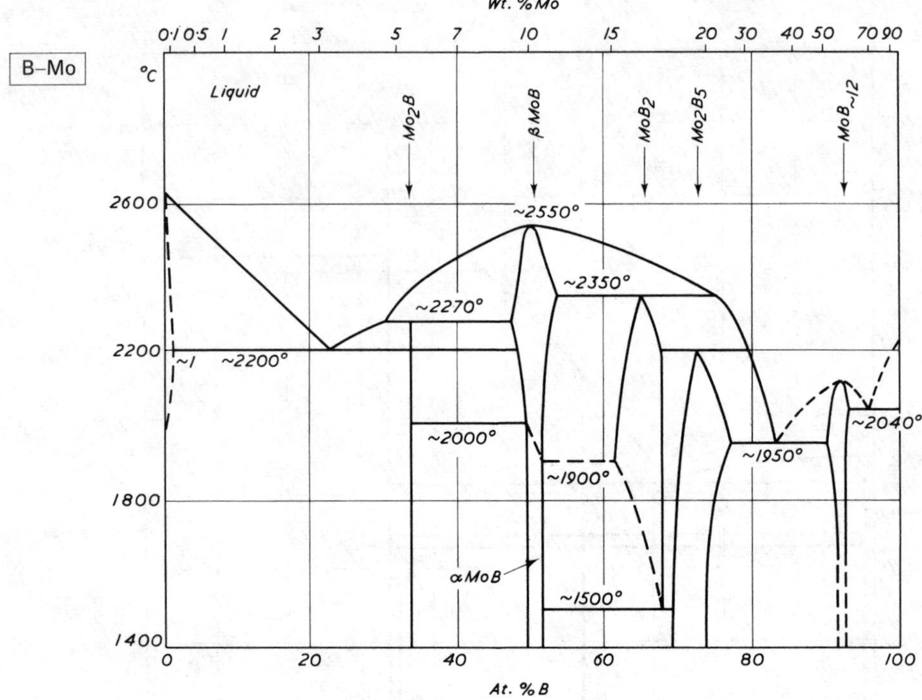
Wt. % Mo
0·1 0·5 1 2 3 5 7 10 15 20 30 40 50 70 90
°C
Liquid
Mo_2B
βMoB
MoB_2
Mo_2B_5
$MoB_{\sim 12}$
2600
~2550°
~2350°
~2270°
2200
~1
~2200°
~2000°
~1900°
~1950°
~2040°
1800
αMoB
~1500°
1400
0 20 40 60 80 100
At. % B

B–Nb

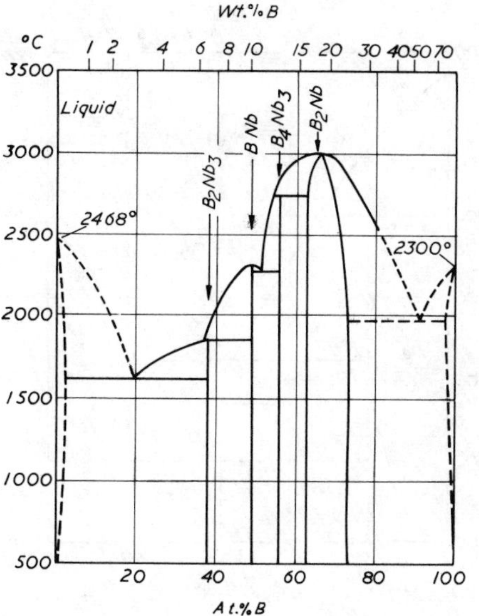
Wt.% B
°C
3500
3000
2500
2000
1500
1000
500
Liquid
B_2Nb_3
B Nb
B_4Nb_3
B_2Nb
2468°
2300°
At.% B

B–Ni

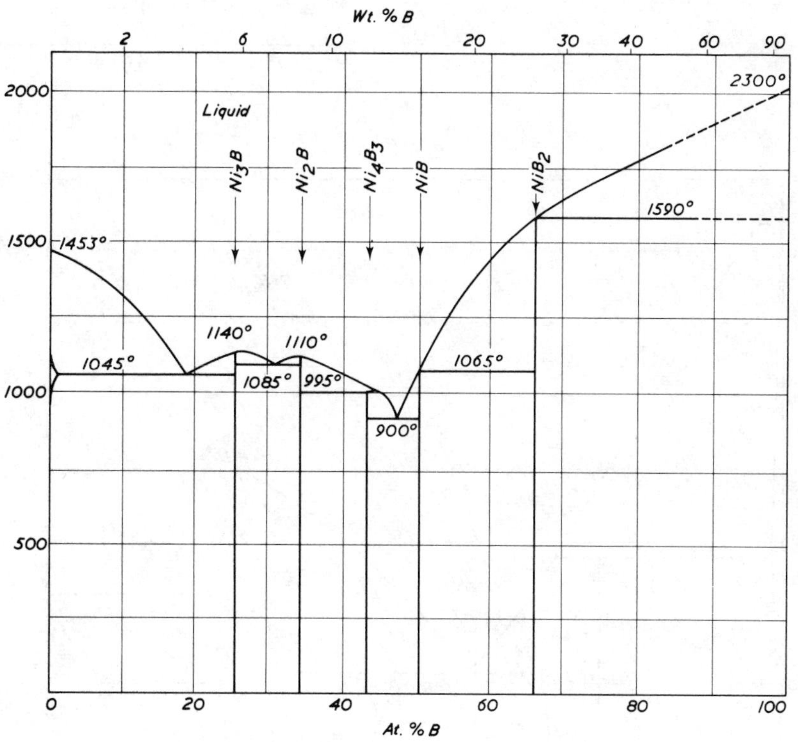
Wt. % B
Liquid
Ni_3B
Ni_2B
Ni_4B_3
NiB
NiB_2
1453°
1045°
1140°
1110°
1085°
995°
900°
1065°
1590°
2300°
At. % B

B–Pd
Wt. % B 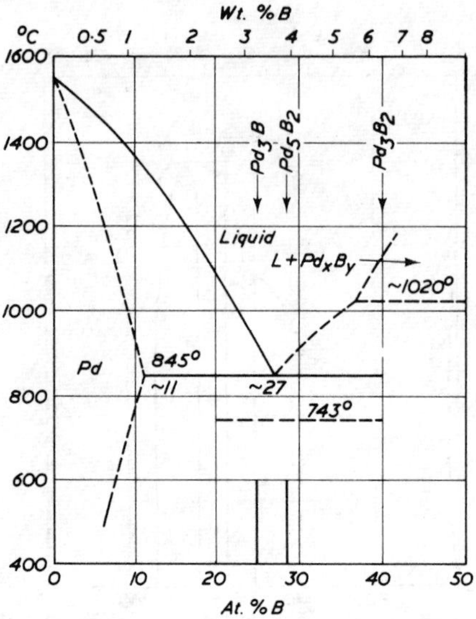
°C
0·5 1 2 3 4 5 6 7 8
1600
1400
1200
1000
800
600
400
Pd_3B
Pd_5B_2
Pd_3B_2
Liquid
$L + Pd_xB_y$
~1020°
845°
Pd
~11
~27
743°
0 10 20 30 40 50
At. % B

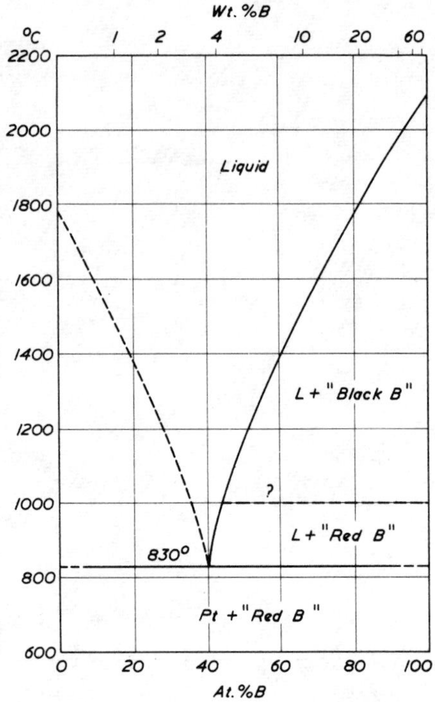

B–Pt
Wt. % B
°C
1 2 4 10 20 60
2200
2000
1800
1600
1400
1200
1000
800
600
Liquid
L + "Black B"
?
L + "Red B"
830°
Pt + "Red B"
0 20 40 60 80 100
At. % B

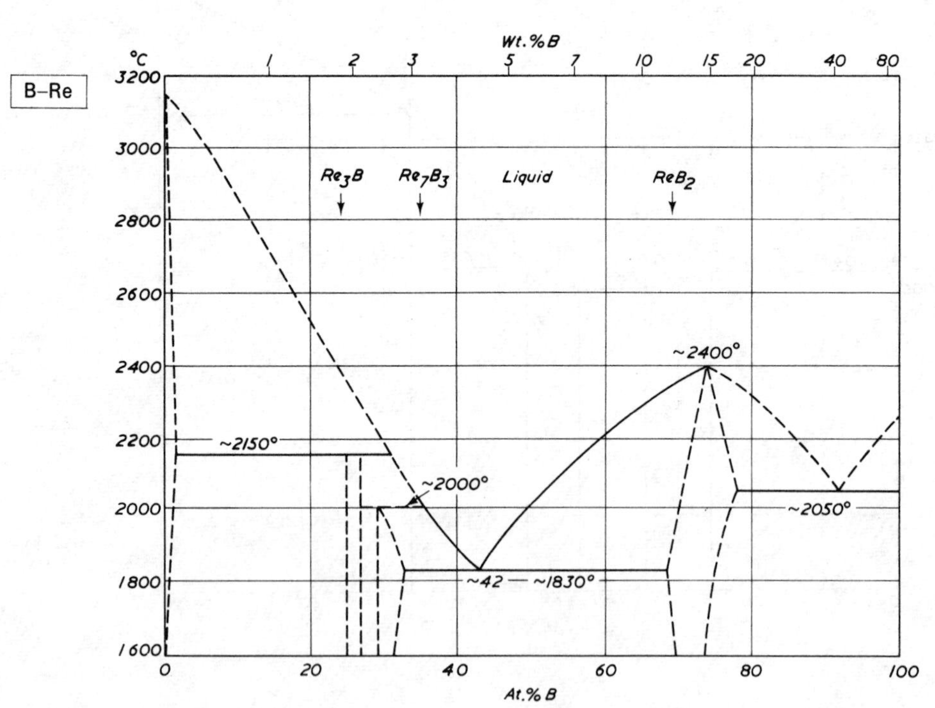
B–Re
Wt. % B
°C
1 2 3 5 7 10 15 20 40 80
3200
3000
2800
2600
2400
2200
2000
1800
1600
Re_3B
Re_7B_3
Liquid
ReB_2
~2400°
~2150°
~2000°
~2050°
~42
~1830°
0 20 40 60 80 100
At. % B

B–Si

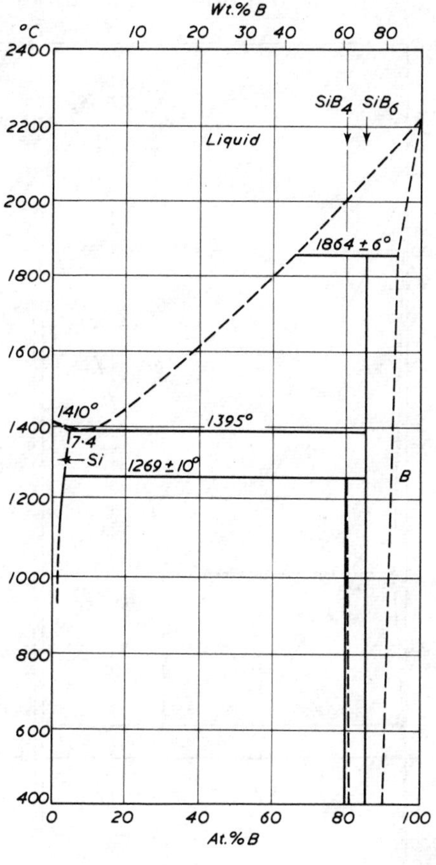

°C
Wt.% B
10
20
30
40
60
80
2400
2200
2000
1800
1600
1400
1200
1000
800
600
400
Liquid
SiB_4
SiB_6
1864 ± 6°
1410°
1395°
7·4
Si
1269 ± 10°
B
0
20
40
60
80
100
At.% B

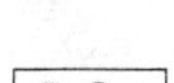

B–Sm

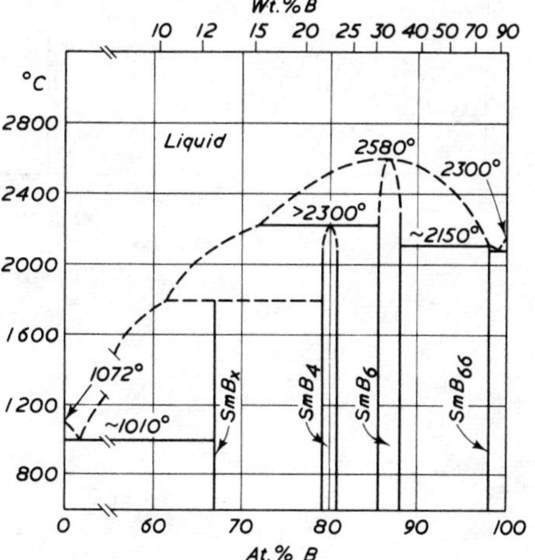

Wt.% B
10
12
15
20
25
30
40
50
70
90
°C
2800
2400
2000
1600
1200
800
Liquid
2580°
2300°
>2300°
~2150°
1072°
~1010°
SmB_x
SmB_4
SmB_6
SmB_{66}
0
60
70
80
90
100
At.% B

B–Ta

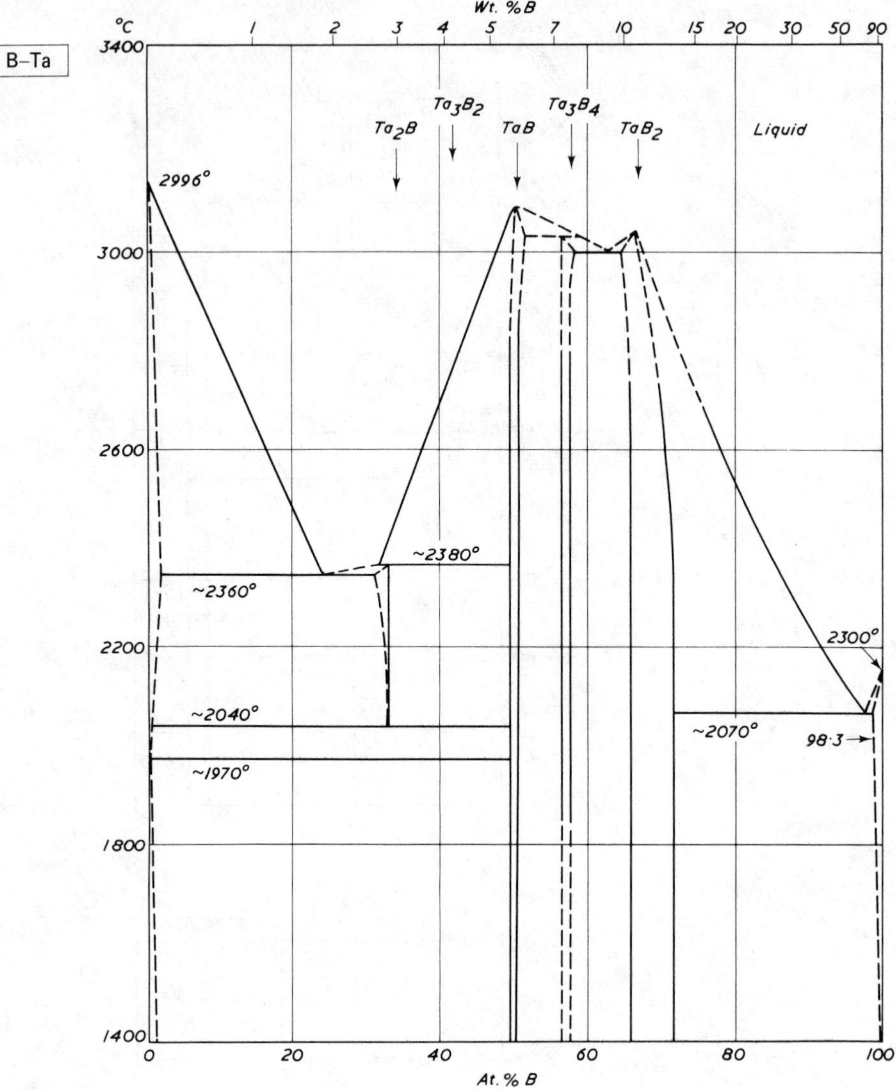

Wt. %B
°C
1
2
3
4
5
7
10
15
20
30
50
90
3400
3000
2600
2200
1800
1400
Ta2B
Ta3B2
TaB
Ta3B4
TaB2
Liquid
2996°
~2380°
~2360°
~2040°
~1970°
2300°
~2070°
98·3
0
20
40
60
80
100
At. % B

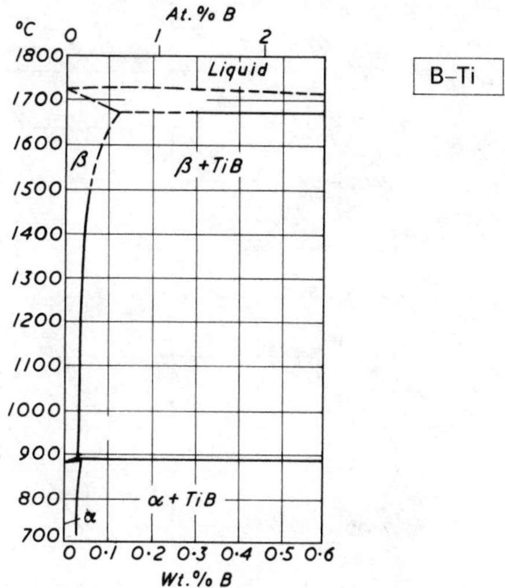
At.% B
°C
Liquid
β
β + TiB
α + TiB
α
Wt.% B

B–Ti

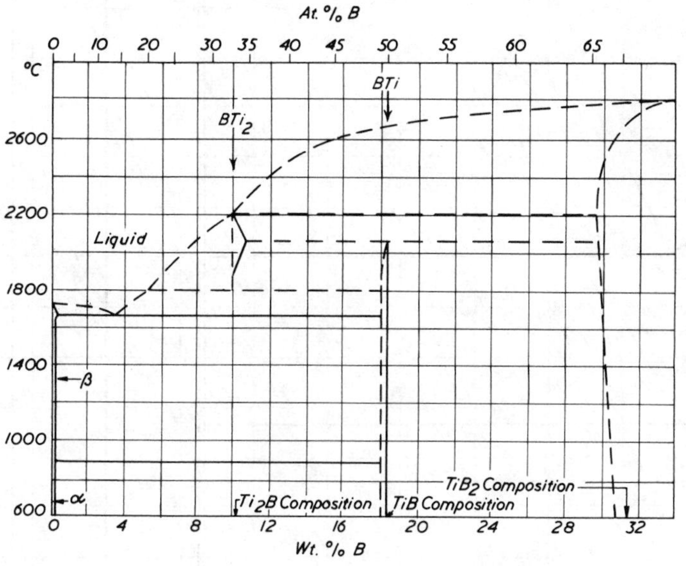
At.% B
°C
BTi
BTi2
Liquid
β
α
Ti2B Composition
TiB Composition
TiB2 Composition
Wt.% B

B–Ti

B–U

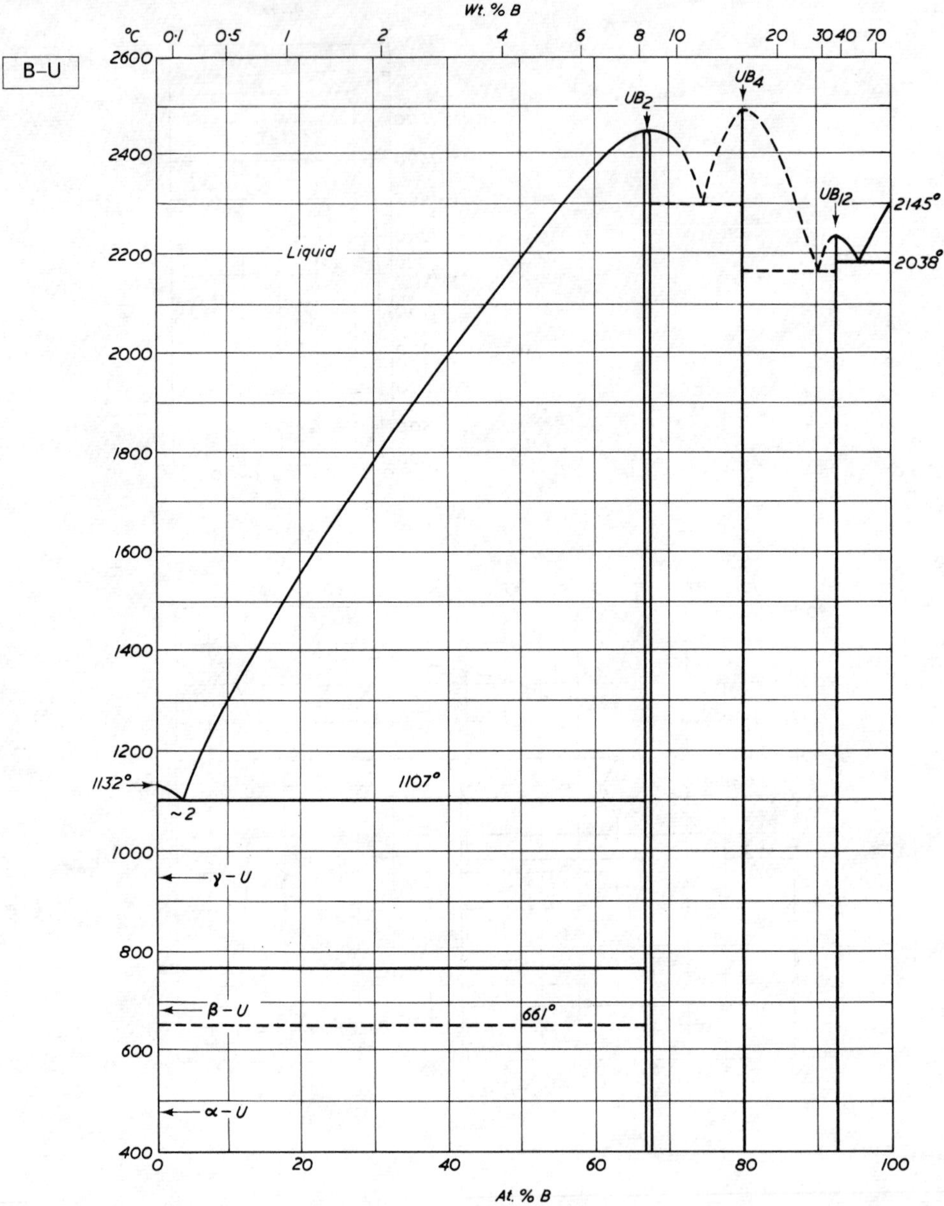

Wt. % B
°C
0·1
0·5
1
2
4
6
8
10
20
30
40
70
2600
2400
2200
2000
1800
1600
1400
1200
1000
800
600
400
UB_2
UB_4
UB_{12}
2145°
2038°
Liquid
1132°
~2
1107°
γ – U
β – U
661°
α – U
0
20
40
60
80
100
At. % B

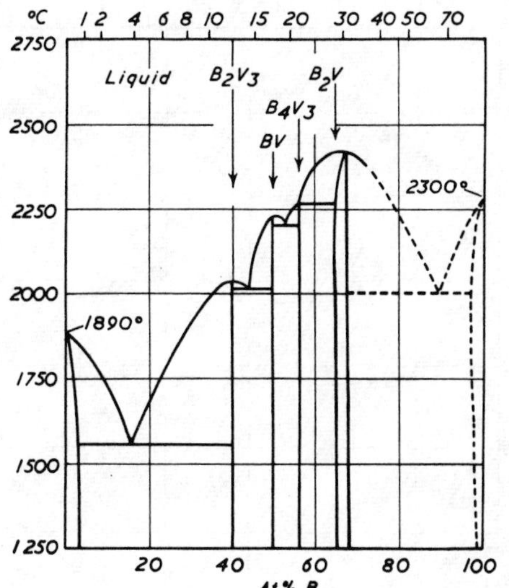
Wt.% B
°C
1 2 4 6 8 10 15 20 30 40 50 70
2750
2500
2250
2000
1750
1500
1250
Liquid
B_2V_3
B_2V
B_4V_3
BV
2300°
1890°
20 40 60 80 100
At.% B

B–V

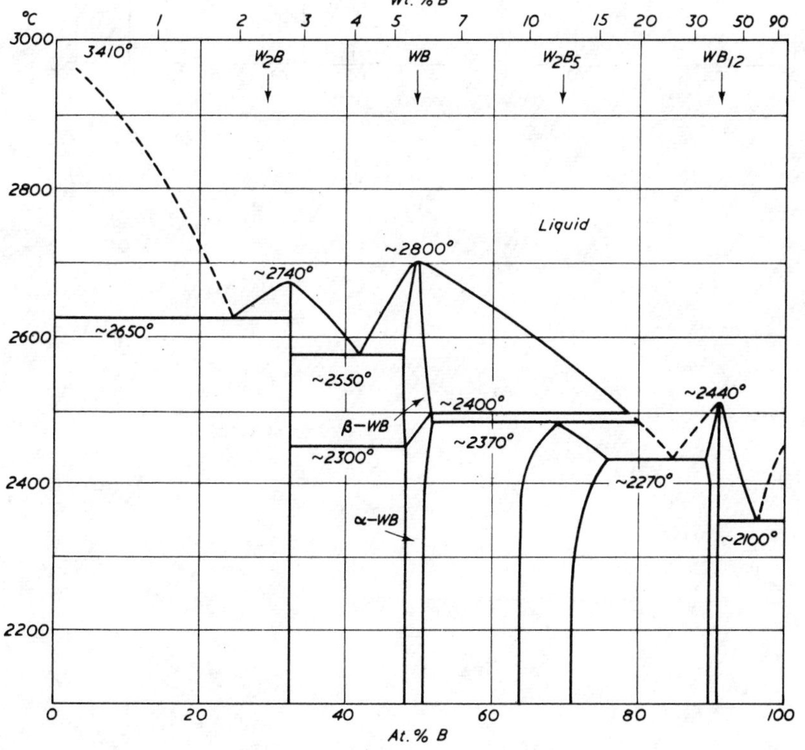
Wt. % B
°C
1 2 3 4 5 7 10 15 20 30 50 90
3000
2800
2600
2400
2200
3410°
W_2B
WB
W_2B_5
WB_{12}
Liquid
~2800°
~2740°
~2650°
~2550°
~2400°
β–WB
~2370°
~2300°
~2440°
~2270°
α–WB
~2100°
0 20 40 60 80 100
At.% B

B–W

B–Y

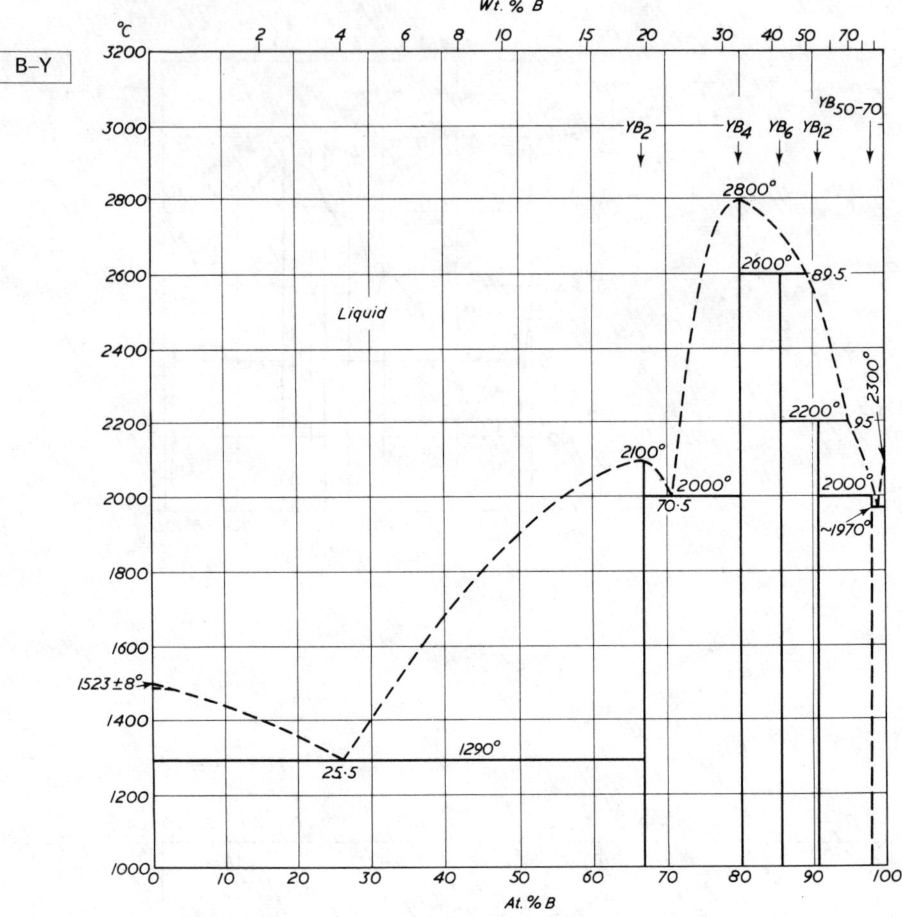
Wt. % B
°C
2
4
6
8
10
15
20
30
40
50
70
3200
3000
2800
2600
2400
2200
2000
1800
1600
1400
1200
1000
YB2
YB4
YB6
YB12
YB50-70
2800°
2600°
89·5
Liquid
2300°
2200°
95
2100°
2000°
70·5
2000°
~1970°
1523±8°
1290°
25·5
0
10
20
30
40
50
60
70
80
90
100
At. % B

B–Zr

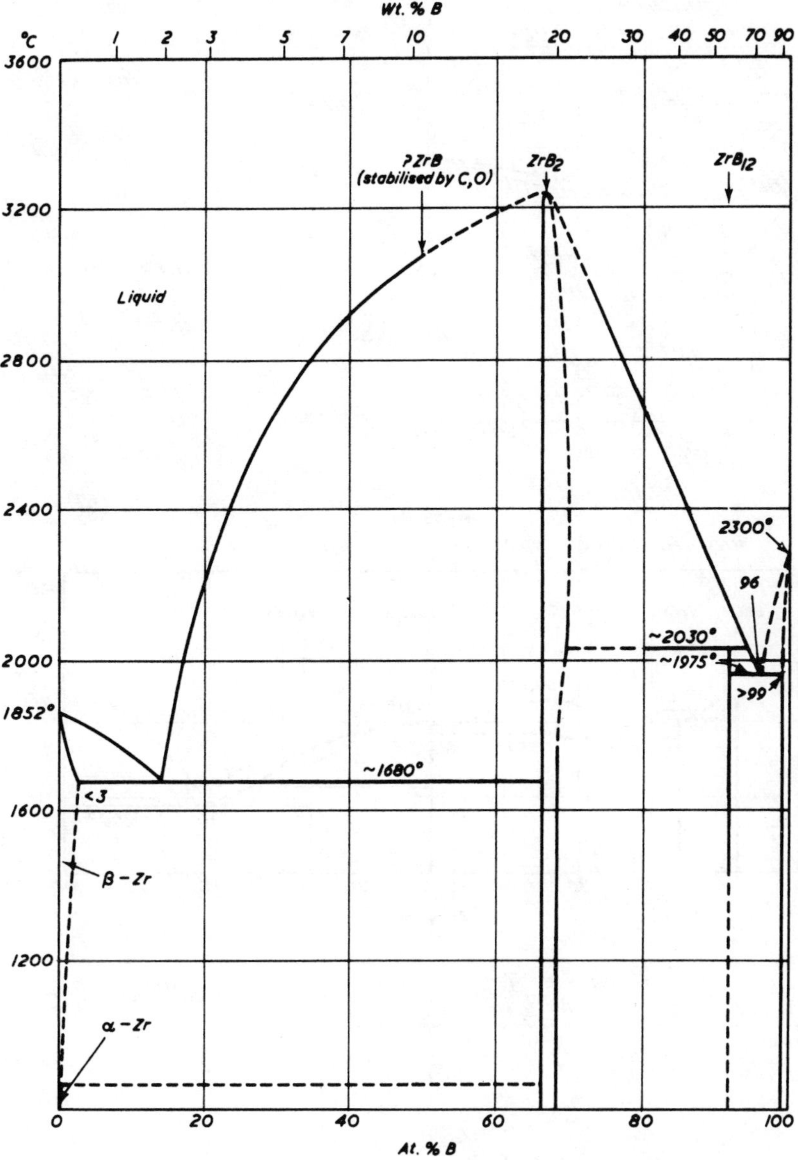

Wt. % B
1
2
3
5
7
10
20
30
40
50
70
90
°C
3600
3200
2800
2400
2000
1852°
1600
1200
?ZrB
(stabilised by C,O)
ZrB_2
ZrB_{12}
Liquid
2300°
96
~2030°
~1975°
>99
~1680°
<3
β-Zr
α-Zr
0
20
40
60
80
100
At. % B

Ba–Ca

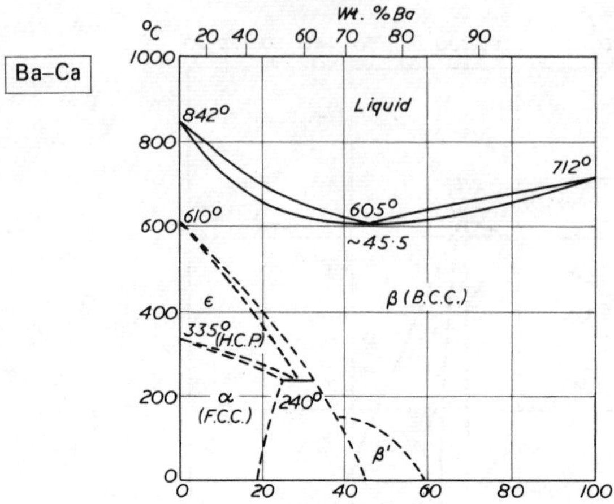

Ba–Cd

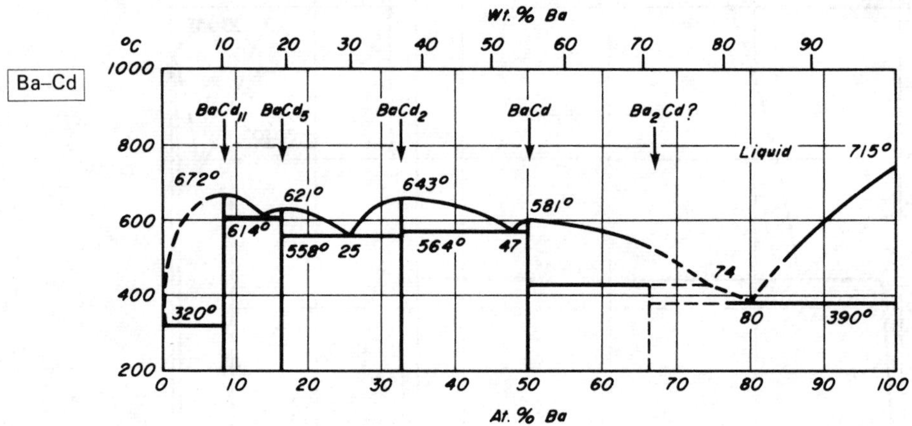

Ba–Cu

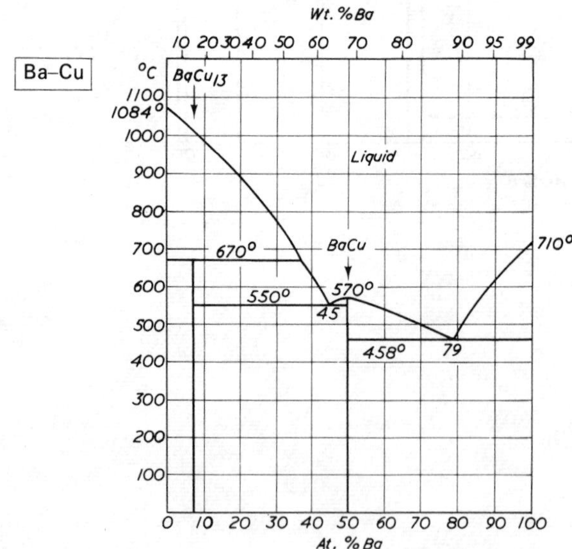

Ba–Ge

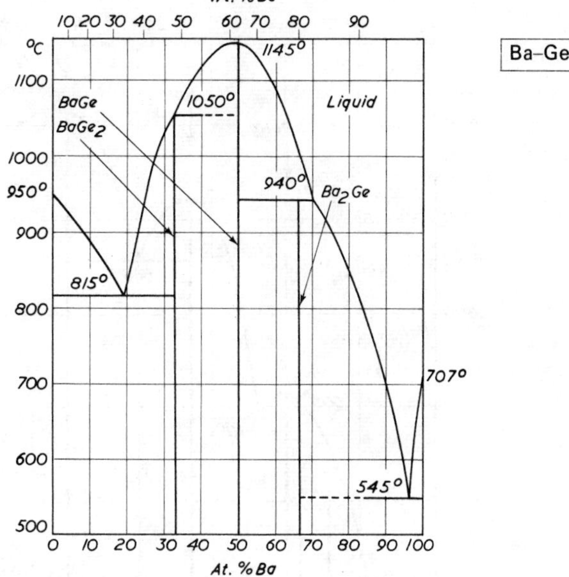
Wt. % Ba
10 20 30 40 50 60 70 80 90
°C
1100
1000
950°
900
800
700
600
500
1145°
1050°
BaGe
BaGe2
Liquid
940°
Ba2Ge
815°
707°
545°
0 10 20 30 40 50 60 70 80 90 100
At. % Ba

Ba–H

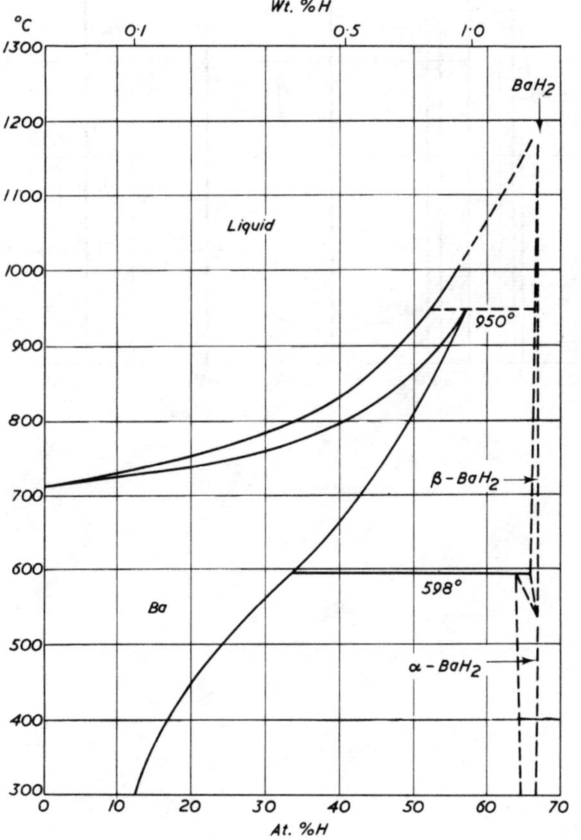
Wt. % H
0·1
0·5
1·0
°C
1300
1200
1100
1000
900
800
700
600
500
400
300
BaH2
Liquid
950°
β - BaH2
598°
Ba
α - BaH2
0 10 20 30 40 50 60 70
At. % H

Ba–Hg

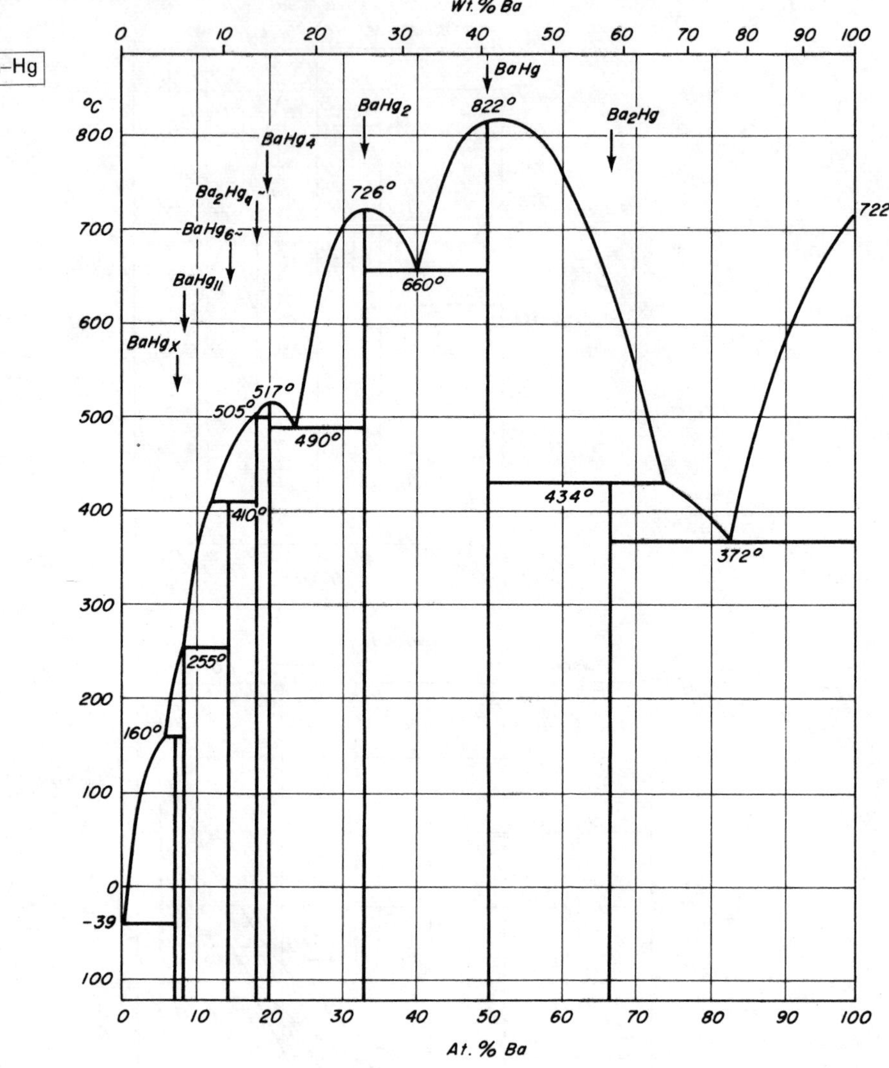

Wt.% Ba
°C
BaHg
822°
BaHg$_2$
Ba$_2$Hg
BaHg$_4$
726°
Ba$_2$Hg$_9$~
BaHg$_6$~
BaHg$_{11}$
660°
722
BaHg$_X$
517°
505°
490°
434°
410°
372°
255°
160°
-39
At.% Ba

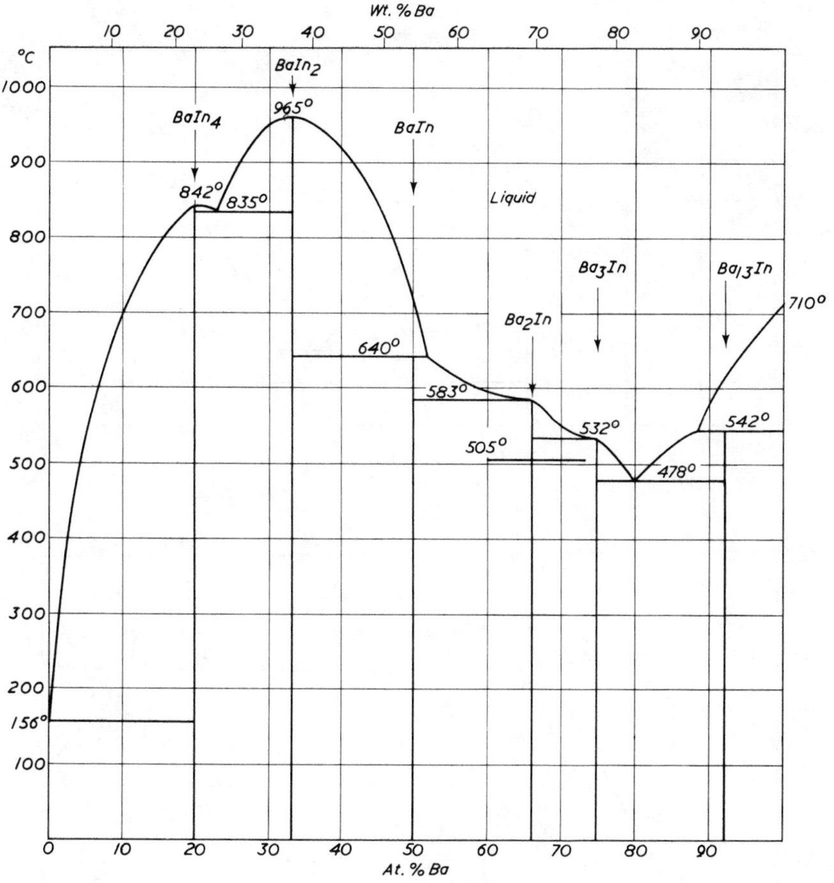

Ba–In

Wt. % Ba
10 20 30 40 50 60 70 80 90
°C
1000
900
800
700
600
500
400
300
200
156°
100
$BaIn_2$
965°
$BaIn_4$
842°
835°
BaIn
Liquid
Ba_3In
$Ba_{13}In$
710°
Ba_2In
640°
583°
532°
542°
505°
478°
0 10 20 30 40 50 60 70 80 90
At. % Ba

Ba–Li

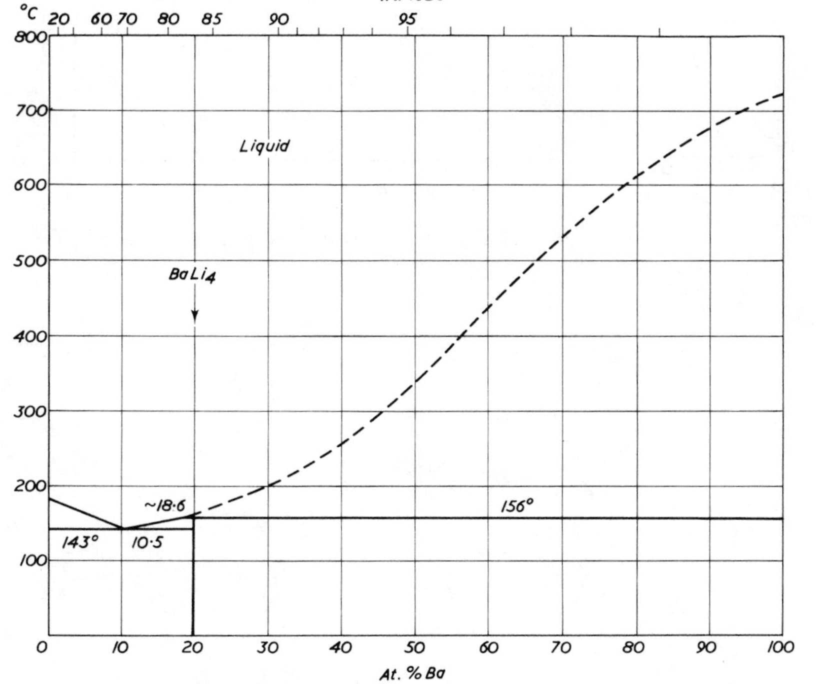

Ba–Mg

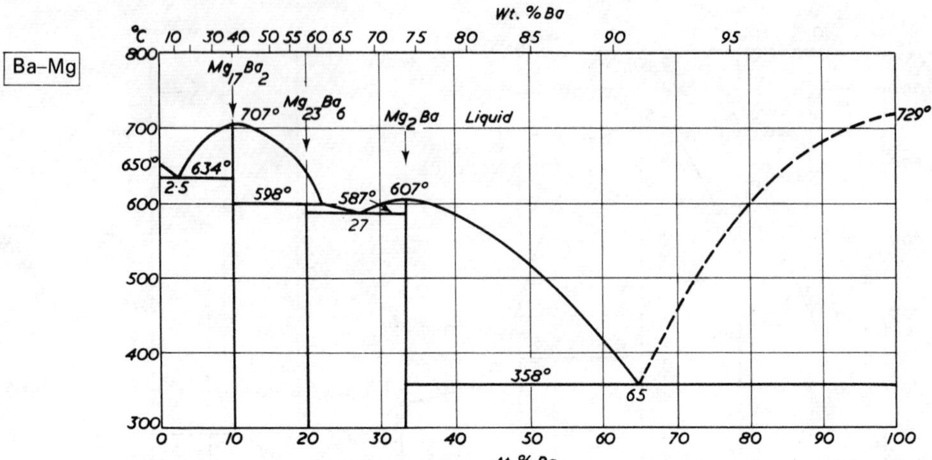
Wt. % Ba
°C
10
30
40
50
55
60
65
70
75
80
85
90
95
800
700
650°
600
500
400
300
Mg17Ba2
707°
Mg23Ba6
Mg2Ba
Liquid
729°
634°
2·5
598°
587°
607°
27
358°
65
0
10
20
30
40
50
60
70
80
90
100
At. % Ba

Ba–Mn

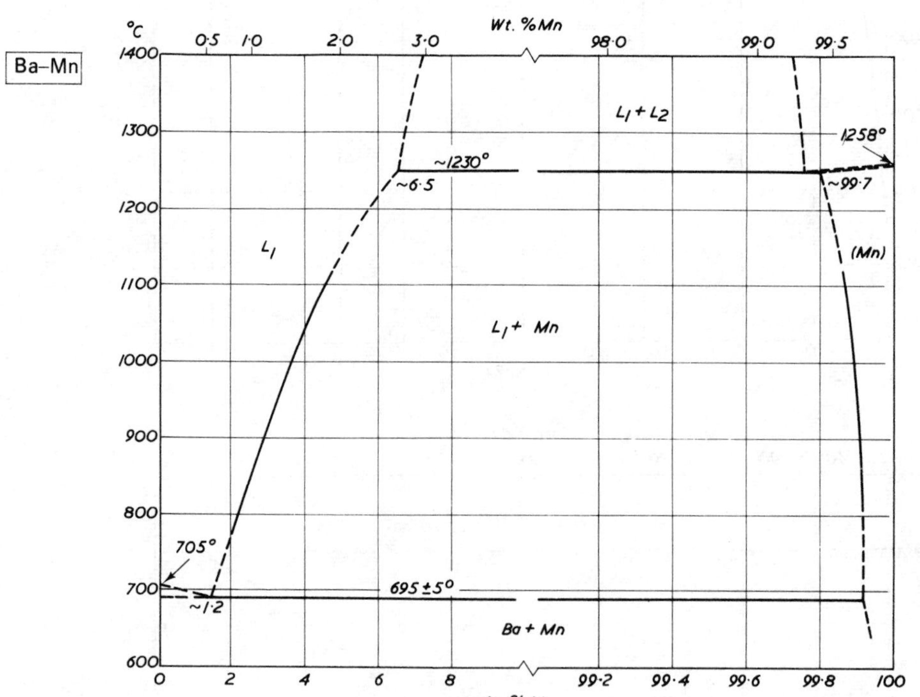
Wt. % Mn
°C
0·5
1·0
2·0
3·0
98·0
99·0
99·5
1400
1300
1200
1100
1000
900
800
700
600
L1 + L2
1258°
~1230°
~6·5
~99·7
L1
(Mn)
L1 + Mn
705°
~1·2
695 ± 5°
Ba + Mn
0
2
4
6
8
99·2
99·4
99·6
99·8
100
At. % Mn

Ba–Na

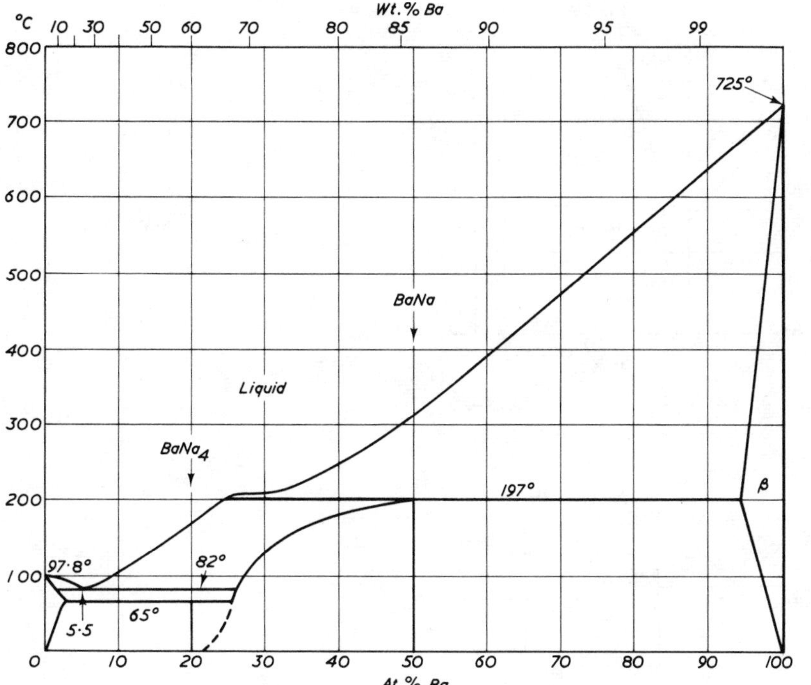
Wt.% Ba
°C
725°
BaNa
Liquid
BaNa4
197°
β
97·8°
82°
65°
5·5
At.% Ba

Ba–Pb

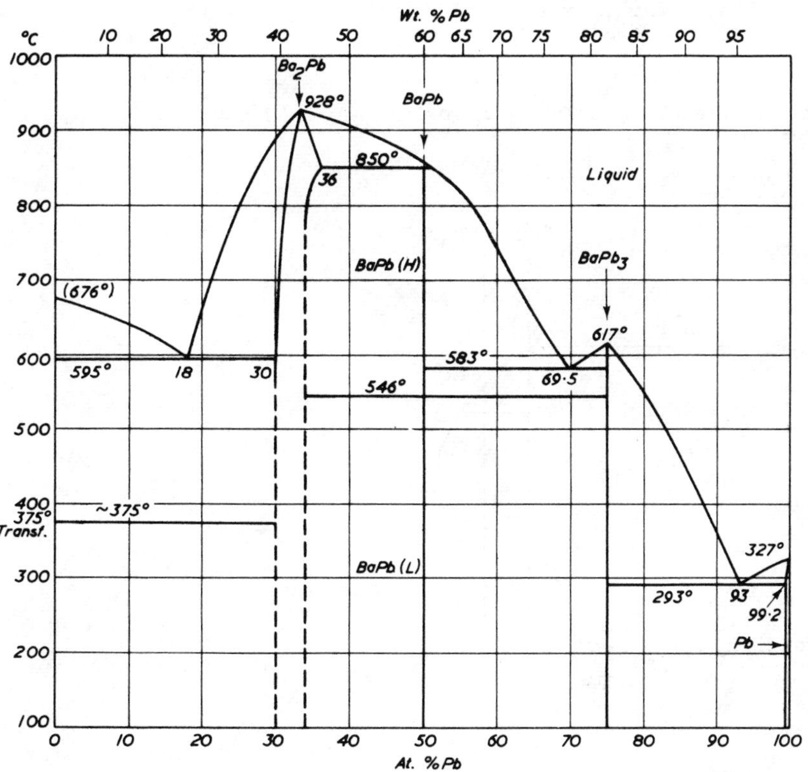
Wt. % Pb
°C
Ba2Pb
928°
BaPb
850°
36
Liquid
BaPb (H)
BaPb3
(676°)
617°
583°
595°
18
30
69·5
546°
~375°
375°
Transf.
BaPb (L)
327°
293°
93
99·2
Pb
At. % Pb

Ba–Pd

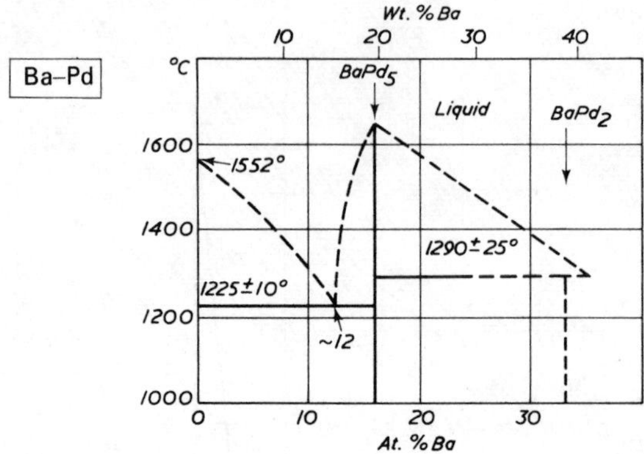

Ba–Pt

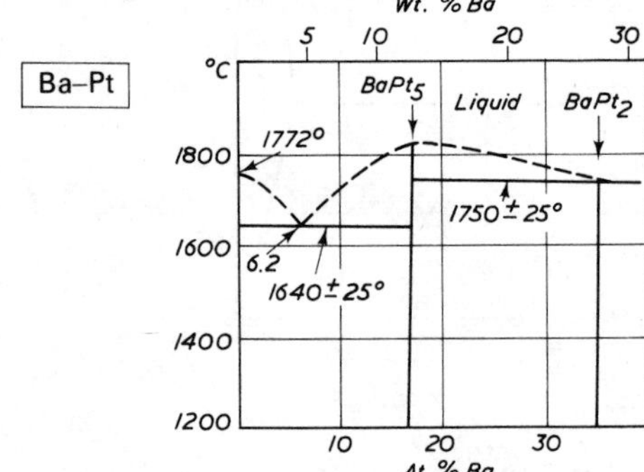

Ba–Si

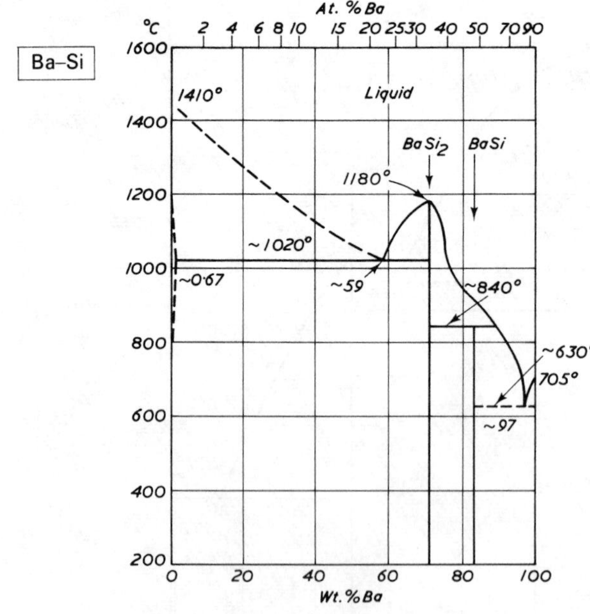

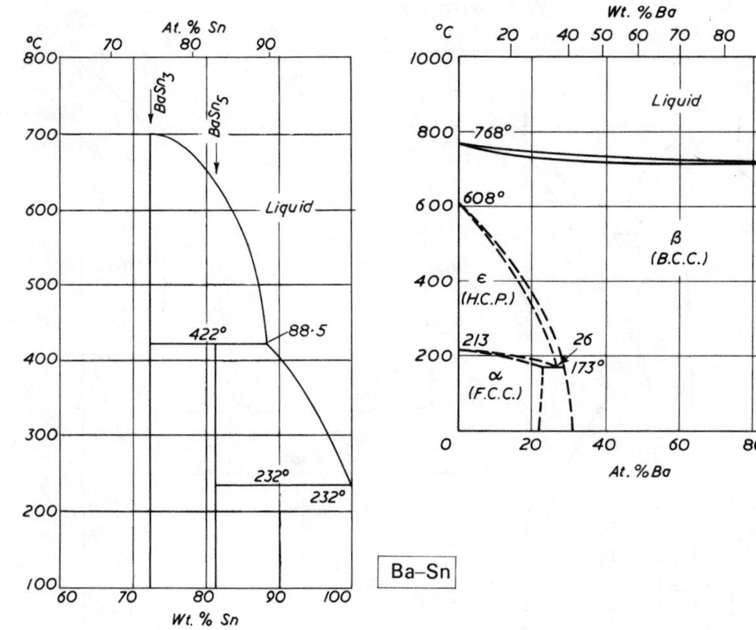

Ba–Sn

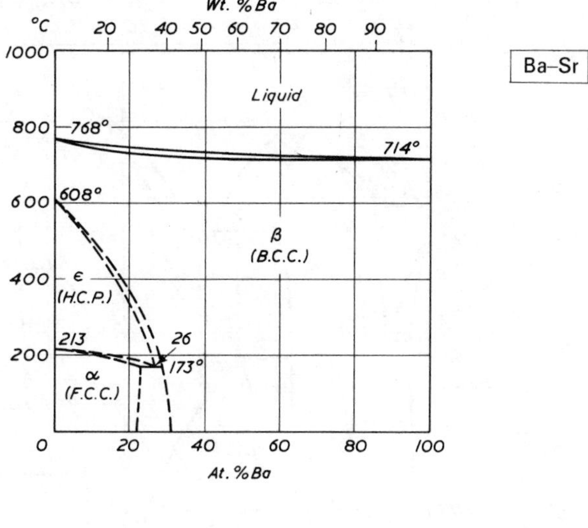

Ba–Sr

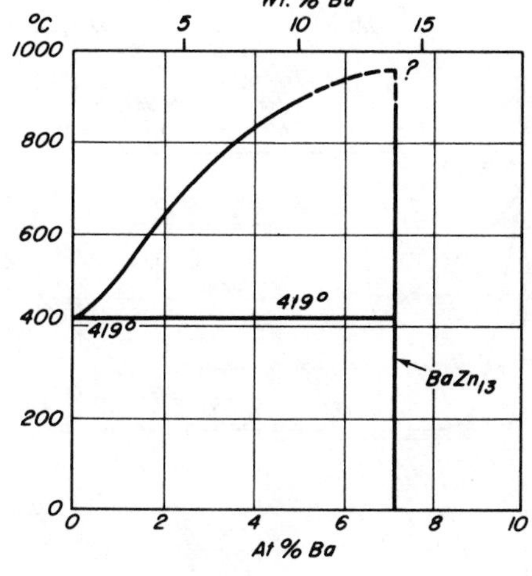

Ba–Zn

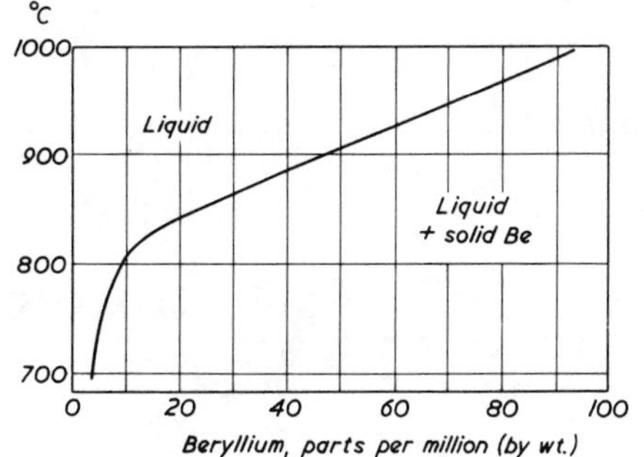

Be–Bi

Be–Co

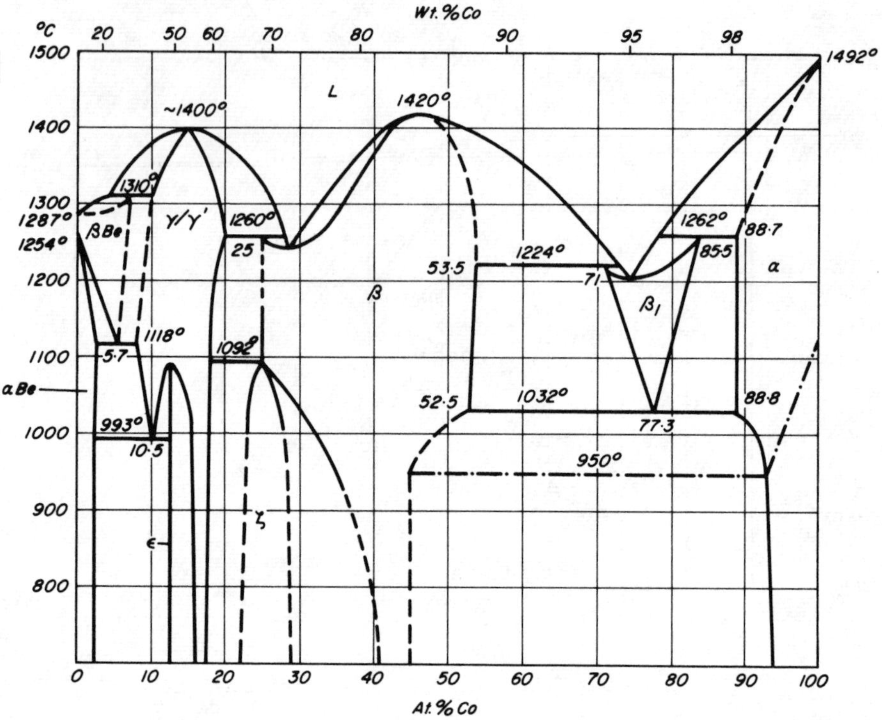

Wt.% Co
20 50 60 70 80 90 95 98
°C
1500
1400
1300
1287°
1254°
1200
1100
αBe
1000
900
800
1492°
L
1420°
~1400°
1310°
βBe
γ/γ'
1260°
25
1262°
88·7
85·5
53·5
1224°
71
α
β
β1
1118°
5·7
1092°
993°
10·5
52·5
1032°
88·8
77·3
950°
ζ
ε
0 10 20 30 40 50 60 70 80 90 100
At.% Co

Be–Cr

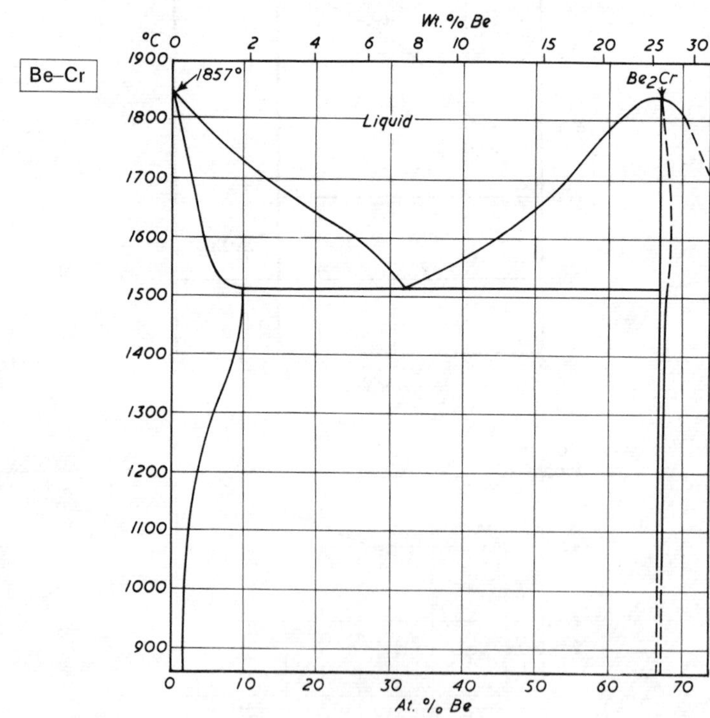

Wt.% Be
0 2 4 6 8 10 15 20 25 30
°C
1900
1800
1700
1600
1500
1400
1300
1200
1100
1000
900
1857°
Liquid
Be_2Cr
0 10 20 30 40 50 60 70
At. % Be

Be–Cu

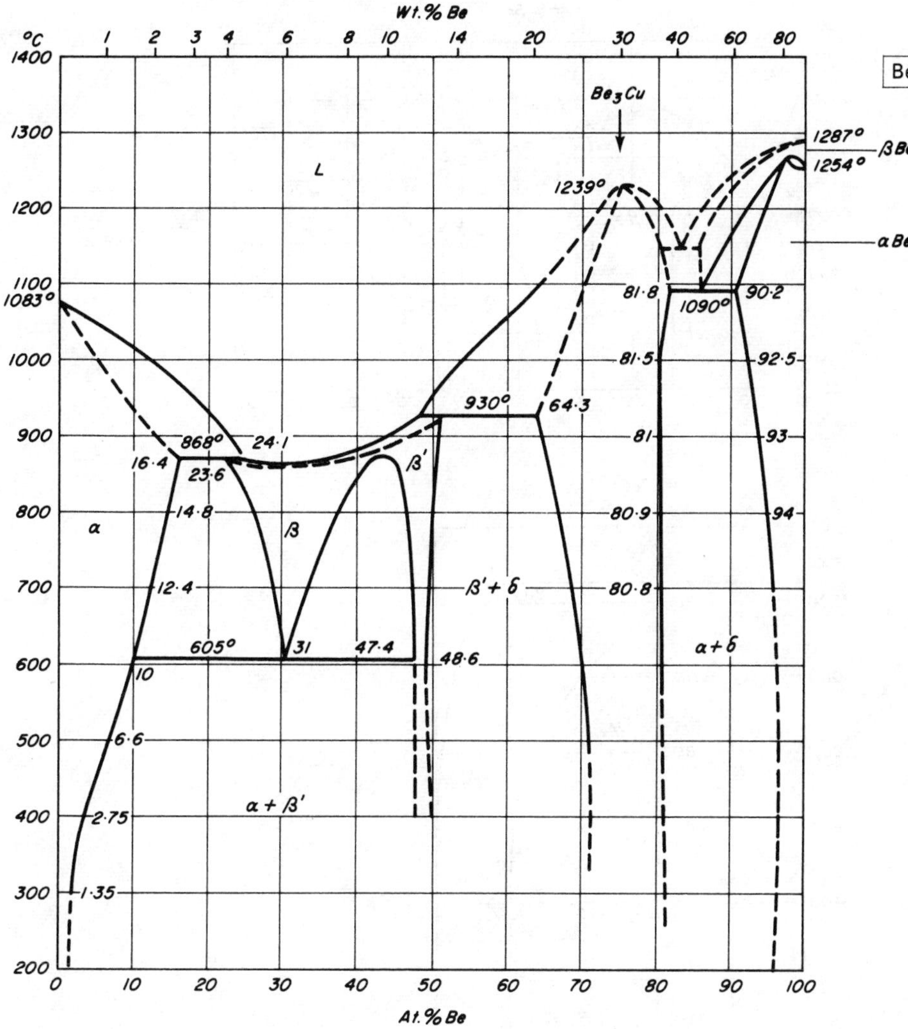

Wt.% Be
°C
1 2 3 4 6 8 10 14 20 30 40 60 80
1400
1300
1200
1100
1083°
1000
900
800
700
600
500
400
300
200
Be₃Cu
L
1287°
βBe
1254°
1239°
αBe
81·8
1090°
90·2
81·5
92·5
930°
64·3
868°
24·1
81
93
16·4
23·6
β′
14·8
80·9
94
α
β
12·4
β′+δ
80·8
605°
31
47·4
α+δ
10
48·6
6·6
2·75
α+β′
1·35
0 10 20 30 40 50 60 70 80 90 100
At.% Be

Be–Cu

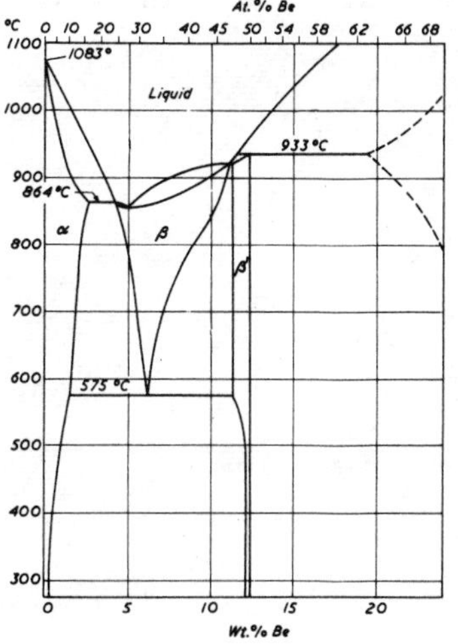

At.% Be
°C
0 10 20 30 40 45 50 54 58 62 66 68
1100
1083°
Liquid
1000
933°C
900
864°C
α
β
800
β′
700
600
575°C
500
400
300
0 5 10 15 20
Wt.% Be

Be–Fe

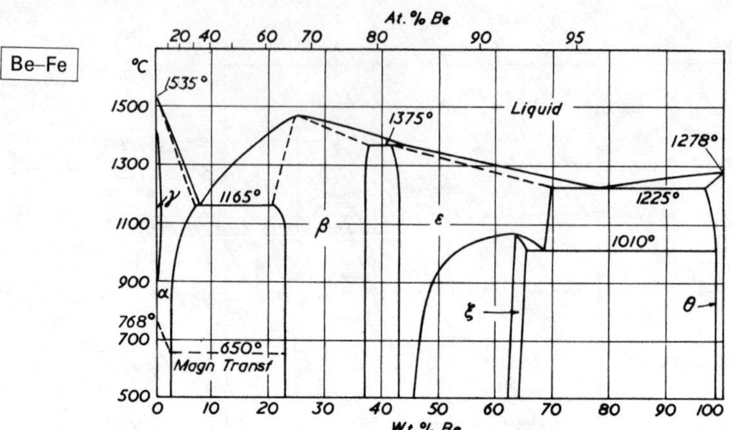
At.% Be
20 40 60 70 80 90 95
°C
1535°
1500
1300
1100
900
768°
700
500
Liquid
1375°
1278°
1165°
1225°
1010°
γ
α
β
ε
ξ
θ
650°
Magn Transf
0 10 20 30 40 50 60 70 80 90 100
Wt.% Be

Be–Ga

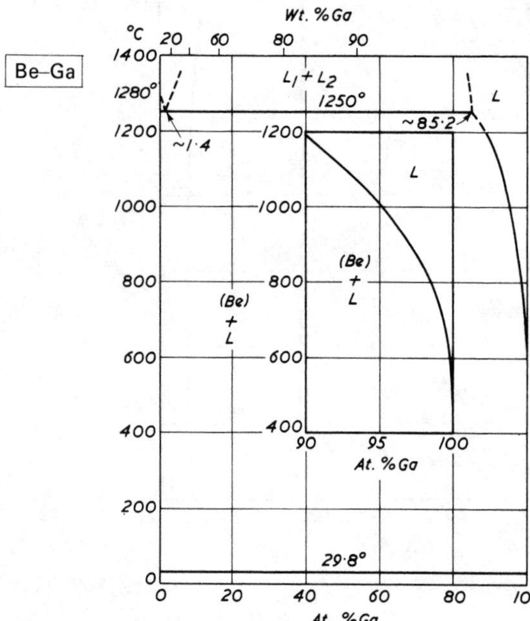
Wt. %Ga
°C 20 60 80 90
1400
1280°
1200
1000
800
600
400
200
0
$L_1 + L_2$
1250°
~1·4
~85·2
L
(Be) + L
29·8°
90 95 100
At. %Ga
0 20 40 60 80 100
At. %Ga

Be–Hf

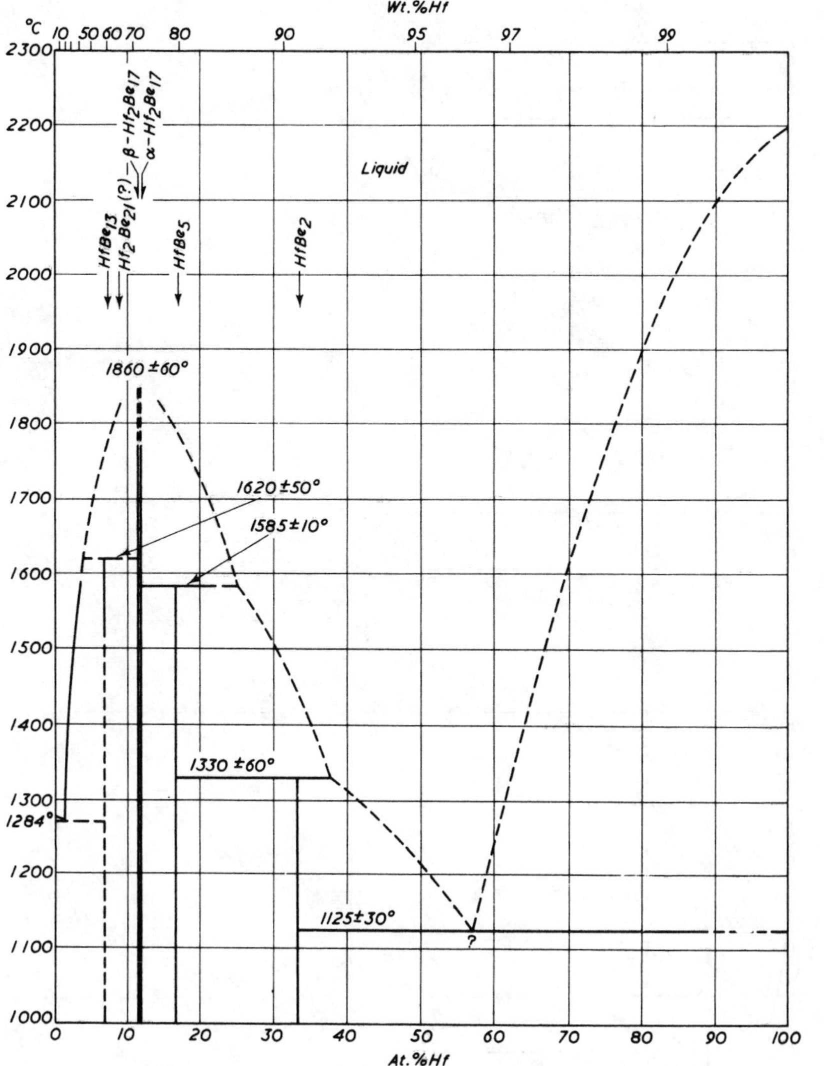

Wt.%Hf
°C
10 50 60 70 80 90 95 97 99
2300
2200
2100
2000
1900
1800
1700
1600
1500
1400
1300
1284°
1200
1100
1000
Liquid
HfBe13
Hf2Be21(?)
β-Hf2Be17
α-Hf2Be17
HfBe5
HfBe2
1860 ±60°
1620 ±50°
1585±10°
1330 ±60°
1125±30°
?
0 10 20 30 40 50 60 70 80 90 100
At.%Hf

Be–In

Wt. %In
°C 20 60 80 90
1400
1200
1000
800
600
400
200
0
$L_1 + L_2$
1280°
(Be)+L
1200
1000
800
600
400
L
(Be) + L
0·002
99·99 100
At. %In
156·4°
20 40 60 80 100
At. %In

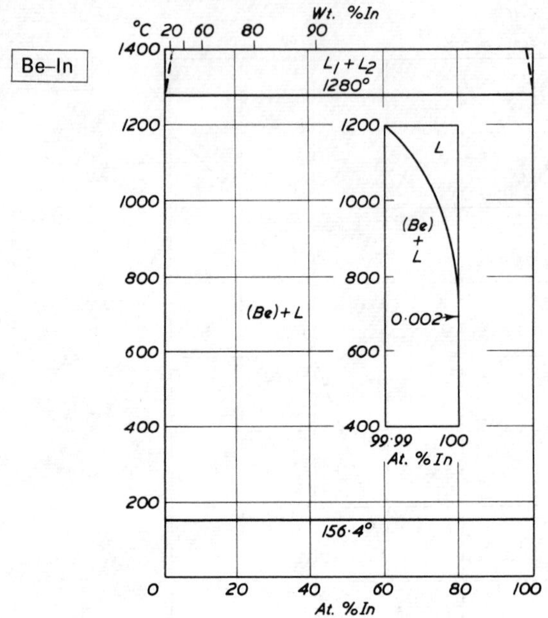

Be–Nb

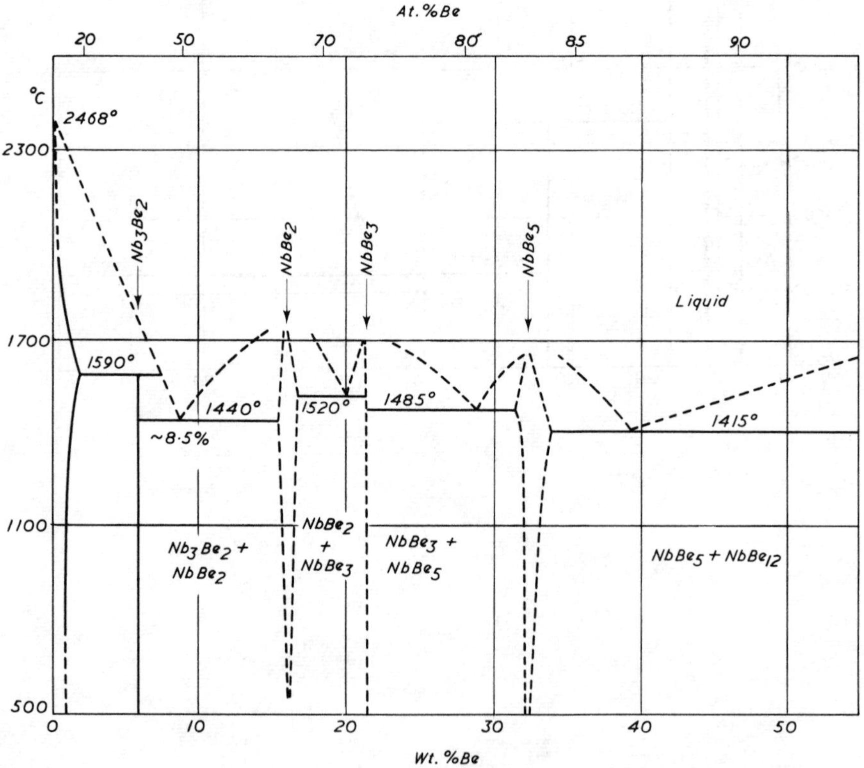

At.%Be
20 50 70 80 85 90
°C
2468°
2300
1700
1100
500
Nb_3Be_2
$NbBe_2$
$NbBe_3$
$NbBe_5$
Liquid
1590°
1440°
1520°
1485°
1415°
~8·5%
Nb_3Be_2 + $NbBe_2$
$NbBe_2$ + $NbBe_3$
$NbBe_3$ + $NbBe_5$
$NbBe_5$ + $NbBe_{12}$
0 10 20 30 40 50
Wt. %Be

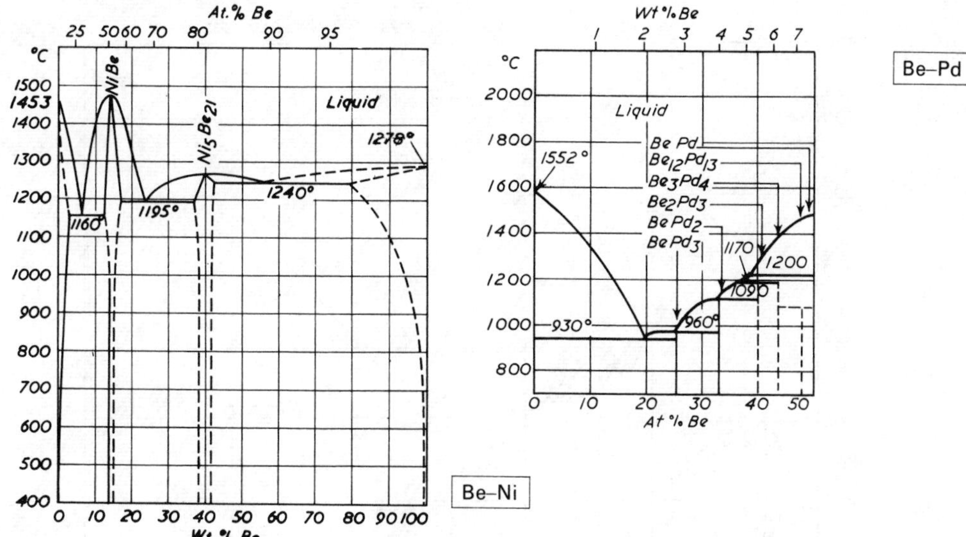
At.% Be
25 50 60 70 80 90 95
°C
1500
1453
1400
1300
1200
1100
1000
900
800
700
600
500
400
NiBe
Liquid
1278°
Ni_5Be_{21}
1160°
1195°
1240°
0 10 20 30 40 50 60 70 80 90 100
Wt.% Be
Be–Ni
Wt % Be
1 2 3 4 5 6 7
°C
2000
1800
1600
1400
1200
1000
800
Liquid
1552°
Be Pd
$Be_{12}Pd_{13}$
Be_3Pd_4
Be_2Pd_3
$BePd_2$
$BePd_3$
1170
1200
1090
930°
960°
0 10 20 30 40 50
At % Be
Be–Pd

Be–Pu

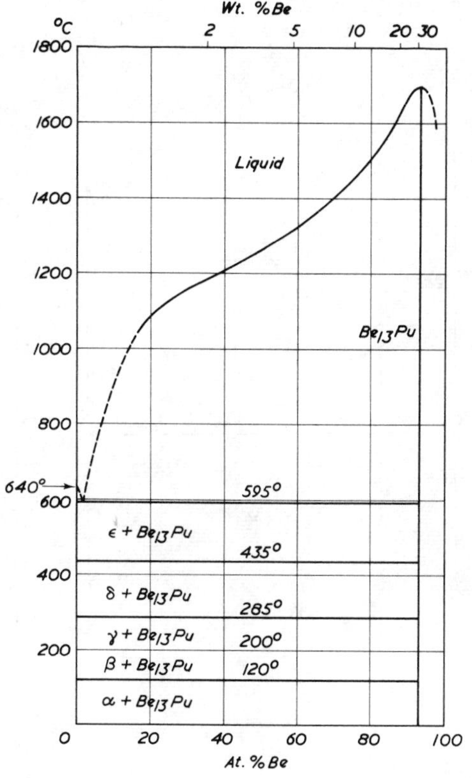
Wt. % Be
2 5 10 20 30
°C
1800
1600
1400
1200
1000
800
600
400
200
0
Liquid
$Be_{13}Pu$
640°
595°
$\epsilon + Be_{13}Pu$
435°
$\delta + Be_{13}Pu$
285°
$\gamma + Be_{13}Pu$
200°
$\beta + Be_{13}Pu$
120°
$\alpha + Be_{13}Pu$
20 40 60 80 100
At. % Be

Be–Ru

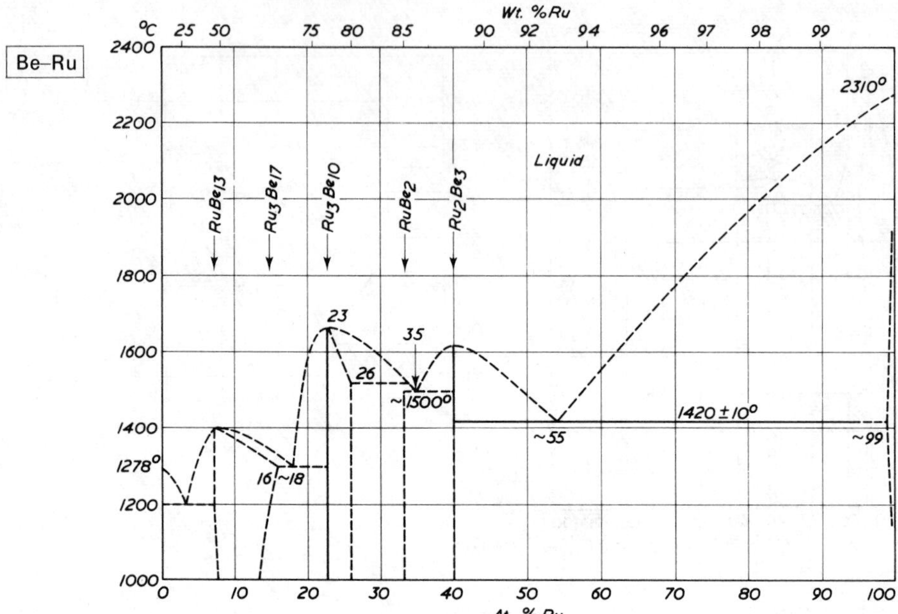

Be–Si

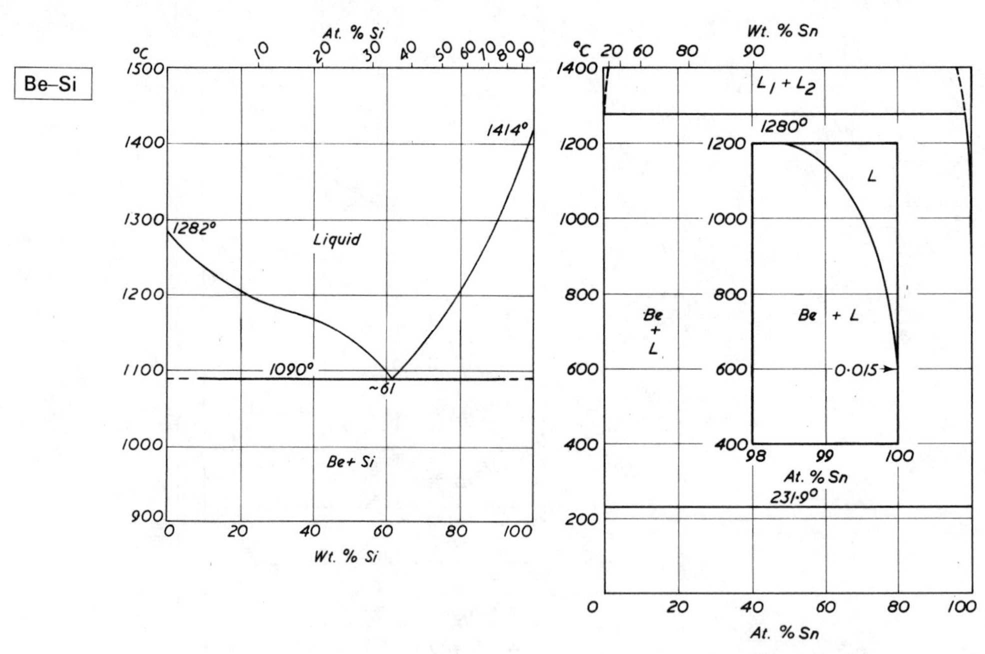

Be–Sn

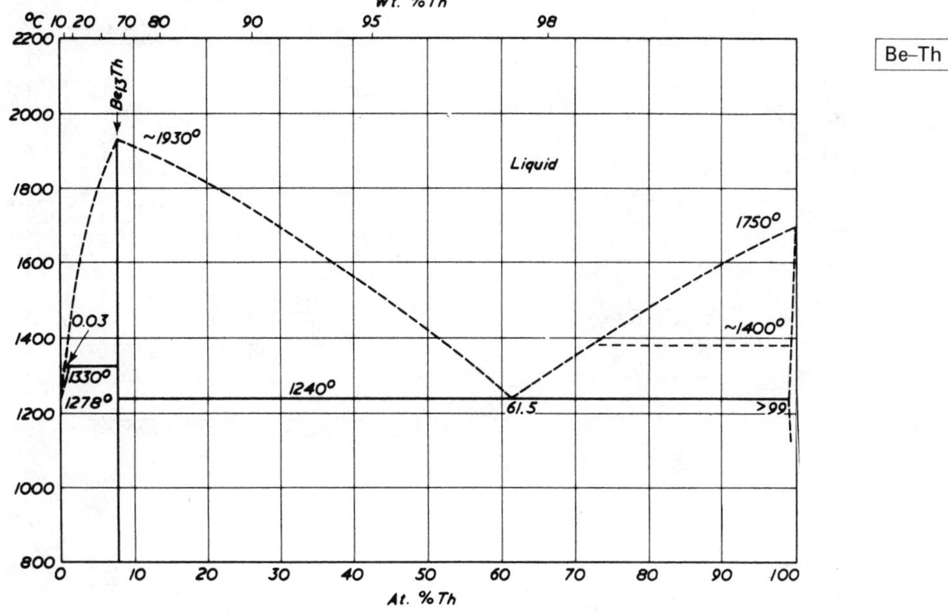

Wt. %Th
°C
10
20
70
80
90
95
98
2200
2000
1800
1600
1400
1200
1000
800
$Be_{13}Th$
~1930°
Liquid
1750°
0.03
~1400°
1330°
1278°
1240°
61.5
>99
0
10
20
30
40
50
60
70
80
90
100
At. %Th

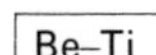

Be–Th

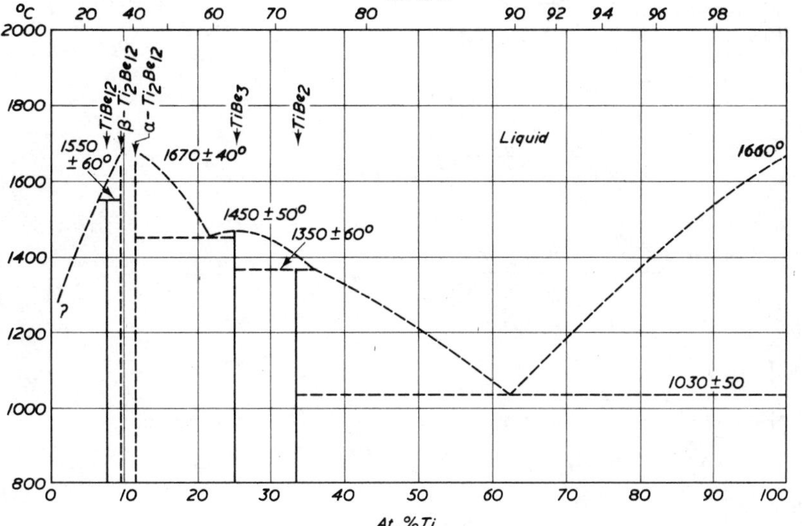

Be–Ti
Wt. %Ti
°C
20
40
60
70
80
90
92
94
96
98
2000
1800
1600
1400
1200
1000
800
$TiBe_{12}$
β- Ti_2Be_{12}
α- Ti_2Be_{12}
$TiBe_3$
$TiBe_2$
Liquid
1550
± 60°
1670 ± 40°
1660°
1450 ± 50°
1350 ± 60°
?
1030 ± 50
0
10
20
30
40
50
60
70
80
90
100
At. %Ti

Be–U

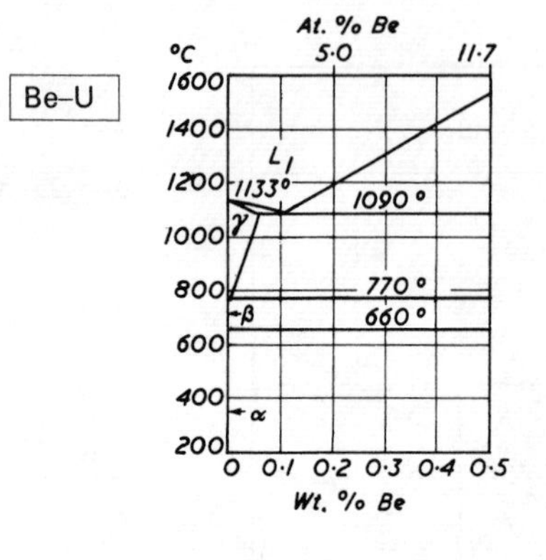

Be–U

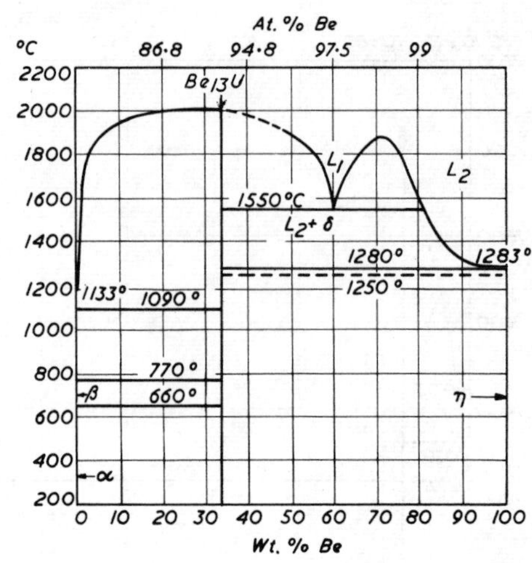

Be–W

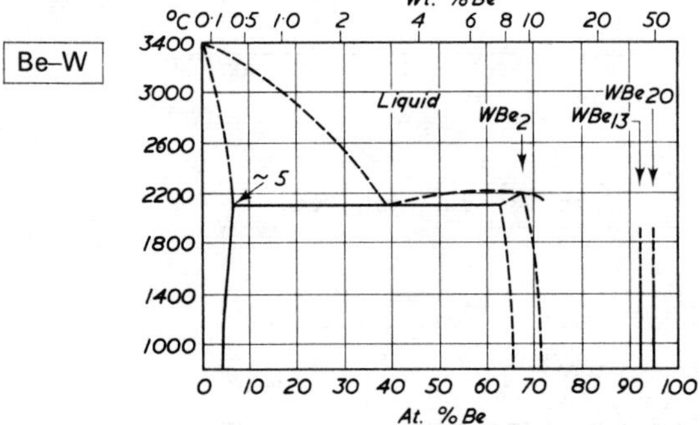

Be–Y

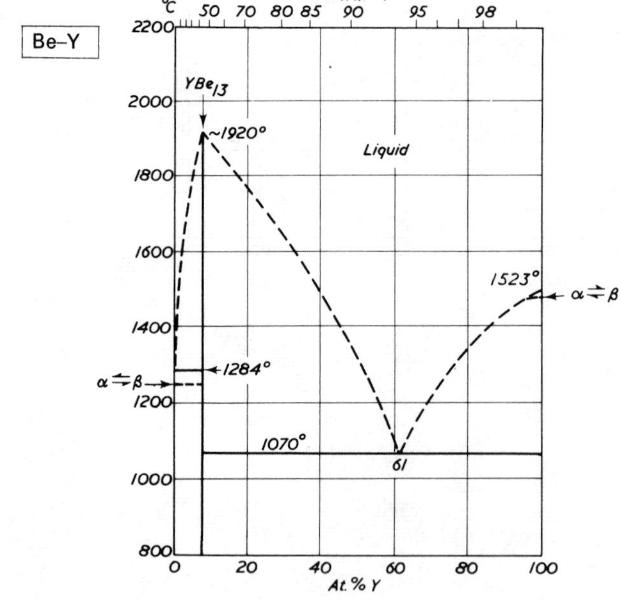

Be–Zr

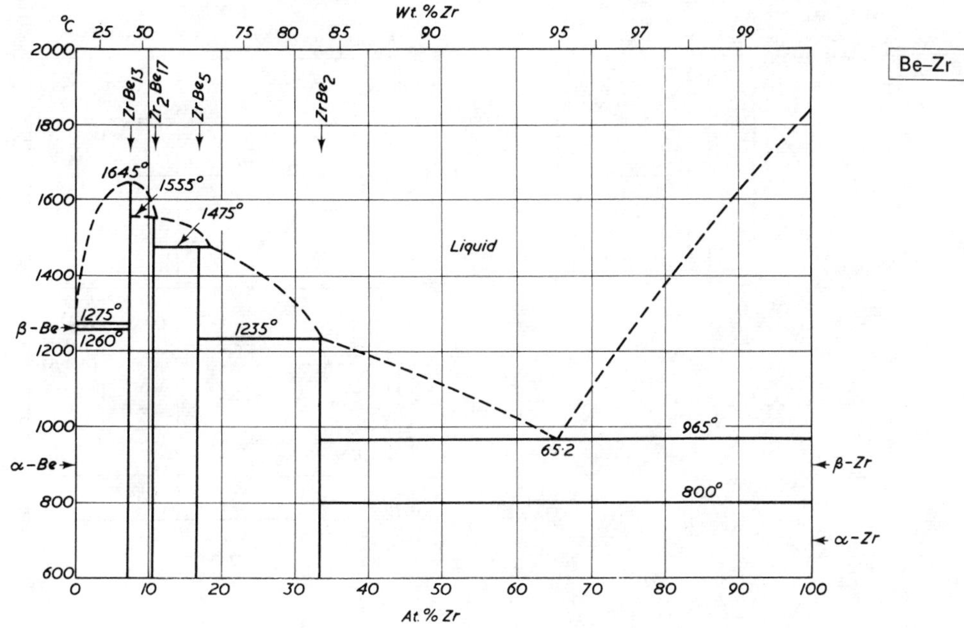

Bi–Ca

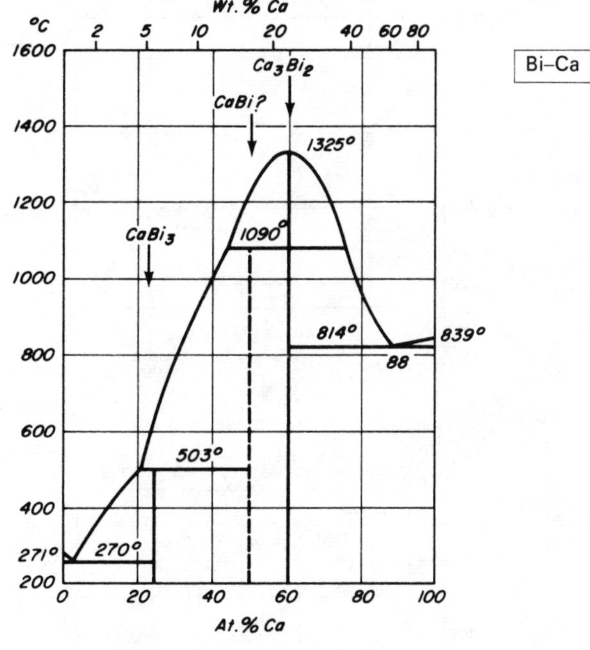

Bi–Cd

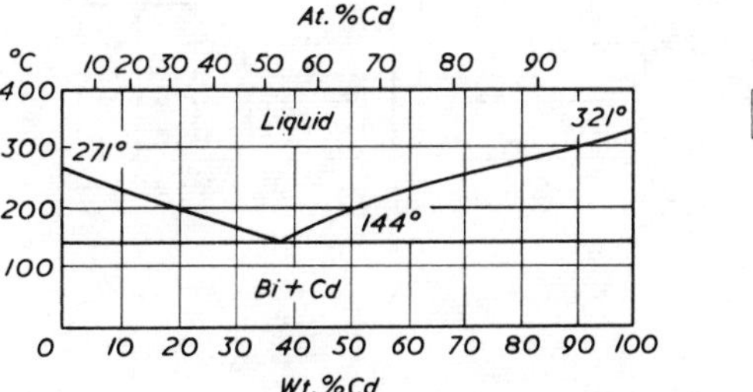

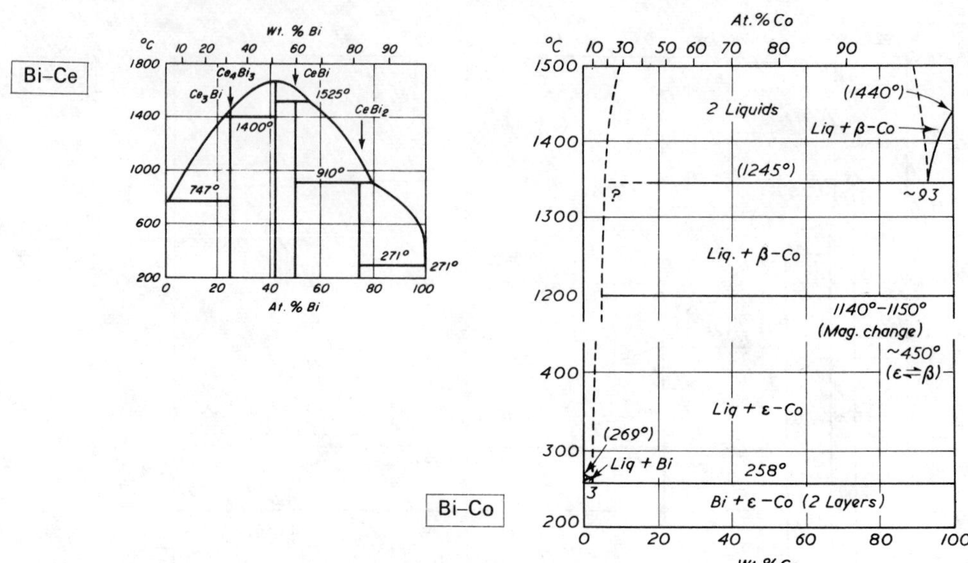
Bi–Ce
Wt. % Bi
°C
1800
1400
1000
600
200
Ce4Bi3
CeBi
Ce3Bi
1525°
CeBi2
1400°
910°
747°
271°
271°
At. % Bi
Bi–Co
At.% Co
°C
1500
1400
1300
1200
400
300
200
2 Liquids
(1440°)
Liq + β-Co
(1245°)
?
~93
Liq. + β-Co
1140°–1150°
(Mag. change)
~450°
(ε ⇌ β)
Liq + ε-Co
(269°)
Liq + Bi
258°
3
Bi + ε-Co (2 Layers)
Wt.%Co

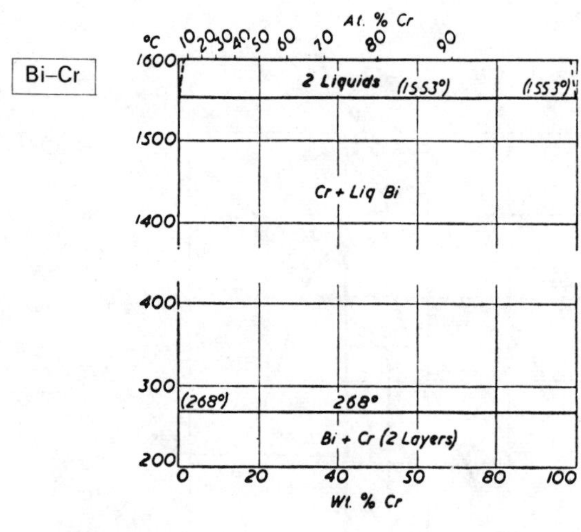
Bi–Cr
At. % Cr
°C
1600
1500
1400
400
300
200
2 Liquids
(1553°)
(1553°)
Cr + Liq Bi
(268°)
268°
Bi + Cr (2 Layers)
Wt. % Cr

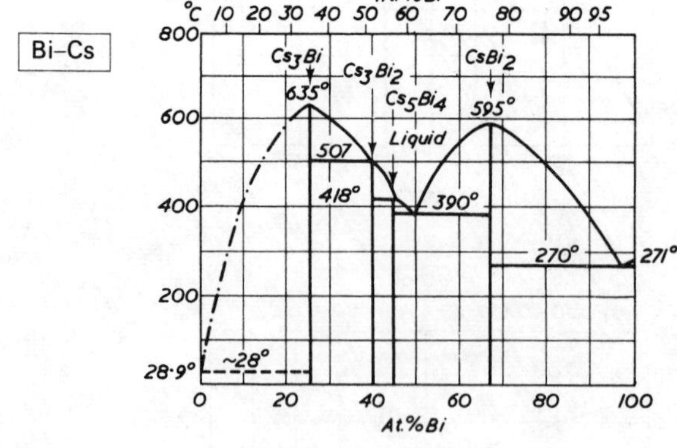
Bi–Cs
Wt. % Bi
°C
800
600
400
200
28·9°
Cs3Bi
Cs3Bi2
Cs5Bi4
CsBi2
635°
595°
Liquid
507
418°
390°
270°
271°
~28°
At.% Bi

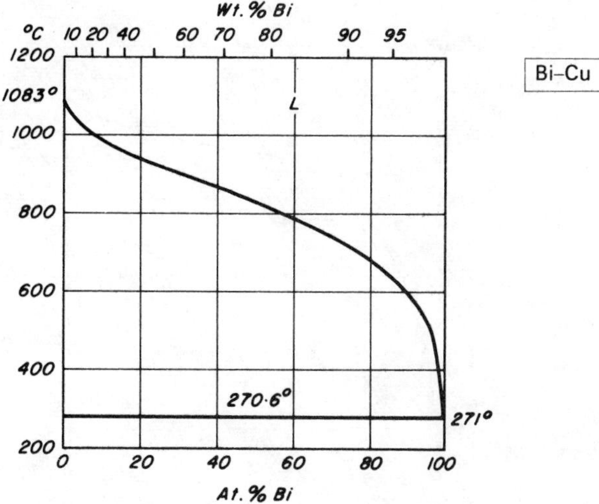
Bi–Cu
Wt. % Bi
10 20 40 60 70 80 90 95
°C
1200
1083°
1000
800
600
400
200
L
270·6°
271°
0 20 40 60 80 100
At. % Bi

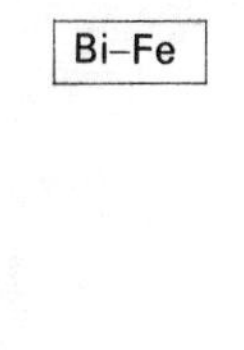
Bi–Fe

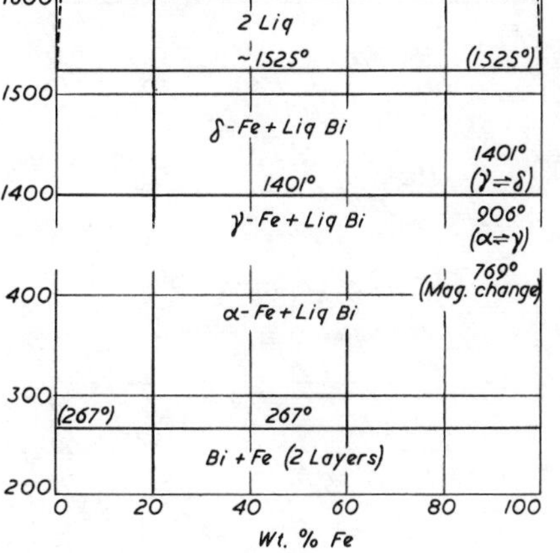
At. % Fe
10 30 50 60 70 80 90
°C
1600
1500
1400
400
300
200
2 Liq
~1525°
(1525°)
δ-Fe + Liq Bi
1401°
1401°
(γ⇌δ)
γ-Fe + Liq Bi
906°
(α⇌γ)
769°
(Mag. change)
α-Fe + Liq Bi
(267°)
267°
Bi + Fe (2 Layers)
0 20 40 60 80 100
Wt. % Fe

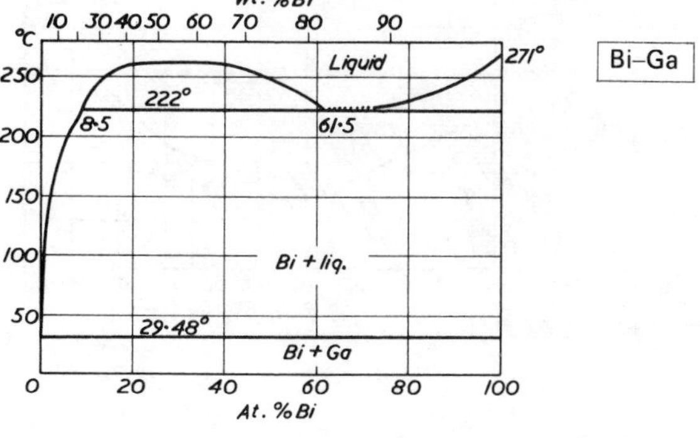
Bi–Ga
Wt. % Bi
10 30 40 50 60 70 80 90
°C
250
200
150
100
50
Liquid
271°
222°
8·5
61·5
Bi + liq.
29·48°
Bi + Ga
0 20 40 60 80 100
At. % Bi

Bi–Ge

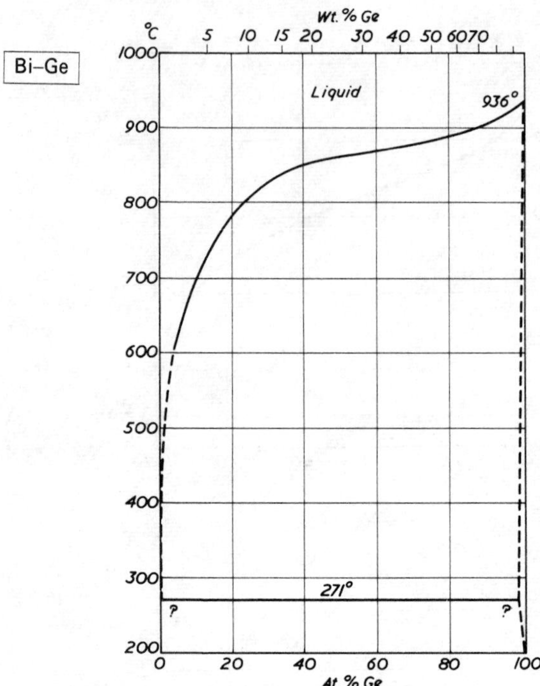

Bi–Hg

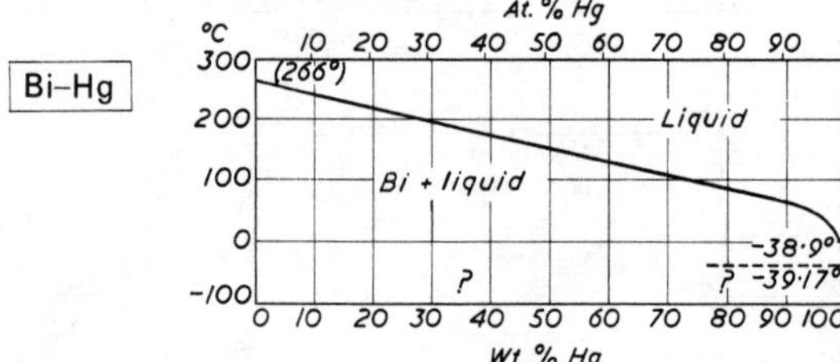

Bi–In

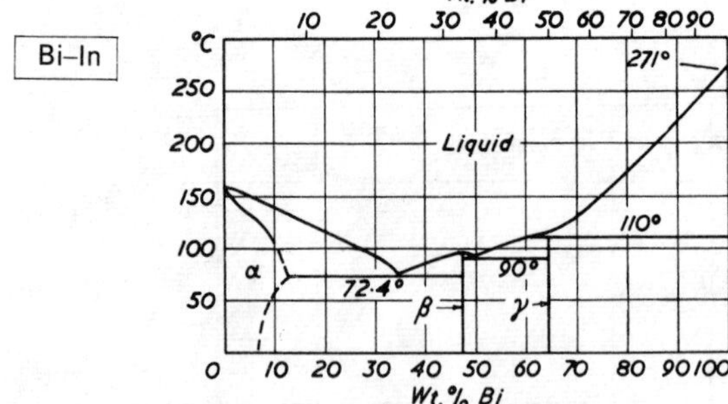

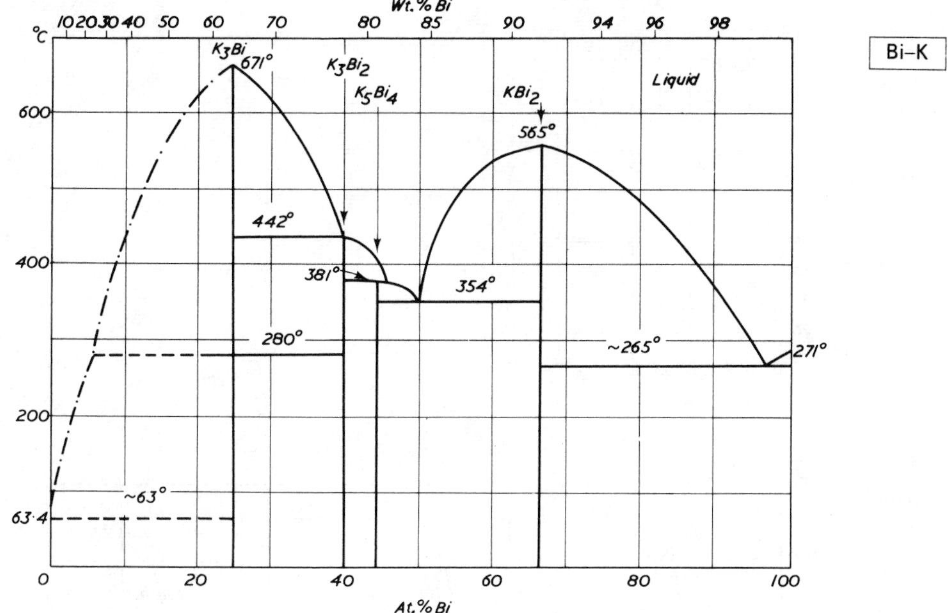

Bi–K

Wt.% Bi
10 20 30 40 50 60 70 80 90
°C
2000
1800
1600
1400
1200
1000
920°
868°
800
600
400
200
0 10 20 30 40 50 60 70 80 90 100
At. % Bi
La_4Bi_3
La_5Bi_3
LaBi
1670°
1615°
La_2Bi
1350°
1252°
$LaBi_2$
Liquid
932°
271°

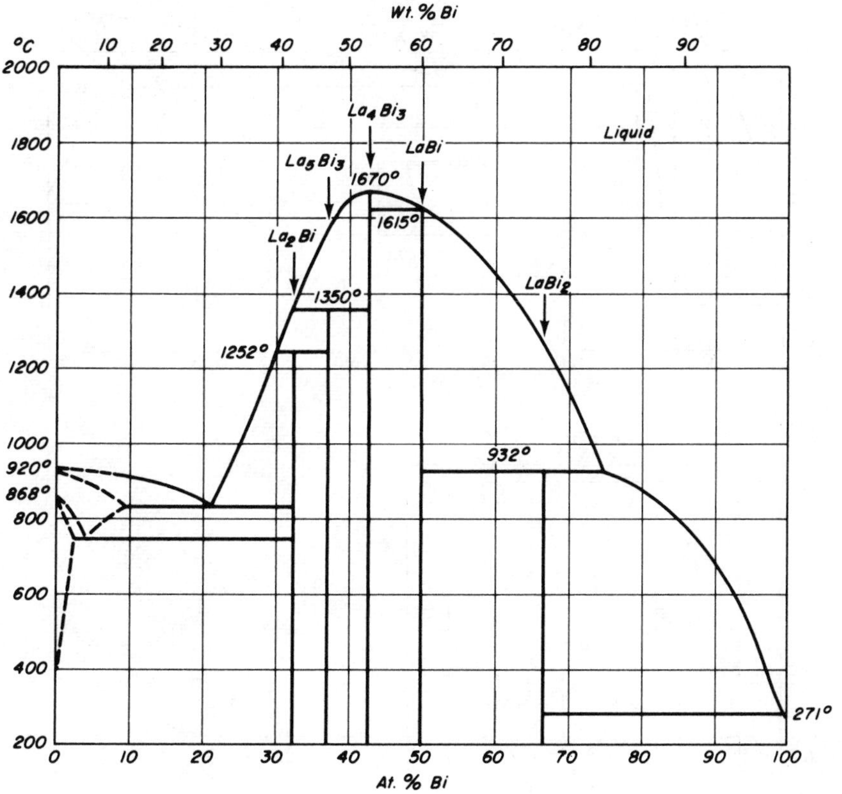

Bi–La

Bi–Li

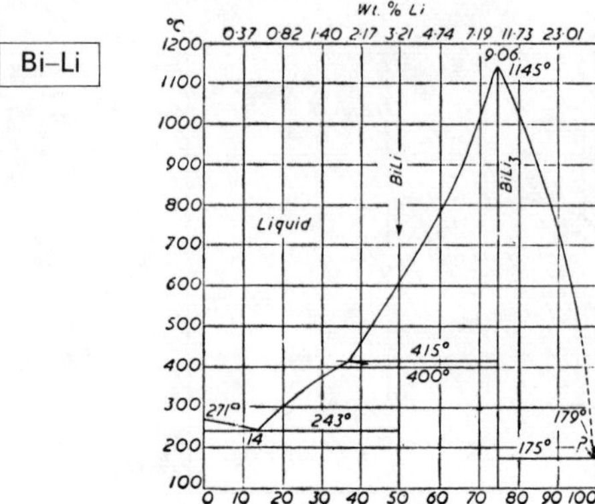

Bi–Mg

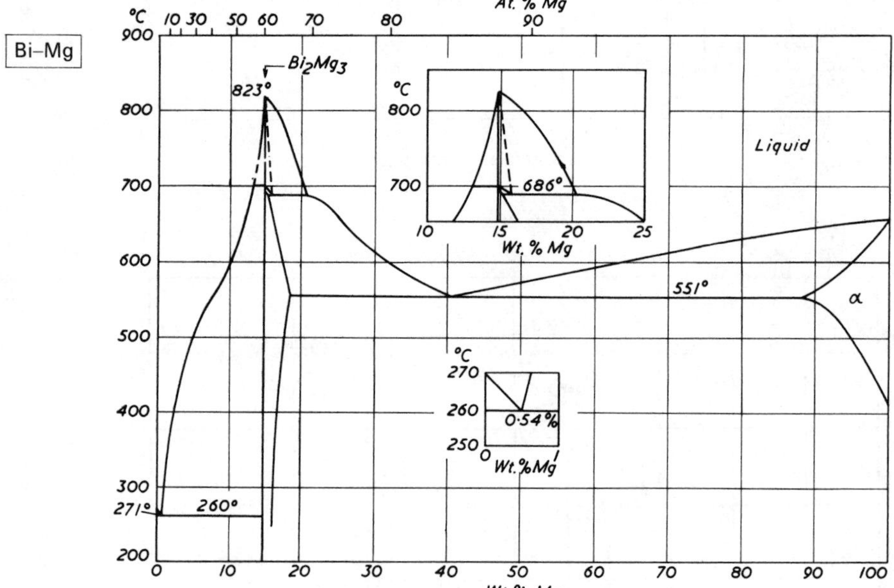

Bi–Mn

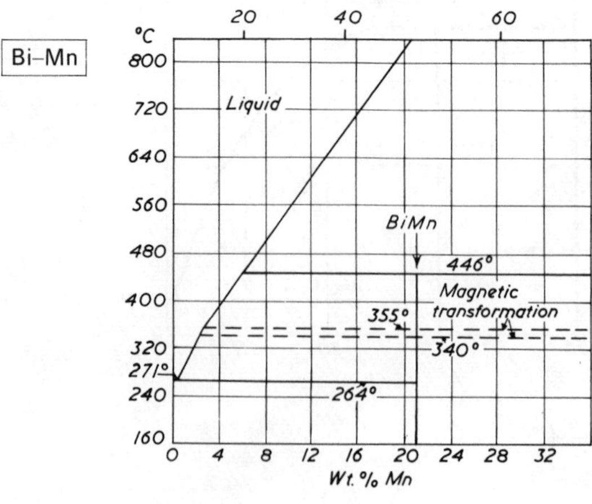

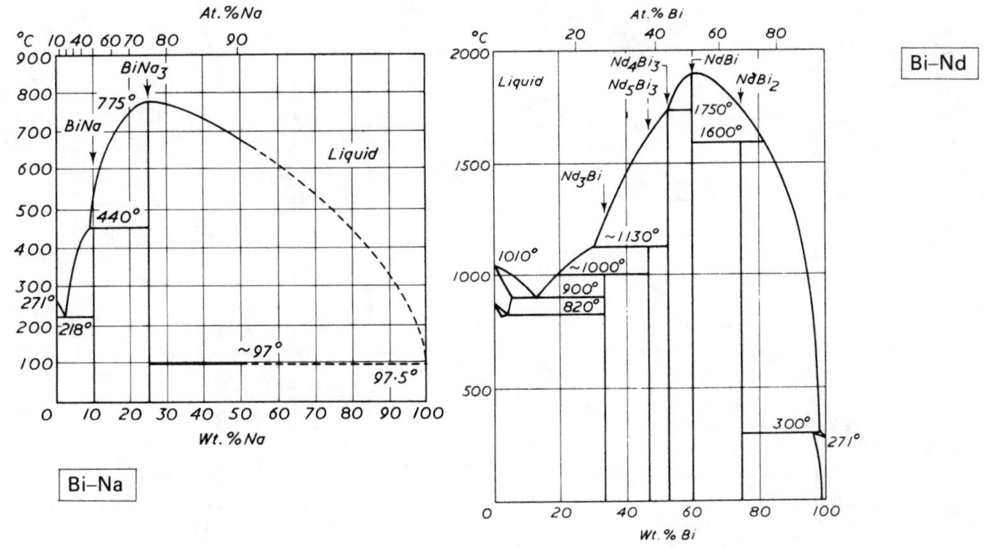
At.%Na
°C
10 40 60 70 80 90
BiNa3
775°
BiNa
Liquid
440°
271°
218°
~97°
97·5°
Wt.%Na
Bi–Na
At.% Bi
20 40 60 80
Liquid
Nd4Bi3
Nd5Bi3
NdBi
NdBi2
1750°
1600°
Nd3Bi
~1130°
1010°
~1000°
900°
820°
300°
271°
Wt.% Bi
Bi–Nd

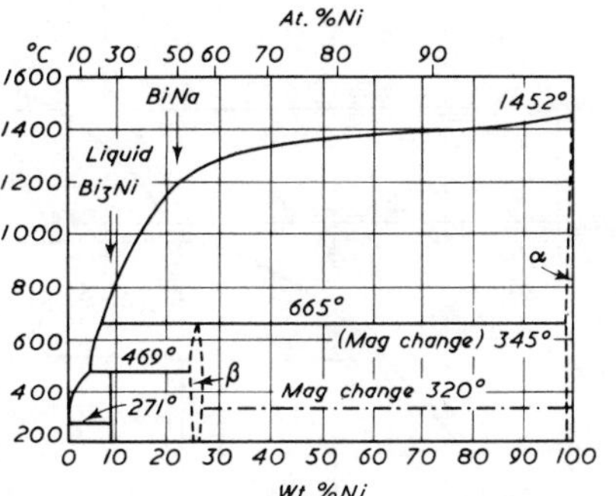
At.%Ni
°C
10 30 50 60 70 80 90
BiNa
1452°
Liquid
Bi3Ni
α
665°
469°
(Mag change) 345°
β
Mag change 320°
271°
Wt.%Ni

Bi–Ni

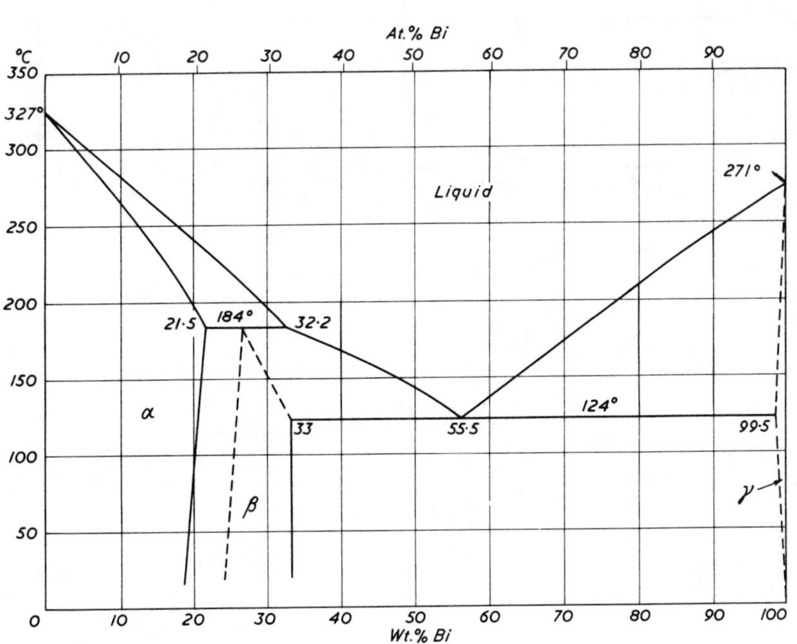
At.% Bi
°C
10 20 30 40 50 60 70 80 90
327°
Liquid
271°
21·5
184°
32·2
α
124°
33
55·5
99·5
β
γ
Wt.% Bi

Bi–Pb

Bi–Pd

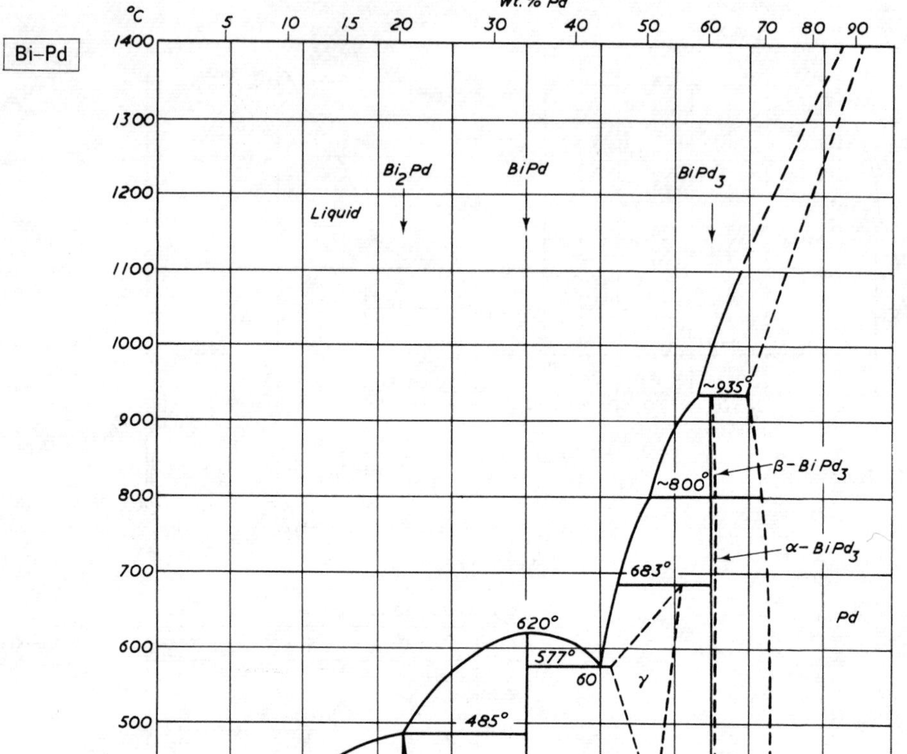
Wt. % Pd
°C
Bi2Pd
BiPd
BiPd3
Liquid
~935°
β-BiPd3
~800°
α-BiPd3
683°
Pd
620°
577°
60
γ
485°
β-Bi2Pd
380°
<400°
16·4
α-Bi2Pd
271°
256°
5·8
At. % Pd

Bi–Pt

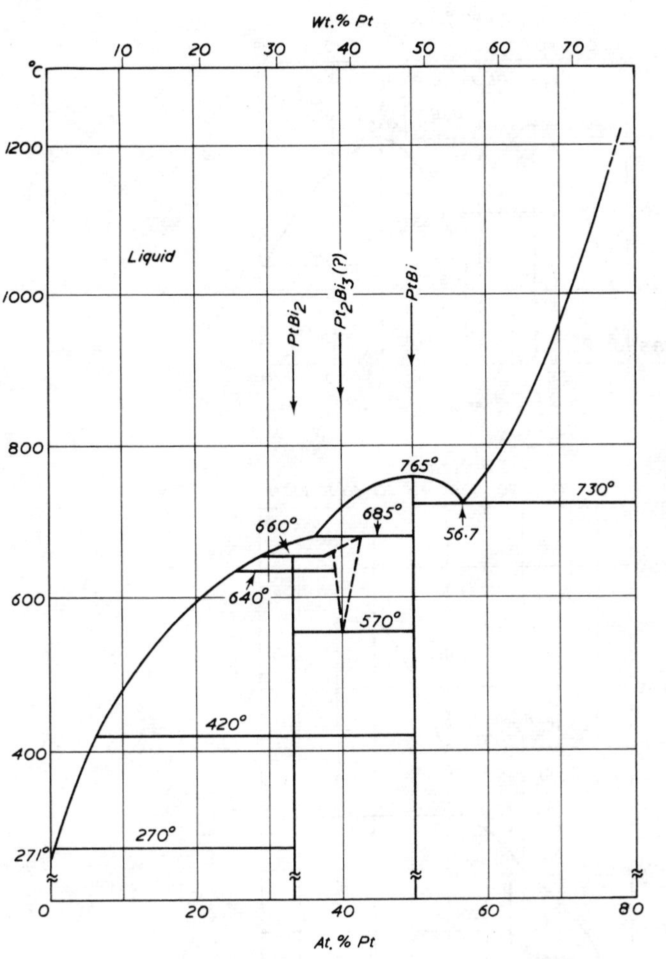
Wt.% Pt
10
20
30
40
50
60
70
°C
1200
1000
800
600
400
Liquid
PtBi2
Pt2Bi3 (?)
PtBi
765°
730°
685°
660°
56·7
640°
570°
420°
270°
271°
0
20
40
60
80
At.% Pt

Bi–Pu

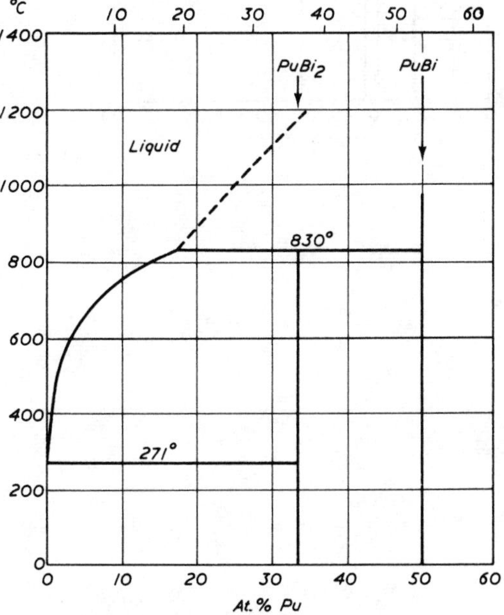
Wt. % Pu
10
20
30
40
50
60
°C
1400
1200
1000
800
600
400
200
0
PuBi2
PuBi
Liquid
830°
271°
0
10
20
30
40
50
60
At.% Pu

Bi–Rb

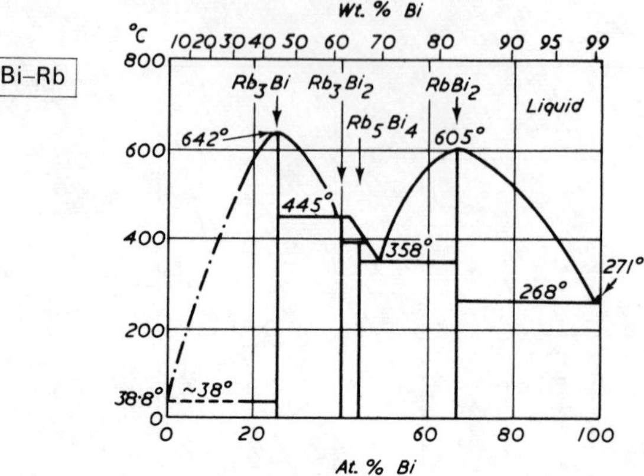
Wt. % Bi
°C
10 20 30 40 50 60 70 80 90 95 99
Rb3Bi
Rb3Bi2
RbBi2
Liquid
Rb5Bi4
642°
605°
445°
358°
271°
268°
~38°
38·8°
At. % Bi

Bi–Rh

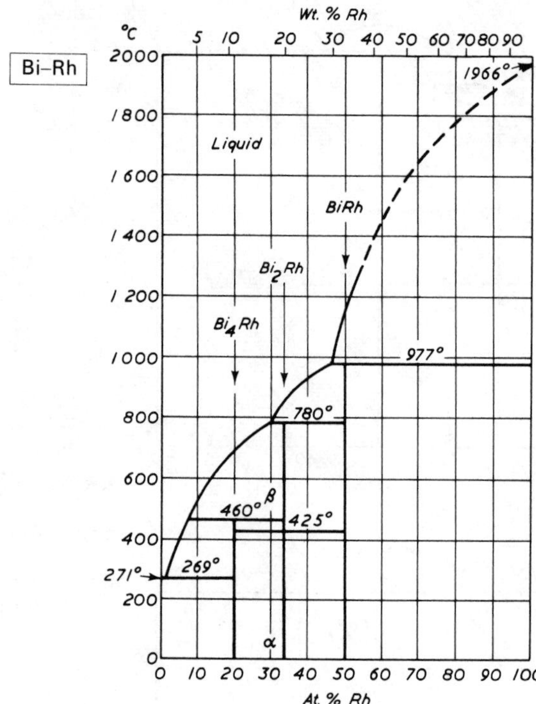
Wt. % Rh
°C
5 10 20 30 40 50 60 70 80 90
1966°
Liquid
BiRh
Bi2Rh
Bi4Rh
977°
780°
β
460°
425°
271°
269°
α
At. % Rh

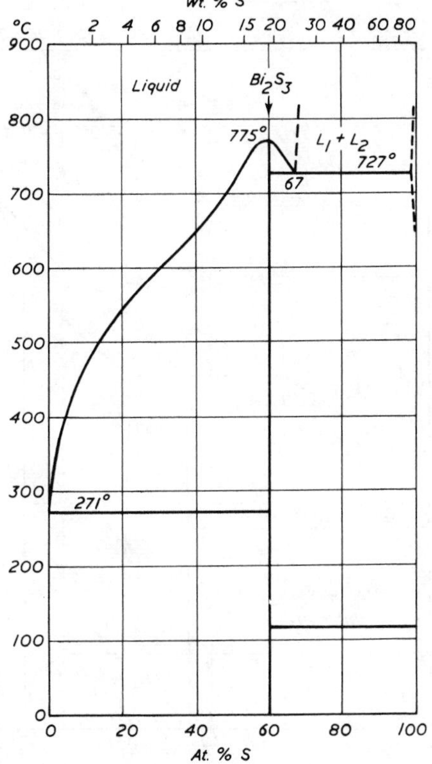
Wt. % S
°C
2 4 6 8 10 15 20 30 40 60 80
900
800
700
600
500
400
300
200
100
0
Liquid
Bi_2S_3
775°
$L_1 + L_2$
727°
67
271°
0 20 40 60 80 100
At. % S

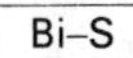
Bi–S

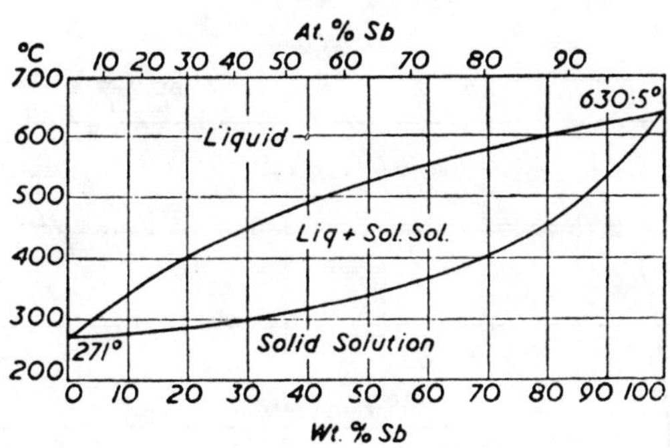
At. % Sb
°C
10 20 30 40 50 60 70 80 90
700
600
500
400
300
200
630·5°
Liquid
Liq + Sol. Sol.
Solid Solution
271°
0 10 20 30 40 50 60 70 80 90 100
Wt. % Sb

Bi–Sb

Bi–Se

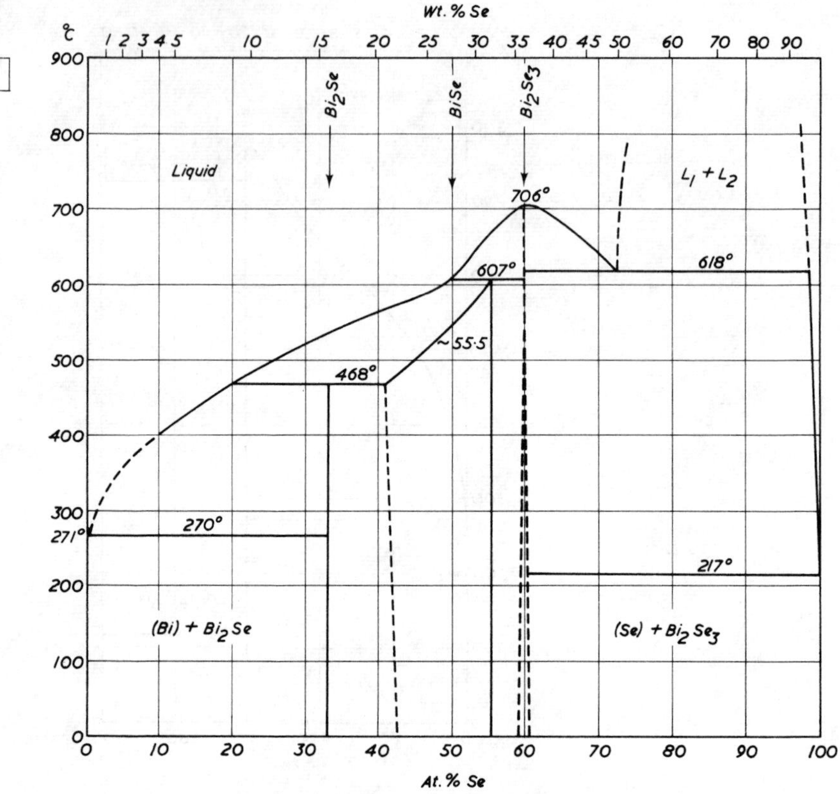
Wt. % Se
°C
1 2 3 4 5 10 15 20 25 30 35 40 45 50 60 70 80 90
900
800
700
600
500
400
300
271°
200
100
0
Bi_2Se
BiSe
Bi_2Se_3
Liquid
$L_1 + L_2$
706°
607°
618°
~55·5
468°
270°
217°
(Bi) + Bi_2Se
(Se) + Bi_2Se_3
0 10 20 30 40 50 60 70 80 90 100
At. % Se

Bi–Si

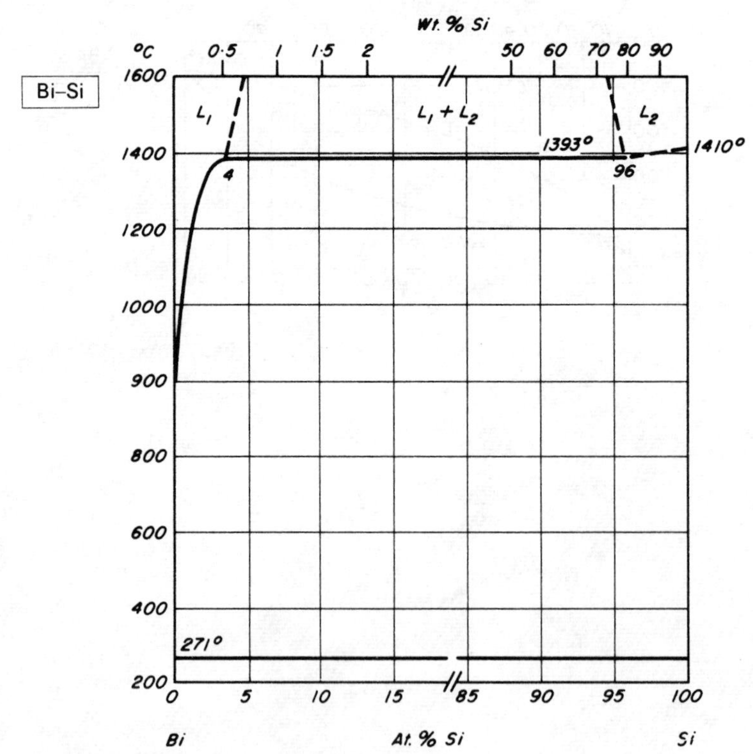
Wt. % Si
°C
0·5 1 1·5 2 50 60 70 80 90
1600
1400
1200
1000
900
800
600
400
200
L_1
$L_1 + L_2$
L_2
1393°
1410°
4
96
271°
0 5 10 15 85 90 95 100
Bi
At. % Si
Si

Bi–Sn

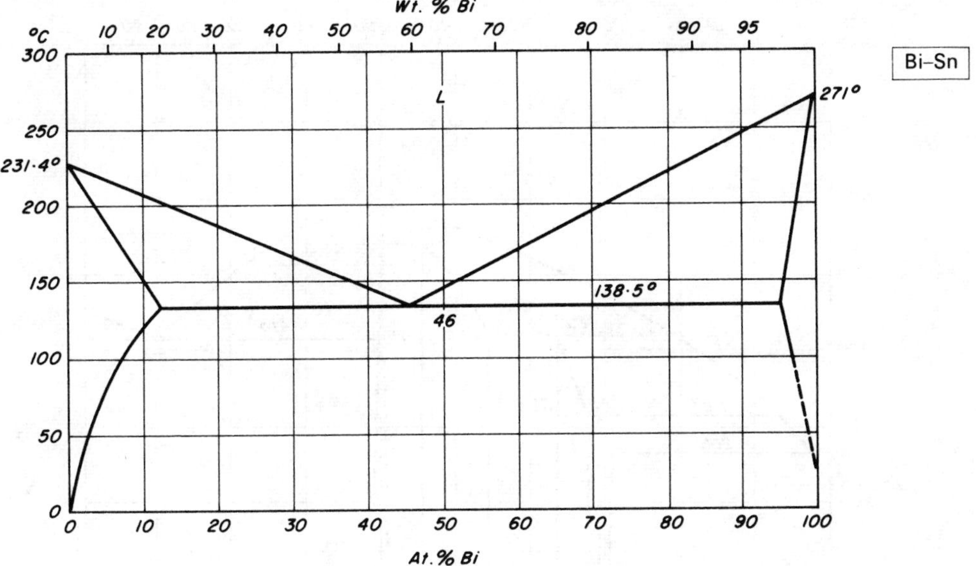
Wt. % Bi
10 20 30 40 50 60 70 80 90 95
°C
300
250
200
150
100
50
0
231·4°
L
271°
138·5°
46
0 10 20 30 40 50 60 70 80 90 100
At.% Bi

Bi–Sr

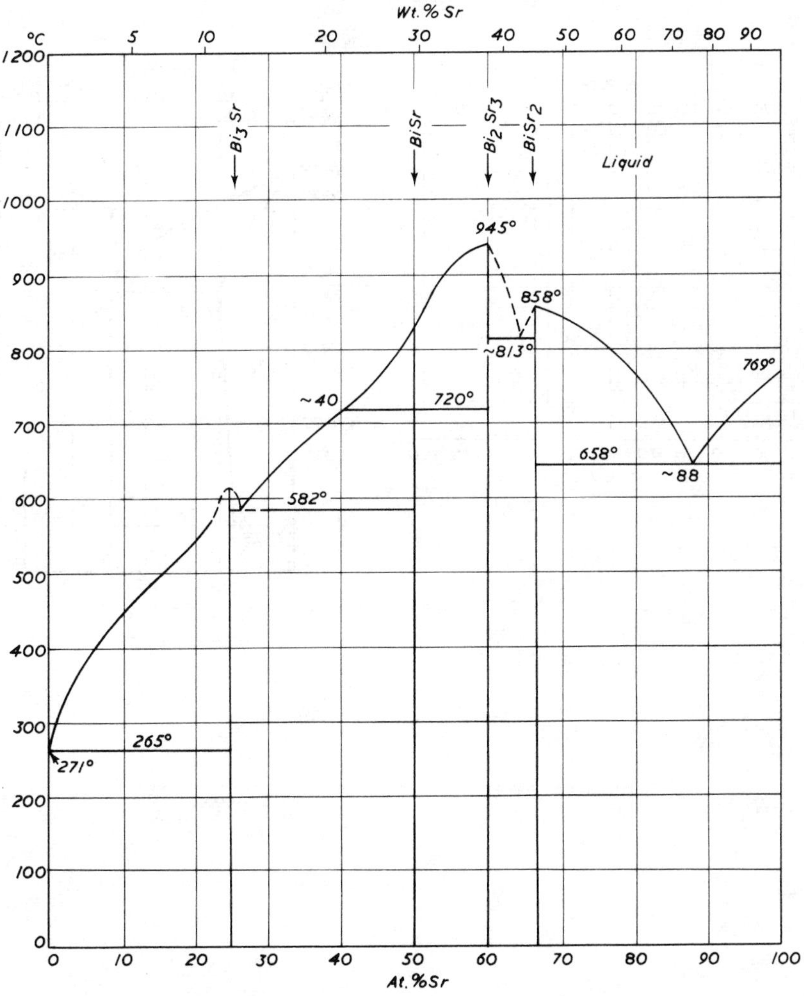

Wt. % Sr
5 10 20 30 40 50 60 70 80 90
°C
1200
1100
1000
900
800
700
600
500
400
300
200
100
0
Bi_3Sr
BiSr
Bi_2Sr_3
$BiSr_2$
Liquid
945°
858°
~813°
769°
~40
720°
658°
~88
582°
265°
271°
0 10 20 30 40 50 60 70 80 90 100
At.%Sr

Bi–Te

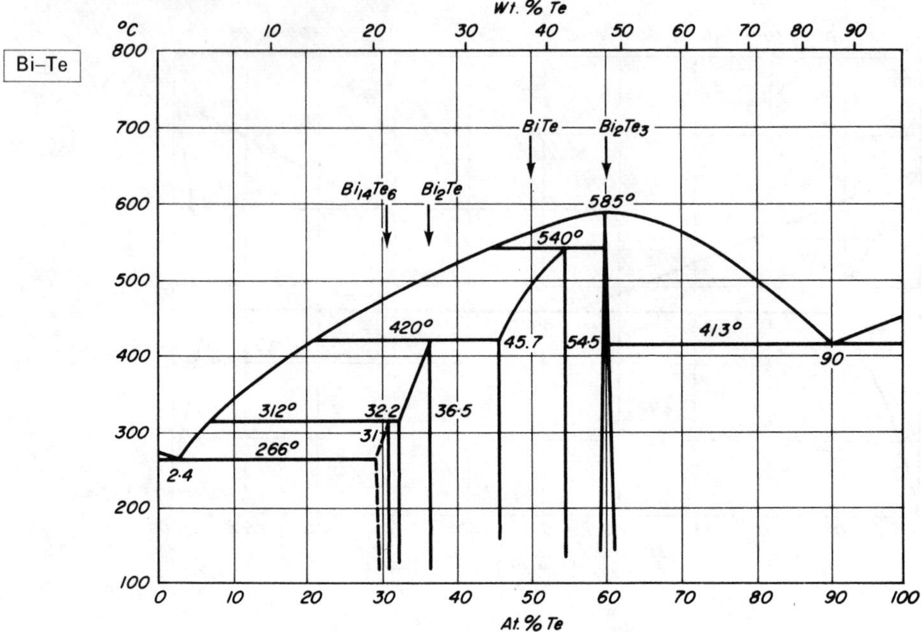

Bi–Th

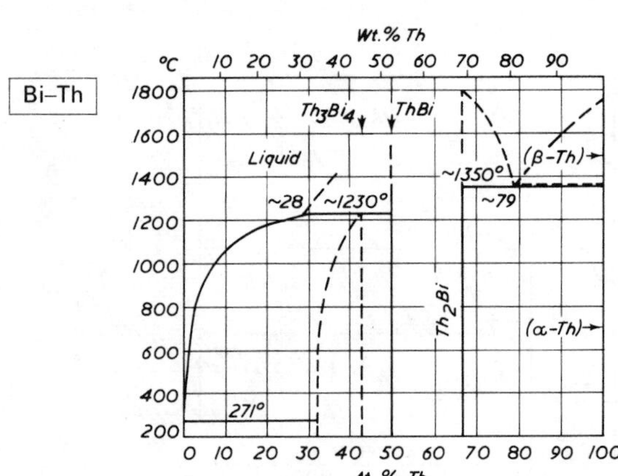

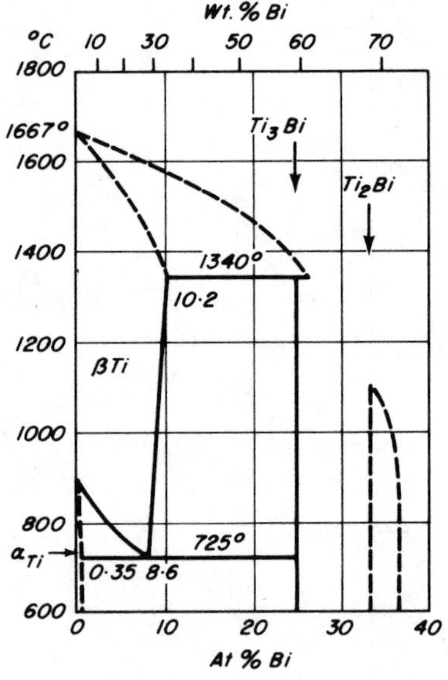

Bi–Ti

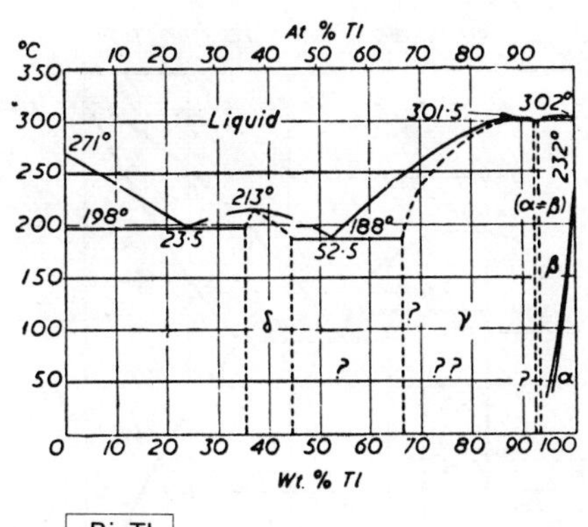

Bi–Tl

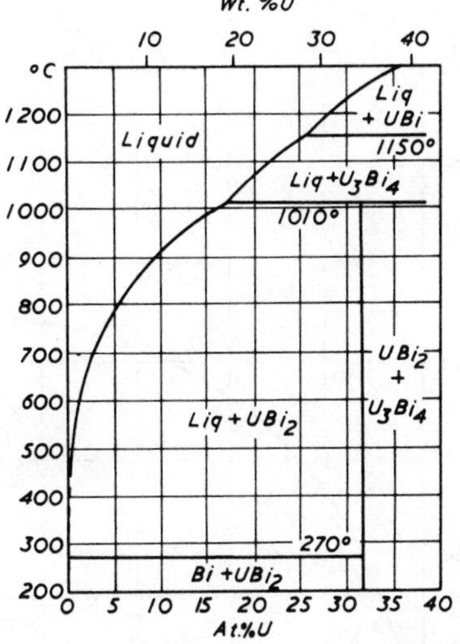

Bi–U

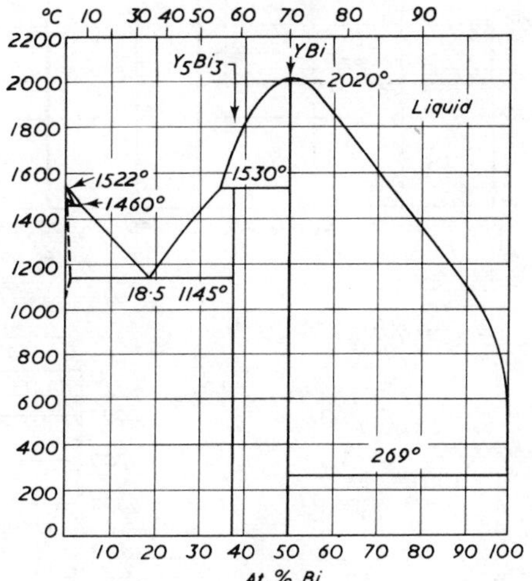

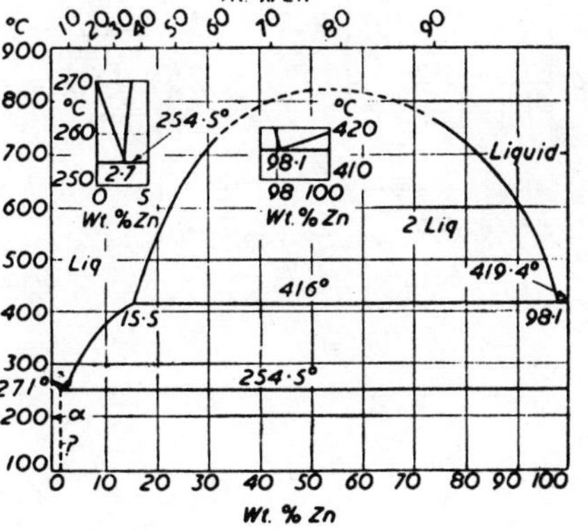

Bi–Zn

Bi–Zr

Wt. % Zr
°C
ZrBi2
ZrBi
Zr3Bi2
Zr2Bi
1495°
1340°
1300°
1290°
(β)
900°
714°
271·3°
(β)+
Zr4Bi
~900°
(α)+
Zr4Bi
(β)
(α)
a/o Zr
Wt. % Zr

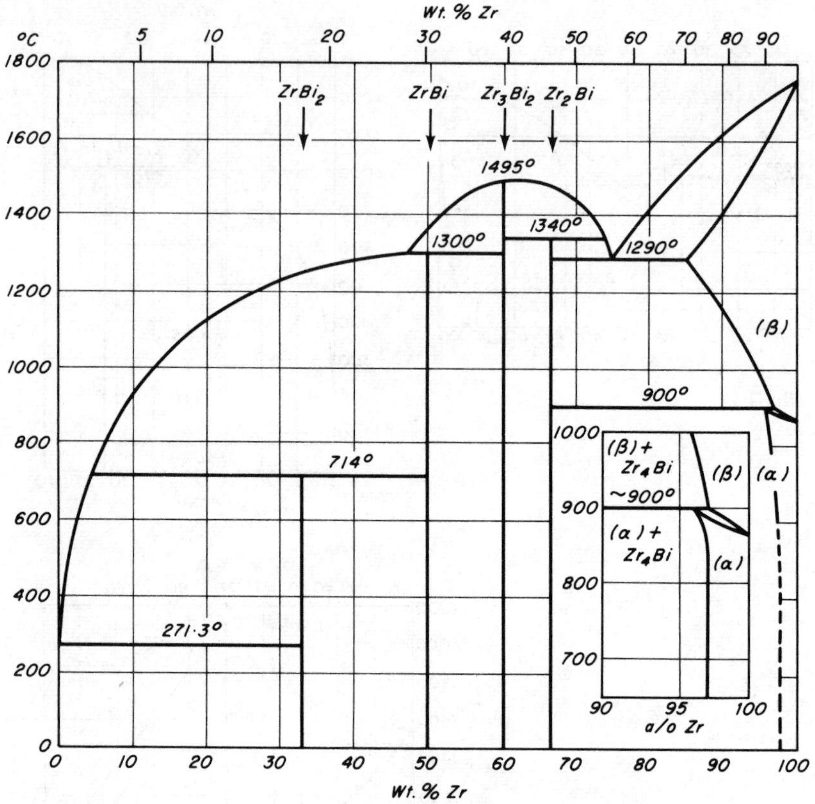

C–Co

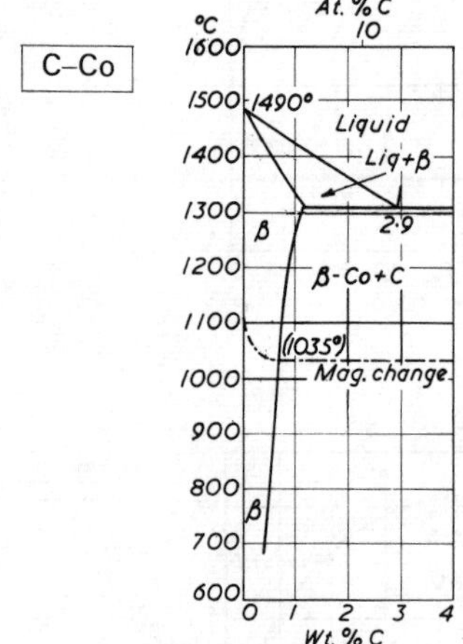

At. % C
°C
1490°
Liquid
Liq+β
2·9
β
β-Co+C
(1035°)
Mag. change
β
Wt. % C

C–Cr

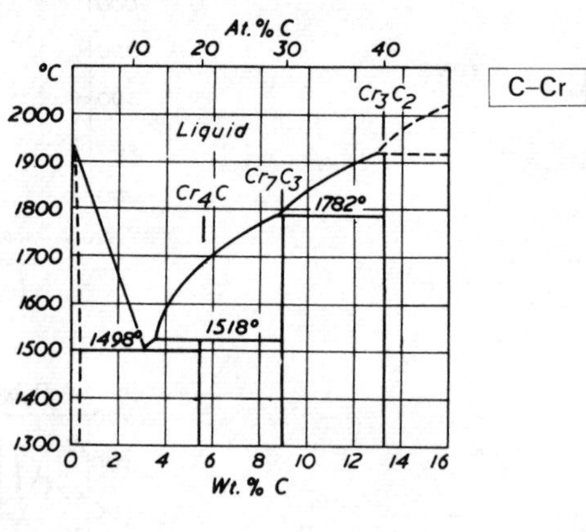

At. % C
°C
Liquid
Cr3C2
Cr7C3
Cr4C
1782°
1518°
1498°
Wt. % C

C–Cu

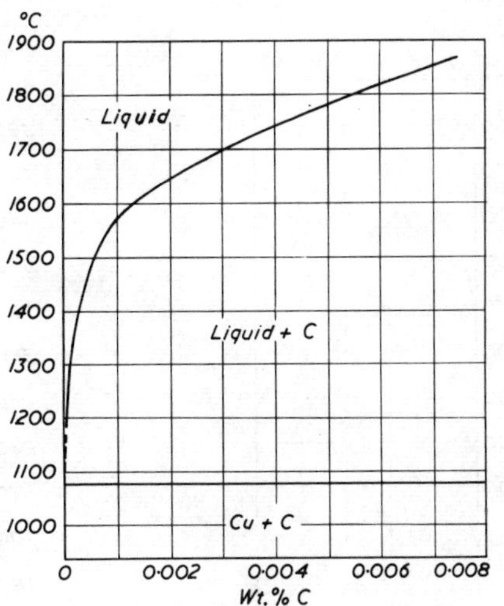
°C
1900
1800
1700
1600
1500
1400
1300
1200
1100
1000
Liquid
Liquid + C
Cu + C
0
0·002
0·004
0·006
0·008
Wt.% C

C–Fe
At.% C 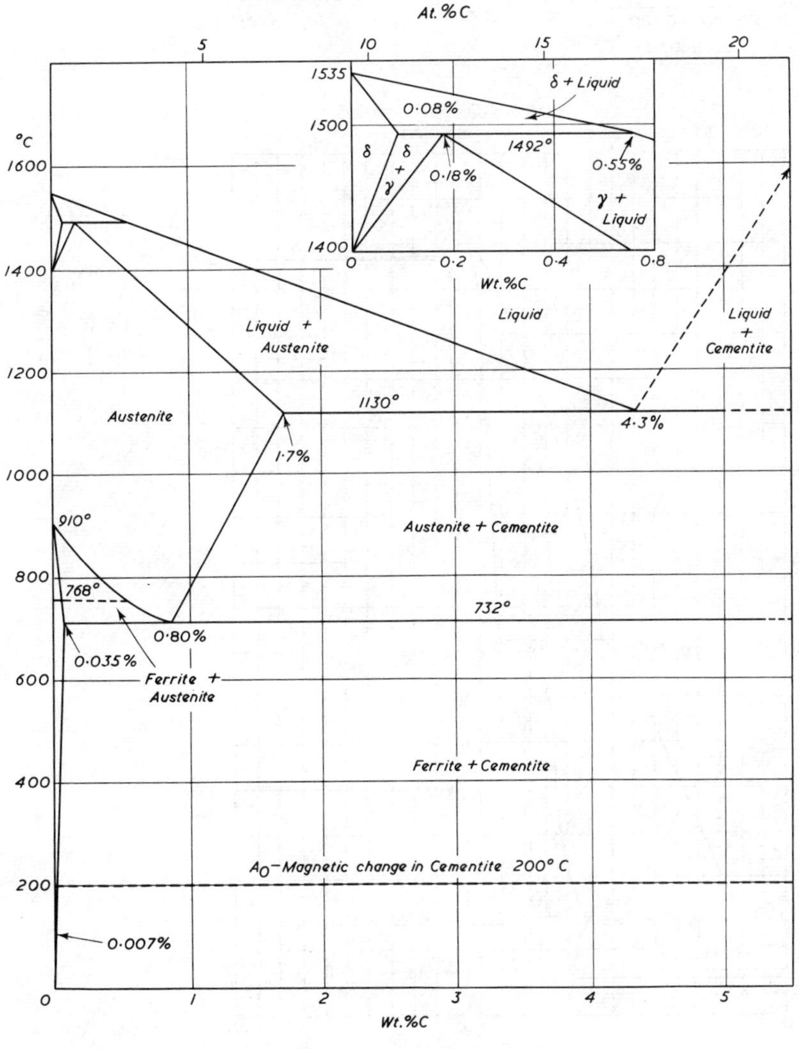
5
10
15
20
°C
1600
1400
1200
1000
800
600
400
200
1535
1500
1400
δ + Liquid
0·08%
1492°
δ
δ + γ
0·18%
0·55%
γ + Liquid
0
0·2
0·4
0·8
Wt.%C
Liquid + Austenite
Liquid
Liquid + Cementite
Austenite
1130°
1·7%
4·3%
910°
Austenite + Cementite
768°
732°
0·80%
0·035%
Ferrite + Austenite
Ferrite + Cementite
A_0–Magnetic change in Cementite 200° C
0·007%
0
1
2
3
4
5
Wt.%C

C–Ge

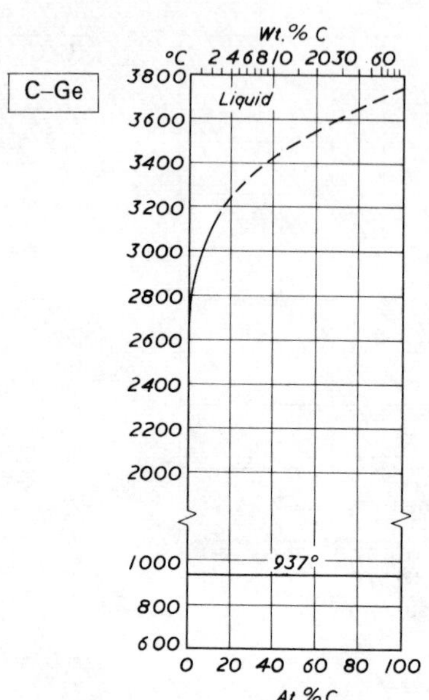

C–Hf

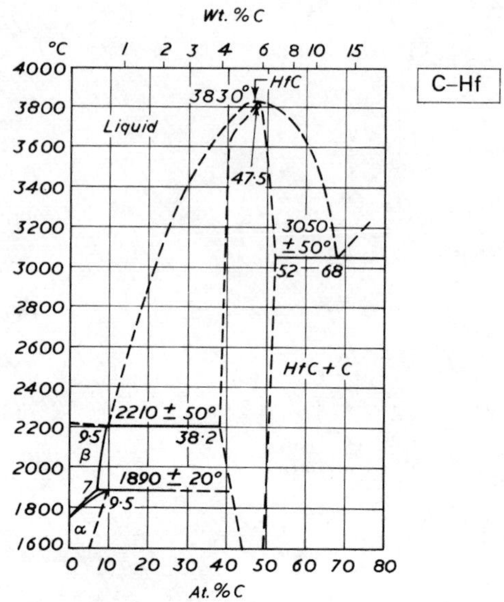

C–La

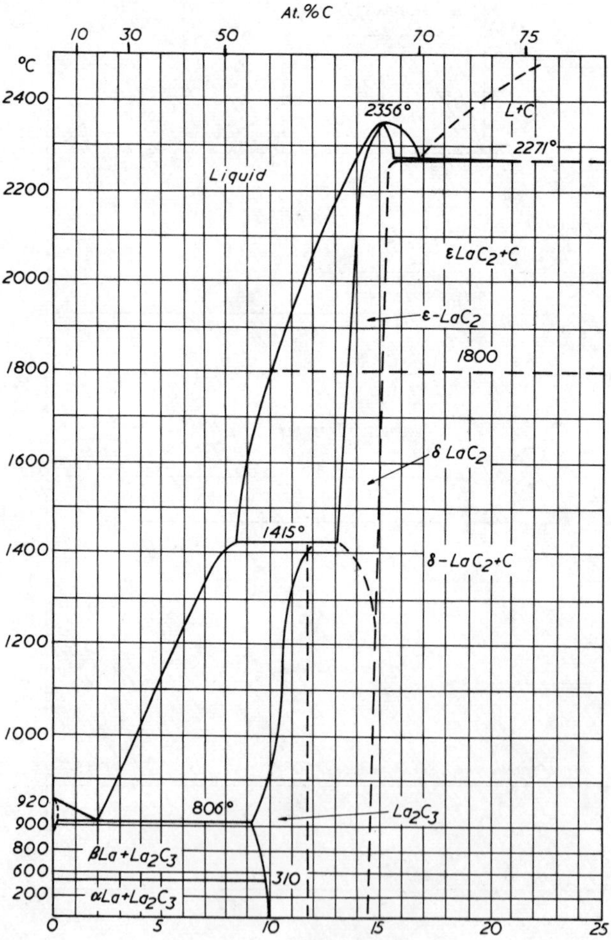

At.% C

°C

2356°

L+C

2271°

Liquid

εLaC₂+C

ε-LaC₂

1800

δ LaC₂

1415°

δ-LaC₂+C

920

806°

La₂C₃

βLa+La₂C₃

310

αLa+La₂C₃

Wt.%C

C–La

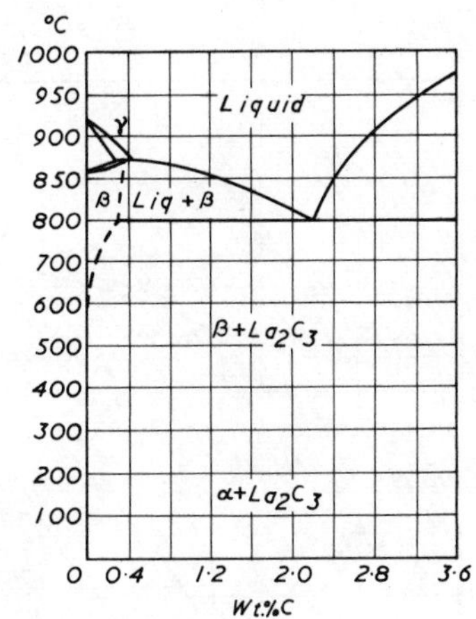
°C
Liquid
γ
β
Liq + β
β+La2C3
α+La2C3
Wt.%C

C–Li

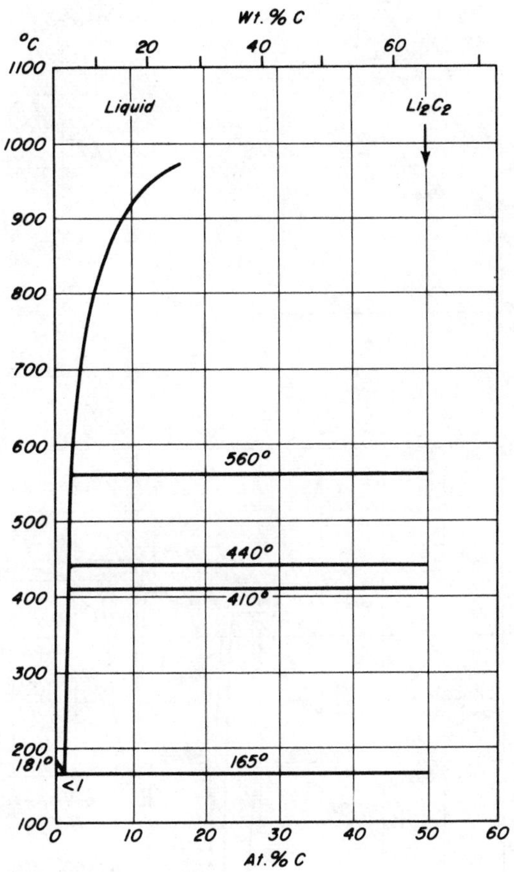
Wt. % C
°C
Liquid
Li2C2
560°
440°
410°
165°
181°
<1
At. % C

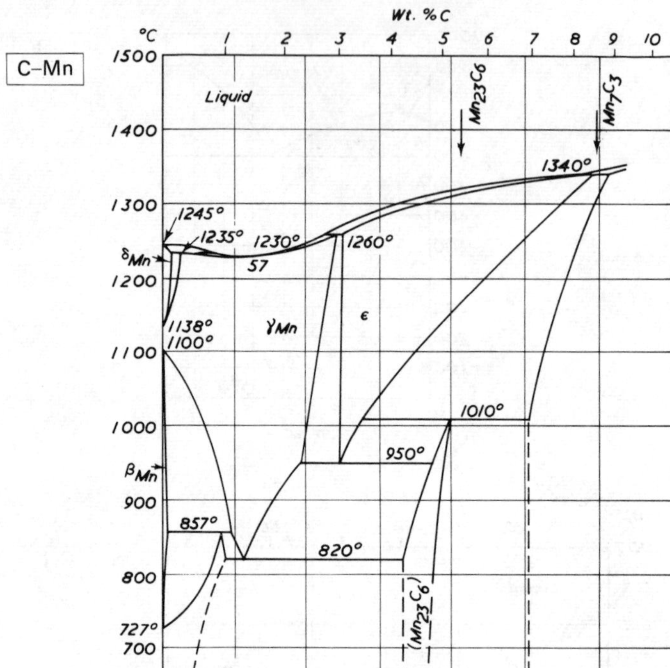
C–Mn
Wt. % C
°C
Liquid
$Mn_{23}C_6$
Mn_7C_3
1340°
1245°
1235°
1230°
57
1260°
δMn
γMn
ε
1138°
1100°
1010°
950°
βMn
857°
820°
$(Mn_{23}C_6)$
727°
αMn
At. % C

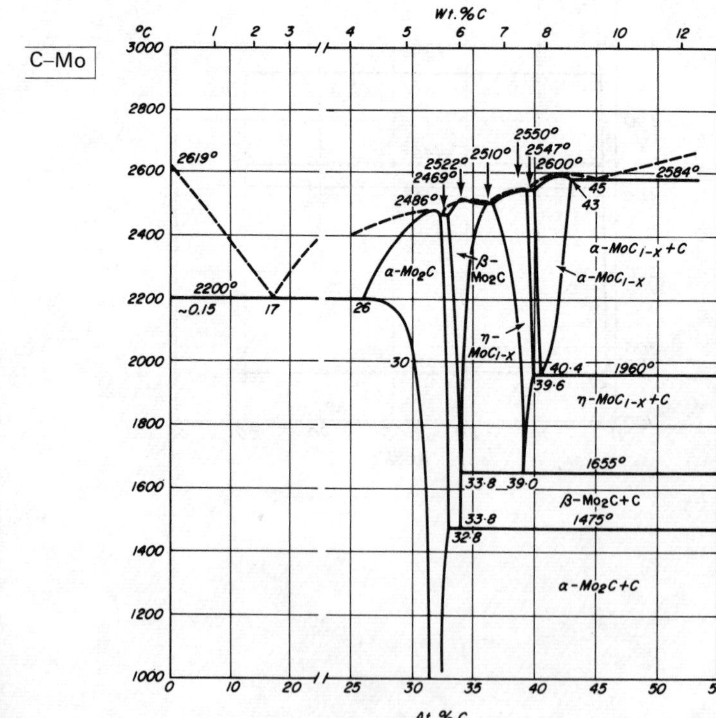
C–Mo
Wt. % C
°C
2619°
2550°
2547°
2522°
2510°
2469°
2600°
2584°
2486°
45
43
α-MoC1−x + C
β-Mo2C
α-Mo2C
α-MoC1−x
2200°
~0.15
17
26
η-MoC1−x
30
40·4
1960°
39·6
η-MoC1−x + C
1655°
33·8
39·0
β-Mo2C + C
33·8
1475°
32·8
α-Mo2C + C
At. % C

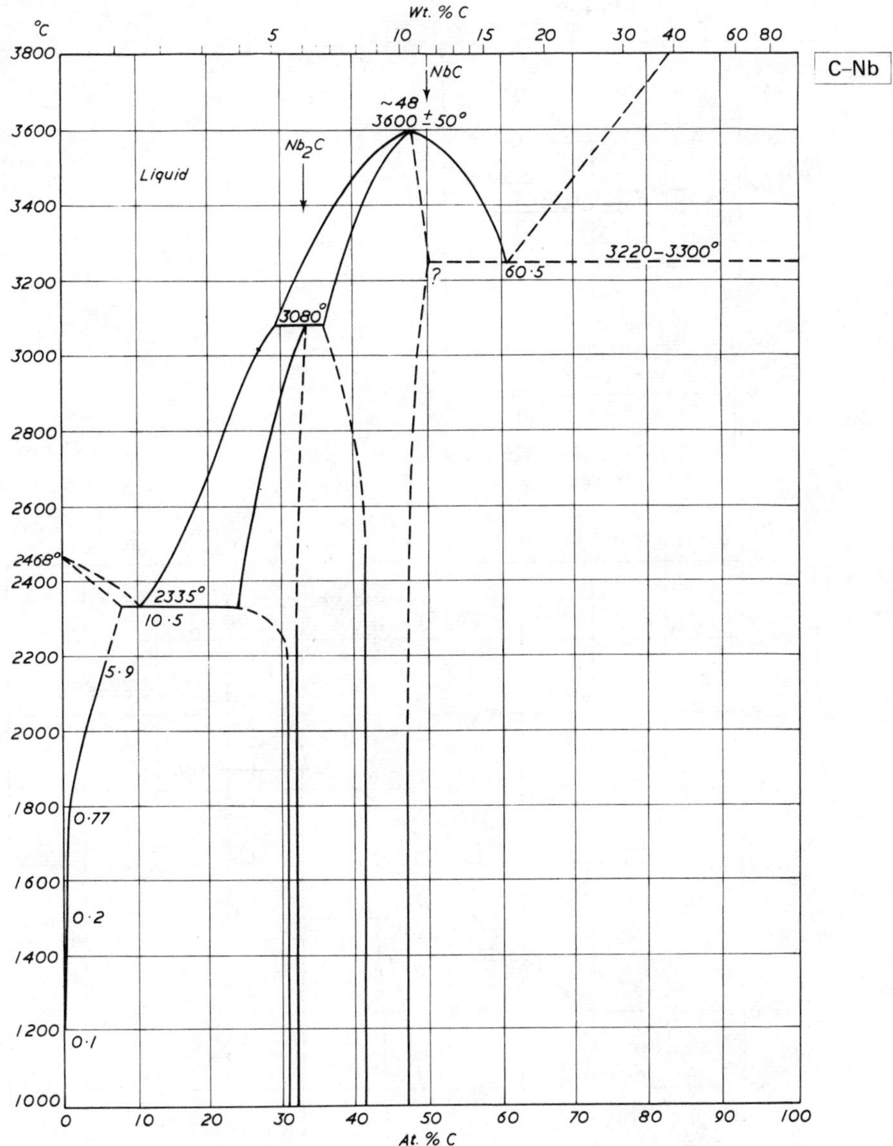
C–Nb
Wt. % C
5
10
15
20
30
40
60
80
°C
3800
3600
3400
3200
3000
2800
2600
2468°
2400
2200
2000
1800
1600
1400
1200
1000
NbC
~48
3600 ± 50°
Nb_2C
Liquid
3220–3300°
?
60·5
3080°
2335°
10·5
5·9
0·77
0·2
0·1
0
10
20
30
40
50
60
70
80
90
100
At. % C

C–Ni

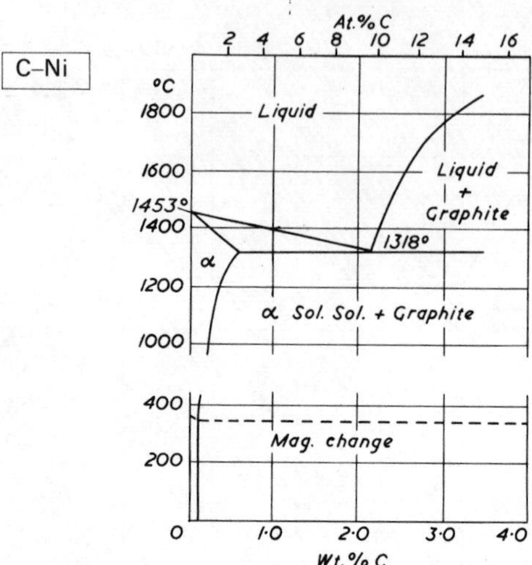

At.% C
2 4 6 8 10 12 14 16
°C
1800
1600
1453°
1400
1200
1000
400
200
0
Liquid
Liquid + Graphite
1318°
α
α Sol. Sol. + Graphite
Mag. change
1·0 2·0 3·0 4·0
Wt.% C

C–Pu

Wt.% C
0·2 0·5 1·0 2 4 6 8 10 20
°C
2600
2400
2200
2000
1800
1600
1400
1200
1000
800
600
400
200
0
Pu_3C_2
PuC
Pu_2C_3
PuC_2
Liquid
2250±20°
2050±20°
1750±20°
1654±20°
630±2°
575°
480°
451°
319°
206°
122°
0 10 20 30 40 50 60 70 80 90 100
At. % C

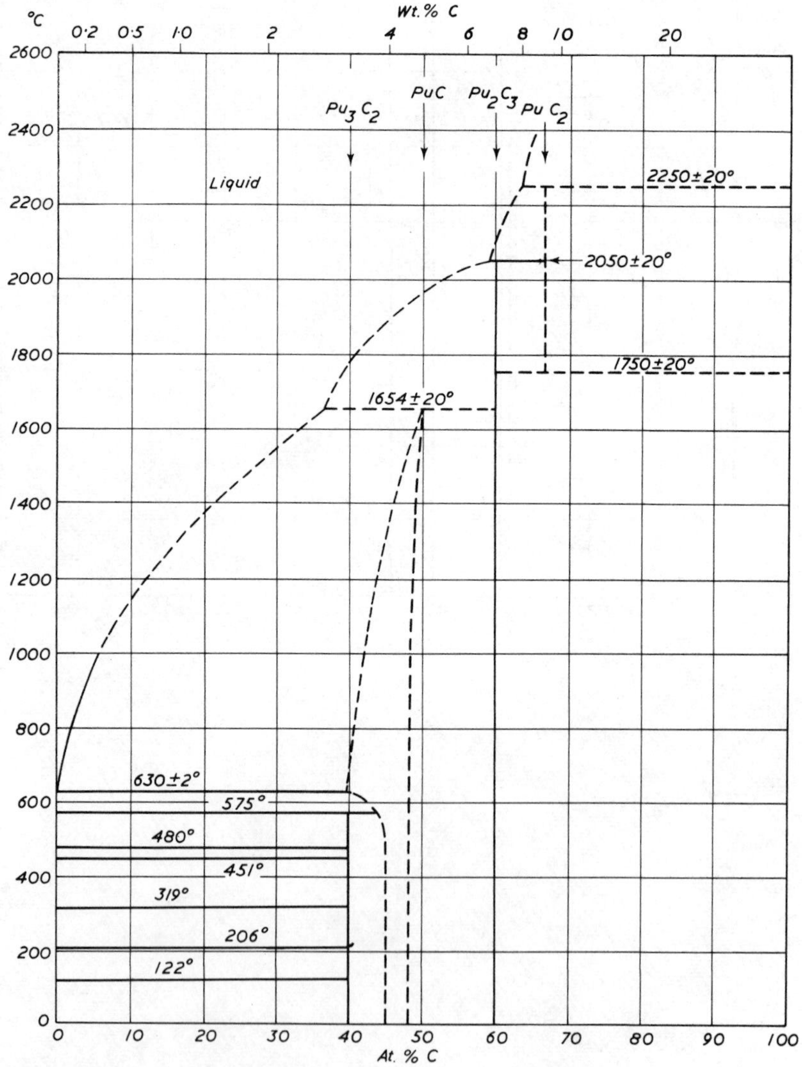

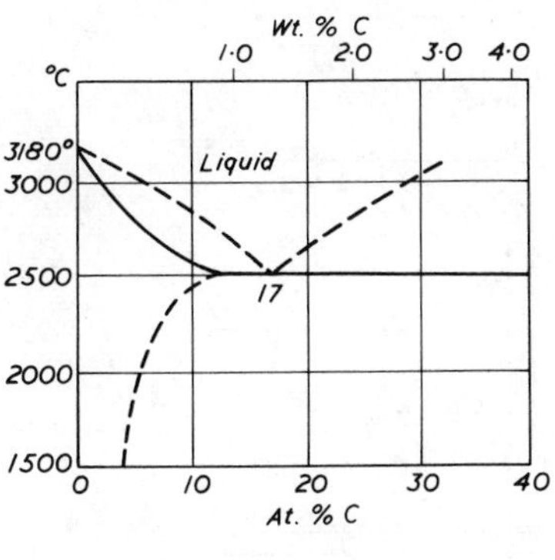
Wt. % C
1·0
2·0
3·0
4·0
°C
3180°
3000
2500
2000
1500
Liquid
17
0
10
20
30
40
At. % C

C–Re

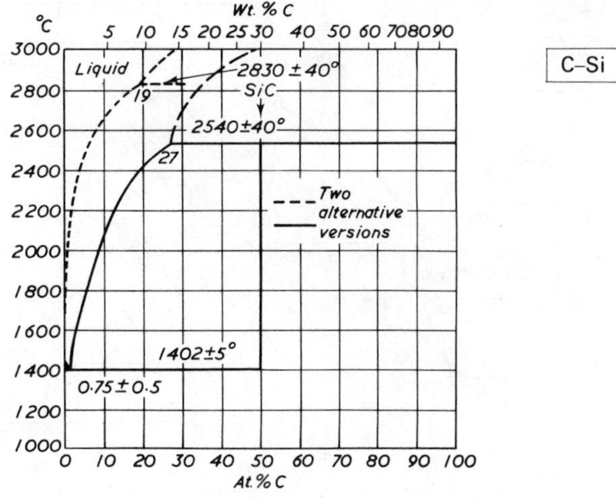
Wt. % C
°C
5 10 15 20 25 30 40 50 60 70 80 90
Liquid
2830 ± 40°
19
SiC
2540±40°
27
Two
alternative
versions
1402±5°
0·75 ± 0·5
3000
2800
2600
2400
2200
2000
1800
1600
1400
1200
1000
0 10 20 30 40 50 60 70 80 90 100
At. % C

C–Si

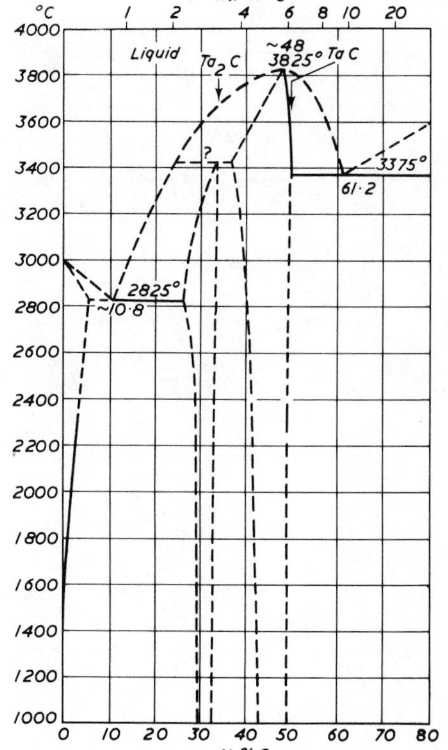
Wt. % C
°C
1 2 4 6 8 10 20
Liquid
Ta_2C
~48
3825°
TaC
?
3375°
61·2
2825°
~10·8
4000
3800
3600
3400
3200
3000
2800
2600
2400
2200
2000
1800
1600
1400
1200
1000
0 10 20 30 40 50 60 70 80
At. % C

C–Ta

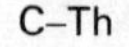
C–Th

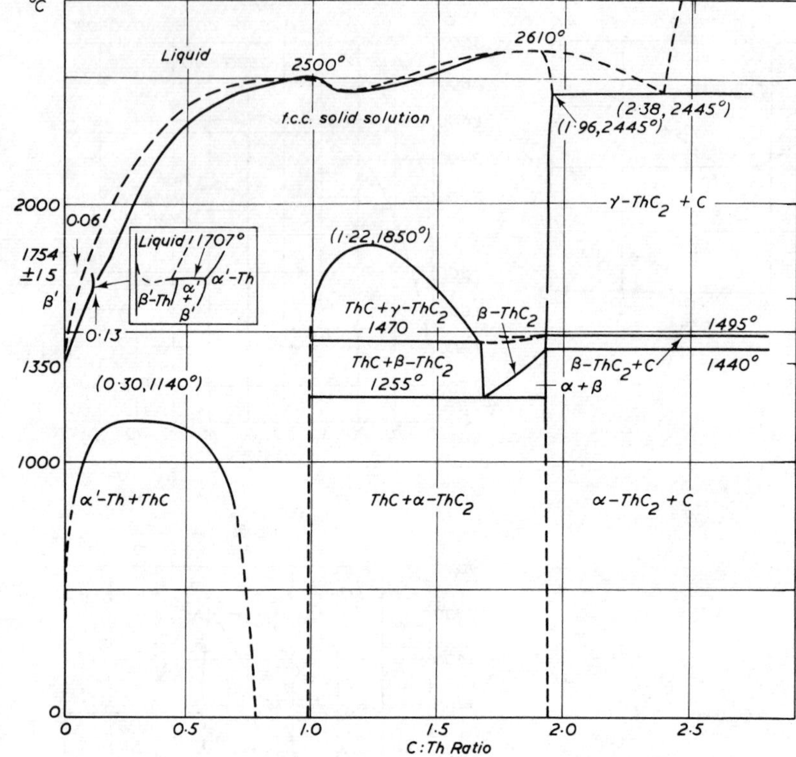

°C
Liquid
2500°
2610°
f.c.c. solid solution
(2·38, 2445°)
(1·96,2445°)
2000
γ-ThC2 + C
0·06
1754
±15
β'
Liquid
1707°
α'-Th
β'-Th
α' + β'
0·13
1350
(1·22,1850°)
ThC+γ-ThC2
1470
β-ThC2
1495°
ThC+β-ThC2
1255°
β-ThC2+C
1440°
α+β
(0·30,1140°)
1000
α'-Th+ThC
ThC+α-ThC2
α-ThC2 + C
0
0·5
1·0
1·5
2·0
2·5
C:Th Ratio

C–Ti

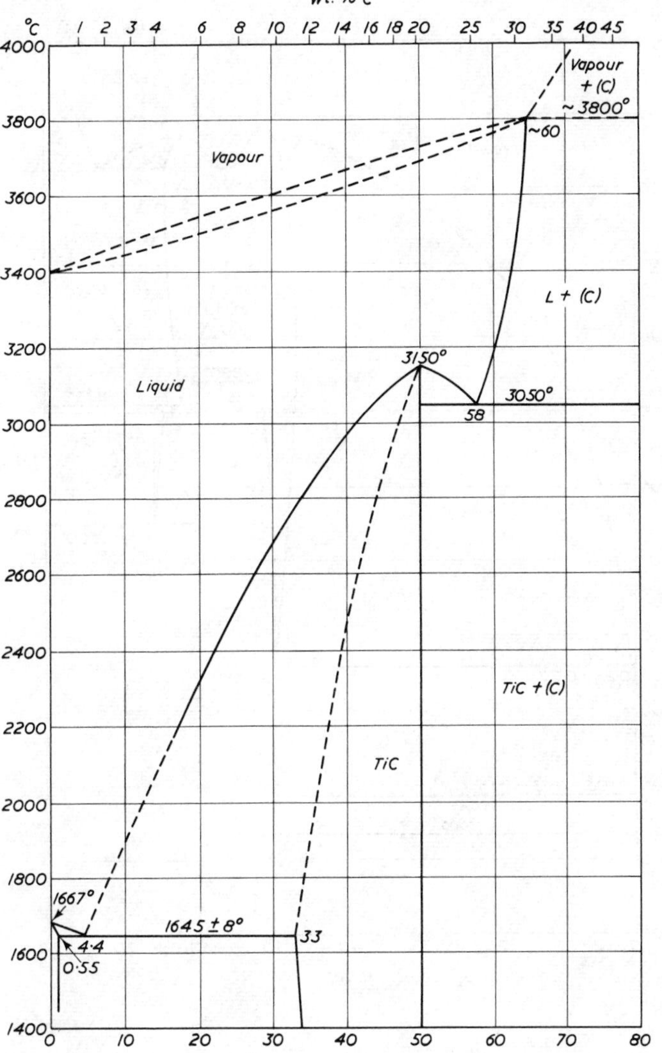
Wt. % C
°C
Vapour
+ (C)
~ 3800°
~60
Vapour
L + (C)
3150°
Liquid
3050°
58
TiC + (C)
TiC
1667°
1645 ± 8°
33
4·4
0·55
At. % C

C–U

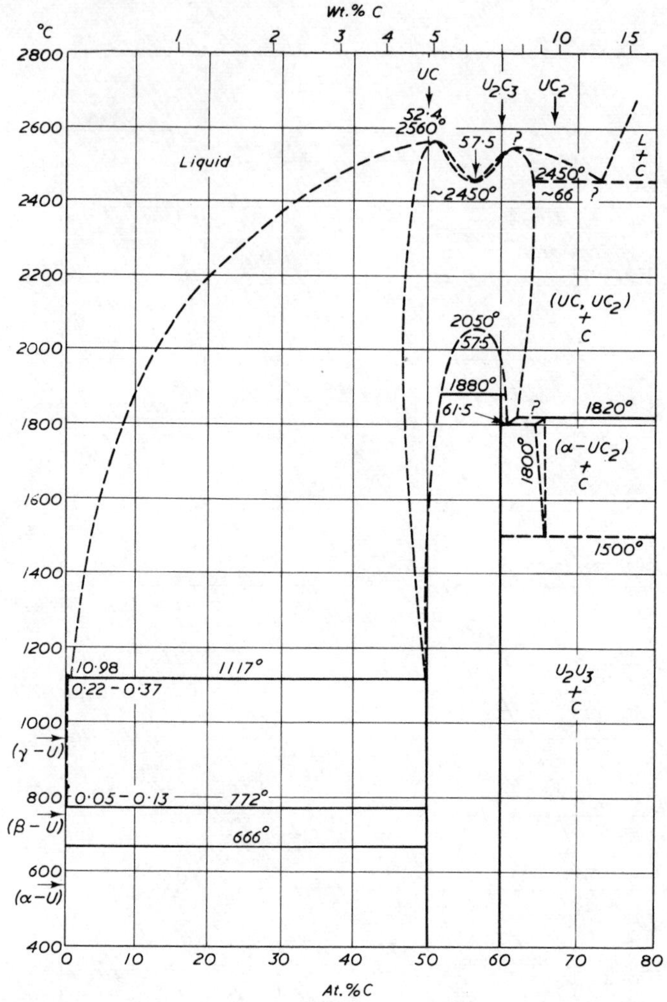

Wt.% C
1 2 3 4 5 10 15
°C
2800
2600
2400
2200
2000
1800
1600
1400
1200
1000
800
600
400
UC
U_2C_3
UC_2
52·4
2560°
57·5
?
L + C
Liquid
~2450°
2450°
~66
?
(UC, UC_2) + C
2050°
57·5
1880°
61·5
?
1820°
1800°
$(\alpha-UC_2)$ + C
1500°
10·98
1117°
0·22 – 0·37
$(\gamma-U)$
0·05 – 0·13
772°
$(\beta-U)$
666°
$(\alpha-U)$
U_2U_3 + C
0 10 20 30 40 50 60 70 80
At.%C

C–V

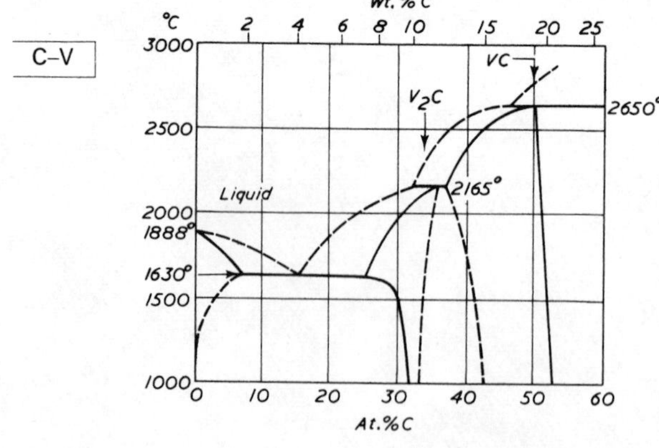

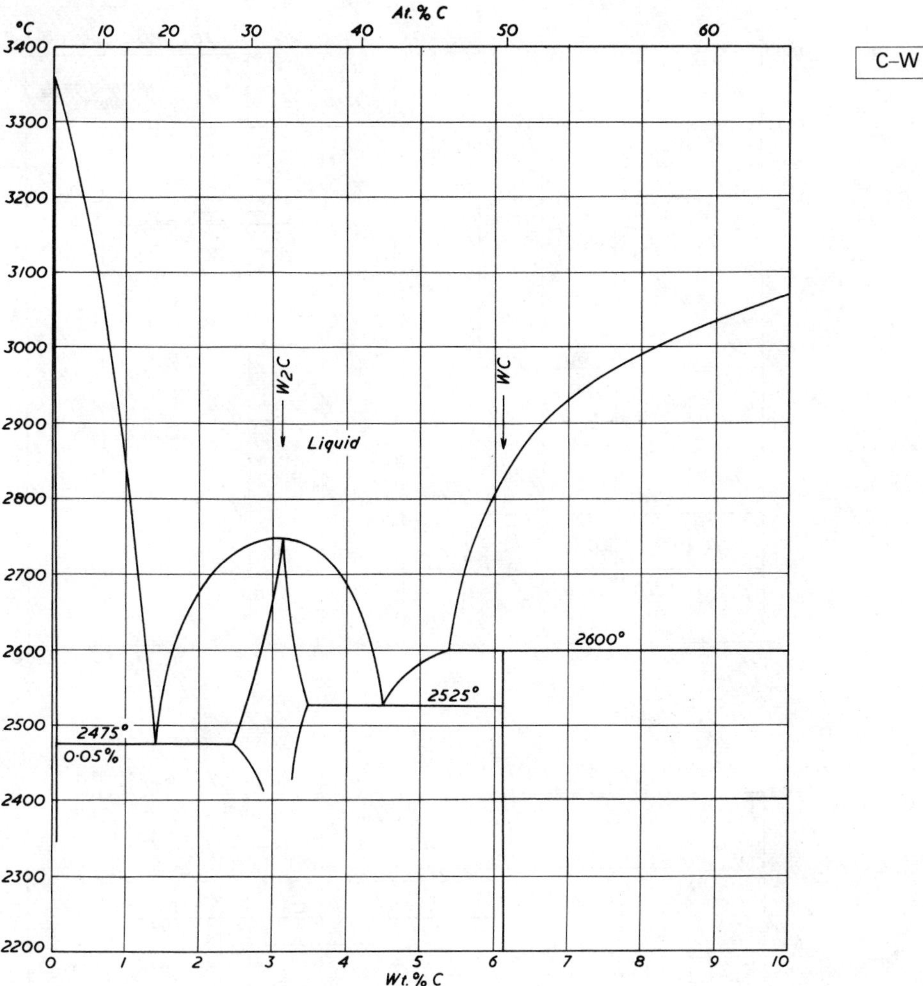

C–W
At. % C
°C
10
20
30
40
50
60
3400
3300
3200
3100
3000
2900
2800
2700
2600
2500
2400
2300
2200
W2C
WC
Liquid
2600°
2525°
2475°
0·05%
0
1
2
3
4
5
6
7
8
9
10
Wt. % C

C–Zr

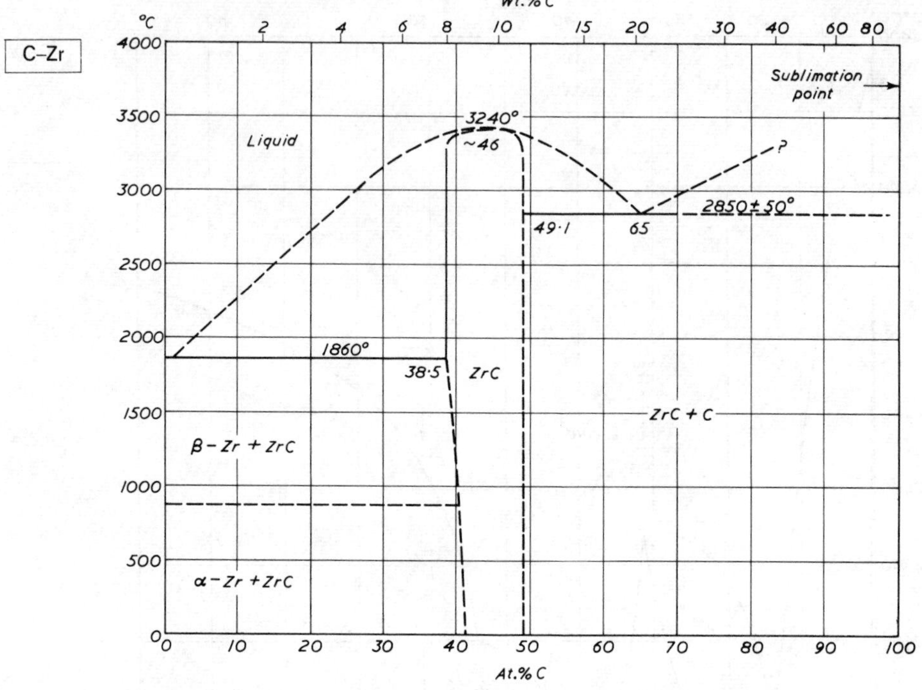
Wt.% C
°C
2
4
6
8
10
15
20
30
40
60
80
4000
3500
3000
2500
2000
1500
1000
500
0
Sublimation point
3240°
~46
Liquid
?
2850±50°
49·1
65
1860°
38·5
ZrC
ZrC + C
β – Zr + ZrC
α – Zr + ZrC
0
10
20
30
40
50
60
70
80
90
100
At.% C

Ca–Cd

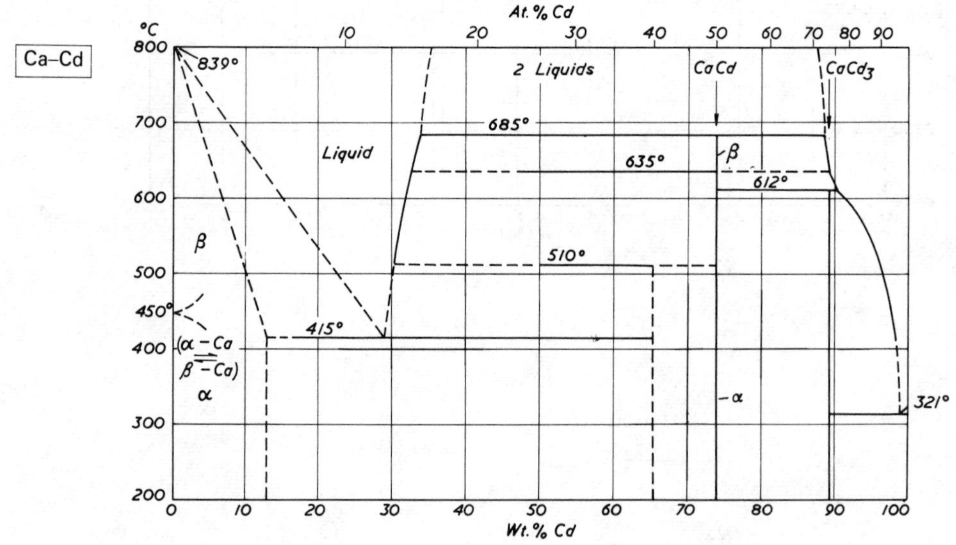
At.% Cd
°C
10
20
30
40
50
60
70
80
90
800
700
600
500
450°
400
300
200
839°
2 Liquids
CaCd
CaCd3
685°
Liquid
β
635°
612°
β
510°
415°
(α – Ca ⇌ β – Ca)
α
– α
321°
0
10
20
30
40
50
60
70
80
90
100
Wt.% Cd

Ca–Co

Wt. % Ca
1 2 3 4 5 6
°C
700
600
500
467°
400
300
200
α
ε
626°
7·9 (5·5)
6·3 (4·4)
0
5
10
Co At. % Ca

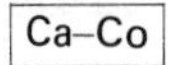

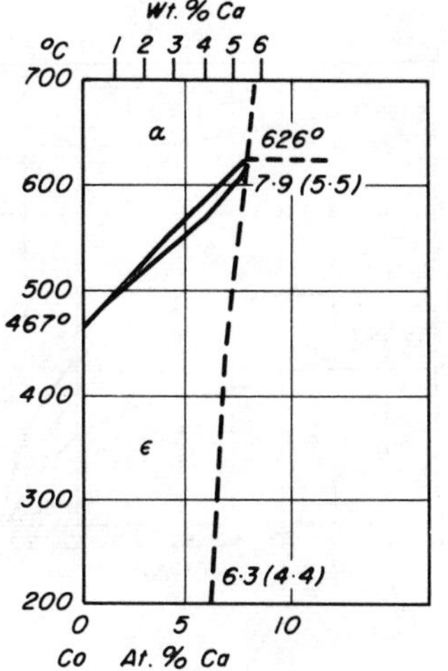

Ca–Cu

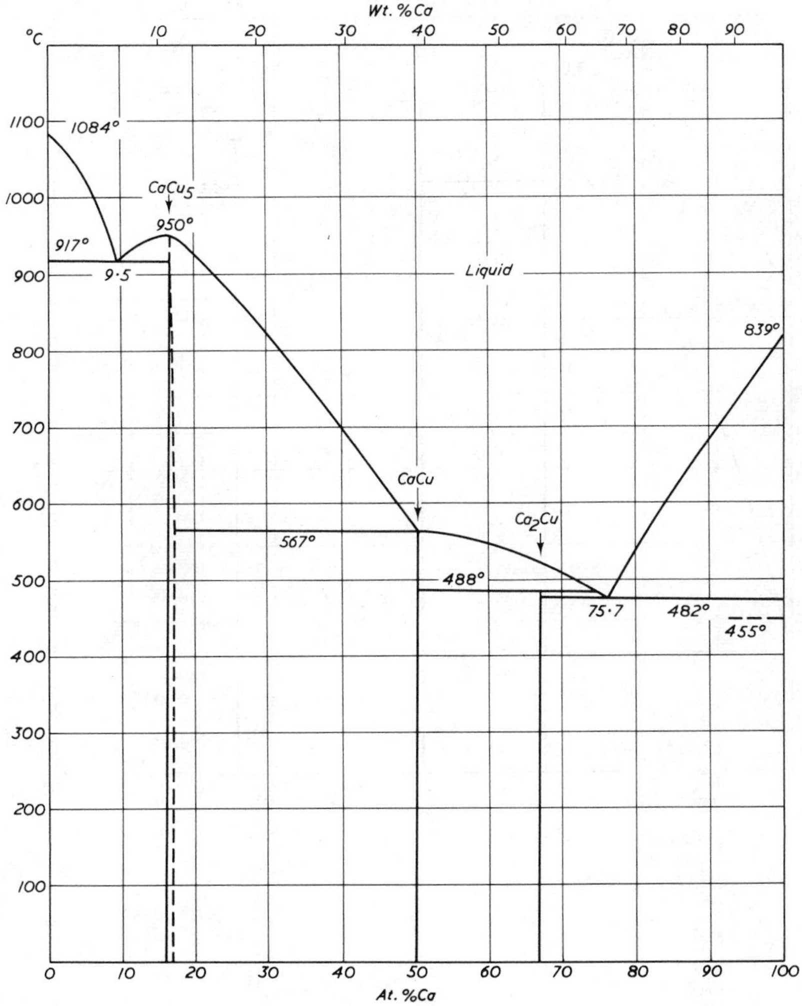

Wt. % Ca
10 20 30 40 50 60 70 80 90
°C
1100
1000
900
800
700
600
500
400
300
200
100
1084°
CaCu5
950°
917°
9·5
Liquid
839°
CaCu
Ca2Cu
567°
488°
75·7
482°
455°
0 10 20 30 40 50 60 70 80 90 100
At. % Ca

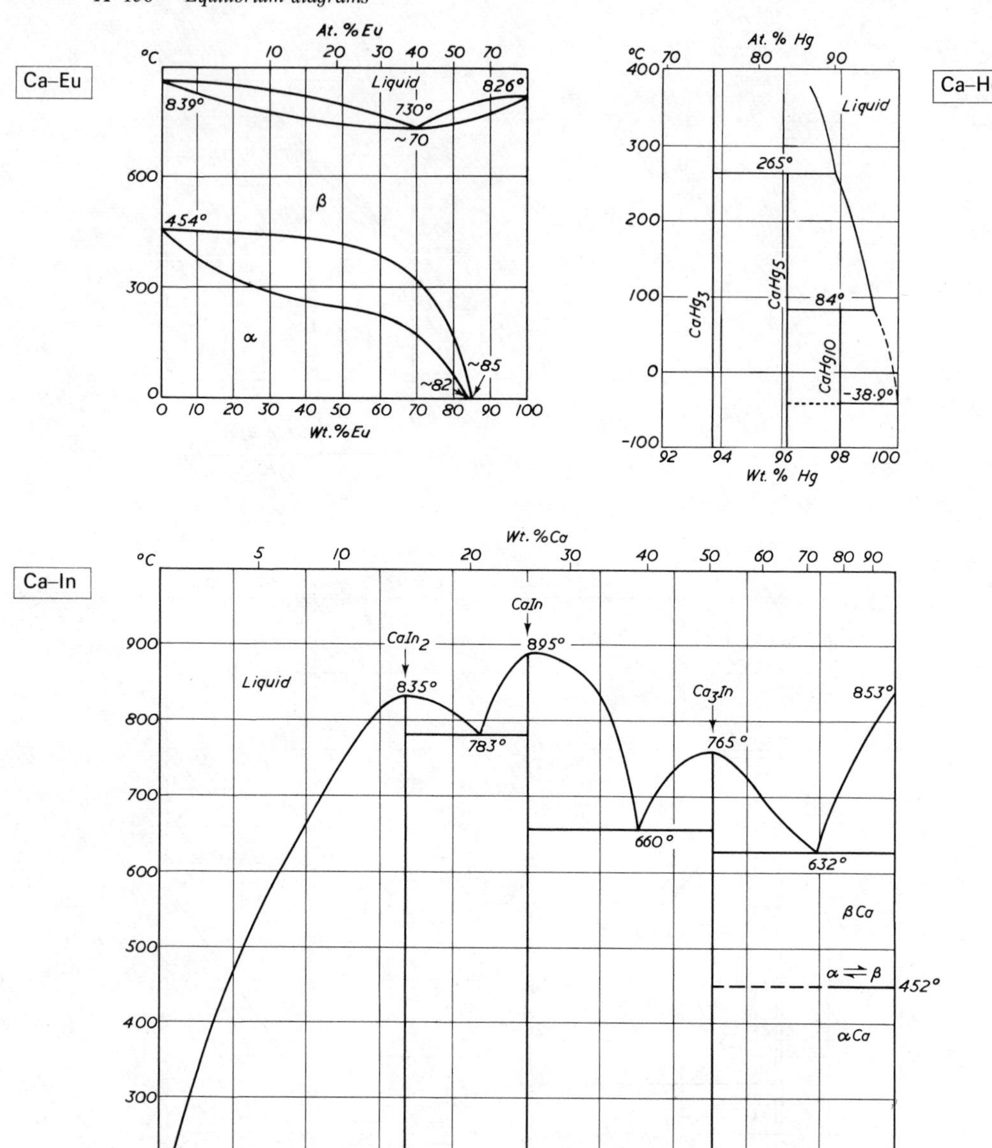
Ca–Eu
At. % Eu
°C
10 20 30 40 50 70
Liquid
826°
839°
730°
~70
600
β
454°
300
α
~85
~82
0
0 10 20 30 40 50 60 70 80 90 100
Wt.% Eu
Ca–Hg
At. % Hg
°C 70 80 90
400
Liquid
300
265°
200
100
CaHg3
CaHg5
84°
CaHg10
0
-38·9°
-100
92 94 96 98 100
Wt. % Hg
Ca–In
Wt. % Ca
°C
5 10 20 30 40 50 60 70 80 90
CaIn
CaIn2
895°
900
Liquid
835°
Ca3In
853°
800
783°
765°
700
660°
632°
600
βCa
500
α ⇌ β
452°
400
αCa
300
200
156°
100
0 10 20 30 40 50 60 70 80 90 100
At.% Ca

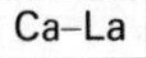
Ca–La

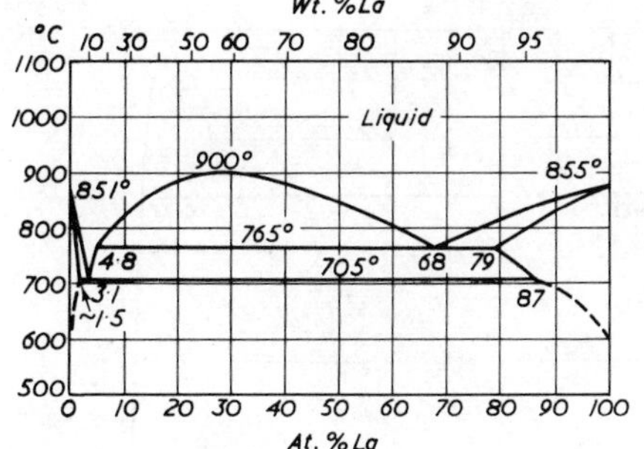
Wt. %La
°C
10 30 50 60 70 80 90 95
1100
1000
900
800
700
600
500
Liquid
900°
855°
851°
765°
4·8
705°
68
79
3·1
87
~1·5
0 10 20 30 40 50 60 70 80 90 100
At. %La

Ca–Li

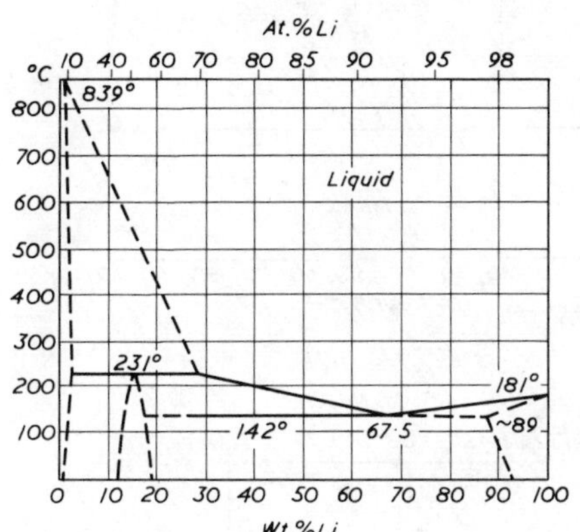
At.%Li
°C
10 40 60 70 80 85 90 95 98
839°
800
700
600
500
400
300
200
100
Liquid
231°
181°
142°
67·5
~89
0 10 20 30 40 50 60 70 80 90 100
Wt.%Li

Ca–Mg

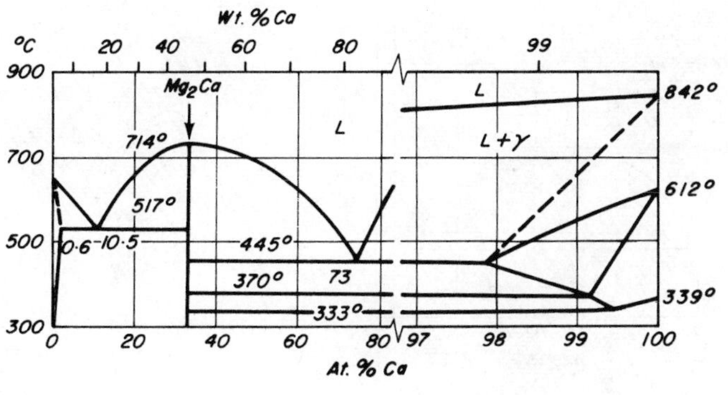
Wt. % Ca
°C
20 40 60 80 99
900
700
500
300
Mg_2Ca
L
L
L+γ
842°
714°
612°
517°
0·6
10·5
445°
370°
73
333°
339°
0 20 40 60 80 97 98 99 100
At. % Ca

Ca–Mn

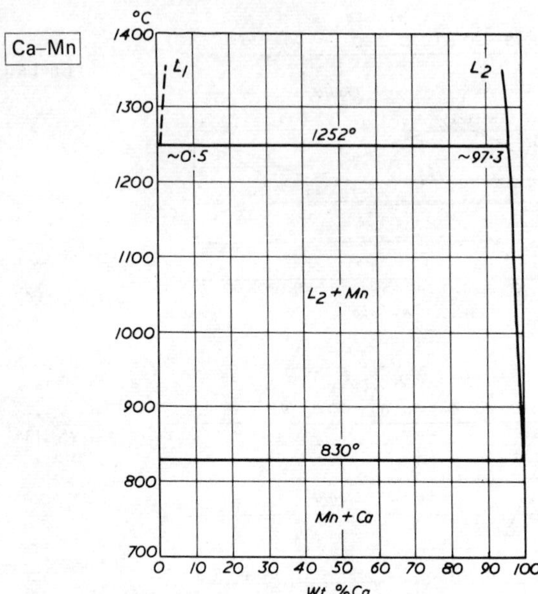

Ca–N

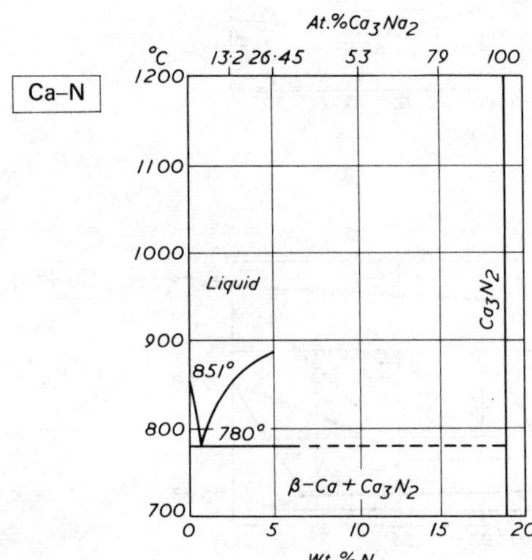

Ca–Na

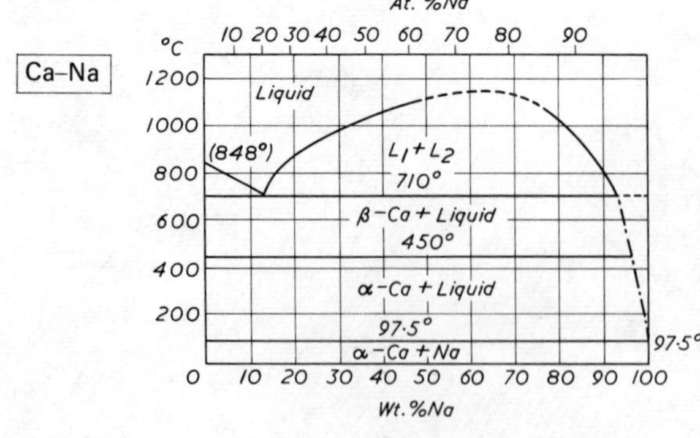

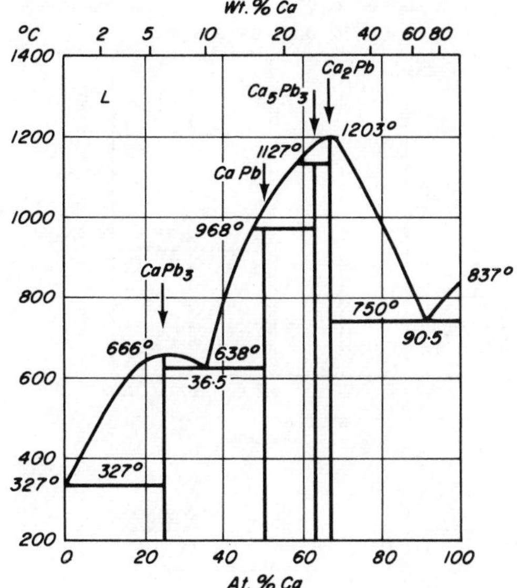
Wt. % Ca
°C
2 5 10 20 40 60 80
1400
1200
1000
800
600
400
327°
200
L
Ca2Pb
Ca5Pb3
1203°
1127°
Ca Pb
968°
CaPb3
837°
750°
90·5
666°
638°
36·5
327°
0 20 40 60 80 100
At. % Ca

Ca–Pb

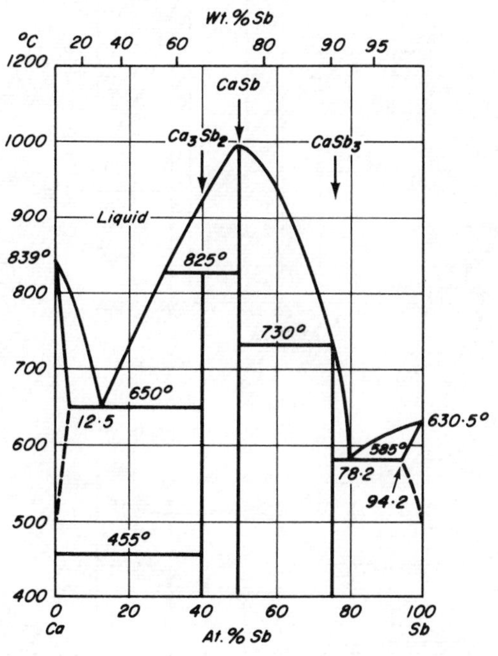
Wt. % Sb
°C
20 40 60 80 90 95
1200
1000
900
839°
800
700
600
500
400
CaSb
Ca3Sb2
CaSb3
Liquid
825°
730°
650°
12·5
630·5°
585°
78·2
94·2
455°
0 20 40 60 80 100
Ca
Sb
At. % Sb

Ca–Sb

Ca–Si

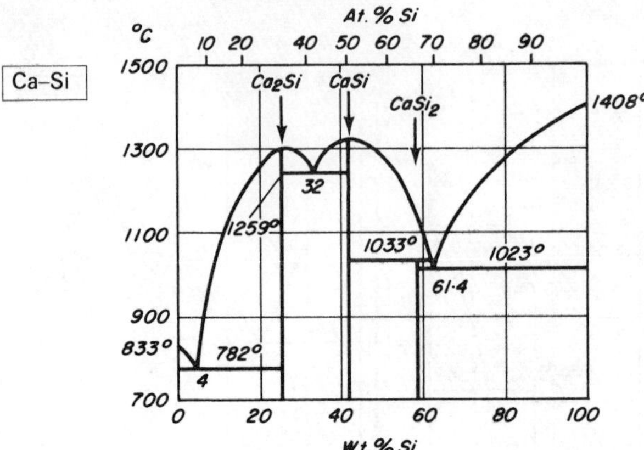

Ca–Sn

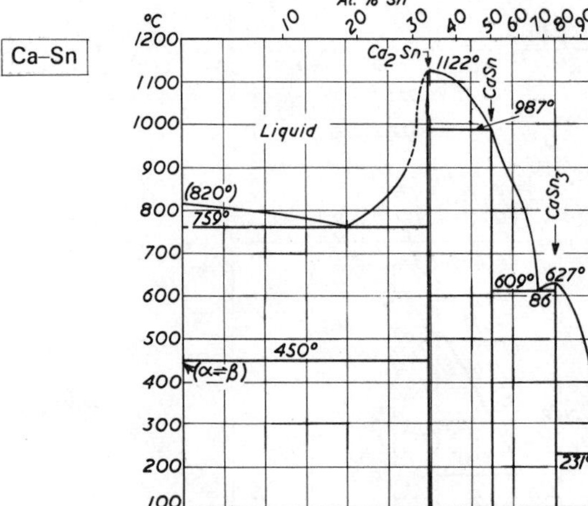

Ca–Sr

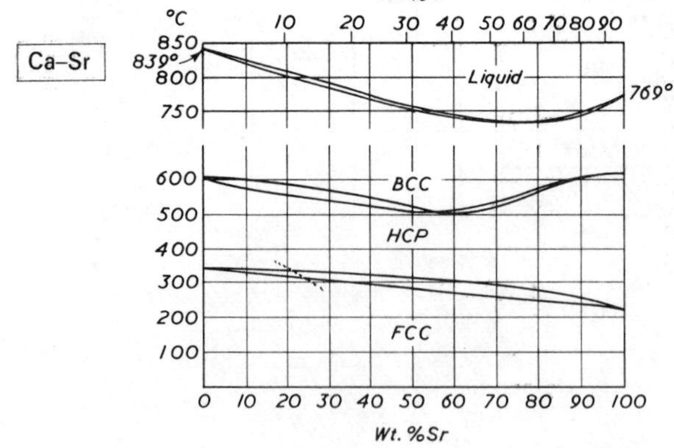

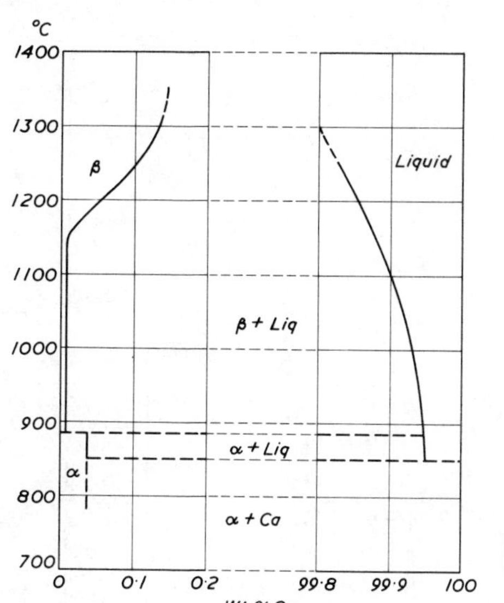

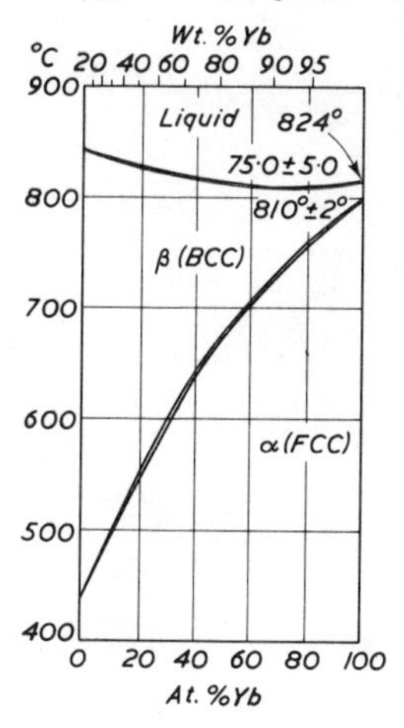

Ca–Yb

Ca–Ti

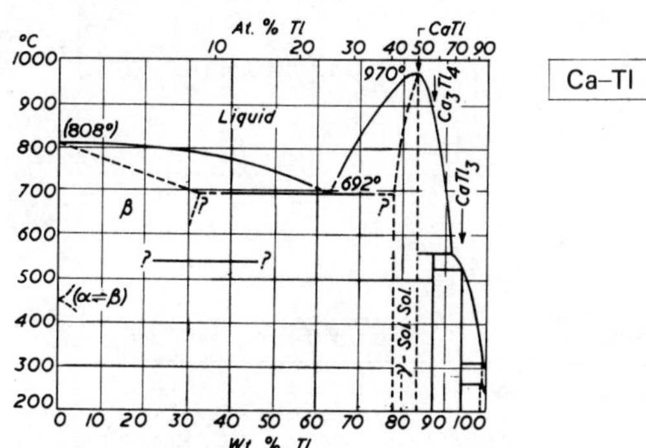

Ca–Tl

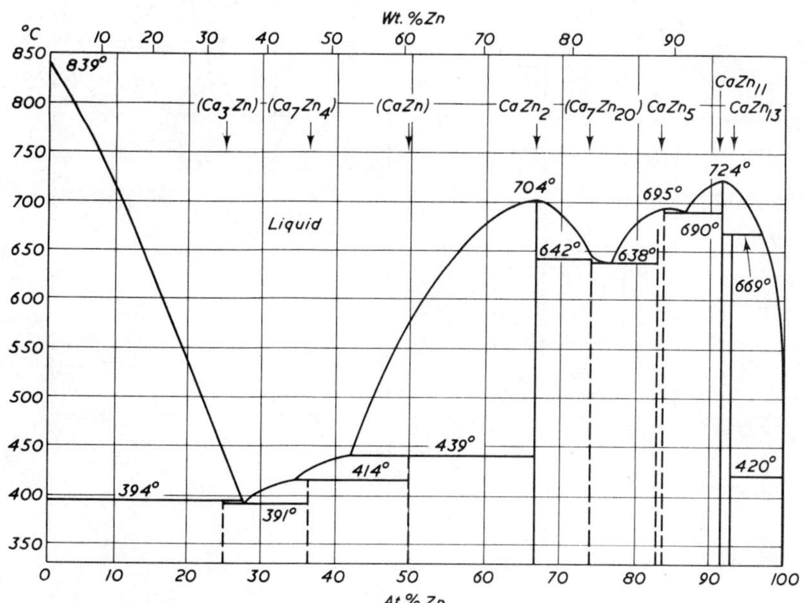

Ca–Zn

Cd–Cu

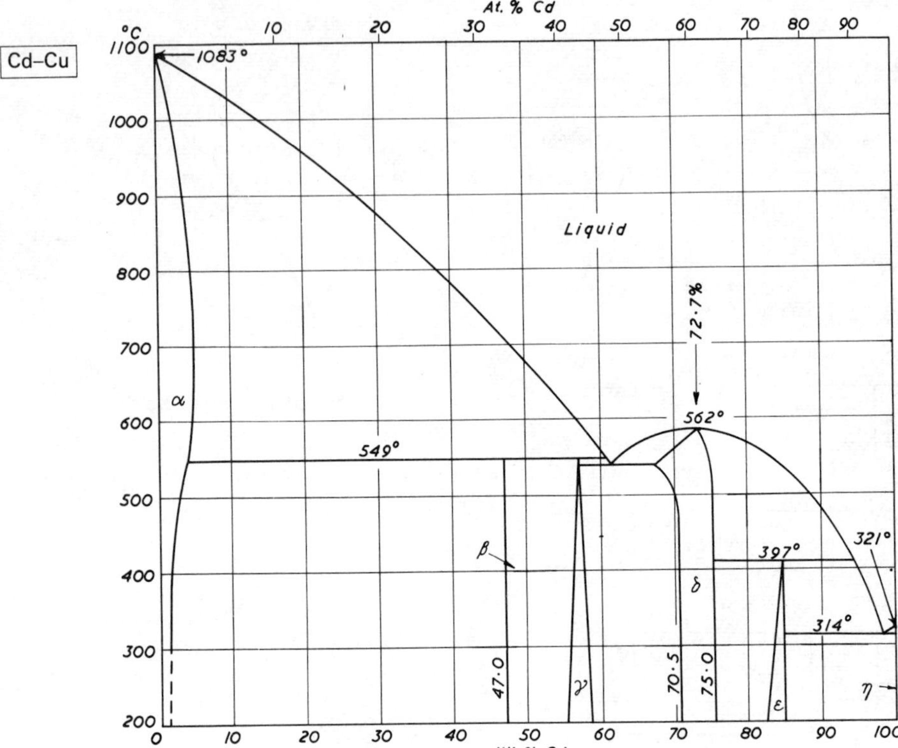
At. % Cd
°C
1083°
Liquid
72·7%
562°
549°
α
β
γ
δ
ε
η
397°
321°
314°
47·0
70·5
75·0
Wt. % Cd

Cd–Eu

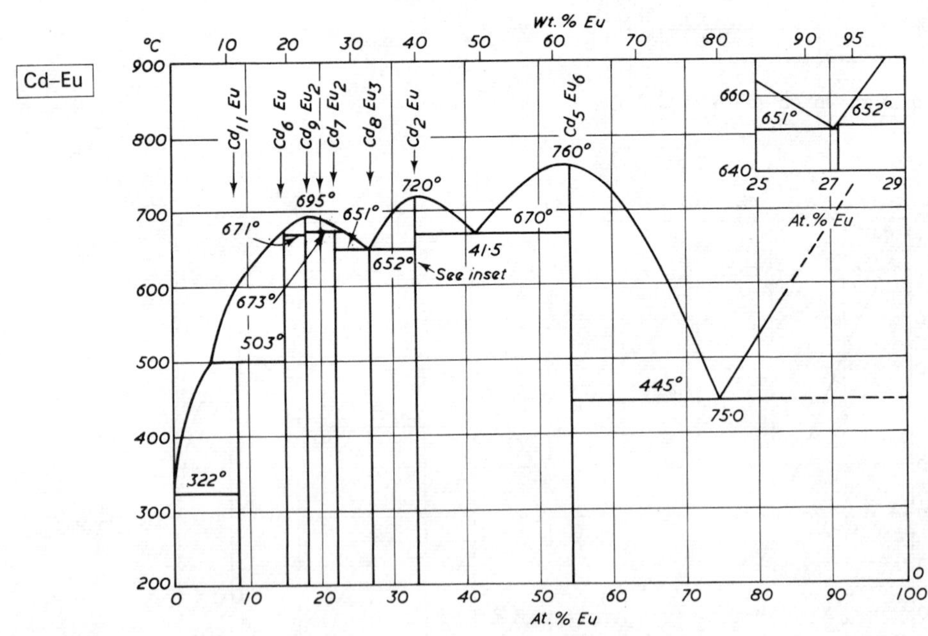
Wt. % Eu
°C
$Cd_{11}Eu$
Cd_6Eu
Cd_9Eu_2
Cd_7Eu_2
Cd_8Eu_3
Cd_2Eu
Cd_5Eu_6
760°
720°
695°
671°
651°
670°
673°
652°
41·5
See inset
503°
322°
445°
75·0
651°
652°
At. % Eu

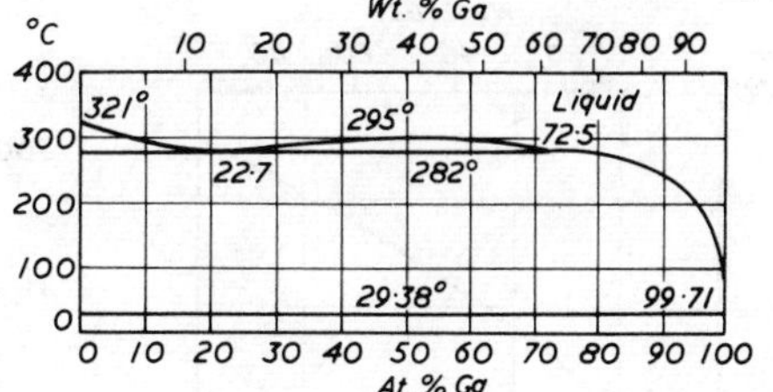

Cd–Ga

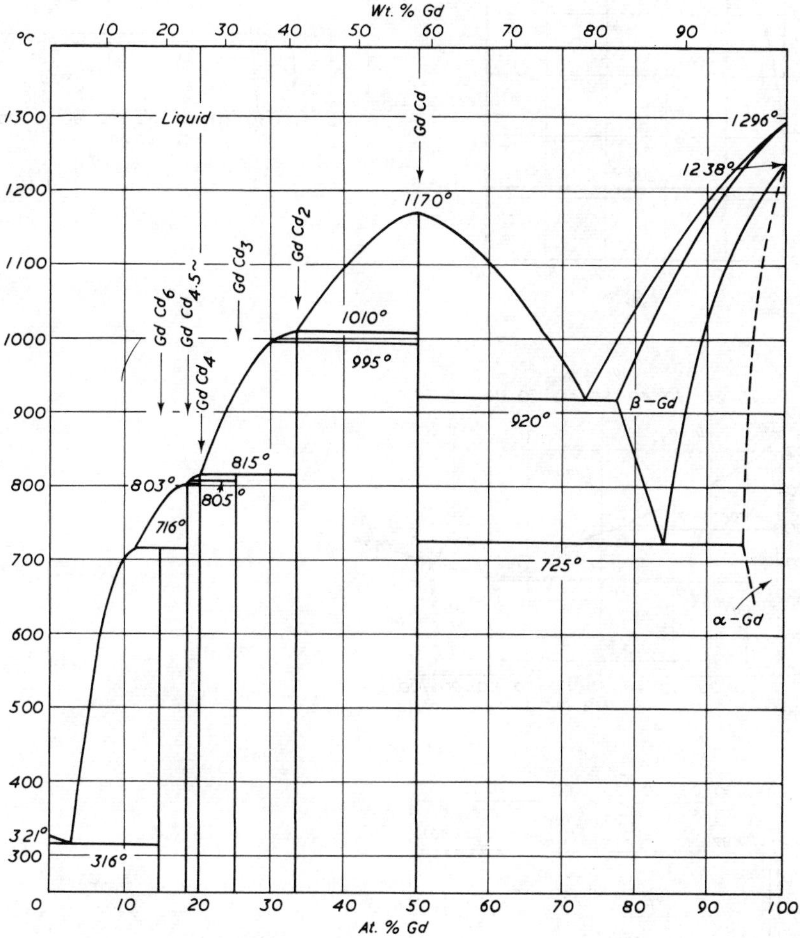

Cd–Gd

Cd–Ge

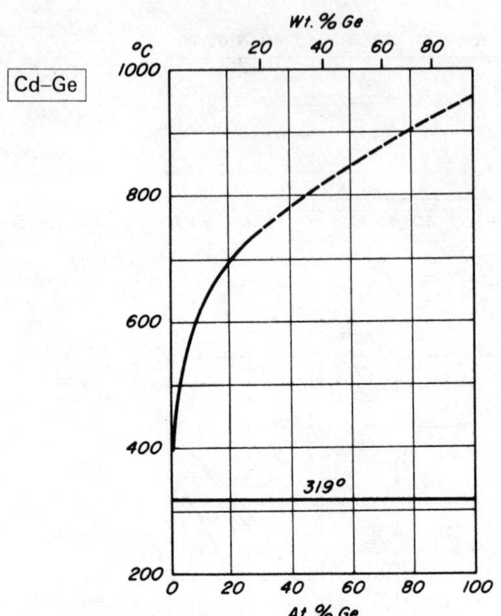

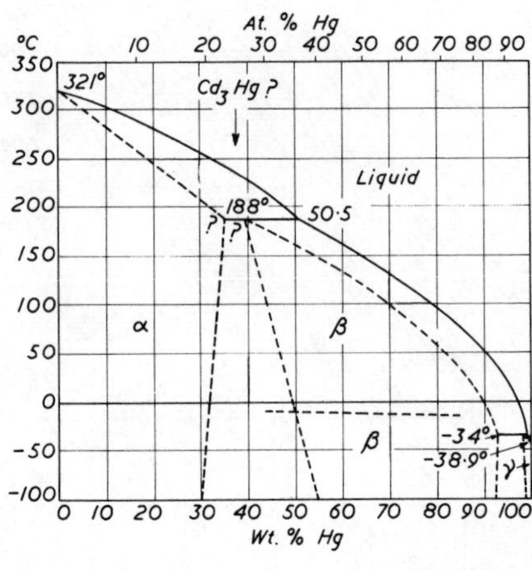

Cd–Hg

Cd–In

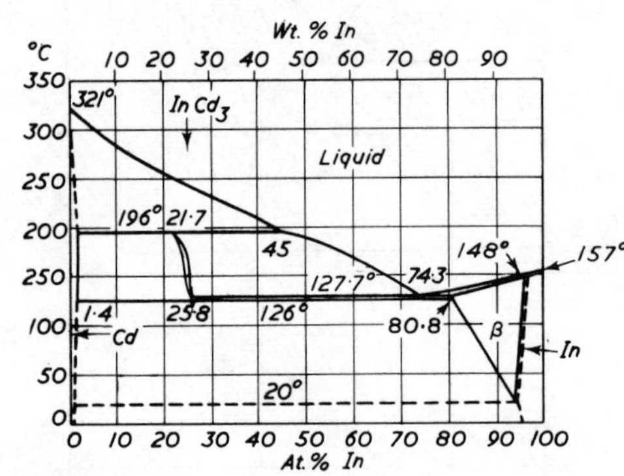

Cd–K

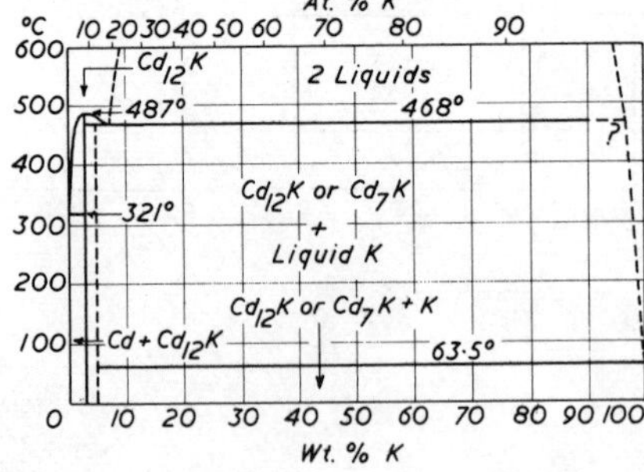

Cd–La

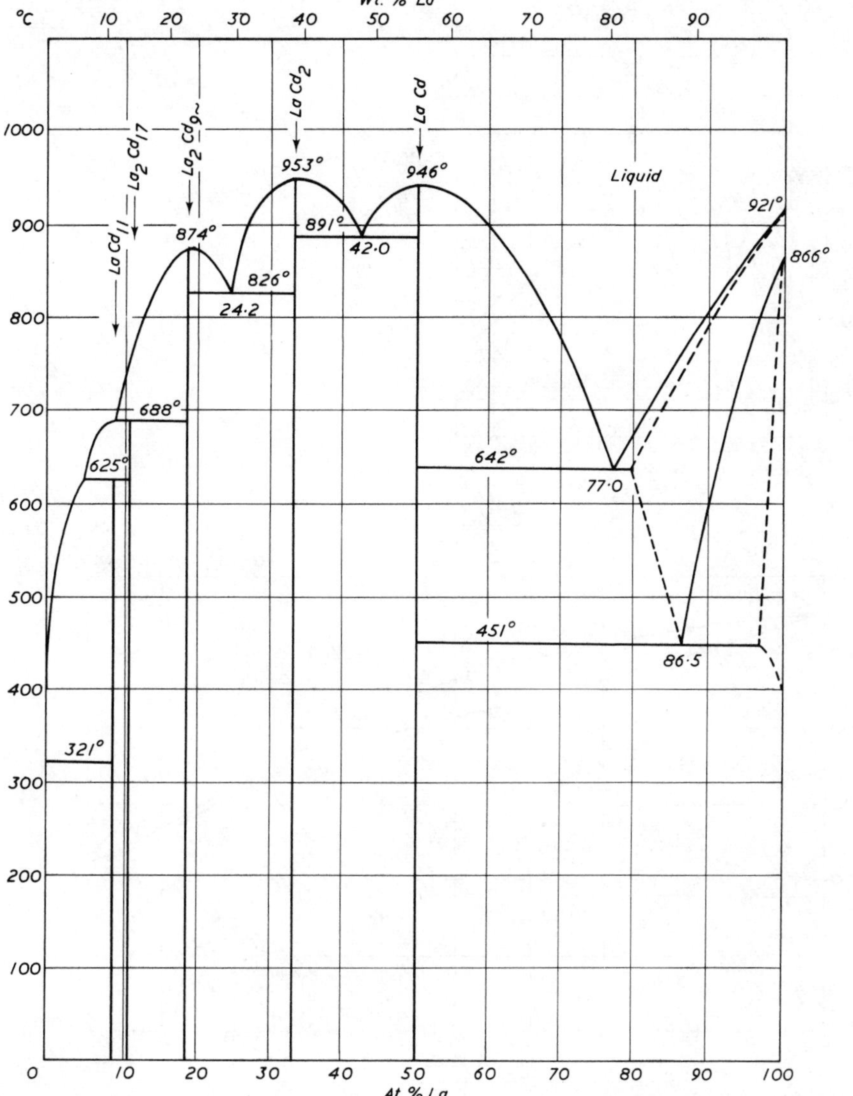
°C
Wt. % La
10
20
30
40
50
60
70
80
90
LaCd11
La2Cd17
La2Cd9~
LaCd2
LaCd
1000
900
800
700
600
500
400
300
200
100
953°
946°
Liquid
921°
874°
891°
42·0
866°
826°
24·2
688°
625°
642°
77·0
451°
86·5
321°
0
10
20
30
40
50
60
70
80
90
100
At. % La

Cd–Li

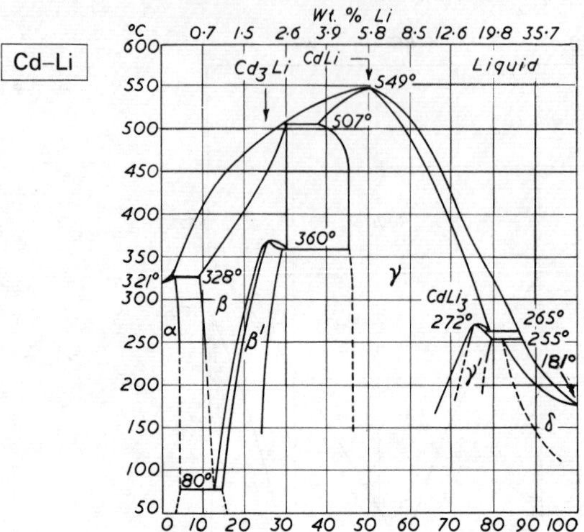

Cd–Mg

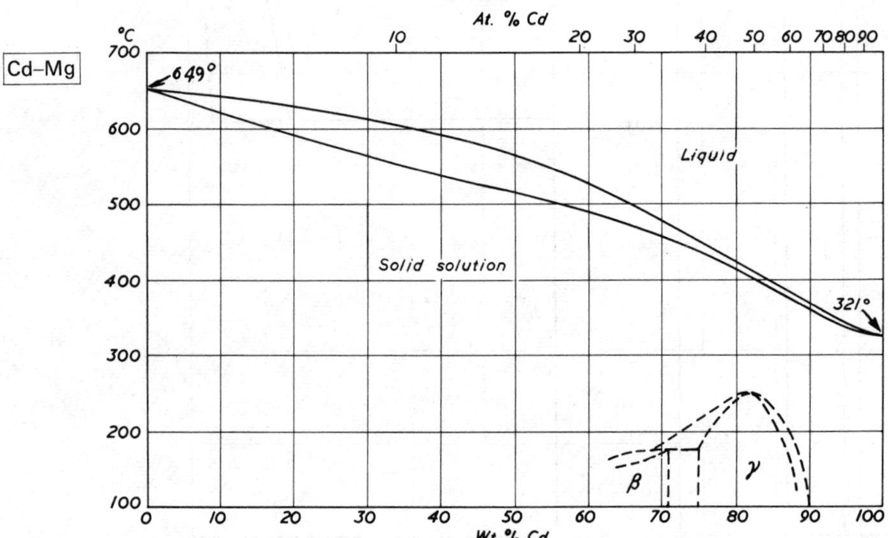

Cd–Na

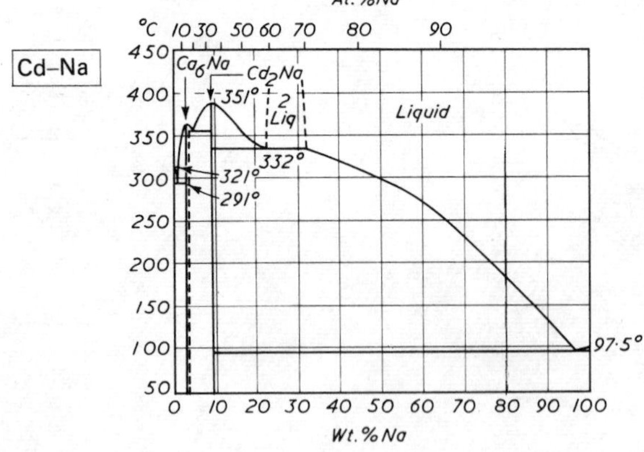

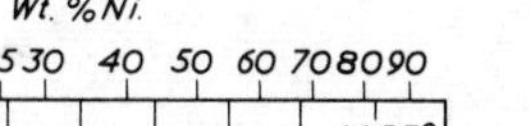

Cd–Ni

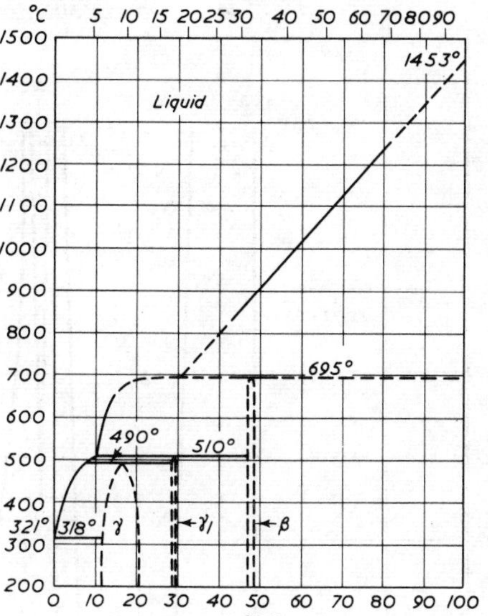

Cd–Np

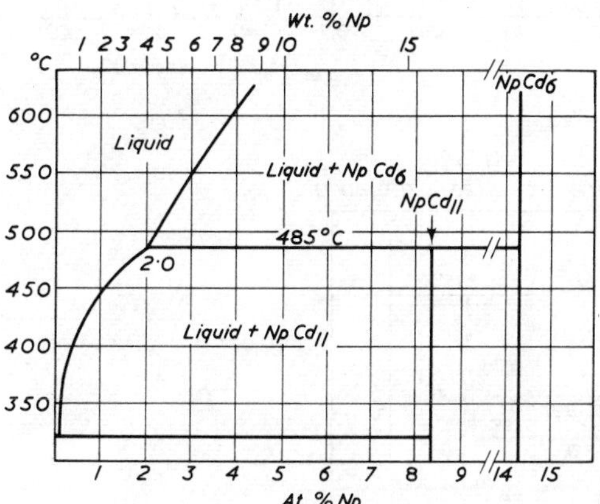

Cd–Pb

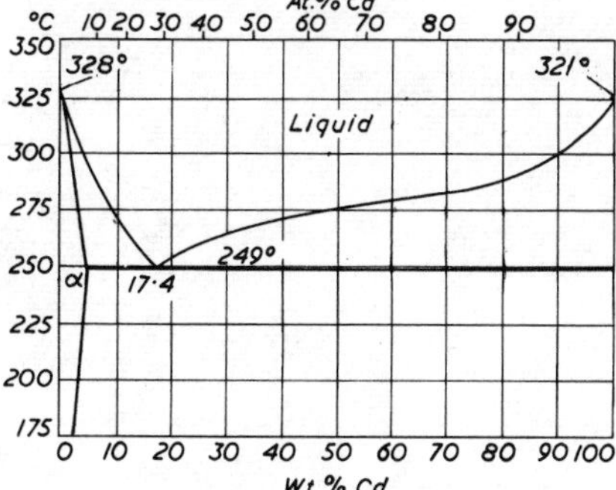

Cd–Pd

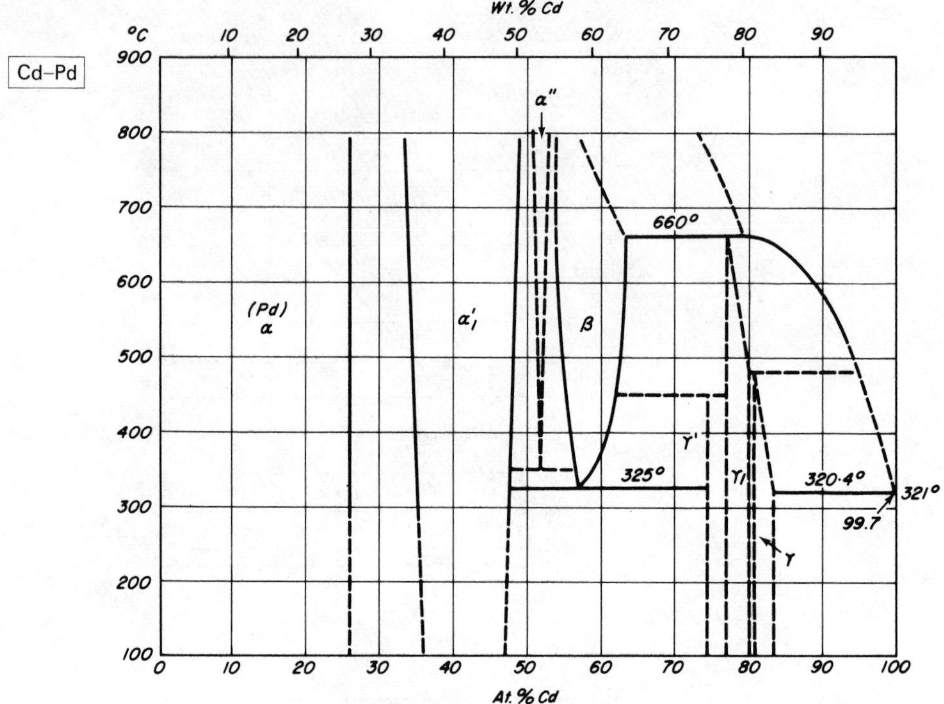

Wt.% Cd
°C
900
800
700
600
500
400
300
200
100
10
20
30
40
50
60
70
80
90
100
α''
660°
(Pd)
α
α'₁
β
γ'
325°
γ₁
320·4°
321°
99.7
γ
At.% Cd

Cd–Pt

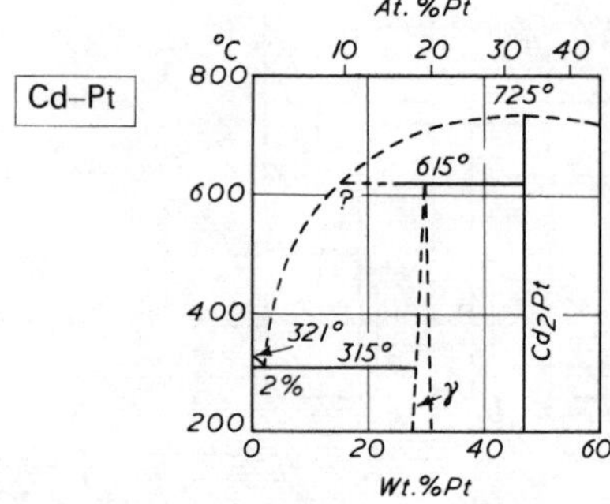

At.%Pt
°C
800
600
400
200
725°
615°
?
321°
315°
2%
γ
Cd₂Pt
Wt.%Pt

Cd–Pu

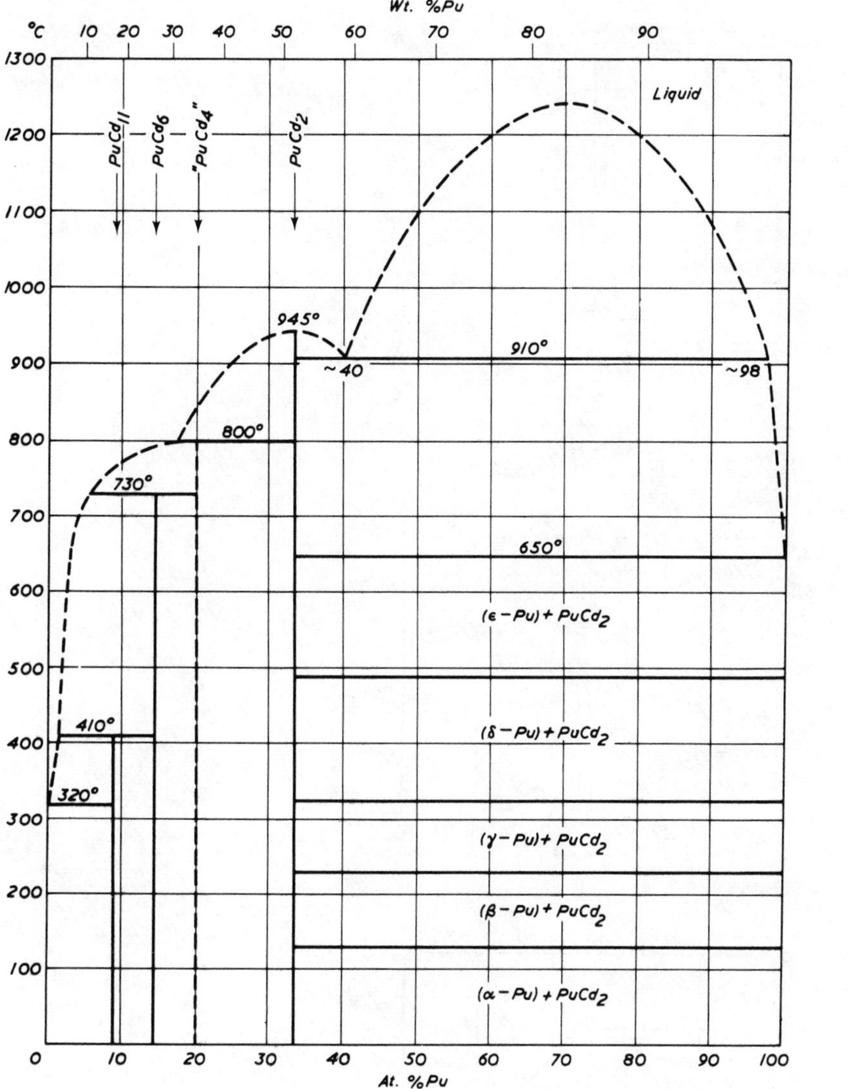
Wt. %Pu
°C
10 20 30 40 50 60 70 80 90
1300
1200
1100
1000
900
800
700
600
500
400
300
200
100
0
Liquid
PuCd11
PuCd6
"PuCd4"
PuCd2
945°
910°
~40
~98
800°
730°
650°
(ε – Pu) + PuCd2
(δ – Pu) + PuCd2
410°
320°
(γ – Pu) + PuCd2
(β – Pu) + PuCd2
(α – Pu) + PuCd2
10 20 30 40 50 60 70 80 90 100
At. %Pu

Cd–S

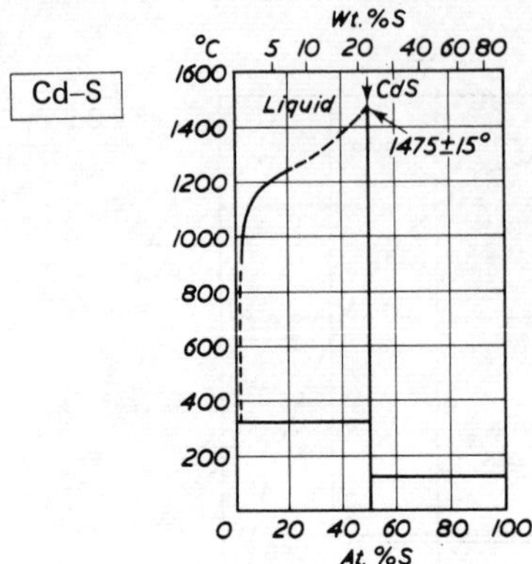

Cd–Sb

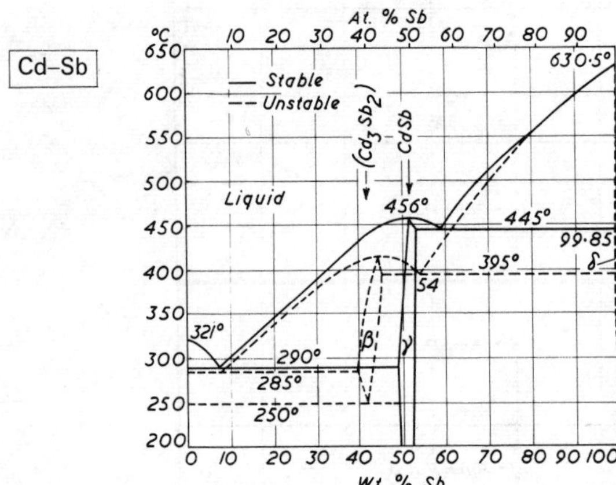

Cd–Se

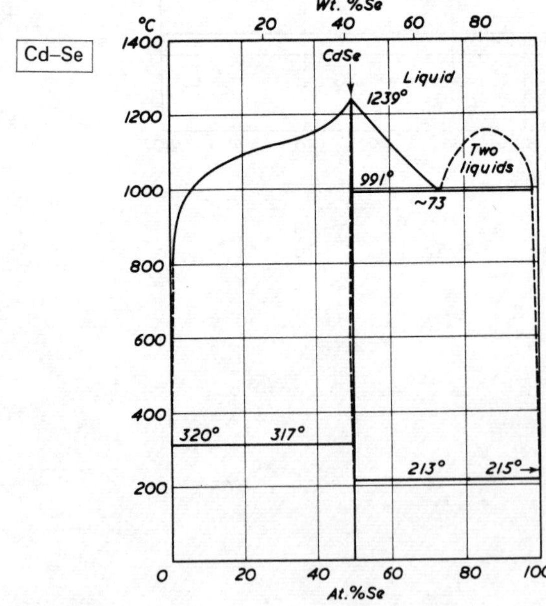

Cd–Sm

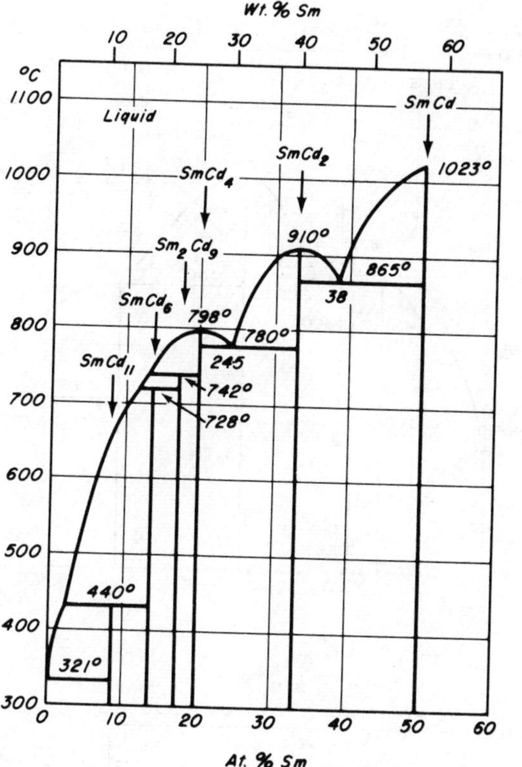
Wt. % Sm
10 20 30 40 50 60
°C
1100
1000
900
800
700
600
500
400
300
Liquid
SmCd
SmCd$_2$
1023°
SmCd$_4$
910°
865°
38
Sm$_2$Cd$_9$
SmCd$_6$
798°
780°
SmCd$_{11}$
245
742°
728°
440°
321°
0 10 20 30 40 50 60
At. % Sm

Cd–Sn

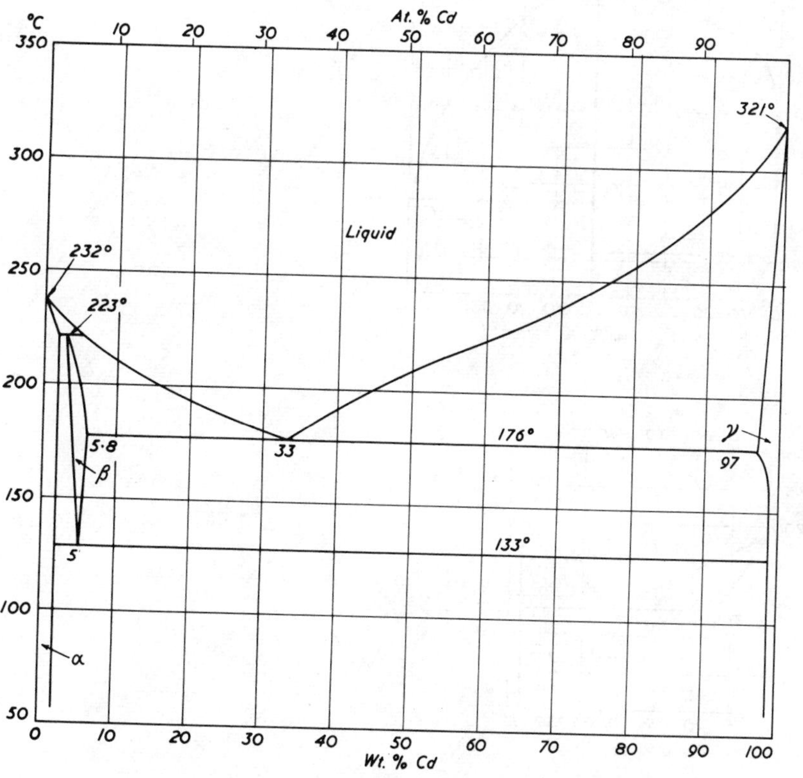
At. % Cd
10 20 30 40 50 60 70 80 90
°C
350
300
250
200
150
100
50
321°
Liquid
232°
223°
5·8
β
33
176°
γ
97
5
133°
α
0 10 20 30 40 50 60 70 80 90 100
Wt. % Cd

Cd–Sr

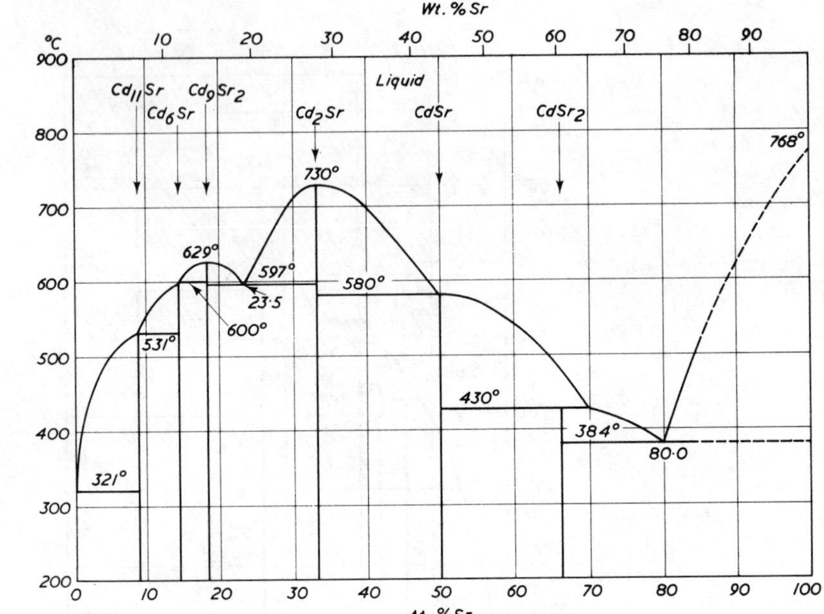

Cd–Te

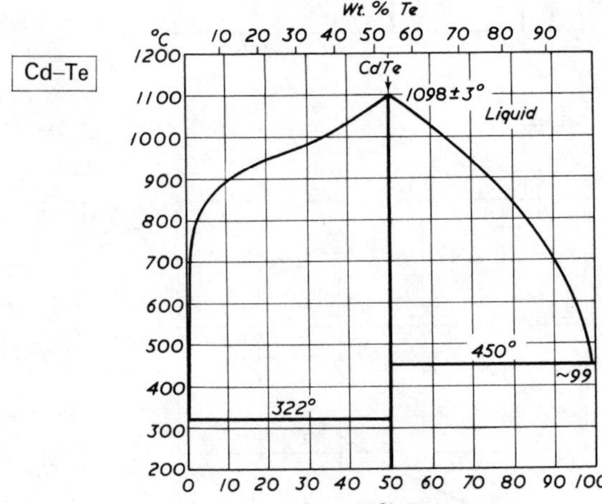

Cd–Ti

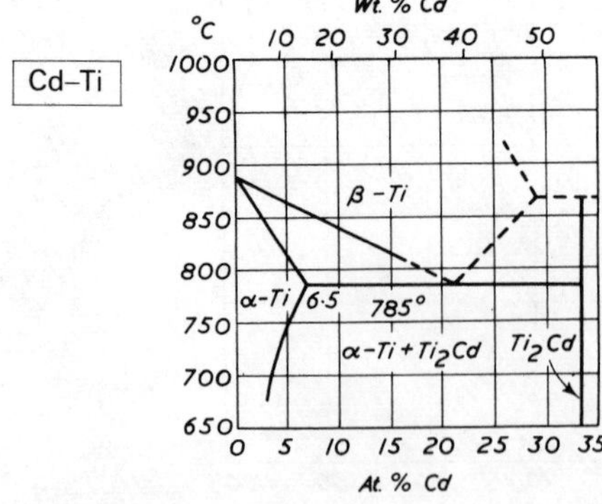

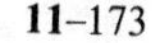

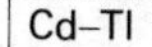

Cd–Tl

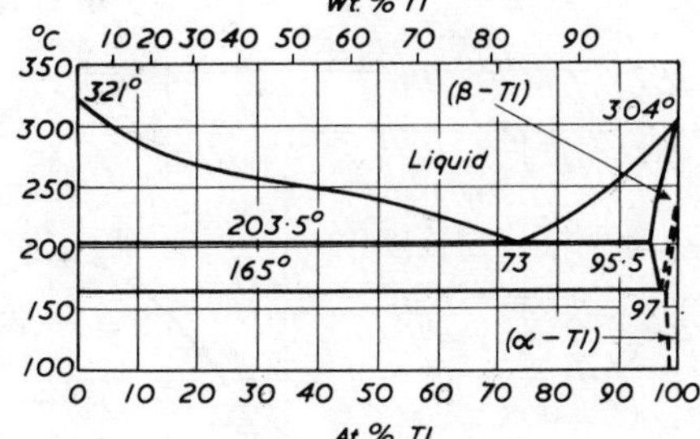
Wt. % Tl
°C
10 20 30 40 50 60 70 80 90
350
300
250
200
150
100
321°
(β – Tl)
304°
Liquid
203·5°
165°
73
95·5
97
(α – Tl)
0 10 20 30 40 50 60 70 80 90 100
At. % Tl

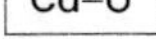

Cd–U

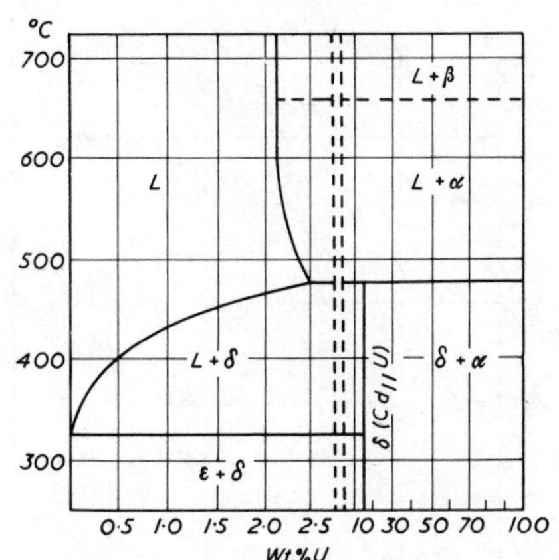
°C
700
600
500
400
300
L + β
L
L + α
L + δ
δ + α
δ (Cd11U)
ε + δ
0·5 1·0 1·5 2·0 2·5 10 30 50 70 100
Wt.%U

Cd–Y

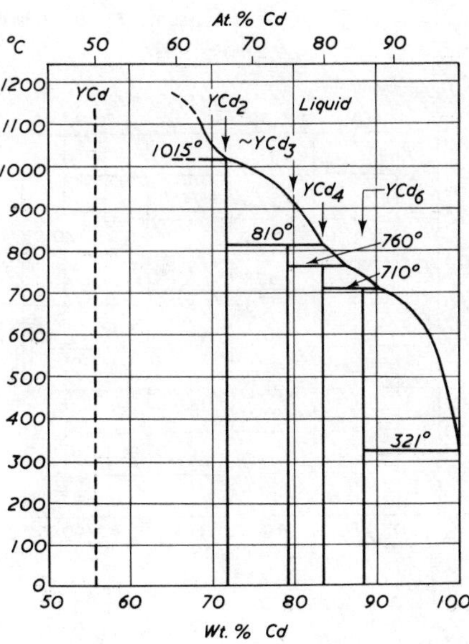
At. % Cd
°C
50 60 70 80 90
1200
1100
1000
900
800
700
600
500
400
300
200
100
0
YCd
YCd2
Liquid
1015°
~YCd3
YCd4
YCd6
810°
760°
710°
321°
50 60 70 80 90 100
Wt. % Cd

Cd–Yb

Wt. % Cd
10
20
30
40
50
60
70
80
90
C
900
800
700
600
500
400
300
200
100
0
Liquid
824°
736°
696°
YbCd
796°
Yb3Cd8
YbCd3·6
YbCd5·7
YbCd6
YbCd2
703°
680°
655°
654°
648°
627°
614°
636°
467°
321°
317°
0
10
20
30
40
50
60
70
80
90
100
At. % Cd

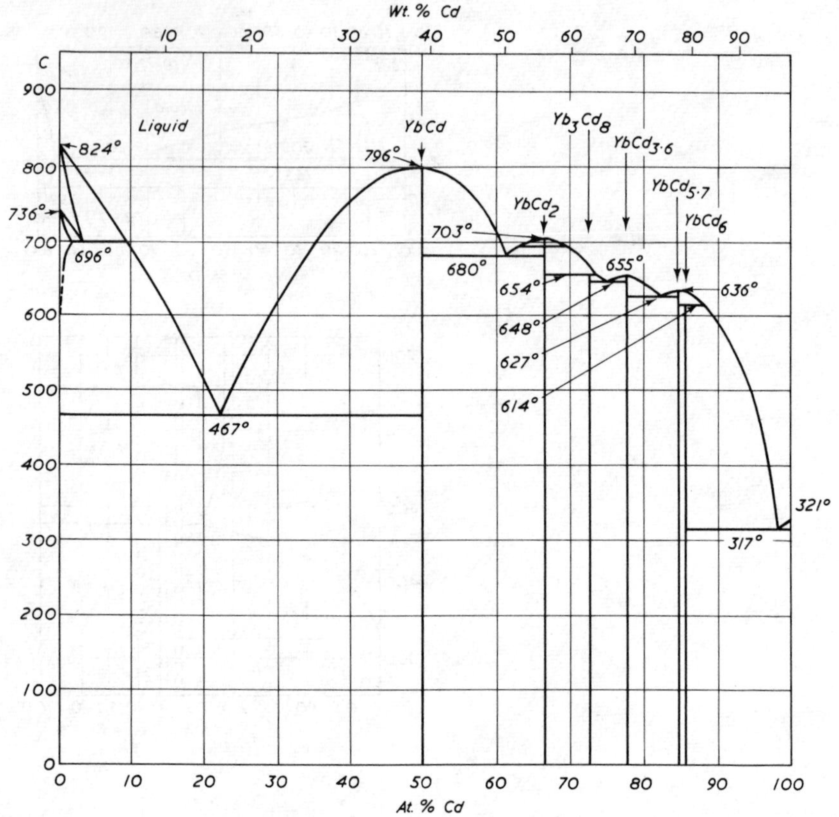

Cd–Zn

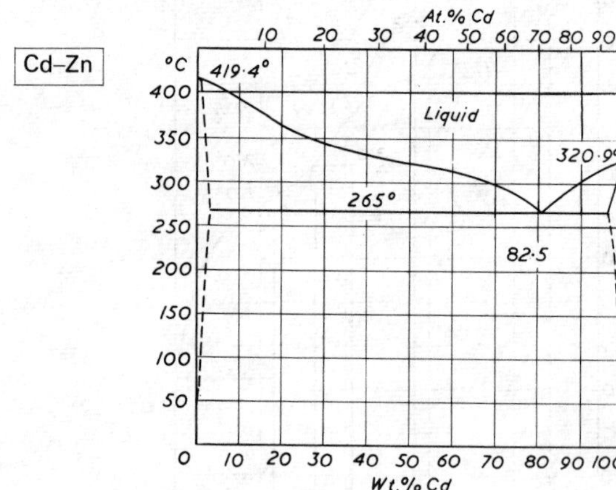

At.% Cd
10
20
30
40
50
60
70
80
90
°C
400
350
300
250
200
150
100
50
419·4°
Liquid
320·9°
265°
82·5
0
10
20
30
40
50
60
70
80
90
100
Wt.% Cd

Cd–Zr

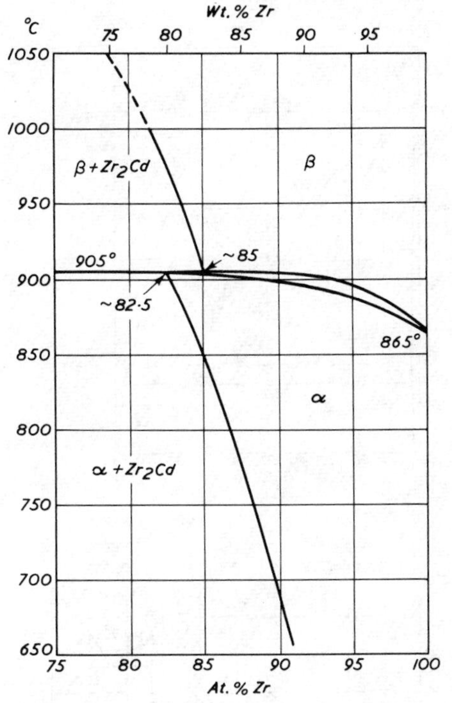
Wt.% Zr
°C
75
80
85
90
95
1050
1000
950
900
850
800
750
700
650
β+Zr2Cd
β
905°
~85
~82·5
865°
α
α+Zr2Cd
75
80
85
90
95
100
At.% Zr

Ce–Co

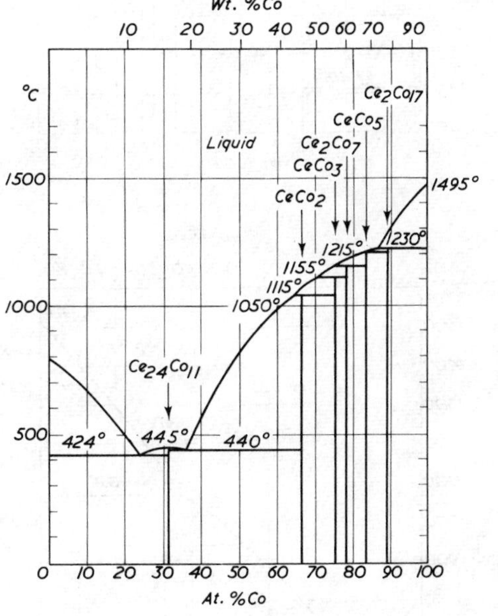
Wt. %Co
10
20
30
40
50
60
70
90
°C
Liquid
Ce2Co17
CeCo5
Ce2Co7
CeCo3
CeCo2
1500
1495°
1215°
1230°
1155°
1115°
1050°
1000
Ce24Co11
500
424°
445°
440°
0
10
20
30
40
50
60
70
80
90
100
At. %Co

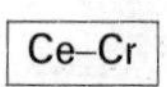

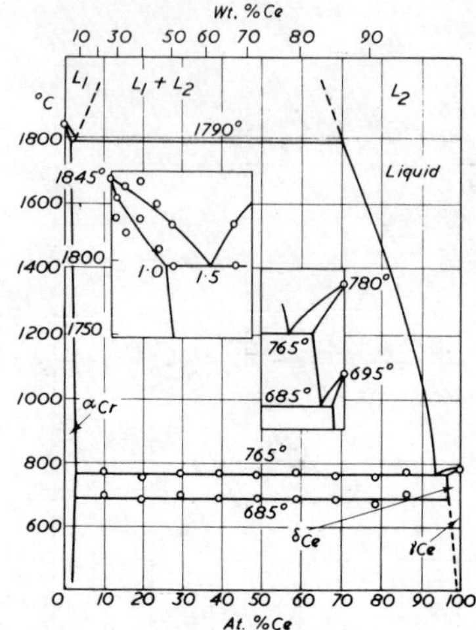
Wt. %Ce
°C
L1
L1 + L2
L2
1790°
Liquid
1845°
1800
1·0
1·5
1750
780°
765°
695°
685°
αCr
765°
685°
δCe
γCe
At. %Ce

Ce–Cu

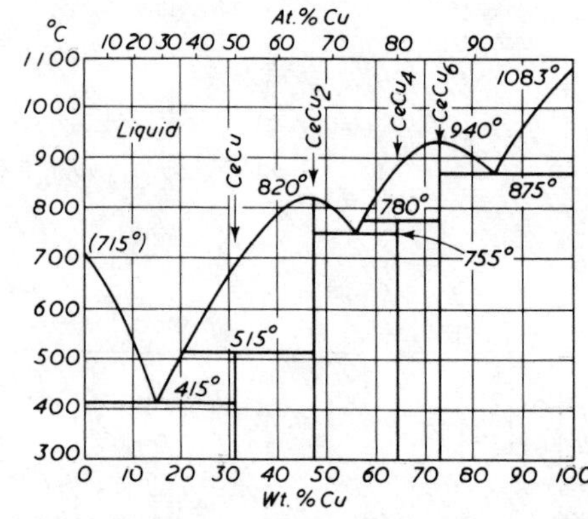
°C
At.% Cu
Liquid
CeCu
CeCu2
CeCu4
CeCu6
1083°
940°
820°
875°
780°
755°
(715°)
515°
415°
Wt. % Cu

Ce–Fe

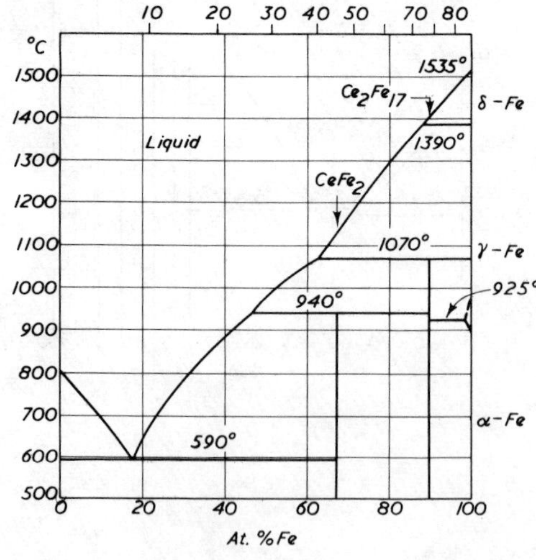
Wt. %Fe
°C
1535°
Ce2Fe17
δ-Fe
Liquid
1390°
CeFe2
1070°
γ-Fe
925°
940°
α-Fe
590°
At. %Fe

Ce–Ge

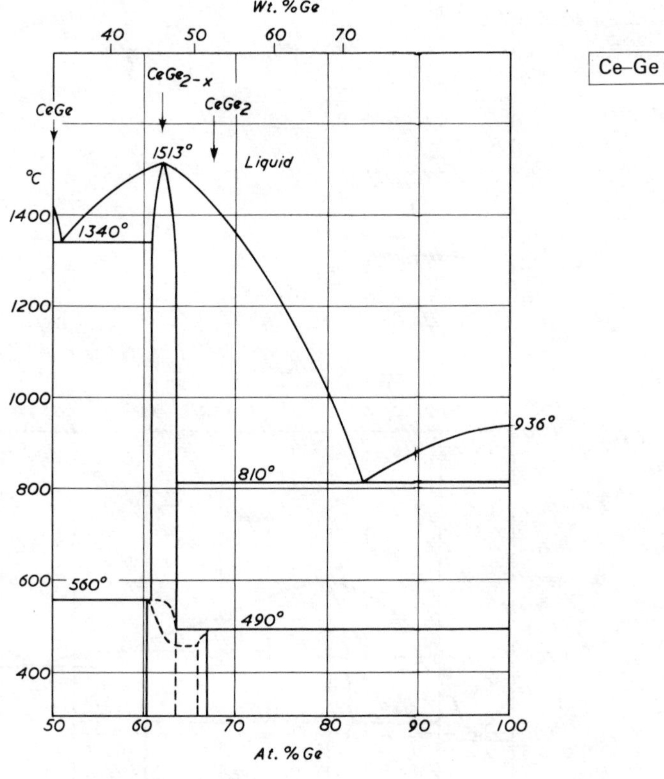

Wt. % Ge
40
50
60
70
CeGe
CeGe2-x
CeGe2
1513°
Liquid
°C
1400
1340°
1200
1000
936°
810°
800
600
560°
490°
400
50
60
70
80
90
100
At. % Ge

Ce–In

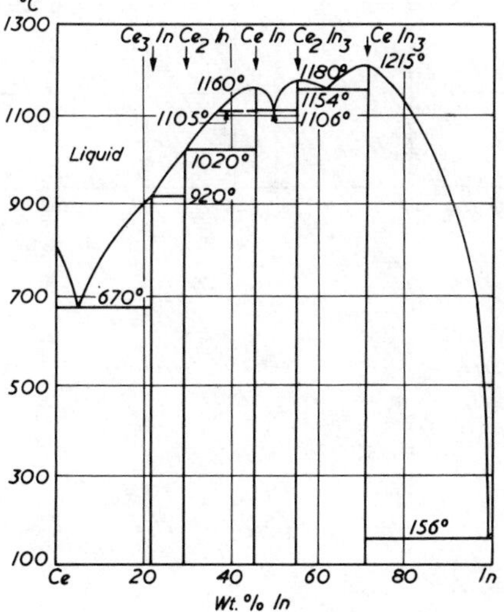

°C
1300
Ce3 In
Ce2 In
Ce In
Ce2 In3
Ce In3
1180°
1215°
1160°
1154°
1100
1105°
1106°
Liquid
1020°
900
920°
700
670°
500
300
156°
100
Ce
20
40
60
80
In
Wt. % In

Ce–La

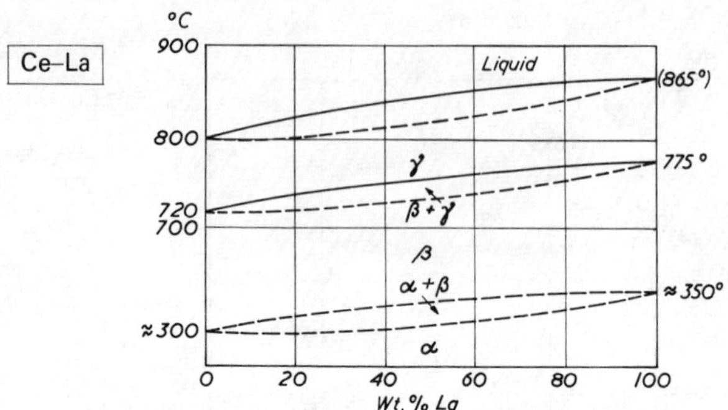
°C
900
800
720
700
≈300
Liquid
(865°)
775°
γ
β+γ
β
α+β
≈350°
α
0
20
40
60
80
100
Wt.% La

Ce–Mg

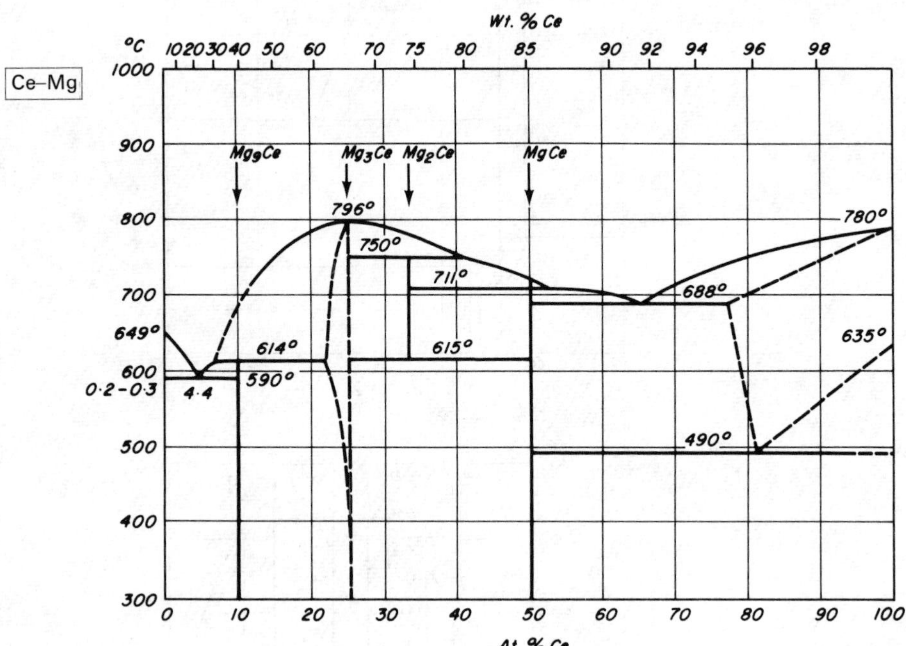
Wt.% Ce
°C
10 20 30 40 50 60 70 75 80 85 90 92 94 96 98
1000
900
800
700
600
500
400
300
Mg9Ce
Mg3Ce
Mg2Ce
MgCe
796°
750°
711°
780°
688°
649°
614°
615°
635°
590°
0·2–0·3
4·4
490°
0
10
20
30
40
50
60
70
80
90
100
At.% Ce

Ce–Mn

Wt. % Ce
°C
10 20 30 40 50 60 70 80 90
1300
1200
1138°
1100
1000
900
800
700
600
500
1244°
L₁
δMn
γMn
L₁+L₂
L₂
998°
18·5(36·7)
32(54·6)
βMn
727°
αMn
622°
86(94)
638°
84(93)
791°
δCe
718°
γCe
0 10 20 30 40 50 60 70 80 90 100
At. % Ce

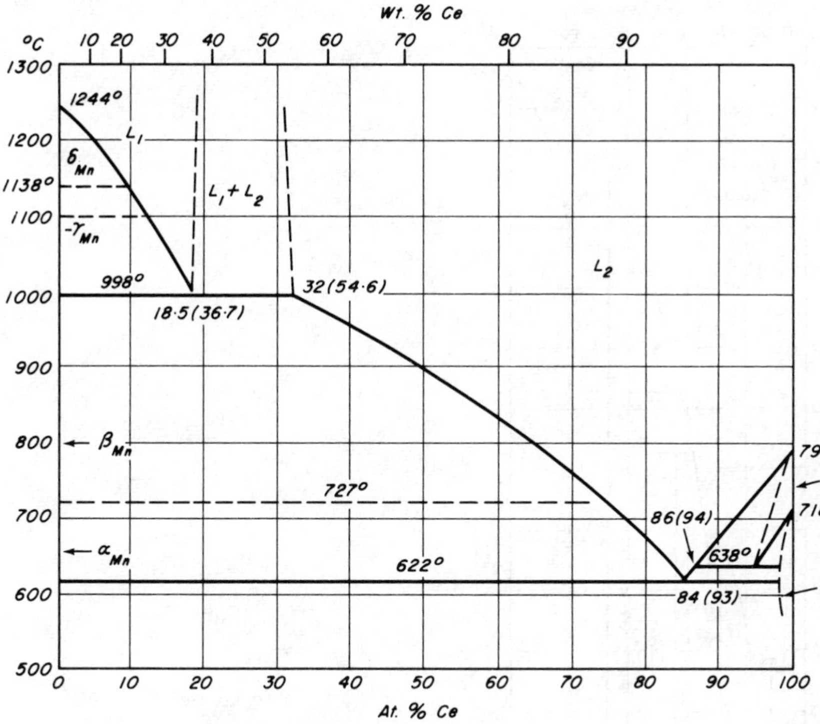

Ce–Nb

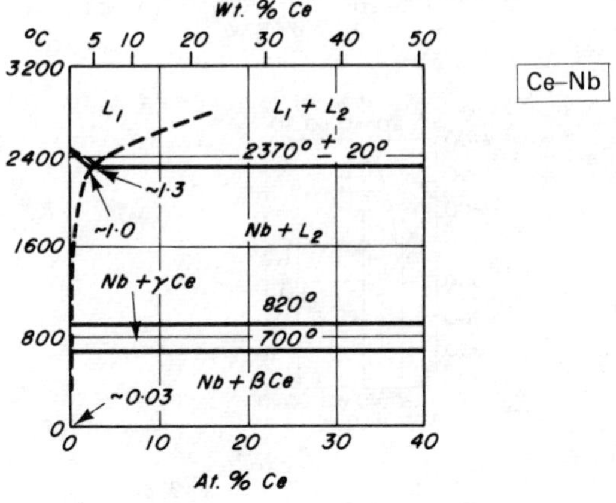

Wt. % Ce
°C
5 10 20 30 40 50
3200
2400
1600
800
0
L₁
L₁ + L₂
2370° ± 20°
~1·3
~1·0
Nb + L₂
Nb + γCe
820°
700°
Nb + βCe
~0·03
0 10 20 30 40
At. % Ce

Ce–Ni

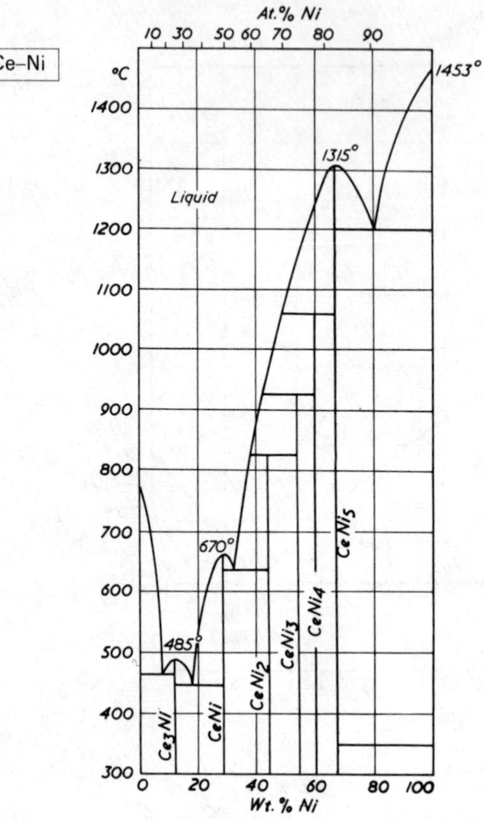

Ce–Pb

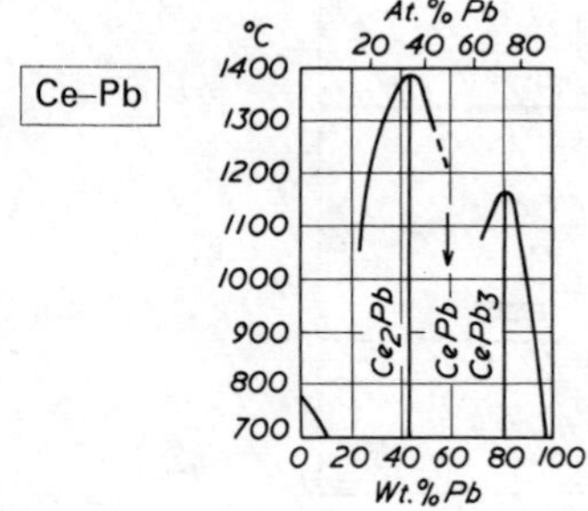

Ce–Pd

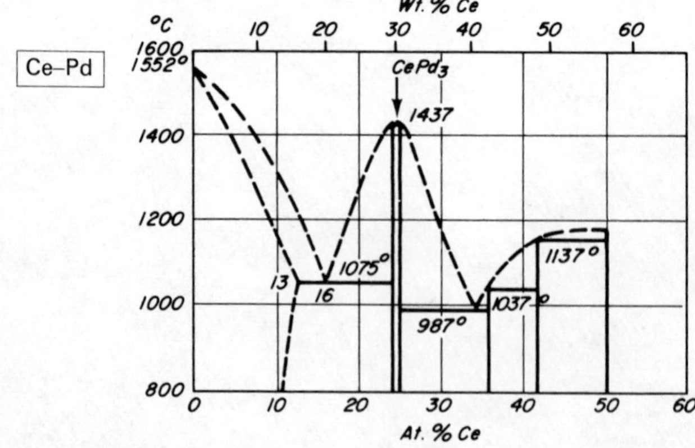

Ce–Pu

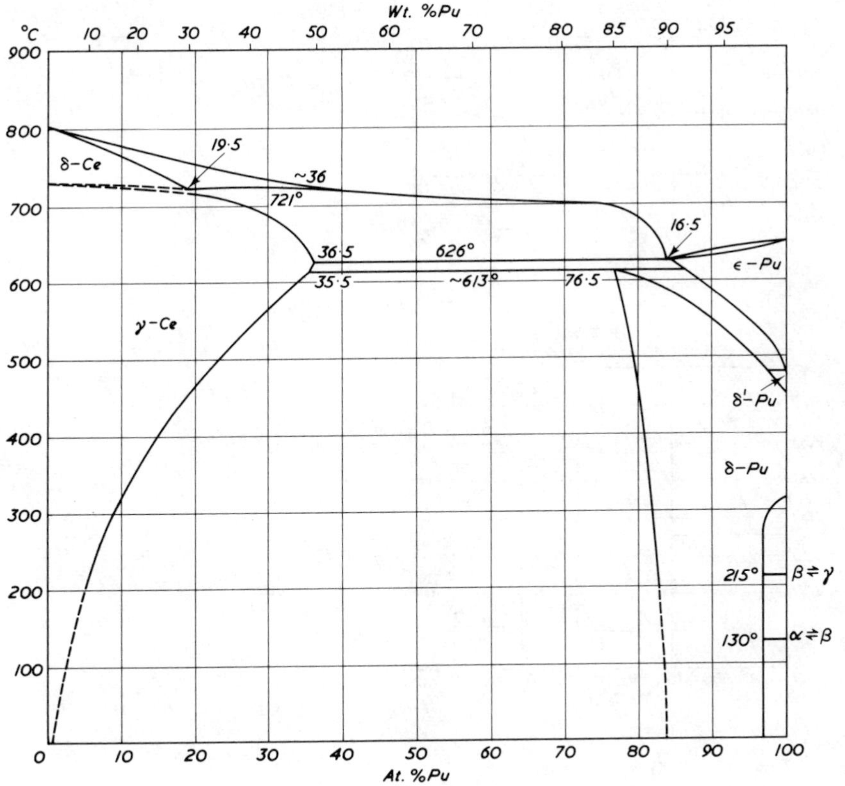

Wt. %Pu
°C
10
20
30
40
50
60
70
80
85
90
95
900
800
700
600
500
400
300
200
100
0
10
20
30
40
50
60
70
80
90
100
At. %Pu
δ-Ce
19·5
~36
721°
16·5
36·5
626°
ε-Pu
35·5
~613°
76·5
γ-Ce
δ'-Pu
δ-Pu
215°
β⇌γ
130°
α⇌β

Ce–Ru

°C
2500
2310°
2000
1500
1000
500
Ru2Ce
Liquid
~1570°
ε
β
798°
645°
32
γ
0
20
40
60
80
100
Wt.% Ce

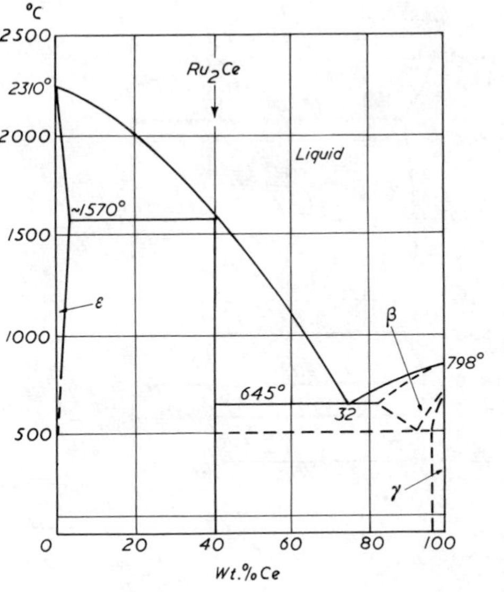

Ce–Si

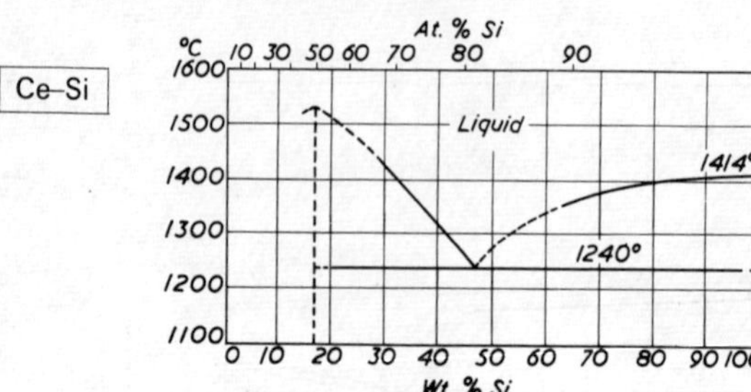

Ce–Sn

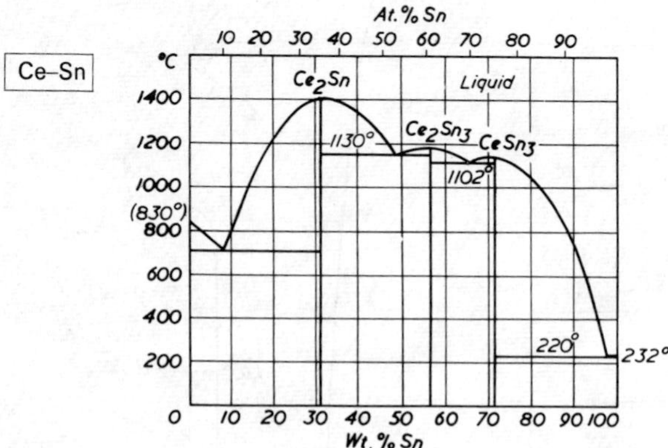

Ce–Ti

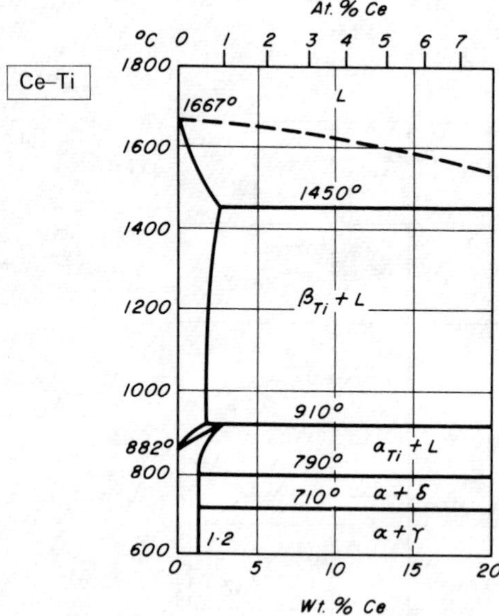

Ce–Th

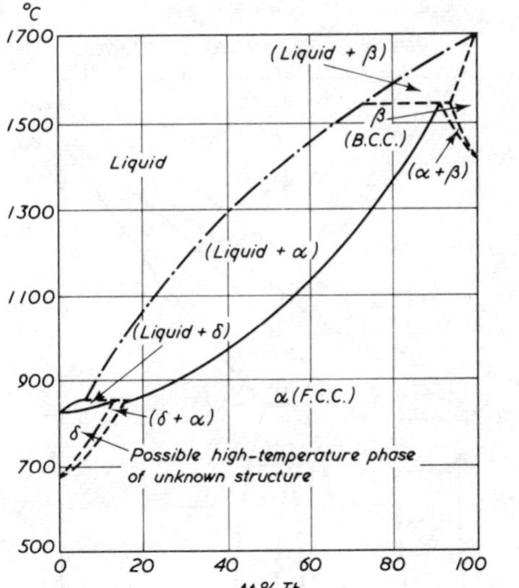

Ce–Tl

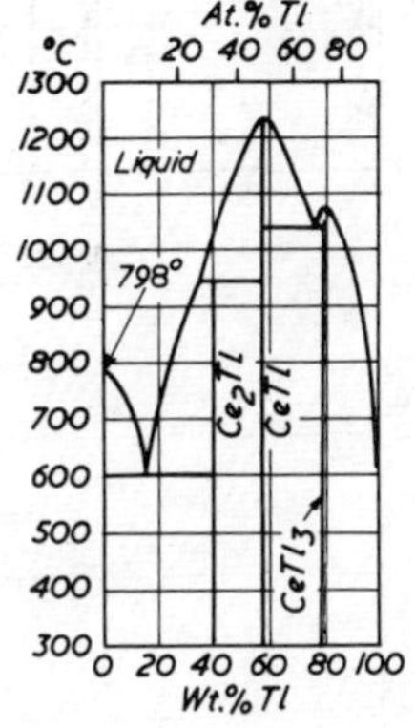

Ce–U

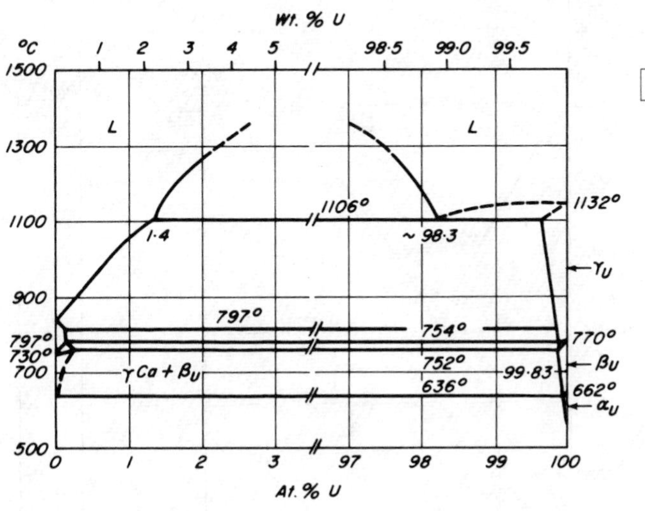

Ce–Y

Wt. % Y
°C
10 20 30 40 50 60 70 80 90
1600
1400
1200
1000
800
600
400
200
0
1520°
1485°
CeY
L
804°
~33
850°
~52
α_Y
730°
700°
γ_{Ce}
β_{Ce}
0 20 40 60 80 100
At. % Y

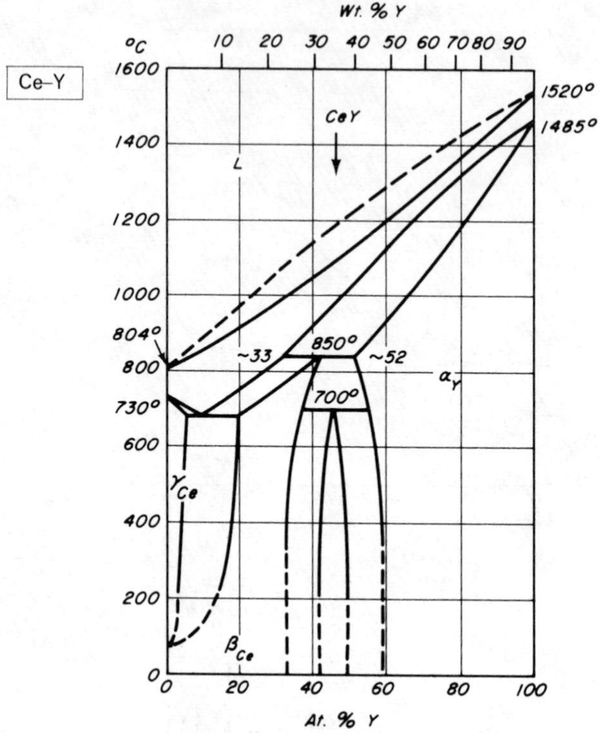

Ce–Zn

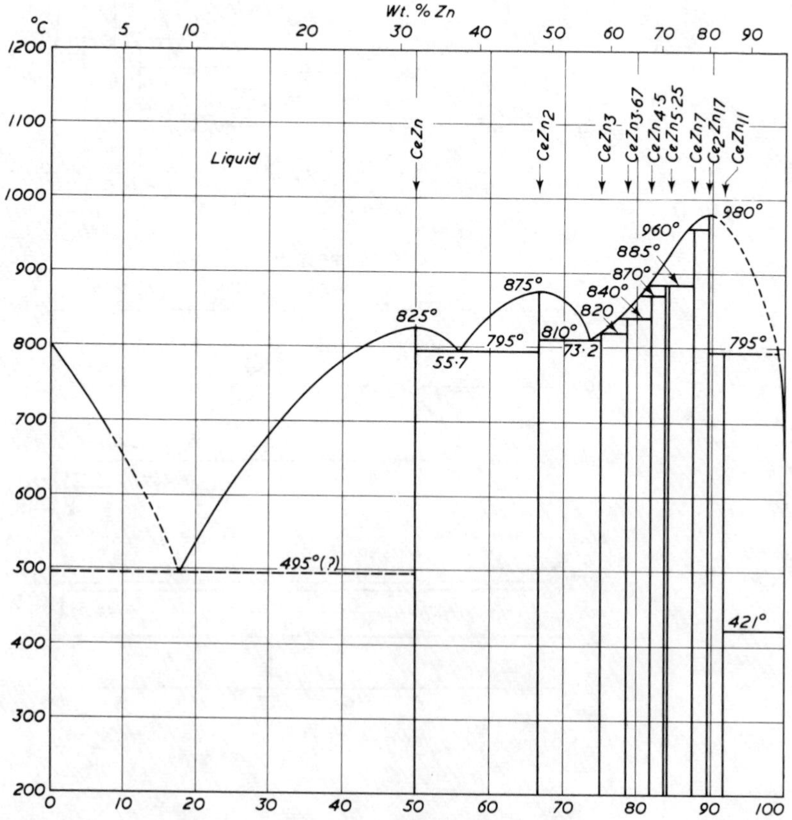

Wt. % Zn
°C
5 10 20 30 40 50 60 70 80 90
1200
1100
1000
900
800
700
600
500
400
300
200
Liquid
CeZn
$CeZn_2$
$CeZn_3$
$CeZn_{3.67}$
$CeZn_{4.5}$
$CeZn_{5.25}$
$CeZn_7$
Ce_2Zn_{17}
$CeZn_{11}$
980°
960°
885°
870°
840°
820
875°
825°
810°
795°
73·2
55·7
795°
495°(?)
421°
0 10 20 30 40 50 60 70 80 90 100
At.%Zn

Co–Cr

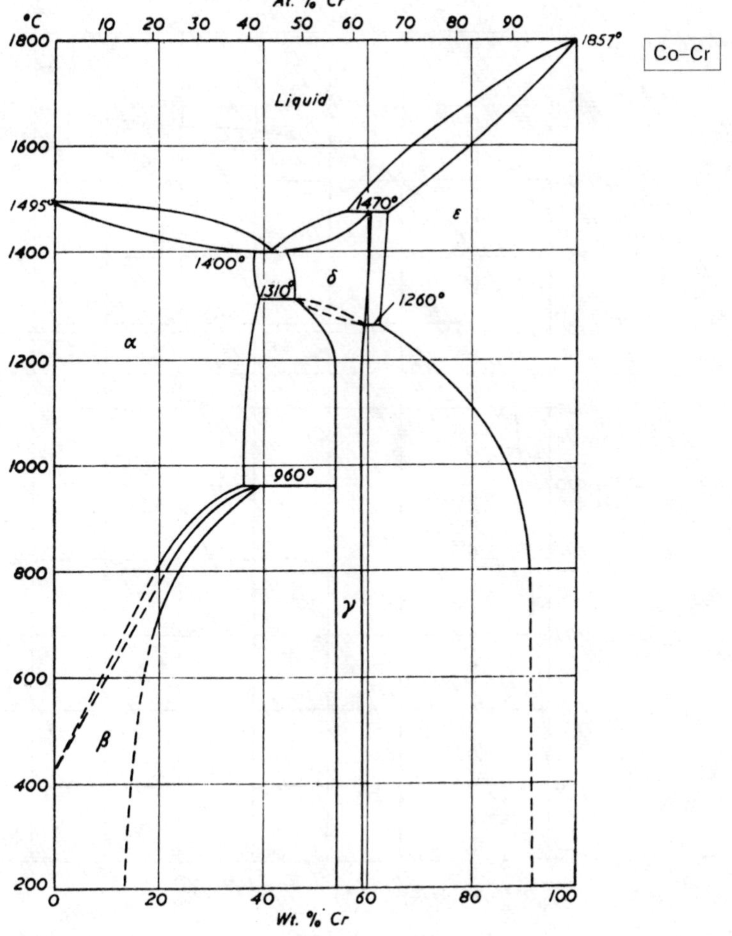
At. % Cr
°C
10
20
30
40
50
60
70
80
90
1800
1857°
Liquid
1600
1495°
1470°
ε
1400
1400°
δ
1310°
1260°
α
1200
1000
960°
800
γ
600
β
400
200
0
20
40
60
80
100
Wt. % Cr

Co–Cu

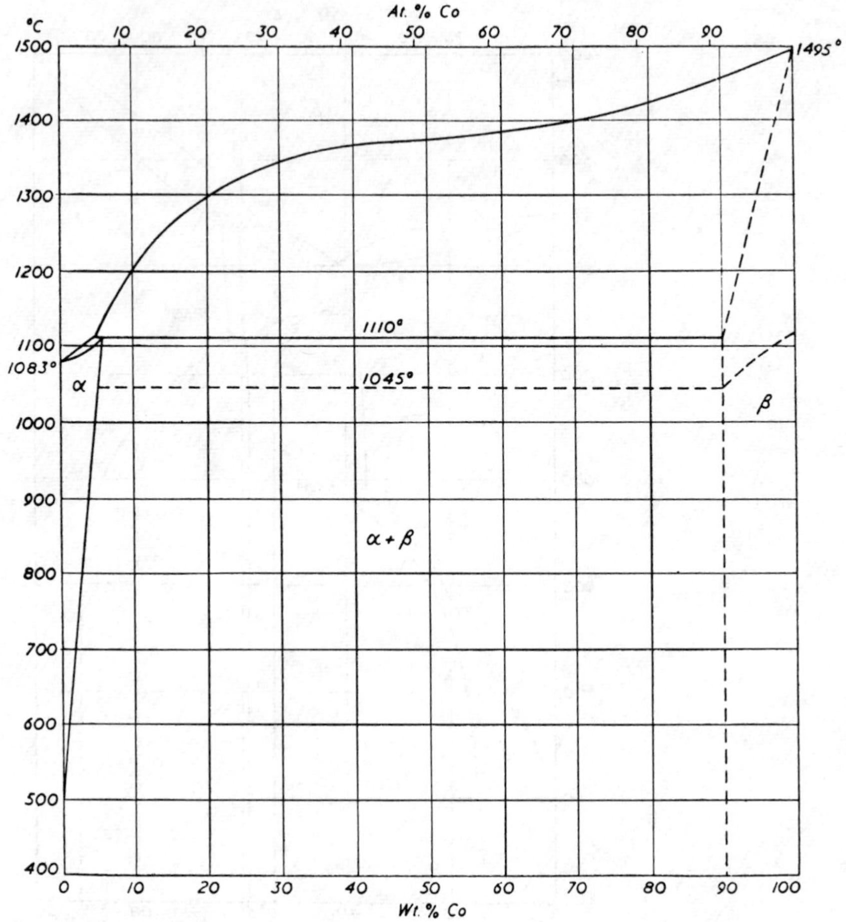

At. % Co
°C
1495°
1110°
1083°
1045°
α
β
α + β
Wt. % Co

Co–Er

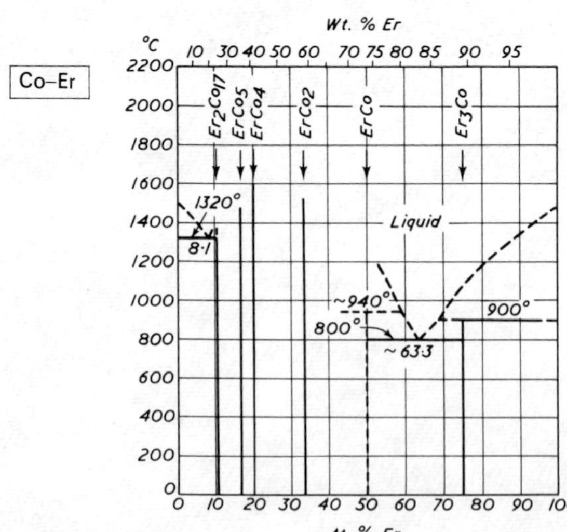

Co–Fe

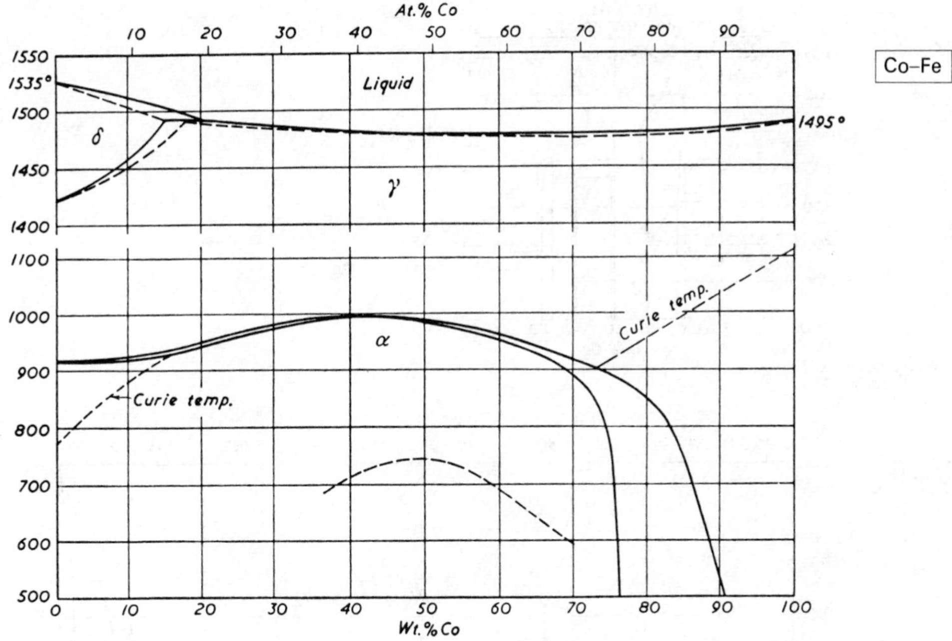

Co–Gd

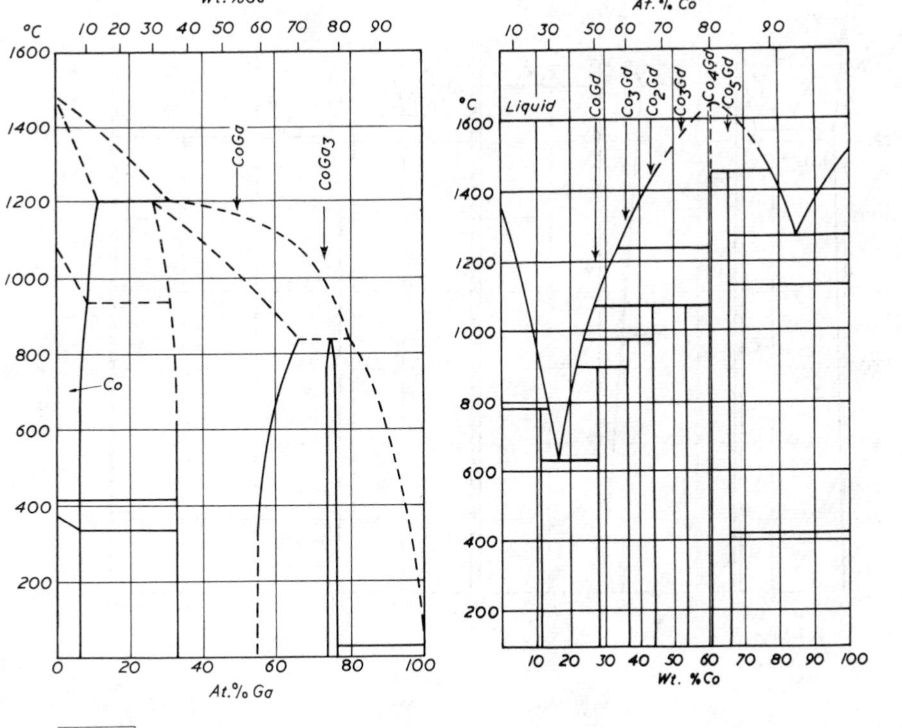

Co–Ga

Co–Ge

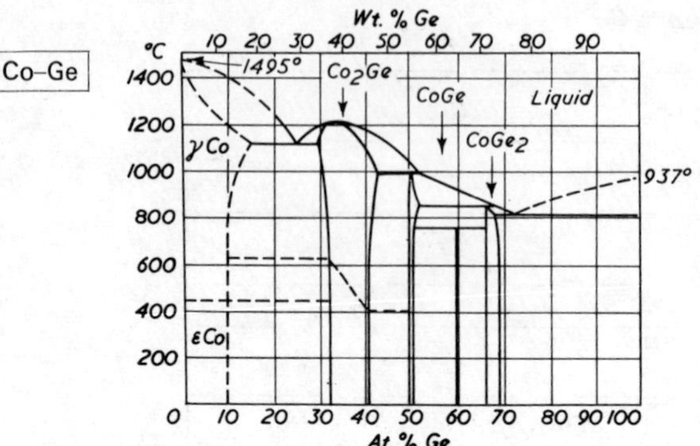
Wt. % Ge
10 20 30 40 50 60 70 80 90
°C
1400
1200
1000
800
600
400
200
0
1495°
Co_2Ge
CoGe
$CoGe_2$
Liquid
γCo
εCo
937°
0 10 20 30 40 50 60 70 80 90 100
At. % Ge

Co–Hf

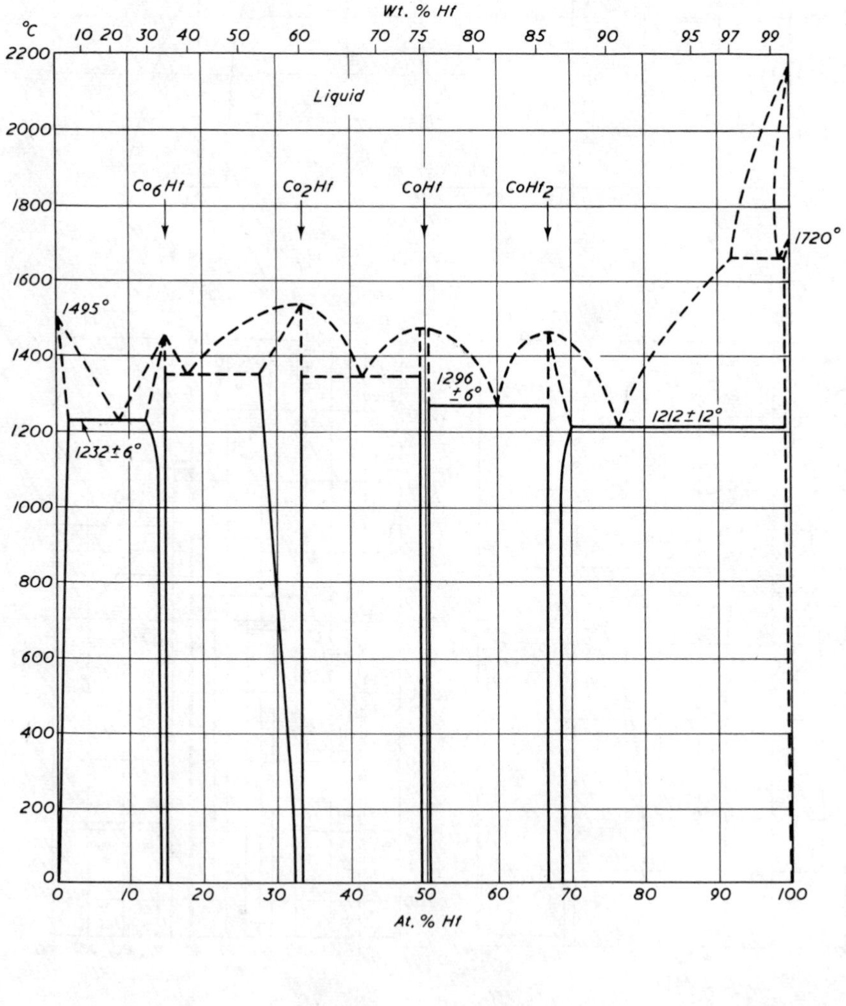

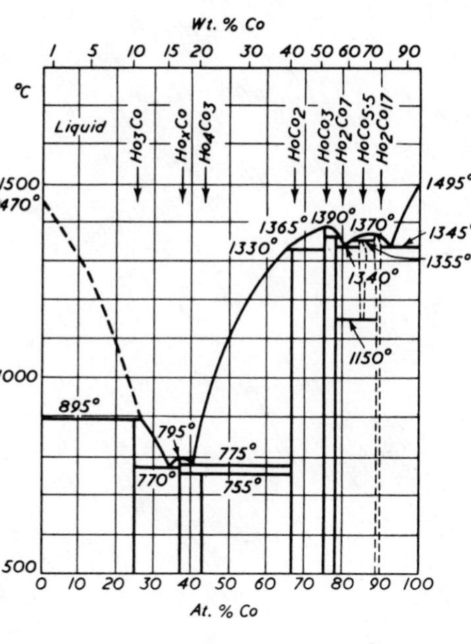

Co–Ho

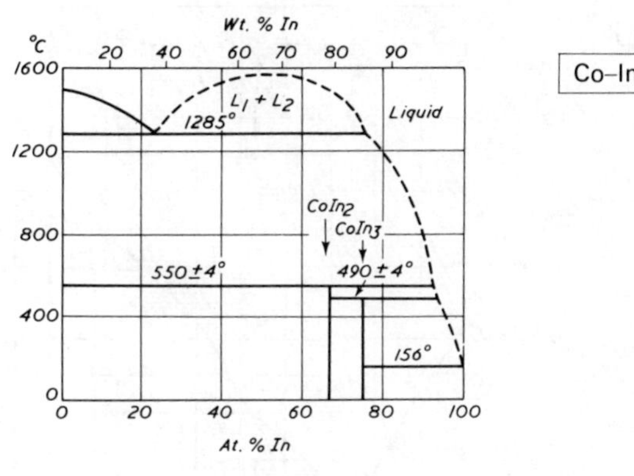

Co–In

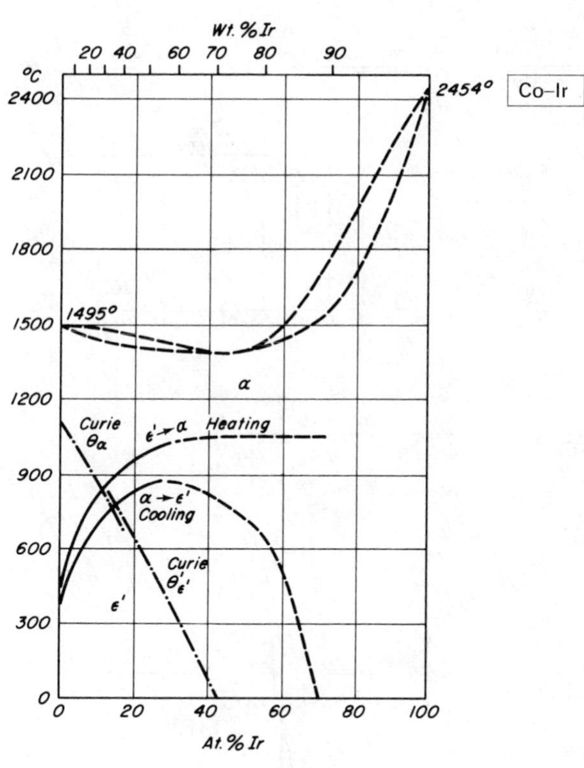

Co–Ir

Co–La

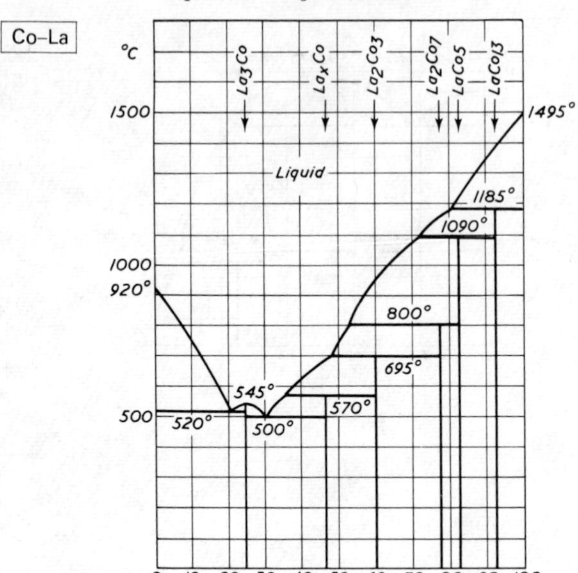

Co–Mg

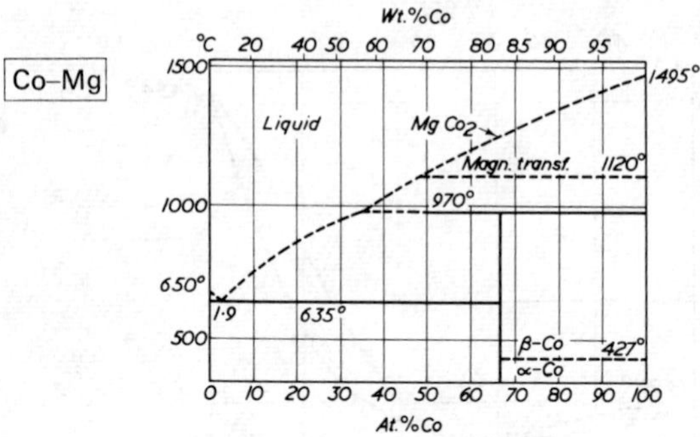

Co–Mn

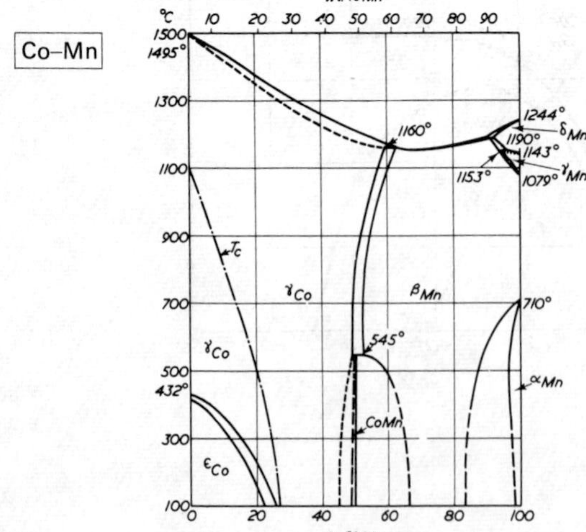

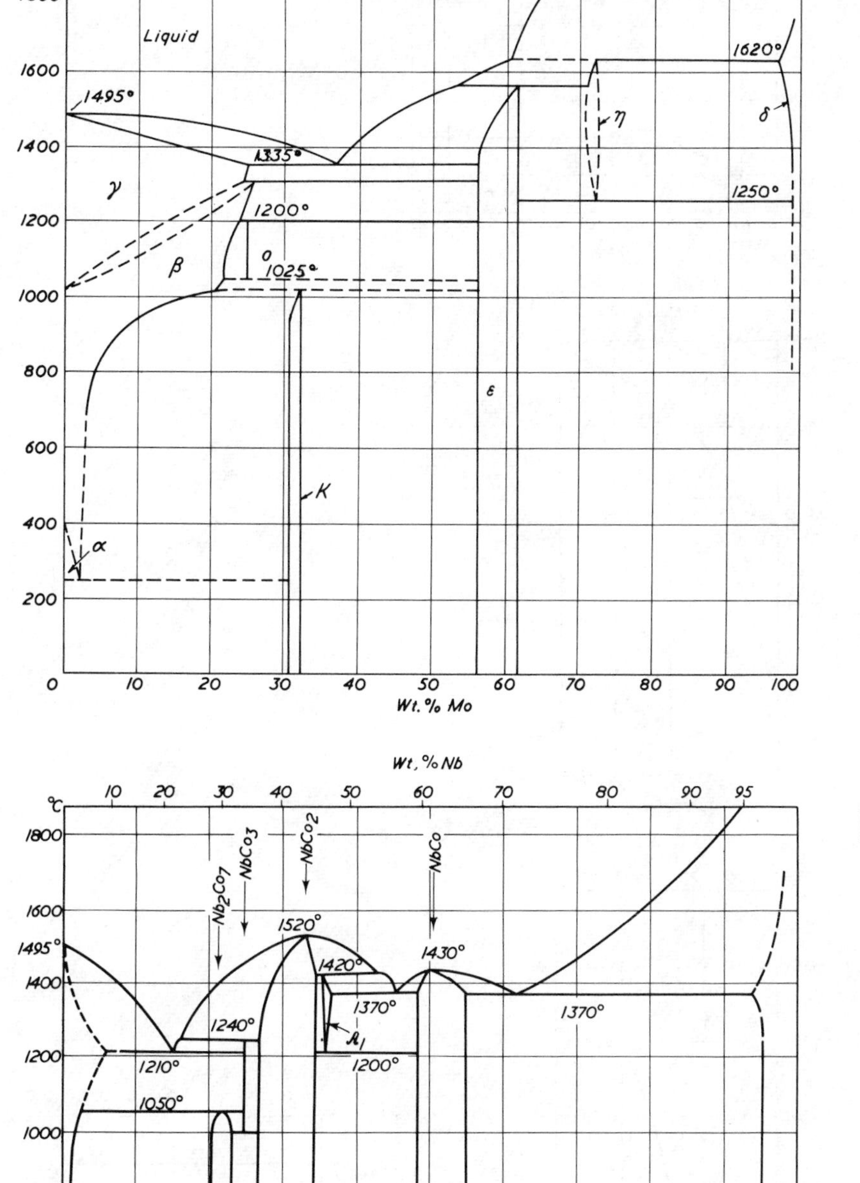
Co–Mo
At. % Mo
°C
Liquid
1495°
1335°
1200°
1025°
1620°
1250°
γ
β
σ
η
δ
ε
K
α
Wt. % Mo
Co–Nb
Wt, % Nb
°C
Nb_2Co_7
$NbCo_3$
$NbCo_2$
NbCo
1520°
1495°
1420°
1430°
1370°
1240°
1210°
1200°
1050°
k_1
α ⇄ β
At. % Nb

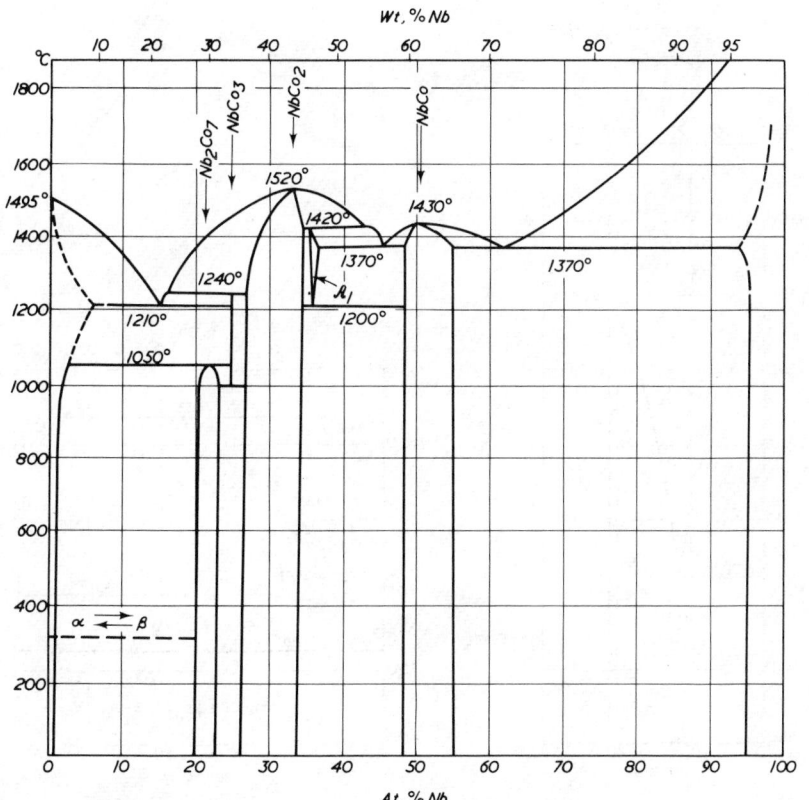

Co–Ni

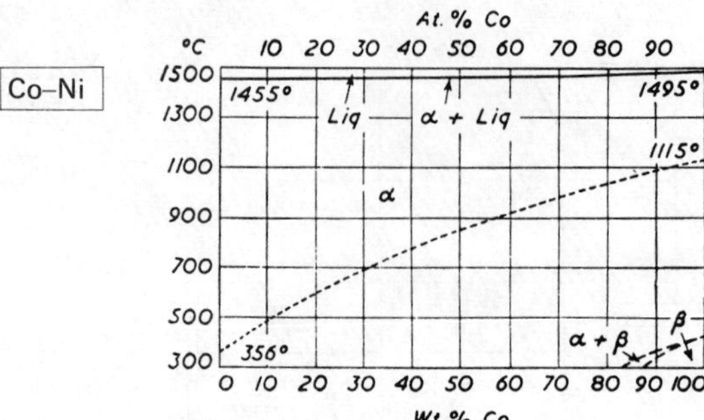

Co–Os

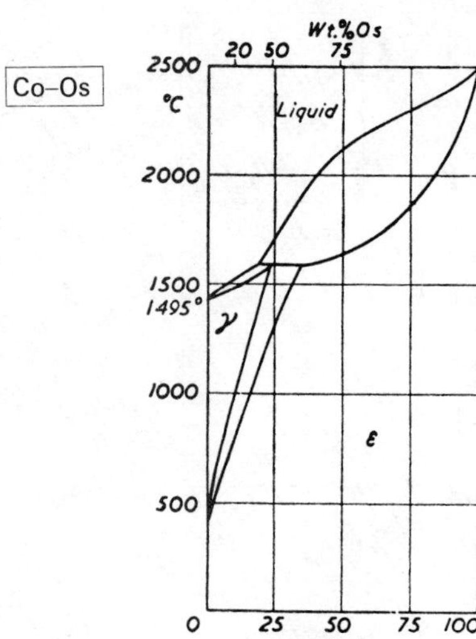

Co–P

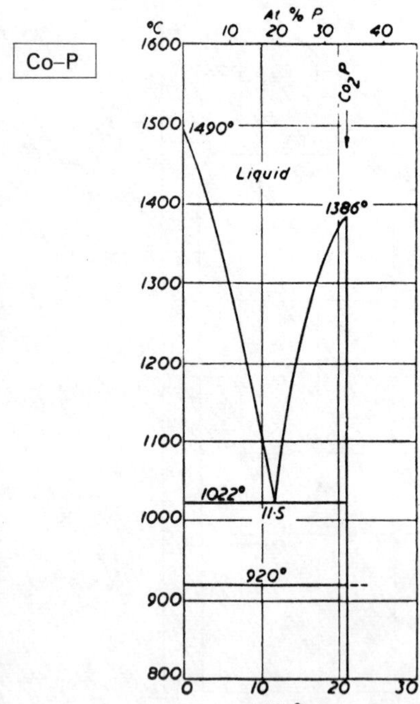

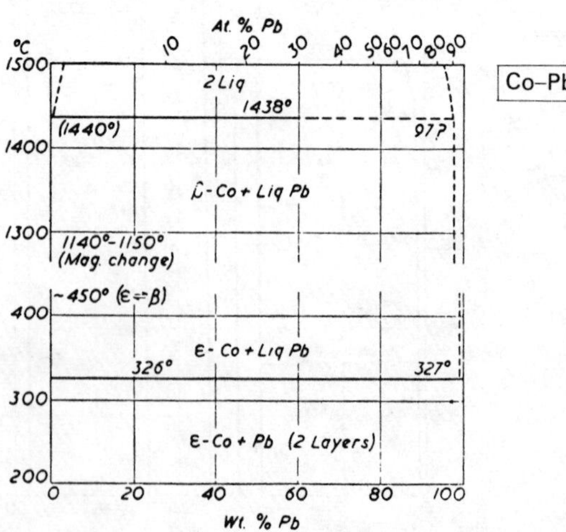

Co–Pb

Co–Pd

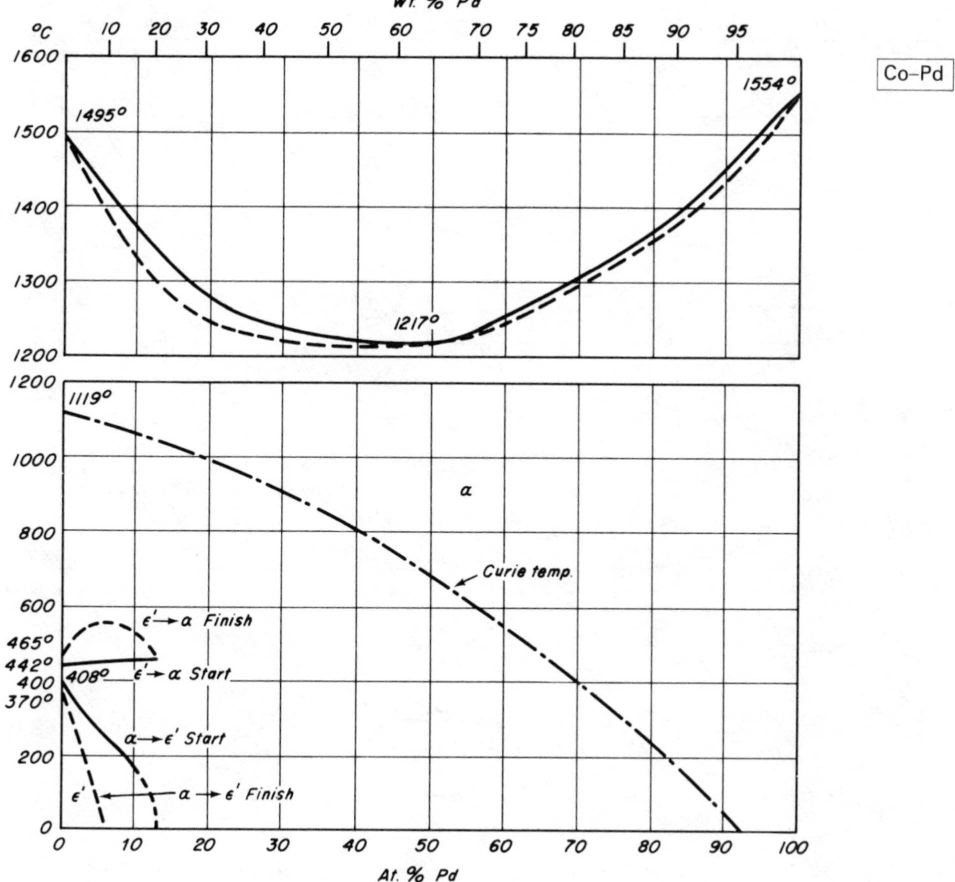
Wt. % Pd
°C
10
20
30
40
50
60
70
75
80
85
90
95
1600
1500
1400
1300
1200
1495°
1554°
1217°
1200
1000
800
600
400
200
0
1119°
α
Curie temp.
465°
442°
370°
408°
ε′→α Finish
ε′→α Start
α→ε′ Start
α→ε′ Finish
ε′
0
10
20
30
40
50
60
70
80
90
100
At. % Pd

Co–Pt

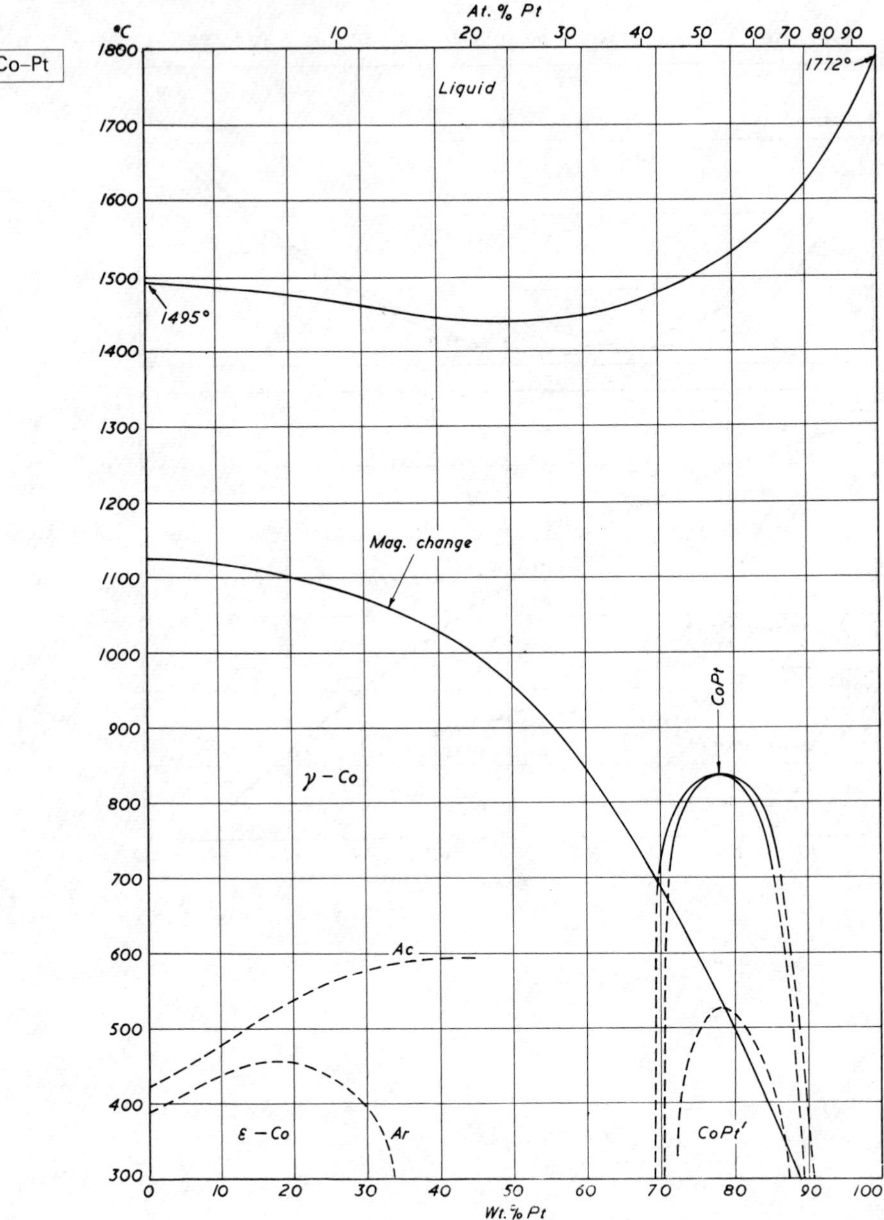
At. % Pt
°C
Liquid
1772°
1495°
Mag. change
CoPt
γ – Co
Ac
Ar
ε – Co
CoPt′
Wt. % Pt

Co–Pu

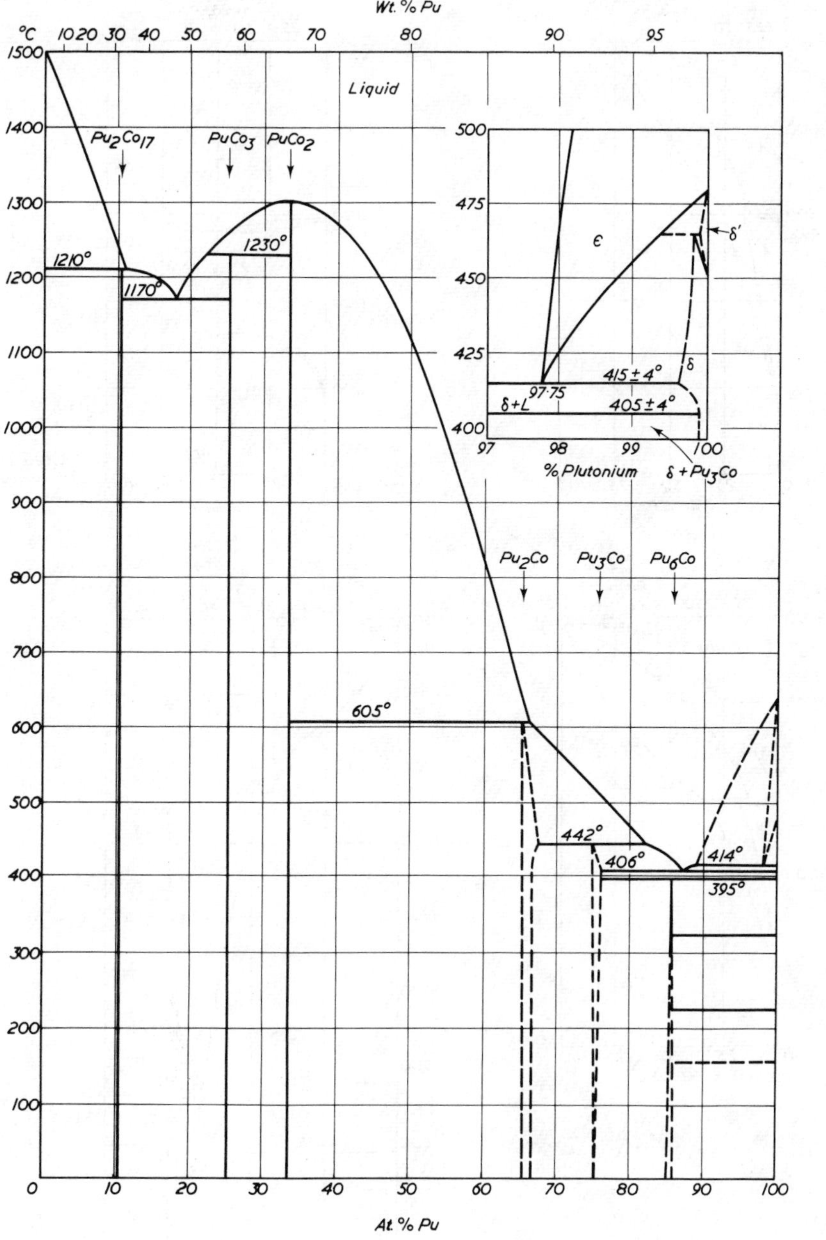

Wt.% Pu
At.% Pu
°C
Liquid
Pu_2Co_{17}
$PuCo_3$
$PuCo_2$
1210°
1170°
1230°
605°
Pu_2Co
Pu_3Co
Pu_6Co
442°
406°
414°
395°
ε
δ'
δ
415 ± 4°
97·75
δ+L
405 ± 4°
% Plutonium
δ + Pu_3Co

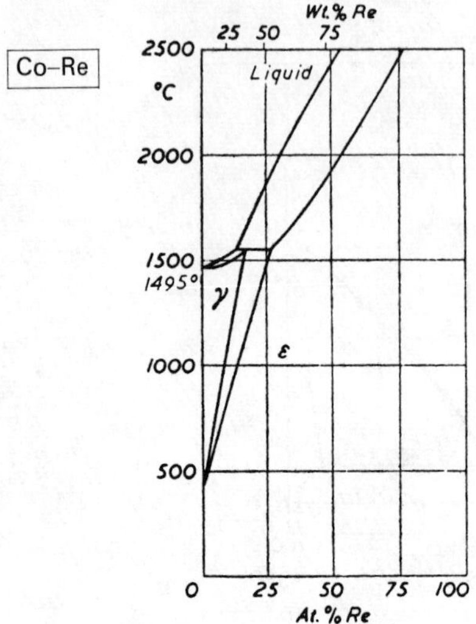
Co–Re
Wt.% Re
25 50 75
2500
°C
Liquid
2000
1500
1495°
γ
ε
1000
500
0 25 50 75 100
At.% Re

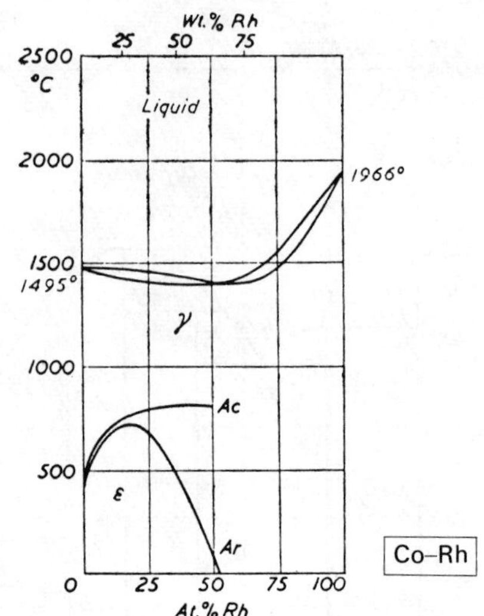
Wt.% Rh
25 50 75
2500
°C
Liquid
2000
1966°
1500
1495°
γ
1000
Ac
500
ε
Ar
0 25 50 75 100
At.% Rh
Co–Rh

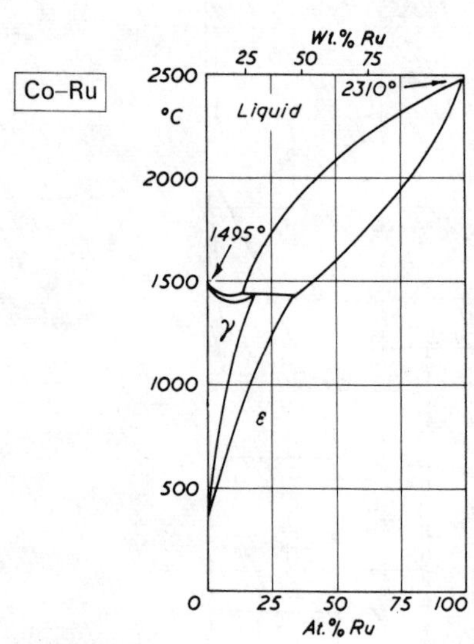
Co–Ru
Wt.% Ru
25 50 75
2500
°C
Liquid
2310°
2000
1495°
1500
γ
1000
ε
500
0 25 50 75 100
At.% Ru

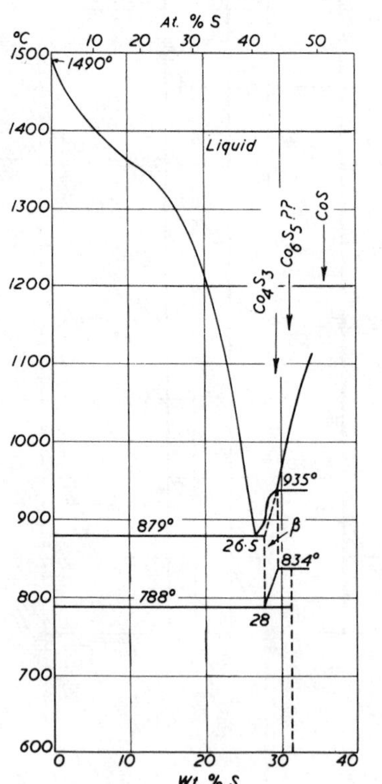
At. % S
°C
10 20 30 40 50
1500
1490°
1400
Liquid
1300
1200
1100
1000
900
800
700
600
Co4S3
Co6S5 ??
CoS
935°
879°
26·5
β
834°
788°
28
0 10 20 30 40
Wt. % S
Co–S

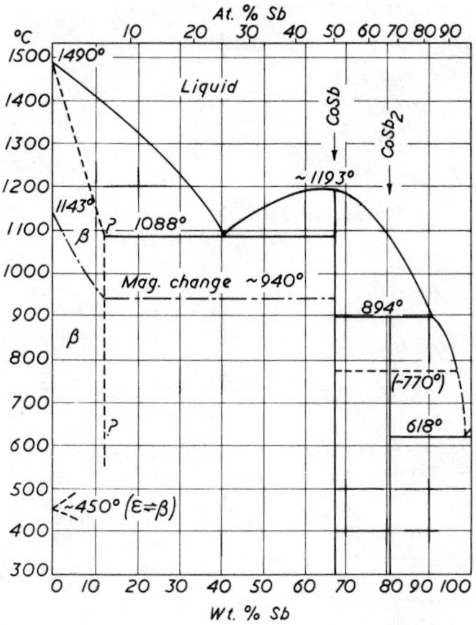

Co–Se

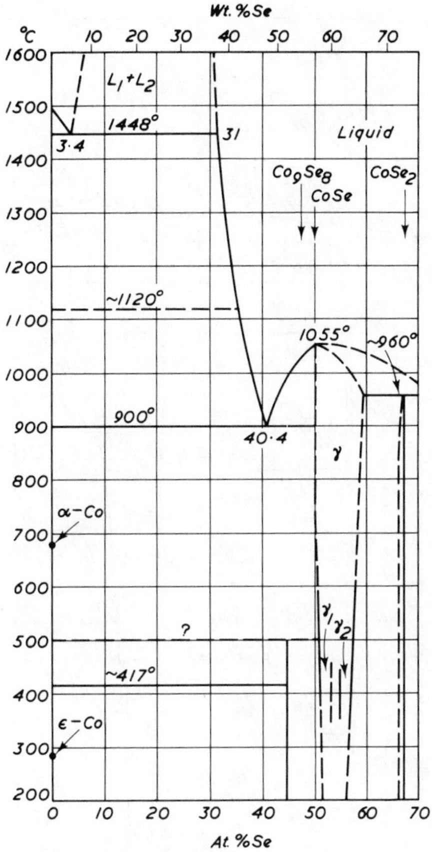

Co–Sb

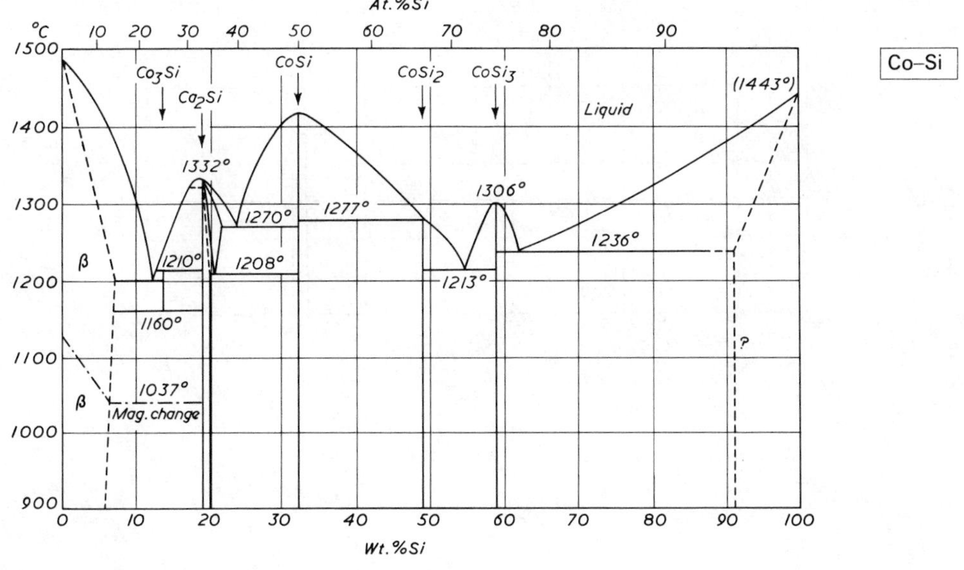

Co–Si

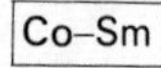

°C

1400

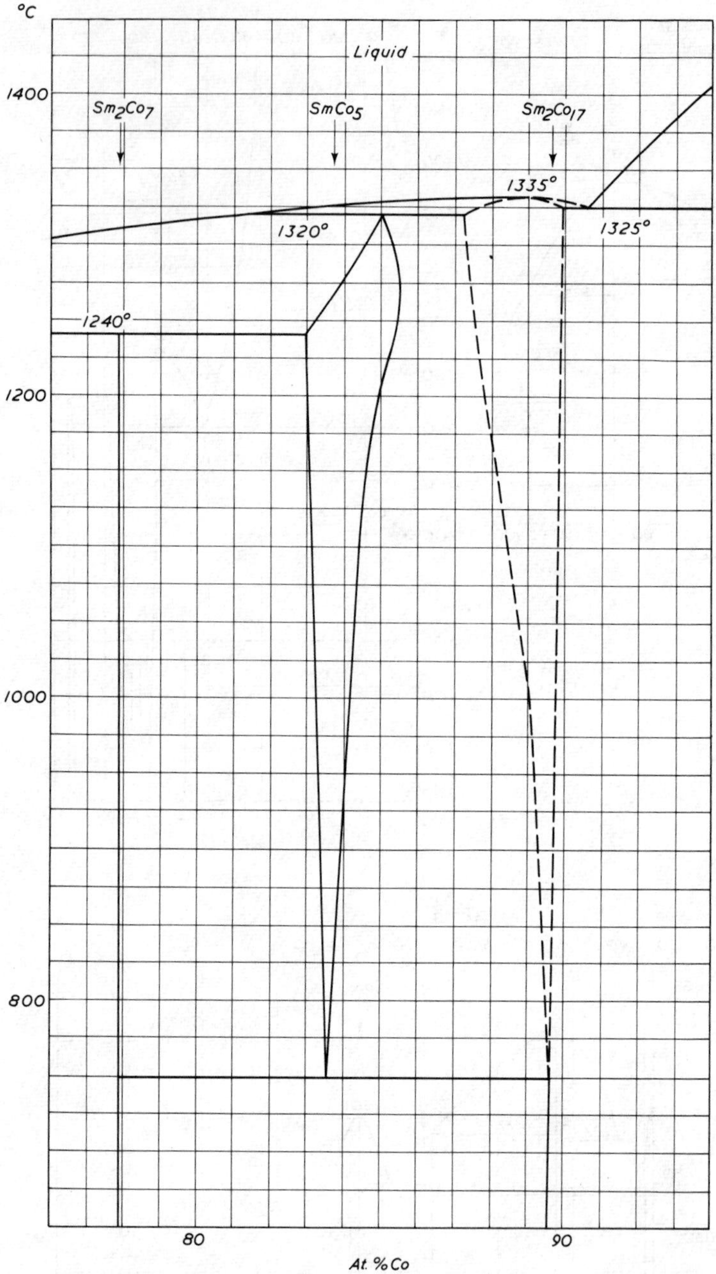

1200

1000

800

Liquid

Sm_2Co_7

$SmCo_5$

Sm_2Co_{17}

1335°

1320°

1325°

1240°

80

90

At. % Co

At. % Co
°C
Liquid
1495°
βCo
γ
γ'
αCo
0·8
Wt. % Co

Co–Sn

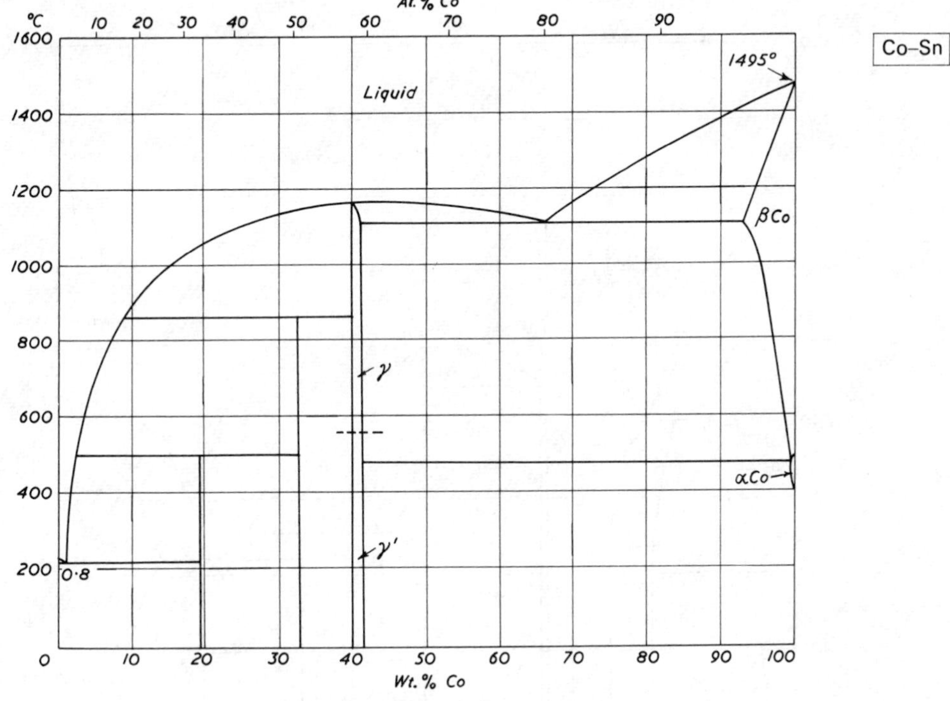

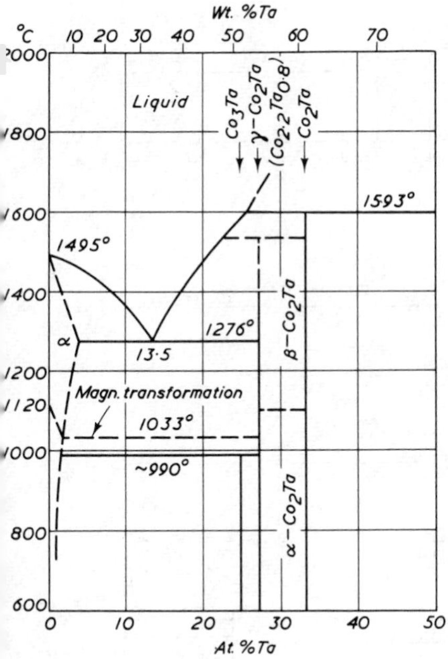

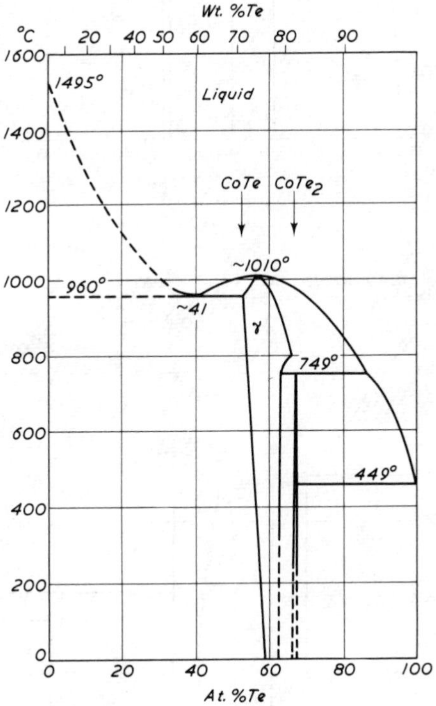

Wt. %Ta
°C
Liquid
Co3Ta
γ–Co2Ta
(Co2.2Ta0.8)
Co2Ta
1593°
1495°
β–Co2Ta
α
1276°
13·5
Magn. transformation
1033°
~990°
α–Co2Ta
At. %Ta

Co–Ta

Wt. %Te
°C
1495°
Liquid
CoTe
CoTe2
~1010°
960°
~41
γ
749°
449°
At. %Te

Co–Te

Co–Ti

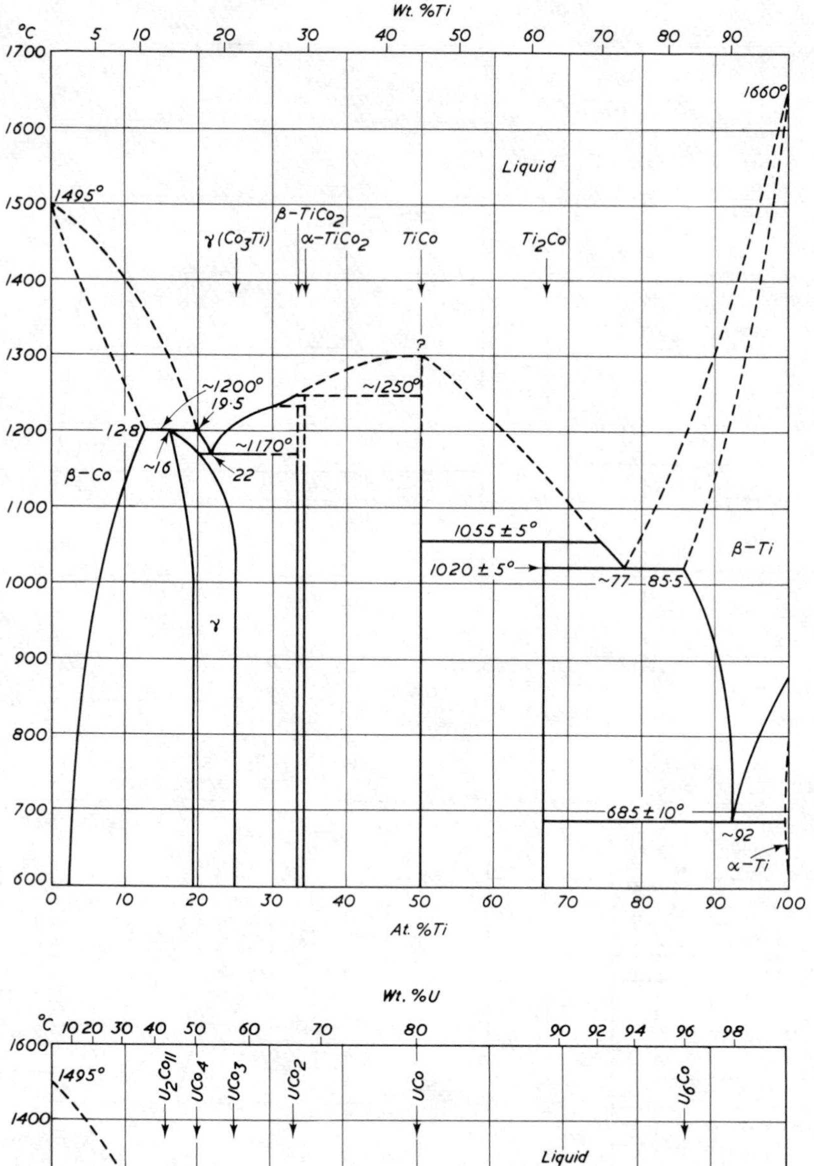

Wt. %Ti
°C
Liquid
1495°
1660°
β-TiCo2
γ(Co3Ti)
α-TiCo2
TiCo
Ti2Co
?
~1200°
19·5
~1250°
12·8
~1170°
~16
22
β-Co
1055 ± 5°
1020 ± 5°
~77
85·5
β-Ti
γ
685 ± 10°
~92
α-Ti
At. %Ti

Co–U

Wt. %U
°C
1495°
U2Co11
UCo4
UCo3
UCo2
UCo
U6Co
Liquid
1132°
(γ-U)
β-U
α-U
At. %U

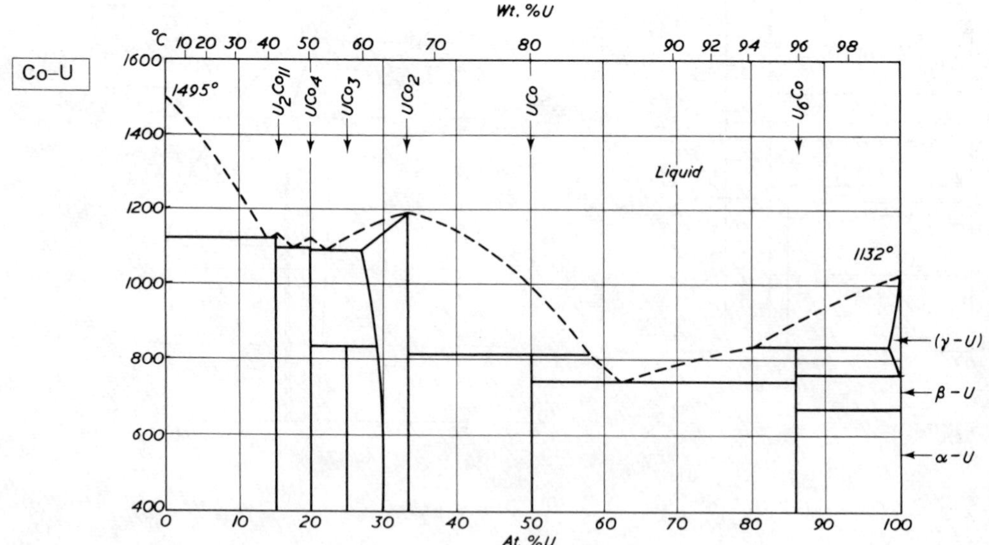

Co–V

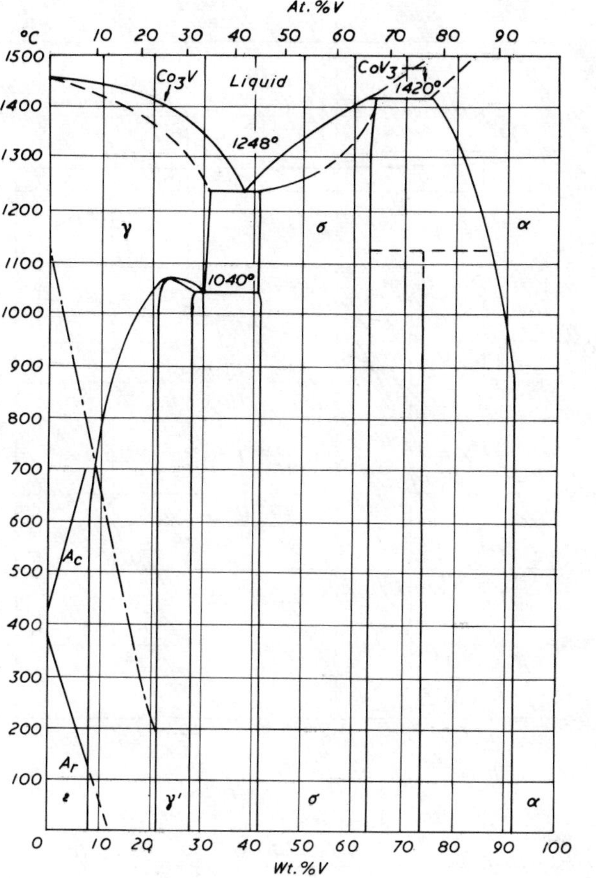
At. %V
°C
10
20
30
40
50
60
70
80
90
1500
1400
1300
1200
1100
1000
900
800
700
600
500
400
300
200
100
0
Co3V
Liquid
CoV3
1420°
1248°
γ
σ
α
1040°
Ac
Ar
ε
γ'
σ
α
10
20
30
40
50
60
70
80
90
100
Wt. %V

Co–W

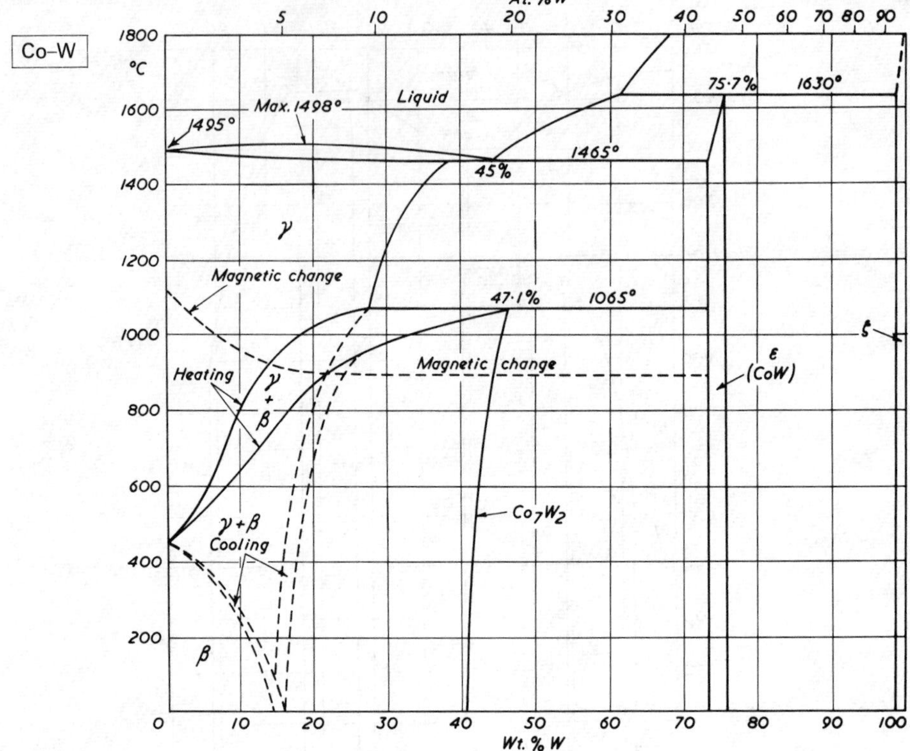
At.% W
5
10
20
30
40
50
60
70
80
90
1800
°C
1600
1400
1200
1000
800
600
400
200
0
Liquid
Max. 1498°
1495°
75·7%
1630°
1465°
45%
γ
Magnetic change
47·1%
1065°
ξ
ε
(CoW)
Magnetic change
Heating
γ
+
β
γ + β
Cooling
Co_7W_2
β
10
20
30
40
50
60
70
80
90
100
Wt. % W

Co–Y

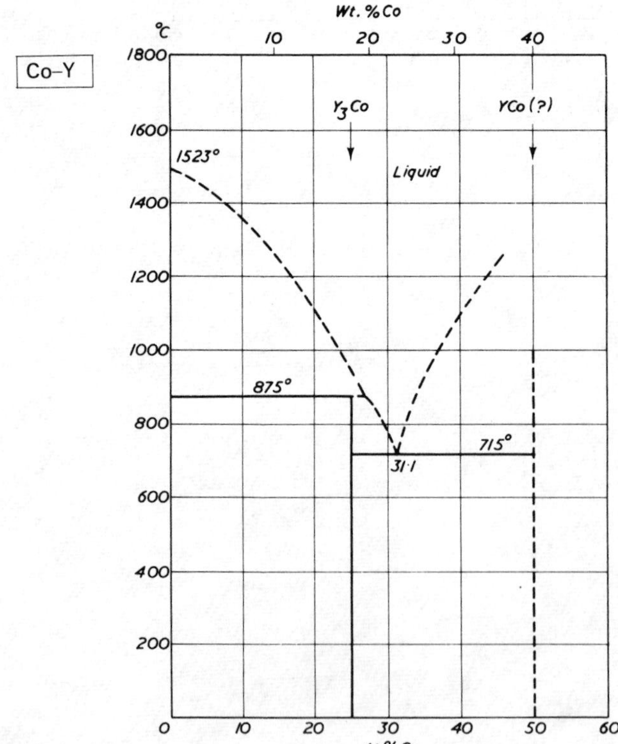
Wt. % Co
10
20
30
40
°C
1800
1600
1400
1200
1000
800
600
400
200
0
Y_3Co
YCo (?)
1523°
Liquid
875°
715°
31·1
10
20
30
40
50
60
At. % Co

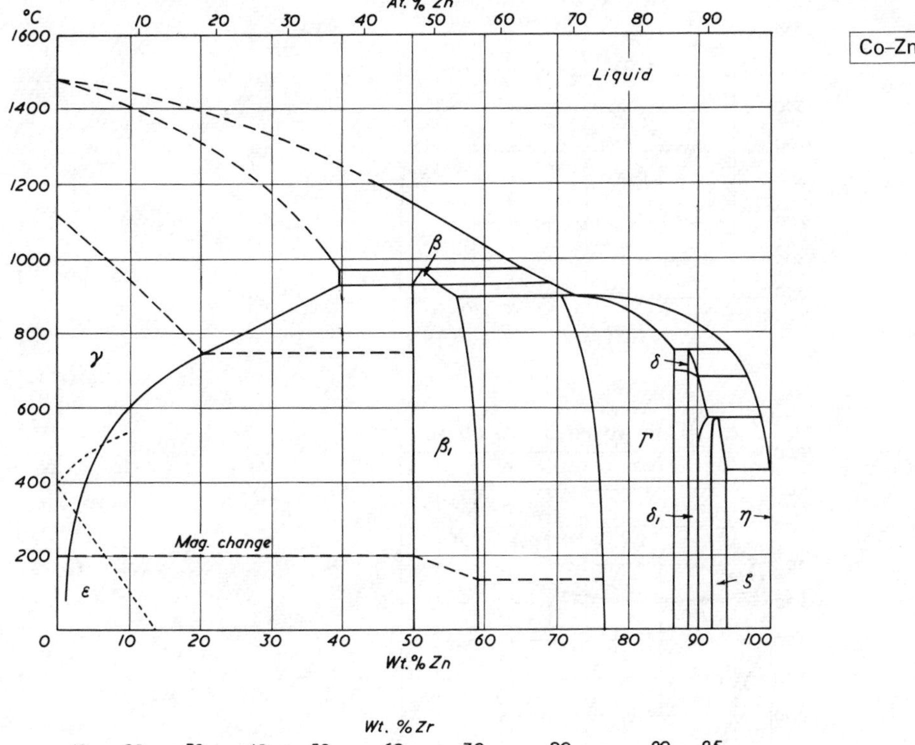

Co–Zn

At. % Zn
°C
1600
1400
1200
1000
800
600
400
200
0
Liquid
β
γ
β₁
Γ
δ
δ₁
η
ζ
ε
Mag. change
Wt.% Zn

Co–Zr

Wt. % Zr
°C
1600
1200
800
400
1495°
1852°
1560°
1270°
1360°
1370°
1260°
1240°
1040°
1090°
980°
940°
830°
Zr2Co11
Zr6Co23
ZrCo2
ZrCo
Zr2Co
Zr3Co
βCo
α ⇄ β
αCo
βZr
αZr
At. % Zr

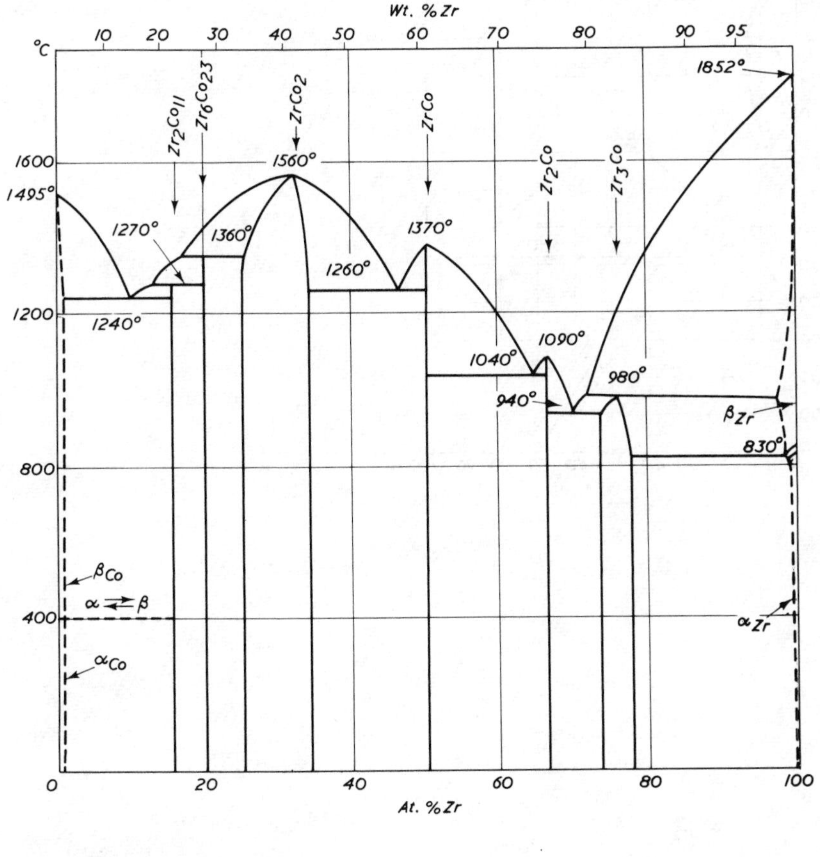

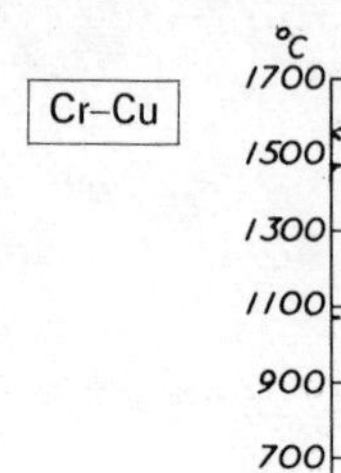
Cr–Cu

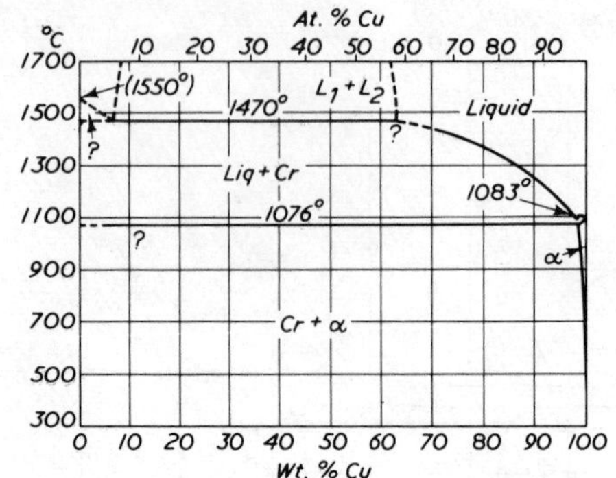
At. % Cu
°C
(1550°)
1470°
$L_1 + L_2$
Liquid
?
Liq + Cr
1083°
1076°
α
Cr + α
Wt. % Cu

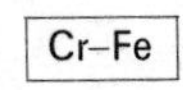
Cr–Fe

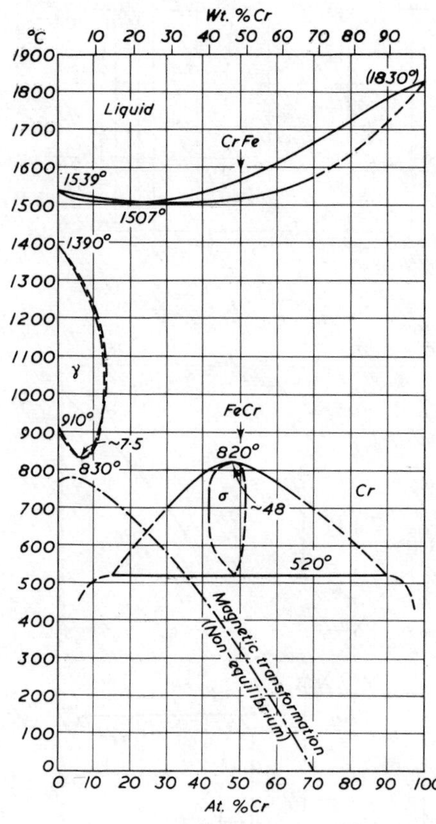
Wt. % Cr
°C
(1830°)
Liquid
CrFe
1539°
1507°
1390°
γ
910°
~7·5
830°
FeCr
820°
σ
~48
Cr
520°
Magnetic transformation
(Non-equilibrium)
At. % Cr

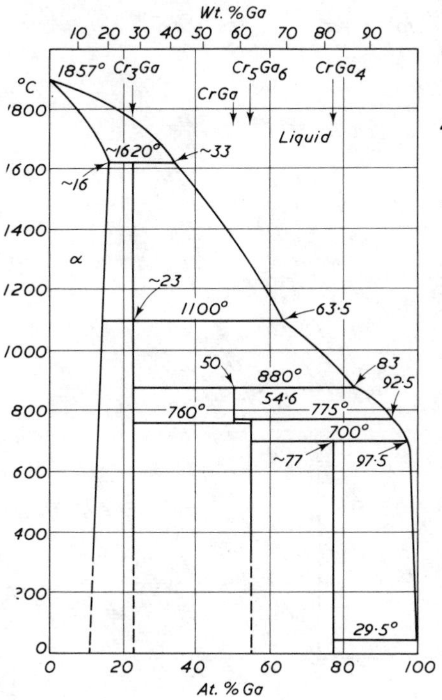

Cr–Gd

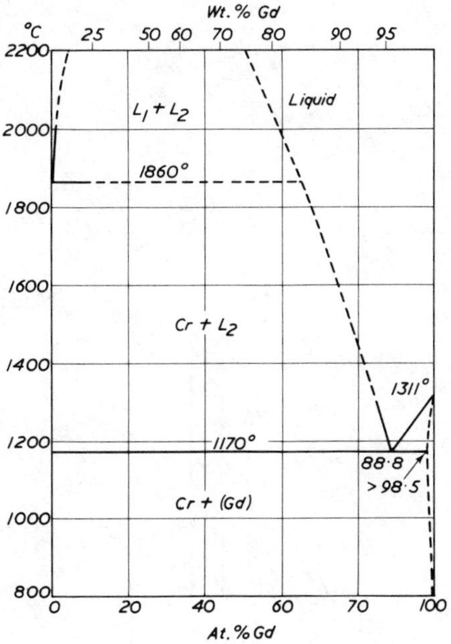

Cr–Ga

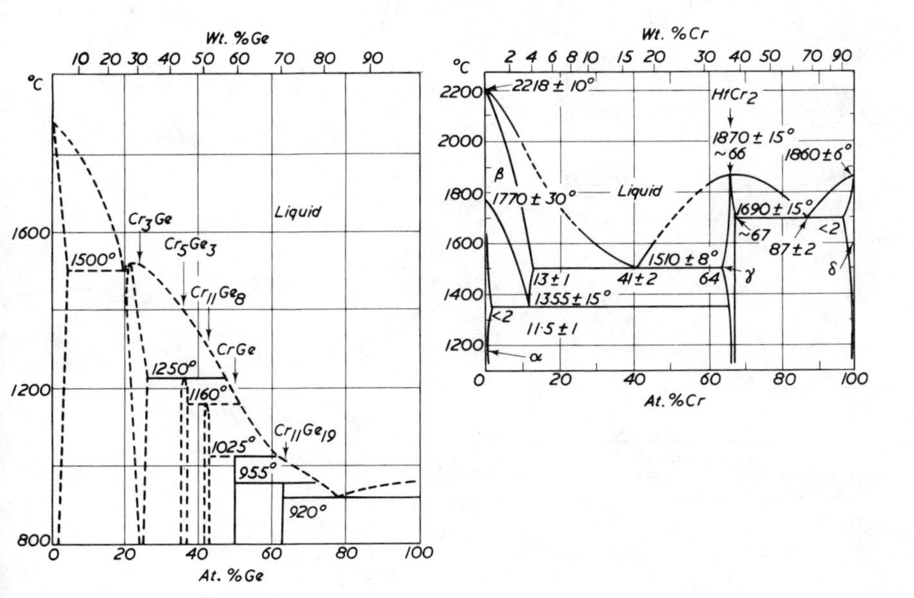

Cr–Ge

Cr–Hf

Cr–Ir

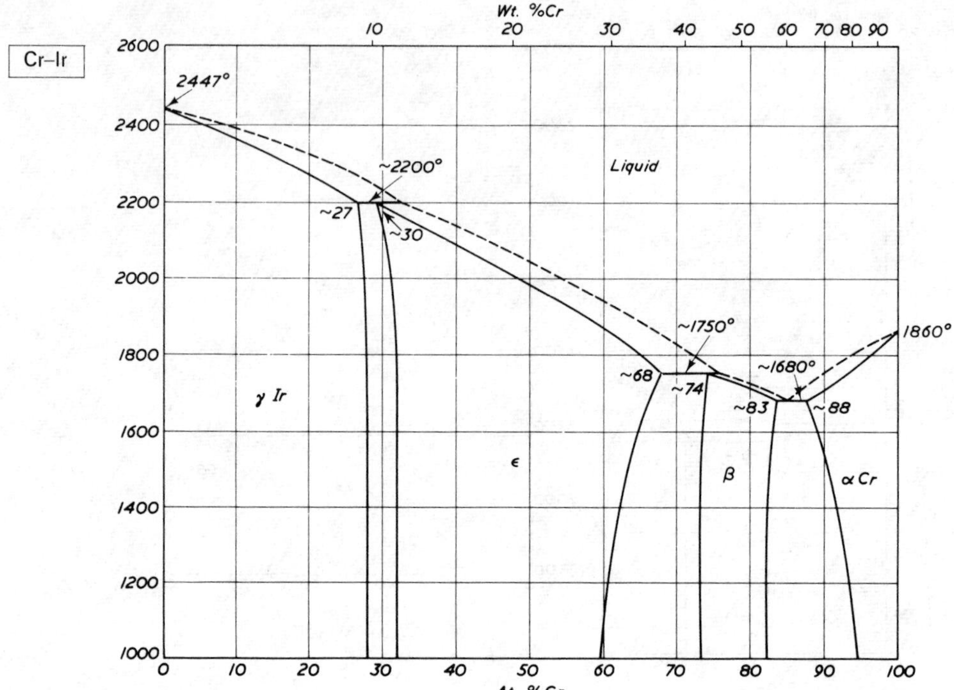
Wt. %Cr
10
20
30
40
50
60
70
80
90
2600
2400
2200
2000
1800
1600
1400
1200
1000
2447°
~2200°
Liquid
~27
~30
~1750°
1860°
~1680°
~68
~74
~83
~88
γ Ir
ε
β
α Cr
0
10
20
30
40
50
60
70
80
90
100
At. %Cr

Cr–La

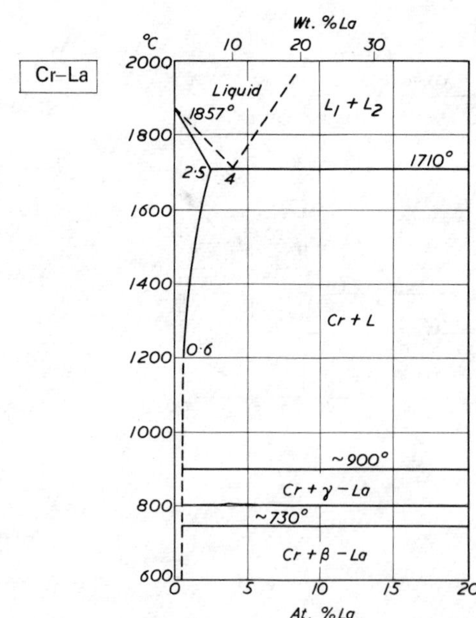
Wt. %La
°C
10
20
30
2000
1800
1600
1400
1200
1000
800
600
Liquid
1857°
$L_1 + L_2$
1710°
2·5
4
Cr + L
0·6
~900°
Cr + γ − La
~730°
Cr + β − La
0
5
10
15
20
At. %La

Cr–Mn

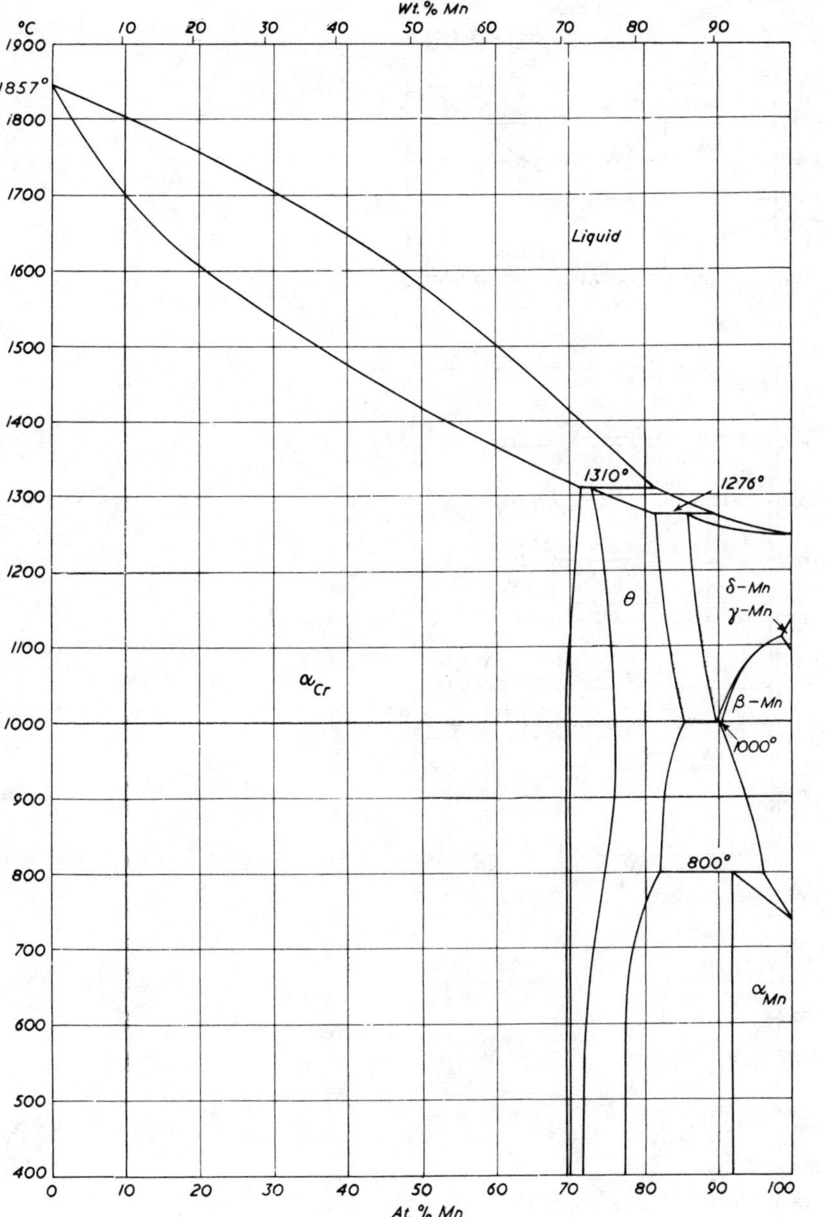

Wt.% Mn
°C
1857°
Liquid
1310°
1276°
θ
δ-Mn
γ-Mn
α_{Cr}
β-Mn
1000°
800°
α_{Mn}
At. % Mn

Cr–Mo

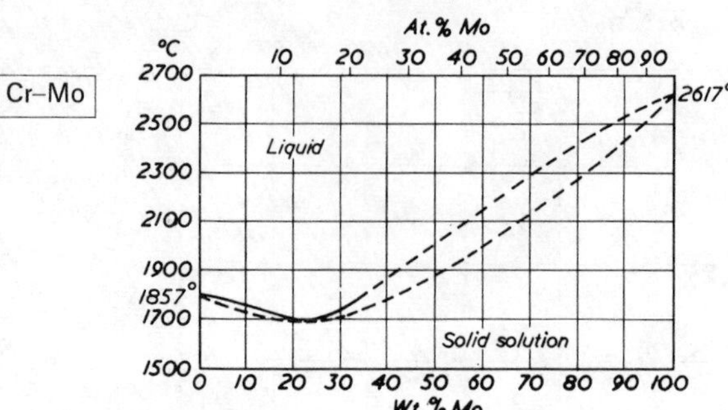

Cr–Nd

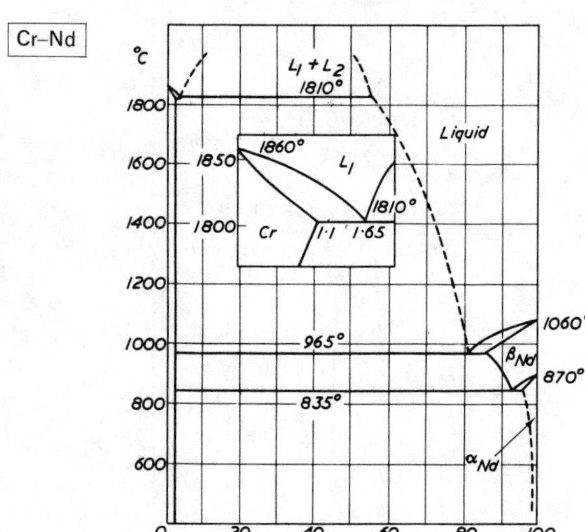

Cr–Ni

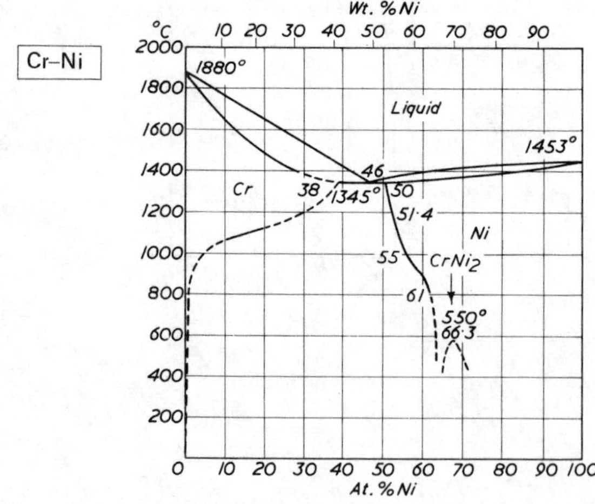

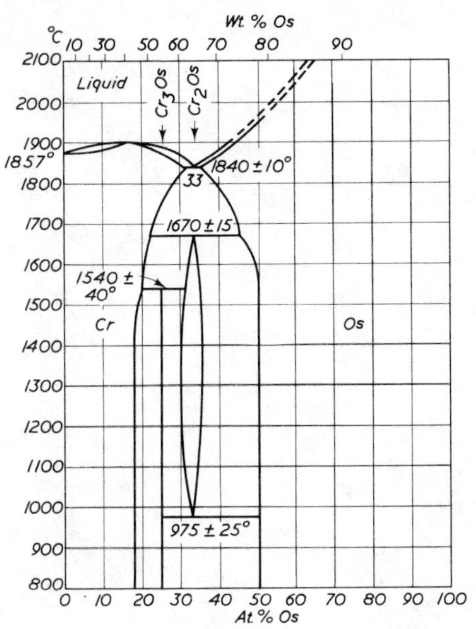

Cr–Os

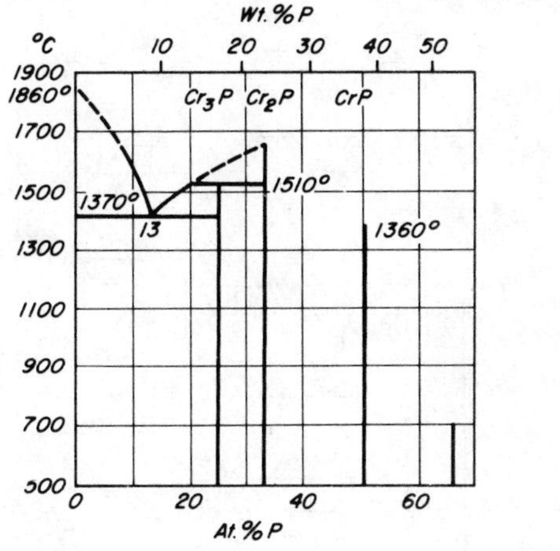

Cr–P

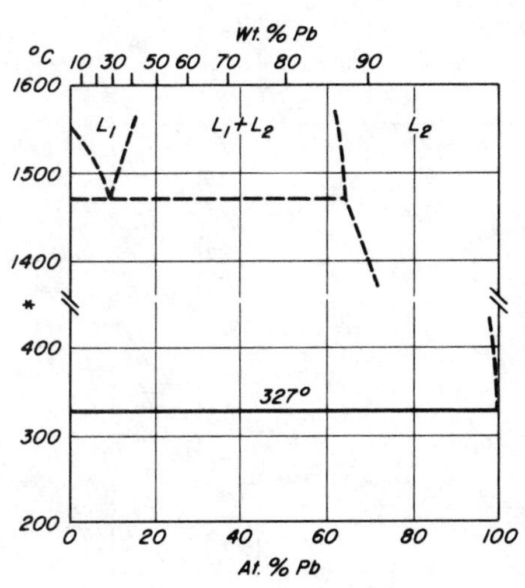

*Sol Ct in liq Pb 908/1210°C

$\log_{10}$ (At % Cr) = 4·315 − 6717/T°K

M.p. Cr=1860°C

Cr–Pb

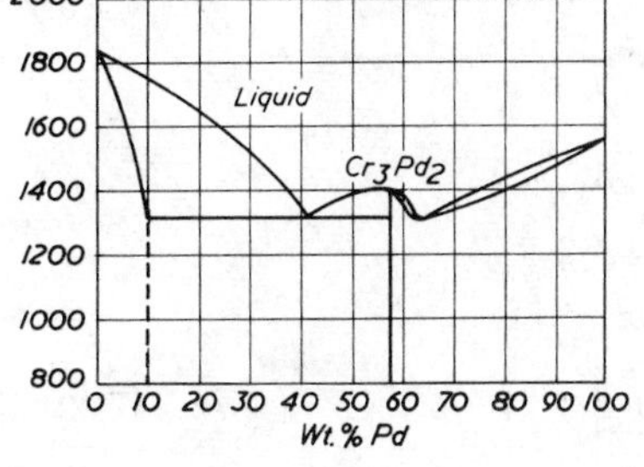

Cr–Pd

Cr–Pr

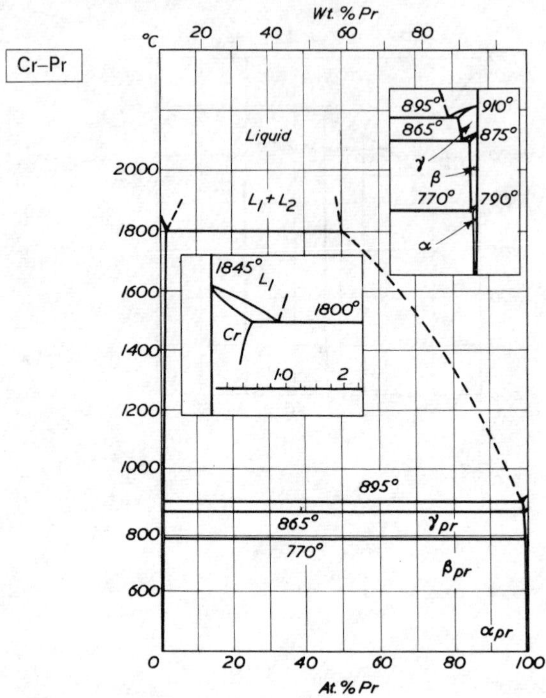
Wt. % Pr
°C
20
40
60
80
Liquid
$L_1 + L_2$
2000
1800
1600
1400
1200
1000
800
600
1845°
L_1
1800°
Cr
1·0
2
895°
910°
865°
875°
γ
β
770°
790°
α
895°
865°
γ_{Pr}
770°
β_{Pr}
α_{Pr}
0
20
40
60
80
100
At. % Pr

Cr–Pt

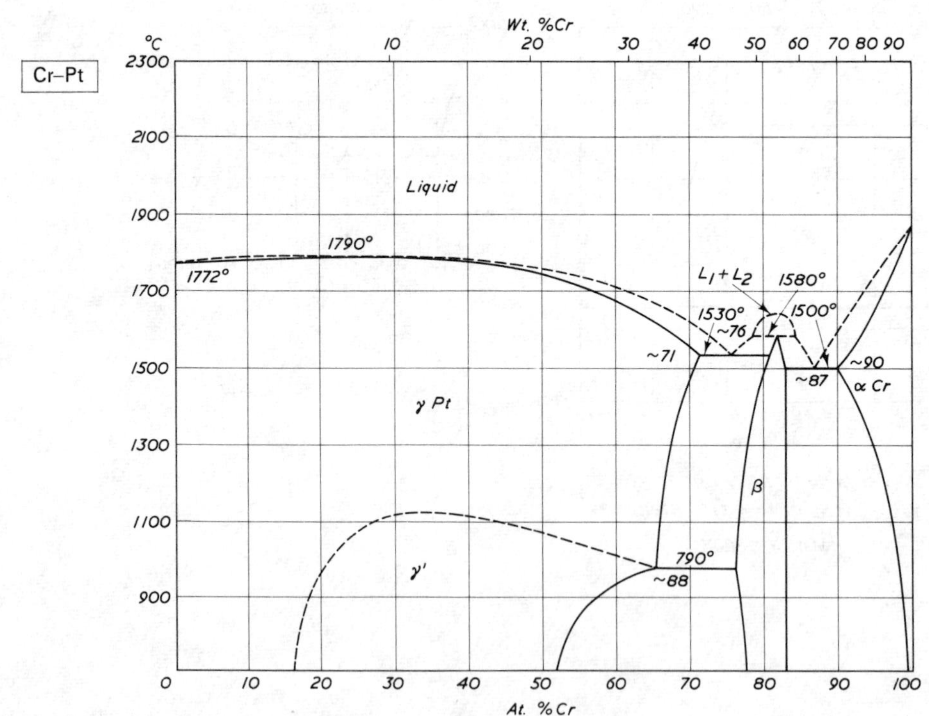
Wt. %Cr
°C
2300
10
20
30
40
50
60
70
80
90
2100
1900
1700
1500
1300
1100
900
Liquid
1790°
1772°
$L_1 + L_2$
1580°
1530°
~76
1500°
~71
~87
~90
α Cr
γ Pt
β
γ'
790°
~88
0
10
20
30
40
50
60
70
80
90
100
At. %Cr

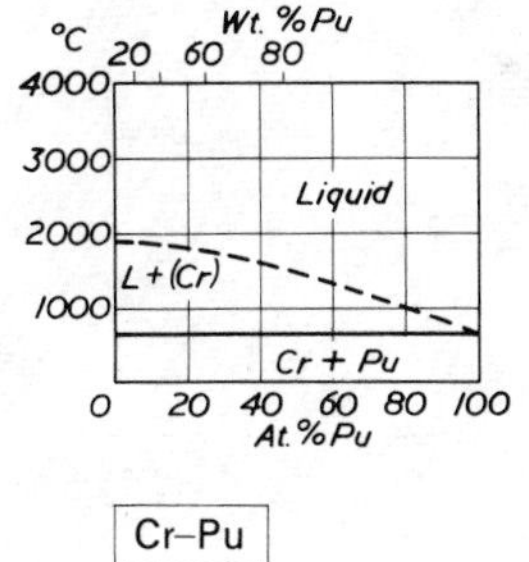
°C
Wt. % Pu
20 60 80
4000
3000
2000
1000
Liquid
L + (Cr)
Cr + Pu
0 20 40 60 80 100
At. % Pu

Cr–Pu

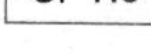
Cr–Re

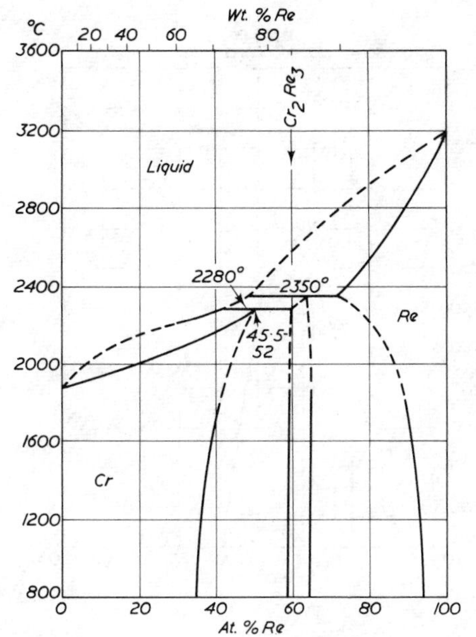
°C
Wt. % Re
20 40 60 80
3600
3200
2800
2400
2000
1600
1200
800
Cr_2Re_3
Liquid
2280°
2350°
45·5
52
Re
Cr
0 20 40 60 80 100
At. % Re

Cr–Rh

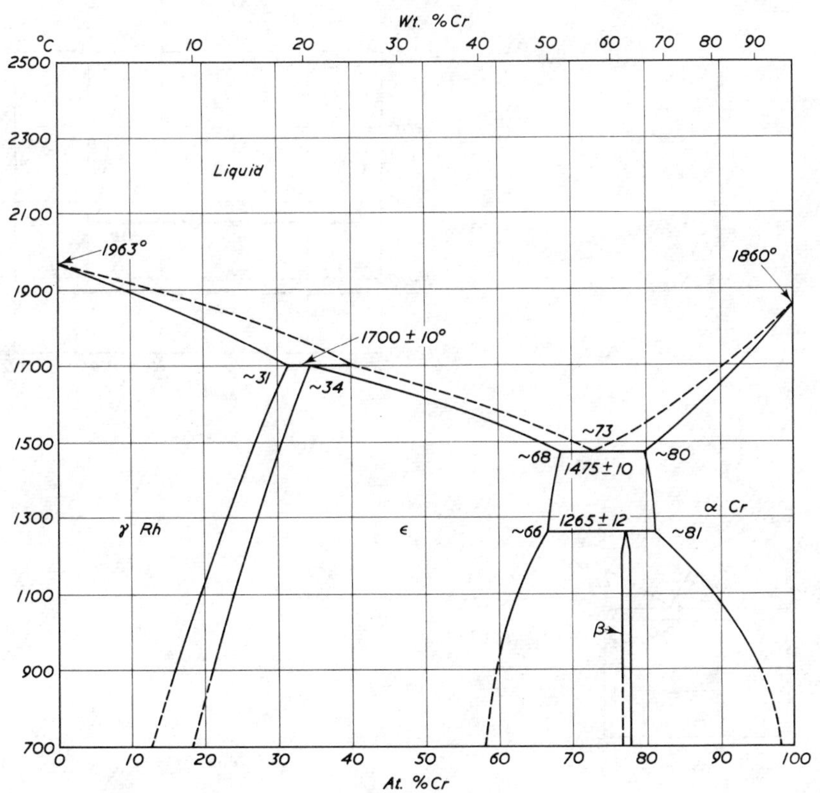
°C
Wt. % Cr
10 20 30 40 50 60 70 80 90
2500
2300
2100
1900
1700
1500
1300
1100
900
700
Liquid
1963°
1860°
1700 ± 10°
~31
~34
~73
~68
~80
1475 ± 10
1265 ± 12
~66
~81
α Cr
γ Rh
ε
β
0 10 20 30 40 50 60 70 80 90 100
At. % Cr

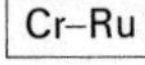
Cr–Ru

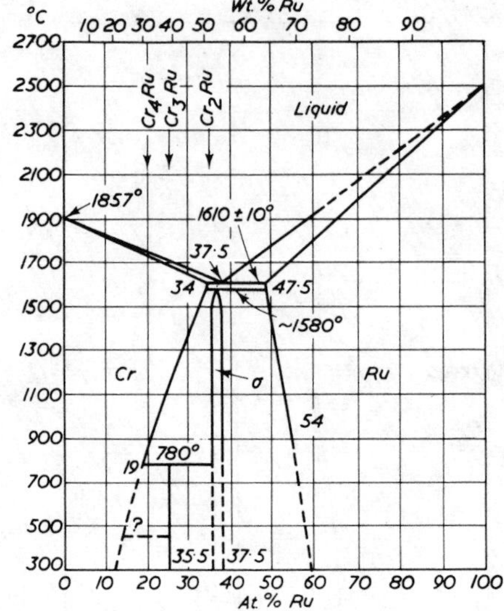

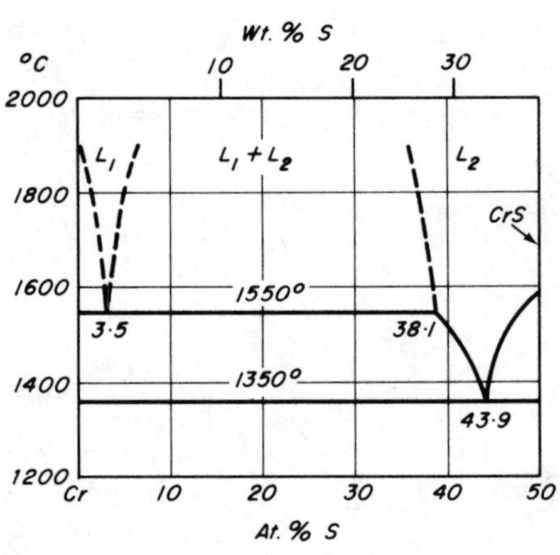

Cr–S

Cr–Sc

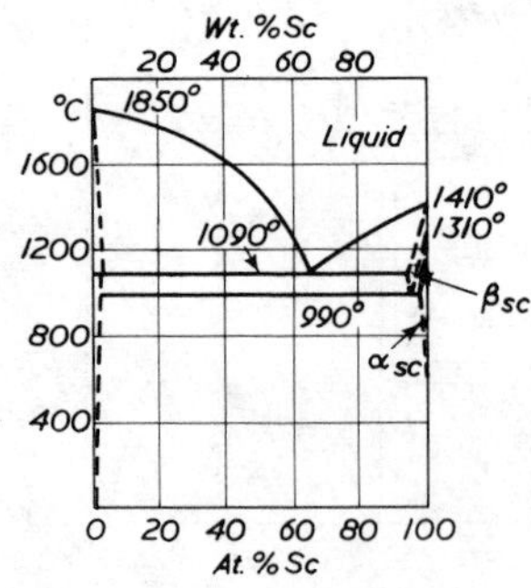

Cr–Si

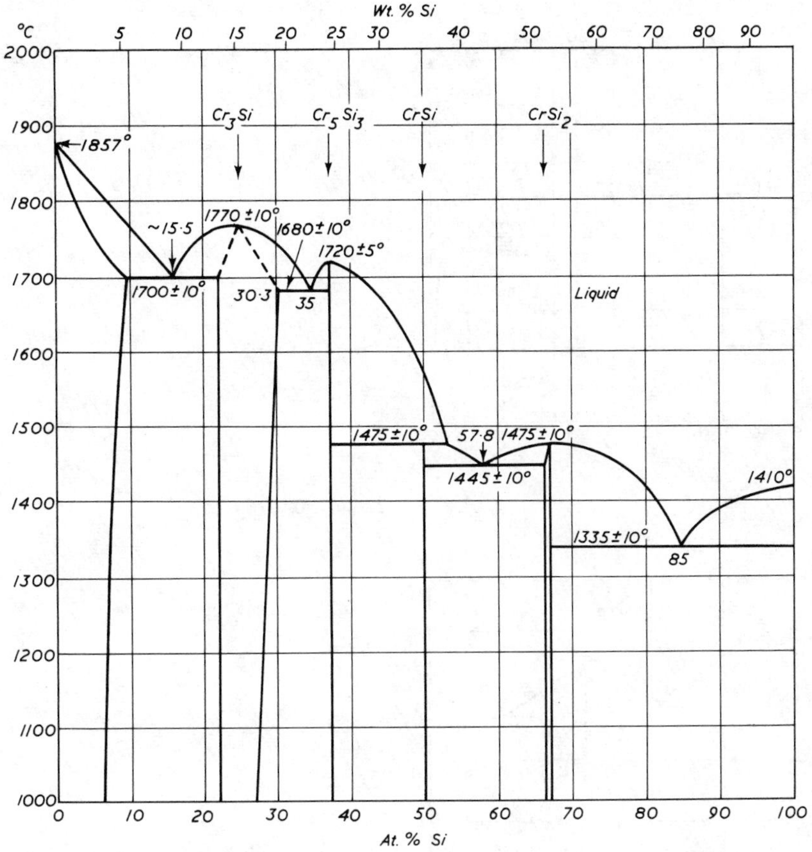

Wt. % Si

°C

Cr$_3$Si Cr$_5$Si$_3$ CrSi CrSi$_2$

←1857°

~15·5

1770 ±10°

1680 ±10°

1720 ±5°

1700 ±10°

30·3

35

Liquid

1475 ±10°

57·8

1475 ±10°

1445 ±10°

1410°

1335 ±10°

85

At. % Si

Cr–Sn

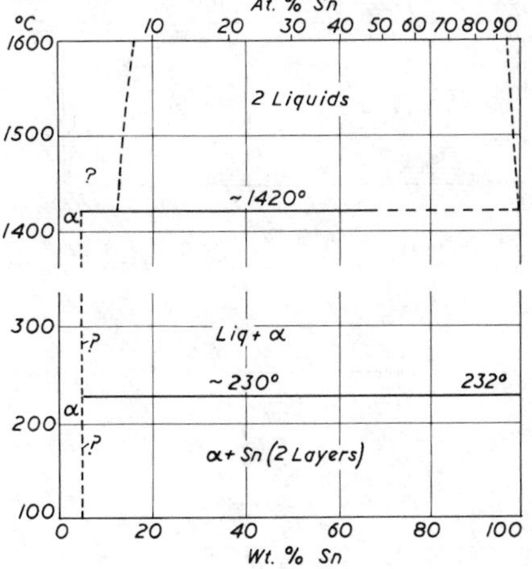

Cr–Ta

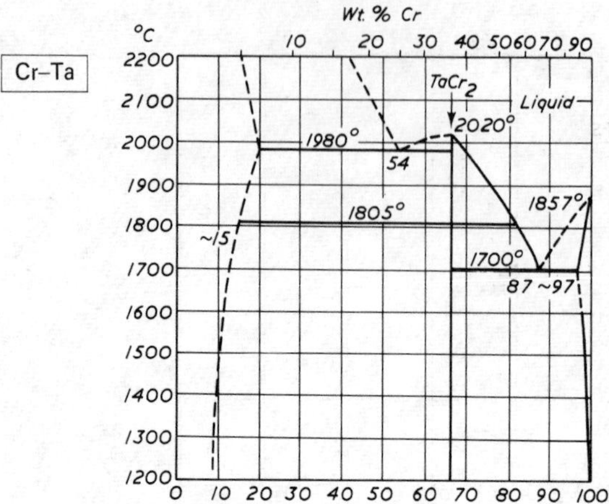

Cr–Tb

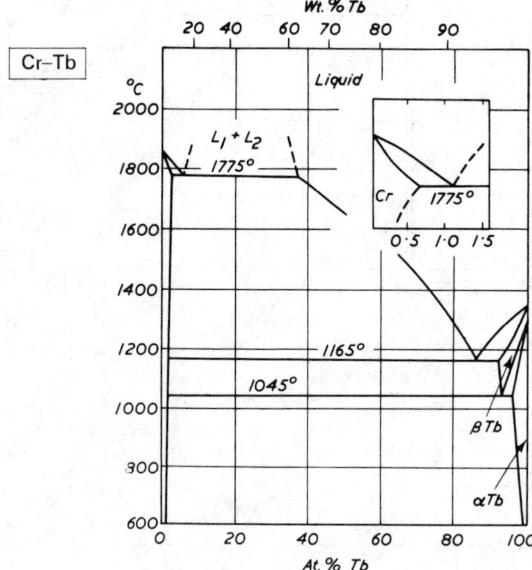

Cr–Ti

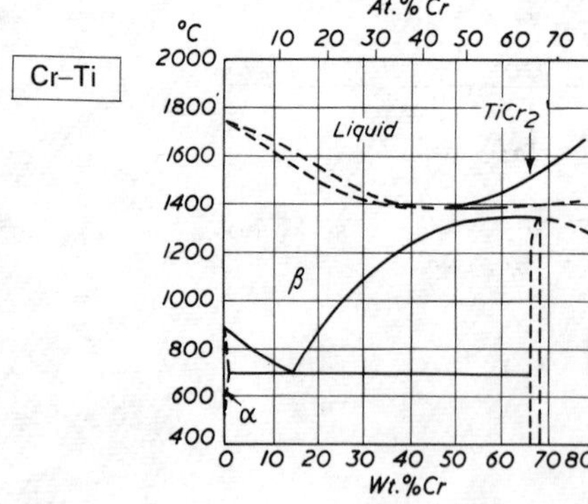

Cr–U

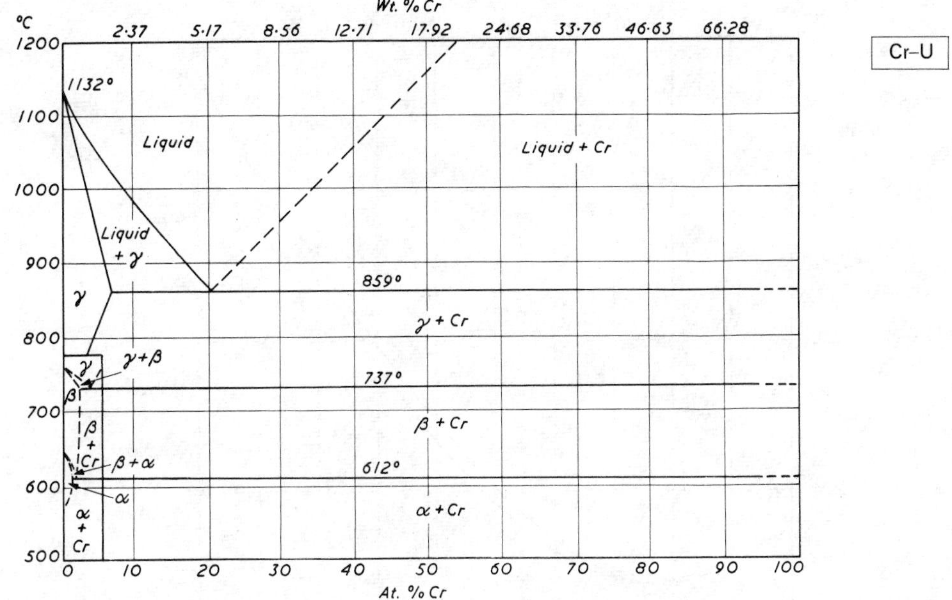
Wt. % Cr
°C
2·37
5·17
8·56
12·71
17·92
24·68
33·76
46·63
66·28
1200
1100
1000
900
800
700
600
500
1132°
Liquid
Liquid + Cr
Liquid + γ
γ
859°
γ + Cr
γ + β
737°
β + Cr
β + Cr
β + α
612°
α
α + Cr
α + Cr
0
10
20
30
40
50
60
70
80
90
100
At. % Cr

Cr–V

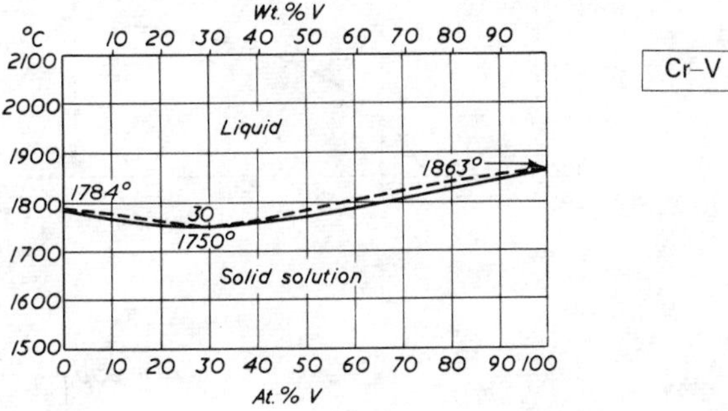
Wt.% V
°C
10 20 30 40 50 60 70 80 90
2100
2000
1900
1800
1700
1600
1500
Liquid
1863°
1784°
30
1750°
Solid solution
0 10 20 30 40 50 60 70 80 90 100
At.% V

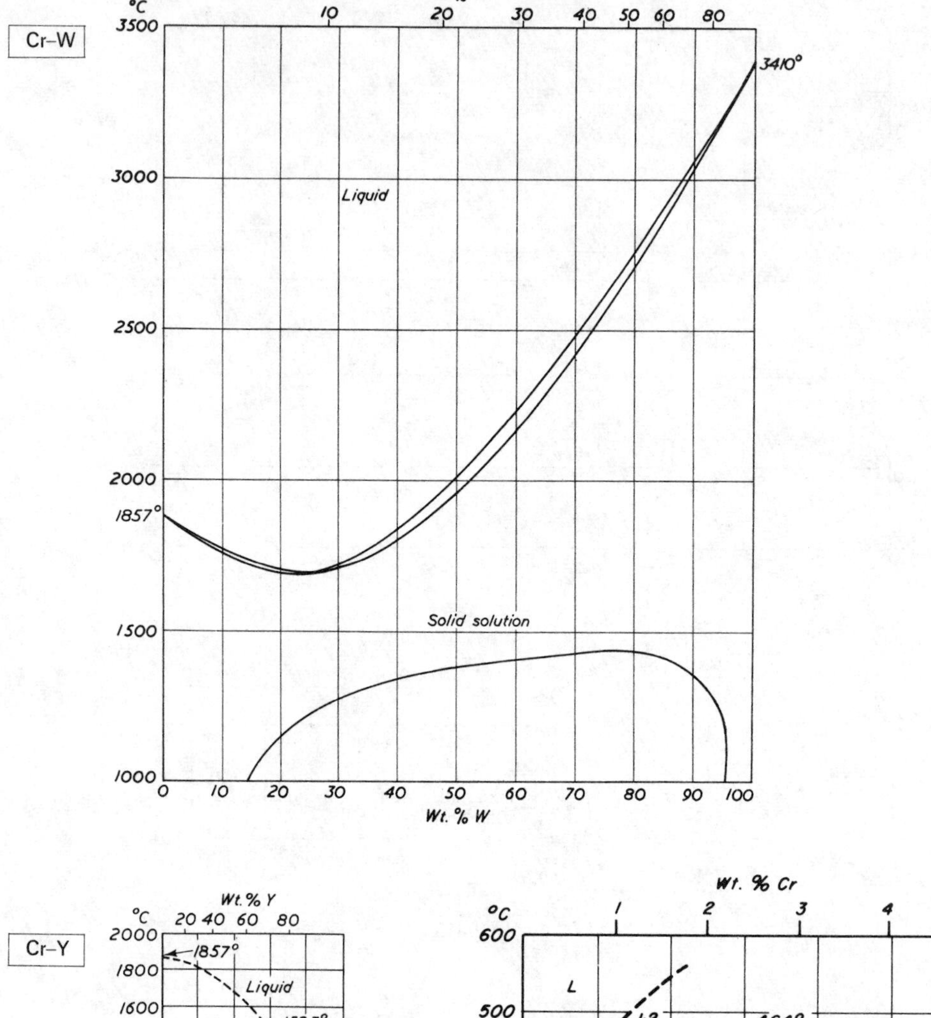
Cr–W
At. % W
10
20
30
40
50
60
80
°C
3500
3000
2500
2000
1857°
1500
1000
3410°
Liquid
Solid solution
0
10
20
30
40
50
60
70
80
90
100
Wt. % W

Cr–Y

Cr–Zn

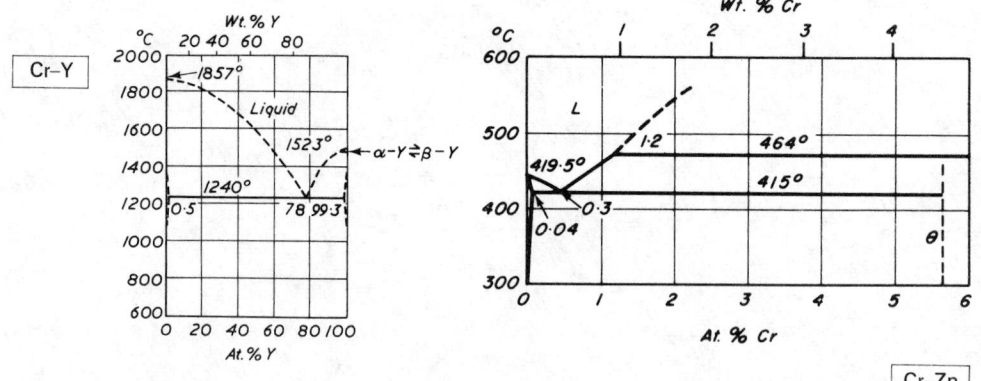

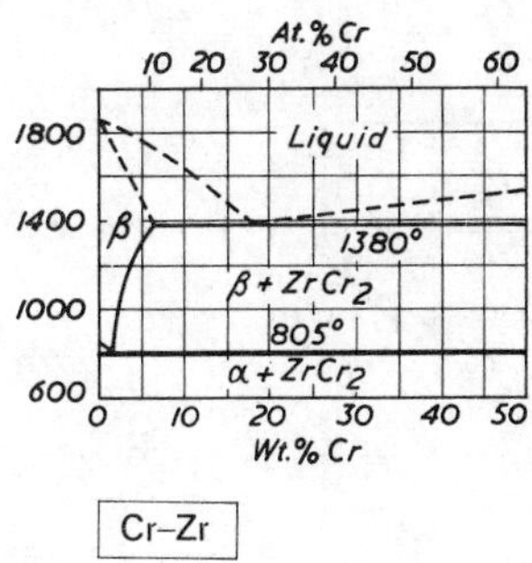

Cr–Zr

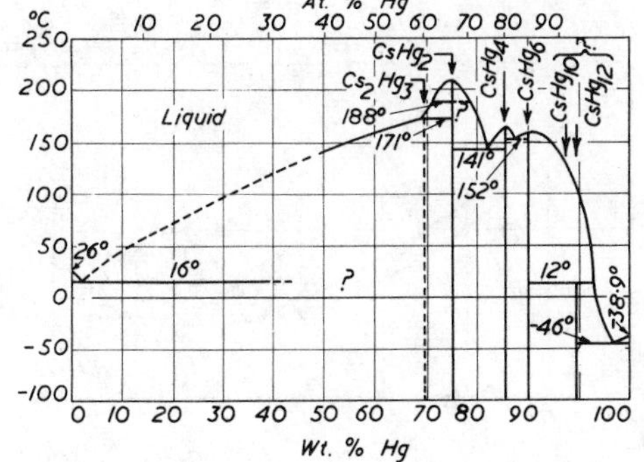

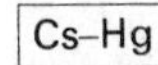

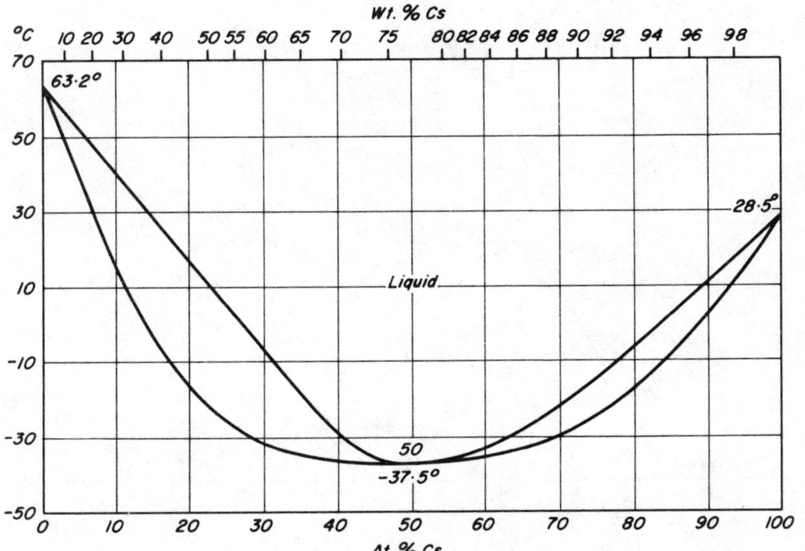

Cs–K

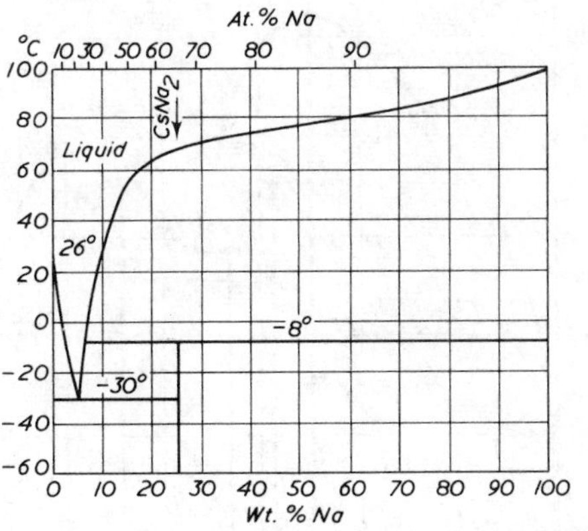

Cs–Na

Cs–Rb

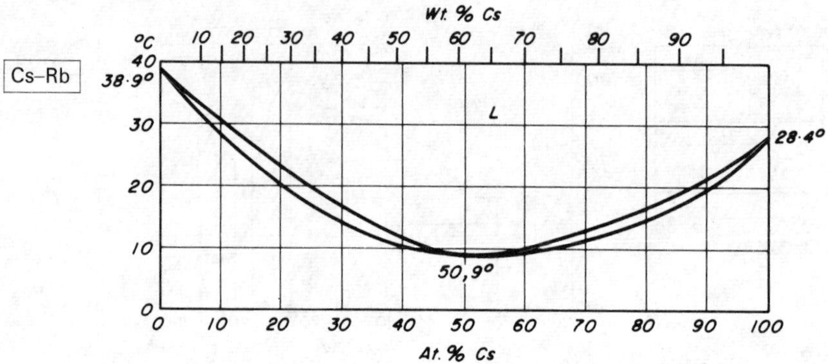

Cs–S

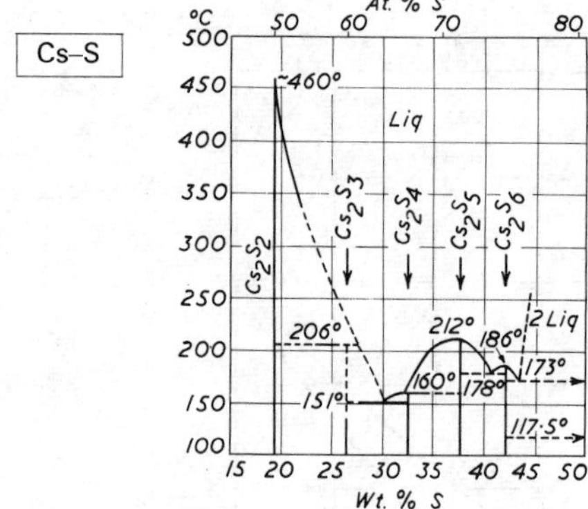

Cs–Sb

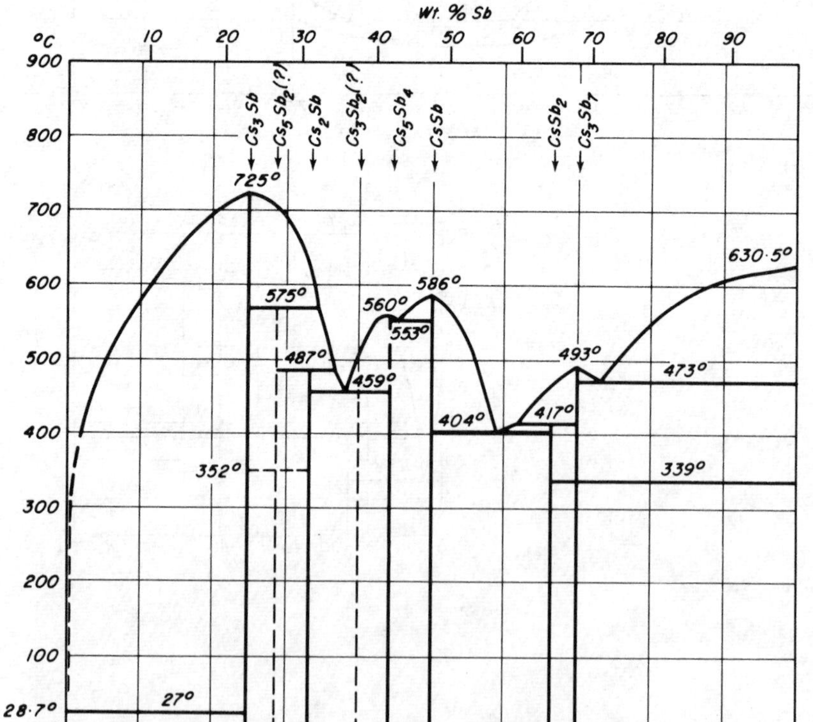

Cu–Er

°C
1600
1522°
1400
1200
1000
800
600
Liquid
ErCu
ErCu2
ErCux
ErCuy
ErCu5
1055°
940°
1010°
1083°
935°
880°
905°
885°
895°
0
20
40
60
80
100
At.%Cu

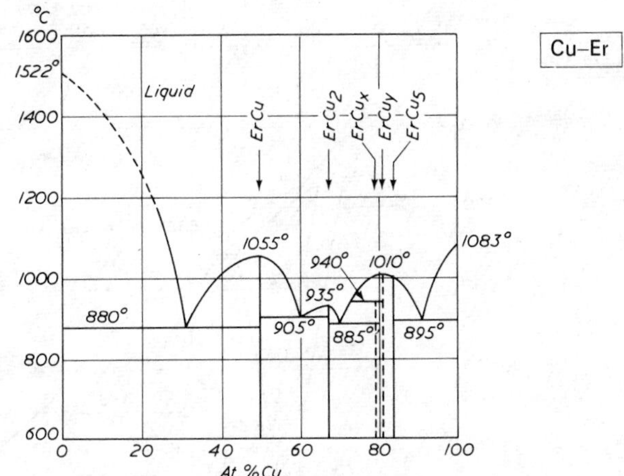

Cu–Fe

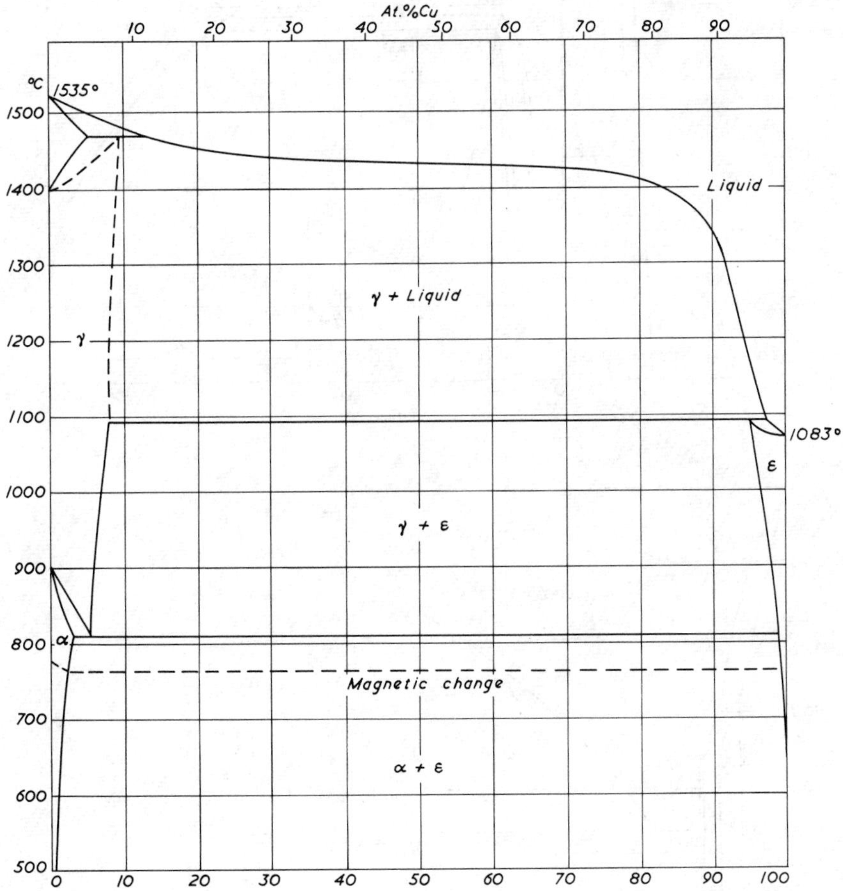

At.%Cu
10
20
30
40
50
60
70
80
90
°C
1535°
1500
1400
1300
1200
1100
1000
900
800
700
600
500
Liquid
γ + Liquid
γ
1083°
ε
γ + ε
α
Magnetic change
α + ε
0
10
20
30
40
50
60
70
80
90
100
Wt.% Cu

Cu–Ga

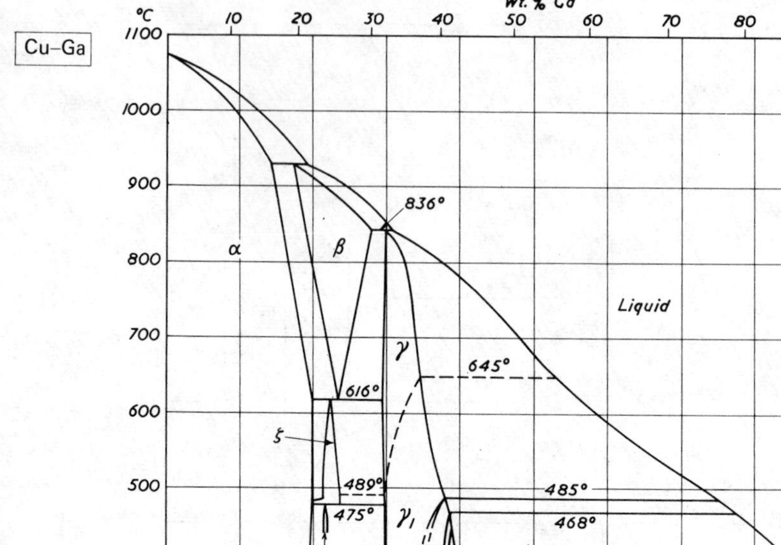
Wt. % Ga
°C
Liquid
836°
645°
616°
489°
475°
485°
468°
254°
29·6°
α
β
γ
ζ
$ζ_1$
$γ_1$
$γ_2$
$γ_3$
θ
At. % Ga

Cu–Ge

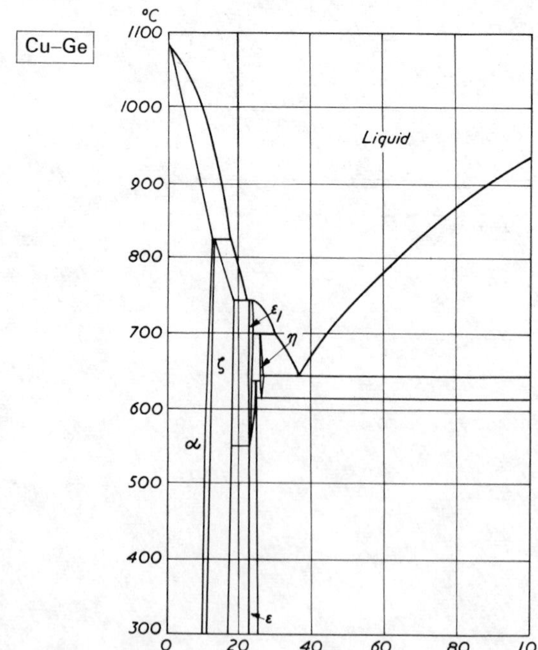
°C
Liquid
$ε_1$
η
ζ
α
ε
At. % Ge

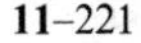

Cu–Hg

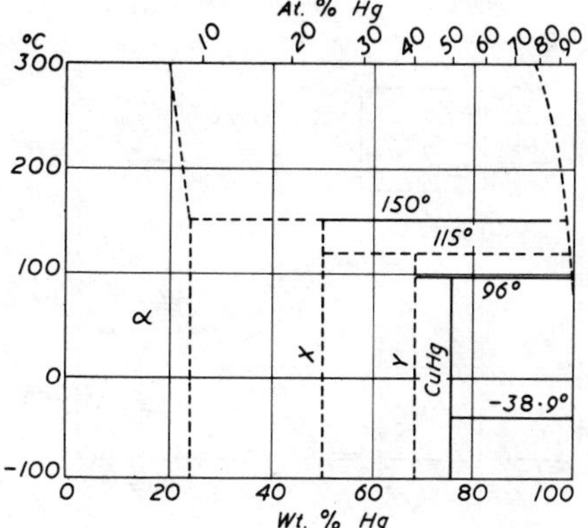
At. % Hg
10
20
30
40
50
60
70
80
90
°C
300
200
100
0
-100
150°
115°
96°
α
X
Y
CuHg
-38·9°
0
20
40
60
80
100
Wt. % Hg

Cu–In

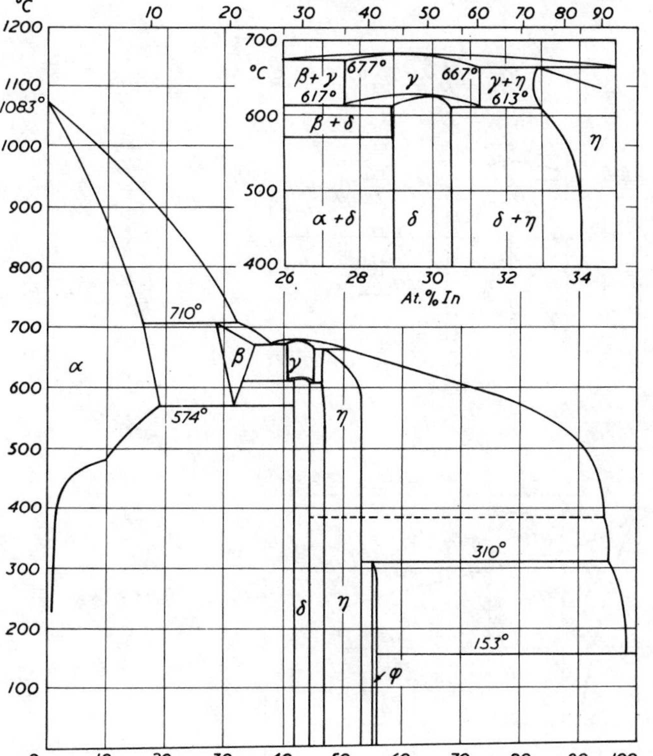
At. % In
10
20
30
40
50
60
70
80
90
°C
1200
1100
1083°
1000
900
800
700
600
500
400
300
200
100
0
700
°C
677°
β+γ
γ
667°
γ+η
617°
613°
600
β+δ
η
500
α+δ
δ
δ+η
400
26
28
30
32
34
At. % In
710°
α
β
γ
η
574°
310°
δ
η
153°
φ
10
20
30
40
50
60
70
80
90
100
Wt. % In

Cu–Ir

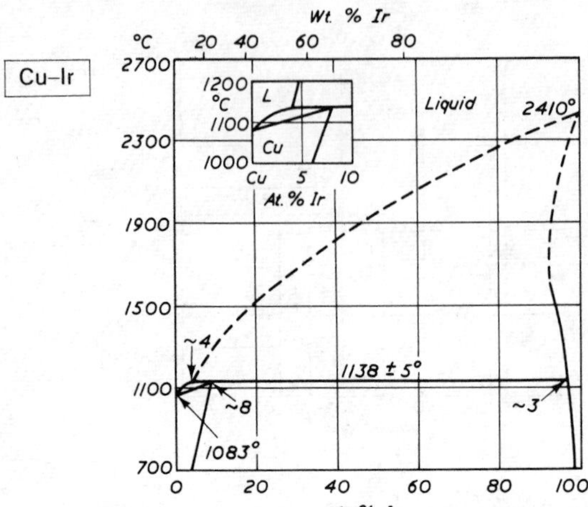

Cu–La

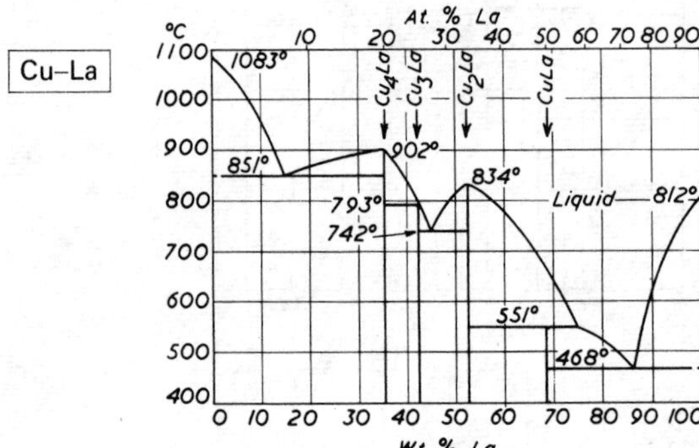

Cu–Li

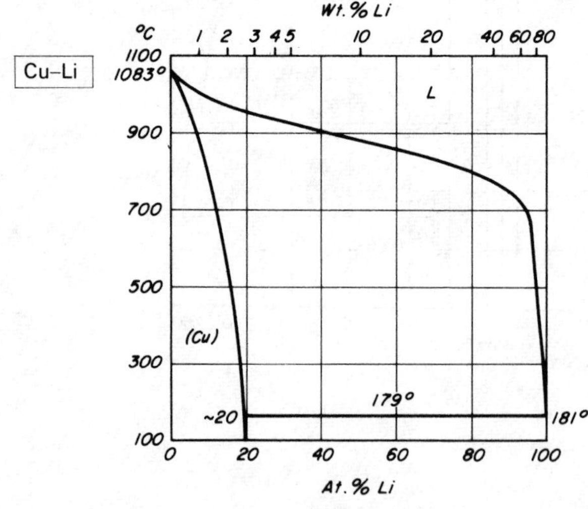

Cu–Mg

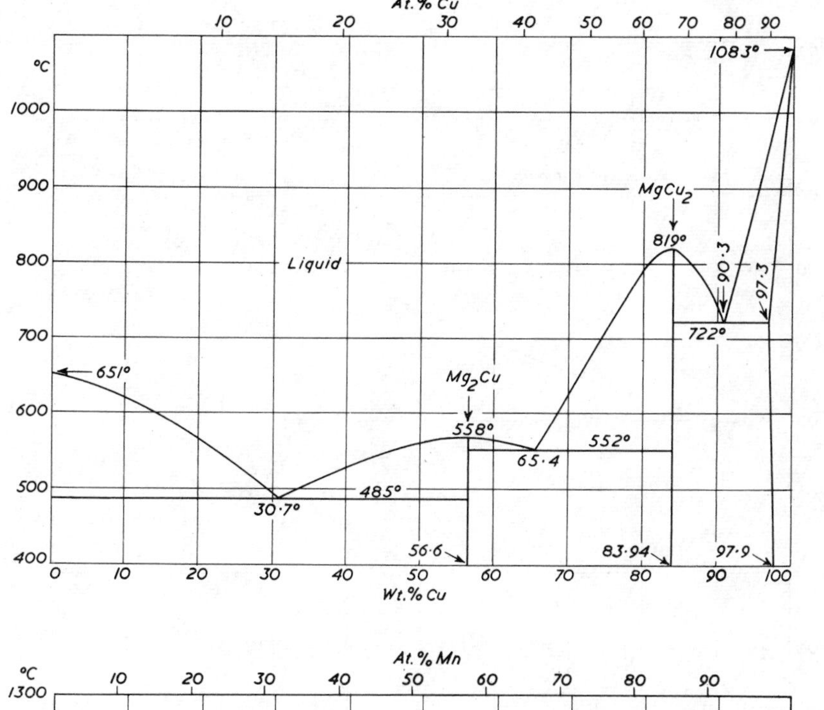
At.% Cu
10
20
30
40
50
60
70
80
90
°C
1000
900
800
700
600
500
400
1083°
MgCu2
819°
90·3
97·3
Liquid
722°
651°
Mg2Cu
558°
552°
65·4
485°
30·7°
56·6
83·94
97·9
0
10
20
30
40
50
60
70
80
90
100
Wt.% Cu

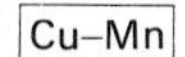
Cu–Mn

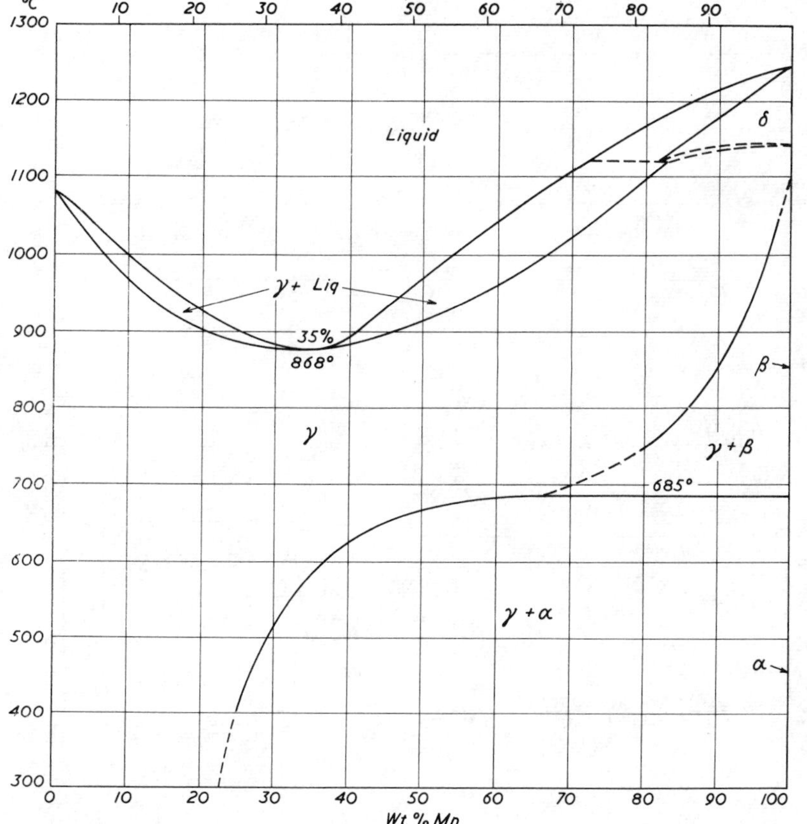
At.% Mn
10
20
30
40
50
60
70
80
90
°C
1300
1200
1100
1000
900
800
700
600
500
400
300
Liquid
δ
γ+ Liq
35%
868°
β
γ
γ+β
685°
γ+α
α
0
10
20
30
40
50
60
70
80
90
100
Wt.% Mn

Cu–Ni

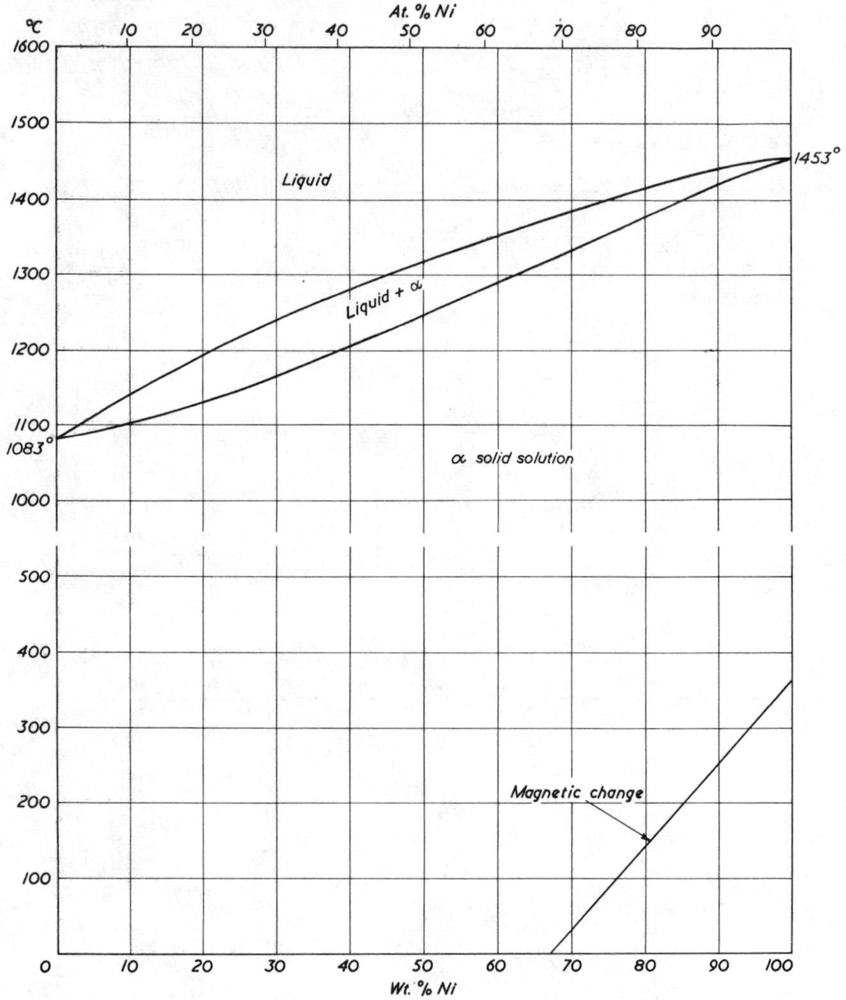

At. % Ni

Liquid

1453°

Liquid + α

1083°

α solid solution

Magnetic change

Wt. % Ni

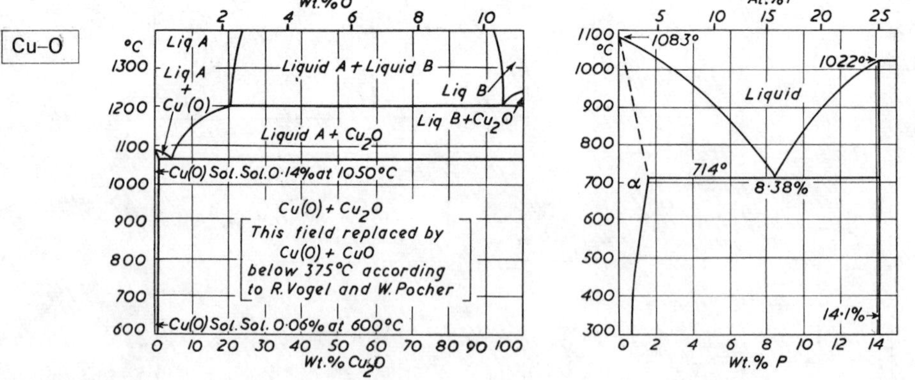

Cu–Pb

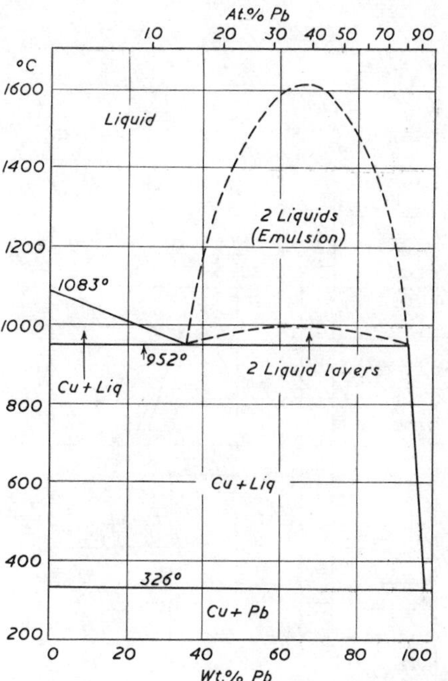
At.% Pb
10
20
30
40
50
70
90
°C
1600
1400
1200
1000
800
600
400
200
Liquid
2 Liquids
(Emulsion)
1083°
952°
Cu + Liq
2 Liquid layers
Cu + Liq
326°
Cu + Pb
0
20
40
60
80
100
Wt.% Pb

Cu–Pd

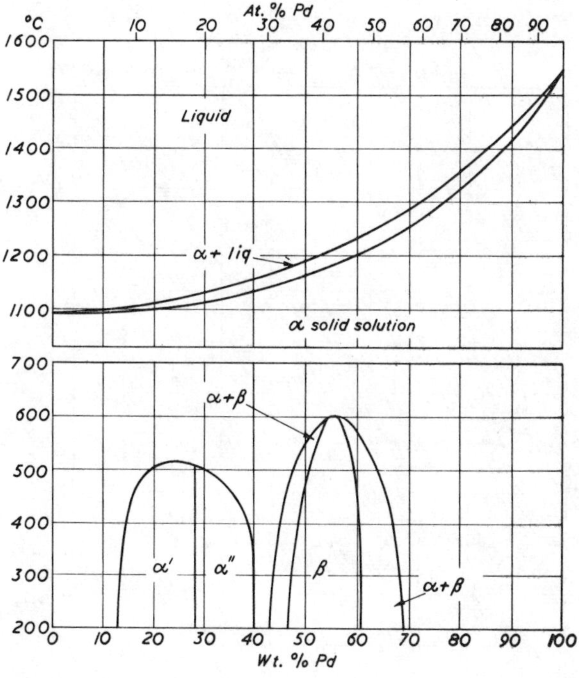
At. % Pd
10
20
30
40
50
60
70
80
90
°C
1600
1500
1400
1300
1200
1100
700
600
500
400
300
200
Liquid
α + liq
α solid solution
α+β
α′
α″
β
α+β
0
10
20
30
40
50
60
70
80
90
100
Wt. % Pd

Cu–Pr

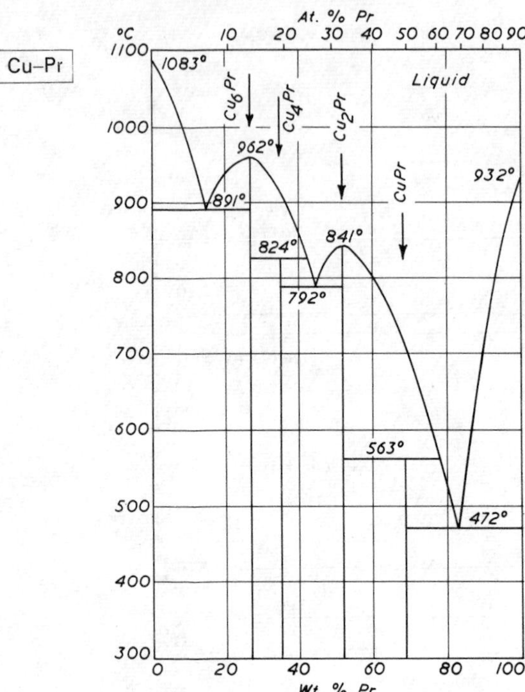
At. % Pr
°C
1083°
Liquid
Cu_6Pr
Cu_4Pr
Cu_2Pr
CuPr
962°
932°
891°
824°
841°
792°
563°
472°
Wt. % Pr

Cu–Pt

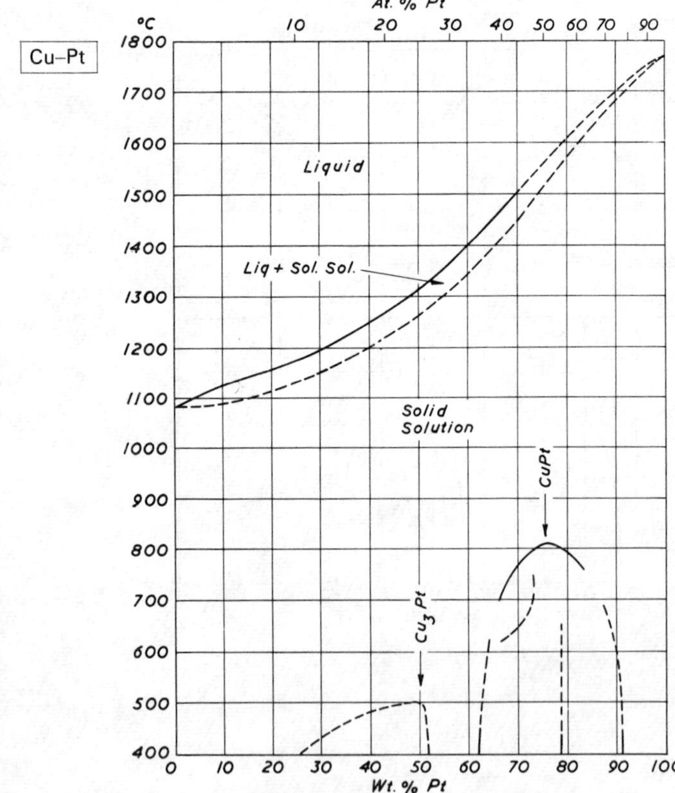
At. % Pt
°C
Liquid
Liq + Sol. Sol.
Solid Solution
CuPt
Cu_3Pt
Wt. % Pt

Cu–Pu

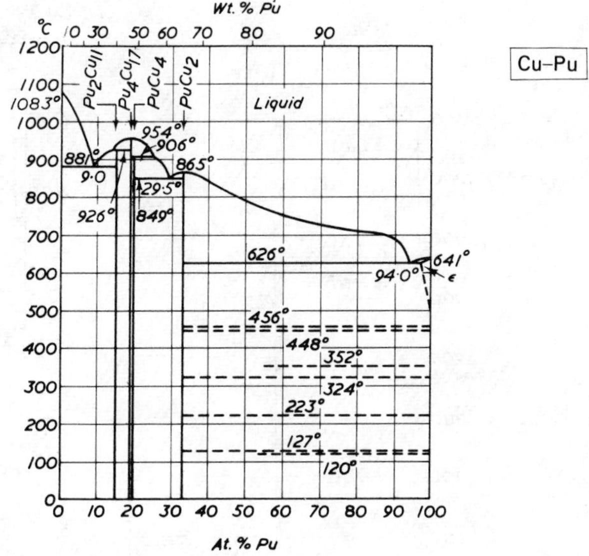
Wt. % Pu
10 30 50 60 70 80 90
°C
1200
1100
1083°
1000
900
800
700
600
500
400
300
200
100
0
Pu_2Cu_{11}
Pu_4Cu_{17}
$PuCu_4$
$PuCu_2$
Liquid
954°
906°
881°
865°
9·0
29·5°
926°
849°
626°
641°
94·0°
ε
456°
448°
352°
324°
223°
127°
120°
0 10 20 30 40 50 60 70 80 90 100
At. % Pu

Cu–Rh

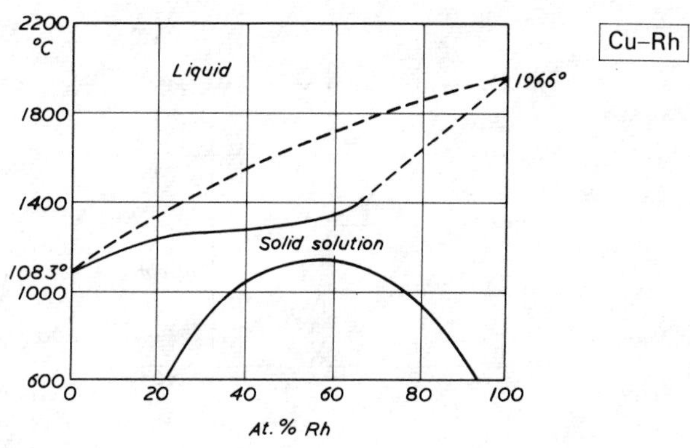
2200
°C
Liquid
1966°
1800
1400
Solid solution
1083°
1000
600
0 20 40 60 80 100
At. % Rh

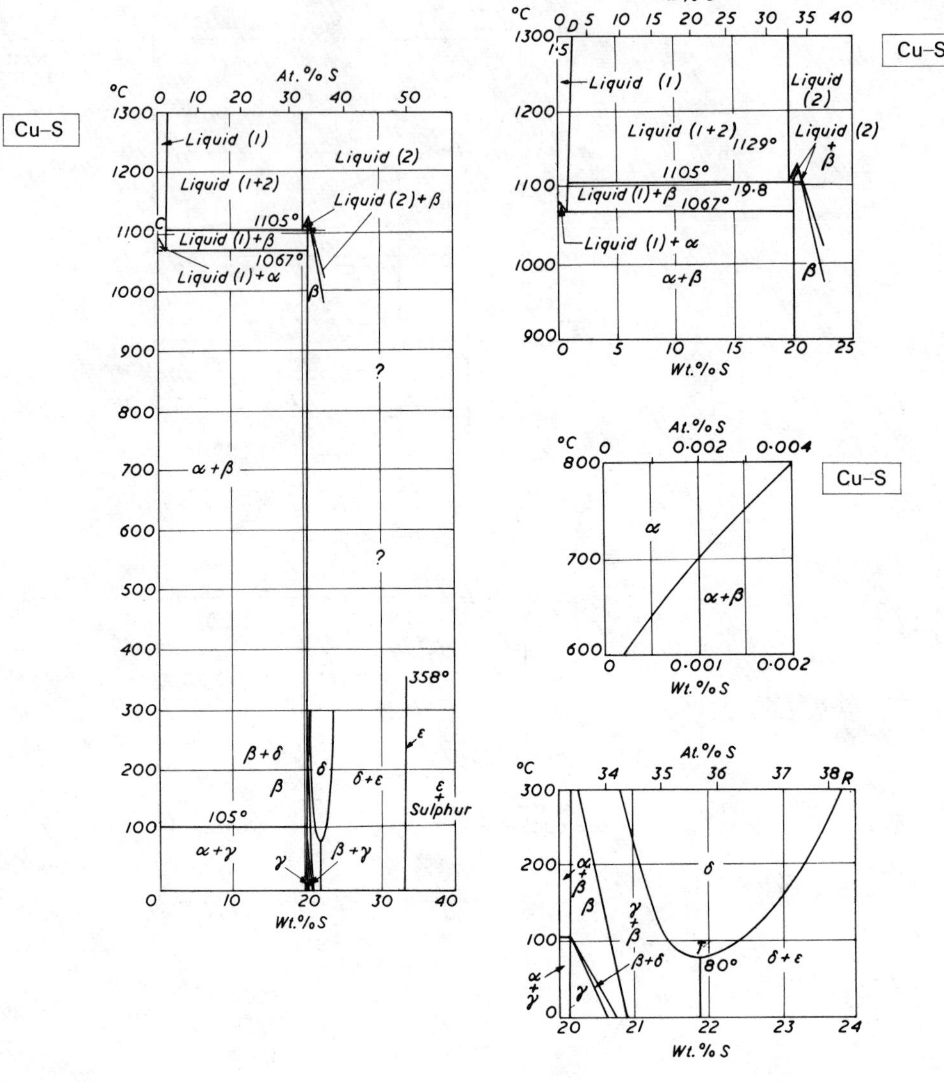
Cu–S
At.% S
°C
Liquid (1)
Liquid (2)
Liquid (1+2)
Liquid (2)+β
1105°
Liquid (1)+β
1067°
Liquid (1)+α
β
?
α+β
358°
ε
β+δ
δ
δ+ε
ε + Sulphur
105°
α+γ
γ
β+γ
Wt.% S
Cu–S
At.% S
°C
1·5
Liquid (1)
Liquid (2)
Liquid (1+2)
1129°
Liquid (2) + β
1105°
Liquid (1)+β
19·8
1067°
Liquid (1) + α
α+β
β
Wt.% S
Cu–S
At.% S
°C
α
α+β
Wt.% S
At.% S
°C
α + β
β
δ
γ + β
β+δ
80°
δ + ε
α + γ
γ
Wt.% S
Cu–S

Cu–Sb

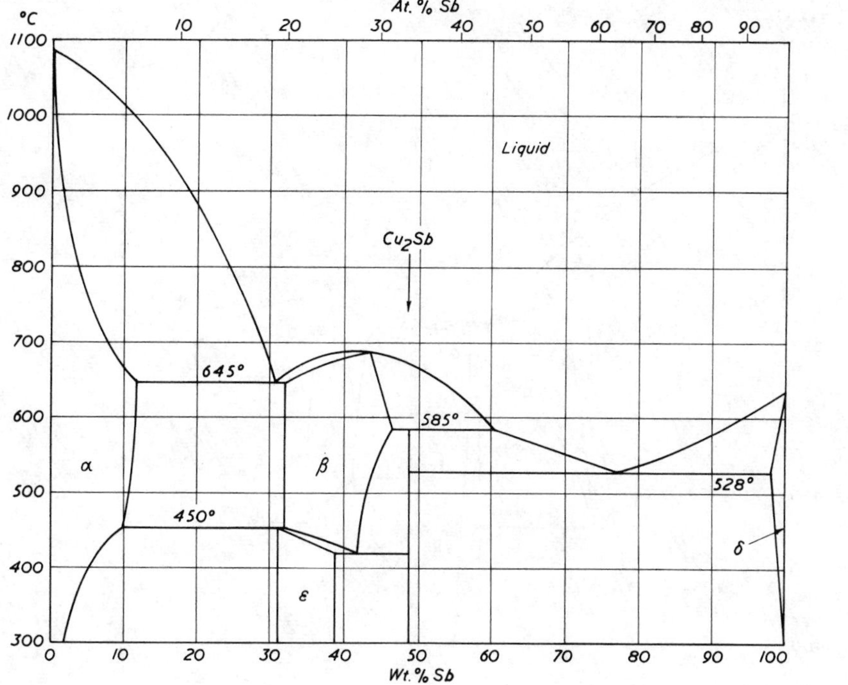
At. % Sb
°C
Liquid
Cu_2Sb
645°
585°
528°
450°
α
β
ε
δ
Wt. % Sb

Cu–Se

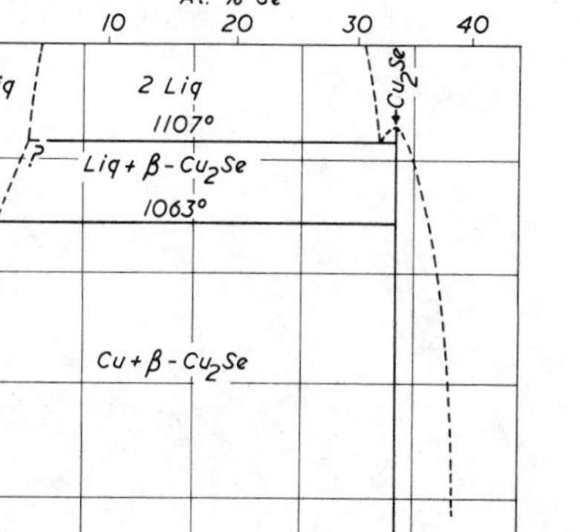
At. % Se
°C
Liq
2 Liq
1107°
Cu_2Se
?
Liq + β-Cu_2Se
1063°
Cu + β-Cu_2Se
Wt. % Se

Cu–Si

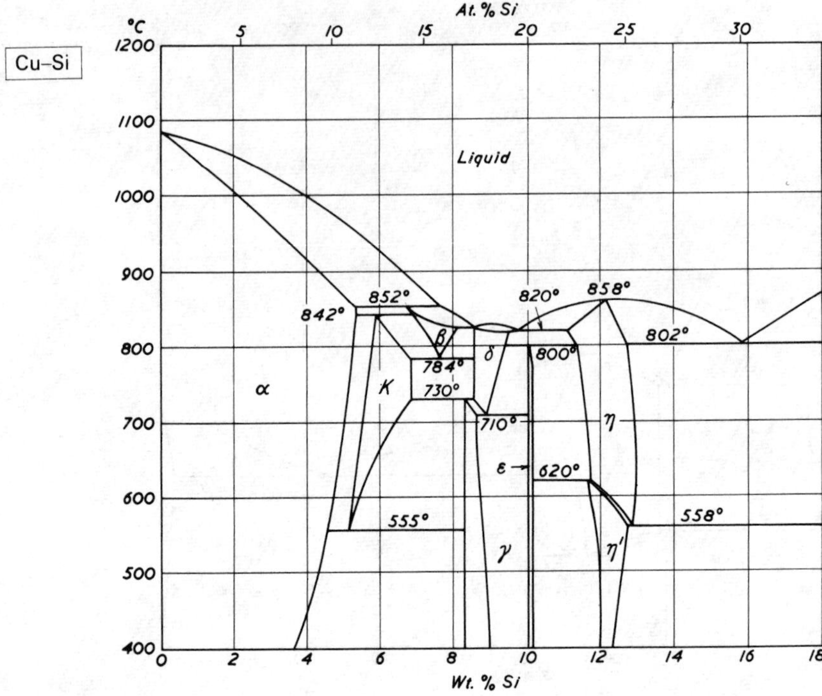
At. % Si
°C
5
10
15
20
25
30
1200
1100
1000
900
800
700
600
500
400
Liquid
852°
842°
820°
858°
802°
800°
784°
730°
710°
620°
555°
558°
α
K
β
δ
γ
ε
η
η'
0
2
4
6
8
10
12
14
16
18
Wt. % Si

Cu–Sn

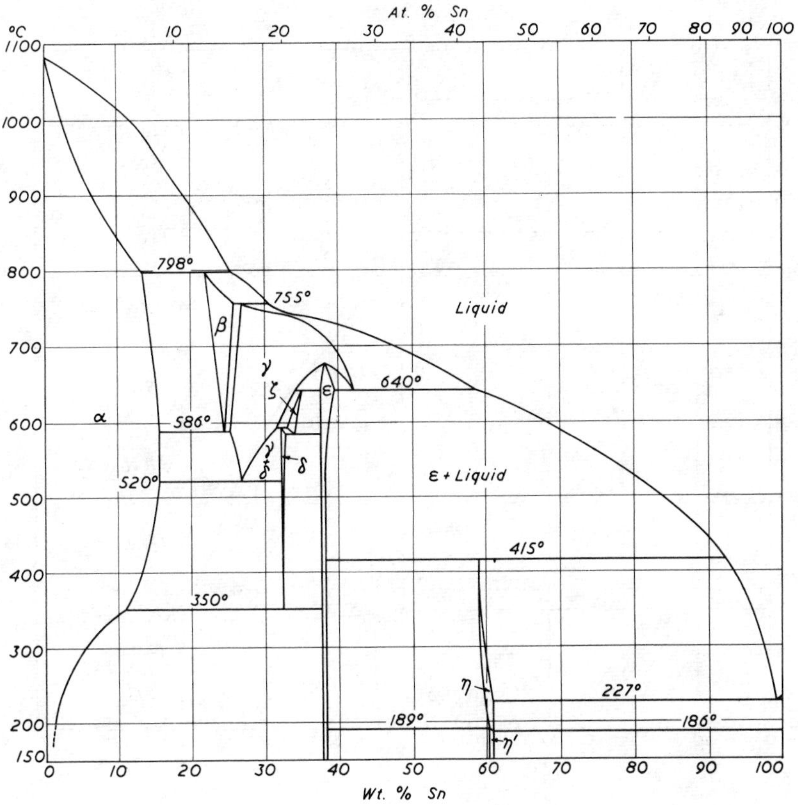

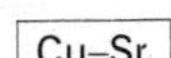

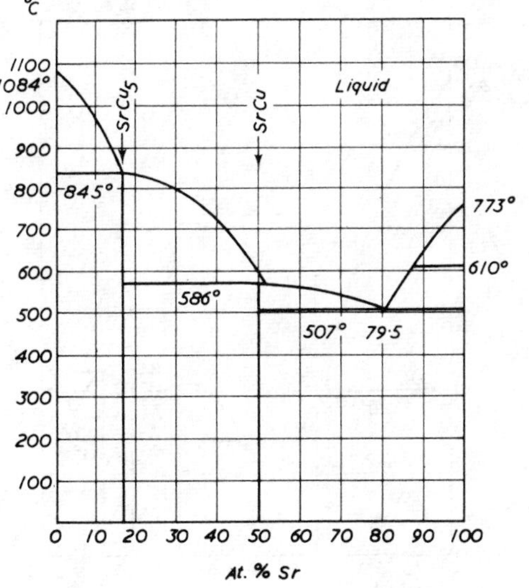
°C
1100
1084°
1000
900
800
700
600
500
400
300
200
100
SrCu5
SrCu
Liquid
845°
773°
610°
586°
507°
79·5
0
10
20
30
40
50
60
70
80
90
100
At. % Sr

Cu–Te

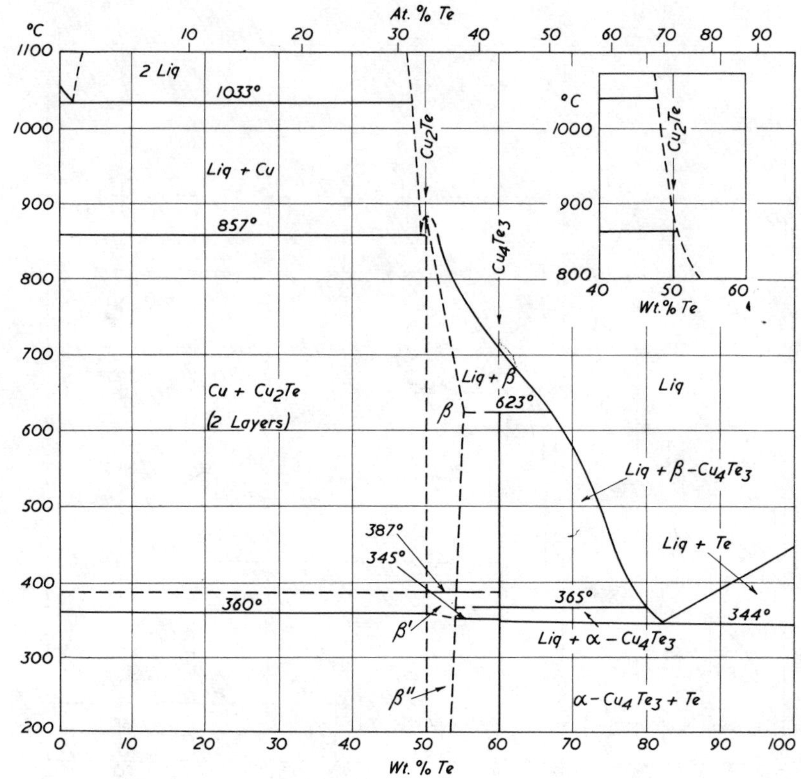

At. % Te
°C
2 Liq
1033°
Liq + Cu
857°
Cu_2Te
Cu_4Te_3
Liq + β
623°
β
Liq
Cu + Cu_2Te
(2 Layers)
Liq + β-Cu_4Te_3
387°
345°
Liq + Te
365°
360°
344°
β′
Liq + α − Cu_4Te_3
β″
α-Cu_4Te_3 + Te
Wt. % Te

Cu–Th

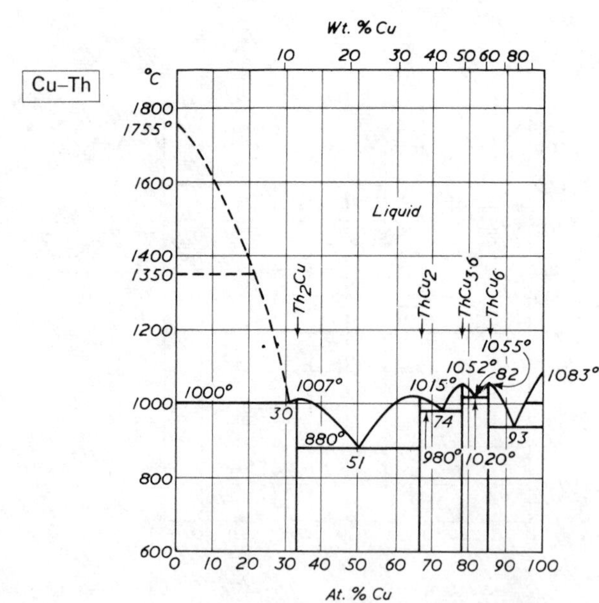

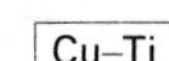
Cu–Ti

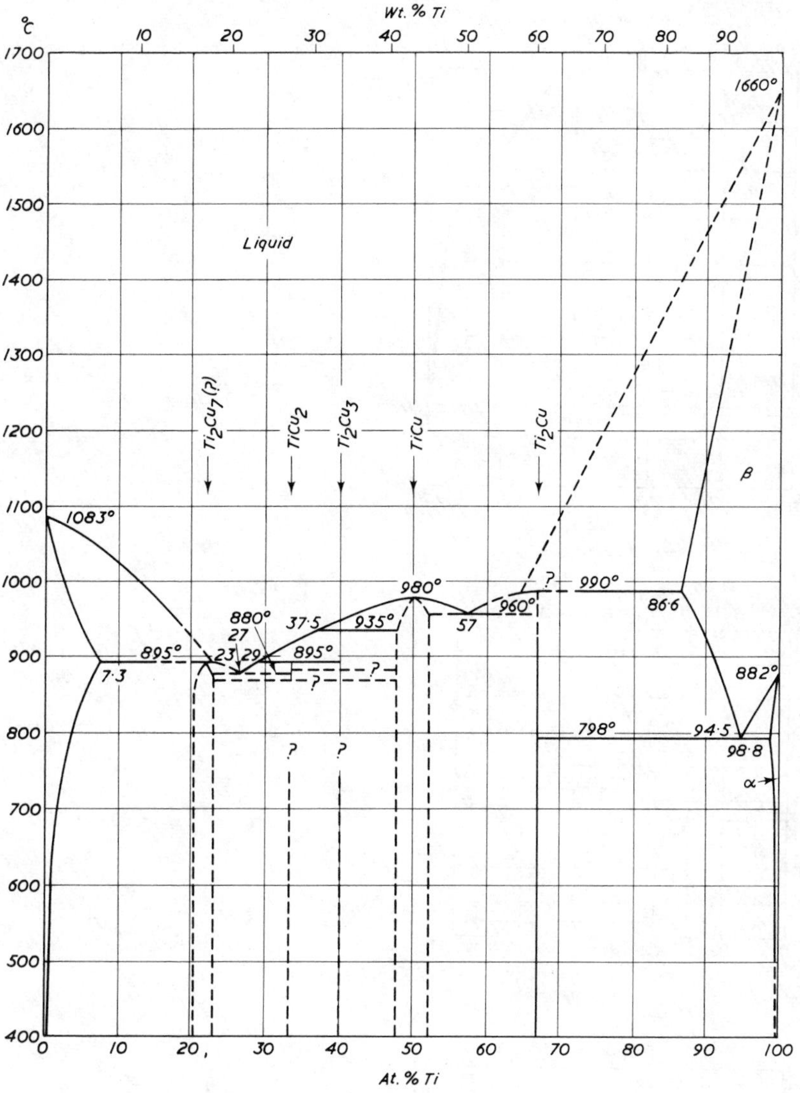

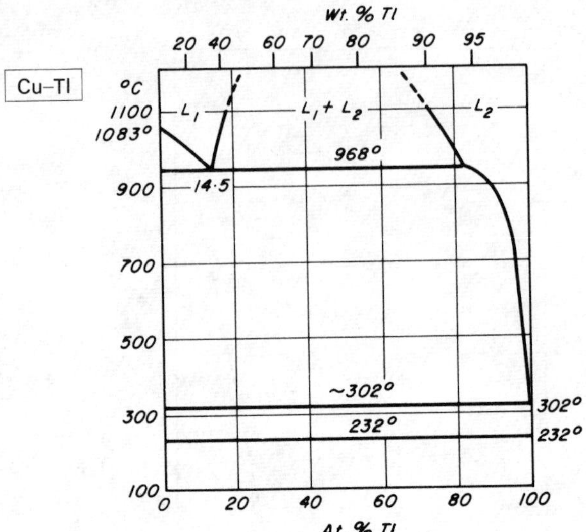
Cu–Tl
Wt. % Tl
20 40 60 70 80 90 95
°C
1100
1083°
900
700
500
300
100
L1
L1 + L2
L2
968°
14·5
~302°
302°
232°
232°
0 20 40 60 80 100
At. % Tl

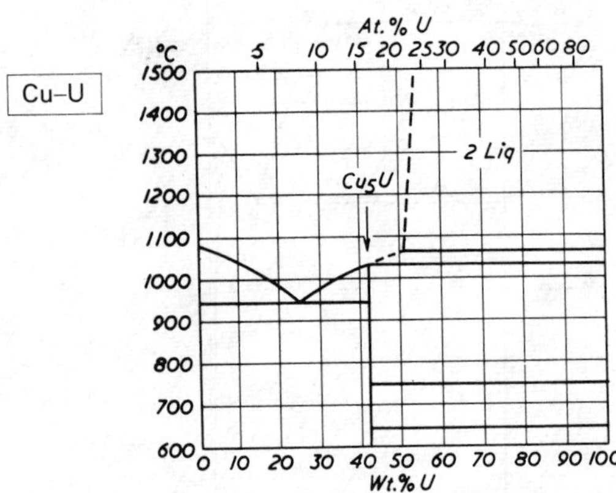
Cu–U
At.% U
5 10 15 20 25 30 40 50 60 80
°C
1500
1400
1300
1200
1100
1000
900
800
700
600
2 Liq
Cu5U
0 10 20 30 40 50 60 70 80 90 100
Wt.% U

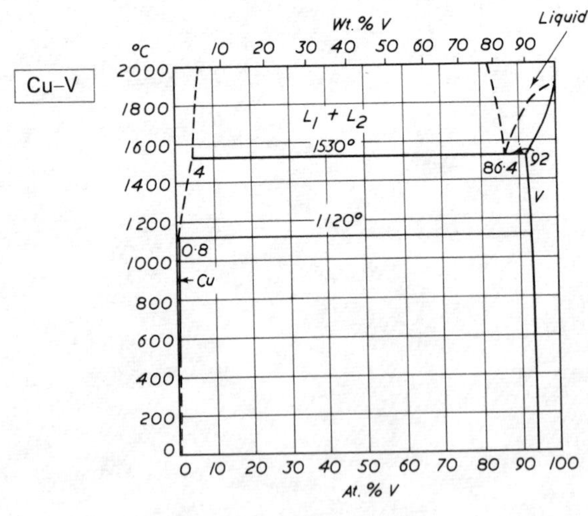
Cu–V
Wt. % V
10 20 30 40 50 60 70 80 90
Liquid
°C
2000
1800
1600
1400
1200
1000
800
600
400
200
0
L1 + L2
1530°
4
86·4
92
V
1120°
0·8
Cu
0 10 20 30 40 50 60 70 80 90 100
At. % V

Cu–Y

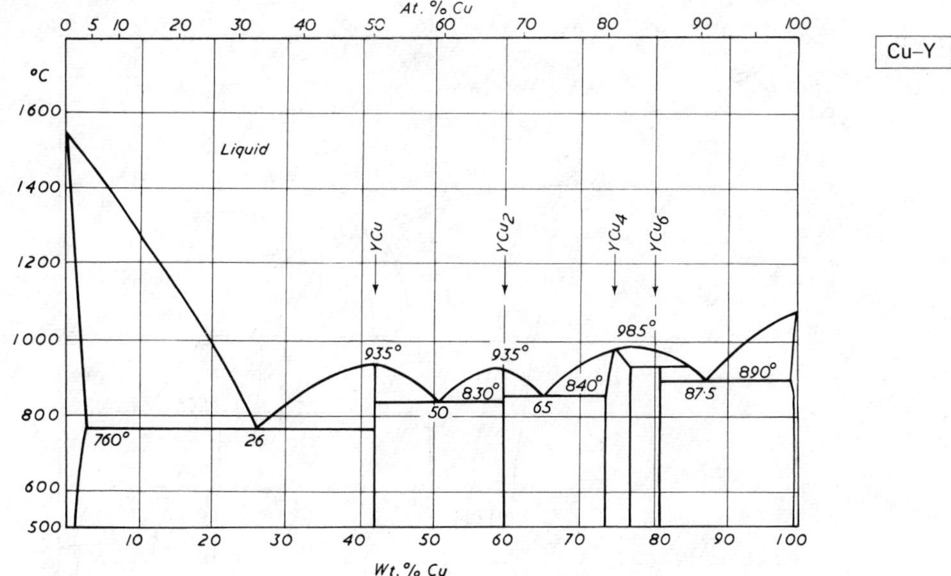
At. % Cu
0 5 10 20 30 40 50 60 70 80 90 100
°C
1600
1400
1200
1000
800
600
500
Liquid
YCu
YCu_2
YCu_4
YCu_6
935°
935°
985°
830°
840°
890°
760°
26
50
65
87·5
10 20 30 40 50 60 70 80 90 100
Wt. % Cu

Cu–Yb

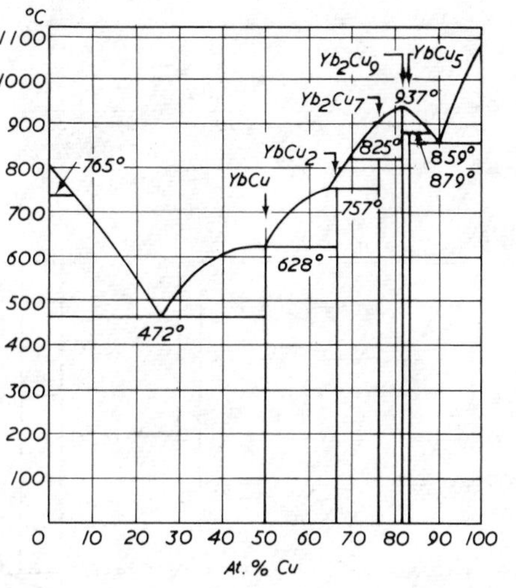
°C
1100
1000
900
800
700
600
500
400
300
200
100
Yb_2Cu_9
$YbCu_5$
Yb_2Cu_7
937°
825°
859°
879°
$YbCu_2$
765°
YbCu
757°
628°
472°
0 10 20 30 40 50 60 70 80 90 100
At. % Cu

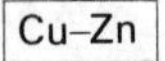
Cu–Zn

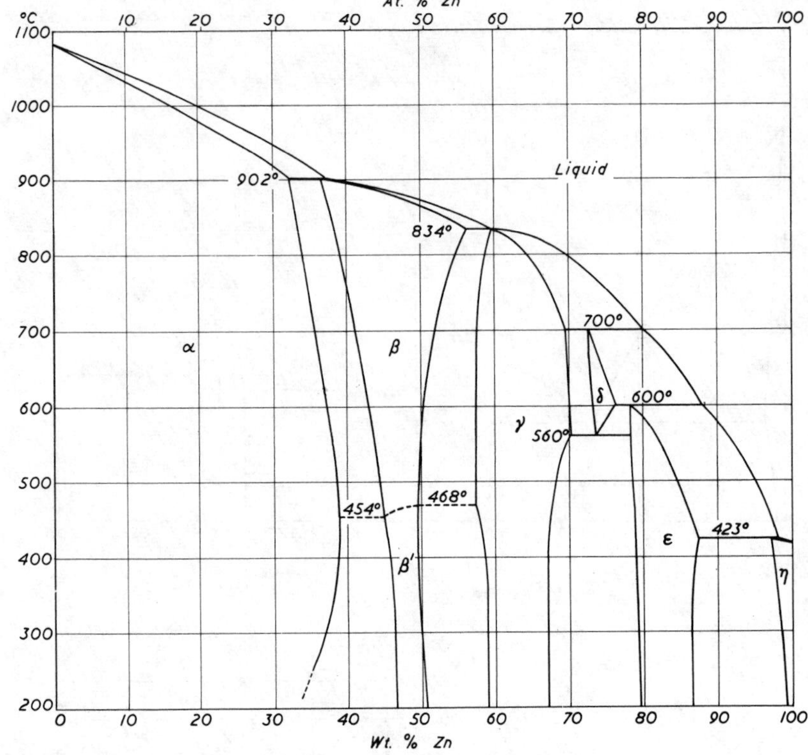
At. % Zn
°C
Liquid
902°
834°
700°
600°
560°
468°
454°
423°
α
β
β′
γ
δ
ε
η
Wt. % Zn

Cu–Zr

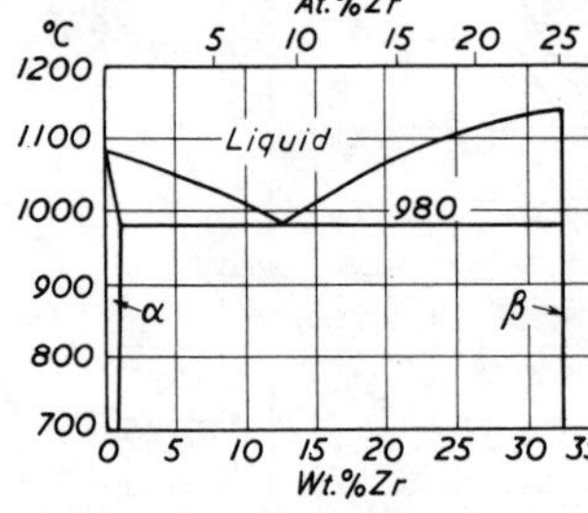
At.%Zr
°C
Liquid
980
α
β
Wt.%Zr

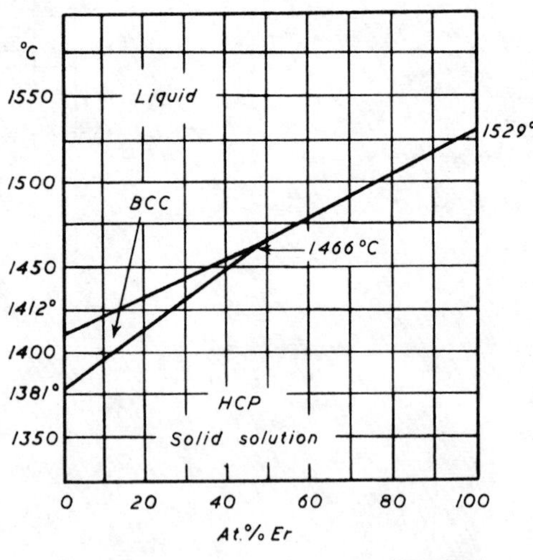
°C
Liquid
1529°
BCC
1466 °C
1412°
1381°
HCP
Solid solution
At.% Er

Dy–Er

Dy–Fe

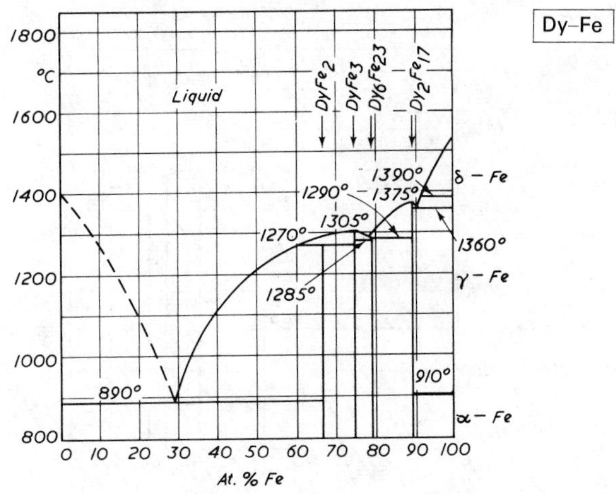

Dy–Ho

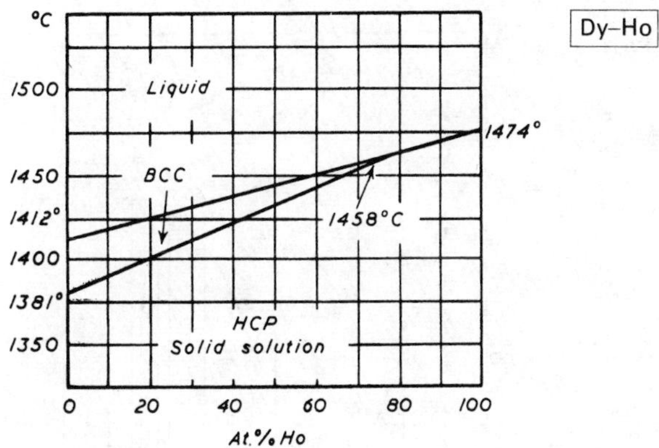

Dy–Zr

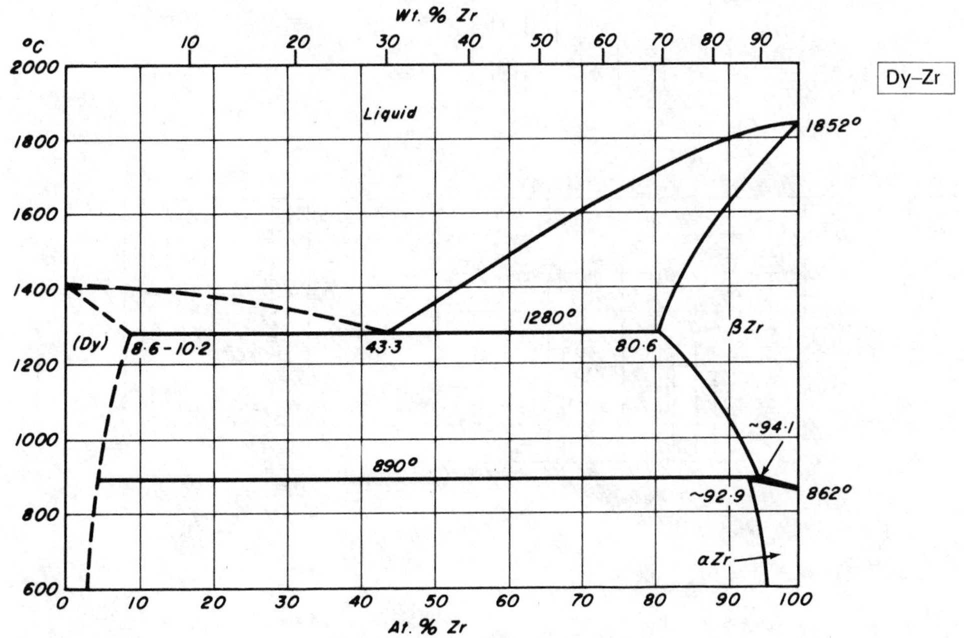

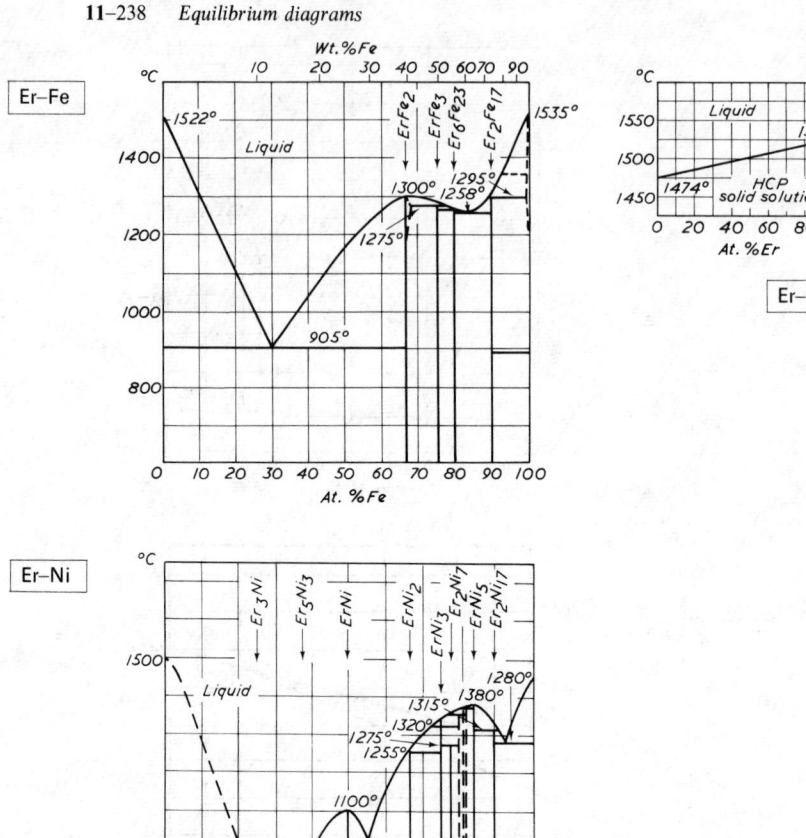

Er–Fe

Wt. % Fe
°C
1522°
Liquid
$ErFe_2$ $ErFe_3$ Er_6Fe_{23} Er_2Fe_{17}
1535°
1300° 1295° 1258°
1275°
905°
At. % Fe

°C
Liquid
1529°
1474°
HCP solid solution
At. % Er

Er–Ho

Er–Ni

°C
Er_3Ni Er_5Ni_3 ErNi $ErNi_2$ $ErNi_3$ Er_2Ni_7 $ErNi_5$ Er_2Ni_{17}
Liquid
1280°
1380°
1315°
1320°
1275°
1255°
1100°
1025°
845°
800°
765°
At. % Ni

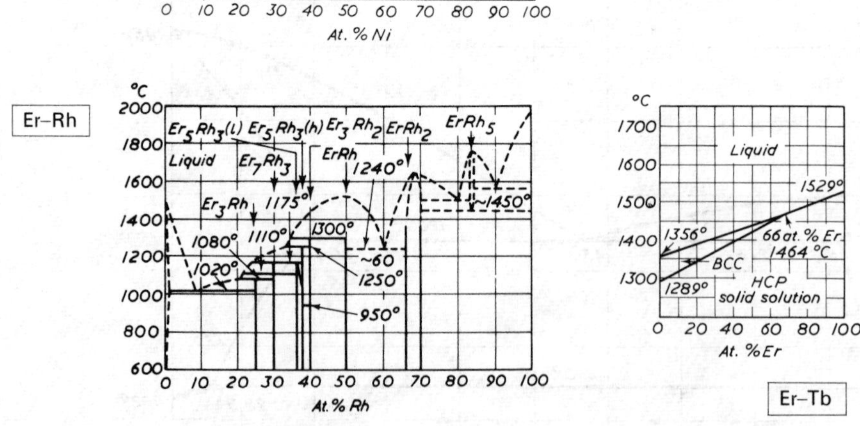

Er–Ti

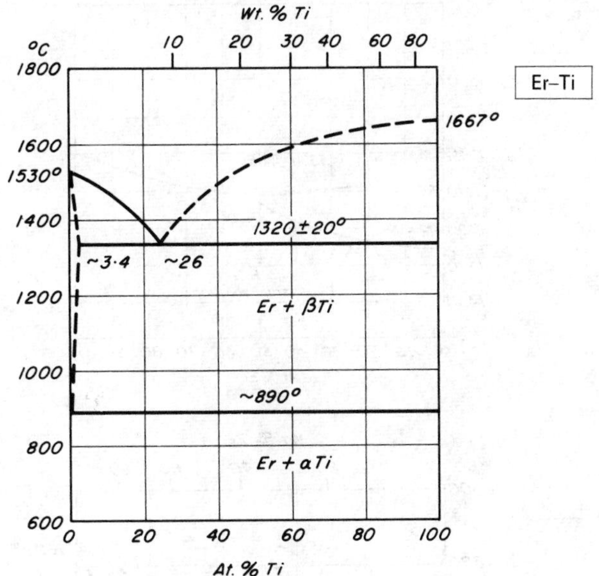
Wt. % Ti
10
20
30
40
60
80
°C
1800
1600
1530°
1400
1200
1000
800
600
1667°
1320±20°
~3·4
~26
Er + βTi
~890°
Er + αTi
0
20
40
60
80
100
At. % Ti

Er–V

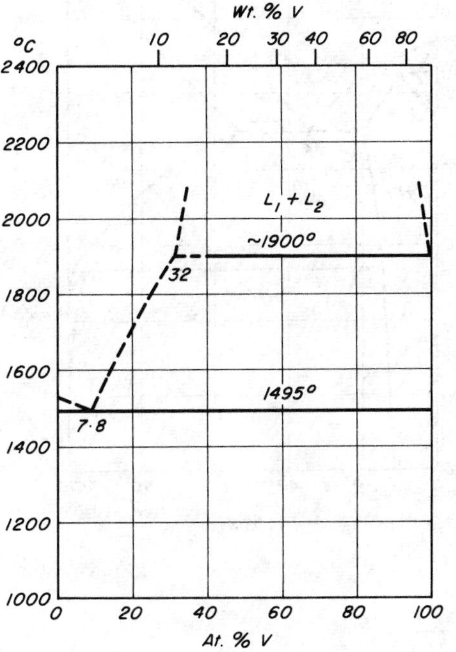
Wt. % V
10
20
30
40
60
80
°C
2400
2200
2000
1800
1600
1400
1200
1000
$L_1 + L_2$
~1900°
32
1495°
7·8
0
20
40
60
80
100
At. % V

Er–Y

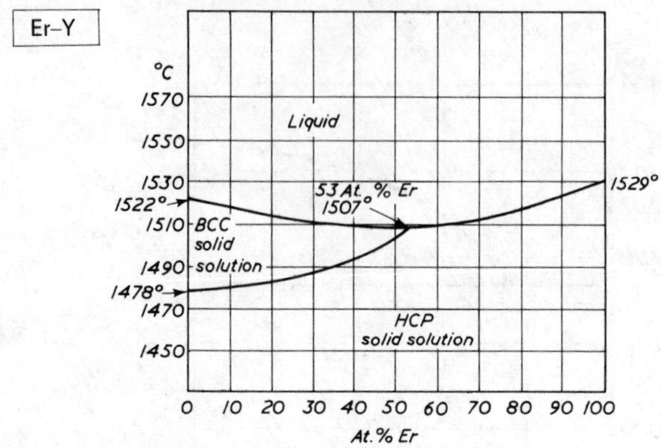
°C
1570
1550
1530
1510
1490
1470
1450
Liquid
1522°
1529°
53 At.% Er
1507°
BCC
solid
solution
1478°
HCP
solid solution
0 10 20 30 40 50 60 70 80 90 100
At.% Er

Er–Zr

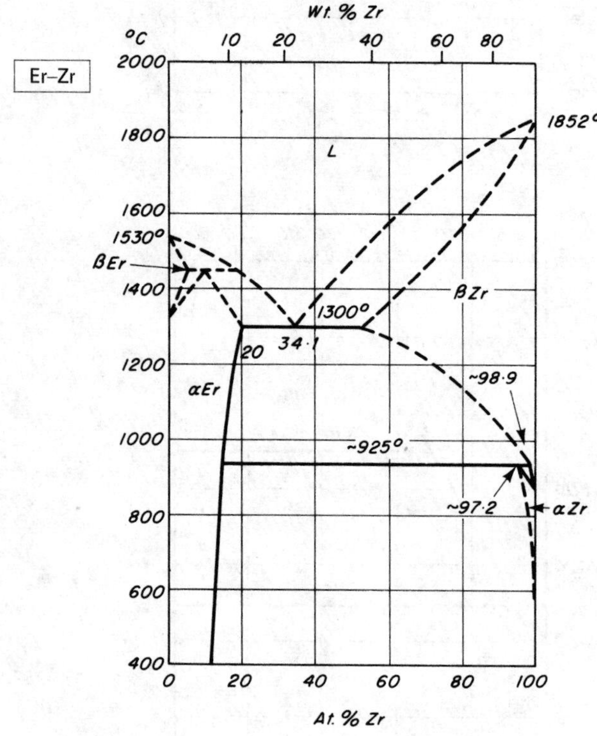
Wt.% Zr
10 20 40 60 80
°C
2000
1800
1600
1400
1200
1000
800
600
400
1852°
L
1530°
βEr
βZr
1300°
34·1
20
αEr
~98·9
~925°
~97·2
αZr
0 20 40 60 80 100
At.% Zr

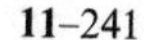
Eu–In

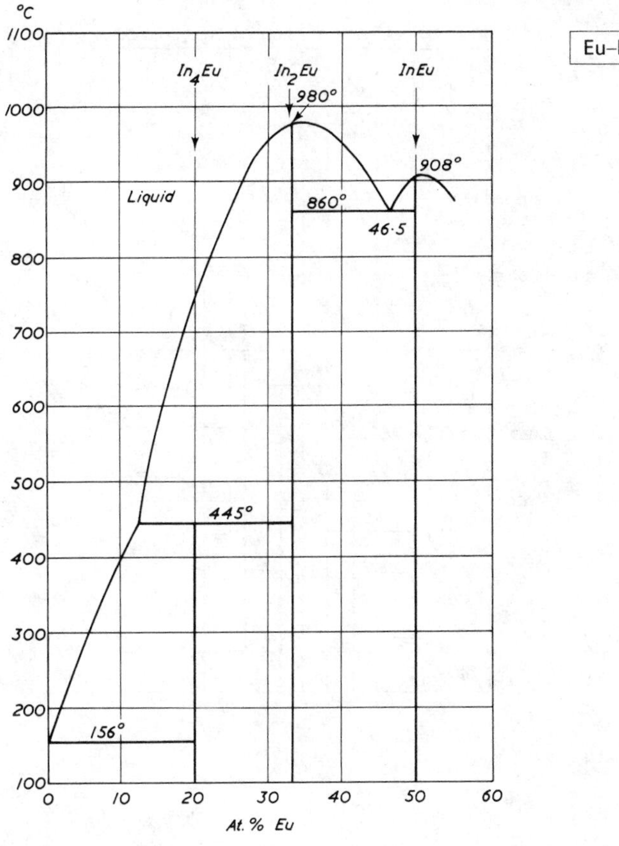
°C
In$_4$Eu
In$_2$Eu
InEu
980°
908°
Liquid
860°
46·5
445°
156°
At. % Eu

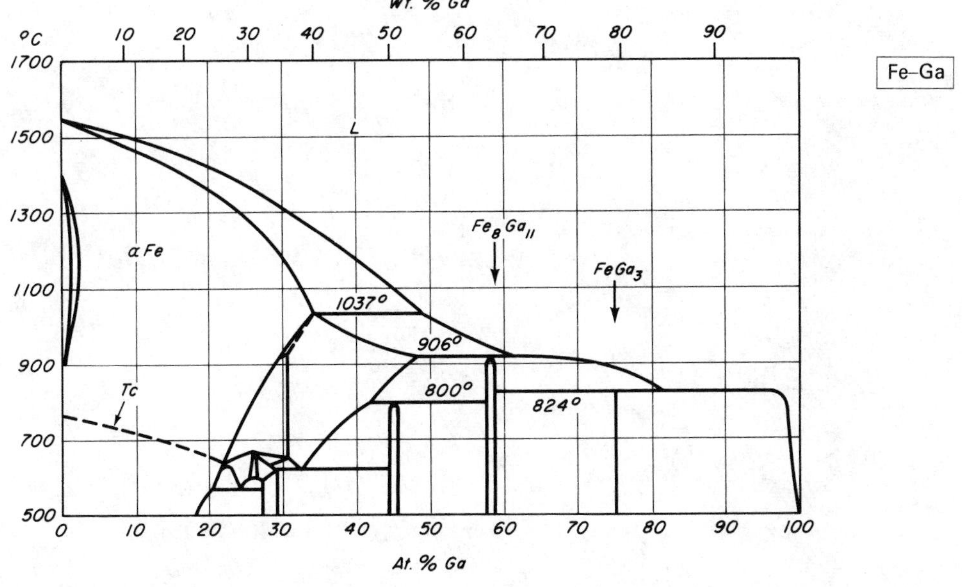
Fe–Ga
Wt. % Ga
°C
L
αFe
Fe$_8$Ga$_{11}$
FeGa$_3$
1037°
906°
800°
824°
Tc
At. % Ga

Fe–Gd

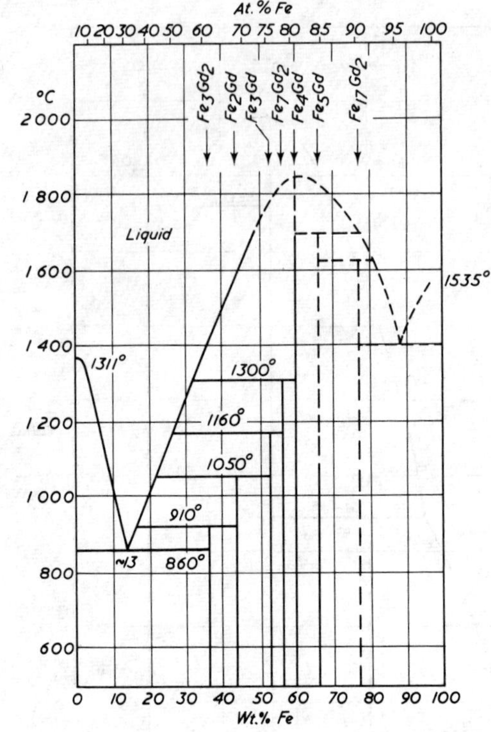

At.% Fe
10 20 30 40 50 60 70 75 80 85 90 95 100
°C
2000
1800
1600
1400
1200
1000
800
600
Fe_3Gd_2
Fe_2Gd
Fe_3Gd
Fe_7Gd_2
Fe_4Gd
Fe_5Gd
$Fe_{17}Gd_2$
Liquid
1535°
1311°
1300°
1160°
1050°
910°
~13
860°
0 10 20 30 40 50 60 70 80 90 100
Wt.% Fe

Fe–Ge

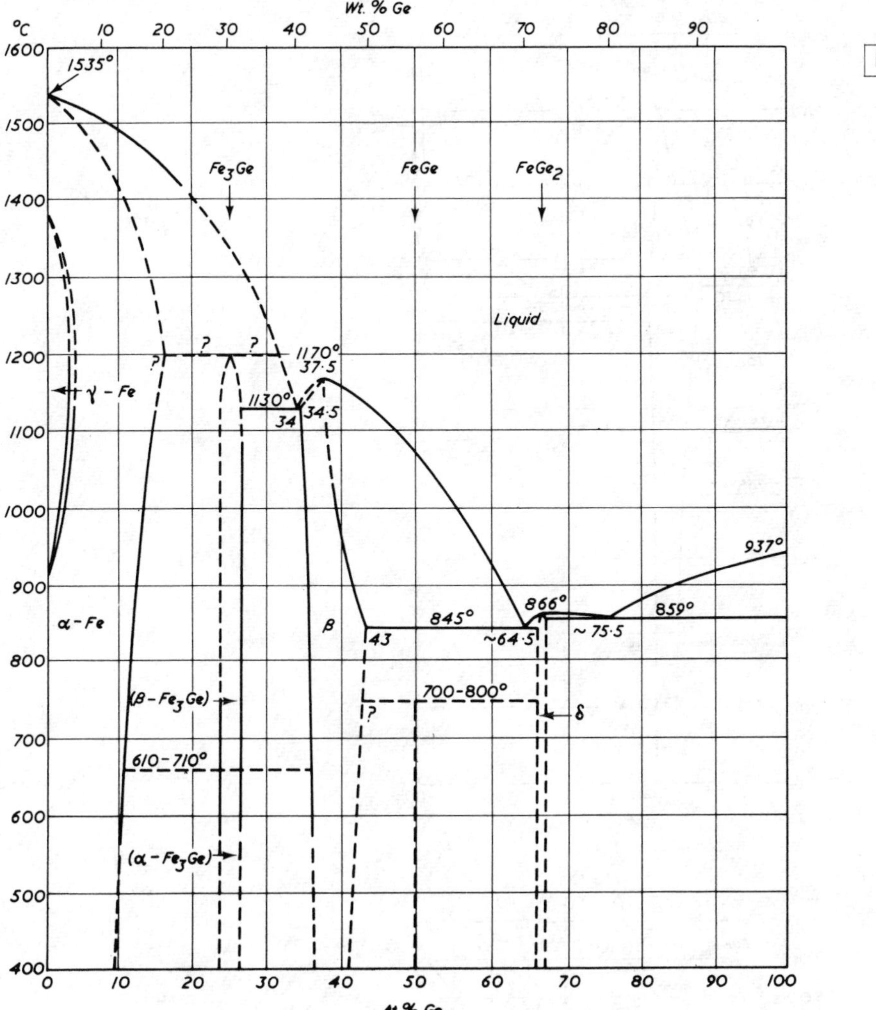

Wt. % Ge
°C
1535°
Fe$_3$Ge
FeGe
FeGe$_2$
Liquid
1170°
37·5
1130°
34
34·5
γ - Fe
937°
866°
859°
845°
~64·5
~75·5
43
α - Fe
β
(β - Fe$_3$Ge)
700-800°
δ
610-710°
(α - Fe$_3$Ge)
At. % Ge

Fe–Hf

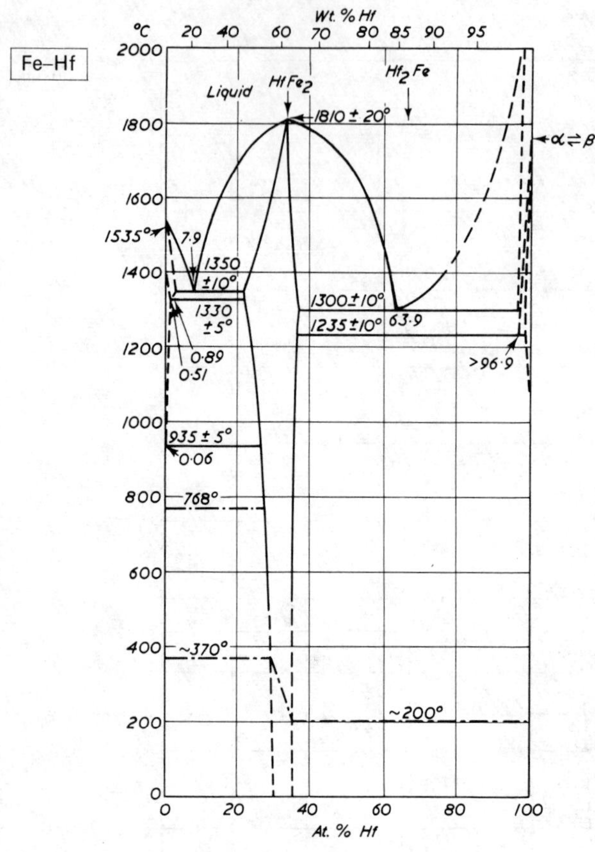
Wt. % Hf
°C
20 40 60 70 80 85 90 95
Liquid
HfFe2
Hf2Fe
1810 ± 20°
α⇌β
1535°
7.9
1350 ±10°
1330 ±5°
0.89
0.51
1300±10°
1235±10°
63.9
>96.9
935 ± 5°
0.06
768°
~370°
~200°
At. % Hf

Fe–Ho

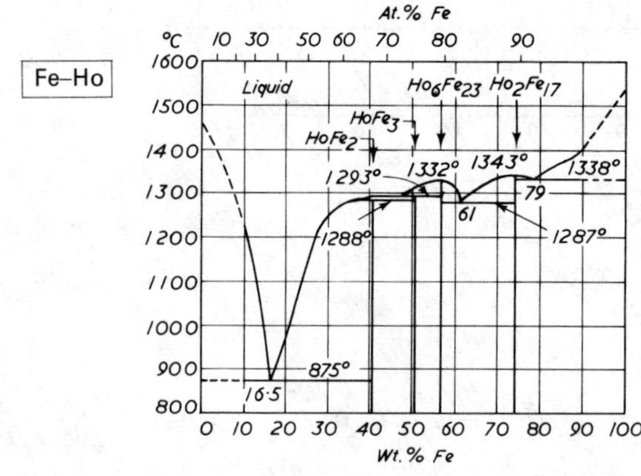
At. % Fe
°C
10 30 50 60 70 80 90
Liquid
Ho6Fe23
Ho2Fe17
HoFe3
HoFe2
1293°
1332°
1343°
1338°
79
61
1288°
1287°
875°
16.5
Wt. % Fe

Fe–In

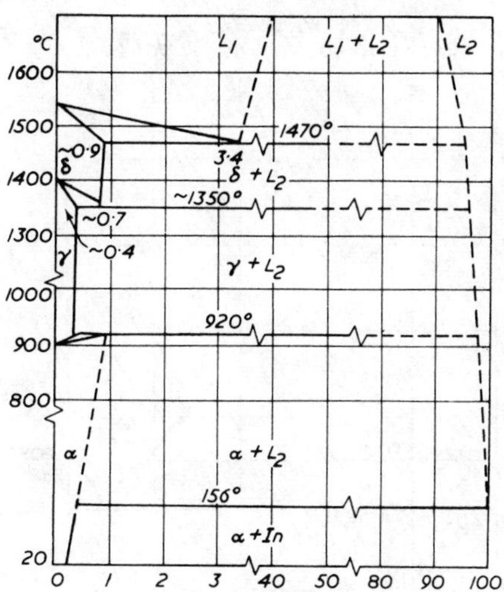
°C
1600
1500
1400
1300
1000
900
800
20
L1
L1 + L2
L2
1470°
~0·9
3·4
δ
δ + L2
~1350°
~0·7
~0·4
γ
γ + L2
920°
α
α + L2
156°
α + In
0 1 2 3 40 50 80 90 100
At. % In

Fe–Ir

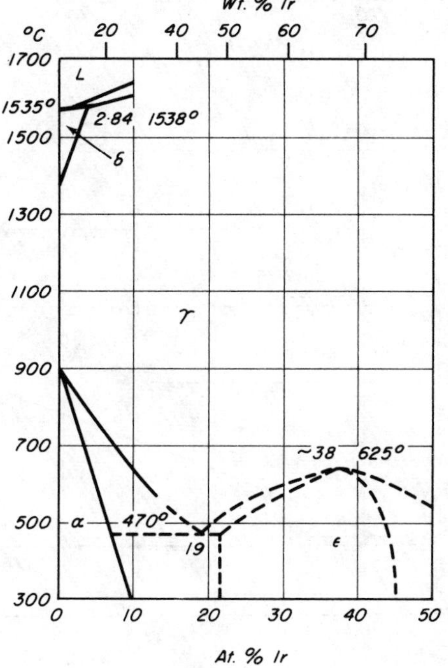
Wt. % Ir
20 40 50 60 70
°C
1700
1535°
1500
1300
1100
900
700
500
300
L
2·84
1538°
δ
γ
~38
625°
α
470°
19
ε
0 10 20 30 40 50
At. % Ir

Fe–La

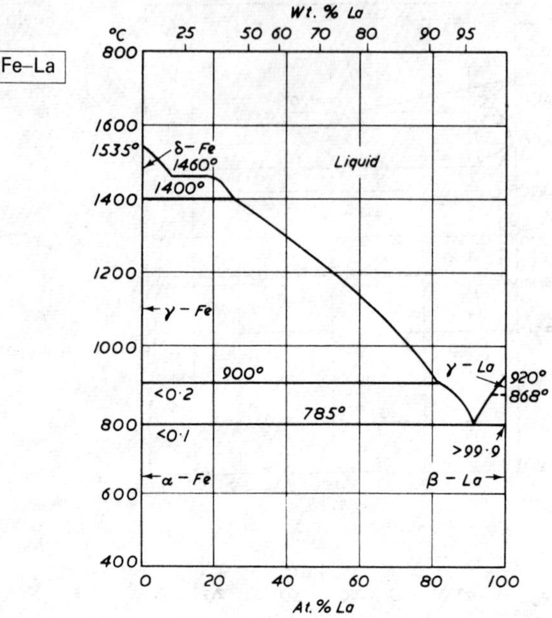
Wt. % La
°C
25
50 60 70 80 90 95
1800
1600
1535°
δ-Fe
1460°
Liquid
1400°
1400
1200
γ-Fe
1000
900°
γ-La
920°
868°
<0·2
785°
800
<0·1
>99·9
α-Fe
β-La
600
400
0
20
40
60
80
100
At. % La

Fe–Mg

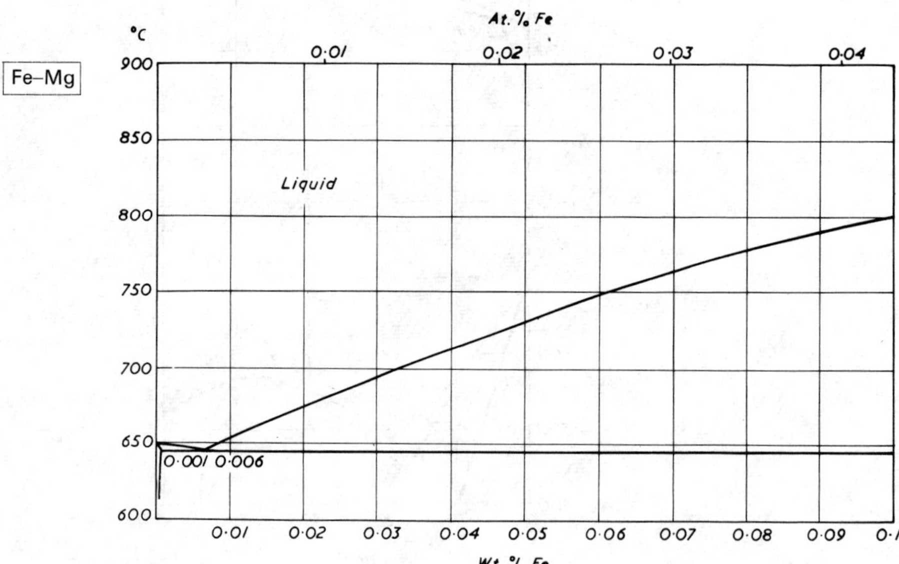
At. % Fe
°C
0·01
0·02
0·03
0·04
900
850
Liquid
800
750
700
650
0·001 0·006
600
0·01
0·02
0·03
0·04
0·05
0·06
0·07
0·08
0·09
0·1
Wt. % Fe

Fe–Mn

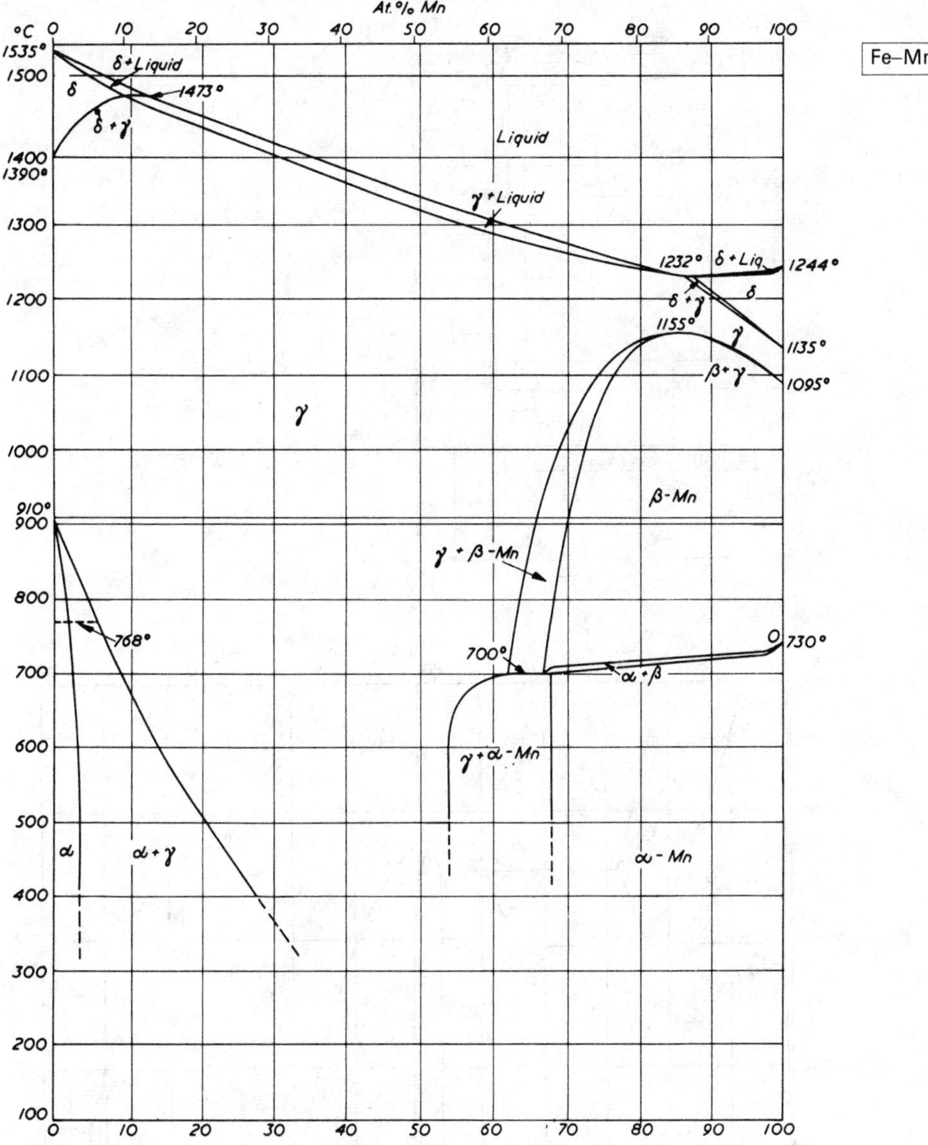
At.% Mn
°C
1535°
1500
1400
1390°
1300
1200
1100
1000
910°
900
800
700
600
500
400
300
200
100
0
10
20
30
40
50
60
70
80
90
100
δ + Liquid
δ
1473°
δ + γ
Liquid
γ + Liquid
1232°
δ + Liq.
1244°
δ + γ
δ
1155°
γ
1135°
β + γ
1095°
γ
β-Mn
γ + β-Mn
768°
700°
730°
α + β
γ + α-Mn
α
α + γ
α-Mn
Wt.% Mn

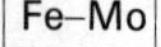
Fe–Mo

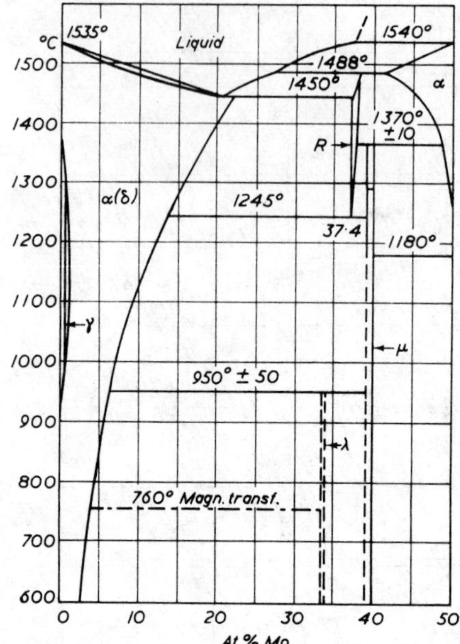
°C
1535°
Liquid
1540°
1488°
1450°
α
1370°
±10
R
α(δ)
1245°
37·4
1180°
γ
μ
950° ± 50
λ
760° Magn. transf.
At.% Mo

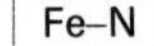
Fe–N

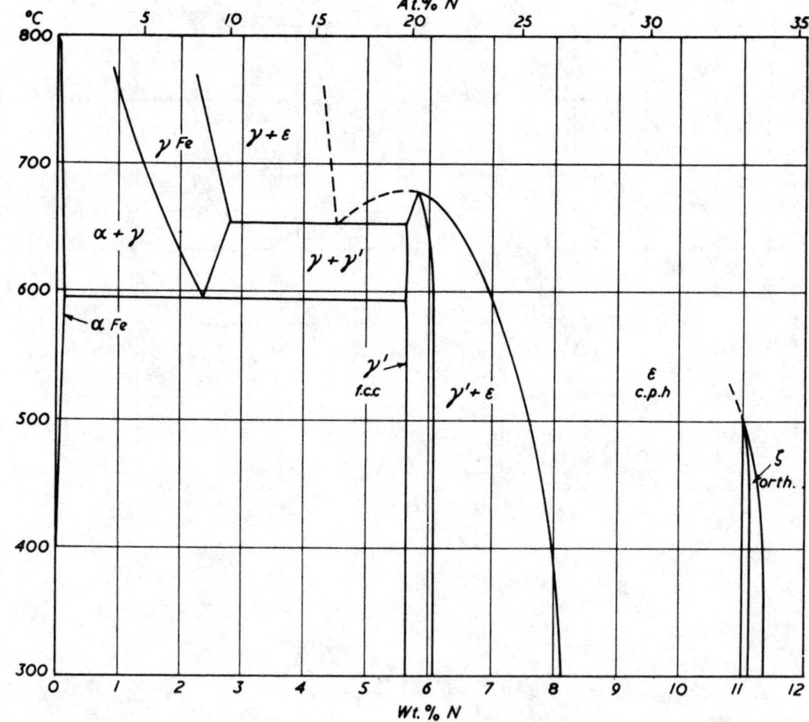
At.% N
°C
γ Fe
γ + ε
α + γ
γ + γ'
α Fe
γ'
f.c.c
γ' + ε
ε
c.p.h
ζ
orth.
Wt.% N

Fe–Nb

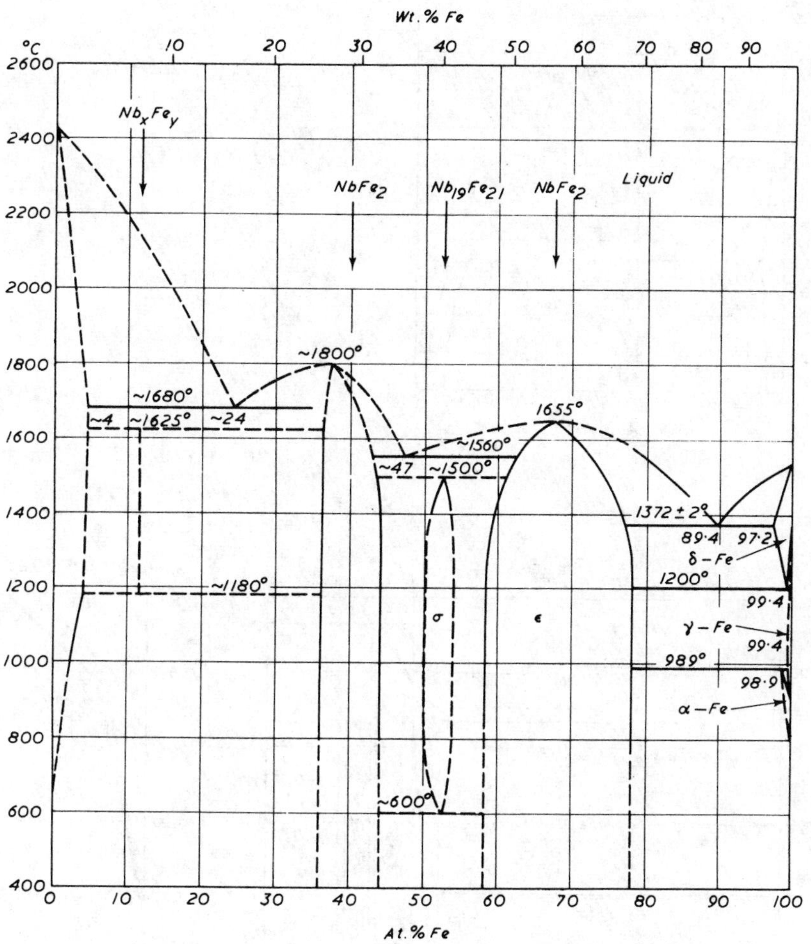
Wt. % Fe
At. % Fe
°C
Nb_xFe_y
$NbFe_2$
$Nb_{19}Fe_{21}$
$NbFe_2$
Liquid
~1800°
~1680°
~4
~1625°
~24
1655°
~1560°
~47
~1500°
~1180°
1372 ± 2°
89·4
97·2
δ – Fe
1200°
99·4
γ – Fe
99·4
989°
98·9
α – Fe
σ
ε
~600°

Fe–Ni

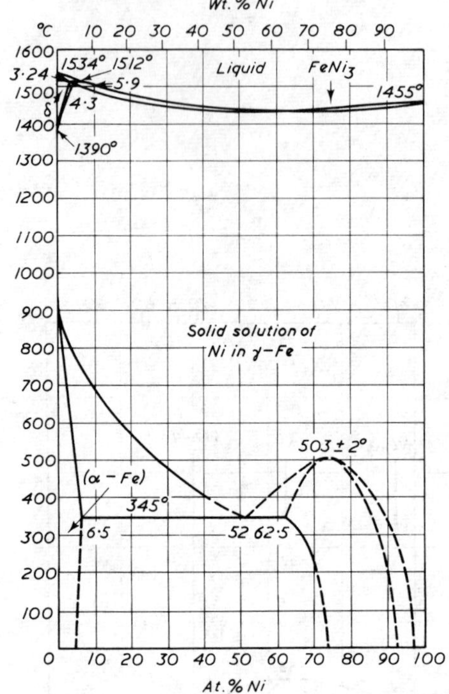

Fe–O

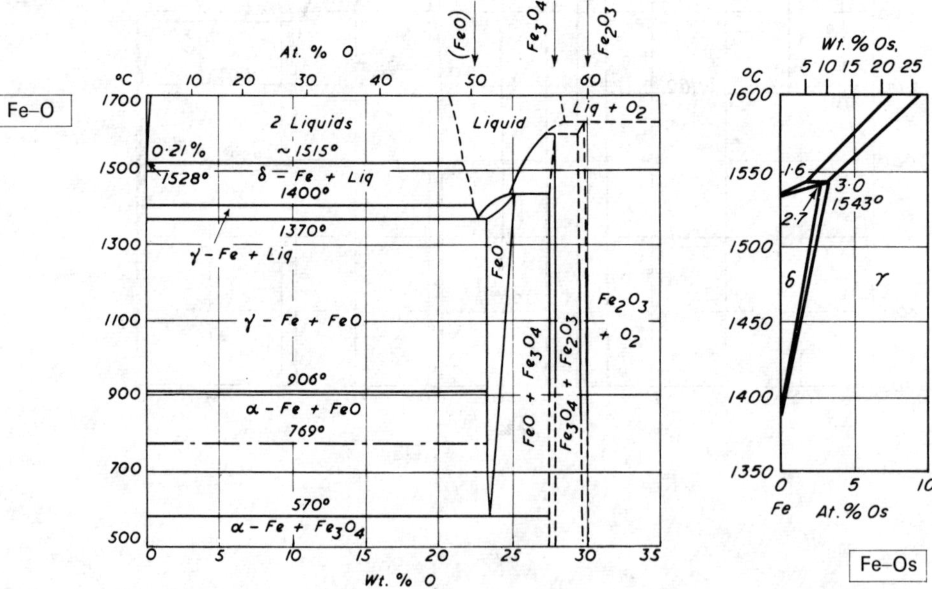

Fe–Os

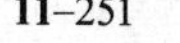
Fe–P

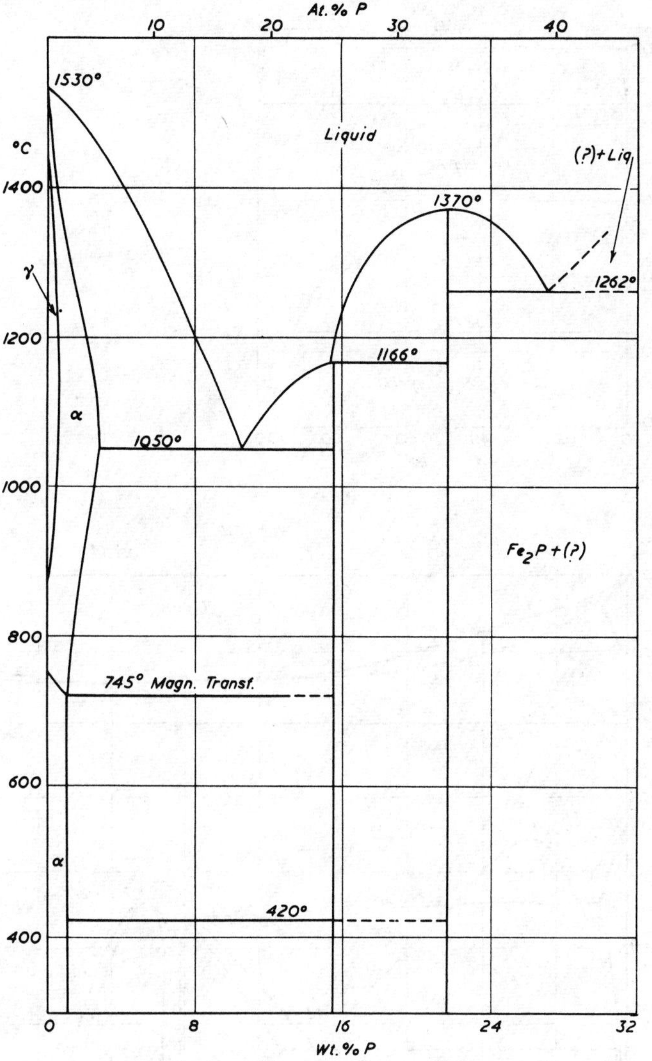
At.% P
10
20
30
40
1530°
Liquid
°C
(?)+Liq
1400
1370°
γ
1262°
1200
1166°
α
1050°
1000
Fe_2P+(?)
800
745° Magn. Transf.
600
α
420°
400
0
8
16
24
32
Wt.% P

Fe–Pb

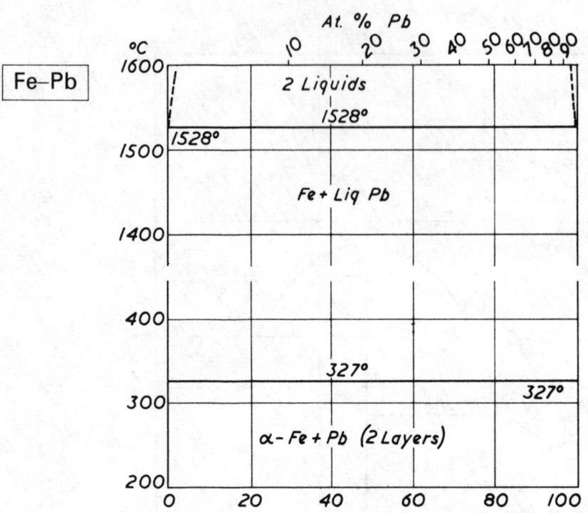

At. % Pb
°C
1600
1500
1400
400
300
200
2 Liquids
1528°
1528°
Fe + Liq Pb
327°
327°
α - Fe + Pb (2 Layers)
0
20
40
60
80
100
Wt. % Pb

Fe–Pd

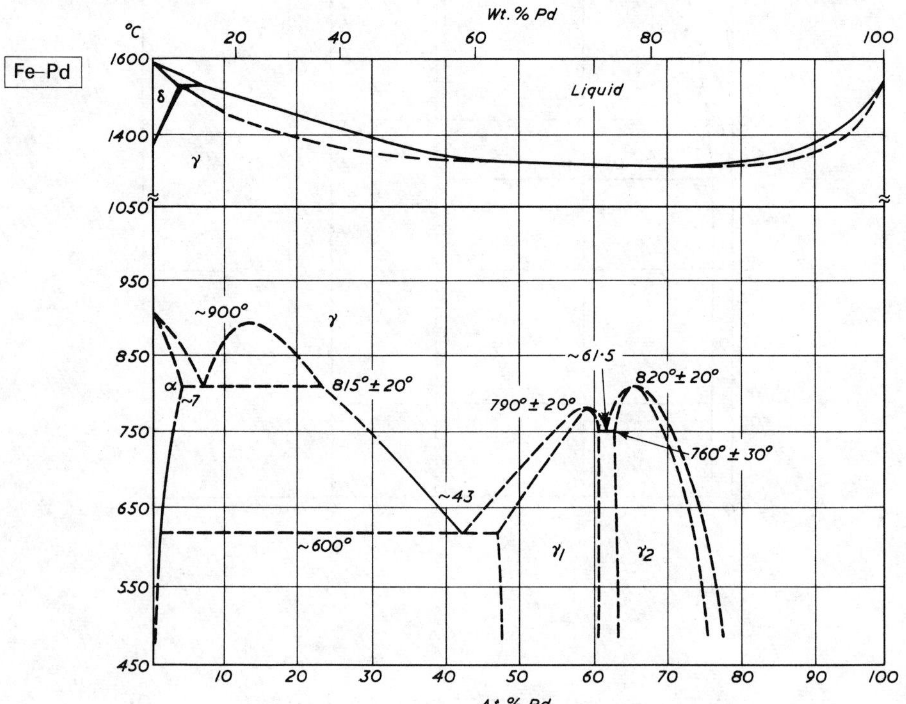

Wt. % Pd
20
40
60
80
100
°C
1600
1400
1050
950
850
750
650
550
450
δ
γ
Liquid
~900°
γ
α
~7
815° ± 20°
~61·5
820° ± 20°
790° ± 20°
760° ± 30°
~43
~600°
γ_1
γ_2
10
20
30
40
50
60
70
80
90
100
At. % Pd

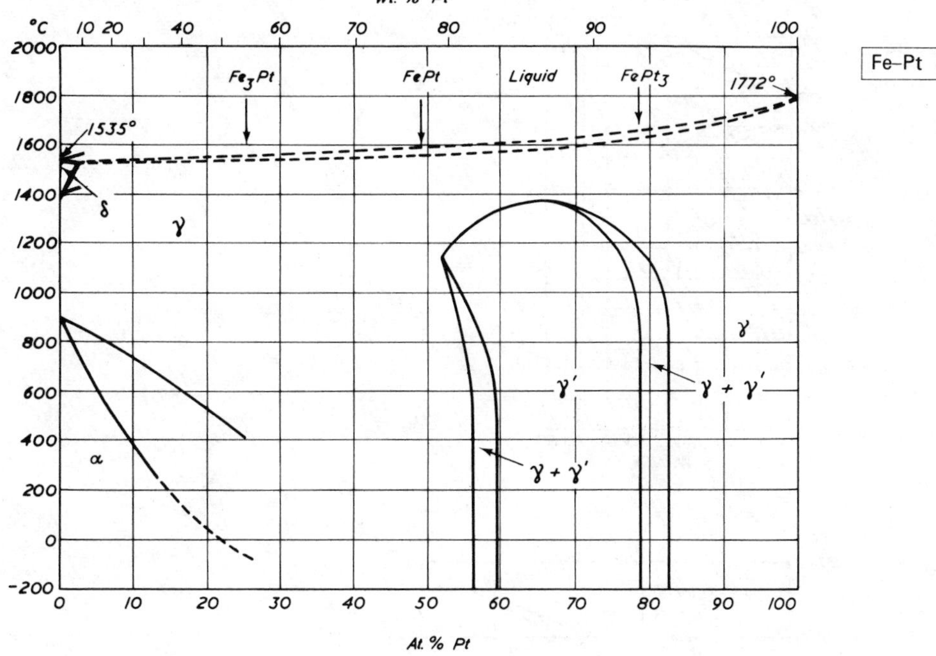
Fe–Pt
Wt. % Pt
°C
10 20 40 60 70 80 90 100
2000
1800
1600
1400
1200
1000
800
600
400
200
0
-200
Fe3Pt
FePt
Liquid
FePt3
1772°
1535°
δ
γ
γ'
γ + γ'
γ + γ'
γ
α
0 10 20 30 40 50 60 70 80 90 100
At. % Pt

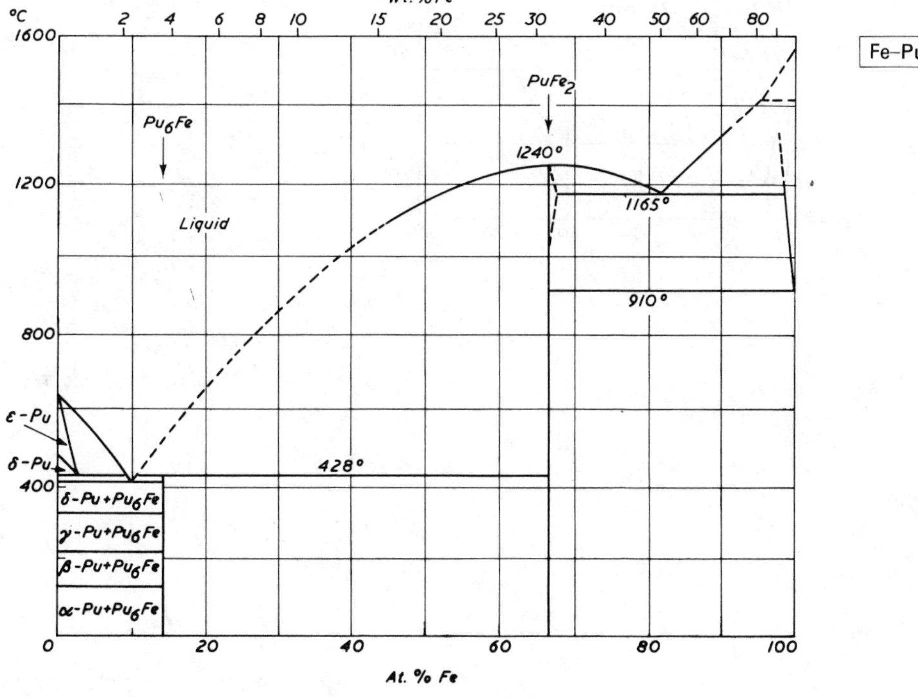
Fe–Pu
Wt. % Fe
°C
2 4 6 8 10 15 20 25 30 40 50 60 80
1600
1200
800
400
PuFe2
Pu6Fe
1240°
Liquid
1165°
910°
ε-Pu
δ-Pu
428°
δ-Pu+Pu6Fe
γ-Pu+Pu6Fe
β-Pu+Pu6Fe
α-Pu+Pu6Fe
0 20 40 60 80 100
At. % Fe

Fe–Re

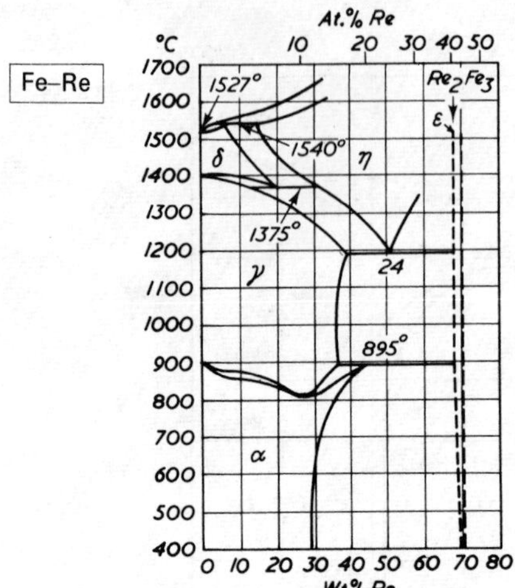
At.% Re
°C
10
20
30
40
50
1527°
Re2Fe3
1540°
δ
η
ε
1375°
24
γ
895°
α
Wt.% Re

Fe–Ru

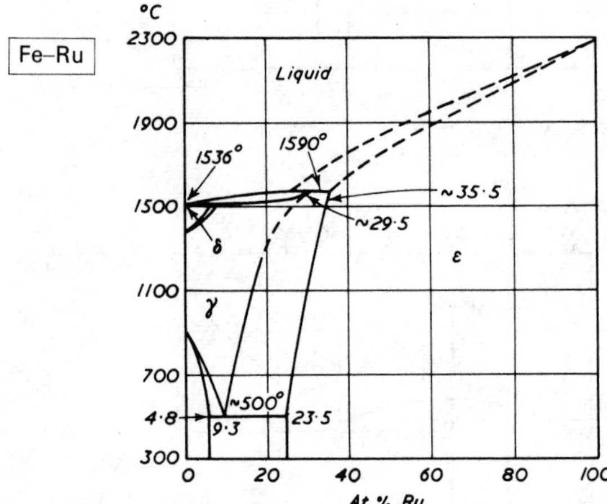
°C
Liquid
1536°
1590°
~35·5
~29·5
δ
ε
γ
~500°
4·8
9·3
23·5
At.% Ru

Fe–S

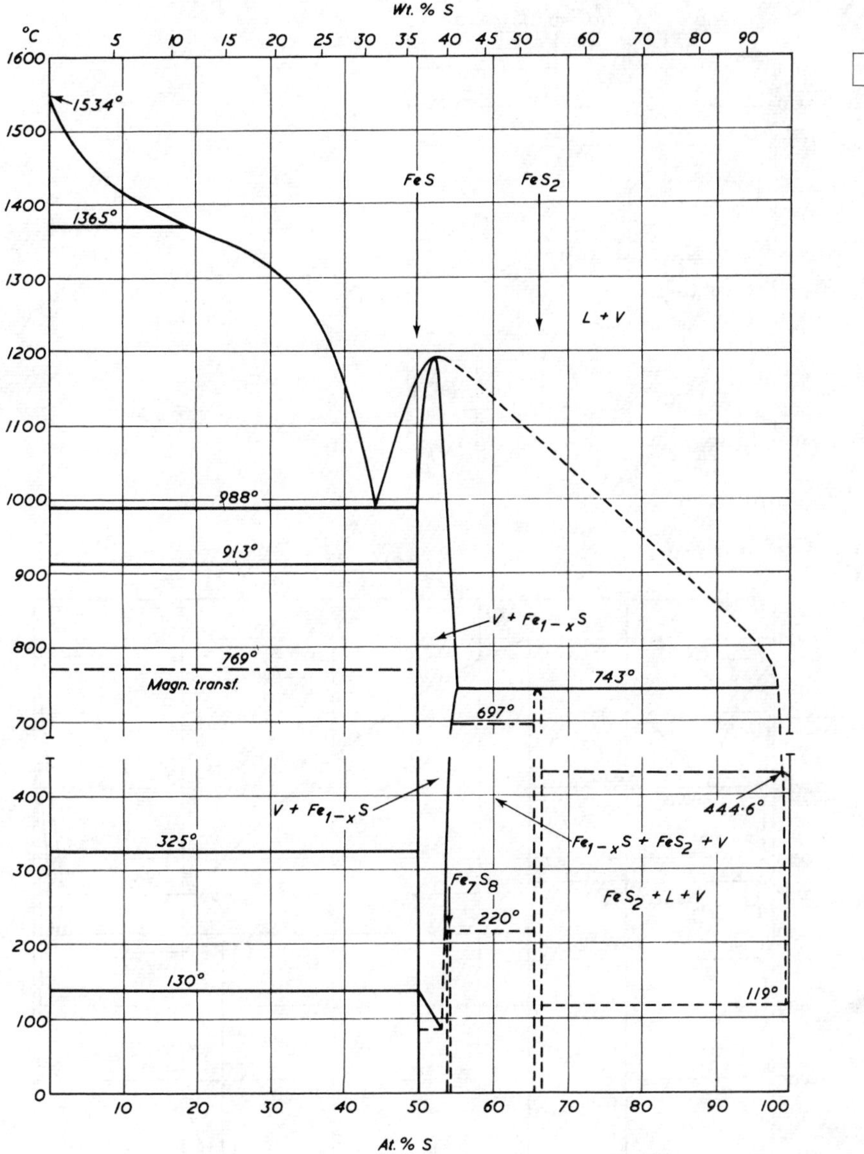
Wt. % S
At. % S
°C
1534°
1365°
988°
913°
769°
Magn. transf.
FeS
FeS2
L + V
V + Fe1–xS
743°
697°
444·6°
Fe1–xS + FeS2 + V
325°
Fe7S8
FeS2 + L + V
220°
130°
119°

Fe–Sb

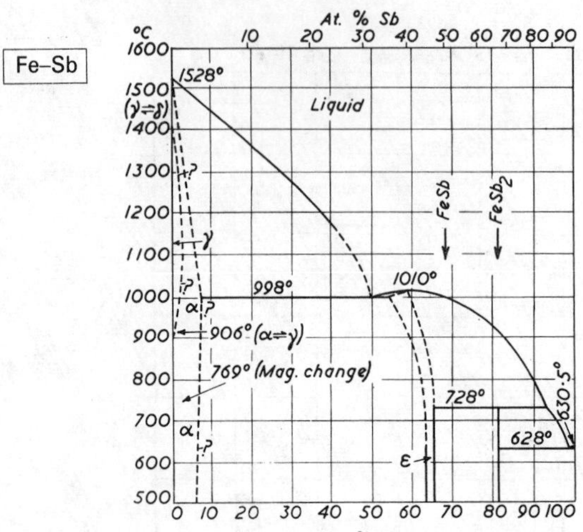
At. % Sb
10 20 30 40 50 60 70 80 90
°C
1600
1500
1400
1300
1200
1100
1000
900
800
700
600
500
1528°
(γ⇌δ)
Liquid
FeSb
FeSb2
γ
998°
1010°
α
906° (α⇌γ)
769° (Mag. change)
728°
630·5°
628°
α
ε
0 10 20 30 40 50 60 70 80 90 100
Wt. % Sb

Fe–Sc

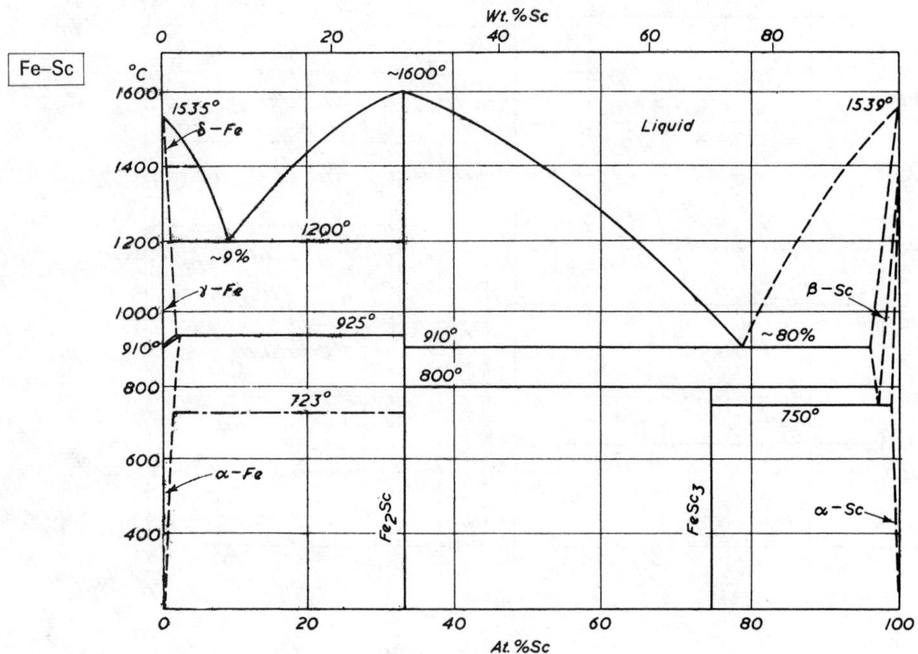
Wt.%Sc
0 20 40 60 80
°C
1600
1400
1200
1000
910°
800
600
400
~1600°
1535°
δ-Fe
1539°
Liquid
1200°
~9%
γ-Fe
β-Sc
925°
910°
~80%
800°
723°
750°
α-Fe
Fe2Sc
FeSc3
α-Sc
0 20 40 60 80 100
At.%Sc

Fe–Si

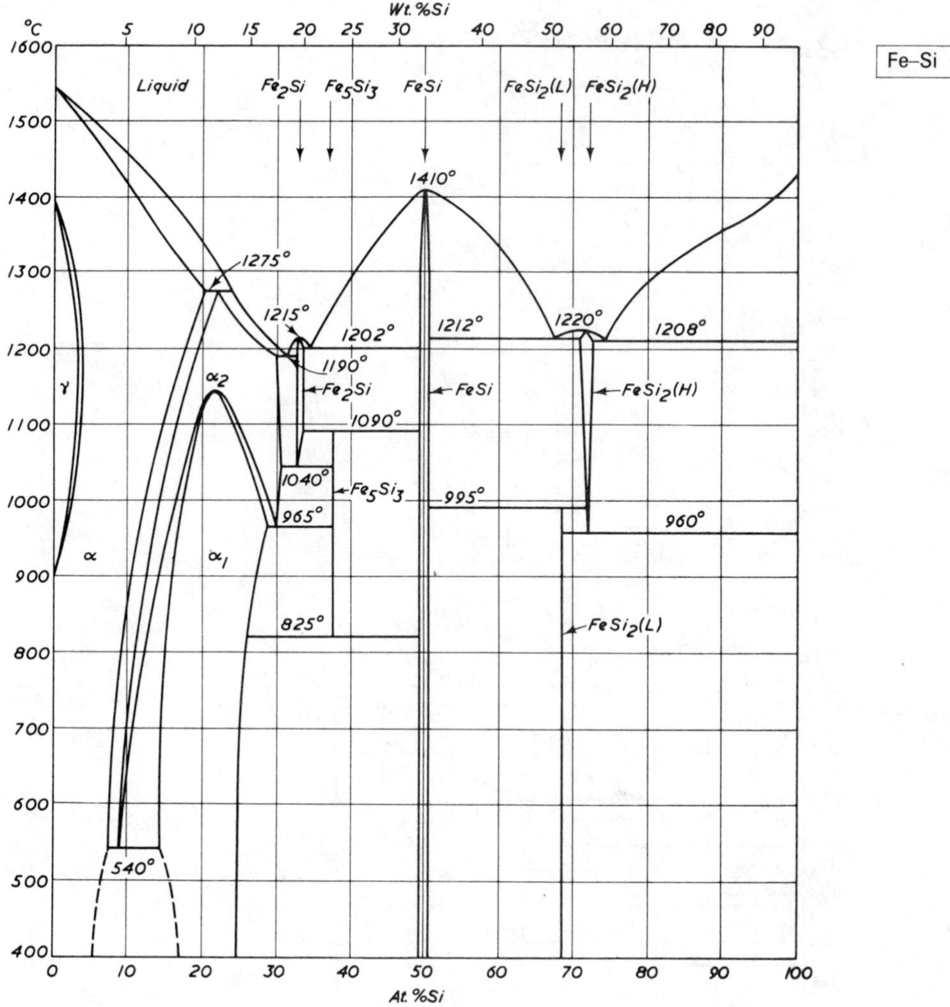

Wt.%Si
°C
5 10 15 20 25 30 40 50 60 70 80 90
1600
1500
1400
1300
1200
1100
1000
900
800
700
600
500
400
Liquid
Fe2Si
Fe5Si3
FeSi
FeSi2(L)
FeSi2(H)
1410°
1275°
1215°
1202°
1212°
1220°
1208°
1190°
α2
γ
Fe2Si
FeSi
FeSi2(H)
1090°
1040°
Fe5Si3
995°
965°
960°
α
α1
825°
FeSi2(L)
540°
0 10 20 30 40 50 60 70 80 90 100
At.%Si

Fe–Sm

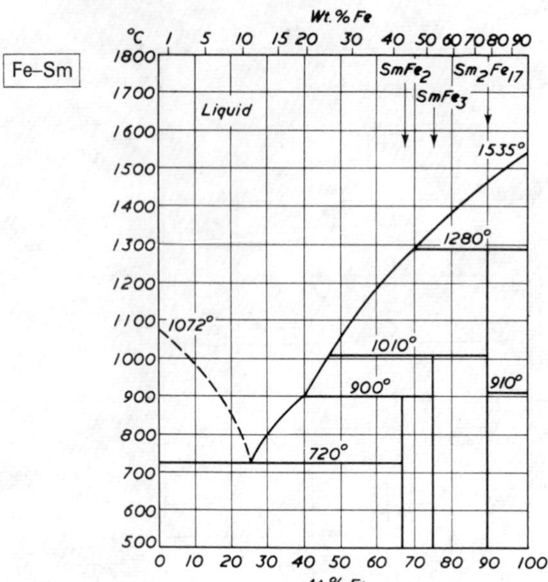
Wt.% Fe
°C
1 5 10 15 20 30 40 50 60 70 80 90
SmFe2
Sm2Fe17
SmFe3
Liquid
1535°
1280°
1072°
1010°
900°
910°
720°
1800
1700
1600
1500
1400
1300
1200
1100
1000
900
800
700
600
500
0 10 20 30 40 50 60 70 80 90 100
At.% Fe

Fe–Sn

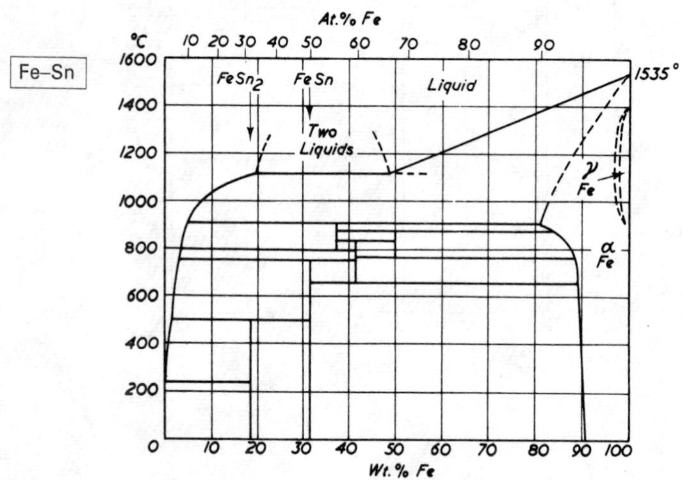
At.% Fe
°C
10 20 30 40 50 60 70 80 90
FeSn2
FeSn
Liquid
1535°
Two
Liquids
γ
Fe
α
Fe
1600
1400
1200
1000
800
600
400
200
0
10 20 30 40 50 60 70 80 90 100
Wt.% Fe

Fe–Ta

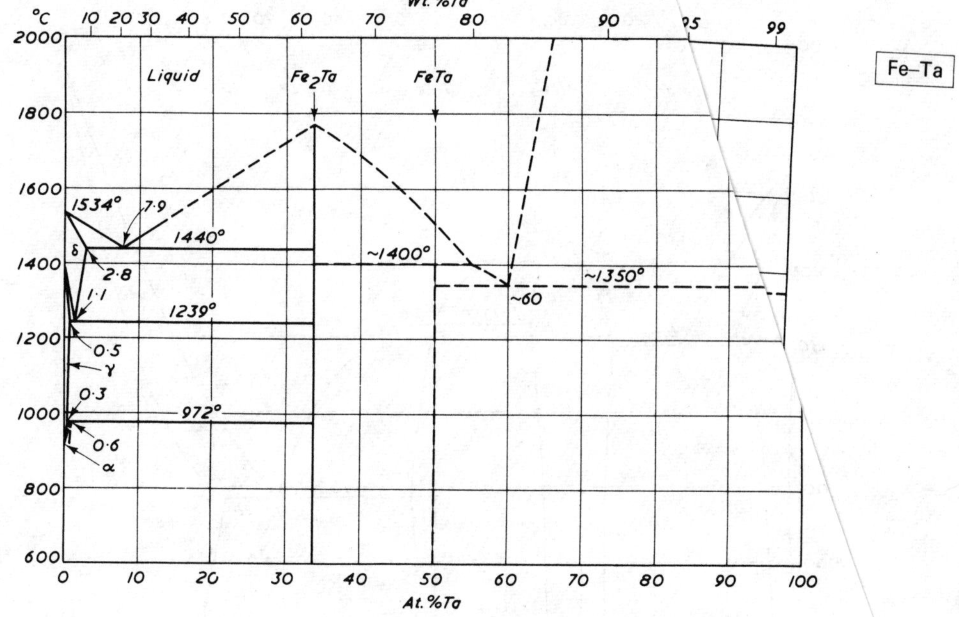
Wt. %Ta
°C
2000
1800
1600
1400
1200
1000
800
600
Liquid
Fe_2Ta
FeTa
1534°
7·9
1440°
δ
2·8
1·1
1239°
0·5
γ
0·3
972°
0·6
α
~1400°
~1350°
~60
At. %Ta

Fe–Ti

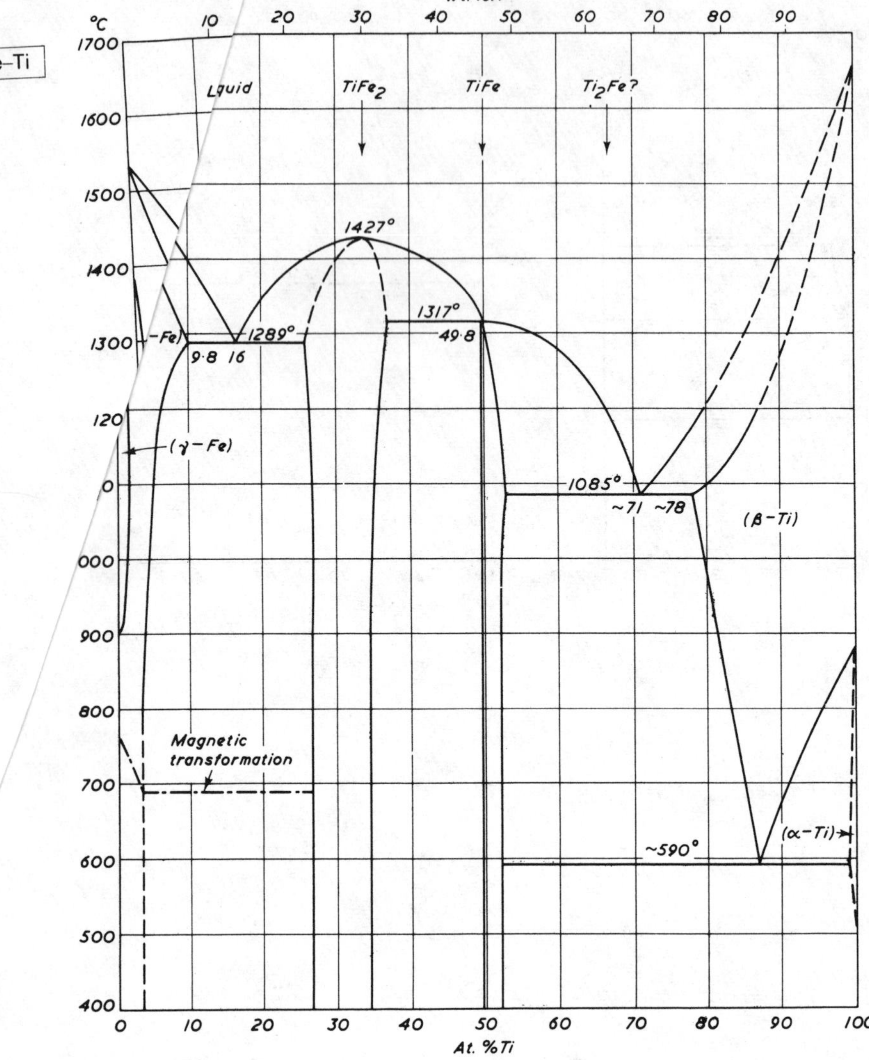

Wt. %Ti
10
20
30
40
50
60
70
80
90
°C
1700
1600
1500
1400
1300
120
0
000
900
800
700
600
500
400
Liquid
$TiFe_2$
TiFe
Ti_2Fe?
1427°
1317°
1289°
9·8
16
49·8
-Fe)
(γ-Fe)
1085°
~71
~78
(β-Ti)
Magnetic
transformation
~590°
(α-Ti)
0
10
20
30
40
50
60
70
80
90
100
At. %Ti

Fe–U

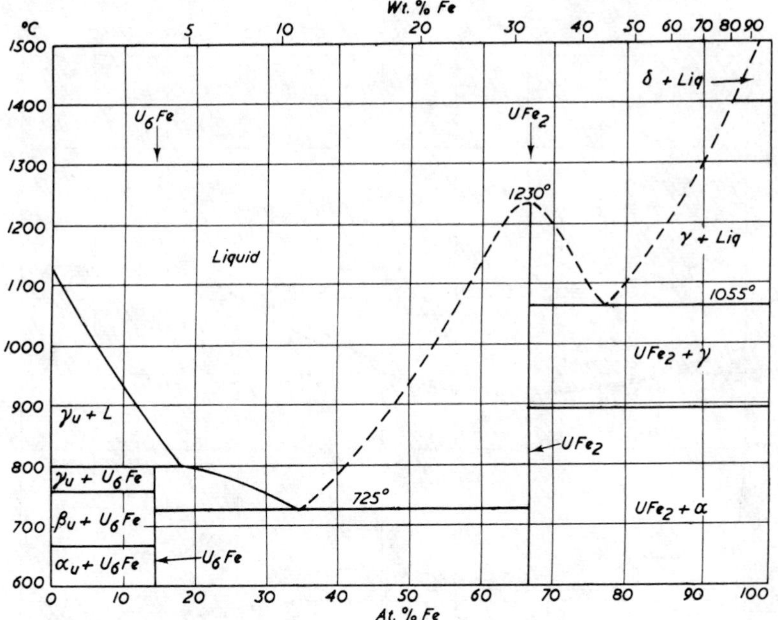
Wt. % Fe
5
10
20
30
40
50
60
70
80
90
°C
1500
1400
1300
1200
1100
1000
900
800
700
600
δ + Liq
U_6Fe
UFe_2
1230°
Liquid
γ + Liq
1055°
UFe_2 + γ
$γ_u$ + L
UFe_2
$γ_u$ + U_6Fe
725°
$β_u$ + U_6Fe
UFe_2 + α
$α_u$ + U_6Fe
U_6Fe
0
10
20
30
40
50
60
70
80
90
100
At. % Fe

Fe–V

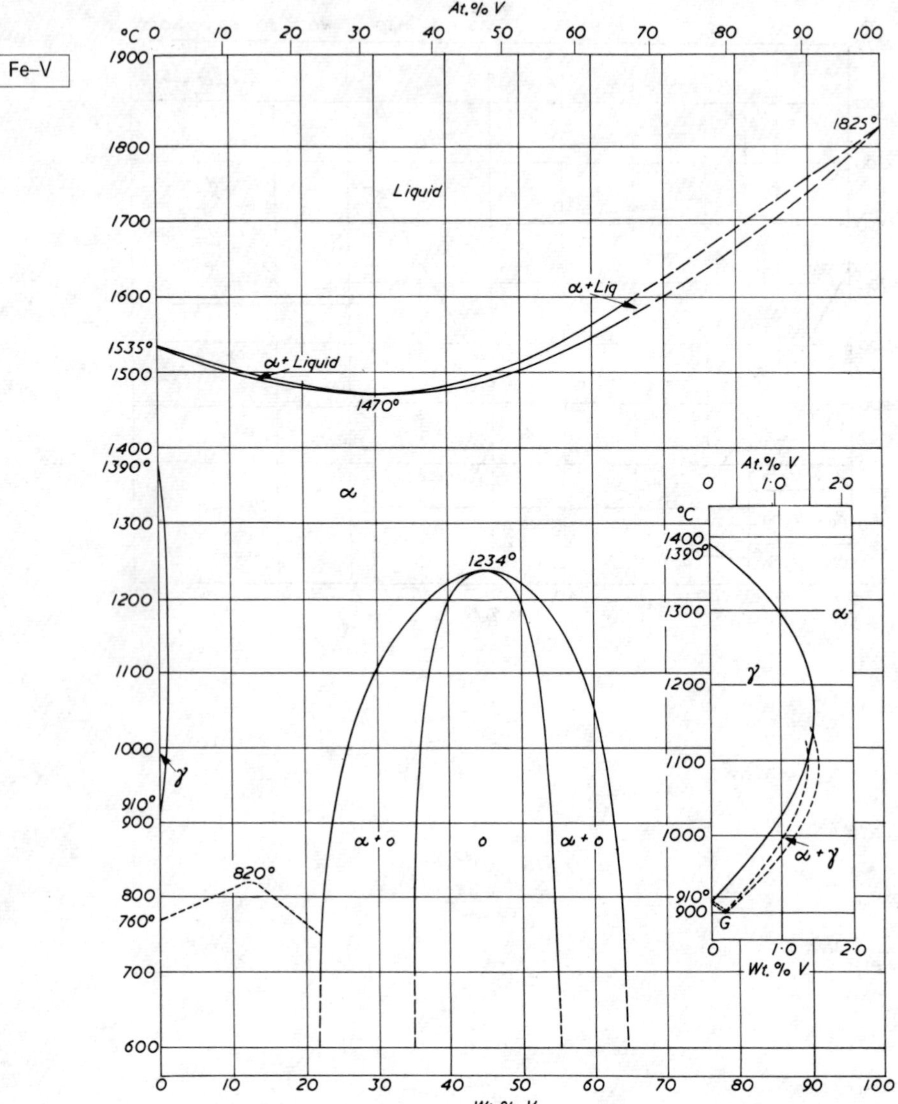
At.% V
°C
Liquid
1825°
α+Liq
1535°
α+ Liquid
1470°
1390°
α
1234°
910°
γ
820°
760°
α + σ
σ
α + σ
Wt. % V
At.% V
1390°
α
γ
α + γ
910°
G
Wt. % V

Fe–W

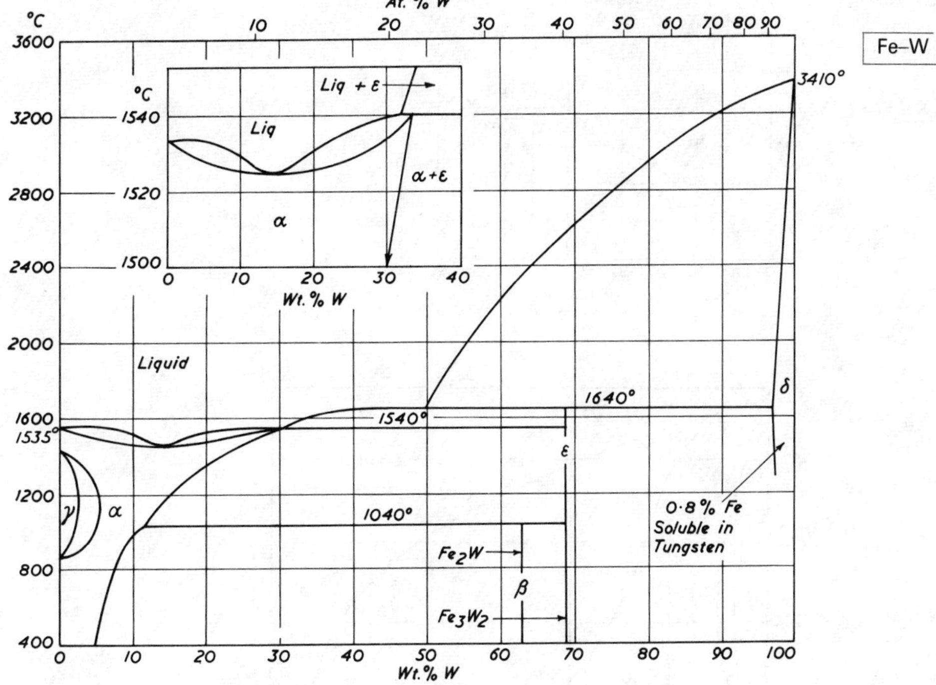
At. % W
°C
3600
3200
2800
2400
2000
1600
1535°
1200
800
400
10
20
30
40
50
60
70
80
90
°C
1540
1520
1500
Liq + ε
Liq
α+ε
α
0
10
20
30
40
Wt. % W
3410°
Liquid
δ
1640°
1540°
ε
γ
α
1040°
0·8 % Fe
Soluble in
Tungsten
Fe2W
β
Fe3W2
0
10
20
30
40
50
60
70
80
90
100
Wt. % W

Fe–Y

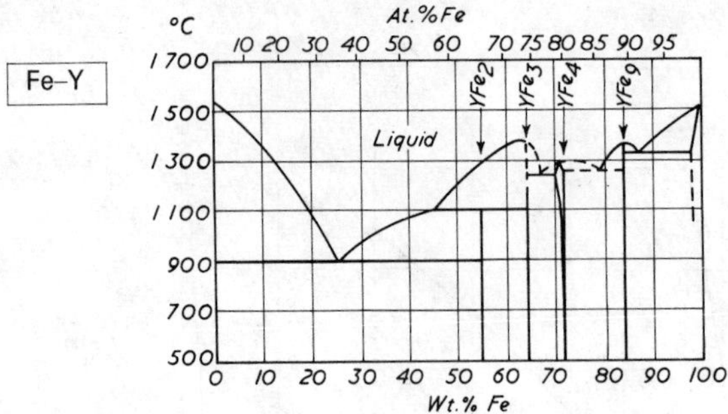

°C
At.%Fe
10 20 30 40 50 60 70 75 80 85 90 95
1700
1500
1300
1100
900
700
500
Liquid
YFe2
YFe3
YFe4
YFe9
0 10 20 30 40 50 60 70 80 90 100
Wt.% Fe

Fe–Zn

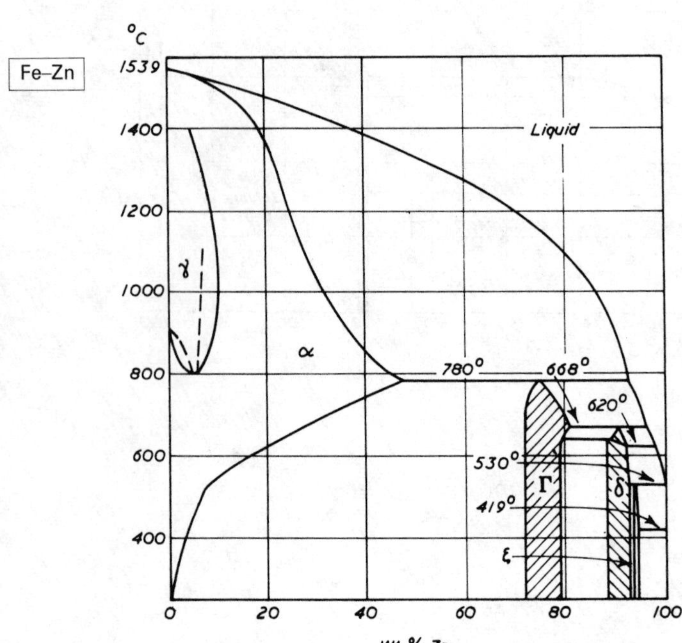

°C
1539
1400
1200
1000
800
600
400
Liquid
γ
α
780°
668°
620°
530°
Γ
δ
419°
ξ
0 20 40 60 80 100
Wt.% Zn

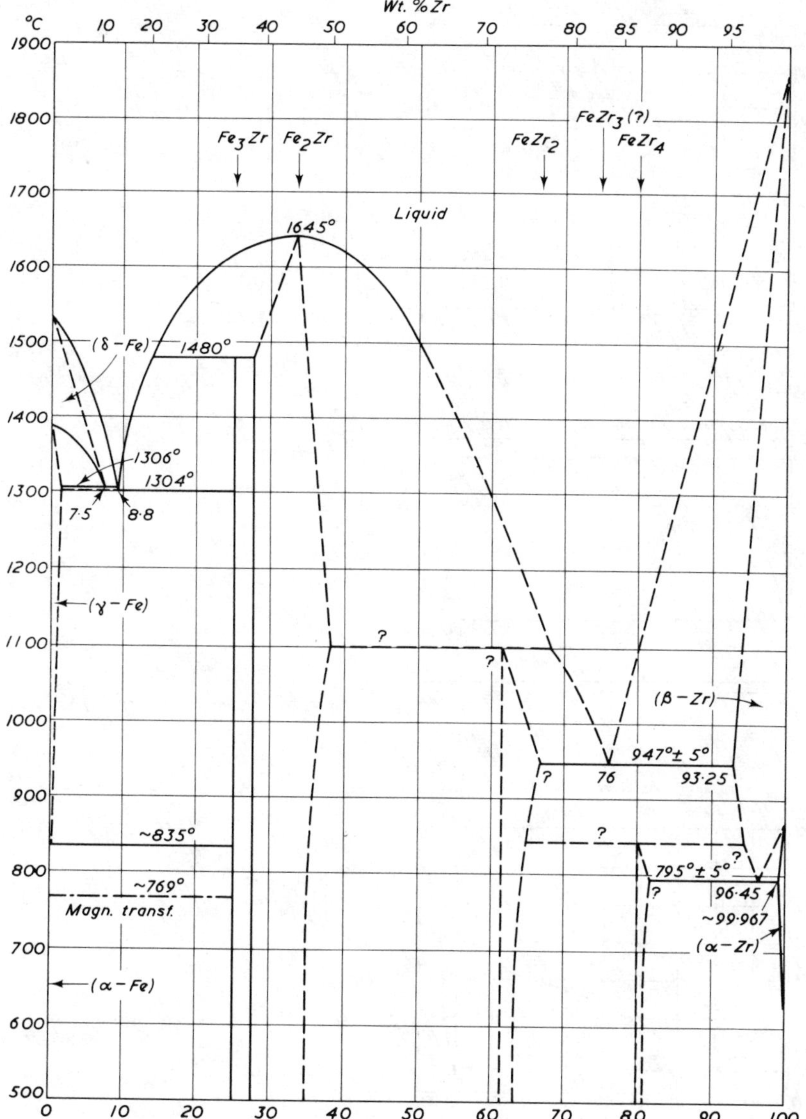

Wt. %Zr
°C
Fe3Zr
Fe2Zr
FeZr2
FeZr3(?)
FeZr4
Liquid
1645°
(δ-Fe)
1480°
1306°
1304°
7·5
8·8
(γ-Fe)
?
?
(β-Zr)
947°± 5°
?
76
93·25
~835°
?
795°± 5°
?
~769°
?
96·45
Magn. transf.
~99·967
(α-Zr)
(α-Fe)
At. %Zr

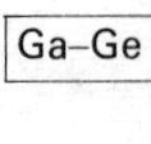

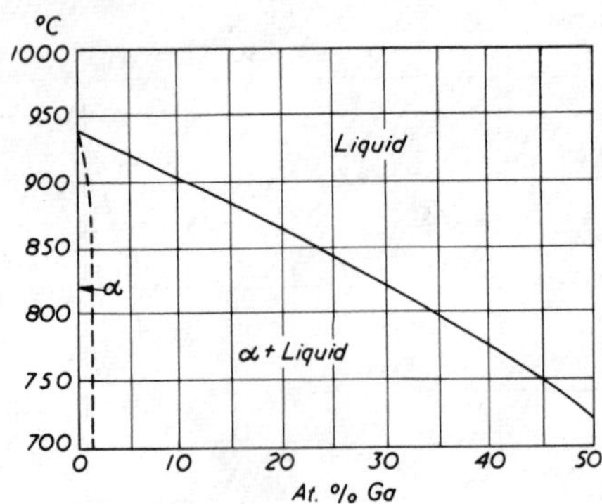

Ga–Hg

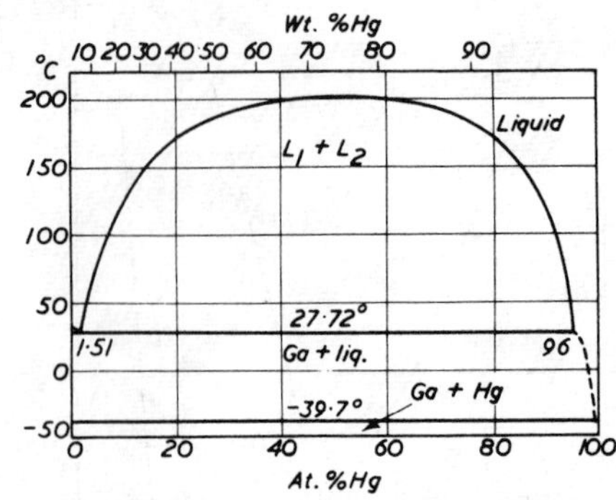

Ga–In

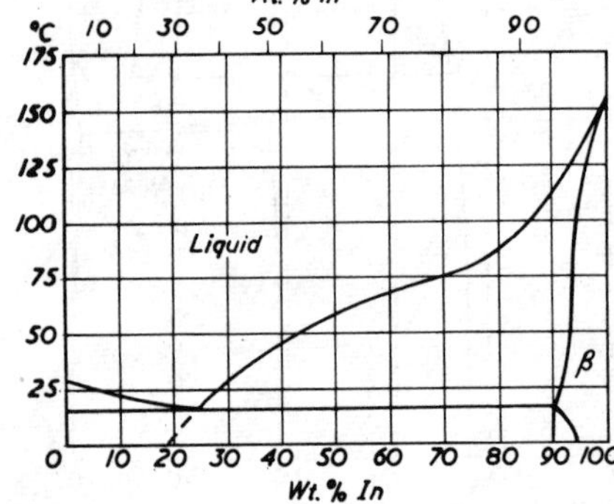

Ga–K

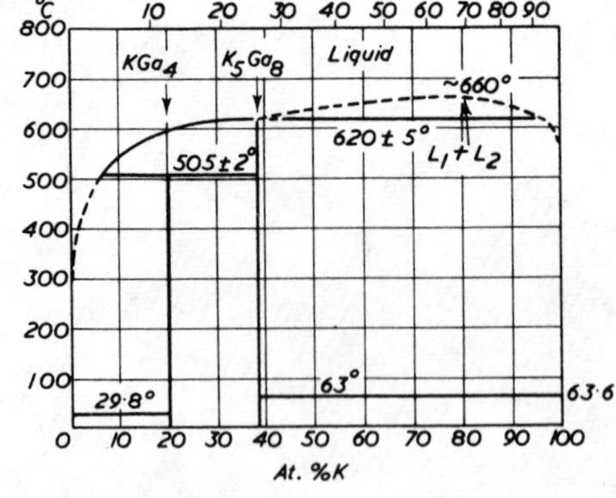

Ga–Mg

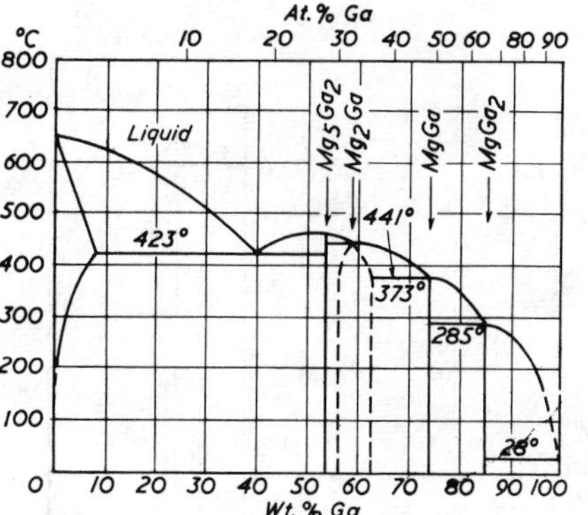
At.% Ga
°C
Liquid
Mg5Ga2
Mg2Ga
MgGa
MgGa2
423°
441°
373°
285°
28°
Wt.% Ga

Ga–Mn

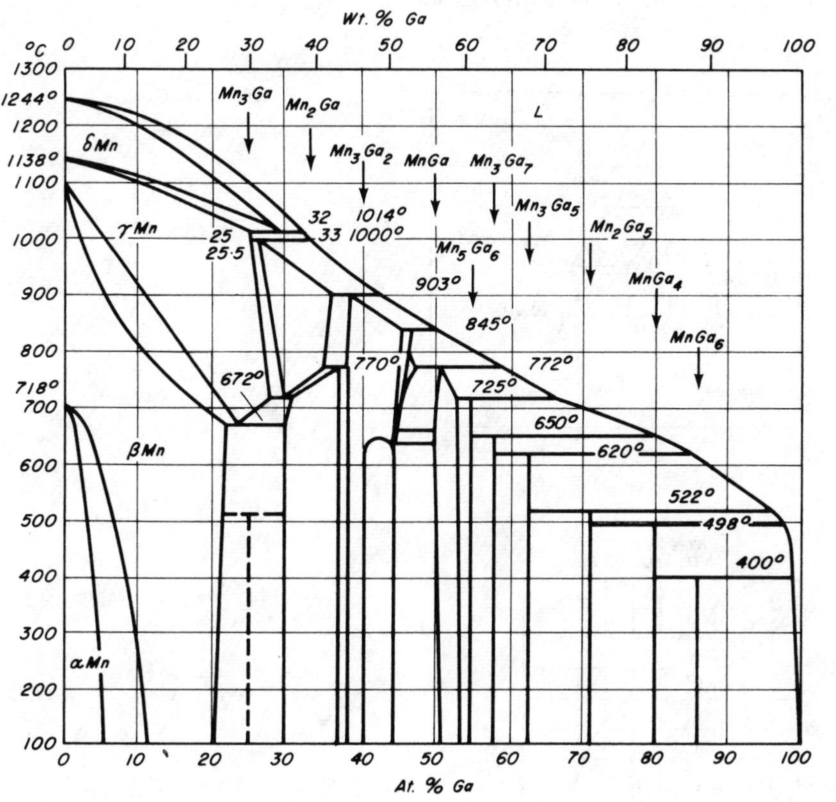
Wt. % Ga
°C
1244°
1138°
718°
δMn
γMn
βMn
αMn
L
Mn3Ga
Mn2Ga
Mn3Ga2
MnGa
Mn3Ga7
Mn3Ga5
Mn5Ga6
Mn2Ga5
MnGa4
MnGa6
32
33
25
25·5
1014°
1000°
903°
845°
770°
772°
725°
672°
650°
620°
522°
498°
400°
At. % Ga

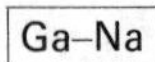
Ga–Na

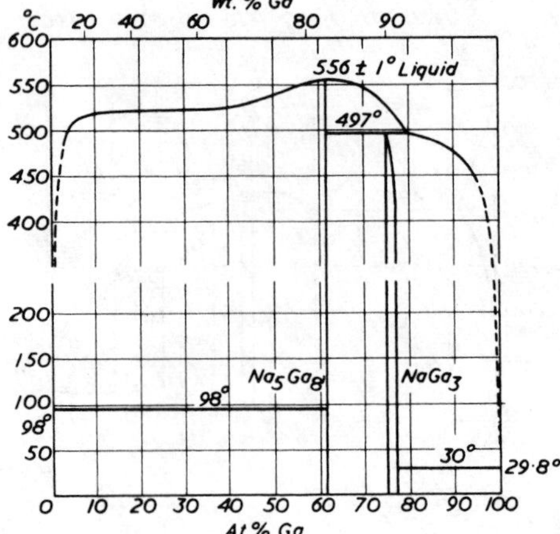
Wt. % Ga
°C
556 ± 1° Liquid
497°
Na5Ga8
NaGa3
98°
30°
29·8°
At.% Ga

Ga–Nb

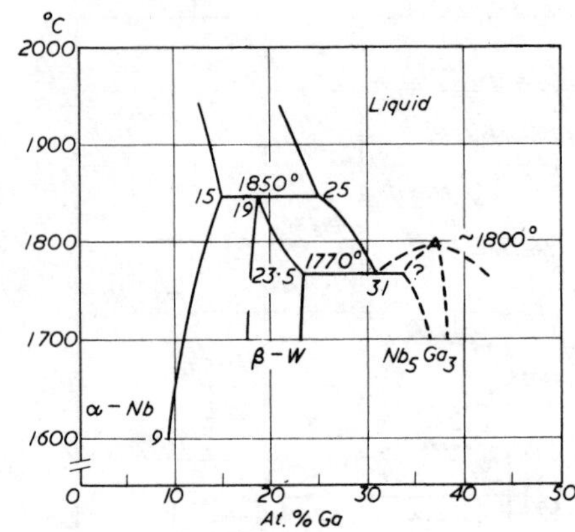
°C
Liquid
1850°
1770°
~1800°
β – W
Nb5Ga3
α – Nb
At. % Ga

Ga–Ni

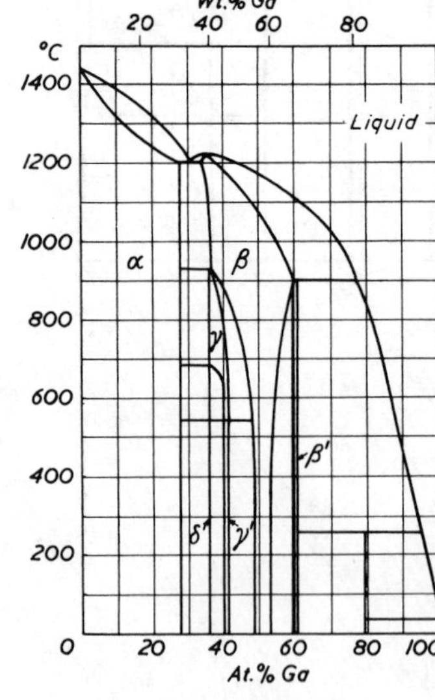
Wt.% Ga
°C
Liquid
α
β
γ
β'
δ
γ'
At.% Ga

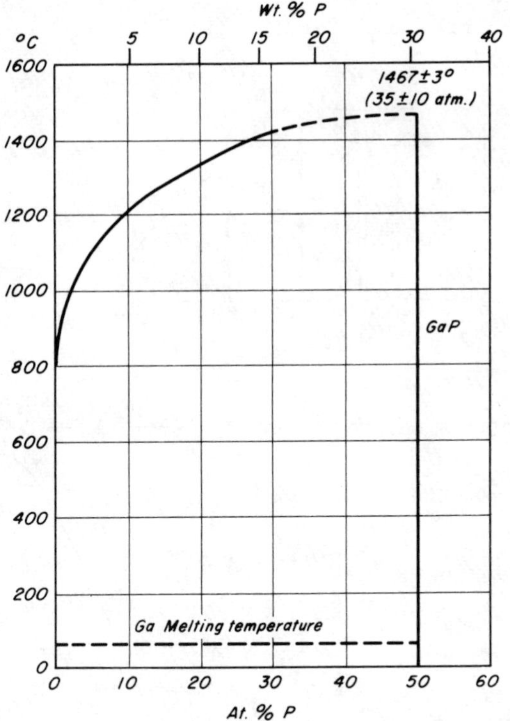
Wt. % P
°C
1467±3°
(35±10 atm.)
GaP
Ga Melting temperature
At. % P

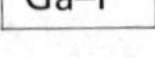
Ga–P

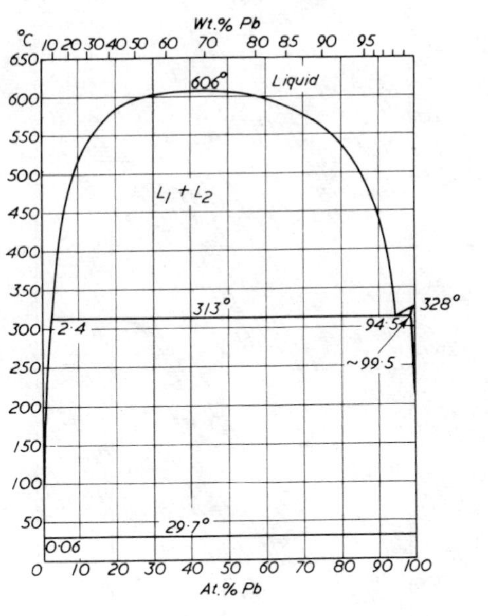
Wt.% Pb
°C
606°
Liquid
$L_1 + L_2$
313°
2·4
94·5
328°
~99·5
29·7°
0·06
At.% Pb

Ga–Pb

Ga–Pd

°C
1400
1200
1000
800
600
400
200
Liquid
PdGa
Pd5
Ga3
Pd2Ga
Pd3Ga
Pd3Ga7
PdGa5
0 10 20 30 40 50 60 70 80 90 100
At. % Ga

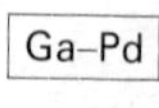

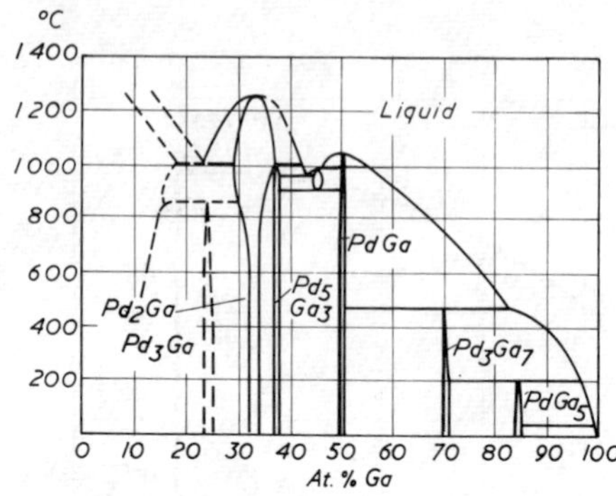

Ga–Pr

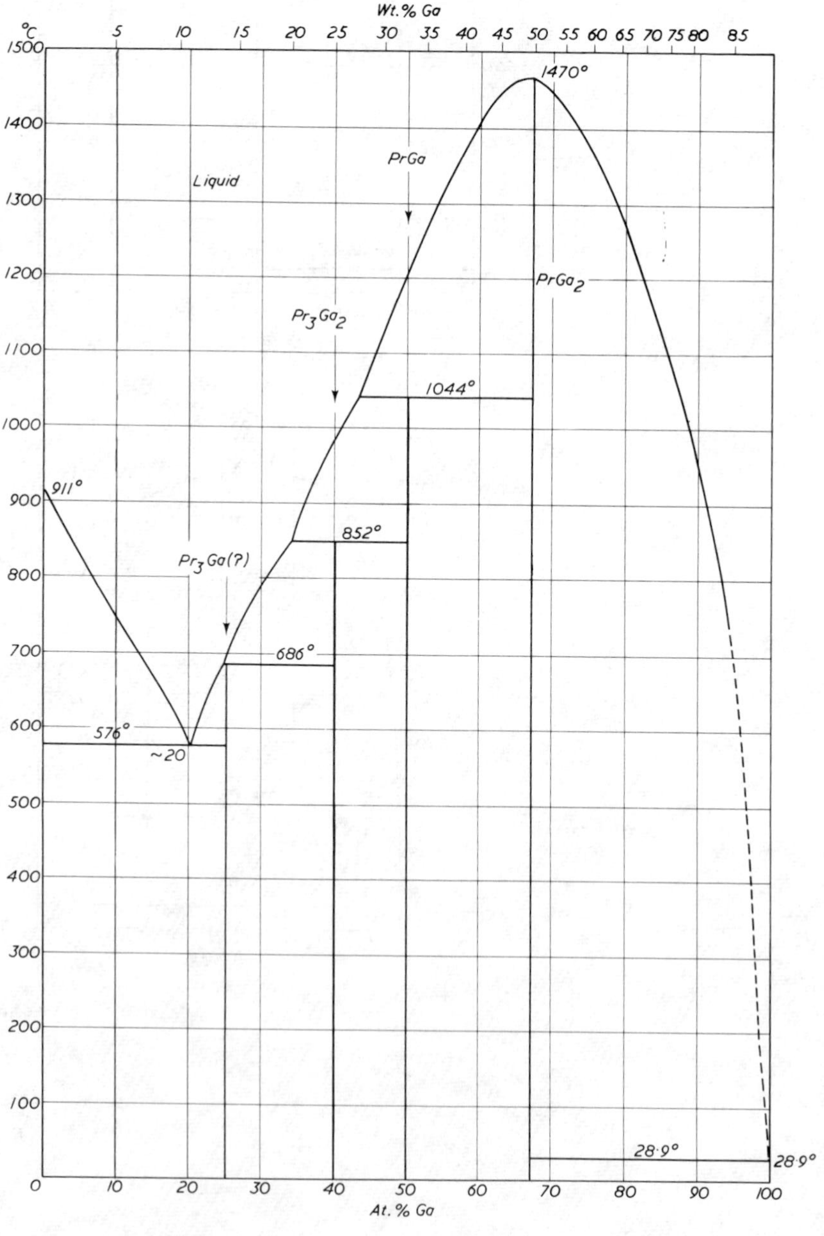

Wt. % Ga
5 10 15 20 25 30 35 40 45 50 55 60 65 70 75 80 85
°C
1500
1400
1300
1200
1100
1000
900
800
700
600
500
400
300
200
100
0
1470°
PrGa
Liquid
PrGa2
Pr3Ga2
1044°
911°
852°
Pr3Ga(?)
686°
576°
~20
28·9°
28·9°
0 10 20 30 40 50 60 70 80 90 100
At. % Ga

Ga–Pu

Ga–Sb

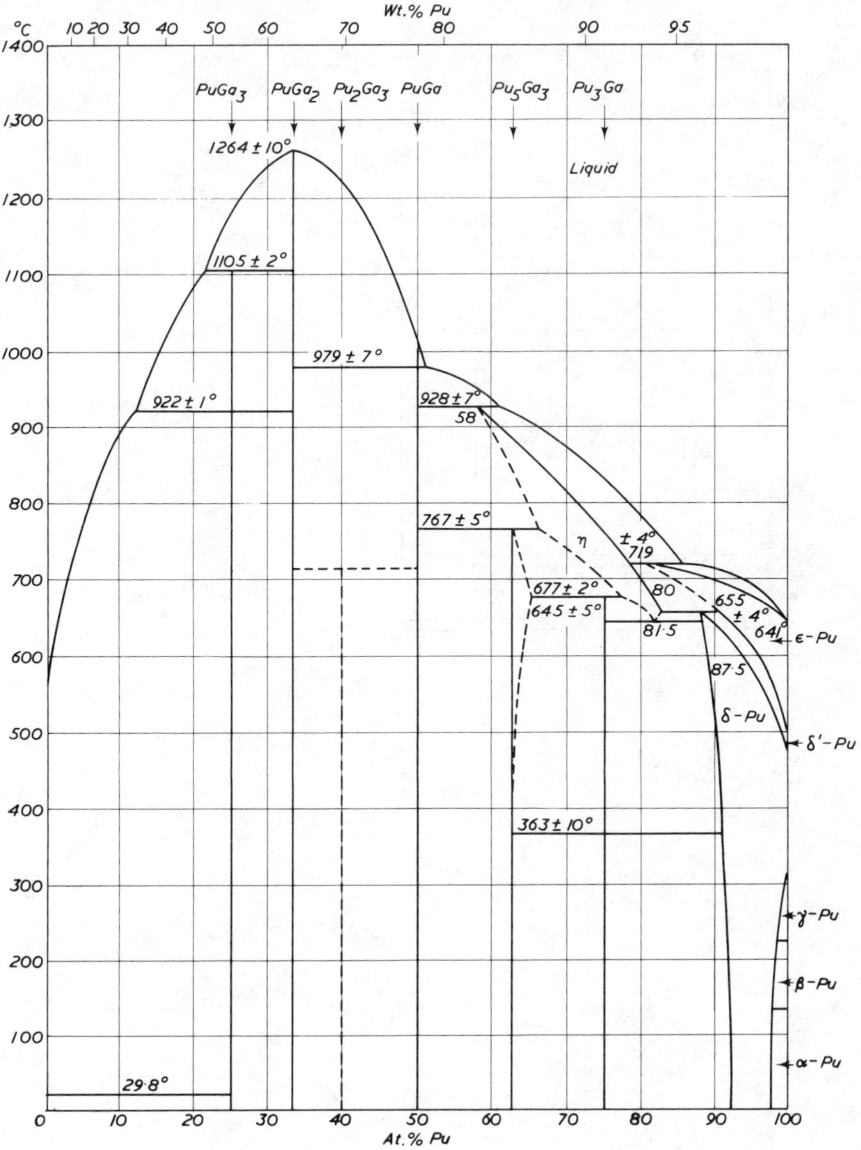

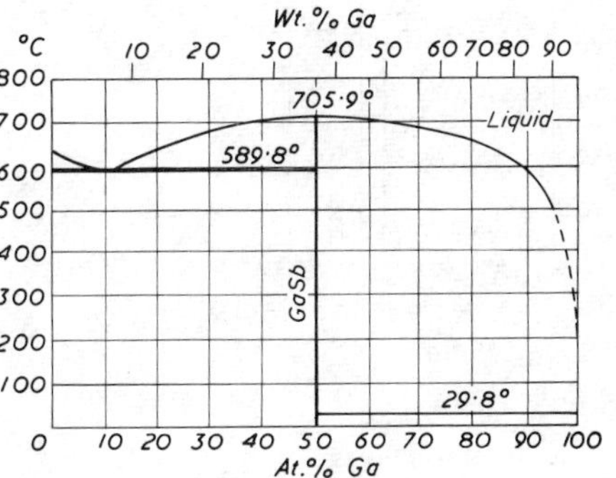
Wt.% Ga
°C
705·9°
Liquid
589·8°
GaSb
29·8°
At.% Ga

Ga–Si

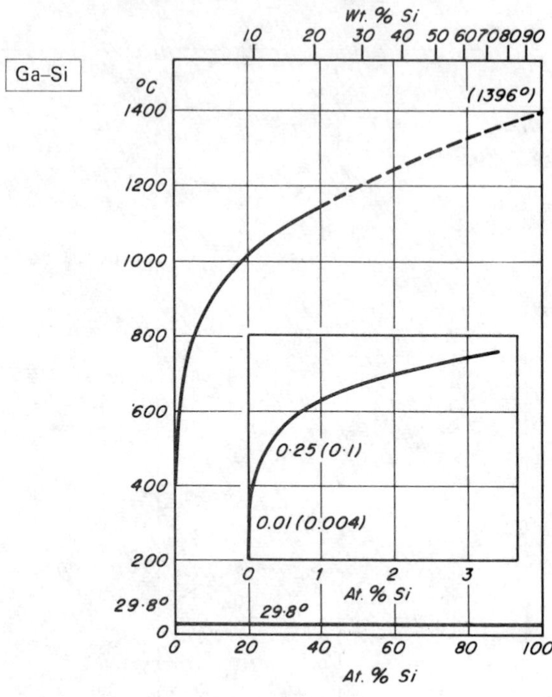

Ga–Sn

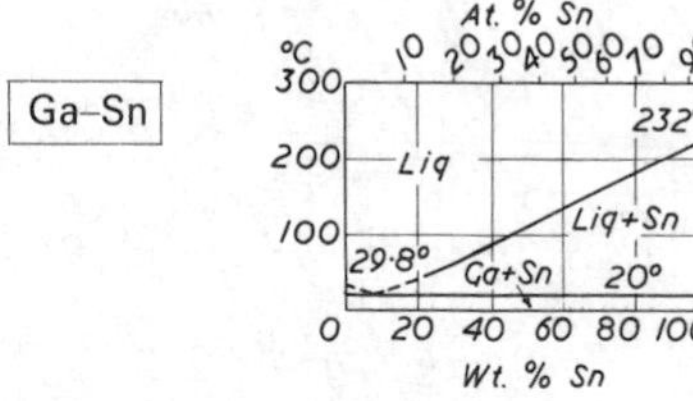

Ga–Te

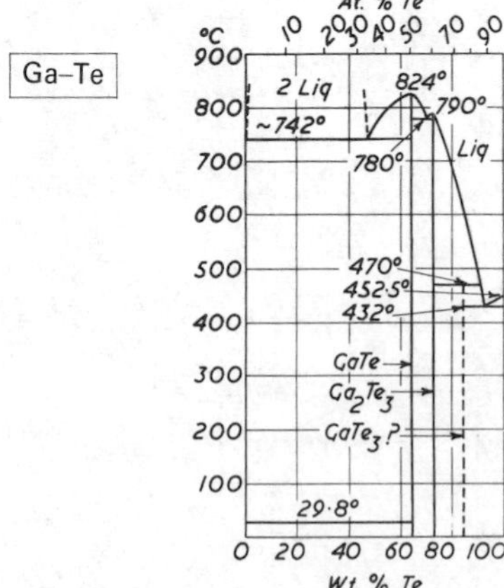

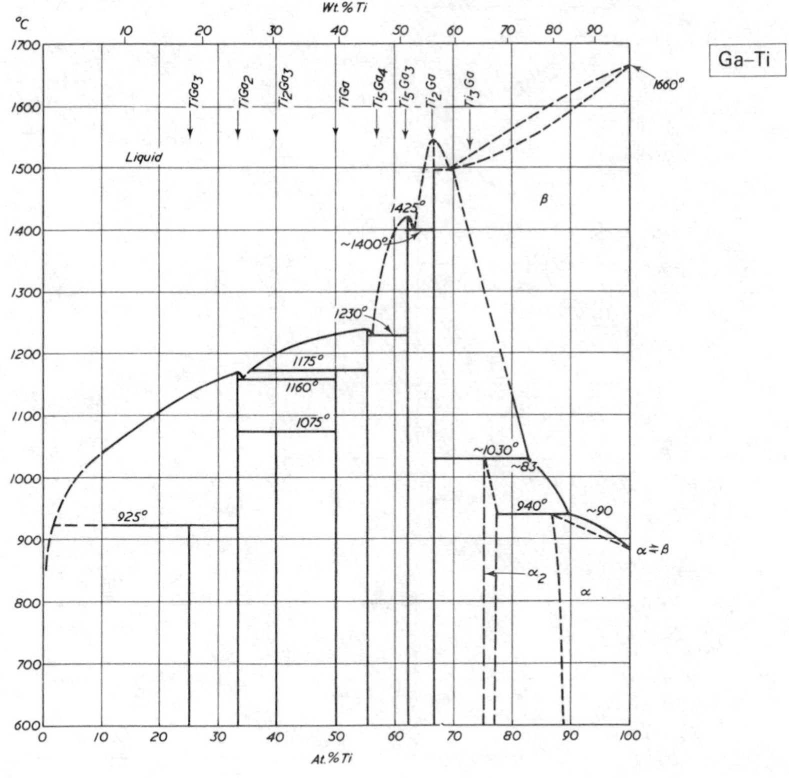
Ga–Ti
Wt.% Ti
°C
At.% Ti
Liquid
TiGa3
TiGa2
Ti2Ga3
TiGa
Ti5Ga4
Ti5Ga3
Ti2Ga
Ti3Ga
1660°
β
1425°
~1400°
1230°
1175°
1160°
1075°
~1030°
~83
940°
~90
925°
α⇌β
α2
α

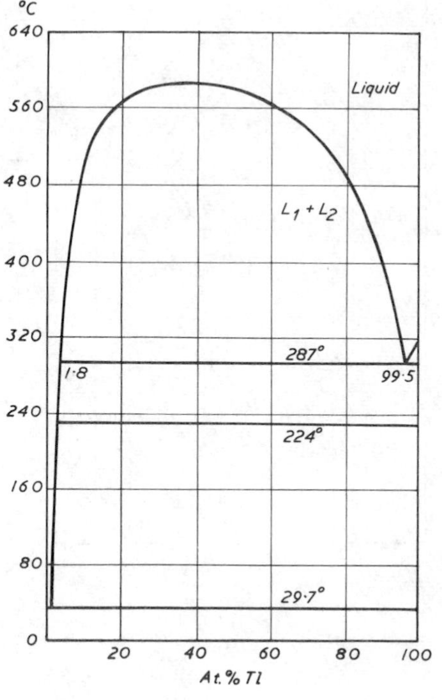
Ga–Tl
°C
Liquid
L1 + L2
287°
1·8
99·5
224°
29·7°
At.% Tl

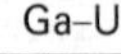
Ga–U

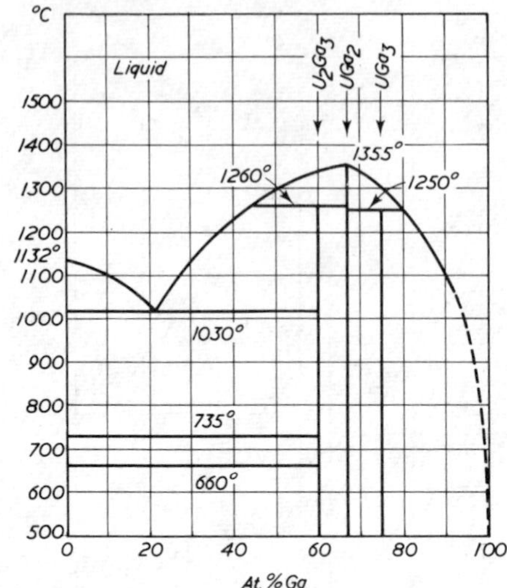
°C
Liquid
U_2Ga_3
UGa_2
UGa_3
1355°
1260°
1250°
1132°
1030°
735°
660°
1500
1400
1300
1200
1100
1000
900
800
700
600
500
0
20
40
60
80
100
At. % Ga

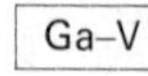
Ga–V

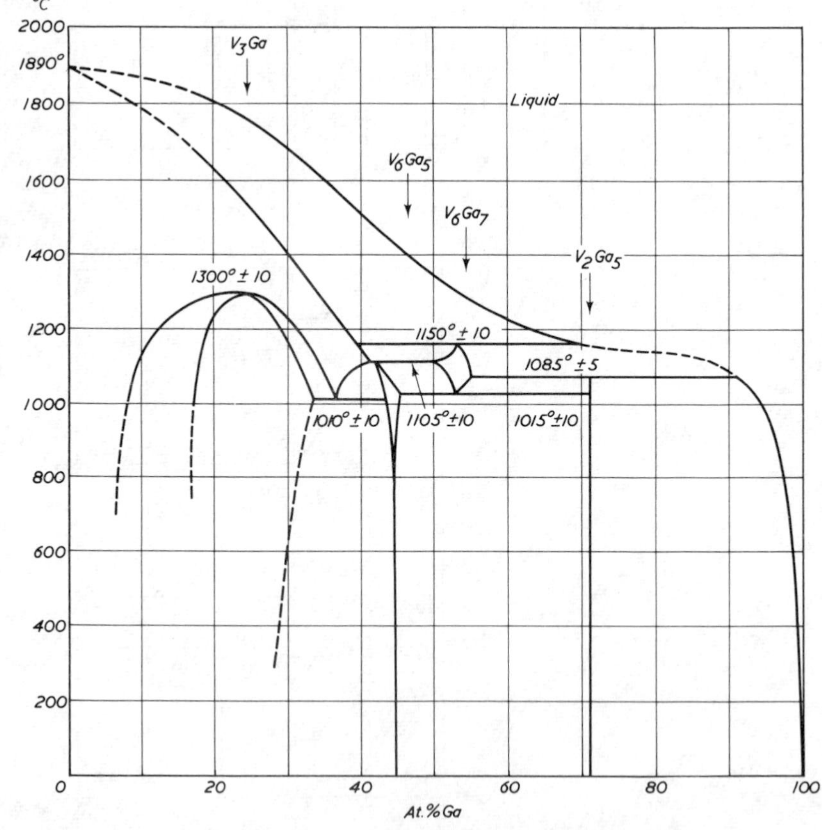
°C
2000
1890°
V_3Ga
Liquid
V_6Ga_5
V_6Ga_7
V_2Ga_5
1300° ± 10
1150° ± 10
1085° ± 5
1010° ± 10
1105° ± 10
1015° ± 10
1800
1600
1400
1200
1000
800
600
400
200
0
20
40
60
80
100
At. % Ga

Ga–Zn

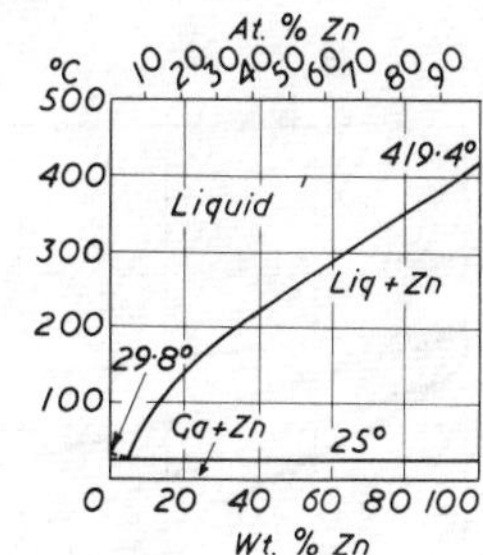

At. % Zn
10 20 30 40 50 60 70 80 90
°C
500
400
300
200
100
0
419·4°
Liquid
Liq + Zn
29·8°
Ga+Zn
25°
0 20 40 60 80 100
Wt. % Zn

Gd–Mn

Wt. % Gd
10 20 30 40 50 60 70 80 90 100
°C
GdMn12
GdMn5 (?)
Gd6Mn23
GdMn2
βGd ⇌ αGd
L ⇌ δMn
1122°
1085°
1080°
1050°
950°
830°
64·7 (84·0)
αMn ⇌ βMn
Magn. Transf.
194±3°
1400
1300
1200
1100
1000
900
800
700
600
500
400
300
200
100
0
0 10 20 30 40 50 60 70 80 90 100
At. % Gd

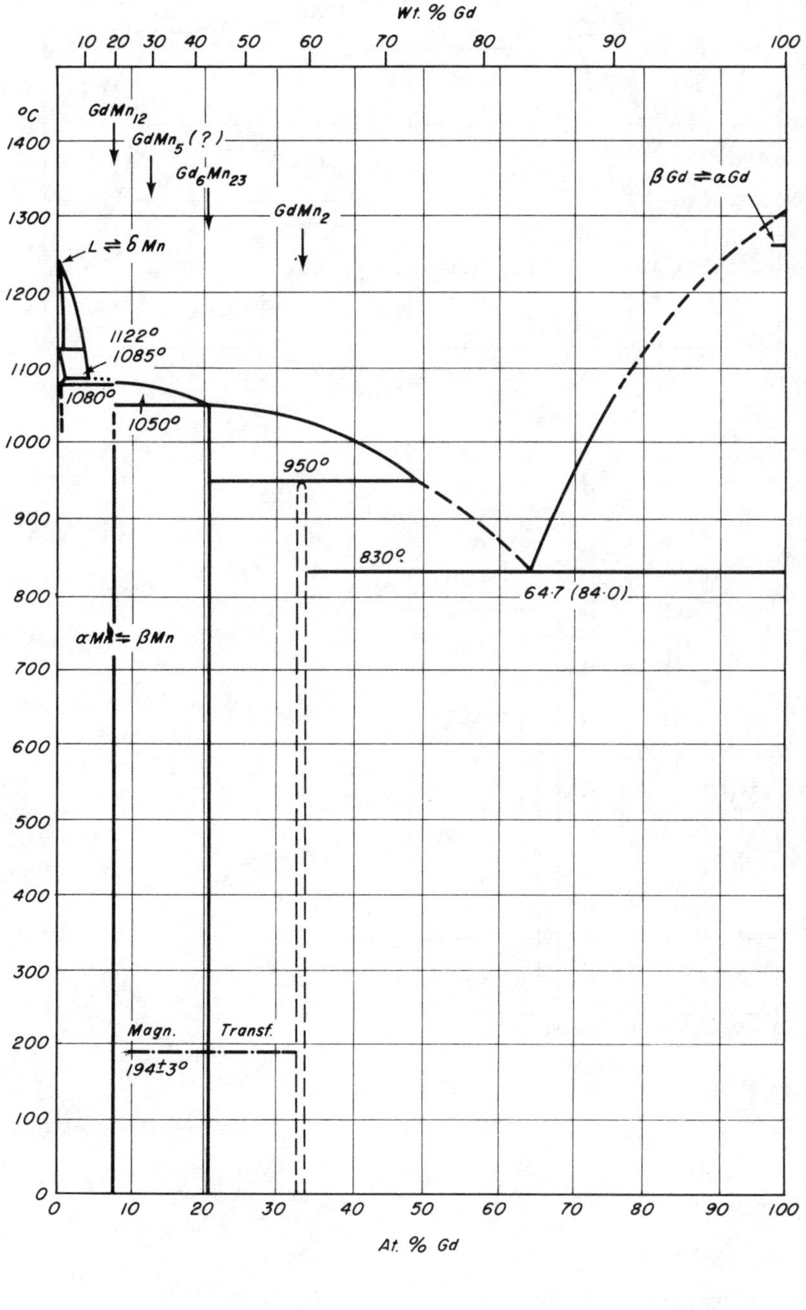

Gd–La

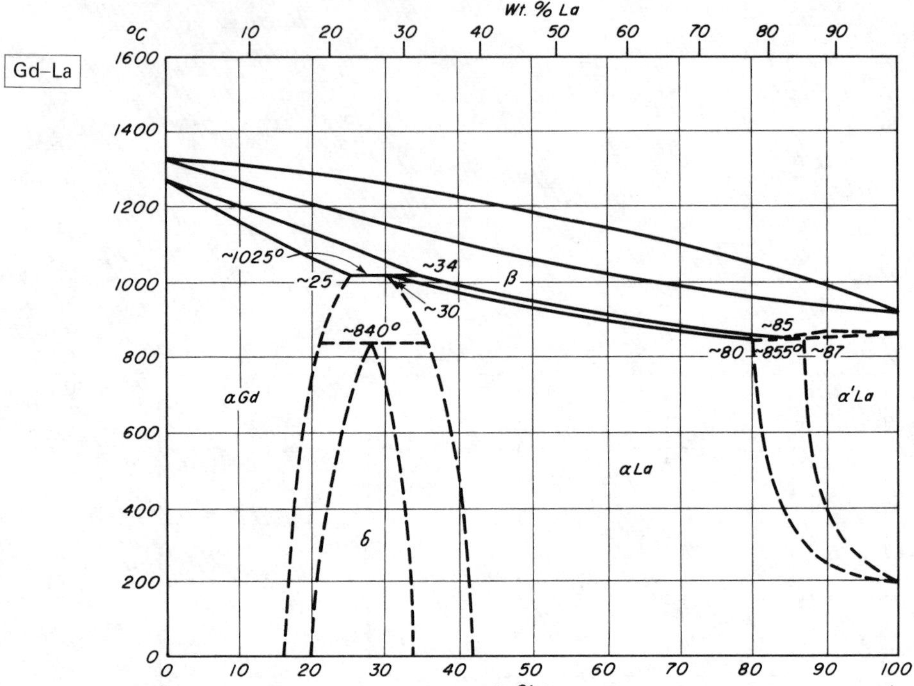
Wt. % La
°C
~1025°
~25
~34
β
~30
~840°
~85
~80
~855°
~87
αGd
α'La
αLa
δ
Gd
At. % La
La

Gd–Ni

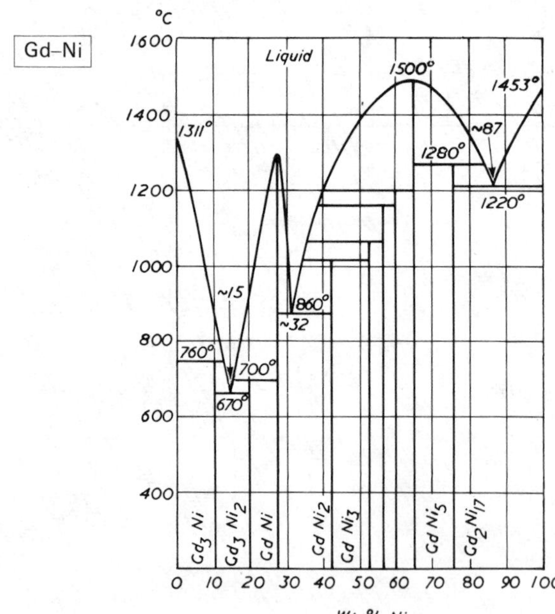
°C
Liquid
1500°
1453°
1311°
~87
1280°
1220°
~15
860°
~32
760°
700°
670°
Gd_3Ni
Gd_3Ni_2
GdNi
$GdNi_2$
$GdNi_3$
$GdNi_5$
Gd_2Ni_{17}
Wt. % Ni

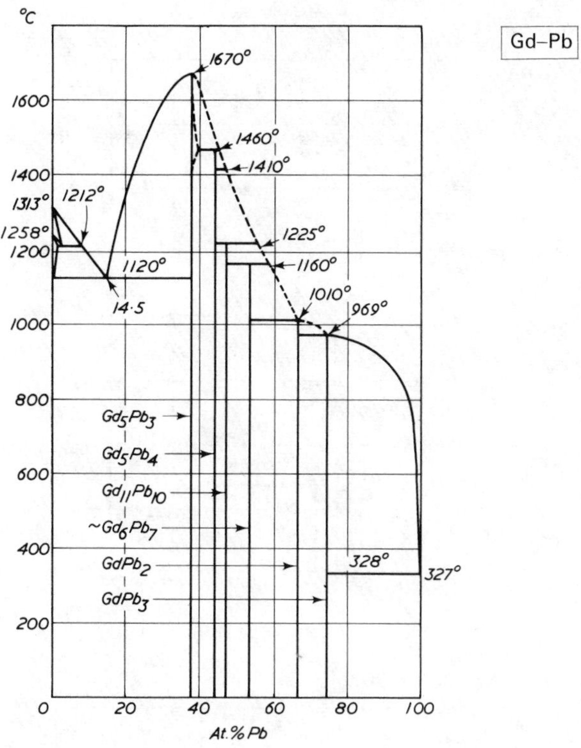
Gd–Pb
°C
1670°
1460°
1410°
1313°
1212°
1258°
1225°
1120°
1160°
14·5
1010°
969°
Gd$_5$Pb$_3$
Gd$_5$Pb$_4$
Gd$_{11}$Pb$_{10}$
~Gd$_6$Pb$_7$
GdPb$_2$
GdPb$_3$
328°
327°
At.% Pb

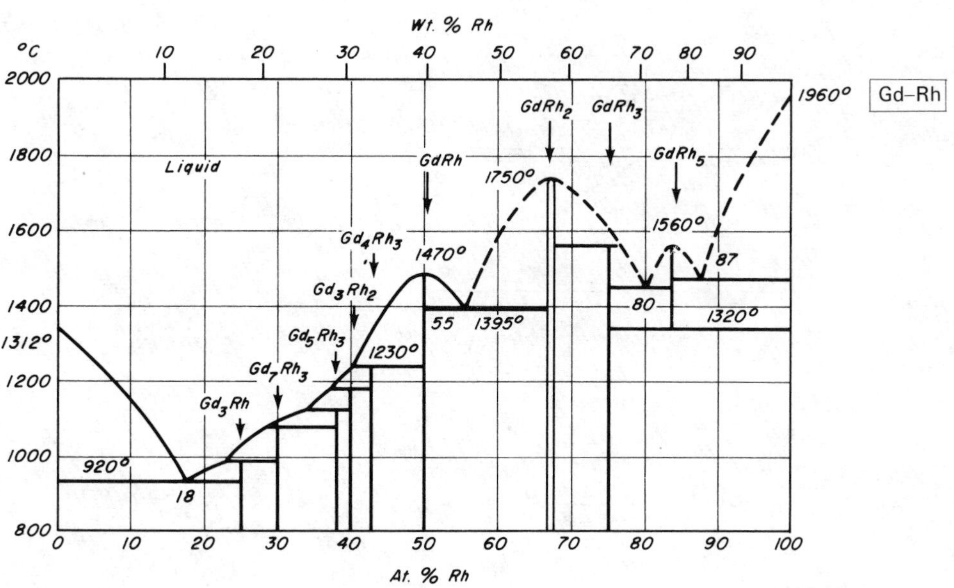
Gd–Rh
Wt. % Rh
°C
1960°
Liquid
GdRh
GdRh$_2$
GdRh$_3$
GdRh$_5$
1750°
1560°
Gd$_4$Rh$_3$
1470°
87
Gd$_3$Rh$_2$
80
55
1395°
1320°
1312°
Gd$_5$Rh$_3$
1230°
Gd$_7$Rh$_3$
Gd$_3$Rh
920°
18
At. % Rh

Gd–Ru

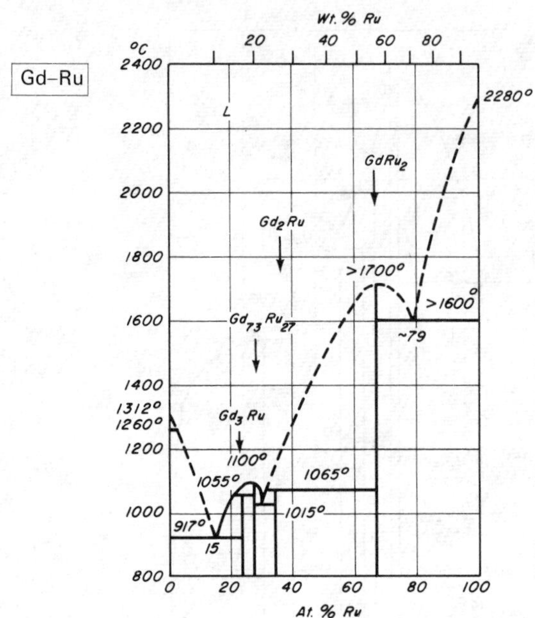

Gd–Sc

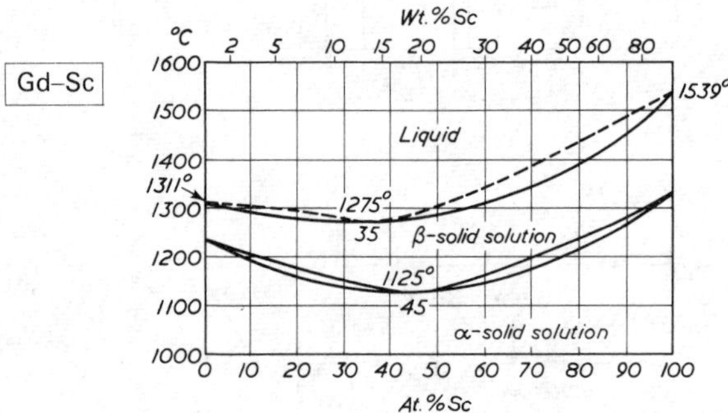

Gd–Ti

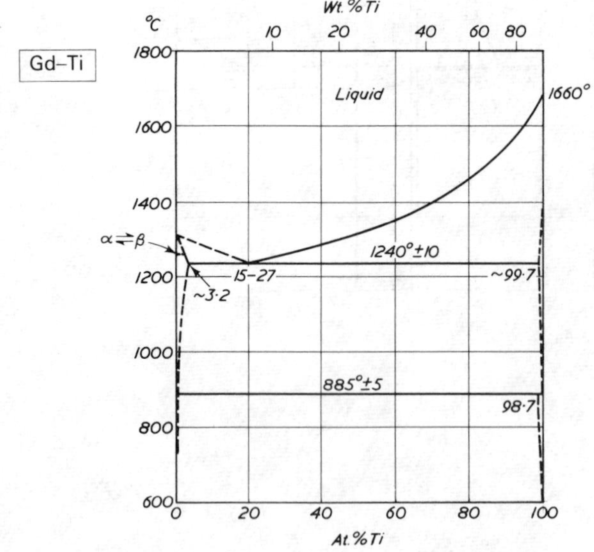

Gd–Y

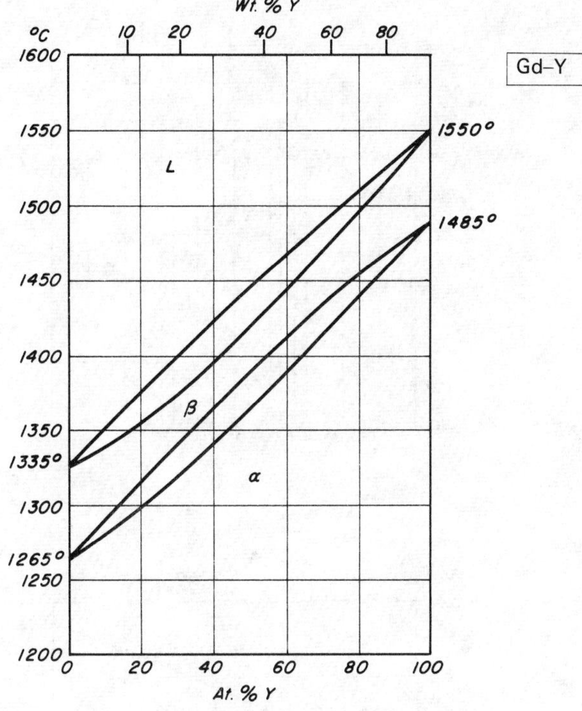
Wt. % Y
10
20
40
60
80
°C
1600
1550
1500
1450
1400
1350
1335°
1300
1265°
1250
1200
1550°
1485°
L
β
α
0
20
40
60
80
100
At. % Y

Gd–Zr

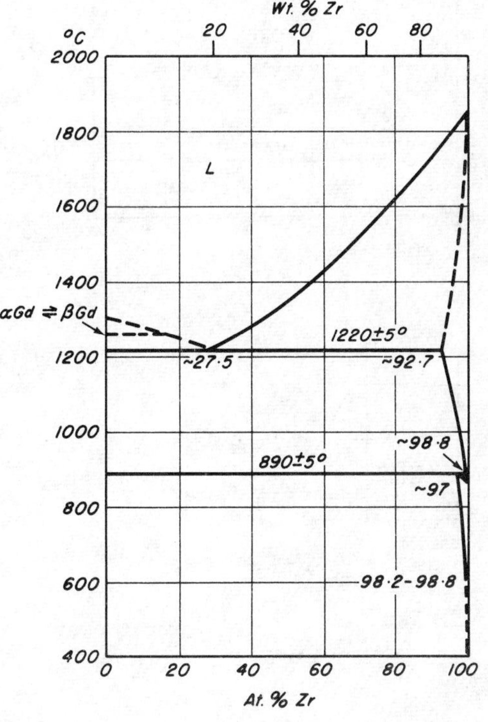
Wt. % Zr
20
40
60
80
°C
2000
1800
1600
1400
1200
1000
800
600
400
L
αGd ⇌ βGd
1220±5°
~27·5
≈92·7
~98·8
890±5°
~97
98·2–98·8
0
20
40
60
80
100
At. % Zr

Ge–Hf

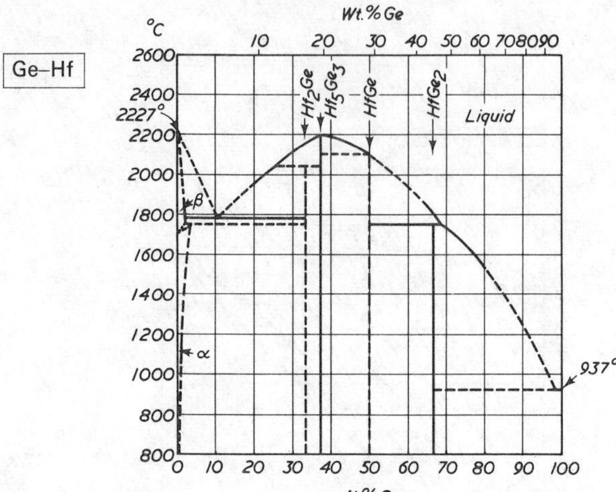

Ge–In

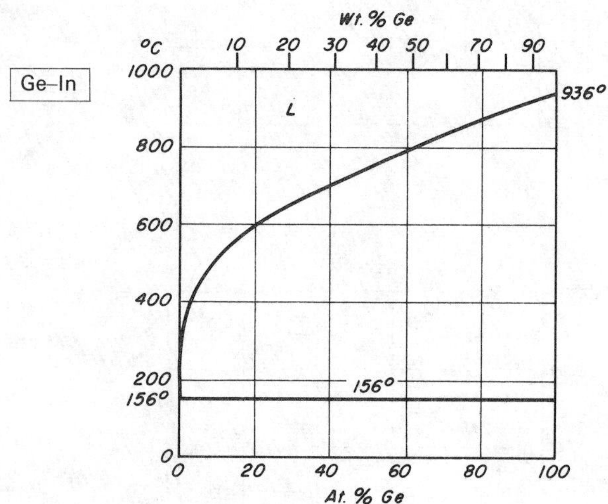

Ge–Mg

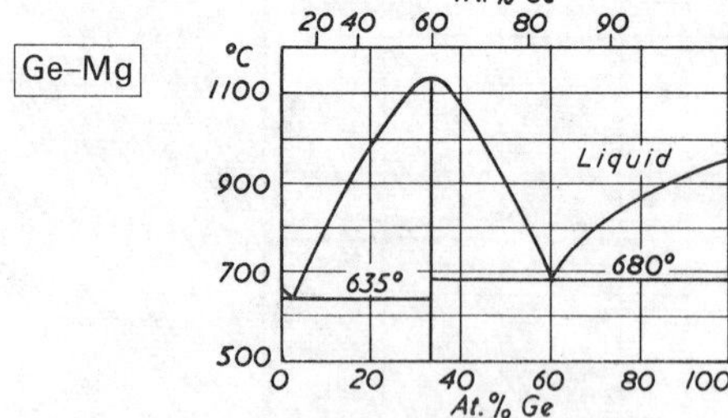

Ge–Mn

Wt. % Ge
°C
L
ε
x
Mn3Ge2
930°
965°
ζ
η
953°
966°
1244°
δ Mn
γ Mn
β Mn
α Mn
918°
790°
796°
720°
57
936°
xxxx Magnetic transformation
At. % Ge

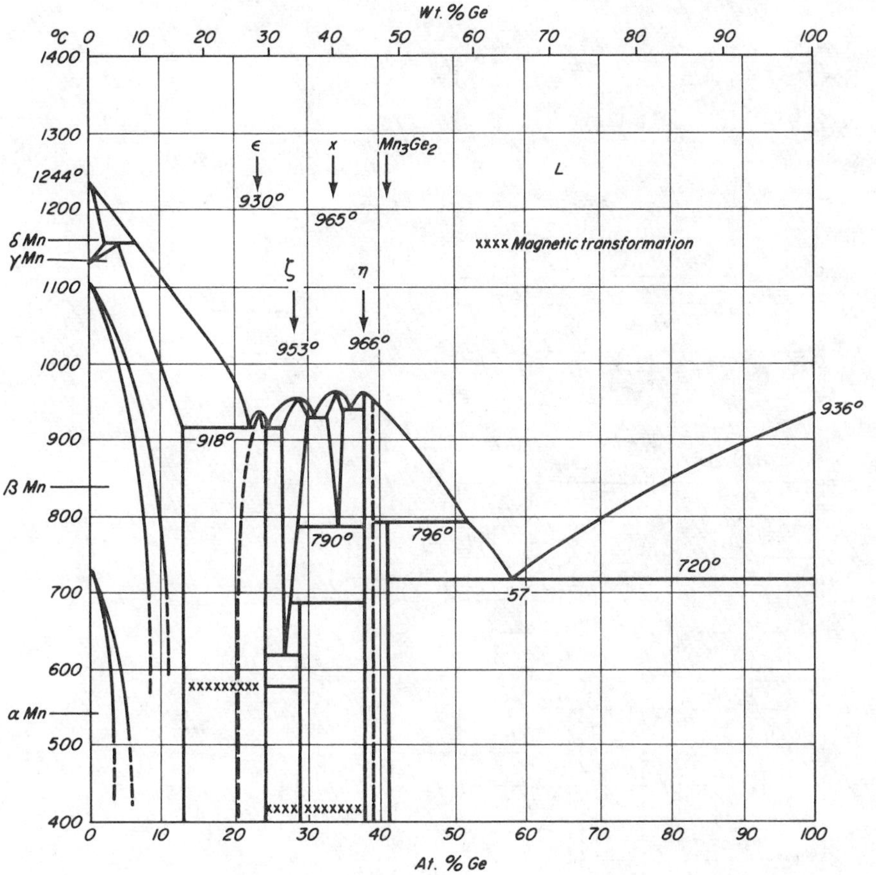

Ge–Mo

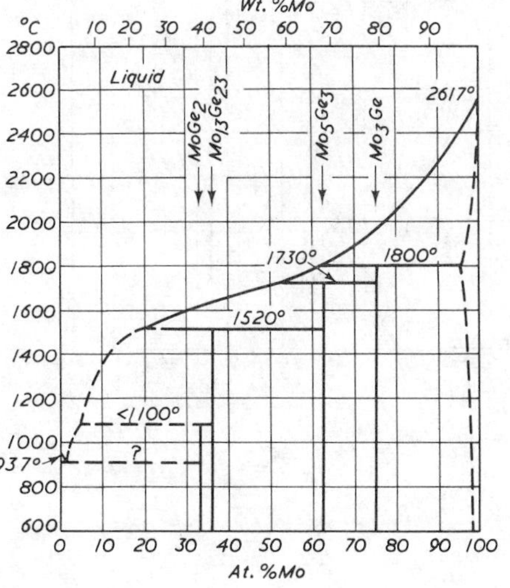

Wt. %Mo
°C
Liquid
MoGe2
Mo13Ge23
Mo5Ge3
Mo3Ge
2617°
1730°
1800°
1520°
<1100°
?
937°
At. %Mo

Ge–Pb

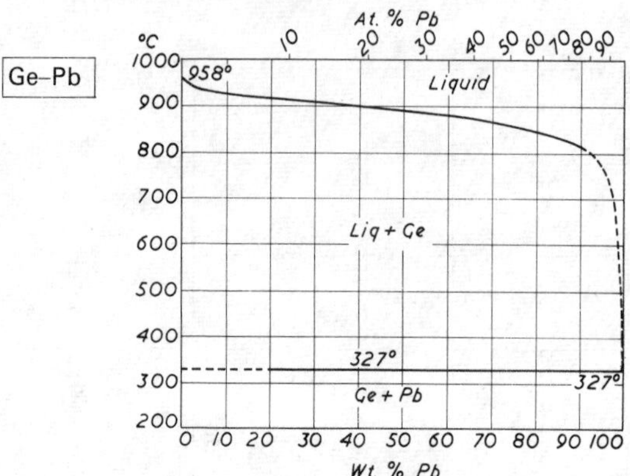
At. % Pb
°C
958°
Liquid
Liq + Ge
327°
327°
Ge + Pb
Wt. % Pb

Ge–Rh

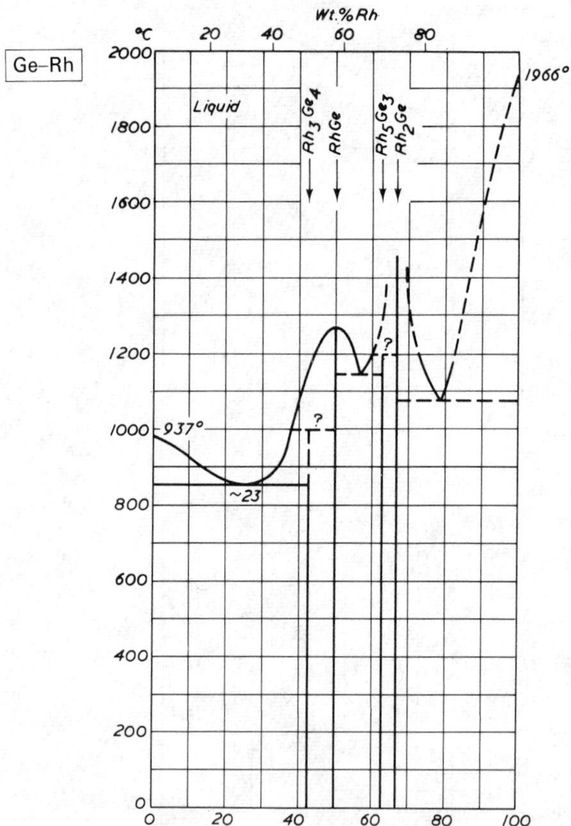
Wt.%Rh
°C
1966°
Liquid
Rh3Ge4
RhGe
Rh5Ge3
Rh2Ge
937°
~23
At.%Rh

°C
2300
1900
1500
1100
700
Liquid
Ge_3Ru_2
Ge Ru
$Ge+Ge_2Ru_2$
GeRu+Ru
0
20
40
60
80
100
At.% Ru

Ge–Ru

Wt.%S
°C
5 10 15 20 30 40 50 60 70 90
1200
1100
1000
900
800
700
600
500
400
300
200
100
0
956°
GeS
GeS_2
L_1+L_2
Liquid
920°±5°
840°±5°
665°±5°
650°±5°
590°±5°
600°
~60
Boiling point
120°±5°
0 10 20 30 40 50 60 70 80 90 100
At.%S

Ge–S

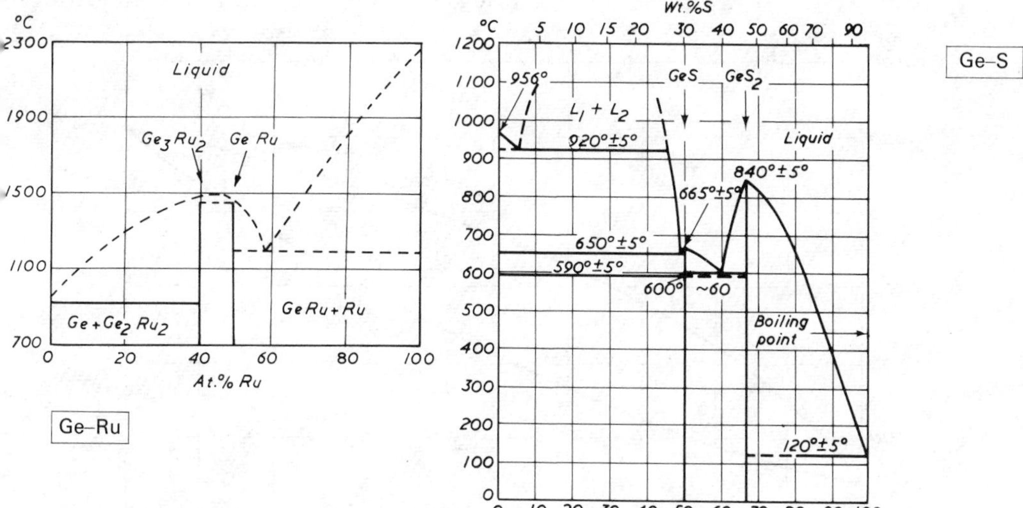

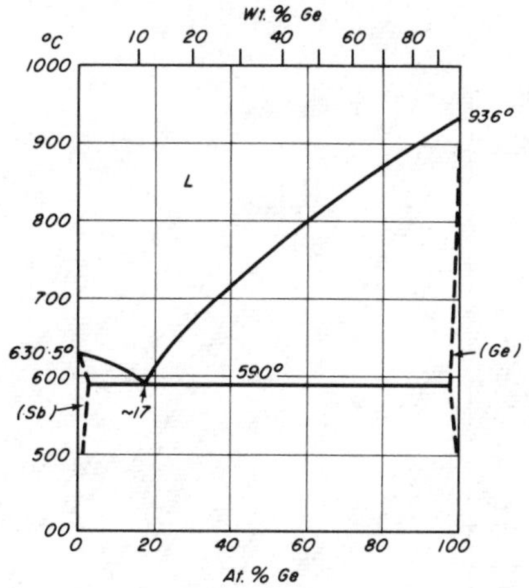

Wt.% Ge
10 20 40 60 80
°C
1000
900
800
700
600
500
00
936°
L
630·5°
590°
(Ge)
(Sb)
~17
0 20 40 60 80 100
At.% Ge

Ge–Sb

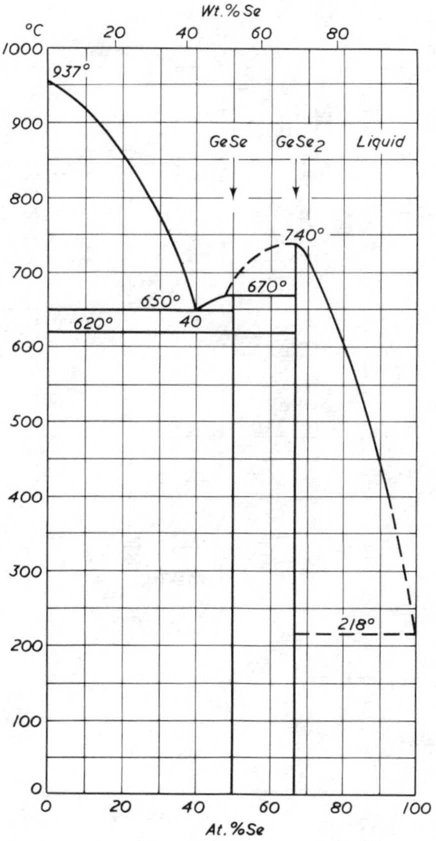

Wt.% Se
20 40 60 80
°C
1000
900
800
700
600
500
400
300
200
100
0
937°
GeSe
$GeSe_2$
Liquid
740°
670°
650°
620°
40
218°
0 20 40 60 80 100
At.%Se

Ge–Se

Ge–Si

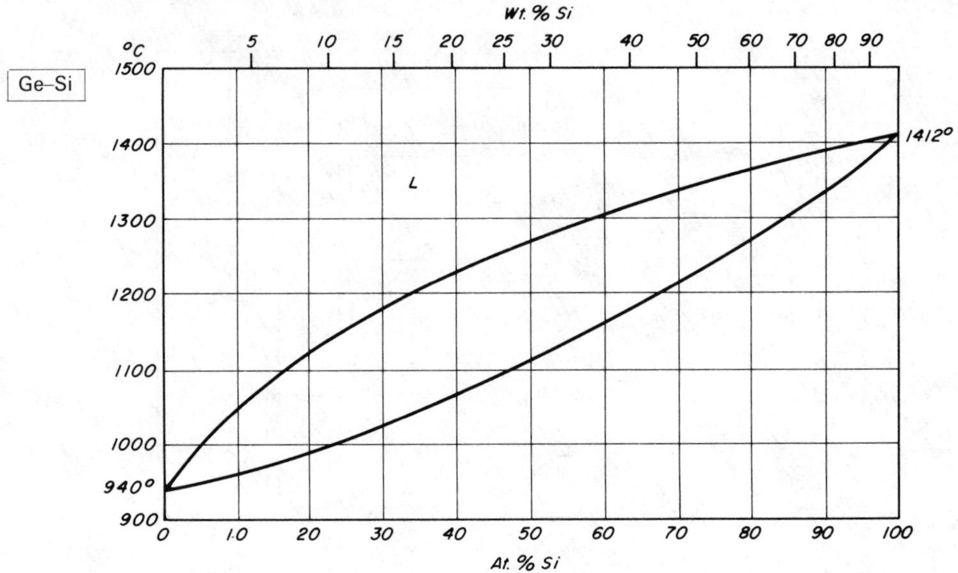
Wt. % Si
5
10
15
20
25
30
40
50
60
70
80
90
°C
1500
1400
1300
1200
1100
1000
940°
900
1412°
L
0
1.0
20
30
40
50
60
70
80
90
100
At. % Si

Ge–Sn

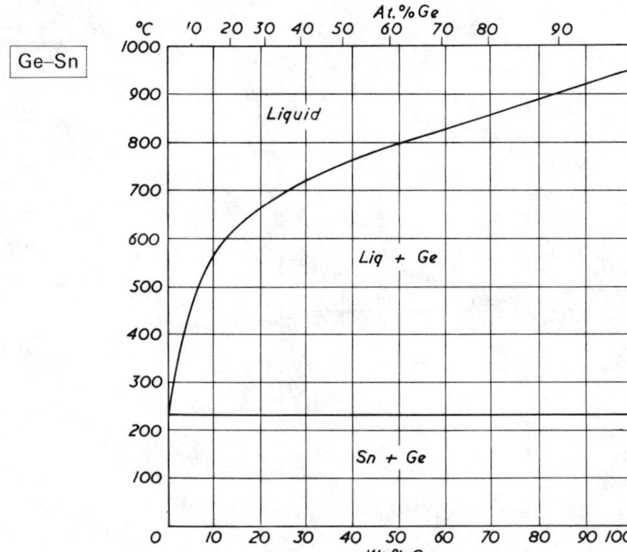
At. % Ge
°C
10
20
30
40
50
60
70
80
90
1000
900
800
700
600
500
400
300
200
100
0
Liquid
Liq + Ge
Sn + Ge
10
20
30
40
50
60
70
80
90
100
Wt. % Ge

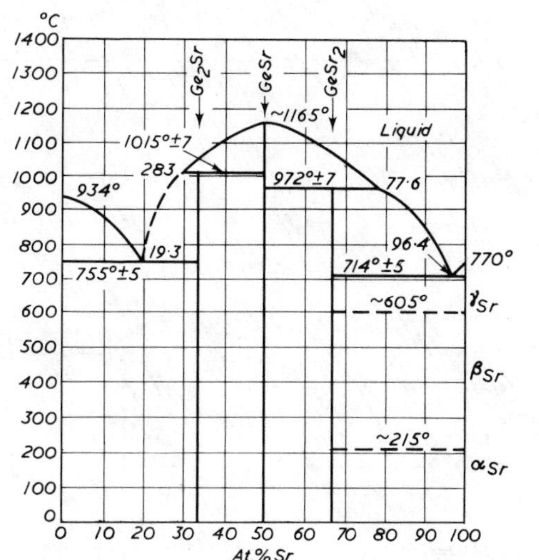
°C
Ge2Sr
GeSr
GeSr2
~1165°
1015°±7
283
934°
972°±7
77·6
Liquid
19·3
755°±5
96·4
714°±5
770°
~605°
γSr
βSr
~215°
αSr
At.% Sr

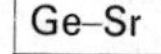
Ge–Sr

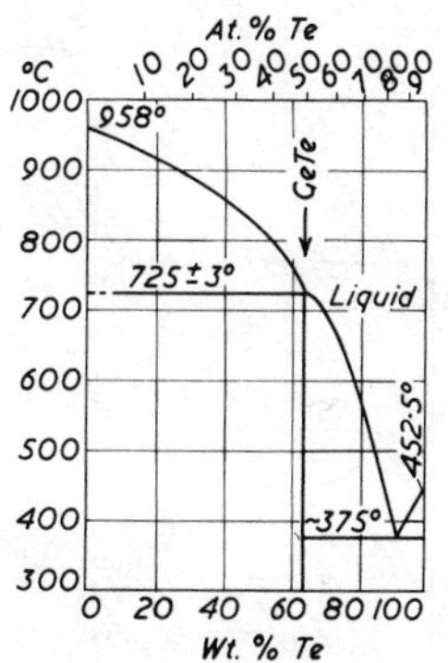
At.% Te
°C
958°
GeTe
725±3°
Liquid
452·5°
~375°
Wt. % Te

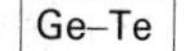
Ge–Te

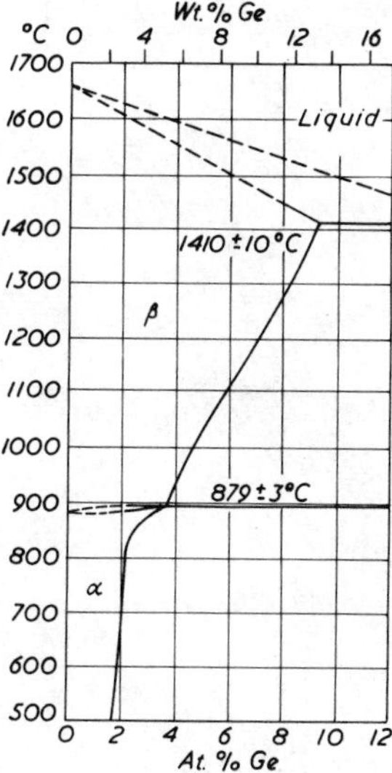
Wt.% Ge
°C
Liquid
1410±10°C
β
879±3°C
α
At. % Ge

Ge–Ti

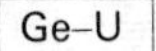

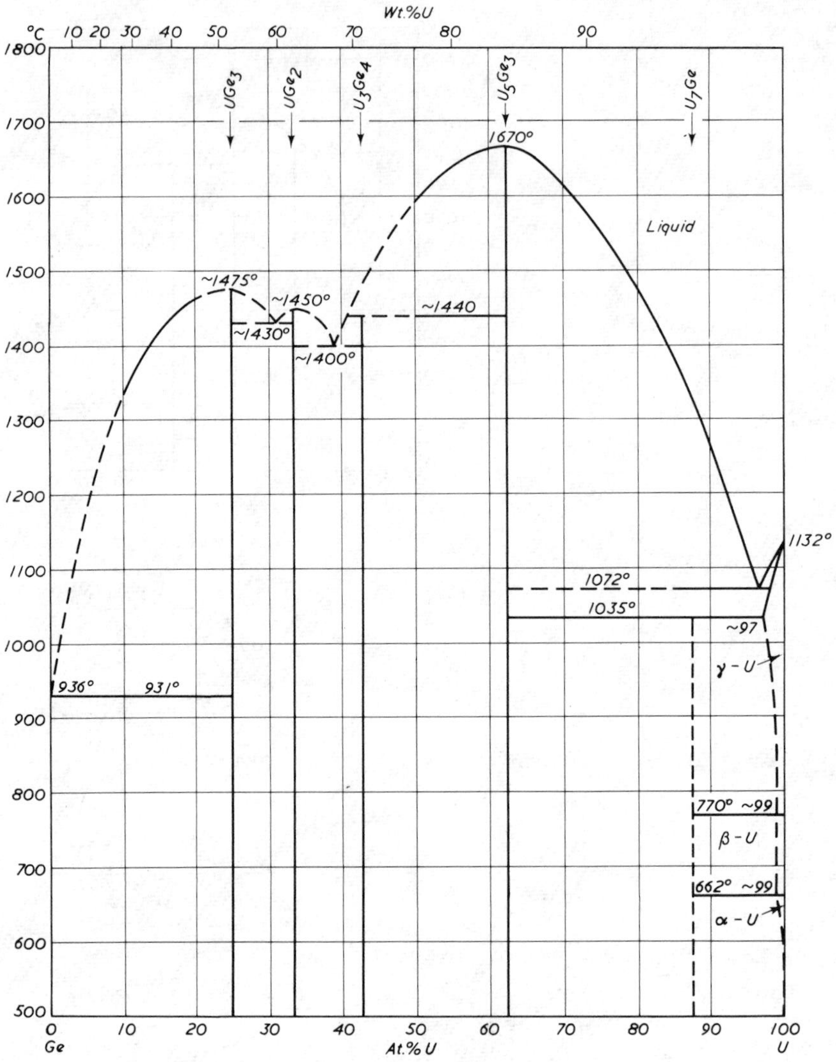

Wt.%U
°C
10 20 30 40 50 60 70 80 90
1800
1700
1600
1500
1400
1300
1200
1100
1000
900
800
700
600
500
UGe3
UGe2
U3Ge4
U5Ge3
U7Ge
1670°
Liquid
~1475°
~1450°
~1440
~1430°
~1400°
1132°
1072°
1035°
~97
γ-U
936°
931°
770° ~99
β-U
662° ~99
α-U
0 10 20 30 40 50 60 70 80 90 100
Ge
At.%U
U

Ge–Y

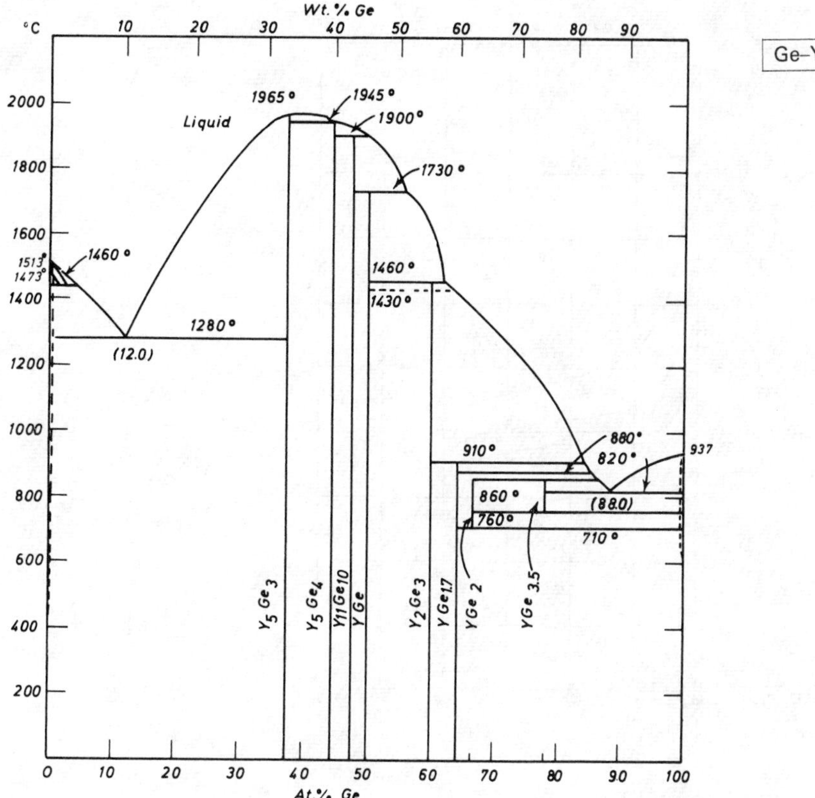

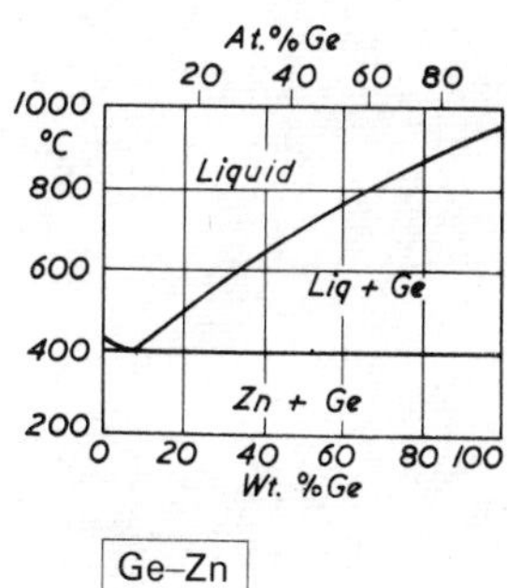

Ge–Zn

Ge–Zr

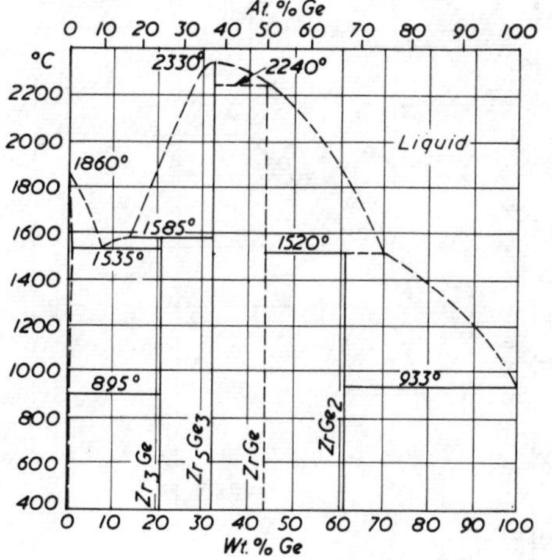

H–Nb

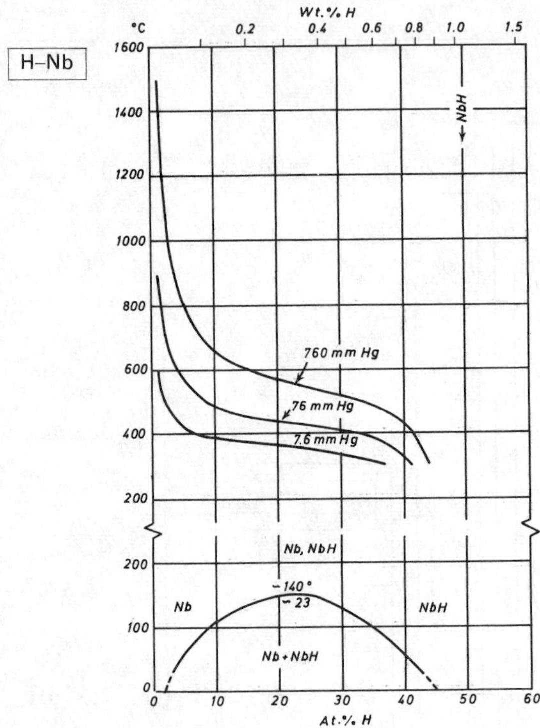

Wt.% H
0.2
0.4
0.6
0.8
1.0
1.5
°C
1600
1400
1200
1000
800
600
400
200
100
0
NbH
760 mm Hg
76 mm Hg
7.6 mm Hg
Nb, NbH
~140°
~23
Nb
NbH
Nb+NbH
0
10
20
30
40
50
60
At.% H

H–Sr

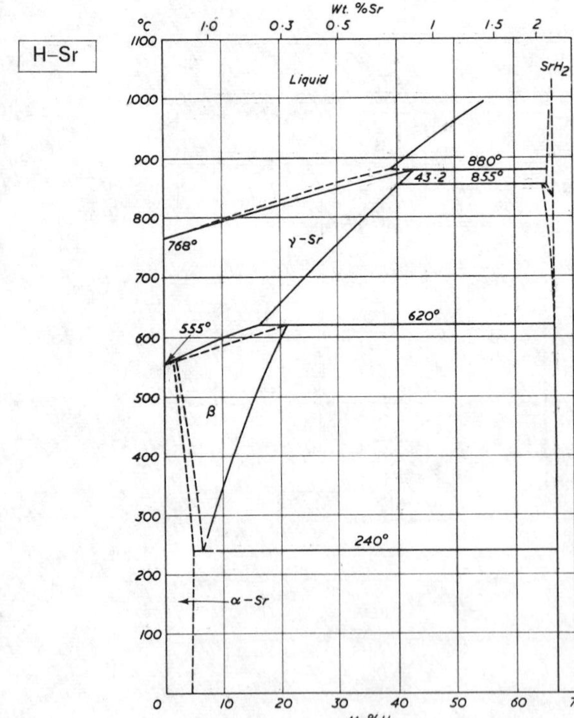

Wt. %Sr
1·0
0·3
0·5
1
1·5
2
°C
1100
1000
900
800
700
600
500
400
300
200
100
Liquid
SrH_2
880°
43·2
855°
768°
γ-Sr
620°
555°
β
240°
α-Sr
0
10
20
30
40
50
60
70
At. %H

H–Ti

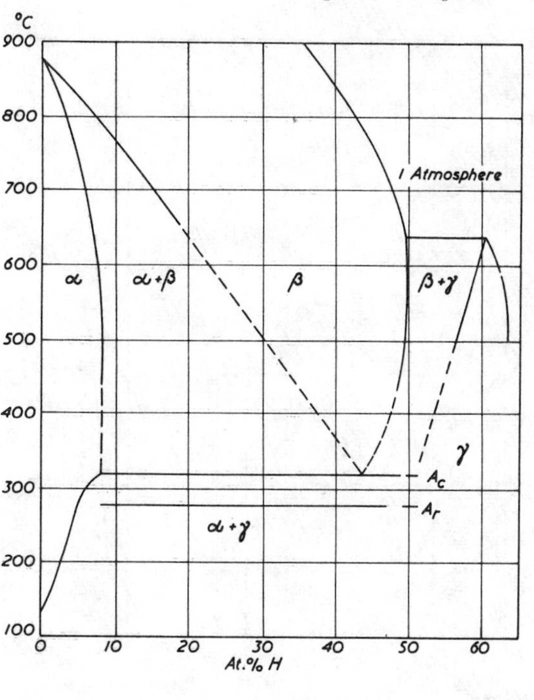
°C
900
800
700
600
500
400
300
200
100
0
10
20
30
40
50
60
At.% H
1 Atmosphere
α
α+β
β
β+γ
γ
A_c
A_r
α+γ

H–Zr

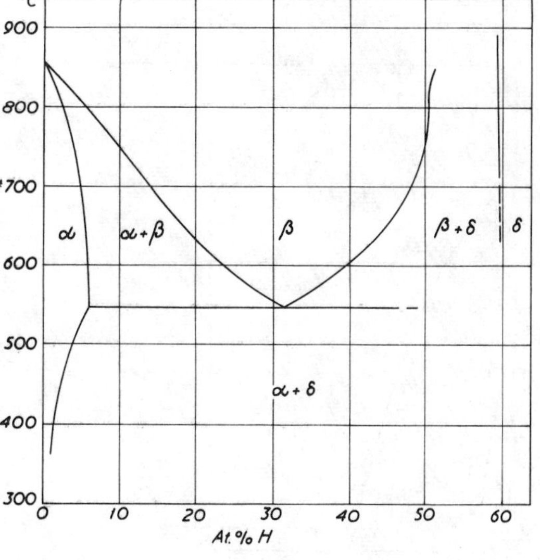
°C
900
800
700
600
500
400
300
0
10
20
30
40
50
60
At.% H
α
α+β
β
β+δ
δ
α+δ

Hf–Ir

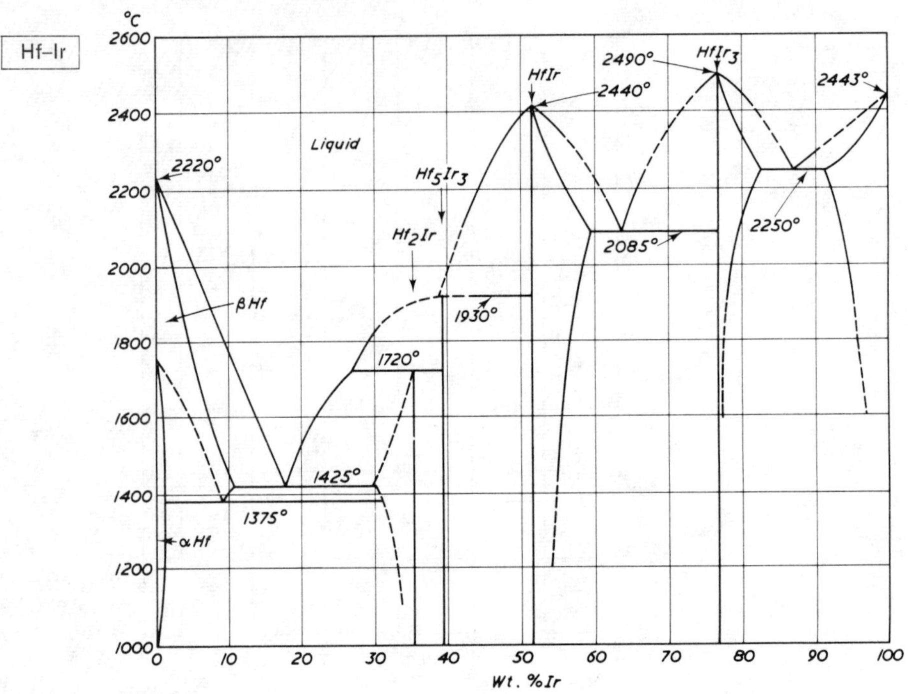

°C
2600
2400
2200
2000
1800
1600
1400
1200
1000
Liquid
2220°
βHf
αHf
1375°
1425°
1720°
Hf_2Ir
Hf_5Ir_3
1930°
HfIr
2440°
2490°
$HfIr_3$
2085°
2443°
2250°
0
10
20
30
40
50
60
70
80
90
100
Wt. %Ir

Hf–Mn

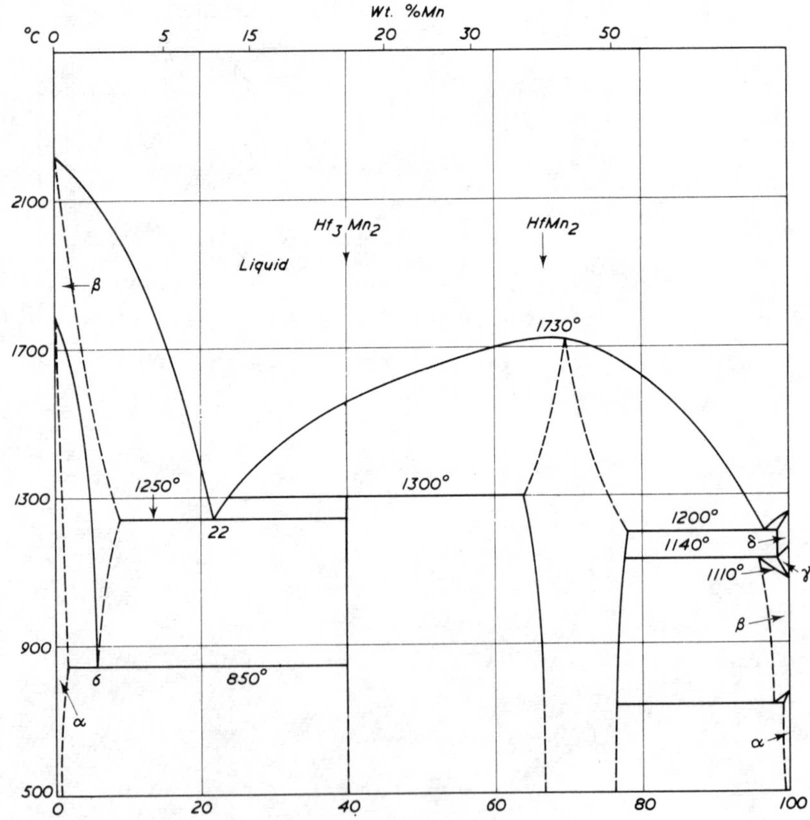

Wt. %Mn
0
5
15
20
30
50
°C
2100
1700
1300
900
500
Liquid
β
α
Hf_3Mn_2
$HfMn_2$
1730°
1250°
22
1300°
6
850°
1200°
1140°
1110°
δ
γ
β
α
0
20
40
60
80
100
At. %Mn

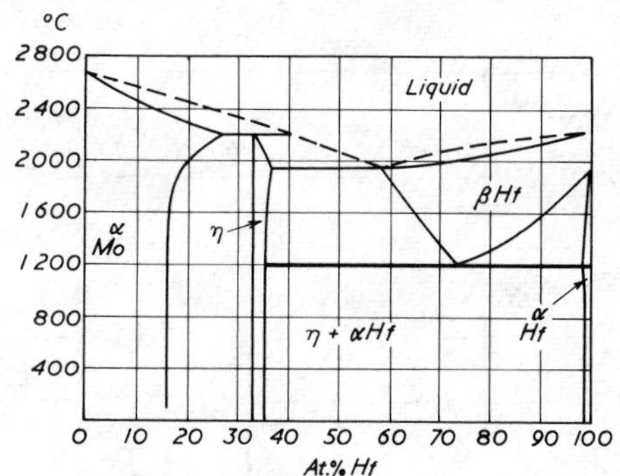
°C
2800
2400
2000
1600
1200
800
400
Liquid
βHf
α Mo
η
η + αHf
α Hf
0 10 20 30 40 50 60 70 80 90 100
At.% Hf

Hf–Mo

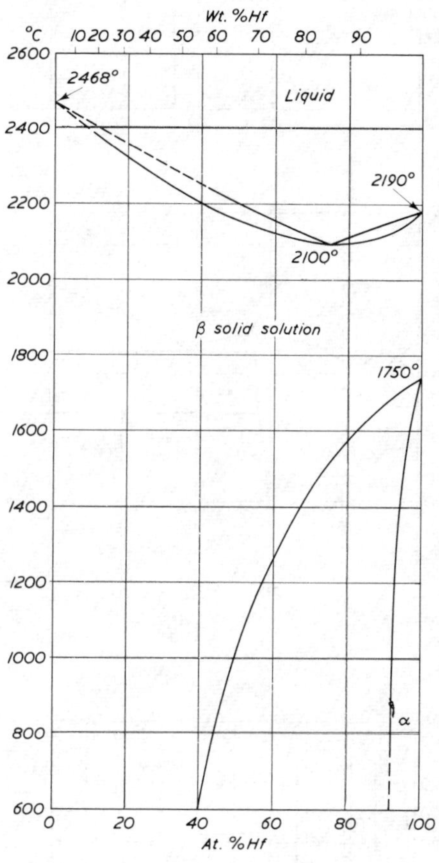
Wt. %Hf
10 20 30 40 50 60 70 80 90
°C
2600
2400
2200
2000
1800
1600
1400
1200
1000
800
600
2468°
Liquid
2190°
2100°
β solid solution
1750°
α
0 20 40 60 80 100
At. %Hf

Hf–Nb

Hf–Ni

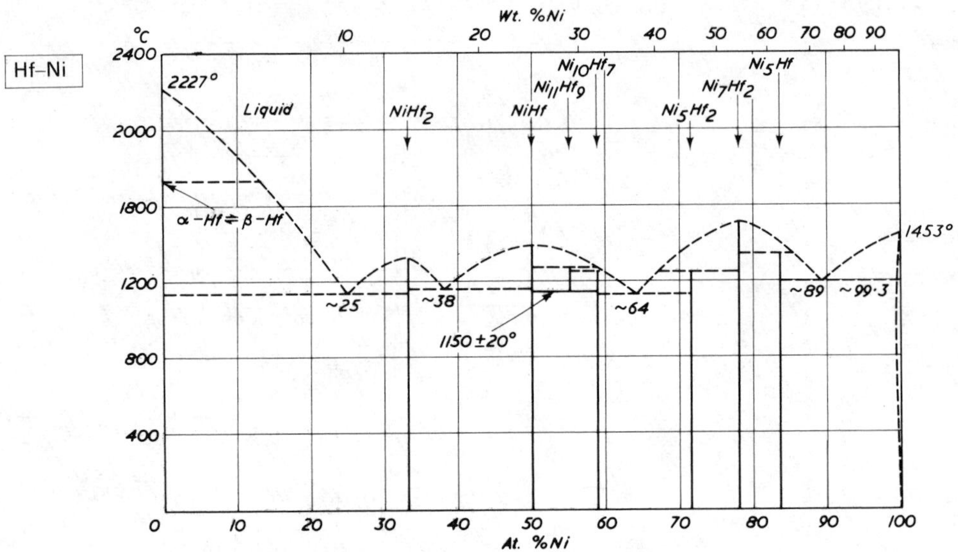

Hf–O

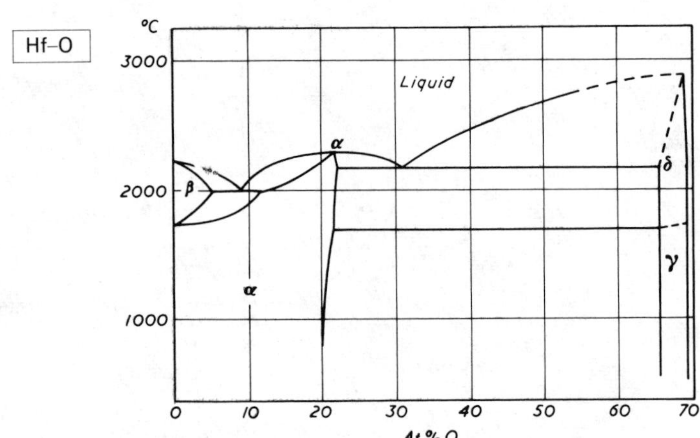

Hf–Re

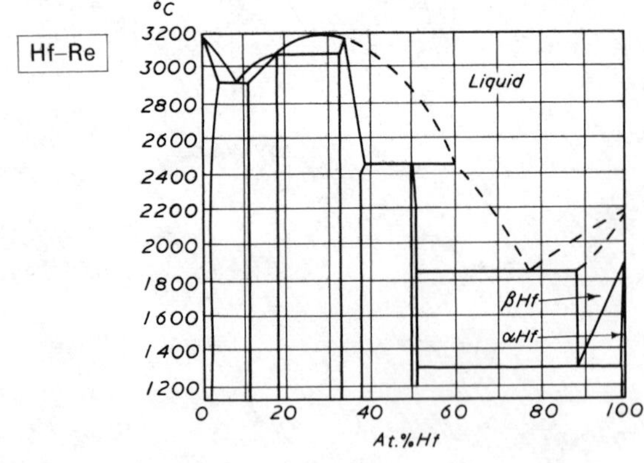

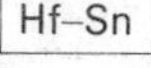
Hf–Sn

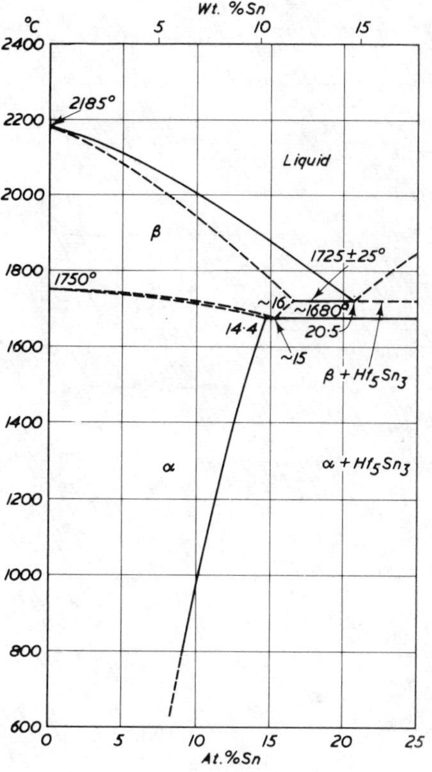
Wt. %Sn
°C
2185°
Liquid
β
1725±25°
1750°
~16
~1680°
14·4
20·5
~15
β + Hf_5Sn_3
α
α + Hf_5Sn_3
At.%Sn

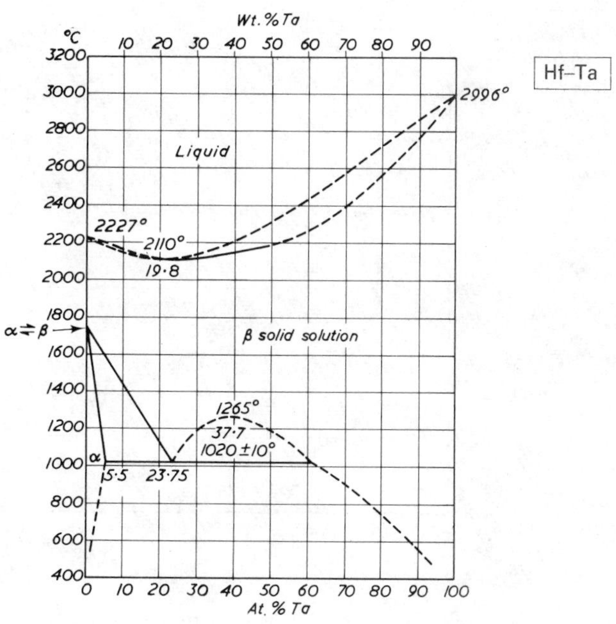
Hf–Ta
Wt.%Ta
°C
2996°
Liquid
2227°
2110°
19·8
α⇌β
β solid solution
1265°
37·7
1020±10°
α
5·5
23·75
At.% Ta

Hf–Th

Wt. %Th
°C
10 20 30 40 50 60 70 80 90
2200
2100
2000
1900
1800
1700
1600
1500
1400
1300
1200
1100
1000
2227°
Liquid
1750°
α⇌β
1600°
1450°
α⇌β
~1·5
68·4
83
1295°
~1·5
88·5
0 10 20 30 40 50 60 70 80 90 100
At. %Th

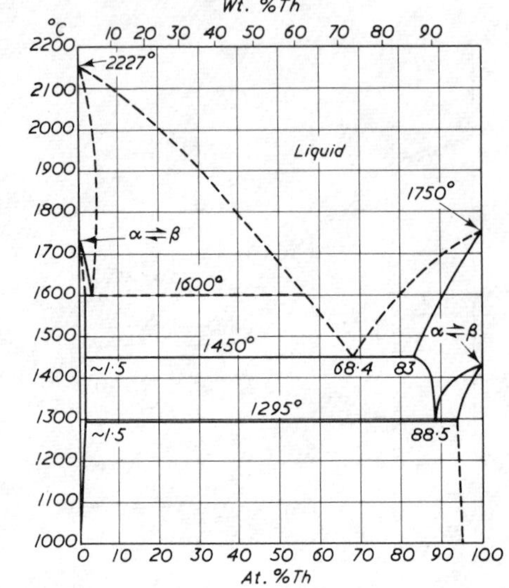

Hf–Ti

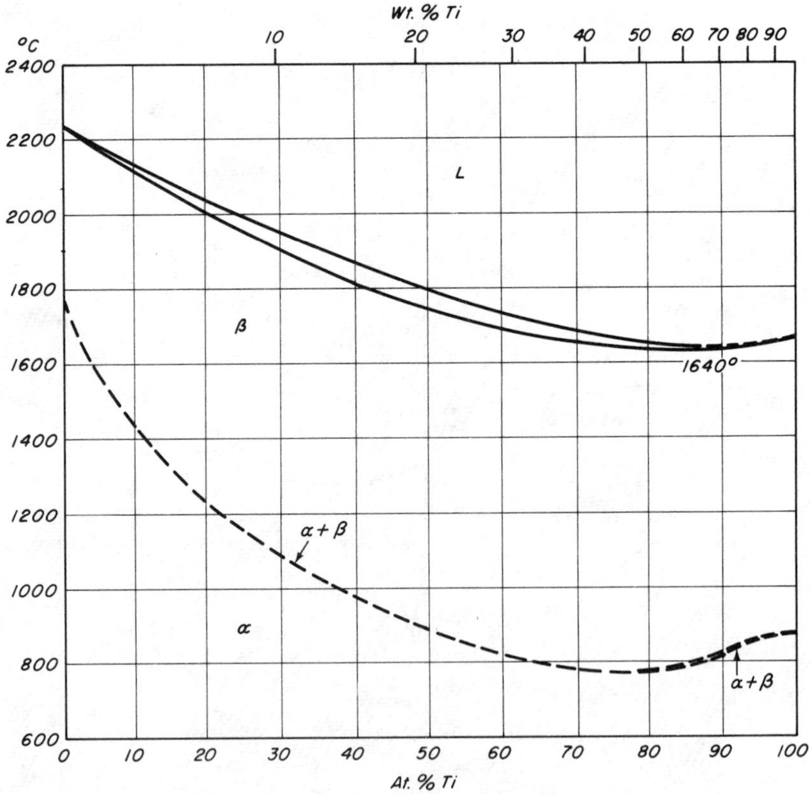

Wt. % Ti
10 20 30 40 50 60 70 80 90
°C
2400
2200
2000
1800
1600
1400
1200
1000
800
600
L
β
1640°
α + β
α
α + β
0 10 20 30 40 50 60 70 80 90 100
At. % Ti

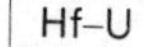
Hf–U

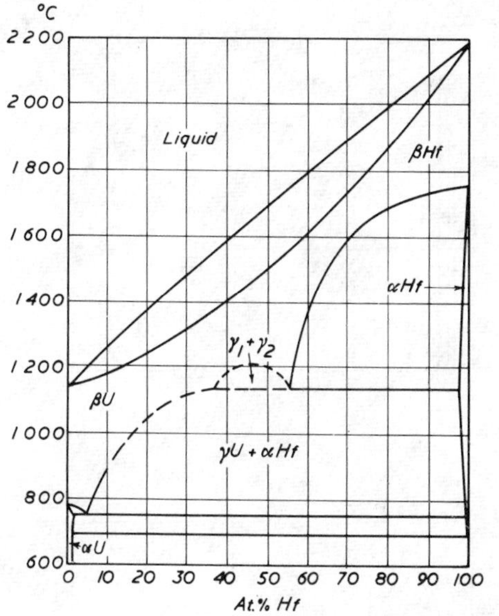
°C
2200
2000
1800
1600
1400
1200
1000
800
600
Liquid
βHf
αHf
γ1+γ2
βU
γU + αHf
αU
0 10 20 30 40 50 60 70 80 90 100
At.% Hf

Hf–W

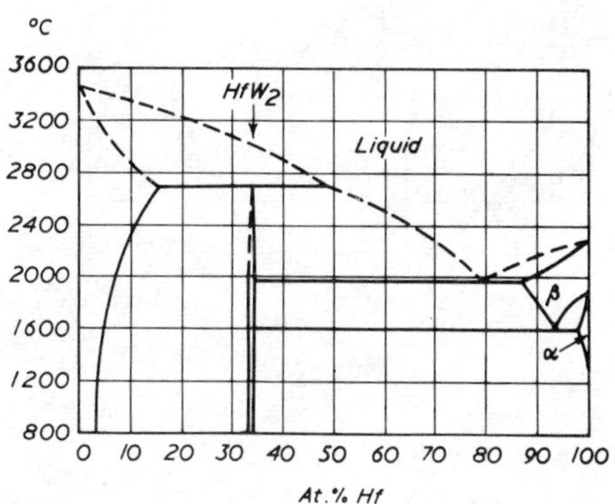
°C
3600
3200
2800
2400
2000
1600
1200
800
HfW2
Liquid
β
α
0 10 20 30 40 50 60 70 80 90 100
At.% Hf

Hg–In

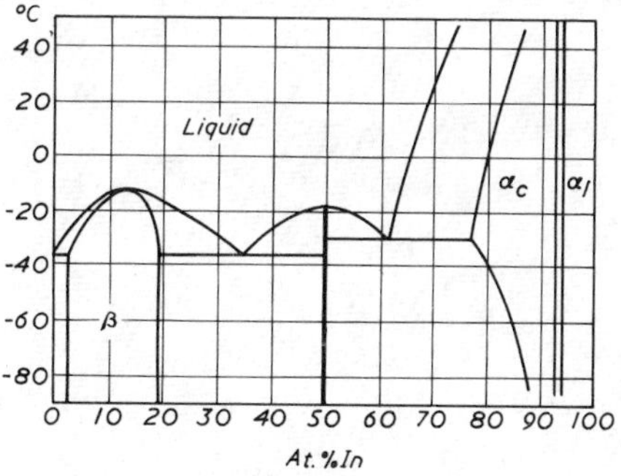
°C
40
20
0
-20
-40
-60
-80
Liquid
β
αc
αl
0 10 20 30 40 50 60 70 80 90 100
At.% In

Hg–K

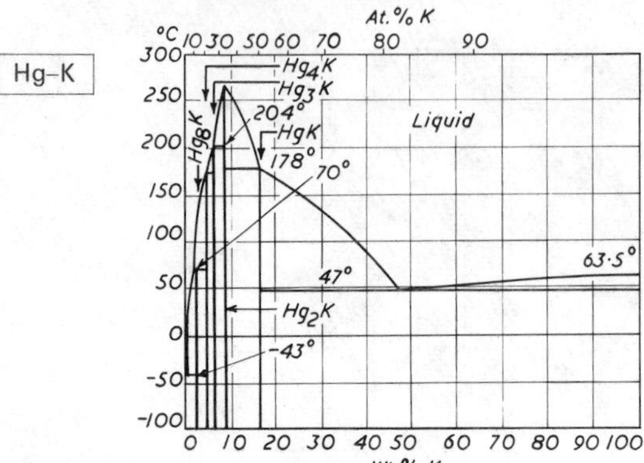
At.% K
°C
Hg4K
Hg3K
Hg8K
204°
HgK
178°
70°
Liquid
63·5°
47°
Hg2K
−43°
Wt.% K

Hg–Li

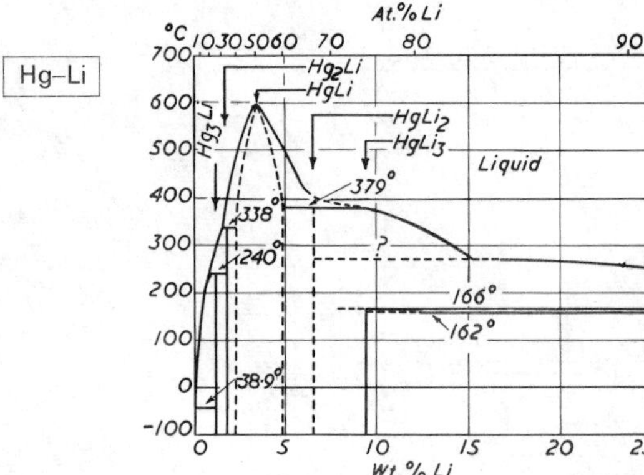
At.% Li
°C
Hg2Li
HgLi
Hg3Li
HgLi2
HgLi3
Liquid
379°
338°
240°
?
166°
162°
38·9°
Wt.% Li

Hg–Mg

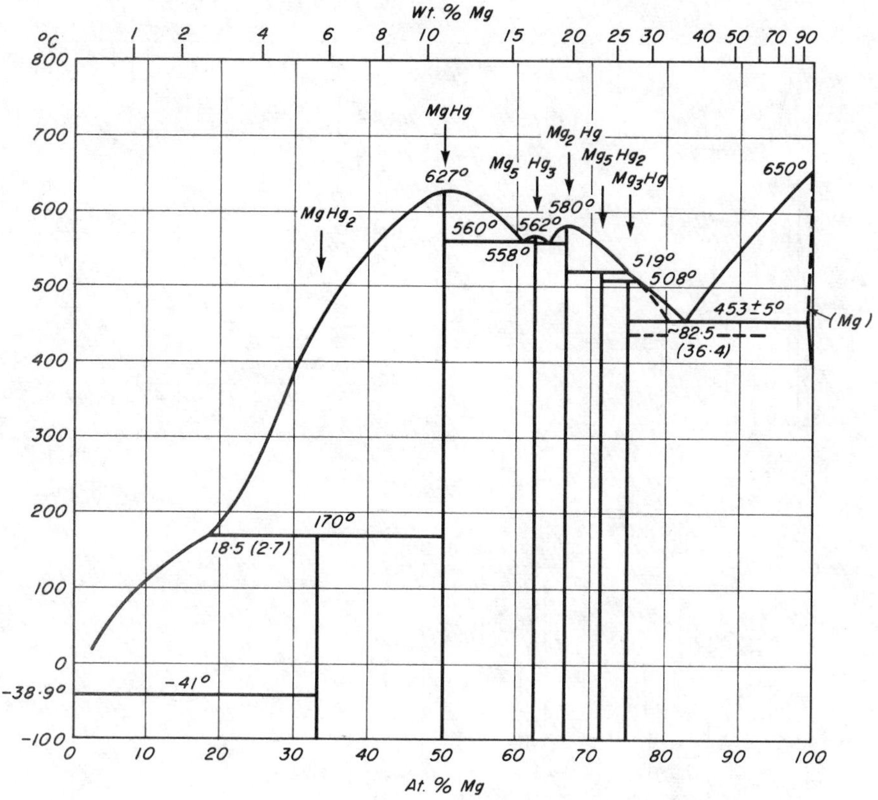

Wt. % Mg
1
2
4
6
8
10
15
20
25
30
40
50
70
90
°C
800
700
600
500
400
300
200
100
0
-38·9°
-100
MgHg
627°
Mg5 Hg3
Mg2 Hg
Mg5 Hg2
Mg3Hg
650°
MgHg2
560°
562°
580°
558°
519°
508°
453±5°
(Mg)
~82·5
(36·4)
170°
18·5 (2·7)
-41°
0
10
20
30
40
50
60
70
80
90
100
At. % Mg

Hg–Mn

Wt. % Hg
98
98·5
99
99·2
99·4
99·6
99·8
100
°C
500
400
300
200
100
Liquid
93
94
95
96
97
98
99
100
At. % Hg

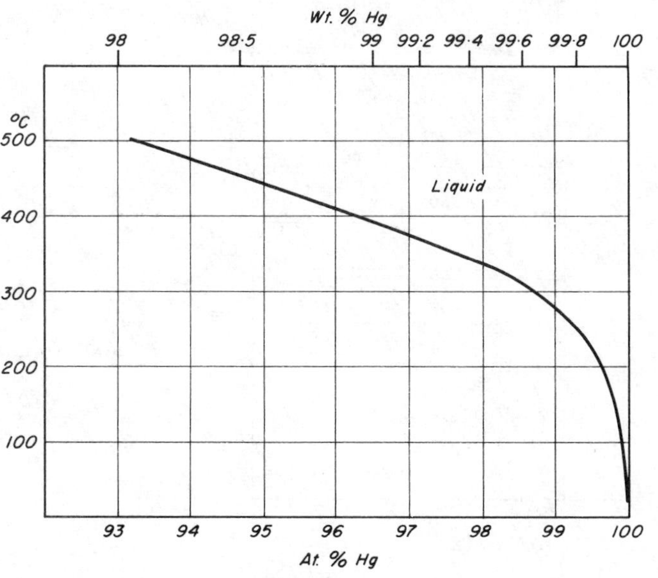

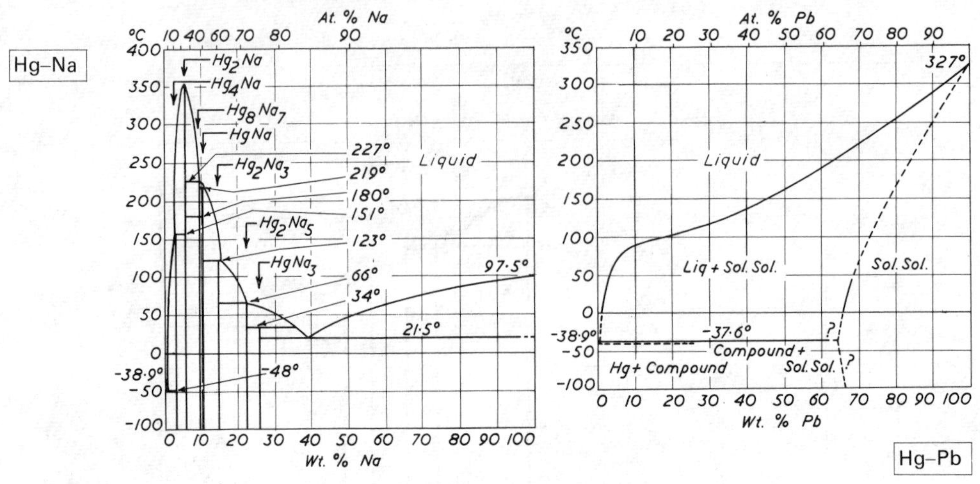
Hg–Na
At. % Na
Wt. % Na
Hg2Na
Hg4Na
Hg8Na7
HgNa
Hg2Na3
Hg2Na5
HgNa3
227°
219°
180°
151°
123°
66°
34°
97·5°
21·5°
-38·9°
-48°
Liquid
Hg–Pb
At. % Pb
Wt. % Pb
327°
Liquid
Liq + Sol. Sol.
Sol. Sol.
-38·9°
-37·6°
Compound +
Sol. Sol.
Hg + Compound

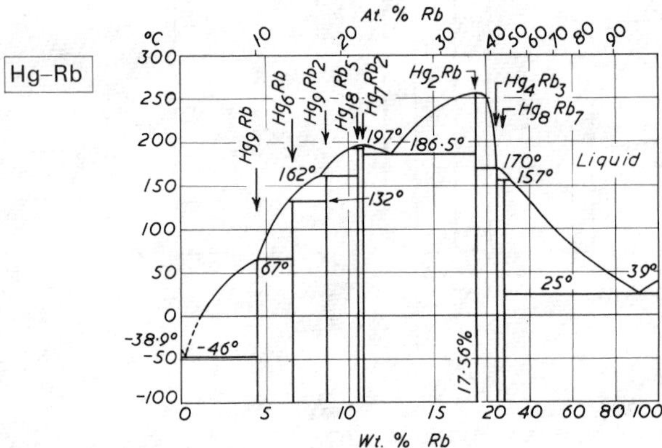
Hg–Rb
At. % Rb
Wt. % Rb
Hg9Rb
Hg6Rb
Hg9Rb2
Hg18Rb5
Hg7Rb2
Hg2Rb
Hg4Rb3
Hg8Rb7
197°
186·5°
162°
132°
170°
157°
Liquid
67°
25°
39°
-38·9°
-46°
17·56%

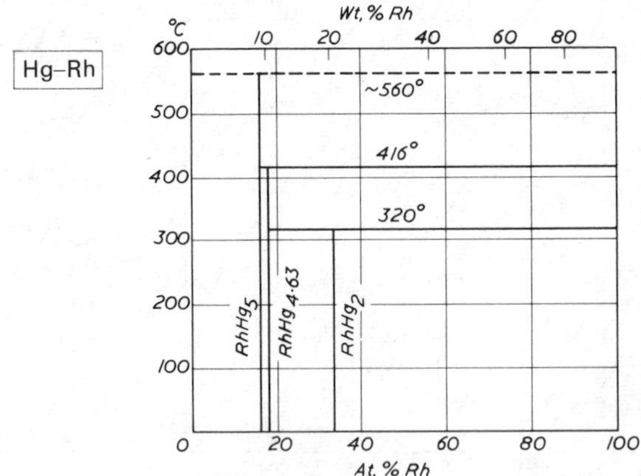
Hg–Rh
Wt, % Rh
At, % Rh
~560°
416°
320°
RhHg5
RhHg4·63
RhHg2

Hg–Sb

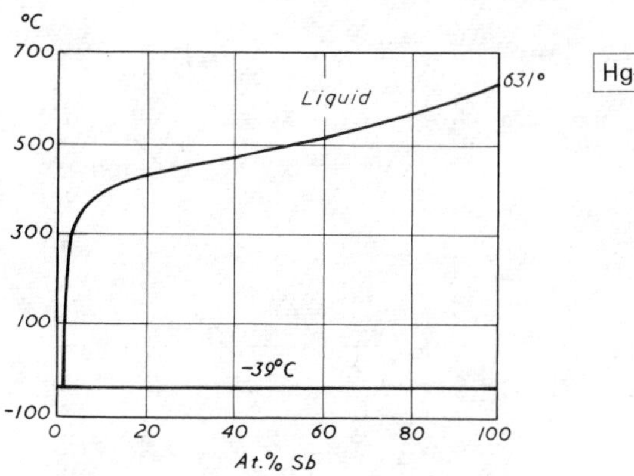

Hg–Sn

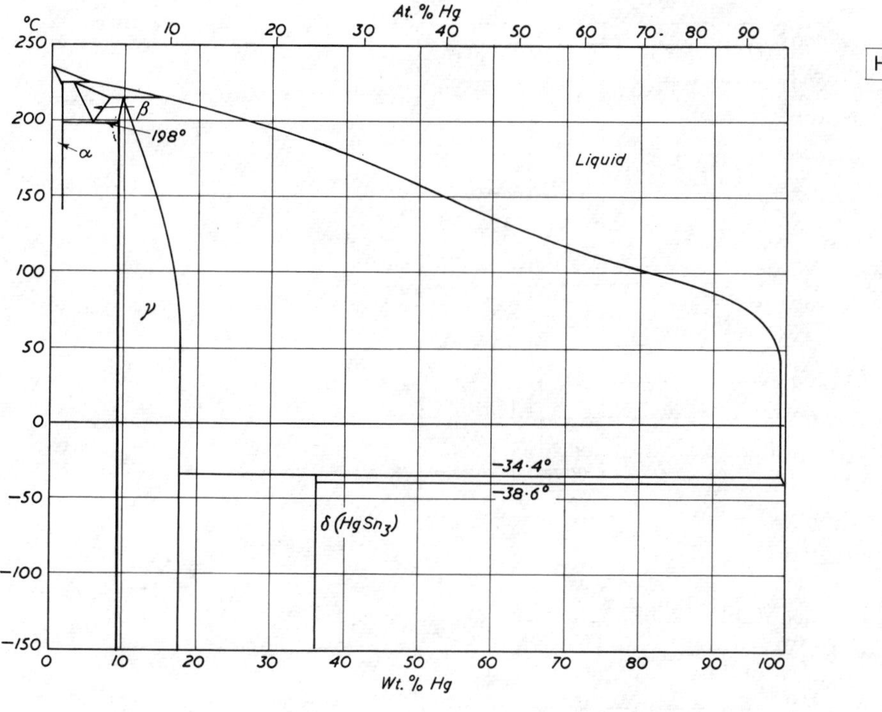

Hg–Te

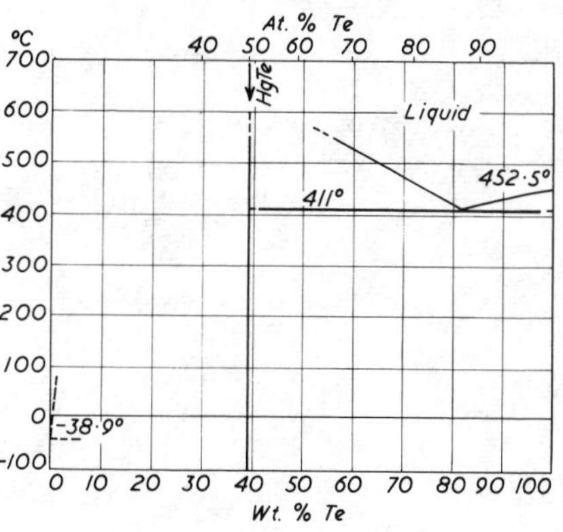

Hg–Tl

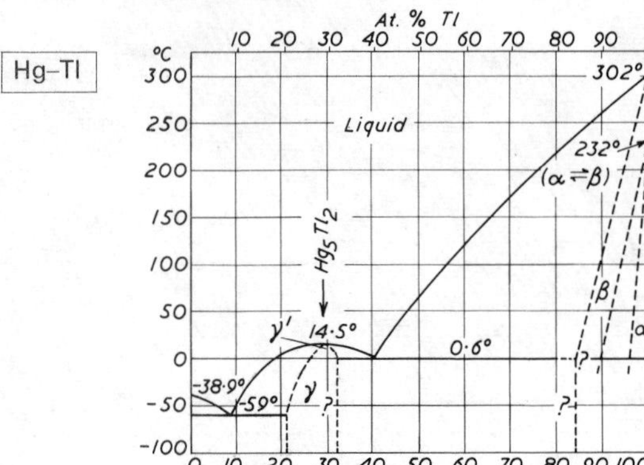

Hg–U

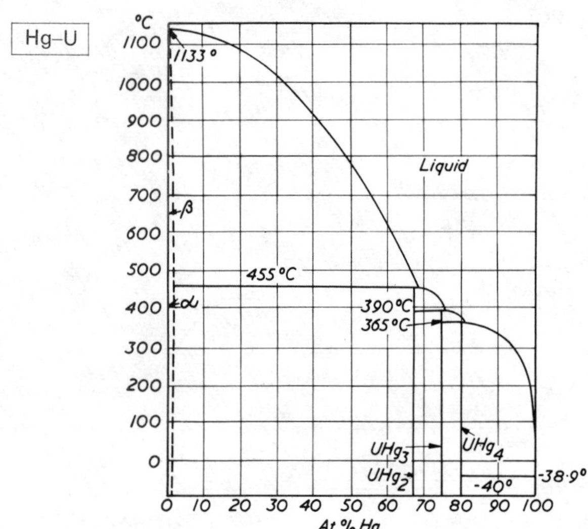

Hg–Zn

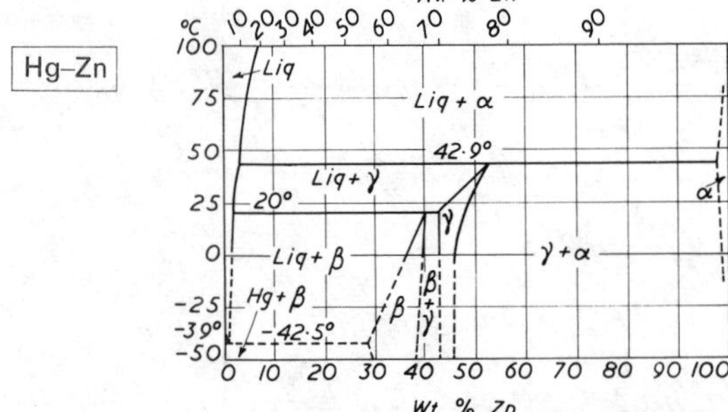

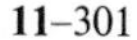
Ho–Tb

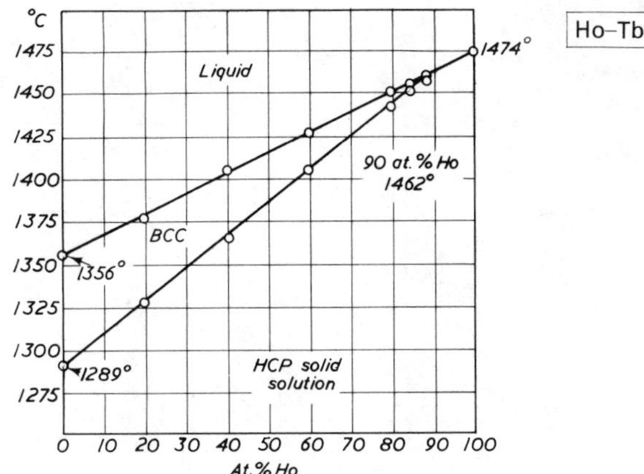
°C
1475
1450
1425
1400
1375
1350
1325
1300
1275
Liquid
1474°
90 at.% Ho
1462°
BCC
1356°
1289°
HCP solid
solution
0 10 20 30 40 50 60 70 80 90 100
At.% Ho

In–Li

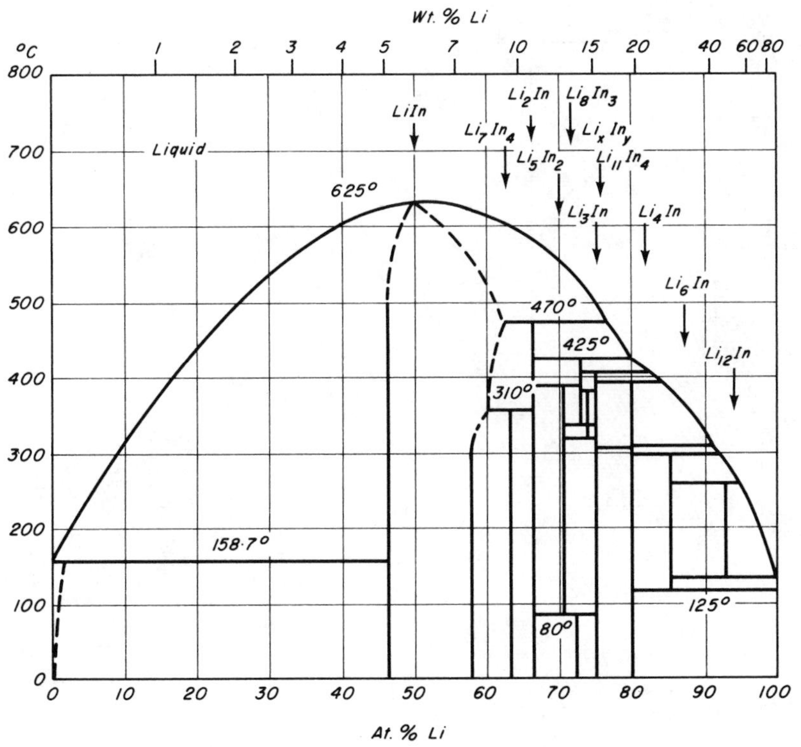
Wt. % Li
1 2 3 4 5 7 10 15 20 40 60 80
°C
800
700
600
500
400
300
200
100
0
Liquid
LiIn
Li_2In
Li_8In_3
Li_7In_4
Li_xIn_y
Li_5In_2
$Li_{11}In_4$
Li_3In
Li_4In
Li_6In
$Li_{12}In$
625°
470°
425°
310°
158·7°
125°
80°
0 10 20 30 40 50 60 70 80 90 100
At. % Li

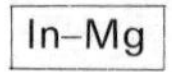
In–Mg

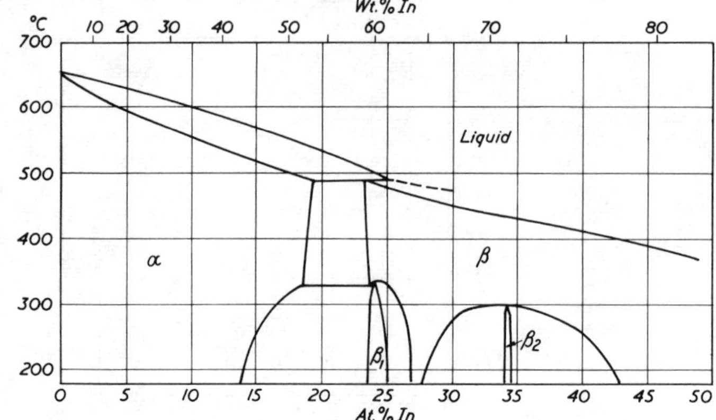
Wt.% In
°C
10 20 30 40 50 60 70 80
700
600
500
400
300
200
Liquid
α
β
β_1
β_2
0 5 10 15 20 25 30 35 40 45 50
At.% In

In–Mn

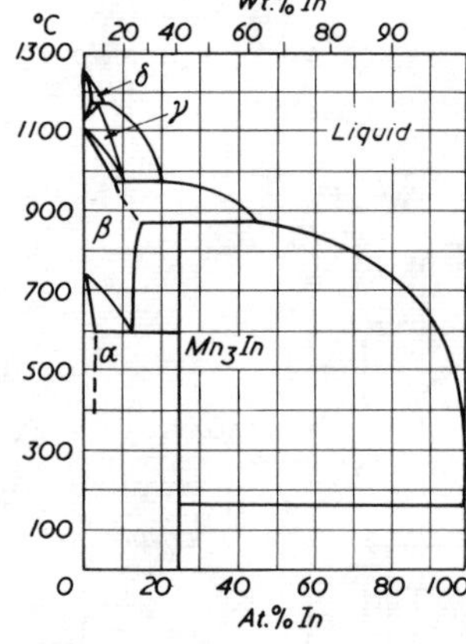
Wt.% In
°C
20 40 60 80 90
1300
1100
900
700
500
300
100
0
δ
γ
Liquid
β
α
Mn_3In
20 40 60 80 100
At.% In

In–Ni

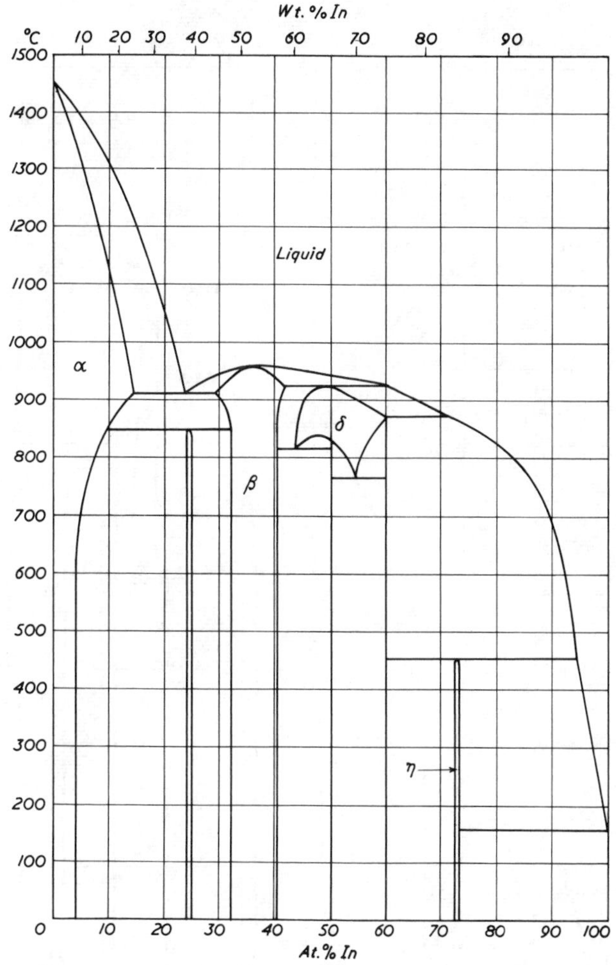

In–Pb

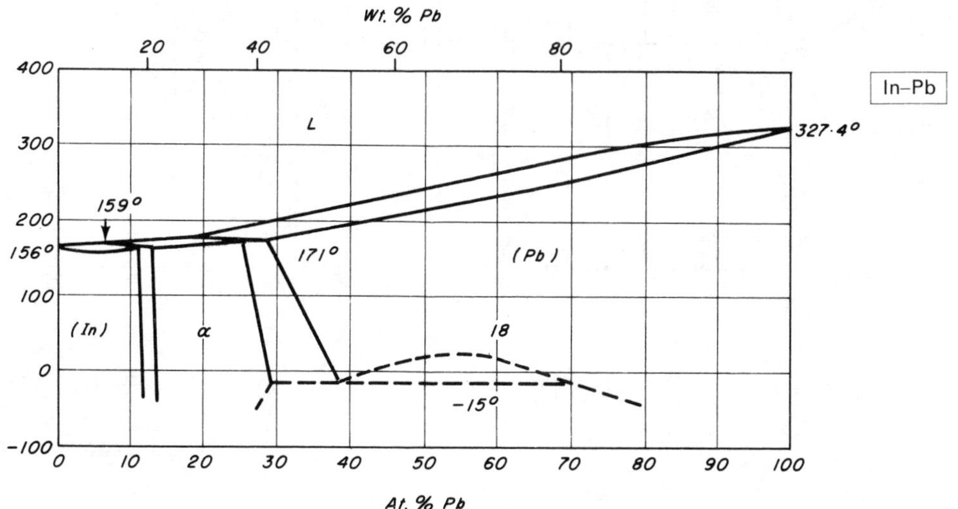
Wt. % Pb
20
40
60
80
400
300
200
100
0
−100
L
327·4°
159°
156°
171°
(Pb)
(In)
α
18
−15°
0
10
20
30
40
50
60
70
80
90
100
At. % Pb

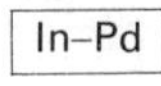

Wt.% Pd

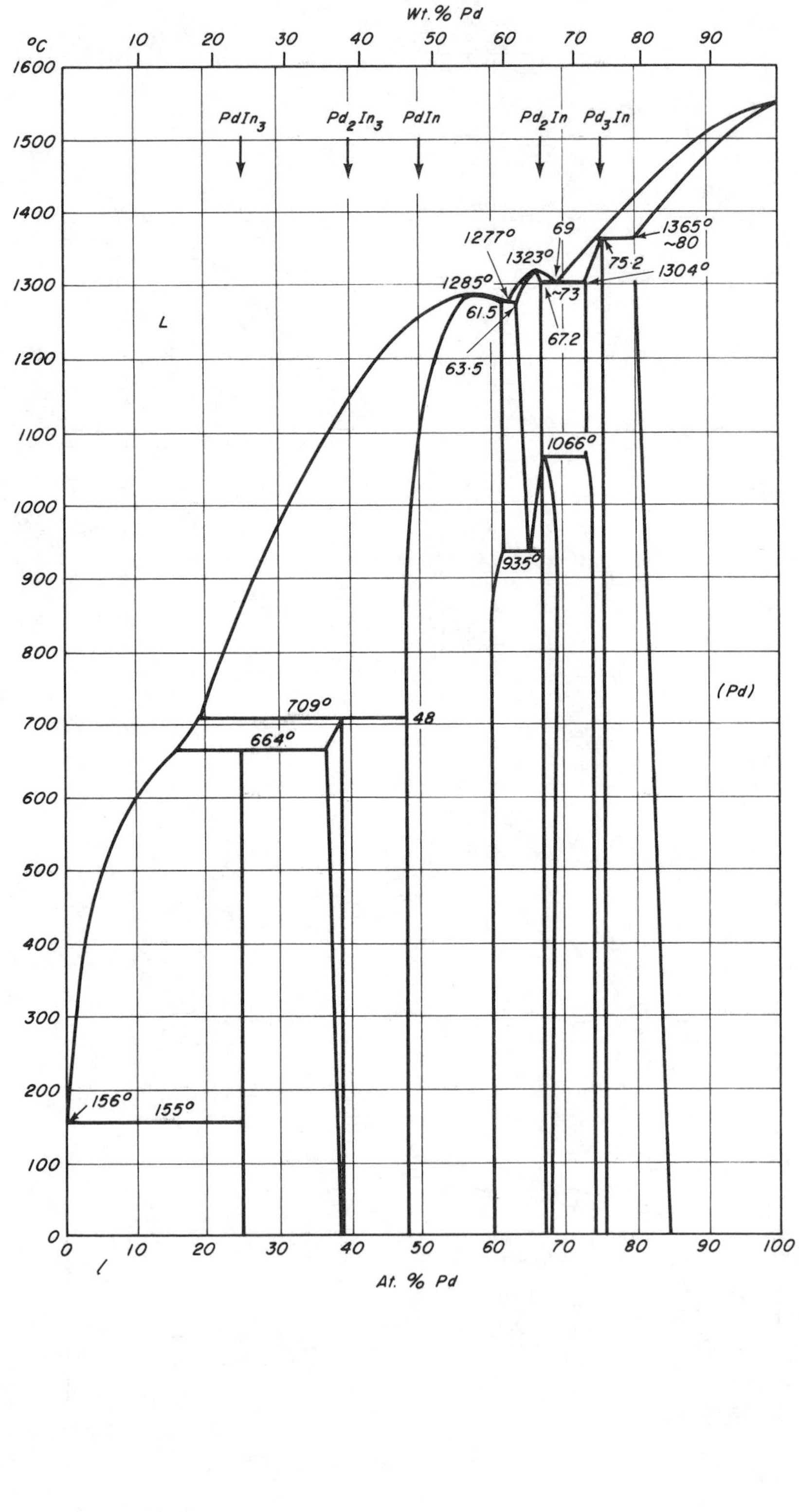

°C

PdIn$_3$ Pd$_2$In$_3$ PdIn Pd$_2$In Pd$_3$In

L

1277° 69 1365° ~80 1323° 75·2 1304° 1285° ~73 61.5 67.2 63·5

1066°

935°

(Pd)

709° 48

664°

156° 155°

At. % Pd

In–Pr

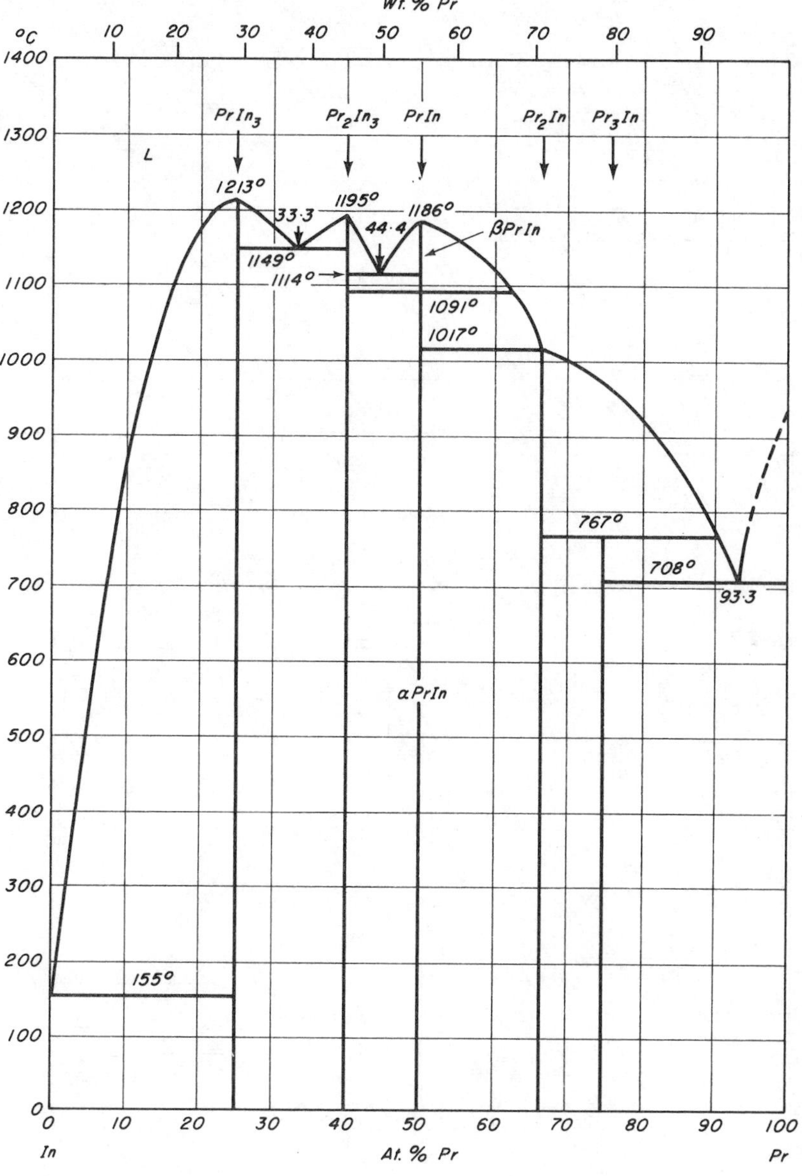
Wt. % Pr
°C
10
20
30
40
50
60
70
80
90
1400
1300
1200
1100
1000
900
800
700
600
500
400
300
200
100
0
L
PrIn3
Pr2In3
PrIn
Pr2In
Pr3In
1213°
33·3
1195°
1186°
44·4
βPrIn
1149°
1114°
1091°
1017°
767°
708°
93·3
αPrIn
155°
0
10
20
30
40
50
60
70
80
90
100
In
At. % Pr
Pr

In–Sb

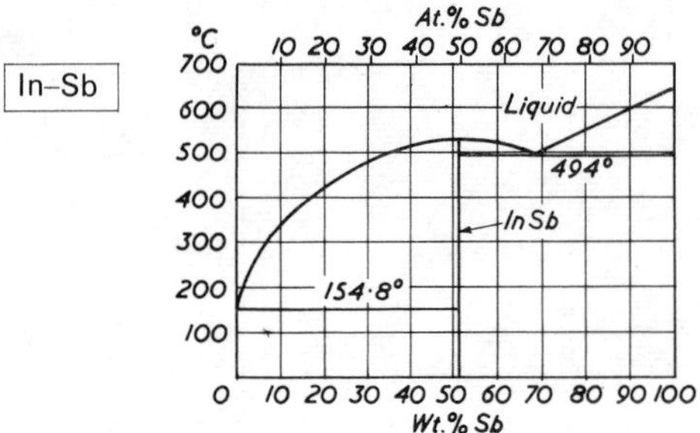

In–Se

Wt. %Se
°C
5 10 15 20 30 40 50 60 70 80 90
1000
900
800
700
600
500
400
300
200
100
In_5Se_6
In_2Se
InSe
In_2Se_3
900
Liquid
$L_3 + L_4$
$\delta - In_2Se_3$
?
~770°
660±10°
750±10°
$\gamma - In_2Se_3$
660±10°
650±10°
620°
51
$L_1 + L_2$
520°
550±10°
540±10°
$\beta - In_5Se_6$
$\beta - In_2Se_3$
$\alpha - In_5Se_6$
225°
227°
195°
155°
$\alpha - In_2Se_3$
0 10 20 30 40 50 60 70 80 90 100
At. %Se

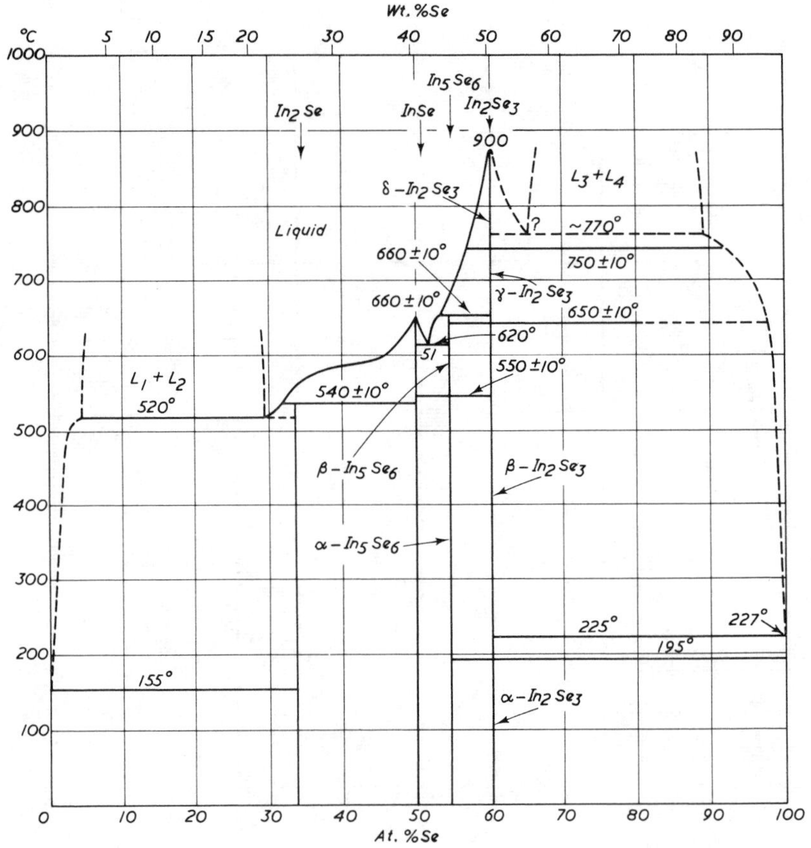

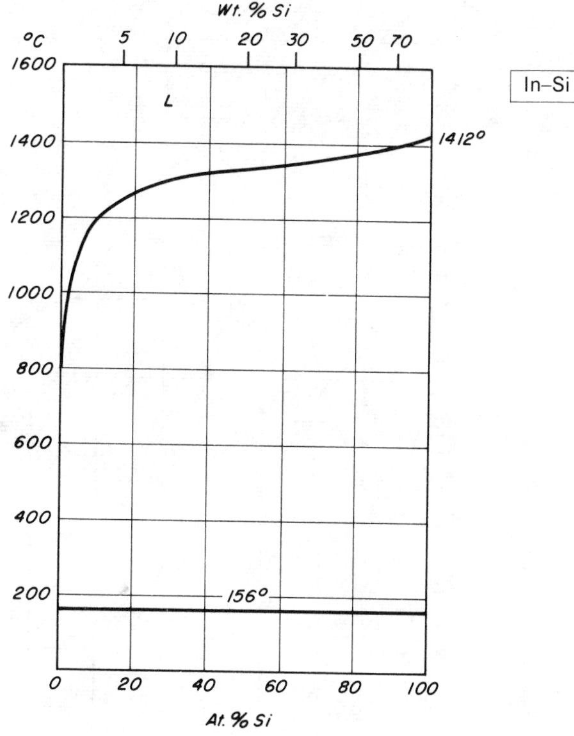
Wt. % Si
°C
5
10
20
30
50
70
1600
L
1400
1412°
1200
1000
800
600
400
200
156°
0
20
40
60
80
100
At. % Si

In–Si

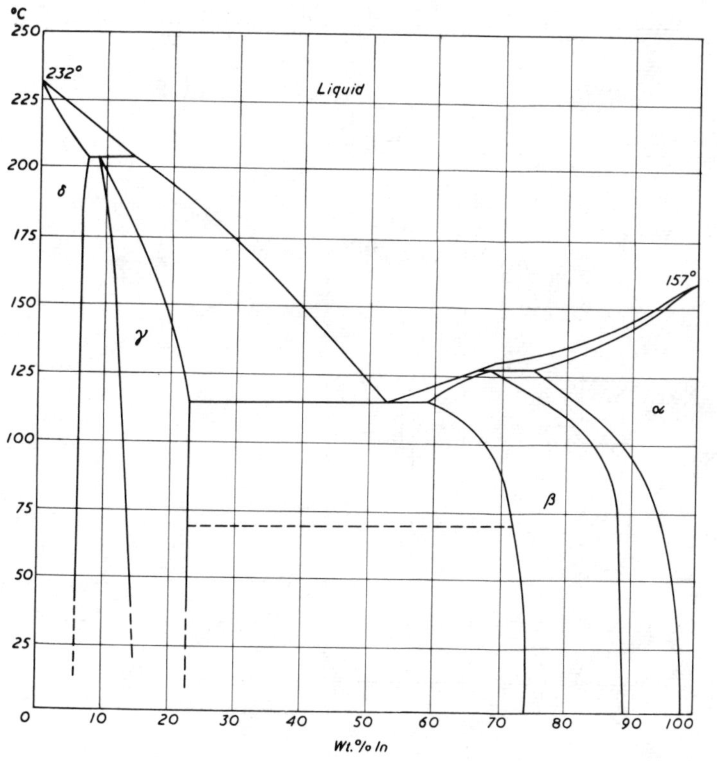
°C
250
232°
225
Liquid
200
δ
175
150
157°
γ
125
100
α
75
β
50
25
0
10
20
30
40
50
60
70
80
90
100
Wt.% In

In–Sn

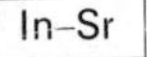
In–Sr

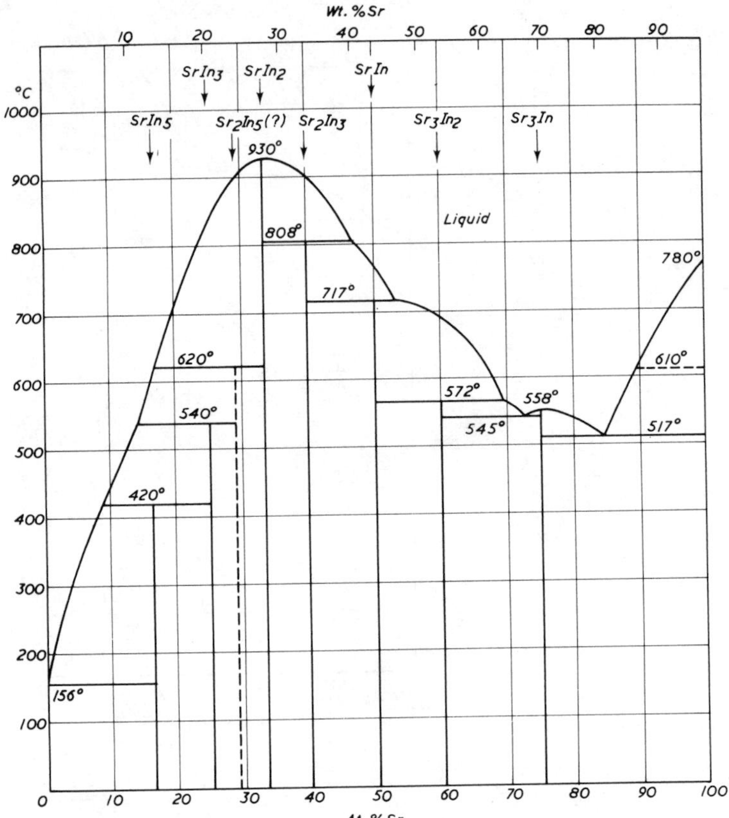
Wt. % Sr
10
20
30
40
50
60
70
80
90
°C
SrIn3
SrIn2
SrIn
SrIn5
Sr2In5(?)
Sr2In3
Sr3In2
Sr3In
930°
808°
Liquid
780°
717°
620°
610°
572°
558°
540°
545°
517°
420°
156°
1000
900
800
700
600
500
400
300
200
100
0
100
At. % Sr

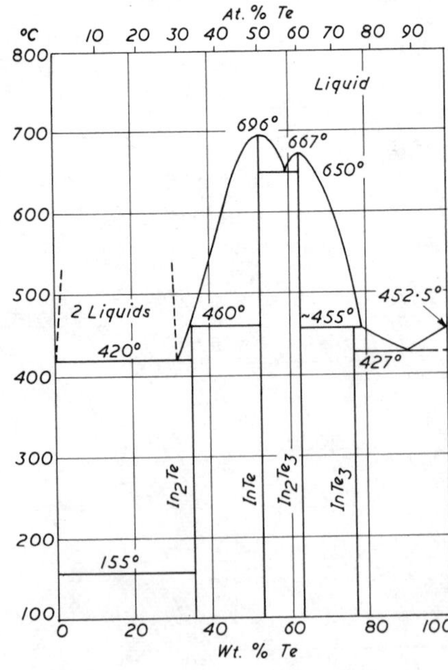
In–Te
At. % Te
°C
10 20 30 40 50 60 70 80 90
800
700
600
500
400
300
200
100
Liquid
696°
667°
650°
2 Liquids
460°
~455°
452·5°
420°
427°
In2Te
InTe
In2Te3
InTe3
155°
0 20 40 60 80 100
Wt. % Te

In–Ti

Wt. %Ti
°C
In_4Ti_3
β-Ti
796 ± 5°
42·8
?
α_2
α-Ti
156 ± 1°
At. %Ti

In–Tl

Wt. %Tl
°C
Liquid
303°
235°
156·4°
171·5°
53
~43
BCC
(β-Tl)
Face-centred
tetragonal
FCC
α
HCP
(α-Tl)
30°
59·9
82·5
87·5
At. %Tl

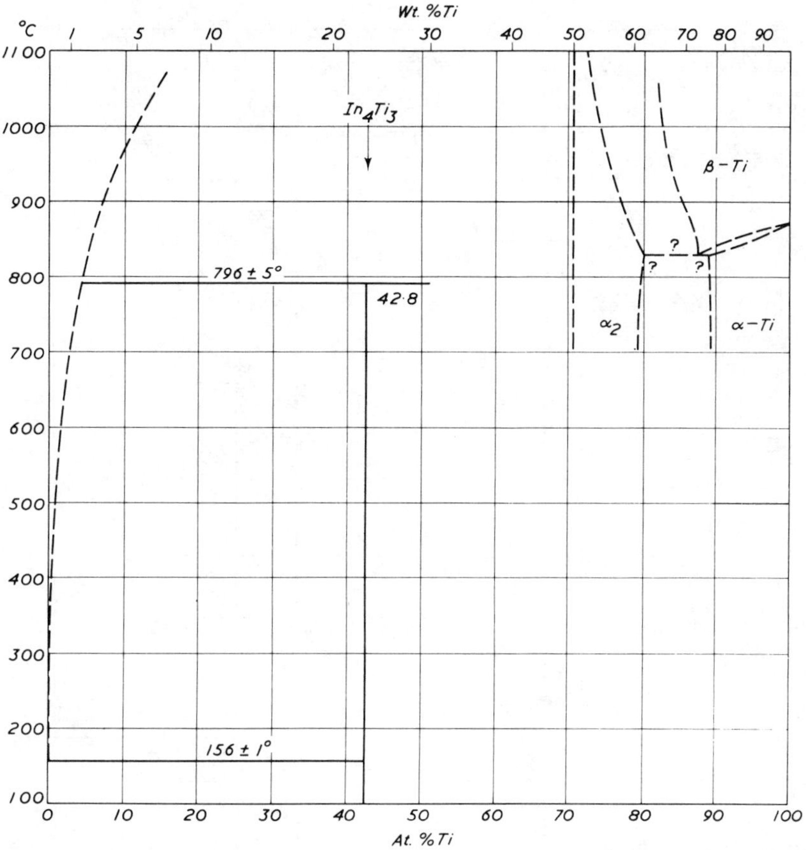

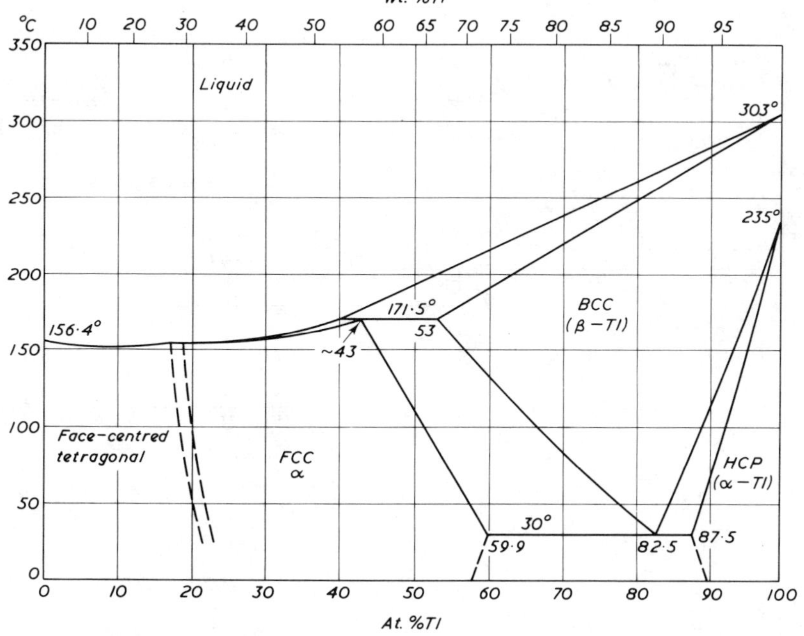

In–Yb

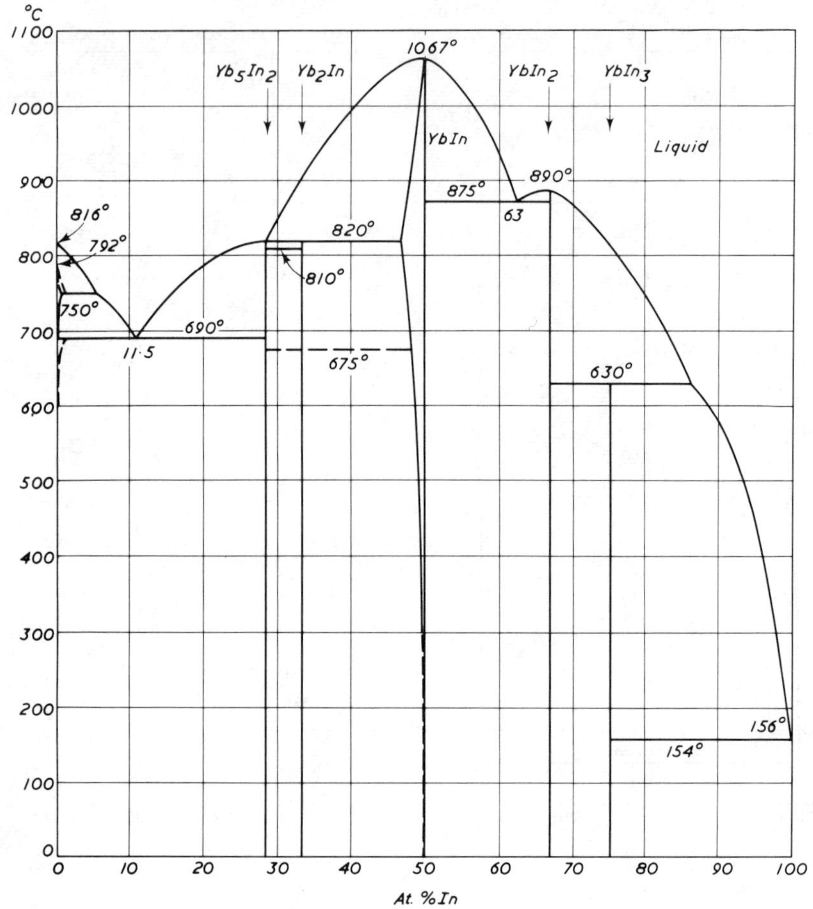

°C
1100
1000
900
800
700
600
500
400
300
200
100
0

Yb_5In_2 Yb_2In 1067° $YbIn_2$ $YbIn_3$

YbIn

Liquid

816° 792° 750° 690° 11·5 820° 810° 675° 875° 63 890° 630° 156° 154°

0 10 20 30 40 50 60 70 80 90 100

At. % In

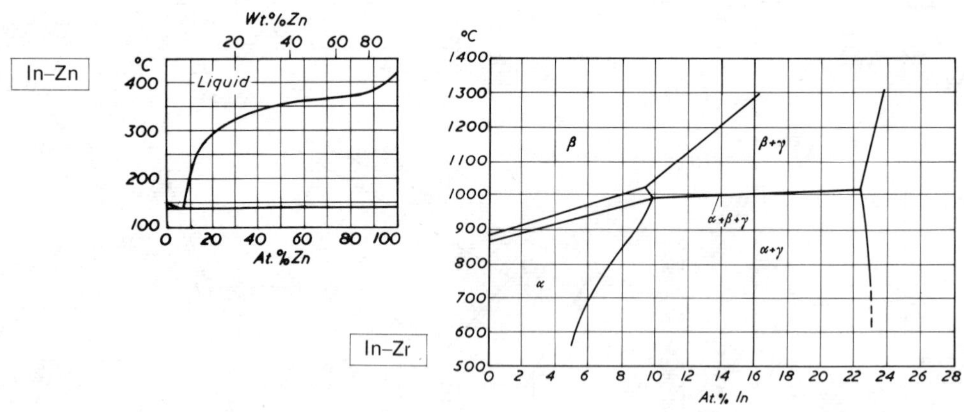

Ir–Mn

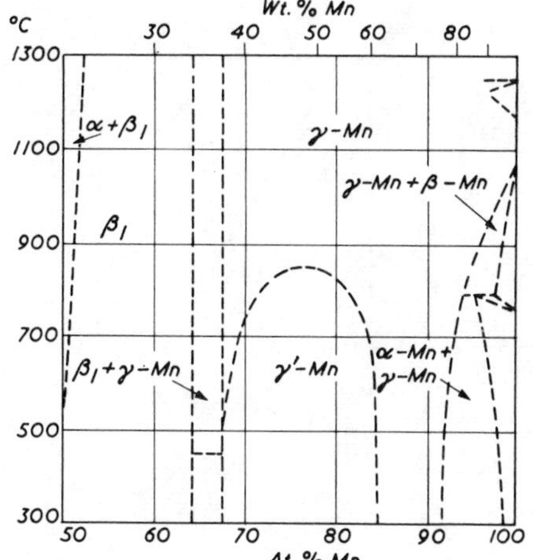
Wt. % Mn
°C
30
40
50
60
80
1300
1100
900
700
500
300
α+β₁
γ-Mn
γ-Mn+β-Mn
β₁
β₁+γ-Mn
γ'-Mn
α-Mn+γ-Mn
50
60
70
80
90
100
At. % Mn

Ir–Mo

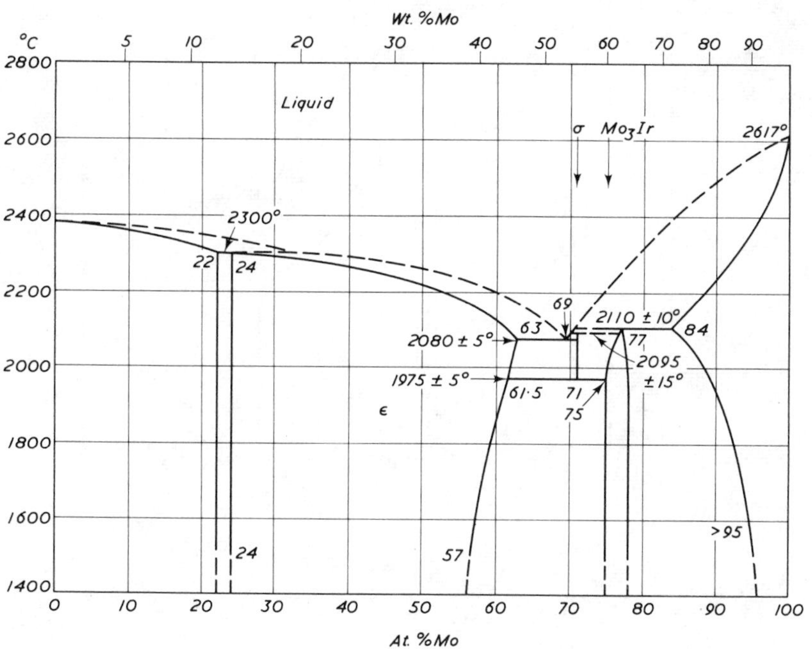
Wt. %Mo
°C
2800
2600
2400
2200
2000
1800
1600
1400
Liquid
σ
Mo₃Ir
2617°
2300°
22
24
69
2110 ± 10°
84
63
2080 ± 5°
77
2095
± 15°
1975 ± 5°
61·5
71
75
ϵ
>95
57
24
0
10
20
30
40
50
60
70
80
90
100
At. %Mo

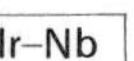
Ir–Nb

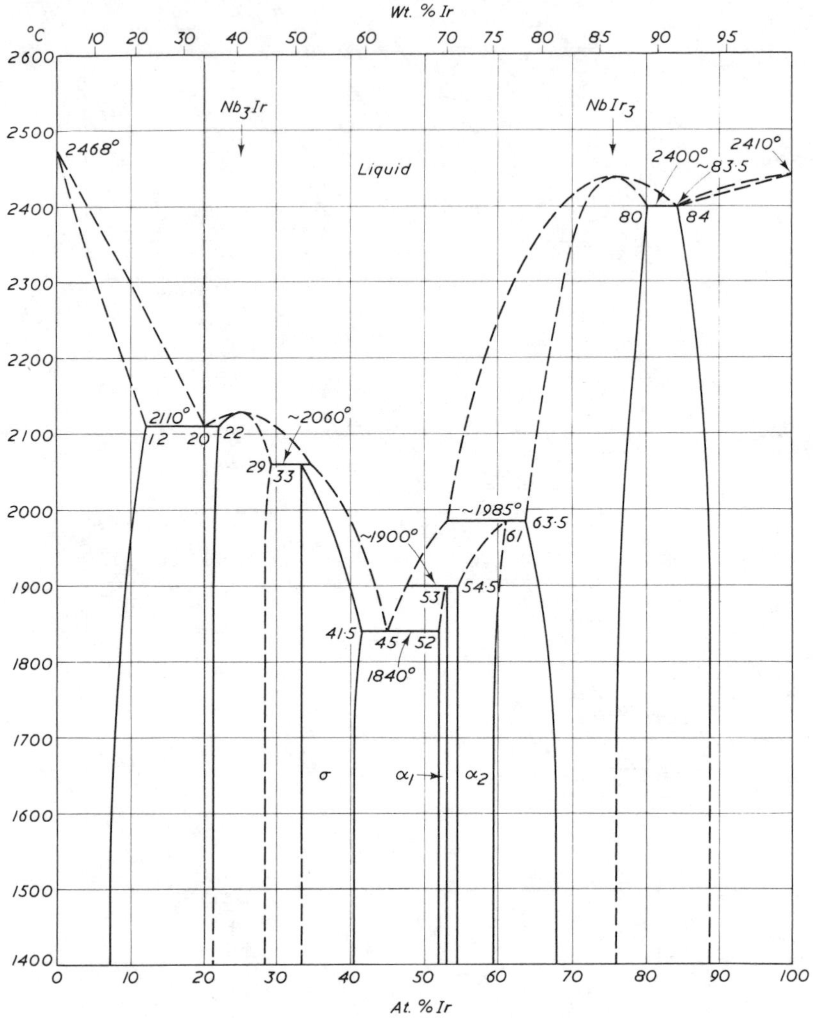

Wt. % Ir
°C
10 20 30 40 50 60 70 75 80 85 90 95
2600
2500
2400
2300
2200
2100
2000
1900
1800
1700
1600
1500
1400
Nb3Ir
NbIr3
Liquid
2468°
2400°
2410°
~83·5
80
84
2110°
12
20
22
~2060°
29
33
~1985°
63·5
61
~1900°
53
54·5
41·5
45
52
1840°
σ
α1
α2
0 10 20 30 40 50 60 70 80 90 100
At. % Ir

Ir–Os

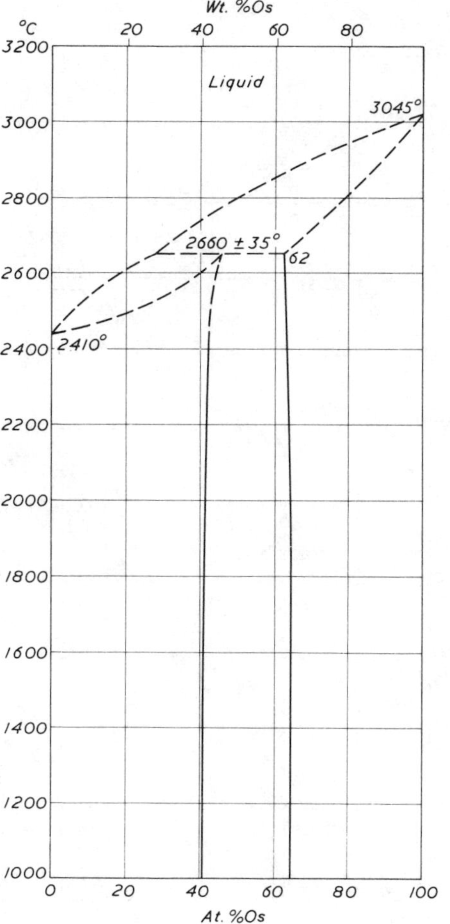
Wt. %Os
°C
20
40
60
80
3200
3000
2800
2600
2400
2200
2000
1800
1600
1400
1200
1000
Liquid
3045°
2660 ± 35°
62
2410°
0
20
40
60
80
100
At. %Os

Ir–Pd

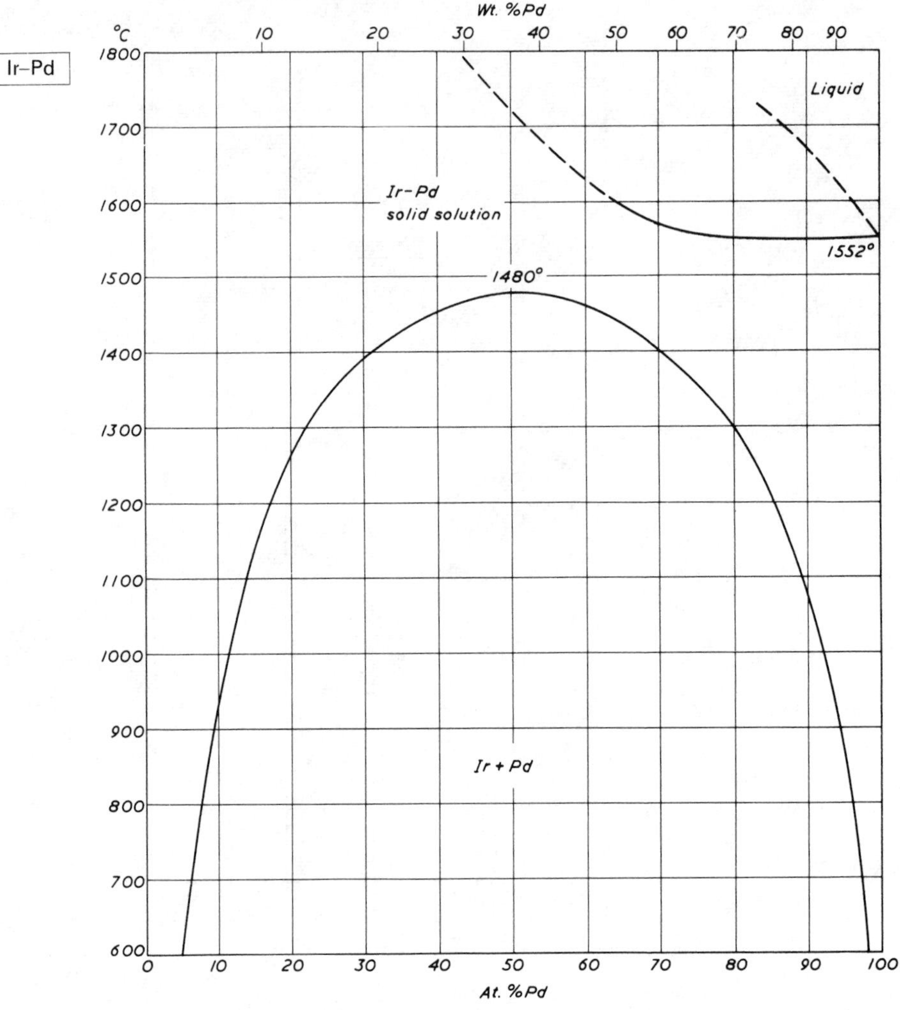
Wt. %Pd
°C
10
20
30
40
50
60
70
80
90
1800
1700
1600
1500
1400
1300
1200
1100
1000
900
800
700
600
Liquid
Ir–Pd
solid solution
1552°
1480°
Ir + Pd
0
10
20
30
40
50
60
70
80
90
100
At. %Pd

Ir–Pt

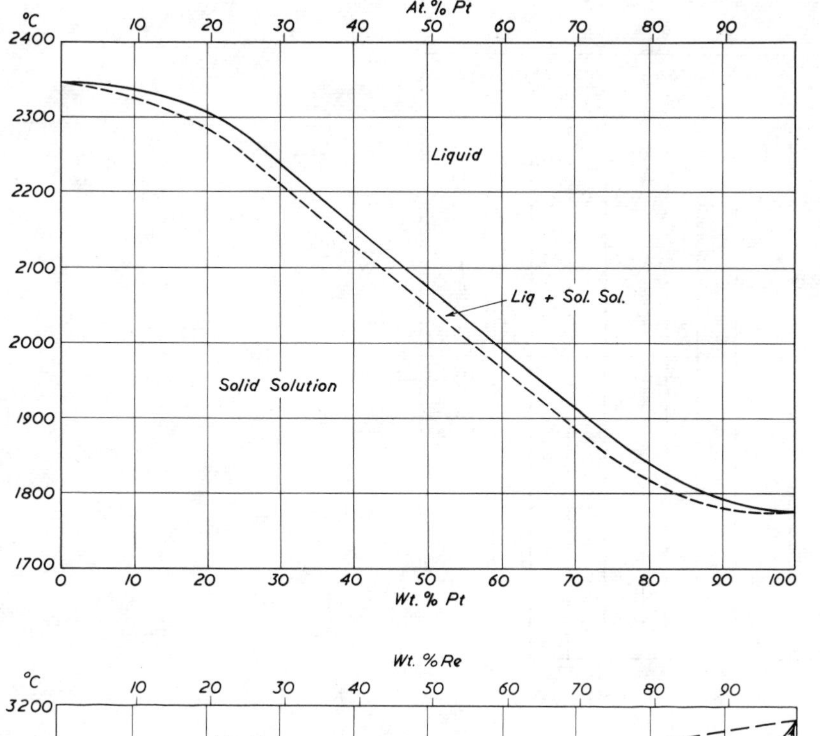

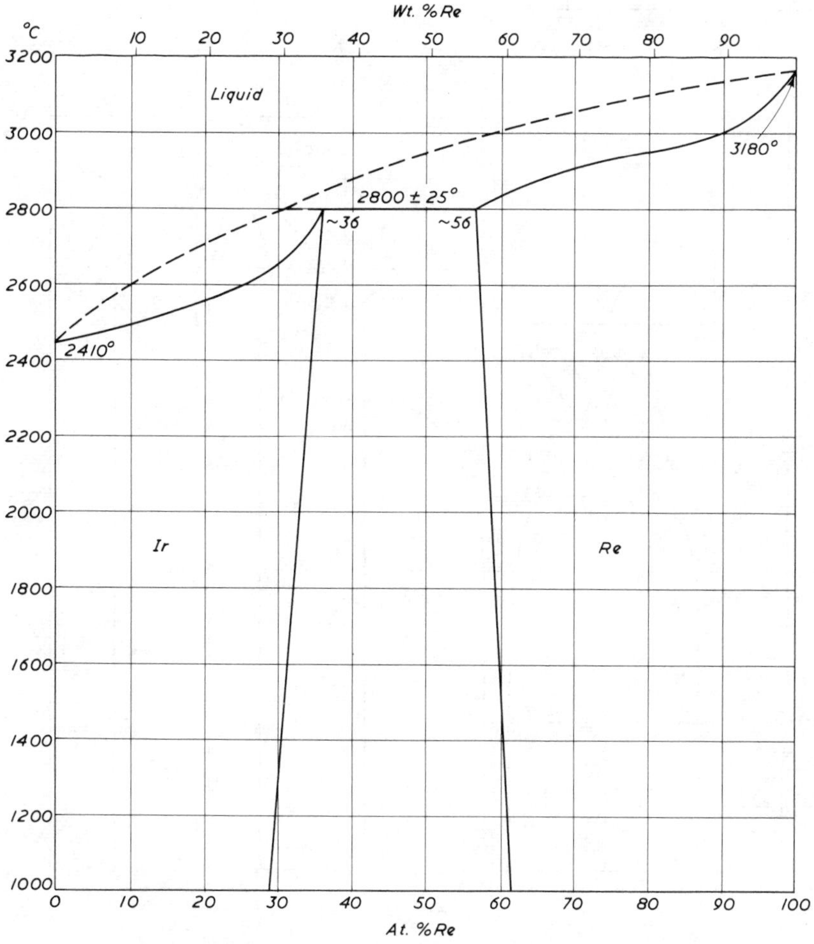

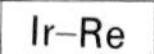

Wt. % Re

10 20 30 40 50 60 70 80 90

°C

3200 3000 2800 2600 2400 2200 2000 1800 1600 1400 1200 1000

Liquid

2800 ± 25°

~36

~56

3180°

2410°

Ir

Re

0 10 20 30 40 50 60 70 80 90 100

At. % Re

Ir–Th

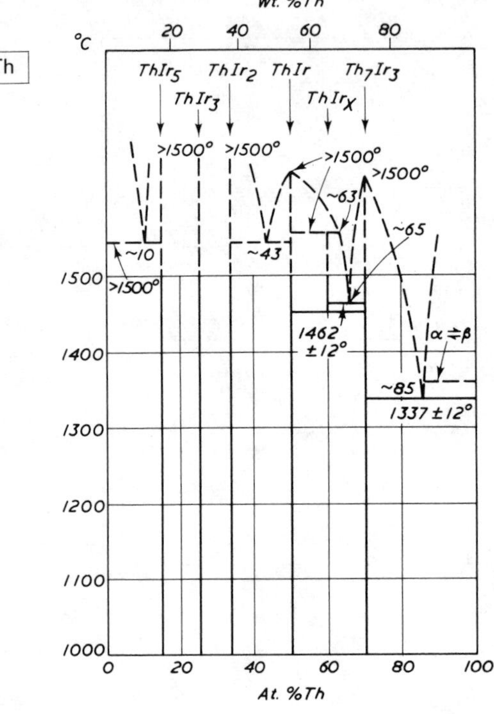

Wt. %Th
20
40
60
80
°C
ThIr5
ThIr3
ThIr2
ThIr
ThIrX
Th7Ir3
>1500°
>1500°
>1500°
>1500°
~63
~65
~10
~43
>1500°
1500
1400
1300
1200
1100
1000
1462
±12°
α⇌β
~85
1337±12°
0
20
40
60
80
100
At. %Th

Ir–Ti

°C
1000
800
600
β
α
0
4
8
At.%Ir
Liquid
2450°
2315°
2120°
2000°
2100
1700
1650°
δ
δ
ε
η
1300
β
900
α
500
0
20
40
60
80
100
At. %Ir

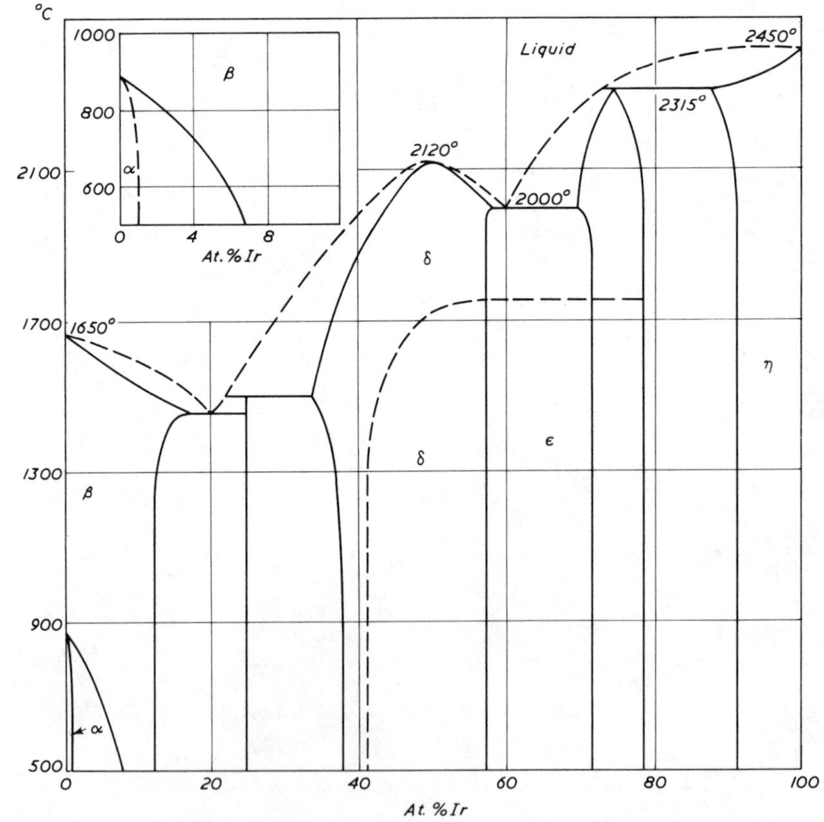

Ir–W

Wt. %W
°C
10 20 30 40 50 60 70 80 90
3400
3200
3000
2800
2600
2400
2200
2000
1800
1600
1400
1200
1000
Liquid
3410°
2540 ± 25°
~74
~90
~75
2490°
~58
~66
~70
2460 ± 25°
2305 ± 25°
2378°
~19
~22
~21
σ
ε
57
75
1810 ± 20°
96
0 10 20 30 40 50 60 70 80 90 100
At. %W

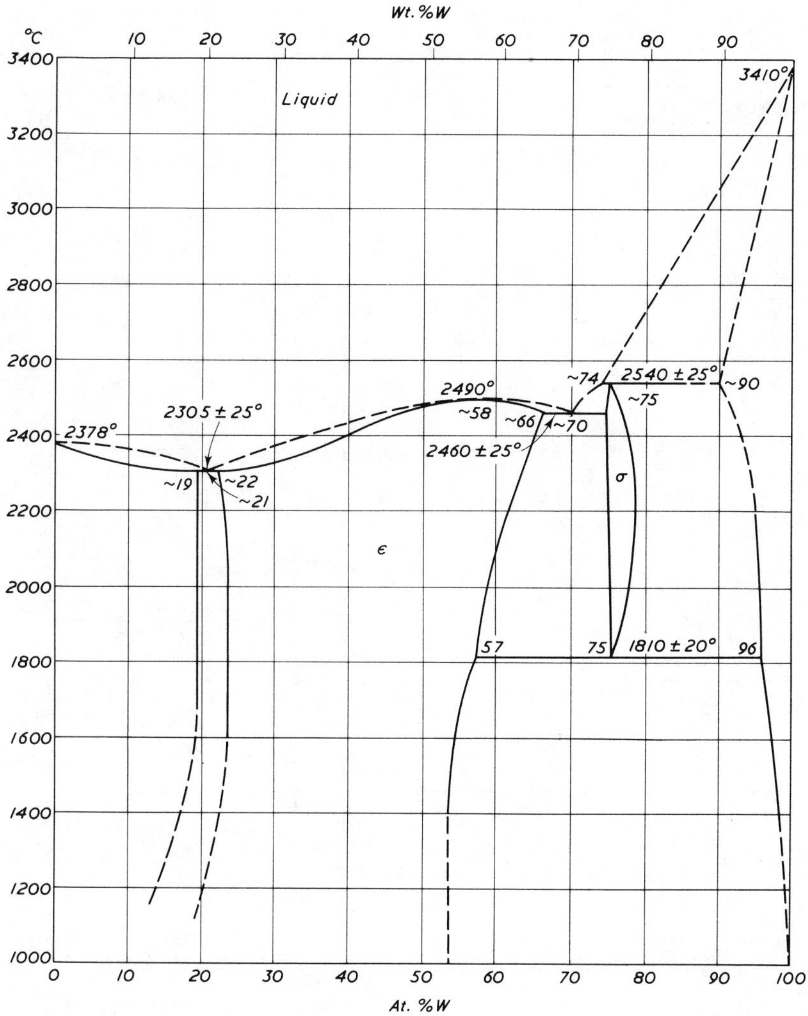

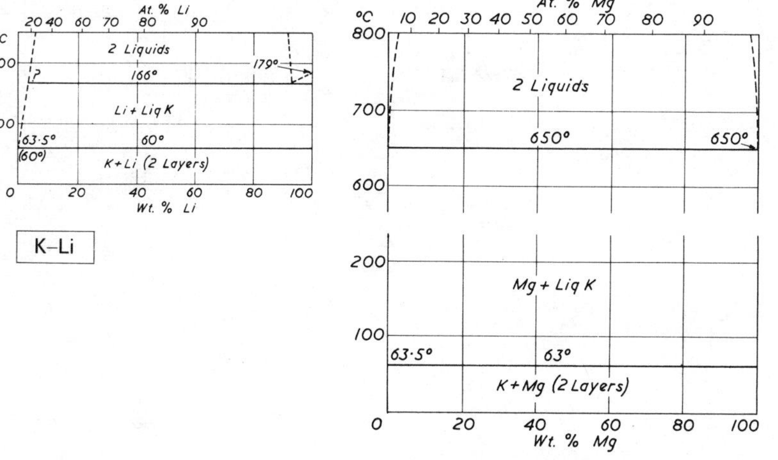

At. % Li
20 40 60 70 80 90
°C
200
100
2 Liquids
179°
?
166°
Li + Liq K
63·5°
(60°)
60°
K + Li (2 Layers)
0 20 40 60 80 100
Wt. % Li

K–Li

K–Mg

At. % Mg
°C
10 20 30 40 50 60 70 80 90
800
700
600
2 Liquids
650°
650°
200
100
Mg + Liq K
63·5°
63°
K + Mg (2 Layers)
0 20 40 60 80 100
Wt. % Mg

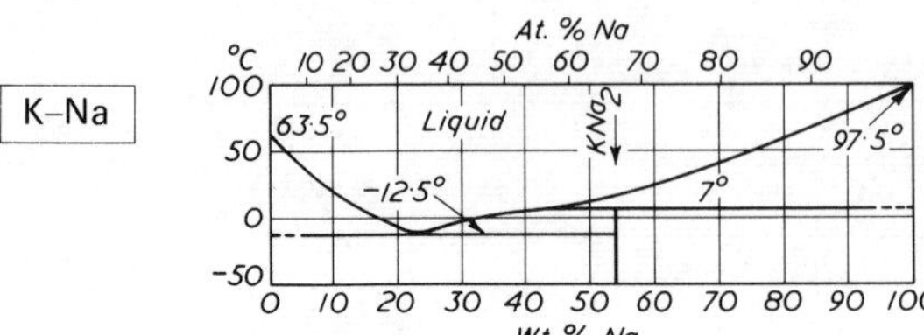
K–Na
At. % Na
°C
Liquid
KNa_2
63·5°
−12·5°
7°
97·5°
Wt % Na

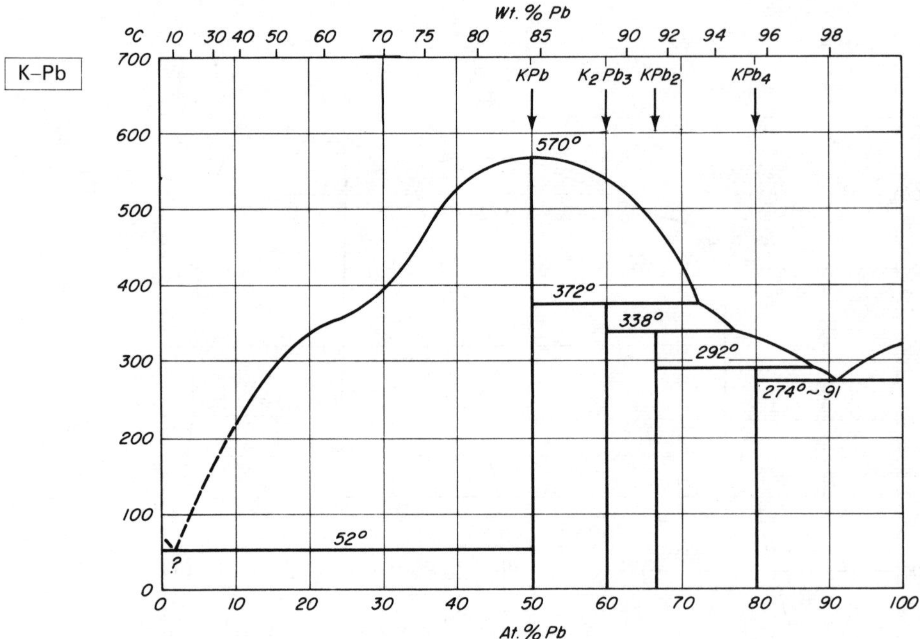
K–Pb
Wt. % Pb
°C
KPb
K_2Pb_3
KPb_2
KPb_4
570°
372°
338°
292°
274°~91
52°
?
At. % Pb

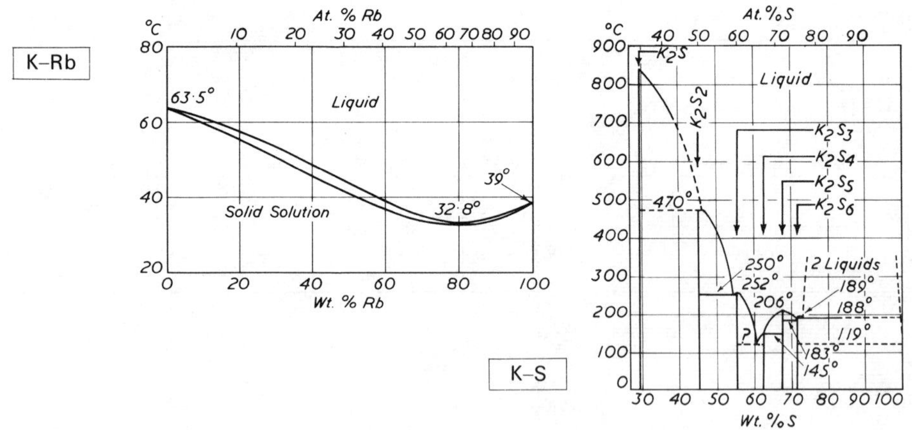
K–Rb
At. % Rb
°C
63·5°
Liquid
39°
Solid Solution
32·8°
Wt. % Rb
K–S
At. % S
°C
K_2S
Liquid
K_2S_2
K_2S_3
K_2S_4
K_2S_5
K_2S_6
470°
250°
252°
206°
2 Liquids
189°
188°
119°
?
183°
145°
Wt. % S

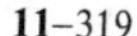

K–Sb

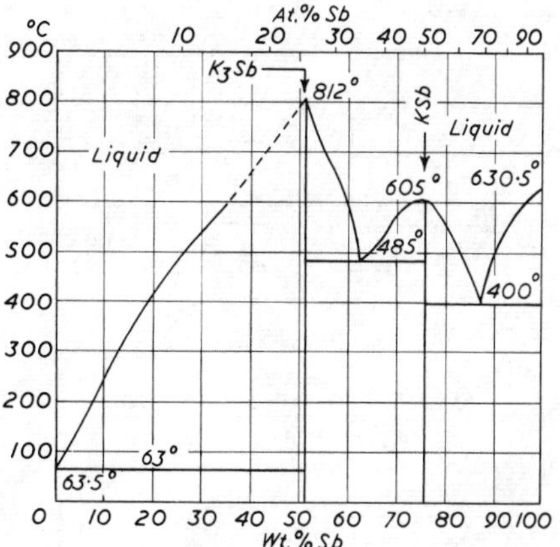
°C
At.% Sb
10 20 30 40 50 70 90
900
800
700
600
500
400
300
200
100
0
K_3Sb
812°
KSb
Liquid
Liquid
605°
630·5°
485°
400°
63°
63·5°
10 20 30 40 50 60 70 80 90 100
Wt.% Sb

K–Sn

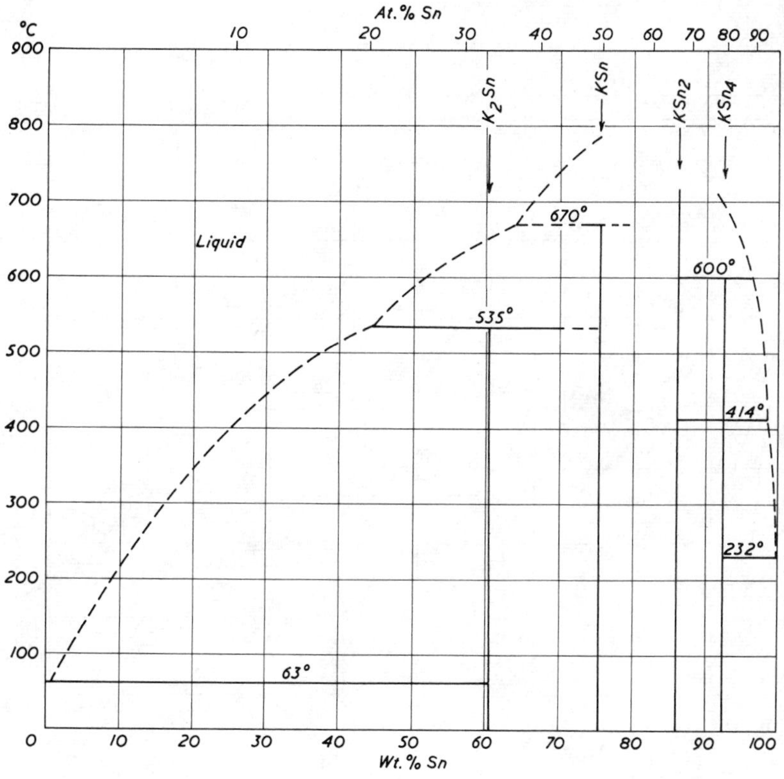
°C
At.% Sn
10 20 30 40 50 60 70 80 90
900
800
700
600
500
400
300
200
100
0
K_2Sn
KSn
KSn_2
KSn_4
Liquid
670°
600°
535°
414°
232°
63°
10 20 30 40 50 60 70 80 90 100
Wt.% Sn

K–Tl

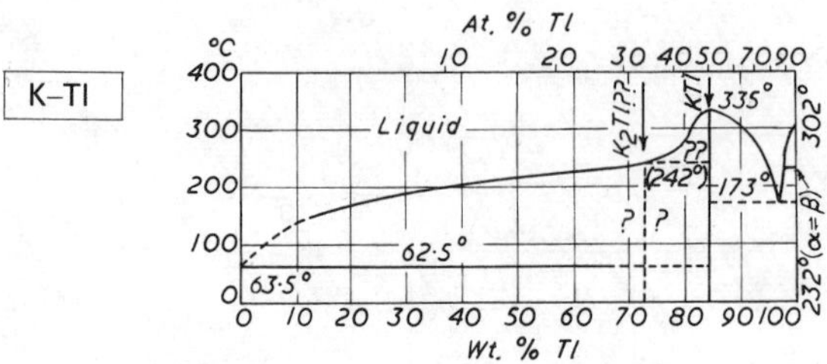
At. % Tl
°C
400
300
200
100
0
10
20
30
40
50
70
90
Liquid
K_2Tl ??
KTl
335°
??
(242°)
173°
302°
? ?
62·5°
63·5°
232°(α⇌β)
0 10 20 30 40 50 60 70 80 90 100
Wt. % Tl

K–Zn

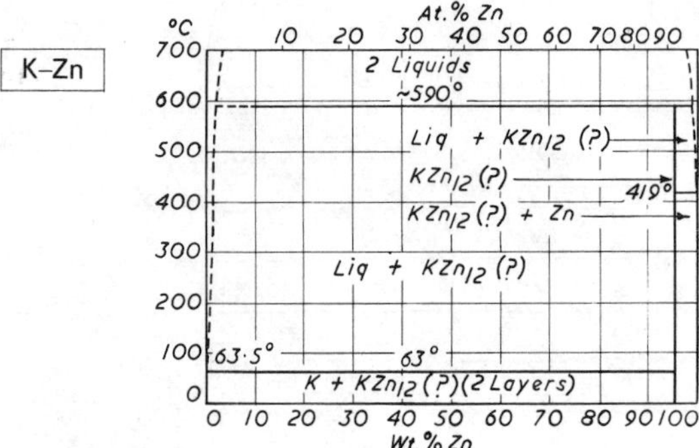
At.% Zn
°C
700
600
500
400
300
200
100
0
10 20 30 40 50 60 70 80 90
2 Liquids
~590°
Liq + KZn_{12} (?)
KZn_{12} (?)
419°
KZn_{12} (?) + Zn
Liq + KZn_{12} (?)
63·5°
63°
K + KZn_{12} (?)(2 Layers)
0 10 20 30 40 50 60 70 80 90 100
Wt.% Zn

La–Mg

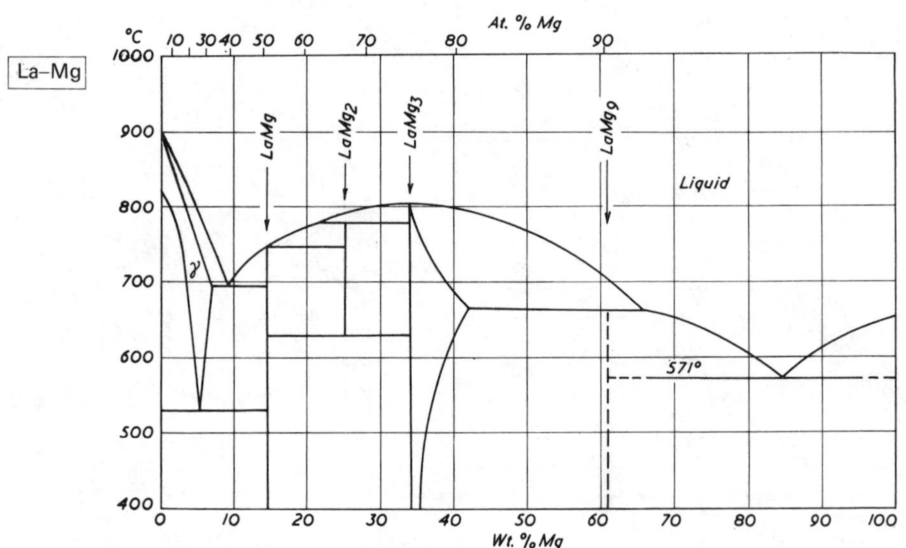
At. % Mg
°C
1000
900
800
700
600
500
400
10 30 40 50 60 70 80 90
LaMg
$LaMg_2$
$LaMg_3$
$LaMg_9$
Liquid
γ
571°
0 10 20 30 40 50 60 70 80 90 100
Wt. % Mg

La–Mn

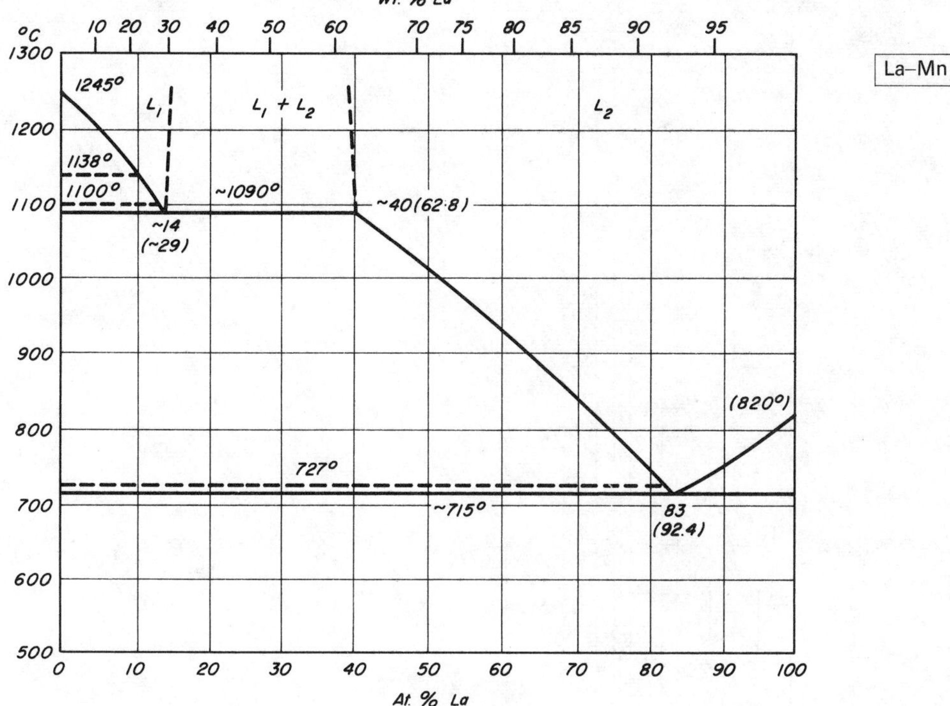

Wt. % La
°C
10 20 30 40 50 60 70 75 80 85 90 95
1300
1200
1100
1000
900
800
700
600
500
1245°
L_1
$L_1 + L_2$
L_2
1138°
1100°
~1090°
~40(62·8)
~14
(~29)
(820°)
727°
~715°
83
(92.4)
0 10 20 30 40 50 60 70 80 90 100
At. % La

La–Nd

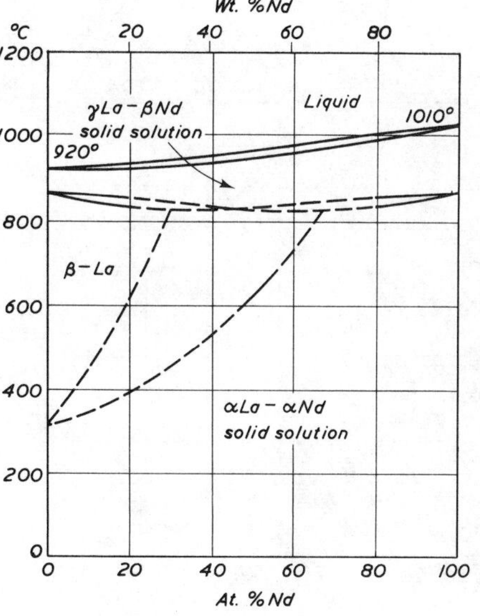

Wt. %Nd
°C
20 40 60 80
1200
1000
800
600
400
200
0
γLa – βNd
solid solution
Liquid
1010°
920°
β–La
αLa – αNd
solid solution
0 20 40 60 80 100
At. %Nd

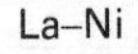
La–Ni

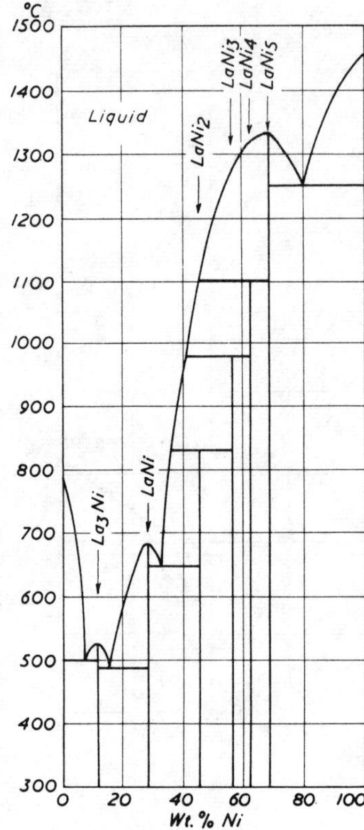
°C
Liquid
La3Ni
LaNi
LaNi2
LaNi3
LaNi4
LaNi5
Wt. % Ni

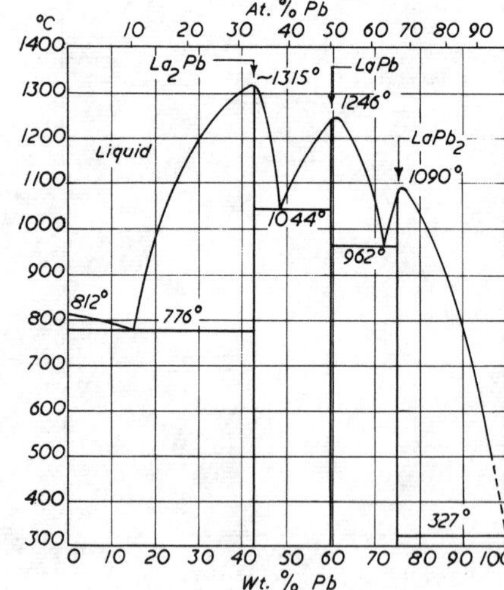
La–Pb
°C
At. % Pb
Liquid
La2Pb
~1315°
LaPb
1246°
LaPb2
1090°
1044°
962°
812°
776°
327°
Wt. % Pb

La–Rh

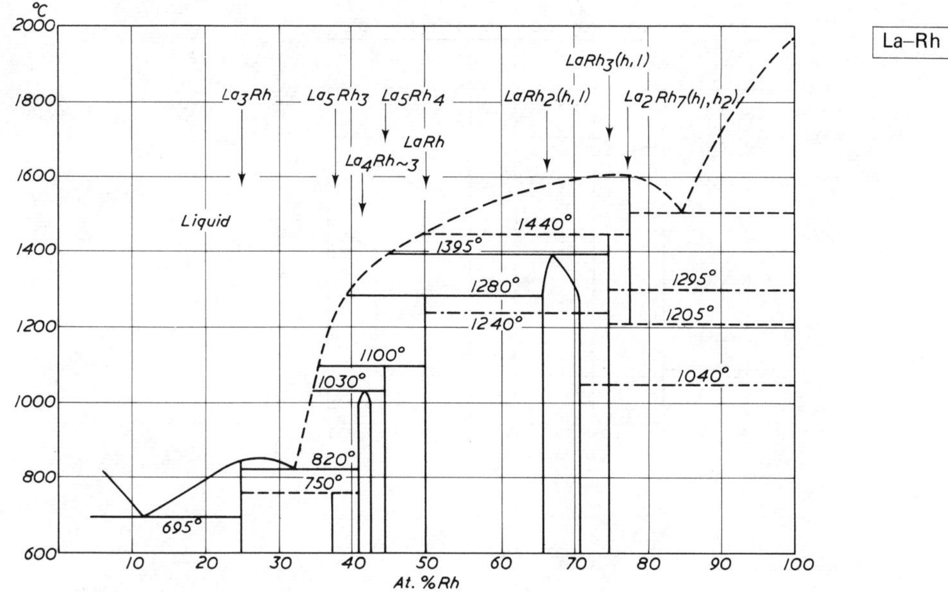
°C
2000
1800
1600
1400
1200
1000
800
600
La3Rh
La5Rh3
La5Rh4
LaRh
La4Rh~3
LaRh2(h, l)
LaRh3(h, l)
La2Rh7(h1, h2)
Liquid
1440°
1395°
1280°
1240°
1295°
1205°
1100°
1030°
1040°
820°
750°
695°
10
20
30
40
50
60
70
80
90
100
At. %Rh

La–Sb

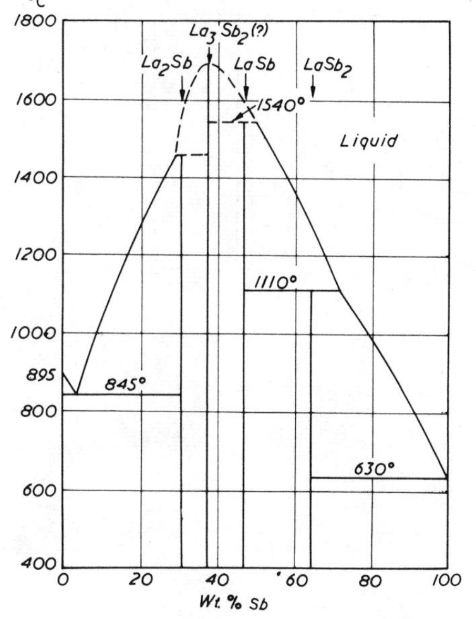
°C
1800
1600
1400
1200
1000
895
800
600
400
La3Sb2(?)
La2Sb
LaSb
LaSb2
1540°
Liquid
1110°
845°
630°
0
20
40
60
80
100
Wt. % Sb

La–Sn

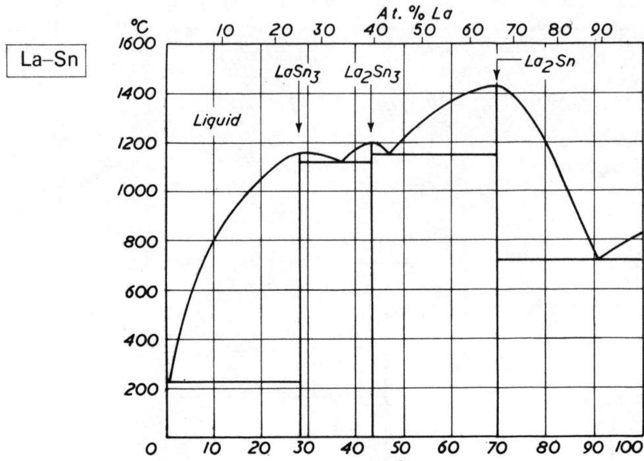
At. % La
10 20 30 40 50 60 70 80 90
°C
1600
1400
1200
1000
800
600
400
200
0
LaSn3
La2Sn3
La2Sn
Liquid
10 20 30 40 50 60 70 80 90 100
Wt. % La

La–Tl

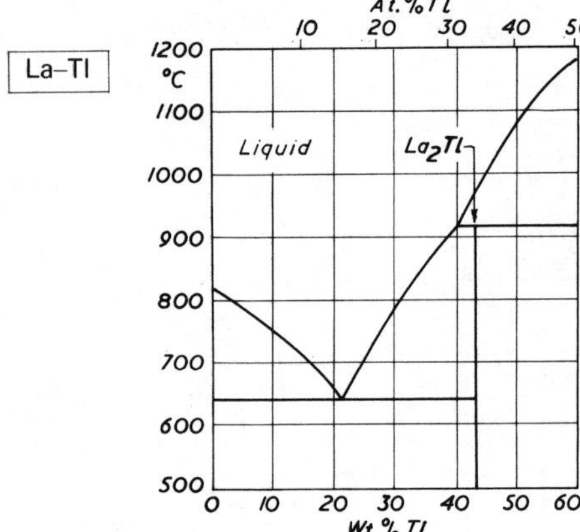
At. % Tl
10 20 30 40 50
1200
°C
1100
1000
900
800
700
600
500
Liquid
La2Tl
0 10 20 30 40 50 60
Wt. % Tl

La–V

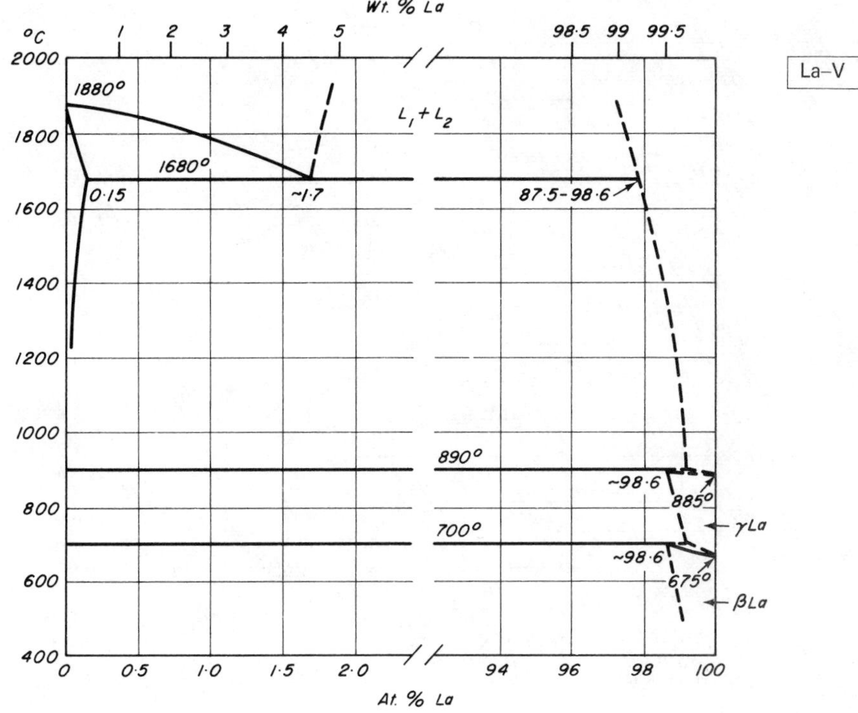
Wt. % La
°C
1880°
L_1+L_2
1680°
0·15
~1·7
87·5–98·6
890°
~98·6
885°
γLa
700°
~98·6
675°
βLa
At. % La

Li–Mg

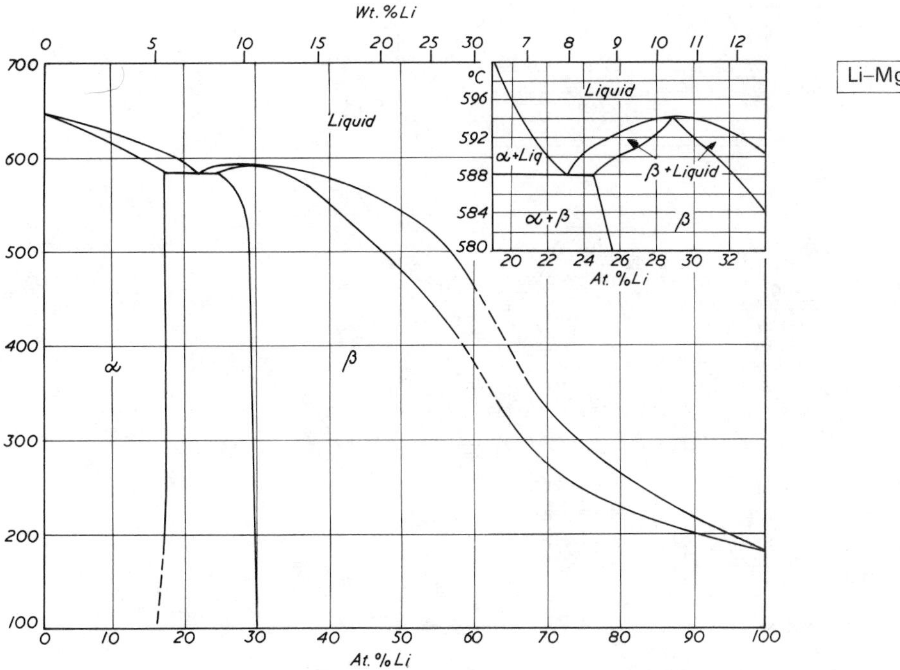
Wt. %Li
Liquid
α
β
°C
Liquid
α+Liq
β+Liquid
α+β
β
At. %Li
At. %Li

Li–Mn

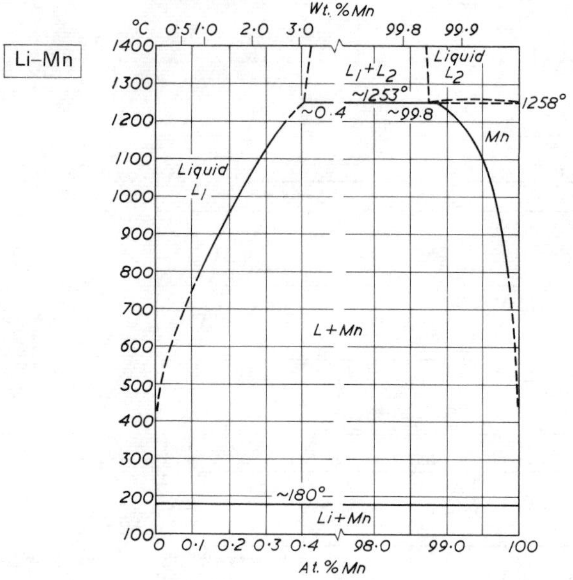

Li–Na

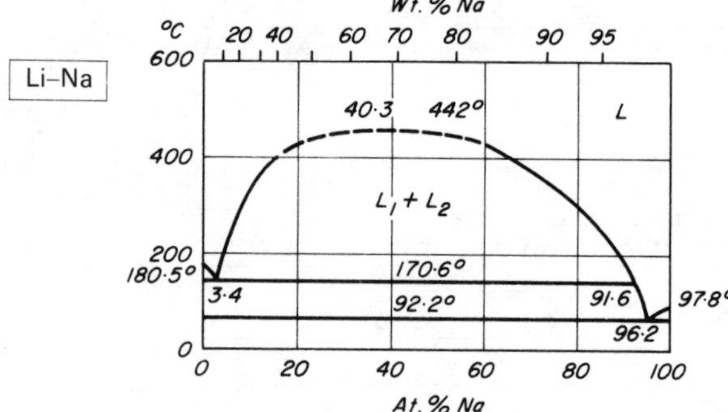

Li–Pb

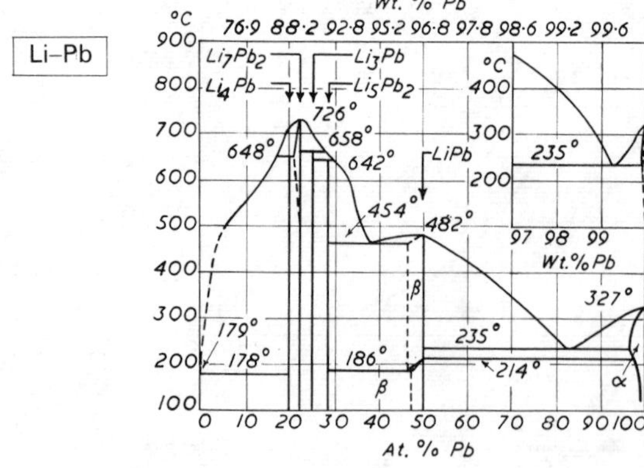

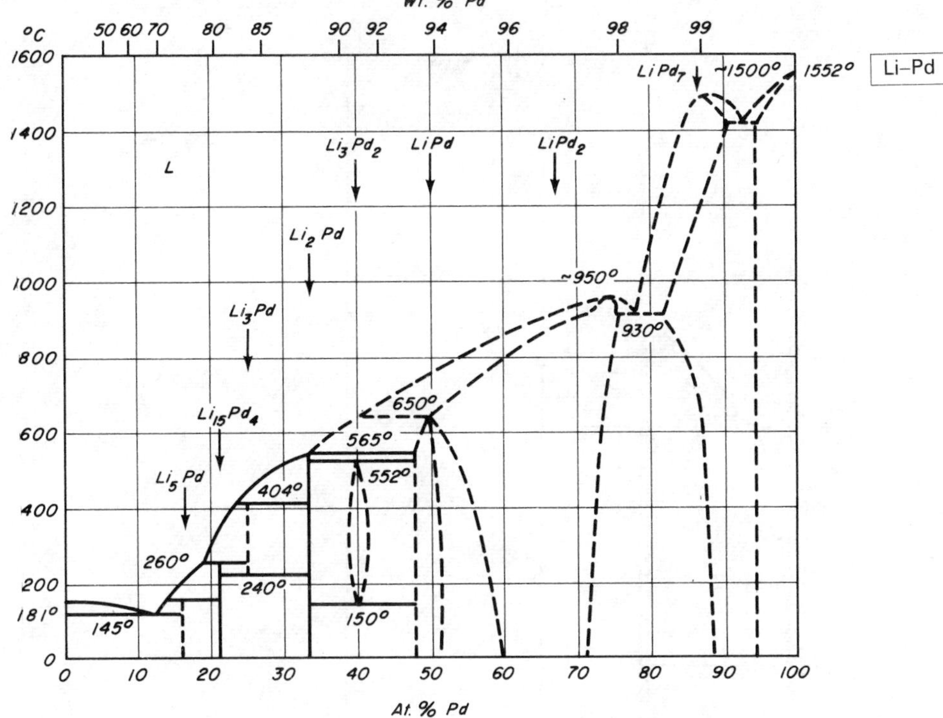
Wt. % Pd
°C
50 60 70 80 85 90 92 94 96 98 99
1600
1400
1200
1000
800
600
400
200
181°
0
L
Li_3Pd_2
LiPd
$LiPd_2$
$LiPd_7$
~1500°
1552°
Li_2Pd
Li_3Pd
~950°
930°
650°
$Li_{15}Pd_4$
565°
552°
Li_5Pd
404°
260°
240°
145°
150°
0 10 20 30 40 50 60 70 80 90 100
At. % Pd

Li–Pd

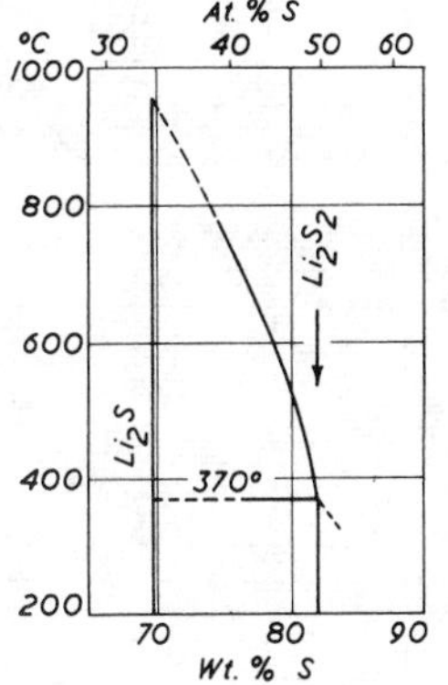
At. % S
°C 30 40 50 60
1000
800
600
400
200
Li_2S_2
Li_2S
370°
70 80 90
Wt. % S

Li–S

Li–Si

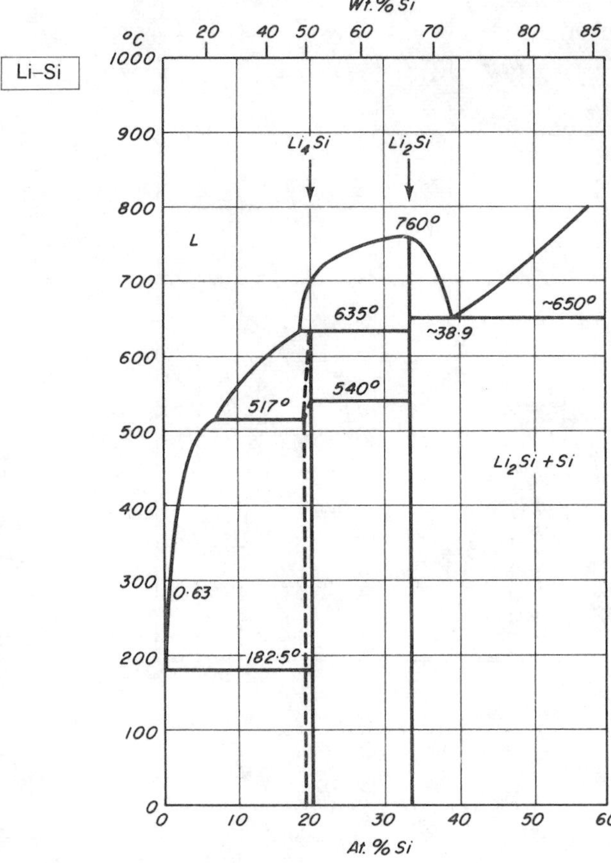
Wt.% Si
°C
20
40
50
60
70
80
85
1000
900
800
700
600
500
400
300
200
100
0
Li_4Si
Li_2Si
760°
L
635°
~650°
~38·9
540°
517°
Li_2Si+Si
0·63
182·5°
0
10
20
30
40
50
60
At.% Si

Li–Sn

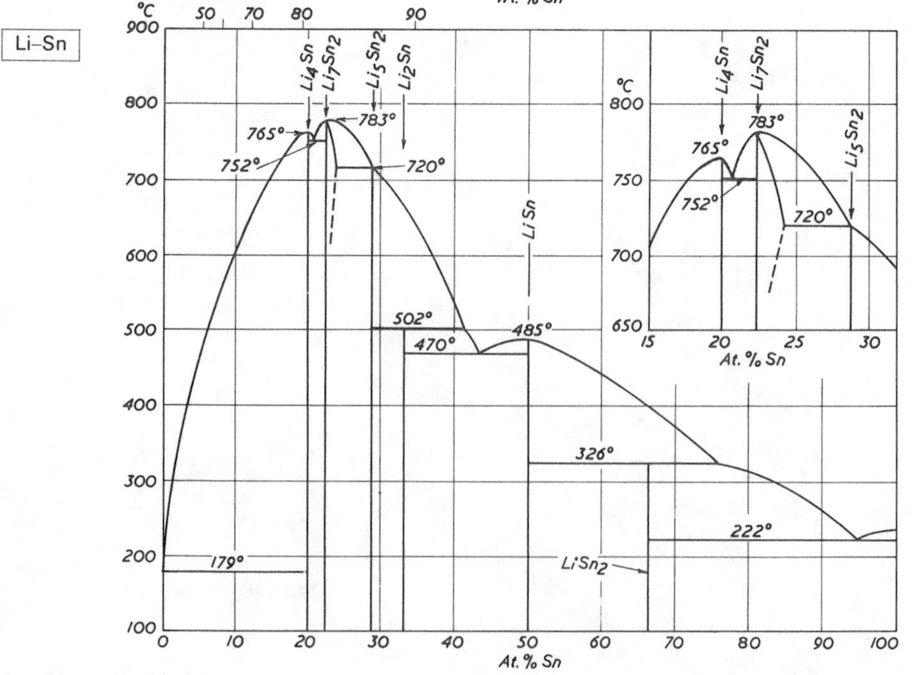
Wt. % Sn
°C
50
70
80
90
900
800
700
600
500
400
300
200
100
Li_4Sn
Li_7Sn_2
Li_5Sn_2
Li_2Sn
765°
783°
752°
720°
LiSn
502°
485°
470°
326°
222°
179°
$LiSn_2$
0
10
20
30
40
50
60
70
80
90
100
At. % Sn
°C
800
750
700
650
15
20
25
30
At. % Sn

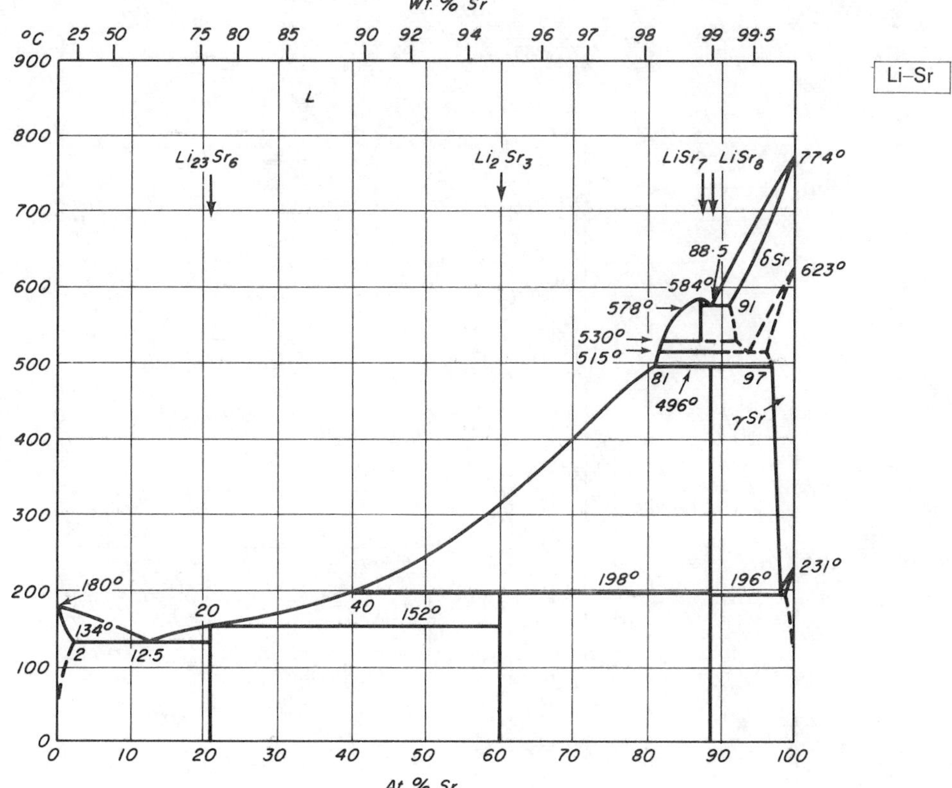
Wt. % Sr
°C
25 50 75 80 85 90 92 94 96 97 98 99 99·5
900
800
700
600
500
400
300
200
100
0
L
$Li_{23}Sr_6$
Li_2Sr_3
$LiSr_7$
$LiSr_8$
774°
88·5
δSr
623°
584°
578°
91
530°
515°
81
97
496°
γSr
231°
198°
196°
180°
134°
20
40
152°
2
12·5
0 10 20 30 40 50 60 70 80 90 100
At. % Sr

Li–Tl

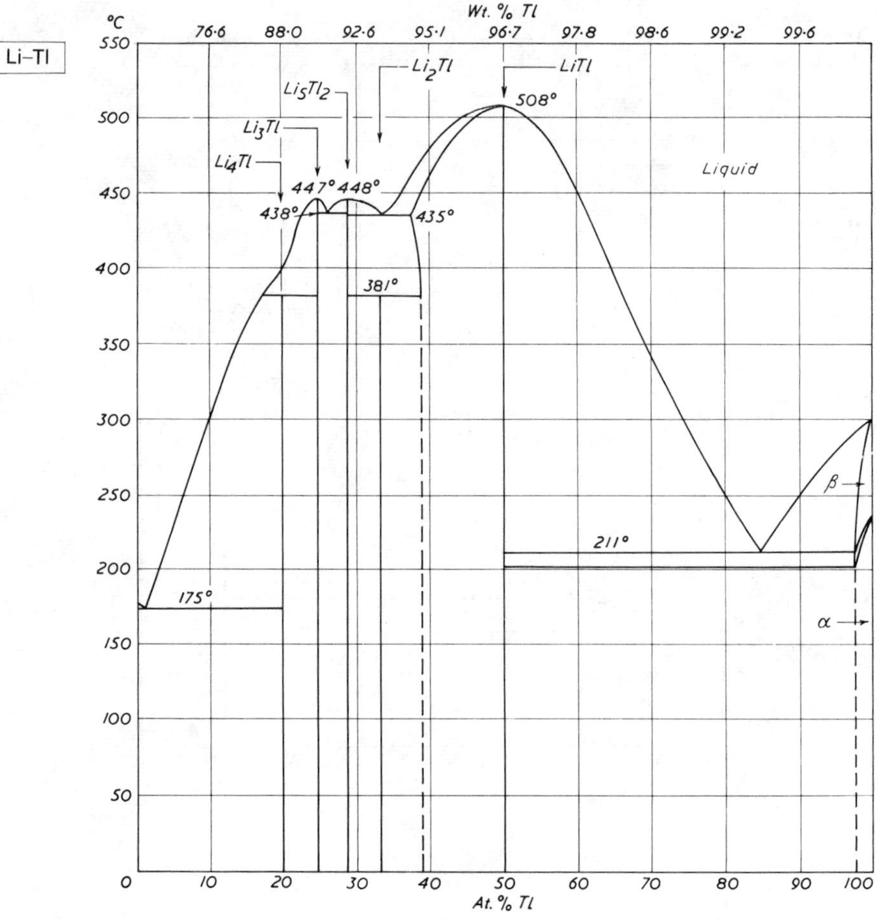
Wt. % Tl
76·6
88·0
92·6
95·1
96·7
97·8
98·6
99·2
99·6
°C
550
500
450
400
350
300
250
200
150
100
50
0
Li5Tl2
Li2Tl
LiTl
Li3Tl
Li4Tl
508°
447° 448°
438°
435°
381°
Liquid
β
211°
175°
α
10
20
30
40
50
60
70
80
90
100
At. % Tl

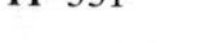

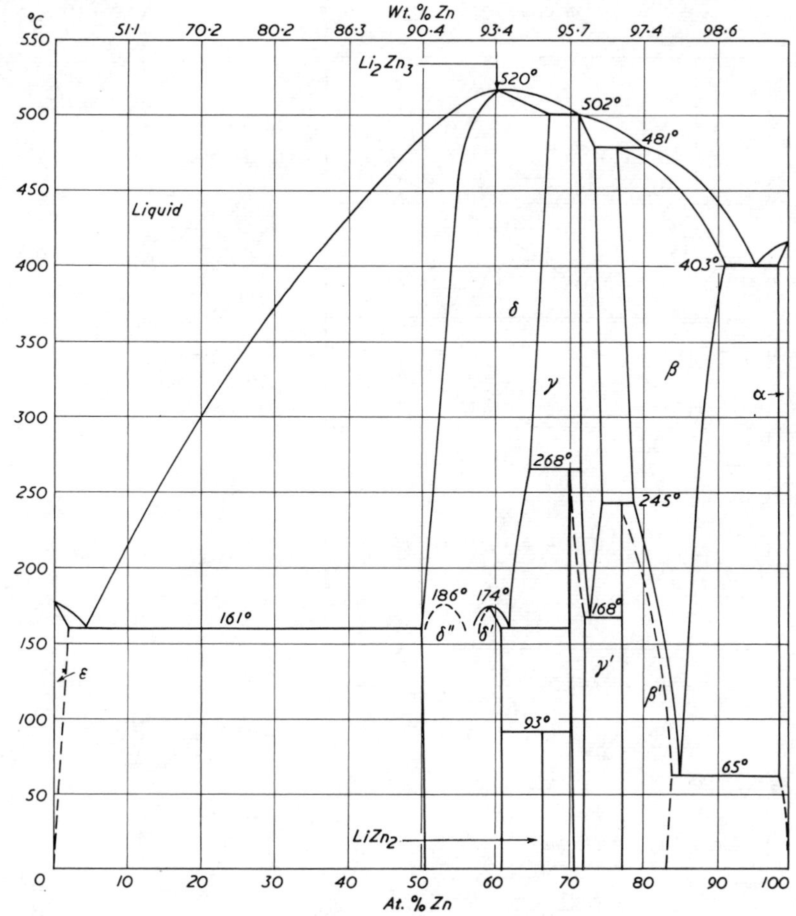

Wt. % Zn
51·1 70·2 80·2 86·3 90·4 93·4 95·7 97·4 98·6
°C
550 500 450 400 350 300 250 200 150 100 50 0
Li_2Zn_3
520° 502° 481° 403° 268° 245° 186° 174° 168° 161° 93° 65°
Liquid
δ γ β α δ″ δ′ γ′ β′ ε
$LiZn_2$
10 20 30 40 50 60 70 80 90 100
At. % Zn

Mg–Mn

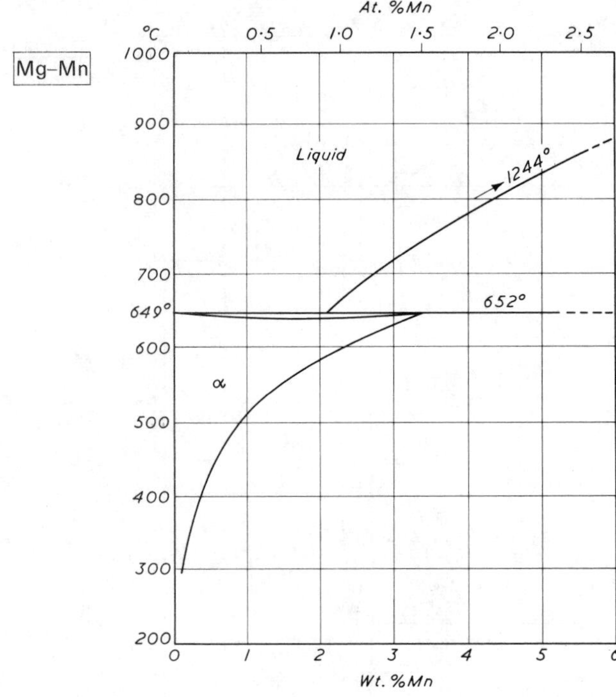
At. %Mn
°C
0·5
1·0
1·5
2·0
2·5
1000
900
800
700
649°
600
500
400
300
200
Liquid
1244°
652°
α
0
1
2
3
4
5
6
Wt. %Mn

Mg–Na

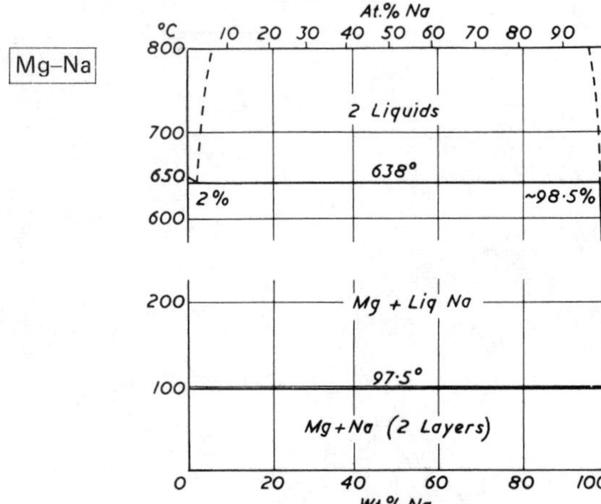
At.% Na
°C
10 20 30 40 50 60 70 80 90
800
700
650
600
200
100
2 Liquids
638°
2%
~98·5%
Mg + Liq Na
97·5°
Mg+Na (2 Layers)
0
20
40
60
80
100
Wt.% Na

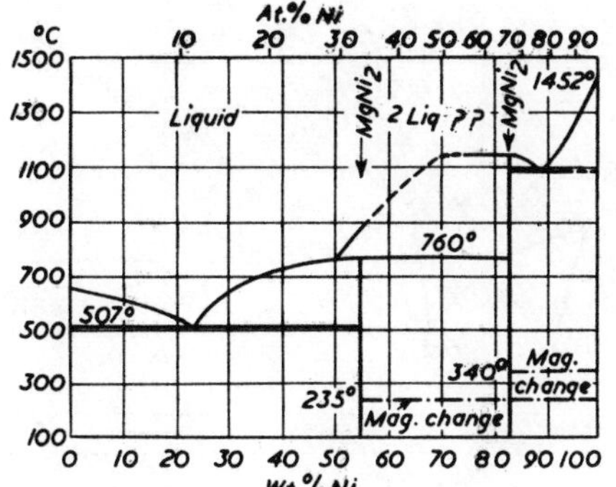

Mg–Ni

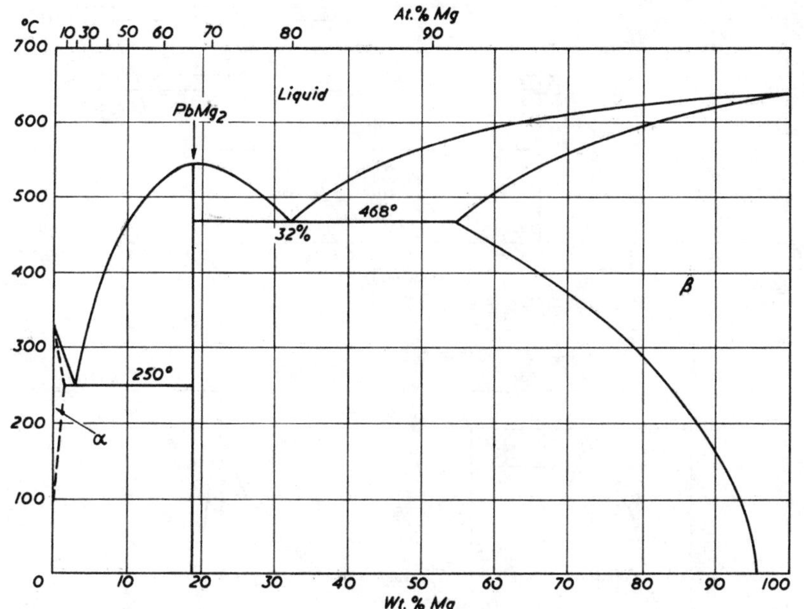

Mg–Pb

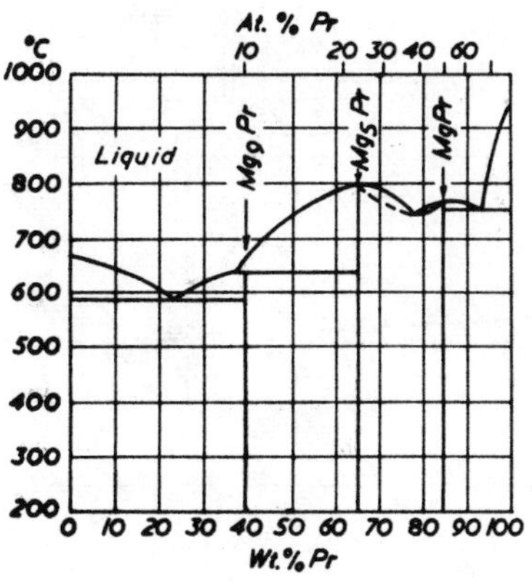

Mg–Pr

Mg–Pu

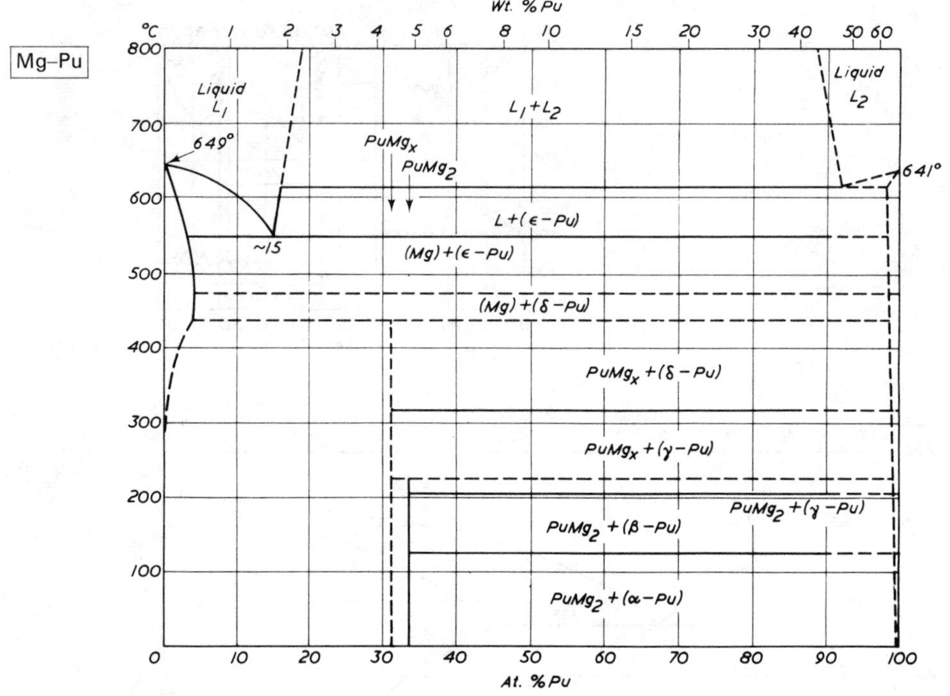
Wt. % Pu
°C
800
700
600
500
400
300
200
100
0
1
2
3
4
5
6
8
10
15
20
30
40
50
60
Liquid
L1
L1+L2
Liquid
L2
649°
PuMgx
PuMg2
641°
~15
L+(ε−Pu)
(Mg)+(ε−Pu)
(Mg)+(δ−Pu)
PuMgx+(δ−Pu)
PuMgx+(γ−Pu)
PuMg2+(γ−Pu)
PuMg2+(β−Pu)
PuMg2+(α−Pu)
10
20
30
40
50
60
70
80
90
100
At. % Pu

Mg–Sb

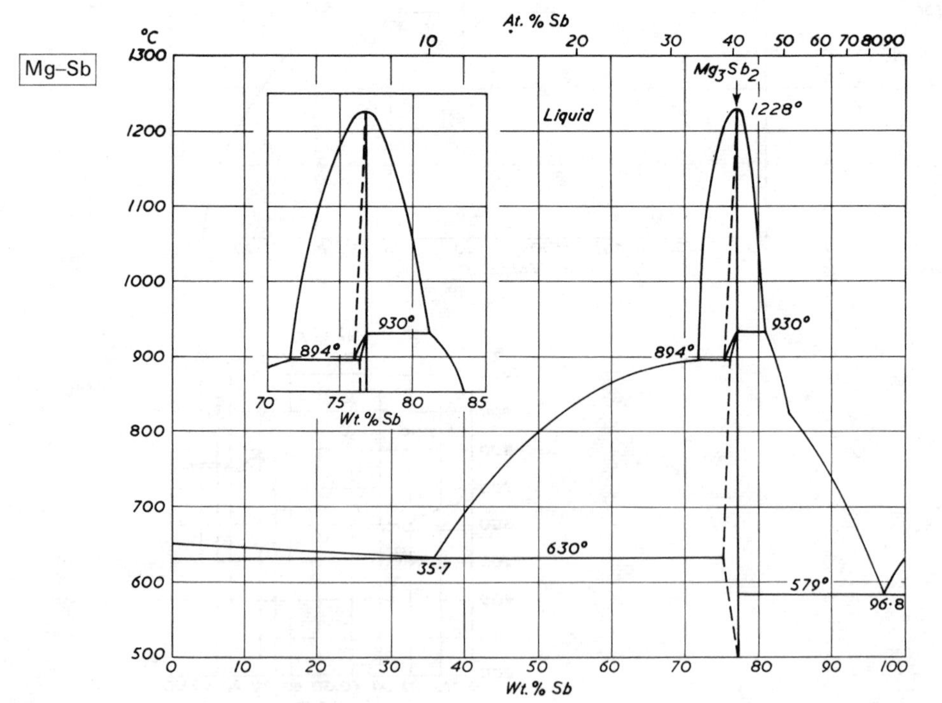
At. % Sb
°C
1300
1200
1100
1000
900
800
700
600
500
10
20
30
40
50
60
70
80
90
Mg3Sb2
1228°
Liquid
930°
894°
930°
894°
70
75
80
85
Wt. % Sb
630°
35·7
579°
96·8
0
10
20
30
40
50
60
70
80
90
100
Wt. % Sb

Mg–Sc

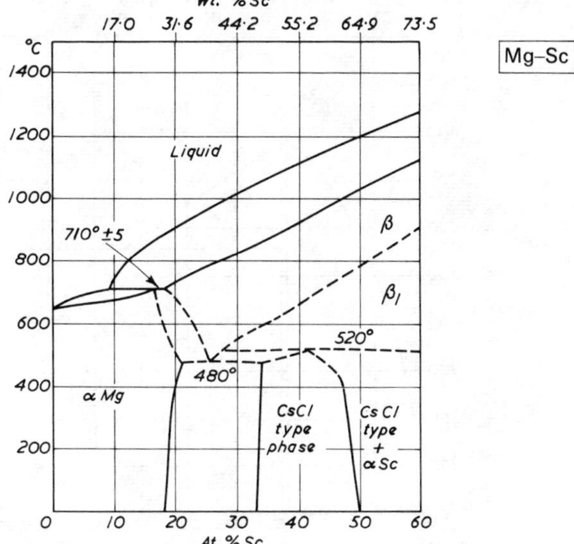

Wt. % Sc
17·0
31·6
44·2
55·2
64·9
73·5
°C
1400
1200
1000
800
600
400
200
Liquid
710° ± 5
β
β1
520°
480°
α Mg
CsCl
type
phase
Cs Cl
type
+
α Sc
0
10
20
30
40
50
60
At. % Sc

Mg–Si

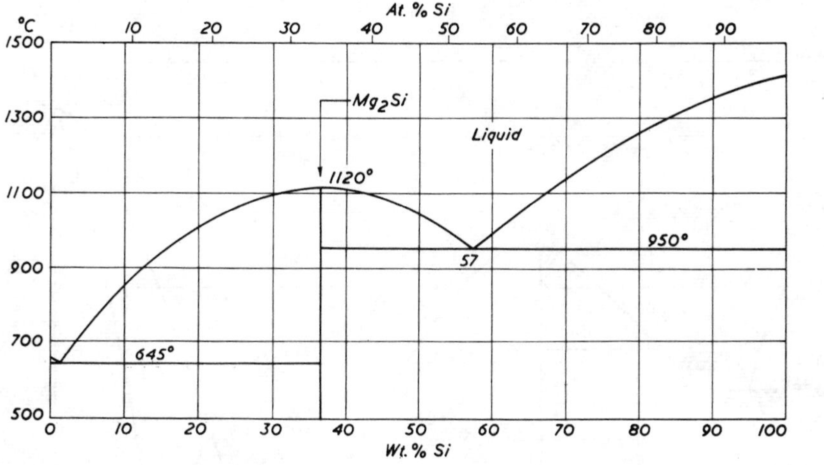

At. % Si
10
20
30
40
50
60
70
80
90
°C
1500
1300
1100
900
700
500
Mg2Si
Liquid
1120°
950°
57
645°
0
10
20
30
40
50
60
70
80
90
100
Wt. % Si

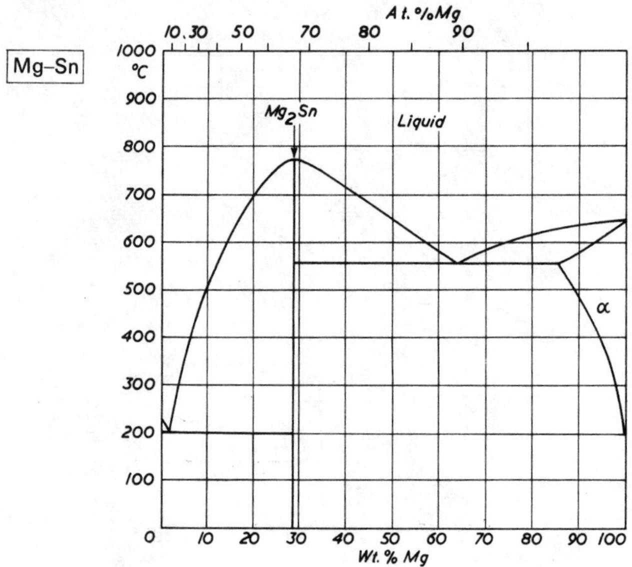
Mg–Sn
At.%Mg
10, 30 50 70 80 90
1000 °C
900
800
700
600
500
400
300
200
100
0
Mg$_2$Sn
Liquid
α
10 20 30 40 50 60 70 80 90 100
Wt.% Mg

Mg–Sr
Wt.% Sr 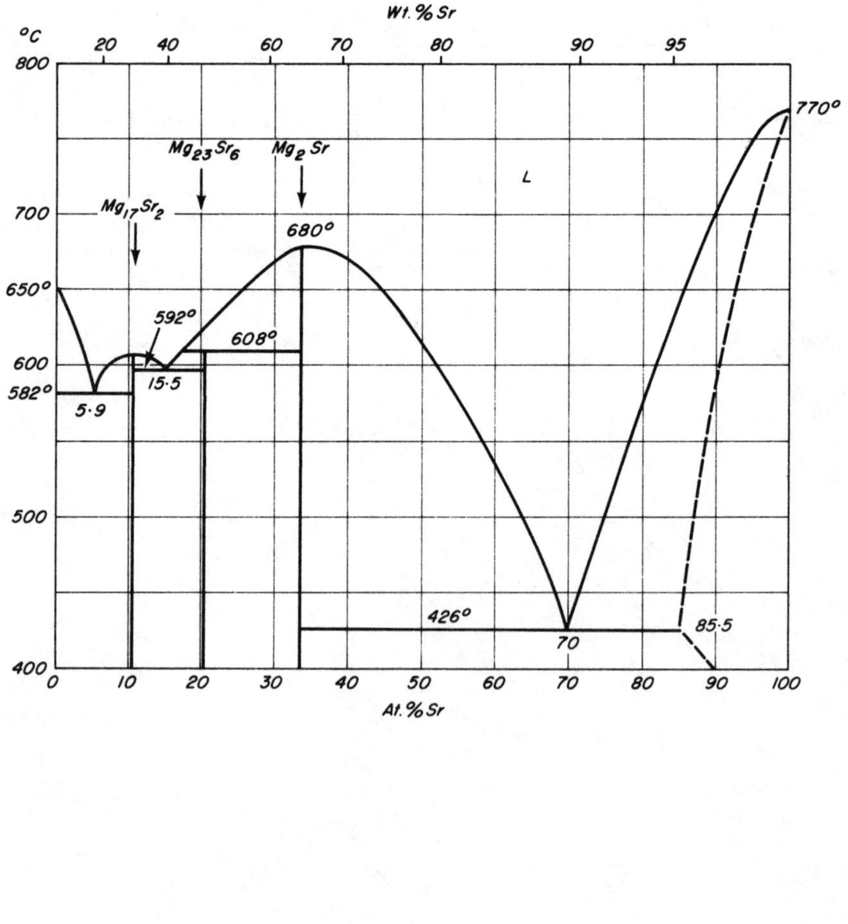
20 40 60 70 80 90 95
°C
800
700
650°
600
582°
500
400
Mg$_{17}$Sr$_2$
Mg$_{23}$Sr$_6$
Mg$_2$Sr
L
770°
680°
592°
608°
5·9
15·5
426°
70
85·5
0 10 20 30 40 50 60 70 80 90 100
At.% Sr

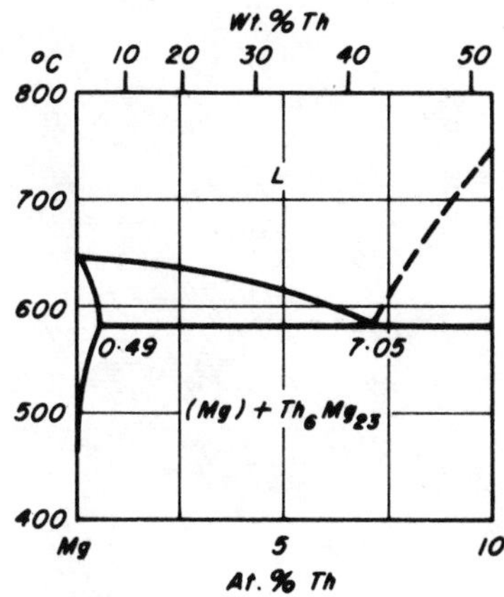

Mg–Th

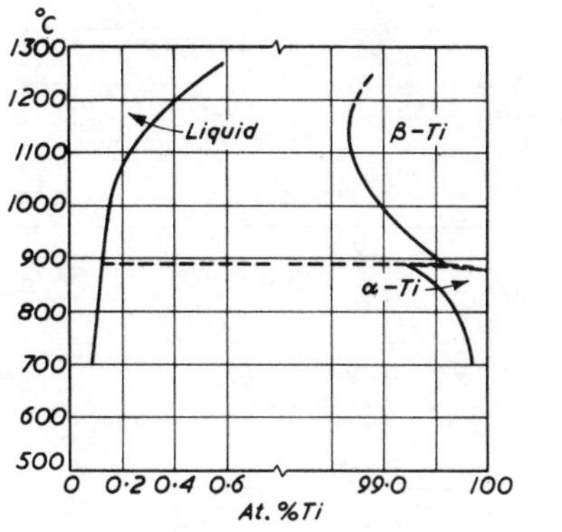

Mg–Ti

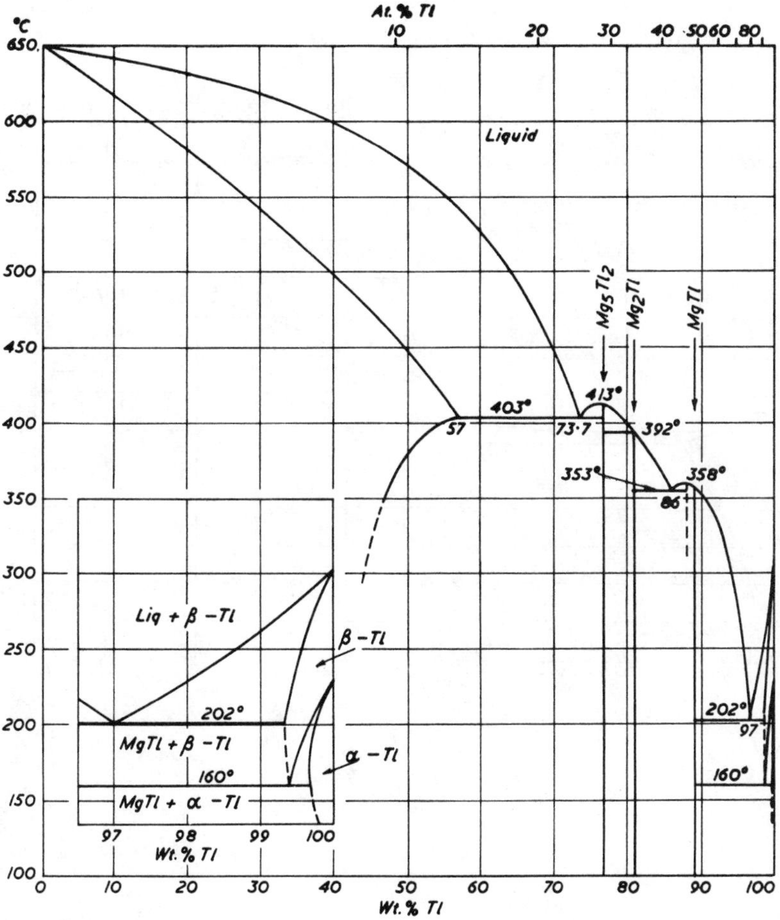

Mg–Tl

Mg–U

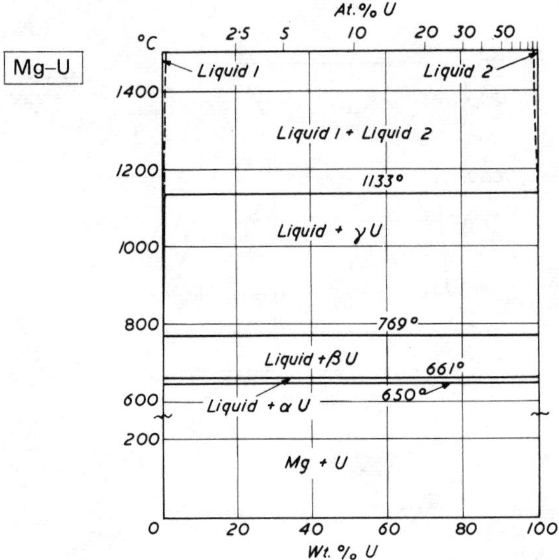
At.% U
°C
2·5
5
10
20
30
50
Liquid 1
Liquid 2
1400
Liquid 1 + Liquid 2
1200
1133°
Liquid + γU
1000
800
769°
Liquid + βU
661°
600
650°
Liquid + α U
200
Mg + U
0
20
40
60
80
100
Wt. % U

Mg–Y

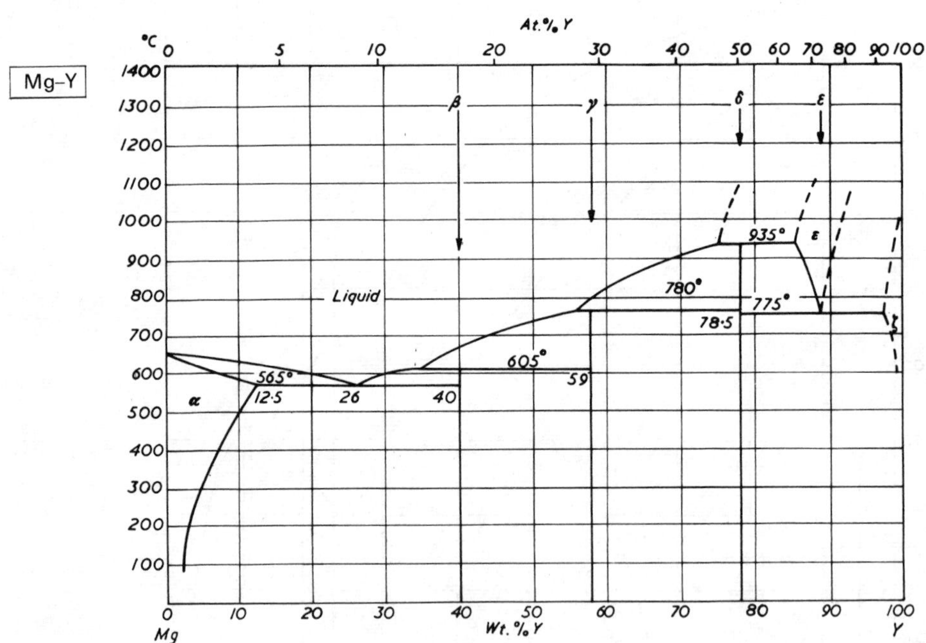
At.% Y
°C
1400
1300
1200
1100
1000
900
800
700
600
500
400
300
200
100
β
γ
δ
ε
935°
780°
775°
78·5
Liquid
605°
59
565°
12·5
26
40
α
ζ
Mg
Y
Wt. % Y

Mg–Zn

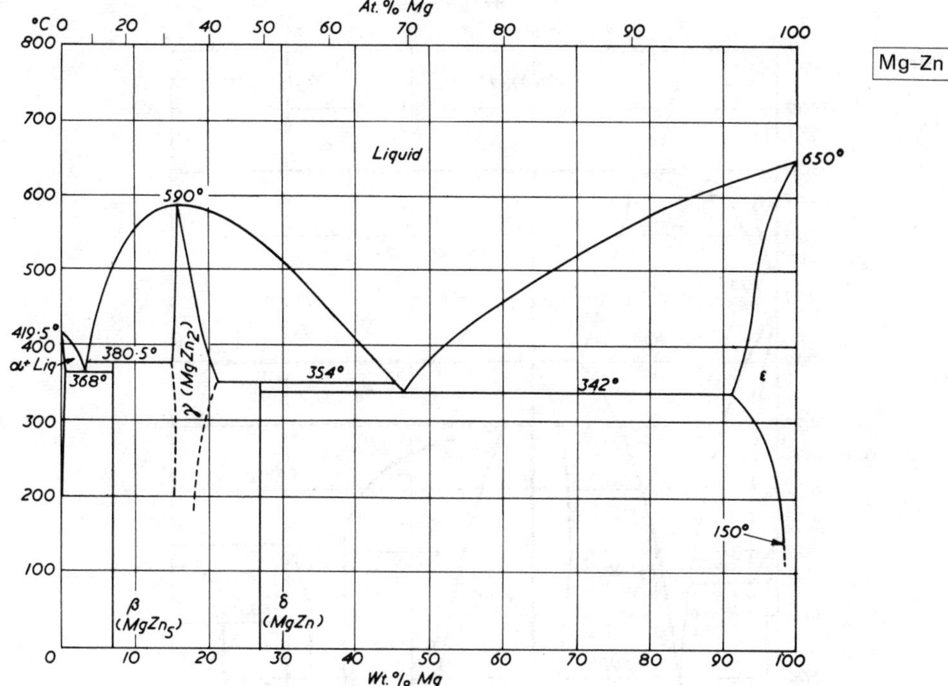

At. % Mg
°C
Liquid
590°
650°
419·5°
α+ Liq
380·5°
368°
γ (MgZn$_2$)
354°
342°
ε
150°
β (MgZn$_5$)
δ (MgZn)
Wt. % Mg

Mg–Zr

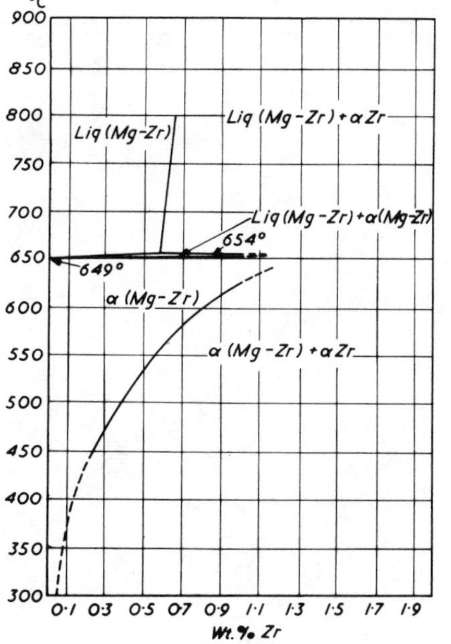

°C
Liq (Mg–Zr)
Liq (Mg–Zr) + α Zr
Liq (Mg–Zr) + α (Mg–Zr)
654°
649°
α (Mg–Zr)
α (Mg–Zr) + α Zr
Wt. % Zr

Mn–N

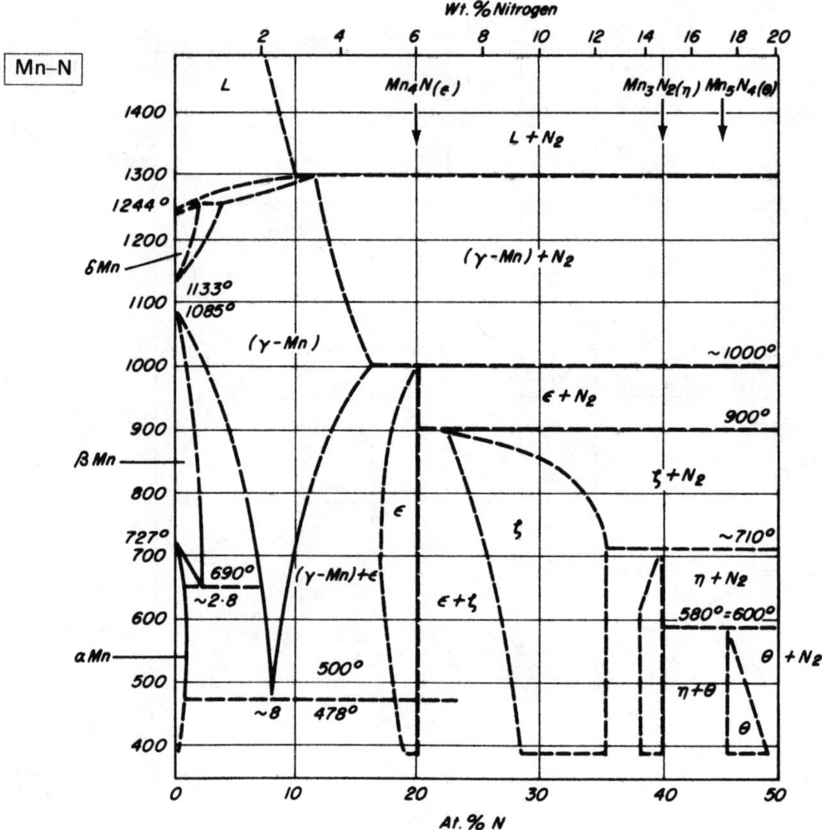

Wt.% Nitrogen
2 4 6 8 10 12 14 16 18 20
L
Mn4N(ε)
Mn3N2(η)
Mn5N4(θ)
L + N2
1400
1300
1244°
1200
δMn
(γ-Mn) + N2
1133°
1100
1085°
(γ-Mn)
~1000°
1000
ε + N2
900
900°
βMn
ζ + N2
800
ε
ζ
727°
700
~710°
690°
(γ-Mn)+ε
η + N2
~2·8
600
ε+ζ
580°=600°
αMn
θ + N2
500°
500
η+θ
~8
478°
θ
400
0 10 20 30 40 50
At.% N

Mn–Nb

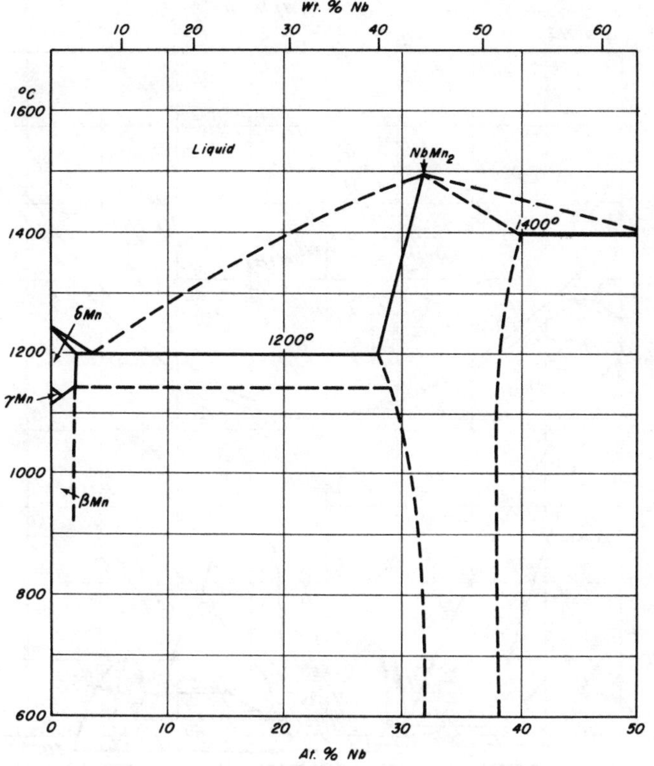
Wt. % Nb
10
20
30
40
50
60
°C
1600
Liquid
NbMn2
1400°
1400
δMn
1200°
1200
γMn
1000
βMn
800
600
0
10
20
30
40
50
At. % Nb

Mn–Nd

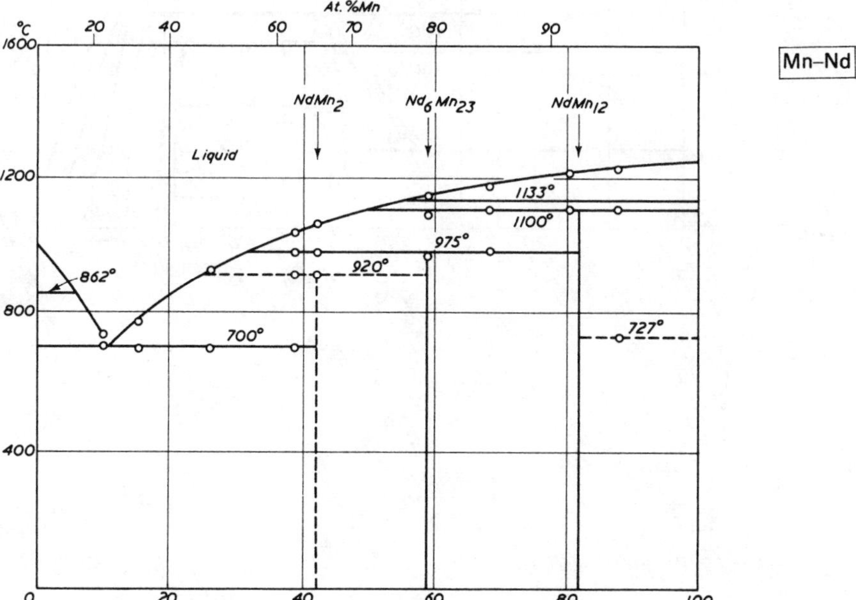
At. %Mn
°C
20
40
60
70
80
90
1600
NdMn2
Nd6Mn23
NdMn12
Liquid
1200
1133°
1100°
975°
862°
920°
800
700°
727°
400
0
20
40
60
80
100
Wt. %Mn

Mn–Ni

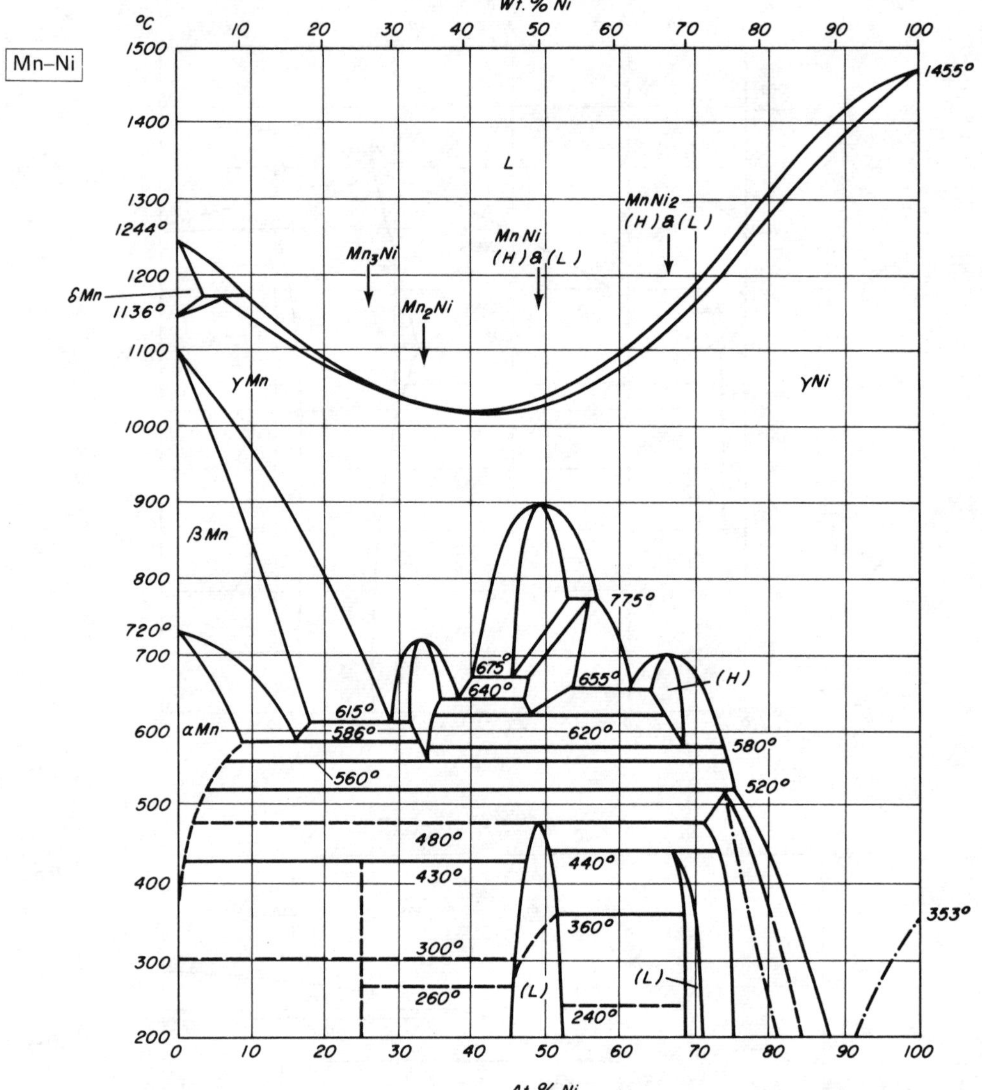

Wt. % Ni
°C
10
20
30
40
50
60
70
80
90
100
1500
1400
1300
1200
1100
1000
900
800
700
600
500
400
300
200
1455°
L
$MnNi_2$
(H) & (L)
MnNi
(H) & (L)
Mn_3Ni
Mn_2Ni
1244°
δMn
1136°
γMn
γNi
βMn
775°
720°
675°
640°
655°
(H)
615°
586°
620°
αMn
580°
560°
520°
480°
440°
430°
360°
353°
300°
260°
(L)
(L)
240°
0
At. % Ni

Mn–O

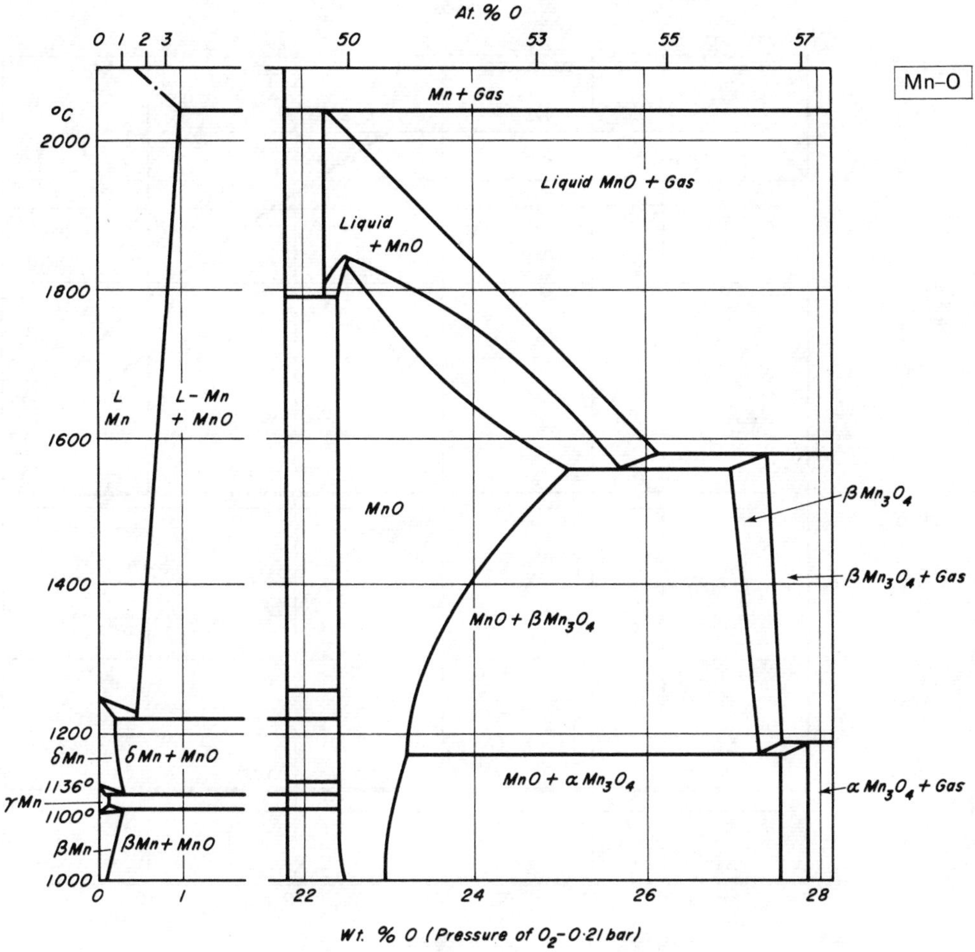

At. % O

0 1 2 3 50 53 55 57

°C

2000 1800 1600 1400 1200 1000

Mn + Gas

Liquid MnO + Gas

Liquid + MnO

L Mn

L – Mn + MnO

MnO

βMn_3O_4

βMn_3O_4 + Gas

MnO + βMn_3O_4

δMn

δMn + MnO

1136°

γMn

1100°

βMn

βMn + MnO

MnO + αMn_3O_4

αMn_3O_4 + Gas

0 1 22 24 26 28

Wt. % O (Pressure of O_2–0·21 bar)

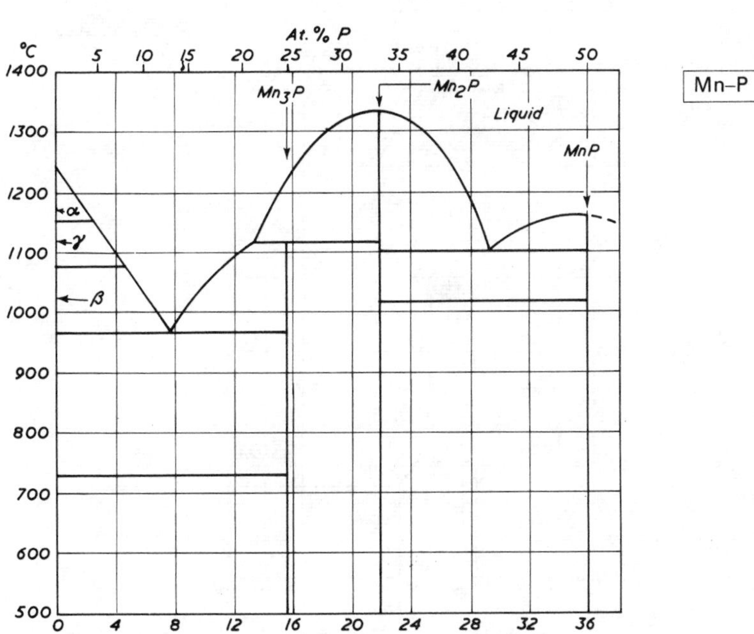

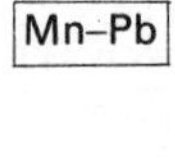
Mn–Pb
Wt. % Pb 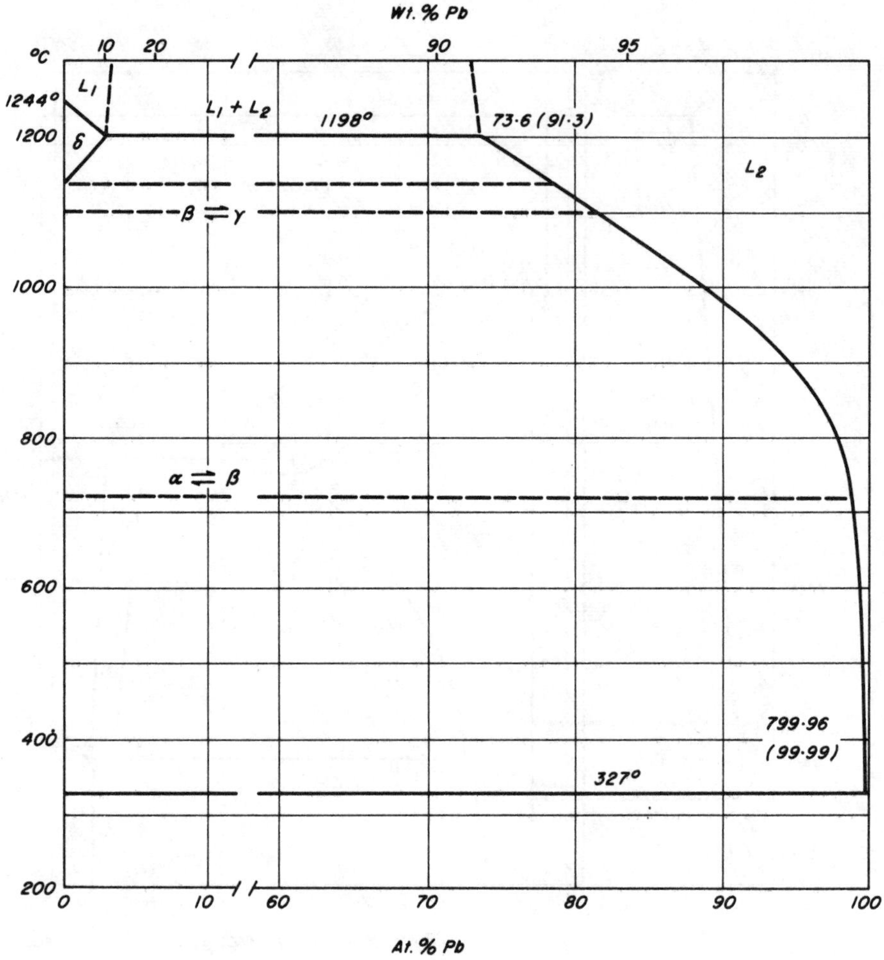
10
20
90
95
°C
L_1
1244°
$L_1 + L_2$
1198°
73·6 (91·3)
1200
δ
L_2
β ⇌ γ
1000
800
α ⇌ β
600
400
799·96
(99·99)
327°
200
0
10
60
70
80
90
100
At. % Pb

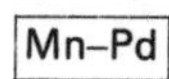
Mn–Pd

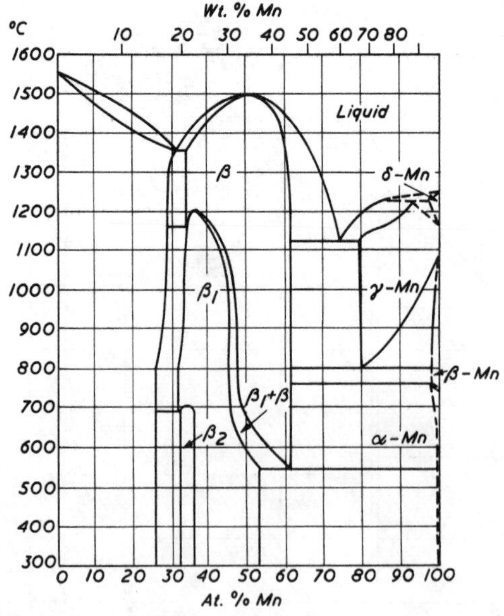
Wt. % Mn
10
20
30
40
50
60
70
80
°C
1600
1500
1400
1300
1200
1100
1000
900
800
700
600
500
400
300
Liquid
β
δ-Mn
β_1
γ-Mn
β-Mn
$\beta_1+\beta$
β_2
α-Mn
0
10
20
30
40
50
60
70
80
90
100
At. % Mn

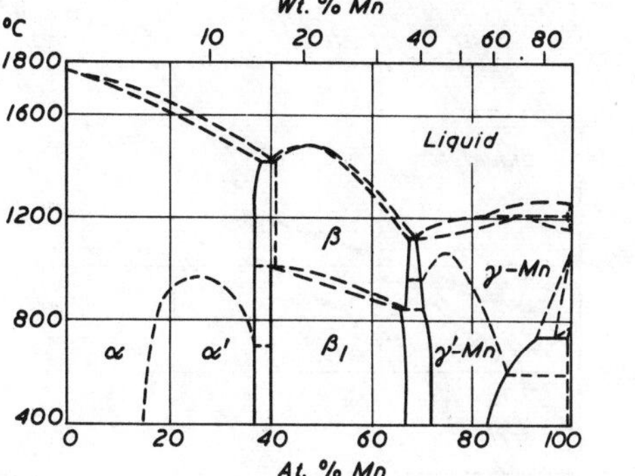
Wt. % Mn
°C
10
20
40
60
80
1800
1600
1200
800
400
Liquid
β
γ-Mn
β1
α
α'
γ'-Mn
0
20
40
60
80
100
At. % Mn

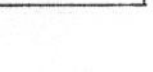
Mn–Pt

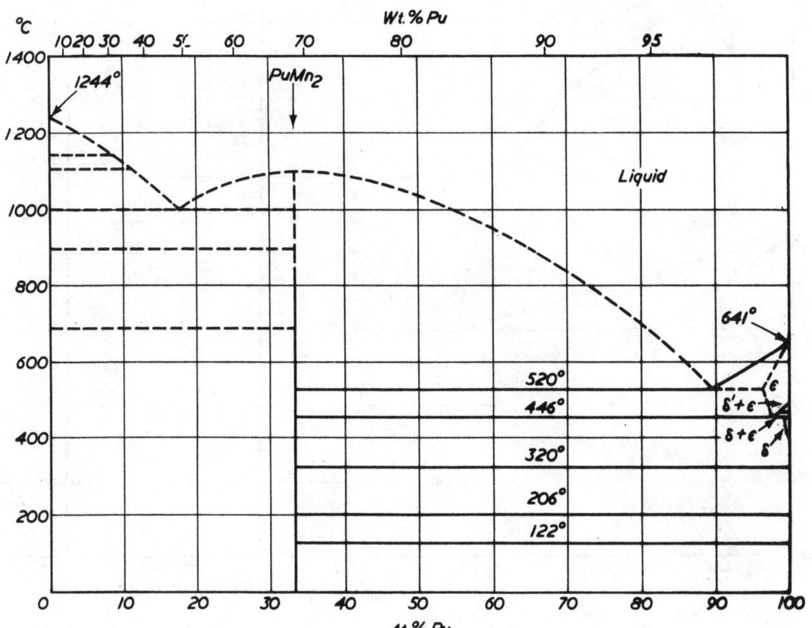
°C
Wt.% Pu
10
20
30
40
50
60
70
80
90
95
1400
1244°
PuMn2
1200
1000
800
600
400
200
Liquid
641°
520°
446°
320°
206°
122°
ε
δ'+ε
δ+ε
δ
0
10
20
30
40
50
60
70
80
90
100
At.% Pu

Mn–Pu

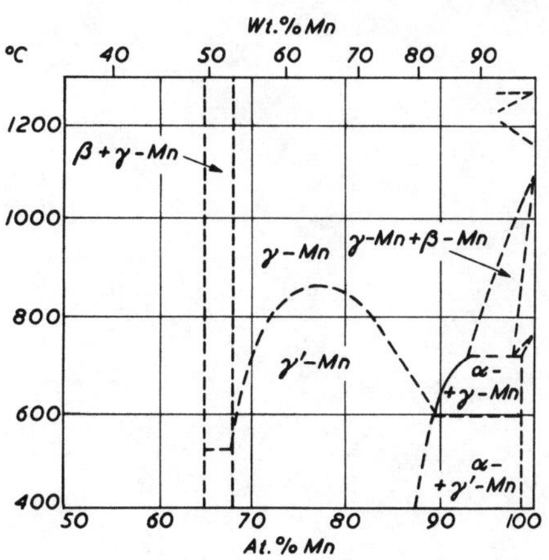
Wt.% Mn
°C
40
50
60
70
80
90
1200
β+γ-Mn
1000
γ-Mn
γ-Mn+β-Mn
800
γ'-Mn
600
α-
+γ-Mn
α-
+γ'-Mn
400
50
60
70
80
90
100
At.% Mn

Mn–Rh

Mn–Ru

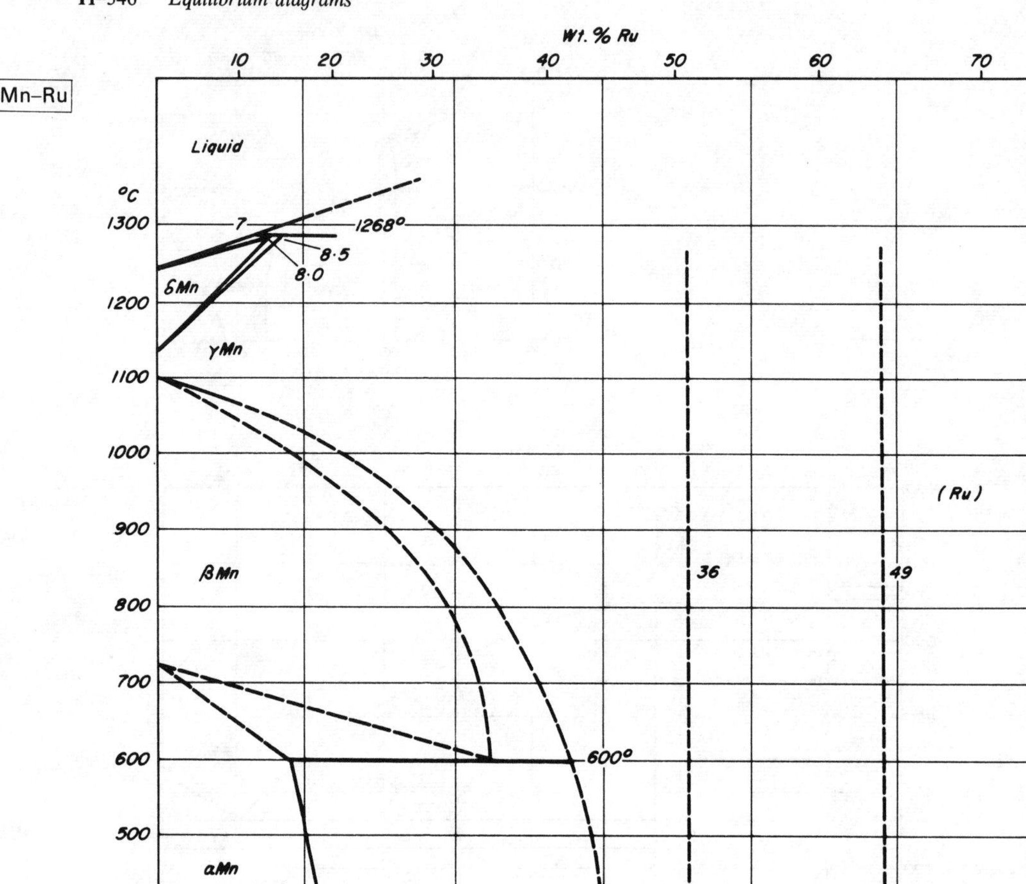
Wt. % Ru
10
20
30
40
50
60
70
Liquid
°C
1300
7
1268°
8·5
8·0
δMn
1200
γMn
1100
1000
(Ru)
900
βMn
36
49
800
700
600
600°
500
αMn
400
300
0
10
20
30
40
50
At. % Ru

Mn–S

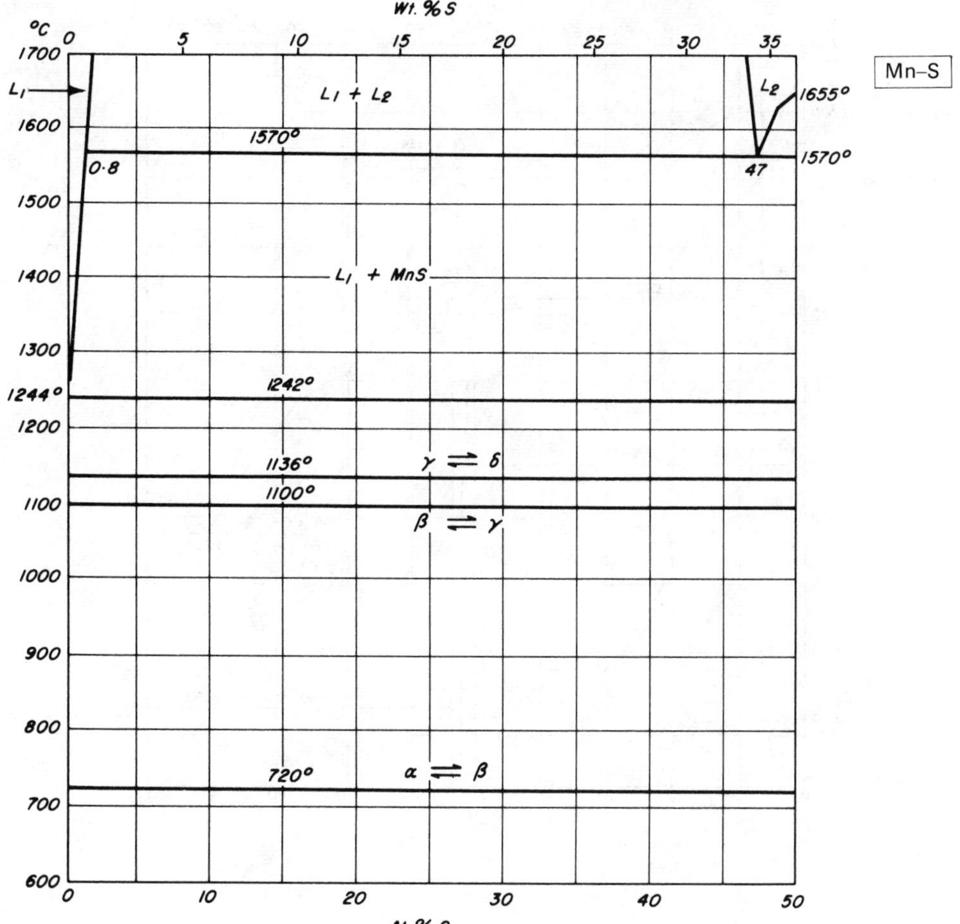
Wt. % S
°C
1700
1600
1500
1400
1300
1244°
1200
1100
1000
900
800
700
600
0
5
10
15
20
25
30
35
L_1
$L_1 + L_2$
L_2
1655°
1570°
0·8
47
1570°
L_1 + MnS
1242°
1136°
$\gamma \rightleftharpoons \delta$
1100°
$\beta \rightleftharpoons \gamma$
720°
$\alpha \rightleftharpoons \beta$
0
10
20
30
40
50
At. % S

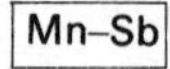
Mn–Sb

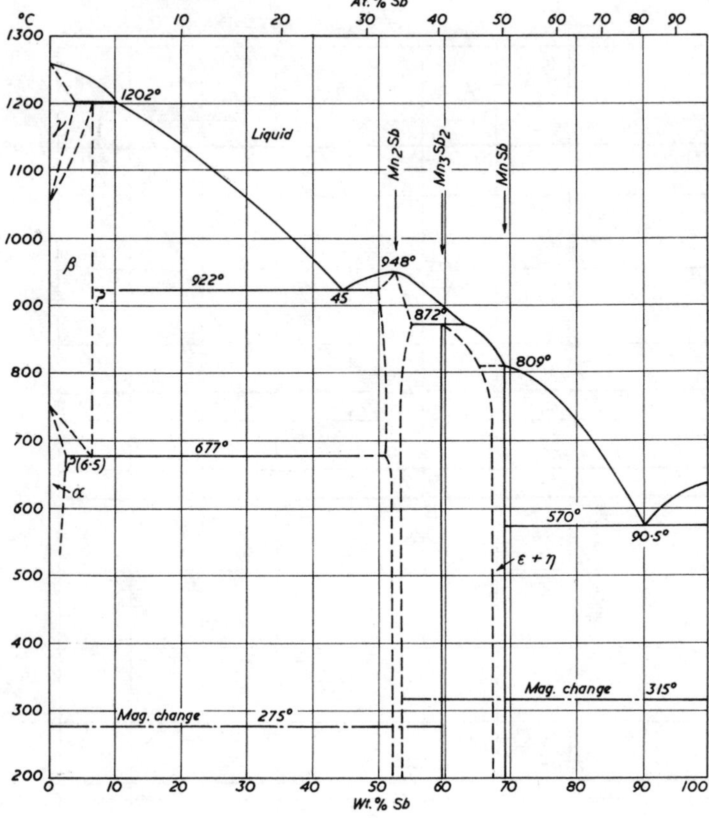
At.% Sb
°C
Liquid
1202°
γ
β
922°
45
948°
Mn2Sb
Mn3Sb2
MnSb
872°
809°
677°
(6·5)
α
570°
90·5°
ε + η
Mag. change
315°
275°
Wt.% Sb

Mn–Si

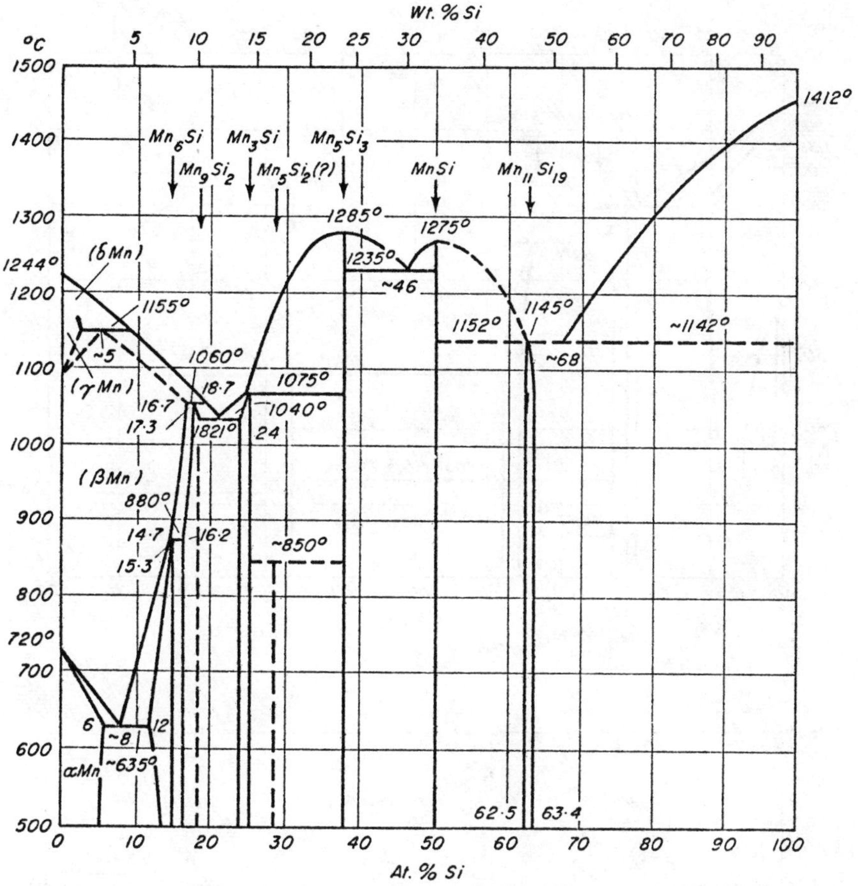

Wt. % Si

°C

1412°

Mn_6Si Mn_9Si_2 Mn_3Si Mn_5Si_2(?) Mn_5Si_3 MnSi $Mn_{11}Si_{19}$

1285° 1275° 1235° ~46 (δMn) 1244° 1155° 1152° 1145° ~1142° ~5 ~68 (γMn) 1060° 18·7 1075° 16·7 17·3 1040° 1821° 24 (βMn) 880° 14·7 16·2 15·3 ~850° 720° 6 ~8 12 αMn ~635° 62·5 63·4

At. % Si

Mn–Sm

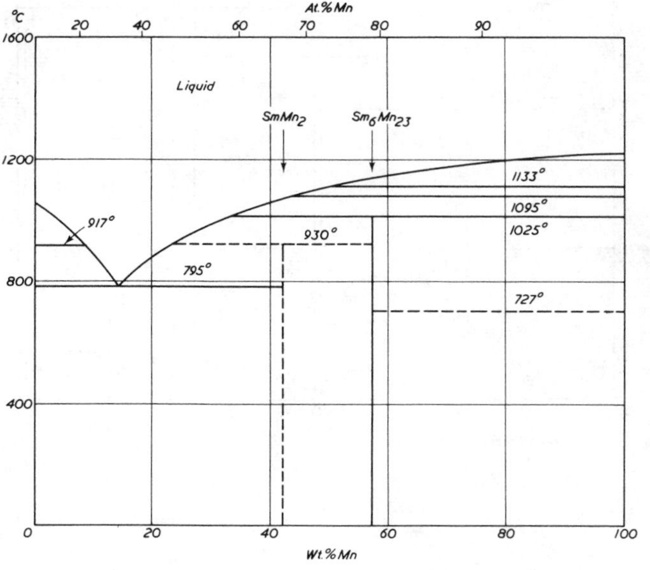

Mn–Sn

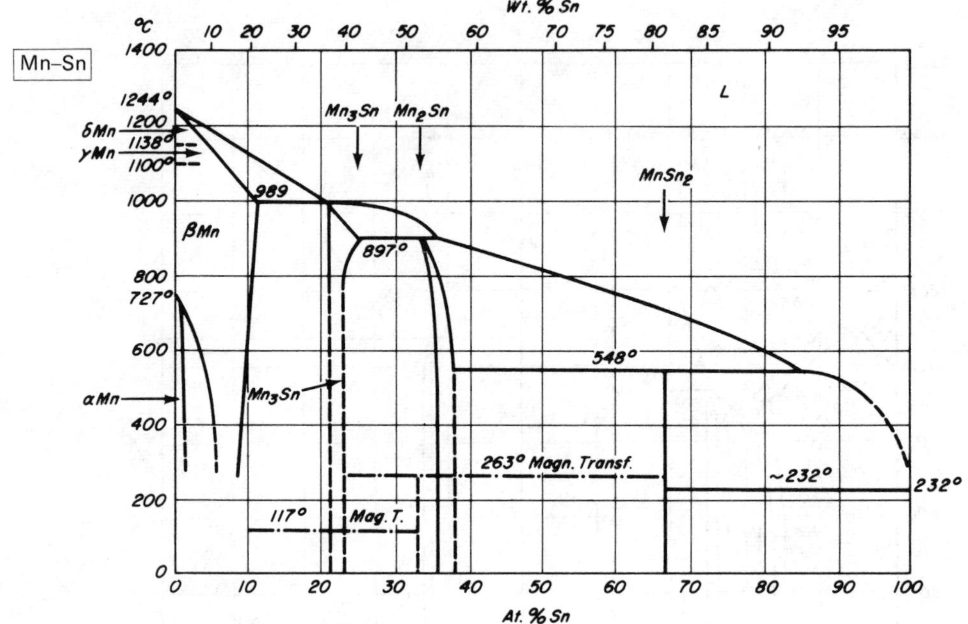
Wt. % Sn
°C
At. % Sn
L
Mn3Sn
Mn2Sn
MnSn2
1244°
δMn
γMn
1138°
1100°
989
βMn
897°
727°
548°
αMn
Mn3Sn
263° Magn. Transf.
~232°
232°
117°
Mag. T.

Mn–Tb

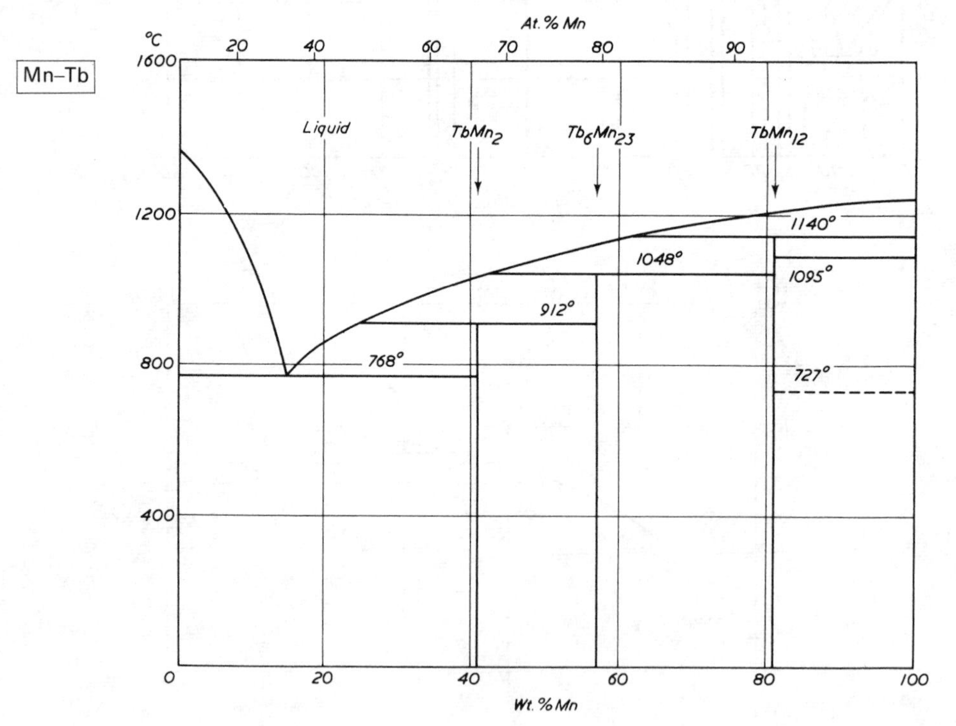
At. % Mn
°C
Liquid
TbMn2
Tb6Mn23
TbMn12
1140°
1048°
1095°
912°
768°
727°
Wt. % Mn

Mn–Ti

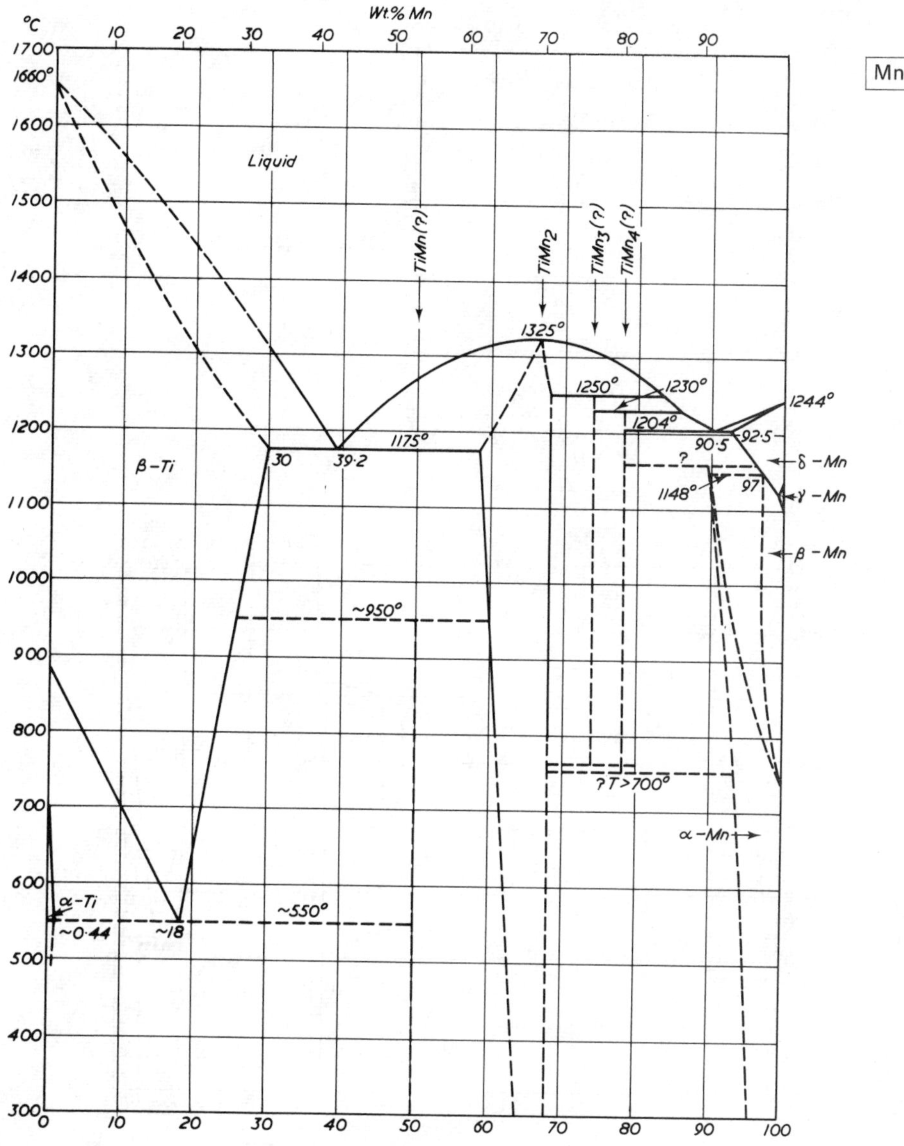
Wt.% Mn
°C
1660°
Liquid
TiMn(?)
TiMn2
TiMn3(?)
TiMn4(?)
1325°
1250°
1230°
1244°
1204°
1175°
30
39·2
90·5
92·5
β-Ti
?
1148°
97
δ-Mn
γ-Mn
β-Mn
~950°
?T>700°
α-Mn
α-Ti
~550°
~0·44
~18
At.%Mn

Mn–Tl

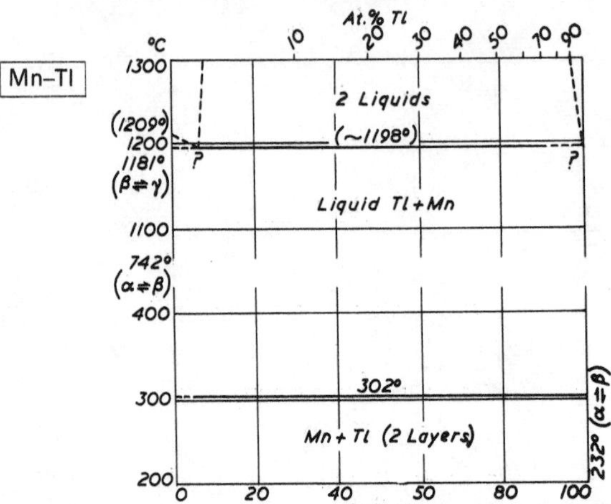
At.% Tl
°C
2 Liquids
(1209°)
(~1198°)
1181°
(β⇌γ)
Liquid Tl+Mn
742°
(α⇌β)
302°
Mn+Tl (2 Layers)
232° (α⇌β)
Wt.% Tl

Mn–U

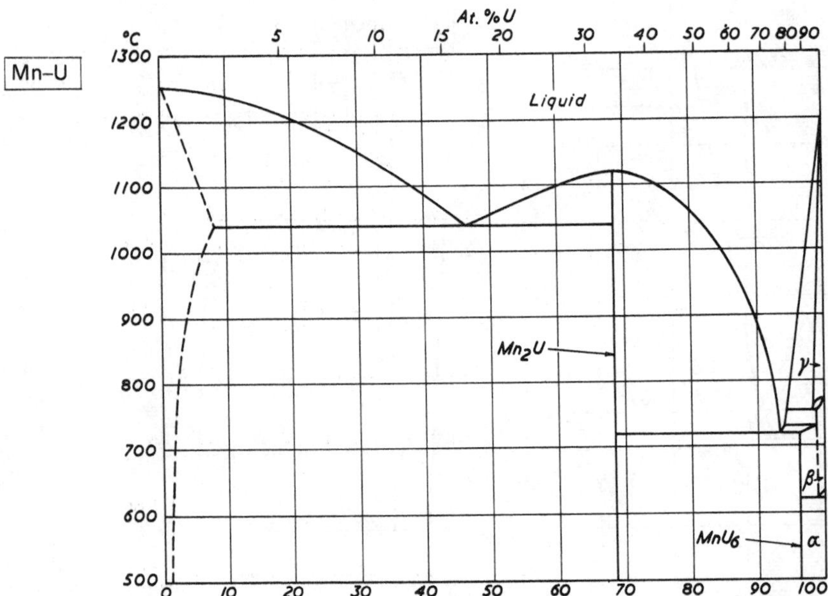
At. % U
°C
Liquid
Mn2U
γ
β
MnU6
α
Wt. % U

Mn–V

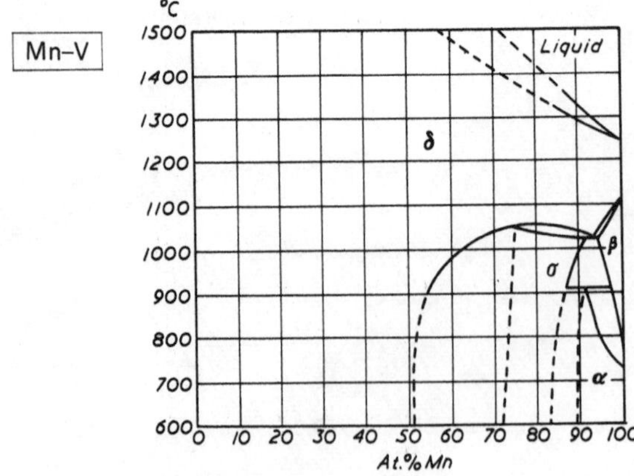
°C
Liquid
δ
σ
β
α
At.% Mn

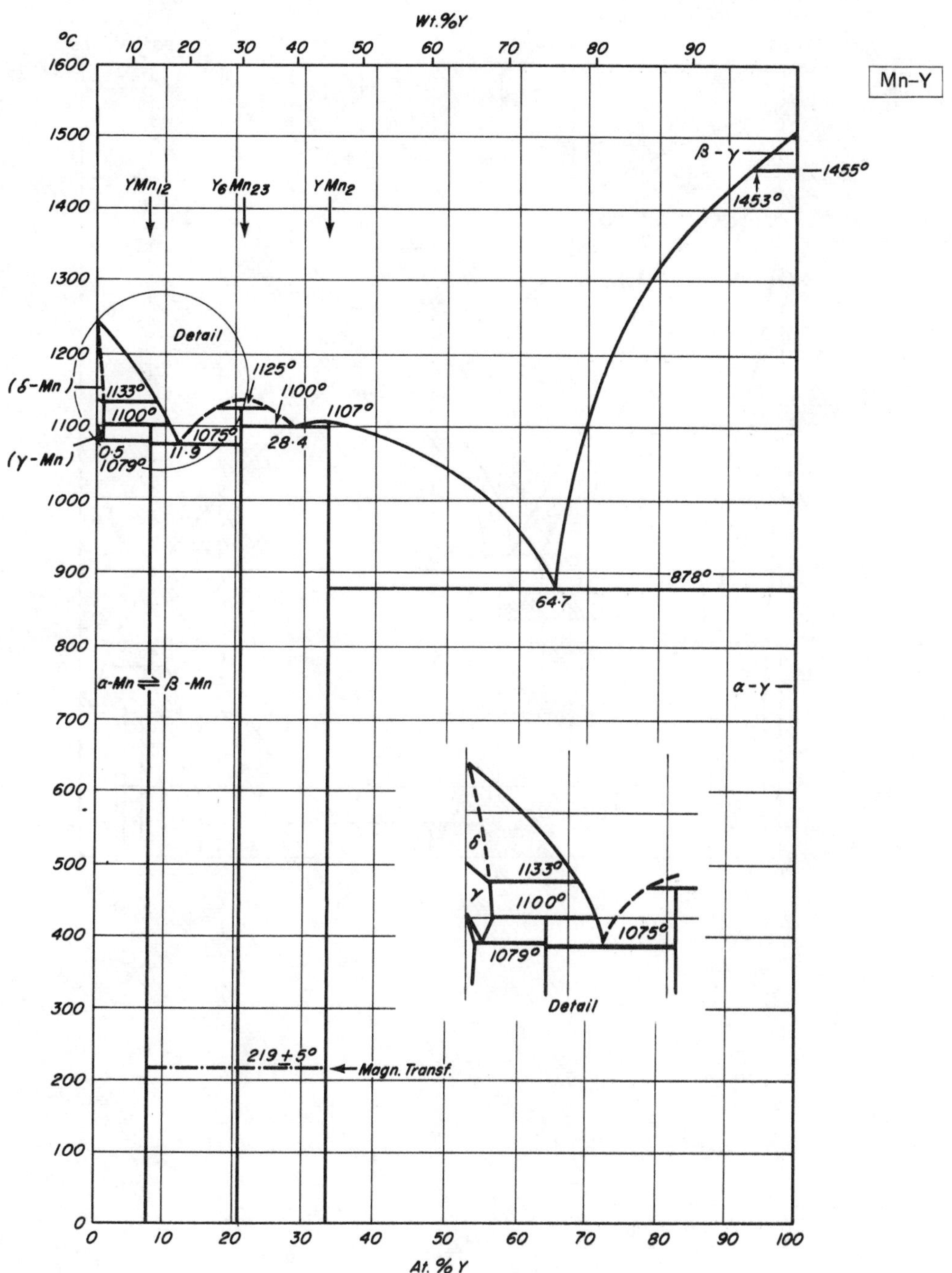
Mn–Y
Wt.%Y
°C
At. % Y
YMn12
Y6Mn23
YMn2
Detail
(δ-Mn)
(γ-Mn)
1133°
1100°
0.5
1079°
11.9
1075°
1125°
1100°
28.4
1107°
64.7
878°
β - γ
1453°
1455°
α-Mn ⇌ β-Mn
α - γ
219 ± 5°
Magn. Transf.
δ
γ
1133°
1100°
1075°
1079°
Detail

Mn–Zn

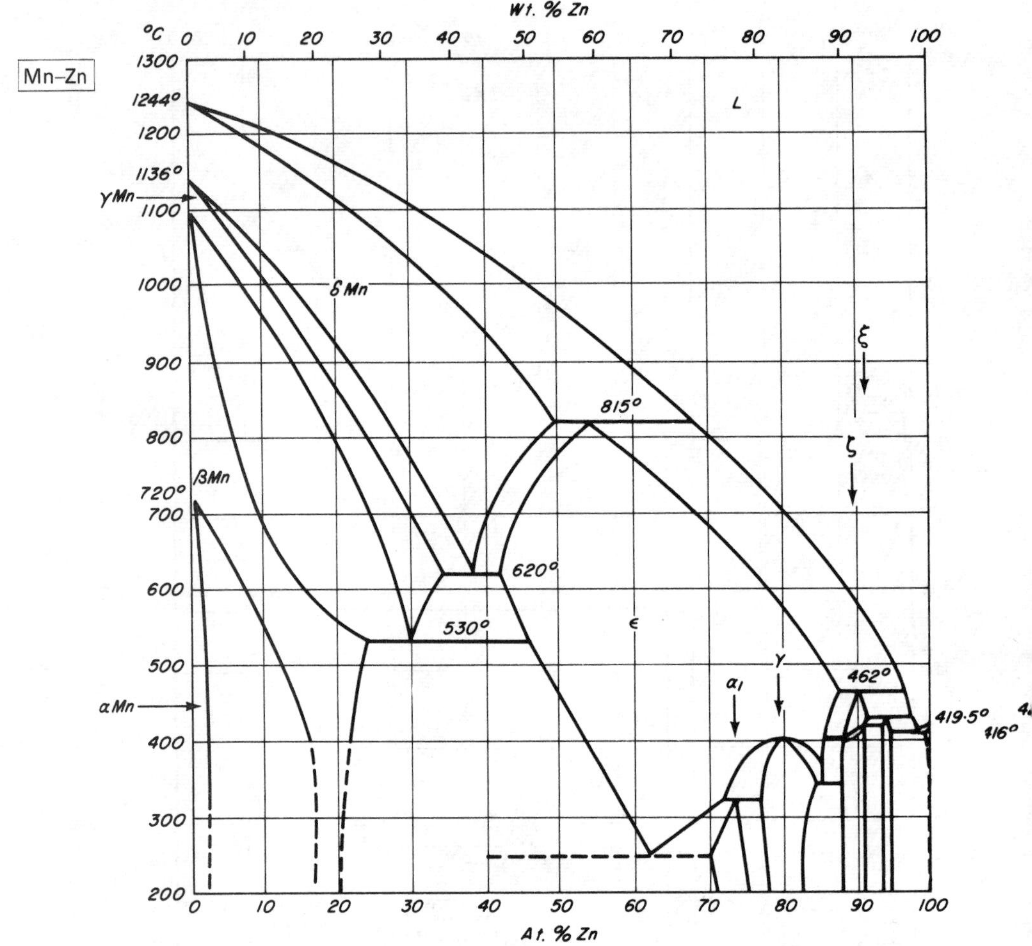

see also detail p. 11–355

Mn–Zn

detail

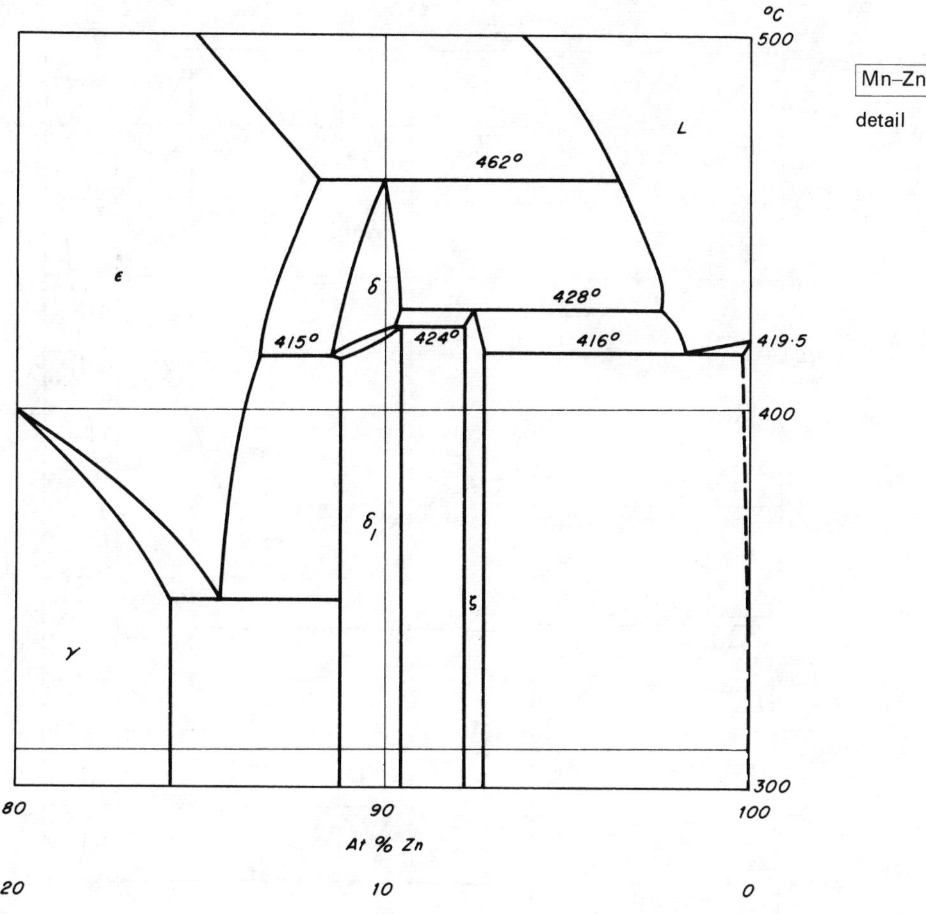
°C
500
400
300
419·5
L
462°
428°
424°
416°
415°
ε
δ
δ_1
ζ
γ
80
90
100
At % Zn
20
10
0
At % Mn

Mn–Zr

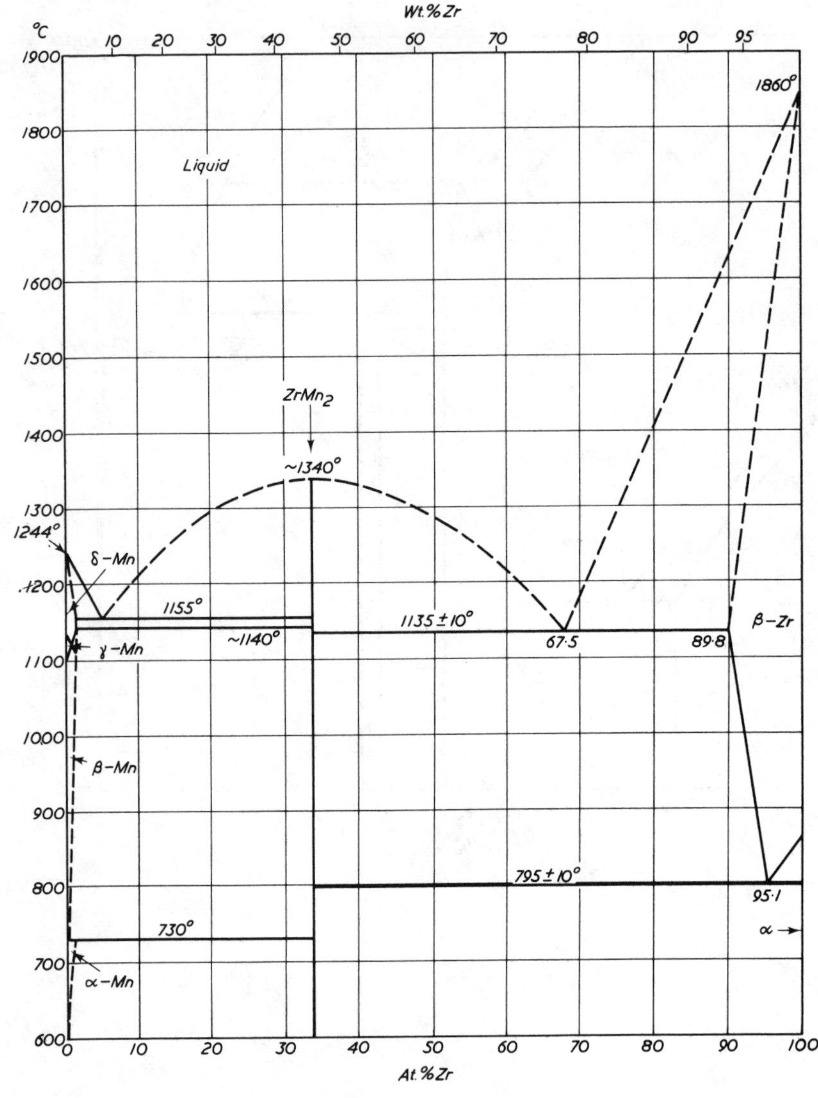

Wt.% Zr
°C
10 20 30 40 50 60 70 80 90 95
1900
1800
1700
1600
1500
1400
1300
1200
1100
1000
900
800
700
600
1860°
Liquid
ZrMn₂
~1340°
1244°
δ-Mn
1155°
1135 ± 10°
~1140°
67·5
89·8
β-Zr
γ-Mn
β-Mn
795 ± 10°
95·1
730°
α
α-Mn
0 10 20 30 40 50 60 70 80 90 100
At.% Zr

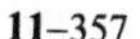

Mo–Ni

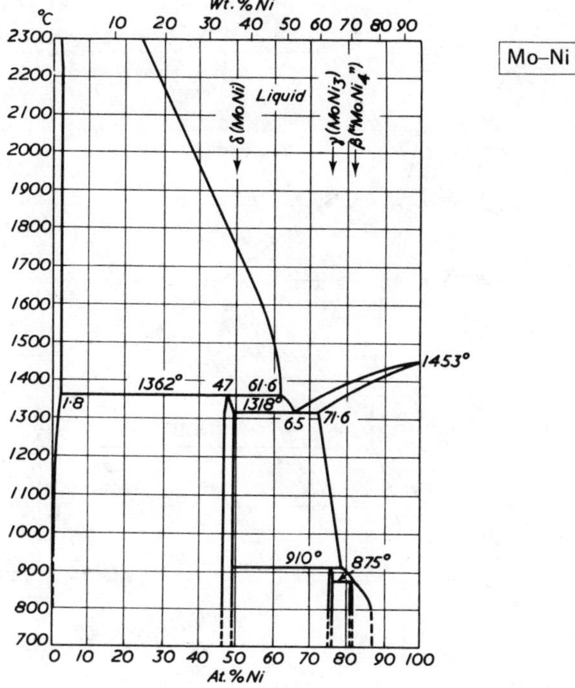

Wt. % Ni
10
20
30
40
50
60
70
80
90
°C
2300
2200
2100
2000
1900
1800
1700
1600
1500
1400
1300
1200
1100
1000
900
800
700
Liquid
δ (MoNi)
γ (MoNi3)
β ("MoNi4")
1453°
1362°
47
61·6
1·8
1318°
65
71·6
910°
875°
0
10
20
30
40
50
60
70
80
90
100
At. % Ni

Mo–Os

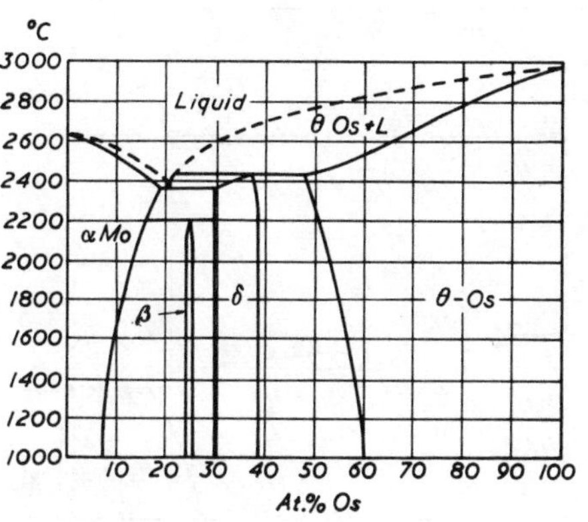

°C
3000
2800
2600
2400
2200
2000
1800
1600
1400
1200
1000
Liquid
θ Os + L
α Mo
β
δ
θ-Os
10
20
30
40
50
60
70
80
90
100
At. % Os

Mo–Pd

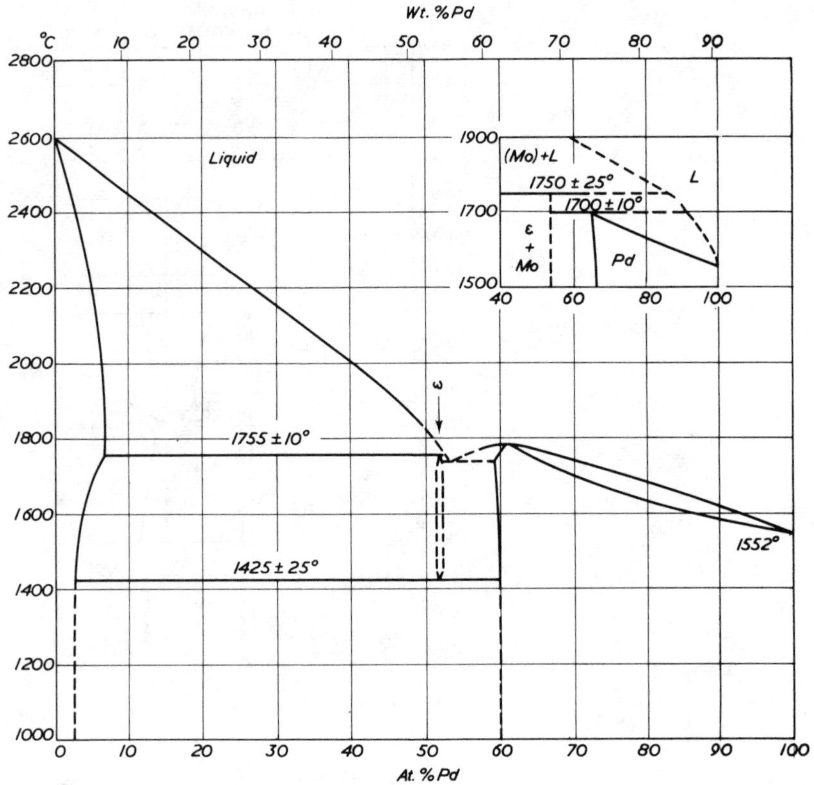

Wt. % Pd
°C
Liquid
(Mo)+L
L
1750 ± 25°
1700 ± 10°
ε + Mo
Pd
ω
1755 ± 10°
1425 ± 25°
1552°
At. % Pd

Mo–Re

°C
Liquid
Mo
σ
σ + Re
σ + Mo
χ + σ
χ
χ + Re
χ + Mo
At.% Re

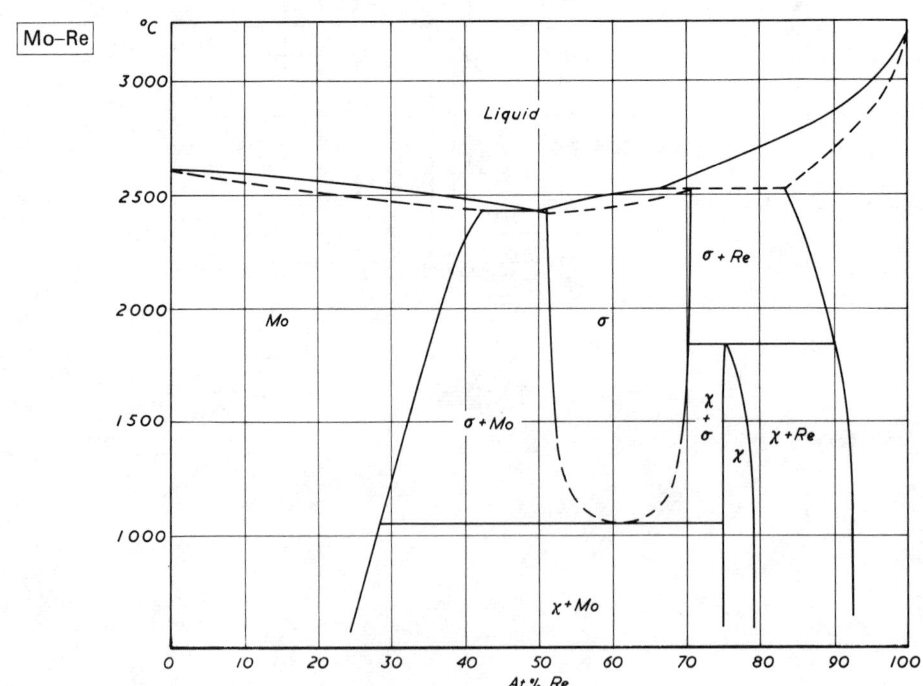

Mo–Rh

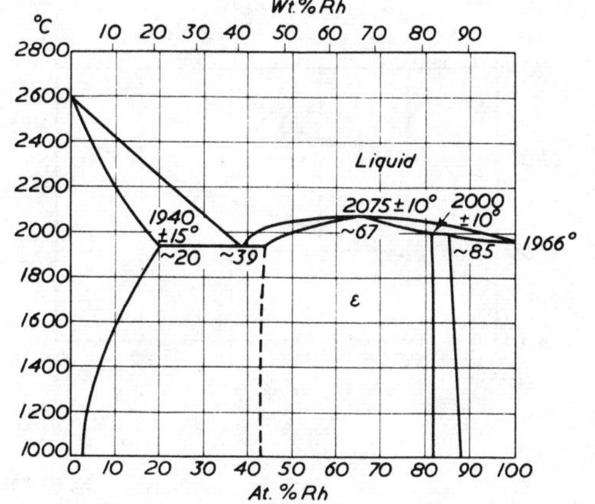

Mo–Ru

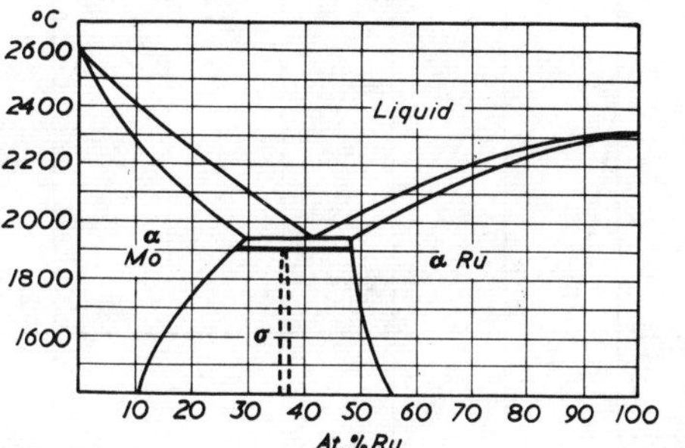

Mo–Si

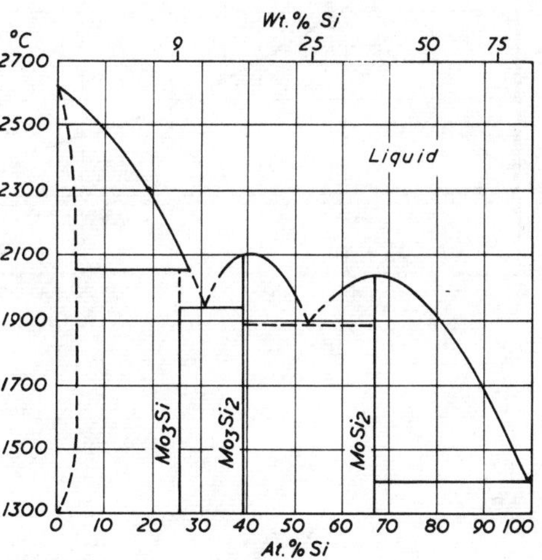

Mo–Ta

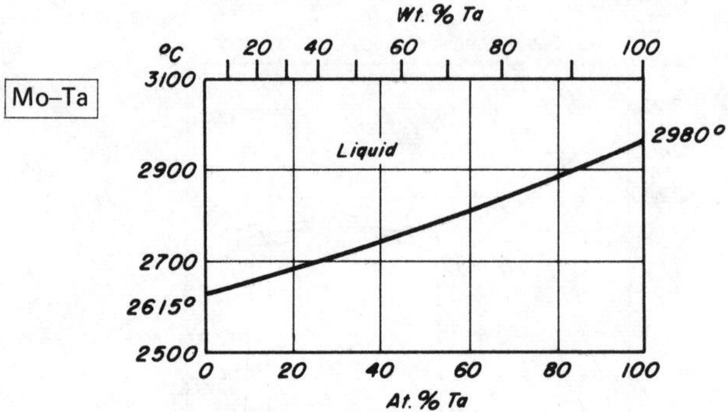

Wt. % Ta
°C
20
40
60
80
100
3100
Liquid
2980°
2900
2700
2615°
2500
0
20
40
60
80
100
At. % Ta

Mo–Th

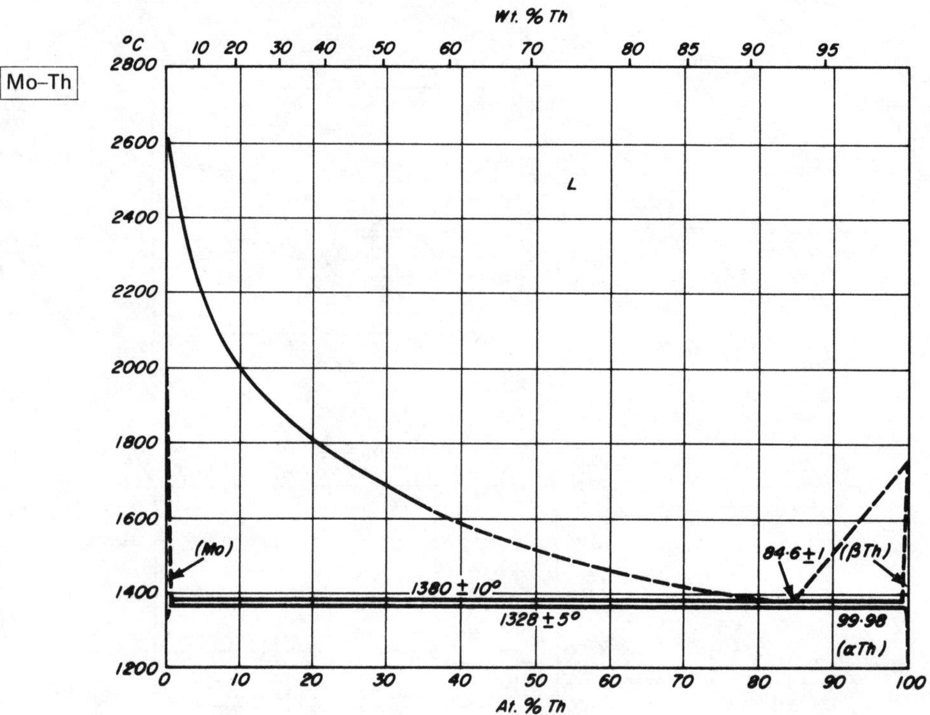

Wt. % Th
°C
10
20
30
40
50
60
70
80
85
90
95
2800
2600
L
2400
2200
2000
1800
1600
(Mo)
84·6±1
(βTh)
1400
1380±10°
1328±5°
99·98
(αTh)
1200
0
10
20
30
40
50
60
70
80
90
100
At. % Th

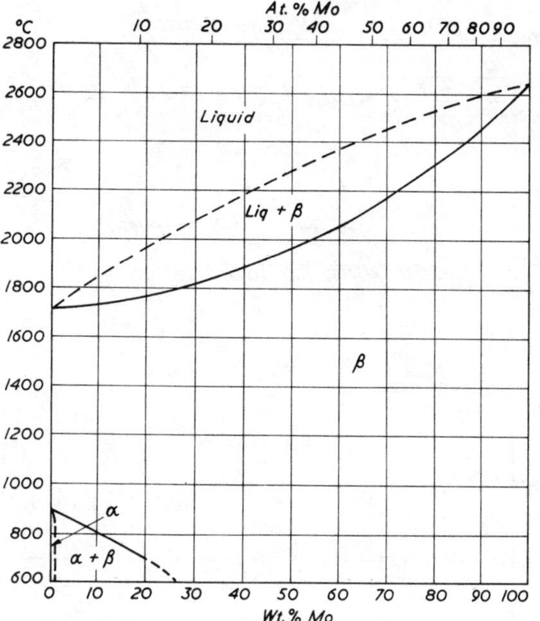

At. % Mo
°C
Liquid
Liq + β
β
α
α + β
Wt. % Mo

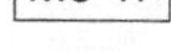

Mo–Ti

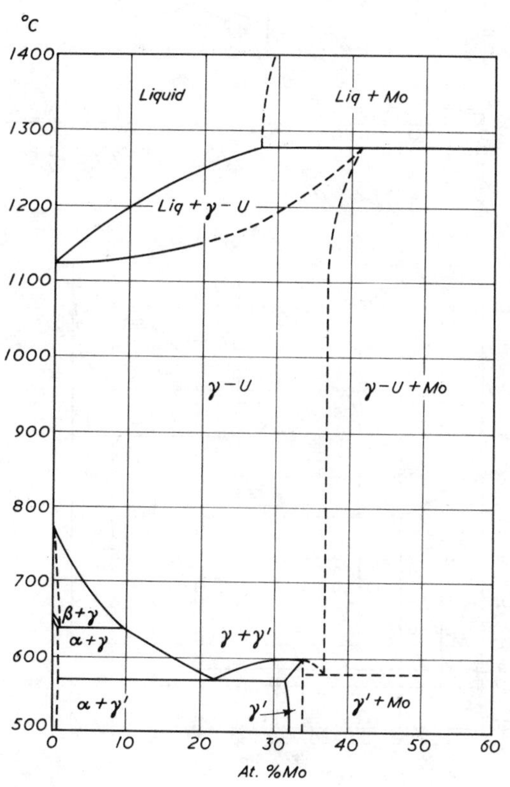

Mo–U
°C
Liquid
Liq + Mo
Liq + γ − U
γ − U
γ − U + Mo
β+γ
α + γ
γ + γ′
α + γ′
γ′
γ′ + Mo
At. % Mo

Mo–V

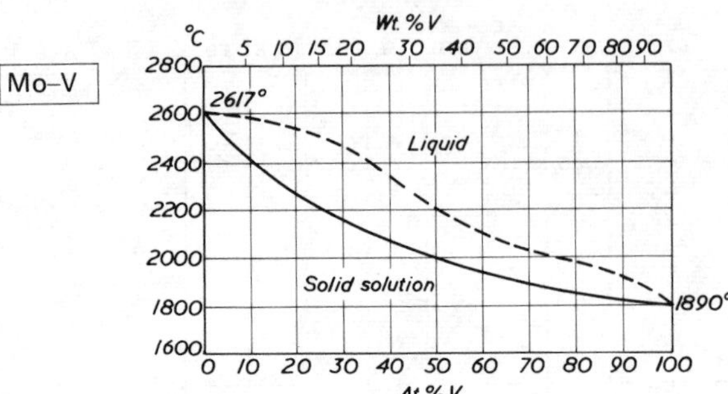
Wt. % V
°C
5 10 15 20 30 40 50 60 70 80 90
2800
2600
2400
2200
2000
1800
1600
2617°
Liquid
Solid solution
1890°
0 10 20 30 40 50 60 70 80 90 100
At. % V

Mo–W

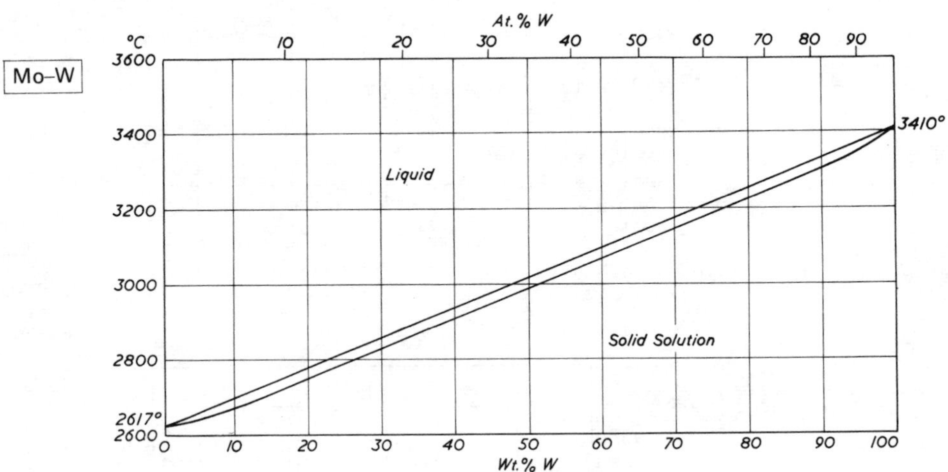
At. % W
°C
10 20 30 40 50 60 70 80 90
3600
3400
3200
3000
2800
2617°
2600
3410°
Liquid
Solid Solution
0 10 20 30 40 50 60 70 80 90 100
Wt. % W

Mo–Zr

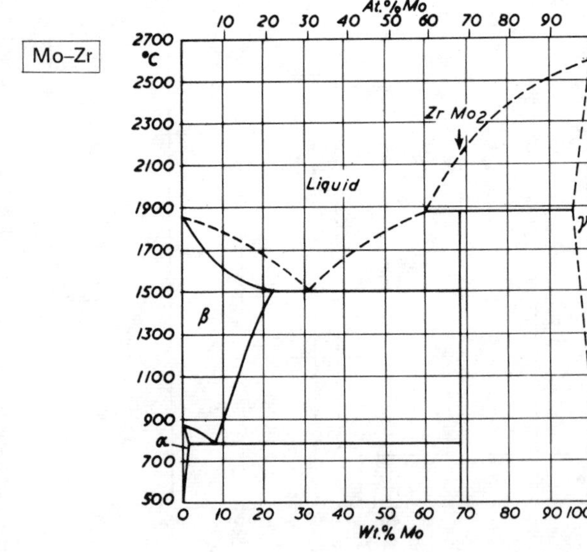
At.% Mo
10 20 30 40 50 60 70 80 90
2700
°C
2500
2300
2100
1900
1700
1500
1300
1100
900
α
700
500
Zr Mo2
Liquid
γ
β
0 10 20 30 40 50 60 70 80 90 100
Wt.% Mo

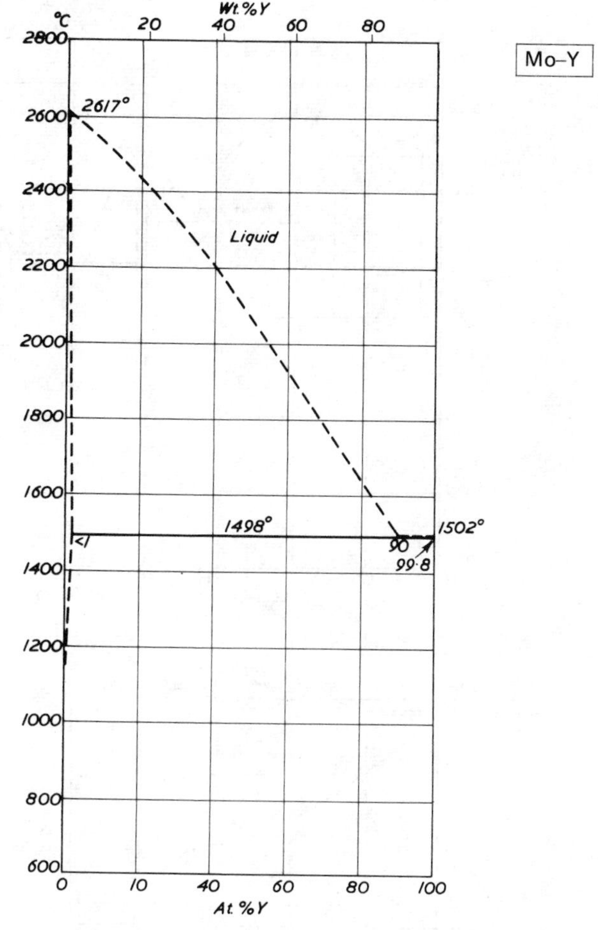
Mo–Y
°C
Wt.%Y
20
40
60
80
2800
2600
2400
2200
2000
1800
1600
1400
1200
1000
800
600
2617°
Liquid
1498°
1502°
<1
90
99·8
0
10
40
60
80
100
At.%Y

N–Ti

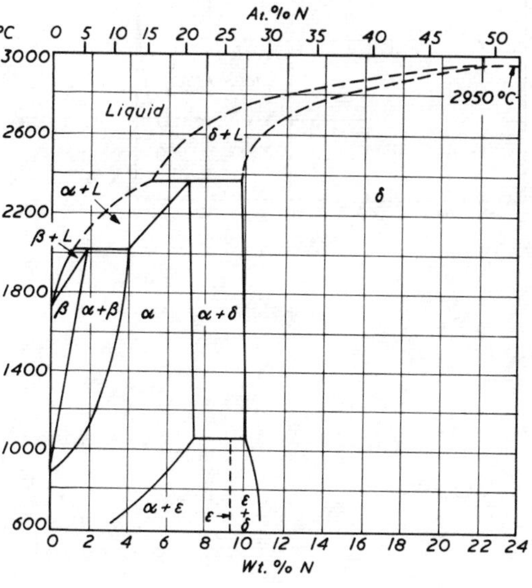
At.% N
°C
0 5 10 15 20 25 30 35 40 45 50
3000
2600
2200
1800
1400
1000
600
2950 °C
Liquid
δ+L
α+L
β+L
δ
β
α+β
α
α+δ
α+ε
ε
ε+δ
0 2 4 6 8 10 12 14 16 18 20 22 24
Wt.% N

Na–Pb

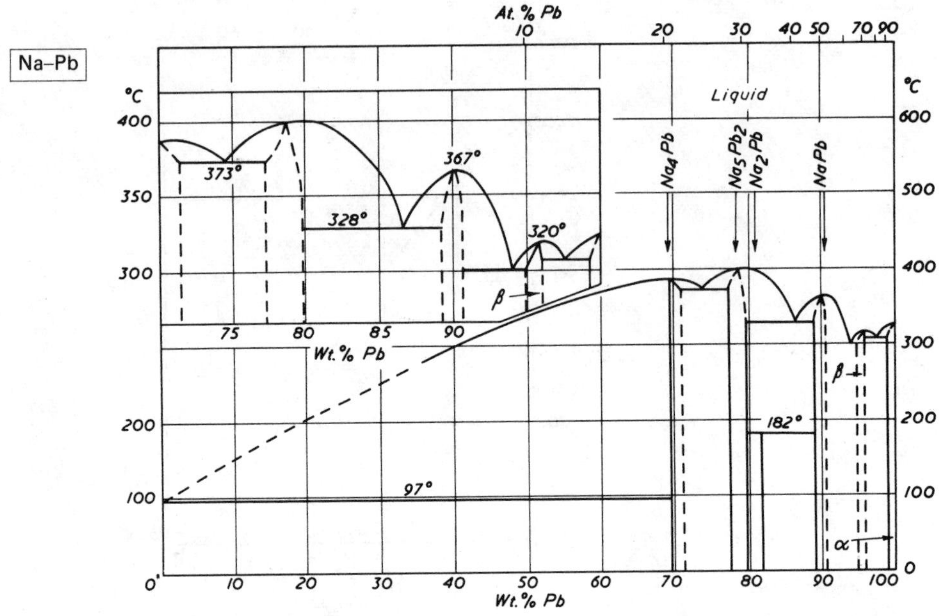

Na–Rb

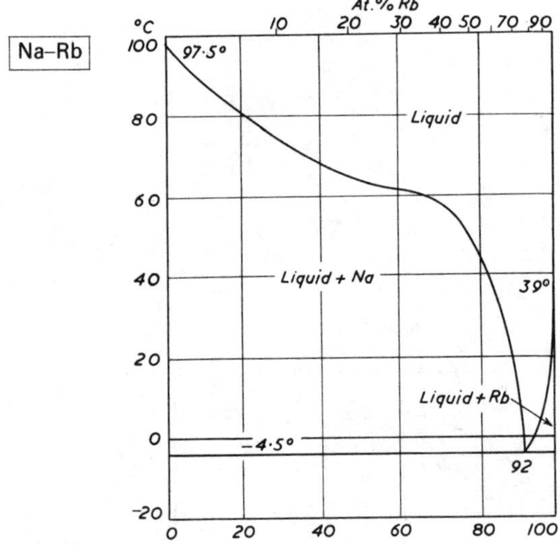

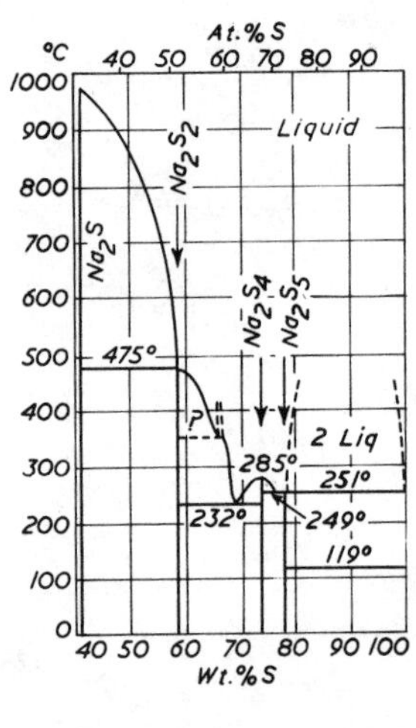

Na–S

Na–Sb

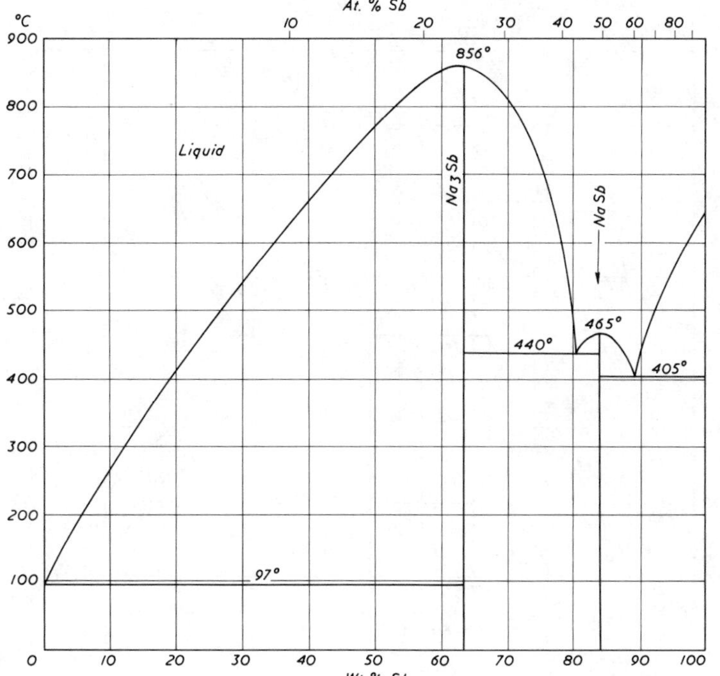
At. % Sb
10
20
30
40
50
60
80
°C
900
800
700
600
500
400
300
200
100
0
856°
Liquid
Na_3Sb
NaSb
465°
440°
405°
97°
10
20
30
40
50
60
70
80
90
100
Wt. % Sb

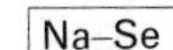
Na–Se

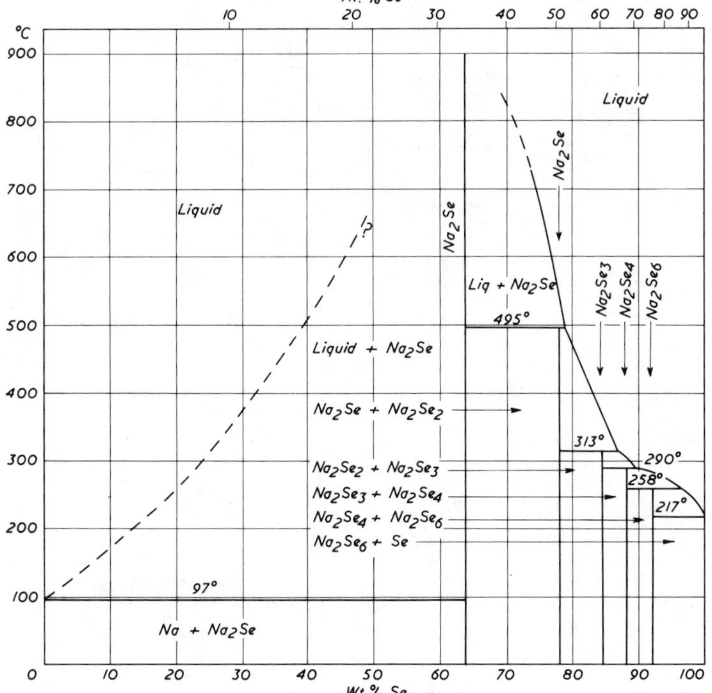
At. % Se
10
20
30
40
50
60
70
80
90
°C
900
800
700
600
500
400
300
200
100
0
Liquid
Liquid
Na_2Se
Na_2Se
?
Liq + Na_2Se
495°
Na_2Se_3
Na_2Se_4
Na_2Se_6
Liquid + Na_2Se
Na_2Se + Na_2Se_2
313°
290°
258°
217°
Na_2Se_2 + Na_2Se_3
Na_2Se_3 + Na_2Se_4
Na_2Se_4 + Na_2Se_6
Na_2Se_6 + Se
97°
Na + Na_2Se
10
20
30
40
50
60
70
80
90
100
Wt. % Se

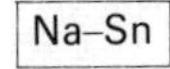

Na–Sn

At. % Na
°C
10 30 40 50 60 70 80 85 90 95
600
550
500
450
400
350
300
250
200
150
100
50
0
10 20 30 40 50 60 70 80 90 100
Wt. % Na
Liq + α NaSn
Liq + β NaSn
Liquid
β NaSn + Na4Sn3
Na4Sn3 + Na2Sn
Liq + Na2Sn
Na2Sn + Na4Sn
Liq + β NaSn
L + NaSn2
NaSn + NaSn2
L + NaSn3
Na2Sn + Na3Sn
Na3Sn + Na4Sn
Liq + Na4Sn
NaSn4 + β NaSn
NaSn3 + NaSn2
Sn + NaSn6
NaSn4 + NaSn3
NaSn6 + NaSn4
Na4Sn + Na

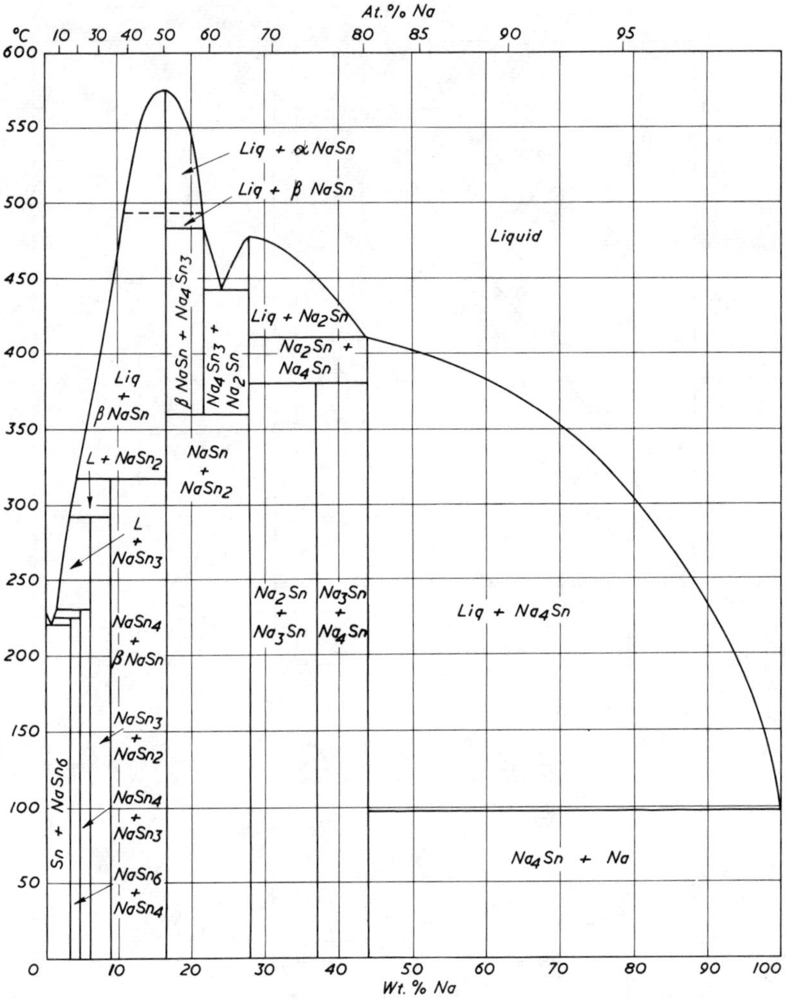

Na–Sr

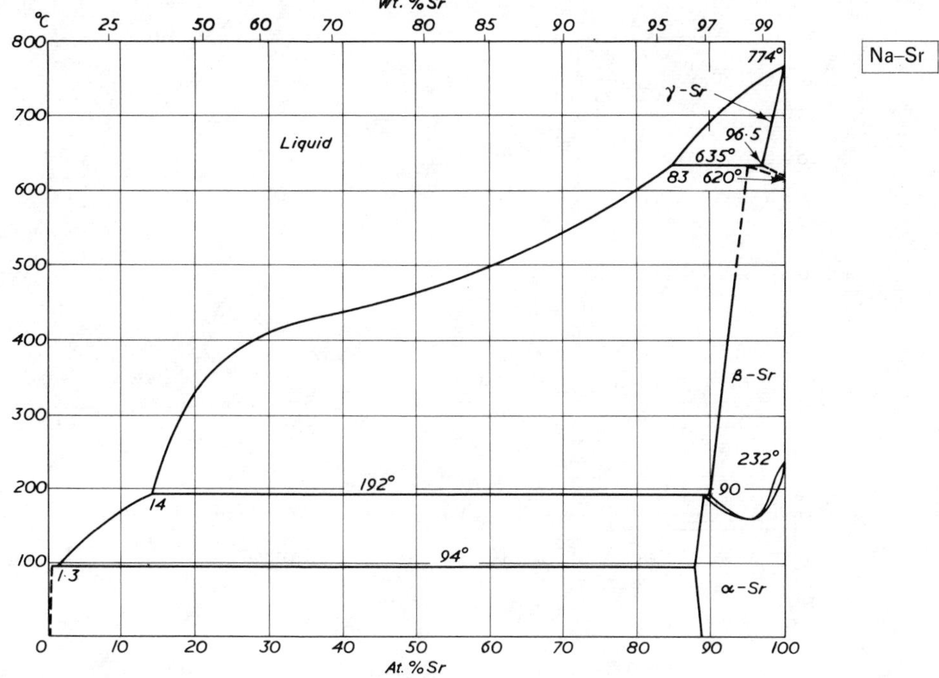

Wt. % Sr
25
50
60
70
80
85
90
95
97
99
°C
774°
γ-Sr
96·5
635°
83
620°
Liquid
β-Sr
232°
192°
14
90
94°
1·3
α-Sr
At. % Sr

Na–Te

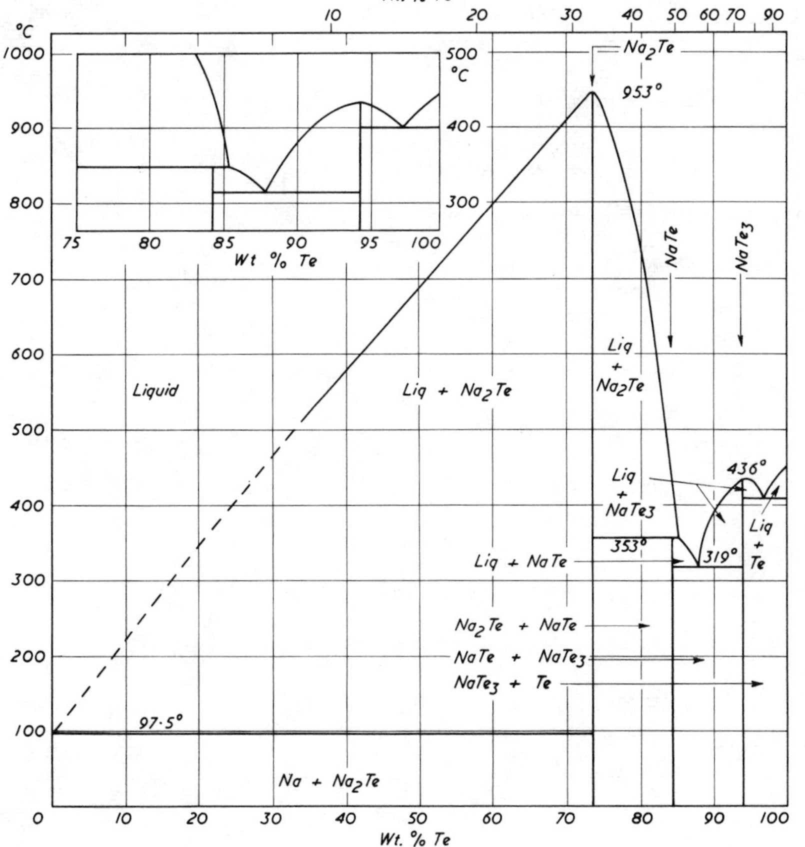

At. % Te
°C
Wt % Te
Na_2Te
953°
NaTe
$NaTe_3$
Liquid
Liq + Na_2Te
Liq + Na_2Te
Liq + $NaTe_3$
436°
Liq + Te
353°
319°
Liq + NaTe
Na_2Te + NaTe
NaTe + $NaTe_3$
$NaTe_3$ + Te
97·5°
Na + Na_2Te
Wt. % Te

Na–Th

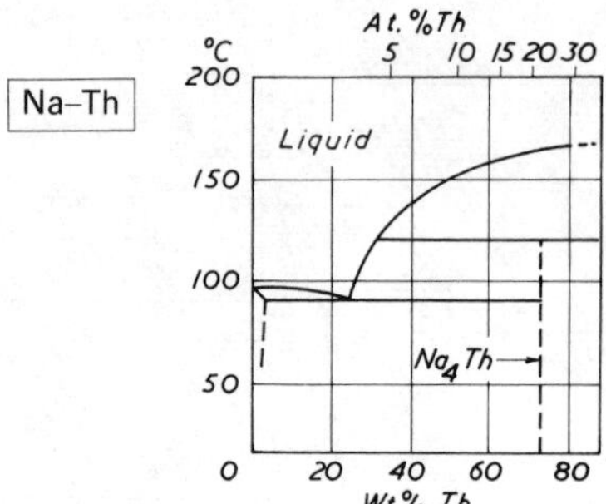

Na–Tl

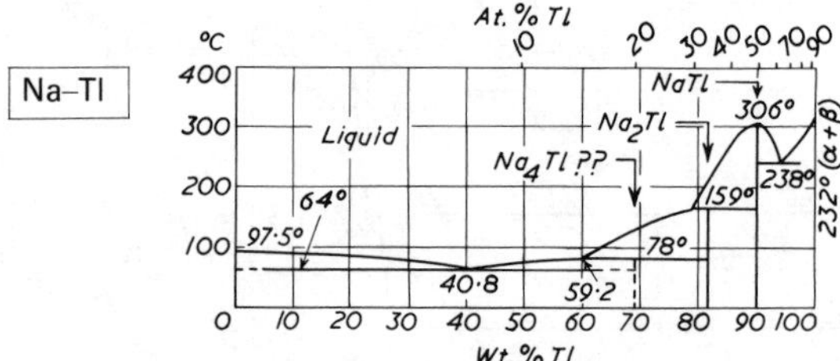

Na–Zn

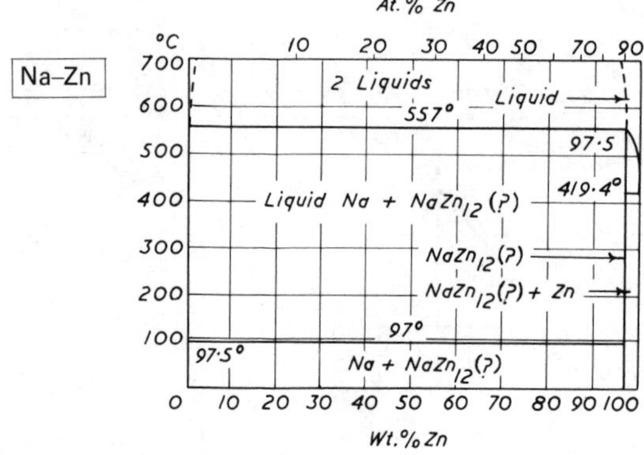

Nb–Ni

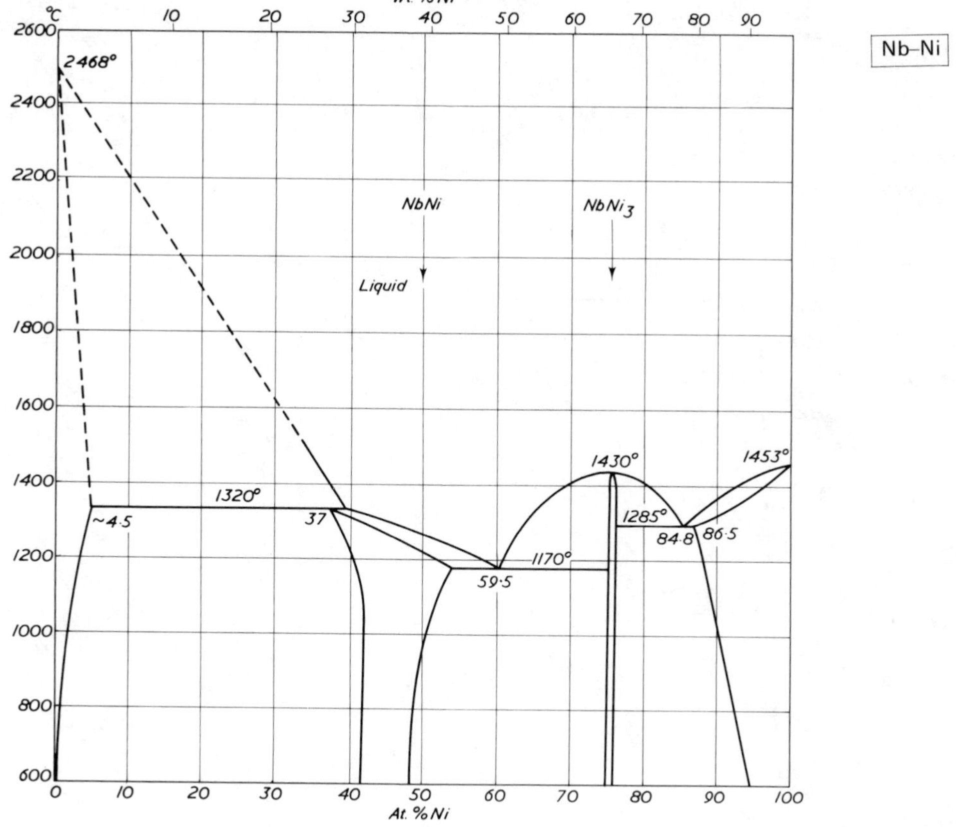

Wt. % Ni
10
20
30
40
50
60
70
80
90
°C
2600
2400
2200
2000
1800
1600
1400
1200
1000
800
600
2468°
NbNi
NbNi3
Liquid
1430°
1453°
1320°
~4·5
37
1285°
84·8
86·5
1170°
59·5
0
10
20
30
40
50
60
70
80
90
100
At. % Ni

Nb–O

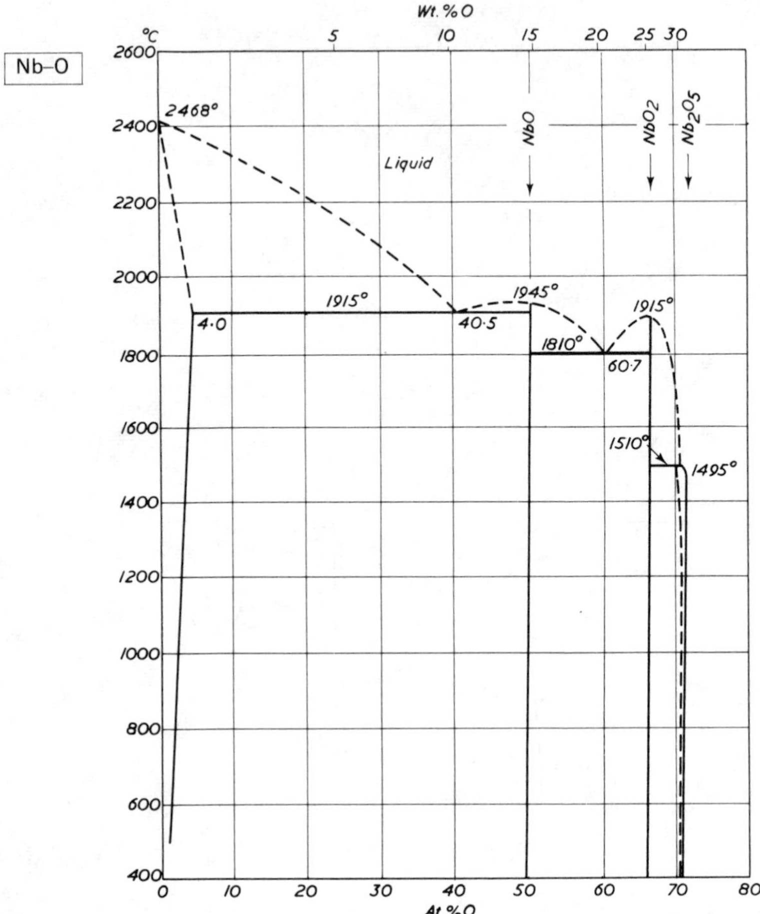
Wt.%O
°C
5
10
15
20
25
30
2600
2400
2200
2000
1800
1600
1400
1200
1000
800
600
400
2468°
Liquid
NbO
NbO_2
Nb_2O_5
1915°
1945°
1915°
4·0
40·5
1810°
60·7
1510°
1495°
0
10
20
30
40
50
60
70
80
At.%O

Nb–Re

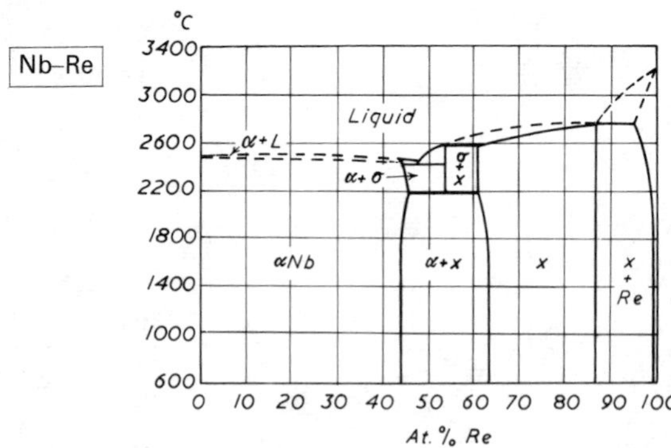
°C
3400
3000
2600
2200
1800
1400
1000
600
Liquid
α+L
α+σ
σ + χ
αNb
α+χ
χ
χ + Re
0
10
20
30
40
50
60
70
80
90
100
At.% Re

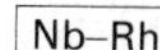

Wt. %Rh

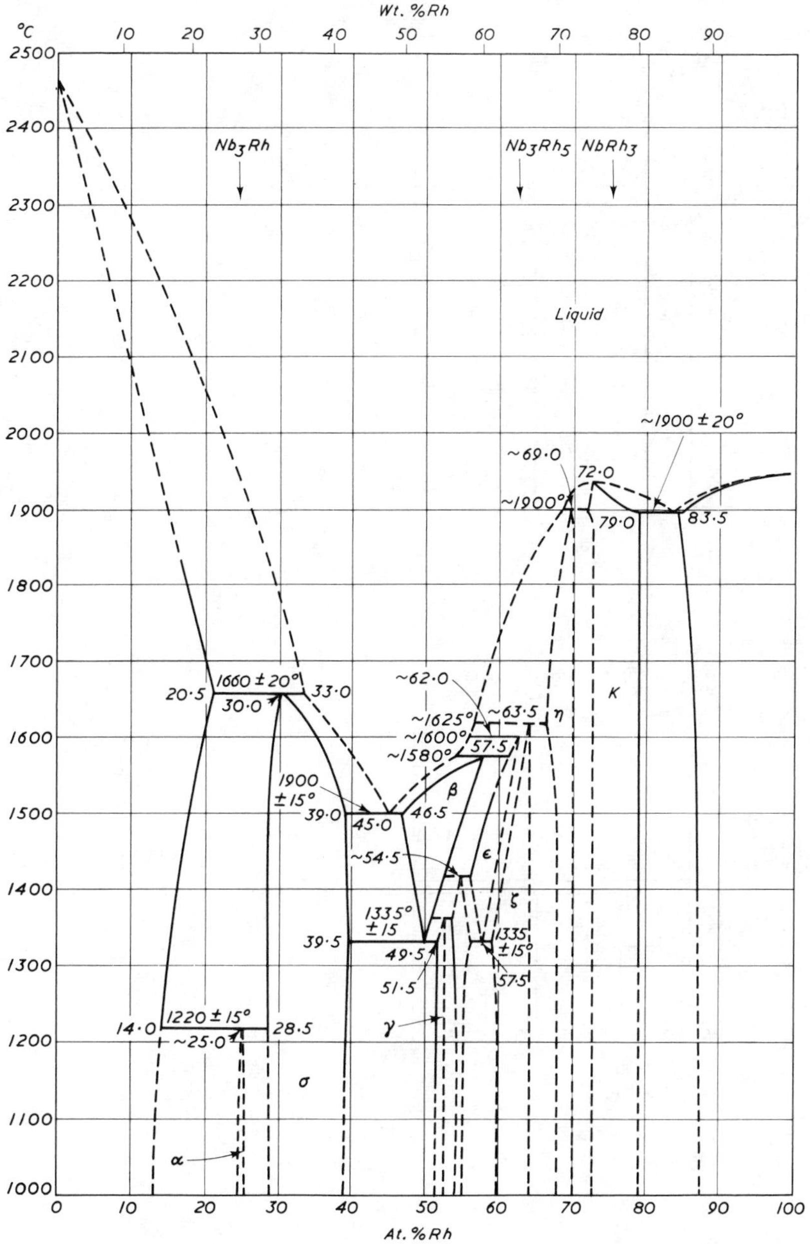

°C

Nb$_3$Rh

Nb$_3$Rh$_5$

NbRh$_3$

Liquid

~1900 ± 20°

~69·0

72·0

~1900°

79·0

83·5

20·5

1660 ± 20°

33·0

30·0

~62·0

~1625°

~63·5

η

K

~1600°

57·5

~1580°

1900 ± 15°

39·0

45·0

46·5

β

~54·5

ε

ζ

1335° ± 15

39·5

49·5

1335 ± 15°

51·5

57·5

14·0

1220 ± 15°

28·5

~25·0

γ

σ

α

At. %Rh

Nb–Si

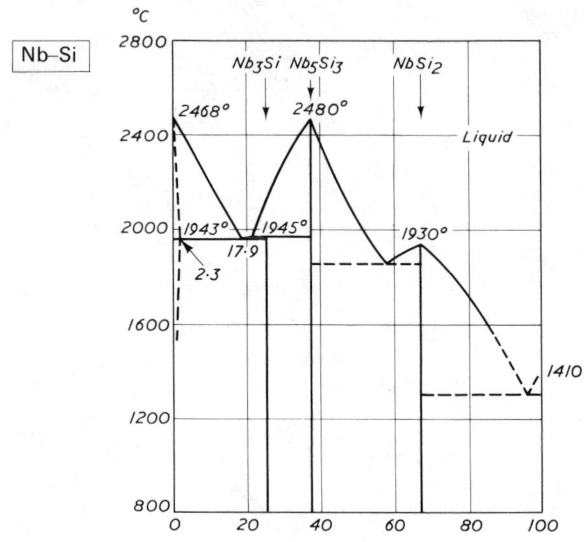
°C
Nb3Si
Nb5Si3
NbSi2
2468°
2480°
Liquid
1943°
1945°
17·9
2·3
1930°
1410°
At.%Si

Nb–Sn

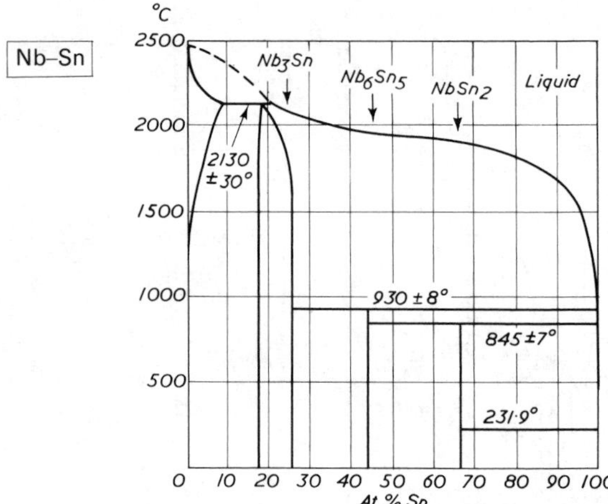
°C
Nb3Sn
Nb6Sn5
NbSn2
Liquid
2130
±30°
930 ± 8°
845 ±7°
231·9°
At.% Sn

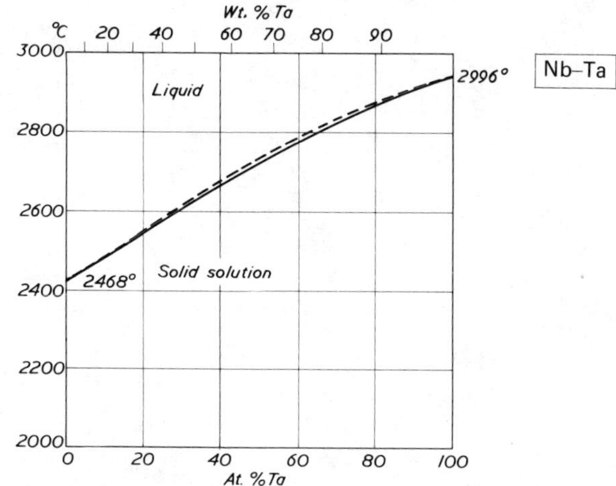

Nb–Ta

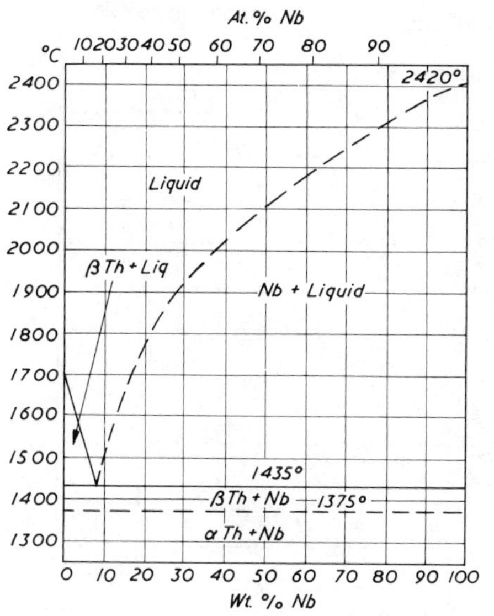

Nb–Th

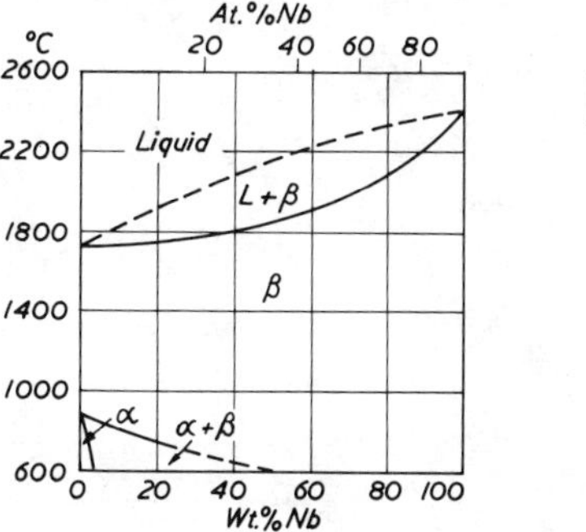

Nb–Ti

Nb–V

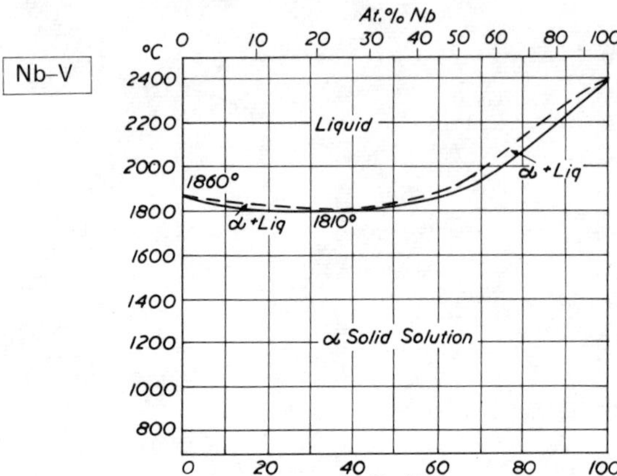
At.% Nb
°C
Liquid
1860°
α + Liq
α + Liq
1810°
α Solid Solution
Wt.% Nb

Nb–Y

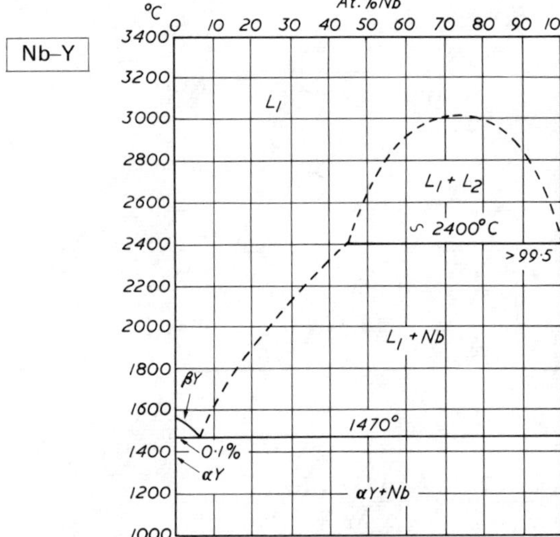
At.%Nb
°C
L_1
$L_1 + L_2$
~ 2400°C
>99·5
L_1 + Nb
βY
1470°
0·1%
αY
αY+Nb
Wt.%Nb

Nb–Zr

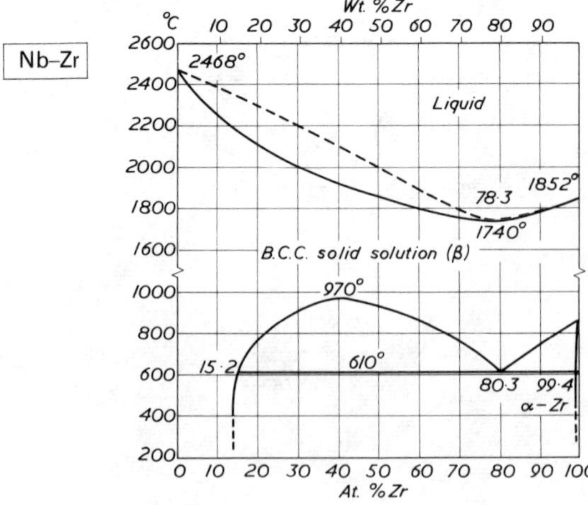
Wt. %Zr
°C
2468°
Liquid
1852°
78·3
1740°
B.C.C. solid solution (β)
970°
15·2
610°
80·3
99·4
α – Zr
At. %Zr

Nd–Pr

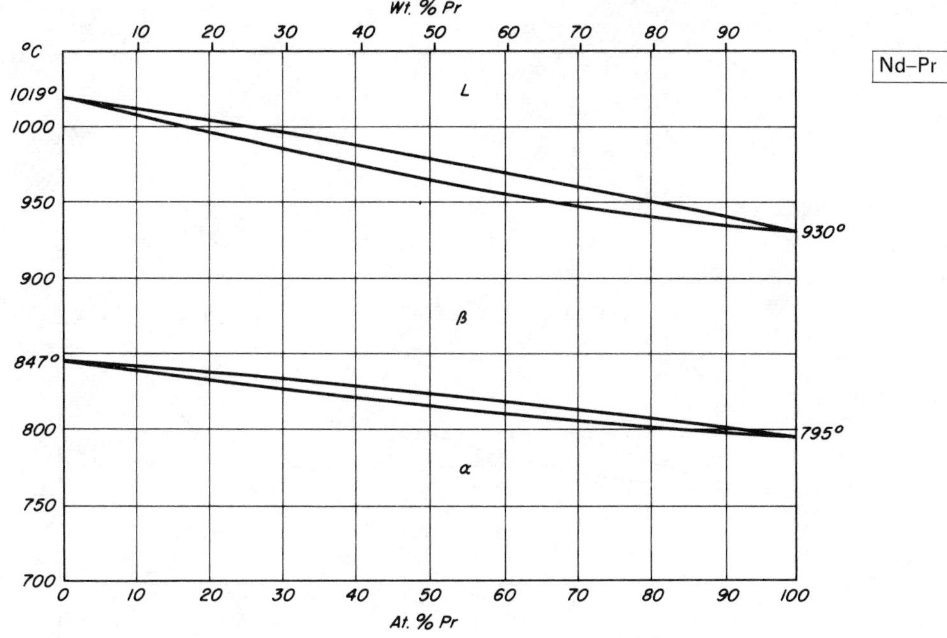
Wt. % Pr
°C
1019°
930°
847°
795°
L
β
α
At. % Pr

Nd–Rh

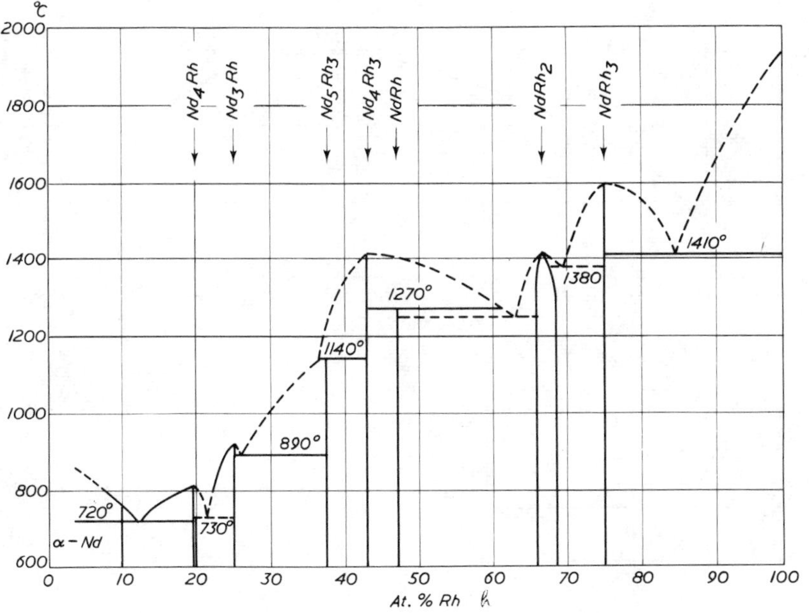
°C
Nd4Rh
Nd3Rh
Nd5Rh3
Nd4Rh3
NdRh
NdRh2
NdRh3
1410°
1380
1270°
1140°
890°
720°
730°
α - Nd
At. % Rh

Nd–Sc

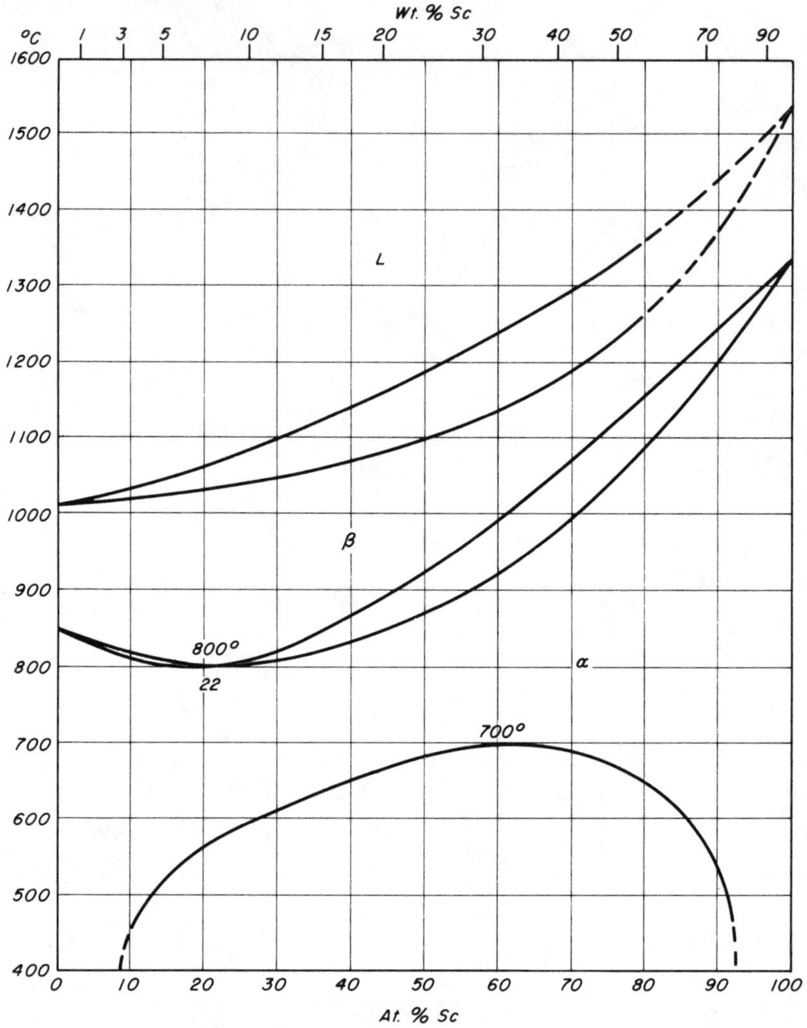

Nd–Ti

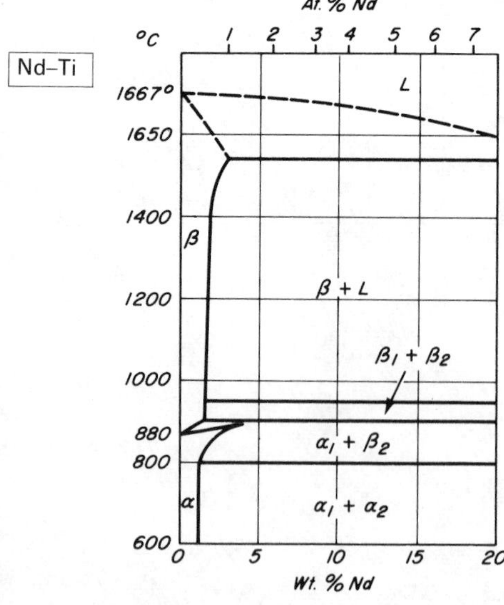
At.% Nd
°C
1 2 3 4 5 6 7
1667°
1650
1400
1200
1000
880
800
600
L
β
β + L
$β_1 + β_2$
$α_1 + β_2$
$α_1 + α_2$
α
0 5 10 15 20
Wt.% Nd

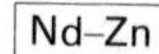
Nd–Zn

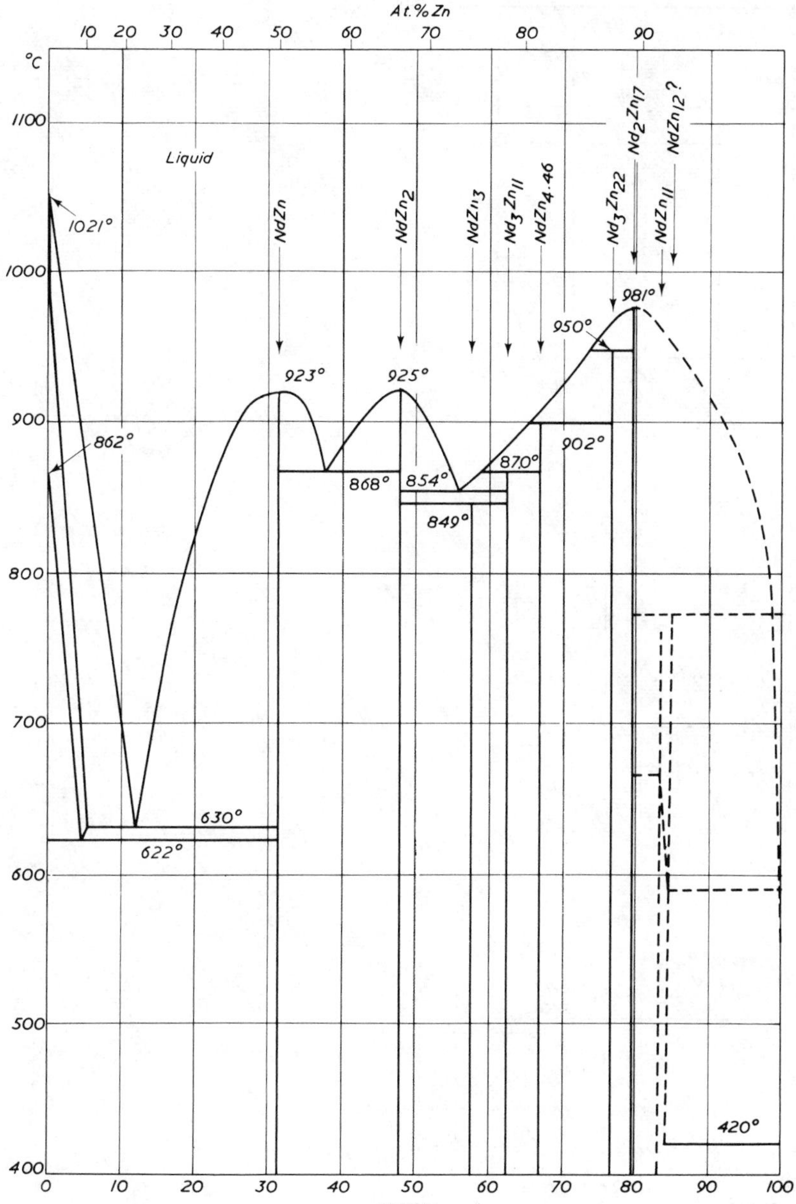

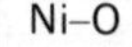
Ni–O

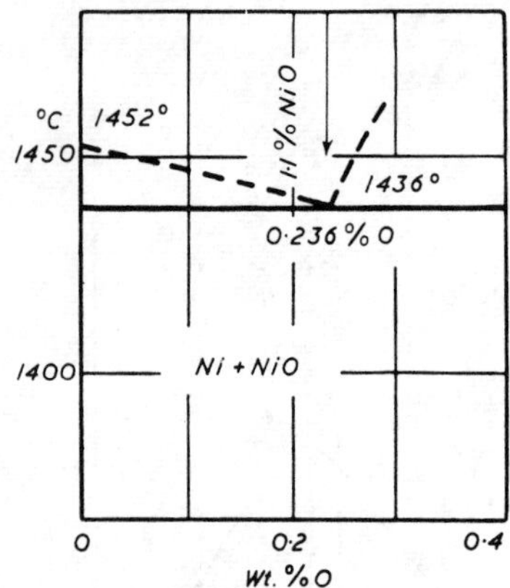
°C
1450
1400
1452°
1·1 % NiO
1436°
0·236 % O
Ni + NiO
0
0·2
0·4
Wt. % O

Ni–P

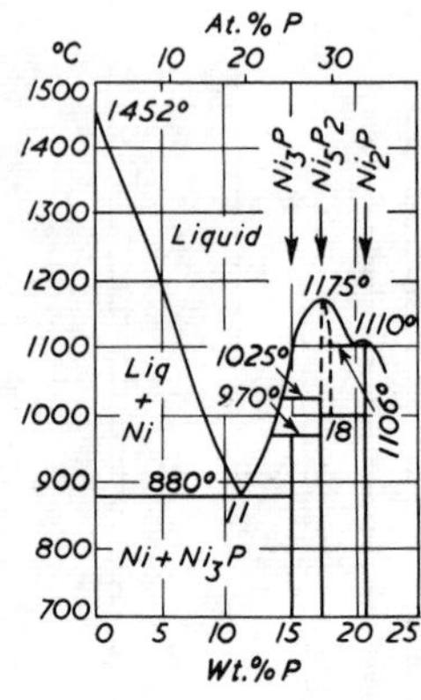
At. % P
°C
10 20 30
1500
1400
1300
1200
1100
1000
900
800
700
1452°
Ni3P
Ni5P2
Ni2P
Liquid
1175°
1110°
1025°
970°
Liq + Ni
18
1106°
880°
11
Ni + Ni3P
0 5 10 15 20 25
Wt. % P

Ni–Pb

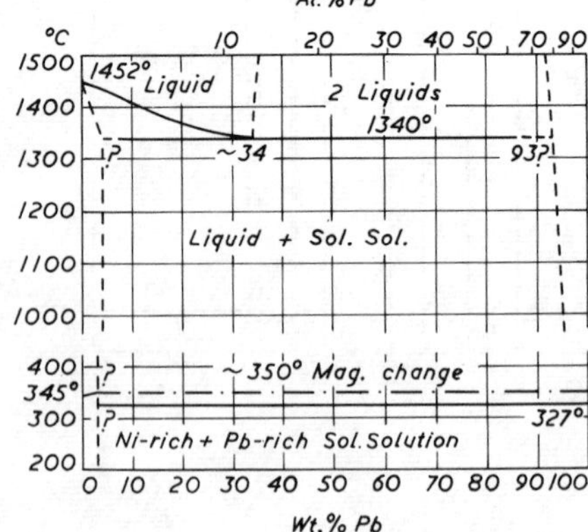
At. % Pb
°C
10 20 30 40 50 70 90
1500
1400
1300
1200
1100
1000
400
345°
300
200
1452°
Liquid
2 Liquids
1340°
?
~34
93?
Liquid + Sol. Sol.
~350° Mag. change
327°
Ni-rich + Pb-rich Sol. Solution
0 10 20 30 40 50 60 70 80 90 100
Wt. % Pb

Ni–Pd

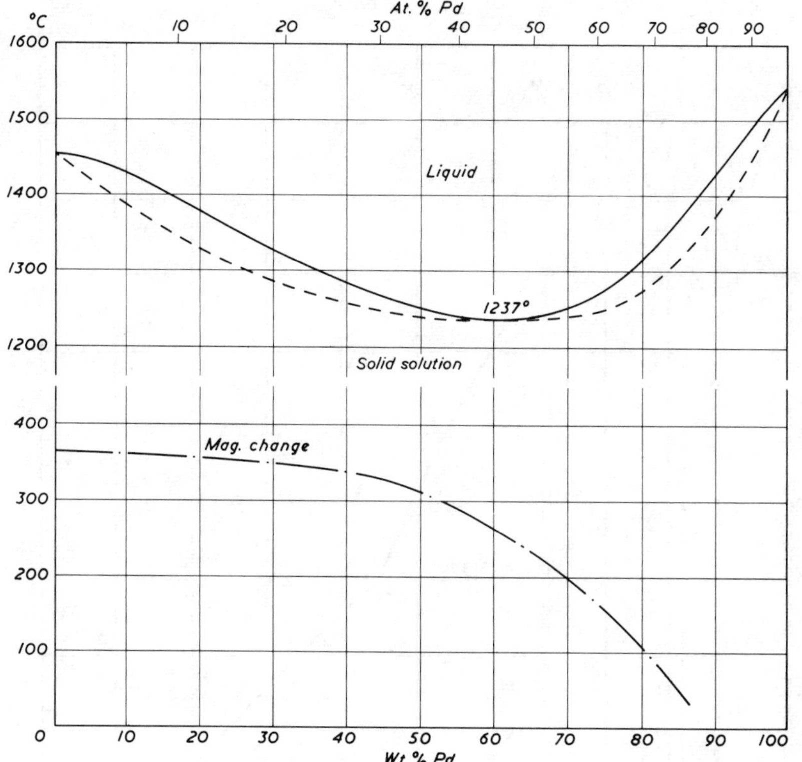

At.% Pd
°C
Liquid
1237°
Solid solution
Mag. change
Wt.% Pd

Ni–Pr

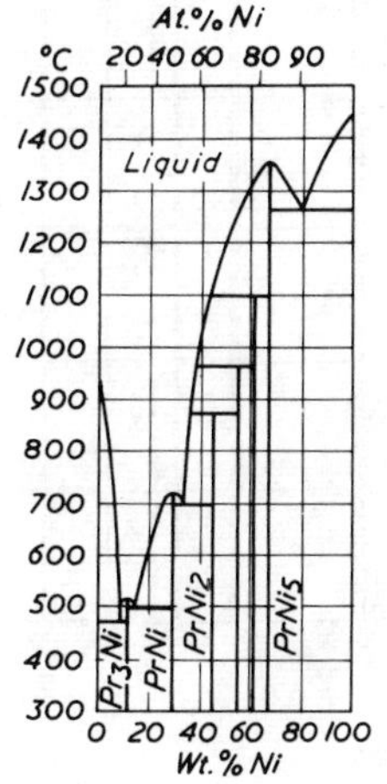

At.% Ni
°C
Liquid
Pr_3Ni
PrNi
$PrNi_2$
$PrNi_5$
Wt.% Ni

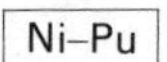
Ni–Pu
Wt.% Pu 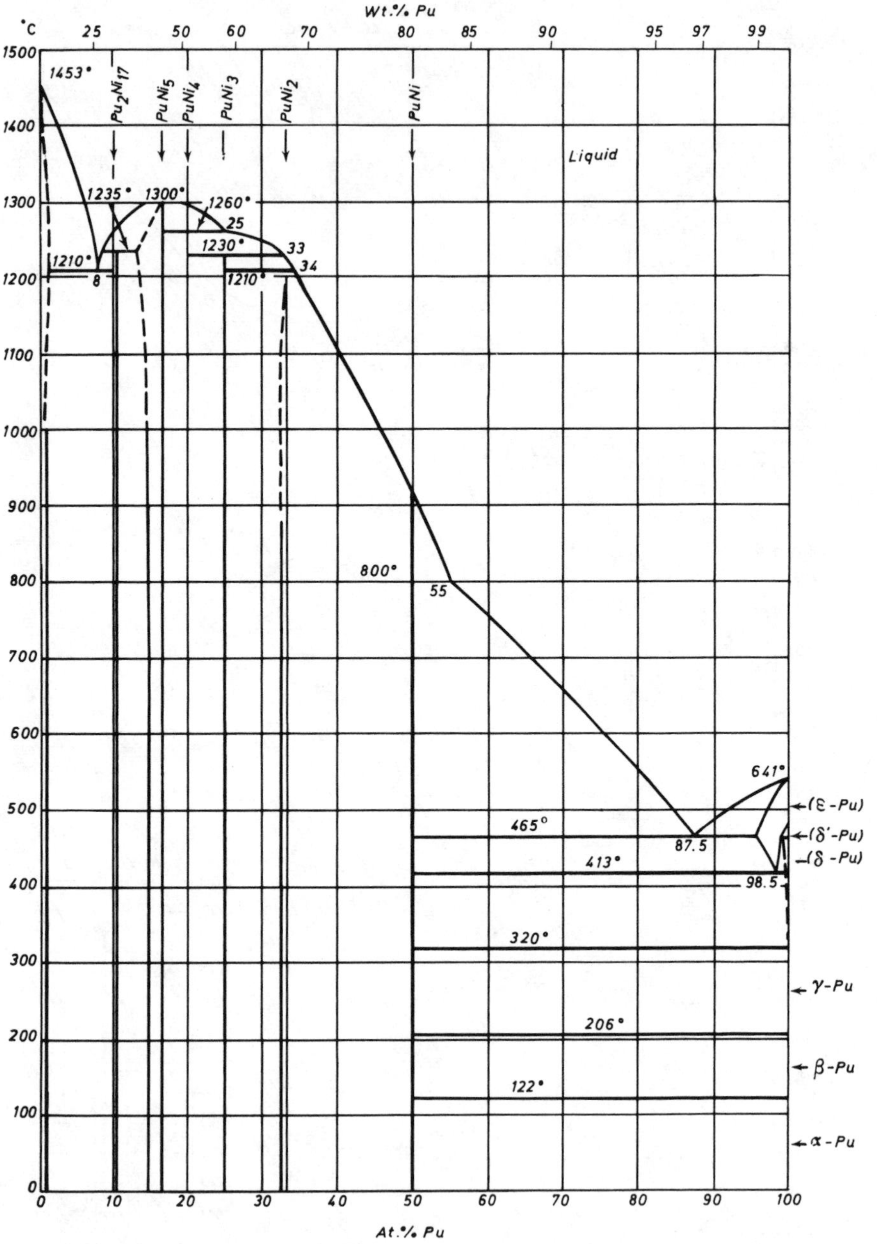
25 50 60 70 80 85 90 95 97 99
°C
1500 1400 1300 1200 1100 1000 900 800 700 600 500 400 300 200 100 0
1453°
Pu_2Ni_{17}
$PuNi_5$
$PuNi_4$
$PuNi_3$
$PuNi_2$
PuNi
Liquid
1235° 1300° 1260°
25
1230° 33
1210° 8
1210° 34
800°
55
641°
465°
87.5
413°
98.5
320°
206°
122°
(ε-Pu)
(δ'-Pu)
(δ-Pu)
γ-Pu
β-Pu
α-Pu
0 10 20 30 40 50 60 70 80 90 100
At.% Pu

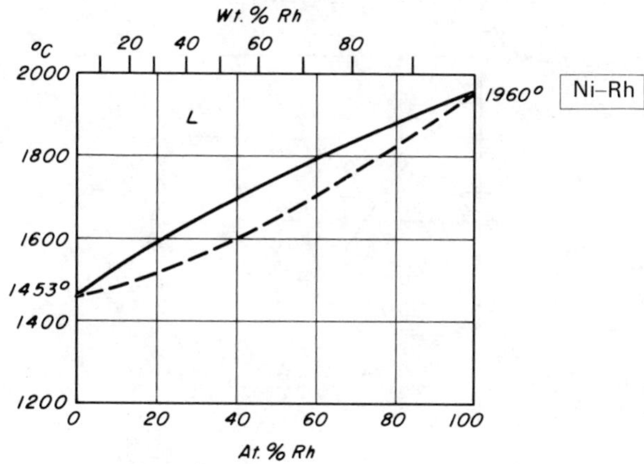
Wt.% Rh
°C
20
40
60
80
2000
1800
1600
1453°
1400
1200
L
1960°
0
20
40
60
80
100
At.% Rh

Ni–Rh

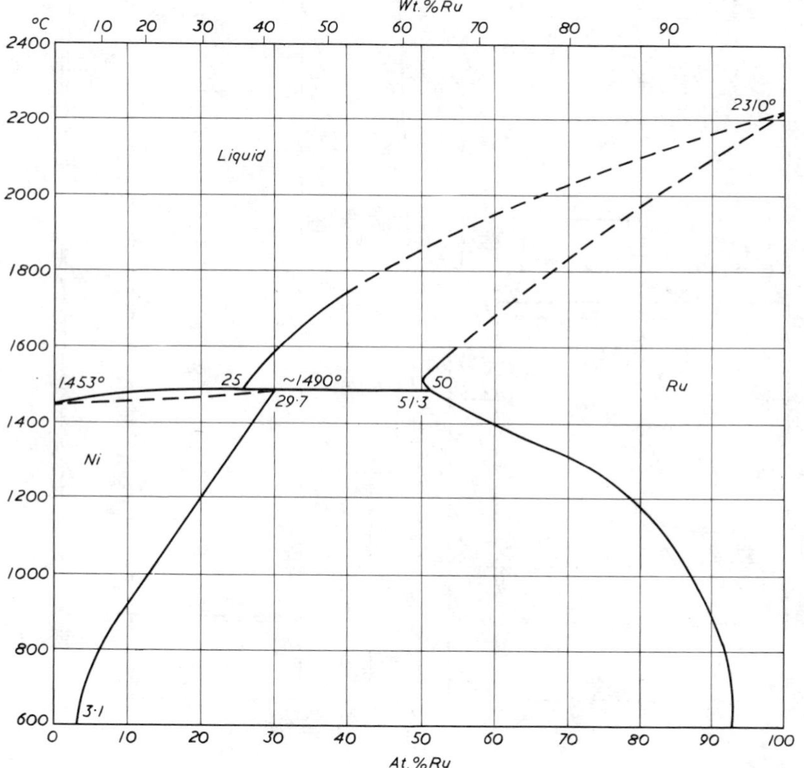
Wt.%Ru
°C
10
20
30
40
50
60
70
80
90
2400
2200
2000
1800
1600
1400
1200
1000
800
600
Liquid
2310°
1453°
25
~1490°
50
29·7
51·3
Ru
Ni
3·1
0
10
20
30
40
50
60
70
80
90
100
At.%Ru

Ni–Ru

Ni–S

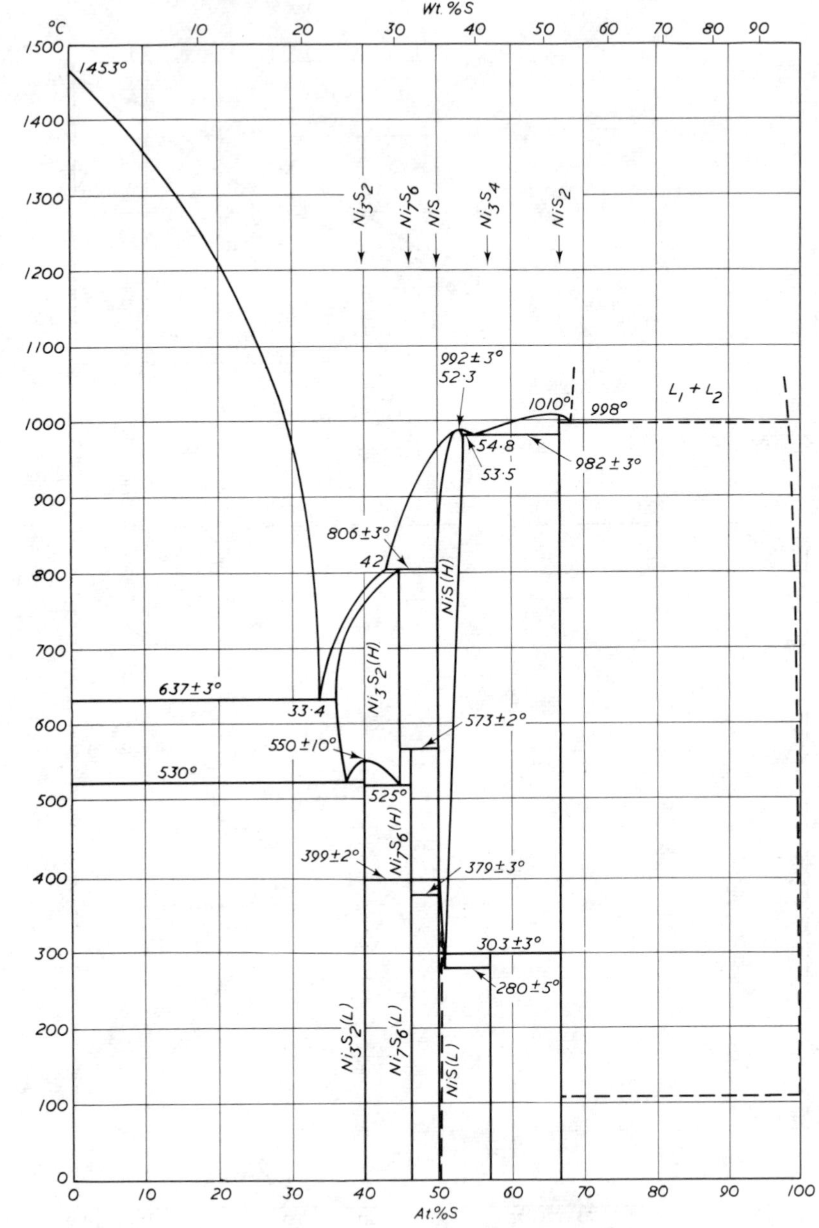
Wt.%S
°C
1453°
Ni3S2
Ni7S6
NiS
Ni3S4
NiS2
992±3°
52·3
1010°
998°
L1 + L2
54·8
53·5
982±3°
806±3°
42
NiS(H)
Ni3S2(H)
637±3°
33·4
573±2°
550±10°
530°
525°
Ni7S6(H)
399±2°
379±3°
303±3°
280±5°
Ni3S2(L)
Ni7S6(L)
NiS(L)
At.%S

Ni–Sb

At.% Sb
°C
10 20 30 40 50 60 70 80 90
1500
1400
1300
1200
1100
1000
900
800
700
600
500
1452°
Liquid
Ni_4Sb
Ni_5Sb_2
NiSb
Ni_2Sb_3 or $NiSb_2$
1075°
1162°
~1160°
1100
35
52·4
β
γ
α
680°
616°
613°
~97
630·5°
β′
0 10 20 30 40 50 60 70 80 90 100
Wt.% Sb

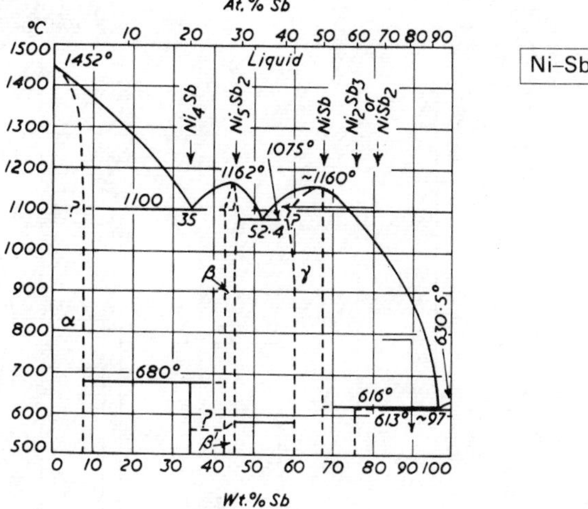

Ni–Si

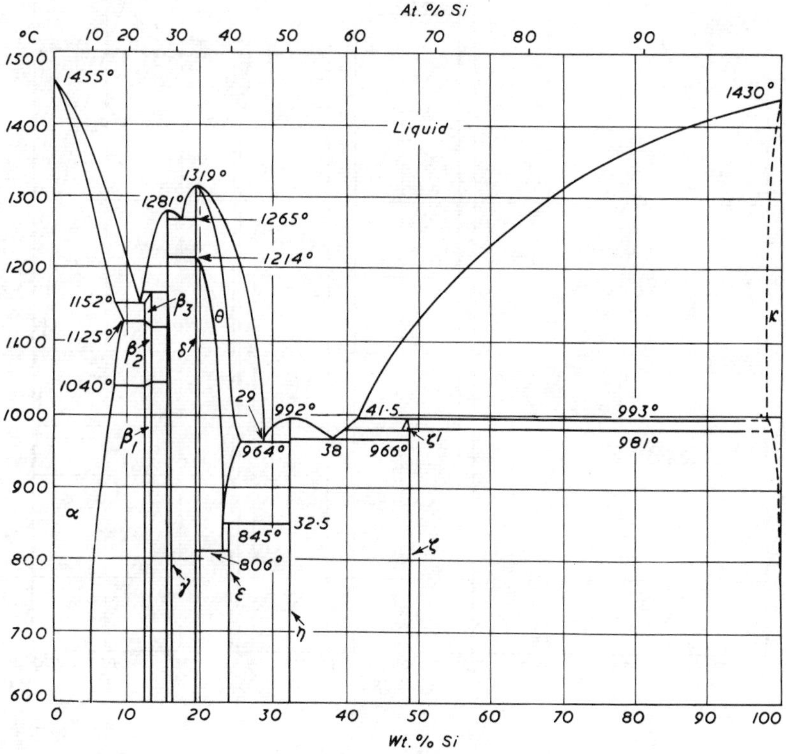

At. % Si
°C
10 20 30 40 50 60 70 80 90
1500
1400
1300
1200
1100
1000
900
800
700
600
1455°
Liquid
1430°
1319°
1281°
1265°
1214°
1152°
1125°
1040°
$β_3$
$β_2$
$β_1$
θ
δ
κ
29
992°
41·5
993°
964°
38
966°
ζ′
981°
α
32·5
845°
806°
ζ
γ
ε
η
0 10 20 30 40 50 60 70 80 90 100
Wt. % Si

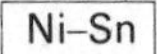
Ni–Sn
At. % Sn 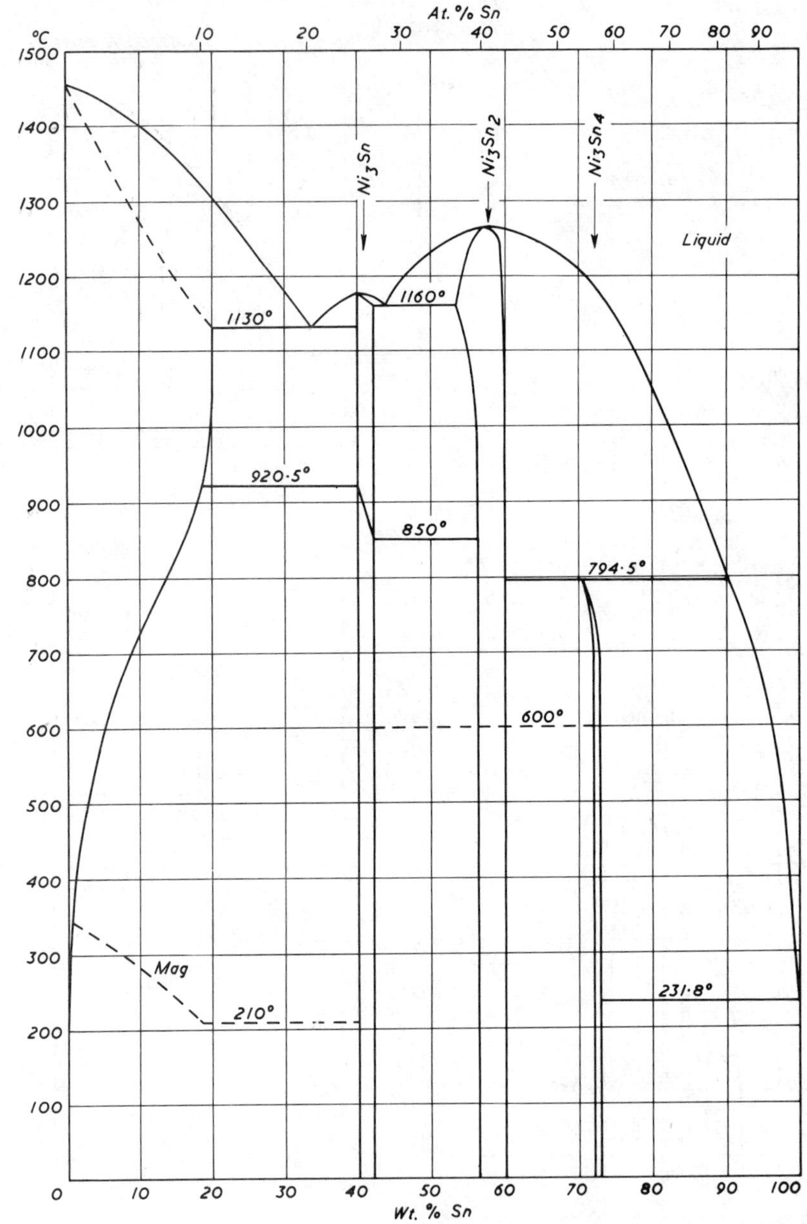
10 20 30 40 50 60 70 80 90
°C
1500
1400
1300
1200
1100
1000
900
800
700
600
500
400
300
200
100
0
Ni_3Sn
Ni_3Sn_2
Ni_3Sn_4
Liquid
1130°
1160°
920·5°
850°
794·5°
600°
Mag
210°
231·8°
10 20 30 40 50 60 70 80 90 100
Wt. % Sn

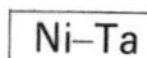
Ni–Ta

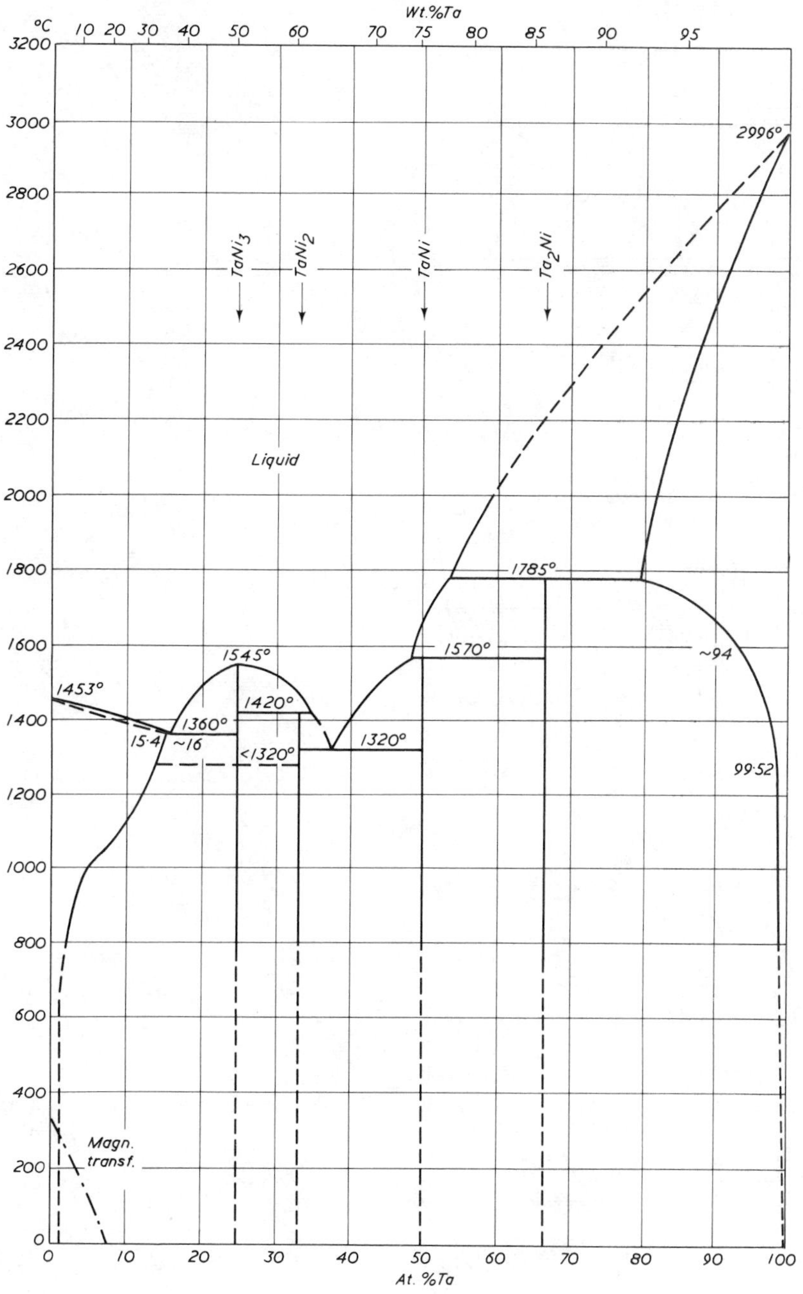

°C
Wt.%Ta
10 20 30 40 50 60 70 75 80 85 90 95
3200
3000
2800
2600
2400
2200
2000
1800
1600
1400
1200
1000
800
600
400
200
0
2996°
TaNi$_3$
TaNi$_2$
TaNi
Ta$_2$Ni
Liquid
1785°
1570°
~94
1545°
1453°
1420°
1360°
15·4
~16
<1320°
1320°
99·52
Magn. transf.
0 10 20 30 40 50 60 70 80 90 100
At. %Ta

Ni–Th

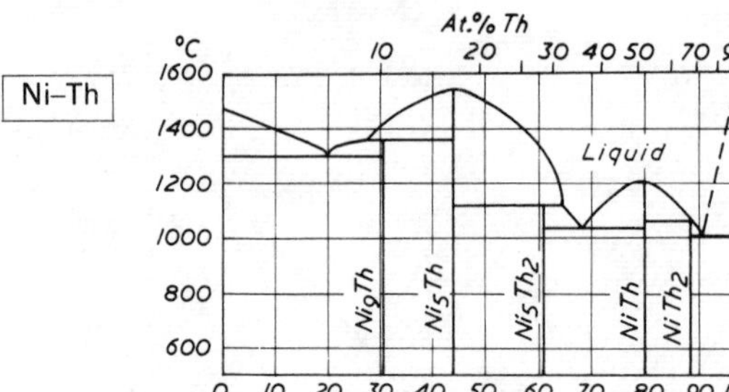

Ni–Ti

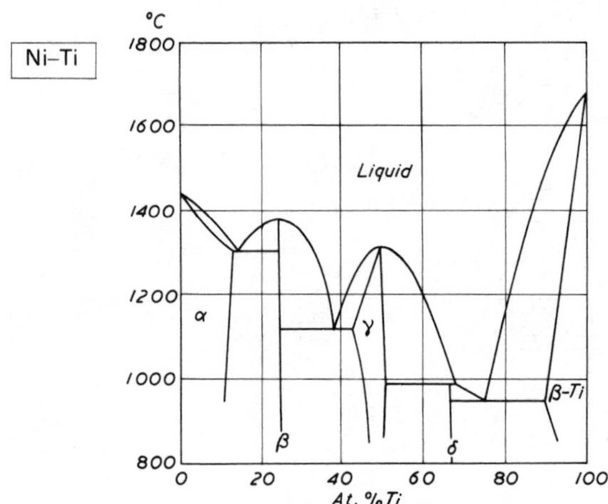

Ni–Tl

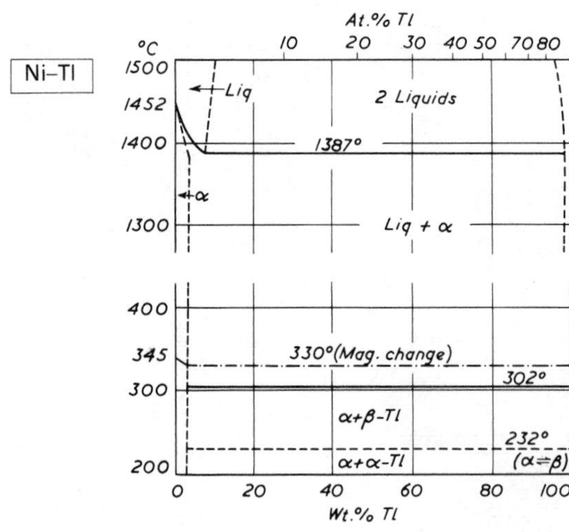

Ni–U

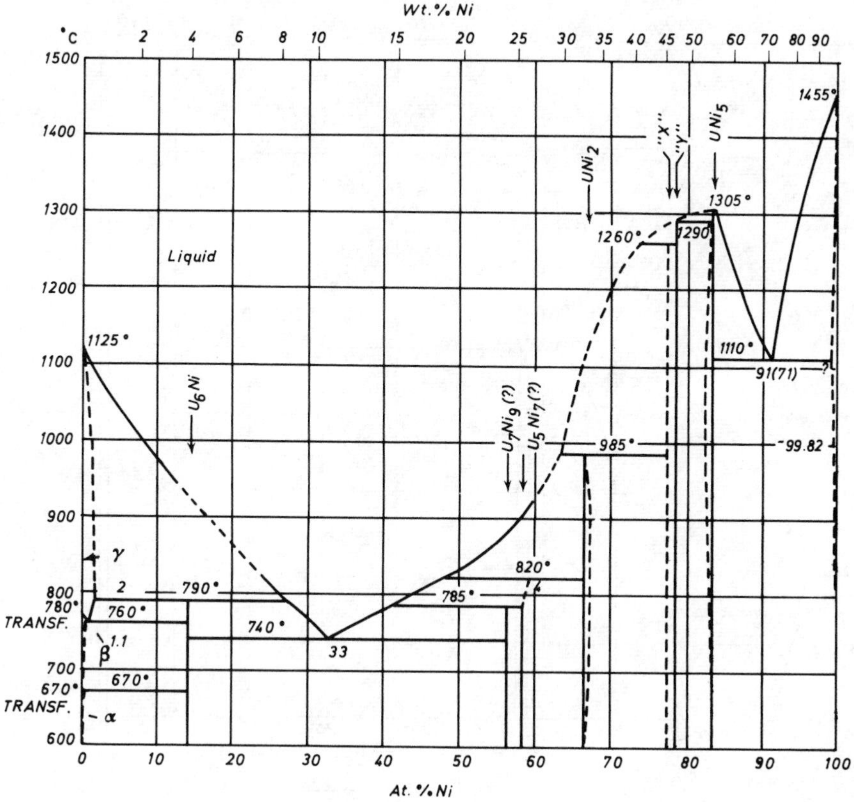

Ni–V

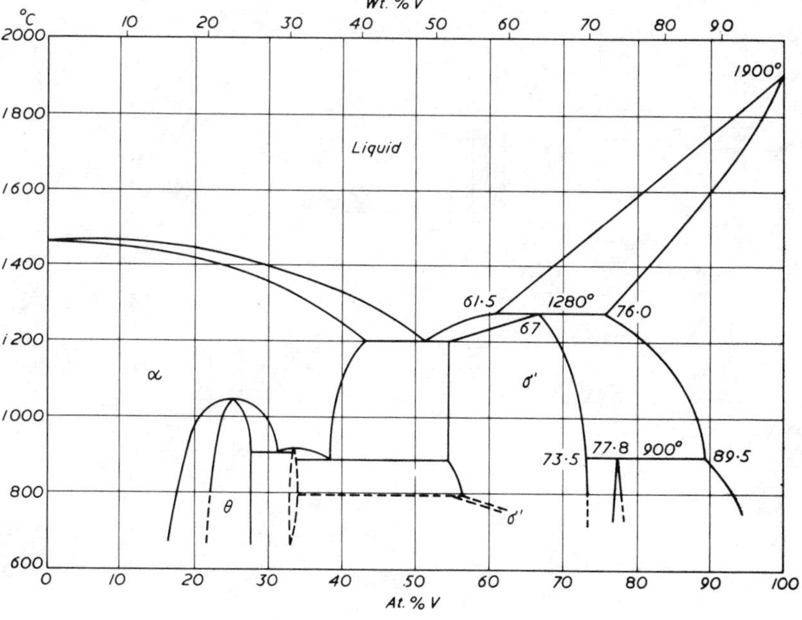
Wt. % V
°C
10
20
30
40
50
60
70
80
90
2000
1800
1600
1400
1200
1000
800
600
Liquid
1900°
61·5
1280°
76·0
67
α
σ'
θ
73·5
77·8
900°
89·5
σ'
0
10
20
30
40
50
60
70
80
90
100
At. % V

Ni–W

At. % W
°C
Liquid
Max. 1505° ± 5
45%
1495° ± 5
99%
γ
43%
970° ± 10
α
β
Magnetic change
Wt. % W

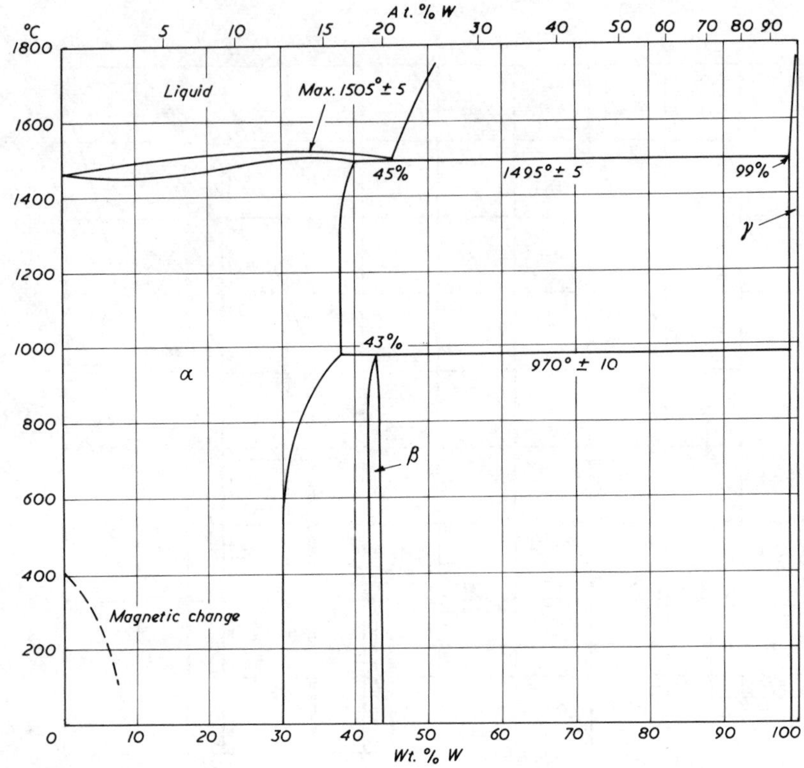

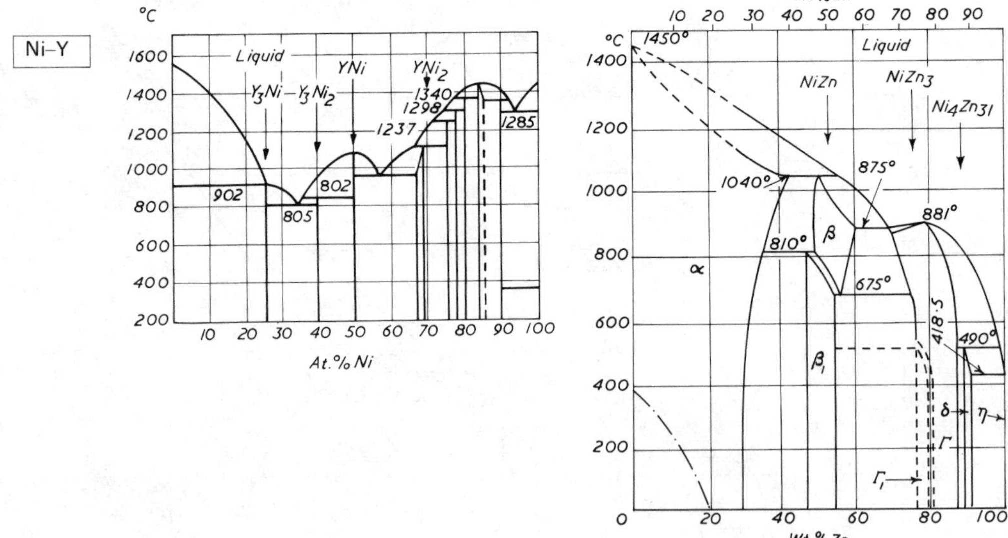

Ni–Y

°C
Liquid
Y_3Ni – Y_3Ni_2
YNi
YNi_2
1340
1298
1237
1285
902
805
802
At. % Ni

At. % Zn
°C
1450°
Liquid
NiZn
$NiZn_3$
Ni_4Zn_{31}
875°
1040°
881°
β
810°
α
675°
418·5
490°
$β_1$
δ
η
Γ
$Γ_1$
Wt. % Zn

Ni–Zn

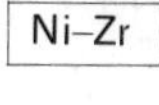
Ni–Zr
Wt. %Zr 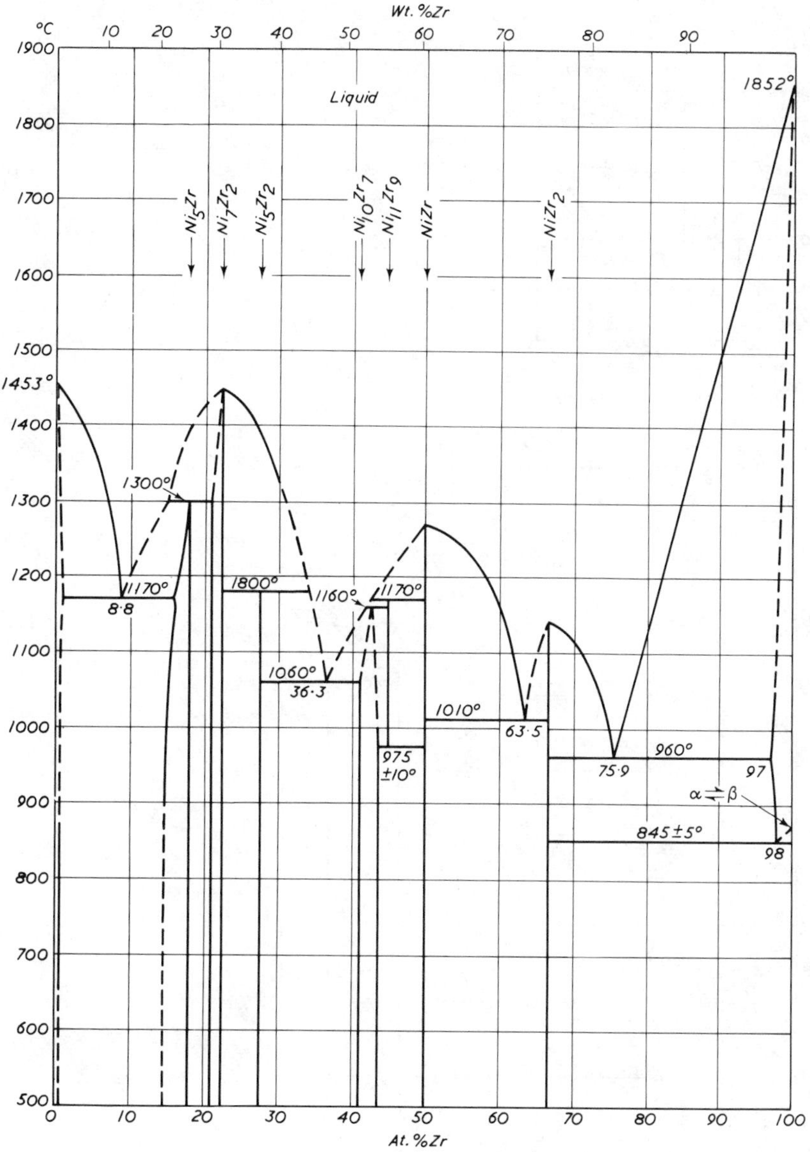
°C
Liquid
Ni_5Zr
Ni_7Zr_2
Ni_5Zr_2
$Ni_{10}Zr_7$
$Ni_{11}Zr_9$
NiZr
$NiZr_2$
1453°
1852°
1300°
1170°
8·8
1800°
1060°
36·3
1160°
1170°
975 ±10°
1010°
63·5
960°
75·9
97
α ⇄ β
845±5°
98
At. %Zr

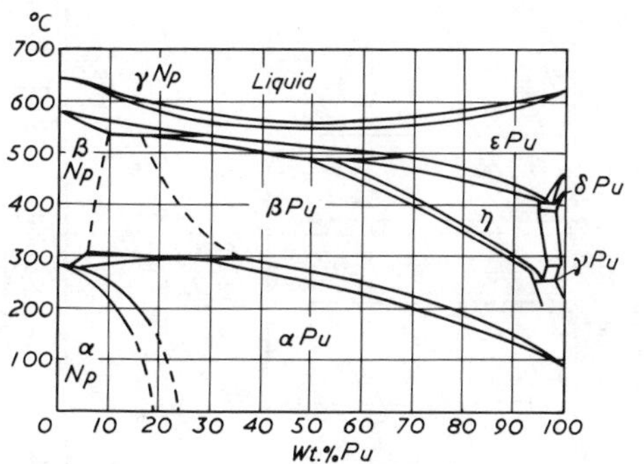
Np–Pu
°C
γ Np
Liquid
β Np
ε Pu
δ Pu
β Pu
η
γ Pu
α Np
α Pu
Wt.% Pu

Np–U

°C
1200
1000
800
600
400
200
0
Liquid
γ-Np + γU
βU
βNp
δ
α-U
αNp
0 10 20 30 40 50 60 70 80 90 100
Wt. %U

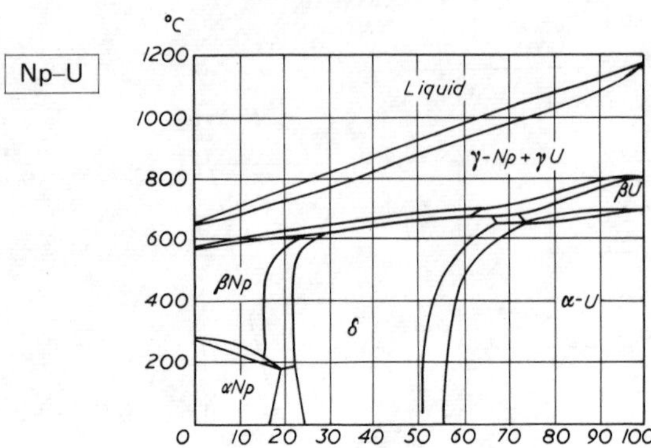

O–Pu

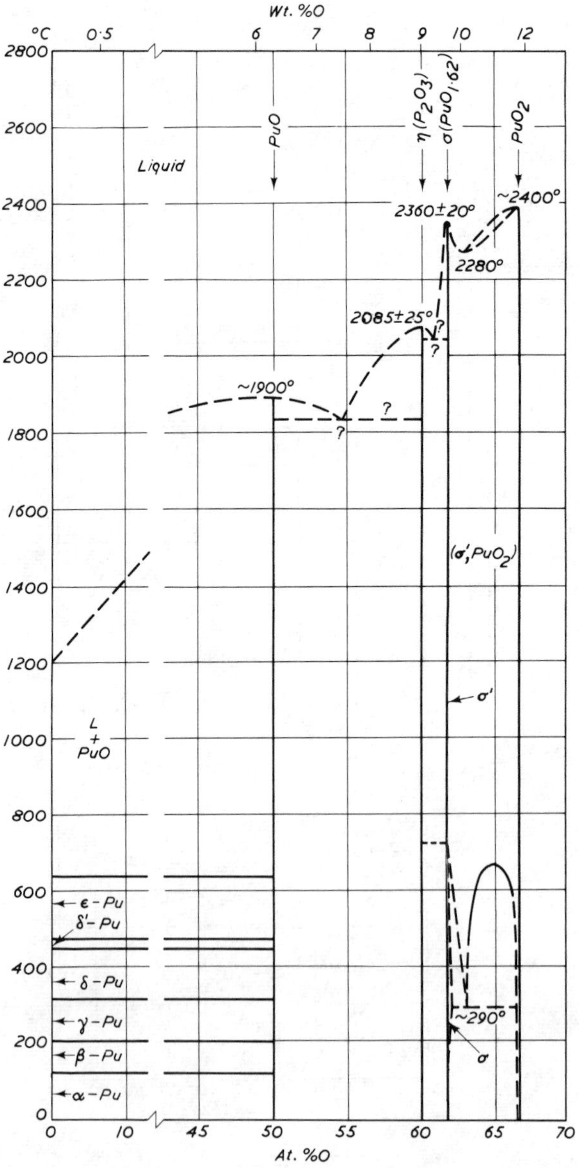

Wt. %O
°C
0·5 6 7 8 9 10 12
2800
2600
2400
2200
2000
1800
1600
1400
1200
1000
800
600
400
200
0
PuO
η(P2O3)
σ(PuO1·62)
PuO2
Liquid
2360±20°
~2400°
2280°
2085±25°
?
~1900°
(σ', PuO2)
σ'
L + PuO
ε-Pu
δ'-Pu
δ-Pu
γ-Pu
β-Pu
α-Pu
~290°
σ
0 10 45 50 55 60 65 70
At. %O

O–Ti

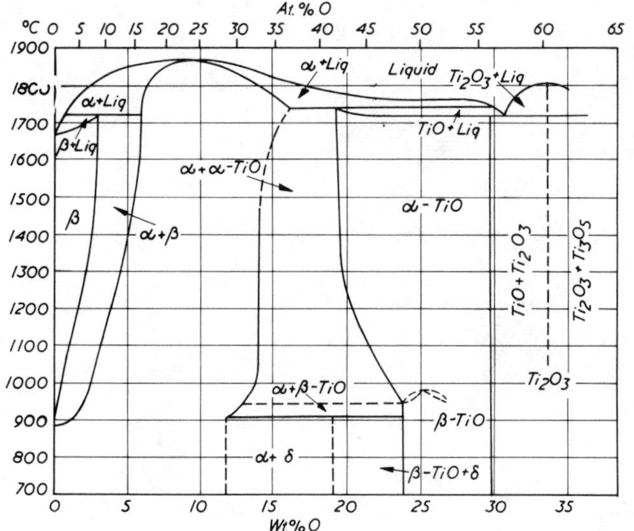

At.% O
°C
Wt.% O
Liquid
α+Liq
β+Liq
Ti2O3+Liq
TiO+Liq
α+α-TiO
α-TiO
β
α+β
TiO+Ti2O3
Ti2O3+Ti3O5
α+β-TiO
Ti2O3
β-TiO
α+δ
β-TiO+δ

O–V

Wt. %O
°C
Liquid
1790 ± 10°
1665 ± 5°
1640 ± 5°
1185 ± 10°
α
β
δ
γ
510 ± 5°
α'
At. %O

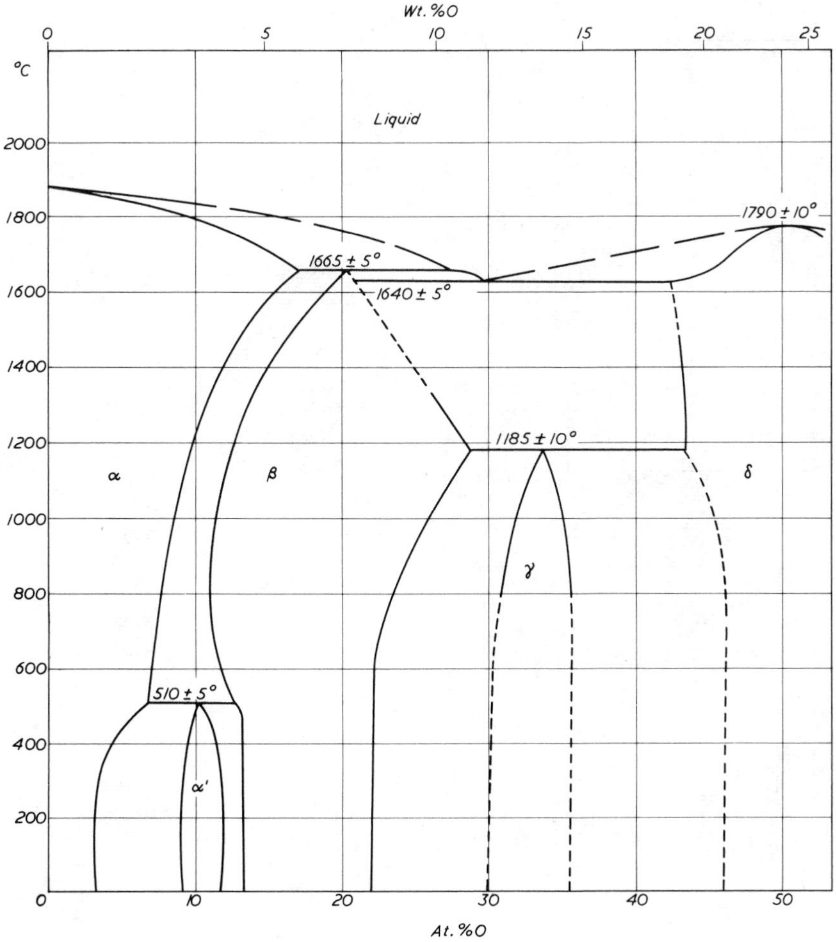

Os–Ru

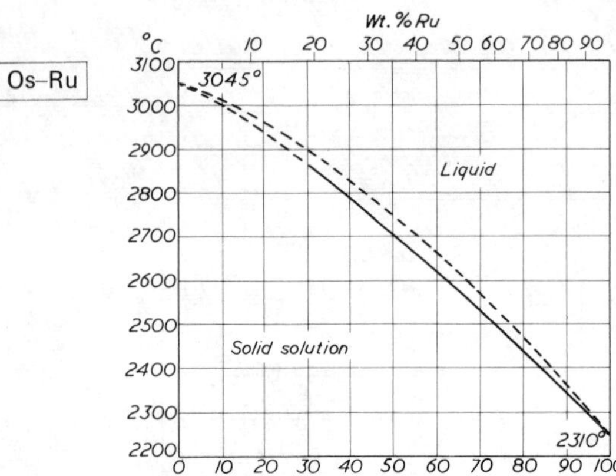
Wt. % Ru
°C
3045°
Liquid
Solid solution
2310°
At. % Ru

Os–U

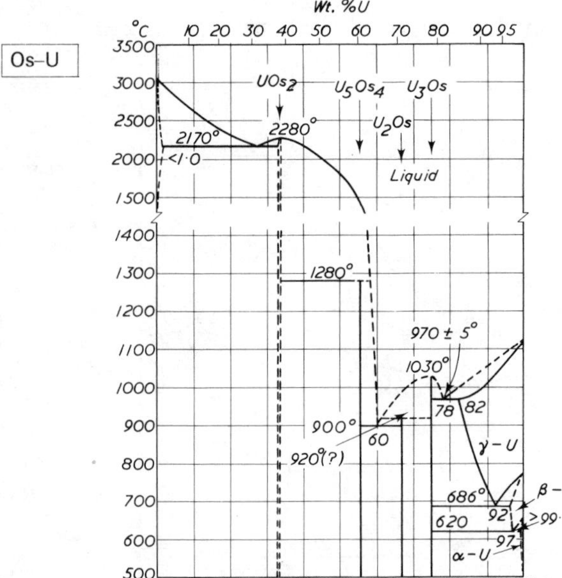
Wt. % U
°C
UOs2
U5Os4
U3Os
U2Os
2280°
2170°
<1·0
Liquid
1280°
970 ± 5°
1030°
78
82
900°
60
920°(?)
γ-U
686°
β-U
92
>99·5
620
97
α-U
At. % U

Os–W

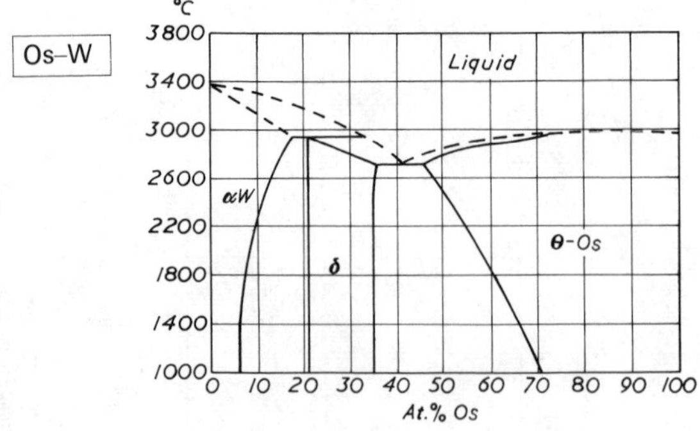
°C
Liquid
αW
δ
θ-Os
At. % Os

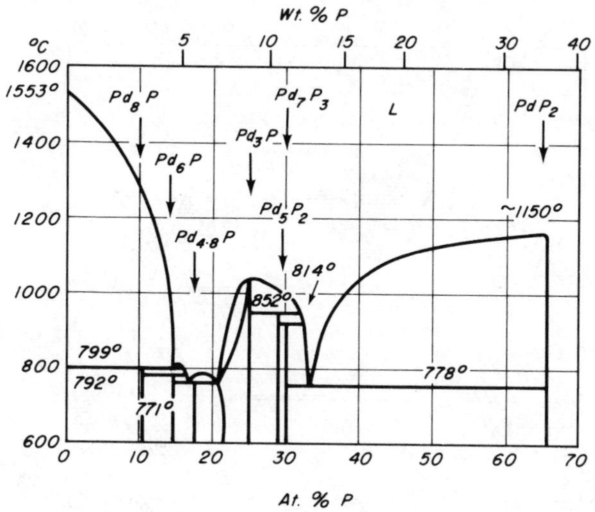
Wt. % P
°C
1553°
Pd$_8$P
Pd$_7$P$_3$
L
PdP$_2$
Pd$_3$P
Pd$_6$P
Pd$_5$P$_2$
~1150°
Pd$_{4\cdot8}$P
814°
852°
799°
792°
771°
778°
At. % P

P–Pd

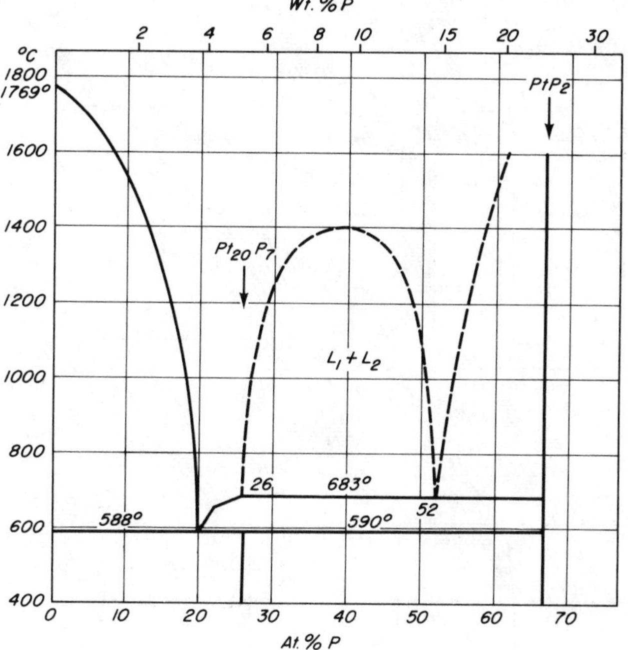
Wt. % P
°C
1769°
PtP$_2$
Pt$_{20}$P$_7$
L$_1$ + L$_2$
26
683°
52
588°
590°
At. % P

P–Pt

P–Si

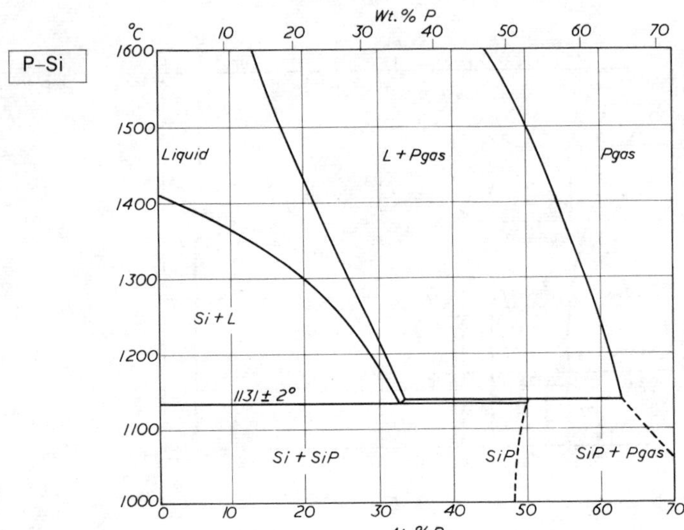
Wt. % P
°C
Liquid
L + Pgas
Pgas
Si + L
1131 ± 2°
Si + SiP
SiP
SiP + Pgas
At. % P

P–Sn

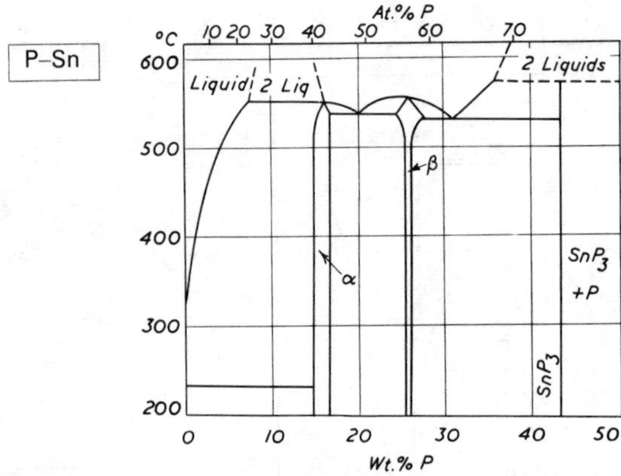
At.% P
°C
Liquid
2 Liq
2 Liquids
β
α
SnP_3
+P
SnP_3
Wt.% P

P–Tl

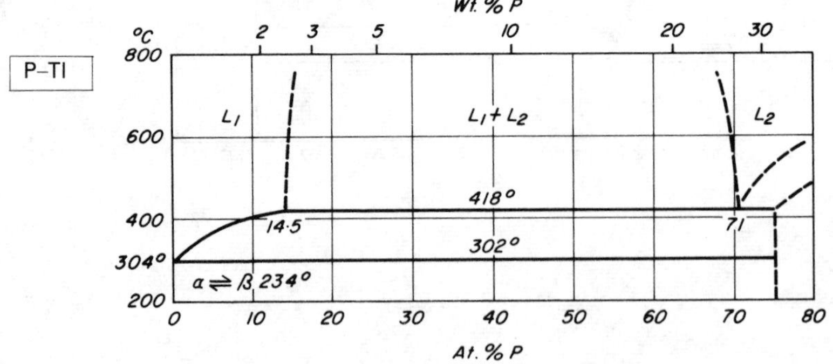
Wt. % P
°C
L_1
$L_1 + L_2$
L_2
418°
14·5
71
302°
304°
α ⇌ β 234°
At. % P

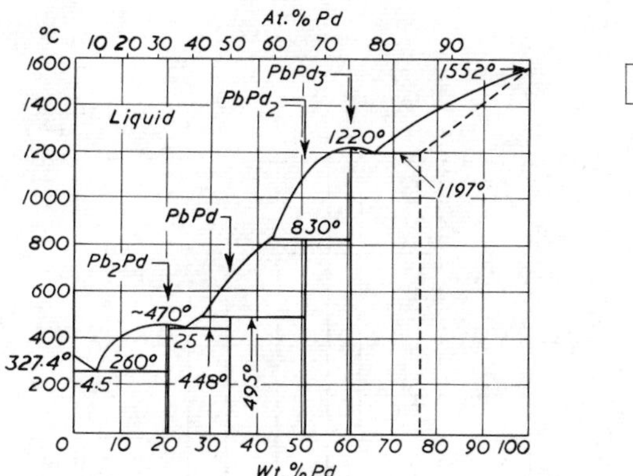

Pb–Pd

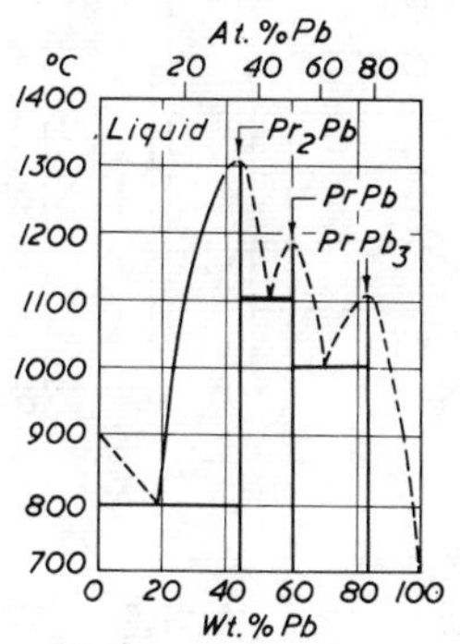

Pb–Pr

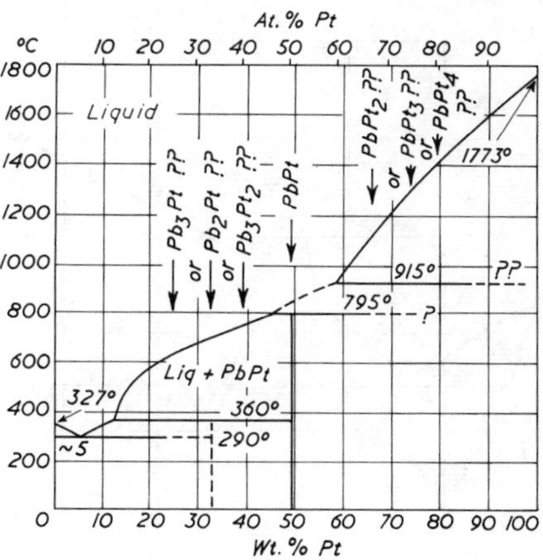

Pb–Pt

Pb–Pu

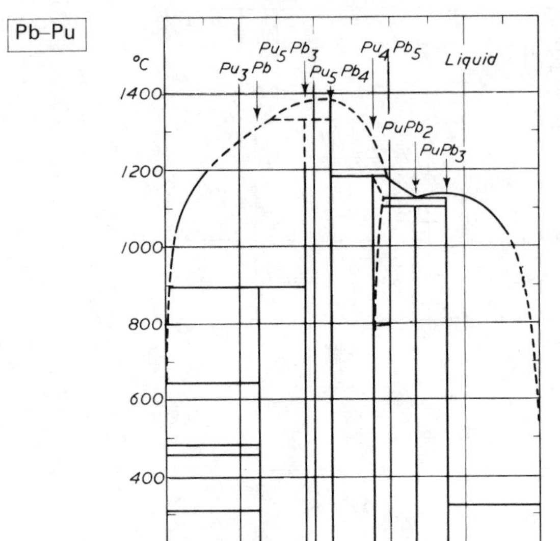
°C
Pu3Pb
Pu5Pb3
Pu5Pb4
Pu4Pb5
Liquid
PuPb2
PuPb3
1400
1200
1000
800
600
400
200
0
20
40
60
80
100
At.% Pb

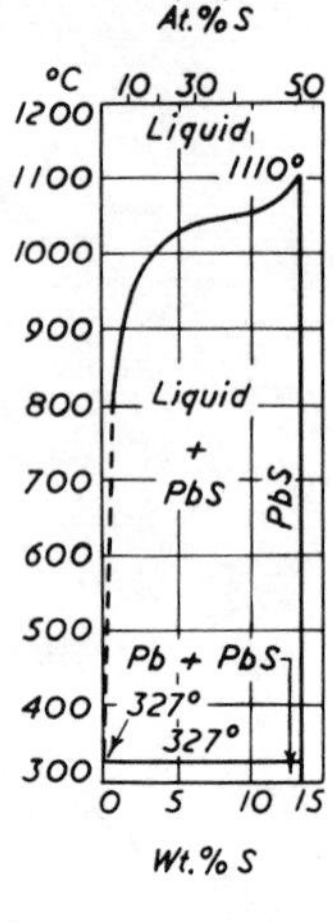
At.% S
°C
10
30
50
1200
1100
1000
900
800
700
600
500
400
300
Liquid
1110°
Liquid
+
PbS
PbS
Pb + PbS
327°
327°
0
5
10
15
Wt.% S

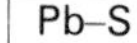
Pb–S

Pb–Sb

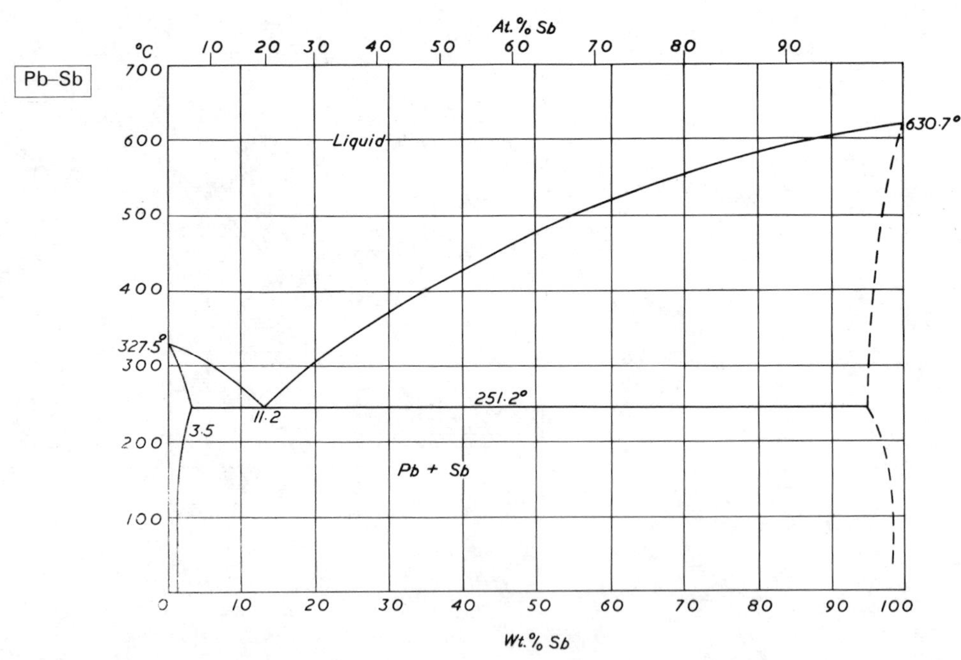
At.% Sb
°C
10
20
30
40
50
60
70
80
90
700
600
500
400
327.5°
300
200
100
Liquid
630.7°
251.2°
3.5
11.2
Pb + Sb
0
10
20
30
40
50
60
70
80
90
100
Wt.% Sb

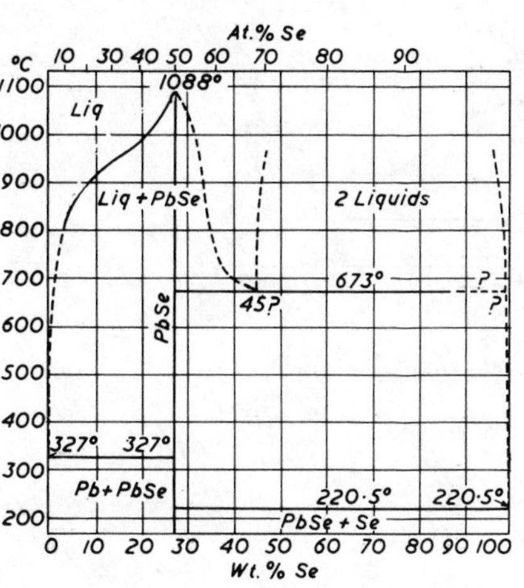

Pb–Se

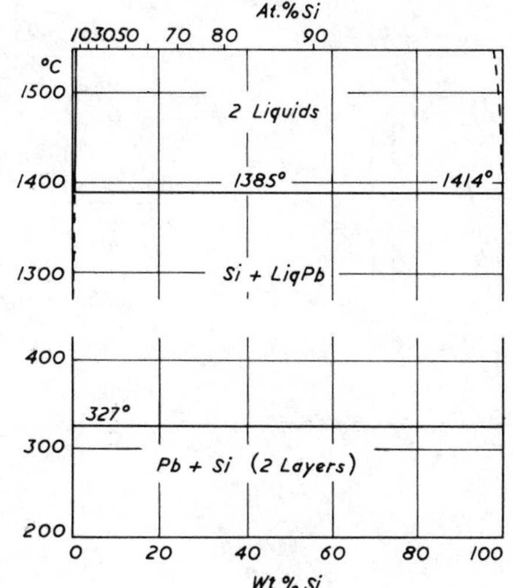

Pb–Si

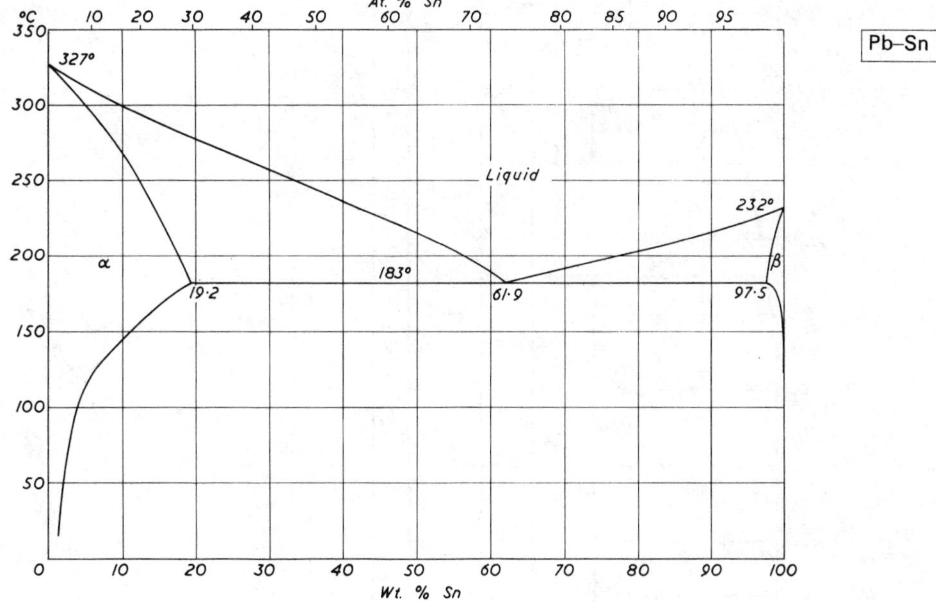

Pb–Sn

Pb–Sr

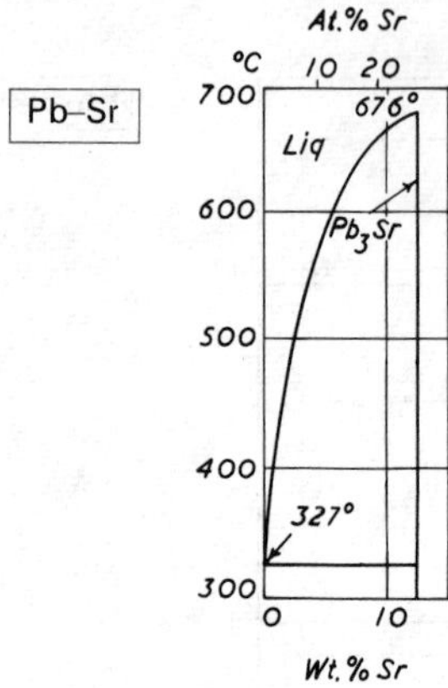

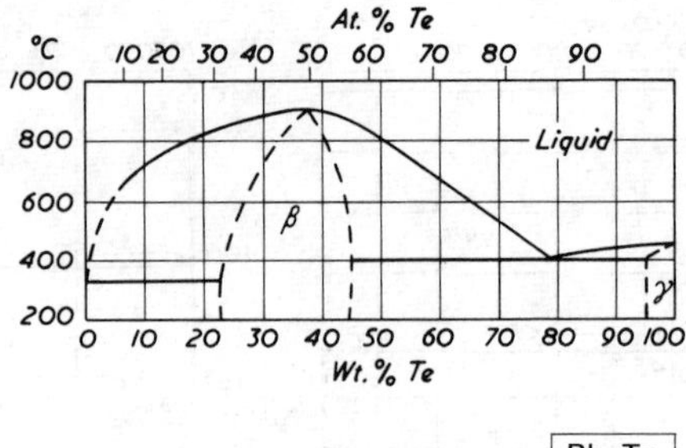

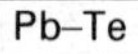
Pb–Te

Pb–Th

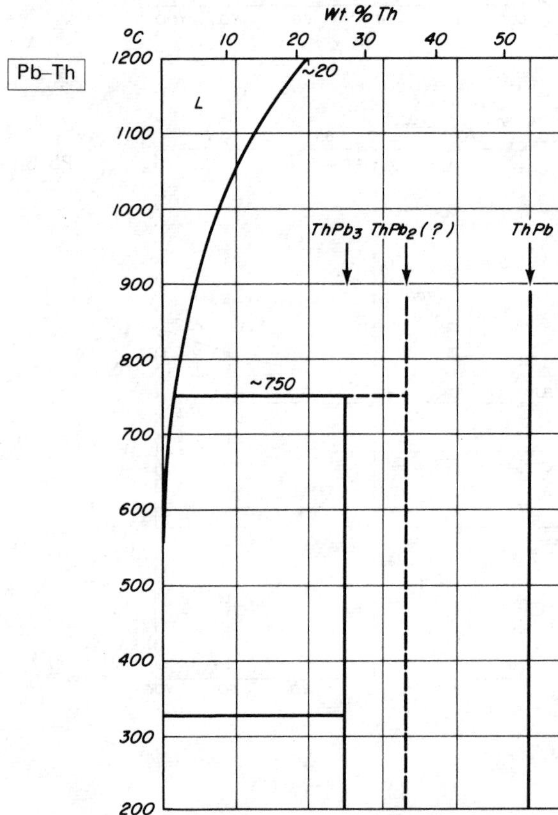

Pb–Ti

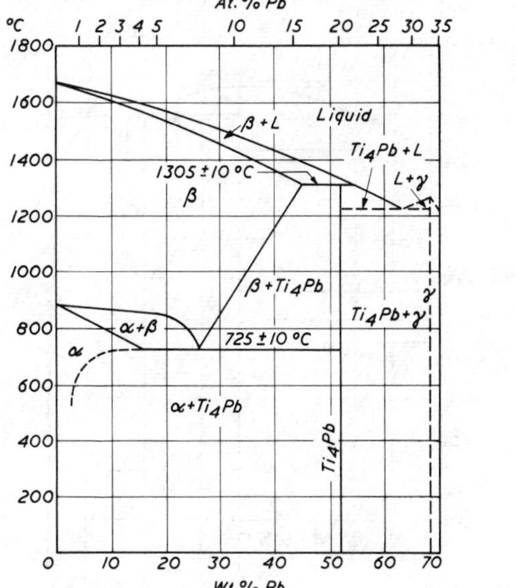
At.% Pb
1 2 3 4 5 10 15 20 25 30 35
°C
1800
1600
1400
1200
1000
800
600
400
200
0
Liquid
β + L
Ti_4Pb + L
L+γ
1305 ± 10 °C
β
β + Ti_4Pb
γ
Ti_4Pb+γ
α+β
725 ± 10 °C
α
α+Ti_4Pb
Ti_4Pb
10 20 30 40 50 60 70
Wt.% Pb

Pb–Tl

At.% Tl
10 20 30 40 50 60 70 80 90
°C
400
350
300
250
200
150
100
50
0
Liquid
327°
380°
310°
(302°)
β
232°(α⇌β)
δ
γ
?
(140°)
α
10 20 30 40 50 60 70 80 90 100
Wt.% Tl

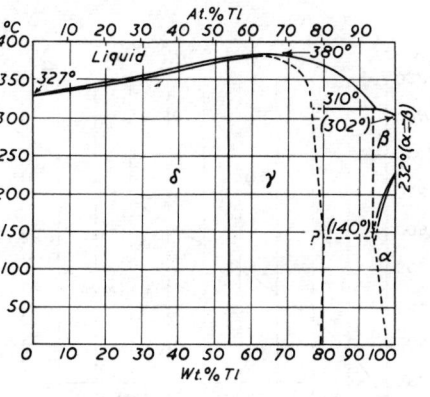

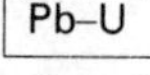
Pb–U

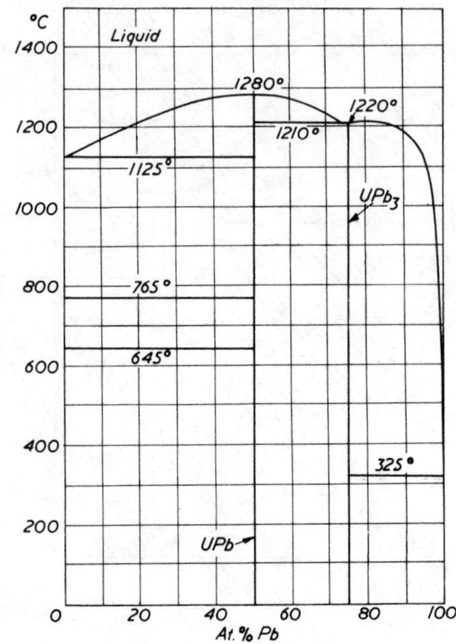
°C
Liquid
1280°
1220°
1210°
1125°
UPb3
765°
645°
325°
UPb
At.% Pb

Pb–W

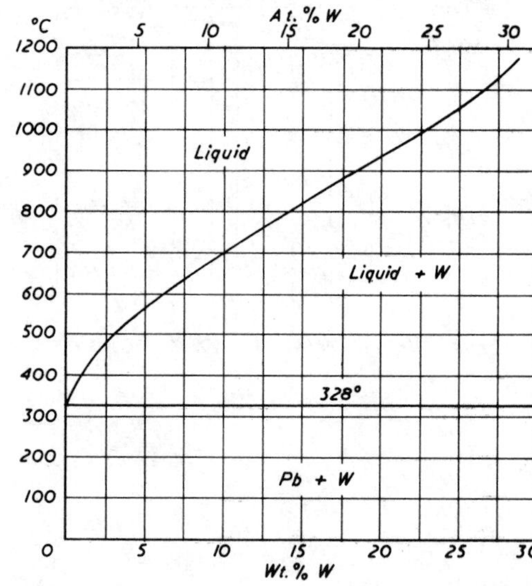
°C
At.% W
Liquid
Liquid + W
328°
Pb + W
Wt.% W

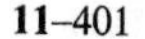

Pb–Zn

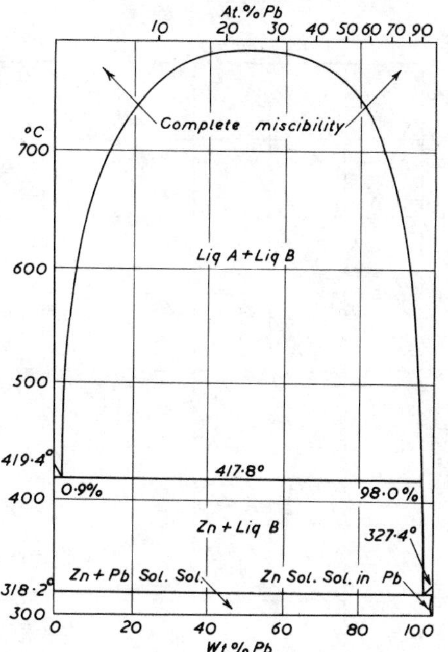
At.% Pb
10
20
30
40
50
60
70
90
°C
700
600
500
419·4°
400
318·2°
300
Complete miscibility
Liq A + Liq B
417·8°
0·9%
98·0%
Zn + Liq B
327·4°
Zn + Pb Sol. Sol.
Zn Sol. Sol. in Pb
0
20
40
60
80
100
Wt.% Pb

Pd–Rh

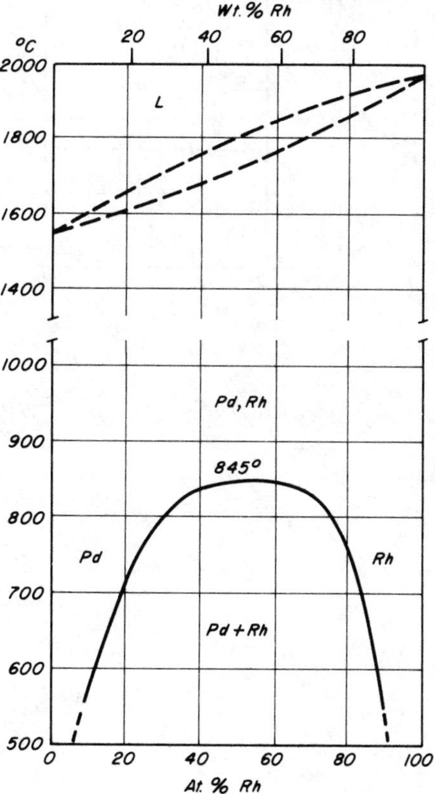
Wt.% Rh
20
40
60
80
°C
2000
1800
1600
1400
1000
900
800
700
600
500
L
Pd, Rh
845°
Pd
Rh
Pd + Rh
0
20
40
60
80
100
At. % Rh

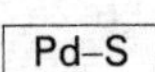
Pd–S

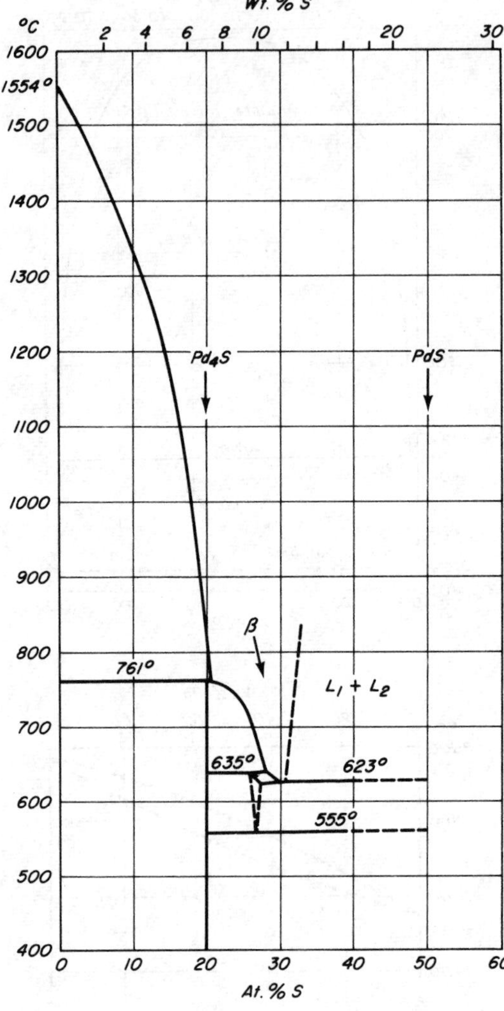
Wt. % S
2
4
6
8
10
20
30
°C
1600
1554°
1500
1400
1300
1200
1100
1000
900
800
700
600
500
400
Pd_4S
PdS
β
761°
$L_1 + L_2$
635°
623°
555°
0
10
20
30
40
50
60
At. % S

Pd–Sb

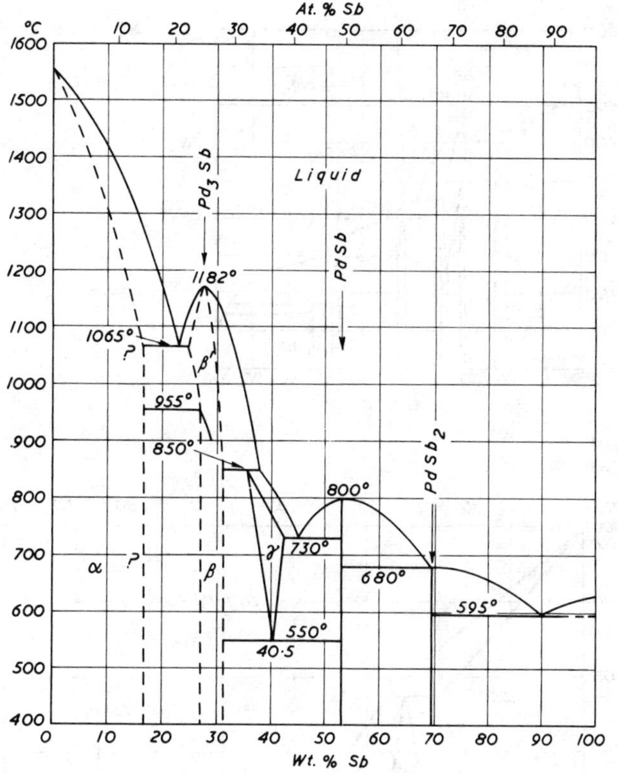
At. % Sb
°C
Wt. % Sb
Liquid
Pd_3Sb
PdSb
$PdSb_2$
1182°
1065°
955°
850°
800°
730°
680°
595°
550°
40·5
α
β
β'
γ'
?

Pd–Si

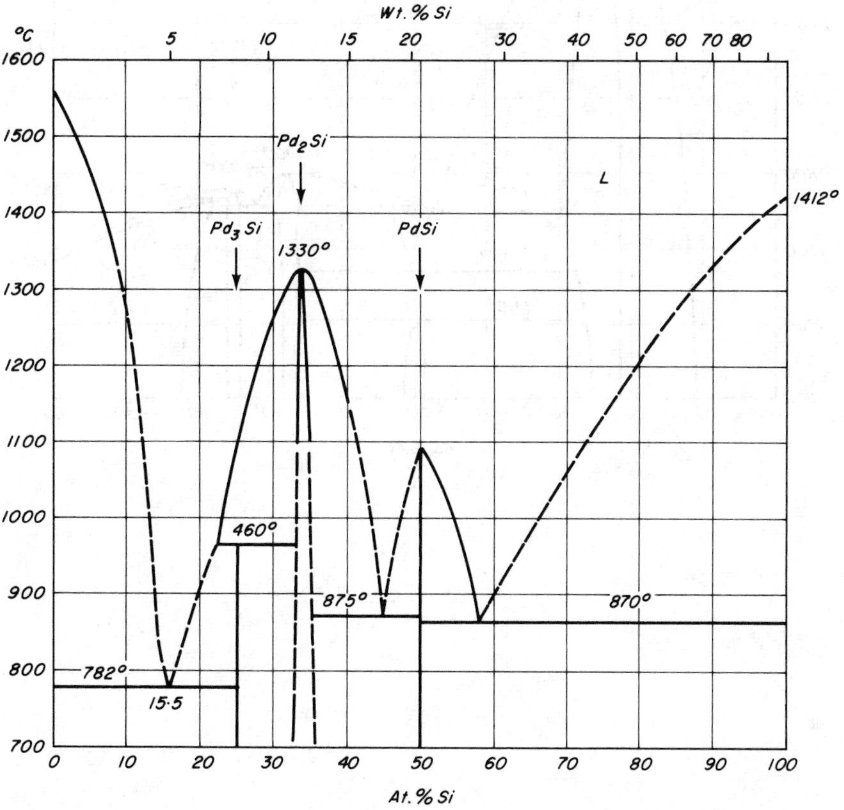

Pd–Sn

°C
1800
1600
1400
1200
1000
800
600
400
200
0
Pd3Sn
Pd2Sn
Pd3Sn2
PdSn
PdSn2
PdSn3
PdSn4
Liquid
10 20 30 40 50 60 70 80 90 100
At. % Sn

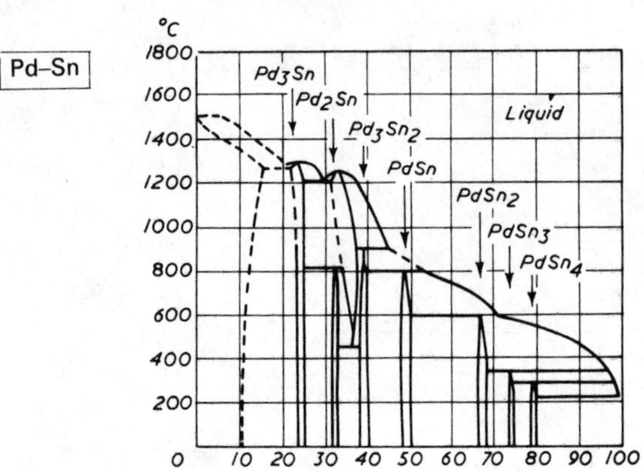

Pd–Ta

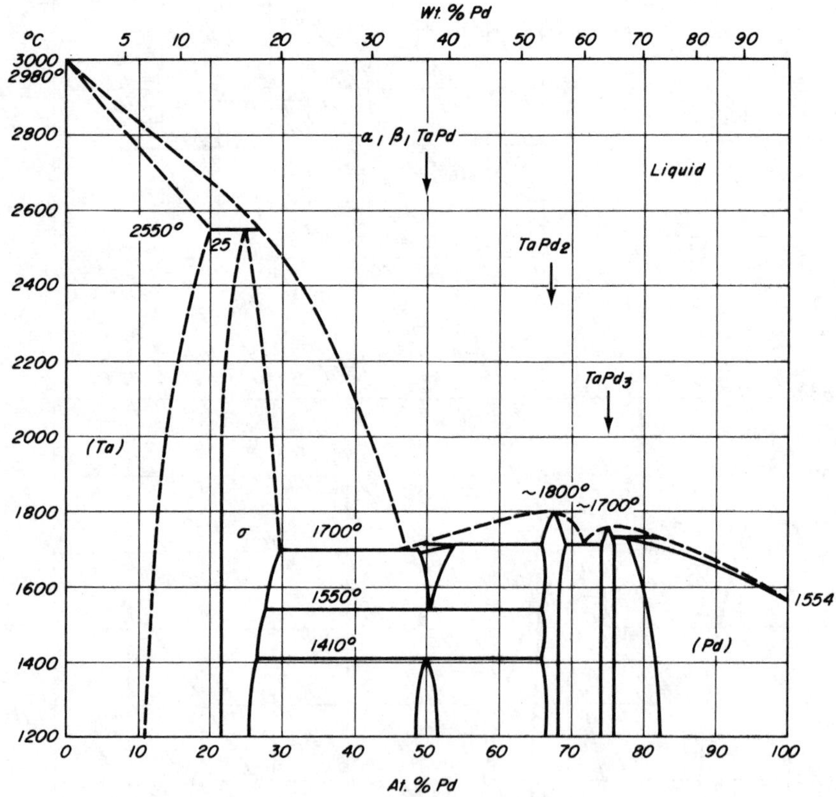

Wt. % Pd
5 10 20 30 40 50 60 70 80 90
°C
3000
2980°
2800
2600
2400
2200
2000
1800
1600
1400
1200
α, β, TaPd
Liquid
2550°
25
TaPd2
TaPd3
(Ta)
~1800°
~1700°
σ
1700°
1550°
1410°
1554°
(Pd)
0 10 20 30 40 50 60 70 80 90 100
At. % Pd

Pd–Te

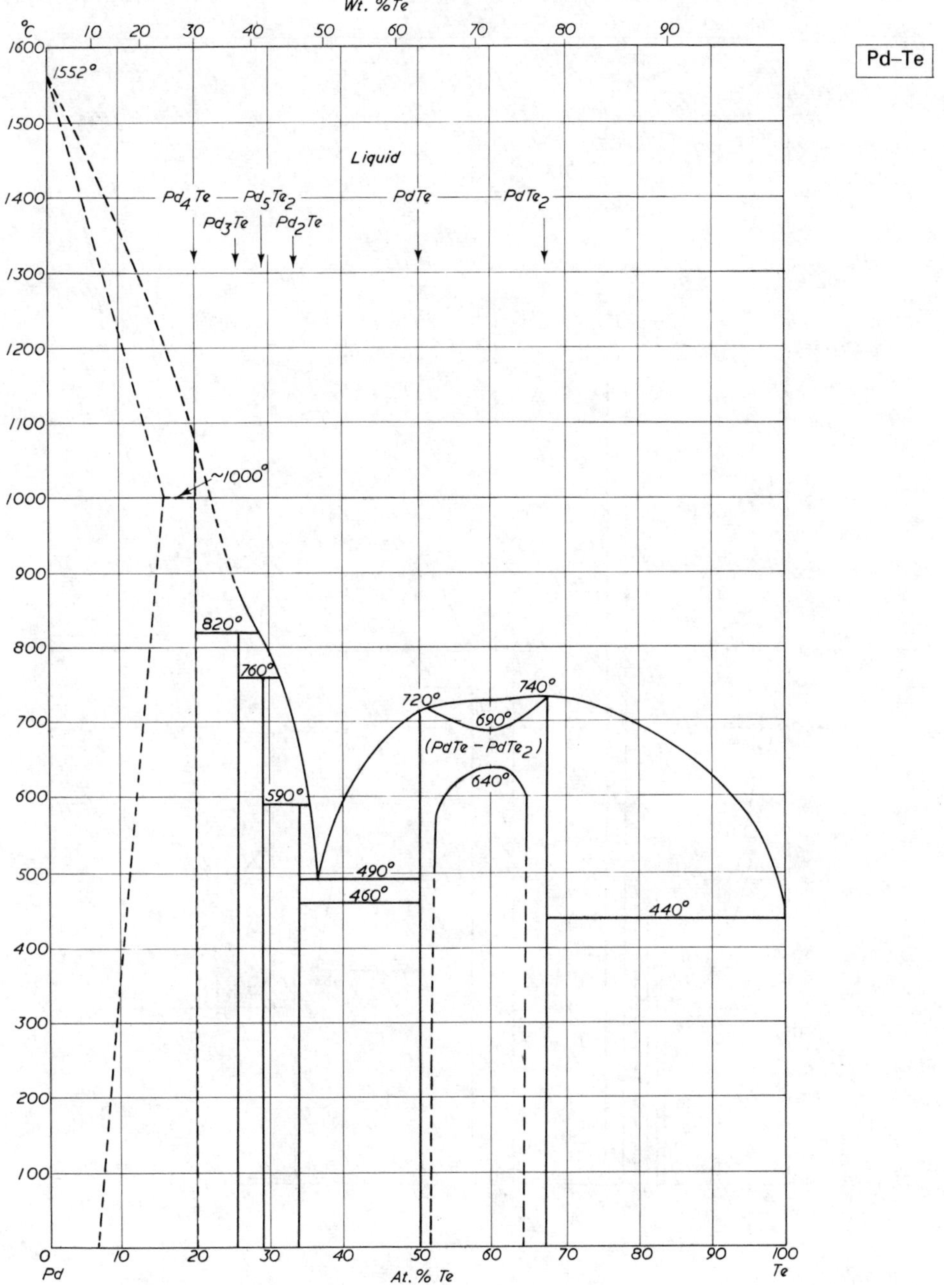

Wt. %Te
°C
1552°
Liquid
Pd_4Te
Pd_5Te_2
PdTe
$PdTe_2$
Pd_3Te
Pd_2Te
~1000°
820°
760°
740°
720°
690°
($PdTe - PdTe_2$)
640°
590°
490°
460°
440°
Pd
At. % Te
Te

Pd–Th

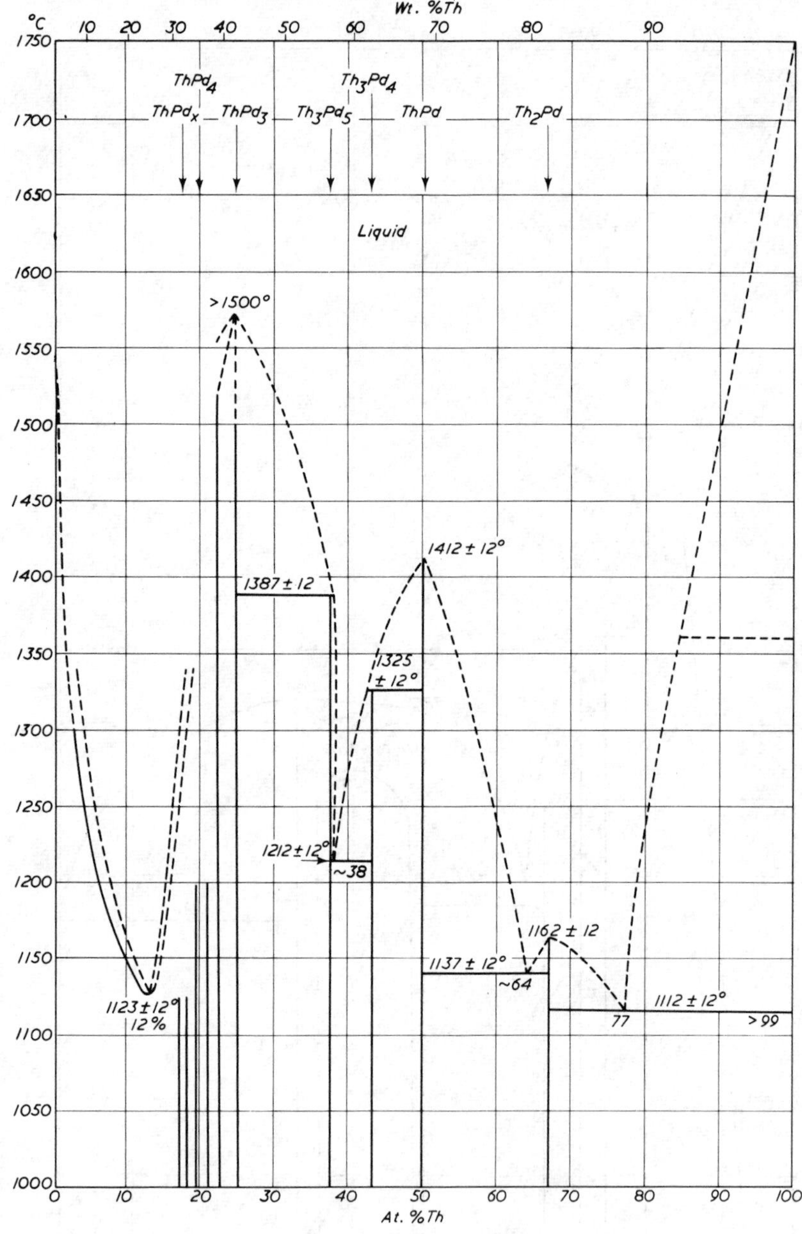
Wt. %Th
°C
10
20
30
40
50
60
70
80
90
1750
1700
1650
1600
1550
1500
1450
1400
1350
1300
1250
1200
1150
1100
1050
1000
ThPd4
ThPdx
ThPd3
Th3Pd4
Th3Pd5
ThPd
Th2Pd
Liquid
>1500°
1412 ± 12°
1387 ± 12
1325
± 12°
1212 ± 12°
~38
1162 ± 12
1137 ± 12°
~64
1123 ± 12°
12 %
1112 ± 12°
77
>99
0
10
20
30
40
50
60
70
80
90
100
At. %Th

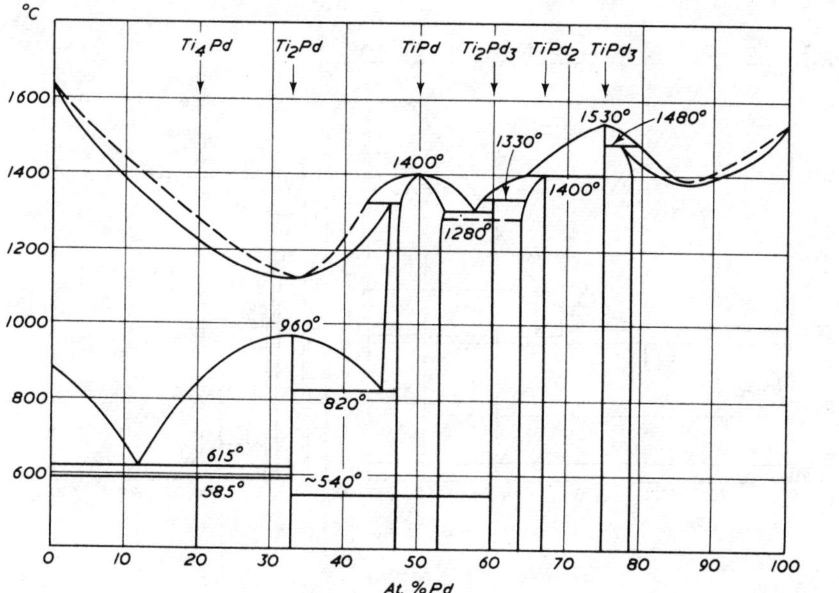
°C
1600
1400
1200
1000
800
600
Ti4Pd
Ti2Pd
TiPd
Ti2Pd3
TiPd2
TiPd3
1530°
1480°
1330°
1400°
1400°
1280°
960°
820°
615°
~540°
585°
0
10
20
30
40
50
60
70
80
90
100
At. %Pd

Pd–Ti

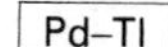

Pd–Tl

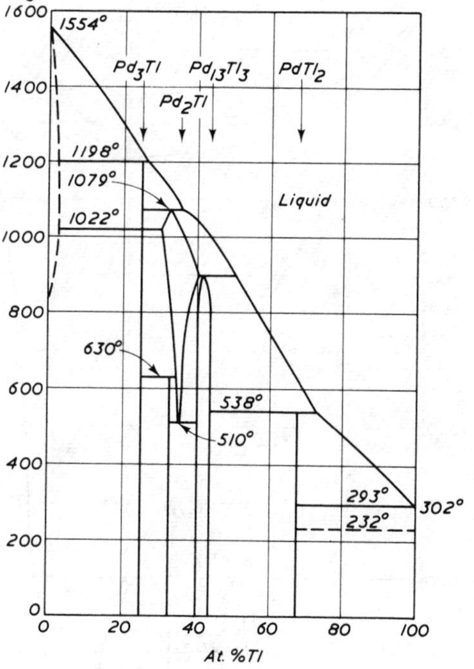
°C
1600
1554°
1400
1200
1000
800
600
400
200
0
Pd3Tl
Pd13Tl3
Pd2Tl
PdTl2
1198°
1079°
1022°
Liquid
630°
538°
510°
293°
302°
232°
20
40
60
80
100
At. %Tl

Pd–U

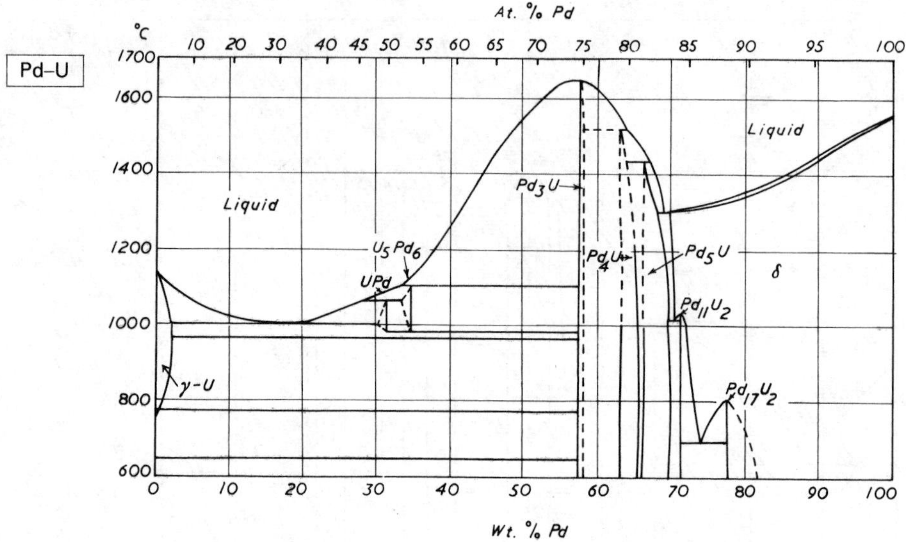

Pd–V

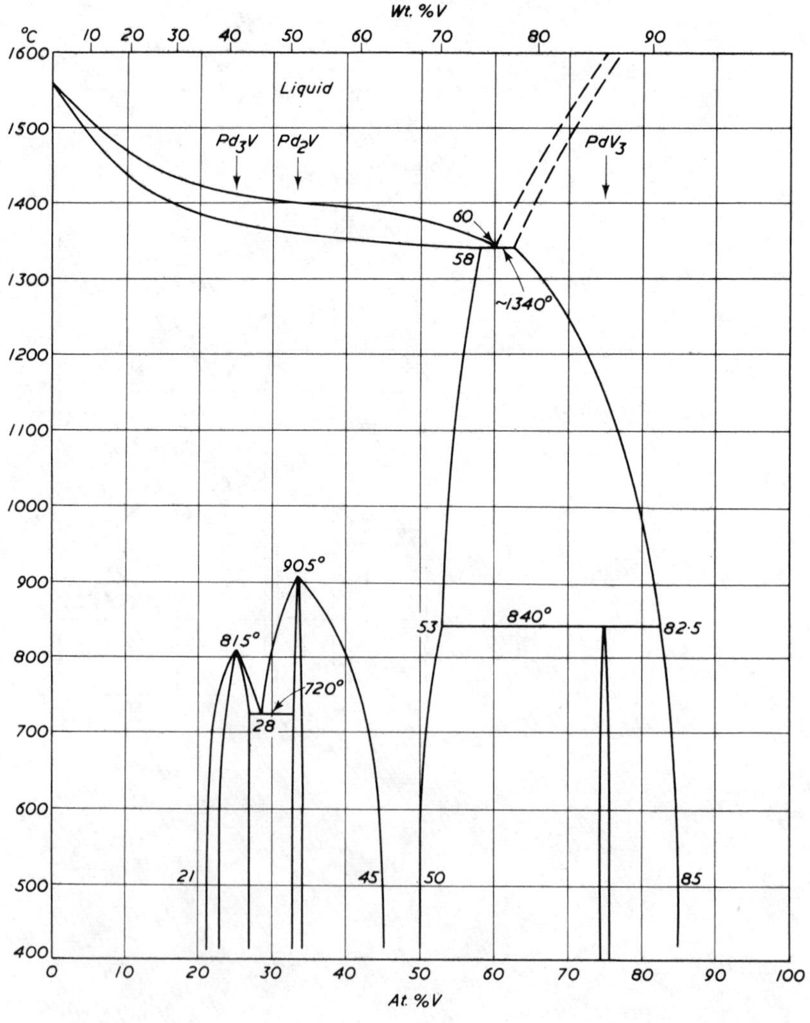

Wt. % V
°C
Liquid
Pd_3V
Pd_2V
PdV_3
60
58
~1340°
905°
815°
720°
28
840°
53
82·5
21
45
50
85
At. % V

Pd–W

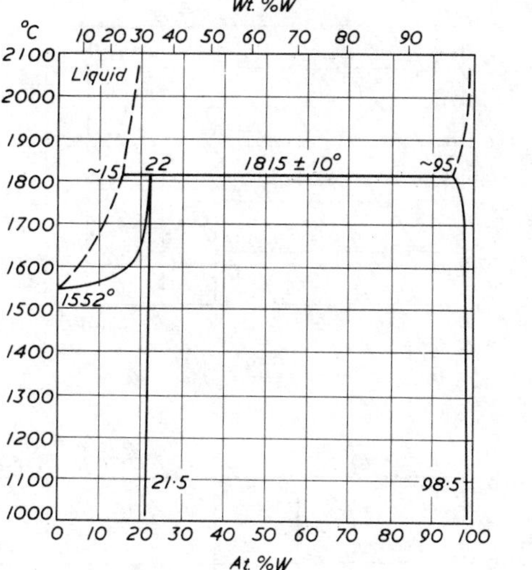

Pd–Zn

Wt. % Zn

°C

10 20 30 40 50 60 70 80 90

1600 1500 1400 1300 1200 1100 1000 900 800 700 600 500 400 300

Liquid

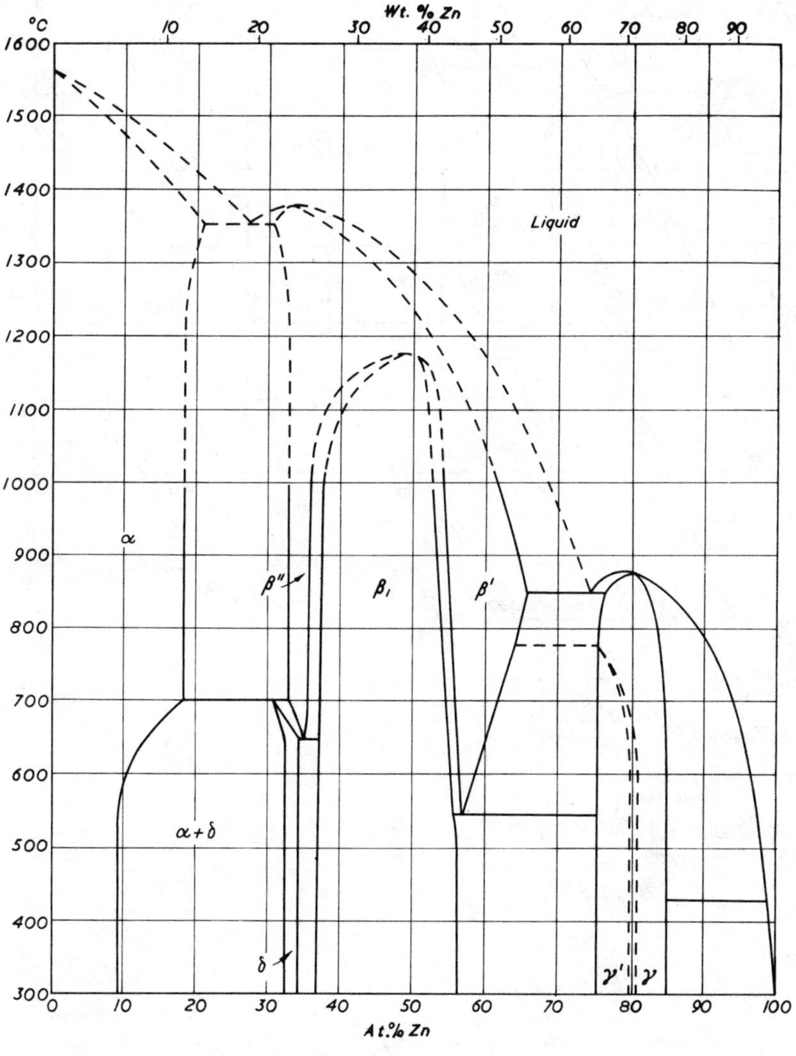

α

β″

$β_1$

β′

α+δ

δ

γ′ γ

0 10 20 30 40 50 60 70 80 90 100

At.% Zn

Pd–Zr

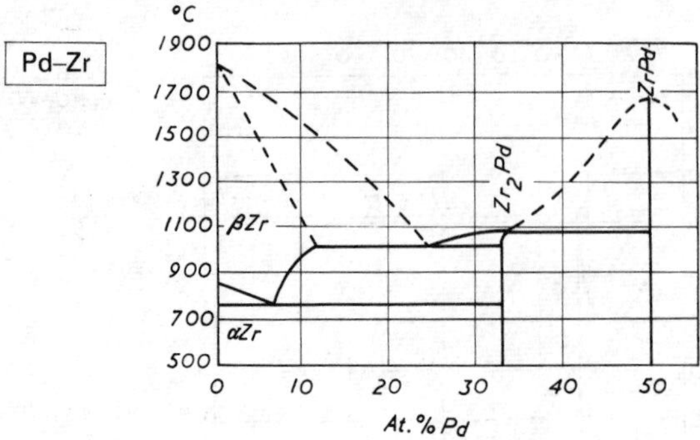

Pr–Sn

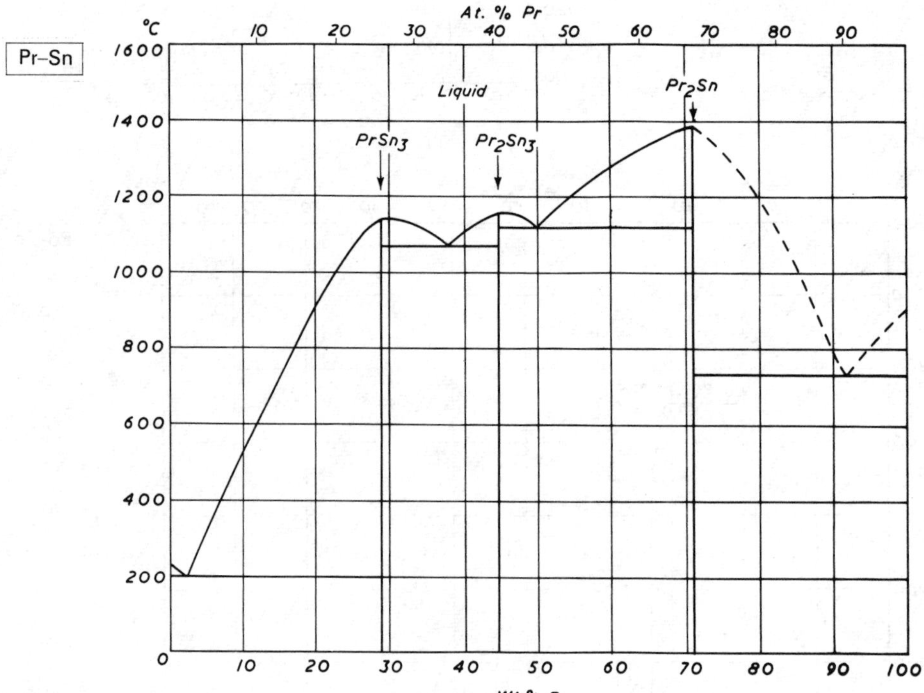

Pr–Tl

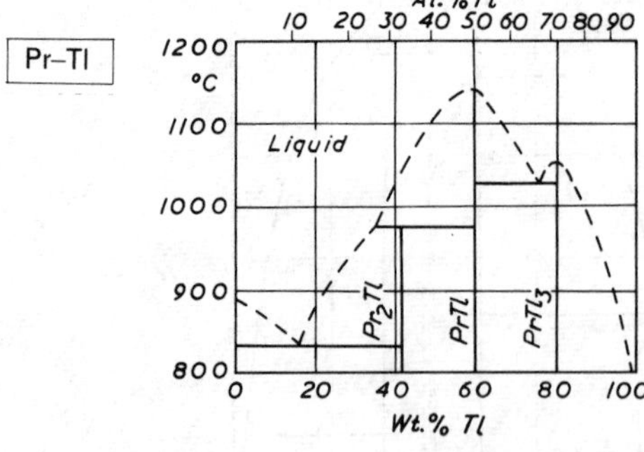

Pr–Zn

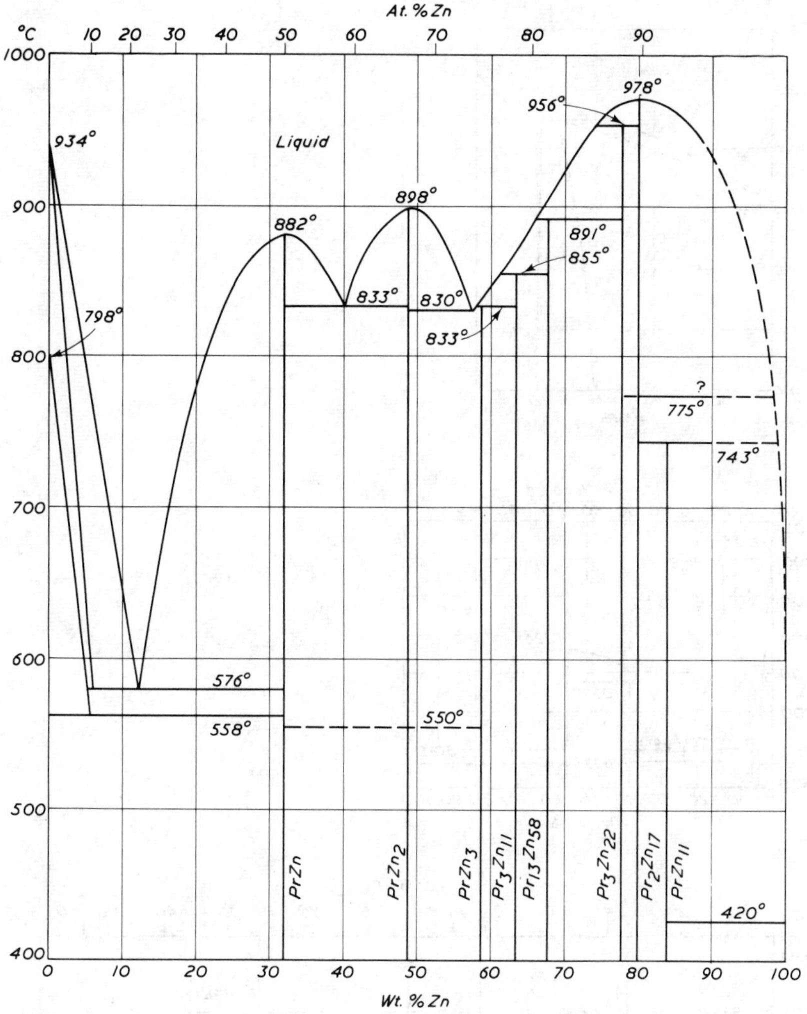

At. % Zn
°C
10
20
30
40
50
60
70
80
90
1000
900
800
700
600
500
400
Liquid
934°
798°
882°
898°
833°
830°
833°
956°
978°
891°
855°
775°
?
743°
576°
558°
550°
420°
PrZn
PrZn2
PrZn3
Pr3Zn11
Pr13Zn58
Pr3Zn22
Pr2Zn17
PrZn11
0
10
20
30
40
50
60
70
80
90
100
Wt. % Zn

Pt–Rh

At.% Rh
°C
10 20 30 40 50 60 70 80 90
2000
1900
1800
1700
1600
Liquid
1966°
1773°
Solid solution
0 10 20 30 40 50 60 70 80 90 100
Wt.% Rh

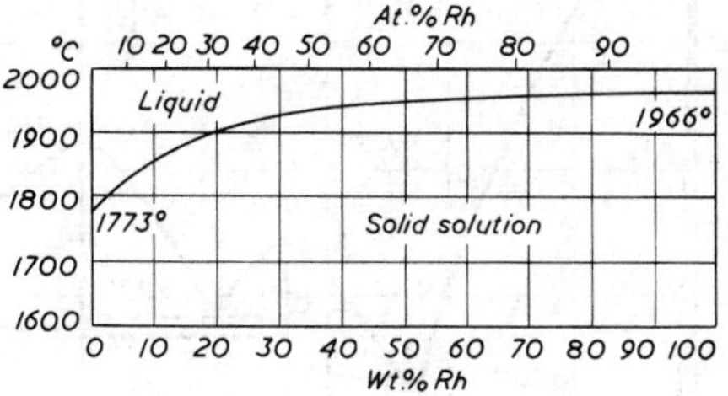

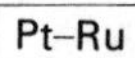
Pt–Ru

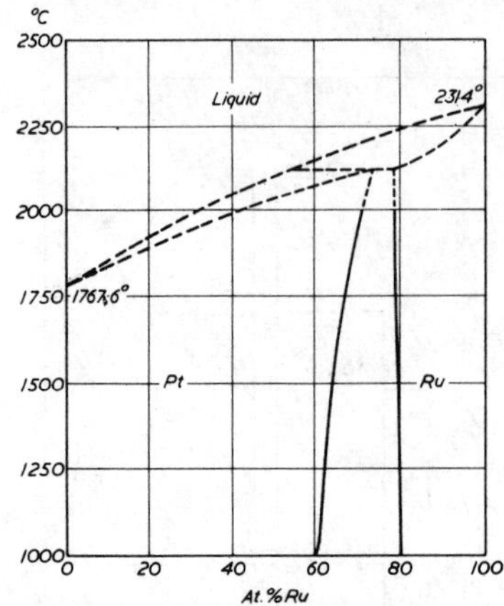
°C
Liquid
2314°
1767.6°
Pt
Ru
At.%Ru

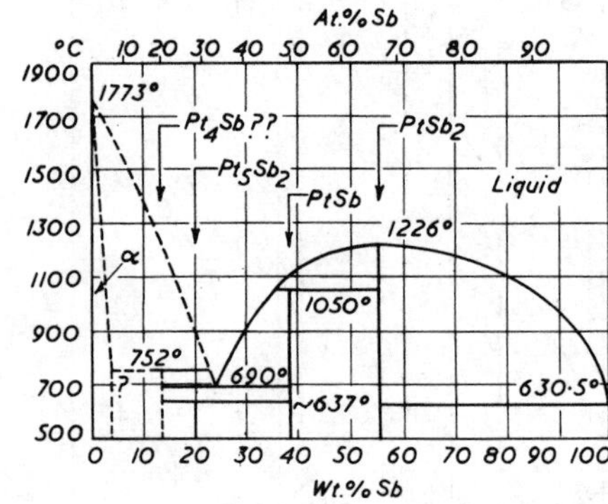
At.% Sb
Wt.% Sb
1773°
Pt4Sb??
Pt5Sb2
PtSb
PtSb2
Liquid
1226°
α
1050°
752°
690°
~637°
630·5°
?

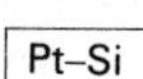
Pt–Si

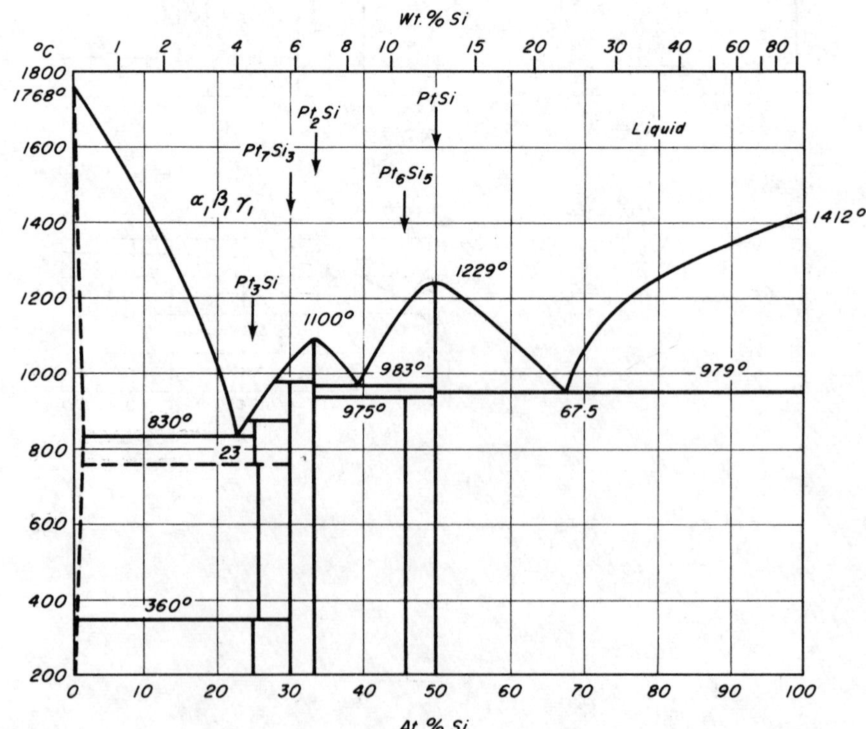
Wt.% Si
°C
1768°
PtSi
Pt2Si
Pt7Si3
Liquid
Pt6Si5
α1 β1 γ1
1412°
1229°
Pt3Si
1100°
983°
979°
975°
67·5
830°
23
360°
At. % Si

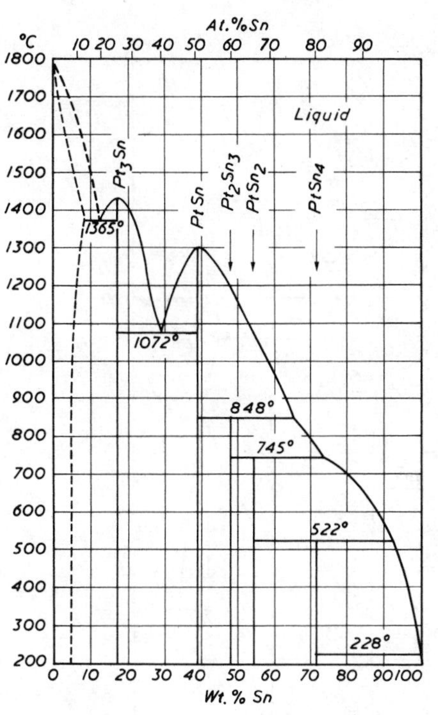

Pt–Sn

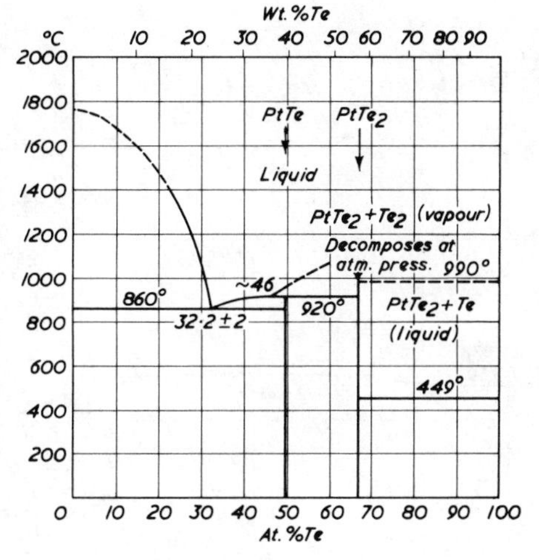

Pt–Te

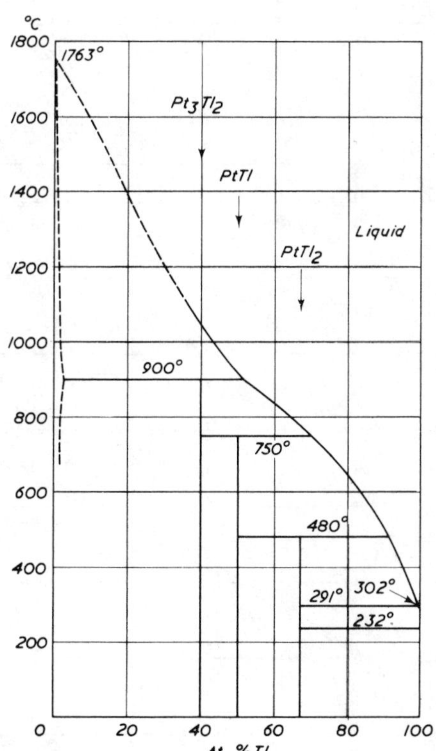

Pt–Tl

Pt–V

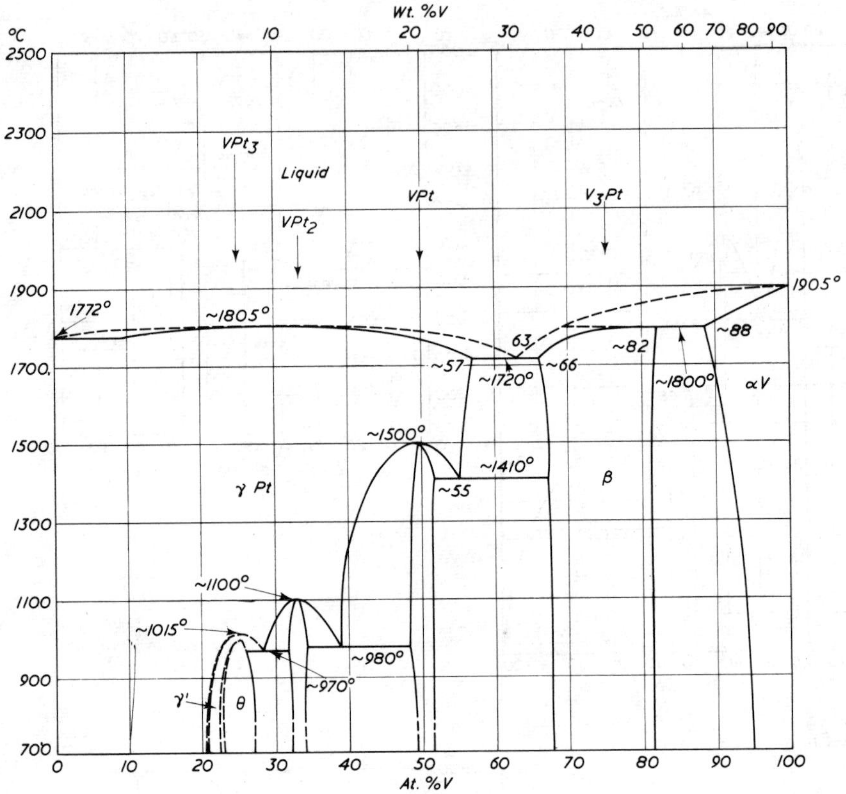

Pt–W

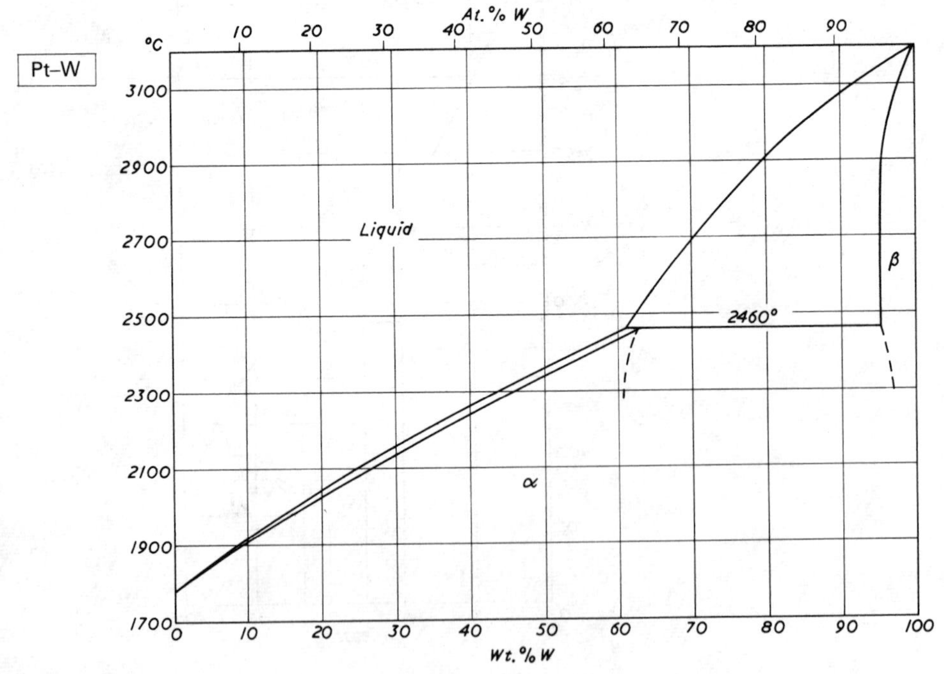

Pt–Zr

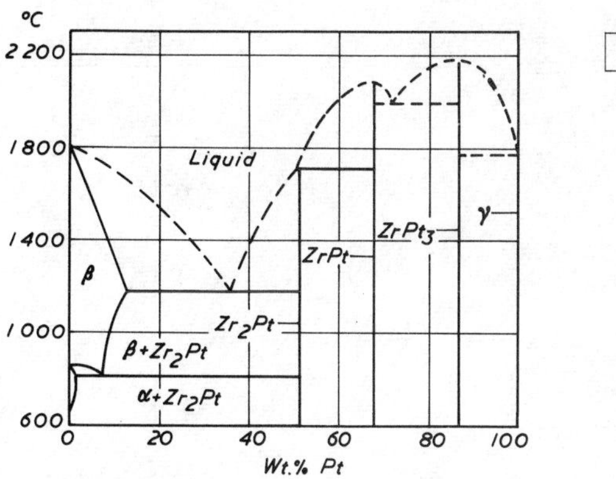
°C
2200
1800
1400
1000
600
Liquid
β
Zr2Pt
β+Zr2Pt
α+Zr2Pt
ZrPt
ZrPt3
γ
0
20
40
60
80
100
Wt.% Pt

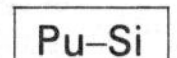
Pu–Si

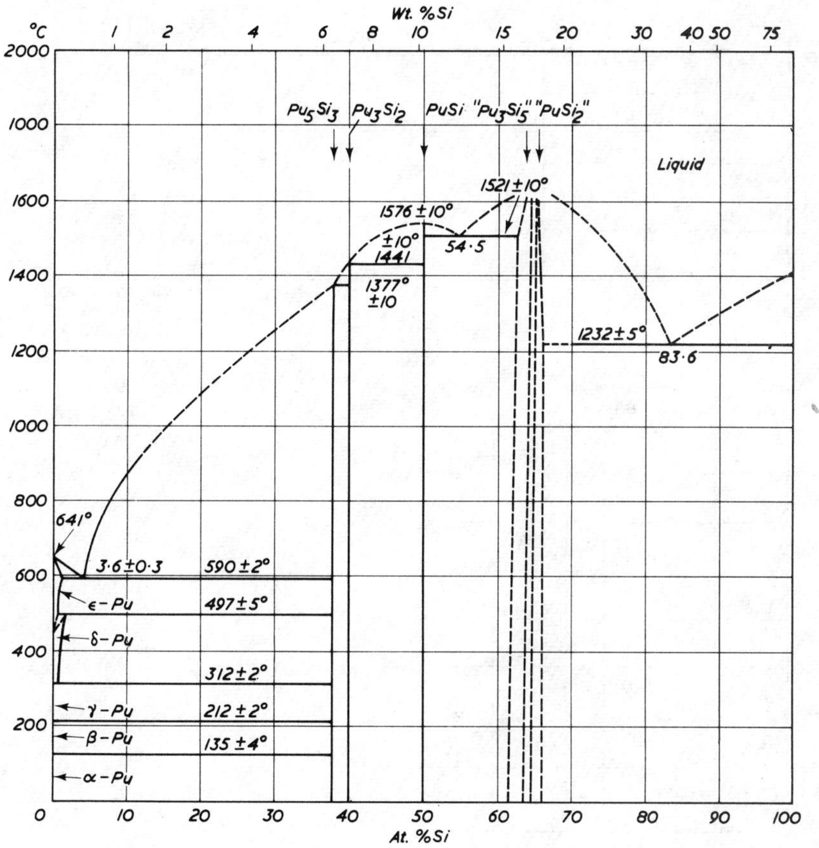
Wt. %Si
°C
2000
1000
1600
1400
1200
1000
800
600
400
200
0
Pu5Si3
Pu3Si2
PuSi
"Pu3Si5"
"PuSi2"
Liquid
1521±10°
1576±10°
±10°
1441
54·5
1377°
±10
1232±5°
83·6
641°
3·6±0·3
590±2°
ε-Pu
497±5°
δ-Pu
312±2°
γ-Pu
212±2°
β-Pu
135±4°
α-Pu
At. %Si

Pu–Th

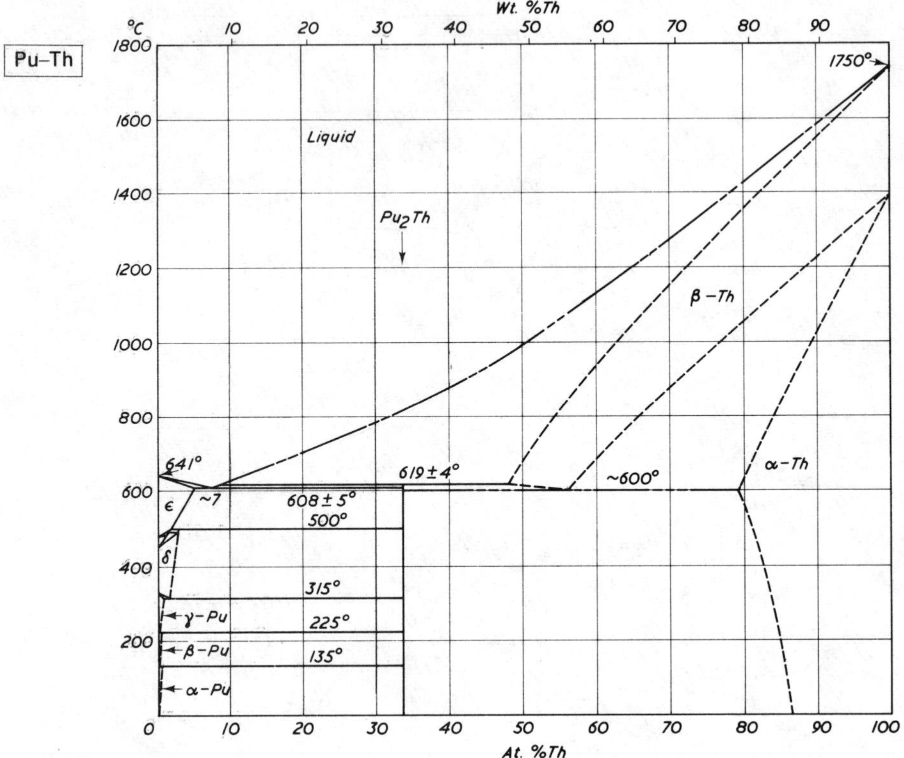
Wt. %Th
°C
Liquid
Pu_2Th
β-Th
α-Th
1750°
641°
619±4°
~600°
~7
608±5°
500°
315°
225°
135°
ε
δ
γ-Pu
β-Pu
α-Pu
At. %Th

Pu–Ti

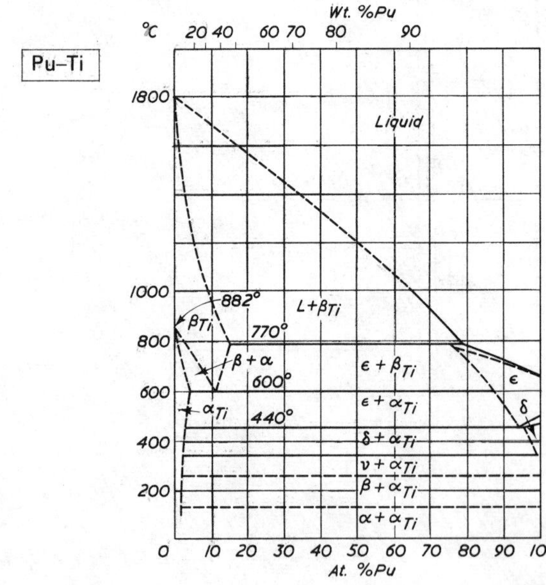
Wt. %Pu
°C
Liquid
882°
$L+\beta_{Ti}$
β_{Ti}
770°
β+α
600°
α_{Ti}
440°
$\epsilon+\beta_{Ti}$
$\epsilon+\alpha_{Ti}$
$\delta+\alpha_{Ti}$
$\nu+\alpha_{Ti}$
$\beta+\alpha_{Ti}$
$\alpha+\alpha_{Ti}$
ε
δ
At. %Pu

Pu–U

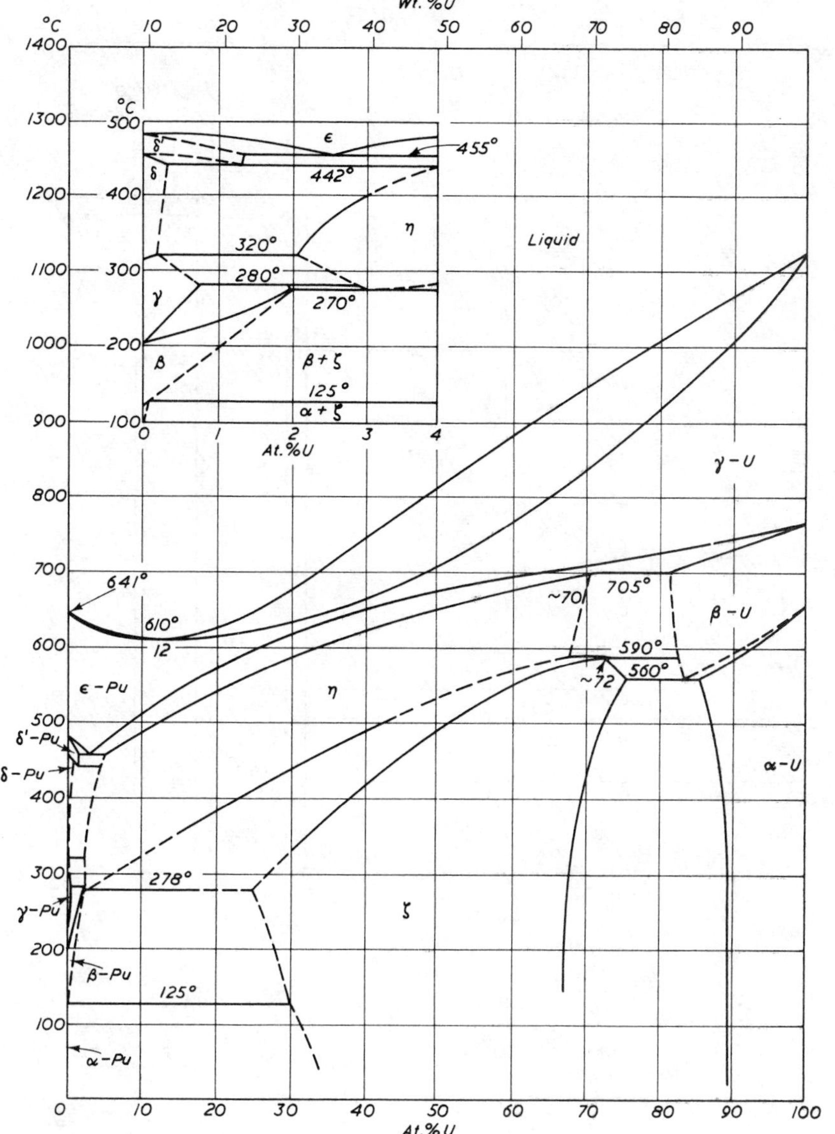

Wt. %U
°C
At.%U
Liquid
641°
610°
12
705°
~70
590°
560°
~72
ε-Pu
δ'-Pu
δ-Pu
γ-Pu
β-Pu
α-Pu
η
ζ
278°
125°
γ-U
β-U
α-U
ε
455°
442°
δ'
δ
320°
280°
270°
γ
β
β+ζ
125°
α+ζ

Pu–Zn

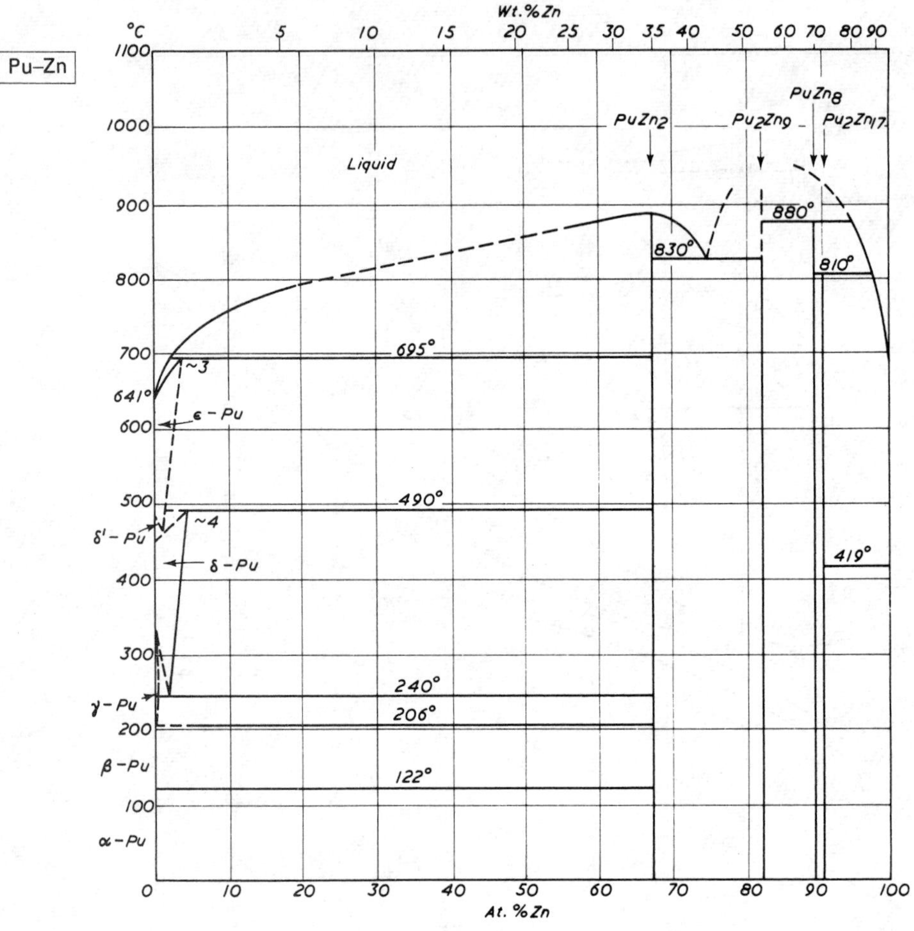

Wt.% Zn
°C
Liquid
PuZn2
Pu2Zn9
PuZn8
Pu2Zn17
880°
830°
810°
695°
~3
641°
ε-Pu
490°
~4
δ'-Pu
δ-Pu
419°
240°
γ-Pu
206°
β-Pu
122°
α-Pu
At. % Zn

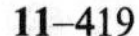

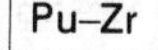
Pu–Zr

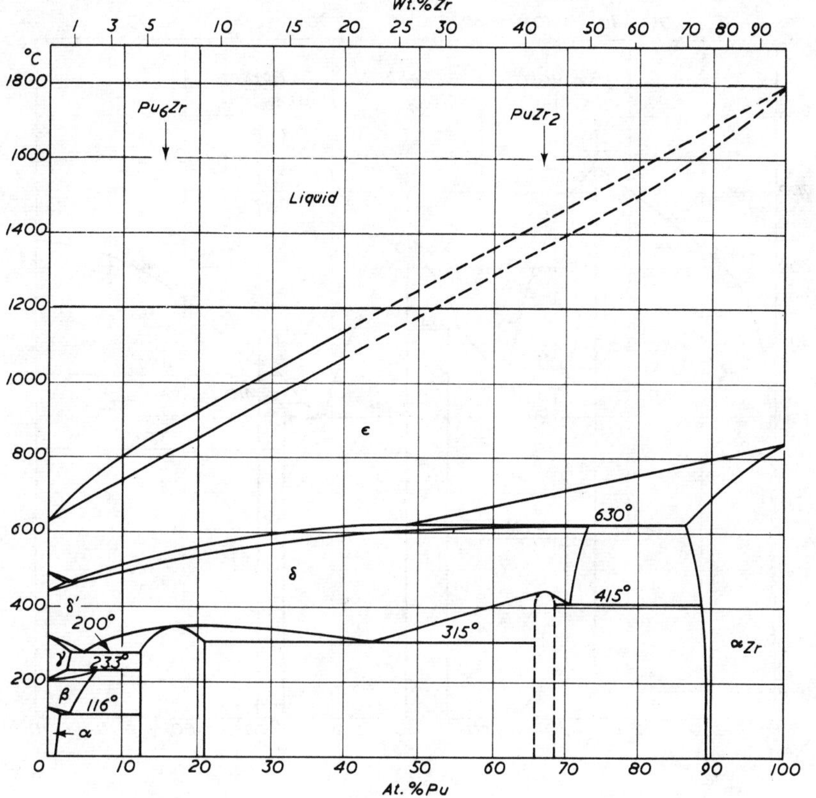
Wt.% Zr
°C
Pu6Zr
PuZr2
Liquid
ε
δ
δ'
γ
β
α
630°
415°
315°
200°
233°
116°
αZr
At. % Pu

Rb–S

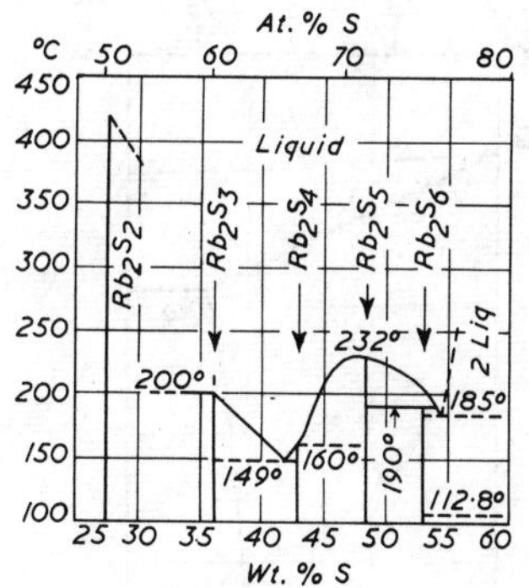
At. % S
°C
Liquid
Rb2S2
Rb2S3
Rb2S4
Rb2S5
Rb2S6
2 Liq
232°
200°
185°
149°
160°
190°
112·8°
Wt. % S

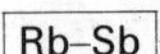
Rb–Sb

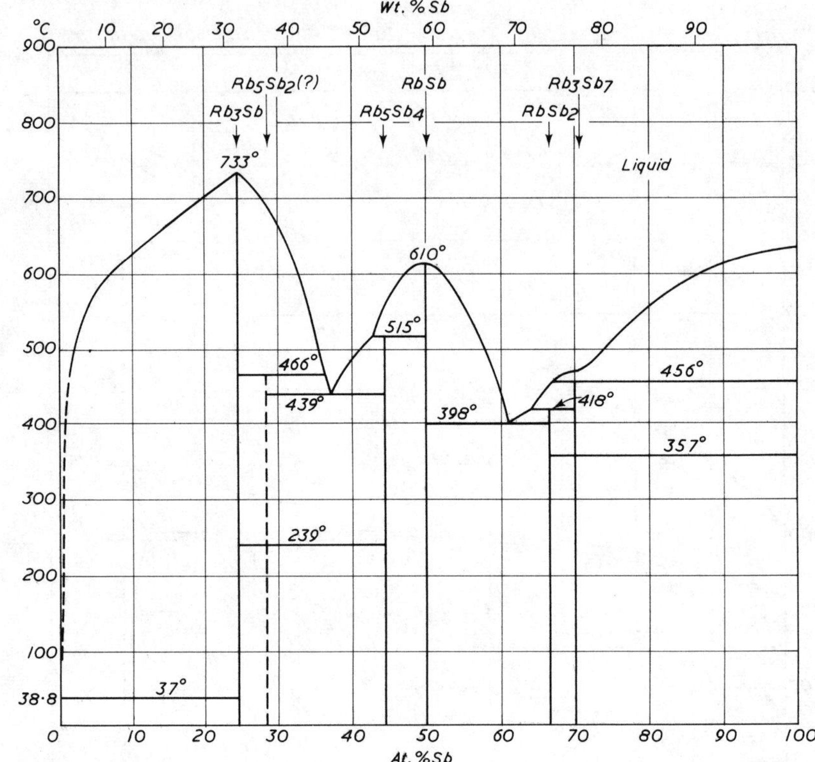
Wt. % Sb
°C
Rb5Sb2(?)
Rb3Sb
RbSb
Rb5Sb4
Rb3Sb7
RbSb2
733°
Liquid
610°
515°
466°
439°
398°
418°
456°
357°
239°
37°
38·8
At. % Sb

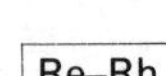
Re–Rh

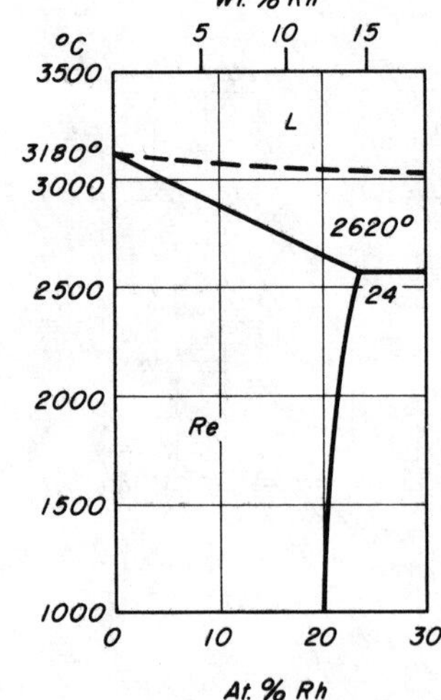
Wt. % Rh
°C
L
3180°
2620°
24
Re
At. % Rh

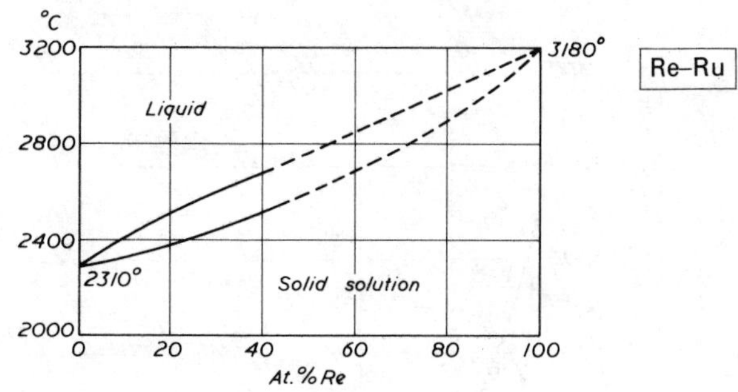

Re–Ru

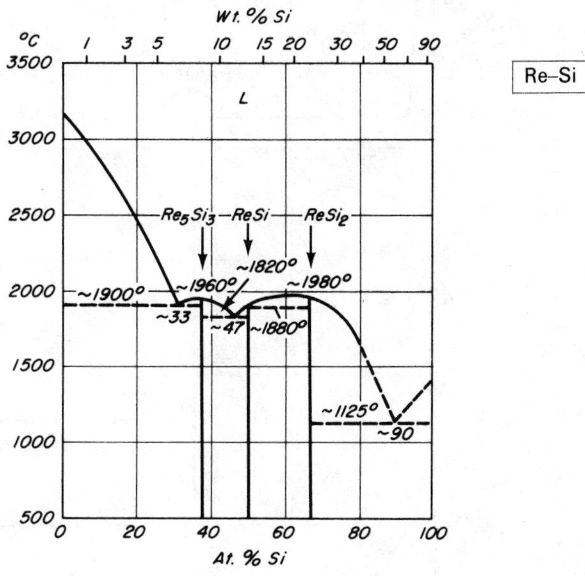

Re–Si

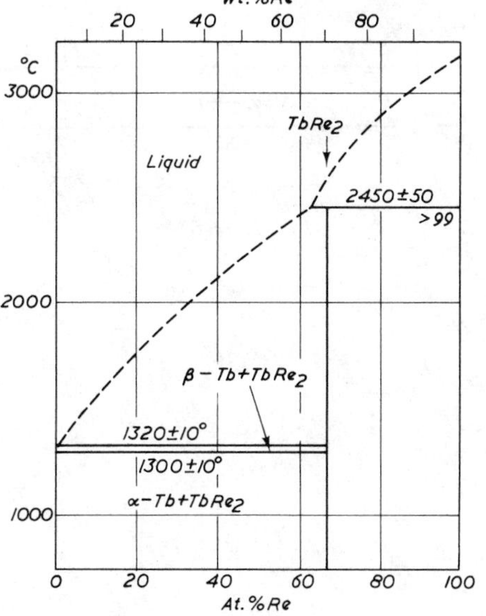

Re–Tb

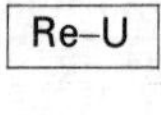
Re–U

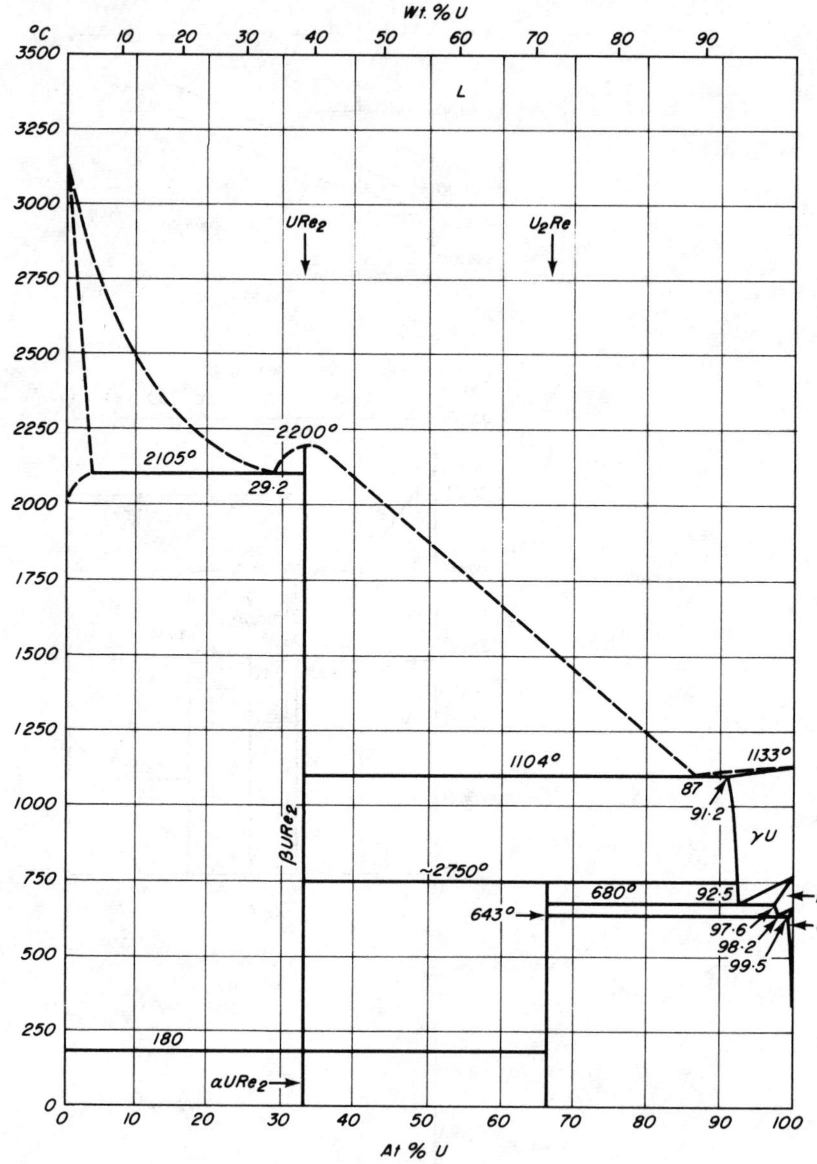

Wt. % U
°C
10
20
30
40
50
60
70
80
90
L
3500
3250
3000
2750
2500
2250
2000
1750
1500
1250
1000
750
500
250
0
URe_2
U_2Re
2200°
2105°
29·2
1104°
1133°
87
91·2
γU
βURe₂
~2750°
680°
92·5
643°
βU
αU
97·6
98·2
99·5
180
αURe₂
0
10
20
30
40
50
60
70
80
90
100
At % U

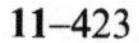

Re–V

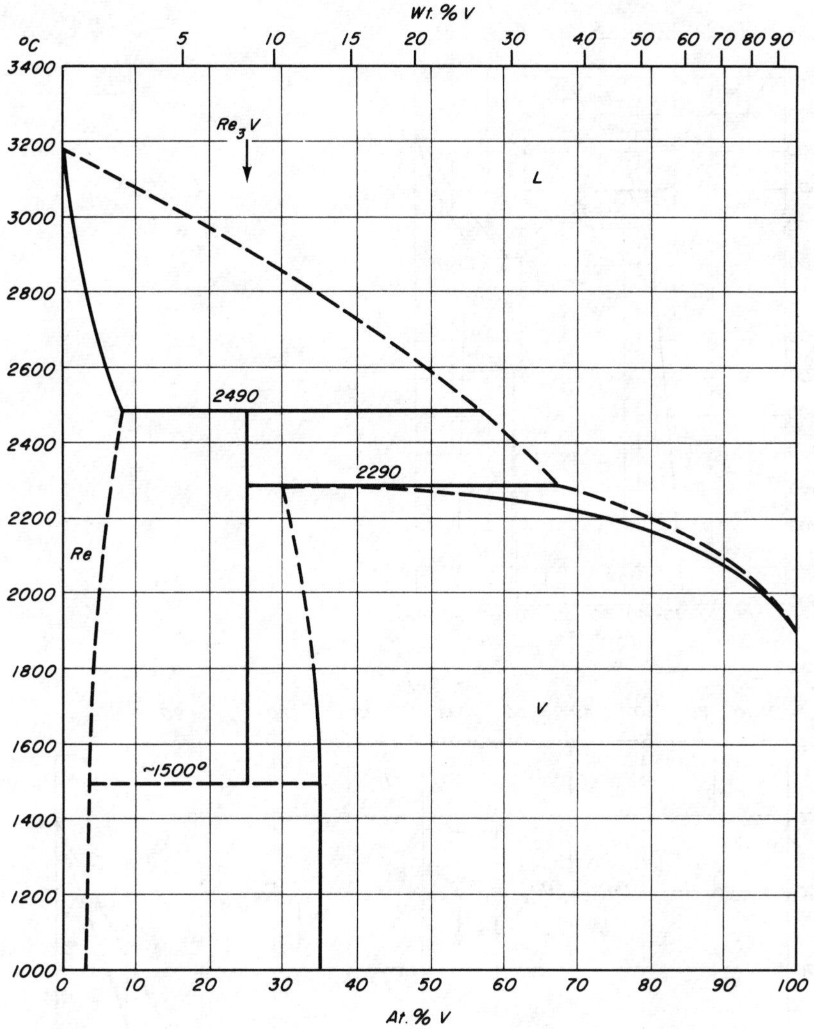

Wt. % V

5 10 15 20 30 40 50 60 70 80 90

°C

3400 3200 3000 2800 2600 2400 2200 2000 1800 1600 1400 1200 1000

Re_3V

L

2490

2290

Re

V

~1500°

0 10 20 30 40 50 60 70 80 90 100

At. % V

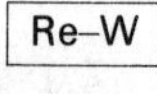
Re–W

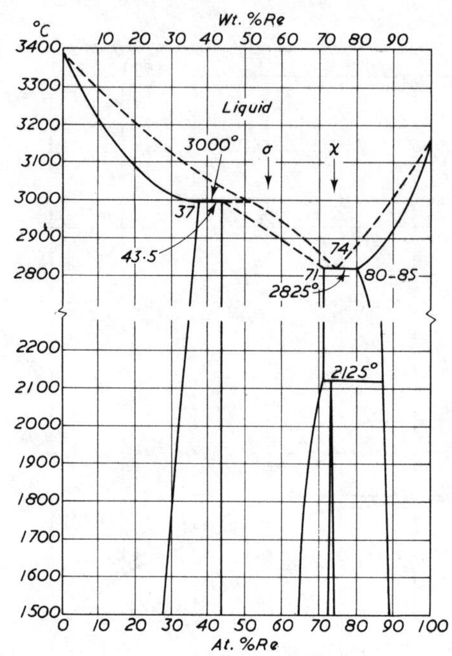
°C
Wt. %Re
10 20 30 40 50 60 70 80 90
3400
3300
3200
3100
3000
2900
2800
2200
2100
2000
1900
1800
1700
1600
1500
Liquid
3000°
σ
χ
37
43·5
74
71
80–85
2825°
2125°
0 10 20 30 40 50 60 70 80 90 100
At. %Re

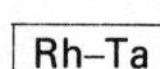
Rh–Ta
Wt. % Ta 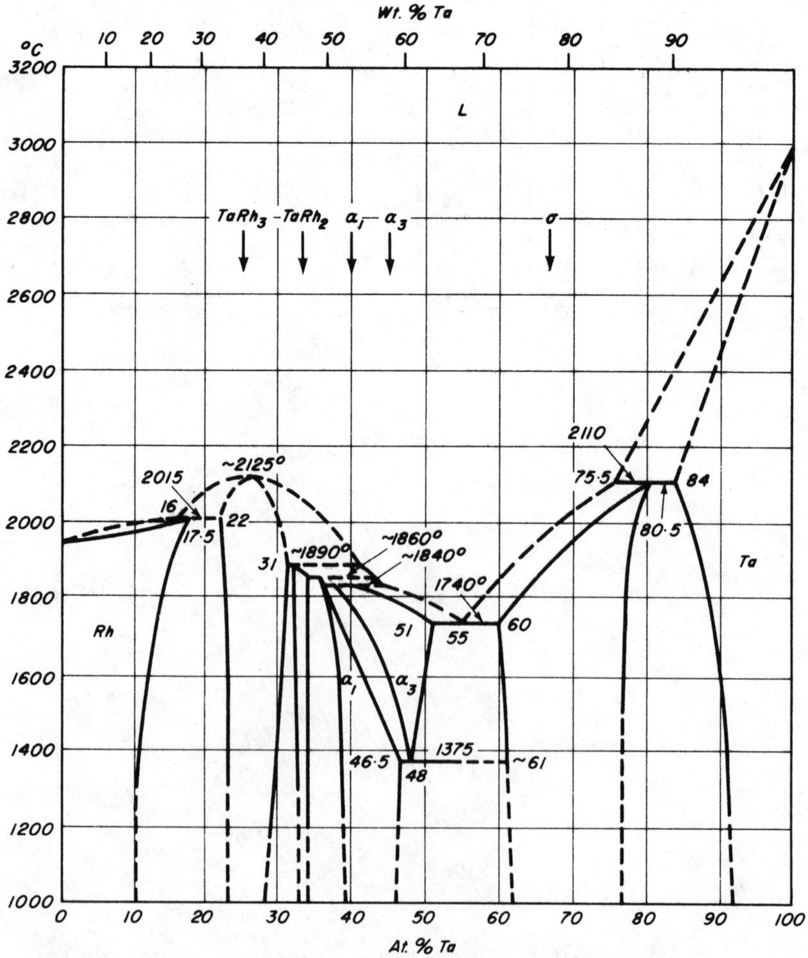
10 20 30 40 50 60 70 80 90
°C
3200
3000
2800
2600
2400
2200
2000
1800
1600
1400
1200
1000
L
TaRh3 -TaRh2 α1 α3
σ
2015
~2125°
16
17·5
22
31
~1890°
~1860°
~1840°
1740°
51
55
60
2110
75·5
84
80·5
Ta
Rh
α1
α3
46·5
48
1375
~61
0 10 20 30 40 50 60 70 80 90 100
At. % Ta

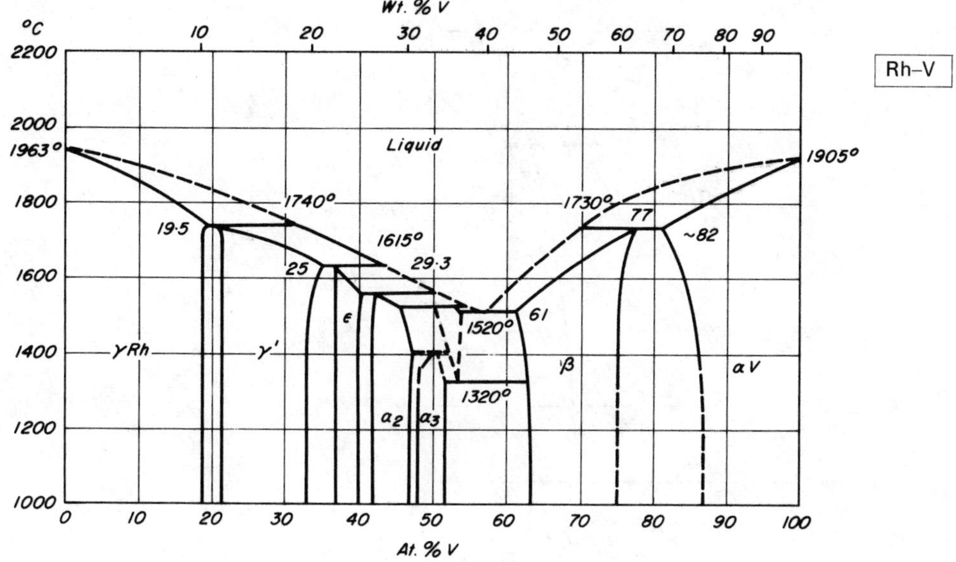

Rh–W

Wt. % W

°C

L

2240

2100

~17

~20

~19

~58

~67

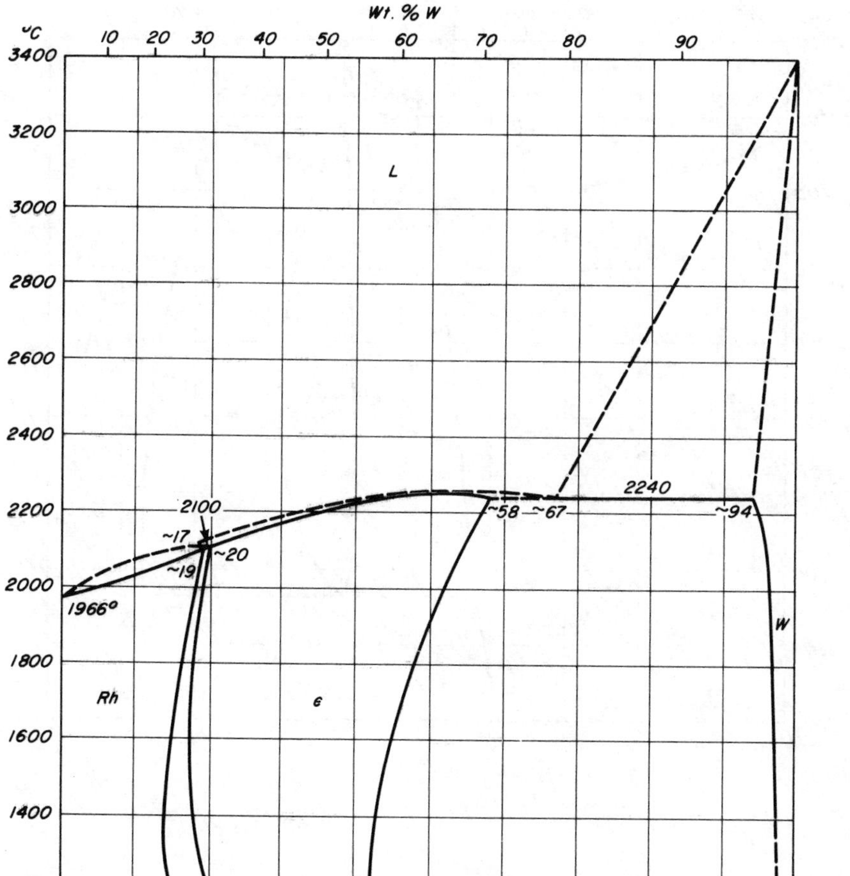

~94

1966°

W

Rh

ε

At. % W

Ru–Pt

°C
2500
2250
2000
1750
1500
1250
1000
Liquid
2314°
1767·6°
Pt
Ru
0
20
40
60
80
100
At.% Ru

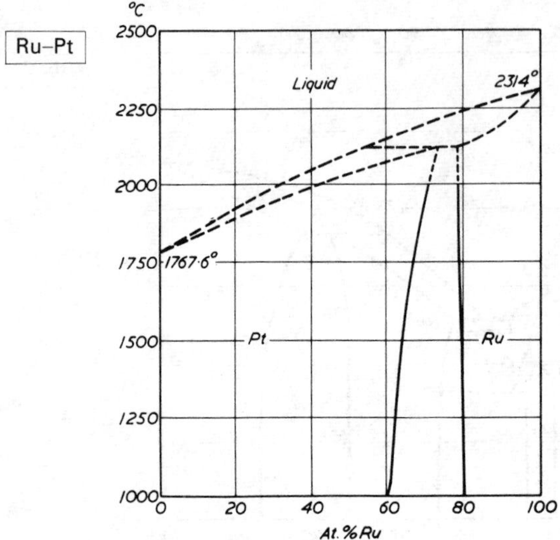

Ru–W

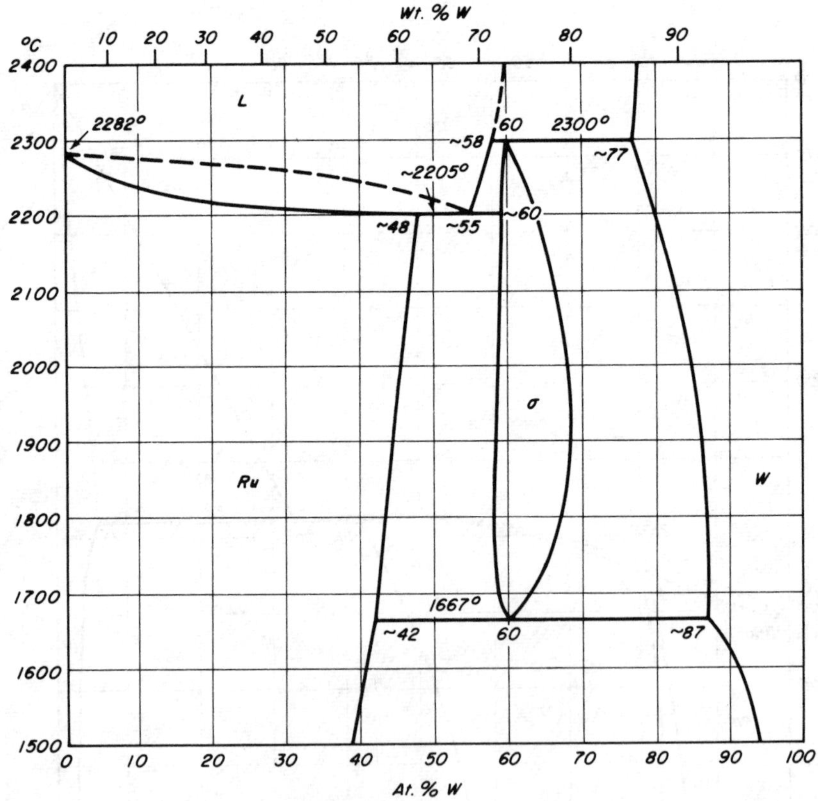

Wt. % W
10
20
30
40
50
60
70
80
90
°C
2400
2300
2200
2100
2000
1900
1800
1700
1600
1500
L
2282°
~58
60
2300°
~77
~2205°
~48
~55
~60
σ
Ru
W
1667°
~42
60
~87
0
10
20
30
40
50
60
70
80
90
100
At. % W

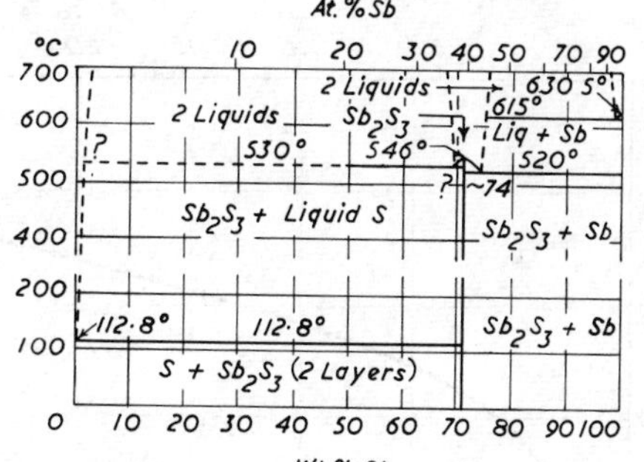
At. % Sb
°C
2 Liquids
630.5°
615°
Liq + Sb
2 Liquids
Sb_2S_3
530°
546°
520°
~74
Sb_2S_3 + Liquid S
Sb_2S_3 + Sb
112·8°
112·8°
Sb_2S_3 + Sb
S + Sb_2S_3 (2 Layers)
Wt. % Sb

S–Sb

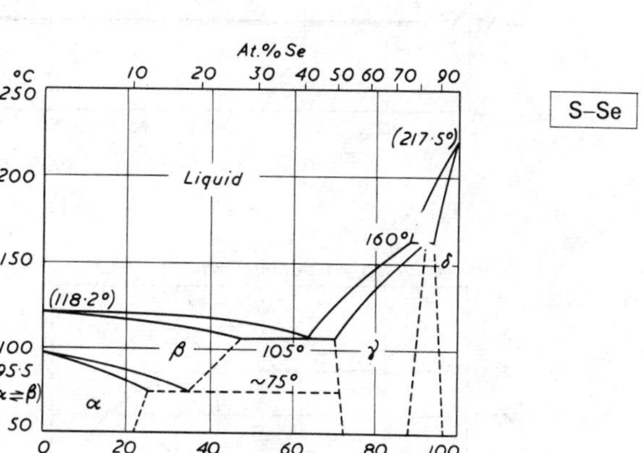
At. % Se
°C
(217·5°)
Liquid
160°
δ
(118·2°)
β
105°
γ
95·5
(α⇌β)
~75°
α
Wt. % Se

S–Se

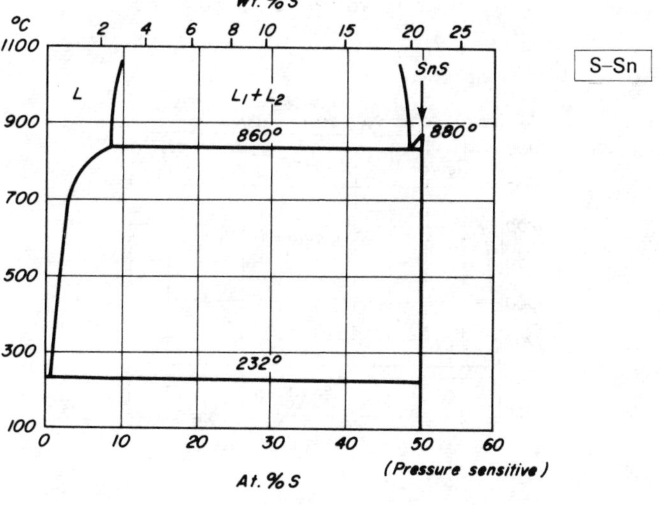
Wt. % S
°C
SnS
L
$L_1 + L_2$
860°
880°
232°
(Pressure sensitive)
At. % S

S–Sn

S–Te

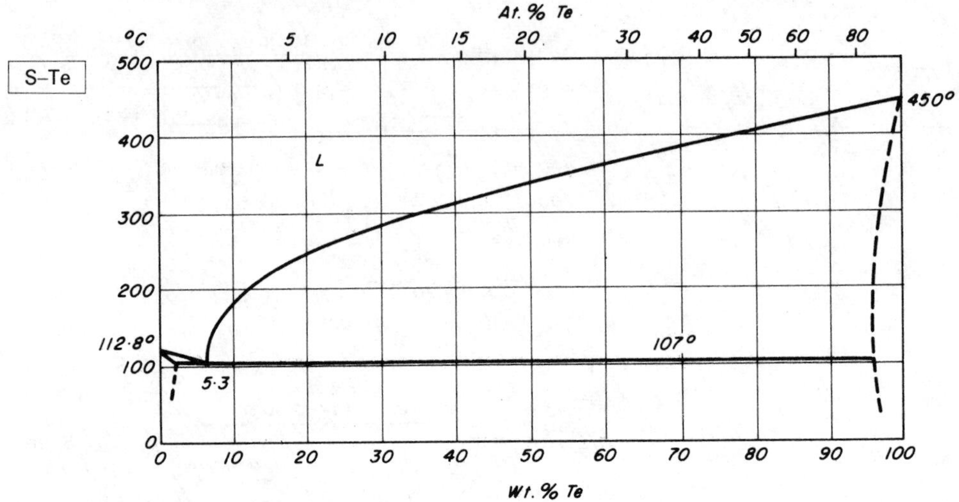

Sb–Se

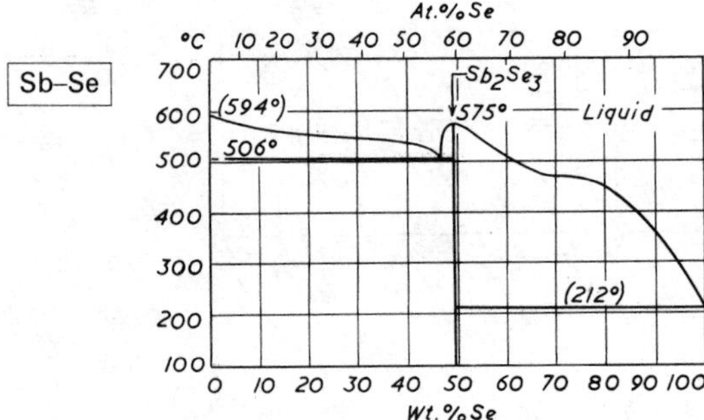

Sb–Si

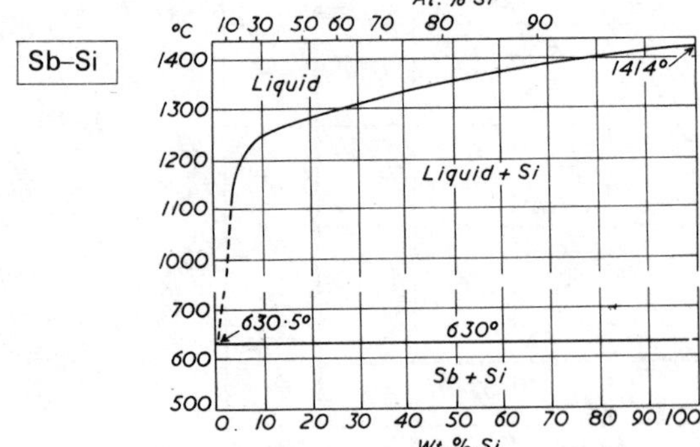

Sb–Sn

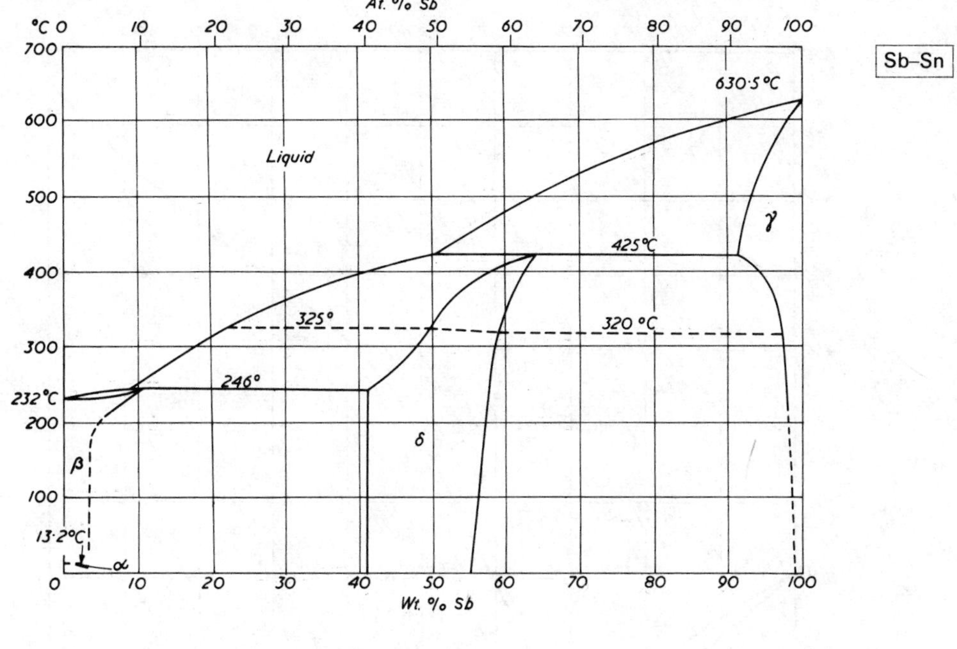
At. % Sb
Wt. % Sb
Liquid
630·5 °C
425 °C
325°
320 °C
246°
232 °C
13·2 °C
β
α
δ
γ

Sb–Te

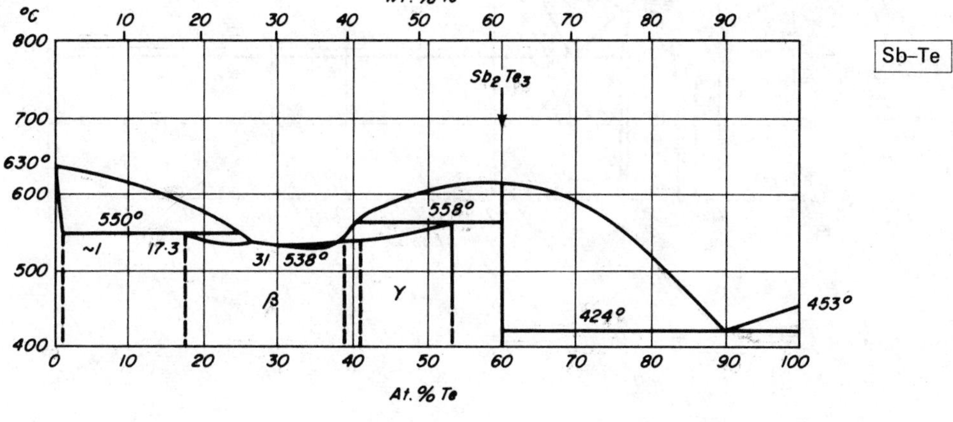
Wt. % Te
At. % Te
Sb_2Te_3
630°
550°
558°
~1
17·3
31
538°
β
γ
424°
453°

Sb–Tl

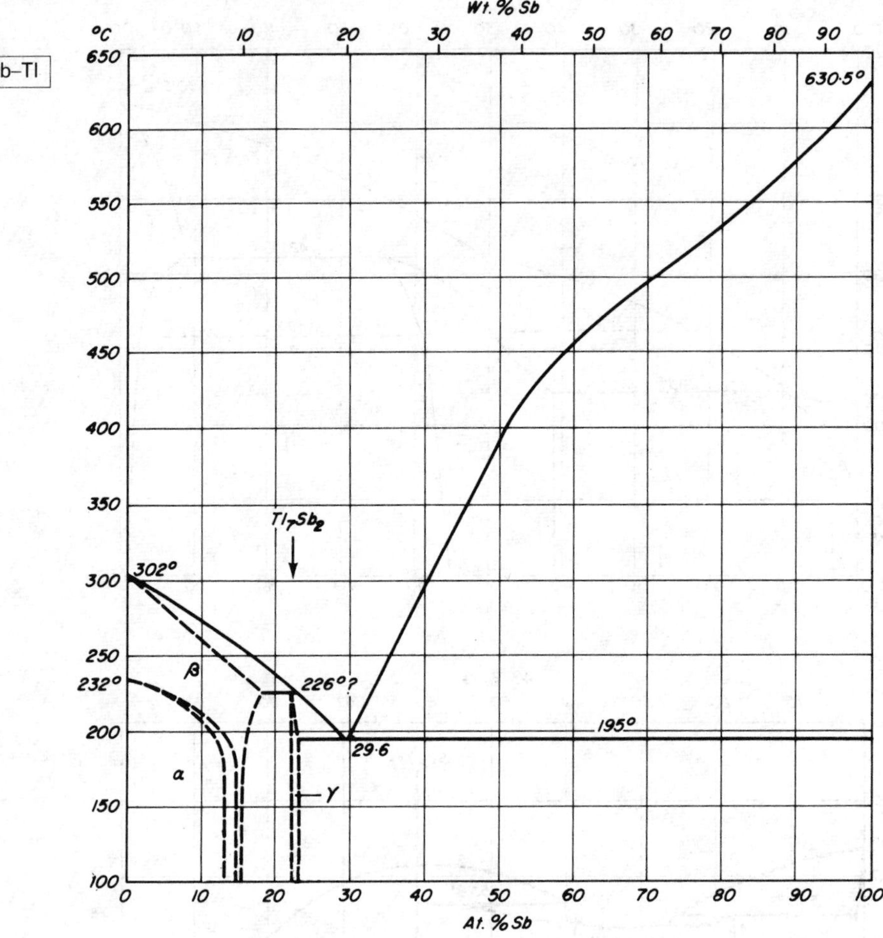

Wt. % Sb
°C
10
20
30
40
50
60
70
80
90
650
600
550
500
450
400
350
300
250
232°
200
150
100
630·5°
Tl7Sb2
302°
β
226°?
195°
29·6
α
γ
0
10
20
30
40
50
60
70
80
90
100
At. % Sb

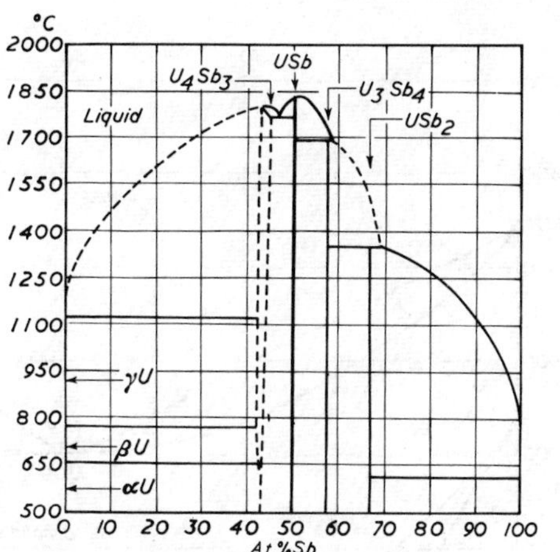
°C
USb
U4Sb3
U3Sb4
USb2
Liquid
γU
βU
αU
At.%Sb

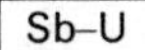
Sb–U

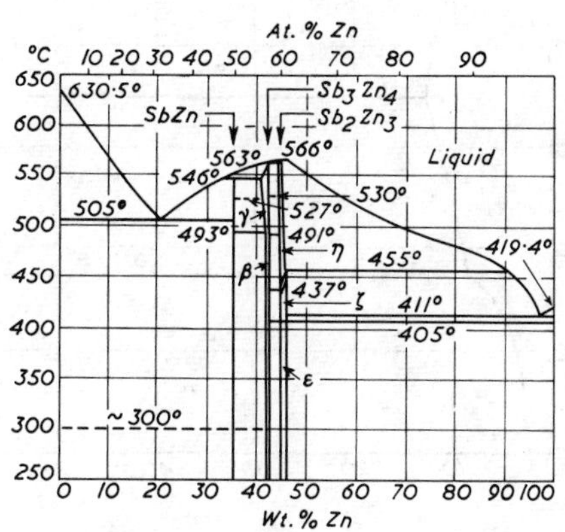
At. % Zn
°C
630·5°
SbZn
Sb3Zn4
Sb2Zn3
563°
566°
546°
Liquid
530°
505°
527°
493°
491°
η
455°
419·4°
β
437°
ζ
411°
405°
ε
~300°
Wt.% Zn

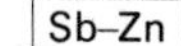
Sb–Zn

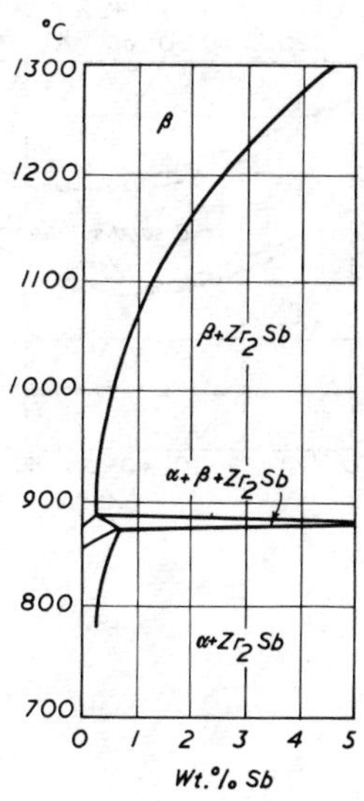
°C
β
β+Zr2Sb
α+β+Zr2Sb
α+Zr2Sb
Wt.% Sb

Sb–Zr

Sc–Ti

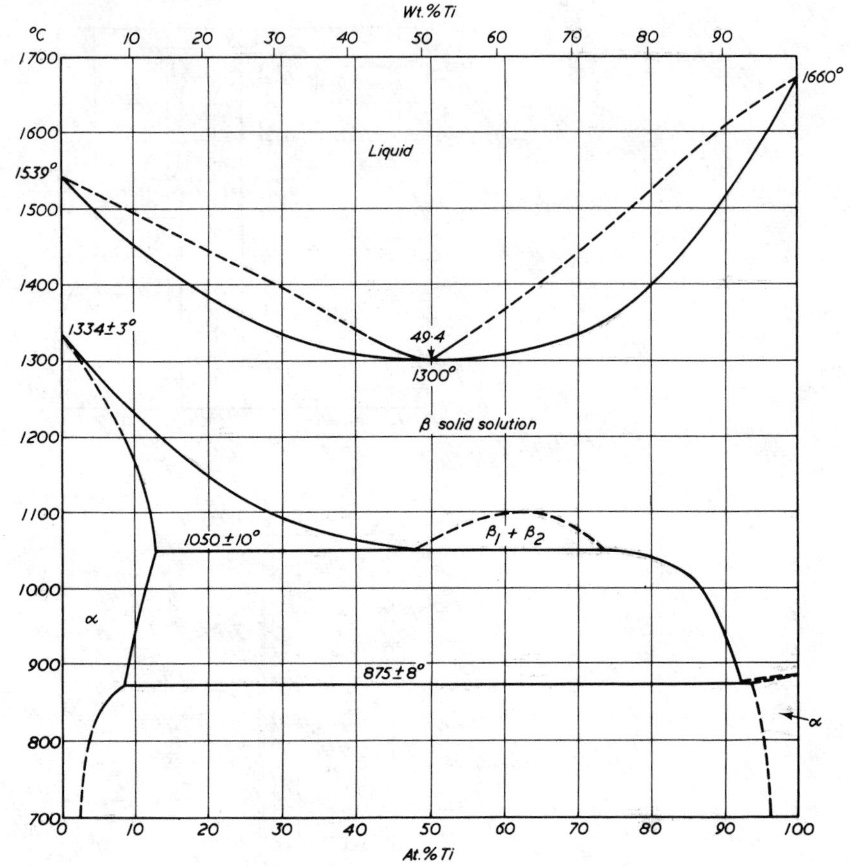

Wt.% Ti
°C
Liquid
1539°
1660°
1334±3°
49·4
1300°
β solid solution
1050±10°
β1 + β2
α
875±8°
α
At.% Ti

Sc–Y

Wt.% Y
°C
1539°
1523°
1365°
50
β solid solution
1175°
43
α solid solution
At.% Y

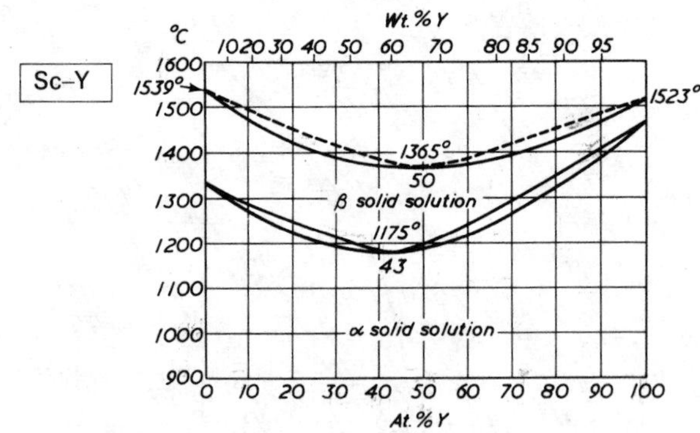

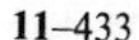

Sc–Zr

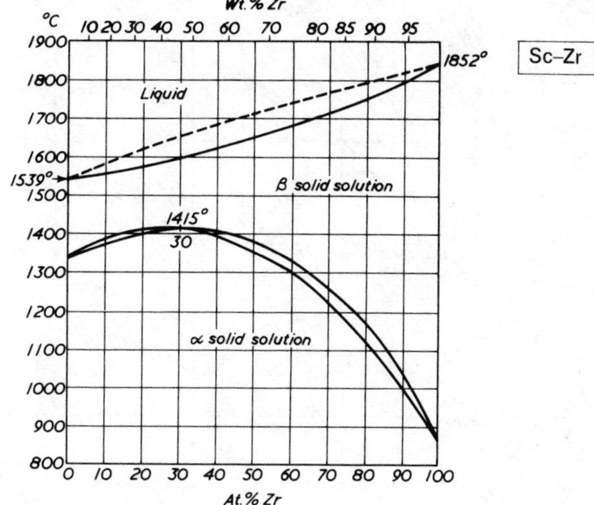
Wt.% Zr
10 20 30 40 50 60 70 80 85 90 95
°C
1900
1800
1700
1600
1500
1400
1300
1200
1100
1000
900
800
1852°
Liquid
1539°
β solid solution
1415°
30
α solid solution
0 10 20 30 40 50 60 70 80 90 100
At.% Zr

Se–Sn

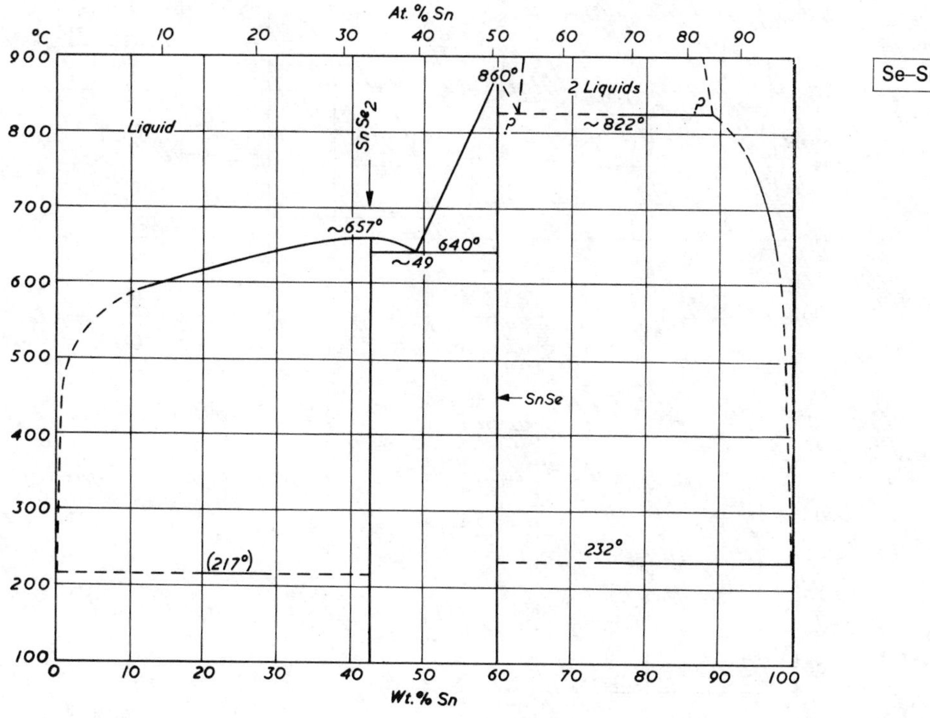
At. % Sn
10 20 30 40 50 60 70 80 90
°C
900
800
700
600
500
400
300
200
100
Liquid
$SnSe_2$
860°
2 Liquids
?
~822°
~657°
640°
~49
SnSe
(217°)
232°
0 10 20 30 40 50 60 70 80 90 100
Wt.% Sn

Se–Te

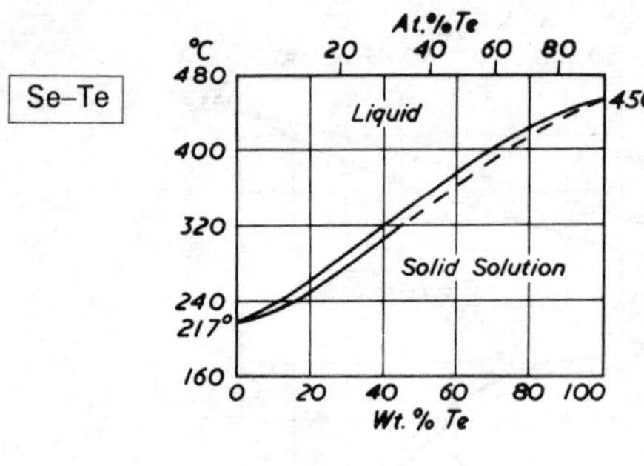

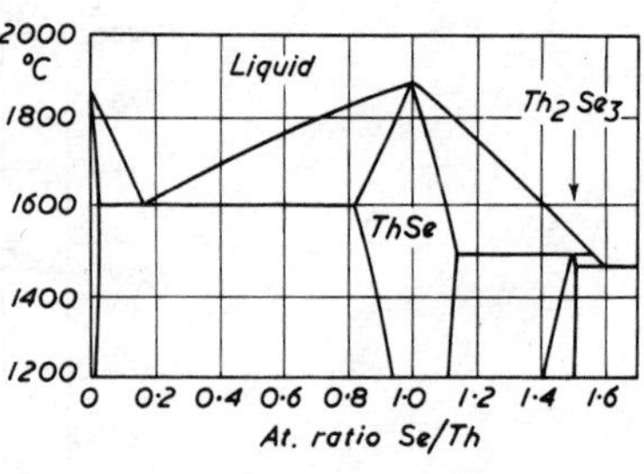

Se–Th

Se–Tl

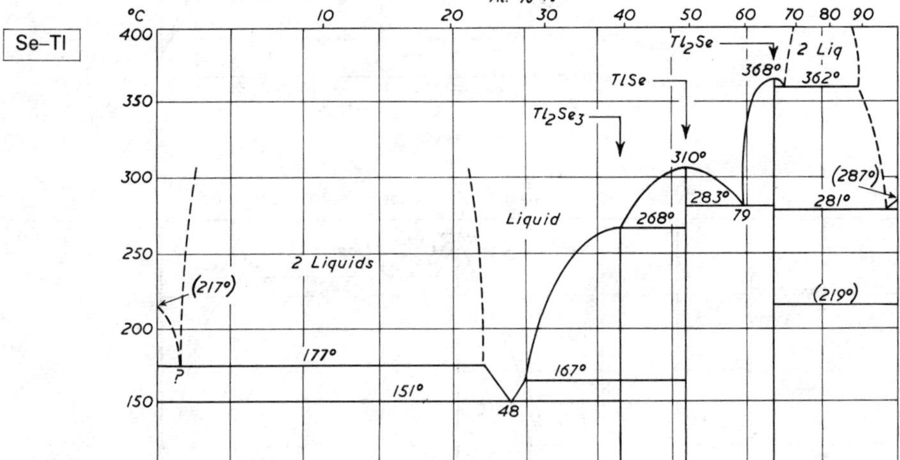

Si–Sn

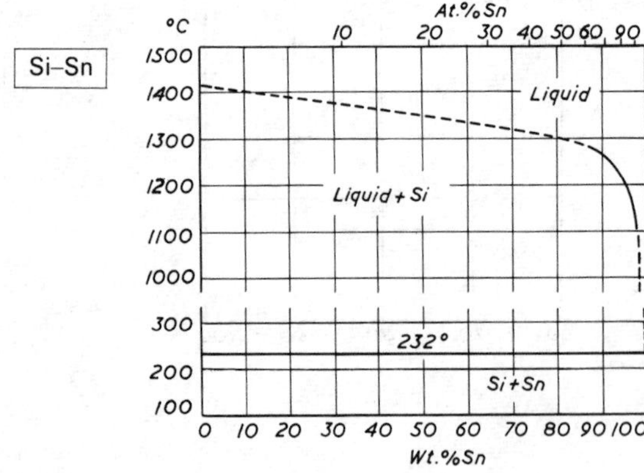

Si–Sr

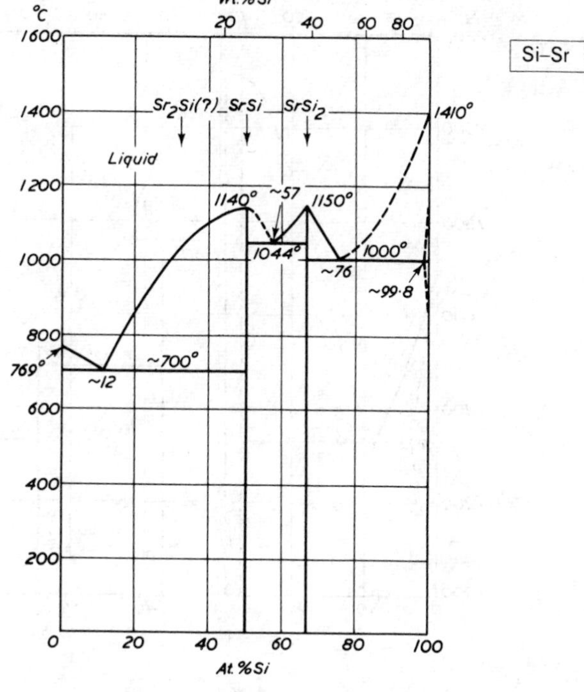
Wt.% Si
20
40
60
80
°C
1600
1400
1200
1000
800
600
400
200
Sr2Si(?)
SrSi
SrSi2
1410°
Liquid
1140°
~57
1150°
1044°
1000°
~76
~99·8
769°
~700°
~12
0
20
40
60
80
100
At.% Si

Si–Th

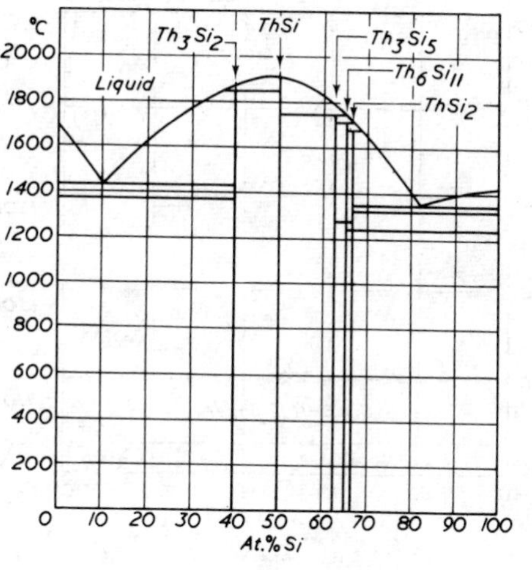
°C
2000
1800
1600
1400
1200
1000
800
600
400
200
Th3Si2
ThSi
Th3Si5
Th6Si11
ThSi2
Liquid
0
10
20
30
40
50
60
70
80
90
100
At.% Si

Si–Ti

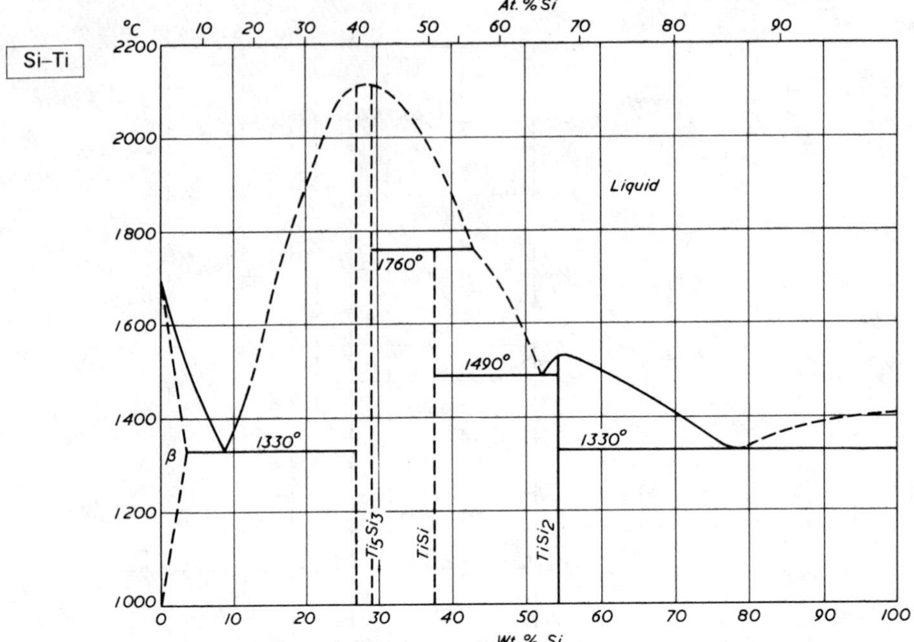

Si–Tl

Si–U

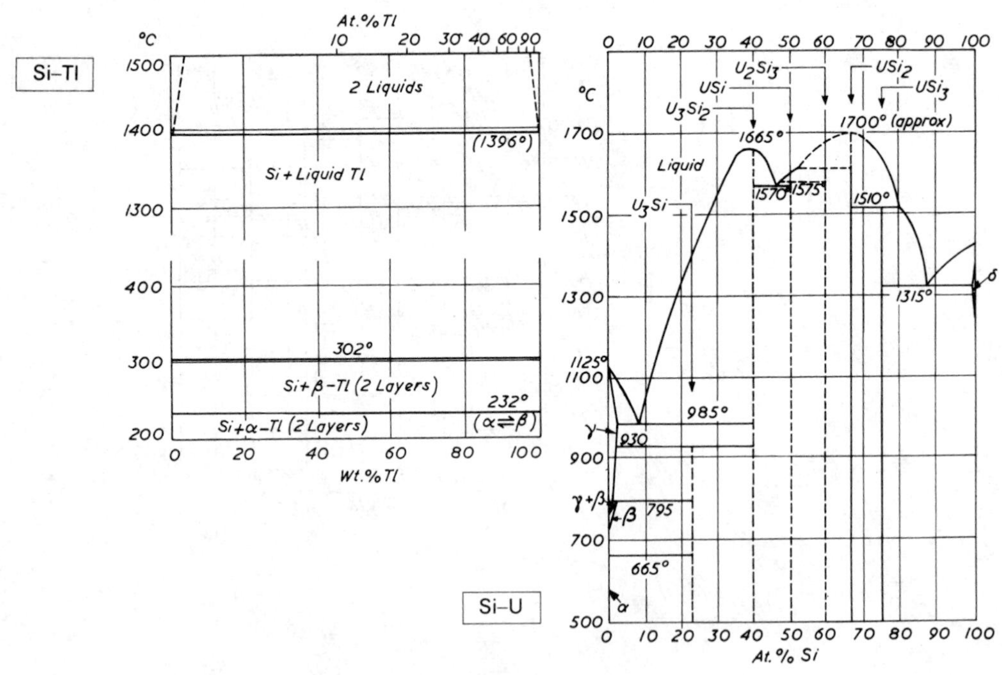

Si–V

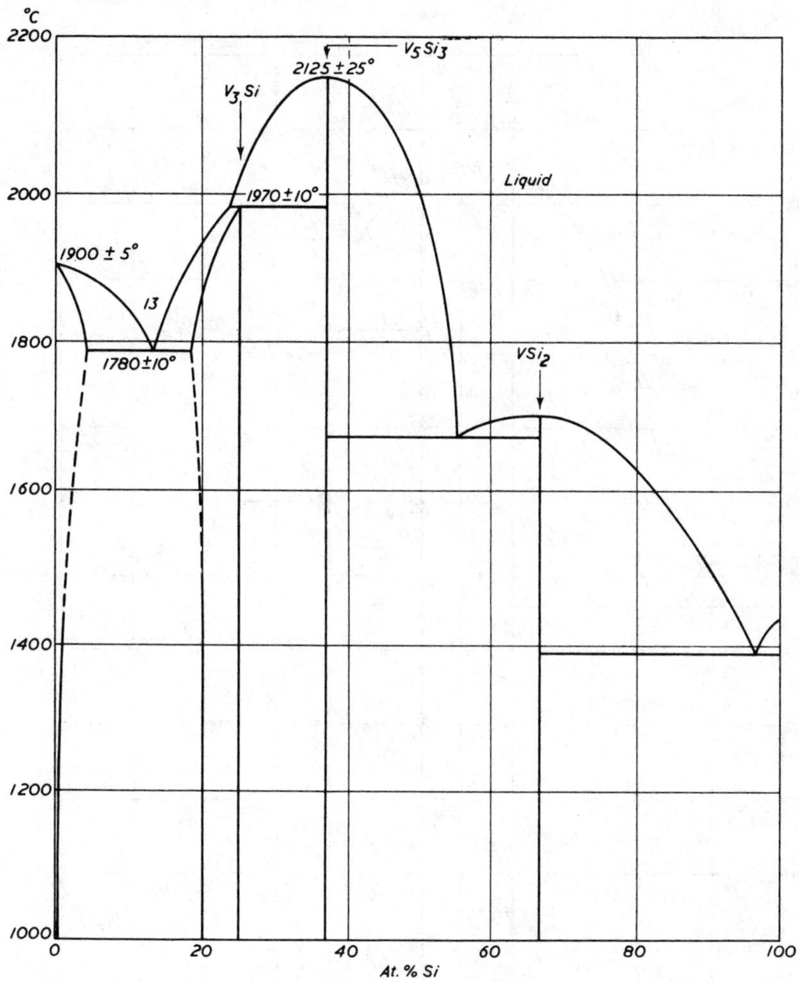

°C
2200
2000
1800
1600
1400
1200
1000
0
20
40
60
80
100
At. % Si
V_5Si_3
2125 ± 25°
V_3Si
Liquid
1970 ± 10°
1900 ± 5°
13
1780 ± 10°
VSi_2

Si–W

Wt. % Si
5
15
50
°C
2700
2500
2300
2100
1900
1700
1500
1300
Liquid
W_3Si_2
WSi_2
0
20
40
60
80
100
At. % Si

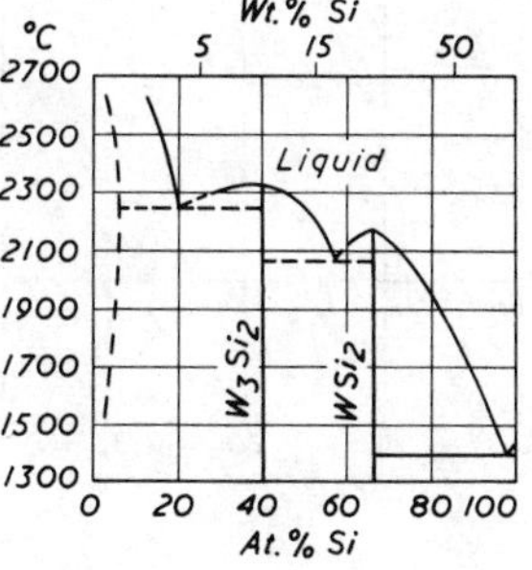

Si–Y

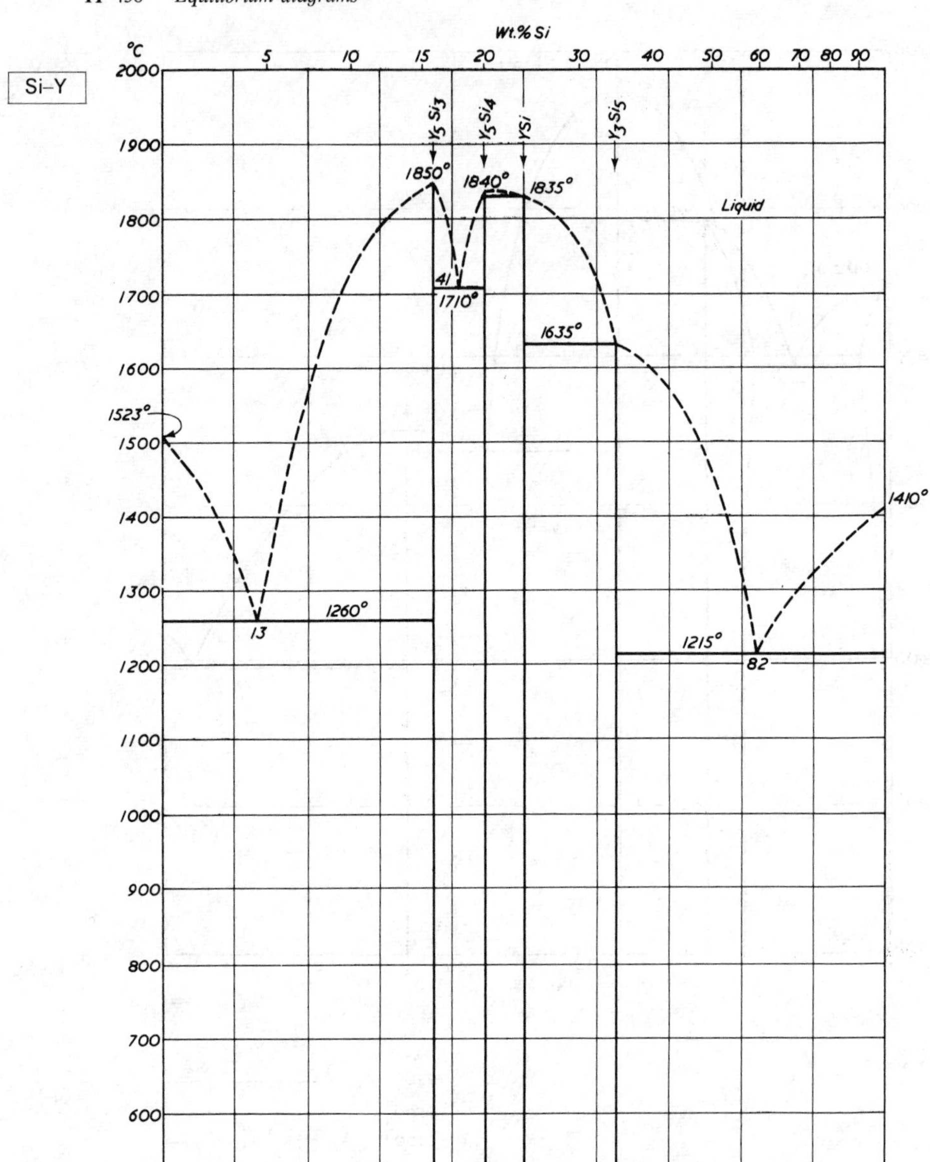

Wt.% Si
°C
2000
1900
1800
1700
1600
1500
1400
1300
1200
1100
1000
900
800
700
600
500
400
5
10
15
20
30
40
50
60
70
80
90
Y5Si3
Y5Si4
YSi
Y3Si5
1850°
1840°
1835°
Liquid
41
1710°
1635°
1523°
1410°
1260°
13
1215°
82
Y3Si5 ⇌ Y3Si5
450±50°
0
10
20
30
40
50
60
70
80
90
100
At.% Si

Si–Zn

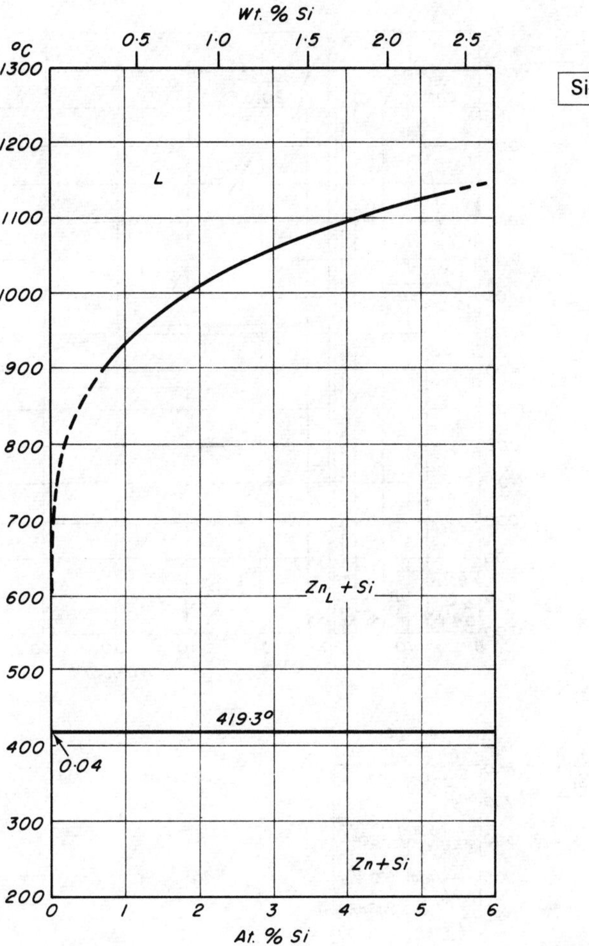
Wt. % Si
0·5
1·0
1·5
2·0
2·5
°C
1300
1200
1100
1000
900
800
700
600
500
400
300
200
L
Zn_L + Si
419·3°
0·04
Zn+Si
0
1
2
3
4
5
6
At. % Si

Si–Zr

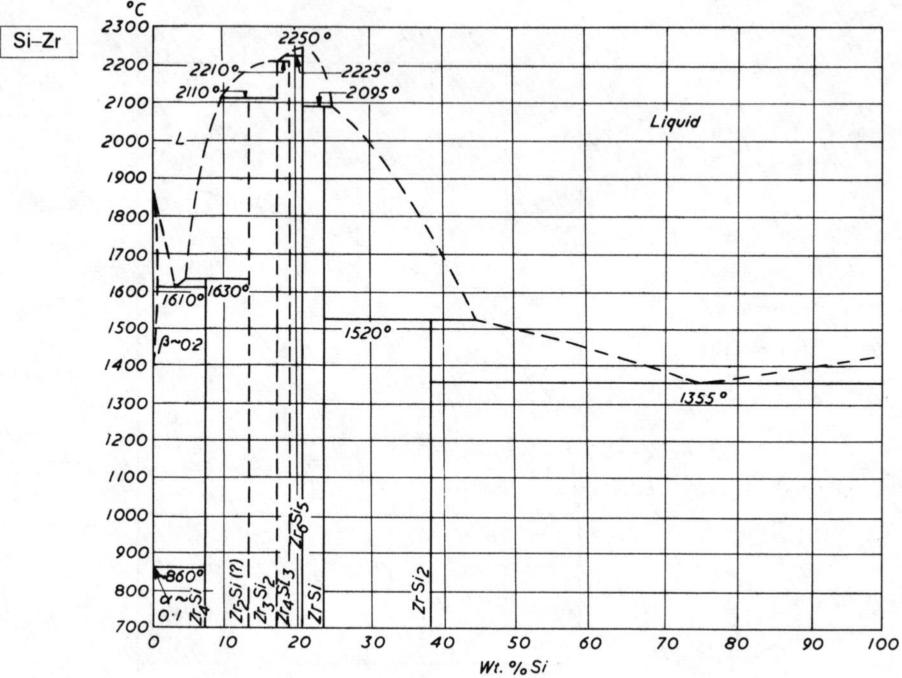

Sn–Sr

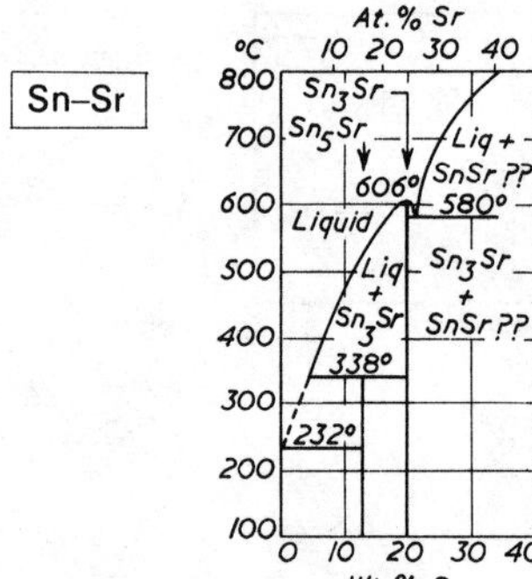

Sn–Te

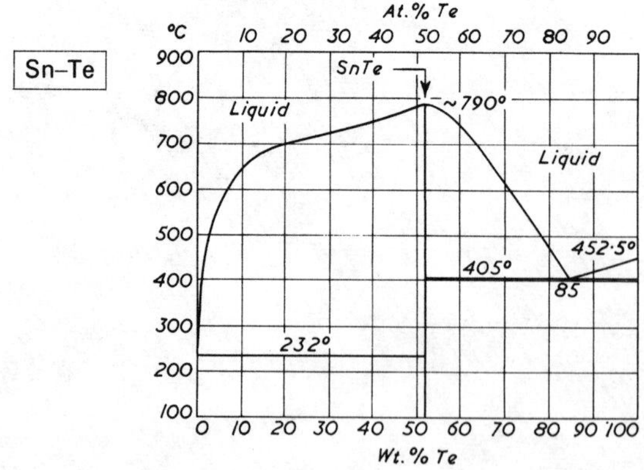

Sn–Ti

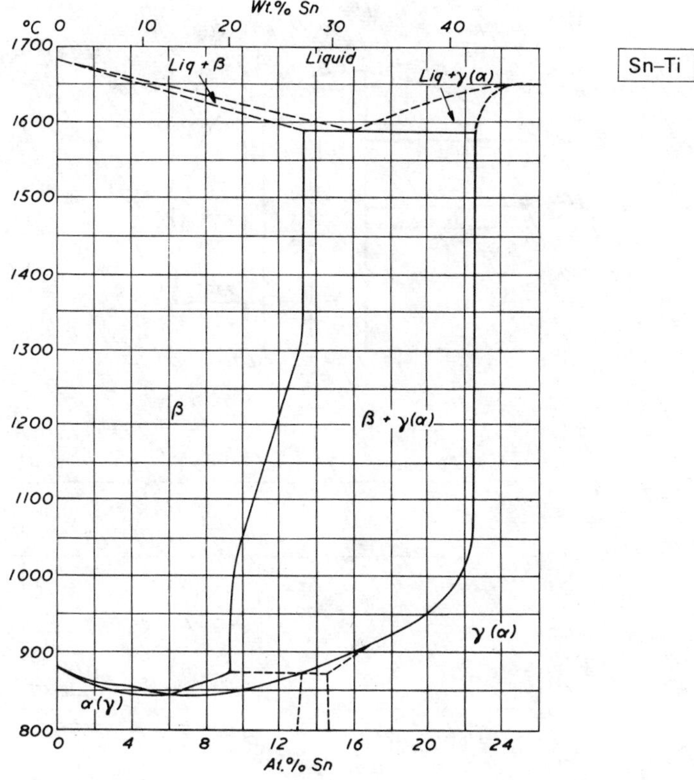
Wt.% Sn
°C
Liquid
Liq + β
Liq + γ(α)
β
β + γ(α)
γ (α)
α(γ)
At.% Sn

Sn–Tl

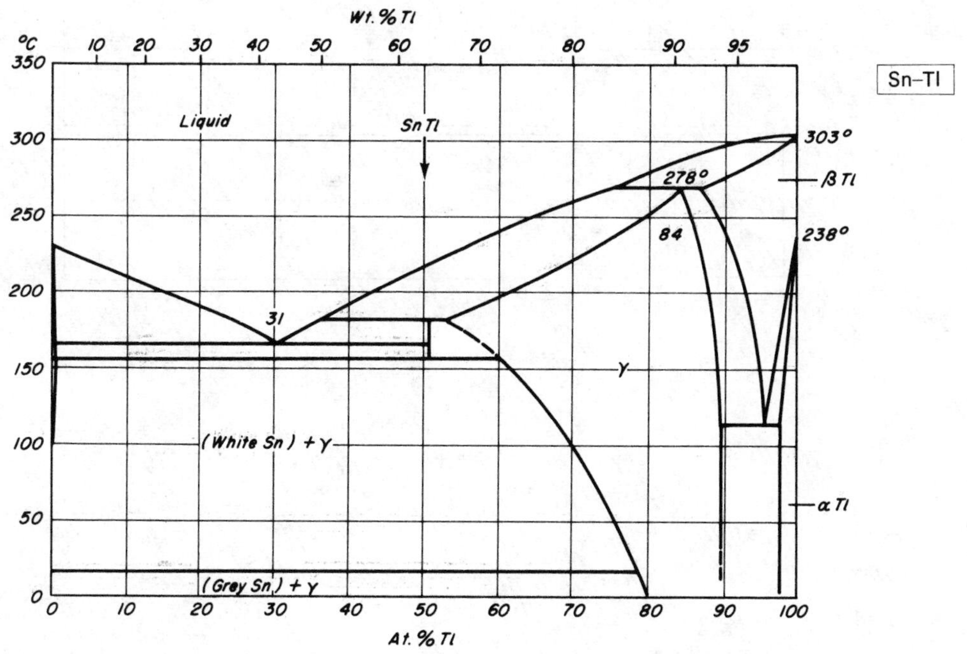
Wt.% Tl
°C
Liquid
SnTl
303°
278°
βTl
84
238°
31
γ
(White Sn) + γ
αTl
(Grey Sn) + γ
At.% Tl

Sn–U

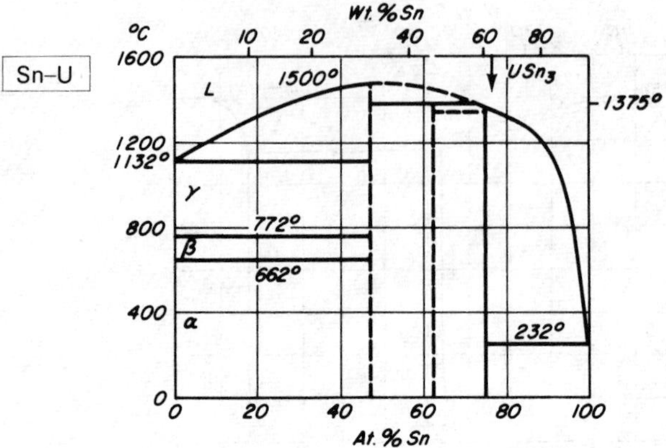
Wt.% Sn
°C
10
20
40
60
80
1600
1200
1132°
800
400
0
L
1500°
USn3
1375°
γ
772°
β
662°
α
232°
0
20
40
60
80
100
At.% Sn

Sn–Zn

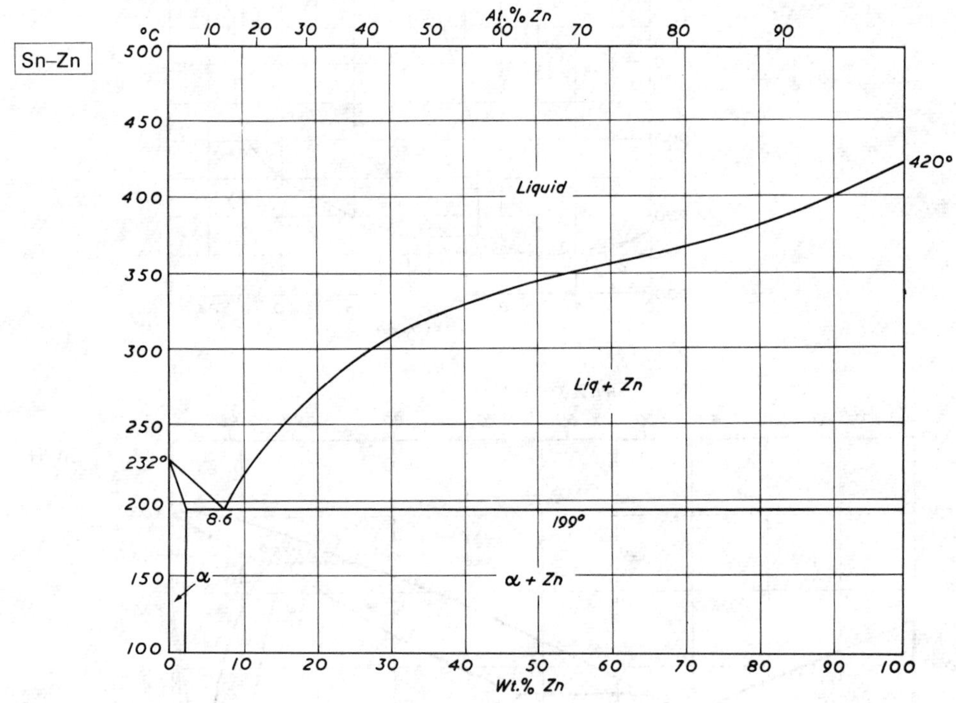
At.% Zn
°C
10
20
30
40
50
60
70
80
90
500
450
400
350
300
250
232°
200
150
100
420°
Liquid
Liq + Zn
8.6
199°
α
α + Zn
0
10
20
30
40
50
60
70
80
90
100
Wt.% Zn

Sn–Zr

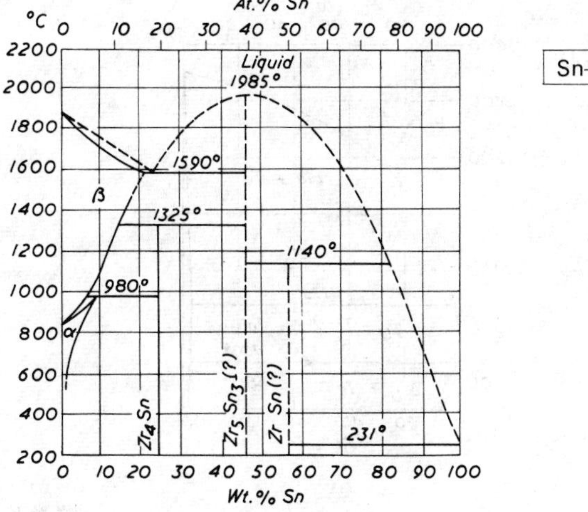

Ta–Th

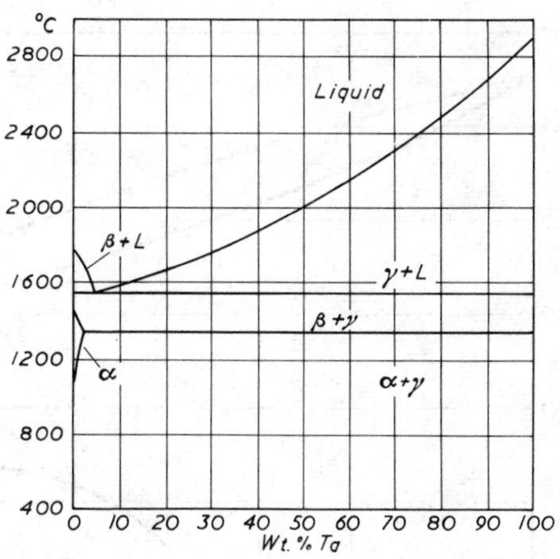

Ta–Ti

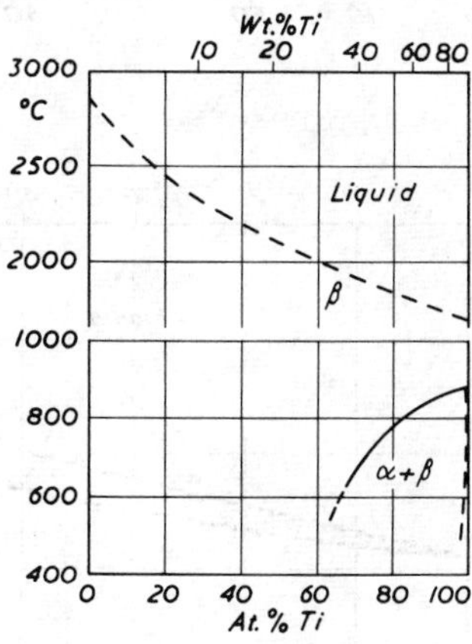

Ta–U

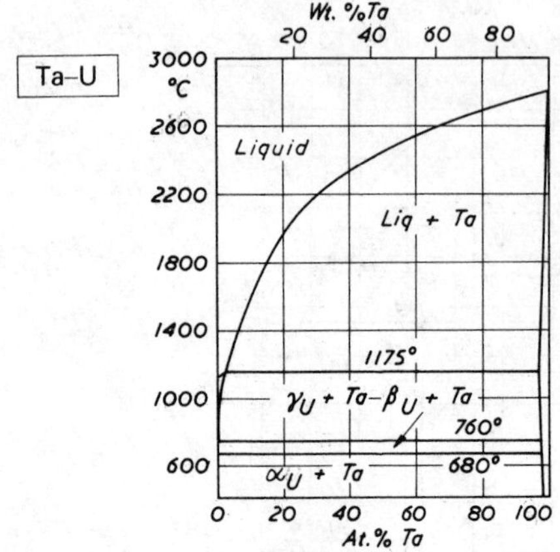

Wt. %Ta
20
40
60
80
3000
°C
2600
2200
1800
1400
1000
600
Liquid
Liq + Ta
1175°
γ_U + Ta–β_U + Ta
760°
680°
α_U + Ta
0
20
40
60
80
100
At.% Ta

Ta–V

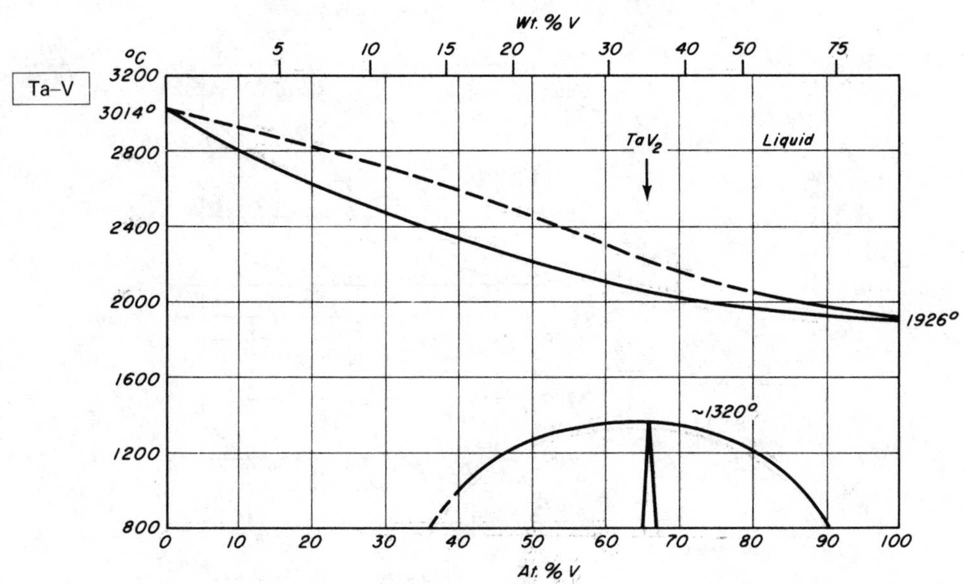

Wt. % V
5
10
15
20
30
40
50
75
°C
3200
3014°
2800
2400
2000
1600
1200
800
TaV_2
Liquid
1926°
~1320°
0
10
20
30
40
50
60
70
80
90
100
At. % V

Ta–W

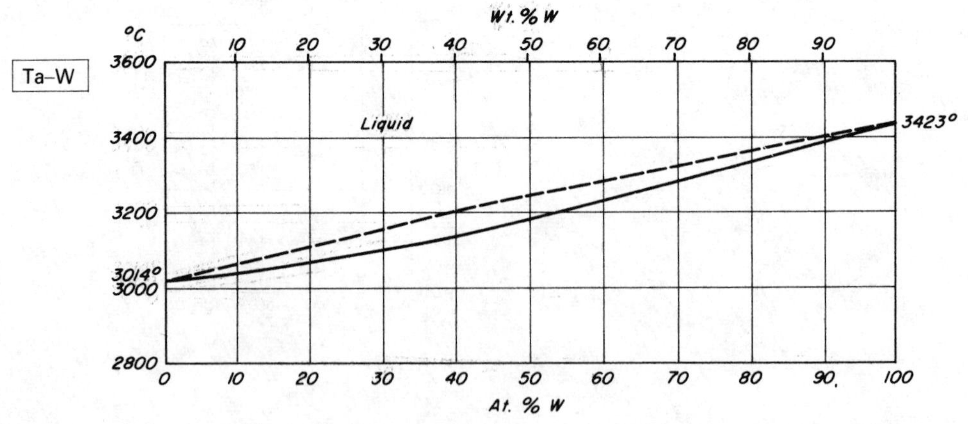

Wt. % W
10
20
30
40
50
60
70
80
90
°C
3600
3400
3200
3014°
3000
2800
Liquid
3423°
0
10
20
30
40
50
60
70
80
90
100
At. % W

Ta–Y

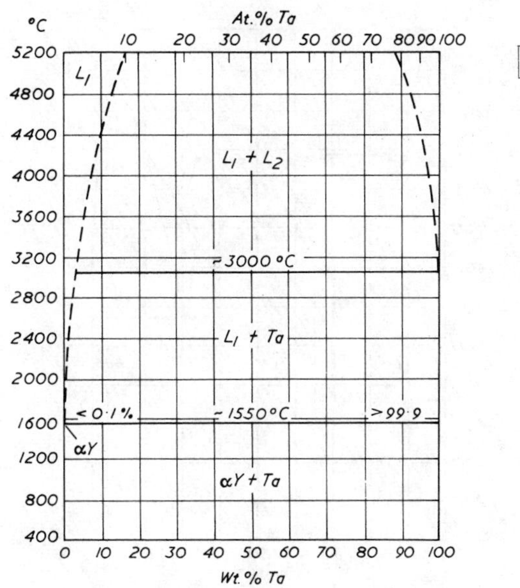
°C
At.% Ta
10 20 30 40 50 60 70 80 90 100
5200
4800
4400
4000
3600
3200
2800
2400
2000
1600
1200
800
400
L_1
$L_1 + L_2$
≈3000 °C
L_1 + Ta
< 0·1 %
~1550 °C
>99·9
αY
αY + Ta
0 10 20 30 40 50 60 70 80 90 100
Wt.% Ta

Ta–Zr

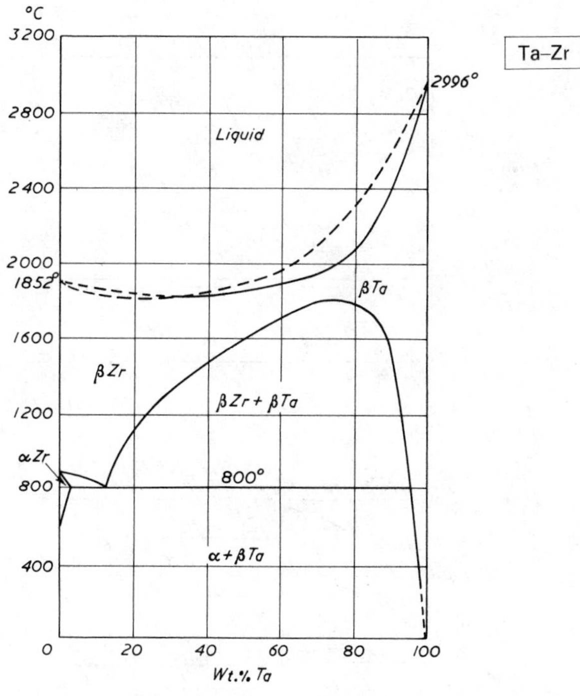
°C
3200
2800
2400
2000
1852°
1600
1200
αZr
800
400
2996°
Liquid
βTa
βZr
βZr + βTa
800°
α + βTa
0 20 40 60 80 100
Wt.% Ta

Tc–V

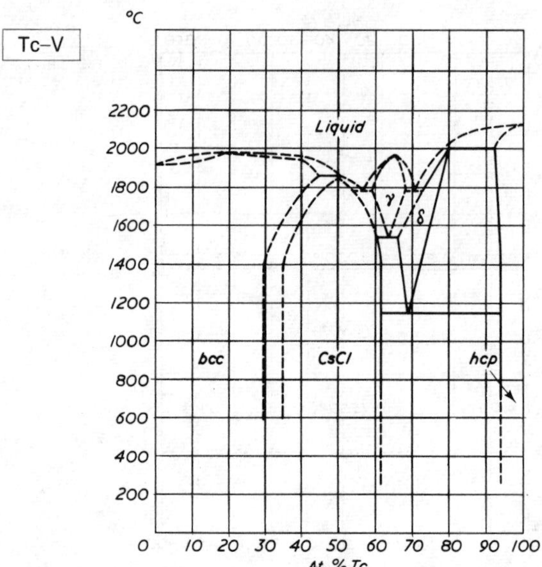

Te–Tl

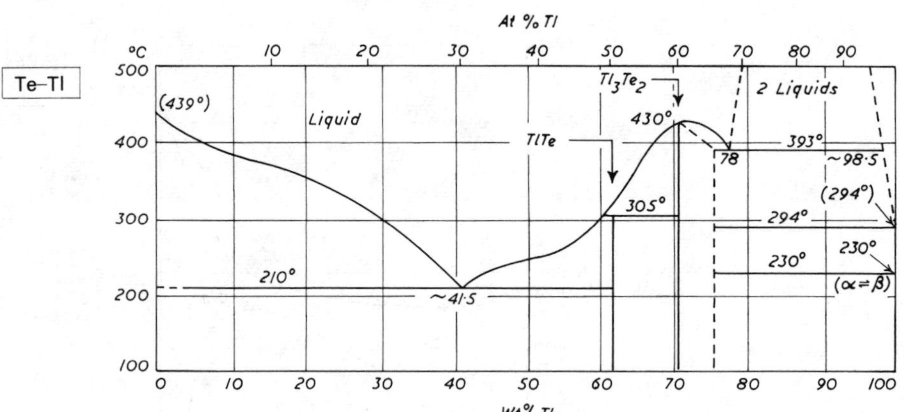

Te–Zn

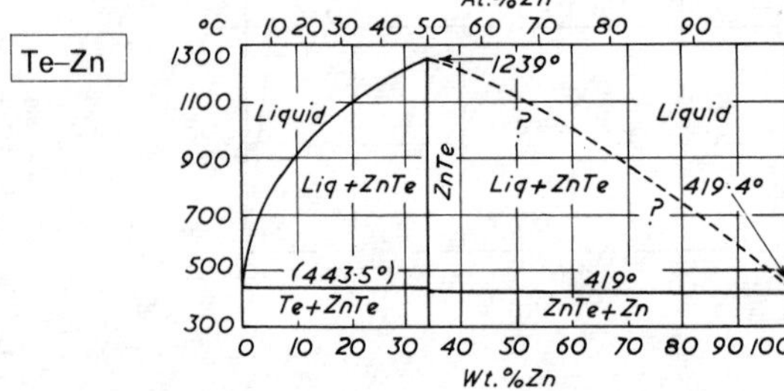

Th–Ti

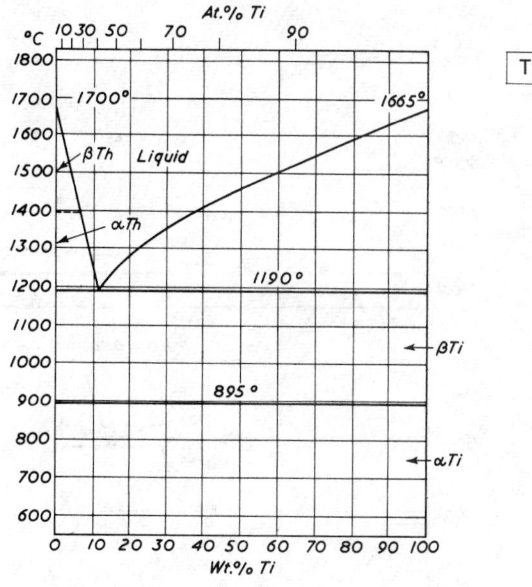
At.% Ti
°C
10 30 50 70 90
1800
1700
1600
1500
1400
1300
1200
1100
1000
900
800
700
600
1700°
1665°
βTh
Liquid
αTh
1190°
βTi
895°
αTi
0 10 20 30 40 50 60 70 80 90 100
Wt.% Ti

Th–U

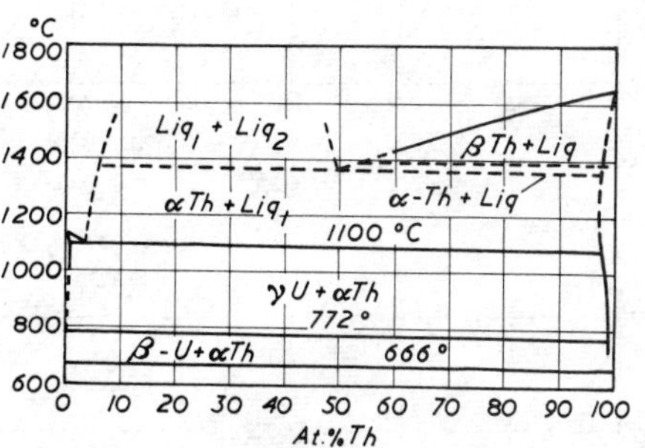
°C
1800
1600
1400
1200
1000
800
600
Liq₁ + Liq₂
βTh + Liq
αTh + Liq₁
α-Th + Liq
1100 °C
γU + αTh
772°
β-U + αTh
666°
0 10 20 30 40 50 60 70 80 90 100
At.% Th

Th–Y

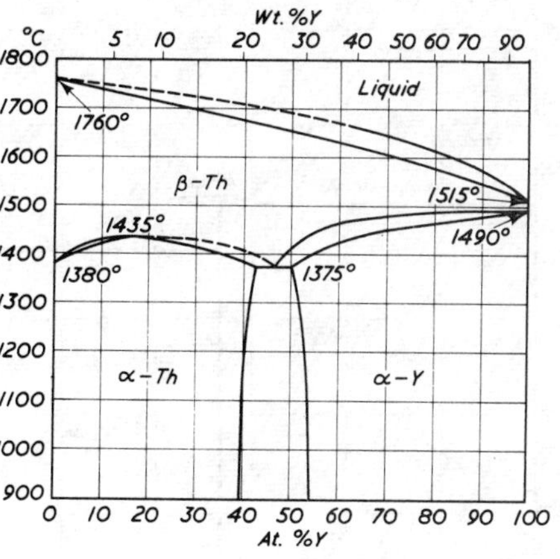
Wt.% Y
°C
5 10 20 30 40 50 60 70 90
1800
1700
1600
1500
1400
1300
1200
1100
1000
900
Liquid
1760°
β-Th
1515°
1435°
1490°
1380°
1375°
α-Th
α-Y
0 10 20 30 40 50 60 70 80 90 100
At. % Y

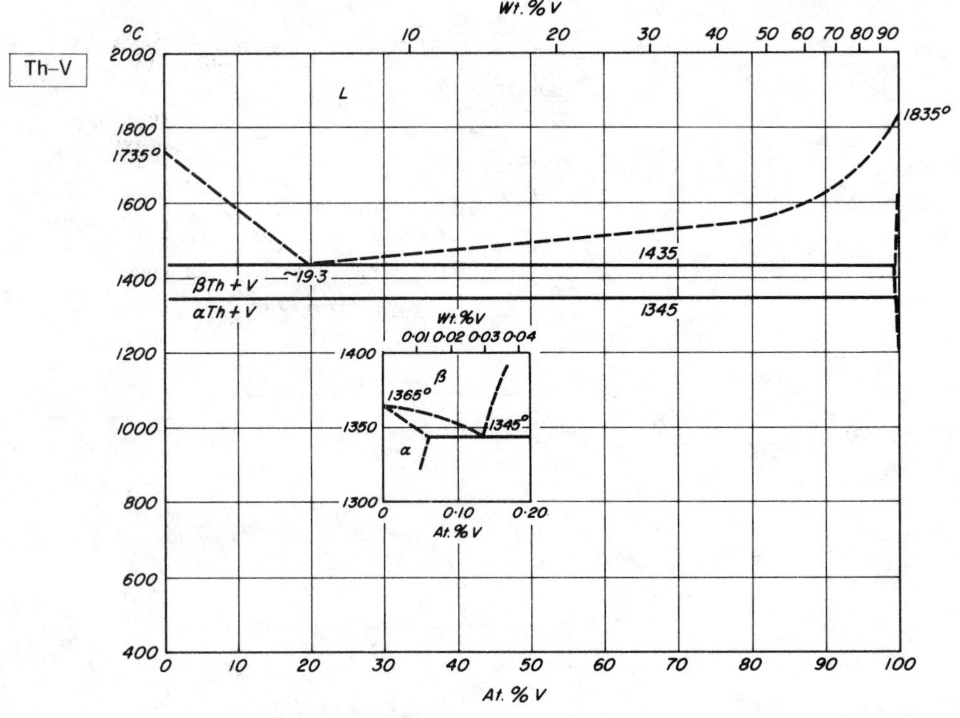
Th–V
Wt.% V
°C
L
1835°
1735°
1435
~19·3
βTh + V
αTh + V
1345
Wt.%V
0·01 0·02 0·03 0·04
β
1365°
1345°
α
At.% V
At. % V

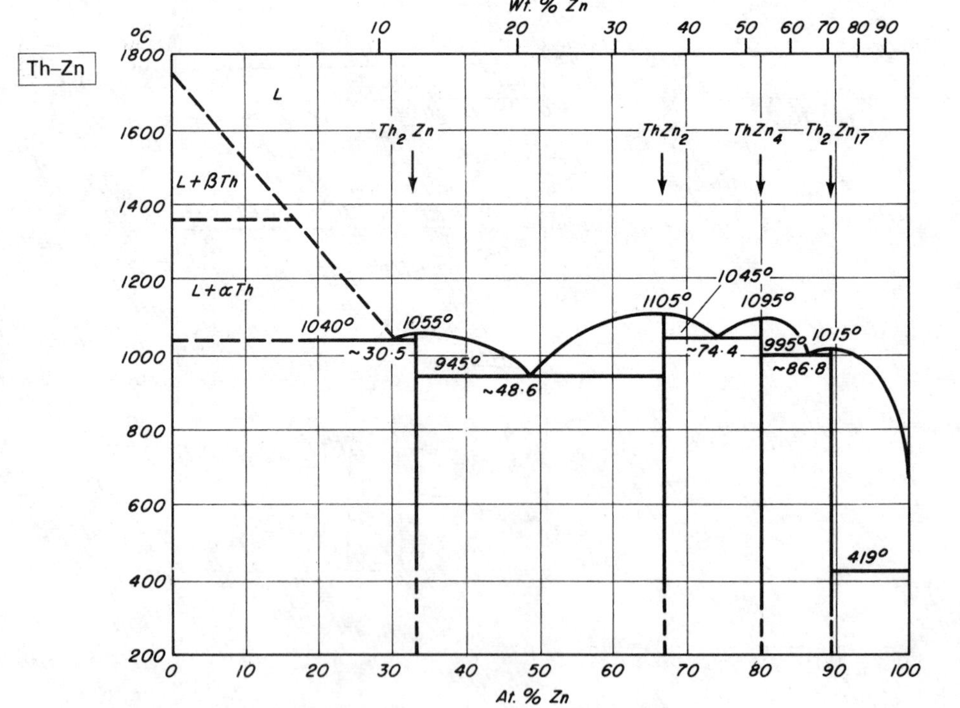
Th–Zn
Wt. % Zn
°C
L
Th2Zn
ThZn2
ThZn4
Th2Zn17
L + βTh
L + αTh
1045°
1105°
1095°
1040°
1055°
995°
1015°
~30·5
945°
~74·4
~86·8
~48·6
419°
At. % Zn

Th–Zr

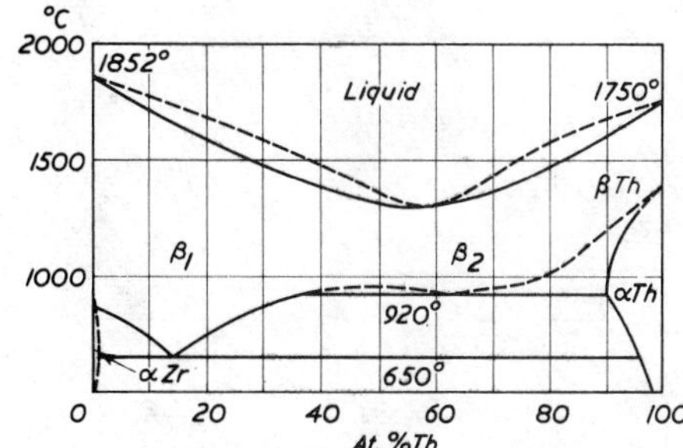
°C
2000
1500
1000
1852°
Liquid
1750°
βTh
β_1
β_2
αTh
920°
αZr
650°
0 20 40 60 80 100
At. %Th

Ti–U

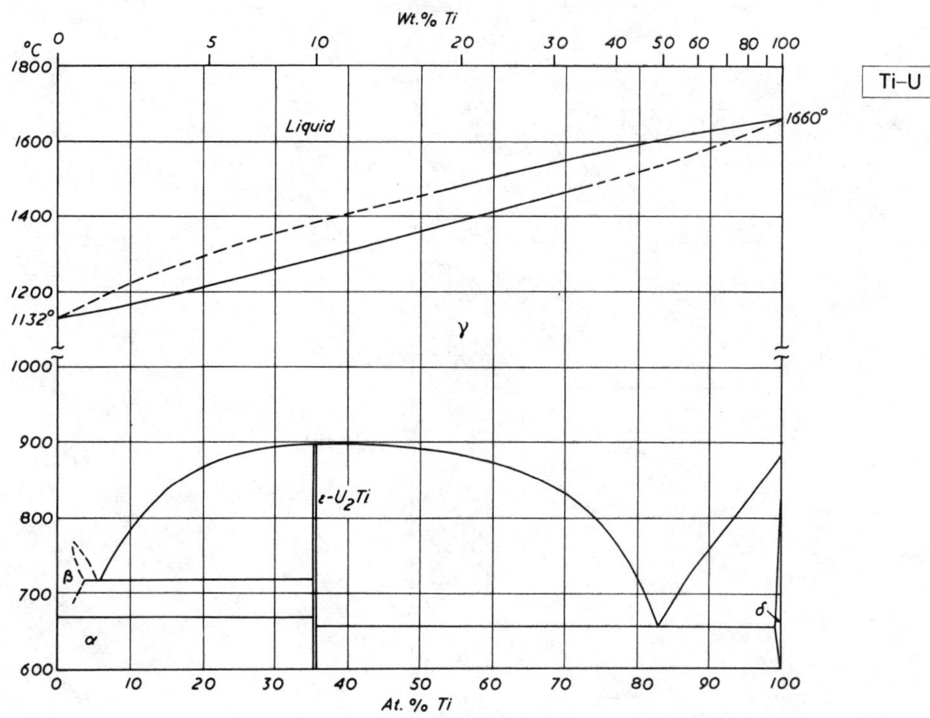
Wt.% Ti
0 5 10 20 30 40 50 60 80 100
°C
1800
1600
1400
1200
1132°
1000
900
800
700
600
Liquid
1660°
γ
ε-U_2Ti
β
α
δ
0 10 20 30 40 50 60 70 80 90 100
At. % Ti

Ti–V

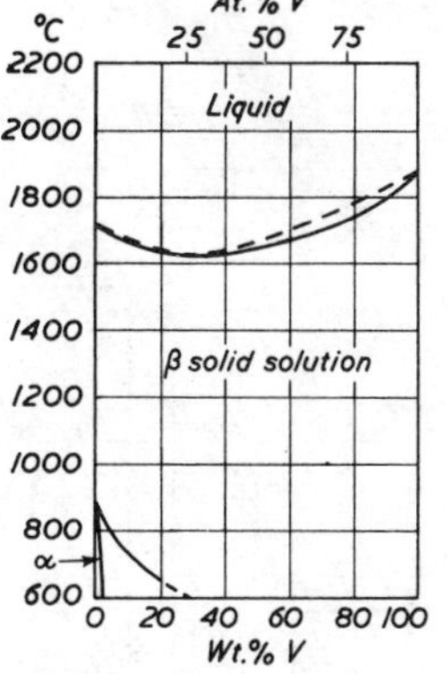
At.% V
25 50 75
°C
2200
2000
1800
1600
1400
1200
1000
800
600
Liquid
β solid solution
α
0 20 40 60 80 100
Wt.% V

Ti–W

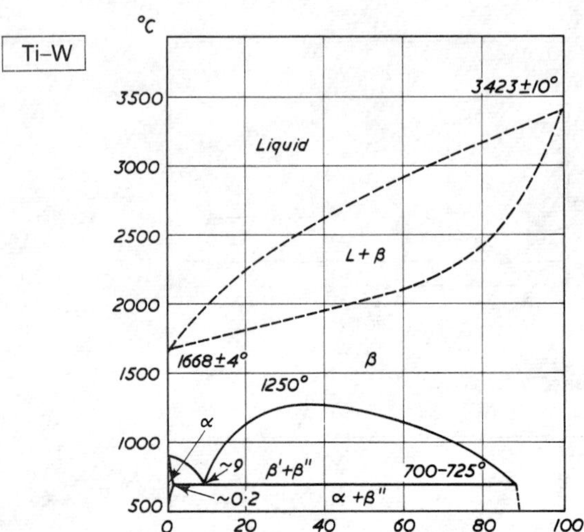
°C
3423±10°
Liquid
L+β
1668±4°
β
1250°
α
~9
β′+β″
700–725°
~0·2
α+β″
At. %W

Ti–Y

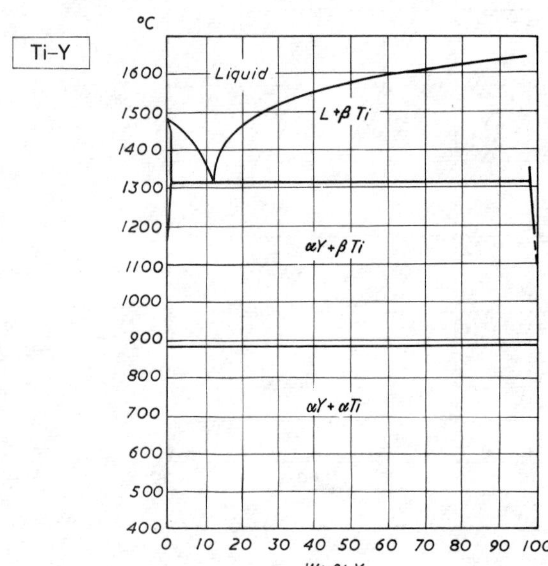
°C
Liquid
L+β Ti
αY+β Ti
αY+αTi
Wt. %Y

Ti–Zn

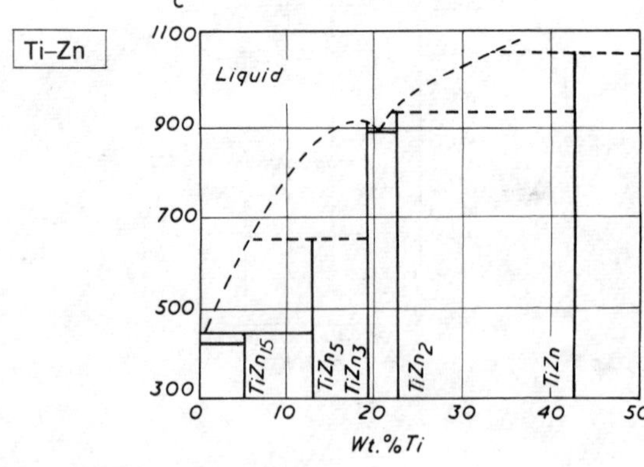
°C
Liquid
$TiZn_{15}$
$TiZn_5$
$TiZn_3$
$TiZn_2$
TiZn
Wt.%Ti

Ti–Zr

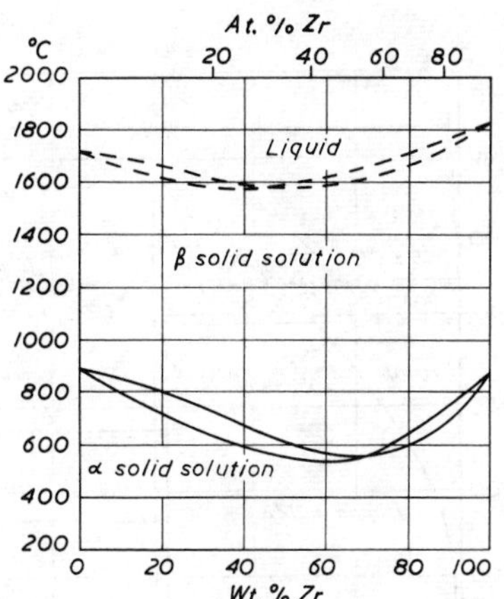

Tl–Zn

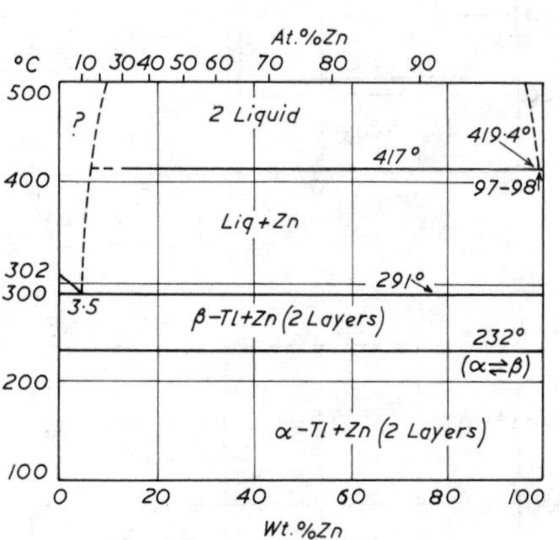

U–W

U–V

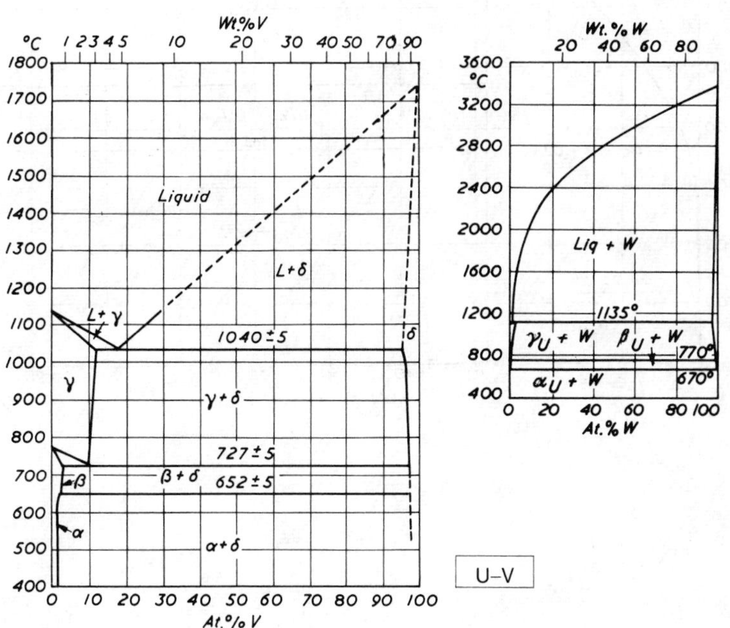

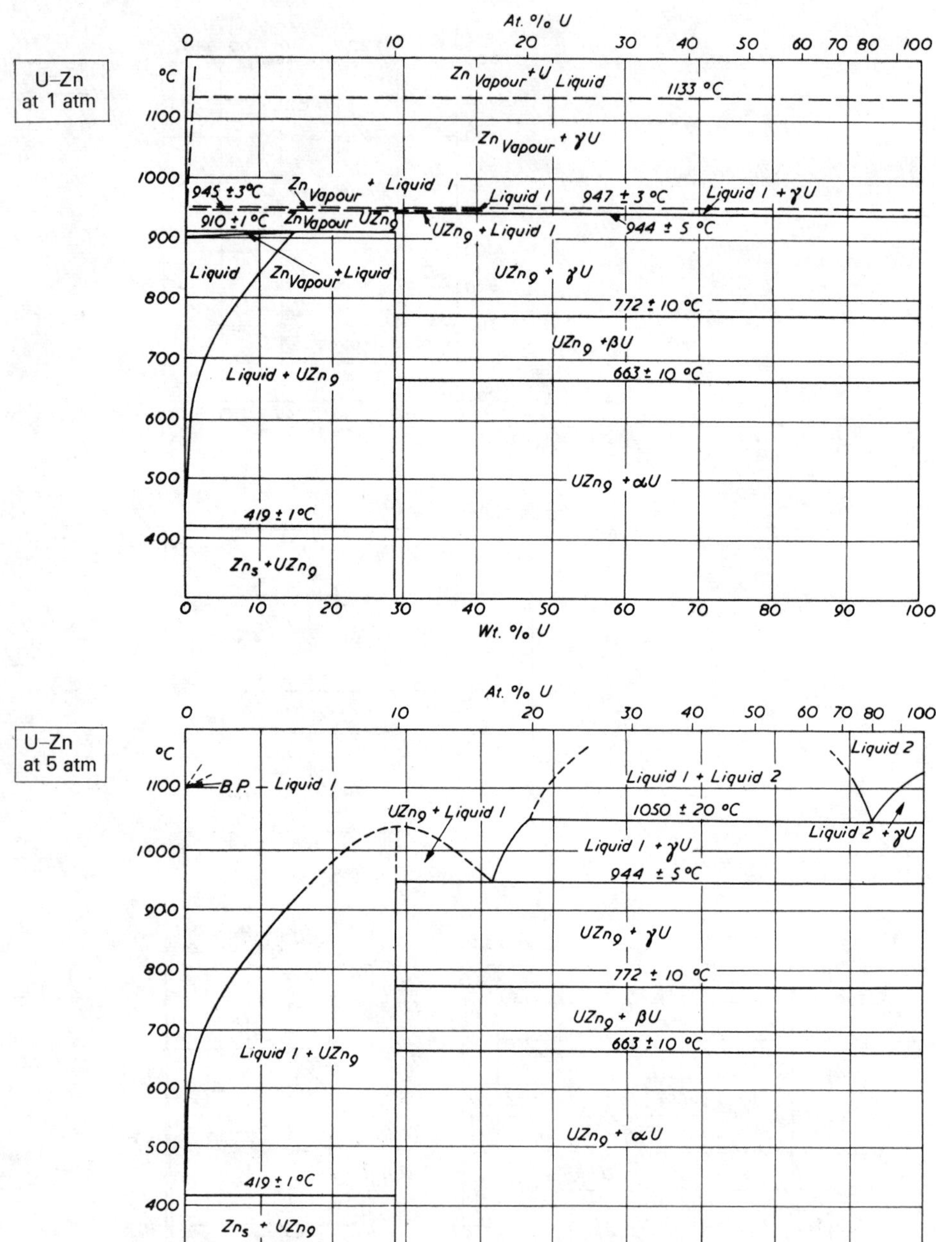
U–Zn at 1 atm
At. % U
Wt. % U
°C
Zn Vapour + U Liquid
1133 °C
Zn Vapour + γU
945 ± 3°C
Zn Vapour + Liquid 1
Liquid 1
947 ± 3 °C
Liquid 1 + γU
910 ± 1°C
Zn Vapour UZn9
UZn9 + Liquid 1
944 ± 5 °C
Liquid
Zn Vapour + Liquid
UZn9 + γU
772 ± 10 °C
UZn9 + βU
663 ± 10 °C
Liquid + UZn9
UZn9 + αU
419 ± 1°C
Zns + UZn9
U–Zn at 5 atm
At. % U
Wt. % U
°C
B.P.
Liquid 1
Liquid 2
Liquid 1 + Liquid 2
UZn9 + Liquid 1
1050 ± 20 °C
Liquid 2 + γU
Liquid 1 + γU
944 ± 5°C
UZn9 + γU
772 ± 10 °C
UZn9 + βU
663 ± 10 °C
Liquid 1 + UZn9
UZn9 + αU
419 ± 1°C
Zns + UZn9

U–Zr

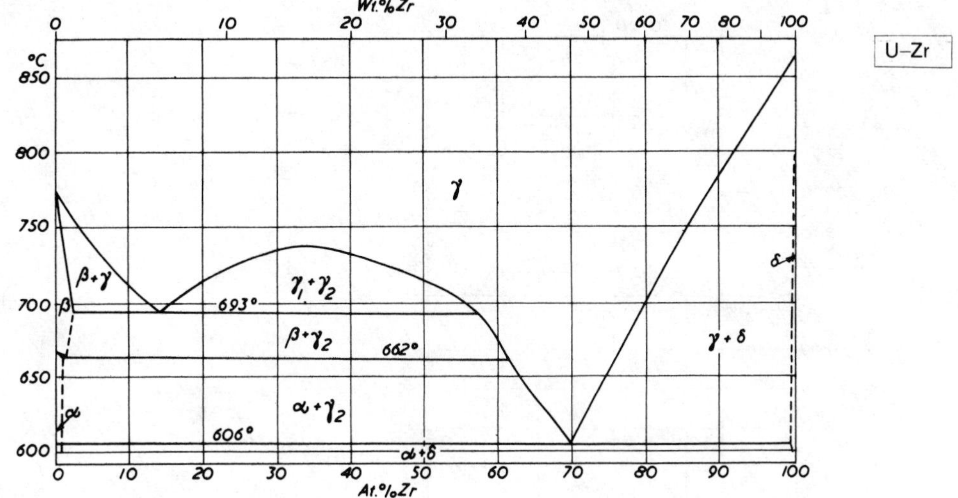
Wt.% Zr
0
10
20
30
40
50
60
70
80
100
°C
850
800
750
700
650
600
γ
β+γ
β
$γ_1+γ_2$
693°
$β+γ_2$
662°
$α+γ_2$
α
606°
α+δ
γ+δ
δ
0
10
20
30
40
50
60
70
80
90
100
At.% Zr

V–W

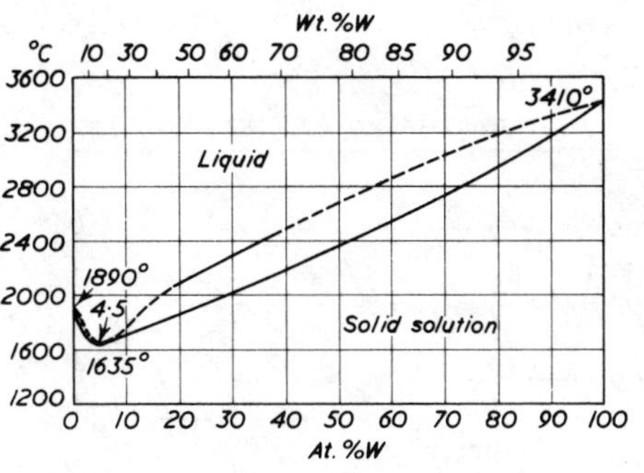
Wt.%W
°C
10
30
50
60
70
80
85
90
95
3600
3200
2800
2400
2000
1600
1200
3410°
Liquid
1890°
4·5
1635°
Solid solution
0
10
20
30
40
50
60
70
80
90
100
At.%W

V–Y

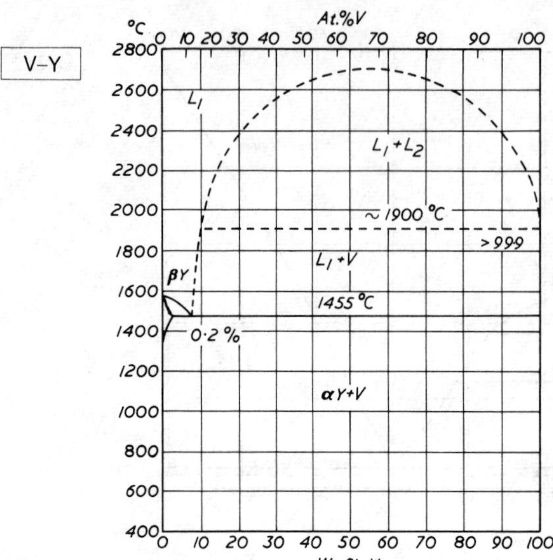

°C
At.%V
0 10 20 30 40 50 60 70 80 90 100
2800
2600
2400
2200
2000
1800
1600
1400
1200
1000
800
600
400
L_1
$L_1 + L_2$
~1900 °C
>99·9
L_1+V
βY
1455 °C
0·2 %
αY+V
0 10 20 30 40 50 60 70 80 90 100
Wt.% V

V–Zr

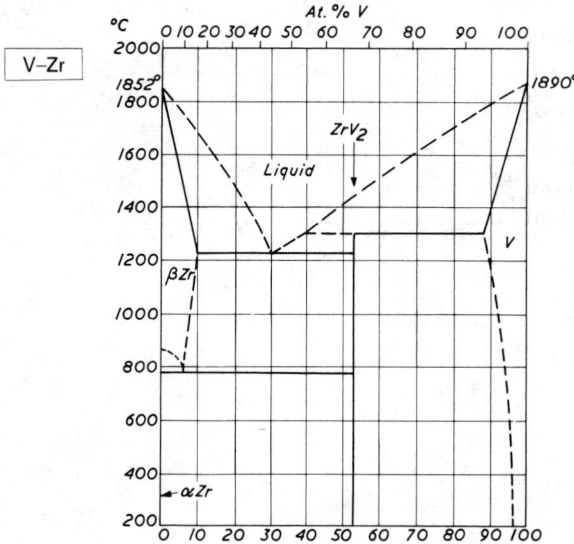

°C
At. % V
0 10 20 30 40 50 60 70 80 90 100
2000
1852°
1800
1890°
1600
1400
1200
1000
800
600
400
200
ZrV_2
Liquid
V
βZr
αZr
0 10 20 30 40 50 60 70 80 90 100
Wt. % V

W–Zr

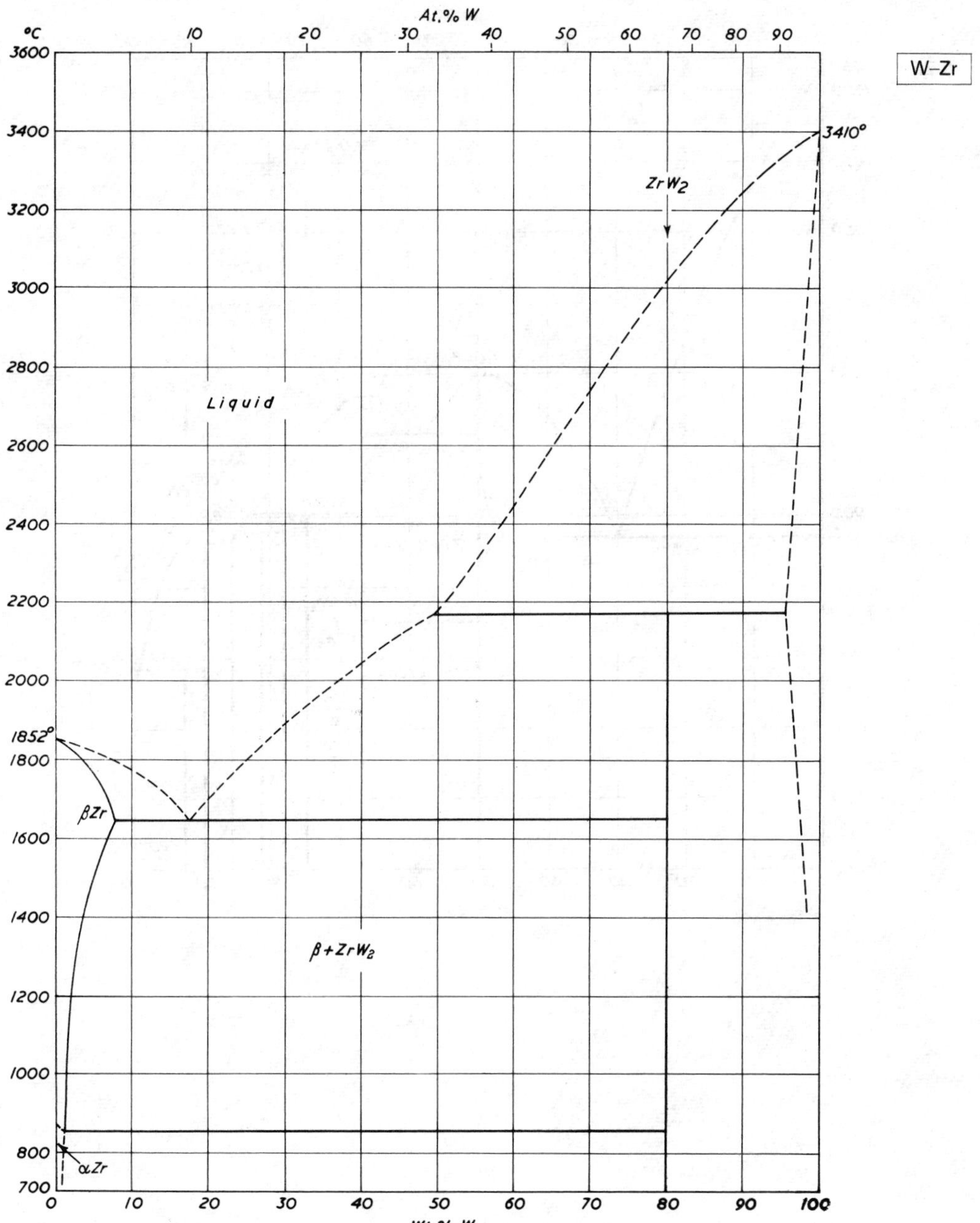
At.% W
°C
10
20
30
40
50
60
70
80
90
3600
3400
3200
3000
2800
2600
2400
2200
2000
1852°
1800
1600
1400
1200
1000
800
700
3410°
ZrW2
Liquid
βZr
β+ZrW2
αZr
0
10
20
30
40
50
60
70
80
90
100
Wt.% W

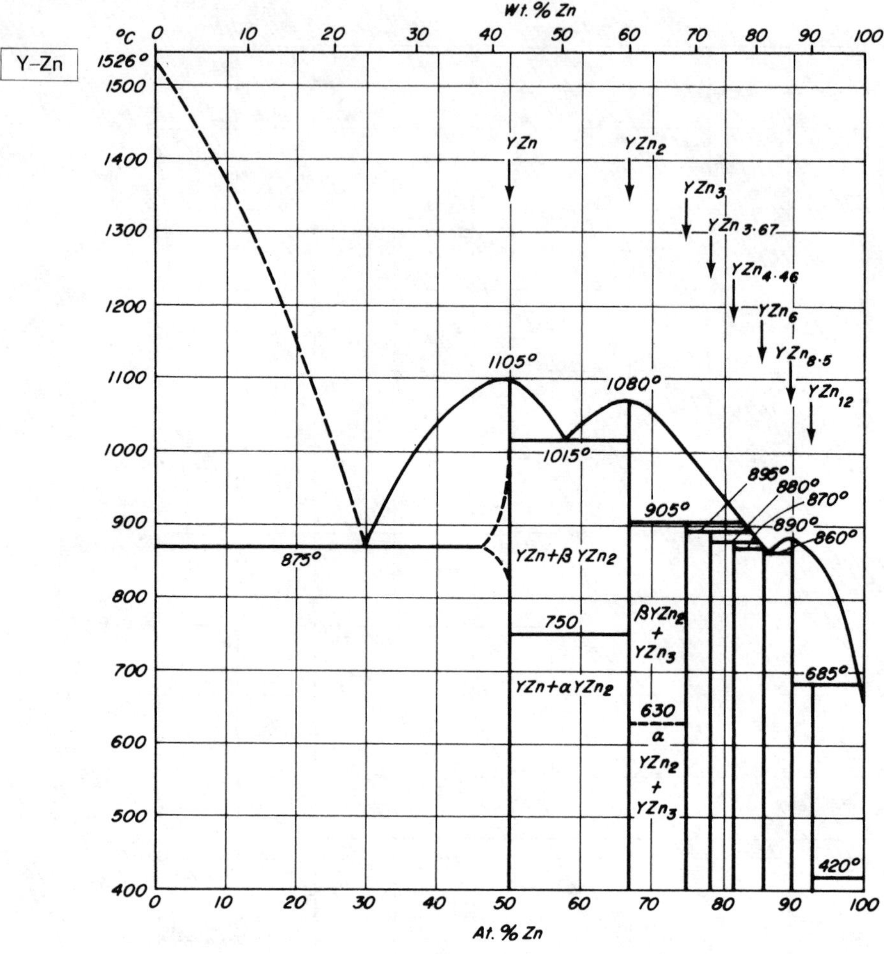
Y–Zn
Wt. % Zn
°C
1526°
875°
1105°
1080°
1015°
905°
895°
880°
870°
890°
860°
750
630
685°
420°
YZn
YZn2
YZn3
YZn3.67
YZn4.46
YZn6
YZn8.5
YZn12
YZn+β YZn2
YZn+α YZn2
βYZn2 + YZn3
α YZn2 + YZn3
At. % Zn

Y–Zr

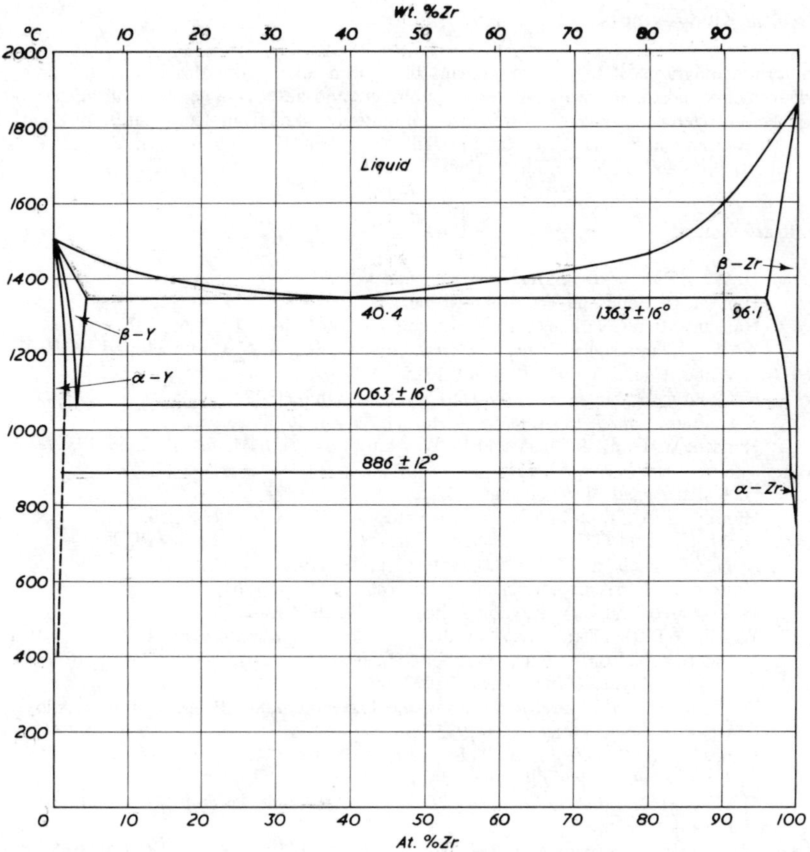

Wt. %Zr
°C
Liquid
β-Zr
40·4
1363 ± 16°
96·1
β-Y
α-Y
1063 ± 16°
886 ± 12°
α-Zr
At. %Zr

Zn–Zr

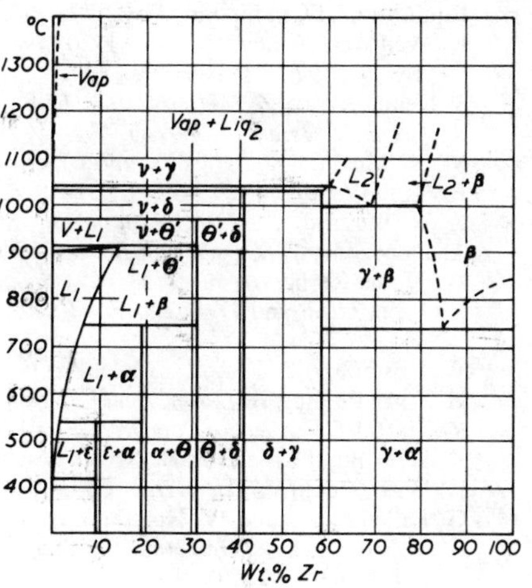
°C
Vap
Vap + Liq2
ν+γ
ν+δ
L2
L2+β
V+L1
ν+θ′
θ′+δ
L1+θ′
β
L1
L1+β
γ+β
L1+α
L1+ε
ε+α
α+θ
θ+δ
δ+γ
γ+α
Wt.% Zr

11.3 Acknowledgements

The equilibrium diagrams of the binary systems have been taken from the following sources. Acknowledgements and thanks are made to the publishers and authors concerned. In the case of systems where no reference is given, use was made of material taken from 'Constitution of Binary Alloys' by M. Hansen and K. Anderko, McGraw-Hill, New York, 1958, and from the Supplements by R. P. Elliott, 1965 and by F. A. Shunk, 1969.

11.3.1 Binary systems

Ag–Al	H. W. L. Phillips, *Inst. Met. Ann. Eq. Diag. No.* 21.
Ag–As	Hansen; G. Eade and W. Hume-Rothery, *Z. Metallk.*, 1959, **50**, 123
Ag–Bi	Hansen; M. W. Nathans and M. Leider, *J. Phys. Chem.*, 1962, **66**, 2012
Ag–Ca	W. A. Alexander *et al.*, *Can. J. Chem.*, 1969, **47**, 611; A. N. Campbell and W. H. W. Wood, *Can. J. Chem.*, 1971, **49**, 1315
Ag–Cd	F. C. Kracek, 'Metals Handbook', Cleveland, Ohio, 1948
Ag–Dy	S. Delfino *et al.*, *J. less-common Metals*, 1976, **44**, 267
Ag–Ga	Hansen; W. Hume-Rothery and K. W. Andrews, *J. Inst. Metals*, 1942, **68**, 133; *Idem*, *Z. Metallk.*, 1959, **50**, 661
Ag–Gd	G. Kiessler *et al.*, *J. less-common Metals*, 1972, **26**, 293
Ag–Ge	Hansen; E. A. Owen and V. W. Rowlands, *J. Inst. Metals*, 1940, **66**, 371
Ag–Hg	H. M. Day and C. H. Mathewson, *Trans. Am. Min. Engrs*, 1938, **128**, 261
Ag–In	A. N. Campbell *et al.*, *Can. J. Chem.*, 1970, **48**, 1703
Ag–La	Hansen; K. A. Gschneider *et al.*, *Met. Trans.*, 1970, **1**, 1961
Ag–Li	W. G. Freeth and G. V. Raynor, *J. Inst. Metals*, 1953/4, **82**, 569
Ag–Mg	W. Hume-Rothery and K. W. Andrews, *J. Inst. Metals*, 1943, **69**, 488; G. F. Sager and B. J. Nelson, 'Metals Handbook', Cleveland, Ohio, 1948; J. L. Haughton and R. J. M. Payne, *J. Inst. Metals*, 1937, **60**, 351
Ag–Mn	R. Hultgren *et al.*, 'Selected Values of the Thermodynamic Props. of Binary Alloys', American Society for Metals, 1973
Ag–Na	Hansen; J. R. Weeks, *Trans. Quart. ASM*, 1969, **62**, 304
Ag–P	R. Vogel *et al.*, *Z. Metallk.*, 1959, **50**, 130
Ag–Pr	G. Canneri, *Metallurgia ital.*, 1934, **26**, 794; K. A. Gschneider *et al.*, *Met. Trans.*, 1970, **1**, 1961
Ag–Sb	Hansen; P. W. Reynolds and W. Hume-Rothery, *J. Inst. Metals*, 1937, **60**, 365
Ag–Sc	I. Stapf *et al.*, *J. less-common Metals*, 1975, **39**(2), 219
Ag–Sn	Equilibrium Data for Sn Alloys, *Tin Res. Inst.*, 1949
Ag–Sr	F. Weibke, *Z. Anorg, Chem.*, 1930, **193**, 297; T. Heumann and N. Harmsen, *Z. Metallk.*, 1970, **61**, 906
Ag–Tb	S. Delfino *et al.*, *Z. Metallk.*, 1976, **67**(6), 392
Ag–Th	E. Raub, *Z. Metallk.*, 1949, **40**, 431
Ag–Ti	M. K. McQuillan, *J. Inst. Metals.*, 1959/60, **88**, 235
Ag–Tl	E. Raub, *Z. Metallk.*, 1949, **40**, 432
Ag–U	R. W. Buzzard *et al.*, *J. Res. natn. Bur. Stand.*, 1954, **52**, 149
Ag–Yb	A. Palenzona, *J. less-common Metals*, 1970, **21**, 443
Ag–Zn	W. Hume-Rothery *et al.*, *Proc. Roy. Soc.*, 1941, **A 177**, 149
Ag–Zr	J. O. Betterton and D. S. Easton, *Trans. Am. Inst. Min. Engrs*, 1958, **12**, 470
Al–As	W. Köster and B. Thoma, *Z. Metallk.*, 1955, **46**, 291
Al–Be	H. W. L. Phillips, *Inst. Met. Ann. Eq. Diag. No.* 19
Al–Bi	R. Martin-Garin *et al.*, *Compte Rendu*, [C], 1966, **262**, 335
Al–Ce	K. H. J. Buschow and J. H. N. van Vucht, *Z. Metallk.*, 1966, **57**, 162
Al–Co	H. W. L. Phillips, *Inst. Met. Ann. Eq. Diag. No.* 20
Al–Cr	G. Falkenhagen and W. Hofmann, *Z. Metallk.*, 1950, **41**, 191
Al–Cu	G. V. Raynor, *Inst. Met. Ann. Eq. Diag. No.* 4; R. P. Jewitt and D. J. Mack, *J. Inst. Met.*, 1963/4, **92**, 59
Al–Dy	F. Casteels, *J. less-common Metals*, 1967, **12**, 210
Al–Er	K. H. J. Buschow and J. H. N. van Vucht, *Z. Metallk.*, 1965, **56**, 9
Al–Fe	H. W. L. Phillips, *Inst. Met. Ann. Eq. Diag., No.* 13
Al–Gd	K. H. J. Buschow, *J. less-common Metals*, 1965, **9**, 453.
Al–Hf	E. M. Savitskii *et al.*, *Russ. Met.*, 1970, (1), 107

Al–In B. Predel, *Z. Metallk.*, 1965, **56**, 791

Al–La K. H. J. Buschow, *Philips Res. Reports*, 1965, **20**, 337.

Al–Mg G. V. Raynor, *Inst. Met. Ann. Eq. Diag. No.* 5; J. B. Clark and F. N. Rhines, *Trans. Am. Inst. Min. Engrs.*, 1957, **209**, 6

Al–Mn T. Godecke and W. Köster, *Z. Metallk.*, 1971, **62**(1), 727

Al–Nb C. E. Lundin and A. S. Yamamoto, *Trans. Am. Inst. Min. Engrs*, 1966, **236**, 863

Al–Nd K. H. J. Buschow, *J. less-common Metals*, 1965, **9**, 452

Al–Ni W. O. Alexander and N. B. Vaughan, *J. Inst. Metals*, 1937, **61**, 250; H. Groeber and V. Hauk, *Z. Metallk.*, 1950, **41**, 283; H. W. L. Phillips, *Inst. Met. Ann. Eq. Diag. No.* 18

Al–Pb E. Scheil and E. Jahn, *Z. Metallk.*, 1949, **40**, 319

Al–Pr K. H. J. Buschow and J. H. N. van Vucht, *Z. Metallk.*, 1966, **57**, 162

Al–Pt R. Huch and W. Klemm, *Z. anorg. allg. Chem.*, 1964, **329**, 123

Al–Pu F. H. Ellinger *et al.*, *J. nucl. Mater.*, 1962, **5**, 165

Al–Ru W. Obrowski, *Metall*, 1963, **17**, 108

Al–Sb H. W. L. Phillips, *Inst. Met. Ann. Eq. Diag. No.* 15

Al–Sc O. P. Naumkin *et al.*, *Metally Splavȳ Élektrotekh.*, 1965, (4), 176

Al–Si H. W. L. Phillips, *Inst. Met. Ann. Eq. Diag. No.* 16

Al–Sm F. Casteels, *J. less-common Metals*, 1967, **12**, 210

Al–Sn R. C. Dorward, *Metal Trans. A*, 1976, **7A**(2), 308

Al–Sr G. Bruzzone and F. Merlo, *J. less-common Metals*, 1975, **39**(1), 1

Al–Th J. R. Murray, *J. Inst. Metals*, 1958/9, **87**, 349

Al–Ti E. Ence and H. Margolin, *Trans. Am. Inst. Min. Engrs*, 1961, **221**, 151; F. A. Crossley, *Trans. Am. Inst. Min. Engrs*, 1966, **236**, 1174; *see also* E. S. Bumps *et al.*, *J. Metals*, 1952, 609; H. W. L. Phillips, *Inst. Met. Ann. Eq. Diag. No.* 22, D. Clark *et al.*, *J. Inst. Metals.*, 1962/3 **91**, 197

Al–U P. R. Roy, *J. Nuc. Mat.*, 1964, **11**(1), 59

Al–V R. Flükiger *et al.*, *J. less-common Metals*, 1975, **40**(1), 103

Al–W W. D. Clark, *J. Inst. Metals*, 1940, **66**, 271.

Al–Y C. E. Lundin and D. T. Klodt, *Trans. AIM*, 1961, **54**, 168

Al–Yb A. Palenzona, *J. less-common Metals*, 1972, **29**, 289

Al–Zn G. V. Raynor, *Inst. Met. Ann. Diag. No.* 1; G. R. Goldak and J. G. Parr, *J. Inst Met.*, 1963/4, **92**, 230

Al–Zr W. L. Fink and L. A. Wiley, *Trans. Am. Inst. Min. Engrs*, 1939, **133**, 69; D. J. McPherson and M. Hansen, *Trans. Am. Soc. Metals*, 1954, **46**, 354

Am–Pu F. H. Ellinger *et al.*, *J. nucl. Mater.*, 1966, **20**, 83

As–Au B. Gather and R. Blacknik, *Z. Metallk.*, 1976, **67**(3), 168

As–Co Hansen; W. Köster and W. Mulfinger, *Z. Metallk.*, 1938, **30**, 348

As–Eu S. Ono *et al.*, *J. less-common Metals*, 1971, **25**, 287

As–Ga W. Köster and B. Thoma, *Z. Metallk.*, 1955, **46**, 291

As–In T. S. Liu and E. A. Peretti, *Trans. Am. Soc. Metals*, 1953, **45**, 677

As–K F. W. Dorn *et al.*, *Z. anorg. allg. Chem.*, 1961, **309**, 204

As–Sn E. A. Peretti and J. K. Paulsen, *J. less-common Metals*, 1969, **17**, 283

As–Te J. R. Eifert and E. A. Peretti, *J. Mater. Sci.*, 1968, **3**, 293

Au–B W. Obrowski, *Naturwiss*, 1961, **48**, 428; F. Wald and R. W. Stormont, *J. less-common Metals*, 1965, **9**, 423

Au–Bi M. W. Nathans and M. Leider, *J. phys. Chem.*, 1962, **66**, 2012

Au–Cd P. J. Durrant, *J. Inst. Metals*, 1929, **41**, 139; V. G. Rivlin *et al.*, *Acta met.*, 1962, **11**, 1143

Au–Ce L. Rolla *et al.*, *Z. Metallk.*, 1943, **35**, 29

Au–Co E. Raub and P. Walter, *Z. Metallk.*, 1950, **41**, 234

Au–Cr E. Raub, *Z. Metallk.*, 1960, **51**, 290

Au–Cs G. Kienast *et al.*, *Z. anorg. allg. Chem.*, 1961, **310**, 143

Au–Cu H. E. Bennett, *J. Inst. Metals*, 1962/3, **91**, 158; Hansen

Au–Dy O. D. McMasters and K. A. Gschneidner, *J. less-common Metals*, 1973, **30**, 325

Au–Er P. E. Rider *et al.*, *Trans. Am. Inst. Min. Engrs*, 1965, **233**, 1488

Au–Ga C. J. Cooke and W. Hume-Rothery, *J. less-common Metals*, 1966, **10**, 42

Au–Gd P. E. Rider *et al.*, *Trans. Am. Inst. Min. Engrs*, 1965, **233**, 1488

Au–Ge R. I. Jaffee *et al.* *Trans. Am. Inst. Min. Engrs*, 1945, **161**, 366

Au–Hg	C. Rolfe and W. Hume-Rothery, *J. less-common Metals*, 1967, **13**, 1
Au–In	S. E. R. Hiscocks and W. Hume-Rothery, *Proc. Roy. Soc.*, 1964, **A 282**, 318
Au–K	G. Kienast and J. Verma, *Z. anorg. allg. Chem.*, 1961, **310**, 143
Au–Li	G. Kienast and J. Verma, *Z. anorg. allg. Chem.*, 1961, **310**, 143
Au–Lu	P. E. Rider *et al.*, *Trans. Am. Inst. Min. Engrs*, 1965, **233**, 1488
Au–Mn	E. Raub, *Z. Metallk.*, 1949, **40**, 359; E. Raub *et al.*, *Z. Metallk.*, 1953, **44**, 312
Au–Na	G. Kienast *et al.*, *Z. anorg. allg. Chem.*, 1961, **310**, 143
Au–P	R. Vogel *et al.*, *Z. Metallk.*, 1959, **50**, 130
Au–Pt	A. S. Darling *et al.*, *J. Inst. Metals*, 1952/3, **81**, 125
Au–Rb	G. Kienast *et al.*, *Z. anorg. allg. Chem.*, 1961, **310**, 143
Au–S	R. Vogel and R. Gebhardt, *Z. Metallk.*, 1961, **52**, 318
Au–Sc	P. E. Rider *et al.*, *Trans. Am. Inst. Min. Engrs*, 1965, **233**, 1488
Au–Se	R. Vogel and R. Gebhardt, *Z. Metallk.*, 1961, **52**, 318; R. Rabenau *et al.*, *J. less-common Metals*, 1971, **24** (3), 291
Au–Si	R. P. Anantatmula *et al.*, *J. Electron. Mat.*, 1975, **4** (3), 445
Au–Sn	Equilibrium Data for Sn Alloys, *Tin Res. Inst.*, 1949
Au–Sr	M. Feller-Kniepmaier and T. Heumann, *Z. Metallk.*, 1960, **51**, 404
Au–Tb	P. E. Rider *et al.*, *Trans. Am. Inst. Min. Engrs*, 1965, **233**, 1488
Au–Tm	P. E. Rider *et al.*, *Trans. Am. Inst. Min. Engrs*, 1965, **233**, 1488
Au–U	R. W. Buzzard and J. J. Park, *J. Res. natn. Bur. Stand.*, 1954, **53**, 291
Au–V	R. Flükiger *et al.*, *J. less-common Metals*, 1975, **40**, 103
Au–Yb	A. Iandelli and A. Palenzona, *J. less-common Metals*, 1969, **18**, 221; P. E. Rider *et al.*, *Trans. Am. Inst. Min. Engrs*, 1965, **233**, 1488
B–Cr	K. I. Portnoi *et al.*, *Poroshk. Metall.*, 1969, No. 4, 51
B–Eu	E. M. Savitskii *et al.*, *Neorg. Materialy*, 1971, **7**, 617
B–Ge	L. R. Bidwell, *J. less-common Metals*, 1970, **20**, 19
B–Hf	K. I. Portnoi *et al.*, *Neorg. Materialy*, 1971, **7**, 1987
B–Mn	G. Pradelli and C. Gianoglio, *La Metallurgia Italiana*, 1974, **12**, 659
B–Mo	K. I. Portnoi *et al.*, *Metally. Splavȳ Élektrotekh.*, 1967, (4), 171
B–Nb	H. Novotny *et al.*, *Z. Metallk.*, 1959, **50**, 417
B–Ni	A. S. Sobolev and T. F. Fedorov, *Inorg. Materials*, 1967, **433**, 643; S. Omori *et al.*, *J. Jap. Soc. Powder Met.*, 1971, (4), 132
B–Re	K. I. Portnoi and V. M. Romashov, *Poroshk. Metall.*, 1968, (2) 41
B–Sm	G. I. Solovyev and K. E. Spear, *J. Am. Ceram. Soc.*, 1972, **55** (9), 475
B–Ta	K. I. Portnoi *et al.*, *Poroshk. Metall.*, 1971, No. 11, (107), 89; E. Rudy and I. Progulski, *Planseeber Pulvermetall*, 1967, **15**, 13; *see also* Elliott for alternative version
B–U	R. W. Mar, *J. Am. Ceram. Soc.*, 1975, **50** (3–4), 145
B–W	K. I. Portnoi *et al.*, *Poroshk. Metall.*, 1967, No. 5 (53), 75
B–Zr	K. I. Portnoi *et al.*, *Poroshk. Metall.*, 1970, No. 7 (91), 68
Ba–Cd	R. T. Dirstine, *J. less-common Metals*, 1975, **39** (1), 181
Ba–Cu	G. Bruzzone, *J. less-common Metals*, 1971, **25**, 361
Ba–Ge	V. G. Andrianov *et al.*, *Neorg. Materialy*, 1966, **2**, 2064
Ba–Hg	G. Bruzzone and F. Merlo, *J. less-common Metals*, 1975, **39**, 271.
Ba–In	G. Bruzzone, *J. less-common Metals*, 1966, **11**, 249
Ba–Na	F. A. Kanda *et al.*, *J. phys. Chem.*, 1965, **69**, 3867
Ba–Pd	E. M. Savitskii *et al.*, *Metally Splavȳ Élektrotekh.*, 1970, (6), 143
Ba–Pt	E. M. Savitskii *et al.*, *Metally Splavȳ Élektrotekh.*, 1971, (1), 157
Be–Bi	G. W. Horsley and J. T. Maskrey, *J. Inst. Metals*, 1957/8, **86**, 401
Be–Co	F. Aldinger and S. Jönsson, *Z. Metallk.*, 1977, **68** (5), 362
Be–Cr	A. R. Edwards and S. T. M. Johnston, *J. Inst. Met.*, 1955/6, **84**, 313
Be–Fe	R. J. Teitel and M. Cohen, *Trans. Am. Inst. Min. Engrs*, 1949, **185**, 285
Be–Nb	A. T. Grigoriev and I. I. Raevsky, *Russ. Met.*, 1968, (5), 134
Be–Ni	E. Jahn, *Z. Metallk.*, 1949, **40**, 399
Be–Pd	O. Winkler, *Z. Metallk.*, 1938, **30**, 162
Be–U	R. W. Buzzard, *J. Res. natn. Bur. Stand.*, 1953, **50**, 63
Be–W	H. J. Goldschmidt and W. M. Ham, *J. less-common Metals*, 1966, **10** (1), 57
Bi–Ca	S. Hoesel, *Z. Phsik. Chem.* (*Leipzig*), 1962, **219**, 205

Bi–Co R. Damm *et al.*, *Z. Metallk.*, 1962, **53**, 196
Bi–Cu M. W. Nathans and M. Leider, *J. Phys. Chem.*, 1962, **66**, 2013
Bi–Cs G. Gnutzmann and W. Klemm, *Z. anorg. allg. Chem.*, 1961, **309**, 181
Bi–Ga B. Predel, *Z. Phys. Chem.*, 1960, **24**, 206
Bi–In O. H. Henry and E. L. Baldwick, *Trans. Am. Inst. Min. Engrs*, 1947, **171**, 389
Bi–K G. Gnutzmann and W. Klemm, *Z. anorg. allg. Chem.*, 1961, **309**, 181
Bi–La K. Nomura *et al.*, *J. less-common Metals*, 1977, **52**, 259
Bi–Mn A. U. Seybolt *et al.*, *Trans. Am. Inst. Min. Engrs*, 1956, **206**, 606 and 1406–8
Bi–Nd G. F. Kobzenko *et al.*, *Neorg. Materialy*, 1971, **7**, 1438
Bi–Pb G. O. Hiers, 'Metals Handbook', Cleveland, Ohio, 1948
Bi–Pt N. N. Zhuravlev *et al.*, *Physics Metals Metallogr.*, 1962, **13**, 51
Bi–Rb G. Gnutzmann and W. Klemm, *Z. anorg. allg. Chem.*, 1961, **309**, 181
Bi–Rh R. G. Ross and W. Hume-Rothery, *J. less-common Metals*, 1962, **4**, 454
Bi–Sn W. Oelsen and K. F. Golucke, *Arch. Eisenhuttenw.*, 1958, **29**, 689
Bi–U P. Cotterill and H. J. Axon, *J. Inst. Metals*, 1958/9, **87**, 159
Bi–Y F. A. Schmidt *et al.*, *J. less-common Metals*, 1969, **18**, 215

C–Cr D. S. Bloom and N. J. Grant, *Trans. Am. Inst. Min. Engrs*, 1950, **188**, 141
C–Cu M. B. Bever and C. F. Floe, *Trans. Am. Inst. Min. Engrs*, 1946, **166**, 128
C–La F. H. Spedding *et al.*, *Trans. Am. Inst. Min. Engrs*, 1959, **215**, 192
C–Re A. I. Evstyukhin *et al.*, *Met. i Metalloved. Chistikh Metallov, Moscow*, 1963, 149
C–Th R. Benz and P. L. Stone, *High Temp. Sci.*, 1969, **1**, 114
C–V E. K. Storms and R. J. McNeal, *J. phys. Chem.*, 1962, **66**, 1401

Ca–Cu G. Bruzzone, *J. less-common Metals*, 1971, **25**, 361
Ca–Eu V. F. Stroganova *et al.*, *Russ. Met.*, 1969, (6), 91
Ca–In G. Bruzzone, *J. less-common Metals*, 1966, (11), 249
Ca–Li M. R. Woolfson, *Trans. Am. Soc. Metals*, 1957, **49**, 794
Ca–Mn I. Obinata *et al.*, *Metallwiss. Technik*, 1963, (12), 1205
Ca–Pb G. Bruzzone and F. Merlo, *J. less-common Metals*, 1976, **48**, 103
Ca–Sb Z. U. Niyazora *et al.*, *Izv. Akad. SSSR Nedrg. Mater.*, 1976, **12** (7), 1293
Ca–Sr J. C. Schottmiller *et al.*, *J. phys. Chem.*, 1958, **62**, 1446
Ca–Si E. Schürmann, *et al.*, *Arch. Eisenhuttenw.*, 1975, **45** (6), 367
Ca–Ti I. Obinata *et al.*, *Trans. Am. Soc. Metals*, 1960, **52**, 1072
Ca–Yb S. D. Soderquist and F. X. Kayser, *J. less-common Metals*, 1968, **16**, 361
Ca–Zn A. F. Messing *et al.*, *Trans. Am. Soc. Metals*, 1963, **56**, 345

Cd–Cu Hansen; E. Raub, *Metallforschung*, 1947, **2**, 120
Cd–Eu W. Köster and J. Meixner, *Z. Metallk.*, 1965, **56**, 695
Cd–Gd G. Bruzzone *et al.*, *J. less-common Metals*, 1971, **25**, 295
Cd–La G. Bruzzone and F. Merlo, *J. less-common Metals*, 1973, **30**, 303
Cd–Mg W. Hume-Rothery and G. V. Raynor, *Proc. Roy. Soc.*, 1940, **A 174**, 471
Cd–Np M. Krumpelt *et al.*, *J. less-common Metals*, 1969, **18**, 35
Cd–Sm G. Bruzzone and M. L. Fornasini, *J. less-common Metals*, 1974, **37**, 289
Cd–Sn D. Hansen and W. T. Pell-Walpole, *J. Inst. Metals*, 1936, **59**, 281
Cd–Sr W. Köster and J. Meixner, *Z. Metallk.*, 1965, **56**, 695
Cd–Ti W. M. Robertson, *Met. Trans*, 1972, **3**, 1443
Cd–U A. E. Martin *et al.*, *Trans. Am. Inst. Min. Engrs*, 1961, **221**, 789
Cd–Y E. Ryba *et al.*, *J. less-common Metals*, 1969, **18**, 49
Cd–Yb A. Palenzona, *J. less-common Metals*, 1971, **25**, 367

Ce–Co R. Vogel, *Z. Metallk.*, 1946, **37**, 98
Ce–Cr V. M. Svechnikov *et al.*, *Dopov. Akad. Nauk. ukr. RSR*, 1969, (4), 354
Ce–Fe J. O. Jepson and P. Duwez, *Trans. Am. Soc. Metals*, 1955, **47**, 543
Ce–Ge K. H. J. Buschow and J. S. Wieringen, *Phys. Status. Solidi*, 1970, **42**, 231
Ce–In R. Vogel and H. Klose, *Z. Metallk.*, 1954, **45**, 633
Ce–La R. Vogel and H. Klose, *Z. Metallk.*, 1954, **45**, 633
Ce–Mn A. Landelli, *Atti Accad. Nazl. Lincei. Rend.*, 1952, **13**, 265; and B. J. Thamer, *J. less-common Metals*, 1964, **7**, 341 and 1965, **8**, 215
Ce–Pb L. Rolla *et al.*, *Z. Metallk.*, 1943, **35**, 29
Ce–Pd J. R. Thomson, *J. less-common Metals*, 1967, **13**, 307; and D. Rossi *et al.*, *J. less-common Metals*, 1975, **40**, 345

Ce–Ru	W. Obrowski, *Z. Metallk.*, 1962, **53**, 736
Ce–Sn	Equilibrium Data for Sn Alloys, *Tin Res. Inst.*, 1949
Ce–Th	R. T. Weiner *et al.*, *J. Inst. Metals*, 1957/8, **86**, 185
Ce–Ti	E. M. Savitskii and G. S. Burkhanov, *J. less-common Metals*, 1962, **4**, 301
Ce–Tl	L. Rolla *et al.*, *Z. Metallk.*, 1943, **35**, 29
Co–Cr	A. R. Elsea *et al.*, *Trans. Am. Inst. Min. Engrs*, 1949, **180**, 579
Co–Cu	C. S. Smith, 'Metals Handbook', Cleveland, Ohio, 1948
Co–Fe	W. C. Ellis and E. S. Greiner, *Trans. Am. Soc. Metals*, 1941, **29**, 415
Co–Ga	K. Schubert *et al.*, *Z. Metallk.*, 1959, **50**, 534
Co–Gd	V. F. Novy *et al.*, *Trans Am. Inst. Min. Engrs*, 1961, **221**, 588
Co–Ge	H. Pfisterer and K. Schubert, *Z. Metallk.*, 1949, **40**, 379
Co–Ho	K. H. J. Buschow and A. S. Van Der Goot, *J. less-common Metals*, 1969, **19**, 153
Co–In	J. D. Schöbel and H. Stadelmaier, *Z. Metallk.*, 1970, **61**, 342
Co–La	K. H. J. Buschow and W. A. J. J. Velge, *J. less-common Metals*, 1967, **13**, 11
Co–Mg	J. F. Smith and M. J. Smith, *Trans. Am. Soc. Metals*, 1964, **57**, 337
Co–Mn	Hansen and Elliott; *see also* K. Tsioplakis and T. Gödecke, *Z. Metallk.*, 1971, **62**, 680 and H. Masumoto *et al.*, *Nippon Kink. Gakk.*, 1969, **33**, 999
Co–Mo	T. F. J. Quinn and W. Hume-Rothery, *J. less-common Metals*, 1963, **5**, 314
Co–Nb	S. K. Bataleva, *Vest. Mosk. Univ. Khim.*, 1970, (4), 432
Co–Ni	C. E. Lacy, 'Metals Handbook', Cleveland, Ohio, 1948
Co–Os	W. Köster and E. Horn, *Z. Metallk.*, 1952, **43**, 444
Co–Pt	W. Köster, *Z. Metallk.*, 1949, **40**, 431
Co–Re	W. Köster and E. Horn, *Z. Metallk.*, 1952, **43**, 444
Co–Rh	W. Köster and E. Horn, *Z. Metallk.*, 1952, **43**, 444
Co–Ru	W. Köster and E. Horn, *Z. Metallk.*, 1952, **43**, 444
Co–Sm	K. H. J. Buschow and F. J. N. den Broeder, to be published
Co–Sn	Equilibrium Data for Sn Alloys, *Tin Res. Inst.*, 1949
Co–Ta	W. Köster and W. Mulfinger, *Z. Metallk.*, 1938, **30**, 348; M. Karchinsky and R. W. Fountain, *Trans. Am. Inst. Min. Engrs*, 1959, **215**, 1033
Co–V	W. Köster and H. Schmid, *Z. Metallk.*, 1955, **46**, 195
Co–W	S. Takeda, *Sci. Rep. Tôhoku Univ.*, 1936, Honda Anniv. Vol., p. 864.
Co–Zn	J. Schramm, *Z. Metallk.*, 1941, **33**, 46
Co–Zr	S. K. Bataleva, *Vest. Mosk. Univ., Khim.*, 1970, **11** (5), 557
Cr–Ga	J. D. Bornand and P. Feschotte, *J. less-common Metals*, 1972, **29**, 81
Cr–Ge	V. L. Zagryazhskii *et al.*, *Poroshk. Metall*, 1966, **8**, 55
Cr–Hf	E. Rudy and S. T. Windisch, *J. less-common Metals*, 1968, **15**, 13
Cr–Ir	R. M. Waterstrat and R. C. Manuszewski, *J. less-common Metals*, 1973, **32** (1), 79
Cr–Mn	S. J. Carlile *et al.*, *J. Inst. Metals*, 1949/50, **76**, 169; A. Hellawell and W. Hume-Rothery, *Phil. Trans. Roy. Soc.*, **A 249**, 1957, 417
Cr–Mo	O. Kubaschewski and A. Schneider, *Z. Elektrochem.*, 1942, **48**, 671
Cr–Nd	V. G. Ivanchenko *et al.*, *Dopov. Akad. Nauk. ukr. RSR*, 1969, (A), (1), 61
Cr–P	R. Vogel and G. W. Kasten Arch. Eisenhuttenw., 1939, **12**, 387 and *J less common Metals*, 1962, **4**, 496
Cr–Pb	G. Hindricks, *JISI*, 1943, **148**, 428 and T. Alden *et al.*, *Trans. AIME*, 1958, **212**, 15
Cr–Pd	G. Grube and R. Knabe, *Z. Elektrochem.*, 1936, **42**, 793
Cr–Pr	V. G. Ivanchenko *et al.*, *Dopov. Akad. Nauk. ukr. RSR*, 1969. (A), (8), 748
Cr–Pt	R. M. Waterstrat, *Met. Trans.*, 1973, **4**, 1585
Cr–Rh	R. M. Waterstrat and R. C. Manuszewski, *J. less-common Metals*, 1973, **32**, 331
Cr–Sc	V. M. Svechnikov *et al.*, *Dopov. Akad. Nauk. ukr. RSR*, 1972, (A), (34), 266
Cr–Tb	V. G. Ivanchenko *et al.*, *Dopov. Akad. Nauk. ukr. RSR*, 1970, (A), (8), 758
Cr–Ti	F. B. Cuff *et al.*, *J. Metals*, 1952, 848
Cr–U	A. H. Daane and A. S. Wilson, *Trans. Am. Inst. Min. Engrs*, 1955, **203**, 1219
Cr–W	W. Trzebiatowski *et al.*, *Analyt. Chem.*, 1947, **19**, 93; H. T. Greenaway, *J. Inst. Metals*, 1951/2, **80**, 589
Cr–Zr	E. T. Hayes *et al.*, *J. Metals*, 1952, 304
Cu–Er	K. H. J. Buschow, *Philips Res. Reports*, 1970, **25**, 227
Cu–Ga	W. Hume-Rothery *et al.*, *Phil. Trans. Roy. Soc.*, 1934, **A 233**, 1; J. O. Betterton and W. Hume-Rothery, *J. Inst. Metals*, 1951/2, **80**, 459
Cu–Ge	W. Hume-Rothery *et al.*, *J. Inst. Metals*, 1940, **66**, 221; J. Reynolds and W. Hume-Rothery, *J. Inst. Metals*, 1956/7, **85**, 120

Cu–In R. O. Jones and E. A. Owen, *J. Inst. Metals*, 1953/4, **82**, 445
Cu–Ir E. Raub and G. Röschel, *Z. Metallk*, 1969, **60**, 142
Cu–Mn B. M. Loring, 'Metals Handbook', Cleveland, Ohio, 1948
Cu–O F. N. Rhines and C. H. Mathewson, *Trans. Am. Inst. Min. Engrs*, 1934, **111**, 339
Cu–Pb G. C. Holder, 'Metals Handbook', Cleveland, Ohio, 1939
Cu–Pd F. W. Jones and C. Sykes, *J. Inst. Metals*, 1939, **65**, 422
Cu–Rh Ch. Raub *et al.*, *Metall*, 1972, **25**, 761
Cu–S J. Nutting, *Inst. Met. Ann. Eq Diag. No.* 24
Cu–Sb Hansen; J. C. Mertz and C. H. Mathewson, *Trans. Am. Inst. Min. Engrs*, 1937, **124**, 68
Cu–Si C. S. Smith, 'Metals Handbook'. Cleveland, Ohio, 1948
Cu–Sn G. V. Raynor, *Inst. Met. Ann. Diag. No.* 2
Cu–Sr G. Bruzzone, *J. less-common Metals*, 1971, **25**, 361–366
Cu–Th R. J. Schiltz *et al.*, *J. less-common Metals*, 1971, **25**, 175
Cu–U H. A. Wilhelm and O. N. Carlson, *Trans. Am. Soc. Metals*, 1950, **42**, 1311
Cu–Y R. F. Domegala *et al.*, *Trans. Am. Soc. Metals*, 1961, **53**, 137
Cu–Yb A. Iandelli and A. Palenzona, *J. less-common Metals*, 1971, **25**, 333
Cu–Zn P. Chiotti *et al.*, *US Energy Com.* 13930, 1960, **78**
Cu–Zr H. L. Burghoff, 'Metals Handbook', Cleveland, Ohio, 1948

Dy–Er F. H. Spedding *et al.*, *J. less-common Metals*, 1973, **31** (1), 1
Dy–Ho F. H. Spedding *et al.*, *J. less-common Metals*, 1973, **31** (1), 1

Er–Fe A. Meyer, *J. less-common Metals*, 1969, **18**, 41
Er–Ho F. H. Spedding *et al.*, *J. less-common Metals*, 1973, **31**, 1
Er–Ni K. H. J. Buschow, *J. less-common Metals*, 1968, **16**, 45
Er–Rh R. H. Ghassem and A. Raman, *Met. Trans.*, 1973, **4**, 745
Er–Tb F. H. Spedding *et al.*, *J. less-common Metals*, 1973, **31**, 1
Er–Y F. H. Spedding *et al.*, *J. less-common Metals*, 1973, **31**, 1

Eu–In W. Köster and J. Meixner, *Z. Metallk*, 1965, **56**, 695

Fe–Ga W. Köster and T. Gödecke, *Z. Metallk.*, 1977, **68** (10), 661
Fe–Gd V. F. Novy *et al.*, *Trans. Am. Inst. Min. Engrs*, 1961, **221**, 580
Fe–Ho G. J. Roe and T. J. O'Keefe, *Met. Trans.*, 1970, **1**, 2565
Fe–In C. Dasarathy, *Trans. Am. Inst. Min. Engrs*, 1969, **245**, 1838
Fe–Mg A. S. Yue, *J. Inst. Metals*, 1962/3, **91**, 166
Fe–Mn A. Hellawell, *Inst. Met. Ann. Eq. Diag. No.* 26
Fe–Mo Hansen; A. K. Sinha *et al.*, *J. Iron Steel Inst.*, 1967, **205**, 191
Fe–N K. H. Jack, *Acta Cryst.*, 1952, **5**, 404
Fe–Pd W. S. Gibson and W. Hume-Rothery, *J. Iron Steel Inst.*, 1958, **189**, 243; E. Raub *et al.*, *Z. Metallk*, 1963, **54**, 549
Fe–Pt A. Kussmann *et al.*, *Z. Metallk*, 1950, **41**, 470; R. A. Buckley and W. Hume-Rothery, *J. Iron Steel Inst.*, 1959, **193**, 61
Fe–Pu P. G. Marsden *et al.*, *J. Inst. Metals*, 1957/8, **86**, 166
Fe–Re H. Eggers, *Mitt. K.-Wilhelm-Inst. Eisenforsch. Düsseld.*, 1938, **20**, 147
Fe–Ru W. S. Gibson and W. Hume-Rothery, *J. Iron Steel Inst.*, 1958, **189**, 243; E. Raub and W. Plate, *Z. Metallk*, 1960, **51**, 477
Fe–S Hansen; V. P. Buistrov *et al.*, *Tsvet. Metally*, 1971, (6), 21
Fe–Sc O. P. Naumkin *et al.*, *Russ. Met.*, 1969, (3), 125
Fe–Si W. Köster and T. Gödecke, *Z. Metallk*, 1968, **59**, 602
Fe–Sm K. H. J. Buschow, *J. less-common Metals*, 1971, **25**, 131
Fe–Sn Equilibrium Data for Sn Alloys, *Tin Res. Inst.*, 1949
Fe–Ta A. K. Sinha and W. Hume-Rothery, *J. Iron Steel Inst.*, 1967, **205**, 671; Elliott
Fe–U J. D. Grogan, *J. Inst. Met.*, 1950, **77**, 571
Fe–V A. Hellawell, *Inst. Met. Ann. Eq. Diag. No.* 27
Fe–Y R. F. Domegala *et al.*, *Trans. Am. Soc. Metals*, 1961, **53**, 137
Fe–Zn S. Budurov *et al.*, *Z. Metallk*, 1972, **63**, 348 and G. F. Bastin *et al.*, *Z. Metallk.*, 1974, **65** (10), 656

Ga–Ge E. S. Greiner and P. Breidt, *Trans. Am. Inst. Min. Engrs*, 1955, **203**, 187
Ga–Hg Bruno Predel, *Z. Phys. Chem.*, 1960, **24**, 206
Ga–In J. P. Denny *et al.*, *J. Metals*, 1952, 39
Ga–Mg H. Gröber and V. Hauk, *Z. Metallk.*, 1950, **41**, (6), 191

Ga–Mn E. Wachtel and K. J. Nier, *Z. Metallk.*, 1965, **56** (11), 779
Ga–Nb R. E. Miller *et al.*, *Solid State Commun.*, 1971, **9** (20), 1769; *see also* L. L. Oden and R. E. Siemens, *J. less-common Metals*, 1968, **14**, 33, and V. V. Baron *et al.*, *Russ. J. Inorg. Chem.*, 1964, **9** (9), 1172 for alternative versions
Ga–Ni E. Hellner, *Z. Metallk.*, 1950, **41**, 480
Ga–Pd K. Schubert *et al.*, *Z. Metallk.*, 1959, **50**, 534
Ga–Sb I. G. Greenfield and R. L. Smith, *Trans. Am. Inst. Min. Engrs*, 1955, **203**, 351
Ga–U K. H. J. Buschow, *J. less-common Metals*, 1973, **31** (1), 165
Ga–V J. H. N. van Vucht *et al.*, *Philips Res. Reports*, 1964, **19**, 407

Gd–Ni V. F. Novy *et al.*, *Trans. Am. Inst. Min. Engrs*, 1961, **221**, 585
Gd–Pb J. T. Demel and K. A. Gschneider, *J. nucl. Mater.*, 1969, **29**, 11
Gd–Rh O. Leobich and E. Raub, *J. less-common Metals*, 1976, **46** (1), 1
Gd–Ru O. Leobich and E. Raub, *J. less-common Metals*, 1976, **46** (1), 7
Gd–Sc B. J. Beauchy and A. H. Daane, *J. less-common Metals*, 1964, **6**, 322

Ge–Mg 'Gmelins Handbuch der anorganischen Chemie, Mg. A. 4', 8th edn, 1952
Ge–Mn E. Wachtel and E. T. Henig, *Z. Metallk.*, 1969, **60** (3), 243 and *J. less-common Metals*, 1970, **21**, 223
Ge–Mo P. Stecher *et al.*, *Monatsh. Chem.*, 1963, **94**, 1154
Ge–Rh N. N. Zhuravlev and G. S. Zhdanov, *Kristallogra*, 1956, **1**, 205
Ge–Ru E. Raub and W. Fritzsche, *Z. Metallk.*, 1962, **53**, 779
Ge–Sn Equilibrium Data for Sn Alloys, *Tin Res. Inst.*, 1949
Ge–Sr R. L. Sharkey, *J. less-common Metals*, 1970, **20**, 113
Ge–Ti M. K. McQuillan, *J. Inst. Metals*, 1954/5, **83**, 485
Ge–Y F. A. Schmidt *et al.*, *J. less-common Metals*, 1972, **26**, 53
Ge–Zn E. Gebhardt, *Z. Metallk.*, 1942, **34**, 255
Ge–Zr O. N. Carlson *et al.*, *Trans. Am. Soc. Metals*, 1956, **48**, 843

H–Sr D. Petersen and R. P. Colburn, *J. phys. Chem.*, 1966, **70**, 468
H–Ti G. A. Lenning *et al.*, *Trans. Am. Inst. Min. Engrs*, 1954, **200**, 367
H–Zr C. E. Ellis and A. D. McQuillan, *J. Inst. Metals*, 1956/7, **85**, 89

Hf–Ir M. I. Copeland and D. Goodrich, *J. less-common Metals*, 1969, **18**, 347
Hf–Mn A. K. Shufrin and G. P. Dmitrievna, *Dopov. Akad. Nauk. ukr. RSR*, 1969 (1), 67
Hf–Mo A. Taylor *et al.*, *J. less-common Metals*, 1961, **3**, 265
Hf–Ni M. E. Kirkpatrick and W. L. Larsen, *Trans. Am. Soc. Metals*, 1961, **54**, 580
Hf–O E. Rudy and P. Stecher, *J. less-common Metals*, 1963, **5**, 75
Hf–Ta L. L. Oden *et al.*, *U.S. Bur. Min. Rep. Invest.* 6521, 1964
Hf–U D. T. Peterson and D. J. Beernstein, *Trans. Am. Soc. Metals*, 1959, **52**, 158
Hf–W B. C. Giessen *et al.*, *Trans. Am. Inst. Min. Engrs.* 1962, **224**, 60; A. Braun and E. Rudy, *Z. Metallk.*, 1960, **51**, 362

Hg–Mn G. Jangg and H. Palman, *Z. Metallk.*, 1963, **54**, 364
Hg–Rh G. Jangg *et al.*, *Z. Metallk.*, 1967, **58**, 724
Hg–Sb G. Jangg *et al.*, *Z. Metallk.*, 1962, **53**, 313
Hg–Sn M. L. Gayler, *J. Inst. Metals*, 1937, **60**, 381
Hg–U B. R. T. Frost, *J. Inst. Metals*, 1953/4, **82**, 456

Ho–Tb F. H. Spedding *et al.*, *J. less-common Metals*, 1973, **31** (1), 1

In–Li W. A. Alexander *et al.*, *Can. J. Chem.*, 1976, **54** (7), 1052
In–Mg G. V. Raynor, *Trans. Faraday Soc.*, 1948, **44**, 15
In–Mn U. Zwicker, *Z. Metallk.*, 1950, **41**, 400
In–Ni E. Hellner, *Z. Metallk.*, 1950, **41**, 402
In–Pb T. Heumann and B. Predel, *Z. Metallk.*, 1966, **57**, 50
In–Si C. D. Thurmond and M. Kowalchik, *Bell System Tech. J.*, 1960, **39**, 169
In–Sn J. C. Blade and E. C. Ellwood, *J. Inst. Metals*, 1956/7, **85**, 30
In–Sr G. Bruzzone, *J. less-common Metals*, 1966, **11**, 249
In–Yb O. D. McMasters, *J. less-common Metals*, 1971, **23**, 253
In–Zn S. Valentiner, *Z. Metallk.*, 1943, **35**, 250
In–Zr J. O. Betterton and W. K. Noya, *Trans. Am. Inst. Min. Engrs*, 1958, **212**, 340

Ir–Mn	E. Raub and W. Mahler, *Z. Metallk.*, 1955, **46**, 282
Ir–Ti	V. N. Eremenko and T. D. Shtepa, *Russ. Met.*, 1970, (6), 127
La–Mg	R. Vogel and T. Heumann, *Z. Metallk.*, 1946, **37**, 1
La–Mn	L. Rolla and A. Landelli, *Ber. Deut. Chem. Gess.*, 1942, **75**, 2091
La–Ni	R. Vogel, *Z. Metallk.*, 1946, **37**, 98
La–Rh	P. P. Singh and A. Raman, *Trans. Am. Inst. Min. Engrs*, 1969, **245** (7), 1561
La–Sn	Equilibrium Data for Sn Alloys, *Tin Res. Inst.*, 1949
La–Tl	L. Rolla *et al.*, *Z. Metallk.*, 1943, **35**, 29
Li–Mg	W. Hume-Rothery *et al.*, *J. Inst. Metals*, 1946, **72**, 538; W. E. Freeth and G. V. Raynor, *J. Inst. Metals*, 1953/4, **82**, 575
Li–Pb	O. Loebich and Ch. J. Raub, *J. less-common Metals*, 1977, **55**, 67
Lu–Pb	K. A. Gschneider *et al.*, *J. less-common Metals*, 1969, **19**, 337
Mg–Mn	W. R. D. Jones, *Inst. Met. Ann. Eq. Diag. No.* 28
Mg–Pd	Hansen; P. Vosskühler, *Z. Metallk.*, 1939, **31**, 109
Mg–Pr	L. Rolla *et al.*, *Z. Metallk.*, 1943, **35**, 29
Mg–Sc	B. J. Beaudry and A. H. Daane, *J. less-common Metals*, 1969, **18**, 305
Mg–Sn	Equilibrium Data for Sn Alloys, *Tin Res. Inst.*, 1949
Mg–U	P. Chiotti *et al.*, *Trans. Am. Inst. Min. Engrs.* 1956, **206**, 562
Mg–Y	E. D. Gibson and O. N. Carlson, *Trans. Am. Soc. Metals*, 1969, **52**, 1084; D. Mizer and J. A. Clark, *Trans. Am. Inst. Min. Engrs*, 1961, **221**, 207
Mg–Zn	W. R. D. Jones, *Inst. Met. Ann. Eq. Diag. No.* 29
Mg–Zr	J. H. Schaum and H. C. Burnett, *J. Res. natn. Bur. Stand.*, 1952, **49**, 155
Mn–N	'Diagrammy Sov. Metall. Sis'., 1967, Moscow
Mn–Nb	A. Hellawell, *J. less-common Metals*, 1959, **1**, 343 and V. M. Svechnikov, *Metallofiz.*, 1976, (**64**), 24
Mn–Nd	H. Kirchmayr and W. Lugscheider, *Z. Metallk.*, 1970, **61** (1), 22
Mn–Ni	K. E. Tsiuplakis and E. Kneller, *Z. Metallk.*, 1969, **60** (5), 433
Mn–O	G. Trömel *et al.*, *Erzmetall*, 1976, **29** (5), 234
Mn–P	J. Berak and T. Heumann, *Z. Metallk.*, 1950, **41**, 21
Mn–Pb	'Manganese Phase Diagrams', The Manganese Centre, 1980
Mn–Pd	E. Raub and W. Mahler, *Z. Metallk.*, 1954, **45**, 430
Mn–Pt	E. Raub and W. Mahler, *Z. Metallk.*, 1955, **46**, 282
Mn–Rh	E. Raub and W. Mahler, *Z. Metallk.*, 1955, **46**, 282
Mn–Ru	E. Raub and W. Mahler, *Z. Metallk.*, 1955, **46**, 282; and A. Hellawell, *J. less-common Metals*, 1959, **1**, 343
Mn–S	L. Staffansson, *Met. Trans. B*, 1976, **7B**, 131
Mn–Si	'Manganese Phase Diagrams'. The Manganese Centre, 1980
Mn–Sm	H. Kirchmayr and W. Lugscheider, *Z. Metallk.*, 1970, **61** (1), 22
Mn–Sn	'Manganese Phase Diagrams', The Manganese Centre, 1980
Mn–Tb	H. Kirchmayr and W. Lugscheider, *Z. Metallk.*, 1970, **61** (1), 22
Mn–U	H. A. Wilhelm and O. N. Carlson, *Trans. Am. Soc. Metals*, 1950, **42**, 1311
Mn–V	R. M. Waterstrat, *Trans. Am. Inst. Min. Engrs*, 1962, **224**, 240
Mn–Y	R. L. Myklebust and A. H. Daane, *Trans. Am. Inst. Min. Engrs*, 1962, **224**, 354
Mn–Zn	O. Romer and E. Wachtel, *Z. Metallk.*, 1971, **62** (11), 820
Mo–Ni	R. E. W. Casselton and W. Hume-Rothery, *J. less-common Metals*, 1964, **7**, 212
Mo–Os	A. Taylor *et al.*, *J. less-common Metals*, 1962, **4**, 436
Mo–Pd	E. Anderson, *J. less-common Metals*, 1964, **6**, 81
Mo–Re	A. G. Knapton, *J. Inst. Metals*, 1958/9, **87**, 62
Mo–Ru	E. Anderson and W. Hume-Rothery, *J. less-common Metals*, 1960, **2**, 443
Mo–Si	R. Kieffer and E. Cerwenka, *Z. Metallk.*, 1952, **43**, 101
Mo–Ti	M. Hansen *et al.*, *J. Metals*, 1951, 881
Mo–U	P. C. L. Pfeil, *J. Inst. Metals*, 1950, **77**, 553; F. G. Streets and J. J. Stobo, *J. Inst. Metals*, 1963/4, **92**, 171
Mo–W	W. P. Sykes, 'Metals Handbook', Cleveland, Ohio, 1948
Mo–Zr	R. F. Domegala *et al.*, *Trans. Am. Inst. Min. Engrs*, 1953, **197**, 73

N–Ti A. E. Palty *et al.*, *Trans. Am. Soc. Metals*, 1954, **46**, 312

Na–Sn Equilibrium Data for Sn Alloys, *Tin Res. Inst.*, 1949
Na–Th G. Grube and L. Botzenhardt, *Z. Elektrochem.*, 1942, **48**, 418

Nb–Re A. G. Knapton, *J. less-common Metals*, 1959, **1**, 480; B. C. Giessen *et al.*, *Trans. Am. Inst. Min. Engrs*, 1961, **221**, 1009
Nb–Si D. A. Deardorff *et al.*, *J. less-common Metals*, 1969, **18**, 11
Nb–Sn J. P. Charlesworth *et al.*, *J. Mater. Sci.*, 1970, **5**, 580
Nb–Th O. N. Charlson *et al.*, *Trans. Am. Inst. Min. Engrs*, 1956, **206**, 132
Nb–Ti M. Hansen *et al.*, *J. Metals*, 1951, 881
Nb–U P. C. L. Pfeil *et al.*, *J. Inst. Metals*, 1958/9, **87**, 204; B. A. Rogers *et al.*, *Trans. Am. Inst. Min. Engrs.* 1958, **212**, 387
Nb–V H. A. Wilhelm *et al.*, *Trans. Am. Inst. Min. Engrs*, 1954, **200**, 915
Nb–Y C. E. Lundin and D. T. Klodt, *J. Inst. Metals*, 1961/2, **90**, 341

Nd–Rh P. P. Singh and A. Raman, *Met. Trans.*, 1970, **1**, 236
Nd–Ti E. M. Savitski and G. B. Burkanov, *J. less-common Metals*, 1962, **4**, 301
Nd–Zn J. T. Mason and P. Chiotti, *Met. Trans.*, 1972, **3**, 2851

Ni–Pr R. Vogel, *Z. Metallk.*, 1946, **37**, 98
Ni–Pt A. Kussmann and H. E. Steinwehr, *Z. Metallk.*, **40**, (7), 263
Ni–Rh E. Raub and E. Röschel, *Z. Metallk.*, 1970, **61**, 113
Ni–Si E. N. Skinner, 'Metals Handbook', Cleveland, Ohio, 1948
Ni–Sn K. Schubert and E. Jahn, *Z. Metallk.*, 1949, **40**, (8), 319
Ni–Th J. R. Thomson, *J. less-common Metals*, 1972, **29**, 183
Ni–Ti D. M. Poole and W. Hume-Rothery, *J. Inst. Metals*, 1954/5, **82**, 473
Ni–V W. B. Pearson and W. Hume-Rothery, *J. Inst. Metals*, 1951/2, **79**, 643; L. R. Stevens and O. N. Charlson, *Met. Trans.*, 1970, **1**, 1267
Ni–W F. H. Ellinger and W. P. Sykes, *Trans. Am. Soc. Metals*, 1940, **28**, 619
Ni–Y B. J. Beaudry and A. H. Daane, *Trans. Am. Inst. Min. Engrs*, 1960, **218**, 854; R. F. Domegala *et al.*, *Trans. Am. Soc. Metals*, 1961, **53**, 137

Np–Pu P. G. Mardon *et al.*, *J. less-common Metals*, 1961, **3**, 281
Np–U P. G. Mardon *et al.*, *J. less-common Metals*, 1959, **1**, 467

O–Ti T. H. Schofield and A. E. Bacon, *J. Inst. Metals*, 1955/6, **84**, 47
O–V D. G. Alexander and O. N. Carlson, *Trans. Am. Soc. Metals*, 1971, **2**, 2805

Os–W A. Taylor *et al.*, *J. less-common Metals*, 1961, **3**, 333

P–Pd L. O. Gullman, *J. less-common Metals*, 1966, **11**, 157
P–Pt S. Rundquist, *Acta Chem. Scand.*, 1961, **15**, 451

Pb–Pd V. C. Marcotte, *Met. Trans. B*, 1977, **8B**, 185
Pb–Pr L. Rolla *et al.*, *Z. Metallk.*, 1943, **35**, 29
Pb–Pu D. H. Wood *et al.*, *J. nucl. Mater*, 1969, **32**, (2), 193
Pb–Sb J. B. Clark and C. W. F. T. Pistorius, *J. less-common Metals*, 1975, **42**, 59
Pb–Sn G. V. Raynor, *Inst. Met. Ann. Eq. Diag. No.* 6
Pb–Te G. O. Hiers, 'Metals Handbook', Cleveland, Ohio, 1948
Pb–Ti P. Farrar and H. Margolin, *Trans. Am. Inst. Min. Engrs*, 1955, **203**, 101
Pb–U B. R. T. Frost and J. T. Maskrey, *J. Inst. Metals*, 1953/4, **82**, 171
Pb–W S. Inouye, *Mem. Coll. Sci. Kyoto Univ.*, 1920, **4**, 43
Pb–Zn E. A. Anderson and J. L. Rodda, 'Metals Handbook', Cleveland, Ohio, 1939

Pd–Si R. H. Willens *et al.*, *J. Metals*, 1964, **16**, 92
Pd–Ta R. M. Waterstrat *et al.*, *Met. Trans. A*, 1978, **9A**, 643
Pd–Ti V. N. Eremenko and T. D. Shtepa, *Poroshk. Metall.*, 1972, (3), 75
Pd–Tl S. Bhan *et al.*, *J. less-common Metals*, 1968, **16**, 415
Pd–U J. A. Catterall *et al.*, *J. Inst. Metals*, 1956/7, **85**, 63; G. P. Pells, *J. Inst. Metals*, 1963/4, **92**, 416
Pd–Zn W. Köster and U. Zwicker, *Heraus Festschrift*, 1951, 76

Pd–Zr K. Anderko, *Z. Metallk.*, 1959, **60**, 681

Pr–Sn Equilibrium Data for Sn Alloys, *Tin Res. Inst.*, 1949
Pr–Tl L. Rolla *et al.*, *Z. Metallk.*, 1943, **35**, 29
Pr–Zn J. T. Mason and P. Chiotti, *Met. Trans.*, 1970, **1**, 2119

Pt–Ru J. M. Hutchinson, *Plantin Metals Rev.*, 1972, **16**, (3), 88
Pt–Sn K. Schubert and E. Jahn, *Z. Metallk.*, 1949, **40**, (10), 399
Pt–Te M. L. Gimpl *et al.*, *Trans. Am. Soc. Metals*, 1963, **56**, 209
Pt–Tl S. Bhan *et al.*, *J. less-common Metals*, 1968, **16**, 415
Pt–V R. M. Waterstrat, *Met. Trans.*, 1973, **4**, 455
Pt–W R. I. Jaffee and H. P. Nielsen, *Trans. Am. Inst. Min. Engrs*, 1949, **180**, 603
Pt–Zr E. G. Kendall *et al.*, *Trans. Am. Inst. Min. Engrs*, 1961, **221**, 445

Pu–Th D. M. Poole *et al.*, *J. Inst. Met.*, 1957/8, **86**, 172
Pu–Ti A. Languille, *Mém. Scient. Revue. Métall.*, 1971, **68**, (6), 435
Pu–Zn F. H. Ellinger *et al.*, 'Extractive and Physical Metallurgy of Plutonium and its Alloys', Interscience, N.Y., 1960, p. 169 et seq.
Pu–Zr J. A. C. Marples, *J. less-common Metals*, 1960, **2**, 331

Rb–Sb F. W. Dorn and W. Klemm, *Z. anorg. allg. Chem.*, 1961, **309**, 189

Re–Ru E. Rudy *et al.*, *Z. Metallk.*, 1962, **53**, 90
Re–Tb E. M. Savitskii and O. Kh. Khamidov, *Russ. Met.*, 1968, (6), 108

Rh–V R. M. Waterstrat and R. C. Manuszewski, *J. less-common Metals*, 1977, **52**, 293

Sb–Sn E. C. Ellwood, *Inst. Met. Ann. Eq. Diag., No.* 23
Sb–U B. J. Bauchy and A. H. Daane, *Trans. Am. Inst. Min. Engrs*, 1959, **215**, 199
Sb–Zr J. O. Betterton and W. M. Spicer, *Trans. Am. Inst. Min. Engrs*, 1958, **212**, 456

Se–Te E. Grison, *J. Chem. Phys.*, 1951, **19**, (9), 1109
Se–Th R. W. M. D'Eye *et al.*, *J. Chem. Soc.*, 1952, 2555, 143; A. Brown and J. J. Norreys, *J. Inst. Metals*, 1960/1, **89**, 238

Si–Ti M. Hansen *et al.*, *Trans. Am. Soc. Metals*, 1952, **44**, 518
Si–U A. Kaufmann *et al.*, *Trans. Am. Inst. Min. Engrs*, 1957, **209**, 23
Si–V H. A. C. M. Bruning, *Philips Res Reports*, 1967, **22**, (4), 349
Si–W R. Kieffer *et al.*, *Z. Metallk.*, 1952, **43**, 284
Si–Zr C. E. Lundin *et al.*, *Trans. Am. Soc. Metals*, 1953, **45**, 901

Sn–Ti M. K. McQuillan, *J. Inst. Metals*, 1955/6, **84**, 307
Sn–Tl Oska Prefect., *Univ. Bull.*, 1966, **15**, 137
Sn–Zr D. J. McPherson and M. Hansen, *Trans. Am. Soc. Metals*, 1953, **45**, 915

Ta–Th O. D. MacMasters and W. D. Larsen, *J. less-common Metals*, 1961, **3**, 312
Ta–Ti D. Summers Smith, *J. Inst. Metals*, 1952/3, **81**, 73
Ta–U C. H. Schramm *et al.*, *Trans. Am. Inst. Min. Engrs*, 1950, **188**, 195
Ta–V AFML–TR–65–2, Part V–Compendium of Phase Diagram Data, (June 1969), 113
Ta–W AFML–TR–65–2, Part V–Compendium of Phase Diagram Data, (June 1969), 144
Ta–Y C. E. Lundin and D. T. Klodt, *J. Inst. Metals*, 1961/2, **90**, 341
Ta–Zr D. E. Williams *et al.*, *Trans. Am. Inst. Min. Engrs*, 1962, **224**, 751

Tc–V C. C. Koch and G. R. Love, *J. less-common Metals*, 1968, **15**, 43

Th–Ti O. N. Carlson *et al.*, *Trans. Am. Inst. Min. Engrs*, 1956, **206**, 132
Th–U J. R. Murray, *J. Inst. Metals*, 1958/9, **87**, 94
Th–Y T. Eash and O. N. Carlson, *Trans. Am. Soc. Metals*, 1959, **52**, 1097, 301
Th–Zn P. Chiotti and K. J. Gill, *Trans. Am. Inst. Min. Engrs*, 1961, **221**, 573

Ti–U A. G. Knapton, *J. Inst. Metals*, 1954/5, **83**, 497
Ti–V H. N. Aderstedt *et al.*, *J. Am. Chem. Soc.*, 1952, **44**, 990

Ti–W E. Rudy and S. T. Windisch, *Trans. Am. Inst. Min. Engrs*, 1968, **242**, 953
Ti–Y D. W. Bau, *Trans. Am. Soc. Metals*, 1961, **53**, 1
Ti–Zn W. Heine and U. Zwicker, *Z. Metallk.*, 1962, **53**, 380

U–Nb H. A. Saller and F. A. Rough, *J. Metals*, 1953, 545
U–W C. H. Schramm *et al.*, *Trans. Am. Inst. Min. Engrs*, 1950, **188**, 195
U–Zn P. Chiotti *et al.*, *Trans. Am. Inst. Min. Engrs*, 1957, **209**, 51
U–Zr D. Summers-Smith, *J. Inst. Metals*, 1954/5, **82**, 277

V–Y C. E. Lundin and D. T. Klodt, *J. Inst. Metals*, 1961/2, **90**, 341
V–Zr J. T. Williams, *Trans. Am. Inst. Min. Engrs*, 1955, **203**, 345

W–Zr R. F. Domegala *et al.*, *Trans. Am. Inst. Min. Engrs*, 1953, **197**, 73

Y–Zn J. T. Mason and P. Chiotti, *Met. Trans. A*, 1976, **7A**, 289

Zn–Zr P. Chiotti and G. R. Kilp, *Trans. Am. Inst. Min. Engrs*, 1959, **215**, 892

11.3.2 Ternary systems and higher systems

No diagrams or references to these systems are included. The literature is now very large. For a recent and comprehensive bibliography of multicomponent systems the reader is referred to A. Prince, 'Multicomponent Alloy Constitution Bibliography 1955–1973', London Metals Society 1978; and 'Multicomponent Alloy Constitution Bibliography 1974–1977', London Metals Society, 1981.

12 Gas–metal systems

12.1 The solution of gases in metals

The gases which can be found in solution in measurable quantities in metals are the diatomic gases hydrogen, nitrogen and oxygen and also the noble gases in Group 8 of the Periodic Table.

12.1.1 Dilute solutions of diatomic gases

A diatomic gas dissociates on solution so that equilibrium between the solute and the gas phase is written:

$$X_2\ (\text{gas}) \rightleftharpoons 2X\ (\text{dissolved in metal})$$

In systems where no gas–metal compounds are formed and where also the solute does not contribute to the stability of the metallic phase, solutions are usually so dilute that Henry's law applies. If, also, the ideal gas laws are assumed for the gas phase and the standard states selected are (1) the infinitely dilute atomic fraction for the solute and (2) the pure element at a standard pressure $p^{\ominus}$ for the gas, the equilibrium constant is given by:

$$\mathrm{K} = \frac{a_x^2}{a_{x_2}} = \frac{N_x^2 p^{\ominus}}{p} \tag{12.1}$$

where a_x and a_{x_2} are the activities of the solute and the diatomic gas, N_x is the atomic fraction of solute and p is the pressure of the gas. In a form rearranged to express the proportionality between the solute concentration, C, and the square root of the gas pressure, p, equation (12.1) is known as Sievert's relation:

$$C = S\left(\frac{p}{p^{\ominus}}\right)^{1/2} \tag{12.2}$$

where S is the solute concentration in equilibrium with the standard pressure, $p^{\ominus}$.

Values for solubilities in dilute solutions are usually presented as solute concentrations in equilibrium with the pure gas at one atmosphere pressure (101 325 Pa). The units of concentration in common use are cm^3 of gas measured at one atmosphere pressure and 273 K per 100 g of metal for hydrogen and mass % for nitrogen and oxygen.

Application of the Van't Hoff isochore gives the variation with temperature of the equilibrium constant at constant pressure and hence also of the corresponding equilibrium solute concentration:

$$\left(\frac{\mathrm{dln}\ K}{\mathrm{d}T}\right)_p = \frac{\Delta H^{\ominus}}{RT^2}$$

where H° is the standard enthalpy of solution of the gas. Substituting for K from equation (12.1) gives:

$$\left(\frac{\mathrm{dln}\ N_x}{\mathrm{d}T}\right)_p = \frac{\Delta H^{\ominus}}{2RT^2}$$

For a small range of temperature it is permissible to disregard the temperature-dependence of $\Delta H^{\ominus}$ so that integration then yields:

$$\log\left(\frac{N_1}{N_2}\right)=\frac{-\Delta H^{\ominus}}{2.303\times 2R}\left(\frac{1}{T_1}-\frac{1}{T_2}\right) \tag{12.3}$$

where N_1 and N_2 are the atomic fractions of solute in equilibrium at temperatures T_1 and T_2 with the same gas pressure. Standard enthalpies of solution $\Delta H^{\ominus}$ can be evaluated from experimental results fitted to equation (12.3), bearing in mind Kubaschewski's reminder[1] that thermochemical data derived from the temperature-dependence of equilibrium constants are uncertain.

12.1.2 Complex gas–metal systems

In systems where the solute contributes to the stability of the metallic phase or where near stoichiometric compounds separate, the information of interest extends to high solute concentrations where there is deviation from Henry's law and to equilibria between condensed phases. If complete, this information is most conveniently presented graphically as isotherms, i.e. expressing composition as a function of equilibrium pressure for selected temperatures. Regions where two condensed phases coexist in the system are manifest by pressure invariance between the compositions of the conjugate phases. For some systems, the available experimental results are limited to values for solute concentrations in equilibrium at selected temperatures with either a second phase or the gas at a selected pressure.

12.1.3 Solutions of hydrogen

Table 12.1 gives selected values of dissolved hydrogen concentrations in equilibrium with hydrogen at one atmosphere pressure for the metals silver, aluminium, cobalt, chromium, copper, iron, magnesium, manganese, molybdenum, nickel, lead, platinum, silicon, tin, zinc and some of their alloys. The solutions are dilute and form endothermically from the gas phase, so that the solubility of hydrogen increases with temperature (except for α-manganese).

The alkali and alkaline-earth metals form hydrides with non-metallic characteristics in which bonding is of predominantly ionic character with the hydrogen present as H^- anions. Table 12.2 gives hydrogen concentrations in equilibrium with the corresponding hydrides for barium, calcium, magnesium, sodium and strontium.

Certain of the metals in the transition, rare earth and actinide series have the remarkable ability to take hydrogen into interstitial solid solution until the atomic fraction of hydrogen approaches simple stoichiometric ratios but without loss of the metallic character of the phase so formed. Additional phases may also appear at higher hydrogen concentrations. The solute hydrogen increases the stability of the metallic phase, the solution process is exothermic and thus for a given hydrogen activity (i.e. for constant pressure of hydrogen in the gas phase) the stability decreases with temperature. Figures 12.1 to 12.11 present information for the systems formed by hydrogen with cerium, niobium, neodymium, palladium, praseodymium, plutonium, tantalum, thorium, titanium, vanadium and zirconium. Tables 12.3 and 12.4 give data for uranium and hafnium.

Table 12.1 HYDROGEN SOLUTIONS IN EQUILIBRIUM WITH GASEOUS HYDROGEN AT ATMOSPHERIC PRESSURE

Solvent metal	*Hydrogen concentration, S at T °C* cm³ *of gas at atmosphere pressure and* 0 °C *per* 100 g *of metal*							*References*
Ag	*T* 400	500	600	700	800	900	960.5	16
(solid)	*S* 0.056	0.11	0.18	0.23	0.33	0.43	0.52	
Al	*T* 350	400	500	600	660			18
(solid)	*S* 0.001 2	0.002 8	0.011	0.030	0.050			also 17
(liquid)	*T* 660	700	750	800	900	1 000		17, 19
	S 0.69	0.92	1.23	1.67	2.75	4.15		also 18

Table 12.1 HYDROGEN SOLUTIONS IN EQUILIBRIUM WITH GASEOUS HYDROGEN AT ATMOSPHERIC PRESSURE—*continued*

Solvent metal			*Hydrogen concentration, S at T °C* cm³ *of gas at atmospheric pressure and* 0 °C *per* 100 g *of metal*									*References*
Al–Cu	*T*		700	800	900	1 000						
(liquid)	*S*	2% Cu	0.75	1.35	2.45	3.80						19
		4% Cu	0.65	1.15	2.25	3.45						19
		8% Cu	0.50	0.95	1.85	2.95						19
		16% Cu	0.40	0.80	1.35	2.30						19
		32% Cu	0.35	0.65	1.15	1.80						19
Al–Si	*T*		700	800	900	1 000						
(liquid)	*S*	2% Si	0.75	1.50	2.50	3.90						19
		4% Si	0.70	1.35	2.35	3.75						19
		8% Si	0.60	1.25	2.25	3.60						19
		16% Si	0.50	1.23	2.15	2.37						19
Co	*T*	600	700	800	900	1 000	1 100	1 200	1 300	1 400	1 492	
(solid)	*S*	0.9	1.22	1.85	2.51	3.30	4.31	5.40	6.7	7.75	8.65	20, 21
(liquid)	*T*	1 500	1 550	1 600	1 650							
	S	20.5	21.6	23.2	24.3							21, 22
Co–Fe		*See* Fe–Co										
Co–Ni		*See* Ni–Co										
Cr	*T*	400	500	600	700	800	900	1 000	1 100	1 200		24
(solid)	*S*	0.2	0.3	0.4	0.6	1.0	1.7	2.6	3.7	5.4		also 23
Cr–Fe		*See* Fe–Cr										
Cu	*T*	300	400	500	600	700	800	900	1 000	1 083		25, 27
(solid)	*S*	0.01	0.05	0.11	0.21	0.27	0.53	0.89	1.34	1.9		also 21, 26
(liquid)	*T*	1 083	1 100	1 150	1 200	1 250	1 300	1 350	1 400			27
	S	5.1	5.4	6.3	7.2	8.3	9.2	10.4	11.8			also 21, 26, 28
Cu–Ag	*T*		1 225	1 225	1 225	1 225	1 225					
(liquid)	% Ag		0	10	20	30	50					
	S		8.6	7.9	7.0	5.9	4.3					7
Cu–Al	*T*		700	800	900	1 000	1 050					
(solid)	*S*	1.43% Al	—	—	0.70	1.15	1.40					27
		3.3% Al	—	0.35	0.70	1.05	1.25					27
		5.77% Al	—	0.10	0.30	0.60	—					27
		6.84% Al	0.10	0.15	0.35	0.65	—					27
		8.1% Al	0.10	0.15	0.35	0.75	—					27
Cu–Al	*T*		1 100	1 150	1 200	1 300	1 400					
(liquid)	*S*	1.45% Al	5.1	5.8	6.6	—	—					27
		3.3% Al	4.4	5.1	5.7	—	—					27
		5.77% Al	3.3	3.9	4.5	—	—					27
		6.84% Al	3.0	3.5	4.1	—	—					27
		8.1% Al	2.7	3.2	3.5	4.5	5.9					27 also 7
Cu–Au	*T*		1 225	1 225	1 225	1 225						
(liquid)		% Au	0	10	30	50						
	S		8.6	7.9	5.6	3.4						7
Cu–Ni	*T*		1 225	1 225	1 225							
(liquid)		% Ni	0	10	20	*See also* Ni–Cu						
	S		8.6	13	17							7
Cu–Pt	*T*		1 225	1 225	1 225							
(liquid)		% Pt	0	10	20							
	S		8.6	9.6	9.7							7

Table 12.1 HYDROGEN SOLUTIONS IN EQUILIBRIUM WITH GASEOUS HYDROGEN AT ATMOSPHERIC PRESSURE—*continued*

Solvent metal			*Hydrogen concentration, S at T* °C cm³ *of gas at atmospheric pressure and* 0 °C *per* 100 g *of metal*										*References*
Cu–Sn	*T*		1 000	1 100	1 200	1 300							
(liquid)	*S*	5.9% Sn	—	4.80	6.28	7.81							28
		11.5% Sn	3.09	4.11	5.35	6.85							28
		21.7% Sn	2.11	2.97	3.94	5.10							28
		40.2% Sn	0.53	0.94	1.50	—							28
		54.8% Sn	0.50	0.76	1.15	1.61							28 also 7
Cu–Zn	*T*		500	600	700	800	875*						
(solid α)	*S*	33% Zn	0.010	0.023	0.044	0.076	0.35						29
			(* α+β region)										
Fe	*T*		200	300	400	500	600	700	800	900			24
(solid α)	*S*		—	0.1	0.2	0.6	1.2	1.7	2.4	3.0			also 30, 31, 32
	T		900	1 000	1 100	1 200	1 250	1 350	1 400				24
(solid γ)	*S*		4.7	5.4	6.4	7.4	8.4	—	9.3				also 21, 31, 33
Fe	*T*		1 400	1 450									24
(solid δ)	*S*		6.1	6.4									also 21, 31, 33
(liquid)	*T*		1 535	1 550	1 600	1 700							59, 60, 61
	S		24.5	25.5	26.5	29.5							21 also 22, 31, 33 34, 35, 36
Fe–Co	*T*		1 600	1 600	1 600	1 600	1 600						
(liquid)		% Co	0	20	40	60	80						
	S		29.8	23.8	20.7	18.2	21.5						22
Fe–Cr	*T*		400	600	700	800	850	900	1 000	1 100	1 200		
(solid)	*S*	4% Cr	0.2	0.9	1.4	1.9	2.2	4.4	5.3	6.3	7.4		24
		10% Cr	0.2	0.7	1.2	2.1	5.0	6.0	7.1	8.3	9.3		24
		20% Cr	0.2	0.5	1.8	3.0	3.7	4.3	4.9	5.6	6.6		24
		44% Cr	0.2	0.5	1.2	2.3	3.5	4.3	5.4	6.2	6.9		24
		77% Cr	0.2	0.5	0.8	1.2	1.5	2.0	3.1	4.4	5.9		24 also 33
Fe–Cr	*T*		1 600	1 600	1 600	1 600	1 600						
(liquid)		% Cr	9.4	18.8	28.5	38.3	48.2						
	S		25	25	25	24	23						33
Fe–Nb	*T*		1 560	1 685									
(liquid)	*S*	*5.01% Nb	24.6	28.3									36
		8.97% Nb	29.2	31.7									36
		15.12% Nb	41.7	43.8									36
			*Estimates from results for equilibrium at 2 900 Pa										
Fe–Ni	*T*		300	400	500	600	700	800	900	1 000	1 100	1 200	
(solid)	*S*	3.3% Ni	—	0.3	0.7	1.3	2.0	3.9	5.0	6.2	7.3	8.4	24
		5.5% Ni	0.2	0.4	0.8	1.4	2.1	4.1	5.2	6.3	7.3	8.4	24
		11.6% Ni	0.4	0.6	1.3	2.1	3.2	4.3	5.3	6.4	7.5	8.5	24
		21.0% Ni	0.5	0.7	1.4	2.2	3.3	4.3	5.4	6.5	7.6	8.8	24
		32.1% Ni	0.6	0.9	1.5	2.4	3.3	4.4	5.5	6.7	7.8	9.0	24
		53.2% Ni	0.8	1.2	1.8	2.7	3.7	4.7	5.8	6.9	8.1	9.3	24
		62.4% Ni	1.3	1.7	2.4	3.3	4.1	5.0	6.1	7.3	8.4	9.6	24
		72.4% Ni	1.6	1.8	2.6	3.4	4.3	5.2	6.3	7.4	8.7	9.9	24
		84.8% Ni	1.8	2.2	2.7	3.7	4.8	6.3	7.6	9.2	10.7	12.4	24 also 21, 33
(liquid)	*T*		1 550	1 600	1 700								
	S	20% Ni	26.5	27.5	31								21
		40% Ni	28.5	30	32.5								21
		60% Ni	31.5	33	36								21
		80% Ni	36	38	41								21 also 22, 33

Table 12.1 HYDROGEN SOLUTIONS IN EQUILIBRIUM WITH GASEOUS HYDROGEN AT ATMOSPHERIC PRESSURE—*continued*

Solvent metal			*Hydrogen concentration, S at T °C* cm^3 *of gas at atmospheric pressure and* 0 °C *per* 100 g *of metal*										*References*
Fe–Si	*T*		1 350	1 400	1 500	1 550	1 650						
(liquid)	*S*	1.78% Si	—	—	25.5	27.5	31.6						35
		11.0% Si	11.5	12.5	14.4	15.4	17.4						35
		21.7% Si	7.0	7.6	8.9	9.5	10.7						35
		31.5% Si	—	6.0	6.5	6.8	7.3						35
		39.1% Si	—	7.0	8.5	9.1	10.3						35
		45.7% Si	9.3	9.9	11.1	11.8	13.1						35
		51.5% Si	12.7	13.1	14.0	14.4	15.3						35
		63.7% Si	20.9	21.4	22.4	22.9	23.8						35
Fe–Ti	*T*		1 560	1 685									
(liquid)	*S*	*0.18% Ti	24.3	26.7									36
		0.45% Ti	25.6	28.2									36
		0.70% Ti	28.2	30.6									36
		2.67% Ti	—	45.8									36
		3.14% Ti	50.0	52.9									36
			*Estimated from results for equilibrium at 2 900 Pa										
Fe–Cr–Ni	*T*		400	500	600	700	800	900	1 000	1 100	1 200		
(solid)	*S*	17.5% Cr + 9% Ni	0.2	0.8	1.8	2.4	3.4	4.9	6.4	7.7	8.8		24
		4.6% Cr + 10% Ni	0.6	1.0	1.7	2.8	3.6	4.4	5.5	6.6	8.1		24
Mg	*T*		650										
(solid)	*S*		15–30 (results unreliable)										38, 39, 41
(liquid)	*T*		675	725	775								41
	S		46	60	63	(results unreliable)							also 38, 39, 40
Mn	*T*		25	100	200	300	400	500	600			Marked hysteresis during transition in range 600 < *T* < 800	43
(solid α)	*S*		21.6	19.9	17.2	14.5	12.4	11.4	11.4				also 42
(solid β and γ)	*T*		800	850	900	950	1 000	1 050	1 100	1 125		Hysteresis during transition in range 1 125 < *T* < 1 150	43
	S		28.6	29.3	30.1	31.4	32.8	34.2	40.0	42.2			also 42
Mn	*T*		1 150	1 175	1 200	1 225	1 243						
(solid δ)	*S*		41.1	41.7	42.8	44.4	46.6						43 also 42
(liquid)	*T*		1 243	1 250	1 275	1 300							
	S		46.6	50.0	58.3	60.2							43 also 42
Mo	*T*		500	600	700	800	900	1 000	1 100	1 200			
(solid)	*S*		0.8	1.3	1.7	2.2	1.8	1.2	0.8	0.5			23
Ni	*T*		300	400	500	600	700	800	900				
(solid)	*S*		2.0	2.5	3.3	4.3	5.6	7.0	8.5				21, 24 also 26, 32, 33
	T		1 000	1 100	1 200	1 300	1 400	1 453					
	S		10.0	11.5	13.0	14.0	16.5	18					21, 24 also 26, 32, 33
(liquid)	*T*		1 453	1 500	1 600	1 700							
	S		41	43	48	51							21 also 22, 26, 33, 34

Table 12.1 HYDROGEN SOLUTIONS IN EQUILIBRIUM WITH GASEOUS HYDROGEN AT ATMOSPHERIC PRESSURE—*continued*

Solvent metal		*Hydrogen concentration, S at T °C* cm³ *of gas at atmospheric pressure and* 0 °C *per* 100 g *of metal*											*References*
Ni–Co (solid)	*T*	1 400	1 400	1 400	1 400								
	% Co	20	40	60	80								
	S	14	11.5	10.0	9.0								21
(liquid)	*T*	1 500	1 600	1 700									
	S 20% Co	37.5	40.5	43.5									21
	40% Co	31.5	34.5	36.5									21
	60% Co	28	30	33.5									21
	80% Co	23.6	26	29									21 also 22
Ni–Cu (liquid)	*T*	1 500	1 600										
	S 20% Cu	41	44.5										21
	40% Cu	37	40										21
	60% Cu	31.5	34.5										21
	80% Cu	23.5	25.5										21
Ni–Fe	*See* Fe–Ni												
Pb (liquid)	*T*	420	500	600	700	800	900	References 45 and 46 report					
	S	—	0.11	0.25	0.45	0.80	1.25	no detected solution for $T<600$					44 also 45, 46
Pb–Ca (liquid)		No solution detected in lead with 0.16 or 0.24% Ca at 420 °C											45
Pb–Mg (liquid)	*T*	500	500	500	500	500	500	500	500	500	500	500	
	% Mg	1.0	2.1	3.1	5.5	6.5	7.6	7.8	8.1	8.5	9.0	17.5	
	S	0.18	0.22	0.45	0.87	1.7	1.5	1.8	1.7	2.4	1.9	4.8	45
Pt (solid)	*T*	400	800	1 000	1 100	1 200	1 300						
	S	0.07	0.10	0.19	0.34	0.54	0.80						47
Si (solid)	*T*	1 200											
	S	0.001 6											48
Sn (liquid)	*T*	1 000	1 100	1 200	1 300								
	S	0.04	0.09	0.21	0.36	(results unreliable)							28
Zn (liquid)	*T*	516											
	S	<0.002											46

Table 12.2 HYDROGEN SOLUTIONS IN EQUILIBRIUM WITH ALKALI AND ALKALINE EARTH METAL HYDRIDES

Solvent metal	*Hydride*		*Atom fraction of hydrogen,* N_H*, and hydride dissociation pressure,* P(Pa) *at temperature,* T °C								*References*
Ba (solid)	BaH_2	*T*	400	500	600	700					
		N_H	0.18	0.24	0.35	0.50					49
		P	—	—	—	—					
Ca (liquid)	CaH_2	*T*	780	800	830	860					50
		N_H	0.26	0.29	0.31	0.32					also 51, 52
		P	1.87×10^3	3.63×10^3	6.27×10^3	1.08×10^4					
Mg (solid)	MgH_2	*T*	440	470	510	560					53
		N_H	0.020	0.031	0.034	0.093					also 54 an
		P	3.90×10^6	6.55×10^6	1.19×10^7	2.36×10^7					Table 1
Na (liquid)	NaH	*T*	250	300	315	330	350	375	400	425	55, 56
		N_H	9.7×10^{-5}	5.1×10^{-4}	1.2×10^{-3}	2.4×10^{-3}	3.0×10^{-3}	5.8×10^{-3}	1.4×10^{-2}	2.8×10^{-2}	also 58
		P	—	—	—	—	—	—	—	—	

'able 12.2 HYDROGEN SOLUTIONS IN EQUILIBRIUM WITH ALKALI AND ALKALINE EARTH METAL YDRIDES—*continued*

olvent etal	*Hydride*		*Atom fraction of hydrogen, N_H, and hydride dissociation pressure, P(Pa) at temperature, T °C*								*References*
r solid :+γ)	SrH_2		αSr		γSr⟶						
		T	212		263	365	456	497	562	595	57
		N_H	0.048		0.062	0.096	0.12	0.14	0.19	0.19	
		P	—		—	—	—	—	—	—	
r solid β)	SrH_2	T	668	740	802	810					57
		N_H	0.21	0.26	0.36	0.40					
		P	—	—	—	—					

ıble 12.3 HYDROGEN SOLUTIONS IN URANIUM EQUILIBRATED WITH GASEOUS HYDROGEN AND WITH URANIUM YDRIDE
ſallet and Trzeciak[62], also reference 63)

ılvent ıase	*S, mass fraction of hydrogen in uranium equilibrated at T °C with hydrogen gas at atmospheric pressure* *Sp, mass fraction of hydrogen in uranium equilibrated at T °C with uranium hydride, UH_3* *P, UH_3 dissociation pressure* (pascals) *at T °C*								
	T	100	200	300	400	432	500	600	662
	S	—	—	—	—	1.6×10^{-6}	1.8×10^{-6}	2.0×10^{-6}	2.2×10^{-6}
olid α)	P	0.2	60	3×10^{3}	5×10^{4}	1.0×10^{5}	3.7×10^{5}	1.81×10^{6}	4.02×10^{6}
	Sp	6×10^{-10}	2×10^{-8}	2×10^{-7}	1.1×10^{-6}	1.6×10^{-6}	3.5×10^{-6}	8.6×10^{-6}	1.35×10^{-5}
	T	662	700	725	750	769			
	S	7.8×10^{-6}	8.5×10^{-6}	9.0×10^{-6}	—	9.7×10^{-6}			
olid β)	P	4.02×10^{6}	6.25×10^{6}	8.24×10^{6}	1.05×10^{7}	1.31×10^{7}			
	Sp	4.9×10^{-5}	6.8×10^{-5}	8.1×10^{-5}	9.7×10^{-5}	1.11×10^{-4}			
	T	769	800	900	1 000	1 100	1 129		
	S	1.47×10^{-5}	1.50×10^{-5}	1.56×10^{-5}	1.62×10^{-5}	1.67×10^{-5}	1.69×10^{-5}		
olid γ)	P	1.31×10^{7}	2.16×10^{7}	4.07×10^{7}	8.04×10^{7}	1.49×10^{8}	1.77×10^{8}		
	Sp	1.68×10^{-4}	1.95×10^{-4}	3.12×10^{-4}	4.57×10^{-4}	6.43×10^{-4}	7.02×10^{-4}		
	T	1 129	1 200	1 300	1 400				
	S	2.81×10^{-5}	2.93×10^{-5}	3.11×10^{-5}	3.27×10^{-5}				
quid)	P	1.77×10^{8}	2.48×10^{8}	3.94×10^{8}	5.96×10^{8}				
	Sp	1.17×10^{-3}	1.45×10^{-3}	1.94×10^{-3}	2.52×10^{-3}				

able 12.4 HYDROGEN SOLUTIONS IN HAFNIUM EQUILIBRATED WITH ASEOUS HYDROGEN AT ATMOSPHERIC PRESSURE
:spagno, Azou and Bastien[70,71], also reference 69)

emperature °C	100	300	500	700	900	950	1 000	1 050	1 100
tom ratio H/Hf	1.80	1.78	1.60	1.38	0.88	0.40	0.09	0.06	0.05

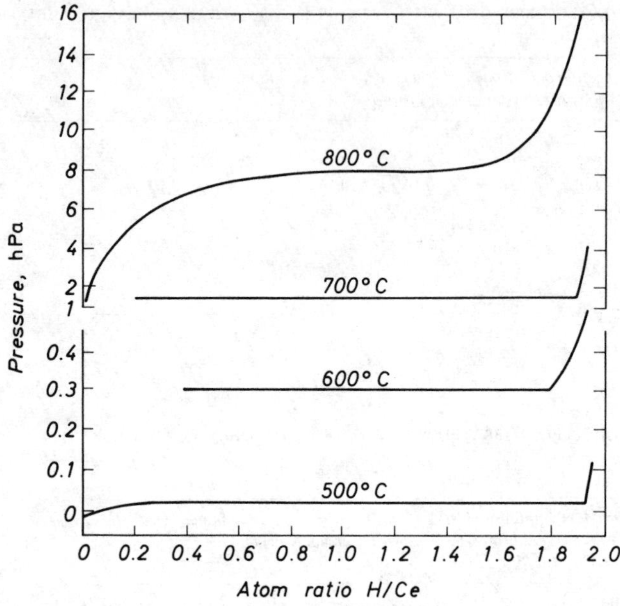

Figure 12.1 *The cerium–hydrogen system (Mulford and Holley.*[64] *See also Sieverts et al.*[65–67] *Ivanov and Stomakhin*[68] *and Edwards and Velekis*[69]*)*

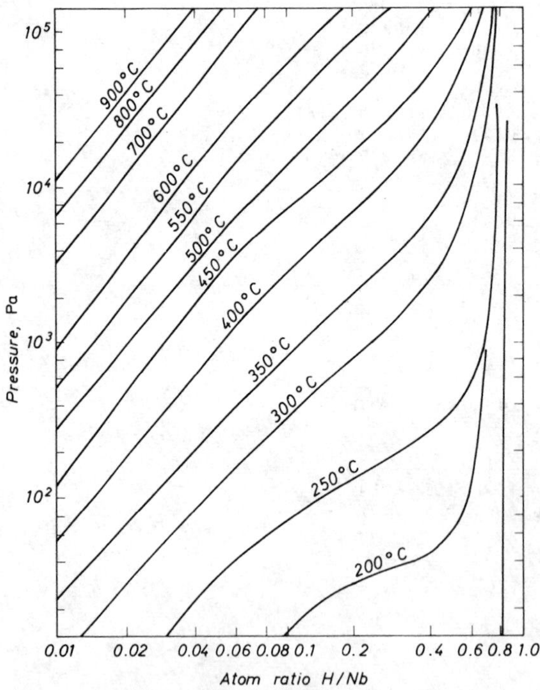

Figure 12.2 *The niobium–hydrogen system (Albrecht, Goode and Mallet.*[73,74] *See also Komjathy*[75] *and Walter and Chandler*[72]*)*

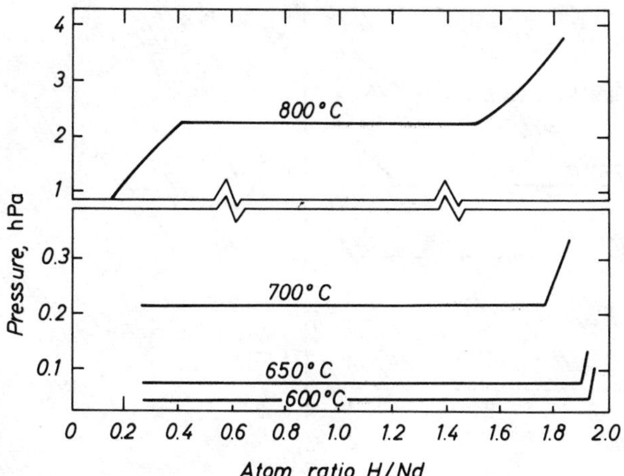

Figure 12.3 *The neodymium–hydrogen system (Mulford and Holley.[64] See also Sieverts and Roell[76])*

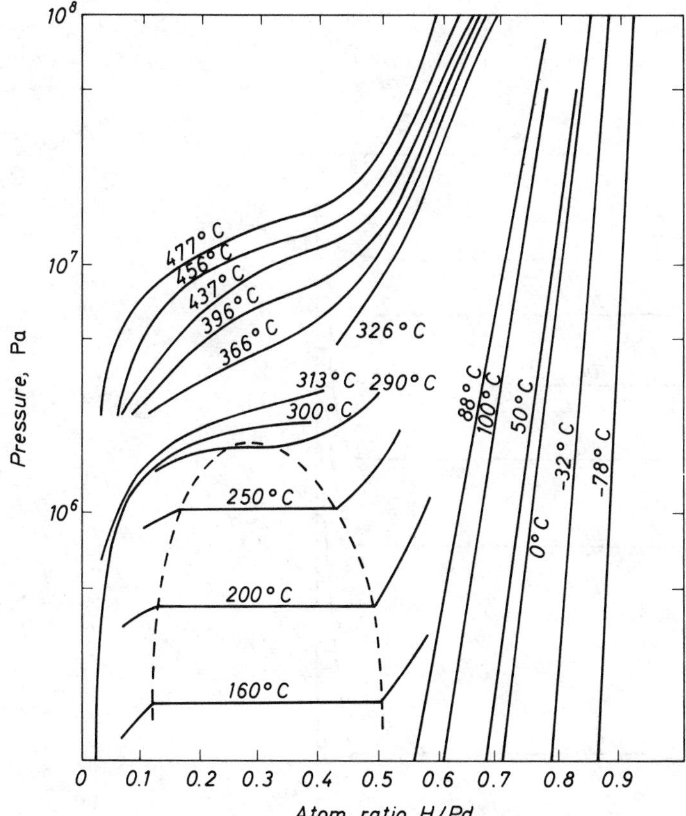

Figure 12.4 *The palladium–hydrogen system (Levine and Weal,[77] Gillespie et al.[78,79] and Perminov.[80] See also Everett and Nordon,[81] Flanagan,[82] Nakhutin and Sutyagina,[83] Mitacek and Aston,[84] Carson et al.,[85] Karpova and Tverdovsky,[86] Vert et al.[87] and Maeland and Flanagan[88])*

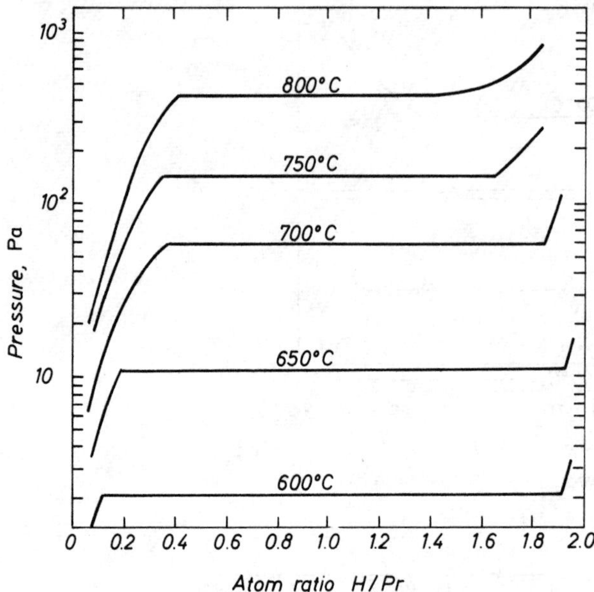

Figure 12.5 *The praseodymium–hydrogen system (Mulford and Holley.[64] See also Sieverts and Roell[76])*

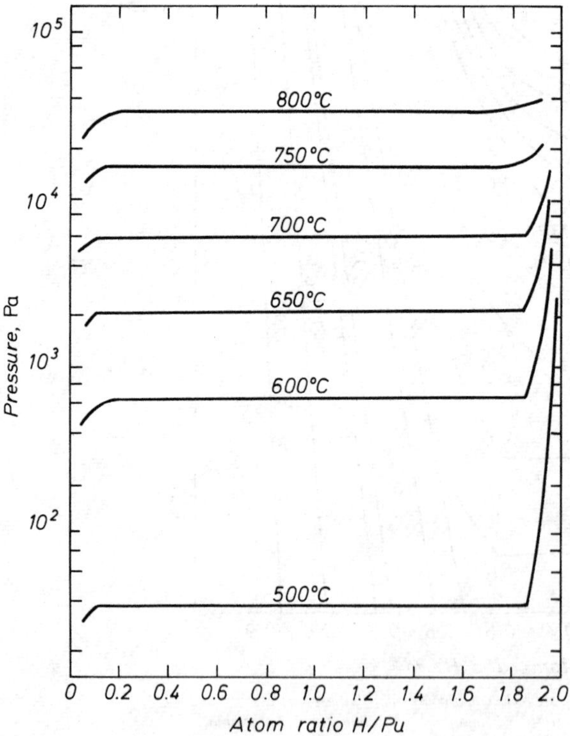

Figure 12.6 *The plutonium–hydrogen system (Mulford and Sturdy[89])*

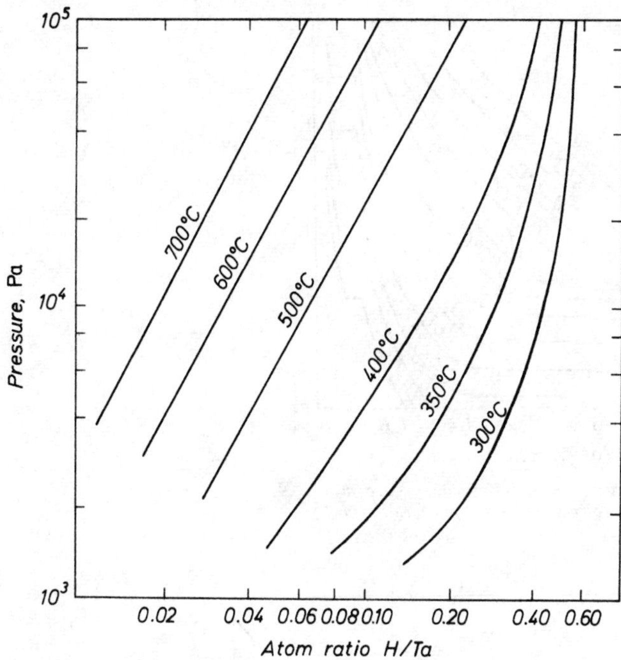

Figure 12.7 *The tantalum–hydrogen system (Mallet and Koehl.*[92] *See also Kofstad et al.*[93] *and Pedersen et al.*[94]*)*

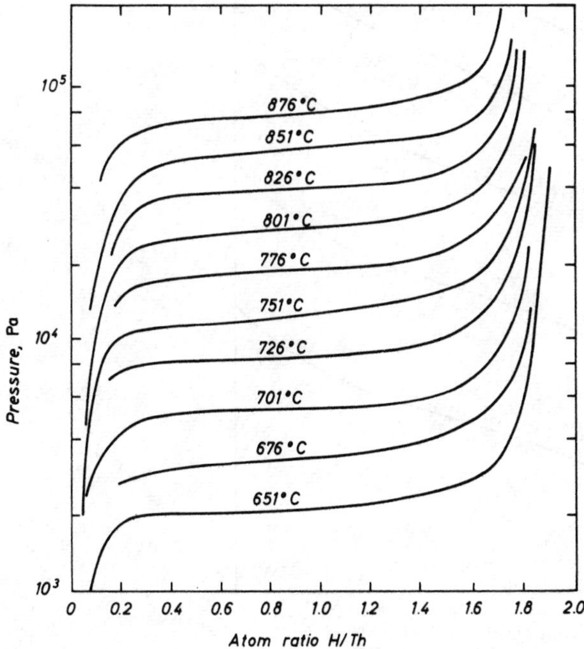

Figure 12.8 *The thorium–hydrogen system (Mallet and Campbell.*[95] *See also Peterson and Westlake*[96]*)*

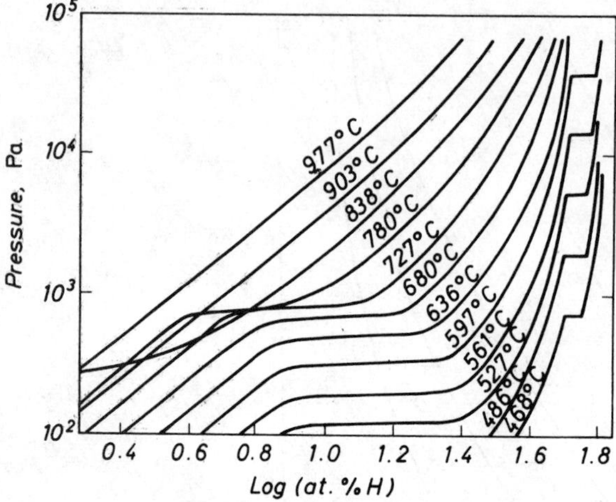

Figure 12.9 *The titanium–hydrogen systems (McQuillan.[97] See also Lenning et al.,[98] McQuillan,[99] Samsonov and Antonova[100] and Krylov[101])*

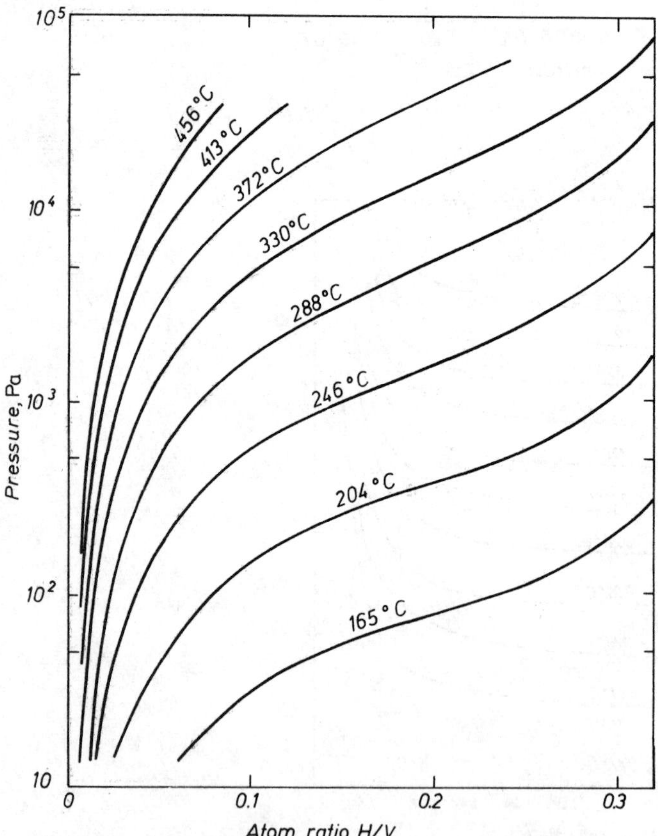

Figure 12.10 *The vanadium–hydrogen system (Kofstad and Wallace.[102] See also Brauer and Schnell[103] and Maeland[104])*

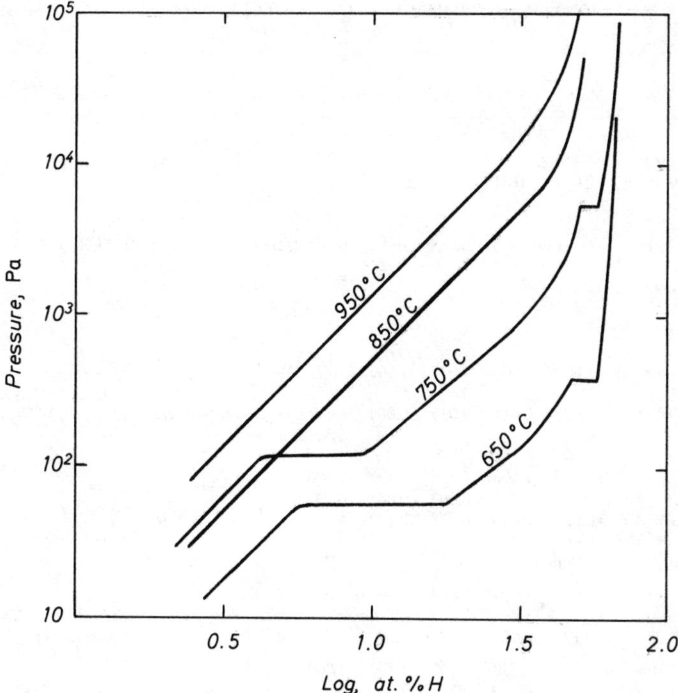

Figure 12.11 *The zirconium–hydrogen system (Private communication from McQuillan based on reference 105. See also Motz,[106] Schwartz and Mallet,[107] Gulbransen and Andrew,[108] Edwards et al.,[109] Mallet and Albrecht,[110] Espagno et al.,[111] Libowitz,[112] La Grange et al.,[113] Slattery,[114] Singh and Gordon Parr,[115] and Katz and Berger[116])*

12.1.4 Solutions of nitrogen

Table 12.5 gives values of dissolved nitrogen concentration in equilibrium with nitrogen at one atmosphere pressure for the metals iron, cobalt, chromium, molybdenum, manganese, nickel, silicon and some of their alloys. The solutions are dilute and the solution process is endothermic, the solubility increasing with temperature.

Table 12.6 gives values for nitrogen concentrations in iron and chromium in equilibrium with nitrides, measured by methods including internal friction and calorimetry.

In the solid metals, the solute atoms are assumed to occupy interstitial sites, only a small proportion of the available sites being filled. If iron is cold-worked, the nitrogen solubility is enhanced by additional solute sites at lattice defects thereby introduced into the metal (*see* references 117 and 118).

Some transition metals dissolve nitrogen exothermically to form concentrated interstitial solid solutions of metallic character analogous to the corresponding hydrogen solutions. Figures 12.12 and 12.13 present information for the systems niobium–nitrogen and tantalum–nitrogen.

Table 12.5 NITROGEN SOLUTIONS IN EQUILIBRIUM WITH GASEOUS NITROGEN AT ATMOSPHERIC PRESSURE

Solvent metal		*Mass % of nitrogen at temperature, T °C*						*References*
Fe	*T*	700	800	900				
(solid α)	*Mass %*	1.5×10^{-3}	2.3×10^{-3}	3.3×10^{-3}				120 also 119
(solid γ)	*T*	900	1 000	1 100	1 200	1 300	1 400	
	Mass %	0.028	0.025	0.024	0.023	0.022	0.021	120, 121, 122, 145

Table 12.5 NITROGEN SOLUTIONS IN EQUILIBRIUM WITH GASEOUS NITROGEN AT ATMOSPHERIC PRESSURE—*continued*

Solvent metal		*Mass % of nitrogen at temperature, T °C*							*References*
Fe									
(solid δ)	*T*	1 400	1 450	1 500	1 535				
	Mass %	0.010 1	0.011 1	0.012 1	0.012 9				121
(liquid)	*T*	1 600							
	Mass %	0.044	(average of 21 independent determinations)						22, 121, 124–147
Fe–Al	*T*	1 600	1 600						
(liquid)	% Al	0.25	0.5						140
	Mass %	0.044	0.044						
Fe–C	*T*	1 600	1 600	1 600	1 600	1 600			
(liquid)	% C	1	2	3	4	5			125, 128, 131, 132,
	Mass %	0.030	0.022	0.015	0.011	0.004	(averages of independent determinations)		139, 140, 144
Fe–Cr	*T*		1 000	1 100	1 200	1 300	1 400		
(solid γ)	*Mass %*	4.76% Cr	0.102	0.079	0.063	0.051	0.034*		146
		8.67% Cr	0.286	0.193	0.138	0.102	—		
		14.10% Cr	0.96	0.48	0.26	—	—		
			* δ phase						
(liquid)	*T*	1 600	1 600	1 600	1 600	1 600	1 600	1 600	
	% Cr	5	10	15	20	40	60	70	127, 128, 133, 136,
	Mass %	0.07	0.12	0.18	0.29	1.00	2.3	3.5	140 also 143
	T	1 700	1 700	1 700	1 700	1 700	1 700	1 700	
	% Cr	10	20	40	50	70	80	90	
	Mass %	0.10	0.22	0.75	1.16	2.6	3.5	4.5	133
Fe–Co	*T*	1 600	1 600	1 600	1 600	1 600	1 600		
(liquid)	% Co	5	10	20	40	60	80		22, 137, 140, 144
	Mass %	0.040	0.035	0.028	0.023	0.013	0.010		
Fe–Cu	*T*	1 600	1 600	1 600	1 600	1 600			
(liquid)	% Cu	2	4	6	8	10			
	Mass %	0.044	0.042	0.042	0.040	0.039			137, 140, 144
Fe–Mn'	*T*		1 050	1 200	1 300				
(solid γ)	*Mass %*	0.43% Mn	0.025	—	—				119
		0.53% Mn	0.024	—	—				
		1.46% Mn	0.025	—	0.020				
		12.98% Mn	0.066	0.046	—				
	T	1 550	1 550	1 550	1 550	1 550	1 550		
(liquid)	% Mn	5	10	20	40	60	80		
	Mass %	0.044	0.052	0.087	0.22	0.43	0.80		141, 142
	T	1 600	1 600	1 600	1 600	1 600			
	% Mn	1	2	5	10	20			
	Mass %	0.046	0.048	0.060	0.074	0.098			133, 140 also 143
Fe–Mo	*T*	1 600	1 600	1 600	1 600	1 600			
(liquid)	% Mo	2	4	6	8	10			135, 137, 140, 144
	Mass %	0.045	0.047	0.051	0.054	0.057			also 143
Fe–Nb	*T*	1 600	1 600	1 600	1 600				
(liquid)	% Nb	2	4	6	8				
	Mass %	0.060	0.085	0.11	0.16				140, also 143
Fe–Ni	*T*		918	999	1 217				
(solid)	*Mass %*	1.01% Ni	0.028 3	0.025 3	0.021 5				
		3.98% Ni	0.025 2	0.021 9	0.018 7				
		8.11% Ni	0.020 5	0.018 6	0.015 6				145 also 143
		15.46% Ni	0.013 2	0.012 5	0.011 1				
		26.8% Ni	0.006 1	0.006 2	0.006 1				
		40.7% Ni	0.001 9	0.002 3	0.002 7				

Table 12.5 NITROGEN SOLUTIONS IN EQUILIBRIUM WITH GASEOUS NITROGEN AT ATMOSPHERIC PRESSURE—*continued*

Solvent metal		*Mass % of nitrogen at temperature, T °C*									*References*
Fe–Ni (liquid)	*T*	1 600	1 600	1 600	1 600	1 600	1 600				137
	% Ni	1	2	5	10	25	50				131, 132, 133, 135,
	Mass %	0.044	0.043	0.039	0.033	0.017	0.006 7				136, 140, 144
											also 143
Fe–O (liquid)	Effect of up to 0.2% O is slight										140
Fe–S (liquid)	Effect of up to 0.3% S is slight										137, 140, 144
Fe–Si (solid α)	*T*		700	800	900	1 000	1 100				
	Mass %	2.83% Si	0.001 3	0.001 6	0.001 9	0.002 1	0.002 4				120
(solid γ)	*T*		1 000	1 050	1 100	1 200	1 300	1 350			
	Mass %	0.20% Si	—	0.024	—	0.022	—	0.020			
		0.58% Si	—	0.022	—	0.020	—	0.018			122, 147
		0.90% Si	0.024	—	0.022	0.021	0.020	—			
		1.26% Si	0.023	—	0.022	0.021	0.019	—			
(liquid)	*T*	1 600	1 600	1 600	1 600	1 600					
	% Si	2	4	6	8	10					126, 140, 144
	Mass %	0.035	0.026	0.020	0.013	0.010					also 143
Fe–Sn (liquid)	*T*	1 600	1 600	1 600	1 600	1 600					
	% Sn	2	4	6	8	10					
	Mass %	0.044	0.042	0.041	0.041	0.040					137, 140
Fe–Ta (liquid)	*T*	1 600	1 600	1 600	1 600	1 600					
	% Ta	2	4	6	8	10					
	Mass %	0.052	0.060	0.072	0.084	0.098					140
Fe–V (liquid)	*T*		1 600	1 700	1 800	1 900					
	Mass %	1% V	0.059	0.059	0.059	0.059					
		2% V	0.074	0.071	0.070	0.067					
		3% V	0.088	0.084	0.081	0.079					135, 140
		5% V	—	0.129	0.120	0.110					
		10% V	—	0.315	0.280	0.237					
Fe–W (liquid)	*T*	1 600	1 600	1 600	1 600	1 600	1 600	1 600			
	% W	2	4	6	8	10	12	14			
	Mass %	0.044	0.044	0.045	0.045	0.046	0.046	0.046			140
Fe–Cr–Ni (liquid)	*T*	1 600	1 600	1 600	1 600	1 600	1 600	1 600			
	% Cr	8.9	13.4	4.0	23.0	33.3	24.7	50.0			
	% Ni	6.5	13.4	29.5	6.0	33.1	50.6	25.0			136
	Mass %	0.083	0.255	0.041	0.316	0.476	0.762	1.35			
		For additional data *see* reference 136									
Fe–Mo–V	*T*	1 700	1 700	1 700	1 700	1 700	1 700	1 700	1 700	1 700	
	% Mo	1	3	5	1	3	5	1	3	5	
	% V	1	1	1	2	2	2	3	3	3	
	Mass %	0.054	0.058	0.063	0.069	0.071	0.077	0.089	0.093	0.099	135
		For additional data *see* reference 135									
Co (solid)	Not detectable for *T* 1 200										149, 150
(liquid)	*T*	1 600									
	Mass %	0.004 7									151
Co alloys (liquid)	*T*	1 600	1 600	1 600	1 600	1 600					
	alloying element	0.8% Al	3.0% Cr	6.0% Cu	1.0% Fe	6.0% Mo					
	Mass %	0.004 4	0.006 4	0.005 3	0.004 6	0.005 3					151
	T	1 600	1 600	1 600	1 600	1 600	1 600				
	alloying element	5.5% Nb	3.5% Ni	1.8% Si	6.0% Ta	1.0% V	6.0% W				
	Mass %	0.009 8	0.003 9	0.002 8	0.007 2	0.005 9	0.005 3				151

Table 12.5 NITROGEN SOLUTIONS IN EQUILIBRIUM WITH GASEOUS NITROGEN AT ATMOSPHERIC PRESSURE—*continued*

Solvent metal		*Mass % of nitrogen at temperature, T °C*							*References*
Co–Fe	*See* Fe–Co								
Cr	*T*	1 600	1 650	1 700	1 725	1 750			
(liquid)	*Mass %*	4.08	3.90	3.84	3.76	3.54			154 also 136, 153
Cr–Ni	*See* Ni–Cr								
Cr–Si	*T*		1 600	1 650	1 700	1 750			
(liquid)	*Mass %*	1.5% Si	3.83	3.68	3.54	3.08			
		7.5% Si	1.98	1.89	1.72	1.68			154
		10.0% Si	0.84	0.74	0.69	0.62			also 153
		20.0% Si	0.33	0.30	0.28	0.26			
Mo	*T*	950	1 000	1 050	1 100	1 150			
(solid)	*Mass %*	0.85	0.56	0.41	0.33	0.26			155
Mn	*T*	1 245	1 300	1 400	1 500	1 600	1 700		
(liquid)	*Mass %*	3.4	2.8	1.9	1.6	1.1	1.0		146, 156
Mn–Fe	*See* Fe–Mn								
Ni	*T*	1 600							22, 136, 137, 144
(liquid)	*Mass %*	$<$0.002 5							
Ni–Cr	*T*	1 600	1 600	1 600	1 600	1 600	1 600	1 600	
(liquid)	% Cr	10	20	30	40	50	60	70	136
	Mass %	0.016	0.068	0.17	0.44	1.0	1.7	2.6	
Ni–Fe	*See* Fe–Ni								
Si	*T*	1 420							
(liquid)	*Mass %*	~0.01							129
U	Some phase relationships discussed in references 171 and 172								

Table 12.6 NITROGEN SOLUTIONS IN IRON AND CHROMIUM IN EQUILIBRIUM WITH NITRIDES

Solvent	*Phase in equilibrium*	*Method of determination*		*Mass % of nitrogen at temperature, T °C*								*Refer-ences*
Pure α Fe	Fe_4N	Calorimetry	*T*	200	240	300	330	400	450	575		
			Mass %	0.008	0.014	0.018	0.026	0.043	0.059	0.097		160
		Invariant pressure	*T*	450	500	550	590					
			Mass %	0.033	—	0.070	0.10					161
				—	0.06	—	—					117
		Internal friction	*T*	250	300	350	400	450	500	575	585*	
			Mass %	0.005	0.010	0.015	0.025	0.035	0.050	0.075	—	162
				—	0.008 4	—	0.025	—	0.055	—	0.095	163
				* eutectoid temperature								
		See also criticism in reference 164 and reply in reference 165										
	Fe_8N	Internal friction	*T*	20	100	150	200	250	300	400		
			Mass %	—	—	0.003 5	0.010	0.020	0.040	—		162
				1.4×10^{-5}	5.2×10^{-4}	—	0.008 8	—	0.055	0.20		163
	N_2 at 1 atm (for comparison)	Internal friction	*T*	500	585	700	800	900				
			Mass %	9.0×10^{-4}	0.001 4	0.002 4	0.003 3	0.004 5				163
				See also Table 12.8								
Fe–2.8% Si	Unidentified nitride	Internal friction	*T*	300	400	500	600	700	800	900	1 000	
			Mass %	0.001 0	0.001 5	0.002 4	0.004 0	0.006 1	0.010	0.014	0.019	166
Cr (solid)	Cr_2N		*T*	1 100	1 200	1 300	1 400					
			Mass %	0.04	0.09	0.14	0.26					152
			P_{Cr_2N} Pa	2×10^2	9×10^2	2.5×10^3	5.8×10^3.					

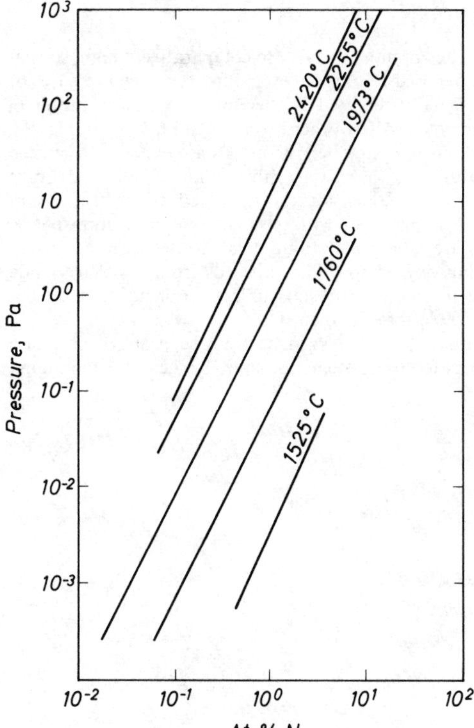

Figure 12.12 *The niobium-nitrogen system (Cost and Wert[168] See also Pemsler[167])*

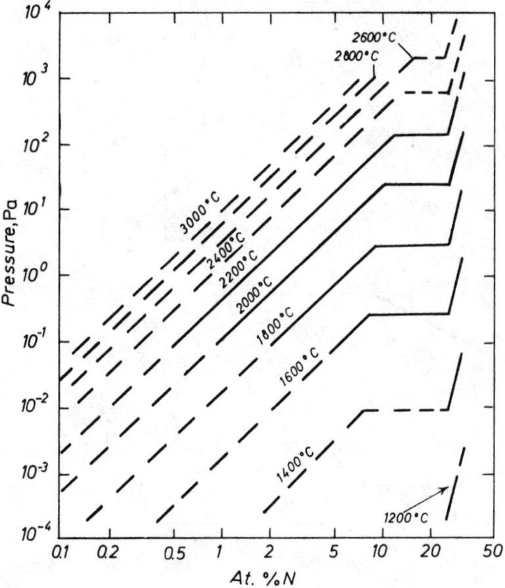

Figure 12.13 *The tantalum–nitrogen system (Gebhardt et al.[169] See also Gebhardt et al.[170] and Pemsler[167])*

12.1.5 Solutions of oxygen

The free energies of formation of the lowest oxides of most metals are comparatively high, so that an oxide film is formed when these metals are exposed to oxygen, except for very low pressures or very high temperatures. The solubility data usually required is therefore the concentration of dissolved oxygen in equilibrium with the oxide phase. A few other metals (Os, Pt, Rh, Au, Hg, Pd, Ru and Ir) form less stable oxides so that no film is present when the metals are exposed to oxygen at atmospheric pressure at elevated temperature. Of these, only silver and palladium dissolve appreciable quantities of oxygen. Table 12.7 gives values for the dissolved concentrations of oxygen in equilibrium with the lowest oxide of the metal or with gaseous oxygen at atmospheric pressure as appropriate. These concentrations are often small and difficult to measure. The usual method of establishing equilibrium by allowing oxygen to diffuse inwards from a surface oxide phase or from the gas phase is liable to lead to erroneous results unless the metal is free from traces of impurities which form oxides more stable than its own oxide.

Some transition metals can dissolve large quantities of oxygen before a separate oxide phase appears. Figures 12.14 and 12.15 give isotherms for the systems niobium–oxygen and tantalum–oxygen.

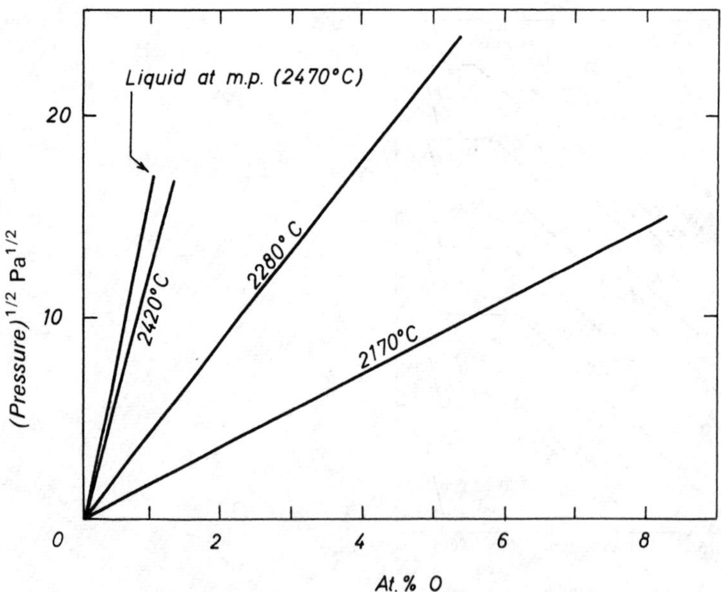

Figure 12.14 *The niobium–oxygen system (Pemsler.[167] See also Elliot[210] and Fromm[208])*

Table 12.7 OXYGEN SOLUTIONS IN EQUILIBRIUM WITH OXIDES OR WITH GASEOUS OXYGEN AT ATMOSPHERIC PRESSURE

Solvent metal	Phase in equilibrium		Mass % of oxygen at temperature, T °C							References
Ag	O_2 (gas at	*T*	300	400	500	600	700	800	900	
(solid)	atmospheric pressure)	*Mass %*	3.0×10^{-5}	1.4×10^{-4}	4.4×10^{-4}	1.07×10^{-3}	2.16×10^{-3}	3.81×10^{-3}	6.14×10^{-3}	174 also 173
(liquid)	O_2 (gas at	*T*	973	1 024	1 075	1 125				
	atmospheric pressure)	*Mass %*	0.305	0.295	0.277	0.264				173
Co	CoO	*T*	600	700	810	875				
(solid α)	(solid)	*Mass %*	0.6×10^{-2}	0.9×10^{-2}	1.6×10^{-2}	2.05×10^{-2}				175
(solid β)	CoO	*T*	875	945	1 000	1 200				
	(solid)	*Mass %*	0.58×10^{-2}	0.7×10^{-2}	0.8×10^{-2}	1.3×10^{-2}				175
(liquid)	CoO	*T*	1 550	1 600	1 650					
	(solid)	*Mass %*	0.13	0.16	0.23					178
Cr	Cr_2O_3	*T*	1 350							
(solid)	(solid)	*Mass %*	approximately 0.03%							179
Cu	Cu_2O	*T*	600	700	800	900	950	1 000	1 050	
(solid)	(solid)	*Mass %*	1.6×10^{-3}	1.7×10^{-3}	2.1×10^{-3}	2.7×10^{-3}	3.4×10^{-3}	4.6×10^{-3}	7.7×10^{-3}	181 also 180, 176, 177
Fe	FeO	*T*	<1 500							
(solid)	(solid)	*Mass %*	<0.009							182, 189
(liquid)	FeO	*T*	1 550	1 600	1 650	1 700				
	(liquid)	*Mass %*	0.18	0.23	0.28	0.34				183, 184, 185, 186, 187, 188, 267, 189, 268
Hf		*See* references 190 and 191								
K	K_2O	*T*	100	150	200	250	300			
(liquid)	(solid)	*Mass %*	0.10	0.17	0.27	0.41	0.60			192
Li	Li_2O	*T*	250	400						
(liquid)	(solid)	*Mass %*	0.0109	0.066						193
Mo		*See* reference 194 for condensed-phase relationships								194
Na	Na_2O	*T*	100	200	300	400	500	550		
(liquid)	(solid)	*Mass %*	0.002	0.005	0.010	0.018	0.047	0.08		192

Table 12.7 OXYGEN SOLUTIONS IN EQUILIBRIUM WITH OXIDES OR WITH GASEOUS OXYGEN AT ATMOSPHERIC PRESSURE—*continued*

Solvent metal	*Phase in equilibrium*	*Mass % of oxygen at temperature, T °C*						*References*
Nb		*See* Figure 12.14 and references 167, 205, 206, 207 and 210						
Ni (solid)	NiO (solid)	*T*	600	800	1 000	1 200		
		Mass %	0.020	0.019	0.014	0.012		195
(liquid)	NiO (liquid)	*T*	1 450	1 500	1 550	1 600	1 650	
		Mass %	0.28	0.46	0.72	1.10	1.66	199
See also references 178, 197 and 198 for activities of O in Ni–Fe and Ni–Co alloys								also 196, 197, 198
Pb (liquid)	PbO (solid)	*T*	350	450	550			
		Mass %	5.4×10^{-4}	8.6×10^{-4}	13.2×10^{-4}			200
Pd (solid)	O_2 (gas at atmospheric pressure)	*T*	1 200					
		Mass %	<0.05					201
Pu		*See* reference 209 for plutonium–oxygen phase diagram						
Rh (solid)	O_2 (gas at atmosphere pressure)	Slight solubility						201
Si (solid)	SiO_2 (solid)	*T*	1 000	1 100	1 200	1 300	1 412	
		Mass %	2.8×10^{-4}	5.3×10^{-4}	9.1×10^{-4}	1.5×10^{-3}	2.2×10^{-3}	202, 203
Sn (liquid)	SnO_2 (solid)	*T*	536	600	700	751		
		Mass %	0.000 18	0.000 55	0.002 8	0.004 9		204
Ta		*See* Figure 12.15 and references 208 and 210–220						
Ti		*See* references 221–229						
U		*See* references 230–232						
V		*See* references 233–236						
Y		*See* reference 237						
Zr		*See* references 238–242						

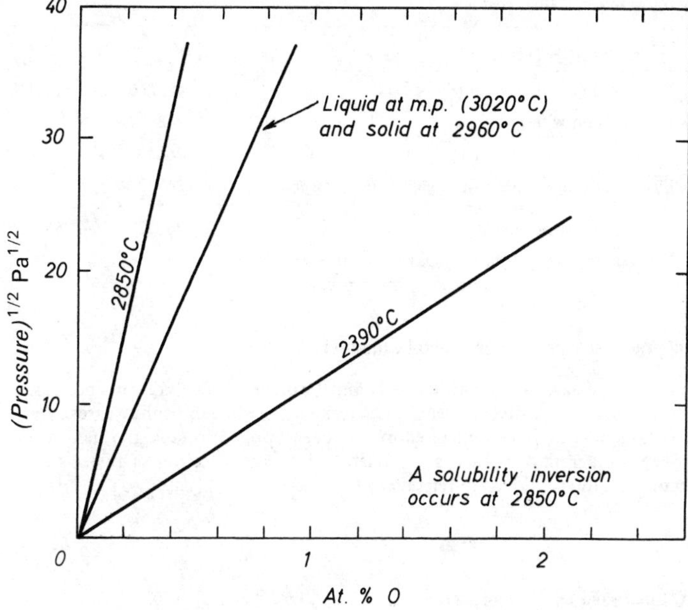

Figure 12.15 *The tantalum–oxygen system (Pemsler.[167] See also Gebhardt et al.,[205,212–218] Powers and Doyle,[219] Marcotte and Larsen,[206] Meussner and Carpenter[207] and Fromm[208])*

12.1.6 Solutions of the noble gases

The solubilities in metals of the noble gases in Group 8 of the Periodic Table are so small that the quantities which dissolve by equilibrating metals with the pure gases are difficult to detect. For example, Kubaschewski's theoretical argument[1] predicts that at 600 °C, only 3.5×10^{-10} atomic fraction of xenon will dissolve in liquid bismuth equilibrated with xenon gas at one atmosphere pressure. However, significant quantities of the noble gases can be inserted into metal lattices by very energetic processes such as nuclear fission or bombardment with accelerated ions. Examples of solutions produced in this manner are given in Table 12.8. Fuller information is given in a review by Blackburn.[14]

Table 12.8 SOLUTIONS OF NOBLE GASES

Solvent	*Solutes*	*Method of introducing solute*	*References*
Ag	Ar, Kr, Xe	Electric discharge on metal in gas at low pressure	243, 244, 246, 247, 249
	Ar	Injection of accelerated ions	245
	Kr	Equilibration with gas phase	248
Al	Ar	Injection of accelerated ions	245
	He	Injection of cyclotron-accelerated α-particles	250, 251, 252
Al–Li	He	Radioactive decay of neutron-irradiation products	253, 254
Al–U	Kr	Radioactive decay of neutron-irradiation products	255
Au	Ar	Injection of accelerated ions	245
Be	He	Radioactive decay of neutron-irradiation products	256, 257
	He	Injection of cyclotron-accelerated α-particles	258
Cu	He	Injection of cyclotron-accelerated α-particles	250, 259
Ge	He	Equilibration with gas phase	48, 260
Pb	Ar	Injection of accelerated ions	245
	K	Equilibration with gas phase	248

Table 12.8 SOLUTIONS OF NOBLE GASES—*continued*

Solvent	*Solutes*	*Method of introducing solute*	*References*
Si	He	Equilibration with gas phase	48, 260
Sn	Kr	Equilibration with gas phase	248
Ti	He		261
U	Ar, Kr, Xe	Electric discharge on metal in gas at low pressure	243, 244
	Kr	Injection of accelerated ions	245
	Kr, Xe	Nuclear fission of solvent	262, 263, 264, 265, 266
Zr	Ar, Kr, Xe	Electric discharge on metal in gas at low pressure	243, 244

12.1.7 Theoretical and practical aspects of gas–metal equilibria

The equilibria between metals and gases are of a wide variety and the practical effects of absorbed gases in metals during industrial processes are diverse, usually deleterious and often difficult to assess. As a result, a vast amount of practical and theoretical effort has been applied in studying gas–metal interactions using numerous different approaches, as illustrated by the selection of reviews and papers of general or theoretical interest given in references 1–15.

REFERENCES

1. O. Kubaschewski, A. Cibula and D. C. Moore, 'Gases and Metals', Iliffe, London, 1970.
2. J. D. Fast, 'Interaction of Metals and Gases', Philips Technical Library, Eindhoven, 1965.
3. R. M. Barrer, *Discuss. Faraday Soc.*, 1948, No. 4, 68.
4. O. Kubaschewski, *Z. Electrochem.*, 1938, **44**/2, 152.
5. J. W. McBain, 'Sorption of Gases by Solids', London, 1932.
6. A. Nikuradse and R. Ulbricht, 'Das Zweistoffsystem Gas-Metal', Munich, 1950.
7. A. Sieverts, *Z. Metallk.*, 1929, **21**, 37.
8. E. Fromm and E. Gebhardt, 'Gases and Carbon in Metals'. Springer-Verlag, Berlin, 1976.
9. C. R. Cupp, *Prog. Metal Phys.*, 1954, **4**, 105.
10. D. P. Smith, 'Hydrogen in Metals', Chicago, 1948.
11. D. E. J. Talbot, 'Effects of Hydrogen in Aluminium, Magnesium, Copper and their Alloys', *International Met. Reviews*, 1975, **20**, 166.
12. P. Cotterill, *Prog. Mater. Sci.*, 1961, **9**, 205.
13. R. Fowler and C. J. Smithells, *Proc. R. Soc.*, 1937, **160**, 37.
14. R. Blackburn, 'Inert Gases in Metals', *Met. Reviews*, 1966, **11**, 159.
15. O. Kubaschewski and B. E. Hopkins, 'Oxidation of Metals and Alloys', 2nd Edition, Butterworth, London, 1962.
16. E. W. Steacie and F. M. G. Johnson, *Proc. R. Soc.*, 1928, A, **117**, 662.
17. C. E. Ransley and H. Neufeld, *J. Inst. Metals*, 1948, **74**, 599.
18. W. Eichenauer, K. Hattenbach and A. Pebler, *Z. Metallk.*, 1961, **52**, 682.
19. W. R. Opie and N. J. Grant, *Trans. AIMME*, 1950, **188**, 1237.
20. A. Sieverts and H. Hagen, *Z. phys. Chem.*, 1934, **169**, 237.
21. H. Schenck and K. W. Lange, *Arch. Eisenhütt Wes.*, 1966, **37**, 739.
22. T. Busch and R. A. Dodd, *Trans. Met. Soc. AIMME*, 1960, **218**, 488.
23. E. Martin, *Arch. Eisenhütt Wes.*, 1929/30, **3**, 407.
24. L. Luckmeyer-Hasse and H. Schenk, *Arch. Eisenhütt Wes.*, 1932–33, **6**, 209.
25. W. Eichenauer and A. Pebbler, *Z. Metallk.*, 1957, **48**, 373.
26. A. Sieverts, *Z. phys. Chem.*, 1911, **77**, 611.
27. P. Röntgen and F. Möller, *Metallwirt., Metallwiss., Metalltech.*, 1934, **13**, 81, 97.
28. M. B. Bever and C. F. Floe, *Trans. AIMME*, 1944, **156**, 149.
29. R. Eborall and A. J. Swain, *J. Inst. Metals*, 1952–53, **81**, 497.
30. W. Eichenauer, H. Kunzig and A. Pebler, *Z. Metallk.*, 1958, **49**, 220.
31. A. Sieverts, G. Zapf and H. Moritz, *Z. phys. Chem.*, 1938, **183**, 19.
32. M. H. Armbruster, *J. Amer. chem. Soc.*, 1943, **65**, 1043.
33. F. de Kazinczy and O. Lindberg, *Jernkont. Annlr*, 1960, **144**, 288.
34. H. Schenck and H. Wünsch, *Arch. Eisenhütt Wes.*, 1961, **32**, 779.
35. H. Liang, M. B. Bever and C. F. Floe, *Trans. AIMME*, 1946, **167**, 395.
36. M. M. Karnaukhov and A. N. Morazov, *Izv. Akad. Nauk*, (Otdelenie Tekh. Nauk), December 1948, 1845.
37. D. J. Carney, J. Chipman and N. J. Grant, *Trans. AIMME*, 1950, **188**, 404.
38. H. Winterhager, *Alumin.-Arch.*, 1938, **12**, 7.
39. R. S. Busk and E. G. Bobalek, *Trans. AIMME*, 1947, **171**, 261.
40. F. Sauerwald, *Z. anorg. allg. Chem.*, 1949, **258**, 27.
41. J. Koenman and A. G. Metcalf, *Trans. Amer. Soc. Metals*, 1959, **51**, 1072.

42. A. Sieverts and H. Moritz, *Z. phys. Chem.*, 1938, **180/4**, 249.
43. E. V. Potter and H. C. Lukens, *Trans. AIMME*, 1947, **171**, 401.
44. W. R. Opie and N. J. Grant, *Trans. AIMME*, 1951, **191**, 244.
45. W. Mannchen and M. Bauman, *Z. Metallk.*, 1955, **9**, 686.
46. W. Hofmann and J. Maatsch, *Z. Metallk.*, 1956, **47**, 89.
47. A. Sieverts, *Ber. dt. chem. Ges.*, 1912, **45**, 221.
48. A. van Wieringen and N. Warmoltz, *Physica*, 1956, **22**, 849.
49. D. T. Peterson and M. Indig, *J. Am. chem. Soc.*, 1960, **82**, 5645.
50. W. D. Treadwell and J. Stecher, *Helv. chim. Acta*, 1953, **36**, 1820.
51. W. C. Johnson, M. F. Stubbs, A. E. Sidwell and A. Pechukas, *J. Am. chem. Soc.*, 1939, **61**, 318.
52. D. T. Peterson and V. G. Fattore, *J. phys. Chem.*, 1961, **65**(11), 2062.
53. J. F. Stampfer, C. E. Holley and J. F. Shuttle, *J. Am. chem. Soc.*, 1960, **82**, 3504.
54. J. A. Kenneley, J. W. Varwig and H. W. Myers, *J. phys. Chem.*, 1960, **64**(5), 703.
55. C. C. Addison, R. J. Pulham and R. J. Roy, *J. chem. Soc.*, 1965, 116.
56. D. D. Williams, J. A. Grand and R. R. Miller, *J. phys. Chem.*, 1957, **61** 379.
57. D. T. Peterson and R. P. Colburn, *J. phys. Chem.*, 1966, **70**, 468.
58. M. D. Banus, J. J. McSharry and E. A. Sullivan, *J. Am. chem. Soc.*, 1955, **77**, 2007.
59. T. Bagshaw and A. Mitchell, *J. Iron Steel Inst.*, 1967, **205**, 769.
60. D. J. Carney, J. Chipman and N. J. Grant, *Trans. AIMME*, 1950, **207**, 597.
61. M. Weinstein and J. F. Elliot, *Trans. Met. Soc. AIMME*, 1963, **227**, 382.
62. M. W. Mallet and M. J. Trzeciak, *Trans. Am. Soc. Metals*, 1958, **50**, 981.
63. H. C. Mattrow, *J. phys. Chem.*, 1955, **59**, 93.
64. R. N. R. Mulford and C. E. Holley, *J. phys. Chem.*, 1955, **59**, 1222.
65. A. Sieverts and G. Muller-Goldegg, *Z. anorg. allg. Chem.*, 1923, **131**, 65.
66. A. Sieverts and E. Roell, *Z. anorg. allg. Chem.*, 1925, **146**, 149.
67. A. Sieverts and A. Gotta, *Z. anorg. allg. Chem.*, 1928, **172**, 1.
68. E. G. Ivanov, A. Ya. Stomakhin, G. M. Medveda and A. F. Filippov, *Chernaya Metallurgiya*, 1966, **5**, 69.
69. R. K. Edwards and E. Veleckis, *J. phys. Chem.*, 1962, **66**, 1657.
70. L. Espagno, P. Azou and P. Bastien, *C. r. hebd. Séanc. Acad. Sci., Paris*, 1960, **250**, 4352.
71. L. Espagno, P. Azou and P. Bastien, *Mém. scient. Rev. Metall.*, 1962, **59**, 182.
72. R. J. Walter and W. T. Chandler, *Trans. AIMME*, 1965, **233**, 762.
73. W. M. Albrecht, W. D. Goode and M. W. Mallet, *J. electrochem. Soc.*, 1959, **106**, 981.
74. W. M. Albrecht, M. W. Mallet and W. D. Goode, *J. electrochem. Soc.*, 1958, **105**, 219.
75. S. Komjathy, *J. less-common Metals*, 1960, **2**, 466.
76. A. Sieverts and E. Roell, *Z. anorg. allg. Chem.*, 1926, **150**, 261.
77. P. L. Levine and K. E. Weal, *Trans. Faraday Soc.*, 1960, **56**, 357.
78. L. J. Gillespie and F. P. Hall, *J. Am. chem. Soc.*, 1926, **48**, 1207.
79. L. J. Gillespie and L. S. Galstaun, *J. Am. chem. Soc.*, 1936, **58**, 2565.
80. T. S. Perminov, A. A. Orlov and A. N. Frumkin, *Dokl. Akad. Nauk, SSSR*, 1952, **84**, 749.
81. D. H. Everett and P. Nordon, *Proc. R. Soc.*, 1960A, **259**, 341.
82. T. B. Flanagan, *J. phys. Chem.*, 1961, **65**(2), 280.
83. I. E. Nakhutin and E. I. Sutyagina, *Fizica Metall.*, 1959, **7**, 459.
84. P. Nitacek and J. G. Aston, *J. Am. chem. Soc.*, 1963, **85**(2), 137.
85. A. W. Carson, T. B. Flanagan and F. A. Lewis, *Trans. Faraday Soc.*, 1960, **56**, 1332 and 371.
86. R. A. Karpova and I. P. Tverdovsky, *Zh. fiz. Khim.*, 1959, **33**, 1393.
87. Zh. L. Vert, I. P. Tverdovsky and I. A. Mosevich, *Zh. fiz Khim.*, 1965, **39**, 1061.
88. A. Maeland and T. B. Flanagan, *Platin. Metals Rev.*, 1966, **10**, 20.
89. R. N. R. Mulford and G. Sturdy, *J. Am. chem. Soc.*, 1955, **77**, 3449.
90. M. L. Lieberman and P. G. Wahlbeck, *J. phys. Chem.*, 1965, **69**, 3973.
91. J. F. Stampfer, U.S. Atomic Energy Commission Rep., 1966 (LA-3473).
92. M. W. Mallet and B. G. Khoehl, *J. electrochem. Soc.*, 1962, **109**, 611 and 968.
93. P. Kofstad. W. E. Wallace and L. J. Hyvönen, *J. Am. chem. Soc.*, 1959, **81**, 5015.
94. B. Pedersen, T. Krogdahl and O. E. Stokkeland, *J. chem. Phys.*, 1969, **42**, 72.
95. M. W. Mallet and I. E. Campbell, *J. Am. chem. Soc.*, 1951, **73**, 4850.
96. D. T. Peterson and D. G. Westlake, *Trans. AIMME*, 1959, **215**, 445.
97. A. D. McQuillan, *Proc. R. Soc.*, 1951, **204**, 309.
98. G. A. Lenning, C. M. Craighead and R. I. Jaffee, *Trans. AIMME*, 1954, **200**, 367.
99. A. D. McQuillan, *J. Inst. Metals*, 1951, **79**, 73.
100. G. V. Samsonov, and M. M. Antonova, *Ukr. khim. Zh.*, 1966, **32**, 555.
101. B. S. Krylov, *Izvest. Akad. Nauk, SSSR Metally*, 1966, **2**, 144.
102. P. Kofstad and W. E. Wallace, *J. Am. chem. Soc.*, 1959, **81**, 5019.
103. G. Brauer and W. D. Schnell, *J. less-common Metals*, 1964, **6**, 326.
104. A. J. Maeland, *J. phys. Chem.*, 1964, **68**, 2197.
105. C. E. Ells and A. D. McQuillan, *J. Inst. Metals*, 1956–57, **85**, 89.
106. J. Motz, *Z. Metallk.*, 1962, **53**, 770.
107. C. M. Schwartz and M. W. Mallet, *Trans. Am. Soc. Metals*, 1954, **46**, 640.
108. E. A. Gulbransen and K. F. Andrew, *J. electrochem. Soc.*, 1954, **101**, 474, and *J. Metals*, 1955, **7**, 136.
109. R. F. Edwards, P. Leversque and D. Cubicciotti, *J. Am. chem. Soc.*, 1955, **77**, 1307.
110. M. W. Mallet and W. M. Albrecht, *J. electrochem. Soc.*, 1957, **104**, 142.
111. L. Espagno, P. Azou and P. Bastien, *Mém. scient. Rev. Metall.*, 1960, **57**, 254.

112. G. G. Libowitz, *J. nucl. Mater.*, 1962, **5**, 228.
113. L. D. La Grange, L. J. Dijkstra, J. M. Dixon and U. Merten, *J. phys. Chem.*, 1959, **63**, 2035.
114. G. F. Slattery, *J. Inst. Metals*, 1967, **95**, 43.
115. K. P. Singh and J. Gordon Parr, *Trans. Faraday Soc.*, 1963, **59**, 2248.
116. O. M. Katz and J. A. Berger, *Trans. Met. Soc. AIMME*, 1965, **233**, 1005 and 1014.
117. H. A. Wriedt and L. S. Darken, *Trans. Met. Soc. AIMME*, 1965, **233**, 111.
118. H. A. Wriedt and L. S. Darken, *ibid.*, 1965, **233**, 122.
119. A. Sieverts and G. Zapf, *Z. phys. Chem.*, 1935, **174**, 359.
120. N. S. Corney and E. T. Turkdogan, *J. Iron Steel Inst.*, 1955, **180**, 344.
121. A. Sieverts, G. Zapf and H. Moritz, *Z. phys. Chem.*, 1938, **183**, 19.
122. L. S. Darken, R. P. Smith and C. W. Filer, *Trans. AIMME*, 1951, **191**, 1174.
123. I. N. Milinskaya and I. A. Tomilin, *Dokl. Akad. Nauk, SSSR*, 1967, **174**, 135.
124. J. Chipman and D. Murphy, *Trans. AIMME*, 1935, **116**, 179.
125. L. Eklund, *Jernkont. Annlr.*, 1939, **123**, 545.
126. J. C. Vaughan and J. Chipman, *Trans. AIMME*, 1940, **140**, 224.
127. R. M. Brick and J. A. Creevy, *Metals Tech.*, *AIMME*, Tech. Pub. No. 1165, April 10; 1940.
128. T. Kootz, *Arch. EisenhüttWes.*, 1941, **15**, 77.
129. C. R. Taylor and J. Chipman, *Trans. AIMME*, 1943, **154**, 228.
130. W. M. Karnaukoy and A. M. Marozov, *Bull. Acad. Sci. URSS Classe sci. tech.*, 1947, **735**, Brutcher Transl., no. 2029.
131. T. Saito, *Sci. Rep. Res. Insts., Tôhoku Univ.*, Ser. A., 1949, **1**, 411.
132. T. Saito, *ibid.*, 419.
133. H. Wentrup and O. Reif, *Arch. EisenhüttWes.*, 1949, **20**, 359.
134. Y. Kasamatu and S. Matóba, *Technology Rep., Tôhoku Univ.*, 1957, **22**, No. 1.
135. V. Kashyap and N. Parlee, *Trans. AIMME*, 1958, **212**, 86.
136. J. Humbert and J. F. Elliot, *Trans. Met. Soc. AIMME*, 1960, **218**, 1076.
137. H. Schenck, M. Frohberg and H. Graf, *Arch. EisenhüttWes.*, 1958, **29**, 673.
138. V. P. Fedotov and A. M. Samarin, *Dokl. Akad. Nauk, SSSR*, 1958, **122**, 597.
139. S. Maekawa and Y. Nakagawa, Tetsu-to-Hagané, *Abstr.*, 1959, **45**, 255.
140. R. D. Pehlke and J. F. Elliott, *Trans. Met. Soc. AIMME*, 1960, **218**, 1088.
141. S. Z. Beer, *Trans. Met. Soc. AIMME*, 1961, **221**, 2.
142. R. A. Dodd and N. A. Gokcen, *Trans. Met. Soc. AIMME*, 1961, **221**, 233.
143. P. H. Turnock and R. D. Pehlke, *Trans. Met. Soc. AIMME*, 1966, **236**, 1540.
144. H. Schenck, M. Frohberg and H. Graf, *Arch. EisenhüttWes.*, 1959, **30**, 533.
145. H. A. Wriedt and O. D. Gonzalez, *Trans. Met. Soc. AIMME*, 1961, **221**, 532.
146. E. T. Turkdogan and S. Ignatowicz, *J. Iron Steel Inst.*, 1958, **188**, 242.
147. E. T. Turkdogan and S. Ignatowicz, *J. Iron Steel Inst.*, 1957, **185**, 200.
148. E. T. Turkdogan, S. Ignatowicz and J. Pearson, *J. Iron Steel Inst.*, 1955, **181**, 227.
149. A. Sieverts and H. Hagen, *Z. phys. Chem.*, 1934, **169**, 337.
150. R. Juza and W. Sachsze, *Z. anorg. allg. Chem.*, 1945, **253**, 95.
151. R. G. Blossey and R. D. Pehlke, *Trans. Met. Soc. AIMME*, 1966, **236**, 28.
152. A. U. Seybolt, and R. A. Oriani, *Trans. AIMME*, 1956, **206**, 556.
153. V. M. Berezhiani and B. M. Mirianashvili, *Trudy Inst. Metall., Tbilisi*, 1965, **14**, 163.
154. V. S. Mozgovoy and A. M. Samarin, *Dokl. Akad. Nauk, SSSR*, 1950, **74**, 729.
155. A. Sieverts and H. Brunig, *Arch. EisenhüttWes.*, 1933, **7**, 641.
156. N. A. Gokcen, *Trans. Met. Soc. AIMME*, 1961, **221**, 200.
157. F. Lihl, P. Ettmayer and A. Kutzelnigg, *Z. Metallk.*, 1962, **53**, 715.
158. V. P. Perepelkin, *Chernaya Metallurgia*, 1966, **3**, 88.
159. W. Kaiser and C. D. Thurmond, *J. appl. Phys.*, 1959, **30**, 427.
160. G. Borelius, S. Berglund and O. Avsan, *Ark. Fys.*, 1950, **2**, 551.
161. V. G. Pararjpe, M. Cohen, M. B. Bever and C. F. Floe, *Trans. AIMME*, 1950, **188**, 261.
162. L. J. Dijkstra, *Trans. AIMME*, 1949, **185**, 252.
163. J. D. Fast and M. B. Verrijp, *J. Iron Steel Inst.*, 1955, **180**, 337.
164. H. U. Aström and G. Borelius, *Acta Met.*, 1954, **2**, 547.
165. J. D. Fast and M. B. Verrijp, *Acta Met.*, 1955, **3**, 203.
166. D. A. Leak, W. R. Thomas and G. M. Leak, *Acta Met.*, 1956, **3**, 501.
167. J. P. Pemsler, *J. electrochem. Soc.*, 1961, **108**, 744.
168. J. R. Cost and C. A. Wert, *Acta Met.*, 1963, **11**, 231.
169. E. Gebhardt, H. D. Seghezzi and E. Fromm, *Z. Metallk.*, 1961, **52**, 464.
170. E. Gebhardt, H. D. Seghezzi and W. Dürrschnabel, *Z. Metallk.*, 1958, **49**, 577.
171. F. Anselin, *J. nucl. Mater.*, 1963, **10**, 301.
172. R. Benz and M. G. Bowman, *J. Am. chem. Soc.*, 1966, **88**, 264.
173. E. W. R. Steacie and F. M. G. Johnson, *Proc. R. Soc.*, 1926, **A112**, 542.
174. W. Eichenauer and G. Muller, *Z. Metallk.*, 1962, **53**, 321.
175. A. U. Seybolt and C. H. Mathewson, *Trans. AIMME*, 1935, **117**, 156.
176. N. G. Schmaal and E. Minzl, *Z. phys. Chem.*, 1965, **314**, 142.
177. W. Hofmann and M. Klein, *Z. Metallk.*, 1966, **57**, 385.
178. V. V. Averin, A. Yu. Polyakov and A. M. Samarin, *Izvest. Akad. Nauk, SSSR (Tekhn.)*, 1957, **8**, 120.
179. D. Caplan and A. A. Burr, *Trans. AIMME.*, 1955, **203**, 1052.
180. F. N. Rhines and C. H. Mathewson, *Trans. AIMME*, 1934, **III**, 337.

181. A. Phillips and E. N. Skinner, *Trans. AIMME*, 1941, **143**, 301.
182. R. Sifferlen, *C. r. hebd. Séanc. Acad. Sci, Paris*, 1957, **244**, 1192.
183. L. S. Darken and R. W. Gurry, *J. Am. chem. Soc.*, 1947, **68**, 798.
184. H. Schenke and E. Steinmetz, *Arch. EisenhüttWes.*, 1967, **38**, 813.
185. T. Fuwa and J. Chipman, *Trans. Met. Soc. AIMME*, 1960, **218**, 887.
186. T. P. Floridis and J. Chipman, *Trans. AIMME*, 1958, **212**, 549.
187. E. S. Tankins, N. A. Gokcen and G. R. Bolton, *Trans. Met. Soc. AIMME*, 1964, **230**, 820.
188. J. Skala, M. Kase and M. Mandl, *Hutn. Listy*, 1962, **17**, 841.
189. J. H. Swisher and E. T. Turkdogan, *Trans. Met. Soc. AIMME*, 1967, **239**, 427.
190. R. F. Dogmagala and R. Ruh, *Trans. Q. Am. Soc. Metals*, 1965, **58**, 164.
191. E. Rudy and P. Stecher, *J. less-common Metals*, 1963, **1**, 78.
192. D. D. Williams, J. A. Grand and R. R. Miller, *J. phys. Chem.*, 1959, **63**, 68.
193. E. E. Hoffman, *Amer. Soc. Test. Mat. Symposium on Newer Materials*, 1959, **1960**, 195.
194. B. Phillips and L. L. Y. Chang, *Trans. Met. Soc. AIMME*, 1965, **233**, 1433.
195. A. U. Seybolt, Dissertation Yale University, 1936.
196. P. D. Merica and R. G. Waltenberg, *Trans. AIMME*, 1925, **71**, 715.
197. H. A. Wriedt and J. Chipman, *Trans. AIMME*, 1955, **203**, 477.
198. A. M. Samarin and V. P. Fedotov, *Izv. Akad Nauk, SSSR* (*Tekhn.*), 1956, **6**, 119.
199. J. E. Bowers, *J. Inst. Metals*, 1961–62, **90**, 321.
200. K. W. Groshelm-Krisko, W. Hoffmann and H. Hanemann, *Z. Metallk.*, 1944, **36**, 91.
201. E. Raub and N. Plate, *Z. Metallk.*, 1957, **48**, 529.
202. H. J. Hrostowski and R. H. Kaiser, *Phys. Chem. Solids*, 1959, **9**, 214.
203. W. Kaiser and P. H. Keck, *J. appl. Phys.*, 1957, **28**, 1427.
204. T. N. Belford and C. B. Alcock, *Trans. Faraday Soc.*, 1965, **61**, 443.
205. E. Gebhardt and R. Rothenbacher, *Z. Metallk.*, 1963, **54**, 623.
206. V. C. Marcotte and W. L. Larsen, *J. less-common Metals*, 1966, **4**, 229.
207. R. A. Meussner and C. D. Carpenter, *Corros. Sci.*, 1967, **2**, 115.
208. F. Fromm, *Z. Metallk.*, 1966, **57**, 540.
209. E. R. Gardner, T. L. Markins and R. S. Street, *J. inorg. nucl. Chem.*, 1965, **27**, 541.
210. R. P. Elliot, Amer. Soc. Metals Reprint, 1959, (143).
211. E. Gebhardt and H. D. Seghezzi, *Z. Metallk.*, 1959, **50**, 521.
212. E. Gebhardt and H. D. Seghezzi, *ibid.*, 1950, **50**, 248.
213. E. Gebhardt and H. D. Seghezzi, *ibid.*, 1955, **46**, 560.
214. E. Gebhardt and H. D. Seghezzi, *ibid.*, 1957, **48**, 430.
215. E. Gebhardt and H. D. Seghezzi, *ibid.*, 1957, **48**, 503.
216. E. Gebhardt and H. D. Seghezzi, *ibid.*, 1957, **48**, 559.
217. E. Gebhardt and Preisendanz, *Plansee Proc.*, 1955, 254.
218. E. Gebhardt and H. D. Seghezzi, *ibid.*, 1959, 280.
219. R. J. Powers and M. V. Doyle, *Trans. Met. Soc. AIMME*, 1959, **215**, 655.
220. M. Hoch and D. B. Bulrymowicz, *Trans. Met. Soc. AIMME*, 1964, **230**, 186.
221. I. Z. Kornilov and V. V. Glazova, *Izv. Akad. Nauk, SSSR Metally*, 1965, **1**, 189.
222. V. V. Glazova, *Dokl. Akad. Nauk, SSSR*, 1965, **164**, 567.
223. B. A. Bolachev, V. A. Livanov and A. A. Bukhanova, *Sov. J. non-ferrous Metals*, 1966, **3**, 94.
224. P. Kofstad, P. B. Anderson and O. J. Krudtaa, *J. less-common Metals*, 1961, **3**, 89.
225. F. Ehrlich, *Z. anorg. Chem.*, 1941, **24**, 53.
226. M. K. McQuillan, *Corros. Anti-Corrosion*, 1962, **10**, 361.
227. T. Hurlen, *J. inst. Metals*, 1960–61, **89**, 128.
228. M. T. Hepworth and W. B. Sample, *Trans. Met. Soc. AIMME*, 1962, **224**, 875.
229. M. T. Hepworth and R. Schuhmann, *Trans. Met. Soc. AIMME*, 1962, **224**, 928.
230. J. Besson, P. L. Blum and J. P. Morlevat, *C. r. hebd. Séanc. Acad. Sci., Paris*, 1965, **260**, 3390.
231. R. A. Smith, US Atomic Energy Commission Rep., 1966 (BMZ–17SS).
232. A. E. Martin and R. K. Edwards, *J. phys. Chem.*, 1965, **69**, 1788.
233. N. P. Allen, O. Kubaschewski and O. V. Goldbeck, *J. electrochem. Soc.*, 1951, **98**, 417.
234. W. Rostoker and A. S. Yamamoto, *Trans. Am. Soc. Metals*, 1955, **47**, 1002.
235. M. A. Gurevich and B. F. Ormont, *Zh. neorg. Khim.*, 1957, **2**, 1566, 2581.
236. M. A. Gurevich and B. F. Ormont, *ibid.*, 1958, **3**, 403.
237. R. C. Tucker, E. D. Gibson and O. N. Carlson, *International Symposium on Compounds of Interest in Nuclear Reactor Technology*, Colorado, USA, 1964, and US Atomic Energy Commission Rep., 1964 (IS–812).
238. H. J. de Boer and J. D. Fast, *Recl. Trav. chim. Pays-Bas Belg.*, 1936, **55**, 449.
239. O. Kubaschewski and W. A. Dench, *J. Inst. Metals*, 1955–56, **84**, 440.
240. B. Holmberg and T. Dagerhamn, *Acta Chem. Scand.*, 1961, **15**, 915.
241. E. Gebhardt, H. D. Seghezzi and W. Durrschnabel, *J. nucl. Mater.*, 1961, **4**, 241, 255 and 269.
242. V. C. Marcotte, W. L. Larsen and D. E. Williams, *J. less-common Metals*, 1964, **5**, 373.
243. G. Brebec, V. Levy and Y. Adda, *C. r. hebd. Séanc. Acad. Sci., Paris*, 1961, **252**, 722.
244. V. Levy *et al.*, *C. r. hebd. Séanc. Acad. Sci., Paris*, 1961, **252**, 876.
245. C. W. Tucker and F. J. Norton, *J. nucl. Mater.*, 1960, **2**, 329.
246. A. D. Le Claire and A. H. Rowe, *Revue Metall., Paris*, 1955, **52**, 94.
247. J. M. Tobin, *Acta Met.*, 1957, **5**, 398.
248. G. W. Johnson and R. Shuttleworth, *Phil. Mag.*, 1959, **4**, 957.

249. J. M. Tobin, *Acta Met.*, 1959, **7**, 701.
250. R. S. Barnes, *Phil. Mag.*, 1960, **5**, 635.
251. C. E. Ells and C. E. Evans, Atomic Energy of Canada Ltd., Rep. 1959 (CR Met–863).
252. C. E. Ells, *J. nucl. Mater.*, 1962, **5**, 147.
253. G. T. Murray, *J. appl. Phys.*, 1961, **32**, 1045.
254. D. W. Lillie, *Trans. Met. Soc. A.I.M.M.E.*, 1960, **218**, 270.
255. M. B. Reynolds, *Nucl. Sci. Engng.*, 1958, **3**, 428.
256. C. E. Ells and E. C. W. Perryman, *J. nucl. Mater.*, 1959, **1**, 73.
257. V. Levy, *Bull. Inform. Sci. Tech.*, 1962, **62**, 56.
258. R. S. Barnes and G. B. Redding, *J. nucl. Energy*, A, 1959, **10**, 32.
259. R. S. Barnes, G. B. Redding and A. H. Cottrell, *Phil. Mag.*, 1958, **3**, 97.
260. A. van Wieringen, Symposium: 'La Diffusion dans les Metaux', 1957, 107.
261. A. M. Rodin and V. V. Surenyants, *Fizika Metall.*, 1960, **10**, 216.
262. M. B. Reynolds, *Nucl. Sci. Engng.*, 1956, **1**, 374.
263. J. F. Walker, UK Atomic Energy Authority Publ. 1959 (IGR–TN/W–1046).
264. F. J. Norton, *J. nucl. Mater.*, 1960, **2**, 350.
265. D. L. Gray, US Atomic Energy Commission Rep. 1960 (HW–62639).
266. N. R. Chellew and R. K. Steunenberg, *Nucl. Sci. Engng.*, 1962, **14**, 1.
267. H. Sawamura and S. Matoba, Sub. comm. for Phys. Chem. of Steelmaking, 19th Comm. 3rd Div. Jap. Soc. for promotion of Sci., July 4th, 1961.
268. Y. Sato, K. Suzuki, Y. Omori and K. Sanbongi, Tetsu-to-Hagané, *Abstr.*, 1968, **54**, 330.

13 Diffusion in metals

13.1 Introduction

In an isotropic medium the diffusion coefficient D^i of species i is defined through Fick's first law,

$$J^i = -D^i \text{ grad } c^i \tag{13.1}$$

J^i is the instantaneous net flux of species i, or diffusion current per unit area, and grad c^i is the gradient of the concentration c^i of i. If J and c are measured in terms of the same unit of quantity (e.g. J in g cm^{-2} s^{-1}, c in g cm^{-3}), D has the dimensions (L^2T^{-1}). It is usually expressed as cm^2 s^{-1}. Generally, D depends on the concentration.

That matter is to be conserved at each point leads to Fick's second law,

$$\frac{\partial c^i}{\partial t} = \text{div } (D^i \text{ grad } c^i) \tag{13.2}$$

giving the rate of the change of concentration with time to which diffusion gives rise.

The fluxes J^i are referred, at least for practical purposes, to axes fixed in the volume of the sample; but volume changes which take place as a result of diffusion lead to some ambiguity in the definition of such axes. Means have been proposed[4, 11] for avoiding this by using axes scaled to the volume changes, but little use is made of these and it is more usual in accurate work to restrict the range of concentration employed so that volume changes are small or negligible.

When the concentration varies along only one direction, say the x axis, (13.1) and (13.2) become

$$J^i = -D^i \frac{\partial c^i}{\partial x} \tag{13.3}$$

$$\frac{\partial c^i}{\partial t} = \frac{\partial}{\partial x}\left(D^i \frac{\partial c^i}{\partial x}\right) \tag{13.4}$$

If, furthermore, D is independent of composition, and so also of position in the sample, (13.4) becomes

$$\frac{\partial c^i}{dt} = D^i \frac{\partial^2 c^i}{\partial x^2} \tag{13.5}$$

In anisotropic media diffusion rates vary with direction. In general, the diffusion flux is in the same direction as grad c *only* when grad c is along one of a set of orthogonal axes known as the 'principal axes of diffusion'. (These always coincide with axes of crystallographic symmetry so there is no difficulty in identifying them, except in cases of symmetry lower than orthorhombic.) For diffusion along principal axes equations like (13.3) may still be written.

$$\left.\begin{aligned} J^i_x &= -D^i_x(\partial c^i/\partial x) \\ J^i_y &= -D^i_y(\partial c^i/\partial y) \\ J^i_z &= -D^i_z(\partial c^i/\partial z) \end{aligned}\right\}$$

D_x, D_y and D_z are called 'principal coefficients of diffusion'.

In general grad c and J are not in the same direction. However, if l, m, n are the direction cosines of grad c then a diffusion coefficient for this direction may be defined as the ratio of the component of J along (l, m, n), divided by grad c. This is

$$D_{lmn}=l^2D_x+m^2D_y+n^2D_z \tag{13.6}$$

Thus anisotropic diffusion can be completely described in terms of the three principal diffusion coefficients. In uniaxial crystals (tetragonal, trigonal, hexagonal) symmetry dictates, if the z axis is the unique axis, that $D_x=D_y$. Thus D is the same for all directions perpendicular to the unique axis and is often denoted $D_\perp$. D_z is then denoted as $D_{\parallel}$. (13.6) may then be written

$$D_\theta=\sin^2\theta\cdot D_\perp+\cos^2\theta\cdot D_{\parallel} \tag{13.7}$$

where $\cos\theta\equiv n$.

Equations (13.4) and (13.5) still hold for anisotropic diffusion, with D given by (13.6) and (13.7).

Equation (13.1) provides a formal definition of a diffusion coefficient as the ratio of J^i to grad c^i. It also assumes that J^i is determined only by grad c^i. In the very large majority of diffusion measurements that have been made this holds true so that the above simple equations provide an adequate description of the diffusion process taking place. Such measurements are of three main types and these are discussed first and the nature of the diffusion coefficients they entail. They are:

1. Measurements which entail diffusion under a chemical concentration gradient (Chemical Diffusion Measurements—Table 13.4).

   (i) Diffusion of a single interstitial solute into a pure metal.

   (ii) Interdiffusion of two metals which form substitutional solid solutions (or interdiffusion between two alloys of the two metals).

2. Measurements which entail diffusion in essentially chemically homogeneous systems. These are possible through the use of radioactive tracers.

The diffusion of an interstitial solute in a pure metal [1(i)] is described by a single equation like (13.1) and the D has a simple and well-defined physical significance as describing diffusion of solute relative to the solvent lattice.

The same is true for the D for diffusion into a metal or alloy of any radioactive tracer. The methods employed (*see* below) require such extremely small amounts and gradients of tracer that the system remains chemically homogeneous during diffusion. Any diffusion of other constituents is altogether negligible so that D refers simply to the diffusion of the tracer species relative to the solvent lattice.

For the interdiffusion of two metals or alloys [1(ii)] the situation is a little less simple. There would appear to be two diffusion coefficients required, one for each species, but *referred to volume fixed axes* these are equal because grad $c_1=-$grad c_2 and J_1 must be equal and opposite to J_2. Again a single equation like (13.1) suffices to describe the diffusion process and the single D refers to the diffusion rate of either species relative to these axes. It is called the *chemical interdiffusion coefficient* and usually denoted $\tilde{D}$ (Table 13.4).

For many practical purposes $\tilde{D}$ is an adequate measure of the diffusion behaviour of a binary substitutional system. But of more fundamental physical interest are the rates of diffusion of the two species relative to local lattice planes. It is well established that generally these rates are not equal in magnitude. There is therefore a net total flux of atoms across any lattice plane, and if the density of lattice sites is to be conserved each plane in the diffusion zone must shift to compensate for this imbalance of the fluxes across it. At the same time lattice sites are created on one side of the sample and eliminated at the other, processes which are achieved by the creation and annihilation of vacancies. This shift of lattice planes, known as the Kirkendall effect, is observed experimentally as a movement of inert markers, usually fine insoluble wires, incorporated into the sample before diffusion. It is clear, then, that diffusion occurs on a lattice which locally is moving relative to the axes with respect to which $\tilde{D}$ was calculated. To provide a more complete description of binary substitutional diffusion it is therefore necessary to introduce diffusion coefficients D_A and D_B to describe diffusion of the two species relative to lattice planes. It is easy to show that these are related to D by the equation

$$\tilde{D}=N_AD_B+N_BD_A \tag{13.8}$$

where N_A and N_B are the fractional concentrations of A and B. D_A and D_B, which are of more direct physical interest than $\tilde{D}$, are known as the *intrinsic* or *partial chemical diffusion coefficients*.

The velocity v of a marker is given by

$$v=(D_A-D_B)\partial N_A/\partial x, \tag{13.9}$$

where $\partial N_A/\partial x$ is the concentration gradient at the marker; so in principle D_A and D_B can be calculated separately when $\tilde{D}$ and v have been measured. In practice this is done usually only for markers placed at the original interface between the two interdiffusing metals or alloys: in this case a measurement of the *displacement* x_m of the marker after time t allows v to be obtained simply, for $v=x_m/2t$.

Equations (13.8) and (13.9) assume no net volume change and a compensation of the flux difference which is complete and which occurs by bulk motion along only the diffusion direction. These conditions are rarely met fully in practice, as is seen from the occurrence often of lateral changes in dimensions and of a porosity in the *side* of the diffusion zone suffering a net loss of atoms. This porosity, attributed to vacancies precipitating instead of being eliminated at sinks, suggests abnormal vacancy concentrations may be present in the diffusion zone. Because it is diffcult to take into account the effect these abnormal conditions in the diffusion zone may have on the calculated values of $\tilde{D}$ and v, and hence on D_A and D_B, chemical interdiffusion experiments may provide results of limited accuracy and, for theoretical purposes, of limited significance: their effect is of course smaller the smaller the concentration gradients employed.

By contrast, radioactive tracer methods altogether avoid these difficulties and uncertainties associated with diffusion in a chemical gradient, and so are preferred in any investigation with a theoretical objective. They have the further advantage that the diffusion coefficients of the several species of an alloy can be determined separately and directly, rather than through any composite coefficient like $\tilde{D}$. These are referred to as *tracer diffusion coefficients* (Table 13.3) and will be denoted D_A^*, D_B^* etc. to distinguish them from the partial chemical diffusion coefficients D_A and D_B determined by chemical diffusion methods.†

Results on the diffusion coefficient D_A^* in very dilute alloys AB containing small concentrations C_B of B are frequently represented in terms of enhancement factors b_1, b_2, etc., in the equation

$$D_A^*(C_B)=D_A^*(C_B=0)\,(1+b_1C_B+b_2C_B^2+\ldots) \tag{13.9a}$$

$D_A^*(C_B=0)$ is of course just the self-diffusion coefficient of pure A.

Except at vanishingly small concentrations of A, D_A and D_A^* differ fundamentally because the presence of the chemical concentration gradient under which D_A is measured imposes on the otherwise random motion of the atoms a bias, which makes atoms jump preferentially in one direction along the concentration gradient. Simple thermodynamic considerations lead to the relation

$$D_A=D_A^*\left(1+\frac{\partial \ln \gamma_A}{\partial \ln N_A}\right) \tag{13.10}$$

between a partial chemical D_A and the corresponding tracer D_A^* measured at the same concentration. γ_A is the activity coefficient of A. In a binary system the bracket term is the same for both species (Gibbs–Duhem relation). Thus

$$\frac{D_A}{D_A^*}=\frac{D_B}{D_B^*} \tag{13.11}$$

(13.10) and (13.11) are approximate forms of more elaborate theoretical expressions, but are reasonably well obeyed experimentally.

When D_A^* is measured at the extremely small concentrations of A that tracer methods permit (by diffusion of tracer into pure metal B) it is called the *tracer impurity diffusion coefficient* of A in B (Table 13.2). Such coefficients are of especial theoretical interest because of the particularly simple type of diffusion they describe (*see* footnote).

Finally, tracer methods are used as the commonest means of measuring *self-diffusion coefficients* in pure metals (Table 13.1, 13.5 and 13.6). By self-diffusion is meant of course the diffusion of a species in the pure lattice of its own kind.

For chemical diffusion in systems of more than two components, equation (13.1) and those following are inadequate. Experimentally it is found that when three or more components are present a concentration gradient of one species can lead to a diffusion flow of another, even if this is distributed homogeneously to start with. To cater for such cases Fick's first law is generalized by writing

$$J_i=\sum_{j=1}^{N} D_{ij}\frac{\partial c_j}{\partial x} \qquad (j=1,\,2\,\ldots\,N) \tag{13.12}$$

† D_A^* and D_B^* are sometimes referred to as the *self-diffusion coefficients of the alloy*. This is a perfectly acceptable alternative terminology. But there is a tendency nowadays to employ the term 'self' to the extent of describing tracer impurity diffusion coefficients (*v.i*) as impurity self-diffusion coefficients. This latter term is ambiguous and misleading and its use is to be discouraged.

But if there are n interstitial and N–n substitutional components, and if the J_i are referred to volume-fixed axes then the relations

$$\sum_{j=n+1}^{N} J_i=0 \quad \text{and} \quad \sum_{j=n+1}^{N} \partial c_j/\partial x=0$$

allow (13.12) to be rewritten

$$J_i=\sum_{j=1}^{N-1} D_{ij}\frac{\partial c_j}{\partial x} \qquad (j=1, 2 \ldots N-1) \tag{13.13}$$

so that $(N-1)^2$ coefficients suffice to describe the diffusion behaviour. The analogue of Fick's second law is

$$\frac{\partial c_i}{\partial t}=\sum_{j=1}^{N-1}\frac{\partial}{\partial x}\left(D_{ij}\frac{\partial c_j}{\partial x}\right) \tag{13.14}$$

These equations have been applied to a few ternary systems.

It is possible to show from the principles of irreversible thermodynamics that not all the D_{ij} are independent and that a total of only $N(N-1)/2$ coefficients are in fact sufficient to describe diffusion in an N-component system. No measurements in metals have employed this reduced scheme of coefficients, for to do so requires a knowledge of the thermodynamic properties of the system that is rarely available.

13.2 Methods of measuring *D*

13.2.1 Steady-state methods

These are based directly on Fick's first law. The usual procedure is to maintain concentrations of diffusant on the opposite sides of a sample, which is usually a thin sheet or a thin-walled tube, and to measure the resulting steady rate of flow J. This is generally practicable only when the diffusing element is a gas or can be supplied to and removed from the sample through a vapour phase. If the surface concentrations c_1 and c_2 in equilibrium with the ambient atmospheres are known, an average D over the concentration range is, for a sheet of thickness t for example, simply $\bar{D}=Jt/(c_1-c_2)$ (Method Ib). Alternatively, if the steady concentration distribution across the sample is determined, $D(c)$ may be calculated from $D=J(\partial c/\partial x)$ (Method Ia).

D may also be calculated from measurements of the time required to reach a steady state (Method Ic).

These methods are used for measuring D only for interstitial solute diffusion: the Kirkendall effect complicates any attempt to apply it reliably to substitutional diffusion.

13.2.2 Non-steady-state methods

The change in the concentration distribution in a sample as a result diffusion is measured and D deduced from a solution of Fick's second law [equations (13.2), (13.4), (13.5) or (13.14) appropriate to the conditions of the experiment. There are three common types of experimental arrangement, two of which are usually employed in chemical diffusion coefficient measurements, the third in measurements of tracer diffusion coefficients.

(i) DIFFUSION COUPLE METHOD

Two metals, or two different homogeneous alloys of concentrations c_1 and c_2, are brought into intimate contact across a plane interface, say by welding. Diffusion is allowed to take place by annealing at a constant temperature for a time t. The distribution of concentration in the sample is then determined in some convenient manner, often by removal and subsequent analysis of a succession of thin layers cut parallel to the initial interface. It is usually arranged that the two halves of the couple be sufficiently thick that the diffusion zone does not extend to either end.

D generally varies with concentration, but no analytic solutions of (13.4) are available so recourse is had to a graphical method of analysis known as the Matano-Boltzmann method. The concentration c is plotted against x and $D(c)$ determined graphically from.

$$D(c)=(2t\cdot \partial c/\partial x)^{-1}\int_{c}^{c_1} x\, dc \quad \text{[Method IIa(i)]} \tag{13.15}$$

The origin of x is located by the condition

$$\int_{c_1}^{c_2} x\, dc=0$$

and this may be shown to coincide, under ideal conditions, with the *initial* position of the interface between the two members of the couple. Thus it is $\tilde{D}$ which is measured in substitutional diffusion. Markers inserted at the interface locate its *final* position after diffusion. It has already been mentioned that measuring their displacement x_m from $x=0$ allows the partial diffusion coefficients to be calculated.

If D varies little in the range c_1 to c_2, and this is often so if the range is sufficiently restricted, equation (13.5) may be used, the solution of which for this case is

$$\frac{c-c_2}{c_1-c_2}=\tfrac{1}{2}\left\{1-erf\left[\frac{x}{2\sqrt{(Dt)}}\right]\right\} \tag{13.16}$$

With $x=0$ defined as before, D can then be calculated directly by a 'least squares' fit of the $c\sim x$ data to this or other appropriate equations [Method IIa(ii)].

Occasionally, the diffusion couple method is used to measure self-diffusion coefficients, one half of the couple being normal metal, the other enriched in one of its active or normal isotopes. It may also be used to measure diffusion coefficients in liquids (Shear-cell method).

With analytic solutions, like (13.16) D can be calculated by measuring c at one position only. This is sometimes done but it is not to be expected that values derived in this way will be as reliable as when derived from a complete $c\sim x$ curve (Method IIb).

The concentration range in a diffusion couple may span any number of phase regions in the equilibrium diagram of the system; the diffusion zone then consists of phase layers with concentration discontinuities across each boundary between two layers. In such cases equation (13.15) [Method IIa(i)] is still applicable. If D is assumed constant, analytic solutions are available and with these it is sometimes possible (Method IIc) to determine D from measurements only of the rates of movement of one or more phase boundaries and knowledge of the equilibrium concentrations at the boundaries.[5, 5a]

(ii) IN-DIFFUSION AND OUT-DIFFUSION METHODS

Material is allowed to diffuse into, or out of, an initially homogeneous sample of concentration c_1 under the condition that the concentration at the surface is maintained at a constant and known value c_0 by being exposed to a constant ambient atmosphere. c_1 is usually zero for in-diffusion experiments and so is c_0 for out-diffusion experiments.

D may be calculated from a measurement either of the total amount of material taken up by or lost from the sample (Method IIIb), or of the concentration distribution within the sample after diffusion (Method IIIa). The first method gives an average D over the range c_1 to c_0. For the second, equation (13.15) can be used again to give $D(c)$ or, if D is constant, it may be calculated from an appropriate analytic solution.

When the loss (or gain) of material from the sample entails the movement of a phase boundary, D can again be calculated from the rate of movement (Method IIIc). This method has been mostly used for interstitial solute diffusion, but also occasionally for substitutional diffusion measurements in systems with a sufficiently volatile component. A disadvantage of it is that conditions at the surface may not always be under adequate control so that c_0 is either ill-defined or not constant or both, with consequent uncertainty in D.

A common method of measuring liquid self-diffusion rates employs a type of out-diffusion method. A capillary tube, closed at one end and containing activated material, is immersed open-end uppermost in a large bath of inactive material. After the diffusion anneal the depleted activity content of the capillary is determined, and D calculated on the assumption that diffusion of the active species out of the tube is subject to zero concentration being maintained at the exit.

For determining the concentration distribution $c(x)$ in any of the above chemical diffusion methods a wide variety of techniques has been employed, including chemical and spectrographic analysis, X-ray and electron diffraction, electron microprobe analysis, X-ray absorption, microhardness measurements and so forth. In this edition of the tables the method of analysis is not recorded for it is probably of less importance in assessing the reliability of a result than other features of the experimental procedure.

(iii) THIN LAYER METHODS

These are used now almost exclusively for the measurement of self and of tracer D's. A very thin layer of radioactive diffusant, of total amount g per unit area, is deposited on a plane surface of the sample, usually by evaporation or electrodeposition. After diffusion for time t the concentration at a distance x from the surface is

$$c(x)=\frac{g}{(\pi Dt)^{1/2}}\exp\left(-\frac{x^2}{4Dt}\right) \tag{13.17}$$

provided the layer thickness is very much less than $(Dt)^{1/2}$. This condition is easy to satisfy because extremely small quantities suffice for studying the diffusion on account of the very high sensitivity of methods of detecting and measuring radioactive substances. For the same reason there is a negligible change in the chemical composition of the sample so D is constant and equation (13.5), of which (13.17), is the solution for this case, is applicable.

After diffusion the activity of each of a series of slices cut from the sample may be determined and D calculated from the slope $(=1/4Dt)$ of the linear plot of log activity in each slice against x^2 [Method IVa(i)]. Alternatively, such a plot may be constructed from intensity measurements made on an autoradiograph of a single section cut along or obliquely to the diffusion direction [Method IVa(ii)].

Another method is to calculate D from measurements made, after the removal of each slice, of the residual activity emanating from each newly exposed surface *of the sample.* [Residual activity method; Method IVb.]

Or, D may be determined by comparing the total activity from the surface $x=0$ after diffusion with the original activity at $t=0$ (surface decrease, Method IVc).

Methods IVb and IVc require an integration of equation (13.17). They are generally regarded as less reliable in principle than Method IVa because they obviously necessitate also a knowledge of the absorption characteristics of the radiation concerned. In addition Method IVc is particularly susceptible to errors arising from possible oxidation and from evaporation losses of the deposited material.

A recent development has been the use of the electron-microprobe to measure even impurity diffusion coefficients: instruments are now available with a sensitivity adequate to monitor diffusion from deposited layers of *inactive* diffusant thin enough to meet the requirements for use of equation (13.17) [Method IVa(iii)].

13.2.3 Indirect methods, not based on Fick's laws

In addition to macroscopic diffusion there are a number of other phenomena in solids which depend for their occurrence on the thermally activated motion of atoms. From suitable measurements made on some of these phenomena it is possible to determine a D. The more important of these are:

1. Internal friction due to a stress-induced redistribution of atoms in interstitial solution in metals (Snoek effect and Gorsky effect, Method Va.)
2. A similar phenomenon occurring in substitutional solid solution and due, it is believed, to stress-induced changes in short range order (Zener effect, Method Va.)
3. Phenomena associated with nuclear magnetic resonance absorption, especially the 'diffusional narrowing' of resonance lines and a contribution, arising from atomic mobility, to the spin-lattice relaxation time T_1 (Method Vb).
4. Some magnetic relaxation phenomena in ferromagnetic substances (Method Vc).
5. The sintering of metal powder particles or wires (Method Vd).

1, 2, 3 and 4 are associated with atomic motion over only a few atomic distances, and so have the advantage of providing measurements of D at temperatures lower than are often practicable by conventional methods. Since in every case measurements are made in homogeneous material the diffusion coefficients obtained are of the nature of tracer rather than chemical diffusion coefficients.

Most measurements of D are conducted at a series of temperatures so as to provide values of the constants A and Q occurring in the Arrhenius equation

$$D = A \exp(-Q/RT) \tag{13.18}$$

which usually describes very well the observed temperature dependence.* A is called the 'frequency factor' and Q the activation energy. Wherever possible, experimental measurements are reported in the tables in terms of A and Q alone.

Experiments may be made by any of the above methods either with single crystal or polycrystalline material. With polycrystals there is, in addition to diffusion through the grains (volume diffusion), diffusion at a more rapid rate locally through the disordered regions of grain boundaries. This can, however, be reduced to a negligible proportion of the whole by using large grain material and by working at relatively high temperatures because, since $Q_{gb} < Q_v$, grain boundary diffusion rates increase less rapidly with temperature than do volume diffusion rates. Obviously single crystals are to be preferred in accurate measurements of what is intended to be volume diffusion but even in their case there may be, at too low temperatures, a contribution to D from diffusion along dislocations. Measured values of D will then tend to be above the values expected from an extrapolation of the high temperature date using (13.18), and when they do so to a noticeable extent are often discarded in estimating Q and A.

From measurements of the concentration distribution around a grain boundary—usually in a bicrystal into which material diffuses parallel to the boundary—a product $D'\delta$ may be deduced.[12] D' is the coefficient for diffusion *in* the boundary of width δ, δ is an uncertain quantity but all results quoted in Table 13.5 give values for A_{gb} calculated assuming $\delta = 5.0 \times 10^{-8}$cm. $D'\delta$ is found to depend on the orientation of the boundary and on the direction of diffusion within it.

13.3 Mechanisms of diffusion

Most theoretical discussions of diffusion are concerned with an understanding of A and Q rather than of D itself. On the basis of theoretical calculations of Q for various possible mechanisms of diffusion and comparison with observed values, it has been supposed for some time that in metals atoms diffuse substitutionally by thermally activated jumps into vacant lattice sites, i.e. by the 'vacancy mechanism'. This has comparatively recently been very convincingly confirmed, at least for f.c.c. metals, by thermal expansion and quenching experiments. While the same mechanism is usually thought to operate in most metal structures, there is considerable doubt at present whether this is in fact true for a number of so-called 'anomalous b.c.c. metals'—β-Ti, β-Zr, β-Hf, β-Pr, γ-U and δ-Ce—or at least whether the vacancy mechanism is the only one operating in their case. It is also believed that the noble metals and other low-valent solutes (Group II), plus the later transition elements, may dissolve interstitially, at least in part, and diffuse by an interstitial-type process in the alkali metals, in the high-valent Group III and IV elements and also in the *early* members of each of the transition groups, the lanthanide series and the aclinide series of elements. This belief stems from the anomalously very large diffusion rates of these solutes in these solvents.[13]

REFERENCES

Textbooks

1. P. G. Shewmon, 'Diffusion in Solids', McGraw-Hill, New York, 1963.
2. W. Seith and T. Heumann, 'Diffusion in Metallen', 2nd edn, Springer-Verlag, Berlin, 1955.
3. K. Hauffe, 'Reaktionen in und an festen Stoffe', Springer-Verlag, Berlin, 1955.
4. J. Crank, 'The Mathematics of Diffusion', Clarendon Press, Oxford, 2nd edn, 1975.
5. W. Jost, 'Diffusion in Solids, Liquids and Gases', 2nd edn, Academic Press, New York, 1964.

5a. Y. Adda and J. Philibert, 'La Diffusion dans les Métaux', Presse Universitaire, Paris, 1966.

Reviews

6. A. D. Le Claire, *Progr. Metal Phys.*, 1949, **1**, 306; 1953, **4**, 265.
7. C. E. Birchenall, *Metall. Rev.*, 1958, **3**, 235.
8. R. E. Howard and A. B. Lidiard, *Rep. Progr. Phys.*, 1964, **XXVII**, 246.
9. C. Tomizuka, 'Methods of Experimental Physics' (edited by K. Lark-Horowitz and V. A. Johnson), 6A, 364, Academic Press, New York, 1959.
10. D. Lazarus, *Solid St. Phys.*, 1960, **10**.

10a. N. L. Peterson, 'Diffusion in Metals', 1968. *Solid St. Physics*, **22**, Academic Press.

10b. 'Diffusion Data', a quarterly review published by Diffusion Information Centre, *Trans. Tech. SA*, Switzerland

*This is often true even of $\tilde{D}$, because Q_A and Q_B for the partial diffusion coefficients do not seem to differ very much.

Papers

11. M. Cohen, C. Wagner and J. E. Reynolds, *Trans, AIME*, 1953, **197**, 1534.
12. A. D. Le Claire, *Br. J. appl. Phys.*, 1963, **14**, 351.
13. A. D. Le Claire, *J. Nuclear Mat.*, 1978, **69**/**70**, 70.

Summary of methods for measuring *D*

STEADY-STATE METHOD with

I. (a) Measurement of concentration distribution within the sample or, Ia
(b) Average gradient calculated from c_1 and c_2 as deduced from equilibrium data or, Ib
(c) Time-delay method (measurement of time to reach steady state) Ic

NON-STEADY METHODS

II. *Diffusion couple methods*
(a) With determination of $c \sim x$ curve and
(i) Use of Matano Boltzmann analysis to give $D(c)$ IIa(i)
(ii) When it is evident (or assumed) that D is effectively constant, calculation of D from an analytic solution IIa(ii)
(iii) When it is evident that D is *not* constant and an analytic solution is used to calculate a D corresponding to each value of c—giving an approximate $D(c)$ IIa(iii)
(b) D calculated from a single concentration measurement IIb
(c) D calculated from an analytic solution, assuming D constant, using measurements of rate of movement of phase boundaries and knowledge of equilibrium concentrations on the boundaries IIc

III. *In-diffusion and out-diffusion methods* in-(i) out-(ii)
(a) D calculated from $c \sim x$ curves IIIa
(b) D calculated from total gain or loss, or rate thereof IIIb
(c) D calculated from rate of phase boundary movement IIIc

IV. *Thin layer methods*
(a) With measurement of $c \sim x$ curve
(i) By sectioning and counting – using radioactive diffusant IVa(i)
(ii) By autoradiography IVa(ii)
(iii) By electron-microprobe—using non-radioactive diffusant IVa(iii)
(b) Residual activity method – using radioactive diffusant IVb
(c) Surface decrease method IVc

V. *Indirect methods*
(a) By internal friction Va
(b) By nuclear magnetic resonance Vb
(c) By ferromagnetic relaxation Vc
(d) By sintering Vd

Notes on the tables

1. All measurements are reported whenever possible in terms of A and Q (see equation 13.18). A in cm^2s^{-1}: Q in kcal mol^{-1}: ($R = 1.987$ cal $mol^{-1}K^{-1}$ 1 eV = 23 kcal mol^{-1}). Where errors are quoted these are authors' estimates.
2. The temperature range' is the range over which measurements were used to calculate A and Q. Extrapolation too far outside this range may not in some cases give reliable values for D.
3. All alloy concentrations are in atomic percentages unless otherwise stated. Purity of material is as quoted and is presumably in weight percentages, although this is not always stated explicitly in papers.
4. s.c. = single crystals; p.c. = polycrystals.
5. In Table 13.4 a single concentration denotes the concentration at which $D(c)$ was determined. Two concentrations separated by a hyphen denote the range of concentration over which measurements were made. Where this is followed by a single D value, or a single set of A and Q values, it is also the concentration range over which these values are averages.
6. Bold type in Table 13.4. This is used: (1) To indicate the species to which the D's, or A and Q values, refer in cases where there might be ambiguity–usually for interstitial solid solutions. Where there is no bold type the data refer to the interdiffusion coefficients of the first two substitutional species. (2) To indicate which component was used in the vapour phase in experiments employing methods I and III.
7. Where several measurements exist an attempt has been made to select what appear to be the most reliable one or two. Mostly these are later measurements and references to earlier work can usually be found by consulting the references quoted.

Table 13.1 SELF-DIFFUSION IN SOLID ELEMENTS

Element	*A*	*Q*	*Temp. range °C*	*Method*	*Ref.*
Group IA					
Li	$0.125^{+0.024}_{-0.020}$	12.673 ± 0.148	35/178	IVa(i), p.c., normal Li ($\approx 8\%$ Li^6)[(a)]	1 and 2
	0.39 ± 0.02	13.49 ± 0.07	70/170	IIa(ii)	3
Na	$A_1 = 0.72 \pm 0.05$[(b)] $A_2 = (57 \pm 4)10^{-4}$	$Q_1 = 11.5 \pm 0.5$ $Q_2 = 8.53 \pm 0.2$	−78/98	IVa(i), p.c. Na^{22}	4
K	0.16 ± 0.1	9.36 ± 0.05	−52/62	IVa(i), p.c. K^{42}	5
Rb	0.23	9.4	−23/40	Vb	6
Group IB					
Cu	0.78	50.4 ± 0.2	698/1 061	IVa(i), s.c., 99.999%, Cu^{64}	7
	0.35	48.66			
(*b*)	$A_1 = 0.10$ $A_2 = 2.0$	$Q_1 = 47.04$ $Q_2 = 55.81$	300/1 061	IVa(i), s.c., 99.999%, Cu^{64}	7 and 8
Ag	0.67	45.2	640/955	IVa(i), s.c., 99.999%	10
	0.235	42.89			
(*b*)	$A_1 = 0.027$ $A_2 = 3.0$	$Q_1 = 40.08$ $Q_2 = 49.74$	277/955	IVa(i), s.c., 99.999%	9 and 10 11
Au	0.091 ± 0.001	41.70 ± 0.3	704/1 048	IVa(i), p.c., 99.95%	12
Group IIA					
Be$\parallel c$ $\perp c$	0.62 ± 0.15 0.52 ± 0.15	39.4 ± 0.7 37.6 ± 0.7	563/1 070	IVb, Be^7, s.c.	13
Mg$\parallel c$ $\perp c$	1.0 1.5	32.2 32.5	468/635	IVa(i), Mg^{28}, s.c., 99.9+%	14
$\parallel c$ $\perp c$	1.78 1.75	33.2 33.0	500/630	IVa(i) and IVb, Mg^{28}, s.c., 99.99%	15
Ca	8.3	38.5	500/800	IVb, Ca^{45}, p.c., 99.95%	16
Group IIB					
Zn$\parallel c$ $\perp c$	0.13 ± 0.01 0.18 ± 0.01	21.9 ± 0.15 23.0 ± 0.11	240/418	IVa(i), $Zn^{65/69}$, s.c., 99.999	17
Cd$\parallel c$ $\perp c$	0.118 0.183	18.61 ± 0.12 19.59 ± 0.12	147/327	IVa(i), Cd^{109}, s.c., 99.999	18
Group IIIA					
Y$\parallel c$ $\perp c$	0.82 5.2	60.3 67.1	900/1 300	IVb, Y^{91}, s.c.	19
β·La	$1.5^{+1.2}_{-0.7}$	45.1 ± 1.2	660/840(β)	IVa(i), La^{140}, p.c., 99.97%	20
γ·La	$0.013^{+0.039}_{-0.009}$	24.5 ± 3.0	867/897	IVa(i), p.c., La^{140}	88
	0.11	29.9	878/910	IVa(i), p.c., 99.85%, La^{140}	89
γ·Ce	$5{\cdot}5 \; {}^{+1.3}_{-1.1} \times 10^{-1}$	36.6 ± 0.4 21.5 ± 0.7	528/692(γ) 719/771(δ)	IVa(i), p.c., 99.9%	21
δ·Ce	$1.2 \; {}^{+0.5}_{-0.4} \times 10^{-2}$	29.4 ± 1.1	(β)	IVa(i). Pr^{142}, p.c., 99.97%	22
β·Pr	$8.7 \; {}^{+5.6}_{-3.4} \times 10^{-2}$	72.05 ± 0.65 72.27 ± 0.36	1 202/1 411	IVa(i), Er^{169}, s.c.	23
Er$\parallel c$	$3.71 \; {}^{+0.87}_{-0.71}$				
$\perp c$	$4.51 \; {}^{+0.55}_{-0.49}$				
Eu	1.0	34.4 ± 0.7	498/801	IVa(i); p.c., Eu^{152}	90
β·Gd	0.01	32.7 ± 0.4	1276/1308	IVa(i); p.c., Gd^{159}	90
α·Yb γ·Yb	0.034 0.12	35.06 28.9	550/710 725/810	IVa(i); p.c.; 99.5%, Y^{169}	91
Group IIIB					
Al	1.71 [(c)]	34.0	450/650	IVa(i), Al^{26}, p.c., 99.9%	24
	2.25	34.5	300/650	IVa(i), p.c.	25

Table 13.1 SELF-DIFFUSION IN SOLID ELEMENTS—*continued*

Element	*A*	*Q*	*Temp. range* °C	*Method*	*Ref.*
Ga		$D \times 10^3$			
		$=5.3 \pm 0.8$	9.8	IVa(i), Ga^{72}, s.c./p.c., 99.999 9%	30
		$=5.3 \pm 1.1$	20.0		
		$=7.8 \pm 3.0$	25.0		
		$=9.3 \pm 1.2$	27.5		
		$=42 \pm 11$	29.7		
In $\parallel c$	2.7	18.7 ± 0.3	44/144	IVa(i), In^{144}, s.c., 99.995%	31
$\perp c$	3.7	18.7 ± 0.3			
α·Tl $\parallel c$	0.4	22.9 ± 0.5	150/225(α)	IVa(i), Tl^{204}, s.c., 99.9+%	32
$\perp c$	0.4	22.6 ± 1.0			
β·Tl	0.7	20.0 ± 0.5	235/275(β)		
Group IVB					
C (Graphite)	0.4–14.1	163 ± 12[e]	2 185/2 347	IIIb, C^{14}, natural crystals	33
Si	1460	115.8	1045/1390	IVa(i)., s.c., Si^{31}	34
Ge	7.8 ± 3.4	68.50 ± 0.96	766/928	IVa(i), Ge^{71}, s.c.	36
	10.8 ± 2.4	69.40 ± 0.44	731/916	IVb, Ge^{71}, s.c.	37
Sn $\parallel c$	8.2 ± 6	25.6 ± 0.8	178/222	IVa(i), Sn^{113}, s.c., 99.998%	38
$\perp c$	1.4 ± 0.5	23.3 ± 0.5			
$\parallel c$	7.7 ± 3	25.6 ± 1.0	160/228	IVa(i), Sn^{113}, s.c., 99.999%	39
$\perp c$	10.7 ± 1.0	25.1 ± 0.8			
Pb	0.995 ± 0.2	25.65 ± 0.25	200/323	IVa(i), Pb^{210}, s.c., 99.999 9%	40 and 41
Group VB					
P	$(3.6 \pm 1.4)10^9$	27.4 ± 0.5	22/43	IVa(i), P^{33} s.c.	87
Sb $\parallel c$	56 ± 20	48.0 ± 0.5	500/630	IVa(i), Sb^{124}, s.c., 99.999 9%	43
$\perp c$	0.10 ± 0.02	35.8 ± 0.3			
	1.05	39.5 ± 1.4	473/583	IVb, Sb^{124}, p.c., 99.9+%	44
Bi		See footnote *f*			45
Group VIB					
S $\parallel c$	—	~ 69.3[g]	80/95	IVa(i), S^{35}, s.c.	47
$\perp c$	2.10^{17}	51.4			
Se $\parallel c$	0.2	27.6 ± 2.5	152/215	IVb, Se^{75}, s.c.	48
$\perp c$	100	32.2 ± 2.1			
Te $\parallel c$	130	40.25 ± 1.15	305/400	IVa(i), Te^{127m}, s.c., 99.999 9%	49
$\perp c$	39 100	46.69 ± 2.30			
Group IVA					
α·Ti	8.6×10^{-6}	35.9	690/880(α)	IVb, Ti^{44}, p.c., 99.99%	50
β·Ti	1.9×10^{-3}	36.5 ± 0.49	900/1 580(β)	IVb, Ti^{44}, p.c., 99.9%	51
	$A_1 = 3.58 \times 10^{-4}$ (b)	$Q_1 = 31.2$	898/1 540(β)	IVa(i), Ti^{44}, p.c., 99.9%	52
	$A_2 = 1.09$	$Q_2 = 60.0$			
α·Ti	$2.1 \cdot 10^{-7}$	27.0 ± 2.9	740/857(α)	IVb, Zr^{95}, p.c., 99.99%	50
	5.6×10^{-4}	45.5	750/850	IVa(i), p.c., Zr^{95}	63
β·Zr	$A_1 = 8.5 \times 10^{-5}$	$Q_1 = 27.70$	900/1750	IVa(i), p.c., 99.94, Zr^{95}	53 and 35
(b)	$A_2 = 1.34$	$Q_2 = 65.20$			
α·Hf $\parallel c$	0.86	88.4 ± 3.2	1 220/1 610(α)	IVa(i), Hf^{181}, s.c., Hf+2.1% Zr	56
$\perp c$	0.28	83.2 ± 4.8			
	7.3×10^{-6}	41.6 ± 2.3	924/1 483(α)	IVb, $Hf^{175/181}$. p.c.,	50
β·Hf	4.8×10^{-3}	43.8 ± 2.2	1 785/2 160(β)	Hf+2.7% Zr	51
α·Th	395	71.6 ± 2.3	690/910(α)	IVc, Th^{208}, p.c.	54
β·Th	$10^4 - 10^6$	99.0 ± 7.0	1 450/1 550(β)	Vd, s.c., 99.9%	55
Group VA					
V	0.36 ± 0.02	73.65 ± 0.15	880/1 356	IVa(i), V^{48}, s.c./p.c., 99.99%	57
	214 ± 20	94.14 ± 0.33	1 356/1 833		
Nb	$A_1 = 8 \times 10^{-3}$	$Q_1 = 83.5$	1080/2420	IVa(i), s.c.; >99.92, Nb^{95}	58
(*b*)	$A_2 = 3.7$	$Q_2 = 104 \cdot 7$			
Ta	0.124	98.7	1 250/2 220	IVa(i), Ta^{182}, p.c.	59

Table 13.1 SELF-DIFFUSION IN SOLID ELEMENTS—*continued*

Element	*A*	*Q*	*Temp. range°C*	*Method*	*Ref.*
Group VIA					
Cr	970	104±1	1100/1820	IVa(i), s.c., 99.995%, Cr,^{51} Cr^{48}	60
Mo (b)	A_1=0.126 A_2=139	Q_1=104.5±5 Q_2=131.2±5	1090/2500	IVa(i), s.c., Mo^{99}	61
W (b)	A_1=0.04 A_2=46	Q_1=125.7±12 Q_2=159.1±14	1430/3140	IVa(i), s.c., 99.999%, W^{187}	62
α·U	$2.0 \cdot 10^{-3}$	40.0	580/650	IIa(ii), Enriched U^{234}, p.c.	64
	‖110 ‖010	‖001			
D=	36.7 ≤1.5	42.9	652.9 'Mosaic' crystal	IVa(i), U$^{235/233}$ s.c.	65
	19.5 ≤1	19.5 $\times 10^{-14}$	625.5 'Mosaic' crystal		
	6 ≤0.35	7.5	587.4 'Mosaic' crystal		
	3.67 ≤0.35	3.84	587.4 'Perfect' crystal		
β·U	1.35×10^{-2}	42.0	700/755	IIa(ii), Enriched U^{234}, p.c.	66
	2.8×10^{-3}	44.2	690/750	IVb, U^{235}, p.c.	67
γ·U	1.19×10^{-3} (h)	26.7	803/1069	Various, p.c.	68 and 69
	0.11×10^{-3}	36.0	850/1050	IVb,(i), p.c., 99.76%	70
	0.12	54.5	830/1080	(j)	
Group VIII					
α·Fe	*See* Figure 13.1 (k)		510/770 (Ferromagnetic)	Various	71, 92 and 93
	2.01	57.5	770/884 (Paramagnetic)	IVb, s.c. and p.c., Fe^{59}	72
	1.67 (l)	61.3±0.9	809/901 (Paramagnetic)	IVa(i), s.c., 99.999%, Fe^{59}	74
γ·Fe	$0.49^{+0.38}_{-0.29}$ (m)	67.86±1.45	1170/1361	IVa(i), Fe^{59}, p.c., 99.98%	75
δ·Fe	2.01	57.5	1428/1492	IVa(i), Fe$^{55/59}$, s.c./p.c.	72
	6.8	61.7	1407/1515	IVa(ii), Fe^{55}, 99.998%, p.c.	77
Co	0.23	64.0	1120/1370	IVb, Co^{60}, p.c., 99.999%	76
	$0.35^{+0.33}_{-0.17}$	65.0	1015/1300	IVa(i), Co^{60}, p.c., 99.45%	78
Ni	1.27	67.2	870/1404	Various (n)	76, 79 and 80
	A_1=0.92 (b) A_2=37	66.4 85.3	540/1400	IVa(i) and b, s.c., Ni^{63}	79 and 94
Pd	$0.205^{+0.05}_{-0.04}$	63.6±0.65	1050/1500	IVa(i), Pd^{103}, s.c., 99.999%	81
Pt	0.33	68.08±1.4	1325/1600	IVa(i), Pt^{195m}, p.c., 99.99%	82
	0.22±0.03	66.47±0.9	1250/1725	IVc, Pt^{195m}, p.c., 99.999%	83
ε·Pu	2.2×10^{-2}	18.5	500/612	IIa(ii), p.c., Pu^{240}	85
	4.5×10^{-3}	16.0	492/613	IVa(i), p.c., Pu^{239}	86
δ·Pu	5.32×10^{32}	141.0	457/477		
δ·Pu	4.5×10^{-3}	23.8	350/440	IIa(ii), p.c., Pu^{238}	84
	5.17×10^{-1}	30.2	321/442	IVa(i), p.c., Pu^{239}	86
γ·Pu	3.8×10^{-1}	28.3	211/291		
β·Pu	1.69×10^{-2}	25.8	136/181		

Notes:

(*a*) Reference 2 reports measurements of self-diffusion in Li of different isotopic compositions, from which are derived values of D for Li^6 diffusing in Li^7 and of Li^7 diffusing in Li^6.

(*b*) The log $D \sim 1/T$ plot shows distinct curvature and is expressed as the sum of two Arrhenius terms. $D = A_1 \exp(-Q_1/RT) + A_2 \exp(-Q_2/RT)$.

(*c*) The Q values are confirmed by creep measurements (26) but other indirect methods indicate lower values of Q at low temperatures e.g. NMR (27), annealing of voids (28)—Q=28.75 and 30.13 respectively. The collected results are discussed in reference 29.

(*d*) There is no clear evidence of marked anisotropy.

(*e*) Assumed in analysis of results that D in direction perpendicular to basal plane is negligible.

(*f*) See reference 45 for account of the highly anomalous self-diffusion behaviour of Bi and reasons for believing the often quoted results of Seith (reference 46) to be very suspect.

(*g*) Orthorhombic S crystals, $D_{\|c}$ is 2 to 4 times less than $D_{\perp c}$.

(*h*) The results quoted from (68) are from a least squares fit to the three measurements of reference 69.

(*i*) Samples pre-annealed at diffusion temperature.

(*j*) Samples all pre-annealed at 1080°C.

(*k*) The temperature dependance of *D* in the ferromagnetic phase is not well established, especially near T_c. *See* Figure 13.1 and references 71, 72, 73, 92 and 93.

(*l*) The authors report a strong dependence of *A* on impurity content.

(*m*) These values are very close indeed to the values $A=0.5$ $Q=68.0$ previously chosen by Badia (76) as best representing the combined results of a number of other investigations over the range 1050/1400 °C.

(*n*) $A=1.27$, $Q=67.2$ are given in reference 76 as the best fit to the combined results of references 79 and 80.

(*o*) Measurements with P^{32} (reference 42) show larger *D*'s due to radiation enhancement from the more energetic P^{32} radiation.

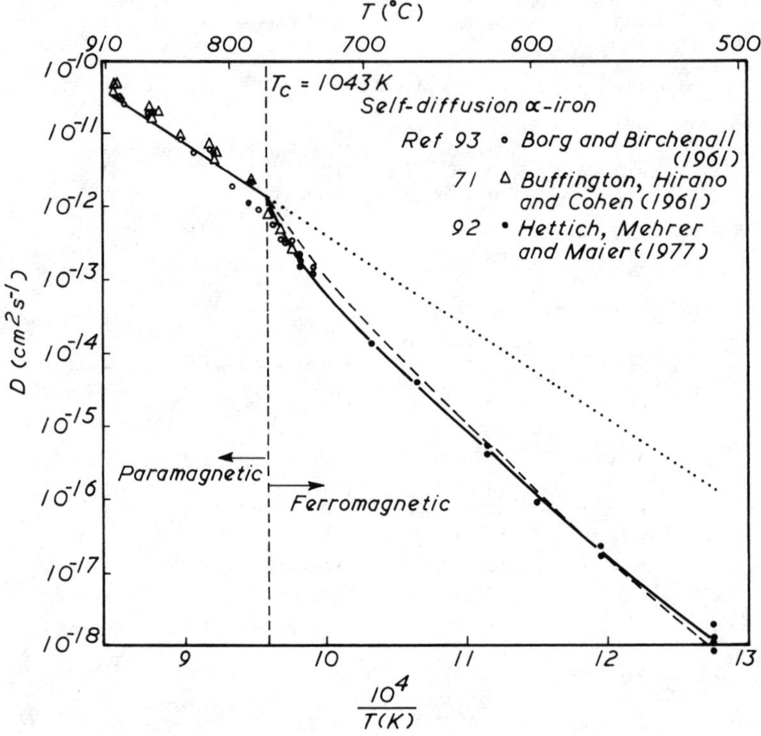

Figure 13.1 *Self-diffusion in αFe, above and below the Curie point*[92]

REFERENCES TO TABLE 13.1

1. A. Ott, J. N. Mundy, L. Löwenberg and A. Lodding, *Z. Naturf.*, 1968, **23A**, 627.
2. A. Lodding, J. N. Mundy and A. Ott, *Phys. Status Solidi*, 1970, **38**, 559.
3. A. N. Naumov and G. Ya. Ryskin, *Soviet Phys. tech. Phys.*, 1959, **4**, 162.
4. J. N. Mundy, *Phys. Rev. B.*, 1971, **3**, 2431.
5. J. N. Mundy, T. E. Miller and R. J. Porte, *Phys. Rev. B*, 1971, **3**, 2445.
6. D. F. Holcomb and R. E. Norberg, *Phys. Rev.*, 1955, **98**, 1074.
7. S. J. Rothman and N. L. Peterson, *Phys. Status Solidi*, 1969, **35**, 305.
8. K. Maier, *Phys. Status Solidi* (*a*), 1977, **44**, 567.
9. N. Q. Lam, S. J. Rothman, H. Mehrer and L. J. Nowicki, *Phys. Status Solidi* (*b*), 1973, **57**, 225.
10. S. J. Rothman, N. L. Peterson and J. T. Robinson, *Phys. Status Solidi*, 1970, **39**, 635.
11. H. M. Morrison, *Phil. Mag.*, 1975, **31**, 243.
12. S. M. Makin, A. H. Rowe and A. D. Le Claire, *Proc. phys. Soc.*, 1957, **2370**, 545.
13. J. M. Dupouy, J. Mathie and Y. Adda, *Mem. scient. Revue Metall.*, 1966, **63**, 481.
14. P. G. Shewmon, *J. Metals, N. Y.*, 1956, **8**, 918.
15. J. Combronde and G. Brebec, *Acta Met.*, 1971, **19**, 1393.
16. L. V. Pavlinov, A. M. Gladyshev and V. N. Bykov, *Fizica Metall.*, 1968, **26**, 823.
17. N. L. Peterson and S. J. Rothman, *Phys. Rev.*, 1967, **163**, 645.
18. C. Mao, *Phys. Rev.*, 1972, **5**, 4693.
19. D. S. Gorny and R. M. Altovski, *Fizica Metall.*, 1970, **30**, 85.

20. M. P. Dariel, G. Erez and G. M. J. Schmidt, *Phil. Mag.*, 1969, **19**, 1053.
21. M. P. Dariel, D. Dayan and A. Languille, *Phys. Rev.*, 1971, **B4**, 4348.
22. M. P. Dariel, G. Erez and G. M. J. Schmidt, *Phil. Mag.*, 1969, **19**, 1045.
23. F. H. Spedding and K. Shiba, *J. chem. Phys.*, 1972, **57**, 612.
24. T. S. Lundy and J. F. Murdoch, *J. appl. Phys.*, 1962, **53**, 1671.
25. M. Beyeler, *Thèse-Paris*, 1968; *J. Phys. (Fr.)*, 1968, **29**, 345.
26. S. L. Robinson and O. D. Sherby, *Phys. Status Solidi*, 1970, **al**, K199.
27. F. Y. Fradin and T. J. Rowland, *Appl. Phys. Lett.*, 1967, **6**, 207.
28. T. E. Volin and R. W. Balluffi, *Phys. Status Solidi*, 1968, **25**, 163.
29. A. Seeger, D. Wolf and H. Mehrer, *Phys. Status Solidi*, 1971. **B48**, 481.
30. A. C. Carter and C. G. Wilson, *Br. J. appl. Phys.*, 1968, **1**, 515.
31. J. E. Dickey, *Acta Met.*, 1959, **7**, 350.
32. G. A. Shirn, *Acta Met.*, 1955, **3**, 87.
33. M. A. Kanter, *Phys. Rev.*, 1957, **107**, 655.
34. H. J. Mayer, H. Mehrer and K. Maier, 'Rad. Effects in Semiconductors', 1976 (Inst. of Phys. Conference Series, 31), pp. 186–193.
35. G. V. Kidson, *Can. J. Phys.*, 1963, **41**, 1563.
36. H. Letaw, W. M. Portney and L. Slifkin, *Phys. Rev.*, 1956, **102**, 636.
37. H. Widmer and G. R. Gunther-Mohr, *Helv. Phys. Acta*, 1961, **34**, 635.
38. J. D. Meakin and E. Klokholm, *Trans. Met. Soc. AIME*, 1960, **218**, 463.
39. C. Coston and N. H. Nachtrieb, *J. Phys. Chem.*, 1964, **68**, 2219.
40. J. W. Miller, *Phys. Rev.*, 1969, **181**, 1095.
41. H. A. Resing and N. H. Nachtrieb, *J. Phys. Chem. Solids*, 1961, **21**, 40.
42. N. H. Nachtrieb and G. S. Handler, *J. chem. Phys.*, 1955, **23**, 1569.
43. H. Cordes and K. Kim, *J. appl. Phys.*, 1966, **37**, 2181; *Z. Naturf.*, 1965, **20a**, 1197.
44. A. Hässner and R. Hässner, *Phys. Status Solidi*, 1965, **11**, 575.
45. W. P. Ellis and N. H. Nachtrieb, *J. appl. Phys.*, 1969, **40**, 472.
46. W. Seith, *Z. Electrochem.*, 1933, **39**, 538.
47. E. M. Hampton and J. N. Sherwood, *Phil. Mag.*, 1974, **29**, 763.
48. P. Brätter and H. Gobrecht, *Phys. Status Solidi*, 1970, **37**, 869.
49. R. N. Ghoshtagore, *Phys. Rev.*, 1967, **155**, 598.
50. F. Dyment and C. M. Libanati, *J. Mater. Sci.*, 1968, **3**, 349.
51. N. E. Walsöe de Reca and C. M. Libanati, *Acta Met.*, 1968, **16**, 1297.
52. J. F. Murdock, T. S. Lundy and E. E. Stansbury, *Acta Met.*, 1964, **12**, 1033.
53. J. I. Federer and T. S. Lundy, *Trans. Met. Soc. AIME*, 1963, **227**, 592.
54. F. Schmitz and M. Fock, *J. nucl. Mater.*, 1967, **21**, 317.
55. C. J. Meechan, *2nd Nuclear and Eng. Sci. Conf.*, Phila. Pa., 1957, Paper No. 57, NESC.-7.
56. B. E. Davis and W. D. McMullen, *Acta Met.*, 1972, **20**, 593.
57. R. F. Peart, *J. Phys. Chem. Solids*, 1965, **26**, 1853.
58. R. E. Einziger, J. N. Mundy and H. A. Hoff, *Phys. Rev.*, 1978, **B17**, 440.
59. R. E. Pawel and T. S. Lundy, *J. Phys. Chem. Solids*, 1965, **26**, 937.
60. J. N. Mundy, C. W. Tse and W. D. McFall, *Phys. Rev.* 1976, **B13**, 2349.
61. K. Maier, H. Mehrer and G. Réin, *Z. Metallk.* 1979, **70**, 271.
62. J. N. Mundy, S. J. Rothman, N. Q. Lam, H. A. Hoff and L. J. Nowicki, *Phys. Rev.*, 1978, **B18**, 6566.
63. P. Flubacher, Report EIR Berichte, No. 49 (Switzerland), 1963.
64. Y. Adda and A. Kirianenko, *J. nucl. Mater.*, 1962, **6**, 130.
65. S. J. Rothman, R. Bastar, J. J. Hines and D. Rokop, *Trans. Met. Soc. AIME*, 1966, **236**, 897.
66. Y. Adda, A. Kirianenko and C. Mairy, *J. nucl. Mater.*, 1959, **1**, 300.
67. G. B. Federov, E. A. Smirnov and S. S. Moiseenko, *Met. Metalloved. Chist. Metal.*, 1968, No. 7, 124.
68. N. L. Peterson and S. J. Rothman, *Phys. Rev.*, 1964, **3A**, A842.
69. A. Bochvar, V. Kuznetsova and V. Sergeev, *Trans. 2nd Geneva Conf. on Peaceful Uses of At. Energy*, 1958, **VI**, 68.
    Y. Adda and A. Kirianenko, *J. nucl. Mater.*, 1959, **1**, 120.
    S. J. Rothman, L. T. Lloyd and A. L. Harkness, *Trans. AIME*, 1960, **218**, 605.
70. G. B. Federov and E. A. Smirnov, *Met. Metalloved. Chist. Metal*, 1967, No. 6, 181.
71. F. S. Buffington, K. Hirano and M. Cohen, *Acta Met.*, 1961, **9**, 434.
72. D. W. James and G. M. Leak, *Phil. Mag.*, 1966, **14**, 701.
73. R. J. Borg and D. Y. F. Lai, *Phil. Mag.*, 1968, **18**, 55.
74. V. Irmer and M. Feller-Kniepmeier, *Phil. Mag.*, 1972, **25**, 1345.
75. Th. Heumann and R. Imm, *J. Phys. Chem. Solids.*, 1968, **29**, 1613.
76. M. Badia, Thesis Nancy (France), 1969.
    M. Badia and A. Vignes, *Acta Met.*, 1969, **17**, 177.
77. D. Graham and D. H. Tomlin, *Phil. Mag.*, 1963, **8**, 1581.
78. W. Lange, A. Hässner and K. Siebov, *Isotopentechnic*, 1962, **2**, 42.
79. H. Bakker, *Phys. Status Solidi*, 1968, **28**, 569.
80. A. Y. Shinyayev, *Physics Metals Metallogr.*, 1963, **15**, 100.
    J. E. Reynolds, B. L. Averbach and M. Cohen, *Acta Met.*, 1957, **5**, 29.
    R. E. Hoffmann, F. W. Pickus and R. A. Ward, *Trans. Met. Soc. AIME*, 1956, **206**, 483.
    J. R. MacEwan, J. V. MacEwan and L. Yaffe, *Can. J. Chem.*, 1959, **37**, 1629.
    K. Monma, *Nippon kink Gakk.*, 1964, **28**, 188.

81. N. L. Peterson, *Phys. Rev.*, 1964, **136**, 568.
82. G. V. Kidson and R. Ross, *Proc. 1st UNESCO Conf.*, 'Radioisotopes in *Sci. Res.*', Pergamon Press, 1958, **1**, 185.
83. F. Cattaneo, E. Germagnoli and F. Grasso, *Phil. Mag.*, 1962, **7**, 1373.
84. R. E. Tate and E. M. Cramer, *Trans. Met. Soc. AIME*, 1964, **230**, 639.
85. M. Dupuy and D. Calais, *Trans. Met. Soc. AIME*, 1968, **242**, 1679.
86. W. Z. Wade, D. W. Short, J. C. Walden and J. W. Magana, *Met. Trans. A*, 1978, **9A**, 965.
87. E. M. Hampton, P. McKay and J. N. Sherwood, *Phil. Mag.*, 1974, **30**, 853.
88. M. P. Dariel, *Phil. Mag.*, 1973, **28**, 915.
89. A. Languille, D. Calais and B. Coqblin, *J. Phys. Chem. Solids*, 1974, **35**, 1461.
90. M. Fromont and G. Marbach, *J. Phys. Chem. Solids*, 1977, **38**, 27.
91. M. Fromont, A. Languille and D. Calais, *J. Phys. Chem. Solids*, 1974, **35**, 1367.
92. G. Hettich, H. Mehrer and K. Maier, *Scripta Met.*, 1977, **11**, 795
93. R. J. Borg and C. E. Birchenall, *Trans. Met. Soc. AIME*, 1960, **218**, 980.
94. K. Maier, H. Mehrer, E. Lessmann and W. Schüle, *Phys. Status Solidii*, 1976, **78**, 689.

Table 13.2 TRACER IMPURITY DIFFUSION COEFFICIENTS

In Ag

Element	*A*	*Q*	*Temp. range* °C	*Method*			*Ref.*
Cd^{115}	0.44 ± 0.05	41.7 ± 0.21	592–937	IVa(i)	s.c.	99.99%	10
In^{114}	0.41 ± 0.04	40.63 ± 0.20	612–936	IVa(i)	s.c.	99.99%	10
Sn^{113}	0.25 ± 0.03	39.30 ± 0.20	592–937	IVa(i)	s.c.	99.99%	10
Sb^{124}	0.169 ± 0.003	38.32	468–942	IVa(i)	s.c.	99.99%	11
Pd^{103}	$9.57^{+1.63}_{-1.37}$	56.75 ± 0.30	735–940	IVa(i)	s.c.	99.999%	9
Pt	6.0	56.9 ± 0.65	650–950	IV	p.c.	—	210
$Ru^{103/106}$	180 ± 70	65.8	793–945	IVa(i)	s.c.	99.99%	12
Cu^{64}	1.23 ± 0.25	46.1 ± 0.9	716–945	IVa(i)	s.c.	99.99%	13
Zn^{65}	0.54 ± 0.05	41.7 ± 0.2	643–924	IVa(i)	s.c.	99.99%	14
Ge^{71}	0.084 ± 0.042	36.5 ± 1.5	670–850	IVa(i)	p.c.	—	15
As	0.042	35.72 ± 0.62	640–940	IVa(iii)	p.c.	99.98%	107
Ni^{63}	21.9 ± 4.7	54.77 ± 0.50	748–951	IVa(i)	s.c.	99.99%	16
	15	51.9	630–930	IVa(i)	s.c.		212
Co^{60}	1.9	48.75	700–940	IVa(i)	s.c.	99.999%	106
Fe^{59}	2.6	49.0	800–932	IVa(i)	s.c.	99.999%	106
Au^{198}	0.85 ± 0.09	48.28 ± 0.25	718–925	IVa(i)	s.c.	99.99%	18
Hg^{203}	0.079 ± 0.008	38.1 ± 0.2	653–948	IVa(i)	s.c.	99.99%	13
Tl^{204}	0.15 ± 0.08	37.9 ± 1.5	644–801	IVa(i)	p.c.	—	15
Pb^{210}	0.22	38.1 ± 2	700–825	IVa(i)	p.c.	—	19
S^{35}	1.65	40.0	600–900	IVb	s.c.	99.999%	108
Al	0.13	38.1 ± 0.6	600–950	IVa(iii)	p.c.	—	211
Te^{125}	$0.47^{+0.17}_{-0.13}$	38.90 ± 0.69	770–940	IVa(i) and b	p.c.	—	109

In Al

Element	*A*	*Q*	*Temp. range* °C	*Method*			*Ref.*
Cu^{64}	0.647	32.27 ± 0.27	433–652	IVa(i)	s.c.	99.999%	26
Ag^{110}	0.118	27.83 ± 0.14	371–605	IVa(i)	s.c.	99.999%	26
	0.13 ± 0.001	28.0 ± 0.014	342–610	IVa(i)	s.c.	99.999%	27
Au^{198}	0.131	27.79 ± 0.24	368–655	IVa(i)	s.c.	99.999%	26
	0.077 ± 0.005	27.0 ± 0.11	423–609	IVa(i)	s.c.	99.999%	27
Mg	0.0623	27.44	325–650	IVb	p.c.	99.999%	215
	1.24	31.15	394–655	IVa(i)	s.c.	99.999%	214
Zn^{65}	0.259	28.86 ± 0.13	357–653	IVa(i)	s.c.	99.999%	26
	1.1 ± 0.4	30.9 ± 0.6	405–654	IVa(i)	p.c.	99.999%	25
Cd^{115m}	1.04 ± 0.17	29.7 ± 0.26	441–634	IVa(i)	s.c.	99.999%	27
Hg	15.3	33.9 ± 1.3	445–590	IVa(i)	p.c.	—	213
Ga^{72}	0.490	29.24 ± 0.14	406–653	IVa(i)	s.c.	99.999%	26
In^{114}	0.123	27.6	400–600	IVa(i) and b	p.c.	99.999%	28
	1.16	29.21	442–656	IVa(i)	s.c.	99.999%	29
Tl	116	36.5 ± 2.3	460–590	IVa(i)	p.c.	—	213

Table 13.2 TRACER IMPURITY DIFFUSION COEFFICIENTS—*continued*

Element	*A*	*Q*	*Temp. range* °C	*Method*			*Ref.*
In Al (cont.)							
Si	2.48	32.75	480–620	IIa	p.c.	99.999%	215
Ge^{71}	0.481	28.98 ± 0.21	401–654	IVa(i)	s.c.	99.999%	26
Sn^{113}	0.245	28.5	400–600	IVa(i) and b	p.c.	99.999%	28
Pb	50	34.8 ± 3.1	500–603	IVa(i)	p.c.	—	213
Sb^{124}	0.09	29.08	448–620	IVb	p.c.	99.995%	130
Cr^{51}	$\sim 5.10^4$	~ 58	586–649	IVa(i)	s.c.	99.999%	26
	2.4×10^3	61	500–650	—	—	—	131
U	$0.1^{+3}_{-0.09}$	28^{+6}_{-4}	525–625	IVa(i)	p.c.	99.995%	132
$Mn^{55/56}$	104	50.4	460–660	IVa(i)	s.c. and p.c.	99.999%	133
Fe^{59}	135 ± 68	46.0 ± 1.4	550–640	IVa(i)	s.c.	99.999%	27
	9.1×10^5	61.64	520–660	IVa(i)	s.c. and p.c.	99.995%	134
Co^{60}	464	41.74 ± 0.41	422–654	IVa(i)	s.c.	99.999%	26
	250	41.63	400–640	IVa(i) and b	p.c.	99.995%	135
Ni	4.4	34.82	470–650	IV	p.c.	99.995%	256
Zr	728	57.9	531–640	IVb	p.c.	99.999%	215
In Au							
Hg^{203}	$0.116^{+0.13}_{-0.06}$	37.38 ± 1.60	600–1 027	IIIa	p.c.	99.994%	20
Pt^{195}	$7.6^{+4.7}_{-2.9}$	60.90 ± 1.2	900–1 056	IVa(i)	p.c. and s.c.	99.98%	21
Ni^{63}	0.034 ± 0.007	42.0 ± 0.4	702–988	IVb	p.c.	99.93%	22
Co^{60}	0.068 ± 0.014	41.6 ± 0.4	702–948	IVb	p.c.	99.93%	22
Fe^{59}	0.082 ± 0.016	41.6 ± 0.4	754–948	IVb	p.c.	99.93%	22
Cu	0.105	40.65	700–906	IVa(iii)	p.c.	99.99%	17
Ag^{110}	0.072 ± 0.08	40.20 ± 0.25	699–1 008	IVa(i)	s.c.	99.99%	18
Al	0.052	34.3 ± 0.55	500–950	IVa(iii)	p.c.	—	216
In	$(7.5 \pm 0.9) \times 10^{-2}$	36.7 ± 0.26	700–1 000	IIa(ii)	p.c.	99.999%	110
Sn	$(4.12 \pm 0.5) \times 10^{-2}$	34.2 ± 0.28	700–1 000	IIa(ii)	p.c.	99.999%	110
Sb	$(1.14 \pm 0.19) \times 10^{-2}$	30.90 ± 0.38	730–1 005	IIa(ii)	p.c.	99.999%	111
In Be							
Cu $\parallel c$	0.38	47.3 }	700–1 000	II	s.c.	—	217 and 224
$\perp c$	0.42	46.0 }					
Ag^{110}	6.2 ± 1.6	46.1 ± 0.9	650–910	IVb	p.c.	99.85%	82
$Ag^{110} \parallel c$	0.41 ± 0.15	39.1 ± 1.6 }	656–897	IVb	s.c.	99.85%	82
$\perp c$	1.98 ± 0.7	45.7 ± 1.8 }					
Al^{26}	1.0	40.2 ± 4.3	795–1 085	IVb	p.c.	—	218
Fe^{59}	0.53 ± 0.2	51.8 ± 1.1	700–1 076	IVb	p.c.	99.85%	82
Ni^{63}	0.2 ± 0.12	58 ± 2.2	800–1 250	IVb	p.c.		125
V^{48}	29	58 ± 4.0	900–150 }				
Nb^{95}	2×10^4	85.9 ± 8.1	1 045–1 240 }	IVb	p.c.	—	219
Ce^{141}	3.1×10^2	72.5 ± 4.5	950–1 240 }				
In Ca							
C^{14}	2.7×10^{-3}	23.3	500–800	IVb	p.c. }		
Fe^{59}	3.2×10^{-5}	29.8	550–800	IVc	p.c. }	99.95%	142
Ni^{63}	1.0×10^{-5}	28.9	550–800	IVb	p.c. }		
U	1.1×10^{-5}	34.8	500–700	IVb	p.c. }		

Table 13.2 TRACER IMPURITY DIFFUSION COEFFICIENTS—*continued*

Element	*A*	*Q*	*Temp. range* °C	*Method*			*Ref.*
In Cd							
$Ag^{110}\parallel c$	1.41	24.64 ± 0.48	200–310				
$\perp c$	0.68	25.07 ± 0.65					
$Au^{195}\parallel c$	1.41	25.47 ± 0.08	175–310				
$\perp c$	3.16	26.43 ± 0.22					
$Zn^{65/69}\parallel c$	0.13	18.03 ± 0.25	155–320	IVa(i)	s.c.	99.999%	129
$\perp c$	0.084	18.02 ± 0.24					
$In^{114m}\parallel c$	0.101	17.45 ± 0.19	160–300				
$\perp c$	0.090	16.94 ± 0.15					
$Hg\parallel$ and $\perp c$	0.212	18.78 ± 0.06	155–300				
$Pb^{210}\parallel c$	0.060	16.46 ± 0.44	240–298	IVa(i)	s.c.	99.999%	225
$\perp c$	0.071	15.71 ± 0.19					
In Ce							
Au^{190}	$(4.4\pm0.7)\times10^{-3}$	14.9 ± 0.3	550–700(γ)				
	$\left(9.5^{+8.6}_{-5.2}\right)\times10^{-2}$	20.5 ± 1.6	726–774(δ)				
Ag^{110}	$\left(2.5^{+0.7}_{-0.6}\right)\times10^{-2}$	21.1 ± 0.5	580–695(γ)	IVa(i)	p.c.	—	209
	$\left(1.2^{+0.8}_{-0.5}\right)\times10^{-1}$	22.2 ± 1.0	723–776(δ)				
Fe^{59}	1.7×10^{-2}	11.9 ± 0.5	(γ)	IVa(i)	p.c.		220
	2×10^{-3}	7.7 ± 0.6	(δ)				
La^{110}	3.8×10^{-2}	24.5	725–775 (δ)	IVa(i)	p.c.		221
In Co							
Fe	0.11	60.5	950–1 370(γ)	IIa	p.c.	99.999%	39,34 and 43
Ni^{63}	0.4	67.4	1 130–1 370(γ)	IIa	p.c.	99.999%	34
Pt^{193m}	0.65	66.7 ± 5.4	1 081–1 208	IVb	p.c.	99.99%	222
Zn^{65}	0.12	63.7	808–T_c (Ferromagnetic)	IVb	s.c.	—	223
	0.08	60.8	T_c–1 300 (Paramagnetic)				
In Cr							
Fe^{55}	$0.47^{+1.63}_{-0.36}$	79.3 ± 4.8	1 245–1 413	IVb	p.c.		61
Mo^{99}	2.7×10^{-3}	58	1 100–1 420	IVb	p.c.		46
C^{14}	9×10^{-3}	26.5	1 200–1 500	IVb	p.c.		44
In Cu							
Zn^{65}	0.34 ± 0.04	45.6 ± 0.09	605–1 049	IVa(i)	s.c.		1
	$0.73^{+0.15}_{-0.12}$	47.5 ± 0.45	890–1 075	IVa(i)	p.c.	99.99%	99
Ga^{67}	$0.523^{+0.038}_{-0.034}$	46.02 ± 0.17	880–1 080	IVa(i)	p.c.	99.99%	100
Ge^{68}	$0.315^{+0.049}_{-0.042}$	44.31 ± 0.35	840–1 050	IVa(i)	p.c.	99.99%	100
	$0.397^{+0.037}_{-0.033}$	44.76 ± 0.20	650–1 015	IVa(i)	s.c.	99.998%	101
As^{73}	$0.202^{+0.041}_{-0.034}$	$42.13+0.44$	810–1 075	IVa(i)	p.c.	99.99%	102

Table 13.2 TRACER IMPURITY DIFFUSION COEFFICIENTS—*continued*

Element	*A*	*Q*	*Temp. range* °C	*Method*			*Ref.*
In Cu (cont.)							
Ni^{63}	3.8 ± 0.2	56.8 ± 0.1	695–1 061	IVa(i)	s.c.	99.99%	3
	2.7 ± 0.35	56.5 ± 0.28	742–1 076	IVa(i)	s.c.	99.998%	4
Co^{60}	5.7 ± 0.34	55.2 ± 1.1	700–950	IVa(i)	s.c.		5
	1.93 ± 0.06	54.1 ± 0.08	843–1 076	IVa(i)	s.c.	99.998%	4
Fe^{59}	1.4 ± 0.28	51.8 ± 0.47	830–1 074	IVa(i)	s.c.	99.998%	4
	1.01 ± 0.23	50.95 ± 0.46	716–1 056	IVa(i)	s.c.	99.998%	6
Mn^{54}	1.02	47.8	698–980	IVb	p.c.		103
V^{48}	2.48	51.39	680–1 070	IVb	p.c.		103
Cr^{51}	0.337	46.61	726–1 065	IVb	p.c.		103
Ag^{110}	0.61	46.60	455–1 064	IVb		99.99%	228
Cd^{115}	0.935 ± 0.27	45.7 ± 0.9	725–950	IVa(i)	s.c.	99.98%	7
In^{114m}	1.87	46.9 ± 0.8	798–1 081	IVa(i)			226
Sn^{113}	0.842	44.95 ± 0.1	737–1 047	IVa(i)	s.c.		227
Sb^{124}	0.34 ± 0.12	42.0 ± 0.7	600–1 002	IVa(i)	s.c.	99.99%	8
Pd^{103}	$1.71^{+0.23}_{-0.21}$	54.37 ± 0.3	807–1 056	IVa(i)	s.c.	99.999%	9
Ru^{103}	8.5	61.5	948–1 062	IVa(i)	s.c.	99.999%	229
Ir^{192}	10.6	66.01	911–1 030	IVa(i)	s.c.	99.99%	230
Nb^{95}	2.04 ± 0.6	60.06 ± 1.2	807–906	IVb	p.c.	99.999%	104
Au^{198}	0.69	49.7		IVa(i)	s.c.		2
Au^{195}	0.897	50.75 ± 0.24	811–1 068	IVa(i)	s.c.	99.99%	231
Hg^{203}	0.35	44.0		IVa(i)	s.c.		2
Tl	0.71	43.3	785–996	IVa(i)	s.c.	99.999%	36
Pb^{210}	0.862	43.56 ± 0.25	733–952	IVa(i)	s.c.	99.99%	231
Bi^{207}	0.766	42.53 ± 0.20	800–1 075	IVa(i)	s.c.	99.99%	231
S^{35}	23 ± 7	49.2 ± 0.7	800–1 000	IVb	s.c.	99.999%	105

In Fe

Element	*A*	*Q*	*Temp. range* °C	*Method*			*Ref.*
Be^{7}	5.34	52.1	800–1 500 (α and δ)	IVb	p.c.	99.9%	148
	0.1	57.6	1 100–1 350 (γ)	IVa(ii)	p.c.	99.9%	149
C^{14}	6.2×10^{-3}	19.2	350–850 (α)	IVb	p.c.	99.93%	58
	0.1	32.4	900–1 060 (γ)				
S^{35}	1.7	53.0	950–1 250 (γ)	IVb	p.c.		150
V^{48}	$0.25^{+0.1}_{-0.08}$	63.1 ± 2.2	1 120–1 380 (γ)				
Cr^{51}	$8.52^{+3.20}_{-2.33}$	59.9 ± 1.6	800–880 (α)	IVb	p.c.	99.98%	151
	$10.80^{+3.35}_{-2.56}$	69.7 ± 1.7	950–1 400 (γ)				
Mn^{54}	$1.49^{+1.0}_{-0.60}$	55.8 ± 2.5	700–760 (Ferrom. α)				
	$0.35^{+0.31}_{-0.17}$	52.5 ± 2.3	800–900 (Param. α)	IVb	p.c.	99.97%	152
	$0.16^{+0.06}_{-0.05}$	62.5 ± 1.0	920–1 280 (γ)				
	0.78[e]	60.1	809–901 (Param. α)	IVa(i)	s.c.	99.99%	153
Co^{60}	118[a]	68.3 ± 2.0	800–904 (Param. α)	IVb	p.c.	99.999%	40
	9.5	62.3 ± 1.0	830–888 (Param. α)	IVa(i)	p.c.	99.97%	55
	1.0	72.1	1 140–1 360 (γ)	IIa	p.c.	99.999%	34 and 42
	6.38 ± 0.8	61.4 ± 0.9	767–1 521 (α and δ)	IVa(i) and IVb	s.c. and p.c.	99.95%	154
	7.19[d]	62.2	683–726 (Ferrom. α)	IVb	p.c.		

Table 13.2 TRACER IMPURITY DIFFUSION COEFFICIENTS—*continued*

Element	*A*	*Q*	*Temp. range* °C	*Method*			*Ref.*
In Fe (cont.)							
Ni^{63}	9.9[(a)]	61.9±2.0	800–900 (Param. α)	IVb	p.c.	99.999%	40
	1.3±0.33	56.1±1.1[(b)]	810–900 (Param. α)	IVb	s.c. and p.c.		
	1.4	58.7[(c)]	600–680 (Ferrom. α)	IVb and c	s.c.	99.97%	41
	0.77±0.2	67.0±0.7	930–1 050 (γ)	IVb	s.c.		
	3.0	75.0	1 140–1 400 (γ)	IIa and IVa(iii)		99.999%	34 and 40
Pt^{193m}	2.7	70.7±2.5	960–1 260 (γ)	IVb	p.c.		222
Cu	300	67.8	772–880 (α)	IVa(iii)	s.c.	99.999%	155
	0.19	65.1	925–1 050 (γ)	IVa(iii)	p.c.	99.999%	155
Nb^{95}	530	82.3	1 160–1 290 (γ)		p.c.	99.95%	157
Ag^{110}	1.95×10^3	69.0	748–888 (α)	IVa(i)	p.c.		158
Sb	1.11×10^3	66.6	800–900 (α)	IVb	p.c.		159
Sn^{113}	5.4	55.5	700–760 (Ferrom. α)	IVb	p.c.		208
	2.4	53.0	800–910 (Param. α)				
Hf^{181}	$3.6^{+1.6}_{-1.1}\times10^3$	97.3±2.6	1 110–1 360 (γ)	IVb	p.c.	99.98%	151
Au^{198}	31.0[(a)]	62.4±1.2	800–900 (Param. α)	IVb	p.c.	99.999%	40
U	7×10^{-5}	31.8	950–1 075 (γ)		p.c.		160

(*a*) At $T<800$ °C D becomes increasingly less than would be calculated from these A and Q, due to the onset of ferromagnetism. *See* Figure 13.2(*a*) (Ni), 13.2(*b*) (Co), 13.2(*c*) (Au).
(*b*) Reference 40 criticizes the estimated error in this Q value and re-estimates it to be ±7 kcal mol^{-1}.
(*c*) The authors report a smooth transition in the values of D from about 800 °C at the bottom end of the linear Arrhenius range in the paramagnetic region, to about 700 °C at the top end of what they represent as a linear Arrhenius ferromagnetic range.
(*d*) Reference 161 discusses the differences between the results in the α-range of references 154 and 40.
(*e*) Reference 153 reports a strong dependence of A on impurity content of the Fe.

In Hf

Co	$5.3^{+1.8}_{-1.3}\times10^{-3}$	22.80±0.86	900–1 900 (α)	IVb	p.c.		235
Cr	$0.14^{+0.07}_{-0.05}$	51.1±1.3	833–1 525(α+β)				

In In

$Ag^{110m}\parallel c$	0.11	11.5±0.3	20–140	IVa(i)	s.c.	99.99%	94
$\perp c$	0.52	12.8±0.3					
Au^{198}	9×10^{-3}	6.7±0.9	20–140	IVa(i)	s.c.[(a)]	99.99%	94
Tl	0.049	15.5±0.78	49–155	IVa(i)	p.c.	99.99%	53

(*a*) For Au, randomly oriented s.c.'s were used.

In K

Au^{198}	$(1.29\pm0.61)10^{-3}$	3.23±0.29	6–53	IVa(i)	p.c.	99.95%	123
Na^{22}	$(5.8\pm1.6)10^{-2}$	7.45±0.17	0–62	IVa(i)	p.c.	99.95%	121
Rb^{86}	$(9.0\pm2.8)10^{-2}$	8.78±0.18	0–60	IVa(i)	p.c.	99.95%	124

In La

Au	$2.2^{+0.8}_{-0.6}\times10^{-2}$	18.1±0.6	600–800	IVa(i)	p.c.	99.97%	197
Ce^{141}	1.8×10^{-2}	25.0	866–897 (γLa)	IVa(i)	p.c.		156

Table 13.2 TRACER IMPURITY DIFFUSION COEFFICIENTS—*continued*

Element	*A*	*Q*	*Temp. range* °C	*Method*			*Ref.*
In Li							
Cu^{64}	0.47 ± 0.11	9.22 ± 0.22	50–121	IVa(i)	p.c.	99.98%	113
Ag^{110m}	0.37 ± 0.13	12.83 ± 0.25	67–161	IVa(i)	p.c.		119
Au^{195}	0.21 ± 0.08	10.99 ± 0.18	46–153	IVa(i)	p.c.		118
Na^{22}	0.41 ± 0.09	12.61 ± 0.15	52–176	IVa(i)	p.c.		116
Zn	$0.57^{+0.31}_{-0.20}$	12.98 ± 0.24	57–173	IVa(i)	p.c.	99.98%	117
Cd^{115m}	0.62	15.00 ± 0.53	82–176	IVa(i)	p.c.	99.98%	112
Hg	1.04	14.18 ± 0.47	58–173	IVa(i)	p.c.	99.98%	112
Ga^{72}	0.21	12.91 ± 0.32	116–173	IVa(i)	p.c.	99.98%	112
In^{114m}	0.39 ± 0.25	15.87 ± 0.36	75–170	IVa(i)	p.c.		115
Sn^{113}	0.62	15.00 ± 1.20	107–174	IVa(i)	p.c.	99.95%	114
Pb	1.6×10^{2}	25.21 ± 1.2	128–170	IVa(i)	p.c.	99.95%	114
Sb	1.6×10^{10}	41.5 ± 4.5	140–176	IVa(i)	p.c.	99.95%	114
Bi	5.3×10^{13}	47.3 ± 3.4	140–177	IVa(i)	p.c.	99.95%	114

Element	*A*	*Q*	*Temp. range* °C	*Method*			*Ref.*
In Mg							
$Ag^{110m} \parallel c$	3.62	31.8 ± 0.67	479–640				
$\perp c$	17.9	35.4 ± 0.35					
$Cd^{109} \parallel c$	1.29	33.6 ± 0.4	460–625				
$\perp c$	0.46	31.7 ± 0.4		IVa(i)	s.c.	99.99%	126
$In^{114m} \parallel c$	1.75	34.25 ± 0.7	474–633				
$\perp c$	1.88	34.0 ± 1.0					
$Sn^{113} \parallel c$ (a)	4.27	35.8 ± 0.8	475–630				
$Sb^{124} \parallel c$	2.57	32.8 ± 0.8	508–623				
$\perp c$	3.27	33.0 ± 0.4					
Zn^{65}	0.41 ± 0.08	28.6 ± 0.5	467–620	IVa(i)	p.c.	99.985%	127

(*a*) $D_{\perp}/D_{\parallel} = 1$ at 629.3 °C and $= 1.13$ at 585.2 °C.

Element	*A*	*Q*	*Temp. range* °C	*Method*			*Ref.*
In Mo							
Cr^{59}	1.88	81.8 ± 4.8	1 000–1 150		s.c.	99.8%	164
Fe^{59}	0.15	82.7 ± 1.9	1 000–1 350	IVb	p.c.	99.96%	232
Co^{60}	18^{+20}_{-9}	106.7 ± 3.8	1 850–2 330(a)	IVa(i)	s.c. and p.c.	99.98%	66
Nb^{95}	14^{+14}_{-7}	108 ± 3.2	1 900–2 275				
W^{185}	1.7	110.0	1 700–2 260	IVa(i)	p.c.		67 and 168
Re^{186}	0.097	94.7	1 700–2 100	IVa(i)	p.c.	—	60
U^{235}	7.6×10^{-3}	76.4	1 500–2 000	IVb	p.c.		165
	1.3×10^{-6}	75.6	1 800–2 100		p.c.		57
S^{35}	320	101.0	2 220–2 470	IVa(i)	s.c.	99.97%	166
P^{32}	0.19	80.5	2 000–2 220	IVa(i)	s.c.	99.97	167
C^{14}	2.8×10^{-4}	34.2	1 100–1 200	IVb	p.c.		89

(*a*) Samples annealed *in vacuo*. *D* is reported to be lower when samples are annealed in argon[66, 68], e.g. $D = 50 \exp(-118\,000/RT)$.

Element	*A*	*Q*	*Temp. range* °C	*Method*			*Ref.*
In Na							
Ag^{110}	$(15 \pm 9)10^{-3}$	5.11 ± 0.25	25–78	IVa(i)	p.c.		122
Au^{198}	$(3.34 \pm 1.0)10^{-4}$	2.21 ± 0.2	1–77	IVa(i)	p.c.		120
K^{42}	0.08 ± 0.018	8.43 ± 0.14	0–92	IVa(i)	p.c.		121
Rb^{86}	0.15 ± 0.04	8.49 ± 0.15	−1–86	IVa(i)	p.c.		121
Cd^{115}	0.37 ± 0.13	9.76 ± 0.22	0–90	IVa(i)	p.c.		122
In^{114}	1.79 ± 1.07	11.64 ± 0.35	20–90	IVa(i)	p.c.		122
Sn^{113}	0.54 ± 0.37	10.49 ± 0.46	43–90	IVa(i)	p.c.		122
Tl^{204}	0.52 ± 0.10	10.18 ± 0.12		IVa(i)	p.c.		122

Table 13.2 TRACER IMPURITY DIFFUSION COEFFICIENTS—*continued*

Element	*A*	*Q*	*Temp. range* °C	*Method*			*Ref.*
In Nb							
Ti^{44}	0.4	88.5	1 625–2 075	IVa(iii)	p.c.		169
	$0.099^{+0.1}_{-0.05}$	86.9 ± 2.2	994–1 492	IVa(i)	s.c.		170
V^{48}	0.47	90	1 625–2 075	IVa(iii)	p.c.		169
	2.21	85.0	1 000–1 400	IVb	s.c.		171
Cr^{51}	$0.3^{+0.28}_{-0.14}$	83.5 ± 1.8	953–1 435	IVa(i)	s.c.	99.92%	172
Fe^{55}	1.5	77.7 ± 1.2	1 390–2 100	IVa(ii)	p.c.	99.74%	68
	0.14	70.3 ± 1.4	1 390–1 895	IVa(i), (iii), IVb	s.c. and p.c.	99.9%	233
Co^{60}	0.042	61.43 ± 2.06	1 074–1 900	IVa(i)	s.c.		234
	0.11	65.6 ± 1.2	1 307–1 647	}			
Ni	0.077	63.1 ± 0.8	1 160–1 773	} IVa(i), (iii), IVb	s.c. and p.c.	99.9%	233
Cu	$D = 3.71 \times 10^{-10}$		1 556	}			
	$D = 1.02 \times 10^{-9}$		1 636	}			
		$Q \simeq 72$		}			
Zr	0.47	87	1 582–2 084	IVa(iii)	p.c.		169
Mo	92	122	1 725–2 182	IVa(iii)	p.c.		169
Ta^{182}	$1.0^{+0.9}_{-0.5}$	99.3 ± 2.4	1 100–2 073	IVa(i)	s.c. and p.c.	99.76%	173
W^{185}	7×10^{4}	156	1 902–2 170	IVa(iii)	p.c.		169
	0.5×10^{-3}	91.7	1 800–2 200	IVb	p.c.	99.8%	174
U^{235}	8.9×10^{-2}	76.8	1 500–2 000	IVb	p.c.	99.55%	165
Y^{91}	1.5×10^{-3}	55.6	1 200–1 600	IVb	s.c.	99.8%	175
Sn^{113}	$0.14^{+0.1}_{-0.05}$	78.9 ± 2.2	1 850–2 390	IVa	p.c.	99.85%	66 and 66a
C^{14}	0.033	37.9	930–1 800	IVa(i)	p.c.	99.5%	176
P^{32}	5.1×10^{-2}	51.5	1 300–1 800	IVa(i)	s.c.	99.9%	167
S^{35}	2.6×10^{3}	73.1	1 370–1 770	IVa(i)	s.c.	99.6%	177
In Nd							
Fe^{59}	4.6×10^{-3}	12.2 ± 0.4	αNd }	IVa(i)	p.c.		220
	1×10^{-2}	13.6 ± 1.0	βNd }				
In Ni							
Cu^{64}	$0.57^{+0.61}_{-0.29}$	61.7 ± 2.2	1 050–1 360	IVa(i)	p.c.	99.95%	35
	0.27 ± 0.06	61.02 ± 0.6	775–1 050	IVa(iii)	s.c.	99.999%	45
Co^{60}	0.59	64.4	850–1 370	IIa and IVb	p.c.	99.999%	33 and 34
Fe	0.22	60.4	950–1 370	IIa and IVb	p.c.	99.999%	34 and 48
Cr^{51}	$1.10^{+0.9}_{-0.5}$	65.1 ± 1.9	1 100–1 270	IVa(i)	p.c.	99.95%	47
V	0.87	66.5	800–1 300	IVb	p.c.	99.99%	143
Mo^{99}	1.6×10^{-3}	51.0	1 100–1 420	IVb	p.c.	—	46
W^{185}	$2.0^{+0.8}_{-0.6}$	71.5 ± 1.1	1 100–1 300	IVa(i)	p.c.	99.95%	59
	1.13	71.0	1 100–1 275	IVa(i)	—	99.9%	49
Pt^{193m}	2.5	68.5 ± 5.2	1 081–1 208	IVb	p.c.	99.99%	222
Au^{198}	2.0	65.0	900–1 100	IVa(ii)	p.c.	99.986%	37
Ce^{144}	0.66 ± 0.18	60.80 ± 0.81	700–1 100 }	IVa(i)	p.c.	99.99%	144
Nd^{147}	0.44 ± 0.13	59.82 ± 0.83	700–1 100 }				
U	$1.0^{+2.5}_{-0.9}$	$56.4^{+3.0}_{-6.0}$	975–1 075	IVa(ii)	p.c.		145
Pu	0.64 ± 0.5	51.0 ± 5	1 025–1 125	IIa(i)	p.c.	99.997%	146
Be^{7}	0.019	46.2	1 020–1 400	IVb	p.c.	99.9%	147
C^{14}	0.1	33.0	600–900	IVb	p.c.	—	50
S^{35}	1.4	52.3	800–1 225	IVa	s.c.		236

Table 13.2 TRACER IMPURITY DIFFUSION COEFFICIENTS—*continued*

Element	*A*	*Q*	*Temp. range* °C	*Method*			*Ref.*
In Ni (cont.)							
Ag	8.94	66.73±0.56	1 200–1 400	IVa(i)	s.c.	99.98	237
Sb^{125}	3.85	63.06±0.25	930–1 400	IVa(i)	s.c.	99.98	52
In	6.78	64.6±0.13	1 000–1 390	IVa(i)	s.c.	99.98	238
Sn^{113}	0.83	58.0	700–1 350	—	p.c.	99.8%	90
	30.0	65.5±0.5	1 273–1 483	IVb	p.c.	—	203
In Pb							
Cu^{64}	$(7.9\pm2.0)10^{-3}$	8.02±0.40	220–320	IVa(i)	p.c. and s.c.		85
Ag^{110}	$(4.6\pm1.0)10^{-2}$	14.44±0.50	120–320				
Au^{195}	4.1×10^{-3}	9.35±0.07	94–325	IVa(i)	s.c.	99.999%	31
Au^{198}	8.7×10^{-3}	10.0	190–320	IVa(i)	s.c.	99.999%	51
Pd^{109}	$(3.4\pm0.6)10^{-3}$	8.46	200–320	IVa(i)	s.c.	99.999%	239
Ni	$(9.4\pm3)10^{-3}$	10.63±0.37	210–320	IVa(i)	s.c.	99.999%	240
Co	9.10^{-3}	11.09	110–300	IVa(i)	p.c.	99.999%	241
Zn^{65}	$(1.65\pm0.26)10^{-2}$	11.42±0.16	180–300	IVa(i)	s.c.	99.999%	242
Cd^{115m}	0.41±0.10	21.23±0.2	150–320	IVa(i)	s.c.	99.999 9%	93
Hg^{203}	1.05±0.24	22.7±0.2	193–300	IVa(i)	s.c.		141
In	33	26.79±0.76	160–200	IVa(iii)	s.c.	99.999%	139
Tl^{204}	0.511±0.2	24.33±0.44	206–323	IVa(i)	p.c.	99.99%	32
Bi^{210}	$D=2.66\times10^{-10}$		290.4				
	$D=1.006\times10^{-9}$		322.7				
Sn^{113}	0.41±0.26	23.75±0.69	240–325	IVa(i)	s.c.	99.999%	243
Sb^{124}	0.29	22.20	188–315	IVa(i)	s.c.	99.999 9%	140
In Pr							
Fe^{59}	$2.1^{+0.5}_{-0.4}\times10^{-3}$	9.4±0.4	(α)	IVa(i)	p.c.		220
	$4^{+1.8}_{-1.3}\times10^{-3}$	10.4±0.4	(β)				
Co^{60}	$4.7^{+2.3}_{-1.6}\times10^{-2}$	16.4±0.8	660–790 (α)	IVa(i)	p.c.	99.93%	200
Cu^{64}	$8.4^{+2.5}_{-1.9}\times10^{-2}$	18.1±0.5	653–786 (α)	IVa(i)	p.c.	99.9%	199
	$5.7^{+2.1}_{-1.5}\times10^{-2}$	17.8±0.7	813–914 (β)				
Ag^{110}	$0.14^{+0.04}_{-0.03}$	25.4±0.5	(α)	IVa(i)	p.c.	99.93%	200
	$3.2^{+1.1}_{-0.8}\times10^{-2}$	21.5±0.7	(β)				
Au^{198}	$4.3^{+1.6}_{-1.1}\times10^{-2}$	19.7±0.6	650–780 (α)	IVa(i)	p.c.	99.93%	200
	$3.3^{+1.1}_{-0.8}\times10^{-2}$	20.1±0.6	800–910 (β)				
In^{114}	$9.6^{+10.2}_{-5.0}\times10^{-2}$	28.9±1.6	(β)	IVa(i)	p.c.	99.96%	201
La^{140}	$1.8^{+1.1}_{-0.7}\times10^{-2}$	25.7±1.1	(β)				
Ho^{166}	$9.5^{+4.3}_{-2.9}\times10^{-3}$	26.3±0.8	(β)				
$Zn^{65, 69}$	$0.18^{+0.09}_{-0.06}$	24.8±0.9	603–766 (α)	IVa(i)	p.c.		202
	$0.63^{+0.32}_{-0.21}$	27.0±0.9	822–921 (β)				
In Pt							
Au^{199}	1.3×10^{-1}	60.2±1.2	580–992	IVa	s.c.		245
Ag	1.3×10^{-1}	61.7±0.9	1 200–1 600	IVa(iii)	p.c.	99.99%	244
Al	1.3×10^{-3}	46.3±1.7	1 100–1 600				
Fe	2.5×10^{-2}	58.2±1.5	1 100–1 400				

Table 13.2 TRACER IMPURITY DIFFUSION COEFFICIENTS—*continued*

Element	*A*	*Q*	*Temp. range* °C	*Method*			*Ref.*
In Pt (cont.)							
Co^{57}	19.6[a]	74.2	900–1 050				
	$A_1 = 156$	$Q_1 = 80.1$		IVc	p.c.	99.9–99.99%	204
	$A_2 = 1.08 \times 10^{-9}$	$Q_2 = 20.9$	750–1 050				

(*a*) The log $D \sim 1/T$ plot is curved and best fitted over the whole temperature range by the two Arrhenius expressions. A_2 and Q_2 probably represent dislocation diffusion.

Element	*A*	*Q*	*Temp. range* °C	*Method*			*Ref.*
In Pu							
Co	1.2×10^{-2}	12.7 ± 3.6	344–426 (δ-Pu)	IVa(i)	p.c.		246
	1.4×10^{-3}	9.9 ± 0.6	484–621 (ε-Pu)	IVa(i)	p.c.		253
Au^{198}	5.7×10^{-5}	10.3 ± 1.2	480–640 (ε-Pu)				
Ag^{110m}	4.9×10^{-5}	9.6 ± 0.8	480–640 (ε-Pu)	IVa(i)	p.c.		247
Cu	1×10^{-3}	12.3 ± 2.0	500–580 (ε-Pu)	IIc	p.c.		247
In Se							
S^{35} ∥ *c*	1.10×10^{-5}	13.8	60–90	IVb	s.c	—	205
⊥ *c*	1.70×10^{3}	26.5					
Tl^{204}	2.0×10^{-3}	16.6	60–150	IVc	p.c.	—	206
In Sn							
Cu^{64}∥*c*	$D \approx 2 \times 10^{-6}$		25	IVa(i)	s.c.	—	95
⊥*c*	2.4×10^{-3}	7.9 ± 1.0	140–220				
Ag^{110}∥*c*	7.1×10^{-3}	12.3 ± 0.4					
⊥*c*	0.18	18.4 ± 0.5	135–225	IVa(i)	s.c.		87
Au^{198}∥*c*	5.8×10^{-3}	11.0 ± 0.4					
⊥*c*	0.16	17.7 ± 0.5					
Zn^{65}∥*c*	1.10×10^{-2}	$12.0 + 0.4$	135–223	IVa(i)	s.c.	99.999%	198
⊥*c*	8.4	21.3 ± 0.2					
In^{114}∥*c*	12.2 ± 2.5	25.6 ± 0.5	180–221	IVa(i)	s.c.	99.998%	30
⊥*c*	34.1 ± 6.5	25.8 ± 0.5					
Tl^{204}	$(1.2 \pm 0.04)10^{-3}$	14.7 ± 0.6	137–216	IVa(ii)	p.c.	99.999%	138
Sb^{124}∥*c*	$71^{+22}_{-17.1}$	29.0 ± 0.3	190–226	IVa(i)	s.c.		198
⊥*c*	73^{+36}_{-24}	29.4 ± 0.4					
Co^{60}	5.5 ± 0.5	22.0 ± 2.0	140–217	IVb	s.c. and p.c.	99.999%	54
Hg^{203}∥*c*	$7.5^{+6.4}_{-3.5}$	25.3 ± 0.6	174–226	IVa(i)	s.c.	99.999 9%	136
⊥*c*	30^{+20}_{-12}	26.8 ± 0.5					
Cd∥*c*	220	28.2 ± 0.6	190–225	IVa(i)	s.c.		198
⊥*c*	120	27.6 ± 0.5					
In Ta							
C^{14}	0.012	40.3	1 450–2 200	IVa(i)	p.c.	—	194
S^{35}	100	70.0	1 970–2 110	IVb	p.c.	99.0%	162
Nb^{95}	0.23	98.7	921–2 484	IVa(i)	s.c.	99.7%	77
Fe^{59}	0.505	71.4	930–1 240	—	p.c.	—	76
Mo^{99}	1.8×10^{-3}	81.0	1 750–2 220	IVb	p.c.		44
Y^{91}	0.12	72.2	1 200–1 500	IVb	s.c.	99.8/99.9	175
U^{235}	7.6×10^{-5}	84.4	1 600–2 150				57
In Te							
Se^{75}	2.6×10^{-2}	28.6	320–440	IVa(i)	p.c.	—	81
Hg^{203}	3.4×10^{-5}	18.7	270–440				

Table 13.2 TRACER IMPURITY DIFFUSION COEFFICIENTS—*continued*

Element	*A*	*Q*	*Temp. range* °C	*Method*			*Ref.*
In Th							
Fe	5×10^{-3}	19.3	α-Th				
	4×10^{-3}	17.1	β-Th				
Co	5×10^{-4}	13.2	α-Th				
	4×10^{-3}	15.6	β-Th	IIa(ii)	p.c.	99.95%	248
Ni	4×10^{-3}	18.6	α-Th				
	4×10^{-4}	9.1	β-Th				
V	1.9×10^{-2}	31.0					
Nb	5×10^{-1}	48.2	1 370–1 665	IIa(ii)	p.c.	99.95%	249
Ta	5.7×10^{-1}	50.3	β-Th				
Pa^{231}	1.26×10^{-2}	74.7 ± 2.8	690–910 (α)	IVc	p.c.	—	207
U^{233}	2.21	79.3 ± 0.9					
In Ti							
α-Ti							
Ni^{63}	$D=2.1\pm0.2\times10^{-8}$		868				
	$D=2.3\pm0.2\times10^{-8}$		844				
	$D=7.6\pm1.1\times10^{-8}$		786	IVa(i)	p.c.		180
	$D=1.8\pm0.2\times10^{-9}$		698				
	$D=6.7\pm0.7\times10^{-10}$		639				
C^{14}	$(7.9\pm4.5)10^{-4}$	30.5 ± 1.1	600–800	IVb	p.c.	—	83
β-Ti							
Be^{7}	0.80	40.2	915–1 300	IVb	p.c.	99.96%	187
P^{32}	$A_1=5^{(a)}$	$Q_1=56.5$	945–1 600	IVa	p.c.	99.7–99.9%	65
	$A_2=3.62\times10^{-3}$	$Q_2=24.1$					
Sc^{46}	4×10^{-3}	32.4	940–1 570	IVa(i)	p.c.	99.95%	84
V^{48}	$A_1=3.4^{(a)}$	$Q_1=61.5$	900–1 540	IVa(i)	p.c.	99.9%	63 and 65
	$A_2=1.0\times10^{-3}$	$Q_2=34.7$					
Cr^{51}	$A_1=14.0^{(a)}$	$Q_1=65.5$	970–1 650	IVa(ii)			
	$A_2=7.4\times10^{-3}$	$Q_2=36.6$					
Mn^{54}	$A_1=12.0^{(a)}$	$Q_1=64.5$	930–1 650	IVa(i)			
	$A_2=7.6\times10^{-3}$	$Q_2=34.3$					
Fe^{55}	$A_1=15.0^{(a)}$	$Q_1=60.7$	920–1 650	IVa(ii)	p.c.	99.7–99.9%	64 and 65
	$A_2=8.0\times10^{-3}$	$Q_2=30.0$					
Co^{60}	$A_1=16.0^{(a)}$	$Q_1=61.3$	910–1 650	IVa(i)			
	$A_2=13.0\times10^{-3}$	$Q_2=30.9$					
Ni^{63}	$A_1=20.0^{(a)}$	$Q_1=60.0$	930–1 650	IVa(ii)			
	$A_2=17.0\times10^{-3}$	$Q_2=31.6$					
Cu	$A_1=11.3^{(a)}$	$Q_1=60.2$	960–1 460	IVa(iii)	p.c.	'Iodide'	88
	$A_2=2.1\times10^{-3}$	$Q_2=29.2$					
Zr^{95}	4.7×10^{-3}	35.4	920–1 500	IVb	p.c.	98.94	70
Nb^{95}	$A_1=9.5^{(a)}$	$Q_1=69.5$	1 000–1 650	IVa(ii)			
	$A_2=1.3\times10^{-3}$	$Q_2=34.9$			p.c.	99.7–99.9%	64 and 65
Mo^{99}	$A_1=3.6^{(a)}$	$Q_1=65.0$	900–1 650	IVa(i)			
	$A_2=0.7\times10^{-3}$	$Q_2=36.9$					
Ag^{110}	3×10^{-3}	43.0	940–1 590	IVa(i)	p.c.	99.95%	84
Sn^{113}	$A_1=9.5^{(a)}$	$Q_1=69.2$	950–1 600	IVa	p.c.	99.7–99.9%	65
	$A_2=0.38\times10^{-3}$	$Q_2=31.6$					
Ta^{182}	$A_1=13^{(a)}$	$Q_1=74.0$	914–1 600	IVa(i)	p.c.		92
	$A_2=3\times10^{-4}$	$Q_2=33.5$					
W^{185}	3.6×10^{-3}	43.9	900–1 250	IVb	p.c.		70
U^{235}	5.1×10^{-4}	29.3	900–1 400	IVb	p.c.		195
	2×10^{-3}	32.9	915–1 025		p.c.		160
Pu	10^{-6}	15.3	900–1 100		p.c.		197

(*a*) The Arrhenius plots of diffusion in β-Ti frequently show a distinct positive curvature and results are often represented as $D=A^1\exp(-Q_1/RT)+A_2\exp(-Q_2/RT)$.

Table 13.2 TRACER IMPURITY DIFFUSION COEFFICIENTS—*continued*

Element	*A*	*Q*	*Temp. range* °C	*Method*			*Ref.*
In Tl							
b.c.c. Tl							
Ag^{110m}	4.2×10^{-2}	11.9 ± 0.4	240–300	IVa(i)	p.c.	99.999 9%	97
Au^{198}	5.2×10^{-4}	6.0 ± 0.4					
h.c.p. Tl							
$Ag^{110m}\parallel c$	2.7×10^{-2}	11.2 ± 0.4	115–220	IVa(i)	s.c.	99.999 9%	97
$\perp c$	3.8×10^{-2}	11.8 ± 0.4					
$Au^{198}\parallel c$	2.0×10^{-5}	2.8 ± 0.4	90–220				
$\perp c$	5.3×10^{-4}	5.2 ± 0.4					
In U							
In γ-U							
Cu^{64}	$1.96^{+0.35}_{-0.30}\times10^{-3}$	24.06 ± 0.40	786–1 040	IVa(i)	p.c.	99.99%	78
Ni^{63}	$5.36^{+0.86}_{-0.74}\times10^{-4}$	15.66 ± 0.35	786–1 040				
Co^{60}	$3.51^{+0.95}_{-0.75}\times10^{-4}$	12.57 ± 0.58	784–990				
Fe^{59}	$2.69^{+0.43}_{-0.37}\times10^{-4}$	12.01 ± 0.34	787–990				
Mn^{54}	$1.81^{+1.95}_{-0.94}\times10^{-4}$	13.88 ± 1.66	787–939				
Cr^{51}	$5.47^{+1.04}_{-0.87}\times10^{-3}$	24.46 ± 0.43	797–1 038				
Nb^{95}	$4.87^{+1.18}_{-0.94}\times10^{-2}$	39.65 ± 0.50	790–1 103				
Au	$(4.86\pm1.3)\times10^{-3}$	30.4 ± 0.520	784–1 007	IVa(i)	p.c.	99.99%	79
In β-U							
Co	$1.5\pm0.2\times10^{-2}$	27.45 ± 0.15	691–763	IVa(i)	p.c.	99.98%	193
Cr^{51}	$D=3.6\times10^{-10(b)}$		670				
	$D=1.07\times10^{-9}$		740				
	$D=1.77\times10^{-9}$		748.2	IVa(i)	p.c.	99.993%	80
Fe^{59}	$D=8.71\times10^{-9(a)}$		701				
	$D=2.6\times10^{-8(a)}$		760				

(*a*) The mean of two values.
(*b*) Values also reported at 5 other temperatures between 670° and 740°, but results too scattered for Arrhenius representation.

Element	*A*	*Q*	*Temp. range* °C	*Method*			*Ref.*
In V							
Co^{60}	$1.12^{+0.41}_{-0.30}$	70.45 ± 1.05	1 025–1 853	IVa(i)	s.c.		250
Ta^{182}	$2.44^{+1.87}_{-1.06}$	71.98 ± 1.92	1 100–1 800	IVa(i)			251
Ti	0.10	68.08	1 100–1 350	IVa(i)	p.c.	99.98%	252
	34.1	86.92	1 350–1 803				
Cr	$\left(9.54^{+13.7}_{-5.62}\right)10^{-3}$	64.6 ± 2.3	960–1 200	IVb	p.c.	99.8%	62
Fe^{59}	0.373 ± 0.051	71.0 ± 0.18	960–1 345	IVa(i)	s.c.		56
	274 ± 15	92.17 ± 0.11	1 415–1 817				
U^{235}	10^{-4}	61.4	1 100–1 500				
C^{14}	$(4.9\pm2.3)10^{-3}$	27.3 ± 0.96	845–1 130	IVa(i)	p.c.		38
P^{32}	2.45×10^{-2}	49.8	1 200–1 450	IVa(i)	p.c.	99.8%	163
S^{35}	3.1×10^{-2}	34.0	1 320–1 520	IVb	p.c.	99.8%	162

Table 13.2 TRACER IMPURITY DIFFUSION COEFFICIENTS—*continued*

Element	*A*	*Q*	*Temp range* °C	*Method*			*Ref.*
In W							
C^{14}	$(3.45 \pm 0.12)10^{-3}$	37.8 ± 0.14	1 500–1 800	IVb	s.c.	99.994%	190
	0.3	49.6	1 250–1 450	IVb	p.c.		89
P	26.8	121.8 ± 7.4	1 880–2 180	IVa(i)	s.c.	99.99%	54
Fe^{59}	11.5	66.0	930–1 240	—	p.c.	—	76
Y^{91}	6.7×10^{-3}	68.1	1 200–1 600	IVb	s.c.	99.8–99.9%	175
Nb^{95}	3.01 ± 0.1	137.6 ± 0.7	1 305–2 367	IVa(i)	p.c.	—	192
Mo^{99}	0.05	121	2 000–2 400	IV	p.c.	—	191
Ta^{182}	3.05 ± 0.2	139.9 ± 1.3	1 305–2 375	IVa(i)	p.c.	—	192
Re^{186}	19.5	141	2 100–2 400	IV	p.c.	—	191
$^{183/4}$	275 ± 110	162.8 ± 2.5	2 666–3 228	IVa(i)	s.c.	99.99%	75
U	2×10^{-3}	103.5	1 700–2 200	—	p.c.	—	57
In Y							
Fe	1.8×10^{-2}	20.0	900–1 330 (α)		p.c.	99+%	44
Ag	5.4×10^{-3}	18.0	905–1 180 (α)				
C^{14}	$(1.7 \pm 0.6) \times 10^{-3}$	58.0 ± 17.0	1 000–1 460	IVb	p.c.		96
In Zn							
$Ni^{63} \parallel c$	$8.1^{+32}_{-6.5}$	32.60 ± 1.98	290–391	IVa(i)	s.c.	99.999%	91
$\perp c$	$0.43^{+0.43}_{-0.21}$	28.98 ± 0.85					
$Cu^{64} \parallel c$	2.22 ± 0.57	29.53 ± 0.29	338–415	IVa(i)	s.c.	99.999%	86
$\perp c$	2.00 ± 0.54	29.92 ± 0.30					
$Ga^{72} \parallel c$	0.016 ± 0.001	18.40 ± 0.06	240–403	IVa(i)	s.c.	99.999%	86
$\perp c$	0.018 ± 0.001	18.15 ± 0.08					
$Ag^{110} \parallel c$	0.32 ± 0.02	26.0 ± 0.1	271–413	IVa(i)	s.c.	99.999%	23
$\perp c$	0.45 ± 0.07	27.6 ± 0.2					
$Cd^{115m} \parallel c$	0.114 ± 0.008	20.54 ± 0.08	224–416	IVa(i)	s.c.	99.999%	24
$\perp c$	0.117 ± 0.003	20.42 ± 0.03					
$In^{114} \parallel c$	0.062 ± 0.008	19.1 ± 0.1	171–395	IVa(i)	s.c.	99.999%	23
$\perp c$	0.14 ± 0.02	19.6 ± 0.1					
$Sn^{113} \parallel c$	0.15 ± 0.07	19.4 ± 0.5	298–400	IVa(i)	s.c.		128
$\perp c$	0.13 ± 0.03	18.4 ± 0.2					
$Au^{198} \parallel c$	0.97 ± 0.22	29.7 ± 0.3	315–415	IVa(i)	s.c.	99.999%	24
$\perp c$	0.29 ± 0.12	29.7 ± 0.5	347–415				
$Hg^{203} \parallel c$	0.056 ± 0.002	19.70 ± 0.05	260–413	IVa(i)	s.c.	99.999%	98
$\perp c$	0.073 ± 0.006	20.18 ± 0.09					
C^{14}	$(1.0 \pm 0.3) \times 10^{-5}$	$12.0 + 2.0$	166–383	IVb	p.c.		96
In Zr							
V	1.12×10^{-8}	22.9	600–850(α)	IVb	p.c.	99.84%	171
	7.59×10^{-3}	45.8	870–1 200(β)				
	0.32	57.2	1 200–1 400(β)				
Cr^{51}	4.9×10^{-3}	30.1	623–832(α)	IVb	p.c.[(a)]	99.5% and 99.999%	178
	4.17×10^{-3}	32.0	900–1 200(β)	IVb	p.c.	99.7%	70
Mn^{54}	2.4×10^{-3}	30.2	620–830(α)	IVb	p.c.[(a)]	99.5% and 99.999	179
	5.6×10^{-3}	33.6	960–1 160(β)				
Fe^{59}	2.5×10^{-2}	48.0	(α)	IV	p.c.		74
	4.0×10^{-2}	30.0	(β)				
	9.1×10^{-3}	27.0	900–1 400(β)	IVb	p.c.	99.7%	70
	$D = 3.7 \times 10^{-8}$		700 (α)	IVa(i)	p.c.		180
	$D = 3.5 \times 10^{-7}$		798 (α)				

Table 13.2 TRACER IMPURITY DIFFUSION COEFFICIENT—*continued*

Element	*A*	*Q*	*Temp. range* °C		*Method*		*Ref.*
InZr (*cont.*)							
Co^{60}	$D=5.8$ to 0.74×10^{-8}		583(α)	IVa(i)	s.c.[(b)]		180
	$(3.26\pm2.77)10^{-3}$	21.82 ± 0.9	920–1 600(β)	IVa(i)	p.c.	99.99%	182
Ni^{63}	$D=1.2\times10^{-7}$		698 } (α)	IVa(i)	s.c.[(b)]		180
	$D=4.0\times10^{-7}$		750 } (α)				
	$D=9.0\times10^{-7}$		801 } (α)				
	$D=8.0\times10^{-7}$		830 } (α)				
Cu^{64} ∥c	0.40	35.5 }	615–859 (α-Zr)	IVa(i)	s.c.	>99.95%	137
⊥c	0.25	36.9 }					
Nb^{95}	6.6×10^{-6}	31.5	740–857(α)	IVb	p.c.	99.99%	183
	$9\times10^{-6}(T/1\,136)^{18.1}$	$25.1+35.5\times(T-1\,136)$	880–1 750(β)	IVa(i)	s.c.	99.94%	69
Mo^{99}	6.22×10^{-8}	24.76	600–850(α)	IVb	p.c.		184
	$A_1=1.99\times10^{-4}$	$Q_1=35.2$ }	900–1 600(β)	IVb	p.c.		185
	$A_2=2.63$[(c)]	$Q_2=68.3$ }					
Ta^{182}	100	70.0	700–800(α)	IVb	p.c.	99.6%	73
W^{185}	0.41	55.8	900–1 250(β)	IVc	p.c.	99.7%	70
U^{235}	$A_1=0.36$ [(c)]	58.0 }	950–1 500 (β-Zr)	IVa	p.c.		72
	$A_2=3\times10^{-6}$	19.7 }					
Ce^{141}	$A_1=3.16\times10^{-2}$	$Q_1=41.4$ }	800–1 600(β)	IVb	p.c.		185
	$A_2=42.17$[(c)]	$Q_2=74.1$ }					
Ag^{110}	5.1×10^{-3}	44.7	764–847 (α-Zr)	IVa(i)	p.c.	99.99%	254
	5.7×10^{-4}	32.7	951–1 190 (β-Zr)				
Au^{198}	$D=1.3\times10^{-11}$		840(α)	IVa(i)	s.c.	99.83%	181
Rb	$1.17\pm0.73\times10^{-2}$	61.0 ± 0.3	760–863(α) }		p.c.		186
	$8.80\pm2.90\times10^{-4}$	36.7 ± 0.6	880–1 030(β) }				
Be^{7}	0.33	31.9	(α-Zr) }	IVa(i)	p.c.	99.99%	255 and 257
	8.33×10^{-2}	31.8	(β-Zr) }				
Zn^{65}	$D=2.8\times10^{-11}$		826(α)	IVa(i)	s.c.	99.93	181
Sn^{113}	1.0×10^{-8}	22.0	700– (α) }	IVb	p.c.	99–99.7%	71
	5×10^{-3}	39.0	–1 250(β) }				
C^{14}	2×10^{-3}	36.23	600–850(α) }	IVb	p.c.		188
	8.9×10^{-2}	31.80	870–1 250(β) }				
P^{32}	0.33	33.3	950–1 200(β)	IVa(i)	p.c.	99.94	163
S^{35}	8.9	44.2	597–807	IVa(i)	p.c.	99.94	189

(*a*) *D* is apparently independent of impurity content in this range.
(*b*) Measurements on s.c. of unspecified orientation.
(*c*) *D* is given as $D=A_1 \exp -Q_1/RT+A_2 \exp -Q_2/RT$.

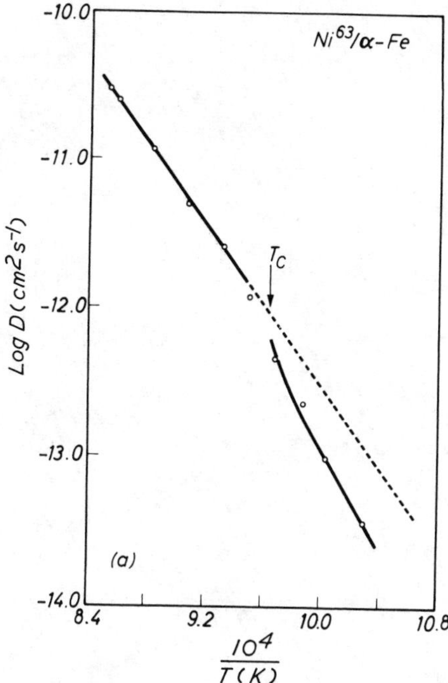

Figure 13.2(a) *Tracer diffusion of* Ni^{63} *into* αFe^{40}

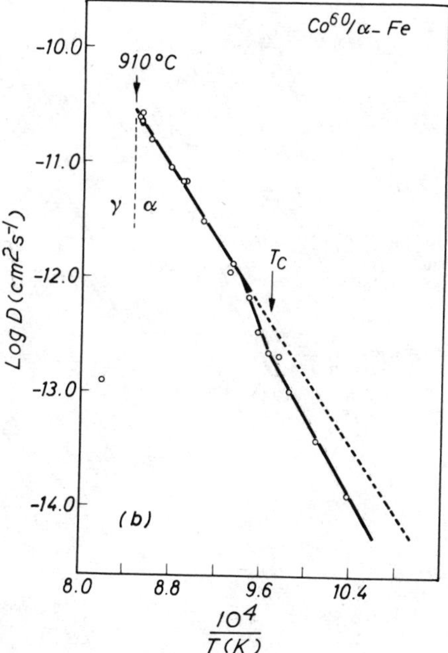

Figure 13.2(b) *Tracer diffusion of* Co^{60} *into* αFe^{40}

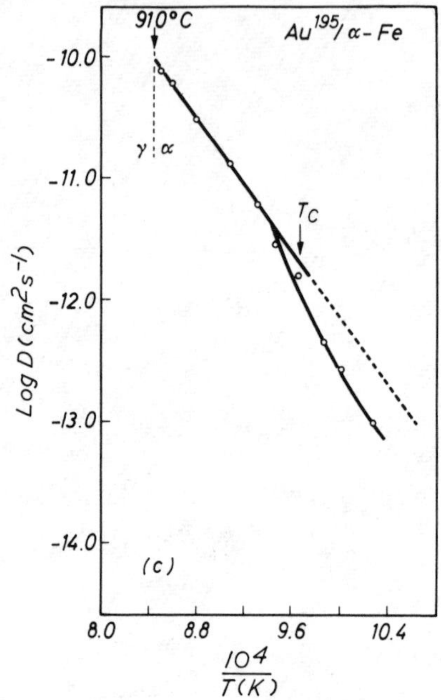

Figure 13.2(c) *Tracer diffusion of* Au^{195} *into* αFe^{40}

REFERENCES TO TABLE 13.2

1. J. Hino, C. Tomizuka and C. Wert, *Acta Met.*, 1957, **5**, 41.
2. C. Tomizuka. Quoted from D. Lazarus 'Solid State Physics' (1960) Vol. 10.
3. A. Ikushima, *J. phys. Soc. Japan*, 1959, **14**, 1636.
4. C. A. Mackliet, *Phys. Rev.*, 1958, **109**, 1964.
5. M. Sakamoto, *J. phys. Soc. Japan*, 1958, **13**, 845.
6. J. G. Mullen, *Phys. Rev.*, 1961, **121**, 1649.
7. T. Hirone, N. Kanitomi, M. Sakamoto and H. Yamaki, *J. phys. Soc. Japan*, 1958, **13**, 838.
8. M. C. Inman and L. W. Barr, *Acta Met.*, 1960, **8**, 112.
9. N. L. Peterson, *Phys. Rev.*, 1963, **132**, 2471.
10. C. T. Tomizuka and L. Slifkin, *Phys. Rev.*, 1954, **96**, 610.
11. E. Sonder, L. Slifkin and C. T. Tomizuka, *Phys. Rev.*, 1954, **93**, 970.
12. C. B. Pierce and D. Lazarus, *Phys. Rev.*, 1959, **114**, 686.
13. A. Sawatzky and F. E. Jaumot Jr., *J. Metals, N. Y.*, 1957, **9**, 1207.
14. A. Sawatzky and F. E. Jaumot Jr., *Phys. Rev.*, 1955, **100**, 1627.
15. R. E. Hoffmann, *Acta Met.*, 1958, **6**, 95.
16. T. Hirone, S. Miura and T. Suzuoka, *J. phys. Soc. Japan*, 1961, **16**, 2456.
17. A. Vignes and J. P. Haeussler, *Men. scient. Revue Metall.*, 1966, **63**, 1091.
18. W. C. Mallard, A. B. Gardner, R. F. Bass and L. M. Slifkin, *Phys. Rev.*, 1963, **129**, 617.
19. R. E. Hoffmann, D. Turnbull and E. W. Hart, *Acta Met.*, 1955, **3**, 417.
20. A. J. Mortlock and A. H. Rowe, *Phil. Mag.*, 1965, **11**, 1157.
21. A. J. Mortlock, A. H. Rowe and A. D. Le Claire, *Phil. Mag.*, 1960, **5**, 803.
22. D. N. Duhl, K. Hirano and M. Cohen, *Acta Met.*, 1963, **11**, 1.
23. J. H. Rosolowski, *Phys. Rev.*, 1961, **124**, 1828.
24. P. B. Ghate, *Phys. Rev.*, 1963, **131**, 174.
25. J. E. Hilliard, B. L. Averbach and M. Cohen, *Acta Met.*, 1959, **7**, 86.
26. N. L. Peterson and S. J. Rothman, *Phys. Rev.*, 1970, **B1**, 3264.
27. W. B. Alexander and L. M. Slifkin, *Phys. Rev.*, 1970, **B1**, 3274.
28. M. S. Anand and R. P. Agarwala, *Phys. Status Solidi*, 1970, **A1**, 41K.
29. G. M. Hood and R. S. Schultz, *Phys. Rev.*, 1971, **B4**, 2339.
30. A. Sawatsky, *J. appl. Phys.*, 1958, **29**, 1303.
31. A. Ascoli, *J. Inst. Metals*, 1961, **89**, 218.
32. H. A. Resing and N. H. Nachtrieb, *J. phys. Chem. Solids*, 1961, **21**, 40.

33. K. Hirano, R. P. Agarwala, B. L. Averbach and E. Cohen, *J. appl. Phys.*, 1962, **33**, 3049.
34. M. Badia and A. Vignes, *Acta Met.*, 1969, **17**, 177.
35. K. Monma, H. Suto and H. Oikawa, *Nippon Kink. Gakk.*, 1964, **28**, 192.
36. S. Komura and N. Kunitomi, *J. phys. Soc. Japan*, 1963, **18**, (Supp. II), 208.
37. A. D. Kurtz, B. L. Averbach and M. Cohen, *Acta Met.*, 1955, **3**, 442.
38. R. Son, S. Ihara, M. Miyake and T. Sano, *Nippon Kink. Gakk.*, 1969, **33**, 1.
39. H. W. Mead and C. E. Birchenall, *J. Metals*, 1955, **7**, 994.
40. R. J. Borg and D. Y. F. Lai, *Acta Met.*, 1963, **11**, 861.
41. K. Hirano, M. Cohen and B. L. Averbach, *Acta Met.*, 1961, **9**, 440.
42. T. Suzuoka, *Trans. Japan Inst. Metals*, 1961, **2**, 176.
43. M. Aucouturier and P. Lacombe, *Acta Met.*, 1965, **13**, 125.
44. J. E. Murphy, G. H. Adams and W. N. Cathey, *Met. Trans. A*, 1975, **6a**, 343.
45. H. Helfmeier and M. Feller-Kniepmeier, *J. appl. Phys.*, 1970, **41**, 3202.
46. P. L. Gruzin, S. V. Zemskii and I. B. Rodina, 'Metallurgy and Metallography of Pure Metals, No. 4, Moscow, 1963, 243, (AERE Trans. 1032).
47. K. Monma, H. Suto and H. Oikawa, *Nippon Kink Gakk.*, 1964, **28**, 188.
48. P. Guiraldenq. *Metaux. Corros. Inds.*, 1964, **39**, 347.
49. H. W. Allison and G. E. Moore, *J. appl. Phys.*, 1958, **29**, 842.
50. P. L. Gruzin, Yu. A. Polikarpov and G. B. Federov, *Physics Metals Metallogr., N.Y.*, 1957, **4**, 94.
51. G. V. Kidson, *Phil. Mag.*, 1966, **13**, 247.
52. A. B. Vladimirov *et al.*, *Fiz. Met. Metalloved*, 1976, **41**, 429.
53. R. E. Eckert and H. G. Drickamer, *J. chem. Phys.*, 1952, **20**, 13.
54. V. P. Iovkov, A. S. Panov and A. V. Ryabenko, *Izv. Akad. Nauk SSSR, Met.*, 1978, **1**, 78.
55. K. Sato, *Trans. Japan Ins. Metals*, 1964, **5**, 91.
56. M. G. Coleman, C. A. Wert and R. F. Peart, *Phys. Rev.*, 1968, **175**, 788.
57. G. B. Federov *et al.*, *Atom Energ. (USSR)*, 1971, **31**, 516.
58. P. L. Gruzin, V. G. Kostogonov and P. A. Platanov, *Dokl. Akad. Nauk SSSR.*, 1955, **100**, 1069.
59. K. Monma, H. Suto and H. Oikawa, *Nippon Kink Gakk.*, 1964, **28**, 197.
60. M. B. Bronfin, 'Diffusion Processes, Structure and Properties of Metals' (Moscow 1964) Translation—Consultants Bureau, New York; 1965, 24.
61. R. A. Wolf and H. W. Paxton, *Trans. Met. Soc. AIME*, 1964, **230**, 1426.
62. R. A. Wolf, Thesis Carnegie Inst. Tech., 1962.
63. J. F. Murdock, T. S. Lundy and E. E. Stansbury, *Acta Met.*, 1964, **12**, 1033.
64. R. B. Gibbs, D. Graham and D. H. Tomlin, *Phil. Mag.*, 1963, **8**, 1269.
65. J. Askill and G. B. Gibbs, *Phys. Status Solidi*, 1965, **11**, 557.
66. J. Askill, Thesis Reading, 1964, and 'Diffusion in B.C.C. Metals' (American Society for Metals 1965) 247.

66a. J. Askill, *Phys. Status Solidi*, 1965, **9**, K167.

67. S. Z., Bokshtein, M. B. Bronfin and S. T. Kishkin, 'Diffusion Processes Structure and Properties of Metals' (Moscow 1964). Translation—Consultants Bureau New York; 1965, 16.
68. R. F. Peart, D. Graham and D. H. Tomlin, *Acta Met.*, 1962, **10**, 519.
69. J. I. Federer and T. S. Lundy, *Trans. Met. Soc. AIME*, 1963, **227**, 592.
70. L. V. Pavlinov, *Fizica Metall.*, 1967, **24**, 272.
71. P. L. Gruzin, V. S. Emelyanov, G. G. Ryabora and G. B. Federov, *Geneva Conference Proceedings*, 1959, **19**, 187.
72. G. B. Federov, E. A. Smirnov, F. I. Zhomev, F. I. Gusev and S. A. Paraev, *Met. Metalloved. Chist. Metal.*, 1971, **9**, 30.
73. E. V. Borisov, Y. G. Godin, P. L. Gruzin, A. I. Eustyukhin and V. S. Emelyanov, *Met. i Met. Izdatel Ak. Nauk. SSSR, Moscow*, 1959 and in Translation NP–TR–448.
74. A. M. Blinkin and V. V. Vorobiov, *Ukr. Fiz. Zh.*, 1964, **9**, 91.
75. R. L. Andelin, J. D. Knight and M. Kahn, *Trans. Met. Soc. AIME*, 1965, **233**, 19.
76. Y. P. Vasil'ev, I. F. Kamardin, V. I. Skatskii, S. G. Chermomorchenko and G. N. Schuppe, *Trudy. Stred. Gos Univ. in u. i. Lenina*, 1955, **65**, 47.
77. R. E. Pawel and T. S. Lundy, *J. Phys. Chem. Solids*, 1965, **26**, 937.
78. N. L. Peterson and S. J. Rothman, *Phys. Rev.*, 1964, **136***A*, 842.
79. S. J. Rothman, *J. nucl. Mater.*, 1961, **3**, 77.
80. S. J. Rothman, N. L. Peterson and S. A. Moore, *J. nucl. Mater.*, 1962, **7**, 212.
81. Sh. Merlanov and A. A. Kulier, *Soviet Phys. solid St.*, 1962, **4**, 394.
82. M. C. Naik, J. M. Dupouy and Y. Adda, *Revue Metall., Paris*, 1966, **63**, 488.
83. A. I. Nakanechnikov and L. V. Pavlinov, *Izv. Akad. Nauk SSSR Metal*, 1972, **No. 2**, 213.
84. J. Askill, *Phys. Status Solidi*, 1971, **B43**, 1K.
85. B. Dyson, T. R. Anthony and D. Turnbull, *J. appl. Phys.*, 1966, **37** (6), 2370.
86. A. P. Batra and H. B. Huntington, *Phys. Rev.*, 1966, **145**, 542.
87. B. Dyson, *J. appl. Phys.*, 1966, **37**, 2375.
88. O. Caloni, A. Ferrari and P. M. Strocchi, *Electrochem. Metall.*, 1969, **4**, 45.
89. L. N. Aleksandrov and V. Ya. Shchelkonogov, *Sov Powder Met. & Metal Ceram.*, 1964, **4**, 288.
90. S. Z. Bokshtein, S. T. Kishkin and L. M. Moroz, 'Inv. of Structure of Metals by Isotopes, (Moscow 1959); **1961**, Aec. tr., U505.
91. A. J. Mortlock and P. M. Ewens, *Phys. Rev.*, 1967, **156**, 814.
92. J. Askill, *Phys. Status Solidi*, 1966, **16**, 63K.
93. J. W. Miller, *Phys. Rev.*, 1969, **181**, 1095.

94. T. R. Anthony and D. Turnbull, *Phys. Rev.*, 1966, **151** (2), 495.
95. B. F. Dyson, T. R. Anthony and D. Turnbull, *J. appl. Phys.*, 1967, **38** (8), 3408.
96. R. M. Dubovstev, V. S. Dôtov, T. I. Miroshnichenko and N. A. Nikolaev, *Fiz. Met. Metalloved*, 1976, **42**, 1314.
97. T. R. Anthony, B. F. Dyson and D. Turnbull, *J. appl. Phys.*, 1968, **39**, (3), 1391.
98. A. P. Batra and H. B. Huntington, *Phys. Rev.*, 1967, **154**, 569.
99. S. M. Klotsman, *et al.*, *Fizica Metall.*, 1969, **28**, 1025.
100. S. M. Klotsman, *et al.*, *Fizica Metall.*, 1971, **31**, 429.
101. F. D. Reinke and C. E. Dahlstrom, *Phil. Mag.*, 1970, **22**, 57.
102. S. M. Klotsman, *et al.*, *Fizica. Metall.*, 1970, **29**, 803.
103. K. Hoshino, Y. Iijima and K-I. Hirano, *Met. Trans A*, 1977, **8A**, 469.
104. M. C. Savena and B. D. Sharma, *Trans. Indian Inst. Metals*, 1970, **23**, 16.
105. F. Moya, G. E. Moya and F. Cabanes-Brouty, *Phys. Status Solidi*, 1969, **35**, 893.
106. J. Bernardini, and J. Cabane, *Acta Met.*, 1973, **21**, 1561.
107. T. Hehenkamp and R. Wübbenhorst, *Z. Metallk.*, 1975, **66**, 275.
108. N. Barbouth, J. Oudar and J. Cabane, *C. r. hebd. Séanc. Acad. Sci., Paris*, 1967, **C264**, 1029.
109. V. N. Kaigorodov, *et al.*, *Fiz. Met. Metallov.*, 1969, **28** (i), 120.
110. K. Dreyer, Chr. Herzig and Th. Heumann, 'Atomic Transport in Solids and Liquids' (Proc. Europhy. Conf.) (*Verlag Z. Naturforsch.* 1971), 237.
111. Ch. Herzig and Th. Heumann, *Z. Naturforsch*, 1972, **27***a*, 613.
112. A. Ott, *Z. Naturforsch*, 1970, **25***a*, 1477.
113. A. Ott, *J. appl. Phys.*, 1969, **40**, 2395.
114. A. Ott, D. Lazarus and A. Lodding, *Phys. Rev.*, 1969, **188**, 1088.
115. A. Ott, *Z. Naturforsch.*, 1968, **23***a*, 2126.
116. J. N. Mundy, A. Ott and L. Lowenberg, *Z. Naturforsch.*, 1967, **22***a*, 2113.
117. J. N. Mundy, A. Ott, L. Lowenberg and A. Lodding, *Phys. Status Solidi*, 1968, **35**, 359.
118. A. Ott, *Z. Naturforsch.*, 1968, **23***a*, 1683.
119. A. Ott and A. Norden-Ott, *Z. Naturforsch.*, 1968, **23***a*, 473.
120. L. W. Barr, J. N. Mundy and F. A. Smith, *Phil. Mag.*, 1969, **20**, 389.
121. L. W. Barr, J. N. Mundy and F. A. Smith, *Phil. Mag.*, 1967, **16**, 1139.
122. L. W. Barr and F. A. Smith, to be published.
123. F. A. Smith and L. W. Barr, *Phil Mag.*, 1970, **21**, 633.
124. F. A. Smith and L. W. Barr, *Phil. Mag.*, 1969, **20**, 205.
125. V. M. Ananyn, *et al.*, *At. Energ.*, 1970, **29**, 220.
126. J. Combronde and G. Brebec, *Acta Met.*, 1972, **20**, 37.
127. K. Lal, Report CEA (Saclay) R3136, 1967.
128. J. S. Warford and H. B. Huntington, *Phys. Rev., B.*, 1970, **1**, 1867.
129. Chih-wen. Mao, *Phys. Rev., B*, 1972, **5**, 4693.
130. S. Badrinarayanan and H. B. Mathur, *Intl. J. appl. Radiat. Isotopes*, 1968, **19**, 353.
131. W. G. Fricke, *Aluminium* (Am. Soc. Metals, Ohio 1967), **1**, 118.
132. J. J. Blechet, A. Van Craeynest and D. Calais, *J. nucl. Mater.*, 1968, **27**, 112.
133. G. M. Hood and R. J. Schultz, *Phil. Mag.*, 1971, **23**, 1479.
134. G. M. Hood, *Phil. Mag.*, 1970, **21**, 305.
135. M. S. Anand and R. P. Agarwala, *Phil. Mag.*, 1972, **26**, 297.
136. W. K. Warburton, *Phys. Rev.*, 1972, **B6**, 2161.
137. G. M. Hood and R. J. Schultz, *Phys. Rev. B*, 1975, **B11**, 3780.
138. L. Bartha and T. Szalay, *Int. J. appl. Radiat. Isotopes*, 1969, **20**, 825.
139. J. Kucera and K. Stransky, *Can. Met. Q.*, 1969, **8**, 91.
140. S. Nishikawa and K. Tsumuraya, *Phil. Mag.*, 1972, **26**, 941.
141. W. K. Warburton, *Phys. Rev.*, 1973, **B15**.
142. L. V. Pavlinov, A. M. Gladyshev and Y. N. Bykov, *Fizica Metall.*, 1968, **26**, 946.
143. S. P. Murarka, M. S. Anand and R. P. Agarwala, *Acta Met.*, 1968, **16**, 69.
144. A. R. Paul and R. P. Agarwala, *Metal Trans.*, 1971, **2**, 2691.
145. J. P. Zanghi, A. Van Craeynest and D. Calais, *J. nucl. Mater.*, 1971, **39**, 133.
146. J. J. Blechet, A. Van Craeynest and D. Calais, *J. nucl. Mater.*, 1968, **28**, 177.
147. G. V. Grigorev and L. V. Pavlinov, *Fizica Metall.*, 1968, **25**, 836.
148. G. V. Grigorev and L. V. Pavlinov, *Fizica Metall.*, 1968, **26**, 946.
149. G. V. Grigorev and L. V. Pavlinov, *Fizica Metall.*, 1968, **25**, 377.
150. A. Hoshino and T. Araki, *Trans. Nat. Res. Inst. Metals*, 1971, **13**, 99.
151. A. W. Bowen and G. M. Leak, *Met. Trans.*, 1970, **6**, 1965.
152. K. Nohara and K. Hirano, *Proc. Int. Conf. Sci. Tech. Iron and Steel*, 1970, **7**, 11.
153. V. Irmer and Feller-Kniepmeier, *J. Phys. Chem. Solids*, 1972, **33**, 2141.
154. D. W. James and G. M. Leak, *Phil. Mag.*, 1966, **14**, 701.
155. G. Salje and M. Feller-Kniepmeier, *J. appd. Phys.*, 1977, **48**, 1833.
156. M. Fromont, *J. Phys. (Paris) Lett.*, 1976, **37**, 117.
157. B. Sparke, D. W. James and G. M. Leak, *J. Iron Steel Inst.*, 1965, **203**, 152.
158. Bondy and V. Levy, *C. r. hebd. Séanc. Acad. Sci., Paris*, 1971, **C272**, 81.
159. G. Bruggeman and J. Roberts, *J. Metals.*, 1968, **20**, 54.
160. F. De Keroulas, J. Morey and Y. Quère, *J. nucl. Mater.*, 1967, **22**, 276.
161. R. J. Borg and D. Y. F. Lai, *Phil. Mag.*, 1968, **18**, 55.

162. B. A. Vandyshev and A. S. Panov, *Izv. Akad. Nauk SSSR*, 1969, **1**, 244.
163. B. A. Vandyshev and A. S. Panov, *Izv. Akad Nauk SSSR*, 1970, **2**, 231.
164. L. M. Mulyakaev, G. U. Scherbedinskii and G. N. Dubinin, *Metallov. term. Obrab. Metall.*, 1971, **8**, 45.
165. L. U. Pavlinov, A. I. Makonechnikov and V. N. Bykov, *Soviet J. Atom Energy*, 1965, **19**, 1495.
166. B. A. Vandyshev and A. S. Panov, *Fizica Metall.*, 1968, **25**, 321.
167. B. A. Vandyshev and A. S. Panov, *Fizica Metall.*, 1968, **26**, 517.
168. J. Askill, *Phys. Status Solidi*, 1967, **23**, K21.
169. F. Roux and A. Vignes, *Rev. Phys. appl.* (*Fr.*), 1970, **5**, 393.
170. J. Pelleg, *Phil. Mag.*, 1970, **21**, 735.
171. R. P. Agarwala, S. P. Murarka and M. S. Anand, *Acta Met.*, 1968, **16**, 61.
172. J. Pelleg, *Phil. Mag.*, 1969, **19**, 25.
173. T. S. Lundy, *et al.*, *Trans. Met. Soc. AIME*, 1965, **233**, 1533.
174. G. B. Federov, F. J. Zhomov and E. A. Smirnov, *Metall. Metalloved. Chist. Metal*, 1969, **8**, 145.
175. D. S. Gornyi and R. M. Altovski, *Fizica Metall.*, 1971, **31**, 781.
176. P. Son, *et al.*, *J. Japan Inst. Metals*, 1967, **31**, 998.
177. B. A. Vandyshev and A. S. Panov, *Izv. Acad. Nauk SSSR.*, 1968, **1**, 206.
178. R. Tendler and C. F. Varotto, *J. nucl. Mater.*, 1972, **44**, 99.
179. R. Tendler and C. F. Varotto, *J. nucl. Mater.* 1973, **46**, 107.
180. G. M. Hood and R. J. Schultz, *Phil Mag.*, 1972, **26**, 329.
181. G. M. Hood, 'Diffusion Processes', **1970**, Vol. 1, New York, Gordon and Breach, p. 361.
182. G. V. Kidson and G. J. Young, *Phil Mag.*, 1969, **20**, 1057.
183. F. Dyment and C. M. Libanati, *J. Mater. Sci.*, 1968, **3**, 349.
184. R. P. Agarwala and A. R. Paul, *Proc. nucl. Rad. Chem. Symp.*, 1967, **3**, 542.
185. A. R. Paul, *et al.*, *Int. Conf. Vac. and Insls. in Metals*, Julich, Sept. 1968, 1, 105.
186. E. Ch. Schwegler, *Intl. J. Mass Spectr. Ion Phys.*, 1968, **1**, 191.
187. L. V. Pavlinov, G. V. Grigorev and G. O. Gromyko, *Izv. Akad. Nauk. SSSR Metal*, 1969, **3**, 207.
188. R. P. Agarwala and A. R. Paul, *J. Nucl. Mat.*, 1975, **58**, 25.
189. B. A. Vandyshev, A. S. Panov and P. L. Gruzin, *Fizica Metall.*, 1967, **23**, 908.
190. A. Shepela, *J. less-common Metals*, 1972, **26**, 33.
191. L. M. Larikov, V. M. Tyshkevich and L. F. Chorna, *Ukr. Fiz. Zh.*, 1967, **12**, 983.
192. R. E. Pawel and T. S. Lundy, *Acta Met.*, 1969, **17**, 979.
193. M. P. Dariel, M. Blumenfeld and G. Kimmel, *J. appl. Phys.*, 1970, **41**, 1480.
194. P. Son, *et al.*, *J. Japan Inst. Met.*, 1966, **30**, 1137.
195. L. V. Pavlinov, *Fizica Metall*, 1970, **30**, 800.
196. A. Languille, *Mém. Scient. Revue Metall.*, 1971, **68**, 435.
197. M. Dariel, G. Erez and G. M. J. Schmidt, *Phil. Mag.*, 1969, **19**, 1053.
198. F. H. Huang and H. B. Huntington, *Phys. Rev.*, 1974, **9B**, 1479.
199. M. Dariel, *J. appl. Phys.*, 1971, **42**, 2251.
200. M. Dariel, G. Erez and G. M. J. Schmidt, *J. appl. Phys.*, 1969, **40**, 2746.
201. M. Dariel, G. Erez and G. M. J. Schmidt, *Phil. Mag.*, 1969, **19**, 1045.
202. M. P. Dariel, *Phil. Mag.*, 1970, **22**, 563.
203. P. P. Kuzmenko and G. Grinevich, *Fizica Metall.*, 1967, **24**, 424.
204. A. Kucera and T. Zemcik, *Can. Met. Q.*, 1968, **7**, 73.
205. P. Brätter and H. Gobrecht, *Phys. Status Solidi*, 1970, **41**, 631.
206. P. Brätter, H. Gobrecht and D. Wobig, *Phys. Status Solidi*, 1972, **11**, 589.
207. F. Schmitz and M. Fock, *J. nucl. Mater.*, 1967, **21**, 353.
208. D. Treheux, *et al.*, *C. r. hebd. Séanc. Acad. Sci.*, *Paris*, Series C. 1972, **274**, 1260.
209. M. P. Dariel, D. Dayan and D. Calais, *Phys. Status Solidi*, 1972, **A10**, 113.
210. R. L. Fogel'son, Ya. Y. Ugai and I. A. Akimova, *Fiz. Met. Metalloved*, 1975, **39**, 447.
211. R. L. Fogel'son, Ya. Y. Ugai and I. A. Akimova, *Izv. Vyssh. Uchebn Zaved.*, *Tsvetn Metall.*, 1975, **2**, 142.
212. J. Ladet, J. Bernardini and F. Cabane-Brouty, *Sci. Met.*, 1976, **10**, 195.
213. F. Sawayanagi and R. R. Hasiguti, *J. Jap. Inst. Met.*, 1978, **42**, 1155.
214. S. J. Rothman, N. L. Peterson, L. J. Nowicki and L. C. Robinson, *Phys. Status Solidii B*, 1974, **63**, K29.
215. K. Hirano and S. Fujikawa, *J. Nuclear Mat.*, 1978, **69/70**, 564.
216. R. L. Fogel'son and N. N. Trofimova, *Izv. Vyssh. Uchebn. Zaved.*, *Tsvetz. Metall.*, 1978, **4**, 152.
217. S. M. Meyers and S. T. Picraux., *Phys. Rev.*, 1974, **9**, 3953.
218. V. P. Gladkov, A. V. Svetlov, D. M. Skorov, V. I. Tenishev and A. N. Shabalin, *At. Energ.*, 1976, **40**, 257.
219. V. M. Anan'in, V. P. Gladkov, A. V. Svetlov and D. M. Skorov, *At. Energ.*, 1976, **40**, 256.
220. M. P. Dariel, *Acta Met.*, 1975, **23**, 473.
221. M. P. Dariel, *Phil. Mag.*, 1973, **28**, 915.
222. B. Million and J. Kucera, *Kovove Mater.*, 1973, **11**, 300.
223. A. Bristoti and A. R. Wazzan, *Rev. Bras. Fis.*, 1974, **4** 1.
224. J. M. Dupouy, J. Mathie and Y. Adda., Proc. Int Conf. Metallurgy of Be, Grenoble, p. 159, 1965.
225. D. C. Yeh, L. A. Acuna and H. B. Huntington, *Phys. Rev*, 1981, **23**, 1771.
226. G. Krautheim, A. Neidhardt and V. Reinhold, *Krist. Techn.*, 1978, **13**, 1335.
227. V. A. Gorbachev *et al.*, *Fiz. Metal. Metalloved.*, 1973, **35**, 889.
228. G. Barreau, G. Brunnel, G. Ciceron and P. Lacombe, *C. R. Acad. Sci* (*Paris*), 1970. **270**, 516.
229. J. Bernardini and J. Cabane, *Acta Met.*, 1973, **21**, 1561.
230. S. M. Klotsman *et al.*, *Fiz. Met. Metalloved*, 1978, **45**, 1104.
231. V. A. Gorbachev *et al.*, *Fiz. Met. Metalloved*, 1977, **44**, 214.

232. K. Nohara and K. Hirano, *Nippon Kinzoku Gakkaischi*, 1973, **37**, 731.
233. D. Ablitzer, *Phil. Mag.*, 1977, **35**, 1239.
234. J. Pelleg, *Phil. Mag.*, 1976, **33**, 165.
235. F. Dyment, *J. Nuclear Mat.*, 1976, **61**, 271.
236. A. B. Vladimirov *et al.*, *Fiz. Metal. Metalloved.*, 1975, **39**, 319.
237. A. B. Vladimirov *et al.*, *Fiz. Metal. Metalloved*, 1978, **45**, 1015.
238. A. B. Vladimirov *et al.*, *Fiz. Metal. Metalloved*, 1978, **45**, 1301.
239. D. L. Decker, C. T. Candland and H. B. Vanfleet, *Phys. Rev.*, 1975, **B11**, 4885.
240. C. T. Candland and H. B. Vanfleet, *Phys. Rev.*, 1973, **B7**, 575.
241. K. Kusunaki and S. Nishikawa, *Scripta Met.*, 1978, **12**, 615.
242. D. L. Decker *et al.*, *Phys. Rev.*, 1977, **B15**, 507.
243. D. L. Decker, J. D. Weiss and H. B. Vanfleet, *Phys. Rev.*, 1977, **B16**, 2392.
244. D. Bergner and K. Schwarz, *Neue Huette*, 1978, **23**, 210.
245. G. Rein, H. Mehrer and K. Maier, *Phys. Status Solidii*, 1978, **A45**, 253.
246. C. Charissoux, D. Calais and G. Gallet, *J. Phys. Chem. Sol.*, 1975, **36**, 981.
247. C. Charissoux and D. Calais, *J. Nucl. Mat.*, 1976, **61**, 317.
248. W. N. Weins and O. N. Carlson, *J. Less Common Met.*, 1979, **66**, 99.
249. F. A. Schmidt, R. J. Conzemius and O. N. Carlson, *J. Less Common Met.*, 1978, **56**, 53.
250. J. Pelleg, *Phil. Mag.*, 1975, **32**, 593.
251. J. Pelleg and M. Herman, *Phil. Mag.*, 1977, **35**, 349.
252. J. F. Murdock and C. J. McHargue, *Acta Met.*, 1968, **16**, 493.
253. C. Charissoux and D. Calais, *J. Nuclear Mat.*, 1975, **57**, 45.
254. R. Tendler and C. F. Varotto, *J. Nucl. Mat.*, 1974, **54**, 212.
255. R. Tendler, J. Abriata and C. F. Varotto, *J. Nucl. Mat.*, 1976, **59**, 215.
256. G. Erdélyi *et al.*, *Phil. Mag.*, 1978, **38**, 445.
257. L. V. Pavlinov, G. V. Grigoriev, and G. O. Gromyko *Izv. Akad Nauk SSSR Metal.*, 1969, **3**, 207.

Table 13.3 DIFFUSION IN HOMOGENEOUS ALLOYS

Tracer diffusion—Self-diffusion of alloys

Element 1 (*purity*) At.%	*Element 2* (*purity*) At.%	A_1^*	Q_1^*	A_2^*	Q_2^*	*Temp. range* °C	*Ref.*
Ag	Al						
(—)	(—)		IVa(i)	p.c.			
	2.05	0.25	42.5	—	—	700–850	1
	9.47	0.83	42.9	—	—		
	14.1	0.73	41.2	—	—		
(99.98)	(99.996)		IVc	p.c.			
0–9.4	—	0.39	28.9	(indep. of conc.)		400–595	69
Ag	Au						
(99.99)	(99.99)		IVa(i)	s.c.			
100	0	0.49	44.47	0.85	48.28	630–1 010	2
	8	0.52	44.79	0.82	48.30		
	17	0.32	44.05	0.48	47.30		
	35	0.23	43.54	0.35	46.67		
	50	0.19	43.41	0.21	45.28		
	66	0.11	41.73	0.17	44.51		
	83	0.09	41.02	0.12	43.05		
	94	0.072	40.26	0.09	42.08		
0	100	0.072	40.2	0.09	41.7		
Ag	Cd						
(99.99)	(99.999)		IVa(i)	s.c. and p.c.			
100	0	0.44	44.27	0.44	41.69	500–900	3
	6.50	0.31	42.61	0.33	40.48		
	13.60	0.23	40.96	0.22	38.61		
	27.5	—	—	0.25	35.95		
	28.0	0.16	37.25	—	—		
			IVa(i)	p.c.			
	31	—	~36	—	~35.2	560–680	74
	34	—	~36[a]	—	~35.1		
	37	—	~36	—	~34.9		

Table 13.3 DIFFUSION IN HOMOGENEOUS ALLOYS—*continued*

Element 1 (*purity*) At.%	*Element 2* (*purity*) At.%	A_1^*	Q_1^*	A_2^*	Q_2^*	*Temo. range* °C	*Ref.*
Ag	Cu						
(—)	(—)		IVa(i)	p.c.			
	1.75	0.66	44.8	—	—	700–890	1
	4.16	1.84	46.6	—	—		
	6.56	0.51	43.5	—	—		
Ag	Ge						
(—)	(—)		IVa(i) and IVb	p.c.			
	1.50	0.55	44.0	D_{Ge} greater than for infinite dilution by ~15%At.% Ge		700–850	1 and 7
	3.00	1.59	45.3				
	4.30	1.89	44.5				
	5.43	2.18	44.2				
Ag	In						
(99.99)	(99.99)		IVa(i)	s.c. and p.c.			
100	0	0.44	44.27	0.41	40.80	500–900	3
	4.40	0.36	42.67	—	—		
	4.70	—	—	0.45	40.30		
	12.40	—	—	0.57	38.39		
	12.60	0.12	37.40	—	—		
	16.60	—	—	0.57	36.61		
	16.70	0.18	36.27	—	—		
Ag	Mg						
(99.95)	(99.95)		IVb	p.c.			
	45.8	1.53	41.3	—	—	500–700	12
	49.8	0.28	40.6	—	—		
	52.0	0.134	38.0	—	—		
(99.98)	(99.9+)		IVa(i)	p.c.			
	41.10	0.095	33.2	—	—	500–700	13
	43.60	0.15	35.3	—	—		
	48.48	0.37	39.5	—	—		
	48.72	0.39	39.7	—	—		
	52.82	0.33	36.7	—	—		
	57.15	0.051	28.7	—	—	500–600	
	60.88	$D^*_{Ag}=4.37\times10^{-9}$ at 500.5 °C					
Ag	Pb						
(—)	(—)		IVa(i)	p.c.			
	0.21	0.22	42.5	—	—	700–800	9 and 1
	0·25	—	—	0.22	37.8		
	0.52	—	—	0.38	38.7		
	0.71	0.89	44.7	—	—		
	1.30	0.70	43.5	—	—		
	1.32	—	—	0.46	38.5		
(—)	(99.999 9)		IVa(i)	s.c.			
0.09	—	D^*_{Pb}/D^*_{PurePb}	=1.115	(b=136.8)	300.3		63
0.18	—		=1.25				
Ag	Pd						
('Spec. Pure')			IVa(i)	p.c.			
	0–21.8	$027e^{-8.2c}$	43.7	—$^{(b)}$	—	715–942	4
	0–20.4	—$^{(b)}$	—	$12.5e^{-7.5c}$	57.2	850–900	5
Ag	Sb						
(99.99)	(99.99)		IVa(i)	s.c.			
	0.53	0.38	43.5	(*c*)	—	550—900	6
	0.89	0.30	42.6	—	—		
	1.42	0.275	42.0	—	—		

(*a*) Activation energies read from graph in Reference 74.
(*b*) c=conc. of Pd.
(*c*) In 0.7 At. % Sb alloys, D_{Sb}^* same as D in pure Ag.
In 2.8 At. % Sb alloys, D_{Sb}^* ~20% greater than D in pure Ag.

Table 13.3 DIFFUSION IN HOMOGENEOUS ALLOYS—*continued*

Element 1 (*purity*) At.%	*Element 2* (*purity*) At.%	A_1^*	Q_1^*	A_2^*	Q_2^*	*Temp range* °C	*Ref.*
Ag	Sn						
(—)	(—)			p.c.			
	0.18	0.13	41.7	—	—		
	0.48	0.13	40.9	—	—		24
	0.91	0.17	40.5	—	—		
	0.97	0.28	39.4	—	—	700–850	25
	2.8	0.1	38.6	—	—		
	4.56	0.23	39.7	—	—		24
	5.1	0.2	38.6	—	—		25
	7.45	0.16	37.0	—	—		
(—)	(—)		p.c. IVb Ag^{110m}, Sn^{113}				
75	25	1.03×10^{-3}	22.4	4.01×10^{-4}	22.8	200–400	128
(—)	(—)		p.c. and s.c. IVa Sn^{113}, Ag^{110}				
100	~0	1.0	45.7	0.17	38.4		
	1.7	—	—	0.125	37.45		
	0.108	0.13	41.3	—	—		
	0.8	0.12	40.2	—	—	620–800	129
	3.0	0.085	38.4	—	—		
	4.7	0.07	37.5	—	—		
	6.0	0.07	37.0	—	—		
Ag	Tl						
(—)	(—)		IVa(i) and IVb p.c.				
100	0	0.724	45.5	0.15	37.9		
	1.1	0.42	43.5	0.72	40.4	640–800	7
	2.6	0.35	41.9	0.57	39.4		
	5.5	0.10	37.6	—	—		
Ag	Zn						
(99.999)	(99.999)		IVa(i)	s.c.			
0.00	—‖*c*	0.31	26.11	0.13	21.90		71 and
	⊥*c*	0.45	27.60	0.18	23.00		23 of
0.35	—‖*c*	—	—	—	—		Table 2
	⊥*c*	0.49	27.65	—	—		
0.57	—‖*c*	—	—	0.14	21.95		
	⊥*c*	—	—	0.22	23.12		
0.68	—‖*c*	0.35	26.09	—	—	320–415	70
	⊥*c*	—	—	—	—		
0.89	—‖*c*	0.42	26.29	—	—		
	⊥*c*	0.69	28.01	—	—		
1.40	—‖*c*	—	—	0.17	22.07		
	⊥*c*	—	—	0.26	23.13		
(99.999)	(99.999)		IVa(i)	s.c.			
—	0	$D_1=1.43\times10^{-10}$		—	—	747	
		$=17.2\times10^{-10}$		—	—	880	
—	1.10	$=1.63\times10^{-10}$		—	—	747	
		$=19.9\times10^{-10}$		—	—	880	
—	2.08	$=1.80\times10^{-10}$		—	—	747	71
		$=21.8\times10^{-10}$		—	—	880	
—	3.10	$=2.0\times10^{-10}$		—	—	747	
		$=24.1\times10^{-10}$		—	—	880	
—	4.05	$=2.28\times10^{-10}$		—	—	747	
		$=26.2\times10^{-10}$		—	—	880	
(99.99)	(99.999)		IVa(i)	p.c.			
85	15		~36	0.11	36.0		10
70	30	0.29	35.99	0.46	35.2	500–700	8
(Merck.)	(99.999)		IIb	p.c.			
52.4	47.6	4.55×10^{-3}	17.6	—	—	400–610	11

Table 13.3 DIFFUSION IN HOMOGENEOUS ALLOYS—*continued*

Element 1 (*purity*) At.%	*Element 2* (*purity*) At.%	A_1^*	Q_1^*	A_2^*	Q_2^*	*Temp. range* °C	*Ref.*
Al	Co						
(99.9)	(Carbonyl)		IVc	p.c.			
	10	—	—	2.65	67.5	1 040–1 220	14
42		—	—	1.84×10^2	85	1 000–1 200	
49		—	—	333×10^2	102	1 100–1 300	15
50.7		—	—	0.013×10^2	65	1 000–1 200	
	49/57		Individual D_{Co}^* values plotted at 1 250				16
Al	Cu						
(99.999)	(—)		IVa(i)	s.c.			
—	1.0	—	—		$D = 4.02 \times 10^{-10}$	489	72
					$D = 7.92 \times 10^{-9}$	608	
( )	( )		IVb	p.c.			
0	—	—	—	0.43	48.5		
2.80	—	—	—	0.46	48.0		
5.50	—	—	—	0.30	47.0		
8.83	—	—	—	0.46	47.1	800–1 040	73
11.7	—	—	—	0.61	47.2		
14.5	—	—	—	4.2	51.1		
(99.994)	( )		Vb	p.c.			
—	0	0.10	30.5	—	—	~330–460	127
	0.15	6×10^{-4}	24.1	—	—		
Al	Fe						
(—)	(—)		IVa	p.c.			
		Diff. of Co					
3.47		0.1	53.0	3.2	59.0		
7.95		1.9	56.0	4.5	60.0		
13.5		6.8	58.5	0.4	52.0		
20.6		22.0	60.0	32.0	63.0		
23.6		27.0	60.0	27.0	62.5	—	17
35.5		210.0	67.0	—	—		
42.0		580.0	71.0	—	—		
47.3		6 300.0	79.0	—	—		
52.0		148.0	67.0	60.0	66.0		
( )	( )		IVb	p.c.			
5.7	—	—	—	3.7	58.7	850–1 185	75
( )	( )		IVa(i)	p.c.			
'AlFe$_3$'		1.74×10^2	50.5	(Individual Fe		550–630	
'AlFe'		8.5×10^{-5}	25.1	diffusion coefficients		550–630	76
Al_2Fe		3.99×10^3	55.5	are reported in		520–630	
Al_3Fe		8.9×10^{-4}	28.0	reference 77)		550–600	
Al	Mg						
(99.994)	( )		Vb	p.c.			
	0	0.10	30.5	—	—		
	2.2	0.93	30.6	—	—	~330–460	127
	5.5	0.21	27.7	—	—		
Al	Ni						
(—)	(—)		IVc	p.c.			
		Diff. of Co^{60}					
	47.3	4.7×10^{-2}	56.6	—	—		
	48.5	9.3×10^{-2}	59.9	—	—		
	49.4	4.4×10^{-3}	52.5	—	—	1 050–1 350	18
	50.7	57.7	80.6	—	—		
	53.1	2.6	67.6	—	—		
	55.1	7.2×10^{-3}	47.1	—	—		

Table 13.3 DIFFUSION IN HOMOGENEOUS ALLOYS—*continued*

Element 1 (*purity*) At.%	*Element 2* (*purity*) At.%	A_1^*	Q_1^*	A_2^*	Q_2^*	*Temp. range* °C	*Ref.*
Al	Ni						
( )	( )		IVb	p.c.			
		Diff. of In^{114m}					
—	48.3	—	—	1.2×10^{-4}	42.5		
—	48.6	1.29×10^{-3}	56.6	1.04×10^{-3}	50.1		
—	49.0	—	—	5.3×10^{-4}	47.9		
—	49.2	—	—	0.23	65.9		
—	50.0	3.98×10^{-3}	58.0	4.461	73.4	1 000–1 350	78 & 130
—	53.2	—	—	0.63	65.5		
—	54.5	—	—	0.15	59.8		
—	58.0	1.83×10^{-4}	40.6	0.035	51.7		
—	58.5	—	—	0.096	60.6		
—	58.7	—	—	0.725	59.8		
—	73.20	—	—	3.11	71.73		
—	74.71	—	—	1.00	72.40	900–1 300	106
—	76.20	—	—	4.41	73.16		
Al	Ni Ti						
( )	( ) ( )						
19.92	— 5.13	—	—	0.055	62.79		
20.83	— 5.01	—	—	0.039	62.02	900–1 300	106
21.84	— 5.13	—	—	0.085	63.74		
Al	Zn						
(99.99)	(99.99)		IVa(i)	p.c.	Zn^{65}		
	0	—	—	0.27	28.75		
	1.16	—	—	0.25	28.43		
(At.%)	1.73	—	—	0.18	27.88		
	2.15	—	—	0.22	28.09	340–620	149
	2.80	—	—	0.22	27.97		
	3.29	—	—	0.24	28.06		
	3.76	—	—	0.23	27.93		
(99.99)	(99.999)		IVa(i)	p.c.			
100	0	—	—	1.1	30.9	400–650	
	4.33	—	—	0.35	28.3	360–610	
	9.23	—	—	0.20	27.0	360–575	
	16.7	—	—	0.10	25.0	360–525	19
	36.9	—	—	0.16	24.1	360–450	
	49.4	—	—	0.048	21.9	325–440	
	62.9	—	—	0.12	20.0	325–405	
0	100.0	—	—	0.031	20.5		
(99.994)	( )		Vb	p.c.			
	0	0.10	30.5	—	—		
	8	0.087	29.3	—	—	~330–460	127
	11.5	0.035	28.0	—	—		
	18.8	0.025	27.1	—	—		
Al	Zn	Mg					
(99.99)	(99.99)	( )		IVa(i)	p.c.	Zn^{65}	
	0	1.11	—	—	0.34	29.03	
	0.71	1.12	—	—	0.25	28.48	
	1.49	1.13	—	—	0.23	28.20	
	2.11	1.14	—	—	0.31	28.57	
(At.%)	0	2.77	—	—	0.24	28.43	420–575 149
	0.71	2.80	—	—	0.27	28.43	
	1.48	2.83	—	—	0.21	27.93	
	2.11	2.86	—	—	0.17	27.51	
	0	4.43	—	—	0.28	28.34	
	0	6.63	—	—	0.50	28.94	

Table 13.3 DIFFUSION IN HOMOGENEOUS ALLOYS—*continued*

Element 1 (*purity* At.%	*Element 2* (*purity*) At.%	A_1^*	Q_1^*	A_2^*	Q_2^*	*Temp. range* °C	*Ref.*
Au	Cd						
(99.99)	(99.95)		IVa(i)	s.c.			
50	50	0.17	27.9	0.23	28.0	300–590[(a)]	20
(99.999+)	(99.999+)		IVa(i)	s.c.			
	47.5	0.23	28.1	1.36	31.0	350–600	
	49.0	0.61	30.0	1.50	31.2	440–550	21
	50.5	0.12	26.2	0.22	27.1	440–550	

(*a*) Between 590° and the m.p. at 626° there is marked upward curvature in the Arrhenius Plot.

Element 1 (*purity* At.%	*Element 2* (*purity*) At.%	A_1^*	Q_1^*	A_2^*	Q_2^*	*Temp. range* °C	*Ref.*
Au	Cu						
( )	( )		IVc	p.c.			
25	75	6.5×10^{-3}	38.2	—	—	550–900 (Disord.)	81
25	75	Diffusion of Co^{57} 4.2×10^{-2}	46.0			650–900	89
Au	Ni						
(99.96)	(99.99)		IVa(ii)	p.c.	IVa(i)		
100	0	0.26	45.3	0.30	46.0		
	10	—	—	0.80	47.6		
	20	0.05	40.2	0.82	47.8	Variable	
	35	0.063	42.7	—	—	(For Ni,	22(Au)
	36	—	—	1.10	49.1	50–75°	
	50	0.091	43.4	0.09	44.5	temp.	
	65	0.51	48.8	0.005	39.6	ranges	23(Ni)
	80	1.1	60.5	0.05	49.2	only)	
	90	—	—	0.04	51.1		
0		2.0	65.0	0.40	63.8	—	
Au	Pb						
( )	(99.999 9)		IVa(i)	s.c.			
0.04	—		$D^*_{Pb}/D^*_{Pure\ Pb}=2.2$	($b_1=5\,726$)		199.4	
0.015	—		=0.55				
0.04	—		=1.6				
			=1.75[(a)]				
0.06	—		=2.85	($b_1=4\,312$)		215.2	63
0.08	—		=3.8				
			=4.05[(a)]				

(*a*) Samples pre-annealed before diffusion.

Element 1 (*purity* At.%	*Element 2* (*purity*) At.%	A_1^*	Q_1^*	A_2^*	Q_2^*	*Temp. range* °C	*Ref.*
Au	Zn						
( )	( )		IVa(i)	s.c.			
—	49	0.19	31.9	0.84	34.6		
—	50	0.33	33.1	1.93	35.4	428–650 (Ordered β′)	79
—	51	0.016	27.0	0.047	27.5		
Be	Ni						
( )	( )		IVb	p.c.			
—	1.68	—	—	0.41	59.0	900–1 100	80
—	7.9	—	—	0.23	45.0		
C	Cr Fe						
	( ) ( )		IVb	p.c.			
	0.5	45×10^{-3}	26.5	—	—	1 400–1 600	85
	1.0	20×10^{-3}	24.0	—	—		

Table 13.3 DIFFUSION IN HOMOGENEOUS ALLOYS—*continued*

Element 1 (*purity*) At.%	*Element 2* (*purity*) At.%		A_1^*	Q_1^*	A_2^*	Q_2^*	*Temp. range* °C	*Ref.*
C	Cr	Ta						
( )	( )	( )		IVb	p.c.			
—	—	1.0	5×10^{-3}	28.8	—	—	1 300–1 500	85
C	Fe							
(—)	(—)			Va and Vc	p.c.			
V. small	100		0.008	19.8[(a)]	—	—	−50 to +150	26
(—)	(—)			IVa(i)	p.c.			
0			—	—	0.44	67.0		
1.15			—	—	0.052	59.0		
2.46			—	—	0.015	54.0	1 000–1 300	28
3.39			—	—	0.021	54.0		
4.97			—	—	0.029	53.8		
6.21			—	—	0.050	53.8		
C	Fe	Ni						
(—)	(—)	(—)		IVb	p.c.			
c%	(γ)	20	—	—	$18.10^{-0.92c}$	$75–6c$	800–1 300	29
c%		25	—	—	$71.10^{-0.65c}$	$79–5c$	1 050–1 330	

(*a*) These values also represent the best fit to the original (reference 26) and later measurements when plotted altogether. *See* reference 27.

Element 1 (*purity*) At.%	*Element 2* (*purity*) At.%		A_1^*	Q_1^*	A_2^*	Q_2^*	*Temp. range* °C	*Ref.*
C	Fe	Si						
	(99.97)		V(a) p.c.					
0.104	(α)	5.5	D_c^* 'virtually' the same as in Si-free Fe				26–70	67
C	Nb							
	(99.4)			Va	p.c.			
V. small	~100		0.004 0	33.02	—	—	130–280	30
C	Ta							
(—)	(—)			Va	p.c.			
V. small	~100		0.006 1	38.51	—	—	190–360	30
C	V							
	(—)			Va	p.c.			
V. small	~100		0.004 5	27.29	—	—	60–165	30
Cd	Mg							
(—)	(—)			IVb	s.c.			
75	25		11.2×10^{-5}	12.4 (Ord.)	—	—	54–90	31
75	25		0.074	16.7 (Disord.)	—	—	95–200	
25	75		4.10^{-5}	16.3 (Ord.)	—	—	120–155	32
25	75		1.2×10^{-6}	12.8 (Disord.)	—	—	155–281	
Cd	Pb							
( )	(99.999 9)			IVa(i)	s.c.			
				Effect of Cd on D_{Pb}^*				
			(*See* eq. 13.9a)	b_1	b_2	b_3		
				45.197	438.1	—	198.7	
0 to 5				30.028	407.7	44 286	248.2	52
				19.138	1 070.2	—	300.5	
Co	Cr							
(—)	(—)			IVb	p.c.			
	4		0.67	65.8	—	—	1 100 – 1 350	33
	7		56.3	79.3	—	—		
Co	Cr	Ni						
(—)	(—)	(—)		IVb	p.c.			
	9	26	6.3	72.1	—	—	1 100 – 1 350	33
	18	26	0.4	64.2	—	—		

Table 13.3 DIFFUSION IN HOMOGENEOUS ALLOYS—*continued*

Element 1 (*purity*) At.%	*Element 2* (*purity*) At.%	A_1^*	Q_1^*	A_2^*	Q_2^*	*Temp. range* °C	*Ref.*
Co	Fe						
( )	( )		IVa(i)	p.c.			
		1.33	69.4	1.26	68.5	1 012–1 164(γ)	34
50	50	2.0	60.0	0.25	55.0	795–945(α)	
		—	133	—	133	655–722(CsCl)	
( )	( )		IVb	p.c.			
—	21	0.54	65.0	—	—	1 100–1 300	33
( )	( )		IVa(iii)	p.c.			
			Diffusion of Ni				
—	0		$D_{Ni}=13.10^{-12}$			1 136	82
—	30.2		$=11.5$				
—	68.5		$=8–8.8$				
—	88.3		$=5$				
—	100		$=6$				
( )	( )		IVb	p.c.	(F = Ferromagnetic)		
0	(bcc)	1.83	55.9	—	—	800–899	107
	(fcc)	0.77	63.3	—	—	950–1 360	
3	(bcc)	9.17	63.6	—	—	630–750(F)	
6.8	(bcc)	0.469	44.7	—	—	630–800(F)	
		5.72×10^{-5}	35.0	—	—	880–920	
	(fcc)	0.109	77.9	—	—	1 010–1 310	
28.6	(bcc)	1.25×10^{-3}	47.3	—	—	630–810(F)	
	(fcc)	3.36×10^{-2}	63.6	—	—	1 060–1 310	
49.6	(bcc)	6.59×10^{-2}	59.0	—	—	750–850(F)	
	(fcc)	0.154	83.5	—	—	1 060–1 310	
67.2	(bcc)	6.04×10^{-3}	45.6	—	—	630–880(F)	
	(fcc)	3.15×10^{-2}	63.3	—	—	1 060–1 310	
89.6	(fcc)	6.44×10^{-2}	60.0	—	—	800–1 010(F)	
		1.61×10^{-2}	55.9	—	—	1 060–1 310	
100	(fcc)	0.50	65.4	—	—	772–1 048(F)	
		0.17	62.2	—	—	1 192–1 297	
(99.8)	(99.9)	IVb	p.c.	Co^{60}			
0		0.029	59.1	—	—	960–1 220	131
6		20.54	76.8	—	—		
10		15.65	73.8	—	—		
15		1.98	69.2	—	—		
20		0.31	62.5	—	—		
Co	Fe Ti						
( )	( ) ( )		IVb	p.c.			
	15 4	0.008	51.2	—	—	1 100–1 200	33
Co	Ga						
(99.998)	(99.999 9)	IVa(i)	p.c.	Co^{60}			
45.2		97.6	66.65	—	—		132
48.0		917	72.3	—	—		
50.0		555	70.59	—	—		
56.0		259	66.92	—	—		
60.0		69	62.07	—	—		
Co	Mn						
( )	( )	IVb	p.c.	Mn^{54}			
100	~0	—	—	3.15×10^{-2}	55.5	860–T_c	133
		—	—	1.1×10^{-2}	52.0	T_c–1 246	
—	5.22	—	—	1.38	64.1	868–T_c	
		—	—	0.501	61.3	T_c–1 200	
—	10.24	—	—	1.36	62.9	903–1 148	

Table 13.3 DIFFUSION IN HOMOGENEOUS ALLOYS—*continued*

Element 1 (*purity*) At.%	*Element 2* (*purity*) At.%	A_1^*	Q_1^*	A_2^*	Q_2^*	*Temp. range* °C	*Ref.*
Co	Ni						
(99.5)	(99.98)		IVb	p.c.		Ferromag.[a]	
100	0	0.5	65.4	0.34	64.3	772–1 048	
	11	0.61	67.1	0.46	65.6	864–1 048	
	20	5.96	73.5	1.66	69.7	845–1 048	
	30	1.16	68.6	2.01	68.5	772–899	
	51	0.096	61.5	0.36	63.6	701–819	
						Paramag.[a]	35
100	0	0.17	62.2	0.10	60.2	1 192–1 297	
	11	0.21	63.6	0.17	62.7	1 144–1 297	
	20	2.42	71.0	0.41	65.7	1 090–1 246	
	30	0.78	67.0	0.67	64.8	1 050–1 250	
	51	0.12	60.2	0.21	60.6	90–1 190	
	100	0.75	64.7	1.70	68.1	700–1 190	
(99.4)	(99.8)		IVb	p.c.			
(a) (b)							
99.5	—	1.66	68.7	3.35	71.0	1 047–1 311(Co) 1 220–1 420(Ni)	
93.8		7.4	72.9	5.40	72.2	1 210–1 370	
89.0		2.52	69.3	6.42	72.6		
80.6		0.99	65.9	1.89	68.2	1 160–1 410	
72.6		0.70	64.6	1.60	67.2		
49.4		0.18	59.8	0.25	61.1	1 100–1 350	
43.3		0.52	62.6	0.69	63.8		83
21.1		0.33	61.1	0.45	62.6	1 160–1 410	
10.8		0.66	63.0	1.47	66.0		
4.3		0.49	62.4	2.86	68.0	1 210–1 370	
0.03		1.11	64.9	1.39	65.9	1 150–1 390	
( )	( )		IVb	p.c.			
		Diffusion of Fe^{59}					
50	50	125	76.5			1 000–1 250	84
	(99.4%)	Diffusion of C^{14}			IVb	p.c.	
5.25		0.4	37.0	—	—	600–900	36

(*a*) According to reference 64 measurements in pure Co, in Co+11% Ni and Co+20% Ni alloys do not convincingly demonstrate a difference in A^* and Q^* values for the paramagnetic and ferromagnetic regions.
(*b*) All alloys contain 0.1–0.2 Mg.

Element 1 (*purity*) At.%	*Element 2* (*purity*) At.%	A_1^*	Q_1^*	A_2^*	Q_2^*	*Temp. range* °C	*Ref.*
Co	Mn Ni						
( )	( ) ( )		IVb	p.c.			
19.5	20.3 60.0	0.86	60.5	—	—	1 020–1 120	
40.6	20.45 38.9	0.22	57.5	—	—	1 020–1 160	14
59.7	20.6 19.5	0.05	54.5	—	—	1 060–1 160	
Co	Ti						
(99.97)	(99.97)		IVa(i)	p.c.			
1.6		—	—	1.26×10^{-3}	34.7	960–1 504	
3.3		—	—	1.58×10^{-3}	33.5	893–1 344	121
4.9		—	—	1.41×10^{-2}	38.3	803–1 300	
7.4		—	—	2.50×10^{-2}	38.9	803–1 211	
Co	Zr						
( )	( )		IVa(i)	p.c.			
		$D^*_{Co}/D^*_{Co\text{ in pure Zr}}$		$D^*_{Zr}/D^*_{Pure\ Zr}$			
0.395	—	0.900		0.948			
0.747	—	—		1.322 8			
1.22	—	0.861		1.668		993	118
1.61	—	—		2.143 3			
1.995	—	0.820		2.223 9			

Table 13.3 DIFFUSION IN HOMOGENEOUS ALLOYS—*continued*

Element 1 (*purity*) At.%	*Element 2* (*purity*) At.%	A_1^*	Q_1^*	A_2^*	Q_2^*	*Temp. range* °C	*Ref.*
Cr	Fe						
(99.9)	(99.9)		IVb	p.c.			
9.13		—	—	9.27	55.1	575–726(Ferro.α)	86
		—	—	0.42	52.4	777–825(Para.α)	
		—	—	0.12	56.7	900–1 040(γ)	87
15.22		—	—	1.25	54.1	595–677(Ferro.α)	
		—	—	0.27	51.5	726–777(Para.α)	86
19.75		—	—	0.65	51.9	575–646(Ferro.α)	
		—	—	0.18	49.7	690–825(Para.α)	
( )		( )	IVb	p.c.			
2		3.21	58.4	—	—		
6		1.21	56.7	—	—		
13		0.64	55.4	—	—	800–1 400	37
16		0.19	52.1	—	—		
19		0.18	51.8	—	—		
15		2.0	57.0	—	—	—	88
( )	( )		IVb	p.c.	Fe^{59}		
0		—	—	4.4	60.5		
0.87	(wt.%)	—	—	4.3	60.4		
1.43		—	—	4.2	60.3		
3.09		—	—	4.0	60.1	Paramag.	134
5.05		—	—	3.7	59.7	α phase	
6.68		—	—	3.6	59.5		
8.18		—	—	3.3	59.2		
11.83		—	—	2.9	58.5		
(≥98)	(≥98)						
26		0.156	48.4	—	—		
51		40.0	70.0	—	—	950–1 320	38
69		24.6	75.5	—	—		
('Electrolytic')			IVa(i)	p.c.			
27		—	—	0.195	50.4		
51		—	—	249	74.6	1 040–1 400	39
84		0.376	64.2	146	81.9		
( )	( )		IVb	p.c.			
		Diffusion of Ni^{63}					
2.02	—	0.356	65.58		fcc		
5.73		0.282	65.22		fcc		
9.47		0.12	63.38		fcc		95
14.1		0.016	56.71		bcc		
20.1		0.007	53.59		bcc		
Cr	Fe Ni						
( )	( ) ( )		IVa(i)	p.c.			
19.9	(a) 24.7	0.19	58.8	1.74	67.9	840–1 290	
			(Ni) IVb	p.c.			102
		4.06	67.5	—	—	840–1 290	
( )	( ) ( )		IVc	p.c.			
17.5	(b) 11.33	—	—	0.58	67.1	808–1 200	41
( )	( ) ( )		IVa	Cr^{51}, Fe^{59}, Ni^{63}	p.c.		
17	12	0.13	63.1	0.37	66.8	600–1 300	135
				$A_{Ni}=8.8$	$Q_{Ni}=60.0$		
( )	( ) ( )		IVa(i)	p.c.			
17	12	6.3×10^{-2}	58.1	—	—	750–1 200	136
( )	( ) ( )		IVb	p.c.			
19	— 10	1.4	72	2.5×10^{-3}	52		
	30	1.4	72	1.0	66.5		
	(wt.%) 45	2×10^{-4}	96	1.2	61.0	900–1 200	137
	55	2.8	71	4.5	61.0		
	65	2.6	71	1.0	65.0		
	75	7.2×10^{-2}	60.5	4.10^2	82.0		
		Diff. of Ni					

(*a*) Plus 0.55 Nb, 0.74 Mn, 0.69 Si, 0.03 Ti.
(*b*) Plus 0.85 Nb, 1.3 Mn, 0.7 Si, 0.1 Mo.

Table 13.3 DIFFUSION IN HOMOGENEOUS ALLOYS—*continued*

Element 1 (*purity*) At.%	*Element 2* (*purity*) At.%	A_1^*	Q_1^*	A_2^*	Q_2^*	*Temp. range* °C	*Ref.*
Cr	Fe Ni						
( )	( ) (99.8)		IVb p.c.	Diffusion of C^{14}			
0.74	0.52 —	—	—	0.1	34.0	600–900	36
4.65	0.36 —	—	—	0.5	37.0	500–900	36
18	— 8			6.18	44.61	450–1 200	103
Cr	N						
( )	( )		Va	p.c.			
—	10–80 ppm	—	—	0.016	27.5	65–190	112
Cr	Ni						
('Electrolytic')			IVb	p.c.			
4.9		—	—	0.15	60.5	950–1 250	122
6.35		0.01	50.5	—	—	950–1 250	122
9.93		—	—	0.039	57.0	950–1 250	122
11.69		0.037	54.7	—	—	950–1 250	122
19.7		0.01	58.0	0.006 3	52.6	950–1 250	122
( )	(99.95)		IVa(i)	p.c.			
0		1.1	65.1	1.9	68.0	1 100–1 270 (D_{Cr}^*)	42
10		1.4	66.5	3.3	70.2	1 040–1 275 (D_{Ni}^*)	42
20		1.9	67.7	1.6	68.5	1 040–1 404 (Ni Self D)	42
( )	( )		IVa(i)	p.c.	IVa(ii)		
35(wt%)		0.2	58.6	4.10^{-3}	49.0	950–1 360	68
80(wt%)		0.28	62.0	—	—	1 150–1 415	68
( )	( )		IVa	p.c.			
25(wt%)		1.12	54.8	—	—	1 000–1 350	109
Cr	Ti						
	(99.4)	p.c.	IVa(ii)				
10.0		0.02	40.2	—	—	925–1 180	44
18.0		0.09	44.5	—	—	925–1 180	44
Cr	Zr						
(99.99)	(99.9)		IV	p.c.	Cr^{51}		
1.5	(wt%)	0.19	45.7	—	—	960–1 100(β)	49
Cu	Fe						
(99.998 5)	( )		IVa(i)	p.c. and s.c.			
	0.2		$D_{Cu}^*/D_{Cu\ Pure}$ = 0.99			1 023	120
	0.5		= 0.98			1 023	120
	1.38		= 0.91			1 020	120
	1.44		= 0.90	($b_1 = -5 \pm 1.5$)		1 078	120
	145		= 0.94			1 020	120
	1.82		= 0.91			992	120
	2.40		= 0.90			1 020	120
Cu	Ni						
(99.99)	(99.95)	p.c.	IVa(i)				
0		0.57	61.7	1.9	68.0	1 050–1 360	43
13.0		1.5	63.0	35.0	74.9	1 050–1 360	43
45.4		2.3	60.3	17.0	66.8	985–1 210	43
78.5		1.9	55.3	0.063	49.7	860–1 113	43
100.0		0.33	48.2	1.7	55.3	860–1 070	43
Cu	Ni Zn						
( )	( ) ( )		IVa(i)	p.c. (Diffusion of Zn^{65})			
100	— —	—	—	0.24	45.1	800–1 040	90
89.9	— 10.1	—	—	0.64	45.6	749–979	90
79.5	— 20.5	—	—	0.35	42.2	748–940	90
69.8	— 30.2	—	—	0.32	39.3	700–902	90
90.7	9.3 —	—	—	0.36	47.8	795–1 040	90
80.4	9.3 10.3	—	—	0.49	46.8	750–1 005	90
70.2	9.3 20.5	—	—	1.41	46.9	750–975	90
60.1	9.1 30.8	—	—	0.39	41.4	700–900	90

Table 13.3 DIFFUSION IN HOMOGENEOUS ALLOYS—*continued*

Element 1 (*purity*) At.%	*Element 2* (*purity*) At.%	A_1^*	Q_1^*	A_2^*	Q_2^*	*Temp. range* °C	*Ref.*
Cu	Ni Zn						
81.8	18.2 —	—	—	0.89	51.3	795–1 005	90
70.8	18.8 10.4	—	—	0.36	47.7	800–1 040	90
60.6	18.6 20.8	—	— Diffusion of Zn^{65}	0.73	44.7	748–940	90
71.4	28.6 —	—	—	1.37	54.1	870–1 080	90
61.2	28.2 10.6	—	—	1.44	52.6	855–1 040	90
50.8	28.2 21.0	—	—	1.17	49.9	800–1 005	90
40.7	27.9 31.4	—	—	1.13	47.4	760–976	90
Cu	Sb						
( )	( )		IVc	p.c.			
		D^*_{Cu}		D^*_{Sb}			
(δ)	19.4	4.10^{-10}		3.10^{-11}		390	61
(χ)	24.4	7×10^{-9}		3.1×10^{-10}		390	61
(γ)	33.4	$\sim10^{-9}$		2.7×10^{-9}		390	61
( )	( )		IVb	p.c.			
(β)	21	8.57×10^{-3}	10.46	—	—	520–630	91
(β)	25	1.99×10^{-3}	7.26	—	—	520–630	91
(β)	29	5.80×10^{-3}	5.80	—	—	520–630	91
Cu	Sn						
(99.999)	(99.999)		IVa(i)	p.c.			
(δ)	20.5	4.7	30.9	2.4×10^{3}	49.7	440–575	92
(γ)	18.0	1.4×10^{-2}	17.8	0.33	29.2	600–725	92
(γ)	19.8	3.6×10^{-3}	20.2	9.2×10^{-2}	27.1	600–725	92
Cu	Zn						
('Spec. P')	('Spec. P')	s.c.	IVa(i)				
(α)	31	0.34	41.9	0.73	40.7	580–905	45
	(99.99)	s.c.	IVa(i)				
(β)	45.65 to 48.1	0.011	22.04	0.003 5	18.78	Disord. 497–817	46
(β)	45.65 to 48.1	180	37.09[(a)]	78.10^{3}	44.23	Ord. 380–450	46
(β)	45.65 to 48.1	80	36.02[(a)]	163.0	36.3	Ord. 264–380	46
(β)	45.65 to 48.1	Diffusion of Sb 0.08	23.5			Disord. 498–594	46
(—)	(—)			p.c. IVa(i)			
(β)	46.7	0.020	23.42	0.022	22.04	Disord. 500–800	47
(β)	46.7	0.80	31.0[(b)]	1.0	32.0	Ord. around 300	47
(—)	(—)	p.c.	IVa(i)				
(β)	47.2	Diffusion of Ag 0.014	21.9[(c)]	Diffusion of Co 0.47	26.9	Disord. 470–700	48
( )	( )		IVa(i)	p.c.			
		D^*_{Cu}		D^*_{Zn}			
—	0	2.71×10^{-10}		9.66×10^{-10}		894.4	93
		6.70		22.7		946.7	93
—	0.62	2.81		—		894.4	93
—	1.09	7.42		—		946.4	93
—	2.17	3.12		—		894.4	93
—	2.68	3.23		—		894.4	93
—	2.99	8.33		—		946.7	93
—	4.06	3.71		—		894.4	93
—	4.13	9.23				946.7	93

(*a*) Arrhenius plots in the ordered region are curved. The values of A and Q reported describe straight line approximations to the data over the temperature ranges indicated.

(*b*) Ditto. The values of A and Q given here refer to the data at the lower end of the ordered temperature range investigated, viz: 300 °C.

(*c*) Values of D^* for the ordered region are shown only in graphical form in reference 48.

Table 13.3 DIFFUSION IN HOMOGENEOUS ALLOYS—*continued*

Element 1 (*purity*) At.%	*Element 2* (*purity*) At.%	A_1^*	Q_1^*	A_2^*	Q_2^*	*Temp. range* °C	*Ref.*
Cu	Zn						
(Electrolytic)	(99.99)		IVb	s.c.			
0		—	—	$\perp c$ 1.62	26	(d)	
0		—	—	$\parallel c$ 0.013	19		
0.2		—	—	$\perp c$ 3.2	26		
0.2		—	—	$\parallel c$ 0.021	19		
0.3		—	—	$\perp c$ 3.4	26		
0.3		—	—	$\parallel c$ 0.025	19	550–630	56
0.4		—	—	$\perp c$ 3.4	26		
0.4		—	—	$\parallel c$ 0.029	19		
0.5		—	—	$\perp c$ 3.5	26		
0.5		—	—	$\parallel c$ 0.035	20		

(*d*) Additions of 0.5 at % Al to this range of CuZn alloys have a negligible effect on $Q_{Zn}{}^*$ but decrease $A_\perp$ by ~20% and A by ~40%.

Element 1 (*purity*) At.%	*Element 2* (*purity*) At.%	A_1^*	Q_1^*	A_2^*	Q_2^*	*Temp. range* °C	*Ref.*
Fe	Ge						
( )	( )		IVa	p.c.			
	4.8	4.8	58.0	—	—	900–1 200	75
Fe	Mn						
(99.97)	(99.94)		IVb	p.c.			
At.%	1.04	9×10^{-2}	63.4	5.5×10^{-2}	59.6		
	2.03	1.05×10^{-1}	62.8	2.0×10^{-2}	56.2		
	2.97	5.8×10^{-2}	61.1	9.6×10^{-3}	53.1		
	4.90	6.6×10^{-2}	60.9	1.7×10^{-2}	54.8	Fe 990–1 300	
	7.04	1.1×10^{-1}	62.6	7.2×10^{-2}	59.3	Mn 710–1 300	
	10.41	3.5×10^{-1}	65.8	2.9×10^{-1}	63.7		98
	18.15	6.4×10^{-1}	67.4	1.9×10^{-1}	62.5		
	25.5	8.5×10^{-1}	66.3	1.2×10^{-1}	60.1		
	33.98	6.0×10^{-1}	66.1	7.3×10^{-2}	57.8		
( )	( )		IVb	p.c.			
	wt.%	Diffusion of Ni^{63}					
	0.42	0.495	67.46				
	1.26	0.364	66.83			fcc range	95
	4.60	0.144	64.61				
	9.7	0.132	63.90				
Fe	Mo						
( )	( )		IVb	p.c. Fe^{59}			
	0.54	15.5	62.9	—	—		
(wt.%)	1.06	23.6	61.4	—	—	680–900 Paramag.	141
	1.50	28.5	63.6	—	—		
	2.50	47.7	63.1	—	—		
Fe	N						
( )	( )		Va	p.c.			
~100	V. small	—	—	0.005	18.4	−20 – +160	51
Fe	Ni						
( )	( )		IVb	p.c.			
	0.2	—	—	1.09	69.33		
	0.55	—	—	1.09	69.33		
wt.%	2.29	—	—	0.593	67.3	γ sol. sol. range	95
	9.21	—	—	0.497	66.4		
	19.34	—	—	0.409	65.41		
(Electrol.)	(Electrol.)		IVc	p.c.			
	5.8	—	—	2.11	73.5	1 160–1 390	50
	14.88	—	—	5.0	75.6		
( )	( )		IVb	p.c.			
	20	18	75	—	—	800–1 300	29
	25	71	79	—	—	1 050–1 330	
( )	( )		IVa(i)	p.c.			
	50	—	—	66	77.1	1 013–1 235	96
	80	—	—	4.8	68.4		

Table 13.3 DIFFUSION IN HOMOGENEOUS ALLOYS—*continued*

Element 1 (*purity*) At.%	*Element 2* (*purity*) At.%	A_1^*	Q_1^*	A_2^*	Q_2^*	*Temp. range* °C	*Ref.*
( )	( )		IVa(i)	s.c.			
0 to 1.46	—	—	—	D^*_{Ni} nearly independent of composition		1 226 and 1326	97
Fe	Pd						
(99.97)	(99.95)		IVa(i) Fe^{59}	IVb Pd^{113}	p.c.		
0		0.18	62.1	0.04	59.6		
10		0.91	65.9	0.37	64.1		
20		0.91	64.4	0.79	64.8		
30	(At.%)	0.60	62.0	0.73	63.7		
40		0.69	61.8	0.79	63.6		
45		0.79	62.1	0.67	63.0		
50		0.95	62.7	0.70	63.2	1 100–1 250	142
60		0.95	63.1	1.05	64.7		
70		0.66	62.7	1.66	66.7		
80		0.93	64.9	1.84	68.0		
90		0.79	66.3	0.70	66.6		
100		0.41	67.1	0.41	67.1		
Fe	Pt						
( )	( )		IVb p.c. Pt^{193m}				
	0	—	—	2.7	70.7 ± 2.5		
	15	—	—	1.1	63.1 ± 8.2		
	20	—	—	0.34	63.3 ± 7.8		
	25	—	—	1.17	63.4 ± 11.4		
	30	—	—	0.28	63.1 ± 80		
	34	—	—	0.15	63.1 ± 5.6	780–1 420	138
	40	—	—	1.3	69.2 ± 7.8		
	45	—	—	1.13	67.9 ± 1.4		
	50	—	—	2.1	69.8 ± 2.9		
	55	—	—	0.85	68.5 ± 1.7		
	60	—	—	0.34	67.1 ± 5.3		
Fe	Sb						
( )	( )		IVa	p.c.			
	2.5	0.51	51.8	—	—	896–1 097	75
Fe	Si						
(99.999)	(99.99)		IVa and b	p.c.			
	0	1.39	56.5	—	—		
	4.7	1.63	55.5	—	—		
	6.3	1.65	54.9	—	—		
	8.2	0.50	50.9	—	—	800–1 300	100
	11.3	1.11	51.1	—	—		
	11.7	1.46	51.7	—	—		
( )	( )		IVb	p.c.			
	2.9	0.44	52.2	—	—	967–1 416	123
( )	( )		IVa(i) s.c.	Fe^{59}			
	7.64	1.38	54.5	—	—	900–1 100	139
(At.%)	11.1	0.63	50.7	—	—		
Fe	Ti						
( )	( )		IVa	p.c.			
	2	2.8	57.8	—	—	900–1 200	75
( )	( )		IVa(i)	p.c.			
	2	0.56	51.7	—	—		
	4	0.27	48.9	—	—	1 000–1 400	99
	6	0.40	49.9	—	—		
( )	( )		IVa(i)	p.c.			
	33.32	11.7	75.0	—	—	850–1 100	125
(*a*)	50.0	0.135	57.2	—	—		
(Spec. P.)	(99.7)	IVa(ii)	p.c.	Diffusion of Nb^{95}			
5		9.2×10^{-2}	39.6	1.82×10^{-3}	34.9	850–1 260	
10		2.14	48.6	2.9×10^{-2}	41.3	850–1 200	53
15		52.5	58.1	9.9	56	850–1 100	

(*a*) Reference 125 also reports data on $D_{Fe}{}^*$ in two-phase alloys within the range 18 to 50 At.% Ti.

Table 13.3 DIFFUSION IN HOMOGENEOUS ALLOYS—*continued*

Element 1 (*purity*) At.%	*Element 2* (*purity*) At.%	A_1^*	Q_1^*	A_2^*	Q_2^*	*Temp. range* °C	*Ref.*
Fe	V						
(—)	(—)		IVb	p.c.			
(γ)	0.53	1.46	68.9	—	—		
(γ)	1.09	0.53	66.2	—	—	1 100–1 300	54
(α)	2.11	0.10	61.5	—	—		
(—)	(—)		IVa(i)	p.c.		(Paramag.)	
	18.0	7.0	61.7	3.9	58.5	880–1 200	55
(99.98)	(—)		IVb	p.c.			
		Diffusion of Hf[181]					
—	2.1	1.31	69.3	3.92	57.6	1 000–1 400	94
(99.999)	( )		IVa	p.c.			
—	1.8	1.4	56.6	—	—	900–1 500	75
—	5.3	1.87	57.4	—	—	900–1 193	
( )	( )		IVb	p.c.			
	2	—	—	3.92	57.6		
	5	—	—	3.00	57.0		
	9	—	—	2.28	56.4	1 000–1 450	99
	14	—	—	2.12	56.5		
	19	—	—	1.66	55.9		
( )	( )	Diffusion of Cr[51]					
	1.7	2.0	57.0	—	—	—	88
Ga	Ni						
( )	( )	IV	p.c.	Ni[63]	Ga[67]		
	47.28	0.123	45.8	0.0029	34.2		
(At.%)	48.76	0.787	50.0	0.0126	36.9		
	49.33	—	—	0.0130	37.3		
	50.01	0.0010	35.0	0.0936	41.2	810–1 111	140
	50.45	—	—	0.1353	41.9		
	50.73	0.0122	39.8	0.0174	37.8		
	51.01	0.109	45.3	0.0107	36.7		
	52.4	5.143	53.05	0.0121	36.8		
Ga	Pu						
( )	( )		IVa(i)	p.c.			
1.0(wt)	—	—	—	76.4	36.3	340–508(δ)	114
		—	—	6.98×10^{-4}	13.4	574–644(ε)	
K	Na						
(99.97)	(99.999 5)		IVa(i)	p.c.			
		—	—	$D^*_{Na} \times 10^9$			
0				1.34			
0.13				1.38			
0.14				1.44			
0.19				1.54			
0.33				1.58		0	
0.49				1.81			
0.56				1.79			
0.67				1.90			
1.03				2.12			108
				$D^*_{Na} \times 10^8$			
0				2 16			
0.13				2.28			
0.18				2.26			
0.38				2.39		50	
0.55				2.61			
0.73				2.86			
1.25				3.40			
Mn	Ti						
(99.97)	(99.97)		IVa(i)	p.c.			
9.7		—	—	1.9×10^{-2}	40.9	860–1 450	
13.3		—	—	2.06×10^{-2}	41.1	800–1 350	121
17.9		—	—	2.60×10^{-2}	42.1	797–1 300	
20.6		—	—	5.47×10^{-1}	49.6	810–1 250	

Table 13.3 DIFFUSION IN HOMOGENEOUS ALLOYS—*continued*

Element 1 (*purity*) At.%	*Element 2* (*purity*) At.%	A_1^*	Q_1^*	A_2^*	Q_2^*	*Temp. range* °C	*Ref.*
Mn	Zr						
( )	( )	IV	p.c.	Mn^{54}	Zr^{95}		
0		5.38×10^{-3}	33.6	0.31×10^{-4}	25.2		
0.5		2.92×10^{-3}	32.4	0.71×10^{-4}	26.9		
1.0	(At.%)	1.38×10^{-3}	30.9	1.2×10^{-4}	27.9	900–1 200	143
1.5		4.6×10^{-4}	28.7	2.17×10^{-4}	29.1		
2.0		8.0×10^{-5}	25.0	3.36×10^{-4}	29.95		
Mo	Ni						
( )	( )		IVa	p.c.			
8		1.31	54.9	2.55	56.4	1 000–1 400	
16		1.3	54.6	0.63	52.2	950–1 350	
18		0.45	52.2	0.34	52.2	950–1 350	126
20		0.25	50.3	0.19	48.8	950–1 300	
23		0.20	49.5	0.12	47.4	1 100–1 300	
(99.4)	(99.4)	IVb	p.c.	Diffusion of C^{14}			
2.94		—	—	1.0	38	600–900	36
Mo	Ti						
(99.98)	(99.62)	IVb	p.c.	Diffusion of U^{235}			
5				0.26	49.3		
10				1.45	56.7		
15				2.88	59.6	900–1 400	116
20				2.69	64.6		
25				6.03	69.0		
30				33.0	75.0		
Mo	U						
(—)	(—)		IIa(ii)	p.c.			
0		—	—	1.8×10^{-3}	27.5	800–1 040	57
10		—	—	2.5×10^{-3}	33.0		
Mo	W						
( )	( )		IVa	p.c.			
	0.1	142	112.0	0.008 5	71.0	1 800–2 400	
	15	265	106.0	1.4	73.0		
	20	146	102.0	1.7	74.6	1 400–2 400	
	25	47	95.0	2.2	77.0		
	35	28	92.0	6.9	85.0	1 500–2 400	
	50	12	88.0	14	95.0	1 700–2 500	104
	65	1.3	86.0	16	102.0	1 800–2 600	
	75	0.2	84.5	20	116.0		
	80	0.11	82.0	22	119.0	1 800–2 800	
	85	0.08	80.0	25	122.0		
	99.9	0.002 5	78.0	24	130.0	2 200–2 800	
( )	( )		IVa	p.c.			
44	—	0.17	107.0	—	—	1 900–2 400	105
55	—	0.12	103.0	—	—		
Nb	Ni						
( )	( )		IVb s.c.	Ni^{63} Nb^{95}			
25	75	2.4×10^2	107.0	—	—	1 270–1 350	144
		—	—	0.18	72.8	1 090–1 370	
N	Nb						
	(99.4)	p.c.	Va				
V. small	~100	0.008 6	34.92	—	—	150–300	30
N	Ta						
( )	( )	p.c.	Va				
V. small	~100	0.005 6	37.84	—	—	185–300	30
N	V						
( )	( )	p.c.	Va				
V. small	~100	0.009 2	34.06	—	—	135–280	30

Table 13.3 DIFFUSION IN HOMOGENEOUS ALLOYS—*continued*

Element 1 (*purity*) At.%	*Element 2* (*purity*) At.%	A_1^*	Q_1^*	A_2^*	Q_2^*	*Temp. range* °C	*Ref.*
Nb	Ni						
(99.9)	(99.99)		IVa(i)	p.c.			
1.2		—	—	0.12	60.8	1 030–1 230	113
8		—	—	0.20	62.2		
10		—	—	1.80	67.1		
( )	( )	p.c.	Va				
Nb	**O**						
( )	( )		Va	p.c.			
~100	Small	—	—	5.6×10^{-3}	26.2	23–1 545	145
( )	( )		IV	p.c. O^{18}			
~100	~0	—	—	3.7×10^{-3}	25.4	550–1 100	146
	1–1.5	—	—	4.55×10^{-3}	25.9	600–1 100	
Nb	Ti						
(99.6)	(99.7)	p.c.	IVa(ii)	Diffusion of Fe^{55}			
5		1.2×10^{-4}	29.9	7.9×10^{-3}	33.1	850–1 300	53
10		5.8×10^{-4}	36.1	11.5×10^{-3}	34.9		
15		1.5×10^{-3}	39.3	7.9×10^{-3}	35.0		
31		9×10^{-3}	50.0	—	—	1 050–1 800	58
54		8×10^{-2}	64.0	—$^{(a)}$	—	1 200–1 800	
66		0.1	72.0	—	—	1 500–2 000	
89		1.0	91.0	—	—	1 700–2 200	
Nb	Ti						
( )	(99.97)		IVa(i) p.c.	Ti^{44}	Nb^{95}		
	100	2.9×10^{-4}	31.03	4.54×10^{-4}	31.30	950–1 511	147
(At.%)	94.6	1.79×10^{-3}	38.22	1.27×10^{-3}	35.61		
	80.4	1.18×10^{-2}	47.33	3.15×10^{-3}	41.96		
	64.3	2.98×10^{-1}	61.73	2.51×10^{-1}	59.02		
( )	( )	IVb	p.c.	Diffusion of U^{235}			
5				0.022	40.7	900–1 400	116
10				0.035	43.5		
15				0.11	48.0		
20				0.26	51.5		
25				0.25	54.0		
30				0.42	56.3		
Nb	U						
( )	( )		IVb	p.c.			
0		1.2×10^{-5}	28.5	1.1×10^{-4}	36.0	950–1 050	110
5				3.2×10^{-6}	34		
10				3.5×10^{-6}	35.2		
20				10^{-5}	39.6		
35		3.1×10^{-2}	68.8	1.25×10^{-4}	53.2	1 050–1 220	
50		3.1×10^{-2}	72.7	2.5×10^{-4}	57.2	1 200–1 450	
65		7.6×10^{-2}	77.5	4.0×10^{-3}	65.6	1 200–1 500	
80		1.1	91.5	6.3×10^{-4}	67.0	1 550–1 800	
90		5.2	100.6	2.5×10^{-3}	73.0	1 700–1 900	
100		0.91	100.6	6.5×10^{-6}	76.7	1 800–2 000	
( )	( )		IIa(ii)	p.c.			
10	—	—	—	1.66×10^{-4}	28.2	800–1 040	57
Nb	Zr						
(99.99)	(99.9)		IV p.c.	Nb^{95}			
2.3	wt.%	1.63×10^{-3}	38.8	—	—	900–1 160	49
Nb	Zr Mo						
( )	( ) ( )		IVb	p.c.			
5	95 —	4.6×10^{-2}	50	4.1×10^{-3}	40.5	1 200–1 500	119
10	90 —	4.4×10^{-1}	55	2.8×10^{-2}	47		
15	85 —	5.5×10^{-1}	58	8.6×10^{-2}	50.5		
2.5	— 2.5	7.6×10^{-2}	51	9.0×10^{-3}	43		
5	— 5	3.1×10^{-1}	56	5.4×10^{-2}	49.5		
7.5	— 7.5	1.1	60	2.8×10^{-1}	55		
—	95 5	—	—	6.4×10^{-3}	42		
—	90 10	—	—	6.6×10^{-2}	49		
—	85 15	—	—	2.10^{-1}	53.5		

(*a*) Values read from graphically plotted results.

Table 13.3 DIFFUSION IN HOMOGENEOUS ALLOYS—*continued*

Element 1 (*purity*) At.%	*Element 2* (*purity*) At.%	A_1^*	Q_1^*	A_2^*	Q_2^*	*Temp. range* °C	*Ref.*
Nb	Zr **O**						
( )	( ) ( )		IV	p.c.	O^{18}		
	0.5	—	—	1.83×10^{-3}	26.3	600–1 100	146
wt.%	0.8	—	—	1.07×10^{-3}	25.9	600–1 100	
	1.0	—	—	4.11×10^{-4}	24.7	600–1 100	
				(O Diffusion)			
Ni	Sb						
( )	( )		IVb	p.c.			
71.7		5.4×10^{-4}	14.7	—	—	600–1 000 °C	101
72.9(β)		4.9×10^{-4}	14.8	—	—		
73.7		6.3×10^{-4}	15.8	—	—		
75.0		6.9×10^{-4}	16.5	—	—		
Ni	Ti						
( )	( )		IVa(i)	p.c.			
	2.44	2.24	68.6	—	—	929–1 228	124
	4.86	1.51	67.5	—	—		
	7.24	0.91	66.2	—	—		
	9.63	6.6	70.7	—	—		
	12.68	3.0	76.2	—	—		
	25	6.9×10^{-3}	92.9	—	—		
Ni	W						
(99.95)			IVa(i)	p.c.			
	0	1.9	68.0	2.0	71.5	1 096–1 395 (D_{Ni})	66
	1.7	30.0	76.5	2.2	73.1		
	5.3	58.0	80.6	17.0	80.5	1 100–1 295 (D_W)	
	9.2	1.1	70.3	1.4	74.5		
O	Ta						
	(99.9)	p.c.	Va				
V. small	100	0.004 4	25.45	—	—	40–160	30
( )	( )	p.c.	Va				
V. small	~100	1.9×10^{-2}	27.4			150–350	111
O	V						
	(—)	p.c.	Va				
V. small	100	0.013 0	29.01	—	—	70–190	30
Pb	Tl						
(99.99)	(99.99)		IVa(i)	p.c. (except 62.6%)			
100	0	1.372	26.06	0.511	24.33	206–323	59
	5.21	1.108	25.75	0.364	23.89		
	10.27	0.880	25.45	0.361	23.83		
	20.2	0.647	25.05	0.353	23.78		
	34.6	0.367	24.53	0.193	23.12		
	50.3	0.231	24.44	0.091	22.52		
	62.4	0.393	25.64	0.101	22.93		
	62.6(s.c.)	0.287	25.29	0.126	23.20		
	74.5	0.691	26.83	0.194	23.86		
	76.2	0.862	27.13	0.330	24.48		
	81.8	2.575	28.24	0.957	25.37		
	87.1	17.0	29.71	1.20	25.53		
Pu	Zr						
( )	( )		IVb p.c.	Pu^{240}			
At.%	40	0.04	29.65	—	—	640–840	148
	10		$D_1 = 1.05 \times 10^{-7}$	—	—	650	
Sn	Zr						
	(99.6)		IVb	p.c.			
0		—	—	5.9×10^{-2}	52	650–827	60
1.0		—	—	5.0	62		
1.85		—	—	2.1×10^{-3}	75	740–827	
2.75		—	—	10	64		

Table 13.3 DIFFUSION IN HOMOGENEOUS ALLOYS—*continued*

Element 1 (*purity*) At.%	*Element 2* (*purity*) At.%	A_1^*	Q_1^*	A_2^*	Q_2^*	*Temp. range* °C	*Ref.*
Ti	V						
(—)	(99.98)		IVa(i)	p.c.			
0						900–1800	115
10						900–1800	115
20						900–1800	115
30		Curved Arrhenius plots *See* Figure 13.3a		Curved Arrhenius plots *See* Figure 13.3b		900–1800	115
40						900–1800	115
50		The lines of Figures 13.3a and 13.3b are drawn at 10 at.% intervals of composition as shown in column 1				900–1800	115
60						900–1800	115
70						900–1800	115
80						900–1800	115
90						900–1800	115
( )	( )	IVb	p.c.	Diffusion of U^{235}			
	5			0.034	41.2	900–1400	116
	10			0.063	43.5	900–1400	116
	15			0.047	43.5	900–1400	116
	20			0.0096	39.3	900–1400	116
	25			0.025	42.5	900–1400	116
	30			0.089	46.7	900–1400	116
	40			0.12	49.4	900–1400	116
	50			0.65	56.3	900–1400	116
	60			2.7	69.0	900–1400	116
	70			5 900	87.2	900–1400	116
U	Zr						
(—)	(—)		IIa(ii)	p.c.			
	10	1.26×10^{-4}	22	—	—	800–1 040	57
( )	( )		IIa(ii)	p.c.	IVa(i)		
	95	$D_U^* = 1.5\times10^{-9}$	—	$D_{Zr}^* = 3.2\times10^{-9}$	—	1 000	62
		$D_U^* = 1.85\times10^{-9}$	—	$D_{Zr}^* = 4.2\times10^{-9}$	—	1 050	62
( )	( )		IVa(i)	p.c.			
0		5.7×10^{-4}	30.5			900–1 065	117
11		7.5×10^{-4}	33.9	0.12	49.1	900–1 065	117
27		3.65×10^{-3}	38.4	2.8×10^{-2}	45.5	900–1 065	117
39		7.10^{-3}	40.3	3.9×10^{-4}	35.0	900–1 065	117
59		8.96	58.6	2.4×10^{-5}	28.3	900–1 065	117
78				3.8×10^{-6}	23.7	900–1 065	117
85				7.5×10^{-7}	20.0	900–1 065	117
100				1.6×10^{-7}	16.3	900–1 065	117

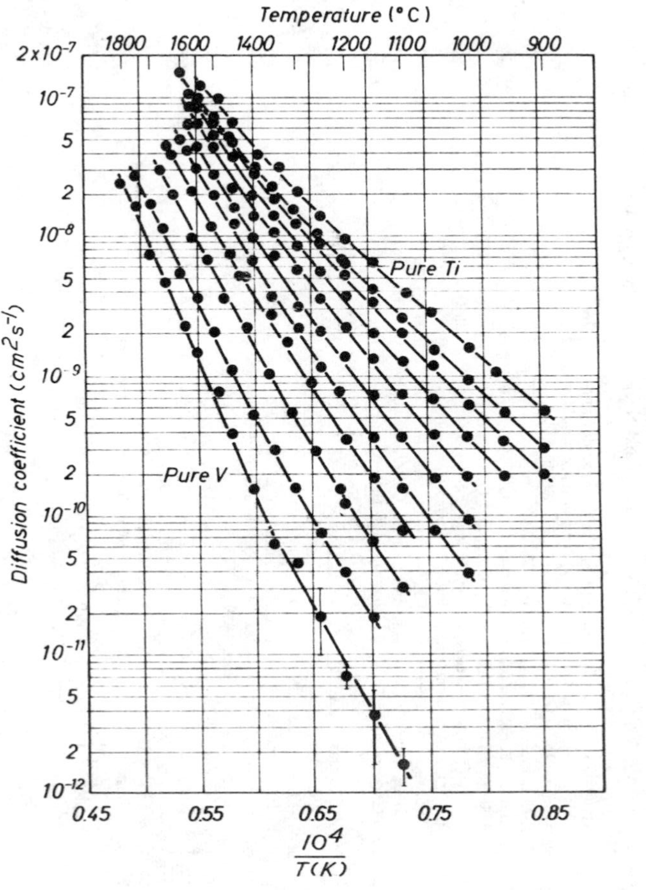

Figure 13.3a *Temperature dependence of diffusion of* Ti^{44} *in titanium–vanadium alloys*[115]

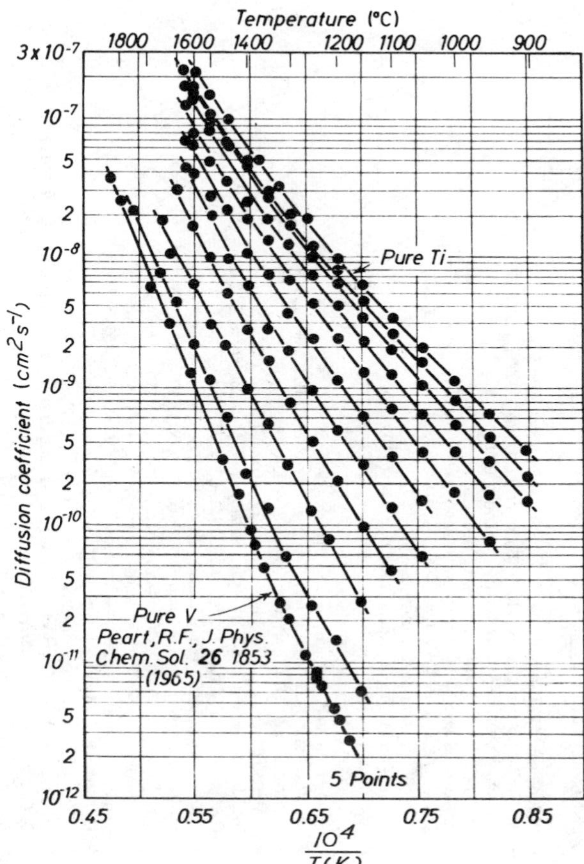

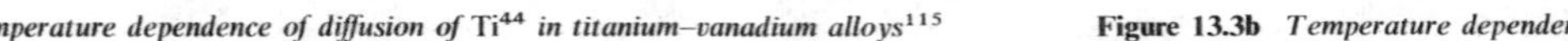
Figure 13.3b *Temperature dependence of diffusion of* V^{48} *in titanium–vanadium alloys*[115]

REFERENCES TO TABLE 13.3

1. R. E. Hoffman, D. Turnbull and E. W. Hart, *Acta metall.*, 1955, **3**, 417.
2. W. C. Mallard, A. B. Gardner, R. F. Bass and L. M. Slifkin, *Phys. Rev.*, 1963, **129**, 617.
3. A. Schoen, *Ph. D. Thesis*, University of Illinois, 1958.
4. N. H. Nachtrieb, J. Petit and J. Wehrenberg, *J. chem. Phys.*, 1957, **26**, 106.
5. R. L. Rowland and N. H. Nachtrieb, *J. phys. Chem.*, 1963, **67**, 2817.
6. E. Sonder, *Phys. Rev.*, 1955, **100**, 1662.
7. R. E. Hoffman, *Acta metall.*, 1958, **6**, 95.
8. D. Lazarus and C. T. Tomizuka, *Phys. Rev.*, 1956, **103**, 1155.
9. R. E. Hoffman and D. Turnbull, *J. appl. Phys.*, 1952, **23**, 1409.
10. C. T. Tomizuka. Unpublished data.
11. T. Heumann and P. Lohman, *Z. Electrochem*, 1955, **59**, 849.
12. W. C. Hagel and J. H. Westbrook, *Trans. metall., Soc. AIME*, 1961, **221**, 951.
13. H. A. Domian and H. I. Aaronson, *ibid.*, 1964, **230**, 44.
14. S. D. Gertsricken and I. Y. Dekhtyar, *Proc. 1955 Geneva Conf.*, 1955, **15**, 99.
15. S. D. Gertsricken and I. Y. Dekhtyar, *Fizika metall. Metallov.*, 1956, **3**, 242.
16. F. C. Nix and F. E. Jaumot, *Phys. Rev.*, 1951, **83**, 1275.
17. S. D. Gertsricken *et al.*, *Issled. zharpr. Splav.*, 1958, **3**, 68.
18. A. E. Berkowitz, F. E. Jaumot and F. C. Nix, *Phys. Rev.*, 1954, **95**, 1185.
19. J. E. Hilliard, B. L. Averbach and M. Cohen, *Acta metall.*, 1959, **7**, 86.
20. H. B. Huntington, N. C. Miller and V. Nerses, *ibid.*, 1961, **9**, 749.
21. D. Gupta, D. Lazarus and D. S. Liebermann, *Phys. Rev.*, 1967, **153**, 863.
22. A. D. Kurtz, B. L. Averbach and M. Cohen, *Acta metall.*, 1955, **3**, 442.
23. J. E. Reynolds, B. L. Averbach and M. Cohen, *ibid.*, 1957, **5**, 29.
24. M. Yanitskaya, A. A. Zhukhavitskii and S. Z. Bokstein, *Dokl. Akad. Nawk SSSR*, 1957, **112**, 720.
25. S. D. Gertsricken and T. K. Yatsenko, *Vop. Fiz.*, 1957, **8**, 101.
26. C. A. Wert and C. Zener, *Phys. Rev.*, 1949, **76**, 1169.
27. R. P. Smith, *Trans. metall. Soc., AIME*, 1962, **224**, 105.
28. H. W. Mead and C. E. Birchenall, *J. Metals.*, 1956, **8**, 1336.
29. P. L. Gruzin and E. V. Kuznetsov, *Dokl. Akad. Nauk SSSR*, 1953, **93**, 808.
30. R. W. Powers and M. V. Doyle, *J. appl. Phys.*, 1959, **30**, 514.
31. B. Khomka, *Acta Physica Polonica*, 1963, **24**, 669.
32. B. Khomka, *Nukleonika*, 1963, **8**, 185.
33. P. L. Gruzin and B. M. Noskov, 'Problems of Metallography and Physics of Metals,' **4**, 509 (Moscow, 1955) and Aec. tr 2924, p. 355.
34. S. G. Fishman, D. Gupta and D. S. Liebermann, *Phys. Rev.*, 1970, **B2**, 1451.
35. K. Hirano, R. P. Agarwala, B. L. Averbach and M. Cohen, *J. appl. Phys.*, 1962, **33**, 3049.
36. D. L. Gruzin, Yu. A. Polikarpov and G. B. Federov, *Fizika metall. Metallov.*, 1957, **4**, 94.
37. A. W. Bowen and G. M. Leak, *Metal Trans.*, 1970, **1**, 2767.
38. H. W. Paxton and T. Kunitake, *Trans. metall Soc. A.I.M.E.*, 1960, **218**, 1003.
39. L. I. Ivanov and N. P. Ivanchev, *Izv. Akad. Nauk. SSSR, Otdel, Tekhn. Nauk.*, 1958, **8**, 15.
40. A. Ya Shinyayev, Conference on Uses of Isotopes and Nuclear Radiations, *Met. Metallogr.*, 1958, p. 299, Moscow.
41. V. Linnenbom, M. Tetenbaum and C. Cheek, *J. appl. Phys.*, 1955, **26**, 932.
42. K. Monma, H. Suto and H. Oikawa, *Nippon Kink. Gakk.*, 1964, **28**, 188.
43. K. Monma, H. Suto and H. Oikawa, *ibid*, 1964, **28**, 192.
44. A. J. Mortlock and D. H. Tomlin, *Phil. Mag.*, 1959, **4**, 628.
45. J. Hino, C. Tomizuka and C. Wert, *Acta Metall.*, 1957, **5**, 41.
46. A. B. Kuper, D. Lazarus, J. R. Manning and C. T. Tomizuka, *Phys. Rev.*, 1956, **104**, 1536.
47. P. Camagni, *Proc. 2nd Geneva Conf. Atomic Energy*, P/1365, Vol. 20, Geneva, 1958.
48. C. Bassani, P. Camagni and S. Pace, *Il Nuovo Cim.*, 1961, **19**, 393.
49. G. P. Tiwari, M. C. Saxena and R. V. Patil, *Trans. Ind. Inst. Met.*, 1973, **26**, 55.
50. J. R. MacEwan, J. U. MacEwan and L. Yaffe, *Can. J. Chem.*, 1959, **37**, 1629.
51. Results given are from a 'best line' through the internal friction data of $(-20° - +32°)$C. Wert, *J. appl. Phys.*, 1950, **21**, 1196; $(9.5° - 21.5°)$ J. D. Fast and M. B. Verrijp, *J. Iron St. Inst.*, 1954, **176**, 24; $(18.5° - 59°)$ W. R. Thomas and G. M. Leak, *Phil. Mag.*, 1954, **45**, 656; $(91° - 161°)$ L. Guillet and P. Gence, *J. Iron St. Inst.*, 1957, **186**, 223.
52. J. W. Miller, *Phys. Rev.*, 1969, **181**, 1095.
53. R. F. Peart and D. H. Tomlin, *Acta metall.*, 1962, **10**, 123.
54. M. S. Zelinski, B. M. Moskov, P. V. Pavlov and E. V. Shitov, *Physics Metals Metallogr.*, 1959, **8** (5), 79 and *Fiz. Metall. Metallov.*, 1959, **8**, 725.
55. J. Stanley and C. Wert, *J. appl. Phys.*, 1961, **32**, 267.
56. I. A. Naskidashvili, *Soobshoheniya Akad. Nauk. Gauzin, SSSR*, 1955, **16**, 509.
57. Y. Adda and A. Kirianenko, *J. nucl. Mater.*, 1962, **6**, 135.
58. G. B. Gibbs, D. Graham and D. H. Tomlin, *Phil. Mag.*, 1963, **8**, 1269.
59. H. A. Resing and N. H. Nachtrieb, *Physics Chem. Solids*, 1961, **21**, 40.
60. V. S. Lyashenko, V. N. Bykov and L. V. Pavlinov, *Physics Metals Metallogr.*, 1959, (3) **8**, 40.
61. T. Heumann and F. Heinemann, *Z. Electrochem.*, 1956, **60**, 1160.
62. Y. Adda, C. Mairy and J. M. Andreu, *Revue Métall., Paris*, 1960, **57**, 549.

63. J. W. Miller, *Phys. Rev.*, 1970, **B2**, 1624.
64. R. J. Borg, *J. appl. Phys.*, 1963, **34**, 1562.
65. S. D. Gertsricken and I. Ya. Dekhtyar, *Proc. 1955 Geneva Conf. Peaceful Uses of Atomic Energy*, 1955, **15**, 124.
66. K. Monma, H. Suto and H. Oikawa, *Nippon Kink. Gakk.*, 1964, **28**, 197.
67. D. A. Leak and G. M. Leak, *J. Iron St. Inst.*, 1958, **189**, 256.
68. J. Askill, *Phys. Status Solidi*, 1971, **a8**, 587.
69. Th. Heumann and H. Böhmer, *J. Phys. Chem. Solids*, 1968, **29**, 237.
70. C. J. Santoro, *Phys. Rev.*, 1969, **179**, 593.
71. S. J. Rothman and N. L. Peterson, *Phys. Rev.*, 1967, **154**, 552.
72. W. B. Alexander and L. M. Slifkin, *Phys. Rev.*, 1970, **B1**, 3274.
73. J. Kucera and B. Million, *Metal Trans.*, 1970, **1**, 2599.
74. A. B. Gardner, R. L. Sanders and L. M. Slifkin, *Phys. Status Solidi*, 1968, **30**, 93.
75. D. Y. F. Lai and R. J. Borg, USAEC Rept. UCRL 50314, 1967.
76. L. N. Larikov, *Avtom. Svarka*, 1971, **24**, 71.
77. L. N. Larikov, *et al.*, *Prot. Coat. Metals*, 1970, **3**, 91.
78. G. F. Hancock and B. R. McDonnell, *Phys. Status Solidi*, 1971, **4**, 143.
79. D. Gupta and D. S. Liebermann, *Phys. Rev. B*, 1971, **4**, 1070.
80. V. M. Ananin, *et al.*, *Atom. Energ. USSR*, 1970, **29**, 220.
81. S. Benci, *et al.*, *J. Phys. Chem. Solids*, 1965, **26**, 687.
82. M. Badia and A. Vignes, *C. r. hebd. Séanc. Acad. Sci., Paris*, 1967, **264C**, 858.
83. A Hässner and W. Lange, *Phys. Status Solidi*, 1965, **8**, 77.
84. P. Guiraldenq and P. Poyet, *C. r. hebd. Séanc. Acad. Sci., Paris*, 1970, **C270**, 2116.
85. E. V. Borisov, D. L. Gruzin and S. V. Zemskii, *Prot. Coat. Metals*, 1968, **2**, 104.
86. S. P. Ray and B. D. Sharma, *Acta Met.*, 1968, **16**, 981.
87. S. P. Ray and B. D. Sharma, *Trans. Indian Inst. Met*, 1970, **23**, 77.
88. L. V. Pavlinov, E. A. Isadzanov and V. P. Smirnov, *Fizica Metall.*, 1968, **25**, 836.
89. S. Benci, G. Gasparrini and T. Rosso, *Phys. Letters.*, 1967, **24**, 418.
90. Oikawa, *et al.*, *ASM Trans. Quart.*, 1968, **61**, 354.
91. Th. Heumann, H. Meiners and H. Stüer, *Z. Naturforsch.*, 1970, **25a**, 1883.
92. R. Ebeling and H. Wever, *Z. Metall.*, 1968, **53**, 222.
93. N. L. Peterson and S. J. Rothman, *Phys. Rev.*, 1970, **B2**, 1540.
94. A. W. Bowen and G. M. Leak, *Metal Trans.*, 1970, **1**, 1695.
95. G. F. Hancock and G. M. Leak, *Met. Sci. J.*, 1967, **1**, 33.
96. A. Ya. Shinyaev. *Izv. Akad. Nauk SSSR, Metl.*, 1969, **4**, 182.
97. H. Bakker, J, Backus and F. Waals, *Phys. Status Solidi*, 1971, **B45**, 633.
98. K. Nohara and K. Hirano, *Proc. Int. Conf. Science Tech. Iron and Steel*, Tokyo, 1970, Sect. 6 1267; and *J. Jap. Inst. Met.*, 1973, **37**, 51.
99. A. W. Bowen and G. M. Leak, *Metal Trans.*, 1970, **1**, 2767.
100. D. Y. F. Lai and R. J. Borg, UCRL Rept. No 50516, 1968; *J. appl. Phys.*, 1970, **41**, 5193.
101. Th. Heumann and H. Stüer, *Phys. Status Solidi*, 1966, **15**, 1966.
102. A. F. Smith and G. B. Gibbs, *Metal Sci. J.*, 1969, **3**, 93; 1968, **2**, 47.
103. R. P. Agarwala, *et al.*, *J. Nucl. Mater.*, 1970, **36**, 41.
104. I. N. Frantsevich, *et al.*, *J. Phys. Chem. Solids*, 1969, **30**, 947.
105. Larikov, *et al.*, *Ukr. Fiz. Zh.*, 1967, **12**, 983.
106. G. F. Hancock, *Phys. Status Solidi*, 1971, **A7**, 535.
107. K. Hirano and M. Cohen, *Trans. Jap. Inst. Met.*, 1972, **13**, 96.
108. J. N. Mundy and W. D. McFall, *Phys. Rev.*, 1972, **B5**, 2835.
109. D. F. Kalinovich, I. I. Kovenskii and M. D. Smolin, *Fizika Iverd. Tela*, 1971, **13**, 2813.
110. G. B. Federov, E. A. Smirnov and V. N. Gusev, *Atomn. Energ.*, 1972, **32**, 11.
111. G. Canneli and L. Verdini, *Nuovo Cim.*, 1969, **B59**, 19.
112. M. J. Klein, *J. appl. Phys.*, 1967, **38**, 167.
113. A. Ya. Shinyaev, *Izv. Akad. Nauk SSSR Metal.*, 1968, **No. 1**, 203.
114. W. Z. Wade, *J. nucl. Mater.*, 1971, **38**, 292.
115. J. F. Murdock and C. J. McHargue, *Acta Met.*, 1968, **16**, 493.
116. L. V. Pavlinov, *Fiz. Metall.*, 1970, **30**, 379.
117. G. B. Federov, E. A. Smirnov and F. I. Zhomov, *Met. Metalloved. Chist. Metal.*, 1968, **No. 7**, 116.
118. G. V. Kidson and J. S. Kirkaldy, *Phil. Mag.*, 1969, **20**, 1057.
119. G. B. Federov, E. A. Smirnov and S. M. Novikov, *Met. Metalloved. Chist. Metal*, 1969, **No. 8**, 41.
120. J. L. Bocquet, *Acta Met.*, 1972, **20**, 1347.
121. E. Santos and F. Dyment, *Phil. Mag.*, 1975, **31**, 809.
122. G. B. Federov, E. A. Smirnov and F. I. Zhomov, *Met. Metalloved. Chist. Metal.*, 1963, **No. 4**, 110.
123. B. Mills, G. K. Walker and G. M. Leak, *Phil. Mag.*, 1965, **12**, 939.
124. A. Ya. Shinyaev, *Phys. Met. Metallogr.*, 1966, **21**, 76.
125. A. Ya. Shinyaev, *Izv. Akad. Nauk SSSR Metal*, 1971, **No. 5**, 210.
126. I. N. Frantsevich, Kalinovich, I. I. Kovenski and M. D. Smolin, 'Atomic Transp. in Solids and Liquids' (Verlag Z. Naturforsch, Tubingen, 1971). p. 68.
127. Stoebe, *et al.*, *Acta Met.*, 1965, **13**, 701.
128. T. Okabe, R. F. Hochman and M. E. McLain, *J. Biomed. Mater. Res.*, 1974, **8**, 381.
129. P. Gas and J. Bernardini, *Scr. Met.*, 1978, **12**, 367.

130. A. Lutze-Birk and H. Jacobi, *Scr. Met.*, 1975, **9**, 761.
131. G. Henry, G. Barreau and G. Cizeron, *C. R. Hebd. Acad. Sci. Sevie C*, 1975, **280**, 1007.
132. N. A. Stolwijk, T. Spruijt, M. A. Hoetjes-Eijkel and H. Bakker, *Phys. Stat. Sol. A*, 1977, **42**, 537.
133. Y. Iijima and K. I. Hirano, *Phil. Mag.*, 1977, **35**, 229.
134. J. Kuceta *et al.*, *Acta Met.*, 1974, **22**, 135.
135. R. A. Perkins, R. A. Padgett and N. K. Tunali, *Met. Trans.*, 1973, **4**, 1665, 2535.
136. A. F. Smith, *Metal Sci*, 1975, **9**, 375.
137. P. Guiraldenq and P. Poyet, *Mem. Sci. Rev. Met.*, 1973, **70**, 715.
138. J. Kucera and B. Million, *Phys. Stat. Sol. A*, 1975, **31**, 275.
139. H. V. M. Mirani *et al.*, *Phys. Stat. Sol. A*, 1975, **29**, 115.
140. A. T. Donaldson and R. D. Rawlings, *Acta Met.*, 1976, **24**, 285.
141. J. Ruzickova and B. Million, *Kovove Mater.*, 1977, **15**, 140.
142. J. Fillon and D. Calais, *J. Phys. Chem. Solids*, 1977, **38**, 81.
143. D. D. Pruthi, M. S. Anand and R. P. Agarwala, *Phil. Mag.*, 1979, **39**, 173.
144. Y. Muramatsu, *Trans. Natl. Res. Inst. Met.*, 1975, **17**, 21.
145. F. J. M. Boratto and R. E. Reed-Hill, *Scr. Met.*, 1977, **11**, 709.
146. R. A. Perkins and R. A. Padgett, *Acta Met.*, 1977, **25**, 1221.
147. A. E. Pontau and D. Lazarus, *Phys. Rev. B*, 1979, **B19**, 4027.
148. J. P. Zanghi and D. Calais, *J. Nuclear Met.*, 1976, **60**, 145.
149. D. Beke, I. Gödeny, F. J. Kedves and G. Groma, *Acta Met.*, 1977, **25**, 539.

Table 13.4 CHEMICAL DIFFUSION COEFFICIENT MEASUREMENTS

Element 1 At.%	*Element 2* At.%	*A*	*Q*	*D*	*Temp. range* °C	*Method*	*Ref.*
A	Ag						
V. small	~100	0.12	33.6	—	600–800	IIIb(ii)	1
A	Mg						
V. small	~100	10^4	52	—	330–540	IIIb(ii)	104
Ag	Al						
0.5		0.21	28.8				
1.0		0.30	29.5				
1.5		0.33	29.8	A few partial D_{Ag}'s			
2.0		0.55	30.8	and D_{Al}'s calculated.	500–595	IIa(i)	2
2.5		0.78	31.4	Very roughly			
3.0		1.50	32.5	$\tilde{D}=D_{Ag}\sim 2D_{Al}$			
3.5	—	3.0	33.7				
6.5	—	11.0	37.0				
8.5		16.0	38.0				
Ag	Au						
0–8.77		0.0242	37.0		806–1017	IIa(ii)	3
45.0				4.7×10^{-9}			
54.0				4.1×10^{-9}			
64.0				3.7×10^{-9}	940	IIIa(i)	4
75.0				2.8×10^{-9}			
85.0				1.9×10^{-9}			
50.8		0.14	41.7		763–965	IIa(i)	5
Ag	Cd						
	0–5	4.7×10^{-3}	31.3		650–895	IIa(ii) and IIb(ii)	6
	0–25			Figure 13.4	627, 727 and 780	IIa(i)	51
Ag	Cu						
0–3		0.012	35.6	—	717–867	IIa(ii)	6
	0–2	0.52	43.9	—	750–800	IIa(ii)	156
Ag	Ga						
	1.9–9.5	0.42	38.9	—	600–940	IIb	219
Ag	**H**						
	0–sol. limit	2.82×10^{-3} (*D* indep. of conc.)	7.5	—	338–600	IIIb(ii)	7
Ag	**Kr**						
~100	V. small	1.05	35.0	—	500–800	IIIb(ii)	8

Table 13.4 CHEMICAL DIFFUSION COEFFICIENT MEASUREMENTS—*continued*

Element 1 At.%	*Element 2* At.%	*A*	*Q*	*D*	*Temp. range* °C	*Method*	*Ref.*
Ag	Mn						
—	0–8.5	0.18	42.9	*D* indep. of conc.	576–933	IIa	151
Ag	**Ne**						
~100	V. small	2.5	59.5	—	800–940	IIIb(ii)	105
Ag	**O**						
	0–sol. limit	3.66×10^{-3}	11.0	—	412–862	IIIb(ii)	9
Ag	Pb						
0–0.12		7.4×10^{-2}	15.2	—	220–285	IIa(ii)	10
Ag	Pd						
50	50	1.5×10^{-6}	24.6	—	600–900	II	152
Ag	**S**						
	Sol. soln	2.34×10^{-3}	26.3	—	604–752		220
Ag	**Xe**						
~100	V. small	0.036	37.5	—	500–800	IIIb(ii)	13
Ag	**Zn**						
(*β*)	50	0.0164	16.5	$D_{Zn}\sim3$ to $4D_{Ag}$	400–610	IIc, IIa(i)	11
	40–55			Figure 13.5			
(*α*)	26.8			8.7×10^{-9}	at 700	IIIa(i)	12
	27.6			2.45×10^{-9}	at 650		
				$D_{Zn}\sim1.5$–$2.2D_{Ag}$			
(*ε*)	74–87	*See* Figure 13.15			310–400	IIa(i)	221
Al	Be						
	0.015	52	39.0	—	500–635	IIa(iii) and IIc	14
	0.022	126	40.3				
	0.03	550	43.1				
Al	Cu						
	0–0.215	0.29	31.12	—	505–635	IIa(ii)	15
	0–~2	0.18	30.10	—	502–538	IIc	154
	(sol. sol. range)						
0		0.131	44.24 (*a*)	—	712–997	IIa(i)	153
2		0.231	44.84 (*a*)	—			
4		0.287	44.80 (*a*)	—			
6		0.364	44.71 (*a*)	—			
8		0.588	45.23 (*a*)	—			
10		1.033	46.44 (*a*)				
12		1.293	45.73 (*a*)				
~25	(*β*)	0.19	27.5	—	646–750	IIc	17
	(γ_2)	0.85	32.5	—	400–535	IIc	155
	(*δ*)	2.1	33.0				
	(ζ_2)	1.6×10^{6}	55.1	—			
	(η_2)	2.2	35.5	—			
	(*θ*)	0.56	30.5				

(*a*) $\log A=-0.8649+0.0829\,C_{Al}$, $Q=44.27+0.14\,C_{Al}$ least squares fit to the data of reference 153.

Element 1 At.%	*Element 2* At.%	*A*	*Q*	*D*	*Temp. range* °C	*Method*	*Ref.*
Al	Fe						
9		2.7×10^{-1}	44.9	—	920–1 210 (Disordered)	IIa	18
17		3.6×10^{-3}	34.0	—			
25		7×10^{-2}	39.9	—			
33		7.3×10^{-3}	31.0	—	1 100–1 210 (Disordered)		
33		2.1×10^{4}	69.4	—	800–1 000 (Ordered)		
41		1.5×10^{5}	72.9	—	800–920		
0–~52	(*α*)	30.1	56.0	—	950–1 100	IIa(ii)	120
Al	**H**						
	0–sol. limit	0.11	9.78	—	360–600	IIIb(ii)	19

Table 13.4 CHEMICAL DIFFUSION COEFFICIENT MEASUREMENTS—*continued*

Element 1 At.%	*Element 2* At.%	*A*	*Q*	*D*	*Temp. range* °C	*Method*	*Ref.*
Al	**He**						
~100	V. small	3.0	36.5	—	—	IIIb(ii)	105
Al	Li						
	0–sol. limit	4.5	33.3	—	417–597	IIa(ii)	20
Al	Mg						
	0.27–2.2	Results scattered. Most fall in the band of Figure 13.8 for this conc. range			400–550	—	21
	0	1	31.0	— }	250–440 }	IIa	157
	0–10	4.4	33.5	— }			
0–20		12	34.3	—	350–420 }		
β phase		2.4×10^{-4}	13.6	— }	325–425	IIc	213
γ phase		9.9×10^{-1}	28.1	— }			
Al	Mn						
	0.02–0.15	—	—	Figure 13.9	600–650	IIa(iii)	21
Al	Na						
	0–0.002	1.1	32.0	—	550–650	IIIb(i)	22
Al	Nb						
33	67	2×10^{-3}	55	— }	1 200–1 500	IIa	222
25	75	2.5	87.5	— }			
Al	Ni						
0–0.7		1.87	64.0	—	1 100–1 280	IIa(ii)	23
Al	Pu						
3–9.1	(δ)	2.25×10^{-4}	25.5	—	350–517	IIa(ii)	186
Al	Si						
	0–0.5	0.346	29.6	—	344–631	IIa(ii)	224
	0–0.5	2.02	32.5 }		480–620	IIa(i)	24
		$A_{Si}=3.95$	$G_{Si}=33.46$ }				
		$A_{Al}=5.07$	$Q_{Al}=34.18$ }				
	0–0.7	—	—	Figure 13.10	450–580	IIa(iii)	21 and 25
Al	Ti						
2.0 }	(β)	{ 1.4×10^{-5}	21.9	$\tilde{D}$ increases linearly with *c*	983–1 250 }	IIa(i)	26
12.0 }		{ 9.0×10^{-5}	25.5				
10.0	(α)	1.6×10^{-5}	23.7		834–900 }		
3.8	(β)			{ $D_{Al}=14.11\times10^{-9}$, $D_{Ti}=4.61\times10^{-9}$ }	1 250 }		
Al	Zn						
		$\tilde{D}$ @ 330°	360°	400° 440°	485°	540°	
	~0	1.84	3.98	12.7 49.2	149	610	
	9.0	1.95	4.85	19.3 69.6	174	610	
	18.1	3.64	6.12	20.0 74.8	2.2	—	27
	37.6	—	1.10	— 51.1	—	—	
		All *D*'s $\times10^{-11}$			330–540	IIa(i)	
To very good approximation		$D_{Zn}=\tilde{D}/c_{Al}$ (c_{Al} = fractional at. conc.)					
Al	Zr						
8		9.2×10^{-3}	40.4	— }	1 100–1 300 }	IIa(i)	223
10		2.3×10^{-2}	42.8	— }			
12		5.2×10^{-2}	45.2	— }			
14		7.6×10^{-2}	45.9	— }			
16		8×10^{-2}	45.9	—	1 200–1 300 }		
	Al_3Zr_5	9.2	67.6	—	1 100–1 300 }		
	Al_2Zr_3	3.4	65.0	—	1 000–1 300 }		
	Al_3Zr_4	1.6×10^{5}	91.3	—	1 000–1 300 }		
As	Fe						
0.6–4.6		4.3	52.5	— (α)	950–	IIa(i)	255
		0.58	58.9	— (γ)	–1 380		
Au	Cu						
	10–90	—	—	Figures 13.6 and 13.7	733–857	IIa(i)	160
'Au Cu'	50	{ 2.36×10^{-6}	13.63	(Disordered)	550–700 }	IIa(i)	167
		{ 7.94×10^{-8}	10.72	(Ordered)	300–450 }		

Table 13.4 CHEMICAL DIFFUSION COEFFICIENT MEASUREMENTS—*continued*

Element 1 At.%	*Element 2* At.%	*A*	*Q*	*D*	*Temp. range* °C	*Method*	*Ref.*
Au	Fe						
	0–18.3	1.16×10^{-4}	24.4	—	750–1 000	IIa(ii)	28
Au	H						
	0–sol. limit	5.6×10^{-4}	5.64	—	500–940	IIIb(ii)	29
Au	In						
				$\tilde{D}$ at 142° $\tilde{D}$ at 151° ($\times 10^{-12}$)			
3				0.47 24.0			
33	(Au In_2)			7.0 29.0			
50	(Au In)			2.6 6.6	—	IIa(i)	30
69	(Au_9 In_4)			5.8 9.8	(Values of D_{In} listed,		
80	(Au_4 In)			0.49 0.68	calculated assuming		
91				0.24 0.28	$D_{Au} \ll D_{In}$)		
Au	Ni						
	2	4.3×10^{-2}	41.4				
	10	3.9×10^{-2}	41.5				
	20	9.5×10^{-2}	43.7				
	30	7.8×10^{-2}	43.9				
	40	2.2×10^{-2}	42.0				
	50	1.3×10^{-2}	42.4				
	55	5.9×10^{-3}	41.7	—	850–975	IIa(i)	31
	60	6.8×10^{-2}	48.8				
	65	1.4×10^{3}	73.0				
	70	2.0×10^{7}	96.2				
	75	6.2×10^{8}	105.0				
	98	1.8×10^{4}	85.2				
Au	Pb						
0–~0.09	0	0.35	14.0	—	113—300	IIa(ii)	32
Au	Pd						
	0–17.1	1.13×10^{-3}	37.4	—	727—970	IIb	33
~50		3.2×10^{-4}	36.5	—	600—1 050	IIa	152
Au	Pt						
98		0.62	54.6				
96		1.0	56.0				
94		0.73	55.4				
92		0.67	55.4				
90		0.60	55.2				
88		0.58	55.2				
86		0.53	55.1	—	925–1 055	IIa(i)	34
84		0.52	55.2				
82		0.47	55.1				
80		0.43	55.0				
2–8		0.37	62.6				
95		$A_{Au}=0.32$	$Q_{Au}=45.6$				
		$A_{Pt}=0.09$	$Q_{Pt}=54$				
B	Fe						
0–0.009 5	(α)	10^{6}	62.0	—	700–835	IIIa(ii)	35
0–0.02	(γ)	0.002	21.0	—	950–1 300	IIIa(ii)	36
	(Analysis of 'Fe' for both experiments (in wt%) 0.0038B, 0.43C, 1.64Mn, 0.02P, 0.019S, 0.37Si, 0.04Cr, 0.01Ni, 0.01Mn)						
Ba	H	4×10^{-3}	4.54	—	200–620	IIa(ii)	250
Ba	U						
0–sol. limit	(γ)	0.112	40.8	—	850–1 040	IIc	85
Be	Cu						
0–~15	(α)	0.19	41.5	—	550–884		
~33	(β)	0.084	27.5	—	650–884	IIa(i)	37
~48	(γ(β′))	0.054	31.0	—	550–884		
~75	(δ)	0.001 2	33.0	—	550–884		
~33	(β)	$A_{Be}=0.035$	$Q_{Be}=29$				
		$A_{Cu}=0.045$	$Q_{Cu}=25$				

Table 13.4 CHEMICAL DIFFUSION COEFFICIENT MEASUREMENTS—*continued*

Element 1 At.%	*Element 2* At.%	*A*	*Q*	*D*	*Temp. range* °C	*Method*	*Ref.*
Be	Fe						
	0–0.2	1.0	54.0	—	800–1 100	IIa(i)	38
		Very little variation of $\tilde{D}$ with c					
Be	H						
		2.3×10^{-7}	4.4	—	200–1 000	IIIb	86
Be	Mg						
0–sol. limit		8.06	37.49	—	500–600	IIIa(i)	212
Bi	Pb						
0–2.0		0.018	18.4	—	220–285	IIa(i)	39
C	Co						
Sol. sol. range		0.53	38.5	—	950–1 050	IIIb(i)	158
~0.1–0 (wt%)		0.31	36.7	—	800–1 400	IIIa(i)	216
C	Co Fe						
	78.5	0.472	37.5	—	850–1 100	IIIb(ii)	41
0.48	89.4	0.442	37.5	—	850–1 112		
0–0.7 (wt%)	0 (γ)	0.04+0.08 c	31.35		1 000–1 200	IIa(i)	146
	5.8	0.04+0.08 c	30.5				
	10.6	0.03+0.1 c	29.9				
	20.2	0.03+0.06 c	28.85				
			(c=wt%C)				
C	Fe						
0–0.1	(α)	3.94×10^{-3}	19.16	—	–40–350	Various	215
0–0.1	(α)	$\tilde{D}(c)=0.008\exp(-19.8/RT)+2.2\exp(-29.3/RT)$			–40–845	Various	40 and 41
0.46	(γ)	0.668	37.46	—	925–1 100	IIIb(ii)	41
1.0	(γ)	0.36	36.0	—	750–1 300	IIa(i)	42
2.0		0.27	34.5				
3.0		0.20	33.3				
4.0		0.14	31.9				
5.0		0.10	30.1				
6.0		0.07	28.5				
C	Fe Mn						
0.4 (wt%)	(γ) 0	0.07	31.35	—	1 000–1 200	IIa(i)	148
	1.0	0.08	31.60				
	12.1	0.19	33.9				
	19.2	0.41	36.1				
C	Fe Ni						
0–0.7 (wt%)	(γ) 0	0.04+0.08 c	31.35		1 000–1 200	IIa(i)	147
	3.9	0.03+0.1 c	31.0				
	9.2	0.03+0.1 c	30.55				
	17.3	0.02+0.1 c	29.8				
			(c=wt%C)				
0–0.1 wt% (γ)	23	0.322	35.38	—	850–1 100	IIIa(i)	56
	40	0.20	33.4				
	60.4	0.296	34.67				
	70	0.372	35.46				
	80	0.680	37.25				
	100	0.366	35.67				
C	Fe Si						
Sol. soln range	(γ) 0–2.35	—	—	Figure 13.23	880–950	IIc	145
C	Mo						
Sol. soln range		0.034	41.0	—	1 780–1 970	IIIa(i)	198

Table 13.4 CHEMICAL DIFFUSION COEFFICIENT MEASUREMENTS—*continued*

Element 1 At.%	*Element 2* At.%	*A*	*Q*	*D*	*Temp. range* °*C*	*Method*	*Ref.*
C	Nb						
0.3–1.0		1.8×10^{-2}	38.0	—	1 600–2 120	IIIb(ii)	128
C	Ni						
0–sol. limit		2.48	40.2	—	730–1 020	IIIb(ii)	43
0–0.1 wt%		0.366	35.67	—	850–1 100	IIIa(i)	56
C	Re						
2 or 3–0		0.1	53.0	—	1 230–1 730	IIIb(ii)	205
C	Ta						
		6.7×10^{-3}	38.6	—	190–2 680	Various	249
C	Th						
~7.2		—	~38	7.4×10^{-9} 2.8×10^{-8} 5.7×10^{-8}	1 000 1 100 1 200	IIIa(i)	44
100–400 p.p.m.	(*β*)	0.022	27.0	—	1 440–1 680	IIa(ii)	159
C	Ti						
0.14–sol. limit	(*α*)	5.06	43.5	$\tilde{D}$ indep. of conc.	736–835	IIc	45
		6×10^{-3}	22.6	—	1 340–1 600	IIa(ii)	252
C	V	8.8×10^{-3}	27.79	—	60–1 825	Various	247
C	Zr						
	(*β*)	0.004 8	26.7	—	900–1 260	IIa	209
Cd	Cu						
0–0.5		3.5×10^{-3}	29.2	—	500–850	IIa(ii)	6
Cd	Pb						
0–1.0		1.85×10^{-3}	15.4	—	167–252	II	46
Cd	**Hg**						
	0–4.0	2.57	19.6	—	156–202	III(ii)	46
Ce	Mg						
Sol. soln range in Mg		450	42.0	—	500–598	IIc	210 and 174
Ce	Pu						
3.74–7.17	(*δ*)	1.31×10^{-2}	29.6	—	403–528	IIa(ii)	187
Ce	U						
0–sol. limit	(*γ*)	3.92	66.4	—	800–1 000	IIc	85
Co	Cr						
	0–15.2	0.084	60.6	—	1 000–1 300	IIa(ii)	47
	0–40	0.443	63.6	—	1 000–1 370	IIa(iii)	48
		(*D* reported f(*c*), but concentration dependence very slight)					
Co	Cu						
0.1		0.6	51.0	—	800–1 073	IIa(i)	256
2	(wt%)	5.7	58.0	—			
Co	Fe						
Sol. soln range 0–100		—	—	Figure 13.11	1 136–1 356	IIa(i)	160
10		1.5×10^{-3}	52.3	—	1 000–1 400	IIa(i)	242
20		2.9×10^{-3}	51.4	—			
30		4.4×10^{-3}	50.7	—			
40		5.8×10^{-3}	51.7	—			
50		7.0×10^{-3}	51.4	—			
60		8.8×10^{-3}	51.9	—			
70		11.5×10^{-3}	52.1	—			
80		12.0×10^{-3}	52.1	—			
90		13.1×10^{-3}	52.3	—			

Table 13.4 CHEMICAL DIFFUSION COEFFICIENT MEASUREMENTS—*continued*

Element 1 At.%	*Element 2* At.%	*A*	*Q*	*D*	*Temp. range* °C	*Method*	*Ref*
Co	**H**						
	5 Atm.	8.3×10^{-3}	11.8	—	400–550(α)	Ic	195
		3.4×10^{-2}	13.8	—	200–400(ε)		
Co	Mn						
	5	7.79	70.75	(Ferro)	860–Tc	IIa(i)	225
	5	0.781	65.25	(Para)	T_c–1 150		
	10	3.07	67.88		860–1 150		
	20	0.70	61.42				
	30	0.721	59.27				
	40	0.627	57.6				
	33	$A_1 = 0.22$	$Q_1 = 62.86$				
		$A_2 = 0.98$	$Q_2 = 54.73$				
Co	Mo						
	0–10	0.231	62.8	—	1 000–1 300	IIa(ii)	47
	0–15	2.48	70.4	—	800–1 300	IIa(i)	226
Co	Ni						
	10	1.76	71.3	—	1 115–1 315	IIa(i)	161
	20	1.61	70.5				
	30	1.89	70.5				
	40	1.50	69.2				
	50	0.480	65.3				
	60	0.166	61.6				
	70	0.140	60.7				
	80	0.725	65.3				
	90	1.94	68.0				
0–100		Figure 13.12	—	—	1 136–1 356	IIa(i)	160
Co	Pd						
10				13.7×10^{-11}	1 150°C	IIa(i)	144
20				25.6			
30				55.0			
40				120			
50				145			
60				120			
70				70.7			
80				28.2			
90				13.9			
Co	Pt						
	0–100		Figure 13.13	—	1 125–1 300	IIa(i)	218
Co	Ti						
	4–8	15	67.0	—	900–1 140	II	227
(Co_3Ti)	21	5.3×10^{-2}	40.0	—			
(Co_2Ti)	30–32	0.28	52.0	—			
(Co Ti)	46–50	4.4×10^{-4}	41.4	—			
βTi	90–95	67	49.5	—	700–850		
Co	V						
	0–17	0.021	53.0	—	1 100–1 300	IIa(ii)	47
Co	W						
	0.5	0.008	56.9	—	1 100–1 300	IIa(ii)	47
Cr	Fe						
10–20	(α)	1.48	54.9	$\tilde{D}$ indep. of *c* (In the range 10–1%, $\tilde{D}$ increases by ~30%) $D_{Cr} \sim 1.5 D_{Fe}$	823–1 440	IIa(i)	49

Table 13.4 CHEMICAL DIFFUSION COEFFICIENT MEASUREMENTS—*continued*

Element 1 At. %	*Element 2* At. %	*A*	*Q*	*D*	*Temp. range* °C	*Method*	*Ref.*
37				80×10^{-10}			
42				56×10^{-10}			
52				27.8×10^{-10}			
62	(α)			11.7×10^{-10}	1 250	IIa(i)	50
72				7.1×10^{-10}			
81				4.9×10^{-10}			
91				4.64×10^{-10}			
0–71	(γ)	0.001 2	52.2	—	900–1 200	IIa(ii)	47
Cr	N Ti						
—	1 Atm. press. 0	0.009 6	28.5	—			
—	1 Atm. press. 0.5	0.009 0	28.2	—	1 000–1 400	Internal oxidation method	196
—	1 Atm. press. 3.0	0.005 4	26.6				
—	1 Atm. press. 5.0	0.002 5	24.0				
Cr	Nb						
(γ)	2	34.0	97.9	—	1 100–1 624		
Cr_2Nb	34	26.0	94.0	—	1 251–1 624		
Cr_2Nb	38[a]	24.0	94.8[b]			IIa(i)	140
(α)	90	2.7	92.0	—	1 100–1 624		
(α)	95	0.31	86.0				

(*a*) The compound Cr_2Nb is reported to have a 5% solubility range.
(*b*) Arrhenius plots not always linear. $\tilde{Q}$ and $\tilde{A}$ derived from measurements at the higher temperatures.

Element 1 At. %	*Element 2* At. %	*A*	*Q*	*D*	*Temp. range* °C	*Method*	*Ref.*
Cr	Ni						
0–12		0.604	61.44	—	1 000–1 300	IIa(ii)	47
Cr	Ti						
9	(β)	—	—	3.6×10^{-9}; $D_{Ti} = 2.8 \times 10^{-9}$; $D_{Cr} = 3.7 \times 10^{-9}$	985	IIa(i)	53
Cr	U						
0–sol. limit	(γ)	0.7	34	—	900–1 000	IIc	54
Cu	Fe	A_{Cu}	Q_{Cu}	A_{Fe} Q_{Fe}			
ε Sol. soln		6.1	64.0	2.7 63.5	900–1 050	IIa(ii)	193
γ Sol. soln		3.6	65.5	8.9 75.0			
Cu	Ga						
	0–3	0.58	46.3		700–1 050	IIb	219
	2.5	3×10^{-4}	32	—			
	4.9	1.8×10^{-3}	34	—			
	7.6	1.6×10^{-2}	37	—			
	10.3	1.8×10^{-1}	40	—	500–700	IIa(i)	228
	13.1	1.3×10^{-1}	37.5	—			
	15.9	8×10^{-2}	35	—			
Cu	H						
	(H) sol. limit	11.31×10^{-3}	9.286	—			
	(D) sol. limit	7.30×10^{-3}	8.794	—	447–927	IIIb(iii)	162
	(T) sol. limit	6.12×10^{-3}	8.717	—			
Cu	Mn						
	0–28	0.58	42.4	*D* indep. of conc.	640–820	IIa	163
Cu	Mn						
	0	1.02	47.8	—			
	10	5.66	49.5	—			
	20	1.75×10	50.4	—			
	30	4.62×10	45.9	—			
	40	9.22×10	49.95	—			
	50	2.11×10^{2}	57.6	—			
	72	1.40×10^{6}	87.5	—	748–930	IIa(i)	243
	76	2.42×10^{7}	95.6	—			
	80	2.78×10^{7}	97.5	—			
	82	2.11×10^{7}	97.5	—			
	84	7.12×10^{6}	94.9	—			
	86	6.06×10^{6}	94.9	—			

Table 13.4 CHEMICAL DIFFUSION COEFFICIENT MEASUREMENTS—*continued*

Element 1. At. %	*Element 2* At. %	*A*	*Q*	*D*	*Temp. range* °C	*Method*	*Ref.*
Cu	Ni						
~0	~100	0.4	61.6				
	0–100			Figure 13.14	765–1 066	IIa(i)	55
~100	~0	1.4	54.5				
Cu	**O**						
		1.76×10^{-2}	16.0	—	800–1 030	I (Electrochem. method)	197
Cu	Pd						
	50	0.48	53.5	—	800–1 050	IIa	152
	0–100	—	—	Figure 13.24	931–1 061	IIa(i)	160
Cu	Pt						
0–13.9		0.049	55.7	—	1 040–1 400	IIa(ii)	28
	Small	0.67	55.7	—	750–1 075	IIb	244
Cu	Si						
	0	0.037	40.0				
	4	0.41	48.2	—	700–800	IIa(i)	59
	8	18.6	53.8				
	α: sol. soln range	11.4	47.84	—	665–775	IIc	164
Cu	Sn						
(α)	0–7	$2\times10^{-2}\times10^{0.133C_{Sn}}$ (c_{Sn}=at.%Sn)	37.28	$D_{Sn}>D_{Cu}$	727–827	II	16
(δ)	20.5	D_{Cu}, D_{Sn}	Figure 13.16	7.7×10^{-10}	428	IIc	60
				3.6×10^{-9}	441		
				1.0×10^{-8}	458		
				1.1×10^{-7}	507		
				1.65×10^{-7}	545		
				1.3×10^{-7}	572		
(γ)	15–22	D_{Cu}, D_{Sn}	Figure 13.18	Figure 13.17	706.5	IIa(i)	
(ε)		1.43×10^{-4}	16.9	—	190–220	II	229
(η)		1.55×10^{-4}	15.5	—			
Cu	Ti						
	0	0.693	46.8	—			
	0.5	0.934	47.6	—	700–1 010	IIa(i)	230
	1.0	1.41	48.8	—			
	1.5	1.92	49.7	—			
Cu	**Zn**						
(α)	1	0.056	40.0				
	5	0.062	40.0	—	780–915		
	10	0.083	39.5				
	16	0.095	38.0			IIa(i)	61
	20	0.09	36.5				
	25	0.031	32.5	—	724–915		
	28	0.016	29.7				
		D_{Cu} and D_{Zn}—Figures 13.19, 13.20 and 13.21					
(α)	10	0.13	40.8				
	15	0.21	40.8				
	20	0.36	40.8	—	700–910	IIIa(i)	62
	28	1.7	41.3				
	28 D_{Zn}	2.1	41.2				
	28 D_{Cu}	0.81	42.7				
(β)	46	—	—	D_{Zn}/D_{Cu}=2.4–3.6	600–800		
(β)	48	0.006 9	18.8	(Disordered)	480–710	IIa	165
		1 440	36.0	(Ordered)	318–447		
(β)	44–48	0.018–0.013	19.9–18.2	—	500–800	IIa(i)	63

Table 13.4 CHEMICAL DIFFUSION COEFFICIENT MEASUREMENTS—*continued*

Element 1 At.%	*Element 2* At.%	*A*	*Q*	*D*	*Temp. range* °C	*Method*	*Ref.*
	59	2.45×10^{-2}	23.4				
	60	2.44×10^{-2}	22.9				
	61	1.71×10^{-2}	21.9				
	62	2.45×10^{-2}	21.9	—	375–650		
(γ)	63	1.14×10^{-2}	20.1				
	64	0.99×10^{-2}	19.3				
	65	0.62×10^{-2}	17.7			IIIa(i)	64
	65.5	0.19×10^{-2}	15.5	—	425–650		
	66.5	0.28×10^{-2}	15.3	—	525–650		
				D_{Zn}/D_{Cu}			
	65–66	—	—	9.4	375–475		
	67	—	—	11.4	525		
(γ)	68	—	—	8.6	575		
	68	—	—	5.7	650		
(ε)	79–86	See Figure 13.33		—	250–400	IIa(i)	251
Fe	**H**						
(α)	0–0.08	1.4×10^{-3}	3.2		200–780	IIIa(ii)	65
(α)	0–sol. limit	0.93×10^{-3}	2.7	Indep. of *c*	200–774	IIIa(ii)	7
	At $T<\sim200$ °C $\tilde{D}$ is usually less than expected from extrapolation of higher T results and apparently depends on sample history. *See* reference 65						
(γ)		0.011	9.95	—	—	Ib	66
Fe	Mn						
	5	5.95	75.0	—			
	10	3.04	72.4	—			
	15	3.37	72.9	—			
	20	2.89	71.9	—			
	25	2.83	70.9	—			
(γ)	30	2.53	70.5	—	1 010–1 250		
	35	3.12	71.4	—			
	40	2.44	70.4	—			
	45	2.17	69.5	—			
	50	1.96	68.3	—		IIa(i)	166
	55	2.04	69.1	—			
	36.4	—	—	$D_{Fe}=3.48\times10^{-12}$, $D_{Mn}=6.22\times10^{-12}$	1 010		
	38.0	—	—	$D_{Fe}=5.29\times10^{-12}$, $D_{Mn}=1.55\times10^{-11}$	1 090		
				$D_{Fe}=9.07\times10^{-11}$, $D_{Mn}=2.42\times10^{-10}$	1 250		
	41.2	—	—	$D_{Fe}=3.36\times10^{-11}$, $D_{Mn}=8.04\times10^{-11}$	1 170		
	5	7.2×10^{-2}	59.9	—			
	10	1.75×10^{-2}	63.0	—			
	15	3×10^{-1}	64.4	—	850–1 300	IIa(i)	231
	20	1.63×10^{-1}	62.6	—			
	25	7.2×10^{-2}	59.4	—			
	30	1.20×10^{-1}	60.0	—			
Fe	Mn, C (wt%)						
	4, 0.02	0.57	66.2	—	1 050–1 450		
(γ)	14, 0.02	0.54	65.4			IIa(i) and (ii)	67
	4, 1.25	0.51	61.2	—	1 000–1 250		
	14, 1.25	0.52	61.0				
		Empirically $\tilde{D}$ may be represented (±20%) over the ranges 0–20% Mn, 0–1.5 wt% C by $\tilde{D}=(0.486+0.011\text{ wt\% Mn})(1+2.53\text{ wt\% C})\exp(-66\,000/RT)$					
(γ)	0–60	—	—	Figure 13.22	1 200	IIa(i)	67

Table 13.4 CHEMICAL DIFFUSION COEFFICIENT MEASUREMENTS—*continued*

Element 1 At.%	*Element 2* At.%	*A*	*Q*	*D*	*Temp. range* °C	*Method*	*Ref.*
Fe	Mo						
(γ)	0–0.59	0.068	59.0	—	1 150–1 260	IIa(ii)	89
		(Addition of 0.4 wt% C increases *A* to 0.091)					
(α)	1.9–3.6	3.467	57.7	—	930–1 260	IIa(ii)	
	α sol. soln range	10	60	—	790–1 185	IIc	52
Fe	N						
($\alpha+\delta$)	0–sol. in eq. with 0.95 atm. N_2	7.8×10^{-3}	18.9	—	500–850 and 1 410–1 470	IIIa(i) and IIIb(ii)	68 69
(γ)	0–sol. in eq. with 0.95 atm. N_2	0.91	40.26	—	950–1 350	IIIa(i) and IIIb(ii)	70 and 71 69
Fe	Ni						
(α)	1	—	—	2.9×10^{-14} 2.75×10^{-12}	700 800	IIa(i)	141
(γ)	10	5.3	76.2	—	— (1 000–1 290)	IIa(i)	141
	20	8.9	76.0	—	—		
	30	15.0	75.9	—	—		
	40	24.5	75.8	—	—		
	50	41.5	75.7	—	1 000–1 290		
	60	58.5	75.7	—	—		
	70	38.5	73.4	$D_{Fe}>D_{Ni}$ for $c<$ ~60% Ni	—		
	80	44.5	73.8	$D_{Fe}<D_{Ni}$ for $c>$ ~60% Ni			
	90	49.5	74.8	—	—		
		Effect of a pressure of 40 kb is to decrease $\tilde{D}\gamma$ by one order of magnitude					
	0–100	Figure 13.27			1 130–1 356	IIa(i)	160
	31–33	$A_{Fe}=3.6$ $Q_{Fe}=68.5$	$A_{Ni}=1.6$ $Q_{Ni}=72.5$	— —			
Fe	**Ni**	C (wt-%)					
(γ) 4	0.03	0.44	67.7	—	1 100–1 450	IIa(i) and (ii)	72
16	0.03	0.51	67.3				
4	0.6	0.46	65.5	—	1 050–1 300		
16	0.6	0.42	64.5				
		Empirically, $\tilde{D}$ may be represented ($\pm$20%) over the ranges 0–20 Ni, 0–1.5 wt% C by $\tilde{D}=(0.344\pm0.012$ wt% Ni$)\times(1+2.3$ wt% C$)\exp(-67\,500/RT)$					
Fe	**O**						
[a]	Sol. soln range	5.75	40.4	—	γ range	Internal oxidation	168
	Sol. soln range	3.7×10^{-2}	23.4	—	α and δ range		
[b]	Sol. soln range	0.4	39.9	—	700–850(α)	Internal oxidation	169
Fe	**P**						
(α and δ)	Sol. soln range	~2.9	~55	—	850–875 and 1 410–1 458	IIa(i)	79
(γ)	Sol. soln range	28.3	69.8	—	1 250–1 350		

[a] Fe+0.1% Al. [b] Fe+0.07% Si.

Table 13.4 CHEMICAL DIFFUSION COEFFICIENT MEASUREMENTS—*continued*

Element 1 At.%	*Element 2* At.%	*A*	*Q*	*D*	*Temp. range* °C	*Method*	*Ref.*
Fe	Pd						
10		—	55.6	—			
20		—	54.5	—			
30		—	54.4	—			
40		—	54.4	—			
50		—	55.5	$D_{Fe} > D_{Pd}$	1 100–1 250	IIa(i)	253
60		—	57.5				
70		—	57.9	—			
80		—	64.7	—			
90		—	65.4	—			
Fe	S						
(α)	Sol. soln range	1.68	48.9	—	750–900	IIIa(i)	73
(α and δ)	Sol. soln range	1.35	48.4	—	750–900 and 1 400–1 450	IIa(i)	73 and 74
(γ)	Sol. soln range	2.42	53.4	—	1 200–1 352	IIa(i)	74
Fe	S Si						
(α)	Sol. soln range 6.2	2.68	49.7	—	900–1 300	IIIa(i)	73
Fe	Sb						
	0	2.6×10^3	63.1	—			
	1	6.4×10^2	59.5	—			
	2	2.2×10^2	56.7	(a)—	780–950	IIa(i)	254
	3	9.3×10	54.5	—			
	4	4.3×10	52.6	—			

(*a*) *D* is a little lower (≈ ×2) at temperatures below the magnetic transformation at 765°C.

Element 1 At.%	*Element 2* At.%	*A*	*Q*	*D*	*Temp. range* °C	*Method*	*Ref.*
Fe	Si						
(α)	4.5 × 7.1	0.44	48.0	—	1 095–1 350	IIa(ii)	75
(α)	0–4.21[a]	$0.735 \times (1+0.124c_{Si})$	52.53	—	900–1 400	IIa(i)	169
	8.35	1.82	51.43	—			
	8.69	1.87	51.37	—			
	9.04	1.77	51.12				
(α)	9.38	1.62	50.82	—	900–1 100	IIa(i)	171
	9.73	1.52	50.58	—			
	10.07	1.55	50.55	—			
	10.41	1.66	50.66	—			
	~0	8	59.5	(Extrapolated)			
	4	17	59.5	—			
(α)	5	17	59.1		800–1 400	IIa(i)	170
	8	35	59.4				
(α Ferro.)	3–12	500	68.5	—	400–600 (Ferromagnetic)	—	172
(γ)	0–2	—	—	4×10^{-10}	1 206	IIa(ii)	75
		—	—	1.7×10^{-9}	1 293		

(*a*) + 1.4%V to stabilize the α-phase.

Element 1 At.%	*Element 2* At.%	*A*	*Q*	*D*	*Temp. range* °C	*Method*	*Ref.*
Fe	Sn						
				9.7×10^{-10}	950		
	0–100	—	~46	2×10^{-9}	1 000	II	81
				3.9×10^{-9}	1 050		
				7.6×10^{-9}	1 100		
Fe	Ti						
(α)	~0.7–3.0	3.15	59.2	—	1 075–1 225	IIc	77
(γ)	0~0.7	0.15	60.0	—	1 075–1 225	IIa(i)	
	2	68	62.4	—	700–1 300		
5	(β)	0.60	45.0	—	900–1 300	IIa	76
10	(β)	0.77	46.1	—	700–1 300		
15	(β)	3.6	51.2	—	700–1 050		

Table 13.4 CHEMICAL DIFFUSION COEFFICIENT MEASUREMENTS— *continued*

Element 1 At.%	*Element 2* At.%	*A*	*Q*	*D*	*Temp. range* °C	*Method*	*Ref.*
Fe	U						
0–sol. limit	(γ)	1.3	32.0	—	790–1 000	IIc	54
Fe	V						
(α) atm.	0.7	0.61	63.8	—	950–1 250	IIa(i)	78
	5.0	3.9	56.9				
	10.0	1.1	53.5				
	15.0	0.70	52.5				
	20.0	0.71	52.8				
	25.0	0.63	52.8				
	30.0	0.59	53.1				
	10.0(1, 2 and 40 kb pressure)			Figure 13.25			
(γ)	0.7(1, 20 and 40 kb)			Figure 13.26	950–1 300		
(γ)	2.0(40 kb)	—	—	5.8×10^{-12}	1 100		
				2.5×10^{-11}	1 162		
				2.9×10^{-11}	1 201		
				1.1×10^{-10}	1 275		
				2.6×10^{-10}	1 350		
(γ)	3.0(40 kb)	—	—	6.3×10^{-12}	1 100		
				2.0×10^{-11}	1 162		
				3.2×10^{-11}	1 201		
				1.6×10^{-10}	1 275		
				1.4×10^{-10}	1 292		
				2.8×10^{-10}	1 350		
Fe	**W**						
0–0.13		11.5	142		1 927–2 527	IIIb(ii)	82
	0–1.3	—	—	3.7×10^{-10}	1 280	IIa	83
	0–1.2	—	—	2.4×10^{-9}	1 330		
	0–3.4	—	—	1.0×10^{-9}	1 330		
Fe	Zn						
8.5	(δ₁)	2.97×10^{-3}	15.4		468–525	IIa(i)	232
9.0		7.88×10^{-3}	17.3				
9.5		1.39×10^{-2}	19.2				
10.0		5.53×10^{-3}	19.0				
10.5		2.82×10^{-3}	19.2				
11.0		1.02×10^{-3}	18.6				
11.5		5.81×10^{-4}	17.7				
12.0		6.98×10^{-4}	17.9				
12.5		1.26×10^{-3}	18.1				
13.0		8.21×10^{-3}	18.7				
21.5–22.5	(Γ_1)	2.04×10^{-4}	19.2				
31–32	(Γ)	1.05×10^{-3}	22.0				
0–6	(ζ)	2.28×10^{-2}	19.9		240–360	IIa	233
Ga	Pu						
3–7.9	(δ)	1.3	37.35	($\tilde{D}$ indep. of conc.)	350–517	IIa(ii)	184
0.48–2.6	(ε)	5.3×10^{-4}	13.2	—	560–640	IIa(i)	185
2.5–6.5	(δ)	0.098	33	—	400–534	IIa(i)	214
Ga	Ti						
Sol. soln range	(α)	4.4×10^{-4}	43.4	—	600–860	IIa and c	191
25	(Ti_3 Ga)	7.4×10^{-5}	43.8	—			
Ge	**H**						
	Range of *c* in eq. with H_2 gas over p, range 10–76 cm Hg	2.72×10^{-3}	8.7	—	800–910	Ic	84
Ge	**He**						
~100	V. small	6.1×10^{-3}	16.0	—	795–872	Ic	106

Table 13.4 CHEMICAL DIFFUSION COEFFICIENT MEASUREMENTS— *continued*

Element 1 At.%	*Element 2* At.%	*A*	*Q*	*D*	*Temp. range* °C	*Method*	*Ref.*
H	Mo	1×10^{-2}	14.0		900–1 500	Various	194
H 0.2–4.3	Nb	5×10^{-4}	2.445[a]	—	0–300	Va	80
H	Ni	6.44×10^{-3}	9.61	—	24–1 120	Various	90
H	Sol. soln range	7.04×10^{-3}	9.434	—	400–1 000	IIIb(ii)	162
D	Sol. soln range	5.27×10^{-3}	9.243	—	400–1 000	IIIb(ii)	162
T	Sol. soln range	4.32×10^{-3}	9.102		400–1 000	IIIb(ii)	162
H 0–~2	Pd	1.3×10^{-4}	5.0	—	231–334(?)	Ic	91
		4.3×10^{-3}	5.62	—	200–700	Ib	92 and 93
		Permeation of H through Pd very much affected by sample history, — contamination, etc. See references 91–95					
H 500 torr	Pt	6×10^{-3}	5.9	$(D_H/D_D=1.16)$	600–900	Ic	203
H 0–$\sim10^{-8}$	Si	9.4×10^{-3}	11.0	—	1 090–1 200	Ic	106
H 0.2–4.3	Ta	4.4×10^{-4}	3.23[a]	—	−20–300	Va	80
H Sol. soln range	Th	2.92×10^{-3}	9.75	—	300–900	IIIa(i) and IIIb(ii)	96
H Sol. soln range	Ti (α)	1.8×10^{-2}	12.38	—	500–824	IIIa(i)	97
Sol. soln range	(β)	1.95×10^{-3}	6.64	—	600–1 000	IIIb(i)	97
H (n.s.)	U (α)	0.0195	11.1	—	390–630	IIIb(ii)	98
	(β)	—	—	7.3×10^{-5}	700	IIIb(ii)	98
	(β)	—	—	1.0×10^{-4}	724	IIIb(ii)	98
(n.s)	(β)	3.3×10^{-4}	3.6	—	698–750	IIIb(ii)	99
	(γ)	1.5×10^{-3}	11.4	—	800–970	IIIb(ii)	99
H 0.2–4.3	V	3.5×10^{-4}	1.15[a]	—	0–300	Va	80
H 10^{-8}–600 torr	W	4.1×10^{-3}	9.0	—	830–2 130	IIIb(ii)	206

(*a*) True diffusion coefficients. Apparent *D*'s may be much smaller, with higher *Q* and *A*, due to surface effects. *See*, for example, references 87, 176, and 173.

Element 1 At.%	*Element 2* At.%	*A*	*Q*	*D*	*Temp. range* °C	*Method*	*Ref.*
H	Zr (α)	4.15×10^{-3}	9.47	—	450–700	IIa(i)	100
40–200 p.p.m.	(α)[a]	7.0×10^{-3}	10.65	$(D_{\parallel c}\approx 2D_{\perp c})$	275–700	IIa(ii)	101
Sol. soln range	(α Zircalloy 2)	2.17×10^{-3}	8.38	—	260–560	IIIa(i)	102

(*a*) No significant difference in results for Zircalloy 2 and Zircalloy 4 from those for pure Zr. Values quoted are means of all measurements.

Element 1 At.%	*Element 2* At.%	*A*	*Q*	*D*	*Temp. range* °C	*Method*	*Ref.*
0–41	(β)	5.32×10^{-3}	8.32	(*D* indep. of *c*)	760–1 010	III(i)	103
	(β)	7.37×10^{-3}	8.54	—	870–1 100	IIa(i)	100
He 0–$\sim6.10^{-9}$	Mg	60.0	36.0	—	400–575	IIIb(ii)	104

Table 13.4 CHEMICAL DIFFUSION COEFFICIENT MEASUREMENTS—*continued*

Element 1 At.%	*Element 2* At.%	*A*	*Q*	*D*	*Temp. range* °C	*Method*	*Ref.*
He 0–~4.10^{-10}	Si	0.11	29	—	1 170–1 207	Ic	106
He 0–0.13	Th (α)	10^{-3}–10^{-4}	38	—	900–1 450	IIIb(ii)	107
He 16	Ti (α)	1.1×10^{-9}	16.1	—	615–720	IIIb(ii)	117
Hf	**O** Sol. soln range, in eq. with oxide	0.66	50.8	—	500–1 050	IIc and IIIb(i)	108
Hf	W 2–10	7.2×10^{-3}	47.2	—	1 550–1 900	IIa(i)	234
Hf	Zr 0–100	*See* Figure 13.31		—	850–1 500	IIa(i)	235
Hg 0–4	Pb	0.35	19	—	177–197	IIa(i)	46
In 0–1	Pb	—	—	2.3×10^{-10}	285	IIa(ii)	46
0–3				3.5×10^{-11}	252		
				3.5×10^{-10}	320		
20–60	—	*See* Figure 13.32		—	115–173	IIa(i)	245
Ir	W						
3	(α)	3.9×10^{4}	170[a]				
24	(σ)	2.4×10^{-4}	60.8				
50	(ϵ)	15.0	120.4	—	1 300–2 110	IIa(i)	140
60	(ϵ)	15.0	120.4				
90	(β)	1.1×10^{3}	145.4				

(*a*) Arrhenius plots non-linear. *Q* and *A* calculated from measurements at the highest temperatures.

Element 1 At.%	*Element 2* At.%	*A*	*Q*	*D*	*Temp. range* °C	*Method*	*Ref.*
La Sol. soln range in La	Mg	0.022	24.4	—	540–598	IIc	210 and 174
La Sol. soln range	U (γ)	117.0	55.7	—	850–1 090	IIc	109
Li V. small	Si	2.5×10^{-3}	15.1	—	800 and 1 350	IIIb(ii)	110
Li n.s., but probably small	W	5.0	41.5	—	1 090–1 230	IIIb(ii)	111
Li Sol. soln range	Zr	0.73	33.7	—	775–850	IIa(ii)	143
Mg 0–<1	Ni	0.44	56	—	1 050–1 300	IIa(ii)	112
Mg	Pb						
0.26		—	—	$2.5 \times 3.7 \times 10^{-10}$	250	IIa(iii)	
1.0		—	—	6.9×10^{-10}			
2.0		—	—	8.6×10^{-10}	250	IIa(i)	
3.0		—	—	1.1×10^{-9}			113
4.1		—	—	$6.4-7.8 \times 10^{-10}$	250	IIa(iii)	
0.26		—	—	9.4×10^{-10}	270		
1.9		—	—	1.2×10^{-10}	220	IIa(iii)	
				1.3×10^{-10}	270		

Table 13.4 CHEMICAL DIFFUSION COEFFICIENT MEASUREMENTS—*continued*

Element 1 At.%	*Element 2* At.%	*A*	*Q*	*D*	*Temp. range* °C	*Method*	*Ref.*
Mg	Pu						
			$\tilde{D}$ at T=420°	475°	534°		
	0.01	in units 10^{-11}	6.1	25.0	130	IIa(i)	118
	0.56		3.5	11.3	46.3		
	1.12		3.1	9.3	49.7		
	1.7		2.1	13.0	23.5		
Mg	U						
	0.025	—	—	1.2×10^{-11}	400	IIa(i)	118
				3.3×10^{-11}	500		
Mn	Ni						
0–4		7.5	67.1	—	1 100–1 300	IIa(ii)	23
Mn	Ti						
8	(β)	1.10^{-3}	35.2	—	830–1 190	IIa(i) and IIIa(ii)	26
		(Very small dep. of $\tilde{D}$ on c in range 2–13%)					
Mo	N						
	Sol. soln range	4.3×10^{-3}	26.0	—	1 300–2 000	IIIb(ii)	199
	Supersat. sol. soln	3×10^{-3}	27.6	—	1 500–2 000	IIIb(ii)	200
Mo	Nb						
~0		1.10^{3}	132	D_{Nb}~ 3–6 × D_{Mo}	1 800–2 165	IIa(i)	115
50		1.10^{3}	137				
~100		1.10^{3}	138				
20		13.5	102.4	—	1 400–2 375	IIa(i)	211
40		3.8	98.7	—			
60		2.1	98.1	—			
80		0.052	82.5	—			
20–80		1.5	95.4	(Average representation of values above)			
Mo	Ni						
0–0.93		3.0	68.9	—	1 150–1 400	IIa(ii)	112
0–9		0.853	64.4	—	1 000–1 300	IIa(ii)	47
Mo	**O**						
		3×10^{-2}	31.0	—	~200	V	58
Mo	Pd						
	61	5.5×10^{-5}	45.0	(D_{pd}≈ 10–20 × D_{Mo})[(a)]	1 000–1 600	IIa(i)	175
	66	4.0×10^{-5}	39.5				
	71	5.0×10^{-5}	42.5				
	75	2.4×10^{-4}	47.9	—			
	80	1.6×10^{-3}	52.2	—			
	85	1.6×10^{-2}	60.5	—			
	90	9.0×10^{-1}	70.0	—			
	95	1.4×10^{-1}	67.5	—			

(*a*) At the original composition in pure Mo/pure Nb couples.

Element 1 At.%	*Element 2* At.%	*A*	*Q*	*D*	*Temp. range* °C	*Method*	*Ref.*
Mo	Ta						
	'Ta rich'	4.68×10^{-5}	60.0	—	1 900–2 300	IIa	190
	'Mo rich'	4.16×10^{-5}	56.0	—			
Mo	Ti						
0		$\sim2\times10^{-2}$	47	—	1 210–1 600	IIa(i)	115
10		$\sim2\times10^{-2}$	50	D_{Ti}/D_{Mo}			
20	(β)	$\sim1\times10^{-2}$	52	~3 at 1 600° ~13 at 820°			
30		$\sim9\times10^{-2}$	63	—			
40		$\sim10^{-2}$	61	—			
(*a*)							

(*a*) Results are reported in reference 115 for the whole composition range (0–100%), but for >40% Mo values vary greatly with type of couple used–incremental or pure metals.

Table 13.4 CHEMICAL DIFFUSION COEFFICIENT MEASUREMENTS—*continued*

Element 1 At. %	*Element 2* At. %	*A*	*Q*	*D*	*Temp. range* °C	*Method*	*Ref.*
0–10	(β)	1.3×10^{-4}	33.1	—	900–1 300		
Sol. soln range	(α)	3.5×10^{-8}	28.4	—	600–800	IIa(ii)	114
Mo	U						
2		2.2	47.5				
4		0.58	45.8				
6		20.0	53.0				
8		16.0	55.0				
10		28.0	56.8				
12	(γ)	3.2	52.2	—	850–1 050		
16		0.096	45.7				
20		3.10^{-3}	39.4			IIa(i)	116
24		4.5×10^{-4}	38.5				
26		2.1×10^{-4}	34.0				
				D_* D_{Mo}			
6.0		—	—	3.4×10^{-9} 5.2×10^{-10}	850		
				1.4×10^{-8} 2.1×10^{-9}	950		
8.0		—	—	1.6×10^{-8} 5.0×10^{-9}	1 000		
10.0		—	—	3.4×10^{-8} 1.3×10^{-8}	1 050		
Mo	W (wt%)						
	10	4.48	117.2				
	20	2.41	114.9	—			
	30	0.64	109.6	—			
	40	0.48	109.3	—			
	50	0.30	107.7	—	2 000–2 500	IIa(i)	119
	60	0.17	105.4	—			
	70	0.14	104.7	—			
	80	0.08	102.7	—			
	90	0.05	100.8	—			
Mo	Zr						
	0–10	1.6	107.3	—	1 650–1 835	IIa(ii)	
	0–10	—	—	1.3×10^{-11} to 3.7×10^{-11}	1 835	IIa(i)	115
(Mo_2Zr)	33⅓	1.10^{-3}	55.6	—	820–1 445	IIc	
N	Nb						
Sol. soln range		0.061	38.8	—	800–1 600	IIIa(i)	121
N	Re						
		1.4×10^{-1}	36.7	—	1 300–1 900	IIIb(ii)	57
N	Th						
Sol. soln range	(α)	2.1×10^{-3}	22.5	—	845–1 490	IIIb(i)	122
50–400p.p.m.	(β)	3.2×10^{-3}	17	—	1 450–1 715	IIa(ii)	159
N	Ti						
Conc. range at diff. temp	(α)	0.012	45.25	—	900–1 570	IIIc(i)	123
Sol. soln range	(β)	0.035	33.8	—	900–1 570	IIIa(i)	
Composition ranges at	(α)	0.2	57	—			
diff. temps	(δ)	90.0	57	—	1 350–1 700	IIc	177
N	V						
		4.17×10^{-2}	35.46	—	60–1 825	Various	247
N	W						
0–300 torr.		2.4×10^{-3}	28.4	—	1 400–2 200	IIIb(ii)	199
1–25 torr.		2.37×10^{-3}	35.9	—∼	1 000–1 800	IIIb(ii)	207

Table 13.4 CHEMICAL DIFFUSION COEFFICIENT MEASUREMENTS—*continued*

Element 1 At.%	*Element 2* At.%	*A*	*Q*	*D*	*Temp. range* °C	*Method*	*Ref.*
N	Zr						
Comp. range at diff. temp.	(α)	0.3	57.0	—	1 350–1 750	IIc	177
Comp. range at diff. temp.	(α)	0.15	54.1	—	650–850	IIIa(i)	178
Sol. soln range	(β)	0.015	30.7	—	920–1 640	IIIa(i)	124
Comp. range at diff. temp.	(δ)	0.06	60.0	—	1 350–1 700	IIc	177
N	Zr Hf						
Sol. soln range	(β) 1.8–2.2	0.003	33.6	—	900–1 600	IIIa(i)	125
N	Zr Sn (wt%)						
Sol. soln range	(β) 1.8	0.011	31.4	—	1 165–1 640		
	2.6	0.014	30.9	—	1 100–1 530	IIIa(i)	126
	5.0	0.011	29.4	—	1 050–1 490		
Nb	Ta						
	10	1.1×10^{-2}	82.0	—			
	20	1.34×10^{-2}	84.0	—			
	25	1.0×10^{-2}	82.0	—			
	35	9.3×10^{-3}	83.0	—			
	40	1.0×10^{-2}	84.0	—			
	45	1.56×10^{-2}	86.0	—	2 000–2 380	IIa(i)	246
	55	1.26×10^{-2}	87.0	—			
	60	2.0×10^{-2}	89.0	—			
	65	3.16×10^{-2}	92.0	—			
	70	4.4×10^{-2}	94.0	—			
	75	5.6×10^{-2}	95.0	—			
	90	1.24×10^{-1}	99.0	—			
Nb	Ti	(a)					
	0	2.5×10^{-3}	70				
	20	2.5×10^{-3}	63				
	40	3.2×10^{-3}	57				
	60	3.8×10^{-3}	50	$D_{Ti} \sim 2 \times D_{Nb}$	1 000–1 590	IIa(i)	115
	80	3.8×10^{-3}	44				
	100	3.8×10^{-3}	40				

(*a*) Values taken from smoothed plots of *A* and *Q* against composition.

Element 1 At.%	*Element 2* At.%	*A*	*Q*	*D*	*Temp. range* °C	*Method*	*Ref.*
Nb	U(γ)	(a)					
	2	2.8×10^{7}	148.9				
	12	2.3×10^{7}	144.2	—~	1 500–1 650		
	18	9.6×10^{6}	140.0				
	22	0.091	73.5	—	1 400–1 600		
	28	0.113	72.9				
	38	0.149	72.8	—	1 300–1 500		
	46	0.064	68.0	—	1 150–1 400		
	54	0.45	69.9	—	1 150–1 350		
	62	0.84	68.5				
	68	1.94	69.7	—	1 075–1 300	IIa(i)	127
	74	0.82	60.4	—	950–1 175		
	78	1.16	60.3				
	82	1.19×10^{-4}	33.4	—	892–1 125		
	93	1.63×10^{-4}	30.2	—	693–1 025		
	97	2.31×10^{-4}	29.9				
	97	D_U— 3.82×10^{-3}	29.8	—	693—1 025		
		D_{Nb}— 7.1×10^{-3}	39.3				

(*a*) This is a representative selection from a larger table of values in reference 127.

Table 13.4 CHEMICAL DIFFUSION COEFFICIENT MEASUREMENTS—*continued*

Element 1 At. %	*Element 2* At. %	*A*	*Q*	*D*		*Temp. range* °C	*Method*	*Ref.*
Nb	U(γ)							
	4	—	—	$D_{Nb} \sim 30 \times D_U$				
	10–100	—	—	$D_U > D_{Nb}$				
Nb	V	(a)						
	0	1.6×10^{-2}	98					
	20	1.95×10^{-2}	82					
	40	2.3×10^{-2}	70					
	60	2.8×10^{-2}	64	$D_V \sim 3\text{–}5 \times D_{Nb}$		1 405–1 750	IIa(i)	150
	80	3.3×10^{-2}	63					
	100	3.8×10^{-2}	63					
Nb	W (wt%)							
	10	81.45	105	—				
	20	22.2	100	—				
	30	1.97	89.9	—				
	40	1.4×10^{-2}	66.9	—				
	50	7.4×10^{-3}	65.0	—		2 000–2 400	IIa(i)	119
	60	3×10^{-3}	60.9	—				
	70	1.8×10^{-3}	59.1	—				
	80	1.0×10^{-3}	56.4	—				
	90	6.0×10^{-4}	54.5	—				
Nb	Zr							
5		—	52	—				
30		—	59	—		900–1 600	IIa	179
95		—	79.5	—				
0		$\sim 10^{-2}$	~47	—				
20		$\sim 4 \times 10^{-2}$	~50	—				
40		$\sim 10^{-1}$	~61	—				
60	(a)	$\sim 3 \times 10^{-1}$	~72	—		1 445–1 690	IIa(i)	115
80		~2	~83	—				
100		~10	~93	—				

(*a*) Values taken from smoothed plots of *A* and *Q* against composition.

Element 1 At. %	*Element 2* At. %	*A*	*Q*	*D*		*Temp. range* °C	*Method*	*Ref.*
Ni	**O**							
	(a)	7.9×10^4	73.9	(s.c)		800–1 200	Internal oxidation method	201
	(b)	9.5×10^4	74.4	(p.c.)		900–1 300		
		5.8	69.7	—		680–830	—	202
Ni	Pb							
0–3		~0.66	25.3	—		285–320	II	46
Ni	Pd		$\tilde{D} \times 10^{11}$					
		859°	*950°*	*1 019°*	*1 150°*			
10		0.68	1.30	1.79	18.6			
20		1.03	2.01	3.37	30.0			
30		1.52	3.54	7.39	70.4			
40		1.72	5.37	11.7	98.2			
50		1.97	6.61	16.1	158		IIa(i)	144
60		1.40	5.42	15.6	131			
70		0.84	3.06	11.3	104			
80		0.56	1.90	5.58	58.7			
90		0.50	1.25	2.30	35.0			
Ni	Pt							
0–14.9		7.9×10^{-4}	43.1	—		1 043–1 401	IIa(ii)	28
	0–100		Figure 13.28			950–1 300	IIa(i)	218
Ni	Si							
	0–<1	1.5	61.7	—		1 120–1 300	IIa(ii)	112
Ni	Ti							
	0–0.9	0.86	61.4	—		1 100–1 300	IIa(ii)	23

(a) Ni + 0.58%Si. s.c.
(b) Ni + 0.48%Si. p.c.

Table 13.4 CHEMICAL DIFFUSION COEFFICIENT MEASUREMENTS—*continued*

Element 1 At.%	*Element 2* At.%	*A*	*Q*	*D*	*Temp. range* °C	*Method*	*Ref.*
Ni	Ta						
Ta_2Ni		2.6×10^{-3}	55.1	—	1 150–1 300	II	236
TaNi		2.1	73.3	—			
$TaNi_2$		0.1	59.8	—			
$TaNi_3$		1.7×10^{-5}	31.9	—			
$TaNi_8$		0.9×10^{-2}	79.8	—			
Ni	U						
0–sol. limit	(γ)	2 500	46.0	—	850–1 000	IIc	54
Ni	V						
	0–16.5	0.287	59.2	—	1 100–1 300	IIa(ii)	47
Ni	W						
	0–1.5	11.1	76.8	—	1 150–1 290	IIa(ii)	23
	0–5	0.86	70.4	—	1 100–1 300	IIa(ii)	47
	1	2.24	72.37	—	1 000–1 316	IIa(i)	180
	2	2.16	72.48	—			
	3	2.11	72.57	—			
	4	2.07	72.65	—			
	5	2.04	72.73	—			
	6	2.01	72.81	—			
	7	1.98	72.89	—			
	8	1.95	72.98				
	9	1.94	73.07	—			
	10	1.92	73.17	—			
	11	1.90	73.26	—			
	12	1.89	73.37	—			
Ni	Zn						
	5–95	$1.05 \times 10^3 \times \exp(-0.142C_{Ni})$	43	—	600–1 000	IIa(i)	181
		A_{Ni} 8.3×10^{-2}	Q_{Ni} 43.5				
		A_{Zn} 0.176	Q_{Zn} 48.5				
δ		7.1×10^{-2}	20.3	—	210–600	II	237
γ′		1.2×10^{-1}	21.7	—			
γ‴		3.0×10^{-2}	23.1	—			
O	Pt						
Sol. soln range		9.3	78.0	—	1 435–1 504	Ic	204
O	Ta						
0–1.13		0.015	26.7	—	700–1 400	IIa(ii)	129
O	Th						
25–220 p.p.m.	(β)	1.3×10^{-3}	11	—	1 440–1 700	IIa(ii)	159
	(α)	1.3×10^2	49		1 000–1 200	IIa(ii)	248
O	Ti						
Sol. soln range	(α)	0.778	48.6	—	932–1 142	IIc	130
Sol. soln range	(β)	1.6	48.2	—	950–1 414	IIIa(i)	123
O	V						
		2.46×10^{-2}	29.5	—	60–1 825	Various	247
O	Zr						
Sol. soln range	(α)	1.32	48.2	—	290–1 500	Various	131
	(α)	0.0661	44.0	—	290–650		
		16.5	54.7[a]	—	650–1 500		

(*a*) Reference 131 reviews all published data. Quoted *A*'s and *Q*'s are 'best mean values'. Data slightly better represented by two Arrhenius expressions, above and below 650 °C.

Table 13.4 CHEMICAL DIFFUSION COEFFICIENT MEASUREMENTS—*continued*

Element 1 At.%	*Element 2* At.%	*A*	*Q*	*D*	*Temp. range* °C	*Method*	*Ref.*
Sol. Soln range	(β)	0.977	41.0	—	1 050–1 200	III	183
Sol. soln range	(α) (Zircalloy)	0.196	41.0	—	1 000–1 500	IIIc(i)	132
Sol. soln range	(β) (Zircalloy)	0.045 3	28.2	—	1 000–1 500	IIIa(i)	132
Pb	Sn						
	0–2	4.0	23.8	—	245–300	IIa(ii)	10
	Sol. soln range		24.0	$\tilde{D}$ increases with conc. of Sn	170 and 181	IIa(i)	135
Pb	Tl						
	0–2	0.025	19.4	—	220–285	IIa(ii)	29
	0.53	1.03	24.6	Almost independent of conc.	260–315	IIa(i)	113
Pd	Ti						
	(β)	1.26×10^{-3}	31.5	—	700–1 000		
	(γ)	1.6×10^{-8}	10.7	—			
	(δ)	3.6×10^{-4}	30.9	—	700–900		
	(ϵ)	1.6×10^{-6}	20.1	—			
	(η)	6.4×10^{-4}	34.7	—		IIa(i)	238
				$D_{Ti}=3.54\times10^{-9}$	900		
				$D_{Pd}=1.32\times10^{-9}$	900		
				$D_{Ti}/D_{Pd}\simeq1$	800		
Pt	W		(b)				
2	(β)	3.1×10^{2}	139.0				
50	(γ)(a)	4.7×10^{-3}	83.6	—	1 300–1 743		
55		3.3×10^{-3}	82.1				
65	(ε)(a)	4.4×10^{-2}	92.0	—	1 473–1 743	IIa(i)	140
77		1.8×10^{-2}	78.0				
80	(α)	1.2×10^{-2}	75.4	—	1 300–1 743		
85		1.3×10^{-2}	74.2		1 300–1 700		

(*a*) The γ and ε are two new phases observed during the diffusion experiments and not previously reported.
(*b*) Arrhenius plots not always linear. Q and A derived from measurements at higher temperatures.

Element 1 At.%	*Element 2* At.%	*A*	*Q*	*D*	*Temp. range* °C	*Method*	*Ref.*
Pu	Ti						
2	(β)	9.4×10^{-4}	29.6	—	900–1 100	IIa(ii)	217
15	(β)	2.3×10^{-3}	30.5	—			
Pu	U						
1.75		0.14×10^{-7}	13.4				
3.50		0.15×10^{-7}	13.7				
5.25		0.18×10^{-7}	14.1	Probably a			
7.0		0.28×10^{-7}	15.2	significant			
8.75	(α)	0.44×10^{-7}	16.3	contribution to	410–540	IIa(i)	134
10.50		0.88×10^{-7}	17.9	$\tilde{D}$ from g.b.			
12.25		1.18×10^{-7}	18.8	diffusion			
14.0		2.0×10^{-7}	20.0				
15.75		2.57×10^{-7}	20.6				
Pu	Zr						
20		7×10^{-1}	44.0	—	750–900		
30		1×10^{-2}	34.5	—	700–850		
40	($\varepsilon\beta$)	1.5×10^{-3}	28.5	—	700–900	IIa(i)	189
50		2.5×10^{-4}	23.5	—	700–870		
60		9×10^{-5}	18.5	—	700–870		
(δ)	4.2–11.4	5.89×10^{-6}	20.0	—	351–475	IIa	188
0.115		0.1	54.0	—			
	(α)				700–800	IIa(ii)	192
1.15		11.1	65.0	—			
		A_{Zr}	Q_{Zr}	A_{Pu}	Q_{Pu}		
20		8×10^{-1}	45.0	6×10^{-1}	44.0	Same T	
30	($\varepsilon\beta$)	7.5	49.0	4×10^{-1}	42.0	range as	189
40		2×10^{-1}	40.0	1×10^{-3}	27.0	for $\tilde{D}$	
50		1.5×10^{-4}	29.0	3×10^{-2}	28.0		

Table 13.4 CHEMICAL DIFFUSION COEFFICIENT MEASUREMENTS—*continued*

Element 1 At.%	*Element 2* At.%	*A*	*Q*	*D*	*Temp. range* °C	*Method*	*Ref.*
Rh	W						
3	(α)	1.3×10^{-6}	58.0				
60	(ε)	1.5×10^{-6}	41.7	—	1 300–1 800	IIa(i)	140
70	(ε)	3.1×10^{-6}	43.4				
90	(β)	2.5×10^{-6}	41.6				
Ru	W		(a)				
5	(α)	5.5×10^{-3}	93.5	—	1 300–2 025		
39	(σ)	1.2×10^{-5}	61.0	—	1 785–2 025	IIa(i)	140
70	(β)	1.8×10^{-5}	49.5	—	1 300–2 025		
90	(β)	1.0×10^{-5}	57.2				

(*a*) Arrhenius plots not always linear. *Q* and *A* derived from measurements at highest temperatures.

Element 1 At.%	*Element 2* At.%	*A*	*Q*	*D*	*Temp. range* °C	*Method*	*Ref.*
S	Ni						
~0.01		2.3×10^{6}	90.0	—	1 000–1 200	IIIa(ii)	142
Si	U						
Sol. soln range	(γ)	20	45.0	—	850–1 050	IIc	54
Sn	Ti						
1.0	(β)	8.4×10^{-7}	15.3	Increases linearly with C	1 000–1 250	IIa(i)	26
8.0		2.7×10^{-4}	29.8		1 090–1 250		
2.0	(β)	—	—	$D_{Sn} = 9.18 \times 10^{-9}$; $D_{Ti} = 2.65 \times 10^{-9}$	1 250		
Sn	Zr						
Sol. soln range	(α)	3.10^{-4}	22.0	—	600–850	IIa(ii)	136
0–3.9	(β)	6.9×10^{-4}	36.0	—	1 100–1 300	IIa(ii)	
Sr	U						
Sol. soln range	(γ)	2.38×10^{-3}	47.0	—	800–1 000	IIc	109
Ta	W						
	'Ta rich'	1.78	119.0	—	2 100–2 500	IIa	190
	'W rich'	4.16×10^{-2}	100.0	—			
Ti	U						
10.0	(γ)	11.10^{-3}	36.6	—	950–1 075	IIa(i)	137
20.0		1.4×10^{-3}	33.0				
30.0		1.6×10^{-3}	34.8				
40.0		4.0×10^{-3}	38.4				
50.0		9.5×10^{-3}	42.0				
60.0		2.6×10^{-3}	39.4				
70.0		2.6×10^{-3}	39.4				
80.0		2.2×10^{-3}	37.5				
90.0		1.1×10^{-3}	33.8				
95.0		0.46×10^{-3}	30.2				
				D_{Ti} D_U			
16.5		—		5.8×10^{-9} 2.2×10^{-8}	1 075		
18.0		—		1.2×10^{-9} 4.7×10^{-8}	950		
				2.9×10^{-9} 9.5×10^{-9}	1 000		
				4.1×10^{-9} 1.6×10^{-8}	1 050		
16.5–18		$Q_U = 38.5$;	$Q_{Ti} = 40.0$				
Ti	V						
(α)	Sol. soln range	—	—	3.91×10^{-15}	600	IIa(ii)	114
				4.7×10^{-15}	700		
(β)	0–10	1.25×10^{-2}	41.4	—	900–1 300	IIa(i)	

Table 13.4 CHEMICAL DIFFUSION COEFFICIENT MEASUREMENTS—*continued*

Element 1 At.%	*Element 2* At.%	*A*	*Q*	*D*	*Temp. range* °C	*Method*	*Ref.*
(β)	2.0	6.0×10^{-3}	39.6	Dep. on *c* in range 2–12% v. slight	900–1 250	IIa(i)	26
(β)	3.5	—	—	$D_{Ti}=1.31\times10^{-9}$, $D_{V}=14.9\times10^{-9}$	1 250		
10		8.3×10^{-4}	47.2	—	650–1 050	IIa(i)	240
20		1.5×10^{-3}	47.6	—			
30		4.4×10^{-3}	48.7	—			
40		1.3×10^{-2}	49.4	—			
50		2.4×10^{-2}	48.6	—			
60		1.1×10^{-2}	44.6	—			
70		8.1×10^{-4}	36.6	—			
80		4.1×10^{-4}	33.4	—			
90		1.6×10^{-4}	29.6	—			
Ti	Zr						
(α)	0–10	1.7×10^{-12}	11.8	—	600–800	IIa(ii)	114
(β)	0–10	1.8×10^{-2}	40.1	—	900–1 300	IIa(ii)	114
10		1.4×10^{-2}	39.3	—	830–1 050	IIa(i)	239
25		3.3×10^{-3}	35.0	—	830–1 050		
		5×10^{-7}	15.7	—	650–830		
40		2.7×10^{-3}	34.1	—	830–1 050		
		1.2×10^{-6}	17.1	—	650–830		
50		2.4×10^{-3}	33.6	—	830–1 050		
		1.7×10^{-6}	17.7	—	650–830		
65		1.6×10^{-3}	32.6	—	830–1 050		
		2.0×10^{-6}	18.0	—	650–830		
80		1.5×10^{-3}	32.2	—	830–1 050		
		2.2×10^{-6}	18.0	—	650–830		
90		1.3×10^{-3}	31.5	—	830–1 050		
50.5				$D_{Zr}=5.1\times10^{-10}$, $D_{Ti}=3.2\times10^{-10}$	800 °C		
U	W						
'low'		1.8×10^{-2}	93.0	—	1 970–2 730	IIIb	241
U	Xe						
(β)	$0–10^{-6}$	9×10^{-7}	23	—	700–750	IIIb(ii)	149
(γ)	$0–10^{-6}$	10^{8}	98	—	810–1 060		
U	Zr						
(γ)	10	9.5×10^{-4}	32.0	—	950–1 075	IIa(i)	138
	20	1.3×10^{-4}	28.6				
	30	0.35×10^{-4}	26.3				
	40	0.4×10^{-4}	27.4				
	50	0.8×10^{-4}	29.7				
	60	0.63×10^{-4}	29.7				
	70	0.55×10^{-4}	29.7				
	80	3.2×10^{-4}	34.3				
	90	78×10^{-4}	41.0				
	95	870×10^{-4}	47.0				
	10–95			D_U and D_{Zr} Figures 13.29 and 13.30	950–1 040	IIa(i)	139

REFERENCES TO TABLE 13.4

1. A. D. Le Claire and A. H. Rowe, *Rev. Métall.*, 1955, **52**, 94.
2. Th. Heumann and S. Dittrich, *Z. Electrochem*, 1957, **61**, 1138.
3. H. Ebert and G. Trommsdorf, *ibid.*, 1950, **54**, 294.
4. R. W. Baluffi and L. L. Seigle, *J. appl. Phys.*, 1954, **25**, 607.
5. W. A. Johnson, *Trans. AIME*, 1942, **147**, 331.
6. O. Kubaschewski, *Trans. Faraday Soc.*, 1950, **46**, 713.
7. W. Eichenauer, H, Kunzi and A. Pebler, *Z, Metallk.*, 1958, **49**, 220.
8. J. M. Tobin, *Acta, metall.*, 1957, **5**, 398.
9. W. Eichenauer and G. Müller, *Z. Metallk.*, 1962, **53**, 321; 1962, **53**, 700.
10 W. Seith and J. G. Laird, *ibid.*, 1932, **24**, 193
11. T. Heumann and P. Lohmann, *Z. Electrochem.*, 1955, **59**, 849.
12. A. G. Guy, *Trans. Metall. Soc., AIME*, 1959, **215**, 279.
13. J. M Tobin, *Acta metall.*, 1959, **7**, 7101.
14. H. Buckle and J. Descamps, *Rev. Métall.*, 1951, **48**, 569.
15. J. B. Murphy, *Acta metall.*, **9**, 563
16. H. Oikawa and A. Hosoi, *Scr. Met.*, 1975, **9**, 823.
17. M. K. Asundia and D. R. F. West, *J. Inst. Metals*, 1964, **92**, 428.
18. K. Hirano and A. Hishunima, *Nippon Kink. Gakk.*, 1968, **32**, 516.
19. W. Eichenauer, K. Hattenbach and A. Pebler, *Z. Metallk.*, 1961, **52**, 682.
20. L. P. Costas, *USA Rep.*, TID-16676, 1962.
21. H. Bückle, *Z. Electrochem.*, 1943, **49**, 238.
22. C. E. Ransley and H. Neufeld, *J. Inst. Metals*, 1950, **78**, 25.
23. R. A. Swalin and A. Martin, *J. metal. Trans. AIME*, 1956, **206**, 567.
24. S. Fujikawa, K. Hirano and Y. Fukushima, *Met. Trans. A*, 1978, **9A**, 1811.
25. R. F. Mehl, F. N. Rhines and K. A. von den Steiner, *Metals Alloys*, 1941, **13**, 41.
26. D. Goold, *J. Inst. Metals*, 1960, **88**, 444.
27. J. E. Hilliard, B. L. Averbach and M. Cohen, *Acta metall.*, 1959, **7**, 86.
28. O. Kubaschewski and H. Ebert, *Z. Electrochem*, 1944, **50**, 138.
29. W. Eichenauer and D. Liebscher, *Z. Naturforsch.*, 1962, **17a**, 355.
30. G. W. Powell and J. D. Braun, *Trans. metall, Soc. AIME*, 1964, **230**, 694.
31. J. E. Reynolds, B. L. Averbach and M. Cohen, *Acta metall.*, 1957, **5**, 29.
32. W. Seith and K. Etzold, *Z. Electrochem.*, 1934, **40**, 829; 1935, **41**, 122.
33. W. Jost, *Z. phys. Chem.*, 1933, **B21**, 158.
34. A. Bolk, *Acta metall.*, 1961, **9**, 643.
35. P. E. Busby and C. Wells, *J. Metals*, 1954, **6**, 972.
36. P. E. Busby, M. E. Warga and C. Wells, *ibid.*, 1953, **5**, 1463.
37. R. Reinbach and F. Krietsh, *Z. Metallk.*, 1963, **54**, 173.
38. R. Le Hazif, G. Donze, J. M. Dupouy and Y. Adda, *Mem. Sci Rev. Met.*, 1964, **LXI**, 467.
39. W. Seith and F. G. Laird, *Z. Metallk.*, 1932, **24**, 193.
40. C. G. Homan, *Acta metall.*, 1964, **12**, 1071.
41. R. P. Smith, *Trans. metall. Soc. AIME*, 1964, **230**, 476.
42. C. Wells, W. Batz and R. F. Mehl, *Trans. Am. Inst. min. Engrs*, 1950, **188**, 553.
43. J. J. Lander, H. E. Kern and A. L. Beach, *J. appl. Phys.*, 1952, **23**, 1305.
44. D. T. Peterson, *Trans. Am. Soc. Metals*, 1961, **53**, 765.
45. F. C. Wagner, E. J. Burcur and M. A. Steinberg, *ibid.*, 1956, **48**, 742.
46. W. Seith, E. Hofer and H. Etzold, *Z. Electrochem.*, 1934, **40**, 332.
47. A. Davin, V. Leroy, D. Coutsouradis and L. Habraken, *Rev. Metall.*, 1963, **60**, 275; *Cobalt*, June 1963, **19**.
48. J. W. Weeton, *Trans. Am. Soc. Metals*, 1952, **44**, 436.
49. T. Heumann and H. Bohmer, *Arch. Eisenhütt Wes.*, 1960, **31**, 749.
50. H. W. Paxton and E. J. Pasierb, *Trans. metall. Soc. AIME*, 1960, **218**, 794.
51. J. R. Manning, *Phys. Rev.*, 1959, **116**, 69.
52. J. P. Pivot, A. Van Craeynest and D. Calais, *J. nucl. Mater.*, 1969, **31**, 342.
53. R. F. Peart and D. H. Tomlin, *J. Phys. Chem. Solids*, 1962, **23**, 1169.
54. M. Mossé, V. Levy and Y. Adda, *C. R. Acad., Sci. Paris*, 1960, **250**, 3171.
55. G. Brunel, G. Cizeron and P. Lacombe, *C. r. hebd. Séanc. Acad. Sci., Paris*, 1969, **269C**, 895.
56. R. P. Smith, *Trans. Met. Soc. AIME*, 1966, **236**, 1224.
57. H. Jehn, K. Hahloch and E. Fromm, *J. less-common Metals*, 1972, **27**, 98.
58. V. I. Baranova *et al.*, *Fiz Khim. Obrab. Mater*, 1968, **2**, 61.
59. F. N. Rhines and R. F. Mehl, *Trans. Am. Inst. min. Engrs*, 1938, **128**, 185.
60. E. Starke and H. Wever, *Z. Metallk.*, 1964, **55**, 107.
61. G. T. Horne and R. F. Mehl. *Trans. Am. Inst. min. Engrs*, 1955, **203**, 88.
62. R. Resnick and R. W. Balluffi, *ibid.*, 1955, **203**, 1004.
63. U. S. Landergren, C. E. Birchenall and R. F. Mehl, *ibid.*, 1956, **206**, 73.
64. R. F. Mehl and C. F. Lutz, *Trans. metall. Soc. AIME*, 1961, **221**, 561.
65. E. W. Johnson and M. L. Hill, *ibid.*, 1960, **218**, 1104.
66. W. Geller and T. H. Sun, *Arch. Eisenhütt Wess.*, 1950, **21**, 423.
67. C. Wells and R. F. Mehl, *Trans. Am. Inst. min. Engrs.*, 1941, **145**, 315.
68. P. Grieveson and E. T. Turkdogan, *Trans. metall. Soc. AIME.*, 1964, **230**, 1604.
69. J. D. Fast and M. B. Verrijp, *J. Iron S. Inst.*, 1954, **176**, 24.

70. P. Grieveson and E. T. Turkdogan, *Trans. metall. Soc. AIME*, 1964, **230**, 411.
71. L. S. Darken, R. P. Smith and E. W. Filer, *Trans. Am. Inst. min. Engrs*, 1951, **191**, 1174.
72. C. Wells and R. F. Mehl, *ibid.*, 1941, **145**, 129.
73. N. G. Ainslie and A. E. Seybolt, *J. Iron St. Inst.*, 1960, **194**, 341.
74. G. Seibel, *C.R. Acad. Sci. Paris*, 1962, **255**, 3182; *Mem. Sci. Rev. Metall.*, 1964, **61**, 413.
75. W. Baltz, H. W. Mead and C. E. Birchenall, *Trans. metall. Soc. AIME*, 1952, **194**, 1070.
76. K. Hirano and Y. Ipposhi, *Nippon Kink Gakk.*, 1968, **32**, 815.
77. S. H. Moll and R. E. Ogilvie, *Trans metall. Soc. AIME*, 1959, **215**, 613.
78. R. E. Hannemann, R. E. Ogilvie and H. C. Gates, *Trans. metall. Soc. AIME*, 1965, **233**, 691.
79. G. Seibel, *C. R. Acad. Sci. Paris*, 1963, **256**, 4661; *Mem. Sci. Rév. Met.*, 1964, **61**, 413.
80. G. Schaumann, J, Völkl and G. Alefeld, *Phys. Stat. Sol.* 1970, **42**, 401.
81. C. O. Bannister and W. D. J. Jones, *J. Iron St. Inst.*, 1931, **124**, 71.
82. J. A. M. van Liempt, *Rec. Trav. Chim. Pays Bas*, 1945, **64**, 239.
83. G. Grube and K. Schneider, *Z. anorg. Chem.*, 1927, **168**, 17.
84. R. C, Frank and J. E. Thomas, *J. Phys. Chem. Solids*, 1960, **16**, 144.
85. J. Tournier, Rep. CEA-R-2446, October 1964.
86. P. M. S. Jones and R. Gibson, *Rept AWRE*, 0-2/67, 1967.
87. W. M. Albrecht, W. D. Goode and M. W. Mallet, *J. Electrochem. Soc.*, 1959, **106**, 981.
88. M. L. Hill and E. W. Johnson, *Acta metall.*, 1955, **3**, 566.
89. J. L. Ham, *Trans. Am Soc. Metals*, 1945, **35**, 331.
90. W. M. Robertson, *Z. f. Metallk.*, 1973, **64**, 436.
91. M. van Sway and C. E. Birchenall, *Trans. metall. Soc. AIME*, 1960, **218**, 285.
92. W. D. Davis, *US Rep.* K.A.P.L. 1227, October, 1954.
93. O. M. Katz and E. A. Gulbransen, *Rev, Sci. Inst.*, 1960, **31**, 615.
94. W. D. Davis, *US Rep.* K.A.P.L. 1375, April 1955.
95. O. N. Salmon, D. Randall and E. A. Wilk, K.A.P.L., 1674, November 1956; K.A.P.L. 984, May 1954.
96. D. T. Peterson and D. G. Westlake, *J. phys. Chem.*, 1960, **64**, 649.
97. R. J. Wasilewski and G. L. Kehl, *Metallurgica*, 1954, **50**, 225.
98. M. W. Mallet and M. J. Trzeciak, *Trans. Am. Soc, Metals*, 1958, **50**, 981.
99. H. W. Meyers, J. W. Varwig, J. L. Marshall, L. G. Weber and J. E. Kenelley, *U.S.A.E.C. Rep.* MCW-1439, December 1959.
100. M. Someno, *Nippon Kink. Gakk.*, 1960, **24**, 249.
101. J. J. Kearns, *J. nucl. Mater.*, 1972, **43**, 330.
102. A. Sawatzky, *J. nucl. Mater.*, 1960, **2**, 62.
103. V. L. Gelezunas, *J. Electrochem Soc.*, 1963, **110**, 779.
104. H. R. Glyde, *Phil. Mag.*, 1965, **12**, 919.
105. H. R. Glyde, *J. nucl. Mater.*, 1967, **23**, 75.
106. A. van Wieringen and N. Warmoltz, *Physica*, 1956, 22, 849.
107. A. Andrew, C. R. Davidson and L. E. Glasgow, *US Rep.* NAA-SR-1598, 1956.
108. J. P. Pemsler, *J. Electrochem. Soc.*, 1964, **111**, 1185.
109. Y. Adda, V. Levy, Z. Hadari and J. Tournier, *Rev. Métall.*, 1959, **57**, 278.
110. E. M. Pell. *Phys. Rev.*, 1960, **119**, 1014.
111. H. M. Love and G. M. McCracken, *Can. J. Phys.*, 1963, **41**, 83.
112. R. A. Swalin, A. Martin and R. Olsen, *Trans. Am Inst. min. Engrs*, 1957, **209**, 936.
113. W. Seith and J. Herrmann, *Z. Electrochem.*, 1940, **46**, 213.
114. R. P. Elliot, *US Rep.* AD.290336, March 1962.
115. C. S. Hartley, J. E. Steedly and L. D. Parsons, *US Rep.* ML-TDR-64-316, December 1964, and 'Diffusion in B.C.C. Metals', *Am. Soc. Met*, 1965, p. 35.
116. Y. Adda and J. Philibert, *C.R. Acad. Sci. Paris*, 1958, **246**, 113; Rep C.E.A.-880, March 1958.
117. A. M. Rodin and V. V. Surenyants, *Phys, Metals Metallogr.*, 1960, **10**, (2), 58.
118. D. Calais, M. Beyeler, M. Mouchnino, A. van Craeynest and Y. Adda, *C. r. hebd. Séanc. Acad. Sci. Paris*, 1963, **257**, 1285.
119. E. P. Nechiporenko *et al.*, *Fizica Metall.*, 1971, **32**, 89.
120. P. Gröbner, *Hutnické listy*, 1955, **10**, 200.
121. W. M. Albrecht and W. D. Goode, *US Rep.* BM1-1360, 1959.
122. A. F. Gerds and M. W. Mallett, *J. Elecrochem. Soc.*, 1954, **101**, 175.
123. R. J. Wasilewski and G. L. Kehl, *J. Inst. Metals*, 1954, **83**, 94.
124. M. W. Mallett. J. Belle and B. B. Cleland, *J. Electrochem. Soc.*, 1954, **101**, 1.
125. M. W. Mallett, E. M. Baroody, H. R. Nelson and C. A. Papp, *J. Electrochem. Soc.*, 1953, **100**, 103.
126. M. W. Mallett, J. Belle and B. B. Cleland, *US Rep.* BM1-829, May 1953.
127. N. L. Peterson and R. E. Ogilvie, *Trans. metall. Soc. AIME*, 1963, **227**, 1083.
128. G. Hoerz and K. Lindenmaier, *Z. Metallk.*, 1972, **63**, 240.
129. E. Gebhardt, H. D. Seghezzi and A. Stegherr, *ibid.*, 1957, **48**, 624.
130. C. J. Rosa, *Met. Trans.*, 1970, **1**, 2517.
131. I. G. Ritchie and A. Atrens, *J. Nucl. Mat*; 1977, **67**, 254.
132. M. W. Mallett, M. W. Albrecht and P. R. Wilson, *ibid.*, 1959, **106**, 181.
133. G. Béranger, *C. r. hebd. Séanc. Acad. Sci. Paris*, 1964, **259**, 4663.
134. M. Dupuy and D. Calais, *Mem. Sci Met.*, 1965, **LXII**, 721.
135. H. Cordus and M. Kukuk, *Z. anorg. Allgem. Chemie*, 1960, **306**, 121.
136. R. Resnick and R. Balluffii, *US Rep.* S.E.P. 118, August, 1953.

137. Y. Adda and J. Philibert, *Acta metall.*, 1960, **8**, 700.
138. Y. Adda, J. Philibert and Faraggi, *Rev. Métall.*, 1957, **54**, 597.
139. Y. Adda, C. Mairy and J. L. Andreu, *ibid.*, 1960, **57**, 550.
140. E. J. Rapperport, V. Merses and M. F. Smith, *US Rep.* ML-TDR-64-61, March 1964.
141. J. I. Goldstein, R. E. Hanneman and R. E. Ogilvie, *Trans. metall. Soc. AIME*, 1965, **233**, 812.
142. I. Pfeiffer, *Z. Metallk.*, 1955, **46**, 516.
143. L. S. DeLuca, *US Rep.* KAPL-M-LSD-1, August, 1960.
144. I. B. Borovski, I. D. Marchukova and Yu. E. Ugaste, *Fizika Metall.*, 1966, **22**, 849.
145. M. A. Krishtal, *Dokl. Akad. Nauk. SSSR.* 1953, **92**, 951 and Nsf-tr 223.
146. M. Blanter, *Zhur. Tech. Phys. SSSR*, 1950, **20**, 1001.
147. M. Blanter, *ibid.*, 1950, **20**, 217.
148. M. Blanter, *ibid.*, 1951, **21**, 818.
149. M. B. Peraillon, V. Levy and M. Y. Adda, *Comm.* to 1964 Autumn Meeting, Société Française de Métallurgie.
150. R. C. Reiss, C. S. Hartley and J. E. Steedly, *J. less-common Metals*, 1965, **9**, 309.
151. R. S. Barclay and P. Niessen, *Amer. Soc. Met. Qt.*, 1969, **62**, 721.
152. O. Neukman, *Galvanotechnick*, 1970, **61**, 626.
153. H. Oikawa, T. Obara and S. Karashima, *Met. Trans.*, 1970, **1**, 2969.
154. J. R. Cahoon, *Metal Trans.*, 1972, **3**, 1324.
155. Y. Funamizu and K. Watanabe, *Trans. Japan Inst. Metals*, 1971, **12**, 147.
156. J. R. Cahoon and W. V. Youdelis, *Trans. Met. Soc. AIME*, 1967, **239**, 127.
157. G. Moreau, J. A. Carnet and D. Calais, *J. nuci. Mater.*, 1971, **38**, 197.
158. H. Lafitau, *C. r. hebd. Séanc. Acad. Sci., Paris*, 1968, **C267**, 132.
159. D. T. Peterson and T. Carnahan. *Trans. Met. Soc. AIME*, 1969, **245**, 213.
160. M. Badia, *Thesis*, Univ. of Nancy (France), 1969.
161. Y. Iijima and K. Hirano, *Nippon Kink. Gakk.*, 1971, **35**, 511.
162. L. Katz, M. Guinan and R. J. Borg, *Phys. Rev.*, 1971, **B4**, 330.
163. O. Caloni and A. Ferrari, *Z. Metallk.*, 1967, **58**, 892.
164. H. I. Aaronson, H. A. Domain and A. D. Brailsford, *Trans AIME*, 1968, **242**, 738.
165. Yu. E. Ugaste and V. N. Pimenov, *Fizica Metall.*, 1971, **31**, 363.
166. A. Tsuji and K. Yamanaka, *Nippon Kink. Gakk.*, 1970, **34**, 486.
167. M. Khobaib and K. P. Gupta, *Sci. Met.*, 1970, **4**, 605.
168. J. H. Swisher and E. T. Turkdogan, *Trans. Met Soc. AIME*, 1967, **239**, 426.
169. R. J. Borg and D. Y. F. Lai, *J. appl. Phys.*, 1970, **41**, 5193.
170. A. Vignes, *Trans. 2nd Nat. Conf. Electron Microprobe Analysis*, Boston, 1967, Paper No. 20.
171. H. V. M. Mirani and P. Maaskant, *Phys. Status Solidi*, 1972, **A14**, 521.
172. P. E. Brommer and H. A. 't Hooft, *Phys. Letters.*, 1967, **26A**, 52.
173. G. L. Holleck, *J. Phys. Chem.*, 1970, **24**, 1957.
174. K. Lal, *C.E.A.* (*France*), Rept. No. CEA-R. 3136, 1967.
175. W. Zaiss, S. Steeb and T. Krabichler, *Z. Metallk.*, 1972, **63**, 180.
176. T. O. Ogurtani, *Met. Trans.*, 1971, **2**, 3035.
177. V. S. Eremeev, Yu. M. Ivanov and A. S. Panov, *Izv. Akad. Nauk. SSSR Metal*, 1969, **4**, 262.
178. C. J. Rosa and W. W. Smeltzer, *Electrochem. Technol.*, 1966, **4**, 149.
179. G. N. Ronami *et al.*, *Vestn. Mos. Univ. ser. 3*, 1970, **11**, 251.
180. J. M. Walsh and M. J. Donachie, *Met. Sci. J.*, 1969, **3**, 68.
181. M. Andreani, P. Azou and P. Bastien, *Mém. Scient. Revue Métall.*, 1969, **66**, 21.
182. C. J. Rosa, *Metal Trans.*, 1970, **1**, 2617.
183. J. Debuigne, *Métaux Corros. Inds*, 1967, **No. 501**, 186.
184. G. R. Edwards, R. E. Tate and E. A. Hakkila, *J. nucl. Mater*, 1968, **25**, 304.
185. M. R. Harvey *et al.*, *J. less-common Metals*, 1971, **23**, 446.
186. R. E. Tate, G. R. Edwards and E. A. Hakkila, *J. nucl. Mater.*, 1969, **29**, 154.
187. M. R. Harvey, A. L. Rafalski and D. H. Riefenberg, *Trans. ASM*, 1968, **61**, 629.
188. M. R. Harvey, A. L. Rafalski and D. H. Riefenberg, *Trans. ASM*, 1969, **62**, 1014.
189. C. Remy, M. Dupuy and D. Calais, *J. nucl. Mater.*, 1970, **34**, 46.
190. A. N. Ivanov, G. B. Krasilnikova and B. S. Mitin, *Fizica Metall.*, 1970, **12**, 291.
191. C. E. Shamblen and C. J. Rosa, *Metal Trans.*, 1971, **2**, 1925.
192. J. C. Lautier, A. van Craeynest and D. Calais, *J. nucl. Mater.*, 1967, **23**, 111.
193. M. A. Krishtal *et al.*, *Fiz. Khim. Obrab. Mater.*, 1971, **No. 3**, 109.
194. R. Gibala and C. A. Wert, Rpt. COO-1676-3, 1967.
195. G. R. Caskey, R. G. Derrick and M. R. Louthan, *Scr. Met.*, 1974, **8**, 481.
196. J. L. Arnold and W. C. Hagel, *Metal Trans.*, 1972, **3**, 1471.
197. R. L. Pastorek and R. A. Rapp, *Trans. Met. Soc. AIME*, 1969, **245**, 1711.
198. P. S. Rudman, *Trans. A.I.M.E.*, 1967, **239**, 1949.
199. H. Jehn and E. Fromm, *J. less-common Metals*, 1970, **21**, 333.
200. J. H. Evans and B. L. Eyre, *Acta Met.*, **17**, 1109.
201. R. Barlow and P. J. Grundy, *J. Mater. Sci.*, 1969, **4**, 797.
202. A. Messner, R. Benson and J. E. Dorn, *Trans. ASM*, 1961, **53**, 227.
203. Y. Ebisuzaki, W. J. Kass and M. O'Keefe, *J. chem. Phys.*, 1968, **49**, 3329.
204. L. R. Velho and R. W. Bartlett, *Met. Trans.*, 1972, **3**, 65.

205. R. Ducros and P. Le Groff, *C. r. hebd. Séanc., Acad. Sci., Paris*, 1968, **267C**, 704.
206. R. Frauenfelder, *J. Vac. Sci. & Technol.*, 1969, **6**, 388.
207. R. L. Wagner, *Metal Trans.*, 1970 **1**, 3365.
208. R. Frauenfelder, *J. chem. Phys.*, 1968, **48**, 3966.
209. L. V. Pavlinov and V. H. Bykov, *Fizica Metall.*, 1965, **19**, 397.
210. K. Lal and V. Levy, *C. r. hebd. Séanc. Acad. Sci., Paris*, 1966, **262C**, 107.
211. B. S. Wyatt and B. B. Argent, *J. less-common Metals*, 1966, **11**, 259.
212. V. F. Yerks, V. F. Zelensky and V. S. Krasnorutskiy, *Physics Metals Metallogr., N.Y.*, 1966, **22**, 112.
213. F. Funamizu and K. Watanabe, *Trans. Japan Inst. Met.*, 1972, **13**, 278.
214. A. L. Rafalski, M. R. Harvey and D. H. Riefenberg, *Trans. Quarterly*, 1967, **60**, 721.
215. A. E. Lord and D. N. Beshers, *Acta Met.*, 1966, **14**, 1659.
216. Th. Hehenkamp, *Acta Met.*, 1966, **14**, 887.
217. A. Languille, *Mem. Scient. Revue Métall.*, 1971, **68**, 435.
218. I. B. Borovskiy, I. D. Marchukova and Yu. E. Ugaste, *Fizica Metall.*, 1967, **24**, 436.
219. I. B. Borovskiy, I. D. Marchukova and Yu. E. Ugaste, *Izv. Vyssh. Uchebn. Zaved. Tsvetn. Metall.*, 1977, **1**, 172.
220. K. Fueki, K. Ota and K. Kishio, *Bull. Chem. Soc. Jpn*, 1978, **51**, 3067.
221. T. Shimozaki and M. Onishi, *J. Jap. Inst. Met.* 1978, **42**, 1083.
222. V. N. Agafonov *et al.*, *Vestn. Mosk. Univ. Khim.*, 1975, **16**, 121.
223. A. Gukelberger and S. Steeb, *Z. f. Metallk.*, 1978, **69**, 255.
224. D. Bergner and E. Cyrener, *Neue Huette*, 1973, **18**, 356.
225. Y. Iijima, O. Taguchi and K. I. Hirano, *Met. Trans. A.*, 1977, **8A**, 991.
226. C. P. Heijwegen and G. D. Rieck, *Acta Met.*, 1974, **22**, 1269.
227. P. J. M. Van der Straten *et al.*, *Z. Metallk.*, 1976, **67**, 152.
228. M. Wilhelm, *Z. Naturforsch. A.*, 1974, **29**, 733.
229. M. Onishi and H. Fujibuchi, *Trans. Jap. Inst. Met.*, 1975, **16**, 539.
230. Y. Iijima, K. Hoshino and K. Hirano, *Met. Trans. A.*, 1977, **8A**, 997.
231. K. Nohara and K. Hirano, *J. Jap. Inst. Met.*, 1973, **37**, 51.
232. Y. Wakamatsu, K. Samura and M. Onishi, *J. Jap. Inst. Met.*, 1977, **41**, 664.
233. Y. Wakamatsu, M. Onishi and H. Miura, *J. Jap. Inst. Met.*, 1975, **39**, 903.
234. B. A. Dainyak and V. I. Kostikov, *Izv. Vyssh. Uchebn. Zaved. Chern. Metall.*, 1976, 11, 15.
235. E. A. Balakir *et al.*, *Izv. Vyssh. Uchebn. Zaved. Tsvetn. Metall.*, 1975, 4, 162.
236. V. N. Pimenov, Y. E. Ugaste, K. A. Akkushkarova, *Izv. Akad. Nauk SSSR*, 1977, **1**, 184.
237. T. Shimozaki and M. Onishi, *J. Japan Inst. Met.*, 1978, **42**, 402.
238. P. Lamparter, T. Krabichler and S. Steeb, *Z. Metallk.*, 1973, **64**, 720.
239. A. Brunch and S. Steeb, *Z. Naturforsch. A.*, 1974, **29**, 1319.
240. A. Brunch and S. Steeb., *High Temp. High Press.*, 1974, **6**, 155.
241. E. C. Schwegler, *Intl. J. Mass Spec: Ion Physics*, 1968, **1**, 191.
242. T. Ustad and H. Sorum, *Phys. Stat. Sal.*, 1973, **A20**, 285.
243. Y. Iijima, K. I. Hirano and K. Sato, *Trans. Jap. Inst. Met.*, 1977, **18**, 835.
244. R. L. Fogelson, Y. A. Ugai and A. V. Pokoev, *Fiz. Met. Metalloved*, 1972, **33**, 1102.
245. D. R. Campbell, K. N. Tu and R. E. Robinson, *Acta Met.*, 1976, **24**, 609.
246. A. Ya Shinyayev and N. I. Kopaleishvili, *Fiz. Met. Metalloved.*, 1974, **38**, 222.
247. F. A. Schmidt and J. C. Warner, *J. Less Comm. Met.*, 1972, **26**, 325.
248. D. T. Peterson, *Trans. A.I.M.E.*, 1961, **221**, 924.
249. F. A. Schmidt and O. N. Carlson, *J. Less Comm. Met.*, 1972, **26**, 247.
250. D. T. Peterson and C. C. Hammerberg, *J. Less Comm. Met.*, 1968, **16**, 457.
251. Y. F. Funamizu and K. Watanabe, *Trans. Jap. Inst. Met.*, 1976, **17**, 59.
252. O. N. Carlson, F. A. Schmidt and R. R. Lichtenberg, *Met. Trans.*, 1975, **6A**, 725.
253. J. P. Gomez, C. Remy and D. Calais, *Mem Sci. Rév. Metall.*, 1973, **70**, 597.
254. K. Nishida, H. Murohashi and T. Yamamato, *Trans. Jap. Inst. Met.*, 1979, **20**, 269.
255. B. I. Bozic and R. J. Lucic, *J. Mat. Sci.*, 1976, **11**, 887.
256. F. J. Bruni and J. W. Christian, *Acta Met.*, 1973, **21**, 385.

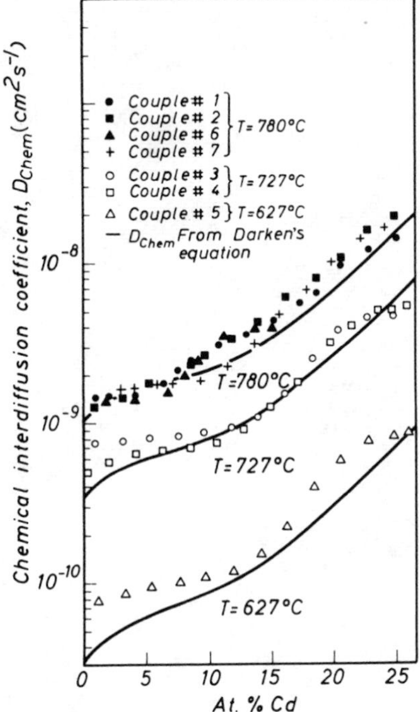

Figure 13.4 *Chemical diffusion in coefficients in* Ag Cd *alloys*[51] *Table 13.4*

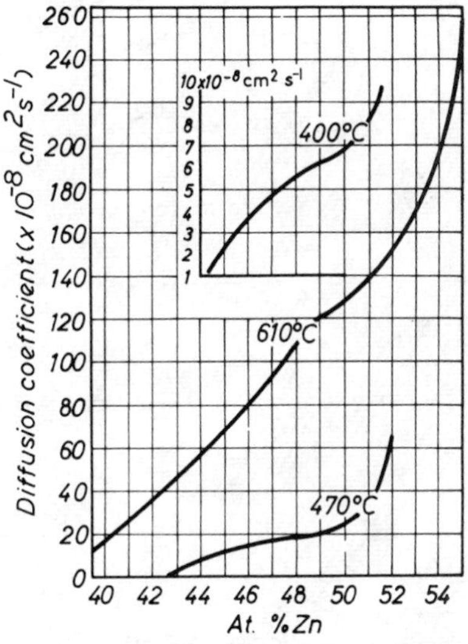

Figure 13.5 *Chemical diffusion coefficients in* Ag Zn *alloys*[11] *Table 13.4*

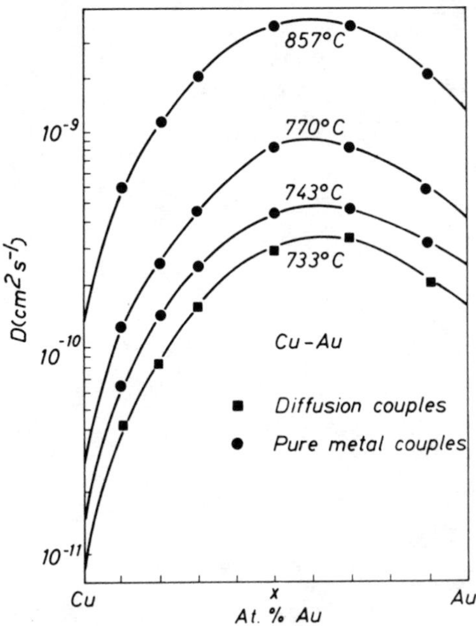

Figure 13.6 *Chemical diffusion coefficients in* Cu Au *alloys as a function of composition*[160]

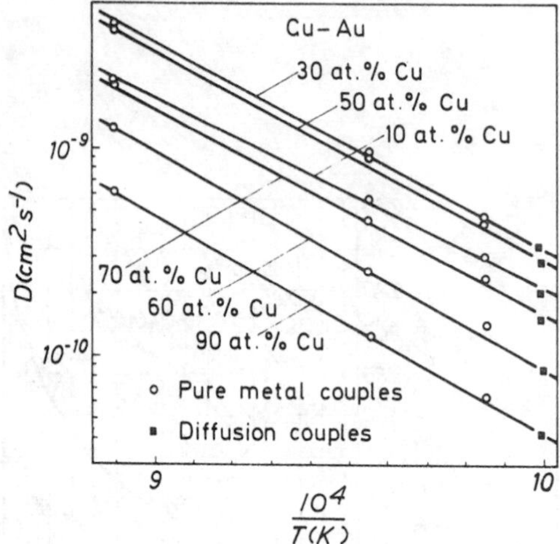

Figure 13.7 *Chemical diffusion in* Cu-Au *as a function of* T[160]

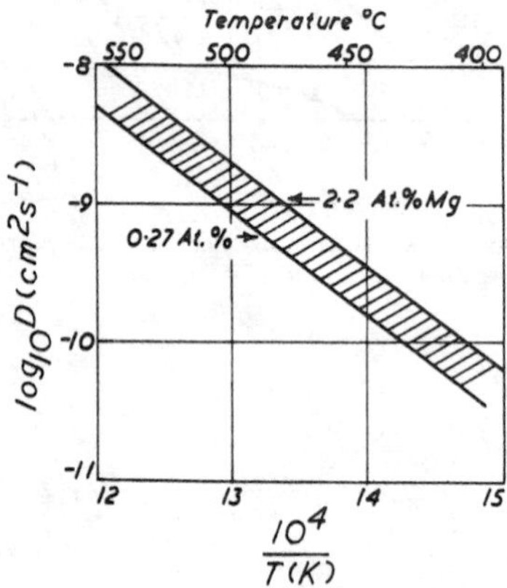

Figure 13.8 *Chemical diffusion in* Al Mg *alloys*[21]

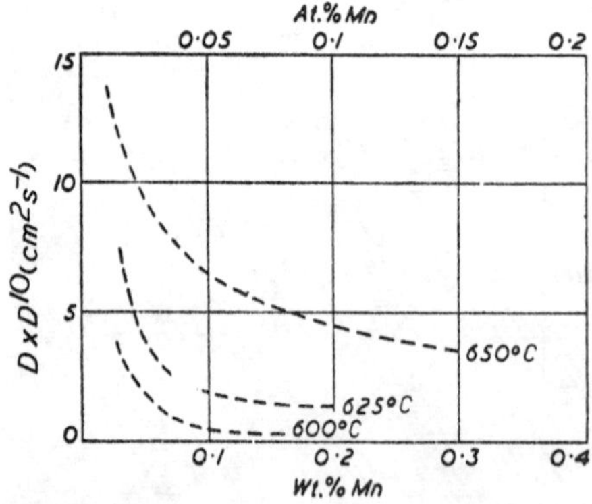

Figure 13.9 *Chemical diffusion in* Al Mn *alloys*[21]

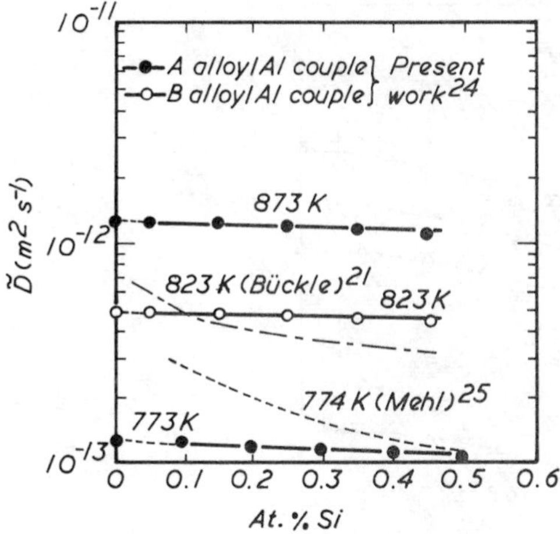

Figure 13.10 *Chemical diffusion in* Al Si *alloys*[21,25]

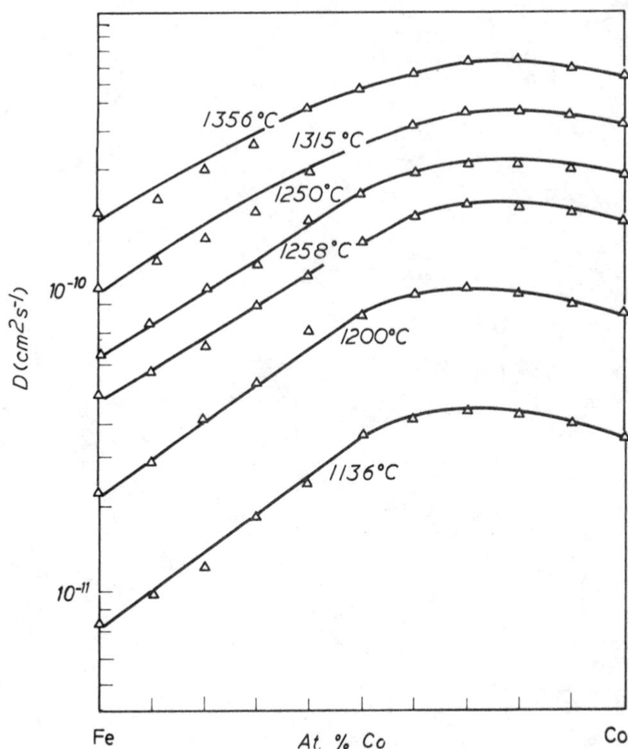

Figure 13.11 *Interdiffusion in the system* Co-Fe[160]

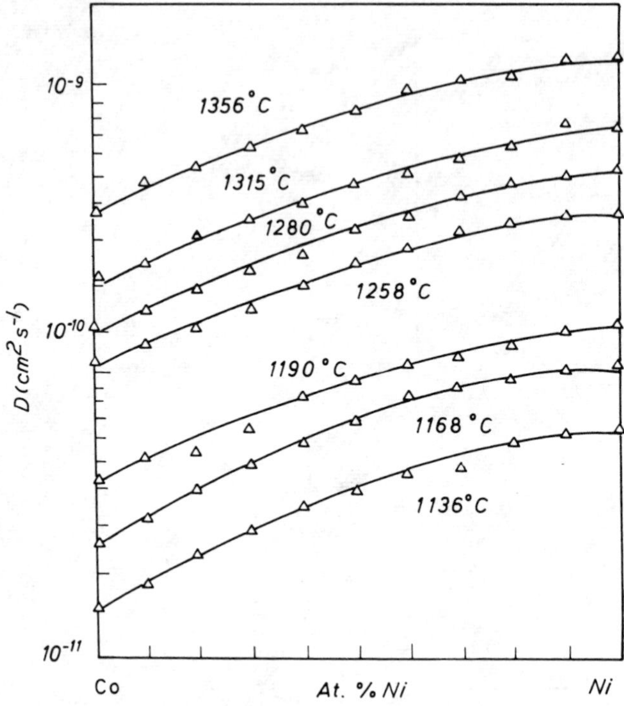

Figure 13.12 *Interdiffusion in the system* Co-Ni[160]

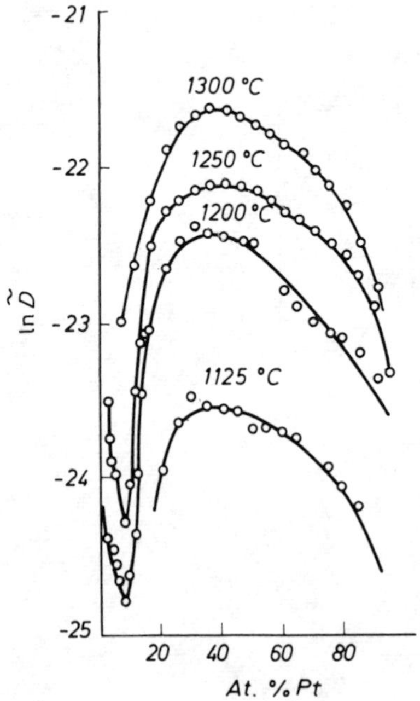

Figure 13.13 *Interdiffusion in* Co-Pt[218]

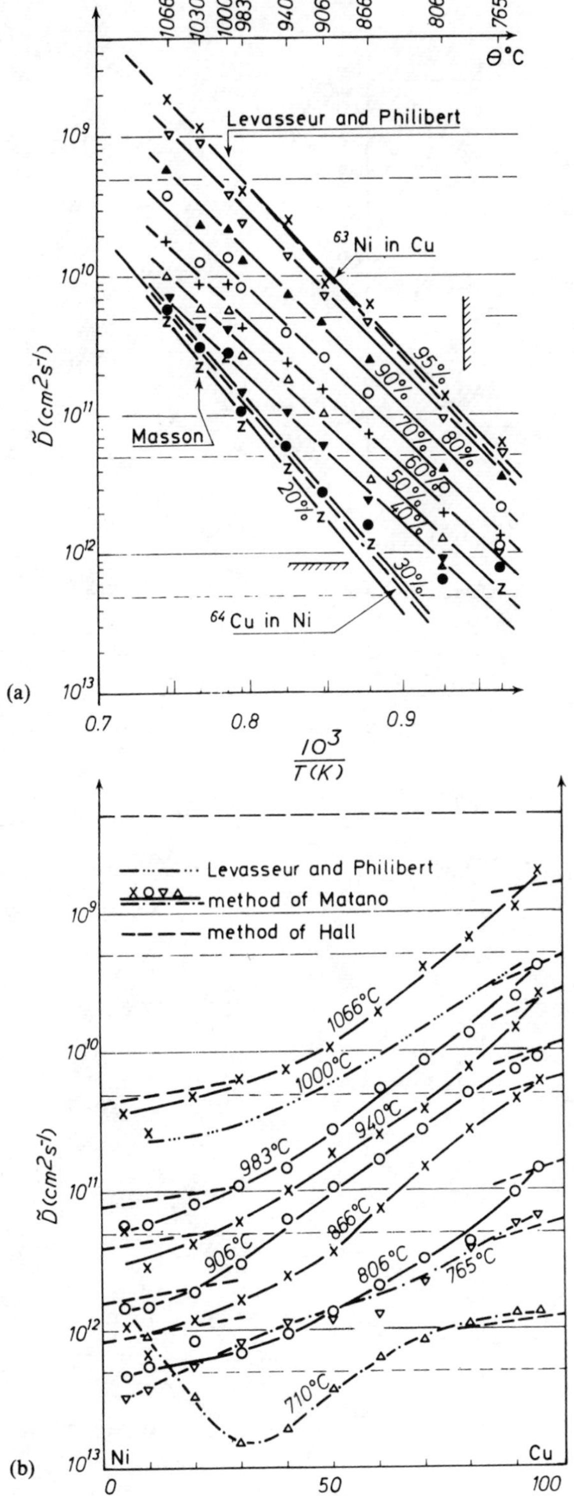

Figure 13.14 *Interdiffusion in* Cu–Ni[55] *(a) as function of* I/T; (b) *As a function of composition*

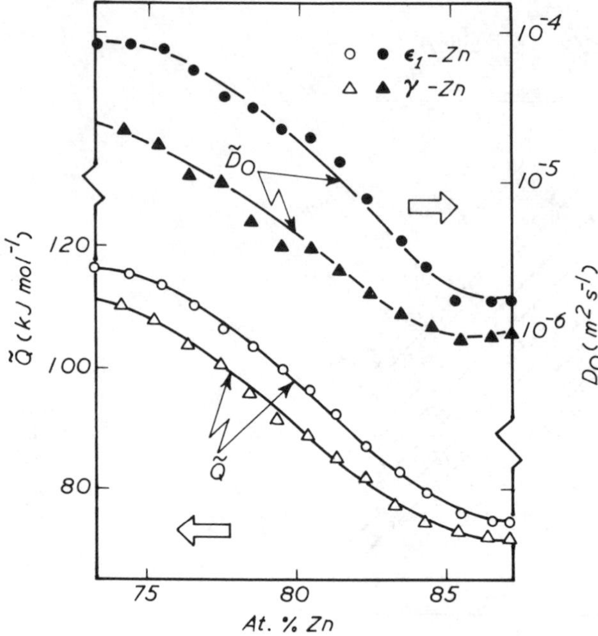

Figure 13.15 *Chemical diffusion* CuZn *alloys*[221].

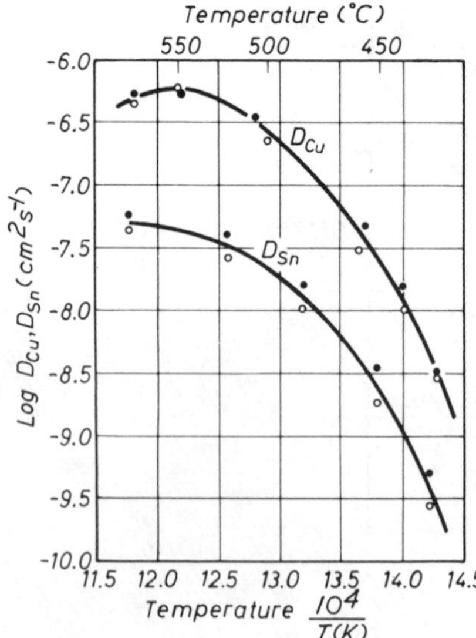

Figure 13.16 *Partial diffusion coefficients for* Cu *and* Sn *in δ-phase* CuSn *alloys*[60]

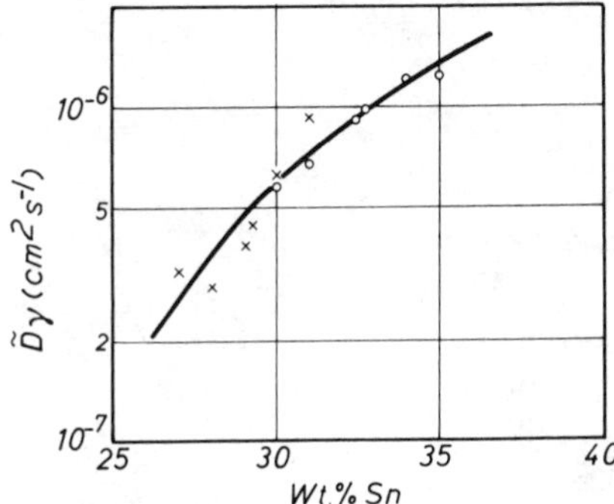

Figure 13.17 *Chemical diffusion coefficients for γ-phase of* Cu-Sn *systems at* 706.5 °C[60]

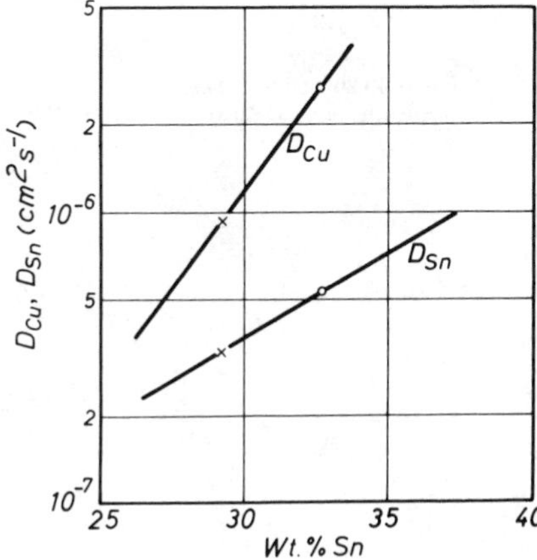

Figure 13.18 *Partial diffusion coefficients for* Cu *and* Sn *in γ* Cu Sn *alloys at* 706.5 °C

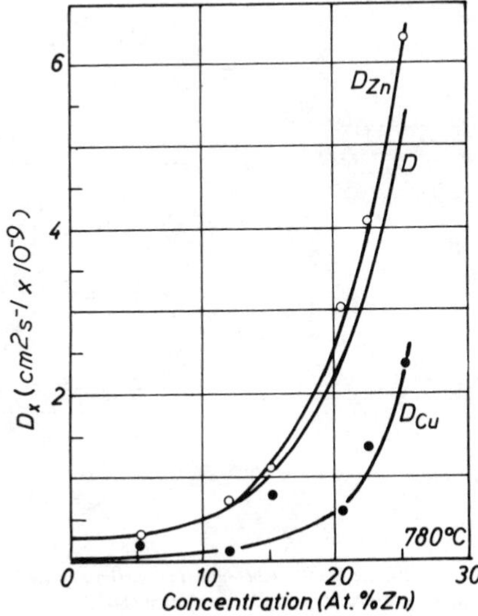

Figure 13.19 *Chemical and partial diffusion coefficients in α-*Cu-Zn *system at* 780 °C[61]

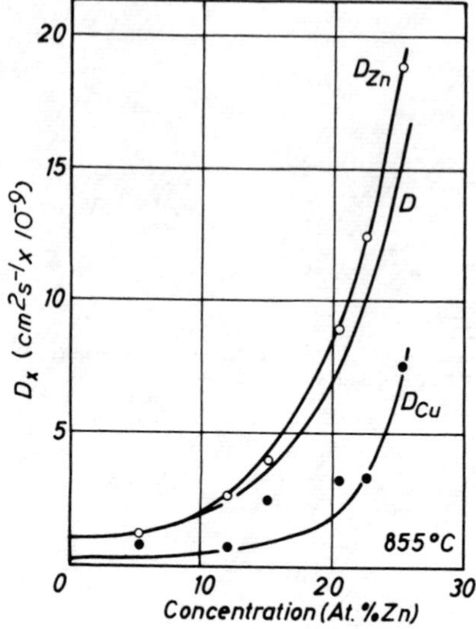

Figure 13.20 *Chemical and partial duffusion coefficients in α-Cu-Zn systems at* 855 °C[61]

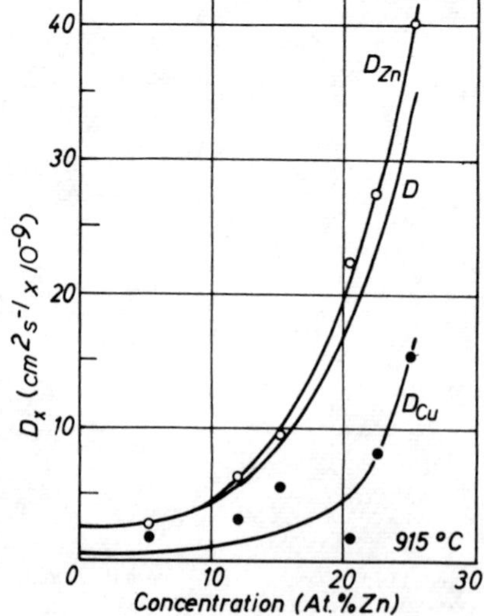

Figure 13.21 *Chemical and partial diffusion coefficients in α-Cu-Zn system at* 915 °C [61]

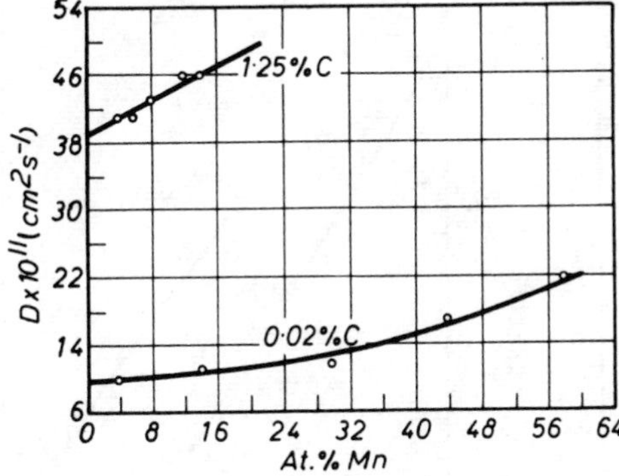

Figure 13.22 *Chemical diffusion coefficients in* γ-Fe-Mn *alloys with* 0.02 *and* 1.25 wt% C. *Temperature* 1 200 °C[67]

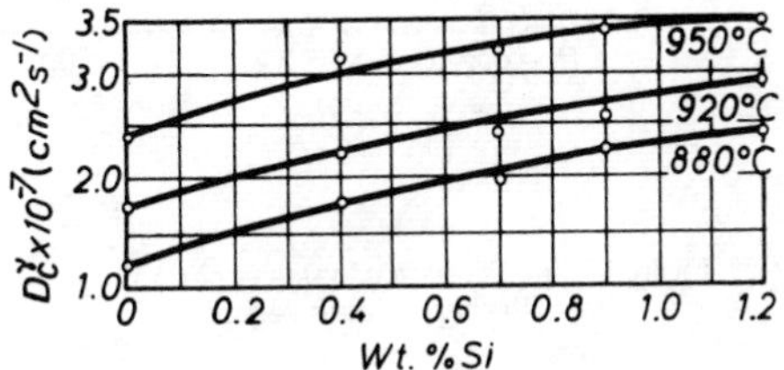

Figure 13.23 *Effect of* Si *content on chemical diffusion of* C *in* γ Fe[145]

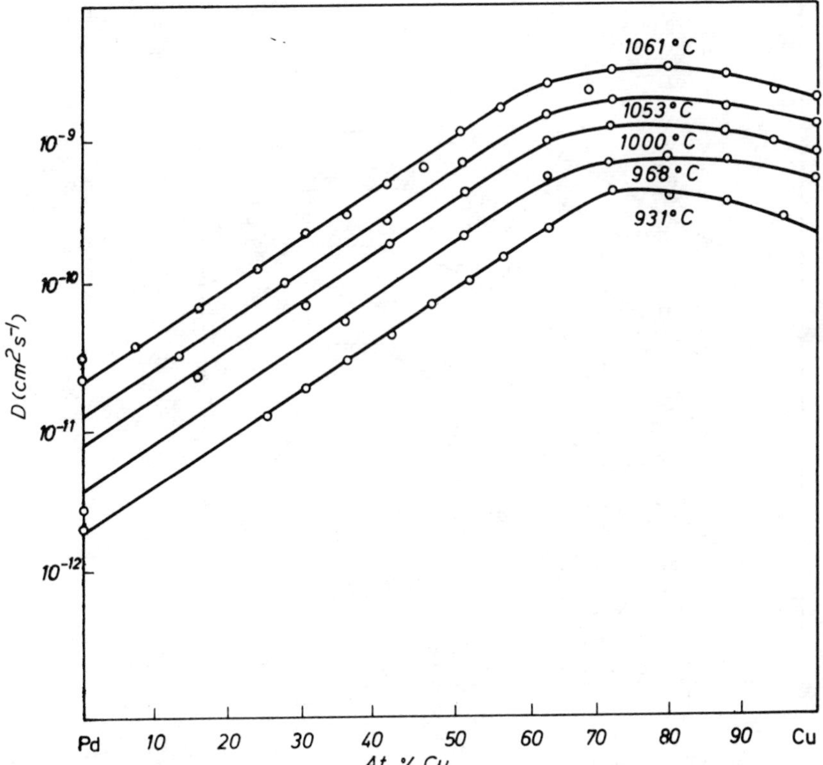

Figure 13.24 *Interdiffusion in the* Cu-Pd *system*[160]

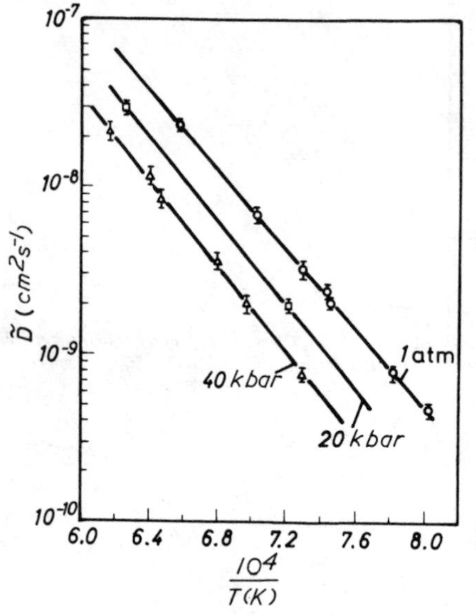

Figure 13.25 *Chemical diffusion in* α-Fe V 10% *alloy at* 1, 20 *and* 24 kbar *pressure*

Figure 13.26 *Chemical diffusion in* γ Fe V 0.7% *alloy at* 1, 20 *and* 40 kbar *pressure*[78]

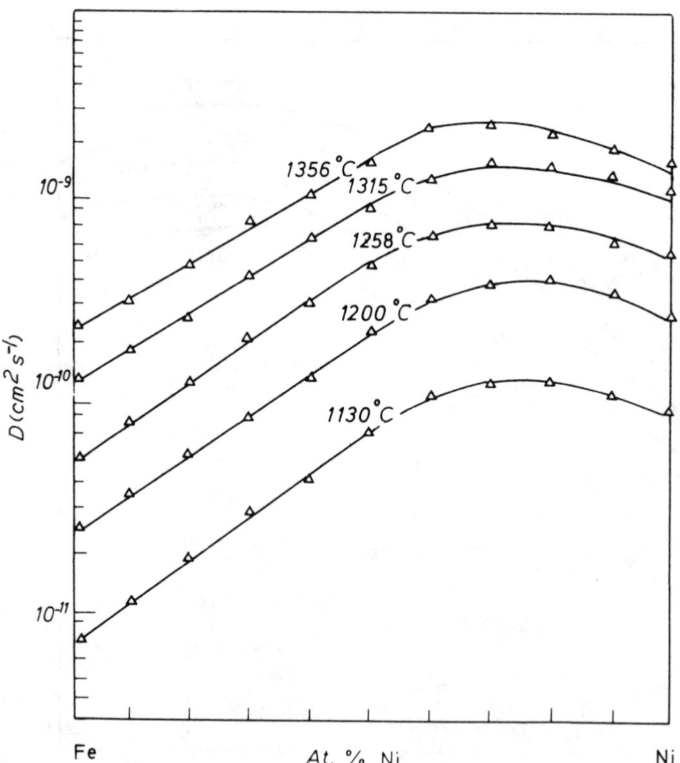

Figure 13.27 *Chemical diffusion coefficients in the* Fe–Ni *system*[160]

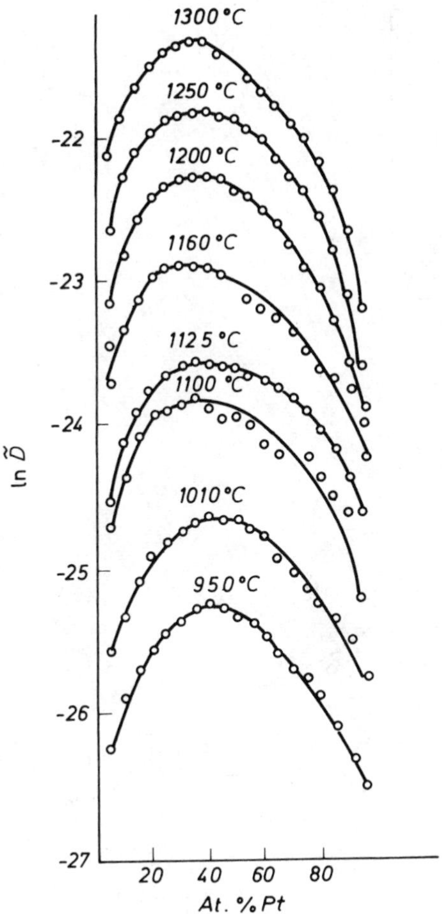

Figure 13.28 *Chemical diffusion coefficients in the* Ni–Pt *system*[218]

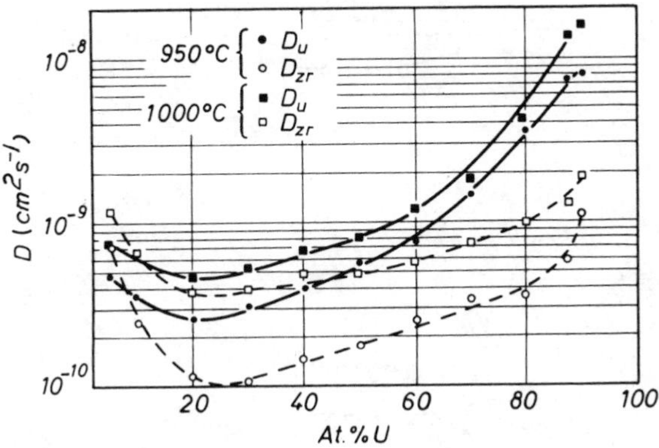

Figure 13.29 *Partial diffusion coefficients* D_u *and* D_{Zr} *in* γ U–Zr *alloys*[139]

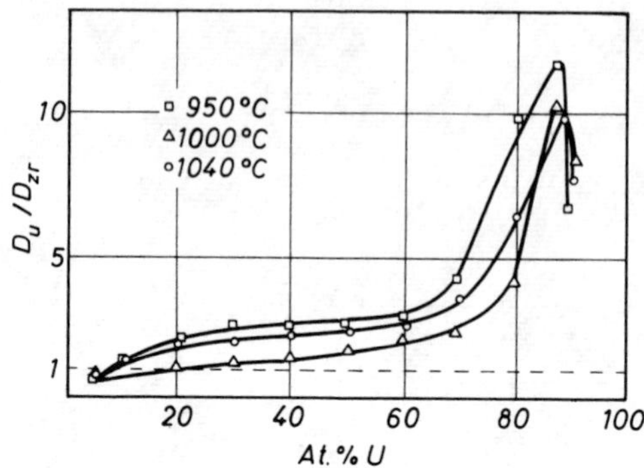

Figure 13.30 *Ratio of partial diffusion coefficients in* γ U–Zr *alloys*[139]

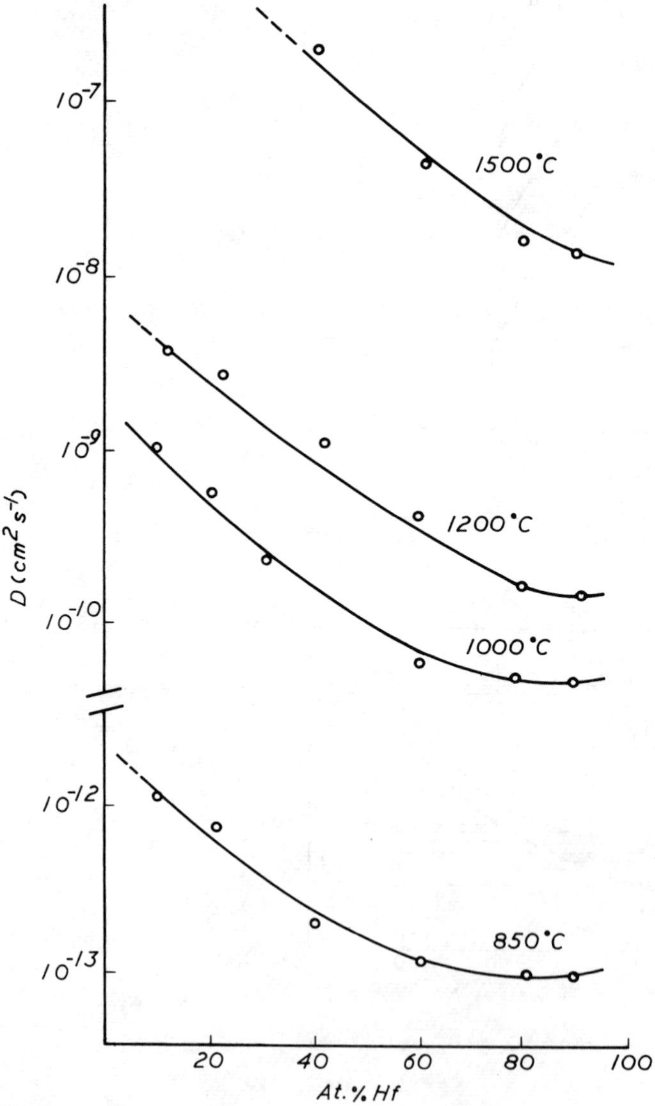

Figure 13.31a *Chemical diffusion in* Hf–Zr *alloys*

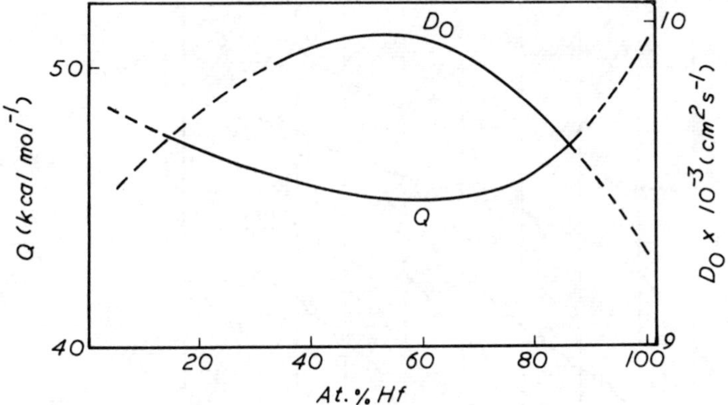

Figure 13.31b *A and Q for chemical diffusion in* Hf–Zr *alloys*[235]

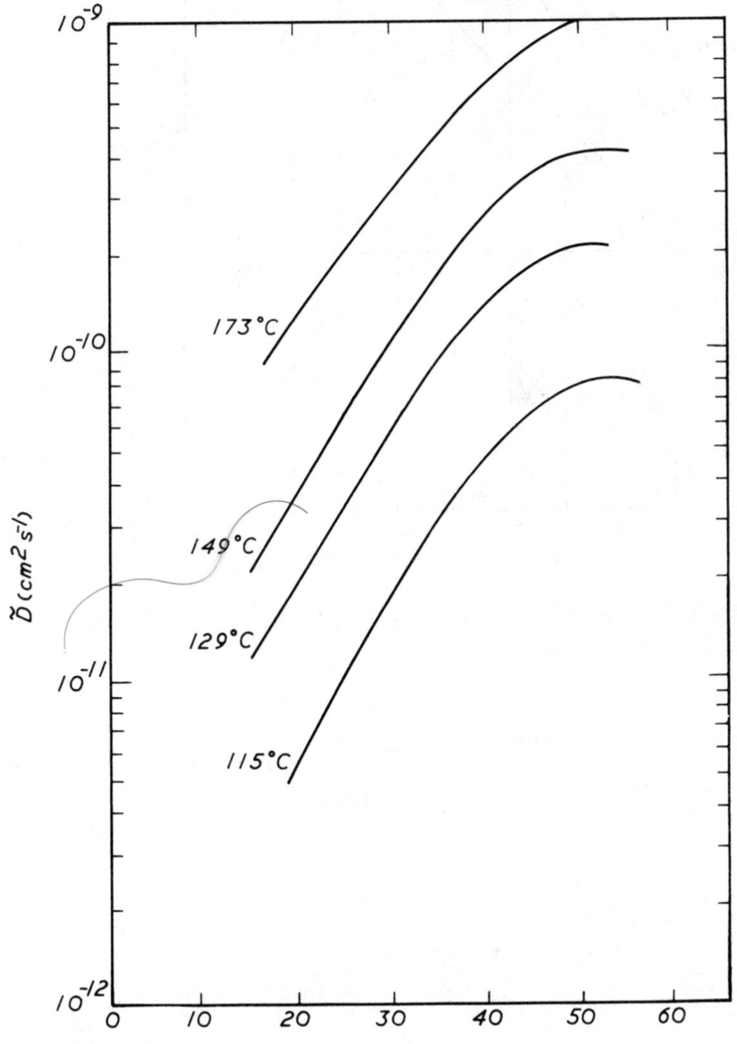

Figure 13.32 *Chemical diffusion in* In–Pb *alloys*[245]

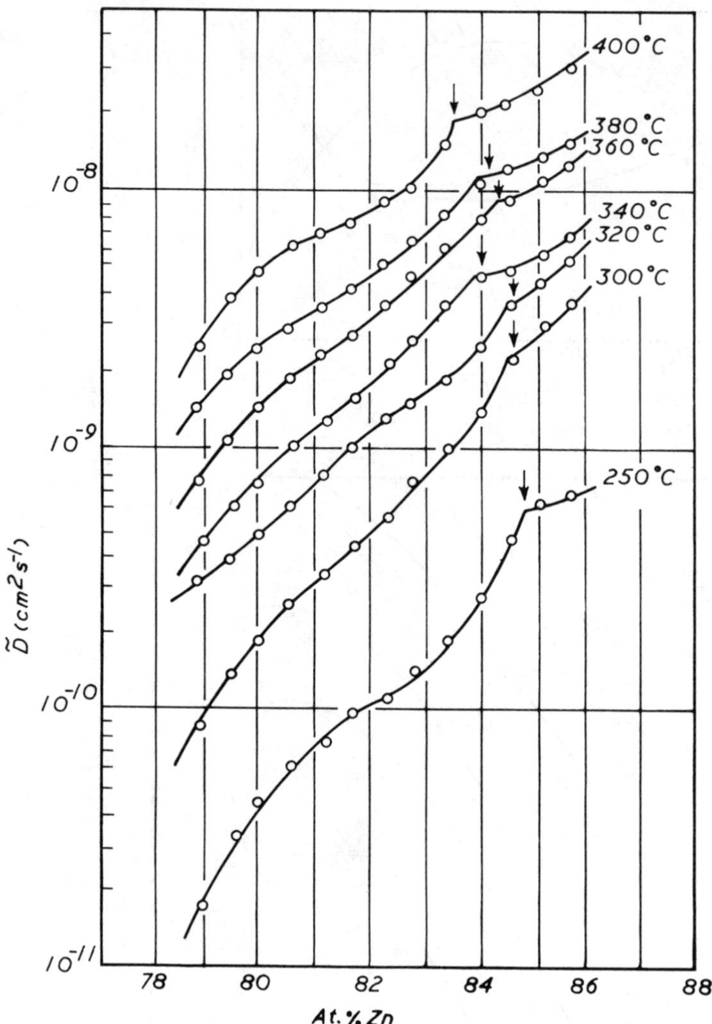

Figure 13.33 *Chemical diffusion in* ε Cu–Zn *alloys*[251]

Table 13.5 GRAIN BOUNDARY SELF-DIFFUSION

Element	A_{gb}	Q_{gb}	*Temp. range* °C	*Method*	*Ref.*
Ag	0.12 0.03	21.5 20.2	350–480	99.999% 99.97% p.c. (F)	1
	0.026	18.4 ± 0.6	350–555	p.c. (W)*	17
	6.8×10^{-4}	11.8 ± 0.7	400–555	⟨100⟩ 16° *tilt* b.c. b'dies (W)*	17
	—	~30	350–498	⟨112⟩ 18° *tilt* b.c. b'dies (W)	17
	140×10^{-3} (D_P)	19.7	400–525	⟨100⟩ 9 to 28° *tilt* b.c. b'dies 99.98% (F)	2

Table 13.5 GRAIN BOUNDARY SELF-DIFFUSION—*continued*

Element	A_{gb}	Q_{gb}	*Temp. range* °C	*Method*	*Ref.*
	3.2×10^{-3} (D_P)	18.0	420–520	⟨100⟩, ⟨110⟩, ⟨111⟩ 2 to 30° *twist* b.c. b'dies, 99.999% (F)	3
		D_{gb} is anisotropic within a boundary			4
Au	6.2×10^{-3}	20.2 ± 0.5	367–444	99.999+, p.c. (W and S)	29
Zn	0.22	14.3 ± 0.2	75–160	99.999% p.c. (F)	5
	0.38	14.6 ± 0.2		99.97% p.c. (F)	
	—	12.3			6
Cd	1.0	13.0 ± 1.2	50–110	99.5% p.c. (F)	7
	$0.67^{+1.0}_{-0.3}$	11.0 ± 0.66	51–135	99.999 5% p.c. (S)	18
Sn	$(6.44^{+5.9}_{-4.9}) \times 10^{-2}$	9.55 ± 0.7	40–115	99.99% p.c. (F)	8
Pb	0.81	15.7	214–260	p.c. (F)	9
		D_{gb} is anisotropic within a boundary			10
	—	4.70 ± 0.41	120–220	⟨100⟩ 30° *tilt* b.c. b'dy	25
	—	4.10 ± 0.84		⟨100⟩ 20° *tilt* b.c. b'dy	
	—	5.75 ± 1.86		⟨100⟩ 14° *tilt* b.c. b'dy	
	—	9.12 ± 2.88		⟨100⟩ 10° *tilt* b.c. b'dy	
	—	8.00 ± 2.56		⟨100⟩ 10° *twist* b.c. b'dy	
				99.999+ (F)	
Sb	$5.87^{+4.07}_{-2.41}$	23.1 ± 0.77	370–548	99.9% p.c. (W)	20
	$3.00^{+2.26}_{-1.29}$	22.2 ± 0.84	385–568	99.999 9% p.c. (W)	
Te^{127m}	0.22	19.1 ± 1.6	280–390	g. b'dies p.c. 99.999 9% (F)*	19
	1.89	21.6 ± 2.30	275–380	screw dislocations s.c. 99.999 9% (F)*	
	0.88×10^{-2}	15.4 ± 1.15	253–401	edge dislocations s.c. 99.999 9% (F)*	
Cr	—	46.0	1 000–1 350	p.c.	11
Fe					
(α)	13.0	40.0	580–660	99.7% p.c. (F)	12
	2.5	40.0	530–650	99.96% p.c. (F)	
	22.4	41.5	501–682	99.95% + p.c. (F)*	26
(γ)	5.0	39.0	923–1 016	99.7% p.c. (W)	13
	3.4	39.0	918–1 014	99.99% p.c. (W)	
	1.54	38.0	949–1 159	99.95% + p.c. (F)*	26
Co	0.15	39 ± 0.5	—	p.c.	14
Ni	$(1.75^{+2.1}_{-1.0}) 10^{-2}$	28.2 ± 2.0	850–1 100	99.85% p.c. (F)	15
	—	~29.9	700–1 100	⟨100⟩ 5° *tilt* b'dies (W)	16 and 28
	9.2×10^{-4}	24.9		⟨100⟩ 10° *tilt* b'dies (W)	
	8.0×10^{-3}	25.6		⟨100⟩ 20° *tilt* b'dies (W)	
	—	26.0 ± 1.5		⟨100⟩ 20–45° *tilt* b'dies (F)	
	0.44	40.7 ± 1.1	600–970	99.97% (W) ⟨112⟩ {111} 10° b.c. *tilt* b'dies (edge dislocations)	24
	0.52	44.9 ± 4.4		⟨111⟩ {111} 10° b.c. twist b'dies (screw dislocations)	
	20 (D_P)	36.8	500–600	99.999% edge dislocations in bent s.c.'s	27
W^{185}	6.66 ± 0.30	92.0 ± 2.0	1 400–2 200	99.992% (F)	21
		Q the same for sub-boundary $(<10^\circ)$ diffusion			
U	3.0×10^2	45.8	690–750	(β) p.c.	22
	2.0×10^2	42.7	850–1 050	(γ) p.c.	23

All values of A_{gb} are calculated assuming a grain boundary width $\delta = 5 \times 10^{-8}$ cm.
p.c. = polycrystals. b.c. = bicrystals. s.c. = single crystal.
Values of A_{gb}, and to a lesser extent of Q_{gb}, depend on the mathematical methods used to analyse the results:
(F) indicates results calculated from Fisher's solution of g.b. diffusion;
(W) indicates results calculated from Whipple's solution of g.b. diffusion:
(S) indicates results calculated from Suzuoka's solution;
( )* indicates the results are reported calculated from more than one solution, in addition to the one listed;
(D_p) signifies that A_{gb} refers to dislocation pipe diffusion—*see* references 2 and 27.

REFERENCES TO TABLE 13.5

1. R. E. Hoffman and D. Turnbull, *J. appl. Phys.*, 1951, **22**, 634.
2. D. Turnbull and R. E. Hoffman, *Acta metall.*, 1954, **2**, 419.
3. G. Love and P. G. Shewmon, *ibid.*, 1963, **11**, 899.
4. R. E. Hoffman, *ibid.*, 1956, **4**, 97.
5. E. S. Wajda, *ibid.*, 1954, **2**, 184.
6. F. Voigtmann, *Thesis*, Dresden, 1961.
7. E. S. Wajda, G. A. Shirn and H. B. Huntington, *Acta metall.*, 1955, **3**, 39.
8. W. Lange and D. Bergner, *Phys. Stat. Sol.*, 1962, **2**, 1410.
9. B. Okkerse, *Acta metall.*, 1954, **2**, 551.
10. B. Okkerse, T. J. Tiedema and W. G. Burgers, *ibid.*, 1955, **3**, 300.
11. S. Z. Bokstein, S. T. Kishkin and L. M. Moroz, *UNESCO Int. Conf. Rad. Isotopes and Sci. Res.*, 1957, Pap. 193.
12. C. Leymonie and P. Lacombe, *Mém. scient. Révue Metall.*, 1960, **57**, 285.
13. P. Guiraldenq and P. Lacombe, *Acta metall.*, 1965, **13**, 51.
14. S. D. Gerzricken, T. K. Yatsenko and L. Slastnikov, *Vaprosy fiziki metallov i metallovedeniva, Kiev*, 1959, **9**, 154.
15. W. Lange, A. Hässner and G. Mischer, *Phys. Stat. Sol.*, 1964, **5**, 63.
16. W. R. Upthegrove and M. J. Sinnott, *Trans. Am. Soc. Metals*, 1958, **50**, 1031.
17. J. T. Robinson and N. L. Peterson, *Surf. Sci.*, 1972, **31**, 586.
18. F. Guenther, A. Haessner and L. Oppermann, *Isotopenpraxis*, 1969, **5**, 461.
19. R. N. Goshtagore, *Phys. Rev.*, 1967, **155**, 603.
20. A. Haessner and G. Voigt, *Z. Metallk.*, 1968, **59**, 559.
21. K. G. Kreider and G. Bruggeman, *Trans. Met. Soc. AIME*, 1967, **239**, 1222.
22. G. B. Federov, E. A. Smirnov and S. S. Moiseenko, *Met. Metalloved Chist, Metall.*, 1968, **No. 7**, 124.
23. G. B. Federov and E. A. Smirnov, *Met. Metalloved. Chist. Metallov*, 1967, **No. 6**, 181.
24. R. F. Cannon and J. P. Stark, *J. appl. Phys.*, 1969, **40**, 4366.
25. J. P. Stark and W. R. Upthegrove, *Trans. ASM*, 1966, **59**, 479.
26. D. W. James and G. M. Leak, *Phil. Mag.*, 1965, **12**, 491.
27. M. Wuttig and H. K. Birnbaum, *Phys. Rev.*, 1966, **147**, 495.
28. R. F. Cannon and J. P. Stark, *J. appl. Phys.*, 1969, **40**, 4361.
29. D. Gupta, *J. appl. Phys.*, 1973, **44**, 4455.

Table 13.6 SELF-DIFFUSION IN LIQUID METALS[22]

Element	*A*	*Q*	*Temp. range* °C	*Method*	*Ref.*
Li	$(14.4 \pm 0.7)\,10^{-4}$	2.87 ± 0.07	192–450	IIIb(ii) Capillary	1 and 2
Na^{24}	$(8.6 \pm 0.9)\,10^{-4}$	2.22 ± 0.08	102–290	IVa(i) Capillary	2
K^{42}	$(7.6 \pm 0.5)\,10^{-4}$	2.02 ± 0.06	82–290	IVa(i) Capillary	2
Rb	$(6.6 \pm 1.1)\,10^{-4}$	1.98 ± 0.16	57–230	Electrotransport method	2 and 3
Cs	4.8×10^{-4}	1.86	50–200	Calculated	2 and 4
Cu^{64}	$(14.6 \pm 1.0)\,10^{-4}$	9.71 ± 0.71	1 140–1 270	IIIb(ii) Capillary	5
Ag^{110}	$(5.8 \pm 1.4)\,10^{-4}$	7.66 ± 0.67	970–1 300	IIIb(ii) Capillary	6
Zn^{65}	8.2×10^{-4}	5.09 ± 0.8	450–610	IIIb(ii) Capillary	7
	12×10^{-4}	5.60 ± 0.6	430–610	IIIb(ii) Capillary	8
Cd	$(3.62 \pm 0.07)\,10^{-4}$	3.31 ± 0.39	330–510	IIIb(ii) Capillary	9
Hg	1.40×10^{-4}	1.22 ± 0.08	−35 – 260	IIa(ii) Shear cell	10
	1.63×10^{-4} (a)	1.33 ± 0.08	0–295	IIIb(ii) Capillary	11
Ga^{72}	0.75×10^{-4}	0.88 ± 0.09	7–77	IVa(i) Capillary	12 and 13
	4.13×10^{-4}	2.01 ± 0.03	77–400		
In	$(2.89 \pm 0.25)\,10^{-4}$	2.43 ± 0.05	170–750	IIIb(ii) Capillary	14
Tl	9.45×10^{-4}	4.4	Estimated from viscosity		15
Sn^{113}	1.11×10^{-4}	1.59 ± 0.45	260–430	IIIb(ii) Capillary	16
	11.2×10^{-4}	4.57 ± 0.62	430–950		
	3.18×10^{-4} (a)	2.77	230–920	IIa(ii) Shear cell	17
	$(7.3 \pm 0.7) \times 10^{-4}$	3.78 ± 0.13	250–1 250	IIa(ii) Shear cell	21
Pb	$(2.37 \pm 0.12)\,10^{-4}$	5.90 ± 1.16	350–510	IIIb(ii) Capillary	9
Bi	$(0.83 \pm 0.11)\,10^{-4}$	2.51 ± 0.13	267–707	IIIb(ii) Capillary	18
Fe	10^{-2}	15.7	(c) 1 337–1 407	IIIb(ii) Capillary	19
	0.43×10^{-2}	12.2	(d) 1 237–1 357		
Te	13.6×10^{-4}	5.66 ± 0.25	437–600	IVb(ii) Capillary	20

(*a*) Average values. *A* and *Q* both increase with temperature.
(*b*) Supercooled.
(*c*) Fe + 2.5% C.
(*d*) Fe + 4.6% C.

REFERENCES TO TABLE 13.6

1. A. Ott and A. Lodding, *Z. Naturforsch.*, 1965, **20a**, 1578.
2. S. J. Larsson, C. Roxbergh and A. Lodding, *Phys. Chem. Liquids*, 1972, **3**, 137.
3. A. Nordén and A. Lodding, *Z. Naturforsch.*, 1968, **22a**, 215.
4. A. Lodding, *Z. Naturforsch.*, 1972, **27a**, 873.
5. J. Henderson and L. Young, *Trans. met. Soc. AIME*, 1961, **221**, 72.
6. V. G. Leak and R. A. Swalin, *Trans. met. Soc. AIME*, 1964, **230**, 426.
7. N. H. Nachtrieb, E. Fraga and C. Wahl, *J. phys. Chem.*, 1963, **67**, 2353.
8. W. Lange, W. Pippel and F. Bendel, *Z. phys. Chem.*, 1959, **212**, 238.
9. M. Mirshamshi, A. Cosgarea and W. Upthegrove, *Trans. metall. Soc., AIME*, 1966, **236**, 122.
10. E. F. Broome and H. A. Walls, *Trans. Met. Soc. AIME*, 1968, **242**, 2177.
11. R. E. Meyer, *J. phys. Chem.*, 1961, **65**, 567.
12. S. Larsson *et al.*, *Z. Naturforsch.*, 1970, **25a**, 1472.
13. E. F. Broome and H. A. Walls, *Trans. Met. Soc. AIME*, 1969, **245**, 739.
14. A. Lodding, *Z. Naturforsch.*, 1956, **11a**, 200.
15. J. A. Cahill and A. V. Grosse, *J. phys. Chem.*, 1965, **69**, 518.
16. C. H. Ma and R. A. Swalin, *J. chem. Phys.*, 1962, **36**, 3014.
17. K. G. Davis and P. Fryzuk, *J. appl. Phys.*, 1968, **39**, 4848.
18. N. Petrescu and M. Petrescu, *Revue Roum. Chim.*, 1970, **15**, 189.
19. L. Yang, M. T. Simnad and G. Derge, *Trans. Met. Soc. AIME*, 1956, **206**, 1577.
20. L. Nicoloiu, L. Ganovici and I. Ganovici, *Rev. Roum. Chim.*, 1970, **15**, 1713.
21. S. N. Kryukov, E. A. Soldatov and V. I. Irkov, *Vestn. Mosk. Univ., Ser. 2*, 1971, **12**, 332.
22. See A. Lodding, Diffusion in Liquid Metals' Diffusion and Defect Monograph Series, *Trans. Tech.* publication, Switzerland. In Preparation.

14 General physical properties

14.1 The physical properties of pure metals

Many physical properties depend on the purity and physical state (annealed, hard drawn, cast, etc.) of the metal. The data in Tables 14.1 and 14.2 refer to metals in the highest state of purity available, and are sufficiently accurate for most purposes. The reader should, however, consult the references before accepting the values quoted as applying to a particular sample.

Table 14.1 THE PHYSICAL PROPERTIES OF PURE METALS AT NORMAL TEMPERATURES

Metal	*Melting point* °C	*Boiling point* °C	*Density at* 20 °C $g cm^{-3}$‡	*Thermal conductivity* 0–100 °C $W m^{-1} K^{-1}$	*Mean specific heat* 0–100 °C $J kg^{-1} K^{-1}$	*Resistivity at* 20 °C μΩcm	*Temp. coeff. of resistivity* 0–100 °C $10^{-3} K^{-1}$	*Coefficient of expansion* 0–100 °C $10^{-6} K^{-1}$
Aluminium	660.37	2 520	2.70	238	917	2.67	4.5	23.5
Antimony	630.74	1 590	6.68	23.8	209	40.1	5.1	8–11
Arsenic	(817)	616	5.727	—	331	33.3	—	5.6
Barium	729	2 130	3.5	—	285	60 (0 °C)	—	18
Beryllium	1 287	2 470	1.848	194	2052	3.3	9.0	12
Bismuth	271.442	1 564	9.80	9	124.8	117	4.6	13.4
Cadmium	321.108	767	8.64	103	233.2	7.3	4.3	31
Caesium	28.5	670	1.87	36.1(s)	234	20	4.8	97
Calcium	839	1 484	1.54	125	624	3.7	4.57	22
Cerium	798	3 430	6.75	11.9	188	85.4	8.7	8
Chromium	1 860	2 680	7.1	91.3	461	13.2	2.14	6.5
Cobalt	1 494	2 930	8.9	96	427	6.34	6.6	12.5
Copper	1 084.8	2 560	8.96	397	386.0	1.694	4.3	17.0
Dysprosium	1 500	(2 630)	8.536	10.0	173	91	1.19	8.6
Erbium	1 530	(2 600)	9.051	9.6	166	86	2.01	9.2
Gadolinium	1 350	(3 000)	α7.895 β7.80	8.8	298	134	0.9/1.76	6.4
Gallium	29.7	2 205	5.91	41.0(s)	377	*	—	18.3
Germanium	937	2 830	5.32	56.4	310	$\sim 89 \times 10^3$	—	5.75
Gold	1 064.43	2 860	19.3[2]	315.5	130	2.20	4.0	14.1
Hafnium	2 227	4 600	13.1	22.9	147	32.2	4.4	6.0
Holmium	1 461	2 600	8.803	—	164	94	1.71	9.5**
Indium	156.4	2 070	7.3	80.0	243	8.8	5.2	24.8
Iridium	2 447	4 390	22.4	146.9	130.6	5.1	4.5	6.8
Iron	1 536	2 860	7.87	78.2	456	10.1	6.5	12.1
Lanthanum	920	(3 420)	α6.174 β6.186 γ5.97	13.8	200	57	2.18	4.9
Lead	327.502	1 750	11.68	34.9	129.8	20.6	4.2	29.0
Lithium	181	1 342	0.534	76.1	3 517	9.29	4.35	56
Lutetium	1 652	3 327	9.842	—	154	68	—	125**
Magnesium	649	1 090	1.74	155.5	1 038	4.2	4.25	26.0
Manganese	1 244	2 060	7.4	7.8	486	160(α)	—	23

* *See* 'Electrical properties',
** at 400 °C,
‡ Densities of higher allotropes not at 20 °C.

Table 14.1 THE PHYSICAL PROPERTIES OF PURE METALS AT NORMAL TEMPERATURES—*continued*

Metal	*Melting point* °C	*Boiling point* °C	*Density at 20 °C* $g cm^{-3}$‡	*Thermal conductivity 0–100 °C* $W m^{-1} K^{-1}$	*Mean specific heat 0–100 °C* $J kg^{-1} K^{-1}$	*Resistivity at 20 °C* μΩcm	*Temp. coeff. of resistivity 0–100 °C* $10^{-3} K^{-1}$	*Coefficient of expansion 0–100 °C* $10^{-6} K^{-1}$
Mercury	−38.87	357	13.546	8.65	138	95.9	1.0	61
Molybdenum	2615	4610	10.2	137	251	5.7	4.35	5.1
Neodymium	1024	(3060)	α7.004 β6.80	13.0	209	64	1.64	6.7
Nickel	1455	2915	8.9	88.5	452	6.9	6.8	13.3
Niobium	2467	4740	8.6	54.1	268	16.0	2.6	7.2
Osmium	3030	5000	22.5	86.9	130	8.8	4.1	4.57
Palladium	1554	2960	12.0	75.2	247	10.8	4.2	11.0
Platinum	1769.9	3830	21.45	73.4	134.4	10.58	3.92	9.0
Plutonium	640	3235	19.84	8.4	142	146.5	—	55
Polonium	246	965	—	—	—	—	—	—
Potassium	63.2	759	0.86	104(s)	754	6.8	5.7	83
Praeseodymium	932	(3020)	α6.782 β6.64	11.7	192	68	1.71	4.8
Radium	700	1500	5	—	—	—	—	—
Rhenium	3180	5690	21.0	47.6	138	18.7	4.5	6.6
Rhodium	1966	3700	12.4	148	243	4.7	4.4	8.5
Rubidium	38.8	688	1.53	58.3(s)	356	12.1	4.8	9.0
Ruthenium	2310	4120	12.2	116.3	234	7.7	4.1	9.6
Samarium	1072	1803	7.536 7.40	—	181	92	1.48	—
Scandium	1538	(2870)	2.99	—	558	66	—	12
Selenium	220.5	685	4.79	—	339	12	—	37
Silicon	1412	3270	2.34	138.5	729	10^3–10^6	—	7.6
Silver	961.93	2163	10.5	425	234	1.63	4.1	19.1
Sodium	97.8	883	0.97	128	1227	4.7	5.5	71
Strontium	770	1375	2.6	—	737	23 (0 °C)	—	100
Tantalum	2980	5370	16.6	57.55	142	13.5	3.5	6.5
Terbium	1356	(2500)	8.272	—	172	116	—	7.0
Tellurium	450	988	6.24	3.8	134	1.6×10^5 (0 °C)	— —	1.7 ∥c axis 27.5 ⊥c axis
Thallium	304	1473	11.85	45.5	130	16.6	5.2	30
Thorium	1755	4290	11.5	49.2	100	14	4.0	11.2
Thulium	1543	1727	9.322	—	160	90	1.95	11.6**
Tin	231.968	2625	7.3	73.2	226	12.6	4.6	23.5
Titanium	1667	3285	4.5	21.6	528	54	3.8	8.9
Tungsten	3387	5555	19.3	174	138	5.4	4.8	4.5
Uranium	1132	4400	19.05(α) 18.89(β)	28	117	27	3.4	‡
Vanadium	1902	3410	6.1	31.6	498	19.6	3.9	8.3
Ytterbium	824	1427	6.977 6.54	—	145	28	1.30	25.0
Yttrium	1520	3300	4.478 4.25	10.2	309	53	2.71	10.8**
Zinc	419.58	911	7.14	119.5	394	5.96	4.2	31
Zirconium	1852	4400	6.49	22.6	289	44	4.4	5.9

(s) = solid * *See* 'Electrical properties.' ‡ α-Uranium 23 ∥ *a* axis, −3.5 ∥ *b* axis, 17 ∥ *c* axis } 25–300 °C β-Uranium 4.6 ∥ *c* axis, 23 ⊥ *c* axis } 20–720 °C

** At 400 °C. ‡ Densities of higher allotropes not at 20 °C.

Rare Earths and Rare Metals ( ).

Melting and boiling points (1) *see also* 'Thermochemical data' p. **8**–1. Electrical resistivity (2, 3) *see also* 'Electrical properties,' p. **19**–1. Specific heat (4, 5) Thermal conductivity (6).

Table 14.2 THE PHYSICAL PROPERTIES OF PURE METALS AT ELEVATED TEMPERATURES

Metal	Temperature t°C	Coefficient of expansion 20–t°C $10^{-6}\,K^{-1}$	Resistivity at t°C μΩ cm	Thermal conductivity at t°C $W\,m^{-1}\,K^{-1}$	Specific heat at t°C $J\,kg^{-1}\,K^{-1}$	References
Aluminium	20	—	2.67		900	7, 8, 9
	100	23.9	3.55		938	
	200	24.3	4.78	238	984	
	300	25.3	5.99		1 030	
	400	26.49	7.30		1 076	
Antimony	20	—	40.1	18.0	205	7, 10, 6
	100	8.4–11.0	59	16.7	214	
	500	9.7–11.6	154	19.7	239	
Beryllium	20	—	3.3	180	1 976	11
	100	12	5.3	152	2 081	
	200	13	10.5	130.2	(2 215)	
	300	14.5	11.1	117.7	(2 353)	
	500	16	21.8	103.0	(2 621)	
	700	17	26	85.8	(2 889)	
Bismuth	20	—	117	8.0	121	7, 12
	100	13.4	156	7.5	130	
	250	—	260	7.5	147	
Cadmium	20	—	7.3	84	230	7, 13, 14
	100	31.8	9.6	87.9	239	
	300	(38)	18.0	104.7	260	
Chromium	20	—	13.2	91.3	444	7, 15, 16
	100	6.6	18 (152 °C)	—	490	
	400	8.4	31 (407 °C)	76.2 (426 °C)	582	
	700	9.4	47 (652 °C)	67.4 (760 °C)	649	
Cobalt	20	—	5.68	—	434	42, 45
	100	12.3	9.30	—	453	
	200	13.1	13.88	—	478	
	300	13.6	19.78	—	502	
	400	14.0	26.56	—	527	
	600	—	40.2	—	575	
	800	—	58.6	—	716	
	1 000	—	77.4	—	800	
	1 200	—	91.9	—	883	
Copper	20	—	1.694	394	385	7, 17, 16, 18
	100	17.1	—	394	389	
	200	17.2	2.93	389	402	
	500	18.3	4.6(497 °C)	341(538 °C)	(427)	
	1 000	20.3	8.1(977 °C)	244(1 037 °C)	(473)	
Gold	20	—	2.2	293	126	7
	100	14.2	2.8	293	130	
	500	15.2	6.8	—	142	
	900	16.7	11.8	—	151	
Hafnium	20	—	35.5	(22.2)	144	43, 44, 48
	100	—	46.5	22.0	148	
	200	—	60.3	21.5	152	
	400	6.3	84.4	20.7	160	
	1 000	6.1	—	—	185	
	1 400	6.0	—	—	—	
	1 800	5.9	—	—	—	
Iridium	20	—	5.1	148(0 °C)	130	19
	100	6.8	6.8	143	134	
	500	7.2	15.1	—	142	
	1 000	7.8	—	—	159	

Table 14.2 THE PHYSICAL PROPERTIES OF PURE METALS AT ELEVATED TEMPERATURES—*continued*

Metal	*Temperature* t °C	*Coefficient of expansion* 20–t °C 10^{-6} K^{-1}	*Resistivity at* t °C μΩ cm	*Thermal conductivity at* t °C W m^{-1} K^{-1}	*Specific heat at* t °C J kg^{-1} K^{-1}	*References*
Iron	20	—	10.1	73.3	444	7, 20
	100	12.2	14.7	68.2	477	
	200	12.9	22.6	61.5	523	
	400	13.8	43.1	48.6	611	
	600	14.5	69.8	38.9	699	
	800	14.6	105.5	29.7	791	
Lead	20	—	20.6	34.8	130	7, 6, 11
	100	29.1	27.0	33.5	134	
	200	30.0	36.0	31.4	134	
	300	31.3	50	29.7	138	
Magnesium	20	—	4.2	167	1 022	7
	100	26.1	5.6	167	1 063	
	200	27.0	7.2	163	1 110	
	400	28.9	12.1	130	1 197	
Molybdenum	20	—	5.7	142	247	7, 21, 22, 23
	100	5.2	7.6	138	260	
	500	5.7	17.6	121	285	
	1 000	5.75	31	105	310	
	1 500	6.51	46	84	339 (mean)	
	2 500	—	77	—	—	
Nickel	20	—	6.9	88	435	24
	100	13.3	10.3	82.9	477	
	200	13.9	15.8	73.3	528	
	300	14.4	22.5	63.6	578	
	400	14.8	30.6	59.5	519	
	500	15.2	34.2	62.0	535	
	900	16.3	45.5	—	595	
Niobium	20	—	14.6	—	268	42, 45
	200	7.19	25.0	56.5	271	
	400	7.39	36.6	60.7	284	
	600	7.56	48.1	65.3	292	
	800	7.72	59.7	—	301	
	1 000	7.88	71.3	—	310	
Palladium	20	—	10.8	75	243	19
	100	11.1	13.8	74	247	
	500	12.4	27.5	—	268	
	1 000	13.6	40	—	297	
Platinum	20	—	10.58	72	134	7, 19, 23, 25
	100	9.1	13.6	72	134	
	500	9.6	27.9	—	147	
	1 000	10.2	43.1	67	159	
	1 500	11.31	55.4	63	176	
Plutonium	20 $\alpha\to\alpha$	47	145.8	(8.4)	131	42, 45
	100 $\alpha\to\alpha$	203	141.6	—	138	
	200 $\alpha\to\beta$	173	107.8	—	145	
	300 $\alpha\to\gamma$	181	107.4	—	153	
	400 $\alpha\to\delta$	109	100.7	—	154	
	500 $\alpha\to\varepsilon$	101	110.6	—	144	
Rhenium	20	12.4 ‖-axis	18.7	48	134	26, 16, 27
	100	4.7 ⊥-axis	25		138	
	2 500	7.29 (2 000 °C)	132	—	209 (2 527 °C)	
Rhodium	20	—	4.7	149	243	19
	100	8.5	6.2	147	255	
	500	9.8	14.6	—	289	
	1 000	10.8	—	—	331	

Table 14.2 THE PHYSICAL PROPERTIES OF PURE METALS AT ELEVATED TEMPERATURES—*continued*

Metal	*Temperature* t°C	*Coefficient of expansion* 20–t°C $10^{-6}\,K^{-1}$	*Resistivity at* t°C μΩ cm	*Thermal conductivity at* t°C $W\,m^{-1}\,K^{-1}$	*Specific heat at* t°C $J\,kg^{-1}\,K^{-1}$	*References*
Silver	20	—	1.63	419	234	7, 28
	100	19.6	2.1	419	222	
	500	20.6	4.7	(377)	(230)	
	900	22.4	7.6	—	(243)	
Tantalum	20	—	13.5	57	138	7, 29, 30, 31, 32
	100	6.5	17.2	54	142	
	500	6.6	35	—	151	
	1 500	—	71	—	167	
	2 500	—	102	—	234(2 727 °C)	
Thallium	20	—	16.6	46	134	33, 6
	100	30	—	45	138	
	200	—	—	45	142	
Tin	20	—	12.6	65	222	7, 34
	100	23.8	15.8	63	239	
	200	24.2	23.0	60	260	
Titanium	20	—	54	16	519	35, 36
	100	8.8	70	15	540	
	200	9.1	88	15	569	
	400	9.4	119	14	619	
	600	9.7	152	13	636	
	800	9.9	165	(13)	682	
Tungsten	20	—	5.4	167	134	37, 33 38
	100	4.5	7.3	159	138	
	500	4.6	18	121	142	
	1 000	4.6	33	111	151	
	2 000	5.4	65	93	—	
	3 000	6.6	100	—	—	
Uranium	20 α	—	30	27	116	Expansion anisotropic 42, 45
	600 α	—	59	38	186	
	700 β	—	55.5	40	176	
	800 γ	—	54	42.3	160	
Vanadium	20	—	24.8	—	492	42, 45
	100	8.3	31.5	31	505	
	500	9.6	—	36.8	570	
	700	—	—	35.2	603	
	900	10.4	—	—	636	
Zinc	20	—	5.96	113	389	39, 13, 40, 41, 6
	100	31	7.8	109	402	
	200	33	11.0	105	414	
	300	34	13.0	101	431	
	400	—	16.5	96	444	

REFERENCES TO TABLES 14.1 AND 14.2

1. W. Slough, *Private Communication*, Chemical Standards Division, NPL, 1972.
2. M. J. Swan, *Private Communication*, Electrical Science Division, NPL, 1972.
3. G. W. C. Kaye and T. H. Laby, 'Tables of Physical and Chemical Constants', Longmans, London, 1966.
4. 'Thermophysical Properties of Matter', TPRC Data Series, Volume 4.
5. R. J. Corrnecini and J. Gniewek, Nat. Bureau of Stds. Monograph; 1960.
6. 'Thermophysical Properties of Matter', TPRC Series Volume 1; 1970.
7. US Bur. Stds. Circular C447, 'Mechanical Properties of Metals and Alloys', Washington, 1952.
8. R. Hase, R. Heierberg and W. Walkenhorst, *Aluminium*, 1940, **22**, 631.
9. T. G. Peason and H. W. L. Phillips, *Met. Rev.*, 1957, **2**, 305.
10. H. Tsutsumi, *Sci. Rep. Tôhoku Univ.* (1), 1918, **7**, 100.
11. R. W. Powell, *Phil. Mag.*, 1953, **44**, 657.

12. E. F. Northrup and V. A. Suydam, *J. Franklin Inst.*, 1913, **175**, 160.
13. Saldau, *Z. Metallogr.*, 1915, **7**, 5.
14. S. Grabe and E. J. Evans, *Phil. Mag.*, 1935, **19**, 773.
15. R. W. Powell and R. P. Tye, *J. Inst. Metals*, 1957, **85**, 185.
16. C. F. Lucks and H. W. Deem, ASTM Special Tech. Pubn., 1958, No. 227.
17. C. S. Smith and E. W. Palmer, *Trans. AIMME*, 1935, **117**, 225.
18. C. J. Meechan and R. R. Eggleston, *Acta Met.*, 1954, **2**, 680.
19. R. F. Vines, 'The Platinum Metals and their Alloys', New York, 1941.
20. BISRA, 'Physical Constants of Some Commercial Steels at Elevated Temperatures', London, 1953.
21. C. Zwikker, *Physica*, 1927, **7**, 73.
22. E. P. Mikol, US Atomic Energy Comm. Publ. ORNL–1131, 1952.
23. O. H. Kirkorian, UCRL–6132, TID–4500, Sept. 1960, USA.
24. Mond Nichel Co. Ltd., *Nickel Bull.* 1951, **24**, 1.
25. K. S. Krishnan and S. C. Jain, *Br. J. appl. Phys.*, 1954, **5**, 426.
26. C. Agte, H. Alterthum, *et al.*, *Z. anorg. Chem.*, 1931, **196**, 129.
27. R. E. Taylor and R. A. Finch, US Atomic Energy Com. Rep., 1961 NAA–SR–6034.
28. US Bur. Min. Circular C412, 'Silver, its Properties and Industrial Uses', Washington, 1936.
29. L. Malter and D. B. Langmuir, *Phys. Rev.*, 1939, **55**, 743.
30. M. Cox, *Phys. Rev.*, 1943, **64**, 241.
31. I. B. Fieldhouse, *et al.*, WADC Tech. Rep. 55–495, 1956.
32. N. S. Rasor and J. D. McClelland, WADC Tech. Rep. 56–400, 1957.
33. A. E. van Arkel, 'Reine Metalle', Berlin, 1939.
34. Int. Tin R. and D. Council. Tech. Publ. B1, 1937.
35. E. S. Greiner and W. C. Ellis, Metals Tech., Sept., 1948.
36. L. Silverman, *J. Metals, N.Y.*, 1953, May, p. 631.
37. C. J. Smithells, 'Tungsten', London, 1952.
38. V. S. Gurnenyuk and V. V. Lebeden, *Fizica Metall.*, 1961, **11**, 29.
39. F. L. Uffelmann, *Phil. Mag.*, (7), 1930, **10**, 633.
40. Lees, *Phil. Trans. R. Soc.*, 1908, **A208**, 432.
41. Dewar and Fleming, *Phil. Mag.*, 1893, **36**, 271.
42. C. A. Hampel, 'Rare Metals Handbook', Chapman & Hall, London, 1961.
43. H. K. Adenstedt, *Trans. A.S.M.*, 1952, **44**, 949.
44. R. P. Cox *et al.*, *Ind. Eng. Chem.*, 1958, **50**, 141.
45. Thermochemical Data Section. *Met. Ref. Book.*
46. R. W. Powell, R. P. Tye and M. J. Woodman, *J. less Common Metals*, 1967, **12**, 1.

14.2 The physical properties of liquid metals

Table 14.3a THE PHYSICAL PROPERTIES OF LIQUID METALS
Density, surface tension and viscosity

The values of the properties listed in Table 14.3a are the weighted mean of the experimental values due to the investigators listed, usually in order of the reliance placed on their results, in the column of references. The references are grouped, and relate to density, surface tension and viscosity in that order. The properties given in parentheses are estimates.

14.2.1 Density

The variation of the density D of most liquid metals with temperature t is found to be well represented by a linear equation

$$D = D_0 + (t - t_0)\,(\mathrm{d}D/\mathrm{d}t)$$

where D_0 is the density of the liquid metal at its melting point t_0.

14.2.2 Surface tension

The surface tension γ of most liquids can be represented over temperature ranges of usual interest by the linear equation

$$\gamma = \gamma_0 + (t - t_0)\,(\mathrm{d}\gamma/\mathrm{d}t)$$

14.2.3 Viscosity

For most liquid metals the variation of viscosity η with temperature T(K) may be written

$$\eta = \eta_0 \exp (E/RT)$$

where η_0 and E are constants, and are given in the table, and R is the gas constant, 8.3144 J K^{-1} mol^{-1}.

Table 14.3a THE PHYSICAL PROPERTIES OF LIQUID METALS

	Temp	*Density*		*Surface tension*		*Viscosity*			
Metal	t_0 °C	D_0 g cm^{-3}	dD/dt mg cm^{-3} K^{-1}	γ_0 mN m^{-1}	dγ/dt mN m^{-1} K^{-1}	η_{mp} mN s m^{-2}	η_0 mN s m^{-2}	E kJ mol^{-1}	*References*
Ag	960.7	9.346	−0.907	903	−0.16	3.88	0.4532	22.2	1, 26, 27/86/90, 91
Al	660.0	2.385	−0.28	914	−0.35	1.30	0.1492	16.5	2–6/86/90, 91
As	817	5.22	−0.535	—	—	—	—	—	7/–/–
Au	1 063	17.36	−1.5	1 140	−0.52	5.0	1.132	15.9	9, 8/86/90, 91
B	2 077	2.08	—	1 070	—	—	—	—	10/86/–
Ba	727	3.321	−0.526	224	−0.095	—	—	—	11, 12/89/–
Be	1 283	1.690	−0.1162	1 390	(−0.29)	—	—	—	13/86/–
Bi	271	10.068	−1.33	378	−0.07	1.80	0.4458	6.45	14/86/90, 91
Ca	865	1.365	−0.221	361	−0.10	1.22	0.0651	27.2	12/86/90, 91
Cd	321	8.02	−1.16	570	−0.26	2.28	0.3001	10.9	15, 16, 17, 18, 77/86/90, 91
Ce	804	6.685	−0.227	740	−0.33	2.88	—	—	19–22/86/59
Co	1 493	7.76	−0.988	1 873	−0.49	4.18	0.2550	44.4	23, 24, 25/86/90, 91
Cr	1 875	6.28	−0.30	1 700	(−0.32)	—	—	—	25, 29/86/–
Cs	28.6	1.854	−0.6381	69	−0.047	0.68	0.1022	4.81	30/30, 84/90, 91
Cu	1 083	8.000	—0.801	1 285	−0.13	4.0	0.3009	30.5	31, 32, 33, 26, 34, 8/73, 34, 87, 88/90, 91
Fe	1 536	7.015	−0.883	1 872	−0.49	5.5	0.3699	41.4	36, 23, 25, 33, 37/86/90, 91
Fr	18	(2.35)	(−0.792)	(62)	(−0.044)	(0.765)	—	—	38/38/38
Ga	29.8	6.09	−0.60	718	−0.10	2.04	0.4359	4.00	39, 40/86/90, 91
Gd	1 312	(7.14)	—	810	−0.16	—	—	—	86/86/–
Ge	934	5.60	−0.625	621	−0.26	0.73	—	—	41, 42, 28, 13, 43 /86/94
Hf	1 943	11.1	—	1 630	(−0.21)	—	—	—	86/86–
Hg	−38.87	13.691	−2.436	498	−0.20	2.10	0.5565	2.51	45, 46/86/90, 91
	0	13.5951	average						
	20	13.5459	over						
			0–100 °C						
	100	13.3515							
In	156.6	7.023	−0.6798	556	−0.09	1.89	0.3020	6.65	47/86/90, 91
Ir	2 443	(20.0)	—	2 250	(−0.31)	—	—	—	48/86/–
K	63.5	0.8270	−0.2285	111.0	−0.0625	0.51	0.1340	5.02	30/30/90, 91
La	930	5.955	−0.237	720	−0.32	2.45	—	—	59/85/59
Li	180.5	0.525	−0.1863	395	−0.150	0.57	0.1456	5.56	30/30/90, 91
Mg	651	1.590	−0.2647	559	−0.35	1.25	0.0245	30.5	49–51/86/90, 91
Mn	1 241	5.73	−0.7	1 090	−0.2	—	—	—	52/53/–
Mo	2 607	(9.34)	—	2 250	(−0.30)	—	—	—	48, 54/86/–
Na	96.5	0.927	−0.2361	195	−0.0895	0.68	0.1525	5.24	30/30/90, 91
Nb	2 468	(7.83)	—	1 900	(−0.24)	—	—	—	48/86/–
Nd	1 024	6.688	−0.528	689	−0.09	—	—	—	22/86/–
Ni	1 454	7.905	−1.160	1 778	−0.38	4.90	0.1663	50.2	36, 25, 55, 37, 56, 23, 34/86/90, 91
Os	2 727	(20.1)	—	2 500	(−0.33)	—	—	—	48/86/–

Table 14.3a THE PHYSICAL PROPERTIES OF LIQUID METALS—*continued*

	Temp	*Density*		*Surface tension*		*Viscosity*			
Metal	t_0 °C	D_0 g cm^{-3}	dD/dt mg cm^{-3} K^{-1}	γ_0 mN m^{-1}	dγ/dt mN m^{-1} K^{-1}	η_{mp} mN s m^{-2}	η_0 mN s m^{-2}	E kJ mol^{-1}	*References*
P	44	—	—	52	—	1.71	—	—	–/86/95
Pb	327	10.678	−1.317 4	468	−0.13	2.65	0.463 6	8.61	57, 58, 15/86/90, 91
Pd	1 552	10.49	−1.266	1 500	(−0.22)	—	—	—	26/86/–
Pr	935	6.611	−0.240	—	—	2.80	—	—	59, 20/–/59
Pt	1 769	19	−2.9	1 800	(−0.17)	—	—	—	26, 60/86/–
Pu	640	16.64	−1.450	550	(−0.10)	6.0	1.089	5.59	61, 62/86/92
Rb	38.9	1.437	−0.486	83	−0.052	0.67	0.094 0	5.15	30/30, 84/90, 91
Re	3 158	(18.8)	—	2 700	(−0.34)	—	—	—	48, 54/86/–
Rh	1 966	10.8	—	2 000	(−0.30)	—	—	—	48/63/86/–
Ru	2 427	(10.9)	—	2 250	(−0.31)	—	—	—	48/86/–
S	119	1.819	−0.800	61	−0.07	~12	—	—	64/86/95
Sb	630.5	6.483	−0.565*	367	−0.05	1.22	0.081 2	22.0	65, 13/86/90, 91
Se	217	3.989	−1.44	106	−0.1	24.8	—	—	66–69/86/94
Si	1 410	2.51	−0.32	865	(−0.13)	0.94	—	—	14, 34, 70/86/94
Sn	232	7.000	−0.612 7	544	−0.07	1.85	0.538 2	—	71/86/90, 91
Sr	770	2.48	—	303	−0.10	—	—	—	72/86/–
Ta	2 977	(15.0)	—	2 150	(−0.25)	—	—	—	48/86/–
Te	451	5.71	−0.360	180	(−0.06)	~2.14	—	—	27, 73, 43, 74/86/94
Th	1 691	10.5	—	978	(−0.14)	—	—	—	72/86/–
Ti	1 685	4.11	−0.702	1 650	(−0.26)	5.2	—	—	75/86/93
Tl	302	11.280	−1.43	464	−0.08	2.64	0.298 3	10.5	76–78/86/90, 91
U	1 133	17.90	−1.031	1 550	−0.14	6.5	0.484 8	30.4	79/96/90, 91
V	1 912	5.7	—	1 950	(−0.31)	—	—	—	75/48/86
W	3 377	(17.6)	—	2 500	(−0.29)	—	—	—	48, 54, 80/86/–
Yb	824	—	—	—	—	1.07	—	—	–/–/94
Zn	419	6.575	−1.10	782	−0.17	3.85	0.413 1	12.7	73, 81, 16, 82, 83/86/90, 91
Zr	1 850	(5.8)	—	1 480	(−0.20)	8.0	—	—	44, 48, 75/86/93

* For antimony the simple linear equation is adequate to about 1 000 °C. However, the results for all the temperatures within the liquid range are better represented by the equation

$$D = 6.596 + 2.022 \times 10^{-4}\,T - 3.629 \times 10^{-7}\,T^2$$

where T is in degrees K (ref. 65).

REFERENCES TO TABLE 14.3a

1. A. D. Kirshenbaum, J. A. Cahill and A. V. Grosse, *J. inorg. nucl. Chem.*, 1962, **24**, 333.
2. J. D. Edwards and T. A. Moorman, *Chem. metall. Engng*, 1921, **24**, 61.
3. E. Gebhardt, M. Becker and S. Dorner, *Z. Metallk.*, 1953, **44**, 510.
4. G. D. Ayushina, E. S. Levin and P. V. Gel'd, *Zh. Fiz. Khim.*, 1969, **43**, 2756.
5. V. N. Naidich and Yu. V. Eremenko, *Fizika Metall.*, 1961, **6**, 62.
6. W. J. Coy and R. S. Mateer, *Trans. Q. ASM*, 1965, **58**, 99.
7. P. J. McGonigal and A. V. Grosse, *J. phys. Chem.*, 1963, **67**, 924.
8. E. Gebhardt and G. Wörwag, *Z. Metallk.*, 1951, **42**, 111, 358.
9. E. Gebhardt and S. Dorner, *Z. Metallk.*, 1951, **42**, 353.
10. F. N. Tavadze *et. al.*, *Dokl. Akad, Nauk SSSR*, 1963, **150**, 544; *Studii Cerc. Met.*, 1965, **10**, 49.
11. C. C. Addison and R. J. Pulham, *J. chem. Soc.*, 1962, 3873.
12. A. V. Grosse and P. J. McGonigal, *J. phys. Chem.*, 1964, **68**, 414.
13. A. V. Grosse and J. A. Cahill, *Trans. Q. ASM*, 1964, **57**, 739.
14. L. D. Lucas and G. Urbain, *C. r., hebd. Séanc. Acad. Sci., Paris*, 1962, **255**, 2414.
15. H. T. Greenaway, *J. Inst. Metals*, 1947–48, **74**, 133.
16. T. R. Hogness, *J. Am. chem. Soc.*, 1921, **43**, 1621.

17. A. Stauffer, Thesis, Gottingen, Dec. 1952.
18. H. J. Fisher and A. Phillips, *J. Metals*, 1954, **6**, 1060.
19. J. F. Eichelberger, Mound Lab. Rep. 1113, 1961, p8.
20. J. F. Eichelberger, Mound Lab. Rep. 1118, 1961, p12.
21. R. H. Perkins, L. A. Geoffrion and J. C. Biery, *Trans. AIME*, 1965, **233**, 1703.
22. W. G. Rohr, *J. less-common Metals*, 1966, **10**, 389.
23. L. D. Lucas and M. P. Pascal, *C. r. hebd. Séanc. Acad. Sci., Paris*, 1960, **250**, 1850.
24. A. D. Kirshenbaum and J. A. Cahill, *Trans. Q. ASM.*, 1963, **56**, 281.
25. T. Saito and Y. Sakuma, *J. Japan Inst. Met.*, 1967, **31**, 1140.
26. L. D. Lucas, *C. r. hebd. Séanc. Acad. Sci., Paris*, 1961, **253**, 2526.
27. M. Wobst and R. Rentzsch, *Z. phys. Chem.* (Leipzig), 1969, **240**, 36.
28. E. S. Levin *et. al.*, 'Poverkhinio Yavleniya Rasplavakh', 1968, 191–202, Edited by V. N. Eremenko, Naukova Dumka, Kiev, USSR.
29. V. N. Eremenko and Ya V. Naidich, *Izv. Akad. Nauk SSR.*, 1959, **2**, 111.
30. J. Freund, 'Thermophysical and Nuclear Parameters of Molten Li, Na, K, Rb and Cs', Inst. Kerntech. Tech. Univ. Berlin, (13), 184; **1969**.
31. J. A. Cahill and A. D. Kirshenbaum, *J. phys. Chem.*, 1962, **66**, 1080.
32. A. E. El-Mehairy and R. G. Ward, *Trans. AIME*, 1963, **227**, 1226.
33. M. G. Froberg and R. Weber, *Arch. Eisenhütt Wes.*, 1964, **35**, 877.
34. R. F. A. Freeman, *Private Communication*, Fulmer Research Institute, 1973.
35. E. Gebhardt, M. Becker and S. Dorner, *Aluminium*, 1955, **31**, 315.
36. A. V. Grosse and A. D. Kirshenbaum, *J. inorg. nucl. Chem.*, 1963, **25**, 331.
37. S. I. Papel, L. M. Shergin, and B. V. Tsarevskii, *Zh. Fiz. Khim.*, 1969, **43**, 2365.
38. Yu. P. Os'minin, *Zh. Fiz. Khim.*, 1969, **45**, 2610.
39. V. I. Nizenko, L. I. Sklyarenko and V. N. Eremenko, *Ukr. Khim. Zhur.*, 1965, **31**, 559.
40. W. H. Hoather, *Pol. phys. Soc.*, 1936, **48**, 699.
41. F. N. Tavadze *et al.*, 'Vap. Metalloved. Korroz. Metal', 1968, 11–18, Edited by F. N. Tavadze, *Iz. Metsniereba*, Tbilisi, USSR.
42. F. N. Tavadze *et al.*, 'Poverkhinio Yavleniya Rasplavakh', 1968, 159–162, Edited by V. N. Eremenko; *Naukova Dumka*, Kiev, USSR.
43. A. Klemm *et al.*, *Mh. Chem.*, 1952, **83**, 629.
44. A. W. Peterson, H. Kedesdy, P. H. Keck and E. Schwarz, *J. appl. Phys.*, 1958, **28**, 213.
45. A. V. Grosse, US Atomic Energy Comm. Contract AT 30–1–2082, May 1965.
46. A. H. Cook, *Phil. Trans. R. Soc.*, A, 1961, **254**, (1038), 125.
47. P. J. McGonigal, J. A. Cahill and A. D. Kirshenbaum, *J. inorg. nucl. Chem.*, 1962, **24**, 1012.
48. B. C. Allen, *Trans. AIME*, 1963, **227**, 1175; 1964, **230**, 1357.
49. P. J. McGonigal, A. D. Kirshenbaum and A. V. Grosse, *J. phys. Chem.*, 1962, **66**, 737.
50. E. Pelzel and F. Sauerwald, *Z. Metallk.*, 1941, **33**, 229.
51. E. Gebhardt, M. Becker and E. Trägner, *Z. Metallk.*, 1955, **46**, 90.
52. S. I. Popel, B. V. Tsarevskii and N. K. Dzhemilev, *Fizica Metall.*, 1964, **18**, 468.
53. N. A. Vatolin and O. A. Esin, *Fizica Metall.*, 1963, **16**, 933.
54. A. I. Pekarev, *Izv. Vÿssh ucheb. Zaved.*, 1963, **6**, 111.
55. S. Y. Shiraishi and R. G. Ward, *Can. Met. Q.*, 1964, **3**, 117.
56. V. N. Eremenko and V. I. Nishenko, *Ukr. Khim. Zhur.*, 1964, **30**, 125.
57. A. D. Kirshenbaum, J. A. Cahill and A. V. Grosse, *J. inorg. nucl. Chem.*, 1961, **22**, 33.
58. S. W. Strauss, L. E. Richards and B. F. Brown, *Nucl. Sci. Engng.*, 1960, **7**, 442.
59. L. J. Wittenberg, D. Ofte and W. G. Rohr, 'Rare Earth Research', 2, p. 257, K. S. Vorres, Gordon and Breach, NY, 1964.
60. Yu. V. Eremenko and V. N. Naidich, *Izv. Akad. Nauk. SSSR*, 1959, **6**, 129.
61. C. E. Olsen, T. A. Sandenaur and C. C. Herrick, Los Alamos Lab. Rep. LA–2358, 1959.
62. L. V. Jones, D. Ofte, W. G. Rohr and L. J. Wittenberg, *Trans. ASM*, 1962, **55**, 819.
63. V. N. Eremenko and Yu. V. Naidich, *Izv. Akad. Nauk. SSSR*, 1961, (6), 100.
64. Y. Ono and S. Matsushima, *Sci. Rep. Res. Insts. Tôhoku Univ.*, 1957, **9**, 309.
65. A. D. Kirshenbaum and J. A. Cahill, *Trans. ASM*, 1962, **55**, 849.
66. L. D. Lucas and G. Urbain, *C. r. hebd. Séanc. Acad. Sci., Paris*, 1962, **254**, 1622.
67. S. Shirai *et al.*, *J. chem. Soc. Japan*, 1963, **84**, 968.
68. S. Dobinsky and J. Weselowsky, *Bull. int. Acad. Pol. Sci. Lett.*, 1936, A. 446.
69. A. N. Campbell and S. Epstein, *J. Am. chem. Soc.*, 1942, **64**, 2679.
70. V. P. Elyutin, V. I. Kostikov and V. Ya. Levin, *Izv. Vÿssh. uchet. Zaved.*, 1970, **13**, 53.
71. A. D. Kirshenbaum and J. A. Cahill, *Trans. ASM*, 1962, **55**, 844.
72. J. F. Elliott and M. Gleiser, 'Thermochemistry for Steelmaking', Addison-Wesley, London, 1960.
73. L. D. Lucas, *Mém. Scient. Revue Métall.*, 1964, **61**, 1.
74. C. S. Smith and D. P. Spitzer, *J. phys. Chem.*, 1962, **66**, 946.
75. M. Maurach, *Trans. Indian Inst. Met.*, 1961, **14**, 209.
76. V. N. Eremenko, Yu. V. Naidich and G. P. Khilya, 'Poverkh. Yavleniya Rasplavakh', 1968, p. 165.
77. A. Schneider and G. Heymer, *Z. anorg. allg. Chem.*, 1956, **286**, 118.
78. F. A. Kanda and D. V. Keller, US Energy Comm. Repts. TID 15863; 1961; TID 20849; 1964.
79. A. V. Grosse, J. A. Cahill and A. D. Kirshenbaum, *J. Am. chem. Soc.*, 1961, **83**, 4665.
80. A. Calverley, *Proc. phys. Soc.*, 1957, **70B**, 1040.
81. E. Ubelacker and L. D. Lucas, *C. r. hebd. Séanc. Acad. Sci., Paris*, 1962, **254**, 1622.

82. Y. Matuyana, *Sci. Rep. Tôhoku Univ.*, 1929, **18**, 737.
83. C. M. Saeger and E. J. Ash, *Bur. Stand. J. Res.*, 1932, **8**, 37.
84. A. A. Kiriyanenko and A. N. Solov'en, *Teplofiz. Vys. Temp.*, 1970, **8**, 537.
85. G. R. Pulliam and E. S. Fitzsimmons, ISC–659, 1955.
86. B. C. Allen, 'Liquid Metals', S. Z. Beer, Dekker; 1972.
87. B. C. Allen and W. D. Kingery, *Trans AIME*, 1959, **215**, 30.
88. G. Metzger, *Z. phys. Chem.*, 1959, **211**, 1.
89. C. C. Addison, J. M. Coldrey and R. J. Pulham, *J. chem. Soc.*, 1963, 1227.
90. L. J. Wittenberg and D. Ofte, 'Techniques of Metals Research', Vol. **4**, 193, Interscience, 1970.
91. R. T. Beyer and E. M. Ring, 'Liquid Metals', S. Z. Beer, Dekker, p450; 1972.
92. L. V. Jones, D. Ofte, W. G. Rohr and L. J. Wittenberg, *Trans. ASM*, 1962, **55**, 819.
93. V. P. Elyutin, M. A. Maurakh and I. A. Penkov, *Izv. Vuz. Chem. Met.*, 1965, 128.
94. V. M. Glazov, S. N. Chizhevskaya and N. N. Glagoleva, 'Liquid Semiconductors', Plenum Press, New York, 1969.
95. R. C. Weast, 'Handbook of Chemistry and Physics', 52nd edn., Chemical Rubber Co., 1971.
96. J. A. Cahill and A. D. Kirshenbaum, *J. inorg. nucl. Chem.*, 1965, **27**, 73.

Table 14.3b THE PHYSICAL PROPERTIES OF LIQUID METALS
Specific heat, thermal conductivity, and electrical resistivity

Element	*Temperature* °C	*Specific heat* $J g^{-1} K^{-1}$	*Thermal conductivity* $W m^{-1} K^{-1}$	*Electrical resistivity* μΩ m	*Reference*
Ag	960.7	0.283	174.8	0.172 5	1
	1 000	0.283	176.5	0.176 0	
	1 100	0.283	180.8	0.184 5	
	1 200	0.283	185.1	0.193 5	
	1 300	0.283	189.3	0.202 3	
	1 400	—	193.5	0.211 1	
Al	660	1.08	94.03	0.242 5	1
	700	1.08	95.37	0.248 3	
	800	1.08	98.71	0.263 0	
	900	1.08	102.05	0.277 7	
	1 000	—	105.35	0.292 4	
As	817	—	—	(2.10)	1
Au	1 063	(0.149)	104.44	0.312 5	1, 21
	1 100	(0.149)	105.44	0.318 0	
	1 200	(0.149)	108.15	0.331 5	
	1 300	(0.149)	110.84	0.348 1	
	1 400	(0.149)	113.53	0.363 1	
B	2 077	2.91	—	(2.10)	1, 21
Ba	727	0.228	—	1.33	2, 21
Be	1 283	3.48	—	(0.45)	1, 21
Bi	271	0.146	17.1	1.290	3, 4, 8, 21
	300	0.143	15.5		
	400	0.147 5	15.5		
	500	0.137 5	15.5		
	600	0.133 6	15.5		
	700	—	15.5		
Ca	865	(0.775)	—	(0.250)	1, 21
Cd	321	0.264	42	0.337	2, 5, 6
	400	0.264	47	0.343 0	
	500	0.264	54	0.351 0	
	600	0.264	(61)	0.360 7	
Ce	804	0.25	—	1.268	15, 21
	1 000	0.25	—	1.294	
	1 200	0.25	—	1.310	
Co	1 493	(0.59)	—	1.02	2, 3, 21

Table 14.3b THE PHYSICAL PROPERTIES OF LIQUID METALS—*continued*

Element	*Temperature* °C	*Specific heat* $J g^{-1} k^{-1}$	*Thermal conductivity* $W m^{-1} K^{-1}$	*Electrical resistivity* μΩ m	*Reference*
Cr	(1 903)	(0.78)	—	(0.316)	1, 21
Cs	28.6	0.28	19.7	0.370	1, 7
	100	0.265	20.2	0.450	
	200	0.240	20.8	0.565	
	400	0.21	20.2	0.810	
	600	0.22	18.3	1.125	
	800	0.25	16.1	1.570	
	1 600	—	4.0	—	
Cu	1 083	0.495	165.6	0.200	1, 21
	1 100	0.495	166.1	0.202	
	1 200	0.495	170.1	0.212	
	1 400	0.495	176.3	0.233	
	1 600	0.495	180.4	0.253	
Fe	1 536	(0.795)	—	1.386	9, 10, 21
Fr	18	(0.142)	—	(0.87)	1, 11
	700	(0.134)	—		
Ga	29.8	0.398	(25.5)	0.26	4, 8, 9
	100	0.398	30.0	0.27	12, 13,
	200	0.398	35.0	0.28	21
	300	0.398	(39.2)	0.30	
Gd	(1 350)	(0.213)	—	(0.278)	1, 21
Ge	934	(0.404)	—	0.672	8, 9, 14,
	1 000	(0.404)	—	0.727	21
Hf	(2 227)	—	—	(2.18)	1
Hg	−38.87	(0.142)	6.78	0.905	1, 8, 21
	0	(0.142)	7.61	0.940	
	20	0.139	8.03	0.957	
	100	0.137	9.47	1.033	
	500	(0.137)	12.67	1.600	
	1 000	—	8.86	3.77	
	1 460 (critical temp.)		~0.000 4	~1 000	
Ho	1 500	(0.203)	—	(1.93)	1, 21
In	156.6	(0.259)	(42)	0.323 0	16, 21
	200	(0.259)		0.333 9	
	400	(0.259)		0.436 1	
	600	(0.259)		(0.513 1)	
K	63.5	0.820	53.0	0.136 5	1, 7
	100	0.810	51.7	0.154	
	200	0.790	47.7	0.215	
	500	0.761	37.8	0.444	
	1 000	(0.838)	24.4	(0.110)	
	1 500	—	15.5		
La	930	(0.057 5)	(21.0)	1.38	17, 21
	1 000	(0.057 5)	—	1.43	
	1 100	(0.057 5)	—	1.50	
	1 200	(0.057 5)	—	1.56	
Li	180.5	4.370	46.4	0.240	1, 7, 9
	200	4.357	47.2	—	
	400	4.215	53.8	—	
	600	4.165	57.5	—	
	800	4.148	58.6	—	
	1 000	4.147	58.4	—	
	1 600	(4.36)	52.0	—	

Table 14.3b THE PHYSICAL PROPERTIES OF LIQUID METALS—*continued*

Element	*Temperature* °C	*Specific heat* $J g^{-1} K^{-1}$	*Thermal conductivity* $W m^{-1} K^{-1}$	*Electrical resistivity* μΩ m	*Reference*
Mg	650	(1.36)	78	0.274	4, 5, 21
	700	(1.36)	81	0.277	
	800	(1.36)	88	0.282	
	1 000	(1.36)	100	—	
Mn	124	0.838	—	0.40	1, 9, 21
Mo	2 607	0.57	—	(0.605)	1, 21
Na	97	1.386	89.7	0.096 4	1, 7
	100	1.385	89.6	0.099	
	200	1.340	82.5	0.134	
	400	1.278	71.6	0.224	
	600	1.255	62.4	0.326	
	800	1.270	53.7	0.469	
	1 000	(1.316)	45.8	—	
	1 200	(1.405)	38.8	—	
Nb	2 468	—	—	(1.05)	1
Nd	1 024	0.232	—	(1.26)	1, 21
Ni	1 454	0.620	—	0.850	9, 21
P (black)	44	—	—	(2.70)	1
Pb	327	0.152	15.4	0.948 5	5, 16
	400	0.144	16.6	0.986 3	
	500	0.137	18.2	1.034 4	
	600	0.135	19.9	(1.082 5)	
	800	—	—	(1.169)	
	1 000	—	—	(1.263)	
Po	254	—	—	(3.98)	1
Pr	935	(0.238)	—	(1.38)	1, 21
Pt	1 770	(0.178)	—	(0.73)	1, 21
Pu	640	—	—	(1.33)	1
Ra	960	(0.136)	—	(1.71)	1, 21
Rb	38.8	0.398	33.4	0.228 3	1, 7
	100	0.383	33.4	0.273 0	
	200	0.364	31.6	0.366 5	
	500	0.348	26.1	0.689 0	
	1 000	(0.378)	17.0	(1.71)	
	1 500	—	8.0	(5.32)	
Re	3 158	—	—	(1.45)	1
Ru	2 427	—	—	(0.84)	1
S	119	0.984	—	$>10^{10}$	2, 21
Sb	630.5	(0.258)	21.8	1.135	4, 9, 21
	700	(0.258)	21.3	1.154	
	800	(0.258)	20.9	1.181	
	1 000	(0.258)	—	1.235	
Sc	1 539	(0.745)	—	(1.31)	1, 21
Se	217	0.445	0.3	$\sim 10^6$	2, 18, 21
Si	1 410	1.04	—	0.75	1, 2, 14, 21
	1 500	1.04	—	0.82	
	1 600	1.04	—	0.86	
Sm	1 072	(0.223)	—	(1.90)	1, 21
Sn	232	0.250	30.0	0.472 0	5, 16, 21
	300	0.242	31.4	0.490 6	
	400	0.241	33.4	0.517 1	
	500	(0.24)	35.4	0.543 5	
	1 000	(0.26)	—	(0.670)	

Table 14.3b THE PHYSICAL PROPERTIES OF LIQUID METALS—*continued*

Element	*Temperature* °C	*Specific heat* $J g^{-1} K^{-1}$	*Thermal conductivity* $W m^{-1} K^{-1}$	*Electrical resistivity* μΩ m	*Reference*
Sr	770	0.354	—	(0.58)	1, 21
Ta	(2 996)	—	—	(1.18)	1
Tb	1 365	—	—	(2.44)	1
Te	450	(0.295)	2.5	5.50	2, 9, 14
	500	(0.295)	3.0	4.80	19, 21
	600	(0.295)	4.1	4.30	
	800	(0.295)	(6.2)	(3.9)	
	1 000	(0.295)	—	(3.8)	
Ti	1 685	(0.700)	—	(1.72)	1, 21
Tl	303	0.149	(24.6)	0.731	9, 21
	400	0.149		0.759	
	500	0.149	—	0.788	
Tm	1 600	—	—	(1.88)	1
U	1 133	(0.161)	—	0.636	20, 21
	1 200	(0.161)	—	0.653	
	1 300	(0.161)	—	(0.678)	
V	1 912	(0.780)	—	(0.71)	1, 21
W	3 377	—	—	(1.27)	1
Y	1 530	(0.377)	—	(1.04)	1, 21
Yb	824	—	—	(1.64)	1
Zn	419.5	0.481	49.5	0.374	5, 9
	500	0.481	54.1	0.368	
	600	0.481	59.9	0.363	
	800	0.481	(60.7)	0.367	
Zr	1 850	(0.367)	—	(1.53)	1, 21

REFERENCES TO TABLE 14.3b

1. A. V. Grosse, *Revue Hautes Temp. & Refrac.*, 1966, **3**, 115.
2. R. S. Allgaier, *Phys. Rev.*, 1969, **185**, 227.
3. G. V. Samsonov, A. D. Panasyuk and E. M. Dudnik, *Izv. Vuz. Tsvet. Met.*, 1969, **12**, 110.
4. D. S. Viswanath and B. C. Mathur, *Met. Trans.*, 1972, **3**, 1769.
5. C. Y. Ho, R. W. Powell and P. E. Liley, NSRDS-NBS 16th February 1968.
6. G. R. B. Elliott, C. C. Herrick, *et al.*, *High Temp. Sci.*, 1969, **1**, 58.
7. J. Freund, 'Thermophysical and Nuclear Parameters of Molten Li, Na, K, Rb and Cs', Inst. Kerntech. Tech. Univ. Berlin, 1969, (13), 184.
8. G. Bush and J. Tieche, *Phys. condens. Matter*, 1963, **1**, 78.
9. J. Wilson, *Met. Rev.*, 1965, **10**, 381.
10. I. A. Pavars, B. A. Baum and P. V. Gel'd, *Zh. fiz. Khim.*, 1969, **43**, 2744.
11. Yu. P. Os'minin, *Zh. fiz. Khim.*, 1969, **43**, 2610.
12. H. J. Guentherodt and H. U. Kuenzi, *Phys. condens. Matter*, 1969, **10**, 285.
13. M. J. Duggin, *Phys. Lett.*, A, 1969, **29**, 470.
14. V. M. Glazov, S. N. Chizhevskaya and N. N. Glagoleva, 'Liquid Semiconductors', Plenum Press, New York, 1969.
15. G. Busch *et al.*, *Phys. Lett.*, A, 1970, **31**, 191.
16. H. A. Davies and J. S. Leach, *Phys. Chem, Liquids*, 1970, **2**, 1.
17. G. M. Kreig, R. B. Genter, and A. V. Grosse, *Inorg. nucl. Chem. Lett.*, 1969, **5**, 819.
18. B. M. Mogilevsky and A. Ph. Chudnovsky, *Proc. Internat. Conf. Phys. Semiconductors* (Moscow), 1968, **2**, 1241.
19. J. C. Perron, *Phys. Lett.*, A, 1970, **32**, 169.
20. G. Busch, H. J. Guentherodt and H. V. Kuenzi, *Phys. Lett.*, A, 1970, **32**, 376.
21. D. R. Stull, and G. C. Sinke, 'Thermodynamic Properties of the Elements', *Amer. chem. Soc.*, 1956; K. K. Kelley and E. G. King, 'Contributions to Data on Theoretical Metallurgy', *XIV*, *U.S. Bur. Mines Bull.* 592; R. Hultgren et al., 'Selected Values of Thermodynamic Properties,' Wiley, 1963.

14.3 The physical properties of aluminium and aluminium alloys

Table 14.4a THE PHYSICAL PROPERTIES OF ALUMINIUM AND ALUMINIUM ALLOYS AT NORMAL TEMPERATURES

Sand Cast

Material	*Nominal composition* %		*Density* $g\,cm^{-3}$	*Coefficient of expansion* 20–100°C $10^{-6}K^{-1}$	*Thermal conductivity* 100°C $W\,m^{-1}K^{-1}$	*Resistivity* $\mu\Omega\,m$	*Modulus of elasticity* MPa × 10^3
Al	Al	99.5	2.70	24.0	218	3.0	69
	Al	99.0	2.70	24.0	209	3.1	—
Al–Cu	Cu	4.5	2.75	22.5	180	3.6	71
	Cu	8	2.83	22.5	138	4.7	—
	Cu	12	2.93	22.5	130	4.9	—
Al–Mg	Mg	3.75	2.66	22.0	134	5.1	—
	Mg	5	2.65	23.0	130	5.6	—
	Mg	10	2.57	25.0	88	8.6	71
Al–Si	Si	5	2.67	21.0	159	4.1	71
	Si	11.5	2.65	20.0	142	4.6	—
Al–Si–Cu	Si	10	2.74	20.0	100	6.6	71
	Cu	1.5					
	Si	4.5	2.76	21.0	134	4.9	71
	Cu	3					
Al–Si–Cu–Mg*	Si	17	2.73	18.0	134	8.6	88
	Cu	4.5					
	Mg	0.5					
Al–Cu–Mg–Ni (Yalloy)	Cu	4	2.78	22.5	126	5.2	71
	Mg	1.5					
	Ni	2					
Al–Cu–Fe–Mg	Cu	10	2.88	22.0	138	4.7	71
	Fe	1.25					
	Mg	0.25					
Al–Si–Cu–Mg–Ni (Lo-Ex)	Si	12	2.71	19.0	121	5.3	71
	Cu	1					
	Mg	1					
	Ni	2					
	Si	23	2.65	16.5	107	—	88
	Cu	1					
	Mg	1					
	Ni	1					

* Die cast.

Table 14.4b THE PHYSICAL PROPERTIES OF ALUMINIUM AND ALUMINIUM ALLOYS AT NORMAL TEMPERATURES

Wrought

Material	*Nominal composition* %		*Condition**		*Density* $g\,cm^{-3}$	*Coefficient of expansion* 20–100°C $10^{-6}K^{-1}$	*Thermal conductivity* 100°C $W\,m^{-1}K^{-1}$	*Resistivity* $\mu\Omega\,cm$	*Temp. coeff. of resistance* 20–100°C	*Modulus of elasticity* MPa × 10^3
Al	Al	99.992	Sheet	O			239	2.68	0.0042	69
				H8	2.70	23.5	234	2.70	0.0042	69
			Extruded				239	2.68	0.0042	69
	Al	99.8	Sheet	O			234	2.74	0.0042	69
				H8	2.70	23.5	230	2.76	0.0042	69
			Extruded				230	2.79	0.0041	69
	Al	99.5	Sheet	O			230	2.80	0.0041	69
				H8	2.71	23.5	230	2.82	0.0041	69
			Extruded				226	2.85	0.0041	69

Table 14.4b THE PHYSICAL PROPERTIES OF ALUMINIUM AND ALUMINIUM ALLOYS AT NORMAL TEMPERATURES—*continued*

Wrought

Material	*Nominal composition %*	*Condition**		*Density* g cm^{-3}	*Coefficient of expansion* 20–100°C 10^{-6} K^{-1}	*Thermal conductivity* 100°C W m^{-1} K^{-1}	*Resistivity* μΩ cm	*Temp. coeff. of resistance* 20–100°C	*Modulus of elasticity* MPa × 10^3
	Al 99	Sheet	O			226	2.87	0.0040	69
			H8	2.71	23.5	226	2.89	0.0040	69
		Extruded				226	2.86	0.0040	69
Al–Mn	Mn 1.25	Sheet	O						
			H2						
			H4	2.74	23.0	180	3.9	0.003 0	69
			H6						
			H8						
		Extruded				151	4.8	0.002 4	—
Al–Mg	Mg 2.25	Sheet	O			159	4.5	0.002 6	70
			H8	2.69	23.5	155	4.7	0.0025	—
		Extruded				147	4.9	0.002 3	—
	Mg 3.5	Sheet	O			142	5.3	0.002 1	70
			H4	2.67	23.5	138	5.4	0.002 1	—
		Extruded				134	5.7	0.001 9	—
	Mg 4.5	Sheet	O	2.66	23.5	130	6.0	0.001 9	—
	Mn 0.7		H4			126	6.1	0.001 9	—
Al–Li	Li 2.0	Sheet	TF	2.56	—	—	—	—	77
Al–Mg–Li	Mg 3.0	Sheet	TF	2.52	—	—	—	—	79
	Li 2.0								
Al–Li–Mg	Li 3.0	Sheet	TF	2.46	—	—	—	—	84
	Mg 2.0								
Al–Mg–Si	Mg 0.5	Extruded	TB	2.70	23.0	193	3.5	0.003 3	71
	Si 0.5		TF			201	3.3	0.003 5	—
	Mg 1.0	Sheet	TB			188	3.6	0.003 3	69
	Si 1.0		TF	2.69	23.0	193	3.4	0.003 5	—
		Extruded	TB			180	3.8	0.003 1	—
			TF			180	3.8	0.003 1	—
Al–Cu–Mg–Si	Cu 4.0	Sheet	TF	2.80	22.5	147	5.0	0.002 3	73
(Duralumin)	Mg 0.6								
	Si 0.4								
	Mn 0.6								
	Cu 4.5	Sheet	TB	2.81	22.5	147	5.2	0.002 2	73
	Mg 0.5		TF			159	4.5	0.002 6	—
	Si 0.75								
	Mn 0.75								
Al–Cu–Mg–Ni	Cu 4.0	Forgings	TF	2.78	22.5	151	4.9	0.002 3	72
(Yalloy)	Mg 1.5								
	Ni 2.0								
Al–Si–Cu–Mg	Si 12.0	Forgings	TF	2.66	19.5	151	4.9	0.002 3	79
(Lo–Ex)	Cu 1.0								
	Mg 1.0								
	Ni 1.0								
Al–Zn–Mg	Zn 10.0	Forgings		2.91	23.5	151	4.9	0.002 3	—
	Cu 1.0								
	Mn 0.7								
	Mg 0.4								
Al–Zn–Mg–Cu	Zn 5.7	Extrusion	TF	2.80	23.5	130	5.7	0.002 0	72
	Mg 2.6								
	Cu 1.6								
	Cr 0.25								

*O = Annealed.
H2 = Quarter hard.
H4 = Half hard.
H6 = Three-quarters hard.
H8 = Hard.
TB = Solution treated and naturally aged.
TF = Solution treated and artificially aged.

14.4 The physical properties of copper and copper alloys

Table 14.5 THE PHYSICAL PROPERTIES OF COPPER AND COPPER ALLOYS AT NORMAL TEMPERATURES

Material	*Composition* %	*Density* g cm^{-3}	*Melting point or liquidus* °C	*Coefficient of expansion* 25–300°C 10^{-6} K^{-1}	*Electrical conductivity* 20°C %IACS*	*Thermal conductivity* W m^{-1} K^{-1}	*Refs.*
OFHC copper	Cu 99.99+	8.94	1 083	17.7	101.5	399	1
Tough pitch HC copper	O_2 0.03	8.92	1 083	17.7	101.5	397	2, 3, 4
Deoxidized non-arsenical copper	P 0.005–0.012	8.94	1 083	17.7	85–96	341–395	5
	P 0.013–0.050	8.94	1 083	17.7	70–90	298–372	5
Deoxidized arsenical copper	P 0.03 As 0.35	8.94	1 082	17.4	45	177	2
Silver bearing copper	O_2 0.02 Ag 0.05	8.92	1 079	17.7	101	397	2, 3
Tellurium copper	Cu 99.5 Te 0.5	8.94	1 082	17.7	98	382	2
Chromium copper	Cu 99.4 Cr 0.6	8.89	1 081	17	45(1) 82(2)	167 188	6
Beryllium copper	Be 1.85 Co 0.25	8.25	1 000	17	17(1) 23(2)	84 105	7
	Be 0.5 Co 2.5	8.75	1 060	17	23(1) 47(2)	126 210	7
Cadmium copper	Cu 99.2 Cd 0.8	8.94	1 080	17	85	376	6
Sulphur copper	Cu 99.65 S 0.35	8.92	1 075	17	95	373	5
Cap copper	Cu 95 Zn 5	8.85	1 065	18.1	56	234	5
Gilding metals	Cu 90 Zn 10	8.80	1 040	18.2	44	188	5
	Cu 85 Zn 15	8.75	1 020	18.7	37	159	5
	Cu 80 Zn 20	8.65	1 000	19.1	32	138	5
Brass	Cu 70 Zn 30	8.55	965	19.9	28	121	5
	Cu 67 Zn 33	8.50	940	20.2	27	121	5
	Cu 63 Zn 37	8.45	920	20.5	26	125	5
	Cu 60 Zn 40	8.40	900	20.8	28	126	5
Aluminium brass	Cu 76 Zn 22 Al 2	8.35	1 010	18.5	23	101	5
Naval brass	Cu 62 Zn 37 Sn 1	8.40	915	21.2	26	117	5
Free cutting brass	Cu 58 Zn 39 Pb 3	8.50	900	20.9	26	109	3, 5
Hot stamping brass	Cu 58 Zn 40 Pb 2	8.45	910	20.9	26	109	3, 5

Table 14.5 THE PHYSICAL PROPERTIES OF COPPER AND COPPER ALLOYS AT NORMAL TEMPERATURES—*continued*

Material	*Composition* %	*Density* $g\,cm^{-3}$	*Melting point or liquidus* °C	*Coefficient of expansion* 25–300 °C $10^{-6}\,K^{-1}$	*Electrical conductivity* 20 °C %IACS*	*Thermal conductivity* $W\,m^{-1}\,K^{-1}$	*Refs.*
High tensile brass	Cu 54–62 Others 7 max. Zinc—balance	8.3–8.4	990 approx.	21 approx.	20–25	88–109	
Nickel silver 10%	Cu 62 Ni 10 Zn 28	8.60	1 010	16.4	8.31	37	8, 9
12%	Cu 62 Ni 12 Zn 26	8.64	1 025	16.2	7.71	30	8, 9
15%	Cu 62 Ni 15 Zn 23	8.69	1 060	16.2	7.01	27	8, 9
18%	Cu 62 Ni 18 Zn 20	8.72	1 100	16.0	6.3	28	8, 9
25%	Cu 62 Ni 25 Zn 13	8.82	1 160	17.0	5.1	21	8, 9
Phosphor bronze	Sn 3.5 P 0.12	8.85	1 070	18.8	18.8	85	5, 10
	Sn 5 P 0.09	8.85	1 060	18.0	16.8	75	5, 10
	Sn 7 P 0.12	8.80	1 050	18.5	14.0	67	5, 10
	Sn 8 P 0.05	8.80	1 040	18.0	14.0	63	10
Cupro-nickel	Ni 5.5 Fe 1.2 Mn 0.5	8.94	1 121	17.5	12.5	67	11
	Ni 10.5 Fe 1.5 Mn 0.75	8.94	1 150	17.1	8.0	42	11
	Ni 31.0 Fe 1.0 Mn 1.0	8.90	1 238	16.6	4.5	21	11
Silicon bronze	Si 3 Mn 1	8.52	1 028	18.0	8.1	50	2
Aluminium bronze	Cu 95 Al 5	8.15	1 065	18.0	17.7	85	2
	Cu 92 Al 8	7.75	1 041	17.8	14.8	80	2
	Cu 85.5, Al 9.5 Fe 3.0, Mn 1.0 Ni 1.0	7.57	1 040	17.0	13	62	2

* The International Annealed Copper Standard is material of which the resistance of a wire 1 metre in length and weighing 1 gram is 0.15328 ohm at 20 °C. 100% IACS at 20 °C = 58.00 $MS\,m^{-1}$.
(1) Solution heat treated.
(2) Fully heat treated (to maximum hardness).

REFERENCES TO TABLE 14.5

1. OFHC Copper—Technical Information, American Metal Climax Inc., 1969.
2. R. A. Wilkins and E. S. Bunn, 'Copper and Copper Base Alloys', New York, 1943.
3. C. S. Smith, *Trans AIMME*, 1930, **89**, 84.
4. C. S. Smith, *Trans. AIMME*, 1931, **93**, 176.
5. Copper Development Association, Copper and Copper Alloy Data Sheets, 1968.
6. Copper Development Association, High Conductivity Copper Alloys, 1968.
7. Copper Development Association, Beryllium Copper, 1962.
8. M. Cook, *J. Inst. Metals*, 1936, **58**, 151.
9. International Nickel Limited, Nickel Silver Engineering Properties, 1970.
10. M. Cook and W. G. Tallis, *J. Inst. Metals*, 1941, **67**, 49.
11. International Nickel Limited, Cupronickel Engineering Properties, 1970.

14.5 The physical properties of magnesium and magnesium alloys

Table 14.6 THE PHYSICAL PROPERTIES OF SOME MAGNESIUM AND MAGNESIUM ALLOYS AT NORMAL TEMPERATURE

Material		*Nominal composition†* %	*Condition*	*Density at 20 °C* $g\,cm^{-3}$	*Melting point* °C *Sol.*	*Liq.*	*Coeff. of thermal expansion 20–200 °C* $10^{-6}\,K^{-1}$	*Thermal conductivity* $W\,m^{-1}\,K^{-1}$	*Electrical resisitivity* μΩ cm	*Specific heat 20–200 °C* $J\,kg^{-1}\,K^{-1}$	*Weldability by argon arc process‡*	*Relative damping capacity§*
Pure Mag		Mg 99.97	X	1.74	650		27.0	167	3.9	1 050	A	
Mg–Mn	(MN70)	Mn 0.75 approx.	X	1.75	650	651	26.9	146	5	1 050	A	
	(AM503)	Mn 1.5	X	1.76	650	651	26.9	142	5.0	1.050	A	C
Mg–Al	AL80	Al 0.75 approx. Be 0.005	X	1.75	630	640	26.5	117	6	1 050	A	
Mg–Al–Zn	(AZ31)	Al 3 Zn 1	X	1.78	575	630	26.0	(84)	10.0	1 050	A	
	(A8)	Al 8 Zn 0.5	AC AC TB	1.81 1.81	475*	600	27.2 27.2	84 84	13.4 —	1 000 1 000	A	C
	(AZ91)	Al 9.5 Zn 0.5	AC AC TB AC TF	1.83 1.83 1.83	470*	595	27.0 27.0 27.0	84 84 84	14.1 — —	1 000 1 000 1 000	A	C
	(AZM)	Al 6 Zn 1	X	1.80	510	610	27.3	79	14.3	1 000	A	
	(AZ855)	Al 8 Zn 0.5	X	1.80	475*	600	27.2	79	14.3	1 000	A	
Mg–Zn–Mn	(ZM21)	Zn 2 Mn 1	X	1.78			27.0			—	A	
Mg–Zn–Zr	(ZW1)	Zn 1.3 Zr 0.6	X	1.80	625	645	27.0	134	5.3	1 000	A	A
	(ZW3)	Zn 3 Zr 0.6	X	1.80	600	635	27.0	125	5.5	960	C	
	(Z5Z)	Zn 4.5 Zr 0.7	AC TF	1.81	560	640	27.3	113	6.6	960	C	
	(ZW6)	Zn 5.5 Zr 0.6	XTE	1.83	530	630	26.0	117	6.0	1 050	C	

Mg–RE–Zn–Zr	(ZRE1)	RE Zn Zr	2.7 2.2 0.7	AC TE	1.80	545	640	26.8	100	7.3	1 050	A	B
	(RZ5)	Zn RE Zr	4.0 1.2 0.7	AC TE	1.84	510	640	27.1	113	6.8	960	B	
	(ZE63)	Zn RE Zr	6 2.5 0.7	AC TF	1.87	515	630	27.0	109	5.6	960	A	
Mg–Th–Zn–Zr	(ZTY)	Th Zn Zr	0.8 0.5 0.6	X	1.76	600	645	26.4	121	6.3	960	A	
	(ZT1)	Th Zn Zr	3.0 2.2 0.7	AC TE	1.83	550	647	26.7	105	7.2	960	A	(B)
	(TZ6)	Zn Th Zr	5.5 1.8 0.7	AC TE	1.87	500	630	27.6	113	6.6	960	B	
Mg–Ag–RE–Zr	(QE22)	Ag RE Zr	2.5 2.0¶ 0.6	AC TF	1.82	550	640	26.7	113	6.85	1000	A	
Mg–Ag–RE–Th–Zr	(QH21)	Ag RE Th Zr	2.5 1.0¶ 1.0 0.7	AC TF	1.82	540	640	26.7	113	6.85	1 005	A	—
Mg–Zr	(ZA)	Zr	0.6	AC	1.75	650	651	27.0	(146)	(4.5)	1 050	A	A

AC Sand cast.
TB Solution heat treated.
TE Precipitation heat treated.
TF Fully heat treated.
† Mg–Al type alloys normally contain 0.2–0.4% Mn to improve corrosion resistance.

X Extruded, rolled or forged.
RE Cerium mischmetal containing approx. 50% Ce.
* Non-equilibrium solidus 420 °C.
() Estimated value.
¶ Mischmetal enriched in neodymium.

‡ Weldability rating:
A Fully weldable.
B Weldable.
C Not recommended where fusion welding is involved.

§ Damping capacity rating:
A Outstanding.
B Equivalent to cast iron.
C Inferior to cast iron but better than Al–base cast alloys.

14.6 The physical properties of nickel and nickel alloys

Table 14.7 THE PHYSICAL PROPERTIES OF WROUGHT NICKEL AND SOME HIGH NICKEL ALLOYS AT ROOM TEMPERATURE

*Alloy**	*Nominal composition* %		*Density* $g\,cm^{-3}$	*Coefficient of expansion* 20–100°C $10^{-6}\,K^{-1}$	*Specific heat* $J\,kg^{-1}\,K^{-1}$	*Thermal conductivity* $W\,m^{-1}\,K^{-1}$	*Electrical resistivity* $\mu\Omega\,cm$
Nickel	99.4 Ni		8.89	13.3	456	74.9	9.5
Monel† alloy 400	30 Cu 1.5 Fe 1.0 Mn		8.83	13.9	423	21.7	51.0
Monel alloy K-500	29 Cu 2.8 Al 0.5 Ti		8.46	13.7	419	17.4	61.4
Cupro-nickel	55 Cu		8.88	14.9	421	19.5	52.0
Inconel† alloy 600	16 Cr 6 Fe		8.42	13.3	460	14.8	103
Inconel alloy 625	22 Cr 4 Nb 9 Mo	0.3 Ti 0.3 Al	8.44	12.8	410	9.8	129
Inconel alloy 718	19 Cr 5 Nb 3 Mo	0.9 Ti 0.5 Al	8.19	13.0	427	11.2	125
Inconel alloy X-750	15 Cr 7 Fe 2.5 Ti	0.6 Al 0.8 Nb	8.25	12.6	425	12.0	122
Incoloy† alloy 800 Incoloy alloy 800 H‡	45 Fe 21 Cr 0.4 Ti	0.4 Al	7.95	14.2	460	11.5	93
Incoloy alloy 825	32 Fe 21 Cr 3 Mo	2 Cu 1.0 Ti	8.14	14.0	441	11.1	113
Incoloy alloy DS	40 Fe 18 Cr 2 Si		7.91	14.2	450	12.0	108
Ni Span† alloy C-902	47 Fe 5.5 Cr 2.5 Ti	0.5 Al	8.10	7.6	502	12.1	101
Hastelloy B 2	28 Mo		9.22	10.3	373	11.1	137
Hastelloy C 4	16 Mo 16 Cr		8.64	10.8	406	10.1	125
Hastelloy alloy X	9 Mo 21 Cr 18 Fe		8.23	13.8	485	9.1	118
Nimonic† alloy 75	20 Cr 0.4 Ti		8.37	11.0	461	11.7	102
Nimonic alloy 80A	20 Cr 2.0 Ti 1.5 Al		8.19	12.7	460	11.2	117
Nimonic alloy 81	30 Cr 1.8 Ti 1.0 Al		8.06	11.1	461	10.9	127

Table 14.7 THE PHYSICAL PROPERTIES OF WROUGHT NICKEL AND SOME HIGH NICKEL ALLOYS AT ROOM TEMPERATURE—*continued*

*Alloy**	*Nominal composition* %		*Density* $g\,cm^{-3}$	*Coefficient of expansion* 20–100°C $10^{-6}\,K^{-1}$	*Specific heat* $J\,kg^{-1}\,K^{-1}$	*Thermal conductivity* $W\,m^{-1}\,K^{-1}$	*Electrical resistivity* $\mu\Omega\,cm$
Nimonic alloy 90	20 Cr 17 Co 2.4 Ti	1.4 Al	8.18	12.7	445	11.5	114
Nimonic alloy 105	15 Cr 20 Co 5 Mo	5 Al 1.2 Ti	8.01	12.2	419	10.9	131
Nimonic alloy 115	14 Cr 13 Co 3 Mo	5 Al 4 Ti	7.85	12.0	444	10.6	139
Nimonic alloy 263	20 Cr 20 Co 6 Mo	2 Ti 0.5 Al	8.36	11.1	461	11.7	115
Nimonic alloy 901	13 Cr 35 Fe 6 Mo	3 Ti	8.16	13.5	419	—	—
Nimonic alloy PE16	16 Cr 32 Fe 3 Mo	1.0 Ti 1.0 Al	8.02	11.3	544	11.7	110

* Where trade marks apply to the name of an alloy there may be materials of similar composition available from other producers who may or may not use the same suffix along with their own trade names. The suffix alone e.g. Alloy 800 is sometimes used as a descriptive term for the type of alloy but trade marks can be used only by the registered user of the mark.
† Registered Trade Mark.
‡ A variant on alloy 800 having controlled carbon and heat treatment to give significantly improved creep-rupture strength.

14.7 The physical properties of titanium and titanium alloys

Table 14.8 PHYSICAL PROPERTIES OF TITANIUM AND TITANIUM ALLOYS AT NORMAL TEMPERATURES

Material IMI designation	*Nominal composition* %		*Density* $g\,cm^{-3}$	*Coefficient of expansion* 20–100°C $10^{-6}\,K^{-1}$	*Thermal con-ductivity* 20–100°C $W\,m^{-1}\,K^{-1}$	*Resistivity* 20°C $\mu\Omega\,cm$	*Temp. coefficient of resistivity* 20–100°C $\mu\Omega\,cm\,K^{-1}$	*Specific heat* 50°C $J\,kg^{-1}\,K^{-1}$	*Magnetic suscept.* 10^{-6} cgs *units* g^{-1}
CP Titanium	Commercially pure		4.51	7.6	16	48.2	0.002 2	528	+3.4
IMI 230	Cu	2.5	4.56	9.0	13	70	0.002 6	—	—
IMI 260/261	Pd	0.2	4.52	7.6	16	48.2	0.002 2	528	—
IMI 315	Al Mn	2.0 2.0	4.51	6.7	8.4	101.5	0.000 3	460	+4.1
IMI 317	Al Sn	5.0 2.5	4.46	7.9	6.3	163	0.000 6	470	+3.2
IMI 318	Al V	6.0 4.0	4.42	8.0	5.8	168	0.000 4	610	+3.3

Table 14.8 PHYSICAL PROPERTIES OF TITANIUM AND TITANIUM ALLOYS AT NORMAL TEMPERATURES —*continued*

Material IMI designation	*Nominal composition %*	*Density* g cm^{-3}	*Coefficient of expansion* 20–100 °C 10^{-6} K^{-1}	*Thermal conductivity* 20–100 °C W m^{-1} K^{-1}	*Resistivity* 20 °C μΩ cm	*Temp. coefficient of resistivity* 20–100 °C μΩ cm K^{-1}	*Specific heat* 50 °C J kg^{-1} K^{-1}	*Magnetic suscept.* 10^{-6} *cgs units* g^{-1}
IMI 550	Al 4.0 Mo 4.0 Sn 2.0 Si 0.5	4.60	8.8	7.9	159	0.000 4	—	—
IMI 551	Al 4.0 Mo 4.0 Sn 4.0 Si 0.5	4.62	8.4	5.7	170	0.000 3	400	+3.1
IMI 679	Sn 11.0 Zr 5.0 Al 2.25 Mo 1.0 Si 0.2	4.84	8.0	7.1	163	0.000 4	—	—
IMI 680	Sn 11.0 Mo 4.0 Al 2.25 Si 0.2	4.86	8.9	7.5	165	0.000 3	—	—
IMI 685	Al 6.0 Zr 5.0 Mo 0.5 Si 0.25	4.45	9.8	4.8	167	0.000 4	—	—
IMI 829	Al 5.5 Sn 3.5 Zr 3.0 Nb 1.0 Mo 0.3 Si 0.3	4.53	9.45	7.8	—	—	530	—

14.8 The physical properties of zinc and zinc alloys

Table 14.9 PHYSICAL PROPERTIES OF ZINC AND ZINC ALLOYS

Material	*Nominal composition*	*Density* g cm^{-3}	*Coefficient of expansion* 10^{-6} K^{-1}	*Thermal conductivity* W m^{-1} K^{-1}	*Electrical conductivity* %IACS 20 °C	*Condition*	*Melting point (liquidus)* °C
Zn Polycrystalline	99.993% Zn	7.13 (25 °C)	39.7 (20–250 °C)	113	28.27	Cast	419.46
ZnAlMg BS1004A	4% Al 0.04% Mg	6.7	27 (20–100 °C)	113	27	Pressure die cast	387
ZnAlCuMg BS1004B	4% Al 1% Cu 0.04% Mg	6.7	27 (20–100 °C)	109	26	Pressure die cast	388
ZnAlCuMg ILZRO 12 (ZA12)	11% Al 1% Cu 0.02% Mg	6.0	28 (20–100 °C)	115	28.3	Chill cast	432
ZA27	27% Al 2.3% Cu 0.015% Mg	5.0	26 (20–100 °C)	123	29.7	Chill cast	487

14.9 The physical properties of zirconium alloys

Table 14.10 PHYSICAL PROPERTIES OF ZIRCONIUM ALLOY

Alloy	*Composition* %	*Density* $g\,cm^{-3}$	*Thermal cond. at 25°C* $W\,m^{-1}\,K^{-1}$	*Coefficient of expansion* 20–100°C 10^{-6}
Zirconium 10	Commercially pure	6.50	21.1	5.04
Zirconium 30	Cu 0.55 Mo 0.55	6.55	25.3	5.93
Zircalloy II	Sn 1.5 Fe 0.12 Cr 0.10 Ni 0.05	6.55	12.3	5.67

See also *Table 26.36* page **26**–52.

14.10 The physical properties of steels

Table 14.11 PHYSICAL PROPERTIES OF STEELS

Material and condition Composition %	*Temperature* °C	*Specific gravity* $g\,cm^{-3}$	*Thermal properties (see Notes)* *Specific heat* $J\,kg^{-1}\,K^{-1}$	*Coefficient of thermal expansion* $10^{-6}\,K^{-1}$	*Thermal conductivity* $W\,m^{-1}\,K^{-1}$	*Electrical resistivity* μΩ cm
		Carbon steels				
C 0.06, Mn 0.4	RT	7.87	—	—	65.3	12.0
	100		48.2	12.62	60.3	17.8
	200		520	13.08	54.9	25.2
Annealed	400		595	13.83	45.2	44.8
	600		754	14.65	36.4	72.5
	800		875	14.72	28.5	107.3
	1 000		—	13.79	27.6	116.0
C 0.08, Mn 0.31	RT	7.86	—	—	59.5	13.2
	100		482	12.19	57.8	19.0
	200		523	12.99	53.2	26.3
Annealed	400		595	13.91	45.6	45.8
	600		741	14.68	36.8	73.4
	800		960	14.79	28.5	108.1
	1 000		—	13.49	27.6	116.5
C 0.23, Mn 0.6 (En 3, 060A22)	RT	7.86	—	—	51.9	15.9
	100		486	12.18	51.1	21.9
	200		520	12.66	49.0	29.2
Annealed	400		599	13.47	42.7	48.7
	600		749	14.41	35.6	75.8
	800		950	12.64	26.0	109.4
	1 000		—	13.37	27.2	116.7

Notes:
1. Where *specific heats* are quoted at temperatures above RT the values have been determined over a range of 50°C up to the temperature quoted.
2. *Coefficients of expansion* are mean values from RT up to the temperature quoted.
3. *Electrical resistivity* values are uncorrected for dimensional changes of the specimen with temperature. Original dimensions as at RT.

Table 14.11 PHYSICAL PROPERTIES OF STEELS—*continued*

Material and condition Composition %	*Temperature* °C	*Specific gravity* g cm^{-3}	*Thermal properties (see Notes)* *Specific heat* J kg^{-1} K^{-1}	*Coefficient of thermal expansion* 10^{-6} K^{-1}	*Thermal conductivity* W m^{-1} K^{-1}	*Electrical resistivity* μΩ cm
C 0.42, Mn 0.64 } En 8, 060A42	RT	7.85	—	—	51.9	16.0
	100		486	11.21	50.7	22.1
	200		515	12.14	48.2	29.6
Annealed	400		586	13.58	41.9	49.3
	600		708	14.58	33.9	76.6
	800		624	11.84	24.7	111.1
	1 000		—	13.59	26.8	122.6
C 0.80, Mn 0.32	RT	7.85	—	—	47.8	17.0
	100		490	11.11	48.2	23.2
	200		532	11.72	45.2	30.8
Annealed	400		607	13.15	38.1	50.5
	600		712	14.16	32.7	77.2
	800		616	13.83	24.3	112.9
	1 000		—	15.72	26.8	119.1
C 1.22, Mn 0.35	RT	7.83	—	—	45.2	18.4
	100		486	10.6	44.8	25.2
	200		540	11.25	43.5	33.3
Annealed	400		599	12.88	38.5	54.0
	600		699	14.16	33.5	80.2
	800		649	14.33	23.9	115.2
	1 000		—	16.84	26.0	122.6
C 0.23, Mn 1.51 } En 14, 150M19	RT	7.85	—	—	46.1	19.7
	100		477	11.89	46.1	25.9
	200		511	12.68	44.8	33.3
Annealed	400		590	13.87	39.8	52.3
	600		741	14.72	34.3	78.6
	800		821	12.11	26.4	110.3
	1 000		—	13.67	27.2	117.4
C 0.13, Mn 0.61, Ni 0.12	0	7.84	435			16.3
	100		494			22.6
	200		528			29.6
	400		599			48.2
Annealed	600		754			74.2
	800		833			110.0
	1 000		657			119.4
		Low alloy steels				
C 0.40, Mn 0.67, Ni 0.80 } 1% Ni, En 12	RT	7.85	*—	—	—	21.9
	100		486	11.90	49.4	26.4
	200		507	12.55	46.9	33.4
Hardened 850°C OQ	400		544	13.75	40.6	52.0
Tempered 600°C (1h) OQ	600		586	14.45	34.8	77.5
C 0.37, Mn 1.56, Mo 0.26 } Mn–Mo, En 16, 605A37	RT	7.85	*—	—	—	25.4
	100		456	12.45	48.2	30.6
	200		477	13.20	45.6	39.1
Hardened 845°C OQ	400		532	14.15	39.4	60.0
Tempered 600°C (1h)	600		599	14.80	33.9	88.5

Notes:
* Where a group is marked with an asterisk the Values are a mean from RT up to the temperature quoted.
1. Where *specific heats* are quoted at temperatures above RT the values have been determined over a range of 50°C up to the temperature quoted.
2. *Coefficients of expansion* are mean values from RT up to the temperature quoted.
3. *Electrical resistivity* values are uncorrected for dimensional changes of the specimen with temperature. Original dimensions as at RT.

Table 14.11 PHYSICAL PROPERTIES OF STEELS—*continued*

Material and condition *Composition* %	*Temperature* °C	*Specific gravity* $g\ cm^{-3}$	*Thermal properties (see Notes)* *Specific heat* $J\ kg^{-1}\ K^{-1}$	*Coefficient of thermal expansion* $10^{-6}\ K^{-1}$	*Thermal conductivity* $W\ m^{-1}\ K^{-1}$	*Electrical resistivity* μΩ cm
C 0.37, Mn 1.48, Mo 0.43 — Mn–Mo, En 17, 608M38	RT	7.85	*—	—	—	22.5
	100		482	12.45	45.6	27.2
	200		494	13.00	44.0	34.3
Hardened 850 °C OQ	400		519	13.90	39.4	52.5
Tempered 620 °C (1h) OQ	600		595	14.75	33.9	77.5
C 0.32, Mn 0.69, Cr 1.09 — 1%Cr, En 18B, 530A32	RT	7.84	—	—	48.6	20.0
	100		494	12.16	46.5	25.9
	200		523	12.83	44.4	33.0
Annealed	400		595	13.72	38.5	51.7
	600		741	14.46	31.8	77.8
	800		934	12.13	26.0	110.6
	1 000		—	13.66	28.1	117.7
C 0.39, Mn 0.79, Cr 1.03 — 1% Cr, En 18D, 530A40	RT	7.85	*—	—	—	22.8
	100		452	12.35	44.8	28.1
	200		473	13.05	43.5	35.2
Hardened 850 °C OQ	400		519	14.40	37.7	53.0
Tempered 640 °C (1h) OQ	600		561	15.70	31.4	78.5
C 0.28/0.33, Mn 0.4/0.6, Si 0.2/0.35, Cr 0.8/1.1, Mo 0.15/0.25 — 1% Cr–Mo	0		—		42.7	—
	RT	7.85	—		—	22.3
	100		477		42.7	27.1
	200		515		—	34.2
	300		544		40.6	—
Hardened and tempered	400		595		—	52.9
	500		657		37.3	—
	600		737		—	78.6
	700		825		31.0	—
	800		883		—	110.3
	1 000		—		28.1	117.1
	1 200		—		30.1	122.2
C 0.41, Mn 0.67, Cr 1.01, Mo 0.23 — 1% Cr–Mo, En 19, 708A42	RT	7.83	*—	—	—	22.2
	100		—	12.25	42.7	26.3
	200		473	12.70	42.3	32.6
	400		519	13.70	37.7	47.5
Hardened 850 °C OQ	600		561	14.45	33.1	64.6
Tempered 600 °C (1h) OQ						
C 0.4, Mn 0.4, Cr 1.1, Mo 0.7 — 1% Cr–Mo, En 20B	RT	7.85		—	—	—
	100			12.3	41.9	
	200			12.6	41.9	
	400			13.7	38.9	
Hardened and tempered	600			14.4	32.7	
	800			—	26.0	
C 0.4, Mn 0.6, Cr 3.0, Mo 0.8, V 0.2 — 3% Cr–Mo–V, En 40C, 897M39	RT	7.83		—	—	—
	100			12.5	37.7	
	200			12.9	37.7	
	400			13.5	34.8	
	600			14.0	31.0	
Hardened and tempered						

Notes:

* The values are a mean from RT up to the temperature quoted.

1. Where *specific heats* are quoted at temperatures above RT the values have been determined over a range of 50 °C up to the temperature quoted.
2. *Coefficients of expansion* are mean values from RT up to the temperature quoted.
3. *Electrical resistivity* values are uncorrected for dimensional changes of the specimen with temperature. Original dimensions as at RT.

Table 14.11 PHYSICAL PROPERTIES OF STEELS—*continued*

Material and condition Composition %		Temperature °C	*Specific gravity* g cm^{-3}	*Thermal properties (see Notes)* *Specific heat* J kg^{-1} K^{-1}	*Coefficient of thermal expansion* 10^{-6} K^{-1}	*Thermal conductivity* W m^{-1} K^{-1}	*Electrical resistivity* μΩ cm
C 0.35	Low Ni–Cr–Mo	RT	7.84	—	—	42.7	21.1
Mn 0.59	En 19	100		477	12.67	42.7	27.1
Ni 0.20		200		515	13.11	41.9	34.2
Cr 0.88		400		595	13.82	38.9	52.9
Mo 0.20		600		737	14.55	33.9	78.6
Annealed		800		883	11.92	26.4	110.3
		1 000		—	13.86	28.1	117.1
C 0.23		RT	7.83		—	38.5	35.5
Mn 0.45	3% Cr–W–Mo–V	100			11.9	33.6	39.0
Si 0.45		200			12.4	33.1	46.2
Cr 2.87		400			13.1	30.6	63.0
W 0.59		600			13.6	29.3	85.4
Mo 0.51		800			14.1	28.9	—
V 0.77							
Hardened and tempered							
C 0.32	3% Ni	RT	7.85	—	—	36.4	25.9
Mn 0.55	En 21	100		482	11.20	37.7	32.0
Ni 3.47		200		523	11.80	38.9	39.0
Annealed		400		590	12.90	36.8	56.7
		600		749	13.87	32.7	81.4
		800		604	11.10	25.1	112.2
		1 000		—	13.29	27.6	118.0
C 0.33	3% Ni–Cr	RT	7.85	—	—	34.3	25.6
Mn 0.50	En 23	100		494	11.36	36.0	31.7
Ni 3.4		200		523	12.29	36.8	38.7
Cr 0.8		400		599	13.18	36.4	56.7
Hardened and		600		775	13.72	31.8	81.7
tempered		800		557	10.69	26.0	111.5
		1 000		—	13.11	27.6	117.8
C 0.41	1½% Ni–Cr–Mo	RT	7.84		—		24.8
Ni 1.43	En 24	100			—		29.8
Cr 1.07	817M40	200			12.40		36.7
Mo 0.26		400			13.60		55.2
Hardened 830°C OQ		600			14.30		79.7
Tempered 630°C (1h) OQ							
C 0.32	2½% Ni–Cr–Mo	RT	7.85		—		27.7
Ni 2.60	En 25	100			—		32.1
Cr 0.67	826M31	200			11.55		38.7
Mo 0.51		400			13.10		57.3
Hardened 830°C OQ		600			13.85		82.5
Tempered 650°C OQ							
C 0.34	3% Ni–Cr–Mo	RT	7.86	—	—	33.1	27.7
Mn 0.54	En 27	100		486	11.63	33.9	33.7
Ni 3.53		200		523	12.12	35.2	40.6
Cr 0.76		400		607	13.12	35.6	58.2
Mo 0.39		600		770	13.79	30.6	82.5
Hardened and tempered		800		636	10.67	26.8	111.4
		1 000		—	12.96	28.5	117.6

Notes:

1. Where *specific heats* are quoted at temperatures above RT, the values have been determined over a range of 50°C up to the temperature quoted.
2. *Coefficients of expansion* are mean values from RT up to the temperature quoted.
3. *Electrical resistivity* values are uncorrected for dimensional changes of the specimen with temperature. Original dimensions as at RT.

Table 14.11 PHYSICAL PROPERTIES OF STEELS—*continued*

Material and condition Composition %	*Temperature* °C	*Specific gravity* g cm^{-3}	*Thermal properties (see Notes)* *Specific heat* J kg^{-1} K^{-1}	*Coefficient of thermal expansion* 10^{-6} K^{-1}	*Thermal conductivity* W m^{-1} K^{-1}	*Electrical resistivity* μΩ cm
C 0.29 } $4\frac{1}{4}$% Ni–Cr	RT	7.83		—	—	37.0
Ni 4.23 } En 30A	100			10.55	27.6	41.6
Cr 1.26 }	200			12.00	29.7	49.3
Hardened 820 °C AC						
Tempered 250 °C (1h)						
C 0.18 } 2% Ni–Mo	RT	7.85		—		24.9
Ni 1.76 } En 34	100			12.50		29.6
Mo 0.20 } 665A17	200			13.10		37.1
Blank carburized 920 °C						
Hardened 800 °C OQ						
C 0.15 } $4\frac{1}{4}$% Ni–Cr–Mo	RT	7.85		—		36.3
Ni 4.25 } En 39B	100			11.30		40.1
Cr 1.18 } 835A15	200			12.55		46.7
Mo 0.20 }						
Blank carburized 920 °C						
Hardened 800 °C OQ						
C 0.39 } Low alloy steel	RT	7.86		—		24.7
Mn 1.35 } En 100	100			12.00		28.2
Ni 0.65 } 945M38	200			12.75		34.0
Cr 0.48 }	400			14.00		52.0
Mo 0.17 }	600			14.75		74.7
Hardened 850 °C OQ						
Tempered 620 °C (1h) OQ						
C 0.39 } Low Ni–Cr–Mo	RT	7.84		—		24.8
Ni 1.39 } En 110	100			12.00		29.2
Cr 1.02 } 816M40	200			12.65		35.6
Mo 0.14 }	400			13.65		54.0
Hardened 840 °C OQ	600			14.30		78.0
Tempered 650 °C (1h) OQ						
C 0.17 } $\frac{3}{4}$% Ni–Cr	RT	7.85		—		29.1
Ni 0.86 } En 351	100			12.80		34.2
Cr 0.71 } 635A14	200			13.10		41.1
Blank carburized 910 °C						
Hardened 820 °C OQ						
C 0.17 } $1\frac{1}{4}$% Ni–Cr–Mo	RT	7.87		—		31.8
Ni 1.25 } En 353	100			11.30		36.6
Cr 1.02 } 815A16	200			12.45		43.2
Mo 0.15 }						
Blank carburized 910 °C						
Hardened 810 °C OQ						
C 0.16 } 2% Ni–Cr–Mo	RT	7.84		—		34.5
Ni 2.00 } En 355	100			11.80		39.2
Cr 1.50 } 822A17	200			12.30		45.7
Mo 0.20 }						
Blank carburized 910 °C						
Hardened 810 °C OQ						

Notes:

1. Where *specific heats* are quoted at temperatures above RT, the values have been determined over a range of 50 °C up to the temperature quoted.
2. *Coefficients of expansion* are mean values from RT up to the temperature quoted.
3. *Electrical resistivity* values are uncorrected for dimensional changes of the specimen with temperature. Original dimensions as at RT.

Table 14.11 PHYSICAL PROPERTIES OF STEELS—*continued*

Material and condition Composition %	*Temperature* °C	*Specific gravity* g cm^{-3}	*Thermal properties (see Notes)* *Specific heat* J kg^{-1} K^{-1}	*Coefficient of thermal expansion* 10^{-6} K^{-1}	*Thermal conductivity* W m^{-1} K^{-1}	*Electrical resistivity* μΩ cm
C 0.48, Mn 0.90, Si 1.98, Cu 0.64 — 2% Si–Cu	RT	7.73	—	—	25.1	41.9
Annealed	100		498	11.19	28.5	47.0
	200		523	12.21	30.1	52.9
	400		603	13.35		68.5
	600		749	14.09		91.1
	800		528	13.59		117.3
	1 000		—	14.54		122.3
C 0.05, Mn 0.3, Si 0.7, B 1.96, Al 0.03 — 2% B	0			—		24.9
Hot worked	RT	7.72	461	—		—
	100			10.0		30.9
	200			11.0		38.7
	400			11.9		57.4
	600			11.8		81.9
	800			13.3		—
C 0.10, Mn 0.14, Si 0.43, B 4.2, Al 0.53 — 4% B	0			—		39.9
As cast	RT	7.40	523	—		—
	100			9.5		50.6
	200			—		61.5
	300			10.4		72.3
	400			—		83.3
	500			11.2		—
	600			—		106.5
	700			11.8		—
	800			—		129.4
	1 000			13.0		—
Typically C 0.10, Mo 0.5, B 0.004 — ½% Mo–B 'Fortiweld'	RT	7.86	*440	12.00	46.1	20.0
Normalized and stress-relieved 600 °C	100		465	12.55	45.2	24.5
	200		494	13.25	44.4	31.0
	400		557	14.30	41.5	48.5
	600		632	15.10	36.9	74.5
	700		674	15.40	35.2	88.0
C 0.10, Si 1.0 max, Cr 4.0/6.0 — 5% Cr AISI 502	30	7.7		—	36.0	
Annealed	100			11.0	—	
	200			11.6	35.2	
	400			12.6	—	
	600			13.3	—	
	800			—	26.8	
	1 200			—	26.8	
High alloy steels						
C 0.45, Mn 0.5, Si 3.5, Cr 3.5 — 3% Cr–3% Si	RT	7.6		—	22.2	80
Hardened and tempered	100			13.0	—	
	300			13.0	—	
	500			13.0	—	
	700			14.0	—	
	900			—	31.4	

Notes:

* The values are a mean from RT up to the temperature quoted.
1. Where *specific heats* are quoted at temperatures above RT, the values have been determined over a range of 50 °C up to the temperature quoted.
2. *Coefficients of expansion* are mean values from RT up to the temperature quoted.
3. *Electrical resistivity* values are uncorrected for dimensional changes of the specimen with temperature. Original dimensions as at RT.

Table 14.11 PHYSICAL PROPERTIES OF STEELS—*continued*

Material and condition *Composition* %	*Temperature* °C	*Specific gravity* g cm^{-3}	*Thermal properties (see Notes)* *Specific heat* J kg^{-1} K^{-1}	*Coefficient of thermal expansion* 10^{-6} K^{-1}	*Thermal conductivity* W m^{-1} K^{-1}	*Electrical resistivity* μΩ cm
C 0.45 — 8% Cr–3% Si	RT	7.6		—	22.2	80.0
Mn 0.5 — En 52	100			13.0	—	—
Cr 8.0 — 401S45	300			13.0	—	110.0
Si 3.4	500			13.0	—	
Hardened and tempered	700			14.0	—	
	900			—	31.4	
C 0.40 — 11% Cr	RT	7.75		—	23.5	60.0
Mn 0.3	100			10.0	—	—
Cr 11.5	300			11.0	—	—
Hardened and tempered	500			12.0	—	—
	700			12.0	—	—
	750			—	24.3	119.0
C 0.12 — 9% Cr–Mo	RT	7.78	*402	11.15	26.0	49.9
Cr 9.0	100		427	11.30	26.4	55.5
Mo 1.0	200		461	11.60	26.8	63.0
Normalized and tempered	400		528	12.10	27.6	79.5
	600		595	12.65	26.8	97.5
	700		624	12.85	26.8	106.5
C 0.20 — 11% Cr–Mo–V–Nb	RT	7.75		—		
Mn 0.4	100			9.3		
Cr 11.0	200			10.9		
Mo 0.5	400			11.5		
V 0.7	600			12.1		
Nb 0.15	800			12.2		
Hardened and tempered						
C 0.13 — 13% Cr	RT	7.74	—	—	26.8	48.6
Mn 0.25 — En 56B	100		473	10.13	27.6	58.4
Cr 12.95 — 420S29	200		515	10.66	27.6	67.9
Ni 0.14	400		607	11.54	27.6	85.4
Annealed	600		779	12.15	26.4	102.1
	800		691	12.56	25.1	116.0
	1 000		—	11.70	27.6	117.0
C 0.07 — 17% Cr	RT	7.7		—	21.8	62.0
Mn 0.8	100			10.0		
Cr 17.0	200			11.0		
Annealed	300			12.0		
C 0.06 — 21% Cr	RT	7.76	482	—	21.8	62.0
Mn 0.8	100			10.0		
Cr 21.0	300			11.0		
Annealed	500			11.0		
	700			12.0		
	900			13.0		
C 0.22 — 30% Cr–Ni	RT	7.90		—	12.6	80.0
Cr 30.4	100			10.0		
Ni 0.26						
Hardened and tempered						

Notes:

* The values are a mean from RT up to the temperature quoted.

1. Where *specific heats* are quoted at temperatures above RT, the values have been determined over a range of 50 °C up to the temperature quoted.
2. *Coefficients of expansion* are mean values from RT, up to the temperature quoted.
3. *Electrical resistivity* values are uncorrected for dimensional changes of the specimen with temperature. Original dimensions as at RT.

Table 14.11 PHYSICAL PROPERTIES OF STEELS—*continued*

Material and condition Composition %		*Temperature* °C	*Specific gravity* g cm^{-3}	*Thermal properties (see Notes)* *Specific heat* J kg^{-1} K^{-1}	*Coefficient of thermal expansion* 10^{-6} K^{-1}	*Thermal conductivity* W m^{-1} K^{-1}	*Electrical resistivity* μΩcm
C 1.22	13% Mn	RT	7.87	—	—	13.0	66.5
Mn 13.0		100		519	18.01	14.6	75.7
1050°C Air-cooled		200		565	19.37	16.3	84.7
		400		607	21.71	19.3	100.4
		600		704	19.86	21.8	110.0
		800		649	21.86	23.5	120.4
		1000		673	23.13	25.5	127.5
C 0.28	28% Ni	RT	8.16	—	—	12.6	82.9
Mn 0.89		100		502	13.73	14.7	89.1
Ni 28.4		200		519	15.28	16.3	94.7
950°C, WQ		400		540	17.02	18.9	103.9
		600		586	17.82	22.2	111.2
		800		586	18.28	25.1	116.5
		1000		599	18.83	27.6	120.6
C 0.10	12% Cr–4% Al	RT	7.42	502	—	—	122
Mn 0.60	AISI 406	100			11.0	25.1	125
Cr 12.0		300			12.0	—	129
Al 4.5		500			12.0	28.5	—
Softened		600			—	—	136
		700			13.0		—
		850			—		141
C 0.72	4% Cr–18% W	RT	8.69	—	—	24.3	40.6
Mn 0.25		100		410	11.23	26.0	47.2
Ni 0.07		200		435	11.71	27.2	54.4
Cr 4.26		400		502	12.20	28.5	71.8
W 18.5		600		599	12.62	27.2	92.2
Annealed 830°C		800		716	12.97	26.0	115.2
		1000		—	12.44	27.6	120.9
C 0.16	16% Cr–Ni	RT	7.7	—	—	18.8	72.0
Mn 0.2	En 57	100		482	10	—	—
Ni 2.5	431S29	300			11	—	—
Cr 16.5		500			12	24.3	103.0
Softened							
C 0.08	18% Cr–8% Ni	RT	7.92	—	—	15.9	69.4
Mn 0.3/0.5	En 58A	100		511	14.82	16.3	77.6
Ni 8	302S25	200		532	16.47	17.2	85.0
Cr 18/20		400		569	17.61	20.1	97.6
1100°C WQ		600		649	18.43	23.9	107.2
		800		641	19.03	26.8	114.1
		1000		—	—	28.1	119.6
C 0.12	18% Cr–11% Ni	RT	7.9		—	15.9	72
Mn 1.5	(Nb stabilized)	100			16.0	—	
Ni 11.0	En 58G	300			18.0	17.2	
Cr 17.5	347S17	500			18.0	18.8	
Nb 1.2		700			19.0	20.1	
Softened							

Notes:

1. Where *specific heats* are quoted at temperatures above RT, the values have been determined over a range of 50°C up to the temperature quoted.
2. *Coefficients of expansion* are mean values from RT, up to the temperature quoted.
3. *Electrical resistivity* values are uncorrected for dimensional changes of the specimen with temperature. Original dimensions as at RT.

Table 14.11 PHYSICAL PROPERTIES OF STEELS—*continued*

Material and condition Composition %		*Temperature* °C	*Specific gravity* g cm^{-3}	*Thermal properties (see Notes)* *Specific heat* J kg^{-1} K^{-1}	*Coefficient of thermal expansion* 10^{-6} K^{-1}	*Thermal conductivity* W m^{-1} K^{-1}	*Electrical resistivity* µΩ cm
C 0.22		RT	7.72		—		82
Mn 0.6	20% Cr–8% Ni	100			15.0		
Cr 20.0	(Ti stabilized)	300			15.0		
Ni 8.5		500			16.0		
Ti 1.2		700			17.0		
Softened		900			18.0		
C 0.15		RT	7.92		—	—	
Mn 0.8	19% Cr–14% Ni	100			17.0	15.1	
Ni 14	(Nb stabilized)	200			17.2	16.8	
Cr 19		400			17.6	20.1	
Nb 1.7		600			18.6	24.3	
Softened							
C 0.30	18% Cr–8% Ni–W	RT	7.8			13	85
Mn 0.6	En 55	100			16.0		
Si 1.5		300			17.0		
Ni 8		500			17.0		
Cr 20		700			18.0		
W 4		900			18.0		
Softened		1 050			—	29	125
C 0.12	18% Cr–8% Ni–Al	RT	7.67		—	18.0	85
Mn 0.3	(Ti stabilized)	100			15		
Ni 8.5		300			15		
Cr 18.5		500			15		
Ti 0.8		700			16		
Al 1.4		900			17	26.0	125
Normalized and tempered							
C 0.10		RT	8.01	490	—	15.5	70
Mn 0.3	12% Cr–12% Ni	100			18	16.8	77
Ni 12.5	En 58D						
Cr 12.5							
Softened							
C 0.10		RT	7.94	*477	14.80	12.6	74.1
Mn 6.0	15/10/6/1	100		494	15.70	13.8	80.0
Cr 15.0	Cr–Ni–Mn–Mo	200		511	16.75	15.4	86.7
Ni 10.0		400		536	18.25	18.8	99.4
Mo 1.0		600		557	18.95	21.8	108.4
Solution treated 1 100 °C		700		565	19.30	23.0	114.4
C 0.27		RT	8.08		—	12.1	97
Mn 1.25	11% Cr–36% Ni	100			14	—	—
Ni 36		300			15	—	—
Cr 11		500			16	18.4	117
Softened							
C 0.1		RT	7.5		10	15.9	88
Mn 1.3	30% Cr–Ni	200			11		
Si 1.2		800			13		
Ni 1.8		1 000			13		
Cr 29.0		1 100				26.4	126
Softened							

Notes:

* The values are a mean from RT up to the temperature quoted.

1. Where *specific heats* are quoted at temperatures above RT, the values have been determined over a range of 50 °C up to the temperature quoted.
2. *Coefficients of expansion* are mean values from RT up to the temperature quoted.
3. *Electrical resistivity* values are uncorrected for dimensional changes of the specimen with temperature. Original dimensions as at RT.

Table 14.11 PHYSICAL PROPERTIES OF STEELS—*continued*

Material and condition *Composition* %	*Temperature* °C	*Specific gravity* g cm^{-3}	*Thermal properties (see Notes)* *Specific heat* J kg^{-1} K^{-1}	*Coefficient of thermal expansion* 10^{-6} K^{-1}	*Thermal conductivity* W m^{-1} K^{-1}	*Electrical resistivity* μΩ cm
C 0.1 (14% Cr–63% Ni)	RT	8.1			12.6	105
Mn 1.0	100			12.0		
Ni 63	200			12.5		
Cr 14	400			13.5		
Softened	600			14.5		
	800			15.5		
	1 000			16.5	28.9	110
C 0.30 (17% Cr–17% Ni–Mo –Co–Nb)	RT	8.0		—	12.6	93.8
Mn 3.0	100			15		
Ni 17.5	300			16		
Cr 16.5	500			16		
Mo 3.0	700			17		
Nb 2.5						
Co 7.0						
Softened						
C 0.4 (13% Cr–13% Ni–W –Nb)	RT	8.03		—		
Mn 0.9	100			16.8		
Si 1.4	200			17.3		
Ni 13.0	400			18.3		
Cr 13.0	600			18.9		
W 2.3	800			19.3		
Nb 0.9						
Normalized						
C 0.4 (13% Cr–13% Ni–W –Mo–Co–Nb)	RT	8.13		—	—	
Mn 0.8	100			15.6	13.4	
Si 1.0	200			15.8	17.2	
Ni 13	400			16.9	18.8	
Cr 13	600			17.3	22.2	
W 2.5	800			18.0	25.5	
Mo 2.0						
Nb 3.0						
Co 10.0						
Solution treated						
C 0.27 (20% Cr–10% Ni –46% Co)	RT	8.26		—	—	
Mn 0.77	100			14.8	14.7	
Ni 10.5	200			15.0	16.3	
Cr 19.1	400			15.2	19.7	
Mo 2.2	600			15.9	23.0	
Nb 1.4	800			16.8	26.0	
V 3.0						
Co 46.6						
Solution treated and aged						
Cast steels						
C 0.11 (Plain carbon B.S. 1617A)	100			12.2	48.6	19.5
Mn 0.35	200			12.6		
A 950 °C, N 950 °C	300			13.2		
	400			13.6		
	500			13.9		
	600			14.2		

Notes:

1. Where *specific heats* are quoted at temperatures above RT, the values have been determined over a range of 50 °C up to the temperature quoted.
2. *Coefficients of expansion* are mean values from RT up to the temperature quoted.
3. *Electrical resistivity* values are uncorrected for dimensional changes of the specimen with temperature. Original dimensions as at RT.

Table 14.11 PHYSICAL PROPERTIES OF STEELS—*continued*

Material and condition Composition %		Temperature °C	*Specific gravity* g cm^{-3}	*Thermal properties (see Notes)* *Specific heat* J kg^{-1} K^{-1}	*Coefficient of thermal expansion* 10^{-6} K^{-1}	*Thermal conductivity* W m^{-1} K^{-1}	*Electrical resistivity* μΩ cm
C 0.4, Mn 0.5	Plain carbon BS 1760 A 900 °C, OQ 830 °C T 650 °C	100			11.8	42.3	23.5
		200			12.4		
		300			12.8		
		400			13.3		
		500			13.7		
		600			14.2		
C 0.17, Mn 0.74, Mo 0.50	Carbon Mo BS 1398 A 920 °C, SR 650 °C	100			12.4		24.2
		200			12.8		
		300			13.1		
		400			13.4		
		500			13.8		
		600			14.2		
C 0.25, Mn 1.55	1½% Mn BS 1456 A A 950 °C, WQ 910 °C, T 660 °C	100			13.2		
		200			13.3		
		300			13.7		
		400			14.1		
		500			14.7		
		600			15.2		
C 0.29, Cr 1.80, Ni 0.46, Mo 0.52	1½% Cr Ni Mo BS 1458 OQ 900 °C, T 660 °C	100			12.5		27.6
		200			12.7		
		300			13.0		
		400			13.4		
		500			13.9		
		600			14.4		
C 0.34, Ni 2.82, Cr 0.74, Mo 0.42	2½% Ni Cr Mo BS 1459 OQ 850 °C, T 640 °C	100			12.0	39.4	27.3
		200			12.3		
		300			12.6		
		400			13.0		
		500			13.5		
		600			13.9		
C 0.24, Cr 3.23, Mo 0.51	3% Cr Mo BS 1461 OQ 900 °C, T 690 °C	100			12.2		
		200			12.4		
		300			12.7		
		400			12.9		
		500			13.3		
		600			13.6		
C 0.1, Cr 4.06, Mo 0.57	5% Cr Mo BS 1462 N 950 °C, T 680 °C	100			11.8		37.1
		200			12.0		
		300			12.3		
		400			12.5		
		500			12.7		
		600			13.0		
C 0.13, Cr 8.29, Mo 1.1, Si 1.1	9% Cr Mo BS 1463 OQ 900 °C, T 690 °C	100			11.9		
		200			11.6		
		300			11.7		
		400			11.7		
		500			11.8		
		600			11.9		

Notes:

1. Where *specific heats* are quoted at temperatures above RT, the values have been determined over a range of 50 °C up to the temperature quoted.
2. *Coefficients of expansion* are mean values from RT up to the temperature quoted.
3. *Electrical resistivity* values are uncorrected for dimensional changes of the specimen with temperature. Original dimensions as at RT.

Table 14.11 PHYSICAL PROPERTIES OF STEELS—*continued*

Material and condition *Composition* %		*Temperature* °C	*Specific gravity* g cm^{-3}	*Thermal properties (see Notes)*			*Electrical resistivity* μΩ cm
				Specific heat J kg^{-1} K^{-1}	*Coefficient of thermal expansion* 10^{-6} K^{-1}	*Thermal conductivity* W m^{-1} K^{-1}	
C 0.27	13% Cr	100			11.5		
Cr 12.1	BS 1630	200			11.8		
Ni 1.07	OQ 930 °C, T 730 °C	300			12.4		
Si 1.16		400			12.6	25.1	
		500			12.7		
		600			12.9		
C 0.47	Carbon Cr	100			12.5		
Cr 0.85	BS 1956 A	200			12.9		
	N 870 °C, T 635 °C	300			13.2		
		400			13.4		
		500			13.5		
		600			13.6		
C 0.1	3½% Ni	100			11.3		
Ni 3.35	BS 1504–503	200			11.9		
	WQ 880 °C, T 650 °C	300			12.2		
		400			12.7		
		500			13.5		
		600			13.6		
C 0.19	1¼% Cr Mo	100			11.8		28.7
Cr 1.13	BS 1504–621	200			12.4		
Mo 0.5	N 920 °C, T 625 °C	300			12.6		
		400			13.3		
		500			13.7		
		600			13.9		
Cast corrosion-resisting steels							
C 0.13	13% Cr	100	7.73	482	11.0	24.7	56
Mn 0.80	BS 1630 A	300			11.0	—	
Cr 12.5		500			12.0	—	
Hardened and tempered		600				27.6	
C 0.25	13% Cr	100	7.75	482	11.0	24.3	57
Mn 0.70	BS 1630 C	300			11.0	—	
Cr 12.5		500			12.0	—	
Hardened and tempered		600			—	26.0	
C 0.07	18% Cr–8% Ni	100	7.93	502	17.0	16.3	72
Si 0.70	BS 1631 A						
Mn 0.80							
Ni 8.5							
Cr 18.0							
Normalised							
C 0.08	18% Cr, 8% Ni Nb	100	7.93	502	17.0	15.9	
Si 1.00	BS 1631 B Nb	300			18.0	—	
Mn 0.50		500			18.0	—	
Ni 9.00		700			19.0	20.1	
Cr 18.0							
Nb 0.9							
Normalised							

Notes:

1. Where *specific heats* are quoted at temperatures above RT, the values have been determined over a range of 50 °C up to the temperature quoted.
2. *Coefficients of expansion* are mean values from RT, up to the temperature quoted.
3. *Electrical resistivity* values are uncorrected for dimensional changes of the specimen with temperature. Original dimensions as at RT.

Table 14.11 PHYSICAL PROPERTIES OF STEELS—*continued*

Material and condition *Composition* %	*Temperature* °C	*Specific gravity* g cm^{-3}	*Specific heat* J kg^{-1} K^{-1}	*Coefficient of thermal expansion* 10^{-6} K^{-1} (*Thermal properties, see Notes*)	*Thermal conductivity* W m^{-1} K^{-1} (*Thermal properties, see Notes*)	*Electrical resistivity* μΩ cm
C 0.12, Si 1.50, Mn 0.80, Ni 9.00, Cr 19.00, Ti 0.6 — 18% Cr, 8% Ni Ti BS 1631 B Ti Normalised	100	7.78	444	17.0	15.5	70
C 0.06, Si 0.70, Mn 0.70, Ni 12.0, Cr 19.0, Mo 3.6 — 19% Cr, 12% Ni, 3½% Mo BS 1632 A Water quenched	RT	7.96	502	—	16.3	
	100			16.0		
C 0.07, Si 1.00, Mn 1.00, Ni 10.5, Cr 18.0, Mo 2.75 — 18% Cr, 10% Ni 2½% Mo BS 1632 B Normalised	RT	7.96	502	—	16.3	73
	100			16.5		
	200			16.9		
	400			17.2		
	500			17.4		
	600			17.9		
	800			19.0		
C 0.08, Si 1.00, Mn 0.50, Ni 10.5, Cr 18.0, Mo 2.75, Nb 0.90 — 18% Cr, 10% Ni, 2½% Mo Nb BS 1632 C Nb Normalised	RT	7.96	502	—	16.3	73
	100			16.0	17.6	
C 0.10, Si 1.50, Mn 0.80, Ni 10.0, Cr 18.0, Mo 2.75, Ti 0.60 — 18% Cr, 10% Ni, 2½% Mo Ti BS 1632 C Ti Normalised	RT	7.78	448	—	15.5	78
	100			17.0		
C 0.06, Si 0.70, Mn 0.60, Ni 8.5, Cr 18.0, Mo 2.5 — 18% Cr, 8% Ni, 2½% Mo BS 1632 D Normalised	RT	7.93	502	—	16.3	
	100			16.0		
Cast heat-resisting steels						
C 0.25, Si 0.70, Mn 0.70, Cr 12.5 — 13% Cr BS 1648 A Hardened and tempered	RT	7.75	482	—	24.3	57
	100			11.0	—	
	300			11.0	—	
	500			12.0	—	
	600			—	26.0	

Notes:

1. Where *specific heats* are quoted at temperatures above RT, the values have been determined over a range of 50°C up to the temperature quoted.
2. *Coefficients of expansion* are mean values from RT, up to the temperature quoted.
3. *Electrical resistivity* values are uncorrected for dimensional changes of the specimen with temperature. Original dimensions as at RT.

Table 14.11 PHYSICAL PROPERTIES OF STEELS—*continued*

Material and condition Composition %		Temperature °C	*Specific gravity* g cm^{-3}	*Thermal properties (see Notes)* *Specific heat* J kg^{-1} K^{-1}	*Coefficient of thermal expansion* 10^{-6} K^{-1}	*Thermal conductivity* W m^{-1} K^{-1}	*Electrical resistivity* μΩ cm
C 0.40	27% Cr	RT	7.63	482	—	20.9	70
Si 0.80	BS 1648 B	100			10.2		
Mn 0.90		200			10.8		
Cr 29.0		400			11.0		
Tempered		600			11.5		
		800			12.4		
		1 000			13.3		
C 1.70	27% Cr	RT	7.63	482	—	20.9	70
Si 0.70	BS 1648 C	100			10.2		
Mn 0.70		200			10.8		
Cr 27.0		400			11.0		
Tempered		600			11.5		
		800			12.4		
		1 000			13.3		
C 0.30	20% Cr 10% Ni	RT	7.74	502	—	—	80
Si 1.50	BS 1648 D	100			—	15.5	
Mn 1.50		500			17.8	—	
Ni 10.0		800			18.5	26.8	
Cr 20.0		1 100			19.6	—	
C 0.35	21% Cr 8% Ni	RT	7.92	435	—	10.9	86
Si 1.50	4% W	100			13.6		
Mn 0.80	BS 1648 D	300			14.5		
Ni 7.0		500			15.4		
Cr 21.0		700			16.5		
W 4.0		900			17.7	26.8	
		1 000			18.3	—	
C 0.20	25% Cr 12% Ni	RT	7.92	544	—	13.8	85
Si 1.20	BS 1648 E	100			16.5		
Mn 1.30		200			16.6		
Ni 13.0		400			16.9		
Cr 25.0		600			17.6		
Normalised		800			18.2		
		1 000			18.7		
C 0.20	25% Cr 12% Ni	RT	7.90	502	—	12.6	87
Si 1.00	3% W	100			15.0		
Mn 0.80	BS 1648 E	300			16.0		
Ni 12.0		500			16.0		
Cr 23.0		700			17.0		
W 3.0		900			19.0		
Normalised		1 000			—	29.3	
C 0.20	25% Cr 20% Ni	RT	7.90	544	—	15.9	90
Si 1.50	BS 1648 F	100			16.5		
Mn 1.00		200			16.9		
Ni 20.0		400			17.5		
Cr 25.0		600			18.3		
Normalised		800			19.2		
		1 000			20.0		

Notes:
1. Where *specific heats* are quoted at temperatures above RT the values have been determined over a range of 50 °C up to the temperature quoted.
2. *Coefficients of expansion* are mean values from RT up to the temperature quoted.
3. *Electrical resistivity* values are uncorrected for dimensional changes of the specimen with temperature. Original dimensions as at RT.

Table 14.11 PHYSICAL PROPERTIES OF STEELS—*continued*

Material and condition		*Temperature* °C	*Specific gravity* g cm^{-3}	*Thermal properties (see Notes)* *Specific heat* J kg^{-1} K^{-1}	*Coefficient of thermal expansion* 10^{-6} K^{-1}	*Thermal conductivity* W m^{-1} K^{-1}	*Electrical resistivity* μΩ cm
Composition %							
C 0.35	25% Ni 15% Cr BS 1648 G	RT	7.90	502	—	12.6	88
Si 0.90		100			15.0		
Mn 0.75		300			16.0		
Ni 25.0		500			17.0		
Cr 15.0		700			17.0		
		900			18.0		
		1 000				29.3	
C 0.50	35% Ni 15% Cr	RT	7.93	460	—	—	100
Si 2.00		100			—	13.4	
Mn 1.50		500			16.0		
Ni 35.0		800			16.5		
Cr 15.0		1 100			17.6		
Cast							
C 0.50	40% Ni 20% Cr BS 1648 H	RT	8.02	460	—	—	105
Si 2.0		100			—	13.4	
Mn 1.50		500			16.0	—	
Ni 40.0		800			16.4	23.9	
Cr 20.0		1 100			17.4	—	
Cast							
C 0.50	60% Ni 15% Cr BS 1648 K	RT	8.12	460	—	—	108
Si 2.00		100			—	13.4	
Mn 1.50		500			14.2	—	
Ni 60.0		800			15.3	23.0	
Cr 15.0		1 100			16.5	—	
Cast							

Notes:

1. Where *specific heats* are quoted at temperatures above RT, the values have been determined over a range of 50°C up to the temperature quoted.
2. *Coefficients of expansion* are mean values from RT, up to the temperature quoted.
3. *Electrical resistivity* values are uncorrected for dimensional changes of the specimen with temperature. Original dimensions as at RT.

REFERENCES TO TABLE 14.11

1. 'Metals Handbook', 4th edn.
2. J. Woolman and R. A. Mottram, 'Mechanical and Physical Properties of BS En Steels (BS 970, 1950), Pergamon Press.
3. Sundry technical information issued by industrial organizations e.g. British Steel Corporation, Mond Nickel Co. Ltd.

Table 14.12 SOME LOW TEMPERATURE THERMAL PROPERTIES OF A SELECTION OF STEELS

There is particular interest in the thermal properties (especially the thermal expansion) of steels used under conditions well below normal atmospheric temperature, and available information is set out below in respect of some such steels.

Material and condition *Analyses %*	*Temperature* °C	*Coefficient of thermal expansion* $10^{-6} K^{-1}$	*Thermal conductivity* $W m^{-1} K^{-1}$
Typically	−200	−9.5	16.0
C 0.09, Ni 9 — 9 Ni	−150	−9.7	19.5
	−100	−9.9	23.0
Double normalized and tempered	−50	−10.2	26.5
	RT	10.5	29.5
	100	11.0	32.0
	200	11.7	34.0
	300	12.3	34.5
C 0.42, Ni 1.58, Cr 1.19, Mo 0.24 — 1½ Ni–Cr–Mo	−150	−10.4	
	−100	−11.2	
	−50	−11.8	
Hardened 840°C OQ; Tempered 650°C (1 h)/AC	RT	12.1	
C 0.12, Mo 0.54, B 0.003 — ½ Mo–B	−150	−10.5	
	−100	−11.2	
	−50	−11.8	
Hardened 960°C OQ; Tempered 700°C (½ h) AC	RT	12.2	
C 0.27, Cr 3.14, Mo 0.49 — 3 Cr–Mo	−150	−9.8	
	−100	−10.3	
	−50	−10.8	
Hardened 900°C OQ; Tempered 650°C (1 h) AC	RT	11.5	
C 0.09, Mn 6.23, Ni 9.88, Cr 14.88, Mo 1.01, V 0.28, B 0.003, Nb 0.94 — 15 Cr–10 Ni–6 Mn–Mo–V–B–Nb	−150	−14.7	
	−100	−15.3	
	−50	−15.7	
1150°C AC	RT	16.2	
C 0.13, Ni 4.16, Cr 1.23, Mo 0.19 — 4½ Ni–Cr–Mo	−150	−9.4	
	−100	−9.8	
	−50	−10.2	
Blank carburized 890°C AC; 820°C (¼ h), transferred to 580°C (½ h) OQ	RT	10.8	
C 0.41, Cr 3.14, Mo 0.97, V 0.20 — 3 Cr–1 Mo–V	−150	−9.7	
	−100	−10.1	
	−50	−10.5	
Hardened 930°C OQ; Tempered 700 (½ h) AC	RT	11.2	
C 0.12, Ni 3.10, Cr 0.91, Mo 0.16 — 3½ Ni–Cr–Mo	−150	−10.1	
	−100	−10.6	
	−50	−11.0	
Blank carburized 910°C AC; Hardened 840°C (¼ h) OQ; Tempered 760°C OQ	RT	11.6	

Table 14.12 SOME LOW TEMPERATURE THERMAL PROPERTIES OF A SELECTION OF STEELS—*continued*

Material and condition *Analyses %*	*Temperature* °C	*Coefficient of thermal expansion* $10^{-6}K^{-1}$	*Thermal conductivity* $W\,m^{-1}\,K^{-1}$
C 0.99, Cr 1.47 } 1 C–1½ Cr	−150	−9.6	
Hardened 850°C AC	−100	−10.6	
Tempered 650°C (½h) AC	−50	−11.6	
	RT	12.3	
C 0.17, Ni 1.74, Cr 0.2, Mo 0.22 } 2 Ni–Mo En 34	−150	−8.5	
Blank carburized 910°C AC	−100	−9.5	
Hardened 870°C OQ	−50	−10.4	
Tempered 770°C OQ	RT	11.3	
C 0.11, Ni 3.04 } 3 Ni	−150	−9.9	
Blank carburized 910°C AC	−100	−10.5	
Hardened 870°C OQ	−50	−11.0	
Tempered 770°C OQ	RT	11.5	

* Thermal expansion values shown for temperatures other than RT are the mean values from RT to that temperature. For RT the instantaneous value is given.

REFERENCE TO TABLE 14.12

1. Sundry technical information issued by British Steel Corporation and Mond Nickel Co. Ltd.

15 Elastic properties, damping capacity and shape memory alloys

15.1 Elastic properties

The elastic properties of a metal reflect the response of the interatomic forces between the atoms concerned to an applied stress. Since the bonding forces vary with crystallographic orientation the elastic properties of metal single crystals may be highly anisotropic. However, polycrystalline metals and alloys with a randomly oriented grain structure behave isotropically. Table 15.1 lists elastic constants for polycrystalline metals and alloys in an isotropic condition. Any preferred orientation or texture resulting from rolling, drawing or extrusion, for example, will result in departures from the listed values to a degree that depends upon the elastic anisotropy of the individual crystals (which may be deduced from the single crystal elastic constants of Tables 15.2 to 15.6 that follow) and the nature and extent of the preferred orientation.

Since the elastic properties are determined by the aggregate response of the interatomic forces between all the atoms in the metal, the presence of small quantities of solute atoms in dilute alloys or their rearrangement by heat treatment will have relatively little effect on the absolute values of their elastic constants. Consequently, the elastic constants of all the plain carbon and low alloy steels will be approximately the same unless some preferred orientation is present. Similarly with Cu-, Al- and Ni- base dilute alloys, etc. In the case of concentrated alloys there may be larger variations in elastic moduli, especially where there is a drastic change in the relative proportions of different phases in a multiphase alloy. In the case of ideal solid solutions the elastic moduli vary linearly with atom fraction. The elastic moduli of non-ideal solid solutions may show positive or negative deviations from linearity. Ordering produces an increase in elastic moduli.

Increase in temperature causes a gradual decrease in elastic moduli. The decrease is fairly linear over wider ranges of temperature but sharply increases in magnitude as the melting point is approached. Discontinuities are observed at structural transformations.

Ferromagnetic materials having a high degree of domain mobility may exhibit considerably higher elastic moduli below the Curie point in the presence of a high magnetic field. The lower elastic moduli in the absence of a magnetic field are due to magnetostrictive dimensional changes caused by stress-induced domain movement.

Table 15.1 ELASTIC CONSTANTS OF POLYCRYSTALLINE METALS AT ROOM TEMPERATURE

Metal	*Young's modulus* GPa	*Rigidity modulus* GPa	*Bulk modulus* GPa	*Poisson's ratio*	*Ref.*
Aluminium	70.6	26.2	75.2	0.345	1
Antimony	54.7	20.7	—	0.25–0.33	2, 3
	77.9	19.3	—	—	4
Barium	12.8	4.86	—	0.28	2
Beryllium	318	156	110	0.02	5
Bismuth	34.0	12.8	—	0.33	2
Brass 70Cu 30Zn	100.6	37.3	111.8	0.35	1
Cadmium	62.6	24.0	51.0	0.30	5
Caesium	1.7	0.65	—	0.295	2
Calcium	19.6	7.9	17.2	0.31	2, 6
Cast Iron—Grey, BS 1452:1977					
Grade 150	100	40	—	0.26	7, 8
Grade 180	109	44	—	0.26	7, 8
Grade 220	120	48	—	0.26	7, 8
Grade 260	128	51	—	0.26	7, 8
Grade 300	135	54	—	0.26	7, 8
Grade 350	140	56	—	0.26	7, 8
Grade 400	145	58	—	0.26	7, 8
—Blackheart malleable BS 310:1972					
Grades B340/12 to B290/6	169	67.6	—	0.26	7, 9
Pearlitic malleable BS 3333:1972					
Grades P4440/7 to P540/5	172	68.8	—	0.26	7, 9
Whiteheart malleable BS 309:1972					
Grades W340/3, W410/4	176	70.4	—	0.26	7, 9
Nodular BS 2789:1973					
Grades 370/17, 420/12	169	66	—	0.275	7, 10
Grades 500/7, 600/3	169–174	65.9	—	0.275	7, 10
Grades 700/2, 800/2 (pearlitic, normalized)	176	68.6	—	0.275	7, 10
pearlite 700/2, 800/2 (hardened, tempered)	172	67.1	—	0.275	7. 10
Cerium	33.5	13.5	—	0.248	11, 12
Chromium	279	115.3	160.2	0.21	1
Cobalt	211	82	181.5	0.32	13, 16
Constantan 45Ni 55Cu	162.4	61.2	156.4	0.327	1
Copper	129.8	48.3	137.8	0.343	1
Cupro-nickel 70Cu 30Ni	144	53.8	—	0.34	14
Duralumin	70.8	26.3	75.4	0.345	1
Gallium	9.81	6.67	—	0.47	2
Germanium	79.9	29.6	—	0.32	2
Gold	78.5	26.0	171	0.42	15, 16
Hafnium	141	56	109	0.26	17, 18
Incoloy 800 20Cr, 32Ni bal Fe	196	73	—	0.334	38
Indium	10.6	3.68	—	0.45	2
Invar 64Fe 36Ni	144	57.2	99.4	0.259	1
Iridium	528	209	371	0.26	16, 19, 20, 21
Iron (pure)	211.4	81.6	169.8	0.293	1
Lanthanum	37.9	14.9	—	0.28	2, 12
Lead	16.1	5.59	45.8	0.44	1
Lithium	4.91	4.24	—	0.36	23, 28
Magnesium	44.7	17.3	35.6	0.291	1
Manganese	191	79.5	—	0.24	2, 24

Table 15.1 ELASTIC CONSTANTS OF POLYCRYSTALLINE METALS AT ROOM TEMPERATURE—*continued*

Metal	*Young's modulus* GPa	*Rigidity modulus* GPa	*Bulk modulus* GPa	*Poisson's ratio*	*Ref.*
Manganese–copper 70Mn 30Cu (high damping alloy)	93	22.4	—	—	25
Molybdenum	324.8	125.6	261.2	0.293	1
Monel 400 63–70Ni, 2 Mn, 2.5Fe, bal Cu	185	66	—	0.32	26
Nickel	199.5	76.0	177.3	0.312	1
Nickel silver 55Cu, 18Ni, 27 Zn	132.5	49.7	132	0.333	1
Nimonic 80A 20Cr, 2.3Ti, 1.8Al, bal Ni (fully heat-treated)	222	85	—	0.31	27
Niobium	104.9	37.5	170.3	0.397	1
Ni–span C902 (constant modulus alloy)	186	66	—	0.41	28
Osmium	559	223	373	0.25	16, 21, 29
Palladium	121	43.6	187	0.39	18, 19, 21, 29
Platinum	170	60.9	276	0.39	16, 19, 29, 30
Plutonium	87.5	34.5	—	0.18	31
Potassium (−190°C)	3.53	1.30 (room temp.)	—	0.35	15
Rhenium	466	181	334	0.26	2, 16 32
Rhodium	379	147	276	0.26	16, 19, 29
Rubidium	2.35	0.91	—	0.30	2
Ruthenium	432	173	286	0.25	18, 21, 29
Selenium	58	—	—	0.447	15
Silicon	113	39.7	—	0.42	2, 33
Silver	82.7	30.3	103.6	0.367	1
Sodium	6.80	2.53	—	0.34	2, 23
Steel—Mild	208–209	81–82	160–169	0.27–0.3	34
0.75C	210	81.1	168.7	0.293	1
0.75C (hardened)	201.4	77.8	165	0.296	1
Tool 0.98C, 1.03 Mn, 0.65 Cr, 1.01 W	211.6	82.2	165.3	0.287	1
Tool 0.98C, 1.03 Mn, 0.65 Cr, 1.01 W (hardened)	203.2	78.5	165.2	0.295	1
Maraging Fe–18Ni 8Co 5Mo	186	72	—	0.30	35
Stainless austenitic (Fe–18Cr,	190–201	74–86	—	0.25–0.29	36
8–10 Ni)	200–206	78–79	—	0.27–0.3	36
Stainless, ferritic (Fe–13Cr)	200–215	80–83	—	0.27–0.3	1, 36
Stainless, martensitic (Fe–13Cr, 0.1–0.3C)	215.3	83.9	166	0.283	1
Stainless, martensitic (Fe–18Cr, 2Ni, 0.2C)					
Strontium	15.7	6.03	12.0	0.28	2, 6
Tantalum	185.7	69.2	196.3	0.342	1
Tellurium	47.1	16.7	—	0.16–0.3	15
Thallium	7.90	2.71	28.5	0.45	2, 6
Thorium	78.3	30.8	54.0	0.26	2, 6
Tin	49.9	18.4	58.2	0.357	1
Titanium	120.2	45.6	108.4	0.361	1
Tungsten	411	160.6	311	0.28	1
Tungsten carbide	534.4	219	319	0.22	1
Uranium	175.8	73.1	97.9	0.20	37
Vanadium	127.6	46.7	158	0.365	1
Yttrium	66.3	25.5	—	0.265	12
Zinc	104.5	41.9	69.4	0.249	1
Zirconium	98	35	89.8	0.38	17, 18

REFERENCES TO TABLE 15.1

1. G. Bradfield, 'Use in Industry of Elasticity Measurements in Metals with the help of Mechanical Vibrations', National Physical Laboratory. Notes on Applied Science No. 30, HMSO, 1964.
2. W. Köster, *Z. Electrochem. Phys. Chem.*, 1943, **49**, 233.
3. W. Köster, *Z. Metall.*, 1948, **39**, 2.
4. 'Metals Handbook', Amer. Soc. Metals, Vol. 1, 1961.
5. D. J. Silversmith and B. L. Averbach, *Phys. Rev.*, 1970, **B1**, 567.
6. S. F. Pugh, *Phil. Mag. Ser. 7*, 1974, **45**, 823.
7. H. T. Angus, 'Cast Irons, Physical and Engineering Properties', Butterworths, London, 1976.
8. 'Engineering Data on Grey Cast Irons', Brit. Cast Iron Res. Assoc., 1977.
9. 'Engineering Data on Malleable Cast Irons', Brit. Cast Iron Res. Assoc., 1974.
10. 'Engineering Data on Nodular Cast Irons', Brit. Cast Iron Res. Assoc., 1974.
11. M. Rosen, *Phys. Rev.*, 1969, **181**, 932.
12. J. F. Smith, C. D. Carlson and F. H. Spedding, *J. Metals*, **9**; *Trans. AIME*, 1957, **209**, 1212.
13. 'Physical and Mechanical Properties of Cobalt', Cobalt Information Centre, Brussels, 1960.
14. 'Cupro-Nickel Alloys, Engineering Properties', Publ. 2969, Inco Europe, London, 1966.
15. Landolt-Börnstein, 'Zahlenwerte und Funktionen', Vol. 2, Part 1, Springer-Verlag, Berlin, 1971.
16. A. S. Darling, *Int. Met. Rev.*, 1973, 91.
17. Private communication, Imperial Metal Industries, Witton, Birmingham.
18. A. S. Darling, *Proc. Inst. Mech. Eng.*, 1965, Pt 3D, **180**, 104.
19. W. Köster, *Z. Metall.*, 1948, **39**, 1.
20. A. Roll and H. Motz, *Z. Metall.*, 1957, **48**, 272.
21. K. H. Schramm, *Z. Metall.*, 1962, **53**, 729.
22. P. W. Bridgman, *Proc. Amer. Acad. Arts Sci.*, 1922, **57**, 41.
23. O. Bender, *Ann. Phys.*, 1939, **34**, 359.
24. M. Rosen, *Phys. Rev.*, 1968, **165**, 357.
25. D. Birchon, *Engineering Mater. and Design*, 1964, **7**, 606.
26. 'Wrought Nickel-Copper Alloys, Engineering Properties', Publ. 7011, Inco Europe, London, 1970.
27. 'Nimonic Alloy 80A', Publ. 3663, Henry Wiggin Ltd., Hereford, 1975.
28. 'Controlled Expansion and Constant Modulus Nickel-Iron Alloys', Publ. 6710, Inco Europe, London, 1967.
29. W. Köster, *Z. Metall.*, 1948, **39**, 111.
30. E. Grüneisen, *Ann. Phys.*, 1908, **25**, 825.
31. 'Plutonium Handbook', Ed. O. J. Wick, Gordon and Breach, New York, 1967, p. 39.
32. T. E. Tietz, B. A. Wilcox and J. W. Wilson, Standford Res. Instit. Calif., Report SU-2436, 1959.
33. R. L. Templin, *Metals and Alloys*, 1932, **3**, 136.
34. J. Woolman and R. A. Mottram, 'The Mechanical and Physical Properties of the British Standard En Steels', Vol. 1, Pergamon, Oxford, 1964.
35. '18% Nickel Maraging Steels', Publ. 4419, Inco Europe, London, 1976.
36. J. Woolman and R. A. Mottram, 'The Mechanical and Physical Properties of the British Standard En Steels', Vol. 3, Pergamon, Oxford, 1969.
37. 'Commercial Uranium', Brit. Nuclear Fuels Ltd., Warrington.
38. 'Incoloy 800', Publ. 3664, Henry Wiggin Ltd., Hereford, 1977.

15.1.1 Elastic compliances and elastic stiffnesses of single crystals

Single crystals are generally anisotropic and therefore require many more constants of proportionality than isotropic materials. The relations between stress and strain are defined by the generalized Hooke's law, which states that the strain components are linear functions of the stress components and vice versa.

That is,

$$\varepsilon_{xx}=S_{11}\sigma_{xx}+S_{12}\sigma_{yy}+S_{13}\sigma_{zz}+S_{14}\sigma_{yz}+S_{15}\sigma_{zx}+S_{16}\sigma_{xy}$$
$$\varepsilon_{yy}=S_{21}\sigma_{xx}+S_{22}\sigma_{yy}+S_{23}\sigma_{zz}+S_{24}\sigma_{yz}+S_{25}\sigma_{zx}+S_{26}\sigma_{xy}$$
$$\cdots\cdots\cdots\cdots\cdots\cdots\cdots\cdots\cdots\cdots$$
$$\varepsilon_{xy}=S_{61}\sigma_{xx}+S_{62}\sigma_{yy}+S_{63}\sigma_{zz}+S_{64}\sigma_{yz}+S_{65}\sigma_{zx}+S_{66}\sigma_{xy}$$

and correspondingly

$$\sigma_{xx}=C_{11}\varepsilon_{xx}+C_{12}\varepsilon_{yy}+C_{13}\varepsilon_{zz}+C_{14}\varepsilon_{yz}+C_{15}\varepsilon_{zx}+C_{16}\varepsilon_{xy}$$
$$\cdots\cdots\cdots\cdots\cdots\cdots\cdots\cdots\cdots\cdots$$
$$\sigma_{xy}=C_{61}\varepsilon_{xx}+C_{62}\varepsilon_{yy}+C_{63}\varepsilon_{zz}+C_{64}\varepsilon_{yz}+C_{65}\varepsilon_{zx}+C_{66}\varepsilon_{xy}$$

where

σ_{xx}, σ_{yy}, σ_{zz} and σ_{yz}, σ_{zx}, σ_{xy} represent normal and shear stresses, respectively;
ε_{xx}, ε_{yy}, ε_{zz} and ε_{yz}, ε_{zx}, ε_{xy} represent normal and shear strains, respectively.

The elastic constants S_{ij} and C_{ij} are called the elastic compliances and elastic stiffnesses, respectively. Many of the constants are equal, the number of independent constants decreasing with increasing crystal symmetry. For example, in the hexagonal system there are five independent constants, while in the cubic system there are only three elastic compliances S_{11}, S_{12}, S_{44} with corresponding elastic stiffnesses C_{11}, C_{12}, C_{44}.

The tensile and shear moduli will vary with orientation in a single crystal of a cubic metal according to

$$\frac{1}{E}=S_{11}-2[(S_{11}-S_{12})-\tfrac{1}{2}S_{44}]\,(l^2m^2+m^2n^2+l^2n^2)$$

$$\frac{1}{G}=S_{44}-2[(S_{11}-S_{12})-\tfrac{1}{2}S_{44}]\,(l^2m^2+m^2n^2+l^2n^2)$$

where l, m, n are the direction cosines of the specimen axis with respect to the crystallographic axes. For an isotropic crystal

$$S_{44}=2(S_{11}-S_{12}) \text{ and } C_{44}=\tfrac{1}{2}(C_{11}-C_{12})$$

hence

$$E=\frac{1}{S_{11}} \text{ and } G=\frac{1}{S_{44}}$$

Therefore, the degree of anisotropy is conveniently specified by

$$\frac{2(S_{11}-S_{12})}{S_{44}} \text{ or } \frac{(C_{11}-C_{12})}{2C_{44}}$$

15.1.2 Principal elastic compliances and elastic stiffnesses at room temperature

The units are TPa^{-1} for S_{ij} (elastic compliances) and GPa for C_{ij} (elastic stiffnesses).

Table 15.2 CUBIC SYSTEMS (3 CONSTANTS)

Metal	S_{11}	S_{44}	S_{12}	C_{11}	C_{44}	C_{12}	*Ref.*
Ag	22.9	22.1	−9.8	123	45.3	92.0	1
Al	16.0	35.3	−5.8	108	28.3	62.0	1, 2
Au	23.4	23.8	−10.7	190	42.3	161	1
Ca	94.0	83.0	−31.0	16.0	12.0	8.0	3
Cr	3.08	9.98	−0.49	346	100	66.0	1
Cs (78 K)	1 676	676	−762	2.49	2.06	1.48	4
Cu	15.0	13.3	−6.3	169	75.3	122	1
Fe	7.67	8.57	−2.83	230	117	135	1
Ge	9.73	14.9	−2.64	129	67.1	48.0	1
Ir	2.24	3.72	−0.67	600	270	260	5
	2.28	3.90	−0.67	580	256	242	6
K	1 215	531	−558	3.71	1.88	3.15	7
	1 339	526	−620	3.69	1.90	3.18	8
Li	315	104	−144	13.4	9.60	11.3	9
Mo	2.71	9.0	−0.74	459	111	168	1
Na	549	233	−250	7.59	4.30	6.33	1
Nb	6.56	35.2	−2.29	245	28.4	132	1, 10, 11

Table 15.2 CUBIC SYSTEMS (3 CONSTANTS)—*continued*

Metal	S_{11}	S_{44}	S_{12}	C_{11}	C_{44}	C_{12}	*Ref.*
Ni (zerofield)	7.67	8.23	−2.93	247	122	153	1, 12
Ni (saturation field)	7.45	8.08	−2.82	249	124	152	1
Pb	93.7	68.0	−43.0	48.8	14.8	41.4	1
Pd	13.7	14.0	−6.0	224	71.6	173	1
Pt	7.35	13.1	−3.08	347	76.5	251	13
Rb	1 330	625	−600	2.96	1.60	2.44	14
Si	7.74	12.6	−2.16	165	79.2	64	1
Sr	148	174	−60	14.7	5.74	9.9	15
Ta	6.89	12.1	−2.57	262	82.6	156	1, 16
Th	27.2	20.9	−10.7	75.3	47.8	48.9	17
	27.4	22.0	−10.9	77.0	45.5	50.9	18
Tl	101	91	−46	40.8	11.0	34.0	19
V	6.76	23.2	−2.32	230	43.2	120	1
W	2.49	6.35	−0.70	517	157	203	1

Table 15.3 HEXAGONAL SYSTEMS (5 CONSTANTS)

Metal		11	33	44	12	13	*Ref.*
Be	S	3.45	2.87	6.16	−0.28	−0.05	1
	C	292	349	163	24	6	1
Cd	S	12.2	33.8	51.1	−1.2	−8.9	1
	C	116	50.9	19.6	42	41	1
Co	S	5.11	3.69	14.1	−2.37	−0.94	1
	C	295	335	71.0	159	111	1
Dy	S	16.0	14.5	41.2	−4.6	−3.2	1
	C	74.0	78.6	24.3	25.5	21.8	1
Er	S	14.1	13.2	36.4	−4.2	−2.6	1
	C	84.1	84.7	27.4	29.4	22.6	1
Gd	S	18.3	16.1	48.3	−5.7	−3.8	20
	S	18.0	16.1	48.1	−5.7	−3.6	21
	C	66.7	71.9	20.7	25.0	21.3	20
	C	67.8	71.2	20.8	25.6	20.7	21
Hf	S	7.16	6.13	18.0	−2.48	−1.57	22
	C	181	197	55.7	77	66	22
Ho	S	15.3	14.0	38.6	−4.3	−2.9	1
	C	76.5	79.6	25.9	25.6	21.0	1
Mg	S	22.0	19.7	60.9	−7.8	−5.0	1
	C	59.3	61.5	16.4	25.7	21.4	1
Nd	S	23.7	18.5	66.5	−9.50	−3.90	23, 24
	C	54.8	60.9	15.0	24.6	16.6	23, 24
Pr	S	26.6	19.3	73.6	−11.3	−3.80	25, 26
	C	49.4	57.4	13.6	23.0	14.3	25, 26
Re	S	2.11	1.70	6.21	−0.80	−0.40	1
	C	616	683	161	273	206	1
Ru	S	2.09	1.82	5.53	−0.58	−0.41	20
	C	563	624	181	188	168	20
Sc	S	12.5	10.6	36.1	−4.30	−2.20	27
	C	99.3	107	27.7	39.7	29.4	27
Tb	S	17.4	15.6	46.0	−5.2	−3.60	21, 28
	C	69.2	74.4	21.8	25.0	21.8	21, 28
Tl	S	104	31.1	139	−83.0	−11.6	29, 30
	C	41.9	54.9	7.20	36.6	29.9	29, 30
Ti	S	9.69	6.86	21.5	−4.71	−1.82	1
	C	160	181	46.5	90.0	66.0	1
Y	S	15.4	14.4	41.1	−5.10	−2.70	31
	C	77.9	76.9	24.3	29.2	20.0	31
Zn	S	8.22	27.7	25.3	−0.60	−7.0	1, 32
	C	165	61.8	39.6	31.1	50.0	1
Zr	S	10.1	8.0	30.1	−4.0	−2.4	1
	C	144	166	33.4	74	67	1

Table 15.4 TRIGONAL SYSTEMS (6 CONSTANTS)

Metal		11	33	44	12	13	14	*Ref.*
As	S	30.6	140	45.0	20.5	−56.0	1.7	33
	C	130	58.7	22.5	30.3	64.3	−3.7	33
Bi	S	25.7	41.1	113	−7.8	−11.2	−21.4	1, 34
	C	62.3	37.0	11.5	23.1	23.4	7.3	1, 34
B	S	—	—	—	—	—	—	—
	C	467	473	198	241	—	15.1	35
Hg (83 K)	S	154	45.0	151	−119	−21	−100	36
	C	36.0	50.5	12.9	28.9	30.3	4.7	36
Sb	S	16.0	29.6	39.1	−6.1	−6.0	−12.4	1
	C	101	44.8	39.6	31.4	27.0	22.1	1
Se	S	131	41	112	−13	−40	56	1
	C	18.6	76.1	14.8	7.3	25.2	5.6	1
Te	S	53.4	24.3	52.1	−16.1	−13.6	26.7	1
	C	34.4	70.8	32.7	9.0	24.9	13.1	1

Table 15.5 TETRAGONAL SYSTEMS (6 CONSTANTS)

Metal		11	33	44	66	12	13	*Ref.*
In	S	149	199	154	83	−44	−96	1
	C	45.2	44.9	6.52	12.0	40	41.2	1
Sn	S	42.4	14.8	45.6	42.1	−32.4	−4.3	1
	C	73.2	90.6	21.9	23.8	59.8	39.1	1

Table 15.6 ORTHORHOMBIC SYSTEMS (9 CONSTANTS)

Metal		11	12	33	44	55	66	12	13	23	*Ref.*
Ga	S	12.2	14.0	8.49	28.6	23.9	24.8	−4.4	−1.7	−2.4	1
	C	100	90.2	135	35.0	41.8	40.3	37.0	33.0	31.0	1
U	S	4.91	6.73	4.79	8.04	13.6	13.4	−1.19	0.08	−2.61	37
	C	215	199	267	124	73.4	74.3	46.5	21.8	108	37

REFERENCES TO TABLES 15.2 TO 15.6

1. Landolt-Börnstein, 'Numerical Data and Functional Relationships in Science and Technology', New Series, Group III, Vol. *2*, Berlin, Springer-Verlag, 1979.
2. C. Gault, P. Boch, A. Dauger, *Phys. Stat. Solidi*, 1977, **a43**, 625.
3. M. Taut and H. Eschrig, *Phys. Stat. Solidi*, 1976, **b73**, 151.
4. F. J. Kollarits and T. Trivisonno, *J. Phys. Chem. Solids*, 1968, **29**, 2133.
5. H. G. Purwins, H. Hieber and J. Labusch, *Phys. Stat. Solidi*, 1965, **11**, k63.
6. R. E. Macfarlane, J. A. Rayne and C. K. Jones, *Phys. Letters*, 1966, **20**, 234.
7. P. A. Smith and C. S. Smith, *J. Phys. Chem. Solids*, 1965, **26**, 279.
8. G. Fritsch and H. Bube, *Phys. Stat. Solidi*, 1975, **a30**, 571.
9. H. C. Nash and C. S. Smith, *J. Phys. Chem. Solids*, 1959, **9**, 113.
10. E. Walker and M. Peter, *J. appl. Phys.*, 1977, **48**, 2820.
11. D. M. Schlader and J. F. Smith, *J. appl. Phys.*, 1977, **48**, 5062.
12. K. Salama and J. A. Alers, *Phys. Stat. Solidi*, 1977, **a41**, 241.
13. R. E. Macfarlane, J. A. Rayne and C. K. Jones, *Phys. Letters*, 1965, **18**, 91.
14. C. A. Roberts and R. Meister, *J. Phys. Chem. Solids*, 1966, **27**, 1401.
15. S. S. Mathur and P. N. Gupta, *Acustica*, 1974, **31**, 114.
16. W. L. Stewart *et al.*, *J. appl. Phys.*, 1977, **48**, 75.
17. P. E. Armstrong, O. N. Carlson and J. F. Smith, *J. appl. Phys.*, 1959, **30**, 36.
18. J. D. Greiner, D. T. Peterson and J. F. Smith, *J. appl. Phys.*, 1977, **48**, 3357.
19. M. S. Shepard and J. F. Smith, *Acta Met.*, 1967, **15**, 357.
20. E. S. Fisher and D. Dever, *Trans. Met. Soc. AIME*, 1967, **239**, 48.
21. S. B. Palmer, E. W. Lee and M. N. Islam, *Proc. roy. Soc.*, 1974, **A338**, 341.

22. E. S. Fisher and C. J. Renken, *Phys. Rev.*, 1964, **135A**, 482.
23. J. D. Greiner *et al.*, *J. appl. Phys.*, 1976, **47**, 3427.
24. J. T. Lenkkeri and S. B. Palmer, *J. Phys.*, 1977, **F7**, 15.
25. J. D. Greiner *et al.*, *J. appl. Phys.*, 1973, **44**, 3862.
26. S. B. Palmer and C. Isci, *Physica*, 1977, **86–88**, 45.
27. E. S. Fisher and D. Dever, *Proc. Rare Earth Res. Conf.*, Coronado Calif., 1968, Vol. 7, p. 237.
28. K. Salama, F. R. Brotzen and P. L. Donoho, *J. appl. Phys.*, 1972, **43**, 3254.
29. R. Weil and A. W. Lawson, *Phys. Rev.*, 1966, **141**, 452.
30. R. W. Ferris, M. L. Shepherd and J. F. Smith, *J. appl. Phys.*, 1963, **34**, 768.
31. J. F. Smith and J. A. Gjevre, *J. appl. Phys.*, 1960, **31**, 645.
32. D. P. Singh, S. Singh and S. Chendra, *Ind. J. Phys.*, 1977, **A51**, 97.
33. N. G. Pace and G. A. Saunders, *J. Phys. Chem. Solids*, 1971, **32**, 1585.
34. A. M. Lichnowski and G. A. Saunders, *J. Phys.*, 1977, **C10**, 3243.
35. I. M. Silvestrova *et al.*, *Mater. Res. Bull.*, 1974, **9**, 1101.
36. H. B. Huntington, 'Solid State Physics', (ed. F. Seitz and D. Turnbull), New York, Academic Press, 1958, Vol. 7, p. 213.
37. E. S. Fisher and H. J. McSkimin, *J. appl. Phys.*, 1958, **29**, 1473.

15.2 Damping capacity

The damping capacity of a metal measures its ability to dissipate elastic strain energy. The existence of this property implies that Hooke's law is not obeyed even at stresses well below the conventional elastic limit. In a perfectly elastic solid in vibration, stress and strain are always in phase and no energy is dissipated.

There are two important types of damping: anelastic and hysteretic.

In an anelastic solid there is a lag between the application of stress and the attainment of the resulting equilibrium strain; unless the stress changes exceedingly slowly. Processes with this characteristic give rise to an energy loss that reaches a peak at a critical frequency of vibration.

An hysteretic solid has a stress–strain curve on loading that does not coincide with that on unloading. The area between the two curves is proportional to the energy loss and does not vary with the frequency with which the load cycle is traversed but changes in a complex fashion with peak stress. Damping from this class of mechanism is often high and since it does not vary with frequency is of particular interest to the engineer since it can contribute to vibration and noise reduction and can limit the intensity of vibrational stress under resonant conditions and thus minimize fatigue failure.

Table 15.7 lists the specific damping capacity of a number of commercial alloys including some of very high damping that might be of interest to vibration engineers. In all cases the damping is predominantly of the hysteretic type.

Table 15.7 THE SPECIFIC DAMPING CAPACITY OF COMMERCIAL ALLOYS AT ROOM TEMPERATURE

The specific damping capacity which is normally measured on solid cylinders stressed in torsion is defined as the ratio of the vibrational strain energy dissipated during one cycle of vibration to the vibrational strain energy at the beginning of the cycle.

Alloy	*Composition %*	*Specific damping capacity %*	*Surface shear stress* MPa
Cast irons			
High carbon inoculated flake iron	2.5% C, 1.9% Si, 1.0% Mn, 20.7% Ni, 1.9% Cr, 0.13% P	19.3	34.5
Spun cast iron	3.54% C, 3.39% G.C., 1.9% Si, 0.4% Mn, 0.38% P	10.8	34.5
Non-inoculated flake iron	3.3% C, 2.2% Si, 0.5% Mn, 0.14% P, 0.03% S	8.5	34.5
Inoculated flake iron	3.3% C, 2.2% Si, 0.5% Mn, 0.14% P, 0.03% S	7.3	34.5
Austenitic flake graphite	2.5% C, 1.9% Si, 1% Mn, 20.7% Ni, 1.9% Cr, 0.03% P, 0.03% S	7.1	34.5
Alloyed flake graphite	3.14% C, 2% Si, 0.6% Mn, 0.7% Ni, 0.4% Mo, 0.14% P, 0.03% S	5.3	34.5

Table 15.7 THE SPECIFIC DAMPING CAPACITY OF COMMERCIAL ALLOYS AT ROOM TEMPERATURE—*continued*

Alloy	*Composition %*	*Specific damping capacity %*	*Surface shear stress* MPa
Nickel-copper austenitic flake	2.55% C, 1.9% Si, 1.25% Mn, 15.2% Ni, 7.3% Cu, 2% Cr, 0.03% P, 0.04% S	3.9	34.5
Undercooled flake graphite titanium/CO_2 treated	3.27% C, 2.2% Si, 0.6% Mn, 0.35% Ti, 0.14% P, 0.03% S	3.9	34.5
Annealed ferritic nodular	3.7% C, 1.8% Si, 0.4% Mn, 0.76% Ni, 0.06% Mg, 0.03% P, 0.01% S, <0.003% Ce	2.8	34.5
Pearlitic malleable	BS 3333/1961 Grade B.33/4	1.6	34.5
Blackheart malleable	BS 310/1958 Grade B.22/14	1.5	34.5
As cast pearlitic nodular	3.66% C, 1.8% Si, 0.4% Mn, 0.76% Ni, 0.06% Mg, 0.03% P, 0.01% S, <0.003% Ce	1.4	34.5
Steels			
BS 970 070M20 En3	0.17% C mild steel, normalized	1.5	34.5
BS 1407 (silver steel)	Spherodized	0.8	34.5
BS 1407 (silver steel)	Water quenched 800°C	0.5	34.5
BS 1407 (silver steel)	Water quenched 800°C aged 100°C $1\frac{1}{2}$ h	0.2	34.5
BS 970 653M31 En23T	3% Ni, 1% Cr, 0.3% C	0.8	34.5
BS 970 503M40 En12Q	1% Ni, 0.4% C	0.3	34.5
BS 970 709M40 En19U	1% Cr, 0.3% Mo, 0.4% C	0.15	34.5
BS 3S62	12% Cr, 0.2% C. Quenched tempered to 225 BHN	3.8	34.5
BS 970 321S20 En58B	18% Cr, 8% Ni, 0.6% Ti, 0.1% C. Solution treated 1 050°C water quenched	1.8	34.5
NMC	0.62% C, 3.86% Cr, 8.6% Ni, 7.3% Mn. Solution treated 1 050°C, water quenched	0.7	34.5
BS 970 302S25 En58A	18% Cr, 8% Ni, 0.1% C. Solution treated 1 050°C, water quenched	0.3	34.5
Copper alloys			
Hidurel 6	As cast	1.35	34.5
Gunmetal	88% Cu, 10% Zn, 2% Sn	1.0	34.5
Brass (BS 265)	As extruded	0.4	34.5
Hidurel 5	As cast	0.4	34.5
Hidurel 7	As cast	0.25	34.5
High tensile brass	As cast	0.25	34.5
Novoston	—	0.25	34.5
Aluminium alloys			
Duralumin (HE 14)	— —	0.25	34.5
RR57 (DTD 5004 WP)	—	0.20	34.5
RR58 (DTD 5014 WP)	—	0.10	34.5
Hiduminium 100 (SAP)	—	5.0	34.5
Magnesium alloys			
DTD 5005	Mg/Zn/Zr/Th	7.4	20.7
BS 1278	Mg/Zn/Mn	1.6	20.7
DTD 721A	Mg/Zn/Zr	0.65	20.7
Magnesium Elektron MSR Alloy	Mg/Ag/Zr	0.4	20.7
Manganese alloys			
Mn–Cu (quenched from 850°C aged 2 h at 425°C)	90% Mn–10% Cu	21	34.5
	85% Mn–15% Cu	27	34.5
	80% Mn–20% Cu	22	34.5
	70% Mn–30% Cu	42	34.5
	60% Mn–40% Cu	42	34.5
	50% Mn–50% Cu	33	34.5
Nickel alloys			
Ni–Ti (Nitinol)	55% Ni–45% Ti	26	69
T–D Nickel	2.5% Thoria	10.7	69
Mallory No-chat	—	9.4	69

Reference: D. Birchon, *Engineering Materials and Design*, Sept., Oct., 1964.

15.2.1 Anelastic damping

Of interest to the physical metallurgist is the fact that a phase lag between stress and strain can give rise to a peak in energy dissipation or damping as a function of temperature or frequency. Several quite distinct atomic processes have been identified with damping peaks and measurements on these peaks in a wide variety of metals and alloys have been used to give diffusion data and to study precipitation, ordering phenomena and the properties of dislocations, point defects and grain boundaries. Table 15.8 identifies the damping peaks found in a number of pure metals and alloys with the relaxation process thought to be involved and also give an indication of the magnitude of the damping peak height. Detailed information on the specific mechanisms involved can be obtained from the reviews below and the references given for the respective damping peaks. The main types of peak that are observed are as follows. In cold worked pure metals movement of dislocation lines results in a number of low temperature peaks known as Bordoni peaks. Interaction of dislocations with point defects give rise to a further series of unstable peaks at higher temperatures. In alloys the stress-induced redistribution of solute atoms results in two types of peak, the Zener-type peak in substitutional solid solutions and the Snoek-type peak in interstitial solid solutions. The interaction of interstitial solute atoms with substitutional solute atoms gives rise to a modified Snoek peak in ternary alloys. In cold worked alloys the interaction of interstitial solute atoms with dislocations results in the Köster-type peaks at higher temperatures than the Snoek-type peaks. In pure metals and alloys the stress-induced migration of grain boundaries and/or polygonised (sub-grain) boundaries gives rise to a further series of high temperature damping peaks.

In the ideal case, for a relaxation process having a single relaxation time (τ) the logarithmic decrement (δ) will be given by

$$\delta=\pi\Delta\frac{\omega\tau}{1+\omega^2\tau^2}$$

where ω is the angular frequency and Δ is the modulus defect. The relaxation time, being diffusion controlled, varies with temperature according to an Arrhenius equation of the form $\tau=\tau_0 \exp (H/RT)$ where τ_0 is a constant, H is the activation energy controlling the relaxation process and T is the absolute temperature. The condition for maximum damping (δ_p) is that $\omega\tau=1$ and hence

$$\delta_p=\frac{\pi\Delta}{2}(=\pi Q^{-1}\ \mathrm{max})$$

The modulus defect which is a measure of the strength of the relaxation can be highly orientation dependent and therefore the values given in the table below must be interpreted with caution. It must also be noted that for many measured damping peaks a distribution of relaxation times is found to be present. This leads to a broader peak being observed than would be present if a single relaxation time were operative. The decrement will be given by

$$\delta=\pi\sum_i\Delta_i\frac{\omega\tau_i}{1+\omega^2\tau_i^2}$$

where each i refers to a component of the total peak that has the same form as that of a peak arising from a single relaxation time.

This theoretical aspect of the analysis of broad peaks in terms of a spectrum of τ's has been comprehensively dealt with by A. S. Nowick and B. S. Berry in *IBM Journal of Research and Development*, 1961, 5(4), 297–311, 312–20.

REVIEWS

1. C. Zener, 'Elasticity and Anelasticity of Metals', Chicago: Chicago University Press, 1948.
2. K. M. Entwistle, 'Progress in Non-Destructive Testing' (edited by E. G. Stanford, J. H. Fearnon), vol. 2, p. 191, London: Heywood, 1960.
3. D. H. Niblett and J. Wilks, *Adv. Phys.*, 1960, **9**, 1.
4. K. M. Entwistle, *Metall. Rev.*, 1962, **7**, 175.
5. A. S. Nowick and B. S. Berry, 'Anelastic Relaxation in Crystalline Solids', Academic Press, 1972.

In Table 15.8, metals and alloys are listed in alphabetical order with the highest concentration constituent first. The values of the modulus defect are deduced using the equation

$$\Delta = \frac{2\delta_p}{\pi} = 2Q^{-1}_{max}$$

where δ_p is the peak decrement and Q^{-1}_{max} is the peak value of Q^{-1}. These relationships are valid only if the damping arises from a process having a single relaxation time. In most cases a distribution of relaxation times exists and the peaks are broader, but there is only rarely sufficient published data to permit this distribution to be deducted. As many authors do not make it clear which damping units they use, the quoted values of Modulus Defect in the table must not be interpreted too precisely. The aim is to record the existence of peaks and list the suggested mechanism giving rise to them, and the values of Δ serve to indicate the approximate strength of the relaxation. If detailed quantitative data are sought peaks should be analysed in particular cases using the method of Berry and Nowick.

Table 15.8 ANELASTIC DAMPING

Alloy	*Composition and physical condition*	*Peak temp.*	*Frequency* Hz	*Modulus defect*	*Activation energy* kJ mol^{-1}	*Mechanism or type*	*Ref.*
Ag	99.999% Ag	37 K	0.7	2.6×10^{-2}	0.84	Bordoni type	1, 2
	(CW* 16% at 4.2 K)	50 K	0.7	1.2×10^{-2}	0.84	Bordoni type	1, 2
Ag	99.999% Ag single crystal deformed at RT 43.7% [121] 5.4% [111]	~50 K	~600	13×10^{-4}	—	Bordoni type	3
	Single crystals deformed at (5.4% [111]) RT	~50 K	~600	6×10^{-3}	—	Bordoni type	3
Ag	99.99% Ag	173 K	10^3	—	21.3	Point defect/ dislocation interaction	4
	(CW* at RT)	200 K	10^3	—	35.6 etc.	Point defect/dislocation interaction	4
Ag	99.998% Ag (84% area reduction, then annealed at 500 °C for 1 h)	163 °C	1.04	38×10^{-4}	92.0	Grain boundary	5
		356 °C	0.98	39×10^{-4}	177	Grain boundary	5
Ag	99.999% Ag	150 K	1.5	1.3×10^{-2}	92	Grain boundary	6
Ag–Au	25–80 At. % Au	364–398 °C	1.0	2×10^{-3}–2×10^{-2}	175.7–	Zener	7
	42 At. % Au	460 °C	0.36	0.104 max	165.3	Grain boundary	8
Ag–Cd	29 At. % Cd	~230 °C	~1	—	152.3	Zener	9
	39 At. % Cd	—	~1	—	123.0	Zener	9
Ag–Cd	32 At. % Cd	220 °C	0.6	4×10^{-2}	146.9	Zener	10
	0.9–32 At. % Cd	367–452 °C	1.5	9.2×10^{-2}–1.38×10^{-1}	159–188	Grain boundary	6, 11
Ag–In	9.6 At. % In–17.9 At. % In	580–536 K	~1	$<3.5\times10^{-3}$–7×10^{-3}	152.8–130.5	Zener	12
Ag–In	7.5 At. % In–15.6 At. % In	—	~1	6.5×10^{-4}–6×10^{-3}	159 133.1	Zener	9
Ag–In	10.8 At. % In–18.1 At. % In	~500 K	10^{-3}–10^{-1}	$\sim20\times10^{-3}$	146.9–139.7	Zener	13
Ag–In	16 At. % In	270 °C	1.6	2.9×10^{-2}	—	Zener	14
Ag–In	1 At. % In–16 At. % In	355–450 °C	1.5	9.5×10^{-2}–1.27×10^{-1}	172–188	Grain boundary	6, 11
Ag–Sb	6.3 At. % Sb	~510 K	10^{-3}–10^{-1}	$\sim5\times10^{-3}$	136	Zener	15
Ag–Sn	0.93 At. % Sn	200 °C	1.5	1.3×10^{-3}	—	Zener	6, 11
Ag–Sn	8.1 At. % Sn	~550 K	10^{-3}–10^{-1}	$\sim5\times10^{-3}$	131.8	Zener	15
Ag–Sn	0.9 At. % Sn–8.0 At. % Sn	390–440 °C	1.5	1.08×10^{-1}–1.27×10^{-1}	171–184	Grain boundary	6, 11
Ag–Zn	15 At. % Zn–30 At. % Zn	280–232 °C	0.68	1.2×10^{-2}–1.4×10^{-2}	142–136	Zener	16

* Cold worked.

Table 15.8 ANELASTIC DAMPING—*continued*

Alloy	*Composition and physical condition*	*Peak temp.*	*Frequency* Hz	*Modulus defect*	*Activation energy* kJ mol^{-1}	*Mechanism or type*	*Ref.*
Al	99.99% Al (CW* 3% at RT)	24 K	1.2×10^4	1.5×10^{-4}	2.3	Bordoni type	17, 18
Al	99.999% Al deformed at 20 K then at 80 K	70 K	3	—	11.6	Bordoni type	19
		100 K	3	—	18.3	Bordoni type	
		115 K	3	—	—	?	
		155 K	3	—		?	
Al	99.99% Al (CW* 6% at RT)	83 K	1.08×10^3	1.6×10^{-3}	16.3	Bordoni type	20
		119 K	1.06×10^3	2.4×10^{-3}	28.9	Bordoni type	
Al	99.999% Al 2 h at 470 K, then CW* by 0.5% at 77 K	110 K	4×10^{-2}	$\sim 10^{-2}$	24	Hasiguti type	21
		155 K	2×10^3	$\sim 2 \times 10^{-3}$	—	Hasiguti type	
Al	99.999% Al deformed at 85 K by 10^{-4} and cycled 10^2 times	130 K	~1	$\sim 30 \times 10^{-3}$	—	Bordoni type with contribution from impurity–dislocation interactions	22
Al	99.994% [111]	139 K	10^7–	—	4.1	Bordoni type	23
	[100]	153 K	5×10^7	—	6.0	Bordoni type	
	[110]	196 K		—	19.5	Bordoni type	
Al	99.999% Al (CW* 18% at RT)	213 K	10^3	—	38.5	Point defect/dislocation interaction	4
Al	99.6% Al reduced by 69%	270 °C	5×10^{-1}	$\sim 1.2 \times 10^{-1}$	—	Relaxation of stresses by shear deformation and recrystallization	24
Al	99.6% Al reduced by 99.3%	270 °C	5×10^{-1}	$\sim 2 \times 10^{-1}$	—	Relaxation of stresses by shear deformation and recrystallization	24
		400 °C	5×10^{-1}	$\sim 10^{-1}$	—	Associated with grain boundaries	24
Al	99.999% Al deformed by 65%, annealed and deformed again	~300 °C	~1	10^{-3}–10^{-2} depending on CW	48.2	Associated with dislocations	25
Al	99.96% Al area reduced by 75% annealed at 325 °C for 2 h	340 °C	2.32	7.85×10^{-2}	144.3	Grain boundary	5
Al	99.991% Al	275 °C	0.69	1.4×10^{-1}	134	Grain boundary	26
Al	99.999% Al CW* 4% then irradiated by neutrons at 80 K, annealed 360 K	110 K	2.5	1.6×10^{-2}	—	Point defect/dislocation interaction	241
Al–Ag	20% Ag (quenched from 520 °C, aged at 155 °C)	140 °C	0.25	1.2×10^{-2}	105–113	Stress induced change of local degree of precipitation	27
Al–Ag	2.5% Ag–30% Ag annealed	~140–~170 °C	~1	$\sim 5 \times 10^{-4}$ $\sim 8 \times 10^{-3}$	155	Diffusion controlled relaxation of partial dislocations around precipitates	28
Al–Ag	15% Ag (quenched from 200 °C)	160–	0.45	$\sim 4 \times 10^{-3}$	—	Associated with γ	29
		210 °C	0.45	$\sim 4 \times 10^{-3}$	—	Associated with clustering	
Al–Ag	Al + 30% Ag (1) quenched	410 K	—	8×10^{-3}	92	Zener in solid solution	247
	(2) aged at 520 K	420 K	—	14×10^{-3}	110	Zener in ξ phase	247
Al–Cu	4% Cu (quenched from 520 °C reverted at 200 °C)	175 °C	0.9	2×10^{-3}	128(117)	Zener	30, 31
Al–Cu	4% Cu (quenched, reverted, aged at 200 °C for 144 h)	120 °C	0.5	2.4×10^{-3}	92	Stress induced change of shape of precipitate	30

* Cold worked.

Table 15.8 ANELASTIC DAMPING—*continued*

Alloy	*Composition and physical condition*	*Peak temp.*	*Frequency* Hz	*Modulus defect*	*Activation energy* kJ mol^{-1}	*Mechanism or type*	*Ref.*
Al–Cu–Mg–Si	Quenched from 500 °C, aged at 50 °C	20 °C	2×10^3	6×10^{-4} max	56.5	Stress induced ordering of complex atom group	32
Al–Fe	Al–0.25% Fe (annealed at 600 °C for 6 h)	~280 °C	20–	~5×10^{-4}	117	?	33
		~310 °C	2×10^2	~2×10^{-3}	167	?	33
Al–Fe	Al–0.06% Fe	~280 °C	~1	Depends on grain size	~140	Grain boundary	34
Al–Fe	Al–0.16% Fe–0.5% Fe	310–360 °C	~1	Depends on grain size	~140	Grain boundary	34
		440–480 °C	~1		200	Related to relaxation of stresses and precipitation of Fe on grain boundaries	34
Al–Mg–Si	Al–0.6% Mg–0.6% Si (quenched from 480 °C and aged at 230 °C for 26½ h)	215 °C	2.25	5.5×10^{-2}	126	Stress induced diffusion of solutes (?)	35
Al–Mg	2% Mg	RT	2.5×10^4	1.2×10^{-5}	—	Thermoelastic	36
Al–Mg	5.45% Mg (quenched from 500 °C, aged at 250 °C)	165 °C	1.5	7×10^{-4}	116.3	Zener	37
Al–Mg	7.5% Mg (annealed at 400 °C then quenched)	~40 °C	~2	—	66.9	Relaxation of solute clusters	38
		~80 °C	~2	—	—	Zener	
		~120 °C	~2	—	102.5	Zener	
		~227 °C	~2	—	125.5	?	
Al–Mg	7.5% Mg (annealed at 400 °C, cooled slowly)	~203 °C	~2	—	125.5		
Al–Mg	0.93% Mg–12.1% Mg	~100–~200 °C	0.4–98	10^{-4}–2×10^{-3}	135.0	Zener	39
Al–Zn	3.7% Zn (quenched from 450 °C to RT)	21 °C	0.26	~2×10^{-3}	53	Stress induced ordering of zinc atom–vacancy complexes	40
Au	99.999% Au. (CW* 16% at 4.2 K)	43 K	0.5	1.4×10^{-1}	9.62	Bordoni type	1, 41
		65 K	0.5	1.6×10^{-1}	18.4	Bordoni type	1, 41
		77 K	0.5	1.6×10^{-1}	18.4	Bordoni type	1, 41
Au	99.999% Au (annealed at 1 170 K for 4 h, then CW* 3% at 77 K)	120 K	4×10^{-2}–	~10^{-4}	—	Hasiguti type	21
		180 K	2×10^3	~5×10^{-3}	35.7	Hasiguti type	21
		210 K		~10^{-4}	58 177 (two stages)	Hasiguti type	21
Au	99.999% Au (CW* 16% at 70 K)	130 K	4.0	4×10^{-3}	21.3	Hasiguti type	42
		190 K	4.0	2.2×10^{-2}	32.84	Hasiguti type	42
		210 K	4.0	6×10^{-3}	34.7	Hasiguti type	42
Au	99.999% Au (quenched) from 700 °C)	160 K	~1	~2×10^{-3}	68(?)	Associated with dislocations	43
		230 K		~2×10^{-3}	—	Associated with dislocations	43
Au	99.999% Au (quenched from 800 °C)	~290 K	~1	~4×10^{-3}	—	Associated with dislocations	43
Au	99.999% (quenched from 1 000 °C)	210 K	~1	~4×10^{-4}	57.7	Stress induced reorientation of divacancies	44
		220 K	~1	—	—	Hasiguti type	44
Au	99.999 9% Au (quenched from 1 000 °C)	0 °C	~10	7×10^{-4}–8×10^{-3}	62.7	Stress induced reorientation of divacancies	45
Au	99.99% Au (CW*, annealed at 600 °C)	330 °C	0.7	Depends on grain size	141.4	Grain boundary	46
Au	99.999 8% Au (CW* 36% annealed, 650–870 °C°)	238 °C	1.0	4.4×10^{-2}	144.3	Grain boundary	47
		404 °C	1.0	3.2×10^{-2}	242.7	Associated with grain boundaries	47

* Cold worked.

Table 15.8 ANELASTIC DAMPING—*continued*

Alloy	*Composition and physical condition*	*Peak temp.*	*Frequency* Hz	*Modulus defect*	*Activation energy* kJ mol^{-1}	*Mechanism or type*	*Ref.*
Au	99.999 95% Au (annealed at 900 °C)	~400 °C	~1	$\sim 10^{-1}$	435	Sliding at grain boundaries	48
Au–Ag–Zn	Au–42 At. % Ag–15 At. % Zn	260 °C	0.7	0.11	146	Zener	49
Au–Cu	10 At. % Cu–90 At. % Cu	326–392 °C	1.0	3.8×10^{-3}–0.57	114.6–165.3	Zener	50, 48
Au–Cu	10 At. % Cu–90 At. % Cu	552–753 °C	1.0	0.14 0.76	201.7–342.3	Adsorption of solute atoms on grain boundaries	48
Au–Cu	10 At. % Cu–90 At. % Cu	175–250 °C	1.0	5×10^{-2}–2×10^{-2}	—	Grain boundaries	48
Au–Cu–Zn	42 At. % Cu–15 At. % Zn	380 °C	0.5	0.14	—	Order–disorder peak	49
Au–Cu–Zn	21 At. % Cu–17 At. % Zn	300 °C	0.5	0.50	—	Order–disorder peak	49
Au–Cu–Zn	63 At. % Cu–17 At. % Zn	340 °C	0.5	2.6×10^{-2}	—	Zener	49
Au–Fe	5% Fe	365 °C–	~1	2.4×10^{-3}	159	Zener	51
		535 °C		1.5×10^{-2}	177.8	Precipitation of Fe	51
Au–Fe	7% Fe	380 °C–	~1	1.9×10^{-2}–	159	Zener	51
		564 °C		1.2×10^{-2}	177.8	Precipitation of Fe	51
Au–Fe	10% Fe–27% Fe	390 °C	~1	3.5×10^{-2}–1.5×10^{-2}	151–151.9	Zener	51
Au–Ni	30 At. % Ni	397 °C	1.0	—	182.0	Zener	52
Au–Ni	Au–30 At. % Ni (quenched)	~380 °C	~0.5	$\sim 10^{-1}$	88.3	Zener, modified by quenched–in vacancies	53
Au–Ni	7.7 At. % Ni–90.8 At. % Ni	397 °C 652 °C	1.0	—	182.0–251.0	Zener	52
Au–Zn	15 At. % Zn	250 °C	0.7	4.2×10^{-2}	218	Zener	49
Au–Zn	Stoichiometric AuZn annealed	260 °C–290 °C	0.2–1	$\sim 10^{-2}$	140	Concerned with short-range order (?)	54
Be		210 K	1.0	—	51.9	Bordoni type	55
		135 °C	1.0	—	101.7	Solute atom/defect interaction	55
Be	98.6% Be (annealed)	213 K	~1	$\sim 3\times10^{-4}$	50.2	Cold work induced line defect interaction (?)	56
		135 °C	~1	$\sim 5\times10^{-4}$	100	Solute interaction with lattice (?)	56
Be–Fe–O	Be–0.4 % Fe + interstitial impurities including oxygen	0.5 °C	1.22	10^{-2}	63.6	Snoek type	57
Cd–Mg	Cd–29.3% Mg	20 °C	0.75	0.26	80	Zener	58
Cd–Mg	5% Mg–30% Mg	20 °C	1.0	0.29 max	79.5	Zener	59
Co	99.23% Co (0.69% N)	215 K (two peaks)	10^3	—	40.6	Bordoni type (?)	4
Co	99.23% Co (0.69% N) CW* at RT	263 K (two peaks)	10^3	—	51.0	Point defect/dislocation interaction	4 4
		297 K	10^3	—	36.8	Twin boundaries (?)	
Co	99.999% Co (quenched)	410 K 600 K	10–5×10^2	$\sim 2\times10^{-3}$	72.3	Movement of divacancies and dislocations	60
Co–C	Heated in C atmosphere at 1 050 °C and quenched	280 °C	1–16×10^5	4×10^{-4}	159	Motion of C atom pairs in lattice	61
Co–Fe–Cr	Co–37.8% Fe–8.7% Cr. In magnetic field of 0.6×10^3 A/M	111 °C	10^3	3×10^{-3}	—	Electron spin redistribution at Curie point	62
Co–Ni	Co–2% Ni	340 °C	1.5–	$\sim 2.5\times10^{-2}$	—	α–β phase	63
		430 °C	1.8×10^4	$\sim 2.8\times10^{-2}$	—	transformation	63

* Cold worked.

Table 15.8 ANELASTIC DAMPING—*continued*

Alloy	*Composition and physical condition*	*Peak temp.*	*Frequency* Hz	*Modulus defect*	*Activation energy* kJ mol^{-1}	*Mechanism or type*	*Ref.*
Co–Ni	Co–23% Ni	150 °C	1.5–	$\sim 2.9 \times 10^{-2}$	—	α–β phase	63
		310 °C	1.8×10^4	$\sim 3.1 \times 10^{-2}$	—	transformation	63
Cr	99.8% Cr	38 °C	—	2×10^{-3}	—	Electron spin redistribution at Neel temp. (40 °C)	64
Cr–N	Cr–0.004 5% N	160 °C	1.0	1.5×10^{-3}	101.7	Snoek type	65
Cr–N	35 ppm N (quenched from 83 °C)	~36 °C	3	$\sim 4 \times 10^{-4}$	115.1	Magneto-mechanical damping	66
Cr–N	(Annealed in NH_3 at 1 150 °C for 48 h)	155 °C	1	up to 2.2×10^{-3}	85.8	Snoek type	67
Cr–Re–N	Cr–35% Re (quenched from 1 000 °C in NH_3 atmosphere)	130 °C	1	$\sim 10^{-3}$	89.1	Snoek type	68
		190 °C	1	$\sim 10^{-3}$	126.4	Snoek type	
Cu	99.999% Cu (CW* at 77 K	—	~1	—	4.34	Niblett–Wilks type	69
		—	~1	—	11.6	Bordoni type	69
Cu	99.999% Cu (single crystal) CW* 5% at 77 K	38 K	1.09×10^4	2×10^{-3}	4.2	Bordoni type	19, 41, 70, 71,
		79 K	1.09×10^4	4×10^{-3}	11.7	Bordoni type	19, 41, 70, 71
Cu	99.999% Cu (single crystal ⟨100⟩ orientation)	80 °C	13	34×10^{-4}	—	Bordoni type	238
Cu	99.999% Cu (single crystal ⟨110⟩ orientation)	70 °C	13	11.4×10^{-4}	—	Bordoni type	238
Cu	Electrolytic Cu (Fatigued, 4×10^5 cycles)	140 K	0.3	$\sim 4 \times 10^{-4}$	—	Dislocation–divacancies interaction	72
		225 K	0.3	$\sim 4 \times 10^{-4}$	—	Dislocation–vacancies interaction	72
		240 K	0.3	$\sim 2 \times 10^{-4}$	—	Dislocation–interstitials interaction	72
		165 K	0.3	$\sim 1 \times 10^{-4}$	—	Dislocation–interstitials interaction	72
Cu	99.999% Cu (deformed 2.5% quenched in liquid He)	30 K	6×10^2	$\sim 10^{-3}$	—	Niblett–Wilks type	73
		70 K	6×10^2	$\sim 3 \times 10^{-3}$	—	Bordoni type	73
		190 K	6×10^2	$\sim 10^{-4}$	—	Hasiguti type	73
Cu	99.999% Cu (CW* 5% at 77 K)	148 K	1.0	4×10^{-3}	31.0	Point defect, dislocation interaction	74
		170 K	1.0	1.2×10^{-2}	33.9	Point defect, dislocation interaction	74
		238 K	1.0	1×10^{-2}	41.4	Rotation of split interstitials (?)	74 75
Cu	99.999% Cu (annealed at 500 °C for 4 h)	215 °C	1.0	2.11×10^{-2}	157	Grain boundary	
Cu	99.99% Cu (area reduced 47%, annealed at 600 °C for 2 h)	216 °C	1.17	1.65×10^{-2}	132	Grain boundary	5
Cu	99.999 9% Cu	416 °C	~1	~0.2	435	Sliding at grain boundaries	48, 76
		735 °C	~1	~0.2	169.5	Sliding at grain boundaries	76
Cu	99.999% Cu	300 °C	5.0	4×10^{-2}	156.9	Grain boundary	75
Cu–Ag	Cu–0.71% Ag	550 °C	1	—	154.7	Grain boundary	237
Cu–Ag	Cu–0.1 At. % Ag CW* 5%	223 K	5×10^3	2.6×10^{-4}	—	Dislocation/silver atom/point defect interaction	250
		283 K	5×10^3	2.5×10^{-4}	—		

* Cold worked.

Table 15.8 ANELASTIC DAMPING—*continued*

Alloy	*Composition and physical condition*	*Peak temp.*	*Frequency* Hz	*Modulus defect*	*Activation energy* kJ mol^{-1}	*Mechanism or type*	*Ref.*
Cu–Al	2% Al–10% Al (Deformed 3% at RT)	145 K	1.5×10^{7}	—	~24	Bordoni type	77
Cu–Al	Cu–16.8 At. % Al	360 °C	0.66	7×10^{-3}	174.9	Zener	78
Cu–Co	(Aged at 575 °C for 3 min)	230 °C	1	7×10^{-4} (depends on ageing)	184.9	Grain boundary	75
Cu–Ga	Cu–16 At. % Ga	330 °C	1.0	2×10^{-2}	—	Zener	33
Cu–Fe	0.5% Fe–10% Fe	320 °C–350 °C	~1	Depends on grain size	159	Grain boundary	79
		480 °C–550 °C	~1	Depends on grain size	209	Connected with precipitation of Fe on Cu grains	79
Cu–Fe	Up to 1.5% Fe (Quenched from 820 °C)	800 °C–850 °C	~1	~10^{-1}	~125	Connected with ageing of alloy	80
Cu–Ni	25% Ni–75% Ni (Quenched from 720 °C to 240 °C)	34 K	~1	—	79.9	Associated with precipitation	81
		150 K	~1	—	111.3	Associated with precipitation	81
Cu–Ni	Cu–3% Ni (Annealed at 1 100 °C)	~580 °C	~1	2.1×10^{-3}	151	Zener	82
Cu–Ni	Cu–45% Ni (Reduced by 90% at RT)	~600 °C	~1	~2×10^{-2}	—	Associated with recrystallization (?)	83
		~800 °C	~1	~5×10^{-2}	208	Associated with dislocations	83
Cu–Ni	5.6 At. % Ni–94.9 At. % Ni	590–726 °C	~1	0.129–0.112	368–264	Grain boundary	84
Cu–Ni–Zn	20 At. % Zn–10 At. % Ni	381 °C	1	5.3×10^{-3}	197	Zener	85
Cu–Ni–Zn	$Cu_2Ni_{1.15}Zn_{0.92}$ single crystal	76 K	6.9	3×10^{-2}	—	Zener	244
		71 K	6.9	1.2×10^{-2}	—	Ordering	244
Cu–Pd	0.01% Pd–0.3% Pd	30–150 K	5×10^{6}	$10^{-4}-10^{-3}$	—	Overdamped resonance of dislocations	86
Cu–Pt	0.01% Pt–0.3% Pd	30–150 K	5×10^{6}	$10^{-4}-10^{-3}$	—	Overdamped resonance of dislocations	86
Cu–Si	Cu + 5.09 wt. % Si	200 °C	3	7.5×10^{-3}	118	Precipitation of K[1]	248
Cu–Sn	3% Sn–9% Sn	490–500 °C	1.8	0.14–0.152	151–205	Grain boundary	87
Cu–Zn	(α) Cu–31% Zn	R.T.	6×10^{3}	1.8×10^{-4}	—	Thermoelastic	88
Cu–Zn	10% Zn–30% Zn	290–350 °C	0.7	$\left(\frac{\textit{At. conc Zn}}{4}\right)$	159–178	Zener	12
Cu–Zn	17.6 At. % Zn–29.4 At. % Zn	657–614 K	1.3	3.5×10^{-3}–9.2×10^{-3}	182.0–161.1	Zener	89
Cu–Zn	Cu–30% Zn	425 °C	0.5	0.12	172(?)	Grain boundary	90
Cu–Zn	(β) Cu–45 At. % Zn (Quenched from 400 °C)	70 °C	0.9	2.6×10^{-3}	69.5	Diffusion of Zn accelerated by vacancies	91
Cu–Zn	(β) Cu–45 At. % Zn	177 °C	0.9	1.4×10^{-3}	130	Stress induced reorientation of Cu atom pairs	91 91
Cu–Zn	(α–β) Cu–43 At. % Zn	285 °C	0.9	8×10^{-3}	159	Stress relaxation at β–α interfaces	91
Cu–Zn	(β–γ) Cu–50 At. % Zn	190 °C	0.9	2.2×10^{-3}	130	Stress relaxation at β–γ interfaces	91
Cu–Zn–Al	77% Cu–89% Cu 5% Zn–20% Zn 2% Al–8% Al	623–672 K	~1	2×10^{-3}–8.5×10^{-3}	—	Zener	92

Table 15.8 ANELASTIC DAMPING—*continued*

Alloy	*Composition and physical condition*	*Peak temp.*	*Frequency* Hz	*Modulus defect*	*Activation energy* kJ mol^{-1}	*Mechanism or type*	*Ref.*
Cu–Zn–Al	Cu–29.5% Zn–2.4% Al	593 K	1	3.5×10^{-2}	150.6	Zener	93
Cu–Zn–Al	Cu–17.0% Zn–9.0% Al	615 K	1	3.27×10^{-2}	163.2	Zener	93
Fe	(Re-electrolytic) (CW* 5% at RT, in magnetic field of 7.5×10^4 A/m)	~50 K	10^5	$\sim1.5\times10^{-3}$	—	Associated with motion of kinks in dislocations	94
Fe	Armco (CW* at RT)	198 K	10^3	—	44.4	Point defect/dislocation interaction	4
Fe	Armco (CW* at RT)	230 K	10^3	—	54.8	Point defect/dislocation interaction	4
Fe	CW* 40% at RT	275 °C	2.9	1.55×10^{-3}	174	Köster type	95
Fe	99.98% Fe ($<5\times10^{-5}$CN)	526 °C	1.03	Depends on grain size	192	Grain boundary	96
Fe	Fe pure (R ratio 1600) irradiated	110 K	1.4	4×10^{-4}	—	Magnetic relaxation of point defects	246
	2×10^{18} cm^{-2} at 20 K	128 K	1.4	13×10^{-4}	—	Magnetic relaxation of point defects	246
		155 K	1.4	55×10^{-4}	—	Relaxation of self-interstitials	246
Fe–Al	Fe–40 At. % Al	180 °C	0.6	3.5×10^{-3}	121	Movement of Al within tetrahedral lattice	97
		320 °C	0.6	6×10^{-3}	163	Movement of Al within tetrahedral lattice	
Fe–Al	Fe–40 At. % Al	440 °C 550 °C	0.6 0.6	1×10^{-3} 5×10^{-4}	184 —	Movement of Al within tetrahedral lattice Zener	97
Fe–Al	Fe–17 At. % Al	520 °C	1.3	1×10^{-2}	234	Zener	98
Fe–Al–C	Fe–19.3 At. % Al–0.01% C (Quenched from 720 °C, aged)	130 °C	1.4	$\sim1\times10^{-3}$	99.6	Stress induced diffusion of C in ordered Fe_3Al lattice	99
		168 °C	1.4	$\sim5\times10^{-3}$	—	(?)	
Fe–Al–C	Fe–0.7 At. % Al–0.01% C	41 °C	1.2	$\sim10^{-2}$	—	Snoek type	99
Fe–Al–C	Fe–9 At. % Al–0.01% C	~100 °C	1.2	$\sim3\times10^{-3}$	—	(?)	99
Fe–B	Fe–0.05% B	79 °C	14.0	—	63	Snoek type	100
Fe–B	Fe–50 ppm B (C impurities)	260 K 50 °C	~1	$\sim10^{-4}$ $\sim3\times10^{-3}$	54 —	Snoek type due to B Snoek type due to C	101
Fe–C	Fe–0.02% C	27 °C	0.27	2×10^{-2}	84.1	Snoek type	102
Fe–C	Fe–0.4% C (Martensite). (Quenched from 850 °C, tempered at 300 °C for 1 h)	~200 °C	~1	$\sim10^{-2}$	84.5	Associated with dislocations	103
Fe–C	Fe–0.02% C (Quenched from 700 °C, CW* 52%, aged)	210 °C	1	$\sim5\times10^{-3}$	—	Dislocation movement between pinning points (precipitates)	104
Fe–C	Fe–0.01% C (CW* 25%)	235 °C	2.20	8×10^{-3}	138	Köster peak	105
Fe–C	—	353 °C	6.65×10^6	$\sim3\times10^{-3}$	80.17	Snoek type	106
Fe–C	Fe–0.022% C	520 °C	1.0	Depends on grain size	347	Grain boundary	107
Fe–Co–Cr	Fe–54% Co–10% Cr	380 K	6×10^3	$\sim5\times10^{-4}$	—	Macro-eddy current peak	108
Fe–Cr	1.2% Cr–43% Cr	526 °C–683 °C	1.0	3×10^{-3}–10^{-2}	215.5–296.2	Grain boundary	109
Fe–Cr	Fe–16% Cr	661 °C	~1	3.18×10^{-2}	233	Grain boundary	110

* Cold worked.

Table 15.8 ANELASTIC DAMPING—*continued*

Alloy	*Composition and physical condition*	*Peak temp.*	*Frequency* Hz	*Modulus defect*	*Activation energy* kJ mol^{-1}	*Mechanism or type*	*Ref.*
Fe–Cr	Fe–22.5% Cr	662 °C	~1	0.55×10^{-2}	222	Grain boundary	110
Fe–Cr	Fe–22.5% Cr	~560 °C	~1	1.1×10^{-2}	222	Zener	111
Fe–Cr–C	Fe–(1.2% Cr–5.2% Cr)–C (Quenched from 750 °C)	~320 K	~1	$\sim 10^{-2}$	75.31–82.72	Snoek type	112
Fe–Cr–N	Fe–4.2% Cr–N	(Several peaks) 266 K–339 K	~1	Up to 2.3×10^{-3}	67.8–92.0	Motion of N atoms in various environments	113
αFe–H	Fe(<0.1 At. % H)	30 K	1.0	8×10^{-4}	—	Snoek type	114
αFe–H	Fe(<0.1 At. % H) (CW* 0.5%)	116 K	1.0	1.6×10^{-3}	—	Köster type	114
Fe–H	Fe–H (aged at 60 °C for 5 h)	~150 K	8×10^4	$\sim 2 \times 10^{-4}$	—	Köster type	115
		~180 K	8×10^4	$\sim 6.5 \times 10^{-4}$	—	Köster type	115
αFe–D	(CW* 0.5%)	35 K	1.0	8×10^{-4}–	—	Snoek type	114
		120 K	1.0	10^{-3}	—	Köster type	114
Fe–Mn–C	Fe–18.5% Mn–0.7% C	250 °C	2.0	1.1×10^{-2}	146	Diffusion of C in fcc Fe–Mn	116
Fe–Mn–C	Fe–15.5% Mn–0.35% C	285 °C	1.01	1.6×10^{-3}	155	Stress induced ordering of C atoms	117
Fe–Mn–N	Fe–2% Mn–N (Quenched from 950 °C)	(Several peaks) 267 K–337 K	~1	Up to 1.5×10^{-2}	66.1–91.6	Motion of N atoms in various environments	113
Fe–Mn–N	Fe–1.6% Mn–N (Heated at 590 °C in NH_3)	7 °C	1	$\sim 4 \times 10^{-3}$	69.0	Snoek type in complex lattice	118
		23 °C	1	$\sim 5 \times 10^{-4}$	77.4	Snoek type in Fe	
		34.5 °C	1	$\sim 2 \times 10^{-3}$	81.6	N atoms jumping from Fe–Mn sites to Fe–Fe sites	
Fe–Mn–N	Fe–2% Mn–0.01% N	135 °C	~1	$\sim 10^{-3}$	—	—	119
αFe–N	—	22 °C	0.89	$\left(\frac{\text{wt. \%N}}{0.63}\right)$	77.8	Snoek type	120
αFe–N	Fe–0.01% N (CW* 25%)	235 °C	2.35	1.3×10^{-2}	138	Köster type	105
Fe–N	Fe–0.05% N (Quenched from 580 °C, area reduced 21%)	250 °C	1	$\sim 5 \times 10^{-3}$	—	Dislocation movement between pinning points (precipitates)	104
Fe–N	—	326 °C	6.65×10^6	$\sim 2 \times 10^{-3}$	76.78	Snoek type	106
Fe–N	Fe–0.018% N	510 °C	1.0	Depends on grain size	315.9	Grain boundary	107
Fe–Ni	Fe–31.5% Ni	415 K	6×10^3	$\sim 10^{-4}$	—	Macro-eddy current peak	108
Fe–Ni	Fe–5% Ni C impurities (CW* 40% at RT)	40 °C	1.2	2×10^{-3}	—	Snoek type	95
		215 °C	2.9	1.15×10^{-3}	135	Köster type	95
Fe–Ni	Fe–4% Ni (CW*, annealed)	500 °C	1.0	0.24	213–259	Grain boundary	121
		750 °C	1.0	0.50	247–289	(?)	121
Fe–Ni–Cr	Fe–26.0% Ni–21.4% Cr (Annealed from 1 000 °C)	629 °C	~1	Depends on grain size	350	Grain boundary	122
		692 °C	~1		284	Grain boundary	122
		769 °C	~1		296	Grain boundary	122
Fe–Ni–N	Fe–2.29% Ni–N	24.5 °C	1.05	1.009×10^{-2}	77.8	Snoek type due to N	113
		40.5 °C	1.05	8.8×10^{-4}	84.1	Snoek type due to C	113
Fe–P	Fe–0.37% C–0.03% P	40 °C	1.1	3.8×10^{-3}	146	Impurity peak due to P	123

* Cold worked.

Table 15.8 ANELASTIC DAMPING—*continued*

Alloy	*Composition and physical condition*	*Peak temp.*	*Frequency* Hz	*Modulus defect*	*Activation energy* kJ mol^{-1}	*Mechanism or type*	*Ref.*
Fe–Si–N	Fe–2.83% Si–0.05% N	22.5 °C	1.26	—	75.3	Snoek type	124
		37 °C	1.26	—	50.2	N atom jump in vicinity of Fe–Si sites	124
		62 °C	1.26	—	—	N atom jump in vicinity of nitride SiN	124
Fe–Ti–N	Fe–0.15% Ti–0.06% N	~120 °C	~1	$\sim10^{-3}$	105	Interstitial N interacting with Ti	125
		~227 °C	~1	$\sim10^{-3}$	130	Interstitial N interacting with Ti	125
Fe–V	Fe–18% V	625 °C	0.73	1.6×10^{-2}	355.6	Zener	126
Fe–V–C	Fe–0.62% V (Carburized at 800 °C for 4 h, quenched)	33 °C	0.95	3.1×10^{-3}	50	Snoek type due to C	127
		82 °C	0.95	6.6×10^{-3}	46	C atom jumps associated with Fe–V sites	127
Fe–V–C	Fe–5.15% V–240 ppm C. (Quenched from 900 °C)	77 °C	451	9.8×10^{-5}	81.0	Interaction between V and C atoms in solution	128
		109 °C	451	6.1×10^{-5}	83.9	Snoek type	
Fe–V–N	Fe–0.51 At. % V–0.58 At. % N	21.5 °C	0.77	2×10^{-2}	79.5	Snoek type	129
		84 °C	0.77	1.4×10^{-2}	—	N atom jump in vicinity of nitride VN	129
Fe–V–N	Fe–0.62% V–N. (In NH_3, 900 °C for 14 h). (C impurities)	24 °C	1	2×10^{-2}	75	Snoek type due to N	127
		35 °C	1	3×10^{-3}	85.8	Snoek type due to C	127
		84 °C	1	3×10^{-3}	75.3	N atom jumps associated with Fe–V sites	127
Ga–As	GaAs (Compound) [111] (with damaged surfaces)	229 K	3.7×10^{4}	4.0×10^{-5}	40.5	Associated with dislocations on surface	130
Ga–Sb	GaSb (Compound) (With damaged surfaces) [100]	225 K	9.46×10^{4}	2.0×10^{-5}	37.6	Associated with dislocations on surface	130
Gd	(Annealed at 900 °C for 2 h)	190 K	~1	$\sim10^{-3}$	—	Micro-eddy current damping	131
		240 K	~1	$\sim3\times10^{-3}$	—	Motion of 90° domain boundaries	131
Ge	Surface damaged by polishing [100]	150 K	$\sim2\times10^{3}$	$\sim2\times10^{-5}$	29	Associated with dislocations	132
	[111]	205 K	$\sim2\times10^{-3}$	$\sim2\times10^{-5}$	29	Associated with dislocations	132
Ge		209 K	3×10^{3}	1.3×10^{-7}	52.1	Associated with dislocations	133
Ge	Surface damaged [111]	228 K	5.49×10^{4}	4.1×10^{-5}	36.6	Associated with dislocations on surface	130
Ge	Zone-refined. (Strained up to 12.2%)	35 °C	$\sim10^{3}$	$\sim1.2\times10^{-4}$	60	Motion of geometrical kinks over jogs on dislocations	134
Ge	Zone-refined. (Strained up to 12.2%, annealed at 550 °C)	180 °C	$\sim10^{3}$	$\sim10^{-5}$	82	Motion of dislocations in region of surface	134
Ge	Single crystal	395 °C	10^{5}	1.8×10^{-5}	77.4	Electronic relaxation	135
		770 °C	10^{5}	4×10^{-7}	173.6	Impurity peak (?)	135
Ge	(CW* at 800 °C)	~450 °C	~1	$\sim10^{-4}$	~175	Bordoni type	136
Ge–Sb	Ge, doped with Sb (CW* to produce screw dislocations)	~80 °C	2.5×10^{3}	$\sim3\times10^{-5}$	14.5	Kink motion	137
		~180 °C	2.5×10^{3}	$\sim5\times10^{-5}$	24.1	Kink motion	137
		~300 °C	2.5×10^{3}	$\sim1.5\times10^{-4}$	106	Kink motion	137

* Cold worked.

Table 15.8 ANELASTIC DAMPING—*continued*

Alloy	*Composition and physical condition*	*Peak temp.*	*Frequency* Hz	*Modulus defect*	*Activation energy* kJ mol^{-1}	*Mechanism or type*	*Ref.*
Ge–Sb	Ge, doped with Sb (CW* to produce Edse dislocations)	~290 °C	2.5×10^3	$\sim 3 \times 10^{-4}$	106	Kink/point defect interaction	137
Ge–Sb	Ge, doped with Sb annealed	~310 °C	2.5×10^3	$\sim 10^{-4}$	106	Dislocation/point defect interaction	137
Ge–Sb	Ge–4×10^{17} atoms Sb/millilitre	216 K	4.1×10^3	2.4×10^{-7}	55.0	Involving dislocations	133
αHf–O	—	490 °C	0.95	—	—	O atom diffusion in vicinity of Zr impurity	138
Hg–Te	Mercury telluride	170–260 K	10^7–3×10^9	—	9.07	Bordoni type	139
In–Sb	InSb compound [111] with damaged surfaces	222 K	5.65×10^4	5.1×10^{-5}	32.8	Associated with dislocations on surface	130
In–Sb	InSb compound with surface damage [211]	231 K	3.54×10^4	3.8×10^{-5}	41.5	Associated with dislocations on surface	130
In–Tl	In–10 At. % Tl	270 K	2	$\sim 5 \times 10^{-4}$	67.4	Zener	140
		390 K	2	$\sim 2 \times 10^{-3}$	111.7	Reorientation of lattice with respect to grain boundaries	140
In–Tl	In–10 At. % Tl	273 K	~1	—	67	Zener	141
		368 K	~1	0.5	29	Motion of twin boundaries	141
Ir	99.96% Ir (Annealed at 1 720 °C)	1 450 °C	2.0	7.5×10^{-2}	339	Grain boundary	142
K	(Nominal purity) annealed at RT cooled to 120 K	160 K	5×10^3	$\sim 10^{-3}$	~125	Stress induced change in size or shape of precipitates	143
K–Rb	20 At. % Rb– 50 At. % Rb	193 K 179.5 K	3.7×10^3 13×10^3	2.25×10^{-4} 1.9×10^{-3}	41.42–40.38	Zener	144
Li–Mg	Li–57 At. % Mg (single crystal)	127 °C	1.38×10^3	1.7×10^{-2}	90.0	Zener	145
Mg	Mg–Small amounts of Al, Zn, Cd, Tl, In	Peaks from 0–300 K	1.5×10^{-2} — 0.75	Complex spectrum	—	Dislocation/dislocation interaction	146
Mg	99.996% Mg	20 K	$\sim 10^4$	$\sim 4 \times 10^{-4}$	—	Associated with grain boundaries (?)	147
		40–240 K	$\sim 10^4$	$\sim 10^{-3}$	—	Dislocation movement in (0001) slip system	147
Mg	99.992% Mg (CW*)	20 K	4×10^4	3×10^{-3}	—	Bordoni type (?)	148, 71
		100 K (very broad)	4×10^4	—	—	Bordoni type	71
Mg	—	37 K	1.5×10^7	—	0.9	Dislocation motion in basal plane	149
		106.5 K	1.5×10^7	—	8.7	Dislocation motion in non-basal plane	149
		155.5 K	1.5×10^7	—	40.5	?	149
Mg	99.99% Mg (CW* at RT)	180 K	10^3	—	28.9	Point defect/dislocation interaction	4
		225 K	10^3	—	42.3	Point defect/dislocation interaction	4
Mg	99.97% Mg	220 °C	0.46	8×10^{-2}	—	Grain boundary	26
Mg	99.999 9% Mg	40 K	1–2	3×10^{-3}	—	Bordoni (B1)	242
	1–2% CW at 10 K	80 K	1–2	1×10^{-2}	—	Bordoni (B2)	242
Mg–Li	Mg–1.88% Li	51 K	1.5×10^7	—	1.74	Dislocation motion in basal plane	149

* Cold worked.

Table 15.8 ANELASTIC DAMPING—*continued*

Alloy	*Composition and physical condition*	*Peak temp.*	*Frequency* Hz	*Modulus defect*	*Activation energy* kJ mol^{-1}	*Mechanism or type*	*Ref.*
Mg–Li (*cont.*)		99.5 K	1.5×10^{7}	—	9.6	Dislocation motion in non-basal plane	149
		154.5 K	1.5×10^{7}	—	126	?	149
Mg–N	Mg–0.0048% N	46 K	1.5×10^{7}	—	1.0	Dislocation motion in basal plane	149
		116 K	1.5×10^{7}	—	22	Dislocation motion in non-basal plane	149
		157 K	1.5×10^{7}	—	35	?	149
Mn–Cu	Mn–12% Cu (Quenched from 925 °C)	268 K	7×10^{2}	3.6×10^{-4}	—	Stress relaxation across twin boundaries	150
Mo	Zone refined (CW* 5% at RT)	80 K	5	3×10^{-3}	17.36	Bordoni type	151
		240 K	5	2.8×10^{-3}	44.4	Bordoni type	151
		285 K	5	2.8×10^{-3}	44.4	Bordoni type	151
Mo	99.9% Mo (CW* 30% at RT)	~110 K	$\sim10^{3}$	$\sim2\times10^{-3}$	—	Associated with dislocations	152
Mo	99.9993% Mo (Deformed 50% at 450 K)	~140 K	$2–2\times10^{7}$	$\sim10^{-2}$	—	Associated with dislocations	153
Mo	99.995% Mo polycrystalline (CW*)	145 K	1.9×10^{3}	$\sim10^{-3}$	—	Dislocation type	154
Mo	High purity Mo R/Ratio 7500–8000	~140 K	500	3×10^{-3}	—	Dislocation motion	245
Mo	Zone refined (CW* 5% at RT)	300 K	0.4	1.4×10^{-2}	—	Köster type	155
Mo	Impure	~100 °C	~1	$\sim3.5\times10^{-3}$	109	Snoek type due to N	156
		~220 °C	~1	$\sim7.2\times10^{-3}$	134	Snoek type due to O	156
		~310 °C	~1	$\sim5.0\times10^{-3}$	163	Snoek type due to C	156
Mo	Zone refined (Annealed at 2300 °C for 20 min)	200–225 °C	~1	$\sim2\times10^{-3}$	126–134	Dislocation type	157
Mo	Zone refined single crystal (annealed at 1800 °C for 40 min)	~800 °C	~1	$\sim2\times10^{-2}$	272	Short range interaction between dislocations and C atom interstitials	158
Mo		820–870 °C	1.0	—	—	Grain boundary	157
Mo	Commercial	1000–1150 °C	~1	$\sim6\times10^{-2}$	372	Grain boundary	159
		1300–1600 °C	~1	~0.13	54	Recrystallization peak	159
Mo	99.98% Mo (Annealed at 1340 °C)	1050 °C	1.3	3.3×10^{-2}	389	Grain boundary	142
Mo	Zone refined, recrystallised, CW* RT, annealed at 1900 °C for 90 min	~1200 °C	5	$\sim10^{-2}$	—	Grain boundary	158
Mo–C	Mo–0.0034% C	225 °C	1.71	4.4×10^{-3}	126–134	Snoek type	157
Nb	99.97% Nb(CW*) (Superconducting type)	3.24 K	8×10^{4}	$\sim2\times10^{-6}$	0.18	Motion of dislocations	160
	(Normal type)	2.08 K	8×10^{4}	$\sim4\times10^{-6}$	0.15	Motion of dislocations	160
Nb	Zone refined (CW* 5% at RT)	11–19 K (Broad)	8.8	10^{-4}	—	Bordoni type (?)	161
Nb	99.9% Nb (CW*)	~30 K	2×10^{4}–1×10^{5}	$\sim2\times10^{-4}$	1.6	Motion of dislocations	162
		~180 K	2×10^{4}–1×10^{5}	$\sim1\times10^{-4}$	24.1	Interaction of dislocations and impurity atoms	162
Nb	High purity single crystal CW at 320 K	190 K	1.3×10^{3}	1.2×10^{-2}	—	Motion of dislocation	239

* Cold worked.

Table 15.8 ANELASTIC DAMPING—*continued*

Alloy	*Composition and physical condition*	*Peak temp.*	*Frequency* Hz	*Modulus defect*	*Activation energy* kJ mol^{-1}	*Mechanism or type*	*Ref.*
Nb	Zone refined (CW* 5% at RT)	130 K (α peaks)	8	6×10^{-4}	23.0	Bordoni type	151, 163, 164
		240 K (β peaks)	7.8	2.1×10^{-3}	45.2	Bordoni type	151, 163
Nb	99% Nb. (CW* 5%, annealed at 70 °C for 2h)	190 K	~1	~10^{-4}	—	Associated with point defect complexes	164
		200 K	~1	~10^{-3}	~95	Associated with point defect complexes	164
		220 K	~1	~10^{-4}	—	?	164
Nb	99.998% Nb (CW* 50% at RT)	220 K	2–2×10^{7}	~7.5×10^{-3}	—	Dislocation damping	153, 165, 166
Nb	Nb–0.005% N–0.005% O (Zone refined)	~580 °C	8×10^{3}	~7.8×10^{-3}	—	Snoek type due to N	167
		~380 °C	8×10^{3}	~3.5×10^{-3}	—	Snoek type due to O	167
		~470 °C	8×10^{3}	~1×10^{-3}	—		167
Nb	(CW*)	~630 °C	8	—	—	Relaxation of ordered cluster of interstitials near dislocations	168
Nb–Al	Nb–0.29% Al (0.004% C, 0.003% N, 0.029% O) (Annealed)	152 °C	1	~9×10^{-3}	—	Snoek type due to N	169
		283 °C	1	~1×10^{-3}	—	Snoek type due to O	169
Nb–Al	Nb–0.29% Al (0.004% C, 0.003% N, >0.029% O) (CW*)	410 °C	1	~1×10^{-3}	—	Dislocation/solute interaction	169
Nb–C	Nb–0.014% C	259 °C	0.57	4×10^{-3}	139.3	Snoek type	170, 171
Nb	High purity single crystal + 180 at ppm H. CW at 320 K	270 K	1.3×10^{3}	3.0×10^{-2}	—	Snoek–Köster type	239
Nb	Nb single crystal R. ratio 2500	α_1 143 K	0.5	10×10^{-3}	29.7	Formation of kink-pairs in non-screw dislocation	249
		α_2 121 K	0.5	6×10^{-3}	19.8	Kink diffusion in screw dislocation	249
Nb–Mo	1.3% Mo–16% Mo (CW* 30% at RT)	180–220 K	2–2×10^{7}	5×10^{-4}–1×10^{-4}	24.1	Type of Bordoni mechanism	153, 166
Nb–Mo–O	Nb–5.3 At. % Mo–0.16 At. % O (Homogenized at 1 200 °C for 15 min)	167 °C	0.6	~10^{-2}	—	Snoek type	172
Nb–N	Nb–0.018% N	274 °C	0.55	1.1×10^{-2}	145.6	Snoek type	170, 171
Nb–N	Nb–0.066% N (CW* 10%)	500 °C	0.31	2.4×10^{-2}	201	Köster type	173
Nb–O	Nb–0.18 At. % O	152 °C	0.6	~10^{-2}	—	Snoek type	172
Nb–O	Nb–0.026% O	168 °C	2.13	1.6×10^{-2}	114.2	Snoek type	174, 171
Nb–O–N	Nb–1.2 At. % O–0.11 At. % N (CW* 34%)	420 K	~1	~5×10^{-3}	—	Segregation of O atoms to dislocations	175
		500 K	~1	~1×10^{-3}	—	Segregation of N atoms to dislocations	175
Nb–Ti	Nb–48% Ti	100 °C	0.6	~7×10^{-3}	100.0	Snoek due to O impurities	176
Nb–Ti	Nb–48% Ti (N atmosphere at 1 200 °C for 1 h)	340 °C	0.6	~10^{-2}	—	Snoek due to N	176

* Cold worked.

Table 15.8 ANELASTIC DAMPING—*continued*

Alloy	*Composition and physical condition*	*Peak temp.*	*Frequency* Hz	*Modulus defect*	*Activation energy* kJ mol^{-1}	*Mechanism or type*	*Ref.*
Nb–V	Nb–0.3% V	~200 K	2–2 × 10^7	2.5 × 10^{-3}	—	Dislocation damping	153
Nb–Zr	Nb–1.0% Zr	~200 K	2–2 × 10^7	1 × 10^{-3}	—	Dislocation damping	153
Nb–Zr	Nb–1% Zr – O + traces of N	~500 °C	5 × 10^4	—	110.9	Snoek due to O in Nb	177
		~500 °C	5 × 10^4	—	111.3	Substitutional–interstitial process involving O	177
		~500 °C	5 × 10^4	—	123.4	Substitutional–interstitial process involving O pairs	177
Nb–Zr	—	~500 °C	5 × 10^4	—	146.4	Snoek due to N in Nb	177
		~500 °C	5 × 10^4	—	147.3	Substitutional–interstitial process involving N	177
Ni	Zone refined (CW*)	138 K	3 × 10^4	2 × 10^{-3}–35 × 10^{-3}	—	Niblett and Wilks type	178
		248 K	3 × 10^4	1 × 10^{-2}–0.10	—	Bordoni type	178
Ni	Zone refined single crystal (CW*)	145–123 K	~2 × 10^4	Up to 1.6 × 10^{-3}	—	Associated with dislocation reactions	179
		223–263 K	~2 × 10^4	Up to 3.0 × 10	—	Bordoni type	179
Ni	99.99% Ni (CW* at RT)	155 K	10^3	—	29.7	Point defect/dislocation interaction	4
		350 K	10^3	—	51.0	Point defect/dislocation interaction	4
		397 K	10^3	—	69.4	Point defect/dislocation interaction	4
Ni	99.9% Ni	70 °C	0.7	—	77.0	Stress induced re-orientation of interstitial Ni atom pairs	180, 181
Ni	99.999 9% Ni	150 °C	1	~0.1	—	Magneto-mechanical damping	182
Ni	99.99% Ni. (Area reduced 90%, annealed at 905 °C for 1 h)	432 °C	1.41	3.4 × 10^{-2}	308	Grain boundary	5
Ni	99.98% Ni	440–460 °C	0.5	0.10	—	Grain boundary	183
		630–720 °C	0.5	0.12	—	Stress relaxation at polygonized boundaries	183
Ni–Al–C	Ni–2% Al–0.5% C (Quenched)	280 °C	0.5	—	—	Diffusion of C in Ni–Al	184
Ni–C	(Quenched)	230 °C	0.5		—	Diffusion of C in Ni	184
Ni–Cr	0.5% Cr–19.5% Cr	530–800 °C	2	Depends on grain size	—	Grain boundary	185
Ni–Cr–C	Ni–20% Cr–1.87% C (Quenched)	250 °C	0.9	—	98.3	Diffusion of C in Ni–Cr	186
Ni–Cu	Ni–20% Cu	~140 °C	1	~10^{-2}	—	Magneto-mechanical damping	182
		(Varies)	1	~10^{-2}	—	Magnetic ordering	182
Ni–Zr	0.1% Zr–0.5% Zr	~200 °C	~1	2 × 10^{-3}–8 × 10^{-4}	—	Magneto-mechanical damping	187
		~450 °C	~1	1 × 10^{-3}–3 × 10^{-4}	—	Grain boundary	187
		600–700 °C	~1	< 1 × 10^{-4}–2 × 10^{-4}	—	'Blocking' peak	187
Pd–H	99.999% Pd (Annealed, electrolytically loaded with H) 40 At. % H–75 At. % H	70–80 K	2.7	~10^{-3}	12.34–16.19	Stress induced ordering of H pairs in β phase	188

* Cold worked.

Table 15.8 ANELASTIC DAMPING—*continued*

Alloy	*Composition and physical condition*	*Peak temp.*	*Frequency* Hz	*Modulus defect*	*Activation energy* kJ mol^{-1}	*Mechanism or type*	*Ref.*
Pd–H	PdH (β phase)	120 K	$\sim 10^3$	$\sim 2\times10^{-3}$	—	Snoek type due to H	189
	(Strained)	$\sim$150 K	$\sim 10^3$	$\sim 5\times10^{-3}$	—	Pinning of dislocations by interstitial impurities	189
Pd–D	99.999% Pd (Annealed, electrolytically loaded with D) 40 At. % D–73 At. % D	78–86 K	2.7	—	15.94–20.71	Stress induced ordering of D pairs in β phase	188
Pt	(Deformed 2.9%, annealed at 1 080 K for 1 h)	$\sim$70 K	5×10^3–6.5×10^4	5×10^{-6}	11.6	Associated with dislocations grouped into sub-grain boundaries	190
Pt	99.999% Pt (CW* 16% at 4.2 K)	125 K	0.8	8×10^{-3}	28.0	Bordoni type	1, 2
Pt	'Pure'	940–1 090 K	10^{-2}–1	—	275	Recrystallisation peak	191
Pu–Al	Pu–5 At. % Al	$\sim$65 K	10^7	—	—	Co-operative electron transition	192
Pu–Ce	Pu–6 At. % Ce	$\sim$65 K	10^7	—	—	Co-operative electron transition	192
Re	—	1 400 K	1.1	4×10^{-2}	586	Grain boundary	193
Si	Surface damaged [100] by polishing	160 K	$\sim 2\times10^{-5}$	$\sim 2\times10^{-5}$	29	Associated with dislocations	132
	[111]	200 K	$\sim 2\times10^3$	$\sim 4\times10^{-5}$	29	Associated with dislocations	132
Si	Pure n-type [111] (Quenched from 1 000°C to 77 K)	689 K	1.2×10^3	$\sim 2\times10^{-4}$	130	Migration of O–vacancy complex	194
Si	Single crystal	655°C	10^5	1.4×10^{-5}	134.7	Electronic relaxation	195, 196
Si–Cu	Si (Cu doped)	398 K	1.2×10^3	$\sim 3\times10^{-5}$	68	Migration of interstitial Cu	194
		626 K	1.2×10^3	$\sim 2\times10^{-4}$	96	Precipitation of Cu	194
Si–Li–B	Si–Li–0.01% B	210°C	1.74×10^4	6.5×10^{-5}	80.0	Reorientation of Li^+B^- pairs	197
Si–O	Single crystal	1 030°C	10^5	6×10^{-4}	246.0	Stress-induced diffusion of interstitial oxygen	195, 196
Sn	99.99% Sn	80°C	3×10^2	3×10^{-2}	79.5	Grain boundary	198
Ta	99.99%Ta (CW* 7%)	24.6 K	2.26×10^4	3.1×10^{-4}	3.7	Bordoni type	199
Ta	Zone refined (CW* 12% at RT)	124 K (α peaks)	0.8	9×10^{-4}	24.3	Bordoni type	151
		202 K (β peaks)	0.8	1.3×10^{-3}	41.4	Bordoni type	151
Ta	(Single crystal ⟨111⟩ orientation) CW* 3.1%	170 K	17	19×10^{-3}	—	Snoek–Köster type	239
Ta	(Single crystal)	150 K	0.7	—	4×10^{-3}	Dislocation kink formation	240
Ta	—	1 100°C	0.65	7×10^{-2}	418	Grain boundary	193
Ta	99.89% (Annealed at 1 700°C)	1 230°C	1.0	5.6×10^{-2}	406	Grain boundary	142
Ta–C	Ta–0.1% C	338°C	0.55	2.6×10^{-2}	161.1	Snoek type	200, 171
Ta–H	99.9% Ta (CW* 1.7% H charged)	27 K	2×10^4–10^5	$\sim 5\times10^{-5}$	1.6	Associated with dislocations	162
Ta–H	—	100 K	1.75×10^2	2.4×10^{-3}–2.7×10^{-3}	—	Bordoni type (?)	201, 162
		190 K	1.75×10^2	—	34.7	Stress induced ordering in the Ta_2H phase	201, 162

* Cold worked.

Table 15.8 ANELASTIC DAMPING—*continued*

Alloy	*Composition and physical condition*	*Peak temp.*	*Frequency* Hz	*Modulus defect*	*Activation energy* kJ mol^{-1}	*Mechanism or type*	*Ref.*
Ta–H	8.5 At. % H–42.4 At. % H	1 °C	1.75×10^2	6×10^{-3}–1.6×10^{-2}	52.3	Stress induced ordering in Ta_2H phase	201
		36 °C	1.75×10^2	3×10^{-3}–1.9×10^{-2}	—	Long range ordering of H in Ta_2H phase	201
		54 °C	1.75×10^2	2.6×10^{-3}–8.5×10^{-3}	—	Short range ordering of H in Ta_2H phase	201
Ta–N	Ta–0.11% N	334 °C	0.6	9×10^{-2}	156.9	Diffusion of interstitial N	202
		362 °C	0.6	9×10^{-2}	167.4	Diffusion of interstitial N atom pairs	202
Ta–O	Ta–0.081% O	137 °C	0.6	8×10^{-2}	104.6	Diffusion of interstitial O	203, 204
		162 °C	0.6	8×10^{-2}	104.6	Diffusion of interstitial O atom pairs	203, 204
Ta–O	Ta–0.01% O (CW* 30%)	340 °C	0.5	4.4×10^{-3}	151	Köster type	205
Ta–Re–N	Ta–(1.3 At. % Re–3.8 At. % Re)–600 ppm N	~340 °C ~380 °C	0.8 0.8	$\sim6\times10^{-3}$ $\sim10^{-3}$	156.9 168.2	Snoek type due to N Snoek type due to N	206
Te	—	~400 K	1.44×10^4	$\sim10^{-4}$	—	Recombination of election–hole pairs	207
Ti	99.7% Ti (CW* at RT)	220 K	10^3	—	33.9	Bordoni type (?)	4
		305 K	10^3	—	42.3	Point defect/dislocation interaction	4
		336 K	10^3	—	51.9	Point defect/dislocation interaction	4
Ti	99.6% Ti	775 °C	1.0	0.38	201	Grain boundary	208
Ti–Al	0.04% Al–0.12% Al	675–725 °C	1.0	2.5×10^{-2}–0.4	218–293	Grain boundary	208
Ti–Au	Ti–0.05% Au	715 °C	1.0	4×10^{-2}	259.4	Grain boundary	208
Ti–Cr	Ti–10 At. % Cr (Quenched from 1 000 °C)	152 K	$\sim10^5$	1.75×10^{-2}	29	Associated with vacancies in 2 co-existing electronic environments	209
Ti–H	Ti–0.15% H	273 K	1.2	1.6×10^{-2}	62.8	Diffusion of interstitial H	210
Ti–Nb	Ti–25 At. % Nb (Quenched from 1 000 °C)	177 K	$\sim10^5$	6.15×10^{-3}	29	Associated with vacancies in 2 co-existing electronic environments	209
Ti–Nb	0.04% Nb–0.12% Nb	625–675 °C	1.0	2.7×10^{-2}–3.7×10^{-2}	213–264	Grain boundary	208
Ti–Ni	Ti–10 At. % Ni (Quenched from 1 000 °C)	152 K	$\sim10^5$	2.15×10^{-3}	29	Associated with vacancies in 2 co-existing electronic environments	209
Ti–Ni	TiNi Intermetallic compound. (Ni–49% Ti). Annealed at 800 °C	203 K	~1	$\sim10^{-2}$	36.6	Dislocation motion (?)	211
		223–313 K	~1	$\sim10^{-2}$	—	Fine structure of 203 K peak	211
Ti–Ni	—	350 °C	~1	$\sim10^{-3}$	—	Impurity effect (?)	211
		600 °C	~1	$\sim2\times10^{-2}$	—	Transition from TiNi (II) to TiNi (I)	211
Ti–O	Ti–2 At. % O	450 °C	1.0	1.2×10^{-2}	200.8	Diffusion of interstitial O in presence of impurities	212, 213,

* Cold worked.

Table 15.8 ANELASTIC DAMPING—*continued*

Alloy	*Composition and physical condition*	*Peak temp.*	*Frequency Hz*	*Modulus defect*	*Activation energy kJ mol^{-1}*	*Mechanism or type*	*Ref.*
Ti–O	Ti–0.6 At. % O	660–650°C	1.0	2.5×10^{-2}	188.3	Grain boundary	214
Ti–V	Ti–20 At. % V (Quenched from 1 000°C)	140 K	$\sim10^5$	2.2×10^{-3}	29	Associated with vacancies in 2 co-existing electronic environments	209
Ti–V	Ti–50 At. % V (Quenched from 1 000°C)	161 K	$\sim10^5$	2×10^{-3}	29	Associated with vacancies in 2 co-existing electronic environments	209
Ti–V	0.02% V–0.12% V	600–700°C	1.0	3×10^{-2}–0.42	230–335	Grain boundary	208
Ti–Zr	0.02% Zr–0.12% Zr	650–700°C	1.0	3.1×10^{-2}–3.6×10^{-2}	251–502	Grain boundary	208
U	99.9% U (CW* at RT)	155 K	10^3	—	23.0	Bordoni type (?)	4
		202 K	10^3	—	42.3	Point defect/dis-location interaction	4
V–C	—	162°C	0.55	6.4×10^{-3}	114.2	Diffusion of interstitial C	215
V–H	99.99% V+600 ppm H	170 K	500	13×10^{-4}	—	Point defect/dis-location interaction	243
V–H	—	18.5 K	2×10^2–10^5	$\sim6\times10^{-5}$	1.1	Dislocation damping	216
		250 K	~	$\sim10^{-5}$	—	Point defect/dis-location interaction	216
		285 K	~	$\sim10^{-5}$	—	Point defect/dis-location interaction	216
V–H	99.99%+H in soln	203 K	2×10^2–10^5	$\sim7\times10^{-4}$	50	Stress induced ordering in β phase	216
V–H	1.2 At. % H–14.5 At. % H	195 K	75	8×10^{-3}–7.2×10^{-2}	37.87	Diffusion of interstitial H	217
V–N	—	272°C	1.0	2.2×10^{-2}	142.7	Diffusion of interstitial N	218
V–O	—	174°C	0.55	1.2×10^{-2}	122.6	Diffusion of interstitial O	215
W	99.999 8% W	—		—	469	Grain boundary	219
		—		—	46	Movement of vacancies	219
		150 K		—	33	Movement of divacancies	219
W	High purity (CW* 3% at 400°C)	165 K	1.5×10^4	10^{-3}	24.3	Bordoni type	151, 200
W	99.99% W single crystal	~300°C	~1	$\sim2\times10^{-4}$	146	Snoek type of uncertain origin	221
		~400°C	~1	$\sim2\times10^{-4}$	188	Snoek type associated with C impurities	221
W	Commercial purity	1 250°C	0.94	0.28	481–523	Grain boundary	193, 222
W	(CW* at RT)	1 535°C	~70	~0.2	—	Primary re-crystallization peak	223
W	Zone refined	1 600–1 650 K	10^{-2}–0.25	~0.6	477	Recrystallization peak	224
W	Commercial purity	1 900°C	0.35	0.56	619	Grain boundary	222, 225
W	(CW* at RT, annealed at 3 000°C)	~2 000°C	~70	~0.1	—	Grain boundary	223
W–Re	W–20% Re	1 950°C	1.08	$\sim5\times10^{-2}$	510	Grain boundary	223
Zn	99.999% Zn (Compressed 2.4% at RT)	~100 K	$\sim2\times10^7$	—	5.8	Dislocation movements in basal plane	226

* Cold worked.

Table 15.8 ANELASTIC DAMPING—*continued*

Alloy	*Composition and physical condition*	*Peak temp.*	*Frequency* Hz	*Modulus defect*	*Activation energy* kJ mol^{-1}	*Mechanism or type*	*Ref.*
Zn (*cont.*)		~170 K	$\sim 2\times 10^7$	—	15.4	Dislocation movements in prismatic plane	226
		~230 K	$\sim 2\times 10^7$	—	19.2	Dislocation movements in pyramidal plane	226
Zn	99.999% Zn (Annealed at 100 °C for 12 h)	383 K	~1	~0.3	95	Grain boundary	227
Zr	99.999% Zr (7 k bar pressure)	80 K	1.5×10^5	$\sim 5\times 10^{-4}$	28	Bordoni type	228
		250 K	1.5×10^5	$\sim 2\times 10^{-3}$	13	(?)	228
Zr	99.9% Zr (CW* at RT)	200–220 K	10^3	9×10^{-4}	33.9(?)	Bordoni type(?)	41, 229
		305 K	10^3	—	42.3	Point defect/dislocation interaction	4
Zr	—	336 K	10^3	—	51.0	Point defect/dislocation interaction	4
Zr	99.9% Zr	600 °C	1.0	4.6×10^{-2}	218–243	Grain boundary	230, 231
		860 °C	1.0	0.28	—	α–β transformation	230, 232
Zr–Cu	Zr–up to 2.5% Cu (Annealed at 900 °C for 1 h, CW* 10%)	~220 K	1.3×10^4	Up to 1.5×10^{-3}	33.8	Bordoni type	233
Zr–Cu–Mo	Zr–0.5% Cu–0.5% Mo (Deformed 10%)	~230 K	1.3×10^4	2×10^{-3}	33.8	Bordoni type	233
Zr–H	Zr–0.89% H	228 K	1.0	4×10^{-3}	48.5	Diffusion of interstitial H	231
Zr–H	Zr–1.28% H	5 °C	1.0	2.4×10^{-2}	71.1	Diffusion of interstitial H atom pairs	231
Zr–H	Zr–1.15% H	~50 °C	$\sim 2\times 10^4$	$\sim 1\times 10^{-3}$	—	Associated with δ and γ phases	234
		~130 °C	$\sim 2\times 10^4$	$\sim 1\times 10^{-2}$	—	Associated with δ and γ phases	234
Zr–H	Zr–0.26% H	230 °C	1.0	5×10^{-3}	—	Associated with ZrH precipitate	231
Zr–Hf–O	Zr–(0.005% Hf–1% Hf)–O	530 °C 540 °C	~1	$\sim 7\times 10^{-3}$ $\sim 10^{-2}$	—	Grain boundary	235
		422 °C	~1	2×10^{-4}–4×10^{-4}	201	Diffusion of O in lattice	235
Zr–Mo	Zr–6 At.% Mo (Quenched from 1 000 °C)	213 K	$\sim 10^5$	1.36×10^{-2}	29	Associated with vacancies in 2 co-existing electronic environments	209
Zr–Nb	Zr–5% Nb	12 K	10^5	$\sim 2\times 10^{-4}$	1.9	?	236
Zr–Nb	5% Nb–25% Nb (CW*)	40 K		$\sim 6\times 10^{-4}$	4.8	Jahn–Teller type	236
		160 K	10^5	Depends on CW	29	Stress induced re-orientation of atoms	236
Zr–Nb	12 At.% Nb–75 At.% Nb (Quenched from 1 000 °C)	163–207 K	$\sim 10^5$	1.5×10^{-4}–5.9×10^{-3}	29	Associated with vacancies in 2 co-existing electronic environments	209
Zr–O	Zr–1.95% O	420 °C	~1	Depends on O conc^n	201	Diffusion of O in lattice	235
Zr–V	Zr–10 At.% V (Quenched from 1 000 °C	193 K	$\sim 10^5$	1.5×10^{-4}	29	Associated with vacancies in 2 co-existing electronic environments	209

* Cold worked.

REFERENCES TO TABLE 15.8

1. S. Okuda, *J. Phys. Soc. Japan*, 1963, **18** (Suppl. 1), 187.
2. S. Okuda, *Appl. Phys. Letters*, 1963, **2**, 163.
3. B. M. Mecs and A. S. Nowick, *Phil. Mag.*, 1968, **17**, (147), 509.
4. R. R. Hasiguti, N. Igata and G. Kamoshita, *Acta Met.*, 1962, **10**, 442.
5. J. N. Cordea and J. W. Spretnak, *Trans. Met. Soc. AIME*, 1966, **236** (12), 1685.
6. L. Rotherham and S. Pearson, *J. Metals*, 1956, **8**, 881, 894.
7. T. J. Turner and G. P. Williams, Jr., *J. phys. Soc., Japan*, 1963, **18** (Suppl. II), 218.
8. T. J. Turner and G. P. Williams, Jr., *Acta Met.*, 1960, **8**, 891.
9. B. Mills, *Phys. Status, Solidii*, 1971, **6a** (1), 55.
10. T. J. Turner and G. P. Williams, Jr., *Acta Met.*, 1962, **10**, 305.
11. S. Pearson, RAE Rep No Met., 1953, **71**.
12. B. N. Finkel'shteyn and K. M. Shtrakhman, *Phys. Met. Metallogr.*, 1964, **18** (4), 132.
13. G. P. Williams, Jr. and T. J. Turner, *Phys. Status Solidii.*, 1968, **26** (2), 645.
14. B. G. Childs and A. D. Le Claire, *Acta Met.*, 1954, **2**, 718.
15. G. P. Williams, Jr. and T. J. Turner, *Phys. Status Solidii*, 1968, **26** (2), 645.
16. A. S. Nowick, *Phys. Rev.*, 1952, **88**, 925.
17. E. Lax and D. H. Filson, *Phys. Rev.*, 1959, **114**, 1273.
18. L. J. Bruner, *Phys. Rev.*, 1960, **118**, 399.
19. J. Völkl, W. Weinländer and J. Carsten, *Phys. Status Solidii*, 1965, **10** (2), 739.
20. W. J. Baxter and J. Wilks, *Acta Met.*, 1963, **11**, 978.
21. W. Benoit, B. Bays, P. A. Grandchamp, B. Vittoz, G. Fantozzi and P. Gobin, *J. Phys. Chem. Solids*, 1970, **31** (8), 1907.
22. J. L. Chevalier, P. Peguin, J. Perez and P. Gobin, *J. Phys.(D)*, 1972, **5** (4), 777.
23. M. Mongy, K. Salama and O. Beckman, *Solid State Commun.*, 1963, **1** (7), 234.
24. A. A. Galkin, O. I. Datsko, V. I. Zaytsev and G. A. Matinin, *Phys. Met. Metallogr.*, 1969, **28** (1), 207.
25. J. Perez and P. Gobin, *Phys. Status Solidi*, 1967, **24** (2), K167.
26. T. S.-Kê *Phys. Rev.*, 1947, **72**, 41.
27. A. C. Damask and A. S. Nowick, *J. appl. Phys.*, 1955, **26**, 1165.
28. G. Schoeck and E. Bisogni, *Phys. Status Solidii*, 1969, **32** (1), 3.
29. R. E. Miner, T. L. Wilson and J. K. Jackson, *Trans. met. Soc., AIME*, 1969, **245** (6), 1375.
30. B. S. Berry and A. S. Nowick, *NACA Tech. Note*, 1958, 4225.
31. I. N. Fitzpatrick, *Ph.D. Thesis*, University of Manchester, 1965.
32. K. M. Entwistle, *J. Inst. Met.*, 1953–1954, **82**, 249.
33. E. A. Attia, *Brit. J. appl. Phys.*, 1967, **18** (9), 1343.
34. B. Ya Pines and A. A. Karmazin, *Phys. Met. Metallogr.*, 1970, **29** (1), 206.
35. K. J. Williams, *Acta Met.*, 1967, **15** (2), 393.
36. R. H. Randall and C. Zener, *Phys. Rev.*, 1940, **58**, 473.
37. W. G. Nilson, *Canad. J. Phys.*, 1961, **39**, 119.
38. B. N. Dey and M. A. Quader, *Canad. J. Phys.*, 1965, **43** (7), 1347.
39. J. Belson, D. Lemercier, P. Moser and P. Vigier, *Phys. Status Solidii*, 1970, **40**, 647.
40. H. Haefner and W. Schneider, *Phys. Status Solidii*, 1971, **4a**, K221.
41. S. Okuda, *J. appl. Phys.*, 1963, **34**, 3107.
42. S. Okuda and R. R. Hasiguti, *Acta Met.*, 1963, **11**, 257.
43. C. H. Neuman, *J. Phys. Chem. Solids*, 1966, **27** (2), 427.
44. S. Okuda and R. R. Hasiguti, *J. phys. Soc. Japan*, 1964, **19** (2), 242.
45. D. G. Franklin and H. K. Birnbaum, *Acta Met.*, 1971, **19** (9), 965.
46. W. Köster, L. Bangert and J. Hafner, *Z. Metall*, 1956, **47**, 224.
47. D. R. Mash and L. D. Hall, *Trans. AIMME*, 1953, **197**, 937.
48. M. E. De Morton and G. M. Leak, *Metal Sci. J.*, 1967, **1**, 166.
49. A. Pirson and C. Wert, *Acta Met.*, 1962, **10**, 299.
50. G. K. Mal'tseva, V. S. Postnikov and V. V. Usanov, *Phys. Met. Metallogr.*, 1963, **16** (2), 120.
51. B. A. Mynard and G. M. Leak, *Phys. Status Solidii*, 1970, **40** (i), 113.
52. C. Ang, J. Sivertson and C. Wert, *Acta Met.*, 1955, **3**, 558.
53. J. R. Cost, *Acta Met.*, 1965, **13** (12), 1263.
54. K. Mukherjee, *J. appl. Phys.*, 1966, **37** (4), 1941.
55. C. Ang and K. T. Kamber, *J. appl. Phys.*, 1963, **34**, 3405.
56. Choh-Yi Ang and K. T. Kamber, *J. appl. Phys.*, 1963, **34** (11), 3405.
57. M. J. Elias and R. Rawlings, *J. Less-common Metals*, 1965, **9** (4), 305.
58. J. Lulay and C. Wert, *Acta Met.*, 1956, **4**, 627.
59. J. Enrietto and C. Wert, *Acta Met.*, 1958, **6**, 130.
60. R. Kamel and K. Z. Botros, *Phys. Status Solidii*, 1965, **12** (1), 399.
61. G. Mah and C. A. Wert, *Trans. met. Soc. AIME*, 1968, **242** (7), 1211.
62. K. P. Belov, G. I. Katayev and R. Z. Levitin, *J. appl. Phys.*, 1960, **31** (Suppl. 1), 1535.
63. V. N. Belko, B. M. Darinskiy, V. S. Postnikov and I. M. Sharshakov, *Phys. Met. Metallogr.*, 1969, **27** (1), 140.
64. M. E. Fine, E. S. Greiner and W. C. Ellis, *J. Metals, N. Y.*, 1951, **191**, 56.
65. M. E. De Morton, *J. appl. Phys.*, 1962, **33**, 2768.

66. M. J. Klein, *J. appl. Phys.*, 1967, **38** (2), 819.
67. M. J. Kelin and A. H. Claver, *Trans. met. Soc., AIME*, 1965, **233** (11), 1771.
68. M. J. Klein, *Trans. Met. Soc., AIME*, 1965, **233** (1), 1943.
69. S. Okuda, *J. appl. Phys.*, 1963, **34** (10), 3107.
70. D. H. Niblett and J. Wilks, *Phil. Mag.*, 1956, **1**, 415.
71. H. S. Sack, *Acta Met.*, 1962, **10**, 455.
72. P. Bajons and B. Weiss, *Scripta Metall.*, 1971, **5**, 511.
73. B. M. Mecs and A. S. Nowick, *Acta Met.*, 1965, **13** (7), 771.
74. M. Koiwa and R. R. Hasiguti, *Acta Met.*, 1963, **11**, 1215.
75. D. T. Peters, J. C. Bisseliches and J. W. Spretnak, *Trans. met. Soc. AIME*, 1964, **230** (3), 530.
76. M. E. De Morton and G. M. Leak, *Acta Met.*, 1966, **14** (9), 1140.
77. H. Kayano, K. Kamigaki and S. Koda, *J. phys. Soc. Japan*, 1967, **23**, (3), 649.
78. C. Y. Li and A. S. Nowick, *Phys. Rev.*, 1956, **103**, 294.
79. A. A. Karmazin and V. I. Startsev, *Phys. Met. Metallogr*, 1970, **29** (6), 191.
80. V. S. Postnikov, S. A. Ammer, A. T. Kosilov and A. M. Belikov, *Phys. Met. Metallogr.*, 1966, **21** (5), 121.
81. B. N. Dey, *Scripta Metall.*, 1968, **2** (9), 501.
82. J. T. A. Roberts and P. Barrand, *Scripta Metall*, 1969, **3** (1), 29.
83. V. S. Postnikov, I. V. Zolotukhin and I. S. Pushkin, *Phys. Met. Metallogr.*, 1968, **26** (4), 147.
84. J. T. A. Roberts, *Metall. Trans.*, 1970, **1** (9), 2487.
85. M. G. Coleman and C. A. Wert, *Trans. Met. Soc. AIME*, 1966, **236** (4), 501.
86. A. Ikushima and T. Kaneda, *Trans. Japan Inst. Metals*, 1968, **9** (Suppl).
87. K. J. Marsh, *Acta Met.*, 1954, **2**, 530.
88. R. H. Randall, F. C. Rose and C. Zener, *Phys. Rev.*, 1939, **56**, 343.
89. K. M. Shtrakhman, *Phys. Met. Metallogr.*, 1967, **24** (3), 116.
90. T. S. Kê, *J. appl. Phys.*, 1948, **19**, 285.
91. L. M. Clareborough, *Acta Met.*, 1957, **5**, 413.
92. K. M. Shtrakhman, Yu. S. Logvinenko, V. F. Grishchenko and Yu. V. Piguzov, *Soviet Phys. solid St.*, 1971, **13** (5), 1238.
93. K. M. Shtrakhman, *Soviet Phys. solid St.*, 1967, **9** (6), 1360.
94. K. Takita and K. Sakamoto, *Scripta Metall.*, 1970, **4** (5), 403.
95. A. I. Surin and M. S. Blanter, *Phys. Met. Metallogr.*, 1970, **29** (1), 199.
96. G. M. Leak, *Proc. phys. Soc., Lond.*, 1961, **78**, 1520.
97. J. Delaplace, J. Hillairet and A. Silvent, *C.r. hebd. Séanc. Acad. Sci., Paris*, 1966 (c), **262** (4), 319.
98. D. B. Fishbach, *Acta Met.*, 1962, **10**, 319.
99. K. Tanaka, *J. phys. Soc. Japan*, 1971, **30** (2), 404.
100. W. R. Thomas and G. M. Leak, *Nature, Lond.*, 1955, **176**, 29.
101. Y. Hayashi and T. Sugeno, *Acta Met.*, 1970, **18** (6), 693.
102. C. A. Wert, *Phys. Rev.*, 1950, **79**, 601.
103. R. Blackwell, *Nature, Lond.*, 1966, **211** (5050), 733.
104. P. Barrand and G. M. Leak, *Acta Met.*, 1964, **12** (10), 1147.
105. K. Kamber, D. Keefer and C. Wert, *Acta Met.*, 1961, **9**, 403.
106. A. E. Lord and D. N. Beshers, *Acta Met.*, 1966, **14** (12), 1659.
107. G. W. Miles and G. M. Leak, *Proc. phys. Soc., Lond.*, 1961, **78**, 1529.
108. T. Maeda, *Japan J. appl. Phys.*, 1971, **10** (10), 1299.
109. P. Barrand, *Acta Met.*, 1966, **14** (10), 1247.
110. P. Barrand, *Metal Sci. J.*, 1967, **1**, 127.
111. P. Barrand, *Metal Sci. J.*, 1967, **1**, 54.
112. C. R. Ward and G. M. Leak, *Metallurgical, ital.*, 1970, **62** (8), 302.
113. I. G. Ritchie and R. Rawlings, *Acta Met.*, 1967, **15** (3), 491.
114. W. R. Heller, *Acta Met.*, 1961, **9**, 600.
115. R. Gibala, *Acta Met.*, 1967, **15** (2), 428.
116. T. S. Kê and C. T. Tsien, *Phys. Met. Metallogr.*, 1957, **4** (2), 78.
117. V. Kandarpa and J. W. Spretnak, *Trans. Met. Soc. AIME*, 1969, **245** (7), 1439.
118. G. J. Couper and R. Kennedy, *J. Iron Steel Inst.*, 1967, **205** (6), 642.
119. E. T. Stephenson, *Metall. Trans.*, 1971, **2** (6), 1613.
120. J. D. Fast and M. B. Verrijp, *J. Iron Steel Inst.*, 1955, **180**, 337.
121. R. S. Lebyedev and V. S. Postnikov, *Phys. Met. Metallogr.*, 1959, **8** (2), 134.
122. D. Siddell and Z. C. Szkopiak, *Metall. Trans.*, 1972, **3** (7), 1907.
123. Yu. V. Grdina, Ye. E. Glikman and Yu. V. Piguzov, *Phys. Met. Metallogr.*, 1966, **21** (4), 90.
124. D. A. Leak, W. R. Thomas and G. M. Leak, *Acta Met.*, 1955, **3**, 501.
125. G. Szabó-Miszenti, *Acta Met.*, 1970, **18** (5), 477.
126. J. Stanley and C. Wert, *J. appl. Phys.*, 1961, **32**, 267.
127. R. M. Jamieson and R. Kennedy, *J. Iron Steel Inst.*, 1966, **204** (2), 1208.
128. H. Sekine, T. Inoue and M. Ogasawara, *Japan. J. appl. Phys.*, 1967, **6** (21), 272.
129. J. D. Fast and J. L. Meijering, *Philips Res. Rep.*, 1953, **8**, 1.
130. W. Hermann, *Solid State Commun.*, 1968, **6** (9), 641.
131. C. F. Burdett, *Phil Mag.*, 1968, **18** (154), 745.
132. B. M. Mecs and A. S. Nowick, *Appl. Phys. Letters*, 1966, **8** (4), 75.
133. A. Zuckerwar and W. Pechhold, *Z. Angew. Phys.*, 1968, **24** (3), 134.

134. K. Ohori and K. Sumino, *Phys. Status Solidii*, 1972(a), **9** (1), 151.
135. P. D. Southgate, *Proc. phys. Soc. Lond.*, 1960, **76**, 385, 398.
136. L. N. Aleksandrov, Yu. N. Golobokov, V. N. Orlov and F. L. '*Edel*' man, *Soviet Phys. solid St.*, 1969, **10** (9), 2269.
137. F. Calzecchi, P. Gondi and S. Mantovani, *J. appl. Phys.*, 1969, **40** (12), 4798.
138. E. Bisogni and C. Wert, US Air Force, Sci. Res. Rep., 1961, Contract AF49(638)672.
139. T. Alper and G. A. Saunders, *Phil. Mag.*, 1969, **20** (164), 225.
140. M. E. De Morton, *Phys. Status Solidii*, 1968, **126**, K73.
141. V. S. Postnikov, I. V. Zolotukhin, V. N. Burmistrov and I. M. Sharshakov, *Phys. Met. Metallogr.*, 1969, **28** (4), 210.
142. M. J. Murray, *J. Less-common Metals*, 1968, **15** (4), 425.
143. T. D. Gulden and J. C. Shyne, *J. Inst. Metals*, 1968, **96** (5), 139.
144. T. D. Gulden and J. C. Shyne, *J. Inst. Metals*, 1968, **96** (5), 143.
145. D. P. Seraphim and A. S. Nowick, *Acta Met.*, 1961, **9**, 85.
146. J. M. Roberts, *Trans. Japan Inst. Metals*, 1968, **9** (Suppl.), 69.
147. R. T. C. Tsui and H. S. Sack, *Acta Met.*, 1967, **15** (11), 1715.
148. H. L. Caswell, *J. appl. Phys.*, 1958, **29**, 1210.
149. S. Koda, K. Kamigaki and H. Kayano, *J. phys. Soc. Japan*, 1963, **18** (Suppl. 1), 195.
150. A. V. Siefert and F. T. Worrel, *J. appl. Phys.*, 1951, **22**, 1257.
151. R. H. Chalmers and J. Schultz, *Acta Met.*, 1962, **10**, 466.
152. H. Mühlbach, *Phys. Status Solidii*, 1969, **36** (1), K33.
153. R. Gibala, M. K. Korenko, M. F. Amateau and T. E. Mitchell, *J. Phys. Chem. Solids*, 1970, **3** (8), 1889.
154. G. Rieu, J. De Fouquet and A. Nadeau, *C.r. hebd. Séanc. Acad. Sci., Paris*, 1970 (c), **270** (3), 287.
155. S. Z. Bokshtein, M. B. Bronfin, *et al.*, *Soviet Phys. solid St.*, 1964, **5** (11), 2253.
156. Yu. V. Piguzov, W. D. Werner and I. Ya. Rzhevskaya, *Phys. Met. Metallogr.*, 1967, **24** (3), 179.
157. R. H. Schnitzel, *Trans. Met. Soc. AIME*, 1964, **230** (3), 609.
158. M. J. Murray, *Phil. Mag.*, 1969, **20** (165), 561.
159. A. A. Belyakov, V. P. Yelyutin and Ye. I. Mozzhukhin, *Phys. Met. Metallogr.*, 1967, **23** (2), 115.
160. E. J. Kramer and C. L. Bauer, *Phys. Rev.*, 1967, **163** (2), 407.
161. J. Schultz, *Bull. Am. phys. Soc.*, 1964, **9**, 214.
162. F. M. Mazzolai and M. Nuovo, *Solid State Commun.*, 1969, **7** (1), 103.
163. J. Filloux, H. Harper and R. H. Chalmers, *Bull. Am. phys. Soc.*, 1964, **9**, 230.
164. M. W. Stanley and Z. C. Szkopiak, *J. Inst. Metals*, 1966, **94** (2), 79.
165. M. F. Amateau, R. Gibala and T. E. Mitchell, *Scripta Metall.*, 1968, **2** (2), 123.
166. M. F. Amateau, T. E. Mitchell and R. Gibala, *Phys. Status Solidii*, 1969, **36** (1), 407.
167. R. A. Hoffman and C. A. Wert, *J. appl. Phys.*, 1966, **37** (1), 237.
168. F. Schlät, *Trans. Japan Inst. Metals*, 1968, **9** (Suppl.), 64.
169. E. Davenport and G. Mah, *Metall. Trans.*, 1970, **1** (5), 1452.
170. R. W. Powers and M. V. Doyle, *J. Metals, N.Y.*, 1957, **9**, 1285.
171. R. W. Powers and M. V. Doyle, *J. appl. Phys.*, 1959, **30**, 514.
172. C. Vercaemer and A. Clauss, *C.r. hebd. Séanc. Acad. Sci., Paris* 1969 (c), **269** (15), 803.
173. D. H. Boone and C. Wert, *J. phys. Soc. Japan*, 1963, **18** (Suppl. 1), 141.
174. C. Y. Ang, *Acta Met.*, 1953, **1**, 123.
175. D. J. Van Ooijen and A. S. Van Der Goot, *Acta Met.*, 1966, **14** (8), 1008.
176. G. Vidal and H. Bibring, *C.r. hebd. Séanc. Acad. Sci., Paris*, 1965, **260** (3), 857.
177. R. E. Miner, D. F. Gibbons and R. Gibala, *Acta Met.*, 1970, **18** (4), 419.
178. A. W. Sommers and D. N. Beshers, *J. appl. Phys.*, 1966, **37** (13), 4603.
179. P. S. Venkatesan and D. N. Beshers, *J. appl. Phys.*, 1970, **41** (1), 42.
180. A. Seeger, P. Schiller and H. Kronmüller, *Phil. Mag.*, 1960, **5**, 853.
181. P. Schiller, H. Kronmüller and A. Seeger, *Acta Met.*, 1962, **10**, 333.
182. J. T. A. Roberts and P. Barrand, *Acta Met.*, 1967, **15** (11), 1685.
183. O. I. Datsko and V. A. Pavlov, *Phys. Met. Metallogr.*, 1958, **6** (5), 122.
184. T. S. Kê, *Acta phys. sin.,1955*, **11** (5), 405.
185. V. N. Gridnev, A. I. Yefimov and N. P. Kushnareva, *Phys. Met. Metallogr.*, 1967, **23** (4), 142.
186. Y. S. Avraamov, L. N. Belyakov and B. G. Livshits, *Phys. Met. Metallogr.*, 1958, **6** (1), 104.
187. V. M. Azhazha, N. P. Bondarenko, M. P. Zeydlits and B. I. Shapoval, *Phys. Met. Metallogr.*, 1970, **29** (2), 101.
188. R. R. Arons, J. Bouman, M. Witzenbeek, P. T. A. Klaase, C. Tuyn, G. Leferink and G. De Vries, *Acta Met.*, 1967, **15** (1), 144.
189. R. R. Arons, C. Tuyn and G. De Vries, *Acta Met.*, 1967, **15** (10), 1673.
190. J. Coremberg and F. M. Mazzolai, *Solid State Commun.*, 1967, **6** (1), 1.
191. V. O. Shestopal, *Phys. Met. Metallogr.*, 1968, **26** (6), 176.
192. M. Rosen, G. Erez and S. Shtrikman, *J. Phys. Chem. Solids*, 1969, **30** (5), 1063.
193. R. Schnitzel, *J. appl. Phys.*, 1959, **30**, 2011.
194. L. N. Aleksandrov, M. I. Zotov, R. Sh. Ibragimov and F. L. 'Edel' man, *Soviet Phys. solid St.*, 1970. **11** (7), 1494.
195. P. D. Southgate, *Proc. phys. Soc. Lond.*, 1957, **70** (B), 804.
196. P. D. Southgate, *Proc. phys. Soc. Lond.*, 1960, **76**, 385, 398.
197. B. S. Berry, *J. Phys. Chem. Solids*, 1970, **13** (8), 1827.

198. L. Rotherham, A. D. N. Smith and G. B. Greenough, *J. Inst. Metals*, 1951, **79**, 439.
199. L. Verdini and L. A. Vienneau, *Canad. J. Phys.*, 1968, **46** (23), 2715.
200. R. W. Powers and M. V. Doyle, *J. appl. Phys.*, 1957, **28**, 255.
201. P. Kofstad and R. A. Butera, *J. appl. Phys.*, 1963, **34**, 1517.
202. R. W. Powers and M. V. Doyle, *Acta Met.*, 1956, **4**, 233.
203. R. W. Powers and M. V. Doyle, *Acta Met.*, 1955, **3**, 135.
204. R. W. Powers and M. V. Doyle, *Trans. AIMME*, 1959, **215**, 655.
205. G. Schoek and M. Mondino, *J. phys. Soc. Japan*, 1963, **18** (Suppl. 1), 149.
206. A. A. Sagues and R. Gibala, *Scripta Metall.*, 1971, **5** (8), 689.
207. G. Arlt and W. Hermann, *Solid State Commun.*, 1969, **7** (1), 75.
208. J. Winter and S. Weinig, *Trans. AIMME*, 1959, **215**, 74.
209. J. E. Doherty and D. F. Gibbons, *Acta Met.*, 1971, **119** (4), 275.
210. W. Köster, L. Bangert and M. Evers, *Z. Metall.*, 1956, **47**, 564.
211. R. R. Hasiguti and K. Iwasaki, *J. appl. Phys.*, 1968, **39** (5), 2182.
212. W. J. Bratina, *Acta Met.*, 1962, **10**, 332.
213. J. N. Pratt, W. J. Bratina and B. Chalmers, *Acta Met.*, 1954, **2**, 203.
214. D. Gupta and S. Weinig, *Acta Met.*, 1962, **10**, 292.
215. R. W. Powers and M. V. Doyle, *Acta Met.*, 1958, **6**, 643.
216. G. Cannelli and F. M. Mazzolai. *J. Phys. Chem. Solids*, 1970, **31** (8), 1913.
217. R. A. Butera and P. Kofstad, *J. appl. Phys.*, 1963, **34**, 2172.
218. R. W. Powers, *Acta Met.*, 1954, **2**, 604.
219. L. N. Aleksandrov and V. S. Mordyuk, *Phys. Met. Metallogr.*, 1966, **21** (1), 101.
220. R. H. Chalmers and J. Schultz, *Phys. Rev. Letters*, 1961, **6**, 273.
221. R. H. Schnitzel, *Trans. Met. Soc. AIME*, 1965, **233** (1), 186.
222. L. H. Aleksandrov, *Phys. Met. Metallogr.*, 1962, **13** (4), 143.
223. I. Berlec, *Metall. Trans*, 1970, **1** (10), 2677.
224. V. O. Shestopal, *Phys. Met. Metallogr.*, 1968, **25** (6), 148.
225. V. P. Yelyutin and A. K. Natanson, *Phys. Met. Metallogr.*, 1963, **15** (5), 89.
226. H. Kayano, *J. phys. Soc. Japan*, 1969, **26** (3), 733.
227. G. Roberts, P. Barrand and G. M. Leak, *Scripta Metall.*, 1969, **3** (6), 409.
228. J. E. Doherty and D. F. Gibbons, *J. appl. Phys.*, 1971, **42** (11), 4502.
229. P. L. Gruzin and A. N. Semenikhin, *Phys. Met. Metallogr.*, 1963, **15** (5), 128.
230. W. J. Bratina and W. C. Winegard, *J. Metals, N.Y.*, 1956, **8**, 186.
231. K. Bungardt and H. Preisendanz, *Z. Metall.*, 1960, **51**, 280.
232. V. Y. Ivanov, B. I. Shapoval and V. M. Amonenko, *Phys. Met. Metallogr.*, 1961, **11** (1), 55.
233. P. Boch, J. Petit, C. Gasc and J. De Fouquet, *C.r. hebd. Séanc. Acad. Sci., Paris*, 1968 (c), **266** (9), 605.
234. H. L. Brown, P. E. Armstrong and C. P. Kempter, *J. Less-common Metals*, 1967, **13** (4), 373.
235. J. L. Gacougnolle, S. Sarrazin and J. De Fouquet, *C.r. hebd. Séanc. Acad. Sci., Paris* 1970 (c), **270** (2), 158.
236. C. W. Nelson, D. F. Gibbons and R. F. Hehemann, *J. appl. Phys.*, 1966, **37** (13), 4677.
237. S. Karashima and K. Saito, *J. Jap. Inst. Metals*, 1973, **37**(3), 326.
238. H. Farman and D. H. Niblett, 'Proc. 3rd Euro. Conf. Int. Frict.', Manchester, 1980, Pergamon Press, p. 7.
239. H. Schulz, U. Rodrian and M. Maul, 'Proc. 3rd Euro. Conf. Int. Frict'., Manchester, 1980, Pergamon Press, p. 19.
240. H. E. Schaeffer, H. Schulz and H. P. Stark, 'Proc. 3rd Euro. Conf. Int. Frict'., Manchester, 1980, Pergamon Press, p. 25.
241. F. Baudraz and R. Gotthardt, 'Proc. 3rd. Euro. Conf. Int. Frict.', Manchester, 1980, Pergamon Press, p. 67.
242. S. M. Seyed Reihani, G. Fantozzi, C. Esnouf and G. Revel, *Scripta Met.*, 1979, **13**(8), 1011.
243. H. Mizubayashi, S. Okuda and M. Daikubara, *Scripta Met.*, 1979 **13**(12), 1131.
244. A De Rooy, P. M. Bronsveld and J. Th M. De Hosson. 'Proc. 3rd. Euro. Conf. Int. Frict.', Manchester, 1980, Pergamon Press, p. 149.
245. J. N. Lomer and C. R. A. Sutton, 'Proc. 3rd Euro. Conf. Int. Frict.', Manchester, 1980, Pergamon Press, p. 199.
246. M. Weller and J. Diehl, 'Proc. 3rd Euro. Conf. Int. Frict.', Manchester, 1980, Pergamon Press, p. 223.
247. R. Schaller and W. Benoit, 'Proc. 3rd Euro. Conf. Int. Frict.', Manchester, 1980, Pergamon Press, p. 311.
248. M. Mondino and R. Gugelmeier, 'Proc. 3rd Euro. Conf. Int. Frict.', Manchester, 1980, Pergamon Press, p. 317.
249. R. Klam, H. Schulz and H. E. Schaeffer, *Acta Met.*, 1979, **278**, 205.
250. K. Iwasaki, K. Lücke and G. Sokolowski, *Acta. Met.*, 1980, **28**, 855.

15.3 Shape memory alloys

15.3.1 Mechanical properties of shape memory alloys

Most shape memory alloys are of compositions at which the structure can change reversibly and reproducibly from one crystalline state to another by a change in temperature or by a change of mechanical stress at temperatures just above the transformation temperature at zero stress.

In shape memory alloys, the change of structure usually occurs over a narrow range of temperatures by means of a self-accommodating martensitic transformation (though this is by no means the only mechanism). If a stressed memory alloy is thermally cycled through its martensitic transformation temperature, the strain–temperature relationship will take the form of a closed hysteresis loop similar in shape to that of ferromagnetic materials. If such an alloy is cooled to below its M_f temperature, and is then deformed plastically, it will recover its original undeformed shape on re-heating to a temperature above its A_s temperature.

On cooling through the transformation to the martensitic state, the temperatures at which the transformation starts and finishes at zero applied stress are denoted by M_s and M_f respectively. On re-heating, the temperatures at which the reverse transformation to the high temperature phase takes place are A_s and A_f respectively. In experiments, these temperatures are detected by thermal or dilatometric analysis or by changes in electrical resistivity. M_d is the highest temperature at which the transformation can be induced by stress, see Figure 15.1.[11]

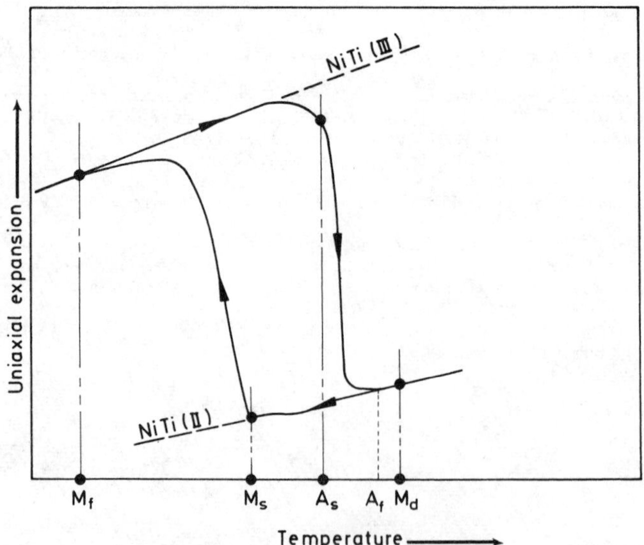

Figure 15.1 *Typical uniaxial dimensional change behaviour for drawn wire.* Ti–55.0% Ni–0.07%C. *After W. B. Cross* et al.[11]

Most data have been obtained for TiNi alloys but increasing research activity is now revealing information about other alloys.

M_s temperature can be varied by changing the composition, e.g. by increasing Ni in TiNi, but the range of M_s values can be extended by partially replacing nickel by iron or cobalt.

Note that if shape memory alloys are cooled under stress, the M_s temperature is raised in direct proportion to the stress.

Table 15.9 COMPOSITIONS AND TRANSFORMATION TEMPERATURES OF SHAPE MEMORY ALLOYS

Alloy composition wt%[(1)]	M_s *temperature at zero stress at* °C	A_s *temperature zero stress* °C	*Maximum shape memory strain* %	*Reference*
Ti–52–56% Ni	~RT			1
Ti–53.5% Ni	98			2
Ti–54.0% Ni	140			2
Ti–54.5% Ni	170	175	8.0	2, 4
Ti–55.0% Ni	140			2
Ti–55.5% Ni	30			2
Ti–56.0% Ni	−25			2
Ti–56.5% Ni	−50			2
Ti–51At.% Ni	28	62		7
TiNi[(2)]	166			3
TiCo	−238			3
Ti Ni_xCo_{1-x}[(3)]	166 where $x=1$ −238 where $x=0$			3
TiFe	~−269			3
ZrRu[(4)]	−233			3
ZrRh[(4)]	380			3
ZrPd[(4)]	727			3
U–9 At.% Nb[(5)]	0	150		5
U–12 At.% Nb[(5)]	0	150		5
U–15 At.% Nb[(5)]	0	60		5
U–18 At.% Nb[(5)]	−196	0		5
Au–28 At.% Cu–46 At.% Zn	−15			6
Au–47.5% Cd	60	70		13
Cu–14.2% Al–4.3% Ni	−20	−15		8
Cu–40% Zn	−70	−120		8
Cu–34.7% Zn–3.0% Sn	−52	−50	4.5 (polycrystal) 8.5 (single crystal)	9
Ti–35% Nb	−175	184	~2.5	10
Ti–51.4% Ni–3.57% Co				
0.3 in. rod	−51	−40	6 to 10	11
0.003 in. foil	−65	−45	6 to 10	11
0.01 in. wire	−73	−51	6 to 10	11
Ti–55.0% Ni–0.07%C				
0.2 in. rod	21	60	6 to 10	11
0.003 in. foil	27	43	6 to 10	11
0.01 in. wire	18	43	6 to 10	11
Ti–54.6% Ni–0.06%C				
0.625 in. rod	43	71	6 to 10	11
0.003 in. foil	38	66	6 to 10	11
0.01 in. wire	32	54	6 to 10	11

(1) Unless otherwise stated.
(2) *See also* Figure 15.2.[3]
(3) *See* Figures 15.3 and 15.4[3].
(4) Non-linear interpolation between these compounds is possible. Relationship is of form $\log_e (M_s K) \infty x$ where x is in the range 0–1 in $Zr–Ru_x–Rh_{1-x}$ etc.
(5) M_s and A_s not accurately determined but are within these temperature ranges.

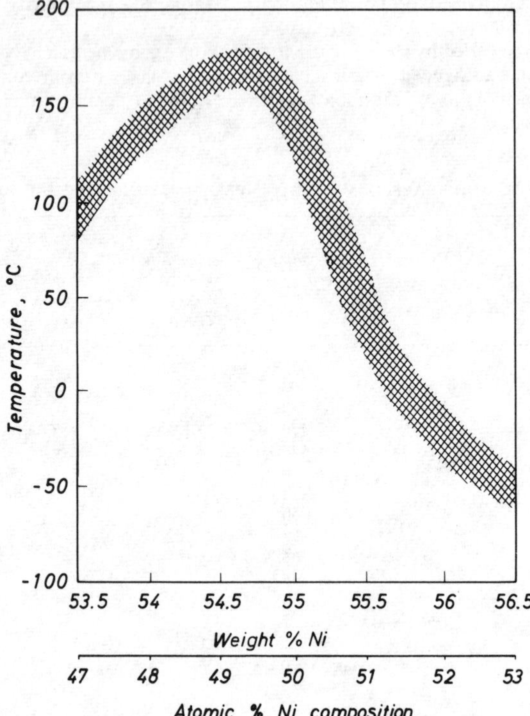

Figure 15.2 *Change of M_s temperature with composition for* TiNi *alloys, after F. E. Wang and W. J. Buehler*[3]

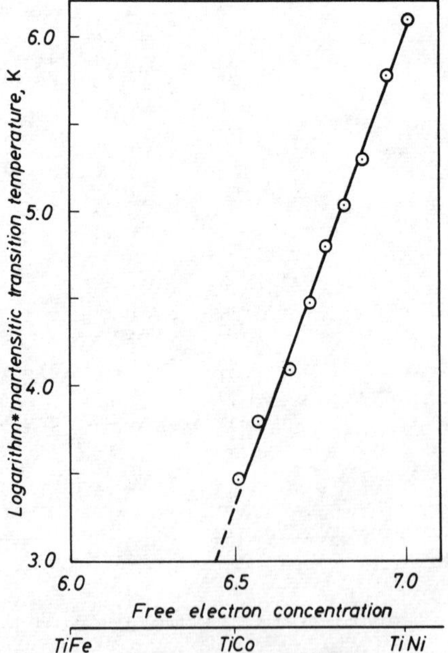

Figure 15.3 *See Table 15.1 after F. E. Wang and W. J. Buehler*[3] *Natural logarithm

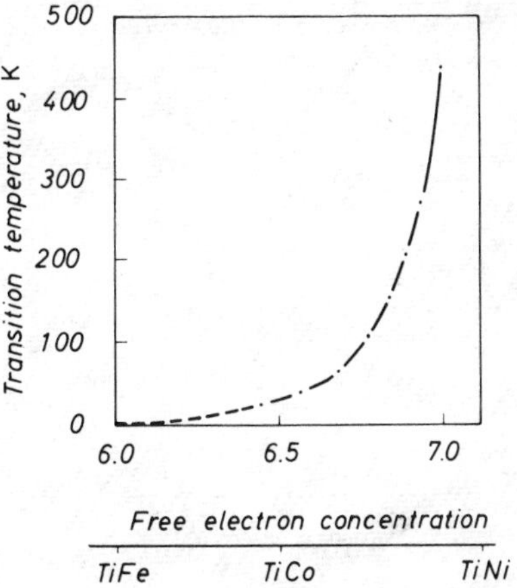

Figure 15.4 *See Table 15.1 after F. E. Wang and W. J. Buehler*[3]

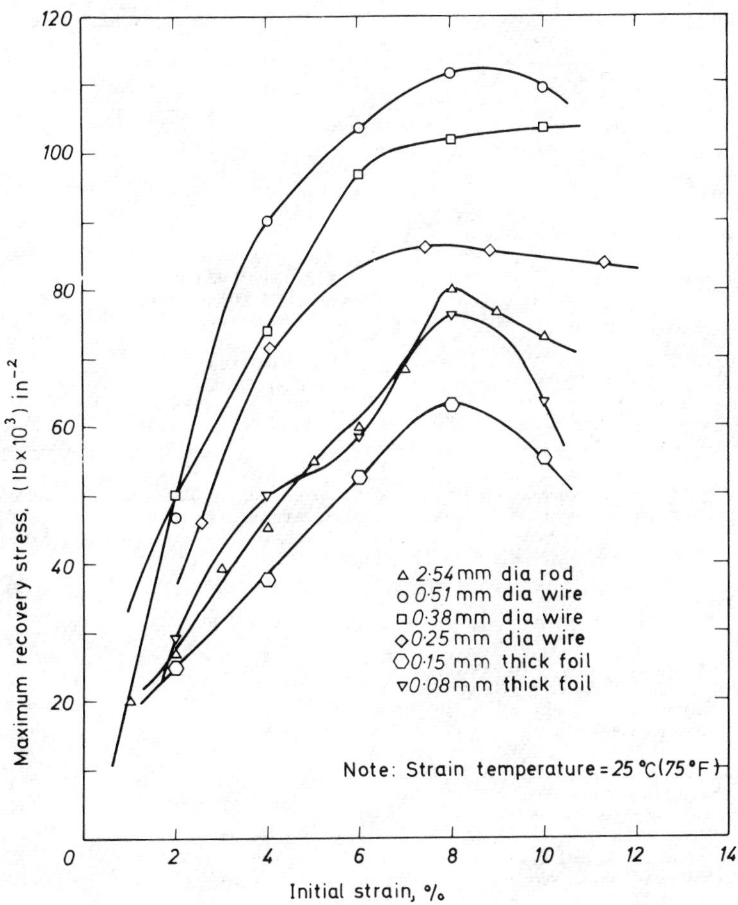

Figure 15.5 *Maximum recovery stress versus initial strain curves for* Ti–55.0% Ni–0.0% *C. See Table 15.2, after W. B. Cross* et al.[11]

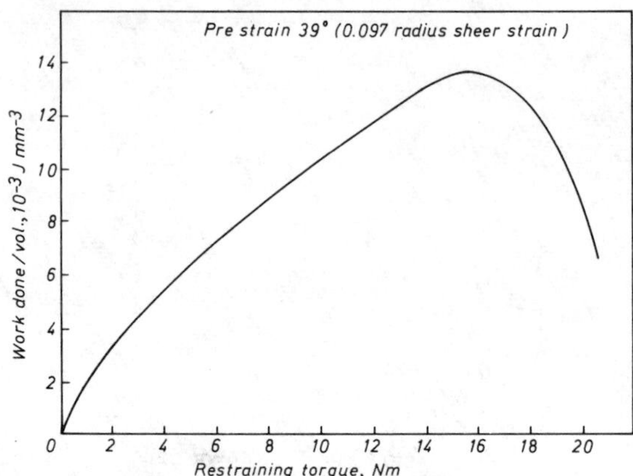

Figure 15.6 *The relationship between the capacity of* TiNi *alloy to do work in torsion and the restraining torque.*[14]

Table 15.10 PHYSICAL AND MECHANICAL PROPERTIES OF A TITANIUM–55% NICKEL SHAPE MEMORY ALLOY

Property	*Value*
Density	6.45 g cm^{-3}
M.p.	1 310°C
Magnetic permeability	<1.002
Electrical resistivity	80 μΩ cm at 20°C 132 μΩ cm at 900°C
Coefficient of thermal expansion (24–900°C)	10.4×10^{-6} °C^{-1}
Mechanical properties at 20°C (i.e. below M_s)	
0.2% yield stress	207 MPa (30 000 lbf in^{-2})
UTS	861 MPa (125 000 lbf in^{-2})
Elongation %	22
Reduction in area %	20
Impact (unnotched)	159J (117 ft lbf) at 20°C 95J (70 ft lbf) at 80°C
Fatigue (rotating beam)	$>25 \times 10^6$ cycles at 483 MPa (70 000 lbf in^{-2})

Note

If recovery is prevented mechanically, the TiNi will exert a stress and is capable of doing work when heated to above the A_s temperature. Samples of Ti–55.0% Ni–0.07%*C* were capable of exerting a stress of up to 758 MPa (110000 lbf in^{-2}) at 171 C (*see* Figure 15.5). The amount of mechanical work which this alloy was capable of doing was 17–20 J cm^{-3} (2500/2900 in lbf in^{-2}) on heating from 24°C to 171°C.

REFERENCES FOR SECTION 15.3

1. W. J. Buehler and R. C. Wiley, US Patent 3 174 851.
2. A. G. Rozner and W. J. Buehler, US Patent 3 351 463.
3. F. E. Wang and W. J. Buehler, US Patent 3 558 369.
4. G. B. Brook, unpublished data.
5. R. J. Jackson, J. F. Boland and J. L. Frankeng, US Patent 3 567 523.
6. N. Nakanishi, *et al.*, *Phys. Letters*, 1971, **37A**, 61.
7. F. E. Wang, *et al.*, *J. appl. Phys.*, 1968, **39**, 2166.
8. K. Otsuka, *et al.*, *Scripta Metall.*, 1972, **6**, 377.
9. J. D. Eisenwasser and L. C. Brown, *Met. Trans. AIME.*, 1972, **3**, 1359.
10. C. Baker, *Met. Sci. J.*, 1971, **5**, 92.
11. W. B. Cross, *et al.*, NASA Report CR—1433, Sept. 1969.
12. H. U. Schuerch, NASA Report CR—1232, Nov. 1968.
13. D. S. Lieberman, T. A. Read and M. S. Wechsler, *J. appl. Phys.*, 1957, **28**, 532.
14. G. B. Brook, *et al.*, Fulmer Research Inst. Rep. No R662/4A, Feb., 1977.

16 Temperature measurement and thermoelectric properties

16.1 Temperature measurement

The unit of the fundamental physical quantity known as thermodynamic temperature, symbol T, is the kelvin, symbol K, defined as the fraction 1/273.16 of the thermodynamic temperature of the triple point of water.*

For historical reasons, connected with the way temperature scales were originally defined, it is common practice to express a temperature in terms of its difference from that of a thermal state 0.01 kelvins lower than the triple point of water. A thermodynamic temperature, T, expressed in this way is known as a Celsius temperature, symbol t, defined by

$$t = T - 273.15 \text{ K}$$

The unit of Celsius temperature is the degree Celsius, symbol °C, which is, by definition, equal in magnitude to the kelvin. A difference of temperature may be expressed in kelvins or degrees Celsius.

The International Practical Temperature Scale of 1968 (IPTS-68) has been constructed in such a way that any temperature measured on it is a close approximation to the numerically corresponding thermodynamic temperature. Moreover, such measurements are easily made and are highly reproducible; in contrast, direct measurements of thermodynamic temperatures are both difficult to make and imprecise. The IPTS-68 uses both International Practical Kelvin Temperatures, symbol T_{68}, and International Practical Celsius Temperatures, symbol t_{68}. The relation between T_{68} and t_{68} is the same as that between T and t, i.e.,

$$t_{68} = T_{68} - 273.15 \text{ K}$$

The units of T_{68} and t_{68} are the kelvin, symbol K, and the degree Celsius, symbol °C; that is, the names of the units are the same as those used for the thermodynamic temperatures T and t.

The second edition of IPTS-68 was adopted by the Fifteenth General Conference on Weights and Measures in 1975.[1] It constitutes only an amendment of the first edition, not a replacement. Any measured temperature, T_{68}, remains unchanged by this amendment of the IPTS-68.

IPTS-68 only extends down to −259.34°C (13.81 K), the triple point of equilibrium hydrogen. For lower temperatures there now exists the '1976 Provisional 0.5 K to 30 K Temperature Scale'.[2]

Table 16.1 FIXED POINTS OF IPTS-68

The defining and secondary fixed points of the IPTS-68 are given below. The defining fixed points are in **heavy type**. t.p. (triple point), b.p. (boiling point), f.p. (freezing point)

Fixed point	*Temperature* °C (*IPTS-68*)
t.p. **equilibrium hydrogen**	−259.34
t.p. normal hydrogen	−259.194
b.p. **equilibrium hydrogen at 33 330.6 Pa**	−256.108

*Thirteenth General Conference of Weights and Measures (1967), Resolutions 3 and 4.

Table 16.1 FIXED POINTS OF IPTS-68—*continued*

Fixed point	*Temperature* °C (*IPTS-68*)
b.p. **equilibrium hydrogen**	–252.87
b.p. normal hydrogen	–252.753
t.p. neon	–248.589
b.p. **neon**	–246.048
t.p. **oxygen**	–218.789
t.p. nitrogen	–210.004
b.p. nitrogen	–195.806
b.p. argon	–185.856
b.p. **oxygen**	–182.962
sublimation point of carbon dioxide	– 78.476
f.p. mercury	– 38.836
ice point	0
t.p. **water**	0.01
t.p. phenoxybenzene	26.87
b.p. **water**	100
t.p. benzoic acid	122.37
f.p. indium	156.634
f.p. **tin**	231.968
f.p. bismuth	271.442
f.p. cadmium	321.108
f.p. lead	327.502
b.p. mercury	356.66
f.p. **zinc**	419.58
b.p. sulphur	444.674
f.p. copper/aluminium eutectic	548.26
f.p. antimony	630.755
f.p. aluminium	660.46
f.p. **silver**	961.93
f.p. **gold**	1 064.43
f.p. copper	1 084.88
f.p. nickel	1 455
f.p. cobalt	1 495
f.p. palladium	1 554
f.p. platinum	1 769
f.p. rhodium	1 963
f.p. alumina, Al_2O_3	2 054
f.p. iridum	2 447
f.p. niobium	2 477
f.p. molybdenum	2 623
f.p. tungsten	3 387

Unless otherwise stated, all boiling points refer to a pressure of 1 standard atmosphere (101 325 Pa). For the relationships between pressure and boiling point refer to reference 1. For details of experimental technique in calibration refer to reference 3. A more extensive list of secondary reference points is given in reference 4.

Table 16.2 THERMAL ELECTROMOTIVE FORCE (millivolts) OF ELEMENTS RELATIVE TO PLATINUM*

A positive sign means that in a simple thermoelectric circuit the element is positive to the platinum at the reference junction (0°C). The e.m.f. generated by any two elements, *A* and *B*, can also be found from this table. It is the algebraic difference, $A_e - B_e$, between the values (A_e and B_e) for the e.m.f. generated by each relative to platinum; a positive sign indicates that *A* is positive to *B* at the reference junction, and a negative sign that *A* is negative to *B*.

Cold junction at 0°C

	Temperature of hot junction °C			
	−200	−100	+100	+200
Aluminium	+0.45	−0.06	+0.42	+ 1.06
Antimony	—	—	+4.89	+10.14
Bismuth	+12.39	+7.54	−7.34	−13.57

* The numerical values given in this table should be taken only as a guide since thermoelectric properties are very sensitive to impurities and state of anneal.

Table 16.2 THERMAL ELECTROMOTIVE FORCE (millivolts) OF ELEMENTS RELATIVE TO PLATINUM*—*continued*

Cold junction at 0° C

	Temperature of hot junction °C			
	−200	−100	+100	+200
Cadmium	− 0.04	−0.31	+0.91	+ 2.32
Caesium	+ 0.22	−0.13	—	—
Calcium	—	—	−0.51	− 1.13
Carbon	—	—	+0.70	+ 1.54
Cerium	—	—	+1.14	+ 2.46
Cobalt	—	—	−1.33	− 3.08
Copper	− 0.19	− 0.37	+ 0.76	+ 1.83
Germanium	−46.00	−26.62	+33.9	+72.4
Gold	− 0.21	− 0.39	+ 0.74	+ 1.77
Indium	—	—	+ 0.69	—
Iridium	− 0.25	− 0.35	+ 0.65	+ 1.49
Iron	− 3.10	− 1.94	+ 1.98	+ 3.69
Lead	+ 0.24	− 0.13	+ 0.44	+ 1.09
Lithium	− 1.12	− 1.00	+ 1.82	—
Magnesium	+ 0.31	− 0.09	+ 0.44	+ 1.10
Mercury	—	—	+ 0.06	+ 0.13
Molybdenum	—	—	+ 1.45	+ 3.19
Nickel	+ 2.28	+ 1.22	− 1.48	− 3.10
Palladium	+ 0.81	+ 0.48	− 0.57	− 1.23
Potassium	+ 1.61	+ 0.78	—	—
Rhodium	− 0.20	− 0.34	+ 0.70	+ 1.61
Rubidium	+ 1.09	+ 0.46	—	—
Silicon	+63.13	+37.17	−41.56	−80.58
Silver	− 0.21	− 0.39	+ 0.74	+ 1.77
Sodium	+ 1.00	+ 0.29	—	—
Tantalum	+ 0.21	− 0.10	+ 0.33	+ 0.93
Thallium	—	—	+ 0.58	+ 1.30
Thorium	—	—	− 0.13	− 0.26
Tin	+ 0.26	− 0.12	+ 0.42	+ 1.07
Tungsten	+ 0.43	− 0.15	+ 1.12	+ 2.62
Zinc	− 0.07	− 0.33	+ 0.76	+ 1.89

Table 16.3 THERMAL ELECTROMOTIVE FORCE (millivolts) OF SOME BINARY ALLOYS RELATIVE TO PLATINUM WITH JUNCTIONS AT 0° AND 100°C*

Metal A: *Metal B:* %*A*	*Lead* *Tin*	*Tin* *Copper*	*Zinc* *Copper*	*Gold* *Silver*	*Gold* *Palladium*	*Nickel* *Copper*	*Tin* *Bismuth*	*Antimony* *Cadmium*	*Antimony* *Bismuth*
0	+0.44	+0.76	+0.76	+0.74	−0.57	+0.76	−7.34	+ 0.90	−7.34
10	+0.44	+0.53	+0.54	+0.55	−0.85	−2.63	+4.00	+ 1.52	−8.82
20	+0.44	+0.56	+0.53	+0.48	−1.25	−3.08	+3.52	+ 2.88	−7.31
30	+0.44	+0.65	+0.54	+0.47	−1.42	−3.54	+2.56	+ 6.4	−5.66
40	+0.45	+0.65	+0.51	+0.47	−1.69	−4.03	+2.10	+12.2	−4.05
50	+0.45	+0.69	+0.54	+0.48	−2.44	−3.64	+1.77	+23.1	−2.51
60	+0.44	+0.72	+0.47	+0.49	−2.97	−3.06	+1.14	+44.4	−1.06
70	+0.44	+0.62	+0.87	+0.49	−2.63	−2.54	+0.95	+21.5	+0.32
80	+0.43	+0.54	+0.66	+0.50	−0.46	−2.49	+0.78	+12.8	+1.79
90	+0.42	+0.48	+0.98	+0.59	−0.05	−1.93	+0.60	+ 8.1	+3.31
100	+0.42	+0.42	+0.76	+0.78	+0.78	−1.48	+0.42	+ 4.89	+4.89

* The numerical values given in these tables should be taken only as a guide since thermoelectric properties are very sensitive to impurities and state of anneal.

Table 16.4 ABSOLUTE THERMOELECTRIC POWER OF PLATINUM

Temperature (K)	300	400	500	600	700	800	900	1 000	1 100	1 200
Thermoelectric power (μVK^{-1})	−5.05	−7.66	−9.69	−11.33	−12.87	−14.38	−15.97	−17.58	−19.03	−20.56

16.2 Thermocouple reference tables

The introduction of the IPTS-68 necessitated the revision of all the thermocouple reference tables then in use. This provided the opportunity for new internal agreement to be reached on common thermocouple reference tables. Tables 16.5 to 16.11 for both noble metal and base metal thermocouples are the result of the new agreement and provide the basis for the new British Standard reference tables for thermocouples, BS 4937 and an identical international standard, IEC 584–1 (1977).

Table 16.5 PLATINUM–10% RHODIUM/PLATINUM THERMOCOUPLE TABLES—TYPE S

Temperature °C (*IPTS-68*) *Reference junction at* 0 °C

Temp.	0	10	20	30	40	50	60	70	80	90	*Temp.*
					e.m.f. μV						
0	0	−53	−103	−150	−194	−236					0
0	0	55	113	173	235	299	365	432	502	573	0
100	645	719	795	872	950	1 029	1 109	1 190	1 273	1 356	100
200	1 440	1 525	1 611	1 698	1 785	1 873	1 962	2 051	2 141	2 232	200
300	2 323	2 414	2 506	2 599	2 692	2 786	2 880	2 974	3 069	3 164	300
400	3 260	3 356	3 452	3 549	3 645	3 743	3 840	3 938	4 036	4 135	400
500	4 234	4 333	4 432	4 532	4 632	4 732	4 832	4 933	5 034	5 136	500
600	5 237	5 339	5 442	5 544	5 648	5 751	5 855	5 960	6 064	6 169	600
700	6 274	6 380	6 486	6 592	6 699	6 805	6 913	7 020	7 128	7 236	700
800	7 345	7 454	7 563	7 672	7 782	7 892	8 003	8 114	8 225	8 336	800
900	8 448	8 560	8 673	8 786	8 899	9 012	9 126	9 240	9 355	9 470	900
1 000	9 585	9 700	9 816	9 932	10 048	10 165	10 282	10 400	10 517	10 635	1 000
1 100	10 754	10 872	10 991	11 110	11 229	11 348	11 467	11 587	11 707	11 827	1 100
1 200	11 947	12 067	12 188	12 308	12 429	12 550	12 671	12 792	12 913	13 034	1 200
1 300	13 155	13 276	13 397	13 519	13 640	13 761	13 883	14 004	14 125	14 247	1 300
1 400	14 368	14 489	14 610	14 731	14 852	14 973	15 094	15 215	15 336	15 456	1 400
1 500	15 576	15 697	15 817	15 937	16 057	16 176	16 296	16 415	16 534	16 653	1 500
1 600	16 771	16 890	17 008	17 125	17 243	17 360	17 477	17 594	17 711	17 826	1 600
1 700	17 942	18 056	18 170	18 282	18 394	18 504	18 612				

Table 16.6 PLATINUM–13% RHODIUM/PLATINUM THERMOCOUPLE TABLES—TYPE R

Temperature °C (*IPTS-68*) *Reference junction at* 0 °C

Temp.	0	10	20	30	40	50	60	70	80	90	*Temp.*
					e.m.f. μV						
0	0	−51	−100	−145	−188	−226					
0	0	54	111	171	232	296	363	431	501	573	0
100	647	723	800	879	959	1 041	1 124	1 208	1 294	1 380	100
200	1 468	1 557	1 647	1 738	1 830	1 923	2 017	2 111	2 207	2 303	200
300	2 400	2 498	2 596	2 695	2 795	2 896	2 997	3 099	3 201	3 304	300
400	3 407	3 511	3 616	3 721	3 826	3 933	4 039	4 146	4 254	4 362	400

Table 16.6 PLATINUM–13% RHODIUM/PLATINUM THERMOCOUPLE TABLES—TYPE R—*continued*

Temperature °C (*IPTS-68*) — *Reference junction at* 0 °C — *e.m.f.* μV

Temp.	0	10	20	30	40	50	60	70	80	90	*Temp.*
500	4 471	4 580	4 689	4 799	4 910	5 021	5 132	5 244	5 356	5 469	500
600	5 582	5 696	5 810	5 925	6 040	6 155	6 272	6 388	6 505	6 623	600
700	6 741	6 860	6 979	7 098	7 218	7 339	7 460	7 582	7 703	7 826	700
800	7 949	8 072	8 196	8 320	8 445	8 570	8 696	8 822	8 949	9 076	800
900	9 203	9 331	9 460	9 589	9 718	9 848	9 978	10 109	10 240	10 371	900
1 000	10 503	10 636	10 768	10 902	11 035	11 170	11 304	11 439	11 574	11 710	1 000
1 100	11 846	11 983	12 119	12 257	12 394	12 532	12 669	12 808	12 946	13 085	1 100
1 200	13 224	13 363	13 502	13 642	13 782	13 922	14 062	14 202	14 343	14 483	1 200
1 300	14 624	14 765	14 906	15 047	15 188	15 329	15 470	15 611	15 752	15 893	1 300
1 400	16 035	16 176	16 317	16 458	16 599	16 741	16 882	17 022	17 163	17 304	1 400
1 500	17 445	17 585	17 726	17 866	18 006	18 146	18 286	18 425	18 564	18 703	1 500
1 600	18 842	18 981	19 119	19 257	19 395	19 533	19 670	19 807	19 944	20 080	1 600
1 700	20 215	20 350	20 483	20 616	20 748	20 878	21 006				

Tables 16.7 PLATINUM–30% RHODIUM/PLATINUM–6% RHODIUM THERMOCOUPLE TABLES–TYPE B

Temperatures °C (*IPTS-68*) — *Reference junction at* 0 °C — *e.m.f.* μV

Temp.	0	10	20	30	40	50	60	70	80	90	*Temp.*
0	0	−2	−3	−2	−0	2	6	11	17	25	0
100	33	43	53	65	78	92	107	123	140	159	100
200	178	199	220	243	266	291	317	344	372	401	200
300	431	462	494	527	561	596	632	669	707	746	300
400	786	827	870	913	957	1 002	1 048	1 095	1 143	1 192	400
500	1 241	1 292	1 344	1 397	1 450	1 505	1 560	1 617	1 674	1 732	500
600	1 791	1 851	1 912	1 974	2 036	2 100	2 164	2 230	2 296	2 363	600
700	2 430	2 499	2 569	2 639	2 710	2 782	2 855	2 928	3 003	3 078	700
800	3 154	3 231	3 308	3 387	3 466	3 546	3 626	3 708	3 790	3 873	800
900	3 957	4 041	4 126	4 212	4 298	4 386	4 474	4 562	4 652	4 742	900
1 000	4 833	4 924	5 016	5 109	5 202	5 297	5 391	5 487	5 583	4 680	1 000
1 100	5 777	5 875	5 973	6 073	6 172	6 273	6 374	6 475	6 577	6 680	1 100
1 200	6 783	6 887	6 991	7 096	7 202	7 308	7 414	7 521	7 628	7 736	1 200
1 300	7 845	7 953	8 063	8 172	8 283	8 393	8 504	8 616	8 727	8 839	1 300
1 400	8 952	9 065	9 178	9 291	9 405	9 519	9 634	9 748	9 863	9 979	1 400
1 500	10 094	10 210	10 325	10 441	10 558	10 674	10 790	10 907	11 024	11 141	1 500
1 600	11 257	11 374	11 491	11 608	11 725	11 842	11 959	12 076	12 193	12 310	1 600
1 700	12 426	12 543	12 659	12 776	12 892	13 008	13 124	13 239	13 354	13 470	1 700
1 800	13 585	13 699	13 814								

Table 16.8 NICKEL–CHROMIUM/COPPER–NICKEL THERMOCOUPLE TABLES—TYPE E (Chrome-Constantan)

Temperatures °C (*IPTS-68*) — *Reference junction at* 0 °C — *e.m.f.* μV

Temp.	0	10	20	30	40	50	60	70	80	90	*Temp.*
−200	−8 824	−9 063	−9 274	−9 455	−9 604	−9 719	−9 797	−9 835			−200
−100	−5 237	−5 680	−6 107	−6 516	−6 907	−7 279	−7 631	−7 963	−8 273	−8 561	−100
0	0	−581	−1 151	−1 709	−2 254	−2 787	−3 306	−3 811	−4 301	−4 777	0

Table 16.8 NICKEL–CHROMIUM/COPPER–NICKEL THERMOCOUPLE TABLES—TYPE E—*continued*
(Chrome-Constantan)

Temperatures °C (*IPTS-68*) *Reference junction at* 0 °C

e.m.f. μV

Temp.	0	10	20	30	40	50	60	70	80	90	*Temp.*
0	0	591	1 192	1 801	2 419	3 047	3 683	4 329	4 983	5 646	0
100	6 317	6 996	7 683	8 377	9 078	9 787	10 501	11 222	11 949	12 681	100
200	13 419	14 161	14 909	15 661	16 417	17 178	17 942	18 710	19 481	20 256	200
300	21 033	21 814	22 597	23 383	24 171	24 961	25 754	26 549	27 345	28 143	300
400	28 943	29 744	30 546	31 350	32 155	32 960	33 767	34 574	35 382	36 190	400
500	36 999	37 808	38 617	39 426	40 236	41 045	41 853	42 662	43 470	44 278	500
600	45 085	45 891	46 697	47 502	48 306	49 109	49 911	50 713	51 513	52 312	600
700	53 110	53 907	54 703	55 498	56 291	57 083	57 873	58 663	59 451	60 237	700
800	61 022	61 806	62 588	63 368	64 147	64 924	65 700	66 473	67 245	68 015	800
900	68 783	69 549	70 313	71 075	71 835	72 593	73 350	74 104	74 857	75 608	900
1 000	76 358										

Table 16.9 IRON COPPER-NICKEL THERMOCOUPLE TABLES—TYPE J
(Iron–Constantan)

Temperatures °C (*IPTS-68*) *Reference junction at* 0 °C

e.m.f. μV

Temp.	0	10	20	30	40	50	60	70	80	90	*Temp.*
−200	−7 890	−8 096									−200
−100	−4 632	−5 036	−5 426	−5 801	−6 159	−6 499	−6 821	−7 122	−7 402	−7 659	−100
0	0	−501	−995	−1 481	−1 960	−2 431	−2 892	−3 344	−3 785	−4 215	0
0	0	507	1 019	1 536	2 058	2 585	3 115	3 649	4 186	4 725	0
100	5 268	5 812	6 359	6 907	7 457	8 008	8 560	9 113	9 667	10 222	100
200	10 777	11 332	11 887	12 442	12 998	13 353	14 108	14 663	15 217	15 771	200
300	16 325	16 879	17 432	17 984	18 537	19 089	19 640	20 192	20 743	21 295	300
400	21 846	22 397	22 949	23 501	24 054	24 607	25 161	25 716	26 272	26 829	400
500	27 388	27 949	28 511	29 075	29 642	30 210	30 782	31 356	31 933	32 513	500
600	33 096	33 683	34 273	34 867	35 464	36 066	36 671	37 280	37 893	38 510	600
700	39 130	39 754	40 382	41 013	41 647	42 283	42 922	43 563	44 207	44 852	700
800	45 498	46 144	46 790	47 434	48 076	48 716	49 354	49 989	50 621	51 249	800
900	51 875	52 496	53 115	53 729	54 341	54 948	55 553	56 155	56 753	57 349	900
1 000	57 942	58 533	59 121	59 708	60 293	60 876	61 459	62 039	62 619	63 199	1 000
1 100	63 777	64 355	64 933	65 510	66 087	66 664	67 240	67 815	68 390	68 964	1 100
1 200	69 536										

Table 16.10 COPPER/COPPER–NICKEL THERMOCOUPLE TABLES—TYPE T
(Copper–Constantan)

Temperatures °C (*IPTS-68*) *Reference junction at* 0 °C

e.m.f. μV

Temp.	0	10	20	30	40	50	60	70	80	90	*Temp.*
−200	−5 603	−5 753	−5 889	−6 007	−6 105	−6 181	−6 232	−6 258			−200
−100	−3 378	−3 656	−3 923	−4 177	−4 419	−4 648	−4 865	−5 069	−5 261	−5 439	−100
0	0	−383	−757	−1 121	−1 475	−1 819	−2 152	−2 475	−2 788	−3 089	0
0	0	391	789	1 196	1 611	2 035	2 467	2 908	3 357	3 813	0
100	4 277	4 749	5 227	5 712	6 204	6 702	7 207	7 718	8 235	8 757	100
200	9 286	9 820	10 360	10 905	11 456	12 011	12 572	13 137	13 707	14 281	200
300	14 860	15 443	16 030	16 621	17 217	17 816	18 420	19 027	19 638	20 252	300
400	20 869										

Table 16.11 NICKEL–CHROMIUM/NICKEL–ALUMINIUM THERMOCOUPLE TABLES—TYPE K (Chromel–Alumel)

Temperatures °C (*IPTS-68*)										*Reference junction at* 0°C	
Temp.	0	10	20	30	40	50	60	70	80	90	*Temp.*
						e.m.f. μV					
−200	−5 891	−6 035	−6 158	−6 262	−6 344	−6 404	−6 441	−6 458			−200
−100	−3 553	−3 852	−4 138	−4 410	−4 669	−4 912	−5 141	−5 354	−5 550	−5 730	−100
0	0	−392	−777	−1 156	−1 527	−1 889	−2 243	2 586	−2 920	−3 242	0
0	0	397	798	1 203	1 611	2 022	2 436	2 850	3 266	3 681	0
100	4 095	4 508	4 919	5 327	5 733	6 137	6 539	6 939	7 338	7 737	100
200	8 137	8 537	8 938	9 341	9 745	10 151	10 560	10 969	11 381	11 793	200
300	12 207	12 623	13 039	13 456	13 874	14 292	14 712	15 132	15 552	15 974	300
400	16 395	16 818	17 241	17 664	18 088	18 513	18 938	19 363	19 788	20 214	400
500	20 640	21 066	21 493	21 919	22 346	22 772	23 198	23 624	24 050	24 476	500
600	24 902	25 327	25 751	26 176	26 599	27 022	27 445	27 867	28 288	28 709	600
700	29 128	29 547	29 965	30 383	30 799	31 214	31 629	32 042	32 455	32 866	700
800	32 277	33 686	34 095	34 502	34 909	35 314	35 718	36 121	36 524	36 925	800
900	37 325	37 724	38 122	38 519	38 915	39 310	39 703	40 096	40 488	40 879	900
1 000	41 269	41 657	42 045	42 432	42 817	43 202	43 585	43 968	44 349	44 729	1 000
1 100	45 108	45 486	45 863	46 238	46 612	46 985	47 356	47 726	48 095	48 462	1 100
1 200	48 828	49 192	49 555	49 916	50 276	50 633	50 990	51 344	51 697	52 049	1 200
1 300	52 398	52 747	53 093	53 439	53 782	54 125	54 466	54 807			

REFERENCES

1. 'The International Practical Temperature Scale of 1968', Amended Edition of 1975. HMSO, London, 1976.
2. 'The 1976 Provisional 0·5K to 30K Temperature Scale', *Metrologia*, 1979, **15**, 65.
3. C. R. Barber, 'The Calibration of Thermometers', HMSO, London, 1971.
4. L. Crovini *et al.*, 'Extended List of Secondary Reference Points', *Metrologia*, 1977, **13**, 197.

17 Radiating properties of metals

The ability of a surface to radiate energy is governed by the material of which the surface is composed and its physical condition. Any attempt, therefore, to place a numerical value on this radiating ability should be related to a definition of the surface condition. It is usual to choose smooth polished surfaces for this purpose and thus arrive at values which are comparable from one metal to another.

A full radiator (black body) provides a standard of comparison for defining the radiating ability of any other body or surface by determining the ratio of the emission of the surface to that of the black body when they are at the same temperature. An examination of the ratios thus obtained shows that the radiating ability of a metal surface varies with wavelength, temperature and angle of emission. The definition of the emissivity, as the ratio is called, must therefore take into account these variations.

DEFINITIONS OF EMISSIVITY

Special emissivity, ε_λ, of a surface is the ratio of the energy radiated over an infinitesimally small wavelength range at wavelength λ in a specified direction, by unit area of the surface, to the energy radiated by unit area of a full radiator at the same temperature.

The emissivity in a direction normal to the surface is most commonly used, but to avoid confusion it should be called the *normal spectral emissivity*.

Total emissivity, ε_t, is the ratio of the total energy radiated in a specified direction by unit area of the surface to the total energy radiated by a full radiator at the same temperature.

The *normal total emissivity* is most commonly employed.

The *absorptivity* (a) and *reflectivity* (r), for an opaque polished surface, are defined as the ratio of the rate of absorption or reflection of energy to the rate of incidence of energy. Since the incident energy must be either reflected or absorbed the sum of the reflectivity and absorptivity must be unity. Since the absorptivity is equal to the emissivity (ε)

$$\varepsilon = a = 1 - r$$

Hence the emissivity may be derived from the reflectivity and this is sometimes more convenient than a direct determination.

The following equation relates the reflectivity to the refractive index, which is a complex quantity given by $(n - jk)$, where n is the real part and k is the complex part and is usually known as the extinction coefficient:

$$r = \frac{n^2 + k^2 + 1 - 2n}{n^2 + k^2 + 1 + 2n}$$

where the extinction coefficient is such that the fraction of light transmitted perpendicularly through a layer d cm thick is exp $(-4\pi dk/\lambda)$, λ being the wavelength in air. Hence

$$\varepsilon = 1 - r = \frac{4n}{n^2 + k^2 + 1 + 2n}$$

The Maxwell electromagnetic theory gives

$$2n^2 = \mu\left[e^2 + \frac{\sigma^2\mu_0^2c^4}{4\pi^2\nu^2}\right]^{1/2} + \mu e$$

$$2k^2 = \mu\left[e^2 + \frac{\sigma^2\mu_0^2c^4}{4\pi^2\nu^2}\right]^{1/2} - \mu e$$

where σ is the electrical conductivity, e the dielectric constant (both at frequency ν), μ_0 the permeability of free space ($= 4\pi \times 10^{-7}$ by definition), μ the relative permeability of the metal (≈ 1), and c the velocity in free space.

It follows that for long wavelengths (ν small), where $\sigma\mu_0 c^2/2\pi\nu$ is large in comparison with e

$$\varepsilon = c_1(\rho/\lambda)^{1/2} \text{ approx.} \tag{17.1}$$

where ρ is the specific resistance in Ωm, and c_1 a constant of value 0.365 when λ is expressed in metres. Further terms may be added to give a better approximation, thus

$$\varepsilon = c_1(\rho/\lambda)^{1/2} + c_2(\rho/\lambda) + c_3(\rho/\lambda)^{3/2} \tag{17.2}$$

where $c_1 = 0.365$, $c_2 = 0.0667$, and $c_3 = 0.0061$.

The equations show that the spectral emissivity of metals should increase as the wavelength decreases and this is in general agreement with experiment. In all cases, however, there is a lack of concordance between the experimental and theoretical values in the visible and ultraviolet (*see* Figure 17.1 for tungsten).

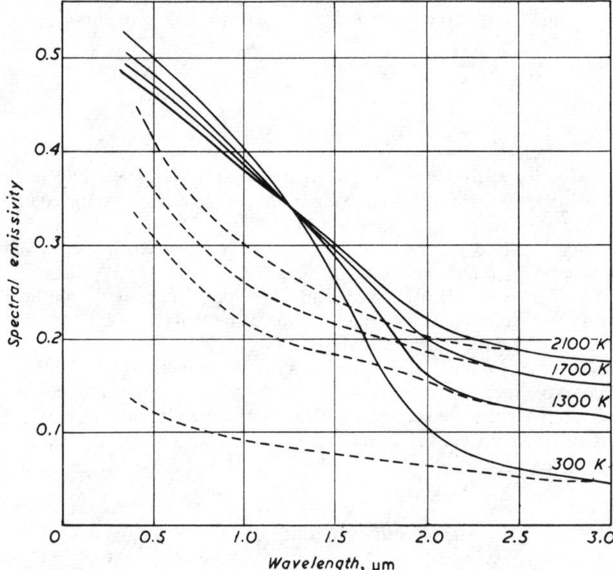

Figure 17.1 *Spectral emissivity of tungsten as a function of wavelength for different temperatures. Dotted lines calculated from equation 17.1. References 1, 2.*

VARIATION OF SPECTRAL EMISSIVITY WITH TEMPERATURE

According to equation 17.1 above, an increase in temperature, since it causes an increase in resistivity, should result in an increase in emissivity. This is generally true in the infra-red region of the spectrum, but in the visible there is often a decrease in emissivity with rise in temperature, as is shown, for example, in Figure 17.1, which gives the variations of spectral emissivity of tungsten with temperature and wavelength. A curious result is shown by these curves: at a wavelength of 1.3 µm there is no variation of emissivity with temperature.

The passage from the solid to the liquid state of a metal does not usually produce a large change in emissivity. The value for platinum, for example, increases by some 15% in the red and even less in the green and violet. For tungsten and molybdenum the change appears to be negligible, but for gold and silver the change in the red is quite marked—from 0.14 to 0.22 for the former and 0.04 to 0.07 for the latter.

VARIATION OF SPECTRAL EMISSIVITY WITH ANGLE OF EMISSION

A radiating surface which obeys Lambert's cosine law gives an intensity of radiation in any given direction which is proportional to the cosine of the angle the direction makes with the normal to the surface. A full radiator obeys the law while metal surfaces do not and since emissivity is defined by comparison with a full radiator, the emissivities must vary with angle of emission. Typical examples of this variation are shown in Figures 17.2 and 17.3.

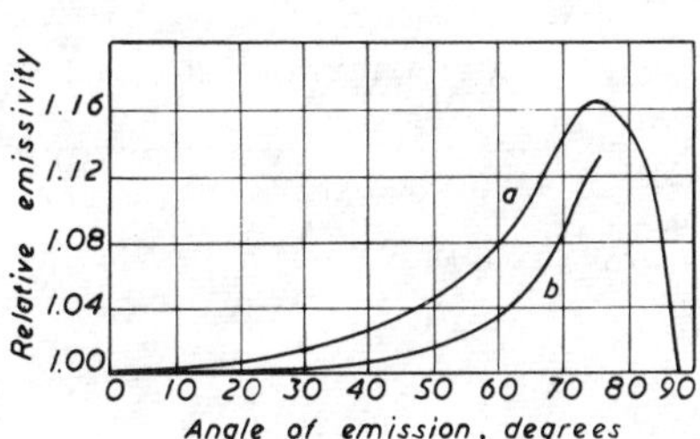

Figure 17.2 *Variation of spectral emissivity of tungsten with angle of emission* (*a*) $\lambda = 0.66\,\mu m$, (*b*) $\lambda = 0.47\,\mu m$. *References 3, 4*

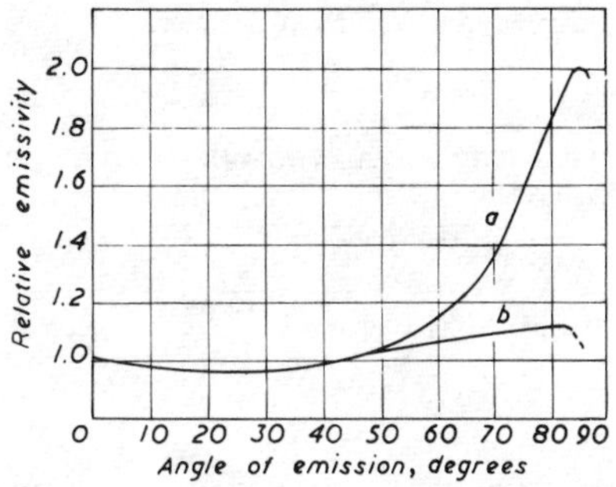

Figure 17.3 *Variation of emissivity of platinum with angle of emission.* (*a*) *total emissivity*, (*b*) *spectral emissivity in the red. Reference 5*

VARIATION OF TOTAL EMISSIVITY WITH TEMPERATURE

The total emissivity in a particular direction may be expressed in terms of the spectral emissivity in the same direction as follows:

$$\varepsilon_t = \frac{\int_0^\infty E_{\lambda T} \varepsilon_\lambda \, d\lambda}{\int_0^\infty E_{\lambda T} \, d\lambda}$$

where $E_{\lambda T}$ is the energy radiated at wavelength λ and temperature T.

Since, however, complete data on the values of ε_λ throughout the spectrum are not available for many metals, the relation cannot be generally employed. On the basis of this formula and equation 17.2 omitting the third term, the following expression may be derived for the total emissivity

$$\varepsilon_t = K_1 \sqrt{(\rho T)} - K_2 \rho T$$

where $K_1 = 5.736$ and $K_2 = 1.769$ if ρ is expressed in ohm m.

The expression is in agreement with experiment in so far as it shows an increase of ε_t with T; numerically, however, it does not represent the facts very well, although there is better agreement at low than at high temperatures. This corresponds with what was noted in connection with spectral emissivities where the agreement between experiment and theory is better at long than at short wavelengths: the energy maximum moves towards the short wavelengths as the temperature increases.

VARIATION OF TOTAL EMISSIVITY WITH ANGLE OF EMISSION

The total emissivity of platinum has been measured for different angles of emission (Figure 17.3) and the variations found to be considerably greater than given by similar measurements of spectral emissivity in the visible region. Since the variations of total emissivity with angle of emission are greater the lower the temperature it may be concluded that the variation of spectral emissivity is greater in the infra-red than in the visible.

POLARIZATION OF RADIATION EMITTED BY A METAL SURFACE

The radiation emitted at an oblique angle by a metal surface is found to be polarized. The fraction of polarized light emitted by tungsten at various angles to the normal is shown in Table 17.1. For a circular tungsten filament the total light emitted in a direction normal to the axis includes about 20% of polarized light. Considerable caution must be exercised in interpreting the results obtained from apparatus employing polarizing components when sighting on a metal surface, as, for example, the Wanner pyrometer.

Table 17.1 POLARIZATION OF LIGHT EMITTED BY TUNGSTEN AS A FUNCTION OF THE ANGLE OF EMISSION[6]

Angle of emission (degrees)	0	30	45	60	75	80	85	90
Fraction of light polarized	0	0.10	0.22	0.46	0.75	0.81	0.90	1.00

TEMPERATURE MEASUREMENT AND EMISSIVITY

Radiation pyrometers, both spectral and total, are normally calibrated in terms of full radiation, and when sighted on a metal surface measure a full radiation temperature. The full radiation temperature, T_r, measured by a total radiation pyrometer is related to the true temperature, T, by the formula

$$T_r = T\sqrt[4]{(\varepsilon_t)}$$

where ϵ_t is the total emissivity of the surface.

The spectral or optical pyrometer measures a luminance temperature T_s which is related to the true temperature by the equation

$$\frac{1}{T} - \frac{1}{T_s} = \frac{\lambda \log_c \varepsilon_\lambda}{C_2}$$

where ε_λ is the spectral emissivity and C_2 is the constant in the Wien equation for distribution of energy in the spectrum and has a value of 1.438 cm K. The corrections to be applied to the two types of pyrometer for various emissivities are shown in Figures 17.4 and 17.5. It will be observed that for the same emissivities the correction is considerably greater for total radiation than for spectral radiation. The difference in correction for a total and optical pyrometer sighted on to the same metal surface is even greater, however, since the spectral emissivity in the visible region for a given temperature is always greater than the total emissivity.

EMISSIVITY VALUES

The values of emissivity given in Tables 17.2 to 17.6 relate, as far as is known, to emission from plane polished or plane unoxidized liquid metal surfaces. The radiation emitted will, of course, be increased considerably by oxidation or roughening of the surface. In any practical application, therefore, the values in the tables must be used with discretion, and where precise measurement is of importance a determination of the emissivity should be made for the surface conditions obtaining. As a rough guide the emissivities for various oxidised surfaces are given in Tables 17.5 and 17.6.

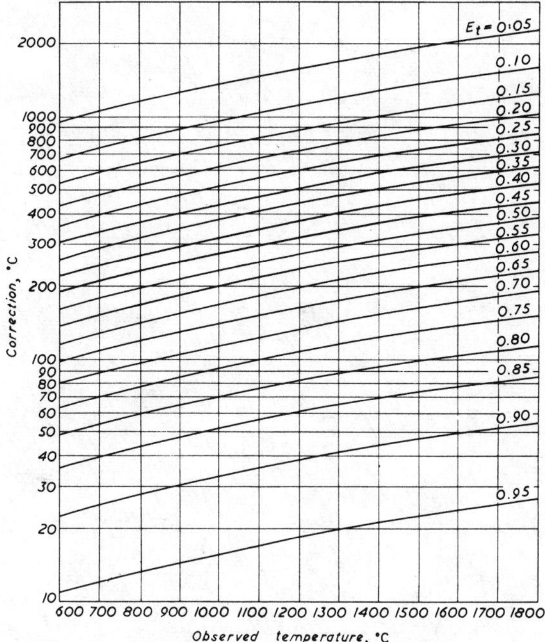

Figure 17.4 *Correction to radiation pyrometer readings for total emissivity*

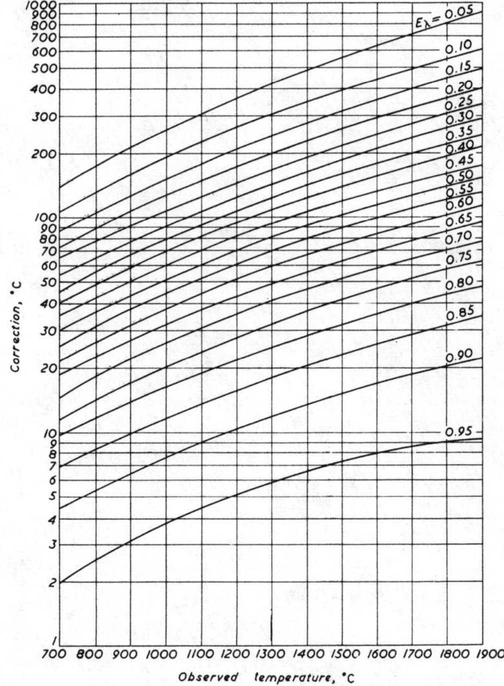

Figure 17.5 *Correction to optical pyrometer readings for spectral emissivity* $\lambda = 0.65\,\mu m$, $c_2 = 1.438\,cm\,K$

Table 17.2 SPECTRAL NORMAL EMISSIVITY OF METALS FOR WAVELENGTH OF 0.65 μm

Metal	*Temperature °C*										*References*
	600	800	1 000	1 200	1 400	1 600	1 800	2 000	2 500	3 000	
Beryllium	—	—	0.37	0.37	—	—	—	—	—	—	7
Chromium	—	—	—	—	0.39	—	—	0.39[m]	—	—	8
Cobalt	—	—	0.33–0.38	0.34–0.37	0.35–0.37	0.37[m]	—	—	—	—	8, 9
Copper	—	0.11	0.10	0.10[m]	0.11[m]	0.12[m]	0.14[m]	—	—	—	10
Erbium	—	—	0.55	0.55	0.55	0.38[m]	—	—	—	—	8
Germanium	—	0.55	—	—	—	—	—	—	—	—	11
Gold	0.16–0.18	0.16–0.19	0.16–0.21	0.13[m]	—	—	—	—	—	—	7
Hafnium	—	—	—	—	0.45	—	—	—	—	—	12
Iridium	—	—	0.36	0.34	0.32	—	—	0.30	—	—	8, 13
Iron	—	0.37	0.36	0.35	0.35	0.37[m]	—	—	—	—	7
Manganese	—	—	—	0.59	0.59[m]	—	—	—	—	—	8
Molybdenum	—	0.37–0.43	0.36–0.42	0.35–0.42	0.34–0.41	0.34–0.41	0.33–0.40	0.32–0.39	0.31–0.37	—	7
Nickel	0.36	0.35	0.34	—	—	—	—	—	—	—	7
Niobium	—	—	0.37	0.37	0.37	0.37	0.37	0.37	0.40[m]	—	14
Osmium	—	—	0.52	0.44	0.40	0.38	0.38	0.38	—	—	15
Palladium	—	0.40	0.37	0.34	0.30	0.37[m]	—	—	—	—	16
Platinum	—	0.29–0.31	0.29–0.31	0.29–0.31	0.29–0.31	0.29–0.31	—	—	—	—	7
Rhenium	—	—	—	—	0.42	0.42	0.41	0.41	0.40	—	17
Rhodium	—	0.25	0.22	0.19	0.18	0.16	—	—	—	—	16
Ruthenium	—	—	0.42	0.35	0.32	0.31	0.31	0.31	—	—	15
Silicon	—	0.63	0.57	0.52	0.46	0.48[m]	—	—	—	—	11
Silver	—	0.055	0.055[m]	—	—	—	—	—	—	—	10
Tantalum	0.47	0.46	0.45	0.44	0.42	0.41	0.40	0.39	0.38	0.36	7
Thorium	—	—	0.38	0.38	0.38	—	—	—	—	—	14
Titanium	—	0.48	0.48	0.48	0.47	—	—	—	—	—	18, 19
Tungsten	—	—	0.46–0.48	0.43–0.48	0.42–0.47	0.42–0.47	0.41–0.47	0.40–0.47	0.38–0.46	0.36–0.45	7
Uranium	—	0.19–0.36	0.19–0.36	0.34[m]	0.34[m]	—	—	—	—	—	7
Yttrium	—	—	—	—	0.35	—	—	—	—	—	8
Zirconium	—	—	0.48	0.45	0.42	0.39	0.36	—	—	—	20
Alloys											
Cast iron	—	0.37	0.37	0.37	0.37	0.40[m]	—	—	—	—	—
Nichrome (in hydrogen)	—	0.35	0.35	0.35	0.35	—	—	—	—	—	7
Steel	—	0.35–0.40	0.32–0.40	0.30–0.40	—	0.37[m]	—	—	—	—	7

[m] Value for molten state.

Table 17.3 SPECTRAL EMISSIVITIES IN THE INFRA-RED OF METALS AT HIGH TEMPERATURES

Metal	*Temperature* °C	*Wavelength* μm 1.0	1.2	1.4	1.5	1.6	1.8	2.0	2.5	3.0	3.5	4.0	4.5	*References*
Cobalt	800	—	0.26	—	—	—	—	0.21	—	—	—	—	—	7
	1 000	—	0.26	—	—	—	—	0.21	—	0.18	—	—	—	7
	1 200	—	0.26	—	—	—	—	0.22	—	0.19	—	—	—	7
Copper	762	—	—	—	0.031	—	—	0.029	—	—	—	0.025	—	21
	901	—	—	—	0.079	—	—	0.065	0.052	0.043	0.038	0.032	—	22
	985	0.049	—	—	0.037	—	0.034	—	0.032	0.031	—	0.030	—	23
Iridium	827	0.229	0.203	0.185	—	0.167	0.152	0.140	—	—	—	—	—	24
	1 227	0.233	0.213	0.194	—	0.180	0.169	0.160	—	—	—	—	—	24
	1 727	0.243	0.228	0.210	—	0.199	0.188	0.180	—	—	—	—	—	24
	2 127	0.247	0.233	0.219	—	0.207	0.199	0.192	—	—	—	—	—	24
Iron	800	—	0.294	—	—	0.264	—	0.237	0.217	—	—	—	—	25
	1 000	—	0.294	—	—	0.267	—	0.245	0.227	—	—	—	—	25
	1 200	—	0.291	—	—	0.300	—	0.252	0.235	—	—	—	—	25
	1 245	0.340	0.316	0.298	0.290	0.282	0.268	0.260	0.248	0.240	0.235	0.225	0.218	22
Molybdenum	1 327	0.335	—	—	0.185	—	—	0.140	—	0.115	—	0.114	—	7
	1 727	0.300	—	—	0.195	—	—	0.170	—	0.155	—	0.145	—	7
	2 527	0.260	—	—	0.210	—	—	0.193	—	0.185	—	0.185	—	7
Nickel	800	—	0.295	0.267	—	0.250	0.230	0.215	—	—	—	—	—	25
	1 000	—	0.293	0.269	—	0.252	0.232	0.219	—	—	—	—	—	25
	1 200	—	0.290	0.271	—	0.253	0.235	0.223	—	—	—	—	—	25
	1 110	—	0.292	0.270	0.250	—	—	0.290	0.205	0.187	0.174	0.162	—	22
Niobium	827	0.345	—	—	0.23	—	—	0.19	—	—	—	—	—	7
	1 227	0.335	—	—	0.25	—	—	0.21	—	—	—	—	—	7
	1 727	0.320	—	—	0.26	—	—	0.23	—	—	—	—	—	7
	2 127	0.315	—	—	0.27	—	—	0.25	—	—	—	—	—	7
Platinum	1 127	—	0.257	—	0.227	—	—	0.193	—	0.151	—	0.130	—	7

Table 17.3 SPECTRAL EMISSIVITIES IN THE INFRA-RED OF METALS AT HIGH TEMPERATURES—*continued*

Metal	*Tempera-ture* °C	*Wavelength* μm 1.0	1.2	1.4	1.5	1.6	1.8	2.0	2.5	3.0	3.5	4.0	4.5	*References*
Rhenium	1 537	0.36	—	—	0.29	—	—	0.25	0.23	—	—	—	—	17
	2 118	0.36	—	—	0.30	—	—	0.27	0.24	—	—	—	—	17
	2 772	0.36	—	—	0.32	—	—	0.29	0.26	—	—	—	—	17
Tantalum	1 427	0.295	—	—	0.220	—	—	0.190	—	0.170	—	0.150	—	7
	1 927	0.310	—	—	0.245	—	—	0.215	—	0.192	—	0.180	—	7
	2 527	0.330	—	—	0.290	—	—	0.270	—	0.240	—	0.230	—	7
Titanium	750	0.490	—	0.510	0.500	—	—	0.455	—	0.525	0.575	0.600	—	26
Tungsten	1 327	0.385	—	—	0.28	—	—	0.21	—	0.13	—	0.095	—	7
	2 127	0.37	—	—	0.292	—	—	0.245	—	0.18	—	0.15	—	7
	2 527	0.36	—	—	0.30	—	—	0.26	—	—	—	—	—	7
Zirconium	1 127	0.46	—	—	0.422	—	—	0.386	0.360	0.348	—	—	—	27
	1 327	0.444	—	—	—	—	—	0.368	—	0.343	—	0.325	—	27
	1 727	0.442	—	—	0.375	—	—	0.357	0.351	0.342	0.330	—	—	27

Table 17.4 SPECTRAL NORMAL EMISSIVITY OF METALS AT ROOM TEMPERATURE
(Derived from reflectivity data by formula $\varepsilon = 1 - r$)

	Wavelength µm							
Metal	10.0	9.0	5.0	3.0	1.0	0.6	0.5	*References*
Aluminium	0.02–0.04	—	0.03–0.08	0.03–0.12	0.08–0.27	—	—	7
Antimony*	—	0.28	0.31	0.35	0.45	0.47	—	28
Bismuth	0.08	—	0.12	0.26	0.72	0.76	0.75	29
Cadmium	—	0.02	0.04	0.07	0.30	—	—	30
Chromium	—	0.08	0.19	0.30	0.43	0.44	0.45	28
Cobalt	—	0.04	0.15	0.23	0.32	—	—	30
Copper	0.021	—	0.024	0.026	0.030	0.080	0.36	7
Gold	0.015	0.015	0.015	0.015	0.020	0.080	0.45	7
Iridium	—	0.04	0.06	0.09	0.22	—	—	30
Iron	—	—	—	—	0.41	0.48	0.49	16
Lead	—	0.06	0.08	—	—	—	—	31
Magnesium*	—	0.07	0.14	0.20	0.26	0.27	0.28	28
Molybdenum	0.15	—	0.16	0.19	0.42	—	—	32
Nickel	—	0.04	0.06	0.12	0.27	—	—	33
Niobium	0.04	—	0.06	0.14	0.29	0.55	—	7
Palladium	—	0.03	0.10	0.12	0.28	0.37	0.42	20, 16, 34
Platinum	0.05	—	0.06	0.11	0.24	0.36	0.40	35
Rhodium	—	0.05	0.07	0.08	0.16	0.21	0.24	28
Silver	0.02	—	0.02	0.02	0.03	0.03	0.03	7
Tantalum	—	0.06	0.07	0.08	0.22	0.55	0.62	30
Tellurium	—	0.22	0.43	0.47	0.50	0.51	—	28
Tin	—	0.14	0.24	0.32	0.46	—	—	30
Titanium	0.05–0.12	—	0.10–0.18	0.25–0.33	0.37–0.49	—	—	7
Tungsten	0.03	—	0.05	0.07	0.40	0.44–0.49	—	7
Vanadium	0.06–0.09	—	0.07–0.11	0.10–0.17	0.36–0.50	0.42–0.57	0.43–0.59	7
Zinc	0.03	—	0.05	0.08	0.50–0.61	0.42–0.58	—	7

* Values for spectral angular (15°) emissivity only.

Table 17.5 SPECTRAL EMISSIVITY OF OXIDIZED METALS FOR WAVELENGTH OF 0.65 µm
(Oxide formed on smooth surfaces)
(For oxides in the form of refractory materials, values of emissivity widely different from those below may be given and will be dependent on the grain size)

Metal	$\varepsilon_{0.65}$	*Metal*	$\varepsilon_{0.65}$
Aluminium	0.30	Uranium	0.30
Beryllium	0.61	Vanadium	0.70
Chromium	0.60	Yttrium	0.60
Cobalt	0.77	Zirconium	0.80
Copper	0.70		
Iron	0.63	*Alloys*	
Magnesium	0.20		
Nickel	0.85	Cast iron	0.70
Niobium	0.71	Nichrome	0.90
Tantalum	0.42	Constantan	0.84
Thorium	0.57	Carbon steel	0.80
Titanium	0.50	Stainless steel	0.85

References 7 and 8.

Table 17.6 TOTAL NORMAL EMISSIVITY OF METALS

Metal	*Temperature °C* 20	100	500	1 000	1 200	1 400	1 600	2 000	2 500	3 000	*References*
Aluminium	—	0.038	0.064	—	—	—	—	—	—	—	36
Beryllium	—	—	—	0.55	0.87	—	—	—	—	—	37
Bismuth	—	0.06	—	—	—	—	—	—	—	—	—
Chromium	—	0.08	0.11–0.14	—	—	—	—	—	—	—	7
Cobalt	—	0.15–0.24	0.34–0.46	—	—	—	—	—	—	—	7
Copper	—	—	0.02	—	0.12[m]	—	—	—	—	—	7
Germanium	—	—	0.54	—	—	—	—	—	—	—	38
Gold	—	0.02	0.02	—	—	—	—	—	—	—	—
Hafnium	—	—	—	—	0.30	0.31	0.32	—	—	—	39
Iron	0.05	0.07	0.14	0.24	—	—	—	—	—	—	7
Lead	—	0.63	—	—	—	—	—	—	—	—	40
Magnesium	—	0.12*	—	—	—	—	—	—	—	—	21
Mercury	—	0.12	—	—	—	—	—	—	—	—	—
Molybdenum	0.065	0.08	0.13	0.19	0.22	0.24	0.27	—	—	—	7
Nickel	—	—	0.09–0.15	0.14–0.22	—	—	—	—	—	—	7
Niobium	—	—	—	0.12	0.14	0.16	0.18	0.21	—	—	41
Palladium	—	—	0.06	0.12	0.15	—	—	—	—	—	16
Platinum	—	—	0.086	0.14	0.16	—	—	—	—	—	7
Rhenium	—	—	—	0.22	0.25	0.27	0.29	—	—	—	42
Rhodium	—	—	0.035	0.07	0.08	0.09	—	—	—	—	16
Silver	—	0.02–0.03	0.02–0.03	—	—	—	—	—	—	—	7
Tantalum	0.03	0.04	0.06	0.11	0.13	0.15	0.18	0.23	0.28	—	7
Tin	—	0.07	—	—	—	—	—	—	—	—	43
Titanium	—	0.11	—	—	—	—	—	—	—	—	44
Tungsten	—	—	0.05	0.11	0.14	0.17	0.19	0.23	0.27	0.30	7
Uranium (α-phase)	—	—	0.33*	—	—	—	—	—	—	—	7
Uranium (γ-phase)	—	—	—	0.29–0.40*	—	—	—	—	—	—	7
Zinc	—	0.07	—	—	—	—	—	—	—	—	45
Zinc (galvanized iron)	—	0.21	—	—	—	—	—	—	—	—	43
Zirconium	—	—	—	0.22	0.25	0.27	—	—	—	—	46
Alloys											
Brass	—	0.059	—	—	—	—	—	—	—	—	43
Cast iron (cleaned)	—	0.21	—	—	—	—	0.29[m]	—	—	—	7
Nichrome	—	—	0.95	0.98	—	—	—	—	—	—	7
Steel (polished)	—	0.13–0.21	0.18–0.26	0.55–0.80	—	—	—	—	—	—	7
Steel (cleaned)	—	0.21–0.38	0.25–0.42	0.50–0.77	—	—	—	—	—	—	7

* Value for total hemispherical emissivity. [m] Value for molten state.

Table 17.7 TOTAL NORMAL EMISSIVITY OF OXIDIZED METALS

The values depend on the degree of oxidation and the grain size.
Unless stated otherwise, the following are results obtained for metals oxidised in general above 600 °C.

Metal	200 °C	400 °C	600 °C	800 °C	1 000 °C	*References*
Alumium	0.11	0.15	0.19	—	—	40
Brass	0.61	0.60	0.59	—	—	40
Chromium	—	0.09	0.14–0.34	—	—	16
Copper (red heat for 30 min)	0.15	0.18	0.23	0.24	—	7
Copper (stably oxidized at 760 °C)	—	0.40–0.50	0.60–0.66	—	—	7
Copper (extreme oxidation)	—	0.88	0.92	—	—	7
Cast iron	0.64	0.71	0.78	—	—	40
Cast iron (strongly oxidized)	0.95	—	—	—	—	—
Iron (red heat for 30 min)	0.45	0.52	0.57	—	—	7
Lead	0.63	—	—	—	—	—
Molybdenum (oxide volatile in vacuum above 540 °C)	—	0.84	—	—	—	7
Monel	0.41	0.44	0.47	—	—	40
Nickel (stably oxidized at 900 °C)	0.15–0.50	0.33–0.51	0.44–0.57	0.49–0.71	—	7
Nimonic (buffed, oxidized at 900 °C)	—	0.46	—	—	—	47
Nimonic (buffed, oxidized at 1 200 °C)	—	0.72	—	—	—	47
Niobium (oxidized and annealed)*	—	—	—	0.74	—	7
Palladium	0.03	0.05	0.076	—	0.124	16
Stainless steel (stably oxidized at high temperature)	—	0.80–0.87	0.84–0.91	0.89–0.95	—	7
Stainless steel (red heat in air for 30 min)	0.12–0.25	0.17–0.30	0.23–0.37	0.30–0.44	~	7
Stainless steel (buffed, stably oxidized at 600 °C)	—	0.41	0.44	0.54	—	7
Stainless steel (polished, oxidized at high temperature)	—	—	0.65–0.70	—	0.73–0.83	7
Stainless steel (shot blasted stably oxidized at 600 °C)	—	0.65	0.67	—	—	7
Tantalum (red heat for 30 min)	0.42	0.42	0.42	—	—	7
Zinc	—	0.11	—	—	—	40

* Value for total hemispherical emissivity.

REFERENCES

1. W. W. Coblentz, *Bull. US Bur. Stand.*, 1918, **14**, 312.
2. W. Weniger and A. H. Pfund, *Phys. Rev.*, 1919, **14**, 427.
3. A. G. Worthing, *Astrophys. J.*, 1912, **26**, 345.
4. A. G. Worthing, *J. opt. Soc. Amer.*, 1926, **13**, 635.
5. E. Bauer and M. Moulin, *J. Phys. Rad.*, 1910, **9**, 468.
6. W. E. Forsythe and A. G. Worthing, *Astrophys. J.*, 1925, **61**, 165.
7. Y. S. Touloukian and D. P. DeWitt (editors), 'Thermophysical Properties of Matter', IFI/Plenum, New York, 1970, Vol. 7, 'Thermal Radiative Properties – Metallic Elements and Alloys'.
8. G. K. Burgess and R. G. Waltenberg, *Bull. US Bur. Stand.*, 1915, **11**, 591.
9. H. B. Wahlin and H. W. Knop, Jr., *Phys. Rev.*, 1948, **74**, 687.
10. C. C. Bidwell, *Phys. Rev.*, 1914, **3**, 439.
11. F. G. Allen, *J. appl. Phys.*, 1957, **28**, 1510.
12. M. L. Shaw, *J. appl. Phys.*, 1966, **37**, 919.
13. O. K. Husmann, *J. appl. Phys.*, 1966, **37**, 4662.
14. L. V. Whitney, *Phys. Rev.*, 1935, **48**, 458.
15. R. W. Douglass and E. F. Adkins, *Trans. Met. Soc. AIME*, 1961, **221**, 248.
16. H. T. Betz, O. H. Olsen, B. D. Schurin and J. C. Morris, WADC-TR-56-222 (Part 2), 1957, 1–184 (AD 202 493).
17. D. T. F. Marple, *J. opt. Soc. Amer.*, 1956, **46**, 490.
18. F. J. Bradshaw, *Proc. phys. Soc.*, 1950, **B63**, 573.
19. H. Seemuller and D. Stark, *Z. Phys.*, 1967, **198**, 201.
20. S. C. Furman and P. A. McManus, USAEC, GEAP-3338, 1960, 1–46.
21. W. G. D. Carpenter and J. H. Sewell, RAE Rpt: CHem-538, 1962, 1–6 (AD 295 648).
22. D. J. Price, *Proc. phys. Soc.*, 1947, **59**, 118.

23. A. E. Anderson, Univ. of Calic., M.S. Thesis, 1962, 1–57.
24. T. B. Barnes, *J. opt. Soc. Amer.*, 1966, **56**, 1546.
25. L. Ward, *Proc. phys. Soc.*, 1956, **69**, 339.
26. J. G. Adams, Northrup Corp., Novair Div., 1962, 1–259 (AD 274 555).
27. J. A. Coffman, G. M. Kibler, T. F. Lyon and B. D. Acchione, WADD-TR-60-646 (Part 2), 1963, 1–183, (AD 297 946).
28. W. W. Coblentz, *Bull. US Bur. Stand.*, 1911, **7**, 197.
29. J. G. Adams, Northrup Space Labs., Hawthorne, Calif., NSL-62-198, 1962, 1–101.
30. W. W. Coblentz, Publ. Carneg. Instn. Wash. No. 65, 1906, p. 91; *Bull. US Bur. Stand.*, 1906, **2**, 457.
31. H. Schmidt and E. Furthmann, *Mitt. K.-Wilhelm.-Inst. Eisenforsch. Dusseld.*, 1928, **10**, 225.
32. R. V. Dunkle and J. T. Gier, Inst. of Eng. Res., Univ. of Calif., Berkeley, Progress Rpt., 1953, 1–73 (AD 16 830).
33. E. Hagen and H. Rubens. *Ann. Phys. Lpz.*, 1900 (4), **1**, 352.
34. E. O. Hulburt, *Astrophys. J.*, 1915, **42**, 203.
35. R. A. Seban, WADD-TR-60-370 (Part 2), 1962, 1–72 (AD 286 863).
36. E. Schmidt, V. A. W. Hauzeitschr and A. G. Erftwerk, *Aluminium*, 1930, **3**, 91.
37. S. Konopken and R. Klemm, NASA-SP-31, 1963, 505–513.
38. V. F. Brekhovskikh, *Inz.-fiz. Zh.*, 1964, **7** (5), 66.
39. D. L. Timrot, V. Yu. Voskresenskii and V. E. Peletskii, *High Temp.*, 1966, **4**, 808.
40. C. P. Randolf and M. J. Overholzer, *Phys. Rev.*, 1913, **2**, 144.
41. G. L. Abbott, WADD-TR-61-94 (Part 3), 1963, 1–30 (AD 435 825) (AD 436 887).
42. G. B. Gaines and C. T. Sims, *J. appl. Phys.*, 1963, **34**, 2922.
43. T. T. Barnes, W. E. Forsythe and E. Q. Adams, *J. opt. Soc. Amer.*, 1947, **37**, 804.
44. J. T. Bevans, J. T. Gier and R. V. Dunkle, *Trans. ASME*, 1958, **80**, 1405.
45. P. F. McDermott, *Rev. Sci. Instrum.*, 1937, **8**, 185.
46. D. L. Timrot and V. E. Peletskii, *High Temp.*, 1965, **3**, 199.
47. A. H. Sully, E. A. Brandes and R. B. Waterhouse, *Br. J. appl. Phys.*, 1952, **3**, 97.

18 Electron emission

Under normal conditions electrons are prevented from leaving a metal by a potential step at the surface. The height of this potential step is called the work function ϕ. Electrons can, however, escape if they are given enough energy. This energy can be supplied in a number of different ways, giving rise to the various types of electron emission.

18.1 Thermionic emission

When a metal is heated, some electrons with energies near the Fermi level are enabled to escape by acquiring extra thermal energy. An adjacent anode carrying a sufficiently positive potential will collect all the electrons emitted, and the saturated emission current will flow. A further increase of anode potential causes a positive field at the metal surface; this lowers the potential barrier slightly and increases the current. The 'zero field' saturated emission current per unit area of the cathode J, is related to the temperature according to the Richardson–Dushman equation.

$$J = AT^2 \exp(-e\phi kT)$$

where A is a constant, e the electronic charge, k Boltzmann's constant and T the absolute temperature.

For a metal, the theoretical value of A is 1.2 MA m^{-2}. In practice ϕ usually has a temperature coefficient and this results in a different value of A. The work function ϕ cannot be calculated reliably, but tends to increase with the density of the metal. The observed values of A and ϕ are shown in Table 18.1 for polycrystalline surfaces of a number of metals.

Table 18.1 THERMIONIC PROPERTIES OF THE ELEMENTS

Element	A kA m^{-2} K^{-2}	ϕ V	*Element*	A kA m^{-2} K^{-2}	ϕ V
Barium	600	2.11	Niobium	1 200	4.19
Beryllium	3 000	3.75	Osmium	1 100 000	5.93
Caesium	1 600	1.81	Palladium	600	4.9
Calcium	600	2.24	Platinum	320	5.32
Carbon	150	4.5	Rhenium	1 200	4.96
Chromium	1 200	3.90	Rhodium	330	4.8
Cobalt	410	4.41	Silicon	80	3.6
Copper	1 200	4.41	Tantalum	1 200	4.25
Hafnium	220	3.60	Thorium	700	3.38
Iridium	1 200	5.27	Titanium	—	3.9
Iron α	260	4.5	Tungsten	600	4.54
Iron γ	15	4.21	Uranium	60	3.27
Molybdenum	550	4.15	Zirconium	3 300	4.12
Nickel	300	4.61			

References: 1, 2, 3, 4 (General); 5 (Hf), 6 (Nb, Ta, Re, Os, Ir); 7 (Be, Cr, Cu).

To calculate the emission from the values of A and ϕ in the table, the emission formula may be written as

$$\log_{10} J = \log_{10} A + 2 \log_{10} T - 5\,040 \phi T^{-1}$$

where J is in $kA\,m^{-2}$, T is in K.

The work function of a metal is lowered when a layer of a more electropositive material is adsorbed on its surface. This increases the thermionic emission, which has a maximum value when the adsorbed layer is approximately monatomic. Table 18.2 shows typical values of thermionic constants for various such surfaces.

Table 18.2 THERMIONIC PROPERTIES OF REFRACTORY METALS WITH ADSORBED ELECTROPOSITIVE LAYERS

Surface	A $kAm^{-2}\,K^{-2}$	ϕ V
Tungsten–barium	15	1.56
Tungsten–caesium	32	1.36
Tungsten–cerium	80	2.71
Tungsten–lanthanum	80	2.71
Tungsten–strontium	—	2.2
Tungsten–thorium	30	2.63
Tungsten–uranium	32	2.84
Tungsten–yttrium	70	2.70
Tungsten–zirconium	50	3.14
Molybdenum–thorium	15	2.59
Tantalum–thorium	15	2.52

Reference 8.

The emission may also be increased by a coating of a refractory metallic compound, usually about 100 μm thick. The thermionic emission follows the usual law but in the case of a semiconductor layer the quantities A and ϕ have a different significance. Table 18.3 shows the

Table 18.3 THERMIONIC PROPERTIES OF REFRACTORY METAL COMPOUNDS

Compound	A $kA\,m^{-2}\,K^{-2}$	ϕ V	*Emission* $kA\,m^{-2}$
TaC	3	3.14	3 at 2 000 K
TiC	250	3.35	63 at 2 000 K
ZrC	3	2.18	40 at 2 000 K
ThC_2	5 500	3.5	40 at 2 000 K
SiC	640	3.5	4 at 2 000 K
UC	330	2.9	50 at 2 000 K
CaB_6	26	2.9	0.12 at 1 670 K
SrB_6	1.4	2.7	0.036 at 1 670 K
BaB_6	160	3.5	0.018 at 1 670 K
LaB_6	290	2.7	7 at 1 670 K
CeB_6	36	2.6	1.7 at 1 670 K
ThB_6	5	2.9	0.022 at 1 670 K
PrB_6	—	3.12	
NdB_6	—	4.6	
ThO_2	50	2.6	20 at 1 900 K
CeO_2	10	2.3	26 at 1 900 K
La_2O_3	9	2.5	8 at 1 900 K
Y_2O_3	10	2.4	13 at 1 900 K
BaO/SrO (oxide cathode)	1–10	1.0	See comments in text

References: 9 (Carbides), 10 (Borides), 11 (Oxides), 12 (UC).

emission constants for various carbides, borides and oxides, and the emission available at a particular operating temperature.

For a thermionic cathode to be technically useful it must have an adequate emission at a temperature where the rate of evaporation is not excessive. This limits the practical cathodes to a relatively small number. Table 18.4 gives the most important of these with their normal maximum operating temperatures and maximum operating emission densities for a generally acceptable life.

Table 18.4 EMISSION AT THE NORMAL MAXIMUM OPERATING TEMPERATURE OF PRACTICAL CATHODES

Cathode	*Operating temperature* K	*Emission* $kA\,m^{-2}$
Tungsten	2 500	3
Tantalum	2 400	8
Rhenium	2 400	0.5
BaO/SrO on nickel d.c.	1 100	10
'Oxide cathode' pulse	1 100	100
BaO/SrO Ni. Matrix type	1 150	20
'L' cathode	1 360	30
Impregnated tungsten	1 350	50
ThO_2 on W or Ir	1 900	10
LaB_6 on Re	1 450	0.5
LaB_6 bulk	1 900	50
Thoriated tungsten	1 900	10

An 'L' cathode consists of a block of porous tungsten, the front emitting surface of which is activated with barium. A reaction between the tungsten and barium oxide at the rear surface produces free barium which diffuses through the porous tungsten.

Impregnated tungsten is a porous tungsten block whose pores are filled with barium calcium aluminate by infiltration in the molten phase at 1 700 °C. The emitting surface is partly the compound and partly barium activated tungsten.

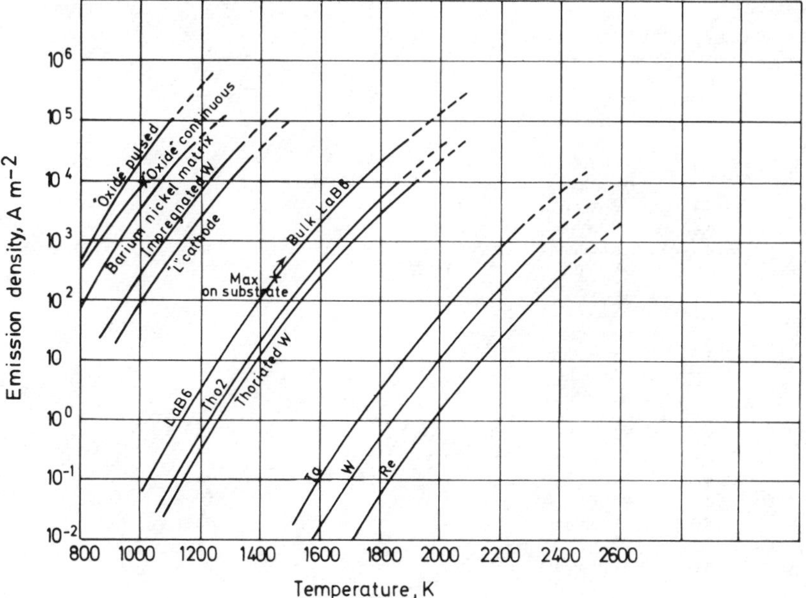

Figure 18.1 *Thermionic emission of practical cathodes as a function of temperature*

The thoriated tungsten is a high density rod or wire containing about 1% thorium oxide. The surface is carburized to form a layer of W_2C. In operation a reaction between the oxide and the carbide produces free thorium, which activates the surface.

Figure 18.1 (reference 13) shows the saturated emissions available from some of these cathodes as a function of temperature. In practice it is usual to operate cathodes at rather less than the saturated emission. The full lines show the region where a useful life is obtainable. The cathode life is usually limited by the evaporation rate and lowering the temperature increases the life considerably. Typically, lowering the temperature of an 'L' or impregnated tungsten cathode by 80 K increases its life by an order.

For an oxide cathode, the continuously drawn emission must be limited to about $10\,kA\,m^{-2}$, but if the current is drawn in microsecond pulses the pulse current may be increased to $100\,kA\,m^{-2}$.

The emissions shown on Figure 18.1 are obtainable only in an environment free from oxidizing gases. In general the partial pressure of gases such as O_2, CO_2, and H_2O should not exceed approximately 10^{-5} Pa. The emission is unaffected by rare gases except when the cathode is bombarded excessively with positive ions; the presence of H_2 may counteract to some extent the effect of oxidizing gases.

18.2 Photoelectric emission

When light of sufficiently high frequency is incident on the surface of a metal, electrons are emitted. In order that electrons shall be emitted with zero velocity from a metal at absolute zero of temperature, the energy of the light photons must equal the energy corresponding to the work function. Thus there is a threshold frequency ν_0 at which $h\nu_0 = e\phi$ where h is Planck's constant. If the frequency of the light is ν, where ν is greater than ν_0, the maximum energy of the emitted electrons is $h\nu - h\nu_0$. At temperatures above absolute zero the threshold is not sharp since light of a frequency less than ν_0 can liberate a small number of electrons. An experiment measuring the energy of photoelectrons emitted for monochromatic light of known energy enables ϕ to be determined fairly accurately, the accuracy improving as the temperature is lowered.

Table 18.5 shows the values of ϕ for a number of metals as determined photoelectrically. The values obtained in this way should correspond to the thermionic values in Table 18.1; discrepancies are probably due to the effect of contamination of the surface, and the photoelectric value is likely to be the more reliable.

Table 18.5 PHOTOELECTRIC WORK FUNCTIONS

Surface	ϕ	*Surface*	ϕ
Aluminium	4.2	Molybdenum	4.2
Antimony	4.1	Nickel	4.9
Arsenic	5.1	Palladium	5.0
Barium	2.5	Platinum	5.3
Beryllium	3.4	Potassium	2.2
Bismuth	4.4	Rhenium	5.0
Boron	4.5	Rhodium	4.6
Cadmium	4.0	Rubidium	2.1
Caesium	1.9	Silicon	4.2
Calcium	2.9	Silver	4.7
Carbon	4.8	Sodium	2.2
Chromium	4.4	Strontium	2.7
Cobalt	4.0	Tantalum	4.1
Copper	4.5	Tellurium	4.8
Gallium	3.9	Thallium	3.8
Germanium	4.8	Thorium	3.5
Gold	4.8	Tin	4.3
Iron	4.4	Titanium	4.1
Iridium	4.6	Tungsten	4.5
Lead	4.0	Uranium	3.6
Lithium	2.4	Zinc	4.3
Manganese	3.8	Zirconium	3.8
Mercury	4.5		

References: 1, 2, 3 (General); 14 (Cu, Ag, Al); 15 (Re).

For practical uses it is required to obtain the maximum photoelectric current for a given light flux. The photoelectric efficiency of a surface may be defined in various ways, the most fundamental being the quantum efficiency Y. This is the ratio of the number of electrons released to the number of incident photons. For clean metals this is very low (10^{-4} approximately) so they are not often used. The efficient photoemitters as used in photocells and photomultipliers are semiconductors with a low effective photon threshold. They are usually formed by combination of one or more alkali metals with an evaporated thin film of antimony (apart from the type consisting of caesium on oxidized silver). Table 18.6 gives the value of maximum photoelectric yields for various such surfaces. This maximum is reached at photon energies 1–1.5 eV above the threshold value, which is also shown.

Table 18.6 PROPERTIES OF EFFICIENT PHOTOELECTRIC EMITTING SURFACE

Surface	*Photoelectric quantum efficiency* Y	*Photon threshold energy* eV
Na_3Sb	0.02	3.1
K_3Sb	0.07	2.6
Rb_3Sb	0.10	2.2
Cs_3Sb	0.25	2.05
NaK_3Sb	0.30	2.0
$CsNaK_3Sb$	0.40	1.55

Reference 16.

The corresponding wavelength of light λ in nm is related to the photon energy $e\phi$ in electron volts by the relationship

$$\lambda = 1.24 \times 10^3 \phi^{-1}$$

18.3 Secondary emission

When electrons (primaries) are incident upon a surface of a solid, electrons (secondaries) are produced which leave the surface in the direction from which the primaries arrive. The total flow of secondaries consists of:

1. Primaries elastically scattered.
2. Primaries reflected inelastically with an energy loss of some tens of volts.
3. True secondaries with an energy independent of the primary energy and a mean value of about 10 eV. Electrons with an energy up to about 50 eV are usually considered to be true secondaries.

The ratio of the total flow of secondaries to that of the primaries is called the secondary emission coefficient δ. As the primary electron energy is increased from zero, δ rises to reach a maximum value δ_{max} for a primary energy V_{max} in the range of 200–1 000 eV for metals, and it falls off more slowly at energies above V_{max}. The shape of the curve relating δ to V is approximately similar for most metals, and a curve normalized to δ_{max} and V_{max} is shown in Figure 18.2.

The values of δ_{max} and V_{max} are shown for most metals in Table 18.7. These values are for clean smooth polycrystalline surfaces. It is impossible to remove oxide films from many metals, such as aluminium or magnesium, by heating. These metals are usually deposited as clean layers either by evaporating in high vacuum or by sputtering. Alternatively the surface of the bulk metal may be cleaned by sputtering in an electric discharge in argon.

It should be noted that the secondary emission is reduced by roughening the surface, as some of the secondaries released in the valleys in the surface may be intercepted by the adjacent high spots. An example of this is carbon; the value of δ_{max} for polished graphite is approximately 1, while that for soot is only about 0.5.

The secondary emission of metal oxides is usually higher than that of metals. Surfaces with high values of δ are used in secondary electron multipliers and are prepared by oxidizing metals containing small quantities, usually about 2%, of magnesium, beryllium or aluminium. Oxidized

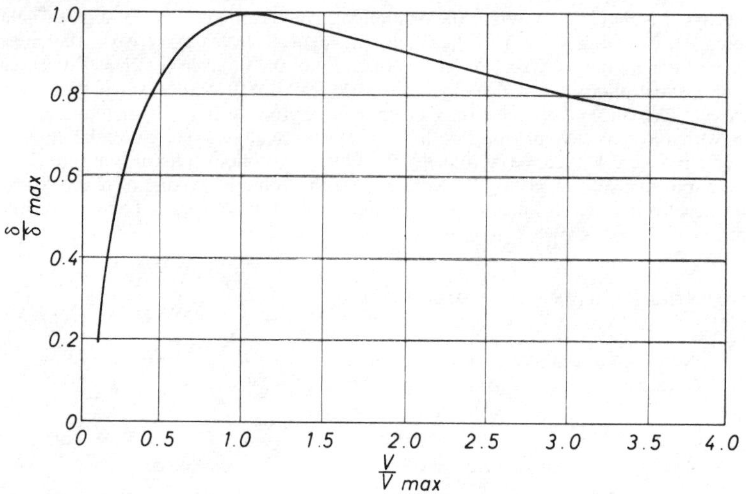

Figure 18.2 *Normalized curve of secondary emission as a function of primary voltage*

Table 18.7 MAXIMUM SECONDARY EMISSION COEFFICIENTS

Element	δ_{max}	V_{max}	*Element*	δ_{max}	V_{max}
Aluminium	0.97	300	Manganese	1.35	200
Antimony	1.30	600	Mercury	1.30	600
Barium	0.85	300	Molybdenum	1.20	350
Beryllium	0.5	200	Nickel	1.35	450
Bismuth	1.15	550	Niobium	1.20	350
Boron	1.2	150	Palladium	1.65	550
Cadmium	1.59	800	Platinum	1.60	720
Calcium	0.60	200	Potassium	0.53	175
Carbon (Graphite)	1.02	300	Rhenium	1.30	800
Caesium	0.72	400	Rubidium	0.90	300
Chromium	1.10	400	Ruthenium	1.40	570
Cobalt	1.35	500	Silicon	1.10	250
Copper	1.28	600	Silver	1.56	800
Dysprosium	0.99	900	Sodium	0.82	300
Erbium	1.05	1 100	Strontium	0.72	400
Gadolinium	1.04	600	Tantalum	1.25	600
Gallium	1.08	600	Terbium	1.02	900
Germanium	1.08	400	Thallium	1.40	800
Gold	1.79	1 000	Thulium	1.05	1 100
Holmium	1.02	900	Thorium	1.10	800
Indium	1.40	500	Tin	1.35	500
Irdium	1.55	700	Titanium	0.90	280
Iron	1.30	200	Tungsten	1.35	650
Lead	1.10	500	Ytterbium	1.04	800
Lithium	0.52	100	Zinc	1.40	800
Magnesium	0.97	275	Zirconium	1.10	350

References: 17; 18 (Pt, Ir, Ru); 19 (Rare earth metals).

metal surfaces with caesium evaporated on to them also have high values of δ and are used in photomultipliers. Table 18.8 shows values of δ obtained from various such surfaces.

Table 18.9 shows the secondary emission from a number of insulating metal compounds either as evaporated films (e) or surface layers on the parent metal (s). These layers must be very thin to avoid accumulating a charge. The secondary electrons originate normally within 10 nm of the surface, so provided films are thicker than this a true value of δ will be obtained.

Table 18.8 SECONDARY EMISSION FROM OXIDIZED ALLOYS AND PHOTOCELL SURFACES

Oxidized alloy	δ_{max}	V_{max}
Ag–Mg	10–16	600
Ag–Be	6	500
Ni–Be	5–10	600
Cu–Be	5	500
Cu–Mg	12.5	900
Cu–Mg–Al	10	700
Photocell surfaces		
Ag–O–Cs	6–10	500
Ni–O–Cs	5.7	500
Ag–O–Rb	5.5	800
Ag–Sb–Cs	8.0	500
Sb–Cs	12	450

Reference 20.

Table 18.9 SECONDARY EMISSION FROM INSULATING METAL COMPOUNDS

Material	δ_{max}	V_{max}
Li F (e)	5.6	—
Na F (e)	5.7	—
NaCl (e)	6–6.8	600
KCl (e)	7.5–8.0	1 500
RbCl (e)	5.8	—
Cs Cl (e)	6.5	—
Na Br (e)	6.25	—
KI (e)	5.6	—
NaI (e)	5.5	—
Ca F_2 (e)	3.2	—
Ba F_2 (e)	4.5	—
Mg F_2 (e)	4.1	410
BeO (s)	3.4–8	200–400
MgO (s)	2.4–17.5	400–1 600
Al_2O_3 (s)	1.5–3.2	350–1 300
Cu_2O (s)	1.19–1.25	440
PbS (s)	1.2	500
MoS_2 (s)	1.10	—
WS_2 (s)	0.96–1.04	—
ZnS (s)	1.8	350
MoO_2 (s)	1.09–1.33	—
Ag_2O (s)	0.90–1.18	—
SiO_2 (e)	2.2	300
Cs_2O (s)	2.3–11	800

Reference 21.

18.4 Auger emission

Auger emission is a byproduct of secondary emission. When the energy distribution of secondaries is examined closely, small peaks can be seen superimposed on the basically smooth curve. These peaks can be made much more obvious by electronic differentiation, and have been shown to originate in Auger transitions, as follows (reference 22). A primary electron removes a secondary electron from an electron shell in an atom. Subsequently another electron in a higher energy shell transfers to the vacancy. The energy thus released is given to a third electron which is emitted. It may be seen that the energy of the third electron is independent of the energy of the primary and is characteristic of three energy levels in the excited surface atom. A large number of these Auger energies have been measured for many elements and they provide identification of the element

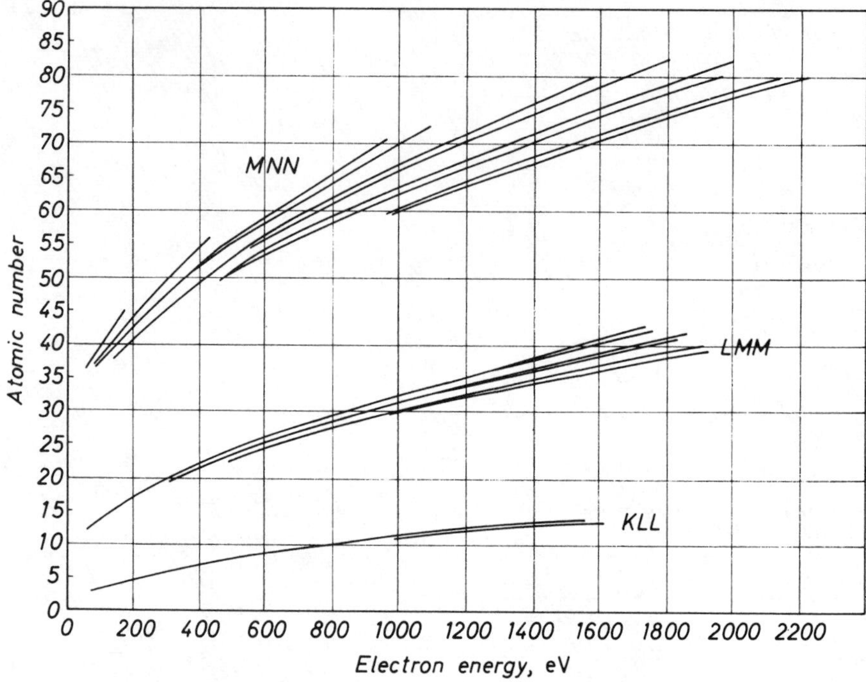

Figure 18.3 *Strongest Auger emission peaks as a function of the atomic number*

concerned. Auger electrons originate within the top 10 nm of the surface and fractions of a monolayer of an element can be detected. This relatively new technique of analysis appears to have a considerable number of possible uses.

The energy of the primary bombarding electrons is usually about 2.5 keV, and the Auger peaks are detected from about 50 eV up to 2 kV. Figure 18.3 shows the Auger energies of the stronger peaks plotted against the atomic number Z. The letters are the electron levels involved in the Auger transition (reference 23).

18.5 Electron emission under positive ion bombardment

When an electrical discharge takes place between two electrodes in a low gas pressure the cathode is bombarded with positive ions and emits electrons. These electrons are essential for maintaining the discharge. The number of electrons released for each arriving ion is usually called γ, the second Townsend coefficient. The coefficient is generally approximately constant for positive ion energies from zero up to about 1 k V. The energy required to release the electron is supplied by neutralization of the positive ion as follows. When the ion is very close to the metal surface the electrostatic is sufficient to extract an electron which neutralizes the positive ion. This releases a photon of energy $(I—\phi)e$, where I is the ionization potential of the gas atom. This photon can then release a photoelectron from the metal provided $(I-\phi)e > \phi e$ or $I > 2\phi$.

The value of γ thus tends to increase with increasing I and decreasing ϕ. Values of γ for various inert gas ions and metals are shown in Table 18.10. At energies above a few keV the value of γ usually increases approximately linearly with energy, the extra electrons being released as a result of kinetic energy transfer.

18.6 Field emission

When a very high positive electric field is applied to the surface of a metal, the potential just outside the metal becomes more positive than the Fermi level in the metal. The work function barrier, instead of being a step becomes very thin, and electrons can ‘tunnel’ through the barrier

Table 18.10 SECOND TOWNSEND COEFFICIENT γ ELECTRONS RELEASED PER POSITIVE ION ARRIVING

Metal	*Ion*	γ	*Ion energy* eV
Tungsten (outgassed)	Ne^+	0.25	0–1 000
Tungsten (outgassed)	He^+	0.24	0–1 000
Tungsten (outgassed)	A^+	0.10	0–1 000
Tungsten (outgassed)	Kr^+	0.05	0–1 000
Tungsten (outgassed)	Xe^+	0.02	0–1 000
Tantalum (outgassed)	A^+	0.02	100
Tantalum (gas covered)	He^+	0.2	500
Tantalum (gas covered)	He^{++}	0.7	500
Molybdenum (outgassed)	He^{++}	0.8	0–1 000
	He^+	0.22	0–1 000
	Ne^{++}	0.7	0–1 000
	Ne^+	0.2	0–1 000
	A^{++}	0.35	0–1 000
	A	0.08	0–1 000
	Kr	0.05	0–1 000
Nickel	He^+	0.7	800
	Ne^+	0.4	800
	A^+	0.1	800

Reference 24

and be emitted. This emission is usually called field emission (sometimes tunnel emission), and the emission density is related to the field by the Fowler–Nordheim law. Table 18.11 shows values of $\log_{10}$ (emission density) for various fields for clean tungsten (ϕ=4.5 V) and also for a barium contaminated surface (ϕ = 2.0 V) and an oxygen contaminated surface (ϕ = 6.3 V).

Table 18.11 FIELD EMISSION FROM TUNGSTEN

ϕ=2.0 eV			ϕ=4.5 eV			ϕ=6.3 eV		
Field 10^9 V m^{-1}	$\log_{10}$ J A m^{-2}	*Current from* 10^{-14} m^2 A	*Field* 10^9 V m^{-1}	$\log_{10}$J A m^{-2}	*Current from* 10^{-14} m^2 A	*Field* 10^9 Vm^{-1}	$\log_{10}$ J A m^{-2}	*Current from* 10^{-14} m^2 A
1.0	6.98	1×10^{-7}	2	0.67	4.7×10^{-14}	2	−8.0	10^{-22}
1.2	8.45	2.8×10^{-6}	3	5.57	3.4×10^{-9}	4	3.12	1.3×10^{-11}
1.4	9.49	3.1×10^{-5}	4	8.06	1.1×10^{-6}	6	7.25	1.8×10^{-7}
1.6	10.23	1.9×10^{-4}	5	9.59	3.4×10^{-5}	8	9.34	2.2×10^{-5}
1.8	10.89	7.8×10^{-4}	6	10.62	4.2×10^{-4}	10	10.66	4.6×10^{-4}
2.0	11.40	2.5×10^{-3}	7	11.36	2.3×10^{-3}	12	11.52	3.3×10^{-3}
2.2	11.82	6.6×10^{-3}	8	11.94	8.8×10^{-3}	14	12.16	1.5×10^{-2}
2.4	12.16	1.5×10^{-2}	9	12.39	2.4×10^{-2}	16	12.65	4.5×10^{-2}
2.6	12.45	2.8×10^{-2}	10	12.76	5.8×10^{-2}	18	13.04	1.1×10^{-1}
			12	13.32	2.1×10^{-2}	20	13.36	2.3×10^{-1}

Reference 25.

High fields which produce appreciable emissions usually result from field concentration at the tips of small projections, spikes or whiskers on metal surfaces. As these usually have submicron tip diameters the emission current in A for an area of 10^{-14} m^2 is also shown in Table 18.11.

The field at the tips of emitting projections is greater than the 'macroscopic' field at an electrode by a factor usually called β. This varies between about 1000 for a rough surface, down to less than 100 for a highly polished and voltage conditioned hard metal surface. When the field emission from a projection reaches an appreciable fraction of an ampere, the projection will melt and vaporize and this may initiate electrical breakdown between the electrodes in vacuum. Alternatively the field emission may heat the anode electrode and release gas or metal vapour which can also initiate a breakdown. From Table 18.11 the necessary tip field is likely to be $2 \times 10^9 - 10^{10}$ V m^{-1}, depending on work function while the macroscopic field will usually be of the order of 1% of this.

Single field emitting sources can be made, consisting of a point 0.1–1.0 μm diameter etched on the end of a refractory metal wire. An anode electrode near the point and carrying a positive potential of a few kV is sufficient to cause field emission. The point usually emits over most of its approximately hemispherical tip. The intensity of the emission varies with direction, as different crystal planes on its surface have different work functions (reference 25). The close packed planes, e.g. the (110) plane of a body centred crystal such as tungsten, have the highest work function and lowest field emission. This effect is shown visually by allowing the electrons to strike a hemispherical fluorescent screen around the point. This arrangement has been used extensively to study surface migration and adsorption phenomena. Measurements can however only be carried out in ultra high vacuum (pressure $< 10^{-7}$ Pa) otherwise positive ions are formed in the gas and these quickly destroy the point by ion bombardment.

REFERENCES

1. C. Herring and M. H. Nichols, *Rev. mod. Phys.*, 1949, **21**, 232.
2. G. Herrman and S. Wagener, 'The Oxide Cathode', Vol. 2. Chapman & Hall; 1951.
3. H. H. Michaelson, *J. appl. Phys.*, 1950, **21**, 536.
4. A. Venema, 'Handbook of Vacuum Physics', Pergamon Press, 1966, **2**, 179–298.
5. D. L. Goldwater and W. E. Danforth, *Phys. Rev.*, 1956, **103**, 871.
6. R. C. Wilson, *J. appl. Phys.*, 1966, **31**, 3170.
7. R. C. Wilson, *J. appl. Phys.*, 1966, **31**, 2265.
8. A. L. Reimann, 'Thermionic Emission', 134, Chapman & Hall; 1934.
9. R. E. Haddad, D. C. Goldwater and F. H. Morgan, *J. appl. Phys.*, 1949, **20**, 886, 1130; ibid., 1951, **22**, 70.
10. J. M. Lafferty, *J. appl. Phys.*, 1951, **22**, 299.
11. D. A. Wright, *Proc. Inst. Elect. Engrs*, 1953, **100**, 125.
12. G. A. Haas and J. T. Jensen, *J. appl. Phys.*, 1960, **31**, 1231.
13. R. O. Jenkins, *Vacuum*, 1969, **19**, 353.
14. E. W. Mitchell and J. W. Mitchell, *Proc. R. Soc.*, 1951, **A210**, 70.
15. R. Levi and G. A. Espersen, *Phys. Rev.*, 1950, **78**, 231.
16. W. E. Spicer and F. Wooten, *Proc. IEEE*, 1963, **51**, 1119.
17. D. J. Gibbons, 'Handbook of Vacuum Physics', Pergamon Press; 1966, **2**, 319.
18. R. O. Jenkins (Unpublished).
19. H. Aspden, London University Ph.D. Thesis, 1968, 166.
20. D. J. Gibbons, 'Handbook of Vacuum Physics', Pergamon Press, 1966, **2**, 375.
21. D. J. Gibbons, 'Handbook of Vacuum Physics', Pergamon Press, 1966, **2**, 336.
22. J. J. Lander, *Phys. Rev.*, 1953, **91**, 1382.
23. P. W. Palmberg, 'Electron Spectroscopy', 838, North-Holland; 1972.
24. S. C. Brown, 'Basic Data of Plasma Physics', 222, Wiley; 1959.
25. R. H. Good and E. W. Müller, 'Encyclopedia of Physics', Springer, 1956, **21**, 188.

19 Electrical properties

19.1 Resistivity

The resistivities of a number of pure metals and alloys are given in Tables 19.1 to 19.4. Resistivity varies with the condition of the material and is sensitive to purity. In general, cold working increases and annealing decreases the resistivity. Common reactive gases such as oxygen, nitrogen and hydrogen may also affect the resistivity, either through selective chemical action with existing metallic impurities (which may even reduce the resistivity) or through solution in the host matrix itself.[1] Thermal cycling through phase transformation, quenching from high temperatures and irradiation, all introduce lattice defects which increase resistivity. These defects include vacancies, dislocations and interstitial atoms. Annealing will promote the movement and eventual removal of these defects. This recovery generally takes place in discrete stages at certain temperatures corresponding to the annealing out of each type of defect. Recovery is complete after treatment at the recrystallization temperature.

Resistivity is often expressed approximately as the sum of the residual resistivity at absolute zero (arising from impurities and lattice defects) and a temperature-dependent intrinsic resistivity (arising from the effect of lattice vibrations upon conduction electrons). The form of temperature dependence is complex, and theories governing it in both solid and liquid metals have been recently discussed in, for example, references 2–15. Over a limited temperature interval, resistivity may be conveniently expressed as a linear relation of the form $\rho_1 = \rho_0\ (1 + \alpha T)$, where T is the interval between two temperatures T_1 and T_0, and α is the temperature coefficient of resistivity (TCR). In Table 19.1, TCR values are given for the temperature range 273–373 K, except for some lower-melting-point elements as indicated.

Table 19.1 RESISTIVITY AND TEMPERATURE COEFFICIENT OF PURE METALS

	Room temperature			*Melting point*			
Element	*Temp.* K	*Resistivity* $10^{-8}\,\Omega$m	*TCR* $10^{-3}\,K^{-1}$	*Temp.* K	*Resistivity* *Solid*	$10^{-8}\,\Omega$m *Liquid*	*Ref.*
Aluminium	293	2.61	4.2	933	—	24.2	7, 16
Americium	300	68.9	—	—	—	—	17
Antimony	293	37.6	5.1	913*	—	113.5	7, 18
Arsenic	293	31	—	—	—	—	19
Barium	300	34.3	5.0	1 002	276	306	20
Beryllium	300	3.76	8.0	—	—	—	20
Bismuth	293	115	4.6	573*	—	128	7, 18
Cadmium	293	6.6	4.3	603*	—	33.7	7, 21
Caesium	300	21.04	5.3†	301.6	21.2	36.9	22
Calcium	300	3.45	3.7	1 113	14.5	33.0	20
Cerium	293	75	0.9	1 068	—	125	23, 24
Chromium	300	12.9	5.9	2 148	—	80	25, 26
Cobalt	293	5.2	6.6	1 766	—	100	26, 27
Copper	293	1.58	4.3	1 356	—	20	26, 27
Dysprosium	293	97	—	—	—	—	23
Erbium	293	80	—	—	—	—	23
Europium	293	116	—	1 099	188	244	23, 24
Francium	300	34	7.2†	300.2	34	55	22

Table 19.1 RESISTIVITY AND TEMPERATURE COEFFICIENT OF PURE METALS—*continued*

	Room temperature			*Melting point*			
Element	*Temp.* K	*Resistivity* $10^{-8}\,\Omega$m	*TCR* $10^{-3}\,K^{-1}$	*Temp.* K	*Resistivity* $10^{-8}\,\Omega$m *Solid*	*Liquid*	*Ref.*
Gadolinium	293	132	—	1 585	—	195	23, 24
Gallium	293	13.65	—	303	—	25.9	28, 29
Gold	293	2.01	4.0	—	—	—	27
Holmium	293	87	—	—	—	—	23
Indium	293	8.0	5.2	430	—	33.1	7, 30
Iridium	300	5.0	4.5	—	—	—	25
Iron	300	9.8	6.5	1 808	—	140	25, 26
Lanthanum	293	62	2.2	1 193	—	140	23, 24
Lead	293	19.3	4.2	673*	—	95	7, 31
Lithium	300	9.55	4.0	454	15.6	24.8	22
Lutetium	293	67	—	1 925	—	224	3, 23
Magnesium	300	4.51	3.7	—	—	—	20
Manganese	293	143.5	0.4	1 517	—	180	26, 32
Mercury	293	94.1	1.0‡	253	—	91	7, 31
Molybdenum	300	5.3	4.35	—	—	—	25
Neodymium	293	64.5	—	—	—	—	23
Nickel	293	6.2	6.8	1 725	—	85	26, 27
Niobium	293	13.27	2.6	—	—	—	33
Osmium	293	8.4	4.1	—	—	—	27
Palladium	300	10.5	4.2	1 825	48	83	25, 34
Platinum	300	10.4	3.9	2 042	—	90	25, 26
Potassium	300	7.47	6.0†	336.4	9.22	13.95	22
Praseodymium	293	66	—	—	—	—	23
Protactinium	298	19.3	—	—	—	—	17
Radium	300	88	6.5	—	—	—	20
Rhenium	293	16.9	4.5	—	—	—	27
Rhodium	293	4.37	4.4	—	—	—	27
Rubidium	300	13.32	6.3†	312.6	14.2	22.5	22
Ruthenium	293	6.7	4.1	—	—	—	27
Samarium	293	86	—	—	—	—	23
Silver	300	1.47	4.1	1 234	—	28.5	26, 27
Sodium	300	4.93	5.3†	371	6.86	9.43	22
Strontium	300	13.5	3.2	1 042	65.6	84.8	20
Tantalum	300	13.1	3.5	—	—	—	25
Terbium	293	107	—	—	—	—	23
Thallium	293	15.0	5.2	576	—	73.1	7, 31
Thorium	293	14.2	4.0	—	—	—	35
Thulium	293	95	—	—	—	—	23
Tin	293	10.1	4.6	—	—	—	36
Titanium	293	39	3.8	1 941	—	400	2, 27
Tungsten	300	5.3	4.8	—	—	—	25
Uranium	293	24.8	2.5	—	—	—	37
Vanadium	300	19.9	3.9	2 163	—	200	25, 26
Ytterbium	293	29	—	1 097	74	109	5, 23
Zinc	293	5.45	4.2	693	—	37.4	7, 38
Zirconium	293	38.9	4.4	—	—	—	27

* Liquid resistivity at temperature above melting point.
† TCR for the interval 250–300 K.
‡ TCR for the liquid phase.

Resistivity values given in Table 19.1 are for bulk material. If the metal is deposited as a thin film, resistivity may deviate from the bulk value, being affected by parameters such as film thickness and grain size. Many recent investigations have been made on these effects (*see* references 39–41).

The resistivity and temperature coefficient of alloys is often dependent upon the method of preparation and heat treatment. The values given in Table 19.2 relate to particular samples and should not be assumed to apply accurately to other samples of similar composition. In systems which exhibit complete mutual solid solubility, there is often a resistivity maximum near 50/50 composition (e.g. Figure 19.1). The transition elements show complex resistivity and TCR behaviour upon alloying. Very low TCRs, which may be of importance industrially, are obtained in some quaternary Ni–Cr alloys after suitable heat treatment. TCR values given here are for the temperature interval 273–373 K.

Theoretical predictions for the resistivity of alloys have recently been discussed in references 42–46.

Table 19.2 RESISTIVITY AND TEMPERATURE COEFFICIENT OF SOME ALLOYS

Alloy		*Nominal composition* wt. %†	*Temperature* K	*Resistivity* 10^{-8} Ωm	*TCR* 10^{-3} K^{-1}
Ag–Au	(normal silver)	Au 0.37	293	1.77	—
		Au 10	293	3.6	—
Ag–Cd–Zn–Cu		Cd 18 Zn 16.5 Cu 15	273	7.0	—
Ag–Cu	(standard silver)	Cu 7.5	293	1.9	—
		Cu 10–50	293	2.0–2.1	—
Ag–Cu–Zn		Cu 25 Zn 15	273	8.3	—
Ag–Mn*		Mn 6	293	14	0.2
		Mn 10	293	27	0.02
		Mn 12	293	33	−0.01
		Mn 16–20	293	42–46	−0.03
Ag–Pd		Pd 5	293	3.8	—
		Pd 10	293	5.8	—
		Pd 20	293	10.1	—
Al–Cu etc.		Cu 6	273	3.1	3.8
	(Duralumin)	Cu 4 Mn 0.6 Mg 0.6	293	5.0–5.3	2.3
		Cu 4.1 Mn 0.5 Mg 1.4 Fe 0.2	273	4.0–4.4	—
Al–Mg etc.		Mg 10	273	8.0	—
		Mg 4.75 Mn 0.63 Fe 0.2 Cr 0.13	273	5.66	—
Al–Mg–Si		Mg 0.5 Si 0.5	293	3.25	3.6
Al–Mn		Mn 1.25	293	3.4–4.4	—
Al–Si		Si 12	273	4.5	—
Al–Zn–Mg etc.		Zn 3.6 Mg 2.55 Mn 0.2 Cr 0.2	273	4.75	—
		Zn 5.6 Cu 1.6 Mg 2.5	273	3.7–5.0	
Au–Ag		Ag 10	273	6.3	1.2
		Ag 33	273	10.8	0.65
Au–Co		Co 2.5	293	32.6	—
Au–Cr–Co		Cr 2.1 Co 0.25	293	39.8	+0.02
		Cr 2.1 Co 0.5	293	44.7	−0.06
		Cr 4.2 Co 0.4	293	57.8	−0.08
Au–Cr–Pd		Cr 2.1 Pd 3.5	293	36.9	0±0.02
		Cr 2.1 Pd 9	293	38.7	0±0.02
Au–Cr–Pt		Cr 2.1 Pt 2	293	34.7	0.07
		Cr 2.1 Pt 6	293	25.6	0.19
Au–Cu–Ag		Cu 78.3 Ag 14.3	273	3.6	1.8
		Cu 26.5 Ag 15.2	273	13.2	0.57
		Cu 15.5 Ag 18.1	273	14.6	0.53
Bi–Sn		Sn 2	273	24.4	—
		Sn 90.5	285	16	—
Bi–Sn–Pb	(Rose's metal)	Sn 23 Pb 28	273	64	2.0
	(Wood's metal)	Sn 12.5 Pb 25	273	52	2.0
Cr–Au[47]		Au 0.6 at %	300	19.2	—
Cu–Al		Al 3	273	8.3	1.0
		Al 10	273	12.6	3.2
Cu–Be		Be 2 (+Ni)	273	6.8–7.4	1.0–1.8
Cu–Mn*		Mn 0.98	273	4.83	—
		Mn 1.49	273	6.66	—
		Mn 4.2	293	17.9	0.25
		Mn 7.4	293	19.7	0.17

* TCR applies over a restricted range at room temperature.
† Unless stated otherwise.

Table 19.2 RESISTIVITY AND TEMPERATURE COEFFICIENT OF SOME ALLOYS—*continued*

Alloy		*Nominal composition* wt. %†	*Temperature* K	*Resistivity* 10^{-8} Ωm	*TCR* 10^{-3} K^{-1}
Cu–Mn–Al*		Mn 9 Al 3	293	38	0.010
		Mn 9 Al 5	293	42	0.012
		Mn 12 Al 3	293	48	−0.005
Cu–Mn–In*		Mn 12 In 1–3	293	42.5	0±0.010
Cu–Mn–Fe*		Mn 23.2 Fe 6.2	273	77	0.01
		Mn 14.76 Fe 0.33	293	53.5	—
		Mn 7.1 Fe 1.9	273	20	0.1
Cu–Mn–Ni*		Mn 72 Ni 10	293	175	1.4
		Mn 24 Ni 3	273	48	−0.03
	(Manganin)	Mn 12 Ni 4	293	44	0.00
Cu–Ni (*see* Figure 19.1)		Ni 45 at %	293	49	0±0.02
		Ni 30 at %	293	36.3	0.05
		Ni 20 at %	293	26.6	0.24
		Ni 10 at %	293	14.1	0.52
Cu–Ni–Fe[48]		Ni 45 Fe 2 at %	300	44	—
		Ni 51 Fe 1 at %	300	48	—
Cu–P		P 0.48	293	8.4	0.84
		P 0.93	293	15.2	0.50
Cu–Sn etc.		Sn 12	293	18	0.5
		Sn 5	273	9.5	—
		Sn 6 Zn 4	288	13.5	—
		Sn 5 Zn 5 Pb 5	273	10.5	—
		Sn 5 Ni 5 Zn 2	273	10.5–14	—
Cu–Zn etc.		Zn 10	293	3.8	—
		Zn 15	273	4.65	—
		Zn 30	273	6.65	1.6
		Zn 40	273	6.81	1.7
		Zn 39 Fe 1 Sn 1	273	7.03	—
	(German silver)	Zn 41 Ni 9 Pb 2	293	30.5	—
	(Admiralty brass)	Zn 27.6 Sn 1	273	6.93	—
Cu–Zn–Ni		Zn 32 Ni 8	273	72	—
		Zn 24 Ni 18	273	30.9	0.04
		Zn 26 Ni 30	273	47.6	0.04
		Zn 25 Ni 14	293	33	0.40
Fe–C etc.	(Mild steel)	C 1 Si 1.8	291	12.0	—
	(Cast iron)	C 3.4	293	66.0	—
Fe–Cr		Cr 20	293	62	—
		Cr 12	293	60	—
Fe–Cr–Ni etc.		Cr 18 Ni 8	293	73	—
		Cr 25 Ni 12	293	87	—
		Cr 25 Ni 20	293	88	—
		Cr 18 Ni 37	293	108	—
		Cr 17.7 Ni 7.4 Al 1.2 Mo 0.7 Si 0.4	273	71–102	—
		Cr 18.4 Ni 9.7 Mn 1.4 Si 0.6	273	70.4	—
		Cr 17.9 Ni 9.8 Mn 1.4 Si 0.6 Ti 0.4	273	73.9	—
		Cr 17.7 Ni 12.2 Mo 2.8 Mn 1.5 Si 0.7	273	76.5	—
		Cr 17.5 Ni 13.1 Mo 2.7 Mn 1.7 Si 0.4	273	71.8	—
		Cr 17.4 Ni 13.3 Mo 2.9 Mn 1.5 Si 0.5	273	77.6	—
		Cr 17 Ni 12.8 Mo 2.7 Mn 1.5 Si 0.4	273	75.0	—

* TCR applies over a restricted range at room temperature.
† Unless stated otherwise.

Table 19.2 RESISTIVITY AND TEMPERATURE COEFFICIENT OF SOME ALLOYS—*continued*

Alloy		*Nominal composition* wt. %†	*Temperature* K	*Resistivity* 10^{-8} Ωm	*TCR* 10^{-3} K^{-1}
Fe–Ni		*See* Figures 19.2 and 19.3	—	—	—
Fe–Ni–Co		Ni 29 Co 17	293	49	3.7
Fe–Si		Si 25	293	45	
		Si 4	293	62	8.0
Fe–Ti etc.		Ti 2.5 C 0.15	293	16	—
Fe–V etc.		V 5 C 1.1	293	121	—
Fe–W etc.		W 20 C 0.2	293	24	—
		W 5 C 0.2	293	20	—
Mg–Al–Zn		Al 9 Zn 2	273	16	—
		Al 4.9 Zn 0.9	273	11.3	—
Na–K liquid[42,50]		K 20 at %	373	28.0	—
		K 40 at %	373	38.2	—
		K 60 at %	373	40.8	—
		K 80 at %	373	33.3	—
Ni–Al etc.		Al 50 at %[49]	300	9.3	—
		Al 2 Mn 2.5 Fe 0.5 Si 1	273	33.3	1.2
		Al 1.6 Si 1.2 Fe 0.1	273	28.1	2.4
Ni–Co–Cr etc.	‡(Udimet 700)	Co 19 Cr 15.2 Mo 5 Al 4.4 Ti 3.4 Fe 0.1	273	131.5	—
Ni–Cr[51]		Cr 5.5 at %	273	41	—
		Cr 11.5 at %	273	70	—
		Cr 27 at %	273	110	—
Ni–Cr etc.	‡(Evanohm)	Cr 20 Al 2.5 Cu 2.5	293	133	0.00
	‡(Karma)	Cr 20 Al 2.5 Fe 2.5	293	134	0.00
		Cr 15 Fe 7	273	98	—
		Cr 20 Fe 5–10	293	109	—
	‡(Hastelloy X)	Cr 22 Fe 20 Mo 9 C 0.15	273	113.8	—
	‡(Inconel X)	Cr 15.4 Fe 6.9 Ti 2.5 Al 0.9	273	124	—
		Cr 9.46 Fe 0.2 Si 0.4	273	70	0.4
Ni–Cu etc.	‡(Monel)	Cu 30 Fe 1.4 Mn 1	273	48	1.9
Ni–Mn		Mn 5	273	18	—
Ni–Mo–Fe		Mo 32 Fe 6	273	135	—
Pb–Sb		Sb 6	273	23	—
Pb–Te–Cu		Te 0.04 Cu 0.06	273	20	—
Pt–Ir		Ir 10	273	24.8	1.3
Pt–Rh		Rh 10	273	18.7	1.66
		Rh 13	273	19.0	1.56
Rh–Co[52]		Co 11 at %	4.2	3.07	—
		Co 20 at %	4.2	4.96	—
		Co 42 at %	4.2	18.98	—
Sn–Pb		Pb 10	288	13.5	—
		Pb 40	273	15	—
		Pb 66.7	288	16	—
Ti–Al–V etc.		Al 6.2 V 4 Fe 0.13	273	167.5	—
Ti–V–Cr–Al etc.		V 13.1 Cr 10.8 Al 3 Fe 0.17	273	149.2	—
Zn–Cu		Cu 1.05	273	6	—
Zn–Al–Cu etc.[53]		Al 4.1 Cu 3.1 Mg 0.05	295	7.2	—

† Unless stated otherwise.
‡ Karma is a trade name of British Driver Harris Co. Ltd.
Inconel and Monel are trade names of Henry Wiggin Co. Ltd.
Udimet is a trade name of Special Metals Corp.
Evanohm is a trade name of Wilbur B. Driver Co.
Hastelloy is a trade name of Cabot Corporation.

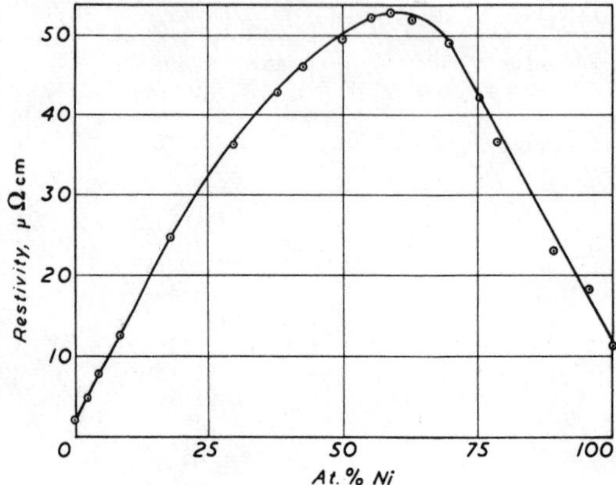

Figure 19.1 *Resistivity of copper-nickel alloys*

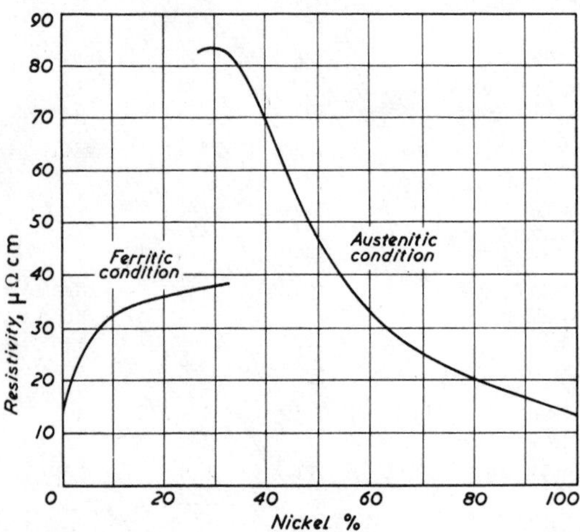

Figure 19.2 *Electrical resistivity of nickel–iron alloys*

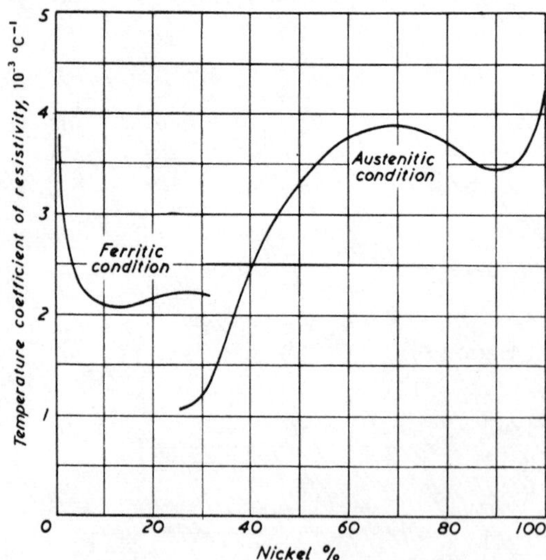

Figure 19.3 *Temperature coefficient of electrical resistivity of nickel–iron alloys*

Dilute copper and aluminium alloys have important applications as electrical conductors. Table 19.3 lists specified resistivity values for certain copper alloys (*see* references 54–56). Table 19.4 shows the effect on the resistivity of EC aluminium (99.45 wt% Al) of the addition of 1 at.% of various solutes. When considering aluminium as an electrical conductor, it is important to achieve strengthening by the addition of suitable solutes without undue reduction in conductivity. The least detrimental additions form intermetallic compounds that remain out of solid solution (*see* references 57 and 57A).

Table 19.3 RESISTIVITY OF SOME SPECIFIC COPPER ALLOYS

Alloy No.	*Composition or description*	*Resistivity at* 293 K 10^{-8} Ωm
101, 102	OFHC	1.55–1.58
110	ETP	1.56–1.65
122	phos. deox.	2.0–2.1
220	10% Zn	3.8–3.9
510	5% Sn	10.5–11.5
706	10% Ni	16.5–19.0

Table 19.4 EFFECT ON THE RESISTIVITY OF EC ALUMINIUM OF SOME ADDITIONAL ELEMENTS

Added element 1 at.%	*Increase in resistivity at* 293 K 10^{-8} Ωm
Antimony	0.238
Cerium	0.049
Gadolinium	0.40
Hafnium	2.20
Lanthanum	0.147
Molybdenum	2.04
Niobium	0.44
Tantalum	1.32
Thorium	0.265
Tungsten	7.4
Yttrium	0.255
Zirconium	0.988

Resistivity of EC aluminium 2.694×10^{-8} Ωm

19.2 Superconductivity

Below a transition temperature T_c some elements, compounds and alloys become superconducting. In this state they exhibit zero electrical resistance and some become perfectly diamagnetic. The application of a magnetic field whether applied externally or generated by a current passing along the superconductor in excess of a critical or limiting value restores the material to its normal resistive state.

Table 19.5 lists the elements which become superconducting when cooled to a sufficiently low temperature together with their transition temperatures and critical fields at 0 K (H_0).[58–60] The critical field H_c at temperature T is closely represented by the relationship

$$H_c = H_0\left[1 - \left(\frac{T}{T_c}\right)^2\right]$$

Table 19.5 TRANSITION TEMPERATURES AND CRITICAL FIELDS FOR SUPERCONDUCTING ELEMENTS

Element		T_c K	H_0 10^4 T	Element	T_c K	H_0 10^4 T
Aluminium		1.18	99	Osmium	0.61	82
Americium		1.06	—	Protactinium	<1.0	—
Beryllium		0.026	—	Rhenium	1.70	201
Cadmium		0.52	30	Ruthenium	0.49	66
Gallium		1.08	51	Tantalum	4.47	830
Hafnium		0.128	—	Technetium	7.8	—
Indium		3.41	283	Thallium	2.38	162
Iridium		0.11	—	Thorium	1.38	—
Lanthanum	α	4.88	—	Tin	3.72	306
	β	6.0	1 600	Titanium	0.4	100
Lead		7.20	803	Tungsten	0.02	—
Mercury	α	4.15	411	Vanadium	5.4	1 310
	β	3.95	340	Zinc	0.85	53
Molybdenum		0.92	—	Zirconium	0.61	47
Niobium		9.25	1 944			

The maximum axial current which a superconducting material of cylindrical form, radius r, will sustain while remaining superconducting is given by

$$I_c = 0.5\, H_c\, r$$

Superconductivity is observed in indium and molybdenum only when these elements are very pure.[60] Traces of magnetic impurity, e.g. Fe > 0.02%, prevent its occurrence.

Table 19.6 CRITICAL TEMPERATURES OF SOME SUPERCONDUCTING ALLOYS AND COMPOUNDS

Alloy or compound	T_c K	Ref.	Alloy or compound	T_c K	Ref.
$Al_{10}V$	1.6	61	Nb_3Rh	2.5	63
AuGa	1.2	62	$NbSi_2$	2.9	67
AuTl	1.92	62	Nb_3Sn	18.3	63
Cr_3Os	4.03	63	Ta_3Ge	8	63
Cr_3Ru	3.43	63	$TaGe_2$	2.7	67
$EuIr_2$	2.6	64	*Ta–42%Ir	6.6	68
$HfGe_2$	2.2	65	$TaSi_2$	4.4	67
$LaGe_2$	2.65	65	Ta_3Sn	6.4	63
La_3In	10.5	64	*Tc–3%Ti	10.2–10.9	70
Mo_3Ge	1.4	63	Ti_3Ir	4.6	63
Mo_3Ir	8.1	63	Ti_3Sb	5.8	63
Mo_3Os	11.68	63	U_6Co	2.5	64
Mo_3Pt	4.56	63	U_6Fe	3.9	64
Mo_3Si	1.3	63	U_6Mn	2.3	64
Mo_2Tc_3	13.5	63	V_3Al	9.6	63
Nb_3Al	18.9	63	V_3Au	3.2	63
Nb_3Au	11	63	V_3Ga	15.4	63
Nb_3Bi	2.25	63	V_3Ge	7	63
Nb_3Ga	20.3	63	V_3In	13.9	63
Nb_5Ga_3	1.35	66	V_3Ir	1.39	63
Nb_3Ge	23	63	V_3Os	5.15	63
$NbGe_2$	16	67	V_3Pb	3.7	63
Nb_3In	8	63	V_3Re_7	9.0	64
Nb_3Ir	1.76	63	V_3Si	17.1	63
*Nb–42%Ir	4–9	68	V_3Sn	4.3	63
*Nb–40% Ir–5%O	10.4–11.7	68	$YbGe_2$	3.8	65
NbN	14.5	64	$ZrGe_2$	8	65
Nb_3P	1.83	69	Zr_2Ir	7.3	71
Nb_3Pt	10	63	$ZrIr_2$	4.1	72

*at.%

A large number of intermetallic compounds, non-stoichiometric alloys and solid solutions exhibit superconductivity. Among the compounds, the sodium chloride structure (e.g. NbN) and the A–15 or 'beta-tungsten' structure (e.g. Nb_3Sn) are two of the more common crystal types. There are currently nearly fifty A–15 compounds known to be superconducting. These alloys and compounds are potentially of considerable technological importance. Table 19.6 lists those with critical temperatures greater than 1 K.

Ternary additions can often lead to increases in critical temperature. Useful results have been discussed in references 68 and 73–77. Increased current-carrying capacity has also been seen to result from energetic particle irradiation.[78]

Superconducting compounds are usually brittle in bulk, and special techniques are required to fabricate them into useful forms. Powder metallurgical techniques may be used, while another typical method results in a composite of multifilamentary superconducting phase in a copper matrix. Superconductors may also be produced as thin films by chemical vapour deposition or sputtering (*see* references 79–83 and chapter 35).

Superconducting properties tend to be dependent upon the method of preparation and heat treatment; this observation is discussed in many of the references quoted.

For overall discussion of superconductivity, its applications, and the prospects of further developments of industrial importance, see references 64 and 84.

REFERENCES

1. B. N. Aleksandrov., *Fiz. Metall.*, 1971, **31**, 1175.
2. J. S. Brown *et al.*, *J. Phys. F.*, 1978, **8** (8), 1703.
3. B. Delley and H. Beck, *J. Phys. F.*, 1979, **9** (3), 505 and 517.
4. H. N. Dunleavy and W. Jones, *J. Phys. F.*, 1978, **8** (7), 1477.
5. H. J. Güntherodt *et al.*, *J. Phys. F.*, 1976, **6** (8), 1513.
6. S. N. Khanna and A. Jain, *J. Phys. F.*, 1977, **7** (12), 2523.
7. F. R. Vukajlovic *et al.*, *Physica*, 1977, **92B** + **C** (1), 66.
8. Y. Waseda *et al.*, *J. Phys. F.*, 1978, **8** (1), 125.
9. S. N. Khanna and A. Jain, *J. Phys. Chem. Solids*, 1977, **38** (5), 447.
10. N. N. Sinha and P. L. Srivastava, *Phys. Status Solidi* (*b*), 1978, **90** (1), 369.
11. T. J. Bastow, *Phys. Lett.*, 1977, **60A** (5), 487.
12. F. J. Ohkawa, *J. Phys. Soc. Jpn*, 1978, **44** (4), 1105.
13. M. Isshiki and K. Igaki, *Trans. Jpn Inst. Met.*, 1978, **19** (8), 431.
14. P. L. Rossiter, *J. Phys. F.*, 1979, **9** (5), 891.
15. J. Ziman, *Phil. Mag.*, 1961, **6**, 1013.
16. British Aluminium Company Ltd. Data Sheet.
17. R. O. A. Hall *et al.*, *J. Low Temp. Phys.*, 1977, **27** (1–2), 305.
18. G. K. White and S. B. Woods, *Phil. Mag.*, 1958, **3**, 342.
19. W. Meissner and B. Voigt, *Ann. Phys.*, 1930, **7**, 892.
20. T. C. Chi, *J. Phys. Chem. Ref. Data*, 1979, **8** (2), 439.
21. S. Gabe and E. G. Evans, *Phil. Mag.*, 1935, **19**, 773.
22. T. C. Chi, *J. Phys. Chem. Ref. Data*, 1979, **8** (2), 339.
23. M. V. Vedernikov *et al.*, *J. less-common Met.*, 1977, **52** (2), 221.
24. H. J. Güntherodt *et al.*, *Phys. Lett.*, 1974, **50A**, 313.
25. G. T. Meaden, 'Electrical Resistance of Metals', Plenum, New York, 1965.
26. K. Hirata *et al.*, *J. Phys. F.*, 1977, **7**, 419.
27. G. K. White and S. B. Woods, *Phil. Trans. R. Soc. London, Ser. A*, 1959, **251**, 273.
28. R. W. Powell, *Proc. Roy. Soc., London, Ser. A*, 1951, **209**, 525.
29. R. N. Lyon, 'Liquid Metals Handbook', 1952. Atomic Energy Commission, Washington DC.
30. R. W. Powell *et al.*, *Phil. Mag.*, 1962, **7**, 1183.
31. E. Grüneisen, *Ergebn. exakt. Naturw.*, 1945, **21**, 50.
32. G. T. Meaden and P. Pelloux-Gervais, *Cryogenics*, 1965, **5**, 227.
33. J. M. Abraham and B. Deviot, *J. less-common Met.*, 1972, **29**, 311.
34. B. C. Dupree *et al.*, *J. Phys. F.*, 1975, **5** (11), L200.
35. P. Haen and G. T. Meaden, *Cryogenics*, 1965, **5**, 194.
36. H. K. Onnes and W. Tuyn, *Proc. Acad. Sci. Amsterdam*, 1923, **25**, 443.
37. G. T. Meaden, *Proc. Roy. Soc., London, Ser. A*, 1963, **276**, 553.
38. W. Tuyn and H. K. Onnes, *Proc. Acad. Sci. Amsterdam*, 1933, **26**, 504.
39. F. Warkusz, *Thin Solid Films*, 1977, **41** (3), 261.
40. F. Warkusz, *J. Phys. D*, 1978, **11** (5), 689.
41. F. Warkusz, *Thin Solid Films*, 1978, **52** (2), 29.
42. L. N. Korochkina *et al.*, *Phys. Met. Metallogr.*, 1975, **40** (2), 1.
43. S. I. Masharov and N. M. Rybalko, *ibid.*, 5.
44. C. A. Rahim and R. D. Barnard, *J. Phys. F*, 1978, **8** (9), 1957.
45. L. V. Meisel and P. J. Cote, *J. Phys. F*, 1977, **7** (12), L321.

46. A. Fert and I. A. Campbell, *J. Phys. F*, 1976, **6** (5), 849.
47. A. Eroglu *et al.*, *Phys. Status Solidi* (*b*), 1978, **87** (1), 287.
48. V. M. Beilin *et al.*, *Phys. Met. Metallogr.*, 1977, **43** (2), 68.
49. Y. Yoshitomi *et al.*, *Solid St. Commun.*, 1976, **20** (8), 741.
50. J. Hennephof *et al.*, *Physica*, 1971, **52**, 279.
51. Y. D. Yao *et al.*, *J. Low Temp. Phys.*, 1975, **21** (3–4), 369.
52. A. Tari, *J. Phys. F*, 1976, **6** (7), 1313.
53. K. Mori and Y. Saito, *Jpn J. Appl. Phys.*, 1976, **15** (10), 1997.
54. Copper Development Association, Standards Handbook.
55. K. Miska, *Mater. Eng.*, 1977, **85** (5), 28.
56. Y. T. Hsu and B. O'Reilly, *J. Met.*, 1977, **29** (12), 21.
57. M. Mujahid and N. N. Engel, *Scr. Metall.*, 1979, **13** (9), 887.
57A. A. Kutner *et al.*, *Aluminium*, 1976, **5**, 322.
58. B. W. Roberts, *J. Phys. Chem. Ref. Data*, 1976, **5**, 581 and NBS. Tech. Note, 983, 1978.
59. E. A. Lynton, 'Superconductivity', Methuen, London 1962.
60. A. C. Rose-Jones, 'Low Temperature Techniques', English Universities Press, London, 1964.
61. T. Claeson, *Commun. Phys.*, 1977, **2** (3), 53.
62. H. R. Khan and C. J. Raub, *Gold Bull.*, 1975, **8** (4), 114.
63. D. Dew-Hughes, *Cryogenics*, 1975, **15**, 435.
64. B. T. Matthias and P. R. Stein, Superconducting Materials, in 'Physics of Modern Materials', Vol. 2, International Atomic Energy Agency, Vienna, 1980.
65. A. K. Ghosh and D. H. Douglass, *Solid State Commun.*, 1977, **23** (4), 223.
66. E. E. Havinga *et al.*, *Phys. Lett.*, 1969, **29A**, 109.
67. C. M. Knoedler and D. H. Donglass, *J. Low Temp. Phys.*, 1979, **37** (1–2), 189.
68. W. L. Johnson and S. J. Poon, *J. less-common Met.*, 1975, **42** (3), 355.
69. J. O. Willis *et al.*, *Phys. Rev. B* (*Solid State*), 1978, **17** (1), 184.
70. C. C. Koch, *J. less-common Met.*, 1976, **44**, 177.
71. D. P. Moiseev *et al.*, *Phys. Met. Metallogr.*, 1975, **39** (6), 33.
72. B. T. Matthias *et al.*, *J. Phys. Chem. Sol.*, 1961, **19**, 130.
73. N. Y. Alekseevskii *et al.*, *Phys. Met. Metallogr.*, 1977, **43** (1), 29.
74. M. Drys and N. Iliew, *J. less-common Met.*, 1976, **44**, 235.
75. C. H. Kopetskii and A. V. Pavlyuchenko, *Phys. Met. Metallogr.*, 1977, **43** (3), 38.
76. R. Somasundaram *et al.*, *J. Appl. Phys.*, 1976, **47** (10), 4656.
77. R. G. Sharma and N. E. Aleksivkii, *J. Phys. D*, 1975, **8** (15), 1783.
78. G. W. Cullen, Proc. Summer Study on Superconducting Devices and Accelerators, Brookhaven National Laboratory, 1968.
79. J. E. Kunzler *et al.*, *Phys. Rev. Lett.*, 1961, **6**, 89.
80. J. P. Harbison and J. Bevk, *J. Appl. Phys.*, 1977, **48** (12), 5180.
81. J. M. E. Harper *et al.*, *J. less-common Met.*, 1975, **43** (1/2), 5.
82. L. A. Pendrys and D. H. Douglass, *Solid State Commun.*, 1976, **18** (2), 177.
83. L. Schultz and R. Bormann, *J. Appl. Phys.*, 1979, **50** (1), 418.
84. T. Luhman and D. Dew-Hughes (eds). 'Metallurgy of Superconducting Materials', Treatise on Materials Science and Technology, Vol. 14, Academic Press, New York, 1979.

20 Steels and alloys with special magnetic properties

20.1 Magnetic definitions

It is desirable to define certain magnetic quantities and to relate these to the practical units used in specifying the characteristics of magnetic materials. In the following definitions, no claim to completeness is made, but fuller treatments may be found in textbooks by Bates[1] and Brailsford.[2]

UNIT MAGNETIC POLE

Forces of attraction and repulsion between magnetic bodies have led to the mathematical concept of the unit magnetic pole by analogy with the electric charge in electrostatic theory. Thus like magnetic poles repel each other and unlike poles attract. The unit pole being that pole which when placed one metre away from a similar pole in vacuo is repelled by a force of $(1/4\pi\mu_0)$ newtons. The magnetic pole is a useful concept when solving problems in magnetostatics, but it must be stressed that a free magnetic pole has no reality as the observed forces arise from magnetic fields caused by moving electric charges. These may be either electric currents, or orbital electrons in the atoms of the magnetic material. The force between two poles of strength m_1 and m_2 is proportional to $m_1 m_2$ and varies inversely as the square of the distance between them.

MAGNETIC FIELD STRENGTH

The magnetic field strength H at a point is a vector quantity measured in both magnitude and direction by the force exerted on a unit pole placed at that point. The unit of magnetic field is the ampere-turn per metre ($A\,m^{-1}$), which is the (axial) field inside a large diameter infinitely long solenoid of 1 turn m^{-1} carrying a current of 1 A.

MAGNETIC MOMENT

A magnetized body has an equal number of like and unlike poles on its surface. The simplest body behaves as a dipole magnet consisting of unlike poles of strength m separated by a distance l. If such a simple magnet is placed with its axis at right angles to the direction of a uniform magnetic field of strength H then it experiences a mechanical couple MH, where M is defined as the magnetic moment, and is equal to ml.

INTENSITY OF MAGNETIZATION

When a magnetic body is placed in a magnetic field it becomes magnetized, that is magnetic poles appear on its surface. The magnitude of this induced magnetization is measured by the intensity of magnetization J. This is a vector quantity and its magnitude equals the magnetic moment per unit volume.

MAGNETIC FLUX DENSITY (OR MAGNETIC INDUCTION)

The magnetic flux density B at a point is a vector quantity arising from the magnetic field and the intensity of magnetization of the medium. It is given by

$$B = \mu_0 H + J$$

where

$$\mu_0 = 4\pi/10^7$$

Total flux ϕ is obtained by integrating B across a given area.

PERMEABILITY

The permeability μ of a medium is the ratio of the magnetic flux B to the magnetic field H producing it. In vacuo or a non-magnetic medium $J=0$ and the permeability is μ_0, known as the permeability of free space. The relative permeability μ_{rel} is the ratio of the magnetic flux density produced in the medium to that produced in a vacuum.

MAGNETIC SUSCEPTIBILITY

From the relation $B=\mu_0 H+J$ we can write $\mu=\mu_0+\mu_0\kappa$ where $\kappa=J/\mu_0 H$. The quantity κ is termed the magnetic susceptibility per unit volume of the material in question.

MASS SUSCEPTIBILITY

The magnetic susceptibility per unit mass, or mass susceptibility, is denoted by χ and is equal to κ/ρ, where ρ is the density of the material.

MAGNETIC SATURATION

Magnetic saturation is said to have been reached when the intensity of magnetization (J) reaches a constant value with increasing magnetizing field H. This is called the saturation intensity (J_s). For ferromagnetic materials where $\kappa \gg 1$ it is conventional to specify the saturation flux density per unit volume (B_s) of the material ($\simeq J_s$) as the fundamental magnetic parameter of the material.

CURIE POINT

The Curie point, or temperature (T_c) is that temperature at which a ferromagnetic material on heating loses its spontaneous magnetization.

MAGNETIC HYSTERESIS

This is a phenomenon by which the magnetization of a material depends not only on the magnetizing field but also on the previous magnetic history. The magnetic flux B always lags behind magnetic field H in a material showing hysteresis. The complete B–H curve, or loop, produced by increasing H from zero to a maximum value, reducing to zero and then repeating with H in the negative sense is termed the hysteresis loop, and the dissipation of energy associated with it, the hysteresis loss. The hysteresis loss per unit volume is given by (area of the loop in $B \times H$) joules.

REMANENCE

The remanence (B_r) is the value of the magnetic flux density remaining in a closed ring of the material following magnetization to saturation in the circumferential direction and subsequent removal of the magnetizing field.

COERCIVE FORCE

The coercive force (H_c) is the reversed value of magnetizing field required to reduce to zero the magnetic flux density in a sample of magnetic material.

$(BH)_{max}$

The energy product of B and H of a magnetic material has a maximum value in the second quadrant of the hysteresis loop. This value $(BH)_{max}$ is directly related to the stored magnetic energy per unit volume of the material.

20.2 Magnetic units and conversion factors

Authors and publishers of current literature on magnetism and magnetic materials have been very slow in adopting SI (system international) units in place of the centimetre gram second (c.g.s.) electromagnetic units previously used. Conversion of magnetic units between the two systems are therefore frequently required, as well as care in using equations which are different for the two systems. The relation between the relevant c.g.s. electromagnetic units and the SI (rationalized m.k.s.) units are given in Table 20.1.

Table 20.1 RELATION BETWEEN SI UNITS AND C.G.S. UNITS OF MAGNETIC QUANTITIES

Quantity	*c.g.s. unit*	*SI unit*	*Conversion factor for c.g.s. to SI units*
Magnetic field H	Oersted (Oe)	Ampere turns per metre ($A m^{-1}$)	$\times 10^3/4\pi$
Magnetic flux ϕ	Maxwell (or line)	Weber (Wb)	$\times 10^{-8}$
Flux density B	Gauss (G)	Tesla (T)	$\times 10^{-4}$
Magnetomotive force	Oersted cm or Gilbert	Ampere-turn	$\times 10/4\pi$
Intensity of magnetization J	e.m.u.	Weber metre^{-2} ($Wb m^{-2}$)	$\times 4\pi \cdot 10^{-4}$
Magnetic moment M	e.m.u.	Weber metre (Wb m)	$\times 4\pi \cdot 10^{-10}$
Pole strength m	e.m.u.	Weber (Wb)	$\times 4\pi \cdot 10^{-8}$
Permeability of vacuum μ_0	1	Henry m^{-1} ($H m^{-1}$)	$\times 4\pi \cdot 10^{-7}$

20.3 Materials with magnetic properties

All substances exhibit magnetic phenomena in the basic forms of either ferromagnetism, paramagnetism, or diamagnetism at any given temperature.

Ferromagnetic materials are characterized by moderate to high permeabilities, which vary with the applied magnetic field, becoming low at very high fields. They also exhibit magnetic hysteresis whereby the intensity of magnetization of the material varies according to whether the applied field is being increased in a positive sense or decreased in a negative sense. A typical hysteresis loop is shown in Figure 20.1 together with some of the more important derived magnetic properties. When the magnetization is cycled continuously round a hysteresis loop, as for example when the applied field arises from an alternating current, there is an energy loss proportional to the area of the included loop. This is termed hysteresis loss, and is measured in joules per metre^3. High hysteresis loss is associated with permanent magnetic properties exhibited by materials commonly termed 'hard' magnetic materials as these almost invariably have hard mechanical properties. Those materials with low hysteresis loss are termed 'soft' and are difficult to magnetize permanently. The ferromagnetic properties of these elements, alloys, and compounds disappear reversibly if heated above the Curie temperature, when they become paramagnetic.

Paramagnetic substances of elongated shape tend to align themselves with an applied magnetic field as do the ferromagnetics, but do not exhibit any permanent magnetic properties when the field is removed as hysteresis is absent. Their magnetic susceptibility is very small but positive, and consequently their relative permeability is only slightly greater than unity. The susceptibility is generally independent of field strength but may decrease with increasing temperature. Among the paramagnetic substances are included many iron salts and rare-earth element salts, palladium and platinum, sodium, potassium, and oxygen, and the ferromagnetics above their Curie point.

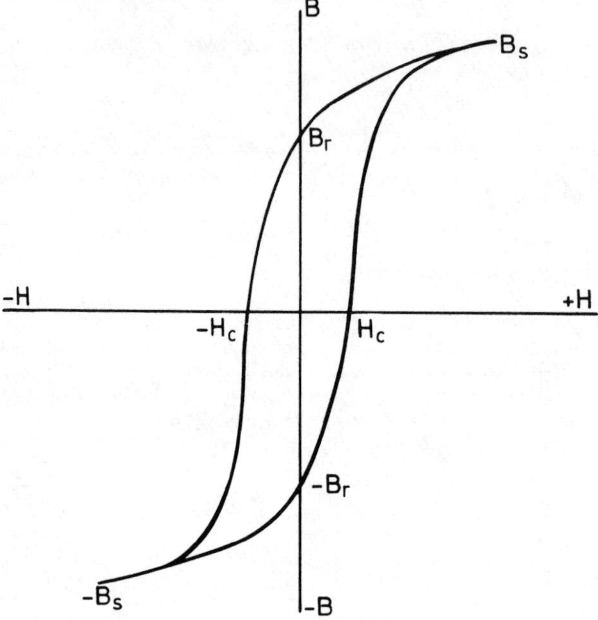

Figure 20.1 *Typical hysteresis loop*

Table 20.2 MASS SUSCEPTIBILITIES OF THE ELEMENTS

Element	*Atomic No.*	*Mass susceptibility* $10^{-8}\,m^3\,kg^{-1}$	*Element*	*Atomic No.*	*Mass susceptibility* $10^{-8}\,m^3\,kg^{-1}$
H	1	−2.49	Sc	21	+9.10
He	2	−0.59	Ti	22	+1.56
			V	23	+1.75
Li	3	+0.62	Cr	24	+3.85
Be	4	−1.25	Mn	25	+14.7
B	5	−0.86	Fe	26	*
C	6	−0.61	Co	27	*
N	7	−1.00	Ni	28	*
O	8	+133			
F	9		Cu	29	−0.11
Ne	10	−0.41	Zn	30	−0.19
			Ga	31	−0.30
Na	11	+0.64	Ge	32	−0.15
Mg	12	+0.69	As	33	−0.39
Al	13	+0.81	Se	34	−0.40
Si	14	−0.16	Br	35	−0.49
P	15	−1.13	Kr	36	−0.49
S	16	−0.61			
Cl	17	−0.71?	Rb	37	+0.26
Ar	18	−0.56	Sr	38	−0.25
			Y	39	+6.62
K	19	+0.65	Zr	40	−0.56
Ca	20	+1.38	Nb	41	+1.88

* Ferromagnetic

Table 20.2 MASS SUSCEPTIBILITIES OF THE ELEMENTS—*continued*

Element	*Atomic No.*	*Mass susceptibility* $10^{-8}\,m^3\,kg^{-1}$	*Element*	*Atomic No.*	*Mass susceptibility* $10^{-8}\,m^3\,kg^{-1}$
Mo	42	+0.05	Tm	69	+189
Tc	43	+3.42	Yb	70	+1.8
Ru	44	+0.63			
Rh	45	+1.39	Lu	71	—
Pd	46	+6.75	Hf	72	+0.53
Ag	47	−0.25	Ta	73	+1.16
Cd	48	−0.23	W	74	+0.35
In	49	−0.14	Re	75	+0.45
Sn	50	−0.31	Os	76	+0.06
Sb	51	−1.09	Ir	77	+0.19
Te	52	−0.39	Pt	78	+1.38
I	53	−0.45			
Xe	54	—	Au	79	−0.19
			Hg	80	−0.21
Cs	55	−0.28	Tl	81	−0.30
Ba	56	+1.12	Pb	82	−0.15
La	57	+1.30	Bi	83	−1.69
Ce	58	+18.8	Po	84	—
Pr	59	+31.2	At	85	—
Nd	60	+45.0	Rn	86	—
Pm	61				
Sm	62	+15.5	Fr	87	—
Eu	63	+275	Ra	88	—
Gd	64	*	Ac	89	—
Tb	65	+1 150	Th	90	+0.14
Dy	66	+795	Pa	91	+3.25
Ho	67	—	U	92	+2.15
Er	68	+330			

* Ferromagnetic

The diamagnetic substances take on a very small magnetization which is opposed to the sense of the applied field. Thus an elongated specimen will align itself transverse to a non-uniform field. The diamagnetics have negative susceptibilities and consequently permeabilities which are very slightly less than unity. Many of the metals and most non-metals are diamagnetic.

It is usual to compare paramagnetic and diamagnetic substances by their susceptibilities rather than by permeability or saturation magnetization as for the ferromagnetic materials. The only materials of commercial interest for their magnetic properties at the present time are the ferromagnetics. These usually contain one or more of the ferromagnetic elements iron, nickel and cobalt. The mass susceptibilities of the elements at room temperature are given in Table 20.2.

20.3.1 Permanent-magnet alloys

Historically the first permanent magnet was the naturally occurring, partially magnetized lodestone or magnetite, a mineral of approximate composition Fe_3O_4, belonging to the spinel group, and widely distributed, especially in igneous rocks. About 1880 hardened carbon steel gave way to low tungsten alloy steels (0.7% C, 6% W), which held the field for many years. Chromium was found in 1917 to function nearly as well as tungsten, and both 6% tungsten and 6% chromium steels fall into the class having a remanence $B_r=0.95$–$1.2\,T$ and a coercive force $H_c=4400$–$5600\,A\,m^{-1}$, with a $(BH)_{max}$ value of the order $2400\,J\,m^{-3}$. (The stored magnetic energy is $(BH)_{max}/2\,J\,m^{-3}$.)

Better steels in the cobalt range followed, culminating in the well-known Honda 35% cobalt steel, which still represents the optimum for martensitic steels ($B_r=0.9\,T$, $H_c=20\,000\,A\,m^{-1}$, $(BH)_{max}=7650\,J\,m^{-3}$). A range of steels containing cobalt, chromium and molybdenum, also appeared and are still in current use.

A new era in permanent-magnet alloys was opened up in 1931 when Mishima in Japan discovered that an alloy of nickel, aluminium and iron possessed a high coercive force ($H_c=32000/48\,000\,A\,m^{-1}$). Subsequent development showed that optimum properties

Table 20.3 SOME DATA ON THE MAGNETIC PROPERTIES OF CAST STEELS AND CAST IRONS

Material	*Condition*	*Flux density* T, $H=$ 4 kA m^{-1}	*Flux density* T, $H=$ 8 kA m^{-1}	*Rel. permeability* Maximum	*Rel. permeability* at H A m^{-1}	*Max. flux density* T, $H=$ 20 kA m^{-1}	*Remanence* T, $H=$ 20 kA m^{-1}	*Coercive force* A m^{-1}, $H=$ 20 kA m^{-1}
Steels								
0.10% carbon steel,[1] Si, 0.33%; Mn, 0.67%	Annealed	1.62	1.74	2 420	216	1.92	0.80	136
	Normalized	1.64	1.76	1 950	248	1.96	0.85	168
	As-cast	1.61	1.74	2 100	208	1.95	0.66	128
0.19% carbon steel,[1] Si, 0.30%; Mn, 0.48%	Annealed	1.60	1.73	2 100	224	1.88	0.87	156
	Normalized	1.57	1.72	1.520	360	1.91	0.90	216
	As-cast	1.55	1.69	1 720	232	1.89	0.73	168
0.34% carbon steel,[1] Si, 0.44%; Mn, 0.55%	Annealed	1.53	1.66	1 200	400	1.85	1.05	296
	Normalized	1.50	1.67	970	575	1.87	1.05	440
	As-cast	1.45	1.65	840	480	1.87	0.85	440
Manganese steel[2] 0.19% C, 0.48% Si, 1.14% Mn	A 925 °C	1.51	1.66	1 300	400	1.86	—	—
0.29% C, 0.29% Si, 1.40% Mn	A 950 °C, N 880 °C, T 600 °C, a.c.	1.50	1.68	650	960	1.78	—	—
0.31% C, 0.52% Si, 1.37% Mn	A 950 °C, OQ 860 °C, T 620 °C, a.c.	1.55	1.65	750	960	1.70	—	—
Chromium-molybdenum steel {0.31% C, 0.69% Mn	A 925 °C, N 880 °C, T 700 °C, a.c.	1.45	1.59	—	—	1.76	—	—
1.16% Cr, 0.39% Mo	A 925 °C, OQ 860 °C, T 650 °C, a.c.	1.54	1.66	—	—	1.81	—	—
						Hysteresis loss J m^{-3} cycle^{-1}		
Cast-irons[3]								
Grey iron T.C. 3.12%, Si 2.2%, Mn 0.67%, P 0.13%	As-cast	0.70	0.86	315	640	2 700	0.41	560
(low phosphorus)	Annealed	0.93	1.07	1 560	200	700	0.44	200
Grey iron T.C. 3.3%, Si 2.04%, Mn 0.52%, P 1.03%	As-cast	0.79	0.97	281	1 040	2 730	0.43	720
(high phosphorus)	Annealed	0.85	1.00	760	320	1 190	0.44	280

Material	Condition							
Whiteheart Malleable TC 1.01%, Si 0.66%, Mn 0.25%	Pearlitic centre	1.32	1.42	730	560	1 490	0.75	360
TC 0.46%, Si 0.66%, Mn 0.25%	Mainly ferritic	1.35	1.61	1 455	280	840	0.74	200
Blackheart Malleable TC 1.26%, Si 0.83%, Mn 0.26%		1.40	—	2 120	176	490	0.62	120
Spheroidal Graphite TC 2.9%, Si 2.61%, Mn 0.72%, Ni 2.18%	Pearlitic	0.93	1.12	290	1 200	3 040	0.54	920
	Annealed, ferritic	1.14	1.26	1 150	320	700	0.46	200
Spheroidal Graphite TC 3.64%, Si 1.41%, Mn 0.3% (nickel free)	Annealed	1.23	1.42	2 060	240	450	0.61	120

Sources of information: 1. 'Effect of Heat Treatment on the Magnetic Properties of Carbon Steel Castings', W. J. Jackson, *J. Iron Steel Inst.*, 1960, **194**, 29.
2. Data submitted by British Steel Foundries.
3. Hillman, M. H. *BCIRA Journal of Research and Development*, 1954, **5**, 188–248.

(B_r = 0.6 T, H_c = 40 000 A m^{-1}, $(BH)_{max}$ = 10 000 J m^{-3}), were obtained with a modified composition containing copper (3–6%).

Simultaneous additions of cobalt and copper to the ternary foundation analysis gave rise to the *Alnico* series of alloys (B_r = 0.75 T, H_c = 40 000 A m^{-1}, $(BH)_{max}$ = 12 800 J m^{-3}), first developed in England at Sheffield.

Directional or anisotropic permanent magnets were obtained by the heat treatment of suitable compositions in a magnetic field. The first experiments were made in Sheffield, England (1938), and continued in Eindhoven, Holland and Schenectady, USA; these culminated in the *Alcomax*, *Ticonal* and *Alnico V* alloys (1940–42), now having along the magnetic axis of the cast magnet properties of a higher order (B_r = 1.25/1.35 T, H_c = 44 000/60 000 A m^{-1}, $(BH)_{max}$ = 36 000/44 000 J m^{-3}).

Columax is the commercial name given to *Alcomax* magnets which have been cooled in such a way during casting that columnar crystals are developed in the same direction as the preferred axis of magnetization. The molten metal is poured into sand moulds on a chilled plate and is allowed to cool mainly in the direction in which the magnetic field is to be applied. The process is therefore only applicable to simple shapes, e.g. rectangular blocks and cylinders. The magnetic properties are lowered as the proportion of crystals not orientated in the preferred direction increases, and are therefore also affected by the presence of cored holes, sharp corners, etc. $(BH)_{max}$ values up to about 60 kJ m^{-3} are available as commercial products in the Columax series. The production of columnar structure is more difficult in the alloys containing titanium as the presence of the latter normally inhibits crystal growth. However, recent discoveries in Sheffield have shown that this difficulty may be overcome by controlled small additions of sulphur or selenium. By this means magnets with $(BH)_{max}$ values up to 80 kJ m^{-3}, with coercivities of 10^5 A m^{-1} or more, are now being produced.

Very high coercivity alloys of the Alnico or Alcomax type have been developed by the inclusion of titanium or niobium. The anisotropy is developed by annealing followed by cooling in a magnetic field. These materials are marketed as Hycomax III or IV and Ticonal X. They have coercivities of 170 kA m^{-1} but relatively low $B_r \sim 0.8$ T, and are used where the magnet length is restricted.

There are many other alloys, particularly those which show precipitation hardening effects, such as copper–nickel–iron and copper–nickel–cobalt alloys, which are classed as permanent-magnet materials. Although their remanence values are low (0.3–0.7 T) they do possess the advantages of being amenable to forging, rolling and stamping, and are machinable. These alloys are manufactured in the USA under the names of *Cunico* and *Cunife*. An alloy known as *Vicalloy*, containing cobalt and vanadium alloyed with iron, also possesses useful properties, and after certain heat treatments and cold working can be rendered anisotropic. All these alloys may be hot worked satisfactorily; they may also be cold worked before the final heat treatment.

Small *Alnico* and *Alcomax* magnets are sometimes produced by the powder metallurgy process, in which case they are compacted to very close tolerances requiring only a minimum of final grinding. The mechanical properties are also improved owing to the finer grain structure. The density of the sintered alloys is about 90% of that of similar cast materials and there is a similar reduction in magnetic properties. Permanent magnets may also be produced from iron and iron/cobalt micropowders, pressed but not sintered, and also from sintered oxides of iron and cobalt.

Even higher coercivities of 370 kA m^{-1} (ref. 1) and $(BH)_{max}$ of 72 kJ m^{-3} have been obtained with platinum–cobalt alloys (Platinax), but since the alloy contains over 70% platinum it is only used where very small compact magnets are required.

Barium ferrite, a new permanent magnet material of approximate composition $BaFe_{12}O_{19}$, was developed originally at Eindhoven under the name of Magnadur and is a sintered oxide ceramic. Barium ferrites are now produced by various manufacturers under different names, in both isotropic and oriented grades. The saturation magnetization and remanence of these materials are low compared with those of the *Alcomax* series, but the coercive force is considerably higher, due to a large crystal anisotropy. It also has a high electrical resistivity.

A very recent development is of very high coercivity magnets consisting of intermetallic compounds of some of the lower molecular weight rare-earth elements and cobalt. These compounds have the composition R Co_5, where R may be yttrium, lanthanum, cerium, praseodymium, or samarium. The crystal structure is hexagonal which gives rise to anisotropy fields between 10 and 20 MA m^{-1}. Samarium–cobalt magnets are now becoming commercially available with energy products $(BH)_{max}$ of up to 160 kJ m^{-3}. These are manufactured by pressing fine particles of the compound in a strongly aligning magnetic field, and sintering to near 100% theoretical density. Temperature stabilization is usually carried out after magnetization by annealing to a temperature slightly above the maximum operating temperature.

Compositions, heat treatments and magnetic properties of selected steels and alloys in common use are summarized in Tables 20.4 and 20.5.

Table 20.4 CHEMICAL COMPOSITIONS OF PERMANENT-MAGNET STEELS AND ALLOYS

Material	*Composition: weight %; bal.* Fe					
Steels	C	Mn	Si	Co	Cr	W
1% Carbon	1.00	0.4	0.2	—	0.5	—
6% Chromium	1.05	0.3	0.15	—	6.0	—
6% Tungsten	0.67	0.3	0.15	0.5	0.4	6.0
6% Cobalt	1.05	0.3	0.15	6.0	9.0	—
9% Cobalt	1.05	0.3	0.15	9.0	9.0	—
15% Cobalt	1.05	0.3	0.15	15.0	9.0	—
35% Cobalt	0.85	0.5	0.15	35.0	6.0	4.0

Alloys (trade names)*	Ni	Co	Al	*Others*
Alni, Reco 1, Alnico III (USA)†	24–30	0	12–14	Sometimes small % Cu, Ti
Alnico, Magloy 5, Reco 160, Alnico II (USA)	16–20	12–14	9–11	Cu, 3–6; small Ti
Hynico II, Reco 2, Alnico XII (USA)	18–21	17–20	8–10	Cu, 2–4; Ti, 4–8 } small Ti
Alcomax II & III, Ticonal G, Magloy 1, Alnico V (USA)	12–15	23–25	8–9	Cu 2–4 + sometimes small % Nb or Ti
Columax, Ticonal GX, Magloy 10, Alnico 5–7, V.D.G. (USA)	13–15	24–25	8	Cu 2–4 + sometimes small % Nb
Alnico 8, Hycomax III & IV, Ticonal X	14–16	32–40	7–8	Cu 3–4, Ti 4–8, sometimes small Nb
Cunife	60% Cu, 20% Ni			
Cunico	35–50% Cu, 30–40% Co, 20–25% Ni			
Vicalloy I	52% Co, 9.5% V			
Vicalloy II	52% Co, 13% V			
Platinax	77% Pt, 23% Co			
Samarium–cobalt	34% Sm, 66% Co.			

* Many other designations are used on the continent.
† The USA designation Alnico, with different subscripts, covers a wide range of alloys.

Table 20.5 PROPERTIES OF PERMANENT-MAGNET MATERIALS

Material	*Magnetic properties* BH_{max} kJ m^{-3}	H_c kA m^{-1}	B_r T	*Heat treatment**	*Density* g cm^{-3}	*Hardness* HV
Carbon steel	1.44	4.8	0.8	SC 740 WQ 780	7.8	940
Chromium steel	2.40	4.8	0.96	AC 750 OQ 840	7.75	775
Tungsten steel	2.40	5.2	1.05	SC 750 WQ 800	8.10	870
Cobalt steels (low and medium cobalt)	3.0–5.2	10.5–14.5	0.72–0.86	AC 1 150 SC 780 AQ 1 000	7.7–7.9	8.00–825
Cobalt steels (high cobalt)	7.2	18.5	0.86	AC 1 150 SC 780 OQ 950	8.15	870
Alni	7.2–9.6	25–50	0.75–0.55	Quench about 1 150 °C and temper at about 550 °C	6.9	
Alnico (low cobalt)	10.4–12.0	36–56	0.75–0.55	Air cool from *ca.* 1 200 °C and temper at about 650 °C	6.9–7.0	

* SC = slow cool AC = air cool OQ = oil quench WQ = water quench Figures are temperatures in °C.

Table 20.5 PROPERTIES OF PERMANENT-MAGNET MATERIALS—*continued*

	Magnetic properties					
Material	BH_{max} kJ m^{-3}	H_c kA m^{-1}	B_r T	*Heat treatment**	*Density* g cm^{-3}	*Hardness* HV
Alnico (medium cobalt)	10.4–17.0	40–50	0.80–0.50	Air cool from *ca.* 1 200 °C and temper at about 550 °C	7.0–7.3	
Hynico	12.8–16.0	60–80	0.58–0.55	Air cool from *ca.* 1 200 °C and temper at about 550 °C	7.2–7.4	570–650
Alcomax II & III	40–44	46–52	1.30–0.90	Cool from high temperature in magnetic field to below about 850 °C, and temper at about 600 °C. Some limitations in shape	7.3	
Alcomax (columnar structure)	56–64	56–62	1.35–1.25	Special casting methods; considerable limitations in size and shape	7.3	
Alnico-8	40–48	110–170	0.75–0.90	Quench, followed by reheat to 800 °C and cool in magnetic field	7.0	—
Cunife	8.0–14.5	32–44	0.55–0.45	Ductile anisotropic alloy; quench from 1 000 °C before cold working and ageing at 650 °C. Value of BH_{max} depends on degree of cold working	8.6	200
Cunico	6.8–8.0	36–56	0.53–0.34	As for Cunife (but no cold working)	8.3	200
Vicalloy I	8.0	24	0.90	Ductile anisotropic; solution treatment 1 200°C + oil quench. Age at 600 °C	8.2	745
Vicalloy II	22.4	36	1.00	As for Vicalloy I. Full properties only attainable with cold work	8.1	—
Platinum–cobalt	75.0	380	0.60	—	—	200
Samarium–cobalt	110–160	560	0.80	Fine particles pressed in magnetic field and subsequently sintered	8.6	—
Barium ferrite, Magnadur, Feroba I, Indox etc.	6.5	16	0.16	Pressed and sintered ceramics	4.8	450
Barium ferrite (oriented) Feroba II, etc.	18–24	72–110	0.30–0.35	Pressed and sintered ceramics. Pressed in magnetic field	4.8	450

* SC = slow cool AC = air cool OQ = oil quench WQ = water quench Figures are temperatures in °C.

20.3.2 Magnetically soft materials

Materials or alloys classed as being magnetically soft are used for a wide range of applications, such as pole cores in electrical machinery, armature stampings, transformer laminations, and other electromagnetic equipment of paramount importance in the electrical, electronic and communication engineering industries. Many special alloys of widely different chemical compositions have been developed to meet the specialised requirements, and each material possesses its own distinguishing characteristics.

Sometimes several conflicting attributes are required in magnetically soft materials for a given application. It is for this reason that a number of different magnetic materials ranging from iron itself through the silicon–iron steels to high nickel–iron alloys, have emerged. Usually, the most important requirement is that the hysteresis loss due to cyclic magnetisation should be as low as possible. The eddy current loss, which is due to the induction of electric currents in the material by the changing magnetic flux, should also be small. Generally the magnetic permeability should be as high as possible, particularly the maximum permeability, and in certain special cases the initial permeability also. Sometimes the requirement is for a high or constant permeability at low field strengths. In general, the saturation intensity should be as high as other conditions will permit. For magnetic temperature compensation there should be a linear relation, over a known range of temperature, between permeability and temperature.

Relevant data concerning a wide selection of materials is given in Tables 20.6 to 20.10 and is discussed in the corresponding sections.

Silicon–iron alloys

For the magnetic circuits of the larger pieces of electrical engineering equipment normally operated at power frequencies of 50 Hz, such as power transformers, alternators and electric motors of all sizes, a magnetic material is essential which will have a low hysteresis, high resistivity and a high value of magnetic saturation. Its production in the form of thin sheet or strip must also be possible, and in view of the big tonnages essential for such purposes it must not be expensive. Silicon–iron alloys are the only ones which meet all these conflicting requirements.

In the very early days of the industry, sheets of puddled iron approximately 0.5 mm thick were used. In 1900 Barrett, Brown and Hadfield carried out a series of tests on the magnetic properties of various iron alloys including silicon, aluminium and manganese, and found that the silicon–iron alloys had the best magnetic properties of the series they tried and were considerably better than the puddled iron in use at that time for magnetic circuits. The silicon content of the alloy they produced was 2.5% and the carbon 0.2–0.3%. The relative magnetic quality of the two types is illustrated by the comparative total watts loss, i.e. hysteresis + eddy currents, of fully annealed sheets 0.5 mm thick, shown below:

Puddled iron	4.0 W kg^{-1} at B_m of 1 T and 50 Hz
Hadfield's silicon steel	2.6 W kg^{-1} at B_m of 1 T and 50 Hz

The manufacture of silicon sheets of this composition was started in England about 1903 and in America about 1907. These sheets were hot rolled in packs in two-high mills and the scale removed by pickling, followed by box annealing in an inert atmosphere. Since that date, the quality of silicon steels has been continuously improved and more has been learned about the effects of silicon on the magnetic properties. These effects are briefly as follows:

1. Silicon precipitates the major portion of the oxygen dissolved in the steel, which reduces hysteresis loss and magnetic ageing.
2. Silicon increases the electrical resistivity and thus reduces the eddy current loss, which is proportional to the resistivity (*see* Figure 20.2).
3. With over 2.5% silicon the alloys are outside the gamma loop on the equilibrium diagram, i.e. the alpha–gamma change point is suppressed (*see* Figure 20.3). Alloys of more than 2.5% silicon can therefore be annealed at high temperatures without recrystallization occurring on cooling. As a result it is easier to produce large grain size in the finished sheet, with a corresponding reduction in hysteresis loss.
4. Silicon precipitates the carbon as graphite if it exceeds about 0.05%. Carbon present as graphite has no effect on hysteresis loss.

Subsequent improvements in hot-rolled silicon sheet were in the main due to:

1. Increase in silicon percentage up to the maximum that could be used without rendering the finished sheets too brittle to shear and punch, namely 4.5%. Another limitation is the effect of silicon on the saturation value, which for iron is 2.2 T. Each 1% of silicon reduces this figure by about 0.5 T because the silicon is not ferromagnetic and reduces the saturation value by dilution.
2. Addition of aluminium to give more complete de-oxidation without increasing brittleness.
3. Reduction in sheet thickness to the optimum economic value for minimum core loss at a frequency of 50 Hz, i.e. 0.35 mm (0.014 in).
4. Reduction of impurities to the minimum economic value. Impurities which increase hysteresis are those which occupy an interstitial position in the crystal lattice, e.g. carbon, oxygen, nitrogen, sulphur.
5. Improvements in the control of annealing equipment and protective atmospheres during final annealing.

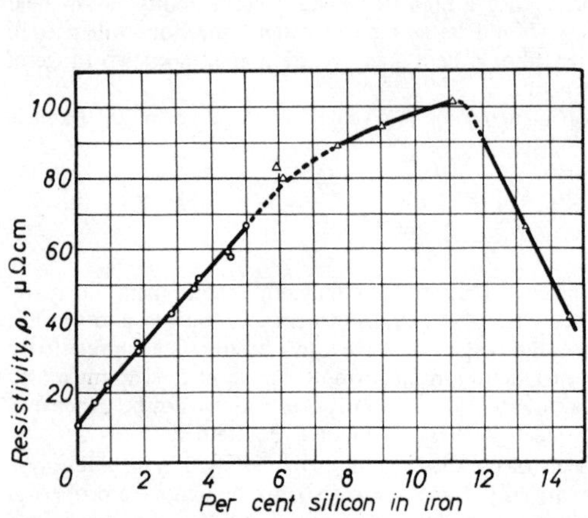

Figure 20.2 *Electrical resistivities of iron-rich iron–silicon alloys*[3]

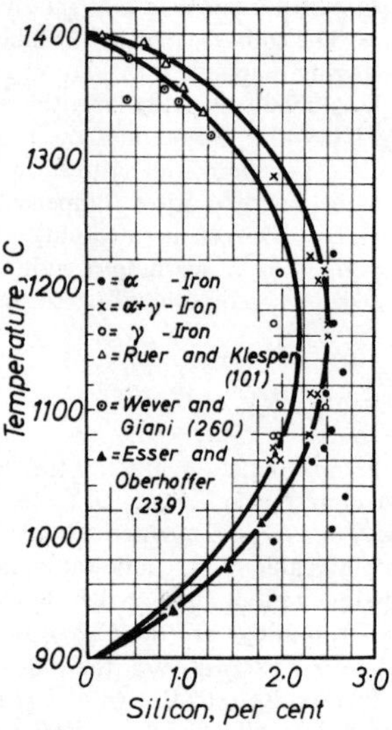

Figure 20.3 *The gamma phase region*[4]

The effect of these improvements on the magnetic quality of hot-rolled sheets is seen in Figure 20.4, which shows the lowest total core loss available with hot-rolled transformer sheet during the last 60 years.

The highest silicon and lowest core loss hot-rolled sheets 0.35 mm (0.014 in) in thickness were used for power transformer cores, but there is a need for cheaper grades of lower silicon content for other uses, such as fractional horse power motors, where the minimum core loss is less important than high saturation needed for maximum torque.

There is, therefore, a whole series of silicon steels rolled with silicon percentages varying from 0.5% up to 4.5%. The various types are shown in Table 20.6, which includes those quoted in the

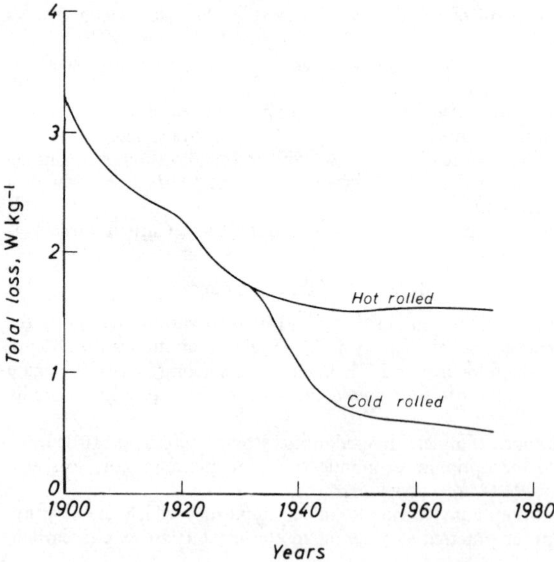

Figure 20.4 *Reduction loss; transformer steel* $B=1.3$ T:$f=50$ Hz

relevant British Standard Specification 601, Part 1: 1959, Part 2: 1961, and the American Iron and Steel Institute Flat Rolled Electrical Steel Manual, Feb. 1964.

The different BSI Grades from 253 to 107 are produced by rolling from alloys of different silicon percentages. The one with the highest core loss, i.e. Grade 253, has approximately 0.5% silicon and the one with the lowest core loss, i.e. Grade 107, has approximately 3.5% silicon. The better transformer grades, i.e. Grade 100 to Grade 74, are normally all rolled from steel of the same composition with 4–4.5% silicon and 0.5% aluminium and are graded, after annealing, on the results of the final core loss test.

Hot rolling in packs to produce silicon steel sheets is being replaced by hot rolling on tandem mills down to about 2.5 mm in thickness, descaling and cold rolling in coils on tandem mills. This can produce a much better surface and the stacking factor is therefore higher. A more uniform gauge can be obtained, die life on punching is better, and the coils can be used on roller feed presses with economy in labour and reduction of scrap.

Cold-rolled dynamo grades with fairly random orientation of the crystal lattice were the first to be produced in this way. It is in fact necessary in the case of sheets for rotating machinery that they should be comparatively isotropic and processing and annealing is arranged with this in mind. In the case of a transformer, however, it is obviously possible to arrange for the flux to be parallel to the direction of rolling of the steel so that, by producing anisotropic strip, advantage can be taken of the fact that the permeability is better and the hysteresis is lower if a single crystal is magnetized with the direction of magnetization parallel to the cube edge of the crystal lattice (*see* Figure 20.5).

Silicon–iron strip for transformers has been produced in large quantities with a very high degree of preferred orientation such that each crystal is almost completely aligned, with the cube edge of the lattice parallel to the direction of rolling. The core losses of this type of steel are much lower than anything produced by hot rolling, although the silicon is limited to a maximum of 3.5% because of the difficulties of cold rolling. The comparative figures of the best available grades of the two types compared at a flux density of 1.3 T, a frequency of 50 Hz, and a thickness of 0.35 mm (0.014 in) are:

Best hot-rolled commercially sold	1.63 W kg^{-1}
Best oriented steel sold	0.66 W kg^{-1}

Originally, oriented cold-rolled steel was used only for small transformers with wound cores in which the magnetic core was a continuously wound ribbon of silicon–iron strip. Today the whole range of power transformers with conventional flat cut cores are made from this type of steel, up to the very largest type of core made so far, i.e. 600 MVA transformers with a core weight of 125 Mg.

Silicon steel sheets used by the electrical engineering industry are of three main types:

1. Cold-rolled grain oriented transformer grades.
2. Cold-rolled non-oriented grades.
3. Hot-rolled transformer and dynamo grades.

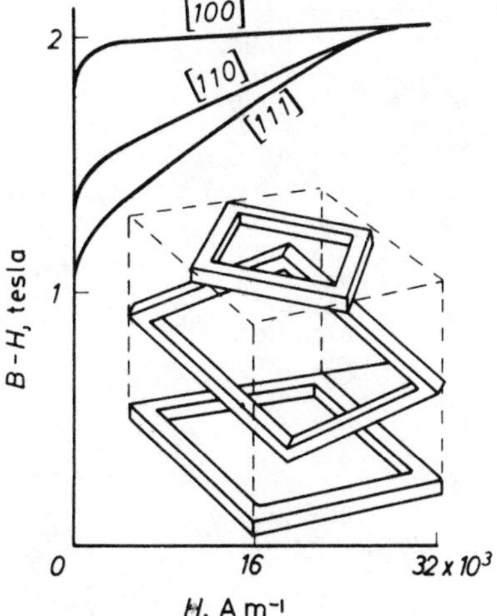

Figure 20.5 *Magnetization curves for 3.85% silicon-iron single crystals at high field strengths in three directions*[5]

These types are subdivided on the basis of total core loss as follows:

1. *Cold-rolled grain oriented transformer grade* is approx. 3.2% silicon and is always graded at a flux density of 1.5 T B_{max} after final annealing. The grades sold are as follows:

BSI grade	*Max. guaranteed core loss B_{max} 1.5 T, frequency 50 Hz, thickness 0.32 mm (0.013 in)* W kg^{-1}
46	1.02
51	1.13
56	1.24

2. *Cold-rolled non-oriented grades* are of different silicon percentages, varying for each grade from 0.5% for those of highest core loss to approx. 3.0% silicon for the best grade with lowest core loss. The core loss test for these qualities is guaranteed at B_m 1.3 T in the UK but normally at B_m 1.0 T abroad. The grades sold are shown below:

BSI Grade	*Approx. Si content %*	*Max. guaranteed core loss* B_{max} 1.3 T, 50 Hz W kg^{-1}	*Thickness* mm (in)
B 100	3.3	2.20	0.35 (0.014)
B 107	3.0	2.36	0.35 (0.014)
A 146	2.5	3.23	0.50 (0.020)
A 170	2.2	3.76	0.50 (0.020)
A 187	2.0	4.14	0.50 (0.020)
A 216	1.0	4.78	0.50 (0.020)
A 253	0.5	5.60	0.50 (0.020)

3. *Hot-rolled non-oriented sheets* are still available in limited quantities with the same silicon percentage and approximately the same magnetic properties as the cold-rolled non-oriented grades but they are rapidly being replaced by the latter type and it is certain that in the near future there will no longer be any hot-mills rolling silicon steels by this method.

The properties of these steels are given in Table 20.6 in order of decreasing total losses.

Table 20.6 SILICON–IRON ALLOYS

Electrical steel grade	*Silicon approx. %*	*Initial relative permea-bility* μ_i	*Max. relative permea-bility* μ_m	*Saturation magnetiza-tion* B_S T	*Total loss* at B_{max} 1.3 T, 50 Hz W kg^{-1}	*Thickness* mm	in	*Coercive force* H_c A m^{-1}	*Specific gravity*	*Resis-tivity* μΩm
Ferrosil 253*	0.5	150	5 000	2.135	2.08	0.35	0.014	80	7.83	0.15
Losil 25*					2.23	0.40	0.016			
BSI A.253					2.38	0.45	0.018			
AISI M.50†					2.53	0.50	0.020			
					2.71	0.55	0.022			
					2.95	0.65	0.025			
Ferrosil 216*	0.7–1.0	250	5 800	2.110	1.76	0.35	0.014	72	7.77	0.15
Losil 22*					1.87	0.40	0.016			
BSI A.216					2.00	0.45	0.018			
AISI M.43†					2.16	0.50	0.020			
					2.32	0.55	0.022			
					2.56	0.65	0.025			
Ferrosil 187*	1.6–2.0	300	6 200	2.060	1.58	0.35	0.014	68	7.71	0.35
Losil 19*					1.68	0.40	0.016			
BSI A.187					1.77	0.45	0.018			
AISI M.36†					1.87	0.50	0.020			
					1.98	0.55	0.022			
					2.13	0.65	0.025			

* British steel Corporation
† USA Standard Grades.

Table 20.6 SILICON–IRON ALLOYS—*continued*

Electrical steel grade	*Silicon approx.* %	*Initial relative permeability* μ_i	*Max. relative permeability* μ_m	*Saturation magnetization* B_S T	*Total loss at* B_{max} 1.3T, 50 Hz W kg^{-1}	*Thickness* mm	in	*Coercive force* H_c A m^{-1}	*Specific gravity*	*Resistivity* μΩm
Ferrosil 170*	2.2	350	6 500	2.050	1.46	0.35	0.014	64	7.68	0.42
Losil 17*					1.54	0.40	0.016			
BSI A.170					1.62	0.45	0.018			
					1.70	0.50	0.020			
					1.78	0.55	0.022			
					1.91	0.65	0.025			
Ferrosil 146*	2.5–2.8	390	6 800	2.020	1.23	0.35	0.014	60	7.63	0.45
Losil 14*					1.33	0.40	0.016			
BSI A. 146					1.40	0.45	0.018			
AISI M. 27†					1.46	0.50	0.020			
					1.54	0.55	0.022			
					1.65	0.65	0.025			
Ferrosil 107*	3.5 hot-	400	7 000	1.985	2.36	0.35	0.014	52	7.62	0.52
Transil 107*	rolled,									
BSI B.107	3.0–3.5									
AISI M.22†	cold-									
	rolled									
Ferrosil 100*	4.0–4.5	450	8 000	1.935–	2.21	0.35	0.014	40	7.55	0.55
Transil 100*	hot-			1.985						
BSI B.100	rolled,									
AISI M.19†	3.0–3.5									
	cold-									
	rolled									
Ferrosil 92*	4.0–4.5	500	9 000	1.935	2.04	0.35	0.014	36	7.50	0.55
Transil 92*										
BSI C.92										
Ferrosil 86*	4.0–4.5	550	9 500	1.935	1.90	0.35	0.014	32	7.50	0.55
Transil 86*										
BSI C.86										
AISI M.17†										
Ferrosil 80*	4.0–4.5	580	10 000	1.935	1.77	0.35	0.014	24	7.50	0.55
Transil 80*										
BSI C.80										
AISI M.15†										
Ferrosil 74*	4.0–4.5	620	11 000	1.935	1.63	0.35	0.014	22	7.50	0.55
Transil 74*										
BSI C.74										
AISI M.14†										
					at B_{max} 1.5T, 50 Hz					
Anisotropic cold-rolled grain oriented types										
Alphasil 62*	3.0–3.25	1 000	37 000	1.975	1.37	0.32	0.013	12	7.65	0.49
Unisil 62*						0.32	0.013			
BSI grade 62						0.32	0.013			
AISI M.8†						0.35	0.014			
Alphasil 56*	3.0–3.25	1 200	46 000	1.975	1.24	0.32	0.013	9.6	7.65	0.49
Unisil 56*					1.24	0.32	0.013			
BSI grade 56					1.24	0.32	0.013			
AISI M.7†					1.22	0.35	0.014			

* British Steel Corporation.
† USA Standard Grades.

Table 20.6 SILICON–IRON ALLOYS—*continued*

Electrical steel grade	*Silicon approx. %*	*Initial relative permeability* μ_i	*Max. relative permeability* μ_m	*Saturation magnetization* B_S T	*Total loss at* B_{max} 1.3T, 50 Hz W kg	*Thickness* mm	in	*Coercive force* H_c A m^{-1}	*Specific gravity*	*Resistivity* μΩm
Alphasil 51*	3.0–3.25	1 500	50 000	1.975	1.13	0.32	0.013	8.0	7.65	0.49
Unisil 51*					1.13	0.32	0.013			
BSI grade 51					1.13	0.32	0.013			
AISI M.6†					1.10	0.35	0.014			
Alphasil 46*	3.0–3.25	1 800	63 500	1.975	1.02	0.32	0.013	6.4	7.65	0.49
Unisil 46*					1.02	0.32	0.013			
BSI grade 46					1.02	0.32	0.013			
AISI M.5†					0.97	0.35	0.014			

* British Steel Corporation.
† USA Standard Grades.

20.3.3 Ferrites

Ferrites are iron oxide compounds usually containing at least one other metallic ion. The magnetic iron oxides Fe_3O_4 and Fe_2O_3, both of which are used in magnetic recording, are also included with the ferrites. The most important of the many ferrite materials now known are sintered ceramics of high resistivity, together with saturation magnetizations up to 0.5T at room temperature. One structural class is the barium hexaferrites with high coercivities used as permanent magnets. The other two structural classes are the spinels and the garnets. The latter materials have the general formula $R_3Fe_5O_{12}$ where R is a rare earth, usually yttrium. These ferrites have low permeabilities and are used in specialized microwave devices having non-reciprocal transmission characteristics.[6]

The spinel-structured ferrites have the general composition MFe_2O_4 where M may be Ni, Mn, Mg, Zn, Cu or Co, or mixtures of these. Many of these may be prepared with high permeability, and by virtue of the high resistivity, which renders eddy current levels negligible, can be used at frequencies up to 10 or 20 MHz as the solid core of inductors or transformers. The losses which do occur are a combination of hysteresis, eddy current, and residual losses which may be separately controlled by composition and processing conditions with due regard to the permeability required and frequency of operation.

The saturation flux density of ferrites is low, making them unsuitable for use in power or high flux transformers. Their use is therefore almost entirely in the electronic and telecommunications industry where they are now largely replacing laminated alloy and powder cores.

Typical properties of the commonly employed manganese–zinc and nickel–zinc ferrites are given in Table 20.7.

Table 20.7 PROPERTIES OF 'INDUCTOR CORE' FERRITES

Type of ferrite	*Typical applications*	*Initial rel. permeability*	*Sat. flux density* T	*Curie temp.* °C	*Freq. range* MHz
Mn Zn	Inductors	1 500	0.38	140	<0.20
Mn Zn	Transformers	6 000	0.42	130	l.f. to 200
Mn Zn	Power transformers	1 500	0.45	220	<0.10
Mn Zn	Rod aerials	700	0.40	210	0.10–2.0
Mn Zn	Low loss inductors	2 200	0.40	190	<0.20
Ni Zn	Transformers	600	0.30	150	0.10–300
Ni Zn	Rod aerials	250	0.33	270	0.5–5.0
Ni Zn	Inductors	35–65	0.25	480	10–40
Ni Zn	Inductors	12–30	0.20	600	20–60
Ni Zn	Power transformers	70–150	0.38	370	2–30

20.3.4 High-permeability materials (nickel–irons, etc.)

The demands of the communications engineer for ferromagnetic core reactors, relays, transformers and loading coils for cables, having minimum size and other unique properties, have been largely met by the development of high permeability alloys of the nickel–iron type. The latent possibilities of these alloys, particularly their high permeabilities at low flux densities (despite their low saturation intensities), were indicated in work carried out as early as 1910. Systematic investigations followed on the nickel–iron alloys, and the effect of nickel on the permeability after an ordinary heat treatment was studied. This early work showed no particular advantage over the iron and iron–silicon alloys, but the ranges of composition over which good magnetic properties could be obtained were established, the optimum saturation indication being in the region of 50% nickel, 50% iron. This led to an intensive study of the 50% nickel, 50% iron alloy for applications in the electronic fields, but for small transformers and loading coils for telephone cables alloys with 70 to 80% nickel proved superior.

Materials in the 70–80% nickel range include the well known *Permalloy* '*C*', made with additions of molybdenum and copper. *Mu-metal* also belongs to this family of alloys and is widely used. Alloys in the nickel range in the region of 50% include the well-known *Radiometal* and *Permalloy B*. Use is also made, as in the case of the silicon–iron alloys, of the effects of preferred orientation produced by heavy cold rolling followed by suitable heat treatment. This gives rise to narrow rectangular hysteresis loops with very high slopes; the properties developed are particularly useful in the construction of reactors, for magnetic amplifiers, vibrating contact rectifiers, and for magnetic switching devices.

Alloys of iron and cobalt are unique and can give up to 13% higher magnetic saturation intensity than pure iron. One of these alloys, containing 50% cobalt, 50% iron, known as *Permendur*, has poor machinability, but with the addition of 2% vanadium is rendered mechanically workable. It is normally used in the form of cold-rolled sheet, finally annealed in hydrogen.

More recently further advances have been made in the improvement of high-permeability materials, mainly by the use of improved techniques of preparation (e.g. vacuum melting) and special heat treatments. The development of Supermalloy, with initial permeability of the order of 100 000, has been followed by Supermumetal, Special Radiometal, Supermendur and others. Production of such materials is, however, costly and on a restricted scale, their use being limited to special applications.

The properties of typical materials of the various types are listed in Table 20.8.

20.3.5 Constant-permeability materials

For certain applications in communication engineering, extreme constancy of effective magnetic permeability is imperative throughout the operating range of flux density. This is particularly important wherever a constant inductance is essential, such as in electrical filters, or in loading coils for telephone lines.

Useful materials for this purpose are the iron–nickel–cobalt alloys containing approximately 30% iron, 45% nickel and 25% cobalt, or alternatively 22.5% iron, 70% nickel and 7.5% cobalt, generally known as the *Perminvar* alloys. Similar but slightly inferior results are also obtained with an alloy containing 50% iron and 50% nickel. This alloy, when given a special heat treatment (i.e. incomplete annealing below 800 °C), is then known as *Conpernik*, the desired constancy is, however, obtained only by considerable sacrifice in value of permeability (*see* Table 20.9).

More recent developments in Germany have produced an iron–nickel series of alloys called *Isoperms*, which contains 40–55% nickel with some alloying element which tends to precipitate under certain conditions of cold-rolling and annealing. Addition elements which have been used with some success are aluminium, copper, beryllium, manganese and titanium, the best being a copper addition of 5–15%. After alternate cold-rolling and heat-treating operations the copper phase is precipitated in such a manner as to produce preferred domain orientation, which results in a low remanence and constancy of permeability (μ_{rel} = 50–60) over a range of magnetizing forces up to 8000 A m^{-1}.

These materials are listed in Table 20.9 in order of decreasing maximum permeability.

Table 20.8 HIGH-PERMEABILITY MATERIALS
(*Approximate typical data*)

Name	*Approximate composition*	*Initial relative permeability*	*Maximum relative permeability*	*Saturation induction* T	*Remanence, B=0.50* T	*Coercive force B=0.50* A m^{-1}	*Hysteresis loss B=0.50* J m^{-3}	*Curie temp.* °C	*Density* g m^{-3}	*Resistivity* μΩm	*Remarks*
Supermalloy, Supermumetal, Ultraperm, Hyperm Max etc.	70–80% Ni usually with small additions of Mo and/or Cu, etc., balance Fe	50 000–100 000	200 000–1 000 000	0.80	0.35–0.55	0.4–0.8	0.5–1.0	300–450	8.8	0.55–0.60	Special melting and/or processing. Heat treatment 1 100–1 300 °C, in dry hydrogen
Permalloy C, Mumetal, 4–79 Permalloy, Hymu 80 M. 1040 Hyperm, etc.		15 000–40 000	50 000–150 000	0.65–0.80	0.25–0.60	1.60–4.0	5–10	300–450	8.8	0.55–0.60	Heat treatment 1 050–1 100 °C, in dry hydrogen
Allegheny 4750, Hipernik, Special Radiometal	*ca.* 50% Ni, sometimes with small additions of Mo, Si, Cu, etc., balance Fe	4 000–4 500	40 000–70 000	1.60	0.40–0.80	4.0–12.0	5–10	400–500	8.2–8.3	0.40–0.45	Special production and/or heat treatment in hydrogen
Permalloy B, Radiometal, 45-Permalloy, Anhyster D, Hyperm 50, etc.	”	2 000–4 000	20 000–40 000	1.50–1.60	0.40–0.80	12–24	10–30	400–500	8.2–8.3	0.45–0.55	Heat treatment 1 000–1 100 °C, in hydrogen
HCR alloy, Orthonol, Deltamax, Hipernik V, Rectimphy, Permenorm 5 000Z, Hyperm 50T, etc.	”	500–1 000	50 000–100 000	1.60	95–98% of satn.	8–12 from satn.	30–70	475	8.2–8.3	0.40	Grain oriented by heavy cold reduction and special heat treatment in hydrogen
Permalloy F	*ca.* 65% Ni	400–2 000	200 000–400 000	1.40	1.33	4 from satn.	22	590	8.4	0.26	Domain oriented by heat treatment in magnetic field in hydrogen
Permalloy B, Rhometal, Anhyster B, Hyperm 36, etc.	*ca.* 35% Ni, sometimes with additions, balance Fe	*ca.* 2 000	*ca.* 7 000	0.90–1.30	0.35	12–24	40–70	180–270	8.1–8.2	0.70–0.90	Heat treatment, 1 000–1 100 °C, in hydrogen
Carbonyl iron Armco iron	99.9% Fe 99.5% Fe	3 000 250	30 000 7 000	2.05–2.15	1.30	8–64*	500*	770 770	7.85 7.85	0.12 0.11	Sintered
V-permendur Supermendur	49% Co, 49% Fe, 2% V 49% Co, 49% Fe, 2% V	700–1 000 —	3 000–6 000 100 000	2.30	1.60†	18–14.5†	1 200*	980 980	8.2 8.15	0.26 0.26	Annealed 800 °C Special melting and heat treatment in magnetic field

* $B=1.0$ † $B=2.0$

Mo-Perminvar	45 Ni, 25 Co, 7.5 Mo, 22.5 Fe	550	3800	1.03	260 ($B=1.2$T)	0.43	52	540	—	0.80	Baked
Perminvar	70 Ni, 7.5 Co, 22.5 Fe	750	3500	1.21	—	0.24	48	—	—	0.16	Annealed
Perminvar	43 Ni, 23 Co, 34 Fe	400	2000	1.55	250 ($B = 1.55$T)	0.30	46	715	—	0.19	Baked. Constant up to $B = 0.08$T. This alloy is anisotropic and if annealed in a magnetic field will give $\mu_{max} = 115000$, $H_c = 8$ A m^{-1}
Compernik	50 Ni, 50 Fe	175	1100	1.51	—	—	—	500	8.25	0.35	Severely cold-worked
Isoperm	40–55 Ni, 60–45 Fe	110	1800	—	—	—	520	—	—	—	Severely cold-worked. μ varies only 10% up to $H = 16000$ A m^{-1}
	36–50 Ni, 9–15 Cu and Fe	—	—								
	40–60 Ni, 3–4 Al and Fe	—	—								

20.3.6 Magnetic powder core materials

The thickness of laminations of ferromagnetic alloy cores must be reduced as the operating frequency is increased above audio range if the increasing loss and reducing permeability caused by eddy currents are to be avoided. However, reduction of thickness of laminations causes increased cost. This may be avoided for many applications up to 200 kHz by the use of magnetic powder cores.

The Permalloys are used in the form of compressed insulated powder cores, the effect of subdivision being to reduce eddy current losses to a minimum. The permeability is also reduced but is made extremely constant, becoming within a small percentage independent of magnetizing force and unaffected by magnetic shock. The properties of such materials are in many ways similar to the constant permeability alloys described in the preceding section. *Sendust*, a brittle alloy of aluminium, silicon and iron, has been similarly used in Japan and Germany. Carbonyl iron powder is used for powder cores mainly at radio frequencies, specially low eddy current losses being obtained on account of the characteristic fine particle size. Some details of these materials are given in Table 20.10.

Table 20.10 MAGNETIC POWDER CORE MATERIALS

			Loss factors $\times 10^6$*			
Material	*Composition*	*Effective initial relative permeability* μ_e	*Eddy current* E	*Hysteresis* $A \times 10^4$	*Residual* C	*Remarks*
Permalloy C	Ni–Fe–Mo	125	2.0	200	4 000	Compressed at high pressures: particles coated with ceramic insulating binder and core heat treated after pressing
Permalloy C	Ni–Fe–Mo	14	0.10	140	2 000	Ditto: particles coated with increased proportion of binder
Carbonyl iron, C-type	Fe 99.95%	50	0.10	500	10 000	High pressure cores
Sendust	Al–Si–Fe	60	0.20	350	12 000	High pressure cores
Carbonyl iron, E-type	Fe 99.0% C 0.6%	12	0.002	50	2 500	Low pressure cores, insulated and bonded with bakelite, etc.
Hydrogen reduced iron	Fe 99.5%	20	0.06	1 000	15 000	Low pressure cores, insulated and bonded with bakelite, etc.

* These are expressed according to the conventional method used in communication engineering, e.g. effective resistance due to core losses $R_e = fL(Ef + AB + C)$.

Where f = frequency in Hz. L = inductance in henries and B = flux density in tesla.

(This analysis is applicable only at low flux densities where for most materials hysteresis loss resistance is proportional to flux density.)

For calculation of loss we have:

Total loss in W cm^{-3} $5B^2R_e/4\pi L$

20.3.7 Magnetic temperature-compensating materials

Alloys having a marked variation of permeability with temperature are used to compensate for errors in electrical instruments caused by changes in the ambient temperature.

The magnetic permeability of iron and iron alloys decreases with temperature (except for low values of the field), and this causes the readings of electrical meters which depend upon the fluxes maintained by a constant voltage or current or by a permanent magnet, to decrease with the ambient temperature. To compensate for such errors it is customary to shunt a certain proportion of the magnetic flux passing through the magnetic circuit of the meter by means of an alloy having an unusually high negative temperature coefficient between 0 and 100 °C, so that, as the ambient temperature increases, the proportion of the shunted flux decreases, causing more flux to pass through the working air gap of the instrument than otherwise would be the case. By correctly proportioning the shunt component it is possible to render the readings almost completely independent of ambient temperature changes.

There are two series of alloys of different composition which may be used for this purpose. First, there are the alloys in the nickel–iron series having approximately 30% nickel, sometimes with additions of chromium, manganese and/or silicon. Secondly, use may be made of the nickel–copper series in the region of 70% nickel, sometimes with small additions of iron and/or manganese. The temperature characteristics of the latter alloys are not so linear and the temperature coefficients are smaller than in the case of the nickel–iron alloys. Details of some alloys are given in Table 20.11.

Table 20.11 THERMOMAGNETIC TEMPERATURE-COMPENSATING MATERIALS

		Magnetic flux density at $H = 8 \times 10^3$ A m^{-1} T					
		Temperature °C					
Material	*Type*	−20	0	20	40	60	*Notes*
R.T.B. Mutemp	Nickel–iron	0.53	0.43	0.31	0.15	—	British
Telcon R 2799 Alloy	Nickel–iron	—	0.42*	0.25*	0.10*	—	British
Carpenter temp. compensator 30 Type 1	Nickel–iron	0.96*	0.89*	0.82*	0.74*	0.68*	American
Hoskin's 567 alloy	Nickel–iron	—	—	—	—	—	American
NHMG alloy	Nickel–iron	—	—	—	—	—	French
Thermoperm	Nickel–iron	0.46	0.35	0.24	0.13	0.03	German
Jae metal	Nickel–copper	0.19	0.17	0.13	0.05	—	British
Calmaloy	Nickel–copper	—	—	—	—	—	American (originally called 'Thermalloy')

* At $H = 4 \times 10^3$ A m^{-1}.

20.3.8 Non-magnetic steels and cast irons

For certain constructional purposes in electrical engineering where magnetic shunting effects have to be avoided there arises a requirement for almost non-magnetic materials possessing certain minimum mechanical properties.

The materials available comprise non-magnetic cast irons, austenitic steels, and certain other sensibly non-magnetic alloys; the choice of materials for any particular purpose is dependent upon the mechanical properties required; e.g. certain intricate stationary parts of electrical machinery would be suitably made from non-magnetic cast iron, while armature straps for high-speed rotors would be made in a high-tensile austenitic alloy steel in order to withstand the dynamic stresses encountered in practice. A high resistivity is sometimes beneficial as well as the possession of virtually non-magnetic properties.

There are two well-known non-magnetic cast irons, one called *Nomag*, a manganese–silicon–nickel–iron with a permeability of 1.03, and a resistivity of 140 microhm cm, and the other *Ni-Resist*, containing lower proportions of manganese and silicon, but a higher proportion of nickel, also a little chromium and copper. These irons with flake graphite are brittle, and can only be used for parts of a simple nature, such as supports which are not subjected to severe mechanical stress. These non-magnetic irons are also made in the spheroidal graphite form e.g. *Nodumag* and *S.G. Ni-Resist*, having the characteristically improved mechanical properties of this type of cast iron.

As examples of austenitic steels, Hadfields 13% manganese steel and the chromium–nickel steels have found considerable use. The stainless steels of the chromium–nickel type are normally non-magnetic but if they are subjected to cold working appreciable magnetic properties may appear. This disadvantage can be overcome by using steels with higher alloy content than the well-known 18/8 type. Besides these materials other non-magnetic austenitic steels have been specially developed to give improved mechanical properties.

Table 20.12 gives details of the non-magnetic cast irons and a number of non-magnetic steels.

Table 20.12 NON-MAGNETIC STEELS AND CAST IRONS

Material	Nominal composition	Relative permeability	Resistivity μΩm	Condition	Mechanical properties: Hardness HB	Proof stress MPa	Max stress MPa	Elongation $L = 4\sqrt{A}$ %	Reduction in area %
Steels									
18/8 Stainless	8% Ni, 18% Cr and various minor constituents	1.005/ 1.03	0.72	Austenized	170	278	618	45	50
18/8 Stainless	8% Ni, 18% Cr and various minor constituents	1.4	—	20% cold reduction	280	—	927	25	—
18/8 Stainless	8% Ni, 18% Cr and various minor constituents	about 10.0	—	50% cold reduction	330	—	1 313	10	—
18/12 Stainless	12% Ni, 18% Cr etc	1.003	0.80	Austenized	160	232	618	60	65
18/12 Stainless	12% Ni, 18% Cr etc.	1.1	—	50% cold reduction	—	—	1 313	—	—
18/12 Stainless	12% Ni, 18% Cr etc.	1.7	—	90% cold reduction	—	—	—	—	—
25/12 Stainless	12% Ni, 25% Cr etc.	1.003	0.90	Austenized	200	309	695	30	35
25/12 Stainless	12% Ni, 25% Cr etc.	1.005	—	90% cold reduction	—	—	—	—	—
Hadfields Mn	13% Mn, 1% Si	1.03	0.71	Water quenched 1 000 °C	—	—	—	—	—
Clyde alloy non-magnetic	8% Ni, 8% Mn, 8% Cr	1.003	0.76	Water quenched 900 °C	—	—	—	—	—
Clyde alloy non-magnetic	8% Ni, 4% Mn, 8% Cr	1.003	—	Water quenched 900 °C	—	571	927	45	45
Firth Vickers NMC	10% Ni, 5% Mn, 4% Cr	1.008/ 1.03	—	—	—	618	927	45	45
Cast irons									
Nomag	7% Mn, 11% Ni, 2.5% Si, 2.7% C	1.03	1.40	Stress relieved	120/150	—	~185	0	—
Nodumag	7% Mn, 11% Ni, 2.5% Si, 2.7% C	1.03	1.10	Stress relieved	120/150	309	432	15	—
Ni-Resist Type 2	Ni 20%, Si 2.5%	1.03	1.70	Stress relieved	150	—	185	0	—
S.G. Ni-Resist Type D-2	Mn 1%, Cr 2%, C 2%	1.03	1.00	Stress relieved	140	232	417	15	—

REFERENCES

1. L. F. Bates, 'Modern Magnetism', 3rd Edn, Cambridge University Press, 1951.
2. F. Brailsford, 'Physical Principles of Magnetism', Van Nostrand, New York, 1966.
3. T. D. Yensen, *Trans. AIEE*, 1915, **34**, 2601.
4. P. Oberhoffer and C. Kreutzer, *Arch. Eisenhüttenw.*, 1928/9 (2), 449.
5. H. J. Williams, *Phys. Rev.*, 1937, **52**, 747.
6. B. Lax and K. J. Button, *Microwave Ferrites and Ferrimagnetics*, McGraw-Hill, New York, 1962.

21 Mechanical testing

21.1 Hardness testing

21.1.1 Brinell hardness

An indenter comprising a hardened steel ball of diameter D mounted in a suitable holder is forced into the material under test under a load F. The diameter of the indentation left in the surface of the material after removal of the load is measured in two directions at right angles. The area of the curved surface of the identation is calculated from the mean diameter, d, the indentation being considered as a segment of a sphere of diameter D. The Brinell hardness is the quotient obtained by dividing the load F, expressed in kilograms-force, by the surface area of the indentation expressed in square millimetres.

Symbols:

F = load in kilograms force (kgf)
D = diameter of ball in millimetres (mm)
d = mean diameter of indentation in millimetres (mm)

$$\text{HB} = \text{Brinell Hardness} = \frac{2F}{\pi D[D - \sqrt{(D^2 - d^2)}]}$$

$$h = \text{depth of indentation in millimetres} = \frac{F}{\pi D(HB)} \text{(mm)}$$

The symbol HB is supplemented by numbers indicating the diameter of the ball used and the load applied. Thus 226 HB 10/3000 indicates that a Brinell hardness of 226 was obtained by using a 10 mm diameter ball with a load of 3000 kgf. If the time of duration of load differs from the standard, a further number is added to show the duration of the load in seconds.

RELATION OF LOAD TO BALL DIAMETER

The choice of load and ball diameter to be used in a Brinell test is determined by two factors:

1. the value of the ratio F/D^2, and
2. the size of the indentation which provides optimum accuracy.

The same value of F/D^2 will, in principle, give the same hardness value for different loads and the load used in practice depends on the nature and the hardness of the material under test. Four standard values of F/D^2, i.e. 30, 10, 5 and 1, have been adopted in the UK. The following combinations of load and diameter are given in BS 240: Part 1: 1962.

Table 21.1 CORRELATED VALUES OF LOAD AND BALL DIAMETER FROM BS 240: PART 1: 1962

Diameter of ball	*Load*							
	$F/D^2=1$		$F/D^2=5$		$F/D^2=10$		$F/D^2=30$	
mm	N	kgf	N	kgf	N	kgf	N	kgf
1	9.80	1	49.03	5	98.01	10	294.2	30
2	39.23	4	196.13	20	392.27	40	1 176.8	120
5	245.17	25	245.17	25	2 451.7	250	7 355	750
10	980.67	100	4 903.3	500	9 806.7	1 000	29 420	3 000

Notes: (1) The specified duration of load is 10–15 seconds.
(2) The maximum depth of indentation shall not exceed $\frac{1}{8}$ of the thickness of the test piece, according to BS 240: Part 1:1962 or $\frac{1}{10}$ of the thickness according to ASTM E10–78.
(3) BS 240 gives values in kgf. Newton values are conversions.
(4) In the USA Specification ASTM E10–78, only F/D^2 values of 30, 15 and 5 with a 10 mm ball are regarded as standard.

21.1.2 Rockwell hardness

An indenter of the standard type comprising a diamond cone or hardened steel ball mounted rigidly in a suitable holder, is forced into the test piece under a preliminary load F_0. When equilibrium has been reached, an indicating device which follows the movement of the indenter and so responds to changes in depth of penetration of the indenter is set to a datum position.

While the preliminary load is still applied, it is augmented by an additional load with resulting increase in penetration of the indenter. When equilibrium has again been reached, the additional load is removed but the preliminary load is maintained.

Removal of the additional load allows a partial recovery, so reducing the depth of penetration. The permanent increase in depth of penetration, e, resulting from application and removal of the additional load is used to deduce the Rockwell hardness number by means of the equation:

$$\mathrm{HR}=E-e$$

where:

HR = Rockwell hardness number
E = a constant depending on the form of indenter:
100 units when a diamond cone indenter is used
130 units when a steel ball indenter is used

A variety of indenters and loads are used to give several scales of hardness, e.g. 60HRC represents a Rockwell hardness of 60 on scale C. The various scales are defined in Table 21.2.

Table 21.2 ROCKWELL HARDNESS SCALES

Scale symbol		*Major load* kN	kgf	*Dial figures*	*Typical applications of scales*
B	$\frac{1}{16}$ in (1.588 mm) ball	0.98	100	Red	Copper alloys, soft steels, aluminium alloys, malleable iron, etc.
C	Diamond	1.47	150	Black	Steel, hard cast irons, pearlitic malleable iron, titanium, deep case hardened steel and other materials harder than HRB100
A	Diamond	0.59	60	Black	Cemented carbides, thin steel, and shallow case-hardened steel
D	Diamond	0.98	100	Black	Thin steel and medium case hardened steel, and pearlitic malleable iron
E	$\frac{1}{8}$ in (3.175 mm) ball	0.98	100	Red	Cast iron, aluminium and magnesium alloys, bearing metals
F	$\frac{1}{16}$ in (1.588 mm) ball	0.59	60	Red	Annealed copper alloys, thin soft sheet metals

Table 21.2 ROCKWELL HARDNESS SCALES—*continued*

Scale symbol		*Major load* kN	kgf	*Dial figures*	*Typical applications of scales*
G	$\frac{1}{16}$ in (1.588 mm) ball	1.47	150	Red	Malleable irons, copper–nickel–zinc and cupro-nickel alloys. Upper limit HRG 92 to avoid possible flattening of ball
H	$\frac{1}{8}$ in (3.175 mm) ball	0.59	60	Red	Aluminium, zinc, lead
K	$\frac{1}{8}$ in (3.175 mm) ball	1.47	150	Red	
L	$\frac{1}{4}$ in (6.350 mm) ball	0.59	60	Red	
M	$\frac{1}{4}$ in (6.350 mm) ball	0.98	100	Red	Bearing metals and other very soft or thin materials. Use smallest ball and heaviest load that does not give anvil effect
P	$\frac{1}{4}$ in (6.350 mm) ball	1.47	150	Red	
R	$\frac{1}{2}$ in (12.70 mm) ball	0.59	60	Red	
S	$\frac{1}{2}$ in (12.70 mm) ball	0.98	100	Red	
V	$\frac{1}{2}$ in (12.70 mm) ball	1.47	150	Red	

Notes:

(1) The preliminary load is 10 kgf for all scales.
(2) The diamond indenter is conical with an included angle of $120 \pm 0.1°$ with a tip rounded to a radius of 0.20 mm.
(3) ASTM E18–79 standard gives values in kgf. Newton values are conversions.
Relevant standards are BS 891 : Part 1 : 1962 in the UK and ASTM E18–79 in the USA.

21.1.3 Rockwell superficial hardness

This test is similar in principle to the Rockwell test, but in order to keep the depth of impression small the minor load is restricted to 3 kg. It is used for thin specimens and for nitrided steels, etc. The following scales are used.

Table 21.3 ROCKWELL SUPERFICIAL HARDNESS SCALES

Scale	*Penetrator*	*Minor load* kgf	N	*Major load* kgf	kN
15-N	Diamond cone	3	29.4	15	0.14
30-N	Diamond cone	3	29.4	30	0.29
45-N	Diamond cone	3	29.4	45	0.44
15-T	$\frac{1}{16}$ in (1.588 mm) steel ball	3	29.4	15	0.14
30-T	$\frac{1}{16}$ in (1.588 mm) steel ball	3	29.4	30	0.29
45-T	$\frac{1}{16}$ in (1.588 mm) steel ball	3	29.4	45	0.44

Relevant standards are BS:4175 Part 1:1967 in the UK and ASTM E18–79 in the USA.

21.1.4 Vickers hardness test

A diamond indenter, in the form of a right pyramid with a square base and an angle of 136° between opposite faces, is forced into the material under a load F. The two diagonals, d_1 and d_2 of the indentation left in the surface of the material after removal of the load are measured and their arithmetic mean d calculated. The area of the sloping surface of the indentation is calculated, the indentation being considered as a right pyramid with a square base of diagonal d and vertex angle of 136°.

The Vickers hardness is the quotient obtained by dividing the load F, expressed in kilograms-force, by the sloping area of the indentation expressed in square millimetres.

Symbols:
F = load in kilograms force (kgf) where 1 kgf = 9.806 65 N
d = arithmetic mean of the two diagonals d_1 and d_2 in millimetres (mm)
HV = Vickers hardness
$= 2F \sin (136°/2) d^2$
$= 1.854F/d^2$

The loads employed vary from 1 to 100 kgf and are maintained from 10 to 15 seconds. Hardness numbers are expressed as 440 HV30 for example where a hardness of 440 is measured using a 30 kgf load.

Relevant standards are BS 427: Part 1:1961 in the UK and ASTM E92–72 in the USA.

21.1.5 Micro-hardness testing

Standards ASTM E384–73

A diamond indenter of specified geometry is forced into the surface of the test piece using a calibrated machine with a test load of 1–1 000 gf to give a micro-indentation, the diagonals of which are subsequently measured by means of a microscope.

Symbols:
HK $= 14\,229\ P_1/d_1^2$
where HK = Knoop hardness
P_1 = load in gf
and d_1 = length of long diagonal of indentation in μm
HV $= 1854.4\ P_1/d_1^2$
where HV = Vickers hardness
P_1 = load in gf
and d_1 = mean length of the diagonals of the indentation in μm

Indenters
Knoop: A highly polished, pointed, rhombic-based pyramidal diamond with included longitudinal edge angles of 172.5° and 130°.
Vickers: A highly polished, pointed, square-based pyramidal diamond with face angles of 136°.

21.1.6 Hardness conversion tables

Table 21.4 only applies to steel of uniform chemical composition and uniform heat treatment, and is not recommended for non-ferrous metals or for case-hardened steels.

Table 21.4 APPROXIMATE CONVERSION OF HARDNESS VALUES
Steels (ASTM E140)

		Rockwell hardness No.			
Vickers hardness No. HV	*Brinell* 3000 kg *load* 10 mm *ball* HB	*C Scale* 150 kg *load diamond cone* HRC	*A scale* 60 kg *load diamond cone* HRA	*Superficial* 30–*N* 10 kg *load diamond cone*	*Scleroscope hardness No.*
100	95	—	43	—	—
120	115	—	46	—	—
140	135	—	50	—	21
160	155	—	53	—	24
180	175	—	56	—	27
200	195	—	58	—	30
220	215	—	60	—	31
240	235	20.3	60.7	41.7	34
260	255	24.0	62.4	45.0	37
280	275	27.1	63.8	47.8	40
300	295	29.8	65.2	50.2	42
320	311	32.2	66.4	52.3	45
340	328	34.4	67.6	54.4	47
360	345	36.6	68.7	56.4	50
380	360	38.8	69.8	58.4	52
400	379	40.8	70.8	60.2	55
420	397	42.7	71.8	61.4	57
440	415	44.5	72.8	63.5	59
460	433	46.1	73.6	64.9	62
480	452	47.7	74.5	66.4	64

Table 21.4 APPROXIMATE CONVERSION OF HARDNESS VALUES—*continued*
Steels (ASTM E140)

		Rockwell hardness No.			
Vickers hardness No. HV	*Brinell** 3 000 kg *load* 10 mm *ball* HB	*C Scale* 150 kg *load diamond cone* HRC	*A scale* 60 kg *load diamond cone* HRA	*Superficial* 30–*N* 10 kg *load diamond cone*	*Scleroscope hardness No.*
500	471	49.1	75.3	67.7	66
520	487	50.5	76.1	69.0	67
540	507	51.7	76.7	70.0	69
560	525	53.0	77.4	71.2	71
580	545	54.1	78.0	72.1	72
600	564	55.2	78.6	73.2	74
620	582	56.3	79.2	74.2	75
640	601	57.3	79.8	75.1	77
660	620	58.3	80.3	75.9	79
680	638	59.2	80.8	76.8	80
700	656	60.1	81.3	77.6	81
720	670	61.0	81.8	78.4	83
740	684	61.8	82.2	79.1	84
760	698	62.5	82.6	79.7	86
780	710	63.3	83.0	80.4	87
800	722	64.0	83.4	81.1	88
820	733	64.7	83.8	81.7	90
840	745	65.3	84.1	82.2	91
860	—	65.9	84.4	82.7	92
880	—	66.4	84.7	83.1	93
900	—	67.0	85.0	83.6	95
920	—	67.5	85.3	84.0	96
940	—	68.0	85.6	84.4	97

* Brinell values greater then 480 are determined with a carbide ball.

Table 21.5 APPROXIMATE CONVERSION OF HARDNESS VALUES
Aluminium and its alloys

		Rockwell superficial HR					*Rockwell superficial* HR		
HV 10 kg	*Brinell** HB	15*T*	30*T*	45*T*	HV 10 kg	*Brinell** HB	15*T*	30*T*	45*T*
20	19.4	2.3	—	—	80	74.6	75.3	44.1	—
24	23.1	18.1	—	—	84	78.3	76.5	46.9	—
28	26.8	29.5	—	—	88	82.0	77.6	49.4	—
32	30.5	38.2	—	—	92	85.7	78.6	51.8	—
36	34.2	45.0	—	—	96	89.4	79.5	54.0	—
40	37.8	50.3	—	—	100	93.0	80.3	55.9	—
44	41.5	54.8	—	—	104	96.7	81.1	57.7	—
48	45.2	58.6	5.2	—	108	100.4	81.8	59.4	—
52	48.9	61.8	12.6	—	112	104.1	82.5	60.9	—
56	52.6	64.5	19.1	—	116	107.8	83.1	62.4	—
60	56.2	66.9	24.5	—	120	111.4	83.7	63.8	39.0
64	59.9	69.0	29.4	—	124	115.1	84.3	65.0	41.2
68	63.6	70.8	33.7	—	128	118.8	84.8	66.2	43.3
72	67.3	72.5	37.6	—	132	122.5	85.3	67.4	45.2
76	71.0	74.0	41.0	—	136	126.2	85.7	68.5	47.1

* The ratio $\frac{\text{load}}{(\text{dia. of ball})^2}$ for the Brinell tests was 5 for Brinell hardnesses of 20–60 and 10 for hardnesses above 60.

Table 21.5 APPROXIMATE CONVERSION OF HARDNESS VALUES—*continued*
Aluminium and its alloys

HV 10 kg	*Brinell** HB	*Rockwell superficial* HR 15*T*	30*T*	45*T*	HV 10 kg	*Brinell** HB	*Rockwell superficial* HR 15*T*	30*T*	45*T*
140	129.8	86.2	69.4	48.8	172	159.3	88.3	75.8	59.6
144	133.5	86.6	70.4	50.4	176	163.0	89.2	76.4	60.7
148	137.2	86.9	71.3	51.9	180	166.6	89.4	77.0	61.7
152	140.9	87.3	72.1	53.3	184	170.3	89.7	77.6	62.7
156	144.6	87.6	72.9	54.7	188	174.0	89.9	78.1	63.6
160	148.2	88.0	73.7	56.0	192	177.7	90.1	78.7	64.5
164	151.9	88.3	74.4	57.3	196	181.4	90.3	79.2	65.4
168	155.6	88.6	75.1	58.4	200	185.0	90.6	79.7	66.2

*The ratio $\frac{\text{load}}{(\text{dia. of ball})^2}$ for the Brinell tests was 5 for Brinell hardnesses of 20–60 and 10 for hardnesses above 60.

Table 21.6 APPROXIMATE CONVERSION OF HARDNESS VALUES
Brass (ASTM E140–65)

HV	*Rockwell hardness No.* *B scale, 100 kg load, $\frac{1}{16}$ in ball*	*F scale, 60 kg load, $\frac{1}{16}$ in ball*	*Rockwell superficial hardness No.* 15-*T scale, 15 kg load, $\frac{1}{16}$ in ball*	30-*T scale, 30 kg load, $\frac{1}{16}$ in ball*	45-*T scale, 45 kg load, $\frac{1}{16}$ in ball*	*Brinell hardness No.* 500 kg *load,* 10 mm *ball*	HV	*Rockwell hardness No.* *B scale, 100 kg load, $\frac{1}{16}$ in ball*	*F scale, 60 kg load, $\frac{1}{16}$ in ball*	*Rockwell superficial hardness No.* 15-*T scale, 15 kg load, $\frac{1}{16}$ in ball*	30-*T scale, 30 kg load, $\frac{1}{16}$ in ball*	45-*T scale, 45 kg load, $\frac{1}{16}$ in ball*	*Brinell hardness No.* 500 kg *load,* 10 mm *ball*
45	—	40.0	—	—	—	42	120	67.0	95.5	—	61.0	41.0	106
47	—	45.0	—	—	—	44	124	69.0	96.5	—	62.5	43.0	110
49	—	49.0	54.5	—	—	46	128	71.0	97.5	—	63.5	45.0	113
52	—	53.5	57.0	—	—	48	132	73.0	98.5	84.5	65.0	46.5	116
56	—	58.8	60.0	15.0	—	52	136	74.5	99.5	85.0	66.0	48.0	120
60	10.0	63.0	62.5	20.5	—	55	140	76.0	100.5	85.5	67.0	50.0	122
64	15.5	66.8	65.0	22.5	—	59	144	77.8	101.5	86.0	68.0	51.5	126
68	21.5	70.0	67.0	30.0	—	62	148	79.0	102.5	—	69.0	53.0	129
72	27.5	73.2	69.0	34.0	—	64	152	80.5	103.0	—	—	54.0	133
76	32.5	76.0	70.5	38.0	4.5	68	156	82.0	104.0	87.0	70.5	55.5	136
80	37.5	78.6	72.0	41.0	10.0	72	160	83.5	—	—	71.5	56.5	139
84	42.0	81.2	73.5	44.0	14.5	76	164	85.0	105.5	—	72.0	58.0	142
88	46.0	83.5	75.0	47.0	19.0	79	168	86.0	106.0	88.0	73.0	59.0	146
92	49.5	85.4	76.5	49.0	23.0	82	172	87.5	106.5	—	73.5	60.0	149
96	53.0	87.2	77.5	51.5	26.5	85	176	88.5	107.0	—	—	61.0	152
100	56.0	89.0	78.5	53.5	29.5	88	180	90.0	107.5	—	75.0	62.0	156
104	58.0	90.5	79.5	55.0	32.0	92	184	91.0	—	—	75.5	63.0	159
108	61.0	92.0	—	57.0	34.5	95	188	92.0	—	89.5	—	64.0	162
112	63.0	93.0	81.0	58.5	37.0	99	192	93.0	—	—	77.0	65.0	166
116	65.0	94.5	82.0	60.0	39.0	103	196	93.5	110.0	90.0	77.5	66.0	169

Note: $\frac{1}{16}$ in ≡ 1.588 mm.

Table 21.7 APPROXIMATE CONVERSION OF HARDNESS VALUES
Hard metals

Vickers diamond hardness, 50 kg *load*	*Rockwell A scale* 60 kg *load: diamond cone*	*Rockwell C scale* 150 kg *load: diamond cone*
1 750	92.4	80.5
1 700	92.0	79.8
1 650	91.7	79.2
1 600	91.3	78.4
1 550	90.9	77.7
1 500	90.5	77.0
1 450	90.1	76.2
1 400	89.7	75.4
1 350	89.3	74.6
1 300	88.9	73.8
1 250	88.5	73.0
1 200	88.1	72.2
1 150	87.6	71.3
1 100	87.0	70.4
1 050	86.4	69.4
1 000	85.7	68.2
950	85.0	66.6
900	84.0	64.6
850	82.8	—

Table 21.8 APPROXIMATE CONVERSION OF HARDNESS VALUES FOR NICKEL AND HIGH NICKEL ALLOYS

Vickers hardness No. HV	*Brinell hardness No.* HB	*Rockwell hardness no.*		
		A. scale	*B scale*	*C scale*
Vickers indenter 1, 5, 10 *and* 30 kgf	10 mm *ball* 3 000 kgf	*Diamond penetrator* 60 kgf	$\frac{1}{16}$ in *ball* 100 kgf	*Diamond penetrator* 150 kgf
77	77	—	30	—
79	79	—	34	—
83	83	—	38	—
87	87	—	42	—
91	91	—	46	—
95	95	—	50	—
100	100	—	54	—
106	106	—	58	—
112	111	39.0	62	—
119	118	41.0	66	—
126	125	43.0	70	—
135	134	45.5	74	—
145	144	47.5	78	—
157	155	50.0	82	—
171	168	52.5	86	—
188	184	55.0	90	—
209	204	57.5	94	—
234	228	60.5	98	20.0
248	241	61.5	100	22.5
285	275	64.5	—	28.5
326	313	67.5	—	34.0
362	346	69.5	—	38.0
404	382	71.5	—	42.0
452	425	73.5	—	46.0
481	450	74.5	—	48.0
513	—	75.5	—	50.0

Note: $\frac{1}{16}$ in ≡ 1.588 mm.

21.2 Tensile testing

21.2.1 Standard test pieces

Table 21.9 DIMENSIONS OF CIRCULAR SECTION TEST PIECES—BRITISH STANDARDS BS 18 : PART 2 : 1971
Gauge length $L_0 = 5.65\sqrt{S_0}$

Cross-sectional area S_0 mm^2	*Diameter d* mm	*Gauge length* L_0 mm	*Minimum parallel length* L_c mm	*Minimum transition radius r* mm
400	22.56	113	124	23.5
200	15.96	80	88	15
150	13.82	69	76	13
100	11.28	56	62	10
50	7.98	40	44	8
25	5.64	28	31	5
12.5	3.99	20	22	4

Notes:

(1) 400 mm^2 cross-sectional area specified in BS 18: Part 2: 1971 for steel specimens. The remaining specimens are common to BS 18: Part 1: 1970 (for non-ferrous specimens) and BS 18: Part 2: 1971.
(2) The minimum transition radius *r* is doubled for cast non-ferrous specimens.

Table 21.10 DIMENSIONS OF CIRCULAR SECTION TEST PIECES—AMERICAN STANDARDS ASTM E8–78
Gauge Length $L_0 = 4d$

Diameter D mm	*Gauge length* L_0 mm	*Minimum parallel length* L_c mm	*Minimum transition radius r* mm
12.50	50	60	10
8.75	35	45	6
6.25	25	32	5
4.00	16	20	4
2.50	10	16	2

Table 21.11 DIMENSIONS OF RECTANGULAR SECTION TEST PIECES–BRITISH STANDARDS BS 18: 1970

Width b mm	*Gauge length* L_0 mm	*Minimum parallel length* L_c mm	*Minimum transition radius r* mm	*Approximate total length* L_t mm
25	100	125	25	300
12.5	50	63	25	200
6	24	30	12	100
3	12	15	6	50

Table 21.12 DIMENSIONS OF RECTANGULAR SECTION TEST PIECES–AMERICAN STANDARDS ASTM E8–78

Width b mm	*Gauge length* L_0 mm	*Minimum parallel length* L_c mm	*Minimum transition radius r* mm	*Approximate total length* L_t mm
40	200	225	25	450
12.5	50	60	13	200
6.25	25	32	6	100

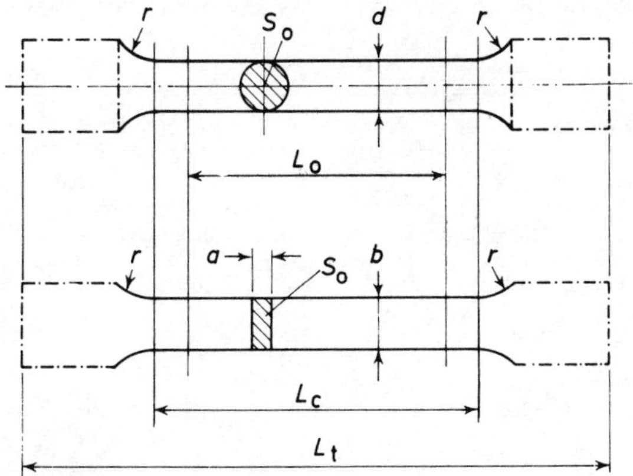

Figure 21.1 *Test pieces of circular and rectangular cross-section*
BS 18: Part 2: 1971

Table 21.13 STANDARD TENSILE TEST PIECES FOR CAST IRON–BRITISH STANDARDS BS: 1452: 1977

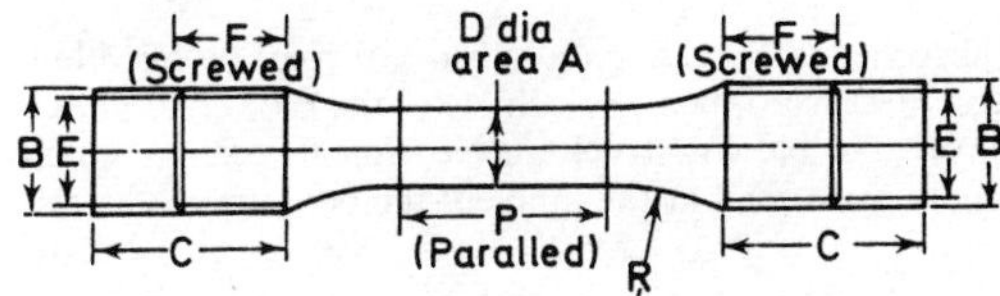

Dimensions of machined tensile test piece

Gauge diameter D		*Minimum parallel length*	*Minimum radius R*		*Plain ends*		*Screwed ends*	
Nominal value	*Machining tolerance**	*P*	*Nominal value*	*Machining tolerance*	*Minimum diameter B*	*Minimum length C*	*Minimum diameter at root of thread E*	*Minimum length F*
mm	mm	mm	mm	mm	mm	mm	mm	mm
20	±0.5	55	25	−0, +5	25	65	25	30

* If it is desired to calculate the tensile strength on the basis of the nominal diameter, the machining tolerance shall be 0.10 mm.

Note: With screwed ends, any form of thread may be used provided that the diameter at the root of the thread is not less than that specified.

Table 21.14 STANDARD TENSILE TEST PIECES FOR CAST IRON–AMERICAN STANDARDS ASTM E8–78

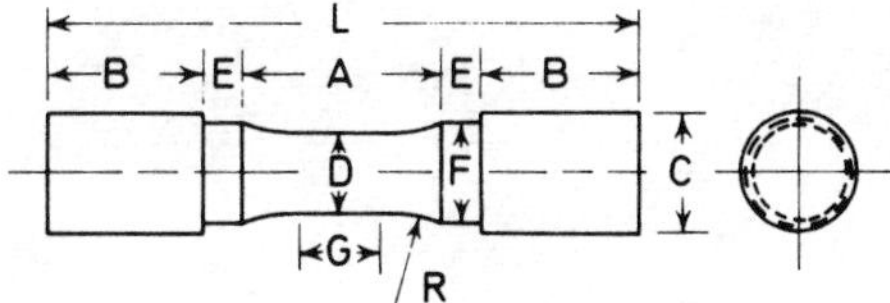

Table 21.14 STANDARD TENSILE TEST PIECES FOR CAST IRON–AMERICAN STANDARDS ASTM E8–78—*continued*

	Dimensions					
	Specimen 1		*Specimen 2*		*Specimen 3*	
	in	mm	in	mm	in	mm
G Length of parallel	Shall be equal to or greater than diameter *D*					
D Diameter	0.500± 0.010	12.5± 0.25	0.750± 0.015	20.0± 0.40	1.25± 0.025	30.0± 0.60
R Radius of fillet, min.	1	25	1	25	2	50
A Length of reduced section, min.	$1\frac{1}{4}$	32	$1\frac{1}{2}$	38	$2\frac{1}{4}$	60
L Overall length, min.	$3\frac{3}{4}$	95	4	100	$6\frac{3}{8}$	160
B Length of end section, approximate	1	25	1	25	$1\frac{3}{4}$	45
C Diameter of end section, approximate	$\frac{3}{4}$	20	$1\frac{1}{8}$	30	$1\frac{7}{8}$	48
E Length of shoulder, min.	$\frac{1}{4}$	6	$\frac{1}{4}$	6	$\frac{5}{16}$	8
F Diameter of shoulder	$\frac{5}{8}\pm\frac{1}{64}$	16.0± 0.40	$\frac{15}{16}\pm\frac{1}{64}$	24.0± 0.40	$1\frac{7}{16}\pm\frac{1}{64}$	36.5± 0.40

Note:

The reduced section and shoulders (dimensions *A*, *D*, *E*, *F*, *G*, and *R*) shall be as shown, but the ends may be of any form to fit the holders of the testing machine in such a way that the load shall be axial. Commonly the ends are threaded and have the dimensions *B* and *C* given above.

21.3 Impact testing of notched bars

21.3.1 Izod test

The test consists of measuring the energy absorbed in breaking a notched test piece by one blow from a striker carried by a pendulum. The test piece is gripped vertically with the root of the notch in the same plane as the upper face of the grips. The blow is struck on the same face as the notch and at a fixed height above it. Tests are usually performed at the ambient temperature of the test house.

TEST PIECES

The standard test pieces are either 10 mm square or 0.45 inches (11.4 mm) diameter cross-section. Complete dimensions are shown in Table 21.15. The notch profile is shown in Figure 21.2.

Table 21.15 DIMENSIONS OF IZOD IMPACT TEST PIECES–BRITISH STANDARD BS 131: Part 1:1961 AMERICAN STANDARD ASTM E23–72

Item	*Square section*		*Circular section*	
	Nominal dimension			
	mm	in *equivalent*	in	mm *equivalent*
Minimum overall length of test piece				
1 Notch	70[1]	2.75	2.8[1]	71
2 Notch	98	3.86	3.9	99
3 Notch	126[2]	4.96	5.0[2]	127
Width	10	0.394 }	0.45 dia.	11.48 dia
Thickness	10	0.394 }		
Root radius of notch	0.25	0.010	0.010	0.25
Maximum depth below notch	8	0.315	0.32	8.1
Distance of plane of symmetry of notch from free end of test piece and from the adjacent notch	28	1.1	1.1	28
Angle of notch	45°		45°	

Notes:

(1) American Standard overall length 75 mm (2.952 in).
(2) American Standard overall length 131 mm (5.157 in).

PRESENTATION OF RESULTS

When standard test pieces are used the following symbols are used in reporting the results of Izod tests:

I for Izod
S for square section
Rs for circular cross-section with straight notch
E.G. I120S: x ft lbf

An energy of x ft lbf was obtained from an Izod test with a striking energy of 120 ft lbf using a square section test piece.

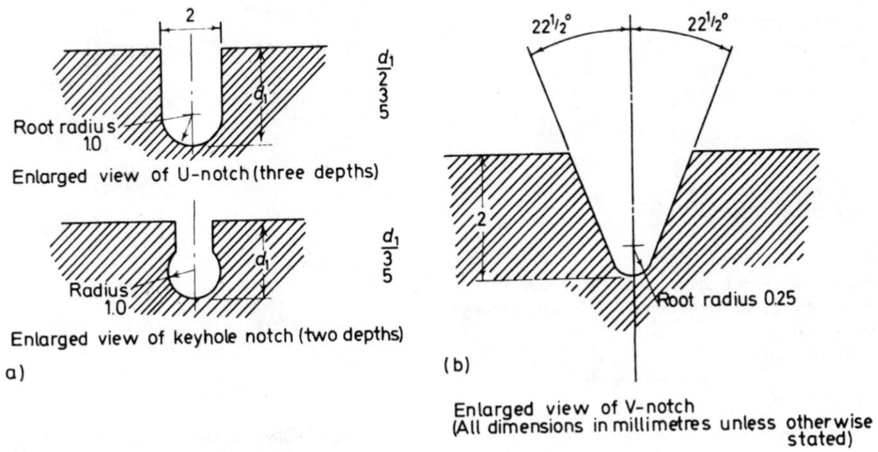

Figure 21.2 *Standard notch profiles for impact test pieces–see Table 21.15 and 21.16*

21.3.2 Charpy test

The test consists of measuring the energy absorbed in breaking by one blow from a pendulum a test piece notched in the middle and supported at each end.

TEST PIECES

The standard test piece has a 10 mm square cross-section with one of the three notch profiles shown in Figure 21.2. The test piece dimensions are shown in Table 21.16 Sub-standard test pieces are used where material thicknesses do not permit full-size specimens. It should be noted that the values obtained from subsidiary specimens cannot be compared with full-size specimens nor can the values obtained from different notches be compared.

PRESENTATION OF RESULTS

The report of the tests should include the following information:
Type of test: Charpy V or U notch; striking energy of the machine; size of test piece if sub-standard.
Nominal depth of notch and form ('U' or 'keyhole').
The energy absorbed (ft lbf) and the test temperature.
e.g. C120V: x ft lbf at y °C

An energy of x ft lbf was recorded from a Charpy V notch specimen tested at y °C with a striking energy of 120 ft lbf.
e.g. C240U2: x ft lbf at y°C
In this case the specimen was a Charpy with a U notch 2 mm deep and the striking energy was 240 ft lbf.

Table 21.16 DIMENSIONS OF CHARPY IMPACT TEST PIECES–BRITISH STANDARDS BS 131 : Part 2 : 1972; BS 131 : Part 3 : 1972; AMERICAN STANDARD ASTM E23–72

	Nominal dimension	
Item	mm	inch *equivalent*
Length	55	2.165
Width standard test piece	10	0.394
subsidiary test piece[1]	7.5	0.295
subsidiary test piece	5.0	0.197
subsidiary test piece	2.5	0.098
Thickness	10	0.394
Root radius of V notch	0.25	0.010
Depth below V notch	8	0.315
Root radius of U notch	1	0.039
Depth below U notch	8 (2 mm notch)[2, 3]	0.315
	7 (3 mm notch)[2, 3]	0.276
	5 (5 mm notch)	0.197
Distance of notch from one end of test piece	27.5	1.083
Angle between plane of symmetry of notch and longitudinal axis of test piece	90°	
Angle of V notch	45°	

Notes:

(1) Additional subsidiary test pieces are permitted by ASTM E23–72 as follows: 5 mm thick with 4 mm depth below the notch and 5, 10 or 20 mm width 3 mm thick with 0.094 in (2.39 mm) depth below notch and 10 mm width.
(2) Not specified in ASTM E23–72.
(3) The 2, 3 and 5 mm notches are sometime referred to as 'Mesnager', 'DVM', and 'Charpy' respectively.

21.4 Plane strain fracture toughness testing

Standards: British BS:5447:1977
American ASTM E399–78

The method involves the loading to failure in tension or three point bend of notched specimens which have been precracked by fatigue loading. The load versus displacement across the notch at the specimen edge is recorded autographically. The load corresponding to a 2% increment of crack extension is established by a specified deviation from the linear portion of the record. The fracture toughness, K_{Ic}, value is calculated from this load by equations which have been established on the basis of elastic stress analysis of the two specimen types.
Stress-intensity factor K_I: A measure of the stress-field intensity near the tip of a crack in a linear elastic body when deformed so that the crack faces are displaced apart, normal to the crack plane (Opening Mode or Mode I deformation). K_I is directly proportional to the applied load and depends on the specimen geometry.
Plain strain fracture toughness K_{Ic}: A material toughness property—the critical value of the stress intensity factor at which rapid propagation of a crack occurs during static loading.

TEST PIECES

The particular type of test piece used depends upon the form, the strength and the toughness of the material under test.

The basis of the test piece size requirements* is that both the crack length, a, and the thickness, B, shall be not less than $2.5[(K_{Ic})/(\sigma_Y)]^2$ where σ_Y is 0.2% proof stress of the material under the conditions of test, i.e. orientation, temperature and loading rate. In the first place, the test piece size has to be based on an approximate estimate of the K_{Ic} of the material, it is better to overestimate the K_{Ic} value initially and subsequently use a more conservative test piece if possible, on the basis of the first test results. Alternatively, the ratio of yield strength to Young's modulus given in Table 21.17 can be used a a guide for selecting the initial test piece size.

Table 21.17 RECOMMENDED THICKNESS AND CRACK LENGTH FOR FRACTURE TOUGHNESS TEST PIECES

	Minimum recommended thickness and crack length	
σ_Y/E	mm	in
up to 0.0050	100	4.0
0.0050–0.0057	75	3.0
0.0057–0.0062	63	2.5
0.0062–0.0065	50	2.0
0.0065–0.0071	38	1.5
0.0071–0.0080	25	1.0
0.0080–0.0095	13	0.5
0.0095 or greater	6.5	0.25

If, however, it is not possible to produce a test piece having both crack length and thickness not less than $2.5\ [(K_{Ic})/(\sigma_Y)]^2$ then it is not possible to obtain a valid K_{Ic} determination according to the standard procedures.

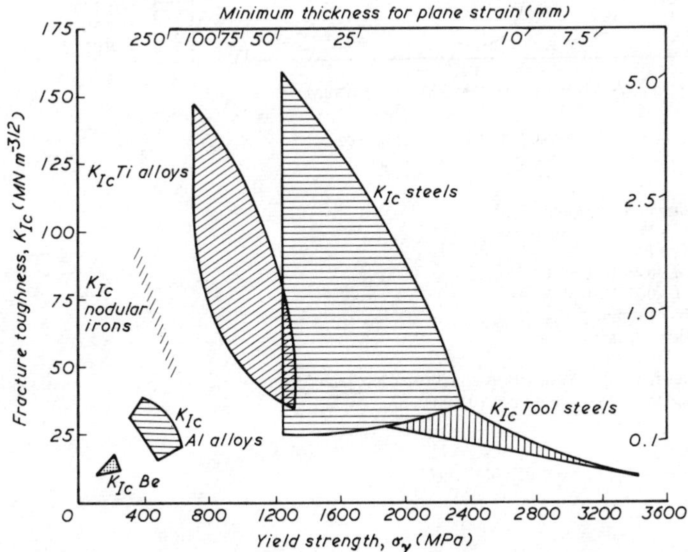

Figure 21.3 *Yield strength versus fracture toughness data for several alloy systems at 20°C. The upper scale of the diagram indicates the minimum thickness requirement for a given strength and toughness by projection of the line joining the origin and data point. (Source: Fulmer Materials Optimizer, Fulmer Laboratories Ltd, Buckinghamshire, UK).*

* For graphical assessment *see* Figure 21.3.

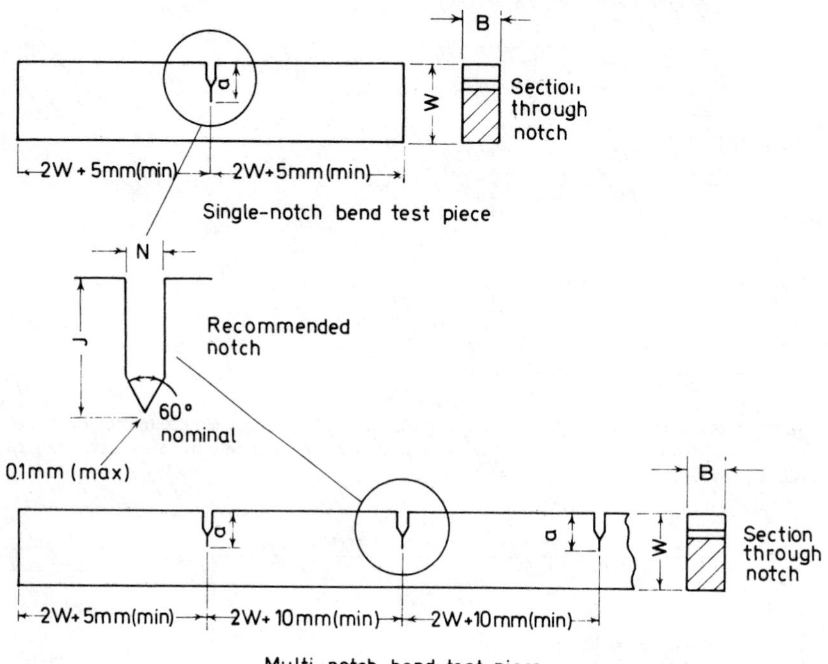

Test piece. Proportion, dimensions and limits

Width	$=W$
Thickness	$=B=\frac{1}{2}W$
Crack length	$a=0.45\ W–0.55\ W$
Notch length	$J=0.25\ W–0.45\ W$

i.e. 0.25 W when $W=13$ mm
0.45 W when $W=150$ mm

Notch width not greater than $\frac{1}{16}$ W (when W is equal to or less than 25 mm N may be up to $1\frac{1}{2}$ mm). Faces parallel and perpendicular to 0.02 mm taper per 10 mm run.

Figure 21.4 *Standard bend test piece from BS: 5447: 1977*

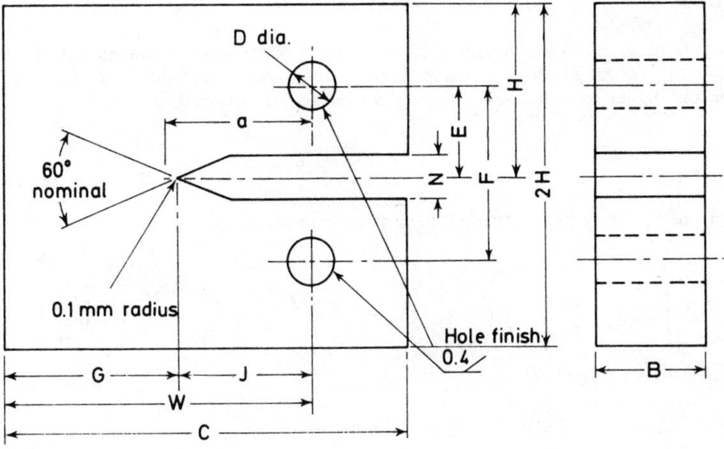

Proportional dimensions and limits

Net width	W
Crack length	$a = 0.45\ W$–$0.55\ W$
$= 0.5\ W$	Thickness
Total width	$C = 1.25\ W$
Hole diameter	$D = 0.25\ W$
Half hole centres	$E = 0.275\ W$
Hole centres	$F = 0.55\ W$
	$G =$ not less than $0.55\ W$
	$H = 0.6\ W$
	$J =$ see Figure 21.4
Notch width	$N =$ not greater than $\frac{1}{16}\ W$

When W is equal to or less than 25 mm N may be up to $1\frac{1}{2}$ mm. Surfaces parallel and perpendicular as applicable to 0.02 mm taper/per 10 mm run.

	Size	*B*	*W*	*C*	*a max.*	*N max.*	*E*	*F*	*H*	*D*	*G min.*
	13	13	26	32.5	14.5	1.5	7.2	14.3	15.6	6.5	14.3
	25	25	50	62.5	27.5	3.1	13.8	27.5	30	12.5	27.5
Hence:	50	50	100	125	55	6.3	27.5	55	60	25	55
	75	75	150	188.5	82.5	9.4	41.3	82.5	90	37.5	82.5
	100	100	200	250	110	12.5	55	110	120	50	110

Finish to be 0.8 unless otherwise specified.
The dimensions given above are for a recommended series of test piece sizes.

Figure 21.5 *Standard tension test piece from BS : 5447 : 1977*

CALCULATION AND INTERPRETATION OF RESULTS

Unlike most other forms of mechanical test, plane strain fracture toughness testing has to be completed and the results analysed before it is known if a valid measurement has been made. It is necessary to calculate a provisional result K_Q, which involves a construction from the test record, and then to determine whether this result is consistent with the test piece size requirements for the proof stress of the material.

OFFSET PROCEDURE

A secant line through the origin is drawn with a slope 5% less than the slope of the tangent to the initial linear part of the record. The force P_Q is defined as the intersection of the secant line with the test record or any higher recorded value of force preceding the intersection.

CALCULATION OF K_Q

K_Q is calculated from P_Q using one of the following relationships.

For the bend test

$$K_Q=\frac{3\,P_QL}{BW^{3/2}}\left[1.93\left(\frac{a}{W}\right)^{1/2}-3.07\left(\frac{a}{W}\right)^{3/2}+14.53\left(\frac{a}{W}\right)^{5/2}-25.11\left(\frac{a}{W}\right)^{7/2}+25.80\left(\frac{a}{W}\right)^{9/2}\right]$$

However, $L=2\,W$, hence $K_Q=P_QY_1/BW^{1/2}$

Values for Y_1 for specific values of a/W are given in Table 21.18.

For tension test piece

$$K_Q=\frac{P_Q}{BW^{1/2}}\left[29.6\left(\frac{a}{W}\right)^{1/2}-185.5\left(\frac{a}{W}\right)^{3/2}+655.7\left(\frac{a}{W}\right)^{5/2}-1017\left(\frac{a}{W}\right)^{7/2}+638.9\left(\frac{a}{W}\right)^{9/2}\right]$$

Hence $K_Q=\dfrac{P_QY_2}{BW^{1/2}}$

Values of Y_2 for specific values of a/W are given in Table 21.19.

CALCULATION OF K_{Ic}

The factor 2.5 $[(K_Q)/(\sigma_Y)]^2$ shall be calculated and if this is less than both the thickness and the crack length of the test piece then K_Q is equal to K_{Ic}. Otherwise, it is necessary to use a larger test piece to determine K_{Ic}, such that both thickness and crack length are not less than 2.5 $[(K_{Ic})/(\sigma_Y)]^2$. The new dimensions can be estimated on the basis of K_Q.

Table 21.18 VALUES OF Y_1 AGAINST a/W FOR BEND TEST PIECE THREE-POINT LOADED OVERALL SPAN TO TEST PIECE WIDTH RATIO 4:1

a/W	0.000	0.001	0.002	0.003	0.004	0.005	0.006	0.007	0.008	0.009
	Y_1 *Stress intensity coefficient*									
0.450	9.10	9.13	9.15	9.18	9.21	9.23	9.26	9.29	9.32	9.35
0.460	9.37	9.40	9.43	9.46	9.49	9.52	9.54	9.57	9.60	9.63
0.470	9.66	9.69	9.72	9.75	9.78	9.81	9.84	9.87	9.90	9.93
0.480	9.96	9.99	10.03	10.06	10.09	10.12	10.15	10.18	10.21	10.25
0.490	10.28	10.31	10.34	10.38	10.41	10.44	10.48	10.51	10.54	10.58
0.500	10.61	10.65	10.68	10.71	10.75	10.78	10.82	10.85	10.89	10.93
0.510	10.96	11.00	11.03	11.07	11.11	11.14	11.18	11.22	11.25	11.29
0.520	11.33	11.37	11.40	11.44	11.48	11.52	11.56	11.60	11.64	11.67
0.530	11.71	11.75	11.79	11.83	11.87	11.91	11.96	12.00	12.04	12.08
0.540	12.12	12.16	12.20	12.25	12.30	12.33	12.37	12.42	12.46	12.50
0.550	12.55									

Table 21.19 VALUES OF Y_2 AGAINST a/W FOR TENSION TEST PIECE

a/W	0.000	0.001	0.002	0.003	0.004	0.005	0.006	0.007	0.008	0.009
	Y_2 *Stress intensity coefficient*									
0.450	8.34	8.36	8.38	8.41	8.43	8.45	8.47	8.50	8.52	8.54
0.460	8.57	8.59	8.61	8.64	8.66	8.69	8.71	8.73	8.76	8.78
0.470	8.81	8.83	8.86	8.88	8.91	8.93	8.96	8.98	9.01	9.03
0.480	9.06	9.09	9.11	9.14	9.16	9.19	9.22	9.24	9.27	9.30
0.490	9.32	9.35	9.38	9.41	9.43	9.46	9.49	9.52	9.55	9.57
0.500	9.60	9.63	9.66	9.69	9.72	9.75	9.78	9.81	9.84	9.87
0.510	9.90	9.93	9.96	9.99	10.02	10.05	10.08	10.11	10.15	10.18
0.520	10.21	10.24	10.27	10.31	10.34	10.37	10.40	10.44	10.47	10.50
0.530	10.54	10.57	10.61	10.64	10.68	10.71	10.75	10.78	10.82	10.85
0.540	10.89	10.92	10.96	11.00	11.03	11.07	11.11	11.15	11.18	11.22
0.550	11.26									

RECORDING OF TEST RESULTS

The following data are usually recorded for each test:

1. Thickness: B in metres.
2. Width: W in metres.
3. Half loading span: L in metres for bend test only.
4. Fatigue precracking.
   (a) The fatigue stress intensity, K_f, during the last 1.25 mm (or 2.5% W whichever is the greater) increment of crack growth. This stress intensity should not exceed 0.7 K_Q.
   (b) The maximum fatigue stress intensity, used to initiate the crack. This stress intensity should not exceed 0.75 K_Q.
   (c) Temperature of the test piece during precracking. When fatigue cracking is conducted at a temperature T_1 and the testing carried out at a temperature T_2, K_f during the final stages of cracking should be less than $[(2\sigma_{Y1})/(3\sigma_{Y2})]\,K_{1C}$ where σ_{Y1} and σ_{Y2} are the proof stresses at the temperatures T_1 and T_2.
   (d) The R ratio, i.e. the ratio of the minimum to the maximum fatigue load. It is recommended that R should not be greater than 0.1.
5. Crack length: a in metres. The crack length is measured from the fractured test piece at 25%, 50% and 75% of the specimen thickness and the average of these three values quoted. The crack is regarded as unsatisfactory if any of these measurements differ by more than 2.5% W or if any two possible crack length measurements differ by more than 5% W.
6. Test temperature.
7. Environment.
8. Loading rate in terms of $\dot{K}$ (change in stress intensity factor per unit time) if outside the standard range of 0.5–2.5 $\mathrm{MN\,m^{-3/2}\,s^{-1}}$.
9. 0.2% proof stress in MPa.
10. K_{Ic}: in $\mathrm{MN\,m^{-3/2}}$. Note $K_{Ic} = K_Q$ if the size requirements are satisfied.

22 Mechanical properties of metals and alloys

The following tables summarize the mechanical properties of the more important industrial metals and alloys.

In the tables of tensile properties at normal temperatures the nominal composition of the alloys is given, followed by the appropriate British and other specification numbers. Most specifications permit considerable latitude in both composition and properties, but the data given in these tables represent typical average values which would be expected from materials of the nominal composition quoted, unless otherwise stated. For design purposes it is essential to consult the appropriate specifications to obtain minimum and maximum values and special conditions where these apply.

The data in the tables referring to properties at elevated and at sub-normal temperatures, and for creep, fatigue and impact strength have been obtained from a more limited number of tests and sometimes from a single example. In these cases the data refer to the particular specimens tested and cannot be relied upon as so generally applicable to other samples of material of the same nominal composition.

22.1 Mechanical properties of aluminium and aluminium alloys

The compositional specifications for wrought aluminium alloys are now internationally agreed throughout Europe, Australia, Japan and the USA. The system involves a four-digit description of the alloy and has been in use as a national standard in the USA since 1957. Registration of wrought alloys is administered by the Aluminum Association in Washington, DC. International agreement on temper designations has not yet been achieved, although the US temper designation has been adopted by the Association of European Aerospace Constructors and, in consequence, is used in the latest BS 'L' series of standards. Nevertheless, the normal UK temper designation is used, where necessary, below.

No comparable rationalization of designations has taken place for cast alloys (Table 22.1).

In the following tables the four-digit system is used, wherever possible, for wrought materials whereas the British Standard designations are generally used for cast alloys.

22.1.1 Alloy designation system for wrought aluminium

The first of the four digits in the designation indicates the alloy group according to the major alloying elements, as follow:

1XXX	aluminium of 99.0% minimum purity and higher
2XXX	copper
3XXX	manganese
4XXX	silicon
5XXX	magnesium
6XXX	magnesium and silicon
7XXX	zinc
8XXX	other element
9XXX	unused

1XXX Group: In this group the last two digits indicate the minimum aluminium percentage. Thus 1099 indicates aluminium with a minimum purity of 99.99%. The second digit indicates modifications in impurity limits or alloying elements, where 0 signifies unalloyed aluminium with natural impurities and the integers 1 to 9, allotted consecutively as needed, indicate special control of limits or additions.

2XXX–8XXX Groups: In these groups the last two digits are simply used to identify the different alloys in the groups and have no special significance. The second digit indicates alloy modifications, zero being allotted to the original alloy.

National variations of existing compositions are indicated by a letter after the numerical designation, allotted in alphabetical sequence, starting with A for the first national variation registered.

Table 22.1 RELATED, OR APPROXIMATELY SIMILAR, SPECIFICATIONS FOR CAST ALLOYS

British	ISO	*American*		*French*	*German*		*USSR*
BS 1490 *or* BS/L *or* DTD	DIS 3522	ASTM B 179–80	SAE 34–383	NF A57–702/3	DIN 1725/2	*Alloy No.*	GOST 2685–63
LM 0	Al 99.5	150.1	—	—	—	—	—
LM 2	Al Si 10 Cu2 Fe	384.1	303	A–S9 U3–Y4	—	—	—
LM 4	Al Si5 Cu3	319.2	326	—	G–Al Si6 Cu4	—	AL 6
LM 5	Al Mg5	514.1	320	A–G6	G–Al Mg5	245	AL 28
LM 6	Al Si12	A413.2	—	—	G–Al Si12	230	—
LM 9	Al Si12 Mg	A360.2	309	—	G–Al Si10 Mg	230/10	AL 4
LM 10	Al Mg10	520.0	324	A–G10–Y4	G–Al Mg10	—	AL 27
LM 12	—	222.1	34	—	—	—	AL 18V
LM 13	Al Si11 Mg Cu	336.1	321	A–S12–UN	—	230/Ni	AL 30
LM 16	Al Si5 Cu1 Mg	355.1	322	—	—	234	AL 5
LM 18	Al Si5	443.0	35	—	—	—	—
LM 20	Al Si12 Cu Fe	A413.1	305	A–S12–Y4	G–Al Si12(Cu)	331	AL 2
LM 21	Al Si6 Cu4 Zn	—	—	A–S5 UZ	G–Al Si6 Cu4	225	AL 16V
LM 22	—	—	—	A–S5 U	—	—	AL 6
LM 24	Al Si8 Cu3 Fe	A380.1	306	A–S9 U3B–Y4	GD–Al Si8 Cu3	333	—
LM 25	Al Si7 Mg	356.1	323	A–S7 G	G–Al Si7 Mg	—	AL 9
LM 26	Al Si9 Cu3 Mg	332.1	332	—	—	—	—
LM 27	Al Si7 Cu2	—	—	—	—	—	—
LM 28	Al Si19 Cu Mg Ni	—	—	—	—	—	—
LM 29	Al Si23 Cu Mg Ni	—	—	—	—	—	AL 26
LM 30	Al Si17 Cu4 Mg	390.0	—	—	—	—	—
4L 35	Al Cu4 Ni2 Mg2	242.2	39	A–U4 NT	—	—	AL 1
2L 99	Al Si7 Mg	A356.2	336	A–S7 GO.3	G–Al Si7 Mg	—	—
L 119	—	—	—	—	—	—	—
L 154/155	—	—	380	—	—	—	AL 20
DTD 5008B	Al Zn5 Mg	712.0	310	A–Z5 G	—	—	—

Table 22.2 ALUMINIUM AND ALUYINIUM ALLOYS—MECHANICAL PROPERTIES AT ROOM TEMPERATURE
Wrought Alloys

Specification	*Nominal composition* %		*Condition*	0.2% *Proof stress* MPa (ton f in^{-2})	*Tensile strength* MPa (ton f in^{-2})	*Elong.* % *on* 50 mm ($\geqslant$ 2.6 mm) or $5.65\sqrt{S_0}$	*Shear strength* MPa (ton f in^{-2})	*Brinell hardness* ($P=5D^2$) kg mm^{-2}	*Fatigue strength (unnotched)* 500 MHz MPa (ton f in^{-2})	*Izod impact* J (ft lbf)	*Remarks*
1099	Al 99.99	Sheet	O	20 (1.4)	55 (3.5)	55	50 (3.3)	15	—	—	Highest quality reflectors
			H4	60 (3.8)	85 (5.5)	20	60 (3.8)	23	—	—	
			H8	85 (5.4)	110 (7.0)	12	70 (1.4)	28	—	—	
1080A	Al 99.8	Sheet	O	25 (1.7)	70 (4.5)	50	60 (3.8)	19	—	—	Domestic trim, chemical plant
			H4	95 (6.2)	100 (6.6)	17	70 (4.5)	29	—	—	
			H8	125 (8.1)	135 (8.9)	11	75 (4.9)	38	—	—	
		Wire	O	—	70 (4.5)	—	60 (3.8)	19	—	—	
			H4	90 (5.9)	105 (6.8)	—	70 (4.6)	30	—	—	
			H8	110–140 (7.1–9.1)	130–160 (8.5–10.5)	—	—	35–41	—	—	
1050A	Al 99.5	Sheet	O	35 (2.3)	80 (5.2)	47	65 (4.2)	21	—	—	General purpose, formable alloy
			H4	105 (6.7)	110 (7.2)	15	75 (4.8)	30	—	—	
			H8	130 (8.4)	145 (9.4)	10	85 (5.5)	40	—	—	
		Bars and sections as extruded		50 (3.3)	75 (5.0)	38	65 (4.2)	22	—	—	
		Rivet stock	H5	125 (8.1)	140 (9.1)	—	—	—	—	—	
		Tubes	O	—	75 (5.0)	—	65 (4.2)	21	—	—	
			H8 < 75 mm	120 (7.7)	125 (8.0)	—	75 (4.8)	—	—	—	
			H8 > 75 mm	110 (7.2)	115 (7.5)	—	70 (4.5)	—	—	—	
		Wire	O	42 (2.7)	75 (4.9)	—	65 (4.2)	21	—	—	
			H4	100 (6.6)	115 (7.5)	—	75 (5.0)	30	—	—	
			H8	115–170	140–195	—	—	38–48	—	—	
				(7.4–11.0)	(9.1–12.6)	—	—	—	—	—	
1350	Al 99.5	Wire	O	28 (1.8)	83 (5.4)	—	55 (3.6)	—	—	—	Electrical conductors
			H4	97 (6.3)	110 (7.1)	—	69 (4.5)	—	—	—	
			H8	165 (10.7)	186 (12.0)	—	103 (6.7)	—	48 (3.1)	—	

Notes: See page **22**–11

Table 22.2 ALUMINIUM AND ALUMINIUM ALLOYS—MECHANICAL PROPERTIES AT ROOM TEMPERATURE—*continued*

Wrought Alloys

Specification	*Nominal composition* %	*Condition*		0.2% *Proof stress* MPa (ton f in^{-2})	*Tensile strength* MPa (ton f in^{-2})	*Elong.* % *on* 50 mm (⩾ 2.6 mm) or $5.65\sqrt{S_0}$	*Shear strength* MPa (ton f in^{-2})	*Brinell hardness* ($P=5D^2$) kg mm^{-2}	*Fatigue strength (unnotched)* 500 MHz MPa (ton f in^{-2})	*Izod impact* J (ft lbf)	*Remarks*
1200	Al 99.0	Sheet	O	35 (2.2)	90 (5.8)	43	70 (44)	22	35 (2.2)	27 (20)	General purpose, slightly higher strength than 1050A
			H2	95 (6.3)	105 (6.8)	20	75 (4.8)	31	40 (2.7)	—	
			H4	115 (7.3)	120 (7.8)	12	80 (5.2)	35	50 (3.1)	31 (23)	
			H6	125 (8.2)	135 (8.8)	11	90 (5.7)	38	60 (4.0)	—	
			H8	145 (9.4)	160 (10.3)	9	95 (6.0)	42	60 (4.0)	26 (19)	
		Bars and sections	as extruded	40 (2.7)	85 (5.5)	38	70 (4.4)	23	45* (3.0*)	27 (20)	
		Tubes	O	—	90 (5.8)	40	70 (4.5)	21	—	—	
			H > 75 mm	128 (8.3)	131 (8.5)	6	100 (6.5)	34	—	—	
			H < 75 mm	120 (7.8)	124 (8.0)	6	95 (6.1)	32	—	—	
2011	Cu 5.5 Bi 0.5 Pb 0.5	Extruded bar	TD 25 mm	295 (19)	340 (22)	14	240 (15.5)	95	—	—	Free machining alloy
			TF 50–75 mm	260 (17)	370 (24)	16	240 (15.5)	100	—	—	
		Wire	TD ⩽ 10 mm	350 (22.5)	365 (23.5)	—	—	—	—	—	
2014A	Cu 4.4 Mg 0.7 Si 0.8 Mn 0.75	Sheet	TB	270 (17.5)	450 (29)	20	260 (17)	115	130* (8.5*)	—	Aircraft applications (cladding when used 1070A)
			TF	430 (28)	480 (31)	10	295 (19)	135	130* (8.5*)	—	
		Clad sheet	TB	250 (16)	425 (27.5)	22	250 (16)	—	95* (6*)	—	
			TF	385 (25)	440 (28.5)	10	260 (17)	—	95* (6*)	—	
		Bars and sections	TB	315 (20.5)	465 (30)	17	—	115	140 (9)	22 (16)	
			TF	465 (30)	500 (32.5)	10	—	135	124 (8)	8 (6)	
		Tubes	TB	310 (20)	425 (27.5)	12	—	115	—	—	
			TF	415 (27)	480 (31)	9	—	135	—	—	
		Wire	TB	340 (22)	445 (29)	15	—	115	—	—	
			TF	425 (27.5)	465 (30)	—	—	135	—	—	
		Rivet stock	TB	340 (22)	450 (29)	—	—	—	—	—	
		Bolt and screw stock	TF	425 (27.5)	460 (30)	—	—	—	—	—	
2024	Cu 4.5 Mg 1.5 Mn 0.6	Plate/sheet/ extrusions	O	75 (4.9)	185 (11.1)	20	125 (8.1)	47	90 (5.8)	—	Aircraft structures
			TB	325 (21.0)	470 (30.4)	20	285 (18.5)	120	140 (9.1)	—	
			TF	395 (25.6)	475 (30.8)	10	—	—	—	—	
2117	Cu 2.5 Si 0.6 Mg 0.4	Sheet	TB	165 (10.7)	295 (19.1)	24	195 (12.6)	70	95 (6.2)	—	Vehicle body sheet

2219	Cu 6	Plate/sheet/ forgings	O	75 (4.9)	170 (11)	18	—	—	—	—	Weldable, creep resistant, high-temperature aerospace applications
	Mn 0.3		TB	185 (11.1)	360 (23.3)	20	—	—	—	—	
	V 0.1		TF	290 (18.8)	415 (26.9)	10	—	—	105 (6.8)	—	
2004	Cu 6	Sheet	O	150 (9.7)	230 (14.9)	15	—	—	100* (6.5*)	—	Superplastically deformable sheet
	Zr 0.4		TF	300 (19.4)	420 (27.2)	12	—	—	150* (9.7*)	—	
2031	Cu 2.3	Forgings	TB	235 (15.2)	355 (23)	22	201 (13)	95	—	—	Aero-engines, missible fins
	Ni 1		TF	340 (22)	420 (27.2)	15	230 (14.9)	120	—	—	
	Mg 0.9										
	Si 0.9										
	Fe 0.9										
2618A	Cu 2.0	Forgings	O	70 (4.5)	170 (11)	20	—	45	85* (5.6*)	—	Aircraft engines
	Mg 1.5		TF	330 (21.5)	430 (28)	8	295 (19)	130	170* (11*)	—	
	Si 0.9										
	Fe 0.9										
	Ni 1.0										
3103	Mn 1.25	Sheet	O	65 (4.1)	110 (7.0)	40	80 (5.25)	30	50 (3.1)	34 (25)	General purpose, holloware, building sheet
			H2	125 (8.2)	130 (8.5)	17	90 (5.75)	40	55 (3.6)	—	
			H4	140 (9.2)	155 (10.0)	11	95 (6.25)	44	60 (4.0)	29 (21)	
			H6	160 (10.4)	180 (11.5)	8	105 (6.75)	47	70 (4.5)	—	
			H8	185 (11.9)	200 (13.0)	7	110 (7.25)	51	70 (4.5)	20 (15)	
		Wire	O	60 (3.9)	115 (7.5)	—	—	30	—	—	
			H4	135 (8.8)	155 (10.0)	—	—	45	—	—	
			H8	170–200 (11.0–13.0)	205–245 (13.3 – 15.9)	—	—	55–65	—	—	
3105	Mn 0.75	Sheet	O	55 (3.6)	115 (7.5)	24	85 (5.5)	—	—	—	Building cladding sheet
	Mg 0.6		H4	150 (9.7)	170 (11)	5	105 (6.8)	—	—	—	
			H8	195 (12.6)	215 (14)	3	115 (7.4)	—	—	—	
3004	Mn 1.2	Sheet	O	70 (4.5)	180 (11.7)	20	110 (7.1)	45	95 (6.2)	—	Sheet metal work, storage tanks
	Mg 1		H4	200 (13)	240 (15.6)	9	125 (8.1)	63	105 (6.8)	—	
			H8	250 (16.2)	285 (18.5)	5	145 (9.4)	77	110 (7.1)	—	
3008	Mn 1.6	Sheet	O	50 (3.2)	120 (7.8)	23	—	—	—	—	Thermally resistant alloy. Vitreous enamelling
	Fe 0.7		H8	270 (17.5)	280 (18.1)	4	—	—	—	—	
	Zr 0.3										
3003 clad with 4343	Mn 1.2	Sheet	O	40 (2.6)	110 (7.1)	30	75 (4.9)	—	—	—	Flux brazing sheet
	Si 7.5		H2	125 (8.1)	130 (8.4)	10	85 (5.6)	—	—	—	
			H4	145 (9.4)	150 (9.8)	8	95 (6.2)	—	—	—	
			H6	170 (11)	175 (11.3)	5	105 (6.8)	—	—	—	

Notes: see page **22**–11

Table 22.2 ALUMINIUM AND ALUMINIUM ALLOYS—MECHANICAL PROPERTIES AT ROOM TEMPERATURE—*continued*
Wrough Alloys

Specification	*Nominal composition* %	*Condition*		0.2% *Proof stress* MPa (ton f in^{-2})	*Tensile strength* MPa (ton f in^{-2})	*Elong.* % *on* 50 mm (≥ 2.6 mm) or $5.65\sqrt{S_0}$	*Shear strength* MPa (ton f in^{-2})	*Brinell hardness* ($P=5D^2$) kg mm^{-2}	*Fatigue strength (unnotched)* 500 MHz MPa (ton f in^{-2})	*Izod impact* J (ft lbf)	*Remarks*
3003 clad with 4004	Mn 1.2 Si 10 Mg 1.5	Sheet	Physical properties		as for 3003 clad with 4343			—	—	—	Vacuum brazing sheet
4032	Si 12.0 Cu 1.0 Mg 1.0 Ni 1.0	Forgings	TF	240 (15.5)	325 (21)	5	—	115	110 (7)	—	Pistons
4043A	Si 5.0	Rolled wire		75 (5.0)	130 (8.3)	20	—	—	—	—	Welding filler wire
4047A	Si 12	Wire	M	189 (12.2)	225 (14.6)	8	—	—	—	—	Brazing rod
5657	Mg 0.8	Sheet	O	40 (2.9)	110 (7.1)	25	75 (4.9)	28	—	—	High base purity, bright trim alloy
			H4	140 (9.1)	160 (10.4)	12	95 (6.2)	40	—	—	
			H8	165 (10.7)	195 (12.7)	7	105 (6.8)	50	—	—	
5005	Mg 0.8	Sheet	O	40 (2.9)	125 (8.1)	25	75 (4.9)	28	—	—	Architectural trim, commercial vehicle trim
			H4	150 (9.7)	160 (10.4)	6	95 (6.2)	—	—	—	
			H8	195 (12.7)	200 (13)	4	110 (7.1)	—	—	—	
5251	Mg 2.25 Mn 0.25	Sheet	O	95 (6.0)	185 (12.0)	22	125 (8.25)	45	110 (7.1)	50 (36)	Sheet metal work
			H4	230 (15.0)	245 (16.0)	7	145 (9.3)	70	125 (8.1)	29 (21)	
			H8	275 (17.9)	285 (18.5)	2	175 (11.2)	80	140 (9.0)	—	
		Bars and sections as extruded		95 (6.0)	185 (12.0)	20	125 (8.0)	45	95* (6.0*)	49 (36)	
		Tubes	O	100 (6.5)	200 (13.0)	20	—	—	—	—	
			H4	230 (14.8)	250 (16.0)	6	—	—	—	—	
			H8	255 (16.5)	270 (17.5)	5	—	—	—	—	
		Wire	O	95 (6.2)	200 (13.0)						
			H8	260–290	280–310	—	—	48	—	—	
				(16.8–18.8)	(18.1–20.1)	—	—	75–85	—	—	
5154A	Mg 3.5 Mn 0.5	Sheet	O	125 (8.0)	240 (15.5)	24	155 (10.2)	55	115 (7.6)	—	Welded structures, storage tanks, salt water service
			H2	245 (15.8)	295 (19.0)	10	175 (11.3)	80	125 (8.0)	—	
			H4	275 (17.9)	310 (20.0)	9	175 (11.4)	95	130 (8.5)	—	
		Bars and sections as extruded		125 (8.0)	230 (15.0)	25	145 (9.5)	55	140* (9.0*)	48 (35)	
		Tubes	O	125 (8.0)	225 (14.5)	20	—	55	—	—	
			H4	[illegible]	[illegible]	7				—	

		Wire	O	125 (8.0)	240 (15.5)	—	—	55	—	—	
			H4	265 (17.1)	295 (19.0)	—	—	90	—	—	
			H8	310 (20.1)	355 (23.0)	—	—	100	—	—	
		Rivet stock	O	125 (8.0)	250 (16.0)	—	—	—	—	—	
			H2	—	290 (18.8)	—	—	—	—	—	
5083	Mg 4.5 Mn 0.7 Cr 0.15	Sheet	O	170 (11.0)	310 (20.0)	21	170 (11.0)	72	—	—	Marine applications, cryogenics, welded pressure vessels.
			H4	290 (18.9)	370 (24.0)	9	210 (13.5)	110	—	—	
		Bars and sections as extruded		180 (11.5)	315 (20.5)	19	180 (11.5)	77	—	—	
5556A	Mg 5	Wire	H4	250 (16.2)	330 (21.4)	12	—	—	—	—	Weld filler wire
5056A	Mg 5.0 Mn 0.5	Wire	O	140 (9.1)	300 (19.4)	—	—	65	—	—	Rivets, bolts, screws
			H4	300 (19.4)	340 (22.0)	—	—	95	—	—	
			H8	340–400 (22.0–26.0)	400–450 (26.0–29.0)	—	—	110–120	—	—	
		Rivet stock	O	140 (9.1)	300 (19.4)	—	—	65	—	—	
			H2	—	350 (22.7)	—	—	—	—	—	
		Bolt and screw stock	H4	300 (19.4)	340 (22.0)	—	—	—	—	—	
6063	Mg 0.5 Si 0.5	Bars and sections	M	85 (5.5)	155 (10)	30	100 (6.5)	35	—	—	Architectural extrusions (fast extruding)
			TB	115 (7.5)	180 (11.5)	30	130 (8.5)	52	60 (4.0)	43 (32)	
			TF	210 (13.5)	245 (16)	20	160 (10.5)	75	70 (4.5)	31 (23)	
		Wire	O	—	115 (7.5)	—	—	—	—	—	
			TB	115 (7.5)	180 (11.5)	—	—	50	—	—	
			TF	195 (12.5)	230 (15)	—	—	70	—	—	
6061	Mg 10 Si 0.6 Cr 0.25 Cu 0.2	Bars and sections	TB	145 (9.5)	230 (15)	20	160 (10.5)	60	—	34 (25)	Intermediate strength extrusion alloy
			TF	280 (18)	310 (20)	13	200 (13.0)	90	—	27 (20)	
		Wire	TH ⩽ 6 mm	310–400 (20–26)	385–430 (25–28)	—	—	—	—	—	
			TH6–10 mm	295–385 (19–25)	380–415 (24.5–27)	—	—	—	—	—	
		Bolt and screw stock	TH	290 (19)	340 (22)	—	—	—	—	—	
6082	Mg 1.0 Si 1.0 Mn 0.5	Sheet	O	60 (4)	125 (8)	26	85 (5.6)	35	—	—	General purpose high duty applications
			TB	155 (10)	245 (16)	21	175 (11.3)	65	—	—	
			TF	285 (18.5)	315 (20.5)	12	205 (13.4)	100	—	—	
		Bars, sections and forgings	TB	160 (10.6)	240 (15.5)	25	180 (11.5)	65	—	41 (30)	
			TF	285 (18.5)	310 (20)	13	215 (14)	100	—	34 (25)	
		Tubes	TB	160 (10.5)	245 (16)	20	—	65	—	—	
			TF	285 (18.5)	325 (21)	10	—	95	—	—	

Notes: see page **22**–11

Table 22.2 ALUMINIUM AND ALUMINIUM ALLOYS—MECHANICAL PROPERTIES AT ROOM TEMPERATURE—*continued*
Wrought Alloys

Specification	*Nominal composition* %		*Condition*	0.2% *Proof stress* MPa (ton f in^{-2})	*Tensile strength* MPa (ton f in^{-2})	*Elong.* % *on* 50 mm ($\geqslant$ 2.6 mm) or $5.65\sqrt{S_0}$	*Shear strength* MPa (ton f in^{-2})	*Brinell hardness* ($P=5D^2$) kg mm^{-2}	*Fatigue strength* (*unnotched*) 500 MHz MPa (ton f in^{-2})	*Izod impact* J (ft lbf)	*Remarks*
6009	Si 0.8 Mg 0.6 Mn 0.5 Cu 0.4	Sheet	TB TF	130 (8.4) 325 (21)	235 (15.2) 345 (22.3)	24 12	205 (13.3) 150 (9.7)	60 —	97 (6.3) 115 (7.4)	— —	Vehicle body sheet
7020	Zn 4.5 Mg 1.2 Zr 0.15	Bars and sections	TB TF	225 (14.5) 310 (20)	340 (22) 370 (24)	18 15	— —	— —	— —	— —	Transportable bridging
7075	Zn 5.6 Mg 2.5 Cu 1.6 Cr 0.25	Sheet/plate/ forgings/ extrusion	O TF (1)	105 (6.8) 505 (32.7) 435 (28.2)	230 (14.9) 570 (37) 505 (32.7)	17 11 13	150 (9.7) 330 (21.4) —	60 150 —	— 160 —	— 7 (5) —	Aircraft structures
7050	Zn 6.2 Mg 2.2 Cu 2.3 Zr 0.12	Thick section plate/ forgings	(2)	455 (29.5)	515 (33.3)	11	—	—	220 (14.2)	—	Low quench sensitivity, high stress corrosion resistance. Aircraft structures
7475	Zn 5.7 Mg 2.2 Cu 1.5 Cr 0.2	Sheet/plate/ forgings	TF(3) (4)	525 (34) —	460 (29.8) —	12 —	— 270 (17.5)	— —	— 220 (14.2)	— —	High base purity. high fracture toughness. Aircraft structures
7016	Zn 4.5 Mg 1.1 Cu 0.75	Extrusions	TF	315 (20.4)	360 (23.3)	12	—	—	—	—	Bright anodized vehicle bumpers
7021	Zn 5.5 Mg 1.5 Cu 0.25 Zr 0.12	Extrusions	O TF	115 (7.4) 395 (25.6)	235 (15.2) 435 (28.2)	16 13	— —	— —	— —	— —	Bumper backing bars
8079	Fe 0.7	Foil	O H8	35 (2.3) 160 (10.4)	95 (6.2) 175 (11.3)	26 2	— —	— —	— —	— —	Domestic foil

Cast alloys											
Al	Al 99.0	Sand cast	M	30 (2)	80 (5)	30	55 (3.5)	25	30* (2*)	19 (14)	High conductivity, high
(LM0)		Chill cast	M	30 (2)	80 (5)	40	55 (3.5)	25	30* (2*)	19 (14)	ductility
Al–Mg	Mg 5.0	Sand cast	M	100 (6.5)	160 (10.5)	6	—	60	45 (3)	8 (6)	Very high corrosion
(LM5)	Mn 0.5	Chill cast	M	100 (6.5)	215 (14)	10	—	65	95 (6)	12 (9)	resistance
(LM10)	Mg 10.0	Sand cast	TB	180 (11.5)	295 (19)	12	230 (15)	85	55 (3.5)	15 (11)	Strength + corrosion
		Chill cast	TB	190 (12.5)	340 (22)	18	230 (15)	95	—	—	resistance
Al–Si	Si 5.0	Sand cast	M	60 (4)	125 (8)	5	90 (6)	40	55 (3.5)	1.5 (1)	Intricate castings
(LM18)		Chill cast	M	70 (4.5)	155 (10)	6	120 (8)	50	85 (5.5)	2.5 (2)	
(LM6) (LM20)	Si 11.5	Sand cast	M	65 (4)	170 (11)	8	110 (7)	55	45* (3*)	4 (3)	Very similar alloys, excellent casting characteristics and corrosion resistance. LM6 has slightly superior corrosion resistance
(LM6) (LM20)		Chill cast	M	75 (5)	215 (14)	10	130 (8.5)	60	60* (4*)	9.5 (7)	
Al-Si-Mg	Si 7	Sand cast	TF	195 (12.6)	240 (15.5)	3	—	—	56 (3.6)	—	Good strength in fairly
	Mg 0.4						—	90	90 (5.8)	—	difficult castings. Cast
(2L99)		Chill cast	TF	210 (13.6)	290 (18.8)	6					vehicle wheels
Al-Cu-Si	Cu 4.2	Sand cast	TB	170 (11)	225 (14.6)	8	—	—	—	—	Aircraft castings
(L154)	Si 1.2	Chill cast	TB	175 (11.3)	280 (18.1)	15	—	—	—	—	
(L155)		Sand cast	TF	215 (13.9)	295 (19.1)	5	—	85	—	—	
		Chill cast	TF	215 (13.9)	320 (20.7)	10	—	90	—	—	
Al-Cu-Si	Cu 3.5	Chill cast	M	110 (7)	200 (13)	3	—	85	—	—	Excellent die casting
(LM24)	Si 8.0	Die cast	M	150 (9.5)	320 (20.5)	2	—	85	—	—	alloy
(LM4)	Cu 3.0	Sand cast	M	90 (6)	155 (10)	3	—	70	70* (4.5*)	0.7 (0.5)	General engineering,
	Si 5.0	Chill cast	M	100 (6.5)	170 (11)	3	—	80	75* (5*)	0.7 (0.5)	particularly sand and
		Sand cast	TF	230 (15)	260 (17)	1	—	105	—	—	permanent mould castings
		Chill cast	TF	260 (17)	330 (21.5)	3	—	110	—	—	
(LM22)	Cu 3.0 Si 5.0 Mn 0.5	Chill cast	TB	115 (7.5)	260 (17)	9	—	75	—	4.5 (3.5)	Good combination of impact resistance and strength
(LM2)	Cu 1.5	Sand cast	M	85 (5.5)	140 (9)	1	—	70	55 (3.5)	—	General purpose die casting alloy
	Si 10.0	Chill cast	M	95 (6)	185 (12)	2	—	80	60 (4)	—	
Al-Cu-Mg	Cu 10.0	Chill cast	M	155 (10)	185 (2)	1	—	85	—	0.9 (0.7)	Castings to withstand high
(LM12)	Mg 0.25	Chill cast	TF	285 (18.5)	310 (20)	—	—	130	60 (4)	—	hydraulic pressure
Al-Cu (L119)	Cu 5 Ni 1.5	Sand cast	TF	200 (12.9)	225 (14.6)	2	—	90		—	Sand castings for elevated temperature service
Al-Zn-Mg (DTD5008B)	Zn 5.3 Mg 0.6 Cr 0.5	Sand cast	TB		220 (14.2)	5	—	—	—	—	Colour anodizing alloy

Notes: see page **22**–11

Table 22.2 ALUMINIUM AND ALUMINIUM ALLOYS—MECHANICAL PROPERTIES AT ROOM TEMPERATURE—*continued*
Cast Alloys

Material (specification)	*Nominal composition* %		*Condition*	0.2% *Proof stress* MPa (ton f in^{-2})	*Tensile strength* MPa (ton f in^{-2})	*Elong.* % *on* 50 mm ($\geqslant$ 2.6 mm) *or* $5.65\sqrt{S_0}$	*Shear strength* MPa (ton f in^{-2})	*Brinell hardness* ($P=5D^2$) kg mm^{-2}	*Fatigue strength (unnotched)* 500 MHz MPa (ton f in^{-2})	*Izod impact* J (ft lbf)	*Remarks*
Al-Cu-Si-Zn	Cu 2 Si 7	Sand cast	M	85 (5.5)	155 (10)	2	—	75	—	—	Versatile general purpose alloy
(LM27)		Chill cast	M	100 (6.5)	180 (11.7)	3	—	80	—	—	
Al-Si-Cu-Mg (LM30)	Si 17.0 Cu 4.5 Mg 0.6	Chill cast	M	160 (10.5)	180 (11.5)	0.5	—	110	—	—	Die castings with high wear resistance, especially automobile cylinder blocks
		Die cast	M	240 (15.5)	275 (18)	1	—	120	—	—	
		Die cast	TS	265 (17)	295 (19)	1	—	—	—	—	
(LM16)	Si 5.0 Cu 1.0 Mg 0.5	Sand cast	TB	130 (8.5)	210 (13.5)	3	200 (15)	80	70 (4.5*)	1.5 (1)	Water-cooled cylinder heads and applications requiring leak-proof castings
		Chill cast	TB	130 (8.5)	245 (16)	6	210 (13.5)	85	85 (5.5*)	2.5 (2)	
		Sand cast	TF	245 (16)	255 (16.5)	1	215 (14)	100	60 (4)	1 (0.75)	
		Chill cast	TF	275 (18)	310 (20)	2	225 (14.5)	110	70 (4.5)	1.5 (1)	
Al-Si-Mg-Mn (LM9)	Si 12.0 Mg 0.4 Mn 0.5	Sand cast	TE	120 (8)	185 (12)	2	120 (8)	70	55* (3.5*)	1.5 (1)	Fluidity, corrosion resistance and high strength. Extensive use for low-pressure castings
		Chill cast	TE	160 (10.5)	255 (16.5)	2.5	160 (10.5)	80	70* (4.5*)	2.5 (2)	
		Sand cast	TF	235 (15)	255 (16.5)	1	200 (13)	100	70* (4.5*)	0.7 (0.5)	
		Chill cast	TF	275 (18)	310 (20)	1	230 (15)	110	85* (5.5*)	1.5 (1)	
(LM25)	Si 7.0 Mg 0.3	Sand cast	M	90 (6)	140 (9)	2.5	—	60	—	—	The most widely used general purpose, high-strength casting alloy
		Chill cast	M	90 (6)	180 (11.5)	4	—	60	—	—	
		Sand cast	TE	135 (8.5)	165 (10.5)	1.5	—	75	55 (3.5)	—	
		Chill cast	TE	165 (10.5)	220 (14)	2.5	—	85	—	—	
		Sand cast	TB7	95 (6)	170 (11)	3	—	65	—	—	
		Chill cast	TB7	100 (6.5)	230 (15)	8	—	70	75 (5)	—	
		Sand cast	TF	225 (14.5)	255 (16.5)	1	—	105	60 (4)		
		Chill cast	TF	240 (15.5)	310 (20)	3	—	105	95 (6)		
Al-Cu-Mg-Ni (Y alloy) (4L35)	Cu 4.0 Mg 1.5 Ni 2.0	Sand cast	TF	220 (14)	235 (15)	1	—	115	80* (5*)	1.5 (1)	Highly stressed components operating at elevated temperatures
		Chill cast	TF	240 (15.5)	290 (19)	2	—	115	110* (7*)	4.5 (3.5)	
Al-Si-Cu-Mg-Zn (LM21)	Si 6.0 Cu 4.0 Mg 0.2 Zn 1.0	Sand cast	M	130 (8.5)	180 (11.5)	1	—	85	—	—	General engineering applications, particularly crankcases
		Chill cast	M	130 (8.5)	200 (13)	2	—	90	—	—	

Al-Si-Cu-	Si 23.0	Sand cast	TE	120 (8)	130 (8.5)	0.3	—	120	—	—	Pistons for high performance
Mg–Ni	Cu 1.0	Chill cast	TE	170 (11)	210 (13.5)	0.3	—	120	—	—	internal combustion engines
(LM29)	Mg 1.0	Sand cast	TF	120 (8)	130 (8.5)	0.3	—	120	—	—	High performance piston alloy
	Ni 1.0	Chill cast	TF	170 (11)	210 (13.5)	0.3	—	120	—	—	
(LM28)	Si 19.0	Chill cast	TE	170 (11)	190 (12.5)	0.5	—	120	—	—	
	Cu 1.5										
	Mg 1.0	Sand cast	TF	120 (8)	130 (8.5)	0.5	—	120	—	—	
	Ni 1.0	Chill cast	TF	170 (11)	200 (13)	0.5	—	120	—	—	
	Si 11.0	Chill cast	TE	—	220 (14)	1	—	105	—	—	Low expansion piston alloy
	Cu 1.0										
(LM13)	Mg 1.0	Sand cast	TF	190 (12.5)	200 (13)	0.5	—	115	85* (5.5*)	—	
	Ni 1.0	Chill cast	TF	280 (18)	290 (19)	1	190 (12.5)	125	100* (6.5*)	1.4 (1)	
	(Lo-Ex)	Sand cast	TF7	140 (9)	150 (9.5)	1	—	75	—	—	
		Chill cast	TF7	200 (13)	210 (13.5)	1	—	75	—	—	
(LM26)	Si 9.0	Chill cast	TE	180 (11.5)	230 (15)	1	—	105	—	1.4 (1)	Piston alloy
	Cu 3.0										
	Mg 1.0										
	Ni 0.7										
Al–Cu–Si–	Cu 2.0	Sand cast	TF	260 (17)	285 (18.5)	1	—	120	80 (5)	—	Aircraft engine
Mg–Fe–Ni	Si 1.5	Chill cast	TF	305 (20)	335 (22)	1	—	125	—	—	castings for elevated
(3L52)	Mg 1.0										temperature service
	Fe 1.0										
	Ni 1.25										
Al–Cu–Si–	Cu 1.5	Sand cast	TE	135 (8.5)	170 (11)	2.5	—	70	—	—	Aircraft engine castings
Fe–Ni–Mg	Si 2.0	Chill cast	TE	150 (9.5)	210 (13.5)	3.5	—	75	—	—	
(3L51)	Fe 1.0										
	Ni 1.4										
	Mg 0.15										

* Fatigue limit for 50×10^6 cycles
M = as manufactured.
O = annealed.
H2, H4, H5, H6 = intermediate tempers
H8 = fully hard temper.

TS = stress relieved only.
TB = solution treated and naturally aged.
TB7 = solution treated and stabilized.
TD = solution treated, cold worked and naturally aged.
TF = solution treated and precipitation treated.
TF7 = full heat treatment plus stabilization.
TH = solution treated, cold work and precipitation treated.
TE = precipitation treated only.

(1) Special temper for maximum stress corrosion resistance (US designation T73).
(2) Special heat treatment for combination of properties (US designation T736).
(3) Special heat treatment for combination of properties (US designation T61).
(4) Special heat treatment for combination of properties (US designation T7351).

Table 22.3 ALUMINIUM AND ALUMINIUM ALLOYS—MECHANICAL PROPERTIES AT ELEVATED TEMPERATURES

Material (*specification*)	*Nominal composition* %	*Condition*	*Temp.* °C	*Time at temp.* h	0.2% *Proof stress* MPa tonf in^{-2}	*Tensile strength* MPa tonf in^{-2}	*Elong.*% *on* 50 mm *or* $5.65\sqrt{S_0}$
Wrought Alloys							
Al	Al 99.95	Rolled rod O	24	—	—	55 (3.6)	61
(1095)			93	—	—	45 (2.8)	63
			203	—	—	25 (1.6)	80
			316	—	—	12 (0.8)	105
			427	—	—	5 (0.3)	131
(1200)	Al 99	O	24	10000	35 (2.2)	90 (5.8)	45
			100	10000	35 (2.2)	75 (4.9)	45
			148	10000	30 (2.0)	60 (3.8)	55
			203	10000	25 (1.6)	40 (2.7)	65
			260	10000	14 (0.9)	30 (1.8)	75
			316	10000	11 (0.7)	17 (1.1)	80
			371	10000	6 (0.4)	14 (0.9)	85
		H4	24	10000	115 (7.6)	125 (8.1)	20
			100	10000	105 (6.7)	110 (7.2)	20
			148	10000	85 (5.4)	90 (5.8)	22
			203	10000	50 (3.1)	65 (4.2)	25
			260	10000	17 (1.1)	30 (1.8)	75
			316	10000	11 (0.7)	17 (1.1)	80
			371	10000	6 (0.4)	14 (0.9)	85
		H8	24	10000	150 (9.8)	165 (10.8)	15
			100	10000	125 (8.1)	150 (9.8)	15
			148	10000	95 (6.3)	125 (8.1)	20
			203	10000	30 (1.8)	40 (2.7)	65
			260	10000	14 (0.9)	30 (1.8)	75
			316	10000	11 (0.7)	17 (1.1)	80
			371	10000	6 (0.4)	14 (0.9)	85
Al–Mn	Mn 1.25	O	24	10000	40 (2.7)	110 (7.1)	40
(3103)			100	10000	37 (2.4)	90 (5.8)	43
			148	10000	34 (2.2)	75 (4.9)	47
			203	10000	30 (2.0)	60 (3.8)	60
			260	10000	25 (1.6)	40 (2.7)	65
			316	10000	17 (1.1)	30 (1.8)	70
			371	10000	14 (0.9)	20 (1.3)	70
		H4	24	10000	145 (9.4)	150 (9.8)	16
			100	10000	130 (8.5)	145 (9.4)	16
			148	10000	110 (7.1)	125 (8.0)	16
			203	10000	60 (4.0)	95 (6.2)	20
			260	10000	30 (1.8)	50 (3.3)	60
			316	10000	17 (1.1)	30 (1.8)	70
			371	10000	14 (0.9)	20 (1.3)	70
		H8	24	10000	185(12.1)	200(12.9)	10
			100	10000	145 (9.4)	180(11.6)	10
			148	10000	110 (7.1)	155(10.2)	11
			203	10000	60 (4.0)	95 (6.2)	18
			260	10000	30 (1.8)	50 (3.3)	60
			316	10000	17 (1.1)	30 (1.8)	70
			371	10000	14 (0.9)	20 (1.3)	70
Al–Mg	Mg 1.4	O	24	10000	55 (3.6)	145 (9.4)	—
(5050)			100	10000	55 (3.6)	145 (9.4)	—
			148	10000	55 (3.6)	130 (8.5)	—
			203	10000	50 (3.3)	95 (6.2)	—
			260	10000	40 (2.7)	60 (4.0)	—
			316	10000	30 (1.8)	40 (2.7)	—
			371	10000	20 (1.3)	30 (1.8)	—

Table 22.3 ALUMINIUM AND ALUMINIUM ALLOYS—MECHANICAL PROPERTIES AT ELEVATED TEMPERATURES—*continued*

Material (*specification*)	*Nominal composition* %	*Condition*	*Temp.* °C	*Time at temp.* h	0.2% *Proof stress* MPa tonf in^{-2}	*Tensile strength* MPa tonf in^{-2}	*Elong.* % *on* 50 mm *or* $5.65\sqrt{S_0}$
Al–Mg (*cont.*)		H4	24	10 000	165(10.7)	195(12.5)	—
			100	10 000	165(10.7)	195(12.5)	—
			148	10 000	150 (9.8)	165(10.7)	—
			203	10 000	50 (3.3)	95 (6.2)	—
			260	10 000	40 (2.7)	60 (4.0)	—
			316	10 000	35 (1.8)	40 (2.7)	—
			371	10 000	20 (1.3)	30 (1.8)	—
		H8	24	10 000	200(12.9)	220(14.2)	—
			100	10 000	200(12.9)	215(13.8)	—
			148	10 000	175(11.2)	180(11.6)	—
			203	10 000	60 (3.8)	95 (6.2)	—
			260	10 000	40 (2.7)	60 (4.0)	—
			316	10 000	35 (1.8)	40 (2.7)	—
			371	10 000	20 (1.3)	30 (1.8)	—
Al–Mg–Cr (5052)	Mg 2.25 Cr 0.25	O	24	10 000	90 (5.8)	195(12.5)	30
			100	10 000	90 (5.8)	190(12.3)	35
			148	10 000	90 (5.8)	165(10.7)	50
			203	10 000	75 (4.9)	125 (8.0)	65
			260	10 000	50 (3.3)	80 (5.3)	80
			316	10 000	35 (2.2)	50 (3.3)	100
			371	10 000	20 (1.3)	35 (2.2)	130
		H4	24	10 000	215(13.8)	260(17.0)	14
			100	10 000	205(13.4)	260(17.0)	16
			148	10 000	185(12.1)	215(13.8)	25
			203	10 000	105 (6.7)	155(10.2)	40
			260	10 000	50 (3.3)	80 (5.3)	80
			316	10 000	35 (2.2)	50 (3.3)	100
			371	10 000	20 (1.3)	35 (2.2)	130
		H8	24	10 000	255(16.5)	290(18.8)	8
			100	10 000	255(16.5)	285(18.3)	9
			148	10 000	200(12.9)	235(15.1)	20
			203	10 000	105 (6.7)	155(10.2)	40
			260	10 000	50 (3.3)	80 (5.3)	80
			316	10 000	35 (2.2)	50 (3.3)	100
			371	10 000	20 (1.3)	35 (2.2)	130
(5154)	Mg 3.5 Cr 0.25	O	24	10 000	125 (8.0)	240(15.6)	25
			100	10 000	125 (8.0)	240(15.6)	30
			148	10 000	125 (8.0)	195(12.5)	40
			203	10 000	95 (6.2)	145 (9.4)	55
			260	10 000	60 (4.0)	110 (7.1)	70
			316	10 000	40 (2.7)	70 (4.5)	100
			371	10 000	30 (1.8)	40 (2.7)	130
		H4	24	10 000	225(14.7)	290(18.8)	12
			100	10 000	220(14.2)	285(18.3)	16
			148	10 000	195(12.5)	235(15.1)	25
			203	10 000	110 (7.1)	175(11.2)	35
			260	10 000	60 (4.0)	110 (7.1)	70
			316	10 000	40 (2.7)	70 (4.5)	100
			371	10 000	30 (1.8)	40 (2.7)	130
		H8	24	10 000	270(17.4)	330(21.5)	8
			100	10 000	255(16.5)	310(20.1)	13
			148	10 000	220(14.2)	270(17.4)	20
			203	10 000	105 (6.7)	155(10.2)	35
			260	10 000	60 (4.0)	110 (7.1)	70
			316	10 000	40 (2.7)	70 (4.5)	100
			371	10 000	30 (1.8)	40 (2.7)	130

Table 22.3 ALUMINIUM AND ALUMINIUM ALLOYS—MECHANICAL PROPERTIES AT ELEVATED TEMPERATURES—*continued*

Material (specification)	*Nominal composition* %	*Condition*		*Temp.* °C	*Time at temp.* h	*0.2% Proof stress* MPa (tonf in^{-2})	*Tensile strength* MPa (tonf in^{-2})	*Elong % on 50 m or* $5.65\sqrt{S_0}$
Al–Mg–Mn (5056A)	Mg 5.0 Mn 0.3	As extruded	M	20	1 000	145 (9.5*)	300(19.5)	25
				50	1 000	145 (9.5)	300(19.5)	27
				100	1 000	145 (9.5)	300(19.5)	32
				150	1 000	135 (8.7)	245(15.8)	45
				200	1 000	111 (7.2)	215(13.8)	56
				250	1 000	75 (5.0)	130 (8.5)	77
				300	1 000	50 (3.2)	95 (6.0)	100
				350	1 000	20 (1.2)	60 (3.9)	140
Al–Mg–Si (6063)	Mg 0.7 Si 0.4		TF	24	10 000	215(13.8)	240(15.6)	18
				100	10 000	195(12.5)	215(13.8)	15
				148	10 000	135 (8.9)	145 (9.4)	20
				203	10 000	45 (2.9)	60 (4.0)	40
				260	10 000	25 (1.6)	30 (2.0)	75
				316	10 000	17 (1.1)	20 (1.3)	80
				371	10 000	14 (0.9)	17 (1.1)	105
(6082)	Mg 0.6 Si 1.0 Cr 0.25		TF	24	10 000	230(18.2)	330(21.5)	17
				100	10 000	270(17.5)	290(18.8)	19
				148	10 000	175(11.2)	185(12.1)	22
				203	10 000	65 (4.2)	80 (5.3)	40
				260	10 000	35 (2.4)	45 (2.9)	50
				316	10 000	30 (2.0)	35 (2.2)	50
				371	10 000	25 (1.6)	30 (1.8)	50
(6061)	Mg 1.0 Si 0.6 Cu 0.25 Cr 0.25		TF	24	10 000	275(17.9)	310(20.1)	17
				100	10 000	260(17.0)	290(18.8)	18
				148	10 000	213(13.8)	235(15.1)	20
				203	10 000	105 (6.7)	130 (8.5)	28
				260	10 000	35 (2.2)	50 (3.3)	60
				316	10 000	17 (1.1)	30 (2.0)	85
				371	10 000	14 (0.9)	20 (1.3)	95
Al–Cu–Mn (2219)	Cu 6 Mn 0.25	Forgings	TF	20	100	230(15*)	385(25)	8
				100	100	—	365(23.6)	—
				150	100	220(14.4)	325(21.0)	—
				200	100	185(12.0)	280(18.0)	—
				250	100	135 (8.6)	205(13.4)	—
				300	100	110 (7.0)	145 (9.5)	—
				350	100	45 (3.0)	70 (4.4)	—
				400	100	20 (1.4)	30 (1.8)	—
Al–Cu–Pb–Bi (2011)	Cu 5.5 Pb 0.5 Bi 0.5		TB	24	10 000	295(19.2)	375(24.3)	15
				100	10 000	235(15.1)	320(20.8)	16
				148	10 000	130 (8.5)	195(12.5)	25
				203	10 000	75 (4.9)	110 (7.1)	35
				260	10 000	30 (1.8)	45 (2.9)	45
				316	10 000	14 (0.9)	25 (1.6)	90
				371	10 000	11 (0.7)	17 (1.1)	125
Al–Cu–Mg–Mn (2017)	Cu 4.0 Mg 0.5 Mn 0.5		TB	24	10 000	275(17.8)	430(27.7)	22
				100	10 000	255(16.5)	385(25.0)	18
				148	10 000	205(13.4)	275(17.8)	16
				203	10 000	115 (7.6)	150 (9.8)	28
				260	10 000	65 (4.2)	80 (5.3)	45
				316	10 000	35 (2.2)	45 (2.9)	95
				371	10 000	25 (1.6)	30 (2.0)	100
(2024)	Cu 4.5 Mg 1.5 Mn 0.6		TB	24	10 000	340(22.0)	470(30.4)	19
				100	10 000	305(19.6)	422(27.3)	17
				148	10 000	245(16.0)	295(19.2)	17
				203	10 000	145 (9.4)	180(11.6)	22
				260	10 000	65 (4.2)	95 (6.2)	45
				316	10 000	35 (2.2)	50 (3.1)	75
				371	10 000	25 (1.6)	35 (2.2)	100

Table 22.3 ALUMINIUM AND ALUMINIUM ALLOYS—MECHANICAL PROPERTIES AT ELEVATED TEMPERATURES—*continued*

Material (specification)	*Nominal composition* %	*Condition*		*Temp.* °C	*Time at temp.* h	0.2% *Proof stress* MPa (tonf in^{-2})	*Tensile strength* MPa (tonf in^{-2})	*Elong.*% *on* 50 mm *or* $5.65\sqrt{S_0}$
Al–Cu–Mg–Si–Mn	Cu 4.4		TF	24	10 000	415(26.8)	485(31.3)	13
(2014)	Mg 0.4			100	10 000	385(25.0)	455(29.5)	14
	Si 0.8			148	10 000	275(17.8)	325(21.0)	15
	Mn 0.8			203	10 000	80 (5.3)	125 (8.0)	35
				260	10 000	60 (3.8)	75 (4.9)	45
				316	10 000	35 (2.2)	45 (2.9)	64
				371	10 000	25 (1.6)	30 (2.0)	20
		Forgings	TF	20	100	415*(27*)	480(31)	10
				100	100	410(26.5)	465(30)	—
				150	100	400(26)	430(28)	—
				200	100	260(17)	295(19)	—
				250	100	85 (5.5)	110 (7.0)	—
				300	100	45 (3.0)	70 (4.5)	—
				350	100	35 (2.4)	50 (3.2)	—
Al–Cu–Mg–Ni	Cu 2.2	Forgings	TF	20	100	325*(21*)	430(28)	8
(2618)	Mg 1.5			150	100	340 (22)	440(26.5)	—
	Ni 1.2			200	100	260(17)	300(19.5)	—
	Fe 1.0			250	100	170(11)	210(13.5)	—
				300	100	70 (4.5)	115 (7.6)	—
				350	100	30 (2.0)	50 (3.2)	—
				400	100	20 (1.2)	30 (2.0)	—
(2031)	Cu 2.2	Forgings	TF	20	100	325*(21*)	430(28)	13
	Mg 1.5			100	100	310(20)	400(26)	—
	Ni 1.2			200	100	255(16.5)	310(20)	—
	Fe 1.0			250	100	110 (7.0)	155(10)	—
	Si 0.8			300	100	45 (3.0)	75 (4.7)	—
				350	100	30 (2.0)	40 (2.5)	—
Al–Si–Cu–Mg–Ni	Si 12.2	Forgings	TF	24	10 000	320(20.6)	380(24.6)	9
(4032)	Cu 0.9			100	10 000	305(19.9)	345(22.3)	9
	Mg 1.1			148	10 000	225(14.7)	255(16.5)	9
	Ni 0.9			203	10 000	60 (4.0)	90 (5.8)	30
				260	10 000	35 (2.4)	55 (3.6)	50
				316	10 000	20 (1.3)	35 (2.2)	70
				371	10 000	14 (0.9)	25 (1.6)	90
Al–Zn–Mg–Cu	Zn 5.6		TF	24	10 000	505 (32.6)	570(37)	11
(7075)	Cu 1.6			100	10 000	430(27.7)	455(29.5)	15
	Mg 2.5			148	10 000	145 (9.4)	175(11.2)	30
	Cr 0.3			203	10 000	80 (5.3)	95 (6.2)	60
				260	10 000	60 (3.8)	75 (4.9)	65
				316	10 000	45 (2.9)	60 (3.8)	80
				371	10 000	30 (2.0)	45 (2.9)	65
Cast Alloys								
Al–Mg	Mg 5.0	Sand cast	M	20	1 000	95 (6.0*)	160(10.5)	4
(LM 5)	Mn 0.5			100	1 000	100 (6.5)	160(10.5)	3
				200	1 000	95 (6.0)	130 (8.5)	3
				300	1 000	55 (3.5)	95 (6.0)	4
				400	1 000	15 (1.0)	30 (2.0)	4
(LM 10)	Mg 10.0	Sand cast	TB	20	1 000	180(11.8*)	340(22.0)	16
				100	1 000	205(13.2)	350(22.6)	10
				150	1 000	154(10.0)	270(17.5)	0
				200	1 000	105 (6.8)	185(12.0)	42
				300	1 000	40 (2.6)	90 (5.9)	85
				400	1 000	11 (0.7)	45 (3.0)	100

Table 22.3 ALUMINIUM AND ALUMINIUM ALLOYS—MECHANICAL PROPERTIES AT ELEVATED TEMPERATURES—*continued*

Material (specification)	*Nominal composition %*	*Condition*		*Temp. °C*	*Time at temp. h*	*0.2% Proof stress* MPa (tonf in^{-2})	*Tensile strength* MPa (tonf in^{-2})	*Elong. % on 50 mm or* $5.65\sqrt{S_0}$
Al–Si (LM 18)	Si 5.0	Pressure die cast	M	24	10 000	110(7.1)	205(13.4)	9
				100	10 000	110(7.1)	175(11.2)	9
				148	10 000	103(6.7)	135(8.9)	10
				203	10 000	80(5.3)	110(7.1)	17
				260	10 000	40(2.7)	55(3.6)	23
(LM 6)	Si 12.0	Pressure die cast	M	24	10 000	145(9.4)	270(17.4)	2
				100	10 000	145(9.4)	225(14.7)	$2\frac{1}{2}$
				148	10 000	125(8.0)	185(12.1)	3
				206	10 000	105(6.7)	150(9.8)	7
				260	10 000	40(2.7)	75(4.9)	13
Al–Si–Cu (LM 4)	Si 5.0, Cu 3.0, Mn 0.5	Sand cast	M	20	1 000	95*(6.0*)	155(10.0)	2
				100	1 000	140(9.0)	180(11.8)	2
				200	1 000	110(7.0)	135(8.8)	2
				300	1 000	40(2.7)	60(4.0)	12
				400	1 000	20(1.3)	30(2.0)	27
Al–Si–Mg (LM 25)	Si 5.0, Mg 0.5	Chill cast	TF	20	1 000	270*(17.6*)	325(21.0)	2
				100	1 000	255(16.6)	290(18.8)	2
				200	1 000	60(4.0)	90(5.8)	25
				300	1 000	25(1.5)	40(2.5)	65
				400	1 000	12(0.8)	25(1.5)	65
Al–Cu–Mg–Ni (4L 35)	Cu 4.0, Mg 1.5, Ni 2.0	Sand cast	TF	20	1 000	200*(13.0*)	275(17.7)	$\frac{1}{2}$
				100	1 000	255(16.6)	325(21.0)	$\frac{1}{2}$
				200	1 000	150(9.8)	135(12.0)	$\frac{1}{2}$
				300	1 000	30(2.0)	55(3.6)	32
				400	1 000	15(1.0)	40(2.5)	60
Al–Si–Ni–Cu–Mg (LM 13)	Si 12.0, Ni 2.5, Cu 1.0, Mg 1.0	Chill cast	TF	20	1 000	275*(17.8*)	285(18.6)	$\frac{1}{2}$
				100	1 000	280(18.0)	320(20.8)	$\frac{1}{2}$
				200	1 000	110(7.2)	165(10.8)	$\frac{1}{2}$
				300	1 000	30(1.8)	60(3.8)	15
				400	1 000	15(1.0)	35(2.2)	25
		Chill cast Special	TF	20	1 000	200*(13.0*)	275(17.8)	1
				100	1 000	195(12.5)	250(16.3)	1
				200	1 000	110(7.2)	170(11.0)	3
				300	1 000	35(2.2)	60(4.0)	15
				400	1 000	15(1.0)	35(2.2)	50

* 0.1% Proof stress.
M = As manufactured.
O = Annealed.
H4 = Half hard.
H8 = Hard.
TB = Solution treated and naturally aged.
TF = Solution treated and artificially aged.

Table 22.4 ALUMINIUM AND ALUMINIUM ALLOYS—MECHANICAL PROPERTIES AT LOW TEMPERATURES

Material (specification)	*Nominal composition %*	*Condition*		*Temp. °C*	*0.2% Proof stress* MPa (tonf in^{-2})	*Tensile strength* MPa (tonf in^{-2})	*Elong. % on 4D or 50 mm*	*Reduction in area %*	*Reference*
Al (1200)	Al 99.0	Rolled and drawn rod	O	24	34(2.2)	90(5.8)	42.5	76.4	
				−28	34(2.2)	95(6.0)	43.0	76.4	1
				−80	37(2.4)	100(6.6)	47.5	77.0	
				−196	43(2.8)	170(11.0)	56	74.4	
			H8	24	140(9.0)	155(9.9)	16	59.8	
				−28	144(9.3)	155(10.2)	15.2	59.4	1
				−80	147(9.5)	165(10.8)	18.0	65.3	
				−196	165(10.8)	225(14.5)	35.2	67.0	

Table 22.4 ALUMINIUM AND ALUMINIUM ALLOYS—MECHANICAL PROPERTIES AT LOW TEMPERATURES—*continued*

Material (specification)	*Nominal composition %*	*Condition*		*Temp. °C*	*0.2% Proof stress* MPa (tonf in^{-2})	*Tensile strength* MPa (tonf in^{-2})	*Elong.* % on 4D or 50 mm	*Reduction in area* %	*Reference*
Al–Mn (3103)	Mn 1.25	Rolled and drawn rod	O	24	40(2.7)	110(7.0)	43.0	80.6	
				−28	40(2.7)	115(7.4)	44.0	80.6	1
				−80	50(3.2)	130(8.5)	45.0	79.9	
				−196	60(3.8)	220(14.3)	48.8	71.2	
		Rolled and drawn rod	H8	24	180(11.7)	195(12.7)	15.0	63.5	
				−28	185(12.0)	205(13.4)	15.0	64.4	1
				−80	195(12.6)	215(14.0)	16.5	66.5	
				−196	220(14.2)	290(18.8)	32.0	62.3	
Al–Mg (5052)	Mg 2.5 Cr 0.25	Rolled and drawn rod	O	24	97(6.3)	199(12.9)	33.2	72.0	
				−28	99(6.4)	201(13.0)	35.8	74.2	1
				−80	97(6.3)	210(13.6)	40.8	76.4	
				−196	115(7.5)	330(21.4)	50.0	69.0	
			H8	24	235(15.2)	275(17.9)	16.6	59.1	
				−28	230(15.0)	280(18.2)	18.3	63.2	1
				−80	236(15.3)	290(18.9)	20.6	64.5	
				−196	275(17.8)	400(25.8)	30.9	57.4	
(5154)	Mg 3.5 Cr 0.25	Sheet	O	26	115(7.6)	240(15.6)	28	66	
				−28	115(7.6)	240(15.6)	32	72	
				−80	115(7.6)	250(16.1)	35	73	
				−196	135(8.9)	350(22.8)	42	60	
			H8	26	275(17.9)	330(21.3)	9	—	
				−80	280(18.2)	340(22.0)	14	—	2
				−196	325(21.0)	455(29.6)	30	—	
				−253	370(24.1)	645(41.7)	35	—	
(5056A)	Mg 5.0 Mn 0.2	Plate	O	20	130(8.4)	290(18.9)	30.5	32.0	
				−75	130(8.4)	290(18.9)	38.2	48.2	1
				−196	145(9.3)	420(27.3)	50.0	36.2	
Al–Mg–Si (6063)	Mg 0.7 Si 0.4	Extrusion	TB	26	90(5.8)	175(11.2)	32	78	
				−28	105(6.7)	190(12.3)	33	75	
				−80	115(7.6)	200(12.9)	36	75	
				−196	115(7.6)	260(16.7)	42	73	
		Extrusion	TF	26	215(13.8)	240(15.6)	16	36	
				−28	220(14.3)	250(16.1)	16	36	
				−80	225(14.7)	260(17.0)	17	38	
				−196	250(16.1)	330(21.4)	21	40	
Al–Mg–Si–Cr (6151)	Mg 0.7 Si 1.0 Cr 0.25	Forging	TF	24	300(19.3)	320(20.8)	15.2	38.8	
				−28	310(20.0)	325(21.1)	12.0	34.0	1
				−80	305(19.7)	330(21.3)	14.9	38.7	
				−196	330(21.5)	385(24.9)	18.3	34.7	
Al–Mg-Si–Cu–Cr (6061)	Mg 1.0 Si 0.6 Cu 0.25 Cr 0.25	Rolled and drawn rod	TF	24	270(17.6)	315(20.5)	21.8	56.4	
				−28	280(18.2)	330(21.5)	21.5	52.5	1
				−80	290(18.8)	345(22.5)	22.5	53.7	
				−196	315(20.5)	425(27.4)	26.5	46.5	
Al–Cu–Mg–Mn (2024)	Cu 4.5 Mg 1.5 Mn 0.6	Rolled and drawn rod	TB	24	300(19.5)	480(31.2)	23.3	31.8	
				−28	305(19.7)	500(32.3)	24.4	33.1	1
				−80	320(20.7)	510(33.2)	25.3	30.8	
				−196	400(26.0)	615(39.8)	26.7	26.3	
		Rolled and drawn rod	TF	24	400(26.0)	500(32.4)	14.5	25.8	
				−28	405(26.1)	502(32.5)	12.7	21.5	1
				−80	415(26.9)	514(33.3)	13.3	22.0	
				−196	460(29.8)	605(39.1)	14.0	19.7	

Table 22.4 ALUMINIUM AND ALUMINIUM ALLOYS—MECHANICAL PROPERTIES AT LOW TEMPERATURES—*continued*

Material (specification)	*Nominal composition %*	*Condition*		*Temp. °C*	*0.2% Proof stress MPa (tonf in^{-2})*	*Tensile strength MPa (tonf in^{-2})*	*Elong. % on 4D or 50 mm*	*Reduction in area %*	*Reference*
Al–Cu–Si–Mg–Mn (2014)	Cu 4.5 Si 0.8 Mg 0.5 Mn 0.8	Rod	TB	26	290(18.8)	430(27.7)	20	28	
				−28	290(18.8)	440(28.6)	22	28	
				−80	302(19.6)	440(28.6)	22	26	
				−196	380(24.6)	545(35.3)	20	20	
		Rod	TF	26	415(26.8)	485(31.3)	13	31	
				−28	415(26.8)	485(31.3)	13	29	1
				−80	420(27.2)	495(32.1)	14	28	
				−196	470(30.4)	565(36.6)	14	26	
		Forging	TF	26	410(26.7)	465(30.1)	12	24	
				−80	460(29.9)	510(32.9)	14	24	2
				−196	530(34.5)	610(39.6)	11	22	
				−253	590(38.2)	715(46.4)	7	22	
Al–Zn–Mg–Cu (7075)	Zn 5.6 Mg 2.5 Cu 1.6	Rolled and drawn rod	TF	24	485(31.4)	560(36.3)	15.0	29.1	
				−28	490(31.8)	570(37.0)	15.3	26.2	1
				−80	505(32.8)	590(38.2)	15.3	23.6	
				−196	570(37.0)	670(43.4)	16.0	20.1	

O = Annealed.
H8 = Fully hard temper.
TB = Solution treated and naturally aged.
TF = Solution treated and precipitation treated.

Table 22.5 ALUMINIUM ALLOYS—CREEP DATA

Material (specification)	*Nominal composition %*	*Condition*		*Temp. °C*	*Stress MPa (tonf in^{-2})*	*Minimum creep rate % per 1 000 h*	*Total extension % in 1 000 h*	*Reference*
Al (1080)	99.8	Sheet	O	20	24.1(1.56)	0.005	0.39	
				20	27.6(1.79)	0.045	1.28	
				80	7.0(0.45)	0.005	0.045	
				80	8.3(0.535)	0.01	0.065	
				250	1.4(0.09)	0.005	0.047	
				250	2.1(0.134)	0.01	0.047	
				250	2.8(0.179)	0.015	0.052	
				250	4.1(0.268)	0.055	0.152	
Al–Mg (5052)	Mg 2.5	Sheet	O	80	45(3)	0.005	0.085	
(LM 5)	Mg 5.6	Cast		100	110(7)	0.055	0.33	3
				100	115(7.5)	0.17	0.57	
				100	125(8)	0.21	1.19	
				200	30(2)	0.08	0.21	
				200	45(3)	0.20	0.39	
				200	60(4)	0.62	0.92	
				300	3.9(0.25)	0.045	0.10	
				300	7.7(0.50)	0.12	0.25	
				300	15(1.0)	0.35	0.60	
(LM 10)	Mg 10	Cast		100	40(2.5)	0.013	0.126	3
				100	55(3.5)	0.022	0.107	
				100	75(5.0)	0.046	0.174	
				150	7.5(0.5)	0.126	0.413	
				150	15(1.0)	0.147	0.647	
				200	7.5(0.5)	0.107	0.341	
				200	15(1.0)	0.273	0.658	

Table 22.5 ALUMINIUM ALLOYS—CREEP DATA—*continued*

Material (specification)	*Nominal composition* %	*Condition*	*Temp* °C	*Stress* MPa (tonf in^{-2})	*Minimum creep rate* % *per* 1 000 h	*Total extension* % *in* 1 000 h	*Reference*
Al–Cu	Cu 4	Cast	205	17(1.1)	0.04	—	5
			205	34(2.2)	0.09	—	
			205	51(3.3)	0.14	—	
			205	70(4.5)	0.69	—	
			315	8.9(0.58)	0.13	—	
			315	13.1(0.85)	0.29	—	
	Cu 10	Cast	205	34(2.2)	0.01	—	5
			205	68(4.4)	0.11	—	
			315	8.9(0.58)	0.12	—	
			315	13.1(0.85)	0.43	—	
			315	17(1.1)	0.99	—	
Al–Si (LM 13)	Si 13 Ni 1.7 Mg 1.3	Sandcast (modified)	100	45(3)	0.016	0.190	3
			100	60(4)	0.06	0.675	
			200	15(1)	0.016	0.096	
			200	23(1.5)	0.054	0.179	
			200	30(2)	0.14	0.432	
			300	3.8(0.25)	0.013	0.026	
			300	7.7(0.50)	0.047	0.098	
			300	15(1.0)	0.223	0.428	
Al–Mn (3103)	Mn 1.25	Extruded rod	200	15(1.0)	0.001	—	8
			200	31(2.0)	0.022	—	
			200	34.8(2.25)	0.040	—	
			200	38.6(2.50)	0.060	—	
			200	42.5(2.75)	0.13	—	
			200	46(3.0)	0.15	—	
			200	54(3.5)	0.73	—	
			300	7.5(0.5)	0.007	—	
			300	15(1.0)	0.39	—	
Al–Cu–Si (2025)	Cu 4 Si 0.8	Extruded TF	150	90(6)	0.03	0.340	3
			150	125(8)	0.045	0.395	
			150	155(10)	0.325	0.722	
			200	30(2)	0.035	0.107	
			200	45(3)	0.1	0.204	
			200	60(4)	0.040	0.700	
			250	15(1)	0.02	0.156	
			250	23(1.5)	0.07	0.176	
			250	30(2)	2.36	—	
Al–Cu–Mg–Mn (2024)	Cu 4.5 Mg 1.5 Mn 0.6	Clad sheet TB	35	415(27)	10.0	—	4
			100	344(22.3)	1.0	—	
			100	385(25)	10.0	—	
			150	276(17.9)	1.0	—	
			150	327(21.2)	10.0	—	
			190	140(9)	1.0	—	
			190	200(13)	10.0	—	
		Clad sheet TF	35	424(27.5)	1.0	—	4
			35	430(28)	10.0	—	
			100	347(22.5)	1.0	—	
			100	363(23.5)	10.0	—	
			150	242(15.7)	1.0	—	
			150	289(18.7)	10.0	—	
			190	117(7.6)	1.0	—	
			190	193(12.5)	10.0	—	

Table 22.5 ALUMINIUM ALLOYS—CREEP DATA—*continued*

Material (specification)	*Nominal composition %*	*Condition*	*Temp* °C	*Stress* MPa (tonf in^{-2})	*Minimum creep rate* % *per* 1 000 h	*Total extension* % *in* 1 000 h	*Reference*
Al–Cu–Mg–Ni (2218)	Cu 4 Mg 1.5 Ni 2.2	Forged TB	100	193(12.5)	0.01	0.394	3
			100	232(15.0)	0.02	0.440	
			100	270(17.5)	0.04	0.835	
			200	77(5.0)	0.028	0.173	
			200	108(7.0)	0.16	0.345	
			300	7(0.5)	0.037	0.078	
			300	15(1.0)	0.5	0.640	
			400	1.5(0.1)	0.05	0.110	
		Cast TB	200	77(5.0)	0.01	0.153	3
			200	116(7.5)	0.08	0.287	
			300	7(0.5)	0.018	0.072	
			300	15(1.0)	0.08	0.151	
			400	1.5(0.1)	0.06	0.132	
Al–Cu–Mg–Zn (7075)	Zn 5.6 Cu 1.6 Mg 2.5	Clad sheet TF	35	430(28)	0.1	—	4
			35	480(31)	1.0	—	
			35	495(32)	10.0	—	
			100	295(19)	0.1	—	
			100	355(23)	1.0	—	
			100	370(24)	10.0	—	
			150	70(4.5)	0.1	—	
			150	170(11)	1.0	—	
			150	245(16)	10.0	—	
			190	45(3)	0.1	—	
			190	75(5)	1.0	—	
			190	125(8)	10.0	—	
Al–Mg–Si–Mn (6351)	Mg 0.7 Si 1.0 Mn 0.6	Extruded rod	100	193(12.5)	0.007	—	8
			100	201(13.0)	0.010	—	
			100	232(15.0)	0.11	—	
			100	255(16.5)	1.6	—	
			150	93(6.0)	0.0087	—	
			150	108(7.0)	0.023	—	
			150	154(10.0)	0.22	—	
			200	31(2.0)	0.011	—	
			200	46(3.0)	0.040	—	
			200	62(4.0)	0.13	—	
			200	77(5.0)	0.28	—	

O = Annealed.
TB = Solution treated and naturally aged, will respond to precipitation treatment.
TF = Solution treated and artificially aged.

Table 22.6 ALUMINIUM ALLOYS—FATIGUE STRENGTH AT VARIOUS TEMPERATURES

Material (specification)	*Nominal composition %*	*Condition*	*Temp.* °C	*Endurance (unnotched)* MPa (tonf in^{-2})	MHz	*Remarks*	*Reference*
Al–Mg (5056)	Mg 5.0	Extruded	−65	184(11.9)	20	Rotating beam	
			−35	164(10.6)			
			+20	133(8.6)			
	Mg 7.0	Extruded rod	−65	182(11.8)	20	Rotating beam	
			−35	178(11.5)			
			+20	173(11.2)			
(LM 10)	Mg 10.0	Sand cast (oil quenched)	20	93(6.0)	30	Rotating beam	9
			150	77(5.0)			
			200	40(2.6)			

Table 22.6 ALUMINIUM ALLOYS—FATIGUE STRENGTH AT VARIOUS TEMPERATURES

Material (specification)	*Nominal composition* %	*Condition*		*Temp.* °C	*Endurance (unnotched)* MPa (tonf in^{-2})	MHz	*Remarks*	*Reference*
Al–Si (LM 6)	Si 12.0	Sand cast (modified)		20 100 200 300	51(3.3) 43(2.8) 35(2.25) 25(1.6)	50	Rotating beam, 24 h at temp.	6
Al–Cu (2219)	Cu 6.0	Forged	TF	20 150 200 250 300 350	117(7.6) 65(4.2) 62(4.0) 46(3.0) 39(2.5) 23(1.5)	120	Reverse bending stresses	9
Al–Si–Cu (LM 22)	Si 4.6 Cu 2.8	Sand cast		20 100 200 300	62(4.0) 54(3.5) 60(3.9) 42(2.7)	50	Rotating beam	6
Al–Cu–Si–Mn (2014)	Cu 4.5 Si 0.8 Mn 0.8	Forgings	TF	148 203 260	65(4.2) 45(2.9) 25(1.6)	100	Rotating beam	
Al–Cu–Mn–Mg (2014)	Cu 4.0 Mn 0.5 Mg 0.5	Extruded rod	TB	25 148 203 260	103(6.7) 93(6.0) 65(4.2) 31(2.0)	500	Rotating beam, 100 days at temp.	7
Al–Cu–Mg–Si–Mn (2014)	Cu 4.4 Mg 0.7 Si 0.8 Mn 0.8	Forgings	TB	20 150 200 250 300	119(7.7) 90(5.8) 62(4.0) 54(3.5) 39(2.5)	120	Reversed bending	9
		Forgings	TF	20 150 200 250 300	130(8.4) 79(5.1) 57(3.7) 39(2.5) 39(2.5)	120	Reversed bending	9
Al–Cu–Mg–Ni (2218)	Cu 4.0 Mg 1.5 Ni 2.0	Forged		20 148 203 260	117(7.6) 103(6.7) 65(4.2) 45(2.9)	500 100 100 100	Rotating beam after prolonged heating	
		Chill cast	TF	20 100 200 300	100(6.5) 105(6.8) 108(7.0) 80(5.2)	50	Rotating beam, 24 h at temp.	6
Al–Ni–Cu	Ni 2.5 Cu 2.2	Forged	TF	20 150 200 250 300 350	113(7.3) 82(5.3) 70(4.5) 59(3.8) 39(2.5) 39(2.5)	120	Reversed bending	9
Al–Si–Cu–Mg–Ni (LM13)	Si 12.0 Cu 1.0 Mg 1.0 Ni 1.0	Chill cast (Lo-Ex)		20 100 200 300	97 (6.3) 107 (6.9) 97 (6.3) 54 (3.5)	50	Rotating beam, 24 h at temp.	6

Table 22.6 ALUMINIUM ALLOYS—FATIGUE STRENGTH AT VARIOUS TEMPERATURES—*continued*

Material (specification)	*Nominal composition %*	*Condition*		*Temp.* °C	*Endurance (unnotched)* MPa (tonf in^{-2})	MHz	*Remarks*	*Reference*
Al–Zn–Mg–Cu (7075)	Zn 5.6 Mg 2.5 Cu 1.6 Cr 0.2	Plate	TF	24 149 204 260	151 (9.8) 83 (5.4) 59 (3.8) 48 (3.1)	500	Reversed bending	—

TB = Solution treated and naturally aged, will respond to precipitation treatment.
TF = Solution treated and artificially aged.

REFERENCES

1. Bogardus, S. W. Steckley and F. M. Howell, N.A.C.A. Technical Note 2082, 1950.
2. 'A Review of Current Literature of Metals at Very Low Temperatures', Battelle Memorial Institute; 1961.
3. J. McKeown and R. D. S. Lushey, *Metallurgia*, 1951, **43**, 15.
4. A. E. Flanigan, L. F. Tedsen and J. E. Dorn, *Trans. Amer. Inst. Min. Met. Eng.*, 1947, **171**, 213.
5. R. R. Kennedy, *Proc. Am. Soc. Test. Mat.*, 1935, **35**, 218.
6. J. McKeown, D. E. Dineen and L. H. Back, *Metallurgia*, 1950, **41**, 393.
7. F. M. Howell and E. S. Howarth, *Proc. Am. Soc. Test Mat.*, 1937, **37**, 206.
8. N. P. Inglis and E. G. Larke, *J. Inst. Mech. Engrs.*, 1959.
9. P. H. Frith, 'Properties of Wrought and Cast Aluminium and Magnesium Alloys at Atmospheric and Elevated Temperatures', HMSO, 1956.

22.2 Mechanical properties of copper and copper alloys

22.2.1 Standard specifications

United Kingdom— British Standards—BS designation

BS 1400	Copper alloy ingots and copper and copper alloy castings
2870	Rolled copper and copper alloys, sheet, strip and coil
2871	Copper and copper alloys—tubes
2872	Copper and copper alloys—forging stock and forgings
2873	Copper and copper alloys—wire
2874	Copper and copper alloys—rods and sections
2875	Copper and copper alloys—plate

International Standards Organisation—ISO designation

ISO/R1190–1	1971 Copper and copper alloys. Parts 1 and 2
ISO 426/I	1973 Wrought copper–zinc alloys. Part 1
ISO 426/II	1973 Wrought copper–zinc alloys. Part 2
ISO 427	1973 Wrought copper–tin alloys
ISO 428	1973 Wrought copper–aluminium alloys
ISO 429	1973 Wrought copper–nickel alloys
ISO 430	1973 Wrought copper–nickel–zinc alloys
ISO 431	1972 Electrolytic tough pitch copper
ISO/R197	1961 Classification of coppers
ISO/R1187	1971 Special wrought copper alloys
ISO/R1190–1	1971 Copper and copper alloys. Parts 1 and 2

Table 22.7 COPPER AND COPPER ALLOYS—TYPICAL MECHANICAL PROPERTIES AT ROOM TEMPERATURE

Condition of material is expressed in accordance with BS Nomenclature, viz:

- O — Material in the annealed condition
- $\frac{1}{4}$H, $\frac{1}{2}$H, H, EH — The various harder tempers produced by cold rolling. For certain of the materials in this schedule, these tempers may be produced by partial annealing
- SH, ESH — Spring hard tempers produced by cold rolling of thinner material
- M — Material in the 'as manufactured' condition. In this schedule confined to hot rolled or extruded material
- W — Material which has been solution heat treated and will respond effectively to precipitation treatment
- W($\frac{1}{4}$H), W($\frac{1}{2}$H), W(H) — Material which has been solution heat treated and subsequently cold worked to various harder tempers
- WP — Material which has been solution heat treated and precipitation treated
- W($\frac{1}{4}$H)P, W($\frac{1}{2}$H)P, W(H)P — Material which has been solution heat treated, cold worked and then precipitation treated

British Standards gives strengths in N mm^{-2} – numerically equal to MPa. 1 N mm^{-2} ≡ 1 MPa.

Material and composition		British Standard specification number	Condition	Limit of proportionality		0.2% Proof stress		UTS		Elongation on 5d or 50 mm	Shear strength		Brinell hardness	Vickers hardness	Modulus of elasticity	
				MPa	(tonf in^{-2})	MPa	(tonf in^{-2})	MPa	(tonf in^{-2})	%	MPa	(tonf in^{-2})	kg mm^{-2}	(10 kilo)	GPa	10^6 lb in^{-2}
OFHC copper*	Cu 99.95%+	BS 2870 C 103	Strip—O	15	1.0	48	3.1	216.	14.0	48	162	10.5	42	51	117	17
	Cu 99.99%+	BS 3839 C110	Strip—$\frac{1}{2}$H	108	7.0	176	11.3	263	17.0	32	170	11.0	82	90		
			Strip—H	154	10.0	265	17.2	314	20.5	16	185	12.0	96	106		
Tough pitch copper $O_2=0.03\%$		BS 2870 C 101	Strip—O	31	2.0	54	3.5	224	14.5	56	162	10.5	49	53		
			Strip—$\frac{1}{2}$H	116	7.5	176	11.3	263	17.0	29	170	11.0	74	88		
			Strip—H	154	10.0	270	17.5	314	20.5	13	185	12.0	87	107		
															117	17
		BS 2873 and BS 2874 C 101	Wire and Rod—O	—	—	—	—	232	15.0	45	—	—	—	—		
			H (Over 5 mm dia.)	—	—	—	—	370	24.0	—	—	—	—	—		
			H (Under 5 mm dia.)	—	—	—	—	448	29.0	—	—	—	—	—		

*'OFHC' is a registered trade mark of Amax Copper Inc.

Table 22.7 COPPER AND COPPER ALLOYS—TYPICAL MECHANICAL PROPERTIES AT ROOM TEMPERATURE—*continued*

Material and composition	*British Standard specification number*	*Condition*	*Limit of proportionality*		*0.2% Proof stress*		*UTS*		*Elongation on 5d or 50 mm*	*Shear strength*		*Brinell hardness*	*Vickers hardness*	*Modulus of elasticity*	
			MPa	(tonf in^{-2})	MPa	(tonf in^{-2})	MPa	(tonf in^{-2})	%	MPa	(tonf in^{-2})	kg mm^{-2}	(10 kilo)	GPa	10^6 lb in^{-2}
Deoxidized non-arsenical copper P 0.04	BS 2875 BS 2871 Pt. 1	Plate M	31	2.0	54	3.5	239	15.5	58	162	10.5	54	58	117	17
	C 106	Tube—½H	—	—	170	11.0	263	17.0	30	—	—	82	90		
Deoxidized arsenical copper P 0.03% As 0.35%	BS 2875 C 107	Plate—M	31	2.0	54	3.5	239	15.5	56	162	10.5	50	54	117	17
Brasses Cu 67 Zn 33	BS 2870 CZ 107	Strip—O	54	3.5	115	7.5	331	21.5	60	247	16.0	65	70	112	16.2
		Strip—½H	154	10.0	278	18.0	388	25.0	35	263	17.0	110	117		
		Strip—H	208	13.5	432	28.0	510	33.0	12	293	19.0	140	150		
Cu 63 Zn 37	BS 2870 CZ 108	Strip—O	70	4.5	131	8.5	331	21.5	55	278	18.0	65	70	108	15.8
		Strip—½H	170	11.0	285	18.5	402	26.0	28	293	19.0	110	117		
		Strip—H	210	13.7	402	26.0	494	32.0	10	309	20.0	136	145		
Cu 60 Zn 40	BS 2874 CZ 123	Rod—½H	—	—	309	20.0	424	27.5	20	300	19.5	120	125	102	14.8
	BS 2875 CZ 123	Plate—M	62	4.0	130	9.0	371	24.0	40	278	18.0	85	90		
Aluminium brass Cu 76 Al 2 Zn 22	BS 2871 CZ 110	Tube—O	—	—	139	9.0	378	24.5	55	269	17.5	75	80	110	16.0
	BS 2875 CZ 110	Plate—M	—	—	154	10.0	393	25.5	50	285	18.5	95	100		
Naval brass Cu 62 Zn 37 Sn 1	BS 2874 CZ 112	Rod—M	93	6.0	170	11.0	408	26.5	35	316	20.5	100	105	103	15.0
		Plate—M	77	5.0	124	8.0	386	25.0	40	285	18.5	90	95		
Free cutting brass Cu 58 Zn 39 Pb 3	BS 2874 CZ 121	Rod—M	—	—	201	13.0	417	27.0	25	309	20.0	105	110	96	13.9
Hot stamping brass Cu 58 Zn 40 Pb 2	BS 2874 CZ 122	Rod—M	—	—	224	14.5	402	26.0	30	316	20.5	95	100	96	13.9
High tensile brass Cu 57.5 Mn 1.5 Sn 0.5 Fe 0.5 Zn 40.0	BS 2874 CZ 115	Rod—M	93	6.0	247	16.0	532	34.5	30	309	20.0	—	—	103	15.0
High tensile brass Cu 58.0 Mn 1.0 Fe 1.0 Sn 1.0 Al 0.5 Zn 38.5	BS 2874 CZ 114	Rod—M	139	9.0	309	20.0	566	36.0	30	340	22.0	—	—	103	15.0

Nickel silvers	BS 2870 NS 103	Strip—O	62	4.0	100	6.5	340	22.0	65	—	—	66	69		
Cu 62 Ni 10 Zn 28		Strip—½H	170	11.0	332	21.5	432	28.0	28	—	—	121	158	121	17.5
		Strip—H	270	17.6	510	33.0	564	36.5	11	—	—	155	177		
Cu 62 Ni 12 Zn 26	BS 2870 NS 104	Strip—O	62	4.0	108	7.0	340	22.0	60	—	—	65	68	124	18.0
		Strip—H	309	20.0	587	38.0	695	45.0	4	—	—	210	220		
Cu 62 Ni 15 Zn 23	BS 2870 NS 105	Strip—O	77	5.0	124	8.0	355	23.0	55	—	—	70	74	121	17.5
		Strip—H	340	22.0	618	40.0	695	45.0	4	—	—	210	220		
Cu 62 Ni 18 Zn 20	BS 2870 NS 106	Strip—O	70	4.5	124	7.5	386	25.0	52	—	—	77	81		
		Strip—½H	91	14.0	386	25.0	486	31.5	21	—	—	138	155	121	17.5
		Strip—H	123	19.0	525	34.0	610	39.5	7	—	—	166	168		
Silver-bearing copper	—	Strip—O	31	2.0	54	3.5	224	14.5	56	162	10.5	55	59		
Ag 0.05%		Strip—½H	—	—	180	11.6	286	18.5	25	178	11.5	75	80	117	17
		Strip—H	—	—	263	17.0	316	20.5	10	185	12.0	90	100		
Tellurium copper	BS 2874 C 109	Rod—O	—	—	54	3.5	232	15.0	40	—	—	49	53	117	17
Te 0.5%		Rod—½H	—	—	232	15.0	278	18.0	15	—	—	80	85		
Chromium copper	—	Rod—W	28	1.8	54	3.5	232	15.0	60	—	—	58	65		
Cr 0.6%		Rod—WP	247	16.0	324	21.0	448	29.0	25	—	—	124	142	—	—
		Rod—W(H)	309	20.0	479	31.0	541	35.0	14	—	—	140	160		
Beryllium copper	BS 2870 CB 101	Strip—W	—	—	224	14.5	479	31.0	47	—	—	100	110		
Be 1.85		Strip—WP	—	—	1 066	69.0	1 205	78.0	7	—	—	350	360	159	18.5
Co 0.25		Strip—W(H)	—	—	1 205	78.0	1 313	85.0	2	—	—	363	380		
Be 0.5	—	Strip—W	—	—	178	11.5	324	21.0	27	—	—	67	75		
Co 2.5		Strip—WP	—	—	618	40.0	757	39.0	11	—	—	205	210	159	18.5
		Strip—WH	—	—	772	50.0	810	52.5	8	—	—	215	220		
Cadmium copper	BS 2873 C 108	Wire—H	—	—	—	—	649	42.0	—	—	—	—	—	—	—
Cd 0.8															
Sulphur copper	BS 2874 C 111	Rod—O			54	3.5	216	14.0	40	—	—	50	55	117	17
S 0.35		Rod—½H			231	15.0	263	17.0	12	—	—	80	85		

Table 22.7 COPPER AND COPPER ALLOYS—TYPICAL MECHANICAL PROPERTIES AT ROOM TEMPERATURE—*continued*

Material and composition	*British Standard specification number*	*Condition*	*Limit of proportionality*		*0.2% Proof stress*		*UTS*		*Elongation on 5d or 50 mm*	*Shear strength*		*Brinell hardness*	*Vickers hardness*	*Modulus of elasticity*	
			MPa	(tonf in^{-2})	MPa	(tonf in^{-2})	MPa	(tonf in^{-2})	%	MPa	(tonf in^{-2})	kg mm^{-2}	(10 kilo)	GPa	10^6 lb in^{-2}
Cap copper		Strip—O	39	2.5	92	6.0	263	17.0	45	192	12.5	65	70		
Cu 95		Strip—½H	124	8.0	223	14.5	308	20.0	30	216	14.0	85	89	124	18
Zn 5		Strip—H	162	10.2	315	20.5	386	25.0	8	232	15.0	105	110		
Gilding metals	BS 2870 CZ 101	Strip—O	39	2.5	100	6.5	262	17.0	60	216	14.0	65	70		
Cu 90		Strip—½H	124	8.0	247	16.0	317	20.5	24	224	14.5	85	90	124	18
Zn 10		Strip—H	173	11.2	324	21.0	386	25.0	8	247	16.0	100	110		
Cu 85	BS 2870 CZ 102	Strip—O	39	2.5	108	7.0	293	19.0	60	224	14.5	65	70		
Zn 15		Strip—½H	142	9.2	254	16.5	340	22.0	28	239	15.5	85	90	121	17.6
		Strip—H	193	12.5	370	24.0	440	28.5	12	285	18.5	125	130		
Cu 80	BS 2870 CZ 103	Strip—O	46	3.0	108	7.0	317	20.5	64	230	14.9	65	70		
Zn 20		Strip—½H	170	11.0	285	18.5	378	24.5	30	247	16.0	100	105	119	17.2
		Strip—H	216	14.0	402	26.0	479	31.0	16	285	18.5	130	140		
Brass	BS 2870 CZ 106	Strip—O	46	3.0	115	7.5	324	21.0	67	230	14.9	65	70		
Cu 70 Zn 30		Strip—½H	154	10.0	270	17.5	378	24.5	40	247	16.0	107	115	115	16.6
		Strip—H	208	13.5	386	25.0	463	30.0	20	254	16.5	132	143		
Nickel silvers	BS 2870 NS 109	Strip—O	77	5.0	124	8.0	386	25.0	50	—	—	75	80	121	17.5
Cu 57 Ni 25 Zn 18		Strip—H	309	20.0	618	40.0	695	45.0	4	—	—	201	210		
Phosphor bronzes	BS 2870 PB 101	Strip—O	54	3.5	124	8.0	324	21.0	50	247	16.0	70	74		
Sn 3.5 P 0.1		Strip—½H	300	19.5	386	25.0	510	33.0	12	340	22.0	160	170	121	17.5
		Strip—H	402	26.0	579	37.5	656	42.5	2	378	24.5	195	205		
Sn 5.0 P 0.1	BS 2870 PB 102	Strip—O	54	3.5	130	8.4	347	22.5	55	254	16.5	71	75		
		Strip—½H	362	23.5	440	28.5	541	35.0	14	347	22.5	165	175	122	17.6
		Strip—H	479	31.0	618	40.0	710	46.0	2	386	25.0	205	215		

Sn 7 P 0.1	BS 2870 PB 103	Strip—O	77	5.0	139	9.0	356	23.0	60	260	16.8	80	84	117	17.0
		Strip—½H	371	24.0	494	32.0	593	38.5	12	371	24.0	175	185		
		Strip—H	494	32.0	687	44.5	741	48.0	5	424	27.5	210	220		
Sn 8 P 0.5	—	Wire—O	—	—	—	—	424	27.5	65	309	20.0	—	—	111	16.2
		Wire—H	—	—	—	—	927	60.0	—	440	28.5	—	—		
Cupro-nickels Ni 5.5 Fe 1.2 Mn 0.5	BS 2871 CN 101	Tube—O	—	—	116	7.5	316	20.5	35	240	15.5	65	70	132	19.2
		Plate—M	—	—	93	6.0	278	18.0	40	201	13.0	60	65		
Ni 10.5 Fe 1.0 Mn 0.75	BS 2871 CN 102	Tube—O	—	—	139	9.0	331	21.5	38	247	16.0	70	74	135	19.6
		Plate—M	—	—	108	7.0	324	21.0	40	240	15.5	65	70		
Ni 31.0 Fe 1.0 Mn 1.0	BS 2871 CN 107	Tube—O	—	—	170	11.0	417	27.0	42	309	20.0	90	95	152	22.0
		Plate—M	—	—	161	10.5	355	23.0	38	278	18.0	90	95		
Silicon bronze Cu 96 Si 3 Mn 1	BS 2870 CS 101	Plate—M	—	—	—	—	410	26.5	55	—	—	—	—	103	15.0
		Rod—O	—	—	77	5.0	362	23.5	60	293	19.0	—	—		
Aluminium bronzes Cu 95 Al 5	BS 2870 CA 101	Plate—M	—	—	147	9.5	378	24.5	50	286	18.5	85	90	126	18.3
Cu 92 Al 8	—	Plate—M	—	—	170	11.0	417	27.0	45	316	20.5	90	95	123	17.9
		Rod—½H	—	—	479	31.0	602	39.0	12	378	24.5	160	170		
Cu 85.5 Al 9.5 Fe 3.0 Mn 1.0 Ni 1.0	BS 2874 CA 103	Rod—H	—	—	417	27.0	757	49.0	18	571	37.0	200	210	131	19.0

Table 22.8 CAST COPPER ALLOYS—TYPICAL MECHANICAL PROPERTIES AT ROOM TEMPERATURE
(Properties vary, dependent on composition, section size and foundry practice)

BS 1400 *designation*	*Material*	*Composition* %	*Condition*	*0.2% Proof stress* N mm^{-2} (tonf in^{-2})	*UTS* N mm^{-2} (tonf in^{-2})	*Elongation on* 50 mm %	*Brinell hardness*
SCB1	Yellow brass	Cu 64 Zn 34 Pb 2	Sand cast	77 (5.0)	232 (15.0)	25	45–60
LG2	Leaded gunmetal or red brass	Cu 85 Pb 5 Sn 5 Zn 5	Sand cast Continuous cast	108 (7.0) 124 (8.0)	216 (14.0) 254 (16.5)	15 20	65–75 75–95
HTB1	High tensile brasses	Cu 56 Al 1.5 Fe 1.5 Mn 1.5 Zn 39.5	Sand cast Centrifugal cast	201 (13.0) 224 (14.5)	494 (32.0) 587 (38.0)	20 22	100–150 100–150
HTB3		Cu 56 Al 5 Fe 2.5 Mn 2.5 Zn 34	Sand cast Centrifugal cast	386 (25.0) 403 (26.0)	710 (46.0) 757 (49.0)	12 15	150–230 150–230
G1	Gunmetal	Cu 88 Sn 10 Zn 2	Sand cast *Continuous cast	139 (9.0) 147 (9.5)	286 (18.5) 317 (20.5)	18 15	50–70 70–90
LB2	Leaded bronze	Cu 80 Sn 10 Pb 10	Sand cast *Continuous cast	108 (7.0) 170 (11.0)	247 (16.0) 293 (19.0)	18 20	65–85 80–90
PB1	Phosphor bronze	Cu 90 Sn 10	Sand cast *Centrifugal cast	139 (9.0) 185 (12.0)	293 (19.0) 370 (24.0)	15 16	70–100 100–150
AB1	Aluminium bronzes	Cu 87 Al 10 Fe 3	Sand cast *Centrifugal cast	201 (13.0) 216 (14.0)	541 (35.0) 571 (37.0)	25 28	90–140 120–160
AB2		Cu 80 Al 10 Fe 5 Ni 5	Sand cast *Centrifugal cast	263 (17.0) 293 (19.0)	649 (42.0) 695 (45.0)	15 15	140–180 140–180
ASB1	Silicon bronze	Cu 95.5 Si 3.5 Mn 1	Sand cast	92 (6.0)	347 (22.5)	25	

* Properties of continuously cast and centifugally cast materials are generally similar.

Table 22.9 COPPER AND COPPER ALLOYS—TYPICAL TENSILE PROPERTIES AT ELEVATED TEMPERATURES
Limit of proportionality or proof stress is reported under a code, viz:

LP = Limit of proportionality
0.1 = 0.1% offset proof stress
0.2 = 0.2% offset proof stress
0.5 = 0.5% offset proof stress

Material	*Composition %*	*Condition (see Table 22.7)*	*Temperature* °C	*Limit of proportionality or proof stress* MPa (tonf in^{-2})	*Code (see above)*	*UTS* MPa (tonf in^{-2})	*Elongation on* 50 mm %	*Remarks*	*Reference*
OFHC* copper	Cu 99.99+	Sheet—O	24	78 (5)	0.2	212 (14)	56.3	Average grain size 0.045 mm	2
			100	77 (5)	0.2	190 (12)	55.4		
			204	70 (4.5)	0.2	(159) (10)	56.9		
Tough-pitch copper	O_2 0.03	Sheet—O	24	68 (4.4)	0.5	214 (14)	57.8	Average grain size 0.043 mm	2
			100	68 (4.4)	0.5	187 (12)	57.4		
			204	67 (4.3)	0.5	157 (10)	56.9		
Deoxidized non-arsenical copper	P=0.04	Sheet—Q	24	65 (4.2)	0.5	210 (14)	53.4	Average grain size 0.044 mm	2
			100	66 (4.3)	0.5	183 (12)	52.5		
			204	57 (3.7)	0.5	161 (10)	52.1		
Deoxidized arsenical copper	As 0.35 P 0.03	Plate—M	20	93 (6.0)	0.1	219 (14)	57	Hot rolled plate	2
			121	93 (6.0)	0.1	204 (13)	57		
			204	83 (5.4)	0.1	184 (12)	52		
Silver-bearing copper	Ag 0.05	Sheet—O	20	48 (3.1)	0.1	227 (15)	53	Average grain size 0.03 mm	2
			300	45 (2.9)	0.1	162 (11)	45		
			500	32 (2.1)	0.1	107 (7)	42		
Tellurium copper	Te–0.5	Rod—H	27	350 (23)	0.2	361 (23)	12.8		2
			260	265 (17)	0.2	266 (17)	4.7		
Chromium copper	Cr—0.6	Rod—O	20	— (—)	—	259 (17)	35	Strain rate 0.1 in min^{-1}	4
			350	— (—)	—	204 (13)	44		
			550	— (—)	—	148 (10)	10		
		Rod—WP	20	— (—)	—	374 (24)	21	Strain rate 0.1 in min^{-1}	4
			350	— (—)	—	298 (19)	10		
			550	— (—)	—	210 (14)	3		

*'OFHC' is a registered trade mark of Amax Copper Inc.

Table 22.9 COPPER AND COPPER ALLOYS—TYPICAL TENSILE PROPERTIES AT ELEVATED TEMPERATURES—*continued*

Material	*Composition* %	*Condition (see Table 22.7)*	*Temperature* °C	*Limit of proportionality or proof stress* MPa (tonf in^{-2})	*Code (see above)*	*UTS* MPa (tonf in^{-2})	*Elongation on* 50 mm %	*Remarks*	*Reference*
Gilding metals	Cu 90 Zn 10	Rod—O	25	— (—)	—	253 (16)	56	—	2
			375	— (—)	—	137 (9)	9	—	
			625	— (—)	—	59 (4)	17	—	
			875	— (—)	—	19 (1)	16	—	
	Cu 85 Zn 15	Plate—O	20	79 (5.1)	0.1	297 (19)	50	—	2
			121	79 (5.1)	0.1	259 (17)	45	—	
			232	74 (4.8)	0.1	238 (15)	36	—	
	Cu 80 Zn 20	Rod—H	23	— (—)	—	557 (36)	13	Cold worked 30%	2
			400	— (—)	—	143 (9)	7		
			850	— (—)	—	10 (0.6)	30		
Brasses	Cu 70 Zn 30	Strip—O	20	99 (6.4)	0.1	332 (22)	64	Grain size 0.03 mm	2
			200	93 (6.0)	0.1	293 (19)	54		
			300	88 (5.7)	0.1	238 (15)	32		
	Cu 63 Zn 37	Strip—O	20	125 (8.1)	0.1	347 (23)	60	Grain size 0.035 mm	2
			200	120 (7.8)	0.1	317 (21)	54		
			300	111 (7.2)	0.1	267 (17)	39		
	Cu 60 Zn 40	Plate—O	20	96 (6.2)	0.2	332 (22)	62	—	2
			121	96 (6.2)	0.2	312 (20)	62		
			204	105 (6.8)	0.2	297 (19)	62		
Aluminium brass	Cu 76 Al 2 Zn 22	Tube—O	20	165 (10.7)	0.2	397 (26)	53.8	Elongation measured on $11.3\sqrt{area}$	2
			200	134 (8.7)	0.2	317 (21)	38.5		
			400	108 (7.0)	0.2	232 (15)	13.3		
Naval brass	Cu 62 Sn 1 Zn 37	Plate—O	21	141 (9.1)	0.1	360 (23)	47	—	2
			121	133 (8.6)	0.1	246 (22)	46	—	
			204	134 (8.7)	0.1	326 (21)	38		

Table 22.9 COPPER AND COPPER ALLOYS—TYPICAL TENSILE PROPERTIES AT ELEVATED TEMPERATURES—*continued*

Material	*Composition* %	*Condition (see Table 22.7)*	*Temperature* °C	*Limit of proportionality or proof stress* MPa (tonf in^{-2})	*Code (see above)*	*UTS* MPa (tonf in^{-2})	*Elongation on* 50 mm %	*Remarks*	*Reference*
Free cutting	Cu 58	Rod—M	21	266	LP	477	22.5	—	2
brass	Zn 39			(17)		(31)			
			482	—	—	76	33.5		
						(5)			
Forging	Cu 59	Rod—O	20	—	—	364	50.2		2
brass	Zn 39					(24)			
	Pb 2		200	—	—	319	42.2		
						(21)			
			400	—	—	163	25.4		
						(11)			
			610	—	—	34	22.5		
						(2)			
	Cu 59.32	Rod—O	27	207	0.5	479	13	Extruded	5
	Zn 35.95			(13)		(31)		and	
	Fe 2.07		232	193	0.5	314	28	annealed	
	Al 1.21			(13)		(20)		816 °C	
	Fb 0.71		427	—	—	208	47	for 20 min	
	Sn 0.67					(13)			
Nickel silver	Cu 75	Rod—O	30	76	LP	347	51.0	—	—
20% Ni	Ni 20			4.9		(23)			
	Zn 5		316	86	LP	310	28.5		
				(5.6)		(20)			
			399	—	—	272	37.0		
						(18)			
Phosphor	Sn 5	Rod—O	17	—	—	337	84	—	2
bronze	P 0.1					(22)			
			260	100	0.2	278	34		
				(6.5)		(18)			
			500	65	0.2	141	6		
				(4.2)		(9.2)			
	Sn 8	Tube—H	20	441	0.2	559	47	Temper	2
	P 0.05			(29)		(36)		hard	
			200	451	0.2	539	37	and stress	
				(29)		(35)		relieved	
			400	304	0.2	345	6		
				(20)		(22)			
Silicon	Si 3	Strip—O	20	104	0.1	371	66	Strain rate	—
bronze	Mn 1			(6.8)		(24)		2 in min^{-1}	
			200	94	0.1	309	54		
				(6.1)		(20)			
			300	92	0.1	276	52		
				(6.0)		(18)			
Cupro-	Ni 5.5	Plate—O	20	158	0.1	301	40	—	2
nickels	Fe 1.2			(10)		(20)			
	Mn 0.5		177	134	0.1	255	38		
				(8.7)		(17)			
			316	133	0.1	230	34		
				(8.6)		(15)			

Table 22.9 COPPER AND COPPER ALLOYS—TYPICAL TENSILE PROPERTIES AT ELEVATED TEMPERATURES—*continued*

Material	*Composition* %	*Condition (see Table 22.7)*	*Temperature* °C	*Limit of proportionality or proof stress* MPa (tonf in^{-2})	*Code (see above)*	*UTS* MPa (tonf in^{-2})	*Elongation on* 50 mm %	*Remarks*	*Reference*
Cupro-nickels (cont.)	Ni 10.5 Fe 1.0 Mn 0.75	Tube—O	20	159 (10)	0.1	371 (24)	35	—	2
			204	147 (9.5)	0.1	329 (21)	28		
			400	139 (9.0)	0.1	287 (19)	18		
	Ni 31.0 Fe 1.0 Mn 1.0	Plate—M	20	114 (7.4)	0.1	369 (24)	50	Hot rolled	2
			232	96 (6.2)	0.1	304 (20)	46		
			371	86 (5.6)	0.1	283 (18)	63		
Aluminium bronze	Cu 89 Al 8 Fe 3	Rod—M	20	294 (19)	0.2	490 (32)	51	Extruded rod elongation measured on 11.3$\sqrt{area}$	2
			300	69 (4.5)	0.2	441 (29)	31		
			500	10 (0.63)	0.2	147 (9.5)	58		
	Cu 81.2 Al 10.1 Ni 4.8 Fe 3.0 Mn 0.8	Rod—M	24	407 (26.4)	0.5	775 (50.2)	20	Extruded rod elongation measured on 4.5 $\sqrt{area}$	
			204	403 (26.1)	0.5	693 (44.9)	13		
			427	167 (10.8)	0.5	244 (15.8)	51		

Table 22.10 COPPER AND COPPER ALLOYS—TYPICAL TENSILE AND IMPACT PROPERTIES AT LOW TEMPERATURES

Limit of proportionality or proof stress is reported under a code, viz:

LP = Limit of proportionality
0.1 = 0.1% offset proof stress
0.2 = 0.2% offset proof stress
0.5 = 0.5% offset proof stress

Impact values are reported either as C = Charpy, V notch test or I = Izod test

Material	Composition %	Condition	Temperature °C	Proof stress MPa (tonf in^{-2})	Proof stress Code (see above)	UTS MPa (tonf in^{-2})	Elongation on 50 mm %	Impact value Joules (ft lbf^{-1})	Impact value Code (see above)	Reference
OFHC copper	Cu 99.99+	Rod—O	Room	75 (4.8)	0.2	222 (14)	86.2	71.1 (52.5)	C	2
			−78	80 (5.2)	0.2	270 (17)	84.5	77.1 (57.0)	C	
			−253	90 (5.9)	0.2	418 (27)	83.0	85.8 (63.5)	C	
Tough pitch copper	O_2 0.03	Rod—O	+20	59 (3.8)	0.1	216 (14)	48.0	58.3 (43.0)	I	2
			−80	69 (4.5)	0.1	266 (17)	47.0	59.6 (44.0)	I	
			−180	79 (5.1)	0.1	351 (23)	57.6	67.7 (50.0)	I	
Beryllium copper	Cu 97.44	Rod—W	+20	171 (11)	0.1	525 (34)	36	55.5 (41)	I	
	Be 2.56		−80	201 (13)	0.1	598 (39)	38	54.2 (40)	I	
			−180	344 (22)	0.1	769 (50)	41	54.2 (40)	I	
		Rod—WP	+20	865 (56)	0.1	1287 (83)	2.6	2.7 (2)	I	6
			−80	1016 (66)	0.1	1388 (90)	0.4	4.1 (3)	I	
			−180	1069 (69)	0.1	1480 (96)	3.0	4.1 (3)	I	
Gilding metals	Cu 90	Rod—O	+22	66 (4.3)	0.2	265 (17)	56 Measured	151.8 (112)	C	2
	Zn 10		−197	91 (5.9)	0.2	381 (25)	86 on 4.52	151.8 (112)	C	
			−269	147 (9.5)	0.2	470 (30)	91 √area	—		
Brasses	Cu 70	Rod—O	+20	194 (13)	0.1	352 (23)	49.4	88.8 (65.5)	I	2
	Zn 30		−80	188 (12)	0.1	394 (26)	59.5	93.5 (69.0)	I	
			−180	204 (13)	0.1	507 (33)	74.6	106.4 (78.5)	I	
	Cu 60	Rod—O	+20	—	—	397 (26)	51.3	41.8 (30.9)	C	
	Zn 40		−78	—	—	421 (27)	53.0	42.0 (31.0)	C	2
			−183	—	—	523 (34)	55.3	40.7 (30.0)	C	
		Rod—H	+20	—	—	549 (36)	19.8	—	—	
		(Cold worked	−78	—	—	571 (37)	21.0	—	—	
		25% reduction)	−183	—	—	669 (43)	24.4	—	—	
Free cutting brass	Cu 58	Rod—H	+20	—	—	559 (36)	27	—	—	
	Zn 39		−80	—	—	598 (39)	27	—	—	2
	Pb 3		−195	—	—	735 (48)	26	—	—	

Table 22.10 COPPER AND COPPER ALLOYS—TYPICAL TENSILE AND IMPACT PROPERTIES AT LOW TEMPERATURES—*continued*

Limit of proportionality or proof stress is reported under a code, viz:

LP = Limit of proportionality
0.1 = 0.1% offset proof stress
0.2 = 0.2% offset proof stress
0.5 = 0.5% offset proof stress

Impact values are reported either as C = Charpy, V notch test or I = Izod test

Material	*Composition* %	*Condition*	*Temperature* °C	*Proof stress* MPa (tonf in^{-2})	*Proof stress* *Code* (*see above*)	*UTS* MPa (tonf in^{-2})	*Elongation on* 50 mm %		*Impact value* Joules (ft lbf^{-1})	*Impact value* *Code* (*see above*)	*Reference*
Forging brass	Cu 58	Rod—O	+ 20	—	—	364 (24)	50.2		21.5 (15.9)	C	
	Zn 40		− 78	—	—	377 (24)	49.8		24.1 (17.8)	C	2
	Pb 2		−183	—	—	475 (31)	50.6		22.6 (16.7)	C	
Phosphor bronzes	Cu 95	Sheet—O	+ 27	155 (10)	0.2	358 (23)	61		—	—	
	Sn 5		− 40	173 (11)	0.2	393 (25)	73		—	—	2
			− 73	180 (12)	0.2	427 (28)	76		—	—	
		Sheet—EH	+ 27	621 (40)	0.2	677 (44)	7		—	—	
			− 40	648 (42)	0.2	703 (46)	9.5		—	—	
			− 73	677 (44)	0.2	738 (48)	11		—	—	
	Cu 92	Rod—H	+ 24	772 (50)	0.2	807 (52)	20	(On 1 in	—	—	
	Sn 8		−196	964 (62)	0.2	986 (64)	30	gauge	—	—	2
			−253	1059 (69)	0.2	1158 (75)	25	length)	—	—	
Silicon bronze	Cu 96	Rod—H	+ 25	—	—	511 (33)	39.8		—	—	
	Si 3	(Cold worked	− 80	—	—	571 (37)	31.7		—	—	6
	Mn 1	42% reduction)	−190	—	—	692 (45)	36.2		—	—	
Nickel silver	Cu 55.15	Sheet—O	+ 20	193 (13)	0.1	519 (34)	33		108.5 (80)	I	
	Ni 30.5		−120	199 (13)	0.1	619 (40)	38		108.5 (80)	I	6
	Zn 14.3		−190	196 (13)	0.1	718 (47)	41		118.0 (87)	I	
Cupro-nickels	Ni 10.5	Rod—O	+ 22	147 (10)	0.5	342 (22)	37	On	154.6 (114)	C	
	Fe 1.0		−197	208 (13)	0.5	569 (37)	50	4.52$\sqrt{area}$	155.9 (115)	C	2
	Mn 0.5		−269	172 (11)	0.5	556 (36)	53		—	—	
	Ni 31.0	Rod—O	+ 22	129 (8)	0.5	398 (26)	47	On	155.9 (115)	C	
	Fe 1.0		−197	218 (14)	0.5	619 (40)	52	4.52$\sqrt{area}$	154.6 (114)	C	2
	Mn 1.0		−253	263 (17)	0.5	715 (46)	51		154.6 (114)	C	
Aluminium bronzes	Cu 92	Rod—O	+ 25	110 (7)	0.2	414 (27)	107	Gauge length	—	—	2
	Al 8		−196	134 (9)	0.2	558 (36)	77	not quoted	—	—	
	Cu 79	Plate—M	+ 20	451 (29)	0.2	745 (48)	12	On	19.3 (14.2)	C	2
	Al 10		−196	588 (38)	0.2	892 (58)	4	5.65$\sqrt{area}$	8.3 (6.1)	C	
	Fe 5										
	Mn 1										
	Ni 5										

Table 22.11 COPPER AND COPPER ALLOYS—FATIGUE PROPERTIES AT ROOM TEMPERATURE
Note: Where the number of cycles is not given the value represents endurance limit.

Material	*Composition* %	*Form and condition*	*Fatigue strength* MPa (tonf in^{-2})	*Number of cycles* 10^6	*Reference*
OFHC copper	Cu 99.99+	Rod—cold worked 29.2%	117 (7.6)	300	2
Tough pitch copper	O_2—0.03	Rod—annealed Grain size—0.040 mm	62 (4.0)	300	2
		Wire—cold worked 37%	107 (6.9)	100	
Deoxidized non-arsenical copper	P = 0.04	Tube—annealed Grain size 0.050 mm	76 (4.9)	20	2
		Strip—cold rolled 60% reduction	128 (8.3)	100	
Silver-bearing copper	Ag—0.03	Strip-cold rolled 50% reduction	103 (6.7)	—	7
Chromium copper	Cr = 0.88 Si = 0.09 Fe = 0.07	Rod—cold drawn 90% reduction	178 (11.5)	300	7
		Rod—cold drawn 90% reduction—heat treated 3 h 400 °C	193 (12.5)	300	
Beryllium coppers	Be = 1.85 Co = 0.25	Strip—solution heat treated	224 (14.5)	100	3
		Strip—rolled 'hard' after solution heat treatment	239 (16)	100	
		Strip—rolled 'hard' after solution heat treatment and then precipitation hardened	284 (18)	100	
	Be = 0.6 Co = 2.5	Strip—rolled '½ hard' after solution treatment	241 (16)	100	
Gilding metal	Cu = 90 Zn = 10	Strip—annealed Grain size 0.030 mm	69 (4.5)	100	2
		Strip—cold worked 21% reduction	110 (7.1)	100	
		Strip—cold worked 37% reduction	114 (7.4)	100	
		Strip—cold worked 60% reduction	124 (8.0)	100	
		Strip—cold worked 68% reduction	138 (8.9)	100	
	Cu = 80 Zn = 20	Strip—annealed Grain size 0.035 mm	97 (6.3)	100	2
		Strip—cold worked 60% reduction	152 (9.8)	100	
Brasses	Cu = 70 Zn = 30	Strip—annealed Grain size 0.025 mm	107 (6.9)	100	2
		Strip—cold rolled 21% reduction	124 (8.0)	100	
		Strip—cold rolled 60% reduction	152 (9.8)	100	
	Cu = 65 Zn = 35	Strip—annealed	104 (6.7)	100	2
		Strip—cold rolled 37.1% reduction	135 (8.7)	100	
		Strip—cold rolled 60.5% reduction	138 (8.9)	100	

Table 22.11 COPPER AND COPPER ALLOYS—FATIGUE PROPERTIES AT ROOM TEMPERATURE—*continued*

Material	*Composition* %	*Form and condition*	*Fatigue strength* MPa (tonf in^{-2})	*Number of cycles* 10^6	*Reference*
Brasses (*cont.*)	Cu = 60 Zn = 40	Rod—annealed Rod—cold worked 25% and stress relieved at 275 °C	148 (9.6) 210 (14)	100 100	2
Aluminium brass	Cu = 76 Zn = 22 Al = 2	Rod—cold worked 20–25% and stress relieved	97 (6.3)	20	2
Naval brass	Cu = 62 Zn = 38 Sn = 1	Rod—hot rolled Rod—cold worked 27% reduction	128 (8.3) 183 (12)	100 100	2
Phosphor bronze	Sn = 5 P = 0.1	Strip—annealed Grain size 0.035 mm Strip—cold rolled 69% reduction	172 (11) 221 (14)	100 100	2
	Sn = 8 P = 0.1	Rod—annealed Grain size 0.020 mm Rod—cold worked 30.1% reduction	221 (14) 234 (15)	1 000 1 000	2
	Sn = 10 P = 0.1	Rod—annealed Grain size—0.065/070 mm Rod—cold worked 30.1% reduction	172 (11) 159 (10)	1 000 1 000	2
Silicon bronze	Cu = 96 Si = 3 Mn = 1	Rod—annealed Rod—hard	130 (8.4) 232 (15)	300 300	7
Cupro-nickels	Ni = 5.5 Fe = 1.0 Mn = 0.5	Rod—annealed Rod—cold worked 25% reduction	131 (8.5) 173 (11)	100 100	2
	Ni = 10.5 Fe = 1.5 Mn = 0.75	Strip—cold rolled Hard temper	145 (9.4)	100	2
	Ni = 31.0 Fe = 1.0 Mn = 1.0	Tube—annealed Tube—cold worked and stress relieved	147 (9.5) 177 (11)	100 100	2
Aluminium bronze	Cu = 92 Al = 8	Rod—lightly worked	203 (13.2)	100	2
	Cu = Rem. Al = 9.7 Fe = 5.1 Ni = 5.3	Rolled rod	323 (20.9)	100	2
Nickel silver	Cu = 55 Ni = 18 Zn = 27	Strip—annealed Strip—cold rolled 70% reduction	114 (7.4) 173 (11)	100 100	—

Table 22.12 COPPER AND COPPER ALLOYS, IMPACT PROPERTIES

Material	*Composition* %	*Form and condition*	*Temperature* °C	*Impact value* *Izod* Joules (ft lbf^{-1})	*Charpy* Joules (ft lbf^{-1})	*Reference*
OFHC copper	Cu 99.99+	Plate—hot rolled	0	—	62.4 (46)	1
			204	—	50.2 (37)	
			316	—	56.9 (42)	
			538	—	44.7 (33)	
			650	—	35.3 (26)	
Tough pitch copper	O_2 0.08	Plate—annealed	20	47.5 (35)	—	8
			200	37.9 (28)	—	
			300	35.3 (26)	—	
			500	31.2 (23)	—	
			600	28.5 (21)	—	
Deoxidized non-arsenical copper	P 0.06	Plate—annealed	20	61.0 (45)	—	8
			200	56.9 (42)	—	
			300	55.6 (41)	—	
			500	42.0 (31)	—	
			600	31.2 (23)	—	
Deoxidized arsenical copper	As 0.36 P 0.07	Plate—annealed	20	62.4 (46)	—	8
			200	56.9 (42)	—	
			400	55.6 (41)	—	
			500	43.4 (32)		
			600	31.2 (23)	—	
Naval brass (American type)	Cu 60.25 Sn 0.75 Zn Bal.	Annealed	20	—	82.4 (60.8)	6
			−50	—	79.9 (58.9)	
			−80	—	84.7 (62.5)	
			−115	—	80.6 (59.2)	
Phosphor bronze	Sn 4 P 0.4	Rod—hard drawn	20	62.4 (46)	—	9
			−41	59.7 (44)	—	
Silicon bronze	Cu 96 Si 3 Mn 1	Rod—annealed	20	—	90.0 (66.4)	6
			−50	—	99.1 (73.1)	
			−80	—	93.8 (69.2)	
			−115	—	87.4 (64.5)	
Aluminium bronzes	Cu 89.6 Al 7.8 Fe 2.6	Rod—extruded annealed and roller straightened	24	—	98.9 (73)	2
			−29	—	105.7 (78)	
			−59	—	103.0 (76)	
			−182	—	94.9 (70)	
	Cu 81.2 Al 10.1 Ni 4.75 Mn 0.8	Rod—extruded annealed and roller straightened	24	—	20.3 (15)	2
			−29	—	19.0 (14)	
			−59	—	17.6 (13)	
			−182	—	12.2 (9)	
Cupro-nickels	Cu 80 Ni 20	Annealed	20	104.4 (77)	—	6
			−80	107.1 (79)	—	
			−120	113.9 (84)	—	
			−180	115.2 (85)	—	
	Cu 70 Ni 30	Annealed	20	—	89.9 (66.3)	6
			−30	—	80.5 (59.4)	
			−50	—	80.5 (59.4)	
			−80	—	79.6 (58.7)	
			−115	—	81.3 (60.0)	

Table 22.12 COPPER AND COPPER ALLOYS, IMPACT PROPERTIES—*continued*

Material	*Composition* %	*Form and condition*	*Temperature* °C	*Impact value* Izod Joules (ft lbf^{-1})	*Impact value* Charpy Joules (ft lbf^{-1})	*Reference*
Nickel silvers	Cu 74.28	Annealed	20	—	91.5 (67.5)	6
	Ni 19.49		−30	—	76.5 (56.4)	
	Zn 5.43		−50			
	Mn 0.80		−80	—	75.7 (55.8)	
			−115	—	71.3 (52.6)	
	Cu 55.15	Annealed	20	—	108.4 (80)	—
	Ni 30.50		−40	—	118.0 (87)	
	Zn 14.3		−120	—	108.5 (80)	
			−180	—	118.0 (87)	

Table 22.13 WROUGHT COPPER AND COPPER ALLOYS, CREEP PROPERTIES

Notes: (1) All values relate to rod or wire products unless specified.
(2) Total extension = Initial extension + total creep.
= Initial extension + intercept + (minimum creep rate × duration)

Materials and composition %	*Condition*	*Test temperature* °C	*Applied stress* MPa (tonf in^{-2})	*Duration* 1 000 h	*Total extension* %	*Intercept* %	*Minimum creep rate in* % *per* 1 000 h	*Remarks*
Oxygen-free copper Cu 99.99+	Annealed grain size 0.025 mm	149	14.5 (0.94)	6.4	0.053	0.024	0.001 7	Reference 2
			21.0 (1.36)	6.5	0.128	0.049	0.007 5	
			31.4 (2.03)	6.0	0.510	0.290	0.023	
			54.8 (3.55)	5.1	2.490	1.560	0.083	
		204	14.2 (0.92)	6.5	0.213	0.054	0.021	
			21.3 (1.38)	6.0	0.580	0.256	0.049	
			28.0 (1.81)	6.5	1.295	0.725	0.078	
			51.0 (3.30)	5.0	4.580	2.670	0.215	
	Hard drawn 84% reduction	149	54.6 (3.53)	6.0	0.102	0.002	0.008 2	
		204	14.5 (0.94)	3.2	0.157	0.090	0.014	
			28.0 (1.81)	6.0	0.422	0.185	0.034 5	
			48.3 (3.13)	6.5	3.80	2.50	0.19	
HC copper O_2 = 0.04		149	14.2 (0.92)	6.4	0.088	0.048	0.003 2	Reference 2
			20.7 (1.34)	6.5	0.257	0.133	0.013	
			41.4 (2.68)	6.5	1.875	1.120	0.057 5	
			55.9 (3.62)	6.5	3.475	1.795	0.088	

Table 22.13 WROUGHT COPPER AND COPPER ALLOYS, CREEP PROPERTIES—*continued*

Materials and composition %	*Condition*	*Test temperature* °C	*Applied stress* MPa (tonf in^{-2})	*Duration* 1 000 h	*Total extension* %	*Intercept* %	*Minimum creep rate in* % *per* 1 000 h	*Remarks*
HC copper $O_2=0.04$ (*cont.*)	Annealed grain size 0.025 mm	260	2.5 (0.16)	6.0	0.084	0.016	0.011	
			7.3 (0.47)	6.5	0.640	0.113	0.079 5	
			13.7 (0.89)	6.5	2.877	0.869	0.306	
	Drawn 84% reduction	149	52.0 (3.37)	6.4	0.118	0.041	0.004 9	
			68.9 (4.46)	6.5	0.167	0.042	0.010	
		204	7.3 (0.47)	6.5	0.064	0.045	0.001 1	
			28.0 (1.81)	6.5	1.080	0.409	0.097	
			49.0 (3.17)	6.5	5.418	2.47	0.44	
Deoxidized non-arsenical copper P = 0.008	Annealed grain size 0.032 mm	204	14.2 (0.92)	6.0	0.078	0.037	0.003 9	Reference 2
			21.0 (1.36)	7.08	0.355	0.164	0.018 5	
			35.1 (2.27)	7.08	1.378	0.660	0.051	
			55.4 (3.59)	6.0	3.334	1.120	0.120	
	Cold worked 84% reduction	204	24.4 (1.58)	7.7	0.126	−0.015 1	0.015 2	Reference 2
			34.7 (2.25)	7.08	0.119	−0.085	0.038	Accelerating creep rate
			62.4 (4.04)	7.08	0.534	−1.110	0.224	Accelerating creep rate
			103.5 (6.70)	0.58	0.169	0.034	0.055	
			103.5 (6.70)	4.2	2.813	−8.630	2.70	Accelerating creep rate
Deoxidized arsenical copper As = 0.35 P = 0.03	Annealed grain size 0.045 mm	149	20.7 (1.34)	6.5	0.055	0.013	<0.000 1	
			35.8 (2.32)	6.4	0.580	0.085	0.002 35	
			54.8 (3.55)	6.5	1.560	0.185	0.008 5	
			70.6 (4.57)	6.4	2.537	0.275	0.019	
		204	21.0 (1.36)	6.0	0.119	0.040	0.002 3	
			38.6 (2.50)	6.0	1.055	0.295	0.014	Reference 2
			49.0 (3.17)	6.0	1.648	0.365	0.022	
		260	14.2 (0.92)	6.86	0.107	0.049	0.005 5	
			24.4 (1.58)	6.86	0.678	0.335	0.024	
			43.1 (2.79)	6.0	2.584	0.700	0.152	

Table 22.13 WROUGHT COPPER AND COPPER ALLOYS, CREEP PROPERTIES—*continued*

Materials and composition %	*Condition*	*Test temperature* °C	*Applied stress* MPa (tonf in^{-2})	*Duration* 1 000 h	*Total extension* %	*Intercept* %	*Minimum creep rate in* % *per* 1 000 h	*Remarks*
Deoxidized arsenical copper As = 0.35 P = 0.03 (*cont.*)	Cold worked 84% reduction	149	85.6 (5.54)	6.4	0.089	0.017	0.000 24	
			138.8 (8.99)	6.5	0.153	0.033	0.000 5	
			207.6 (13.44)	6.5	0.282	0.089	0.001 6	
			275.8 (17.86)	9.45	0.452	0.118	0.007 8	
			325.4 (21.07)	0.5	0.750	0.160	0.42	
			325.4 (21.07)	0.75	0.910	—	—	
		260	10.7 (0.69)	6.0	0.029	0.001 5	0.001 45	
			14.2 (0.92)	6.0	0.075	−0.012	0.012	Accelerating creep rate
			24.4 (1.58)	7.3	0.605	−0.187	0.105	Accelerating creep rate
			34.4 (2.23)	6.5	0.933	−0.672	0.24	Accelerating creep rate
Silver-bearing copper Ag–0.086 O_2–0.02	Annealed strip grain size 0.030 mm	130	137.9 (8.93)	2.4	15.8	14.7	1.05	Reference 2
		175	96.5 (6.25)	2.6	7.0	6.7	0.35	
			137.9 (8.93)	2.4	27.4	23.2	1.85	
		225	41.2 (2.67)	3.0	0.9	0.6	0.02	
			96.5 (6.25)	2.5	10.6	8.0	1.1	
	Cold worked strip 25% reduction	130	55.1 (3.57)	4.75	0.08	0.075	0.002	Reference 2
			96.5 (6.25)	10.2	0.18	0.16	0.004	
			137.9 (8.93)	7.2	0.26	0.24	0.005	
		225	55.1 (3.57)	8.9	0.21	0.14	0.006 4	
			96.5 (6.25)	11.5	0.56	0.38	0.017	
	Cold worked strip 50% reduction	130	55.1 (3.57)	4.55	0.09	0.08	0.001 5	
			96.5 (6.25)	11.4	0.20	0.185	0.001 5	
			137.9 (8.93)	7.25	0.29	0.265	0.004	
		225	55.1 (3.57)	8.9	0.26	0.15	0.011	
			96.5 (6.25)	12.9	0.795	0.335	0.029	
			137.9 (8.93)	3.0	0.825	0.525	0.10	
Tellurium copper Te–0.5	Annealed grain size 0.025 mm	149	21.0 (1.36)	6.0	0.134 4	0.077	0.001 4	Reference 2
			35.8 (2.32)	6.0	0.251 5	0.121	0.007 8	
			59.0 (3.82)	6.0	1.553	0.737	0.022 3	

Table 22.13 WROUGHT COPPER AND COPPER ALLOYS, CREEP PROPERTIES—*continued*

Materials and composition %	*Condition*	*Test temperature* °C	*Applied stress* MPa (tonf in^{-2})	*Duration* 1 000 h	*Total extension* %	*Intercept* %	*Minimum creep rate in* % *per* 1 000 h	*Remarks*
Tellurium copper Te–0.5 (*cont.*)	Cold drawn 37%	149	47.6 (3.08)	6.0	0.089 5	0.034 4	0.000 85	
			68.9 (4.46)	6.0	0.149 4	0.068 1	0.001 9	
			103.5 (6.70)	6.0	0.223 4	0.088	0.005 2	
			137.5 (8.90)	6.0	0.390 5	0.155 5	0.011 5	
			201.3 (13.04)	6.0	1.133	0.479	0.080	
Cadmium copper Cd = 1.03	Cold worked 20% reduction	130	137.9 (8.93)	—	—	—	0.007	Reference 5
		205	55.1 (3.57)	—	—	—	0.011	
			137.9 (8.93)	—	—	—	0.075	
Chromium copper Cr = 0.73	Fully heat treated and drawn	343	68.9 (4.46)	—	—	—	0.008 9	Reference 5
			110.3 (7.14)	—	—	—	0.014	
			137.9 (8.93)	—	—	—	0.038	
Cap copper Cu 95 Zn 5	Cold drawn 51% reduction	200	98.1 (6.35)	3.305	0.123	—	—	Reference 2
			117.7 (7.62)	1.995	0.21	—	—	
			137.3 (8.89)	1.999	0.38	—	—	
			140.2 (9.08)	2.011	0.51	—	—	
Gilding metal Cu 85 Zn 15	Annealed grain size 0.060 mm	149	31.0 (2.01)	4.5	0.031	0.006	0.000 7	Reference 2
			48.0 (3.11)	5.1	0.087	0.024	0.002 9	
			67.6 (4.38)	4.5	1.07	0.44	0.026	
		260	13.9 (0.89)	5.0	0.091	0.026	0.008	
			24.2 (1.57)	5.3	0.241	0.037	0.030	
			34.3 (2.22)	5.0	0.507	0.083	0.073	
			41.3 (2.68)	5.14	0.958	0.180	0.138	
			47.9 (3.10)	1.6	1.100	0.510	0.32	
	Cold worked 84% reduction	149	66.9 (4.33)	4.4	0.096	0.029	0.000 7	
			136.5 (8.84)	5.1	0.197	0.054	0.0002 6	
			273.7 (17.72)	4.5	0.445	0.108	0.011	
			371.0 (24.02)	5.1	0.860	0.250	0.033	

Table 22.13 WROUGHT COPPER AND COPPER ALLOYS, CREEP PROPERTIES—*continued*

Materials and composition %	*Condition*	*Test temperature* °C	*Applied stress* MPa (tonf in^{-2})	*Duration* 1 000 h	*Total extension* %	*Intercept* %	*Minimum creep rate in* % *per* 1 000 h	*Remarks*
Gilding metal Cu 85 Zn 15 (*cont.*)	Cold worked 84% reduction	260	4.2 (0.27)	5.54	0.096	0.038	0.010	
			6.6 (0.43)	5.3	0.270	0.080	0.034	
			13.4 (0.87)	2.95	0.678	0.104	0.19	
			20.5 (1.33)	3.65	2.715	−0.045	0.76	Accelerating creep rate
Brass Cu 70 Zn 30	Annealed grain size 0.022 mm	204	19.3 (1.25)	3.3	0.052	0.016	0.008 1	Reference 2
			39.2 (2.54)	1.6	0.104	0.025	0.036	
			59.2 (3.83)	3.3	0.430	0.035	0.11	
		260	7.7 (0.50)	4.08	0.115	0.014	0.021	
			13.3 (0.86)	3.72	0.357	0.061	0.074	
			18.8 (1.22)	4.08	0.908	0.090	0.195	
			25.3 (1.64)	3.72	1.557	0.147	0.37	
	Cold worked 84% reduction (fine grained)	149	68.9 (4.46)	5.2	0.156	0.058	0.002 5	Reference 2
			135.8 (8.79)	5.23	0.326	0.129	0.005 4	
			275.4 (17.83)	6.85	0.780	0.260	0.026	
			348.8 (22.59)	4.75	1.430	0.540	0.10	
		204	6.9 (0.45)	5.1	0.128	0.084	0.006 7	
			21.2 (1.37)	5.06	0.530	0.180	0.063 6	
			34.7 (2.25)	4.7	1.494	0.213	0.265	
		260	3.4 (0.22)	5.0	0.311	0.121	0.037	
			5.6 (0.36)	5.0	0.970	0.175	0.159	
			10.2 (0.66)	2.62	3.015	0.212	1.07	
Brass Cu 60 Zn 40	Annealed	149	34.4 (2.23)	6.43	0.053 5	0.009 5	0.001 1	Reference 2
			51.3 (3.32)	6.43	0.099	0.029	0.002 3	
			68.9 (4.46)	6.43	0.158	0.039	0.006	
			103.4 (6.70)	6.43	0.313	0.084	0.011 5	
			137.1 (8.88)	6.43	3.580	1.265	0.20	
		204	7.3 (0.47)	7.7	0.048	0.013 2	0.002 9	
			14.2 (0.92)	7.7	0.090	0.025	0.005 8	
			28.0 (1.81)	2.28	0.18	0.010	0.022	
			42.0 (2.72)	7.68	1.975	0.053	0.246	

Table 22.13 WROUGHT COPPER AND COPPER ALLOYS, CREEP PROPERTIES—*continued*

Materials and composition %	*Condition*	*Test temperature* °C	*Applied stress* MPa (tonf in^{-2})	*Duration* 1 000 h	*Total extension* %	*Intercept* %	*Minimum creep rate in* % *per* 1 000 h	*Remarks*
Forging brass Cu 59.32 Zn 35.95 Pb 2.07	Extruded and annealed 816°C for 20 min	149	68.9 (4.46)	—	—	—	0.052	Reference 5
		177	68.9 (4.46)	—	—	—	0.18	
		204	68.9 (4.46)	—	—	—	0.92	
		260	68.9 (4.46)	—	—	—	89.0	
		177	20.7 (1.34)	—	—	—	0.030	
		232	20.7 (1.34)	—	—	—	0.075	
		260	20.7 (1.34)	—	—	—	0.134	
		288	20.7 (1.34)	—	—	—	1.14	
Admiralty brass Cu 70 Zn 29 Sn 1	Annealed grain size 0.055 mm	149	31.4 (2.03)	4.5	0.029	0.013	<0.001	Reference 2
			82.3 (5.33)	3.38	0.090	0.023	0.001 4	
			104.4 (6.76)	4.4	0.134	0.042	0.002 8	
		260	6.6 (0.43)	4.98	0.026	0.003	0.003 7	
			13.6 (0.88)	4.3	0.072	0.031	0.008	
			20.7 (1.34)	4.98	0.149	0.034	0.021	
			31.0 2.01)	4.3	0.306	0.019	0.062	Accelerating creep rate
			42.5 (2.75)	6.3	1.528	−0.35	0.295	Accelerating creep rate
	Cold worked 60% reduction	149	86.5 (5.60)	5.2	0.164	0.066	0.001 6	Reference 2
			103.5 (6.70)	6.5	0.203	0.078	0.002 7	
			138.5 (8.97)	5.4	0.287	0.116	0.005 2	
			208.8 (13.52)	6.5	0.451	0.179	0.008 8	
			280.7 (18.15)	5.2	0.685	0.307	0.018	
			361.5 (22.4)	5.2	1.053	0.385	0.060	
		260	2.0 (0.13)	5.75	0.088	0.052	0.005 5	
			6.8 (0.44)	2.95	0.425	0.090	0.11	
			13.7 (0.89)	5.3	2.481	0.160	0.435	
			20.4 . (1.32)	2.6	2.601	0.340	0.86	
Aluminium brass Cu 76 Zn 22 Al 2	Annealed grain size 0.030 mm	149	40.6 (2.63)	5.5	0.037 5	0.002	0.001 0	Reference 2
			68.3 (4.42)	6.5	0.099	0.024	0 003 8	
			137.1 (8.88)	6.5	0.450	0.146	0.029	
			151.7 (9.82)	9.45	0.979	0.367	0.051	

Table 22.13 WROUGHT COPPER AND COPPER ALLOYS, CREEP PROPERTIES—*continued*

Materials and composition %	*Condition*	*Test temperature* °C	*Applied stress* MPa (tonf in^{-2})	*Duration* 1 000 h	*Total extension* %	*Intercept* %	*Minimum creep rate in* % *per* 1 000 h	*Remarks*
Aluminium brass Cu 76 Zn 22 Al 2 (*cont.*)	Annealed grain size 0.030 mm	260	3.8 (0.24)	6.86	0.097	0.043	0.007 1	
			7.3 (0.47)	6.86	0.298	0.082	0.030	
			10.7 (0.69)	6.5	0.524	0.128	0.059	
	Cold worked 37% reduction	149	132.0 (8.55)	6.4	0.208	0.074	0.002 8	
			207.6 (13.44)	6.5	0.349	0.128	0.006 3	
			275.1 (17.81)	11.1	0.590	0.233	0.010	
			344.7 (22.32)	6.5	1.169	0.498	0.054	
		260	7.3 (0.47)	6.0	0.159	0.081	0.012	
			14.4 (0.92)	6.0	0.577	0.197	0.061	
			21.0 (1.36)	6.0	1.243	0.117	0.184	
			34.7 (2.25)	3.48	2.322	−0.181	0.71	
Phosphor bronze Cu 95 Sn 5 P 0.2	Annealed grain size 0.050 mm	149	31.0 (2.01)	5.62	0.028	0.003	0.000 3	
			68.9 (4.46)	5.62	0.072	0.009	0.000 8	
			104.1 (6.74)	5.62	0.142	0.016	0.001 9	
			117.5 (7.61)	4.75	0.978	0.008	0.009 4	
		260	10.7 (0.69)	5.1	0.039	0.017	0.001 6	
			17.0 (1.10)	5.0	0.066	0.020	0.005	
			34.3 (2.22)	5.8	0.216	0.065	0.018	
			72.7 (4.71)	5.64	1.191	−0.100	0.213	Accelerating creep rate
	Cold worked 84% reduction	149	31.5 (2.04)	5.62	0.045	0.009	0.000 9	Reference 2
			103.0 (6.67)	5.62	0.170	0.050	0.002 6	
			136.5 (8.84)	5.62	0.235	0.075	0.003 7	
			205.4 (13.30)	5.62	0.367	0.112	0.008	
			344.4 (22.30)	4.75	0.726	0.250	0.027	
		260	2.2 (0.14)	5.1	0.113	0.025	0.016 4	
			3.7 (0.24)	5.64	0.285	0.090	0.033 6	
			6.80 (0.44)	8.15	0.797	0.269	0.063 4	
			20.8 (1.35)	5.75	2.511	−0.243	0.47	Accelerating creep rate
Silicon bronze Cu 96 Si 3 Mn 1	Annealed 450°C	204	86.5 (5.60)	—	—	—	0.065	Reference 11
			103.5 (6.70)	—	—	—	0.084	

Table 22.13 WROUGHT COPPER AND COPPER ALLOYS, CREEP PROPERTIES—*continued*

Materials and composition %	*Condition*	*Test temperature* °C	*Applied stress* MPa (tonf in^{-2})	*Duration* 1 000 h	*Total extension* %	*Intercept* %	*Minimum creep rate in* % *per* 1 000 h	*Remarks*
Silicon bronze Cu 96 Si 3 Mn 1 (*cont.*)	Annealed 450 °C	288	34.6 (2.24)	—	—	—	0.035	
			41.4 (2.68)	—	—	—	0.080	
			51.9 (3.36)	—	—	—	0.19	
			69.0 (4.47)	—	—	—	0.65	
Aluminium bronze Cu 95 Al 5	Annealed plate	200	107.8 (6.98)	3.0	0.2	—	0.000 07	—
		300	33.4 (2.16)	2.0	0.2	—	0.000 1	
	Cold drawn (rod)	550	17.3 (1.12)	5.8	7.5	0.035	640	Rupture test results
	Rockwell 92B		34.4 (2.23)	0.52	5.7	0.62	8 340	Reference 2
		600	6.8 (0.44)	85.3	23.5	0.090	79	
			17.3 (1.12)	2.9	8.5	0.316	1 250	
			34.4 (2.23)	0.14	8.0	0.0	23 500	
Aluminium bronze Cu 79 Al 10 Fe 5 Mn 1 Ni 5	Extruded	250	61.8 (4.0)	1.17	0.068	—	<0.004 2	Reference 2
			92.7 (6.0)	1.44	0.123	—	0.008 3	
			139.0 (9.0)	1.44	0.236	—	0.031 7	
			216.2 (14.0)	1.27	0.622	—	0.258 3	
			308.9 (20.0)	1.128	3.67	—	—	
		400	46.3 (3.0)	0.72	0.43	—	0.237 5	
			61.8 (4.0)	0.72	0.76	—	0.541 7	
			77.2 (5.0)	0.72	1.93	—	1.375	
Cupro-nickel Ni 10.5 Fe 1.0 Mn 0.5	Annealed grain size 0.025 mm	149	103.5 (6.70)	6.0	0.870 5	0.1	<0.000 1	Reference 2
			137.9 (8.93)	6.0	2.131	0.242	0.000 16	
			172.4 (11.16)	6.0	4.705	0.163 7	0.000 22	
		260	63.5 (4.11)	6.0	0.090	0.014 3	0.000 61	
			90.7 (5.87)	6.0	0.516	0.253 8	0.001 7	
			126.6 (8.10)	6.0	1.803	0.175 6	0.003 8	
	Cold worked 21% reduction	149	138.2 (8.95)	6.0	0.139 1	0.018 8	<0.000 1	
			206.8 (13.39)	6.0	0.199	0.014 8	0.000 2	
			276.1 (17.88)	6.0	0.277	0.027 6	0.001 4	
			310.3 (20.09)	6.0	0.410	0.061	0.002 4	
			343.2 (22.32)	6.0	0.635	0.164	0.003 5	

Table 22.13 WROUGHT COPPER AND COPPER ALLOYS, CREEP PROPERTIES—*continued*

Materials and composition %	*Condition*	*Test temperature* °C	*Applied stress* MPa (tonf in^{-2})	*Duration* 1 000 h	*Total extension* %	*Intercept* %	*Minimum creep rate in* % *per* 1 000 h	*Remarks*
Cupro-nickel Ni 10.5 Fe 1.0 Mn 0.5 (*cont.*)	Cold worked 21% reduction	260	139.9 (9.06)	6.0	0.198 8	0.057 6	0.002 2	
			207.9 (13.46)	6.0	0.442 5	0.169	0.013 6	
			244.8 (15.85)	4.32	0.607	0.189	0.044	
			244.8 (15.85)	6.0	0.700	0.102	0.061 7	
Cupro-nickel Ni 31 Fe 1 Mn 1	Cold worked and stress relieved	399	124.2 (8.04)	2.5	0.219	—	0.015	Reference 2
			172.4 (11.16)	2.5	0.319	—	0.032	
			206.8 (13.39)	1.5	0.359	—	0.055	
			241.4 (15.63)	1.0	0.490	—	0.17	
			275.8 (17.86)	1.0	0.818	—	0.40	
			310.3 (20.09)	1.0	3.25	—	1.0	
		454	48.3 (3.13)	2.5	0.142	—	0.019	
			96.5 (6.25)	1.5	0.339	—	0.072	
			172.4 (11.16)	1.0	0.993	—	0.61	
			206.8 (13.39)	1.5	9.80	—	2.2	
		510	13.7 (0.89)	1.5	0.096	—	0.032	
			41.4 (2.68)	1.0	0.292	—	0.18	
			124.2 (8.04)	1.5	11.2	—	3.4	
		566	10.3 (0.67)	0.5	0.185	—	0.3	
			68.9 (4.46)	0.5	7.20	—	7	
Nickel silver Cu 74.23 Ni 20.08 Zn 5.08 Mn 0.69	Annealed 650	316	34.4 (2.23)	—	0	—	—	Reference 11
			86.5 (5.60)	—	0.006	—	—	
			103.5 (6.70)	—	0.013	—	—	
			137.5 (8.90)	—	0.034	—	—	
			173.0 (11.20)	—	0.072	—	—	
		399	34.4 (2.23)	—	0	—	—	
			61.8 (4.00)	—	0.015	—	—	
			86.5 (5.60)	—	0.065	—	—	
			103.5 (6.70)	—	0.365	—	—	

Table 22.14 CAST COPPER ALLOYS—CREEP PROPERTIES

Material and composition %	*Test temperature* °C	*Applied stress* MPa (tonf in^{-2})	*Minimum creep rate in* % per 1 000 h	*Reference*
Manganese bronze	149	69 (4.48)	0.118	
Cu 57.14				
Al 0.44	177	69 (4.48)	0.69	
Sn 0.10				11
Pb 0.49	288	21 (1.34)	1.81	
Fe 1.72				
Zn 40.11				
Silicon brass	260	69 (4.46)	0.075	
Cu 81.6				
Zn 13.95	316	21 (1.34)	0.236	
Si 4.40				
Fe 0.05	371	21 (1.34)	0.73	5
		69 (4.46)	35	
Aluminium bronze—sand cast	250	77 (5.0)	<0.146	
Cu 90		131 (8.5)	<0.153	
Al 10		185 (12.0)	<1.44	10
		309 (20.0)	400	
Aluminium bronze—sand cast	250	77 (5.0)	<0.037 5	
Cu 80		131 (8.5)	<0.121	
Al 10		185 (12.0)	<0.321	10
Fe 5		309 (20.0)	<1.85	
Ni 5				
Leaded gun metal	232	55 (3.57)	0.007	
Cu 84.8		76 (4.91)	0.032	5
Pb 4.8		97 (6.25)	0.140	
Sn 5.1	260	41 (2.68)	0.007	
Zn 4.8		45 (2.90)	0.013	
		55 (3.57)	0.040	
Tin bronze	260	103 (6.70)	1.44	
Cu 90		55 (3.57)	0.042	
Sn 10				5
	316	69 (4.46)	2.0	
		55 (3.57)	0.89	
		28 (1.79)	0.225	
		7 (0.45)	0.030	
Admiralty gun metal	204	93 (6.0)	0.039	
Cu 88		120 (7.80)	0.75	
Sn 10				
Zn 2	260	42 (2.70)	0.019 5	
		62 (4.00)	0.023 5	
		76 (4.90)	0.06	12
	316	21 (1.34)	0.011	
		28 (1.79)	0.013	
		35 (2.24)	0.138	

Table 22.15 TENSILE AND CREEP PROPERTIES OF TOUGH PITCH COPPER-SILVER ALLOYS

Silver %	*Cold work* %	*Test temp.* °C	0.2% *Proof stress* MPa	*UTS* MPa	*Elong. on* 50 mm %	*Stress for* 1% *strain in* 10^5 *hours* MPa	*Stress for rupture in* 10^5 *hours* MPa
0.002	0	RT	64	220	49	—	—
	0	100	63	193	50	42	107
	0	150	63	176	50	39	59
	30	RT	287	307	18	—	—
	30	100	262	270	10	—	120
	30	150	244	253	14	—	67

Table 22.15 TENSILE AND CREEP PROPERTIES OF TOUGH PITCH COPPER-SILVER ALLOYS—*continued*

Silver %	*Cold work* %	*Test temp.* °C	0.2% *Proof stress* MPa	*UTS* MPa	*Elong. on* 50 mm %	*Stress for* 1% *strain in* 10^5 *hours* MPa	*Stress for rupture in* 10^5 *hours* MPa
0.034	0	RT	73	225	53	—	—
	0	100	73	197	52	—	142
	0	150	62	182	52	—	112
	10	RT	199	241	35	—	—
	10	100	189	209	34	159	167
	10	150	178	192	36	134	137
	20	RT	291	297	13	—	—
	20	100	260	266	8	—	—
	20	150	244	248	7	—	—
	30	RT	296	303	11	—	—
	30	100	267	277	8	182	189
	30	150	253	260	8	139	147
0.065	0	RT	83	225	50	—	—
	0	100	81	199	52	70	151
	0	150	75	185	50	63	119
	10	RT	212	245	34	—	—
	10	100	204	215	30	168	179
	10	150	190	201	28	142	148
	20	RT	293	296	12	—	—
	20	100	275	278	7	—	—
	20	150	259	261	7	—	—
	30	RT	299	304	10	—	—
	30	100	277	282	7	203	215
	30	150	263	268	7	142	166
0.14	0	RT	67	230	51	—	—
	0	100	66	203	51	74	—
	0	150	69	189	53	65	144
	10	RT	205	249	36	—	—
	10	100	198	219	32	176	192
	10	150	189	203	30	153	166
	20	RT	288	296	12	—	—
	20	100	272	276	8	230	238
	20	150	261	263	7	200	204
	30	RT	328	337	7	—	—
	30	100	308	314	6	253	265
	30	150	290	306	5	198	217
0.32	0	RT	82	232	52	—	—
	0	100	74	206	52	83	—
	0	150	82	194	51	66	—
	10	RT	194	249	41	—	—
	10	100	184	221	36	—	—
	10	150	186	211	35	—	—
	20	RT	278	289	18	—	—
	20	100	267	272	10	—	—
	20	150	255	261	9	—	—
	30	RT	330	336	11	—	—
	30	100	314	317	7	282	280
	30	150	299	308	6	241	249

REFERENCES

1. OFHC Copper—Technical Information, American Metal Climax Inc.; 1969.
2. Copper Development Association, Copper and Copper Data Sheets.
3. Copper Development Association, High Conductivity Copper Alloys, 1968.
4. M. Cook and E. C. Larke, *J. Inst. Loco. Eng.*, 1938, **28**, 609.
5. Elevated Temperature Properties of Copper and Copper Base Alloys. ASTM Spl. Publication No. 181, 1956.
6. C. S. Smith, *Proc. Am. Soc. test. Mater.*, 1939, **39**, 642.
7. A. R. Anderson and C. S. Smith, *Proc. Am. Soc. test. Mater.*, 1941, **41**, 849.
8. M. Cook and E. C. Larke, *J. Inst. Metals*, 1942, **58**, 1.

9. H. W. Gillet, *Proc. Am. Soc., Project 13*, 1941.
10. E. Voce, *Metallurgia*, 1946, **35**, 3.
11. Compilation of Available High Temperature Creep Characteristics of Metals and Alloys. *Amer. Soc. Mech. Engineering*, 1938.
12. G. Chadwick, *J. Am. Soc. Naval Enging*, 1938, **50**, 52.
13. J. E. Bowers and R. D. S. Lushey, *Met. and Mat. Tech.*, 1978, **10**(7), 381.

22.3 Mechanical properties of lead and lead alloys

The mechanical properties of lead and its alloys, particularly the more dilute alloys, are extremely sensitive to variations in composition, grain size, metallurgical history and temperature and rate of testing. They are therefore rarely reproducible with any degree of accuracy, except on the same sample under identical test conditions and, even then, a delay of a few hours between tests may affect the results obtained.

The figures quoted in the following tables should therefore be considered only as *typical* and should not be used where accuracy is necessary. For these same reasons, the materials quoted in the tables are grouped according to their principal uses and the typical values quoted for various properties are merely intended as an indication of their suitability for those uses.

Lead alloys are also dealt with under 'Solders', Chapter 34.

Table 22.16 LEAD AND LEAD ALLOYS—TYPICAL MECHANICAL PROPERTIES

Common name	*Nominal composition* %	*Specification*[1] (BS)	*Hardness VPN*	*Tensile strength*[2] N mm^{-2}	*Fatigue*[3] *strength* N mm^{-2}
A. Lead and lead alloys for chemical applications					
Type A lead	Pb > 99.99	334 Type A	4	16.8 (20°C) 12.1 (60°C)	±3.17 (20°C) ±2.24 (60°C)
Copper lead	Cu 0.05–0.07	334 Type B1	4.5	17.6 (20°C) 15.2 (60°C)	±4.96 (20°C) ±4.69 (60°C)
Tellurium copper lead	Te 0.02–0.05 Cu 0.05–0.07	334 Type B2	6	21.1 (20°C) 18.7 (60°C)	±7.24 (20°C) ±6.55 (60°C)
Silver copper lead	Ag 0.003–0.005 Cu 0.003–0.005	334 Type B3	5	16.4 (20°C) 13.4 (60°C)	±4.48 (20°C) ±3.45 (60°C)
Antimonial lead	Sb 2.5–11.0	334 Type C			
	Sb 4.0		8–12	30.0 (20°C) 25.2 (60°C)	±10.62 (20°C) ±10.04 (60°C)
	Sb 8.0		9–16	37.7 (20°C) 35.9 (60°C)	±14.82 (20°C) ±12.06 (60°C)
Dispersion strengthened lead[4]	PbO 1.5 dispersed phase	—	14	29.4	±13.4
	PbO 4.0 dispersed phase	—	—	35.5	±13.7
B. Lead and lead alloys for building applications					
Milled (rolled) lead sheet[8]	Pb > 99.9	1178[5]	4	15–18[6]	±2.9[7]
Lead pipes (not chemical)	Pb > 99.8	602[5] Comp. I	4	17[6]	±2.9[7]
	Pb 99.25–99.8	602[5] Comp. II	4–4.5	17[6]	±3[7]
Silver copper lead pipes	Ag 0.003–0.005 Cu 0.003–0.005	1085[5]	5	17[6]	±4.2[9]

Table 22.16 LEAD AND LEAD ALLOYS—TYPICAL MECHANICAL PROPERTIES—*continued*

Common name	*Nominal composition* %	*Specification*[1] (BS)	*Hardness VPN*	*Tensile strength*[2] $N\,mm^{-2}$	*Fatigue*[3] *strength* $N\,mm^{-2}$
C. Lead alloys for cable sheathing					
	Sb 0.85	801 Alloy B	6–15 (depending on heat treatment)	31[6]	±8.3[7]
	Cd 0.075 Sn 0.2	801 Alloy ½ C			±3.8[7]
	Sb 0.4 Sn 0.2	801 Alloy E	6	18.5[6]	±6.6[7]

Notes:
(1) Data from BS 334 Revision–Draft for Public Comment.
(2) Rate of testing: 0.4 mm mm^{-1} min^{-1}.
(3) 20×10^6 cycles.
(4) Although no longer in commercial production in the UK, these materials are included for their scientific interest.
(5) This information is correct at time of writing, but the standards listed are shortly due for revision.
(6) Testing rate: 0.1–0.25 mm mm^{-1} min^{-1}.
(7) 10^7 cycles.
(8) Although not included in the current BS 1178, it is now generally recommended to add 0.02–0.06% copper to improve fatigue properties.
(9) 2.5×10^7 cycles.

Table 22.17 IMPORTANT LEAD ALLOYS WITH UNSPECIFIED MECHANICAL PROPERTIES[1]

Main application	*Nominal composition* %	*Specification* (BS)	*Specific use*
Soldering	Sn 64 or 60	219 Grades A, AP, K, KP	Electrical and electronics
	Sn 40, 45, 50, 60	219 Grades G, R, F, KP	Can soldering, general engineering
	Sn 30, 35	219 Grades J, H	Cable sheaths
	Sn 15, 20	219 Grades W, V	Lamps, low service temperatures
	Sn 40 Sb 2.2	219 Grade C	Heat exchangers, general dip soldering
	Sn 30, 32 Sb 1.6, 1.8	219 Grades D, L	Plumbing, wiped joints
	Sn 28, 30 Sb 1.5, 1.6	AU 90 Nos. 28A, 30A	General body soldering
	Sn 25 Sb 1.4	AU 90 No. 25A	Radiator core dipping
	Sn 16, 18, 22 Sb 2.5, 1.0, 1.2	AU 90 Nos. 16AX, 19A, 22A	Body soldering, radiator dipping
	Sn 5.0 Sb 4.0	AU 90 No. 5AX	Hot dip coating, tubular radiator manufacture
	Sn 2.6, 10.2 Sb 5.1, 4.0 As 0.5	AU 90 Nos. 3AX §, 10AX §	Body solders for shallow areas
Printing			
	Sn 2–4 Sb 2–5	—	Electrobacking metal
	Sn 2–5 Sb 10–13	—	Slug casting metal

Table 22.17 IMPORTANT LEAD ALLOYS WITH UNSPECIFIED MECHANICAL PROPERTIES[1]—*continued*

Main application	*Nominal composition %*	*Specification* (BS)	*Specific use*
Printing (cont.)	Sn 5–10 Sb 15	—	Stereotype metal
	Sn 6–13 Sb 15–19	—	Monotype casting metal
	Sn 13–22 Sb 20–28	—	Cast type casting metal

Notes:
(1) In many uses of lead alloys, the strength, hardness, etc. of the alloy are not of prime importance. In general soldering, for example, the geometry of the joint and the materials being joined, are of greater importance than the strength of the bulk solder. In most applications of type metals, fluidity of the liquid alloy, contraction properties, and wear resistance of the solid alloy, are the most important characteristics in formulating an alloy. See also Chapter 34.

BIBLIOGRAPHY

L. I. Goff and G. Hewish, 'Creep Resistant Lead Sheet by the D.M. Process', *3rd Inter. Lead Conf.*, Venice, 1968, Publ'd. LDA, London.
L. I. Goff and R. D. Semmens, 'Lead Products by Continuous Casting', *2nd Inter. Lead Conf.*, Arnhem, 1965, Publ'd. LDA, London.
J. N. Greenwood, *Met. Rev.*, 1961, **6**(**23**), 279–351.
W. Hofmann, 'Lead and Lead Alloys', English trans. of 2nd revised German edition, Springer-Verlag, Berlin.
'Lead Abstracts', 1962–80. LDA, London.
Early Trials.' *Second Inter. Lead Conf.*, Arnhem, 1965.
A. Lloyd and E. R. Newson, 'Dispersion Strengthened Lead Properties and Potentialities in Chemical Plant and some
A. Lloyd and E. R. Newson, 'Dispersion Strengthened Lead—Developments and Applications in the Chemical Industry, *3rd Inter. Lead Conf.*, Venice, 1968, Publ'd. LDA, London.

22.4 Mechanical properties of magnesium and magnesium alloys

Table 22.18 MAGNESIUM AND MAGNESIUM ALLOYS (WROUGHT)—TYPICAL MECHANICAL PROPERTIES AT ROOM TEMPERATURE

			Specifications				*Tension*					*Compression*		
Material	*Nominal* composition* %	*Form*	DTD or BS (Air)	BS (Gen. Eng.)	ASTM	*Elektron*	*Proof stress* 0.2% MPa	tonf in^{-2}	*UTS* MPa	tonf in^{-2}	*Elong.* %	*Proof stress* 0.2% MPa	tonf in^{-2}	*VPN* 30 kg
Mg	Mg 99.9	Sheet, annealed	—				69	4.5	185	12	4	—	—	30–35
		Bar, extruded	—				100	6.5	232	15	6	—	—	35–45
Mg–Mn	Mn 1.5	Sheet	118C	3370–MAG–S–101M	—	AM503	100	6.5	232	15.0	6	—	—	35–45
		Extruded bar (1 in diam.)	142B	3373–MAG–E–101M	M1A–F, B107		162	10.5	263	17.0	7	124	8.0	45–55
		Extruded tube	737A	3373–MAG–E–101 M	M1A, B107		154	10.0	247	16.0	6	—	—	45–55
Mg–Al–Zn	Al 3.0	Sheet, annealed		3370–MAG–S–111O	AZ31, B90	AZ31	131	8.5	232	15.0	13	—	—	50–60
	Zn 1.0	half hard		3370–MAG–S–111M	AZ31, B90		170	11.0	263	17.0	10	100	6.5	55–70
	Mn 0.3	Extruded bar and sections		3373–MAG–E–111M	AZ31, B107		162	10.5	255	16.5	11	93	6.0	50–60
	Al 6.0	Forgings	2L513	3372–MAG–F–121M	AZ61, B91	AZM	183	12.0	293	19.0	8	147	9.5	60–70
	Zn 1.0	Extruded bar and sections	2L512	3373–MAG–E–121M	AZ61, B107		183	12.0	293	19.0	8	147	9.5	55–70
	Mn 0.3	Extruded tube	2L503	3373–MAG–E–121M	AZ61, B107		170	11.0	278	18.0	8	147	9.5	60–70
	Al 8.0	Forgings	88C	—	AZ80A, B91	AZ855	208	13.5	293	19.0	8	185	12.0	65–7[illegible]
	Zn 0.5													
	Mn 0.3													
Mg–Zn–Mn	Zn 2.0	Sheet, annealed	5091	3370–MAG–S–131O		ZM21	131	8.5	232	15.0	13			
	Zn 1.0	half hard	5101	3370–MAG–S–131M			170	11.0	263	17.0	10			
		Extruded bar sections	—	3373–MAG–E–131M			162	10.5	255	16.5	11			
Mg–Zn–Zr	Zn 1.0	Sheet	2L515	3370–MAG–S–141M	—	ZW1	178	11.5	263	17.0	10	154	10.0	55–70
	Zr 0.6	Extruded bar and sections	2L508	3373–MAG–E–141M	—		208	13.5	293	19.0	13	177	11.5	60–75
		Extruded tube	2L509	3373–MAG–E–141M	—		193	12.5	278	18.0	7	—	—	60–75
	Zn 3.0	Sheet	2L504	3370–MAG–S–151M	—	ZW3	185	12.0	270	17.5	8	154	10.0	60–70
	Zr 0.6	Forgings	2L514	3372–MAG–F–151M	—		224	14.5	309	20.0	8	193	12.5	60–80
		Extruded bar and sections (1 in diam.)	2L505	3373–MAG–E–151M	—		239	15.5	309	20.0	18	213	13.8	65–75
	Zn 5.5	Bars and sections												
	Zr 0.6	Heat treated	5041A	3373–MAG–E–161TE	ZK60A–T5, B107–70	ZW6	270	17.5	340	22.0	10	255	16.5	60–80

Material	Composition	Form	DTD	BS	ASTM	Elektron	Proof stress MPa	tonf in^{-2}	UTS MPa	tonf in^{-2}	Elong. %	Compression proof stress MPa	tonf in^{-2}	Hardness
Mg–Th–Zn–Zr	Th 0.8	Extruded bar and sections 5111	—	—	—	ZTY	147	9.5	263	17	18	—	—	50–70
(Creep resistant)	Zn 0.5 Zr 0.6	Forgings 5111	—	—	—		147	9.5	232	15	13	—	—	50–70
Mg–Th–Mn (Creep resistant)	Th 2.0 Mn 0.75	Sheet	—	—	HM21–T8, B90	—	165	10.7	247	16.0	9	179	11.6	—
	Th 3.0 Mn 1.2	Extruded bar and sections	—	—	HM31–T5	—	227	14.7	287	18.6	8	185	12.0	—

Nuclear alloys: Two wrought magnesium alloys (Magnox AL80; Mg 0.75Al–0.005Be and MN70; Mg 0.75Mn) of interest only for their nuclear and high-temperature properties have room-temperature tensile properties similar to those of AM503.

*It is usual to add 0.2–0.4% Mn to alloys containing aluminium to improve corrosion resistance. M = As manufactured. O = Fully annealed. TE = Precipitation treated.

Table 22.19 MAGNESIUM AND MAGNESIUM ALLOYS (CAST) TYPICAL MECHANICAL PROPERTIES ATROOM TEMPERATURE

			Specifications				Tension					Compression		
Material	*Nominal* composition* %	*Condition*	DTD or BS(Air)	BS(Gen. Eng.)	ASTM	*Elektron*	*Proof stress* 0.2% MPa	tonf in^{-2}	MPa	*UTS* tonf in^{-2}	*Elong.* %	*Proof stress* 0.2% MPa	tonf in^{-2}	*Brinell hardness*† *VPN* 30 kg
Mg–Zr	Zr 0.6	AC	—	—	KIA, B80	ZA	51	3.3	185	12.0	2.0	54	3.5	40–50
Mg–Al–Zn	Al 6.0	AC	—	—	AZ63A–F, B80	—	97	6.3	199	12.9	5	97	6.3	50
	Zn 3.0	TB	—	—	AZ63A–T4, B80		97	6.3	275	17.8	10	97	6.3	55
		TF	—	—	AZ63A–T6, B80		131	8.5	275	17.8	5	131	8.5	73
	Al 8.0	AC	—	2970 MAG 1–M		A8	86	5.6	158	10.2	4	86	5.6	50–60
	Zn 0.4	TB	3L122	2970 MAG 1–TB	AZ81A–T4, B80		82	5.3	247	16.0	11	82	5.3	50–60
	Al 9.5	AC	—	2970 MAG 3–M	AZ91C–F, B80	AZ91	93	6.0	154	10.0	2	93	6.0	55–65
	Zn 0.4	TB	3L124	2970 MAG 3–TB	AZ91C–T4, B80		90	5.8	232	15.0	6	90	5.8	55–65
		TF	3L125	2970 MAG 3–TF	AZ91C–T6, B80		127	8.2	239	15.5	2	124	8.0	75–85
		Die cast	—	—	AZ91B–F, B94		111	7.2	216	14.0	3	108	7.0	60–70
	Al 9.0	AC	—	—	AZ92A, B80	—	97	6.3	165	10.7	2	97	6.3	65
	Zn 2.0	TB	—	—	—		97	6.3	275	17.8	8	97	6.3	63
		TF	—	—	—		145	9.4	275	17.8	2	145	9.4	84
Mg–Zn–Zr	Zn 4.5 Zr 0.7	TE	2L127	2970 MAG 4–TE	ZK51A–T5, B80	Z5Z	161	10.4	263	17.0	6	162	10.5	65–75

Table 22.19 MAGNESIUM AND MAGNESIUM ALLOYS (CAST)—TYPICAL MECHANICAL PROPERTIES AT ROOM TEMPERATURE—*continued*

			Specifications				Tension					Compression		
Material	*Nominal* composition* %	*Condition*	DTD or BS(Air)	BS(Gen. Eng.)	ASTM	*Elektron*	*Proof stress* 0.2% MPa	tonf in^{-2}	*UTS* MPa	tonf in^{-2}	*Elong.* %	*Proof stress* 0.2% MPa	tonf in^{-2}	*Brinell hardness*† *VPN* 30kg
Mg–Zn–RE–Zr	Zn 4.0 RE 1.2 Zr 0.7	TE	2L128	2970 MAG 5–TE	ZE41A–T5, B80	RZ5	150	9.7	216	14.0	5	139	9.0	55–75
	Zn 6.0 RE 2.5 Zr 0.7	TF §	5045	—		ZE63	190	12.3	295	19.1	7	190	12.3	70–80
Mg–RE–Zn–Zr (Creep resistant to 250°C)	RE 2.7 Zn 2.2 Zr 0.7	TE	2L126	2970 MAG 6–TE	EZ33A–T5, B80	ZRE1	95	6.2	162	10.5	4.5	93	6.0	50–60
Mg–Th–Zn–Zr (Creep resistant to 350°C)	Th 3.0 Zn 2.2 Zr 0.7	TE	5005A	2970 MAG 8–TE	HZ32A–T5, B80	ZT1	93	6.0	216	14.0	7	93	6.0	50–60
Mg–Zn–Th–Zr	Zn 5.5 Th 1.8 Zr 0.7	TE	5015A	2970 MAG 9–TE	ZH62A–T5, B80	TZ6	167	10.8	270	17.5	8	162	10.5	65–75
Mg–Th–Zr	Th 3.0 Zr 0.7	TF	—	—	HK31A–T6, B80	MTZ	93	6.0	208	13.5	5	93	6.0	50–60
Mg–Ag–RE‡–Zr	Ag 2.5 RE 2.0‡ Zr 0.6	TF	5025A 5035A	—		MSR–A MSR–B	187 204	12.1 13.2	247 260	16.0 16.8	5 3	178 193	11.5 12.5	65–80 65–80
	Ag 2.5 RE 2.0‡ Zr 0.6	TF	5055	—	QE22A–T6, B80	QE22	200	12.9	260	16.8	4	195	12.6	65–80
Mg–Ag–Th–RE‡Zr	Ag 2.5 RE 1.0‡ Th 1.0 Zr 0.7	TF	—	—	QH21A–T6, B80	QH21A	210	13.6	270	17.5	4	200	12.9	65–80

* It is usual to add 0.2–0.4% Mn to alloys containing aluminium to improve corrosion resistance. RE = Cerium mischmetal containing approx. 50% cerium.
† Brinell tests with 500 kg on 10 mm ball for 30 s.
‡ Fractionated rare earth metals: MSR-A contains 1.7%; MSR-B contains 2.5%.
§ Solution heat treated in an atmosphere of hydrogen.
AC = Sand cast. TE = Precipitation heat treated.
TB = Solution heat treated. TF = Fully heat treated.

Table 22.20 MAGNESIUM AND MAGNESIUM ALLOYS (EXCLUDING HIGH TEMPERATURE ALLOYS FOR WHICH SEE TABLE 22.21)—TYPICAL TENSILE PROPERTIES AT ELEVATED TEMPERATURES

					'Short-time' tension†						
Material	*Nominal composition** %		*Form and condition*	*Test temp.* °C	*Young's modulus* GPa	10^6 lbf in^{-2}	*0.2% proof stress* MPa	tonf in^{-2}	*UTS* MPa	tonf in^{-2}	*Elong.* %
Mg	Mg	99.95	Forged	20	45	6.5	—	—	170	11.0	5
				100	—	—	—	—	128	8.3	8
				150	—	—	—	—	93	6.0	16
				200	—	—	—	—	54	3.5	43
Mg–Al–Zn	Al	8.0	Sand cast	20	45	6.5	86	5.6	158	10.2	4
	Zn	0.4		100	34	5.0	76	4.9	154	10.0	5
	(A8)			150	32	4.6	65	4.2	145	9.4	11
				200	25	3.6	62	4.0	100	6.5	20
				250	—	—	—	—	73	4.7	27
			Sand cast	20	45	6.5	82	5.3	247	16.0	11
			and	100	34	5.0	73	4.7	202	13.1	16
			solution	150	33	4.8	65	4.2	154	10.0	21
			treated	200	28	4.0	62	4.0	116	7.5	25
				250	—	—	—	—	85	5.5	21
	(AZ855)		Forged	20	45	6.5	221	14.3	309	20.0	8
				150	—	—	153	9.9	216	14.0	25
				200	—	—	102	6.6	154	10.0	28
	Al	9.5	Sand cast	20	45	6.5	93	6.0	154	10.0	2
	Zn	0.4		100	—	—	—	—	131	8.5	2
	(AZ91)			150	—	—	—	—	122	7.9	6
				200	—	—	—	—	108	7.0	25
				250	—	—	—	—	77	5.0	34
			Sand cast	20	45	6.5	90	5.8	232	15.0	6
			and	100	—	—	—	—	222	14.4	12
			solution	150	—	—	—	—	196	12.7	16
			treated	200	—	—	—	—	139	9.0	20
			Sand cast	20	45	6.5	127	8.2	239	15.5	2
			and	100	40	5.8	91	5.9	232	15.0	6
			fully heat	150	37	5.4	77	5.0	185	12.0	25
			treated	200	28	4.0	62	4.0	133	8.6	34
				250	19	2.7	46	3.0	103	6.7	30
Mg–Zn–Zr	Zn	4.5	Sand cast	20	45	6.5	161	10.4	263	17.0	6
	Zr	0.7	and heat	100	34	5.0	124	8.0	185	12.0	14
	(Z5Z)		treated	150	28	4.0	102	6.6	145	9.4	20
				200	22	3.25	79	5.1	113	7.3	23
				250	19	2.75	57	3.7	85	5.5	20
	Zn	3.0	Extruded	20	45	6.5	255	16.5	309	20.0	18
	Zr	0.6		100	40	5.75	162	10.5	182	11.8	33
	(ZW3)			200	22	3.25	46	3.0	127	8.2	56
				250	12	1.75	11	0.7	100	6.5	71
			Sheet	20	45	6.5	195	12.6	270	17.5	10
				100	40	5.75	120	7.8	165	10.7	33
				150	33	4.75	74	4.8	116	7.5	42
				200	—	—	—	—	76	4.9	51
				250	—	—	—	—	49	3.2	59
Mg–Zn–RE–Zr	Zn	4.0	Sand cast	20	45	6.5	150	9.7	216	14.0	4
	RE	1.2	and heat	100	41	6.0	134	8.7	195	12.6	6
	Zr	0.7	treated	150	40	5.75	120	7.8	167	10.8	19
	(RZ5)			200	38	5.5	99	6.4	131	8.5	29
				250	33	4.75	74	4.8	99	6.4	35

Table 22.20 MAGNESIUM AND MAGNESIUM ALLOYS (EXCLUDING HIGH-TEMPERATURE ALLOYS FOR WHICH SEE TABLE 22.21) TYPICAL TENSILE PROPERTIES AT ELEVATED TEMPERATURES—*continued*

				'Short-time' tension†						
Material	*Nominal composition** %	*Form and condition*	*Test temp.* °C	*Young's modulus* GPa	10^6 lbf in^{-2}	*0.2% proof stress* MPa	tonf in^{-2}	*UTS* MPa	tonf in^{-2}	*Elong.* %
Mg–Zn–Th–Zr	Zn 5.5	Sand cast	20	45	6.5	161	10.4	270	17.5	9
	Th 1.8	and heat	100	34	5.0	134	8.7	224	14.5	22
	Zr 0.7	treated	150	31	4.5	110	7.1	178	11.5	26
	(TZ 6)		200	28	4.0	82	5.3	130	8.4	26
			250	26	3.75	52	3.4	91	5.9	25
Mg–Ag–	Ag 2.5	Sand cast	20	45	6.5	201	13.0	259	16.8	4
RE–Zr	RE(D)2.0	and fully	100	41	6.0	185	12.0	232	15.0	12
(D)	Zr 0.6	heat	150	40	5.75	171	11.1	210	13.6	16
	(QE22)	treated	200	38	5.5	154	10.0	185	12.0	20
			250	34	5.0	102	6.6	142	9.2	27
			300	31	4.5	68	4.4	88	5.7	59
Mg–Ag–Re(D)	Ag 2.5	Sand cast	20	45	6.5	210	13.6	270	17.5	4
Th–Zr	RE(D)1.0	and fully	100	41	5.9	199	12.9	242	15.7	17
‡	Th 1.0	heat	150	40	5.8	190	12.3	224	14.5	20
	Zr 0.6	treated	200	38	5.5	183	11.8	205	13.3	18
	(QH21)		250	37	5.4	167	10.8	185	12.0	19
			300	33	4.8	120	7.8	131	8.5	20

* It is usual to add 0.2–0.4% Mn to alloys containing aluminium to improve corrosion resistance.
† In accordance with BS 1094: 1943; 1 h at temperature and strain rate 0.1–0.25 in in^{-1} min^{-1}.
‡ Tested according to BS 4A4. RE = Cerium mischmetal containing approx. 50% Ce. Re(D) = Neodymium enriched mischmetal.

Table 22.21 HIGH-TEMPERATURE MAGNESIUM ALLOYS—TENSILE PROPERTIES AT ELEVATED TEMPERATURES

				'Short-time' tension†						
Material	*Nominal composition** %	*Form and condition*	*Test temp.* °C	*Young's modulus* GPa	10^6 lbf in^{-2}	*0.2% proof stress* MPa	tonf in^{-2}	*UTS* MPa	tonf in^{-2}	*Elong.* %
Mg–RE–Zn	RE 2.7	Sand cast	20	45	6.5	93	6.0	162	10.5	4.5
	Zn 2.2	and heat	100	40	5.75	79	5.1	150	9.7	11
	Zr 0.7	treated	150	38	5.5	76	4.9	139	9.0	19
	(ZRE1)		200	36	5.25	74	4.8	125	8.1	26
			250	33	4.75	65	4.2	107	6.9	35
			300	28	4.0	48	3.1	85	5.5	51
			350	21	3.0	26	1.7	56	3.6	90
Mg–Th–Zr	Th 3.0	Sand cast	20	45	6.5	93	6.0	208	13.5	4
	Zr 0.7	and fully	100	40	5.75	88	5.7	188	12.2	10
	(HK31)	heat	150	38	5.5	86	5.6	174	11.3	13
	(MTZ)	treated	200	38	5.5	85	5.5	162	10.5	17
			250	36	5.25	83	5.4	150	9.7	20
			300	34	5.0	73	4.7	136	8.8	22
			350	29	4.25	56	3.6	103	6.7	23
Mg–Th–Zn–	Th 3.0	Sand cast	20	45	6.5	93	6.0	216	14.0	9
Zr	Zn 2.2	and heat	100	36	5.25	88	5.7	159	10.3	23
	Zr 0.7	treated	150	34	5.0	79	5.1	131	8.5	27
	(ZT1)		200	33	4.75	65	4.2	108	7.0	33
			250	33	4.75	56	3.6	90	5.8	38
			300	31	4.5	49	3.2	76	4.9	41
			350	28	4.0	45	2.9	63	4.1	34
	Th 0.8	Sheet	20	45	6.5	181	11.7	266	17.2	10
	Zn 0.5		100	41	6.0	179	11.6	224	14.5	10
	Zr 0.6		150	41	6.0	176	11.4	201	13.0	11
	(ZTY)		200	40	5.75	165	10.7	171	11.1	15
			250	40	5.75	124	8.0	134	8.7	20
			300	34	5.0	73	4.7	96	6.2	27
			350	29	4.25	17	1.1	56	3.6	38

Table 22.21 HIGH-TEMPERATURE MAGNESIUM ALLOYS—TENSILE PROPERTIES AT ELEVATED TEMPERATURES —*continued*

Material	*Nominal composition** %	*Form and condition*	*Test temp.* °C	*'Short-time' tension†* *Young's modulus* GPa	10^6 lbf in^{-2}	*0.2% proof stress* MPa	tonf in^{-2}	*UTS* MPa	tonf in^{-2}	*Elong.* %
Mg–Ag–RE(D)–Zr	Ag 2.5 RE(D) 2.0 Zr 0.6 (QE22)	Sand cast and fully heat treated	High strength cast alloys with good elevated temperature properties —for which *see* Table 22.19.							
Mg–Ag–RE(D)–Th–Zr	Ag 2.5 RE(D) 1.0 Th 1.0 Zr 0.6 (QH21)	Sand cast and fully heat treated								

* It is usual to add 0.2–0.4% Mn to alloys containing aluminium to improve corrosion resistance.
† In accordance with BS 1094:1943; 1 h at temperature; strain rate 0.1–0.25 in in^{-1} min^{-1}.
RE = Cerium mischmetal containing approx. 50% Ce. RE(D) = neodymium-enriched mischmetal.

Table 22.22 HIGH-TEMPERATURE MAGNESIUM ALLOYS—LONG-TERM CREEP RESISTANCE

Material	Nominal composition %	Form and Condition	Temp. °C	Time† h	Stress to produce specified creep strains %									
					0.05		0.1		0.2		0.5		1.0	
					MPa	tonf in^{-2}	MPa	tonf in^{-2}	MPa	tonf in^{-2}	MPa	tonf in^{-2}	MPa	tonf in^{-2}
Mg–RE–Zb–Zr	RE 2.7	Sand cast	200	100	52	3.37	66	4.27	71	4.6	—	—	—	—
	Zn 2.2	and heat		500	41	2.65	54	3.50	65	4.21	—	—	—	—
	Zr 0.7	treated		1000	36	2.33	47	3.04	58	3.76	—	—	—	—
	(ZRE1)													
			250	100	23	1.49	28	1.81	32	2.07	36	2.33	—	—
				500	11	0.71	19	1.23	24	1.55	30	1.94	34	2.20
				1000	—	—	14	0.91	20	1.29	26	1.68	30	1.94
			315	100	5.6	0.36	7.4	0.48	8	0.52	—	—	—	—
				500	—	—	5.2	0.34	6.5	0.42	—	—	—	—
				1000	—	—	4.3	0.28	5.6	0.36	—	—	—	—
	Zn 4.0	Sand cast	100	100	—	—	97	6.28	111	7.19	117	7.58	—	—
	RE 1.2	and heat		500	—	—	—	—	106	6.86	117	7.58	—	—
	Zr 0.7	treated		1000	—	—	—	—	103	6.67	116	7.51	—	—
	(RZ5)													
			150	100	77	4.99	86	5.57	97	6.28	101	6.54	107	6.93
				500	—	—	75	4.86	88	5.70	96	6.22	100	6.47
				1000	—	—	70	4.53	83	5.37	91	5.89	97	6.28
			200	100	29	1.88	43	2.78	52	3.37	67	4.35	73	4.73
				500	22	1.42	28	1.81	37	2.40	52	3.37	64	4.14
				1000	20	1.29	23	1.49	31	2.01	43	2.78	53	3.43
			250	100	6.2	0.40	12	0.78	19	1.23	32	2.07	39	2.53
				500	4.3	0.28	6.2	0.40	8.6	0.56	15	0.97	19	1.23
				1000	3.9	0.25	5.4	0.35	6.9	0.45	12	0.78	15	0.97
Mg–Th–Zr	Th 3.0	Sand cast	200	100	31*	2.01*	45*	2.91*	63*	4.08*	97*	6.28*	111*	7.19*
	Zr 0.7	and fully		1000	—	—	—	—	62*	4.01*	100*	6.47*	—	—
	(HK31)	heat												
	(MTZ)	treated	260	100	—	—	28*	1.81*	43*	2.78*	65*	4.21*	—	—
				1000	—	—	—	—	29*	1.88*	45*	2.91*	—	—
			315	100	9.3*	0.60*	14*	0.91*	19*	1.23*	27*	1.75*	32*	2.07*
				1000	—	—	—	—	6.9	0.45*	9.3*	0.60*	—	
Mg–Th–Zn–Zr	Th 0.8	Sheet	250	100		Stress of 46 MPa (3 tonf in^{-2}) produced 0.01% creep strain								
	Zn 0.5			1000		Stress of 46 MPa (3 tonf in^{-2}) produced 0.03% creep strain								
	Zr 0.6													
	(ZTY)													

Table 22.22 HIGH-TEMPERATURE MAGNESIUM ALLOYS—LONG-TERM CREEP RESISTANCE—*continued*

					Stress to produce specified creep strains %									
					0.05		0.1		0.2		0.5		1.0	
Material	*Nominal composition* %	*Form and Condition*	*Temp.* °C	*Time*† h	MPa	tonf in^{-2}	MPa	tonf in^{-2}	MPa	tonf in^{-2}	MPa	tonf in^{-2}	MPa	tonf in^{-2}
Mg–Th–Zn–Zr	Th 3.0	Sand cast	250	100	42	2.72	50	3.24	56	3.63	63	4.08	66	4.27
	Zn 2.2	and heat		500	35	2.27	43	2.78	51	3.30	58	3.76	63	4.08
	Zr 0.7	treated		1 000	31	2.01	39	2.53	48	3.10	56	3.63	61	3.95
	(ZT1)													
			300	100	23	1.49	28	1.81	35	2.27	46	2.98	52	3.37
				500	19	1.23	21	1.36	25	1.62	36	2.33	41	2.65
				1 000	17	1.10	19	1.23	21	1.36	32	2.07	36	2.33
			325	100	14	0.91	19	1.23	24	1.55	29	1.88	36	2.33
				500	12	0.78	13	0.84	16	1.04	21	1.36	25	1.62
				1 000	10	0.65	12	0.78	13	0.84	15	0.97	20	1.29
			350	100	10	0.65	12	0.78	18	1.17	21	1.36	23	1.49
				500	—	—	9	0.58	10	0.65	12	0.78	14	0.91
				1 000	—	—	8	0.52	8	0.52	9	0.58	10	0.65
			375	100	—	—	8	0.52	11	0.71	12	0.78	13	0.84
				500	—	—	—	—	—	—	8	0.52	9	0.58
				1000	—	—	—	—	—	—	—	—	8	0.52
	Zn 5.5	Sand cast	150	100	51	3.30	66	4.27	82	5.31	96	6.22	102	6.60
	Th 1.8	and heat		500	36	2.33	56	3.63	69	4.47	85	5.50	94	6.09
	Zr 0.7	treated		1 000	26	1.68	51	3.30	63	4.08	80	5.18	90	5.83
	(TZ6)													
			200	100	26	1.68	32	2.07	45	2.91	56	3.63	62	4.01
				500	15	0.97	22	1.42	26	1.68	40	2.59	49	3.17
				1 000	11	0.71	17	1.10	20	1.29	31	2.01	40	2.59
Mg–Ag–	Ag 2.5	Sand cast	200	100	55	3.56	74	4.79	88	5.70	—	—	—	—
RE(D)–Zr	RE(D) 2.0	and fully		500	—	—	54	3.50	65	4.21	82	5.31	89	5.76
	Zr 0.6	heat		1 000	—	—	46	2.98	56	3.63	73	4.73	79	5.12
	(QE22)	treated												
			250	100	18	1.17	26	1.68	33	2.14	—	—	—	—
				500	—	—	15	0.97	22	1.42	28	1.81	31	2.01
				1 000	—	—	10	0.65	16	1.04	22	1.42	26	1.68
Mg–Ag–	Ag 2.5	Sand cast	250	100	22	1.42	32	2.07	39	2.53	—	—	—	—
RE(D)–Th–Zr	RE(D) 1.0	and fully		500	—	—	20	1.29	26	1.68	32	2.07	36	2.33
	Th 1.0	heat		1 000	—	—	—	—	21	1.36	26	1.68	30	1.94
	Zr 0.6	treated												
	(QH21)													

* Total strains.
† 4–6 h heating to test temperature followed by 16 h soaking at test temperature.
RE = Cerium mischmetal containing approx. 50% Ce.
RE(D) = Neodymium-enriched mischmetal.

Table 22.23 HIGH-TEMPERATURE MAGNESIUM ALLOYS—SHORT-TERM CREEP RESISTANCE

					Stress to produce specified total strains %											
					0.5		1.0		2.0		5.0		10.0		*Stress to fracture*	
Material	*Nominal composition* %	*Form and condition*	*Temp.* °C	*Time.*† s	MPa	tonf in^{-2}	MPa	tonf in^{-2}	MPa	tonf in^{-2}	MPa	tonf in^{-2}	MPa	tonf in^{-2}	MPa	tonf in^{-2}
Mg–RE–Zn–Zr	RE 2.7	Sand cast	200	30	—	—	—	—	98	6.35	118	7.64	130	8.42	136	8.81
	Zn 2.2	and heat		60	—	—	—	—	97	6.28	117	7.58	128	8.29	134	8.68
	Zr 0.7	treated		600	—	—	—	—	96	6.22	116	7.51	125	8.09	129	8.35
	(ZRE1)															
			250	30	76	4.92	84	5.44	92	5.96	111	7.19	123	7.96	130	8.42
				60	74	4.79	83	5.37	91	5.89	110	7.12	120	7.77	129	8.35
				600	73	4.73	82	5.31	89	5.76	108	6.99	114	7.38	125	8.09
			315	30	52	3.37	59	3.82	73	4.73	80	5.18	85	5.50	90	5.83
				60	51	3.30	58	3.76	69	4.47	76	4.92	83	5.37	88	5.70
				600	42	2.72	49	3.17	56	3.63	62	4.01	68	4.40	73	4.73
	Zn 4.0	Sand cast	200	30	100	6.47	107	6.93	116	7.51	127	8.22	—	—	136	8.81
	RE 1.2	and heat		60	99	6.41	105	6.80	114	7.38	124	8.03	—	—	134	8.68
	Zr 0.7	treated		600	86	5.57	99	6.41	103	6.67	114	7.38	—	—	125	8.09
	(RZ5)															
			250	30	86	5.57	90	5.83	94	6.09	99	6.41	—	—	116	7.51
				60	83	5.37	88	5.70	91	5.89	96	6.22	—	—	113	7.32
				600	71	4.60	76	4.92	81	5.24	86	5.57	—	—	93	6.02
			315	30	62	4.01	69	4.47	76	4.92	79	5.12	83	5.37	86	5.57
				60	59	3.82	66	4.27	73	4.73	76	4.92	79	5.12	82	5.31
				600	48	3.09	53	3.43	59	3.82	64	4.14	67	4.34	69	4.47
Mg–Th–Zr	Th 3.0	Sand cast	250	30	96	6.22	103	6.67	119	7.71	138	8.94	—	—	145	9.39
	Zr 0.7	and fully		60	95	6.15	103	6.67	118	7.64	137	8.87	—	—	145	9.39
	(HK31)	heat		600	94	6.09	102	6.60	117	7.58	137	8.87	—	—	144	9.32
	(MTZ)	treated														
			315	30	80	5.18	88	5.70	103	6.67	117	7.58	—	—	128	8.29
				60	78	5.05	86	5.57	102	6.60	116	7.51	—	—	127	8.22
				600	74	4.79	82	5.31	96	6.22	107	6.93	—	—	120	7.77

Mg–Th–	Th	0.8	Sheet	250	30	—	—	—	—	110	7.12	159	10.30	163	10.55	165	10.68
Zn–Zr	Zn	0.5			60	—	—	—	—	95	6.15	157	10.17	160	10.36	162	10.49
	Zr	0.6			600	—	—	—	—	—	—	145	9.39	149	9.65	151	9.78
	(ZTY)			350	30	20	1.29	32	2.07	48	3.11	80	5.18	93	6.02	102	6.60
					60	18	1.17	28	1.81	40	2.59	67	4.34	82	5.31	98	6.35
					600	—	—	15	0.97	20	1.29	31	2.01	42	2.72	66	4.27
Mg–Th–	Th	3.0	Sand cast	200	30	—	—	—	—	—	—	100	6.47	118	7.64	125	8.09
Zn–Zr	Zn	2.2	and heat		60	—	—	—	—	—	—	96	6.22	114	7.38	123	7.96
	Zr	0.7	treated		600	—	—	—	—	—	—	85	5.50	103	6.67	114	7.38
	(ZT1)			250	30	58	3.76	65	4.21	71	4.60	84	5.44	102	6.60	111	7.19
					60	57	3.69	64	4.14	69	4.47	81	5.24	99	6.41	107	6.93
					600	56	3.63	63	4.08	68	4.40	74	4.79	86	5.57	98	6.35
				315	30	55	3.56	60	3.88	64	4.14	73	4.73	76	4.92	82	5.31
					60	53	3.43	59	3.82	63	4.08	72	4.66	76	4.92	80	5.18
					600	50	3.24	59	3.82	61	3.95	71	4.60	74	4.79	77	4.99
	Zn	5.5	Sand cast	200	30	96	6.22	113	7.32	120	7.77	128	8.29	137	8.87	144	9.32
	Th	1.8	and heat		60	93	6.02	109	7.06	117	7.58	124	8.03	133	8.61	137	8.87
	Zr	0.7	treated		600	83	5.37	90	5.83	102	6.60	110	7.12	114	7.38	119	7.71
	(TZ6)			250	30	70	4.53	77	4.99	85	5.50	96	6.22	99	6.41	107	6.99
					60	65	4.21	74	4.79	80	5.18	90	5.83	94	6.09	99	6.41
					600	56	3.63	60	3.88	66	4.27	74	4.79	77	4.99	82	5.31
				315	30	54	3.50	59	3.82	64	4.14	70	4.53	74	4.79	76	4.92
					60	52	3.37	57	3.69	62	4.01	66	4.27	70	4.53	73	4.73
					600	44	2.85	49	3.17	53	3.43	56	3.63	58	3.76	59	3.82

† 1 h heating to test temperature followed by 1 h soaking at test temperature.
Re = cerium mischmetal containing approx. 50% Ce.

Table 22.24 MAGNESIUM AND MAGNESIUM

Material	*Nominal* composition* %	Condition	‡ *State*	*Test temp.* °C	*Fatigue strength†* 10^5 MPa	10^5 tonf in^{-2}	5×10^5 MPa	5×10^5 tonf in^{-2}
Mg–Mn	Mn 1.5	Extruded	U	20	107	6.9	90	5.8
	(AM503)		N		76	4.9	60	3.9
Mg–Al–Zn	Al 6.0	Extruded	U	20	161	10.4	139	9.0
	Zn 1.0		N		127	8.25	110	7.1
	(AZM)							
	Al 8	Sand cast	U	20	108	7.0	93	6.0
	Zn 0.4		N		107	6.9	80	5.2
	(A8)	Sand cast and	U	20	124	8.0	102	6.6
		solution	N		108	7.0	86	5.6
		treated	U	150	93	6.0	69	4.5
			U	200	71	4.6	52	3.4
	Al 9.5	Sand cast	U	20	114	7.4	91	5.9
	Zn 0.4		N		110	7.1	83	5.4
	(AZ91)	Sand cast and	U	20	124	8.0	93	6.0
		solution	N		103	6.7	82	5.3
		treated						
		Sand cast and	U	20	117	7.6	90	5.8
		fully heat	N		93	6.0	66	4.3
		treated						
Mg–Zn–Zr	Zn 3.0	Extruded	U	20	151	9.8	137	8.9
	Zr 0.6		N		124	8.0	99	6.4
	(ZW3)							
	Zn 4.5	Sand cast and	U	20	111	7.2	86	5.6
	Zr 0.7	heat treated	N		90	5.8	86	5.6
	(Z5Z)							
Mg–Zn–RE–Zr	Zn 4.0	Sand cast and	U	20	124	8.0	99	6.4
	RE 1.2	heat treated	N		108	7.0	93	6.0
	Zr 0.7		U	150	97	6.25	85	5.5
	(RZ5)		U	200	93	6.0	74	4.8
	Zn 2.2	Sand cast and	U	20	100	6.5	82	5.3
	RE 2.7	heat treated	N		77	5.0	59	3.8
	Zr 0.7		U	150	69	4.5	60	3.9
	(ZRE 1)		U	200	68	4.4	59	3.8
			U	250	59	3.8	48	3.1
			U	300	49	3.2	39	2.5
	Zn 6	Sand cast	U	20	144	9.32	131	8.48
	RE 2.5	and fully	N		99	6.41	83	5.37
	Zr 0.6	heat treated**						
	(ZE63)							
Mg–Ag–RE(D)–Zr	Ag 2.5	Sand cast and	U	20	119	7.7	103	6.7
	RE(D) 2.5	fully heat	N		77	5.0	65	4.2
	Zr 0.6	treated	U	200	—	—	—	—
	(MSR–B)		U	250	—	—	77	5.0
Mg–Ag–RE(D)Th–Zr	Ag 2.5	Sand cast and	U	20	135	8.74	114	7.38
	RE(D) 1.0	fully heat	N		86	5.57	72	4.66
	Th 1.0	treated	U	250	108	6.99	76	4.92
	Zr 0.6							
	(QH21)							
Mg–Zn–Th–Zr	Zn 5.5	Sand cast and	U	20	120	7.8	86	5.6
	Th 1.8	heat treated	N		100	6.5	86	5.6
	Zr 0.7							
	(TZ6)							

ALLOYS—FATIGUE AND IMPACT STRENGTHS

at specified cycles								*Impact strength § for single blow fracture*				
10^6		5×10^6		10^7		5×10^7		*Test temp.*	*Unnotched*		*Notched*	
MPa	tonf in^{-2}	MPa	tonf in^{-2}	MPa	tonf in^{-2}	MPa	tonf in^{-2}	°C	J	ft lbf	J	ft lbf
88	5.7	86	5.6	85	5.5	83	5.4	20	12–14	9–10	4–4.5	3–3.5
54	3.5	51	3.3	50	3.25	48	3.1					
133	8.6	125	8.1	124	8.0	120	7.8	20	34–43	25–32	7–9.5	5–7
103	6.7	97	6.3	94	6.1	91	5.9					
90	5.8	88	5.7	88	5.7	86	5.6	20	3–5	2–3.5	1.5–2	1–1.5
73	4.7	66	4.3	65	4.2	63	4.1					
97	6.3	91	5.9	90	5.8	90	5.8	20	18–27	13–20	4.5–7	3.5–5
82	5.3	74	4.8	73	4.7	69	4.5					
66	4.3	59	3.8	57	3.7	57	3.7					
48	3.1	39	2.5	36	2.3	31	2.0	−196	1.5	1.0		
89	5.75	88	5.7	86	5.6	85	5.5	20	1.5–2.0	1–1.5	1–1.5	0.5–1
74	4.8	68	4.4	66	4.3	63	4.1					
93	6.0	93	6.0	93	6.0	—	—	20	7–9.5	5–7	3–4	2–3
80	5.2	79	5.1	79	5.1	77	5.0					
80	5.2	79	5.1	77	5.0	76	4.9	20	3–4	2–3	1–1.5	0.5–1
66	4.3	65	4.2	65	4.2	65	4.2					
134	8.7	128	8.3	127	8.2	124	8.0	20	23–31	17–23	9.5–12	7–9
93	6.0	91	5.9	90	5.8	88	5.7					
85	5.5	82	5.3	80	5.2	77	5.0	20	7–12	5–9	3–4	2–3
85	5.5	82	5.3	80	5.2	77	5.0	−196	0.8	0.6		
97	6.3	97	6.25	96	6.2	94	6.1	20	4–5.5	3–4	1–2	0.5–1.5
91	5.9	88	5.7	86	5.6	83	5.4	−196	0.7	0.5		
80	5.2	73	4.7	69	4.5	65	4.2					
69	4.5	62	4.0	59	3.8	54	3.5					
80	5.2	79	5.1	77	5.0	74	4.8	20	6–7.5	4.5–5.5	1–2	0.5–1.5
54	3.5	52	3.4	52	3.4	51	3.3					
59	3.8	57	3.7	57	3.7	57	3.7					
56	3.6	52	3.4	51	3.3	51	3.3					
45	2.9	43	2.8	43	2.8	42	2.7	−196	0.5	0.3		
37	2.4	37	2.4	36	2.3	34	2.2					
127	8.2	121	7.8	119	7.7	117	7.6	20	12.9–17.6	9–13	2.3–2.7	1.5–2
79	5.1	73	4.7	72	4.7	71	4.6					
103	6.7	103	6.7	102	6.6	100	6.5					
63	4.1	62	4.0	62	4.0	62	4.0					
—	—	90	5.8	88	5.7	86	5.6					
68	4.4	57	3.7	54	3.5	51	3.3					
111	7.2	109	7.1	108	7.0	108	7.0					
69	4.5	64	4.1	63	4.1	62	4.0					
65	4.2	56	3.6	55	3.6	52	3.4					
85	5.5	83	5.4	83	5.4	82	5.3	20	8–11	6–8	1.5–3	1–2
80	5.2	77	5.0	76	4.9	76	4.9					
								−196	0.5	0.4		

Table 22.24 MAGNESIUM AND MAGNESIUM

Material	Nominal* composition %	Condition	‡ State	Test temp. °C	Fatigue strength† 10^5 MPa	10^5 tonf in^{-2}	5×10^5 MPa	5×10^5 tonf in^{-2}
Mg–Th–Zr	Th 3.0	Sand cast and	U	20	—	—	74	4.8
	Zr 0.7	fully heat	N		—	—	48	3.1
	(MTZ)	treated	U	200	—	—	74	4.8
			U	250	80	5.2	63	4.1
Mg–Th–Zn–Zr	Th 0.7	Extruded	U	20	100	6.5	86	5.6
	Zn 0.5		N		73	4.7	52	3.4
	Zr 0.6		U	200	—	—	74	4.8
	(ZTY)		U	250	80	5.2	63	4.1
	Th 3.0	Sand cast and	U	20	97	6.3	82	5.3
	Zn 2.2	heat treated	N		76	4.9	59	3.8
	Zr 0.7		U	200	71	4.6	60	3.9
	(ZT1)		U	250	66	4.3	51	3.3
			U	325	—	—	—	—

* It is usual to add 0.2–0.4% Mn to alloys containing aluminium to improve corrosion resistance.
** Solution heat treated in an atmosphere of hydrogen.
† Wohler rotating beam tests at 2960 c.p.m.
‡ U = Unnotched.
N = Notched. Semi-circular notch of 0.12 cm (0.047 in) radius. Stress concentration factor 1.8.
§ Hounsfield balanced impact test. Notched bar values are equivalent to Izod values.
RE(D) = Neodymium enriched mischmetal

Table 22.25 HEAT TREATMENT OF MAGNESIUM ALLOY CASTINGS

Heat treatment conditions for magnesium sand castings can be varied depending on the particular components and specific properties required. The following are examples of the conditions used for each alloy which will give properties meeting current national and international specifications.

Material		Nominal* composition %	Condition	Time h	Temperature °C
Mg–Al–Zn	(AZ80)	Al 8.0	TB	12–24	400–420
		Zn 0.4			
	(AZ91)	Al 9.5	TB	16–24	400–420
		Zn 0.4			
	(AZ91)		TF	16–24	400–420
					Air cool
				8–16	180–210
Mg–Zn–Zr	(Z5Z)	Zn 4.5	TE	10–20	170–200
		Zr 0.7			
Mg–Zn–RE–Zr	(RZ5)	Zn 4.0	TE	2–4	320–340
		RE 1.2			
		Zr 0.7			Air cool
				10–20	170–200
	(ZRE1)	RE 2.7	TE	10–20	170–200
		Zn 2.2			
		Zr 0.7			
Mg–Th–Zr	(HK31)	Th 3.0	TF	2–4	560–570
		Zr 0.7			Air cool
				10–20	195–205
Mg–Zn–Th–Zr	(TZ6)	Zn 5.5	TE	2–4	320–340
		Th 1.8			Air cool
		Zr 0.7		10–20	170–200

ALLOYS—FATIGUE AND IMPACT STRENGTHS—*continued*

at specified cycles								*Impact strength § for single blow fracture*				
10^6		5×10^6		10^7		5×10^7		*Test temp.*	*Unnotched*		*Notched*	
MPa	tonf in^{-2}	MPa	tonf in^{-2}	MPa	tonf in^{-2}	MPa	tonf in^{-2}	°C	J	ft lbf	J	ft lbf
68	4.4	65	4.2	63	4.1	62	4.0					
40	2.6	36	2.3	34	2.2	32	2.1					
68	4.4	60	3.9	59	3.8	58	3.75					
59	3.8	54	3.5	52	3.4	51	3.3					
83	5.4	79	5.1	76	4.9	74	4.8					
51	3.3	49	3.2	48	3.1	46	3.0					
68	4.4	60	3.9	59	3.8	57	3.7					
59	3.8	54	3.5	52	3.4	51	3.3					
79	5.1	74	4.8	71	4.6	68	4.4	20	7–8	5–6	1.5–3	1–2
56	3.6	51	3.3	49	3.2	48	3.1					
59	3.8	54	3.5	52	3.4	51	3.3					
46	3.0	43	2.8	42	2.7	39	2.5					
43	2.8	37	2.4	34	2.2	29	1.9	−196	0.8	0.6		

Table 22.25 —*continued*

Material		*Nominal* composition* %	*Condition*	*Time* h	*Temperature* °C
	(ZT1)	Th 3.0 Zn 2.2 Zr 0.7	TE	10–20	310–320
Mg–Ag–RE(D)Zr	(QE22)	Ag 2.5 RE(D) 2.0 Zr 0.6	TF	4–12 Water/Oil Quench 8–16	520–530 195–205
	(QH21)	Ag 2.5 RE(D) 1.0 Th 1.0 Zr 0.6	TF	4–12 Water/Oil Quench 12–20	520–530 195–205

* It is usual to add 0.2–0.4% Mn to alloys containing aluminium to improve corrosion resistance.
RE = Cerium mischmetal containing approximately 50% cerium.
RE(D) = Neodymium-enriched mischmetal.
TB = Solution heat treated.
TE = Precipitation heat treated.
TF = Fully heat treated.

Note:
Above 350°C. furnace atmospheres must be inhibited to prevent oxidation of magnesium alloys. This can be achieved either by:
(i) adding $\frac{1}{2}$–1% SO_2 gas to the furnace atmosphere;
or (ii) carrying out the heat treatment in an atmosphere of 100% dry CO_2.

Mechanical properties at subnormal temperatures

At temperatures down to −200°C tensile properties have approximately linear temperature coefficients: proof stress and UTS increase by 0.1–0.2% of the RT value per °C fall in temperature, and elongation falls at the same rate: modulus of elasticity rises approximately 19 MPa (2800 lbf in^{-2}) per °C over the range 0°C to −100°C. No brittle-ductile transitions have been found.

Tests at −70°C have suggested that the magnesium–zinc–zirconium alloys show the best retention of ductility and notched impact resistance at this temperature.

22.5 Mechanical properties of nickel and nickel alloys

Table 22.26 WROUGHT NICKEL AND HIGH NICKEL ALLOYS, STANDARD SPECIFICATIONS* AND DESIGNATIONS

Alloy	*Nominal composition* %		*France* AFNOR	*Germany* DIN	Werkstoff Nr.	*UK* BS *and* DTD	*USA* ASTM	ASME	AMS	AICMA†
Nickel	Ni	99.0	—	17740: Ni 99.2	2.406 6 2.406 8	3072–76: NA 11	B160–163	SB160–163	5553	—
Monel 400	Cu Fe Mn	30 1.5 1.0	—	17743: Ni Cu 30 Fe	2.4360	3072–76: NA13 DTD 10 DTD 192 DTD 196 DTD 200 DTD 204 DTD 477	B127 B163–165 B564	SB127 SB163–165 SB395 SB564	4544 4558 4574 4675 4730	—
Monel K-500	Cu Al Ti	29 2.8 0.5	—	17743: Ni Cu 30 Al	2.437 5	3072–76: NA 18 DTD 487	—	—	4676	—
Inconel 600	Cr Fe	16 6	—	17742: Ni Cr 15 Fe	2.481 6	3072–76: NA 14 DTD 328	B163 B166–168	SB163 SB166–168	5540 5580 5665 5687	—
Inconel 625							B443 B444 B446	SB443 SB444 SB446	5666 5599 5837	—
Inconel X-750	Cr Fe Ti Al Nb	15 7 2.5 0.6 0.8	—	—	—	—	A637	SA637	5542 5582 5598 5667/8/9 5671 5698/9	—
Incoloy 800 } Incoloy 800H }	Fe Cr Ti Al	45 21 0.4 0.4	—	X10 Ni Cr Al Ti 3220	1.487 6	3072–76: NA 15 NA15H	B163 B407–409 B564	SB163 SB407–409 SB564	5766	—

Incoloy 825	Fe Cr Mo Cu Ti	32 21 3 2 1	—	17749: Ni Cr 21 Mo	2.485 8	3072–76: NA 16	B163 B423–425	SB163 SB423–425	—	—
Incoloy DS	Fe Cr Si	40 18 2	— —	X12 Ni Cr Si 361 6	1.486 4	3072–76: NA 17	—	—	—	—
Ni–Span C-902	Fe Cr Ti Al	47 5.5 2.5 0.5	—	—	—	—	—	—	5221 5223 5225	—
Hastelloy B–2	Mo	28	—	—	—	—	B333– B335– B619 B622	SB333 SB335 SB619 SB662	—	—
Hastelloy C–4	Mo Cr	16 16	—	—	—	—	B574 B575 B619 B662	SB574 SB575 SB619 SB662	—	—
Hastelloy X	Mo Cr Fe	9 21 18	NC22FeD	LW2.4665	—	HR6 HR204	B435 B572 B619 B622 B626	—	5536 5587 5754 5588	Ni–P93–HT
Nimonic 75	Cr Ti	20 0.4	NC 20T	17742: Ni Cr 20 Ti LW 2.4630	2.495 1 2.463 0	HR 5 HR 203 HR 403 HR 504	— —	— —	— —	Ni–P91–HT
Nimonic 80A	Cr Ti Al	20 2.0 1.5	NC 20TA	Ni Cr 20 Ti Al	2.463 1	HR 1 HR 201 HR 401 HR 601	A637	—	—	Ni–P95–HT

Table 22.26 WROUGHT NICKEL AND HIGH NICKEL ALLOYS, STANDARD SPECIFICATIONS* AND DESIGNATIONS—*continued*

Alloy	*Nominal composition* %		*France* AFNOR	*Germany* DIN	Werkstoff Nr.	*UK* BS *and* DTD	*USA* ASTM	ASME	AMS	AECMA†
Nimonic 90	Cr Co Ti Al	20 17 2.4 1.4	NCK 20TA	17744: Ni Cr 20 Co 18 Ti LW2.4632	2.496 9 2.963 2	HR 2 HR 202 HR 403 HR 501 HR 502 BS 3075 DTD 5087	—	—	—	Ni–P96–HT
Nimonic 105	Cr Co Mo Al Ti	15 20 5 5 1.2	NK 20CDA	Ni Co 20 Cr 15 Mo Al Ti	2.463 4	HR 3	—	—	—	Ni–P61–HT
Nimonic 115	Cr Co Mo Al Ti	14 13 3 5 4	—	Ni Co 15 Cr 15 Mo Al Ti	2.463 6	HR 4	—	—	—	Ni–P102– HT
Nimonic 263	Cr Co Mo Ti Al	20 20 6 2 0.5	NCK 20D	LW2.4650	2.465 0	HR 10 HR 206 HR 404	—	—	5872	Ni–P105– HT
Nimonic 901	Fe Cr Mo Ti	35 13 6 3	Z8 NC DT 42	Ni Cr 15 Mo Ti LW2.4662	2.466 2	—	—	—	5660 5661	Fe–PA99– HT
Nimonic PE16	Fe Cr Mo Ti Al	32 16 3 1.0 1.0	—	X8 Ni Cr Mo Ti Al 4316	—	HR 11 HR 207	—	—	—	—

* Specifications constantly under review; check for latest issue.
† Association Européenne des Constructeurs de Matériel Aérospatial.

Table 22.27 WROUGHT NICKEL AND HIGH NICKEL ALLOYS, MECHANICAL PROPERTIES[1] AT ROOM TEMPERATURES

Alloy[3]	*Nominal composition* %	*Condition*	0.2% *Proof stress* MPa (tonf in^{-2})	*UTS* MPa (tonf in^{-2})	*Elongation* %	*Brinell hardness* No.	— *Izod impact* J (ft lbf)
Nickel—pure	Ni 99.9	Annealed	60 (4)	310(20)	40	85	160(120)
Nickel—comc.	Ni 99.0	Annealed	150(10)	400(26)	40	100	160(120)
		Hot rolled	200(13)	500(32)	40	120	160(120)
		Cold drawn	480(31)	660(43)	25	190	—
Monel[1] 400	Cu 30	Annealed	230(15)	530(34)	40	125	140(100)
	Fe 1.5	Hot rolled	490(32)	620(40)	35	150	140(100)
	Mn 1.0	Cold drawn	570(37)	700(45)	25	190	110(80)
Monel K-500	Cu 29 Al 2.8	Annealed and aged	600(45)	1 000(65)	30	290	—
	Ti 0.5	Hot rolled and aged	880(57)	1 110(72)	25	310	—
		Cold drawn and aged	850(55)	1 110(72)	20	310	—
Cupro-nickel[4]	Cu 55	Annealed	200(13)	400(26)	45	—	—
		Cold drawn.	450(29)	590(38)	20	—	—
Inconel[1] 600	Cr 16	Annealed	260(17)	620(40)	45	150	160(120)
	Fe 6	Cold drawn	700(45)	880(57)	20	200	110(80)
Inconel 625	Cr 22	Annealed	534(34)	930(60)	45	186	—
	Nb 4 Mo 9	Solution treated	352(23)	810(52)	50	157	—
Inconel 718	Cr 19 Nb 5 Mo 3 Ti 0.9 Al 0.5	Fully heat treated[5]	1 160(75)	1 360(88)	15	410	—
Inconel X-750	Cr 15 Fe 7 Ti 2.5 Al 0.6 Nb 0.8	Fully heat treated	900(58)	1 240(80)	20	280	—
Incoloy[1] 800	Fe 45	Annealed	310(20)	590(38)	45	180	160(120)
	Cr 21 Ti 0.4 Al 0.4	Cold drawn	700(45)	880(57)	20	250	110 (80)
Incoloy 825	Fe 32 Cr 21 Mo 3 Cu 2 Ti 1	Annealed	340(22)	650(42)	40	150	160(120)
Incoloy DS	Fe 40	Annealed	390(25)	700(45)	45	190	140(100)
	Cr 18 Si 2	Cold drawn	850(55)	1 000(65)	10	300	110 (80)
Ni-Span[1] C-902	Fe 47 Cr 5.5 Ti 2.5 Al 0.5	Fully heat treated	770(50)	1 200(78)	25	300	—
Hastelloy B-2	Mo 28	Solution treated	526(34)	955(62)	53	235	—
Hastelloy C-4	Mo 16 Cr 16	Solution treated	416(27)	768(50)	52	184	—
Hastelloy C-276	Mo 16 Cr 15 W 4 Fe 5 C 0.02	Solution treated	390(25)	740(48)	45	200	160(120)

Table 22.27 WROUGHT NICKEL AND HIGH NICKEL ALLOYS, MECHANICAL PROPERTIES[1] AT ROOM TEMPERATURES—*continued*

Alloy[3]	*Nominal composition* %	*Condition*	0.2% *Proof stress* MPa (tonf in^{-2})	*UTS* MPa (tonf in^{-2})	*Elongation* %	*Brinell hardness* No.	*Izod Impact* J (ft lbf)
Hastelloy X	Mo 9 Cr 21 Fe 18	Solution treated	350(23)	800(52)	45	175	75 (55)
Nickel-chromium types[6]	Cr 20 Si 1	Annealed	310(20)	700(45)	40	160	—
Nimonic[1] 75	Cr 20 Ti 0.4	Annealed	350(23)	800(52)	42	180	110 (80)
			0.1% PS				
Nimonic 80A	Cr 20 Ti 2.0 Al 1.5	Fully heat treated	800(50)	1 250(81)	20	370	70 (50)
Nimonic 90	Cr 20 Co 17 Ti 2.4 Al 1.4	Fully heat treated	800(52)	1 250(81)	25	380	70 (50)
Nimonic 105	Cr 15 Co 20 Mo 5 Al 5 Ti 1.2	Fully heat treated	800(52)	1 160(75)	12	380	16 (12)
Nimonic 115	Cr 14 Co 13 Mo 3 Al 5 Ti 4	Fully heat treated	830(54)	1 250(81)	25	400	—
Nimonic 263	Cr 20 Co 20 Mo 6 Ti 2 Al 0.5	Fully heat treated	600(38)	1 000(65)	40	320	—
Nimonic 901	Fe 35 Cr 13 Mo 6 Ti 3	Fully heat treated	480(31)	1 050(68)	30	300	—
Nimonic PE16	Fe 32 Cr 16 Mo 3 Ti 1.0 Al 1.0	Fully heat treated	460(30)	850(55)	30	280	—
Super alloys—see(7)							

(1) Registered Trade Mark
(2) All values in the tables in this section are representative and not to be considered as minima or maxima not to be used for specification purposes. Conversion figures have all been rounded off and are not to be taken as precise equivalents.
(3) Where trade marks apply to the name of an alloy there may be materials of similar composition available from other producers who may or may not use the same suffix along with their own trade names. The suffix alone, e.g. Alloy 800 is sometimes used as a descriptive term for the type of alloy but trade marks can be used only by the registered user of the mark.
(4) Other copper–nickel alloys, cupro-nickels, will be found listed under copper base alloys.
(5) Fully heat-treated usually implies a solution treatment followed by some form of precipitation hardening cycle. The precise heat-treatment and the associated properties may be varied to suit particular applications or according to the form of the alloy as forging, bar, sheet or welded fabrication.
(6) There are many nickel chromium alloys with and without iron used for electrical resistance heating and for general high-temperature applications in furnaces and heat-treatment plant.
(7) There are many compositions for complex alloys which give creep-resistance or other high-temperature properties. These are referred to as 'superalloys', most do contain nickel, many are nickel base but there is no systematic classification.

Table 22.28 WROUGHT NICKEL AND HIGH NICKEL ALLOYS, SHORT-TIME HIGH-TEMPERATURE TENSILE PROPERTIES

Alloy	*Nominal composition* %	*Condition*	*Test temperature* °C	0.2% *proof stress* MPa (tonf in^{-2})	*UTS* MPa (tonf in^{-2})	*Elongation* %
Nickel	Ni 99.0	Hot rolled	20	170(11)	490(32)	50
			200	150(10)	540(35)	50
			400	140 (9)	540(35)	50
			600	110 (7)	250(16)	60
			800	—	170(11)	60
Monel* 400	Cu 30	Hot rolled	20	230(15)	560(36)	45
	Fe 1.5		200	200(13)	540(35)	50
	Mn 1.0		400	220(14)	460(30)	52
			600	120 (8)	260(17)	30
			800	77 (5)	120 (8)	54
Monel* K-500	Cu 29	Fully heat treated	20	340(22)	680(44)	45
	Al 2.8		200	290(19)	650(42)	40
	Ti 0.5		400	260(17)	600(39)	30
			600	290(19)	460(30)	5
			800	—	185(12)	30
Inconel* 600	Cr 16	Hot rolled	20	250(16)	590(38)	50
	Fe 6		400	185(12)	560(36)	50
			600	150(10)	530(34)	10
			800	—	250(16)	20
			1 000	—	110 (7)	50
Inconel X-750	Cr 15	Fully heat treated	20	620(40)	1 110(72)	24
	Fe 7		400	590(38)	1 000(65)	28
	Ti 2.5		650	280(18)	830(54)	9
	Al 0.6		800	310(20)	370(24)	22
	Nb 0.8		800	100 (6)	170(11)	90
Incoloy* 800	Fe 45	Annealed	20	300(19)	600(39)	45
	Cr 21		600	210(14)	440(29)	40
	Ti 0.4					
	Al 0.4					
Incoloy DS	Fe 40	Annealed	20	300(19)	700(45)	45
	Cr 18		600	—	460(30)	50
	Si 2		800	—	150(10)	90
			1 000	—	75 (5)	100
Hastelloy* B-2	Mo 28	Solution treated	200	451(29)	885(57)	50
			320	426(28)	864(56)	49
			430	418(27)	866(56)	51
Hastelloy C-4	Mo 16	Solution-treated	200	403(26)	706(46)	49
	Cr 16		320	371(29)	675(44)	52
			430	320(21)	656(42)	64
Hastelloy X	Mo 9	Solution treated	20	360(23)	790(51)	45
	Cr 21		600	280(18)	620(40)	40
	Fe 18		800	230(15)	410(26)	37
			1 000	110 (7)	160(10)	45
Nimonic* 75	Cr 20	Annealed	20	420(27)	800(52)	35
	Ti 0.4		400	400(26)	740(48)	30
			600	310(20)	590(38)	30
			800	110 (7)	220(14)	80

* Registered trade marks

Table 22.28 WROUGHT NICKEL AND HIGH NICKEL ALLOYS, SHORT-TIME HIGH-TEMPERATURE TENSILE PROPERTIES—*continued*

Alloy	*Nominal composition* %	*Condition*	*Test temperature* °C	0.2% *proof stress* MPa (tonf in^{-2})	*UTS* MPa (tonf in^{-2})	*Elongation* %
Nimonic* 80A	Cr 20	Fully heat	20	740(48)	1 240(80)	24
	Ti 2.0	treated	400	680(44)	1 150(74)	26
	Al 1.5		600	620(40)	1 080(70)	20
			800	490(32)	620(40)	24
Nimonic 90	Cr 20	Fully heat	20	750(49)	1 175(76)	30
	Co 17	treated	600	680(44)	1 030(67)	26
	Ti 2.4		800	530(34)	900(58)	18
	Al 1.4		1 000	48 (3)	76 (5)	130
Nimonic 105	Cr 15	Fully heat	20	780(50)	1 140(74)	22
	Co 20	treated	600	720(47)	1 040(67)	25
	Mo 5		800	680(44)	810(53)	25
	Al 5		1 000	150(10)	175(11)	42
	Ti 1.2					
Nimonic 115	Cr 14	Fully heat	20	860(56)	1 230(80)	27
	Co 13	treated	600	790(51)	1 100(71)	20
	Mo 3		800	760(49)	1 020(66)	19
	Al 5		1 000	200(13)	420(27)	26
	Ti 4					
				0.1% PS		
Nimonic 263	Cr 20	Fully heat	20	570(37)	970(63)	40
	Co 20	treated	600	460(30)	790(51)	41
	Mo 6		800	390(25)	560(36)	20
	Ti 2		1 000	60 (4)	110 (7)	65
	Al 0.5					
Nimonic PE 16	Cr 16	Fully heat	20	490(32)	880(57)	37
	Fe 32	treated	600	450(29)	730(47)	27
	Mo 3		800	290(19)	390(25)	53
	Ti 1.0		1 000	46 (3)	92 (6)	100
	Al 1.0					

* Registered trade mark.

Table 22.29 WROUGHT NICKEL AND HIGH NICKEL ALLOYS, CRYOGENIC PROPERTIES

Alloy	*Nominal composition* %	*Condition*	*Test temp.* °C	0.2% *Proof stress* MPa (tonf in^{-2})	*UTS* MPa (tonf in^{-2})	*Elongation* %	*Charpy impact* J (ft lbf)	*Notched fatigue strength** MPa (tonf in^{-2})
Nickel	Ni 99.0	Annealed	20	160 (10)	500 (32)	48	230 (170)	—
			−75	170 (11)	560 (36)	58	230 (170)	120 (8)
			−200	230 (15)	710 (46)	54	230 (170)	140 (9)
Monel* 400	Cu 30	Annealed	20	220 (14)	540 (35)	52	260 (190)	—
	Fe 1.5							
	Mn 1.0		−180	340 (22)	800 (52)	50	240 (180)	—

* 10^6 cycles Kt = 3.1. * Registered trade mark

Table 22.29 WROUGHT NICKEL AND HIGH NICKEL ALLOYS, CRYOGENIC PROPERTIES—*continued*

Alloy	*Nominal composition* %	*Condition*	*Test temp.* °C	0.2% *Proof stress* MPa (tonf in^{-2})	*UTS* MPa (tonf in^{-2})	*Elongation* %	*Charpy impact* J ft lbf	*Notched fatigue strength** MPa (tonf in^{-2})
Monel** K-500	Cu 29, Al 2.8, Ti 0.5	Fully heat treated	20	670 (43)	1 060 (69)	22	74 (55)	—
			−80	740 (48)	1 140 (74)	24	68 (50)	300 (19)
			−200	830 (54)	1 260 (82)	30	—	330 (21)
			−255	940 (61)	1 380 (89)	28	—	330 (21)
Inconel** 600	Cr 16, Fe 6	Annealed	20	230 (15)	650 (42)	37	240 (180)	280 (18)
			−80	290 (19)	730 (47)	40	210 (150)	280 (18)
			−190	—	—	—	190 (140)	300 (19)
Inconel 718	Cr 19, Nb 5, Mo 3, Ti 0.9, Al 0.5	Fully heat treated	20	1 400 (91)	1 500 (97)	—	—	—
			−80	1 400 (91)	1 600 (104)	—	—	—
			−190	1 600 (104)	1 800 (117)	—	—	—
Inconel X-750	Cr 15, Fe 7, Ti 2.5, Al 0.6, Nb 0.8	Fully heat treated	20	700 (45)	1 210 (78)	24	—	—
			−75	790 (51)	1 290 (84)	23	—	410 (27)
			−200	810 (52)	1 440 (93)	19	—	440 (29)
			−255	900 (58)	1 450 (94)	16	—	460 (30)
Ni-Span** C-902	Fe 47, Cr 5.5, Ti 2.5, Al 0.5	Fully heat treated	20	760 (49)	1 210 (78)	30	24 (18)	320 (21)
			−130	860 (56)	1 410 (91)	28	23 (17)	310 (20)
			−255	1 000 (65)	1 690 (110)	25	23 (17)	390 (25)

* 10^6 cycles 14=3.1 ** Registered trade marks

Table 22.30 WROUGHT NICKEL AND HIGH NICKEL ALLOYS, FATIGUE PROPERTIES

Alloy	*Nominal composition* %	*Condition*	*Test temperature* °C	*Average endurance limit Rotating beam tests* 10^8 cycles MPa (tonf in^{-2})
Nickel	Ni 99.0	Annealed	20	230(15)
		Cold drawn	20	340(22)
Monel**400	Cu 30, Fe 1.5, Mn 1.0	Annealed	20	230(15)
		Cold drawn*	20	300(19)
Monel K-500	Cu 29, Al 2.8, Ti 0.5	Annealed	20	260(17)
		Cold drawn and aged	20	320(21)

* Stress equalized 3 h at 260 °C ** Registered trade mark.

Table 22.30 WROUGHT NICKEL AND HIGH NICKEL ALLOYS, FATIGUE PROPERTIES—*continued*

Alloy	*Nominal composition* %	*Condition*	*Test temperature* °C	*Average endurance limit Rotating beam tests* 10^8 cycles MPa (tonf in^{-2})		
Inconel** 600	Cr 16 Fe 6	Annealed	20 650 850	270(17) 180(12) 75 (5)		
Inconel 718	Cr 19 Nb 5 Mo 3 Ti 0.9 Al 0.5	Fully heat treated	20 650	620(40) 500(32)		
Inconel X-750	Cr 15 Fe 7 Ti 2.5 Al 0.6 Nb 0.8	Fully heat treated	20 700	280(18) 340(22)		
Incoloy** 800	Fe 45 Cr 21 Ti 0.4	Annealed	20 500 850	290(19) 260(17) 95 (6)		
				0 ± stress. For lives of		
				MPa (tonf in^{-2})	h	cycles
Nimonic** 75	Cr 20 Ti 0.4	Annealed	20 750	260(17) 190(12)	65 300	10×10^6 10×10^6
Nimonic 90	Cr 20 Co 17 Ti 2.4 Al 1.4	Fully heat treated	750 870 	260(17) 240(16) 140 (9) 110(7)	300 1 000 300 1 000	36×10^6 120×10^6 36×10^6 120×10^6
Nimonic 105	Cr 15 Co 20 Mo 5 Al 5 Ti 1.2	Fully heat treated	20 750 870 980 	350(23) 250(16) 260(17) 250(16) 240(16) 190(12) 170(11) 124 (8)	50 500 50 500 50 500 50 500	3×10^7 3×10^8 3×10^7 3×10^8 3×10^7 3×10^8 3×10^7 3×10^8

* Stress equalized 3h at 260 °C. ** Registered trade marks.

Table 22.31 WROUGHT NICKEL AND HIGH NICKEL ALLOYS, CREEP PROPERTIES

				Stress MPa (tonf in^{-2}) *for creep extension of*							
				0.1%			0.2%			1%	*Rupture*
Alloy	*Nominal composition* %	*Condition*	*Test temp.* °C	300 h	1 000 h	10 000 h	300 h	1 000 h	10 000 h	10 000 h	10 000 h
Nickel	Ni 99.0	Cold drawn	400	—	—	77 (5)	—	—	—	220 (14.5)	—
Monel* 400	Cu 30 Fe 1.5 Mn 1.0	Hot rolled	400	—	—	140 (9)	—	—	—	215 (14)	—
Monel K-500	Cu 29 Al 2.8 Ti 0.5	Cold drawn and aged	400	—	—	355 (23)	—	—	—	—	—

* Registered trade mark.

Table 22.31 WROUGHT NICKEL AND HIGH NICKEL ALLOYS, CREEP PROPERTIES—*continued*

Alloy	*Nominal composition* %	*Condition*	*Test temp.* °C	*Stress* MPa (tonf in^{-2}) *for creep extension of* 0.1% 300 h	0.1% 1 000 h	0.1% 10 000 h	0.2% 300 h	0.2% 1 000 h	0.2% 10 000 h	1% 10 000 h	*Rupture* 10 000 h
Inconel* 600	Cr 16 Fe 6	Hot rolled	400	—	—	340 (22)	—	—	—	—	—
			600	—	—	38 (2.5)	—	—	—	77 (5)	—
Incoloy* 800H	Fe 45 Cr 21 Tl 0.4 Al 0.4	Solution treated	650	—	115 (7.4)	90 (5.8)	—	—	—	—	121 (7.8)
			870	—	24 (1.6)	20 (1.2)	—	—	—	—	24 (1.6)
			980	—	7 (0.45)	3.8 (0.25)	—	—	—	—	8.3 (0.54)
Inconel X-750	Cr 15 Fe 7 Ti 2.5 Al 0.6 Nb 0.8	Fully heat treated	650	—	—	—	—	420 (27)	—	—	—
			815	—	—	—	—	110 (7)	—	—	—
Nimonic* 75	Cr 20 Ti 0.4	Annealed	650	56 (3.6)	—	—	—	62 (4.0)	51 (3.3)	—	—
			750	25 (1.6)	—	—	—	29 (1.9)	25 (1.6)	—	—
Nimonic 80A	Cr 20 Ti 2.0 Al 1.5	Fully heat treated	650	390 (25)	185 (12.0)	—	460 (30)	390 (24.8)	260 (16.6)	0.5% 280 (18.0)	280 (18.2)
			750	170 (11.0)	120 (7.8)	51 (3.3)	200 (13.0)	150 (9.8)	66 (4.3)	90 (5.9)	85 (5.5)
			815	93 (6.0)	58 (3.8)	23 (1.5)	108 (7.0)	73 (4.7)	26 (1.7)	29 (1.9)	34 (2.2)
Nimonic 90	Cr 20 Co 17 Ti 2.4 Al 1.4	Fully heat treated	650	400 (26)	348 (22.5)	250 (160)	450 (29)	400 (26)	310 (20)	325 (21)	340 (22)
			750	185 (12.0)	137 (8.9)	65 (4.2)	220 (14.2)	175 (11.3)	85 (5.5)	110 (7.1)	135 (8.7)
			815	85 (5.5)	57 (3.7)	28 (1.8)	108 (7.0)	77 (5.0)	31 (2.0)	38 (2.5)	54 (3.5)
			870	49 (3.2)	29 (1.9)	15 (1.0)	60 (3.9)	37 (2.4)	17 (1.1)	18 (1.2)	28 (1.8)
Nimonic 105	Cr 15 Co 20 Mo 5 Al 5 Ti 1.2	Fully heat treated	750	285 (18.5)	230 (15.0)	—	325 (21)	280 (18.0)	193 (12.5)	—	208 (13.5)
			815	162 (10.5)	119 (7.7)	—	193 (12.5)	148 (9.6)	93 (6.0)	102 (6.6)	120 (7.8)
			870	86 (5.6)	57 (3.7)	—	108 (7.0)	83 (5.4)	34 (2.2)	45 (2.9)	57 (3.7)
			940	31 (2.0)	22 (1.4)	—	48 (3.1)	31 (2.0)	12 (0.8)	22 (1.4)	35 (2.3)
Nimonic 115	Cr 14 Co 13 Mo 3 Al 5 Ti 4	Fully heat treated	750	355 (23)	310 (20)	—	385 (25)	340 (22)	—	—	250 (16)
			850	154 (10.0)	116 (7.5)	51 (3.3)	222 (14.4)	137 (8.9)	53 (3.5)	74 (4.8)	108 (7.0)
			950	51 (3.3)	34 (2.2)	14 (0.9)	63 (4.1)	43 (2.8)	15 (1.0)	17 (1.1)	52 (3.4)
			1 000	22 (1.4)	15 (1.0)	—	29 (1.9)	19 (1.2)	6 (0.4)	8 (0.5)	—

* Registered trade mark.

BIBLIOGRAPHY

'Nickel and Its Alloys', US Department of Commerce NBS Monograph 106, Washington, USA; 1968.

W. Betteridge and J. Heslop, 'The Nimonic Alloys', Edward Arnold, London; 1974.

22.6 Mechanical properties of titanium and titanium alloys

Table 22.32 TITANIUM AND TITANIUM ALLOYS. CORRESPONDING GRADES OR SPECIFICATIONS

IMI designation	*UK British Standards (Aerospace series) and Min. of Def. DTD series**	*France* AIR–9182, 9183, 9184	*Germany BWB series†*	*AECMA recom-mendations*	*USA AMS series‡*
IMI 115	BS TA 1, DTD 5013	T-35	3.7024	Ti-POI	
IMI 125	BS TA 2, 3, 4, 5	T-40	3.7034	Ti-PO2	AMS 4902, 4941, 4942, 4951
IMI 130	DTD 5023, 5273, 5283, 5293	T-50			AMS 4900
IMI 155 IMI 160	BS TA 6 BS TA 7, 8, 9	T-60	3.7064	Ti-PO4	AMS 4901 AMS 4921
IMI 230	BS TA 21, 22, 23, 24 BS TA 52-55, 58	T-U2·		Ti-P11	
IMI 315	DTD 5043				
IMI 317	BS TA 14, 15, 16, 17	T-A5E		Ti-P65	AMS 4909, 4910, 4926, 4924, 4953, 4966
IMI 318	BS TA 10, 11, 12, 13, 28, 56	T-A6V	3.7164	Ti-P63	AMS 4911, 4928, 4934, 4935, 4954, 4965, 4967
IMI 550	BS TA 45–51, 57	T-A4DE	3.7184	Ti-P68	
IMI 551	BS TA 38–42				
IMI 679	BS TA 18–20, 25–27				AMS 4974
IMI 680	DTD 5213	T-E11DA			
IMI 685	BS TA 43, 44	T-A6ZD	3.7154	Ti-P67	

*UK BS 3531 Part 1 (Metal Implants in Bone Surgery), and Draft British Standard for Lining of Vessels and Equipment for Chemical Processes, Part 9, also refer.
†Germany DIN 17850, 17860, 17862, 17863, 17864 (3.7025/35/55/65), and TUV 230–1–68 Group I, II, III and IV also refer.
‡USA MIL–T–9011, 9046, 9047, 14577, 46038, 46077, 05–10737 and ASTM B265–69, B338–65, B348–59T, B367–61T, B381–61T, B382–61T also refer.

Table 22.33 PURE TITANIUM. TYPICAL MECHANICAL PROPERTIES AT ROOM TEMPERATURE

*Designation**	*Grade*	*Condition*	0.2% *proof stress* MPa (tonf in^{-2})	*Tensile strength* MPa (tonf in^{-2})	*Elongation %* *on* 50 mm	*Elongation %* *on* 5D	*Red. in area* %	*Specification bend radius* 180° bend <1.83 mm	*Specification bend radius* 180° bend <3.25 mm	*Mod. of elasticity* GPa (10^6 lbf in^{-2})	*Mod. of rigidity* GPa (10^6 lbf in^{-2})
Iodide	Pure, 60 HV		103 (6.7)	241 (15.6)	55		80				
IMI 115	Commercially pure	Annealed sheet	255 (16)	370 (24)	33			lt	2t		
		Annealed rod	220 (14)	370 (24)		40	70				
		Annealed wire		390 (25)	38						
IMI 125	Commercially pure	Annealed sheet	340 (22)	460 (30)	30			1½t	2t		
		Annealed rod	305 (20)	460 (30)		28	57				
		Annealed tube	325 (21)	480 (31)	35						
IMI 130	Commercially pure	Annealed sheet	420 (27)	540 (35)	25			2t⁺	2½t		
		Annealed rod	360 (23)	540 (35)		24	48			105 (15)	38 (5.5)
		Annealed wire		550 (36)	24						
		Hard-drawn wire		700 (45)	11.5						
IMI 155	Commercially pure	Annealed sheet	540 (35)	640(41)	24			2½t	3t		
IMI 160	Commercially pure	Annealed rod	500 (32)	670 (43)		23	46				
		Annealed wire		690 (45)	24						

* IMI Nomenclature. † Up to 1.63 mm.

Table 22.34 TITANIUM ALLOYS. TYPICAL MECHANICAL PROPERTIES AT ROOM TEMPERATURE

*Designation**	*Nominal composition* %	*Condition*	0.2% *proof stress* MPa (tonf in^{-2})	*Tensile strength* MPa (tonf in^{-2})	*Elongation %* *on* 50 mm	*Elongation %* *on* 5D	*Red. in area* %	*Specification bend radius* 180°	*Mod. of elasticity* GPa (10^6 lbf in^{-2})	*Mod. of rigidity* GPa (10^6 lbf in^{-2})
IMI 230	Cu 2.5	Annealed sheet	520 (34)	620 (40)	24			2t (0.5–3 mm)	125 (17.5)	
		Aged sheet	670 (43)	770 (50)	20			2t (typical)		
		Annealed rod	500 (33)	630 (41)		27	45			
		Aged rod	580 (38)	740 (48)		22	41		125	
IMI 260	Pd 0.2	Similar to commercially pure Titanium 115								
IMI 261	Pd 0.2	Similar to commercially pure Titanium 125								
IMI 315	Al 2.0, Mn 2.0	Annealed rod	590 (38)	720 (47)		21	50		120 (17)	

Table 22.34 TITANIUM ALLOYS. TYPICAL MECHANICAL PROPERTIES AT ROOM TEMPERATURE — *continued*

*Designation**	*Nominal composition* %	*Condition*	0.2% *proof stress* MPa (tonf in^{-2})	*Tensile strength* MPa (tonf in^{-2})	*Elongation* % *on* 50 mm	*Elongation* % *on* 5D	*Red. in area* %	*Specification bend radius* 180°	*Mod. of elasticity* GPa (10^6 bf in^{-2})	*Mod. of rigidity* GPa (10^6 lbf in^{-2})
IMI 317	Al 5.0 Sn 2.5	Annealed sheet	820 (53)	860 (56)	16			4t (<2 mm) 4½t (⩽3 mm)		
		Annealed rod	930(60)	1 000 (65)		15	37		120 (17)	
IMI 318	Al 6.0 V 4.0	Annealed sheet	1 110 (72)	1 160 (75)	10			5t (⩽3.25 mm)		
		Annealed rod	990 (64)	1 050 (68)		15	40		106 (15.3)	46 (6.7)
		Aged rod (fastener stock)	1 050 (68)	1 140 (74)		15	40			
		Hard-drawn wire		1 410 (91)	4					
IMI 550	Al 4.0 Mo 4.0 Sn 2.0 Si 0.5	F.h.t. rod	1 070 (69)	1 200 (78)		14	42		116 (17)	
IMI 551	Al 4.0 Mo 4.0 Sn 4.0 Si 0.5	F.h.t. rod	1 140 (74)	1 300 (84)		12	40		113 (16.5)	43 (6.2)
IMI 679	Sn 11.0 Zr 5.0 Al 2.25 Mo 1.0 Si 0.2	Quenched and aged rod	1 080 (70)	1 230 (79)		11	40		108 (15.6)	46 (6.7)
		Air-cooled and aged rod	1 000(65)	1 120 (72)		13	45			
IMI 680	Sn 11.0 Mo 4.0 Al 2.25 Si 0.2	Quenched and aged rod	1 200 (78)	1 350 (87)		12	37		115(16.7)	
		Furnace-cooled and aged rod	1 080 (70)	1 160 (75)		14	47			
IMI 685	Al 6.0 Zr 5.0 Mo 0.5 Si 0.25	F.h.t. rod	920 (60)	1 020 (66)		11	22		124 (18)	47 (6.7)
IMI 829	Al 5.5 Sn 3.5 Zr 3.0 Nb 1.0 Mo 0.3 Si 0.3	F.h.t. rod	848 (55)	965 (63)		12	22		120 (17)	

*IMI Nomenclature.

Table 22.35 COMMERCIALLY PURE TITANIUM SHEET, TYPICAL VARIATION OF PROPERTIES WITH TEMPERATURE

*Designation**	*Temperature* °C	0.2% *proof stress*		*Tensile strength*		*Elongation on* 50 mm	*Mod. of elasticity*		*Transformation temperature*
		MPa	tonf in^{-2}	MPa	tonf in^{-2}	%	GPa	10^6 lbf in^{-2}	°C
IMI 115	−196	442	28.6	641	41.5	34			$\alpha/\alpha+\beta$
	−100	306	19.8	444	28.8	34			865
	20	207	13.4	337	21.8	40			
	100	168	10.9	296	19.2	43			
	200	99	6.4	218	14.1	38			
	300	53	3.4	167	10.8	47			
	400	42	2.7	131	8.5	52			
	450	36	2.3	120	7.8	49			
IMI 125	20	334	21.6	479	31.0	31			
	100	250	16.2	397	25.7	32			
	200	184	11.9	300	19.4	40			
	300	142	9.2	232	15.0	45			
	400	127	8.2	190	12.3	38			
	450	119	7.7	175	11.3	35			
IMI 130	−196	730	47.3	855	55.4	28			
	−100	590	38.2	737	47.7	28			
	20	394	25.5	547	35.4	28	108	15.7	
	100	315	20.4	462	29.9	29	99	14.3	$\alpha+\beta/\beta$
	200	205	13.3	331	21.4	37	91	13.2	915
	300	139	9.0	247	16.0	40	83	12.1	
	400	102	6.6	199	12.9	34	65	9.4	
	450	93	6.0	182	11.8	28			
	500						46	6.6	
IMI 155	20	460	29.7	625	40.5	25			
	100	372	24.1	537	34.8	26			
	200	219	14.2	386	24.9	32			
	300	151	9.8	281	18.2	36			
	400	110	7.1	221	14.3	33			
	450	96	6.2	202	13.1	26			

* IMI nomenclature.

Table 22.36 TITANIUM ALLOYS, TYPICAL VARIATION OF PROPERTIES WITH TEMPERATURE

Designation	*Nominal composition* %	*Condition*	*Temperature* °C	0.2% *Proof stress*		*Tensile strength*		*Elongation* %		*Red. in area* %	*Mod. of elasticity*		*Transformation temperature* °C
				MPa	tonf in^{-2}	MPa	tonf in^{-2}	*on* 50 mm	*on* 5D		GPa	10^6 lbf in^{-2}	
IMI 230	Cu 2.5	S.h.t. (trans.)	20	500	32.4	605	39.2	24					$\alpha/\alpha+\beta$
			100	410	26.5	540	35.0	29					790
			200	310	20.1	450	29.1	33					
			300	270	17.5	410	26.5	31					$\alpha+\beta/\beta$
			400	250	16.2	380	24.6	30					895 ± 10
			500	220	14.2	380	24.6	33					
		Aged sheet (trans.)	20	622	40.3	761	49.3	24					
			100	553	35.8	704	45.6	23					
			200	471	30.5	635	41.1	26					
			300	457	29.6	607	39.3	23					
			400	429	27.8	573	37.1	19					
			500	357	23.1	468	30.3	21					
		Aged rod	20	638	41.3	795	51.5		22	40	107	15.5	
			100	601	38.9	761	49.3		21	39	100	14.5	
			200	507	32.8	687	44.5		23	45	92	13.3	
			300	496	32.1	658	42.6		20	50	85	12.3	
			400	415	26.9	592	38.3		21	53	78	11.3	
			500	361	23.4	491	31.8		27	57	71	10.3	
IMI 260	Pd 0.2	Similar to IMI 115											
IMI 262	Pd 0.2	Similar to IMI 125											
IMI 315	Al 2.0	Annealed rod	20	618	40	757	49		18	41	110	16	$\alpha+\beta/\beta$
	Mn 2.0		100	510	33	649	42		21	46	107	15.5	915 ± 20
			200	386	25	525	34		22	48	97	14	
			300	293	19	432	28		19	50	86	12.5	
			400	278	18	417	27		18	56	76	11	
			500	201	13	340	22		22	72	62	9	
IMI 317	Al 5.0	Annealed rod	20	822	53.2	919	59.5		18	39	112	16.3	$\alpha/\alpha+\beta$
	Sn 2.5		100	692	44.8	798	51.7		19	40	109	15.8	950
			200	494	32.0	638	41.3		18	44	105	15.3	
			300	415	26.9	576	37.3		19	42	89	12.9	$\alpha+\beta/\beta$
			400	374	24.2	522	33.8		18	41	84	12.4	1025 ± 20
			500	346	22.4	485	31.5		21	57	81	11.8	

IMI 318	Al	6.0	Annealed rod	−196	1 560	101.0	1 675	108.5	6	29			α+β/β
	V	4.0		−100	1 165	75.4	1 265	81.9	12	33			1 000±15
				20	970	62.8	1 040	67.3	15	38	106	15.3	
				100	825	53.4	920	59.6	17	43	102	14.8	
				200	710	46.0	815	55.7	18	49	96	13.9	
				300	645	41.7	750	48.6	18	56	90	13.1	
				400	580	37.6	700	45.3	18	63	85	12.3	
				500	450	29.1	605	39.2	26	72	79	11.5	
				600	125	8.1	265	17.2	58	85			
				700	40	2.6	135	8.7	127	94			
			Heat-treated rod (fastener stock)	20	1 035	67	1 145	74	14				
				100	925	60	1 035	67	15				
				200	805	52	925	60	16				
				300	710	46	850	55	16				
				400	635	41	805	52	18				
				500	540	35	695	45	25				
IMI 550	Al	4.0	F.h.t rod	20	1 081	69.1	1 220	79.4	15	49	116	16.8	α+β/β
	Mo	4.0		100	965	62.1	1 130	73.2	15	49	112	16.2	980±10
	Sn	2.0		200	805	50.2	960	62.6	16	60	106	15.4	
	Si	0.5		300	700	45.3	900	58.3	16	55	101	14.6	
				400	655	42.4	835	54.0	17	60	95	13.8	
				500	585	38.0	780	50.4	19	68	90	13.1	
				600	310	20.8	585	37.8	26	83	85	12.3	
IMI 551	Al	4.0	F.h.t. rod	20	1 250	80.8	1 390	90.0	10	27	113	16.4	α+β/β
	Mo	4.0		100	1 125	72.8	1 300	84.2	11	29	108	15.7	1 050±15
	Sn	4.0		200	925	59.9	1 145	74.0	14	38	103	14.9	
	Si	0.5		300	815	52.8	1 045	67.8	15	38	98	14.2	
				400	745	48.4	970	62.8	14	41	93	13.5	
				500	670	43.5	920	59.6	18	55	88	12.8	
				600	460	29.8	755	49.0	27	65	81	11.7	

Table 22.36 TITANIUM ALLOYS, TYPICAL VARIATION OF PROPERTIES WITH TEMPERATURE—*continued*

Designation	*Nominal composition* %	*Condition*	*Temperature* °C	*0.2% Proof stress* MPa	tonf in^{-2}	*Tensile strength* MPa	tonf in^{-2}	*Elongation %* *on* 50 mm	*on* 5D	*Red. in area* %	*Mod. of elasticity* GPa	10^6 lbf in^{-2}	*Transformation temperature* °C
IMI 679	Sn 11.0	Quenched and	20	1 050	68	1 230	79.5		10	37			$\alpha+\beta/\beta$
	Zr 5.0	aged rod	100	940	61	1 145	74		11	43			950±10
	Al 2.25		200	820	53	1 020	66		12	45			
	Mo 1.0		300	740	48	990	64		11	46			
	Si 0.2		400	710	46	940	61		11	46			
			450	680	44	910	59		11	46			
		Air-cooled and	20	1 020	66	1 095	71		14	41	108	15.7	
		aged rod	100	895	58	995	64.5		16	47	103	14.9	
			200	770	50	900	58.5		16	49	99	14.4	
			300	695	45	865	56		14	49	94	13.6	
			400	665	43	850	55		14	48	90	13.1	
			500	600	39	795	51.5		15	48	85	12.3	
IMI 680	Sn 11.0	Quenched and	20	1 180	76.5	1 330	86		12	43	106	15.4	$\alpha+\beta/\beta$
	Mo 4.0	aged rod	100	1 020	66	1 190	77		14	49	100	14.5	945±15
	Al 2.25		200	905	58.5	1 105	71.5		15	53	96	13.9	
	Si 0.2		300	835	54	1 075	69.5		15	56	94	13.6	
			400	805	52	1 020	66		14	57	90	13.1	
			450	725	47	975	63		13	54	88	12.8	
		Furnace-cooled	−196	1 630	105.5	1 730	112		$8\frac{1}{2}$	36			
		and aged rod	−100	1 280	83	1 380	89.5		10	43			
			20	1 030	67	1 130	73		15	49			
IMI 685	Al 6.0	F.h.t. rod	−196	1 480	96	1 560	101		6	13			$\alpha+\beta/\beta$
	Zr 5.0		−100	1 140	74	1 270	82		10	18			1 020±10
	Mo 0.5		20	890	57.5	1 030	67		12	22	124	18.0	
	Si 0.25		100	800	52	935	60.5		13	22	120	17.4	
			200	720	46.5	850	55		15	24	114	16.5	
			300	650	42	800	52		16	27	108	15.7	
			400	595	38.5	750	48.5		18	31	102	14.8	
			500	535	34.5	695	45		19	37	95	13.8	
IMI 829	Al 5.5	F.h.t. rod	20	895	57.9	1 028	66.5		$10\frac{1}{2}$	22	119	17.3	$\alpha+\beta/\beta$
	Sn 3.5		200	622	40.2	792	51.2		$14\frac{1}{2}$	28	110	16.0	1 015±15
	Zr 3.0		500	501	32.4	665	43.0		15	36	93	13.5	
	Nb 1.0		540	487	31.5	653	42.2		16	42	91	13.2	
	Mo 0.3		600	457	29.6	634	41.0		14	38	88	12.8	
	Si 0.3												

Table 22.37 COMMERCIALLY PURE TITANIUM—TYPICAL CREEP PROPERTIES

IMI designation	*Temperature* °C	*Stress* MPa (tonf in^{-2}) *to produce* 0.1% *plastic strain in*		
		1 000 h	10 000 h	100 000 h
IMI 130	20	288(18.6)	270(17.5)	207(13.4)
	50	243(15.7)	221(14.3)	165(10.7)
	100	179(11.6)	165(10.7)	119 (7.7)
	150	140 (9.1)	133 (8.6)	96 (6.2)
	200	113 (7.3)	116 (7.5)	77 (5.0)
	250	96 (6.2)	101 (6.5)	66 (4.3)
	300	87 (5.6)	83 (5.4)	55 (3.6)
IMI 155	20	309(20.0)	278(18.0)	260(16.8)
	50	252(16.3)	232(15.0)	213(13.8)
	100	188(12.2)	170(11.0)	157(10.2)
	150	145 (9.4)	131 (8.5)	122 (7.9)
	200	116 (7.5)	108 (7.0)	104 (6.7)
	250	102 (6.6)	97 (6.3)	94 (6.1)
	300	93 (6.0)	90 (5.8)	86 (5.6)

Table 22.38 TITANIUM ALLOYS—TYPICAL CREEP PROPERTIES

IMI designation	*Nominal composition* %		*Condition*	*Temperature* °C	*Stress* MPa* *to produce* 0.1% *total plastic strain in*			
					100 h	300 h	500 h	1 000 h
IMI 230	Cu	2.5	Aged sheet	200	435(28.1)	—	—	—
				300	375(24.3)	—	—	—
				400	220(14.2)	—	—	—
				450	109 (7.1)	—	—	—
			Annealed sheet	20	360(23.3)	—	—	—
				100	279(18.1)	—	—	—
				200	235(15.2)	—	—	—
				300	202(13.1)	—	—	—
				400	125 (8.1)	—	—	—
IMI 317	Al	5.0	Annealed rod	20	633(41.0)	608(39.4)	—	593(38.4)
	Sn	2.5		100	474(30.7)	463(30.0)	—	458(29.7)
				200	370(24.0)	—	—	370(24.0)
				300	359(23.2)	—	—	359(23.2)
				400	337(21.8)	—	—	337(21.8)
				500	162(10.5)	119 (7.7)	—	88(5.7)
IMI 318	Al	60	Annealed rod	20	832(53.9)	818(53.0)	—	788(51.0)
	V	4.0		100	704(45.6)	680(44.0)	—	676(43.8)
				200	638(41.3)	636(41.2)	—	635(41.1)
				300	576(37.3)	568(36.8)	—	—
				400	287(18.6)	144 (9.3)	—	102 (6.6)
				500	32 (2.1)	18 (1.2)	—	—
IMI 550	Al	4.0	Fully heat-treated bar	300	724(46.8)	718(46.5)	—	710(46.0)
	Mo	4.0		400	551(35.7)	519(33.6)	—	471(30.5)
	Sn	2.0		450	254(16.5)	174(11.3)	—	101 (6.5)
	Si	0.5		500	82 (5.3)	51 (3.3)	—	31 (2.0)
IMI 551	Al	4.0	Fully heat-treated rod	400	621(40.2)	575(37.2)	540(34.9)	501(32.4)
	Mo	4.0		450	307(19.9)	217(14.1)	—	—
	Sn	4.0						
	Si	0.5						
IMI 679	Sn	11.0	Air-cooled and aged rod	20	896(58.0)	880(57.0)	—	880(57.0)
	Zr	5.0		150	703(45.5)	695(45.0)	—	672(43.5)
	Al	2.25		300	664(43.0)	664(43.0)	—	649(42.0)
	Mo	1.0		400	579(37.5)	571(37.0)	—	526(34.0)
	Si	0.2		450	448(29.0)	386(25.0)	—	247(16.0)
				500	131 (8.5)	93 (6.0)	—	62 (4.0)

Table 22.38 TITANIUM ALLOYS—TYPICAL CREEP PROPERTIES—*continued*

IMI designation	*Nominal composition %*	*Condition*	*Temperature °C*	*Stress* MPa* *to produce* 0.1% *total plastic strain in* 100 h	300 h	500 h	1 000 h
IMI 680	Sn 11.0	Quenched and	20	1 127(73.0)	1 112(72.0)	—	—
	Mo 4.0	aged rod	150	945(61.2)	942(61.0)	—	—
	Al 2.25		200	862(55.8)	856(55.5)	—	—
	Si 0.2		300	804(52.0)	788(51.0)	—	—
			400	555(36.0)	540(35.0)	—	—
			450	298(19.3)	209(13.5)	—	—
			500	88 (5.7)	51 (3.3)	—	—
		Furnace-cooled	300	570(37.0)	—	—	—
		and aged rod	350	540(35.0)	—	—	—
			400	490(32.0)	—	—	—
IMI 685	Al 6.0	Heat-treated	200	599(38.8)	—	592(38.3)	589(38.1)
	Zr 5.0	forgings	300	551(35.7)	—	541(35.0)	535(34.6)
	Mo 0.5		400	497(32.1)	—	480(31.1)	462(29.9)
	Si 0.25		450	461(29.8)	—	431(27.9)	426(27.6)
			500	408(26.4)	—	340(22.0)	—
IMI 829	Al 5.5	Fully heat	450	478(30.9)	—	—	—
	Sn 3.5	treated rod	500	420(27.2)	—	—	—
	Zr 3.0		550	300(19.4)	—	—	—
	Nb 1.0		600	130 (8.4)	—	—	—
	Mo 0.3						
	Si 0.3						

*Figures in parentheses are the stresses in tonf in^{-2}.

Table 22.39 TITANIUM AND TITANIUM ALLOYS—TYPICAL FATIGUE PROPERTIES

IMI designation	*Nominal composition %*	*Condition*	*Temperature °C*	*Tensile strength* MPa (tonf in^{-2})	*Details of test*	*Endurance limit for* 10^7 cycles (*except where stated*) MPa (tonf in^{-2})
					Rotating bend	
IMI 115	Commercial purity	Annealed rod	Room	354(22.9)	Smooth $K_t=1$	±193(±12.5)
				354(22.9)	Notched $K_t=3$	±123(± 8.0)
					Rotating bend	
IMI 125	Commercial purity	Annealed rod	Room	417(27.0)	Smooth $K_t=1$	±232(±15.0)
				417(27.0)	Notched $K_t=3$	±154(±10.0)
					Rotating bend	
IMI 130	Commercial purity	Annealed rod	Room	550(35.6)	Smooth $K_t=1$	±270(±17.5)
				550(35.6)	Notched $K_t=2$	±170(±11.0)
				550(35.6)	Notched $K_t=3.3$	±170(±11.0)
					Direct stress (Zero mean)	
				550(35.6)	Smooth $K_t=1$	±263(±17.0)
				550(35.6)	Notched $K_t=1.5$	±247(±16.0)
				550(35.6)	Notched $K_t=2$	±170(±11.0)
				550(35.6)	Notched $K_t=3.3$	±116(± 7.5)
				589(38.0)	Smooth $K_t=1$	±278(±18.0)
				589(38.0)	Notched $K_t=2$	±147(± 9.5)
				589(38.0)	Notched $K_t=3$	±123(± 8.0)
				589(38.0)	Notched $K_t=4$	±116(± 7.5)
					Direct stress (Zero mean)	
IMI160	Commercial purity	Annealed rod	Room	674(43.5)	Smooth $K_t=1$	±376(±24.3)

Table 22.39 TITANIUM AND TITANIUM ALLOYS—TYPICAL FATIGUE PROPERTIES—*continued*

IMI designation	*Nominal composition* %	*Condition*	*Temperature* °C	*Tensile strength* MPa (tonf in^{-2})	*Details of test*	*Endurance limit for* 10^7 *cycles (except where stated)* MPa (tonf in^{-2})
IMI 230	Cu 2.5	Annealed sheet	Room	564(36.5)	Reversed bend	±390(±25.0)
		Aged sheet	Room	772(50.0)	Reversed bend	±490(±31.5)
					Direct stress (Zero minimum)	
		Aged sheet	Room	761(49.3)	Smooth $K_t=1$	0→560(0→36.0)
					Rotating bend	
		Annealed rod	Room	598(38.7)	Smooth $K_t=1$	±370(±24.0)
			400		Smooth $K_t=1$	±150(±10.0)
					Direct stress (zero mean)	
		Annealed rod	Room	638(41.3)	Smooth $K_t=1$	±280(±18.0)
					Rotating bend	
		Aged rod	Room	700(45.3)	Smooth $K_t=1$	±450(±29.0)
			400	—	Smooth $K_t=1$	±290(±19.0)
					Direct stress (Zero mean)	
		Aged rod	Room	792(51.3)	Smooth $K_t=1$	±470(±30.0)
					Notched $K_t=3.3$	±200(±13.0)
IMI 260	Pd 0.2	Similar to IMI 115				
IMI 262	Pd 0.2	Similar to IMI 125			Rotating bend	Limits for this alloy 10^8 cycles
IMI 317	Al 5.0 Sn 2.5	Annealed rod	Room	—	Smooth $K_t=1.0$	±371(±24.0)
					Notched $K_t=2.0$	±263(±17.0)
					Notched $K_t=3.3$	±239(±15.5)
					Direct stress (Zero mean	
					Smooth $K_t=1.0$	±433(±28.0)
					Notched $K_t=1.5$	±278(±18.0)
					Notched $K_t=2.0$	±201(±13.0)
					Notched $K_t=3.3$	±154(±10.0)
					Rotating bend	
IMI 318	Al 6.0 V 4.0	Annealed rod	Room	960(62.5)	Smooth $K_t=1$	±470(±30.4)
				960(62.5)	Notched $K_t=2.7$	±230(±14.9)
					Direct stress (Zero minimum)	
				1 015(65.6)	Smooth $K_t=1$	0→750(0→48.6)
				1 015(65.6)	Notched $K_t=3$	0→325(0→21.1)
					Direct stress (Zero minimum)	
IMI 550	Al 4.0 Mo 4.0 Sn 2.0 Si 0.5	Fully heat-treated rod	Room	1 180(76.8)	Smooth $K_t=1$	0→850(0→55.0)
				1 180(76.8)	Notched $K_t=3$	0→350(0→22.7)
					Rotating bend	
					Smooth $K_t=1$	±587(±38.0)
					Notched $K_t=2.4$	±394(±25.5)

Table 22.39 TITANIUM AND TITANIUM ALLOYS—TYPICAL FATIGUE PROPERTIES—*continued*

IMI designation	*Nominal composition %*	*Condition*	*Temperature °C*	*Tensile strength* MPa ($tonf\,in^{-2}$)	*Details of test*	*Endurance limit for* 10^7 cycles *(except where stated)* MPa ($tonf\,in^{-2}$)
					Rotating bend	
IMI 551	Al 4.0	Fully heat-treated rod	Room	—	Smooth $K_t=1$	±750(±48.6)
	Mo 4.0			—	Notched $K_t=3.2$	±430(±27.8)
	Sn 4.0					
	Si 0.5					
					Rotating bend	
IMI 679	Sn 11.0	Air-cooled and aged rod	Room	—	Smooth $K_t=1.0$	±641*(±41.5)*
	Zr 5.0		200	—	Smooth $K_t=1.0$	±510*(±33.0)*
	Al 2.25		400	—	Smooth $K_t=1.0$	±510*(±33.0)*
	Mo 1.0		450	—	Smooth $K_t=1.0$	±556(±36.0)
	Si 0.2		500	—	Smooth $K_t=1.0$	±495(±32.0)
					Rotating bend	(Limits for 2×10^7 cycles)
IMI 680	Sn 11.0	Quenched and aged rod	Room	1 272(82.4)	Smooth $K_t=1$	±710(±46.0)
	Mo 4.0			1 272(82.4)	Notched $K_t=2$	±340(±22.0)
	Al 2.25			1 272(82.4)	Notched $K_t=3.3$	±293(±19.0)
	Si 0.2					
					Direct stress (Zero mean)	(Limits for 2×10^7 cycles)
			Room	1 272(82.4)	Smooth $K_t=1$	±695(±45.0)
					Notched $K_t=2$	±371(±24.0)
					Notched $K_t=3.3$	±232(±15.0)
					Rotating bend	(Limits for 10^8 cycles)
			Room	—	Smooth $K_t=1$	±648(±42.0)
			200	—	Smooth $K_t=1$	±495(±32.0)
			400	—	Smooth $K_t=1$	±479(±31.0)
					Direct stress (Zero mean)	
		Furnace-cooled rod	Room	1 100(70.0)	Smooth $K_t = 1$	±680(±43.0)
					Direct stress (Zero mean)	
IMI 685	Al 6.0	Fully heat-treated rod	20	—	Smooth $K_t = 1$	±440(±28.5)
	Zr 5.0		450	—	Smooth $K_t = 1$	±300(±19.4)
	Mo 0.5		520	—	Smooth $K_t = 1$	±260(±16.8)
	Si 0.25					
					Direct stress (Zero minimum)	
			450	—	Smooth $K_t = 1$	0→475(0→30.7)
			520	—	Smooth $K_t = 1$	0→425(0→27.5)
					Direct stress (Zero minimum)	
		Fully heat-treated forging	Room	—	Smooth $K_t = 1$	0→640(0→41.4)
			Room	—	Notched $K_t = 3.5$	0→220(0→14.2)
			475	—	Smooth $K_t = 1$	0→460(0→29.8)
			475	—	Notched $K_t = 3.5$	0→210(0→13.6)
					Direct stress (Zero minimum)	
IMI 829	Al 5.5	Fully heat-treated rod	Room		Smooth $K_t = 1$	0→550(35.6)
	Sn 3.5				Notched $K_t = 3$	0→260(16.8)
	Zr 3.0					
	Nb 1.0					
	Mo 0.3					
	Si 0.3					

* Limits for 10^8 cycles.

Table 22.40 IZOD IMPACT PROPERTIES OF TITANIUM AND TITANIUM ALLOYS

IMI designation	*Nominal composition %*		*Condition*	*Izod value* Joules (ft lbf)*							
				−196 °C	−78 °C	20 °C	100 °C	200 °C	300 °C	400 °C	500 °C
IMI 130†	Commer- cially pure		Annealed rod	—	62.4 (46)	61.0 (45)	62.4 (46)	72 (53)	82 (60½)	84 (62)	82 (60½)
IMI 317	Sn Al	5.0 2.5	Annealed rod	17.6 (13)	20.3 (15)	27.1 (20)	35.2 (26)	52.8 (39)	63.7 (47)	70.5 (52)	71.8 (53)
IMI 318	Al V	6.0 4.0	Annealed rod	13.5 (10)	14.9 (11)	20.3 (15)	25.7 (19)	40.6 (30)	65.0 (48)	83.5 (63)	92.0 (68)
IMI 550	Al Mo Sn Si	4.0 4.0 2.0 0.5	Fully heat-treated rod	—	—	19.0 (14)	—	—	—	—	—

IMI designation	*Nominal composition %*		*Condition*	*Charpy value* Joules (ft lbf)							
				−196 °C	−78 °C	20 °C	100 °C	200 °C	300 °C	400 °C	500 °C
IMI 551	Al Mo Sn Si	4.0 4.0 4.0 0.5	Fully heat-treated rod	13.5 (10)	19 (14)	20.3 (15)	21.7 (16)	24.4 (18)	26.5 (19½)	28.5 (21)	31.2 (23)
IMI 679	Sn Zr Al Mo Si	11.0 5.0 2.25 1.0 0.2	Air-cooled and aged	10.8 (8)	13.5 (10)	14.9 (11)	16.3 (12)	19 (14½)	25 (18½)	30 (22)	33.9 (25)
IMI 680	Sn Mo Al Si	11.0 4.0 2.25 0.2	Quenched and aged rod	8.1 (6)	8.8 (6½)	10.8 (8)	12.2 (9)	14.9 (11)	17.6 (13)	20.3 (15)	25.7 (19)
IMI 685	Al Zr Mo Si	6.0 5.0 0.5 0.25	Fully heat-treated rod	31.2 (23)	39.3 (29)	43.4 (32)	—	—	—	—	—

* BSS 131 (1) 0.45 in diameter straight notched test pieces.
† Izod values of commercial purity titanium are appreciably affected by variation in hydrogen content within commercial limits (0.008% maximum) in Ti 130 rod.

22.7 Mechanical properties of zinc and zinc alloys

Table 22.41 MECHANICAL PROPERTIES OF ZINC ALLOYS AT ROOM TEMPERATURE

	Composition						*Properties*		
Zinc	Unalloyed zinc is generally used only in the wrought form. The data here are some typical values								
	Zn min	Pb	Cd max	Sn	Fe	*Total* Pb + Cd + Sn + Fe + Cu (max)		*Parallel to rolling direction*	*Across rolling direction*
BS 3436 Zn 1	99.99	0.003	0.003	0.001	—	0.01	Zn 1		
Zn 2	99.95	0.03	0.02	0.001	0.01	0.05	Tensile strength		
Zn 3	99.5	0.35	0.15	0.001	0.03	0.5	MPa	120	150
Zn 4	98.5	1.35	0.15	0.02	0.04	1.5	Elongation %	60–80	40–60
							Hardness Vickers	30	

Table 22.41 MECHANICAL PROPERTIES OF ZINC ALLOYS AT ROOM TEMPERATURE—*continued*

	Composition						*Properties*		
Zinc	Unalloyed zinc is generally used only in the wrought form. The data here are some typical values								
	Zn min	Pb	Cd max	Sn	Fe	*Total* Pb+ Cd+Sn+ Fe+Cu (max)		*Parallel to rolling direction*	*Across rolling direction*
Zinc-copper-titanium	Cu	Ti				*Balance*			
	0.14	0.1–0.15				zinc of 99.995% purity	Tensile strength MPa Elongation % Hardness (Brinell)	 180 35 40–45	 216 20

These alloys are used principally in pressure die castings and other castings. Properties are given in Table 26.36 page **26**–52. Casting and foundry data

22.8 Mechanical properties of zirconium and zirconium alloys

Table 22.42 MECHANICAL PROPERTIES OF ZIRCONIUM ALLOYS AT ROOM TEMPERATURE

Material	*Nominal composition*	*Condition*	0.1% *proof stress* MPa (10^3 lbf in^{-2})	*UTS* MPa (10^3 lbf in^{-2})	*Elongation* %	*Macro-hardness* HV	*Reference*
Zr (ex iodide)	>99.9% purity Impurities in ppm by wt $O_2$65, $N_2$15, $H_2$12, Hf 35, Ni 20	Crystal bar, cold rolled and vacuum annealed 2 h at 750 °C	100–130 (14.5–18.9)	170–210 (24.7–30.5)	40–45 on 1 in gauge	85–100	1
Zr (ex sponge)	>99.6% purity Impurities in ppm by wt $O_2$1300, $N_2$80, $H_2$20, Hf 400, Ni 40	Sheet material, cold rolled and vacuum annealed 2 h at 750 °C	250–310 (36.3–45.0) in r.d.	350–390 (50.8–56.6) in r.d.	23–31 on 1 in gauge	195–215	1
Zircalloy 2	Sn 1.2–1.7% Fe 0.07–0.2% Cr 0.05–0.15% Ni 0.03–0.08% Zr–remainder	Plate materials cold rolled and annealed 1 h at 750 °C	340† (49.3) in r.d. 490 (71.1) in t.d.	450 (65.3) in r.d. 520 (75.4) in t.d.	29 in r.d. 23 in t.d. on 1 in gauge	205–220*	2 1
Zirconium 30*	Cu 0.46–0.66% Mo 0.50–0.60% Zr–remainder	Sheet and strip	220–320 (31.9–46.4) in r.d.	470–550 (68.2–79.8) in r.d.	20–31 on 2 in gauge	130–180	3
Zr/2½% Nb	Nb 2.55% Impurities in ppm by wt $O_2$1050, $N_2$20, $H_2$10, Hf 70 N<40	Plate, hot rolled at 750 °C later annealed ½ h at 700 °C	470 (68.2) for forged product	590 (85.6) for forged product 410–500 ppm O_2 600–2 500 ppm O_2	24 gauge un- specified‡	150–500 ppm O_2 230–2500 ppm O_2	4 3

* Imperial Metal Industries Nomenclature.

Table 22.43 MECHANICAL PROPERTIES OF ZIRCONIUM ALLOYS AT ELEVATED TEMPERATURES

Material	*Nominal composition*	*Test temperature* °C	0.1% *proof stress* MN m^{-2} (10^3 lbf in^{-2})	*UTS* MN m^{-2} (10^3 lbf in^{-2})	*Elongation* %	*Reference*
Zr (ex sponge)	>99.6% purity Impurities in ppm by wt $O_2$1300, $N_2$80, $H_2$20, Hf 400, Ni 40	371	50† (7.3)	110 (15.0)	57	3
Zircalloy 2	Sn 1.2–1.7% Fe 0.07–0.2% Cr 0.05–0.15% Ni 0.03–0.08% Zr–remainder	300	92–126 (13.4–18.3)	210–260 (30.5–37.7)	33–44 on 2 in gauge	3
Zirconium 30	Cu 0.46–0.66% Mo 0.50–0.60% Zr–remainder	300	160 (23.2)	250 (36.3)	34 on $5.65\sqrt{A}$ gauge	3
Zr/2½% Nb	Nb 2.55% Impurities in ppm by wt $O_2$1050, $N_2$20,$H_2$10, Hf70, Ni<40	300	210 (30.5)	340 (49.3) 280–500 ppm O_2* 350–2 500 ppm O_2*	33 gauge unspecified	3 4*

r.d. rolling direction.
t.d. transverse direction.
* Imperial Metal Industries nomenclature.
† 0.2% proof stress.

REFERENCES TO TABLES 22.42 and 22.43

1. B. J. Gill, Dept. of Metallurgy and Materials Tech., University of Surrey, Guildford, England, 1972.
2. W. Evans and G. W. Parry, *Electrochem. Tech.*, 1966, **4**, 225.
3. Imperial Metal Industries pubn 2 Ed/MK105/33/366.
4. J. Winton and R. A. Murgatroyd, *Electrochem. Tech.*, 1966, **4**, 358.

22.9 Steels

Table 22.44 FORGED OR ROLLED STEELS—ROOM TEMPERATURE LONGITUDINAL MECHANICAL PROPERTIES

Material	British or other standard	Composition % C	Mn	Ni	Cr	Other elements	Condition	Limiting ruling section mm	in	UTS MPa (tonf in^{-2})	Yield stress MPa (tonf in^{-2})	Elong. (gl $5.65\sqrt{S_0}$) %	Izod J (ft lbf)	Remarks
1. *Carbon steels with up to 1.7%* Mn *content, including free cutting steels* (BS 970:Part 1:1972, DIN 1651:1970)														
0.20C Steel	BS 070M20 (En3)	0.16	0.50	—	—	P, S	Normalized	152	6	430 (28)	215 (14)	22	—	General constructional steel suitable for welding.
		0.24	0.90	—	—	0.050 max		254	10	400 (26)	200 (13)	20	—	
							Cold drawn	13	½	530 (34)	385 (25)	12	—	
								76	3	430 (28)	340 (22)	14		
							Hardened and tempered	19	¾	540/690 (35/45)	355 (23)	20	41‡(30)	
0.26 C Steel	BS 070M26	0.22	0.50	—	—	P, S	Normalized	64	2½	490 (32)	245 (16)	20	—	Medium strength engineering steel
		0.30	0.90	—	—	0.050 max		254	10	430 (28)	215 (14)	20	—	
							Cold drawn	13	½	570 (37)	430 (28)	11	—	
								76	3	490 (32)	370 (24)	13	—	
							Hardened and tempered	13	½	620/770 (40/50)	415 (27)	16	34‡ (25)	
								29	1⅛	540/690 (35/45)	355 (23)	20	41‡ (30)	
0.30 C Steel	BS 080M30 (En5)	0.26	0.60	—	—	P, S	Normalized	152	6	490 (32)	245 (16)	20	—	General engineering steels widely used in the bright drawn condition
		0.34	1.00	—	—	0.050 max		254	10	460 (30)	230 (15)	19	—	
							Cold drawn	13	½	600 (39)	450 (29)	10	—	
								76	3	530 (34)	385 (25)	12	—	
							Hardened and tempered	19	¾	620/770 (40/50)	420 (27)	16	34‡ (25)	
								64	2½	540/690 (35/45)	340 (22)	18	34‡ (25)	
0.36 C Steel	BS 080M36	0.32	0.60	—	—	P, S	Normalized	64	2½	540 (35)	280 (18)	16	27‡ (20)	
		0.40	1.00	—	—	0.050 max		254	10	490 (32)	245 (16)	18	—	
							Cold drawn	13	½	620 (40)	480 (31)	9	—	
								76	3	540 (35)	400 (26)	11	—	
							Hardened and tempered	13	½	690/850 (45/55)	465 (30)	16	34‡ (25)	
								29	1⅛	620/770 (40/50)	400 (26)	16	34‡ (25)	

0.40 C Steel	BS 080M40 (En8)	0.36	0.60	—	—	P, S	Normalized	152	6	540 (35)	280 (18)	16	20‡ (15)	Nuts and bolts,
		0.44	1.00	—	—	0.050 max		254	10	510 (33)	245 (16)	17	—	forgings and
							Cold drawn	13	½	650 (42)	510 (33)	8	—	general
								76	3	570 (37)	430 (28)	10	—	engineering
							Hardened	19	¾	690/850 (45/55)	465 (30)	16	34‡ (25)	parts
							and	64	2½	620/770 (40/50)	385 (25)	16	34‡ (25)	
							tempered							
0.46 C Steel	BS 080M46	0.42	0.60	—	—	P, S	Normalized	64	2½	620 (40)	310 (20)	14	—	Nuts and bolts,
		0.50	1.00	—	—	0.050 max		254	10	540 (35)	280 (18)	15	—	forgings and
							Cold drawn	13	½	690 (45)	555 (36)	7		general
								76	3	620 (40)	480 (31)	9	—	engineering
							Hardened	13	½	770/930 (50/60)	525 (34)	14	—	parts
							and	102	4	620/770 (40/50)	370 (24)	16	—	
							tempered							
0.50 C Steel	BS 080M50	0.45	0.60	—	—	P, S	Normalized	152	6	620 (40)	310 (20)	14	—	Gears and
	(En43A)	0.55	1.00	—	—	0.050 max		254	10	570 (37)	280 (18)	14	—	machined parts
							Cold drawn	13	½	730 (47)	585 (38)	8	—	for flame or
								76	3	650 (42)	510 (33)	10	—	induction
							Hardened	13	½	850/1 000 (55/65)	570 (37)	12	—	hardening
							and	63	2½	690/850 (45/55)	430 (28)	14	—	
							tempered							
0.55 C Steel	BS 070M55	0.50	0.50	—	—	P, S	Normalized	64	2½	700 (45)	355 (23)	12	—	General machine
		0.60	0.90	—	—	0.050 max		254	10	600 (39)	310 (20)	13	—	parts requiring
							Cold drawn	13	½	790 (51)	620 (40)	7	—	higher wear
								76	3	710 (46)	570 (37)	9	—	resistance
							Hardened	19	¾	850/1 000 (55/65)	570 (37)	12	—	
							and	102	4	690/850 (45/55)	415 (27)	14	—	
							tempered							
0.19 C–	BS 120M19	0.15	1.00	—	—	P, S	Normalized	102	4	490 (32)	295 (19)	20	34‡ (25)	Armature shafts
1.2 Mn		0.23	1.40	—	—	0.050 max		254	10	460 (30)	260 (17)	19	—	High tensile
Steel							Cold drawn	13	½	600 (39)	450 (29)	11	—	bolts. Lifting
								76	3	530 (34)	385 (25)	12	34‡ (25)	gear chains.
							Hardened	19	¾	690/850 (45/55)	510 (33)	16	34‡ (25)‡	Automotive
							and	102	4	540/690 (35/45)	355 (23)	18	47‡ (35)‡	forgings
							tempered							

‡Grain controlled steel.

Table 22.44 FORGED OR ROLLED STEELS—ROOM TEMPERATURE LONGITUDINAL MECHANICAL PROPERTIES—*continued*

Material	*British or other standard*	Composition % C	Mn	Ni	Cr	*Other elements*	*Condition*	*Limiting ruling section* mm	in	Properties (minima unless otherwise stated) *UTS* MPa (tonf in^{-2})	*Yield stress* MPa (tonf in^{-2})	*Elong.* (gl $5.65\sqrt{S_0}$) %	*Izod* J (ft lbf)	*Remarks*
0.19 C–	BS 150M19	0.15	1.30	—	—	P, S	Normalized	152	6	540 (35)	325 (21)	18	41‡ (30)	Armature shafts.
1.5 Mn	(En14A)	0.23	1.70	—	—	0.050 max		254	10	510 (33)	295 (19)	17	—	High tensile
Steel							Hardened	29	$1\frac{1}{8}$	690/850 (45/55)	510 (33)	16	41‡ (30)	bolts. Lifting
							and	152	6	540/690 (35/45)	340 (22)	18	54‡ (40)	gear chains.
							tempered							Automotive
														forgings
0.28 C–	BS 120M28	0.24	1.00	—	—	P, S	Normalized	152	6	540 (35)	325 (21)	16	34‡ (25)	
1.2 Mn		0.32	1.40	—	—	0.050 max		254	10	530 (34)	310 (20)	17	—	
Steel							Cold drawn	13	$\frac{1}{2}$	650 (42)	510 (33)	8	—	
								76	3	570 (37)	430 (28)	10	—	
							Hardened	29	$1\frac{1}{8}$	690/850 (45/55)	510 (33)	16	34‡ (25)	
							and	102	4	620/770 (40/50)	415 (27)	16	41‡ (30)	
							tempered							
0.28 C–	BS 150M28	0.24	1.30	—	—	P, S	Normalized	152	6	590 (38)	355 (23)	16	34‡ (25)	
1.5 Mn	(En14B)	0.32	1.70	—	—	0.050 max		254	10	560 (36)	325 (21)	16	—	
Steel							Hardened	13	$\frac{1}{2}$	770/930 (50/60)	570 (37)	16	34‡ (25)	
							and	152	6	620/770 (40/50)	400 (26)	16	47‡ (35)	
							tempered							
0.36 C–	BS 120M36	0.32	1.00	—	—	P, S	Normalized	152	6	590 (38)	355 (23)	15	—	General
1.2 Mn	(En15B)	0.40	1.40	—	—	0.050 mm		254	10	570 (37)	340 (22)	16	—	engineering
Steel							Cold drawn	13	$\frac{1}{2}$	690 (45)	555 (36)	7	—	steel
								76	3	620 (40)	480 (31)	9	—	
							Hardened	19	$\frac{3}{4}$	770/930 (50/60)	570 (37)	14	34‡ (25)	
							and	102	4	620/770 (40/50)	415 (27)	18	41‡ (30)	
							tempered							
0.36 C–	BS 150M36	0.32	1.30	—	—	P, S	Normalized	152	6	620 (40)	385 (25)	14	—	
1.5 Mn	(En15)	0.40	1.70	—	—	0.050 max		254	10	600 (39)	355 (23)	15	—	
Steel							Hardened	13	$\frac{1}{2}$	850/1000 (55/65)	635 (41)	12	34‡ (25)	
							and	152	6	620/770 (40/50)	400 (26)	18	47‡ (35)	
							tempered							

Low C Free cutting steel	BS 220M07 (EnlA)	0.15 max	0.90 1.30	— —	— —	S 0.20 0.30	Cold drawn	13	½	460 (30)	Not given	7	—	Free cutting mild steels to be machined in high speed automatic lathes
								76	3	360 (23)	Not given	10	—	
							Hot rolled	102	4	360 (23)	215 (14)	22		
Low C Free cutting steel	BS 230M07	0.15 max	0.90 1.30	— —	— —	S 0.25 0.35	Cold drawn	13	½	460 (30)	Not given	7	—	Free cutting mild steels to be machined in high speed automatic lathes
								76	3	360 (23)	Not given	10	—	
							Hot rolled	102	4	360 (23)	215 (14)	22	—	
Low C Free cutting steel	BS 240M07 (En1B)	0.15 max	1.10 1.50	— —	— —	S 0.30 0.60	Cold drawn	13	½	450 (29)	Not given	7		Free cutting mild steels to be machined in high speed automatic lathes
								76	3	360 (23)	Not given	10	—	
							Hot rolled	64	2½	360 (23)	215 (14)	20	—	
Low C Free cutting steels containing lead	DIN 9SMnPb28	0.14 max	0.90 1.30	— —	— —	S 0.24– 0.32 Pb 0.15 0.30	Cold drawn	16	⅝	510/690 (33/45)	415 (27)	7	—	Free cutting mild steels to be machined in high speed automatic lathes
								64	2½	415/665 (27/43)	310 (20)	9	—	
								102	4	385/635 (25/41)	245 (16)	10	—	
0.28 C Free cutting steel	BS 216M28	0.24 0.32	1.10 1.50	— —	— —	S 0.12– 0.20	Cold drawn	13	½	570 (37)	430 (28)	10	—	Free cutting medium strength engineering steels for machine parts and engine components, etc.
								76	3	490 (32)	370 (24)	12	—	
							Hardened and tempered	19	¾	620/770 (40/50)	430 (28)	18	34‡ (25)	
								64	2½	540/690 (35/45)	355 (23)	20	34‡ (25)	
0.36 C Free cutting steel	BS 212M36 (En8M)	0.32 0.40	1.00 1.40	— —	— —	S 0.12– 0.20	Cold drawn	13	½	620 (40)	480 (31)	7	—	Free cutting medium strength engineering steels for machine parts and engine components, etc.
								76	3	540 (35)	400 (26)	9	—	
							Hardened and tempered	13	½	690/850 (45/55)	495 (32)	16	54‡ (40)	
								102	4	540/690 (35/45)	340 (22)	20	34‡ (25)	

‡ Grain controlled steel

Table 22.44 FORGED OR ROLLED STEELS—ROOM TEMPERATURE LONGITUDINAL MECHANICAL PROPERTIES—*continued*

Material	*British or other standard*	*Composition %* C	Mn	Ni	Cr	*Other elements*	*Condition*	*Limiting ruling section* mm	in	*Properties (minima unless otherwise stated)* *UTS* MPa (tonf in^{-2})	*Yield stress* MPa (tonf in^{-2})	*Elong.* (gl 5.65$\sqrt{S_0}$) %	*Izod* J (ft lbf)	*Remarks*
0.36 C Free cutting steels higher Mn	BS 216M36 (En 15AM)	0.32	1.30	—	—	S 0.12–	Cold worked	16	$\frac{5}{8}$	650 (42)	510 (33)	7	—	Free cutting medium strength engineering steels for machine parts and engine
								76	3	570 (37)	415 (27)	9	—	
		0.40	1.70	—	—	0.20	Hardened and tempered	102	4	540 (35/45)	340 (22)	20	34‡ (25)	
						P 0.060 Si 0.25 max		29	$1\frac{1}{8}$	690/850 (45/55)	480 (31)	16	34‡ (25)	
0.36 C Free cutting steels	BS 225M36	0.32	1.00	—	—	S 0.20–	Cold worked	16	$\frac{5}{8}$	620 (40)	480 (31)	7	—	Free cutting medium strength engineering steels
		0.40	1.40	—	—	0.30		76	3	540 (35)	430 (28)	9	—	
						P 0.060	Hardened and	64	$2\frac{1}{2}$	620/770 (40/50)	400 (26)	18	34‡ (25)	
						Si 0.25 max	tempered	29	$1\frac{1}{8}$	690/850 (45/55)	480 (31)	16	34‡ (25)	
0.44 C Free cutting steels	BS 212M44 (En 8M)	0.40	1.00	—	—	S 0.12–	Hardened and	102	4	620/770 (40/50)	400 (26)	18	34‡ (25)	Free cutting medium strength engineering steels
		0.48	1.40	—	—	0.20								
						P 0.060 Si 0.25 max	tempered	13	$\frac{1}{2}$	770/930 (50/60)	540 (35)	14	27‡ (20)	
0.44 C Free cutting steel	BS 225M44	0.40	1.30	—	—	S 0.20–	Hardened and	13	$\frac{1}{2}$	850/1 000 (55/65)	600 (39)	12	27‡ (20)	Free cutting medium strength engineering steels for machine parts and engine components, etc.
		0.48	1.70	—	—	0.30 P 0.060 Si 0.25 max	tempered	100	4	690/850 (45/55)	450 (29)	16	34‡ (25)	
2. *Low alloy high strength weldable steels* (BS 4360:1979, 1501:Pt 2:1970, 1503:1969, ASTM Standards)														
Mn steel	BS 4360–40B	0.20 max	1.50 max	—	—	—	As rolled	16	$\frac{5}{8}$	400/480 (26/31)	230 (15)	25	27 (20)	Structural steel
Mn steel	BS 4360–40C	0.18 max	1.50 max	—	—	—	As rolled	16	$\frac{5}{8}$	400/480 (26/31)	230 (15)	25	27 (20)	Structural steel

Mn Nb (V)	BS 4360–40D	0.16 max	1.50 max	—	—	Nb and/or V 0.10 max	Normalized	16	$\frac{5}{8}$	400/480 (26/31)	260 (17)	25	41 (30)	Structural steel
								40	$1\frac{1}{2}$	400/480 (26/31)	245 (16)	25	27 (20)	
Mn steel	BS 4360–40E	0.16 max	1.50 max	—	—	P + S 0.04 max	Normalized	16	$\frac{5}{8}$	400/480 (26/31)	260 (17)	25	61 (45)	Structural steel
								40	$1\frac{1}{2}$	400/480 (26/31)	245 (16)	25	47 (34)	
								63	$2\frac{1}{2}$	400/480 (26/31)	240 (15.5)	25	27 (20)	
Mn steel	BS 4360–43A	0.25 max	1.60 max	—	—	— (Cu 0.50 max)	As rolled	16	$\frac{5}{8}$	430/510 (28/33)	245 (16)	22	—	Structural steel
Mn steel	BS 4360–43B	0.22 max	1.50 max	—	—	— (Cu 0.50 max)	As rolled	16	$\frac{5}{8}$	430/510 (28/33)	245 (16)	22	27 (20)	Structural steel
Mn steel	BS 4360–43C	0.19 max	1.50 max	—	—	—	As rolled	16	$\frac{5}{8}$	430/510 (28/33)	245 (16)	22	27 (20)	Structural steel
Mn Nb (V) steel	BS 4360–43D	0.16 max	1.50 max	—	—	Nb and/or V 0.10 max	Normalized	16	$\frac{5}{8}$	430/510 (28/33)	280 (18)	22	41 (30)	Structural steel
								40	$1\frac{1}{2}$	430/510 (28/33)	270 (17.5)	22	27 (20)	
Mn steel	BS 4360–43E	0.16 max	1.50 max	—	—	—	Normalized	16	$\frac{5}{8}$	430/510 (28/33)	280 (18)	22	61 (45)	Structural steel
								40	$1\frac{1}{2}$	430/510 (28/33)	270 (17.5)	22	47 (34)	
								63	$2\frac{1}{2}$	430/510 (28/33)	255 (16.5)	22	27 (20)	
Mn Nb (V) steel	BS 4360–50B	0.20 max	1.50 max	—	—	Nb and/or V 0.10 max	As rolled	16	$\frac{5}{8}$	490/620 (31.5/40)	355 (23.0)	20	—	Structural steel Plate, bars, sections, tubes
		0.20 max	1.50 max	—	—	Nb and/or V 0.10 max (Cu 0.50) max	Normalized	100	4	490/620 (31.5/40)	325 (21)	20	—	
Mn Nb (V) steel	BS 4360–50C	0.20 max	1.50 max	— —	— —	Nb and/or V 0.10 max (Cu 0.50) max	As rolled	16	$\frac{5}{8}$	490/620 (31.5/40)	355 (23)	20	Charpy V 41 (30) at −5 °C	Structural steel. Plate bars, sections, tubes
		0.20 max	1.50 max	— —	— —	Nb and/or V 0.10 max (Cu 0.50) max	Normalized	100	4	490/620 (32/40)	325 (21)	20	27 (20) at −15 °C	

‡ Grain controlled steel.

Table 22.44 FORGED OR ROLLED STEELS—ROOM TEMPERATURE LONGITUDINAL MECHANICAL PROPERTIES—*continued*

Material	*British or other standard*	*Composition %* C	Mn	Ni	Cr	*Other elements*	*Condition*	*Limiting ruling section* mm	in	*Properties (minima unless otherwise stated)* *UTS* MPa (tonf in^{-2})	*Yield stress* MPa (tonf in^{-2})	*Elong.* (gl $5.65\sqrt{S_0}$) %	*Izod* J (ft lbf)	*Remarks*
Mn Nb (V) steel	BS 4360–50D	0.18 max	1.50 max	— —	— —	Nb and/or V 0.10 max	Normalized	16	$\frac{5}{8}$	490/620 (31.5/40)	355 (23)	20	Charpy V 41 (30) at −20 °C	Structural steel. Plate, bars, sections, tubes
							Normalized	63	$2\frac{1}{2}$	490/620 (31.5/40)	340 (22)	20	27 (20) at −30 °C	
Mn Nb (V) Steel	BS 4360–50D1	0.18 max	1.50 max	—	—	Nb and/or V 0.10 max	Normalized	16	$\frac{5}{8}$	490/620 (32/40)	355 (23)	20	41 (30)	Structural steel
								40	$1\frac{1}{2}$	490/620 (32/40)	345 (22.5)	20	27 (20)	
Mn Nb (V) Steel	BS 4360–50E	0.18 max	1.50 max	—	—	Nb and/or V 0.15 max	Normalized	16	$\frac{5}{8}$	490/620 (32/40)	355 (23)	20	47 (35)	Structural steel
								40	$1\frac{1}{2}$	490/620 (32/40)	345 (22.5)	20	41 (30)	
								63	$2\frac{1}{2}$	490/620 (32/40)	340 (22)	20	27 (20)	
Mn Nb (V) Steel	BS 4360–50F	0.16 max	1.50 max	—	—	Nb and/or V 0.10 max	Hardened and tempered	16	$\frac{5}{8}$	490/620 (32/40)	390 (25)	20	41 (30)	Structural steel
								25	1	490/620 (32/40)	390 (25)	20	35 (25)	
								40	$1\frac{1}{2}$	490/620 (32/40)	390 (25)	20	27 (20)	
Mn Nb (V) Steel	BS 4360–55C	0.22 max	1.60 max	— —	— —	Nb 0.10 max and/or	As rolled	16	$\frac{5}{8}$	550/700 (36/45)	450 (29)	19	Charpy V 27 (20) at 0 °C	Structural steel. Plate bars, sections, tubes
						V 0.20 max (Cu 0.50 max) + Al	As rolled	38	$1\frac{1}{2}$	550/700 (36/45)	415 (27)	19		
Mn Nb (V) Steel	BS 4360–55E	0.22 max	1.60 max	— —	— —	Nb 0.10 max and/or	Normalized	16	$\frac{5}{8}$	550/700 (36/45)	450 (29)	19	Charpy V 61 (45) at −20 °C	Structural steel. Plate, bars, sections, tubes
						V 0.20 max (Cu 0.50 max) + Al	Normalized or hardened and tempered	63	$2\frac{1}{2}$	550/700 (36/45)	400 (26)	19	27 (20) at −50 °C	
CrCu Steel	BS 4360 WR 50A, WR 50A1	0.12 max	0.60 max	0.65 max	0.30 1.25	Cu 0.25 0.55 P 0.070	As rolled	12	$\frac{1}{2}$	480 (31)	345 (22.5)	21	Charpy V 27 (20) at 0 °C	Weather resisting plate
						0.150		38	$1\frac{1}{2}$	480 (31)	325 (21)	21	—	

CrCuV Steel	BS 4360 WR 50B WR 50B1	0.10 0.19	0.90 1.25	— —	0.40 0.70	Cu 0.25 0.40 V 0.02 0.10	As rolled	12	$\frac{1}{2}$	480 (31)	345 (22.5)	21	Charpy V 27 (20) at 0 °C	Weather resisting plate
								50	2	480 (31)	340 (22)	21	—	
CrCuV Steel	BS 4360 WR 50C WR 50C1	0.10 0.22	0.90 1.45	— —	0.40 0.70	Cu 0.25 0.40 V 0.02 0.10	Normalized	12	$\frac{1}{2}$	480 (31)	345 (22.5)	21	Charpy V 27 (20) at −15 °C	Weather resisting plate
								50	2	480 (31)	340 (22)	21		
Mn V Steel	ASTM A 242	0.22 max	1.25 max	—	—	V 0.02 min	As rolled	38	$1\frac{1}{2}$	420 (27)	310 (20)	21	—	
Mn V Nb Steel	ASTM A 572	0.22 max	1.35 max	— —	— —	Nb 0.01 min and/or V 0.02 min N 0.015 max	As rolled	19	$\frac{3}{4}$	550 (35.5)	450 (29)	17	—	
Mo B Steel	BS 1501–261 (Fortiweld)	0.10 0.17	0.40 0.80	0.30 max	0.25 —	Mo 0.40– 0.60 B 0.001– 0.005	Normalized	89	$3\frac{1}{2}$	550/670 (36/43)	420 (27)	16 (transv)	Charpy V 20 (15)	Pressure vessel plate
CrMoZr Steel	ASTM A 242	0.15 max	1.00 max	— —	0.40– 0.70	Si 0.60– 0.90 Mo 0.20 max Zr 0.15 max	As rolled	19	$\frac{3}{4}$	480 (31)	340 (22)	20	—	High strength low alloy structural steel
1CrMo Steel	BS 1501–261 (Grade 27)	0.09 0.18	0.40 0.70	0.30 max	0.70– 1.20	Mo 0.45– 0.65	Normalized and tempered	152	6	420/540 (27/35)	285 (18.5)	19 (transv)	—	Pressure vessel plate
1CrMo Steel	BS 1501–620 (Grade 31)	0.12 0.18	0.40 0.70	0.30 max	0.70 1.20	Mo 0.45 0.65	Normalized and tempered	76	3	480/600 (31/39)	340 (22)	18	—	Pressure vessel plate
								152	6	450/570 (29/37)	315 (20)	16	—	
1CrMo Steel	BS 1503–620	0.08 0.15	0.30 0.80	0.40 max	0.70 1.10	Mo 0.45 0.65	Normalized and tempered or hardened and tempered	—	—	420/570 (27/37)	232 (15)	18 (transv)	—	Pressure vessel forgings

Table 22.44 FORGED OR ROLLED STEELS—ROOM TEMPERATURE LONGITUDINAL MECHANICAL PROPERTIES—*continued*

Material	*British or other standard*	*Composition %* C	Mn	Ni	Cr	*Other elements*	*Condition*	*Limiting ruling section* mm	in	*Properties (minima unless otherwise stated)* *UTS* MPa ($tonf\ in^{-2}$)	*Yield stress* MPa ($tonf\ in^{-2}$)	*Elong.* (gl $5.65\sqrt{S_0}$) %	*Izod* J (ft lbf)	*Remarks*
$1\frac{1}{4}$CrMo Steel	BS 1501–621	0.09 0.15	0.40 0.70	0.30 max	1.00 1.50	Mo 0.45–0.65	Normalized and tempered	152 76	6 3	450/570 (29/37) 480/600 (31/39)	315 (20) 340 (22)	16 (transv) 18 (transv)	—	Pressure vessel steel
$1\frac{1}{4}$CrMo Steel	BS 1503–621	0.08 0.17	0.30 0.80	0.40 max	1.00 1.50	Mo 0.45 0.65	Normalized and tempered or quenched and tempered	—	—	460/620 (30/40)	265 (17)	16 (transv)	—	Pressure vessel forgings
$1\frac{3}{4}$CrMoV Steel	ASTM A 517 Grade E (SSS100)	0.12 0.20	0.40 0.70	— —	1.40 2.00	Cu 0.30 0.40 Mo 0.40–0.60 V 0.04 or Ti 0.10 +B	Hardened and tempered	64	$2\frac{1}{2}$	790/930 (51/60)	700 (45)	15	—	Pressure vessel plate
MnCrMoV Steel	BS 1501–271	0.11 0.17	1.00 1.50	0.70 max	0.40 0.70	Mo 0.20–0.28 V 0.04–0.12	Normalized and tempered	150 25	6 1	560/680 (36/44) 590/700 (38/45)	385 (25) 465 (30)	16 (transv) 16 (transv)	Charpy V 41 (30) at 20 °C 27 (20) at 0 °C	Pressure vessel plate
MnCrMoV Steel	BS 1503–271	0.17 max	1.00 1.50	0.30 0.70	0.50 1.00	Mo 0.20–0.35 V 0.05–0.10	Normalized and tempered or hardened and tempered	—	—	560/710 (36/46)	370 (24)	19	—	Pressure vessel forgings
MnNiMo Steel	ASTM A533 Grade B	0.25 max	1.15 1.50	0.40 0.70	— —	Mo 0.45–0.60	Hardened and tempered	102	4	620 (40)	480 (31)	17	—	Pressure vessel plate

MnNiMo Steel	ASTM A508 Class 3	0.15 0.25	1.20 1.50	0.40 0.80	— —	Mo 0.45–0.60 V 0.05 max	Hardened and tempered	—	—	550 (35.5)	340 (22)	18	Charpy V 41 (30) at +4 °C	Pressure vessel forgings
CrMoZr Steel	WG Proprietary steel (N-A-XTRA 70)	0.20 max	0.70– 1.10	— —	0.60– 1.00	Si 0.50–0.90 Mo 0.20–0.60 Zr 0.06–0.12	Hardened and tempered	25	1	790/930 (51/60)	700 (45)	16	—	
NiCuMo Steel	UK Proprietary steel (Nicuage Type 1)	0.06 max	0.40 0.65	0.70 1.00	— —	Cu 1.00–1.30 Nb 0.02 min	As rolled	13	$\frac{1}{2}$	540/630 (35/41)	500/580 (32/38)	25	Charpy V 61–135 (45–100) at −10°C 41–115 (30–85) at −20°C	Structural steel resistant to atmospheric corrosion. Good low temperature toughness
							Aged at 500/570 °C	13	$\frac{1}{2}$	630/740 (41/48)	590/680 (38/44)	20	41–115 (30–85) at −10°C 34–81 (25–60) at −20°C	
NiCuMo Steel	ASTM A588	0.15 max	1.00 max	0.75 max	— —	Mo 0.15–0.25 Cu 0.50–1.00	As rolled	102	4	480 (31)	340 (22)	20	—	Resistant to atmospheric corrosion
NiCrMoVNb Steel	BS 1501–281	0.09 0.15	0.90 1.30	0.70 1.00	0.40 0.70	Mo 0.20–0.28 V 0.04–0.12	Normalized and tempered	152	6	560/680 (36/44)	385 (25)	16 (transv)	Charpy V 68 (50) at −10 °C 27 (20) at −40 °C	Pressure vessel plate
						Nb 0.10 max N 0.015 max		25	1	590/700 (38/45)	465 (30)	16 (transv)		
NiCrMoVB Steel	ASTM A517F (T1)	0.10 0.20	0.60 1.00	0.70 1.00	0.40 0.65	Mo 0.40–0.60 V 0.03–0.08 Cu 0.15–0.50 +B	Hardened and tempered	63	$2\frac{1}{2}$	790/930 (51/60)	700 (45)	16	Charpy V 20 (15) at −45 °C	Pressure vessel plate

Table 22.44 FORGED OR ROLLED STEELS—ROOM TEMPERATURE LONGITUDINAL MECHANICAL PROPERTIES—*continued*

Material	*British or other standard*	*Composition %* C	Mn	Ni	Cr	*Other elements*	*Condition*	*Limiting ruling section* mm	in	*Properties (minima unless otherwise stated)* *UTS* MPa (tonf in^{-2})	*Yield stress* MPa (tonf in^{-2})	*Elong.* (gl $5.65\sqrt{S_0}$) %	*Izod* J (ft lbf)	*Remarks*
NiCrMo Steel	UK Proprietary steel	0.22 max	1.00 1.20	0.70 0.85	0.90 1.10	Mo 0.45–0.60	Hardened and tempered	—	—	660 (43) (Typical values)	510 (33) (Typical values)	20	Charpy V 120 (89)	Pressure vessel forgings
$1\frac{1}{2}$NiCrMoV Steel	BS 1501–282	0.12 0.17	0.90 1.30	1.40 1.60	0.30 0.70	Mo 0.30–0.40 V 0.08–0.12	Double normalized and tempered	152	6	570/700 (37/45)	415 (27)	18 (transv)	Charpy V 81 (60) at −10 °C 27 (20) at −50 °C	Pressure vessel plate
								76	3	590/710 (35/46)	—	18 (transv)		
$2\frac{1}{4}$CrMo Steel	BS 1501–622 Grade 31	0.10 0.15	0.40 0.80	0.30 max	2.00 2.50	Mo 0.90–1.20	Normalized and tempered	152	6	480/600 (31/39)	280 (18)	16 (transv)	—	Pressure vessel plate
$2\frac{1}{4}$CrMo Steel	BS 1501–622 Grade 45	0.13 0.18	0.40 0.80	0.30 max	2.00 2.50	Mo 0.90–1.20	Normalized and tempered	152	6	700/820 (45/53)	550 (36)	15 (transv)	—	Pressure vessel plate
	ASTM A542	0.15 max	0.30 0.60	— —	2.00 2.50	Mo 0.90–1.10	Hardened and tempered	102	4	710/835 (46/54)	590 (38)	16	—	Plate
	BS 1503–622 (Esshete CRM2)	0.08 0.15	0.40 0.70	0.40 max	2.00 2.50	Mo 0.90–1.10	Normalized and tempered or hardened and tempered	—	—	540/700 (35/45)	370 (24)	19	—	Pressure vessel forgings
3CrMo Steel	BS 1503–623 Grade 38	0.22 max	0.30 0.80	0.40 max	2.75 3.50	Mo 0.45–0.60	Normalized and tempered, or hardened and tempered	—	—	590/740 (38/48)	415 (27)	19	—	Pressure vessel forgings

3CrMo Steel	BS 1503–623 Grade 47	0.20 0.30	0.30 0.80	0.40 max	2.75 3.50	Mo	0.45–0.60	Normalized and tempered or hardened and tempered	—	—	730/880 (47/57)	540 (35)	17	—	Pressure vessel forgings
3NiCrMo Steel	ASTM A543 (HY 80)	0.18 max	0.40 —	2.25 3.25	1.00 1.50	Mo V	0.45–0.60 0.03 max	Hardened and tempered	102	4	710/835 (46/54)	590 (38)	17	—	Plate
	ASTM A543 (HY 100)	0.20 max	0.40 —	2.50 3.50	1.30 1.80	Mo V	0.45–0.60 0.03 max	Hardened and tempered	102	4	790/930 (51/60)	700 (45)	16	—	Plate
3.25NiCrMo Steel	ASTM A508 Class 4a	0.23 max	0.20 0.40	2.75 3.90	1.50 2.00	Mo V	0.40–0.60 0.03 max	Hardened and tempered	—	—	790 (51)	700 (45)	16	Charpy V 47 (35) at −30 °C	Pressure vessel forgings
3½NiCrMo Steel	BS 1501–503	0.15 max	0.30 0.80	3.25 3.75	0.30 max	Mo	0.10 max	Normalized and tempered or hardened and tempered	38	1½	450 (29)	265 (17)	20	Charpy V 34 (25) at −80 °C 18 (13) at −100 °C	Pressure vessel plate
5NiCrMoV Steel	ASTM (HY 130)	0.12 max	0.60 0.90	4.75 5.25	0.40 0.70	Mo V	0.30–0.65 0.05–0.10	Hardened and tempered	102	4	990 (64)	900 (58)	15	Charpy V 88 (65) at −20 °C	Plate

3. *Low alloy direct hardening steels including ultra high strength and steels suitable for nitriding*
(BS 970: Part 2: 1970, 4670: 1971, DIN 17200: 1969; 17211: 1970, Stahl Eisen Werkstoffblatt 550–57)

1Ni Steel	BS 503M40 (En12)	0.36 0.44	0.70 1.00	0.70 1.00	— —	— —		Hardened and tempered	22 254	⅞ 10	770/930 (50/60) 620/770 (40/50)	585 (38) 430 (28)	15 17	40 (30) 27 (20) —	Forgings
½Cr Steel	DIN 38Cr2	0.34 0.41	0.50 0.80	— —	0.40 0.60	—		Hardened and tempered	16 102	⅝ 4	790/930 (51/60) 590/740 (38/48)	540 (35) 340 (22)	14 17	— —	

Table 22.44 FORGED OR ROLLED STEELS—ROOM TEMPERATURE LONGITUDINAL MECHANICAL PROPERTIES—*continued*

Material	*British or other standard*	*Composition %* C	Mn	Ni	Cr	*Other elements*	*Condition*	*Limiting ruling section* mm	in	*Properties (minima unless otherwise stated)* *UTS* MPa (tonf in^{-2})	*Yield stress* MPa (tonf in^{-2})	*Elong.* (gl 5.65$\sqrt{S_0}$) %	*Izod* J (ft lbf)	*Remarks*
$\frac{1}{2}$Cr Steel	DIN 46Cr2	0.42 0.50	0.50 0.80	— —	0.40 0.60	— —	Hardened and tempered	16 102	$\frac{5}{8}$ 4	880/1 080 (57/70) 680/835 (44/54)	630 (41) 450 (29)	12 15	— —	
$\frac{3}{4}$Cr Steel	BS 526M60 (En11)	0.55 0.65	0.50 0.80	— —	0.50 0.80	— —	Hardened and tempered	64 102	$2\frac{1}{2}$ 4	1 000/1 160 (65/75) 850/1 000 (55/65)	740 (48) 620 (40)	8 11	— —	
1Cr Steel	DIN 34Cr4	0.30 0.37	0.60 0.90	— —	0.90 1.20	— —	Hardened and tempered	16 102	$\frac{5}{8}$ 4	880/1 080 (57/70) 680/835 (44/54)	680 (44) 465 (30)	12 15	— —	
1Cr Steel	BS 530M40 (En18)	0.36 0.44	0.60 0.90	— —	0.90 1.20	— —	Hardened and tempered	29 102	$1\frac{1}{8}$ 4	850/1 000 (55/65) 690/850 (45/55)	680 (44) 525 (34)	13 17	54 (40) 54 (40)	
$1\frac{1}{2}$MnMo Steel	BS 605M30 (En16D)	0.26 0.34	1.30 1.70	— —	— —	Mo 0.22–0.32	Hardened and tempered	152 19	6 $\frac{3}{4}$	700/850 (45/55) 1 000/1 160 (65/75)	525 (34) 850 (55)	17 12	54 (40) 47 (35)	
$1\frac{1}{2}$MnMo Steel	BS 605M36 (En16)	0.32 0.40	1.30 1.70	— —	— —	Mo 0.22–0.32	Hardened and tempered	254 19	10 $\frac{3}{4}$	700/850 (45/55) 1 000/1 160 (65/75)	445 (32) 850 (55)	15 12	34 (25) 47 (35)	
$1\frac{1}{2}$MnMo Steel	BS 606M36 (En16M)	0.32 0.40	1.30 1.70	— —	— —	Mo 0.22–0.32 S 0.15–0.25	Hardened and tempered	102 29	4 $1\frac{1}{8}$	690/850 (45/55) 850/1 000 (55/65)	525 (34) 680 (44)	15 11	54 (40) 41 (30)	Free cutting steel
$1\frac{1}{2}$MnMo Steel	BS 608M38 (En17)	0.34 0.42	1.30 1.70	— —	— —	Mo 0.40 0.55	Hardened and tempered	254 29	10 $1\frac{1}{8}$	690/850 (45/55) 1 000/1 160(65/75)	445 (32) 850 (55)	15 12	41 (30) 47 (35)	
$1\frac{1}{4}$NiCr Steel	BS 640M40 (En111)	0.36 0.44	0.60 0.90	1.10 1.50	0.50 0.80	— —	Hardened and tempered	152 29	6 $1\frac{1}{8}$	700/850 (45/55) 930/1 080 (60/70)	525 (34) 755 (49)	17 12	54 (40) 47 (35)	

3NiCr Steel	BS 653M31 (En23)	0.27 0.35	0.45 0.70	2.75 3.25	0.90 1.20	— —		Hardened and tempered	152 64	6 $2\frac{1}{2}$	770/930 (50/60) 930/1 080 (60/70)	585 (38) 755 (49)	15 12	54 (40) 47 (35)	
$\frac{3}{4}$CrMo Steel	UK Proprietary	0.53 (Typical)	0.70	—	0.70	Mo	0.40	Hardened and tempered	508	20	805 (52) (Typical values)	540 (35) (Typical values)	14	—	Crankshaft forgings
1CrMo Steel	DIN 25CrMo4	0.22 0.29	0.50 0.80	— —	0.90 1.20	Mo	0.15– 0.30	Hardened and tempered	40 150	$1\frac{1}{2}$ 6	790/930 (51/60) 635/790 (41/51)	585 (38) 415 (27)	14 16	— —	
1CrMo Steel	DIN 34CrMo4	0.30 0.37	0.50 0.80	— —	0.90 1.20	Mo	0.15– 0.30	Hardened and tempered	100 250	4 10	790/930 (51/60) 680/835 (44/54)	550 (36) 465 (30)	14 15	— —	
1CrMo Steel	BS 708M40 (En19A)	0.36 0.44	0.70 1.00	— —	0.90 1.20	Mo	0.15– 0.25	Hardened and tempered	152 29	6 $1\frac{1}{8}$	700/850 (45/55) 930/1 080 (60/70)	525 (34) 755 (49)	17 12	54 (40) 47 (35)	
1CrMo Steel	BS 709M40 (En19)	0.36 0.44	0.70 1.00	0.40 max	0.90 1.50	Mo	0.25– 0.40	Hardened and tempered	254 29	10 $1\frac{1}{8}$	690/850 (45/55) 1000/1160(65/75)	495 (32) 850(55)	15 12	34 (25) 47(35)	
1CrMo Steel	BS 711M40	0.36 0.44	0.60 1.00	0.40 max	0.90 1.50	Mo	0.25– 0.40 max	Hardened and tempered	500 250	20 10	700/850 (45/55) 800/950 (52/62)	500 (32) 600 (39)	12 14	30 (22) 25 (18)	Forgings
3CrMo Steel	BS 722M24 (En40B)	0.20 0.28	0.45 0.70	— —	3.00 3.50	Mo	0.45– 0.65	Hardened and tempered	152 254	6 10	930/1 080 (60/70) 850/1 000 (55/65)	755 (49) 650 (42)	12 13	47 (35) 41 (30)	Nitriding steel
3CrMo Steel	BS 722M29 (En29B)	0.25 0.33	0.45 0.70	0.40 max	3.00 3.50	Mo	0.45 0.65	Hardened and tempered	250 1 000	10 39	900/1 050 (58/68) 650/800 (42/52)	720 (47) 470 (30)	13 13	41 (30) 37 (27)	Nitriding steel forgings
3CrMo Steel	DIN 32CrMo12	0.28 0.35	0.40 0.70	0.30 max	2.80 3.30	Mo	0.30 0.50	Hardened and tempered	150 250	6 10	930/1 130 (60/73) 880/1 080 (57/70)	740 (48) 680 (44)	11 12	— —	Nitriding steel
MnNiMo Steel	BS 785M19 (En13)	0.15 0.23	1.40 1.80	0.40 0.70	— —	Mo	0.15 0.35	Hardened and tempered	152 254	6 10	620/770 (40/50) 620/770 (40/50)	465 (30) 450 (29)	18 16	Charpy V 54 (40) 41 (30)	Forgings

Table 22.44 FORGED OR ROLLED STEELS—ROOM TEMPERATURE LONGITUDINAL MECHANICAL PROPERTIES—*continued*

Material	*British or other standard*	*Composition %* C	Mn	Ni	Cr	*Other elements*	*Condition*	*Limiting ruling section* mm	in	*Properties (minima unless otherwise stated)* *UTS* MPa (tonf in^{-2})	*Yield stress* MPa (tonf in^{-2})	*Elong.* (gl 5.65 $\sqrt{S_0}$) %	*Izod* J (ft lbf)	*Remarks*
1½NiCrMo	BS 816M40	0.36	0.45	1.30	1.00	Mo 0.10	Hardened	254	10	770/930 (50/60)	555 (36)	15	34 (25)	
Steel	(En110)	0.44	0.70	1.70	1.40	0.20	and	29	1⅛	1 000/1 160 (65/75)	850 (55)	12	47 (35)	
							tempered							
1½NiCrMo	BS 817M40	0.36	0.45	1.30	1.00	Mo 0.20	Hardened	254	10	850/1 000 (55/65)	650 (42)	13	41 (30)	
Steel	(En24)	0.44	0.70	1.70	1.40	0.35	and	29	1⅛	1 540 (100)	1 240 (80)	5	11 (8)	
							tempered							
1½NiCrMo	BS 818M40	0.36	0.45	1.30	1.00	Mo 0.20–	Hardened	1 000	39	800/950 (52/62)	610 (40)	14	33 (24)	Forgings
Steel		0.44	0.85	1.80	1.50	0.40	and	250	10	950/1 100 (62/71)	780 (50)	12	41 (30)	
							tempered							
1NiCrMo	DIN 36CrNi	0.32	0.50	0.90	0.90	Mo 0.15	Hardened	100	4	880/1 035 (57/67)	680 (44)	12	—	
Steel	Mo4	0.40	0.80	1.20	1.20	0.30	and	250	10	740/880 (48/57)	540 (35)	14	—	
							tempered							
2NiCrMo	BS 823M30	0.26	0.35	1.80	1.80	Mo 0.30	Hardened	64	2½	1 540 (100)	1 235 (80)	7	14 (10)	
Steel		0.34	0.60	2.20	2.20	0.50	and	254	10	850/1 000 (55/65)	650 (42)	13	41 (30)	
							tempered							
2½NiCrMo	BS 826M31	0.27	0.45	2.30	0.50	Mo 0.45	Hardened	64	2½	1 540 (100)	1 235 (80)	5	11 (8)	
Steel	(En25)	0.35	0.70	2.80	0.80	0.65	and	254	10	850/1 000 (55/65)	650 (42)	13	41 (30)	
							tempered	1 000	39	800/950 (52/62)	610 (40)	14	33 (24)	Forgings
2½NiCrMo	BS 826M40	0.36	0.45	2.30	0.50	Mo 0.45	Hardened	102	4	1 540 (100)	1 235 (80)	7	14 (10)	
Steel	(En26)	0.44	0.70	2.80	0.80	0.65	and	254	10	900/1 050 (58/68)	720 (47)	13	47 (34)	Forgings
							tempered	1 000	39	850/1 000 (55/65)	660 (43)	13	33 (24)	Forgings
3NiCrMo	BS 830M31	0.27	0.45	2.75	0.90	Mo 0.25	Hardened							
Steel	(En27)	0.35	0.70	3.25	1.20	0.35	and	254	10	850/1 000 (55/65)	650 (42)	13	41 (30)	
							tempered	64	2½	1 080/1 240 (70/80)	940 (61)	11	41 (30)	
4NiCrMo	BS 835M30	0.26	0.45	3.90	1.10	Mo 0.20	Hardened	152	6	1 540 (100)	1 235 (80)	7	20 (15)	
Steel	(En30B)	0.34	0.70	4.30	1.40	0.35	and							
							tempered							
													Charpy V	
4NiCrMo	BS S146	0.38	0.40	4.00	1.80	Mo 0.50	Hardened	25	1	1 820 (118)	1 450 (94)	11	26 (19)	Ultra high
Steel	French		(Typical)				and	102	4	1 760 (114)	1 420 (92)	9	20 (15)	strength
										(Typical values)	(Typical values)			aircraft steel

$3\frac{1}{4}$CrMoV Steel	BS 897M39 (En40C) DIN 39Cr MoV139	0.35 0.43	0.45 0.70	— —	3.00 3.50	Mo 0.80 1.10 V 0.15 0.25	Hardened and tempered	64 250 1 000	$2\frac{1}{2}$ 10 39	1 310 (85) 1 100/1 250 (71/81) 850/1 000 (55/65)	1 160 (75) 940 (61) 660 (43)	8 11 13	20 (15) 24 (18) 43 (31)	Nitriding steel Forgings Forgings
5CrMoV	BS BH11 ASTM A579 Grade 41	0.38	0.40 (Typical)	—	5.13	Si 0.92 Mo 1.36 V 0.54	Hardened and tempered	29 108	$1\frac{1}{8}$ $4\frac{1}{4}$	2 040 (132) 2 040 (132) (Typical values)	1 605 (104) 1 605 (104) (Typical values)	9 9	— —	Ultra high strength aircraft steel
1CrMoNiV Steel	ASTM A579 Grade 23 (D6AC)	0.45	0.75 (Typical)	0.55	1.05	Mo 1.00 V 0.10	Hardened and tempered	25	1	1 885 (122) (Typical values)	1 680 (109) (Typical values)	10	Charpy V 20 (15)	Ultra high strength aircraft steel
2NiCrMoV Steel	ASTM A579 Grade 32 (300M)	0.44	0.70 (Typical)	1.85	0.80	Si 1.60 Mo 0.35 V 0.06	Hardened and tempered	25 76	1 3	1 990 (129) 1 930 (125) (Typical values)	1 685 (109) 1 620 (105) (Typical values)	9 9	Charpy V 27 (20) 26 (19)	Ultra high strength aircraft steel
NiCrMoV Steel	UK Proprietary (NCMV)	0.44	0.45 (Typical)	1.80	1.50	Mo 0.90 V 0.25	Hardened and tempered	29	$1\frac{1}{8}$	2 070 (134) (Typical values)	1 700 (110) (Typical values)	10	Charpy V 27 (20)	Ultra high strength aircraft steel
$1\frac{1}{2}$CrAlMo Steel	BS 905 M31 (En41A)	0.27 0.35	0.40 0.65	— —	1.40 1.80	Mo 0.15 0.25 Al 0.90 1.30	Hardened and tempered	64 102	$2\frac{1}{2}$ 4	770/930 (50/60) 690/850 (45/55)	585 (38) 525 (34)	15 17	54 (40) 54 (40)	Nitriding steel §
$1\frac{1}{2}$Cr AlMo Steel	BS 905 M39 (En41B)	0.35 0.43	0.40 0.65	— —	1.40 1.80	Mo 0.15 0.25 Al 0.90 1.30	Hardened and tempered	64 152	$2\frac{1}{2}$ 6	850/1 000 (55/65) 690/850 (45/55)	680 (44) 525 (34)	13 17	47 (35) 54 (40)	Nitriding steel §
$1\frac{1}{2}$CrAlMo Steel	DIN 41CrAlMo7	0.38 0.45	0.50 0.80	— —	1.50 1.80	Mo 0.25 0.40 Al 0.80 1.20	Hardened and tempered	102 152	4 6	930/1 125 (60/73) 835/1 035 (54/67)	740 (48) 635 (41)	12 14	— —	Nitriding steel §
$1\frac{1}{2}$CrNiAlMo Steel	DIN 34CrAlNi Mo7	0.30 0.37	0.40 0.70	0.85 1.15	1.50 1.80	Mo 0.15 0.25 Al 0.80 1.20	Hardened and tempered	254	10	790/990 (51/64)	585 (38)	13	—	Nitriding steel §
$1\frac{1}{2}$MnNiCr Mo Steel	BS 945M38 (En100)	0.34 0.42	1.20 1.60	0.60 0.90	0.40 0.60	Mo 0.15 0.25	Hardened and tempered	29 254	$1\frac{1}{8}$ 10	1 000/1 160 (65/75) 690/850 (45/55)	850 (55) 495 (32)	12 15	47 (35) 34 (25)	

§ Surface hardness after Nitriding 950HV

Table 22.44 FORGED OR ROLLED STEELS—ROOM TEMPERATURE LONGITUDINAL MECHANICAL PROPERTIES—*continued*

Material	*British or other standard*	Composition % C	Mn	Ni	Cr	*Other elements*	*Condition*	Limiting ruling section mm	in	Properties (minima unless otherwise stated) *UTS* MPa ($tonf\,in^{-2}$)	*Yield stress* MPa ($tonf\,in^{-2}$)	*Elong.* (gl $5.65\sqrt{S_0}$) %	*Izod* J (ft lbf)	*Remarks*
$3\frac{1}{4}$NiCrMoV Steel	BS 976 M33	0.28 0.38	0.20 0.60	2.90 3.60	0.90 1.70	Mo 0.45 0.65 V 0.08 0.15	Hardened and tempered	250 1 000	10 39	1 100/1 250 (71/81) 850/1 000 (55/65)	980 (64) 710 (46)	11 13	37 (27) 46 (33)	Forgings Forgings
4. *Carburizing steels* (BS 970: Part 3:1971; DIN 17210:1969)														
$\frac{1}{2}$Cr Steel	BS 523 M15	0.12 0.18	0.30 0.60	— —	0.30 0.60	— —	Oil quenched	19	$\frac{3}{4}$	620 (40)	Not specified	13	34 (25)	
$\frac{3}{4}$Cr Steel	BS 527 M20	0.17 0.23	0.60 0.90	— —	0.60 0.90	— —	Oil quenched	19	$\frac{3}{4}$	770 (50)	Not specified	12	20 (15)	
$1\frac{1}{4}$MnCr Steel	DIN 16MnCr5	0.14 0.19	1.00 1.30	— —	0.80 1.10		Oil quenched	11 30	$\frac{7}{16}$ $1\frac{3}{16}$	880/1 175 (57/76) 790/1 080 (51/70)	630 (41) 590 (38)	9 10	— —	
$1\frac{1}{4}$MnCr Steel	DIN 20MnCr5	0.17 0.22	1.10 1.40	— —	1.00 1.30	— —	Oil quenched	11 30	$\frac{7}{16}$ $1\frac{3}{16}$	1 080/1 375 (70/89) 990/1 280 (64/83)	740 (48) 680 (44)	7 8	— —	
$\frac{3}{4}$NiCr Steel	BS 635 M15 (En351)	0.12 0.18	0.60 0.90	0.70 1.10	0.40 0.80	— —	Oil quenched	19	$\frac{3}{4}$	770 (50)	Not specified	12	27 (20)	
1NiCr Steel	BS 637 M17 (En352)	0.14 0.20	0.60 0.90	0.85 1.25	0.60 1.00	— —	Oil quenched	19	$\frac{3}{4}$	930 (60)	Not specified	10	20 (15)	
$1\frac{1}{2}$NiCr Steel	DIN 15CrNi6	0.12 0.17	0.40 0.60	1.40 1.70	1.40 1.70	— —	Oil quenched	11 30	$\frac{7}{16}$ $1\frac{3}{16}$	960/1 280 (62/83) 880/1 175 (57/76)	680 (44) 635 (41)	8 9	— —	
$3\frac{1}{4}$NiCr Steel	BS 655 M13 (En36A)	0.10 0.16	0.35 0.60	3.00 3.75	0.70 1.00	— —	Oil quenched	19	$\frac{3}{4}$	1 000 (65)	Not specified	9	41 (30)	
4NiCr Steel	BS 659 M15 (En39A)	0.12 0.18	0.25 0.50	3.90 4.30	1.00 1.40	— —	Oil quenched	Test piece size		1 310 (85)	Not specified	8	34 (25)	
$\frac{1}{2}$CrMo Steel	DIN 20MoCr4	0.17 0.22	0.60 0.90	— —	0.30 0.50	Mo 0.40 0.50	Oil quenched	11 30	$\frac{7}{16}$ $1\frac{3}{16}$	880/1 175 (57/76) 790/1 080 (51/70)	635 (41) 590 (38)	9 10	— —	

$\frac{1}{2}$CrMo Steel	DIN 25MoCr4	0.23 0.29	0.60 0.90	— —	0.40 0.60	Mo 0.40 0.50	Oil quenched	11 30	$\frac{7}{16}$ $1\frac{3}{16}$	1 080/1 375 (70/89) 990/1 280 (64/83)	740 (48) 680 (44)	7 8	— —
$1\frac{3}{4}$NiMo Steel	BS 665 M17 (En34)	0.14 0.20	0.35 0.75	1.50 2.00	— —	Mo 0.20 0.30	Oil quenched	19	$\frac{3}{4}$	770 (50)	Not specified	12	41 (30)
$1\frac{3}{4}$NiMo Steel	BS 665 M20	0.17 0.23	0.35 0.75	1.50 2.00	— —	Mo 0.20 0.30	Oil quenched	19	$\frac{3}{4}$	850 (55)	Not specified	11	27 (20)
$1\frac{3}{4}$NiMo Steel	BS 665 M23 (En35)	0.20 0.26	0.35 0.75	1.50 2.00	— —	Mo 0.20 0.30	Oil quenched	19	$\frac{3}{4}$	930 (60)	Not specified	10	16 (12)
$\frac{1}{2}$NiCrMo Steel	BS 805 M17 (En361)	0.14 0.20	0.60 0.95	0.35 0.75	0.35 0.65	Mo 0.15 0.25	Oil quenched	19	$\frac{3}{4}$	770 (50)	Not specified	12	27 (20)
$\frac{1}{2}$NiCrMo Steel	BS 805 M20 (En362)	0.17 0.23	0.60 0.95	0.35 0.75	0.35 0.65	Mo 0.15 0.25	Oil quenched	19	$\frac{3}{4}$	850 (55)	Not specified	11	20 (15)
$\frac{1}{2}$NiCrMo Steel	BS 805 M22	0.19 0.25	0.60 0.95	0.35 0.75	0.35 0.65	Mo 0.15 0.25	Oil quenched	19	$\frac{3}{4}$	930 (60)	Not specified	10	14 (10)
$\frac{1}{2}$NiCrMo Steel	BS 805 M25 (En363)	0.22 0.28	0.60 0.95	0.35 0.75	0.35 0.65	Mo 0.15 0.25	Oil quenched	19	$\frac{3}{4}$	1 000 (65)	Not specified	9	—
$1\frac{1}{2}$NiCrMo Steel	BS 815 M17 (En353)	0.14 0.20	0.60 0.90	1.20 1.70	0.80 1.20	Mo 0.10 0.20	Oil quenched	19	$\frac{3}{4}$	1 080 (70)	Not specified	8	27 (20)
$1\frac{1}{2}$NiCrMo Steel	DIN 17CrNiMo6	0.14 0.19	0.40 0.60	1.40 1.70	1.50 1.80	Mo 0.25 0.35	Oil quenched	11 30	$\frac{7}{16}$ $1\frac{3}{16}$	1 175/1 420 (76/92) 1 080/1 330 (70/86)	835 (54) 790 (51)	7 8	— —
$1\frac{3}{4}$NiCrMo Steel	BS 820 M17 (En354)	0.14 0.20	0.60 0.90	1.50 2.00	0.80 1.20	Mo 0.10 0.20	Oil quenched	19	$\frac{3}{4}$	1 160 (75)	Not specified	8	27 (20)
2NiCrMo Steel	BS 822 M17 (En355)	0.14 0.20	0.40 0.70	1.75 2.25	1.30 1.70	Mo 0.15 0.25	Oil quenched	Test piece size		1 310 (85)	Not specified	8	27 (20)
$3\frac{1}{2}$NiCrMo Steel	BS 832 M13 (En36C)	0.10 0.16	0.35 0.60	3.00 3.75	0.70 1.00	Mo 0.10 0.25	Oil quenched	19	$\frac{3}{4}$	1 080 (70)	Not specified	8	34 (25)
4NiCrMo Steel	BS 835 M15 (En39B)	0.12 0.18	0.25 0.50	3.90 4.30	1.00 1.40	Mo 0.15 0.30	Oil quenched	Test piece size		1 310 (85)	Not specified	8	34 (25)

Table 22.44 FORGED OR ROLLED STEELS—ROOM TEMPERATURE LONGITUDINAL MECHANICAL PROPERTIES—*continued*

Material	*British or other standard*	*Composition limits %* C	Mn	Ni	Cr	*Other elements*	*Condition*	*Limiting ruling section* mm	in	*Properties (minima unless otherwise stated)* *UTS* MPa (tonf in^{-2})	*Yield stress* MPa (tonf in^{-2})	*Elong.* (gl 5.65$\sqrt{S_0}$) %	*Izod* J (ft lbf)	*Remarks*
5. *High alloy steels*														
Stainless, heat resisting and valve steels														
BS 970: Part 4:1970: 1449:Part 2; 1975, and Aircraft steels DIN 17224:1968; 17440:1967, Stahl-Eisen Werkstoffblatt 390:61, 400:60, 470:60, 670:69, AICMA (Association International des Constructeurs de Material Aerospacial) Standards														
Ferritic stainless and heat resisting steels														
12Cr Steel	BS 409 S17	0.09 max	1.00 max	0.70 max	10.5 12.5	Ti 5C 0.70	Softened	—	—	420 (27)	245 (16)	20	—	P:Sh:St*
12Cr Steel	BS S61	0.12 max	1.00 max	1.00 max	11.5 13.5	—	Softened	—	—	540/700 (35/45)	355 (23)	20	34 (25)	—
13Cr Steel	BS 403 S17	0.08 max	1.00 max	0.50 max	12.0 14.0	—	Softened	—	—	420 (27)	280 (18)	20	—	B:F:P:Sh:St*
13CrSi Steel	DIN X10CrSi13	0.12 max	1.00 max	— —	12.0 14.0	Si 1.9 2.4	Softened	—	—	540/680 (35/44)	340 (22)	15	—	Scaling resistant up to 950 °C in air
13CrAl Steel	BS 405 S17	0.08 max	1.00 max	0.50 max	12.0 14.0	Al 0.10 0.30	Softened	—	—	420 (27)	245 (16)	20	—	P:Sh:St*
13CrAl Steel	DIN X10CrA113	0.12 max	1.00 max	— —	12.0 14.0	Al 0.7 1.2	Softened	—	—	495/635 (32/41)	295 (19)	15	—	Scaling resistant up to 950 °C in air
17Cr Steel	BS 430 S15 (En60)	0.10 max	1.00 max	0.50 max	16.0 18.0	— —	Softened	64	$2\frac{1}{2}$	430 (28)	280 (18)	20	—	B:F:P:Sh:St*
17CrTi Steel	DIN X8CrTi17	0.10 max	1.00 max	— —	16.0 18.0	Ti 7× C min	Softened	16	$\frac{5}{8}$	450/585 (29/38)	265 (17)	20	—	
17CrNb Steel	DIN X8CrNb17	0.10 max	1.00 max	— —	16.0 18.0	Nb 12× C min	Softened	16	$\frac{5}{8}$	450/585 (29/38)	265 (17)	20	—	
17CrMo Steel	BS 434 S19	0.10 max	1.00 max	0.50 max	16.0 18.0	Mo 0.90 1.30	Softened	—	—	430 (28)	245 (16)	20	—	Sh:St*

17CrMo Steel	DIN X6CrMo17	0.07 max	1.00 max	— —	16.0 17.5	Mo 0.90 1.20	Softened	16	$\frac{5}{8}$	450/635 (29/41)	295 (19)	25	—	
17CrMoTi Steel	DIN X8CrMoTi17	0.10 max	1.00 max	— —	16.0 18.0	Mo 1.50 2.00 7 × Ti C min	Softened	—	—	495/635 (32/41)	295 (19)	20	—	Used in the chemical and textile industries
18CrSi Steel	DIN X10CrSi18	0.12 max	1.00 max	— —	17.0 19.0	Si 1.90 2.40	Softened	—	—	540/680 (35/44)	340 (22)	15	—	Scaling resistant up to 1 050 °C in air
18CrAl Steel	DIN X10CrA118	0.12 max	1.00 max	— —	17.0 19.0	Al 0.70 1.20	Softened	—	—	495/635 (32/41)	295 (19)	12	—	Scaling resistant up to 1 050 °C in air
20Cr Steel	BS 442 S19	0.10 max	1.00 max	0.50 max	18.0 22.0	—	Softened	—	—	430 (28)	245 (16)	20	—	Sh:St*
24CrAl Steel	DIN X10CrA124	0.12 max	1.00 max		23.0 25.0	Al 1.20 1.70	Softened	—	—	495/635 (32/41)	295 (19)	10	—	Scaling resistant up to 1 200 °C
29CrSi Steel	DIN X10CrSi29	0.12 max	1.00 max	— —	28.0 31.0	Si 1.00 2.00	Softened	— —	— —	540/680 (35/44)	385 (25)	12	—	Scaling resistant up to 1 150 °C in air
16CrNiMo Nb Steel	UK (FV702)	0.03	0.60 (Typical)	2.5	16.0	Si 0.50 Mo 1.00 Nb 0.50	Softened	19	$\frac{3}{4}$	705 (45.5) (Typical values)	500 (32.5) (Typical values)	22	—	Resistant to stress corrosion
Martensitic stainless and heat resisting steels														
13Cr Steel	BS 410 S21 (En56A)	0.09 0.15	1.00 max	1.00 max	11.5 13.5	— —	Hardened and	152	6	540/690 (35/45)	370 (24)	20	34 (25)	B:F:P:Sh:St*
							tempered	64	$2\frac{1}{2}$	690/850 (45/55)	525 (34)	15	34 (25)	
13CrS Steel	BS 416S21 (En56AM)	0.09 0.15	1.50 max	1.00 max	11.5 13.5	(Mo 0.60 max) S 0.15	Hardened and	152	6	540/690 (35/45)	370 (24)	15	34 (25)	B:F* Free cutting steel
						0.30	tempered	64	$2\frac{1}{2}$	690/850 (45/55)	525 (34)	11	27 (20)	
13CrSe steel	BS 416S41 (En 56AM)	0.09 0.15	1.50 max	1.00 max	11.5 13.5	Mo 0.60 max Se 0.15 0.30	Hardened and tempered	152	6	540/690 (35/45)	370 (24)	15	34 (25)	B:F* Free cutting steel

* B = Bars. F = Forgings. P = Plates. Sh = Sheets. St = Strips.

Table 22.44 FORGED OR ROLLED STEELS—ROOM TEMPERATURE LONGITUDINAL MECHANICAL PROPERTIES—*continued*

Material	*British or other standard*	Composition limits % C	Mn	Ni	Cr	*Other elements*	*Condition*	*Limiting ruling section* mm	in	Properties (minima unless otherwise stated) *UTS* MPa (tonf in^{-2})	*Yield stress* MPa (tonf in^{-2})	*Elong.* (gl $5.65\sqrt{S_0}$) %	*Izod* J (ft lbf)	*Remarks*
13Cr Steel	BS 420S29 (En56B)	0.14 0.20	1.00 max	1.00 max	11.5 13.5	— —	Hardened and tempered	152 29	6 1⅛	690/850 (45/55) 770/930 (50/60)	525 (34) 585 (38)	15 13	27 (20) 27 (20)	B:F*
13CrS Steel	BS 416 S29 (En56BM)	0.14 0.20	1.50 max	1.00 max	11.5 13.5	(Mo 0.60 max) S 0.15 0.30	Hardened and tempered	152 29	6 1⅛	700/850 (45/55) 770/930 (50/60)	520 (34) 585 (38)	11 10	27 (20) 14 (10)	B:F* Free cutting steel
13Cr Steel	BS 3S62	0.18 0.25	1.00 max	1.00 max	12.0 14.0	— —	Hardened and tempered	64 152	2½ 6	700/850 (45/55) 700/850 (45/55)	525 (34) 525 (34)	15 15	34 (25) 27 (20)	— —
13Cr Steel	BS2S124	0.15 0.25	1.50 max	1.00 max	12.0 14.0	Mo 0.60 max Zr 0.60 max Mo 1.00 max +Zr S 0.15 0.40	Hardened and tempered	152	6	690/850 (45/55)	450 (29)	11	27 (20)	Free machining steel
13Cr Steel	BS 420S37 (En56C)	0.20 0.28	1.00 max	1.00 max	12.0 14.0		Hardened and tempered	152	6	770/930 (50/60)	585 (38)	13	13.5 (10)	B:F*
13CrS Steel	BS 416S37 (En56CM)	0.20 0.28	1.50 max	1.00 max	12.0 14.0	(Mo 0.60 max) S 0.15 0.30	Hardened and tempered	152	6	770/930 (50/60)	585 (38)	10	13.5 (10)	B:F* Free cutting steel
13Cr Steel	BS 420S45 (En 56D)	0.28 0.36	1.00 max	1.00 max	12.0 14.0	— —	Hardened and tempered	152	6	770/930 (50/60)	585 (38)	13	13.5 (10)	B:F:Sh:St*
17CrNi Steel	BS 431S29 (En57) 4S80	0.12 0.20	1.00 max	2.0 3.0	15.0 18.0	— —	Hardened and tempered	152 64	6 2½	850/1 005 (55/65) 880/1 080 (57/70)	680 (44) 690 (45)	11 12	20 (15) 34 (25)	B:F*

17CrNiS Steel	BS 441S29	0.12 0.20	1.50 max	2.0 3.0	15.0 18.0	(Mo 0.60 max) S 0.15 0.30	Hardened and tempered	64	$2\frac{1}{2}$	850/1 005 (55/65)	680 (44)	8	20 (15)	B:F* Free cutting steel
17CrNiS Steel	BS S137	0.12 0.20	1.50 max	2.0 3.0	15.0 18.0	(Mo 0.60 max) S 0.15 0.30	Hardened and tempered	70	$2\frac{3}{4}$	880/1 080 (57/70)	690 (45)	11	—	Primarily intended for nuts: free machining
17CrNiSe Steel	BS 441S49	0.12 0.20	1.50 max	2.00 3.00	15.0 18.0	(Mo 0.60 max) Se 0.15 0.30	Hardened and tempered	64	$2\frac{1}{2}$	850/1 000 (55/65)	680 (44)	8	20 (15)	B:F*
12CrMo Steel	DIN X19CrMo12.1	0.15 0.23	0.30 0.80	0.80 max	11.0 12.5	Mo 0.80 1.20	Hardened and tempered	152	6	680/835 (44/54)	495 (32)	16	—	B:F* Creep resistant
12CrMoV Steel	DIN X20CrMoV12.1	0.17 0.23	0.30 0.80	0.30 0.80	11.0 12.5	Mo 0.80 1.20 V 0.25 0.35	Hardened and tempered	19	$\frac{3}{4}$	680/835 (44/54)	495 (32)	16	—	P:Sh:St:T* Creep resistant Resistant to high pressure H_2
12CrMoV Steel	DIN X22CrMoV12.1	0.20 0.26	0.30 0.80	0.30 0.80	11.0 12.5	Mo 0.80 1.20 V 0.25 0.35	Hardened and tempered	152	6	790/925 (51/60)	590 (38)	14	—	B:F* Creep resistant
12CrMoVNb Steel	AICME Fe-PM36 (UK FV448 Jethete 160)	0.11 0.19	0.20 1.25	0.50 1.20	10.0 12.0	Mo 0.40 1.00 V 0.10 0.70 Nb 0.10 0.60 +N	Hardened and tempered	530 dia 89 thick disc	21 dia $3\frac{1}{2}$ thick disc	975 (63) (Typical tangential values)	835 (54)	16	—	Gas turbine components
12CrNiMoV Steel	AICME Fe-PM37 (UK FV566 (+Nb) Jethete M152 Jethete M154)	0.08 0.15	0.50 0.90	2.0 3.0	11.0 12.5	Mo 1.50 2.00 V 0.25 0.40 +N	Hardened and tempered	127 sq	5 sq	1 065 (69) (Typical longitudinal values)	850 (55)	18	77 (57)	Gas turbine components

* B = Bars. F = Forgings. P = Plates. Sh = Sheets. St = Strips. T = Tubes.

Table 22.44 FORGED OR ROLLED STEELS—ROOM TEMPERATURE LONGITUDINAL MECHANICAL PROPERTIES—*continued*

Material	*British or other standard*	*Composition limits* Composition % C	Mn	Ni	Cr	*Other elements*	*Condition*	*Limiting ruling section* mm	in	*Properties (minima unless otherwise stated)* *UTS* MPa ($tonf\ in^{-2}$)	*Yield stress* MPa ($tonf\ in^{-2}$)	*Elong.* (gl $5.65\sqrt{S_0}$) %	*Izod* J (ft lbf)	*Remarks*
12CrCoMoV Nb Steel	AICME Fe-PM38 (UK FV535)	0.05 0.12	0.20 1.35	0.20 1.20	9.8 11.5	Co 5.00 7.50 Mo 0.50 1.10 V 0.10 0.60 Nb 0.20 0.60 +B	Hardened and tempered	560 dia 102 thick	22 dia 4 thick	1 065 (69) (Typical tangential values)	925 (60)	17	—	Gas turbine components
Austenitic stainless and heat resisting steels														
12/12CrNi Steel	DIN X8CrNi 12 12 UK (FV DDQ)	0.10 max	2.00 max	11.50 13.50	12.00 14.00	— —	Softened	—	—	495/635 (32/41)	200 (13)	50	—	Spoons and forks
14/10CrNi CuMoTi Steel	UK (FV467)	0.20	0.90	9.50 (Typical)	14.00	Mo 2.00 Cu 2.50 Ti 0.80	Precipitation hardened	—	—	680 (44) (Typical values)	300 (19.5) (Typical values)	52	135 (100)	B:F* Creep resistant
17/7CrNi Steel	BS 301S21	0.15 max	0.50 2.00	6.00 8.00	16.00 18.00	— —	Softened	—	—	540 (35)	215 (14)	35	—	Sh:St*. Also used as cold rolled strip or cold drawn wire for springs
18/8CrNi Steel	BS S205	0.15 max	0.50 2.00	7.50 9.00	17.0 19.0	— —	Low temperature Heat treatment			1 350/1 550 (87/100) (Typical values)	—	—	—	Cold drawn and heat treated wire and springs
18/9CrNi Steel	BS 304S15 (En58E)	0.06 max	0.50 2.00	8.00 11.00	17.50 19.00	— —	Softened Cold drawn	— 19 44	— $\frac{3}{4}$ $1\frac{3}{4}$	460 (30) 860 (56) 650 (42)	170 (11) 695 (45) 310 (20)	40 12 28	— — —	B:F:P:Sh:St*
18/9CrNiN Steel	UK (Hi-proof 304)	0.06 max	2.00 max	8.00 11.00	17.50 19.00	N 0.15 0.25	Softened	102	4	585 (38)	295 (19)	35	—	B:P*

18/9CrNi Steel	BS 302S17	0.08 max	0.50 2.00	8.00 11.00	17.00 19.00	— —		Softened	—	—	510(33)	210(13.5)	40	—	Sh:St*
18/9CrNiTi Steel	BS S129	0.08 max	0.50 2.00	8.00 11.00	17.00 19.00	Mo Ti	0.70 max 5 × C 0.8	Softened	152	6	540(35)	210(13.5)	35	69(50)	
18/9CrNiNb Steel	BS S130	0.08 max	0.50 2.00	8.00 11.00	17.00 19.00	Mo Nb	0.70 max 10 × C 1.1	Softened	152	6	540(35)	210(13.5)	35	69(50)	
18/9CrNiTi Steel	BS 321S12 (En58B, 58C)	0.08 max	0.50 2.00	9.00 12.00	17.00 19.00	Ti	5 × C/0.70	Softened Cold drawn	— 19 44	— $\frac{3}{4}$ $1\frac{3}{4}$	490(32) 860(56) 650(42)	195(12.5) 695(45) 310(20)	40 12 28	— — —	B:F:P:Sh:St*
18/9CrNiNb Steel	BS 347S17 (En58F, 58G)	0.08 max	0.50 2.00	9.00 12.00	17.00 19.00	Nb	10 × C/1.0	Softened	—	—	510(33)	210(13.5)	40	—	B:F:P:Sh:St*
18/9CrNiNb N Steel	UK (Hi-proof 347)	0.08 max	2.00 max	9.0 12.0	17.0 19.0	Nb N	10 × C/1.0 0.15 0.25	Softened	102	4	650(42)	340(22)	35	—	B:P*
18/9CrNi Steel	BS 302S25 (En58A)	0.12 max	0.50 2.00	8.0 11.0	17.0 19.0	— —		Softened Cold drawn	— 19 44	— $\frac{3}{4}$ $1\frac{3}{4}$	510(33) 860(56) 650(42)	210(13.5) 695(45) 310(20)	40 12 28	— — —	B:F:Sh:St*
18/9CrNiS Steel	BS 303S21 (En58M)	0.12 max	1.00 2.00	8.0 11.0	17.0 19.0	S	0.15 0.30	Softened Cold drawn	— 19 44	— $\frac{3}{4}$ $1\frac{3}{4}$	510(33) 860(56) 650(42)	210(13.5) 695(45) 310(20)	40 12 28	— — —	B:F* Free cutting
18/9CrNiSe Steel	BS 303S41 (En58M)	0.12 max	1.00 2.00	8.00 11.00	17.0 19.0	Se	0.15 3.00	Softened	—	—	510(33)	210(13.5)	40	—	B:F* Free cutting
18/9CrNiTi Steel	BS 321S20 (En58B, 58C)	0.12 max	0.50 2.00	8.0 11.0	17.0 19.0	Ti	5 × C/0.90	Softened Cold drawn	— 19 44	— $\frac{3}{4}$ $1\frac{3}{4}$	510(33) 860(56) 650(42)	210(13.5) 695(45) 310(20)	40 12 28	— — —	B:F* Scale resistant up to 800 °C in air
18/9CrNi TiS Steel	BS 325S21 (En58M)	0.12 max	1.00 2.00	8.0 11.0	17.0 19.0	Ti S	5 × C/0.90 0.15 0.30	Softened Cold drawn	— 19 44	— $\frac{3}{4}$ $1\frac{3}{4}$	510(33) 860(56) 650(42)	210(13.5) 695(45) 310(20)	40 12 28	— — —	B:F* Free cutting

* B = Bars. F = Forgings. P = Plates. Sh = Sheets. St = Strips. T = Tubes.

Table 22.44 FORGED OR ROLLED STEELS—ROOM TEMPERATURE LONGITUDINAL MECHANICAL PROPERTIES—*continued*

Material	*British or other standard*	*Composition %* C	Mn	Ni	Cr	*Other elements*		*Condition*	*Limiting ruling section* mm	in	*Properties (minima unless otherwise stated)* *UTS* MPa ($tonf\,in^{-2}$)	*Yield stress* MPa ($tonf\,in^{-2}$)	*Elong.* (gl $5.65\sqrt{S_0}$) %	*Izod* J (ft lbf)	*Remarks*
18/10CrNi Steel	BS S536	0.03	0.50 2.00	9.0 12.0	17.5 19.0	— —		Softened	—	—	500/700 (32/45)	190 (12.5)	40	—	Sh:St*
18/10CrNi Steel	BS 304S12	0.03 max	0.50 2.00	9.0 12.0	17.5 19.0	— —		Softened	—	—	460 (30)	170 (11)	40	—	B:F:P:Sh:St*
18/10CrNi N Steel	UK (Hi-proof 304L)	0.03 max	2.00 max	9.00 12.0	17.5 19.0	N	0.15 0.25	Softened	102	4	585 (38)	295 (19)	35	—	B:F*
18/10CrNi Steel	BS 304S16 (En58E)	0.06 max	0.50 2.00	9.0 11.0	17.5 19.0	—		Softened	—	—	510 (33)	210 (13.5)	40	—	Sh:St*
18/10CrNiTi Steel	BS S524, S526	0.08	0.50 2.00	9.0 11.0	17.0 19.0	Ti	5 × C/0.70	Softened Cold rolled	— —	— —	540 (35) 800/1 110 (52/71)	210 (13.5) 640 (41)	30 15	— —	Sh:St*
18/10CrNiNb Steel	BS S525, S527	0.08	0.50 2.00	9.0 11.0	17.0 19.0	Nb	10 × C/1.0	Softened Cold rolled	— —	— —	540 (35) 800/1 110 (52/71)	210 (13.5) 640 (41)	30 15	— —	Sh:St*
18/11CrNi Steel	BS 305S19	0.10 max	0.50 2.00	11.0 13.0	17.0 19.0	— —		Softened	—	—	460 (30)	170 (11)	40	—	Sh:St*
17/10CrNi Mo Steel	BS 315S16 (En58H)	0.07 max	0.50 2.00	9.0 11.0	16.5 18.5	Mo	1.25 1.75	Softened	—	—	460 (30)	170 (11)	40	—	B:F:Sh:St*
17/10CrNi MoNb Steel	UK (FV548)	0.08	1.00	11.5 (Typical)	16.5	Mo Nb	1.5 1.0	Softened	—	—	600 (39) (Typical values)	225 (14.5) (Typical values)	55	135 (100)	B:F* Creep resistant
17/11CrNi MoSe Steel	BS 326S36	0.08 max	0.50 2.00	10.0 13.0	16.5 18.5	Mo Se	2.25 3.00 0.15 0.30	Softened	—	—	510 (33)	210 (13.5)	40	—	B:F* Free cutting
17/12CrNi Mo Steel	BS 316S12	0.03 max	0.50 2.00	11.0 14.0	16.5 18.5	Mo	2.25 3.00	Softened	—	—	460 (30)	170 (11)	40	—	B:F:P:Sh:St*

17/12CrNi MoN Steel	UK (Hi-proof 316L)	0.03 max	2.00 max	11.0 14.0	16.5 18.5	Mo N	2.25 3.00 0.15 0.25	Softened	102	4	620 (40)	315 (20.5)	35	—	B:F*
17/12CrNi Mo Steel	BS 316S16 (En58J)	0.07 max	0.50 2.00	10.0 13.0	16.5 18.5	Mo	2.25 3.00	Softened Cold drawn	— 19 44	— $\frac{3}{4}$ $1\frac{3}{4}$	460 (30) 860 (56) 650 (42)	170 (11) 695 (45) 310 (20)	40 12 28	— — —	B:F:P:Sh:St* Also used as cold rolled strip or cold drawn wire for springs
17/12CrNi MoN Steel	UK (Hi-proof 316)	0.07 max	2.00 max	10.00 13.0	16.5 18.5	Mo N	2.25 3.00 0.15 0.25	Softened	102	4	620 (40)	315 (20.5)	35	—	B:F*
17/12CrNi MoTi Steel	BS 320S17 (En58J-Ti)	0.08 max	0.50 2.00	11.0 14.0	16.5 18.5	Mo Ti	2.25 3.00 4 × C/0.60	Softened	—	—	490 (32)	195 (12.5)	40	—	B:F:P:Sh:St*
18/12 CrNiMo Nb Steel	BS (En58J-Nb) DIN X10CrNiMoNb 18 12	0.10 max	2.00 max	12.0 14.5	16.5 18.5	Mo Nb	2.5 3.0 8 × C min	Softened	64	$2\frac{1}{2}$	495/740 (32/48)	225 (14.5)	40	—	B:F:P:Sh:St*
16/13CrNi Nb Steel	DIN X8CrNiNb 16 13	0.04 0.10	1.5 max	12.0 14.0	15.0 17.0	Nb	10 × C/1.2	Softened	152	6	510/680 (33/44)	200 (13)	35	—	Creep resistant
16/13CrNi MoVNb Steel	DIN X8CrNiMo VNb 16 13	0.04 0.10	1.5 max	12.5 14.5	15.5 17.5	Mo V Nb +N	1.1 1.5 0.60 0.85 10 × C/1.2	Precipitation hardened	152	6	540/740 (35/48)	255 (16.5)	30	—	Creep resistant. Resistant to high pressure hydrogen
16/16 CrNiMo Nb Steel	DIN X8CrNi MoNb 16 16	0.04 0.10	1.50 max	15.5 17.5	15.5 17.5	Mo Nb	1.6 2.0 10 × C/1.2	Softened	152	6	525/680 (34/44)	215 (14)	35	—	Creep resistant

* B = Bars. F = Forgings. P = Plates. Sh = Sheets. St = Strips. T = Tubes.

Table 22.44 FORGED OR ROLLED STEELS—ROOM TEMPERATURE LONGITUDINAL MECHANICAL PROPERTIES—*continued*

		Composition %							*Limiting ruling section*		*Properties (minima unless otherwise stated)*				
Material	*British or other standard*	C	Mn	Ni	Cr	*Other elements*		*Condition*	mm	in	*UTS* MPa (tonf in^{-2})	*Yield stress* MPa (tonf in^{-2})	*Elong.* (gl $5.65\sqrt{S_0}$) %	*Izod* J (ft lbf)	*Remarks*
16/16CrNiW Nb Steel	DIN X6CrNiW Nb 16 16	0.04 0.10	1.50 max	15.5 17.5	15.5 17.5	W Nb +N	2.5 3.5 10 × C/1.2	Softened	152	6	540/740 (35/48)	255 (16.5)	30	—	Creep resistant
16/16CrNi MoNbB Steel	DIN X8CrNiMo NbB 16 16	0.04 0.10	1.50 max	15.5 17.5	15.5 17.5	Mo Nb B	1.6 2.0 10 × C/1.2 0.05 0.10	Precipitation hardened	152	6	540/740 (35/48)	280 (18)	30	—	Nuclear reactor components
18/15CrNi Mo Steel	BS 317S12	0.03 max	0.50 2.00	14.0 17.0	17.5 19.5	Mo	3.0 4.0	Softened	—	—	460 (30)	170 (11)	40	—	B:F:P:Sh:St*
18/13CrNi Mo Steel	BS 317S16	0.06 max	0.50 2.00	12.0 15.0	17.5 19.5	Mo	3.0 4.0	Softened	—	—	460 (30)	170 (11)	40	—	B:F:P:Sh:St*
20/18NiCr MoCuNb Steel	DIN X5CrNiMo CuNb 18 18	0.07 max	2.00 max	19.0 21.0	16.5 18.5	Mo Cu Nb	2.0 2.5 1.8 2.2 8 × C min	Softened	—	—	495/740 (32/48)	225 (14.5)	40	—	Resistant to sulphuric acid attack
20/12CrNiSi Steel	DIN X15CrNiSi 20 12	0.20 max	2.00 max	11.0 13.0	19.0 21.0	Si	1.8 2.3	Softened	—	—	585/740 (38/48)	295 (19)	40	—	Resistant to scaling up to 1050 °C in air
23/14CrNi Steel	BS 309S24, S522	0.15 max	0.50 2.00	13.0 16.0	22.0 25.0	—		Softened	—	—	540 (35)	215 (14)	40	—	P:Sh:St*
23/14CrNiTi Steel	BS S125, S528	0.15 max	0.50 2.00	13.0 16.0	22.0 25.0	Ti	5 × C/0.90	Softened	152	6	540 (35)	215 (14)	28	68 (50)	B:F:Sh:St*
23/14CrNi Nb Steel	BS S126, S529	0.15 max	0.50 2.00	13.0 16.0	22.0 25.0	Nb	10 × C/1.40	Softened	152	6	540 (35)	215 (14)	28	68 (50)	B:F:Sh:St*

24/17CrNi Steel	BS 312S24, S523	0.15 max	0.50 2.00	16.0 19.0	23.0 26.0	—		Softened	—	—	540(35)	215(14)	40	—	P:Sh:St*
24/17CrNiTi Steel	BS S127, S530	0.15 max	0.50 2.00	16.0 19.0	23.0 26.0	Ti	5 × C/0.90	Softened	152	6	540(35)	215(14)	28	68(50)	B:F:Sh:St*
24/17CrNi Nb Steel	BS S128, S531	0.15 max	0.50 2.00	16.0 19.0	23.0 26.0	Nb	10 × C/1.40	Softened	152	6	540(35)	215(14)	28	68(50)	B:F:Sh:St*
24/20CrNi Steel	BS 310S24	0.15 max	0.50 2.00	19.0 22.0	23.0 26.0	— —		Softened	—	—	540(35)	215(14)	40	—	B:F:P:Sh:St* Resistant to scaling up to 1 050 °C in air
25/15NiCr TiMoVA1B Steel	AICMA FE-PA92HT (UK FV559 ASTM A286)	0.08 max	1.00 2.00	24.0 27.0	13.5 16.0	Si Ti Mo V Al +B	0.4 1.0 1.9 2.3 1.0 1.5 0.1 0.5 0.35 max	Precipitation hardened	—	—	900(58)	590(38)	20	—	Gas turbine components. Also suitable for use at sub-zero temperatures
25/20CrNiSi Steel	DIN X15CrNiSi 25 20	0.20 max	2.00 max	19.0 21.0	24.0 26.0	Si	1.8 2.3	Softened	—	—	590/740 (38/48)	295(19)	40	—	Scaling resistant up to 1 200 °C in air
25/25 CrNiMoTi Steel	DIN X5CrNiMoTi 25 25	0.06 max	2.00 max	24.0 26.0	24.0 26.0	Mo Ti	2.0 2.5 5 × C min	Softened	—	—	500/740(32/48)	225(14.5)	35	—	Used in the chemical industry
36/16NiCrSi Steel	DIN X12NiCrSi 36 16	0.15 max	2.00 max	34.0 37.0	15.0 17.0	Si	1.5 2.0	Softened	—	—	540/740 (35/48)	260(17)	40	—	Resistant to scaling up to 1 100 °C in air
42/12 NiCrMo TiCoAl B Steel	AICMA FE-PA99-HT	0.10 max	0.50 max	40.0 45.0	11.0 14.0	Mo Ti Co Al +B	5.0 6.5 2.6 3.1 1.0 max 0.35 max	Precipitation hardened	—	—	1 130(73)	820(53)	10	—	Gas turbine components

* B = Bars. F = Forgings. P = Plates. Sh = Sheets. St = Strips. T = Tubes.

Table 22.44 FORGED OR ROLLED STEELS—ROOM TEMPERATURE LONGITUDINAL MECHANICAL PROPERTIES—*continued*

Material	*British or other standard*	*Composition %* C	Mn	Ni	Cr	*Other elements*	*Condition*	*Limiting ruling section* mm	in	*Properties (minima unless otherwise stated)* *UTS* MPa (tonf in^{-2})	*Yield stress* MPa (tonf in^{-2})	*Elong.* (gl $5.65\sqrt{S_0}$) %	*Izod* J (ft lbf)	*Remarks*
20/20/20 CrNiCoMo WNbN Steel	AICMA FE-PA91-HT	0.08 0.16	1.00 2.00	19.0 21.0	20.0 22.5	Co 18.5 21.0 Mo 2.5 3.5 W 2.0 3.0 Nb 0.75 1.25 N 0.10 0.20	Precipitation hardened	—	—	680/960(44/62)	340(22)	30	—	Gas turbine components
12/11/6 CrNiMn Steel	DIN X15CrNiMn 12 10	0.05 0.20	5.5 6.5	9.0 11.0	10.5 12.5	— —	Softened Cold formed	— —	— —	500/650(32/42) 630/835(41/54)	220(14) 500(32)	45 35	— —	
15/10/6 CrNiMn- MoNbVB Steel	UK (Esshete 1250)	0.15 max	5.5 7.0	9.0 11.0	14.0 16.0	Mo 0.8 1.2 Nb 0.75 1.25 V 0.15 0.40 +B	Softened	—	—	495(32)	180(11.5)	30	—	Pressure vessel plate, super- heater tube, steam piping. Creep resistant
17/6/4 CrMnMiN Steel	AISI 201	0.15 max	5.5 7.5	3.5 5.5	16.0 18.0	N 0.25 max	Softened	—	—	790 (51) (Typical values)	380 (24.5) (Typical values)	48	—	
17/8/5 CrMnNiN steel	BS284S16	0.07 max	7.00 10.00	4.0 6.5	16.5 18.5	N 0.15 0.25	Softened	—	—	630(41)	300(19.5)	40	—	P:Sh:St*
18/9/5 CrMnNiN Steel	AISI 202 DIN X8 CrMnNi 18 8	0.15 max	7.5 10.0	4.0 6.0	17.0 19.0	N 0.25 max	Softened	—	—	725(47) (Typical values)	380(24.5) (Typical values)	48	—	
18/8/6 CrMnNiN Steel	DIN X7CrMnNiN 18 8	0.07	8.9	5.9 (Typical)	18.0	Mo 0.30 N 0.22	Softened	—	—	718(46.5) (Typical values)	370(24) (Typical values)	55	—	Used in the chemical industry

18/10/9 CrMnNi- MoN Steel	DIN X3CrMnNi- MoN 18 10	0.04	11.0	9.0 (Typical)	19.0	Mo N	2.35 0.30	Softened	—	—	820(53)	430(28)	38	—	Used in the chemical industry
18/12/2 MnCrNi- Mo Steel	DIN X12MnCr 18 12	0.15 max	17.0 19.0	1.5 2.5	11.0 13.0	Mo P	0.30 0.80 0.08 max	Softened Cold formed	— —	— —	630/790(41/51) 790/990(51/64)	295(19) 500(32.5)	45 25	— —	Used in the cutlery industry
Austenitic-ferritic stainless and heat resisting steels															
25/5CrNi Steel	Swedish (UHB 45) AISI 327	0.10 —	Not given (Typical)	4.8	25.5	—		Softened	—	—	630(41) (Typical values)	460(30) (Typical values)	30	—	Resistant to scaling up to 1 075°C in air
25/4CrNiSi Steel	DIN X20CrNiSi 25 4	0.15 0.25	2.0 max	3.5 5.5	24.0 26.0	Si	0.8 1.3	Softened	—	—	590/740(38/48)	390(25)	26	—	Resistant to scaling up to 1 100°C in air
25/CrNiMo Steel	Swedish (UHB44) AISI 329	0.08	Not given (Typical)	5.3	25.0	Mo	1.50	Softened	—	—	660(42.5) (Typical values)	500(32.5) (Typical values)	30	—	Used in the chemical industry. Scaling resistant up to 1 075°C in air
Valve steels (BS 970: Part 4:1970, Stahl Eisen Werkstoffblatt 490–52)															
9CrSi Steel	BS401S45 (En52)	0.40 0.50	0.30 0.75	0.50 max	7.5 9.5	Si	3.00 3.75	Hardened and tempered	—	—	925(60) (Typical values)	680(44) (Typical values)	22	—	
14/14CrNiW Steel	BS331S40 (En54)	0.35 0.50	0.50 1.00	12.0 15.0	12.0 15.0	Si W	1.0/2.0 2.0/3.0	Softened	—	—	895(58) (Typical values)	510(33) (Typical values)	23	—	
14/14CrNi WMo Steel	BS331S42, S111 (En54A)	0.37 0.47	0.50 1.00	13.0 15.0	13.0 15.0	Si W Mo	1.0/2.0 2.2/3.0 0.40/0.70	Softened	—	—	—	—	—	20(15)	
18/9 CrNiWSi Steel	DIN X45CrNiW 189 (Similar to BS En55)	0.40 0.50	0.80 1.50	8.00 10.0	17.0 19.0	Si W	2.0 3.0 0.8 1.2	Age hardened	—	—	790/990(51/64)	390(25)	25	—	Resistant to lead oxide corrosion
20CrNiSi Steel	BS443S65 (En59)	0.75 0.85	0.30 0.75	1.20 1.70	19.0 21.0	Si	1.75 2.25	Hardened and tempered	—	—	895/1 050(58/68)	710(46)	7	—	

* B = Bars. F = Forgings. P = Plates. Sh = Sheets. St = Strips. T = Tubes.

Table 22.44 FORGED OR ROLLED STEELS—ROOM TEMPERATURE LONGITUDINAL MECHANICAL PROPERTIES—*continued*

Material	*British or other standard*	*Composition %* C	Mn	Ni	Cr	*Other elements*	*Condition*	*Limiting ruling section* mm	in	*Properties (minima unless otherwise stated)* *UTS* MPa (tonf in^{-2})	*Yield stress* MPa (tonf in^{-2})	*Elong.* (gl 5.65$\sqrt{S_0}$) %	*Izod* J (ft lbf)	*Remarks*
21/11CrNiSi Steel	BS381S34	0.15 0.25	1.50 max	10.5 12.5	20.0 22.0	Si 0.75 1.25 N 0.15 0.30	Precipitation hardened	—	—	—	—	—	—	Resistant to lead oxide corrosion
21/9/4 CrMnNiN Steel	BS349S52	0.48 0.58	8.0 10.0	3.25 4.50	20.0 22.0	N 0.38 0.50 C+N 0.90 min	Precipitation hardened	—	—	1 050(68) (Typical values)	620(40) (Typical values)	9	—	Resistant to lead oxide corrosion
21/9/4 CrMnNi-NS Steel	BS349S54	0.48 0.58	8.0 10.0	3.25 4.50	20.0 22.0	N 0.38 0.50 S 0.035 0.080 C+N 0.90 min	Precipitation hardened	—	—	1 050(68) (Typical values)	620(40) (Typical values)	8	—	Free cutting steel
21/9/4 CrMnNiNNb Steel	BS352S52	0.48 0.58	8.00 10.00	3.25 4.50	20.0 22.0	N 0.38 0.50 Nb 2.00 3.00 C+N 0.90 min		—	—	—	—	—	—	
21/9/4 CrMnNiNNb S steel	BS352S54	0.48 0.58	8.00 10.00	3.25 4.50	20.0 22.0	N 0.38 0.50 Nb 2.00 3.00 S 0.035 0.080	—	—	—	—	—	—	—	Free cutting steel
23/5 CrNiMo Steel	DIN X45CrNi Mo23 5	0.40 0.50	0.90 1.20	4.5 5.5	22.0 24.0	Si 1.0 1.3 Mo 2.5 3.0	Precipitation hardened	—	—	1 235(80) (HRC40)	—	—	—	

High-strength stainless steels (ASTM A579-67, DIN 17224:1968)
Semi-Austenitic steels (*Typical compositions and mechanical properties*)

17/7CrNiAl Steel	AISI 631 ASTM A579 Grade 62 (17/7PH) DINX7CrNiAl 177	0.07	0.8	7.0	17.0	Al	1.2	Precipitation hardened	—	—	1 360 (88)	1 250 (81)	10	—	Sheet and strip. Also used as cold-rolled strip or cold-drawn wire for springs at temperatures up to 350°C
15/7CrNi-MoAl Steel	AISI 632 ASTM A579 Grade 63 (PH15/7Mo)	0.07	0.8	7.1	15.1	Mo Al	2.2 1.2	Precipitation hardened	—	—	1 515 (98)	1 420 (92)	9	—	Pressure vessels, springs
14/5 CrNiCu MoNb Steel	BS S145 (FV520B)	0.07	1.00	5.8	14.5	Mo Cu Nb	1.6 1.6 0.30	Precipitation hardened			1 470(95)	1 030(66)	10	20(15)	Bars and forgings for aircraft parts
14/8CrNiMo Al Steel	USA (PHI4/8Mo)	0.04	0.6	8.3	15.1	Mo Al +N	2.2 1.2	Precipitation hardened	—	—	1 570(102)	1 450(94)	8	—	Pressure vessels, aircraft parts
16/4CrNi-MoN Steel	AISI 633 (AM 350)	0.10	0.8	4.25	16.5	Mo N	2.75 0.10	Precipitation hardened	—	—	1 405(91)	1 175(76)	11	Charpy V 20(15)	Valves, piping, aircraft parts
15/4CrNi MoN Steel	AISI 634 ASTM A579 Grade 64 (AM 355)	0.13	0.8	4.25	15.5	Mo N	2.75 0.10	Precipitation hardened	—	—	1 480(96)	1 250(81)	11	Charpy V 20(15)	Aircraft parts, valves, turbine parts
15/5CrNi-CuMoTi Steel	BS S533 (FV520S)	0.05	1.3	5.5	16.0	Mo Cu Ti	1.8 2.00 0.10	Precipitation hardened	—	—	1 370(89)	980(64)	11	—	Sheet and strip for aircraft parts
Martensitic steels															
17/4CrNiCu Nb Steel	AISI 630 ASTM A579 Grade 61 (17/4PH)	0.04	0.8	4.3	16.0	Cu Nb	3.3 0.27	Precipitation hardened	—	—	1 360(88)	1 265(82)	12	—	Gears, springs, cutlery, aircraft parts, turbine components

Table 22.44 FORGED OR ROLLED SHEETS—ROOM TEMPERATURE LONGITUDINAL MECHANICAL PROPERTIES—*continued*

		Composition %						*Limiting ruling section*		*Properties (minima unless otherwise stated)*				
Material	*British or other standard*	C	Mn	Ni	Cr	*Other elements*	*Condition*	mm	in	*UTS* MPa (tonf in^{-2})	*Yield stress* MPa (tonf in^{-2})	*Elong.* (gl $5.65\sqrt{S_0}$) %	*Izod* J (ft lbf)	*Remarks*
15/5CrNi CuNb Steel	USA(15/5PH)	0.04	0.8	4.6	15.0	Cu 3.3 Nb 0.27	Precipitation hardened	—	—	1 360(88)	1 265(82)	12	—	Gears, cams, cutlery, aircraft parts
13/8CrNi-MoAl Steel	USA(PH13/8Mo)	0.03	0.10 max	8.2	12.8	Mo 2.2 Al 1.1 +N	Precipitation hardened	76	3	1 465(95)	1 310(85)	12	—	Aircraft parts, fasteners, shafts
12/8CrNi-CuTiNb Steel	USA(Custom 455)	0.03	0.25	8.5	11.7	Cu 2.2 Ti 1.2 Nb 0.3	Precipitation hardened	102	4	1 420(92)	1 345(87)	10	Charpy V 16(12)	Fasteners, springs, pressure vessels, valve parts, forgings
14/13CrCoV Steel	USA(AFC77)	0.15	0.20	—	14.5	Co 13.0 Mo 5.0 V 0.40	Precipitation hardened	—	—	1 630(106)	1 295(84)	10	—	Die casting dies, glass moulds
14/15CrCo-NiMoTi Steel	USA(AM367)	0.02	Not given	3.5	14.0	Co 15.0 Mo 2.0 Ti 0.4	Precipitation hardened	—	—	1 500(97)	1 465(95)	11	—	Bars, forgings, sheet and strip
12/12CrCo-NiMoTiNb Steel	UK(D70)	0.02	Not given	4.0	12.0	Co 12.0 Mo 4.0 Ti 0.40 Nb 0.10 Al 0.10 +Zr, B	Precipitation hardened	—	—	1 650(107)	1 600(104)	9.5	16(12)	Bars, forgings, sheet and strip
12/8/5CrNi-CoMoTi Steel	WG(Ultrafort 401)	0.01	Not given	8.1	12.5	Co 5.2 Mo 2.0 Ti 0.80 +Al, Zr, B	Precipitation hardened	—	—	1 650(107)	1 570(102)	11	—	Pressure vessels, springs, fasteners, aircraft parts, extrusion and stamping tools

Maraging nickel steels (ASTM A579-67)
Typical compositions

12/5NiCr-MoAlTi Steel	ASTM A579 Grade 75	0.02	Not given	12.0	5.0	Mo 3.0 Al 0.40 Ti 0.20	Precipitation hardened	—	—	1 310(85) (Typical values)	1 250(81) (Typical values)	15	Charpy V 74(55)	Gears, fasteners, shafts, rocket and missile cases, aircraft parts, plastic mould dies, die holders, die casting die inserts, extrusion tools
18/8NiCo-MoTi Steel	ASTM A579 Grade 71	0.01	Not given	18.0	—	Co 8.5 Mo 3.0 Ti 0.20 Al 0.10 +Ca, Zr, B	Precipitation hardened	—	—	1 390/1 540(90/100)	1 330/1 480(86/96)	9/13	Charpy V 34/68 (25/50)	Gears, fasteners, shafts, rocket and missile cases, aircraft parts, plastic mould dies, die holders, die casting die inserts, extrusion tools
18/8NiCo MoTi Steel	ASTM A579 Grade 72	0.01	Not given	18.0	—	Co 8.0 Mo 5.0 Ti 0.40 Al 0.10 +Ca, Zr, B	Precipitation hardened	—	—	1 680/1 910(109/124)	1 630/1 820 (106/118)	8/10	Charpy V 20/41 (15/30)	Gears, fasteners, shafts, rocket and missile cases, aircraft parts, plastic mould dies, die holders, die casting die inserts, estrusion tools
18/8NiCo-MoTi Steel	ASTM A579 Grade 73	0.01	Not given	18.0	—	Co 9.0 Mo 5.0 Ti 0.60 Al 0.10 +Ca, Zr, B	Precipitation hardened	—	—	1 820/2 130(118/138)	1 790/2 090 (116/135)	6/9	Charpy V 14/27 (10/20)	Gears, fasteners, shafts, rocket and missile cases, aircraft parts, plastic mould dies, die holders, die casting die inserts, extrusion tools

Table 22.44 FORGED OR ROLLED STEELS—ROOM TEMPERATURE LONGITUDINAL MECHANICAL PROPERTIES—*continued*

Material	*British or other standard*	*Composition %* C	Mn	Ni	Cr	*Other elements*	*Condition*	*Limiting ruling section* mm	in	*Properties (minima unless otherwise stated)* *UTS* MPa (tonf in^{-2})	*Yield stress* MPa (tonf in^{-2})	*Elong.* (gl $5.65\sqrt{S_0}$) %	*Izod* J (ft lbf)	*Remarks*
Other high alloy steels (BS 1501: Part 2:1970, 1503:1969, Stahl Eisen Werkstoffblatt 390-61, ASTM 579-67)														
12Mn Steel	DIN X120Mn12	1.1 1.3	11.5 13.5	— —	— —	P 0.1 max	Softened	—	—	790/1 080(51/70)	340(22)	40	—	Non-magnetic
18Mn Steel	DIN X35Mn18	0.30 0.40	17.0 19.0	— —	— —	P 0.1 max	Softened	—	—	680/930(44/60)	250(16)	30	—	Non-magnetic
18MnCr Steel	DIN X40MnCr18	0.30 0.50	17.0 19.0	— —	3.0 3.5	P 0.1 max	Softened Cold formed	— —	— —	740/930(48/60) 990/1 175(64/76)	295(19) 880(57)	45 20	— —	Non-magnetic end bells for alternator rotors
23MnCr Steel	DIN X40Mn-Cr23	0.30 0.50	21.0 24.0	— —	3.0 3.5	P 0.1 max	Softened	—	—	680/880(44/57)	310(20)	45	—	Non-magnetic
9Ni Steel	BS 1501-509, 510, 1503-509	0.10 max	0.30 0.80	8.5 9.75	0.30 max	Mo 0.20 max	Double normalised and tempered, or, hardened and tempered	—	—	700(45)	525(34)	18	Charpy V 68(50) at −100°C 47(35) at −160°C 34(25) at −196°C	Pressure vessel plates 50 mm (2 in) max thick and forgings for sub-zero applications
13Cr Steel	UK (Silver Fox 67)	0.60 0.70	0.50 1.0	— —	12.0 13.5	— —	Annealed Lightly cold rolled	— —	— —	770(50) 1 080(70)	500(32) 1 050(68)	18 3	— —	Razor and surgical biades, general cutting tools
9/4NiCoCr-MoV Steel	USA(HP9-4-20)	0.20	0.30 (Typical)	9.2	0.8	Co 4.5 Mo 1.0 V 0.10	Hardened and tempered	—	—	1 420(92) (Typical values)	1 235(80) (Typical values)	14	Charpy V 74(55)	Rocket motor cases, seamless tubing, shafts, pressure vessels, piping
9/4NiCo CrMoV Steel	ASTM A579 Grade 81 (HP9-4-25)	0.25	0.25 (Typical)	8.0	0.45	Co 4.0 Mo 0.45 V 0.10	Hardened and tempered	152	6	1 375(89) (Typical values)	1 265(82) (Typical values)	13	Charpy V 47(35)	Rocket motor cases, seamless tubing, shafts, pressure vessels, piping

9/4NiCoCr MoV Steel	ASTM A579 Grade 82 (HP9-4-30)	0.30	0.25 (Typical)	7.5	1.0	Co 4.5 Mo 1.0 V 0.10	Hardened and tempered	126	5	1 570(102) (Typical values)	1 375(89) (Typical values)	11 —	Charpy V 34(25)	Armour plate aircraft parts
9/4NiCo-MoV Steel	ASTM A579 Grade 83 (HP9-4-45)	0.45	0.25 (Typical)	7.5	0.30	Co 4.0 Mo 0.30 V 0.10	Salt bath quenched	76	3	1 850(120)	1 540(100)	9	Charpy V 27(20)	Aircraft parts, fasteners, connecting rods, valve-spring wires
							Hardened, subzero cooled and tempered	76	3	1 990(129) (Typical values)	1 710(111) (Typical values)	7	20(15)	
6. *Spring steels* (BS 970: Part 5: 1972, DIN 17221: 1955, 17222: 1955, 17225: 1955)														
Carbon steels (*cold rolled strip except for 060 A96*)														
0.53C Steel	BS080A52 (En43) DIN Ck53	0.50 0.55	0.70 0.90	— —	— —	Si 0.10 0.35	Hardened and tempered	—	—	1 175/1 390(76/90)	1 035(67)	7	—	Laminated springs
0.67C Steel	BS080A67 (En43E) DIN Ck67	0.65 0.70	0.70 0.90	— —	— —	Si 0.10 0.35	Hardened and tempered	—	—	1 375/1 620(89/105)	1 280(83)	6	—	Laminated springs
0.72C Steel	BS070A72 (En42)	0.70 0.75	0.60 0.80	— —	— —	Si 0.10 0.35	Hardened and tempered	—	—	—	—	—	—	Laminated springs
0.78C Steel	BS070A78 (En42) DIN Mk75	0.75 0.82	0.60 0.80	— —	— —	Si 0.10 0.35	Hardened and tempered	—	—	1 575/1 775(102/115)	1 465(95)	6	—	Laminated springs
0.96C Steel	BS 060A96 (En44) DIN Mk101	0.93 1.00	0.50 0.70	— —	— —	Si 0.10 0.35	Hardened and tempered	—	—	1 760/2 315(114/150)	1 670(108)	5	—	Coil springs up to 25 mm (1 in) bar dia
Manganese and manganese silicon steels														
2Mn Steel	DIN 50Mn7	0.45	1.80 (Typical)	—	—	—	Hardened and tempered	—	—	1 175/1 390(76/90)	1 030(67)	7	—	Laminated springs
$1\frac{3}{4}$Si Steel	DIN 46Si7	0.42 0.50	0.50 0.80	— —	— —	Si 1.5 1.8	Hardened and tempered	—	—	1 280/1 465(83/95)	1 080(70)	6	—	Elliptic or helical springs, laminated springs

Table 22.44 FORGED OR ROLLED STEELS—ROOM TEMPERATURE LONGITUDINAL MECHANICAL PROPERTIES—*continued*

		Composition %						*Limiting ruling section*		*Properties (minima unless otherwise stated)*				
Material	*British or other standard*	C	Mn	Ni	Cr	*Other elements*	*Condition*	mm	in	*UTS* MPa (tonf in^{-2})	*Yield stress* MPa (tonf in^{-2})	*Elong.* (gl $5.65\sqrt{S_0}$) %	*Izod* J (ft lbf)	*Remarks*
2Si Steel	BS250A53 (En45) DIN 51Si7	0.50 0.57	0.70 1.00	— —	— —	Si 1.70 2.10	Hardened and tempered	—	—	1 280/1 465(83/95)	1 080(70)	6	—	Laminated springs, coil springs up to 25 mm (1 in) bar dia.
2Si Steel	BS250A58 (En45A) DIN 55Si7	0.55 0.62	0.70 1.00	— —	— —	Si 1.70 2.10	Hardened and tempered	—	—	1 280/1 465(83/95)	1 080(70)	6	—	Laminated springs, coil springs up to 25 mm (1 in) bar dia.
2Si Steel	BS250A61 (En45A) DIN 66Si7	0.58 0.65	0.70 1.00	— —	— —	Si 1.70 2.10	Hardened and tempered	—	—	1 375/1 540(89/100)	1 175(76)	6	—	Laminated springs, coil springs over 25 mm (1 in) bar dia
1 SiMn Steel	DIN 60SiMn5	0.55 0.65	0.90 1.10	— —	— —	Si 1.0 1.3	Hardened and tempered	—	—	1 330/1 540(86/100)	1 030(67)	6	—	Laminated springs, ring springs
Low alloy steels (hot rolled)														
1¾ MnV Steel	DIN 50MnV7	0.50	1.70 (Typical)	—	—	0.10	Hardened and tempered	—	—	1 280/1 465(83/95)	1 130(73)	7	—	Laminated springs, coil springs up to 25 mm (1 in) bar dia
¾Cr Steel	BS527A60 (En48)	0.55 0.65	0.70 1.00	— —	0.60 0.90	Si 0.10 0.35	Hardened and tempered	—	—	—	—	—	—	Laminated springs
1¼ SiCr Steel	DIN67 SiCr5	0.62 0.72	0.40 0.60	— —	0.40 0.60	Si 1.2 1.4	Hardened and tempered	—	—	1 465/1 690(95/109)	1330(86)	5	—	Coil springs and valve springs. Used up to 300°C

1CrV Steel	BS735A50 (En47) DIN 50CrV4	0.46 0.54	0.60 0.90	— —	0.80 1.10	V 0.15 min Si 0.10 0.35	Hardened and tempered	—	—	1 330/1 540(86/100)	1 175(76)	6	—	Laminated springs, coil springs. Torsion bars up to 40 mm ($1\frac{9}{16}$in) dia. Used up to 300°C
1CrV Steel	DIN 58CrV4	0.55 0.62	0.80 1.10	— —	0.90 1.20	V 0.07 0.12	Hardened and tempered	—	—	1 465/1 690(95/109)	1 330(86)	6	—	Torsion bars over 40 mm ($1\frac{9}{16}$in) bar dia
1CrMo Steel	DIN 50CrMo4	0.50	0.90 (Typical)	—	1.0	Mo 0.20	Hardened and tempered	—	—	1 330/1 540(86/100)	1 175(76)	6	—	Laminated springs, coil springs
$\frac{1}{2}$NiCrMo Steel	BS805A60	0.55 0.65	0.70 1.00	0.40 0.70	0.40 0.60	Mo 0.15 0.25 Si 0.10 0.35	Hardened and tempered	—	—	—	—	—	—	Laminated springs, coil springs
2SiCrMo Steel	BS925A60	0.55 0.65	0.70 1.00	— —	0.20 0.40	Si 1.70 2.10 Mo 0.20 0.30	Hardened and tempered	—	—	—	—	—	—	Laminated springs, coil springs
1CrMoV Steel	DIN 51CrMoV4	0.51	0.90 (Typical)	—	1.0	Mo 0.20 V 0.10	Hardened and tempered	—	—	1 465/1 690(95/109)	1 330(86)	6	—	Torsion bars over 40 mm ($1\frac{9}{16}$in) bar dia
$1\frac{1}{2}$CrMoV Steel	DIN45CrMoV67	0.40 0.50	0.60 0.80	— —	1.3 1.5	Mo 0.65 V 0.75 0.25 0.35	Hardened and tempered	—	—	1 390/1 690(90/109)	Not given	Not given	—	Used up to 450°C
$2\frac{1}{4}$CrWV Steel	DIN 30WCrV-179	0.25 0.35	0.20 0.40	— —	2.2 2.5	W 4.0 4.5 V 0.50 0.70	Hardened and tempered	—	—	1 390/1 690(90/109)	Not given	Not given	—	Used up to 500°C
High alloy steel														
8WCrMoV Steel	DIN 65WMo348	0.63 0.68	0.30 appr	— —	3.5 4.0	W 8.0 9.0 Mo 0.80 0.90 V 0.60 0.80	Hardened and tempered	—	—	1 390/1 690(90/109)	Not given	Not given	—	Used up to 550°C

Table 22.45 TYPICAL HOT TENSILE PROPERTIES OF FORGED OR ROLLED STEELS IN THE LONGITUDINAL DIRECTION

		Typical composition %						*Tensile properties*				
Material	*British or other standards*	C	Mn	Ni	Cr	*Other elements*	*Condition*	*Temperature* °C	*UTS* MPa ($tonf\,in^{-2}$)	*Yield stress* MPa ($tonf\,in^{-2}$)	*Elong.* (gl $5.65\sqrt{S_0}$) %	*RA* %
1. *Carbon steels*												
Armco Iron	—	0.02	0.03	—	—	—	Normalized	RT	340 (22)	185 (12)	39	69
								200	448 (29)	185 (12)	23	54
								400	309 (20)	Not given	15	67
0.15C Steel	BS 040A12 (En2B)	0.13	0.50	—	—	—	Normalized	RT	417 (27)	247(16)	33	Not given
								200	463(30)	232(15)	23	Not given
								400	386 (25)	193 (12.5)	33	Not given
								500	309 (20)	Not given	33	Not given
0.20C Steel	BS 070M20 (En3)	0.20	0.70	—	—	—	Normalized	RT	448 (29)	263 (17)	30	62
								200	463 (30)	224 (14.5)	24	54
								400	371 (24)	185 (12)	31	67
0.25C Steel	BS 070M26 (En4)	0.26	0.70	—	—	—	Normalized	RT	479 (31)	247 (16)	30	Not given
								200	510 (33)	232 (15)	22	Not given
								400	448 (29)	201 (13)	27	Not given
								500	340 (22)	185 (12)	28	Not given
0.35C Steel	BS 080M36	0.36	0.80	—	—	—	Normalized	RT	602 (39)	309 (20)	26	Not given
								200	633 (41)	309 (20)	17	Not given
								400	587 (38)	216 (14)	25	Not given
								500	432 (28)	Not given	28	Not given
0.40C Steel	BS 080M40 (En8)	0.40	0.80	—	—	—	Normalized	RT	602 (39)	340 (22)	29	53
								250	633 (41)	293 (19)	19	41
								450	479 (31)	Not given	31	59

2. *Low alloy weldable steels*

$\frac{1}{2}$ Mo Steel	Similar to DIN 15Mo3	0.17	0.60	—	—	Mo	0.60	Normalized	RT	510 (33)	317 (20.5)	33	69
									400	510 (33)	232 (15)	30	73
									500	417 (27)	224 (14.5)	29	77
									600	293 (19)	185 (12)	34	77
$\frac{1}{2}$MoV Steel	—	0.12	0.60	—	—	Mo	0.50	Normalized	RT	587 (38)	394 (25.5)	23	76
						V	0.30	and	200	556 (36)	386 (25)	22	73
								tempered	400	479 (31)	340 (22)	25	77
									600	278 (18)	232 (15)	32	81
1CrMo Steel	BS 1501–620	0.15	0.70	—	0.95	Mo	0.50	Normalized	RT	479 (31)	286 (18.5)	21	68
								and	400	595 (38.5)	263 (17)	23	57
								tempered	500	525 (34)	232 (15)	24	65
									600	340 (22)	216 (14)	27	82
$2\frac{1}{4}$CrMo Steel	BS 1501–622	0.13	0.60	—	2.20	Mo	1.10	Normalized	RT	641 (41.5)	417 (27)	18	75
								and	535	479 (31)	402 (26)	28	79
								tempered	610	309 (20)	263 (17)	47	89

3. *Low alloy direct hardening steels*

$\frac{3}{4}$MoV Steel	—	0.18	0.50	—	—	Mo	0.70	Hardened	RT	896 (58)	811 (52.5)	20	68
						V	0.20	and	400	741 (48)	680 (44)	21	62
								tempered	600	494 (32)	456 (29.5)	24	73
1CrMo Steel	—	0.40	0.40	—	1.10	Mo	0.70	Hardened	RT	1 004 (65)	903 (58.5)	19	38
								and	300	942 (61)	857 (55.5)	Not given	Not given
								tempered	400	842 (54.5)	741 (48)	20	44
									600	479 (31)	324 (21)	28	64
3CrMo Steel	BS 722M24 (En40B)	0.24	0.60	—	3.30	Mo	0.55	Hardened	RT	757 (49)	556 (36)	20	Not given
								and	400	664 (43)	510 (33)	21	Not given
								tempered	450	618 (40)	479 (31)	25	Not given
									500	556 (36)	463 (30)	27	Not given

Table 22.45 TYPICAL HOT TENSILE PROPERTIES OF FORGED OR ROLLED STEELS IN THE LONGITUDINAL DIRECTION—*continued*

Material	*British or other standards*	*Typical composition %* C	Mn	Ni	Cr	*Other elements*	*Condition*	*Tensile properties* *Temperature* °C	*UTS* MPa (tonf in^{-2})	*Yield stress* MPa (tonf in^{-2})	*Elong.* (gl $5.65\sqrt{S_0}$) %	*RA* %
3CrMoV Steel	—	0.28	0.40	—	3.30	Mo 0.50	Hardened	RT	788 (51)	649 (42)	21	63
						V 0.20	and	400	664 (43)	510 (33)	20	76
							tempered	600	510 (33)	386 (25)	24	76
3CrVMoW Steel	—	0.23	0.30	—	2.70	V 0.70	Hardened	RT	988 (64)	903 (58.5)	17	40
						Mo 0.50	and	200	896 (58)	826 (53.5)	18	38
						W 0.50	tempered	400	811 (52.5)	764 (49.5)	16	36
								600	571 (37)	510 (33)	16	33
								700	355 (23)	301 (19.5)	21	36
$3\frac{1}{2}$Ni Steel	BS (En22)	0.34	0.60	3.50	—	—	Hardened	RT	788 (51)	571 (37)	21	Not given
							and	200	741 (48)	525 (34)	23	Not given
							tempered	400	726 (47)	510 (33)	22	Not given
								500	448 (29)	340 (22)	24	Not given
$3\frac{1}{2}$ NiCr Steel	—	0.30	0.50	3.50	0.60	—	Hardened	RT	896 (58)	757 (49)	20	Not given
							and	200	849 (55)	710 (46)	17	Not given
							tempered	400	849 (55)	726 (47)	16	Not given
								500	695 (45)	602 (39)	19	Not given
$2\frac{1}{2}$NiCrMo Steel	BS 826M31 (En25)	0.31	0.55	2.70	0.60	Mo 0.55	Hardened	RT	988 (64)	880 (57)	21	Not given
							and	200	942 (61)	834 (54)	17	Not given
							tempered	400	911 (59)	834 (54)	17	Not given
								500	741 (48)	664 (43)	19	Not given
4NiCrMo Steel	—	0.30	0.30	4.10	2.00	Mo 0.30	Hardened	RT	927 (60)	664 (43)	20	61
							and	400	757 (49)	571 (37)	19	61
							tempered	600	355 (23)	170 (11)	40	88
3NiCrMoV Steel	—	0.29	0.40	3.10	0.70	Mo 0.50	Hardened	RT	788 (51)	649 (42)	20	60
						V 0.20	and	400	633 (41)	494 (32)	19	64
							tempered	600	386 (25)	232 (15)	27	84

4NiCrMoV Steel	—	0.36	0.60	3.90	2.20	Mo V	0.50 0.20	Hardened and tempered	RT	927 (60)	695 (45)	17	56
									400	772 (50)	587 (38)	17	55
									550	525 (34)	293 (19)	28	77
4. High alloy steels													
8CrSi Valve steel	BS 401S45 (En52)	0.45	0.50	—	8.50	Si	3.50	Hardened and tempered	RT	942 (61)	718 (46.5)	23	53
									400	726 (47)	471 (30.5)	24	60
									600	270 (17.5)	178 (11.5)	52	93
13Cr Steel	BS 410S21 (En56A)	0.12	0.50	—	13.00	—		Hardened and tempered	RT	610 (39.5)	417 (27)	29	72
									200	502 (32.5)	363 (23.5)	28	76
									400	463 (30)	332 (21.5)	24	74
									600	309 (20)	178 (11.5)	43	90
13Cr Steel	BS 420S37 (En56C)	0.24	0.50	—	13.0	—		Hardened and tempered	RT	703 (45.5)	541 (35)	23	60
									200	595 (38.5)	456 (29.5)	23	62
									400	541 (35)	432 (28)	18	58
									600	293 (19)	293 (19)	38	80
13Cr Steel	DIN X40Cr13	0.40	0.50	—	12.0	—		Hardened and tempered	RT	772 (50)	541 (35)	17	Not given
									400	687 (44.5)	Not given	14	49
									600	347 (22.5)	Not given	26	67
11CrVMoNb Steel	—	0.20	0.40	—	11.0	V Mo Nb	0.70 0.50 0.20	Hardened and tempered	RT	958 (62)	849 (55)	15	Not given
									400	795 (51.5)	741 (48)	10	Not given
									600	525 (34)	479 (31)	21	Not given
									650	448 (29)	378 (24.5)	23	Not given
11CrNiMoNbV Steel	—	0.20	0.80	1.20	10.5	Mo Nb V	0.70 0.60 0.20	Hardened and tempered	RT	1 081 (70)	942 (61)	16	46
									400	803 (52)	741 (48)	16	48
									600	587 (38)	541 (35)	28	73
									700	363 (23.5)	324 (21)	30	81
20Cr Steel	BS 442S19	0.06	0.80	—	21.0	—		Softened	RT	541 (35)	409 (26.5)	30	60
									400	432 (28)	317 (20.5)	21	58
									600	216 (14)	208 (13.5)	45	79
14/8CrNiMoAl Steel	Similar to USA PH-13/8Mo	0.03	0.80	8.50	14.5	Mo A	2.30 1.10	Precipitation hardened	RT	1 683 (109)	1 591 (103)	6	Not given
									31.5	1 375 (89)	1 236 (80)	4	Not given
									535	896 (58)	726 (47)	16	Not given

Table 22.45 TYPICAL HOT TENSILE PROPERTIES OF FORGED OR ROLLED STEELS IN THE LONGITUDINAL DIRECTION—*continued*

Material	*British or other standards*	*Typical composition %* C	Mn	Ni	Cr	*Other elements*	*Condition*	*Tensile properties* *Temperature* °C	*UTS* MPa ($tonf\,in^{-2}$)	*Yield stress* MPa ($tonf\,in^{-2}$)	*Elong.* (gl $5.65\sqrt{S_0}$) %	*RA* %
18/8CrNi Steel	BS 304S15 (En58E)	0.06	0.80	10.00	18.0	—	Softened	RT	571 (37)	216 (14)	52	70
								600	324 (21)	93 (6)	35	57
18/8CrNiTi Steel	BS 321S20 (En58B)	0.10	0.80	8.50	18.0	Ti 0.60	Softened	RT	656 (42.5)	255 (16.5)	46	68
								400	463 (30)	193 (12.5)	36	65
								600	378 (24.5)	162 (10.5)	31	67
18/11CrNiNb Steel	Similar to BS 347S17 (En58G)	0.08	1.50	11.0	17.5	Nb 1.20	Softened	RT	633 (41)	263 (17)	50	65
								400	432 (28)	178 (11.5)	33	64
								600	378 (24.5)	162 (10.5)	35	62
								750	239 (15.5)	Not given	42	73
17/13CrNiMo Steel	BS 316S16 (En58J)	0.07	1.40	12.50	16.50	Mo 2.80	Softened	RT	710 (46)	247 (16)	51	76
								600	463 (30)	124 (8)	43	62
								800	216 (14)	108 (7)	47	62
14/14CrNiW Valve steel	BS 331S40 (En54)	0.42	0.70	14.00	14.00	W 2.50 Si 1.50	Softened	RT	973 (63)	556 (36)	22	24
								400	656 (42.5)	456 (29.5)	20	33
								600	525 (34)	301 (19.5)	24	55
								800	263 (17)	139 (9)	35	74
20/8CrNiW Valve steel	BS (En55)	0.30	0.80	7.50	20.00	W 4.00 Si 1.30	Softened	RT	834 (54)	463 (30)	33	36
								400	656 (42.5)	371 (24)	26	43
								600	510 (33)	263 (17)	30	41
								800	247 (16)	Not given	43	55
23/12CrNiW Valve steel	BS (En55)	0.20	0.40	11.50	23.00	W 3.00 Si 1.60	Softened	RT	718 (46.5)	402 (26)	28	34
								400	610 (39.5)	301 (19.5)	30	42
								600	456 (29.5)	247 (16)	29	36
								800	232 (15)	Not given	41	41
25/12CrNi Steel	—	0.10	1.50	12.00	25.00	—	Softened	RT	656 (42.5)	340 (22)	40	70
								400	556 (36)	293 (19)	37	58
								600	432 (28)	239 (15.5)	30	55
								800	170 (11)	139 (9)	Not given	Not given

24/20CrNi Steel	BS 310S24	0.14	0.80	21.00	24.00	Si	1.00	Softened	RT	649 (42)	317 (20.5)	43	63
									400	556 (36)	201 (13)	37	63
									600	463 (30)	193 (12.5)	28	42
									800	193 (12.5)	Not given	40	42
18/18/7 CrNiCoMo-CuTi Steel	—	0.21	0.80	17.50	17.00	Co	7.00	Precipitation hardened	RT	687 (44.5)	409 (26.5)	32	43
						Mo	2.50		400	579 (37.5)	355 (23)	28	41
						Cu	2.50		600	510 (33)	355 (23)	22	29
						Ti	0.70		800	332 (21.5)	293 (19)	20	29
20/20/20 CrNiCoMo-WNb Steel	AICMA FE-PA91-HT	0.12	1.50	20.00	21.00	Co	20.00	Precipitation hardened	RT	811 (52.5)	394 (25.5)	35	48
						Mo	3.00		400	734 (47.5)	386 (25)	30	47
						W	2.00		600	602 (39)	324 (21)	29	45
						Nb	1.00		800	340 (22)	208 (13.5)	29	42
35/15NiCr Steel	—	0.10	1.00	35.00	15.00	—		Softened	RT	571 (37)	293 (19)	41	70
									400	463 (30)	224 (14.5)	28	42
									600	355 (23)	170 (11)	22	27
									800	170 (11)	139 (9)	16	20

Table 22.46 TYPICAL FATIGUE STRENGTH OF STEELS ON SMOOTH SPECIMENS AT ROOM AND HIGHER TEMPERATURES

		Typical composition %						*Fatigue limit for 10^7 cycles of stress at temperatures indicated,* MPa (tonf in^{-2})								
Material	*British or other standards*	C	Mn	Ni	Cr	*Other elements*	*Condition*	RT	100°C	200°C	300°C	400°C	500°C	600°C	650°C	*Remarks*
1. *Forged or rolled steels tested in longitudinal direction* *Carbon steels with up to 1.5% Mn content*																
Armco Iron	—	0.02	0.03	—	—	—	As rolled	±185 (±12)	±170 (±11)	±178 (±11.5)	±232 (±15)	±178 (±11.5)	±116 (±7.5)	—	—	—
0.20C Steel	BS070 M20 (En3)	0.20	0.70	—	—	—	Normalized	±193 (±12.5)	±193 (±12.5)	±193 (±12.5)	±247 (±16)	±232 (±15)	±154 (±10)	—	—	—
0.25C Steel	BS070 M26 (En4)	0.26	0.70	—	—	—	Normalized	±201 (±13)	±193 (±12.5)	±193 (±12.5)	±247 (±16)	±263 (±17)	±185 (±12)	—	—	—
0.30C Steel	BS080 M30 (En5)	0.30	0.80	—	—	—	Normalized	±232 (±15)	—	—	—	—	—	—	—	—
0.40C Steel	BS080 M40 (En8)	0.40	0.80	—	—	—	Hardened and tempered	±278 (±18)	—	—	—	—	—	—	—	—
0.55C Steel	BS070 M55 (En9)	0.55	0.65	—	—	—	Hardened and tempered	±293 (±19)	—	—	—	—	—	—	—	—
CMn Steel	BS150 M19 (En14A)	0.28	1.50	—	—	—	Normalized	±278 (±18)	—	—	—	—	—	—	—	—
Low alloy weldable steel																
½Mo Steel	—	0.14	0.43	—	—	Mo 0.50	Normalized	±317 (±20.5)	—	—	±402 (±26)	±371 (±24)	±278 (±18)	—	—	—
Low alloy direct hardening steel																
2NiCrMo Steel	—	0.27	Not given	2.0	0.90	Mo 0.40	Hardened and tempered	±432 (±28)	±432 (±28)	±432 (±28)	±440 (±28.5)	±432 (±28)	±247 (±16)	—	—	—
2¾CrMoVW Steel	—	0.23	0.30	—	2.70	Mo 0.50 V 0.70 W 0.50	Hardened and tempered	±440 (±28.5)	—	—	—	—	±324 (±21)	±247 (±16)	—	4×10^7 cycles

3½Ni Steel	BS(En22)	0.40	0.80	3.50	—		Hardened and tempered	±525 (±34)	±533 (±34.5)	±556 (±36)	±556 (±36)	±517 (±33.5)	±394 (±25.5)	—	—	Rotating bending
3CrMo Steel	BS722M24 (En40B)	0.24	0.60	—	3.30	Mo 0.55	Hardened and tempered	±293 (±19)	—	—	—	—	—	—	—	—
3NiCr Steel	BS653M31 (En23)	0.31	0.60	3.0	0.90		Hardened and tempered	±432 (±28)	—	—	—	—	—	—	—	—
3NiCrMoV Steel	Similar to BS976M33	0.29	0.40	3.1	0.70	Mo 0.50 V 0.20	Hardened and tempered	±486 (±31.5)	—	—	—	—	—	—	—	—
4NiCrMo Steel	Similar to BS835M30 (En30B)	0.30	0.50	3.9	1.20	Mo 0.35	Hardened and tempered	±525 (±34)	±432 (±28)	—	—	—	—	—	—	—
4½NiCrMo Steel	—	0.20	0.57	4.65	1.40	Mo 0.58	Hardened and tempered	±571 (±37)	±448 (±29)	—	—	—	—	—	—	—
High alloy steels																
13Cr Steel	BS410S21 (En56A)	0.12	0.50	—	13.0	—	Hardened and tempered	±340 (±22)	—	—	—	—	—	—	—	—
13Cr Steel	BS420S37 (En56C)	0.24	0.50	—	13.0	—	Hardened and tempered	±402 (±26)	±394 (±25.5)	±386 (±25)	±355 (±23)	±309 (±20)	±201 (±13)	—	—	—
17CrNi Steel	BS431S29 (En57)	0.16	0.50	2.50	16.50	—	Hardened and tempered	±371 (±24)	—	—	—	—	—	—	—	—
18/10CrNi Steel	BS304S15 (En58E)	0.06	0.80	10.0	18.0	—	Softened	±263 (±17)	—	—	—	—	—	—	—	—
18/9CrNi Steel	BS302S17	0.07	1.50	9.5	18.0	—	Softened	±278 (±18)	—	—	—	±216 (±14)	±216 (±14)	—	±208 (±13.5)	—
18/10CrNi Steel	BS302S25 (En58A)	0.12	1.50	9.7	18.5	—	Softened	±293 (±19)	—	—	—	±255 (±16.5)	±263 (±17)	—	±216 (±14)	—
18/8CrNiTi Steel	BS321S20 (En58B)	0.10	0.80	8.5	18.0	Ti 0.60	Softened	±270 (±17.5)	—	—	—	—	—	—	—	—

Table 22.46 TYPICAL FATIGUE STRENGTH OF STEELS ON SMOOTH SPECIMENS AT ROOM AND HIGHER TEMPERATURES—*continued*

		Typical composition %						*Fatigue limit for* 10^7 *cycles of stress at temperatures indicated,* MPa (tonf in^{-2})								
Material	*British or other standards*	C	Mn	Ni	Cr	*Other elements*	*Condition*	RT	100°C	200°C	300°C	400°C	500°C	600°C	650°C	*Remarks*
18/10CrNi-Nb Steel	Similar to BS347S17 (En58G)	0.08	1.50	11.0	17.5	Nb 1.20	Softened	±301 (±19.5)	—	—	—	—	—	±208 (±13.5)	±178 (±11.5)	—
18/9CrNiMo Steel	BS315S16 (En58H)	0.07	1.00	9.5	18.0	Mo 1.25	Softened	±270 (±17.5)	—	—	—	—	—	—	—	—
18/8CrNiMo Steel	—	0.07	0.50	8.0	18.0	Mo 2.70	Softened	±270 (±17.5)	—	—	—	—	—	—	—	—
18/18CrNi-MoCuTi Steel	—	0.07	0.80	18.0	18.0	Mo 3.70 Cu 2.40 Ti 0.60	Precipitation hardened	±263 (±17)	—	—	—	—	—	—	—	—
2. *Steel castings*																
Carbon steel castings	BS592 Grade B	0.32	0.77	—	—	—	Annealed	±229 (±14.8)	—	—	—	—	—	—	—	—
							Normalized and tempered	±258 (±16.7)								
$1\frac{1}{2}$Mn Steel Castings	BS1456 Grade B1	0.31	1.60	—	—	—	Normalized and tempered	±334 (±21.6)	—	—	—	—	—	—	—	—
	BS1456 Grade B2	0.31	1.60	—	—	—	Hardened and tempered	±403 (±26.1)	—	—	—	—	—	—	—	—
$\frac{1}{2}$NiCrMo Steel castings	BS1458 Grade A	0.36	0.89	0.38	0.59	Mo 0.32	Hardened and tempered	±372 (±24.1)	—	—	—	—	—	—	—	—
$1\frac{3}{4}$NiCrMo Steel castings	BS1458 Grade C	0.34	0.60	1.74	0.70	Mo 0.30	Hardened and tempered	±534 (±34.6)	—	—	—	—	—	—	—	—

Table 22.47 TYPICAL HOT CREEP AND RUPTURE PROPERTIES OF FORGED OR ROLLED STEELS (BS1501:Part 2:1970, DIN 17175:1959, 17240:1959, Stahl-Eisen Werkstoffblatt 550–57, 670–69)

		Typical composition %							*Stress to produce 1.0% strain*		*Stress to cause rupture*	
Material	*British or other standards*	C	Mn	Ni	Cr	*Other elements*	*Condition*	*Temperature* °C	MPa ($tonf\,in^{-2}$) in 10 000 h	MPa ($tonf\,in^{-2}$) in 100 000 h	MPa ($tonf\,in^{-2}$) in 10 000 h	MPa ($tonf\,in^{-2}$) in 100 000 h
1. *Carbon steels*												
0.15C Steel	BS 040A12	0.13	0.50	—	—	—	Normalized	400	136 (8.8)	96 (6.2)	192(12.4)	133 (8.6)
	DIN St35.8							450	80 (5.2)	49 (3.2)	113 (7.3)	68 (4.4)
								500	39 (2.5)	20 (1.3)	54 (3.5)	29 (1.9)
0.20C Steel	BS 070M20	0.20	0.70	—	—	—	Normalized	400	136 (8.8)	96 (6.2)	192(12.4)	133 (8.6)
	DIN St45.8							450	80 (5.2)	49 (3.2)	113 (7.3)	68 (4.4)
								500	39 (2.5)	20 (1.3)	54 (3.5)	29 (1.9)
0.35C Steel	BS 080M36	0.36	0.80	—	—	—	Hardened and	400	147 (9.5)	99 (6.4)	193(12.5)	137 (8.9)
	DIN Ck35						tempered	450	80 (5.2)	49 (3.2)	113 (7.3)	68 (4.4)
								500	39 (2.5)	22 (1.4)	54 (3.5)	34 (2.2)
0.45C Steel	BS 080M46	0.46	0.80	—	—	—	Hardened and	400	147 (9.5)	99 (6.4)	193(12.5)	137 (8.9)
	DIN Ck45						tempered	450	80 (5.2)	49 (3.2)	113 (7.3)	68 (4.4)
								500	39 (2.5)	22 (1.4)	54 (3.5)	34 (2.2)
2. *Low alloy weldable steels*												
½Mo Steel	DIN 15Mo3	0.16	0.65	—	—	Mo 0.30	Normalized	450	216(14.0)	167(10.8)	304(19.7)	246(15.9)
								500	133 (8.6)	74 (4.8)	176(11.4)	93 (6.0)
								530	85 (5.5)	37 (2.4)	105 (6.8)	48 (3.1)
1CrMo Steel	BS1501–620	0.15	0.70	—	0.95	Mo 0.50	Normalized	450	247(16.0)	193(12.5)	371(24.0)	286(18.5)
	DIN 13CrMo44						and	500	162(10.5)	100 (6.5)	232(15.0)	139 (9.0)
							tempered	560	63 (4.1)	31 (2.0)	93 (6.0)	40 (2.6)
1¼CrMo Steel	BS1501–621	0.12	0.55	—	1.25	Mo 0.55	Normalized	450	309(20.0)*	224(14.5)*	432(28.0)	314(20.3)
							and	500	181(11.7)*	119 (7.7)*	258(16.7)	168(10.9)
							tempered	560	69 (4.5)*	31 (2.0)*	100 (6.5)	40 (2.6)
2¼CrMo Steel	BS1501-622	0.13	0.60	—	2.20	Mo 1.10	Normalized	500	148 (9.6)	103 (6.7)	216(14.0)	147 (9.5)
	DIN 10CrMo910						and	550	83 (5.4)	53 (3.4)	116 (7.5)	77 (5.0)
							tempered	580	57 (3.7)	31 (2.0)	77 (5.0)	45 (2.9)

* Estimated

Table 22.47 TYPICAL HOT CREEP AND RUPTURE PROPERTIES OF FORGED OR ROLLED STEELS (BS 1501:Part 2:1970, DIN 17175:1959, 17240:1959, Stahl-Eisen Werkstoffblatt 550–57, 670–69)—*continued*

Material	*British or other standards*	*Typical composition %* C	Mn	Ni	Cr	*Other elements*	*Condition*	*Temperature* °C	*Stress to produce 1.0% strain* MPa (tonf in^{-2}) in 10 000 h	*Stress to produce 1.0% strain* MPa (tonf in^{-2}) in 100 000 h	*Stress to cause rupture* MPa (tonf in^{-2}) in 10 000 h	*Stress to cause rupture* MPa (tonf in^{-2}) in 100 000 h
1¼MnCrMoV	BS1501-271	0.14	1.25	—	0.55	Mo 0.24	Normalized	450	263(17.0)*	216(14.0)*	371(24.0)	309(20.0)
Steel						V 0.08	and	500	170(11.0)*	100 (6.5)*	235(15.2)	139 (9.0)
							tempered	550	74 (4.8)*	28 (1.8)*	102 (6.6)	39 (2.5)
1NiCrMoV	BS1501-281	0.12	1.10	0.85	0.55	Mo 0.24	Normalized	450	216(14.0)*	178(11.5)*	297(19.2)	247(16.0)
Steel						V 0.08	and	500	139 (9.0)*	85 (5.5)*	187(12.1)	111 (7.2)
							tempered	550	62 (4.0)*	23 (1.5)*	82 (5.3)	31 (2.0)
1½NiCrMoV	BS1501-282	0.15	1.10	1.50	0.50	Mo 0.35	Normalized	450	216(14.0)*	178(11.5)*	297(19.2)	247(16.0)
Steel						V 0.10	and	500	139 (9.0)*	85 (5.5)*	187(12.1)	111 (7.2)
							tempered	550	62 (4.0)*	23 (1.5)*	182 (5.3)	31 (2.0)
3. *Low alloy direct hardening steels*												
1CrMo Steel	DIN 24CrMo5	0.24	0.65	—	1.00	Mo 0.25	Hardened	450	227(14.7)	171(11.1)	310(20.1)	225(14.6)
							and	500	139 (9.0)	93 (6.0)	176(11.4)	117 (7.6)
							tempered	550	62 (4.0)	25 (1.6)	79 (5.1)	37 (2.4)
1½CrMoV	DIN 24CrMoV	0.24	0.45	—	1.35	Mo 0.55	Hardened	450	303(19.6)	239(15.5)	405(26.2)	321(20.8)
Steel	55					V 0.20	and	500	193(12.5)	128 (8.3)	256(16.6)	184(11.9)
							tempered	550	93 (6.0)	56 (3.6)	139 (9.0)	74 (4.8)
1½CrMoV	DIN 21CrMoV	0.21	0.40	—	1.35	Mo 1.10	Hardened	450	340(22.0)	276(17.9)	423(27.4)	349(22.6)
Steel	511					V 0.30	and	500	230(14.9)	165(10.7)	303(19.6)	212(13.7)
							tempered	550	120 (7.8)	65 (4.2)	156(10.1)	93 (6.0)
2½CrMoV	DIN	0.30	0.55	—	2.50	Mo 0.20	Hardened	450	246(15.9)	187(12.1)	324(21.0)	264(17.1)
Steel	30CrMoV92					V 0.15	and	500	158(10.2)	99 (6.4)	216(14.0)	137 (8.9)
							tempered	550	74 (4.8)	36 (2.3)	99 (6.4)	46 (3.0)
4. *High alloy steels*												
9CrMo Steel	—	0.12	0.44	—	9.50	Mo 0.95	Hardened	590	43 (2.8)	14 (0.9)	59 (3.8)*	19 (1.2)*
							and	650	19 (1.2)	6 (0.4)	26 (1.7)	8(0.55)*
							tempered					

12CrMo Steel	DIN X19CrMo12 1	0.19	0.55	—	12.0	Mo 1.00	Hardened and tempered	500	241(15.6)	187(12.1)	300(19.4)	235(15.2)
								550	137 (8.9)	93 (6.0)	176(11.4)	117 (7.6)
								600	59 (3.8)	34 (2.2)	83 (5.4)	46 (3.0)
12CrMoV Steel	DIN X22CrMoV12 1	0.22	0.55	—	12.00	Mo 1.00 V 0.30	Hardened and tempered	500	289(18.7)	221(14.3)	338(21.9)	275(17.8)
								550	165(10.7)	108 (7.0)	212(13.7)	137 (8.9)
								600	79 (5.1)	45 (2.9)	103 (6.7)	59 (3.8)
18/9CrNi Steel	BS304S15 (En58E)	0.06	0.80	10.00	18.0	—	Softened	600	85 (5.5)	46 (3.0)	116 (7.5)*	62 (4.0)*
								650	57 (3.7)	29 (1.9)	77 (5.0)*	39 (2.5)*
								700	39 (2.5)	17 (1.1)	54 (3.5)*	23 (1.5)*
16/13CrNiNb Steel	DIN X8CrNiNb 16 13	0.07	1.00	13.0	16.00	Nb 10 × C/1.2	Softened	600	113 (7.3)	79 (5.1)	158(10.2)	108 (7.0)
								650	79 (5.1)	49 (3.2)	103 (6.7)	63 (4.1)
								700	49 (3.2)	26 (1.7)	63 (4.1)	34 (2.2)
								750	34 (2.2)	15 (1.0)	45 (2.9)	20 (1.3)
17/13CrNiMo Steel	BS316S16 (En58J)	0.07	1.40	12.50	16.50	Mo 2.80	Softened	600	139 (9.0)	77 (5.0)	185(12.0)*	108 (7.0)*
								650	100 (6.5)	46 (3.0)	131 (8.5)*	62 (4.0)*
								700	63 (4.1)	29 (1.9)	85 (5.5)*	39 (2.5)*
								750	43 (2.8)	20 (1.3)	56 (3.6)*	26 (1.7)*
16/16 CrNiMoNb Steel	DIN X8CrNiMoNb 16 16	0.07	1.00	16.50	16.50	Mo 1.80 Nb 10 × C/1.2	Softened	600	158(10.2)	108 (7.0)	225(14.6)	151 (9.8)
								650	108 (7.0)	63 (4.1)	137 (8.9)	83 (5.4)
								700	63 (4.1)	34 (2.2)	83 (5.4)	45 (2.9)
								750	42 (2.7)	15 (1.0)	54 (3.5)	20 (1.3)
16/16 CrNiWNb Steel	DIN X6CrNiWNb 16 16	0.06	1.00	16.50	16.50	W 3.00 Nb 10 × C/1.2 N 0.10	Softened	600	167(10.8)	117 (7.6)	235(15.2)	162(10.5)
								650	113 (7.3)	68 (4.4)	147 (9.5)	90 (5.8)
								700	68 (4.4)	39 (2.5)	91 (5.9)	49 (3.2)
								750	39 (2.5)	17 (1.1)	54 (3.5)	23 (1.5)
25/13CrNi Steel	—	0.06	1.55	13.40	24.9	—	Softened	650	69 (4.5)	39 (2.5)	93 (6.0)*	54 (3.5)*
								700	48 (3.1)	17 (1.1)	62 (4.0)*	23 (1.5)*
24/20CrNi Steel	BS 310S24	0.14	0.80	21.00	24.0	Si 1.00	Softened	600	154(10.0)	77 (5.0)	224(14.5)*	103 (6.7)*
								650	103 (6.7)	56 (3.6)	139 (9.0)*	74 (4.8)*
								700	65 (4.2)	34 (2.2)	86 (5.6)*	45 (2.9)*

*Estimated

Table 22.48 MECHANICAL PROPERTIES OF FORGED OR ROLLED STEELS AT SUBZERO TEMPERATURES IN THE LONGITUDINAL DIRECTION

		Typical composition %							*Typical properties*					
													Impact	
Material	*British or other standards*	C	Mn	Ni	Cr	*Other elements*	*Conditions*	*Temperature* °C	*UTS* MPa (tonf in^{-2})	*Yield stress* MPa (tonf in^{-2})	*Elong.* (gl $5.65\sqrt{S_0}$) %	*RA* %	*Test spec.*	J (ft lbf)
1. *Carbon steels*														
Armco iron	—	0.02	0.03	—	—	—	Not given	RT	309 (20)	Not given	Not given	73	Izod	106 (78)
								−75	432 (28)	293 (19)	33	72	Izod	5.5(4)
								−120	525 (34)	463 (30)	15	68	—	—
0.1C Steel	BS 040A10 (En2A)	0.10	0.45	—	—	—	As rolled	RT	463 (30)	340 (22)	27	72	Charpy	80 (59)
								−30	—	—	—	—	Charpy	60 (44)
								−45	—	—	—	—	Charpy	8 (6)
								−160	649 (42)	556 (36)	24	62	—	—
0.15C Steel	BS 040A12 (En2B)	0.13	0.50	—	—	—	Normalized	RT	463 (30)	355 (23)	28	67	Charpy	163(120)*
								−65	—	—	—	—	Charpy	9.0(6.5)*
								−160	710 (46)	571 (37)	30	58	—	—
0.20C Steel	BS 070M20 (En3)	0.20	0.70	—	—	—	Normalized	RT	430 (28)	215 (14)	21	65	Charpy	99 (73)*
								−20	—	—	—	—	Charpy	13.5(10)*
								−80	—	—	—	—	Charpy	2.7(2)*
								−160	726 (47)	587 (38)	Not given	56	—	—
0.25C Steel	BS 0/0M26 (En4)	0.26	0.70	—	—	—	Normalized	RT	494 (32)	247 (16)	20	Not given	—	—
							Cold drawn	RT	571 (37)	432 (28)	11	Not given	Izod	27 (20)
								−40	757 (49)	Not given	11	Not given	Izod	3.4(2.5)
0.35C Steel	BS 080M36	0.36	0.80	—	—	—	Normalized	RT	541 (35)	278 (18)	27	62	Charpy	110 (81)*
								−55	—	—	—	—	Charpy	30 (22)*
								−80	—	—	—	—	Charpy	11 (8)*
								−100	741 (48)	571 (37)	28	57	—	—
0.40C Steel	BS 080M40 (En8)	0.40	0.80	—	—	—	Hardened and tempered	RT	772 (50)	463(30)	22	Not given	—	—
								−185	1 112 (72)	Not given	9	Not given	—	—
0.45C Steel	BS 080M46	0.46	0.80	—	—	—	Normalized	RT	618 (40)	309 (20)	15	Not given	Charpy	22 (16)*
								−80	896 (58)	Not given	14	Not given	Charpy	11 (8)*
							Hardened and tempered	RT	927 (60)	525 (34)	14	Not given	Charpy	68 (50)*
								−40	—	—	—	—	Charpy	36.5(27)*
								−80	1 050 (68)	Not given	12	Not given	Charpy	24.5(18)*

0.50C Steel	BS 080M50 (En9) (En43A)	0.50	0.80	—	—	—	Normalized	RT	618 (40)	309 (20)	14	Not given	—	—
								−30	803 (52)	Not given	12	Not given	—	—
2. *Low alloy weldable steels*														
MnCrMoV Steel	BS1501–271	0.14	1.40	—	0.70	Mo 0.28 V 0.10	Normalized and tempered	RT	649 (42)	463 (30)	18	Not given	Charpy V Izod	41 (30) 68 (50)
								0	—	—	—	—	Charpy V Izod	27 (20) 51.5(38)
								−60	—	—	—	—	Izod	16 (12)
1NiCrMoV Steel	BS 1501–281	0.12	1.10	0.85	0.55	Mo 0.24 V 0.08 Nb 0.10	Normalized and tempered	RT	649 (42)	417 (27)	18	Not given	—	—
								−10	—	—	—	—	Charpy V	68 (50)
								−20	—	—	—	—	Charpy V	54 (40)
								−40	—	—	—	—	Charpy V	27 (20)
$1\frac{1}{2}$NiCrMoV Steel	BS 1501–282	0.15	1.10	1.50	0.50	Mo 0.35 V 0.10	Double normalized and tempered	RT	649 (42)	432 (28)	18	Not given	—	—
								−10	—	—	—	—	Charpy V	81 (60)
								−20	—	—	—	—	Charpy V	68 (50)
								−40	—	—	—	—	Charpy V	41 (30)
								−50	—	—	—	—	Charpy V	27 (20)
$3\frac{1}{2}$Ni Steel	BS 1501–503	0.15	0.55	3.50	—	—	Normalized and tempered	RT	448 (29)	263 (17)	20	Not given	—	—
								−80	—	—	—	—	Charpy V	34 (25)
								−100	—	—	—	—	Charpy V	17.5(13)
5Ni Steel	DIN 12Ni19	0.13	0.40	5.10	—	—	Hardened and tempered	RT	710 (46)	Not given	22	74	—	—
								−150	1 050 (68)	Not given	22	57	—	—
								−195	1 205 (78)	Not given	18	50	—	—
3. *Low alloy direct hardening steels*														
1CrV Steel	—	0.29	0.70	—	1.00	V 0.20	Hardened and tempered	RT	896 (58)	Not given	13	Not given	—	—
								−75	1 035 (67)	Not given	13	Not given	—	—
$\frac{3}{4}$CrMo Steel	—	0.31	0.65	—	0.70	Mo 0.20	Hardened and tempered	RT	880 (57)	Not given	17	Not given	Izod	110 (81)
								−40	1 035 (67)	Not given	17	Not given	Izod	109 (80)
1CrMo Steel	BS 709M40 (En19)	0.40	0.80	—	1.10	Mo 0.30	Hardened and tempered	RT	849 (55)	494 (32)	15	70	Charpy	141 (104)*
								−70	—	—	—	—	Charpy	84 (62)*
								−185	1 251 (81)	1 143 (74)	19	47	—	—

* Charpy specimen—Izod notch V = Charpy specimen—V notch.

Table 22.48 MECHANICAL PROPERTIES OF FORGED OR ROLLED STEELS AT SUBZERO TEMPERATURES IN THE LONGITUDINAL DIRECTION—*continued*

		Typical composition %						*Typical properties*						
Material	*British or other standards*	C	Mn	Ni	Cr	*Other elements*	*Condition*	*Temperature* °C	*UTS* MPa (tonf in^{-2})	*Yield stress* MPa (tonf in^{-2})	*Elong.* (gl $5.65\sqrt{S_0}$) %	*RA* %	*Impact* *Test spec.*	*Impact* J (ft lbf)
$1\frac{1}{4}$NiCr Steel	BS 640M40 (En111)	0.40	0.80	1.30	0.65	—	Hardened and tempered	RT	1 004 (65)	680 (44)	13	60	Izod	54 (40)
								−185	1 699 (110)	Not given	4.5	48	—	—
1NiCrMo Steel	—	0.47	0.80	1.10	1.00	Mo 0.20	Hardened and tempered	RT	1 081 (70)	Not given	12	Not given	—	—
								−76	1 220 (79)	Not given	11	Not given	—	—
$2\frac{1}{2}$NiCrMo Steel	BS 826M31 (En25)	0.31	0.55	2.70	0.60	Mo 0.55	Hardened and tempered	RT	1 004 (65)	741 (48)	13	65	Izod	34 (25)
								−60	1 127 (73)	988 (64)	12	63	—	—
								−180	1 390 (90)	1 266 (82)	12	63	—	—
$2\frac{1}{2}$NiCrMo Steel	—	0.34	0.50	2.30	1.90	Mo 0.40	Hardened and tempered	RT	1 158 (75)	1 019 (66)	15	65	—	—
								−70	1 251 (81)	1 127 (73)	15	63	—	—
								−185	1 560 (101)	1 390 (90)	17	62	—	—
$3\frac{1}{4}$Ni Steel	BS (En21)	0.34	0.70	3.30	—	—	Hardened and tempered	RT	927 (60)	865 (56)	15	64	Charpy	109 (80)*
								−75	1 066 (69)	988 (64)	14	60	Charpy	38 (28)*
3NiCr Steel	BS 653M31 (En23)	0.31	0.60	0.90	0.90	—	Hardened and tempered	RT	1 004 (65)	757 (49)	12	60	Charpy	80 (59)
								−95	—	—	—	—	Charpy	61 (45)
								−185	1 683 (109)	1 668 (108)	5	49	—	—
$4\frac{1}{2}$NiCr Steel	BS (En30A)	0.30	0.50	4.40	1.40		Hardened and tempered	RT	865 (56)	Not given	24	Not given	—	—
								−30	1 019 (66)	Not given	26	Not given	—	—
								−90	1 050 (68)	Not given	23	Not given	—	—
4NiCrMo Steel	BS 835M30 (En30B)	0.30	0.50	3.90	1.20	Mo 0.35	Hardened and tempered	RT	1 544 (100)	1 235 (80)	17	54	Charpy	34 (25)*
								−80	1 761 (114)	1 266 (82)	17	58	—	—
								−100	—	—	—	—	Charpy	20.5 (15)*
								−196	2 039 (132)	1 544 (100)	17	50	—	—
4. *High alloy steels*														
9Ni Steel	BS 1501–509, 1503–509 DIN X8Ni9 ASTM A353	0.10	0.55	9.25	—	—	Hardened and tempered	RT	695 (45)	525 (34)	18	Not given	Charpy	76 (56)
								−100	—	—	—	—	Charpy V	68 (50)
								−160	—	—	—	—	Charpy V	47 (35)
								−196	1 158 (75)	911 (59)	18	Not given	Charpy	64 (47)
													Charpy V	34 (25)

13Cr Steel	BS 420S45 (En56D)	0.32	0.40	—	13.5	—	Hardened and tempered	RT	927 (60)	587 (38)	13	37	—	—
								−185	1 792 (116)	1 452 (94)	3	4	—	—
17Cr Steel	BS 430S15 (En60)	0.09	0.45	—	17.0	—	Softened	RT	432 (28)	278 (18)	25	71	Charpy	61 (45)*
								−20	—	—	—	—	Charpy	34 (25)*
								−60	—	—	—	—	Charpy	6.8 (5)*
								−185	1 004 (65)	849 (55)	12	14	—	—
20CrCu Steel	—	0.25	0.50	—	20.0	Cu 1.0	Softened	RT	633 (41)	355 (23)	21	59	—	—
								−185	680 (44)	618 (40)	Not given	Not given	—	—
17CrNi Steel	BS 431S29 (En57)	0.14	Not given	2.50	17.5	—	Hardened and tempered	RT	1 004 (65)	680 (44)	21	Not given	—	—
								−90	1 050 (68)	Not given	18	Not given	—	—
18/10CrNi Steel	BS 302S25 (En58A) DIN X12Cr-Ni 18.9	0.12	0.50	9.7	18.5	—	Softened	RT	510 (33)	216 (14)	47	74	Izod	155(114)U
								−60	942 (61)	340(22)	52	74	Izod	146(108)U
								−120	1 236 (80)	494 (32)	43	63	Izod	146(108)U
								−180	1 544 (100)	510 (33)	40	55	Izod	146(108)U
18/8Cr NiMo Steel		0.05	0.50	8.20	18.0	Mo 2.70	Softened	RT	710 (46)	Not given	58	Not given	—	—
								−30	896 (58)	Not given	51	Not given	—	—
								−55	1 050 (68)	Not given	42	Not given	—	—
								−90	1 174 (76)	Not given	35	Not given	—	—
14/14CrNi Steel	BS (En58D)	0.11	Not given	14.0	14.2	—	Softened	RT	602 (39)	232 (15)	44	76	Izod	159 (117)U
								−65	772 (50)	293 (19)	51	76	Izod	160 (118)U
								−120	958 (62)	417 (27)	49	72	Izod	159 (117)U
								−180	1 328 (86)	819 (53)	45	60	Izod	157 (116)U
25/20CrNi Steel	BS 310S24	0.14	0.80	21.0	24.0	Si 1.00	Softened	RT	649 (42)	317 (20.5)	43	Not given	Charpy	203 (150)*
								−30	741 (48)	Not given	52	Not given	—	—
								−50	—	—	—	—	Charpy	190 (140)*
								−90	849 (55)	Not given	51	Not given	—	—
								−150	—	—	—	—	Charpy	122 (90)*
26Ni Steel	—	0.40	1.50	26.0	—	—	Softened	RT	695 (45)	371 (24)	33	52	—	—
								−185	1 421 (92)	1 097 (71)	10	10	—	—
36Ni Steel	—	0.16	0.90	36.0	—	—	Softened	RT	556 (36)	355 (23)	28	58	—	—
								−250	988 (64)	880 (57)	18	60	—	—

*Charpy specimen—Izod notch U = Unbroken V = Charpy specimen—V notch U = unbroken.

Table 22.49 TOOL STEELS AND THEIR USES
(BS 4659:1971, Stahl-Eisen Werkstoffblatt 150:1971, 200:1969, 250:1970, 320:1969)

Material	*British or other standards*	*Chemical composition limits %* C	Si	Mn	Cr	Mo	W	V	Co	*Other elements*	*Typical uses*
1. *Carbon tool steels*											
0.8C Steel	DIN C80W1	0.75 0.85	0.10 0.25	0.10 0.25	— —	— —	— —	— —	— —	— —	Cold heading dies, chisels, mandrels, punches, rivet snaps, vice jaws, drifts
0.9C Steel	BS BW1A	0.85 0.95	0.30 max.	0.35 max	0.15 max	0.10 max	— —	— —	— —	Ni 0.20 max	Gauges, chisels, punches, shear blades
1C Steel	BS BW1B	0.95 1.10	0.30 max	0.35 max	0.15 max	0.10 max	— —	— —	— —	Ni 0.20 max	Large drills, reamers, cutters, cold heading tools, shear blades, woodworking tools, lathe centres
1.2C Steel	BS BW1C	1.10 1.30	0.30 max	0.35 max	0.15 max	0.10 max	— —	— —	— —	Ni 0.20 max	Twist drills, cutters, reamers, taps, files, screw gauges, woodworking tools
1CV Steel	BS BW2	0.95 1.10	0.30 max	0.35 max	0.15 max	0.10 max	— —	0.15 0.35	— —	Ni 0.20 max	Large drills, reamers, cutters, cold heading tools, shear blades, woodworking tools, lathe centres, when fine grained shallow case is required with increased toughness
2. *Hot work tool steels*											
$1\frac{3}{4}$ NiCr-MoV Steel	DIN 55Ni-CrMoV6 Similar to BS No5 die steel	0.55	0.30	0.60	0.70	0.30 (Typical)	—	0.10	—	Ni 1.70	Solid dies for drop-hammer or press forging
3CrMoV Steel	BS BH10	0.30 0.40	1.10 max	0.40 max	2.80 3.20	2.65 2.95	— —	0.30 0.50	— —	— —	Mandrels, hot extrusion and forging dies, punches, die inserts, gripper and header dies, hot shears, aluminium die casting dies
3CrMoV Co Steel	BS BH10A	0.30 0.40	1.10 max	0.40 max	2.80 3.20	2.65 2.95	— —	0.30 1.10	2.80 3.20	— —	Closed die forging tools
5CrMoV Steel	BS BH11	0.32 0.42	0.85 1.15	0.40 max	4.75 5.25	1.25 1.75	— —	0.30 0.50	— —	— —	Die casting dies and die inserts for light alloys, punches, piercing tools, mandrels, forging dies and die inserts, ejector pins, hot extrusion dies, sleeves, slides
5CrMoV Steel	BS BH13	0.32 0.42	0.85 1.15	0.40 max	4.75 5.25	1.25 1.75	— —	0.90 1.10	— —	— —	As for BH11, when higher performance is required

Table 22.49 TOOL STEELS AND THEIR USES—*continued*
(BS 4659:1971, Stahl-Eisen Werkstoffblatt 150:1971, 200:1969, 250:1970, 320:1969)

Material	*British or other standards*	*Chemical composition limits %* C	Si	Mn	Cr	Mo	W	V	Co	*Other elements*	*Typical uses*
5CrMo-WV Steel	BS BH12	0.30 0.40	0.85 1.15	0.40 max	4.75 5.25	1.25 1.75	1.25 1.75	0.50 max	— —	— —	Extrusion dies, gripper and header dies, forging die inserts, punches, mandrels, sleeves
4WCrV Steel	DIN X30-WCrV53	0.30	0.20	0.30	2.40	— (Typical)	4.30	0.60	—	—	Mandrels, hot extrusion and forging dies, punches, die inserts, gripper and header dies, hot shears, aluminium die casting dies
4WCrCo-VMo Steel	BS BH19	0.35 0.45	0.40 max	0.40 max	4.00 4.50	0.45 max	4.00 4.50	2.00 2.40	4.00 4.50	— —	Extrusion dies and die inserts, forging die inserts, punches and mandrels when the highest performance is required
9WCr-MoV Steel	BS BH21	0.25 0.35	0.40 max	0.40 max	2.25 3.25	0.60 max	8.50 10.00	0.40 max	— —	— —	Mandrels, hot blanking dies, hot punches, extrusion dies and die casting dies for brass, piercer points, gripper dies, hot headers–when high performance is required
9WCrMo VNi Steel	BS BH21A	0.20 0.30	0.40 max	0.40 max	2.25 3.25	0.60 max	8.50 10.00	0.50 max	— —	Ni 2.0 2.5	Hot extrusion tools
18WCrMo VCo Steel	BS BH26	0.50 0.60	0.40 max	0.40 max	3.75 4.50	0.60 max	17.50 18.50	1.00 1.50	0.60 max	— —	Closed die forging tools
3. *Shock resisting tool steels*											
$2\frac{1}{4}$WCrV Steel	BS BS1	0.45 0.55	0.70 1.00	0.30 0.70	1.20 1.70	— —	2.00 2.50	0.10 0.30	— —	— —	Bolt header dies, chipping and caulking chisels, concrete drills, forming dies, grippers, mandrels, punches, pneumatic tools, beading tools, scarfing tools, swaging dies, shear blades
1SiMoV Steel	BS BS2	0.45 0.55	0.90 1.20	0.30 0.50	— —	0.30 0.60	— —	0.10 0.30	— —	— —	Hand and pneumatic chisels, forming tools, ejector pins, mandrels, shear blades, spindles, stamps, tool shanks
2SiCrV Steel	BS BS5	0.50 0.60	1.60 2.10	0.60 0.80	— —	0.30 0.60	— —	0.10 0.30	— —	— —	Hand and pneumatic chisels, forming tools, ejector pins, mandrels, shear blades, spindles, stamps, tool shanks, lathe collets, bending dies, punches, rotary shears

Table 22.49 TOOL STEELS AND THEIR USES—*continued*
(BS 4659:1971, Stahl-Eisen Werkstoffblatt 150:1971, 200:1969, 250:1970, 320:1969)

Material	*British or other standards*	*Chemical composition limits %* C	Si	Mn	Cr	Mo	W	V	Co	*Other elements*	*Typical uses*
4. *Cold work tool steels*											
12CrV Steel	BS BD3	1.90 2.30	0.60 max	0.60 max	12.0 13.0	— —	— —	0.50 max	— —	— —	Blanking dies, cold forming dies, thread rolling dies, shear blades, slitter knives, forming rolls, seaming rolls, burnishing tools, punches, gauges, crimping dies, swaging dies
12CrMoV Steel	BS BD2	1.40 1.60	0.60 max	0.60 max	11.50 12.50	0.70 1.20	— —	0.25 1.00	— —	— —	Blanking dies, cold forming dies, drawing dies, thread rolling dies, shear blades, slitter knives, forming rolls, burnishing tools, punches, gauges, knurling tools, lathe centres, broaches, cold extrusion dies, mandrels, swaging dies
13CrMoV Steel	BS BD2A	1.60 1.90	0.60 max	0.60 max	12.00 13.00	0.70 0.80	— —	0.25 1.00	— —	— —	Coining dies
5CrMoV Steel	BS BA2	0.95 1.05	0.40 max	0.30 0.70	4.75 5.25	0.90 1.10	— —	0.15 0.40	— —	— —	Thread rolling dies, extrusion dies, trimming, blanking and coining dies, mandrels, shear blades, spinning rolls, forming rolls, gauges, beading dies, burnishing tools, embossing dies, plastic moulds, stamping dies, bushes, punches, liners for brick moulds
2MnMo-Cr Steel	BS BA6	0.65 0.75	0.40 max	1.80 2.10	0.85 1.15	1.20 1.60	— —	— —	— —	— —	Blanking, forming, coining and trimming dies, punches, shear blades, spindles, mandrels, plastic moulds
1¼MnCr-WV Steel	BS BO1	0.85 1.00	0.40 max	1.10 1.35	0.40 0.60	— —	0.40 0.60	0.25 max	— —	— —	Blanking, drawing and trimming dies, plastic moulds, paper slitters, shear blades, taps, reamers, gauges, jigs, bending and forming dies, bushes, punches
1¾MnV Steel	BS BO2	0.85 0.95	0.40 max	1.50 1.80	— —	— —	— —	0.25 max	— —	— —	Blanking, stamping, trimming and forming dies and punches, threading dies, taps, reamers, gauges, plugs, jigs, broaches, circular cutters and saws, bushes

Table 22.49 TOOL STEELS AND THEIR USES—*continued*
(BS 4659:1971, Stahl-Eisen Werkstoffblatt 150:1971, 200:1969, 250:1970, 320:1969)

Material	*British or other standards*	*Chemical composition limits %* C	Si	Mn	Cr	Mo	W	V	Co	*Other elements*	*Typical uses*
$1\frac{1}{2}$CrV Steel	BS BL3	0.95 1.05	0.40 max	0.40 max	1.30 1.50	— —	— —	0.10 0.30	— —	— —	Mandrels, cold rolls, ball bearings, precision gauges, reamers, broaches, taps, drills, thread rolling dies, files
$1\frac{1}{2}$WCrV Steel	BS BF1	1.15 1.35	0.40 max	0.40 max	0.25 0.50	— —	1.30 1.60	0.30 max	— —	— —	Taps, broaches, reamers, drills
3V Steel	DIN 145-V33	1.45	0.30	0.40	—	— (Typical)	—	3.30	—	—	Cold heading dies and punches
5. *High speed tool steels*											
3-3-2 WMoV Steel	DIN S3-3-2	0.95 1.03	0.40 max	0.40 max	3.80 4.50	2.50 2.80	2.70 3.00	2.20 2.50	— —	— —	Hacksaw and circular saw blades for metal cutting
2-9-1 WMoV Steel	BS BM1	0.75 0.85	0.40 max	0.40 max	3.75 4.50	8.00 9.00	1.00 2.00	1.00 1.25	0.60 max	— —	Drills, taps, reamers, milling cutters, hobs, punches, lathe and planer tools, form cutters, saws, chasers, broaches, routers, woodworking tools
2-9-2 WMoV Co Steel	BS BM34	0.85 0.95	0.40 max	0.40 max	3.75 4.50	8.00 9.00	1.70 2.20	1.75 2.05	7.75 8.75	— —	Cutting tools
6-5-2 WMoV Steel	BS BM2	0.80 0.90	0.40 max	0.40 max	3.75 4.50	4.75 5.50	6.00 6.75	1.75 2.05	0.60 max	— —	Drills, taps, reamers, milling cutters, hobs, form cutters, saws, lathe and planer tools, chasers, broaches, boring tools, cold forming tools, e.g. punches for cold extrusion, cold heading die inserts
6-5-4 WMoV Steel	BS BM4	1.25 1.40	0.40 max	0.40 max	3.75 4.50	4.25 5.00	5.75 6.50	3.75 4.25	0.60 max	— —	Heavy duty broaches, reamers, milling cutters, chasers, form cutters, lathe and planer tools, blanking dies and punches, swaging dies
6-5-4 WMoV Co Steel	BS BM15	1.45 1.60	0.40 max	0.40 max	4.50 5.00	2.75 3.25	6.25 7.00	4.75 5.25	4.50 5.50	— —	Cutting tools
6-5-2-5 WMo-VCo Steel	DIN S6-5-2-5	0.88 0.96	0.40 max	0.40 max	3.80 4.50	4.70 5.20	6.00 6.70	1.70 2.00	4.50 5.00	— —	Milling cutters, twist drills and taps, particularly suitable for intermittent cutting and drilling

Table 22.49 TOOL STEELS AND THEIR USES—*continued*
(BS 4658:1971, Stahl-Eisen Werkstoffblatt 150:1971, 200:1969, 250:1970, 320:1969)

Material	*British or other standards*	*Chemical composition limits %* C	Si	Mn	Cr	Mo	W	V	Co	*Other elements*	*Typical uses*
2-9-1-8 WMo VCo Steel	BS BM42	1.00 1.10	0.40 max	0.40 max	3.50 4.25	9.00 10.00	1.00 2.00	1.00 1.30	7.50 8.50	— —	Heavy duty drills, reamers, form cutters, lathe tools, hobs, broaches, milling cutters, twist drills
18-0-1 WV Steel	BS BT1	0.70 0.80	0.40 max	0.40 max	3.75 4.50	0.70 max	17.5 18.5	1.00 1.25	0.60 max	— —	Drills, taps, reamers, hobs, lathe and planer tools, broaches, burnishing dies, chasers, form cutters, milling cutters
18-0-2 WV Steel	BS BT2	0.75 0.85	0.40 max	0.40 max	3.75 4.50	0.70 max	17.50 18.50	1.75 2.05	0.60 max	— —	Cutting tools
18-1-1-5 WMo-VCo Steel	BS BT4	0.70 0.80	0.40 max	0.40 max	3.75 4.50	1.00 max	17.5 18.5	1.00 1.25	4.50 5.50	— —	As for BT1, when higher performance is required
19-1-2-10 WMoV Co Steel	BS BT5	0.75 0.80	0.40 max	0.40 max	3.75 4.50	1.00 max	18.50 19.50	1.75 2.05	9.00 10.00	— —	Cutting tools
20-1-2-12 WMo-VCo Steel	BS BT6	0.75 0.85	0.40 max	0.40 max	3.75 4.50	1.00 max	20.0 21.0	1.25 1.75	11.25 12.25	— —	Heavy duty lathes and planer tools, drills, cut-off tools, milling cutters, hobs
12-1-5-5 WMo-VCo Steel	BS BT15	1.40 1.60	0.40 max	0.40 max	4.25 5.00	1.00 max	12.0 13.0	4.75 5.25	4.50 5.50	— —	Heavy duty form cutters, milling cutters, broaches, blanking dies, lathe and planer tools
14-1 WV Steel	BS BT21	0.60 0.70	0.40 max	0.40 max	3.50 4.25	0.70 max	13.50 14.50	0.40 0.60	0.60 max	— —	Cutting tools
22-1 WV Steel	BS BT20	0.75 0.85	0.40 max	0.40 max	4.25 5.00	1.00 max	21.00 22.50	1.40 1.60	0.60 max	— —	Cutting tools
9-3-3-9 WMo-VCo Steel	BS BT42	1.25 1.40	0.40 max	0.40 max	3.75 4.50	2.75 3.50	8.50 9.50	2.75 3.25	9.00 10.00	— —	Heavy duty milling cutters and form cutters when high performance is required

Table 22.50 TOOL STEEL HEAT TREATMENTS AND HARDNESS
(BS 4659:1971, Stahl-Eisen Werkstoffblatt 150:1971, 200:1969, 250:1970, 320:1969)

Material	*British or other standard*	*Annealing* °C	*Hardening* °C	*Quenching medium*	*Tempering* °C	*Typical Rockwell hardness limits*	*Remarks*
1. *Carbon tool steels*							
0.8C Steel	DIN C80W1	680/710	780/810	Water or brine	150/350	64/50	The wear resistance increases but the toughness decreases with increasing carbon content
0.9C Steel	BS BW1A	740/770	770/790	Water or brine	150/350	64/50	
1C Steel	BS BW1B	740/770	770/790	Water or brine	150/350	64/50	
1.2C Steel	BS BW1C	740/770	760/780	Water or brine	150/350	64/50	
1CV Steel	BS BW2	740/770	780/800	Water or brine	150/350	64/50	The V addition reduces the case depth, and raises the resistance to cracking
2. *Hot work tool steel*							
$1\frac{3}{4}$NiCr-MoV Steel	DIN 55NiCr MoV6 Similar to BS No. 5 Die Steel	650/700	830/870	Oil	500/600	44/34	Tough and wear resistant
3CrMoV Steel	BS BH10	850/870	1 000/1 060	Oil or air	530/650	54/39	Wear resistant and fairly tough. Resistant to softening on tempering
3CrMoV Co Steel	BS BH10A	850/870	1 000/1 060	Air or oil	530/650	54/39	Wear resistant
5CrMoV Steel	BS BH11	850/870	1 000/1 030	Air or oil	530/650	54/38	Wear resistant and tough
5CrMoV Steel	BS BH13	850/870	1 020/1 060	Air or oil	530/650	55/42	More resistant to wear and softening on tempering than BH11
5CrMo-WV Steel	BS BH12	850/870	1 000/1 030	Air or oil	530/650	55/39	Wear resistant and tough
4WCrV Steel	DIN X30WCrV53	750/800	1 060/1 100	Oil, air or salt bath of 500/550 °C temperature	530/650	53/42	Less tough but more resistant to softening on tempering than BH11. Not suitable for water cooling in service
4WCrCo-VMo Steel	BS BH19	850/870	1 150/1 200	Air or oil	530/650	58/44	Highly resistant to wear and softening on tempering. Not suitable for water cooling in service
9WCr-MoV Steel	BS BH21	870/890	1 100/1 180	Oil or air	560/675	54/36	Very resistant to softening on tempering. Not suitable for water cooling in service

Table 22.50 TOOL STEEL HEAT TREATMENTS AND HARDNESS—*continued*
(BS 4659:1971, Stahl-Eisen Werkstoffblatt 150:1971, 200:1969, 250:1970, 320:1969)

Material	*British or other standard*	*Annealing* °C	*Hardening* °C	*Hardening* *Quenching medium*	*Tempering* °C	*Typical Rockwell hardness limits*	*Remarks*
9WCrMo V Steel	BS BH21A	870/890	1 100/1 170	Oil or air	560/675	55/40	Resistant to softening
18WCrMo V Steel	BS BH26	870/890	1 180/1 260	Oil or air	550/570	65/60	Resistant to softening
3. *Shock resisting tool steels*							
$2\frac{1}{4}$WCrV Steel	BS BS1	790/820	870/950	Oil	200/650	56/40	Tough and wear resistant
1SiMoV Steel	BS BS2	790/820	850/900	Water or oil	175/425	56/50	Tough and wear resistant
2SiMoV Steel	BS BS5	790/820	870/920	Oil	175/425	58/50	More resistant to wear, but less tough, than BS2
4. *Cold work tool steels*							
12CrV Steel	BS BD3	850/870	950/1 000	Air or oil	200/540	62.5/54	Highly resistant to wear and abrasion. Resistant to corrosion and softening on tempering. Very little distortion after air hardening
12CrMoV Steel	BS BD2	850/870	980/1 030	Air or oil	200/540	61.5/54	Similar to, but tougher than, BD3
13CrMoV Steel	BS BD24	850/870	980/1 030	Oil or air	200/540	60/55	Tough and wear resistant
5CrMoV Steel	BS BA2	850/870	950/980	Air	175/540	61.5/57	Tougher, but less corrosion and wear resistant than BD2. Very little distortion after hardening
2MnMo-Cr Steel	BS BA6	730/750	830/850	Air	150/250	61/56	Less resistant to wear and softening on tempering than BD2. Very little distortion after hardening
$1\frac{1}{4}$MnCr-WV Steel	BS BO1	760/780	780/820	Oil	175/250	61.5/57	Hard and wear resistant. Little distortion after hardening
$1\frac{3}{4}$MnV Steel	BS BO2	760/780	760/780	Oil	175/250	61.5/57	Hard and wear resistant. Little distortion after hardening
$1\frac{1}{2}$CrV Steel	BS BL3	790/810	820/840 790/810	Oil Water	175/320	62.5/56	Hard and wear resistant
$1\frac{1}{2}$WCrV Steel	BS BF1	780/800	780/800	Oil or water	175/250	62.5/60	Hard and wear resistant

Table 22.50 TOOL STEEL HEAT TREATMENTS AND HARDNESS—*continued*
(BS 4659:1971, Stahl-Eisen Werkstoffblatt 150:1971, 200:1969, 250:1970, 320:1969)

Material	*British or other standard*	*Annealing* °C	*Hardening* °C	*Hardening* *Quenching medium*	*Tempering* °C	*Typical Rockwell Hardness limits*	*Remarks*
3V Steel	DIN 145V33	720/760	800/920	Water	100/250	65/60	Very hard and wear resistant. The case depth is controlled by the hardening temperature
5. *High speed steels*							
3-3-2 WMoV Steel	DIN S3-3-2	800/840	1 180/1 220	Oil, air or salt bath	520/550*	62/64	Steel for special applications, e.g. metal cutting saws
2-8-1 WMoV Steel	BS BM1	850/870	1 200/1 220	Oil, air or salt bath	530/550*	65/63	General purpose cutting tools
2-9-2 WMoV Steel	BS BM34	870/900	1 215/1 235	Oil, air or salt bath	530/550*	65/63	General purpose cutting tools
6-5-2 WMoV Steel	BS BM2	850/870	1 210/1 230	Oil, air or salt bath	550/570*	65.5/63	General purpose cutting tools
6-5-4 WMoV Steel	BS BM4	850/870	1 200/1 220	Oil, air or salt bath	540/560*	66/64	Steel for special applications when very high resistance to wear is required
6-5-4 WMoV Co Steel	BS BM15	870/900	1 210/1 230	Oil, air or salt bath	540/560*	66/64	High wear resistance
6-5-2-5 WMo-VCo Steel	DIN S6-5-2-5	800/840	1 210/1 240	Oil, air or salt bath	540/570†	66/64	Steel of high hot hardness, particularly suitable for intermittent cutting and drilling
2-9-1-8 WMo-VCo Steel	BS BM42	870/900	1 180/1 200	Oil, air or salt bath	520/540†	68/66	Steel for special applications when very high hardness is required
18-0-1 WV Steel	BS BT1	870/890	1 270/1 290	Oil, air or salt bath	550/570*	65/63	General purpose cutting tools
18-0-2 WV Steel	BSBT2	870/890	1270/1 290	Oil, air or salt bath	550/570*	65/63	General purpose cutting tools
18-1-1-5 WMo-VCo Steel	BS BT4	880/900	1 280/1 300	Oil, air or salt bath	550/570†	66/64	Steel of high hot hardness, is used at higher speed for heavier cuts than is BT1
20-1-2-12 WMo-VCo Steel	BS BT6	880/900	1 290/1 310	Oil, air or salt bath	550/570†	66.5/65	Steel for special applications when the highest hot hardness is required

* Double tempering.
† Treble tempering.

Table 22.50 TOOL STEEL HEAT TREATMENTS AND HARDNESS—*continued*
(BS 4659:1971. Stahl-Eisen Werkstoffblatt 150:1971, 200:1969, 250:1970, 320:1969)

Material	*British or other standard*	*Annealing* °C	*Hardening* °C	*Quenching medium*	*Tempering* °C	*Typical Rockwell Hardness limits*	*Remarks*
12-1-5-5 WMo-VCo Steel	BS BT15	870/890	1 230/1 250	Oil, air or salt bath	550/570†	67.5/66	Steel for special applications when the highest wear resistance is required
9-3-3-9 WMo-VCo Steel	BS BT42	850/870	1 200/1 240	Oil, air or salt bath	550/570†	68.5/67	Steel for special applications when both the highest wear resistance and highest hot hardness are required

* Double tempering.
† Treble tempering.

22.10 Mechanical properties of cast irons and cast steels

Table 22.51 CAST IRONS AND CAST STEELS—MECHANICAL PROPERTIES AT ROOM TEMPERATURE

		Composition limits %					*Properties (minima unless otherwise stated)*						
											Impact values		
Material	*British or other standards*	C	Si	Mn	*Other elements*	*Condition*	*Brinell hardness* HB	*UTS* MPa (tonf in^{-2})	*Yield stress*† MPa (tonf in^{-2})	*Elong.* (gl $5.65\sqrt{S_0}$) %	*Specimen*	J (ft lbf)	*Remarks*
1. *Iron castings*													
Grey iron or flake graphite iron castings (BS 1452: 1977)													
Grey iron	BS Grade 150	—	—	—	—	As cast	—	150 (10)*	98 (6)	—	Izod unnotched, 20 mm (0.8 in) diameter	8/13 (6/10)	—
Grey iron	BS Grade 180	—	—	—	—	As cast	—	180 (12)*	117 (8)	—	Izod unnotched, 20 mm (0.8 in) diameter	8/13 (6/10)	—
Grey iron	BS Grade 220	—	—	—	—	As cast	—	220 (14)*	146 (9)	—	Izod unnotched, 20 mm (0.8 in) diameter	8/16 (6/12)	—
Grey iron	BS Grade 260	—	—	—	—	As cast	—	260 (17)*	169 (11)	—	Izod unnotched, 20 mm (0.8 in) diameter	13/23 (10/17)	—
Grey iron	BS Grade 300	—	—	—	—	As cast	—	300 (19)*	195 (13)	—	Izod unnotched, 20 mm (0.8 in) diameter	16/31 (12/23)	—
Grey iron	BS Grade 350	—	—	—	—	As cast	—	350 (23)*	228 (15)	—	Izod unnotched, 20 mm (0.8 in) diameter	24/47 (18/35)	—

* Determined on 20 mm diameter bar. † 0.1% proof stress.

Table 22.51 CAST IRONS AND CAST STEELS—MECHANICAL PROPERTIES AT ROOM TEMPERATURE—*continued*

		Composition limits %						*Properties (minima unless otherwise stated)*					
											Impact values		
Material	*British or other standards*	C	Si	Mn	*Other elements*	*Condition*	*Brinell hardness* HB	*UTS* MPa (tonf in^{-2})	*Yield stress*† MPa (tonf in^{-2})	*Elong.* (gl $5.65\sqrt{S_0}$) %	*Specimen*	J (ft lbf)	*Remarks*
Grey iron	BS Grade 400	—	—	—	—	As cast	—	400 (26)*	260 (17)	—	Izod unnotched, 20 mm (0.8 in) diameter	24/47 (18/35)	—
Malleable iron castings (BS 3333:1972, DIN 1692:1963)													
Malleable iron	DIN GTS-35	2.30 2.70	0.80 1.50	0.55 max	— —	As treated	150 max	340 (22)	190 (12.5)	10	—	—	Ferritic
Malleable iron	BS P440/7	—	—	—	—	As treated	149/197	440 (28.5)	270 (17.5)	7	—	—	Pearlitic–ferritic
Malleable iron	BS P510/4	2.35 2.50	1.00 1.50	0.28 0.48	P 0.06 0.12 S 0.16 0.20	As treated	170/229	510 (33)	310 (20)	4	—	—	Pearlitic
Malleable iron	BS P540/5	2.35 2.50	1.00 1.50	0.28 0.48	P 0.06 0.12 S 0.16 0.20	As treated	179/229	540 (35)	340 (22)	5	—	—	Pearlitic
Malleable iron	BS P570/3	2.35 2.50	1.00 1.50	0.28 0.48	P 0.06 0.12 S 0.16 0.20	As treated	197/241	570 (37)	420 (27)	3	—	—	Pearlitic
Malleable iron	DIN GTS-65	2.35 2.50	1.00 1.50	0.28 0.48	P 0.06 0.12 S 0.16 0.20	As treated	210/250	630 (41)	420 (27)	3	—	—	Pearlitic
Malleable iron	BS P690/2	2.35 2.50	1.00 1.50	0.28 0.48	P 0.06 0.12 S 0.16 0.20	Hardened and tempered	241/285	690 (44.5)	540 (35)	2	—	—	Tempered martensitic

Spheroidal or nodular graphite or ductile iron castings (BS 2789: 1973)													
Nodular iron	BS 370/17	3.40 4.00	2.00 2.75	0.20 0.60	Mg 0.02 0.07	Annealed	179 max	370 (24)	230 (15)	17	Izod	13 (10)	Ferritic
Nodular iron	BS 420/12	3.40 4.00	2.00 2.75	0.20 0.60	Ni 1.00 max Mg 0.02 0.07	Annealed	201 max	420 (27)	250 (16)	12	Izod	8 (6)	Ferritic
Nodular iron	BS 500/7	3.40 4.00	2.00 2.75	0.20 0.60	Ni 1.00 max Mg 0.02 0.07	Normalized and tempered	170/241	500 (32)	310 (20)	7	—	—	Ferritic–pearlitic
Nodular iron	BS 600/3	3.30 3.80	2.00 3.00	0.20 0.50	Ni 1.00 max Mg 0.02 0.07	Normalized and tempered	192/269	600 (39)	350 (23)	3	—	—	Ferritic–pearlitic
Nodular iron	BS 700/2	3.30 3.80	2.00 3.00	0.20 0.50	Ni 1.00 max Mg 0.02 0.07	Normalized and tempered	229/302	700 (45)	400 (26)	2	—	—	Pearlitic
Nodular iron	BS 800/2	3.40 3.80	2.00 2.75	0.30 0.60	No 2.00 max Mo 1.00 max Mg 0.02 0.07	Normalized and tempered, or hardened and tempered	248/352	800 (52)	460 (30)	2	—	—	Pearlitic or tempered martensitic
Austenitic iron castings (BS 3468: 1974, DIN 1694: 1966) *Austenitic cast irons with flake graphite*													
13/7NiMn Iron	DIN GGL BS L-NiMn 13 7	3.00 max	1.50 3.00	6.00 7.00	Ni 12.0 14.0 Cr 0.2 max Cu 0.5 max	As cast	150 max	140 (9)	—	—	—	—	Non-magnetic

* Determined on 20 mm diameter bar. † 0.1% proof stress.

Table 22.51 CAST IRONS AND CAST STEELS—MECHANICAL PROPERTIES AT ROOM TEMPERATURE—*continued*

		Composition limits %					*Properties (minima unless otherwise stated)*						
											Impact values		
Material	*British or other standards*	C	Si	Mn	*Other elements*	*Condition*	*Brinell hardness* HB	*UTS* MPa (tonf in^{-2})	*Yield stress*† MPa (tonf in^{-2})	*Elong.* (gl $5.65\sqrt{S_0}$) %	*Specimen*	J (ft lbf)	*Remarks*
15/6/2 NiCuCr Iron	BS L-NiCuCr 15 6 2	3.00 max	1.00 2.80	0.50 1.50	Ni 13.50 17.50 Cu 5.50 7.50 Cr 1.00 2.50	As cast	200 max	170 (11)	—	2	Izod	2.7/5.4 (2/4)	Resistant to corrosion by weak acids. Heat resistant—non-magnetic when Cr content is low
15/6/3 NiCuCr Iron	BS L-NiCuCr 15 6 3	3.00 max	1.00 2.80	0.50 1.50	Ni 13.50 17.50 Cu 5.50 7.50 Cr 2.50 3.50	As cast	250 max	190 (12)	—	1–2	Izod	2.7/5.4 (2/4)	More resistant to corrosion and erosion than is Grade A
20/2 NiCr Iron	BS L-NiCr 20 2	3.00 max	1.00 2.80	0.50 1.50	Ni 18.0 22.0 Cr 1.00 2.50	As cast	215 max	170 (11)	—	2–3	Izod	2.7/5.4 (2/4)	Resistant to corrosion by alkalis. Heat resistant. Non-magnetic when Cr is low
20/3 NiCr Iron	BS L-NiCr 20 3	3.00 max	1.00 2.80	0.50 1.50	Ni 18.0 22.0 Cr 2.50 3.50	As cast	250 max	190 (12)	—	1–2	Izod	2.7/5.4 (2/4)	More resistant to heat and erosion than is Grade A
20/5/3 NiSiCr Iron	BS L-NiSiCr 20 5 3	2.50 max	4.50 5.50	0.50 1.50	Ni 18.0 22.0 Cr 1.50 4.50	As cast	250 max	190 (12)	—	2–3	Izod	5.4/8.1 (4/6)	Resistant to corrosion even by weak sulphuric acid. Highly resistant to heat and scaling

30/3 NiCr Iron	BS L-NiCr 30 3	2.50 max	1.00 2.00	0.50 1.50	Ni 28.0 32.0 Cr 2.50 3.50	As cast	215 max	190 (12)	—	1–3	Izod	2.7/5.4 (2/4)	Highly resistant to heat up to 800 °C particularly to thermal shock. Resistant to erosion in wet steam, slurries etc.
30/5/5 NiSiCr Iron	BS L-NiSiCr 30 5 5	2.50 max	5.00 6.00	0.50 1.50	Ni 29.0 32.0 Cr 4.50 5.50	As cast	210 max	170 (11)	—	—	—	—	Particularly resistant to erosion corrosion and heat
35Ni Iron	BS L-Ni 35	2.40 max	1.00 2.00	0.50 1.50	Ni 34.0 36.0	As cast	140 max	120 (8)	—	1–3	—	—	Resistant to thermal shock
Austenitic cast iron with nodular graphite													
13/7 Ni-Mn Iron	BS S-NiMn 137	3.00 max	2.00 3.00	6.00 7.00	Ni 12.0 14.0	As cast	170 max	390 (25)	210 (14)	15	Charpy V	15/27.5(11/20)	Non-magnetic
22Ni Iron	BS S-Ni 22	3.00 max	1.00 3.00	1.50 2.50	Ni 21.0 24.0 P 0.08 max	As cast	170 max	370 (24)	170 (11)	20	Charpy V	20/33 (15/24)	Tough and ductile at temp. down to −100 °C. Non-magnetic
23/4 Ni-Mn Iron	BS S-NiMn 23 4	2.60 max	1.50 2.50	4.00 4.50	Ni 22.0 24.0	As cast	180 max	440 (28)	210 (14)	25	Charpy V	24 (18)	Tough and ductile at temp. down to −196 °C. Non-magnetic
20/2 NiCr Iron	BS S-NiCr 20 2	3.00 max	1.50 3.00	0.50 1.50	Ni 18.0 22.0 Cr 1.00 2.50 P 0.08 max	As cast	200 max	370 (24)	210 (14)	7	Charpy V	13.5/27.5 (10/20)	Non-magnetic when Cr content is low, resistant to corrosion by alkalis. Heat resistant

†0.1% proof stress.

Table 22.51 CAST IRONS AND CAST STEELS—MECHANICAL PROPERTIES AT ROOM TEMPERATURE—*continued*

		Composition limits %					*Properties (minima unless otherwise stated)*						
Material	*British or other standards*	C	Si	Mn	*Other elements*	*Condition*	*Brinell hardness* HB	*UTS* MPa (tonf in^{-2})	*Yield stress*† MPa (tonf in^{-2})	*Elong.* (gl $5.65\sqrt{S_0}$) %	*Impact values* *Specimen*	*Impact values* J (ft lbf)	*Remarks*
20/3 NiCr Iron	BS S-NiCr 20 3	3.00 max	1.50 3.00	0.50 1.50	Ni 18.0 22.0 Cr 2.50 3.50 P 0.08 max	As cast	255 max	390 (25)	210 (14)	7	Charpy V	12 (9)	More resistant to heat and erosion than is Grade A
20/5/2 NiSiCr Iron	BS S-NiSiCr 20 5 2	3.0 max	4.50 5.50	0.50 1.50	Ni 18.0 22.0 Cr 1.00 2.50 P 0.08 max	As cast	230 max	370 (24)	210 (14)	10	Charpy V	14.9 (11)	Resistant to corrosion even by weak sulphuric acid. Highly resistant to heat and scaling
30/3 NiCr Iron	BS S-NiCr 30 3	2.6 max	1.50 3.00	0.50 1.50	Ni 28.0 32.0 Cr 2.50 3.50 P 0.08 max	As cast	200 max	370 (24)	210 (14)	7	Charpy V	8.5 (6)	Highly resistant to heat, particularly to thermal shock. up to 800 °C Resistant to erosion in wet steam slurries, etc. High hot strength, especially with 1% Mo addition
30/1 NiCr Iron	BS S-NiCr 30 1	2.6 max	1.50 3.00	0.50 1.50	Ni 28.0 32.0 Cr 1.00 1.50 P 0.08 max	As cast	190 max	370 (24)	210 (14)	13	Charpy V	17 (12.5)	Similar to S-NiCr 30 3: good bearing properties
30/5/5 NiSrCr Iron	BS NiSiCr 30 5 5	2.6 max	5.00 6.00	0.50 1.50	Ni 28.0 32.0	As cast	250 max	390(25)	240 (16)	1	Charpy V	3.9/5.9 (3/4)	Particularly resistant to corrosion, erosion and heat

					Cr 4.50 5.50 P 0.08 max								
35 Ni Iron	BS S-Ni 35	2.4 max	1.50 3.00	0.50 1.50	Ni 34.0 36.0 Cr 0.20 max P 0.08 max	As cast	180 max	370 (24)	210 (14)	20	Charpy V	20.5 (15)	Low coefficient of thermal expansion. Resistant to thermal shock
35/3 NiCr Iron	BS S-NiCr 35 3	2.4 max	1.50 3.00	0.50 1.50	Ni 34.0 36.0 Cr 2.00 3.00 P 0.08 max	As cast	190 max	370 (24)	210 (14)	7	Charpy V	7.0 (5)	Low coefficient of thermal expansion. High hot strength especially with 1% Mo addition
Other corrosion and heat resistant iron castings													
14 Si Iron	BS Si14	1.00 max	14.25 15.25	0.50 max	P 0.25 max Mo 3.50 max	Annealed	—	90/125 (6/8)	—	—	Charpy unnotched	2.7/5.4 (2/4)	Resistant to corrosive acids
High Cr Iron	—	1.20 2.50	0.50 2.50	0.30 1.00	Cr 20.0 35.0	As cast	—	200/620 (13/40)	—	—	Charpy unnotched	27/47 (20/35)	Resistant to oxidizing acids and to scaling up to about 1 100 °C
Wear resistant white iron castings													
Unalloyed White Iron	—	2.80 3.60	0.50 1.30	0.40 0.90	—	As cast	—	140/340 (9/22)	—	—	Charpy unnotched	4/13 (3/10)	—
$3\frac{1}{2}$ NiCr White Iron	USA (Ni Hard)	2.80 3.60	0.40 0.70	0.20 0.70	Ni 2.50 4.75 Cr 1.20 1.35	As cast	—	280/510 (18/33)	—	—	Charpy unnotched	27/74 (20/55)	—

†0.1% proof stress.

Table 22.51 CAST IRONS AND CAST STEELS—MECHANICAL PROPERTIES AT ROOM TEMPERATURE—*continued*

		Composition limits %						*Properties (minima unless otherwise stated)*					
											Impact values		
Material	*British or other standards*	C	Si	Mn	*Other elements*	*Condition*	*Brinell hardness* HB	*UTS* MPa (tonf in^{-2})	*Yield stress*† MPa (tonf in^{-2})	*Elong.* (gl $5.65\sqrt{S_0}$) %	*Specimen*	J (ft lbf)	*Remarks*
2. *Steel castings* (BS 1504:1976, 3100:1976)													
Carbon steel castings with up to 1.6% manganese content													
C Steel	BS 3100 AM1	0.15 max	0.60 max	0.50 max	S, P each 0.05 max	Annealed or normalized	—	340/430 (22/28)	185 (12)	22	—	—	High magnetic permeability
C Steel	BS 3100 AM2	0.25 max	0.60 max	0.50 max	S, P each 0.05 max	Annealed or normalized	—	400/490 (26/32)	215 (14)	22	—	—	High magnetic permeability
C Steel	BS 3100 A1	0.25 max	0.60 max	0.90 max	S, P each 0.06 max	Annealed or normalized	—	430 (28)	230 (15)	22	Charpy V	25 (18)	General purpose engineering castings
C Steel	BS 3100 A2	0.35 max	0.60 max	1.00 max	S, P each 0.06 max	Annealed or normalized	—	490 (32)	260 (17)	18	Charpy V	20 (15)	General purpose engineering castings
C Steel	BS 3100 A3	0.45 max	0.60 max	1.00 max	S, P each 0.06 max	Normalized and tempered	—	540 (35)	295 (19)	14	Charpy V	18 (13)	General purpose engineering castings, higher strength and wear resistance
C Steel	BS 3100 AW2	0.40 0.50	0.60 max	1.00 max	S, P each 0.05 max	Normalized and tempered	—	620 (40)	325 (21)	12	—	—	Wear resistant. Suitable for surface hardening
C Steel	BS 3100 AW3	0.50 0.60	0.60 max	1.00 max	S, P each 0.05 max	Normalized and tempered	—	690 (45)	370 (24)	8	—	—	Wear resistant. Suitable for surface hardening
CMn Steel	BS 3100 AL1	0.20 max	0.60 max	1.10 max	S, P each 0.04 max	Normalized and tempered	—	430 (28)	230 (15)	22	Charpy V	20 (15)	—

C Mn Steel	BS 3100 A4	0.18 0.25	0.60 max	1.20 1.60	S, P each 0.05 max	Normalized, normalized and tempered, or hardened and tempered	152/207	540/690 (35/45)	320 (21)	16	Charpy V	30 (22)	—
C Mn Steel	BS 3100 A6	0.25 0.33	0.60 max	1.20 1.60	S, P each 0.05 max	Hardened and tempered	201/255	690/850 (45/55)	495 (32)	13	Charpy V	25 (18)	Thickness 63 mm ($2\frac{1}{2}$ in) max
	A5					Normalized and tempered	179/229	620/770 (40/50)	370 (24)	13	Charpy V	25 (18)	Thickness 100 mm (4 in) max
Low alloy steel castings for direct hardening (Stahl-Eisen Werkstoffblatt 510: 1962)													
1 CrMo Steel	BS 3100 BT1 DIN GS-42CrMo4	0.38 0.45	0.50 max	0.80 max	Cr 0.80 1.20 Mo 0.20 0.30 S, P each 0.05 max	Hardened and tempered	201/255	690/850 (45/55)	495 (32)	11	Charpy V	35 (26)	Thickness 100 mm (4 in) max
$1\frac{1}{2}$ Cr-MoV Steel	BS 3100 BT2 DIN GS-30 CrMoV 6 4	0.27 0.34	0.50 max	0.80 max	Cr 1.30 1.70 Mo 0.30 0.50 V 0.05 0.15 S, P each 0.04 max	Hardened and tempered	248/302	850/1 000 (55/65)	585 (38)	8	Charpy V	25 (18)	Thickness 100 mm (4 in) max
$1\frac{1}{2}$ Cr-MoV Steel	BS 3100 BT3 DIN GS-30 CrMoV 6 4	0.27 0.34	0.50 max	0.80 max	Cr 1.30 1.70 Mo 0.30 0.50 V 0.05 0.15 S, P each 0.03 max	Hardened and tempered	293/341	1 000/1 160 (65/75)	695 (45)	6	Charpy V	20 (15)	Thickness 100 mm (4 in) max

† 0.1% proof stress.

Table 22.51 CAST IRONS AND CAST STEELS—MECHANICAL PROPERTIES AT ROOM TEMPERATURE—*continued*

		Composition limits %						*Properties (minima unless otherwise stated)*					
											Impact values		
Material	*British or other standards*	C	Si	Mn	*Other elements*	*Condition*	*Brinell hardness* HB	*UTS* MPa (tonf in^{-2})	*Yield stress*† MPa (tonf in^{-2})	*Elong.* (gl $5.65\sqrt{S_0}$) %	*Specimen*	J (ft lbf)	*Remarks*
Steel castings for carburizing													
C Steel	BS 3100 AW1	0.10 0.18	0.60 max	0.60 1.00	S, P each 0.05 max	Water Quenced	—	460 (30)	—	12	Charpy V	25 (18)	29 mm (1⅛ in) Test bar
3½ NiCr-Mo Steel	BS 3100 BW1	0.12 0.18	0.60 max	0.30 0.60	Ni 3.00 3.75 Cr 0.60 1.10 Mo 0.15 0.25 S, P each 0.04 max	Oil quenched	—	1 000 (65)	—	7	Charpy V	20 (15)	29 mm (1⅛ in) Test bar
Low alloy creep resisting steel castings (DIN 17245:1967)													
½ Mo Steel	BS 3100 B1 DIN GS-22Mo4	0.20 max	0.20 0.60	0.50 1.00	Mo 0.45 0.65 S, P each 0.05 max	Normalized and tempered or hardened and tempered		460(30)	260 (17)	18	Charpy V	20 (15)	Steam turbine parts
1¼ CrMo Steel	BS 3100 B2 DIN GS-17CrMo5 5	0.20 max	0.60 max	0.50 0.80	Cr 1.00 1.50 Mo 0.45 0.65 S, P each 0.05 max	Normalized and tempered or hardened and tempered		480 (31)	280 (18)	17	Charpy V	30(22)	Steam turbine parts
2½ CrMo Steel	BS 3100 B3	0.18 max	0.60 max	0.40 0.70	Cr 2.00 2.75 Mo 0.90 1.20 S, P each 0.05 max	Normalized and tempered or hardened and temper[illegible]	—	540 (35)	325 (21)	17	Charpy V	25(18)	Steam turbine parts. Also used in the oil and gas industries

½ MoCrV Steel	BS 3100 B7	0.10 0.15	0.45 max	0.40 0.70	Mo 0.40 0.60 Cr 0.30 0.50 V 0.22 0.30 S, P each 0.05 max	Normalized and tempered or hardened and tempered	—	510 (33)	295 (19)	17	—	—	Steam turbine parts
1¼ Cr-MoV Steel	DIN GS-17CrMoV 5 11	0.15 0.20	0.30 0.50	0.50 0.80	Cr 1.20 1.50 Mo 0.90 1.10 V 0.20 0.30 S, P each 0.04 max	Hardened and tempered	—	590 (38)	430 (28)	15	—	—	Steam turbine parts
3 CrMo Steel	BS 3100 B4	0.25 max	0.75 max	0.30 0.70	Cr 2.50 3.50 Mo 0.35 0.60 S, P each 0.04 max	Hardened and tempered	179/229	620 (40)	370 (24)	13	Charpy V	25 (18)	Suitable for nitriding. Used in the oil and gas industries. Resistant to high pressure hydrogen
5 CrMo Steel	BS 3100 B5	0.20 max	0.75 max	0.40 0.70	Cr 4.00 6.00 Mo 0.45 0.65 S, P each 0.04 max	Hardened and tempered	179/229	620 (40)	420 (27)	13	Charpy V	25 (18)	Used in the oil and gas industries. Resistant to high pressure hydrogen
Low-alloy wear resisting steel castings													
1 Cr Steel	BS 3100 BW2	0.45 0.55	0.75 max	0.50 1.00	Cr 0.80 1.20 S, P each 0.06 max	Normalized and tempered	201/255	700 (45)	—	7	—	—	—
	BW3					Hardened and tempered	293	990 (64)	—	—	—	—	—

† 0.1% proof stress.

Table 22.51 CAST IRONS AND CAST STEELS—MECHANICAL PROPERTIES AT ROOM TEMPERATURE—*continued*

		Composition limits %						*Properties (minima unless otherwise stated)*					
											Impact values		
Material	*British or other standards*	C	Si	Mn	*Other elements*	*Condition*	*Brinell hardness* HB	*UTS* MPa (tonf in^{-2})	*Yield stress*† MPa (tonf in^{-2})	*Elong.* (gl $5.65\sqrt{S_0}$) %	*Specimen*	J (ft lbf)	*Remarks*
1CrMo Steel	BS 3100 BW4	0.55 0.65	0.75 max	0.50 1.00	Cr 0.80 1.50 Mo 0.20 0.40 S, P each 0.06 max	Hardened and tempered	341	1 145 (74)	—	—	—	—	—
$1\frac{1}{4}$ MnSi-CrNi Steel	Austria CHRONIT CMN 100	0.70	1.00 (Typical)	1.30	Cr 1.00 Ni 1.00	Hardened and tempered	390	1 310 (85)	—	—	—	—	—
Low alloy steel castings for use at sub-zero temperatures (Stahl-Eisen Werkstoffblatt 685:1968)													
$\frac{1}{2}$ Mo Steel	BS 3100 BL1	0.20 max	0.60 max	1.00 max	Mo 0.45 0.65 S, P each 0.04 max	Hardened and tempered	—	460 (30)	260 (17)	18	Charpy V at −50 °C	20 (15)	—
1 CrMo Steel	DIN GS-26 CrMo4	0.22 0.29	0.30 0.50	0.50 0.80	Cr 0.80 1.20 Mo 0.20 0.30 S 0.025 max P 0.030 max	Hardened and tempered	—	540/700 (35/45)	340 (22)	16	—	—	—
$1\frac{1}{2}$ Ni Steel	DIN GS-10 Ni 6	0.06 0.12	0.30 0.50	0.50 0.80	Ni 1.30 1.80 S, P each 0.025 max	Hardened and tempered	—	390/540 (25/35)	250 (16)	18	—	—	—

$3\frac{1}{2}$ Ni Steel	BS 3100 BL2	0.12 max	0.60 max	0.80 max	Ni 3.00 4.00 S, P each 0.030 max	Hardened and tempered	—	460 (30)	280 (18)	20	Charpy V at −60 °C	20 (15)	
5 Ni Steel	DIN GS-10Ni 19	0.06 0.12	0.30 0.50	0.50 0.80	Ni 4.50 5.50 S, P each 0.025 max	Hardened and tempered	—	500/700 (32/45)	340 (22)	17	—	—	—
Stainless and heat resisting steel castings (DIN 17245:1967, 17445:1969, Stahl-Eisen Werkstoffblatt 410:70, 471:60, 595:70, ASTM Standards) *Martensitic steel castings*													
12 Cr Steel	BS 3100 410 C21	0.15 max	1.00 max	1.00 max	Cr 11.50 13.50 Ni 1.00 max S, P each 0.04 max	Hardened and tempered	152/207	540 (35)	370 (24)	15	—	—	—
12 Cr Steel	ASTM A 296 CA-15	0.15 max	1.50 max	1.00 max	Cr 11.50 14.00 Ni 1.00 max Mo 0.50 max S, P each 0.04 max	Hardened and tempered	183	620 (40)	450 (29)	16	—	—	—
12 Cr Steel	BS 3100 420 C29	0.20 max	1.00 max	1.00 max	Cr 11.50 13.50 Ni 1.00 max S, P each 0.04 max	Hardened and tempered	201/255	690 (45)	465 (30)	11	—	—	—

† 0.1% proof stress.

Table 22.51 CAST IRONS AND CAST STEELS—MECHANICAL PROPERTIES AT ROOM TEMPERATURE—*continued*

		Composition limits %					*Properties (minima unless otherwise stated)*						
											Impact values		
Material	*British or other standards*	C	Si	Mn	*Other elements*	*Condition*	*Brinell hardness* HB	*UTS* MPa ($tonf\ in^{-2}$)	*Yield stress*† MPa ($tonf\ in^{-2}$)	*Elong.* (gl $5.65\sqrt{S_0}$) %	*Specimen*	J (ft lbf)	*Remarks*
17 CrNi Steel	DIN G-X22-CrNi17	0.20 0.27	1.00 max	1.00 max	Cr 16.00 18.00; Ni 1.00 2.00	Hardened and tempered	230/300	790(51)	590(38)	14	—	—	—
12 CrMo-NiV Steel	DIN G-X22CrMoV 121	0.20 0.26	0.20 0.40	0.50 0.70	Cr 11.30 12.20; Mo 1.00 1.20; Ni 0.70 1.00; V 0.25 0.35; S 0.030 max; P 0.045 max	Hardened and tempered	—	700(45)	590(38)	15	—	—	Steam and gas turbine parts. Also used in the oil and gas industries
13/4 CrNi Steel	DIN G-X5CrNi-134 USA (CA-6NM)	0.07 max	1.00 max	1.00 max	Cr 12.00 13.50; Ni 3.50 5.00; Mo 1.00 max; S, P each 0.04 max	Hardened and tempered	—	770(50)	590(38)	14	—	—	Turbine and compressor parts in hydro-electric plants
Austenitic steels													
18/8 CrNi Steel	BS3100 304C15	0.08 max	1.50 max	2.00 max	Cr 17.00 21.00; Ni 8.00 min; S. P each 0.04 max	Softened	—	480(31)	240(16)	26	Charpy V at −196°C	41 (30)	Suitable for use at sub-zero temperatures

18/9 CrNiNb Steel	BS3100 347C17	0.08 max	1.50 max	2.00 max	Cr 17.00 21.00 Ni 8.50 min. Nb 8 × C/1.00 S, P each 0.04 max	Softened	—	480(31)	240(16)	22	Charpy V at −196°C	20(15)	
18/8 CrNi Steel	BS3100 304C12	0.03 max	1.50 max	2.00 max	Cr 17.00 21.00 Ni 8.00 min S, P each 0.04 max	Softened	—	430(28)	215(14)	26	Charpy V at −196°C	34(25)	Suitable for use at sub-zero temperatures
18/8 CrNi Steel	BS3100 302C25	0.12 max	1.50 max	2.00 max	Cr 17.00 21.00 Ni 8.00 min S, P each 0.04 max	Softened	—	480(31)	240(16)	26	—	—	Resistant to scaling up to 850°C in air
22/9 CrNiSi Steel	DIN G-X40CrNiSi 22.9 Similar to BS3100 309C30	0.30 0.50	1.00 2.50	1.50 max	Cr 21.00 23.00 Ni 9.00 11.00 S 0.03 max P 0.045 max	Softened	—	450(29)	—	12	—	—	Resistant to scaling up to 1 000°C in air
25/12 CrNiN Steel	BS3100 309C32	0.20 0.45	1.50 max	2.50 max	Cr 24.00 28.00 Ni 11.00 14.00 Mo 1.50 max N 0.20 max S. P each 0.04 max	Aged at 760°C	—	560(36)	—	3	—	—	—

†0.1% proof stress.

Table 22.51 CAST IRONS AND CAST STEELS—MECHANICAL PROPERTIES AT ROOM TEMPERATURE—*continued*

		Composition limits %					*Properties (minima unless otherwise stated)*						
Material	*British or other standards*	C	Si	Mn	*Other elements*	*Condition*	*Brinell hardness* HB	*UTS* MPa (tonf in^{-2})	*Yield stress*† MPa (tonf in^{-2})	*Elong.* (gl $5.65\sqrt{S_0}$) %	*Impact values* *Specimen*	*Impact values* J (ft lbf)	*Remarks*
25/12 CrNi Steel	BS3100 309C35 DIN G-X35CrNi 25 12	0.20 0.50	1.50 max	2.00 max	Cr 24.00 28.00 Ni 11.00 14.00 Mo 1.50 max S, P each 0.04 max	As cast	—	510(33)	—	7	—	—	Used in the oil and gas industries
						Softened	—	450(29)	230(15)	10	—	—	
25/12 CrNiSi Steel	DIN G-X35CrNiSi 25 12	0.20 0.50	1.50 2.50	1.50 max	Cr 24.00 26.00 Ni 11.00 14.00 S 0.030 max P 0.045 max	Softened	—	450(29)	—	10	—	—	Resistant to scaling up to 1 100 °C in air
25/13 CrNi Steel	DIN G-X15CrNi 2513	0.10 0.20	1.50 max	2.00 max	Cr 24.00 26.00 Ni 12.00 14.00 S 0.030 max P 0.045 max	Softened	—	430(28)	200(13)	25	—	—	Used in the oil and gas industries
26/14 CrNiSi Steel	DIN G-X40CrNiSi 26 14	0.20 0.50	1.50 2.50	1.50 max	Cr 25.00 28.00 Ni 13.00 16.00 S 0.030 max P 0.045 max	Softened	—	450(29)	—	8	—	—	Resistant to scaling up to 1 150 °C in air

25/20 CrNi Steel	BS3100 310C40 DIN G-X35CrNi25 20	0.30 0.50	1.50 max	2.00 max	Cr 24.00 27.00 Ni 19.00 22.00 Mo 1.50 max S. P each 0.04 max	As cast	—	450 (29)	—	7	—	—	Used in the oil and gas industries
25/20 CrNiSi Steel	DIN G-X40CrNiSi 25 20	0.20 0.50	1.50 2.50	1.50 max	Cr 24.00 27.00 Ni 19.00 21.00 S 0.030 max P 0.045 max	Softened	—	450(29)	—	8	—	—	Resistant to scaling up to 1 150°C in air.
25/20 CrNi Steel	DIN G-X15CrNi 25 20	0.10 0.20	1.50 max	2.00 max	Cr 24.00 26.00 Ni 19.00 21.00 S 0.030 max P 0.045 max	Softened	—	430(28)	200(13)	25	—	—	Used in the oil and gas industries
35/15 NiCr Steel	BS3100 330C11	0.35 0.55	1.50 max	2.00 max	Ni 33.00 37.00 Cr 13.00 17.00 Mo 0.50 max S. P each 0.04 max	As cast	—	450(29)	—	3	—	—	
35/20 NiCr Steel	DIN G-X35NiCr 35 20	0.25 0.45	1.50 max	2.00 max	Ni 33.00 35.00 Cr 20.00 22.00 S 0.03 max P 0.045 max	Softened	—	450(29)	220(14)	10	—	—	Used in the oil and gas industries

† 0.1% proof stress.

Table 22.51 CAST IRONS AND CAST STEELS—MECHANICAL PROPERTIES AT ROOM TEMPERATURE—*continued*

		Composition limits %					*Properties (minima unless otherwise stated)*						
											Impact values		
Material	*British or other standards*	C	Si	Mn	*Other elements*	*Condition*	*Brinell hardness* HB	*UTS* MPa (tonf in^{-2})	*Yield stress*† MPa (tonf in^{-2})	*Elong.* (gl $5.65\sqrt{S_0}$) %	*Specimen*	J (ft lbf)	*Remarks*
35/25 NiCr Steel	DIN G-X35NiCr 35 25	0.25 0.45	1.50 max	2.00 max	Ni 33.00 35.00 Cr 24.00 26.00 S 0.03 max P 0.045 max	Softened	—	450(29)	230(15)	10	—	—	Used in the oil and gas industries
36/16 NiCrSi Steel	BS 3100 331C40 DIN G-X40NiCrSi 36 16	0.35 0.55	1.00 2.50	1.50 max	Ni 36.00 40.00 Cr 16.00 20.00 Mo 0.50 max S 0.030 max P 0.045 max	As cast	—	450(29)	—	3	—	—	Resistant to scaling up to 1 100 °C in air
						Softened	—	390(25)	—	8	—	—	
19/10 CrNiMo Steel	BS 3100 317C16	0.08 max	1.50 max	2.00 max	Cr 17.00 21.00 Ni 10.00 min Mo 3.00 4.00 S, P each 0.04 max	Softened	—	480(31)	240(16)	22	—	—	Used in the paper and chemical industries

18/10 CrNiMo Steel	BS 3100 316C16	0.08 max	1.50 max	2.00 max	Cr 17.00 21.00 Ni 10.00 min Mo 2.00 3.00 S, P each 0.04 max.	Softened	—	480(31)	240(16)	26	Charpy V at −196 °C	34(25)	Suitable for use at sub-zero temperatures
18/10 CrNi-MoNb Steel	BS 3100 318C17	0.08 max	1.50 max	2.00 max	Cr 17.00 21.00 Ni 10.00 min Mo 2.00 3.00 Nb 8 × C/ 1.00 S, P each 0.04 max	Softened	—	480(31)	240(16)	18	—	—	—
18/9 CrNiMo Steel	BS 3100 316C71	0.08 max	1.50 max	2.00 max	Cr 17.00 21.00 Ni 8.00 min Mo 2.00 3.00 S, P each 0.04 max	Softened	—	510(33)	260(17)	26	Charpy V at −196 °C	34(25)	Suitable for use at sub-zero temperatures
18/8 CrNiMo Steel	BS 3100 315C16	0.08 max	1.50 max	2.00 max	Cr 17.00 21.00 Ni 8.00 min Mo 1.00 1.75 S, P each 0.04 max	Softened	—	480(31)	240(16)	26	Charpy V at −196 °C	34(25)	Suitable for use at sub-zero temperatures

†0.1% proof stress.

Table 22.51 CAST IRONS AND CAST STEELS—MECHANICAL PROPERTIES AT ROOM TEMPERATURE—*continued*

Material	*British or other standards*	*Composition limits %* C	Si	Mn	*Other elements*	*Condition*	*Brinell hardness* HB	*Properties (minima unless otherwise stated)* *UTS* MPa (tonf in^{-2})	*Yield stress*† MPa (tonf in^{-2})	*Elong.* (gl $5.65\sqrt{S_0}$) %	*Impact values* *Specimen*	J (ft lbf)	*Remarks*
18/10 CrNiMo Steel	BS 3100 316C12	0.03 max	1.50 max	2.00 max	Cr 17.00 21.00 Ni 10.00 min Mo 2.00 3.00 S, P each 0.04 max	Softened	—	430 (28)	215 (14)	26	Charpy V at −196 °C	41 (30)	Suitable for use at sub-zero temperatures
17/13 CrNiMo Steel	DIN G-X6CrNiMo 17 13	0.07 max	1.00 max	2.00 max	Cr 16.00 18.00 Ni 12.50 14.50 Mo 4.00 5.00 S, P each 0.04 max.	Softened	—	390 (25)	180 (11.5)	15	—	—	Used in the chemical industry
16/16 CrNi MoNb Steel	DIN G-X8CrNi-MoNb 16 16	0.08 max	1.50 max	1.50 max	Cr 15.00 18.00 Ni 14.00 17.00 Mo 1.70 2.20 Nb 10 × C/ 1.20 S 0.03 max P 0.045 max	Softened	—	480 (31)	200 (13)	17	—	—	High creep resistant gas turbine parts

20/18 NiCrMo CuNb Steel	DIN G-X7CrNiMo CuNb 2018	0.08 max	1.50 max	2.00 max	Ni 19.0 21.0 Cr 16.5 18.5 Mo 2.00 2.50 Cu 1.80 2.40 Nb 8 × C min S 0.03 max P 0.045 max	Softened	—	450 (29)	180 (11.5)	15	—	—	Resistant to corrosion by sulphuric acid solutions
25/20 Ni-CrMoCu Nb Steel	DIN G-X7NiCrMo-CuNb 25 20	0.08 max	1.50 max	2.00 max	Ni 24.00 26.00 Cr 19.00 21.00 Mo 2.50 3.50 Cu 1.50 2.50 Nb 8 × C min S 0.03 max P 0.045 max	Softened	—	450 (29)	200 (13)	15	—	—	Resistant to corrosion by sulphuric acid solutions
Ferritic—austenitic steel castings													
27/4 CrNi Steel	DIN G-X40CrNi 274	0.30 0.50	1.00 2.00	1.50 max	Cr 26.00 28.00 Ni 3.50 5.50 S 0.03 max P 0.045 max	Softened	—	500 (32)	—	4	—	—	Furnace parts resistant to sulphur-containing gases, and scaling up to 1 150 °C in air

† 0.1% proof stress.

Table 22.51 CAST IRONS AND CAST STEELS—MECHANICAL PROPERTIES AT ROOM TEMPERATURE—*continued*

		Composition limits %						*Properties (minima unless otherwise stated)*					
											Impact values		
Material	*British or other standards*	C	Si	Mn	*Other elements*	*Condition*	*Brinell hardness* HB	*UTS* MPa (tonf in^{-2})	*Yield stress*† MPa (tonf in^{-2})	*Elong.* (gl $5.65\sqrt{S_0}$) %	*Specimen*	J (ft lbf)	*Remarks*
26/5 Cr-NiCuMo Steel	ASTM CD-4MCu	0.04 max	1.00 max	1.00 max	Cr 25.00 27.00 Ni 4.75 6.00 Cu 2.75 3.25 Mo 1.75 2.25 S, P each 0.04 max	Precipitation hardened	290/320	930/1 080 (60/70)	700/835 (45/54)	9/22	Charpy U	16 (12)	Resistant to stress corrosion. Used in naval construction and ordnance
High strength stainless steel castings													
17/4 CrNiCu Steel	ASTM CB-7Cu (17.4 PH)	0.05 max	1.00 max	1.00 max	Cr 15.50 17.00 Ni 3.50 4.50 Cu 2.30 3.30 S, P each 0.04 max	Precipitation hardened	340	1 145 (74) (Typical)	1 050 (68)	12	Charpy U	31 (23)	Martensitic steel, aircraft parts
15/5 CrNiCu Steel	UK (FV 520 B)	0.07	0.70 (typical)	0.50	Cr 15.00 Ni 5.00 Cu 2.00 Mo 1.00 Nb 0.50	Precipitation hardened	375	1 265 (82)	(typical) 1 050 (68)	10	—	—	Martensitic steel

15/4 CrNi-MoN Steel	USA AM 355	0.10 0.15	0.50 max	0.50 1.25	Cr 15.00 16.00 Ni 4.00 5.00 Mo 2.50 3.25 N 0.07 0.13 S, P each 0.04 max	Precipitation hardened	430/460	1 435/1 540 (93/100)	1 020/1 205 (66/78)	9/17	—	—	Semi-austenitic steels. Aircraft parts
Other high alloy steel castings (Stahl Eisen Werkstoffblatt 390:1961, ASTM Standards)													
12Mn Steel	BS 3100 BW10 DIN G-X120Mn12	1.00 1.25	1.00 max	11.00 min	S 0.06 max P 0.07 max	Softened	—	590 (38)	295 (19)	25	—	—	Non-magnetic, wear resistant
13MnMo Steel (low carbon)	ASTM A 128 Grade E1	0.70 1.10	0.40 0.80	12.00 14.00	Mo 0.90 1.10 Cr 0.70 max P 0.05 max	Softened	185/200	590/760 (38/49)	340/370 (22/24)	26/35	—	—	Non-magnetic, wear resistant. For castings of 125 to 250 mm (5–10 in) thickness
13MnMo Steel (high carbon)	ASTM A 128 Grade E1	1.10 1.40	0.40 0.70	12.00 14.00	Mo 0.60 1.20 Cr 0.70 max P 0.05 max	Softened	200/210	620/760 (40/49)	370/420 (24/27)	17/26	—	—	Non-magnetic, wear resistant. For castings of less than 125 mm (5 in) thickness
8/8/6 MnCrNi Steel	DIN G-X25Mn-CrNi 8 8 6	0.22 0.28	0.70 max	7.50 9.50	Cr 7.00 8.50 Ni 5.00 6.50 S 0.03 max P 0.045 max	Softened	—	500 (32)	220 (14)	30	—	—	Non-magnetic

† 0.1% proof stress.

Table 22.51 CAST IRONS AND CAST STEELS—MECHANICAL PROPERTIES AT ROOM TEMPERATURE—*continued*

		Composition limits %					*Properties (minima unless otherwise stated)*						
											Impact values		
Material	*British or other standards*	C	Si	Mn	*Other elements*	*Condition*	*Brinell hardness* HB	*UTS* MPa (tonf in^{-2})	*Yield stress*† MPa (tonf in^{-2})	*Elong.* (gl $5.65\sqrt{S_0}$) %	*Specimen*	J (ft lbf)	*Remarks*
8/8/5 MnNi-CrV Steel	DIN G-X45MnNi CrV 8 8 5	0.40 0.50	1.00 max	7.50 9.50	Ni 7.00 8.50 Cr 4.50 6.00 V 1.00 1.50 S 0.03 max P 0.045 max	Softened Age hardened	— —	500(32) 630(41)	340(22) 540(35)	10 3	— —	— —	Non-magnetic
9CrMo Steel	BS 3100 B6	0.20 max	1.00 max	0.30 0.70	Cr 8.00 10.00 Mo 0.90 1.20 S, P each 0.04 max	Hardened and tempered	179/229	620(40)	420(27)	13	—	—	Used in the oil and gas industries. Resistant to high pressure hydrogen.

†0.1% proof stress.

Table 22.52 TYPICAL HOT TENSILE PROPERTIES OF IRON AND STEEL CASTINGS

		Typical composition %						*Mechanical properties*			
Material	*British or other standards*	*Total* C	Si	Mn	*Other elements*	*Condition*	*Temperature* °C	*UTS* MPa ($tonf\,in^{-2}$)	*Yield stress* MPa ($tonf\,in^{-2}$)	*Elong.* (gl $5.65\sqrt{S_0}$) %	*RA* %
1. *Iron castings*											
Grey iron	BS 1452 Grade 220	3.29	1.27	0.28	—	As cast	RT	236 (15.3)	Not given	0.50	Not given
							370	242 (15.7)	Not given	0.50	Not given
							450	229 (14.8)	Not given	0.75	Not given
Heat resisting grey iron		2.39	5.72	0.67	—	As cast	RT	219 (14.2)	Not given	0.50	Not given
							370	210 (13.6)	Not given	0.50	Not given
							450	219 (14.2)	Not given	0.75	Not given
							540	136 (8.8)	Not given	0.75	Not given
Austenitic iron	Similar to BS L–NiSiCr 20 5 3	1.75	5.84	0.68	Ni 17.72 Cr 2.10	As cast	RT	190 (12.3)	Not given	Not given	2.5
							450	128 (8.3)	Not given	Not given	1.5
							538	119 (7.7)	Not given	Not given	1.5
Malleable iron	DIN 1692 GTS-35	2.78	0.91	0.47	—	As treated	RT	340 (22.0)	210 (13.6)	10.0	Not given
							320	267 (17.3)	176 (11.4)	5.5	Not given
							430	229 (14.8)	159 (10.3)	5.5	Not given
							540	111 (7.2)	100 (6.5)	10.0	Not given
2. *Steel castings* *Carbon steel castings*											
C Steel	BS 3100 AM1	0.14	0.29	0.45	—	Annealed	RT	385 (24.9)	235 (15.2)	22.0	36.0
							100	352 (22.8)	187 (12.1)	19.8	35.6
							200	400 (25.9)	187 (12.1)	10.5	22.8
							300	414 (26.8)	131 (8.5)	13.5	21.0
							400	372 (24.1)	137 (8.9)	21.6	26.5
							500	201 (13)	97 (6.3)	25.5	43.2
C Steel	BS 3100 A2	0.28	0.35	0.66	—	Normalized and tempered	RT	524 (33.9)	289 (18.7)	29	54
							100	486 (31.5)	283 (18.3)	24.5	48
							200	480 (31.1)	240 (15.5)	21.5	43
							300	463 (30.0)	201 (13.0)	21	44
							400	448 (29.0)	190 (12.3)	23.5	58
							500	324 (21.0)	162 (10.5)	32	74

Table 22.52 TYPICAL HOT TENSILE PROPERTIES OF IRON AND STEEL CASTINGS—*continued*

		Typical composition %						*Mechanical properties*			
Material	*British or other standards*	*Total C*	Si	Mn	*Other elements*	*Condition*	*Temperature* °C	*UTS* MPa ($tonf\,in^{-2}$)	*Yield stress* MPa ($tonf\,in^{-2}$)	*Elong.* (gl $5.65\sqrt{S_0}$) %	*RA* %
Low alloy steel castings for direct hardening											
2NiCr Steel	BS 3100	0.35	Not given	0.80	Ni 2.00	Hardened	RT	703 (45.5)	571 (37)	19	50
	BT1				Cr 0.70	and	100	672 (43.5)	571 (37)	17	Not given
						tempered	300	703 (45.5)	479 (31)	10	Not given
							500	332 (21.5)	286 (18.5)	24	Not given
							600	170 (11.0)	255 (16.5)	28	Not given
Low alloy creep resisting castings											
$\frac{1}{3}$Mo Steel	Similar to	0.22	Not given	0.57	Mo 0.32	Normalized	RT	470 (30.4)	278 (18)	21	43
	BS 3100					and	100	434 (28.1)	255 (16.5)	12	38
	B1					tempered	200	558 (36.1)	249 (16.1)	12.8	38
							300	524 (33.9)	241 (15.6)	18.5	42
							400	472 (30.6)	235 (15.2)	18.5	42
$\frac{3}{4}$Mo Steel	Similar to	0.22	Not given	Not given	Mo 0.80	Normalized	RT	537 (34.8)	324 (21)	22	Not given
	BS 3100					and	400	456 (29.5)	193 (12.5)	17	Not given
	B1					tempered	500	358 (23.2)	193 (12.5)	17	Not given
$\frac{3}{4}$NiCrMo Steel	—	0.13	0.37	0.51	Ni 0.77	Normalized	RT	541 (35)	340 (22)	28	55
					Cr 0.53	and	510	386 (25)	216 (14)	30	69
					Mo 0.51	tempered	540	347 (22.5)	216 (14)	33	75
							570	324 (21)	201 (13)	36	76.5
$\frac{3}{4}$NiMoCr Steel	—	0.12	0.28	0.40	Mo 1.08	Normalized	RT	556 (36)	355 (23)	27	63
					Ni 0.81	and	540	386 (25)	247 (16)	27	74
					Cr 0.57	tempered	570	355 (23)	232 (15)	27	78
							590	309 (20)	216 (14)	21	79

$1\frac{1}{4}$CrMo Steel	BS 3100 B2	0.11	0.39	0.59	Cr 1.30 Mo 0.54	Normalized and tempered	RT	540 (35)	365 (23.6)	28	55
							510	378 (24.5)	170 (11)	27	65
							540	340 (22)	178 (11.5)	29	70
							570	309 (20)	162 (10.5)	35	77
$2\frac{1}{2}$CrMo Steel	Similar to BS 3100 B4	0.23	1.10	0.72	Cr 2.36 Mo 0.61	Hardened and tempered	RT	641 (41.5)	402 (26)	18	45
							540	425 (27.5)	378 (24.5)	24	70
							650	216 (14)	185 (12)	42	76
5CrMo Steel	Similar to BS 3100 B5	0.22	0.36	0.57	Cr 5.46 Ni 0.51 Mo 0.50	Hardened and tempered	RT	687 (44.5)	494 (32)	17.5	40
							430	502 (32.5)	378 (24.5)	17	49
							540	386 (25)	301 (19.5)	26	75
							650	247 (16)	69 (4.5)	25	81
High alloy steel castings											
9CrMoAl Steel	Similar to BS 3100 B6	0.22	0.84	1.05	Cr 9.09 Mo 1.56 Al 0.16	Hardened and tempered	RT	752 (48.7)	539 (34.9)	20	60
							540	405 (26.2)	349 (22.6)	37	76
							590	304 (19.7)	249 (16.1)	41	83
							650	196 (12.7)	154 (10)	61	90
19/10CrNiNb Steel	BS 3100 347C17	0.04	0.94	0.84	Cr 19.50 Ni 10.80 Nb 0.75	Softened	RT	510 (33)	278 (18)	35	52
							150	425 (27.5)	219 (14.2)	31	51
							290	397 (25.7)	173 (11.2)	29	49
							430	375 (24.3)	151 (9.8)	28	45
							540	361 (23.4)	142 (9.2)	28	46
							650	306 (19.8)	145 (9.4)	26	46
18/13CrNiMo Steel	BS 3100 316C16	0.08	1.03	0.50	Cr 18.10 Ni 13.70 Mo 3.00	Softened	RT	503 (32.6)	253 (16.4)	37	45
							700	269 (17.4)	110 (7.1)	42	Not given
							740	263 (17)	110 (7.1)	33	58

Table 22.53 TYPICAL HOT CREEP AND RUPTURE PROPERTIES OF IRON AND STEEL CASTINGS (DIN 17245:1967), Stahl-Eisen Werkstoffblatt 471:1960, 595:1970)

		Typical composition %						*Stress to produce 1.0% strain* MPa (tonf in^{-2})			*Stress to cause rupture* MPa (tonf in^{-2})	
Material	*British or other standards*	*Total C*	Si	Mn	*Other elements*	*Condition*	*Tempera-ture* °C	*in* 1 000 h	*in* 10 000 h	*in* 100 000 h	*in* 10 000 h	*in* 100 000 h
1. *Iron castings*												
Grey Iron	BS 1452 Grade 220	3.29	1.27	0.28	—	As cast	450	100 (6.5)	Not given	Not given	Not given	Not given
Heat resisting Grey Iron	—	2.39	5.72	0.67	—	As cast	540	60 (3.9)	Not given	Not given	Not given	Not given
Malleable Iron	DIN 1692 GTS-35	2.78	0.91	0.47	—	As treated	430	111 (7.2)	Not given	Not given	Not given	Not given
Austenitic Iron	Similar to BSL-NiSiCr 20 5 3	1.75	5.84	0.68	Ni 17.72	As cast	450	63 (4.1)	Not given	Not given	Not given	Not given
					Cr 2.10		540	22 (1.4)	Not given	Not given	Not given	Not given
2 *Steel castings*												
Carbon steel castings												
C Steel	DIN GS-C25	0.20	0.40	0.60	—	Hardened	400	Not given	128 (8.3)	88 (5.7)	162 (10.5)	128 (8.3)
						and	450	Not given	68 (4.4)	45 (2.9)	88 (5.7)	63 (4.1)
						tempered	500	Not given	34 (2.2)	20 (1.3)	49 (3.2)	29 (1.9)
Low alloy steel castings for direct hardening												
MnCrMo Steel	—	0.28	Not given	1.35	Cr 0.80	Hardened	430	Not given	358 (23.2)	Not given	479 (31.0)*	Not given
					Mo 0.35	and tempered	540	Not given	63 (4.1)	Not given	93 (6.0)*	Not given
1NiCrMo Steel	—	0.30	Not given	0.60	Ni 1.15	Hardened	430	Not given	337 (21.8)	Not given	463 (20.0)*	Not given
					Cr 0.65	and	480	Not given	193 (12.5)	Not given	263 (17.0)*	Not given
					Mo 0.45	tempered	540	Not given	43 (2.8)	Not given	62 (4.0)*	Not given
2NiCrMo Steel	—	0.40	0.28	0.57	Ni 2.03	Hardened	290	Not given	420 (27.2)	Not given	556 (36.0)*	Not given
					Cr 0.82	and	430	Not given	275 (17.8)	Not given	386 (25.0)*	Not given
					Mo 0.23	tempered	540	Not given	39 (2.5)	Not given	56 (3.6)*	Not given

5CrMo Steel	—	0.27	0.60	0.58	Cr	5.14	Hardened	480	Not given	113 (7.3)	Not given	160 (10.4)	Not given
					Mo	0.61	and	540	Not given	59 (3.8)	Not given	83 (5.4)	Not given
							tempered						
5CrW Steel	—	0.28	0.36	0.42	Cr	5.10	Hardened	480	Not given	117 (7.6)	Not given	162 (10.5)*	Not given
					W	0.79	and	540	Not given	42 (2.7)	Not given	59 (3.8)*	Not given
							tempered	590	Not given	19 (1.2)	Not given	26 (1.7)*	Not given
Low alloy creep resisting steel castings													
½Mo Steel	BS 3100	0.20	0.35	0.70	Mo	0.45	Hardened	450	Not given	204 (13.2)	131 (8.5)	284 (18.4)	196 (12.7)
	B1						and	500	Not given	128 (8.3)	63 (4.1)	162 (10.5)	88 (5.7)
	DIN GS-22Mo4						tempered	550	Not given	49 (3.2)	17 (1.1)	68 (4.4)	25 (1.6)
1¼CrMo Steel	BS 3100	0.17	0.40	0.65	Cr	1.25	Hardened	450	Not given	219 (14.2)	168 (10.9)	295 (19.1)	225 (14.6)
	B2				Mo	0.50	and	500	Not given	148 (9.6)	91 (5.9)	187 (12.1)	128 (8.3)
	DIN GS-17CrMo55						tempered	550	Not given	74 (4.8)	34 (2.2)	99 (6.4)	49 (3.2)
2¼CrMo Steel	BS 3100	0.16	0.40	0.55	Cr	2.30	Normalized	450	Not given	Not given	196 (12.7)	Not given	255 (16.5)
	B3				Mo	1.00	and	500	Not given	Not given	108 (7.0)	Not given	151 (9.8)
							tempered	550	Not given	Not given	43 (2.8)	Not given	63 (4.1)
½MoCrV Steel	BS 3100	0.14	0.40	0.60	Mo	0.60	Normalized	450	Not given	Not given	221 (14.3)	Not given	280 (18.1)
	B7				Cr	0.40	and	500	Not given	Not given	133 (8.6)	Not given	171 (11.1)
					V	0.25	tempered	550	Not given	Not given	59 (3.8)	Not given	79 (5.1)
1¼CrMoV	DIN	0.17	0.40	0.65	Cr	1.30	Hardened	500	Not given	205 (13.3)	147 (9.5)	259 (16.8)	192 (12.4)
Steel	GS-17CrMoV				Mo	1.00	and	550	Not given	122 (7.9)	63 (4.1)	158 (10.2)	93 (6.0)
	5 11				V	0.25	tempered	580	Not given	74 (4.8)	34 (2.2)	103 (6.7)	49 (3.2)
High alloy steel castings													
12CrMoNiV	DIN	0.22	0.30	0.60	Cr	12.0	Hardened	500	Not given	216 (14.0)	162 (10.5)	270 (17.5)	205 (13.3)
Steel	G-X22CrMoV				Mo	1.10	and	550	Not given	131 (8.5)	91 (5.9)	167 (10.8)	117 (7.6)
	12 1				Ni	0.85	tempered	600	Not given	66 (4.3)	34 (2.2)	83 (5.4)	49 (3.2)
					V	0.30							
18/9CrNi	BS 3100	0.07	0.65	0.50	Cr	18.2	Softened	540	Not given	100 (6.5)	Not given	139 (9.0)*	Not given
Steel	304C15				Ni	9.6		650	Not given	54 (3.5)	Not given	76 (4.9)*	Not given
22/9CrNiSi	BS 1648	0.40	2.00	1.00	Cr	22.0	Softened	600	88 (5.7)	Not given	Not given	Not given	Not given
Steel	Grade E				Ni	10.0		700	39 (2.5)	Not given	Not given	Not given	Not given
	DIN							800	17 (1.1)	Not given	Not given	Not given	Not given
	G-X40CrNiSi 22 9												

* Estimated

Table 22.53 TYPICAL HOT CREEP AND RUPTURE PROPERTIES OF IRON AND STEEL CASTINGS (DIN 17245: 1967), Stahl-Eisen Werkstoffblatt 471: 1960, 595: 1970)—*continued*

Material	British or other standards	Typical composition %				Condition	Temperature °C	Stress to produce 1.0% strain MPa (tonf in^{-2})			Stress to cause rupture MPa (tonf in^{-2})	
		Total C	Si	Mn	Other elements			in 1000 h	in 10 000 h	in 100 000 h	in 10 000 h	in 100 000 h
25/12CrNi	BS 3100	0.35	1.00	1.50	Cr 25.0	Softened	600	Not given	54 (3.5)*	Not given	77 (5.0)	Not given
Steel	309C35				Ni 12.0		700	Not given	32 (2.1)*	Not given	46 (3.0)	Not given
	DIN						800	Not given	12 (0.8)*	Not given	17 (1.1)	Not given
	G-X35CrNi											
	25 12											
25/13CrNi	DIN	0.15	1.00	1.50	Cr 25.0	Softened	600	Not given	49 (3.2)*	Not given	69 (4.5)	Not given
Steel	G-X15CrNi				Ni 13.0		700	Not given	31 (2.0)*	Not given	45 (2.9)	Not given
	25 12						800	Not given	9 (0.6)*	Not given	12 (0.8)	Not given
25/20CrNi	BS 3100	0.35	1.00	1.50	Cr 25.0	Softened	700	Not given	49 (3.2)*	Not given	69 (4.5)	Not given
Steel	310C40				Ni 20.0		800	Not given	23 (1.5)*	Not given	32 (2.1)	Not given
	DIN						900	Not given	14 (0.9)*	Not given	20 (1.3)	Not given
	G-X35CrNi											
	25 20											
25/20CrNi	DIN G-X15	0.15	1.00	1.50	Cr 25.0	Softened	700	Not given	36 (2.3)*	Not given	49 (3.2)	Not given
Steel	CrNi				Ni 20.0		800	Not given	11 (0.7)*	Not given	15 (1.0)	Not given
	25 20											
35/20NiCr	DIN	0.35	1.00	1.50	Ni 35.0	Softened	700	Not given	53 (3.4)*	Not given	74 (4.8)	Not given
Steel	G-X35NiCr				Cr 21.0		800	Not given	25 (1.6)*	Not given	36 (2.3)	Not given
	35 20						900	Not given	17 (1.1)*	Not given	23 (1.5)	Not given
35/25NiCr	DIN	0.35	1.00	1.50	Ni 35.0	Softened	700	Not given	53 (3.4)*	Not given	74 (4.8)	Not given
Steel	G-X35NiCr				Cr 25.0		800	Not given	25 (1.6)*	Not given	36 (2.3)	Not given
	35 25						900	Not given	17 (1.1)*	Not given	23 (1.5)	Not given
36/16NiCrSi	BS 3100	0.40	1.50	1.00	Ni 37.0	Softened	600	99 (6.4)	Not given	Not given	Not given	Not given
Steel	331C40				Cr 17.0		700	49 (3.2)	Not given	Not given	Not given	Not given
	DIN						800	29 (1.9)	Not given	Not given	Not given	Not given
	G-X40NiCrSi						900	14 (0.9)	Not given	Not given	Not given	Not given
	36 16											

* Estimated

Table 22.54 CAST IRONS AND CAST STEELS—MECHANICAL PROPERTIES AT SUB-ZERO TEMPERATURES (BS 3100: 19[illegible], DIN 1694: 19[illegible], Stahl-Eisen Werkstoffblatt 685: 19[illegible])

		Typical Composition %						*Typical properties*				
Material	*British or other standards*	*Total C*	Si	Mn	*Other Elements*	*Condition*	*Temperature* °C	*UTS* MPa (tonf in^{-2})	*Yield stress* MPa (tonf in^{-2})	*Elong.* (gl $5.65\sqrt{S_0}$)	*RA* %	*Impact Charpy V* J (ft.lbf)
1. *Iron castings*												
Nodular Cast Iron	BS 500/7	3.62	3.11	0.49	Ce 0.054	As cast	RT	530 (34.3)	355(23.0)	7.5	Not given	22.8 (16.8)
							−15	—	—	—	—	11.5 (8.5)
							−50	—	—	—	—	9.5 (7.0)
							−80	—	—	—	—	4.1 (3.0)
Austenitic Cast Iron	DIN 1694 GGG-NiMn 23 4	2.40	2.25	4.20	Ni 23.0	As cast	RT	440 (28.5)	201 (13.0)	35.0	32.0	—
							−100	494 (32.0)	293 (19.0)	40.0	37.0	—
							−150	525 (34.0)	340 (22.0)	38.0	35.0	—
							−196	618 (40.0)	440 (28.5)	27.0	25.0	—
2. *Steel castings*												
Carbon steel castings with up to 1.2% Mn content												
0.15C Steel		0.15	0.38	0.75	—	Normalized	RT	448 (29)	239 (15.5)	23.0	Not given	69 (51)
							−20	—	—	—	—	34 (25)
							−40	—	—	—	—	24 (17.5)
							−60	—	—	—	—	22 (16)
							−100	—	—	—	—	7 (5)
CMn Steel	BS 3100 A4	0.19	0.28	1.20	—	Hardened and tempered	RT	564 (36.5)	409 (26.5)	21.0	42.0	68 (50)
							−20	—	—	—	—	58 (43)
							−40	—	—	—	—	34 (25)
							−60	—	—	—	—	20 (14.5)
Low alloy steel castings particularly suitable for use at sub-zero temperatures												
½Mo Steel	BS 3100 BL1	0.17	0.40	0.80	Cr 0.55	Hardened and tempered	RT	479 (31.0)	332 (21.5)	19.0	Not given	—
							−50	—	—	—	—	24 (17.5)
1 Ni Steel	Similar to DIN GS-10Ni6	0.12	0.30	0.60	Ni 1.10	Hardened and tempered	RT	417 (27.0)	286 (18.5)	27.0	63.0	99 (73)
							−45	—	—	—	—	31 (23)
							−70	—	—	—	—	11 (8)
3½ Ni Steel	BS 4242 Grade C	0.09	0.32	0.55	Ni 3.96	Hardened and tempered	RT	494 (32.0)	355 (23.0)	22.0	Not given	54 (40)
							−20	—	—	—	—	41 (30)
							−60	—	—	—	—	34 (25)
							−80	—	—	—	—	19 (14)
							−100	—	—	—	—	14 (10.5)

Table 22.54 CAST IRONS AND CAST STEELS—MECHANICAL PROPERTIES AT SUB-ZERO TEMPERATURES (BS 3100: 1967, DIN 1694: 1966, Stahl-Eisen Werkstoffblatt 685: 1968)—*continued*

		Typical Composition %						*Typical properties*				
Material	*British or other standards*	*Total* C	Si	Mn	*Other Elements*	*Condition*	*Temperature* °C	*UTS* MPa (tonf in^{-2})	*Yield stress* MPa (tonf in^{-2})	*Elong.* (gl 5.65$\sqrt{S_0}$)	*RA* %	*Impact Charpy V* J (ft, lbf)
$3\frac{1}{2}$ NiMo Steel	—	0.11	0.36	0.65	Ni 3.45 Mo 0.27	Hardened and tempered	RT	602 (39.0)	432 (28.0)	23.0	59.0	96 (71)
							−70	—	—	—	—	41 (30)
							−100	—	—	—	—	24 (17.5)
							−130	—	—	—	—	15 (11)
5 Ni Steel	DIN GS-10Ni19	0.11	0.30	0.60	Ni 5.00	Normalized and tempered	RT	625 (40.5)	332 (21.5)	23.0	46.0	75 (55.5)
							−45	—	—	—	—	73 (54)
							−70	—	—	—	—	23 (17)
							−100	—	—	—	—	8 (6)
Low alloy steel casting for direct hardening												
$3\frac{1}{2}$ Ni Steel	BS 3100 BTI	0.30	Not given	0.70	Ni 3.50	Hardened and tempered	RT	695 (45.0)	525 (34.0)	23.0	66.0	Not given
							−184	1 027 (66.5)	919 (59.5)	13.0	17.0	Not given
High alloy steel castings												
12Cr Steel	BS 3100 410 C21	0.15	0.43	0.38	Cr 12.0 Ni 0.60	Hardened and tempered	RT	687 (44.5)	541 (35.0)	20.0	Not given	41 (30)
							0	—	—	—	—	34 (25)
							−10	—	—	—	—	24 (17.5)
							−20	—	—	—	—	20 (14.5)
18/8 CrNi Steel	BS 3100 302C25	0.11	0.70	0.90	Cr 18.6 Ni 8.50	Softened	RT	517 (33.5)	247 (16.0)	50.0	53.0	122 (90)
							0	—	—	—	—	111 (82)
							−20	—	—	—	—	88 (65)
							−50	—	—	—	—	71 (52.5)
							−120	—	—	—	—	61 (45)
							−196	—	—	—	—	34 (25)
18/10 CrNiNb Steel	BS 3100 347C17	0.10	0.74	0.83	Cr 18.40 Ni 10.00 Nb 0.75	Softened	RT	556 (36.0)	247 (16.0)	30.0	30.0	47 (34.5)
							−20	—	—	—	—	38 (28)
							−50	—	—	—	—	33 (24.5)
							−120	—	—	—	—	27 (20)
							−196	—	—	—	—	20(14.5)

Weld metal	*Type*	*Composition limits %* C	Si	Mn	S	P	*Other elements*	*Typical mechanical properties* *UTS* MPa (tonf in^{-2})	*Yield stress* MPa (tonf in^{-2})	*Elong.* (gl $5.65\sqrt{S_0}$) %	*RA* %	*Impact values* *Charpy** J (ft lbf)	*Izod* J (ft lbf)	*Applications*
1. *Carbon steels with up to 1.50% Mn content*														
Carbon steel	Cellulosic	0.10	0.27	0.56 (Typical)	0.02	0.02	—	495 (32.0)	405 (26.2)	32.0	66.0	—	68 (50)	Joining low carbon steels
Carbon steel	Rutile	0.07	0.13	0.37 (Typical)	0.02	0.03	—	510 (33.0)	445 (28.8)	29.0	63.0	—	95 (70)	Joining low carbon steels
Carbon steel	Basic (low hydrogen)	0.08	0.60	0.50 (Typical)	0.02	0.02	—	510 (33.0)	400 (25.9)	31.0	75.0	175 (129) V	—	Joining low carbon, low alloy steels
CMn steel	Basic (low hydrogen)	0.08	0.60	1.50 (Typical)	0.02	0.02	—	570 (36.9)	495 (32.0)	31.0	70.0	165 (122) V	—	Joining low alloy steels of 35–40 tsi tensile strengths
2. *Low alloy steels*														
½Mo steel	Basic (low hydrogen)	0.07	0.30	1.25 (Typical)	0.02	0.03	Mo 0.60	695 (45.0)	570 (36.9)	24.0	65.0	—	102 (75)	Joining low alloy steels of 40–45 tsi tensile strengths
1½NiMoV steel	Basic (low hydrogen)	0.07	0.40	0.90 (Typical)	0.02	0.03	Ni 1.60 Mo 0.80 V 0.25	850 (55.0)	710 (46.0)	17.0	62.0	110 (81) K	—	Joining air hardening steels
½Mo steel	Rutile	0.07	0.09	0.56 (Typical)	0.03	0.03	Mo 0.57	495 (32.0)	415 (26.9)	29.0	60.0	—	69 (51)	Joining creep resisting ½Mo steels
½MoCr steel	Basic (low hydrogen)	0.08 max	0.50 max	1.00 max	0.03 max	0.03 max	Mo 0.50 0.60 Cr 0.25 0.50	550 (35.6)	Not given	28.5	55.0	—	—	Joining creep resisting ½Mo steels
½MoCrV steel	Basic (low hydrogen)	0.08 max	0.50 max	1.00 max	0.03 max	0.03 max	Mo 0.50 0.60 Cr 0.25 0.50 V 0.22 0.30	640 (41.4)	Not given	25.5	55.0	—	—	Joining creep resisting low CrMoV steels

* V = V notch. K = keyhole notch

Table 22.55 STEEL WELD METAL—MECHANICAL PROPERTIES AT ROOM TEMPERATURE (DIN 17440: 1967, Stahl-Eisen Werkstoffblatt 400: 1960, 470: 1960, 880: 1959)—*continued*

		Composition limits %						*Typical mechanical properties*						
												Impact values		
Weld metal	*Type*	C	Si	Mn	S	P	*Other elements*	*UTS* MPa (tonf in^{-2})	*Yield stress* MPa (tonf in^{-2})	*Elong.* (gl $5.65\sqrt{S_0}$) %	*RA* %	*Charpy** J (ft lbf)	*Izod* J (ft lbf)	*Applications*
1CrMo steel	Basic (low hydrogen)	0.08 max	0.50 max	1.00 max	0.03 max	0.03 max	Cr 0.80 1.20 Mo 0.45 0.65	695 (45.0)	520 (33.7)	22.5	55.0	—	—	Joining creep resisting 1CrMo steels
2¼CrMo steel	Basic (low hydrogen)	0.08 max	0.50 max	1.00 max	0.03 max	0.03 max	Cr 2.00 2.50 Mo 0.80 1.15	770 (49.9)	655 (42.4)	17.5	45.0	—	—	Joining creep resisting 2¼CrMo steels
3. *High alloy steels* *Stainless and heat resisting steels*														
14Cr steel	Ferritic	0.10 max	0.75 max	1.50 max	0.03 max	0.03 max	Cr 13.50 15.50	—	—	—	—	—	—	Over-laying non-stabilized ferritic stainless steels
18Cr steel	Ferritic	0.10 max	1.50 max	1.50 max	0.03 max	0.03 max	Cr 16.50 18.50	480 (31.1)	—	20.0	—	—	—	Over-laying non-stabilized ferritic stainless steels
17CrNb steel	Ferritic stabilized	0.10 max	1.50 max	1.50 max	0.03 max	0.03 max	Cr 16.00 18.00 Nb 12×C min	—	—	—	—	—	—	Over-laying stabilized ferritic stainless steels
14Cr steel	Martensitic	0.30 0.40	0.30 0.50	0.50 0.70	0.03 max	0.03 max	Cr 13.00 15.00	—	—	—	—	—	—	Over-laying martensitic stainless steels
19/9CrNi steel	Austenitic	0.06 max	1.50 max	1.50 max	0.02 max	0.025 max	Cr 18.00 20.00 Ni 8.50 10.50	620 (40.1)	385 (24.9)	35.0	55.0	—	—	Joining non-stabilized 18/8 CrNi austenitic stainless steels
19/10CrNi steel	Austenitic	0.025 max	1.50 max	1.50 max	0.02 max	0.025 max	Cr 18.00 20.00 Ni 9.00	—	—	—	—	—	—	Joining non-stabilized 18/8 CrNi austenitic

19/9 CrNiNb steel	Austenitic stabilized	0.07 max	2.00 max	1.50 max	0.02 max	0.025 max	Cr 18.00–20.00 Ni 8.00–10.00 Nb 12 × C min	635 (41.1)	385 (24.9)	39.0	55.0	—	—	Joining stabilized 18/8CrNi austenitic stainless steels
19/11 CrNiMo steel	Austenitic	0.06 max	1.50 max	1.50 max	0.02 max	0.025 max	Cr 18.00–20.00 Ni 10.00–12.00 Mo 2.50–3.00	620 (40.1)	370 (24.0)	39.0	45.0	—	—	Joining non-stabilized 18/10 CrNiMo austenitic stainless steels
19/11 CrNiMo steel	Austenitic	0.025 max	1.50 max	1.50 max	0.02 max	0.025 max	Cr 17.00–19.00 Ni 10.00–13.00 Mo 2.50–3.00	—	—	—	—	—	—	Joining non-stabilized 18/10 CrNiMo austenitic stainless steels
19/11 CrNiMoNb steel	Austenitic stabilized	0.07 max	2.00 max	1.50 max	0.02 max	0.025 max	Cr 18.00–20.00 Ni 10.00–13.00 Mo 2.50–3.00 Nb 12 × C min	680 (44.0)	415 (26.9)	37.0	45.0	—	—	Joining stabilized 18/10 CrNiMo austenitic stainless steels
20/18 NiCrMo-CuNb steel	Austenitic stabilized	0.08 max	1.50 max	1.50 max	0.02 max	0.025 max	Ni 20.00–22.00 Cr 17.50–19.50 Mo 2.00–2.50 Cu 1.80–2.20 Nb 12 × C min	—	—	—	—	—	—	Joining Cu containing stabilized 18/18 CrNiMo austenitic stainless steels
25/15 CrNiMo steel	Austenitic	0.20 max	1.50 max	2.00 max	0.02 max	0.025 max	Cr 25.00 Ni 14.50 Mo 2.00 (Typical)	—	—	—	—	—	—	Joining non-stabilized 25/15 CrNiMo austenitic stainless steels

* V = V notch, K = keyhole notch.

Table 22.55 STEEL WELD METAL—MECHANICAL PROPERTIES AT ROOM TEMPERATURE (DIN 17440: 1967, Stahl-Eisen Werkstoffblatt 400: 1960, 470: 1960, 880: 1959)—*continued*

		Composition limits %						*Typical mechanical properties*				*Impact values*		
Weld metal	*Type*	C	Si	Mn	S	P	*Other elements*	*UTS* MPa ($tonf\,in^{-2}$)	*Yield stress* MPa ($tonf\,in^{-2}$)	*Elong.* (gl $5.65\sqrt{S_0}$) %	*RA* %	*Charpy** J (ft lbf)	*Izod* J (ft lbf)	*Applications*
25/20 CrNi steel	Austenitic	0.15 max	1.50 max	1.50 max	0.02 max	0.025 max	Cr 24.00 26.00 Ni 19.00 21.00	—	—	—	—	—	—	Joining non-stabilized 25/20 CrNiSi austenitic heat resisting steels
25/25 CrNiMoNb steel	Austenitic stabilized	0.10 max	1.50 max	1.50 max	0.02 max	0.025 max	Cr 25.00 27.00 Ni 24.00 26.00 Mo 2.00 2.50 Nb 12 × C min	—	—	—	—	—	—	Joining stabilized 25/25 CrNiMo austenitic stainless steels
36/18 NiCr steel	Austenitic	0.20 max	2.00 max	2.00 max	0.02 max	0.025 max	Ni 36.00 40.00 Cr 17.00 19.00	—	—	—	—	—	—	Joining non-stabilized 36/18 CrNiSi austenitic heat resisting steels
18/8/6 CrNiMn steel	Austenitic	0.15 max	1.50 max	5.50 7.50	0.02 max	0.035 max	Cr 17.00 20.00 Ni 7.50 9.50	—	—	—	—	—	—	Joining non-stabilized CrNiMn austenitic stainless steels
Wear resisting steels														
14Mn steel	Austenitic	1.00 1.25	0.35 0.50	13.50 14.50	0.02 max	0.08 max	—	—	—	—	—	—	—	Over-laying wear resisting austenitic Mn steels
18Mn steel	Austenitic	0.80 1.00	0.60 1.00	17.00 18.00	0.02 max	0.08 max	—	—	—	—	—	—	—	Over-laying wear resisting austenitic Mn steels
9CrSi steel	Martensitic	0.40 0.50	2.80 3.30	0.30 0.50	0.025 max	0.025 max	Cr 8.00 10.00	—	—	—	—	—	—	Overlaying 9CrSi valve steels

* V = V notch. K = keyhole notch.

22.11 Other metals of industrial importance

In Table 22.56 are summarized the mechanical properties of metals of industrial importance not included in previous tables. Unless otherwise indicated, the data refer, as far as possible, to the purest metals available.

Table 22.56 MECHANICAL PROPERTIES OF OTHER METALS OF INDUSTRIAL IMPORTANCE

Metal	*Condition*	*Purity* %	*Condition*	*Proof stress* MPa (10^3 lbf in^{-2})	*UTS* MPa (10^3 lbf in^{-2})	*Elong.* %	*Hardness* HV	*Impact strength* J (ft lbf)	*Refs.*[1]
Ag	Soft	99.9	Annealed 600°C	—	172 (25)	50	25	5 (4)	1
	Hard	99.9	Cold worked, 70% reduction	—	330 (48)	4	95	—	1
Au	Soft	99.99	Annealed 450°C	—	130 (19)	40–50	20–30	—	2
	Hard	99.99	60% reduction	205 (30)	220 (32)	4	60	—	3, 4
Be	Soft	99.5	Hot pressed	240 (35)	310 (45)	2	(150)	1[2] (0.8)	5
	Hard	99.5	Forged	345 (50)	550 (80)	10	(200)	3–5[2] (2–4)	5
Ca	Soft	—	Cast	14 (2)	55 (8)	55	17HB	—	6
	Hard	—	Rolled	84.5 (12.3)	115 (16.8)	7	—	—	7
Co	Soft	99.9	Annealed	345–485 (50–70)	760 (110)	7–20	170	—	—
	Hard	99.9	25% reduction	—	1 135 165	0.5	320	—	—
Cr	V. pure, soft	99.99	Iodide reduced; arc-cast, swaged, annealed 600°C	—	415 (60)	44	230	—	8
	Soft	99.95[3]	Electrolytic; arc-cast, hot worked recrystallized above 900°C	—	103 (15)	0	130	—	8
	Hard	99.95[3]	Drawn ~ 500°C, 80–90% reduction	—	689 (100)	10	220	—	8
Hf	Soft	—	Iodide; hot rolled, 0–925°C; cold rolled, 5%; annealed 800°C for 1 h	240 (35)	445 (64, 5)	43	(150–180)	—	9
	Hard	—	Iodide: extruded	365 (53)	745 (108)	20–27	—	—	7
Ir	Soft	—	Annealed	—	550–1 100 (80–160)	—	200–300	—	—
	Hard	—	Hot forged and swaged 800–1 500°C	—	1 200 (175)	13	650	—	10

Table 22.56 MECHANICAL PROPERTIES OF OTHER METALS OF INDUSTRIAL IMPORTANCE—*continued*

Metal	*Condition*	*Purity* %	*Condition*	*Proof stress* MPa (10^3 lbf in^{-2})	*UTS* MPa (10^3 lbf in^{-2})	*Elong.* %	*Hardness* HV	*Impact strength* J (ft lbf)	*Refs.*[1]
Mo	V. pure, soft	99.99	Electron beam melted: hot rolled; recrystallized at 1 100°C for 1 h	345Y (50Y)	435 (63)	5–25	—	—	8
	Soft	99.95[3]	Arc-cast; hot worked and recrystallized ~1 200°C	415–450Y (60–65Y)	485–550 (70–80)	30–40	200	—	8
	Hard	99.95[3]	Rolled at about 1 000°C	550Y (80Y)	620–690 (90–100)	10–20	250	—	8
Nb	V. pure, soft	99.97	Electron beam melted, swaged recrystallized ~1 100°C	170 (25)	240 (35)	50	65	122–230 (90–170)	8, 11, 12
	Soft	99.9[3]	Arc-melted; cold forged; recrystallized for 4 h at 1 200°C	240 (35)	330 (48)	50	115	16–115[1] (12–85)[1]	8
	Hard	99.9[3]	Cold rolled ~90%	550 (80)	585 (85)	5	160	—	8, 11
Np	Soft	99.1	Cast	—	—	—	355	—	13
Os	Soft	—	Wrought and annealed	—	—	—	350	—	6
	Hard	—	Cold worked 7%	—	—	—	1 000	—	4
Pd	Soft	99.9	Wrought and annealed	34, 5 (5)	140–195 (20–28)	24–40	40	—	4
	Hard	99.9	Wrought and cold worked	205 (30)	325 (47)	15	100	—	4
Pt	Soft	99.9	Wrought and annealed	14–35 (2–5)	125–145 (18–21)	30–40	40	—	4
	Hard	99.9	Wrought and cold worked ~50%	185 (27)	195–205 (28–30)	3	100	—	4
Pu	Soft	99.9	Cast (αPu)	205–310 (30–45)	310–550 (45–80)	1	260–270	3 (2)	14
Re	Soft	99.9	Sintered, swaged and annealed	315 (46)	1 125 (164)	24	280	—	8
	Hard	99.9	Cold rolled sheet, 30% reduction	2 150 (311)	2 225 (322)	2	700	—	8
Rh	Soft	—	Annealed, 900°C	69–275 (10–40)	690–760 (100–110)	20–40	120	—	15
	Hard	—	Cold rolled	—	1 380–2 070 (200–300)	—	300	—	4, 7, 15
Ru	Soft	99.9	Wrought and annealed at 1 500°C	372 (54)	495 (72)	3	350	—	15
	Hard	99.9	Wrought and cold worked 10%	—	—	—	750	—	4
Sc	Soft	—	—	—	—	—	78HB	—	16
	Hard	—	Cold swaged 22%	—	—	—	136HB	—	16

Table 22.56 MECHANICAL PROPERTIES OF OTHER METALS OF INDUSTRIAL IMPORTANCE—*continued*

Metal	*Condition*	*Purity* %	*Condition*	*Proof stress* MPa (10^3 lbfin^{-2})	*UTS* MPa (10^3 lbfin^{-2})	*Elong.* %	*Hardness* HV	*Impact strength* J (ft lbf)	*Refs.*[1]
Ta	V. pure, soft	99.98	Electron beam melted; cold rolled annealed 1 200°C	180 (26)	205 (30)	35–45	70–80	—	8
	Soft	99.95[3]	Recrystallized sheet	310–380 (45–55)	310–485 (45–70)	25–40	90	—	8,11
	Hard	99.95[3]	Cold worked 95%	705 (102)	760 (10)	3	200	—	11
Th	Soft	~99.95	Iodide; wrought and annealed	48 (7)	115 (17)	36	38	41 (30)	7
	Hard	~99.95	Iodide, cold rolled	295 (43)	305 (44)	6	70	—	17
U	Soft	99.95	Coarse grained, cast	190 (28)	385 (56)	4	187	19 (14)	18
	Hard	99.95	Fine grained—rolled at 550°C —quenched from 722°C	250 (36)	580 (84)	9	250	15 (11)	18
V	V. pure soft	99.95	Iodide reduced: arc-melted, swaged and recrystallized	103 (15)	190 (27, 5)	39	55	—	8
	Soft	99.8[3]	Calcium reduced: arc-melted, cold worked recrystallized	170–450 (25–65)	260–585 (38–70)	28–34	80	10–136 (7–100)[4]	8
	Hard	99.8[3]	Cold worked	515–690 (75–100)	530–730 (77–110)	1–3	150	—	Several
W	Soft	99.95	Sintered rod; swaged and annealed, 1 590 °C	550 (80)	550–620 (80–90)	0	360	—	8, 11
	Hard	99.95	Sintered, cold worked	—	1920 (280)	0	500	—	11
Y	Soft	—	—	57Y (8.3Y)	130 (19)	(25)	30–60 HB	24 (18)	16
	Hard	—	50% cold swaged	375Y (54Y)	455 (66)	2	100–140HB	—	16

[1] Where data exist from more than one source, ranges of values are given or, alternatively, if the values are similar a probable average is given.
[2] Unnotched Charpy.
[3] Principal impurities are interstitial O, C, N.
[4] Properties very sensitive to impurity content.
Y ='Yield stress'.
( ) Interpolated from another source.
HB = Brinell.
HV = Vickers PN.

REFERENCES TO TABLE 22.56

1. A. Butts and C. D. Coxe (Eds.), 'Silver', Van Nostrand, Princeton, 1967.
2. E. M. Wise, 'Gold', Van Nostrand, Princeton, 1964.
3. Engelhard Industries Inc., Newark, N.J.
4. Materials Selector, *Mater. Engng*, 1972, **76**, No. 4.
5. T. W. Farthing and J. R. Leech, *Proc. Inst. mech. Engrs*, 1956–66, **180**, Part 3D.
6. A. H. Everts and G. D. Bagley, *J. electrochem. Soc.*, 1948, **93**, 265.
7. C. A. Hampel (Ed.), 'Rare Metals Handbook', 2nd Edn, Reinhold, New York, 1961.
8. T. E. Tietz and J. W. Wilson, 'Behaviour and Properties of Refractory Metals', Arnold, London, 1965.

9. USAEC, 'Reactor Handbook Materials—General Properties', McGraw-Hill, New York, 1955.
10. W. Betteridge, 5th Plansee Seminar, Reutte, 1964, paper 27.
11. R. Syne, 'Handbook of Properties of Nb, Mo, Ta and W', NATO, 1965.
12. 'Gmelins Handbuch der Anorganischen Chemie—Niob. Teil A', Verlag Chemie, Weinheim, 1969.
13. V. W. Eldred and G. C. Curtis, *Nature, Lond.*, 1957, **179**, 910.
14. E. Grison, W. B. H. Lord and R. D. Fowler (Eds.), 'Plutonium 1960', Cleaver-Hume, London, 1961.
15. International Nickel Ltd. Research Laboratories.
16. E. V. Kleber and B. Love, 'Technology of Scandium, Yttrium and Rare Earth Metals', Pergamon Press, Oxford, 1963.
17. A. B. Schwope, G. T. Muelenkamp and L. L. Marsh, BMI, 1952, 784.
18. J. H. Gittus, 'Uranium', Butterworths, London, 1963.

22.12 Bearing alloys

Bearing alloys have to satisfy a number of criteria, depending on their application, of which the following are the more important.

WEAR RESISTANCE

The bearing must resist wear and more importantly must not cause wear of the mating surface. The stronger, harder bearing materials must therefore be operated against a hardened mating surface.

EMBEDDABILITY

When dirt or detritus from associated parts is present in the lubricant supply the bearing must minimize damage to the mating surface. Soft bearing materials can absorb more dirt than hard materials.

CONFORMABILITY

The misalignments inherent in mechanisms under load require an ability in the bearing to conform to the geometry of the mating surface. Softness and conformability in a bearing material also facilitate the creation of a hydrodynamic film of lubricant over the bearing surface under sparse lubrication conditions.

STRENGTH

Under static loads the bearing must have a yield strength greater than the applied load. Under dynamic fatigue or shock loads, a combination of high yield strength and ductility is required. The relative fatigue strengths of various bearing alloys can be seen from Figure 22.1 and mechanical properties from Table 22.56.

CORROSION RESISTANCE

The bearing material must resist corrosion by the atmosphere or by the lubricant.

THERMAL EXPANSION

The expansion coefficient of the bearing material must be sufficiently close to that of the housing for there to be no significant loss of interference fit at bearing operating temperatures.

It will be apparent that a bearing material is a compromise between the conflicting requirements for softness and strength, and the optimum compromise depends on the application. High effective strength from a relatively weak bearing material may be achieved if the material is present as a thin lining on a steel backing, and the majority of bearings have this construction. The copper-based engine bearing linings and the stronger aluminium alloys are overlay plated with a thin layer of lead–indium. This soft, partly sacrifical layer provides surface embeddability and conformability, and is thin enough (0.01–0.03 mm) to have good load carrying capacity.

Table 22.57 MECHANICAL PROPERTIES OF BEARING ALLOYS

	Nominal composition and manufacturing process						*Mechanical properties*			
Alloy	Al	Cu	Pb	Sb	Sn		*Proof stress* MPa (tonf in^{-2})	*UTS* MPa (tonf in^{-2})	*Elongation* %	*Hardness* HV5
Copper base alloys	—	Rem	10	—	10	Sintered steel backed	249 (16.0)	303 (19.5)	5	120
	—	Rem	22	—	4.5	Sintered steel backed	81 (5.2)	121 (7.8)	5	46
	—	Rem	25	—	1.5	Sintered steel backed	75 (4.8)	112 (7.2)	5	45
	—	Rem	30	—	—	Sintered steel backed	62 (4.0)	93 (6.0)	5	34
	—	Rem	—	—	10	P 0.5% max. Continuously cast	233 (15)	420 (27)	16	120
	—	Rem	9.5	—	5	Continuously cast	155 (10)	280 (18)	15	70
	—	Rem	10	—	10	Sand cast	101 (6.5)	233 (15)	10	75
	—	Rem	20	—	5	Continuously cast	124 (8.0)	233 (15)	10	70
	—	Rem	26	—	2	Rotary lined on steel	62 (4.0)	110 (7)	3	48
Aluminium base alloys	Rem	1	—	—	20	Continuously cast steel backed	42 (2.7)	120 (8.0)	25	37
	Rem	1	—	—	6	Ni 1.0% Continuously cast steel backed	50 (3.2)	140 (9.0)	28	45
	Rem	—	—	—	40	Continuously cast steel backed	40 (2.6)	67 (4.3)	26	27
	Rem	1	—	—	—	Si 10.6% Continuously cast steel backed	87 (5.6)	194 (12.5)	17	56
	Rem	1	—	—	6	Ni 1.0% Cold worked 4% Continuously cast	140 (9)	147 (9.5)	22	50
	Rem	1.5	—	—	6	Ni 1.4% Mg 0.9% Si 0.5% Continuously cast	83 (5.4)	207 (13.4)	10	78
	Rem	1	0.9	—	—	Zn 5% Si 1.5%	—	216 (14)	15	64
Whitemetals tin base	—	3.3	—	7.5	Rem	Cast on steel strip	65 (4.2)	76 (4.9)	10	27
	—	3.3	—	7.5	Rem	Rotary lined on steel	39 (2.5)	70 (4.5)	15	31
	—	3	—	7.7	Rem	Cd 1.25% Rotary lined on steel	76 (4.9)	90 (5.8)	8	32
Whitemetals lead base	—	0.5	Rem	15	1	As 1% Cast on steel strip and annealed	25 (1.6)	70 (4.5)	6	17
	—	—	Rem	10	6	Cast on steel strip and annealed	30 (2.0)	42 (2.7)	33	16
	—	0.5	Rem	10	10	Rotary lined on steel	60 (3.9)	73 (4.7)	8	25

Rem = Remainder. Properties are typical values.

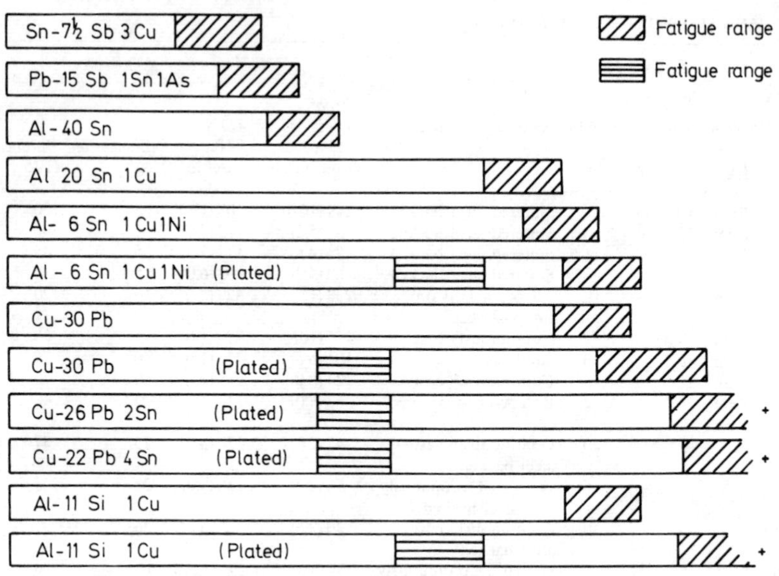

Figure 22.1 *Relative fatigue strength of bearing alloys. Back of bearing temperatures from* 70°C *to* 100°C

BIBLIOGRAPHY—BEARING ALLOYS

'Copper Alloy Bearing Materials', Copper Development Assoc., tech. Note TN9, 1972.
R. Booser, 'Selecting Sleeve Bearing Materials', *Mater. Des. Engng.*, 1958, **48**, 119.
P. G. Forester, 'Bearing Materials', *Met. Rev.*, 1960, **5**, 507.
M. C. Shaw and F. Macks, 'Analysis and Lubrication of Bearings', New York, McGraw-Hill, 1949.
M. J. Neale, 'Tribology Handbook', Butterworths, London, 1973.
European Tribology Conf. London, *Inst. Mech. Engrs*, Sept. 1973.

23 Hardmetals

Hardmetals are sintered powder-metallurgical compacts consisting of minute, ultra-hard particles of refractory compounds (Table 23.1) 'cemented' by a tough, but comparatively soft, metallic matrix. In most grades, the hard particles—typically 1–5 μm in diameter—are tungsten monocarbide (WC) or are based on tungsten carbide, while the 'cement' or binder is metallic cobalt. More complex compositions include a variety of other carbides, occasionally also refractory oxides or nitrides, and may also have alternative binders such as nickel or a nickel–molybdenum alloy.

More than 99% of commercial production (Tables 23.2 to 23.4) is based on either tungsten or titanium carbides, with or without additives of the carbides of tantalum, niobium, chromium, vanadium, hafnium and/or molybdenum. Sintered hardmetals based on other carbides, or on borides, nitrides or silicides, or with other additives, have largely disappeared. The more recent developments in compacted diamond and cubic boron nitride, and in oxide-type cutting ceramics—which are not strictly hardmetals—have, however, created some new interest in the more 'exotic' hardmetals.

Metallurgically, hardmetals are generally pseudo-eutectics, with the carbide particles and binder

Table 23.1 PROPERTIES OF COMPOUNDS USED IN HARDMETALS AND FOR REFRACTORY COATINGS ON HARDMETALS

Compound	*Melting point approx* °C	*Micro-hardness* HV	*Density* g cm^{-3}
Cr_3C_2	1 895±25	1 300	6.68
HfC	3 930±55	2 900	12.7
Mo_2C	2 690±50	1 500	9.2
NbC	3 500±75	2 400	7.85
TaC	3 915±55	2 150	14.53
TiC	3 065±15	3 200	4.93
VC	2 730±85	2 600	5.81
WC	2 780	2 400	15.77
W_2C	2 730±15	3 000	17.34
ZrC	3 440±20	3 000	6.73
Al_2O_3	2 050±20	3 100	3.97
HfO_2	2 840±50	—	9.68
NbO	1 945	—	6.27
TiO_2	1 840±15	—	3.84
ZrO_2	2 770±85	—	5.73
CrN	1 500		6.1
HfN	3 310	2 600	14.0
NbN	2 050	—	7.3
TaN	3 090	—	14.1
TiN	2 930	2 100	5.43
VN	2 050	—	6.1
ZrN	2 980	1 600	7.35

Table 23.2 TYPICAL HARDMETAL GRADES FOR METAL-CUTTING (COMPOSITIONS AND NOMINAL PROPERTIES MAY DIFFER GREATLY BETWEEN DIFFERENT MANUFACTURERS)

Designation		*Composition*, wt%						*Property*		
ISO application code	*US industry code*	WC	TiC	Ta(Nb)C	Mo_2C	Co	Ni	*Density* $g\,cm^{-3}$	*Hardness* HV	*Transverse rupture strength* $N\,mm^{-2}$
P01	C8	—	80	—	10	—	10	5.8	1 900	850
P01	C8	50	35	7	—	6	—	8.5	1 900	1 100
P05	C7	78	16	—	—	6	—	11.4	1 820	1 300
P10	C7	69	15	8	—	8	—	11.5	1 740	1 400
P15	C6	78	12	3	—	7	—	11.7	1 660	1 500
P20	C6	79	8	5	—	8	—	12.1	1 580	1 600
P25	C6	82	6	4	—	8	—	12.9	1 530	1 700
P30	C5	84	5	2	—	9	—	13.3	1 490	1 850
P40	C5	85	5	—	—	10	—	13.4	1 420	1 950
P50	—	78	3	3	—	16	—	13.1	1 250	2 300
M10	—	85	5	4	—	6	—	13.4	1 590	1 800
M20	—	82	5	5	—	8	—	13.3	1 540	1 900
M30	—	86	4	—	—	10	—	13.6	1 440	2 000
M40	—	84	4	2	—	10	—	14.0	1 380	2 100
					—		—			
K01	C4	97	—	—	—	3	—	15.2	1 850	1 450
K05	C4	95	—	1	—	4	—	15.0	1 780	1 550
K10	C3	92	—	2	—	6	—	14.9	1 730	1 700
K20	C2	94	—	—	—	6	—	14.8	1 650	1 950
K30	C1	91	—	—	—	9	—	14.4	1 400	2 250
K40	C1	89	—	—	—	11	—	14.1	1 320	2 500

Table 23.3 TYPICAL MECHANICAL PROPERTIES OF REPRESENTATIVE GRADES OF HARDMETAL USED IN WEAR- AND CORROSION-RESISTING APPLICATIONS

Composition wt%	Grain size, μm	Density $g\,cm^{-3}$	Hardness HV	Hardness HRA	Transverse rupture strength $N\,mm^{-2}$	Compressive properties: Ultimate strength $N\,mm^{-2}$	Elastic limit $N\,mm^{-2}$	Young's modulus $kN\,mm^{-2}$	Poisson's ratio	Ductility %	Charpy impact value J	Abrasion resistance $(vol.\ loss)^{-1}$
97WC/3Co	1	15.3	1 850	93.0	1 600	4 300	3 500	640	0.24	0.6	1.1	100
94WC/6Co	1	15.0	1 680	920	2 000	4 700	1 950	650	0.28	0.8	1.4	55
94WC/6Co	2	15.0	1 550	91.0	2 200	4 800	1 450	640	0.26	1.0	1.4	25
91WC/9Co	2	14.7	1 380	89.7	2 350	4 150	950	610	0.22	1.5	1.4	15
87WC/13Co	3.5	14.2	1 250	88.5	2 700	3 650	550	550	0.25	1.9	1.9	6
84WC/16Co	3.5	13.8	1 100	87.0	2 900	3 800	650	530	0.22	2.7	2.8	5
80WC/20Co	3.5	13.4	950	85.5	2 750	3 500	520	500	0.23	3.0	3.0	4
75WC/25Co	3.5	13.0	800	85.0	2 600	3 100	380	480	0.25	>3.5	3.1	3
$83Cr_3C_2$/15Ni/2W	4	7.1	1 370	89.5	700	3 500	900	350	0.28	1.1	0.4	3
79WC/4TiC/8TaC/9Co	2	12.5	1 580	91.3	1 800	3 650	700	550	0.22	2.0	1.1	13
72WC/8TiC/11TaC/9Co	2	12.5	1 600	91.5	1 700	4 350	800	560	0.24	0.8	0.9	12

Table 23.4 TYPICAL THERMAL AND ELECTRICAL PROPERTIES OF REPRESENTATIVE HARDMETALS

Composition wt%	Grain size μm	Thermal conductivity, $W\,m^{-1}K^{-1}$: 50°C	100°C	150°C	200°C	300°C	400°C	500°C	600°C	Thermal expansion coefficient $10^{-6}K^{-1}$, from 20°C to: 200°C	400°C	600°C	800°C	1 000°C	Electrical resistivity μΩ cm
97WC/3Co	1	125	125	120	120	—	—	—	—	4.0	4.2	4.5	4.7	4.9	17
94WC/3Co	1	120	120	115	115	—	—	—	—	4.6	4.8	5.1	5.4	5.5	17
94WC/6Co	2	120	120	110	100	90	85	80	—	4.3	4.8	5.0	5.4	5.6	17.5
91WC/9Co	2	110	110	110	100	—	—	—	—	4.9	5.2	5.4	5.8	5.9	16
87WC/13Co	3.5	110	110	100	100	—	—	—	—	5.4	5.5	5.9	6.2	6.4	17
84WC/16Co	3.5	90	80	80	80	80	80	80	80	5.8	6.0	6.2	6.8	7.1	19
80WC/20Co	3.5	80	75	75	75	—	—	—	—	6.0	6.3	6.5	7.1	7.4	18.5
75WC/25Co	3.5	70	70	70	75	—	—	—	—	6.3	6.5	7.3	7.4	7.7	18
$83Cr_3C_2$/15N/2W	4	11	12	12.5	13	14	15	17	—	9.1	9.9	10.7	11.3	12.2	85
79WC/4TiC/8TaC/9Co	2	60	60	60	55	55	50	50	50	5.2	5.8	6.2	6.6	6.9	30
72WC/8TiC/11TaC/9Co	2	55	50	50	50	50	50	45	45	5.7	6.0	6.4	6.5	6.8	35

each representing one of the eutectic 'elements'. Nevertheless, even with simple systems such as WC/Co, a slight departure from stoichiometric composition causes problems, through entry into three-phase metastable zones of the tertiary system. The result, depending on whether there is a deficiency or excess of carbon, is either the formation of a brittle double carbide ('eta phase') of tungsten and cobalt, or the presence of free graphite, which has a rather less embrittling effect.

Some important properties are extremely structure-dependent, with interdiffusion between carbide phases related closely to the mode of carburization. Thus, with the more highly alloyed grades, a variety of double, triple and multiple carbide phases may occur, depending on the degree of mutual solubility, the starting ingredients, their individual grain sizes, and the thermal and chemical histories of the various constituents. Departure from stoichiometric composition is less important with such grades, and may even be advantageous with respect to certain key characteristics.

During liquid-phase sintering (Table 23.5), fine particles dissolve preferentially during the heating cycle, while precipitation tends to take place on the largest particles during cooling. Thus any differentiation in grain size in the starting materials tends to result in deleterious grain growth during sintering, unless small amounts of specific grain-growth inhibitors—such as TaC, VC or Cr_3C_2—are included in the composition.

Grain size is one of the primary factors in determining the toughness, hardness and magnetic coercivity (used for quality control—see Table 23.6) of a hardmetal: others include binder and carbide alloy content, porosity and departure from stoichiometric composition. In general, grades with fine grain size (around 1 μm) are hardest but least shock-resistant, while coarse grades (above 3 μm) are much tougher but considerably less resistant to abrasive wear. Exceptionally fine-grained carbides, however, with average grain size significantly below 1 μm, combine excellent hardness with remarkable toughness.

Because of the many variables of composition and manufacture, there are no international standards—and no worthwhile national standards—which in any way prescribe compositions or properties. Instead, the ISO recommendation 513 (Table 23.7) and related national standards relate purely to *applications*, relying on manufacturers to produce roughly similar compositions to meet a particular requirement. Thus a nickel–molybdenum-bonded titanium carbide and a cobalt-bonded tungsten–tantalum–hafnium carbide could each be given the ISO designation 'P01'.

It follows that specific compositions and properties can relate only to a particular manufacturer, or to a number of manufacturers whose compositions and process conditions happen to be closely similar for a particular grade or grades. The more complex the alloy, the more diverse in properties and performance are the compositionally similar products of different manufacturers.

Table 23.5 TYPICAL SINTERING TEMPERATURES FOR REPRESENTATIVE HARDMETALS

Composition wt%					*Sintering temperature*
WC	TiC	Ta(Nb)C	Cr_3C_2	Co	°C
94	—	—	—	6	1 540
91	—	—	—	9	1 480
89	—	—	—	11	1 460
87	—	—	—	13	1 450
80	—	—	—	20	1 400
75	—	—	—	25	1 380
70	—	—	—	30	1 350
96.5	—	—	0.5	3	1 640
95	—	—	0.5	4.5	1 620
93.5	—	—	0.5	6	1 560
90.5	—	—	0.5	9	1 500
85.5	7	3.5	—	4	1 640
81.5	7	3.5	—	8	1 560
80	14	—	—	6	1 620
84	10	—	—	6	1 600
87	7	—	—	6	1 590
87	5	—	—	8	1 550
66	25	—	—	9	1 620

Table 23.6 VARIATION OF MAGNETIC COERCIVITY OF SINTERED WC/Co HARDMETALS WITH COBALT CONTENT AND WC GRAIN SIZE C ADDITIVES AND OTHER FACTORS ALSO AFFECT THIS PROPERTY)

Composition wt%		*Coercivity*, kA m^{-1} *with WC grain size indicated*			
WC	Co	1 μm	2 μm	3 μm	4 μm
96	4	24	18	15.5	—
94	6	21	15	12	11
92	8	19	12.5	10	8.5
90	10	17.5	10.5	8	6.5
85	15	14	7.5	5.5	—
80	20	12	6	—	—
75	25	10.5	—	—	—

23.1 Alloying effects

No large-scale and cost-effective substitute has yet been found for WC/Co hardmetals where simple wear resistance is a primary requirement. This requirement includes such applications as wire-drawing dies and the cutting of most non-ferrous metals and non-metallics. Grain-refining additives improve wear resistance and, where extreme heat is a factor, the incorporation of larger amounts of TaC is beneficial.

In the machining of steels, WC/Co hardmetals suffer from an unusual and potentially catastrophic kind of wear called 'cratering', caused by decomposition and diffusion of the carbide constituents into the steel chip behind the cutting edge. TiC, TaC, NbC and HfC have considerably greater stability, and correspondingly greater crater resistance, than WC under this circumstance. Hardmetals containing these carbides are therefore used almost universally in machining steels. The need to harmonize crater resistance with optimum grain structure, toughness, hardness, wear resistance and other properties helps to account for the wide variety of proprietary compositions.

In recent years, small additions of ruthenium have been found to increase still further the toughness and shock resistance of complex steel-cutting hardmetals.

23.2 Coatings

By applying coatings to appropriate substrates, it is possible to combine the extreme crater resistance of selected pure carbides, nitrides and oxides with the toughness and other mass properties of WC/Co and low-alloy hardmetals. The method almost invariably used is chemical vapour deposition (see chapter 35), though a number of physical methods have also been attempted with varying success. Titanium carbide, derived from the reaction of titanium tetrachloride with methane, was initially employed, to be followed by titanium nitride and carbonitride, hafnium carbide and nitride, aluminium and zirconium oxides, and multi-layered combinations of these and other compounds.

Because of the absence of a metallic binder, pure vapour-deposited coatings are even more abrasion resistant than the equivalent conventional hardmetal. They may, however, fail at the coating/substrate interface because of differential thermal expansion, brittle eta-phase interlayers or 'lamination wear' (caused by the accumulation of dislocations or microcracks below the surface). Thus the 'traditional' multiplicity of hardmetal compositions and structures has become even more complex through the proliferation of chemical-vapour deposited coatings.

23.3 Hot isostatic pressing

Although the porosity of conventionally sintered carbide is extremely low, it can be reduced almost to zero by a subsequent operation of hot isostatic pressing, or 'HIPing'. Toughness and shock resistance are greatly increased and the possibility of rejection of certain large and expensive components at a late stage of grinding or polishing virtually eliminated. Regularly 'HIPed' components include Sendzimir foil rolls, mining-tool inserts, drawing dies and other large abrasion-resisting parts.

Table 23.7 ISO CLASSIFICATION OF CARBIDES ACCORDING TO USE

Main groups of chip removal		*Groups of application*				*Direction of increase in characteristic*	
Symbol	*Broad categories of material to be machined*	*Distinguishing colours*	*Designation*	*Material to be machined*	*Use and working conditions*	*of cut*	*of carbide*
P	Ferrous metals with long chips	BLUE	P 01	Steel, steel castings	Finish turning and boring, high cutting speeds, small chip section, accuracy of dimensions and fine finish, vibration-free operation	↑ Increasing speed; ↓ Increasing feed	↑ Wear resistance; ↓ Toughness
			P 10	Steel, steel castings	Turning, copying, threading and milling, high cutting speeds, small or medium chip sections		
			P 20	Steel, steel castings Malleable cast iron with long chips	Turning, copying, milling, medium cutting speeds and chip sections, planing with small chip sections		
			P 30	Steel, steel castings Malleable cast iron with long chips	Turning, milling, planing, medium or low cutting speeds, medium or large chip sections, and machining in unfavourable conditions*		
			P 40	Steel Steel castings with sand inclusion and cavities	Turning, planing, slotting, low cutting speeds, large chip sections, with the possibility of large cutting angles for machining in unfavourable conditions* and work on automatic machines		
			P 50	Steel Steel castings of medium or low tensile strength, with sand inclusion and cavities	For operations demanding very tough carbides: turning, planning slotting, low cutting speeds, large chip sections, with the possibility. of large cutting angles for machining in unfavourable conditions* and work on automatic machines		

M	Ferrous metals with long or short chips and non-ferrous metals	YELLOW	M 10	Steel, steel castings, manganese steel Grey cast iron, alloy cast iron	Turning, medium or high cutting speeds. Small or medium chip sections	←Increasing speed— —Increasing feed→ ←Wear resistance— —Toughness→
			M 20	Steel, steel castings, austenitic or manganese steel, grey cast iron	Turning, milling. Medium cutting speeds and chip sections	
			M 30	Steel, steel castings, austenitic steel, grey cast iron, high temperature resistant alloys	Turning, milling, planing. Medium cutting speeds, medium or large chip sections	
			M 40	Mild free cutting steel, low tensile steel Non-ferrous metals and light alloys	Turning, parting off, particularly on automatic machines	
K	Ferrous metals with short chips non-ferrous metals and non-metallic materials	RED	K 01	Very hard grey cast iron, chilled castings of over 85 Shore, high silicon aluminium alloys, hardened steel, highly abrasive plastics, hard cardboard, ceramics	Turning, finish turning, boring, milling, scraping	←Increasing speed— —Increasing feed→ ←Wear resistance— —Toughness→
			K 10	Grey cast iron over 220 Brinell, malleable cast iron with short chips, hardened steel, silicon aluminium alloys, copper alloys, plastics, glass, hard rubber, hard cardboard, porcelain, stone.	Turning, milling, drilling, boring, broaching, scraping	
			K 20	Grey cast iron up to 220 Brinell, non-ferrous metals: copper, brass, aluminium	Turning, milling, planing, boring, broaching, demanding very tough carbide	
			K 30	Low hardness grey cast iron, low tensile steel, compressed wood	Turning, milling, planing, slotting, for machining in unfavourable conditions* and with the possibility of large cutting angles	
			K 40	Soft wood or hard wood Non-ferrous metals	Turning, milling, planing, slotting, for machining in unfavourable conditions* and with the possibility of large cutting angles	

* Raw material or components in shapes which are awkward to machine: casting or forging skins, variable hardness etc., variable depth of cut, interrupted cut, work subject to vibrations.

Reproduced from ISO recommendation 513 by permission of the British Standards Institution, 2 Park Street, London, W1A 2BS.

MAIN REFERENCES FOR TABLES

Table 23.1: P. Schwarzkopf and R. Kieffer, 'Hartstoffe und Hartmetalle', Vienna, 1953.
American Ceramic Society, 'Engineering Properties of Selected Ceramic Materials', Columbus, Ohio, 1966.

Table 23.2: K. J. A. Brookes, 'World Directory and Handbook of Hardmetals', 2nd edn, London, 1979.

Tables 23.3 to 23.6: K. J. A. Brookes, Analysis of published and privately communicated data.

Table 23.7: International Standards Organisation, Recommendation 513 (by permission of British Standards Organisation, London).

24 Lubricants

24.1 Introduction

Lubricants minimize friction and wear in rubbing contacts. They may also prevent rusting and with liquid lubricants remove heat. Lubricants may be solid, such as graphite, molybdenum disulphide, polytetrafluoroethylene and talc; or gaseous, commonly air; but the principal lubricants are liquids such as mineral oil, or the semi-solid greases formed from liquids by the use of thickening agents.

24.1.1 Hydrodynamic lubrication

A viscous fluid interposed between two solid surfaces moving in very close proximity to one another becomes pressurized and holds those surfaces apart, often against considerable loads, provided there is a position of closest approach of the surfaces towards which the net flow of fluid is directed. The effect is called hydrodynamic lubrication.

The condition that the net lubricant flow must be towards the position of closest approach of the surfaces is fulfilled in journal bearings by the journal automatically assuming an eccentric position in the bearing. It is not fulfilled if the surfaces are absolutely parallel to one another, but if the bearing surface can tilt appropriately as in the Michel tilting pad bearing, considerable loads can be carried. The amount of tilt required is almost imperceptible: approximately 1 in 10 000. If a position of closest approach can be developed by slight elastic or thermal distortion considerable fluid pressures can be developed between apparently parallel surface.

For very high pressures over tiny contact areas films as thin as 10 nm have been measured.

24.2 Friction, wear and boundary lubrication

In hydrodynamic lubrication friction is purely viscous and is directly dependent on the area of the film, the rate of shear and the viscosity of the lubricant. Coefficients of friction are as low as 0.001–0.003 and wear is negligible. When the speed of sliding and the oil viscosity are insufficient for the viscous film to carry the load, contact occurs between asperities on the opposing surfaces. Friction then rises and wear ensues. Where viscous effects are absent or negligible, lubrication is independent of the nominal area of contact and is said to be of boundary type. Where boundary and viscous effects occur together, the conditions are said to be those of 'mixed' lubrication.

When sliding speeds are very low indeed a 'stick-slip' or jerky motion arises due in part to the elastic response of the drive and in part to the coefficient of static friction exceeding the coefficient of dynamic friction. This undesirable effect can be suppressed by the use of special lubricants containing fatty acids, acid phosphates, or similar materials able to react with the metal surfaces to produce soft, easily-sheared layers of soap which considerably reduce the coefficient of static friction.

For moderate sliding speed and load, boundary lubrication can exist with only the oxide film being worn away and replaced at a tolerable rate of wear. Under these conditions lubricants which produce low friction do not necessarily produce low wear. For most machine bearings, boundary lubrication is inadequate, since the oxide film wears away very rapidly and direct metal to metal contact occurs. The surface asperities weld together momentarily and are broken apart again to produce wear debris largely composed of metallic particles. This is called adhesive wear, or scuffing.

Boundary lubrication is the most important mode in the chipless-forming type of metal working operation since the local pressures have to be high enough to exceed the yield strength of the metal. In such operations the surface area of the workpiece is being enlarged and new, easily-weldable areas are being created. For severe operations, therefore, solid lubricants are used which can resist high pressures and are capable of extension to cover the new areas. Examples of such lubricants are waxes, graphite, molybdenum disulphide, talc, whiting and pastes of dry soap and fats. For more severe operations such as tube and bar drawing thick layers of lubricants may be built up before the operation by, for example, phosphating, baking on a coating lime, treating with special inorganic salts, or by coating with soft metals such as lead.

24.3 Characteristics of lubricating oils

24.3.1 Viscosity

Viscosity is probably the most important property of a lubricating oil or grease. For most fluids in laminar flow there is a linear relationship between the shear stress and the rate of shear. The constant of proportionality is the viscosity or, more specifically, the dynamic viscosity to distinguish it from the kinematic viscosity which is given by the ratio of viscosity to density.

Dynamic viscosity may be defined as the shear stress necessary to move a flat surface at unit speed over a parallel surface unit distance apart when the intervening space is filled with the fluid. In SI units, dynamic viscosity is expressed in $N\,s\,m^{-2}$, but the centipoise or $nN\,s\,m^{-2}$ is at present commonly used. Direct measurements are relatively few. Usually dynamic viscosities are derived from measurement of density and kinematic viscosity, which is easily measured in calibrated glass capillary tubes using gravity flow. The units of kinematic viscosity are $m^2\,s^{-1}$ but centistokes ($1 \times 10^{-6}\,m^2\,s^{-1}$) are most commonly used. Obsolescent units such as Redwood and Saybolt seconds and Engler degrees derived originally from short efflux tube instruments are still to be found, but nowadays they are derived from measurement of kinematic viscosity with which they have almost linear relationships.

With increase of temperature, the viscosity of gases rises while that of liquids falls. A moderately viscous mineral oil falls from $1\,N\,s\,m^{-2}$ at 5°C to $0.01\,N\,s\,m^{-2}$ at 100°C. This property is commonly expressed as the Kinematic Viscosity Index (KVI) measured over the range 37.8–98.9°C (100–210°F) and expressed in values usually ranging between 0 and 100. However, a more rational Dynamic Viscosity Index, which may be measured over any convenient range, has been proposed. Oils with a KVI of 35 and below are known as Low Viscosity Index (LVI) oils, those between 35 and 80 are Medium Viscosity Index (MVI) oils, those between 80 and 110 are High Viscosity Index (HVI) oils and those over about 110 are Very High Viscosity Index (VHVI) oils.

With minerals oils and oils of similar molecular weight, viscosity is independent of the rate of shear, exept possibly under very high pressures. Shear rates in elastohydrodynamic lubrication (or EHL) are said to be 'Newtonian'. Oils of very high molecular weight, such as silicones, exhibit a reduction in viscosity at quite moderate rates of shear and are said to be 'non-Newtonian'. While such loss in viscosity may be only temporary, there may also be permanent loss due to mechanical breakdown of very high molecular weight polymer molecules.

The lubricant viscosity usable in practice depends on the type of machine involved. Low viscosity oils are used for high speeds and high viscosities for low speeds. Very approximately $3\,N\,s\,m^{-2}$ is the maximum viscosity at which machines can be started up, while 0.002 is the minimum for maintaining hydrodynamic lubrication under running conditions.

24.3.2 Boundary lubrication properties

This property of a lubricant is essentially its ability to produce on one or both of the rubbing surfaces a layer of adequate thickness of coherent and adherent, low shear strength material which will minimize metal-to-metal contact and reduce friction. Lubricants exemplifying this property under very mild, slow speed conditions are the fatty acids which produce soap layers on metals with active oxide surfaces, such as copper.

Under such mild conditions the property is often known as 'lubricity'. Under more severe conditions of load, speed and temperature, the property is known as 'Extreme Pressure' or 'EP'. As an example, free sulphur is added to cutting oils to prevent excessive tool wear with tough steels.

24.3.3 Chemical stability

Lubricants are usually required to have an effective life of some hundreds or thousands of hours during which its necessary properties are sensibly unaltered. They may, however, change in whole or part by thermal decomposition, oxidation or by hydrolysis. Thermal decomposition followed by polymerization results in the formation of materials of very high molecular weight, especially insoluble coke-like substances, as well as low molecular weight materials including gases. The viscosity and flash point of the lubricant therefore generally drops. Similar materials are also formed during oxidation, but in addition highly oxygenated species including lacquers and organic acids are formed and the viscosity generally increases. This viscosity increase, the amount of insoluble material and the increase in organic acidity, are conveniently used as expressions of the degree of oxidation. Lubricants based on esters and lubricants containing esters or salts as components, e.g. as additives, are subject to hydrolysis and here again acidity, perhaps with a corrosion test for a sensitive metal such as lead, are used as criteria of stability.

Water-containing lubricants require particularly clean working conditions as contamination may lead to bacterial attack and thus to unpleasant odour, corrosion, and reduced effectiveness. Systems should be prepared and regularly cleaned by flushing with 5% solutions of caustic soda, detergent solutions or both. Biostats or biocides may also be helpful.

24.3.4 Physical properties

The thermal capacity of an oil is particularly important as in many cases the flow of oil is used to remove heat. Thermal capacity varies from around $2000\,J\,kg^{-1}K$ for mineral oils to around 1 500 for silicones and triaryl phosphate esters, which compare with $4200\,J\,kg^{-1}K$ for water.

In general, lubricants must not evaporate rapidly at the highest temperatures of usage. At well above their maximum usable temperatures, mineral oils and similar flammable oils reach their flash points at which, in particular apparatus, the evaporation is sufficient to reach the lower explosive limit with air. High volatility is, however, occasionally desirable. In some metalworking operations, working is followed by an annealing operation and any lubricant remaining must evaporate away cleanly.

At low temperatures paraffinic mineral oils reach a lower limit of usage at the pour point. At about $-7\,°C$ a separate wax phase comes out of solution from the rest of the oil and at about $-10\,°C$ sufficient needle-like crystals form to block the flow.

24.4 Mineral oils

The most important type of oil, the mineral oils, are made from petroleum. They are abundant, are comparatively low in cost and are available in a wide range of viscosities. The fatty oils are inferior in all respects except boundary friction properties which, where required, can easily be provided by the incorporation of a low concentration of fatty material. The various synthetic oils are of growing importance and find effective use in extreme applications where some particular property or properties justifies their high price but, in general, their scale of use is very limited.

There are, broadly speaking, two types of mineral oil: paraffinic, having comparatively long alkyl side chains in the molecule and consequently high KVI, together with a 'wax' pour point; and naphthenic, having comparatively short side chains, low to medium VI and a 'viscosity' pour point. With both types a wide range of viscosity grades can be made. Low viscosity mineral oils are frequently known as 'spindle oils' from their early use in textile machinery while the more viscous oils are known as 'cylinder oils' from their use in steam engines. Grades between these two extremes are called 'light machine' and 'heavy machine' oils. These basic grades are produced by refining and intermediate grades are produced by blending.

The wide range of viscosities available is indicated in Figure 24.1 for paraffinic oils of 95 KVI to BS 4231, in which each grade is designated by its mid-point viscosity at 37.8 °C (100 °F) expressed in centistokes ($mm^2\,s^{-1}$). Figure 24.1 also indicates in its margins the important classification system of the Society of Automotive Engineers (SAE), the 'W' or winter grades of which are classified in dynamic units, the remainder in Saybolt seconds.

Oxidation stability may be significantly increased by the use of antioxidants as shown in

Figure 24.2, which indicates the life obtainable for a moderate degree of deterioration of a well refined paraffinic oil.

Oxidized oils tend to be corrosive to active metals such as cadmium, zinc and, especially, lead but there is no difficulty in selecting suitable materials of construction since resistant paints and elastomers have been developed over many years.

Typical physical properties for mineral oils are given in Table 24.1.

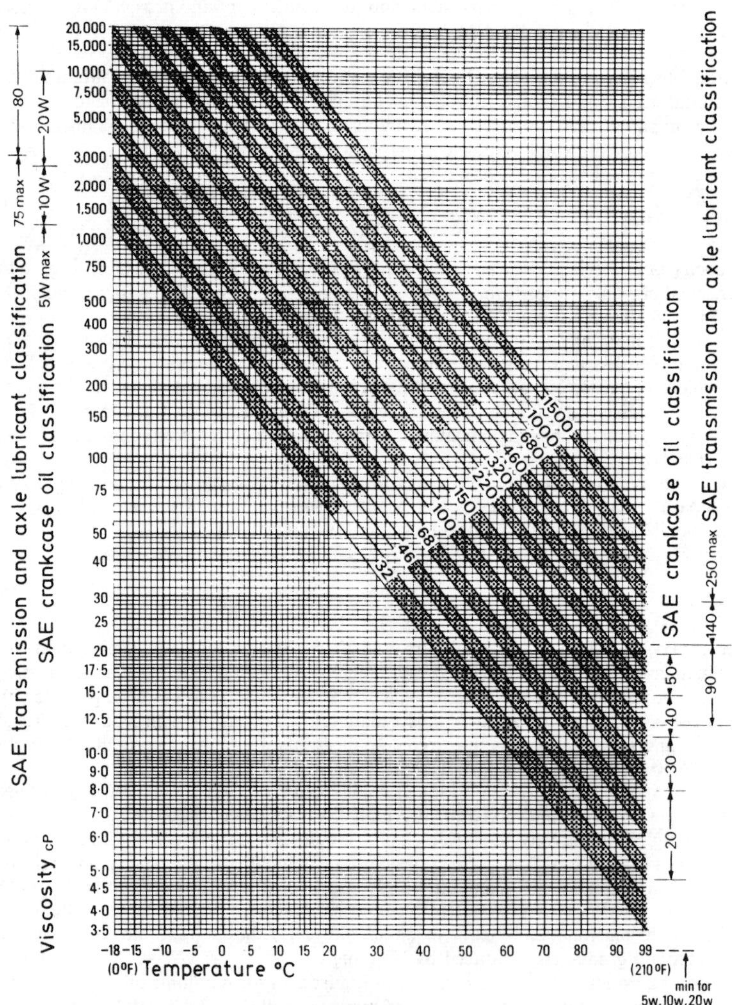

Figure 24.1 BS 4231 *ranges for* 95 K VI *oils with* SAE *limits*
NB: Such oils normally have a pour point of approximately −10 °F, but the lines have been extrapolated to permit comparison with the SAE grades

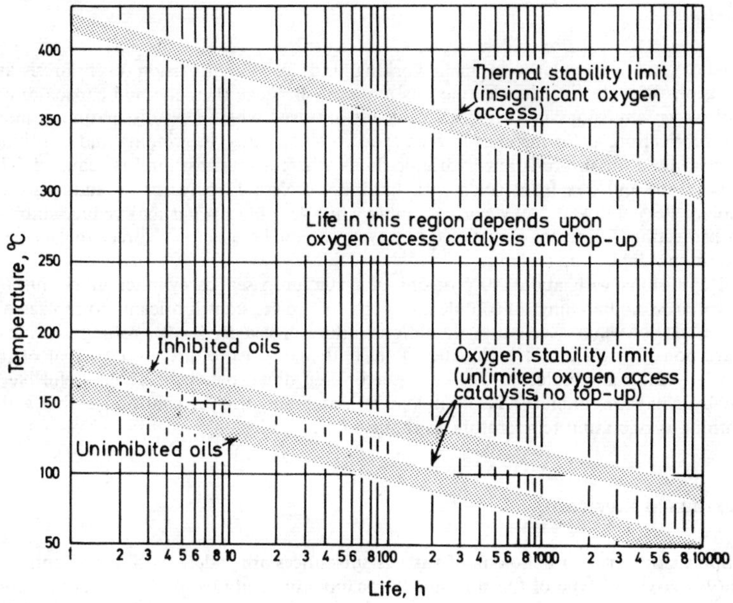

Figure 24.2 *Approximate life of well-refined mineral oils*

Table 24.1 TYPICAL PHYSICAL PROPERTIES OF HIGHLY REFINED MINERAL OILS

	Naphthenic oils			*Paraffinic oils*		
	Spindle	*Light machine*	*Heavy machine*	*Light machine*	*Heavy machine*	*Bright stock*
Density at 25 °C	0.862	0.880	0.897	0.862	0.875	0.891
Viscosity ($mN\,s\,m^{-2}$) at 30 °C	18.6	45.0	171	42.0	153	810
60 °C	6.3	12.0	31	13.5	34	135
100 °C	2.4	3.9	7.5	4.3	9.1	27
Dynamic viscosity index	92	68	38	109	96	96
Pour point, °C	−43	−40	−29	−9	−9	−9
Pressure–viscosity coefficient ($m^2\,N^{-1} \times 10^8$) at						
30 °C	2.0	2.8	2.8	2.2	2.4	3.4
60 °C	1.6	2.0	2.3	1.9	2.1	2.8
100 °C	1.3	1.6	1.8	1.4	1.6	2.2
Isentropic secant bulk modulus at 35 $MN\,m^{-2}$ at						
30 °C	—	—	—	198	206	—
60 °C	—	—	—	172	177	—
100 °C	—	—	—	141	149	—
Thermal capacity ($J\,kg^{-1}\,°C$) at						
30 °C	1 880	1 860	1 850	1 960	1 910	1 880
60 °C	1 990	1 960	1 910	2 020	2 010	1 990
100 °C	2 120	2 100	2 080	2 170	2 150	2 120
Thermal conductivity $W\,m\,m^{-2}$ at						
30 °C	0.132	0.130	0.128	0.133	0.131	0.128
60 °C	0.131	0.128	0.126	0.131	0.129	0.126
100 °C	0.127	0.125	0.123	0.127	0.126	0.123
Temperature (°C) for vapour pressure of 0.001 mmHg	35	60	95	95	110	125
Flash point, open, °C	163	175	210	227	257	300

24.5 Emulsions

For many applications, particularly in metalworking and for the provision of fire-resistant lubricants, emulsions of water and lubricating oils are used. Because their thermal capacities are close to that of water, emulsions of 1–10% oil in water are used where heat dissipation is more important than lubrication, as for instance in high speed metal cutting, in grinding and in rolling. Emulsions are also used for fire-resistant hydraulic fluids where a low cost fluid is required. The water should be clean and free from acids with hardness preferably between 15 and 50 p.p.m. $CaCO_3$ equivalent. Very soft water may cause foaming while very hard water may reduce stability and corrosion protection. Oil/water emulsions, however, cannot be used as a direct replacement for oil in conventional hydraulic systems.

Water in oil emulsions with about 40% deionized water are used as cylinder lubricants for reciprocating compressors handling oil-soluble gases and as fire-resistant lubricants to replace oil, however, the presence of water accelerates the fatigue failure of heavily loaded rolling bearings.

Emulsions are non-Newtonian. At high rates of shear their viscosities are close to that of the base oil, which is usually a spindle oil. They are not broken down by shear. Their useful life is largely limited by emulsion stability. To limit loss of water and loss of emulsion stability the maximum continuous operating temperature is about 65 °C.

24.6 Water-based lubricants

Aqueous solutions, either true or colloidal, of various substances are widely used as coolants and lubricants. The water-glycol type of fire-resistant lubricants are solutions of 50–65% polyglycols, sometimes including ethylene glycol. Two or three grades between 0.03 and 0.07 N s m^{-2} at 38 °C (100 °F) are usually available. Their VIs are very high and pour points very low, e.g. −40 °C. They are completely shear stable but operating temperatures are usually limited to 65 °C in order to avoid excessive loss of water.

Solutions of corrosion inhibitors and load carrying additives are used for high speed cutting and grinding operations.

24.7 Synthetic oils

A large number of synthetic lubricants have been described but only a few are of commercial importance.

24.7.1 Diesters

These were developed principally for aviation gas turbines because of their very high VI, and low pour point gives them a wide range of usage while the susceptibility to antioxidants allows them to work continuously at 120–150 °C with bearing temperatures up to 250 °C.

The load carrying capacity of the diesters in spur gear rigs is about twice as high as that of equiviscous mineral oil; under boundary conditions they appear to be equal to mineral oil while in heavily loaded ball and roller bearings the diesters appear to provide somewhat superior protection against surface fatigue.

24.7.2 Neopentyl polyol esters

This group, based on neopentyl alcohols and mixed aliphatic acids, has significantly better thermal stability and the possibility of rather higher viscosity (0.002–0.008 N s m^{-2}) than the diesters. These esters are suitable for continuous operation at 200 °C with hot spots of 275 °C but at these high operating temperatures it is necessary to avoid the use of cadmium, magnesium and silver because of corrosion and to use silicone and fluorinated types of elastomers.

The load-carrying properties of the neopentyl esters are better than those of the diesters, because of their higher viscosity. They apparently do not promote surface fatigue pitting in heavily loaded ball and roller bearings.

Table 24.2 TYPICAL DATA ON FINISHED SYNTHETIC BASED OILS

	Organic acid esters							*Silicones*		
Base oil	*di-(2-ethyl-hexyl) sebacate*	*di(iso-octyl) azelate*	*Mixed C_3–C_6 penta-erythritol ester*	*Mixed C_4–C_{10} dipenta-erythritol ester*	*Triaryl phosphate ester*	*Fluorocarbon*	*Polyglycol*	*Poly-dimethyl (1 000cS at 25 °C)*	*Medium phenyl*	*Chloro-phenyl*
Density at 25 °C	0.911	0.911	1.00	1.02	1.13	1.95	1.02	0.97	1.07	1.01
Viscosity, N s m^{-2}, at 30 °C	0.016	0.016 5	0.032	0.087	0.087	0.280	0.220	0.140	0.170	0.046
60 °C	0.006 5	0.006 77	0.012	0.025	0.019 5	0.038 5	0.070	0.083	0.060	0.027
100 °C	0.002 87	0.003 01	0.005	0.083	0.005 3	0.010 3	0.022 5	0.045	0.022 5	0.015 6
DVI	145	141	144	132	0	−27	164	200	175	197
Pour point, °C	−60	< −65	−60	−50	−18	−15	−25	−55	−50	< −73
Pressure-viscosity coefficient* (× 10^{-8}), m^2 N^{-1}, at										
30 °C	1.40	—	—	—	3.3	4.4	1.76	1.81	2.0	—
60 °C	1.28	1.38	1.8†	2.4†	2.0	3.7	1.43	1.81	1.9	2.0
100 °C	1.05	1.18	1.2†	1.6†	1.2	2.83	1.22	1.94	1.9	2.0
Thermal capacity, 1 kg °C, at 30 °C	—	—	—	—	1 510	1 340	1 870	—	—	1 490
60 °C	1 960	1 960	—	—	1 610	—	1 970	1 550	—	1 550
100 °C	2 100	2 100	—	—	1 700	—	2 100	—	—	1 670
Thermal conductivity at 30 °C	0.154	0.151	—	—	0.127	—	0.150	0.162	—	0.150
60 °C	0.149	0.148	—	—	0.127	—	0.148	0.159	0.146	0.145
100 °C	0.142	0.144	—	—	0.126	—	0.146	0.155	—	0.140
Temperature for vapour pressure of 0.001 mmHg, °C	117	93	—	—	125	35	—	177	145	—
Flash point, open, °C	230	243	260	300	—	none	277	316	—	> 273
Approximate cost relative to mineral oil	5	5	10	15	5	300	5	30	60	70

* Average value over pressure range 0–5000 lbf in^{-2} (0–34.5 M Pa). † Estimated

24.7.3 Triaryl phosphate esters

These esters find their greatest use as fire-resistant hydraulic fluids in diecasting machines, and like machines, as well as in the governor systems of large steam turbines. Viscosities range from 0.0036 to 0.008 $N s m^{-2}$ at 100 °C. Their oxidation stability is good, but thermal degradation is catalysed by steel. This limits their maximum operating temperature to about 120 °C. The fire-resistance is on a par with that of the water-in-oil emulsions, and is generally adequate for industrial purposes.

The hydrolytic stability of these esters is rather poor and appears to be catalysed by acid impurities or similar substances developed by initial hydrolysis. If the acidity can be kept low by filtration through fullers earth the degradation can be very greatly retarded. Hydrolysis may result in some corrosion of aluminium and steel but the phosphate esters are not corrosive to cadmium, zinc or other common metals.

Triaryl phosphate esters tend to damage conventional rubber elastomers and therefore butyl, silicone or fluorinated types are preferred. Ordinary paints are also affected and those based on epoxy resins should be used.

These esters are good lubricants under both boundary and hydrodynamic (including elasto-hydrodynamic) conditions but they promote fatigue pitting of rolling element bearings and, for a life equivalent to that when using mineral oils, a 20% reduction in load may be required.

Toxicity is largely related to the amount of ortho-tolyl isomer in the oil which is accordingly kept to a low value. Particular attention should be paid to personal cleanliness and good ventilation wherever people come into contact with these oils.

24.7.4. Fluorocarbons

Usually these are polymers of trifluorovinyl chloride, the terminal groups being fluorine. The range of oils with pour points below 20 °C is only 0.002–0.004 $N s m^{-2}$. Densities and volatilities are unusually high. They are exceptionally stable to strong oxidizing agents such as fuming nitric acid, hydrogen peroxide, etc. Thermally, they are completely stable below 300 °C and the degradation at high temperatures is depolymerization so that carbonaceous deposits are not formed.

Fluorocarbons are non-corrosive to metals. Load carrying capacity under boundary conditions is rather better than that of equiviscous mineral oil, but they may have a lower protection against fatigue failure of heavily loaded rolling bearings.

24.7.5 Polyglycols

These oils are also known as polyalkylene glycols, polyoxyalkylenes, glycols and polyethers. The water soluble types are mainly polyethylene oxides and have high pour points and very high viscosity indices, while the mainly polypropylene oxides are water insoluble with low pour points and somewhat lower viscosity indices. A wide viscosity range is covered: from 0.008 to 19.5 $N s m^{-2}$ at 38°C.

Polyglycols are very responsive to oxidation inhibitors and when inhibited are much more stable to oxidation than mineral oils. At about 250 °C polyglycols exhibit rapid thermal decomposition, but as the products of decomposition are volatile they do not form deposits. Polyglycols are not corrosive to the usual metals, but since even the water-insoluble grades are slightly hygroscopic rust inhibited grades are preferred wherever moisture may enter the oil.

The water-soluble types have important uses as components of water-glycol type fire-resistant lubricants and automotive brake fluids and are very good lubricants under hydrodynamic and elastohydrodynamic conditions. Under boundary conditions they are not very good but may be provided with suitable properties by the addition of small amounts of long-chain fatty acids.

Typical physical properties of these synthetic lubricating oils are given in Table 24.2.

24.8 Greases

24.8.1 Composition

The standard definition of a lubricating grease is 'A solid to semi-fluid product of dispersion of a thickening agent in a liquid lubricant. Other ingredients imparting special properties may be included.' The most common types of thickener are calcium and lithium metal soaps. Bentonite is used for high temperatures, above about 140 °C. Esterified silica, vat dyestuffs and urea compounds are used for the most specialised applications.

The fatty acids of the metal soaps also influence the properties of the grease. Mixed acids from tallow, stearic acid and hydroxy stearic acid are probably the most widely used. Complex soaps formed by the co-crystallization of two compounds permit operation at high temperatures.

A grease is usually 80–90% liquid lubricant, commonly low and medium viscosity mineral oil but high viscosity residual oils are used for high temperatures and low speeds. For special purposes, synthetic oils are used.

Additives are commonly used in greases for particular purposes as follows:

Solid lubricants	e.g. graphite, molybdenum disulphide for heavily loaded low speed applications where lubrication will be mainly of the boundary type.
Antioxidants	To prevent rapid oxidation during storage and in use.
Metal passivators	To reduce catalytic oxidation of the grease by cuprous-metals, e.g. in the cages of rolling element bearings. To prevent rusting, particularly of rolling element bearings, during storage and use.
Extreme pressure	To prevent scuffing and wear under boundary lubrication conditions, particularly those arising temporarily from shock loads.

24.8.2 Properties

The essential property of a grease is that it possesses a yield stress up to which it only deforms elastically and above which it flow plastically. When flow commences the ratio shear stress/rate of shear decreases smoothly until at shear rates in the region of $10^6\,s^{-1}$ it closely approaches that ratio for the liquid phase of the grease, i.e. its viscosity. Above the yield stress greases are non-Newtonian liquids, and at any point the ratio shear stress/rate of shear is called its 'apparent viscosity', which is, in effect, the viscosity a Newtonian fluid would have if it exhibited the same shear stress at the same shear rate.

The significance of these properties, in relation to plain bearings is that under stationary conditions grease tends to remain in place in clearance spaces and at the ends of bearings. Thus lubricant is available immediately the machine starts up again, and grease clinging to the ends of bearings acts as a seal to exclude dirt.

Semi-fluid greases of negligible yield stress reduce leakage from gearboxes by virtue of their very high apparent viscosity at low rates of shear. They also permit feeding through long narrow bore piping, particularly at sub-zero temperatures.

The yield stress of a grease is not easily measured and for production quality control and other ordinary purposes the worked Penetration (IP 50/69, ASTM D217–C8) i.e. the depth in mm of the penetration in the grease of a special metal cone under its own weight, is used. The National Lubricating Grease Institute has classified greases according to their consistency after a specified amount of mechanical working as follows:

Table 24.3 NLGI GREASE CLASSIFICATION

NLGI grade	*Worked penetration range*
000	445–475
00	400–430
0	355–385
1	310–340
2	265–295
3	220–250
4	175–205
5	130–160
6	85–115

No. 2 grade is popular since it combines satisfactory yield properties with easy pumpability, but where there are extreme vibration and shock loads a No 3 grade is preferred. Grades more fluid than No. 0 or stiffer than No. 3 are not normally used for roller bearings.

Important properties of the thickener structure are temperature stability, resistance to water and mechanical stability. Table 24.4 lists the various types of thickeners and indicates the extent to which they have these properties

Table 24.4 COMPARISON OF GREASE THICKENERS

Thickener type	*Temperature stability*	*Water resistance*	*Mechanical stability*
Calcium soap	Low	Excellent	Excellent
Sodium soap	Good	Poor	Fair
Lithium soap	Very good	Very good	Very good
Modified clay	Excellent	Good	Good

Synthetic oils are used in place of mineral oils for the liquid phase where their special advantages outweigh their greater cost. Diesters are particularly useful where low volatility and good performance at low temperature are needed, e.g. in aircraft bearings. Polyglycols are used where good oxidation stability and good lubrication between steel and bronze are required, also for special cases where the liquid phase is required to evaporate at high temperature without passing through deposit-forming decomposition stages. Silicones are used where good stability is required at high temperatures without the conditions of load and speed being at all severe. Fluorocarbons are, however, preferred in spite of their very high cost where maximum resistance to oxidation is required, e.g. from contact with liquid oxygen or ozone.

24.9 Oil additives

Plain mineral oils are used in many units and systems for the lubrication of bearings, gears and other mechanisms where their oxidation stability, operating temperature range, ability to prevent wear, etc. are adequate. The addition of fatty oils improves boundary lubrication properties at the expense of oxidation stability and demulsibility, but over the last 30 years oil-soluble chemical compounds called 'additives' have been developed which improve or confer a wide range of properties. The functions required of these 'additives' gives them their common names as indicated in Table 24.5.

Table 24.5 TYPES OF ADDITIVE

Main type	*Function and sub-types*
Acid neutralizers	Neutralise contaminating strong acids formed for example by combustion of high sulphur fuels or, less often, by decomposition of active EP additives
Anti-foam	Reduce surface foam
Antioxidants	Reduce oxidation. Various types are: oxidation inhibitors, retarders; anti-catalyst metal deactivators, metal passivators
Anti-rust	Reduces rusting of ferrous surfaces swept by oil
Antiwear agents	Reduce wear and prevent scuffing of rubbing surfaces under steady load operating conditions, nature of film uncertain
Corrosion inhibitors	Type (1) Reduces corrosion of lead. Type (2) Reduces corrosion of cuprous metals
Detergents	Reduce or prevent deposits formed at high temperatures, e.g. in i.c. engines.
Dispersants	Prevent deposition of sludge by dispersing a finely divided suspension of insoluble material formed at low temperature
Emulsifiers	Form emulsions either water in oil or oil in water according to type
Extreme pressure	Prevent scuffing of rubbing surfaces under shock load operating conditions mainly by formation of inorganic surface films
Lubricity	Reduce friction under boundary lubrication condition, increase load carrying capacity especially where limited by frictional temperature rise, by formation of organic surface film. Examples are fatty acids and their esters
Pour point depressant	Reduce pour point of paraffinic oils
Tackiness	Reduces loss of oil by gravity, e.g. from vertical sliding surfaces, or centrifugal forces
Viscosity index improvers	Reduce the decrease in viscosity due to increase of temperature

24.9.1 Machinery lubricants

As shown in Tables 24.6 and 24.7 below, additives and oils are combined in various ways to provide the performance required. It must be emphasized, however, that indiscriminate mixing can produce undesirable interactions. Indeed some additives may be included in a blend simply to overcome problems caused by other additives.

24.9.2 Cutting oils

Factors entering into the selection of cutting oils are: the material of the workpiece; the speed and nature of the operation; whether cooling is more important than lubrication; and the compatibility of the cutting oil with the machine tool.

Table 24.7 gives a very general system of lubricant selection.

Table 24.6 TYPES OF OIL REQUIRED FOR VARIOUS TYPES OF MACHINERY

Type of machinery	*Usual base oil type*	*Usual additives*	*Special requirements*
Food processing	Medicinal white oil	None	Safety in case of ingestion
Plain roll-neck bearings of rolling mills	HVI	None	Best demulsibility
Oil hydraulic	HVI down to −20 °C MVIN below	Antioxidant Anti-rust Anti-wear Pour point depressant VI Improver Anti-foam	Minimum viscosity change with temperature Minimum wear of steel/steel
Steam and gas turbines	HVI or MVIN distillates	Antioxidant Anti-rust	Ready separation from water, good oxidation stability
Steam engine cylinders	Unrefined or refined residual or high viscosity distillates	None or fatty oil	Maintenance of oil film on hot surfaces, resistance to washing away by wet steam
Air compressor cylinders	HVI or MVIN distillates	Antioxidant, anti-rust	Low deposit formation tendency
Gears (steel/steel)	HVI or MVIN	Anti-wear EP antioxidant Anti-foam Pour point depressant	Protection against wear and scuffing
Gears (steel/bronze)	HVI	Oiliness, tackiness	Maintains smooth sliding at very low speeds. Keeps film on vertical surfaces
Hermetically sealed refrigerators	MVIN	None	Good thermal stability, miscibility with refrigerant, low floc point
Diesel engines	HVI or MVIN	Detergent Dispersant Antioxidant Acid-neutralizer Anti-foam Anti-wear Corrosion inhibitor	Vary with type of engine thus affecting additive combination

Table 24.7 CHIP-FORMING METALWORKING LUBRICANTS

Type of lubricant	*Base lubricant*	*Additive*	*Remarks*
Soluble oil (oil-in-water emulsion)	LVI oil	Emulsifiers Rust inhibitors	With 20–50 parts water For grinding and light cutting operation where cooling and absence of fuming important

Table 24.7 CHIP-FORMING METAL WORKING LUBRICANTS—*continued*

Type of lubricant	*Base lubricant*	*Additive*	*Remarks*
Aqueous cutting solution	Water	Rust inhibitors	As for soluble oil
Inactive EP cutting oil	HVI	Mild EP but no free sulphur	For cutting yellow metal and other non-ferrous alloys where good lubrication without staining required
Active EP cutting oil	HVI	Mild EP and free sulphur	For heavy cuts on tough steel

24.9.3 Lubricants for chipless-forming

Lubricants for chipless-forming probably present a greater range of diversity than any other branch of lubrication. Table 24.8 gives the types of lubricants used in drawing, stamping and pressing, and Table 24.9 gives lubricants used in rolling.

Table 24.8 TYPES OF LUBRICANTS FOR DRAWING, STAMPING AND PRESSING

Metal	*Lubricant in order of severity of operation*
Steel	Mineral oils of medium to heavy viscosities Fatty oil/mineral oil blends Soap solutions Soap/fat pastes Baked-on-lime coatings Soft metals, e.g. lead
Brass and copper	Dilute soap solutions Light mineral oil Soap/fat pastes with solid lubricants Dried on soap
Magnesium	Colloidal graphite in low volatile mineral oils Graphite in volatile solvents
Aluminium	Oil-in-water emulsions Mineral oils, viscosity increasing with severity Mineral oil with 10–15% fatty oil

24.9.4 Rolling oils

Lubrication is often required in metal rolling but excessive lubricity causes 'lack of bite' or roll slippage. Lubrication affects surface finish, ease of application, removal, and uniformity of adherence to the surface.

Table 24.9 ROLLING OILS

Metal	*Lubricant in order of severity of operation*
Steel	Oil-in-water emulsions Mineral oil/fatty oil blends and with lubricity and EP additives Palm oil
Brass and copper	Oil-in-water emulsion Mineral oil Mineral oil with lubricity additive
Aluminium	Mineral oils of viscosity from 40 mN s m^{-2} at 20 °C to 50 mN s m^{-2} at 40 °C with lubricity additives

BIBLIOGRAPHY

F. P. Bowden and D. Tabor, 'The Friction and Lubrication of Solids', Parts I and II, Oxford, 1954, 1964.
E. L. H. Bastian, 'Metalworking Lubricants', McGraw-Hill, New York, 1951.
'Lubrication and Wear: Fundamentals and Application to Design', *Proc. Inst. mech. Engrs*, 1967–68, 182, Part 3A.
'A Glossary of Petroleum Terms', Inst. Petroleum, London, 1961.

25 Friction and wear

The friction and wear characteristics of materials are not intrinsic properties but, rather, depend on a large number of variables including the physical, chemical and mechanical properties of the material and surfaces and the environment.

25.1 Friction

25.1.1 Friction of unlubricated surfaces

DEFINITION

The friction between two bodies is generally defined as the force which acts between them at their surface of contact so as to resist their sliding on one another. The frictional force F is the force required to initiate or maintain motion. If W is the normal reaction of one body on the other, the coefficient of friction μ is defined as $\mu = F/W$.

STATIC AND KINETIC FRICTION

If the force to initiate motion of one of the bodies is F_s and the force to maintain its motion at a given speed is F_k, there is a corresponding coefficient of static friction $\mu_s = F_s/W$ and a coefficient of kinetic friction $\mu_k = F_k/W$. In some cases these coefficients are approximately equal; in most cases $\mu_s > \mu_k$ and there is a tendency for intermittent or 'stick-slip' motion to occur.

BASIC LAWS OF FRICTION

The two basic laws of friction, which are valid over a wide range of experimental conditions, state that:

1. The frictional force F between solid bodies is proportional to the normal force between the surfaces, i.e. μ is independent of W.
2. The frictional force F is independent of the apparent area of contact.

25.1.2 Friction of unlubricated materials

When clean metal surfaces are placed in contact they do not touch over the whole of their apparent area of contact. The load is supported by surface irregularities (asperities) which deform plastically as the load is applied. The area of real contact is approximately proportional to the load and almost independent of the size and geometry of the surfaces.[1] This is also the case when asperity contact is primarily elastic,[2] which may occur with well run-in surfaces, particularly in the presence of a lubricant or surface oxide films. The limiting values to the true area of contact[3] for a wide range of practical situations are W/p and $10W/p$, where W is normal load and p is plastic flow pressure of the asperities, of the same order as the indentation hardness of the material.

For very clean surfaces strong adhesion occurs at regions of real contact, a part of which may be due to cold-welding, and these junctions must be sheared if sliding is to take place. Thus, it is almost impossible to slide such surfaces in a vacuum and complete seizure often occurs as shown in Table 25.1. However, if the surfaces are contaminated the adhesion is much weaker because the formation of strong junctions is inhibited. For example, hydrogen or nitrogen atmospheres have little effect on in-vacuo coefficients of friction, but the smallest trace of oxygen or water vapour produces a profound reduction in friction (Table 25.1). A further reduction in the coefficient of friction often occurs at high sliding speeds, particularly at speeds sufficient to produce local hot-spots and surface melting,[4] e.g. ice at 0.1 m s^{-1} or steel at 500 m s^{-1} for which μ may be less than 0.1.

Table 25.1 STATIC FRICTION OF METALS (SPECTROSCOPICALLY PURE) IN VACUUM (OUTGASSED) AND IN AIR (UNLUBRICATED)

	Metals											
Conditions	Ag	Al	Co	Cr	Cu	Fe	In	Mg	Mo	Ni	Pb	Pt
μ_s metal on itself in vacuo	S	S	0.6	1.5	S	1.5	S	0.8	1.1	2.4	S	4
μ_s metal on itself in air	1.4	1.3	0.3	0.4	1.3	1.0	2	0.5	0.9	0.7	1.5	1.3

S signifies gross seizure ($\mu = 10$).

Friction values for metal couples in air depend on a number of factors. Principal ones are the tendency for formation of oxide films, the degree of deformation in sliding, the ability of oxide films to survive sliding contact and the tendency for transfer of material from one surface to the other. Table 25.2 shows the relative hardnesses of some common metals and their oxides and the load (for a spherical slider on polished surfaces) at which appreciable metallic contact occurs. Thus, the oxide on copper is not easily penetrated, whereas the very hard aluminium oxide on the soft aluminium substrate is readily shattered during sliding. Thick oxide films, such as produced by anodizing aluminium, may be more protective because sliding deformation can be restricted entirely to the oxide. Similarly, with very hard metal substrates, such as chromium, the surface deformation may be so small that the oxide is never ruptured.

Table 25.2 BREAKDOWN OF OXIDE FILMS PRODUCED DURING SLIDING

	Vickers hardness (kg mm^{-1})		*Load* (g) *at which appreciable metallic contact occurs*
Metal	*Metal*	*Oxide*	
Gold	20	—	0
Silver	26	—	0.003
Tin	5	1 650	0.02
Aluminium	15	1 800	0.2
Zinc	35	200	0.5
Copper	40	130	1
Iron	120	150	10
Chromium plate	800	—	Never

Reference: 4.

The static coefficients of friction of a number of metals and alloys on steel are shown in Table 25.3. Of particular note are the values for indium and lead, which are the same as those for sliding on themselves (see Table 25.1). Pick-up occurs on the steel surface such that the sliding couple becomes the metal on itself.

Static friction of ferrous materials is shown in Table 25.4. The data illustrate the effect of increasing hardness on reducing friction through greater support of the surface oxide, the effect of second phases such as carbides and graphite in reducing adhesion of junctions, and the effect of the very thin oxide coating on austenitic stainless steel which is easily ruptured in sliding leading to a high coefficient of friction.

Table 25.3 STATIC FRICTION OF UNLUBRICATED METALS AND ALLOYS (PREPARED GREASE FREE) ON STEEL

The results quoted are for sliders of pure metals and alloys sliding over 0.13% C, 3.42% Ni, normalized steel. The results on mild steel are essentially the same.

Metal or alloy	μ_s	*Metal or alloy*	μ_s
Aluminium (pure)	0.6	Molybdenum (pure)	0.5
Aluminium bronze	0.45	Nickel (pure)	0.5
Brass (Cu 70, Zn 30)	0.5	Phosphor-bronze	0.35
Cast iron	0.4	Silver	0.5
Chromium (pure)	0.5	Steel (0.13 C, 3.42 Ni)	0.8
Constantan	0.4	Tin (pure)	0.9
Copper (pure)	0.8	White metal (tin-base):	0.8
Copper–lead (dendritic: Pb 20)	0.2	(Sb 6.4, Cu 4.2, Ni 0.1, Sn 89.2)	
Copper–lead (non-dendritic: Pb 27)	0.28	White metal (lead-base):	0.5
Indium (pure)	2	(Sb 15, Cu 0.5, Sn 6, Pb 78.5)	
Lead (pure)	1.5	Wood's alloy	0.7

Table 25.4 STATIC FRICTION OF UNLUBRICATED FERROUS MATERIALS ON THEMSELVES

Alloy	*VPN* kg mm^{-1}	μ_s	*Alloy*	*VPN* kg mm^{-1}	μ_s
Pure iron (cold-welded)	150	1–1.2	Ball race steel (Hoffman)	900	0.7–0.7
Normalized steel (C 0.13, Ni 3.42)	170	0.7–0.8	Tool steel (C 0.8, containing carbides)	900	0.3–0.4
Austenitic steel (Cr 18, Ni 8)	200	1	Chromium plate (hard bright)	1 000	0.6
Cast iron (pearlitic)	200	0.3–0.4			

Table 25.5 FRICTION OF VERY HARD SOLIDS

(a) Bonded tungsten carbide (cobalt binder) slider.

Material	μ_s
Tungsten carbide	*c.* 0.2
Aluminium oxide	*c.* 0.25
Copper	*c.* 0.4
Cadmium	0.8–1
Iron	0.4–0.8
Cobalt	0.3

(b) Hard solids sliding on themselves.[4,18]

	Coefficient of friction μ_s		
		Outgassed and measured in vacuo	
Material	*In air at* 20°C	20–1 000°C	*Comments*
Aluminium oxide	0.2	—	
Boron carbide	0.2	0.9	Rises rapidly above 1 800°C
Silicon carbide	0.2	0.6	
Silicon nitride	0.2	—	
Titanium carbide	0.15	1.0	Rises rapidly above 1 200°C
Titanium monoxide	0.2	0.6	
Titanium sesquioxide	0.3	0.7	
Tungsten carbide	0.15	0.6	Rises rapidly above 1 000°C

Very hard solids often have low coefficients when sliding on themselves or other materials because of the limited surface deformation that occurs during sliding (Table 25.5).

Similarly, very low coefficients of friction may be obtained by plating hard metal substrates with thin soft metal films (Table 25.6). The substrate supports the load while sliding occurs within the soft film. Typical film thicknesses are 1 to 10 μm.

Table 25.6 FRICTION OF THIN METALLIC FILMS[1]
(Sliding on a 6 mm diameter steel sphere)

	Coefficient of static friction μ_s			
Load g	*Indium film on steel*	*Indium film on silver*	*Lead film on copper*	*Copper film on steel*
4 000	0.08	0.1	0.18	0.3
8 000	0.04	0.07	0.12	0.2

The friction of many materials is little affected by high or low temperatures (see Table 25.7). Exceptions are when the plastic flow pressure changes significantly or when oxide films become very much thicker.

Table 25.7 FRICTION OF MATERIALS SLIDING ON THEMSELVES AT LOW AND HIGH TEMPERATURES[19,20]

	Coefficient of friction μ					
	Low temperatures in gaseous medium			*High temperatures in air*		
Material	4K	77K	295K	315°C	650°C	980°C
Aluminium	1.52	1.45	1.49	—	—	—
Austenitic stainless steel	0.26	0.35	0.99	—	—	—
Carbon–graphite	—	—	—	0.18	—	—
Copper	0.81	0.78	0.70	—	—	—
Iron	0.97	0.84	0.75	—	—	—
Nickel	1.06	1.10	1.12	—	—	—
Silicon nitride	—	—	—	0.60	0.28	0.48
Tool steel (15Mo15Co)	—	—	—	—	0.30	0.26
Zinc	0.43	0.39	0.52	—	—	—

The friction behaviour of polymers differs from that of metals in three respects. First, the coefficient of friction tends to decrease with increasing load; it also tends to decrease if the geometric contact area is decreased. Second, if the surfaces are left in contact under load the area of true contact may increase with time because of creep and the starting friction may be correspondingly larger. Thirdly, the friction may show changes with speed which reflect the visco-elastic properties of the polymer but the most marked changes occur as a result of frictional heating. Even at speeds of only a few m s^{-1} the friction of unlubricated polymers can rise to very high values. On the other hand at extremely high speeds the friction may fall again because of the formation of a molten lubricating film.

The main effect of speed of sliding is the generation of high local temperatures produced by frictional heating at the regions of real contact. Local hot-spots may produce phase changes or alloy formation at or near the sliding interface, they may produce local melting and they may greatly change the rate of surface oxidation. At speeds of a few m s^{-1} these effects are not as marked as at very high speeds (see Table 25.10) but they may still be significant. In general the kinetic friction at moderate speeds is of the same order as the static friction (compare previous tables) but is usually somewhat smaller. Results in Table 25.9 are for stationary sliders rubbing on a mild-steel disc rotating at a few m s^{-1}. The materials are grouped in descending order of friction. At very high sliding speeds the friction generally falls off because of the formation of a very thin molten surface layer which acts as a lubricant film.[4] Although this is, broadly speaking, the main trend other factors may considerably change the behaviour. For example, with steel sliding on diamond the friction first diminishes and then increases, because at higher speeds the steel is transferred to the diamond so that the sliding resembles that of steel on steel. In some cases the metals may fragment at these very high speeds particularly if they are of limited ductility. Again, if appreciable melting occurs the friction may rise at high speeds because of the viscous resistance of the liquid interface: this occurs with bismuth.

Table 25.8 FRICTION OF STEEL ON POLYMERS: ROOM TEMPERATURE, LOW SLIDING SPEEDS[4,7,8]

Material	*Condition*	μ
Nylon	Dry	0.4
Nylon	Wet	0.15
Perspex (Plexiglass)	Dry	0.5
PVC	Dry	0.5
Polystyrene	Dry	0.5
Low density polythene (no plasticizer)	Dry or wet	0.4
Low density polythene (with plasticizer)	Dry or wet	0.1
High density polythene (no plasticizer)	Dry or wet	0.15
Soft wood	Natural	0.25
Lignum vitae	Natural	0.1
PTFE (low speeds)	Dry or wet	0.06
PTFE (high speeds)	Dry or wet	0.3
Filled PTFE (15% glass fibre)	Dry	0.12
Filled PTFE (15% graphite)	Dry	0.09
Filled PTFE (60% bronze)	Dry	0.09
Rubber (polyurethane)	Dry	1.6
Rubber (isoprene)	Dry	3–10
Rubber (isoprene)	Wet (water–alcohol solution)	2–4

Table 25.9 KINETIC FRICTION OF UNLUBRICATED MATERIALS SLIDING ON MILD STEEL AT SPEEDS OF A FEW m s^{-1}

Slider	μ_k
Nickel, mild-steel	0.55–0.65
Aluminium, brass (70:30), cadmium, magnesium	0.4–0.5
Chromium (hard plate), steel (hard)	0.4
Copper, copper–cadmium alloy	0.3–0.35
Bearing alloys:	
Tin-base	0.46
Lead-base	0.34
Phosphor-bronze	0.34
Copper-lead (Pb 20)	0.18
Non-metals:	
Brake materials	0.4
Garnet	0.4
Carbon	0.2
Bakelite	0.13
Diamond	0.08

Table 25.10 KINETIC FRICTION OF UNLUBRICATED METALS AT VERY HIGH SLIDING SPEEDS (UP TO 600 m s^{-1}) SLIDING ON A SPHERE OF BALL-BEARING STEEL

	Duration of expt.	*Coefficient of friction* μ_k			
Surface	s	9 m s^{-1}	45 m s^{-1}	225 m s^{-1}	450 m s^{-1}
Bismuth	1–10	0.25	0.1	0.05	—
Lead	1–10	0.8	0.6	0.2	0.12
Cadmium	1–10	0.3	0.25	0.15	0.1
Copper	1–10	>1.5	1.5	0.7	0.25
Molybdenum	1–10	1	0.8	0.3	0.2
Tungsten	1–10	0.5	0.4	0.2	0.2
Diamond	1–10	0.06	0.05	0.1	≈0.1

Reference: 4.

25.1.3 Friction of lubricated surfaces

DEFINITIONS

When moving surfaces are separated by a relatively thick film of lubricant the resistance to motion is due entirely to the viscosity of the interposed layer. The friction is extremely low (μ=0.001 ~ .0001) and there is no wear of the solid surfaces. These are the conditions of hydrodynamic lubrication under which bearings operate in the ideal case. If the pressures are too high or the sliding speeds too low the hydrodynamic film becomes so thin that it may be less than the height of the surface irregularities. The asperities then rub on one another and are separated by films only one or two molecular layers thick. The friction under these conditions ($\mu \approx 0.05$ to 0.15) is much higher than for ideal hydrodynamic lubrication and some wear of the surfaces occurs. This type of lubricated sliding is called 'boundary' lubrication.[10] The friction does not depend on the viscosity of the lubricant, but on a more elusive property sometimes called 'oiliness'. Under boundary conditions as for unlubricated surfaces the frictional resistance is proportional to the load and independent of the size of the surfaces.

In certain circumstances a further type of lubrication, known as elastohydrodynamic lubrication, may obtain. It arises in the following way.[11,12,13,14] Under conditions of severe loading the moving surfaces may undergo appreciable elastic deformation: this not only changes the geometry of the surfaces, it also implies that very high pressures are exerted on the oil film. The main effect of this is to produce a prodigious increase in the viscosity of the oil. For example at contact pressures of 30, 60, 100 $kg\,mm^{-2}$ (such as may occur between gear teeth of hardened steel) the viscosity of a simple mineral oil is increased by 200, 40 000 and 1 000 000 fold respectively. Thus the harder the surfaces are pressed together the more difficult it is to extrude the lubricant. Consequently effective lubrication may obtain under conditions where it would normally be expected to break down.

In general, elastohydrodynamic lubrication becomes effective when the oil film thickness is of the order of 10^{-1}–1 µm. This is very much thicker than the boundary film (1–10 nm) but it is very small in engineering terms. Consequently for practical exploitation of elastohydrodynamic lubrication the surfaces must be very smooth and carefully aligned.

25.1.4 Boundary lubrication

Boundary lubricants function by interposing between the sliding surfaces a thin film which can reduce metallic interaction and which is, in itself, easily sheared. The latter criterion restricts boundary lubricants almost exclusively to long chain organic compounds, e.g. paraffins, alcohols, esters, fatty acids and waxes. Radioactive tracer experiments show that while a good boundary lubricant may reduce the friction by a factor of about 20 (from $\mu \approx 1$ to $\mu \approx 0.05$) it may reduce the metallic transfer by a factor of 20 000 or more. Under these conditions the metallic junctions contribute very little to the frictional resistance: the friction is due almost entirely to the force required to shear the lubricant film itself. For this reason two good boundary lubricants may give indistinguishable coefficients of friction, but one may easily give 20 times as much metallic transfer (i.e. wear) as the other. Thus with good boundary lubricants the friction may be an inadequate indication of the effectiveness of the lubricant.

Most boundary lubricants are used as additives, dissolved as a few per cent in a mineral oil:

Table 25.11 LUBRICATION OF STEEL* SURFACES BY PARAFFINS, ALCOHOLS AND FATTY ACIDS STATIC FRICTION

Lubricant	*Length of chain*	μ_s	*Lubricant*	*Length of chain*	μ_s
(a) Paraffins			(c) Fatty acids		
Decane	C_{10}	0.23	Valeric	C_5	0.17
Cetane	C_{16}	0.16	Capric	C_{10}	0.11
Triacontane	C_{30}	0.11	Lauric	C_{12}	0.11
(b) Alcohols			Palmitic	C_{16}	0.1
Octyl	C_8	0.23	Stearic	C_{18}	0.1
Decyl	C_{10}	0.16			
Cetyl	C_{16}	0.1			

*C 0.13, Ni 3.42.

they provide lubrication by adsorbing from solution on to the surfaces. As the temperature is raised the film may dissolve into the superincumbent fluid and lubrication may become ineffective at temperatures appreciably below the melting point of the film itself. The breakdown temperature depends on solubility and concentration, as well as on speed, load, and surface roughness.[15]

With more protracted heating, oxidation of the lubricant occurs and the behaviour is now determined by the properties of the oxidation products themselves. In the early stages these may be beneficial but later they lead to polymerization, gumming and the formation of other deleterious products.

Table 25.12 STATIC FRICTION OF VARIOUS METALS (SPECTROSCOPICALLY PURE) LUBRICATED WITH 1% SOLUTION OF LAURIC ACID (M.P. 44°C) IN PARAFFIN OIL AT ROOM TEMPERATURE

	Coefficient of friction μ_s	
Metal	*Unlubricated*	*Lubricated*
Aluminium	1.3	0.3
Cadmium	0.5	0.05
Chromium	0.4	0.34
Copper	1.4	0.10
Iron	1.0	0.15
Magnesium	0.5	0.10
Nickel	0.7	0.3
Platinum	1.3	0.25
Silver	1.4	0.55

Reference: 4.

Table 25.13 LUBRICATION OF STEEL SURFACES BY VARIOUS LUBRICANTS. STATIC FRICTION

	Static friction μ_s			*Static friction* μ_s	
Lubricant	20 °C	100 °C	*Lubricant*	20 °C	100 °C
None	0.58	—	*Mineral oils*		
Vegetable oils			Light machine	0.16	0.19
Castor	0.095	0.105	Thick gear	0.125	0.15
Rape	0.105	0.105	Solvent refined	0.15	0.2
Olive	0.105	0.105	Heavy motor	0.195	0.205
Coconut	0.08	0.08	BP paraffin	0.18	0.22
			Extreme pressure	0.09 ~ 0.1	0.09 ~ 0.1
Animal oils			Graphited oil	0.13	0.15
Sperm	0.10	0.10	Oleic acid	0.08	0.08
Pale whale	0.095	0.095	Trichlorethylene	0.33	—
Neatsfoot	0.095	0.095	Alcohol	0.43	—
Lard	0.085	0.085	Benzene	0.48	—
			Glycerine	0.2	0.25

Table 25.14 LUBRICATION OF METALS ON STEEL. STATIC FRICTION

Bearing surface	*Rape oil* μ_s	*Castor oil* μ_s	*Mineral oil* μ_s	*Long chain fatty acids* μ_s
Hard steel (axle steel)	0.14	0.12	0.16	0.09
Cast iron	0.10	0.13	0.21	—
Gun metal	0.15	0.16	0.21	—
Bronze	0.12	0.12	0.16	—
Pure lead	—	—	0.5	0.22
Lead-base white metal (Sb 15, Cu 0.5, Sn 6, Pb 78.5)	—	—	0.1	0.08
Pure tin	—	—	0.6	0.21
Tin-base white metal (Sb 6.5, Cu 4.2, Ni 0.1, Sn 89.2)	—	—	0.11	0.07
Sintered bronze	—	—	0.13	—
Brass (Cu 70, Zn 30)	—	0.11	0.19	0.13

25.1.5 Extreme pressure (EP) lubricants

Even the best boundary lubricants (e.g. long-chain acids or soaps) cease to provide any lubrication above about 200 °C. Since localized hot-spots of very much higher temperature are often reached in running mechanisms it is necessary to use surface films that have a high melting point and which, as far as possible, possess a low shear strength. One obvious method is to coat the metal with a thin film of a softer metal. These films are effective up to their melting point but are gradually worn away with repeated sliding. Other materials which are very effective are listed in Table 25.15.

Table 25.15 FRICTION OF METALS LUBRICATED WITH CERTAIN PROTECTIVE FILMS

Protective film	*Coefficient of friction* μ_s	*Temperature up to which lubrication is effective*
PTFE (Teflon)	0.05	~320 °C
Graphite	0.07–0.13	~600 °C
Molybdenum disulphide	0.07–0.1	~800 °C

Another approach is to form a protective film *in situ* by chemical attack, a small quantity of a suitable reactive compound being added to the lubricating oil. The most common materials are additives containing sulphur or chlorine or both. Phosphates are also used. The additive must not be too reactive, otherwise excessive corrosion will occur. The results in Table 25.16 are based on laboratory experiments in which metal surfaces were exposed to H_2S or HCl vapour and the frictional properties of the surface examined. The results show that the films formed by H_2S give a higher friction than those formed by HCl: however in the latter case the films decompose in the presence of water to liberate HCl and for this reason chlorine additives are less commonly used than sulphur additives.

The detailed behaviour of commercial additives depends not only on the reactivity of the metal and the chemical nature of the additive but also on the type of carrier fluid used (e.g. aromatic, naphthenic, paraffinic). Further the chemical reactions which occur are far more complicated than originally supposed. With sulphurized additives oxide formation appears to be at least as important as sulphide formation. With phosphates the surface reaction is still the subject of dispute.

Table 25.16 EFFECT OF SULPHIDE AND CHLORIDE FILMS ON FRICTION OF METALS

	Coefficient of friction (μ_s)				
		Sulphide films		*Chloride films*	
Metal	*Clean*	*Dry*	*Covered with lubricating oil*	*Dry*	*Covered with lubricating oil*
Cadmium on cadmium	0.5	—	—	0.3	0.15
Copper on copper	1.4	0.3	0.2	0.3	0.25
Silver on silver	1.4	0.4	0.2	—	—
Steel on steel (0.13 C, 3.42 Ni)	0.8	0.2	0.05	0.15	0.05

Reference: 1.

The differences in friction are not very marked showing that the friction is a very poor criterion of the effectiveness of an EP lubricant. Marked differences in seizure-preventing properties are often

Table 25.17 KINETIC FRICTION, INITIAL SEIZURE LOADS AND WELD LOADS OF BALL BEARING STEEL SURFACES LUBRICATED WITH TYPICAL EP ADDITIVES.[16] FOUR BALL MACHINE. FRICTION MEASUREMENTS AT 10 kg LOAD

Lubricant		*Coefficient of friction*	*Initial seizure load*	*Weld load*
Base oil	*Additive*	μ_s	kg	kg
Mineral oil	None	0.09	~45	~120
Mineral oil	Zinc di-secbutyl thio-phosphate (10% wt)	0.09	80	230
Mineral oil	Sulphurized sperm oil (5% wt)	0.095	80	250
	Sulphur	—	65	340
	Chlorinated additive (1% wt)	0.085	85	310
Mineral oil	Tributyl phosphate (1% wt)	—	80	150
Paraffin oil	Tributyl phosphate (1% wt)	—	40	125
Mineral oil	Tricresyl phosphate (1% wt)	—	75	140
Paraffin oil	Tricresyl phosphate (1% wt)	—	40	110

accompanied by almost indistinguishable coefficients of friction. The last four lines of the table also show that EP effectiveness depends to some extent on the nature of the base oil.

25.2 Wear

DEFINITIONS

Wear is the progressive loss of substance from the operating surface of a body occurring as a result of relative motion at the surface. Wear is usually detrimental, but in mild form may be beneficial, e.g. during the running-in of engineering surfaces. The major types of wear are abrasive wear, adhesive wear, erosive wear and fretting. Abrasive wear is wear by displacement of material caused by hard protuberances or particles. Adhesive wear is, strictly, wear by transference of material from one surface to another due to the process of solid-phase welding. Adhesive wear is often used, loosely, to describe other metal-to-metal wear mechanisms, including the removal of particles detached by fatigue arising from cyclic contact stresses and in which no adhesion occurs. Erosive wear is loss of material from a solid surface due to relative motion in contact with a fluid which contains solid particles or collapsing vapour bubbles. Fretting is a wear phenomenon occurring between two surfaces having oscillatory motion of small amplitude and is used, frequently, to include fretting corrosion, in which a chemical reaction predominates.

25.2.1 Abrasive wear

Abrasive wear rates and relative wear resistance (defined as wear of a reference material divided by wear of a test material) vary considerably for abrasives of different hardness, size and shape. Wear rates increase approximately linearly with increasing applied load per unit area up to loads at which extensive failure of the abrasive occurs. Figure 25.1 shows the major effect of relative hardness of the worn surface and abrasive on volume wear rate. Thus, relative wear rates in practice may vary over a wide range, as shown in Tables 25.18 to 25.22. Bulk properties of materials are very approximate guides only to abrasive wear resistance, but wear resistance generally increases as the material bulk hardness increases, except when material is hardened by prior plastic deformation.

25.2.2 Adhesive wear

Metal-to-metal wear involves the contact and interaction of asperities on two surfaces. Local stresses at asperities may be high even when applied loads are low. Adhesive wear is promoted by two major factors:

1. The tendency for different materials to form solid solutions or intermetallic compounds with one another. Thus, material combinations of different crystal structure and chemical properties tend to have lower wear rates and friction. Figure 25.2 illustrates the tendency of metal couples to adhere together.
2. The cleanliness of the surface. Cleaner surfaces are more likely to bond together. Surfaces having a thick oxide film have low wear. Stainless steels and nickel alloys, that do not form thick oxides, have poor adhesive wear resistance.

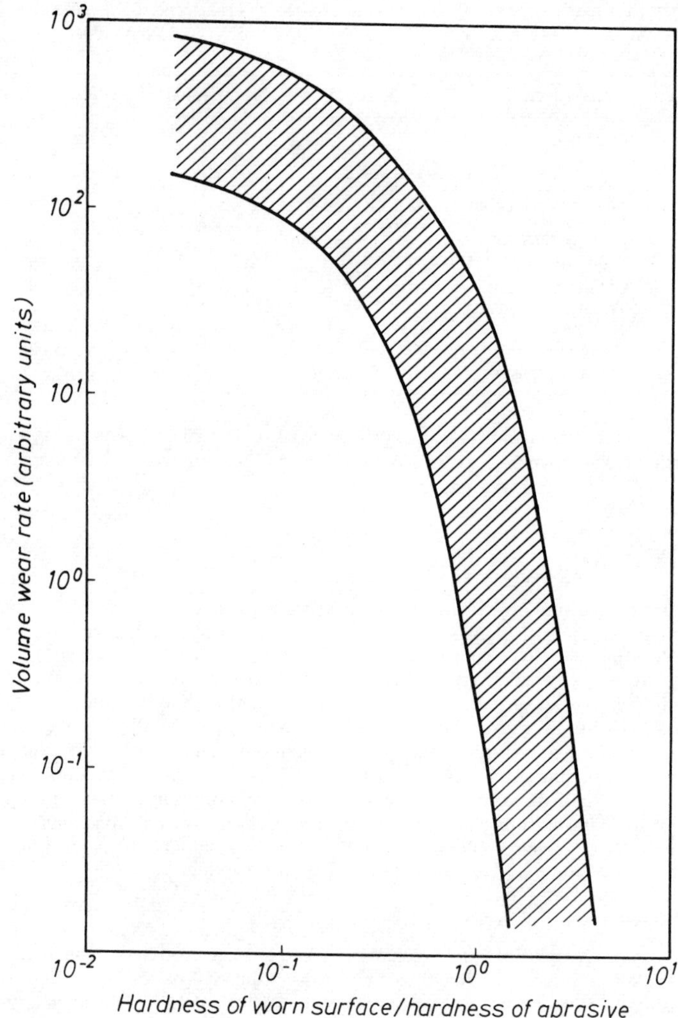

Figure 25.1 *Effect of abrasive hardness on wear rate of metallic materials and ceramics worn on* 80–400 μm *commercial bonded abrasives under an applied stress of* 1 MN m^{-2}.[21,22,23] *(Reproduced from The Fulmer Materials Optimizer by permission of Fulmer Research Institute Ltd.).*

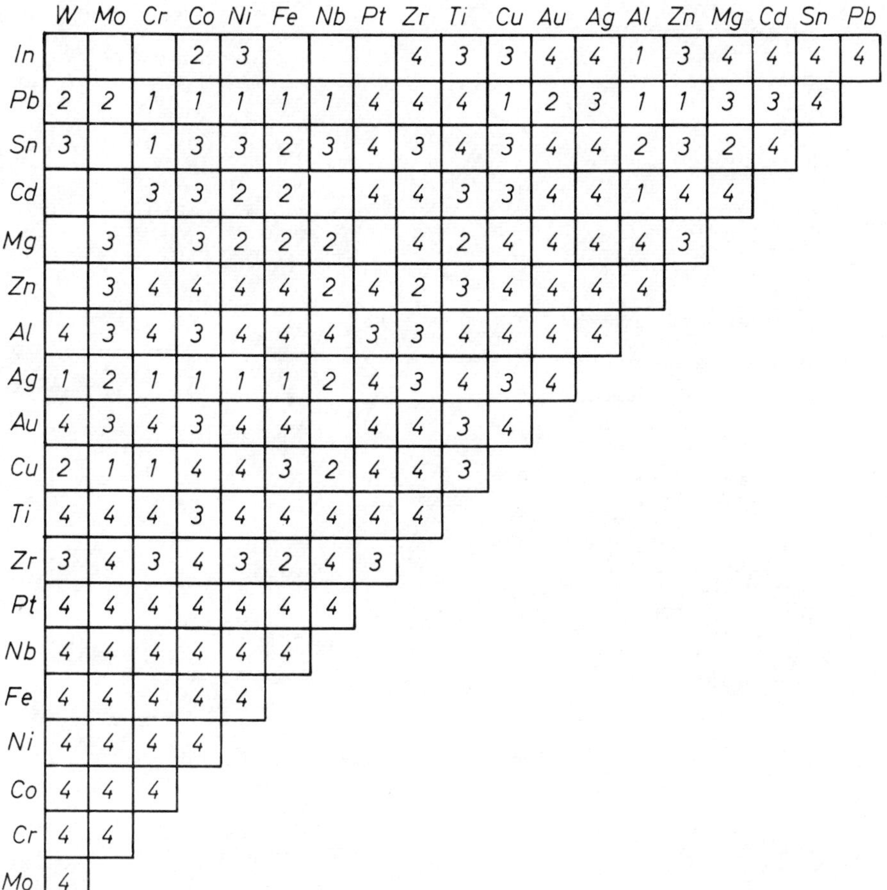

	W	Mo	Cr	Co	Ni	Fe	Nb	Pt	Zr	Ti	Cu	Au	Ag	Al	Zn	Mg	Cd	Sn	Pb
In				2	3				4	3	3	4	4	1	3	4	4	4	4
Pb	2	2	1	1	1	1	1	4	4	4	1	2	3	1	1	3	3	4	
Sn	3		1	3	3	2	3	4	3	4	3	4	4	2	3	2	4		
Cd			3	3	2	2		4	4	3	3	4	4	1	4	4			
Mg		3		3	2	2	2		4	2	4	4	4	4	3				
Zn		3	4	4	4	4	2	4	2	3	4	4	4	4					
Al	4	3	4	3	4	4	4	3	3	4	4	4	4						
Ag	1	2	1	1	1	1	2	4	3	4	3	4							
Au	4	3	4	3	4	4		4	4	3	4								
Cu	2	1	1	4	4	3	2	4	4	3									
Ti	4	4	4	3	4	4	4	4	4										
Zr	3	4	3	4	3	2	4	3											
Pt	4	4	4	4	4	4	4												
Nb	4	4	4	4	4	4													
Fe	4	4	4	4	4														
Ni	4	4	4	4															
Co	4	4	4																
Cr	4	4																	
Mo	4																		

Figure 25.2 *Tendency of metal couples to adhere together. 1 represents the greatest resistance and thus the best combination for wear. 4 represents the least resistance and thus the worst combination for wear. 2 and 3 represent intermediate resistance.*

In metal-to-metal wear, two forms of wear debris are often observed; at very low and very high loads the debris is mainly oxide, but at intermediate loads it is metallic. The transition from oxidative to metallic wear is accompanied by a rapid increase in wear rate. The transition load varies for different materials, microstructures, sliding speed and environment. Thus, wear rates of materials vary by several orders of magnitude (Table 25.23).

Surface treatments are often beneficial in metal-to-metal wear, through a change in surface chemistry, an increase in surface hardness, a change in surface structure or a change in surface topography. Certain coatings are beneficial during running-in, e.g. phosphating and sulphidized coatings, causing metal asperity separation and adherence of lubricant films. Tables 25.24 and 25.25 show the performance of a number of coated and uncoated metal pairs.

25.2.3 Erosive wear

Erosive wear due to the impact of a stream of solid particles is dependent on the size, hardness, velocity and angle of impact of the particles. Wear rate generally increases rapidly with increasing particle size and hardness and impact velocity. For strong and tough materials the maximum wear rate occurs at an impact angle of about 30°, but for hard and brittle materials it occurs at an impact angle of about 90° and for tough and elastic materials at an impact angle close to 0° (Figure 25.3). Thus, material ranking order changes occur for different erosive wear environments.

Table 25.18a COMMONLY USED MATERI

Type of material	Typical commercially available materials	Sliding wear by coke	Wear by blast furnace sinter* sliding	Wear by blast furnace sinter* impact	Wear of ball mill media/grinding quartz ores	Wear by flint stone sand loam agricultural soil	Wear in laboratory jaw crusher, siliceous ores	Wear on commerc bonded 384μm flint abrasive
		Wear rates relative to 0.4% C low alloy steel quenched and tempered to about 500 Vickers hardness						
Cast Irons	Low alloy 2.5–2.8% C, ~800 Vickers	1.2	0.8	1.2	~1.0	0.3	—	0.6
	Heat treated nodular graphite, 700 Vickers	1.0	0.15	0.5	—	1.5	—	—
	15/3 Cr/Mo Martensitic	—	—	—	0.8	—	0.04–0.3	0.7–0.8
	High Cr, 25–30%, Martensitic	0.55–0.8	0.15	0.3	0.8	0.3	0.08–0.6	~0.5
	Ni-hard type (3% C, 4% Ni 2% Cr)	0.6	0.07	0.3	~1.0	0.4	0.08	0.6
Steel	0.4% C, low alloy, ~500 Vickers	1.0	1.0	1.0	1.0	1.0	1.0	1.0
Cast And Rolled Steels	0.8% C, ~800 Vickers	—	—	—	~1.0	0.5	—	0.55
	0.3% C, 0.6% Mn, 1.5% Cr, 0.75% Ni, 0.4% Mo, ~450 Vickers	0.3	—	—	—	—	—	—
	0.2% C, 1.2% Mn, 1.3% Cr, 0.25% Mo, 350 Vickers	1.2	0.85	—	—	—	—	—
	2% C, 12% Cr, ~700 Vickers	—	—	—	0.9	0.5	—	0.55
	1% C, 6% Mn, Cr/Mo, austenitic	—	—	—	~1.0	—	0.3–1.3	—
	1% C, 12–14% Mn, austenitic	0.7–1.1	—	1.1	1.2	0.9	0.35–1.4	0.8
Hard Facings (*See also* I.D-MSC) and Table 25.20	3–5% C, 20–30% Cr, Co/Mo/V/W/B Mn/Nl ferrous alloys, manual arc deposited	0.45–0.8	0.09	0.6	—	0.25–0.4	0.7	0.25–0.7
	Tungsten carbide/ferrous matrix, arc or gas tubular rods	—	0.2	—	—	~0.3	—	0.35
	3.5% C, 33% Cr, 13% W, Co alloy	—	—	—	—	0.2–0.5	—	0.45–0.8
	1% C, 1% Fe, 26% Cr, 4% Si, 3.5% B, Ni alloy	—	—	—	—	~0.3	—	0.85

* The Sinter was produced from foreign ore with ASTM $\frac{1}{4}$-strength index of about 47.
Reproduced by courtesy of Fulmer Research Institute Ltd.

OR ABRASIVE WEAR RESISTANCE[22–30]

ase and onvenience f eplacement	*Typical fields of application*	*Remarks*
sually convenient with good esign to facilitate replacement	Cast irons are very suitable materials to resist medium to high stress abrasive wear due to their good wear resistance and reasonable cost. At very severe levels of impact abrasion, however, inadequate toughness can be a problem and only materials of the work-hardening type should be used. Also cheaper materials may be preferred due to the excessively high wear rates involved	These materials have the merit that a combination of strength i.e. toughness and hardness, may be readily obtained by varying the alloying method of manufacture, and treatment; thus giving suitable combinations of these properties to suit a particular application and wear situation. Various techniques of surface hardening can also be employed to improve resistance to abrasive types of wear. Other products are sintered metals and metal coatings, e.g. Cr plate and sprayed coatings
sually convenient with good esign to facilitate replacement	Due to the very large quantity production involved, steels tend to be comparatively cheap. Thus steels with low wear rates become a competitive materials choice. Their main application lies in hardened steels to resist medium stress abrasion as very low wear rates can be obtained. Austenitic manganese steels can be used in more severe situations due to their work-hardening capability	
eplacement can be difficult if plied *in situ*. These materials e often chosen because hard eld may be built up and orn away several times to its tal depth under severe wear tuations	For medium and high stress abrasion hardfacings give low wear rates generally, and so are used in many situations to resist abrasive wear, e.g. excavator teeth and other earth moving applications	

Table 25.18b COMMONLY USED MATERIALS FOR ABRASIVE WEAR RESISTANCE

Type of material	Typical commercially available materials	Wear rates relative to 0.4%C low alloy steel quenched and tempered to about 500 vickers hardness: Sliding wear by coke	Sliding wear by blast furnace sinter	Impact wear by blast furnace sinter	Wear on commercial bonded 384 μm flint abrasive	Ease and convenience of replacment	Typical fields of application	Remarks
Ceramics	Fusion cast 50% Al_2O_3, 32% ZrO_2, 16% SiO_2	0.1–0.2	~0.2	0.6	—	Convenient if ceramic is bolted in place. Less convenient if ceramic is fixed by adhesive or cement as long curing times may lead to unacceptably long down-times	Possible to achieve very high hardness but brittleness tends to be a problem. Most suitable to resist low stress abrasion by low density materials and powders	—
	Sintered 95–99% Al_2O_3	—	~0.2	~0.7	0.04–0.3			
	Reaction Bonded SiC	—	6.9	—	~0.02			
	Cast basalt	0.9	0.9	—	~3			
	Tungsten carbide/6% Co	0.07	—	—	0.007			
Glass	Plate glass	4.5	—	—	22	Used in sheet form where transparency is required	Glass is brittle and so it is only used at the lowest levels of abrasive wear	—
Concretes	Aluminous cement-based concrete with proprietary aggregates	3.5	15	—	—	Long curing times can lead to unacceptably long down-times. Can be messy and difficult under dirty conditions	Useful to resist wear of irregularly shaped components and when abrasion is of low to medium stress	Also useful in large flat areas, especially when curing time is no real problem, e.g. aircraft hanger flooring, etc.
	As above with 2% by volume 25 × 0.4 mm diam, wire fibres	—	—	2	—			
	Concrete tile — 6 mm wear resistant surface	—	6.7	—	—			Easily castable
Elastomers	Wear resistant rubbers, 55°–70° shore hardness	7.8	—	—	—	Bonded and bolted. Sticking with adhesive can be difficult under dirty conditions	Very useful to resist impact abrasion – most wear resistant at 90° impact angles. Softer types of rubber are used for low stress impact abrasion. Resilient rubber for more severe impact	Bonding of rubber to component is a very large problem in high stress abrasive wear. Good anti-sticking properties and low density
	65° shore hardness rubber with saw tooth surface profile	15.1	—	—	—			
Plastics	Polyurethane	18.5	2.7	—	—	Usually used in sheet form. Difficult to bond plastic to component. Solid moulded components are superior but are limited to small sizes	Low coefficient of friction, good anti-sticking properties. Best for low stress abrasion by fine particles. Resin bonded aggregates are trowellable and so are useful to resist wear or irregularly shaped components	Composite plastics are only as tough as their bonding matrix and therefore find more applications where low stress abrasion by powders or small particles takes place
	High density polyethylene	15.5–31	—	—	—			
	Epoxy resin based PTFE	40	—	—	—			
	Calcined bauxite filled epoxy resin	11	—	—	—			

Reproduced by courtesy of Fulmer Research Institute Ltd.

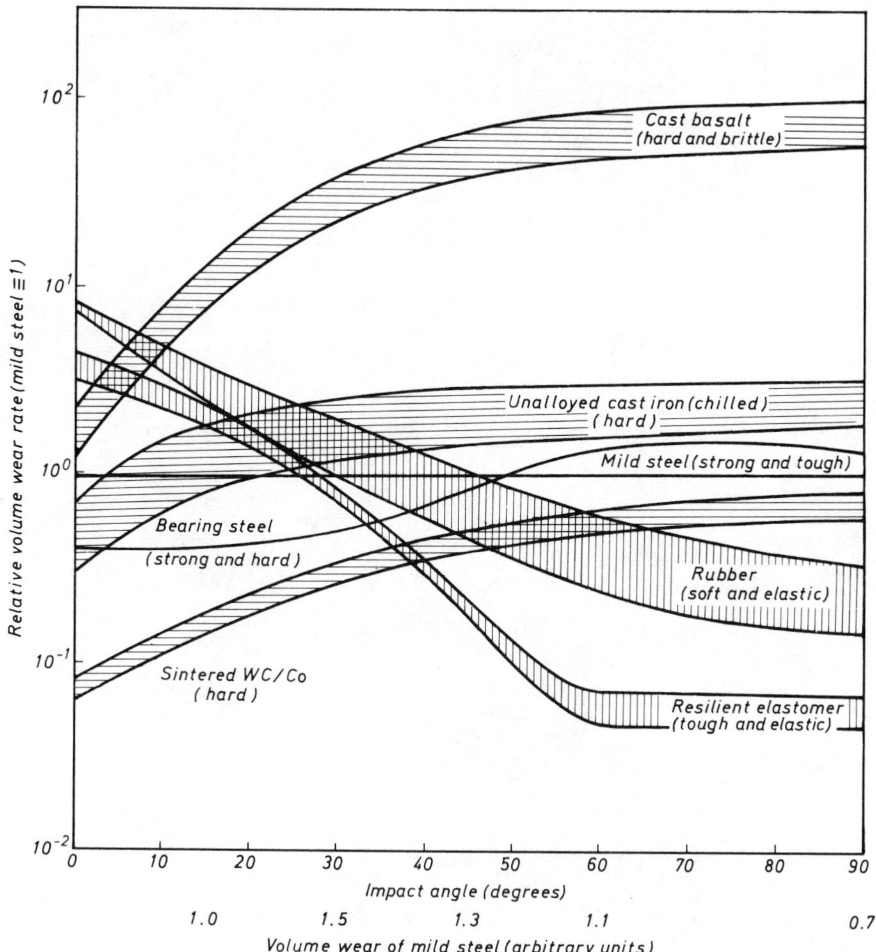

Figure 25.3 *Effect of impact angle on erosion wear of materials impacted with dry 0.2–1.5* mn *quartz.*[36] (*Reproduced from The Fulmer Materials Optimizer by permission of Fulmer Research Institute Ltd.*).

The performance of materials in erosion by sandy water and in pneumatic conveying are given in Table 25.26.

In cavitation, vapour bubbles formed at low pressure collapse in high pressure regions. Cavitation erosion is wear resulting from localized high impact stresses when bubbles collapse at or close to a surface. The cavitation erosion resistance of a range of materials is given in Table 25.27.

25.2.4 Fretting wear

Fretting wear occurs when two contacting surfaces are subject to very small oscillatory slip (of no more than 150 μm). Damage occurs when oxide films are disrupted locally, and may proceed by continuous formation and removal of the oxide, by the abrasive action of the oxide or by localized formation and failure of metal-to-metal adhesive bonds. The rate of fretting wear is normally very low—about 0.1 mg per 10^6 cycles, per MN m^{-2} normal load, per μm amplitude of slip for mild steel. However, localized cyclic stresses may enhance fatigue crack initiation, causing up to 80% reduction in fatigue strength.

Fretting damage is reduced by eliminating slip (by increasing the contact pressure or separating the surfaces) by lubrication (to separate surfaces and wash away debris) and by surface treatments such as electrodeposits of soft metals or chemical conversion coatings of phosphate and sulphidized coatings on steels and anodized coatings on aluminium alloys.

Table 25.19 COMPARISONS OF RELATIVE WEAR RATES OF FERROUS MATERIALS[22,26,31]

		Wear rates relative to 0.4% C, $1\frac{1}{2}$% Ni/Cr/Mo steel, quenched and tempered to 500 kg mm^{-2} (Vickers)												
		Practical wear environments							*Laboratory wear environments*					
		Agricultural soils			*Quartz/feldspar Mo ores*				*Commercial bonded abrasive discs*				*Rubber wheel*	
Material	*Vickers hardness* kg mm^{-2}	*Pumice*	*Stone free sand*	*Iron-stone loam/sand*	*Ball mill*	*Slusher scraper*	*Screen rods*	*Mine car wheels*	84 μm *Corundum* 1 MN m^{-2}	84 μm *Flint* 1 MN m^{-2}	384 μm *Flint* 1 MN m^{-2}	84 μm *Glass* 1 MN m^{-2}	*Dry quartz sand low stress*	*Wet quartz sand high stress*
0.4% C, $1\frac{1}{2}$% Ni/Cr/Mo steel	500	1.0	1.0	1.0	1.0	1.0	1.0	1.0	1.0	1.0	1.0	1.0	1.0	1.0
0.4% C steel	500	—	—	0.95	—	—	—	—	1.0	0.93	—	0.76	—	—
0.8% C steel	800	0.06	0.43	0.57	—	—	—	—	0.65	0.49	0.56	<0.01	—	—
0.95% C steel	550	—	—	—	—	—	0.77	—	—	0.85	—	0.34	0.81	0.95
2% C, 12% Cr steel	700	—	—	0.52	0.91	—	—	—	0.57	0.09	0.56	—	—	—
1% C, 12% Mn steel	210	0.40	0.83	0.92	1.2	0.83	—	0.59–0.83	0.72	0.63	0.79	0.07	—	—
18/8 Cr/Ni stainless	150	—	—	1.9	—	—	—	—	0.92	0.91	—	1.7	—	—
3% C chilled iron	600	—	—	0.43	1.7	—	—	2.0	0.65	0.23	0.63	<0.01	0.29	1.1
3% C, 30% Cr white iron	700	0.06	0.10	0.44	0.83	—	—	—	0.47	<0.01	0.44	—	0.07	0.51
15% Cr, 3% Mo white iron	900	—	—	—	0.77	—	—	—	—	0.26	0.67	—	—	—
Ni-hard type iron	700	—	—	0.58	1.0	—	—	—	0.66	0.17	0.67	—	0.17	0.47

Reproduced by courtesy of Fulmer Research Institute Ltd.

Table 25.20 RELATIVE WEAR RATES OF HARDFACINGS[29,31]

Material	Flint clay soil	Commercial bonded abrasives: 34 μm flint 1 MN m^{-2}	Commercial bonded abrasives: 384 μm flint 1 MN m^{-2}	Rubber wheel test: Dry and low stress	Rubber wheel test: Wet sand high stress
	Wear rates relative to 0.4% C, 1½% Ni/Cr/Mo steel at 500 kg mm^2 Vickers hardness				
Tubular Fe/70% tungsten carbide, arc weld	0.29	0.04	0.35	—	—
Ni alloy/40% tungsten carbide, fusion spray	0.18	0.04	0.26	—	—
3.5%, 33% Cr austenitic iron, arc weld	0.24	0.08	0.59	~0.05	~0.95
High C/Cr martensitic iron arc weld	0.44	0.10	0.42	~0.05	~0.87
0.8% C, 3% Ni, 5% Cr, 12% Mn austenitic steel, arc weld	1.01	0.78	0.89	—	—
0.9% C, 4.5% Cr, 7.5% Mo, 1.6% V, 2% W, 1.5% Si, 1.3% Mn martensitic steel, arc weld	0.59	0.79	0.74	~0.29	~0.70
0.95% C, 26% Cr, 4% Si, 3.5% B Ni alloy, fusion spray	0.32	0.12	0.85	—	—
3.5% C, 33% Cr, 13% W Co alloy, gas weld	0.48	0.07	0.78	~0.08	~0.63
97.6% Al_2O_3, 2.5% TiO_2 plasma spray	—	1.64	4.33	—	—
82% Cr, 18% B paste, fused to substrate by gas weld	—	<0.01	0.03	—	—

Reproduced by courtesy of Fulmer Research Institute Ltd.

Table 25.21 RELATIVE WEAR RATES OF CERAMICS[23,32]

Material	Vickers hardness GN m^{-2}	Fracture toughness K_c MN m$^{-3/2}$	Commercial bonded flint abrasives: 84 μm flint 1 MN m^{-2}	384 μm flint 1 MN m^{-2}	84 μm corundum 1 MN m^{-2}	84 μm SiC 1 MN m^{-2}	Diamond sawing
			Wear rates relative to a sintered 95% alumina (bracketed figures relative to 0.4% C, 1½% Ni/Co/Mo steel at 500 kg mm^{-2} Vickers hardness)				
95% sintered Al_2O_3	12–13	4–6	1.0 (0.024)	1.0 (0.099)	1.0 (0.39)	1.0 (0.81)	1.0
Hot pressed Si_3N_4	~16	5–9	0.32	0.11	—	0.56	1.1
Reaction bonded Si_3N_4	~7	4–5	3.8	8.8	—	2.3	—
Reaction bonded SiC	16–20	7–9	0.15	0.11	—	0.41	—
Hot pressed B_4C	~30	6–9.5	0.21	0.06	—	0.04	0.65
Hot pressed Al_2O_3	14–20	5–7.5	0.3	0.15	—	0.27	—
97.5% sintered Al_2O_3	~15	~8	1.0	0.3	—	0.58	—
99.7% sintered Al_2O_3	~12	~4	1.8	1.4	—	1.2	—
Cast basalt	~5	~2	26	26	—	3	—
Soda-lime glass	4–5	2–4	84	168	100	4.5	—
Sintered TiO_2	~7	~2.5	16	12	—	4.5	—
ZrO_2	15	~2.5	—	—	—	—	1.3
Spinel	16	~1.7	—	—	—	—	2.9
MgO	7	~2.2	—	—	—	—	5.6

Reproduced by courtesy of Fulmer Research Institute Ltd.

Table 25.22 RELATIVE WEAR RATES OF CARBIDE COMPOSITES[23,33–35]

Composition WC%	TiC+TaC%	*Binder%*	*Carbide grain size* μm	*Vickers hardness* $GN\,m^{-2}$	*Fracture toughness* K_c $MN\,m^{-3/2}$	*Wear rates relative to 4–5% Co composites* — *Rock drilling*: *Percussive, granite*	*Rock drilling*: *Rotary sandstone*	*Laboratory wear tests* — *Loose abrasive water slurry*: 0.2–0.5 mm Al_2O_3	*Loose abrasive water slurry*: 0.1–0.5 mm SiC/Al_2O_3	*Commercial bonded abrasives*: 84 μm *flint* $1\,MN\,m^{-2}$	*Commercial bonded abrasives*: 384 μm *flint* $1\,MN\,m^{-2}$	*Commercial bonded abrasives*: 84 μm SiC 1 MN
96		4 Co	1–2	~18	~7	(1.0)	1.0	1.0				
96		4 Co	2–3	~16	~10		2.5					
95.5	0.5	4 Co		~17					1.0			
93	2	5 Co	1–2	~14	~7					1.0	1.0	1.0
94		6 Co	1–2	~16	~10	~1.7	1.7	2.0	~1.7			
94		6 Co	2–3	~15	~13		5.0	2.5		1.6	1.4	1.6
93	1	6 Co	2–3	~13	~7					2.0	1.1	1.5
92	2	6 Co		~16.5					1.9			
72	22	6 Co		~17					5.3			
93		7 Co		~14.5					3.3–4.2			
90	3	7 Co		~17					0.5			
81.5	12	7.5 Co		~16.5					3.5			
92		8 Co	2–3	~15	~12	5.0				2.6	1.9	1.3
91		9 Co	1–2	~15	~9	4.0	3.3	5.0				
91		9 Co	2–3	~13	~13		6.7	10.0				
84.5	6.5	9 Co		~15					6.3			
71	20	9 Co		~16					6.3			
50	40	10 Co		~17					7.3			
89		11 Co		~12		10.0			6.0–7.0			
87		13 Co		~11					8.7			
86	1	13 Co		~13					7.9			
85		15 Co	1–2	~11	18		10.0		~11.0			
85		15 Co	2–3	~11	19		25.0	10.0				
69	15	16 Co		~13					13.0			
80		20 Co		~10.5					18.0			
	80	10 Mo/10Ni		~17					9.0			
75		25 Co		~9					20.0			
70		30 Co		~8.5					25.0			
0.4% C 1½Ni/Cr/Mo steel				5					850–1 500	530	150	6.3
2% C, 12–14% Cr tool steel				~7					140	48	84	~4

Reproduced by courtesy of Fulmer Research Institute Ltd.

Table 25.23 WEAR RATES OF SOME COMMON ENGINEERING MATERIALS IN UNLUBRICATED SLIDING AT 1.8 $m\,s^{-1}$ AND A LOAD OF 400 g

Material	*Wear rate* $mm^3\,mm^{-1}$
Mild steel on itself	1.57×10^{-11}
60/40 leaded brass on hardened tool steel	2.4×10^{-12}
PTFE on hardened tool steel	2.0×10^{-13}
Stellite on hardened tool steel	3.2×10^{-14}
Ferritic stainless steel on hardened tool steel	2.7×10^{-14}
Polyethylene on hardened tool steel	3.0×10^{-15}
Tungsten carbide composite on itself	2.0×10^{-16}

Table 25.24 WEAR PERFORMANCE OF COATED AND UNCOATED FERROUS MATERIALS

		Rolling slip (Amsler)						*Sliding*									
		without impact				*with impact*		*Lubricated* 0.1 $m s^{-1}$		*Unlubricated* *Low Load*							
		Lubricated	*Unlubricated*			*Lubricated*											
		200 N *Load* *V. Itself*	50 N *Load* *V. Itself*	500 N *Load* *V. Itself*	500 N *Load* *V. Normalized steel*	500 N *Load* *V. Itself*	2000 N *Load* *V. Itself*	*Medium Load* *V. Itself*	*High Load* *V. Itself*	0.1 $m s^{-1}$ *V. Itself*	0.1 $m s^{-1}$ *V. Normalized steel*	1.5 $m s^{-1}$ *V. Itself*	1.5 $m s^{-1}$ *V. Normalized steel*	10 $m s^{-1}$ *V. Itself*	10 $m s^{-1}$ *V. Normalized steel*
Normalized steels	0.15% C	2													
	0.45% C	4	3	2						4 / 4		3 / 3		6 / 5	
	0.9% C	5													
Hardened steels	0.45% C		4	2	3			7	5	5				6 / 5	
	0.9% C	2													
Pearlitic cast iron		5													
Low carbon steel	Carbo-nitrided	3													
	Carburized	2	4			6	5					4 / 5	4 / 4	6 / 4	
	Nitro-Carburized		4	1	3							4 / 4	4 / 4	5 / 5	5 / 5

Table 25.24 WEAR PERFORMANCE OF COATED AND UNCOATED FERROUS MATERIALS—*continued*

		Rolling slip (Amsler)						*Sliding*									
		without impact				*with impact*		*Lubricated* 0.1 m s^{-1}		*Unlubricated* Low Load							
		Lub-ricated	*Unlubricated*			*Lubricated*											
		200 N *Load* *V. Itself*	50 N *Load* *V. Itself*	500 N *Load* *V. Itself*	500 N *Load* *V. Normalized steel*	500 N *Load* *V. Itself*	2000 N *Load* *V. Itself*	*Medium Load* *V. Itself*	*High Load* *V. Itself*	0.1 m s^{-1} *V. Itself*	0.1 m s^{-1} *V. Normalized steel*	1.5 m s^{-1} *V. Itself*	1.5 m s^{-1} *V. Normalized steel*	10 m s^{-1} *V. Itself*	10 m s^{-1} *V. Normalized steel*
Low carbon steel–*cont.*	Nitrided	3								5					
Alloy steels	Nitrided					5	5	7	5						
	Boronized		4	1	3	4	2	8	6	6		5 / 6	4 / 4	5 / 6	6 / 6
	Vanadized					5	4		5	6					
Surface treated normalized medium carbon steel	Tufftride		4	1	2						4 / 4	4 / 4	4 / 4	5 / 5	5 / 5
	Sulfinuz	6	2	1	1						4 / 4	4 / 4	4 / 5	5 / 5	5 / 5
	Noskuff	5	4	3	3						4 / 4	4 / 6	4 / 6	4 / 6	4 / 5
	Sulph BT	5	4	1	1						4 / 4		3 / 3	5 / 5	4 / 5

Table 25.24 WEAR PERFORMANCE OF COATED AND UNCOATED FERROUS MATERIALS—*continued*

		Rolling slip (Amsler)						*Sliding*								
			without impact			*with impact*		*Lubricated 0.1 m s^{-1}*		*Unlubricated Low Load*						
		Lub-ricated	*Unlubricated*			*Lubricated*										
		200 N Load *V. Itself*	50 N Load *V. Itself*	500 N Load *V. Itself*	500 N Load *V. Normalized steel*	500 N Load *V. Itself*	2000 N Load *V. Itself*	*Medium Load V. Itself*	*High Load V. Itself*	0.1 m s^{-1} *V. Itself*	0.1 m s^{-1} *V. Normalized steel*	1.5 m s^{-1} *V. Itself*	1.5 m s^{-1} *V. Normalized steel*	10 m s^{-1} *V. Itself*	10 m s^{-1} *V. Normalized steel*
Surface treated normalized medium carbon steel–*cont.*	Sursulf		3	1	2						A 4 / B 4	A 5 / B 4	A 4 / B 4	A 5 / B 5	A 6 / B 5
	Stanal		2	1	1						A 4 / B 4	A 4 / B 4	A 5 / B 4	A 5 / B 4	A 6 / B 4
	Forez		3	2	1							A 4 / B 4	A 5 / B 4	A 6 / B 4	A 6 / B 4
	Phosphate	2													

Key: Numerals represent orders of magnitude of wear, i.e. $1 \equiv 10^{-1}$ to 9×10^{-1}, $6 \equiv 10^{-6}$ to 9×10^{-6} in units of volume (mm^3) distance travelled (m). For sliding A (upper right) / B (lower left) of diagonal: A ≡ sweeping element. B ≡ swept element.

*Normalized 0.45% C steel sweeping element.

Table 25.25 WEAR IN UNLUBRICATED SLIDING (ROTATIONAL) (0.4 m s^{-1}, 300 kN m^{-2} APPLIED LOAD)

Swept element		Sweeping Element: *Steel annealed*	*Grey cast iron*	*0.15% C Steel carburized*	*Tin bronze*	*Brass*
0.15% C steel	Normalized	3.4 / 3.4				
0.15% C steel	Carburized	2.4 / 2.3	2.3 / 1.5		1.6 / 1.8	>4.0 / 1.0
0.15% C steel	Carburized plus sulph BT	2.4 / 1.6		2.2 / 1.9		
0.15% C steel	Carburized plus hard chrome		2.0 / 1.0		2.0 / 1.0	
0.15% C steel	Nitrided	2.2 / 2	1.0 / 1.9	1.3 / 1.8	0.3 / 1.6	>4.0 / 1.7
0.15% C steel	Boronized	2.4 / 1.0	2.0 / 0.7	1.4 / 1.4	2.1 / 0.8	>4.0 / 1.0
Tool steel	Hardened	2.5 / 2		2.2 / 1.3		
Tool steel	Titanium carbide (CVD)	1.5 / 0.9	1.5 / 1.2	0.9 / 0.9		
Mo flame sprayed				0.7 / 0.7	1.8 / 0.7	1.9 / 1.0
Al_2O_3 plasma sprayed					2.8 / 1.8	2.1 / 1.0
WC/Co plasma sprayed		1.5 / 1.7	1.3 / 1.0	0.7 / 0.7	1.1 / 0	3.5 / 1.5

Key: Numerical values ≡ LOG_{10} relative wear rate.

(Cell: A upper right, B lower left)

A ≡ wear of sweeping element.

B ≡ wear of swept element.

Table 25.26 RELATIVE RESISTANCE OF VARIOUS MATERIALS TO EROSIVE WEAR[37,38]

Materials	*Relative erosive wear resistance* — *In sandy water*	*Relative erosive wear resistance* — *Pneumatic conveying of minerals*
Ceramics	2–150	—
Aluminas	—	6.5–10
Weld overlays (hard facings)	1.1–20	—
Cast irons	0.5–6	1.2–10
Quenched and tempered steels	—	1.9–5
Rolled and forged steels	1–6	—
Mild steel	1	1
Cast steels	1–2.5	—
Titanium	1	—
Wrought copper alloys	0.3–1	—
Cast copper alloys	0.4–0.9	—
Aluminium alloys	0.1–0.5	—
Plastics	0.04–0.3	—
Elastomers	0.04–0.09	0.12–0.45
Hard Concrete	—	0.11–0.16

Table 25.27 RESISTANCE OF VARIOUS MATERIALS TO CAVITATION EROSION

Range of materials	*Typically used materials* — *Material type*	*Typically used materials* — *Specification*	*High intensity relative cavitation erosion resistance (Aluminium bronze = 1)*	*Remarks*
Cast bronzes	Aluminium bronze	BS1400 AB2	1.3	Aluminium bronze also has good corrosion resistance in sea water, but is more difficult to cast than Novoston bronze
	Aluminium bronze	BS1400 AB1	1.0 (standard)	
	Novoston bronze	BS1400 CMA1	0.5	
	Gunmetal	BS1400 G1	0.22	
	High tensile brass	BS1400 HTB1	0.18	
	Phosphor bronze	BS1400 PB3	0.15	
	Nickel gunmetal	—	0.15	
	Leaded gunmetal	BS1400 LG3	0.14	
Wrought bronzes	Manganese bronze	—	0.2	—
	Everdur A	BS2872 CZ101	0.16	
	Aluminium brass	BS2872 CZ110	0.15	
	Free turning brass	BS2872 CZ121	0.14	
	Naval brass	BS2872 CZ112	0.12	
	Hot stamping brass	BS2872 CZ122	0.11	
	Tellurium copper	BS2872 C 109	0.03	
Nickel alloys	Monel K-500 (aged)	BS3076 NA18	0.7	—
	S Monel	BS3071 NA3	0.5	
	Monel K-500 (cold drawn)	BS3076 NA18	0.3	
	Nimonic 75	DTD 703A	0.22	
	Nickel (hard)	BS3076 NA11	0.21	
	Nickel (annealed)	BS3076 NA11	0.2	
Wrought aluminium alloys	Al 4% Cu solution treated and aged	BS1476 HE15	0.08	
	Al 5% Mg As rolled	BS1476 NE6M	0.04	Low density
	Al 3.5% Mg As rolled	BS1476 NE5M	0.03	
Titanium alloys	Ti-2.25 Al-4 Mo-0.255C-11Sn	DTD 5213	1.0	Excellent corrosion resistance
	Ti CP with medium interstitial content	DTD 5273	0.7	Low density. High cost
Cast irons and cast steels	Martensitic stainless	Hardened	1.8	Stainless steels are reasonably corrosion resistant. Martensitic stainless steels are less corrosion resistant than austenitic stainless steels
	Austenitic stainless	Annealed	0.8	
	Ferritic stainless	Annealed	0.77	
	SG cast iron	Tempered	0.7	
	Martensitic stainless	Tempered	0.5	
	Alloy cast iron (D2 Ni-resist)		0.2	

Table 25.27 RESISTANCE OF VARIOUS MATERIALS TO CAVITATION EROSION —*continued*

Range of materials	*Typically used materials* — *Material type*	*Specification*	*High intensity relative cavitation erosion resistance (Aluminium bronze = 1)*	*Remarks*
Wrought steels	HT low alloy	Hardened En24	2.5	
	Martensitic stainless	Tempered En57	0.67	
	Martensitic stainless	Tempered En56	0.62	
	Austenitic stainless	Annealed En58J	0.5	
	Mild steel	En1A	0.2	
Stellites (cobalt alloys)	Stellite 12 (cast) – 29% Cr 9%W 1%C		19.3	Most resistant to high cavitation erosion
	Stellite 4 (cast) – 31% Cr 14%W 1%C		15.0	
	Stellite 6 (cast) – 26% Cr 5%W 1%C		5.3	
	Stellite 6 (wrought)		3.4	
	Stellite 7 (cast) – 26% Cr 6%W 0.4%C		2.5	
Plastics	Nylons, acetals, high impact polyethylenes		1.7–3.0	Values of cavitation erosion resistance fall sharply at elevated temperatures. Generally, plastics with high impact strengths have good cavitation erosion resistance. Very suitable for mass manufacture of small components, e.g. boat propellers. Corrosion resistant
	Polyethylenes, Polypropylenes		0.9–1.6	
	PVC, PTFE		0.15–3.5	
	Chlorinated polyether		0.04–0.1	
	Perspex		0.01–0.035	
Rubbers	Bonded rubber coatings, Neoprene		–	Rubber requires careful preparation and application. Rubber gives satisfactory resistance to mild cavitation erosion, and where relative fluid velocities are low. Failure usually occurs at bond to components at higher fluid velocities

Reproduced by courtesy of Fulmer Research Institute Ltd.

REFERENCES

1. F. P. Bowden and D. Tabor, 'Friction and Lubrication of Solids', Pt. I, revised reprint, Clarendon Press, Oxford, 1954.
2. J. A. Greenwood and J. B. P. Williamson, *Proc. R. Soc.*, 1966, **A295**, 300.
3. D. Tabor, Review article on 'Friction, Lubrication and Wear' in *Surface and Colloid Science* (Ed. Egon Matijevic), **5**, 245–312, Wiley-Interscience, New York, 1972.
4. F. P. Bowden and D. Tabor, 'Friction and Lubrication of Solids', Pt. II, Clarendon Press, Oxford, 1964.
5. F. P. Bowden and A. E. Hanwell, *Nature, Lond.*, 1964, **201**, 1279.
6. S. C. Cohen and D. Tabor, *Proc. R. Soc.*, 1966, **A291**, 186.
7. B. J. Briscoe, V. Mustafaev and D. Tabor, *Wear*, 1972, **19**, 389.
8. Mrs. C. M. Pooley and D. Tabor, *Proc. R. Soc.*, 1972, **A329**, 251–274.
9. W. G. Beare and F. P. Bowden, *Phil. Trans.*, 1935, **A234**, 329.
10. Sir W. B. Hardy, 'Collected Scientific Papers', 1936, Cambridge University Press, 1936.
11. A. W. Crook, *Phil. Trans.*, 1958, **A250**, 387; *ibid.*, 1961, **A254**, 237.
12. J. F. Archard and M. T. Kirk, *Proc. R. Soc.*, 1961, **A261**, 532.
13. D. Dowson and G. R. Higginson, 'Elastohydrodynamic Lubrication', Pergamon Press, Oxford, 1966.
14. A. Cameron, 'Basic Lubrication Theory', Longman, Harlow, 1971.
15. W. Hirst and J. V. Stafford, *Proc. Inst. mech. Engrs*, 1972, **186**, 179.
16. H. Naylor, Private Communication.

17. J. F. Archard, *J. appl. Phys.*, 1953, **24**, 981.
18. M. B. Peterson and S. F. Murray, *Metals Eng. Quart.*, 1967, May 22.
19. E. F. Finkin, J. Calabrese and M. B. Peterson, *J. Amer. Soc. Lubr. Engrs.*, 1973, **29**, 197.
20. B. V. Elkonin, *Wear*, 1980, **61**, 169.
21. G. K. Nathan and W. J. D. Jones, *Proc. I. Mech. E.*, 1966/67, **181**, 215.
22. R. C. D. Richardson, *Wear*, 1968, **11**, 245.
23. M. A. Moore and F. S. King, *Wear*, 1980, **60**, 123.
24. H. Hocke, *Iron & Steel Int.*, 1977, Dec., 361.
25. F. Borik and L. W. G. Scholz, *J. Materials*, 1971, **6**, 590.
26. T. E. Norman and E. R. Hall, ASTM STP No. 446, 1969.
27. W. Fairhurst and K. Rohrig, *Foundry Trade J.*, 1974.
28. R. C. D. Richardson, *J. Agric. Engng. Res.*, 1967, **12**, 22.
29. M. A. Moore, *J. Agric. Engng. Res.*, 1975, **20**, 167.
30. G. G. Brown and J. D. Watson, *East Asian Iron & Steel Conf.*, 1977.
31. H. S. Avery, SAE off-Highway Vehicle Meeting, Milwaukee, 1975.
32. A. G. Evans and T. R. Wilshaw, *Acta Met.*, 1976, **24**, 939.
33. E. Cuboni, *Metallurgia Italiana*, 1969, **12**, 593.
34. J. Larsen-Badse, *Powder Met.*, 1973, **16**(31), 1.
35. H. Feld and P. Walter, *Z. Werkstoffkde.*, 1976, **7**, 300.
36. K. Wellinger and H. Uetz, *Jernkont. Ann.*, 1963, **147**, 845.
37. W. A. Stauffer, *Schinzer Archiv. für Angewandte Wissenschaft und Technik*, 1958, **24**, 3.
38. E. Olsen, *Bulk-Storage Movement Control*, 1976, Jan/Feb., 48.

26 Casting alloys and foundry data

26.1 Casting techniques*

Table 26.1 TECHNIQUES USING AN EXPENDABLE MOULD AND PATTERN

	Investment casting				*Full mould*
Equivalent terminology	Lost wax casting Precision casting				Cavity-less casting
Pattern material	Metal, Rubber, Plastic } die Wax, thermoplastics, frozen mercury				Foamed polystyrene or polyurethane
Mould material	Silica base ceramic Slurry + stucco				Silica sand
Binder	Ethyl silicate Magnesium or aluminium phosphate				Usually Na_2SiO_3 or resin (Furane) Can be used unbonded
Size range (kg)	Up to 4.5 except frozen Hg for which up to 45				Up to 5 000 in iron
Scope	Complex shapes and fine detail				One-off castings, e.g. jigs and press tools, complex shapes
Alloys cast	High temp alloys	Steel	Cu base	Al base	Mainly grey iron castings
Surface finish (CLA)	25	25–125	50–85	50–125	No better than 300
Minimum section for length					
25 mm	1.5	1.0	1.5	12	
150 mm	1.5	1.5	1.5	12	
Dimensional tolerance† ±mm					
25 mm	0.13	0.13	0.13	0.6	
150 mm	0.38	0.25	0.18	2.5	
Labour cost per kg	High				Low
Equipment cost per kg	High				Low
Mould material cost	High				Polystyrene pattern approximate cost one third that of wood. Low for unbonded. For other *see* Table 26.2.

* Report No. 187 *PERA*, March 1969.
R. A. Flinn, 'Fundamentals of Metal Casting', Addison–Wesley, New York, 1963.
† The following additional tolerances should be added if dimension chosen crosses parting line:
Sand ±0.38 Permanent ±0.38 Centrifugal ±0.25
Shell ±0.25 Plaster ±0.25 Shaw ±0.10 to 0.15
for a 25 mm casting.

Table 26.2 TECHNIQUES USING AN EXPENDABLE MOULD

	Ceramic mould casting	*Plaster mould casting*	*Shell mould casting*			*Sand casting*	
Equivalent terminology	Shaw process Ceramic shell process	'Antioch' process Gypsum moulding	'C' process Cronig process			—	
Pattern material	Wood, plastic, metal	Wood, plastic, metal	Metal			Wood, plastic, metal	
Mould material	Molochite, sillimanite, mullite	Calcium sulphate	Silica or zircon sand			Silica, zircon, chromite, olivine, sand	
Binder	Ethyl silicate plus gelling agent, e.g. 50/50 NH_4OH/ H_2O	Water (autoclave in saturated steam atmosphere at 0.1 N mm^{-2} (15 psi) for 6–8 h-stand at amb. temp 14 h)	Phenol-formaldehyde resin (Novolac) + hexamethylenetetramine + heat (from pattern at *c.* 260 °C)			Many variants, e.g. clay, CO_2-silicate, cement, resin self set, silicate self set, etc.	
Size range (kg)	Usually up to 180	All sizes	Up to 136			Green sand up to 500, dry sand 500 and above, others to 500 and above	
Scope	Complex shapes and fine detail	Limited by pattern withdrawal. Intricate shapes, e.g. wave guides	Better surface finish and accuracy than conventional sand castings			Limited only by pattern work	
Alloys cast	Most types: Mg base, Al base, Cu base, steels	Mainly aluminium alloys and some Mg and Cu base	Carbon and alloy steels. Manganese steel. Some stainless and HR steels	Grey, SG and malleable irons	Al, Cu and Mg base alloys	All common foundry alloys	Steel Cast-irons Cu base Al base
Surface finish (CLA)	120 to 180, with fine zircon facing 80 or better	30–40	100–300	100–250	100	300–1 000	
Minimum section for length							
25mm	2.5 2.5 3.0	Al 1.4 Cu 1.1	4.6	3.2	Al 1.6 Cu 2.3	6.3 3.2	2.3 3.2
150 mm	3.2–4.0 3.2 3.2–4.0	1.4 1.4	6.4	3.5	3.1 3.1	6.3 5.1	3.6 3.6
Dimensional tolerance ± mm							
25mm	0.18 0.18 0.18	0.13 0.13	0.25	0.25	0.13 0.13	1.52 0.76	0.38 0.38
150mm	0.30 0.25 0.25	0.18 0.25	0.76	0.76	0.33 0.33	2.54 1.27	0.38 0.38
Labour cost per kg	High	Medium	Low			Low	
Equipment cost per kg	Moderately high	Moderately high	Medium to high			Low	
Mould material cost	Moderately high	Medium	Medium			Low	

Table 26.3 TECHNIQUES USING A NON-EXPENDABLE MOULD (DIE)

	Permanent mould casting (Gravity-die casting)	*Low-pressure die casting*	*High-Pressure die casting (including 'Acurad')*	*Centrifugal casting*	*Continuous casting*
Equivalent terminology	Gravity-die casting chill casting	—	Pressure-die casting Cold-chamber (Al-base) Hot-chamber (Zn base)	'Spun-cast'	'Semi'-continuous (in non-ferrous)
Pattern material	—	—	—	Conventional wood	—
Mould material	Heat-resisting cast iron (expendable cores sometimes used)	Heat-resisting cast-iron (expendable cores sometimes used)	Heat-resistant steel die	Metal die Conventional sand mould	Water-cooled metal die
Binder	—	—	—	—	—
Size range (kg)	Usually 0.45–25 in aluminium but can be up to 250	Up to 11.5 in Al-alloy	Up to 2.25 in brass, 20 in Mg. 34 in Zn, 45 in Al	Pipes up to 1.2m dia. × 11 m long. Rings up to 15.2m dia.	Casting speeds vary between 0.04 and 0.11 m s^{-1}
Scope	Casting design must allow for removal from die	Complex shapes and pressure tight castings	Restricted internal shapes. Die withdrawal	Mainly grey iron pipes and cylinder liners and SG iron pipes and other symmetrical annular shapes	Semi-finished simple shapes (strip, rod, bar) in non-ferrous and grey iron and steel
Alloys cast	All common non-ferrous foundry alloys + some grey iron	Almost exclusive to Al-alloys but steel, cast in USA using graphite mould	Mainly Zn and Al base but some brass and Mg base, Sn and Pb	Highly alloyed steels, grey and SG irons. Most Cu base	Alloy and plain carbon steels, Al and Cu base
	Zn base Cu base Mg base Cast-iron Al base	Aluminium	Zn base Al base Mg base Cu base		
Surface finish (CLA)	100–250	40–100	40–100	Varies widely according to die used and metal cast. Typical range 100–500	100–200
Minimum section for length				Varies widely according to die and metal cast. Typical range 1.8–7.6	Down to 7.6
25 mm	2.5 2.5 6.4	1.3–6.4	1.3 0.8		
150 mm	4.1 5.0 10.2		2.0 1.5		
Dimensional tolerance* ± mm				Dependent on type of die. Usually in region of ± 0.64	Usually within 0.13–25 mm
25 mm	0.18 0.38 0.76	0.10	0.10 0.10		
150 mm	0.25 0.53 0.76	0.36	2.00 0.51		
Labour cost per kg	Low	Low	Low	Moderately low	Moderate
Equipment cost per kg	High	High	High	High	High

* The following additional tolerances should be added if dimension chosen crosses parting line:
Sand ± 0.38 Parmanent ± 0.38 Centrifugal ± 0.25
Shell ± 0.25 Plaster ± 0.25 Shaw ± 0.10 to 0.15
for a 25 mm casting.

Table 26.4 MOULDING AND CORE-MAKING MATERIALS.† CLASSIFICATION SPECIFICATION AND PROPERTIES

(a) Typical chemical analyses of moulding and coremaking sands

Origin	SiO_2	Al_2O_3	Fe_2O_3	CaO	Na_2O + K_2O	*Loss on ignition*	*AFS GFN**	*AFS clay content*	*Remarks*
Naturally bonded sands									
Dullatur 40	90.20	5.74	0.91	0.07	1.45	1.56	40	16.4	For steel and heavy iron
Levenseat No. 9	82.63	9.28	1.65	0.31	0.22	4.45	55	20.3	For steel
Dursand High Bond	80.98	9.64	2.55	0.22	1.99	3.32	73	15.7	For steel
Weatherill	80.76	8.90	2.06	0.10	3.36	3.18	76	14.0	For steel
Bramcote	87.50	4.41	2.51	1.00	1.49	2.55	75	8.4	For iron and non-ferrous
Pickering	89.70	4.44	1.36	0.08	1.70	2.00	100	21.1	(Yellow) light iron and non-ferrous
Swynnerton	84.10	7.69	1.91	0.11	3.27	1.80	115	11.4	For iron and non-ferrous
Mansfield Red	82.50	4.96	1.27	2.39 [MgO 1.56]	2.30	2.15	137	13.6	Non-ferrous and some iron
Erith loam	83.80	6.92	2.91	0.52 [MgO 0.83]	2.63	1.95	158	13.2	Iron and non-ferrous
Washed Silica Sands									
Biddulph + 36 mesh	97.65	0.38	0.65	0.10	0.03	0.37	24	—	Sub-angular/angular
Arnold No. 19 (dried)	98.76	0.34	trace	0.11	0.30	0.00	29	—	Rounded/sub-angular
Garside dried No. 21	98.01	0.43	0.90	0.17	0.37	0.00	36	—	Sub-angular
	97.91	1.13	0.50	0.11	0.72	0.21	44	—	Sub-angular
Erith silica	98.97	0.03	0.65	0.05	0.04	0.06	56	—	Sub-angular
Kings Lynn 60 (4F)	98.50	0.78	0.19	0.03	0.35	0.30	66	—	Sub-angular
Redhill 65 (F)	99.50	0.14	0.06	0.00	0.08	0.14	66	—	Sub-angular
New Windsor Rose	97.00	1.50	0.54	0.22	0.93	0.45	68	—	Sub-angular
Chelford Fine (95)	94.70	2.73	0.20	0.10	1.76	0.40	98	—	Sub-angular
Kings Lynn SS (100)	98.40	0.82	0.15	0.03	0.37	0.25	107	—	Sub-angular
Redhill H (110)	99.00	0.27	0.12	<0.01	0.04	0.35	105	—	Sub-angular/angular
Ryarsh	97.50	0.87	0.40	0.03	0.31	0.56	129	—	Sub-angular/angular
Non-siliceous sands									
		ZrO_2		TiO_2					
Zircoruf	33.19	64.90	0.14	0.64	—	0.16	70	—	Zirconium silicate
AMA zircon	32.00	63.40	2.40	0.08	—	0.23	110	—	Zirconium silicate
		Al_2O_3		CaO	MgO				
Olivine No. 2	41.35	0.60	6.25	1.51	48.75	0.33	59	—	Forsterite/Fayalite
Olivine No. 3½	41.76	0.80	6.15	0.84	49.40	0.43	96	—	Forsterite/Fayalite
				Cr_2O_3					
FW chromite	3.10	15.60	26.50	42.00	10.40	under nitrogen 0.23		—	Chromite and other spinels
Fine grade chromite	3.01	12.20	21.50	49.50	12.30	under nitrogen 1.35	111	—	Chromite and other spinels

(b) Typical mechanical analyses of moulding and core-making sands

Origin	*Wt.% Retained on British Standard Sieve Mesh No. ‡ (typical)* 16	22	30	44	60	100	150	200	*Thro'* 200	*Specific § surface* $cm^2\ g^{-1}$
Naturally bonded sands										
Dullatur 40	—	31.8	23.2	16.7	10.3	8.4	3.6	2.0	4.0	—
Levenseat No. 9	0.9	5.3	12.0	15.1	13.9	23.4	5.1	1.4	2.4	122
Dursand High Bond	22.9	2.8	5.4	8.8	13.8	25.1	9.8	8.8	6.0	190

*For comparison purposes only.
†'Data Sheets on Moulding Materials', SCRATA, 3rd Ed. 1972.
‡ For aperture sizes *see* BS410, 1969–*see* Introductory Tables, Table 2.9.
§ Included to give some indication of binder requirement, particularly liquid types.

Table 26.4 MOULDING AND CORE-MAKING MATERIALS:† CLASSIFICATION SPECIFICATION AND PROPERTIES—*continued*

(b) Typical mechanical analyses of moulding and core-making sands—*continued*

Origin	*Wt.% Retained on British Standard Sieve Mesh No.* ‡ *(typical)*									*Specific*§ *surface* $cm^2\ g^{-1}$
	16	22	30	44	60	100	150	200	*Thro'* 200	
Weatherill	12.4	4.8	5.2	8.5	15.1	29.8	12.9	5.3	5.0	173
Bramcote	—	0.0	0.5	3.0	16.0	45.0	22.0	7.5	6.0	—
Pickering	—	1.9	0.8	0.9	1.5	17.5	57.1	14.2	6.1	—
Swynnerton	—	0.5	0.5	2.2	6.7	26.0	25.0	15.3	23.8	—
Mansfield Red	0.4	0.4	0.4	0.8	1.6	4.4	30.4	22.9	25.1	—
Erith loam	1.0	0.6	0.6	0.4	0.4	1.0	15.5	28.8	38.5	—
Washed silica sands										
Biddulph + 36 mesh	6.7	28.4	40.1	17.3	4.5	1.4	0.6	0.3	0.2	63
Arnold No. 19 (dried)	0.0	0.6	38.5	47.5	10.9	1.6	0.2	0.0	0.0	66
Garside dried No. 21	—	0.6	14.4	49.5	27.8	6.7	0.7	0.1	0.1	75
Chelford W.S. (50)	0.1	1.2	6.4	23.6	44.0	23.8	0.7	0.0	0.0	93
Erith silica	0.0	0.0	0.1	3.6	43.2	50.4	2.2	0.3	0.2	128
Kings Lynn 60 (4F)	0.0	0.0	1.0	7.5	24.4	40.6	13.1	2.3	0.6	162
Redhill 65 (F)	0.3	1.4	4.1	8.5	18.7	42.3	20.2	1.7	0.8	160
New Windsor Rose	0.0	0.1	0.5	3.4	20.9	58.8	14.5	1.3	0.4	133
Chelford Fine (95)	0.0	0.0	0.1	0.3	1.7	49.1	33.2	10.4	4.7	206
Kings Lynn SS (100)	0.0	0.0	0.1	0.2	0.6	16.4	63.6	16.6	2.2	220
Redhill H (110)	—	0.0	0.1	0.4	2.2	22.0	41.6	15.3	17.5	260
Ryarsh	0.0	0.0	0.0	0.2	0.2	1.3	59.0	29.1	7.1	320
Non-siliceous sands										
Zircoruf	—	—	—	0.2	16.0	70.7	12.8	0.1	0.2	73
AMA, zircon	—	—	—	—	0.0	6.8	72.6	19.5	1.0	113
Olivine No. 2	5.6	8.8	13.1	18.7	19.9	19.4	6.0	2.4	5.7	161
Olivine No. $3\frac{1}{2}$	0.2	2.1	5.9	15.1	18.7	24.2	12.4	6.1	14.6	274
FW, chromite	0.3	1.0	4.5	13.7	22.8	32.7	15.1	6.4	3.8	106
Fine grade chromite	—	0.1	0.7	6.0	14.0	28.0	23.5	15.5	12.3	187

† Data Sheets on Moulding Materials', SCRATA, 3rd Ed. 1972.
‡ For aperture sizes *see* BS410, 1969–*see* Introductory Tables, Table 2.9.
§ Included to give some indication of binder requirement, particularly liquid types.

Table 26.5 TYPICAL PROPERTIES OF SOME SODIUM SILICATES USED IN THE CARBON-DIOXIDE, SELF-SETTING SILICATE AND FLUID-SAND PROCESSES*

Grade	*Weight ratio* $SiO_2:Na_2O$	*Molecular ratio* $SiO_2:Na_2O$	*Typical analyses % by weight*			*Specific gravity at 20 °C/20*			*Approximate viscosity at 20 °C* cP
			Na_2O	SiO_2	H_2O	*SG*	*Degrees Twaddell*	*Degrees Baumé*	
C112	2.0	2.05	15.2	30.4	54.4	1.56	112	51.8	850
C125	2.0	2.05	16.6	33.2	50.2	1.625	125	55.8	4 500
E100	2.21	2.28	13.2	29.2	57.6	1.50	100	48.1	220
H100	2.4	2.50	12.7	30.8	56.5	1.50	100	48.1	310
H112	2.4	2.5	13.7	33.3	53.0	1.56	112	51.8	2 500
M75	2.9	3.0	9.2	26.8	64.0	1.38	75	39.4	100

* K. E. L. Nicholas, 'The CO_2-Silicate Process in Foundries', BCIRA, 1972.

Table 26.6 TYPES OF MATERIAL USED IN GREEN AND DRY SAND MOULDING*

Class	*Material*	*Approximate AFS/GFN*
A	Naturally bonded sands of medium grain size (Dullatur 40 and other 'rotten rocks')	40
B	Naturally bonded sands of fine grain size (Bramcote)	80
C	Naturally bonded sands of very fine grain size (Mansfield Red)	Over 130
D	Silica sand of medium grain size (Arnold No. 19, Garside No. 21)	30–40
E	Silica sand of fine grain size (Redhill 65, Windsor Rose)	60–70
F	Silica sand of very fine grain size (Ryarsh, Redhill H)	>100
G	Old sand renovated for further use. Grain fineness depends on original sand	
H	Added clay bond—bentonite	
I	Fuller's earth or other intermediate bonds with proprietory trade names	
J	Fire clay	
K	Organic binders—molasses, sulphite lye solution	
L	Dextrine and other cereal binders	
M	Special additions—coal dust (increasing fines and decreasing quantities for thinner castings), blacklead (graphite), blacking	
N	Sulphur	
P	Boric acid	
Q	Fluoride salts, e.g. ammonium bifluoride NH_4HF_2, ammonium silicofluoride $(NH_4)_2SiF_6$, ammonium borofluoride NH_4BF_4, ammonium fluoride NH_4F	
R	Crushed materials such as 'grog' firebrick or ganister	

* W. H. Salmon and E. N. Simons, 'Foundry Practice', Pitman, London, 1966.

Table 26.7 TYPICAL MIXTURES USED IN GREEN AND DRY SAND MOULDING*

The figures refer to weight percentages and the letters to the materials shown in Table 26.6

Nature of material	*Sand*	*Clay*	*Additions*	*Added moisture %*	*Remarks*
Cast-iron					
Green sand:					
Thin castings	60G 37C		3M	6	NB: Coal dust not used on castings for vitreous enamelling
Medium castings	76G 20B		4M	6	
Thick castings	66G 28B		6M	7	
Synthetic green sand:					
Thin castings	91G	3H	6M	$3\frac{1}{2}$	
Thick castings	54G 36E	4I	6M	$3\frac{1}{2}$	
Non-ferrous					
Copper alloys	70C 30F			7	
Phosphor bronze–dry sand	65G 25C		10M	7	
Aluminium alloys	90G 10C			$6\frac{1}{2}$	+5% P for high Mg alloys
Magnesium alloys	88E	4H	6N, 1P	4	
Steel†					
Synthetic green sand	73E 25D	$1\frac{1}{2}$H	$\frac{1}{2}$L	$3\frac{1}{2}$	
Synthetic dry sand	85G 14D	$\frac{1}{2}$H	$\frac{1}{2}$L	5	
Naturally bonded dry sand	75A 25D			7	NB: For 'Compo' stove to dull red-heat. Highly permeable refractory mixture for very large castings
Compo-dried mould	84R	15J	1M	8	
Loam	40B 40R	10J	10M	15–20	

* W. H. Salmon and E. N. Simons, 'Foundry Practice', Pitman, London, 1966.
† Refractory sand essential for steel casting.

Table 26.8 TYPICAL STANDARD AFS TESTS ON CLAY-BONDED MOULDING SAND MIXTURES*

Casting	*Moisture* %	*AFS green permeability number*	*Compressive strength* 10^4 Pa	
			Green	*Dry*
Cast iron				
Stove plate	9.0	10	4.1	13.8
Radiators	7.0	35	3.4	34.5
Cylinder blocks	6.5	80	4.8	31.0
Average castings	7.0	35	4.1	41.4
Synthetic–thin	3.5	70	5.5	41.4
Non-ferrous copper base				
Small	6.0	25	4.8	28.0
Large	5.5	60	6.9	41.4
Aluminium				
Small castings	7.0	20	3.4	13.8
Large castings	7.0	35	3.4	20.7
Magnesium	4.0	80–150	5.5	69.0
Steel				
Green sand	3.5	180	5.5	48.3
Dry sand	7.5	120	6.2	103
Compo	7.5	40	8.3	172

* W. H. Salmon and E. N. Simons, 'Foundary Practice', Pitman, London, 1960.

Note: The tensile strength of sands is less than half the compressive strength.

Table 26.9 TYPICAL MIXTURES FOR THE CO_2-PROCESS AND TYPICAL PROPERTIES OF STANDARD AFS COMPACTS*

Sand	*Silicate* Weight ratio $SiO_2:Na_2O$	°TW	% *Added*	*Gassing time* s	*Compression strength* (k Pa) *As gassed*	*After* 24 h *storage*	
Chelford 50	2:1	112	4	60	1 050	2 680	
Redhill 65	2:1	112	4	60	860	2 140	
Erith (56)	2:1	112	4	60	550	2 300	
King's Lynn (95)	2:1	112	4	60	1 700	—	
						Gassing time s	*After 24 h storage*
Windsor	2.0:1	112	4	40	1 340	30	5 600
Rose	2.4:1	112	4	40	2 040	24	1 790
(68)	2.9:1	100	6	40	1 620	18	1 035

* K. E. L. Nicholas, 'The CO_2-Silicate Process in Foundries', BCIRA, 1963.

CO_2 flow-rate in all cases 2.5/l min^{-1}

Table 26.10 SUMMARY OF

Method	*Binder*	*Equipment*
Natural sand	Clay	Wood, metal, or resin core boxes
Synthetic sand core	Clay, usually bentonite	Metal, wood, or resin core boxes
Loam	Clay	Generally strickled on spindle or barrel
Cereals	Starch and corn products Dextrose	Wood, metal, or resin core boxes
Oil	Wide variety including linseed, cotton-seed and fish oils	Wood, metal, or resin core boxes
Cellulose	Water-soluble cellulose ethers and melamine resins	Wood, metal, or resin core boxes
CO_2	Carbon dioxide and sodium silicate plus breakdown agent, e.g. dextrose monohydrate or molasses	Wood, metal, or resin core boxes
Self-setting silicate	Sodium silicate plus one of: (a) Dicalcium silicate (b) Ferro-silicon (c) Organic esters (acetins or glycol diacetates)	Wood, metal, or resin core boxes. Special box paints preferred for (a) and (b)
Air set (no bake)	Oils, usually heat-treated Tung linseed and alkyd resin modified. Catalysts—sodium perborate, 'metal driers', e.g. cobalt naphthenate, and isocyanates	Wood, metal, or resin core boxes
Cold box	Resin (Novolak) isocyanate MDI plus trimethylamine/ air vapour	Gas dispensers Fume extraction Wood, metal, or resin core boxes
Air set (no bake)	(a) Urea-formaldehyde/furfuryl alcohol plus phosphoric acid (b) Phenol–formaldehyde/furfuryl alcohol or (c) Furfuryl alcohol polymer both plus paratoluene sulphonic acid (b) and (c) are nitrogen free	Wood, metal, or resin core boxes
Fascold	Resin (polyurethane) + catalyst (isocyanate) Chemical polymerization at room temperature	Need special core-blowing machine (sand + resin to mix with sand + catalyst) Wood, Metal, Resin } core boxes
Hot box	Phenolic furfuryl or other resin and catalyst (NH_4^+ salts or p.t.s.a.)	Metal core boxes on machine with heating and cooling cycle
Shell	Phenol formaldehyde resin Novalac type plus hexamine catalyst originally. Now usually 'precoated' or 'resin-coated' sands used	Metal core boxes on machine with heating and cooling cycle
Shaw	Ehyl silicate	Accurate and highly finished metal core boxes
Plaster	Gypsum	Accurate and highly finished metal core boxes
SO_2 process	UFFA or Furane polymer resin with peroxide mix	Gas dispensers. Fume extraction. Wood, metal, or resin core boxes

CORE BINDING PROCESSES

Process	*Applications and limitations*
Cores usually baked to develop the clay bond	Suitable for medium and large simple cores for jobbing foundries
Strength developed by baking clay bond	General use for jobbing foundry for all sizes of core. High green strength, good breakdown
Hand-made cores shaped by strickling on straw rope base, clay bond developed by baking	Suitable only for jobbing foundries for simple round and cylindrical cores
Bond developed by baking	Suitable for small and medium cores in jobbing repetition foundries
Oil oxidized by baking to develop bond	For general applications in all foundries but mostly on small and medium work. Poor green strength, and accuracy therefore difficult to maintain
Cores stoved to develop bond	Suitable for jobbing or repetition work. Good breakdown, finish and low gas content
No baking. Core gassed with CO_2 in core box	Suitable for jobbing or repetition work. Fast—accurate—stoving cores, low gas evolution. Main difficulties on breakdown, improved using breakdown agent, but storage life reduced
Silicate cures by either dehydration or for chemical gelling in 0.5–4 h	Suitable for jobbing foundry. Medium to large cores. Good finish. No fume. Breakdown little better than CO_2
Air drying process of 0.5–4 h duration plus 1–2 h baking	Suitable for jobbing foundry and large cores. Good finish, easy breakdown, good green strength, high accuracy
Almost instantaneous hardening in box. Gas passage, e.g. 3s Air purge, e.g. 5s	Sand air must be dry. Good dimensional accuracy. Knockout excellent. Efficient ventilation required. Cores of intermediate size in batches too small to warrant use of shell process
Resin cures after addition of acid in 10–45 min. No baking	Suitable for jobbing foundry and large cores. Good finish, easy breakdown. Good green strength. High accuracy
Both resin and catalyst liquid. Cured in 30–45s Usable 30–60 min after strip	Machine operations are automatic Mass production of small cores—similar sizes to hot-box (Usually for iron-foundry)
Resin cures after addition of catalyst in 10–40s baking at 180–300°C	Suitable only for large quantity repetition work and generally small cores.
Resin cured by heating cycle of 30–120s at 200–400°C	Suitable for large quantity repetition work. Good finish, accuracy and breakdown. High permeability due to shell construction
Conversion of ethyl silicate to silica bond by hydrolysis	For accurate high-temperature metal application where surface finish and accuracy are of prime importance
Hydration of gypsum subsequently dried	For accurate low-temperature applications where surface finish and accuracy are of prime importance
Almost instantaneous hardening by SO_2 gas	Good dimensional accuracy. Knockout excellent. Not sensitive to water

Table 26.11 TYPICAL MIXTURES AND PROPERTIES OF SOME CORE-MAKING PROCESSES*

Process	*Sand* %	*Binder* %	*Hardener* %	*Compression strength* 10^4 Pa
Self-setting silicate	89 Coarse-medium grain silica	6.0 Na_2SiO_3(2.2 ratio) 1.0 water	4 Ca_2SiO_4	After 24 h 17.2
Cold curing oil	97.5 Clean dry silica	2.0 Oil	Isocyanate 0.5 (25% oil content)	After (h) 1 2 3 24 6.2 14.8 21.9 50.0
Cold curing resin	97.5 Clean dry silica	2.0 Resin	H_3PO_4 or p. t.s.a. 0.8 (40% resin content)	After (h) 1 2 3 24 17.2 27.6 42.8 62.0
Cold-set stoving oil	97.9 Clean dry silica	2.0 Oil	Sodium perborate 0.1 (5% of oil content)	After (h) 1 2 3 Baked 1.6 10.3 13.8 96.5 Tensile strength 10^4 Pa (cured at 220 °C)
Hot box	97.5 Clean dry silica	2.0 Resin†	H_3PO_4 or p. t.s.a. 0.5 (25% of resin content)	After (s) 10 20 30 40 60 30.3 37.2 38.5 39.2 31.0
Oil-bonded (linseed type)	95 Clean dry silica (AFS 55)	Oil 10 Cereal 1.5 Water 2.5	Heat	After (min) 30 45 60 90 17.9 20.3 21.0 22.0
Resin bonded (types (b) and (d) below)	95 Clean dry silica (AFS 55)	Resin 1.25 Cereal 1.50 Water 2.25	Heat	After (min) 10 20 30 45 17.2 29.0 30.0 26.2

* 'Foseco Foundryman's Handbook, 7th Ed. Pergamon Press, Oxford, 1964
† Can be (*a*) Urea formaldehyde–furfuryl alcohol. (*b*) Phenol–formaldehyde (*c*) Urea–phenolformaldehyde. (*d*) Urea–formaldehyde.

26.2 Patterns—crucibles—fluxing

Table 26.12 CONTRACTION ALLOWANCES OF COMMON ALLOYS (SAND CAST)*

Alloy	*Pattern size* m	(in)	*Type of construction*	*Contraction allowance Traditional*	%
Grey cast iron (*see* Note 2)	Up to 0.61	(2.4)	Open	1 in 96	1.04
	0.64–1.22	(25–48)	Open	1 in 120	0.83
	Over 1.22	(Over 48)	Open	1 in 144	0.69
	Up to 0.61	(24)	Cored	1 in 96	1.04
	0.64–1.22	(25–48)	Cored	1 in 120	0.83
	Over 1.22	(Over 48)	Cored	1 in 144	0.69
Cast steel	Up to 0.61	(24)	Open	1 in 48	2.08
	0.64–1.83	(25–72)	Open	1 in 64	1.56
	Over 1.83	(Over 72)	Open	1 in 77	1.30
	Up to 0.46	(18)	Cored	1 in 48	2.08
	0.48–1.22	(19–48)	Cored	1 in 64	1.56
	1.24–1.68	(49–66)	Cored	1 in 77	1.30
	Over 1.68	(Over 66)	Cored	1 in 96	1.04
Manganese steel	—	—	—	1 in 38 decreasing to 1 in 64 for long castings	2.63 1.56

Table 26.12 CONTRACTION ALLOWANCES OF COMMON ALLOYS (SAND CAST)*—*continued*

Alloy	*Pattern size* m (in)		*Type of construction*	*Contraction allowance Traditional*	%
Malleable cast iron	—		Section thickness		
(*see* Note 3)			mm (in)		
			1.6($\frac{1}{16}$)	1 in 70	1.43
			3.2 ($\frac{1}{8}$)	1 in 77	1.30
			4.8 ($\frac{3}{16}$)	1 in 80	1.25
			6.4 ($\frac{1}{4}$)	1 in 85	1.18
			9.5 ($\frac{3}{8}$)	1 in 96	1.04
			12.7 ($\frac{1}{2}$)	1 in 110	0.91
			15.9 ($\frac{5}{8}$)	1 in 128	0.78
			19.1 ($\frac{3}{4}$)	1 in 152	0.66
			22.2 ($\frac{7}{8}$)	1 in 256	0.39
			25.4 (1)	1 in 384	0.26
Pearlitic nodular iron, as cast	—		—	1 in 120 to 1 in 180	0.83 to 0.56
Pearlitic nodular iron, heat-treated to ferritic	—		—	1 in 120	0.83
Nodular iron (thin section) containing carbide as cast and annealed	—		—	1 in 120 contraction to 1 in 240 expansion	0.83 0.42
Aluminium alloys	Up to 1.22	(48)	Open	1 in 77	1.30
	1.24–1.83	(49–72)	Open	1 in 85	1.18
	Over 1.83	(Over 72)	Open	1 in 96	1.04
	Up to 0.61	(24)	Cored	1 in 77	1.30
	0.64–1.22	(25–48)	Cored	1 in 85 to 1 in 96	1.20 to 1.04
	Over 1.22	(Over 48)	Cored	1 in 96 to 1 in 192	1.04 to 0.52
Magnesium alloys	Up to 1.22	(48)	Open	1 in 70	1.43
(*see* Note 4)	Over 1.22	(Over 48)	Open	1 in 77	1.30
	Up to 0.61	(24)	Cored	1 in 77	1.30
	Over 0.61	(Over 24)	Cored	1 in 77 to 1 in 96	1.30 to 1.04
Brass	—		—	1 in 64	1.56
Bronze	—		—	1 in 96 to 1 in 48	1.04 to 2.08
Nickel alloys	—		—	1 in 48	2.08
Everdur (silicon bronze)	—		—	1 in 64	1.56
PMG (silicon bronze)	—		—	1 in 96 to 1 in 64	1.04 to 1.56
Manganese bronze	—		—	1 in 120 to 1 in 64	0.83 to 1.56
Aluminium bronze	—		—	1 in 43	2.32
Zinc alloys	—		—	1 in 85	1.18

* From 'Cast Metals Handbook', American Foundrymen's Society.

Notes

(1) Contraction varies with the casting design, type of metal, pouring temperature, and mould or core resistance. It may be necessary to use several different contraction allowances for the various dimensions of a single pattern.

(2) Standard pattern maker's allowance for common grey iron is 1 in 96. For higher strength alloys and white cast irons, the contraction allowance averages 1 in 77.

(3) Contractions shown for malleable irons are net values, e.g. white iron castings which shrink 1 in 48 when cast, expand 1 in 96 during the anneal, giving a net contraction of 1 in 96.

(4) Contraction varies with the alloy. Average values are:

A8, AZ91, C	1 in 85.
Z5Z, RZ5, TZ6, MSR–A, MSR–B	1 in 77
ZREI, ZT1	1 in 64.

Table 26.13 WEIGHTS OF CASTINGS FROM PATTERN WEIGHTS
Multiply pattern weight by the factor shown for the type of wood/metal and the alloys. For wooden patterns, allowance must be made for any metal reinforcement in the pattern.

Pattern material \ *Metal cast*	*Cast steel*	*Cast iron*	*Yellow brass*	*Gun metal or bronze*	*Zinc*	*Copper*	*Aluminium*	*Mag-nesium*
Mahogany	9.5	8.5	9.5	10.0	8.2	10.1	3.1	2.05
White pine	16.3	14.7	16.5	17.3	14.3	17.5	5.3	3.5
Yellow pine	14.4	13.1	14.7	15.4	12.7	15.6	4.7	3.1
Oak	10.4	9.4	10.5	10.8	9.1	11.2	3.4	2.24
Aluminium alloys	2.87	2.56	2.87	2.98	2.45	3.08	0.93	0.61
Brass	0.98	0.84	0.95	0.99	0.81	1.04	0.31	0.21
Iron	1.09	0.97	1.09	1.13	0.93	1.17	0.35	0.23
White metal	1.08	0.96	1.08	1.11	0.92	1.15	0.34	0.23

Note:
Allowance must also be made for cores and coreprints. For round cores and prints multiply the square of the diameter by the length of the core and prints in inches, and the product by 0.014. This will give the weight of the white pine core in pounds to be deducted from the weight of the pattern.

26.2.1 Pattern materials

(1) Wood: Yellow pine (for limited production) and mahogany (for production patterns) represent the most usual practice.
(2) Metal: (a) Cast iron.
(b) Brass.
(c) Aluminium alloy (e.g. LM6, LM4).
(d) White metal (only to a limited extent because of softness and easily worn away by moulding sand. Used, for example, for lining stripping plates).

Notes:
(1) For small patterns, an alloy of 40% Sn, 18% Sb, 40% Pb, 2% Bi, is used, shrinkage being negligible. Similarly, use is made of a modified 'Wood's metal' (50% Bi, 26.7% Pb, 10% Cd, 13.3% Sn) for making master patterns. It is completely molten at about 70°C and can readily be poured at the temperature of boiling water into wood or plaster moulds.
(2) Contraction allowance for 'master' patterns for casting metal patterns is contraction allowance for casting material, plus contraction allowance for metal pattern material.
(3) Plaster of Paris and proprietory cements: mainly for odd-side impressions.
(4) Plastics: epoxy resins, often reinforced with fibreglass.

Table 26.14 RECOMMENDED STANDARD COLOURS FOR PATTERNS

Part of pattern		*Colour*
As-cast surfaces which are to be left unmachined		Red or orange
Surfaces which are to be machined		Yellow
Core prints for unmachined openings and end prints	Periphery	Black
	Ends	Black
Core prints for machined openings	'A' periphery	Yellow stripes on black
	'B' ends	Black
Pattern joint (split patterns)	'A' cored section	Black
	'B' metal section	Clear varnish
Touch core	Cored shape	Black
	Legend	'Touch'
Seats of and for loose pieces and loose core prints		Green
Stop offs		Diagonal black stripes with clear varnish
Chilled surfaces	Outlined in	Black
	Legend	'Chill'

Table 26.15 GUIDE TO MACHINING ALLOWANCE ADDITIONS TO PATTERN CONTRACTION ALLOWANCES*

Casting alloy	*Pattern size* m(in)	*Bore* mm(in)	*Finish* mm(in)
Cast iron	Up to 0.3 (12)	3.12 (0.125)	2.38 (0.09)
	0.3–0.6 (13–24)	4.76 (0.187 5)	3.18 (0.125)
	0.6–1.1 (25–42)	6.35 (0.25)	4.76 (0.187 5)
	1.1–1.5 (43–60)	7.94 (0.312 5)	6.35 (0.25)
	1.5–2.0 (61–80)	9.53 (0.375)	7.94 (0.312 5)
	2.0–3.0 (80–120)	11.1 (0.437 5)	9.53 (0.375)
Cast steel	Up to 0.3 (12)	4.76 (0.187 5)	3.18 (0.125)
	0.3–0.6 (13–24)	6.35 (0.25)	4.76 (0.187 5)
	0.6–1.1 (25–42)	7.94 (0.312 5)	7.94 (0.312 5)
	1.1–1.5 (43–60)	9.53 (0.375)	9.53 (0.375)
	1.5–2.0 (61–80)	12.7 (0.5)	11.1 (0.437 5)
	2.0–3.0 (80–120)	15.9 (0.625)	12.7 (0.5)
Malleable iron	Up to 0.15 (6)	1.58 (0.062 5)	1.58 (0.062 5)
	0.15–0.23 (6–9)	2.38 (0.093 75)	1.58 (0.062 5)
	0.23–0.30 (9.12)	2.38 (0.093 75)	2.38 (0.093 75)
	0.3–0.6 (12–24)	3.97 (0.156 25)	3.18 (0.125)
	0.6–0.9 (24–35)	4.76 (0.187 5)	4.76 (0.187 5)
Brass, bronze and aluminium alloy castings	Up to 0.3 (12)	2.38 (0.093 75)	1 58 (0.062 5)
	0.3–0.6 (13–24)	4.76 (0.187 5)	3.18 (0.125)
	0.6–0.9 (25–36)	4.76 (0.187 5)	3.97 (0.156 25)

* 'Pattern Makers Manual', American Foundrymen's Society. *Note:* Above pattern sizes quoted, need special instructions.

Table 26.16 MATERIALS FOR PARTING POWDERS AND LIQUIDS

Powder	*Moisture proofer*
Pulverized limestone*	
Precipitated calcium carbonate*	Paraffin wax
Fireclay grog	
Sillimanite	Calcium stearate
Phosphate rock	
Bone ash	Aluminium stearate
Walnut shells	
Ptfe (from aerosol spray)	
Lycopodium powder (very expensive, for art casting only)	
Aluminium powder (as in aluminium paint)	
Liquid—paraffin plus small addition Colza oil	

* React unsatisfactory with chemical binders such as acid-catalysed resin binders

MOULD DRESSING, POWDERS, PAINTS

Table 26.17 SURFACE POWDERS FOR GREEN SAND MOULDS

Non-ferrous	Flour, soapstone, talc (french chalk)
Cast iron	Plumbago, black lead, bituminous coal dust

Material dusted on to mould surface, rubbed in by hand or sleeked with a smooth tool.

Table 26.18 MATERIALS FOR MOULD DRESSINGS

Filler material	*Suspending agent*	*Binder*	*Carrier*
Silica flour	(*a*) *For water base*	Core oil	Water
Zircon flour	Western bentonite	Dextrine	Iso-propyl alcohol ($<5\% H_2O$)
Graphite	China clay	Fire clay	Carbon tetrachloride
Olivine	Carboxy-methyl-cellulose	Linseed oil	Toluene
Mica	Sodium alginates	Phenolic resin	Petroleum-ether
Molochite	(*b*) *For spirit base*	Vinsol resin	Methylated spirits
Magnesite	Sodium bentonite	Polyvinyl acetate	
Chamotte	Bentone	Molasses	
Talc			
Plumbago			
Carbon black			

Tables 26.19 MIXES FOR WATER–BASED MOULD DRESSINGS*

Alloy type	*Mixture components* %				*Added water* %
Copper base	20A	6.6B	6.6C		66.6
High lead or phosphor bronze	11.5D	23B	8.5C		57
Cores for heavy section bronze		34E			66
Aluminium	22A	11F	11C		56
Cast-iron					
Blackwash	22G	4 H or C			74
Blackwash	21G	4L	4H		71
Heavy section	20J	6.6L	6.6B		66.6
Heavy section	25K	6L	3H	3M	63
Steel					
Light section	30K	1.5L	4.5M		64
Facing mix	45J	1.5D	4H		49.5
Manganese steel	42.5N	5L	2.5H		50
Zircon to 1.8Sp.G. (65 °CBé)	60P	4L	1Q		35

* W. H. Salmon and E. N. Simons, 'Foundry Practice', Pitman, London, 1966.

A. Talc.
B. Plumbago.
C. Molasses.
D. China clay.
E. Sodium silicate.
F. Whitening.
G. Blackening.
H. Dextrine.
J. Chamotte (200 mesh).
K. Silica flour.
L. Bentonite.
M. Core oil.
N. Magnesite.
P. Zircon (200 mesh).
Q. Core cream.

Table 26.20 TYPICAL 'SPIRIT–BASE' OR 'FLASH-OFF' DRESSINGS

Component	*Polar solvent carrier*		*Non-polar solvent carrier*	
	Description	*p.b.w.**	*Description*	*p.b.w.**
Refractory filler	Zircon flour (200 mesh)	69.0	Zircon flour (300 mesh)	29.0
Suspension agent	Bentone 18C + pure	2.8	Bentone 34 + Ind.	1.0
	grade toluene	1.5	meths (64 o.p.)	0.4
Resin binder	Novolac type	1.3	Cumarone/Indene type	2.0
Carrier	Iso-propyl alcohol	25.0	White spirit (SBP6)	31.6

* p.b.w.—parts by weight.

Table 26.21 MOULD COATINGS FOR SPECIAL APPLICATIONS

Cast metal	*Coating constituent(s)*	*Purpose*	*For further information consult*
Aluminium base alloys	Acetylene soot	Increased fluidity,	—
	Hexachlorethane-containing wash	Increased fluidity + refined grain	—
Copper base alloys	325 mesh molochite Powdered vinsol resin 99% Iso-propyl alcohol	Prevention of 'steam' reaction at metal–mould interface resulting in globular sub-surface porosity	BNFMTC
Phosphor-bronzes and gunmetals	Finely divided Al–Mg alloy suspended in a solvent (e.g. toluene)	Prevention of pinhole porosity	BNFMTC
Grey iron	Finely divided bismuth	Suppresses micro-shrinkage porosity in 'hot spots'	BCIRA
Grey iron	Finely divided tellurium	Local chilling for additional hardness	BCIRA
Malleable irons	Zinc/ground coke or cobalt, cobalt/iron oxide	Prevents hot-checking on white-iron castings and modifies dendrite structure	BCIRA

BNFMTC—British Non-Ferrous Metals Technology Centre. BCIRA—British Cast Iron Research Association.

26.2.2. Crucibles and melting vessels

Refractory crucibles were originally produced from clay alone, but since the beginning of the century most melting crucibles have incorporated up to 30% graphite, bonded with clay, to improve thermal shock resistance, erosion from fluxes and conductivity. Additions of silicon carbide are also made where higher duty is required. The life of crucibles has been improved over recent years by development of glazes which protect the crucible against oxidation or 'perishing'.

Carbon-bonded silicon carbide crucibles made from silicon carbide and graphite bonded with tar, which on firing produces a carbon bond, have superior thermal shock resistance and improved performance against flux attack and oxidation under certain conditions. 80% of crucibles manufactured in recent years are produced from carbon-bonded silicon carbide. They are generally more expensive than clay-graphite crucibles.

TYPES OF CRUCIBLES

Three types of crucibles are in general use for non-ferrous work. These are:

1. *Crucibles used in pit or 'pull-out' furnaces*
   A and C shape which have to be removed from the furnace for metal to be poured.
   These vary in size from a few kilograms to greater than 1000 kg brass capacity, though the latter sizes are not generally used on account of the difficulty of handling them. In oil- and gas-fired furnaces the crucible rests on a small stand or stool—of similar material to the crucible, and a ring or cylinder to accommodate and protect protruding solid metal is sometimes placed on the top of the crucible. The chief value of the pit furnace lies in its flexibility and low capital cost and it is ideal for small or medium-sized foundries where relatively small quantities of a variety of metals or alloys are required.
2. *Crucibles used in tilting furnaces*
   For convenience in pouring or where the size of crucible is such it cannot be readily handled, it is mounted permanently in the furnace, which is tilted to discharge its molten contents.
   Crucibles varying from 70 kg to greater than 760 kg brass capacity are in general use, though larger crucibles and furnaces are now available with capacities of up to 1520 kg brass (590 kg aluminium)
3. *Maintaining crucibles*
   Crucibles or basins for holding molten metal for periodic casting or for process work are extensively used for purposes such as die casting.
   Those used for die casting are usually of basin shape and are wide in proportion to their height. They are made in sizes up to 600 kg aluminium capacity for aluminium die casting, or 135 to 180 kg brass capacity for aluminium bronze or brass die casting.
   Silicon carbide basins are particularly suited to maintaining molten aluminium alloys for die casting, as neither aluminium nor aluminium oxide attack the crucible material.

26.2.3 Iron and steel crucibles—fluxing

Steel crucibles are normally used for the melting of magnesium alloys. Low carbon steels are used and nickel and cobalt should be absent from the steel because of the adverse effect of even small traces of these elements on the corrosion resistance of magnesium alloys.

The crucibles may be of a pressed construction or made up of welded boiler plate.

For very large melts, cast-steel (~0.4% carbon) crucibles are used.

Although cast-iron crucibles were often used in holding furnaces for aluminium alloy die casting, they appear to have fallen out of favour and are now rarely found in use.

Where they are still used a refractory wash of similar nature and composition to the following:

Finely ground whitening	9 wt.%
Water	90 wt.%
Waterglass (hot conc. soln)	1 wt.%

applied daily will minimize iron contamination.

Silicon carbide or plumbago basins are used in preference to cast-iron pots nowadays, for aluminium melting.

Table 26.22 COMMON FLUXING AND INOCULATION

Alloy group	*Cover fluxes*	*Degassing and/or cleansing fluxes and treatments*
Most aluminium alloys (except Al–Mg, *see* below)	(a) NaCl+NaF (or cryolite)+KCl+small amounts of $SO_4^{2-}+NO_3^-$ (b) Na_2SiF_6+NaCl ↓ also used for metal recovery from drosses (c) Reverberatory furnaces (typical) $CaCl_2$ 50% NaCl 30% KCl 20% (d) Rotary furnaces (typical) NaCl 45% KCl 45% Cryolite 10% } m. pt 607°C	(a) Inert gases N_2, He, Ar [Response time slowest] [Use N_2 at < 700°C or → AlN formation Use N_2 at ≤ 670°C with Al-Mg alloys or → Mg_3N_2 formation] (b) Inert active gas mixtures e.g. $90N_2/10Cl_2$ [Response time reasonably fast] (c) Active gases Cl_2, F_2 or compounds containing them, e.g. hexachlorethane C_2Cl_6 and Freon 12 CCl_2F_2 [Response time fast] (d) Metal chlorides $ZnCl_2$ BCl_3 $CuCl_2$ $MnCl_2$ $TiCl_2$ } Careful selection to prevent contamination All hygroscopic
Aluminium–magnesium alloys (3–10% Mg)	Similar to magnesium alloys. Mixtures of chlorides and fluorides, e.g. $MgCl_2+KCl+MgF_2$ or CaF_2 *NB*: No sodium compounds of any description if present → embrittlement, hot-shortness	Groups (b) and (c) above (20% Mg loss at 690°C 90% Mg loss at 710°C, 50% Mg loss with C_2Cl_6 regardless of temperature) Also used—K_2ZrF_6, Ti sponge
Magnesium alloys		Degas with (a) C_2Cl_6 (b) Cl_2 (c) N_2 'Flux bubbled' [$MgCl_2$ rich flux on crucible bottom 650–680°C]
(a) Mg Mg–Al Mg–Mn	$CaCl_2$ 40% NaCl 30% KCl 20% $MgCl_2$ 10% } 'Melrasal Z'	$MgCl_2$ 35% $CaCl_2$ 15% NcCl 10% KCl 10% CaF_2 20% MgO 10% } Inspissating flux 'Melrasal E' $CaCl_2$ 19% NaCl 11% KCl 7.5% $BaCl_2$ 37.5% MgF_2 25% } For holding furnaces $BaCl_2$ addition to increase density
	'Melrasal UE' all-purpose flux (melting, refining, covering) made up of 1 part 'Z' flux to 6 parts 'E' flux	
(b) Mg–Zr	$BaCl_2$ 39%, $MgCl_2$ 30%, KCl 10%, BaF_2 21% $BaCl_2$ 28%, $CaCl_2$ 26%, KCl 10%, NaCl 16%, MgF_2 20% } for alloys containing rare earths $BaCl_2$ 40% $MgCl_2$ 35% KCl 10% MgF_2 15% } for alloys containing thorium	Zr acts as its own degasser $Zr+2[H]Mg \rightarrow ZrH_2$ ↓
Copper-base alloys HC Copper	Charcoal	Degas N_2 only if necessary Deoxidation: 15% phosphor-copper (equiv. to 0.005% P) followed by CaB_2, Li (0.003–0.01%) or 10% Be–Cu (0.005–0.02% Be)

PRACTICES FOR VARIOUS ALLOYS

Special purpose fluxes and inoculants

(1) Grain refinement
 (a) Combined additions of K_2TiF_6
 +
 KBF_4
 +
 (sometimes) C_2Cl_6 for better efficiency
 KBF_4 used alone if alloy contains Ti near maximum specified
 (b) Using a 'hardener', e.g.: Al–5% Ti–1%B

(2) 'Modification' of Al–Si alloys
 (a) Eutectic or near-eutectic (11–14% Si)
 Salt mixtures containing NaF,
 e.g. 88% NaCl–KCl (1:1)+12%NaF (fuses at 607 °C)
 Vacuum-melted sodium sealed in air-tight Al containers
 or Sr { Sr modification lasts for several hours; Sr modified alloys can be remelted without loss of fine structure }
 (b) Hypereutectic (17–25% Si)
 Red phosphorus introduced as particles (<60 µm) in a mixture of KCl and K_2TiF_6. Effect strengthened if subsequent Cl_2 treatment

Grain refinement with salt mixtures as in 1(a) above

Grain refinement:

Alloy type	*Treatment*	*Degree of grain refinement achieved*
Mg–Al(–Zn–Mn)	Carbon inoculation (C_2Cl_6)	Marked
	Superheat to ≥850 °C	Marked
Mg–Al–Mn(–Zn)	$FeCl_3$	Marked
Mg–Zn(–RE–Mn)	$FeCl_3$, Zn–T% Fe alloy, or NH_3	Very marked
Mg–Mn	Ca + N_2	Mild
	Zr	Increases with falling Mn content

Protection:
(1) 'Dusting' fluxes
 S, NH_4BF_4, NH_4HF_2, NH_4F, H_3, BO_4
 e.g. S + 5 to 30% H_3BO_4
(2) SO_2 atmosphere
 'Purge' mould prior to cast

Grain refinement: Use master alloy (proprietory)
Max gr. ref. at max solubility of Zr (0.6%)

Zr will not grain refine in presence of:	Zr will grain refine in presence of:
Al, Si, Sn, Ni, Fe, Co, Mn, Sb	Zn, Cd, Ce, Ag, Th, Tl, Cu, Bi, Pb, Ca

Table 26.22 COMMON FLUXING AND INOCULATION PRACTICES FOR VARIOUS ALLOYS—*continued*

Alloy group	*Cover fluxes*		*Degassing and/or cleansing fluxes and treatments*	
Gunmetals	Charcoal	C(≥25 mm thick)	Degas with dry N_2 (e.g. 45 kg melts at 5–10 l min^{-1} for about 5 min) or use 'oxidation-reduction' technique	
	Borocalcite	$4CaO \cdot 5B_2O_3 \cdot 9H_2O$		
	Boric oxide	B_2O_3 (crucible attack)		
	Borax	$Na_2B_4O_7 10H_2O$		
	Soda ash	Na_2CO_3 (*desulphuriser*)	*Oxidizing agents*	*Deoxidants*
	Common salt	NaCl	Cupric oxide CuO	15% P–Cu
	Calcium fluoride	CaF_2	Cuprous oxide Cu_2O*	Lithium
	Silica sand	SiO_2	Manganese dioxide } MnO_2†	10% Li–Cu
	Glass (green bottle)	—	Manganese ore } MnO_2†	98% Zn–2% Na
	Barium carbonate	$BaCO_3$ (desulphuriser)	Barium peroxide BaO_2	
	Calcium oxide (lime)	CaO	[*Reduced by H_2 to Cu, does not evolve oxygen †Also removes As, Fe, Si and S]	
Tin bronzes	Typical: Fused borax 30, dry silica sand 50, cupric oxide 20:2.75 kg charge Electric furnace melting: 3 parts lime + 1 part fluorspar			
Brasses and manganese bronzes	Charcoal, borax, greenbottle glass ↓ Virgin melts ↳All scrap melts		None usually practised (Zn, Al, Mn are all deoxidants)	
Aluminium bronzes	Manganese chloride ($MnCl_2$) } Both reduce Cryolite (AlF_3–3NaF) } crucible life (Do not use compounds containing boron)		Degas with dry N_2 or CO_2	
Silicon bronzes	Bottle glass + small % of fluoride salts (Pb free) Can be thickened before skim with silica sand. *NB*: Not charcoal		Degas with dry N_2 or CO_2 Deoxidation not required	
Nickel silvers	Broken glass, 80:20 borax–boracic acid, Borocalcite, all Pb free. *NB*: not charcoal. Thicken for removal with powdered CaO		Deoxidation 0.1% Mn + 0.06% max. Mg (Mg also desulphurizes)	
Cupro-nickels and Monels	As for Ni-Silvers + MnO_2 (1% charge wt), NiO or Cu_2O *NB*: Not charcoal		Deoxidation Si 0.5–2.0% dependent on % Ni Mg 0.03–0.06% dependent on % Ni e.g. for 30% Ni alley 1.2% Mn followed by 0.6% Si followed by 0.03% Mg Can degas with N_2 or dry air	
Cast-irons			Acid cupola: Limestone 28–32 kg tonne^{-1} providing (a) Standard cupola practice (b) Coke ash content–10% (c) $CaCO_3$ is 96% of limestone 'Energized' flux for increased efficiency 21 kg tonne^{-1} (14 kg limestone + 7 kg CaF_2) Basic cupola: 80 to 90 kg tonne^{-1} of flux made up as Limestone 30–40% coke wt Dolomite 10% coke wt * Fluorspar 25% of limestone	

* Can replace fluorspar by 50:50 fluorspar/soda ash.

Special purpose fluxes and inoculants

Grain refinement not usually beneficial with these alloys–encourages 'layer porosity'–although has been achieved with 0.03% Zr+0.02% B, Ti, Co

Grain refinement of manganese bronzes by iron

Fe in alloy acts as a grain-refiner

—

Some grain refinement with Mn

—

Anhydrous soda ash or calcium carbide for desulphurization of acid cupola metal
Calcium silicide: 'Meehanite iron'
Ferro–silicon (+1–2% Al and up to 1% Ca), zirconium etc., } for grain size control (eutectic cells)
75% FeSi (Al and Ca free) and 1–4% Sr } and reduction of 'chill depth'
↓
elimination of 'pinholing' in green sand moulds

Nodularization (spheroidization): sulphur $\leq$0.02%
Magnesium for hypo- and hypereutectic irons introduced as: Mg impregnated coke
Mg–Ni(Si) alloys
Mg–FeSi alloys
Mg–Si alloys

Cerium for hypereutectic irons introduced as mischmetal

26.3 Aluminium casting alloys

Table 26.23 ALUMINIUM–SILICON ALLOYS

Specification BS 1490:1970 *Related British Specifications*	LM6M(Ge) BS L33	LM20M(Ge)	LM9M(SP)	LM9TE(SP)	LM9TE(SP)	LM13TE(SP)	LM13TF(SP)	LM13TF7(SP)
Composition (%) (single figure indicates maximum)								
Copper	0.1	0.4		0.1			0.7–1.5	
Magnesium	0.1	0.2–0.6		0.2			0.8–1.5	
Silicon	10.0–13.0	10.0–13.0		10.0–13.0			10.0–12.0	
Iron	0.6	1.0		0.6			1.0	
Manganese	0.5	0.5		0.3–0.7			0.5	
Nickel	0.1	0.1		0.1			1.5	
Zinc	0.1	0.2		0.1			0.5	
Lead	0.1	0.1		0.1			0.1	
Tin	0.05	0.1		0.05			0.1	
Titanium	0.2	0.2		0.2			0.2	
Other	—	—		—			—	
Properties of material								
Suitability for:								
Sand casting	E	E*		G			G	
Chill casting (gravity die)	E	E		E			G	
Die casting (press die)	G	G		G*			F*	
Strength at elevated temperature	P	P		G			E	
Corrosion resistance	E	G		E			G	
Pressure tightness	E	E		G			F	
Fluidity	E	E		G			G	
Resistance to hot shortness	E	E		E			E	
Machinability	F	F		F			F	
Melting range, °C	565–575	565–575		550–575			525–560	
Casting temperature range, °C	710–740	680–740		690–740			680–760	
Specific gravity	2.65	2.68		2.68			2.70	

Heat treatment								
Solution temperature, °C	—	—	—	—	520–535	—	515–525	515–525
Solution time, h	—	—	—	—	2–8	—	8 (minimum)	8 (minimum)
Quench	—	—	—	—	Cold water	—	Water, 70–80 °C	Water, 70–80 °C
Precipitation temperature, °C	—	—	—	150–170	150–170	160–180	160–180	For pistons:
Precipitation time, h	—	—	—	16 (minimum)	16 (minimum)	4–16	4–16	200–250
Stabilization temperature, °C	—	—	—	—	—	—	—	4–6**
Stabilization time, h	—	—	—	—	—	—	—	
Special properties	Suitable for thin and intricate castings, readily welded	Die casting alloy		Suitable for low-pressure casting High strength and hardness			Low coefficient of expansion Good bearing properties Piston alloy	
Mechanical properties—sand cast—SI units (Imperial units in brackets)								
Tensile stress min., MPa (tonf in^{-2})	160 (10.4)	—	—	170 (11.0)	240 (15.5)	—	170 (11.0)	140 (9.1)
Elongation min. %	5	—	—	1.5	0–1	—	0.5	1
Expected 0.2% proof stress, MPa (tonf in^{-2})	60–70 (3.9–4.5)	—	—	110–130 (7.1–8.4)	220–250 (14.2–16.2)	—	160–190 (10.4–12.3) HB 100–150	130 (8.4) HB 65–85
Mechanical properties—chill cast—SI units (Imperial units in brackets)								
Tensile stress min., MPa (tonf in^{-2})	190 (12.3)	190 (12.3)	190 (12.3)	230 (14.9)	295 (19.1)	210 (13.6)	280 (18.1)	200 (12.9)
Elongation, min. %	7	5	3	2	0–1	1	1	1
Expected 0.2% proof stress, MPa (tonf in^{-2})	70–80 (4.5–5.2)	70–80 (4.5–5.2)	75–85 (4.9–5.5)	150–170 (9.7–11.0)	270–280 (17.5–18.1)	— HB90–120	270–300 (17.5–19.4) HB 100–150	190 (12.3) HB 65–85

*Not normally used in this form.
†If Ti alone is used for grain refinement then Ti ≮ 0.05%.
‡Fully heat-treated.
§Refine with phosphorus—subject to examination under microscope.
**Or for such time to give required BHN.

Notes
Association of Light Alloy Refiners and Smelters Grading:
E—Excellent, F—Fair, G—Good, P—Poor, U—Unsuitable,
(Ge—General purpose alloy; SP—Special purpose alloy as per BS 1490:1970).

Table 26.23 ALUMINIUM-SILICON ALLOYS—*continued*

Specification BS 1490:1970 *Related British Specifications*	LM18M(SP)	LM25M(Ge) LM25TE(Ge) LM25TB7(Ge) LM25TF(Ge)	LM29TE(SP) LM29TF(SP)
Composition % (Single figure indicates maximum)			
Copper	0.1	0.1	0.8–1.3
Magnesium	0.1	0.20–0.60	0.8–1.3
Silicon	4.5–6.0	6.5–7.0	22–25
Iron	0.6	0.5	0.7
Manganese	0.5	0.3	0.6
Nickel	0.1	0.1	0.8–1.3
Zinc	0.1	0.1	0.2
Lead	0.1	0.1	0.1
Tin	0.05	0.05	0.1
Titanium	0.2	0.2†	0.2
Other	—	—	Cr 0.6; Co 0.5, P§
Properties of material			
Suitability for:			
Sand casting	G	G	P
Chill casting (gravity die)	G	E	F
Die casting (press die)	G*	G*	U
Strength at elevated temperature	P	G‡	G
Corrosion resistance	E	E	G
Pressure tightness	E	G	F
Fluidity	G	G	F
Resistance to hot shortness	E	G	E
Machinability	F	F	P
Melting range, °C	565–625	550–615	520–770
Casting temperature range, °C	700–740	680–740	At least 830
Specific gravity	2.69	2.68	2.65

Heat treatment							
Solution temperature, °C	—	—	—	525–545	525–545	—	495–505
Solution time, h	—	—	—	4–12	4–12	—	4
Quench	—	—	—	Water, 70–80 °C	Water, 70–80 °C	—	Air blast
Precipitation temperature, °C	—	—	155–175	—	155–175	185	185
Preciptation time, h	—	—	8–12	—	8–12	To produce HB requirement	8
Stabilization temperature, °C	—	—	—	250	—	—	—
Stabilization time, h	—	—	—	2–4	—	—	—
Special properties	Readily welded		General purpose high-strength casting alloy			More suited to chill (grav. die) casting Piston alloy	
Mechanical properties—sand cast—SI units (Imperial units in brackets)							
Tensile stress min., MPa (tonf in^{-2})	120(7.8)	130(8.4)	150(9.7)	160(10.4)	230(149)	120(7.8)	120(7.8)
Elongation min. %	3	2	1	2.5	0–2	0.3	0.3
Expected 0.2% proof stress, MPa (tonf in^{-2})	55–60(3.6–3.9)	80–100(5.2–6.5)	120–150(7.8–9.7)	80–110(5.2–6.5)	200–250(12.9–16.2)	120(7.8) HB 100–140	120(7.8) HB 100–140
Mechanical properties—chill cast—SI units (Imperial units in brackets)							
Tensile stress min., MPa (tonf in^{-2})	140(9.1)	160(10.4)	190(12.3)	230(14.9)	280(18.1)	190(12.3)	190(12.3)
Elongation min. %	4	3	2	5	2	0.3	0.3
Expected 0.2% proof stress, MPa (tonf in^{-2})	60–70(3.9–4.5)	80–100(5.2–6.5)	130–200(8.4–12.9)	90–110(5.8–7.1)	220–260(14.2–16.8)	170(11.0) HB 100–140	170–190(11.0–12.3) HB 100–140

*Not normally used in this form.
† If Ti alone is used for grain refinement then Ti≮0.05%.
‡ Fully heat-treated.
§ Refine with phosphorus—subject to examination under microscope.
**Or for such time to give required BHN.

Note
E—Excellent. F—Fair. G—Good. P—Poor. U—Unsuitable.
(Ge—General purpose alloy; SP—Special purpose alloy as per BS: 1490:1970).

Table 26.24 ALUMINIUM–SILICON–COPPER ALLOYS

Specification BS 1490; 1970 *Related British Specifications*	LM2M(Ge)	LM4M(Ge)	LM4MTF (Ge)	LM16TB (SP)	LM16TF (SP) 3L78	LM21M(SP)
Composition % (Single figures indicate maximum)						
Copper	0.7–2.5	2.0–4.0		1.0–1.5		3.0–5.0
Magnesium	0.30	0.15		0.4–0.6		0.1–0.3
Silicon	9.0–11.5	4.0–6.0		4.5–5.5		5.0–7.0
Iron	1.0	0.8		0.6		1.0
Manganese	0.5	0.2–0.6		0.5		0.2–0.6
Nickel	0.5	0.3		0.25		0.3
Zinc	2.0	0.5		0.1		2.0
Lead	0.3	0.1		0.1		0.2
Tin	0.2	0.1		0.05		0.1
Titanium	0.2	0.2		0.2*		0.2
Properties of material						
Suitability for:						
Sand casting	G†	G		G		G
Chill casting (gravity die)	G†	G		G		G
Die casting (press die)	E	G		F†		G†
Strength at elevated temp.	G‡	G		G		G
Corrosion resistance	G	G		G		G
Pressure tightness	G	G		G		G
Fluidity	G	G		G		G
Resistance to hot shortness	E	G		G		G
Machinability	F	G		G		G
Melting range, °C	525–570	525–625		550–620		520–615
Casting temperature range, °C	—	700–760		690–760		680–760
Specific gravity	2.74	2.73		2.70		2.81
Heat treatment						
Solution temperature, °C	—	—	505–520	520–530	520–530	—
Solution time, h	—	—	6–16	12 (min)	12 (min)	—
Quench	—	—	Water at 70–80 °C	Water at 70–80 °C	Water at 70–80 °C	—
Precipitation temperature, °C	—	—	150–170	—	160–170	—
Precipitation time, h	—	—	6–18	—	8–10	—
Special properties	Alloy for pressure die castings	General engineering alloy Can tolerate relatively high static loading in TF condition		Pressure tight. High strength alloy in TF condition		Equally suited to all casting processes
Mechanical properties—sand cast—SI units (Imperial units in brackets)						
Tensile stress min. MPa (tonf in^{-2})	—	140(9.1)	230(14.9)	170(11.0)	230(14.9)	150(9.7)
Elongation min. %	—	2	—	2	—	1
Expected 0.2% proof stress, MPa (tonf in^{-2})	—	70–110 (4.5–7.1)	200–250 (12.9–16.2)	120–140 (7.8–9.1)	220–280 (14.2–18.1)	80–140 (5.2–9.1)
Mechanical properties—chill cast—SI units (Imperial units in brackets)						
Tensile strength min. MPa (tonf in^{-2})	150(9.7)	160(10.4)	280(18.1)	230(14.9)	280(18.1)	170(11.0)
Elongation min. %.	1	2	1	3	—	1
Expected 0.2% proof stress, MPa (tonf in^{-2})	90–130 (5.8–8.4)	80–110 (5.2–7.1)	200–300 (12.9–19.4)	140–150 (9.1–9.7)	250–300 (16.2–19.4)	80–140 (5.2–9.1)

* 0.05% min. if Ti alone used for grain refinement.
† Not normally used in this form.
‡ The use of die castings is usually restricted to only moderately elevated temperatures.

Table 26.24 ALUMINIUM–SILICON–COPPER ALLOYS—*continued*

Specification BS 1490: 1970 *Related British Specifications*	LM22TB (SP)	LM24M (Ge)	LM26TE (SP)	LM27M (Ge)	LM30M (SP)	LM30TS (SP)
Composition % (Single figures indicate maximum)						
Copper	2.8–3.8	3.0–4.0	2.0–4.0	1.5–2.5		4.0–5.0
Magnesium	0.05	0.1	0.5–1.5	0.3		0.4–0.7
Silicon	4.0–6.0	7.5–9.5	8.5–10.5	6.0–8.0		16–18
Iron	0.6	1.3	1.2	0.8		1.1
Manganese	0.2–0.6	0.5	0.5	0.2–0.6		0.3
Nickel	0.15	0.5	1.0	0.3		0.1
Zinc	0.15	3.0	1.0	1.0		0.2
Lead	0.1	0.3	0.2	0.2		0.1
Tin	0.05	0.2	0.1	0.1		0.1
Titanium	0.2	0.2	0.2	0.2		0.2
Properties of material						
Suitability for:						
Sand casting	G†	F†	G	G		U
Chill casting (gravity die)	G	F†	G	E		F
Die casting (press die)	G†	E	F†	G†		G
Strength at elevated temp.	G	G‡	E	G		G
Corrosion resistance	G	G	G	G		G
Pressure tightness	G	G	F	G		F
Fluidity	G	G	G	G		G
Resistance to hot shortness	G	G	F	G		F
Machinability	G	F	F	G		P
Melting range, °C	525–625	520–580	520–580	525–605		505–650
Casting temperature range, °C	700–740	—	670–740	680–740		Well above 650 °C
Specific gravity	2.77	2.79	2.76	2.75		2.73
Heat treatment						
Solution temperature, °C	515–530	—	—	—	—	
Solution time, h	6–9	—	—	—	—	
Quench	Water at 70–80 °C	—	—	—	—	*Stress relief*
Precipitation temperature, °C	—	—	200–210	—	—	175–225
Precipitation time, h	—	—	7–9	—	—	8(minimum)
Special properties	Chill casting alloy (grav. die)	Alloy for pressure die castings	Piston alloy, retains strength and hardness at elevated temps.	Excellent castability	Alloy for pressure die casting automobile engine cylinder blocks	
Mechanical properties—sand cast—SI units (Imperial units in brackets)						
Tensile stress min., MPa (tonf in^{-2})	—	—	—	140(9.1)	—	—
Elongation min. %	—	—	—	1	—	—
Expected 0.2% proof stress, MPa (tonf in^{-2})	—	—	—	80–90 (5.2–5.8)	—	—
Mechanical properties—chill cast—SI units (Imperial units in brackets) HB = 90–120						
Tensile strength min., MPa (tonf in^{-2})	245(15.9)	180(11.7)	210(13.6)	160(10.4)	150(9.7)	160(10.4)
Elongation min. %	8	1.5	1	2	0.5	0.5
Expected 0.2% proof stress, MPa (tonf in^{-2})	110–120 (7.1–7.8)	100–120 (6.7–7.7)	160–190 (10.4–12.3)	90–110 (5.8–7.1)	150–200 (9.7–12.9)	160–200 (10.4–12.9)

Note:
E–Excellent. F–Fair. G–Good. P–Poor. U–Unsuitable.
(Ge–General purpose alloy; Sp–Special purpose alloy as per BS 1490:1970).

Table 26.25 ALUMINIUM–COPPER ALLOYS

Specification BS 1490: 1970	LM12M(SP) / LM12TF(SP)*	[LM14-WP]†	[LM11-W] / [LM11-WP]	—	—
Aerospace					
BSL series	— / —	4L35	2L91 / 2L92	—	—
DTD series	— / —	—	— / —	361B	741A
Composition % (Single figures indicate maximum)					
Copper	9.0–11.0	3.5–4.5	4.0–5.0	4.0–5.0	3.5–4.5
Magnesium	0.2–0.4	1.2–1.7	0.10	0.10	1.2–2.5
Silicon	2.5	0.6‡	0.25	0.25	0.5
Iron	1.0	0.6‡	0.25	0.25	0.5
Manganese	0.6	0.6	0.10	0.10	0.1
Nickel	0.5	1.8–2.3	0.10	0.10	0.1
Zinc	0.8	0.1	0.10	0.10	0.1
Lead	0.1	0.05	0.05	0.05	0.1
Tin	0.1	0.05	0.05	0.05	0.05
Titanium	0.2	0.25	0.25	Ti + Nb 0.05–0.30	—
Other	—	—	—	—	Co 0.5–1.0 Nb 0.05–0.3
Properties of material					
Suitability for:					
Sand casting	F	F	F	F	F
Chill casting (gravity die)	G	G	P	P	G
Die casting (press die)	U	U	U	U	—
Strength at elevated temperature	G	E	F	—	—
Corrosion resistance	P	F	F	F	F
Pressure tightness	G	E	P	P	F
Fluidity	F	G	F	F	G
Resistance to hot shortness	G	G	P	P	G
Machinability	E	G	G	G	G
Melting range, °C	525–625	530–640	545–640	540–650	530–640
Casting temperature range, °C	700–760	700–750	680–700	675–750	710–725
Specific gravity	2.94	2.82	2.80	2.80	2.80

Heat treatment							
Solution temp., °C	—	515–520	500–520	525–545	525–545	525–545	495–505
Solution time, h	—	6	6	12–16	12–16	16 (minimum)	10 (minimum)††
Quench	—	Water at 70–80 °C	Boiling water	Water at 70–80 °C	Water at 70–80 °C	Water or oil	Oil at 80–90 °C
Precipitation temperature, °C	—	175–180	95–103§	120–140	120–170	160–70	195–205
Precipitation time, h	—	2 (minimum)	2**	1–2	12–14	8–16	4–5
Special properties	Piston alloy, now superseded by LM13 and LM26. Excellent machinability		Excellent props. at elevated temperatures Grav. die alloy	Good shock resistance		High strength alloy	
Mechanical properties—sand cast—SI units (Imperial units in brackets)							
Tensile stress min., MPa (tonf in^{-2})	—	—	220 (14.2)	220 (14.2)	280 (18.1)	324 (21.0)	263 (17.0)
Elongation %	—	—	—	7	4	—	—
Expected 0.2% proof stress, min., MPa (tonf in^{-2})	—	—	210–240 (13.6–15.5)	165–200 (10.7–12.9)	200–240 (12.9–15.5)	310 (20.1)	250 (16.2)
Mechanical properties—chill cast—SI units (Imperial units in brackets)							
Tensile stress min., MPa (tonf in^{-2})	170	278 (18.0)	280 (18.1)	265 (17.1)	310 (20.1)	402 (26.0)	340 (22.0)
Elongation %	—	—	—	13	9	4	—
Expected 0.2% proof stress, min., MPa (tonf in^{-2})	140–170	139–170 (9.0–11.0) HB 100–150	230–260 (14.9–16.8) HB 100–130	165–200 (10.7–12.9)	200–240 (12.5–15.5)	360 (23.3)	260 (16.8)

* Not included in BS 1490:1970.
†[ ] signifies obsolete specification.
‡ Si + Fe 1.0 max.
§ Or 5 days ageing at room temp.
** Can substitute stabilizing treatment at 200–250 °C if used for pistons.
†† Allow to cool to 480 °C before quench.

Note:
E–Excellent. F–Fair. G–Good. P–Poor. U–Unsuitable.
(Ge–General purpose alloy; SP–Special purpose alloy as per BS 1490:1970).

Table 26.26 MISCELLANEOUS ALUMINIUM ALLOYS

Specification BS 1400: 1970	LM5M(SP)	—	LM10TB(SP)	—
Aerospace				
BSL series	—	—	4L53	L99
DTD series	—	5018A	—	—
Composition % (Single figures indicate maximum)				
Copper	0.1	0.2	0.1	0.1
Magnesium	3.0–6.0	7.4–7.9	9.5–11.0	0.20–0.45
Silicon	0.3	0.25	0.25	6.5–7.5
Iron	0.6	0.35	0.35	0.20
Manganese	0.3–0.7	0.1–0.3	0.10	0.10
Nickel	0.1	0.1	0.10	0.10
Zinc	0.1	0.9–1.4	0.10	0.10
Lead	0.05	0.05	0.05	0.05
Tin	0.1	0.05	0.05	0.05
Titanium	0.2	0.25	0.2†	0.20
Other	—	—	—	—
Properties of material				
Suitability for:				
Sand casting	F	F	F	G
Chill casting (gravity die)	F	F	F	E
Die casting (press die)	F‡	—	F‡	F‡
Strength at elevated temp.	F	F	F	—
Corrosion resistance	E	E	E	E
Pressure tightness	P	P	P	G
Fluidity	F	F	F	G
Resistance to hot shortness	F	G	G	G
Machinability	G	G	G	F
Melting range, °C	580–642	—	450–620	550–615
Casting temperature range, °C	680–740	680–720	680–720	680–740
Specific gravity	2.65	2.64	2.57	2.67
Heat treatment				
Solution temperature, °C	—	425–435§	425–435	535–545
Solution time, h	—	8	8	12
Quench	—	Oil at 160 °C** or boiling water	Oil at no more†† than 160 °C	Water at 65 °C min
Precipitation temp., °C	—	—	—	150–160
Precipitation time, h	—	—	—	4
Special properties	Good corrosion resistance in marine atmospheres	—	Good shock resistance and high corrosion resistance	Excellent castability with good mech. props.
Mechanical properties—sand cast—SI units (Imperial units in brackets)				
Tensile stress min, MPa (tonf in^{-2})	140 (9.1)	278	280 (18.0)	230 (14.9)
Elongation %	3	3	8	2
Expected 0.2% proof stress, min MPa (tonf in^{-2})	90–110 (5.8–7.1)	170 (11.0)	170–190 (11.0–12.3)	185 (12.0)
Mechanical properties—chill cast—SI units (Imperial units in brackets)				
Tensile stress min, MPa (tonf in^{-2})	170 (11.0)	309 (20.0)	310 (20.1)	280 (18.1)
Elongation %	5	10	12	5
Expected 0.2% proof stress, min MPa (tonf in^{-2})	90–120 (5.8–7.8)	170 (11.0)	170–200 (11.0–12.9)	200 (12.9)

* [ ] obsolete.
† 0.05% min, if Ti alone used for grain refinement.
‡ Not normally used in this form.
§ Or 8 h at 435–445 °C then raise to 490–500 °C for further 8 h and quench as in table.
** Do not retain castings in oil for more than 1 h.

Table 26.26 MISCELLANEOUS ALUMINIUM ALLOYS–*continued*

Specification BS 1400:1970	LM28TE(SP)	LM28TF(SP)	[LM23P]*	[LM15WP]*	—
Aerospace					
BSL series	—	—	3L51	3L52	—
DTD series	—	—	—	—	5008B
Composition % (Single figures indicate maximum)					
Copper	1.3–1.8		0.8–2.0	1.3–3.0	0.1
Magnesium	0.8–1.5		0.05–0.2	0.5–1.7	0.5–0.75
Silicon	17–20		1.5–2.8	0.6–2.0	0.25
Iron	0.7		0.8–1.4	0.8–1.4	0.5
Manganese	0.6		0.1	0.1	0.1
Nickel	0.8–1.5		0.8–1.7	0.5–2.0	0.1
Zinc	0.2		0.1	0.1	4.8–5.7
Lead	0.1		0.05	0.05	0.1
Tin	0.1		0.05	0.05	0.05
Titanium	0.2		0.25	0.25	0.15–0.25
Other	Cr 0.6 Co 0.5		—	—	Cr 0.4–0.6
Properties of material					
Suitability for:					
Sand casting	P		G	F	F
Chill casting (gravity die)	F		G	G	P
Die casting (press die)	—		G‡	U	U‡
Strength at elevated temp.	F		G	E	F
Corrosion resistance	G		G	G	E
Pressure tightness	F		G	F	F
Fluidity	F		F	F	F
Resistance to hot shortness	G		G	G	P
Machinability	P		G	G	G
Melting range, °C	520–675		545–635	600–645	572–615
Casting temp. range, °C	$\nless$ 735		680–750	685–755	730–770
Specific gravity	2.68		2.77	2.75	2.81
Heat treatment					
Solution temperature, °C	—	495–505	—	520–540	—
Solution time, h	—	4	—	4	—
Quench	—	Air blast	—	Water at 80–100 °C Oil or air blast	—
Precipitation temp, °C	185	185	150–175	150–180 (195–205)	175–185‡‡ (at least 24 h after cast)
Precipitation time, h	To produce required HB	8	8–24	8–24 (2–5)	10 (at least 24 h after cast)
Special properties	Piston alloy		Aircraft engine castings	High mechanical props. at elevated temps.	Good strength without heat treatment. *See*‡‡
Mechanical properties—sand cast—SI units (Imperial units in brackets)					
Tensile stress min, MPa (tonf in^{-2})	—	120 (7.8)	160 (10.4)	280 (18.1)	216 (14.0)
Elongation %	—	—	2	—	4
Expected 0.2% proof stress, min MPa (tonf in^{-2})	—	— HB 100–140	125 (8.1)	245 (15.9)	150 (9.7)
Mechanical properties—chill cast—SI units (Imperial units in brackets)					
Tensile stress min, MPa (tonf in^{-2})	170 (11.0)	190 (12.3)	200 (13.0)	325 (21.0)	232 (15.0)
Elongation %	—	—	3	—	5
Expected 0.2% proof stress min, MPa (tonf in^{-2})	— HB 90–130	160–190 (10.4–12.3) HB 100–140	140 (19.1)	295 (19.1)	180 (11.7)

†† Can be furnace cooled to 385–395 °C before quench. Do not retain in oil for more than 1 h. Further quench in water or air.
‡‡ Alternative—room temp. age-harden for 3 weeks.

Note:
E–Excellent. F–Fair. G–Good. P–Poor. U–Unsuitable.
(Ge–General purpose alloy; SP–Special purpose alloy as per BS 1490: 1970).

26.4 Copper base casting alloys

Table 26.27 GUNMETALS

Alloy	*Gunmetal* 88/10/2	*Nickel gunmetal* (*as cast*)	*Nickel gunmetal** (*fully heat treated*)
Specification BS 1400:1969	G1	G3	G3WP
Alloy grouping†	C	C	C
Composition % (Single figures indicate maximum)			
Tin	9.5–10.5		6.5–7.5
Zinc	1.75–2.75		1.5–3.0
Lead	1.5		0.10–0.50
Nickel	1.0		5.25–5.75
Silicon	0.02		0.01
Bismuth	0.03		0.02
Aluminium	0.01		0.01
Iron + arsenic + antimony	0.20		0.20
Manganese	—		0.20
Total impurities	0.50		0.50
Properties of material			
Suitability for:			
Sand casting	2§		2
Chill casting	3		3
Die casting (gravity)	3		3
Centrifugal	2		2
Continuous	1		2
Pressure tightness for sand castings:			
Thin sections	2**		1
Thick sections	2		1
Machining	2§§		2
Resistance to corrosion	All these alloys withstand atmospheric natural water and sea water corrosion to a degree depending largely upon the tin content (88/10/2 best)		
Fluidity	Addition of phosphorus in amounts of order of 0.05% for deoxidation gives high fluidity in all cases		
Hot shortness	All alloys are very hot short		
Bearing properties	All alloys suitable for bearing applications—choice dependent on conditions		
Density (g cm^{-3})	8.8		8.8
Casting temperature range, °C			
Under 13 mm section	1 200		1 080–1 200
13–39 mm section	1 170		
Over 40 mm section	1 130		
Mechanical properties—SI units (Imperial units in brackets) in order sand cast, chill cast, continuously cast, centrifugally cast†† (BS 1400)			
Tensile stress min,			
MPa	270, 230, 300, 250	280, —, 340, —	430, —, 430, —
(tonf in^{-2})	(17.5, 14.9, 19.4, 16.2)	(18.1, —, 22.0, —)	(27.8, —, 27.8, —)
0.2% proof stress min			
MPa	130, 130, 140, 130	140, —, 170, —	280, —, 280, —
(tonf in^{-2})	(8.4, 8.4, 9.1, 8.4)	(9.1, —, 11.0, —)	(18.1, —, 18.1, —)
Elongation $5.65\sqrt{S_0}$%	13, 3, 9, 5	16, —, 18, —	3, —, 3, — HB 160 min

* Heat treatment 2 h at 790 ± 10 °C air cool plus 6 h at 320 ± 10 °C air cool.
† Group A alloys in common use; group B special purpose alloys; group C alloys in limited production.
‡ Tin + ½ Nickel content 7.0–8.0%.
§ Grading 1 = Excellent.
2 = Satisfactory.
3 = Possible with special techniques.
4 = Unsuitable.
5 = Not applicable.
** Grading 1 = Suitable
2 = Less suitable.
3 = Unsuitable.
†† Values apply to samples cut from centrifugal castings made in metallic moulds. Min. props. of centrifugal castings made in sand moulds same as for other sand castings.

Table 26.27 GUNMETALS—*continued*

Alloy	*Leaded gunmetal* 83/3/9/5	*Leaded gunmetal* 85/5/5/5	*Leaded gunmetal* 87/7/3/3	*Leaded semi-red brass* 76/3/6/15
Specification				
BS 1400:1969	LG1	LG2	LG4	—
Alloy grouping†	B	A	A	A
Composition % (Single figures indicate maximum)				
Tin	2.0–3.5	4.0–6.0	6.0–8.0‡	2.5–3.5
Zinc	7.0–9.5	4.0–6.0	1.5–3.0	13.0–17.0
Lead	4.0–6.0	4.0–6.0	2.5–3.5	5.25–6.75
Nickel	2.0	2.0	2.0‡	1.0
Silicon	0.02	0.02	0.01	0.01
Bismuth	0.10	0.05	0.05	—
Aluminium	0.01	0.01	0.01	—
Iron + arsenic + antimony	0.75	0.50	0.40	0.55
Manganese	—	—	—	—
Total impurities	1.0	0.80	0.70	—
Properties of material				
Suitability for:				
Sand casting	1	1	1	1
Chill casting	2	2	2	2
Die casting (gravity)	3	3	3	3
Centrifugal	2	1	1	2
Continuous	2	1	1	2
Pressure tightness for sand castings:				
Thin sections	1	1	2	2
Thick sections	2	2	1	2
Machining	1	1	1	2
Resistance to corrosion	All these alloys withstand atmospheric natural water and sea water corrosion to a degree depending largely upon the tin content (88/10/2 best)			
Fluidity	Addition of phosphorus in amounts of order of 0.05% for deoxidation gives high fluidity in all cases			
Hot shortness	All alloys are very hot short			
Bearing properties	All alloys suitable for bearing applications—choice dependent on conditions			
Density (g cm^{-3})	8.8	8.8	8.8	8.77
Casting temperature range °C				
Under 13 mm section	1 180	1 200	1 200	1 150–1 260
13–39 mm section	1 140	1 150	1 160	1 110–1 220
Over 40 mm section	1 100	1 120	1 120	1 066–1 177
Mechanical properties—SI units (Imperial units in brackets) in order sand cast, chill cast, continuously cast, centrifugally cast†† (BS 1400)				
Tensile stress min				
MPa	180, 180, –, –	200, 200, 270, 220	250, 250, 300, 250	Sand cast 172
(tonf in^{-2})	(11.7, 11.7, —, —)	(13.0, 13.0, 17.5, 13.0)	(16.2, 16.2, 19.4, 16.2)	(Sand cast 11.6)
0.2% proof stress min, MPa	80, —, —, —	100, 110, 100, 110	130, 130, 130, 130	0.5% proof stress sand cast 83
(tonf in^{-2})	(5.2, —, —, —)	(6.5, 7.1, 6.5, 7.1)	(8.4, 8.4, 8.4, 8.4)	(Sand cast 5.4)
Elongation $5.65\sqrt{S_0}$%	11.2, —, —	13, 6, 13, 8	16, 5, 13, 6	Sand cast 15 (50.8 mm GL)

§§ Grading 1 = Excellent
2 = Good
3 = Satisfactory with special techniques.
(Comparison between copper alloys rather than with other metals).

Table 26.28 BEARING BRONZES

Alloy	*Phosphor bronze*	*Phosphor bronze for gear blanks*	*Phosphor bronze*	*Phosphor bronze*	*Leaded phosphor bronze*
Specification BS 1400:1969	PB1	PB2	CT1	PB4	LPB1
Alloy grouping†	B	B	B	A	A
Composition % (Single figures indicate maximum unless otherwise stated)					
Tin	10.0 min	11.0–13.0	9.0–11.0	9.5 min	6.5–8.5
Phosphorus	0.50 min	0.15 min	0.15 max	0.40 min	0.30 min
Lead	0.25	0.50	0.25	0.75	2.0–5.0
Zinc	0.05	0.30	0.05	0.50	2.0
Nickel	0.10	0.50	0.25	0.50	1.0
Iron	0.10	0.15	—	—	—
Silicon	0.02	0.02	—	—	—
Aluminium	0.01	0.01	—	—	—
Total impurities	0.60	0.20	0.80	0.50	0.50
Copper	REM	REM	REM	REM	REM
Properties of material					
Suitability for:					
Sand casting	2‡	2	2	2	2
Chill casting	1	1	1	1	1
Die casting (gravity)	3	3	3	3	3
Centrifugal	1	1	1	1	2
Continuous	1	1	1	1	1
Pressure tight sand castings					
Thin sections	3§	3	2	3	2
Thick sections	3	3	3	3	2
Machining	2‡‡	2	2	2	1
Corrosion resistance	All alloys are resistant to natural and sea water				
Fluidity	Good	Good	Good	Good	Good
Hot shortness	All alloys are very hot short				
Density (g cm^{-3})	8.8	8.8	8.8	8.8	8.8
Casting temperature range, °C					
Under 13 mm section	1 120	1 170	1 100	1 120	1 130
13–39 mm section	1 100	1 120	1 070	1 100	1 050
Over 40 mm section	1 040	1 070	1 040	1 060	1 030
Mechanical properties—SI units (Imperial units in brackets) in order sand cast, chill cast, continuously cast, centrifugally cast** BS 1400					
Tensile stress (min)					
MPa	210, 310, 360, 330	220, 270, 310, 280	230, 270, 310, 280	190, 270, 330, 280	190, 220, 270, 230
(tonf in^{-2})	(13.6, 20.1, 23.3, 21.4)	(14.2, 17.5, 20.1, 18.1)	(14.9, 17.5, 20.1, 18.1)	(12.3, 17.5, 21.4, 18.1)	(12.3, 14.2, 17.5, 14.9)
0.2% proof stress (min)					
MPa	130, 170, 170, 170	130, 170, 170, 170	130, 140, 160, 140	100, 140, 160, 140	80, 130, 130, 130
(tonf in^{-2})	(8.4, 11.0, 11.0, 11.0)	(8.4, 11.0, 11.0, 11.0)	(8.4, 9.1, 10.4, 9.1)	(6.5, 9.1, 10.4, 9.1)	(5.2, 8.4, 8.4, 8.4)
Elongation on $5.65\sqrt{S_0}$%	3, 2, 6, 4	5, 3, 5, 3	6, 5, 9, 6	3, 2, 7, 4	3, 2, 5, 4

Alloy	76/9/0/15 *Leaded bronze*	80/10/0/10 *Leaded bronze*	85/5/0/10 *Leaded bronze*	75/5/0/20 *Leaded bronze*
Specification B.S. 1400:1969	LB1	LB2	LB4	LB5
Alloy grouping†	C	A	A	B
Composition % (Single figures indicate maximum unless otherwise stated)				
Tin	8,0–10.0	9.0–11.0	4.0–6.0	4.0–6.0
Phosphorus	0.10	0.10	0.10	0.10
Lead	13.0–17.0	8.5–11.0	8.0–10.0	18.0–23.0
Zinc	1.0	1.0	2.0	1.0
Nickel	2.0	2.0	2.0	2.0
Iron	Sb 0.50	Sb 0.50 Fe 0.15	Sb 0.50	Sb 0.50
Silicon	0.02	0.02	0.02	0.01
Aluminium	—	0.01	—	—
Total impurities	0.30	0.50	0.50	0.30
Copper	REM	REM	REM	REM
Properties of material				
Suitability for:				
Sand casting	3	2	2	3
Chill casting	2	1	1	2
Die casting (gravity)	4	4	4	4
Centrifugal	2	2	2	3
Continuous	1	1	1	3
Pressure tight sand castings				
Thin sections	2	2	2	2
Thick sections	3	2	2	3
Machining	1	1	1	1
Corrosion resistance	All alloys are resistant to natural and sea water			
Fluidity	Good	Good	Good	Good
Hot shortness	All alloys are very hot short			
Density (g cm^{-3})	9.1	9.0	9.0	9.2
Casting temperature range, °C				
Under 13 mm section	1 110	1 130	1 110	1 090
13–39 mm section	1 030	1 080	1 070	1 030
Over 40 mm section	1 010	1 030	1 040	1 010
Mechanical properties—SI units (Imperial units in brackets) in order sand cast, chill cast, continually cast, centrifugally cast** BS 1400				
Tensile stress (min)				
MPa	170, 200, 230, 220	190, 220, 280, 230	160, 200, 230, 220	160, 170, 190, 190
(tonf in^{-2})	(11.0, 13.0, 14.9, 14.2)	(12.3, 14.2, 18.1, 14.9)	(10.4, 13.0, 14.9, 14.2)	(10.4, 11.1, 12.3, 12.3)
0.2% proof stress (min)				
MPa	80, 130, 130, 130	80, 140, 160, 140	60, 80, 130, 80	60, 80, 100, 80
(tonf in^{-2})	(5.2, 8.4, 8.4, 8.4)	(5.2, 9.1, 10.4, 9.1)	(3.9, 5.2, 8.4, 5.2)	(3.9, 5.2, 6.5, 5.2)
Elongation on $5.65\sqrt{S_0}$%	4, 3, 9, 4	5, 3, 6, 5	7, 5, 9, 6	5, 5, 8, 7

* This bronze is now primarily intended for shaped castings rather than bearings.

† *See* footnote † to Table 26.27.

‡ *See* footnote § to Table 26.27.

§ *See* footnote** to Table 26.27.

** *See* footnote †† to Table 26.27.

‡‡ *See* footnote §§ to Table 26.27.

Table 26.29 BRASSES

Alloy	*Brass sand cast*	*Brass sand cast*	*Naval brass sand cast*
Specification BS 1400:1969	SCB1	SCB3	SCB4
Alloy grouping	A	A	C
Composition % (Single figures indicate maximum unless otherwise stated)			
Copper	70.0–80.0	63.0–70.0	60.0–63.0
Lead	2.0–5.0	1.0–3.0	0.5
Tin	1.0–3.0	1.5	1.0–1.5
Iron	0.75	0.75	—
Nickel	1.0	1.0	—
Aluminium	0.01	0.1	0.01
Others	—	—	—
Total impurities	1.0	1.0	0.75
Zinc	REM	REM	REM
Properties of material			
Suitability for:			
Sand casting	1‡	1	1
Chill casting	5	5	5
Die casting (gravity)	5	5	5
Centrifugal	3	3	3
Continuous	2	2	5
Pressure tight castings (sand)			
Thin sections	1§	1¶	1
Thick sections	1	1¶	1
Machining	1§§	1	2
Corrosion resistance	Generally excellent in natural waters, DCB1, DCB3, PCB1, SCB1, SCB3 undergo dezincification in sea water		
Hot shortness	Alpha brasses tend to be hot short. Alpha/beta brasses relatively good		
Density (g cm^{-3})	8.5	8.4	8.3
Casting temperature range, °C			
Under 13 mm section	1 150	1 100	1 100
13–39 mm section	1 100	1 050	1 050
Over 40 mm section	1 070	1 020	1 020
Mechanical properties—SI units (Imperial units in brackets)—sand cast props.** for alloys SCB1, 3, 4, 6—chill cast‡‡ for DCB1 and 3 and PCB1			
Tensile stress (typical)			
MPa	170–200	190–220	250–310
(tonf in^{-2})	(11.0–13.0)	(12.3–14.2)	(16.2–20.1)
0.2% proof stress (typical)			
MPa	80–110	70–110	70–110
(tonf in^{-2})	(5.2–7.1)	(4.5–7.1)	(4.5–7.1)
Elongation on $5.65\sqrt{S_0}$% (typical)	18–40	11–30	18–40

* 0.1% Pb if required.

† Nickel to be counted as copper.

‡, § *See* footnotes § and ** to Table 26.27.

¶ For pressure tight castings Al ≤ 0.02%.

** On separately cast test bars.

†† Values based on 15–40 mm thick sections from castings.

‡‡ Values based on 15–40 mm sections for DCB1, DCB3 and PCB1.

§§ *See* footnote §§ to Table 26.27.

Alloy	*Brass brazable castings*	*Brass die castings*	*Brass for diecasting*	*Brass for pressure diecasting*
Specification BS 1400:1969	SCB6	DCB1*	DCB3†	PCB1
Alloy grouping	A	A	A	A
Composition % (Single figures indicate maximum unless otherwise stated)				
Copper	83.0–86.0	59.0–63.0	58.0–63.0	57.0–60.0
Lead	0.5	0.25	0.5–2.5	0.5–2.5
Tin	—	—	1.0	0.5
Iron	—	—	0.8	0.3
Nickel	—	—	1.0	—
Aluminium	—	0.5	0.2–0.8	0.5
Others	As 0.05–0.20	—	Mn 0.5 Si 0.05	—
Total impurities	1.0 (incl. Pb)	0.75	2.0 (excl. Ni + Pb + Al)	0.5
Zinc	REM	REM	REM	REM

Table 26.29 BRASSES—*continued*

Alloy	*Brass brazable castings*	*Brass die castings*	*Brass for diecasting*	*Brass for pressure diecasting*
Specification BS 1400:1969	SCB6	DCB1*	DCB3†	PCB1
Alloy grouping	A	A	A	A
Properties of material				
Suitability for:				
Sand casting	1	5	5	5
Chill casting	5	1	1	1
Die casting (gravity)	5	1	1	1
Centrifugal	3	2	2	2
Continuous	2	5	3	5
Pressure tight castings (sand)				
Thin sections	1	—	—	—
Thick sections	1	—	—	—
Machining	3	2	1	2
Corrosion resistance	Generally excellent in natural waters. DCB1, DCB3, PCB1, SCB1, SCB3 undergo dezincification in sea water			
Hot shortness	Alpha brasses tend to be hot short. Alpha/beta brasses relatively good			
Density (g cm^{-3})	8.6	8.3	8.3	8.3
Casting temperature range, °C				
Under 13 mm section	1 150	1 050	1 050	Injection temp. 950
13–39 mm section	1 100			
Over 40 mm section	1 070			
Mechanical properties—SI units (Imperial units in brackets)—sand cast props.,**for alloys SCB1, 3, 4, 6—chill cast‡‡ for DCB1 and 3 and PCB1				
Tensile stress (typical)				
MPa	170–190	280–370	300–340	280–370
(tonf in^{-2})	(11.0–12.3)	(18.1–24.0)	(19.4–22.0)	(18.1–24.0)
0.2% proof stress (typical)				
MPa	80–110	90–120	90–120	90–120
(tonf in^{-2})	(5.2–7.1)	(5.8–7.8)	(5.8–7.8)	(5.8–7.8)
Elongation on $5.65\sqrt{S_0}$% (typical)	18–40	23–50	13–40	25–40

Table 26.30 HIGH TENSILE BRASSES
Zinc equivalents

Guillet method		*American method*	
Element	*Coefficient*	*Element*	*Coefficient**
Silicon	10	Silicon	+10
Aluminium	6	Aluminium	+5.0
Tin	2	Tin	+1.0
Lead	1	Lead	—
Iron	0.9	Iron	−0.1
Manganese	0.5	Manganese	−0.5
Nickel	−1.2	Nickel	−2.3
Magnesium	2	Magnesium	+1.0
Zinc equivalent $=\dfrac{A}{B}\times 100$		Zinc equivalent $=\left[100-\dfrac{100\times\%\text{Cu}}{100+\Sigma(\%M\times\text{coeff})}\right]$	

Where
A = sum of (zinc equivalent × % of element of each alloying element) + % zinc
B = A + % copper
M = alloying element
% α-phase = 10 (46.6-zinc equivalent)
In practice, these factors operate with good accuracy, provided that the element considered is not present in amounts greater than 2%.
* These values are Guillet Coefficients minus 1.

Table 26.30 HIGH TENSILE BRASSES—*continued*

Alloy	*Alpha Beta* (460 MPa†) *High tensile brass*	*All Beta* (740 MPa‡) *High tensile brass*
Specification BS 1490: 1969	HTB1§	HTB3
Alloy grouping	B	B
Composition % (Single figures indicate maximum)		
Copper	55.0 min	55.0 min
Manganese	3.0	4.0
Aluminium	0.5–2.5	3.0–6.0
Iron	0.7–2.0¶	1.5–3.25
Tin	1.0	0.20
Nickel	1.0	1.0
Lead	0.50	0.20
Silicon	0.10	0.10
Total impurities	0.20	0.20
Zinc	REM	REM
Properties of material		
Suitability for:		
Sand casting	2††	2
Chill casting	5	5
Gravity die casting	2	4
Centrifugal	2	2
Continuous	5	5
Pressure tight sand castings		
Thin sections	1‡‡	1
Thick sections	1	1
Machining	3**	3
Corrosion resistance	α/β structure–not susceptible to stress corrosion cracking	All β alloy–susceptible to stress corrosion cracking
Density (g cm^{-3})	8.3	7.9
Casting temperature range, °C		
Under 13 mm section	1 060	1 060
13–39 mm section	1 020	1 020
Over 40 mm section	980	980
Mechanical properties—SI units (Imperial units in brackets) in order sand cast, chill cast, continuously cast, centrifugally cast§§		
Tensile stress, min		
MPa	470, 500, —, 500	740, —, —, 740
(tonf in^{-2})	(30.4, 32.4, —, 32.4)	(47.9, —, —, 47.9)
0.2% proof stress, min		
MPa	170, 210, —, 210	400, —, —, 400
(tonf in^{-2})	(11.0, 13.6, —, 13.6)	(25.9, —, —, 25.9)
Elongation on $5.65\sqrt{S_0}$%min	18, 18, —, 20	11, —, —, 13

† Originally known as 30 ton HTB.
‡ Originally known as 48 ton HTB.
§ Micro-structure requirement: 15%α-phase min.
¶ For grain refinement (grain dia. 0.5 mm or less on 29 mm sand cast test bars). Optimum iron for grain refinement approx.–1.1% in HTB 1.
†† For grading *see* footnote § to Table 26.27.
‡‡ For grading *see* footnote** to Table 26.27.
§§ Values apply to samples cut from castings made in metallic moulds.
** *See* footnote §§ to Table 26.27.

Table 26.31 MISCELLANEOUS COPPER-BASE ALLOYS

Alloy	*Aluminium bronze*	*Aluminium bronze*	*Copper manganese aluminium*	*Copper manganese aluminium*
Specification BS 1400:1969	AB1	AB2	CMA1	CMA2
*Alloy grouping**	B	B	B	B
Composition % (Single figures indicate maximum)				
Aluminium	8.5–10.5	8.8–10.0	7.5–8.5	8.5–9.0
Iron	1.5–3.5	4.0–5.5	2.0–4.0	2.0–4.0
Nickel	1.0	4.0–5.5	1.5–4.5	1.5–4.5
Manganese	1.0	1.5	11.0–15.0	11.0–15.0
Zinc	0.50	0.50	0.50	0.50
Silicon	0.25	0.10	0.15	0.15
Tin	0.10	0.10	1.0	1.0
Lead	0.05†	0.05†	0.05	0.05
Phosphorus	—	—	0.05	0.05
Magnesium	0.05	0.05	—	—
Chromium	—	—	—	—
Copper	REM	REM	REM	REM
Total impurities	0.30	0.30	0.30	0.30
Properties of material				
Suitability for:				
Sand casting	2‡	2	2	2
Chill casting	5	5	5	5
Gravity die casting	1	2	2	2
Centrifugal	2	2	2	2
Continuous	3	3	3	3
Pressure tight castings (sand)				
Thin sections	1§	1	1	1
Thick sections	1	1	1	1
Machining	3¶	3	3	3
Corrosion resistance	Very good except under reducing conditions**	Very good except under reducing conditions**	Very good	Very good
Fluidity	Fair	Fair	Good	Good
Strength at elevated temperatures	Very good	Excellent	Very good	Very good
Electrical conductivity % IACS at 15 °C	13	8	3	3
Density ($g\,cm^{-3}$)	7.6	7.6	7.5	7.5
Casting temperature range, °C				
Under 13 mm section	1 250	1 240	1 150	1 150
13–39 mm section	1 200	1 170	1 100	1 100
Over 40 mm section	1 150	1 120	1 050	1 050

Table 26.31 MISCELLANEOUS COPPER-BASE ALLOYS—*continued*

Alloy	*Aluminium bronze*	*Aluminium bronze*	*Copper manganese aluminium*	*Copper manganese aluminium*
Specification BS 1400:1969	AB1	AB2	CMA1	CMA2
*Alloys grouping**	B	B	B	B
Mechanical properties—SI units (Imperial units in brackets) in order sand cast, chill cast, centrifugally cast				
Typical tensile stress				
MPa	500–590, 540–620, 560–650	640–700, 650–740, 670–730	650–730, 670–740, —	740–820, —, —
(tonf in^{-2})	(32.4–38.2, 35.0–40.1,36.3–42.1)	(41.4–45.3, 42.1–47.9, 43.4–47.3)	(42.1–47.3, 43.4–47.9, —)	(47.9–53.1, —, —)
Typical 0.2% proof stress				
MPa	170–200, 200–270, 200–270	250–300, 250–310, 250–310	280–340, 310–370, —	380–470, —, —
(tonf in^{-2})	(11.0–13.0, 13.0–17.5, 13.0–17.5)	(16.2–19.4, 16.2–20.1, 16.2–20.1)	(18.1–22.0, 20.1–24.0, —)	(24.6–30.4, —, —)
Typical elongation $5.65\sqrt{S_0}\%$	18–40, 18–40, 20–30	13–20, 13–20, 13–20	18–35, 27–40, —	9–20, —, —
Typical hardness range HB	90–140, 130–160, 120–160	140–180, 160–190, 140–180	160–210, —, —	220–260, —, —

Alloy	*Nickel silver*	*High-conductivity copper*	*Copper chromium*	*High-strength cupro-nickel*
Specification BS 1400:1969	—	HCC1**	CC1–WP††	—
*Alloy grouping**	—	B	B	B
Composition % (Single figures indicate maximum)				
Aluminium	—	—	—	—
Iron	1.0	—	—	1.0–1.4
Nickel	18–22	—	—	28.0–32.0
Manganese	1.0	—	—	1.0–1.4
Zinc	6.0–10.0	—	—	—
Silicon	0.15	—	—	0.35–0.50§§
Tin	2.5–4.0	—	—	—
Lead	4.0–6.0	—	—	0.01
Phosphorus	0.05	—	—	0.02
Magnesium	—	—	—	—
Chromium	—	—	0.6–1.2	0.3
Copper	REM	B.S. 1035–1037 grade	REM	REM
Total impurities	1.0 (incl. Fe)	—	—	—¶¶

Properties of material				
Suitability for:				
Sand casting	2	2	2	2
Chill casting	—	5	5	2
Gravity die casting	—	3	3	—
Centrifugal	—	2	2	—
Continuous	—	3	3	—
Pressure tight castings (sand)				
Thin sections	—	1	2	1
Thick sections	—	2	2	2
Machining	2	3	3	2
Corrosion resistance	Natural and sea-water applications	Very good	Very good	Excellent
Fluidity	Fair	Good	Fair	Fair
Strength at elevated temperatures	—	Oxidize more readily than the other alloys specified in BS 1400		Good up to 300 °C
Electrical conductivity % IACS at 15 °C	5	90	80	5
Density (g cm^{-3})	8.7	8.9	8.85	—
Casting temperature range, °C				
Under 13 mm section	1 300	1 200	1 230	1 450
13–39 mm section	1 280	1 170	1 190	1 425
Over 40 mm section	1 250	1 130	1 150	1 400
Mechanical properties—SI units (Imperial units in brackets) in order sand cast, chill cast, centrifugally cast				
	Sand cast			Sand cast
Typical tensile stress				
MPa	309	160–190, —, —	270–340, —, —	540–618
(tonf in^{-2})	(20.0)	(10.4–12.3, —, —)	(17.5–22.0, —, —)	(35–40)
Typical 0.2% proof stress	(0.5%)			(0.5%)
MPa	185	—	170–250, —, —	355–386
(tonf in^{-2})	(12.0)	—	(11.0–16.2, —, —)	(23–25)
Typical elongation 5.65$\sqrt{S_0}$%	(4$\sqrt{S_0}$)			(4$\sqrt{S_0}$)
	15.0	23–40, —, —	18–30, —, —	20–25
Typical hardness range HB	80–110	—	Mandatory 100 min	160 (typical)

* *See* footnote † to Table 26.27.
† When castings to be welded lead not to exceed 0.01%
‡ *See* footnote § to Table 26.27.
§ *See* footnote ** to Table 26.27.
¶ *See* footnote §§ to Table 26.27.
** Maximum resistivity 0.019 $\mu\Omega$ m^{-1}.
†† Maximum resistivity 0.022 $\mu\Omega$ m^{-1}.
§§ Can be reduced to 0.2% to improve weldability under severe constraint.
¶¶ Others C 0.04% S 0.02%
Cr 0.30% Nb 1.20–1.40%

26.5 Nickel-base casting alloys

Table 26.32 TYPICAL COMPOSITION

		Chemical					
Alloy	*Trade name*	C	Si	Mn	Cu	Cr	Fe
	(Single figures indicate maximum, (n) indicates nominal composition)						
Nickel	—	0.1–0.3	1.0–2.0	1.0–1.5	0.3	—	1.0
Ni–Cu	Monel†	0.1–0.3	0.5–1.5	0.5–1.5	28–32	—	3.0
Ni–Cu	Monel	0.15	2.5–3.0	0.5–1.5	28–32	—	3.0
Ni–Cu	Monel 'S'	0.15	3.5–4.5	0.5–1.5	28–32	—	3.0
Ni–Fe–Cr	Nichrome† Cronite†	0.8	1.5–2.0	0.5–1.5	—	15–19	9–16
Ni–Mo	Hastelloy† B	0.12	1.0	1.0	—	1.0	4.0–6.0
Ni–Mo–Cr–W	Hastelloy C	0.12	1.0	1.0	—	15.5–17.5	4.5–7.0
Ni–Si	Hastelloy D	0.12	8.5–10.0	0.5–1.25	2.0–4.0	1.0	2.0
Ni–Cr–Fe–Mo	Hastelloy X	0.05–0.15	1.0	1.0	—	20.5–23.0	17.0–20.0
Ni–Cr–Ti–Al	Nimocast† 80	0.07 (n)‡	0.4 (n)	0.4 (n)	—	19.5 (n)	2.0
Ni–Cr–Co–Ti–Al	Nimocast 90	0.07 (n)	0.4 (n)	0.4 (n)	—	19.5 (n)	2.0
Ni–Co–Cr–Mo	Nimocast 242	0.35 (n)	0.4 (n)	0.5	0.2	22.0 (n)	0.75
Ni–Cr–Mo–Nb–W	Nimocast PE10	0.05	0.25 (n)	0.30 (n)	0.2	20.0 (n)	3.0 (n)
Ni–Cr–Mo–Al–W–Nb	Nimocast PD16	0.13 (n)	0.5	0.5	0.5	6.0 (n)	0.5
Ni–Cr–Co–Mo–Ti–Al	Nimocast 263	0.06 (n)	0.4	0.6	0.2	20.0 (n)	0.7
Ni–Cr–Co–Mo–Ti–Al	Alloy IN–100	0.18 (n)	0.2	0.2	0.2	9.5–10.5	0.5
Ni–Cr–Mo–Ti–Al	Alloy IN–162	0.10–0.15	0.3	0.2	—	9.0–11.0	0.5
Ni–Cr–Co–W–Nb	Alloy IN–643	0.40–0.55	1.0	—	—	24.0–26.0	5.0
Ni–Cr–Nb	Alloy IN–657	0.1	—	—	—	48.0–52.0	1.0
Ni–Cr–Mo–Ti–Al–Nb	Alloy 713	0.08–0.20	0.50	0.25	0.50	12.0–14.0	2.5
Ni–Cr–Mo–Ti–Al–Nb	Alloy 713LC	0.03–0.07	0.50	0.25	0.5	11.0–13.0	0.5

* Not to be used for specification purposes.
† Registered Trade Mark.
‡ (n) = nominal composition.

Table 26.33 TYPICAL PROPERTIES

Alloy	*Trade name*	*Casting characteristics*	*Melting range* °C	*Casting temperature* °C	*Nominal density* kg dm^{-3}
Nickel	—	Carbon 0.2–0.3% for better castability	1 360–1 430	1 500–1 600	8.4
Ni–Cu	Monel 401	Somewhat hot short Castability improves as % Si is increased	1 315–1 350	1 500–1 560	8.6

OF SOME NICKEL-BASE CASTING ALLOYS

*composition (%)**							*Standard specification*
Mo	Co	Ti	Al	W	Nb	*Others*	
		(Single figures indicate maximum, (n) indicates nominal composition)					
—	—	—	—	—	—	Mg 0.08–0.12, S 0.05 max, Pb 0.005 max	—
—	—	—	—	—	—	Mg 0.08–0.12, S. 0.05 max, Pb 0.005 max	BS 3071 (1959) NA 1, ASTM 296–71 M35 (2.0% max silicon)
—	—	—	—	—	—	Mg 0.08–0.12, S 0.05 max, Pb 0.005 max	BS 3071 (1959) NA 2
—	—	—	—	—	—	Mg 0.08–0.12, S 0.05 max, Pb 0.005 max	BS 3071 (1959) NA 3
—	—	—	—	—	—	—	—
26–30	2.5	—	—	—	—	V 0.2–0.6, S 0.05 max, Pb 0.005 max	AMS 5396
16–18	2.5	—	—	3.75–5.25	—	S 0.05 max, Pb 0.005 max	AMS 5388C, AMS 5389A
—	1.5	—	—	—	—	—	—
8.0–10.0	0.5–2.5	—	—	0.2–1.0	—	—	AMS 5390
—	—	2.5(n)	1.5(n)	—	—	Ca 0.02 max, Pb 0.005 max	—
—	17(n)	2.5(n)	1.5(n)	—	—	Ca 0.02 max, Pb 0.005 max	—
10.5(n)	10.0(n)	0.3	0.2	—	—	Pb 0.005 max	—
6.0(n)	—	—	—	2.5(n)	7.0(n)	—	AFNOR NC 20 Nb AICMA Ni–C103 HT
2.0(n)	—	0.5	6.0(n)	11.0(n)	1.5(n)	—	—
6.0(n)	20.0(n)	2.0(n)	0.5(n)	—	—	—	—
2.75–3.50	15.0(n)	5.0(n)	5.5(n)	—	—	V 1.0(n)	AFNOR–NK15 CAT. AMS 5397, AICMA–104HT Ni–C
3.5–4.5	—	0.6–1.2	6.2–6.7	2.0(n)	3.0(n)	—	—
0.3–0.8	11.0–13.0	0.01–0.25	—	9.0(n)	2.0(n)	—	—
—	—	—	0.1(n)	—	1.5(n)	—	—
3.8–5.2	—	0.5–1.0	5.5–6.5	—	2.0(n)	—	DIN G–NiCr, 13A16Mo Nb. W. No. 24670, AFNOR NC13 AD-H, AICMA Ni–C98, AMS 5391A
3.8–5.2	—	0.4–1.0	5.5–6.5	—	2.0(n)	—	—

OF SOME NICKEL-BASE CASTING ALLOYS

		Mechanical properties		
Special properties	*Normal† heat treatment*	*Tensile stress* MPa ($tonf\ in^{-2}$)	*0.1% proof stress* MPa ($tonf\ in^{-2}$)	*Elongation* %
Corrosion resistant to alkalis	None	390 (25)	125 (8)	20
Good corrosion resistance to wide range of chemicals	None	495 (32)	155 (10)	25

Table 26.33 TYPICAL PROPERTIES OF SOME NICKEL-BASE CASTING ALLOYS—*continued*

Alloy	*Trade name*	*Casting characteristics*	*Melting range* °C	*Casting temperature* °C	*Nominal density* $kg\,dm^{-3}$
Ni–Cu	Monel 505	Ni–Cu alloys tend to be hot-short	1 290–1 320	1 470–1 530	8.6
Ni–Cu	'S' Monel		1 260–1 290	1 420–1 500	8.5
Ni–Fe–Cr	Nichrome Cronite	Susceptible to hydrogen absorption	1 375–1 425	1 540–1 620	8.2
Ni–Mo	Hastelloy B	Less liable to hydrogen absorption than the Ni–Fe–Cr alloys	1 320–1 350	1 500–1 600	9.2
Ni–Mo–Cr–W	Hastelloy C	Castability–moderate	1 270–1 310	1 500–1 550	8.9
Ni–Si	Hastelloy D	Castability–similar to Monels. Somewhat hot-short	1 110–1 120	1 300–1 400	7.8
Ni–Cr–Fe–Mo	Hastelloy X	Moderate	1 380–1 400	1 480–1 520	8.2
Ni–Cr–Ti–Al	Nimocast 80		1 380–1 400	1 480–1 520	8.2
Ni–Co–Cr–Ti–Al	Nimocast 90	Moderate	1 380–1 400	1 480–1 520	8.2
Ni–Co–Cr–Mo	Nimocast 242		1 370–1 400	1 480–1 520	8.4
Ni–Cr–Mo–Nb–W	Nimocast PE10		1 300–1 330	1 450–1 480	8.6
Ni–Cr–Mo–Al–W–Nb	Nimocast PD16		1 300–1 375	1 450–1 500	8.5
Ni–Cr–Co–Mo–Ti–Al	Nimocast 263	Vacuum cast	1 300–1 350	1 450–1 500	8.4
Ni–Cr–Co–Mo–Ti–Al	Alloy IN–100		1 220–1 300	1 450–1 500	7.8
Ni–Cr–Mo–Ti–Al	Alloy IN–162		1 275–1 300	1 450–1 500	8.1
Ni–Cr–Mo–W–Nb	Alloy IN–643	Moderate	1 450–1 500	1 570–1 600	8.7
Ni–Cr–Nb	Alloy IN–657	Moderate	1 480–1 500	1 530–1 550	8.4
Ni–Cr–Mo–Ti–Al–Nb	Alloy IN–713C	Vacuum cast	1 260–1 290	1 500–1 550	7.9
Ni–Cr–Mo–Ti–Al–Nb	Alloy IN–713LC	Vacuum cast	1 290–1 320	1 640–1 660	7.9

† Heat treatment may be varied to suit specific applications.
‡ 0.2% Proof stress.

		Mechanical properties		
Special properties	*Normal*† *heat treatment*	*Tensile stress* MPa (tonf in^{-2})	*0.1% proof stress* MPa (tonf in^{-2})	*Elongation* %
Similar to above but gall-resisting	Can be precipitation-treated between 350 °C and 450 °C to increase hardness	590 (38)	230 (15)	12
Similar to above but with maximum hardness and gall-resistance		695 (45)	—	Nil
Heat and oxidation resistant. For furnace and heat treatment equipment	None	465 (30)	185 (12)	10
Good corrosion resistance to wide range of chemicals notably hydrochloric acid	Sand castings: air cool from 1120°C Investment castings – none	540 (35)	390 (25)	8
Good corrosion resistance to wide range of chemicals in severe service conditions	Sand castings: air cool from 1200°C Investment castings – none	540 (35)	310 (20)	8
Corrosion resistant to hot concentrated sulphuric acid	Furnace cool from 1050°C and handle with care	770 (5)	—	1
For high temperature service, mainly in gas turbines	Sand castings: air cool from 1200°C Investment castings – none	480 (31)	320 (21)	12
High temperature service: gas turbines and furnace equipment	Solution treat up to 8 h at 1080°C Age up to 16 h at 700°C	770 (50)	525 (34)	14
As for Nimocast 80	As for Nimocast 80	730 (47)	525 (34)	12
Good high temperature thermal shock resistance	None	480 (31)	260 (17)	7
High temperature service in gas turbines, tubochargers, diesel engines	Solution treat for up to 4 h at 1100°C. Age up to 16 h at 750°C. Also used 'as-cast'	710 (46)	560 (36)	10
High temperature service. Gas turbines rotor and stator blades	None	750 (49)	680 (44)	3
As for Nimocast PD16	None	1 000 (65)	650 (42)	—
As for Nimocast PD16	None	900 (58)	850 (55)	7
As above	None	930 (60)	770 (50)	6
High temperature service. Stressed tubes for chemical plant reformers	None	600 (39)	280 (18)	12
High temperature service. Notable resistance to fuel-ash corrosion	Air cool from 1100°C	750 (49)	450 (29)	25
High temperature service. Rotor and stator blades. Integral rotors. Turbochargers	None	865 (56)	740‡ (48)	8
As for alloy IN–713C	None	910 (58)	740‡ (48)	12

26.6 Magnesium alloys

Table 26.34 MAGNESIUM–ALUMINIUM ALLOYS

Grain refined (0.05–0.2 mm *chill cast) then superheated to 850–900 °C or suitably treated with carbon (as hexachlorethane)*
(From: E. F. Emley, 'Principles of Magnesium Technology', Pergamon Press, Oxford, 1966)

Electron designation	A8		A8 (*High purity*)	
Specifications BS 2970:1972	MAG 1M* (GP)†	MAG 1TB* (GP)	MAG2M (SP)†	MAG2TB (SP)
BSS L series	—	3L. 122	—	—
Equivalent DTD	—	—	684A	690A
Composition % (Single figures indicate maximum)				
Aluminium	7.5–9.0		7.5–9.0	
Zinc	0.3–1.0		0.3–1.0	
Manganese	0.15–0.4		0.15–0.7	
Copper	0.15		0.005	
Silicon	0.3		0.01	
Iron	0.05		0.003	
Nickel	0.01		0.001	
Cu + Si + Fe + Ni	0.40		—	
Material properties				
Founding	Good		Good	
Characteristics	Sand and permanent‡ mould		Special melting technique required	
Tendency to hot tearing	Little		Little	
Tendency to micro–porosity	Appreciable		Appreciable	
Castability§	A		A	
Weldability (Ar-Arc process)	Good		Good	
Relative damping capacity¶	C		C	
Strength at elevated temperature**	C		C	
Corrosion resistance	Moderate		Excellent	
Density, g cm^{-3}	1.81		1.81	
Liquidus, °C	600		600	
Solidus, °C	475		475	
Non-equilibrium solidus, °C	420		420	
Casting temperature range, °C	680–800		680–800	
Heat treatment				
Solution:				
Time, h	—	12 (min)	—	12 (min)
Temperature, °C	—	435 (max)‡‡	—	435 (max)‡‡
Cooling	—	Air, oil or water	—	Air, oil or water

Electron designation	AZ91			*C alloy*		
Specifications BS 2970:1972	MAG 3M (GP)	MAG 3TB (GP)	MAG 3TF (GP)	MAG 7M (GP)	MAG 7TB (GP)	MAG 7TF (GP)
BSS L series	—	3L.124	3L.125	—	—	—
Equivalent DTD.	—	—	—	—	—	—
Composition % (Single figures indicate maximum)						
Aluminium		9.0–10.5			7.5–9.5	
Zinc		0.3–1.0			0.3–1.5	
Manganese		0.15–0.4			0.15–0.8	
Copper		0.15			0.35	
Silicon		0.3			0.40	
Iron		0.05			0.05	
Nickel		0.01			0.02	
Cu + Si + Fe + Ni		0.40			0.75	
Material properties						
Founding		Good			Good	
Characteristics		Sand, permanent mould and die (pressure)			Sand, permanent mould and die (pressure)	
Tendency to hot tearing		Little			Little	
Tendency to micro-porosity		Less than MAG 1			Less than MAG 1	
Castability§		A			A	
Weldability (Ar-Arc process)		Good, but some difficulty with die castings			Good, but some difficulty with die castings	
Relative damping capacity¶		C			C	
Strength at elevated temperature**	C	C	B		C	
Corrosion resistance		Moderate			Moderate	
Density, g cm^{-3}		1.83			1.82	
Liquidus, °C		595			600	
Solidus, °C		470			475	
Non-equilibrium solidus, °C		420			420	
Casting temperature range, °C		680–800			600–800	
Heat treatment						
Solution:						
Time, h	—	16 (min)		—	16 (min)	
Temperature, °C	—	435 (max)‡‡		—	435 (max)‡‡	
Cooling	—	Air, oil or water		—	Air, oil or water	

Table 26.34 MAGNESIUM–ALUMINIUM ALLOYS—*continued*

Electron designation	A8		A8 (*High purity*)	
Specifications BS. 2970:1972	MAG 1M* (GP)†	MAG 1TB* (GP)	MAG2M (SP)†	MAG2TB (SP)
BSS L series	—	3L.122	—	—
Equivalent DTD	—	—	684A	690A
Heat treatment–continued				
Precipitation:				
Time, h	—	—	—	—
Temperature, °C	—	—	—	—
Stress relief:				
Time, h	2–4	—	2–4	—
Temperature, °C	250–330	—	250–330	—
Mechanical properties—sand cast—(SI units first, Imperial units following in brackets)				
Tensile strength (min), MPa (tonf in^{-2})	140 (9.1)	200 (13.0)	140 (9.1)	200 (13.0)
0.2% proof stress (min), MPa (tonf in^{-2})	85 (5.5)	80 (5.2)	85 (5.5)	80 (5.2)
Elongation % (min) ($5.65\sqrt{S_0}$)	2	6	2	6
Mechanical properties—chill cast—(SI units first, Imperial units following in brackets)				
Tensile strength (min), MPa (tonf in^{-2})	185 (12.0)	230 (14.9)	185 (12.0)	230 (14.9)
0.2% proof stress (min), MPa (tonf in^{-2})	85 (5.5)	80 (5.2)	85 (5.5)	80 (5.2)
Elongation % (min)($5.65\sqrt{S_0}$)	4	10	4	10
Applications	Automobile road wheels	Good ductility and shock resistance	High-purity alloy—offers good corrosion resistance	

Electron designation	AZ 91		*C alloy*			
Specifications BS 2970:1972	MAG 3M (GP)	MAG 3TB (GP)	MAG 3TF (GP)	MAG 7M (GP)	MAG 7TB (GP)	MAG 7TF (GP)
BSS L series	—	3L.124	3L.125	—	—	—
Equivalent DTD	—	—	—	—	—	—
Heat treatment—continued						
Precipitation:						
Time, h	—	—	8 (min)	—	—	8 (min)
Temperature, °C	—	—	210 (max)	—	—	210 (max)
Stress relief:						
Time, h	2—4	—	—	2—4	—	—
Temperature, °C	250—330	—	—	250—330	—	—
Mechanical properties—sand cast—(SI units first, Imperial units following in brackets)						
Tensile strength (min), MPa (tonf in^{-2})††	125 (8.1)	200 (13.0)	200 (13.0)	125 (8.1)	185 (12.0)	185 (12.0)
0.2% proof stress (min), MPa (tonf in^{-2})	95 (6.2)	85 (5.5)	130 (8.4)	85 (5.5)	80 (5.2)	110 (7.1)
Elongation % (min) ($5.65\sqrt{S_0}$)	—	4	—	—	4	—
Mechanical properties—chill cast—(SI units first, Imperial units following in brackets)						
Tensile strength (min), MPa (tonf in^{-2})	170 (11.0)	215 (13.9)	215 (13.9)	170 (11.0)	215 (13.9)	215 (13.9)
0.2% proof stress (min), MPa (tonf in^{-2})	100 (6.5)	85 (5.5)	130 (8.4)	85 (5.5)	80 (5.2)	110 (7.1)
Elongation % (min) ($5.65\sqrt{S_0}$)	2	5	2	2	5	2
Applications	For pressure tight applications	Increased proof stress after full heat treatment		Principal alloy for commercial usage		

* M—As cast.
TS—Stress relieved only.
TE—Precipitation treated only.
TB—Solution treated only.
TF—Solution and precipitation treated.
† GP General purpose alloy.
SP Special purpose alloy.
‡ Permanent mould = gravity die casting.
§ Ability to fill mould easily. A, B, C, indicate decreasing castability.

¶ Damping capacity ratings.
A = Outstanding; better than grey cast iron.
B = Equivalent to cast-iron.
C = Inferior to cast-iron but better Al-base cast alloys.
** A = Particularly recommended.
B = Suitable but not especially recommended.
C = Not recommended where strength at elev. temps is likely to be an important consideration.
†† 1 MPa = 1 N mm^{-2} = 0.064 75 tonf in^{-2}.
‡‡ Carbon dioxide atmosphere.

Table 26.35 MAGNESIUM–ZIRCONIUM ALLOYS
Inherently fine grained (0.015–0.035 mm chill cast)
(From: E. F. Emley, 'Principles of Magnesium Technology', Pergammon Press, Oxford, 1966)

Electron designation	Z5Z	RZ5	ZRE1
Specifications			
BS 2970:1972	MAG4 TE* (GP)†	MAG5 TE (SP)	MAG6 TE (SP)
BSS L series	2L. 127	2L. 128	2L. 126
Equivalent DTD	—	—	—
Composition % (Single figures indicate maximum)			
Zinc	3.5–5.5	3.5–5.5	0.8–3.0
Silver	—	—	—
Rare earth metals	—	0.75–1.75	2.5–4.0
Thorium	—	—	—
Zirconium	0.4–1.0	0.4–1.0	0.4–1.0
Copper	0.03	0.03	0.03
Nickel	0.005	0.005	0.005
Iron	—	—	—
Silicon	—	—	—
Manganese	—	—	—
Material properties			
Founding characteristics	Good Sand and permanent moulds§	Good	Excellent Sand and permanent mould
Tendency to hot tearing	Marked	Some	Little
Tendency to micro-porosity	Very appreciable	Virtually none	None
Castability¶	B	A	A
Weldability (Ar-Arc Process)	Not recommended	Fair	Very good
Relative damping capacity**	—	—	B
Strength at elevated temperature††	C	B	A
Resistance to creep at elevated temp.	Poor	Fair	Good up to 250 °C
Corrosion resistance	Moderate	Moderate	Moderate
Density, g cm^{-3} (20 °C)	1.81	1.84	1.80
Liquidus, °C	640	640	640
Solidus, °C	560	510	545
Casting temperature range, °C	720–810	720–810	720–810
Heat treatment			
Solution:			
Time, h	—	—	—
Temperature, °C	—	—	—
Cooling	—	—	—
Precipitation:			
Time, h	16	2	8
Temperature, °C	200	330	200
	Air cool	Air cool	Air cool
Post-weld stress relief:			
Time, h	2	Precipitation treatment affords s/relief	10
Temperature, °C	330 to precede precipitation treatment		250 max Air cool
Mechanical properties—sand cast—SI units (Imperial units in brackets)			
Tensile strength min, MPa (tonf in^{-2})	230 (14.9)	200 (13.0)	140 (9.1)
0.2% proof stress min, MPa (tonf in^{-2})	145 (9.4)	135 (8.7)	95 (6.2)
Elongation, % ($5.65\sqrt{S_0}$) min	5	3	3
Mechanical properties—chill cast—SI units (Imperial units in brackets)			
Tensile strength min, MPa (tonf in^{-2})	245 (15.9)	215 (13.9)	155 (10.0)

Table 26.35 MAGNESIUM ZIRCONIUM ALLOYS. *Inherently fine grained—continued*

Electron designation	Z5Z	RZ5	ZRE1
Specifications			
BS 2970:1972	MAG4 TE* (GP)†	MAG5 TE (SP)	MAG6 TE (SP)
BSS L series	2L. 127	2L. 128	2L. 126
Equivalent DTD	—	—	—
0.2% proof stress min, MPa (tonf in^{-2})	145 (9.4)	135 (8.7)	110 (7.1)
Elongation, % ($5.65\sqrt{S_0}$) min	7	4	3
Applications	High strength plus good ductility. Not suitable for spidery complex shapes	For high-strength pressure-tight applications	High degree of pressure tightness at room and elevated temperatures

Electron designation	ZT1	TZ6	ZE63A
Specifications			
BS 2970:1972	MAG8 TE (SP)	MAG9 TE (SP)	—
BSS L series	—	—	—
Equivalent DTD	5005A	5015A	5045
Composition % (Single figures indicate maximum)			
Zinc	1.7–2.5	5.0–6.0	5.5–6.0
Silver	—	—	—
Rare earth metals	0.10	0.20	2.0–3.0
Thorium	2.5–4.0	1.5–2.3	—
Zirconium	0.4–1.0	0.4–1.0	0.4–1.0
Copper	0.03	0.03	0.03
Nickel	0.005	0.005	0.005
Iron	0.01	0.01	0.01
Silicon	0.01	0.01	0.01
Manganese	0.15	0.15	0.15
Material properties			
Founding characteristics	As per MAG7 but more sluggish	Similar to MAG5	Good
Tendency to hot tearing	Little	Very little	Negligible
Tendency to micro-porosity	None	Low	Virtually none
Castability ¶	C	B	A
Weldability (Ar-Arc process)	Very good	Fair	Very good***
Relative damping capacity**	B	—	—
Strength at elevated temperature††	A	B	—
Resistance to creep at elevated temperature	Good up to 350°C	Good for short times up to 350°C	—
Corrosion resistance	Moderate	Moderate	Moderate
Density, g cm^{-3} (20°C)	1.85	1.87	—
Liquidus, °C	645	630	—
Solidus, °C	550	520	516
Casting temperature range, °C	720–810	720–810	720–810
Heat treatment			
Solution:			
Time, h	—	—	30 for 12 mm sctn. 70 for 25 mm sctn.
Temperature, °C	—	—	525†††
Cooling	—	—	Air blast or water spray
Precipitation:			
Time, h	16	2 } followed { 16	48 } or { 72
Temperature, °C	315	330 } by { 200	138 } or { 127
	Air cool	Air cool after each	Air cool

Table 26.35 MAGNESIUM–ZIRCONIUM ALLOYS. *Inherently fine grained—continued*

Electron designation	ZT1	TZ6	ZE63A
Specifications			
BS 2970:1972	MAG8 TE (SP)	MAG9 TE (SP)	—
BSS L series			
Equivalent DTD	5005A	5015A	5045
Post weld stress relief:			
Time, h	2	Pptn. treatment	—
Temperature, °C	350 Air cool	affords stress relief	—
Mechanical properties—sand cast—SI units (Imperial units in brackets)			
Tensile strength min, MPa (tonf in^{-2})	185 (12.0)	255 (16.5)	275 (17.8)
0.2% proof stress min, MPa (tonf in^{-2})	85 (5.5)	155 (10.0)	170 (11.0)
Elongation, % ($5.65\sqrt{S_0}$) min	5	5	5
Mechanical properties—chill cast—SI units (Imperial units in brackets)			
Tensile strength min, MPa (tonf in^{-2})	185 (12.0)	255 (16.5)	Sand
0.2% proof stress min, MPa	85 (5.5)	155 (10.0)	Cast
Elongation, % ($5.65\sqrt{S_0}$) min	5	5	Alloy
Applications	Creep resistant alloy	For heavy duty structural usage	High strength with good ductility and excellent fatigue resistance. Structural parts aircraft, etc.

Electron designation	MSR-A	MSR-B	—	MTZ (HK31)
Specifications				
BS 2970: 1972	—	—	—	—
BSS L series	—	—	—	—
Equivalent DTD	5025A	5035A	5055	—
Composition % (Single figures indicate maximum)				
Zinc	0.2	0.2	0.2	0.3
Silver	2.0–3.0	2.0–3.0	2.0–3.0	—
Rare earth metals	1.2–2.0‡	2.0–3.0‡	1.8–2.5‡	0.1
Thorium	—	—	—	2.5–4.0
Zirconium	0.4–1.0	0.4–1.0	0.4–1.0	0.4–1.0
Copper	0.03	0.03	0.03	0.03
Nickel	0.005	0.005	0.005	0.005
Iron	0.01	0.01	0.01	0.01
Silicon	0.01	0.01	0.01	0.01
Manganese	0.15	0.15	0.15	0.15
Material properties				
Founding characteristics	Good	Good	Good	Less easy to found than MSR types
Tendency to hot tearing	Little	Little	Little	Very little
Tendency to micro-porosity	Slight	Slight	Slight	Negligible
Castability¶	B	B	B	C
Weldability (Ar-Arc Process)	Very good	Very good	Very good	Very good
Relative damping capacity**	B/C	B/C	B/C	—
Strength at elevated temperature††	A	A	A	A
Resistance to creep at elevated temperature	Good up to 200 °C	Good up to 200 °C	Good up to 200 °C	Good to ≯ 350 °C for short time applications

Table 26.35 MAGNESIUM-ZINCONIUM ALLOYS. INHERENTLY FINE GRAINED—*continued*

Electron designation	MSR-A	MSR-B	—	MTZ (HK31)
Specifications				
BS 2970:1972	—	—	—	—
BSS L series	—	—	—	—
Equipment DTD	5025A	5035A	5055	—
Corrosion resistance	Moderate	Moderate	Moderate	Moderate
Density, $g\,cm^{-3}$ (20°C)	1.81	1.82	1.81	1.84
Liquidus, °C	640	640	640	645
Solidus, °C	550	550	550	590
Casting temperature range, °C	720–810	720–810	720–810	720–810
Heat treatment				
Solution:				
Time, h	4 followed 8	4 followed 8	4 followed 8	2
Temperature, °C	500‡‡ by 530‡‡	500‡‡ by 530‡‡	500‡‡ by 530‡‡	565‡‡,¶¶
Cooling	Water or oil	Water or oil	Water or oil	Air cool
Precipitation:				
Time, h	16	16	16	16
Temperature, °C	200	200	200	200
	Air cool	Air cool	Air cool	Air cool
Post-wéld stress relief:				
Time, h	Repeat above	Repeat above	Repeat above	Repeat above
Temperature, °C	cycle	cycle	cycle	cycle
Mechanical properties—sand cast—SI units (Imperial units in brackets)				
Tensile strength min, MPa				
(tonf in^{-2})	240 (15.5)	240 (15.5)	240 (15.5)	200 (13.0)
0.2% proof stress min, MPa				
(tonf in^{-2})	170 (11.0)	185 (12.0)	175 (11.3)	93 (6.0)
Elongation, % ($5.65\sqrt{S_0}$) min	4	2	2	5
Mechanical properties—chill cast—SI units (Imperial units in brackets)				
Tensile strength min, MPa				
(tonf in^{-2})	240 (15.5)	240 (15.5)	240 (15.5)	
0.2% proof stress min, MPa				
(tonf in^{-2})	170 (11.0)	185 (12.1)	175 (11.3)	Usually sand cast
Elongation, % ($5.65\sqrt{S_0}$) min	4	2	2	
Applications	High strength in thick and thin section castings. Elevated temperature (up to 250°C) short time tensile and fatigue props. are the best of Mg casting alloys		Similar to MSRA-B	Superior short time tensile and creep resistance at temperatures around 300°C

* *See* footnote to Table 26.34.
† *See* footnote to Table 26.34.
‡ Neodymium-rich rare earths (others Ce-rich).
¶ *See* footnote to Table 26.34.
** *See* footnote to Table 26.34.
†† *See* footnote to Table 26.34.
‡‡ SO_2 atmosphere.
¶¶ Castings to be loaded into furnace at operating temperature.
*** But only before hydriding treatment.
††† In hydrogen at atmospheric pressure.

26.7 Zinc base casting alloys

Table 26.36 ZINC BASE ALLOYS

Alloy	BS1004A	BS1004B	ILZRO*†–12	No. 2 alloy	ZA27	ILZRO16
Composition % (Single figures indicate maximum unless otherwise stated)						
Aluminium	3.8–4.3	3.8–4.3	10.5–11.5	3.5–4.3	25–28	0.01–0.04
Copper	0.10	0.75–1.25	0.5–1.25	2.5–3.2	2.0–2.5	1.0–1.5
Magnesium	0.03–0.06	0.03–0.06	0.01–0.03	0.03–0.06	0.01–0.02	0.02
Iron	0.10	0.10	0.10	0.075	0.10	0.04
Nickel	0.020	0.020	—	—	—	—
Lead	0.005	0.005	0.005	—	0.004	0.005
Cadmium	0.005	0.005	0.005	Pb + Cd 0.009	0.004	0.004
Tin	0.002	0.002	0.002	0.002	0.002	0.003
Thallium	0.001	0.001	—	—	—	—
Indium	0.000 5	0.000 5	—	—	—	—
Titanium	—	—	—	—	—	0.15–0.25
Chromium	—	—	—	—	—	0.10–0.20
Zinc	Remainder	Remainder	Remainder	Remainder	Remainder	Remainder
					Sand cast	0.10–0.20
Properties						
Density at 21 °C, g cm^{-3}	6.7	6.7	6.0	6.8	5.0	Remainder
Liquidus temperature, °C	387	388	432	390	487	7.1
Solidus temperature, °C	382	379	377	379	375	418
Casting temperature range, °C	393–427	393–427	477–521	395–430	525	416
Specific heat, J kg^{-1} K^{-1} (293–373 K)	418.7	418.7	500	419	—	402
Thermal conductivity, W m^{-1} K	113.1	109.5	116	92.1 to 104.7	123	105
Electrical conductivity, % IACS	27	26	28	25	29.7	20
Electrical resistivity, $\mu\Omega$ cm (at 20 °C)	6.37	6.5	6.8	6.8	—	8.4
Ageing at room temperature	Negligible	Slight growth but no loss of impact	Very slow drop in mech. levels over several years	Impact strength falls fast at 95 °C. Reduction to 30% of initial value after 3 years at ambient temperature marked growth	Little change at 21 °C, loss of UTS at high temperatures	Good creep resistance at high temperatures
Castability	Very good	Slightly better than A	Good. Similar to LM4 (Al-alloy)	Very good	Very good	Good

Property		O	A_1	A_2	A_3		O	A_1	A_2	A_3					
Dimensional changes after casting (mm m^{-1})											Shrinkage 0.005% after 30 days ambient, and 0.03% after 1 000 days. At 95 °C, shrinkage followed by growth, i.e. zero change after 1 000 days	—	Shrinkage 0.005% after 30 days ambient, and 0.015% after 1 000 days, At 95 °C rapid shrinkage then growth to +0.1% after 1 000 days.		
After 5 weeks		−0.32					−0.69								
After 6 months		−0.56					−1.03								
After 5 years		−0.73					−1.36								
After 8 years		−0.79					−1.41								
Dimensional changes after stabilizing heat treatment (mm m^{-1}) (6 h at 100 °C ± 5 °C − air cool)															
After 5 weeks		−0.20					−0.22				—				
After 3 months		−0.30					−0.26				—				
After 2 years		−0.30					−0.37				—				
Mechanical properties—SI units (Imperial units in brackets)											Sand cast		Sand cast	Chill cast	Die cast
Tensile stress, MPa§ (Strain rates 6.33 mm m^{-1} crosshead speed) (tonf in^{-2})		O	A_1	A_2	A_3		O	A_1	A_2	A_3		As cast: 216 to 275 (16.4 to 17.8 tonf)	420	430	230
	C	286	264	247	235	C	335	320	292	255	As cast: 300				
	S	273	264	243	235	S	312	290	—	255					
	[C	18.5	17.1	16.0	15.2]	[C	21.7	20.7	18.9	16.5]	After 200 days at 100 °C: 240				
	S	17.7	17.1	15.7	15.2]	S	20.2	18.8	—	16.5]					
Applications and uses		Extensive use where a large number of strong dimensionally accurate metal components required, e.g. components of cars, domestic appliances, business machines, record players, hydraulic and pneumatic valves, toy models									For prototype, short and long production runs and in plaster moulds. Especially useful for gravity die casting. Used to replace materials requiring extensive fabrication and machining	Pressure die, sand and gravity die cast. For applications where hardness in an advantage. Sheet metal forming dies, moulds for plastics‡, zip fastener sliders	Sand, gravity or pressure die cast. Can be used stressed at up to 150 °C		Creep resistant pressure-die casting alloy
Elongation % on 50 mm (2.0 in)		O	A_1	A_2	A_3		O	A_1	A_2	A_3	As cast: 3 After 200 days at 100 °C: 6	As cast: 0.5–3.0	3–6	—	5
	C	15	25	20	29	C	9	12	14	23					
	S	17	24	19	29	S	10	14	—	23					
Impact strength, J (un-notched specimens)	C	57	58	60	50	C	58	57	56	14	As cast: 8	As cast: 50 (35 ft lbf)	44	—	80
	S	61	56	57	50	S	60	61	—	14	After 200 days at 100 °C: 3.4 (2.5 ft lbf)				
Hardness Brinell 10 FL		82	67	65	48		91	74	74	64	105–125		110–120		80

O—Original value. A_1—After 12 months normal ageing. A_2—After 8 years normal ageing. A_3—After 12 months dry ageing at 95 °C. C—As cast. S—Stabilized.

* ILZRO International Lead Zinc Research Organisation.

† Properties of castings in this alloy little affected by cooling rate.

‡ No. 2 alloy not suitable for resins which release HCl during condensation.

§ 0.2% proof stress ILZRO 12 221 MPa (14.3 tonf in^{-2}). (Expected)—No 2—167–226 MPa (10.8–14.6 tonf in^{-2}).

¶ Zinc to BS 3436 Zn 1.

26.8 Steel castings

26.8.1 Casting characteristics

Casting is usually carried out at 1500–1700 °C, or even higher, according to the type of steel and section of the casting. Highly refractory sands with high resistance to the molten metal stream are essential. Liquid shrinkage is high, so that generous risers are needed and the metal is hot short. Manganese steels, in particular, attack all furnace and ladle refractories, and only olivine sand has adequate resistance to constitute a satisfactory moulding material, tolerably resistant to burn-on.

26.8.2 Heat treatment

HOMOGENIZING

High-temperature treatment intended to reduce interdendritic segregation.

Table 26.37 BRITISH STANDARDS

BS Specification	*Type*	*Chemical composition, %*								
		C	Si *max*	Mn *max*	Ni *max*	Cr *max*	Mo *max*	Cu *max*	S *max*	P *max*
BS 592A	—	0.25 max	0.60	1.00	0.40	0.25	0.15	0.30	0.060	0.060
BS 592B	—	0.35 max	0.60	1.00	—	—	—	—	0.060	0.060
BS 592C	—	0.45 max	0.60	1.00	—	—	—	—	0.060	0.060
BS 1504-101A	For structural purposes	—	—	—	0.40	0.25	0.15	0.40	0.060	0.060
BS 1504–101B	For structural purposes	—	—	—	0.40	0.25	0.15	0.40	0.060	0.060
BS 1504–101C	For structural purposes	—	—	—	0.40	0.25	0.15	0.40	0.060	0.060
BS 1504–161A	For parts under pressure	0.25 max	0.50	0.90	0.40	0.25	0.15	0.40	0.050	0.050
BS 1504–161B	For parts under pressure	0.30 max	0.50	0.90	0.40	0.25	0.15	0.40	0.050	0.050
BS 1617A	High magnetic permeability	0.15 max	0.60	0.50	0.40	0.25	0.15	0.30	0.050	0.050
BS 1617B	High magnetic permeability	0.25 max	0.60	0.50	0.40	0.25	0.15	0.30	0.050	0.050
BS 1760A	For surface hardening	0.40–0.50	0.60	1.00	0.40	0.25	0.15	0.30	0.050	0.050
BS 1760B	For surface hardening	0.50–0.60	0.60	1.00	0.40	0.25	0.15	0.30	0.050	0.050
BS 4239	Case hardening	0.10–0.18	0.60	0.6–1.0	0.40	0.25	0.15	0.30	0.050	0.050

Notes:
BS 1504–101B is comparable with BS 592A. BS 1504–101C is comparable with BS 592C.

* Hardening tests: a test piece ½ in dia. × 3 in (min.) long shall show a hardness of 450 min. for Grade A or 535 (min.) for Grade B after water quenching from 800 to 840 °C.
† For Grades A and B either a bend or an impact test may be specified.
‡ The Izod Impact Test Requirement is mandatory only if specified by the purchaser upon enquiry and order.

ANNEALING

Heating above the critical range (Ac_3) for about 1 h per 25 mm of maximum cross-section and furnace cooling to nearly room temperature. Removes brittleness associated with coarse as-cast grain size or Widmanstätten structure, gives most complete relief of internal stresses, gives minimum yield stress and tensile strength, good ductility but poor impact values. Best magnetic properties in low carbon steels.

SUB-CRITICAL ANNEALING

Heating to a temperature below the Ac_1, and cooling slowly. Used for stress relief after welding, etc., and for softening high alloy steel castings.

NORMALIZING

Similar to annealing except that cooling is in still air. Gives a smaller ferrite/pearlite grain size, higher yield stress, tensile strength and impact, but slightly lower elongation values than

SPECIFICATIONS FOR CARBON STEELS

Usual heat treatment	*Tensile strength*		*Yield stress*		*Elongation*	*Bend test*		*Izod impact*		*Note*
	MPa	tonf in^{-2}	MPa	tonf in^{-2}	%	*Radius of bend*	*Angle min*	Joules	ft lb	
A, A+N N+T, H+T	432 min	28 min	232	15	22	$1\frac{1}{2}$Th	120°	20.3	15†	
A, A+N N+T, H+T	494 min	32 min	263	17	18	$1\frac{1}{2}$Th	90°	20.3	15†	
A, A+N N+T, H+T	541 min	35 min	293	19	14	—	—	13.6	10‡	1
A, N, A+N (N+T)	402–494	26–32	201	13	20	$1\frac{1}{2}$Th	120°	—	—	
A, N, A+N (N+T)	432–541	28–35	216	14	20	$1\frac{1}{2}$Th	120°	—	—	
A, N, A+N (N+T)	541–618	35–40	270	17.5	15	$1\frac{1}{2}$Th	90°	—	—	
A, N, A+N (N+T)	432 min	28 min	216	14	22	$1\frac{1}{2}$Th	120°	—	—	2
A, N, A+N (N+T)	479 min	31 min	247	16	22	$1\frac{1}{2}$Th	90°	20.3	15	
A	340–432	22–28	185	12	22	$1\frac{1}{2}$Th	120°	—	—	
A	402–494	26–32	216	14	22	$1\frac{1}{2}$Th	120°	—	—	3
A	618 min	40 min	324	21	12	—	—	—	—	
A	695 min	45 min	371	24	8	—	—	—	—	4
As Cast A, A+N	494 min	32 min	—	—	12	—	—	27.1	20	

Notes:

1 For ships' marine engines, railway rolling stock and general engineering.
2 Oil and chemical industry applications.
3 For electrical applications.
4 For surface hardening by local heating and quenching.

annealing. Medium and higher carbon steels which harden appreciably are usually tempered subsequently. Normalizing may be preceded by annealing, or a prior normalizing treatment at a higher temperature to break down the cast structure.

QUENCHING (OR HARDENING) AND TEMPERING

Heating above the critical range, but not so high as for annealing or normalizing, and rapidly cooling in water, oil or air blast followed by reheating to a temperature below the critical range until the desired properties are achieved followed by air cooling, furnace cooling (if stress relief is also required) or cooling in water (to minimize temper embrittlement). Gives highest physical properties; high tempering temperatures usually give lower strength but higher elongation and impact values. Normally preceded by annealing.

SURFACE HARDENING

Heating the surface layers by flame or induction to a temperature above the critical range and quenching by water jets following the heat source or by immersion.

Table 26.38 BRITISH STANDARDS

		Chemical composition, % (single figure normally max)								
Specification	*Type*	C	Si	Mn	Ni	Cr	Mo	Cu *max*	S *max*	P *max*
BS 1398A	Carbon molybdenum	0.25	0.20–0.50	0.50–1.00	0.40	0.25	0.40–0.70	0.30	0.050	0.050
BS 1504-240	Carbon molybdenum	0.15–0.25	0.20–0.50	0.50–1.00	0.40	0.25	0.40–0.70	0.40	0.050	0.050
BS 1456A	Pearlitic manganese	0.18–0.25	0.50	1.20–1.60	—	—	—	—	0.050	0.050
BS 1456B$_1$	Pearlitic manganese	0.25–0.33	0.50	1.20–1.60	—	—	—	—	0.050	0.050
BS 1456B$_2$	Pearlitic manganese	0.25–0.33	0.50	1.20–1.60	—	—	—	—	0.050	0.050
BS 1457	Austenitic manganese	1.00–1.25*	1.00	11.0 min.	—	—	—	—	0.060	0.070
BS 1504-503	3½% Nickel	0.15	0.60	0.50–0.80	3.0–4.0	0.3	—	—	0.050	0.050
BS 1398B	1½% Chromium molybdenum	0.20	0.60	0.50–0.80	0.40	1.00–1.50	0.45–0.65	0.30	0.050	0.050
BS 1504-621	1½% Chromium molybdenum	0.20	0.60	0.50–0.80	0.40	1.00–1.50	0.45–0.65	0.40	0.050	0.050
BS 1398C	2¼% Chromium 1% molybdenum	0.18	0.60	0.40–0.70	0.40	2.00–2.75	0.90–1.20	0.30	0.050	0.050
BS 1504-622	2¼% Chromium 1% molybdenum	0.18	0.60	0.40–0.70	0.40	2.00–2.75	0.90–1.20	0.30	0.050	0.050
BS 1461	3% Chromium molybdenum	0.25	0.75	0.30–0.70	0.40	2.50–3.50	0.35–0.60	0.30	0.040	0.040
BS 1504-623	3% Chromium molybdenum	0.25	0.75	0.30–0.70	0.40	2.50–3.50	0.35–0.60	0.40	0.050	0.050
BS 1462	5% Chromium molybdenum	0.20	0.75	0.40–0.70	0.40	4.00–6.00	0.45–0.65	0.30	0.040	0.040
BS 1504-625	5% Chromium molybdenum	0.20	0.75	0.40–0.70	0.40	4.00–6 00	0.45–0.65	0.40	0.040	0.040

26.8.3 Notes to the tables

British Standards for steel castings for general engineering purposes are summarized in BS 3100:1967, and BS 1504:1958 covers steel castings for the chemical, petroleum and allied industries. Tables 26.37, 26.38 and 26.39 summarize these specifications. Tables 26.40 and 26.41 give typical compositions and tensile properties of a selection of cast steels, as some steel specifications do not specify compositions or tensile properties.

The following abbreviations have been used:

A = annealed	WQ = water quenched
N = normalized	T = tempered
H = hardened	S = softened
AH = air hardened	AC = air cooled
OH = oil hardened	Th = thickness (of bend test specimen)
OQ = oil quenched	

SPECIFICATIONS FOR ALLOY STEELS

Usual heat treatment	*Tensile strength min. or range*		*Yield stress*		*Elong. % min*	*Bend test*		*Izod impact*		*Brinell hardness number*	*Note*
	MPa	tonf in^{-2}	MPa	tonf in^{-2}		*Radius of bend*	*Angle min*	*Joules min*	*ft lb min*		
By agreement	463	30	278	18	18	1½Th	120°†	20.3	15†	—	1
A, A+N (A+N+T)	463	30	247	16	20	1½Th	120°†	20.3	15†	—	1
N, N+T, H+T	540–695	35–45	340	22	16	—	—	33.9	25	152–207	
N, N+T, H+T	618–772	40–50	371	24	13	—	—	27.1	20	179–229	
OH+T, WQ+T	541–849	45–55	494	32	13	—	—	27.1	20	201–255	
WQ.1000 °C (min)				As agreed							
A+N+T	448	29	270	17.5	25	1½Th	90°	40.7	30	—	
By agreement	479	31	278	18	17	1½Th	120°†	33.9	25†	—	
A, A+N+T (650 °C max)	479	31.0	278	18	20	1½Th	120°	33.9	25	—	
By agreement	540	35	324	21	17	3Th	120°†	27.1	20†	—	
A, A+N+T (690 °C max)	479	31	278	18	20	3Th	90°	27.1	20	—	
A+AH+T	618–772	40–50	371	24	13	3Th	120°	27.1	20	—	2
A+OH+T	618–772	40–50	371	24	18	3Th	120°	27.1	20	—	2
A+AH+T	618	40	417	27	13	3Th	90°	27.1	20	179–229	2
A+AH+T	618	40	417	27	18	3Th	90°	27.1	20	—	2

Table 26.38 BRITISH STANDARD

Specification	Type	Chemical composition % (single figure normally max)								
		C	Si	Mn	Ni	Cr	Mo	Cu *max*	S *max*	P *max*
BS 1463	9% Chromium molybdenum	0.20	1.00	0.30–0.70	0.40	8.00–10.00	0.90–1.20	0.30	0.040	0.040
BS 1504–629	9% Chromium molybdenum	0.20	1.00	0.30–0.70	0.40	8.00–10.0	0.90–1.20	0.40	0.040	0.040
BS 1630A	13% Chromium	0.15	1.0	1.0	1.0	11.5–13.5	—	—	0.040	0.040
BS 1630B	13% Chromium	0.20	1.0	1.0	1.0	11.5–13.5	—	0.40	0.040	0.040
BS 1504-713	13% Chromium	0.12–0.20	1.25	1.0	1.0	11.5–13.5	0.15	0.40	0.050	0.050
BS 1398D	Chromium–molybdenum–vanadium**	0.15	0.50	0.40–0.80	0.40	0.25–0.50	0.50–0.70	0.30	0.050	0.050
BS 1398E	Chromium–molybdenum–vanadium**	0.15	0.50	0.40–0.80	0.40	0.70–1.0	0.70–1.0	0.30	0.050	0.050
BS 1458A	Higher tensile strength alloy steels	Chemical composition chosen to suit thickness of the casting and severity of the quenching process				—	—	—	0.050	0.050
BS 1458B	Higher tensile strength alloy steels	Chemical composition chosen to suit thickness of the casting and severity of the quenching process				—	—	—	0.040	0.040
BS 1458C	Higher tensile strength alloy steels	Chemical composition chosen to suit thickness of the casting and severity of the quenching process				—	—	—	0.030	0.030
DTD 666	Medium high tensile (927 MPa min)	Chemical composition chosen to suit thickness of the casting and severity of the quenching process				—	—	—	0.020	0.025
DTD 705	High tensile (1158–1266 MPa)	Chemical composition chosen to suit thickness of the casting and severity of the quenching process				—	—	—	0.020	0.025
DTD 5072	1158–1313 MPa investment castings (non-stainless)	Chemical composition chosen to suit thickness of the casting and severity of the quenching process				—	—	—	0.020	0.025
DTD 5172	850–1000 MPa low alloy investment castings	Chemical composition chosen to suit thickness of the casting and severity of the quenching process				—	—	—	—	—
BS 1956A	1% Chromium (abrasion resistance)	0.45–0.55	0.75	0.50–1.00	—	0.80–1.20	—	—	0.060	0.060
BS 1956B	1% Chromium (abrasion resistance)	0.45–0.55	0.75	0.50–1.00	—	0.80–1.20	—	—	0.060	0.060
BS 1956C	1% Chromium–molybdenum (abrasion resisting)	0.55–0.65	0.75	0.50–1.00	—	0.80–1.50	0.20–0.40	—	0.060	0.060
BS 4240	3% Nickel case-hardening	0.10–0.18	0.60	0.30–0.60	2.75–3.50	0.25	0.15	0.30	0.040	0.040
BS 4241	3½% Nickel–chromium–molybdenum (case-hardening)	0.12–0.18	0.60	0.30–0.60	3.0–3.75	0.60–1.10	0.15–0.25	0.30	0.040	0.040

SPECIFICATIONS FOR ALLOY STEELS–*continued*

Usual heat treatment	*Tensile strength min. or range*		*Yield stress*		*Elong.*	*Bend test*		*Izod impact*		*Brinell hardness number*	*Note*
	MPa	tonf in^{-2}	MPa	tonf in^{-2}	% *min*	*Radius of bend*	*Angle min*	Joules *min*	ft lb *min*		
A+AH+T A+OH+T	618	40	417	27	13	3Th	90°	—	—	179–229	
A+AH+T A+OH+T	618	40	417	27	18	3Th	90°	—	—	179–229	
A+AH+T A+OH+T	540	35	371	24	15§	2Th§	120°	—	—	152–207‖	3
	695	45	463	30	11§	—	—	—	—	201–255‖	3
	618	40	448	29	18	2Th	90°	27.1	20	—	
By agreement	510	33	309	20	17	3Th	120°	—	—	—	
By agreement	540	35	309	20	12	3Th	120°	—	—	—	
A+H+T	695–849	45–55	494	32	11	—	—	33.9	25	201–255	
A+H+T	849–1 004	55–65	587	38	8	—	—	27.1	20	248–302	
A+H+T	1 004–1 158	65–75	695	45	6	—	—	20.3	15	293–341	
A+H+T	927	60	695¶	45	12	—	—	40.7	30	269 min	
A+H+T	1 158–1 266	75–82	927¶	60	7	—	—	13.6	10	340 min	
N or A+H+T	1 158–1 313	75–85	927¶	60	7	—	—	13.6	10	240–390	
N or A+H+T	849–1 004	55–65	664¶	43	12	—	—	40.7	30	248–302	
A, A+N+T, A+H+T	695	45	—	—	7	—	—	—	—	201–255‖	4
	—	—	—	—	—	—	—	—	—	293 min‖	4
	—	—	—	—	—	—	—	—	—	341 min‖	4
A or N	695	45	—	—	11	—	—	33.9 min	25 min	—	
A	1 004	65	—	—	7	—	—	20.3 min	15 min	—	

Table 26.38 BRITISH STANDARDS

Specification	*Type*	Chemical composition % (single figure normally max) C	Si	Mn	Ni	Cr	Mo	Cu *max*	S *max*	P *max*
BS 4242A	Ferritic steel for use at low temperatures	0.20	0.60	1.10‡	—	—	—	—	0.040	0.040
BS 4242B	Ferritic steel for use at low temperatures	0.20	0.60	1.00	—	—	0.45–0.65	—	0.040	0.040
BS 4242C	Ferritic steel for use at low temperatures	0.12	0.60	0.80	3.0–4.0	—	—	—	0.030	0.030

Notes:
1 For use at elevated temperatures.
2 Use above 400 °C.
3 Good corrosion resistance.
4 Good wear and abrasion resistance.

* For special applications by agreement between the manufacturer and the purchaser the maximum carbon content may be increased to 1.35%.
† Either a bend or an impact test may be specified.
‡ Manganese/carbon ratio to be greater than 3:1.
§ Free machining grade S = 0.50% max bend test and elongation min does not apply.
|| May be supplied harder than this, tensile and bend tests do not apply.
*¶ 0.1% proof stress.
** BS 1398D V = 0.22–0.30 BS 1398E V = 0.20–0.35.
†† *See* Table 26.38A for impact properties.

Table 26.39 BRITISH STANDARDS SPECIFICATIONS

BS specification	*Type*	Chemical composition* % C *max*	Si *max*	Mn *max*	Ni *min*	Cr	Mo	Nb¶
BS 1631D	Austenitic Cr–Ni	0.12	1.50	2.0	8.0	17.0–21.0	—	—
BS 1504-801	Austenitic Cr–Ni	0.12	2.0	2.0	7.0	17.0 min	—	—
BS 1631A	Austenitic Cr–Ni	0.08	1.50	2.0	8.0	17.0–21.0	—	—
BS 1631B†	Stabilized austenitic Cr–Ni	0.08‡	1.50	2.0	8.5	17.0–21.0	—	8 × C – 1.0
BS 1504-821†	Stabilized austenitic Cr–Ni	0.12	2.00	2.0	8.5	17.0–20.0	—	8 × C – 1.10
BS 1631C	Extra low carbon Cr–Ni	0.03	1.50	2.0	8.0	17.0–21.0	—	—
BS 1632A	Austenitic Cr–Ni–Mo	0.08	1.50	2.0	10.0	18.0–20.0	3.0–4.0	—
BS 1504-846	Austenitic Cr–Ni–Mo	0.08	1.50	2.0	11.0–14.0	18.0–20.0	3.0–4.0	—
BS 1632B	Austenitic Cr–Ni–Mo	0.08	1.50	2.0	10.0	17.0–20.0	2.0–3.0	—
BS 1504-845	Austenitic Cr–Ni–Mo	0.08	1.50	2.0	10.0	16.5–18.5	2.0–3.0	—
BS 1632C†	Stabilized austenitic Cr–Ni–Mo	0.08	1.50	2.0	10.0	17.0–20.0	2.0–3.0	8 × C – 1.0
BS 1504-845 Nb†	Stabilized austenitic Cr–Ni–Mo	0.12	1.50	2.0	10.0	16.5–18.5	2.25–2.75	8 × C – 1.10
BS 1632D	Austenitic Cr–Ni–Mo	0.08	1.50	2.0	8.0	17.0–20.0	2.0–3.0	—
BS 1632E	Austenitic Cr–Ni–Mo	0.08	1.50	2.0	8.0	17.0–20.0	1.0–1.75	—
BS 1632F	Austenitic Cr–Ni–Mo	0.03	1.50	2.0	10.0	17.0–20.0	2.0–3.0	—

SPECIFICATIONS FOR ALLOY STEELS–*continued*

Usual heat treatment	*Tensile strength min. or range*		*Yield stress*		*Elong.*	*Bend test*		*Izod impact*			
	MPa	tonf in^{-2}	MPa	tonf in^{-2}	% *min*	*Radius of bend*	*Angle min*	Joules *min*	ft lb *min*	*Brinell hardne number*	*Note*
Manufacturer's choice	432	28	232	15	22	—	—	††	—	—	
	463	30	278	18	18	—	—	††	—	—	
	463	30	278	18	20	—	—	††	—	—	

Table 26.38A IMPACT PROPERTIES OF BS 4242

Charpy V-notch impact test at temperature of	*Grade A* −40 °C	*Grade B* −50 °C	*Grade C* −60 °C
Minimum average value			
J	20.3	20.3	20.3
ft lbf	15	15	15
Minimum individual value			
J	13.6	13.6	13.6
ft lbf	10	10	10

FOR AUSTENITIC AND HEAT RESISTING STEELS

S *max.*	P *max.*	*Heat treatment*	*Tensile strength min.* MPa	tonf in^{-2}	*Yield or 0.5% proof stress (min.)* MPa	tonf in^{-2}	*Elongation* %	*Bend test* *Radius of bend*	*Angle of bend*	*Impact*
0.040§	0.040	N or W.Q. (1 000–1 100 °C)	479	31	208	13.5	26	—	—	—
0.045	0.045	N or W.Q. (1 000–1 100°C)	463	30	209	13.5	20	1½Th	120°	—
0.040§	0.040	N or W.Q. (1 000–1 100 °C)	479	31	208	13.5	26	—	—	††
0.040§	0.040	N or W.Q. (1 000–1 100 °C)	479	31	208	13.5	22	—	—	††
0.045	0.045	N or W.Q. (1 000–1 100 °C)	463	30	209	13.5	20	1½Th	120°	—
0.040§	0.040	N or W.Q. (1 000–1 100 °C)	432	28	185	12	26	—	—	††
0.040§	0.040	N or W.Q. (1 000–1 150 °C)	479	31	209	13.5	22	—	—	—
0.045	0.045	N or W.Q. (1 000–1 100 °C)	463	30	209	13.5	15	2Th	120°	—
0.040§	0.040	N or W.Q. (1 000–1 150 °C)	479	31	208	13.5	26	—	—	††
0.045	0.045	N or W.Q. (1 000–1 100 °C)	463	30	209	13.5	15	2Th	120°	—
0.040§	0.040	N or W.Q. (1 000–1 150 °C)	463	30	209	13.5	18	—	—	—
0.045	0.045	N or W.Q. (1 000–1 100 °C)	463	30	209	13.5	15	2Th	120°	—
0.040§	0.040	N or W.Q. (1 000–1 150 °C)	510	33	232	15	26	—	—	††
0.040§	0.040	N or W.Q. (1 000–1 150 °C)	432	31	185	13.5	26	—	—	††
0.040§	0.040	N or W.Q. (1 000–1 150 °C)	—	28	—	12	26	—	—	††

Table 26.39 BRITISH STANDARDS SPECIFICATIONS

		*Chemical composition**%						
BS specification	*Type*	C *max*	Si *max*	Mn *max*	Ni *min*	Cr	Mo	Nb¶
BS 1648A	Heat resisting	0.25	2.0	1.0	—	12.0–16.0	—	—
BS 1648B1	Heat resisting	1.0	2.0	1.0	4.0	25.0–30.0	1.0	—
BS 1648B2	Heat resisting	0.5	2.0	2.0	8.0–12.0	25.0–30.0	1.0	—
BS 1648C	Heat resisting	1.0–2.0	2.0	1.0	4.0	25.0–30.0	1.0	—
BS 1648D	Heat resisting	0.4	2.0	2.0	6.0–10.0	17.0–22.0	1.0	—
BS 1648E	Heat resisting	0.5	2.5	2.0	10.0–14.0	22.0–27.0	1.0	—
BS 1648F	Heat resisting	0.5	3.0	2.0	17.0–27.0	22.0–27.0	1.0	—
BS 1648G	Heat resisting	0.5	3.0	2.0	23.0–28.0	17.0–23.0	1.0	—
BS 1648H1	Heat resisting	0.75	3.0	2.0	30.0–40.0	13.0–20.0	1.0	—
BS 1648H2	Heat resisting	0.75	3.0	2.0	36.0–46.0	15.0–25.0	1.0	—
BS 1648K	Heat resisting	0.75	3.0	2.0	55.0–65.0	10.0–20.0	1.0	—
BS 4238EC1**	Close composition high alloy steels	0.20–0.45	1.5	2.5	11.0–14.0	24.0–28.0	0.50	—
BS 4238EC2	Close composition high alloy steels	0.20–0.50	1.5	2.0	11.0–14.0	24.0–28.0	0.50	—
BS 4238FC	Close composition high alloy steels	0.30–0.50	1.5	2.0	19.0–22.0	24.0–27.0	0.50	—
BS 4238H1C	Close composition high alloy steels	0.35–0.55	1.5	2.0	33.0–37.0	13.0–17.0	0.50	—
BS 4238H2C	Close composition high alloy steels	0.35–0.55	1.5	2.0	37.0–41.0	17.0–21.0	0.50	—

* Additional elements may be present in BS 1631, 1632 (W, Cu, V, etc.) and 1648 (W, A1, Nb, etc.).
† Suitable for welding without subsequent heat treatment.
‡ If required with specific impact properties at low temperatures the maximum carbon content shall be 0.06% and the maximum niobium content shall be 0.90%.
§ BS 1631 and BS 1632 free machining grades S—0.5% max and/or other suitable elements may be present. Elongation 12% min. on gauge length $5.65\sqrt{S_0}$.
¶ BS 1631B and BS 1632C titanium in the proportion 5 × C and 0.70% max may be substituted for niobium by agreement.
BS 1504–821 titanium grade, contains titanium in the proportion 4 × C min 0.70% max nickel 7.5% min.
** BS 4238 EC1 N = 0.20% max.
†† *See* Table 26.39A for impact properties.

FOR AUSTENITIC AND HEAT RESISTING STEELS—*continued*

S	P	*Heat treatment*	*Tensile strength min.*		*Yield or 0.5% proof stress (min)*		*Elongation*	*Bend test*		*Impact*
max	*max*		MPa	tonf in^{-2}	MPa	tonf in^{-2}	%	*Radius of bend*	*Angle of bend*	
0.060	0.060	As cast or heat treated at the manufacturer's discretion	—	—	—	—	—	—	—	—
0.060	0.060	As cast or heat treated at the manufacturer's discretion	—	—	—	—	—	—	—	—
0.060	0.060	As cast or heat treated at the manufacturer's discretion	—	—	—	—	—	—	—	—
0.060	0.060	As cast or heat treated at the manufacturer's discretion	—	—	—	—	—	—	—	—
0.060	0.060	As cast or heat treated at the manufacturer's discretion	—	—	—	—	—	—	—	—
0.060	0.060	As cast or heat treated at the manufacturer's discretion	—	—	—	—	—	—	—	—
0.060	0.060	As cast or heat treated at the manufacturer's discretion	—	—	—	—	—	—	—	—
0.060	0.060	As cast or heat treated at the manufacturer's discretion	—	—	—	—	—	—	—	—
0.060	0.060	As cast or heat treated at the manufacturer's discretion	—	—	—	—	—	—	—	—
0.060	0.060	As cast or heat treated at the manufacturer's discretion	—	—	—	—	—	—	—	—
0.060	0.060	As cast or heat treated at the manufacturer's discretion	—	—	—	—	—	—	—	—
0.040	0.040	As cast	556	36	—	—	3	—	—	—
0.040	0.040	As cast	510	33	—	—	3	—	—	—
0.040	0.040	As cast	448	29	—	—	7	—	—	—
0.040	0.040	As cast	448	29	—	—	3	—	—	—
0.040	0.040	As cast	448	29	—	—	3	—	—	—

Table 26.39A CHARPY V-NOTCH IMPACT TEST AT TEMPERATURE OF −196 °C

Specification	*Minimum average value*		*Minimum individual value*	
	Joules	ft lbf	Joules	ft lbf
BS 1631 Grade A	40.7	30	33.9	25
BS 1631 Grade B	20.3	15	13.6	10
BS 1631 Grade C	40.7	30	33.9	25
BS 1632 Grade B	33.9	25	27.1	20
BS 1632 Grade D	33.9	25	27.1	20
BS 1632 Grade E	33.9	25	27.1	20
BS 1632 Grade F	40.7	30	33.9	25

Table 26.40 TYPICAL PROPERTIES
(See notes on

Type	Chemical composition, %							
	C	Si	Mn	Ni	Cr	Other	S	P
C–Mo	0.16	0.35	0.65	—	—	0.53 Mo	0.024	0.024
C–Mo–V	0.12	0.29	0.48	—	—	0.57 Mo, 0.26 V	0.024	0.023
1½% Mn	0.24	0.27	1.46	—	—	—	—	—
1½% Mn–Mo	0.35	—	1.7	—	—	0.35 Mo	—	—
1½% Mn–Mo	0.26	0.28	1.30	—	—	0.25 Mo	—	—
1½% Mn–V	0.35	0.40	1.40	—	—	0.10 V	—	—
4% Ni	0.10	0.2–0.4	0.45–0.75	3.75–4.25	—	—	—	—
2% Ni	0.20	—	0.6–0.9	2.0	—	—	—	—
2½% Ni–Mo	0.3	0.3	0.5	2.5	—	0.4 Mo	—	—
1½% Ni–V	0.28	0.35	1.0	1.5	—	0.1 V	—	—
1% Cr	0.50–0.55	0.35–0.40	0.8–1.0	—	0.9–1.0	—	—	—
¾% Cr–Mo	0.47	0.29	0.66	—	0.72	0.25 Mo	—	—
1¼% Ni–Cr	0.30	—	0.6	1.25	0.6	—	—	—
1½% Ni–Cr	0.40	—	0.6	1.5	0.8	—	—	—
Ni–Cr–Mo	0.23–0.29	0.46–0.55	1.21–1.42	0.59–0.74	0.40–0.52	0.29–0.43 Mo	0.016–0.023	0.024–0.049
Ni–Cr–Mo	0.28–0.35	0.4	0.8–1.0	1.1–1.3	0.5–0.7	0.10–0.20 Mo	—	—
Ni–Cr–Mo	0.25–0.35	0.4	0.8–1.0	1.2–1.5	0.6–0.9	0.25–0.35 Mo	—	—
Ni–Cr–Mo	0.33	0.28	0.61	2.62	0.61	0.47 Mo	0.031	0.042
Ni–Cr–Mo	0.33	0.28	0.61	2.62	0.61	0.47 Mo	0.031	0.042
Cr–V	0.30	—	1.3	—	0.6	0.10 V	—	—
1¼% Cr–Mo	0.99	0.6	0.7	—	1.1	Mo 0.25	—	—
Cr–Mo–V	0.13	0.25	0.47	—	0.82	0.77 Mo, 0.32 V	0.021	0.016
Cr–Mo–V	0.14	0.30	0.43	—	0.93	0.75 Mo, 0.36 V	0.014	0.023
Cr–Mo–V	0.14	0.30	0.43	—	0.93	0.75 Mo, 0.36 V	0.014	0.023
Cr–Mo–V	0.14	0.30	0.43	—	0.93	0.75 Mo, 0.36 V	0.014	0.023
Cr–Mo–V	0.14	0.30	0.43	—	0.93	0.75 Mo, 0.36 V	0.014	0.023
Cr–Mo–V	0.14	0.30	0.43	—	0.93	0.75 Mo, 0.36 V	0.014	0.023
Cu	0.13	—	0.62	—	—	1.0 Cu	—	—
Cu	0.13	—	0.62	—	—	1.0 Cu	—	—
Cu	0.29	—	—	—	—	0.94 Cu	—	—
Cu	0.29	—	—	—	—	0.94 Cu	—	—
Cu–Mo	0.26	—	0.73	—	—	0.36 Mo, 0.88 Cu	—	—
Cu–Mo	0.26	—	0.73	—	—	0.36 Mo, 0.88 Cu	—	—
Low C, Mo–B	0.11	0.32	0.75	0.10	0.07	0.58 Mo‡	0.040	0.042

* On 4d † On 5d ‡ 0.01% Ti. 0.21% Al. 0.001 7% acid soluble B, 0.005% total B, 0.0025% soluble N, 0.0006 5% total N.

Table 26.41 PROPERTIES OF SOME AUSTENITIC
(See notes on

Type	Chemical composition, %									Heat treatment
	C	Si	Mn	Ni	Cr	Mo	Others	S	P	
17% Cr	0.14	0.5	0.6	2.0	17.0	—	—	—	—	H+T
27% Cr	0.43	0.75	0.75	—	26.0	—	0.15 N	0.030	0.030	N or WQ 800 °C
29% Cr	0.30	1.0	0.8	0.25	29.0	—	—	—	—	—
18 Cr/8 Ni	0.07	1.0	1.0	9.0	18.5	—	—	0.030	0.030	N or WQ 1 100 °C
18/8/Nb	0.10	1.25	0.5	9.5	18.0	—	1.0 Nb	—	—	N
18/8/Nb*	0.10	1.25	0.5	9.5	18.0	—	1.0 Nb	0.15	—	N
18/8/Ti	0.12	1.5	0.8	9.0	19.0	—	0.60 Ti	—	—	N 1 100 °C
18/8/3 Mo	0.12	0.62	0.66	8.96	18.34	2.62	—	0.022	0.036	WQ 1 080 °C
18/8/3 Mo*	0.11	0.72	0.62	9.04	18.78	3.27	—	0.26	0.030	WQ 1 080 °C
18/8/3/Nb	0.10	1.25	0.5	9.0	18.0	2.5	1.0 Nb	—	—	N

OF SOME ALLOY STEELS
page 26–57

Heat treatment	*Tensile strength*		*Yield stress*		*Elongation*	*Reduction of area*, %	*Izod impact value*		*Brinell hardness*
	MPa	tonf in^{-2}	MPa	tonf in^{-2}	%		Joules	ft lb	
A 950°, T 650 °C	494	32.0	278	18.0	29	—	—	—	—
A 950°, N 900°, T 680 °C	510	33.0	309	20.0	27	—	—	—	—
NT 520 °C	633	41	432	28	29	48	54.2	40	194
OQ 900°, T 650°	710	46.0	510	33.0	26	58.0	39.3	29	—
N 860°, T 650 °C	559	36.2	395	25.6	26	43.4	—	—	—
N 870°, N 870°, T	710	46.0	525	34	27.5†	57.6	71.6	52.8	—
A, WQ 900°, T 700 °C (WQ)	571–618	37–40	371–402	24–26	23–28	40–50	—	—	159–174
N	520–588	33.7–38.1	313–383	20.3–24.8	25–32†	45–65	—	—	—
N	686–706	44.4–45.7	363–383	23.5–24.8	15–23†	15–32	—	—	—
N 920°, N 815°, T	633	41	479	31	28.5*	59.4	100.7	74.3	—
A, N 860°, T 700 °C	695–849	45–55	355–448	23–29	15–25	—	—	—	212–255
OQ 860°, T 650 °C	822	53.2	681	44.1	20.5	33	33.9–36.6	25–27	—
N 900°, T 670 °C	726	47.0	402	26.0	23	45	59.6	44	—
N 900°, T 670 °C	726	47.0	463	30.0	20	35	35.2	26	—
A 920°, WQ 920°, T 610–670°	676–809	43.8–52.4	482–666	31.2–43.2	16–25	41.4–57	66.4–92.1	49–68	—
A 900°, OQ 870°, T 630 °C	695–849	45–55	494–618	32–40	15–25	35–55	33.9–47.4	25–45	200–255
A 900°, OQ 870°, T 570 °C	849–1 004	55–65	695–772	45–50	15–25	30–50	27.1–54.2	20–40	255–300
OQ 830°, T 620 °C	985	63.8	839	54.3	22.5	55	0	0	315
OQ 830°, T 650 °C	880	57.0	751	48.6	23.5	57	—	—	285
N 900°, T 700 °C	726	47.0	556	36.0	29.0	60	50.1	37	—
HT	1 096	71	618	40	7	—	—	—	—
A 950°, N 1 000°, T 700 °C	587	38.0	417	27.0	27.0	—	—	—	—
A 950 °C	507	32.9	417	20.0	28	—	39.3	29	—
A 950°, T 700 °C	513	33.2	340	22.0	30	—	52.9	39	—
A 950°, N 1 000 °C	803	52.0	—	—	3	—	6.8	5	—
A 950°, N 1 000 °C, T 700 °C	842	54.5	741	48.0	17	—	6.8	9	—
A 950°, N 1 000°, T 750 °C	695	45.0	571	37.0	22	—	12.2	9	—
N 870 °C	483	31.3	344	22.3	31*	51.4	—	—	143
N 870°, T 510 °C	618	40.0	483	31.3	23.5*	49.7	—	—	196
N 815 °C	607	39.3	40.0	25.9	19*	31.2	—	—	179
N 815°, T 540 °C	717	46.4	524	33.9	20.5*	36.4	—	—	207
N 845 °C	703	45.4	412	26.7	17.5*	24.4	—	—	196
N 845°, T 540 °C	717	46.4	545	35.3	22*	41.0	—	—	217
N	718	46.5	452	29.3	23.5	47.1	—	—	—

AND HEAT-RESISTING STEELS
page 26–57

Tensile strength		*Yield stress*		*Elongation*	*Reduction of area*	*Izod*		*Brinell hardness number*	*Melting point* °C *(approx.)*	*Specific gravity*	*Max. recommended service temp.* °C
MPa	tonf in^{-2}	MPa	tonf in^{-2}	%	%	Joules	ft lb				
772	50	618	40	12	—	—	—	229	1 470	7.70	650
479	31	—	—	—	—	—	—	170	—	7.62	1 100
—	—	—	—	—	—	—	—	—	1 480	7.50	—
479	31	208	13.5	40	40	54.2	40	170	1 440	7.90	—
525	34	255	16.5	30	35	—	—	170	1 430	7.93	800
494	32	247	16	20	25	—	—	—	—	—	—
618	40	309	20	18	—	—	—	187	—	7.78	800
559	36.2	338	21.9	54	64.2	130.1	96	—	—	—	—
567	36.7	298	19.3	41	42.3	48.8	36	—	—	—	—
525	34	263	17	25	30	—	—	187	1 420	7.96	—

Table 26.41 PROPERTIES OF SOME AUSTENITIC

Type	Chemical composition, %									Heat treatment
	C	Si	Mn	Ni	Cr	Mo	Others	S	P	
18/8/3/Ti	0.12	1.5	0.8	9.0	19.0	2.5	0.60 Ti	—	—	N 1 100 °C
18/10/Nb	0.10	1.25	0.5	9.5	18.0	—	1.0 Nb	—	—	N
18/10/Nb/Ti	0.11	0.92	2.0	11.0	18.0	—	0.19 Ti, 0.60 Nb	—	—	WQ 1 050 °C
18/10/3 Mo	0.07	1.0	1.0	10.5	19.0	2.75	—	0.030	0.030	N or WQ 1 150 °C
18/10/3/Nb	0.07	1.0	1.0	10.5	19.0	2.75	0.85 Nb	0.030	0.030	N or WQ 1 150 °C
18/18/4/1/4 Cu*	0.10	0.7	0.8	18.0	18.0	3.75	1.0 Nb, 3.5 Cu	0.15	—	As cast
21/8/2 Cu/4 W	0.30	1.5	0.7	9.0	21.0	—	2.0 Cu, 4.0 W	—	—	N 1 100 °C
21 Cr/7 Ni	0.15	1.5	—	7.0	21.0	—	—	—	—	—
23/12/3 W	0.20	1.0	0.8	12.0	23.0	—	3.0 W	—	—	N
21/7/4 W	0.35	1.5	—	7.0	21.0	—	4.0 W	—	—	—
25 Cr/12 Ni	0.20	1.2	1.3	13.0	25.0	—	—	0.030	0.030	N or WQ 1 100 °C
25 Cr/20 Ni	0.20	1.5	1.0	20.0	25.0	—	—	0.030	0.030	N or WQ 1 100 °C
25 Ni/15 Cr	0.35	0.9	0.75	25.0	15.0	—	—	—	—	—
45 Ni/20 Cr	0.60	1.0	0.75	45.0	20.0	—	—	—	—	—
60 Ni/15 Cr	0.30	1.5	—	60	12–20	—	—	—	—	—
CD–4M Cu	0.04 max.	1.0 max.	1.0 max.	4.75–6.0	25–27	1.75–2.25	2.75–3.25 Cu	—	—	As cast 1 h/1 120 °C WQ 1 h/1 120 °C WQ+ 3 h/510 °C 2 h/1 065 °C WQ 2 h/1 065 °C WQ+ 4 h/455 °C
17–4 PH	0.07 max.	1.0 max.	1.0 max.	3.0–5.0	15.25–17.25	—	2.3–3.0 Cu	—	—	1 h/1 040 °C AC+ 1 h/470 °C 1 h/1 040 °C AC+ 1 h/565 °C
FV 520	—	—	—	—	—	—	—	—	—	1 050 °C + 1 h/650 °C

* Free cutting steel. † Charpy V-notch.

Table 26.42 TYPICAL APPLICATIONS OF CARBON AND LOW ALLOY STEELS*

Description	Specification	Main constituents %	Typical uses and remarks
Carbon steel	BS 3100–592 Grade A	C 0.25 max.	General engineering and marine castings where ductility and weldability are important, e.g. crank webs, sternframes, brackets, housings, links, etc. By arrangement, can be supplied suitable for carburizing and cyaniding
	BS 1504–101 Grade A	As necessary	
	BS 3100–592 Grade B	C 0.35 max	For slightly more highly stressed components, e.g. earth-moving machinery and heavy rolling mill equipment, also carbon steel valves
	BS 3100–592 Grade C	C 0.45 max.	For moderate wear resistance applications, e.g. gear wheels of all sizes
	BS 1504–161 Grade A	C 0.25 max.	Pressure castings for valves, etc., for chemical and petroleum industries
	BS 1504–161 Grade B	C 0.30 max.	
Mild steel of high magnetic permeability	BS 3100–1617 Grade A	C 0.15 max.	Electrical applications, e.g. motor yokes, magnet frames, etc.
	BS 3100–1617 Grade B	C 0.25 max	As for Grade A but where magnetic permeability associated with a higher tensile strength is required

AND HEAT-RESISTING STEELS—*continued*

Tensile strength		*Yield stress*		*Elongation*	*Reduction of area*	*Izod*		*Brinell hardness number*	*Melting point* °C *(approx.)*	*Specific gravity*	*Max. recommended service temp.* °C
MPa	tonf in^{-2}	MPa	tonf in^{-2}	%	%	Joules	ft lb				
679	44	309	20	25	—	—	—	212	—	7.78	—
525	34	262	17	25	25	—	—	179	1 430	7.90	—
525	34	247	16	49	41	—	—	—	—	—	—
540	35	208	13.5	35	40	40.7	30	174	—	7.96	800
540	35	208	13.5	35	40	40.7	30	174	—	7.96	800
402	26	185	12	12	16	—	—	163	1 375	8.04	—
695	45	386	25	15	15	—	—	229	—	7.93	—
695	45	309	20	15	—	—	—	207	—	7.81	900
540	35	278	18	28	30	—	—	217	1 400	7.90	—
741	48	468	30.3	15	—	—	—	223	—	7.92	950
540	35	247	16	35	40	40.7	30	179	—	7.94	1 050
587	38	232	15	35	40	40.7	30	179	—	7.90	1 100
494	32	278	18	8	12	—	—	—	1 400	7.90	—
—	—	—	—	—	—	—	—	—	1 370	7.90	—
741	48	340	22	12	—	—	—	200	—	8.16	1 100
834	54	579	37.5	27	41	25.8	19†	262	—	—	—
710	46	587	38	19	—	—	—	250	—	—	—
965	62.5	828	53.6	16	—	—	—	321	—	—	—
772	50	557	36.1	29	48	141	104†	248	—	—	—
973	63	689	44.6	25	38	12.9	9.5†	202	—	—	—
1 172 min.	75.9 min.	965 min.	62.5 min.	6 min.	15 min.	—	—	360–420	—	—	—
896 min.	58 min.	689 min.	44.6 min.	12 min.	15 min.	—	—	269–350	—	—	—
1 056	68.4	1 012	65.5	8	12	—	—	350	—	—	—

Table 26.42 TYPICAL APPLICATIONS OF CARBON AND LOW ALLOY STEELS*—*continued*

Description	*Specification*	*Main constituents* %	*Typical uses and remarks*
Carbon steel for surface hardening	BS 3100–1760 Grade A BS 3100–1760 Grade B	C 0.40–0.50	Castings requiring good wear resistance and suitable for flame hardening, e.g. gear wheels, pinions, etc.
$1\frac{1}{2}$% manganese steel	BS 3100–1456 Grade A	C 0.18–0.25 Mn 1.20–1.70	Applications requiring a medium strength steel with good shock and wear resistance. Thin castings up to $\frac{3}{4}$ in
	BS 3100–1456 Grade B1 and B2	C 0.25–0.33 Mn 1.20–1.70	As for Grade A but suitable for sections up to $1\frac{1}{2}$ in. Axle arms and steel castings for use on exterior of armoured fighting vehicles
	BS 2772 Part 3	C 0.12–0.22 Mn 1.30–1.70	Castings requiring improved resistance to impact shock
Austenitic manganese steel	BS 3100–1457	C 1.0–1.35 Mn 11.0 min.	Applications involving severe wear combined with impact resistance, e.g. crushing machinery, tank track links, etc.
Carbon-molybdenum steel	BS 3100–1398A BS 1504–240	C 0.15–0.25 Mo 0.40–0.70	General engineering purposes where castings are subjected to stress at elevated temperatures, e.g. 424 °C (800 °F), such as turbine components, valves, pumps, steam chests, etc.

Table 26.42 TYPICAL APPLICATIONS OF CARBON AND LOW ALLOY STEELS*—*continued*

Description	*Specification*	*Main constituents %*	*Typical uses and remarks*
1% Chromium steel, (abrasion resisting)	BS 3100–1956 Grades A and B BS 3100:1956 Grade C	C 0.45–0.55 Cr 0.80–1.20 C 0.55–0.65 Cr 0.80–1.50	Excavator shovel teeth and other details of earth-moving equipment, grinding rollers and roller tracks where high resistance to abrasion is required but where impact is not an important consideration
3½% Nickel steel	BS 1504–503	C 0.15 max. Ni 3.0–4.0	Valves and similar pressure castings for chemical and petroleum industries. Suitable for sub-zero applications down to −100 °C
1¼% Chromium–molybdenum steel	BS 1504–621 BS 1398B	C 0.20 max. Cr 1.00–1.50 Mo 0.45–0.65	Valves and similar pressure castings requiring good creep strength at moderately elevated temperatures. *See also* remarks under BS 3100/1461, 1462 and 1463 below
2¼% Chromium 1% molybdenum steel	BS 1504–622 BS 1398C	C. 0.18 max. Cr 2.00–2.75 Mo 0.90–1.20	
3% chromium–moybdenum steel	BS 3100–1461	C 0.25 max. Cr 2.50–3.50 Mo 0.35–0.60	General engineering castings for use up to temperatures of 500 °C and for certain steam-raising applications
	BS 1504–623	C 0.25 max. Cr 2.50–3.50 Mo 0.35–0.60	
5% Chromium–molybdenum steel	BS 3100–1462	C 0.20 max. Cr 4.00–6.00 Mo 0.45–0.65	General engineering castings subjected to stress at elevated temperatures up to 550 °C and requiring some degree of corrosion resistance. Used for valves, impellers, etc.
	BS 1504–625	C 0.20 max. Cr 4.00–6.00 Mo 0.45–0.65	
½% Chromium–½% molybdenum–¼% vanadium	BS 3100–1398D	C 0.15 Cr 0.25–0.50 Mo 0.50–0.70 V 0.22–0.30	Power plant, steampipes, turbine castings, etc.
1% Chromium–1% molybdenum–¼% vanadium	BS 3100–1398E	C 0.15 Cr 0.70–1.00 Mo 0.70–1.00 V 0.22–0.35	Power plant, steampipes, turbine rotors, etc.
45/55 ton alloy steel	BS 3100–1458A	As necessary	Machinery parts requiring high strength and notch ductility, e.g. gears, pinions, excavator teeth and buckets, etc.
55/65 ton alloy steel	BS 3100–1458B	As necessary	As for BS 3100/1458A but where higher strength and hardness is required, e.g. die blocks, grinding rings, dredger parts, brackets, levers, etc.
66/75 ton alloy steel	BS 3100–1458C	As necessary	Similar to BS 1458A and B

* MOD/Min. Tech. Def. Std.01–1 (1969).

Table 26.43 CHARACTERISTICS AND APPLICATIONS OF CORROSION-RESISTING STEEL CASTINGS

Code	*Alloy type and specifications*	*Outstanding features*	*Typical uses*
Cl	13 Cr (0.15% C max.) BS 1630A	Mildly abrasion resistant and more resistant to atmospheric corrosion than low alloy steels; excellent resistance to many organic media in relatively mild service. Capable of being heat treated for high strength and hardness, retains high	In chemical and petroleum and power supply industries for components of high ductility, working under impact loads such as hydraulic press valves, cracking plant equipment, domestic utensils. For valve and pump parts at normal and

Table 26.43 CHARACTERISTICS AND APPLICATIONS OF CORROSION-RESISTING STEEL CASTINGS—*continued*

Code	*Alloy type and specifications*	*Outstanding features*	*Typical uses*
		proportions of strength up to 600 °C and possesses scaling resistance up to 700 °C has low coefficient of expansion	super-heated steam temperatures and furnace burner tips and pilot cones. For mildly corrosive conditions, e.g. atmospheric precipitation, aqueous solutions of salts of organic acids at room temperature, etc.
C2	13 Cr (0.20% C max.) BS 1504–713	Intermediate between C1 and C3	Used where a compromise between C1 and C3 is required
C3	13 Cr (0.30% C max.) BS 1630B	Similar to C1, but greater hardness after heat treatment with little sacrifice in corrosion resistance. Hardness gives useful properties for cutting edge and wear resistance	Choppers, cylinder liners, pump parts, steam turbine parts, mould dies, impellers, paper machinery parts, runners for water turbines, and general engineering work
C4	18 Cr/8 Ni BS 1631A, C and D BS 1504–801	Excellent stability at sub-zero temperatures, used at temperatures down to −225 °C. Low carbon content gives maximum corrosion resistance to strongly oxidizing media, such as boiling nitric acid, sulphates and organic acids. Liable to carbide precipitation between 500 and 800 °C leading to intergranular corrosion	General purpose corrosion-resistant alloy for applications in chemical, pharmaceutical, textile and food processing industries and in oil refineries, handling alcohol, ammonia mercury, soap, sugar, CO_2 and steam. Autoclaves, blast furnace parts, filter press plates, headers, heating coils, spray nozzles, pump and valve parts
C5	18 Cr/8 Ni/Nb BS 1631B BS 1504–821 Nb	Stabilized, so safe to use in temperature range 500–800 °C. Possesses slightly better hot strength stability, resistance to hot nitric acid and knife edge attack in fusion zone of the welds than C6	Used for parts that cannot be heat treated after welding, also parts for use in carbide precipitation range, particularly for aircraft exhaust systems. jet engine parts, shroud assemblies, autoclaves
C6	18 Cr/8 Ni/Ti BS 1631 B, Ti BS 1504–821Ti	Stabilized, recommended for range 599–800 °C, resistance to acid, gas and intercrystalline corrosion. Most successful in majority of applications while being cheaper than C5	For chemical industry equipment, main pipe to waste gas collectors, oil refining equipment, pickling plant plates
C7	19 Cr/12 Ni/3½ Mo BS 1632A BS 1504–486	Higher molybdenum than C8 or C11	Has better resistance to corrosives mentioned under C8
C8	18 Cr/10 Ni/2½ Mo BS 1632B BS 1504–845B	Molybdenum addition gives added corrosion resistance to sulphite liquors and other chemicals and decreases problem of pitting as compared with C4, particularly in contact with chloride, though it is not so resistant to boiling nitric acid at C4. Should have low carbon content or be stabilized to impart resistance to intercrystalline corrosion	Used in textile, paper and chemical industries for valves and pumps handling hot organic acids and fatty acids at high temperatures, chlorides and acid salts, reducing acids, sulphite liquor. For agitators, evaporator parts, jet engine components, spray nozzles, high pressure steam valves. For marine fittings, and use in tropical sea-water, dock and estuary waters
C9	18 Cr/10 Ni/2½ Mo/Nb BS 1632 C. Nb BS 1504–845	Corrosion resistant to additional salt solutions and acids at elevated temperatures	Used for similar applications as C8, particularly in high temperature ranges
C10	18 Cr/10 Ni/2½ Mo/Ti BS 1632 C. Ti	Similar to C9 being corrosion resistant to additional salt solutions as compared to C8. Heat resistant at temperatures up to 800 °C and not subject to intercrystalline corrosion	For components required to resist the effects of sulphuric, sulphurous, phosphoric, formic and acetic acids, also bleaching powder and sulphite liquor
C11	18 Cr/8 Ni/2½ Mo BS 1632D	Has higher strength than C8, and is more easily welded, but has higher magnetic permeability. Corrosion resistance is similar to C8	Similar uses to C8, but slightly less suitable for service at elevated temperatures

Table 26.44 CHARACTERISTICS AND APPLICATION OF HEAT-RESISTING STEEL CASTINGS

Code	*Alloy type and specification*	*Structure*	*Resistance to cyclic heating*	*Resistance to damage by carbon penetration*	*Ability to avoid catalysing hydrocarbon cracking*	*Outstanding features*	*Typical uses*
H1	13 Cr BS 1648A	Martensitic	—	—	—	Limited use for scaling resistance up to 700 °C. Heat treatable to medium-high strengths	Impellers and turbine blades. Steam turbine parts. Moulds and dies. Furnace burners. Internal combustion engines
H2	27 Cr BS 1648B	Ferritic	Poor	Poor	Good	Will resist high sulphur atmospheres when heavy loading is not involved	Sintering and ore roasting furnaces. Grate bars. Retorts. Dampers. Hearths
H3	27 Cr BS 1648C	Ferritic	—	—	—	High abrasion resistance	As for H2
H4	20 Cr/10 Ni BS 1648D	Austenitic	—	—	—	Good oxidation resistance in the range 650–870 °C	Tube supports in oil refineries. Cement mill parts. Conveyor belts. Heat treatment furnace parts
H5	21 Cr/8 Ni/4 W BS 1648D	Austenitic	—	—	—	High strength and oxidation resistance	As for H4, but used where sulphurous atmospheres may be encountered
H6	25 Cr/12 Ni BS 1648E	Austenitic*	Fair	Fair†	—	High strength and oxidation resistance. Can be partially ferritic or fully austenitic. Between 650 and 870 °C, austenitic type must be used	Tube supports in oil refineries. Pots. Heat treatment furnace parts
H7	25 Cr/12 Ni/3 W BS 1648E	Austenitic	—	—	—	High strength values up to 900 °C	Superheater supports. Furnace skids. Heat treatment handling. Furnace racks
H8	25 Cr/20 Ni BS 1648F	Austenitic	Good	Good	Fair	Good hot gas corrosion resistance. Behaviour erratic where thermal shock is involved	Furnace parts. Chains. Radiant heater tubes. Boxes. Muffles. Pots
H9	25 Ni/15 Cr BS 1648G	Austenitic	—	—	—	Good high temperature strength and corrosion resistance	Brazing supports
H10	35 Ni/15 Cr BS 1648	Austenitic	Good	Good	—	Good life when subject to rapid heating and cooling	Glass rolls. Radiant heater tubes. Retorts. Conveyors. Heat treatment furnace parts
H11	40 Ni/20 Cr BS 1648H	Austenitic	Good	Good	—	Good hot strength and corrosion resistance	As for H10, but used where improved resistance to hot gas corrosion is required
H12	60 Ni/15 Cr BS 1648K	Austenitic	Excellent	Excellent	—	Excellent for severe service involving cyclic heating and thermal shock. Resistant to vanadium pentoxide attack	Furnace parts. Quenching apparatus. Molten lead containers

* Depending on composition, can be partially ferritic.
† Resistance is improved when silicon content is 1.5–1.7%.

Table 26.45 COMPOSITION PROPERTIES AND APPLICATIONS OF CARBON AND LOW ALLOY STEELS FOR INVESTMENT CASTINGS†

Steel BS 3146 Part 1 1959	*Element*	*Grade A %* *min*	*Grade A %* *max*	*Grade B %* *min*	*Grade B %* *max*	*Grade C %* *min*	*Grade C %* *max*	*%* *min*	*%* *max*	*Typical applications*
Type CLA 1	Carbon		0.25		0.35		0.45			General purpose engineering steels for low- and medium-strength applications and for structural parts involving welding or cold-working. Suitable for parts which are not heavily stressed but where ductility is an advantage. Used for brackets, housings, links, etc., in, for example, textile, sewing and calculating machinery
	Silicon		0.60		0.60		0.60			
	Manganese		1.00		1.00		1.00			
	Nickel		0.40		0.40		0.40			
	Chromium		0.25		0.25		0.25			
	Molybdenum		0.15		0.15		0.15			
	Copper		0.30		0.30		0.30			
	Sulphur		0.050		0.050		0.050			
	Phosphorus		0.050		0.050		0.050			
Tensile strength:										
tonf in^{-2}		28		32		35				
MPa		432		494		541				
Yield stress or 0.5% proof stress:										
tonf in^{-2}		14		16		17.5				
MPa		216		247		270				
Elongation %		22		20		15				
Angle of bend		120°		90°		no test				
Radius of bend		$1\frac{1}{2}t$*		$1\frac{1}{2}t$*		—				
Hardness		121/174 HB		143/183 HB		163/207 HB				
Type CLA 2	Carbon	0.18	0.25	0.25	0.33					Used for parts where a high yield stress and ductility are required, with a degree of shock resistance, for welded structures subjected to fairly high stress and in general for medium-strength applications where an alloy steel is not justified and for parts subject to temperatures up to 250 °C, e.g. links, levers, light bracket structures, and lightly loaded moving parts
	Silicon	—	0.50	—	0.50					
	Manganese	1.20	1.70	1.20	1.70					
	Nickel	—	0.40	—	0.40					
	Chromium	—	0.25	—	0.25					
	Molybdenum	—	0.15	—	0.15					
	Copper	—	0.30	—	0.50					
	Sulphur	—	0.050	—	0.050					
	Phosphorus	—	0.050	—	0.050					
Tensile strength:										
tonf in^{-2}		35	45	45	55					
MPa		541	695	695	849					
Yield stress or 0.5% proof stress:										
tonf in^{-2} MPa		60% of tensile strength		60% of tensile strength						
Elongation, %		18		15						
Izod impact value, J		34		27						
Hardness HB		152/201	201/255							

*t = Thickness of test piece.

† E. G. Donaldson, 'Investment Casting', *Engineers Digest*, 1962, **23**, 91.

Table 26.45 COMPOSITION PROPERTIES AND APPLICATIONS OF CARBON AND LOW ALLOY STEELS FOR INVESTMENT CASTINGS† —*continued*

Steel BS 3146 Part 1 1959	*Element*	*Grade A %*		*Grade B %*		*Grade C %*		*%*		*Typical applications*
		min	*max*	*min*	*max*	*min*	*max*	*min*	*max*	
Types CLA 3, CLA 4 and CLA 5				Composition not specified						Alloy steels for engineering applications and structural parts requiring medium to high tensile strengths, coupled with a good ductility and shock and fatigue strengths, e.g. airframe and missile parts, power and pneumatic tools, and hydraulic machinery
Type CLA 3										For parts requiring medium tensile strength and subject to medium shock loading
Tensile strength:										
tonf in^{-2}								45	55	
MPa								695	849	
Yield stress or 0.5% proof stress:										
tonf in^{-2}								32		
MPa								494		
Elongation, %								15		
Izod impact value, J								34		
Hardness								201/255 HB		
Type CLA 4										Suitable for components subject to well-lubricated reciprocating loading, but with insufficient surface hardness to give good wear resistance. Suitable for brackets and levers requiring high tensile strength
Tensile strength:										
tonf in^{-2}								55	65	
MPa								849	1 004	
Yield stress or 0.5% proof stress:										
tonf in^{-2}								38		
MPa								587		
Elongation, %								12		
Izod impact value, J								20		
Hardness								248/302 HB		
Type CLA 5 (Grades A and B)										Suitable for structural members subjected to heavy loading, for moderate wear resistance and heavy-duty reciprocating parts, particularly aircraft components Used where load requirements are comparatively high

Tensile strength:						
tonf in^{-2}		60	75			
MPa		927	1 158			
0.1% proof stress:						
tonf in^{-2}		45	60			
MPa		695	927			
Elongation, %		12	7			
Izod impact value, J		41	14			
Hardness		269/321 HB	341/388 HB			
Type CLA 6	Carbon			0.15	0.25	Used for structural parts calling for a medium tensile strength and good ductility, and for parts operating at elevated temperatures
	Silicon			0.20	0.50	
	Manganese			0.50	1.00	
	Nickel			—	0.40	
	Chromium			—	0.25	
	Molybdenum			0.40	0.70	
	Copper			—	0.40	
	Sulphur			—	0.050	
	Phosphorus			—	0.050	
Type CLA 7	Carbon			—	0.25	
	Silicon			—	0.75	
	Manganese			0.30	0.60	
	Nickel			—	0.40	
	Chromium			2.50	3.50	
	Molybdenum			0.35	0.60	
	Copper			—	0.40	
	Sulphur			—	0.050	
	Phosphorus			—	0.050	
Type CLA 6						
Tensile strength:						
tonf in^{-2}				30		
MPa				463		
Yield stress or 0.5% proof stress:						
tonf in^{-2}				16		
MPa				247		
Elongation, %				20		
Angle of bend				120°		
Radius of bend				$1\frac{1}{2}t$*		
Izod impact value, J				20		
Hardness				131/183 HB		

*t = Thickness of test piece.
† E. G. Donaldson, 'Investment Castings', *Engineers' Digest*, 1962, **23**, 91.

Table 26.45 COMPOSITION PROPERTIES AND APPLICATIONS OF CARBON AND LOW ALLOY STEELS FOR INVESTMENT CASTINGS†—*continued*

Steel BS 3146 Part 1 1959	*Element*	*Grade A %*		*Grade B %*		*Grade C %*		*%*		*Typical applications*
		min	*max*	*min*	*max*	*min*	*max*	*min*	*max*	
Type CLA 7										
Tensile strength:										
tonf in^{-2}								40	50	
MPa								618	772	
Yield stress or 0.5% proof stress:										
tonf in^{-2}								24		
MPa								371		
Elongation, %								18		
Angle of bend								120°		
Radius of bend								3*t**		
Izod impact value, J								34		
Hardness								179/223 HB		
Type CLA 8	Carbon							0.35	0.45	Primarily for components where the general requirements are low to medium tensile strength with high local surface hardness. Typical components include pawls, ratchets and triggers
	Silicon							—	0.60	
	Manganese							0.5	0.80	
	Nickel							—	0.40	
	Chromium							—	0.25	
	Molybdenum							—	0.15	
	Copper							—	0.40	
	Sulphur							—	0.050	
	Phosphorus							—	0.050	
Tensile strength										
tonf in^{-2}								35		
MPa								541		
Yield stress										
tonf in^{-2}								17.5		
MPa								270		
Elongation, %								15		
Type CLA 9	Carbon							0.10	0.18	Generally for low-stressed parts requiring high surface hardness, e.g. ratchets, operating levers and parts subject to low reciprocating loading
	Silicon							0.25	0.60	
	Manganese							0.60	1.0	
	Nickel							—	0.40	
	Chromium							—	0.30	
	Molybdenum							—	0.15	

	Copper	—	0.30	
	Sulphur	—	0.050	
	Phosphorus	—	0.050	
Tensile strength:				
tonf in^{-2}		32		
MPa		494		
Elongation, %		20		
Izod impact value, J		27		
Type CLA 10	Carbon	0.10	0.18	High-duty parts subject to wear under reciprocating or intermittent loading. High-speed connecting links and levers, heavy-duty pawls and ratchets are typical applications
	Silicon	0.25	0.60	
	Manganese	0.30	0.60	
	Nickel	2.75	3.50	
	Chromium	—	0.30	
	Molybdenum	—	0.15	
	Copper	—	0.30	
	Sulphur	—	0.050	
	Phosphorus	—	0.050	
Tensile strength:				
tonf in^{-2}		45		
MPa		695		
Elongation, %		18		
Izod impact value, J		41		
Type CLA 11	Carbon	0.20	0.30	For moving members where high hardness is required to resist abrasion or wear, such as crankpins, crankshafts and similar parts
	Silicon	0.25	0.75	
	Manganese	0.30	0.60	
	Nickel	—	0.40	
	Chromium	2.90	3.50	
	Molybdenum	0.40	0.70	
	Copper	—	0.40	
	Sulphur	—	0.045	
	Phosphorus	—	0.045	
Tensile strength:				
tonf in^{-2}		55	65	
MPa		849	1 004	
Yield stress or 0.5% proof stress:				
tonf in^{-2}		38		
MPa		587		
Elongation, %		12		
Izod impact value, J		20		
Hardness		248/302 HB		

* t = Thickness of test piece.
† E. G. Donaldson, 'Investment Castings', *Engineers' Digest*, 1962, **23**, 91.

Table 26.45 COMPOSITION PROPERTIES AND APPLICATIONS OF CARBON AND LOW ALLOY STEELS FOR INVESTMENT CASTINGS†– *continued*

Steel BS 3146 Part 1 1969	*Element*	*Grade A %*		*Grade B %*		*Grade C %*		*%*		*Typical applications*
		min	*max*	*min*	*max*	*min*	*max*	*min*	*max*	
Type CLA 12	Carbon	0.45	0.55	0.45	0.55	0.55	0.65			Hardened parts subject to medium to high stresses. Grade A is useful under moderate conditions of wear, but Grades B and C can be used under heavy-duty conditions where intimate contact with abrasive dust or material exists
	Silicon	—	0.75	—	0.75	—	0.75			
	Manganese	0.50	1.00	0.50	1.00	0.50	1.00			
	Nickel	—	0.40	—	0.40	—	0.40			
	Chromium	0.80	1.20	0.80	1.20	0.80	1.50			
	Molybdenum	—	0.15		0.15	0.20	0.40			
	Copper	—	0.40	—	0.40	—	0.40			
	Sulphur	—	0.050	—	0.050	—	0.050			
	Phosphorus	—	0.050	—	0.050	—	0.050			
Tensile strength:										
tonf in^{-2}		45		—		—				
MPa		695		—		—				
Elongation, %		10		—		—				
Hardness		207 HB		293 HB		341 HB				

† E. G. Donaldson, 'Investment Castings', *Engineers' Digest*, 1962, **23**, 91.

Table 26.46 COMPOSITION, PROPERTIES AND APPLICATIONS OF HIGH ALLOY STEELS NICKEL AND COBALT ALLOYS FOR INVESTMENT CASTINGS†

Alloy BS 3146 Part II 1959	*Element*	*Grade A %*		*Grade B %*		*Grade C %*		*%*		*Typical applications*
		min	*max*	*min*	*max*	*min*	*max*	*min*	*max*	
ANC1-A	Carbon	—	0.15							In chemical, petroleum and power-supply industries, and for components of high ductility, working under impact loads, such as turbine blades, and lightly stressed aircraft and general engineering fittings; also, golf club heads. Used for valves and pump parts for liquids and superheated steam
	Silicon	—	1.25							
	Manganese	—	1.00							
	Nickel	—	1.00							
	Chromium	11.50	13.50							
	Molybdenum	—	0.15							
	Copper	—	0.40							
	Sulphur	—	0.050							
	Phosphorus	—	0.050							

Tensile strength:							
tonf in^{-2}		35					
MPa		541					
Yield stress or 0.5% proof stress:							
tonf in^{-2}		24					
MPa		371					
Elongation, %		20					
Angle of bend		120°					
Radius of bend		2t*					
Hardness		152/207 HB					
ANC1-B	Carbon		0.12	0.20			Suitable for hydraulic, steam and oil-refinery pump and valve parts, turbine blading, and for general engineering, automobile and aircraft fittings. Used for various heat-resistant parts not subjected in service to any great stress, but where scaling resistance is essential such as the tips of gas and fuel-oil burners
	Silicon		—	1.25			
	Manganese		—	1.00			
	Nickel		—	1.00			
	Chromium		11.50	13.50			
	Molybdenum		—	0.15			
	Copper		—	0.40			
	Sulphur		—	0.050			
	Phosphorus		—	0.050			
Tensile strength:							
tonf in^{-2}			40				
MPa			618				
Yield stress or 0.5% proof stress:							
tonf in^{-2}			29				
MPa			448				
Elongation, %			18				
Angle of bend			90°				
Radius of bend			2t*				
Hardness			183/229 HB				
ANC1-C	Carbon				0.20	0.30	Used for cutting blades, particularly surgical and dental equipment, pump and steam-turbine parts, mould dies, impellers, paper-cutting machine parts and general engineering work. An alloy with a higher carbon content is recommended for scalpels and similar surgical implements
	Silicon				—	1.25	
	Manganese				—	1.00	
	Nickel				—	1.00	
	Chromium				11.50	13.50	
	Molybdenum				—	0.15	
	Copper				—	0.40	
	Sulphur				—	0.050	
	Phosphorus				—	0.050	

*t = Thickness of test piece.

† E. G. Donaldson, 'Investment Castings', *Engineers' Digest*, 1962, **23**, 91.

Table 26.46 COMPOSITION, PROPERTIES AND APPLICATIONS OF HIGH ALLOY STEELS NICKEL AND COBALT ALLOYS FOR INVESTMENT CASTINGS†—*continued*

Alloy BS 3146 PART II 1969	*Element*	*Grade A %*		*Grade B %*		*Grade C %*		*%*		*Typical applications*
		min	*max*	*min*	*max*	*min*	*max*	*min*	*max*	
Tensile strength:										
tonf in^{-2}						45				
MPa						695				
Yield stress or 0.5% proof stress:										
tonf in^{-2}						30				
MPa						463				
Elongation, %						15				
Angle of bend						—				
Radius of bend						—				
Hardness						201/255 HB				
ANC2	Carbon							—	0.25	Suitable for highly stressed aircraft and general engineering fittings, in processing rayon and nitric acid manufacture, and in general for corrosive conditions insufficiently drastic to require an austenitic steel. Typical applications include pump and valve parts
	Silicon							0.10	1.0	
	Manganese							—	1.0	
	Nickel							1.0	3.0	
	Chromium							15.5	20.0	
	Sulphur							—	0.045	
	Phosphorus							—	0.045	
Tensile strength:										
tonf in^{-2}								55	65	
MPa								849	1 004	
0.1% proof stress:										
tonf in^{-2}								45		
MPa								695		
Elongation, %								10		
Hardness								248/302 HB		
ANC3-A	Carbon	—	0.12							Suitable for application in the chemical, pharmaceutical, textile, dairy and oil industries. Applications include filter-press plates, headers, heating coils, spray nozzles and pump and valve parts
	Silicon	—	2.00							
	Manganese	—	2.00							
	Nickel	7.5	—							
	Chromium	17.0	—							
	Niobium (columbium)	—	—							
	Sulphur	—	0.045							
	Phosphorus	—	0.045							

ANC5-B	Carbon			—	0.12			Used for parts that cannot be heat treated after welding; also, for parts used in the carbide-precipitation range, particularly for aircraft exhaust systems, jet engines, shroud assemblies, gas turbine rotors and blades, and marine fittings to a certain extent
	Silicon			—	2.00			
	Manganese			—	2.00			
	Nickel			8.5	—			
	Chromium			17.0	20.0			
	Niobium (columbium)			8 × C	1.10			
	Sulphur			—	0.045			
	Phosphorus			—	0.045			
Tensile strength:								
tonf in^{-2}		30		30				
MPa		463		463				
Yield stress or 0.5% proof stress:								
tonf in^{-2}		209		209				
MPa		21		21				
Elongation, %		25		25				
Angle of bend		120°		120°				
Radius of bend		$1\frac{1}{2}t$*		$1\frac{1}{2}t$*				
ANC4-A	Carbon	—	0.08					For fittings employed in the dyeing, textile, paper and chemical industries, for valves and pumps handling hot acids at high temperatures, and chlorides and acid salts. Suitable for agitator and evaporator components and spray nozzles
	Silicon	—	1.50					
	Manganese	—	2.00					
	Nickel	11.0	14.0					
	Chromium	18.0	20.0					
	Molybdenum	3.00	4.0					
	Sulphur	—	0.045					
	Phosphorus	—	0.045					
ANC4-B	Carbon			—	0.08			Applications are similar to those of ANC4-A
	Silicon			—	1.50			
	Manganese			—	2.00			
	Nickel			10.0	—			
	Chromium			16.5	18.5			
	Molybdenum			2.00	3.0			
	Sulphur			—	0.045			
	Phosphorus			—	0.045			
ANC4-C	Carbon					—	0.12	Applications are similar to those of ANC4-A, and also in the high temperature range for jet engine components and high pressure steam valves and components. Most commonly used alloy in this group for investment casting. Popular for marine fittings, pharmaceutical filters, fittings, etc.
	Silicon					—	1.50	
	Manganese					—	2.00	
	Nickel					10.0	—	
	Chromium					16.5	18.5	
	Molybdenum					2.25	2.75	
	Niobium (columbium)					8 × C	1.10	
	Sulphur					—	0.045	
	Phosphorus					—	0.045	

* t = Thickness of test piece.
† E. G. Donaldson, 'Investment Castings', *Engineers' Digest*, 1962, **23**, 91.

Table 26.46 COMPOSITION, PROPERTIES AND APPLICATIONS OF HIGH ALLOY STEELS NICKEL AND COBALT ALLOYS FOR INVESTMENT CASTINGS† —*continued*

Alloy BS 3146 Part II 1959	*Element*	*Grade A %*		*Grade B %*		*Grade C %*		*%*		*Typical applications*
		min	*max*	*min*	*max*	*min*	*max*	*min*	*max*	
Tensile strength:										
tonf in^{-2}		30		30		30				
MPa		463		463		463				
Yield stress or 0.5% proof stress:										
tonf in^{-2}		30		30		30				
MPa		209		209		209				
Elongation, %		15		15		15				
Angle of bend		120°		120°		120°				
Radius of bend		2*t**		2*t**		2*t**				
ANC5-A	Carbon	—	0.5							For furnace and radiant-heater parts, brazing and welding fixtures, jet engine parts
	Silicon	—	3.0							
	Manganese	—	2.0							
	Nickel	17.0	22.0							
	Chromium	22.0	27.0							
ANC5-B	Carbon			—	0.5					Heat-treatment furnace and radiant-heater components. Parts for retorts and conveyors, particularly where improved resistance to hot gas corrosion is required
	Silicon			—	3.0					
	Manganese			—	2.0					
	Nickel			36.0	46.0					
	Chromium			15.0	25.0					
ANC5-C	Carbon					—	0.75			Components for furnaces and quenching apparatus; in salt and lead baths
	Silicon					—	3.0			
	Manganese					—	2.0			
	Nickel					55.0	65.0			
	Chromium					10.0	20.0			
ANC6-A	Carbon	0.15	0.30							Used for oil-refinery components and in heat-treatment and molten-lead installations
	Silicon	0.75	2.0							
	Manganese	—	1.0							
	Nickel	10.0	15.0							
	Chromium	20.0	25.0							
	Tungsten	—	—							
	Sulphur	—	0.045							
	Phosphorus	—	0.045							

ANC6-B	Carbon		0.15	0.30			For blast-furnace parts, components for heat-treatment furnaces and super-heaters, oil and gas burners, stabilizing parts for aircraft gyroscopes, and welding fixtures. The most widely used alloy for heat-resistant applications
	Silicon		0.75	2.0			
	Manganese		—	1.0			
	Nickel		10.0	15.0			
	Chromium		20.0	25.0			
	Tungsten		2.5	3.5			
	Sulphur		—	0.045			
	Phosphorus		—	0.045			
Tensile strength:							
tonf in^{-2}		30	30				
MPa		463	463				
Elongation, %		20	20				
Hardness		250	250				
ANC7	Carbon				0.30	0.40	Suitable for applications such as rotors, turbine blades, discs, high-duty steam-piping fittings and nozzle guide-vane material
	Silicon				0.60	1.60	
	Manganese				0.5	1.0	
	Nickel				12.0	14.0	
	Chromium				18.0	20.0	
	Molybdenum				1.5	2.0	
	Tungsten				2.0	3.0	
	Niobium				2.5	3.5	
	Cobalt				9.0	11.0	
	Sulphur				—	0.045	
	Phosphorus				—	0.045	
Tensile strength:							
tonf in^{-2}					35		
MPa					541		
0.1% proof stress:							
tonf in^{-2}					20		
MPa					309		
ANC8	Carbon				0.05	0.15	Suitable for use in furnace parts and fan hubs
	Silicon				0.20	1.0	
	Manganese				0.20	1.0	
	Nickel				Remainder		
	Chromium				18.0	22.0	
	Titanium				0.2	0.6	
	Aluminium				1.0	0.4	
	Iron				—	5.0	
Tensile strength:							
tonf in^{-2}					27		
MPa					417		
Elongation, %					5		

* t = Thickness of test piece.
† E. G. Donaldson, 'Investment Castings', *Engineers' Digest*, 1962, **23**, 91.

Table 26.46 COMPOSITION, PROPERTIES AND APPLICATIONS OF HIGH ALLOY STEELS NICKEL AND COBALT ALLOYS FOR INVESTMENT CASTINGS†— *continued*

Alloy BS 3146 Part II 1959	*Element*	*Grade A %*		*Grade B %*		*Grade C %*		*%*		*Typical applications*
		min	*max*	*min*	*max*	*min*	*max*	*min*	*max*	
ANC9	Carbon							0.03	0.10	For diesel engine pre-combustion chambers and gas-turbine stator blades
	Silicon							0.20	1.0	
	Manganese							0.20	1.0	
	Nickel							Remainder		
	Chromium							18.0	22.0	
	Titanium							2.20	3.00	
	Aluminium							0.8	1.60	
	Iron							—	5.0	
Tensile strength:										
tonf in^{-2}								37		
MPa								571		
0.1% proof stress:										
tonf in^{-2}								27		
MPa								417		
Elongation, %								5		
ANC10	Carbon							0.05	0.12	For gas-turbine stator blades, and for turbine and turbocharger rotors
	Silicon							0.20	1.0	
	Manganese							0.20	1.0	
	Nickel							Remainder		
	Chromium							18.0	22.0	
	Titanium							2.20	3.00	
	Aluminium							0.80	1.60	
	Cobalt							15.0	20	
	Iron							—	5	
Tensile strength:										
tonf in^{-2}								37		
MPa								571		
0.1% proof stress:										
tonf in^{-2}								27		
MPa								417		
Elongation, %								5		
ANC11	Carbon							0.27	0.40	For gas-turbine stator blades
	Silicon							0.2	0.5	
	Manganese							0.2	0.5	
	Nickel							Remainder		

	Chromium	18.0	23.0	
	Molybdenum	9.5	11.0	
	Titanium	—	0.30	
	Aluminium	—	0.20	
	Cobalt	9.0	11.0	
	Iron	—	1.0	
Tensile strength:				
tonf in^{-2}		27		
MPa		417		
Elongation, %		5		
ANC12	Carbon	—	0.10	For gas-turbine stator blades
	Silicon	—	1.00	
	Manganese	—	1.00	
	Nickel	Remainder		
	Chromium	20.0	23.0	
	Molybdenum	9.0	10.50	
	Titanium	2.20	2.80	
	Aluminium	0.50	0.90	
	Cobalt	—	1.0	
	Iron	—	1.0	
Tensile strength:				
tonf in^{-2}		33		
MPa		510		
Elongation, %		5		
ANC13	Carbon	0.45	0.60	For gas-turbine blades, impellers for exhaust-driven turboblowers, hot metal dies, valve components and boot-and shoe-machinery parts
	Silicon	—	1.5	
	Manganese	—	1.5	
	Nickel	9.5	11.5	
	Chromium	24.0	27.0	
	Cobalt	Remainder		
	Tungsten	6.5	8.5	
	Iron	—	2.0	
Tensile strength:				
tonf in^{-2}		42		
MPa		649		
0.1% proof stress:				
tonf in^{-2}		27		
MPa		417		
Elongation, %		5		

† E. G. Donaldson, 'Investment Castings', *Engineers' Digest*, 1962, **23**, 91.

Table 26.46 COMPOSITION, PROPERTIES AND APPLICATIONS OF HIGH ALLOY STEELS NICKEL AND COBALT ALLOYS FOR INVESTMENT CASTINGS†—*continued*

Alloy BS 3146 Part II 1959	*Element*	*Grade A %*		*Grade B %*		*Grade C %*		%		*Typical applications*
		min	*max*	*min*	*max*	*min*	*max*	*min*	*max*	
ANC14	Carbon							0.2	0.3	For gas-turbine blades, impellers and valve components for high-temperature service
	Silicon							—	1.0	
	Manganese							—	1.0	
	Nickel							1.75	3.75	
	Chromium							25.0	29.0	
	Molybdenum							5.0	6.0	
	Cobalt							Remainder		
	Iron							—	3.0	
Tensile strength:										
tonf in^{-2}								40		
0.1% proof stress:								618		
tonf in^{-2}								28		
MPa								432		
Elongation, %								5		
ANC15	Carbon							0.02	0.12	Suitable for use in components for chemical and petroleum plant and pickling equipment
	Silicon							0.5	1.2	
	Manganese							0.5	1.2	
	Nickel							Remainder		
	Molybdenum							26.0	30.0	
	Iron							4.0	7.0	
	Sulphur							—	0.030	
Tensile strength:										
tonf in^{-2}								30		
MPa								463		
0.1% proof stress										
tonf in^{-2}								15		
MPa								232		
Elongation, %								10		
ANC16	Carbon							0.05	0.15	For components for chemical and petroleum plant
	Silicon							0.5	1.2	
	Manganese							0.5	1.2	
	Nickel							Remainder		
	Chromium							15.5	17.5	
	Molybdenum							16.0	18.0	

	Tungsten					3.75	5.25	
	Iron					4.0	7.0	
	Sulphur					—	0.030	
Tensile strength:								
tonf in^{-2}						28		
MPa						432		
0.1% proof stress:								
tonf in^{-2}						15		
MPa						232		
Elongation, %						5		
ANC17	Carbon					0.05	0.12	For components for chemical and petroleum plant
	Silicon					8.5	10.0	
	Manganese					0.5	1.2	
	Nickel					Remainder		
	Iron					—	2.0	
	Copper					2.0	4.0	
	Sulphur					—	0.030	
Tensile strength		(20 t.s.i. min) 309 MPa min.						
ANC18-A	Carbon	0.1	0.3					For power plant, marine equipment, chemical and process industries, and dye-cleaning and electrical equipment
	Silicon	0.5	1.5					
	Manganese	0.5	1.5					
	Nickel	Remainder						
	Iron	—	3.0					
	Copper	28.0	32.0					
	Magnesium	0.07	0.13					
	Sulphur	—	0.050					
ANC18-B	Carbon			0.05	0.15			Used where a harder material is required under corrosive conditions, e.g. in valve trims and fittings
	Silicon			2.5	3.0			
	Manganese			0.5	1.5			
	Nickel			Remainder				
	Iron				3.0			
	Copper			28.0	32.0			
	Magnesium			0.07	0.13			
	Sulphur			—	0.050			

Table 26.46 COMPOSITION, PROPERTIES AND APPLICATIONS OF HIGH ALLOY STEELS NICKEL AND COBALT ALLOYS FOR INVESTMENT CASTINGS†—*continued*

Alloy BS 3146 Part II 1959	*Element*	*Grade A* %		*Grade B* %		*Grade C* %		%		*Typical applications*
		min	*max*	*min*	*max*	*min*	*max*	*min*	*max*	
ANC18-C	Carbon					0.05	0.15			For impellers, sleeves and valve parts and fittings
	Silicon					3.5	4.5			
	Manganese					0.5	1.5			
	Nickel					Remainder				
	Iron					—	3.0			
	Copper					28.0	32.0			
	Magnesium					0.05	0.13			
	Sulphur					—	0.050			
Tensile strength:										
tonf in^{-2}		20		28		38				
MPa		309		432		587				
0.1% proof stress:										
tonf in^{-2}		7.5		15		32				
MPa		116		232		494				
Elongation, %		15		10		—				

† E. G. Donaldson, 'Investment Castings', *Engineers' Digest*, 1962, **23**, 91.

26.9 Cast irons

26.9.1 Classification of cast irons

Cast irons may be divided into two main groups, comprising the general purpose grades which are used for the majority of engineering applications and the special purpose or alloy cast irons which are used where the operating conditions involve extremes of heat, corrosion or abrasion.

26.9.2 General purpose cast irons

These materials may be further classified into three groups, depending on the graphite form.

British Standards specifications exist for each of these materials. New metric specifications incorporating SI units are available for malleable irons and are in course of preparation for grey and nodular irons. The grade number denotes the tensile strength and the % elongation, e.g. BS 2789 Grade 24/17 has a minimum tensile strength of 371 MPa (24 tonf in^{-2}) and a minimum elongation of 17%. In the case of grey irons, which all fail with an elongation less than 1%, the grade number denotes tensile strength only in test specimens of standard dimensions.

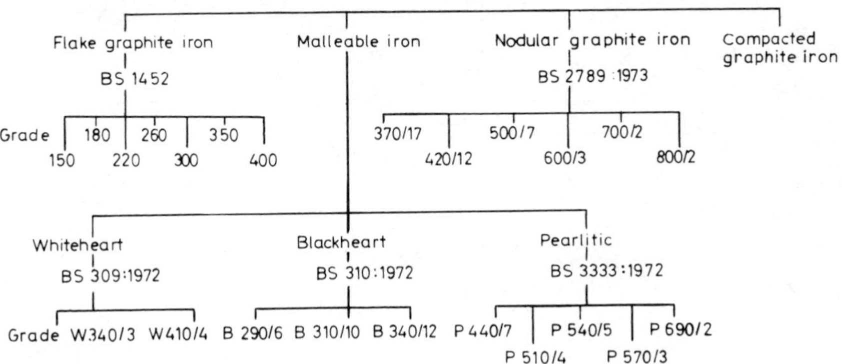

Figure 26.1 *Classification of general purpose cast irons*

GREY IRONS

The grade number denotes the minimum tensile strength requirement on a test piece machined from a 30 mm (1.2 in) diameter as-cast bar and this, rather than analysis, is the basis for specification of cast irons for engineering purposes.

For unalloyed irons the main constituents influencing tensile strength are carbon, silicon and phosphorus. Their combined effect may be expressed as a carbon equivalent value (CEV), where

$$\text{CEV} = \text{total carbon }\% + \frac{\text{phosphorus }\% + \text{silicon }\%}{3}$$

Typical analysis and CEVs are given in Table 26.47.

Table 26.47 TYPICAL ANALYSES OF GREY IRONS

BS grade	150	180	220	260	300	350	400
Total carbon %	3.1–3.4	3.2–3.5	3.2–3.4	3.0–3.2	2.9–3.1	3.1 max.	2.9 max.
Silicon % (final)	2.5–2.8	2.2–2.5	2.0–2.5	1.6–1.9	1.5–1.8	1.4–1.6	1.4–1.6
Manganese %	0.5–0.7	0.5–0.8	0.6–0.8	0.6–0.8	0.5–0.7	0.6–0.75	0.6–0.75
Sulphur % (max.)	0.15	0.15	0.15	0.15	0.12	0.12	0.12
Phosphorus %	0.9–1.2	0.6–0.9	0.1–0.5	0.3 max	0.2 max	0.15 max	0.15 max
Chromium %	—	—	—	—	0.4–0.6	—	—
Nickel %*	—	—	—	—	0.8–1.5	1.5	2.0
Molybdenum %	—	—	—	—	—	0.3–0.5	0.5–0.6
CEV	4.5–4.55	4.3–4.35	4.1–4.2	3.8–3.9	3.5–3.6	—	—

* Copper may be used partially to replace nickel.

Grey irons are section sensitive, their structure and mechanical properties depending on cooling rate in the mould, which in turn depends to a first approximation on section thickness. This effect is shown in Figure 26.2, with tensile and test bar diameter requirements in Table 26.48.

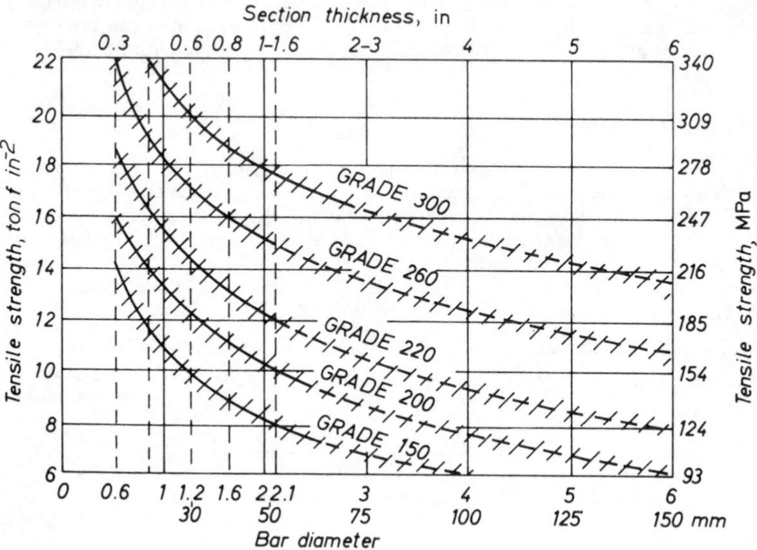

Figure 26.2 *Variation of tensile strength with section thickness*

Table 26.48 SPECIFIED PROPERTIES OF GREY IRONS

Cross-sectional thickness of casting represented	*Diam. of as-cast bar*	*Gauge diam.*	*Grade Minimum tensile strength*						
			10	12	14	17	20	23	26
mm (in)	mm (in)	mm (in)	MPa (tonf in^{-2})	MPa (tonf in^{-2})	MPa (tonf in^{-2})	MPa (tonf in^{-2})	MPa (tonf in^{-2})	MPa (tonf in^{-2})	MPa (tonf in^{-2})
Up to 9.5 (up to $\frac{3}{8}$)	15.2 (0.6)	10.1 (0.399)	170 (11.0)	201 (13.0)	247 (16.0)	294 (19.0)	340 (22.0)	386 (25.0)	432 (28.0)
9.5–19.0 ($\frac{3}{8}$–$\frac{3}{4}$)	22.2 (0.875)	14.3 (0.564)	162 (10.5)	193 (12.5)	232 (15.0)	278 (18.0)	324 (21.0)	371 (24.0)	417 (27.0)
19.0–29.6 ($\frac{3}{4}$–1$\frac{1}{8}$)	30.5 (1.2)	20.3 (0.798)	154 (10.0)	185 (12.0)	216 (14.0)	262 (17.0)	309 (20.0)	355 (23.0)	402 (26.0)
29.6–41.3 (1$\frac{1}{8}$–1$\frac{5}{8}$)	40.6 (1.6)	28.6 (1.128)	147 (9.5)	178 (11.5)	209 (13.5)	247 (16.0)	294 (19.0)	340 (22.0)	386 (25.0)
over 41.3 (over 1$\frac{5}{8}$)	53.3 (2.1)	37.9 (1.493)	139 (9.0)	170 (11.0)	201 (13.0)	232 (15.0)	278 (18.0)	324 (21.0)	371 (24.0)

Excessive rates of cooling lead to the formation of free carbides, or chilling. This effect is more severe with the higher strength irons, and occurs first at free edges of castings, where the cooling rate is a maximum, as shown in Table 26.49.

Table 26.49 CHILLING TENDENCY IN DIFFERENT GRADES OF IRON

BS grade	150	180	220	260	300
Approximate minimum chill free edge					
(in)	0.1–0.2	0.2–0.3	0.25–0.4	0.4–0.65	0.6–1.1
(mm)	2.5–5.0	5.0–7.5	6.3–10.0	10.0–16	15–30

Inoculation of high duty grey irons, from Grade 260 to 400, is carried out using a technique in which a small amount of inoculant, usually containing silicon, is added to the metal immediately before pouring so as to increase the silicon content by 0.3–0.5%. Some commonly used inoculants include ferrosilicon, calcium silicide, SMZ (silicon–manganese–zirconium). Graphite is also a powerful inoculant. Several other proprietary inoculants, most of which are based on one or more of these substances, are also effective. The efficiency of a graphite inoculant depends on high purity, but that of other types depends on the presence of minor constituents such as aluminium, calcium, barium, strontium and cerium.

The tensile strength of hypo-eutectic grey irons (i.e. CEV less than 4.3) can be significantly increased by inoculation, particularly with silicon-containing inoculants. When the CEV is in the range 3.9–4.1 maximum strength is obtained when the inoculant adds 0.2–0.3% silicon, corresponding to an increase of approximately 15 MPa (1 tonf in^{-2}) compared with uninoculated material. With CEV below 3.9 the addition of inoculant sufficient to give a 0.5% increase in silicon content can add up to 60 MPa (4 tonf in^{-2}) to the tensile strength, the effect being progressive with increasing additions up to at least this level.

The chilling tendency of grey irons is also reduced by inoculation, enabling significantly thinner chill free sections than those in Table 26.50 to be cast. The effect is progressive up to at least 0.4% of added silicon.

Table 26.50 ANALYSES OF PIG IRONS

	Typical composition				
	TC %	Si %	Mn %	S % *max*	P %
Hematite	3.7–4.5	0.5–3.5	0.5–1.2	0.05	0.05 max
Low phosphorus	3.8–4.2	1.0–4.5	0.7–1.1	0.05	0.065 max
	3.8–4.2	1.0–4.5	1.1–1.5	0.05	0.065 max
	3.8–4.2	1.0–4.5	1.5–2.0	0.05	0.065 max
Medium phosphorus	3.5–4.0	2.0–3.5	0.8–1.0	0.05	0.5–0.7
High phosphorus	3.3–3.8	2.0–4.5	0.6–0.8	0.05	0.7–1.2
	3.3–3.8	2.0–4.5	0.8–1.2	0.05	0.7–1.2

The chill reducing effect forms the basis for most control tests for inoculation. A sand cast wedge or a small block cast in sand with one face against a metal plate may be made from the metal before and after inoculation. The test pieces are then fractured and the change in depth of chill measured as an assessment of the success of the treatment. Typical results are illustrated in Figure 26.3.

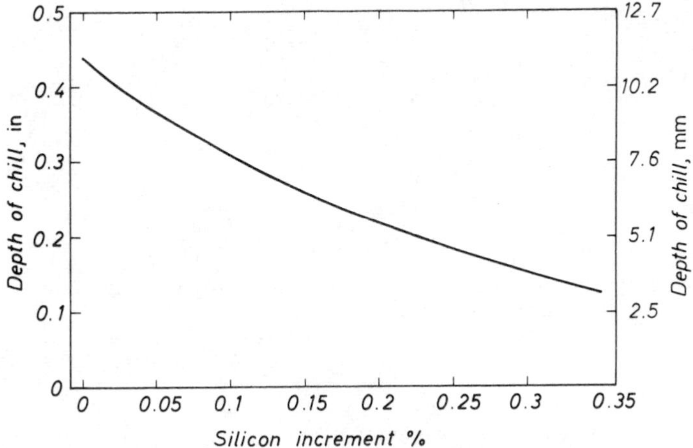

Figure 26.3 *Effect of inoculation on chill depth*

In castings which are susceptible to shrinkage defects inoculation will accentuate these problems, and its use should be restricted only to those castings where chilling or strength considerations make it necessary.

Charge materials for production of grey irons include pig irons, cast iron scrap, steel scrap and ferro alloys.

The main purposes served by pig irons is the provision of carbon, silicon, manganese and, where required, phosphorus and alloying elements. They also limit the sulphur content of cupola melted iron.

Blast furnace pig irons are available in a variety of grades which differ mainly in phosphorus content. Each grade is generally available within different ranges of silicon and manganese contents.

The carbon content will generally be at the lower end of the above ranges when the silicon content is at the upper end of the corresponding range. The silicon content can be specified in increments of 0.25 or 0.50% within the ranges given.

Refined alloy and special pig irons are produced to the customer's specific requirements in a wide range of compositions. They may contain alloying elements if required, within specified ranges of alloy content.

Silvery pig iron (10–12% silicon, 12–14% silicon) provides silicon in a more dilute form than ferrosilicon. It is used to minimise variation in silicon content of cupola melted iron when a substantial quantity of silicon is added, and to reduce the risk of aluminium contamination of the metal which may arise if large additions of ferrosilicon are made.

Special irons, generally of high carbon content, but with a very low content of residual elements are also available, their main use being for the production of nodular graphite iron.

Return scrap is the best source of scrap for re-melting providing this is of known and consistent composition and this should be fully utilised.

Purchased scrap is available in several fairly readily identifiable types as given in Table 26.51.

In comparison with pig iron and cast iron scrap, steel scrap is low in carbon and silicon contents, and the phosphorus and sulphur contents are generally below 0.05%. It is used to lower carbon and silicon contents, particularly in the production of higher strength irons.

Ferro-alloys are used to remedy deficiencies of certain elements, particularly silicon and manganese in the charges.

Table 26.51 COMPOSITION OF SCRAP

	Approximate composition %				
Type of scrap	C	Si	Mn	S	P
Light section scrap, usually less than ¼ in thick	3.2–3.5	2.2–2.8	0.5–0.7	0.10–0.15	1.0–1.5
Textile and machine scrap, generally up to 1½ in average section	3.0–3.3	1.8–2.2	0.5–0.8	0.10–0.15	0.5–1.0
Railway chairs	2.8–3.5	1.5–2.5	Up to 0.5	Up to 0.25	1.0–1.5
Automobile engine scrap	3.1–3.3	2.0–2.2	0.5–0.8	0.08–0.15	<0.2
Ingot mould scrap (heavy section)	3.5–3.8	1.4–1.8	0.5–1.0	0.08	<0.1
Blackheart malleable scrap	2.2–3.1	1.3–1.6	0.3–0.6	0.09–0.25	0.06–0.08
Whiteheart malleable scrap	0.2–2.3	0.5–0.8	0.2–0.3	0.15–0.25	0.06–0.08

FERROSILICON

The two most common grades, in lump form, contain 75–80% and 45–50% of silicon.

Ferromanganese in lump form contains 75–80% manganese.

BRIQUETTES

For relatively small additions of silicon, manganese and chromium, briquettes may be used for convenience and consistency. They contain a fixed amount of the alloy and avoid the necessity of weighing the addition.

Table 26.52 LOSSES OF ALLOYING ELEMENTS

	Melted in cupola	*Added to ladle*
Copper	0 of amount charged	0
Nickel	0 of amount charged	0
Molybdenum	0–5% of amount charged	5% of amount added
Manganese	20–30% of amount charged	5–10% of amount added
Chromium	10% of amount charged	15% of amount added
Silicon	10–15% of amount charged	10% of amount added

Table 26.53 SUMMARY OF STRUCTURAL EFFECTS OF ALLOYING ELEMENTS ON CAST IRON

Element	*% used in pearlitic irons*	*'Chill'*	*Effect on carbides (at high temps.)*	*Effect on graphite structure*	*Effect on combined carbon in pearlite*	*Effect on matrix*
Chromium chill inducing	0.15–0.5	Increases*	Strongly stabilizes	Mildly refines	Increases	Refines pearlite and hardens
Vanadium	0.15–0.3	Increases	Strongly stabilizes	Refines	Increases	
Boron	—	Strongly increases	—	—	—	—
Manganese mildly chill inducing	0.3–1.25	Mildly increases	Stabilizes	Mildly refines	Increases	Refines pearlite and hardens
Molybdenum	0.3–1.0	Mildly increases	About neutral	Strongly refines	Mildly increases	Refines pearlite and strengthens
Tin	0.1–0.2	About neutral	About neutral	About neutral	Increases	
Copper mildly chill restraining	0.5–2.0	Mildly restrains	About neutral	About neutral	Mildly decreases	Hardens
Carbon chill restraining	—	Strongly restrains	Decreases stability	Coarsens	Strongly decreases	Produces ferrite and softens
Silicon	—	Strongly restrains	Decreases stability	Coarsens	Strongly decreases	Produces ferrite and softens
Aluminium	—	Strongly restrains	Decreases stability	Coarsens	Strongly decreases	Produces ferrite and softens
Nickel	0.1–3.0	Restrains†	Mildly decreases stability	Mildly refines	Mildly decreases and stabilizes at eutectoid	Refines pearlite and hardens
Titanium	0.05–0.10	Restrains†	Decreases stability	Strongly refines‡	Decreases	Produces ferrite and softens
Zirconium	0.10–0.30	Restrains†	—	About neutral	—	Produces ferrite and softens

* Chill inducing effect of 1 part of chromium about balances chill restraining effect of $1\frac{1}{2}$ parts silicon or $2\frac{1}{2}$ parts nickel.
† Chill restraining effect of nickel about half that of silicon.
‡ Strong refining action of titanium takes place when small amounts are added, particularly when oxygen is also present.

Table 26.54

Element	*Graphitizing value* (Si = 1)	*Element*	*Graphitizing value* (Si = 1)
Aluminium	0.5	Manganese	−0.25
Copper	0.35	Molybdenum	−0.35
Nickel	0.3–0.4	Chromium	−1.2

Melting range varies widely with composition, the melting temperature falling with increase of carbon content so that low carbon irons must be cast at considerably higher temperatures than high carbon irons. With grey irons containing high phosphorus, melting of the steadite (phosphide eutectic) begins at about 960 °C, giving a relatively long melting range.

FLUIDITY OF CAST IRON

The fluidity of cast iron depends primarily upon composition and pouring temperature. The fluidity, as measured by a special test spiral casting, may be improved by increasing the carbon, silicon or phosphorus contents, or by increasing the pouring temperature. Of these variables, carbon content is the most effective from the point of view of composition, but an increase of 15–20 °C in the pouring temperature improves the fluidity as much as an increase of 0.10% carbon, 0.30% silicon or 0.20% phosphorus.

Table 26.55 DENSITIES AND LIQUIDUS TEMPERATURES OF SOME TYPICAL GREY CAST IRONS (BASED ON 3.8 cm ($1\frac{1}{2}$in) SECTION)

	Iron number						
	1	2	3	4	5	6	7
Density, g cm^{-3} (room temperature)	7.02	7.09	7.26	7.03	7.08	7.27	7.14
Density, g cm^{-3} (liquidus)	6.90	6.94	6.92	6.89	6.89	6.92	6.89
Liquidus temperature, °C	1 150	1 150	1 250	1 150	1 155	1 250	1 195
Liquid contraction per 100 °C(%)	1.1	1.1	1.1	1.1	1.1	1.1	1.1
Composition, %							
Total carbon	3.69	3.67	3.10	3.39	3.27	3.08	2.90
Graphitic carbon	3.53	3.26	2.31	3.20	2.88	2.18	2.68
Silicon	2.87	2.10	1.69	2.86	2.87	1.68	2.88
Phosphorus	0.68	0.46	0.35	0.67	0.59	0.35	0.66
Manganese	0.59	0.54	0.48	0.58	0.52	0.44	0.44
Sulphur	0.03	0.05	0.04	0.03	0.03	0.04	0.03

Table 26.56 HARDNESS AND DENSITY OF MICRO-CONSTITUENTS OF CAST IRON

Constituent	*Density* g cm^{-3}	*Brinell hardness No.*	*Remarks*
Ferrite	7.86	70–75	Iron (electrolytic)
Silico-ferrite	—	88	Containing 0.82% silicon
Silico-ferrite	—	124	Containing 2.28% silicon
Silico-ferrite	—	150	Containing 3.4% silicon
Pearlite	7.846	240	—
Pearlite (silico-ferrite and cementite)	—	200–450	Depending on interlamellar distance
Graphite	2.55	—	—
Ledeburite (massive cementite and sat. austenite)	—	680–840	—
Steadite	7.32	—	—
Manganese sulphide	4.00	—	—
Iron sulphide	5.02	—	—
Divorced pearlite	—	130–150	—
Phosphide eutectic	—	—	400–600 HV
Iron carbide Fe_3C or $(FeCr)_3C$	7.66	550	800–1 200 HV
Iron carbide $(CrFe)_7C_3$	—	—	1 300–1 800 HV

Table 26.57 SPECIFIC GRAVITY OF TYPICAL IRONS OF BRITISH STANDARDS GRADES (BS 1452)

Grade	*Up to* 180	220	260	300	350	400
Specific gravity	6.8–7.15	7.2–7.3	7.25–7.35	7.3–7.4	7.3–7.4	7.4–7.6

Heat treatment of grey cast irons may be carried out to eliminate residual stresses, improve machinability or to increase wear resistance.

Stress relief heat treatment is carried out by slow heating at 50–100 °C per hour to 600 °C ± 10 °C, holding at 600 °C for one hour plus an additional half hour per cm (0.4 in) of maximum section thickness, and cooling at 50–100 °C per hour to below 200 °C followed by air cooling.

Annealing heat treatment is applied when improved machinability is required. There are two distinct sets of conditions in which annealing may be carried out.

1. To break down free carbide or chill formed as a result of failure to match carbon equivalent value to minimum free edge thickness. Annealing must then be carried out at 900 °C for 1–5 h to ensure complete breakdown of carbide, followed by air cooling to ensure a pearlitic matrix. This treatment is carried out as an emergency measure to salvage otherwise unmachinable castings.
2. To provide castings which can be machined at a very high rate. Castings which are fully pearlitic can be annealed to give a mainly ferritic matrix with a hardness of 140–180 HB by holding at 780–820 °C for 1–2 h followed by slow cooling.

Hardening by quenching and tempering is carried out where high surface hardness is required, with corresponding improvement in wear resistance. Hardening temperatures in the range 850–880 °C are used, followed by oil quenching and tempering at 300 °C to reduce internal stresses. Tensile strength is not increased to the same extent as hardness (Figure 26.4) and because of the risk of cracking during quenching, this process is usually restricted to small castings of simple shape.

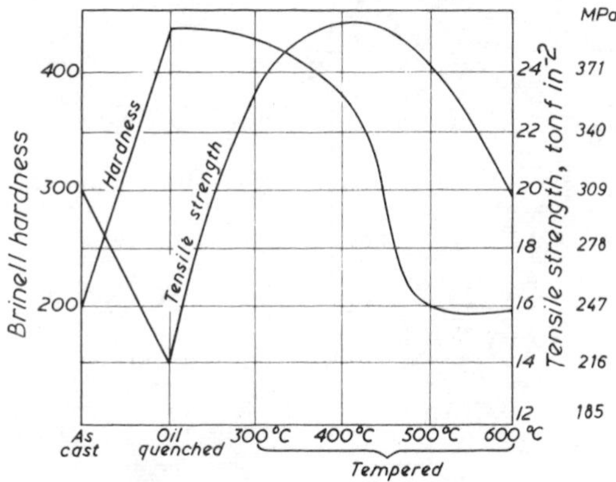

Figure 26.4 *Effect of heat treatment on the strength and hardness of alloy cast iron*

Surface hardening, by flame or induction heating, is widely used to improve the wear resistance of critical surfaces on large castings such as slideways on machine tools. For good response to this treatment the as-cast structure must be fully pearlitic, and the phosphorus content must be below 0.2% to avoid pitting. Hardness in the range 450–500 HV are obtained with depths of 1–3 mm (0.040–0.125 in).

MALLEABLE IRONS

Malleable irons are brittle as cast, their structure consisting of iron carbide in a pearlitic matrix. By suitable heat treatment the carbides are broken down resulting in a structure that consists of graphite aggregates (temper carbon) in a matrix which may be ferritic or pearlitic, depending on composition and heat treatment conditions.

The British Standards (BS 309, 310 and 3333) specify neither composition (apart from maximum phosphorus content of 0.12%) nor heat treatment.

Table 26.58 SPECIFIED PROPERTIES OF MALLEABLE IRONS‡

Type	*Grade*	*Diameter of test bar*†		*Tensile strength*		*0.2% proof stress**		*0.5% proof stress*		*Elongation*	*Hardness range**
		mm	in	MPa	tonf in^{-2}	MPa	tonf in^{-2}	MPa	tonf in^{-2}	%	HB
Whiteheart BS 309: 1972	W410/4	9†	0.394	350	22.6	170	11.0	190	12.3	10	229 max
		12†	0.472	390	25.2	210	13.6	230	14.9	6	229 max
		15†	0.590	410	26.5	220	14.2	250	16.2	4	229 max
	W340/3	9†	0.394	270	17.5	—	—	—	—	7	229 max
		12†	0.472	310	20.1	—	—	—	—	4	229 max
		15†	0.590	340	22.0	—	—	—	—	3	229 max
Blackheart BS 310: 1972	B340/12	15	0.590	340	22.0	190	12.3	200	12.9	12	149 max
	B310/10	15	0.590	310	20.1	180	11.6	190	12.3	10	149 max
	B290/6	15	0.590	290	18.8	—	—	170	11.0	6	149 max
Pearlitic BS 3333: 1972	P690/2	15	0.590	690	44.6	520	33.6	540	34.9	2	241–285
	P570/3	15	0.590	570	36.9	400	25.9	420	27.2	3	197–241
	P540/5	15	0.590	540	34.9	320	20.7	340	22.0	5	179–229
	P510/4	15	0.590	510	33.0	290	18.8	310	20.1	4	170–229
	P440/7	15	0.590	440	28.5	250	16.2	270	17.5	7	149–197

* Not mandatory.
† Separately cast bars.
‡ Diameter to be representative of the important sectional thickness of the casting. Bars are tested in the unmachined condition.

In practice, whiteheart malleable irons to BS 309 have an initial total carbon content of about 3.5% which is reduced to the range 0.25–2.0% by heat treating the castings at 900 °C in an oxidizing environment, and slowly cooling. The required conditions may be produced by packing the castings in an oxidizer (e.g. hematite ore) or by use of a controlled atmosphere furnace. This results in a carbon gradient within the castings, the outer layer being normally ferritic and graphite-free while the core structure consists of temper carbon aggregates in a pearlitic matrix. Small castings of thin section may have a fully decarburized structure throughout, and this is sometimes referred to as a weldable grade of malleable iron.

Blackheart (BS 310) and pearlitic (BS 3333) malleable irons have a lower initial total carbon content in the range 2–3%. Heat treatment in a neutral atmosphere at 850–875 °C followed by slow cooling results in a uniform structure of temper carbon in a ferritic matrix.

Malleable irons to BS 3333 have a structure consisting of temper carbon in a pearlitic matrix. This is produced either by rapid cooling after annealing or by the addition of 0.5% or more manganese. Pearlitic malleable irons have a good response to surface hardening by flame or induction heating, and hardness values of HV 500 can be consistently achieved in production.

NODULAR IRONS

In nodular (spheroidal graphite) irons free graphite is present as spheres or nodules in the as-cast condition. Graphite in this form has a much smaller weakening effect on the matrix than the dispersed graphite flakes in grey ions. Nodular irons therefore have considerably higher strength, ductility, and impact values than grey irons.

Cerium and magnesium additions both produce nodular structures, but the latter has been found to be more adaptable and economical. Both elements are desulphurizers and nodule formation is not possible until the sulphur content has been lowered to about 0.02%. Very small amounts of trace elements, such as 0.003% bismuth, 0.004% antimony, 0.009% lead and 0.12% titanium prevent nodule formation. The effect of these elements is additive, but it can be neutralized by the addition of sufficient cerium to give a residual content of 0.005–0.01%.

Magnesium may be added directly to the ladle as nickel–magnesium, nickel–silicon–magnesium or iron–silicon–magnesium alloy. Higher magnesium recovery is obtained using a plunging technique in which lower density, higher magnesium content additions such as magnesium impregnated coke are held below the liquid metal surface by means of a plunging head. Maximum recovery results from the addition of pure magnesium to the molten iron in a closed pressure-tight converter vessel. Because of equipment costs the use of this latter method is normally restricted to large-scale production.

In all cases the amount of magnesium to be added is given by:

$$\text{Mg} = \frac{\frac{3}{4}\text{ (initial sulphur content)} + \text{residual magnesium content (usually 0.03–0.05\%)}}{\text{expected magnesium recovery}}$$

Nodular irons are inoculated with 0.4–0.8% silicon after nodulizing to refine the structure and minimize chilling.

The carbon content of nodular irons is usually kept above 3.5% in the interests of good castability. Silicon, manganese and phosphorus should be below 2.3%, 0.4% and 0.06% respectively to give maximum ductility and impact value in the ferritic condition.

Nodular irons are slightly more prone to shrinkage defects than grey irons.

Adequate feed metal should be provided and moulds of high rigidity are to be preferred. Running systems should be designed to minimize turbulence, so as to prevent the entrapment of dross which tends to be formed as a result of the magnesium content.

Although nodular irons are much less section sensitive than grey irons, depending on the trace amounts of carbide stablizing elements present, their matrix structures may range from fully pearlitic to completely ferritic, and chilling may occur in sections thinner than 5 mm (0.2 in).

By close control of analysis and inoculation practice nodular irons can be produced in the as-cast condition over a wide range of section thicknesses with any required matrix structure from fully ferritic to fully pearlitic.

Alternatively, the matrix structure of nodular iron castings can be modified by appropriate heat treatments, since the presence of free carbon in the form of graphite enables diffusion of carbon to or from the graphite particles to take place. This is not possible with steels, which contain no free graphite. The effect of variation in matrix structure on mechanical properties is much more pronounced with nodular iron than with flake graphite cast iron, and by heat treatment of an iron of fixed composition foundries can produce castings conforming to the complete range of the grades of BS 2789: 1973.

Table 26.59 SPECIFIED PROPERTIES OF NODULAR IRONS

BS 2789 *grade*	*Tensile strength*‡		*0.2% proof stress**		Elongation	*Typical** hardness	*Impact value*†			
							Average of set of 3 on one sample		*Individual*	
	MPa	tonf in^{-2}	MPa	tonf in^{-2}	%	HB	J	ft lbf	J	ft lbf
SNG 370/17	370	24.0	230	14.9	17	179 max	13.0	10	12.0	9
SNG 420/12	420	27.2	250	16.2	12	201 max	—	—	—	—
SNG 500/7	500	32.3	310	20.1	7	170–241	—	—	—	—
SNG 600/3	600	38.8	350	22.7	3	192–269	—	—	—	—
SNG 700/2	700	45.3	400	25.9	3	229–302	—	—	—	—
SNG 800/2	800	51.8	460	29.8	2	248–352	—	—	—	—

* Not mandatory.
† 10 mm square notched test piece or round test piece to BS 131 Part 1, or Charpy V notch test piece to BS 131 Part 2.
‡ Separately cast bars

Practical heat treatments include:

Annealing Heat to 850–900 °C where the matrix becomes completely austenitic and slow furnace cool at 20–35 °C per hour to below 700 °C. Alternatively, cool more rapidly to 700–720 °C and hold for 4–12 h, followed by air cooling. Ferritic irons produced in this way conform to grades 370/17 and 420/12 of BS 2789: 1973.

Normalizing This is carried out by air cooling from 850 to 900 °C, and produces a mainly pearlitic matrix conforming to grades 700/2 and 800/2 in castings of light and medium section. The use of alloying elements is often necessary to produce a pearlitic matrix in heavier section castings.

Hardened and tempered structures These are produced by oil quenching from 850 to 900 °C and tempering at 550–600 °C. Material conforming to BS 2789 Grade 800/2 is sometimes produced by this method.

Mixed matrix structures Structures intermediate between the annealed and normalized grades have a range of mechanical properties depending on the ratio of ferrite to pearlite. The corresponding grades of BS 2789 are 500/7 and 600/3. In practice these structures are produced by austenitizing at 850–900 °C followed by either controlled rapid cooling at approximately 100 °C per hour through the critical temperature range of 720–800 °C or by rapid air cooling from an appropriate intermediate temperature, e.g. 730 °C, within the critical range.

Stress relieving This is often carried out if required after normalizing by reheating to approximately 600 °C followed by slow cooling, using a similar heat treatment cycle to that used for grey iron castings.

Surface hardening Nodular irons having mainly pearlitic matrices, i.e. Grade 600/3 upwards, have good response to surface hardening by flame or induction heating. In practice, surface hardnesses of 500 HV with total case depths of 0.75–3.0 mm (0.030–0.120 in) are readily obtainable.

26.9.3 Compacted graphite irons

A relatively recent addition to the principal catagories of cast irons. It is characterized by short, stubby graphite flakes. It has properties intermediate between those of grey iron (long flakes) and nodular iron (fully rounded graphite nodules). It is finding uses where thermal shock and thermal cycling resistance are required such as for ingot moulds and dies for the casting of metals.

26.9.4 Special purpose cast irons

The main subdivision of these materials is into graphite-free and graphite-containing irons, the graphite as in the general purpose irons being present in either flake or nodular form.

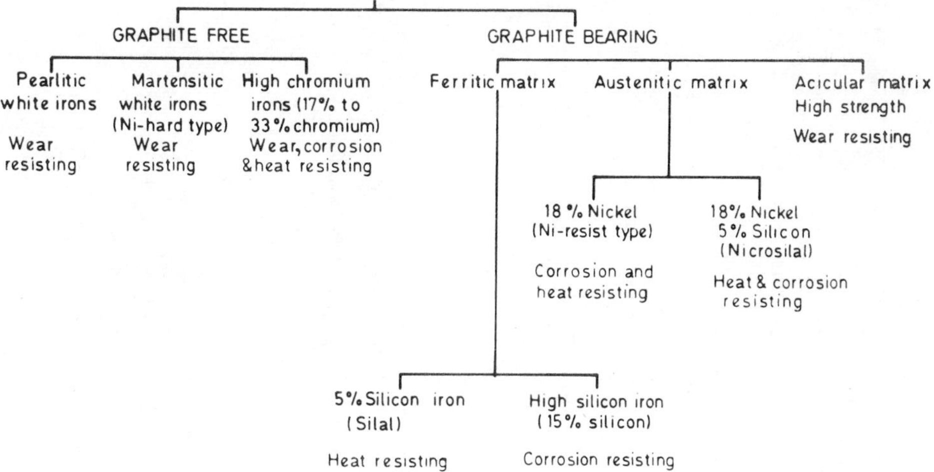

Figure 26.5 *Classification of special purpose cast irons*

British Standards Specifications exist for the austenitic grades (BS 3468: 1974) including the Ni-resist* series and Nicrosilal, and also for high silicon irons (BS 1591: 1949) and abrasion-resistant white irons (BS 4844: 1974).

The austenitic irons conforming to BS 3468 are specified in terms of both analysis and mechanical properties, while BS 4844 specifies analysis and hardness for white irons and BS 1591 specifies analysis only for high silicon irons.

28.9.5 Applications of special purpose cast irons

HIGH TEMPERATURE

Irons in the general engineering group are suitable for applications up to at least 350 °C where long-term dimensional stability is required. There are also many instances of the use of grey irons in the temperature range up to 700 °C provided that appreciable growth and scaling can be tolerated.

For extended life at temperatures up to 850 °C one of the alloyed irons, such as Ni-resist, Silal or Nicrosilal, may be used. The austenitic materials Ni-resist and Nicrosilal have good thermal shock resistance, but Silal is limited to applications where severe temperature gradients are absent. Above 850 °C the most suitable material is 30% chromium iron, which has good oxidation resistance and a useful level of creep strength up to 1050 °C.

CORROSION

Ni-resist, Nicrosilal, high silicon iron and high chromium iron have good corrosion resistance in appropriate media. For example, Ni-resist and Nicrosilal have good resistance to sea water, strong alkalis, inorganic salts and weak acids. These materials are machinable without difficulty. High silicon irons have excellent resistance to sulphuric and nitric acids at all temperatures and concentrations, but have the disadvantage that they are brittle and can only be machined by grinding. High chromium (30%) irons are the only cast irons which can be regarded as stainless so far as atmospheric exposure is concerned. They develop a passive film under oxidizing conditions and their outstanding characteristic is the ability to withstand attack by nitric acid at temperatures up to boiling point and concentrations up to 70%. This grade of iron is machinable without difficulty provided the carbon content is restricted to a maximum of 1.5%.

* Trade name–International Nickel Co. Ltd.

Table 26.60 SPECIFIED ANALYSES OF AUSTENITIC AND HIGH SILICON IRONS*

Type grade	Austenitic											High silicon
	Flake graphite types						Nodular graphite types					
	AUS101		AUS102		AUS104	AUS105	AUS202		AUS203	AUS204	AUS205	
	A	B	A	B	—	—	A	B	—	—	—	—
Carbon%	3.0 max	3.0 max	3.0 max	3.0 max	1.6–2.2	2.6 max	3.0 max	3.0 max	3.0 max	3.0 max	2.6 max	0.35–1.0
Silicon%	1.0–2.8	1.0–2.8	1.0–2.8	1.0–2.8	4.5–5.5	1.0–2.0	1.0–2.8	1.0–2.8	1.0–2.8	4.5–5.5	1.5–2.8	14.25–15.25
Manganese%	1.0–1.5	1.0–1.5	1.0–1.5	1.0–1.5	1.0–1.5	0.4–0.8	0.7–1.5	0.7–1.5	1.8–2.4	1.0–1.5	0.5 max	1.0 max
Phosphorus%	—	—	—	—	—	—	0.08 max	0.08 max	0.08 max	0.08 max	0.08 max	1.0 max
Nickel%	13.5–17.5	13.5–17.5	18.0–22.0	18.0–22.0	18.0–22.0	28.0–32.0	18.0–22.0	18.0–22.0	21.0–24.0	18.0–22.0	28.0–32.0	—
Copper%	5.5–7.5	5.5–7.5	0.5 max	0.5 max	0.5 max	0.5 max	—	—	—	—	—	—
Chromium%	1.0–2.5	2.0–3.5	1.0–2.5	2.0–3.5	1.8–4.5	2.5–3.5	1.0–2.5	2.0–3.5	0.5 max	1.0–2.5	2.5–3.5	—

* These details relate to BS 3468: 1962. This specification was revised in 1974.

Table 26.61 SPECIFIED MECHANICAL PROPERTIES OF AUSTENITIC IRONS†

Property	Flake graphite types						Nodular graphite types				
	AUS101		AUS102		AUS104	AUS105	AUS202		AUS203	AUS204	AUS205
	A	B	A	B	—	—	A	B	—	—	—
Tensile strength, MPa	219	292	219	292	185	170	371	371	371	371	371
Tensile strength tonf in^{-2}	14.2	18.9	14.2	18.9	12.0	11.0	24.0	24.0	24.0	24.0	24.0
0.5% permanent set stress*, MPa	—	—	—	—	—	—	232	232	232	232	232
0.5% permanent set stress*, tonf in^{-2}	—	—	—	—	—	—	15.0	15.0	15.0	15.0	15.0
Elongation,%	2	—	2	—	2	—	8	6	20	10	7
Hardness HB	212 max	248 max	212 max	248 max	248 max	212 max	201 max	255 max	170 max	230 max	201 max
Charpy V notch impact value, J	—	—	—	—	—	—	—	—	27	—	—
Charpy V notch impact value, ft lbf	—	—	—	—	—	—	—	—	20	—	—

* Not mandatory.
† These details relate to BS 3468 : 1962. This specification was revised in 1974.

Table 26.62 TYPICAL ANALYSES OF OTHER SPECIAL PURPOSE CAST IRONS

Type or trade name	C %	Si %	Mn %	S %	P %	Ni %	Cr %	Mo %	*Characteristics*
Silal	2.2	5.5	0.6	0.1	0.1	—	—	—	Heat resistant
Pearlitic white cast iron	2.9	1.2	0.8	0.1	0.1	—	—	—	Wear resistant
Ni-hard*	3.0	0.8	0.8	0.1	0.3	4.5	2.1	—	Wear resistant
High chromium cast irons	1.0	0.4	0.4	0.03	0.05	—	30.0	—	Heat and corrosion resistant
	2.7	0.8	0.8	0.03	0.05	—	27.0	—	Wear and corrosion resistant
	3.0	0.8	0.8	0.03	0.05	—	15.0	3.0	Wear and corrosion resistant
Nomag†	3.0	1.5	7.0	0.03	0.05	11	—	—	Non-magnetic
Low expansion cast iron	2.2	1.5	0.8	0.03	0.05	35	2.0	—	Minimum thermal expansion

* Trade name–International Nickel Ltd.
† Trade name–Ferranti Ltd.

ABRASION

A graphite-free structure is essential for good resistance to abrasion. The irons which meet this requirement, as shown in Figure 26.5 are the unalloyed pearlitic white irons, martensitic white irons of the Ni-hard type and high carbon high chromium irons.

It is difficult to relate wear characteristics determined in a laboratory to practical service conditions, and there is considerable overlapping between the fields of application for these materials. In the absence of experience of similar conditions, evaluation of material must sometimes be made by trial. In general, martensitic white irons have better abrasion resistance than pearlitic white irons. High chromium, high carbon irons are particularly useful where abrasion is combined with impact loading and where abrasive and corrosive conditions exist together. *See* Chapter 25.

27 Refractory materials

Table 27.1 RAW MATERIALS

Raw material	Al_2O_3	CaO	K_2O	Na_2O	MgO	Cr_2O_3	Fe_2O_3	SiO_2	TiO_2	ZrO_2	LOI*	C	S
Andalusite	58.6			5.6		—	—	33.8	—	—	—	—	—
Ball clay (Devon)	20 to 30	0.04	1.2 to 2.7	0.2 to 0.4	0.1 to 0.5	—	0.8 to 1.4	53 to 71	0.9 to 1.6	—	6 to 10	0.1 to 0.7	0.1 to 1.5
Bauxite (Calcined)	84.9	—	0.9	0.9	—	—	3.3	8.0	1.2	—	—	—	—
China clay	38.1	0.1	1.5	0.2	0.2	—	0.7	47.4	—	—	11.9	—	—
Chrome ore (Turkey)	15.0	0.7	—	—	14.0	56.8	(FeO) 12.1	1.5	—	—	—	—	—
Diatomite (Skye)	5.2	1.7	0.1	0.6	1.3	—	3.1	72.1	0.4	—	14.6	—	0.4
Dolomite (Salop)	1.6	31.7	—	—	19.3	—	0.5	1.2	—	—	45.6	—	—
Fireclay	34.9	0.4	1.6		0.9	—	1.9	46.8	1.5	—	11.2	—	—
Flint clay (English)	—	0.3	Trace	0.1	0.2	—	Trace	98.2	—	—	1.4	—	—
Foundry sand (Natural)	8.0	0.1	0.8		0.1	—	0.2	88.4	0.2	—	2.2	—	—
Ganister (Sheffield)	0.9	0.1	—	—	—	—	0.7	96.8	—	—	—	—	—
Kyanite (Kenya)	59.7	0.3	—	—	0.1	—	0.6	37.6	1.1	—	0.6	—	—
Magnesite (Austrian)	0.8	1.8	—	—	88.4	—	6.8	1.4	—	—	—	—	—
Magnesite (Sea water)	0.5	2.2	—	—	93.7	—	1.4	2.1	—	—	—	—	—
Olivine (Norwegian)	0.7	—	—	—	49.0	0.4	6.5	42.0	—	—	1.2	—	—
Quartzite (Welsh)	0.6	0.1	—	—	—	—	0.4	97.8	—	—	—	—	—
Serpentine (Shetland)	2.5	—	—	—	37.6	4.4	(FeO) 6.9	33.2	—	—	15.2	—	—
Silica sand (Pure)	0.5	—	0.1	0.1	—	—	0.1	99.3	—	—	—	—	—
Sillimanite (Australia)	58.8	0.8	0.4	—	0.1	—	2.4	34.5	1.5	—	1.4	—	—
Zircon sand (Australia)	1.2	0.1	—	—	—	0.4	0.4	30.0	2.1	65.4	—	—	—

* Loss on ignition.

Table 27.2 PROPERTIES OF TYPICAL DENSE FIRED REFRACTORY BRICKS

Material	*Chemical analysis* (% by weight) Al_2O_3	CaO	Cr_2O_3	Fe_2O_3	K_2O	MgO	Na_2O	SiO_2	TiO_2	ZrO_2	*Bulk density* g cm^{-3}	*Apparent porosity* %	28 p.s.i. RUL** % *deformation at* °C	*PLC* %††	*Free thermal expansion* %	*Thermal conductivity* Wm^{-1}K^{-1} 500 900 1 300 °C	*Max. service temp.* °C	*Uses*
High Alumina†	87	0.3	—	1.5 to 2.0	—	0.2	—	6.6 to 9.5	2.4 to 3.0	—	2.82 to 2.97	17 to 21	10 1 730	0 to −2 2 h 1 600 °C	0.95 to 1 400 °C	1.3 1.6 1.9	1 800	Arc furnace roof torpedo ladle
Fired dolomite	2	55	—	2	—	37	—	3	—	—	2.7	18	—	0 to 0.8 2 h 1 700 °C	1.4 to 1 000 °C	—	—	Rotary kilns
Firebrick†	38	0.5	<0.1	2.9	0.6	0.55	0.5	56	1.4	—	2.1	18 to 25	5 1 500	−0.7 2 h 1 410 °C	0.5 to 1 000 °C	—	—	Ladles, rotary kilns
Chrome–magnesite‡	13 to 16	0.9 to 1.3	17 to 23	9 to 12	—	46 to 52	—	2.5 to 3.5	—	—	2.9 to 3.1	18 to 24	—	0 to +3 5 h 1 700 °C	1.5 to 1 400 °C	1.8 to 2.2	—	Rotary kilns, electric arc furnaces
Firebrick†	41.7	0.5	—	2.7	0.6	0.5	0.1	52.3	1.7	—	1.9 to 2.0	21 to 26	—	0 to −3 2 h 1 600 °C	0.5 to 1 000 °C	1.3 1.5	—	Glass tank furnaces, blast furnace stack
Magnesite–chrome†	10 to 12	0.8 to 1.1	14 to 18	8 to 10	—	60 to 65	—	1.8 to 2.3	—	—	3 to 3.2	16 to 20	—	−0.5 to +3.0 5 h 1 800 °C		2.2 to 2.7	—	Rotary kilns, electric arc furnaces
Magnesite†	0.2 to 0.4	1.8 to 2.3	—	0.15 to 0.3	—	95.5	—	0.7 to 0.9	—	—	2.9 to 3.0	15 to 19	—	−0.2 to −1.5 5 h 1 800 °C	2.1 to 1 400 °C	3.7 to 4.4	—	Electric arc furnaces, LD backing lining
Mullite†	74.2	—	—	0.7	0.7	0.1	0.3	23.2	0.2	—	2.6 to 2.7	13 to 17	2 1 700	−0.2 2 h 1 700 °C	0.63 to 1 400 °C	1.5 1.7 2.0	1 700	Glass tank furnaces

Silica†	0.6	1.7	—	0.5	0.2	0.2	0.2	96.5	0.1	—	1.7 to 1.8	21 to 25	10 1 680	0 4 h 1 600 °C	1.3 to 1 200 °C	1.3	1.7		1 700	Coke ovens, hot blast stoves
Zircon†	1.2	—	—	0.3	—	—	—	31.9	0.3	64.2	3.7 to 3.9	14 to 18	0 1 700	0 2 h 1 600 °C	0.7 to 1 400 °C	2.6	2.4	2.3	1 700	Glass tank furnaces
Carbon‡	—	—	—	—	—	—	—	—	—	—	1.56 to 1.64	—	0 1 700	−0.5 2 h 1 500 °C	0.55 to 1 000 °C	3.6			—	Blast furnace hearth and bosh
Pitch impregnated fired magnesite*	0.15 to 0.25	1.9 to 2.3	—	0.15 to 0.30	—	96.0 to 97.0	—	0.7 to 0.9	—	—	2.85 to 3.0	14 to 18	—	−0.2 to −1.5 5 h 1 800 °C	2.1 to 1 400 °C	3.7 to 4.4			—	LD, Q-BOP

* Data from Steetley Refractories Ltd, Worksop, Notts.
† Data from Pickford Holland Co., Sheffield.
‡ Typical range into which most products fall.
** Refractoriness under load.
†† Permanent linear change.

Table 27.3 PROPERTIES OF TYPICAL UNFIRED REFRACTORY BRICKS

Material	*Chemical analysis* (% by weight) Al_2O_3	CaO	Fe_2O_3	K_2O	MgO	Na_2O	SiO_2	TiO_2	*Bulk density* g cm^{-3}	*PLC* %**	*Thermal conductivity* Wm^{-1} K^{-1}	*Residual carbon* %	*Loss on ignition* %	*Compressive strength* MN m^{-2} 20 °C	120 °C	180 °C	300 °C	*Uses*
Pitch bonded magnesite*	0.5 to 0.7	1.5 to 3.0	0.5 to 1.0	—	92.5 to 96.5	—	0.8 to 1.3	—	2.9 to 3.1	—	3.9 to 4.6 at 900 °C	1.8 to 2.5	3 to 5	—				LD
Chemically bonded high alumina†	86 to 88	0.3	1.5 to 2.0	0.2	0.2	0.2	6.6 to 9.5	2.4 to 3.0	2.9 to 3.1	+1.5 to +3.5 2 h at 1 600 °C	—	—	—	—				Aluminium melting vessels, electric arc furnaces
Pitch bonded dolomite (tempered)*	0.6 to 0.9	53.0 to 57.0	1.8 to 2.4	—	39.0 to 41.0	—	1.0 to 1.6	—	2.80 to 2.95	—	2.4 to 2.9 at 900 °C	1.8 to 2.2	3.8 to 4.8	21 to 42	2 to 6	1 to 6	1 to 4	BOS converters
Pitch bonded magnesia doloma (tempered)*	0.4 to 0.7	21.0 to 26.0	1.0 to 1.5	—	70.0 to 75.0	—	0.8 to 1.2	—	2.87 to 3.02	—	3.5 to 4.1 at 900 °C	2.5 to 3.2	4.2 to 5.2	20 to 40	2 to 6	2 to 6	1 to 4	BOS converters
Pitch bonded magnesia (tempered)*	0.1 to 0.3	1.8 to 2.8	0.2 to 0.5	—	95.0 to 97.0	—	0.7 to 1.0	—	3.00 to 3.15	—	3.8 to 4.4 at 900 °C	3.0 to 3.6	4.5 to 5.5	20 to 40	5 to 20	3 to 14	2 to 6	BOS converters

* Data from Steetley Refractories Ltd, Worksop, Notts.
† Data from Pickford Holland Co., Sheffield.
** Permanent linear change.

Table 27.4 PROPERTIES OF TYPICAL LIGHTWEIGHT FIRED REFRACTORY BRICKS

Material	*Chemical analysis* (% by weight)									*Bulk density*	*Apparent porosity*	*MOR* at room temp.*	*Thermal expansion*	*Max. service*	*Thermal conductivity* $Wm^{-1} K^{-1}$			*PLC %*	
	Al_2O_3	CaO	B_2O_3	Fe_2O_3	K_2O	MgO	Na_2O	SiO_2	TiO_2	g cm^{-3}	%	MN m^{-2}	%	*temp.*°C	400 °C	600 °C	1 100 °C	at °C	*Uses*
Insulating firebrick	38 to 40	1.0 to 14.0	—	0.4 to 2.0	0.1 to 0.3	—	0.1 to 0.3	45 to 55	1.0 to 1.5	0.6 to 0.9	70 to 75	0.8 to 1.0	0.45 to 0.6 to 1 000 °C	1 300 to 1 400	0.3 to 0.1	0.35 to 0.15	0.4 to 0.25	0 to −1.1 1 300	General insulation
Insulating firebrick	78 to 94	—	—	0.1 to 0.2	0 to 0.1	—	0 to 0.2	—	0 to 0.5	1.2 to 1.4	60 to 65	1 to 3	0.6 to 0.7 to 1 000 °C	1 700	0.3 to 0.5	0.4 to 0.6	0.4 to 0.6	−0.4 to −0.7 1 700	High temperature insulation
Semi-insulating firebrick	30 to 40	0.2 to 1.6	—	1 to 2	0 to 1	0.1 to 0.6	0 to 1	51 to 61	1 to 2	0.6 to 1.0	—	1 to 3	—	1 320	0.25 to 0.35	0.35 to 0.45	0.45 to 0.55	0 to −0.5 1 300	Rotary kilns
Diatomite insulating brick	—	—	—	—	—	—	—	—	—	0.5 to 0.6	72 to 78	—	—	900	0.15 to 0.17	0.18 to 0.19			Low temperature insulation
											Fibre blanket								
Fibre blanket	47 to 65	0 to 0.2	0 to 1.2	0 to 0.3	0.1 to 0.4	0 to 0.1	0.1 to 0.4	38 to 50	0 to 0.1	—	—	—	—	1 300	0.05 to 0.09	0.09 to 0.22	0.11 to 0.45	−2.5 to −4.0 1 260 °C	Low temperature insulation, vacuum degasser gaskets, jointing material, etc.

N.B. Fibre board is similar to fibre blanket but contains a rigidizer. * Modulus of rupture.

27.1.1 Prepared but unshaped refractory materials

Use of these materials is commonly made for installation and/or repair of refractory linings. By definition, unshaped refractory materials are prepared mixtures for use either as delivered or after the addition of an appropriate liquid. This definition covers the refractory cements, mouldable and castable materials, ramming and gunning mixes. One major difference between materials in this group and brick or blocks is the considerably reduced number of joints in a structure, i.e. these materials tend towards a monolithic construction.

27.1.2 Aluminous cements

Hydraulic aluminous cement is manufactured by fusing or sintering a mixture of bauxite and limestone. In general the silica content is kept as low as the raw materials permit and it is preferred that the iron should be in the ferric rather than ferrous condition. Commerical aluminous cements vary somewhat in composition but usually lie in the following ranges: SiO_2 4–7%, Al_2O_3 36–42%, Fe_2O_3 8–12%, FeO 4–8%, TiO_2 2–3% and CaO 36–42%. These cements depend on the presence of calcium aluminates for their properties and by adding various forms of alumina, calcined bauxite, chrome and magnesia, it is possible to produce hydraulic cements with excellent refractory properties.

Calcium aluminate cements have unique properties in that they can be moulded, air-hardened and used directly as high-class refractories. They are relatively slow-setting (1–2 h) but rapid-hardening (24 h). In general, the higher the temperature of firing and within limits the higher the percentage of alumina they contain, the more refractory they become. They are notable for their relatively small shrinkage on heating to 400°C and small expansion on heating to 1350°C (0.5–1.5%). Their resistance to heat is very good and the best products can be used up to 1600°C without softening. Calcium aluminate cements lose strength when heated to 400°C, but then remain unchanged up to 1000°C and gradually recover again at higher temperatures.

27.1.3 Castable materials

These are mixtures of graded refractory aggregate and either a hydraulic cement or a chemical bonding agent. The material is usually supplied dry, and at the appropriate moisture content it may be cast or rammed. Properties of typical materials in this group are shown in Table 27.5. Castable materials are used in monolithic furnace linings, production of special shapes, covers of soaking pits, burner blocks, floors of aluminium holding vessels, cyclones and incinerators.

27.1.4 Mouldable materials

These are mixtures of graded refractory aggregates and plasticizers, usually clay, supplied mixed with water in a workable condition. Chemical bonding agents may also be incorporated. The workability of the material is such that it may be placed by hand malleting. These materials usually have good thermal shock resistance, but a low compressive strength which does, however, improve after the production of a ceramic bond. The thermal conductivity is usually lower than that of the equivalent fired material.

Mouldables are used in many high temperature vessels, ships' boilers, coke ovens and tunnel kilns.

Properties of a typical material from this group are shown in Table 27.5.

27.1.5 Ramming material

This is a mixture of graded refractory aggregate with or without the addition of a plasticizer and with or without water usually supplied at a consistency which requires a mechanical method of application. The material is placed in position by means of hand or pneumatic rammers. It is most important that all the material should be compacted to the same extent (i.e. an even density of packing) and that laminations should be avoided.

Ramming mixes are used in copper production vessels, blast furnaces, LD converters, Q–BOP converters, electric arc furnaces, hot metal mixtures, reheat furnaces, soaking pits and spouts of torpedo ladles.

Properties of a phosphate-bonded ramming mix are given in Table 27.5.

Table 27.5 TYPICAL PROPERTIES OF UNSHAPED REFRACTORIES

Material	*Chemical analysis* (% by weight)* Al_2O_3	CaO	Fe_2O_3	MgO	K_2O	Na_2O	SiO_2	TiO_2	*Bulk density* g cm^{-3} Unfired	Fired	*PLC*%** at °C	*Max service temp.* °C	*Thermal conductivity* Wm^{-1} K^{-1} 400 °C	600 °C	1 100 °C	*Refractoriness*	*MOR*†† MN m^{-1}
Lightweight insulating castable†	30 to 49	9.5 to 12	5 to 8	0 to 1	0 to 1	0 to 1	30 to 50	1 to 2	1.4 to 1.6	1.2 to 1.4	−0.2 to −1.0 1 200 °C	1 300	0.2 to 0.4	0.3 to 0.45		1 350 °C–1 450 °C	—
Dense castable†	96	3	0.1	—	0.05	0.05	0.1	—	2.7 to 3.1	2.6 to 2.9	−1 to +1 1 700 °C	1 800		1.1 to 1.3	1.1 to 2.2	1 800 °C+	—
Ramming mix [Phosphate Bonded]†	83 to 85	0.2	1.6	0.2	0.03	—	9.1 to 10.9	2.7	1.81	—	−1 1 700 °C	1 800	—			1 800 °C+	3.0 at 20°C 14.2 at 1000°C 3.0 at 1400°C
High alumina mouldable†	42 to 70	0 to 10	1 to 5	0 to 0.5	0.2 to 1	0 to 1	25 to 50	0 to 2	—	1.0 to 2.0	−0.5 to +1.5 1 600 °C	1 600		0.35 to 0.6	0.45 to 0.7	—	—

*The chemical constituents of the bonding material are not included: these may include phosphates and organic materials.
**Permanent linear change.
†Data from Pickford Holland Co., Sheffield.
††Modulus of rupture.

27.1.6 Gunning material

Many of the materials which fall into the above groups may be suitably prepared for application by gunning techniques. The suitably prepared material is introduced into a high pressure compressed air line in a specially designed gun and the material is blasted at the desired area to be lined. Some rebounding of material occurs, but by taking care in preparation and application this may be kept to a minimum.

Material applied by gunning techniques is used for general resurfacing of refractory brickwork, recontouring of ladle linings and installation of flue linings.

27.1.7 Design of refractory linings

Essential to the designer of refractory structures is a knowledge of the mechanical properties of materials to be used (e.g. thermal expansion, thermal conductivity, Young's modulus and ultimate strength). Standard test methods, such as those described in BS 1902, do not necessarily give the most useful data. The cold crushing strength of a fired brick is generally much greater than its crushing strength at higher temperatures; rectangular blocks heated from one end do not expand as expected from the free thermal expansion; nor is the Young's modulus of a stressed block the same as that of an unstressed block. These last two points are illustrated in Table 27.6.

For these reasons, great care should be taken to ensure that correct expansion allowances and bricking methods are used. Vessels to which this particularly applies are those in which bricks are suspended, rotated or tilted, e.g. electric arc furnaces, rotary kilns, LD or Q–BOP converters, torpedo ladles or steel ladles.

As a rule of thumb, however, expansion allowances for fired refractories should be approximately half the thermal expansion to the expected temperatures.

In the case of monolithic refractories opinions on expansion allowance vary greatly from no expansion allowance to half the thermal expansion to the expected temperature. With a number of castables there is some evidence to suggest that flexibility of castables in the service environment is sufficient to allow both opinions to be correct.

Expansion allowances may be achieved by placing cardboard, felt, wood or fibre between certain bricks. (N.B. Wood expansion pieces should not be used with silica refractories.)

27.1.8 Physical and chemical properties of pure ceramics

Many ceramics may be used at high temperatures: properties of many of these are given in Table 27.7.

Table 27.6 THERMAL EXPANSION AND YOUNG'S MODULUS FOR FIRED MAGNESITE*

	Temperature °C												
	100	200	300	400	500	600	700	800	900	1 000	1 100	1 200	1 300
Free thermal expansion (%)	0.08	0.24	0.38	0.55	0.70	0.88	1.05	1.21	1.38	1.54	1.71	1.87	—
Thermal expansion measured under thermal gradient heating (%)	0.14	0.29	0.45	0.61	0.76	0.92	1.09	1.25	1.37	1.54	—	—	—
Young's modulus (sonic method) ($GN\ m^{-2}$)	84	83	81	79	78	76	74	73	71	70	69	68	67
Young's modulus measured during restraint of thermal expansion ($GN\ m^{-2}$)	73	28	16	11	9.3	7.8	7.0	6.2	5.0	3.4	1.7	0.81	0.38

* Data published by permission of the British Ceramic Research Association.

Table 27.7 PROPERTIES OF PURE CERAMIC MATERIALS*

Material	*Melting point* °C	*Bulk density* g cm^{-3}	*Thermal conductivity* Wm^{-1} K^{-1} *at temp.* °C	*Thermal expansion coefficient* 10^{-6} *to temp.* °C	*Ultimate stress* MN m^{-2}	*Remarks*
Chromium diboride, CrB_2	2 100	5.2	32 at 20	7.5 20	—	Stable in presence of carbon
Hafnium diboride, HfB_2	3 250	11.2	6.3 at 20	5.3 500 5.5 1 000	—	No reaction with basic slags for 6 minutes at 1 520 °C. Stable in presence of carbon, no reaction with steel for 6 min at 1 620 °C
Tantalum monoboride, TaB	2 340	14.0	—	—	—	Oxidation severe at 1 100–1 400 °C. Unstable in presence of carbon
Tantalum diboride, TaB_2	3 200	12.4	10.9 at 20 13.9 at 200	—	—	Oxidized in air at 800 °C. Stable in presence of carbon
Titanium diboride, TiB_2	2 980	4.5	26 at 20 26 at 200	—	—	Very stable, even in presence of carbon. Oxidation in air severe at 1 100–1 400 °C
Zirconium diboride, ZrB_2	3 060	6.1	23 at 20 23–26 at 200	5.5–6.6 1 000 7.0 1 500	—	No reaction with basic slags at 1 520 °C for 6 min. No reaction with carbon steel at 1 620 °C for 2 h. Stable in presence of carbon. Oxidation in air severe at 1 100–1 400 °C. Stable under inert or reducing conditions to over 2 000 °C
Boron carbide, B_4C	2 350	2.51	29 at 20 84 at 425	4.8 500 5.5 1 000 6.5 2 000 7.1 2 500	2900 in compression at 20°C, 155 in compression at 980°C, 300 MOR at 20°C	Attacked by iron, Thermal shock resistance poor. Resistant to air up to 1000°C
Hafnium monocarbide, HfC	3 890	12.2	—	6.3 500 6.25 1 000	—	Oxidation in air severe at 1 100–1 400 °C. Stable to 2 000 °C in He

**See also* S. J. Burnett, 'Properties of Refractory Materials', UKAEA Research Group, Atomic Energy Research Establishment Harwell, England.

Table 27.7 PROPERTIES OF PURE CERAMIC MATERIALS*—*continued*

Material	*Melting point* °C	*Bulk density* g cm^{-3}	*Thermal conductivity* Wm^{-1}K^{-1} *at temp.* °C	*Thermal expansion coefficient* 10^{-6} *to temp.* °C	*Ultimate stress* MN m^{-2}	*Remarks*
Silicon carbide, SiC	2 700	3.17	42 at 20 21 at 1 000	4.6 500 5.5 1 500 5.9 2 500	550 in compression at 20°C	Quite resistant to oxidation by air up to 1500 °C. Thermal shock resistance very good. Reacts with Fe and MgO in basic slags to give silicides. Reacts with iron. High resistance to acid or neutral slags and coal ash: Used in gasifiers, zinc retorts. Also used as abrasive. Used in electrical heating elements
Tantalum monocarbide, TaC	3 880	14.7	22 at 20	6.3 500 6.7 1 000 8.4 2 500	—	Oxidation in air severe at 1 100–1 400 °C. Useful in He to 3 760 °C
Titanium monocarbide, TiC	3 140	4.25	32 at 20 5.5 at 1 000	7.7 1 000 9.7 2 500	750 to 1 300 in compression at 20 °C	Oxidation in air becomes severe at 1 200 °C. Max useful temp. 3 000 °C in He
Tungsten carbide, WC	2 777	15.7	84 at 20	4.9 1 000 5.8 to 6.1 2 000	—	Oxidation in air severe 500–800 °C useful to 2 000 °C in He. Extremely hard (9 + Mohs. Vickers pyramid 2 400) used in drill tips
Zinconium monocarbide, ZrC	3 540	6.7	21 at 20	6.1 500 6.6 1 000 7.6 2 000	1 640 at 20 °C compression	Oxidation in air becomes severe at 1 100–1 400 °C. Max. useful temp. 2 350 °C in He
Graphite, C	3 650 (sublimes)	1.50 to 2.25	63 to 210 at 20, parallel to grain. 42 to 130 at 20, perpendicular to grain. 47 at 1 300 34 at 2 500	Parallel: 1 to 4 20 Perpendicular: 2.5 to 4.5 20 4.0 to 9.8 1000 5.5 to 11 1 500 compression	Parallel 3.5 to 7.6 at 20 °C Perpendicular 3.5 to 70 at 20°C compression 36 at 2 500°C compression	Thermal shock resistance very good. Not wetted by iron. Resistant to acidic and basic slags. Oxidized in air above 300 °C. Resistant to non-oxidizing gases. May be 'welded' using molybdenum disilicide. Excellent conductor of electricity
Aluminium nitride, AlN	2 230	3.26		4.8 500 5.5 1 000	—	Oxidized by O_2 above 1 000 °C. Unstable in water vapour. Stable in N_2–H_2 mixtures at 1 200–1 600 °C

* *See also* S. J. Burnett, 'Properties of Refractory Materials', UKAEA Research Group, Atomic Energy Research Establishment, Harwell, England.

Table 27.7 PROPERTIES OF PURE CERAMIC MATERIALS*—*continued*

Material	*Melting point* °C	*Bulk density* g cm^{-3}	*Thermal conductivity* Wm^{-1}K^{-1} *at temp.* °C	*Thermal expansion coefficient* 10^{-6} *to temp.* °C	*Ultimate stress* MN m^{-2}	*Remarks*
Boron nitride, BN (hexagonal)	2 730 (sub-limes)	2.1	15 at 20 27 at 1 000 (Perpendicular values are approx. half of these)	Parallel to pressing direction 2.0 1 000 Perpendicular 13.3 1 000	Parallel to pressing direction 310 at 20 °C Compressive perpendi-cular 235 at 20 °C compressive	Thermal shock resistance very good when dry. No reaction with iron at 1 600 °C for 30 min. Oxidation in air severe at 1 100–1 400°C. Stable to 1000°C in O_2. Resists attack by molten metals and glasses. Low coeff. of friction. Fabricated by hot pressing. Machines easily
Trisilicon tetranitride, Si_3N_4	1 900 (sub-limes)	3.2	2.3 to 13 at 20 9.4 at 1 200	α phase 2.1 500 3.7 1 500 β phase 1.5 500 3.1 1 500	—	Thermal shock resistance good. Reacts with iron. Useful to 1 850 °C in reducing or inert conditions. Stable in air to 1 200°C. Resistant to molten glasses, molten Al, Pb, Zn, Sn and to HCl, H_2So_4 and HNO_3. Reacts with molten Cu. Slowly attacked by boiling water. May be partially nitrided, machined, then fully nitrided without great loss of strength. Used' in aluminium handling thermocouple sheaths
Titanium mononitride, TiN	2 900	5.3	29 at 20 8.5 at 1 000	—	—	No reaction with basic slags. Sightly wetted by carbon steel at 1 620 °C. Poor oxidation resistance to O_2 at 600 °C and to CO_2 at 1 200 °C
Zirconium mononitride, ZrN	2 950	7.1	27 at 20 6.7 at 1 000	6.13 450 7.03 680	—	Slight reaction with cast iron at 1 450 °C for 2 h. Oxidation in air severe at 1 100–1 400 °C. Slow hydrolysis in water
Aluminium oxide, Al_2O_3	2 050	3.97	39 at 20 9.2 at 600 5.9 at 1 400 7.1 at 1 800	7.6 500 8.5 1 000 8.9 to 9.1 1 400	2 940 at 20 °C compressive 48 at 1 600 °C compressive 203 at 20 °C shear 23 at 1 500 °C shear	Thermal shock resistance fair. Two polymorphic forms, α and γ. α stable above 450 °C (corundum) and γ metastable under all conditions. Good resistance t(basic and acidic slags

* *See also* S. J. Burnett, 'Properties of Refractory Materials', UKAEA Research Group, Atomic Energy Research Establishment, Harwell, England.

Table 27.7 PROPERTIES OF PURE CERAMIC MATERIALS*—*continued*

Material	*Melting point* °C	*Bulk density* g cm^{-3}	*Thermal conductivity* Wm^{-1} K^{-1} *at temp.* °C	*Thermal expansion coefficient* 10^{-6} *to temp.* °C	*Ultimate stress* MN m^{-2}	*Remarks*
Beryllium oxide, BeO	2 530	3.00	202 at 100 29 at 1 000 15 at 1 700	7.6 500 8.6 to 9.0 1 000 10.3 1 500 11.1 2 000	786 at 20 °C compressive 48 at 1 600 °C compressive	Becomes volatile at 2 100 °C; very poisonous. Not reduced by carbon or hydrogen. Attacked by acids, fluxed by alumina
Calcium oxide, CaO	2 572	3.32	15.5 at 100 8 at 1 000	11.8 500 13.1 1 000 15.3 1 500	—	Thermal shock resistance fair. Poor resistance to attack by slags containing FeO and SiO_2. Subject to hydration
Cerium dioxide, CeO_2	2 600	7.3	—	8.2 500 8.9 1 000	—	Useful in air to 2 400 °C. Not useful in reducing conditions. Subject to hydration
Dichromium trioxide, Cr_2O_3	2 435	5.21	—	8.4 500 8.6 1 000 8.8 1 500	—	Slag resistance good under oxidizing conditions. Less resistant to basic slags
Hafnium dioxide, HfO_2	2 810	9.68	—	Monoclinic 5.5 500 5.8 1 000 6.4 1 700 Tetragonal 1.3 1 700 3.0 2 000	— —	Useful in air to 2 400 °C. Stable in H_2 to 1 925 °C
Magnesium oxide, MgO	2 800	3.58	46 at 20 8.4 at 800 6.3 at 1 400 9.2 at 1 800	12.8 500 13.6 1 000 15.1 1 500 15.9 1 800	83 at 20 °C shear 39 at 1 300 °C shear	Thermal shock resistance poor; can be improved by small amounts of spinel. Resistance to both acidic and basic slags excellent. Limits of usefulness: to 1 600 °C in vacuum; 1 700–1 980 °C in reducing atmosphere; 2 400 °C in air. Melts at 2 680 °C in oxygen-free helium
Silicon dioxide SiO_2	1 710	2.32	1.5 at 20 2.5 at 1 600	α quartz 22.2 575 β quartz 27.8 575 14.6 1 000 Vitreous 0.55 1 000		Thermal shock resistance of vitrified silica is excellent Polymorphic forms of silica are quartz, tridymite and cristobalite. The last two of these are metastable under ordinary conditions. Transition temps: quartz-tridymite 870 °C. Tridymite-cristobalite 1 470 °C. Vitreous silica devitrifies at 1 100 °C

* *See also* S. J. Burnett, 'Properties of Refractory Materials', UKAEA Research Group, Atomic Energy Research Establishment, Harwell, England.

Table 27.7 PROPERTIES OF PURE CERAMIC MATERIALS*—*continued*

Material	*Melting point* °C	*Bulk density* g cm^{-3}	*Thermal conductivity* $Wm^{-1} K^{-1}$ *at temp.* °C	*Thermal expansion coefficient* 10^{-6} *to temp.* °C	*Ultimate stress* MN m^{-2}	*Remarks*
Thorium dioxide ThO_2	3 205	9.7	10 to 15 at 20 2.9 at 1 000 2.5 at 1 400	8.6 500 9.1 1 000 to 9.4 1 400 10.4	1 480 at 20 °C compressive 10 at 1 500 °C compressive 8.3 at 1 300 °C shear	Thermal shock resistance poor. High resistance to basic slags. Reduced by carbon at high temp. Becomes volatile in He at 2 300 °C
Zirconia, ZrO_2	2 690	5.75	2.0 at 25 2.3 at 800 2.7 at 1400	Monoclinic 6.5 500 7.7 1050 Tetragonal 7.9 600 8.3 1400	2070 at 20 °C compressive 102 at 20 °C compressive 19 at 1500 °C compressive	Zirconia phase changes monoclinic → cubic at 1 050°C. Thermal shock resistance influenced by volume change accompanying phase change. Stable in oxidizing atmosphere. Fairly stable in reducing atmosphere. Excellent resistance to basic and acidic slags
Mullite, $3Al_2O_3 . 2SiO_2$	1 830	3.15	7.1 at 25 4.0 at 800 3.8 at 1 400	5.1 to 5.8 1 000 °C	16.6 at 1 100 °C shear	Thermal shock resistance good
Magnesia spinel, $MgO\ Al_2O_3$	2 135	3.51	18 at 25 8 at 600 5.5 at 1 200	8.4 1 000 to 8.6 9.4 1 400	1 370 at 550°C compression 59 at 1 600 °C compression 65 at 20 °C shear 37 at 1 300 °C shear	Resistant to slags containing iron oxide. More resistant than alumina to action of reducing slags. Stable to 1 400 °C in H_2. Thermal shock resistance fair
Zircon, $ZrO_2 . SiO_2$	2 550	4.56	6.1 at 100 4.2 at 800 4.0 at 1 400	3.8 500 4.6 1 000 5.3 1 500	60 at 20 °C shear 16 at 1 300 °C shear	Dissociates above 1 700 °C. Due to action of slags containing FeO, zircon would dissociate in steel-making environments. Thermal shock resistance good
Molybdenum disilicide, $MoSi_2$	2 030	5.95 to 6.24	31.5 at 20 to 200 17 at 1 100	7.8 500 8.5 1 000 9.0 1 500	2 280 at 20 °C compressive	Carbon reduces melting point to 1 870 °C. Corrosion in air becomes severe at 1 700 °C. No attack by O_2 to 1 100 °C. Used in electrical heating elements
Tungsten disilicide, WSi_2	2 165 to 2 180	9.25		7.8 500 8.3 1 000		Corrosion in air severe above 1 950 °C

**See also* S. J. Burnett, 'Properties of Refractory Materials', UKAEA Research Group, Atomic Energy Research Establishment, Harwell, England.

Table 27.8 INDEX OF REFRACTORY STANDARDS

Subject	*BS*	*DIN*	*ASTM*	*PRE*	*ISO*
Abrasion resistance	1902	—	C704–76a	—	—
Acid resistance	—	—	—	R22	—
	—	—	—	R40	—
Alkali attack	—	—	C767–73	—	—
Basic refractories	4982 Part 1	—	—	—	—
	3056	—	—	—	—
	1902	—	—	—	—
Boilers	—	—	C64–72	—	—
Bricks — application	—	1082 Bbl*	—	—	—
Bricks — dimensions	4982	—	C134–70	R20	—
	3056	—	C861–77	R3	—
	2496	—	C909–79	R38	—
	5187	—	C134–70	—	—
Bricks — End arches	—	1082*	—	R3	1145
— dimensions	—	—	—	R36	—
Bricks — rectangular — dimensions	—	1081*	—	R3	R475
Bricks — side arches — dimensions	—	1082*	—	R3	R1145
Bricks — skewbacks — dimensions	—	—	—	R37	—
Carbon monoxide attack	1902	—	C288–78	—	—
Castable refractories	1902	—	C401–77	R25	—
	—	—	C179–72	R27	—
	—	—	C862–77	R28	—
Cement	4550	—	C105–47	—	—
	—	—	C198–76	—	—
	—	—	C606–70	—	—
	—	—	C199–72	—	—
Chemical analysis — alumina refractories	4140	51077	C573–70	—	—
	1902	—	—	—	—
Chemical analysis — Aluminosilicate refractories	1902	51070	C573–70	R24	—
	—	E51083	C575–70	—	—
Chemical analysis — carbon-containing refractories	—	—	C571–70	—	—
Chemical analysis — chrome refractories	1902	51074	C572–70	—	—
Chemical analysis — dolomite	1902	—	C574–71	—	—
Chemical analysis — magnesia refractories	1902	51073	C574–71	R33	—
Chemical analysis — raw materials	—	—	C572–70	—	—
Chemical analysis — sample preparation	—	51062*	—	—	—
Chemical analysis — silicon carbide refractories	—	51075	—	—	—
	—	51076	—	—	—
Chemical analysis — zircon	—	—	C576–70	—	—
Chemical analysis — zirconia	—	—	C705–72	—	—
Chemically-bonded basic bricks	—	51050	—	—	—
Chimneys and flues	4207	1057	—	—	—
Chrome brick	—	—	C455–76	—	—
Chrome–Magnesite brick	—	—	C455–76	—	—
Classification	2973	—	—	R42	1109
Coal tar	616	—	—	R43	—
Coke ovens	999	1089*	—	R44	—
	4966	—	—	—	—
Cold crushing strength	1902	51050	C133–72	R14	—
	—	51067	C93–67	R15	—
Concrete	1881	—	C860–77	—	—
	—	—	C865–77	—	—
	—	—	C862–7	—	—
	—	—	C903–79	—	—
Corrosion resistance	—	V51069	C622–68	R34	—
	—	—	C621–68	—	—
	—	—	C768–73	—	—
	—	—	C874–77	—	—
	—	—	C767–73	—	—
	—	—	C575–70	—	—

*Indicates English translation available.

Table 27.8 INDEX OF REFRACTORY STANDARDS—*continued*

Subject	*BS*	*DIN*	*ASTM*	*PRE*	*ISO*
Creep	—	51053	—	R6	—
Density	—	51050	C914–79	R8	—
	1902	51057	C357–70	R9	—
	—	51065	C134–70	R10	—
	—	—	C830–79	R30	—
	—	—	C20–74	—	—
	—	—	C493–70	—	—
Dimensional tolerances	—	—	—	R23	—
Dolomite	—	—	C468–70	—	—
Drying shrinkage	—	—	C179–72	—	—
Electrical insulation	1598	—	—	—	—
Fibrous products	DD41 (draft standard)	—	—	R41	—
Fireclay	—	—	C105–47	—	—
Fireclay products	4982 Part 2	1089	C63–61	R29	—
	1758	51060	C673–71	—	—
	3056	—	C27–70	—	—
	—	—	C605–72	—	—
Glass melting furnaces	4966	—	—	—	—
	5187	—	—	—	—
Glossary	3446	—	C108–46	—	R836
	—	—	C71–73	—	2246
Grain size	—	51033	—	—	—
Heat transmission	—	—	C108–46	—	—
High alumina	4982 Part 2	—	C673–71	—	—
	3056	—	C27–70	—	—
Hydration	—	—	C492–66	—	—
	—	—	C544–68	—	—
	—	—	C620–70	—	—
	—	—	C456–68	—	—
Incinerators	—	—	C64–72	—	—
Insulating firebrick	—	—	C434–61	R39	2245
	—	—	C155–70	—	—
	—	—	C134–70	—	—
	—	—	C210–68	—	—
	—	—	C134–70	—	—
	—	—	C182–72	—	—
Insulating refractory products	1598	—	—	R39	—
Liquid absorption	—	—	C830–79	—	—
	—	—	C20–74	—	—
Magnesite brick	—	—	C455–76	—	—
Magnesite–chrome brick	—	—	C455–76	—	—
Modulus of rupture	1902	51048	C133–72	R18	—
	—	—	C583–76	R21	—
	—	—	C607–67	—	—
	—	—	C93–67	—	—
	—	—	C606–70	—	—
Moisture content	—	—	C92–76	R11	—
Monolithic linings	4207	—	—	—	—
Mullite	—	—	C467–72	—	—
Non-destructive methods of test	4408	—	—	—	—
Nozzles, fireclay	—	—	C605–72	—	—
Pallets	—	—	—	R 1	—
Permanent linear change	1902	51066	C113–74	R13	2477
	—	—	—	R19	2478
	—	—	C436–70	—	—
	—	—	C605–72	—	—
	—	—	C210–68	—	—
	—	—	C179–72	—	—
Permeability	1902	51058	C577–68	R16	—
	—	51050	—	—	—

*Indicates English translation available

Table 27.8 INDEX OF REFRACTORY STANDARDS—*continued*

Subject	*BS*	*DIN*	*ASTM*	*PRE*	*ISO*
Porosity	1902		C20–74		
			C830–79	R9	
	—	—	C493–70	—	—
Pouring pit	—	—	C435–70	—	—
Pyrometric cone	—	51063	—	—	R1146
Pyrometric cone equivalent	1902	51063	C24–79	—	R528
Ramming mixes	—	—	C673–71	—	—
Refractoriness-under-load	1902	51053	—	R4	R1893
	—	51064	—	—	—
Rotary cement kilns	4982	—	—	R38	—
Sampling	2973	51061	—	—	—
	616	—	—	R7	5022
	1902	—	—	—	—
Sedimentation	—	51033	—	—	—
Sieve analysis	1902	51033	C92–76	—	—
Silica	4966	1089*	C49–57	—	—
	5187	—	C416–70	—	—
	—	—	C575–70	—	—
	—	—	C439–61	—	—
Silicon carbide refractories	—	—	C863–77	—	—
Slag resistance	—	—	C768–73	—	—
	—	—	C874–77	—	—
Specific gravity	1902	—	C830–79	—	—
	—	—	C20–74	—	—
	—	—	C604–79	—	—
	—	—	C135–66	—	—
Strength testing	1902	—	C106–70	—	—
	—	—	C16–77	—	—
	—	—	C546–67	—	—
Tar-bonded refractories	—	—	C831–76	R35	—
Tar-impregnated refractories	—	—	C831–76	R35	—
Test methods	1902	1089	—	—	—
	—	V51046	—	—	—
	—	51048	—	—	—
Thermocouple reference tables	4937	—	—	—	—
Thermal conductivity	1902	V51046	C201–68	R32	—
	—	—	C202–71	—	—
	—	—	C767–73	—	—
	—	—	C182–72	—	—
Thermal expansion	1902	51045	C832–76	—	—
Thermal shock resistance	1902	51068	C38–79	R5	—
	—	E51068	C107–76	—	—
	—	—	C122–76	—	—
	—	—	C439–61	—	—
Unshaped refractory products	1902	51061	C673–71	—	1927
	—	—	C179–72	—	—
	—	—	C491–72	—	—
	—	—	C180–72	—	—
	—	—	C417–72	—	—
	—	—	C181–76	—	—
	—	—	C860–77	—	—
	—	—	C865–77	—	—
	—	—	C862–77	—	—
	—	—	C903–79	—	—
Warpage	1902	—	C154–72	—	—
Young's modulus	—	—	C885–78	—	—
Zircon	—	—	C545–70	—	—

*Indicates English translation available

28 Fuels

28.1 Coal

28.1.1 Analysis and testing of coal

SAMPLING FOR ANALYSIS

In order to be representative, the gross sample is compiled by collecting a number of increments spaced evenly throughout the mass of the consignment of coal and, for the specific method, BS1017:Part 1:1977 should be consulted. The mass of an increment is determined by the maximum size of the coal shown in Table 28.1.

Table 28.1 MINIMUM MASS OF INCREMENT AND SIZE OF ENTRY INTO SAMPLING IMPLEMENT

Nominal upper size of coal, mm	<10	10–25	25–50	50–75	75–100	100–125	125–150	>150
Minimum size of entry into sampling implement, mm	30	75	150	200	250	320	375	>375
Minimum mass of increment kg	0.5	1.5	3.0	4.5	6.0	8.0	14.0	>14

The minimum number of increments required may be obtained from Table 28.2 and is determined by the class of coal and the purpose of the sample. Separate general analysis and total moisture samples may be desirable, e.g. when the coal is very wet, otherwise a common sample can be taken from which both total moisture and general analysis samples are prepared.

Table 28.2 INCREMENTS REQUIRED TO FORM A GROSS SAMPLE

Situation	*Minimum number of increments to be collected from a consignment weighing up to 1000 tonnes*							
	Common sample			*General analysis sample*		*Total moisture sample*		*Size analysis sample*
	Sized coals—dry-cleaned or washed	*Washed smalls (< 50mm)*	*Blended part-treated, untreated, run-of-mine and 'unknown' coals*	*Sized coals—dry-cleaned or washed and unwashed dry coals*	*Blended part-treated, untreated, run-of-mine and 'unknown' coals*	*Sized coals—dry-cleaned or washed and unwashed dry coals*	*Washed smalls (< 50 mm), blended, part-treated, untreated, run-of-mine and 'unknown' coals*	*All coals*
Streams	20	35	35	20	35	20	35	40
Wagons and lorries Barges Sea-going ships (from conveyor during off-loading)	25	35	50	25	50	20	35	40
Sea-going ships (from the hold) Stockpiles	35	35	65	35	65	20	35	40

Reference standards for the precision of measurements on the samples are given in Tables 28.3 and 28.4. The levels of precision represent the 95% probability limits of the deviation of any single value from the true value.

Table 28.3 REFERENCE STANDARDS OF PRECISION FOR MOISTURE AND ASH

True value	*Total moisture*	*Ash (dry basis)*
Below 10%	1% absolute	1% absolute
10% to 20%	0.1 of true value	0.1 of true value
Above 20%	2% absolute	2% absolute

Table 28.4 REFERENCE STANDARDS OF PRECISION FOR SIZE ANALYSIS: PERCENTAGE BETWEEN TWO SIEVES

Per cent in fraction	<5	5–10	10–20	20–30	30–50
Precision % absolute	0.8	1.8	2.7	3.2	3.5

Sampling from a stopped belt is the ideal method for sampling commercial coal and it should be used whenever practicable as the standard against which other methods are checked.

PROXIMATE ANALYSIS Consists of the following determinations (BS 1016: Part 3: 1973):

Total moisture which accounts for the 'free' or adventitious moisture, together with the 'inherent' or original moisture always associated with the coal. Moisture is determined by heating the air-dried coal, ground to pass a 0.2 mm sieve, to 105–110 °C in a vacuum oven or in a stream of nitrogen. The loss in weight is the moisture on the air-dried sample.

Volatile matter is the % loss in weight corrected for moisture when 1 g of the less than 0.2 mm coal is heated in the absence of air to 900 °C.

Ash is determined by placing 1 g of the 0.2 mm coal in a muffle furnace at room temperature, raising the temperature to 500 °C in 30 min, to 815 °C in a further 60–90 min and maintaining this temperature until the residue, which is the ash, is constant in weight.

Fixed carbon is defined as:

100—(moisture + ash + volatile matter)

ULTIMATE ANALYSIS

Requires in addition to ash and moisture determinations (described above) figures for carbon, hydrogen, nitrogen and sulphur.

CALORIFIC VALUE

For practical purposes may be expressed as the number of heat units liberated by the complete combustion of unit weight of coal in a bomb calorimeter. Corrections are made for the formation of nitric and sulphuric acid originally present as nitrogen and sulphur in the coal (BS 1016: Part 5: 1972).

CALCULATION OF CALORIFIC VALUE FROM ULTIMATE ANALYSIS

The calorific value of a fuel may be checked from the ultimate analysis. Usually the calculated value agrees to within 1–2% of the determined value.

In the following formula, which gives the *gross* calorific value (CV), the symbols used give the percentages of: C, carbon; H, hydrogen; O, oxygen; N, nitrogen; S, sulphur.

Grummell Davies:

$$CV=(0.01522\,H+0.937)[C/3+H-(O-S)/8] \qquad MJ\,kg^{-1}$$

Gross and net calorific values All coals contain hydrogen and water, and in the determination of calorific value the water vapour resulting from the combustion of the hydrogen and the vaporization of the original water is condensed to the liquid state. In boiler practice it is not possible to cool the flue gases to a temperature below the dew point and thus the latent heat of condensation of the steam is not recovered.

The *gross* calorific value as determined by the bomb is, therefore, corrected for boiler efficiency work by deducting 2.454 MJ kg^{-1} (1055 Btu lb^{-1}) of water obtained on combustion. The corrected figure is termed the *net* calorific value.

BS SWELLING NUMBER

Provides a means of assessing the tendency of a coal to swell when it is carbonized or used in a combustion appliance. If 1 g of less than 0.2 mm coal is heated in a squat-shaped silica crucible of standard dimensions by a Meker burner (with rich gas) or a Teclu burner with coal gas or a specially designed electric furnace, a coke button of a definite size and shape is produced. By reference to standard profiles a BS swelling number may be assigned to the coal (BS 1016: Part 12: 1959).

ROGA TEST[1]

A mixture of 1 g of less than 0.2 mm coal and 5 g of a standard anthracite is carbonized in a crucible. The resulting coke button is tested in a Roga drum for its resistance to abrasion. From the results obtained the coal coking index (Roga index) is calculated.

GRAY KING COKE TYPE

The caking properties of a coal or blend of coals is assessed by carbonizing in a laboratory assay under standard conditions. The coke residue is classified by comparison with a series of described standard coke types (BS 1016: Part 12: 1959).

AUDIBERT–ARNU DILATOMETER TEST[1]

The test assesses the coking properties of coal or coal blends. A pencil of powdered coal is inserted in a narrow tube and topped by a steel rod which slides in the bore of the tube. The whole is heated at a constant rate. The displacement of the piston is recorded as a function of the temperature. The maximum dilatation and contraction are recorded as a percentage of the original length of the pencil, and the temperatures of the points of softening, maximum dilatation and maximum contraction are noted.

ASH FUSION POINT

Ash fusion point of a coal ash is considered to be a rough guide to its clinkering propensities. The fusion point in a reducing atmosphere is lower by about 40 °C than that in an oxidizing atmosphere.

- Group 1 Fusion temperature 1425–1710 °C, Clinkering troubles absent.
- Group 2 Fusion temperature 1200–1425 °C. Clinkering manageable.
- Group 3 Fusion temperature 1040–1200 °C. Clinkering troubles excessive, unless adequate precautions are taken.

The determination is made on a trilateral pyramid, a cube or a right cylinder prepared in a mould from finely ground ash (BS 1016: Part 15: 1970). The test-piece is heated in an oxidizing or a reducing atmosphere defined as follows:

A reducing atmosphere An atmosphere consisting by volume of 50% hydrogen and 50% carbon dioxide with a tolerance of ±5%.

or

An oxidizing atmosphere An atmosphere consisting of either carbon dioxide or air.

The temperatures determined are:

(a) *Deformation temperature.* The temperature at which the first sign of rounding of the tip of the test specimen occurs.
(b) *Hemisphere temperature.* The temperature at which the height of the specimen is equal to half the base, its shape being approximately hemispherical.
(c) *Flow temperature.* The temperature at which the height of the specimen is equal to one third of that at the hemisphere temperature.

CALCULATION OF THE MINERAL MATTER CONTENT OF COAL

All classification systems are based on coal free from mineral matter and moisture. The analytical data for the coal 'as received' thus require correction in the sense of the following equations. The symbols have the following meanings: W_a, moisture; A, ash; V, volatile matter; C, carbon; H, hydrogen; S, sulphur; N, nitrogen; M_b and M_k, estimates of the mineral matter content; all expressed as percentages on the air-dried basis. Q is the calorific value MJ kg^{-1} (Btu lb^{-1}), on air-dried basis and W_t is the total moisture % on the 'as received' or 'as fired' basis.

(a) To convert 'air dried' to 'as received':

$$\text{Multiply } A, V, C, H, S, N, Q \text{ by } \left[\frac{100-W_t}{100-W_a}\right]$$

(b) To convert 'air dried' to 'dry ash-free':

$$\text{Multiply } V, C, H, S, N, Q \text{ by } \left[\frac{100}{100-(W_a+A)}\right]$$

(c) To convert 'air dried' to mineral matter-free basis:

$$\text{Multiply } C, H, N, Q \text{ by } \left[\frac{100}{100-(W_a+M_k)}\right]$$

The volatile matter on a dry mineral matter-free basis is given by[2]

$$\frac{100(V-c)}{[100-(W_a+M_k)]}$$

According to the analysis available the correction c is given by

$$c = 0.13A + 0.2S_{\text{pyr}} + 0.7CO_2 + 0.7Cl - 0.20$$
$$\text{or } c = 0.13A + 0.2S_{\text{total}} + 0.7CO_2 + 0.7Cl - 0.32$$
$$\text{or } c = 0.13A + 0.2S_{\text{total}} + 0.7CO_2 - 0.12$$

The mineral matter (M_k) is most accurately expressed by a modification of the King, Maries and Crossley[3] formula:

$$M_k = 1.13A + 0.5S_{\text{pyr}} + 0.8CO_2 - 2.8S_{\text{ash}} + 2.8S_{\text{sulph}} + 0.5Cl$$

in which A is the determined ash; CO_2, carbon dioxide; S_{pyr}, pyritic sulphur; S_{ash}, sulphur in ash; S_{sulph}, sulphate sulphur in coal; Cl, chlorine.

Where full analytical data are not available the mineral matter (M_b) may be approximately assessed for British coals by the modified BCURA formula:[4]

$$M_b = 1.1A + 0.53S_{\text{total}} + 0.74CO_2 - 0.32 \simeq 1.15A$$

28.1.2 Classification

Three major classification systems have been devised based on the proximate analysis of coals.

These are the Fuel Research Board/National Coal Board (NCB) classification,[2] the American ASTM classification,[5] and the International Classification[1] of Hard Coals by Type devised by the Economic Commission for Europe (ECE). The Fuel Research Board/NCB classification is described below and outlines are given of the others. The technological characteristics of coals are better defined in terms of their petrographical constituents and a coal classification system on this basis is in prospect.

FUEL RESEARCH BOARD/NCB CLASSIFICATION[2]

Coals are assigned code numbers according to their volatile matter content on a dry, mineral matter-free basis and their caking propensities are assessed by the Gray–King low temperature assay (BS 1016: Part 12: 1959). Clean coal must be used for determining the Gray–King coke type, and if the coal has initially a higher ash content than 10% it is floated at such a specific gravity as will give the maximum yield of coal with not more than 10% of ash.

Using the criterion of volatile matter alone, a first division into the following groups is obtained:

	Volatile matter	*Code no.*
Anthracites	Under 9.1%	100
Low-volatile steam coals	9.1–19.5%	200
Medium-volatile coals	19.6–32.0%	300
High-volatile coals	Over 32%	*See* Table 28.5

In the first three groups, i.e in coals of volatile matter up to 32%, there is a close relationship between volatile matter content and caking properties. Consequently, the effect of subdividing into progressive ranges of volatile matter content is also to produce classes with progressive ranges of caking power. The corresponding Gray–King coke types are given in Table 28.5 as an indication of caking properties.

In the fourth group, i.e. in coals with more than 32% of volatile matter—there is a wide range of caking properties at any given volatile matter content, and subdivision has been made on the basis of the Gray–King coke type. Six ranges of caking properties, listed in Table 28.5, are recognized for these high-volatile coals.

Each of the 400–900 classes can be further subdivided according to volatile-matter content: a 1 in the third figure of the code number indicates that the volatile matter lies between 32.1 and 36.0, and a 2 that it is over 36%. Also, a subdivision is made of the 301 class into 301a and 301b with volatile ranges of 19.6–27.5% and 27.6–32% respectively.

Certain coals have been affected by the heat from nearby igneous intrusions, with the result that their caking properties are generally subnormal compared with those of other coals of similar volatile content. These affected coals are distinguished by the code numbers 201H, 203H, 302H and 303H. They occur mainly in Scotland, but some are found in Durham.

A full list of code numbers, with ranges of volatile contents and caking properties, is given in Table 28.5.

Table 28.5 COAL CLASSIFICATION SYSTEM USED BY NATIONAL COAL BOARD (REVISION OF 1964)

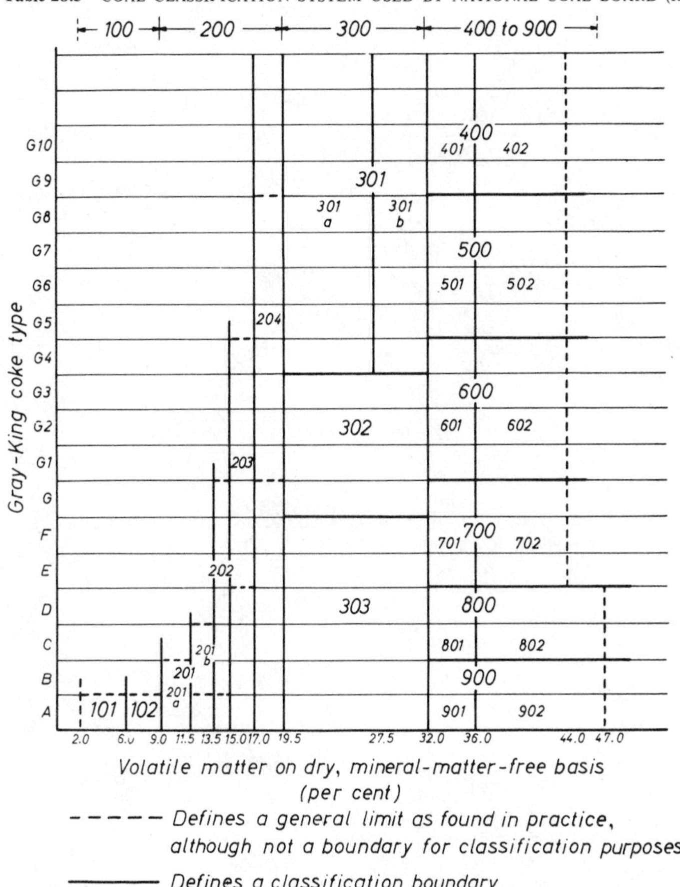

Notes:

(1) Coals that have been affected by igneous intrusions ('heat-altered' coals) occur mainly in classes 100, 200 and 300, and when recognized should be distinguished by adding the suffix H to the coal rank code, e.g. 102H, 201bH.

(2) Coals that have been oxidized by weathering may occur in any class, and when recognized should be distinguished by adding the suffix W to the coal rank code, e.g. 801W.

ASTM CLASSIFICATION OF COALS BY RANK[5]

Coals are classified according to their fixed carbon and calorific value expressed in Btu lb^{-1} on a mineral matter-free basis. The higher rank coals are classified according to fixed carbon on the dry basis; the lower rank coals are classified according to calorific value on the moist basis. Agglomerating character is used to differentiate between certain adjacent groups. There are four classes: I anthracite, II bituminous, III sub-bituminous, IV lignite, each containing a number of named groups. The position of a coal in the scale of rank can be expressed in a condensed form, e.g. (62–146) in which the parentheses signify that the contained numbers are on a mineral matter-free basis. The first number represents the fixed carbon on the dry basis reported to the nearest whole per cent, the second the calorific value expressed as hundreds of Btu lb^{-1} to the nearest hundred.

ECE INTERNATIONAL CLASSIFICATION OF HARD COALS[1]

Coals are first placed in classes according to their volatile matter on the dry ash free basis. Then coals with a volatile matter greater than 33% are placed in classes according to their gross calorific value, Table 28.6.

Table 28.6 DIVISION OF COALS INTO CLASSES (ECE)

Class number	*Volatile matter* % d.a.f.	*Class number*	*Gross calorific value moist ash free* kcal kg^{-1}	(MJ kg^{-1})
1A	3–6.5	6	>7 750	(32.5)
1B	>6.5–10	7	>7 200–7 750	(30.1–32.5)
2	>10–14	8	>6 100–7 200	(25.5–30.1)
3	>14–20	9	>5 700–6 100	(23.9–25.5)
4	>20–28			
5	>28–33			
6–9	>33			

Each class is further subdivided into groups according to their caking properties expressed either by their crucible swelling number or their Roga index, Table 28.7.

Table 28.7 DIVISION OF COAL CLASSES INTO GROUPS (ECE)

Group number	*Crucible swelling number*	*Roga index*
1	0–½	0–5
2	1–2	5–20
3	2½–4	20–45
4	>4	>45

Each group is then subdivided into subgroups according to their coking properties assessed by their maximum dilatation in the Audibert–Arnu dilatometer test or by their Gray–King coke type, Table 28.8.

Table 28.8 DIVISION OF COAL GROUPS INTO SUBGROUPS (ECE)

Subgroup number	*Maximum dilatation*	*Gray–King coke type*
0	Non-softening	A
1	Contraction only	B–D
2	0 and less	E–G
3	>0–50	G1–G4
4	>50–140	G5–G8
5	>140	>G8

A three-digit code number is used to describe the classified coal. The first digit indicates the class, the second digit the group, and the third digit the subgroup.

28.1.3 Physical properties of coal

Table 28.9 PHYSICAL PROPERTIES OF COAL

	Density kg m^{-3}	Bulk density* kg m^{-3} (lb ft^{-3})	Specific heat† kJ kg^{-1} K^{-1}	Coeff. linear thermal exp.‡ 10^{-6} K^{-1} 30 °C	90 °C	220 °C	330 °C
Fusain	—	—	0.88–0.92	—	—	—	—
Bituminous coal	1 250–1 450	600–670 (38–42)	1.00–1.09	33	—	45	60
Anthracite	1 400–1 700	700–790 (44–49)	0.92–0.96	—	—	—	—
Anthracite parallel to bedding plane	—	—	—	—	15	16.5	18
Anthracite perp. to bedding plane	—	—	—	—	27	29	29
Coal ash	—	—	0.67–0.71	—	—	—	—

* The approximate bulk densities refer to dry graded coals loosely packed in large containers. The bulk density is influenced by: (1) size and grading; (2) size of the container; (3) % 'free' moisture in excess of the inherent moisture; (4) shape of particles; (5) method of packing.
† Reference 6. The specific heat of coal increases with increase in volatile content and decrease in the carbon/hydrogen ratio.
‡ Data supplied by British Coal Utilization Research Association.

Methods for predicting the specific heat, enthalpy, and entropy of coal, char, tar and ash as a function of temperature and material composition are presented in reference 7.

The mean thermal conductivity, k, of coking coals between 0 °C and t °C is given approximately by

$$k = 130 + 0.67t + 0.00067t^2 \quad \text{mW m}^{-1}\text{K}^{-1}$$

28.2 Metallurgical cokes

28.2.1 Analysis and testing of coke

SAMPLING FOR ANALYSIS AND SHATTER TEST

Analysis In order to obtain a representative sample cokes are divided into four classes as follows:

Class 1: Large or graded gas coke from which breeze has been removed.
Class 2: Large or graded oven cokes from which breeze has been removed.
Class 3: Gas or oven cokes from which breeze has not been removed.
Class 4: Breeze.

In order to obtain a specified accuracy for any particular determination the number of increments required depends on the type of coke, its moisture, the degree of accuracy chosen and, in many cases, the conditions under which the sampling is to be carried out. The number of increments is independent of the total weight of coke sampled.

Table 28.10 TYPICAL ANALYSES OF SOLID FUELS
The table attempts to give analyses for fuels falling into each of the classes given but it should be understood that they can only be considered as a guide and that wide variations will be encountered among fuels belonging to each class

	Anthracite	*Semi-anthracite*	*Semi-bituminous coals*	*Bituminous coals*							*Lignite*	*Peat*	*Wood*	*Manufactured fuels*		
														Charcoal	*Coke*	*Semi-coke*
Code numbers: NCB	100a	201	204	301a	401	502	601	702	802	902	—	—	—	—	—	—
ECE	120A	221	344	445	545	844	843	832	821	921	—	—	—	—	—	—
Proximate analysis (air-dried basis)																
Moisture	1.0	1.0	1.0	0.9	0.9	1.9	2.0	5.8	8.6	13.8	15.0	20.0	15.0	2.0	2.5	2.5
Volatile matter less moisture	5.0	11.2	17.9	25.9	30.8	34.4	32.7	33.6	34.0	34.7	40.0	50.0	70.0	8.0	1.5	8.5
Fixed carbon	91.0	83.8	77.1	71.3	64.0	56.6	58.5	55.3	52.5	46.9	40.0	25.0	14.5	89.0	88.0	80.0
Ash	3.0	4.0	4.0	1.9	4.3	7.1	6.8	5.3	4.9	4.6	5.0	0.5	0.5	1.0	8.0	9.0
Volatile matter (dry, ash-free)	5.2	11.8	18.8	26.6	32.3	37.8	35.9	37.8	39.4	42.5	50.0	66.7	82.8	8.2	1.7	9.6
Ultimate analysis (air-dried coal)																
Carbon	89.4	86.9	86.0	86.6	83.5	76.7	77.0	74.4	70.0	64.6	55.2	43.1	42.4	90.4	85.1	82.3
Hydrogen	2.9	3.8	4.3	4.8	5.1	4.9	4.8	4.8	4.6	4.4	3.9	4.6	5.1	2.4	0.8	2.7
Nitrogen	1.1	1.2	1.3	1.6	1.5	1.6	1.5	1.5	1.2	1.3	0.7	0.6	0.3	0.8	0.9	0.9
Sulphur	0.9	1.0	1.0	0.8	1.1	2.6	1.5	1.1	0.9	0.6	0.6	1.3	0.3	0.7	0.7	1.1
Oxygen and errors	1.7	2.1	2.4	3.4	3.6	5.2	6.4	7.1	9.8	10.7	19.6	25.4	36.4	2.7	1.6	1.5
*Ultimate analysis (dry, mineral water-free basis)**																
Carbon	93.5	92.0	91.0	89.4	88.8	85.9	85.6	84.3	81.7	79.9	69.0	57.5	50.2	93.2	95.1	93.0
Hydrogen	3.0	4.0	4.5	5.0	5.3	5.4	5.3	5.3	5.4	5.4	4.9	6.1	6.0	2.5	0.9	3.1
Nitrogen	1.2	1.3	1.4	1.7	1.6	1.7	1.7	1.7	1.4	1.6	0.9	0.8	0.4	0.8	1.0	1.0
Sulphur†	0.9	1.1	1.1	0.8	0.8	1.2	1.2	0.8	1.1	0.8	0.7	1.8	0.4	0.7	1.2	1.2
Oxygen and errors	1.4	1.6	2.0	3.1	3.5	5.8	6.2	7.9	10.4	12.3	24.5	33.8	43.0	2.8	1.8	1.7
Caking and swelling tests																
BS swelling number	1	2	7	8	9	8	7	3½	1½	1	—	—	—	—	—	—
Gray–King coke type	A	B	G6	G9	G10	G6	G4	E	C	B	—	—	—	—	—	—
Calorific value MJ kg^{-1} (Btu lb^{-1})																
Air-dried coal	34.24 (14 720)	34.52 (14 840)	34.61 (14 880)	35.00 (15 050)	34.35 (14 770)	31.91 (13 720)	31.82 (13 680)	30.42 (13 080)	28.28 (12 160)	26.05 (11 200)	21.03 (9 040)	16.68 (7 170)	15.75 (6 770)	33.70 (14 500)	30.17 (12 970)	30.38 (13 060)
Dry, mineral matter-free coal	35.80 (15 400)	36.52 (15 700)	36.63 (15 750)	36.66 (15 760)	36.52 (15 700)	35.59 (15 300)	35.54 (15 280)	34.42 (14 800)	33.00 (14 190)	32.19 (13 840)	26.30 (11 300)	22.24 (9 560)	18.60 (8 010)	34.77 (14 950)	33.70 (14 490)	34.33 (14 760)

* Dry, ash-free basis for lignite, peat, wood and manufactured fuels.
† Organic sulphur for coals.

The number of increments of the various classes of coke for ash and moisture determinations required to give an accuracy of ±1% at the 95% probability level is given in Table 28.11 and the weight of the increments in Table 28.12 according to BS 1017: Part 2: 1960.

Table 28.11 NUMBER OF INCREMENTS FOR ACCURACY OF ±1%

Class	*Moisture*		
	3% or less	*3–5%*	*Over 5%*
1	32	32	48
2	48	72	108
3	100	150	225
4	16	16	16

Table 28.12 WEIGHT OF INCREMENT AND SAMPLING IMPLEMENTS

Maximum size of coke	*Sampling implement*
38 mm (1½ in) (i.e. not more than 5% over 1½ in)	1.14 kg (2½ lb) scoop
76 mm (3 in) (i.e. not more than 5% over 3 in)	2.3 kg (5 lb) scoop
101 mm (4 in) (i.e. not more than 5% over 4 in)	4.5 kg (10 lb) scoop
Over 101 mm (4 in)	6.8 kg (15 lb) scoop

Shatter test According to BS 1016: Part 13: 1969 the gross sample of 25 kg for the shatter test should be collected specifically for the test according to BS 1017: Part 2: 1960 and should contain all the sizes over 51 mm (2 in) in approximately the same proportion as are found in the original size analysis.

GENERAL

The qualities of coke which have the most influence on metallurgical practice are purity, hardness and combustibility. The purity of any particular sample is determined by chemical tests, but the physical properties of hardness and combustibility may only be assessed by empirical tests.[8]

CHEMICAL ANALYSIS

Chemical analysis normally includes determinations of water, ash, volatile matter and sulphur. Phosphorus is important in the manufacture of acid pig iron.

The carbon content of a coke is an index of its thermal value and is roughly assessed by subtracting the 'impurities', as determined by chemical analysis, from 100.

OTHER TESTS

The size of coke is specified by the size of the square meshed screen through which it passes or on which it rests, the results being expressed as cumulative percentages on screens of decreasing sizes.

Bulk density of coke This is an indication of the weight of the lump material that will fill a known (large) volume. The cubical container used has a capacity of 0.1 m^3 (2 ft^3), 465 mm (15 in) side internally (BS 1016: Part 13: 1969).

Apparent specific gravity is the ratio of the weight of a given volume of coke to the weight of an equal volume of water (BS 1016: Part 13: 1969).

True specific gravity is the ratio of the weight of a given volume of dry coke passing a 0.2 mm test sieve to the weight of an equal volume of water at the same (atmospheric) temperature (BS 1016: Part 13: 1969).

Porosity may be either 'apparent' or 'total'.

$$\% \text{ apparent porosity} = \left[\frac{\text{volume of open pores}}{\text{volume of coke}}\right] \times 100$$

$$= \left[\frac{W_3 - W_1}{W_3 - W_2}\right] \times 100$$

where W_1 = weight of dried coke, W_2 = weight of coke saturated with water weighed in a tank of cold water, W_3 = weight of coke saturated with water.

It can also be shown that:

$$\% \text{ total porosity} = \left[\frac{\text{real specific gravity} - \text{apparent specific gravity}}{\text{real specific gravity}}\right] \times 100$$

The real specific gravity in the above expression is determined by the specific gravity bottle method on material passing a 0.2 mm sieve, care being taken, however, to boil the coke with water in order to remove air and to saturate it (BS 1016:Part 13:1969).

The apparent specific gravity may be obtained from weighings required by the apparent porosity:

$$\text{apparent specific gravity} = \frac{W_1}{W_3 - W_2}$$

The micum indices of a coke should measure its liability to attrition in the blast furnace. It is determined as follows: 25 kg of coke over 60 mm in size and with less than 5% moisture is placed in a special drum and rotated for 100 revolutions. A size analysis of the coke is then made and the percentages of coke remaining on a 40 mm sieve (M_{40}) and passing through a 10 mm sieve (M_{10}) are normally reported (BS 1016:Part 13:1969).

The 'shatter index' is a measure of the liability of a coke to form breeze during loading, unloading and charging operations.

To determine this index 25 kg of greater than 51 mm (2 in) coke is dropped four times from a special box which is placed 1.83 m (6 ft) above a cast iron or steel plate. The shattered coke is then screened and the average of three tests of the percentages retained on 51 mm (2 in), 38 mm ($1\frac{1}{2}$ in), 25 mm (1 in) and 13 mm ($\frac{1}{2}$ in) square aperture screens are reported as respective shatter indices (BS 1016:Part 13:1969).

The reactivity of coke determines its behaviour towards air or oxygen or the rate with which it reduces carbon dioxide to monoxide.

The 'critical air blast' (CAB), determines the reactivity of coke to air by finding, by trial and error, the minimum rate of blast which will maintain combustion in an ignited bed of 1.2–0.6 mm coke contained in a glass or quartz tube of specified dimensions (BS 1016: Part 13: 1969).

The thermal value of the volatile matter remaining in the coke (volatile therms) is a measure of its ignitability and, indirectly, of its reactivity to air (BS 1016:Part 13:1969).

28.2.2 Properties of metallurgical coke

BLAST FURNACE COKE

A specification for blast furnace coke proposed by the BSC/BISRA iron-making panel of the Iron and Steel Institute (Publication P127:1969) is shown in Table 28.13.

Table 28.13 SPECIFICATION FOR BLAST-FURNACE COKE

Moisture content
This shall not exceed 3%; a mean of 2% is desired
Variation −2 to +3 on single samples
−0.3 to +0.5 on weekly average

Size
The overall size range shall be 19–64 mm (¾–2½ in); this implies pre-crushing of the coke

Shatter index
The 38 mm (1½ in) shatter index shall not be less than 90
Variation ±2 on single samples
±0.3 on weekly average

Micum index
The M_{40} index shall not be less than 75
Variation ±3 on single samples
±0.45 on weekly average

The M_{10} index shall not exceed 7
Variation ±1 on single samples
±0.45 on weekly average

Ash content
The ash content should not exceed 3%
Variation ±1.7 on single samples
±0.27 on weekly average

Sulphur content
The sulphur content should not exceed 0.6%
Variation ±0.17 on single samples
±0.03 on weekly average

A size range of 20–80 mm as charged and with higher ash is more usual. Consistency of the properties of the coke supplied is of the utmost importance.

Table 28.14 PROPERTIES OF COKES

Real density, kg m^{-3}	1 700–2 000
Apparent density, kg m^{-3}	700–1 100
Total porosity, %	36–55
Apparent porosity, %	35–47
Calorific value, MJ kg^{-1}	33.14
Ash, % dry	8–11
Volatile matter	0.6–1.1
Sulphur	0.57–1.4
Phosphorus	0.01–0.14
Critical air blast, l min^{-1}	1.56–2.4

FOUNDRY COKE

The range of properties specified[9] for foundry coke from various plants in Wales and Durham is given in Table 28.15. These supersede the recommendations of TS 47 1959[10].

Table 28.15 SPECIFICATION FOR FOUNDRY COKE

Moisture, % maximum	3.0–5.5
Ash, % maximum	9
Volatile matter, % maximum	0.7–1.0
Sulphur, % maximum	0.85–1.0
Shatter index, 50 mm (2 in) minimum	90
Mean size minimum, mm	102–107
(in)	(4–4.2)
Undersize, not more than 4% less than 50 mm (2 in)	

There are indications[11] that a narrow range of size of coke as charged is more important to the operation of the cupola furnace than the mean size in the size range 40–110 mm and that there is little to be gained by using large coke. Consistency in the properties of the coke supplied is most important.

FORMED COKE

Processes developed to produce formed coke briquettes[12] from weakly coking coals are shown in Table 28.16. Properties of formed cokes used in blast furnace trials are given in Table 28.17.

Table 28.16 CHARACTERISTICS OF FORMED COKE PROCESSES

Process	*Forming*	*Feed*	*Binder*
BBF	Hot briquetting	Any coal	30% caking coal
Consol–BNR	Hot pelletizing	High-volatile coal	Caking coal
Iniex	Briquetting	Low-volatile, non-caking	Pitch
FMC	Briquetting	High-volatile coal	Pitch
Sapoznikov	Hot briquetting	Slightly caking	Caking coal
Guiprokoks	Hot briquetting	Low–medium volatile, weakly caking	High-volatile, weakly caking coal
DKS	Briquetting	Non-caking	Pitch and caking coal

Table 28.17 PROPERTIES OF FORMED COKE BRIQUETTES

	Process and coal source				
Property	*BBF,* Germany*	*BBF,* UK*	*FMC, USA*	*DKS Japan*	*GI, USSR*
Analysis, dry %					
Fixed carbon	81.5	81.2	89.9	80.3	na
Ash	5.5	12.1	5.5	12.6	na
Volatile matter	9.1	6.0	3.9	6.5	1.5
Sulphur	0.9	1.0	0.7	0.5	na
Bulk density, kg m^{-3}	578	622	554	779	na
Strength					
M+40	84	86	—	—	85–90
M+30	—	—	95	—	—
M+20	—	—	—	94	—
M−10	9.3	10.9	5.1	5.6	7–8

* Non-calcined.
na—not available.

BULK DENSITY OF COKE

A graded coke 20–40 mm with normal ash and moisture has a bulk density of 420–480 kg m^{-3}, and run of oven coke 460–510 kg m^{-3}.

SPECIFIC HEAT OF COKE[13]

The relationship between the specific heat of coke and the ash content is a linear one. Values for the mean specific heat between 21° and t°C are given in Table 21.18, where A denotes % ash present in the coke.

Table 28.18 MEAN SPECIFIC HEAT OF COKE

Temperature °C	*Mean specific heat* kJ kg^{-1} K^{-1}
400	1.11–0.001 8 A
500	1.26–0.002 8 A
600	1.36–0.003 4 A
700	1.45–0.004 4 A
800	1.50–0.004 5 A
900	1.56–0.005 4 A
1 000	1.60–0.005 7 A
1 100	1.63–0.005 7 A
1 200	1.66–0.005 7 A
1 300	1.69–0.005 7 A

Table 28.19 THERMAL EXPANSION OF COKE*

	Coefficient of linear thermal expansion 10^{-6} K^{-1}			
Temperature of measurement °C	*Bituminous coal. Strongly coking Carbonization temp.* °C		*Bituminous coal. Weakly coking Carbonization temp.* °C	
	600	1000	600	1000
100	9.0	3.4	6.8	1.8
200	10.0	4.1	7.8	1.8
300	10.5	4.5	8.0	2.3
400	—	4.6	8.0	—

* Data provided by BCURA.

28.3 Gaseous fuels, liquid fuels and energy requirements

28.3.1 Liquid fuels

Liquid fuels are easy to handle, store and control. The two main groups are derived from (1) petroleum, (2) coal carbonization. (In the long term we must expect a full range of liquid fuels manufactured from coal either by gasification and synthesis or by direct routes involving pyrolyses, solvent extraction and hydrogenation.) Distillate fuels contain practically no ash, and residual fuels contain very little ash in comparison with solid fuels. Sulphur in residual fuel oil depends mainly on the source of the crude oil from which it was obtained. High sulphur contents are usually undesirable metallurgically. Slagging troubles occur at over 700 °C owing to the presence of Na_2O, S and V_2O_5 in the fuel. The higher flame emissivity of coal tar fuels is an advantage in high temperature processes.[14]

British Standards specifications[15, 16] for liquid fuels are intended as a guide but more details should be specified for metallurgical use of fuel. Tables 28.20 and 28.21 refer to typical properties of petroleum and tar fuels respectively.

Table 28.20 PETROLEUM LIQUID FUELS

	Kinematic viscosity cSt (10^{-2} m^2 s^{-1})	*I.B.P.* °C	*F.B.P.* °C	*Specific gravity* 15.6°/15.6 °C 60°/60 °F	C %	H %	O+N %	S %	*Ash* %	*Calorific value* MJ kg^{-1} (Btu lb^{-1}) *gross*
	at 37.8 °C (100 °F)									
Primary flash distillate	—	37	72	0.649	83.98	16.0	—	0.02	—	47.9 (20 600)
Primary flash distillate	—	32	163	0.704	84.87	15.0	—	0.03	—	47.1 (20 250)
Kerosine	1–2	160	185	0.78	85.9	14.0	—	0.08	—	46.5 (20 000)
Gas oil	3.3	190	—	0.844	85.3	13.2	—	0.30	0.001	45.4 (19 600)
Gas oil	3.3	190	—	0.833	85.0	13.05	—	0.95	0.001	45.6 (19 500)
	at 82.2 °C (180 °F)		H_2O %							
Light fuel oil	10	—		0.935	85.55	11.50	0.70	2.55	0.02	43.3 (18 600)
Light fuel oil	12.5	—	—	0.99	86.94	11.4	0.5	1.1	0.06	42.9 (18 450)
Medium fuel oil	30	—	0.05	0.967	84.12	11.50	0.8	3.5	0.03	42.6 (18 300)
Medium fuel oil	34	—	0.05	0.968	85.85	11.83	0.8	1.4	0.03	42.98 (18 480)
Heavy fuel oil	50	—	0.1	0.950	85.2	11.70	1.0	1.9	0.1	43.0 (18 500)
Heavy fuel oil	70	—	0.1	0.980	84.5	10.7	0.8	3.8	0.1	42.6 (18 300)
Heavy fuel oil	72	—	0.25	0.939	87.6	11.07	0.9	0.36	0.02	43.7 (18 780)

Table 28.21 COAL TAR FUELS

Type	*CTF* 50	*CTF* 100	*CTF* 200	*CTF* 250	*CTF* 300	*CTF* 400	*Hard pitch*	*Crude tar*
Specific gravity	1.010	1.025	1.145	1.175	1.205	1.245	1.22	1.165
Calorific value, gross								
MJ kg^{-1}	39.66	39.43	38.77	38.59	38.49	37.63	37.22	37.68
(Btu lb^{-1})	(17 050)	(16 950)	(16 670)	(16 590)	(16 550)	(16 180)	(16 000)	(16 200)
Calorific value, net								
MJ kg^{-1}	38.03	37.91	37.45	37.29	37.24	36.49	36.15	36.49
(Btu lb^{-1})	(16 350)	(16 300)	(16 100)	(16 030)	(16 010)	(15 690)	(15 540)	(15 690)
Viscosity								
cSt (10^{-2} m^2 s^{-1})	13	24	300	35 000	25 000	—	—	~300
at temp °C	37.8	37.8	37.8	30	55			37.8
Analysis, %								
C	87.65	88.80	89.36	89.57	89.88	90.42	90.66	90.5
H	7.38	6.90	5.90	5.90	5.73	5.23	4.90	5.4
O (difference)	3.35	2.79	2.49	2.33	2.23	2.01	1.70	1.7
N	0.92	0.84	1.11	1.16	1.22	1.38	1.42	1.13
S	0.66	0.63	0.89	0.84	0.69	0.65	0.86	0.6
Ash	0.04	0.04	0.10	0.20	0.25	0.31	0.46	0.5

(A) CALORIFIC VALUE

In the absence of a bomb calorimeter determination, approximate values may be calculated for petroleum oils, or tars. However, the correlations are separate for each group.

gross calorific value $= 51.91 - 8.79d^2 - \{0.5191 - 0.0879d^2(\%H_2O + \%\ \text{ash} + \%S)\} + 0.0942(\%S)$ MJ kg^{-1}

or $22.320 - 3780d^2 - \{223 - 37.8d^2(\%H_2O + \%\ \text{ash} + \%S)\} + 40.5(\%S)$ Btu lb^{-1}

$\simeq 59.91 - 8.79d^2$ MJ kg^{-1}

or $22\,320 - 3780d^2$ Btu lb^{-1} (reference 17)

net calorific value $= 46.5 + 3.14d - 8.84d^2$ MJ kg^{-1}

or $20\,000 + 1350d - 3800d^2$ Btu lb^{-1} (reference 17)

where d is the specific gravity (relative density) at 15.6 °C (60 °F) for petroleum oils.

The net calorific values are about 2.8 MJ kg^{-1} (1200 Btu lb^{-1}) less than the gross values for distillate fuels down to about 2.3 MJ kg^{-1} (1000 Btu lb^{-1}), less for heavy fuel oils.

For tar fuels:

gross calorific value $= 0.337(\%\ C) + 1.44(\%\ H - \frac{1}{8}\%\ O) + 0.093(\%\ S)$ MJ kg^{-1}

or $145(\%\ C) + 620(\%\ H - \frac{1}{8}\%\ O) + 40(\%\ S)$ Btu lb^{-1} (reference 18)

net calorific value $= 0.75$ (gross CV $+ 10.9$) MJ kg^{-1}

or 0.75 (gross CV $+ 4700$) Btu lb^{-1} (reference 18)

(B) SPECIFIC GRAVITY CORRECTION COEFFICIENTS PER 1 °C

Table 28.22 CORRECTION COEFFICIENTS FOR PETROLEUM PRODUCTS[19]

Specific gravity 15.6/15.6 °C (60/60 °F)	*Correction coeff. per* 1 °C	*Specific gravity* 15.6/15.6 °C (60/60 °F)	*Correction coeff. per* 1 °C
0.605 0–0.613 3	0.001	0.742 5–0.753 7	0.000 79
0.613 4–0.621 9	0.000 99	0.753 8–0.764 9	0.000 77
0.622 0–0.632 0	0.000 97	0.765 0–0.776 0	0.000 76
0.632 1–0.641 9	0.000 95	0.776 1–0.786 9	0.000 74
0.642 0–0.653 0	0.000 94	0.787 0–0.798 8	0.000 72
0.653 1–0.664 9	0.000 92	0.798 9–0.812 4	0.000 70
0.665 0–0.677 5	0.000 90	0.812 5–0.828 3	0.000 68
0.677 6–0.689 9	0.000 88	0.828 4–0.859 9	0.000 67
0.690 0–0.702 5	0.000 86	0.860 0–0.925 0	0.000 65
0.702 6–0.716 6	0.000 85	0.925 1–1.024 9	0.000 63
0.716 7–0.730 0	0.000 83	1.025 0–1.074 9	0.000 61
0.730 1–0.742 4	0.000 81	1.075 0–1.124 9	0.000 59

Table 28.23 CORRECTION COEFFICIENTS FOR COAL TAR FUELS[18]

Type	*Specific gravity correction coeff. per* 1 °C
CTF 50+100	0.000 76
CTF 200	0.000 63
CTF 250	0.000 58
CTF 300	0.000 50
CTF 400	0.000 49

(C) SPECIFIC HEAT

For petroleum oils.[17]

$$\text{specific heat} = \frac{1.69 + 0.0034t}{\sqrt{d}} \text{kJ kg}^{-1} \text{K}^{-1}$$

+2% for naphthenic base crudes
−2% for paraffin base crudes

where

d = specific gravity 15.6/15.6 °C (60/60 °F)
t = temperature, °C

For coal tar fuels.[18]

Specific heat = 1.46 – 1.68 kJ kg^{-1} K^{-1}
(0.35–0.40 Btu lb^{-1} F°$^{-1}$)

(D) THERMAL CONDUCTIVITY

For petroleum oils:[17]

$$K = \frac{0.1184 - 0.0000195T}{d} \text{W m}^{-1} \text{K}^{-1}$$

$$\text{or } \frac{0.821 - 0.000244t}{d} \text{Btu (ft}^2 \text{ h)}^{-1} (\text{°F in}^{-1})^{-1}$$

where
T = temperature, °C
t = temperature, °F

For tar fuels,[18]

K = 0.138–0.147 W m^{-1} K^{-1}
or 0.96–1.02 Btu (ft^2 h)$^{-1}$ (°F in^{-1})$^{-1}$

(E) VISCOSITY

British fuel oil kinematic viscosities are quoted in centistokes (cSt) at a specified temperature (BS 4708: 1971), and, formerly, by seconds Redwood I at 100 °F (Redwood II is approximately 1/10 seconds Redwood I). Coal tar fuels are numbered according to the temperature in °F at which their viscosity is 100 seconds Redwood I, i.e. about 24 cSt at 37.8 °C. Suitable handling temperatures are quoted for each class of fuel.

Table 28.24 VISCOSITY OF LIQUID FUEL

Class	*Kinematic viscosity*		*Storage* °C	*Pumping* °C	*Atomizing temperature* °C
	Temperature °C (°F)	cSt(10^{-2}m^{2}s^{-1})			
Gas oil	37.7 (100)	4	—	—	Ambient
Light fuel oil	82.2 (180)	12.5	10	10	55
Medium fuel oil	82.2 (180)	30	25	30	90
Heavy fuel oil	82.2 (180)	70	35	45	125
CTF 50	37.8 (100)	13	—	—	15
CTF 100	37.8 (100)	24	35	35	38–50
CTF 200	37.8 (100)	300	35	40	80–95
CTF 250	30	35 000	55	75	115–130
CTF 300	55	25 000	85	100	140–155
CTF 400	—	—	135	150	190–205

(F) FLASH POINT

Special precautions are required by statute[20] for liquids with flash points below 32 °C (90 °F). Typical values for fuels are:

	Flash point °C (°F)
Petrol	−40 (−40)
Kerosine	+43 (+110)
All coal tar fuels	over 65 (150)
Gas oil	77 (170)
Light fuel oil	82 (180)
Medium fuel oil	93 (200)
Heavy fuel oil	115 (240)

28.3.2 Gaseous fuels

Table 28.25 shows the properties of most fuel gas constituents.

With the advent of both SI units and natural gas there have been a number of changes in both nomenclature and practice.[21]

Metric standard reference conditions for gas (*st*) The standard reference conditions for gas are now 15 °C and 1 013.25 mbar and *dry*, as defined by the International Gas Union. This is to be compared with the Imperial Standard conditions ISC (or more commonly STP) of 60 °F (15.6 °C), 30 inches Hg (equivalent to 1 013.75 mbar) usually applied to the gas *saturated with water vapour*, and Normal Temperature and Pressure NTP of 0 °C, 760 mm Hg (equivalent to 1 013.25 mbar) also referred to as STP in BS 350:1963.

Pure gaseous fuels are rarely used in metallurgical heating, but the performance of industrial fuels can be predicted from their properties. There is a considerable range in analysis possible with industrial fuels, and the ones given below are considered typical.

(A) BLAST FURNACE GAS

This is the byproduct of iron or ferromanganese production, obtained from the top of the furnace after cooling and suitable dust removal. Injection processes for steam, oil, coal, gas and oxygen tend to vary the typical analyses. Below 3.7 MJ m^{-3} (100 Btu ft^{-3}) calorific value the gas should be enriched[22] or pre-heated before combustion.

The specific gravity relative to air varies from 1 to 1.07.

Table 28.25 PROPERTIES OF CONSTITUENTS OF GASEOUS FUELS

			Gas density (*calc.*)				*Calorific value* Btu ft^{-3}	
Gas	*Formula*	*Specific gravity* (*air* = 1)	kg m^{-3} *NTP dry*	lb ft^{-3} *dry*	*STP sat.*	*Calorific value* MJ m^{-3} (st) *dry*	*STP sat. gross*	*STP sat. net*
Oxygen	O_2	1.104 4	1.428	0.084 57	0.083 93	—	—	—
Atmospheric nitrogen	N_2	0.972 3	1.257	0.074 46	0.074 00	—	—	—
Air	—	1.000 0	1.293	0.076 57	0.076 07	—	—	—
Carbon dioxide	CO_2	1.518 5	1.963	0.116 28	0.115 09	—	—	—
Carbon monoxide	CO	0.966 3	1.250	0.074 00	0.073.54	12.04	318	318
Hydrogen	H_2	0.069 58	0.090	0.005 33	0.006 07	12.12	320	270
Methane	CH_4	0.5533	0.715	0.042 37	0.042 47	37.68	995	895
Ethane	C_2H_6	1.037 1	1.341	0.079 41	0.078 87	65.52	1 730	1 580
Propane	C_3H_8	1.521 0	1.966	0.116 46	0.115 26	93.87	2 479	2 282
Butane	C_4H_{10}	1.935 8	2.503	0.148 23	0.146 48	117.23	3 095	2 848
Acetylene	C_2H_2	0.898 0	1.161	0.068 76	0.068 40	55.00	1 452	1 402
Ethylene	C_2H_4	0.967 5	1.251	0.074 09	0.073 63	59.08	1 560	1 460

Table 28.25 PROPERTIES OF CONSTITUENTS OF GASEOUS FUELS—*continued*

Gas	*Formula*	*Specific gravity (air = 1)*	*Gas density (calc.)* kg m^{-3} *NTP dry*	lb ft^{-3} *dry*	*STP sat.*	*Calorific value* MJ m^{-3} (st) *dry*	*Calorific value* Btu ft^{-3} *STP sat. gross*	*STP sat. net*
Propylene	C_3H_6	1.451 2	1.876	0.111 13	0.110 03	87.11	2 300	2 150
Butylene	C_4H_8	1.935 0	2.501	0.148 17	0.146 43	115.13	3 040	2 840
Benzene	C_6H_6	2.693 8	3.483	0.206 26	0.203 51	141.64	3 740	3 590
Water	H_2O	0.621 8	0.804	—	0.047 61	—	—	—
Hydrogen sulphide	H_2S	1.176 2	1.521	0.090 07	0.089 33	23.86	630	580
Sulphur dioxide	SO_2	2.211 5	2.860	0.169 33	0.167 21	—	—	—

Table 28.26 BLAST FURNACE GAS ANALYSES (MODERN FURNACES WITH OIL INJECTION)

% Analysis by volume CO	CO_2	H_2	N_2	*Calorific value STP sat.* MJ m^{-3}	Btu ft^{-3}
20–23	20–22	3–5	52–55	2.9–3.2	79–86

(B) LIQUEFIED PETROLEUM GAS, LPG

This type of industrial gas is a byproduct of the petroleum industry. These gases are transported in liquid form and vaporized before combustion. They may be used directly, mixed with other gases or distributed as a mixture with air.[23] The liquids have large coefficients of thermal expansion.

Table 28.27 TYPICAL RICH GASEOUS FUELS

	Natural gas	*LPG*[24, 25] *Propane*	*Butane*	*Refinery gas*[26] *Low*	*Medium*	*High*
		% Analysis by volume				
H_2	—	—	—	56	22	0.5
CH_4	86–90	—	—	12	22	22
C_2H_6	2.9–5.3	2	0.5	13	22	25
C_3H_8	0.5–1.3	87	10	13	18	36
C_4H_{10}	0.2–0.3	6	85	2	3	3
C_2H_4	—	2	0.5	1	3	2
C_3H_6	—	2	3	1	7	9
C_4H_8	—	1	1	1	2	2
N_2	1.2–6.8	—	—	1	1	0.5
Specific gravity (air = 1.0)	0.585–0.631	1.528	1.872	0.531	0.876	1.180
Calorific value,						
Gross						
MJ m^{-3} (st)	37.9–39.2	92.7	111.7	36.5	55.1	72.3
Btu ft^{-3}	1 016–1 051	2 486	2 995	977	1 478	1 938
MJ kg^{-1}	—	49.4	48.6	55.8	51.3	49.9
Net						
MJ m^{-3}	—	85.4	102.9	32.5	50.4	66.4
Btu ft^{-3}	—	2 289	2 758	872	1 352	1 780
MJ kg^{-1}	—	45.5	44.8	49.8	46.9	45.8
Liquid specific gravity	—	0.51	0.58	—	—	—
Latent heat of vaporization kJ kg^{-1}	—	426	391	—	—	—
Vapour pressure at 37.8 °C bar (a)	—	16.1	5.86	—	—	—
Liquid coefficient of Cubical expansion per °C		0.0016	0.0011			

(C) NATURAL GAS

Natural gas, supplied from the North Sea gas fields, is a rich hydrocarbon gas, predominantly methane, which is now superseding the previous town gas. Large users have been supplied on an interruptible basis often necessitating the use of dual-fuel burners.[27] Its limits of inflammability are narrow (*see* Table 28.30) and flame speed low when compared with town gas.

(D) PRODUCER GAS

Producer gas[28] is formed by partial oxidation of a solid fuel bed. The process is modified by the introduction of steam (blue water gas) and by thermal cracking of a hydrocarbon fuel such as natural gas, propane, butane, petroleum distillate, gas oil or heavy fuel oil (carburetted water gas).

Normally oxidation of the fuel bed is carried out with air, but total gasification processes also use oxygen. Tar in hot raw producer gas may increase the calorific value to around 74 MJ m^{-3} (200 Btu ft^{-3}).

Table 28.28 ANALYSES OF PRODUCER GAS

Representative of type	*Gas analysis (% volume)*					*Specific gravity (air = 1.00)*	$kJ\,m^{-3}$ *sat. at* 15.6 °C (Btu/SCF *sat.*)	
	CO	CO_2	H_2	CH_4	N_2		*Gross*	*Net*
Producer gas								
Coke, no steam	31.3	2.1	0.5	—	66.1	0.978	3 770 (101.2)	3 763 (101.0)
Coke, steam	29.3	5.6	12.4	—	52.7	0.889	4 952 (132.9)	4 720 (126.7)
Anthracite	24.0	7.5	16.5	1.2	50.8	0.858	5 253 (141.0)	4 900 (131.5)
Bituminous coal	27.0	4.5	14.0	3.0	51.5	0.857	5 980 (160.5)	5 707 (150.5)
Blue water gas								
Coke, steam	43.5	3.5	47.3	0.7	5.1	0.560	11 050 (296.7)	10 150 (272.4)
Fixed bed slagging gasifier	65.5	4.0	29.0	0.4	1.0	0.727	11 380 (305.5)	10 830 (290.6)

(E) REFINERY GAS

Refinery gas may be a byproduct of distillation, cracking or reforming of gas in the petroleum industry. Typical gases are shown in Table 28.27.

(F) TOWN GAS

Town gas used to be made almost exclusively by coal carbonization. Now it can be made of a complex mixture of gases,[29] including coke oven gas, natural gas, liquefied petroleum gas, water gas, refinery gas, reformed hydrocarbon fuels or hydrocarbon fuels diluted with air. To make gas safer the $CO+H_2O=CO_2+H_2$ shift reaction may be used to reduce the quantity of CO present.[30] CO_2 may be removed to increase calorific value and to reduce the specific gravity.

Gases are grouped according to the Wobbe index.

$$WI=\frac{\text{gross calorific value MJ m}^{-3}\text{ (st) dry}}{\sqrt{[\text{specific gravity (air}=1.0)]}}$$

or

$$\frac{\text{gross calorific value Btu ft}^{-3}\text{ STP (ISC) sat.}}{\sqrt{[\text{specific gravity (air}=1.0)]}}$$

Gas group	Wobbe index Btu ft^{-3} STP *sat.*	MJ m^{-3} (st) *dry*
G3	800 ± 40	30.4 ± 1.52
G4	730 ± 30	27.7 ± 1.14
G5	670 ± 30	25.4 ± 1.14
G6	615 ± 25	23.3 ± 0.95
G7	560 ± 30	21.3 ± 1.14
Natural gas	1 335 ± 5%	50.68 ± 5%

Table 28.29 shows the typical properties of some manufactured gaseous fuels.

Table 28.29 PROPERTIES OF SOME MANUFACTURED GASEOUS FUELS

	Horizontal retort	*Inter-mittent vertical chamber*	*Coke oven*	*Con-tinuous vertical retort*	*Low temp.*	*Car-buretted water gas*	*Butane air*	*CWG, CO_2, CO conv. butane enriched*
			% Analysis by volume					
CO	7.20	13.04	7.72	14.06	3.68	31.29	—	4.0
CO_2	1.72	3.01	2.39	3.49	4.89	4.59	—	21.0
H_2	56.07	50.24	48.02	55.99	16.58	36.48	—	49.7
CH_4	26.52	19.11	25.94	17.81	33.52	8.04	—	7.0
N_2	1.88	7.98	9.15	4.35	25.80	8.41	57.67	7.0
O_2	0.14	1.14	0.68	0.18	2.80	0.72	15.33	—
$C_nH_m^*$	4.0	3.10	4.34	2.45	3.04	8.75	—	8.6
	($C_{2.5}H_5$)	($C_{2.5}H_5$)	($C_{2.5}H_5$)	($C_{2.5}H_5$)	(C_4H_8)	($C_{2.5}H_5$)	—	($C_{2.5}H_5$)
C_nH_{2n+2}	2.47	2.38	1.76	1.67	9.69	1.72	27.0	2.7
	(C_2H_6)	(C_2H_6)	(C_2H_6)	($C_{2.5}H_7$)	($C_{2.5}H_7$)	(C_2H_6)	(C_4H_{10})	(C_4H_{10})
Calorific value, Gross								
MJ m^{-3}	21.8	18.4	20.5	18.0	25.9	18.4	31.1	18.2
Btu ft^{-3}	586	494	550	483	695	495	836	488
Net								
MJ m^{-3}	19.5	16.5	18.4	16.1	23.5	16.9	28.6	16.4
Btu ft^{-3}	523	442	493	432	630	455	769	439
Specific gravity	0.375	0.465	0.455	0.422	0.771	0.655	1.252	0.655
Wobbe index								
MJ m^{-3}	35.6	27	30.4	27.7	29.5	27.3	27.8	22.5
Btu ft^{-3}	956	724	816	743	792	734	747	603

* C_nH_m refers to the average analysis of the mixture of remaining hydrocarbons.

For a given piece of equipment the gases within a group as defined by the Wobbe index are interchangeable[31] without burner and nozzle adjustment. However, when changing gaseous fuels, blow-off and flame speed and flash-back must be carefully considered.

(G) ANALYSIS OF FUEL GASES[32]

The following constituents may also be found in fuel gases.

Dust is found in insufficiently cleaned blast furnace and producer gas.

Naphthalene in town gas should be below 23 mg m^{-3} (1 gr per 100 ft^3) in winter and 46 mg m^{-3} (2 gr per 100 ft^3) in summer.

Hydrogen sulphide should not be present in town gas.

7–20 g m^{-3} (300–900 gr per 100 ft^3) may be present in unpurified coal gas.

45 g m^{-3} (2000 gr per 100 ft^3) may be present in refinery gas.

Organic sulphur Up to 1 g m^{-3} (40 gr per 100 ft^3) may be present in town gas. This figure will be lowered by oil washing, benzole stripping, etc. Below 0.02% by weight sulphur is usually present in LPG. Natural gas contains little organic sulphur, 0.3–9.0 ppm v/v and a limit of about 30 mg m^{-3} will be set.

Condensable vapours are chiefly benzene and toluene in carbonization gas, the quantity depending chiefly on temperature of carbonization, volatile matter in coal, and degree of benzole removal.

Tar may be found in producer gas, made from coal.

Water vapour The majority of fuel gases contain water vapour, hence the usual determinations used to assume that the gases were saturated with water vapour at 30 in Hg total pressure (1013.75 mbar) and at 60 °F(15.6 °C). Natural gas properties are quoted on a dry basis (st).

(H) LIMITS OF INFLAMMABILITY

The limits of inflammability[33] of gaseous fuels and vapours are shown approximately in Table 28.30.

Table 28.30 LIMITS OF INFLAMMABILITY

	% By volume in air	
	Lower limit	*Upper limit*
Acetylene	2.5	80
Blast furnace gas	35	74
Butane	1.9	8.5
Carbon monoxide	12.5	74
Carburetted water gas	5.5	36
Coal gas	5.3	32
Coke oven gas	4.4	34
Ethane	3.0	12.5
Hydrogen	4.0	75
Methane	5.3	15
Petrol	1.4	7.6
Producer gas	17	70
Propane	2.2	9.5
Water gas	7.0	72
Commercial butane	1.9	8.5
Commerical propane	2.4	9.5

These limits are affected by direction of propagation, temperature, pressure and inert diluents.

(I) OXYGEN AND PRE-HEAT

Regeneration to about 1100 °C and recuperation to about 800 °C as a mean of pre-heating combustion air are valuable methods of waste heat recovery in large furnaces. The recuperative burner can be employed on smaller furnaces.[34] Lean fuels such as blast furnace and producer gas can be pre-heated to about 500 °C without cracking, to raise the adiabatic flame temperature of the fuel.

Oxygen is available from 200 to 1700 tonnes per day units (700 m^3 t^{-1} STP). In bulk, oxygen must be handled in oil- and grease-free equipment. Pipelines are required to be of stainless or special carbon steel, the oxygen velocity should not exceed 8 m s^{-1}. Non-ferrous valves, flow controllers and water-cooled oxygen-free copper lance tips are desirable for controlling the flow of oxygen in refining processes.

Tonnage oxygen for steelmaking is normally supplied with a purity of 99.5% O_2 for the following purposes:

(a) Oxidation refining processes.
(b) Raising flame temperature.
(c) To obtain extra output from a given furnace system by burning more fuel per unit combustion space.[35,36,37]

(d) Reducing metallic losses, eg. electric steelmaking.
(e) Reducing nitrogen pick-up by metal from combustion air.

Tonnage oxygen for copper smelting, carried out outside the UK, is normally supplied with a purity of 95% O_2.

Use of oxygen lowers the volume of waste gases and hence the heat loss from a furnace as compared with atmospheric air for combustion. Oxygen enrichment of combustion air has a similar effect to increasing its pre-heat temperature.

The adiabatic flame temperature is a guide to the maximum furnace temperature obtainable from a fuel. Its value is lowered by the following conditions.

(a) Dilution with inert gases such as N_2, CO_2.
(b) Dissociation of H_2O and CO_2, particularly over 2000 °C.
(c) Heat loss by radiation.

Table 28.31 FLAME TEMPERATURES OF SOME GASES

	Gross calorific value of fuel		*Flame temperature* °C	
	MJ m^{-3}(st) *dry*	Btu per SCF *sat.*	*In air maximum*	*In oxygen including dissociation*
Acetylene	55.1	1 452	2 325	3 200
Butane	117.5	3 095	1 895	—
Butylene	115.4	3 040	1 930	—
Carbon monoxide	12.07	318	1 950	—
Ethane	65.7	1 730	1 895	—
Ethylene	59.2	1 560	1 975	—
Hydrogen	12.15	320	2 045	2 200
Methane	37.8	995	1 880	—
Propane	94.1	2 479	1 925	2 500
Propylene	87.3	2 300	1 935	—
Coal gas	21.1	560	2 045	2 100
Coal gas	18.0	475	2 045	—
Blue water gas	11.2	295	2 080	—
Producer gas	6.3	165	1 800	—
Producer gas	4.8	128	1 690	—
Blast furnace gas	3.5	92	1 460	—

28.3.3 Fuel material and energy data for various metallurgical processes

The iron and steel industry has much published information,[38] particularly regarding large-scale processes. Non-ferrous processes are in general carried out on a much smaller scale, and comparisons are more difficult to make because of varying conditions of production.

(A) COKE PRODUCTION

Modern coke ovens are compound-fired by coke oven gas, and pre-heated blast furnace gas. Coke breeze is used in sinter production. A surplus of coal tar fuel and coke oven gas is available for other purposes.

Developments[40] to extend the range of coals for producing metallurgical coke or to reduce energy usage or pollution include: coal blending, pre-heat of the coal charge, stamping of the charge, programmed heating, and dry cooling of the hot coke.

Typical figures are shown in Table 28.32.

Table 28.32 ENERGY IN COKE PRODUCTION[40]

Dry coal carbonized	1.000 tonne
Blast furnace coke	690 kg
Coke breeze	60 kg
Oven underfiring fuel	2 500 MJ (24 therms)
Power required	49 MJ (13.8 kWh)
Steam required	620 MJ (300 kg)
Crude tar	43 kg
Crude benzole	11.7 kg
Sulphate of ammonia	8.5 kg
Coke oven gas	6 040 MJ (300 m^3)
Dry cooling of coke saves	1 300 MJ
Coal pre-heating saves	190–330 MJ
Programmed heating of coke ovens saves	10%

Formed-coke processes with calcining, in relation to the blast furnace, probably result in a 5% increase in total energy used per tonne of iron.[40]

(B) ORE PREPARATION

Mixes vary according to materials available and iron quality required but the national average in Table 28.33 will indicate the order of constituents used for sinter for ironmaking.

Table 28.33 AVERAGE MATERIALS USED IN SINTER MAKING

Materials	*Sinter* (kg $tonne^{-1}$)
Home iron ore	273
Imported iron ore	721
Imported manganese ore	2
Pyrites residue	10
Scale	41
Slag	2
Flue dust	4
Iron and steel scrap	2
Other Fe-bearing materials	45
Limestone and dolomite	100
Other burden material	16
Coke and coke breeze	82
Total	1 296
Coke oven and blast furnace gas igniter	
	170 MJ $tonne^{-1}$ (1.6 therm ton^{-1})
	45 MJ m^{-2} bed surface[39]
Power consumed	79 MJ $tonne^{-1}$ (22 kWh)

(C) IRONMAKING

Plants may use home ore, foreign ore or mixed ore; Table 28.34 gives national averages showing the general quantities of materials used.

Table 28.34 AVERAGE MATERIALS USED IN THE UK FOR IRONMAKING, kg TONNE^{-1}

Material	*Steelmaking*	*Foundry*	*Ferro-alloys*	*Total average*
Home iron ore	14	—	—	14
Imported iron ore	331	312	—	328
Manganese ore	6	31	1 775	21
Sinter	1 320	1 292	—	1 311
Limestone and dolomite	39	76	137	41
Scrap	6	53	42	9
Other materials	53	70	55	53
Total	1 771	1 834	2 009	1 777
Coke rate	576	702	1 525	587

The modern blast furnace has a high thermal efficiency. Improvements of the economics of the blast furnace ironmaking route are therefore directed towards reducing the capital investment, hence the trend to larger production units, and utilizing the cheapest sources of energy compatible with good operation.[36] Oxygen injection in the blast increases the production rate, a one percentage point increase in the blast oxygen analysis increases production by 6% with an ultimate limit[41] of 26% oxygen in the blast. Fuel injection lowers the coke rate with coke replacement ratios of 0.95 to 1.5. Fuel oil is used mostly, though gas[43] pulverized coal,[44] coal-oil slurries and coal tar are all proven candidates.[36] Details of operation of modern blast furnaces[42] are shown in Table 28.35.

Table 28.35 MODERN BLAST FURNACE OPERATION[42]

Dry coke kg t^{-1} hot metal	405–440
Blast Nm^3 t^{-1} hot metal	1 020–1 690
Oxygen enrichment Nm^3 t^{-1} hot metal	0–30
Oil injected kg t^{-1} hot metal	40–120
Moisture kg t^{-1} hot metal	12–35
Blast temperature, K	1 330–1 550
Tuyere pressure bar gauge	2–3.4
Top pressure bar gauge	0.3–2.2
Pig iron tonnes per day	2 500–9 900

Direct reduction processes use natural gas or coal to reduce iron ore to sponge iron, suitable for electric furnace melting. Reported energy requirements[45] are in the range 13.8–15.9 GJ tonne^{-1} for the reduction process. The total energy requirement of the direct reduction electric arc furnace route is similar to that of the blast furnace–basic oxygen furnace route (16–20 GJ t^{-1}).

(D) FERROUS CASTINGS

The cupola is largely used in foundries for melting cast iron. The main raw materials are pig iron, cast iron and steel scrap, limestone and coke.

Cold-blast cupolas are being modified to divided blast[46] in which the air is introduced at two levels giving savings of 30% on coke rate and increasing melting rate by 20%. Recuperative hot blast (300–760 °C) cupolas with long daily melting campaigns have total foundry energy requirements 30% less than cold blast or electric melting.[47] Oxygen enrichment[37] of the blast (2%) can benefit quality melting rate and temperature. Electric induction[48,49] melting accounts for 10% of metal melted. Some electric arc furnaces[50] are used for melting awkward shaped scrap. With cold blast cupolas, additions and super-heating in an electric induction furnace (duplexing) can give savings. The cokeless cupola[51] using propane oil or natural gas is of interest in countries with excess of high CV gas. The oil-fired rotary furnace can be economical in a small foundry.[52] Emission of pollutants from foundries will receive increasing attention.[47]

Reduction of metal recycled as foundry returns (Table 28.36) has great potential for reducing the total energy used.

Table 28.36 AVERAGE MATERIALS USED IN IRON FOUNDRIES, kg TONNE^{-1} OF CASTINGS

Scrap	*Pig iron*	*Foundry returns*	*Coke*	*Coal*
595	405	469	264	32

Table 28.37 ENERGY USED IN IRONMAKING PER TONNE OF MOLTEN METAL

Process	*Coke* kg	*Electricity* kWh	*Gas or oil*
Cold blast cupola[47]	~200		
Divided cold blast cupola[46,47]	~140		
Recuperative hot blast cupola[47]	~90		
Cokeless cupola[47,51]			
with gas			34 Nm^3 propane (3.2 GJ)
with oil			80 kg oil (3.4 GJ)
Electric coreless[48]		650	
Induction			
Holding		30	
Super-heating		60–75	
Electric arc[50]		550–625	
Duplexing		50–60	
Holding		50–150	
Rotary furnace[52]			125–216 kg oil
Oil–fired			(5.4–9.3 GJ)

Table 28.38 INDUCTION MELTING FURNACES[53]

	Coreless melting furnaces			
	Production rate		*Average furnace capacity*	
Rating kW	50 Hz kg h^{-1}	150 Hz kg h^{-1}	50 Hz kg	150 Hz kg
200	350	320	—	600
300	550	500	2 100	1 000
400	720	630	2 700	1 200
500	860	770	3 000	1 500
1 000	1 800	1 600	8 000	2 600
2 000	3 600	3 200	15 000	5 000
4 000	7 700	—	27 000	—

	Channel furnaces (50 Hz)		
	Production rate		
Rating kW	*Melting* kg h^{-1}	*Super-heating* kg h^{-1}	*Average capacity* kg
300	800	500	9 000
500	1 400	850	15 000
1 000	2 800	1 800	25 000
2 000	5 900	4 000	75 000
5 000	13 600	9 000	200 000

(E) NON-FERROUS MELTING

Non-ferrous industry usually operates on a smaller scale than the steel industry, hence specific fuel consumptions are higher than for steelmaking despite a lower required melt temperature. Gas-fired furnaces are described and assessed in reference 34.

Table 28.39 TYPICAL NON-FERROUS FUEL CONSUMPTION

Type of furnace	*Capacity* kg (lb)	*Alloy*	*Fuel* MJ tonne^{-1} (therm ton^{-1})
Crucible	50 (100)	Aluminium	23 000 (220)
Bale out	200 (400)	Aluminium	4 000–7 000 (40–70)
Reverberatory	600 (1 200)	Aluminium	2 600–4 000 (25–40)
Reverberatory	2 500 (2.5 ton)	Aluminium 1 tonne h^{-1}	3 200 (31)
Reverberatory	—	Bronze $2\frac{1}{2}$ tonne h^{-1}	3 700 (36)

Mains frequency induction melting furnaces can be used for non-ferrous melting.

Table 28.40 MAINS FREQUENCY INDUCTION MELTING[54]

Capacity kg (lb) *Rating* kW	150 (300) 25		1200 (2640) 120		4000 (8814) 300	
	kg h^{-1} (lb h^{-1})	MJ t^{-1} (kWh ton^{-1})	kg h^{-1} (lb h^{-1})	MJ t^{-1} (kWh ton^{-1})	kg h^{-1} (lb h^{-1})	MJ t^{-1} (kWh ton^{-1})
Copper	100 (220)	885 (250)	340 (750)	1 130 (320)	900 (1 984)	1 150 (325)
Brass	145 (320)	700 (200)	520 (1 150)	780 (220)	1 400 (3 086)	740 (210)
Aluminium	60 (132)	1 600 (450)	295 (650)	1 400 (400)	720 (1 587)	1 400 (400)
Zinc	180 (400)	500 (140)	885 (1 950)	460 (130)	2 200 (4 850)	440 (125)

(F) STEELMAKING[50,55,56,57,58]

The raw materials of steelmaking are principally basic iron, cold or molten, and scrap together with various additions as shown in Table 28.41 for national averages. Production from hot metal and scrap in oxygen converters and from scrap in electric arc furnaces now predominates.

Depending on the proportions of hot metal to scrap and compositions oxygen converters require 45–70 Nm^3 of oxygen per tonne metal.

Arc furnaces ranging in capacity from 10 to 200 tonnes have maximum power requirements at meltdown of 600 to 300 kW t^{-1} respectively.[50] The energy requirement is about 2.3 GJ t^{-1} (650 kWh t^{-1}) of which about 720 MJ t^{-1} (200 kWh t^{-1}) is provided by chemical reactions and combustion of carbon in the metal and electrode carbon, which use about 15 m^3 t^{-1} of oxygen. A reduction of the electrical power input by 400 MJ t^{-1} (110 kWh t^{-1}) can be achieved by preheating the scrap metal charge.

An open hearth furnace melting scrap uses 6.2 GJ tonne^{-1} (60 therms ton^{-1}); with 60% hot metal uses 3.7 GJ tonne^{-1} (36 therms ton^{-1}); with 70% hot metal and 40 m^3 tonne^{-1} of oxygen uses 2.4 GJ tonne^{-1} (23 therms ton^{-1}).

Table 28.41 AVERAGE MATERIAL CONSUMED kg TONNE^{-1} OF STEEL PRODUCED[38]

Process	*Iron*		*Scrap*		*Oxides*	*Finishings*	*Fluxes and fettling materials*	*Total*
	Molten	*Cold*	*Steel*	*Cast iron*				
Oxygen converters	851	15	265	4	12	11	102	1 260
Electric arc	—	56	1 027	33	5	29	57	1 207
Basic open hearth	426	104	532	50	62	5	82	1 261

The ancillary steelmaking energy requirements are as follows:

Lime burning	4700 MJ t^{-1} (45 therm ton^{-1})
Hot metal mixer (inactive)	155 MJ t^{-1} (1.5 therm ton^{-1})
Ladles, stopper rods	145 MJ t^{-1} (1.4 therm ton^{-1}) steel
Refractory repairs, heating up	145 MJ t^{-1} (1.4 therm ton^{-1}) steel
Waste heat boiler power	40 MJ t^{-1} (5 kWh per 1 000 lb) steam
Waste heat recovery	670 kg (1 500 lb) steam per tonne ingot
Cranes	14 MJ (4 kWh) per tonne ingot
Kaldo vessel rotation	7 MJ (2 kWh) per tonne ingot
Waste gas cleaning plant	25 MJ (7 kWh) per tonne ingot
Other steelworks auxiliaries	7 MJ (2 kWh) per tonne ingot
Oxygen production	2–2.5 MJ m^{-3} (16–20 kWh per 1 000 ft^3) oxygen
Arc furnace electrodes	5 kg per ingot tonne
Arc furnace electrode	4 kg per ingot tonne for large furnace using oxygen
Steam for liquid fuel	0.9 kg per litre fuel for pumping, heating, atomising

(G) CONTINUOUS CASTING

Power required, total	43 MJ t^{-1} (12 kWh ton^{-1})
Ladle and tundish heating	155 MJ t^{-1} (1.5 therm ton^{-1})
Yield of slab, bloom, billet	92–98% depending on ladle size

Continuous casting to produce semi-finished products can show considerable energy saving and increased yield compared with conventional, casting, soaking and primary mill practice.

(H) ENERGY REQUIREMENTS OF VARIOUS METAL HEATING PROCESSES[59,60,61,62,63,64]

Ingot soaking pits

Lowest possible	100 MJ t^{-1} (1 therm ton^{-1})
Good practice, hot metal	400 MJ t^{-1} (4 therm ton^{-1}) (3 h soaking)
Good practice, cold metal	2 000 MJ t^{-1} (20 therm ton^{-1}) (15 h soaking)
Electric, short track time	106 MJ (30 kWh) per tonne
Electric, large ingots	320 MJ (90 kWh) per tonne
Electric cold	1 200 MJ (330 kWh) per tonne
Auxiliary power required soaking pits	10 MJ (2.75 kWh) per tonne
Ingot stocking	4.3 MJ (1.2 kWh) per tonne
Scarfing	7 MJ (2 kWh) per tonne

Slab reheating furnaces

	Throughput t h^{-1}	*Nominal metal thickness* mm (in)	MJ t^{-1}	therm ton^{-1}
1 zone	20	8 (3)	1 500	(15)
2 zone	50	13 (5)	2 100–2 500	(20–24)
3 zone	100	20 (8)	2 400–3 300	(23–32)
5 zone	200	30 (12)	3 000–4 100	(29–40)

Hearth loading	350–1 000 kg m^{-2} h^{-1}
Power required	32 MJ t^{-1} (9 kWh ton^{-1})

Batch furnaces

	Hearth loading kg m^{-2} h^{-1} (lb ft^{-2} h^{-1})	*Temperature* °C	*Charge* t h^{-1}	MJ t^{-1} (therm ton^{-1})
Bogie furnace	245 (50)	—	12	2 100 (20)
Normalizing	200–300 (40–60)	1 000	—	1 700 (16)
Batch forging furnace	300–400 (60–80)	—	—	6 000–17 000 (60–160)
Tempering furnace	150–300 (30–40)	500	—	800 (8)
Annealing coils	—	750	11	650 (7)

Billet and bloom reheating furnaces

Continuous billet	1 400 MJ t^{-1} (14 therm ton^{-1})
Continuous bloom reheater	2 500 MJ t^{-1} (25 therm ton^{-1})

(I) ROLLING OF METAL

Table 28.42 ROLLING MILL POWER REQUIRED

	MJ t^{-1} (kWh ton^{-1})	
Slabbing mill	57 (16)	Ingot
Continuous roughing and billet mill	124 (35)	Bloom
Scarfing machine	7 (2)	Bloom
Plate mill	354 (100)	Plate
Roughing mill	57 (16)	Slab
Finishing mill	159 (45)	Strip
Mill auxiliaries	85 (24)	Slab
Descaling	39 (11)	Slab
Universal slab mill	57 (16)	Slab
Strip mill	248 (70)	Strip

Power consumptions for small mills for various sections are not quoted because of variability in reduction and other working conditions.

(J) FLUE GASES

Combustion of most fuels results in the formation of CO_2, SO_2, N_2 and H_2O. Excess air is shown by the presence of O_2 in low-temperature flue gas. Incomplete combustion is shown by the presence of CO and H_2. *See* references 65 and 66 for verification of the accuracy of the analysis of a flue gas and charts illustrating the effect of excess air and waste gas temperatures on the loss of heat up the stack.

(K) ENERGY CONSERVATION

There is much variety in processes and proportions and compositions of materials for the making of iron and steel and the contributions of all the sources of energy need to be calculated when assessing them.[67,68] With the rise in the real cost of energy, a determination of all the energy inputs[47] necessary for a process, i.e. including the energy inputs for capital plant transport materials, etc. gives a useful perspective for long-term planning.

National energy audits[40,47,69,70,71] have clearly shown that substantial reductions in energy usage can be made simply by good housekeeping: checking insulation standards;[72] analysing combustion products and regulating inputs to burners accordingly;[65,66] scheduling furnaces to keep holding times to a minimum; minimizing the recycle of metal, e.g. foundry returns.[47]

About 40% of the heat input to reheating furnaces leaves in the waste gases and can be used to pre-heat the combustion air in a recuperator or to raise steam. Older furnaces can be relined with ceramic fibre insulation and dampers fitted to control air inleakage.

The chemical and sensible heat of waste gases from oxygen converters and arc furnaces can be used to pre-heat scrap or raise steam.[73]

Analyses of the potential for energy conservation in the iron and steel industry are given in references 73 and 74, and in the aluminium and copper industries in references 75 and 76.

Pollutant emission control in an integrated works now uses 45 kWh $tonne^{-1}$ of steel and improved standards, particularly in cokemaking, will increase this to 119 kWh $tonne^{-1}$, which corresponds to 550 − 1200 kWh $tonne^{-1}$ of dust.[77] The total power used in non-electric steelmaking is about 390 kWh $tonne^{-1}$.

Reducing the number of processes in a production system is an effective energy saver.

REFERENCES

1. 'International Classification of Hard Coals by Type', Economic Commission for Europe, E/ECE/247, Geneva, 1956.
2. 'The Coal Classification System Used by the National Coal Board (Revision of 1964)'. The National Coal Board, London, 1964.
3. J. G. King, M. B. Maries and H. E. Crossley, *J. Soc. Chem. Ind.*, 1936, **55**, 2771.
4. R. L. Brown, R. L. Caldwell and F. Fereday, *Fuel*, 1952, **31**, 261.
5. 'Classification of Coals by Rank', Amer. Soc. for Testing Materials ANSI/ASTM D388–77, 1977.
6. S. Coles, *J. Soc. Chem. Ind.*, 1923, **42**, 4351.
7. W. Eisermann *et al.*, 'Estimating Thermodynamic Properties of Coal, Char, Tar and Ash'. *Fuel Process. Tech.*, 1980, **3**, 39.
8. R. A. Mott and R. V. Wheeler, 'The Quality of Coke', London, 1939.
9. H. J. Leyshon. 'The Requirements of Coke for Iron Foundries'. Coke Oven Managers Year Book 1975, p. 216.
10. Report of TS 47., *Brit. Foundryman*, 1959, **52**, 136.
11. J. Gibson, 'Recent Research and Development Work on Foundry Coke', *Brit. Foundryman*, 1973, **66**, 203.
12. E. W. Voice and J. M. Ridgion, 'Changes in Ironmaking Technology in relation to the Availability of Coking Coals'. *Iron and Steel Making*, 1974, **1**, 2.
13. E. Terres and A. Schaller, *Gas Wass.*, 1922, **65**, 761.
14. D. N. Gwyther, 'Coal Tar Fuels in the Open Hearth Furnace', Coke Oven Man. Assoc. Year Book, 1962.
15. Oil Fuels, BS 2869: 1970.
16. Coal Tar Liquid Fuels, BS 1469: 1962.
17. C. S. Cragoe, US Bureau of Standards Misc. Publication No 97, 1929.
18. W. H. Huxtable, 'Coal Tar Fuels' 2nd Edn, Assoc. of Tar Distillers, London, 1961.
19. 'I.P. Petroleum Measurement Tables', Am. Soc. Test. Mat.
20. 'The Highly Flammable Liquids and Liquefied Petroleum Gases Regulations,' 1972, Statutory Instrument No. 917, HMSO.
21. SI Units and conversion factors for use in the British Gas Industry, Gas Council and SBGI, 1971.
22. H. B. Lloyd, C. G. Miles and F. H. Dawes, 'The Use of Naphtha for Blast Furnace Gas Enrichment.' *J. Iron Steel Inst.*, 1966, **204**, 203.
23. I. Carter, 'LPG air Installations,' *J. Inst. Fuel*, 1968, **41**, 366.
24. Specifications for Commercial Propane and Commercial Butane, BS 4250: 1975.
25. A. F. Williams and W. L. Lom, 'Liquefied Petroleum Gases: Guide to Props. Applications and Usage'. Ellis Horwood, London, 1974.
26. J. Burns, 'The Romford Reforming Plant', *Trans. Instn. Gas Engrs.*, 1959, **108**, 1260.
27. 'Interim Code of Practice for Large Gas and Dual Fuel Burners, The Gas Council Report No 764/70, 1971.
28. W. R. Bulcraig, 'Components of Raw Producer Gas', *J. Inst. Fuel*, 1961, **34**, 280.
29. A. L. Roberts, J. H. Towler and B. H. Holland, 'The Hydrocarbon Content of Fuel Gases'. *Trans. Instn. Gas Engrs.*, 1957, **106**, 378.

30. W. B. S. Newling and J. D. F. Marsh, 'The Partial Removal of Oxides of Carbon from Fuel Gases', *Trans. Instn. Gas Engrs.*, 1963, **3**, 143.
31. J. A. Prigg and D. E. Rooke, 'Utilisation Problems and Their Relation to New Methods of Gas Production', *Trans. Instn. Gas Engrs.*, 1963, **3**, 85.
32. Analysis of Fuel Gases, BS 3156: 1959.
33. H. F. Coward and G. W. Jones, 'Limits of Flammability of Gases and Vapours', Bur. of Mines Bulletin No 503, 1952.
34. E. F. Winter, 'The Optimum Industrial Utilization of Gaseous Fuels', *J. Inst. Fuel*, 1978, **51**, 46.
35. T. C. Churcher, 'The Use of Oxygen in Combustion Processes', *J. Inst. Fuel*, 1960, **33**, 73.
36. H. G. Lunn and G. Waterhouse, 'Fuel Oil Injection into Blast Furnaces', *J. Inst. Fuel*, 1976, **49**, 70.
37. S. L. Rowland, 'Recent Experience in the Use of Oxygen in Cupola Operation', *Brit. Foundryman*, 1974, **67**, 187.
38. Iron and Steel Industry Annual Statistics, 1978, B.S.C. for Iron and Steel Statistics Bureau, 1979.
39. A. Bates, 'Modern Sinter Plant Practice', *J. Iron Steel Inst.*, 1970, **208**, 439.
40. P. B. Taylor and K. S. B. Rose, 'The Coke Making Industry', Energy Audit Series 9, Depts. of Energy and Ind. London, 1979.
41. T. Miyashita *et al.*, 'Limits of Oxygen Enrichment and Tuyere Fuel Injection', *Trans. ISIJ*, 1973, **13**.
42. J. G. Peacey and W. G. Davenport, 'The Iron Blast Furnace'. Pergamon, Oxford, 1979.
43. R. Wild, 'Fuel Injection into the Blast Furnace', *J. Iron Steel Inst.*, 1967, **205**, 245.
44. E. M. Summers, L. MacNaughton and J. R. Monson, 'Coal Injection into No. 5 Blast Furnace at Stanton and Staveley Ltd.', *J. Iron Steel Inst.*, 1963, **201**, 666.
45. E. J. Smith and K. P. Hass, 'Present and Future Position of Coal in Steel Tech.', *Ironmaking and Steelmaking*, 1976, **3**, 10.
46. H. J. Leyshon and M. J. Selby, 'Improved Cupola Performance by Correct Distribution of the Blast Supply Between Two Rows of Tuyeres', *Brit. Foundryman*, 1972, **65**, 43.
47. K. S. B. Rose, 'Iron Casting Industry', Energy Audit Series 1, Depts. of Energy and Ind., London, 1979.
48. P. A. Wilson, 'Comparison of Energy Consumption for Cupola, Arc, and Mains Frequency Induction Furnaces for Cast Iron'. *Brit. Foundryman*, 1975, **68**, 173.
49. F. W. Walker, 'Some Experiences in the Use of Vertical Channel Induction Furnaces for Melting Grey Cast Iron'. *Brit. Foundryman*, 1975, **68**, 209.
50. P. A. Wilson, 'Iron Foundry Melting—The Modern Arc Furnace System', *Brit. Foundryman*, 1974, **67**, 250.
51. R. T. Taft, 'The First Twelve Months Operation of a Totally Gas-Fired Cupola', *Brit. Foundryman*, 1972, **65**, 321.
52. W. J. Roscrow, 'Melting and Superheating Iron in Rotary Furnaces'. *Brit. Foundryman*, 1976, **69**, 81.
53. M. J. Severs, 'Electric Melting in Iron Foundries', *Brit. Foundryman*, 1970, **63**, 15.
54. T. R. Brown, 'Mains Frequency Induction Melting Furnaces', *Fndry. Tr. J.*, Nov. 1962, 661.
55. 'Energy Management in Iron and Steel Works', Iron and Steel Inst. pub. P105, 1968.
56. C. R. Hall, 'Four Years of Steelmaking in 200 ton L.D. Furnaces', *J. Iron Steel Inst.*, 1968, **206**, 113.
57. R. S. Howes, 'Electric Steelmaking at Templeborough', *J. Iron Steel Inst.*, 1968, **206**, 205.
58. K. Wakabayashi, 'The Flexibility of L.D. Operation', *Iron Making and Steel Making*, 1976, **3**, 252.
59. K. Donnan and J. J. Blackmore, 'Fuel and Energy Required for Rolling Mill and Reheating Furnaces', *J. Inst. Fuel*, 1962, **35**, 360.
60. 'Plate Mill Furnaces', *J. Iron Steel Inst.*, 1960, **194**, 365.
61. 'Recent Developments in Annealing', *Iron and Steel Inst.*, Special Report No. 79, London, 1963.
62. J. R. Pattison, 'Continuous Reheating Furnaces in The Steel Industry', *J. Inst. Fuel*, 1968, **41**, 345.
63. D. J. Smithson and A. T. Sheridan, 'Energy Use in Mill Areas'. *Ironmaking and Steelmaking*, 1975, **2**, 286.
64. N. Hopkins and K. C. Gandhi, 'Energy Saving as Applied to Large Multizone Pusher and Underfired Walking Beam Furnaces'. *Ironmaking and Steelmaking*, 1975, **2**, 295.
65. J. W. Rose and J. R. Cooper, 'Technical Data on Fuel', 7th Edn, London, 1977.
66. T. F. Hurley and G. R. Stern, 'The NIFES Chart', *J. Inst. Fuel*, 1961, **34**, 393.
67. N. A. Robins, 'Theoretical Energy Requirements for Ironmaking'. *Iron and Steelmaker*, 1976, **2**, 39.
68. H. A. Fine and G. H. Geiger, 'Handbook on Material and Energy Balance Calculations in Metallurgical Processes'. Publ. Met. Soc. AIME, Warrendale, Penn., 1979.
69. 'Energy Use in Copper Sector of the Non-ferrous metals Industry'. IETS No. 11, Dept. of Industry, London, 1979.
70. 'Energy Use in the Lead, Zinc and Other Base Metals Sector', IETS No 12., Dept. of Industry, London, 1979.
71. 'The Zinc and Lead Industries', Energy Audit Series, No. 10. Depts. of Energy and Industry, London, 1980.
72. 'The Economic Thickness of Insulation for hot Pipes', Fuel Efficiency Booklet No 8, Dept. of Energy, London, 1977.
73. W. Montgomery, 'Energy: Availability Conservation and Use in the Steel Industry', *Ironmaking and Steelmaking*, 1975, **2**, 231.
74. K. J. Irvine, 'The Energy Challenge in the Steel Industry'. *Brit. Foundryman*, 1978, **71**, 233.
75. 'The Aluminium Industry', Energy Audit Series No. 6. Depts. of Energy and Industry, London, 1979.
76. P. F. Chapman, 'Energy Conservation and Recycling of Copper and Aluminium', *Metals and Materials*, 1974, **8**, 311.
77. J. A. Harrop and D. J. Smithson, 'Energy Requirements for Pollution Control in the British Iron and Steel Industry'. *Ironmaking and Steelmaking*, 1980, **7**, 196.

29 Heat treatment environments

Controlled atmospheres are used in the heat treatment of metals to ensure that the treated material has the desired surface characteristics.[1,2,3,18] The latter may call for cementation by carbon or nitrogen, decarburization, surface etching or freedom from oxide and other types of surface films, and necessitates control over reactions which take place between the metal charge and the furnace gases, or between the gases themselves. By passing into the furnace chamber a gas of controlled composition to displace the natural atmosphere it is possible either to prevent or to promote reactions at the metal surface, and to eliminate subsequent cleaning processes.

Cementation by carbon and nitrogen and freedom from oxidation may also be achieved by the use of suitable molten salt mixtures as an alternative to controlled atmospheres.

29.1 Chemistry of controlled atmosphere processes

The reactions which take place within the furnace chamber may be classified broadly into three groups:

Group I metal + gas A = metal compound + gas B

e.g. $Fe + H_2O = FeO + H_2$

$Cu + H_2S = CuS + H_2$

Group II alloy_{xy} + gas A = alloy_x + gas B

e.g. C(in Fe) + CO_2 = Fe + 2CO

O(in Cu) + H_2 = Cu + H_2O

Group III gas A + gas B = gas C + gas D (or solid D)

e.g. $CO_2 + H_2 = CO + H_2O$

$CH_4 = 2H_2 + C$

The composition of these systems at equilibrium is determined for the reaction $A + B \rightleftharpoons C + D$ by the equation:

$$\frac{[C][D]}{[A][B]} = K_T$$

where $[A]$, etc. are the activities, usually expressed as concentrations of reactants, and K_T the equilibrium constant at temperature T.

The effect of temperature upon the equilibrium constant is given by the equation*

$$\log_{10} K_T = -\frac{\Delta G}{4.576T}$$

ΔG is the free energy of the reaction in calories at TK. Values of ΔG in joules for various reactions are given in Chapter 8, Table 8.8a-j and for a wider range of reactions, equations of the form

$$\Delta G_T = A + BT\log_{10} T + CT$$

have been tabulated.[15]

* For ΔG in joules, equation becomes $\log_{10} K_T = -\frac{\Delta G}{19.146T}$

29.1.1 Group I

The stability of metal oxides at heat-treating temperatures is an important factor in controlled atmosphere processes. Dissociation pressures of the common metallic oxides are very low at normal heat treating temperatures (Figure 29.1) and deoxidation by treatment in an inert gas or *in vacuo* is not practicable. In the presence of a reducing gas such as hydrogen, oxides may be reduced; this forms the basis of many industrial processes such as the manufacture of certain metal and alloy powders. Where the dissociation pressure is high, the equilibrium concentration of oxidizing gas (e.g. H_2O) in relation to the reducing gas (e.g. H_2) may be quite high, while for oxides of low dissociation pressure, small traces of oxygen-bearing gases may be sufficient to promote their formation (Table 29.1).

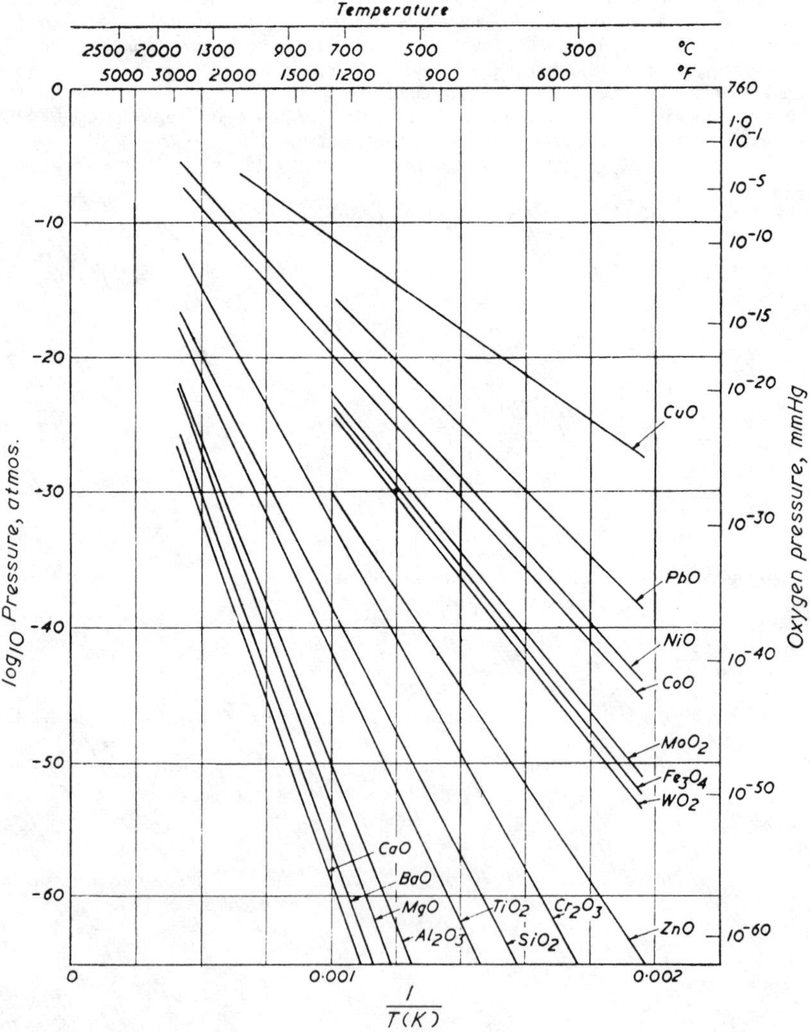

Figure 29.1 *Dissociation pressures of the common metallic oxides at normal heat-treating temperatures*

Table 29.1 OXIDATION OF METALS BY WATER VAPOUR AND BY CARBON DIOXIDE

	For oxidation at 727 °C	
Material	*% water vapour in hydrogen*	*% carbon dioxide in carbon monoxide*
Copper	>99.999	>99.999 9
Lead	99.982	99.989
Nickel	99.63	99.94
Iron	35.5	44.4
Carbon	24.4	31.0
Zinc	0.084	0.152
Chromium	0.001	0.002

29.1.2 Group II

In this type of reaction, the solid reactant is in solution in the parent metal and any excess of the constituent present as a separate phase in the alloy only enters into the reaction after solution in the matrix, e.g. carbon in iron or oxygen in copper.

$$C(\text{in}Fe) + CO_2 \rightleftharpoons Fe + 2CO$$

$$K = \frac{P^2_{CO}}{P_C \times P_{CO_2}}$$

where P_C is dependent upon the concentration of carbon in solution in the iron, and this in turn determines the equilibrium ratio of $P^2_{CO_2}/P_{CO_2}$ (Figure 29.4, p. 29–10). Reactions of this type are important in the heat treatment of carbon steels, where decarburization must be prevented, and in the annealing of tough pitch copper; they form the basis of industrial processes for the carburization and nitriding of steels, and the decarburization of white cast iron in the production of whiteheart malleable.

29.1.3 Group III

This group includes reactions which take place between constituents of the controlled atmosphere, a typical example being the water gas reaction (Table 29.5):

$$CO_2 + H_2 \rightleftharpoons CO + H_2O$$

Iron and nickel are both very active catalysts for the reaction, especially at their normal heat-treating temperatures, and the catalytic activity of metals for this and similar reactions may cause marked roughening of the metal surfaces. Other reactions included in this group are the 'cracking' of hydrocarbon gases and the breakdown of carbon monoxide, both of which may result in sooting. To prevent this the hydrocarbon (or carbon monoxide) is maintained below its equilibrium concentration for the reaction at the temperature of treatment (Table 29.5).

29.2 Types of controlled atmospheres

Controlled atmospheres can be classified according to the source from which they are derived, and the composition of those most widely used is given in Table 29.2a. The fields of application of the various atmospheres overlap to a large extent, and the choice is often determined by economic considerations.

Table 29.2a INDUSTRIAL CONTROLLED ATMOSPHERES

		Approximate composition (volume%)						*Approx. relative cost per unit vol.*		
Atmosphere	*Air/gas ratio*	CO_2	CO	H_2	CH_4	N_2	*Dew point*		*Characteristics*	*Main applications*
Wet hydrogen	—	—	—	97.5–98.0 H_2 2.5–5.0 H_2O	—	—	—	250	Strongly reducing and decarburizing combustible	Decarburizing electrical sheet steel
Dry purified hydrogen	—	—	—	100	—	—	−50 °C	250	Strongly reducing and explosive with air	High-temperature brazing of stainless steel and heat-resisting alloys
Cracked ammonia	—	—	—	75	—	25	−40 °C	100	Strongly reducing and explosive with air	Bright annealing stainless and high carbon steels. Clean annealing brass
Burnt ammonia	2.5/1 to 3.75/1	—	—	0–25	—	100–75	Room* temp. to −50 °C	45	Reducing. Just combustible	Bright annealing carbon steels and non-ferrous metals. Clean annealing brass. Bright hardening on short time heating cycles
Nitrogen from ammonia	3.75/1	—	—	—	—	100	Room* temp. to −50 °C	45	Inert and non-combustible	Bright annealing and bright and clean hardening carbon steels. Annealing silicon steel
Hydrogen from cracked ammonia separated by Ag–Pd membrane	—	—	—	100	—	—	−40 °C	150	Strongly reducing and decarburizing	Heat treatment of pure metals at elevated temperature
Endothermic gas	2.2/1† (for natural gas)	0.25/0.4	20	40	0.5	40	−5 to +5 °C	35	Strongly reducing, toxic and combustible	Bright and clean hardening. Brazing carbon and alloy steels. Carburizing
Partially burnt city gas (natural gas)	6.5/1	4.0	9.5	15	—	71.5	Room* temp. to −50 °C	20	Reducing. Just combustible and toxic	Bright annealing mild steel. Bright brazing and soldering mild steel. Malleablizing of cast iron
Completely burnt city gas (natural gas)	4.5/1† 9/1	10.0	0.5–1.0	0.5–1.0	0.0	87–89	Room* temp. to −50 °C	15	Slightly reducing. Non-combustible	Bright annealing copper and high copper brasses. Copper brazing
Nitrogen from city gas, completely burnt and scrubbed of CO_2 and H_2O	4.5/1†	0.0	0.5–1.0	0.5–1.0	0.0	98–99	−50 °C	12	Slightly reducing. Non-combustible	Bright annealing carbon steels and low alloy steels. Bright and clean hardening carbon steels

* Dew point of wet gases may be reduced if required from room temperature to less than −50 °C by means of suitable drying agents.
‡ Determined by the composition of the gas.

Table 29.2a INDUSTRIAL CONTROLLED ATMOSPHERES—*continued*

Atmosphere	*Air/gas ratio*	*Approximate composition (volume%)* CO_2	CO	H_2	CH_4	N_2	*Dew point*	*Approx. relative cost per unit vol.*	*Character-istics*	*Main applications*
Completely burnt city gas, treated over hot charcoal	4.5/1†	0.0	20.0	2.0	0.0	78.0	Less than −20 °C	10	Just combustible. Reducing and toxic	Bright annealing high carbon steels. Bright and clean hardening carbon steels. Diluent for carburizing atmospheres
Charcoal gas. High temperature external heating	—	—	34.0	—	—	66.0	Less tan 20 °C*	12	Reducing, combustible and toxic	Bright annealing carbon steels. Bright and clean hardening carbon steels. Diluent for carburizing atmospheres
Organic liquids	—	—	—	—	—	—	—	60	—	For carburizing and carbo-nitriding steels
Vacuum	—	—	—	—	—	—	—	(Not comparable)	—	High temperature brazing and heat treatment of reactive metals such as titanium and zirconium; sintering of carbides and high temperature alloys; bright annealing of ferrous and non-ferrous alloys
Byproduct nitrogen from liquid oxygen plant; 0.2% O_2 removed over catalyst with added H_2	—	—	Few p.p.m.	2–5	—	95–98	−40 °C	—	Slightly reducing, otherwise inert and non-explosive	Bright annealing of steel
Burnt city gas scrubbed of CO_2. reformed with steam, scrubbed of CO_2 and H_2O	—	0.1	0.5	2–30	—	Remainder	−20 °C	—	Reducing combustible if H_2 concentration high	Annealing of tin plate

* Dew point of wet gases may be reduced if required from room temperature to less than −50°C by means of suitable drying agents.
† Determined by the composition of the gas.

Table 29.2b

High purity bottled gases	*Dew point*	*Cost per unit volume*	*Characteristics*	*Main applications*
Nitrogen	⟨−70 °C	300	Inert	Clean annealing; hardening
Hydrogen	⟨−70 °C	300	Highly reducing and combustible	Hardening; annealing and brazing pure metals
Argon	⟨−70 °C	3 000	Inert	Titanium treatment
Ammonia	−40 °C	70	Alkaline to toxic	Nitriding; carbonitriding

29.2.1 Ammonia derivatives

Ammonia is the source of cracked and burnt ammonia, and is also a source of hydrogen. Cracked ammonia is produced by the catalytic dissociation of synthetic anhydrous ammonia over a catalyst of nickel maintained at a temperature of about 850 °C. The dissociated gas consists of 75% hydrogen and 25% nitrogen, with less than 0.5% free ammonia. The cracked gas is as dry and as pure as the anhydrous ammonia from which it is obtained and is suitable for the bright heat treatment of metals forming oxides of low dissociation pressures (e.g. stainless steels). Burnt ammonia is the product of reaction of ammonia with air over a catalyst to produce moist nitrogen having an oxygen content below 2 p.p.m. and containing hydrogen in the range of 0.5–25%. Dew point is approximately 5 °C but this may be reduced by drying to −40 to −70 °C. A regenerative system has been developed whereby the burnt ammonia is returned from the furnace to the combustion chamber of the generator, where any free oxygen is neutralized by the necessary amount of fresh cracked ammonia, and the hydrogen level is maintained in the atmosphere returned to the furnace. Atmosphere costs may thereby be reduced appreciably. Hydrogen is produced by passing cracked ammonia under pressure over heated Ag–Pd membranes. The hydrogen separates by diffusion through the membrane and is obtained in a state of extreme purity.

29.2.2 Hydrocarbon gases

The principal feedstock gases for the production of the majority of controlled atmospheres are natural gas and liquified petroleum gases (propane and butane). They are unsatisfactory as furnace atmospheres by themselves mainly due to formation of soot at elevated temperatures; but by the use of controlled combustion in suitable generators, a wide range of atmospheres may be produced containing oxides of carbon, hydrogen, methane, nitrogen and water vapour. In general, sulphur content is very low and is not normally a problem. In special cases, however, it may be removed if necessary. Town's gas, or coal gas, is now only rarely used as a source of controlled atmospheres.

Generators are of two types, depending upon whether the reactions between air and the hydrocarbon gas are predominantly exothermic or endothermic. The former covers the combustible range of rich air/gas mixtures from complete combustion to partial combustion at the practical limit of combustibility of the mixture (approximately 6.5/1 for natural gas).

The generator is of simple construction comprising means of metering accurately the volumes of air and gas, a combustion chamber heated solely by the mixture, a cooler and condensing unit. Means of removal of CO_2 and water vapour are provided where this is beneficial. Occasionally air/kerosene mixtures may be used in specially adapted generators, but only the lean end of the range is used to avoid soot deposition.

The endothermic type generator caters for air/gas mixtures which are not combustible and provides an atmosphere rich in hydrogen and carbon monoxide according to the following simplified equation.

$$2CH_4 + O_2 + 4N_2 = 2CO + 4H_2 + 4N_2$$

An externally heated retort filled with a suitable catalyst is utilized as a reaction chamber for the air/gas mixture. The ratio of the latter and the temperature of the catalyst bed influence the composition of the endogas, but in practice both are controlled to give a near as possible the ideal CO and hydrogen contents. Residual methane, CO_2 and H_2O contents indicate the quality of the product gas as a basis for the various ferrous treatments. Typical composition figures for exothermic and endothermic atmospheres are given in Figure 29.2.

For some processes, e.g. carburizing, bright annealing of copper etc., the air/gas ratio may be controlled from a gas monitor which analyses continuously for CO, CO_2, H_2O or O_2 depending upon the particular process. Industrially, more attention has so far been paid to the control of atmospheres for carburizing.[19]

29.2.3 Charcoal derivatives

The combustion of pure carbon with dry air produces carbon monoxide and carbon dioxide, diluted with nitrogen (Table 29.3), the relative concentrations being determined in the main by the temperature of the carbon. Industrial atmospheres of this type are produced by the combustion of wood charcoal with air, and usually contain a measurable concentration of hydrogen arising from volatiles in the charcoal and reaction between the latter and absorbed water vapour. Control of

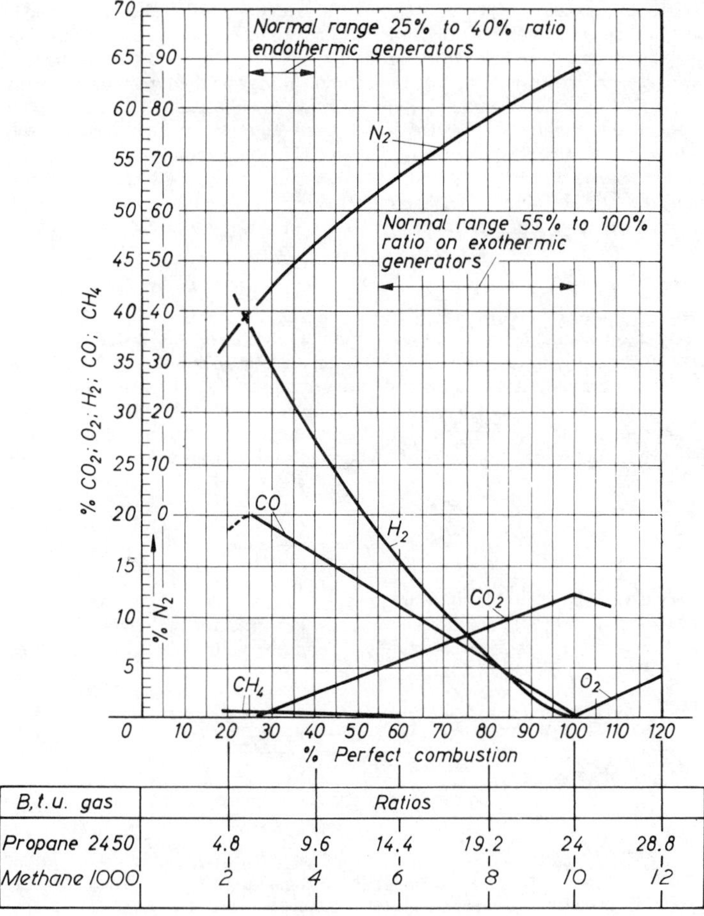

Figure 29.2 *Controlled atmosphere chart. The above percentages of atmosphere constituents are average. Actual analyses will vary slightly according to generator design.*

Table 29.3 COMPOSITION OF ATMOSPHERES DERIVED FROM COMBUSTION OF PURE CARBON WITH DRY AIR (THEORETICAL)

	Composition (volume %)		
Temperature (°C)	CO_2	CO	N_2
500	19.4	2.7	77.9
550	17.5	5.7	76.8
600	14.5	10.7	74.8
650	9.8	18.5	71.7
700	6.1	24.7	69.2
750	3.0	29.8	67.2
800	1.07	32.9	66.0
850	0.66	33.6	65.7
900	0.30	34.3	65.4
950	0.16	34.4	65.4
1 000	0.085	34.5	65.4
1 050	0.025	34.6	65.4
1 100	0.016	34.6	65.4

composition is facilitated by passage of the gases through a secondary bed of heated charcoal, usually superimposed on the main combustion zone. As in producer gas practice, steam or air/steam mixtures may be used in place of air, with a resulting increase in the hydrogen content of the generated atmosphere at the expense of the nitrogen. Charcoal may be used for the processing of atmospheres generated from other sources, to reduce the carbon dioxide and water vapour concentrations to any desired equilibrium level, as in the heat treatment of high carbon steels where decarburization by these gases must be prevented. The temperature of the charcoal determines the degree of conversion of the decarburants. For similar reasons, charcoal may be used as the high-temperature catalyst in the endothermic type generator for the treatment of rich hydrocarbon gas/air mixtures. In practice, charcoal is very little used nowadays.

29.2.4 Nitrogen

High purity nitrogen is readily available for use in heat treatment, both as a gas in bottles and as liquified gas in tanks. Consequently its use in many applications is increasing. It may be used for annealing of many ferrous and non-ferrous metals and alloys either by itself or with additions of hydrogen or hydrocarbon, e.g. nitrogen plus hydrogen for annealing brass, nickel silver; nitrogen plus natural gas for annealing and hardening of medium and high carbon steels; nitrogen and carbon dioxide plus hydrocarbon or alcohol for carburizing;[20, 21] and nitrogen plus ammonia for nitro-carburizing[22] of steels.

29.3 Purification

For certain applications it may be desirable to remove one or more of the gaseous constituents such as carbon dioxide, water vapour and sulphur compounds. Processes for the removal of these constituents from complex gas mixtures are well-established and are summarized in Table 29.4.

Table 29.4 INDUSTRIAL PROCESSES FOR THE PURIFICATION OF CONTROLLED ATMOSPHERES

Gas	*Absorbent*	*Remarks*
Carbon dioxide	Treatment over hot charcoal	Used only for atmospheres already containing CO. Degree of conversion of CO_2 dependent on temperature
	Organic reagents, e.g. solutions of ethanolamine, tetramine, alkacids, etc.	Complete absorption. Solution regenerated
	Under pressure in water	Partial absorption
	Molecular sieves	Absorption to 0.1%
Hydrogen sulphide	Moist bog iron ore, at 50 °C	Complete absorption
	Activated carbon	Complete absorption
	Organic reagents as for CO_2	Complete absorption. Solution regenerated
Sulphur dioxide	Soda ash or lime	Complete absorption
	Activated carbon	Complete absorption
	Organic reagents as for CO_2	Complete absorption. Regeneration difficult
Organic sulphur compounds	Catalytic conversion to SO_2 or H_2S and absorption of these gases	Very small residual equilibrium concentration of organic sulphur compounds
	Activated carbon	Complete absorption
	Organic reagents as for CO_2	Complete absorption. Regeneration difficult
Water vapour	Compression	Partial dehydration only, to about 0 °C
	Refrigeration	Partial dehydration only, to about 5 °C
	Treatment over hot charcoal	As for CO_2 (above)
	Silica gel	Dehydration to about −40 °C
	Activated alumina	Dehydration to about −60 °C
	Molecular sieves	Dehydration to about −90 °C

29.4 Applications to ferrous alloys

LOW CARBON STEEL

The primary reactions to be considered during heat treatment are as follows:

(1) $Fe+H_2O \rightleftharpoons FeO+H_2$ $\qquad K_1=\dfrac{P_{H2}}{P_{H2O}}$

(2) $Fe+CO_2 \rightleftharpoons FeO+CO$ $\qquad K_2=\dfrac{P_{CO}}{P_{CO2}}$

(3) $C(\text{in iron})+2H_2 \rightleftharpoons Fe+CH_4$ $\qquad K_3=\dfrac{P_{CH4}}{P_C \times P_{H_2}^2}$

(4) $C(\text{in iron})+CO_2 \rightleftharpoons Fe+2CO$ $\qquad K_4=\dfrac{P_{CO}^2}{P_C \times P_{CO_2}}$

(5) $C(\text{in iron})+H_2O \rightleftharpoons Fe+H_2+CO$ $\qquad K_5=\dfrac{P_{H2} \times P_{CO}}{P_C \times P_{H2O}}$

(6) $CH_4 \rightleftharpoons 2H_2+C(\text{soot})$ $\qquad K_6=\dfrac{P_{H2}^2}{P_{CH4}}$

(7) $C(\text{soot})+CO_2 \rightleftharpoons 2CO$ $\qquad K_7=\dfrac{P_{CO}^2}{P_{CO2}}$

(8) $CO+H_2O \rightleftharpoons CO_2+H_2$ $\qquad K_8=\dfrac{P_{CO2} \times P_{H2}}{P_{CO} \times P_{H2O}}$

Equilibrium data for reactions 1, 2, 6 to 8 are given in Table 29.5 and for reactions 3 to 5 in Figures 29.3, 29.4 and 29.5. Partially burnt natural gas, propane or butane and, to a lesser extent, burnt ammonia, are the controlled atmospheres most widely used in the heat treatment of steel, including bright normalizing, bright annealing and the descaling of hot rolled strip, the latter usually being combined with a normalizing treatment. In the case of burnt ammonia the reactions

Table 29.5 EQUILIBRIUM CONSTANTS FOR THE ABOVE REACTIONS

t°C	K_1	K_2	K_6	K_7	K_8
400	4.20*	0.33*	0.0707	9.4×10^{-5}	12.7
450	3.52*	0.44*	0.200	6.2×10^{-4}	8.04
500	3.02*	0.56*	0.494	3.9×10^{-3}	5.36
550	2.64*	0.70*	1.09	0.019	3.80
600	2.35	0.84	2.21	0.080	2.79
650	2.11	1.00	4.14	0.285	2.11
700	1.92	1.16	7.27	0.891	1.65
750	1.78	1.33	—	2.49	1.34
800	1.67	1.50	—	6.34	1.12
850	1.58	1.68	—	14.82	0.946
900	1.51	1.85	—	32.25	0.814
950	1.44	2.03	—	65.85	0.710
1000	1.38	2.22	—	127.7	0.622
1050	1.33	2.39	—	233.4	0.556
1100	1.28	2.75	—	410.2	0.499
1150	1.24	2.75	—	692.8	0.451
1200	1.20	2.93	—	1129	0.410
1250	1.17	3.11	—	1784	0.376
1300	1.14	3.29	—	2745	0.346

* Stable oxide at this temperature is Fe_3O_4 not FeO; the equilibrium constants so marked are not as accurate as the others.

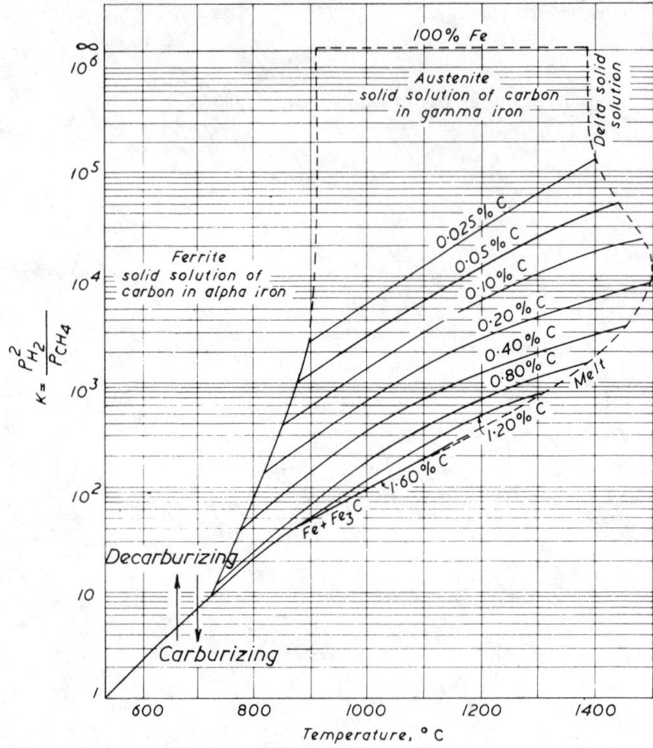

Figure 29.3 *Equilibrium data for the reaction* $Fe+CH_4 \rightleftharpoons C$ *(in iron)*$+2H_2$

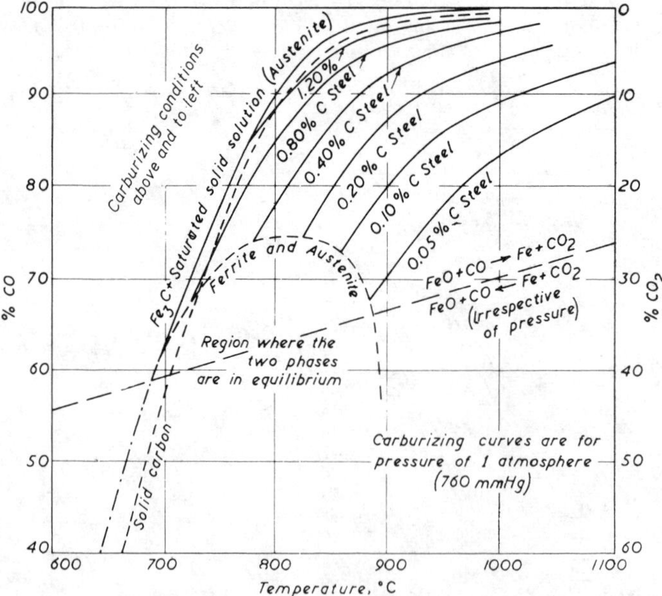

Figure 29.4 *Equilibrium data for the reaction* $Fe+2CO \rightleftharpoons C$ *(in iron)*$+CO_2$. (*After M. I. Becker, Journal I & SI*, 1930, **1**, 354).

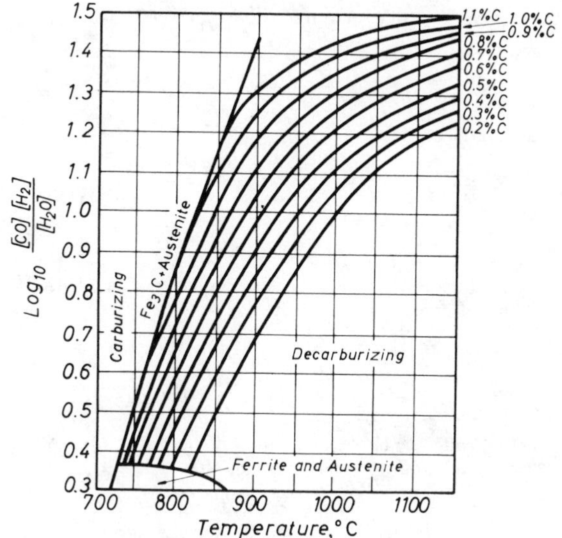

Figure 29.5 *Moisture, carbon content and temperature*[16]

likely to be involved are 1, 3 and 5; in the case of partially burnt hydrocarbon gas all the reactions 1 to 8 may have to be considered. It is advisable to have concentrations of reducing gases in excess of equilibrium for reactions 1 and 2 at the heat treating temperature, to cater for any pick-up of oxidizing gases by the furnace atmosphere. It should be noted that while the equilibrium concentration of carbon dioxide for reaction 2 increases with decreasing temperature, that of water vapour for reaction 1 decreases. Oxidation of steel on cooling is prevented by using a sufficiently rapid cooling rate, by restricting the volume of the controlled atmosphere coming into contact with the steel, or by dehydration of the furnace gases.

The carburizing–decarburizing reactions 3 to 5 are only effective at temperatures near to and above the Ac_3 point, when the solubility of carbon in steel becomes appreciable. For the same reason, soot formation via reactions 6 and 7 is more likely to take place at temperatures below 700 °C, since the equilibrium concentration of carbon monoxide decreases with decreasing temperature, but can be prevented by control both of the cooling rate and the volume of gases contacting the steel surface. Reaction 8 is catalysed by iron and can lead to considerable roughening of the surface if it is appreciably displaced from equilibrium at the heat treating temperature.

Blueing of low carbon steel is preferably carried out in an atmosphere of steam. Chain reactions which may take place in air/steam mixtures below 350 °C lead to hydroxide formation and rusting. In an air furnace, steam should not be introduced below this temperature, but once it is, air should be completely purged before the steel reaches about 400 °C. The final blueing temperature in steam is between 450 and 600 °C, dependent on application.

HIGH CARBON STEEL

The same primary reactions apply to the heat treatment of high carbon steels as to the low carbon type, but the decarburizing reactions 3 to 5 are of most importance. A variety of controlled atmospheres is availabe for industrial use and they are summarized below, the final choice being determined by the quality of product required and whether the heat treating temperature is above or below the Ac_3 point.

Above 680–700 °C the atmospheres for treatment are:

1. High temperature endothermic gas, produced from rich hydrocarbon gas/air mixtures, plus hydrocarbon addition if necessary.
2. Completely burnt hydrocarbon gas, processed over hot charcoal, plus hydrocarbon if necessary.

3. Chemically purified burnt hydrocarbon gas, dry and free from carbon dioxide, plus hydrocarbon if necessary.
4. Nitrogen or nitrogen plus hydrocarbon.
5. Vacuum.

Below 680–700 °C the atmospheres for treatment are:

1. Dry burnt hydrocarbon gas.
2. Dry burnt ammonia.
3. Cracked ammonia.
4. Nitrogen.

Processes have been developed for bright hardening in which the steel is quenched in a blast of non-oxidizing controlled atmospheres, which, if necessary, is refrigerated.

ALLOY STEELS

The nature and concentrations of the alloying elements determine the type of controlled atmosphere which will ensure freedom from oxidation or decarburization during heat treatment.

Austenitic and martensitic stainless steels, etc. If the alloying elements form oxides of low dissociation pressure, and are present in concentrations exceeding about 1%, the controlled atmosphere must be quite free from oxygen-bearing gases, and the choice is limited to cracked ammonia, dry hydrogen, or dry nitrogen/hydrogen mixtures.

Straight nickel steels These can be bright heat treated successfully in atmospheres suitable for their plain carbon steels equivalents, except that the use of a desulphurized atmosphere is advisable. For this reason all alloy steels should be degreased before treatment if surface staining is to be avoided.

Martensitic types of stainless steel Such types are treated in cracked ammonia having a controlled addition of methane or propane to prevent decarburization, and quenched in the controlled atmosphere[5] or oil.

High speed steels are preferably treated in salt bath furnaces, but where the nature and size of the work does not permit this a controlled atmosphere can be used.[6] The formation of an oxide skin is not objectionable, since the tools are normally surface-ground before use, and such a skin may in fact give added protection against decarburization.

29.4.1 Gas carburizing

Atmospheres developed for the heat treatment of carbon steels with freedom from decarburization form the basis of gas carburizing processes.[7] The active carburizing agent is a hydrocarbon gas such as methane, propane or butane, but in order to prevent soot formation, which may inhibit the carburizing reaction, it is necessary to dilute with a non-decarburizing atmosphere. The depth of case is determined by the time and temperature of treatment, and the surface carbon content by the composition of the furnace atmosphere and the temperature (Figure 29.6). The total depth of case is given by $x^2=4Dt$. The effect of temperature on the case depth is given by $x^2/t=A-B/T$. The case may not be sufficiently hardenable over its entire depth. If C_{eff} is the minimum effective carbon content which is sufficiently hardenable, then $x_{eff}/x=C_0-C_{eff}/C_0$ where C_0 is the surface concentration of carbon, equal to the carburizing potential of the atmosphere or the saturation concentration, whichever is the lower. Instruments are now available for rapidly determining the carbon potential of a carburizing atmosphere.

The relationship between the dew point or the carbon dioxide content of the furnace atmosphere and the carbon content of the steel at equilibrium with it has been determined quite accurately over a range of temperatures for a number of different furnace atmospheres, and typical curves for those derived from methane, butane and propane are shown in Figure 29.7. Excessive build-up of carbon in the surface can be removed by a diffusion treatment following carburization. The process lends itself to the direct quenching of the work from the carburizing temperature when carburizing steels of controlled grain size are used. Gas carburizing atmospheres may be summarized as follows:

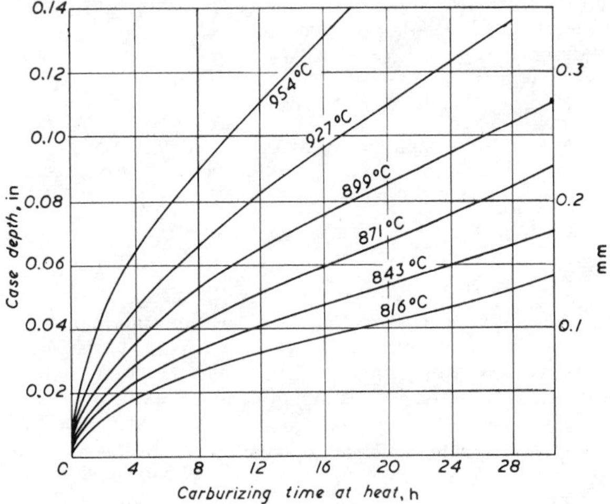

Figure 29.6 *Variation of case depth with time and temperature of treatment*

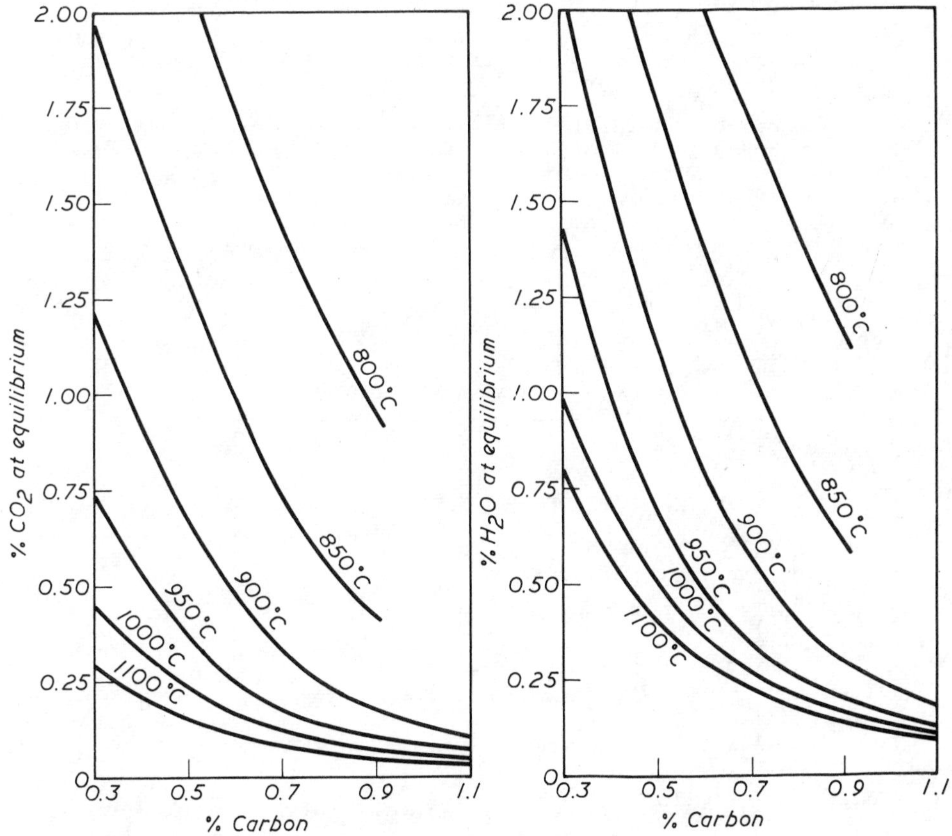

Figure 29.7 *Relationship between carbon potential and the* CO_2 *and* H_2O *contents of an endothermic atmosphere prepared from propane. Composition of atmosphere: 23% CO, 31%* H_2*, traces of* CO_2*,* H_2O *and* CH_4*, balance* N_2

1. High temperature endothermic gas + hydrocarbon gas.
2. Completely burnt hydrocarbon gas, processed over hot charcoal + hydrocarbon gas.
3. Chemically purified burnt hydrocarbon gas + hydrocarbon gas.
4. Drip feed carburizing, using organic liquids.
5. High purity nitrogen plus O_2 or CO_2 or alcohol plus hydrocarbon.

Furnaces for gas carburizing may be pit type or integral quench batch or continuous pushers of gas-tight construction. The process atmosphere may be manually or automatically controlled to a pre-selected 'carbon potential' level by measurement of dew point, CO_2 or oxygen potential.[19]

29.4.2 Nitriding

Nitriding is applied to steels containing one or more alloying elements which form stable nitrides, present in the steel in a highly dispersed state. This is the primary factor contributing to the high surface hardness obtained. The resulting excessive distortion of the crystal lattice slows down considerably the rate of diffusion of the nitrogen, so that long times of treatment are required to produce satisfactory case depths (Figure 29.8). The treatment is carried out in anhydrous ammonia, which partially dissociates at the steel surface to nascent nitrogen and hydrogen. The degree of dissociation is maintained between 15 and 35% by control of the rate of flow of ammonia, the latter being of the order of 20 to 30 ft^3 per hour per 100 ft^2 of surface (61 $l\,h^{-1}$ per m^2 of surface) being nitrided. The nitriding temperature is usually 500–520 °C, depending upon the type of steel being used.

There are three principal groups of structural steels which are nitrided, each giving a different level of surface hardness as follows:

1. 1% Cr steels	500–600 HV
2. 3% Cr steels	800–900 HV.
3. 1.5% Cr, 1% Al, 0.25% Mo steels	950–1100 HV.

Nitrided steels possess a surface layer which is white under the microscope and which is brittle. This layer, probably Fe_4N, has good corrosion resistance, but for maximum wear resistance it should be removed by grinding or chemical means. Attempts to minimize the white layer thickness have been made in which artificially raising the ammonia dissociation level has been achieved by dilution of the ammonia with nitrogen, hydrogen or a mixture of the two. Because of the low temperature of treatment, distortion during nitride hardening is small and can be reduced to

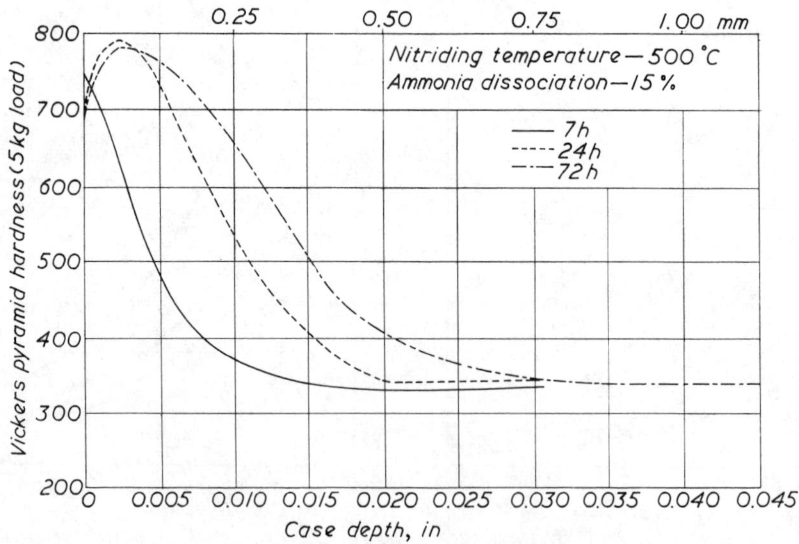

Figure 29.8 *Variation of surface hardness with case depth for different times of treatment*

negligible proportions by giving a pre-stabilizing treatment at about 50 °C higher than the subsequent nitriding temperature to remove machining and other stresses. Degreasing prior to nitriding is essential.

Austenitic valve steels, etc. can be successfully nitrided following a special surface preparation, such as pickling in hot 10% sulphuric acid followed by copper plating 0.000 01 in thick, from a cyanide bath.[8] The nitriding temperature is higher than that used for the more common nitriding steels, usually about 550 °C.

High speed tool steels may be nitrided after the conventional treatment to increase the surface hardness and give improved life.

Certain cast iron alloys, usually aluminium-bearing, can also be nitride-hardened in ammonia at 500–550 °C, developing a surface hardness of 900–950 VPN with a total carbon content of about 2.5%. The nitrided iron has good corrosion and wear resisting properties and is used for cylinders, piston rings, etc.[9]

29.4.3 Carbonitriding

The carbonitriding process which is normally carried out in the temperature range 800–925 °C is one in which simultaneous carburizing and nitriding of the steel occurs. The controlled atmosphere is basically a carburizing atmosphere to which a small addition of ammonia is made.[10] The process temperature and the detailed composition of the atmosphere depend upon the type of steel being treated and the type of case and core properties required.

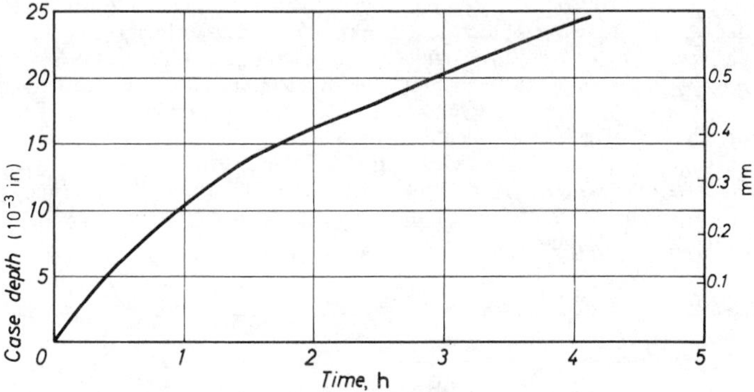

Figure 29.9 *Carbonitriding. Variation of case depth with time at 850 °C*

The concentration of nitrogen in the case decreases with increasing temperature of treatment. In general the process is carried out below the Ac_3 point where a carburizing treatment would result in a shallow, sharply marked case. The simultaneous absorption of carbon and nitrogen, however, results in a greater depth of diffused case than when carbon alone is used and the presence of nitrogen increases the depth of hardenable case (Figure 29.9).

The furnaces used for carbonitriding may be of the batch or continuous type, and are similar in design to those used for gas carburizing. The quenching bath for subsequent hardening can be made an integral part of the furnace, so that the work is maintained in a protective atmosphere up to the point of immersion in the quenching medium. The treatment is most suitable for the production of hard, shallow, wear-resistant cases.

Low temperature nitrocarburizing is an anti-scuffing treatment which may be carried out in nitrogen/ammonia atmospheres, usually at 570 °C followed by oil quenching.[22]

29.4.4 Malleabilizing

The annealing of white iron in the production of whiteheart malleable cast iron is a simultaneous graphitizing and decarburizing treatment. The latter can be carried out in any type of controlled atmosphere containing the necessary concentrations of decarburizing gases, but the most economical process is one in which the castings themselves generate the initial atmosphere by reaction between the air in the furnace chamber and the carbon in the castings.[11] The resulting atmosphere, rich in carbon dioxide, is circulated, and carbon monoxide converted to carbon dioxide by addition of an oxidizing gas such as steam or air. The addition of either of the latter is controlled to leave sufficient reducing gas present to prevent oxidation of the iron while maintaining the maximum possible rate of decarburization. The process is carried out industrially in suitable batch or continuous furnaces at 1 050 °C in an atmosphere containing CO_2 and CO with a ratio of approximately 2.8/1.[17] The depth of decarburization[17] is given approximately by:

$$x^2 = 4D\left(\frac{C_s}{2C_A - C_s}\right)t \quad \text{and the loss in weight by} \quad w = x\left(C_A - \frac{C_s}{2}\right),$$

where C_A equals the total concentration (g cm^{-1}), C_s is the maximum solubility, x is the depth of decarburization, D is the diffusivity constant (g cm^{-2} s^{-1}), and w is the loss in weight (g cm^{-2}) of surface.

In the manufacture of blackheart malleable iron, however, decarburization must be prevented and a completely neutral atmosphere is used, such as dry nitrogen from bulk supply or completely burned hydrocarbon gas, stripped of CO_2 to 0.1% and water vapour to a dew point of −40 °C. Dried, fully burned ammonia is a more expensive but satisfactory alternative.

29.5 Applications to non-ferrous alloys

COPPER AND COPPER ALLOYS

Steam and carbon dioxide are virtually inactive to copper and both are used on an industrial scale. The former has the disadvantage of giving rise to water staining, while the latter is less economical. Burnt hydrocarbon gas is used extensively for bright annealing copper, but the metal is very susceptible to staining by hydrogen sulphide. Organic sulphur and sulphur dioxide do not attack copper, but may be converted to hydrogen sulphide, so that complete desulphurization of the controlled atmosphere is advisable. The form in which sulphur occurs in burnt gases and the ease with which it may be removed depend upon the degree of combustion.[17] Table 29.6 indicates the difference between the relative concentrations of sulphur compounds likely to exist in the partially burnt and in the fully burnt gas. Cracked and burnt ammonia are suitable for copper annealing, and are sulphur free, but the former is little used for economic reasons.

Table 29.6

		Composition %				
Gas	*Type*	CO_2	CO	H_2	H_2O *before cooling*	N_2
A	Partially burnt	5.0	8.0	10.0	14	Remainder
B	Fully burnt	10	1.0	0.5	20	Remainder

		Approximate ratios (calc.) by volume			
Gas	*Type*	SO_2	: H_2S	: COS	: CS_2
A	Partially burnt	1	: 200	: 5.0	: 1.1
B	Fully burnt	1	: 0.01	: 0.000 3	: 7×10^{-5}

Tough pitch copper contains oxygen and treatment in hydrogen-containing atmospheres can result in embrittlement due to the formation of water vapour within the metal; it is necessary to reduce the hydrogen content of the controlled atmosphere to 1% or less if embrittlement is to be avoided. Both burnt ammonia and burnt hydrocarbon gas can be generated to meet the requirements. It should be noted that steam can give rise to embrittlement due to the formation of hydrogen as a result of reaction between the steam and iron of the furnace chamber. A recent development is the open flame furnace, in which copper is bright annealed in the completely burnt products of the fuel used for heating. The fuel is normally desulphurized before use, but any sulphur in the combustion products is present as the harmless dioxide.

Straight copper/tin and copper/nickel alloys can be successfully bright annealed in any of the atmospheres advocated for copper, except that in view of the marked susceptibility of nickel to attack by sulphur in any form it is essential that desulphurization should be absolutely complete. For alloys containing elements which are more readily oxidized (e.g. manganese), the range of suitable atmospheres is restricted to cracked ammonia or dry burnt ammonia.

Copper/zinc alloys A successful industrial process[23] has been developed for the full annealing of brass with zinc content up to at least 37%. Since zinc is readily oxidized by oxygen, CO, CO_2 and water vapour, all of these constituents must be controlled to a minimum and it is beneficial for the atmosphere to contain hydrogen. A typical suitable composition is a 25% hydrogen/nitrogen mixture with the following limits: O_2: 1 v.p.m.; CO_2: 20 v.p.m.; dew point better than −70 °C.

Zinc also has a considerable vapour pressure at annealing temperatures resulting in possible volatilization which may affect the colour and surface finish of the brass. Gases evolved from the alloy may also contribute to staining, while lubricants are a potential source of discolouration. Therefore, temperatures and soak times must be selected with care.

Where finish is less critical, 'flash' annealing may be applied to light gauge material with good results where rapid heating and cooling may be applied. Commercial annealing is often carried out in the cheaper types of controlled atmosphere which produces a light oxide film sufficient to restrict excessive zinc loss.

Gilding metal This contains a comparatively small concentration of zinc and can be bright annealed readily in atmospheres used for the annealing of copper, especially when in the form of coiled strip.

NICKEL AND NICKEL ALLOYS

Nickel oxide has a higher dissociation pressure than iron oxide, and thus at elevated temperatures is not so readily oxidized by carbon dioxide and water vapour (Table 29.7). On the other hand, nickel is an active catalyst for most gas reactions and some care is necessary in the selection of a suitable bright annealing atmosphere. Complete freedom from sulphur in any form is essential, and the relative concentrations of the oxides of carbon in a burnt hydrocarbon type of atmosphere must be carefully controlled to prevent sooting. Cracked and burnt ammonia are suitable atmospheres for bright annealing nickel. Desulphurized burnt hydrocarbon gas, usually dried, is also used, with a CO_2/CO ratio intermediate between those advocated for the bright annealing of copper and of low carbon steel (cf. Table 29.2).

Table 29.7 EQUILIBRIUM FOR THE REACTION:
$NiO+H_2 \rightleftharpoons +H_2O$

Temperature °C	$K=\dfrac{p_{H_2O}}{p_{H_2}}$
450	263
500	258
600	249
650	245
700	243

Nickel iron alloys can be satisfactorily treated in any atmosphere suitable for both nickel and iron individually, but the alloys of nickel containing chromium can be treated only in cracked ammonia, dry hydrogen both with dew point below −50 °C, or vacuum, because of the low dissociation pressure of chromic oxide.

The nickel silvers containing over 15% zinc, like the brasses, are difficult to bright anneal but this may be achieved with special purified atmospheres as for brasses.

LIGHT ALLOYS

The use of controlled atmospheres in the heat treatment of aluminium and of magnesium alloys has yet to be fully investigated. Magnesium alloys are protected against oxidation during solution heat treatment in an atmosphere of air, containing about 1% or less of sulphur dioxide,[13] which results in the formation of an unidentified, transparent film which affords adequate protection against oxidation. Small concentrations of carbon dioxide in air, and a dried, partially burnt hydrocarbon gas, free from oxygen and containing 2% carbon monoxide, 8% carbon dioxide have also been recommended, while a patented process advocates the use of an atmosphere containing water vapour and oxygen.[14] The composition of the annealing atmosphere for aluminium strip may be controlled as that staining of the strip by decomposition of rolling lubricant is reduced to a minimum. Usually, dried lean exothermic gas is used commercially.

Salt baths Molten salt baths have been used for many years, not only on account of their high rate of heat transfer but also for their ability to keep metal surfaces clean. Also a variety of processes[24] may be carried out such as case hardening, direct hardening and tempering, anti-scuffing treatments, descaling, treatment of aluminium alloys, etc. Salt baths may also be used as quenchants for austempering and martempering processes.

There are suitable salt mixtures available for every normal heat treatment temperature from 150 to 1350 °C.[25] They are either neutral to the surface of the metal being treated or are capable of carburizing, nitriding, descaling, etc.

The salts may be heated electrically or by fuel gas up to about 1000 °C, but above this temperature electric heating is the only feasible method.

Salt baths may be divided into three groups:

1. For operating temperature range 150–650 °C.
2. For operating temperature range 750–950 °C.
3. For operating temperature range 950–1350 °C.

GROUP 1 (150–650 °C)

Processes: Tempering, secondary hardening of high-speed steels.
Solution treatment of aluminium alloys.
Austempering, martempering.
Descaling, brazing of aluminium.
Nitriding, nitrocarburizing.

GROUP 2 (750–950 °C)

Processes: Hardening and carburizing of steels.
Brazing.

GROUP 3 (950–1350 °C)

Processes: Hardening of steel tools and dies.
Hardening of high-speed steels.
High temperature brazing.

REFERENCES

1. I. Jenkins, 'Controlled Atmospheres for the Heat of Metals', London, 1946.
2. C. E. Peck, *Metals and Alloys*, 1944, **19**, 593–9; 1945, **22**, 85–91.
3. B. Lustman, *Metal Progress*, 1946, **51**, 850–5.
4. 'Controlled Atmospheres for the Heat-treatment of Metals Using Liquefied Petroleum Gases', Shell-Mex and B.P. Gases Ltd., London.
5. W. E. Mahin and W. C. Troy, 'Controlled Atmospheres', Cleveland, Ohio, 1941.
6. W. A. Schlegel, *Steel*, 1943, **117**, 106.
7. I. Jenkins, 'Gas Carburising', *J. Iron Steel Inst.*, 1946, **154**, 195.
8. B. Jones, *J. Iron Steel Inst.*, 1937, **136**, 169–185; *Carnegie Schol. Mem. Iron Steel Inst.*, 1932, **21**, 39; 1933, **22**, 51; 1934, **23**, 139.
9. J. E. Hurst, *J. Iron Steel Inst.*, 1932, **125**, 223.
10. W. H. Holcroft, *Metal Progress*, 1947, **52**, 380.
11. I. Jenkins, *Metal Treatment*, 1947, **14**, 175.
12. G. W. P. Rengstorff, M. B. Bever and C. F. Floe, *Metal Progress*, 1949, **56**, 651.
13. P. T. Stroup, 'Controlled Atmospheres', Cleveland, Ohio, 1941.
14. F. Allan, *Light Metals*, 1947, **10**, 169–72.
15. O. Kubaschewski and C. B. Alcock, 'Metallurgical Thermochemistry', 5th edn, Pergamon, Oxford, 1979.

16. D. M. Dovey and I. Jenkins, *G. E. C. Journal*, July, 1948.
17. D. M. Dovey, *Atmospheres in Heat Treatment*, Institution of Metallurgists Refresher Course, 1963.
18. L. Fairbank and L. G. W. Palethorpe, 'Controlled Atmospheres for the Heat Treatment of Metals', Iron and Steel Institute special report No. 95, 1966, pp. 52–69.
19. L. G. W. Palethorpe, 'Modern Developments in Atmosphere Control', *Iron and Steel*, 1971, **5**, 326–330.
20. A Cook, 'Nitrogen-based Carbon Controlled Atmosphere — an Alternative to Endogas', *Heat Treatment of Metals*, 1976, **1**, 15.
21. R. G. Bowes, B. J. Sheehy and P. F. Stratton, 'A New Approach to Nitrogen-based Carburising', *Heat Treatment of Metals*, 1979, **3**, 53.
22. W. I. James, 'Practical Experience with Nitrogen-based Nitrocarburising', *Heat Treatment of Metals*, 1979, **1**, 13.
23. P. H. Ebner and J. B. Carrol, 'Bright Annealing of Brass in a Controlled Atmosphere', *Heat Treatment of Metals*, 1978, **4**, 83.
24. Electricity Council, 'Electric Resistance Heating', Electricity and Productivity Series, No. 5, 1966, pp. 118–126.
25. 'Cassel Manual of Heat Treatment and Case Hardening', 7th edn, Sections, 1.23 and 1.2.4. Cassel, London, 1966.

30 Laser metal working

30.1 Introduction

Since the first demonstration of laser action in 1960, vigorous research and development have led to a rapid, sustained growth in the number of laser types, in the output powers produced, and in the scope of their applications. In the relatively specialized field of metal working, it has emerged that only a very few laser types dominate and attention is confined here to neodymium YAG and carbon dioxide lasers. They can be regarded simply as the source of a highly controllable heating beam offering power densities up to 10^6 W cm^{-2} or more.

30.2 Lasers

The three essential components of a laser are the laser medium, the excitation source and the optical resonator (Figure 30.1). The excitation source drives the atoms, ions or molecules of the laser medium to a situation where there is an excess of those at a high energy level over those at a low energy. This inversion of the normal thermodynamic population distribution leads to laser action: an excited member of the medium undergoing a transition from high to low energy will emit a photon, which in turn stimulates further emission, perfectly in phase, and at the same wavelength, from the other excited members of the medium. The radiation is thus rapidly amplified; the role of the optical resonator is to direct and control the radiation by allowing an appropriate fraction to be bled off as a near-parallel beam while the remainder is circulated within the cavity to maintain laser action. The output is monochromatic with high phase coherence which permits focusing down to a spot diameter of the order of the wavelength, yielding the potential for very high power densities.

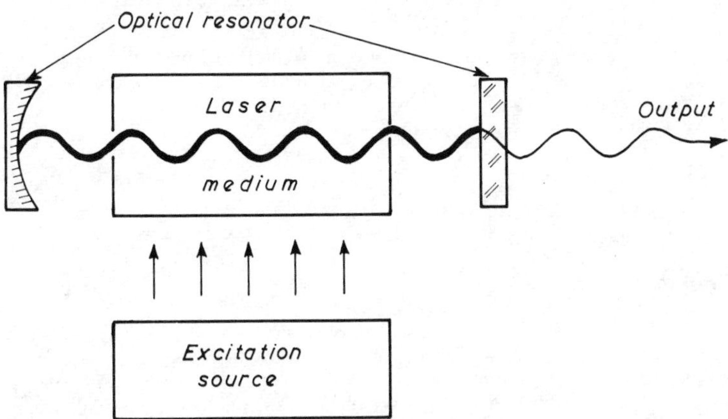

Figure 30.1 *Essentials of a laser*

Table 30.1 summarizes the salient characteristics of neodymium YAG and carbon dioxide lasers.

Table 30.1 CHARACTERISTICS OF NEODYMIUM YAG AND CARBON DIOXIDE LASERS

	NdYAG	CO_2
Relevant energy levels	Electron orbits of neodymium ion in host lattice of yttrium aluminium garnet ($Y_3Al_5O_{12}$)	Stretching of CO_2 molecule
Laser wavelength	1.06 μm	10.6 μm
Laser medium	Solid rod of NdYAG	Gaseous CO_2 (usually with He and N_2, total pressure ~50 mbar)
Excitation	White light excitation from flash tubes beside rod	Collisional excitation by electrons of glow discharge operated in gas mixture
Waste heat removal	Water cool rod	(i) Operate glow discharge in glass tubes and water cool tube walls (ii) Blow gas through discharge and then cool in heat exchanger
Typical operating conditions, including pulsing	c.w. $\lesssim$800 W Pulse lamp 20 Hz 20 J per pulse Q switch 20 kHz 150 W mean	$\lesssim$2 kW c.w. also pulsed discharge $\lesssim$2 kHz 2–10 kW c.w. mainly
Overall efficiency	$\lesssim$4%	$\lesssim$10%
Resonator	Stable or unstable	Stable or unstable
Mirrors	Dielectric	Metal
Lenses	Glass, as for visible light	eg. KC1, Ge, ZnSe
Part of eye most at risk from accidental exposure	Retina	Cornea

30.2.1 Operating conditions

Pulsed or continuous wave (c.w.) operation is possible with both CO_2 and NdYAG lasers. Pulsed operation is of interest because (i) the high peak power in the pulse aids the coupling of beams into reflective metal surfaces, and the ejection of material in hole drilling; (ii) bulk heating of the workpiece is reduced; and (iii) the ability to switch the beam rapidly on and off facilitates a number of high speed processes.

30.2.2 Resonators

The resonators fitted to the above lasers fall into two categories: stable and unstable. These terms refer to a mathematical description of the resonator which will not be attempted here. Figure 30.2 compares typical practical realizations of a stable (a) and an unstable (b) cavity.

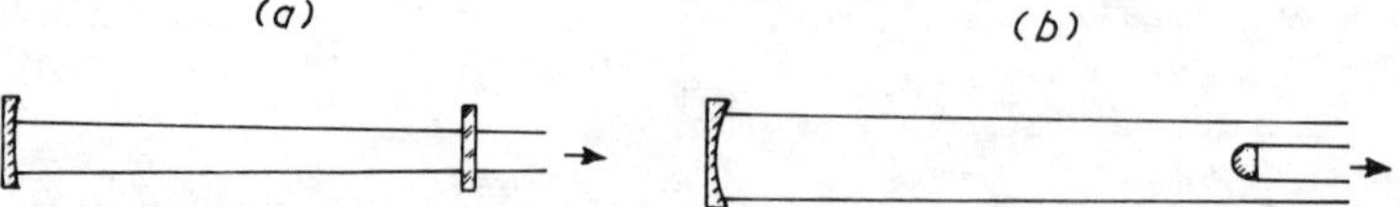

Figure 30.2 *(a) Stable and (b) unstable cavity resonators*

The stable cavity is particularly suited to long lasers of low aperture, while the converse is true of the unstable. The stable cavity yields a laser output (near field) having a gaussian intensity distribution, while the unstable cavity output yields a uniformly illuminated annulus. Focusing the former gives a 'far field' intensity distribution which is still gaussian, and focusing the latter gives an Airy pattern distribution which is strongly peaked on axis but with some power in concentric rings.

30.2.3 Lenses

Beam focusing conditions can be chosen to offer spot sizes and depths of focus to suit particular processes.

For the case of a laser with stable cavity producing zero order mode (Figure 30.3), the beam envelope consists of a waist region (diameter w) and a region bounded by straight lines of divergence α. The lens, focal length F, transforms the near field distribution at waist w_1 to the far field distribution at waist w_2. Further lenses can effect further transformations, the product $w_1\alpha_1 = w_2\alpha_2 = w_n\alpha_n$ being a constant. It can be shown that the diameter of the focused spot

$$w_2 \approx \frac{4}{\pi}\frac{\lambda F}{d}$$

where λ is the laser wavelength, d is the beam diameter at the lens, and F is the focal length of the lens. If depth of focus is defined by a distance z from focus at which the axial power density is 90% of that at focus, then

$$z \approx \pm\frac{\pi}{12}\frac{w_2^2}{\lambda}$$

Thus, short wavelength will give smaller spots and greater depth of focus than long, and short focal length lenses will give smaller spots and more shallow depth of focus than long (provided aberrations do not dominate).

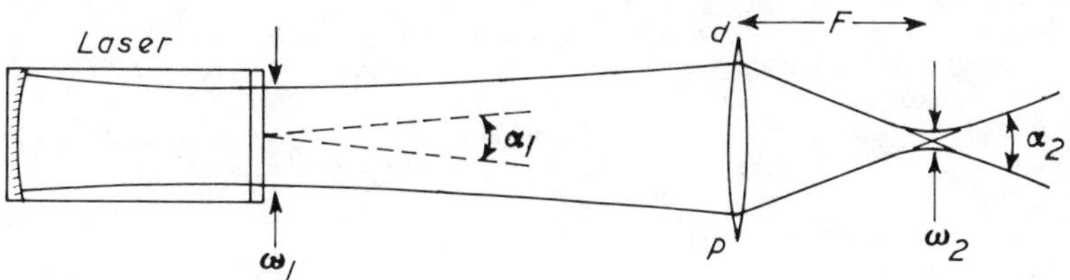

Figure 30.3 *Laser beam focusing*

30.2.4 Safety

Requirements and guidelines for the safe handling of laser power supplies and laser beams are well established. In a production environment, it is normal to place the laser system within a beam-tight enclosure, the opening of access doors breaking an interlock and shutting off the beam. In a research and development environment, where these conditions cannot apply and where there is a finite chance of the operator being exposed to the beam, eyes should be regarded as vulnerable and protected by goggles absorbent to the 10.6 μm or 1.06 μm beam. In metal-working, fumes may be generated and should be treated as in more conventional processing; in high-power laser welding, UV radiation will be emitted by the plasma at the beam interaction point, as in arc welding, and it must be similarly screened.

30.3 Laser welding

30.3.1 Process

The laser beam may be used locally to melt and join appropriate metals, usually by employing focused power densities of about 10^6 W cm^{-2} or greater. The workpiece surface then undergoes melting and vaporization which is sufficiently intense to disrupt the melt to form a capillary or keyhole, thus enabling the beam to penetrate relatively deep into the workpiece.

Using a pulsed laser, a single pulse will create a spot weld and, by suitably moving the workpiece, a pulse train can produce a series of overlapping spots thereby effecting a seam weld. The process is well suited to spot welding of fragile workpieces where resistance electrode clamping is unacceptable, or to seam welding of miniature packages.

With a moving c.w. laser beam, the keyhole translates along the joint line, metal melting ahead and flowing round to solidify behind. The process, like that of electron beam welding but without the vacuum requirement, thus produces welds which are energy efficient, of low shrinkage (because they are narrow) and of low distortion (because they are parallel-sided). Continuous laser welding is particularly suited to autogenous welding, close fit-up of the parts being normally required, although filler material can be added if a gap exists. At higher beam powers, plasma formation at the workpiece becomes important since it can be responsible for the broadening of the weld beads, a reduction in workpiece penetration and (more desirably) a smoothing of both weld surfaces. The plasma is frequently controlled by directing at it a jet of helium (which serves also to prevent oxidation).

A very wide range of metals and alloys (notably ferrous, nickel and titanium based) can be successfully laser welded, and the process is amenable to the welding of dissimilar metals and can cope with a reasonable mismatch of thermal properties. However, materials of high thermal conductivity and optical reflectivity such as gold and copper are difficult to weld, and aluminium and its alloys require careful control.

30.3.2 Performance and applications

It is possible to give only an approximate guide to laser welding performance, since precise figures for penetration and speed depend on power (and energy for a laser pulse), spot size and intensity distribution, position of focus and plasma conditions. However, a range of results in ferrous materials indicate that maximum practical penetrations at c.w. powers of say 0.5, 2 and 5 kW are about 1, 4 and 10 mm, respectively.

Figure 30.4 shows schematically some weld cross-sections representative of a range of different lasers and conditions. The spot weld at (a) is typical of spot welds performed by NdYAG lasers. They are used in production for assembly of filaments and relays. The seam weld at (b) is typical of that produced by NdYAG or CO_2 lasers (pulsed or c.w.), and used for sealing batteries or relay cans, for fabrication of bellows or pen bodies. Welds (c) and (d) indicate the potential of multi-kilowatt CO_2 lasers; they would appear to have scope for significant application in for example the aerospace, automotive and nuclear industries.

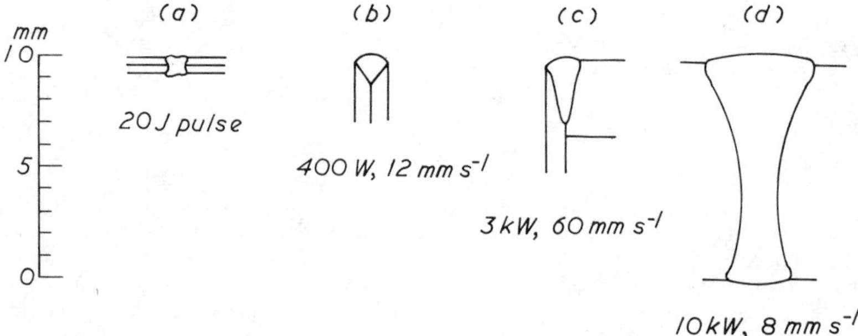

Figure 30.4 *Laser welds*

30.4 Laser drilling and cutting

30.4.1 Process

Metal drilling is most successfully carried out by pulsed solid-state lasers because they can produce focused power densities sufficiently high to promote in the workpiece explosive vapour release, resulting in the ejection of material in the liquid and vapour phase. Ejected matter can disperse in the period between pulses so that beam blocking effects are minimized.

In metal cutting the workpiece is essentially melted by the beam, and a gas jet is used to displace the melt. A gas may be chosen which reacts exothermically with the workpiece, e.g. the use of oxygen (possibly mixed with another gas) when cutting steel gives a useful energy release and the formation of low melting point slag as lubricant.

Laser formed holes and slots are generally characterized by having substantially paralled sides (although with slight taper on entry and exit), a high ratio of depth to width, a very thin recast layer on the wall, a restricted amount of heat conducted into the workpiece, and minimal frozen material adhering to the exit side. A full theoretical understanding of the interaction is still lacking, and indeed empirical optimization of the process is very time-consuming. It is sufficient to note here that high-quality results demand good beam quality; indeed, plane polarization of the beam has been shown to result in different cutting performance in directions parallel and perpendicular to the planes of polarization.

30.4.2 Performance and applications

DRILLING

Table 30.2 indicates the range of parameters encountered in laser hole-drilling.

Table 30.2 LASER HOLE DRILLING

Pulse duration	0.1–1 ms	Solid-state lasers offer higher energy pulses and higher peak powers, but lower repetition rates than gas lasers
Pulse energy	1–50 J	
Repetition rate	2–1 000 Hz	
Hole diameter	0.1–1 mm	Probably more cost-effective to drill diameters >1 mm by other means
Hole depth	$\lesssim$5 mm	

Solid-state lasers are used for the production drilling of precision holes in a wide range of metals and components. Typical of these are holes in jet engine parts, and controlled bleed holes in automotive components.

There are two production applications related to laser drilling. In the first, laser scribing, the amount of material removed from a component surface is sufficient only to render a legible mark; commercially available laser systems use either mask imaging or a dot-matrix to create machine-readable codes or alphanumeric characters on metal components. In the second, laser balancing, lasers are used to balance small rotating components by synchronizing a laser machining pulse with the out-of-balance signal.

CUTTING

For all but the smallest and thinnest components, carbon dioxide, rather than solid-state lasers, will be employed in metal cutting because of the greater ease with which higher mean powers are achieved. In the more difficult materials, such as aluminium, pulsing of the laser output will give an enhanced power to aid coupling. There are still considerable discrepancies between results published by different laser manufacturers, and Table 30.3 gives an indication of cutting performance in mild steel with oxygen assistance.

Table 30.3 LASER CUTTING OF MILD STEEL

Laser power kW	*Maximum* practicable thickness* mm	*Typical thickness* (in* mm) *cut at around* 40 mm s^{-1}
0.5	3	2
3	8	4
5	15	6

* Depends strongly on edge quality required.

Lasers in the power range 0.3–1.0 kW are well established in the profile cutting of metals up to about 6 mm thick. They appear to fill a need in the thickness range 1–3 mm where plasma torch cutting is difficult, mechanical cutting may create distortion, and the volume of production is insufficient to justify manufacture of blanking tools. This thickness range is also covered by NC machine tool punches, which can generally out-perform the laser on straight lines but not on complicated profiles, as evidenced by a current trend for punch manufacturers to offer their equipment with a laser as optional extra.

30.5 Surface treatments

These treatments concern the laser modification of workpiece surfaces to yield superior wear- or corrosion-resistant properties just where necessary. In contrast to cutting and welding, a heating pattern of finite extent is normally required at the work and, depending on shape or uniformity, a defocused beam spot (possibly from a multi-mode laser resonator) or one of several high-speed spot rastering techniques may be used. For convenience, the treatments can be divided into those which do not and those which do involve surface melting.

30.5.1 Solid-state transformation hardening of steels and irons

By use of a diffuse scanning beam, the surface hardening of carbon steels and irons through martensitic transformation can be carried out. The laser can be distinguished from alternative surface hardening techniques by one or more of the following attributes:

1. It characteristically operates as a rapidly scanning source so that overall heat input (and therefore distortion) is minimized, and adequate quenching rates are obtained solely by conduction into the substrate.
2. The beam can be manipulated and directed into bores and conventionally inaccessible regions without the hindrance of supply cables and pipes.
3. Its heating pattern may be rapidly altered to suit the application.

The power densities (usually 10^3–10^4 W cm^{-2}) and scan speeds (or beam dwell times) are chosen to austenitize a relatively shallow case depth (usually ≤ 1 mm) while avoiding surface melting and excessively rapid temperature cycles which may yield unduly hard microstructures or insufficient time for carbon dissolution. The workpiece surface is given a prior coating of zinc or manganese phosphate, or even colloidal graphite, to ensure efficient energy absorption at the surface. For a 0.5 mm case depth, coverage per kW is approximately 65 mm^2 s^{-1} for cast iron and 135 mm^2 s^{-1} for steel (dependent on composition and prior structure).

The majority of applications involve use of CO_2 lasers of around 1 kW and above. Typical of these is the creation of wear-resistant patterns inside cylinder bores, and it is argued that automotive component hardening, featuring high volume production and/or requirements of minimum distortion, is particularly suited to laser processing.

30.5.2 Surface fusion treatments

By use of a higher power density beam, surface melting of substrates occurs and additives may be introduced to produce a surface composition of high hardness or corrosion resistance. Compared with conventional heat sources, the laser offers the potential of reduced distortion, additive usage, and dilution; additionally, faster cooling rates may offer control of microstructure. While not yet significantly established in production, the process is undergoing extensive investigation, mainly at powers of several kilowatts and above, and technical feasibility has been demonstrated in operations such as the deposition of hard-facing alloys on the seats of engine exhaust valves, and in the strengthening of aluminium silicon alloy components by localized enhancement of silicon levels.

Finally, on workpieces subjected to very high power densities for very short times, shallow surface melts occur which experience very fast cooling rates. Rapid scanning of the beam over a component may melt surface layers that are so thin ($\sim$20 μm) that they subsequently self-quench (at rates in excess of 10^5 °C s^{-1}) by conduction to the substrate. In specific metal alloy compositions the resulting product may be a thin non-crystalline 'metallic glass' or, in other alloys and under less extreme conditions, a new alloy phase that may exhibit enhanced wear or corrosion resistance. There is considerable interest in the potential of this as a coating technique, and in its extension to the building up of successive layers of rapidly cooled material, thus creating bulk components which should exhibit the high strength characteristics of rapidly quenched surfaces.

BIBLIOGRAPHY

S. S. Charschan (ed.), 'Lasers in Industry', Van-Nostrand-Reinhold, Princeton, New Jersey, 1972.
W. W. Duley, 'CO_2 Lasers, Effects and Applications', Academic Press, New York, 1976.
J. F. Ready, 'Industrial Applications of Lasers', Academic Press, New York, 1978.
'Laser Focus Buyers' Guide', Advanced Technology Publ., Newton, Massachusetts, annual publication.
BS 4803:1972, 'Guide on Protection of Personnel Against Hazards from Laser Radiation', British Standards Institution. Revised version: 'Radiation Safety of Laser Products and Equipment—Manufacturing Requirements, User's Guide and Classification', 1982.

31 Guide to corrosion control

31.1 Introduction

Metals may be chosen specifically for their resistance to a corrosive environment but in industry, where economic considerations affect the selection of materials, it may be less costly to choose a metal that has a comparatively short life, and carry out regular maintenance or replacement rather than a high initial capital investment in a resistant metal or alloy that will withstand the conditions of corrosion during the lifetime of the plant. There are many examples where either of these two extremes has been the more economic choice, and therefore a wide choice of materials is required. It may be that the most economical decision would be to use coatings (*see* Chapter 35), cathodic protection or control of the environment (inhibitors, etc.). Resistant materials will be listed, where these are available, followed by the less resistant metals and alloys where shorter lifetimes may be tolerated.

31.1.1 Types of corrosion

Corrosion damage to a metal or alloy can be (a) general or uniform corrosion, (b) Localized or pitting corrosion, and may be caused or enhanced by one or more of the following broad classifications:

bimetallic coupling (and dealloying),
crevice corrosion,
erosion corrosion,
stress corrosion cracking,
corrosion fatigue (and fretting corrosion),
hydrogen embrittlement.

31.1.2 Environments which cause corrosion

These may be broadly classified into three main groups:

1. *Natural*
   - Atmosphere — Humid, polluted condensation.
   - Water — Sea water, river, potable, condensation.
   - Soil — Buried structures.
   - Storage — Possible corrosion during storage, transit or erection.
2. *Chemicals*
   - Acids — Sulphuric, hydrochloric, nitric, phosphoric.
   - Alkalis — Sodium and calcium hydroxides.
   - Fertilizers — Nitrogen compounds, ammonia, organic acids.
   - Closed circulating systems — Concentration of river, well or sea water.
   - Manufacture of chemicals — 1000+ varieties.

Process chemistry	— Treatment; e.g. dyeing, pickling, food processing, paper-making.
High temperatures	— Oxidation of metals, exhaust fumes, flue gases.
Wet flue gases	— Combustion systems where gases can condense, i.e. dew point corrosion.

3. *Contact*

Wood	— Some vapours from wood are aggressive.
Polymers	— Some polymers, in contact with metals can cause corrosion.

31.1.3 Accelerating factors

Corrosion rates may be drastically changed when temperature, flow rates, pressures and concentrations of chemicals are varied; corrosion of metals from these parameters can therefore be regarded as an 'add on' factor and under certain conditions of temperature, pressure and flow the corrosion rates may then become excessive. There are many anomalies, e.g. mild steel is not attacked by very high concentrations of H_2SO_4 but rapidly at concentrations below 70% w/v. Copper can withstand sea water but is attacked when the flow rate is excessive.

31.1.4 Measurement of corrosion damage

Corrosion attack is not often uniform and it can be misleading to apply much of the published data which convert weight loss into penetration rate (mm yr^{-1}). For instance, mild steel in sea water corrodes at approximately 0.1 mm yr^{-1}, but pitting can occur up to 0.4 mm yr^{-1} over a relatively small area of the total surface. In the case of intergranular attack at the metal grain boundaries a relatively small rate of attack can cause deep penetration so that whole grains drop out.

Corrosion rates must also be accompanied by the type of attack and in the case of pitting by the probability of finding the deepest pit at a certain depth.

In the case of high temperature oxidation and also for atmospheric corrosion, adherent corrosion products may be produced. The measurement of *weight gain* is then recorded.

Corrosion damage, although not excessive, can be very undesirable or even dangerous. When metals are under tensile or cyclic stresses a small amount of pitting could give rise to stress concentrations that lead ultimately to failure by cracking. Formation of corrosion products in a confined space can lead to 'oxide jacking' where the expansion suffered by the components can cause bursting and distortion. This has been widespread in some forms of concrete reinforcement where relatively mild corrosion can give rise to serious cracking of the nearby concrete.

31.1.5 Chemicals

For data on particular systems the following bibliography may be helpful.

SOURCES OF CORROSION RATE DATA

Three main sources of information:

1. data books;
2. national and international standards;
3. scientific journals and abstract literature.

1. Data books

'Corrosion Guide', E. Rabald, VDI, Düsseldorf, 1969.
'Corrosion Data Survey', G. A. Nelson, NACE, Houston, 1967.
'Werkstoffetabelle', DECHEMA, Frankfurt am Main, 1980.
'Fulmer Materials Selector', Fulmer Research, Slough, 1980.

2. *See* 'Corrosion Prevention Directory' HMSO, for extensive list.

3. Abstract literature

Metal Abstracts, ASM, Met. Soc., London.
Corrosion Abstracts, NACE, Houston, Texas.
Corrosion Profile (Chem. Abstracts), UKCIS, University of Nottingham.

31.2 Bimetallic corrosion

Designers require structures and machines to have metals and alloys of differing mechanical properties in close proximity, e.g. mild steel backed copper bearing surfaces, lightweight aluminium structure on a mild steel base, and many fasteners, rivets, screws etc.

Table 31.1 CORROSION RATE IN mm yr^{-1} FOR BIMETALLIC COUPLING IN SEA WATER OF COMMON STRUCTURAL MATERIALS**

Cathodic member M	M:Al*			M:*Mild steel**			M:*Zinc**		
	10:1	1:1	1:10	10:1	1:1	1:10	10:1	1:1	1:10
Carbon	39	2–10	0.4	7.5	2–8	—	60	33	9
Carbon (Vs H30)	32	2–10	0.5						
Carbon (Vs NS4)	49	2–10							
Lead	0.5	0.15							
Silicon bronze	2.0	0.6		1.7	2.2				
10% Al bronze	2.4	0.6							
NiAl bronze (Vs H30)	2.8	0.7							
HY80	1.6	0.3	0.2	1.3	1.0				
Ni-Resist	0.7	0.06							
Ni Al bronze	2.4	0.15	0.09						
Tin	2.2	0.14							
CN30	1.7	0.1		2.5	0.6				
EN57	2.7	0.15		1.6	0.5				
Monel	2.2	0.12		3.3	1.4				
Mild steel (Vs H30)	3.0	0.16	0.04						
Mild steel	1.7	0.07	0.06				7.4	2.2	
LG4 gunmetal				2.8	1.8	0.04			
Lead				2.0	1.4				
Titanium				2.2	0.6		4.2	1.6	
EN58J				2.7	0.3				
Tin				2.8	0.35				
Aluminium							1.2	0.3	
Copper	4.5	0.66	0.05	6.7	0.3	0.12	14.5	0.5	0.27

* Includes effect of anode/cathode area ratio.
** Based on BS PD6484 (1979).

Table 31.2 CORROSION RATE (mm yr^{-1}) FOR CATHODIC MATERIALS CARBON, TITANIUM AND COPPER COUPLED TO VARIOUS ALLOYS

Anodic member M	*Carbon*:M*			*Titanium*:M*			*Copper*:M*		
	10:1	1:1	1:10	10:1	1:1	1:10	10:1	1:1	1:10
EN57	0.5	0.3		0.08	0.003		0.1	0.01	0.002
HY80	10	2–7					1.4	0.2	0.14
Ferralium 40V	0.007	0.01							
Ni-Resist	5.5	2					0.85	0.27	0.04
Tin	4	0.3		0.25	0.05		4.9	0.35	0.19
2% Al brass	0.4	0.1							
Si Al bronze	0.26	0.16		0.02	0.005				
Monel	0.02	0.02		0.004	0.002	0.002			
Silicon bronze	1.2	0.3	0.01						
Lead	0.5	0.16					1.5	0.46	0.06
Naval brass	0.5	0.22							

* Includes the effect of varying the anode/cathode area.

In these cases, providing the environment is sufficiently conducting, serious acceleration of corrosion may occur. However, compatibility can be achieved.

Table 31.1 gives the accelerating effect on coupling various materials with sea water as the conducting electrolyte. Comparable results may occur with other conducting chemical solutions. For electronic materials with thin metallic coatings in humid conditions, corrosion products may affect performance.

31.2.1 Bimetallic coupling associated with electronic materials

Various noble metals are used to provide good conducting, tarnish-free contacts which are reliable over the life-time of the equipment. These metals are used generally as very thin films (1–5 μm) are usually porous. They are therefore likely to act as a good cathode and induce corrosion even under slightly humid conditions. Thus, the connector can become covered with a thin film of corrosion products which can reduce the performance of the electronic device. Frequently this corrosion effect occurs in handling and in storage. It can also arise with unsatisfactory packaging, and the bimetallic coupling encourages attack.

Table 31.3 contains the combinations which have given satisfactory service and those that have been known to cause corrosion should the environment permit.

Table 31.3 SEVERITY OF GALVANIC CORROSION FROM METALLIC COMBINATIONS
Coatings are shown in brackets: (Ni)Cu = nickel-plated copper; (r.Sn)Cu = reflowed tinned copper; (s.d.)Cu = solder-dipped copper.

Completely satisfactory	*Satisfactory slight corrosion*	*Borderline moderate corrosion*	*Unsatisfactory severe corrosion*
Cu–(Ni)Cu	Cu–(Ag)Cu	(Au)Cu–(s.d.)Cu	Al–Brass
Cu–(Au)Cu	(s.d.)Cu–(Sn)Al	(Sn)Al–(Ni)Cu	Al–Cu
(Sn)Cu–Al	Cu–(Sn)Cu	Al–(s.d.)Al	(Sn)Al–Cu
(Sn)Cu–(Ni)Cu	Cu–(s.d.)Cu		Al–(Ni)Cu
(Sn)Cu–(s.d.)Cu	Cu–(r.Sn)Cu		Al–(Ni)Brass
(Sn)Brass–Al	(Ag)Cu–(Sn)Cu		Al–(Ag)Cu
(s.d.)Cu–(Ni)Cu	(Ag)Cu–(s.d.)Cu		(Sn)Al–(Ag)Cu
(Ni)Cu–(Au)Cu	(Au)Cu–(Sn)Cu		Al–(Au)Cu
(Ni)Cu–(Ag)Cu	Al–(Sn)Al**		(Sn)Al–(Au)Cu
(Au)Cu–(Ag)Cu			
Al–(Sn)Al*			

* No copper undercoat. ** Zincate process.

31.2.2 Dealloying-selective dissolution as a form of bimetallic corrosion

A special case of bimetallic corrosion occurs for certain alloys where the base metal can be preferentially dissolved. Copper–zinc alloys are prone to this corrosion, and dezincification can be a serious corrosion problem resulting from the chemical composition of some natural waters. Information about these water supplies may be obtained from the British Non-Ferrous Metal Association. The corrosion can be reduced by a 1% alloy addition of Sn or by 0.02–0.06% of Sb or P. Copper alloys with resistance to dezincification are given in Table 31.4.

Table 31.4 COPPER ALLOYS RESISTANT TO DEZINCIFICATION

Designation	*Form*	*Brass-type*	*Composition, % by weight*	*β-phase content %*
A	Rod	Special-brass	Cu 61.7, Pb 2.15, Sn 0.02, Al <0.002, Fe 0.02, Ni <0.002, Mn <0.002, As 0.03, Si <0.002, Sb <0.002, Zn rem.	15
B	Sheet	70/30-brass	Cu 69.9, Pb 0.002, Sn <0.002, Al 0.004, Fe 0.012, Ni <0.002, Mn <0.002, As <0.002, Si 0.004, Sb <0.002, Zn rem.	0
C	Rod	Special-brass	Cu 59.8, Pb 0.75, Sn 0.85, Al 0.65, Fe 0.85, Ni <0.005, Mn 1.95, As <0.002, Si 0.030, Sb <0.002, Zn rem.	35

Table 31.4 COPPER ALLOYS RESISTANT TO DEZINCIFICATION—*continued*

Designation	*Form*	*Brass-type*	*Composition, % by weight*	*β-phase content %*
D	Rod	Freeturning brass	Cu 57.5, Pb 2.60, Sn 0.09, Al 0.002, Fe 0.05, Ni 0.010, Mn <0.002, As 0.005, Si <0.002, Sb <0.002, Zn rem.	35
E	Rod	Special-brass	Cu 66.3, Pb 2.2, Sn 0.13, Al <0.002, Fe 0.10, Ni 0.03, Mn 0.58, As 0.05, Si 0.89, Sb <0.002, Zn rem.	5% unidentified phases
F	Press. diecast	Special-brass	Cu 64.9, Pb 2.3, Sn 0.60, Al 1.60, Fe 0.40, Ni 0.17, Mn <0.01, As 0.048, Si 0.40, Sb <0.01, Zn rem.	20
G	Rod	Special-brass	Cu 61.85, Pb 1.95, Sn 0.01, Al <0.002, Fe 0.02, Ni <0.002, Mn <0.002, As 0.024, Si <0.002, Sb <0.002, Zn rem.	1.5
H	Rod	Gunmetal	Cu 84.15, Pb 5.35, Sn 4.6, Al 0.01, Fe 0.03, Ni 0.13, Mn <0.002, As 0.024, Si <0.002, Sb <0.002, Zn rem.	0
I	Tube	70/30-brass	Cu 70.15, Pb <0.002, Sn <0.002, Al <0.005, Fe 0.005, Ni 0.01, Mn <0.002, As 0.034, Si <0.002, Sb <0.002, Zn rem.	0
J	Tube	Al-brass	Cu 76.45, Pb 0.006, Sn 0.008, Al 2.0, Fe 0.03, Ni <0.002, Mn <0.002, As 0.030, Si <0.002, Sb <0.002, Zn rem.	0
K	Chill-cast	Special-brass	Cu 63.8, Pb 0.32, Sn <0.01, Al <0.05, Fe 0.02, Ni 0.01, Mn <0.01, As <0.01, Si 0.21, Sb 0.08, Zn rem.	35

31.3 Crevice corrosion

Many engineering designs place metals together for joining, or create narrow slots or pockets where liquids could be retained. Such crevices include screw threads, nuts, washers, gaskets, some weldments, heat exchanger rolled in tubes, valve packings, etc. In neutral aerated waters there is the strong possibility of corrosion within these crevices, particularly with strongly passive metals in chloride solutions such as sea water. Practically all metals can suffer from this form of attack and the usual remedy is to remove the crevice by careful design of the fit of the components, or by sealing or coating. Table 31.5 gives an order of resistance to crevice corrosion which shows that some metals show good resistance. It is interesting to note that many of the popular stainless steels can be affected to a serious extent by crevice corrosion.

Table 31.5 RESISTANCE TO CREVICE CORROSION

Metal	*Very resistant* →	*Moderate corrosion*		→	*Severe corrosion*
Mild steel					
Low alloy steel		Sea water			
Cupro nickel Cu 10Ni 1.5Fe	Sea water	Neutral solutions			
Cu and Cu/Zn	√				
Titanium and Ti alloys	√ (ambient temp)	95 °C			Halide* and sulphate solns
Stainless steel 10% H_2SO_4 RT	321	316	304	302	13% Cr
10% H_2SO_4 + NaCl	316	321	302	304	16% Cr
Fe 18Cr 13Ni 3Mo 2Si NNG	√				
Fe 18Cr 14Ni 2Mo NiTi	√				
Fe 18Cr 24Ni 3Mo 2Cu	√				
Fe 20Cr 25Ni 5Mo 1.5Cu	√				
Fe 18Cr 10Ni 2.5Mo 2.5Si	√				
Fe 25Cr 2Mo(duplex)	√				
Fe 25Cr 6.5Ni 3Mo(duplex) 0.3 W	√				

Table 31.5 RESISTANCE TO CREVICE CORROSION—*continued*

Metal	*Very resistant*	→	*Moderate corrosion*	→	*Severe corrosion*
Nickel alloys					
Hastalloy 'C'	< 60 °C				
Inconel 625	< 60 °C				
Cobalt alloys					
Vitallium	< 60 °C				

* Ti 0.2% Pt 2% Mo(or 2% Ni) relatively more resistant.

31.4 Corrosion/erosion resistant materials

Flowing electrolytes can increase corrosion rates that are dependent on diffusion. This usually applies to aerated solutions. Corrosion rates in mm yr^{-1} for flowing sea water are given in Table 31.6 and for copper central-heating tube in Table 31.7.

Table 31.6 CORROSION/EROSION IN FLOWING SEA WATER AT AMBIENT TEMPERATURE
Corrosion rate mm yr^{-1}

	Flow rate m s^{-1}		
Metal	*0.3*	*1.2*	*8.0*
Carbon steel	0.16	0.33	1.0
Ni resist.	—	—	0.8
Cast iron	0.2	—	1.25 (7 at 28 °C)
Silicon bronze	0.005	0.08	1.6
Admiralty brass	0.08	0.08	0.7
Aluminium bronze	0.16	—	1.0
Aluminium brass	0.08	—	0.5
Cu, 10% Ni, 1% Fe	0.2	—	0.4
Cu, 30% Ni, 0.05% Fe	0.08	—	0.8
Cu, 30% Ni, 0.5% Fe	0.005	—	0.16
Monel 400	0.005	—	0.005
Stainless steel Type 316	0.004	—	0.005
Hastalloy C	0.005	—	0.05
Titanium	0	—	0
Inconel 625	0	—	0

Table 31.7 EROSION OF COPPER TUBING IN CENTRAL-HEATING SYSTEMS
12 months' tests—corrosion rate in mm yr^{-1}

			Flow rate m s^{-1}			
Temp. °C	*pH*	*Aeration*	1	3	6	12
30	8	Yes	—	—	—	0.03
65	8	Yes	—	—	0.015	0.07
65	8	Yes	—	—	—	0.07
90	8	Yes	0	0.05	0.07	0.07
65	8	No	—	0.05	—	—

31.5 Cavitation

Table 31.8 CAVITATION—HIGH VELOCITY LIQUID FLOW

Highly resistant	*Good resistance*	*Poor resistance*
Stellite 12	Monel 500 5% Si	Aluminium alloys
Stellite 4	Aluminium bronze	Mild steel
Stellite 6	White cast iron	Alloy cast iron
	Silicon iron	Copper
Titanium 99%	High strength Martensitic steel	Silicon bronze
Ti 2.25Al 4Mo 0.2Si 11Sn	Stainless-Martensitic steel EN 5C	Manganese brasses
	Austenitic stainless steel EN 58J	Aluminium brass
		Admiralty brass
		Cu 10Zn

31.6 Corrosion fatigue

This form of failure by cracking is responsible for about 20% of the failures in the chemical process industry.

Corrosion fatigue is the failure of a material caused by cyclic alternating stress in the presence of a corrosive environment. All metals are known to fail at stresses below the yield stress when under cyclic loading. The effect of corrosion is in general to reduce even further the stress level at which failure occurs.

31.6.1 Measurement

It is usual to measure the number of cycles (N) to failure at various applied stress levels (S), the so-called S/N curves. For some metals there is a 'limiting stress level' below which the metal will not fail by cracking. For many metals there is no 'limiting stress level' and for these materials the stress level which leads to failure in a certain number of cycles is recorded, e.g. 10^7 or 10^8 cycles.

31.6.2 Effect of sea water on corrosion fatigue resistance

Table 31.9 gives results for various materials and shows the lowering in strength as a result of cyclic loading. This table is based on laboratory testing and is only a guide. It does, however, demonstrate that account should be taken of the effect of vibration and cyclic loading.

Table 31.9 CORROSION FATIGUE OF ALLOYS IN SEA WATER
Conditions: ambient temperature in flowing sea water (0.6 $m\,s^{-1}$) mean stress zero: 10^8 cycles.

Alloy	*UTS* M Pa	*Corrosion fatigue limit* 10^8 *cycles* M Pa	% *reduction in strength*
Mild steel	400 to 500	14	3.5
Titanium 6% Al–4% V	1060	610	58
Copper alloys			
Monel	1210	179	15
Al bronze	710	152	21
70% Cu–30% Ni	572	62	11
Stainless steels			
304	545	104	19
304L	545	97	18
316	586	97	17
316L	545	90	16
13% Cr–1% Ni (cast)	717	69	10

Table 31.9 CORROSION FATIGUE OF ALLOYS IN SEA WATER—*continued*

Alloy	*UTS* M Pa	*Corrosion fatigue limit* 10^8 *cycles* M Pa	% *reduction in strength*
Nickel alloys			
15% Cr, Mo 3%, Fe 7% (IN–12)	1415	480	34
Inconel 718 19% Cr–3% Mo–18.5% Fe			
Nb + Ta 0.08 mm	1304	414	32
0.4 mm	1030	345	33
Hastalloy C	745	220	30

Table 31.10 EFFECT OF ENVIRONMENT ON CORROSION FATIGUE LIMIT AT 10^7 CYCLES MPa 20°C MEAN ZERO STRESS

Material	*Air*	*Sea water*	*Other corroding solution*
0.17% C steel	218	63	
0.50% C steel	281	70	
15% Cr steel	328	203	
17% Cr–1% Ni steel	422	265	
18% Cr–8% Ni steel Type 304	390	280	10% H_2SO_4 210
Magnesium alloy (2.5% Al)	91	16	
Duralumin (Al–4% Cu)	125	55	
Mild steel	290	3% NaCl 200	De-aerated distilled water 290
		De-aerated 3% NaCl 290	De-aerated distilled water 250
Aluminium zinc magnesium alloy	140	3% NaCl 50	
Shot peened surface	200	3% NaCl 100	

31.7 Stress corrosion cracking

Table 31.11 STRESS CORROSION CRACKING (SCC) IN MILD AND HIGH STRENGTH STEELS

Material	*Conditions affecting stress corrosion cracking*
Mild and low carbon steels	Accelerated SCC by: NaOH, KOH, Na_3PO_4, HNO_3, $NaNO_3$, $Ca(NO_3)_2$ NH_4NO_3, $(NH_4)_2CO_3$, bicarbonates, fuming H_2SO_4, $FeCl_3$. $Fe(Al)_3O_3 + Al_2O_3 + CaO$, ethanolamine + $H_2S + CO_2$, $H_2O + CO + CO_2$ HCN *Hot nitrate solutions:* C0.005–0.19% susceptible 0.09–0.19% C { Moderate plastic deformation increases susceptibility; Large amount of cold work decreases susceptibility Welds with up to 0.26% C in hot nitrates susceptible Welds stress annealed at 600 °C prevents stress corrosion cracking Time to failure decreased as concentration of nitrate increases in boiling solutions. Temperature increase decreases time to failure. Alloying with Cr, Ta, Nb, Ti, Mn reduces susceptibility Inhibitors reduce susceptibility. Cathodic protection in some cases

Table 31–11 STRESS CORROSION CRACKING (SCC) IN MILD AND HIGH STRENGTH STEELS—*continued*

Material	*Conditions affecting stress corrosion cracking*
Mild and low carbon steels—*cont.*	*Caustic solutions:* C0 → 0.02% increases susceptibility C > 0.02% increases time to failure Fully killed steels very susceptible Semi-killed steels slightly susceptible Low C rimming steels very resistant to SCC Temperature increase reduces time to failure. The higher the temperature the lower the concentration at which cracking occurs. Plastic deformation increases SCC. Annealing prevents SCC unless deformation is beyond yield stress
High strength steels (yield > 1000 MPa)	Below 1000 MPa yield stress hardenable steels are generally immune from SCC Above 1000 MPa yield:— *Chloride solutions:* Increase above 0.25% NaCl has little effect on time to fracture by SCC. Temperature has little effect in range 15–80 °C. pH has little effect but above pH 10 susceptibility is reduced. Hydrogen embrittlement and cathodic protection can increase susceptibility to SCC

Table 31.12 CONDITIONS FOR STRESS CORROSION CRACKING IN STAINLESS STEELS

Environment for SCC:	*Chloride solutions*	H_2S *with or without chlorides*	*Caustic solutions*	*Polythionic acid*
Susceptible alloys:	*Types* 304, 316 *martensitics*	*Martensitics ferritics ferritic-austenitics*	*Types* 304, 316	*Types* 304, 316
Temperature °C	> 70	< 60	> 120	Ambient
Chloride content	10 p.p.m. temp. dependent	—	—	—
pH	> 2.0 At lower pH general corrosion for susceptible alloys	pH affects threshold stress	> 12	—
Stress level	Moderate	High	Moderate	Low
Metallurgical condition	All	All	All	Sensitized

Table 31.13 STAINLESS STEEL GRADE SELECTION IN ENVIRONMENTS WITH RISK OF STRESS CORROSION

Solution	*Conditions* Temp. °C	*Conditions* Chloride content	*Recommended grade*
Chloride solutions	≤ 70	Low	304 or 316 depending on risk of pitting
	≤ 70	High	2RE 65,[1] high pitting resistance
	≥ 70	Low	2RE 60[2]
	> 70	High	2RE 65[1]
	> 300	Low	SANICRO 30[3]
H_2S-containing solutions	> 60	None	316
	> 60	Low	2RE 60[2]
	> 60	High	2RE 65[1]
	< 60	None	316
	< 60	Present	2RE 65[1]

(1) 2RE65 Fe 25 Ni, 19.5 Cr, 4.5 Mo, 1.5 Cu
(2) 3RE60 Fe 18.5 Cr, 4.5 Ni, 1.6 Si, 2.5 Mo, 0.02 C
(3) SANICRO 30 Fe–Cr–Ni; small additions of Mn, Cu, Al, Ti.

Table 31.13 STAINLESS STEEL GRADE SELECTION IN ENVIRONMENTS WITH RISK OF STRESS CORROSION—*continued*

Solution	*Conditions* Temp. °C	Chloride content	*Recommended grade*
Caustic solutions	<120	Caustic conc. <20 wt. %	304, 316
		>20 wt. %	SANICRO 30[(3)]
Polythionic acid	—	None	321 or 346 stabilized
	—	Present	SANICRO 30[(3)]

(1) 2RE65 Fe 25 Ni, 19.5 Cr, 4.5 Mo, 1.5 Cu
(2) 3RE60 Fe 18.5 Cr, 4.5 Ni, 1.6 Si, 2.5 Mo, 0.02 C
(3) SANICRO30 Fe–Cr–Ni; small additions of Mn, Cu, Al, Ti.

31.8 Hydrogen embrittlement

Stress corrosion cracking susceptibility can be connected with hydrogen embrittlement. Hydrogen embrittlement may be caused by pick up of hydrogen from melting, welding, electroplating, corrosion reactions or hydrogen gas storage. Table 31.14 gives the order of susceptibility to hydrogen embrittlement of a range of common metals.

Table 3.14 SUSCEPTIBILITY TO HYDROGEN EMBRITTLEMENT

Susceptibility ranking in decreasing susceptibility	*Material*
High susceptibility	High strength steels at high yield strength High strength nickel steels Medium strength and low strength steels Iron–silicon single crystals Cobalt alloys
Low susceptibility	Stainless steels: 310, 304
Little or no susceptibility	Copper alloys
	Stabilized stainless steels Aluminium alloys Molybdenum

Note: Carbon and alloy steels in H_2S atmospheres should be annealed and tempered to a strength less than 800 MPa.

31.9 Fracture toughness under corrosive conditions

Linear elastic fracture mechanics applied to stress corrosion cracking contributes to the quantitative assessment of the effect of crack size and crack growth rate. The crack rate velocity depends uniquely on the plain strain stress intensity, *KI*, and there is a common relationship between these parameters. Tolerable flaw sizes and acceptable slow crack growth rate for many structures can be calculated from this parameter.

31.9.1 Stress corrosion intensity factor *KISCC*

KISCC may be defined as the minimum stress below which subcritical size cracks do not grow. For certain environments and flaw sizes, the stress intensity may be exceeded, if in the specific environment the crack growth rate is not excessive.

$$\text{Critical crack size } C = X\left(\frac{KISCC}{\sigma_y}\right)^2$$

where X is a factor dependent on the crack geometry,
σ_y is the yield stress,
C is maximum allowable crack length if crack growth is to be avoided.

Note: If $KISCC$ is in MPa $m^{1/2}$ and σ_y in MPa, then C is in metres. For stress intensity greater than $KISCC$ a crack will grow to the critical crack size usually at a constant crack velocity, the magnitude of which depends on the chemical environment. If the crack velocity is sufficiently low, the structure may be safe within its design life. Temperature also affects the magnitude of the crack velocity.

See also Chapter 21, 'Mechanical Testing'.

Table 31.15 KIC AND KISCC VALUES
Assume critical crack size $C = 200\ (K/\sigma)^2$, where σ = yield stress, in MPa; C = crack length, in mm.

Alloy	*in air* *KIC* MPa $m^{1/2}$	C mm	*Environment which induces stress corrosion*	*KISCC* MPa $m^{1/2}$	C mm
Copper alloys					
Cu–30% Zn	200	13	NH_4OH pH7	1	0.003
Iron–steel alloys					
Mild steel 0.2 C 0.8 Mn	120	15	10 M NaOH (boiling)	1	0.001
High strength steels					
Steel	36	—	Distilled water	20	—
reinforcing	31	—	$Ca(OH)_2$ (sat.) + NaCl	22	—
bar	31	—	pH 120		
	42	—	above but coupled to Mg	18	—
			above but stress relieve at 430 °C	25	—
Martensite steel			yield stress 1400 MPa		
0.47 C 1.14 Cr			Natural sea water	12–	0.015–
0.82 Mn 0.6 Ni	~100	–1.0	3.3% NaCl	20	0.04
1.0 Mo Fe rem (D6–AC)			Distilled water		
			Hydrazine inhibitor 2%	25	—
High carbon steel					
0.84 C 0.26 Si 0.86 Mn	77	0.5	Tensile strength 1500 MPa	48	0.2
			Distilled water 600 ppm Cl′ + 1300 ppm SO_4''	48	0.2
			Cathodic protection	42	0.16
High strength alloy steel			3.5% NaCl, 20 °C		
			Heat treatment to		
4340 steel	66	0.5	1420 MPa	20	0.05
0.3 C, 0.63 Mn			1000 MPa	94	1.0
0.87 Cr, 0.39 Mo,			H_2S gas pressure 0.5 MPa	33	—
2.29 Ni			H_2S gas pressure 0.35 MPa	30	—
Stainless steels					
13% Cr steel (0.2% C)	60	—	Distilled water 23 °C	17	—
			3.5% NaCl	12	0.008
18% Cr, 8% Ni,	200	75	42% $MgCl_2$–boiling	10	0.18

Ultra high strength steels virtually independent of composition variation, P and S have little effect on *KIC* and *KISCC*.

Table 31.15 KIC AND KISCC VALUES—*continued*

Alloy	*in air* *KIC* $MPa\,m^{1/2}$	*C* mm	*Environment which induces stress corrosion*	*KISCC* $MPa\,m^{1/2}$	*C* mm
Stainless steels—cont.					
16% Cr, 4% Ni, 4% Cu, 1% Mn, 1% Si,			3.5% NaCl		
Nb–Ti–0.3%, 0.07% C.	187	—	tempered at 1150 °C	140	—
	124	—	tempered at 900 °C	87	—
(17-PH alloy)			Cathodic protection		
			by coupling with Zn	33	—
			(900 °C temper) with Al	60	—
			with Mg	29	—
High alloy steels					
18% Mn, 5% Ni	145	—	Hot aqueous halide solutions	8	0.01
13% Co, 10% Ni, 1% Mo, 0.15% C. Prec. hardened	140	1.4	3.5% NaCl at 20 °C		
			at 0.2% proof stress 1660 MPa	19	0.03
			at 0.2% proof stress 1440 MPa	33	0.10
9% Ni, 4% Co,	67	0.4	Martensitic yield at 1650 MPa	15	0.02
0.45% C	100	1.0	Bainitic, yield at 1500 MPa	18	0.04
Aluminium alloys			3.5% NaCl + 0.2 M $Na_2Cr_2O_7$ + 0.07 M Na acetate, + acetic acid to pH 4.0		
Type 2024	18	—	Resistant temper T851	14	—
	25	—	Susceptible, temper T351	12	—
Type 2219	22	—	Resistant, temper T87	22	—
	22	—	Susceptible, temper T37	12	—
Type 7075	24	—	Resistant, temper T 7351	20	—
	20	—	Susceptible, temper T651	4	—
Ag–3Mg–7 Zn	25	—	Aqueous halides at 20 °C	5	0.04
Titanium alloys			NaCl solutions		
Ti–6Al–2Nb 1% Ta + 0.8% Mo	150	9	Heat treated to 700 MPa	116	5
	138	6	Heat treated to 800 MPa	44	0.6
Ti–6Al–1V	60	1.1	0.6 M KCl	20	0.12
Ti–3Al–8V	35	0.2	Methyl alcohol–HCl	6	0.006

REFERENCES TO TABLE 31.15.

H. L. Craig (Ed.), 'Stress corrosion—new approaches', ASTM–STP 610, Philadelphia (1975).
R. W. Staehle *et al.*, 'Stress corrosion cracking and hydrogen embrittlement of iron-based alloys', NACE, 5 (1973).

31.10 Uniform corrosion

Table 31.16 UNIFORM CORROSION* IN INDUSTRIAL ATMOSPHERE

Alloy	*Uniform rate of corrosion* $mm\,yr^{-1}$
Mild steel	0.125
Wrought iron	0.200 (% Si < 0.1%)
Wrought iron	0.100 (% Si > 0.3%)
Copper–bearing mild steel	0.100
Fe, 1% Cr, 0.5% Cu	0.075
Aluminium	0.005
Zinc	0.006
Copper	0.007
70/30 brass	0.020
Nickel	0.010
80 Ni 20 Cr	0.004
18 Cr 8 Ni stainless steel	0.001

* Pitting may take place and penetration can be 2–5 times uniform values.

Table 31.17 UNIFORM CORROSION* IN SPECIAL ATMOSPHERES

Environment	*Rate of uniform corrosion for mild steel* mm yr^{-1}
Domestic kitchens and bathrooms	0.0025–0.001
Laundry	0.0075
Sulphuric acid plant	0.048
Paper mill	0.068
Pickling sheet steel	>0.45

* Pitting may occur and penetration may be 2–5 times uniform rate.

31.11 High temperature oxidation resistance

Oxidation of metals in air produces a relatively thick oxide scale dependent on temperature. Above certain high temperatures the scale becomes excessive and will spall away to give wastage. For close fitting components, such as bolts, values etc. this could cause seizure and cracking.

Table 31.18 HIGH TEMPERATURE OXIDATION OF STEELS

Steel	*Temperature* °C *for oxidation* >0.127 mm yr^{-1}
Plain carbon	580
Fe 0.5% Mo	600
Fe 0.5% Cr, 0.5% Mo	600
Fe 2.25% Cr, 1% Mo	613
Fe 5.0% Cr, 0.5% Mo	607
Fe 9% Cr, 1% Mo	670

Table 31.19 RECOMMENDED TEMPERATURE LIMIT FOR OXIDATION OF VARIOUS ALLOYS IN AIR

Metal	*Temperature limit* °C
Carbon steel	450
½% Mo steel	500
½% Mo 1% Cr steel	550
½% Mo, 5% Cr steel	550
1% Mo, 9% Cr steel	550
12% Cr, Mo, V, steel	575
18% Cr 8% Ni stainless steel	650
12% Cr 8% Ni 1% Nb steel	650
19% Cr, 11% Ni 2% Si steel	1000
23% Cr, 14% Ni steel	1000
23% Cr, 30% Ni steel	1100
80% Ni, 20% Cr	900
60% Ni, 20% Cr 20% Co, Al, Ti	900
55% Ni, 15% Cr 20% Co, 5% Mo 5% Al and Ti	1100

31.12 Contact corrosion

Contact corrosion can be a serious problem in packaging and in electronics. As miniaturization and sophistication of electronic devices have increased, the hazard presented by corrosion is often the limiting factor inhibiting the attainment of expected levels of reliability.

Semiconducting devices, switches and miniaturized VHF circuits are all particularly sensitive to the slightest reaction on critical surfaces, and in devices calling for the highest levels of reliability even the most inert of the phenolic, epoxide and silicone resins are not considered to be fully acceptable; corrosion of electronic assemblies may often be enhanced by migration of ions to sensitive areas under applied potentials, and by local heating effects associated with current flows.

For more information see: P. D. Donavan in L. L. Shreir, 'Corrosion', Vol. 2., Butterworths, London (1976).

Table 31.20 CONTACT CORROSION DANGERS: POLYMERS ETC.

Material	*Severity of corrosion**	*Volatiles evolved and remarks*
Rubbers, elastomers and adhesives		
1. Natural rubber		
(*a*) Non-vulcanized	Slightly corrosive on prolonged exposure	Formic and acetic acid evolved
(*b*) Vulcanized	Slightly–moderately corrosive	Hydrogen sulphide and sulphur dioxide evolved
2. (*a*) Synthetic rubbers	Non-corrosive at ambient condition—most are corrosive above 100 °C	Many are chlorinated and evolve HCl on heating; Hypalon may also emit sulphur dioxide
(*b*) Polysulphide rubbers (cold curing)	Moderately corrosive–very corrosive	Formic acid; the catalysts used are peroxides
3. Silicone polymers	Non-corrosive–very corrosive	Acetic and formic acids. Some single-pack silicone sealants cure by hydrolysis of acetoxy groups releasing acetic acid and are very corrosive; some two-pack formulations evolve formic acid and are corrosive, and others are reputed to be among the most inert polymers
4. Phenol and ureaformaldehyde glues	Slightly corrosive–very corrosive	Formaldehyde, phenol, ammonia and HCl may be evolved. Various acids and salts that yield acids (e.g. formic acid and hydrochloric acid) are used in cold-set formulations. Volatiles evolved during cure may be absorbed by the materials being bonded
Thermoplastics		
1. Polyvinyl chloride (PVC) (and other chlorinated thermoplastics)	Non-corrosive at ambient temperature (but *see* column 3); moderately–very corrosive at 70 °C	Hydrogen chloride (HCl). May become corrosive at ambient temperature if irradiated with UV radiation or in the presence of certain contaminants, e.g. zinc ions
2. Fluorinated thermoplastics (e.g. PTFE)	Non-corrosive at ambient and moderate temperatures; very corrosive above about 350 °C	Decompose to release HF and F_2
3. Nitrocellulose	Slightly–very corrosive	Oxides of nitrogen may be evolved progressively with ageing
4. Nylons		
(*a*) Nylon 6	Corrosive	Acetic acid; formulations frequently contain acetic acid additions as molecular weight regulators
(*b*) Nylon 66	Non-corrosive	

Table 31.20 CONTACT CORROSION DANGERS: POLYMERS ETC.—*continued*

Material	*Severity of corrosion**	*Volatiles evolved and remarks*
5. PVA (polyvinyl acetates and alcohols)	Non-corrosive–very corrosive	Acetic acid released; corrosivity dependent on conditions and formulation (degree of hydrolysis and presence of stabilizers and inhibitors)
6. Cellulose acetate	Slightly corrosive	Acetic acid may be released
7. Polyacetals		
(*a*) Homopolymer	Slightly corrosive at ambient temperature, more corrosive above 40 °C	Acetic acid and formic acid evolved (acetic acid may be used as an end-stopper)
(*b*) Copolymer (formaldehyde and 10% ethylene oxide)	Usually non-corrosive at ambient temperatures, corrosive above 45 °C	Formic acid evolved (if arduous moulding conditions have been used, the polymer may be corrosive at ambient temperatures)
8. Polyolefines, polyesters, polycarbonates, polystyrene, polysulphone, polyphenylene oxide and polymethylmethacrylate	Non-corrosive at ambient temperatures	
Thermosetting resins		
1. Cross-linked polyesters		
(*a*) Cold cured polyesters	MEKP catalyst and cobalt naphthenate accelerator—very corrosive. Other peroxide catalyst systems slightly–moderately corrosive. Irradiation or non-oxidizing catalyst } non-corrosive	Formic and acetic acids evolved. Corrosivity is determined largely by the catalyst used, but is also affected by the formulation, in particular diethylene glycol gives more corrosive resins than does propylene glycol
(*b*) Hot cured polyesters	Non-corrosive–moderately corrosive	

* Refers directly to Zn, Mg and steel.

Defence Guides, DG–3A, 'Prevention of corrosion of zinc and cadmium coatings by vapours from organic materials', HMSO.

Table 31.21 CORROSION BY CONTACT WITH WOOD

Wood	*Classification*	*Typical* pH *values*
Oak	Most corrosive	3.35, 3.45, 3.85, 3.9
Sweet chestnut	Most corrosive	3.4, 3.45, 3.65
Steamed European beech	Moderately corrosive	3.85, 4.2
Birch	Moderately corrosive	4.85, 5.05, 5.35
Douglas fir	Moderately corrosive	3.45, 3.55, 4.15, 4.2
Gahoon	Moderately corrosive	4.2, 4.45, 5.05, 5.2
Teak	Moderately corrosive	4.65, 5.45
Western red cedar	Moderately corrosive	3.45
Parana pine	Least corrosive	5.2 to 8.8
Spruce	Least corrosive	4.0, 4.45
Elm	Least corrosive	6.45, 7.15
African mahogany	Least corrosive	5.1, 5.4, 5.55, 6.65
Walnut	Least corrosive	4.4, 4.55, 4.85, 5.2
Iroko	Least corrosive	5.4, 6.2, 7.25
Ramin	Least corrosive	5.25, 5.35
Obeche	Least corrosive	4.75, 6.75

REFERENCE TO TABLE 31.21

V. R. Gray, *J. Inst. Wood Sci.*, 1958, **1**, 58.

32 Electroplating and metal finishing

The processes and solutions described in this section are intended to give a general guide to surface finishing procedures. To operate these systems on an industrial scale would normally require recourse to one of the Chemical Supply Houses which retail properietary solutions. This particularly applies to electroplating baths containing brighteners.

32.1 Polishing compositions

The following abrasive powders are used for polishing metal.

ALOXITE

Aluminium oxide made by fusing bauxite. Used for cutting down in the same way as emery.

ALUMINA

Certain grades of alumina are used for polishing stainless steel and chromium. The material is generally used in the form of a composition in which the powder is mixed with stearines or other fats.

EMERY POWDER

Used principally in cutting down and for preliminary operations. It is applied to the mop by means of an adhesive, usually glue. Emery powder is an impure aluminium oxide containing about 50–60% Al_2O_3, 30–40% magnetite and small amounts of ferric oxide, silica, chromium, etc. Emery powder should never be used on magnesium or aluminium components because of the adverse effect on corrosion resistance.

TRIPOLI

A calcined diatomaceous earth used for polishing brass, steel and aluminium. It is used generally in the intermediate stages, and is usually compounded with stearines and paraffin wax to make a polishing composition which can be used directly on a mop.

CROCUS POWDER

A polishing composition consisting essentially of ferric oxide, of coarser grade than rouge, used for polishing iron and steel, and also, tin. Usually compounded with stearine and used with a mop or fibre brush.

ROUGE

A high-grade ferric oxide supplied in various degrees of fineness. It can be used in the form of a paste directly on to a soft mop or can be made into a composition with stearine. It is used essentially for finishing to obtain a very high polish on gold, silver, brass, aluminium, etc.

BLACK ROUGE

This consists of black oxide of iron and is sometimes used for finishing operations.

GREEN ROUGE

Chromic oxide used for polishing chromium and stainless steel and can be used either in the form of a composition mixed with stearine or as a paste applied directly to the mop.

VIENNA LIME

Used for making the white finish for polishing nickel, etc. It consists of a calcined dolomite and contains about 60% calcium oxide and 40% magnesia.

CARBORUNDUM

Silicon carbide used for low tensile strength materials, e.g. brass, copper, aluminium, etc. and also brittle metals, such as hard alloys and cast irons.

32.2 Cleaning and pickling processes

VAPOUR DEGREASING

Used to remove excess oil and grease. Components are suspended in a solvent vapour, such as tri- or tetrachloroethylene.
Note: Both vapours are toxic and care should be taken to ensure efficient condensation or extraction of vapours.

EMULSION CLEANING

An emulsion cleaner suitable for most metals can be prepared by diluting the mixture given below with a mixture of equal parts of white spirit and solvent naphtha.

Pine oil	62 g
Oleic acid	10.8 g
Triethanolamine	7.2 g
Ethylene glycol–monobutyl ether	20 g

This is used at room temperature and should be followed by thorough swilling.

Table 32.1 ALKALINE CLEANING SOLUTIONS

Metal to be cleaned	*Composition of solution*			*Temperature*		*Remarks*
		oz gal^{-1}	g l^{-1}	°F	°C	
All common metals other than aluminium and zinc, but including magnesium	Caustic soda (NaOH)	6	37.5	180–200	80–90	For heavy duty
	Soda ash (Na_2CO_3)	4	25.0			
	Tribasic sodium phosphate ($Na_3PO_4.12H_2O$)	1	6.2			
	Wetting agent	$\frac{1}{4}$	1.5			

Table 32.1 ALKALINE CLEANING SOLUTIONS—*continued*

Metal to be cleaned	*Composition of solution*	oz gal^{-1}	g l^{-1}	*Temperature* °F	°C	*Remarks*
	Caustic soda	2	12.5	180–200	80–90	For medium duty
	Soda ash	4	25.0			
	Tribasic sodium phosphate	2	12.5			
	Sodium metasilicate ($Na_2SiO_3.5H_2O$)	2	12.5			
	Wetting agent	$\frac{1}{8}$	0.75			
	Tribasic sodium phosphate	4	25.0	180–200	80–90	For light duty
	Sodium metasilicate	4	25.0			
	Wetting agent	$\frac{1}{8}$	0.75			
Aluminium and zinc	Tribasic sodium phosphate	2	12.5	180–200	80–90	—
	Sodium metasilicate	4	25.0			
	Wetting agent	$\frac{1}{8}$	0.75			
Most common metals	Soda ash	2	12.5	180–200	80–90	Electrolytic cleaner, 6 V Current density 100/A ft^{-2} (10/A dm^{-2}) Article to be cleaned may be made cathode or anode or both alternately
	Tribasic sodium phosphate	4	25.0			
	Wetting agent	$\frac{1}{4}$	1.5			
Most common metals	Soda ash	6	37.5	Room	Room	May be used electrolytically
	Caustic soda	1	6.25			
	Tribasic sodium phosphate	2	12.5			
	Sodium cyanide (NaCN)	2	12.5			
	Sodium metasilicate	1	6.25			
	Wetting agent	$\frac{1}{8}$	0.75			

Table 32.2 PICKLING SOLUTIONS

Metal to be pickled	*Composition of solution*	oz gal^{-1}	g l^{-1}	*Temperature* °F	°C	*Remarks*
Alumimium (wrought)	*For etching* Tribasic sodium phosphate ($Na_3PO_4.12H_2O$)	6	37.5	180–200	80–95	Articles dipped until they gas freely, then swilled, and dipped in nitric acid 1 part by vol. to 1 of water (room temperature)
	For deeper etching Caustic soda (NaOH)	8	56	104–176	40–80	Conditions as above

* Sulphuric acid, pure comcl. grade, s.g. 1.84.

Note:

It is almost universal practice to use an inhibitor in the pickling bath. This ensures dissolution of the scale with practically no attack on the metal. Inhibitors are usually of the long chain amine type and often proprietary materials. Examples are Galvene and Stannine made by ICI and Golpanol by Badische Anilin and Soda Fabrik. As an example of the amount of inhibitor used:
Golpanol B—HCl or H_2SO_4 baths ranging from 5 to 20%; Liquid Golpanol B from about 0.15 to 0.6%; Solid Golpanol about $\frac{1}{4}$ of this.
The inhibitors may also be used in baths for pickling stainless steels.

Table 32.2 PICKLING SOLUTIONS—*continued*

Metal to be pickled	*Composition of solution*			*Temperature*		*Remarks*
		oz gal^{-1}	g l^{-1}	°F	°C	
Aluminium (cast and wrought)	Nitric acid, s.g. 1.42	3 gal	3 l	Room	Room	Articles first cleaned in solvent degreaser. Use polythene or PVC tanks
	Hydrofluoric acid (52%)	1 gal	1 l			
	Bright dip					
	Chromic acid	0.84 oz	5.2 g	195	90	Immerse for 1½ min. Solution has limited life. AR chemicals and deionized or distilled water should be used. USP 2 593 448; 2 593 449)
	Ammonium bifluoride	0.72 oz	4.5 g			
	Cane syrup	0.68 oz	4.2 g			
	Copper nitrate	0.04 oz	0.25 g			
	Nitric acid (s.g. 1.4.)	4.8 oz	30 ml			
	Water (distilled) to	1 gal	1 l			
Aluminium and other non-ferrous metals	*Bright dip*					
	Phosphoric acid (s.g. 1.69)	9.4 gal	9.4 l	195	90	Immerse for several min. Agitate work and solution. Good ventilation necessary. Addition of acetic acid useful with some alloys (BP 659 747)
	Nitric acid (s.g. 1.37)	0.6 gal	0.6 l			
Copper and copper alloys	*To remove scale*					
	Sulphuric acid	1 gal	1 l	150–170	65–75	After pickling articles can be dipped in sodium cyanide: 4 oz gal^{-1} (25 g l^{-1}) to remove tarnish
	Water	4 gal	4 l			
	Or					
	Sulphuric acid*	1 gal	1 l	70–175	20–75	This solution leaves a slight passive film which helps to prevent tarnish
	Sodium dichromate ($Na_2Cr_2O_7.2H_2O$)	12 oz	75 g			
	Water	4 gal	4 l			
	Bright dip					
	Sulphuric acid*	2 gal	2 l	Room	Room	If any scale first dip in spent bright dip. Remove stains by dipping in sodium cyanide 4 oz gal^{-1} (25 g l^{-1})
	Nitric acid	1 gal	1 l			
	Water	1 gal	1 l			
	Hydrochloric acid	0.5 oz	25 ml			
	Matt dip					
	Sulphuric acid*	1 gal	1 l	160–180	70–80	If the finish is too fine add nitric acid. If too coarse add sulphuric acid
	Nitric acid (s.g. 1.42)	1 gal	1 l			
	Zinc oxide (ZnO)	2 lb	200 g			
	Semi-matt dip					
	Sodium dichromate	3 oz	19 g	Room	Room	
	Sulphuric acid	18 oz	114 g			
	Water	1 gal	1 l			
Iron and steel	*Slow pickle to loosen heavy scale*					
	Sulphuric acid	2%	—	Room	Room	Leave for several hours or overnight
	Glue size	0.25%	—			

* Sulphuric acid, pure comcl. grade, s.g. 1.84.

Note:

It is almost universal practice to use an inhibitor in the pickling bath. This ensures dissolution of the scale with practically no attack on the metal. Inhibitors are usually of the long chain amine type and often proprietary materials. Examples are Galvene and Stannine made by ICI and Golpanol by Badische Anilin and Soda Fabrik. As an example of the amount of inhibitor used:
Golpanol B—HCl or H_2SO_4 baths ranging from 5 to 20%; Liquid Golpanol B from about 0.15 to 0.6%; Solid Golpanol about ¼ of this.
The inhibitors may also be used in baths for pickling stainless steels.

Table 32.2 PICKLING SOLUTIONS—*continued*

Metal to be pickled	*Composition of solution*			*Temperature*		*Remarks*
		oz gal^{-1}	g l^{-1}	°F	°C	
Iron and steel *continued*	*To remove scale*					
	Sulphuric acid	10%	—	120–180	50–80	Or hydrochloric acid 10–20%
	Bright dip					
	Oxalic acid crystals	4 oz	25 g	Room	Room	This solution has so far only been used on an experimental basis
	Hydrogen peroxide (100 vol.)	2 oz	13 g			
	Sulphuric acid (10%)	0.02 oz	0.1 g			
	Water to	1 gal	1 l			
	Anode etching					
	Sulphuric acid*	1 gal	1 l	Not above 75	Not above 25	Current density: 200 A ft^{-2} (20 A dm^{-2})
	Water	2 gal	2 l			
	For polished work					
	Sulphuric acid*	—	—	Not above 75	Not above 25	Density must not fall below 1.61 g cm^{-3} or work will be etched
Magnesium and magnesium alloys	*General cleaner*					
	Chromic acid	16–32	100–200	Up to b.p.	Up to b.p.	For removal of oxide films, corrosion products, etc. Should not be used on oily or painted material
	Sulphuric acid pickle					
	Sulphuric acid	4.8	30	Room	Room	Should be used on rough castings or heavy sheet only. Removes approx. 0.002 in 20–30 s
	Nitro-sulphuric pickle					
	Nitric acid	12.8	80	Room	Room	
	Sulphuric acid	3.2	20			
	Bright pickle for wrought products					
	Chromic acid	23	150	Room	Room	Lustrous appearance. Involves metal removal
	Sodium nitrate	4	25			
	Calcium or magnesium fluoride	$\frac{1}{8}$	$\frac{3}{4}$			
	Bright pickle for castings					
	Chromic acid	$37\frac{1}{2}$	235	Room	Room	
	Concentrated nitric acid (70%)	$3\frac{1}{4}$	20			
	Hydrofluoric acid (50%)	1	6.2			

* Sulphuric acid, pure comcl. grade, s.g. 1.84.

Note:

It is almost universal practice to use an inhibitor in the pickling bath. This ensures dissolution of the scale with practically no attack on the metal. Inhibitors are usually of the long chain amine type and often proprietary materials. Examples are Galvene and Stannine made by ICI and Golpanol by Badische Anilin and Soda Fabrik.
As an example of the amount of inhibitor used:
Golpanol B—HCl or H_2SO_4 baths ranging from 5 to 20%; Liquid Golpanol B from about 0.15 to 0.6%; Solid Golpanol about $\frac{1}{4}$ of this.
The inhibitors may also be used in baths for pickling stainless steels.

Table 32.2 PICKLING SOLUTIONS—*continued*

Metal to be pickled	*Composition of solution*	oz gal^{-1}	g l^{-1}	*Temperature* °F	°C	*Remarks*
	Acetic acid	8 approx.	50 approx.	Room	Room	Special purpose pickles
	Citric acid	8 approx.	50 approx.	Room	Room	Special purpose pickles
Stainless steel	*To loosen scale*					
	Sulphuric acid	13–30	80–180	130–160	60–70	Use prior to scale removal treatment, for heavy scales.
	Hydrochloric acid (s.g. 1.16)	6–20	40–120			
	To remove scale					
	Nitric acid (s.g. 1.4)	32	200	130–150	55–65	
	Hydrofluoric acid (52% HF)	6	40			
	Or					
	Sulphuric acid	10	60	Room	Room	
	Hydrofluoric acid	10	60			
	Chromic acid (CrO_3)	10	60			
	Bright pickle					
	Hydrochloric acid (s.g. 1.16)	40	250	140–160	60–70	
	Nitric acid (s.g. 1.4)	3	22			
	White matt finish					
	Ferric sulphate [$Fe_2(SO_4)_3$]	13	80	160–180	70–80	5–15 min
	Hydrofluoric acid (52% HF)	6	40			
Zinc and zinc alloys	*Bright dip*					
	Chromic acid (CrO_3)	40	250	Room	Room	5–30s. If yellow film persists after rinsing dip in sulphuric acid: 1 fl oz per gal (6 ml l^{-1}) and rinse again
	Sodium sulphate (Na_2SO_4)	3	19			

* Sulphuric acid, pure comcl. grade, s.g. 1.84.

Note:

It is almost universal practice to use an inhibitor in the pickling bath. This ensures dissolution of the scale with practically no attack on the metal. Inhibitors are usually of the long chain amine type and often proprietary materials. Examples are Galvene and Stannine made by ICI and Golpanol by Badische Anilin and Soda Fabrik. As an example of the amount of inhibitor used:
Golpanol B—HCl or H_2SO_4 baths ranging from 5 to 20%; Liquid Golpanol B from about 0.15 to 0.6%; Solid Golpanol about $\frac{1}{4}$ of this.
The inhibitors may also be used in baths for pickling stainless steels.

32.3 Anodizing and plating processes

Table 32.3 ANODIZING PROCESSES FOR ALUMINIUM
Good ventilation above the bath and agitation of the bath is advisable in all cases.

Composition of solution			*Temperature*		*Current density* amp ft^{-2} (A dm^{-2})	*Time and voltage*	*Cathodes*	*Vat*	*Hangers*	*Remarks*
	oz gal^{-1}	g l^{-1}	°F	°C						
Chromic acid (CrO_3), chloride content must not exceed 0.2 g l^{-1}, sulphate less than 0.5 g l^{-1} (After Bengough-Stuart)	5–16	30–100	103–108	38–42	Current controlled by voltage. Average 3–4 (0.3–0.4) d.c.	†1–10 min 0–40 V increased in steps of 5 V 5–35 min Maintain at 40 V 3–5 min Increase gradually to 50 V 4–5 min Maintain at 50 V	Tank or stainless steel	Steel (exhausted)	Pure aluminium or titanium	Slight agitation is required. This process cannot be used with alloys containing more than 5% copper
Sulphuric acid (s.g. 1.84)	32	200	60–75	15–24	10–20 (1–2) d.c.	12–18 V 20–40 min	Aluminium or lead plates (tank if lead lined)	Lead lined steel	Pure aluminium or titanium	The current must not exceed 0.2 A l^{-1} of electrolyte
Hard anodizing Hardas process Sulphuric acid	32	200	23–41	−5–+5	25–400 (2.5–40) d.c.	40–120 V	Lead	Lead lined steel	Aluminium or titanium	Agitation required. Gives coating 1–3 thou. thick
Eloxal GX process Oxalic acid $(COOH)_2.2H_2O$	12.8	80	70	20	10–20 (1–2) d.c.	50 V 30–60 min	Vat lining	Lead lined steel	Aluminium or titanium	Oxalic acid processes are more expensive than sulphuric acid anodizing; but coatings are thicker and are coloured.
Eloxal WX process Oxalic acid	12.8	80	75–95	25–35	20–30 (2–3) a.c.	20–60 V 40–60 min	Vat lining	Lead lined steel	Aluminium or titanium	
Integral colour Anodizing Kalcolor process Sulphuric acid	0.8	5	72	22	30 (3) d.c.	25–60 V 20–45 min	Lead	Lead lined steel	Aluminium or titanium	Aluminium level in solution must be maintained between 1.5 and 3 g l^{-1}
Sulphosalicylic acid	16	100								

† Period according to degree of protection. Complete cycle normally 40 min.

Table 32.4 ANODIZING PROCESSES FOR MAGNESIUM ALLOYS

Composition of solution			*Temperature*		*Current density* A ft^{-2} ($A\ dm^{-2}$)	*Time and voltage*	*Cathodes*	*Vat*	*Hangers*	*Remarks*
	oz gal^{-1}	$g\ l^{-1}$	°F	°C						
HAE process										
Potassium hydroxide	19.2	120	<95	<35	12–15 (1.2–1.5)	90 min at 85 V approx. a.c. preferred	Mg alloy for a.c. Mg or steel if d.c. used	Mild steel or rubber lined	Mg alloy	Matt hard, brittle, corrosion resistant, dark brown 25–50 μm thick, abrasion resistant
Aluminium	1.7	10.4								
Potassium fluoride	5.5	34								
Trisodium phosphate	5.5	34								
Potassium manganate (British Patents 777228 and 777229)	3.2	20								
Dow 17 process										
Ammonium bifluoride	39	232	160–180	70–85	5–50 (0.5–5)	10–100 min up to 110 V a.c. or d.c.	Mg alloy for a.c. Mg or steel for d.c.	Mild steel or rubber lined	Mg alloy	Matt dark green, corrosion resistant, 25 μm thick approx., abrasion resistant
Sodium dichromate	16	100								
Phosphoric acid 85% H_3PO_4 (British Patent 762195)	14	88								
Cr 22 process										
Chromic acid	4	25	165–205	75–95	15 (1.5)	12 min 380 V a.c.	—	Mild steel	Mg alloy	Matt dark green, corrosion resistant, 25 μm thick approx.
Hydrofluoric acid (50%)	4	25								
Phosphoric acid H_3PO_4 (85%)	13.5	84								
Ammonia solution	25–30	160–180								
MEL process Fluoride anodize										
Ammonium bifluoride (British Patent 721445)	16	100	<86	<30	5–100 (0.5–10)	30 min 120 V a.c. preferred	Mg alloy for a.c. Mg or steel for d.c.	Rubber lined	Mg alloy	Principally a cleaning process to improve corrosive resistance by dissolving or ejecting cathodic particles from the surface

32.4 Electroplating processes

Table 32.5 PLATING PROCESSES

Metal	Type and composition	Operating conditions									Anodes	Vat	Remarks
				Temperature		Current density		Current efficiency					
		oz gal^{-1}	g l^{-1}	°F	°C	A ft^{-2}	A dm^{-2}	%	Voltage	pH			
Aluminium	Aluminium chloride	65	400	60–140	15–60	20	2	100	—	—	Aluminium	Glass sealed	Operation must be in an atmosphere of nitrogen and the work introduced through a lock. Connections must not spark
	Lithium aluminium hydride	2	13										
	Diethyl ether	Solvent											
Antimony	Antimony oxide (Sb_2O_3)	7	45	130	55	25	2.5	—	—	3.6	Antimony or carbon	Hard rubber or rubber lined	—
	Potassium citrate	20	130										
	Citric acid	24	150										
Brass	Sodium cyanide (NaCN)	7.5	45	75–100	24–40	3–5	0.3–0.5	60–70 (cathode)	2–3	10.5–11.5	Brass (80/20) cast or rolled.	Steel or rubber lined steel	*Brightener:* 2 lb caustic soda in $\frac{1}{2}$ gal of water to which is added 1 lb white arsenic. Use 2–4 fl oz gal^{-1} solution (15–30 ml per 100 l). Free cyanide by analysis
	Copper cyanide (CuCN)	4	26										
	Zinc cyanide ($Zn(CN)_2$)	1.25	7.7										
	Sodium carbonate (Na_2CO_3)	4	26										
	Ammonia (sp. gr. 0.88)	0.2*	1.5*										
	Free cyanide	2.6	17										
Bronze imitation	*Zinc*												
	Sodium cyanide	5	33	Room for light colour, warm for red colour		2–4	0.2–0.4	—	2–3	—	Copper 92% Zinc 8%	Steel	—
	Copper cyanide	4	26										
	Zinc cyanide	0.3	2										
	Rochelle salt	2	13										
	($KNaC_4H_4O_6.4H_2O$)	0.3	2										
	Free cyanide												
	Cadmium	4.5	29	Room		2–5	0.2–0.5	—	2–3	—	Copper	Steel	Cadmium content maintained by addition of small quantities of cadmium oxide dissolved in sodium cyanide
	Sodium cyanide	3	20										
	Copper cyanide	0.25	1.5										
	Cadmium oxide	2	13										
	Sodium carbonate (Na_2CO_3)	1	6.5										
	Free cyanide												

* 0.2 fl oz gal^{-1} or 1.5 ml l^{-1}.

Table 32.5 PLATING PROCESSES

Metal	Type and composition			*Operating conditions*									*Remarks*
				Temperature		*Current density*		*Current efficiency*					
Metal	*Type and composition*	oz gal^{-1}	g l^{-1}	°F	°C	A ft^{-2}	A dm^{-2}	%	*Voltage*	pH	*Anodes*	*Vat*	*Remarks*
Bronze	Potassium cyanide	9	60	150	65	Up to 100	Up to 10	40–50	—	12.5	Copper	Steel	Must be kept free of bivalent tin. Maintain tin content by additions of potassium stannate
	Copper cyanide	4	26										
	Potassium stannate (K_2SnO_3)	5	33										
	Potassium hydroxide	1.5	10										
	Rochelle salt	6	40										
	Free cyanide	3	20										
Cadmium	Sodium cyanide	12–15	75–100	75–90	24–30	10–20	1–2	90	2–3	13	Cast cadmium or cadmium balls in a steel cage	—	*Brightener:* Organics (such as dextrin) or metallic (such as nickel salts)
	Cadmium oxide	3–5	20–33										
	Free cyanide	8–10	52–66										
	Addition agents	0.015–2.4	0.1–15										
Chromium	*Heavy solution*												
	Chromic acid (CrO_3)	72	450	100–120	40–50	120–200	12–20	12–15	4–5	—	Antimonial lead (7%)	Steel lined with antimonial lead 7%	This solution requires reducing: either boil with citric acid 2 oz gal^{-1} (12.5 g l^{-1}) or with tartaric acid 3 oz gal^{-1} (18 g l^{-1}) or with oxalic acid 4 oz gal^{-1} (25 g l^{-1})
	Sulphuric acid	0.72	4.5										
	Light solution												
	Chromic acid	40	250	100–120	40–50	120–200	12–20	12	4–5	—	Antimonial lead (7%)	Steel lined with antimonial lead 7%	Reduction as above: citric acid 1 oz gal^{-1} (6.25 g l^{-1}) or tartaric acid $1\frac{1}{2}$ oz gal^{-1} (9 g l^{-1}) or oxalic acid 2 oz gal^{-1} (12.5 g l^{-1})
	Sulphuric acid	0.4	2.5										
Copper	*Acid*												
	Copper sulphate ($CuSO_4.5H_2O$)	32	200	60–120	16–50	10–200	1–20	95–97	1–3	—	Pure copper	Lead or rubber lined wood or steel	The phenol is sulphonated by heating with its own weight of sulphuric acid to 120° C for 1 h before use. Agitation is necessary for high current density. Constant filtration is advisable
	Sulphuric acid	8.0	50										
	Phenol	0.16	1										

	Through-hole plating												
	Copper sulphate	14–17	88–110	75–90	24–30	10–45	1–45	—	—	—	Copper	Lead or rubber-lined steel	Chloride content serves as a deposit modifier
	Sulphuric acid	27–30	170–190										
	Chloride	> 15 ppm											
	Cyanide (strike)												
	Sodium cyanide (NaCN)	3	19	110–140	45–60	10–30	1–3	10–60	6	11–12	Pure copper rolled or extruded	Steel	Used to deposit thin undercoats for other metals
	Copper cyanide (CuCN)	2	13										
	Sodium carbonate (Na_2CO_3)	2	13										
	Cyanide (high efficiency)												
	Sodium cyanide	13	82	140–180	60–80	10–100	1–10	100	2–4	—	Oxide-free copper sheet	Steel	For rapid plating
	Copper cyanide	10	60										
	Sodium hydroxide	4	26										
	Cyanide (Rochelle)												
	Sodium cyanide	6	37.5	125–160	50–70	20–60	2–6	50–60	6	12.2–12.8	Copper, rolled and annealed	Steel	—
	Copper cyanide	4	26										
	Rochelle salt ($KNaC_4H_4O_6.4H_2O$)	8	50										
	Sodium carbonate	5	30										
	Free cyanide	0.5–1	3–6										
	Pyrophosphate												
	Copper pyrophosphate ($Cu_2P_2O_7.3H_2O$)	11	66	125–140	50–60	10–80	1–8	100	—	8–8.8	Copper	Steel	Commonly used for plating printed circuit boards. Use vigorous agitation
	Potassium pyrophosphate	45	300										
	Ammonium nitrate	1	6										
	Ammonia	0.1	0.6										
Gold	*Hard*												
	Potassium gold cyanide ($KAu(CN)_2$)	2	12	97	35	5–15	0.5–1.5	—	—	3–4.5	Insoluble	—	—
	Citric acid	16	105										
	Phosphoric acid	2	12.5 ml										
	Cobalt (as CoK_2EDTA)	0.16	1										
	Alkaline cyanide												
	Potassium cyanide	5	30	120–150	50–65	1–5	0.1–0.5	100	1.5–2	11	Fine gold (24 carat) or insoluble: stainless steel, platinum or graphite	Enamelled iron	If insoluble anodes are used, solution must be renewed periodically
	Potassium gold cyanide	2	12										
	Potassium carbonate	5	30										
	Dipotassium phosphate (K_2HPO_4)	5	30										

Table 32.5 PLATING PROCESSES—*continued*

Metal	*Type and composition*			*Operating conditions*							*Anodes*	*Vat*	*Remarks*
				Temperature		*Current density*		*Current efficiency*					
		oz gal^{-1}	g l^{-1}	°F	°C	A ft^{-2}	A dm^{-2}	%	*Voltage*	pH			
Indium	(1) Indium fluoborate	38	230	70–90	20–30	50–100	5–10	75	—	1	Part indium, part insoluble	—	Use fluoboric acid to adjust pH
	Boric acid	4.8	30										
	Ammonium fluoborate	7.5	47										
	(2) Indium (as hydroxide)	2.5–5	15–30	70–90	20–30	15–30	1.5–3	50	—	11–12	Steel	—	—
	Potassium cyanide	22–25	140–160										
	Potassium hydroxide	5–6	30–40										
	D-glucose	3–5	20–30										
Iron	Ferrous chloride ($FeCl_2.4H_2O$)	48	300	195	90	Up to 120	Up to 12	—	—	1.2–1.8	Pure iron	Lead or rubber lined	Agitation is desirable for high current densities
	Calcium chloride	50	335										
Lead	Lead fluoborate ($Pb(BF_4)_2$)	40	240	77–100	25–40	5–70	0.5–7	100	—	—	Pure lead free from antimony	Rubber lined steel	—
	Fluoboric acid	10	60										
	Boric acid (H_3BO_3) free	4.5	27										
	Glue	0.03	0.2										
Nickel	*Watts bath*												
	Nickel sulphate ($NiSO_4.6H_2O$)	50	350	110–150	45–65	50	5	95	—	3–4	Cast or rolled Ni (99–100%) bagged	Lead or rubber lined	Agitation desirable for high current densities. Constant filtration desirable. *Wetting agent*. Sod, lauryl sulphate
	Nickel chloride ($NiCl_2.6H_2O$)	7	45										
	Boric acid	6	37										
	Wetting agent	0.015–0.075	0.1–0.5										
	For plating Zinc and Zinc-base alloys												
	Nickel sulphate	12–17	75–112	70–90	20–32	10–30	1–3	—	3–4	5.3–5.8	Nickel (99–100%)	Lead or rubber lined	Agitation can be used
	Sodium sulphate (anhydrous)	12–17	75–112										
	Ammonium chloride	2.4–6	15–37.5										
	Boric acid	2.4	15										

	Sulphamate bath												
	Nickel sulphamate ($Ni(NH_2SO_3)_2$)	48	300	80–140	25–60	20–250	2–25	—	—	3.5–4.2	Nickel (99–100%)	Lead or rubber lined	Air or mechanical agitation
	Boric acid	4.8	30										
	Nickel chloride	1	6										
	Hard nickel												
	Nickel sulphate	28	180	110–140	43–60	20–100	2–10	—	6–8	5.6–5.9	Nickel (99–100%)	—	For building up worn parts
	Ammonium chloride	4	25										
	Boric acid	4.8	30										
	(1) Bright												
	Nickel sulphate	50	330	110–150	45–65	25–100	2.5–10	95	—	3–4	Nickel (99–100%)	Rubber lined steel	Bright nickel plating baths are basically Watts solutions containing brighteners
	Nickel chloride	7	45										
	Boric acid	6	38										
	Sodium naphthalene trisulphonate ($C_{10}H_7(SO_3)_3Na$)	5.6	35										
	(2) Bright (low metal)												
	Nickel sulphate	9.6	60	97–140	35–60	25–100	2.5–10	95	—	3.5–4.2	Nickel	Rubber lined steel	Agitation constant filtration necessary
	Nickel chloride	18	110										
	Boric acid	8	50										
	Black or grey												
	Nickel chloride	12	75	Room temperature		1.5	0.15	—	—	5	Nickel or insoluble	Rubber lined	—
	Ammonium chloride	4.8	30										
	Zinc chloride	4.8	30										
	Sodium thiocyanate (NaCNS)	2.4	15										
Palladium	Palladium (as $Pd(NH_3)_4Br_2$)	4.8	30	120	50	40	4	—	—	9.2	Insoluble	Glass or rubber lined	Can be used to produce thick deposits for electro forming
	Ammonium bromide	7	45										
Platinum	Platinum (as dinitrodiamino platinum)	1.6	10	203	95	70	7	10	2–4	—	Platinum or insoluble	Glass or rubber lined	Solution maintained by addition of platinum salt
	Ammonium nitrate	16	100										
	Sodium nitrite	1.6	10										
	Ammonia (sp. gr. 0.88)	7*	44*										
Rhodium	Rhodium (as sulphate concentrate)	0.32	2	104	40	10–40	1–4	—	3–6	—	Platinum or insoluble	Glass or rubber lined	During plating remove bubbles by cathode agitation
	Sulphuric acid	3.2**	20**										

* 7 fl oz gal^{-1} or 44 ml l^{-1}.
** 3.2 fl oz gal^{-1} or 20 ml l^{-1}.

Table 32.5 PLATING PROCESSES—*continued*

Metal	*Type and composition*	*Operating conditions*									*Anodes*	*Vat*	*Remarks*
				Temperature		*Current density*		*Current efficiency*					
		oz gal^{-1}	g l^{-1}	°F	°C	A ft^{-2}	A dm^{-2}	%	*Voltage*	pH			
Silver	Potassium cyanide	8–12	50–78	70–80	20–27	5–15	0.5–1.5	99–100	<1	—	Fine silver rolled	Lead or rubber lined	Cathode bar may be rocked. *Brightener:* Carbon bisulphide dissolved in silver solution
	Silver cyanide (80% Ag)	5–9	31–56										
	Potassium carbonate	2.5–14	15–90										
	Free cyanide	5.5–8	35–50										
	High speed												
	Silver cyanide (80%)	7–25	44–150	100–120	38–50	5–100	0.5–10	—	—	12	Pure silver (bagged)	Enamelled iron or rubber lined	Agitation is necessary and is usually effected by solution pumpong or cathode movement. Constant filtration advisable
	Potassium cyanide (92%)	11–38	70–240										
	Caustic soda	0.6–4.8	4–30										
	Potassium carbonate	2.5–14	15–90										
	Potassium nitrate	6.4–9.6	40–60										
	Strike solution for non-ferrous metals												
	Silver cyanide	0.7	4.5	Room temperature		15–20	1.5–2.0	—	4–6	—	Silver or steel	Steel or earthenware	—
	Potassium cyanide	13	80										
Tin	*Acid*												
	Sulphuric acid	8	50	68	20	10–100	1–10	~100	0.4–0.8	—	Pure tin baged in terylene	Lead or rubber. lined	Constant filtration advantageous. Periodic filtration essential. Use agitation at higher current densities
	Cresol sulphonic acid ($CH_3.C_6H_3OH.SO_3H$)	6.4	40										
	Stannous sulphate (90% $SnSO_4$)	10	65										
	Gelatin	0.3	2										
	Beta-naphthol	0.16	1										
	Alkaline												
	Caustic soda	1.6	10	167	75	5–30	0.5–3	85	4–6	13	Pure tin, (high speed 1% Al) or insoluble	Steel	Anode must be filmed for uniform dissolution
	Sodium stannate (48% SnO_2)	16	100										
	Immersion (on steel)												
	Stannous sulphate	0.16–0.32	1–2	200–232	90–100	—	—	—	—	—	—	—	Immersion time 5–20 min, work immersed in Monel or stainless steel baskets
	Sulphuric acid	0.8–2.5	5–15										

Tin–Nickel	Stannous chloride ($SnCl_2$)	8	50	154	68	10–30	1–3	100	2–3	2–2.5	Nickel	Rubber lined	Tin content maintained by regular additions of anhydrous stannous chloride
	Nickel chloride	48	300										
	Ammonium bifloride ($NH_4F.HF$)	9	56										
Tin–Lead	Tin (as fluoborate)	4	25	68–86	20–30	5–35	0.5–3.5	95	6–12	—	60/40 tin/lead alloy	Rubber lined	This process gives a 60/40 tin/lead deposit. Proprietary grain refiners are available to replace peptone
	Lead (as fluoborate)	2	12										
	Fluoboric acid	16	100										
	Peptone	0.8	4.5										
	Boric acid	5	30										
Zinc	*Acid*												
	Zinc sulphate ($ZnSO_4.7H_2O$)	60	370	100–130	30–55	100–300	10–30	100	—	3–4.5	Zinc (99.9%)	Lead or rubber lined	Agitation and constant filtration necessary for high current densities
	Sodium chloride	2.5	15										
	Boric acid	3.7	23										
	Aluminium sulphate ($Al_2(SO_4)_3$)	5	31										
	Dextrine (yellow)	2.5	15										
	Cyanide (decorative)												
	Sodium cyanide	8–22	50–140	68–120	20–50	25–150	2.5–15	~85	—	—	Pure Zinc	Steel	—
	Caustic soda	10–20	60–120										
	Zinc oxide (ZnO)	4–9	25–55										
	Sodium carbonate	3.2–20	20–120										
	Cyanide (protective)												
	Sodium cyanide	15–25	90–150	68–120	20–50	25–150	2.5–15	—	—	—	Pure Zinc free from lead	Steel	—
	Caustic soda	15–22	90–140										
	Zinc oxide	9–12	55–75										
	Sodium carbonate	5–12	30–75										
	Zincating												
	(on aluminium) Caustic soda	80	500	77	25	—	—	—	—	—	—	—	Improved adhesion by zincating stripping in 40% HNO_3 and rezincating
	Zinc oxide	16	100										

*7 fl oz gal^{-1} or 44 ml l^{-1}

**3.2 fl oz gal^{-1} or 20 ml l^{-1}

32.5 Plating processes for magnesium alloys

DOW PROCESS (H. K. DELONG)

This process depends on the formation of a zinc immersion coat in a bath of the following composition:

Component	*Concentration* oz gal^{-1}	$g\ l^{-1}$
Tetrasodium pyrophosphate	16	120
Zinc sulphate	5.3	40
Potassium fluoride	1.0	7

The treatment time is 3–5 min at a temperature of 175–185 °F (80–85 °C) with mild agitation. The pH of the bath should be 10–10.4.

The steps of the complete process are:

1. Solvent or vapour degreasing.
2. Hot caustic soda clean or cathodic cleaning in alkaline cleaner.
3. Pickle $\frac{1}{4}$–$1\frac{1}{2}$ min in 1% hydrochloric acid and rinse.
4. Zinc immersion bath as above without drying off from the rinse.
5. Cold rinse and immediately apply copper strike as under.

Component	*Concentration* oz gal^{-1}	$g\ l^{-1}$
Copper cyanide	4.2	26
Potassium cyanide	7.4	46
Potassium carbonate	2.4	15
Potassium hydroxide	1.2	7.5
Potassium fluoride	4.8	30
Free cyanide	1.2	7.5
pH	12.8–13.2	—
Temperature	140 °F (60 °C)	—

CONDITIONS

30–40 A ft^{-2} (3–4 A dm^{-2}) for $\frac{1}{2}$–1 min, reducing to 15–20 A ft^{-2} (1.5–2 A dm^{-2}) for 5 min or longer.

If required, the copper thickness from the above strike can be built up in the usual alkaline or proprietary bright plating baths. Following the above steps, further plating may be carried out in conventional electroplating baths.

ELECTROLESS PLATING ON MAGNESIUM

Deposits of a compound of nickel and phosphorus can be obtained on magnesium alloy components by direct immersion in baths of suitable compositions. Details of the process may be obtained from the inventors, The Dow Chemical Co. Inc., Midland, Michigan, USA.

'GAS PLATING' OF MAGNESIUM (VAPOUR PLATING)

Deposits of various metals on magnesium components (as on other metals) can be produced by heating the article in an atmosphere of a carbonyl or hydride of the metal in question. The process is operated by the Commonwealth Engineering Co. of Ohio Inc., Dayton, Ohio, USA.

32.6 Electroplating process parameters

Table 32.6 AVERAGE CURRENT EFFICIENCIES OF PLATING SOLUTIONS
The figures given below are approximate

	%
Cadmium (oxide)	85–95
Chromium	12–16
Copper (acid)	95–99
Copper (cyanide)	30–60
Copper (Rochelle)	40–65
Gold	70–85
Indium (cyanide)	30–50
Indium (sulphate)	70–90
Iron	90–95
Lead	90–100
Nickel	94–98
Silver	100
Tin (acid)	90–95
Tin (stannate)	70–85
Rhodium	35–40
Zinc (acid)	97–99
Zinc (cyanide)	85–90

Thickness of metal deposited per hour is given in mils (1 mil = 0.001 in) by

$$\frac{\text{CD} \times \text{W} \times \text{CE}}{237 \times \Delta}$$

where

CD = current density in A ft^{-2}
W = g A^{-1} h (Table 32.7)
CE = current efficiency (Table 32.6)
Δ = density of metal deposited

Table 32.7 THEORETICAL RELATIONS OF METAL AND CURRENT

	Metal deposited		*Current required*	
Metal	g A^{-1} h	oz A^{-1} h	A h lb^{-1}	A h kg^{-1}
Aluminium	0.335	0.011 8	1 356	2 989
Antimony (antimonious)	1.515	0.053 4	299	659
Cadmium	2.096	0.073 9	216	476
Chromium (hexavalent)	0.323 5	0.011 4	701	1 545
Cobalt	1.099	0.038 8	413	911
Copper (cuprous)	2.372	0.083 7	191	421
Copper (cupric)	1.186	0.041 8	383	844
Gold (auric)	2.452	0.086 5	185	408
Indium	1.428	0.050 3	318	701
Iron (ferrous)	1.042	0.036 8	435	959
Lead	3.866	0.136 3	117	258
Nickel	1.095	0.038 6	414	913
Palladium	1.990	0.070 2	228	503
Platinum	3.642	0.128 4	125	276
Rhodium	1.920	0.067 7	236	520
Silver	4.025	0.129 4 (Troy)	113 (Avoir.)	350
Tin (stannous)	2.215	0.078 1	205	451
Tin (stannic)	1.108	0.039 1	409	902
Zinc	1.220	0.043 0	372	820

32.7 Miscellaneous coating processes

(1) AUTOCATALYTIC PLATING

Autocatalytic plating is a form of electroless plating in which metal is deposited via a chemical reduction process (as opposed to immersion plating in which thin coatings are formed by electrochemical displacement of the coating metal).

Processes exist for the autocatalytic deposition of a large number of metals, particularly nickel, gold, silver and copper. Basically, the solutions contain a salt of the metal to be deposited and a suitable reducing agent (most commonly hypophosphite, but also hydrazine and boranes etc.). When a metal substrate, which is catalytic to the solution, is introduced into the bath, it becomes covered with a layer of the coating metal which is itself catalytic and thus the process can continue. This mechanism results in an extremely even distribution of deposit on the substrate, i.e. these solutions have a high 'throwing power'.

The most widely used autocatalytic process is nickel–phosphorus; a typical acid bath is as follows:

nickel chloride	4.8 oz gal^{-1}	(30 g l^{-1})
sodium hypophosphite	1.6 oz gal^{-1}	(10 g l^{-1})
sodium glycollate	8.0 oz gal^{-1}	(50 g l^{-1})

The solution is operated at temperatures between 75 and 100 °C, at pH 4–6, giving deposition rates up to 0.6 thou per hour (15 μm h^{-1}). The deposit is an alloy of nickel and phosphorus. containing about 7–10% phosphorus. A useful property of this material is that it can be hardened (typically by heat treating for 1 hour at 400 °C) so as to increase the as-deposited hardness from 400 HV to almost 1 000 HV. Thus, autocatalytic nickel coatings find engineering applications, often as a replacement for hard chromium electrodeposits.

(2) ELECTROPOLISHING

Electropolishing is a method of controlled anodic dissolution of a metal. The solutions used are chosen to have a 'levelling' action. Thus they etch away the asperities of the surface to a greater extent than the recesses and so produce a smooth surface. They are chiefly of use for the examination of metallurgical specimens where it is required to produce a metal surface free from the distortion and damage brought about by mechanical polishing (*see* 'Metallography', p. **10**–8).

(3) ELECTROSTATIC AND ELECTROPHORETIC METHODS OF PAINT APPLICATION

These are methods which have been developed for the economic application of paint to articles of complicated shapes (and often skeleton structure) in large numbers.

When an article like a metal chair is sprayed by a conventional spray gun procedure, much of the spray overshoots the surfaces and is wasted. In the electrostatic method a very high electrical potential is developed between the gun and the article being painted. The droplets of paint assume a charge of opposite sign to the workpiece and are attracted to it. This ensures more uniform coverage, less overspray, and thus greater economy in operation.

Electrophoresis has been utilized in a somewhat similar way. When a direct current is applied to an aqueous emulsion, large dispersed molecules and even oily particles are caused to move towards one of the electrodes. In the electrophoretic process a paint is provided as an aqueous emulsion and current is applied in such a manner that the globules of paint move towards and attach themselves to the object to be painted. The process can be made automatic and continuous and results in very uniform build-up even on points and sharp edges. It is only suitable for large-scale operations but is very economical in paint.

(4) COATING WITH CERAMIC MATERIALS

Just as metals can be sprayed, certain refractory oxides and silicates and the like can be applied by flame gun. Coating thickness and uniformity can be controlled by suitable means and hard, dense coatings can be built up. The coatings resemble biscuit-ware rather than a vitreous glaze, that is, they are absorbent: their chief use is to provide abrasion resistance and to delay heat transfer.

(5) MECHANICAL PLATING

Mechanical plating is a method of plating which utilizes mechanical energy to deposit metal coatings on to metal parts. Parts, glass beads, water, chemicals and metal powder are tumbled together in a barrel at around room temperature to obtain the desired coating.

The process is used primarily to provide ferrous-based parts with sacrificial coatings of zinc, cadmium, and co-deposits with tin. Parts treated by this method are most often fasteners, springs, clips and sintered iron components which are typically handled in bulk.

Parts which have been degreased, descaled, and copper-flashed are tumbled in rubber-lined barrels with water, glass bead impact media, promoter chemicals, and a finely divided powder of the metal to be plated. The promoter chemical serves to clean the metal powder and controls the size of the metal powder agglomerates that are formed. The mechanical energy generated from the barrel's rotation is transmitted through the glass impact media and causes the clean metal powder to be cold welded to the clean metal parts, thereby providing an adherent, metallic coating.

Due to the absence of an impressed current during coating, the process does not produce hydrogen diffusion into the steel substrate. Thus, a post-electroplating bake, in order to preclude hydrogen embrittlement of high strength steel components, is not required for mechanically plated deposits.

32.8 Plating formulae for non-conducting surfaces

(1) METAL POWDERS

The article to be treated may be coated with a metal powder. The best powder for this purpose is a finely ground copper, which is generally sold under the commercial title of bronze powder. This may be applied by mixing it with a cellulose lacquer to which has been added five parts by volume of thinner and spraying it on the object concerned. Alternatively, the object may first be sprayed with lacquer and before it is quite dry may be brushed over with the bronze powder using a soft camel-hair brush. After this treatment the article can be struck over in an acid copper solution.

Waxes (for gramophone records, etc.) may be coated directly by brushing with bronze powder and a soft brush. After brushing with bronze powder they may be treated in the following manner to improve the conductivity and reduce the time of covering in the plating bath:

1. Brush with a soft brush and a 50% mixture of methylated spirit and distilled water.
2. Immerse in a solution containing 30 $g\ l^{-1}$ of sodium cyanide and 6 $g\ l^{-1}$ of silver nitrate.
3. Make up two solutions as follows: (a) Pyrogallic acid 7 $g\ l^{-1}$, citric acid 4 $g\ l^{-1}$ and (b) silver nitrate, 40 $g\ l^{-1}$. Take four parts of solution (a) and one part of solution (b) and mix together and immerse article in this solution for about 10 min.
4. Swill and place in plating bath.

(2) SILVER REDUCTION

A number of articles can be treated by directly reducing silver on the surface. This process is particularly applicable to plastics and glass. Any process which will form a good silver mirror may be used, but the following will be found satisfactory for most purposes.

1. The surface of the article must be very thoroughly cleaned and completely free from grease. Glass and porcelain may be cleaned by using concentrated acid and alkali alternately. Plastics may be treated by brushing or barrelling with a mixture of Vienna lime and pumice, by treating with a suitable solvent or by immersing in a solution of chromic acid.
2. Priming. After cleaning the article is immersed in a 10% stannous chloride solution. Alternately, the solution may be swabbed on to the surface with cotton wool. The article is then thoroughly swilled.
3. The article is then silvered by immersing in a silvering solution, the formula for which is given below. The silvering operation generally takes about 20 min and the temperature must be carefully controlled during this period; usually about 21 °C will be found the most satisfactory. The solution should be slightly agitated during the process.
4. The articles are thoroughly swilled and struck over in a copper tartrate bath. After being coated over with copper any desired plating can be made upon it.

The silvering solution is prepared from the following:

Solution (a) 100 g l^{-1} Rochelle salt.
Solution (b) 10 g l^{-1} silver nitrate.
Solution (c) 200 cm^3 per litre ammonia (sp. gr. 0.880).

Take 100 cm^3 of solution (b) and add solution (c) carefully a little at a time until the precipitate which first forms just redissolves. If too much is added and the solution becomes quite clear, add a few drops of solution (b) until a very faint turbidity is produced. Then add 20 cm^3 of solution (a), thoroughly mix and use immediately.

[N.B. The brown precipitate formed by adding ammonia to silver nitrate is explosive if allowed to dry. Care should be taken therefore to see that this does not happen.]

(3) 'VACUUM METALLIZING'

This process is carried out at less than 0.0007 mmHg pressure and can only be applied to objects which are stable under these conditions. The metal to be deposited is heated until it evaporates and there are several ways in which this is achieved. The vapour recondenses on the first cool surface it encounters, and can be made to form a thick dense coating. The 'throwing power' is very poor since the evaporated metal travels in straight lines and steps must be taken to rotate or manipulate objects exposed to it in order to achieve uniform coating. By controlling the temperature of the work piece the crystal structure of the deposit can be varied.

32.9 Methods of stripping electroplated coatings

CADMIUM OR TIN FROM STEEL

Coatings may be stripped from steel by immersing the article in a solution containing 1 gal (4.5 litres) of hydrochloric acid, 2 oz (57 g) of antimony trioxide and $\frac{1}{2}$ pint (280 ml) of water. After stripping and rinsing the article will probably require wiping to remove smuts.

CHROMIUM

Chromium may be stripped from non-ferrous metals by dissolving it in dilute hydrochloric acid. From steel it is best stripped by making it the anode in a solution of sodium hydroxide. If the base metal is zinc or zinc base diecastings, it is best to strip the chromium by making it the anode in sodium carbonate solution as this will not attack the exposed zinc. The conditions of operation are not critical.

COPPER FROM STEEL

Copper can be stripped from steel by immersing in a solution containing 5 lb gal^{-1} (500 g l^{-1}) of chromic acid and 8 oz gal^{-1} (50 g l^{-1}) of sulphuric acid. This solution will work at room temperature but strips the copper very quickly if heated.

Alternatively, the article can be made anodic (2–6 V) in a solution containing 14 oz gal^{-1} (90 g l^{-1}) sodium cyanide and 2.4 oz gal^{-1} (15 g l^{-1}) sodium hydroxide.

COPPER FROM ZINC AND ZINC BASE DIECASTINGS

Copper may be stripped from these materials by immersing in a solution prepared by dissolving 18 g of sulphur and 250 g of sodium sulphide ($Na_2S.9H_2O$) in a litre of solution. The solution works very rapidly if warmed. The sludge formed on the surface of the object will require removing from time to time by brushing or immersion in a 120 g l^{-1} sodium cyanide solution.

COPPER, NICKEL, ETC., FROM MAGNESIUM

Most metal deposits can be removed from plated magnesium by submitting the component to the fluoride anodizing process. Alternating current is used in a bath of 10% ammonium bifluoride, in which the plating gradually dissolves without affecting the magnesium. About 4–10 V are used until most of the plating has disappeared. Finally, the voltage is raised to 120 to complete the process and cleanse the magnesium from remaining traces of foreign metal. Magnesium hangers must be used with firm connections. The bath should be operated cold. Direct current may be used if the work piece is made the anode using mild steel or magnesium cathodes.

LEAD FROM STEEL, COPPER AND BRASS

Immersion of article in a solution of 95 vol. % glacial acetic acid, 5 vol. % hydrogen peroxide (30 wt %). Dilute solutions may be used although this can lead to pitting of steel.

NICKEL FROM COPPER AND BRASS

Nickel may be stripped from copper by an anodic treatment in either 60 vol. % sulphuric acid or 15 g l^{-1} hydrochloric acid. Care must be taken with the concentration of the acids as this can affect the pitting of the substrates.

NICKEL FROM STEEL

Nickel can be anodically stripped in sulphuric acid, as for nickel from copper. Copper sulphate (30 $g l^{-1}$) or glycerine (30 $g l^{-1}$) can be added to reduce pitting of the steel. An immersion process involves the use of fuming nitric acid (85–95% HNO_3, sp. gr. 1.50) from which water is excluded.

SILVER FROM BRASS

Silver may be stripped from brass by immersing the object in a mixture of 1 vol. conc. nitric acid and 19 vols. conc. sulphuric acid heated to 175 °F (80 °C). The silver is dissolved in a few minutes. The articles should be immediately removed and swilled.

ZINC FROM STEEL

The reagent used for stripping cadmium may be used. Alternatively zinc may be stripped from steel in either warm dilute hydrochloric or sulphuric acid, or 10–15% ammonium nitrate solution, or hot sodium hydroxide solution.

32.10 Conversion coating processes

(1) PHOSPHATING

Phosphating solutions are used to produce corrosion-resistant coatings on ferrous metals and also zinc, cadmium and aluminium. Probably the most important application for these coatings is to act as bases for subsequent painting operations.

Basically, phosphate solutions comprise metal phosphates dissolved in carefully balanced solutions of phosphoric acid. When a clean metal surface is dipped into the solution, the free acid present reacts with the metal, liberating hydrogen and causing the pH of the solution, adjacent to the metal, to rise. This unbalances the solution, resulting in the precipitation of metal phosphates which form a film, chemically bonded to the substrate.

Due to the complexity of modern phosphate solutions, these processes are normally proprietary, examples of which are listed in section 32.11.

There are four main types of phosphate solution: iron, zinc, heavy zinc and manganese, and these produce increasing weights of coating from 30–90 mg ft^{-2} for iron phosphate solutions to 1000–4000 mg ft^{-2} for manganese phosphate solutions.

(2) CHROMATING

Chromating solutions contain hexavalent chromium ions and a mineral acid and are used to increase the corrosion resistance of metals, in particular zinc, cadmium, aluminium and magnesium, by forming a surface layer containing chromium compounds.

The process is usually performed by immersion, although spraying or brushing processes are also used. A wide variety of proprietary solutions are available which produce coatings of different thicknesses. These coatings are often distinguishable by their colour which can vary from clear, to blue, to iridescent yellow and finally to black.

(3) COLOURING OF METALS

The following solutions and operating conditions will produce coloured conversion coatings, as detailed:

(i) Copper and Brass

Black

Copper carbonate	1 lb (454 g)
Ammonia	2 pt (950 ml)
Water	5 pt (2.4 l)

The copper carbonate and ammonia are mixed before adding the water. The solution is operated at 175 °F (80 °C).

The blue black colour may be fixed by dipping in $2\frac{1}{2}\%$ sodium hydroxide solution.

Green

Water	1 gal (USA)	(3.8 l)
Sodium thiosulphate	8 oz	(227g)
Nickel ammonium sulphate	8 oz	(227 g)
or Ferric nitrate	1 oz	(28 g)
Temperature	160 °F	(71 °C)

Brown

Potassium chlorate	$5\frac{1}{2}$ oz	(154 g)
Nickel sulphate	$2\frac{3}{4}$ oz	(77 g)
Copper sulphate	24 oz	(680 g)
Water	1 gal (USA)	(3.8 l)
Temperature	195–212 °F	(90–100 °C)

(ii) Iron and steel

Black

Sodium hydroxide	8 lb	(3.6 kg)
Sodium nitrate	$1\frac{1}{2}$ oz	(42 g)
Sodium dichromate	$1\frac{1}{2}$ oz.	(42 g)
Water	1 gal.	(3.8 l)
Temperature	295°F	(146 °C)

Blue

Ferric chloride	2 oz	(56 g)
Mercuric nitrate	2 oz	(56 g)
Hydrochloric acid	2 oz	(56 g)
Alcohol	8 oz	(227 g)
Water	8 oz	(227 g)

Room temperature. Parts are immersed for 20 minutes, removed and allowed to stand in air for 12 hours. Repeat and then boil in water for 1 hour. Dry, scratch-brush and oil.

Brown

Copper sulphate	3 oz gal^{-1}	(20 g l^{-1})
Mercuric chloride	0.8 oz gal^{-1}	(5 g l^{-1})
Ferric chloride	5 oz gal^{-1}	(30 g l^{-1})
Nitric acid	25 oz gal^{-1}	(150 g l^{-1})
Alcohol	93 fl oz gal^{-1}	(700 ml l^{-1})

Dip in solution, place in hot box at 175 °F (80 °C) for 30 minutes, stand in steam box at 150 °F (65 °C) until coated in red rust, immerse in boiling water to form black oxide, dry and scratch brush. Repeat this operation three times before oiling with linseed oil.

(iii) Stainless steel
Black

Sulphuric acid	180 parts	
Water	200 parts	
Potassium dichromate	50 parts	
Temperature	210 °F (99 °C)	

(iv) Zinc
Black

Ammonium molybdate	4 oz gal^{-1}	(24 g l^{-1})
Ammonia	6 fl oz gal^{-1}	(45 ml l^{-1})

Heat solution to obtain a deep black. Rinse in cold and then hot water; allow to dry and harden.

(v) Aluminium
Black

Potassium permanganate	1.6 oz gal^{-1}	(10 g l^{-1})
Nitric acid	0.5 fl oz gal^{-1}	(4 ml l^{-1})
Copper nitrate	4 oz gal^{-1}	(25 g l^{-1})

Operate at 70 °F (24 °C) for 10 minutes.
Blue

Ferric chloride	60 oz gal^{-1}	(360 g l^{-1})
Potassium ferricyanide	60 oz gal^{-1}	(360 g l^{-1})
Temperature	150 °F	(66 °C)

32.11 Glossary of trade names for coating processes

32.11.1 Wet processes

(1) PHOSPHATE PROCESSES

Processes by which a coating of phosphate is produced on the surface of steel or zinc base alloys by treatment in or with a solution of acid phosphates. For rustproofing, the metal must receive a finishing treatment with paint, varnish, lacquer or oil; examples of typical finishing treatments are given under *Parkerizing* but it should be understood that firms using or marketing other proprietary phosphate processes may apply different designations or use different media for the necessary finishing treatment.

Bonderizing A proprietary phosphate process applied to steel and zinc, marketed by the Pyrene Company (similar to Parkerizing for steel but produces a thinner and less protective coating; synonymous with Parkerizing for zinc alloys).
Coslettizing The original phosphate process for steel, introduced in 1903.
Electro-granodizing A proprietary phosphate process applied to steel, marketed by ICI Ltd. (Paints Division). The chemical action of the solution is assisted by electrolysis.
Granodizing Proprietary phosphate processes applied to steel and zinc marketed by ICI Ltd. (Paints Division).
Lithoform A proprietary phosphate solution applied to zinc, marketed by ICI Ltd. (Paints Division).
Merlizing A proprietary phosphate process applied to steel by the Singer Manufacturing Co.
Parkerizing A proprietary phosphate process applied to steel and zinc, marketed by the Pyrene Company. Examples of subsequent finishing treatments are:

P20	Dewatering black finish
P41	Black shellac finish
P57	Oiled finish
P75	Oiled finish
P96	Oiled finish
SP55	Mineral oil finish

Rovalizing A term applied by International Corrodeless Ltd. to any coating processes marketed by them; includes a phosphate treatment. The term is incomplete without mention of the particular processes referred to.

Walterizing A proprietary phosphate process applied to steel and zinc base alloys, marketed by the Walterisation Co., Ltd.

(2) ALKALINE OXIDATION PROCESSES

Processes by which a black oxide film is formed on steel by treatment in a strongly alkaline solution containing an oxidizing agent. For rustproofing, the metal must receive a finishing treatment which is usually carried out with oil.

Alkablac Blackening processes marketed by M. L. Alkan Ltd.

Black Magic Blackening processes marketed by M. L. Alkan Ltd.

Blakodizing Black chemical finishes on steel, marketed by Tool Treatments Ltd.

Brunofix A proprietary process marketed by Metal Processes Ltd.

Chemag A proprietary process operated by Chemag Metal Colouring Co., Ltd.

Ebonol Range of conversion coating processes for both ferrous and non-ferrous metals, marketed by Enthone/Imasa Ltd.

Jetal An American process marketed by Protective Metal Finishes Ltd.

Pentratol black A proprietary process which originated in USA under the name of Pentrate black, now marketed by Alfred P. Mill.

(3) CHROMATE PROCESSES

Alkrotect Protective chromate treatments, marketed by M. L. Alkan Ltd.

Cronak A patented process for protection of zinc against corrosion licensed by National Smelting Co. Ltd., involving immersion in a solution mainly of acid dichromate.

Enthox Chromating solutions marketed by Enthone/Imasa Ltd.

Kenvert Chromating solutions marketed by the 3M Company Ltd.

(4) ANODIC OXIDATION OF ALUMINIUM AND ITS ALLOYS (ANODIZING)

For protection against corrosion and wear, for decoration, for aiding heat emission and for miscellaneous uses based on the absorptive properties of the oxide film when freshly made; used in conjunction with electrolytic 'polishing' for producing reflectors. Processes involving electrolytic treatment in solution, generally of chromic, sulphuric or oxalic acid with the production of a relatively thick film of oxide.

Alumilite process (sulphuric acid) A proprietary process marketed by Alumilite and Alzak Ltd.

Alzak process An American process for producing reflectors. Operated and marketed in the UK by Alumilite and Alzak Ltd.

Anobrite Bright anodizing process operated by Anobrite Ltd.

Anochrome (chromic acid) An anodic oxidation process operated by London Aluminium Co. Ltd.

Anodolex An anodic oxidation process for the treatment of pistons, operated by Birmingham Aluminium Casting Co.

Bengough–Stuart process (chromic acid) The first anodizing process patented in 1923.

Brytal process A process of producing reflectors introduced by British Aluminium Co. Ltd.

Di-Alumin (sulphuric acid) An anodizing and dyeing process operated by Aluminium Protection Ltd.

Eloxal A generic term used in Germany for anodic oxidation.

Sheppard processes Anodizing processes marketed by British Anodizing Ltd.

(5) IMMERSION PROCESSES FOR THE TREATMENT OF ALUMINIUM ALLOYS

Alumon Zincating process marketed by Enthone/Imasa Ltd.

Bondal Zincating process marketed by W. Canning Ltd.

Decoral Oxidizing process giving electrically conducting coatings capable of being coloured by dyeing. Marketed by Lea Manufacturing Co. Ltd.

MBV process (Modified Bauer Vogel) A process of forming a thin oxide film by immersion in an alkaline solution containing chromates.
Pylumin process A similar process marketed by the Pyrene Company.
Alocrom process A process of priming a thin greenish yellow film in a cold acid solution containing chromates. Marketed by ICI Ltd. (Paints Division).

(6) NON-ELECTROLYTIC PROCESSES

Enplate Electroless nickel plating solutions, marketed by Enthone/Imasa Ltd.
Niklad Electroless nickel plating solutions, marketed by Lea Manufacturing Co. Ltd.
Sylek Electroless nickel-boron plating solutions, marketed by Imasa Ltd.
Transiflo Mechanical plating process, marketed by the 3M Company Ltd.

(7) ELECTROPLATING AND ELECTRODEPOSITION PROCESSES

Achrolyte Tin-cobalt alloy plating process marketed by Udylite/Oxy Metals Ltd.
Abecra 3000 Trivalent chromium plating process marketed by Albright and Wilson Ltd.
Brylanising Zinc coating of wire by the 'Bethanizing' electroplating process of the Bethlehem Steel Co.
Chromonyx Black chromium plating process, often used for coating solar panels. Marketed by Harshaw Chemicals Ltd.
Cromalin process A process of electroplating on aluminium or its alloys operated by Metal Finishes Ltd.
Durionizing A term applied by Durion Ltd. to electrodepositing with hard chromium as applied by them.
Fescolizing A term applied by Fescol Ltd., to any electrodeposition process carried out by them (incomplete without mention of the metal referred to, e.g. Fescolizing in chromium, etc.).
Listard process–Van der Horst process Patented processes of hard chromium plating for protecting the cylinders of internal combustion and other engines from wear. These processes give an oil-retaining surface to the chromium. The processes are operated by Listard Processing Co. Ltd., and British Van der Horst Ltd.
Niron Bright nickel–iron plating process, marketed by Udylite/Oxy Metals Ltd.
Rovalizing A term applied by International Corrodeless Ltd., to any protective coating process used or marketed by them. Incomplete without description of the process referred to (e.g. Roval cadmium, etc., *see also* phosphate processes).
Technichrome process A process operated by Aluminium Protection Co. Ltd. for the deposition of hard chrome on ferrous and non-ferrous metals.
Tryposit A name used by Thomas Try Ltd., to indicate their special process of electrodepositing heavy nickel or hard chromium for engineering purposes.
Zartan Alloy deposit, used as an alternative to decorative chromium. Marketed by M. L. Alkan Ltd.

32.11.2 Dry processes

(1) THERMAL PROCESSES

The processes described below involve heating of the object to be coated in contact with the coating metal (in the form of powder or as a coating to secure inter-penetration) or with a compound of the coating element. Used for coating steel except where otherwise stated.

Aluminizing A process involving spraying with aluminium and heating to cause alloying.
Bower Barff A process of coating steel with a black oxide by heating in contact with steam.
Bruntonizing A process of coating wire with zinc consisting of hot galvanizing followed by drawing—operated by Messrs. Bruntson (Musselburgh), Ltd.
Calorizing A process of coating with aluminium, similar to sherardizing (*see* below) but using aluminium in place of zinc.

Chromizing A process of coating with chromium involving heating to a high temperature in contact either with a vapour of chromous chloride or with metallic chromium.

Corronizing A process of coating steel by electroplating with a duplex coating of nickel followed by either zinc or tin and finally heat treating to diffuse the zinc or tin into the nickel—developed by Messrs. Standard Steel Spring Co., Cooraopolis, Pa.

Follsain Penetral processes A proprietary cementation process operated by Follsain Metals Ltd., involving heating in contact with a mixture of aluminium, metallic oxides and a catalyst.

Galvanizing A process of coating with zinc by dipping pickled steel into molten zinc (electrogalvanizing is sometimes used to mean electroplating with zinc).

Ihrigizing A process of coating with silicon by heating to a high temperature in contact with the vapour of silicon tetrachloride.

Nitriding A process of forming a hard layer on steel involving heating in a suitable atmosphere (usually ammonia vapour) to form a surface layer rich in nitride.

Sherardizing A process of coating with zinc involving heating in contact with zinc powder in revolving drums (introduced by Sherard Cowper Coles).

(2) METAL SPRAYING PROCESSES

A method of coating consisting of projecting a stream of molten metallic particles at high velocity against the surface to be coated. Mainly for protection against corrosion but also used for restoring the dimensions of undersized parts.

Wire or Schoop process A process marketed by Metallization Ltd., and Metallizing Equipment Co. Ltd., in which the coating metal in the form of wire is melted and atomized.

Schori or powder process A process marketed by Schori Metallizing Process Ltd., in which a stream of the powdered coating metal is fed into a flame and blown on to the surface to be coated.

Mellosing or molten metal process A process marketed by Mellowes & Co. Ltd. in which a molten metal is fed into a jet of heated compressed air which serves to atomize it and to project it against the surface to be coated.

Plasma spray process A process in which a very high temperature flame is produced by blowing gas through an electric arc. Metal wire or powder is melted by passage through the flame and is projected by the gas on to the surface to be coated.

Detonation process or flame plating Oxygen, acetylene and the material to be plated, are introduced into a detonation chamber where a spark ignites the mixture. A detonation wave travelling at supersonic speed, forces the powdered material heated to $>3\,500\,^{\circ}C$, on to the substrate. A special building is required for sound insulation. Very high density coatings of refractory materials like tungsten carbide, can be plated by the process.

33 Welding

33.1 Introduction

The 'Glossary of Welding Terms' is intended to explain the meaning of terms common in welding technology and used throughout this section and is not confined to British Standard terminology. For this, the reader is referred to BS 499.

Resistance welding, friction welding, diffusion bonding and fusion welding are considered separately, information on the welding of common materials being included in each section.

A list of British Standards relating to welding is given in Section 33.6.

33.2 Glossary of welding terms

Arc blow In arc welding, the deflection of the arc by magnetic forces induced by the welding current or residual magnetism in the workpiece.

Arc welding A fusion welding process wherein the source of heat is an electric arc.

Argon arc welding A fusion welding process wherein the source of heat is an electric arc struck between an argon shielded non-consumable tungsten electrode and the work. The filler wire, if required, is added separately.

Atomic hydrogen welding A fusion welding process wherein hydrogen is dissociated in an arc struck between two tungsten electrodes afterwards recombining to supply the welding heat. The filler wire is added separately.

Autogenous welding Fusion welding without a filler metal addition, in which the weld metal is provided by melting of the parent material.

Backward welding A fusion welding technique in which the source of heat is directed towards the already deposited weld.

Backing bar A bar of material used for backing up a joint during welding to control penetration at the root and not contiguous with the weld.

Backing strip A strip of metal used for backing up a joint during welding and which may or may not be left on after welding.

Backing run See *Sealing run.*

Backstep welding Welding in which increments of a run are deposited in a direction opposite to the general direction of welding.

Bead-on-plate weld A single run of weld metal deposited on an unbroken surface.

Braze welding A joining process whereby a brazing type of filler is deposited in a prepared joint using a *gas* welding technique.

Bronze welding A joining process whereby a brazing type of filler is deposited in a prepared joint using an *arc* welding technique.

Butt weld A weld between two members lying approximately in the same plane and not overlapping.

Carbon arc welding A fusion welding process wherein the source of heat is an arc struck between the work and a non-consumable carbon electrode. A filler wire may or may not be used. A shielding gas may also be employed.

Carbon dioxide welding A carbon dioxide shielded metal arc welding process using a continuous consumable bare wire electrode.

Chain intermittent fillet welding Two lines of intermittent fillet welding on either side of a joint wherein the fillet welds on one side are opposite those on the other side.
Cold welding Pressure welding at room temperature.
Constant potential power source A power source, the output voltage/ampere characteristic of which is substantially parallel to the current axis. Ideal for self-adjusting arc welding
Controlled arc welding Metal arc welding in which the arc length is kept constant by controlling, from the arc voltage, the rate of feed of the consumable electrode.
Controlled tungsten arc welding Tungsten arc welding in which the arc length is controlled from the arc voltage.
Corner weld An outside weld between two members approximately at right angles forming an L.
Cover glass A clear glass, sometimes gelatin coated, used in welders' helmets and hand shields to protect the filter glass from spatter and fumes.
Crater A depression left in the weld metal at the termination of a run.
Diffusion bonding A joining process wherein the component parts are held together under a pressure too low to cause significant plastic deformation, generally at an elevated temperature. The resultant atomic diffusion causes bonding, sometimes with the formation of a liquid phase.
Dip transfer A mode of particle transfer in gas-shielded metal arc welding wherein use is made of controlled short circuits between electrode and pool.
Downhand weld See *flat position weld.*
Drooping characteristic power source A power source, the output voltage of which falls as current demand increases. Necessary for metal arc welding and controlled-arc welding.
Edge weld A weld between the edges of two or more parallel and faying members.
Electrogas welding A vertical butt welding process using a gas-shielded consumable electrode to deposit metal into a molten pool held in place by moving dams which move upwards as the joint is made.
Electron beam welding A fusion welding process employing a high voltage focussed electron beam.
Electroslag welding A vertical butt welding process in which the filler metal electrode is melted in a rising bath of conducting slag held in position by moving dams.
Explosive welding A pressure welding process employing the energy from the controlled explosion of sheet explosive to effect joining.
Face of weld The exposed surface of a fusion weld on the side from which welding was carried out.
Faying surface That surface of a member which is in contact with another member to which it is to be joined.
Filler wire Metal to be added in making a weld, usually in the form of a bare or flux-coated wire.
Fillet weld A weld of approximately triangular cross-section between two members approximately at right angles to each other.
Filter glass A dark coloured glass used to cut down the radiation from a fusion welding process to assist vision and protect the eyes of the welding operator. *See* BS 679.
Flash butt welding A resistance welding process wherein coalescence is produced by the heat obtained from resistance to the flow of a heavy electric current between two lightly abutting surfaces and by the application of pressure upon attainment of welding temperature.
Flat position weld A fusion weld in which the weld face is approximately horizontal and uppermost.
Forge welding A group of welding processes wherein coalescence is produced by applying pressure or blows to material rendered plastic by heating.
Forward welding A fusion welding technique in which the source of heat is directed ahead of the already deposited weld.
Friction welding A hot pressure welding process in which frictional heat is produced by rotating one component of the joint against a stationary mating surface under slight pressure. An upset force is applied when rotation is stopped. Orbital or oscillating movement may be employed rather than rotation.
Fusion boundary The boundary between weld metal and parent metal in a fusion weld.
Fusion welding All welding processes in which the weld is made by fusion of the parent metal by means of externally applied heat but without hammering or pressure. Filler wire may or may not be added.
Gap The minimum distance at any cross-section between edges, ends or surfaces to be joined.
Gas welding A fusion welding process in which the source of heat is the combustion of gas.
Hammer welding A forge welding process in which the welding pressure is obtained by means of hammer blows.

Heat affected zone The parent material alongside a weld which is heated during the welding operation.
High frequency induction welding A resistance welding process wherein coalescence is produced by the application of pressure to edges heated by the skin effect of high frequency alternating current.
High frequency injection The superimposing of a high frequency voltage on the alternating current used in argon-arc welding to ensure that the arc will restrike at the beginning of electrode positive half-cycles.
Horizontal vertical welding A weld made in such a position that the longitudinal axis of the weld is approximately horizontal and the plane of the weld face is approximately vertical.
Inert gas metal arc welding A group of fusion welding processes in which the source of heat is an electric arc struck between the work and an inert gas shielded, continuous, consumable electrode which acts as filler metal.
Interpass temperature In a multiple pass weld, the lowest temperature of the deposited weld metal before the next pass is started.
Lap weld A fillet weld between two overlapping members.
Laser welding A fusion welding process employing the energy from the excitation of an optical laser.
Leg of a fillet weld The distance from the root of the joint to the toe of a fillet weld.
Manual metal arc welding Metal arc welding using a flux-coated wire or rod electrode under manual control.
Metal arc welding An arc welding process in which the metal electrode forms the filler metal.
MIG (metal inert gas) welding See *Inert gas metal arc welding.*
Nugget The fusion zone of a spot, seam or projection weld.
Overhead position weld A weld made on a surface lying horizontally or at an angle not more than 45° to the horizontal, the weld being made from the underside of the parts joined. Alternatively a weld carried out on the underside of a joint.
Overlap Protrusion of weld metal beyond the bond at the toe of the weld.
Penetration The maximum depth a groove weld extends from its face into a joint, exclusive of reinforcement. In resistance welding, the distance from the interface to the edge of the weld nugget measured on a cross-section through the centre of the weld and normal to the surface.
Plasma arc An arc which is constricted mechanically or magnetically to produce a high heat concentration over a small area.
Plug weld A fusion weld made in a hole formed in one of the parts of a lapped joint to attach the other part.
Positional welding Welding carried out in all positions other than the normal flat position.
Pressure welding A process of joining metals in which the surfaces to be joined are brought into close contact by pressure either cold or heated below the melting point.
Projection welding A resistance welding process in which current and pressure concentration for making a weld is achieved by means of small projections usually raised on one of the workpieces.
Pulsed welding Arc welding, usually gas shielded, in which one or more welding variables, but most commonly current, are periodically varied between two levels.
Reinforcement Weld metal in a butt weld lying outside the plane joining the toes of the weld.
Resistance butt welding A resistance welding process in which the parts to be joined are butted together under pressure while current is allowed to flow until a predetermined temperature is attained and metal at the interface is upset and a weld produced.
Resistance welding A generic term covering those welding processes in which the welding heat is produced by the electrical resistance of the weldment and interfacial contact resistance during passage of the welding current.
Reverse polarity In direct current welding, the arrangement of current supply wherein the work is made the negative pole and the electrode the positive pole of the welding circuit. The term is misleading, and it is preferable to refer to polarity.
Roller spot welding A spot welding process using a machine similar to that used in seam welding. Pressure is applied continuously and current intermittently.
Root (a) The zone in the preparation for V, U, J and bevel butt welds, in the neighbourhood of and including the gap.
(b) The zone between the prepared edges adjacent to a backing strip in an open square butt weld.
(c) The zone, in parts to be fillet welded, in the neighbourhood of the actual or projected intersection of the fusion faces.
Root face That portion of the groove face adjacent to the root and normal to the face of the weld.

Root gap The separation between the members to be joined at the root of a joint.
Run-on and run-off tags Small pieces of metal tacked to a weldment at the beginning and end of welds to facilitate avoidance of undercutting at free edges.
Sealing run A weld bead laid along the back of a groove weld.
Seam welding A resistance welding process in which a continuous weld is produced in overlapped sheets by means of two electrode wheels or between an electrode wheel and an electrode bar. The electrode wheels provide continuous pressure and current flow is intermittent with accurately timed periods.
Self-adjusting arc welding Metal arc welding in which the consumable electrode is fed into the arc at a constant speed with a high current density, any alteration in arc length being corrected by naturally occurring changes in the burn-off rate.
Series welding The making of two spot or seam welds or two or more projection welds simultaneously with electrodes forming a series circuit.
Size of weld (a) Butt weld—The joint penetration.
(b) Fillet weld—The leg length of the largest isosceles right-angled triangle which can be inscribed within the fillet weld cross-section.
Spot welding A resistance welding process in which overlapping parts are welded at one or more spots by means of shaped electrodes which give a high current density at the welding point and maintain mechanical pressure on the weld during and after current flow.
Spray transfer A transfer mode in gas-shielded metal arc welding, wherein metal transfers from the electrode to the weld pool as a stream of droplets.
Staggered intermittent fillet welding Two lines of intermittent fillet welding on either side of a joint wherein the fillet welds on one side are staggered relative to the fillet welds on the other side.
Stitch welding A spot welding process in which the welds overlap.
Straight polarity In direct current welding, the arrangement of current supply wherein the work is made the positive pole and the electrode the negative pole of the welding circuit. The term is misleading, and it is better to refer to actual polarity.
Stringer bead A weld bead deposited without appreciable transverse oscillation of the electrode.
Strip cladding A method of surfacing components by automatic submerged arc or gas-shielded arc welding, using a consumable electrode in the form of a strip rather than wire.
Submerged arc welding Arc welding in which a bare wire consumable electrode is used, the arc being enveloped in a powdered flux, some of which fuses to form a protective slag covering the weld.
Surge injection A means of arc re-ignition whereby a timed uni-directional medium voltage surge is applied across the arc gap. Used in a.c. argon-arc welding.
Tack weld A short weld used in assembly of parts and for preventing distortion during welding.
Thermit welding A fusion welding process where the heat for fusion is obtained from liquid steel resulting from a thermit reaction, the steel so produced being used as the added metal.
Throat of a fillet weld (a) Theoretical. The distance from the beginning of the root of the joint perpendicular to the hypotenuse of the largest right-angled triangle that can be inscribed within the fillet weld cross-section.
(b) Actual. The shortest distance from the root of a fillet weld to its face.
TIG (tungsten inert gas) welding Arc welding using a non-consumable tungsten electrode, surrounded by an inert gas shield. (See *Argon arc welding*.)
Toe of weld The junction between the face of a weld and the parent metal.
Touch welding A metal arc welding technique employing flux coated electrodes of special type, whereby the tip of the rod is rested on the parent metal during welding.
Undercut A groove melted into the parent metal at the toe of a weld and left unfilled by weld metal.
Vertical position weld A weld made in such a position that the longitudinal axis of the weld is approximately vertical.
Weaving The deposition of a weld bead with transverse oscillation of the electrode.
Weldment An assembly whose component parts are joined by welding.
Weld metal That portion of metal which has been molten during welding.

33.3 Resistance welding

There are four main types of resistance welding process: spot welding, projection welding, resistance butt welding and flash butt welding. Seam, stitch and roller spot welding are closely related to spot welding.

Spot welding finds application in the lap joining of sheet material in the range of thickness from 2 × 0.3 mm to 2 × 4 mm and occasionally thicker materials. Stitch and seam welding are used for the manufacture of pressure-tight seams, stitch welding being more suitable for irregularly shaped components and seam welding for long, straight runs or curves with regular or generous radii. Multiple pressure and heating cycles are possible in stitch welding and roller spot welding, but not in seam welding. Roller spot welding is used for producing long, straight rows of spots at higher production rates than are possible in spot welding.

Projection welding may also be used for lap joining of sheets provided the material has sufficient ductility for the production of suitable projections and sufficient strength in the projections to withstand the high loads employed. This process is also used for the production of T-joints and the attachment of studs, nuts, collars and discs to sheet materials. Cross-wire welding is regarded as a form of projection welding.

Resistance butt welding is mainly used for the butt joining of wire and light gauge rods. Heating is more widespread, power consumption is higher and the condition of the end faces is more important than in *flash butt welding*. The latter process is used for butt joining of heavier and more complex sections and bevel joining.

33.3.1 The influence of metallurgical properties on resistance weldability

The properties of most importance which influence weldability, and therefore the selection of suitable welding conditions, are conductivity, expansion characteristics, nature and condition of the surface film, high temperature strength and the structure of the fusion and/or heat affected zone.

High conductivity materials require more current than mild steels to give an equivalent heat input for two reasons, first, due to the increased electrical conductivity and secondly, due to the increased conduction of heat away from the weld area. In practice, high conductivity materials such as aluminium are welded for short times with exceptionally high currents. There is, therefore, a need for more complex and expensive equipment for welding these materials than would be required for mild steel. Splashing or boiling of the weld metal may occur in low conductivity materials, and short times and high currents are used for such cases also.

Coefficient of thermal expansion and shrinkage contraction on solidification are important factors in governing the type of mechanical system used in welding machines. In materials of high thermal expansion and of high thermal conductivity, where shrinkage takes place relatively quickly, it is necessary to provide a high electrode force immediately following the current pulse, in order to avoid shrinkage cavities and cracks.

Consistency of surface condition is extremely important in most resistance welding processes (flash butt welding is an exception), particularly in materials with thick natural oxide films, such as aluminium and magnesium alloys. Cleaning procedures are invariably recommended in such cases. Thin and uniform oxide films, such as those obtained on pickled stainless steels and Nimonic alloys, give consistent welding behaviour, particularly with high electrode forces. Rust, paint and grease influence consistency, and should be removed. Metallic coatings do not interfere with weldability seriously, although consistency may suffer slightly and electrode tip life in spot welding may be reduced.

Materials which have high strength when hot, require high electrode forces in order to maintain a seal around the molten weld metal.

It is important that the metallurgical structure of both weld and heat affected zone should be adequate to withstand the demands made upon the joint in service. The heat affected zone in many alloys is softened, and it is therefore important to limit its extent by the use of short times. Ferritic steels suffer from hardening, which increases with carbon and alloy content. In serious cases this may necessitate some form of post-weld heat treatment to overcome the resulting brittleness.

33.3.2 The resistance welding of various metals and alloys

STEELS

Low carbon steels may be readily resistance welded by all processes, clean deep drawing steel being commonly regarded as excellent in this respect. A guide to the maximum carbon content which can be tolerated in spot and projection welding without excessive hardening is given by the formula

$$C_{max} = 0.1 + 0.012t$$

where t = thickness in millimetres of the thinnest sheet in the combination.

Spot and projection welds in medium carbon and low alloy steels may be made, and it is desirable that a high electrode force be applied after the welding current pulse to prevent cracking followed by a tempering treatment which is most conveniently carried out in the welding machine. The electrode force may be maintained at a high level during tempering or reduced to its original value. Hardening is not such a serious problem in resistance butt or flash welding, since cooling rates are lower; however, butt welding is usually confined to steels that do not form refractory oxides.

Austenitic stainless steels require somewhat lower currents and higher electrode forces in spot and projection welding. In flash butt welding these materials require similar currents to mild steel, but with higher open-circuit voltages and upset forces.

Relative conditions for spot welding mild steel and other materials are summarized in Table 33.2.

Table 33.1 THE RESISTANCE WELDING OF METALS AND ALLOYS—SUITABILITY OF PROCESSES

Material	*Process*				*Material*	*Process*			
	Spot	*Projection*	*Resistance butt*	*Flash butt*		*Spot*	*Projection*	*Resistance butt*	*Flash butt*
Low carbon steel	S	S	S	S	Aluminium bronze	P	P	P	P
Low alloy steel Medium carbon steel	P	P	S	S	Silicon bronze	S	—	S	—
					Cupronickel	S	—	S	—
Austenitic stainless steel	S	S	N	S	Nickel–silver	S	—	S	—
					Magnesium alloys	P	N	N	S
Aluminium and low-strength alloys	P	N	S	S	Nickel and its alloys	S	S	N	S
					Titanium	S	P	N	S
Medium and high-strength aluminium alloys	S	N	P	S	Zirconium	S	S	N	S
Copper	N	N	S	S	Tantalum	S	—	—	—
Beryllium–copper	S	—	S	S	Niobium	S	—	—	—
Gilding metals	N	N	S	N	Molybdenum	P	—	N	S
70/30 and 60/40 brass	S	—	S	N	Tungsten	P	—	N	S
Tin bronze	S	S	S	S					

S = Suitable. N = Not recommended. P = Possible under certain conditions.
— = Information not available.

Table 33.2 RELATIVE CONDITIONS FOR SPOT WELDING MILD STEEL AND OTHER MATERIALS

Material	*Electrode force*	*Weld time*	*Welding current*	*Remarks*
Mild steel	A-70 MPa on tip area	Short B	High C	
Low alloy steels	A × 1	B × 1	C × 1	Post-heating needed
18/8 stainless steels	A × 3	B × 1	C × 0.9	Surface oxides may cause trouble in rare cases
Nimonic alloys	A × 4	B × 1	C × 0.9	(Surface oxides may cause trouble in rare cases)
Nickel	A × 2.5	B × 1	C × 1.5	
Monel, Inconel	A × 4	B × 1	C × 0.9	
Aluminium alloys	A × 2	B × 0.5	C × 4	Surface oxide must be removed
Cupronickel, silicon bronze, 70/30 brass, nickel-silver, etc.	A × 0.8	B × 0.8	C × 1.2 – 2.0	Current depends on conductivity of alloy
Titanium and its alloys	A × 2	B × 1	C × 1	Avoid excessive penetration

Taken from 'Resistance Welding', published by the former British Welding Research Association.

ALUMINIUM AND ALUMINIUM ALLOYS

The resistance weldability of aluminium alloys is governed mainly by their relatively high conductivity and the presence of a tenacious high resistance oxide film. Care is therefore required in surface preparation to produce a consistent contact resistance, and this is normally done by controlled scratch brushing or chemical dipping. The use of paste fluxes is not recommended, due to inconsistent quality. Degreasing alone is occasionally practised in applications where consistency is less important. Cleaning of outer surfaces improves tip life in spot welding.

The need for high currents and short weld times has already been mentioned, and comparative figures are given in Table 33.2.

In this group, pure aluminium is the most difficult material to spot weld, and probably the easiest to resistance butt weld. Strong aluminium alloys clad with pure aluminium do not give the same trouble in spot welding due to their lower conductivity and higher resistance to indentation Generally, spot welding difficulties increase with decreasing parent metal strength and increasing conductivity. All the common sheet materials may be spot welded.

There is little experience with projection welding, the main difficulty being the relatively low strength in compression of the projection. Resistance butt welding is used only for pure aluminium and low strength alloys in small sections, and care with surface preparation is again necessary. Flash butt welding can be used for most aluminium alloys, provided rapid heating and careful control in the time and speed of application of the upset force is applied.

COPPER AND COPPER ALLOYS

Due to its high conductivity, copper is not normally regarded as weldable by the spot and projection welding processes, although very thin copper may be spot welded with molybdenum tipped electrodes using very high currents and short times. The alternative of interposing of shim of relatively high resistance brazing filler metal is often employed, giving a resistance brazed joint rather than a weld. Pure copper may be satisfactorily resistance butt and flash butt welded, and both processes are used in the wire industry.

Cadmium–copper is difficult to resistance weld, and resistance brazing is preferred, but beryllium–copper may be readily spot or butt welded, provided the normal precautions for high conductivity materials are taken.

Brasses increase in weldability with increasing zinc content, and it is difficult to spot weld gilding metals, though 70/30 and particularly 60/40 brass may be welded satisfactorily by this process. Resistance butt welding is frequently used for brass wire, but flash butt welding is not suitable, due to zinc volatilization. Information on current, weld time and upset force compared with mild steel is given in Table 33.2. Certain brasses should be rendered immune from season cracking by a low temperature annealing treatment after welding.

Tin bronzes have relatively low conductivity, and may be readily joined by practically all resistance welding methods. Phosphor bronzes may, however, tend to stick to copper alloy electrodes in spot welding, and plating of the electrodes may be required.

Information on aluminium bronze is meagre, but satisfactory resistance welds of all types have been produced in single phase alloys. Flash butt welding has been used for complex alloys.

Silicon bronzes have low conductivity, and may be satisfactorily spot and resistance butt welded. Weld times and electrode forces are rather lower than for mild steel and currents somewhat higher. It is important that rapid follow up of the electrodes during heating should take place.

Cupronickels and nickel–silvers behave in a similar manner to silicon bronze.

MAGNESIUM AND MAGNESIUM ALLOYS

Magnesium and its alloys may be satisfactorily spot welded for low duty service using equipment similar to that used for aluminium alloys. Reference has been made to the satisfactory weldability of AM503, AZM, AX31, ZW1 and the 3% thorium–0.7% zirconium alloy. Cleaning is important, chemical methods being preferred to mechanical methods, although good results can be obtained by brushing with wire wool. Electrode cleanliness is important if copper contamination of the work-piece is to be avoided. High currents and short weld times are used together with a rapid follow up of the electrodes to give a force somewhat lower than that used for aluminium. Radiused tip electrodes help to form a pressure seal around the weld.

Flash butt welding may also be used for magnesium alloys.

NICKEL AND NICKEL ALLOYS

Nickel, Monel,* Inconel* and the Nimonic* alloys may be satisfactorily spot, projection and flash butt welded.

Conditions for spot welding Nimonic alloys are similar to those for austenitic stainless steel but with somewhat higher electrode forces. Nickel requires higher currents and lower forces although forces are still much higher than for mild steel. Monel and Inconel requires currents and forces similar to those recommended for Nimonic alloys. Comparative figures are given in Table 33.2. Cleanliness is important, sulphur and lead contamination being particularly dangerous. Sticking which may occur when spot welding annealed nickel, may be overcome by silver plating of the electrodes. Cracking may be encountered with some precipitation hardening alloys if electrode pressure is too low. Increased weld time and pressure are beneficial.

Flash butt welding is satisfactory if clamping is firm to prevent arcing, clamping distances are kept short because of the low conductivity and upset forces are high. Open-circuit voltages require to be higher than used for mild steel.

REFRACTORY METALS

Among this group of materials are included titanium, zirconium, tantalum, niobium, molybdenum and tungsten. The problem of contamination by reaction with the atmosphere is not as serious in spot welding as in fusion welding, but the quality of projection welds may be improved by the use of an inert atmosphere.

Titanium may be readily spot or flash butt welded. Conditions for spot welding are similar to mild steel but with higher electrode forces. The high electrical resistance tends to give a large weld with high penetration, and seam welding may need to be done under water to prevent contamination after welding. Flash welding conditions are similar to those used for aluminium alloys. Projection welds have also been made in titanium.

Zirconium in thin gauges may be spot welded using conditions similar to those used for stainless steel. No special cleaning other than degreasing is normally necessary. Projection welding and flash butt welding are also possible.

Tantalum may be spot welded in thin gauges, and welding may be carried out in air if the time is less than one cycle. Larger times require water cooling of the weld area. Surfaces should be degreased and pickled in a sulphuric/chromic acid mixture. Niobium should be treated in a similar way.

Resistance welds in molybdenum and tungsten are often inherently brittle due to the high brittle/ductile transition temperature of the recrystallised and as cast materials. Other problems are contamination of surfaces with electrode materials and contamination of the electrodes themselves. These troubles are minimized by welding under water. Surfaces should be thoroughly cleaned, if possible by grit blasting. Both materials may be resistance brazed readily, using shims of tantalum, zirconium or nickel. Both materials have been satisfactorily flash butt welded.

DISSIMILAR METALS

It is normally easier to make satisfactory joints between dissimilar metals by resistance welding or resistance brazing than by fusion welding or normal brazing techniques, since the problem of fluxing does not arise and techniques may be chosen to minimize the danger of brittle intermetallic phases within the joint. Copper and aluminium, for example, form a series of brittle phases when melted together, but flash butt welding of copper to aluminium is widely practised, since these phases are forced out of the joint when the upset force is applied.

When spot welding dissimilar metals, it may be necessary to use electrodes of differing conductivity against the different parts, i.e. a high conductivity electrode against the lower conductivity material.

Some indication of the spot welding characteristics of dissimilar metals is given in Table 33.3. In some cases where spot welding is not possible due to excessive formation of intermetallics, it is possible to interpose a layer of a third material compatible with both the parts requiring to be joined.

---

* Henry Wiggin & Co. Ltd.

Table 33.3 SPOT WELDING OF DISSIMILAR METALS

		Nimonic	*Monel, Inconel*	*Nickel*	*Phosphor bronze*	*Silicon bronze*	*Nickel–silver*	*25–40% zinc brass*	*10–25% zinc brass*	*Copper*	*Aluminium and alloys*	*18/8 stainless steel*	*Mild steel: Other coatings*	*Mild steel: Tin or zinc coated*	*Mild steel: Scaly*	*Mild steel: Clean*
Mild steel	Clean	G 5 7	P 8	P 8	P 8	P 8	P 8	P 6 8	U 6 8	U 6 8	P 6 8	G 5	P 3 4	G 3 4	P 2	E 1
Mild steel	Scaly				U 2 8	U 2 8	U 2 8				U 2 8 6	U 2	U 2 4 3	U 2 4 3	P 2	
Mild steel	Tin or zinc coated		P 3 8 4	P 3 4	P 3 4	P 3 4	P 3 4	P 3 6 4	U	U	U	G 3 4	P 3 4	G 3 4		
Mild steel	Other coatings		U 3 8 4	U 3 4	U	U	U	U	U	U	U	U	P 3 4			
18/8 stainless steel		G 5	P	P	P	P	P	U	U	U	U	E 5				
Aluminium and alloys		U	U	U	P 5 7	P 5 7		P 5 7	U	U	G 5 7					
Copper		U	U	U	P 6 8 7	P 6 8 7	P 6 8 7	P 6 8 7	U	U						
10–25% zinc brass		U	U	U	U	U	U	U	P							
25–40% zinc brass			P 5 7	P 5 7	P 5 7	P 5 7	P 5 7	G 5 7								
Nickel–silver			G	G	P	P	G									
Silicon bronze			P	P	P	G										
Phosphor bronze			P	P	G											
Nickel			G	G												
Monel, Inconel			G													
Nimonic		G 5 7														

Key: Spot weldability: E Excellent; G Good; P Poor; U Unsatisfactory.
Taken from 'Resistance Welding', published by the former British Welding Research Association.

Notes:

(1) Wide range of welding conditions.
(2) Inconsistent welds of poor strength. Shot blasting or pickling recommended.
(3) Coating thickness should be uniform.
(4) Electrodes should be cleaned frequently to prevent sticking.
(5) High currents and short times preferred.
(6) Thin gauges may be welded with special conditions.
(7) Welding conditions must be accurately controlled.
(8) Low weld strength.

33.4 Friction welding

The most common method of generating frictional heat for welding is to rotate one component relative to the other under an axial load. When the metal temperature at the interface is sufficiently high, the rotation is stopped, and the workpieces are forced together under the original or higher load. Inertia welding is a variant of the process, using the kinetic energy of a flywheel to provide rotation and thermal energy. In friction welding, one component usually has a circular cross-section, but this is not always essential, particularly when orbital rather than rotational differential movement of the workpieces is used.

The principal welding variables are the welding speed, applied heating pressure, forge pressure, and heating duration. For any particular joint configuration, selection of optimum welding conditions is primarily dependent on material strength and thermal conductivity characteristics, although with dissimilar metal joints, it may be necessary to use conditions minimising the formation of brittle phases at the interface.

Welding speed is not generally critical, peripheral velocities of between 0.3 and 7 m s^{-1} being common for a range of materials. Higher welding speeds increase the width of the heat-affected zone (HAZ) and welding time.

Both heating and forge pressures must be selected to achieve uniform heating and maintain the faying surfaces in intimate contact. Typical values for mild steel are 45 MPa and 75 MPa for heating and forging, respectively. With mild steel, doubling the heating pressure increases the power required by 50%. Excessive pressure is to be avoided with alloy steels, since the steep temperature gradient may lead to the formation of hard structures, with reduced ductility. Higher pressures are required for materials of greater hot strength. The heating pressure is particularly important in determining both the torque required to make the weld, and the temperature gradient at the interface. With high conductivity materials, such as copper, it is advantageous to use a low heating pressure to obtain rapid local heating, followed by a high forge pressure. Duration of heating is selected to obtain suitable thermal conditions for the production of a sound joint on forging, without overheating of surrounding material. For mild steel, times of 0.5–5 s are general depending on the joint geometry, lower times being applicable to smaller sections, or higher conductivity materials.

Since it is to some extent self-cleaning, the method is tolerant with respect to surface preparation. The atmosphere is excluded from the weld area, and reactive metals can be joined with no special precautions, while the absence of a liquid phase during welding may avoid problems of porosity, cracking, etc., associated with fusion processes. In fact, if a low melting point liquid area is formed at the interface, it may act as a lubricant, and restrict the development of frictional heat essential for welding. This can arise when welding on to galvanized steel, unless the zinc is removed from the abutting region. Sulphur-bearing free machining steels cause similar difficulties.

Most common alloys are weldable to themselves, while a number of dissimilar metal combinations can be joined, that are normally weldable by other techniques. Table 33.4 illustrates the relative weldability of a range of materials.

DIFFUSION BONDING

In diffusion bonding, contacting surfaces are joined by the simultaneous application of temperature and pressure, using conditions adequate for diffusional processes to occur, but insufficient for macroscopic deformation. As most commonly practised, it is a solid-state process, but a liquid phase may be present stemming either from interdiffusion of dissimilar metals, or from the incorporation of an interlayer into the joint. Interlayers can be applied by vacuum spraying, galvanizing or as powder or foils, and are employed to accelerate diffusion or to prevent the formation of brittle intermetallic phases. Bonding is normally carried out in vacuum, although protective gas shielding can be used. The range of metals and alloys which have been bonded either directly or indirectly via an interlayer is illustrated by Table 33.5, with butt joint properties in Table 33.6. Bonds between metallic and non-metallic components such as glass or ceramics have been made.

The process is governed by four main interrelated variables: pressure, temperature, time and surface condition. A summary of reported bonding conditions is given in Table 33.7.

Table 33.4 WELDABILITY OF DIFFERENT MATERIALS BY FRICTION WELDING*

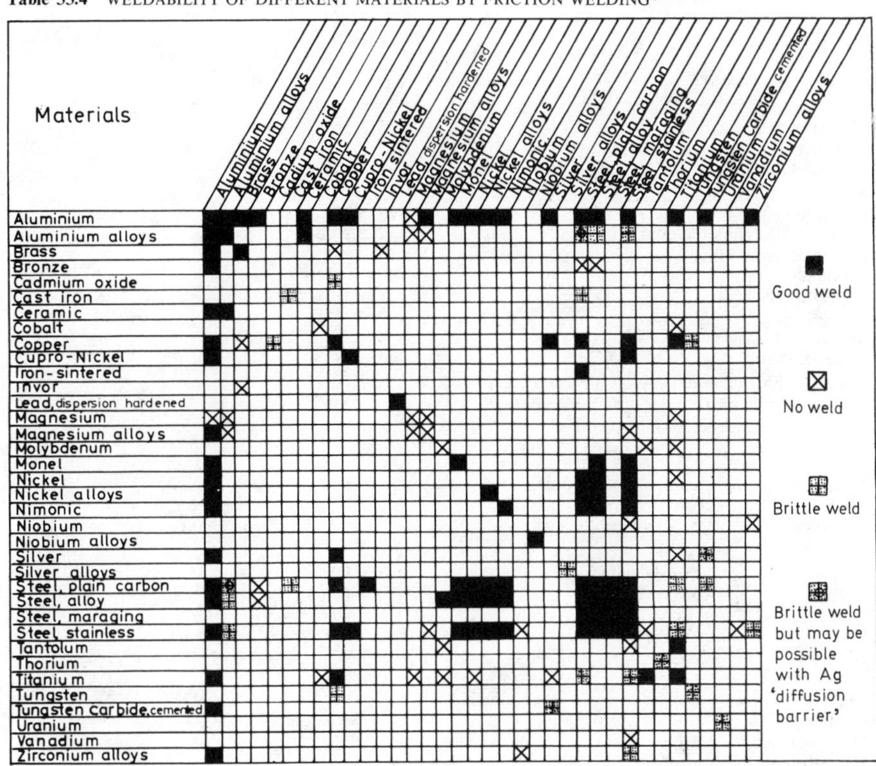

*For further information, *see* C. R. G. Ellis, *Met. Const. & Brit. Weld. J.*, 1970, **2**, 185–188.

Table 33.5 DIFFUSION BONDED COMBINATIONS OF METALS AND ALLOYS WITHOUT AN INTERLAYER (DIRECT) AND WITH AN INTERLAYER (INDIRECT)*

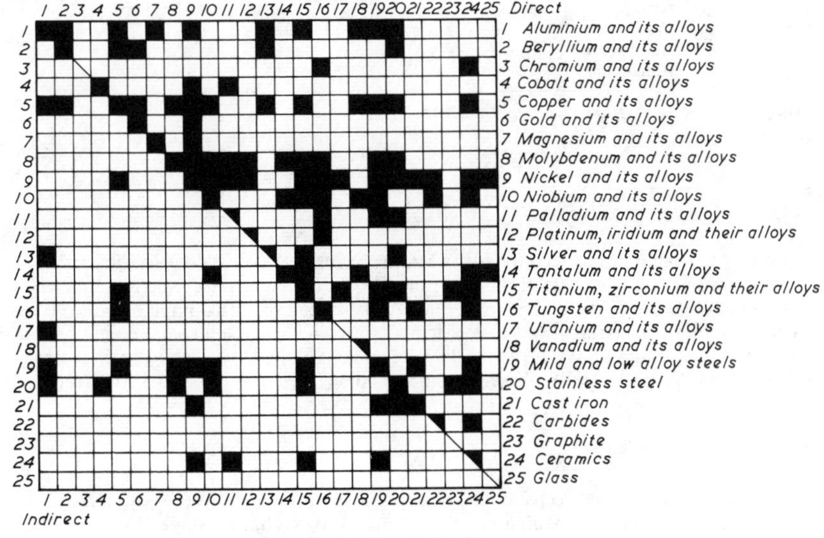

* For further information, *see* P. M. Bartle, *Welding J.*, 1975, **54**, 799–804.

Table 33.6 EXAMPLES OF DIFFUSION BUTT JOINT PROPERTIES*

Material combination	*Joint strength*			*Tensile joint efficiency*,%
	tonf in^{-2}	ksi	MPa	
Mild steel/mild steel	30	66	465	100
High strength steel/high strength steel	75	165	1160	90
Mild steel/cast iron	28	61	430	100
Stainless steel/cast iron	34	75	525	100
Cast iron/cast iron	56	124	865	100
Mild steel/aluminium alloy	10	22	155	50
Aluminium alloy/aluminium alloy	10	22	155	50
Aluminium alloy/copper	9	20	140	70
Copper/nickel	14	31	215	100
Copper/copper	14	31	215	100
Nimonic/Nimonic	59	130	910	90
Stellite/Stellite	60	132	925	90
Titanium alloy/titanium alloy	69	152	1 065	100

* For further information, *see* Table 33.5.

Table 33.7 DIFFUSION BONDING PARAMETERS*

Material I	*Material II*	*Interlay*	*T*, °C	*P*, MPa	*t*, min
Copper	Molybdenum	—	900	7.35	10
Copper	Steel	—	900	4.9	10
Copper	Nickel	—	900	14.7	20
Copper	Copper	—	800–850	4.9–6.9	15–20
Titanium	Nickel	—	800	9.8	10
Titanium	Copper	Molybdenum	950	4.9	30
Titanium	Copper	Niobium	950	4.9	30
Titanium	Copper	—	800	4.9	30
Molybdenum	Molybdenum	Titanium	915	6 860	20
Molybdenum	Steel	—	1 200	4.9	10
Tungsten	Tungsten	Niobium	925	6 860	20
Tantalum	Tantalum	Zirconium	870	—	—
Niobium	Niobium	Zirconium	870	—	—
Zircaloy-2	Zircaloy-2	Copper	1 040	20.6	30–120
Beryllium	Beryllium	63-Ag-27	800	—	30
Copper	Copper	Cu-10 In	—	—	—
Kovar	Kovar	—	1 000–1 110	24.5–19.6	20–25
Steel	Cast iron	—	850–950	14.7	5.7
Steel	Aluminium	—	500	7.35	30
Steel	Aluminium	—	550	4.9	10

*For further information *see* P. Wiesner, *Met. Con. & Brit. Weld. J.*, 1971, **3**, 91–93.

PRESSURE

It is necessary to obtain intimate contact between the faying surfaces, and hence the pressure must be high enough to cause plastic deformation at the tips of abutting surface asperities and to induce creep so that contact over the entire surface area can be achieved. This implies pressures below, but often close to, the material yield stress at temperature and hence solid-state bonding pressures for a range of metals and alloys are typically 5–15 MPa, although higher pressures are required for refractory metals. With liquid phase diffusion bonding, pressures of 0.5–1.5 MPa are normal.

TEMPERATURE

To achieve sufficiently rapid creep for practical purposes, a minimum temperature of 0.7 T_m is normal, T_m being the material melting point in degrees kelvin. Higher temperatures may be used to accelerate bonding.

TIME

The diffusion time is strongly dependent on the temperature. Bonding can occur after only a few minutes, but times of 0.5–2.0 hours are preferred to give complete interface void removal by diffusion, and improved joint ductility.

SURFACE PREPARATION

The rate of bonding depends on the flatness and roughness of the faying surfaces, and a finish of $\sim$0.4 μm is preferred. Removal of thick oxide layers and degreasing prior to bonding are required. although to some extent contaminants and oxides can dissolve in the parent material, especially with iron- and copper-base systems.

33.5 Fusion welding

While gas welding using an oxy-acetylene torch is still frequently used for mild steel and some non-ferrous metals in sheet gauges, arc welding processes account for the greater proportion of welding carried out at the present time.

Metallic arc welding using a coated electrode is the most popular for the manual welding of mild and alloy steels, and electrodes are available for welding cast iron, nickel and its alloys, certain copper alloys and a limited range of aluminium alloys.

A greater number of processes are available for the mechanized welding of mild steel plates. The submerged arc process employs a bare wire electrode and a flux cover continuously supplied via a hopper ahead of the arc. In alternative processes, the flux is applied as a wrapping on the continuously fed electrode or in the core of an electrode formed from folded strip. Other processes using both flux and shielding gas (carbon dioxide) have been developed for mechanized fillet welding.

The gas-shielded processes may be applied to almost all metallurgical materials. They fall into two principal groups, namely those employing a non-consumable electrode, and those wherein the electrode is in the form of a continuous consumable wire.

The argon arc, or TIG, process is the best known of the former type and has been successfully employed for practically all the materials mentioned below, with the possible exception of cast iron. A.c. argon arc welding is normally recommended for aluminium and magnesium alloys and materials forming refractory films, such as aluminium bronze and beryllium copper, while d.c. electrode negative is used for steels, copper and nickel alloys. D.c. nitrogen arc welding is used for copper, and d.c. helium arc welding may be used in circumstances where the high gas cost can be justified in terms of higher welding speeds or increased penetration.

The gas-shielded metal arc processes have been applied to almost as wide a range of materials as the TIG process. The shielding gases used vary with the metals being welded, argon being used for aluminium, magnesium, nickel alloys and refractory metals, argon or nitrogen for pure copper, argon for copper alloys, and argon, argon–oxygen mixtures or carbon dioxide for all types of steel, including stainless steels. D.c. electrode positive is invariably used in present commercial equipment. CO_2 welding is particularly valuable as a low hydrogen process for welding low alloy steels.

Atomic hydrogen and carbon–arc welding are still occasionally used, particularly for mild steel sheet metal welding.

Fusion welding can also be achieved by the electron beam and laser processes. Both processes are applicable to a range of ferrous and non-ferrous alloys, and because of the high energy density, they give a rapid thermal cycle and little distortion. The former method is normally carried out in vacuum and hence is especially appropriate for reactive metals. Laser welding commonly employs helium shielding, this gas having a high ionization potential with reduced energy loss from the laser beam and, for the same reason, helium is also used for out-of-vacuum electron welding.

33.5.1 The fusion welding of metals and alloys—ferrous metals

MILD STEELS

The welding of mild steel may be accomplished by the majority of processes, the choice of method being dependent upon thickness, type and position of joint and to some extent the type of steel employed. The processes most applicable to thin sheet are gas welding, TIG welding, CO_2 shielded metal arc welding, and less commonly, atomic hydrogen welding. The TIG and atomic hydrogen processes may be manual or mechanized.

Joints in thicker sheet and plate will generally employ the manual metal arc, or, if circumstances permit, one of the mechanized processes such as submerged arc or CO_2 welding. The Fusarc process† employing a continuous flux-covered consumable also finds application. Plate of around 100 mm thickness and above may be welded in a single pass using the electroslag process.

The composition of the steel to be welded may limit the choice of process. Rimming steels can give porous welds in gas-shielded welding, and if a mechanized welding process is required, atomic hydrogen, providing a reducing atmosphere, would seem to be the most suitable. Killed steels give better results with argon arc and gas-shielded metal arc welding. High sulphur-free cutting steels are not recommended for welding, since the sulphur gives rise to hot cracking. If such steel has to be welded, basic coated metal arc electrodes should be used.

Gas welding

A neutral flame and no flux is used. The filler material is preferably copper coated to ensure freedom from rust. For general purposes, filler metal to BS 1453: 1972 A1 should be used. This is a low carbon mild steel giving a tensile strength of 330 MPa minimum. For somewhat higher strength A2 filler, containing manganese and silicon, is available; *see* Table 33.8.

Table 33.8 FILLER RODS FOR THE GAS WELDING OF FERRITIC STEELS (BS 1453:1972)

Type	Element wt %						Typical application
	C	Si	Mn	Ni	Cr	Mo	
A1	0.10 max	—	0.60 max	0.25 max	—	—	General purpose mild steel
A2	0.10–0.20	0.10–0.35	1.00–1.60	—	—	—	Mild steel, 420 MPa
A3	0.25–0.30	0.30–0.50	1.30–1.60	0.25 max*	0.25 max*	—	Medium tensile, 480 MPa
A4	0.25–0.35	0.10–0.35	0.35–0.75	2.75–3.25	-0.30	—	Heat treatable deposit
A5	0.35–0.45	0.40–0.70	0.90–1.10	—	0.90–1.20	—	Wearing surfaces
A6	0.15 max	0.25–0.50	0.60–1.50	0.20 max*	0.20 max*	0.45–0.65	Welding ½% Mo steels
A7	0.08–0.15	0.10–0.35	0.80–1.10	—	—	—	Similar to A2 but lower alloy
A32	0.12 max	0.20–0.90	0.40–1.60	—	1.10–1.50	0.45–0.65	Welding 1¼% Cr/½% Mo steels
A33	0.12 max	0.20–0.90	0.40–1.60	—	2.00–2.70	0.90–1.10	Welding 2¼% Cr/1% Mo steels

* If present as a residual element.
Both S and P to be 0.040% max, for A1 to A7, and 0.030% max for A32 and A33.

Metal arc welding

The choice of electrode depends upon the application and the type of welding power source available.

All electrodes supplied to BS 639 : 1976 are coded according to the following system:

1. *Compulsory part:*
   (a) The letter E for a covered manual metal-arc electrode.
   (b) Tensile and yield strength of deposited weld metal.
   (c) Elongation and Charpy impact values of deposited weld metal.
   (d) Type of flux coating.
2. *Optional part:*
   (e) Nominal electrode efficiency.
   (f) Welding positions in which the electrode can be used.
   (g) Recommended current and voltage conditions.
   (h) Whether or not the electrode is hydrogen controlled.

For full details of this system, BS 639 : 1976 should be consulted. The salient coating characteristics are summarized in Table 33.9.

† British Oxygen Co Ltd.

Table 33.9 TYPES OF COATINGS FOR CARBON AND CARBON MANGANESE STEEL ELECTRODES IN BS 639:1976

Coding	*Type of covering*	*Resultant slag*	*Special features*
A	Acid iron oxide-manganese oxide-silica	Inflated honeycombe	High fusion rate and penetration. Normally used in flat position.
AR	Acid-rutile: as A, but with up to 35% titanium oxide	As A, but more fluid	As A
B	Basic lime-fluospar	Dense	Low weld metal hydrogen contents are possible, with high resistance to solidification cracking. Generally suitable for all welding positions
C	Cellulosic	Thin and friable	High fusion rate and penetration. Can normally be used in all welding positions
O	Oxidizing iron oxide	Heavy, solid	Low penetration, with fluid weld pool. Normally used in flat position, and when smooth surface more important than mechanical properties
R	Rutile	Fluid	Smooth arc with little spatter, and usable in all welding positions
RR	Rutile	Fluid, but more viscous than R	Similar to R, but with heavier coating thickness. Good weld surface finish

Automatic processes employing flux

As with manual metal arc welding, the flux type involved is of considerable importance in the submerged arc and Fusarc processes and in techniques employing flux-cored consumables. The choice of flux obtainable with these processes is more limited than is the case with manual metal arc welding but both rutile and basic fluxes (equivalent to Classes R and B in Table 33.9) are available for particular applications, together with more neutral fluxes.

Submerged arc welding fluxes may be in a fused or agglomerated form, depending on the method of manufacture, and have in the past normally been of a neutral or acidic character. More basic fluxes have been developed, which may offer advantages in the mechanical properties of deposited weld metal, although usually with more difficult slag detachment and a more irregular weld surface.

Low carbon wire consumables are used, generally with added manganese and silicon as deoxidants.

Gas-shielded welding processes

Consumables used for gas-shielded welding of steels contain added deoxidants (usually silicon and manganese). Examples for mild steel are given in Table 33.10, A15 and A17. TIG welding is usually carried out using argon shielding for optimum weld pool control, although helium may be used on thicker sections to increase penetration of the weld. With gas-shielded metal arc welding of mild steel, the surrounding gas is generally CO_2 or argon/CO_2 mixtures.

LOW ALLOY STEELS

In general, the application of the various welding processes is similar for both mild and low alloy steels, the principal difference being that the choice of consumables for low alloy steels may be more limited. Manual metal arc and submerged arc welding are most widely used for low alloy steels in industry, consumables for the former process being given in Table 33.11.

It must be appreciated that in the heat-affected zone (HAZ) around a weld, very high cooling rates may be experienced. Particularly with alloy steels, hard HAZ microstructures can be produced, having a high susceptibility to cracking under the influence of residual welding stresses, and hydrogen picked up during the welding operation. The necessity to avoid hydrogen induced cracking is a major factor in the selection of welding conditions for low alloy steels.

Table 33.10 FILLER RODS AND WIRES FOR GAS-SHIELDED ARC WELDING OF FERRITIC STEELS (BS 2901 : PART 1 : 1970)

Type	Element wt %							
	C	Si	Mn	Cr	Mo	Al	S	P
A15*	0.12 max	0.30–0.90	0.90–1.60	—	—	0.04–0.40	0.040 max	0.040 max
A16	0.25–0.30	0.30–0.50	1.30–1.60	—	—	—	0.040 max	0.040 max
A17	0.12 max	0.20–0.50	0.85–1.40	—	—	—	0.040 max	0.040 max
A18	0.12 max	0.70–1.20	0.90–1.60	—	—	—	0.040 max	0.040 max
A19	0.08–0.12	0.30–0.50	1.00–1.30	—	—	0.35–0.75	0.040 max	0.040 max
A30	0.12 max	0.90–0.90	0.40–1.60	—	0.45–0.65	—	0.030 max	0.030 max
A31	0.14 max	0.50–0.90	1.60–2.10	—	0.40–0.60	—	0.030 max	0.030 max
A32	0.12 max	0.20–0.90	0.40–1.60	1.10–1.50	0.45–0.65	—	0.030 max	0.030 max
A33	0.12 max	0.20–0.90	0.40–1.60	2.00–2.70	0.90–1.10	—	0.030 max	0.030 max
A34	0.12 max	0.20–0.90	0.40–1.60	5.00–6.00	0.45–0.65	—	0.030 max	0.030 max

* Titanium or zirconium may be present up to 0.15% max each.

Table 33.11 MANUAL METAL ARC ELECTRODES FOR LOW ALLOY STEELS (BS 2493:1971)

Composition code	Element wt %							
	C	Si	Mn	Ni	Cr	Mo	S	P
MoB*	0.10 max	0.50 max	0.75–1.20	—	—	0.40–0.70	0.035 max	0.030 max
MoC, MoR	0.10 max	0.30 max	0.35 min	—	—	0.40–0.70	0.035 max	0.030 max
1CrMoB*	0.10 max	0.50 max	0.75–1.20	—	1.0–1.5	0.40–0.70	0.035 max	0.030 max
1CrMoR	0.10 max	0.30 max	0.35 min	—	1.0–1.5	0.40–0.70	0.035 max	0.030 max
2CrMoB*	0.10 max	0.50 max	0.75–1.20	—	2.0–2.5	0.90–1.20	0.035 max	0.030 max
2CrMoR	0.10 max	0.30 max	0.35 min	—	2.0–2.5	0.90–1.20	0.035 max	0.030 max
5CrMoB	0.10 max	0.50 max	0.50–1.00	—	4.0–6.0	0.40–0.70	0.035 max	0.030 max
7CrMoB	0.10 max	0.60 max	0.50–1.00	—	6.0–8.0	0.40–0.70	0.035 max	0.030 max
9CrMoB	0.10 max	0.60 max	0.50–1.00	—	8.0–10.0	0.90–1.20	0.035 max	0.030 max
1NiB	0.10 max	0.80 max	1.20 max	0.80–1.10	—	0.35 max	0.030 max	0.030 max
2NiB	0.10 max	0.80 max	1.20 max	2.00–2.75	—	—	0.030 max	0.030 max
3NiB	0.10 max	0.80 max	1.20 max	2.80–3.50	—	—	0.030 max	0.030 max
MnMob	0.10 max	0.80 max	1.20–1.80	—	—	0.25–0.45	0.035 max	0.030 max

*Mn:Si ratio at least 2:1.

HAZ cracking in ferritic steels is influenced by the following inter-related variables:

1. Steel composition and transformation behaviour.
2. Welding process and type of consumable.
3. Energy input of the welding process, and preheating temperature.
4. Joint restraint.

In general, more highly alloyed steels have a greater tendency to cracking, due to their lower transformation temperature and the formation of harder transformation products. In practice, the important aspects of transformation behaviour of a steel are the hardenability of the steel, and the susceptibility of the hardened structure to cracking.

Both hardenability and susceptibility generally increase with increases in carbon and alloying element content. The dominant effect of carbon has been recognized by the empirical derivation of carbon equivalent (CE) formulae which are intended to describe the weldability of a steel. A widely used formula for carbon–manganese structural steels to BS 4360: 1979, is as follows:

$$CE = C + \frac{Mn}{6} + \frac{Cu + Ni}{15} + \frac{Cr + Mo + V}{5}$$

With increased CE, more stringent precautions are necessary to avoid cracking. It should be appreciated that such CE formulae apply only to the particular compositions used in determining them, and are not universally applicable.

The risk of HAZ cracking is highly dependent upon the hydrogen content of the freshly deposited weld metal. With respect to manual metal arc welding, the risk of cracking in a particular steel is lower when low hydrogen basic electrodes are used rather than the common rutile type, particularly if the former electrodes are dried at temperatures above about 300 °C. The submerged-arc and gas-shielded metal-arc processes can produce weld metal deposits having hydrogen contents comparable to or lower than those of basic electrodes, and some relaxation of welding conditions may be possible with these processes, provided that clean dry consumables are used. Typical weld metal hydrogen levels are given in Table 33.12.

Table 33.12 TYPICAL HYDROGEN CONTENTS FOR DIFFERENT STEEL WELD METALS

Weld metal type	*Range of hydrogen contents* (ml per 100 g *of deposited metal*)
BS 639:1976 Classification C	70–100
BS 639:1976 Classification R	20–35
BS 639:1976 Classification B (dried 100–150 °C)	10–15
BS 639:1976 Classification B (baked 350–450 °C)†	3–10
Submerged arc	5*–25
Gas-shielded metal arc (Ar or CO_2)	3*–10
Flux cored CO_2	10–30

* Lower values for clean, dry consumables.
† Consumable manufacturer's advice on drying conditions should be sought.

Thermal conditions prevailing during welding are controlled for one or both of two reasons. First, by control of the cooling rate, it is often possible to determine the transformation product after welding. Secondly, cracking may be avoided by holding the weld area at about 250 °C (when the embrittling effect of hydrogen is negligible) for a sufficient length of time for hydrogen to diffuse away from susceptible regions.

For a given weld geometry, increased heat input during welding will result in a lower cooling rate. In many materials, such reduction in cooling rate will be sufficient to avoid transformation to hard, susceptible microstructures. Increased heat input may be achieved by increasing the welding current or voltage, or by reducing the speed of travel. If cracking cannot be controlled by increasing energy input, preheat may be applied to reduce the cooling rate after welding. The preheat temperature required in a particular case will depend upon the material composition, the joint restraint, and the welding process used, but it will usually be in the range 100–250 °C. In multipass welds, the preheat temperature must be maintained between weld runs as an interpass temperature.

Joint restraint may be defined as resistance to deformation which would relieve contractural welding stresses. Restraint will, in general, increase with increasing thickness of the component parts being joined. Increase in material size will also increase the cooling rate of the weld by affording a larger heat sink. Joint type is significant in determining the number of paths along which heat may be conducted away from the weld, and the effect of joint geometry on cooling conditions can be described by a 'combined thickness' parameter. The combined thickness is the total thickness of material in mm through which heat can flow away from the weld, as indicated in Table 33.13.

When doubt exists regarding the risk of cracking in a given situation, it must be recommended that a procedural trial be undertaken.

It has been found possible with carbon–manganese structural steels to avoid cracking by controlling welding conditions so that the microstructure in the HAZ is not harder than a critical value. When using manual metal arc welding consumables of relatively high hydrogen potential, the HAZ hardness should not exceed 350 HV. With low hydrogen consumables or CO_2 welding, a hardness of 400 HV or even above is appropriate. Figure 33.1 is a nomogram for material of CE determined as above, showing welding conditions such that the critical hardness is not exceeded, and cracking can be avoided. Allowance is made for the use of processes with different hydrogen potentials, and for the combined thickness of the weldment. From the intended welding process, the relevant CE scale is selected. A vertical line is drawn from the CE of the material being considered, to intersect the intended preheat temperature. From this intersection, a line is taken horizontally to the appropriate combined thickness, and then vertically down to determine the

Table 33.13 DETERMINATION OF COMBINED THICKNESS

Type of joint and heat flow	Combined thickness (mm)
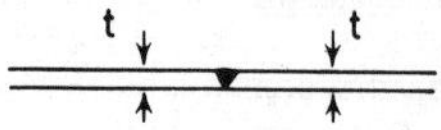 t, t	2t
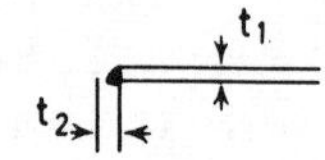 t_1, t_2	$t_1 + t_2$
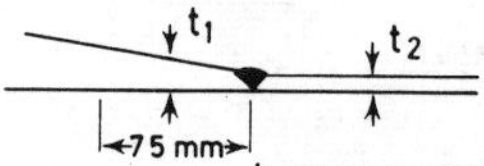 t_1, t_2, 75 mm; t_1 = average thickness over 75 mm	$t_1 + t_2$
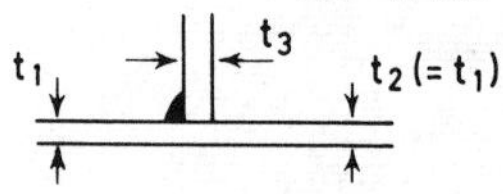 t_1, t_3, $t_2 (= t_1)$	$t_1 + t_2 + t_3$
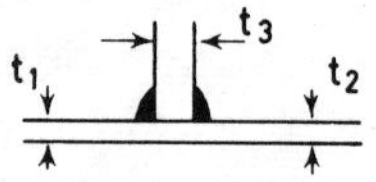 t_1, t_3, t_2; Twin fillet welds made simultaneously	$\frac{1}{2}(t_1 + t_2 + t_3)$

minimum arc energy input necessary during welding to avoid exceeding the critical hardness. Similarly, given a material, thickness and arc energy, the minimum preheat temperature can be determined. Further reference to BS 5135:1974 should be made.

The CE approach for formulating welding procedures is unworkable at CEs above 0.6. With low alloy steels, it is possible to classify susceptibility to HAZ cracking according to their transformation behaviour. Five general classifications are shown in Table 33.14, together with summaries of precautions necessary for avoiding cracking. Appropriate preheat and interpass temperatures for steels in Classes 4 and 5 can be obtained from Figure 33.2. More precise definition of behaviour is not possible since cracking susceptibility is not uniquely related to hardness, while the effect of joint geometry cannot be adequately quantified for these materials.

In general, weld metals used for mild and low alloy steels contain rather less carbon than the parent material, and in consequence are less susceptible to hydrogen-induced cracking than is a HAZ. The factors outlined above for HAZs are nonetheless applicable, and precautions to avoid hydrogen-induced weld metal cracking may be necessary in some cases.

HAZ liquation cracking

Another problem in the welding of ferritic steels is cracking associated with liquation of nonmetallic phases at prior austenite grain boundaries in the HAZ, under the influence of thermal stresses in this area. Susceptibility is dependent mainly on the relative amounts of sulphur, carbon, manganese and phosphorus in the steel and, in plain iron–carbon alloys, the problem becomes significant with carbon contents above about 0.2%. At this carbon content, a manganese to sulphur ratio of at least 20:1 is required to avoid the problem, and with higher carbon contents, this ratio may need to be as high as 30:1 or 50:1.

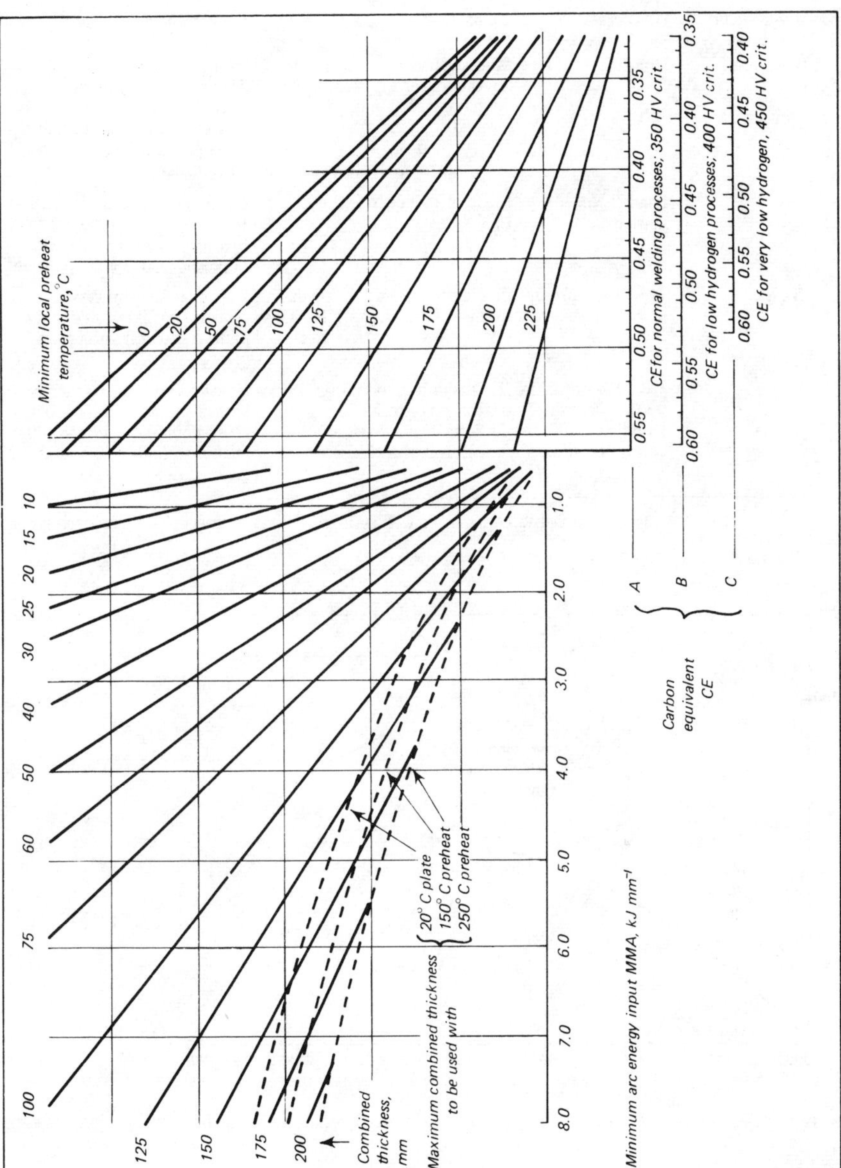

Figure 33.1 *Welding conditions for mild and* C–Mn *steels*

For further information, *see* N. Bailey, *Met. Constr. & Brit. Weld J.*, 1970, **2**, 442.
CE Scale A: Normal welding processes; HAZ hardness restricted to below 350 HV.
CE Scale B: Low hydrogen welding processes; HAZ hardness restricted to below 400 HV.
CE Scale C: Very low hydrogen welding processes; HAZ hardness restricted to below 450 HV.

Table 33.14 CLASSIFICATION OF LOW ALLOY STEELS

Class	*Hardenability*	*Susceptibility to HAZ cracking*	*Examples*	*Precautions in welding*
1	Low	None	Mild steel	None
2	Low	Low	(1) Mild steel with C > 0.15% but < 0.25%, and Mn < 1.0%. (2) C–Mn steels with C < 0.2% and Mn < 1.4%	Thin sections, none; with thicker sections, lower H_2 processes, high heat input and some preheat are all advantageous
3	Low	High (twinned martensite)	Medium carbon steels, e.g. En5, 8, 9*	Production of non-martensitic HAZ more important than low hydrogen. Use high heat inputs, and 250–350 °C preheat
4	High	Low (low carbon martensite)	Low carbon, low alloy high strength steels, e.g. 9% Ni HY80, C–Mn steels in thick sections	Low hydrogen processes, increased heat input and some preheat generally necessary
5	High	High (twinned martensite or bainitic constituents)	(1) Medium carbon alloy steels, e.g. En15 to 30. (2) Some creep resistant steels, e.g. 5 Cr$\frac{1}{2}$Mo 2$\frac{1}{4}$CrlMo	Preheat and interpass temperature in the range 200–350 °C. (May be possible to use 150 °C in thin sections.) Post-weld heat treatment advisable

For further information, *see* T. Boniszewski and R. G. Baker, Proc. Second Commonwealth Welding Conference, Institute of Welding, London. 1965, Paper M5.
* En5 and En8 are equivalent to BS 970 Part 1—080M30 and 080M40.

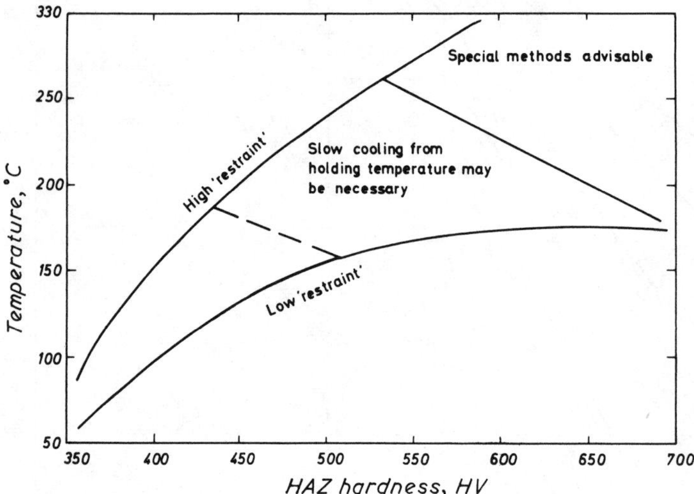

Figure 33.2 *Suggested preheat conditions for procedural tests for steels in Classes 4 and 5 of Table 33.14.*

Lamellar tearing

Lamellar tearing may arise in joints in which the fusion boundary of the weld is parallel to the plate surface, and tensile residual stresses act across the plate thickness. Cracking is generally step-like in character, and is associated with inclusions which are rolled into planar form during plate manufacture. Such inclusions result in reduced ductility in the through-thickness direction of the plate. Lamellar tearing will not normally arise in plate with a through-thickness ductility above 20% reduction in area (RA). Between 10 and 20% RA, it may be encountered in highly restrained cases, such as fully penetrating nozzles. If the ductility is below 10% RA, cracks may form even in relatively lightly restrained T-joints. Lamellar tears have been found in plate of 10 to 175 mm thickness, but are not common below 25 mm thickness. Where experience with a particular material and joint configuration indicates a risk of lamellar tearing, consideration must be given to joint design and welding procedure to reduce the effect of restraint, and to ensure that the fusion boundary of the weld runs across possible planes of weakness in the material.

Stress relief cracking

A number of ferritic steels may be subject to stress relief, or reheat, cracking. Cracking is intergranular, and occurs in the HAZ when restrained joints are given a post-weld heat treatment. The problem is analogous to that in austenitic steels considered below, and arises primarily in alloys showing secondary hardening. Cracking behaviour cannot be predicted at the present time with any precision. Cracking has not been reported in alloys such as $2\frac{1}{4}$Cr/1Mo or $\frac{1}{2}$Mo/B steels at below 75 mm thickness, or in $\frac{1}{2}$Cr/$\frac{1}{2}$Mo/$\frac{1}{4}$V at below 18 mm thickness. The risk of cracking can be reduced by weld dressing to reduce local stress concentrations, and by controlled heat treatment procedures.

HIGH ALLOY HEAT AND CORROSION-RESISTANT STEELS

These fall into three basic categories: the martensitic chromium-containing steels, the higher chromium ferritic steels and the chromium–nickel austenitic steels. With different chromium and nickel contents, materials containing mixtures of the three main microstructural consituents may be produced. The welding behaviour of such materials is generally dependent on the major phase present.

MARTENSITIC STAINLESS STEELS

The principal problem with the martensitic alloys is that of hydrogen-induced cold cracking, as with low alloy steels. Preheating is required, generally to over 200 °C, followed by slow cooling and post-weld heat treatment at 650–750 °C. The likelihood of cracking depends on a number of factors, and procedural trials are normally required to determine the optimum welding conditions in any given instance. Carbon content is of particular importance, and thin plate material of less than 0.1% carbon can often be welded without preheat or post-weld heat treatment; with thicker material or carbon content above 0.2%, both preheat and heat treatment are necessary. At higher carbon levels, it is essential that the welds do not cool to room temperature between completion of welding and application of heat treatment (although the joint should cool sufficiently for complete transformation to martensite to occur in the HAZ, prior to heat treatment), unless sufficient time is given to permit hydrogen diffusion out of the weld area. The martensite transformation finish temperature (M_f) may not be well defined, and in such cases, a suitable cool-out temperature should be defined by procedural tests. Suggested welding conditions for typical alloys are given in Table 33.15. If matching composition consumables are required (e.g. Table 33.16), careful consideration to hydrogen-induced weld metal cracking should be paid, and adequate electrode baking is essential. The use of austenitic consumables is preferred. Although these are of lower strength than the parent material, hydrogen is more soluble in austenite than in martensite, and thus hydrogen tends to remain in the weld metal, reducing the risk of cracking. Manual metal arc welding is generally employed for martensitic stainless steels, with TIG welding finding application for thin gauge material.

Table 33.15 SUGGESTED WELDING CONDITIONS FOR 13% CR STEELS

	Temperature, °C		
Material	*Preheat*	*Interpass*	*Post-weld heat treatment*
13%Cr/0.1%C	250	250–350	700–750
13%Cr/0.2%C	250	250–350	700–750
13%Cr/4%Ni/0.06%C	150	150–250	600–650

Table 33.16 MANUAL METAL ARC WELDING ELECTRODES FOR CR AND CR/NI STEELS (BS 2926:1970)

Composition code	*Element* wt %									
	C	Si	Mn	Cr	Ni	Mo	Nb	W	S	P
13	0.08 max	1.0 max	1.0 max	11.0–13.5	0.60 max				0.030 max	0.040 max
17	0.10 max	1.0 max	1.0 max	15.0–18.0	0.60 max				0.030 max	0.040 max
19.9	0.08 max	1.0 max	0.5–2.5	18.0–21.0	9.0–11.0				0.035 max	0.040 max
19.9.L	0.04 max	1.0 max	0.5–2.5	18.0–21.0	9.0–11.0				0.035 max	0.040 max
19.9.Nb	0.10 max	1.0 max	0.5–2.5	18.0–21.0	9.0–11.0		10 × C – 1.10		0.035 max	0.040 max
19.12.3	0.08 max	1.0 max	0.5–2.5	17.0–20.0	10.0–14.0	2.5–3.5			0.035 max	0.040 max
19.12.3.L	0.04 max	1.0 max	0.5–2.5	17.0–20.0	10.0–14.0	2.5–3.5			0.035 max	0.040 max
19.12.3.Nb	0.10 max	1.0 max	0.5–2.5	17.0–20.0	10.0–14.0	2.5–3.5	10 × C – 1.10		0.035 max	0.040 max
19.13.4	0.08 max	1.0 max	0.5–2.5	17.0–20.0	11.0–15.0	3.5–5.5			0.035 max	0.040 max
19.13.4.L	0.04 max	1.0 max	0.5–2.5	17.0–20.0	11.0–15.0	3.5–5.5			0.035 max	0.040 max
19.13.4.Nb	0.10 max	1.0 max	0.5–2.5	17.0–20.0	11.0–15.0	3.5–5.5	10 × C – 1.10		0.035 max	0.040 max
23.12	0.15 max	1.0 max	0.5–2.5	22.0–25.0	11.0–14.0				0.035 max	0.040 max
23.12.Nb	0.10 max	1.0 max	0.5–2.5	22.0–25.0	11.0–14.0		10 × C – 1.10		0.035 max	0.040 max
23.12.2	0.10 max	1.0 max	0.5–2.5	22.0–25.0	11.0–14.0	2.0–3.0			0.035 max	0.040 max
23.12.W	0.20 max	1.0 max	0.5–2.5	22.0–25.0	11.0–14.0			2.0–4.0	0.035 max	0.040 max
25.20	0.20 max	0.7 max	6.0 max	24.0–28.0	18.0–22.0				0.030 max	0.030 max
25.20.H	0.35–0.45	0.7 max	0.5–2.0	24.0–28.0	18.0–22.0				0.030 max	0.030 max
25.20.Nb	0.12 max	0.7 max	6.0 max	24.0–28.0	18.0–22.0		10 × C – 1.20		0.030 max	0.030 max
17.8.2	0.06–0.10	0.8 max	0.5–2.5	16.5–18.5	8.0–9.5	1.5–2.5			0.030 max	0.030 max
20.9.3	0.10 max	1.0 max	3.0 max	18.5–21.0	8.0–10.0	2.0–4.0			0.035 max	0.040 max

FERRITIC STAINLESS STEELS

Ferritic stainless steels suffer embrittlement on welding due to grain growth and grain boundary martensite formation in the heat affected zone that is heated above 1100 °C. These materials are notch sensitive, and the grain growth raises the ductile–brittle transition temperature to above room temperature, resulting in a brittle joint when the weldment cools out. Preheat to 200 °C and post-heat treatment at 750–850 °C are recommended. These measures do not avoid the deleterious microstructure, but serve to reduce residual welding stresses as far as possible and obtain some HAZ softening. The toughness of matching composition weld metal (e.g. Table 33.16) may be lower than that of the heat affected zone. Thus, austenitic consumables are preferred unless differences in coefficient of expansion between the weld metal and parent material are likely to cause thermal fatigue in service. Manual metal arc and TIG welding are most commonly used for ferritic grades of stainless steel.

Ferritic stainless steels are also sensitive to intercrystalline corrosive attack in the region around a weld that is heated to above 1100 °C. The problem may be avoided by post-weld heat treatment at between 700 and 850 °C. In some circumstances sigma phase may form during heat treatment, causing brittleness below 200 °C. This can be eliminated by annealing at above 850 °C.

The problems of loss of toughness and corrosion resistance in the weld area restrict the industrial use of fusion welded ferritic stainless steel assemblies. These materials can, however, offer good general corrosion resistance, and are considerably more resistant to chloride-induced stress corrosion than are austenitic grades. To avoid the welding problems of conventional ferritic stainless steels, alloys have been developed based either on reduction in interstitial elements, or on balanced chromium–nickel contents so that mixed ferrite–martensite or ferrite–austenite microstructures are obtained. Such materials have been successfully welded in thin gauges for service conditions where a risk of stress corrosion has precluded the use of austenitic steels.

AUSTENITIC STAINLESS STEELS

For practical purposes, austenitic stainless steels can be regarded as among the most weldable materials, and virtually any process can be used to make a joint. The manual metal arc and TIG processes are most common, with MIG and submerged arc welding being used when high deposition rates are required. Manual metal arc and gas-shielded welding consumables are given in Tables 33.16 and 33.17. Oxyacetylene welding is readily carried out with consumables as in Table 33.18, but is not preferred since excessive heating and a carburizing flame can cause weld decay, as considered below. A flux is desirable, and a neutral flame is recommended, oxidizing conditions causing porosity.

Table 33.17 FILLER RODS AND WIRES FOR GAS-SHIELDED ARC WELDING OF AUSTENITIC STAINLESS STEELS (BS 2901: PART 2: 1970)

	Element wt %						
Type	C	Si	Mn	Ni	Cr	Mo	Nb
308S92	0.03 max	0.25–0.60	1.0–2.5	9.0–11.0	19.5–22.0	—	—
308S96	0.08 max	0.25–0.60	1.0–2.5	9.0–11.0	19.5–22.0	—	—
308S93	0.03 max	0.70–1.00	1.5–2.0	9.5–10.5	20.0–21.0	0.30 max	—
347S96	0.08 max	0.25–0.60	1.0–2.5	9.0–11.0	19.0–21.5	—	10 × C–1.0
309S94	0.12 max	0.25–0.60	1.0–2.5	12.0–14.0	23.0–25.0	—	—
311S94	0.12 max	0.25–0.60	1.0–2.5	12.0–14.0	23.0–25.0	—	10 × C–1.3
310S94	0.08–0.15	0.25–0.60	1.0–2.5	20.0–22.5	25.0–28.0	—	—
310S98	0.35–0.45	0.80–1.30	1.0–2.5	20.0–22.5	25.0–28.0	—	—
313S94	0.06–0.13	0.25–0.60	1.0–2.5	20.0–22.5	25.0–28.0	—	10 × C–1.3
316S92	0.03 max	0.25–0.60	1.0–2.5	11.0–14.0	18.0–20.0	2.0–3.0	—
316S96	0.08 max	0.25–0.60	1.0–2.5	11.0–14.0	18.0–20.0	2.0–3.0	—
316S93	0.03 max	0.70–1.00	1.2–1.8	10.0–13.5	18.0–20.0	2.5–3.0	—
317S96	0.08 max	0.25–0.60	1.0–2.5	13.0–15.0	18.5–20.5	3.0–4.0	—
318S96	0.08 max	0.25–0.60	1.0–2.5	11.0–14.0	18.0–20.0	2.0–3.0	10 × C–1.0

S and P each to be 0.030% max.

Table 33.18 FILLER RODS FOR THE GAS WELDING OF AUSTENITIC STAINLESS STEELS (BS 1453:1972)

	Element wt %						
Type	C	Si	Mn	Ni	Cr	Mo	Nb
347S96	0.08 max	0.25–0.60	1.0–2.5	9.0–11.0	19.0–21.5	—	10×C–1.0
309S94	0.12 max	0.25–0.60	1.0–2.5	12.0–14.0	23.0–25.0	—	—
311S94	0.12 max	0.25–0.60	1.0–2.5	12.0–14.0	23.0–25.0	—	10×C–1.3
310S94	0.08–0.15	0.25–0.60	1.0–2.5	20.0–22.5	25.0–28.0	—	—
313S94	0.06–0.13	0.25–0.60	1.0–2.5	20.0–22.5	25.0–28.0	—	10×C–1.3
316S96	0.08 max	0.25–0.60	1.0–2.5	11.0–14.0	18.0–20.0	2.0–3.0	—
318S96	0.08 max	0.25–0.60	1.0–2.5	11.0–14.0	18.0–20.0	2.0–3.0	10×C–1.0

S and P each to be 0.030% max

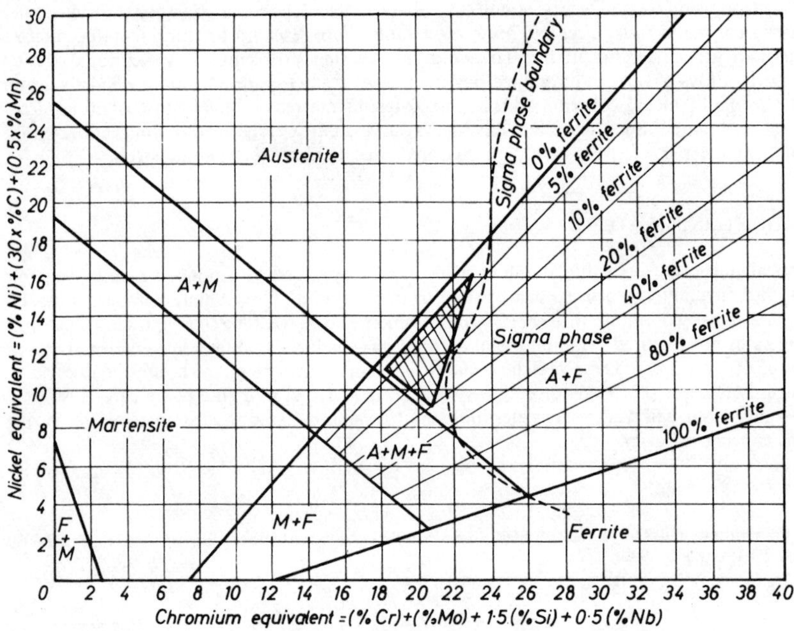

Figure 33.3 *Modified Schaeffler diagram for constitution of stainless steel weld metal.*

The main metallurgical problems with austenitic stainless steels are the avoidance of weld metal and heat affected zone cracking, and maintenance of full corrosion resistance in the weld area.

Fully austenitic weld metals are susceptible to cracking both during solidification and in underlying reheated runs. The problem may be entirely avoided by the use of consumables with composition such that the deposited weld metal contains more than about 3–5% ferrite. A guide to the relationship between composition and structure is given by the Schaeffler diagram, reproduced in Figure 33.3. Dilution from parent material must be taken into account when using the Schaeffler diagram.

The presence of ferrite in the weld metal may promote transformation to the embrittling sigma phase during heat treatment or in service at temperatures between 550 and 850 °C. The compositional range likely to suffer rapid embrittlement is indicated in Figure 33.3. Particularly if sigma formation is a hazard, the preferred weld metal compositions lie in the shaded area.

In certain corrosion-resistant applications, notably urea plant, the presence of ferrite may be undesirable since it can lead to preferential attack taking place on the weld metal. In such cases, low or zero ferrite consumables should be used, and the risk of cracking controlled by minimizing joint restraint and welding with minimum heat input. Increased manganese contents of 3–5% are beneficial.

During the welding cycle, intergranular precipitation of chromium-rich carbides can occur in areas heated to within 500–900 °C. This precipitation causes local loss of chromium from the matrix, with a consequent reduction in corrosion resistance and susceptibility to intercrystalline attack, or 'weld decay'. The problem is normally avoided by the use of material either with carbon contents below 0.03%, or containing strong carbide forming elements such as niobium or titanium. Such 'stabilized' steels are not immune to intercrystalline attack under all circumstances, and heat treatment within the sensitizing range 500–900 °C can induce sensitization to intercrystalline corrosion. If post-weld heat treatment is carried out, the temperature should be above 900 °C. For practical purposes, intercrystalline attack due to arc welding is unlikely to be encountered in unstabilized molybdenum-free material of 0.06% carbon and below, provided that the arc energy per unit length of weld metal is below 2 kJ mm^{-1}, although service in highly oxidizing media should be regarded with caution.

Austenitic steels may suffer HAZ liquation cracking during welding. The problem is minimized by the use of low arc energy welding conditions and with wrought material, by avoiding grain sizes coarser than about ASTM 3–4. In castings, cracking can often be suppressed by using material containing above 5% ferrite. The liquation cracks are of the order of 0.5 mm long, and are not generally significant in service. In welds of high restraint, however, they can form initiation points for 'reheat cracking' during elevated temperature service or post-weld heat treatment. At elevated temperature, intragranular strain-induced precipitation occurs in the HAZ. This causes a loss of creep ductility, and if joint restraint is high enough, intergranular cracking results. All common grades of austenitic stainless steel are susceptible to reheat cracking, with the exception of the 18%Cr/12%Ni/3%Mo types, provided that these do not contain residual carbide forming elements such as niobium or titanium. The risk of reheat cracking is reduced by dressing weld toes to remove liquation cracks, by the use of low hot strength weld metal, and by stress relief at above 950 °C.

High proof stress variants of the common austenitic stainless steels have been developed, based either on solid solution hardening by nitrogen or on 'warm working' by rolling at down to 850 °C to obtain a work hardening effect. These materials can generally be welded with normal consumables with no loss of strength in the weld area. With the nitrogen-bearing steels, excessive dilution of the weld pool by parent material causes a fully austenitic weld metal, with a risk of solidification cracking. High dilution situations, such as the root pass of a butt weld in thick plate, should therefore be regarded with caution. Joint preparation should be such that at least 50% of the molten weld pool is filler material.

CLAD STEELS

These consist of mild or low alloy steel clad with an overlay by rolling, explosively or by weld deposition. A number of overlays such as nickel, Monel, Inconel or stainless steel are available, to fulfil different requirements.

Various welding processes may be used for joining clad plate, although manual metal arc welding is normally employed in view of the range of electrodes available, and the facility of control. The recommended procedure is to prepare the ferritic side of the joint, and weld this conventionally with suitable electrodes, taking care that no cladding material is picked up by the weld metal. The clad side is then chipped out to sound metal to below the depth of the cladding, and welded using suitable filler metal. The choice of consumable for this weld is determined primarily by the necessity to accept dilution from the ferritic substrate material, and give the desired final weld metal composition without the formation of undesirable microstructures. A two-pass technique is usually specified, the first pass employing a consumable tolerant of dilution, and the second pass being intended to deposit weld metal of matching composition to the cladding. Typical consumable compositions are given in Table 33.19.

CAST STEELS

The welding of cast steels presents no special problems additional to those encountered in wrought metal of similar composition. Silicon and manganese contents are usually high, and this has an influence on weldability. Repairs to steel castings are usually subject to the same conditions of restraint as repairs to cast iron, and preheating is often desirable for this reason alone. Plain carbon steels with less than 0.25% C may otherwise be welded without preheat following the procedure in Figure 33.1. For steels containing 0.25–0.50% C, preheat temperatures up to 300 °C may be used, while for steels of carbon content greater than 0.50%, a preheat of 300 °C and a post-weld stress relief at 650 °C should be employed. It may be necessary to use nickel-bearing electrodes when carbon contents are very high. Alternatively, bronze welding may be used.

Preheating, when recommended, should preferably be applied to the whole casting and the figures in Table 33.20 should be regarded as minima. Manual metal arc welding is normally employed, although other methods are possible.

Table 33.19 MANUAL METAL ARC WELDING ELECTRODE COMPOSITIONS FOR WELDING CLAD STEELS ON THE CLADDING SIDE

Cladding	*Electrode*	
	First pass to cover steel	*Remaining passes*
Austenitic stainless steel	25Cr/12Ni	Matching composition to cladding
Chromium stainless steel	25/12	25/12 or matching composition to cladding
Nickel	Nickel 141	Nickel 141
Monel 400	Monel 190	Monel 190
Inconel 600	Inconel 182	Inconel 182

Table 33.20 MANUAL METAL ARC WELDING PROCEDURES FOR A SELECTION OF CAST STEELS

Steel type	*Specification*	*Electrode*	*Preheat °C*	*Post-heat °C*
0.25%C max	BS 592:1967—Grade A	BS 639:1976 E43XXR‡	20–150§	600–650*
0.35%C max	BS 592:1967—Grade B	BS 639:1976 E51XXB	20–150§	600–650
0.45%C max	BS 592:1967—Grade C	BS 639:1976 E51XXB	20–150§	600–650
C/0.5% Mo	BS 1398:1967—Grade A	BS 2493:1971—MoB	150	630–680
2.25%Cr/0.5%Mo	BS 1504–622	BS 2493:1971—2CrMoB	275	640–690†
9%Cr/1%Mo	BS 1463:1967	BS 2926:1970—23.12	250	650–720†
13%Cr	BS 1630:1967—Grade A	BS 2926:1970—23.12	250	680–750†
18%Cr/8%Ni/Nb	BS 1631:1967—Grade B	BS 2926:1970—19.9.Nb	None	None
18%Cr/8%Ni/Mo	BS 1632:1967—Grade B	BS 2926:1970—19.12.3	None	None

* Desirable, but not essential.
† Immediate post-weld heat treatment and special care essential.
‡ Select electrode to match casting properties.
§ *See* Figure 33.1

CAST IRONS

Malleable irons, grey iron, spheroidal graphite cast irons and austenitic cast irons may be welded, provided suitable precautions are taken.

Difficulties are due to lack of ductility in the parent material to accommodate weld shrinkage stresses, the transfer of carbon to the weld metal, resulting in hard, brittle deposits, and hardening in the heat affected zone. In addition, high sulphur contents may result in hot shortness in the weld and subsequent cracking.

Spheroidal graphite irons and other alloy cast irons have increased ductility and impact resistance over normal grey iron, and so may be welded with rather less difficulty. White cast irons can seldom be welded satisfactorily.

Table 33.21 FILLER RODS FOR GAS WELDING CAST IRON (BS 1453:1972)

Type	*Element wt %*						*Applications*
	C	Si	Mn	Ni	S	P	
B1	3.0–3.6	2.8–3.5	0.5–1.0	—	0.15 max	1.5 max	Easy machining
B2	3.0–3.6	2.0–2.5	0.5–1.0	—	0.15 max	1.5 max	Hard (valve seats)
B3	3.0–3.5	2.0–2.5	0.5–1.0	1.25–1.75	0.10 max	0.50 max	Ni cast iron

The choice of process is influenced by the type of component and its composition, gas welding and braze welding being suitable for light components, and metal arc and bronze welding for the heavier types of construction. Filler rods for gas welding are given in Table 33.21. For braze welding, consumables C4, C5, or C6 in Table 33.28 may be used. Types of electrodes for manual metal arc and bronze welding are given in Table 33.22.

Table 33.22 ELECTRODES FOR METAL ARC AND BRONZE WELDING OF CAST IRON

Electrode type	*Applications and remarks*
High nickel	Minimum preheat, easily machined. Not for high sulphur irons
60% nickel, 40% iron	Suitable for spheroidal graphite irons. Moderate machineability
Cast iron (soft iron)	General purpose, preheat essential
Austenitic stainless	Austenitic castings
Phosphor bronze	Not affected by sulphur, poor machineability
Aluminium bronze	Good strength, wear resistance and machineability

To accommodate shrinkage stresses and to minimize hardening in the heat affected zone, preheating to 550 °C and slow cooling is essential unless minimum penetration techniques are employed. In the latter case, repairs in thin sections may be made using high nickel, nickel–iron or bronze electrodes; and in heavy sections, buttering of the edges of the joint with nickel–iron alloy should be followed by welding with soft iron electrodes. If preheating is not employed, minimum heat input is essential, by the use of short weld runs and small diameter electrodes.

33.5.2 Non-ferrous metals

ALUMINIUM AND ALUMINIUM ALLOYS

The main processes for fusion welding this group of materials are the TIG and MIG systems. Manual metal arc and oxyacetylene welding find very limited application, and then only when alternative processes are not available.

The gas-shielded MIG and TIG processes may be used for all the weldable alloys. Sound joints with good mechanical properties can be obtained, as long as weld cleaning is carefully carried out. TIG is suitable for sheet metal work, butt welds up to 6 mm thick and fillet welds where runs are short. It is also valuable in cases where the edge preparation permits autogenous welding to be used. MIG welding is particularly suitable for fillet welding and for the butt welding of material 5 mm thick and above. MIG welding normally employs commercial purity argon as a shielding gas. This gas is also general for TIG welding, although in sections above 6 mm thickness, helium may offer advantages in increased penetration and travel speed.

The majority of aluminium alloys may be welded without difficulty, provided the correct filler wire is used. Recommendations for all processes are given in Table 33.23, and consumable compositions are in Table 33.24. Certain alloys, notably N4 (aluminium–$2\frac{1}{4}$% magnesium), H9 and H30 (aluminium–magnesium silicide type), and H15 (duralumin type) suffer from hot cracking when welded with parent metal fillers and such fillers should be used only under closely controlled conditions of low restraint. The fusion welding of the latter types of alloy is not recommended.

Table 33.23 FILLER WIRES AND ELECTRODES FOR WELDING WROUGHT Al ALLOYS TO BS 1470–1477

Parent material		*Filler wire or electrode*[1]		
Designation	*Type*	*Gas welding*[3]	MIG *or* TIG[3]	*Metal arc*[4]
1	99.99% Al	G1B	G1A (G1B)	99.5% Al
1A	99.8% Al	G1B	G1A G1B)	99.5% Al
1B	99.5% Al	G1B	G1B	99.5% Al
1C	99.0% Al	G1B	G1B	99.5% Al
N3	Al–1.25% Mn	NG3	NG3	Al–Mn
N4[5] N5[5]	Al–Mg	NG6	NG6 (NG61)	—
N51[6]	Al–Mg	NG52	NG52	—
N8	Al–Mg–Mn	NG6	NG61	—
H9[7] H20[7] H30[7]	Al–Mg–Si	NG21 (NG2)	NG21 (NG2)[2] (NG6)	Al–5% Si, Al–10% Si
H12 H15 H16 H18	Al–Cu–Mg–Si	NR[8] (NG2)	NR[8] (NG2)[2]	NR[8] Al–10% Si)

(1) Recommended fillers given first, and alternatives in parentheses.
(2) Fillers to BS 1453:1972.
(3) Filler wires to BS 2901:Part 4:1970.
(4) These are not covered by a British Standard.
(5) These Al–Mg alloys may be susceptible to hot cracking.
(6) BS 4300:Part 8:1969.
(7) NG6 may be used with care to weld these alloys, especially when anodizing is to be carried out, to give a better colour match.
(8) These are not recommended as weldable alloys but NG2 gives the best chance of success.

ALUMINIUM CASTINGS

For welding heat treatable castings, the choice of filler wire should be based upon the composition of the casting itself if post-weld heat treatment is to be employed. If cracking is encountered, a higher alloy content in the filler wire may be required. Alloys containing zinc are generally difficult to weld. For welding LM6, LM9 and LM20, NG2 may be used. NG21 is recommended for LM18 and LM25, while NG6 and NG61 can be used for LM5 and LM10. Parent material is suggested for the remainder of the weldable materials.

DISSIMILAR ALUMINIUM ALLOYS

Recommendations are given in Table 33.25.

Table 33.25 FILLER ALLOYS FOR WELDING DISSIMILAR ALLUMINIUM ALLOYS (GAS SHIELDED WELDING PROCESSES)

Material 1	*Material 2*	*Filler*
1, 1A, 1B, 1C	N3	NG3 (GIB)
1, 1A, 1B, 1C, N3	N4, N5, N51, N8	NG6 (NG61)
N4, N5, N51, N8	H9, H20, H30	NG6 (NG61)
1, 1A, 1B, 1C	H9, H20, H30	NG21 (NG2)
H9, H20, H30	LM2, LM6, LM9	NG2, NG21
N4, N5, N51, N8	LM5, LM10	NG6, NG61

33.5.3 Copper and copper alloys

Copper is produced in several grades, which vary in weldability according to the nature and quantity of the residual elements present. The material has a high thermal conductivity, and heat conduction away from the weld area may restrict the size of molten pool that can be obtained. Preheat is applied to counteract this, particularly with thicker material. Table 33.26 gives an indication of the preheat temperatures for copper and various copper alloys.

Table 33.24 FILLER RODS AND WIRES FOR THE GAS-SHIELDED WELDING OF ALUMINIUM ALLOYS (BS 2901:PART 4:1970)

Type	*Element* wt %									*Notes*
	Al	Cu	Mg	Si	Fe	Mn	Zn	Cr	Ti	
G1A	99.8 min	0.02 max		0.15 max	0.15 max	0.03 max	0.06 max			Cu + Si + Fe + Mn + Zn: 0.2% max Si content should be less than that of Fe
G1B	99.5 min	0.05 max		0.3 max	0.4 max	0.05 max	0.10 max			Cu + Si + Fe + Mn + Zn: 0.5% max Si content should be less than that of Fe
NG21	Remainder	0.10 max	0.2 max	4.5–6.0	0.6 max	0.5 max	0.2 max			
NG3*	Remainder	0.1 max	0.1 max	0.6 max	0.7 max	0.8–1.5	0.2 max	0.2 max	0.2 max	Cr + Ti*: 0.2% max
NG5*	Remainder	0.10 max	3.1–3.9	0.5 max	0.5 max	0.5 max	0.2 max	0.25 max	0.2 max	Mn + Cr: 0.5% max
NG52	Remainder	0.10 max	2.4–3.0			0.50–1.0	0.20 max	0.05–0.20	0.05–0.20	Si + Fe: 0.40% max
NG6*	Remainder	0.10 max	4.5–5.5	0.3 max	0.5 max	0.5 max	0.2 max	0.25 max	0.2 max	Mn + Cr: 0.1–0.5%
NG61	Remainder	0.10 max	5.0–5.5			0.6–1.0	0.2 max	0.05–0.20	0.05–0.20	Si + Fe: 0.40% max

* Ti content can include other grain refining elements.

Welding is normally carried out using the gas shielded processes, and the choice of gas influences the thickness above which preheat is desirable as in Table 33.27. In general, the welding speed and penetration increase with change in shielding gas in the order Ar, He, N_2, and the level of preheat decreases with the same order of gases.

Manual metal arc welding of copper and its alloys is possible, but the gas-shielded processes are preferred. Manual metal arc welding is used mainly when other methods or suitable gas-shielded consumables are not available.

Filler wires for gas welding and gas shielded arc welding of copper and copper alloys are given in Tables 33.28 and 33.29.

Tough pitch copper, which contains residual oxygen, is available in several degrees of purity, and only the high conductivity grades of tough pitch copper should be used for welding. The inert gas-shielded processes are suitable using boron deoxidized copper filler (Table 33.29, C21) where electrical conductivity is important, or silicon–manganese deoxidized filler (Table 33.29, C7). Argon, helium, or mixtures of these gases should be used for shielding. Tough pitch copper may also be bronze welded with silicon bronze or aluminium bronze electrodes, or braze welded with filler to Table 33.28, C2 (silicon brass). The presence of arsenic does not affect weldability. Oxy-acetylene welding is not recommended, due to the risk of 'gassing'.

Phosphorus-deoxidized (PDO) copper may be welded by the oxy-acetylene, TIG or MIG processes. Argon, helium or nitrogen may be used for gas-shielded methods, either separately or mixed.

Table 33.26 SUGGESTED PREHEATING CONDITIONS FOR VARIOUS COPPER ALLOYS USING ARGON SHIELDING

	Preheating temperature, °C	
Material type	*Minimum*	*Maximum*
Copper	300*	530
Silicon bronze	20	65
Phosphor bronze	175	290
70/30 Cu/Ni	20	110
Aluminium bronze	20	150

* 350 °C required for TIG welding.
For further information *see* P. G. F. duPré, *Philips Welding Reporter*, 1972, **8**, 14–26.

Table 33.27 THICKNESS ABOVE WHICH PREHEAT MAY BE REQUIRED—COPPER AND COPPER ALLOYS

	Shielding gas		
Process	*Argon*	*Helium*	*Nitrogen*
TIG	3 mm	6 mm	9 mm
MIG	6 mm	9 mm	12 mm

See Table 33.26 for further information.

Table 33.28 FILLER WIRES FOR GAS WELDING COPPER AND COPPER ALLOYS (BS 1453:1972)

	Element wt %							
Type	Cu	Zn	Pb	Al	Fe	Ni	Mn	Si**
C1	99.85 min†		0.010 max		0.030 max	0.10 max		
C2	57.0–63.0	rem.	0.03 max	0.03 max				0.2–0.5
C2B	56.0–60.0	rem.	0.05 max	0.01 max	0.25–1.2	0.2–0.8	0.01–0.50	0.04–0.15
C2C	56.0–60.0	rem.	0.05 max	0.01 max	0.25–1.2		0.01–0.50	0.04–0.15
C3	59.0–61.0	rem.	0.03 max	0.03 max				
C4	57.0–63.0	rem.	0.03 max	0.03 max	0.1–0.5		0.05–0.25	0.15–0.3
C5	45.0–53.0	rem.	0.03 max	0.03 max.	0.5 max	8.0–11.0	0.5 max	0.15–0.5
C6	41.0–45.0	rem.	0.03 max	0.03 max	0.3 max	14.0–16.0	0.2 max	0.2–0.5

** For other elements *see* next page.

Table 33.28 FILLER WIRES FOR GAS WELDING COPPER AND COPPER ALLOYS (BS 1453:1972)—*continued*

	***Element* wt %						*Total impurities excluding*
Type	Sn	As	Sb	Bi	P	Tl	Ag, Ni, As, P
C1	0.01 max	0.05 max	0.005 max	0.003 0 max	0.015–0.08	0.010 max*	0.060 max
C2	0.5 max						
C2B	0.8–1.1						0.50 max incl. Pb and Al
C2C	0.8–1.1						0.50 max incl. Pb and Al
C3							0.50 max
C4	0.5 max						
C5	0.5 max						
C6	1.0 max						

*Se plus Tl: 0.020% max. †Includes Ag 0.5–1.2%. **For other elements *see* previous page.

The use of nitrogen produces a hotter arc with increased penetration, but with MIG welding, less satisfactory metal transfer may result. Phosphorus content is important, and should be as low as possible to minimize porosity. Phosphorus does not act as an efficient deoxidant in gas-shielded welding, and for this reason filler wires containing additional deoxidants should be used. Autogenous welding is not possible without the risk of porosity, and if welding without filler wire is required, zinc deoxidized (cap) copper can be employed, using the TIG process. Zinc content is relatively unimportant within the range 0.5–3.0% zinc. Oxy-acetylene welds are made using a copper–silver–phosphorus filler rod (Table 33.28, C1) and a flux. Such welds are hot hammered during welding to remove porosity and frequently cold hammered to improve mechanical properties.

Oxygen-free high conductivity copper may be oxy-acetylene welded without gassing, but if TIG or MIG welding is employed, there is a risk of porosity formation which may be overcome by using boron–copper filler wire.

All grades of copper may be bronze welded using the manual metal arc, TIG or MIG processes and fillers of the aluminium, silicon or tin bronze types. The technique involves the use of a wide edge preparation, preferably a fillet and the use of soft arcs to minimize penetration. Deoxidized and oxygen-free copper may also be braze welded using silicon brass (Table 33.28 C2) or manganese bronze (Table 33.28 C4) fillers.

COPPER–ALUMINIUM ALLOYS

Gas welding is not recommended but carbon arc welding using cryolite flux, manual metal arc welding, or preferably the argon or helium TIG welding processes may be used. The iron-bearing single phase alloy Cu/7 Al/3 Fe is normally welded with a nickel-bearing duplex alloy (Table 33.29, C20). Most duplex and complex bronzes are welded with fillers of matching composition, as are the manganese–aluminium bronzes. An important consideration is corrosion resistance, and care should be taken to avoid a combination of manganese–aluminium bronze and the normal single phase and duplex bronzes. The nickel-bearing filler C20 is resistant to dealuminification in all but the most severe environments and is frequently used as a facing deposit in welds in BS 1400: AB2C castings, which are normally made with 10% aluminium fillers (Table 33.29, C13). If a single phase deposit is required, C12 Fe filler may be used for a corrosion resistant layer on top of a more crack resistant filler. If stress corrosion is a problem, a small tin addition may be made to both parent and filler metals.

The welding of the aluminium bronzes should be carried out with as low a heat input as possible. Thus, preheating and high inter-run temperatures should be avoided, as should weaving when depositing filler metal. It is often necessary to give a post-weld thermal treatment to eliminate the risk of stress corrosion cracking.

COPPER–NICKEL ALLOYS

The problems of embrittlement and porosity in welding cupro-nickel may be overcome by using filler wires containing manganese as a desulphurizer and titanium as a deoxidant. Filler wires are available for 90/10, 80/20 and 70/30 cupro-nickels and for 94/5/1 copper–nickel–iron alloy (C16, C17, C18 and C19, Table 33.29). Either argon arc or inert-gas metal arc welding is normally employed, but flux-coated manual metal arc electrodes are available for some alloys.

COPPER–SILICON ALLOYS

Silicon bronzes and 'Everdur' are readily weldable, the inert-gas processes being preferred. Parent metal filler is employed (C9, Table 33.29). Bronze welding is also possible.

COPPER–TIN ALLOYS

Tin bronzes usually contain phosphorus, and their welding behaviour is somewhat similar to phosphorus deoxidized copper, both oxy-acetylene and TIG welds tending to be porous. Filler rods employed are given in Table 33.29 (C10, C11). Bronze welding is preferable.

The weldability of the gunmetals depends upon the lead content. Those containing 0.1% Pb are weldable by the TIG, MIG, manual metal arc and gas welding processes. At 0.1–0.5% Pb, manual metal arc with phosphorus bronze electrodes or bronze welding offers a reasonable chance of success. The leaded gunmetals can be considered to be unweldable although single layer cosmetic repairs may sometimes be made successfully.

Table 33.29 FILLER RODS AND WIRES FOR THE GAS-SHIELDED WELDING OF COPPER AND COPPER ALLOYS (BS 2901:PART 3:1970)

	Element wt %					
Type	Cu	Al	Ti	Fe	Ni	Mn
C7*	98.5 min	0.03 max		0.03 max	0.10 max	0.15–0.35
C8†	99.4 min	0.1–0.3 Al + Ti: 0.25–0.5	0.1–0.3	0.30 max	0.10 max	
C9	remainder	0.03 max		0.10 max	0.10 max	0.75–1.25
C10	93.8 min	0.03 max				
C11	92.3 min	0.03 max				
C12	90.0 min	6.0–7.5		(Fe + Ni + Mn)‡: 1.0–2.5		
C12Fe	89.0 min	6.5–8.5		2.3–3.5		
C13	86.0 min	9.0–11.0		0.75–1.5	1.0 max	1.0 max
C16	remainder	0.03 max	0.20–0.50	1.5–1.8	10.0–12.0	0.5–1.0
C18	66.5 min	0.03 max	0.20–0.50	0.4–1.0	30.0–32.0	0.5–1.5
C20	80.5–85.0	8.0–9.5		1.5–3.5	3.5–5.0	0.5–2.0
C21§	99.8 min			0.030 max		0.020 max
C22	remainder	7.0–8.5		2.0–4.0	1.5–3.0	11.0–14.0

	Element wt %				
Type	Si	Zn	Sn	P	B
C7*	0.20–0.35		1.0 max	0.015 max	
C8†				0.015 max	
C9	2.75–3.25	0.5 max		0.020 max	
C10			4.5–6.0	0.02–0.40	
C11			6.0–7.5	0.02–0.40	
C12	0.10	0.2 max			
C12Fe	0.10 max	0.2 max			
C13	0.10 max	0.2 max			
C16	0.01 max			0.01 max	
C18	0.01 max			0.01 max	
C20	0.10 max	0.2 max			
C21	0.020 max				0.02–0.10
C22					

* These rods and wires are intended for welding Cu using Ar or He as the shielding gas.
† These rods and wires are intended for welding Cu using N_2 as the shielding gas: Ar or He may be used.
‡ Optional elements.
§ British Patent No. 810 233.

Both classes of material have long freezing ranges and are consequently hot short. They are also subject to coring and shrinkage porosity if large molten pools are employed.

COPPER–ZINC ALLOYS

The welding of brass is not easy, due to excessive fume formation, and fume extraction plant may prove necessary. Oxy-acetylene welding using an oxidizing flame and parent metal filler or silicon or manganese–brass or nickel–silver fillers is possible. TIG and MIG welding both employ zinc-free filler alloys in preference to brass, although brass fillers are available for the former process (Table 33.29). For manual metal arc and MIG welding, silicon bronze or aluminium bronze electrodes are employed. The susceptibility of brasses to season cracking should be borne in mind, and stress relief of brass components is desirable.

BERYLLIUM–COPPER

Alternating current TIG welding is preferred. In order to obtain the maximum joint efficiency, welding speed should be high, and post-weld heat treatment is essential. Both the high Be/low Co and high Co/low Be alloys should be welded in the solution heat-treated condition using high Be filler rods. Post-weld ageing will restore most of the strength and hardness. Although possible, consumable electrode methods are not recommended due to the toxicity problem.

CADMIUM–COPPER

Experience is limited, but it is suggested that for gas welding, filler rods to BS 1453 C1 (Table 33.28) should be used. Parent metal filler may be suitable for inert-gas welding. Cadmium is extremely toxic and fusion welding should be attempted only under stringent ventilation control.

CHROMIUM–COPPER

Chromium–copper is heat-treatable and suffers from a loss in mechanical properties in the heat affected zone. It also exhibits a tendency to hot shortness. Bronze welding with aluminium bronze filler has been successfully employed, but joint efficiencies are low after post-weld heat treatment. Joints of low restraint may be made using a d.c. electrode negative TIG technique with helium shielding. The components should be solution treated prior to welding, and subsequently aged.

OTHER COPPER-RICH ALLOYS

Silver–copper may be welded by the inert-gas processes using boron–copper filler wire. Tellurium- and selenium–copper suffer from excessive weld porosity. Leaded copper is hot short, and welding is difficult.

33.5.4 Lead and lead alloys

Lead and its alloys may be welded by the oxy-acetylene process using no flux and a neutral flame. Other fuel gases, such as hydrogen, butane or coal gas, may be used. The edges should be cleaned before welding, and parent metal filler is used. Other processes are little used.

33.5.5 Magnesium alloys

Magnesium alloys are preferably welded by the a.c. TIG process with argon shielding. Normal inert-gas metal arc welding is not suitable for magnesium because of the high deposition rates required. However, progress has been made with the pulsed version of this process. The use of gas welding is confined to the wrought Mg–Mn and Mg–Al–Zn alloys, and should never be used for fillet or lap welds or for welds in zirconium-bearing alloys due to the corrosion hazard from flux residues. Proprietary fluxes are available. Flux removal and chemical cleaning should be followed by chromating.

The argon arc process may be used for welding all the commercially available alloys, with the exception of the wrought Mg–3% Zn–0.7% Zr and Mg–5.5% Zn–0.7% Zr alloys and the cast Mg–4.5% Zn–0.7% Zr alloy.

British Standard filler alloys are listed in Table 33.30. For castings parent metal filler is invariably used. A guide to filler metal selection is given in Tables 33.31, 33.32 and 33.33.

33.5.6 Nickel and nickel alloys

These materials fall into two groups, namely solid solution and precipitation hardening alloys. Nickel alloys are highly sensitive to contamination during welding and to the presence of various minor impurities. To avoid weld cracking and porosity, all surfaces to be welded and filler rods must be grease-free and scrupulously clean. The presence of sulphur, lead and zinc are particularly detrimental, the effects of sulphur being most marked with the chromium-free alloys. Most nickel alloy weld metals are quite fluid and, if possible, components should be welded in the flat position.

It is normally recommended that nickel alloys be stress-free prior to welding. The solid solution hardened grades should be annealed, although a certain amount of cold work is permissible. Post-weld heat treatment is not usually necessary. Incoloy DS* containing silicon is sensitive to hot

Table 33.30 FILLER RODS AND WIRES FOR GAS-SHIELDED WELDING OF MAGNESIUM ALLOYS (BS 2901:PART 4:1970)

	Element wt %						
Type	Zn	Mn	Al	Zr	Cd	*Rare earths*	*Notes*
D1	1.0 max	0.15–0.4	9.0–10.5				Cu + Si + Fe + Ni = 0.40% max
D2	0.03 max	1.0–2.0	0.08 max				
D3	1.5 max	0.15–0.40	5.5–8.5				
D4	0.6–1.4	0.15–0.7	2.3–3.5				
D5	0.75–1.5	0.15 max	0.02 max	0.4–1.0			
D6	1.5–2.5	0.15 max	0.02 max	0.4–1.0	1.5–2.5		Alloys covered by British Patent No. 511 137
D7	3.5–5.0	0.15 max		0.4–1.0		0.75–1.75	
D8	0.8–3.0	0.15 max		0.4–1.0		2.5–4.0	

Table 33.31 FILLER ALLOY SELECTION AND WELDABILITY FOR SIMILAR AND DISSIMILAR WROUGHT MAGNESIUM ALLOYS

Type	*Elektron*	AM503	AZM	AZ31	ZW1	ZW3	ZW6	ZTY
1.5 Mn	AM503	D2 / A	D2 / C	D2 / C	D5 / B	— / D	— / D	— / D
6.0 Al–1.0 Zn	AZM		D3 / A	D3 / A	D5 / C	— / D	— / D	— / D
3.0 Al–1.0 Zn	AZ31			D4 / A	D5 / C	— / D	— / D	— / D
1.3 Zn–0.7 Zr	ZW1				D5 / A	D5 / B	— / D	D5 / A
3.0 Zn–0.7 Zr	ZW3					D6† / D	— / D	D5 / C
5.5 Zn–0.7 Zr	ZW6						— / D	— / D
0.75 Th–0.5 Zn–0.5 Zr	ZTY							D7 / A

Key

Filler (upper part of cell) / Weldability (lower part of cell)

A = Good
B = Fair
C = Possible
D = Not recommended

† For mechanized welds only. D5 may be used for manual welds, but with reduced tensile efficiency.

Note.
The nuclear alloys AL80 and ZR55 are both weldable with parent metal filler.

* Henry Wiggin & Co. Ltd.

Table 33.32 FILLER ALLOY SELECTION AND WELDABILITY FOR DISSIMILAR MAGNESIUM CASTING ALLOYS

Alloy type	*Elektron*	ZRE1	MSR	RZ5	TZ6	ZT1	Z5Z*
8.0 Al–0.4 Zn	A8	ZRE1 / C	MSR / C	ZRE1 / C	ZT1 / C	ZT1 / C	— / D
2.7 RE–2.2 Zn–0.6 Zr	ZRE1		MSR / A	ZRE1 / A	ZT1 / C	ZT1 / C	— / D
1.7–2.5 RE–0.6 Zr–2.5 Ag	MSR			MSR / A	ZT1 / C	ZT1 / C	— / D
1.2 RE–4.0 Zn–0.7 Zr	RZ5				ZT1 / C	ZT1 / C	— / D
1.8 Th–5.5 Zn–0.7 Zr	TZ6					ZT1 / A	— / D
5.0 Th–2.2 Zn–0.7 Zr	ZT1						— / D

For key *see* Table 33.31.
For similar alloys use parent metal filler (weldability A).
* Only simple repairs possible on Z5Z castings.

Table 33.33 FILLER ALLOY SELECTION AND WELDABILITY FOR WELDING WROUGHT TO CAST MAGNESIUM ALLOYS

	Elektron	*Wrought materials*					
		AM503	AZM	AZ31	ZW1	ZW3	ZTY
Casting	A8	D2 / C	D3 / A	D3 / A	— / D	— / D	— / D
Casting	ZRE1	— / D	ZRE1 / C	ZRE1 / C	D5 / A	D5 / B	D7 / C
Casting	MSR	— / D	MSR / C	MSR / C	MSR / A	MSR / B	MSR / C
Casting	RZ5	— / D	ZRE1 / C	ZRE1 / C	D5 / A	D5 / B	D7 / C
Casting	TZ6	— / D	— / D	— / D	D5 / A	D5 / B	D7 / A
Casting	TZ6	— / D	— / D	— / D	D5 / A	D5 / B	D7 / A

For key *see* Table 33.31
Z5Z and ZW6 are not included due to poor weldability.
For alloy types *see* Tables 33.31 and 33.32.

cracking and restraint must be kept to a minimum. Precipitation hardened materials should be solution treated before welding, and aged afterwards. Some materials based on the Ni–Al–Ti precipitation hardening system are susceptible to post-weld heat treatment cracking, depending on the joint restraint, as are austenitic stainless steels. In such cases, a full heat treatment after welding is recommended. HAZ liquation cracking may arise and in certain precipitation hardened alloys, such as Nimonic* 80A, 90 and PK33, restricts the thickness that can be welded to below 5 mm.

TIG welding, both manual and mechanized is the most widely used process for sheet material. A.c. or d.c. may be employed, although the latter is preferred. Most alloys may be welded and suitable fillers are listed in Tables 33.34 and 33.35. Chromium-free grades are susceptible to porosity during autogenous welding, and the addition of filler is recommended so that at least 50% of the weld pool volume is constituted by filler metal. Commercial purity argon is generally used for shielding with additions of up to 10% hydrogen helping to reduce porosity and increase welding speeds. Helium or argon–helium mixtures are applicable and may also be advantageous in increasing welding speed. Nitrogen in the gas or from the atmosphere can cause porosity. Complete coverage of the weld area with shielding gas is essential. The largest gas nozzle possible should be used with the minimum practicable distance from the nozzle to the workpiece, while adequate gas backing must be applied. Many nickel alloys form tenacious oxide films, which tend to be patchy and interfere with weld uniformity. Argon–5% hydrogen mixtures can be used as backing and shielding gases to overcome such problems. Nickel–molybdenum alloys may give a very fluid weld pool, when the TIG process is preferred for positional welding, in view of the facility of control.

For welding nickel and its alloys in heavier gauges, manual metal arc and MIG welding are generally suitable, the welding technique differing little from that practised with austenitic stainless steels. With manual metal arc welding, it is advisable to use small gauge electrodes, and to allow cooling between weld runs. Complete slag removal is imperative, particularly for high temperature service where reactions between the metal and residual slag may cause severe corrosion. Table 33.35 gives typical consumables for some of the more common commercial alloys; d.c. current is essential with positive electrodes. There is no British Standard for manual metal arc electrodes for nickel alloys, but a number of proprietary makes are obtainable.

Argon shielding is preferred with MIG welding, although the addition of up to 20% helium may be of benefit in spreading the weld pool and reducing the incidence of cold laps. Spray, globular dip, or dip transfer conditions have been employed for most nickel alloys, with consumables shown in Tables 33.34 and 33.35. Spray transfer gives the highest deposition rates but may cause weld metal cracking, particularly if the welding conditions are such that a weld bead with a concave surface is provided. Pulsed MIG welding has also been found applicable for nickel alloys, especially for positional welding.

Oxy-acetylene welding is little used for nickel alloys. It is suitable for most solid solution hardened alloys and Monel K-500, although carbon pick-up may reduce corrosion resistance. Nickel and Incoloy DS can be welded without flux while fluxes are available for the Monel, Incoloy, Inconel, Brightray and Hastelloy D alloys. Fluxes used with the chromium-containing alloys must be free from boron, since the presence of boron compounds can cause weld metal

Table 33.34 FILLER RODS AND WIRES FOR GAS SHIELDED WELDING OF NICKEL ALLOYS (BS 2901:PART 5:1970)

	Element wt %						
Type	Ni	Cr	Co	Fe	Mo	Ti	Al**
NA32	93.0 min		1.00 max	1.0 max		2.0–3.50	1.50 max
NA33	62.0–69.0		1.00 max	2.5 max		1.5–3.0	1.25 max
NA34	Remainder	18.0–21.0	1.00 max	0.5 max			
NA35	67.0 min	18.0–22.0	0.10 max	3.0 max		0.75 max	
NA36	Remainder	18.0–21.0	15.0–21.0	3.0 max		1.8–3.0	0.9–2.0
NA37	Remainder	16.0–20.0	12.0–16.0	1.0 max	5.0–9.0	1.5–3.0	1.7–2.5
NA38	Remainder	19.0–21.0	19.0–21.0	0.7 max	5.6–6.1	1.9–2.4	0.3–0.60
NA39	67.0 min	14.0–17.0	1.00 max	0.80 max		2.5–5.0	
NA40	Remainder	20.5–23.0	0.5–2.5	17.0–20.0	8.0–10.0		
NA41	33.0–46.0	19.5–23.5	1.00 max	Remainder	2.5–3.40	0.6–2.20	0.20 max
NA42	42.0–45.0	15.0–18.0	2.00 max	Remainder	2.5–4.0	0.9–1.5	0.9–1.5

** For other elements *see* next page.

* Henry Wiggin & Co. Ltd.

Table 33.34 FILLER RODS AND WIRES FOR GAS SHIELDED WELDING OF NICKEL ALLOYS (BS 2901: PART 5:1970)—*continued*

Type	***Element* wt %				*Notes*
	Mn	C	Si	Cu	
NA32	1.00 max	0.15 max	0.75 max	0.25 max	Other elements 0.5% max
NA33	3.0–4.0	0.15 max	1.25 max	remainder	
NA34	1.2 max	0.26 max	0.50 max	0.20 max	
NA35	2.5–3.50	0.10 max	0.50 max	0.50 max	Ta 0.30% max: (Nb+Ta) 2.0–3.00%
NA36	1.0 max	0.13 max	1.5 max		
NA37	0.5 max	0.07 max	0.5 max	0.2 max	B 0.005% max: Zr 0.06% max
NA38	0.2–0.60	0.04–0.08	0.1–0.40	0.2 max	(Ti+Al) 2.4–2.8%
NA39	2.0–2.75	0.08 max	0.35 max	0.50 max	
NA40	1.0 max	0.15 max	1.0 max		W 0.2–1.0%
NA41	1.0 max	0.05 max	0.50 max	1.5–3.00	Other elements 0.5% max
NA42	0.2 max	0.10 max	0.3 max		B 0.005% max: Zr 0.05% max

** For other elements *see* previous page.

Table 33.35 TYPICAL CONSUMABLES FOR WELDING NICKEL ALLOYS

Parent material	*Welding process*		
	Manual metal arc	*Gas*	*Inert gas shielded*
Nickel	Nickel electrode 141	Nickel 41	NA32 (Nickel 61)
Monel† 400	Monel electrode 190	Monel 40	NA33 (Monel 60)
Monel K–500	Monel electrode 134	Monel 64	Monel 64
Inconel† 600	Inconel electrodes 132 or 182	NA34 (NC80/20)†	NA35 (Inconel 82)
Incoloy† 800	Incoweld† A	NA34 (NC80/20)	NA35 (Inconel 82)
Incoloy DS	Incoweld A	NA34 (NC80/20)	NA34 (NC80/20)
Nimonic† 75	Inconel electrodes 132 or 182	NA34 (NC80/20)	NA34 (NC80/20)
Nimonic 80A			NA36 (Nimonic 90)
Nimonic 90			NA36 (Nimonic 90)
Hastelloy* X			NA40

See Table 33.34 for NA32, 33, 34, 35, 36 and 40.
* Union Carbide U.K. Ltd. † Henry Wiggin & Co. Ltd.

cracking. Complete flux removal after welding is essential. The flame should be slightly reducing for nickel and Monel 400, more reducing with the chromium-containing alloys, and highly reducing for Monel K-500 and Hastelloy D. Filler alloys are given in Table 33.35.

Submerged-arc welding is applicable to various nickel alloys, generally when heavy section material is being joined for corrosion resistant applications. Only a limited range of flux types is available. D.c. is used, with electrode positive being preferred to obtain deeper penetration and reduced risk of slag entrapment.

Electron beam welding can be used for most nickel alloys, and is a common fabrication process for high temperature items such as gas turbine components.

33.5.7 Noble metals

SILVER

Silver is difficult to weld due to its property of oxygen absorption. A slightly reducing flame is used for oxy-acetylene welding with a borax–boric acid flux. Silver–0.5% aluminium rods reduce the tendency to porosity formation. Argon arc welding using d.c. is possible, though slight porosity is usually present. A.c. argon arc welding using aluminium bearing fillers or traces of aluminium powder in the joint preparation is also possible.

GOLD

Gold and gold–copper alloys are readily welded by the gas welding process using parent metal filler and borax fluxes. A reducing flame is recommended for the alloys.

PLATINUM

An oxidizing oxy-hydrogen flame is used without flux. The platinum–rhodium and platinum–iridium alloys may also be welded using this technique.

33.5.8 Refractory metals

The problem common to the welding of this group of materials is their strong affinity for atmospheric and other gases, and the consequent need to avoid contamination, both of the molten pool and the surfaces of the cooling solidified weld bead and heat affected zones. This problem is overcome in two ways, either by provision of a complete argon-filled chamber in which the workpiece and welding head is placed, or by the provision of extended argon shrouds and argon backing.

These materials may all be welded by the argon arc process, and titanium has been successfully welded by the inert-gas metal arc process. The use of this process may eventually be extended to cover other materials in this group.

TITANIUM AND TITANIUM ALLOYS

The α titanium alloys (e.g. commercially pure Ti, Ti–5%Al–2½%Sn) are fully weldable, as are alloys with only small amounts of eutectoid formers or β stabilizers (e.g. Ti–2½% Cu, Ti–8%Al–1%Mo–1%V, Ti–7%Al–2%Nb–1%Ta). As the β stabiliser content rises, the weld zone becomes less ductile and eventually brittle. The Ti–6%Al–4%V alloy represents about the limiting composition for $\alpha\beta$ alloys for good weldability.

In the weldable alloys, TIG welds (d.c., electrode negative) may be made without filler or with parent metal filler depending on thickness. For $\alpha\beta$ alloys, the ductility of the weld metal can be increased by using commercially pure Ti filler, but this will not improve the ductility of the heat affected zone. MIG welding is feasible, but experience is limited. All metal reaching higher than 600–650 °C should be protected by argon.

NIOBIUM AND TANTALUM AND THEIR ALLOYS

The same technique as for titanium should be used, except that argon protection will be required at and above 400 °C.

MOLYBDENUM

Electron beam and TIG (d.c., electrode negative) welds can be made, but the weld zone will be brittle below 300–500 °C. The techniques used for titanium should be applied. The metal produced by sintering gives grossly porous welds, and that made by arc casting should be used.

ZIRCONIUM ALLOYS

Experience is limited to the zirconium–tin and zirconium–niobium alloys, which are used in nuclear engineering. These materials are particularly susceptible to nitrogen contamination, which is harmful to the corrosion resistance. Superficial contamination during welding is removed by pickling. D.c. electrode negative is used, and gas cover should extend to areas in excess of 400 °C.

URANIUM

Uranium may be welded without difficulty by the argon arc process using d.c. electrode negative. It is less sensitive to contamination than titanium.

33.5.9 Zinc and zinc alloys

Zinc and zinc-base castings are difficult to weld. The oxy-acetylene process using a slightly reducing flame and a zinc-ammonium chloride flux is most suitable. Parent metal filler should be used.

33.5.10 Dissimilar metals

Direct fusion welding of dissimilar metals is possible in certain cases, and special techniques have been developed for difficult combinations.

Steels may be joined to nickel alloys, but the joint dilution and selection of filler materials must be made to give either a high nickel–alloy weld or a weld composition falling within the shaded area of the Schaeffler diagram (Figure 33.3).

Combinations of copper and chromium should be avoided, as when welding Monel to chromium bearing steels, and in such cases, nickel-base filler should be interposed.

Steels may be joined to copper and copper alloys by bronze welding techniques, but joining to aluminium involves precoating of the steel with aluminium–silicon brazing alloy, and joint ductility is low.

Copper may be welded direct to nickel, provided a deoxidant (titanium) is present either in the nickel or the filler wire.

Copper to aluminium joints are made by coating the copper with a layer of silver solder (BS 1845: Type AG1 or AG2) and welding by argon arc or inert gas metal arc processes, using aluminium 10% silicon filler.

33.6 British standards relating to welding

General

BS	
499:	Welding terms and symbols. Part 1. 1965. Welding, brazing and thermal cutting glossary. Part 2. 1980 Specifications for Symbols for welding. Part 3. 1965. Terminology of and abbreviations for fusion weld imperfections as revealed by radiography.
499C:1965	Chart of British Standard welding symbols (based on BS 499 Part 2).
679:1959	Filters for use during welding and similar industrial operations.
1542:1960	Equipment for eye, face and neck protection against radiation arising during welding and similar operations.
2653:1955	Protective clothing for welders.
5378:1976	Safety colours and safety designs.

Processes

BS	
693:1960	General requirements of oxy-acetylene welding of mild steel.
1140:1957	General requirements for spot welding of light assemblies in mild steel.
1723:1963	Brazing.
1724:1959	Bronze welding by gas.
1821:1957	Class I oxy-acetylene welding for steel pipelines and pipe assemblies for carrying fluids.
2360:1955	Projection welding of low carbon steel sheet and strip.
2633:1973	Class I arc welding of ferritic steel pipework for carrying fluids.
2640:1955	Class II oxy-acetylene welding of steel pipelines and pipe assemblies for carrying fluids.
2937:1957	General requirements for seam welding in mild steel.
2971:1977	Class II arc welding of carbon steel pipework for carrying fluids.
2996:1958	Projection welding of low carbon wrought steel studs, bosses, bolts, nuts and annular rings.
3019: —	General recommendations for manual inert-gas tungsten-arc welding. Part 1:1958 Wrought aluminium, aluminium alloys and magnesium alloys. Part 2:1960 Austenitic stainless and heat-resisting steels.
3571: —	General recommendations for manual inert-gas metal-arc welding. Part 1:1962 Aluminium and aluminium alloys.
3847:1965	General requirements for mash seam welding in mild steel.
4204:1967	General requirements for the flash welding of steel pipes and tubes for pressure and other high duty applications.
4515:1969	Field welding of carbon steel pipelines.
4570:1970	Fusion welding of steel castings. Part 1 Production, rectification and repair. Part 2 Fabrication welding.

BS	
4677:1971	Class I arc welding of austenitic stainless steel pipework for carrying fluids.
5135:1974	Metal arc welding of carbon and carbon-manganese steels.

Equipment

638: —	Arc welding power sources, equipment and accessories. Part 1:1979 Specification for oil-cooled power sources for manual, semi-automatic and automatic metal-arc welding and for TIG welding. Part 2:1979 Specification for air-cooled power sources for manual metal-arc welding with covered electrodes and for TIG welding. Part 3:1979 Specification for air-cooled power sources for semi-automatic and automatic metal-arc welding. Part 4:1979 Specification for welding cables.
1389:1960	Dimensions of hose connections for welding and cutting equipment.
3065:1965	The rating of resistance welding and resistance heating machines.
3856:1965	Platens for projection welding machines.
4819:1972	Resistance welding water-cooled transformers of the press-package and portable types.
5120:1975	Rubber hose for gas welding and allied processes.
5741:1979	Specification for pressure regulators used in welding, cutting and related processes.

Filler rods and electrodes

639:1976	Covered electrodes for the manual metal-arc welding of carbon and carbon-manganese steels.
807:1955	Spot welding electrodes.
1453:1972	Filler rods and wires for gas welding.
1845:1977	Filler metals for brazing.
2493:1971	Low alloy steel electrodes for manual metal-arc welding.
2901:1970	Filler rods and wires for gas-shielded arc welding. Part 1 Ferritic steels. Part 2 Austenitic stainless steels. Part 3 Copper and copper alloys. Part 4 Aluminium and aluminium alloys and magnesium alloys. Part 5 Nickel and nickel alloys.
2926:1970	Chromium-nickel austenitic and chromium steel electrodes for manual metal-arc welding.
3067:1959	Dimensions of blanks for seam welding wheels.
4165:1971	Electrode wires and fluxes for the submerged arc welding of carbon steel and medium tensile steel.

BS	
4215:1967	Spot welding electrodes and electrode holders.
4577:1970	Materials for resistance welding electrodes and ancillary equipment.
5465:1977	Electrodes wires and fluxes for the submerged arc welding of austenitic stainless steel based on weld metal composition.

Testing and inspection

709:1971	Methods of testing fusion welded joints and weld metal in steel.
1077:1963	Fusion-welded joints in copper.
1138:1943	Test pieces for production control of aluminium alloy spot welds.
1295:1959	Tests for use in the training of welders. Manual metal-arc and oxy-acetylene welding of mild steel.
2600: —	Methods of radiographic examination of fusion-welded butt joints in steels. Part 1:1973 5 mm up to and including 50 mm thick. Part 2:1973 Over 50 mm up to and including 200 mm thick.
2704:1978	Calibration blocks and recommendations for their use in ultrasonic flaw detection.
2910:1973	Methods for radiographic examination of fusion-welded circumferential butt joints in steel pipes.
3451:1973	Methods of testing fusion welds in aluminium and aluminium alloys.
3923: —	Methods for ultrasonic examination of welds. Part 1:1978 Manual examination of fusion butt joints in ferritic steels. Part 2:1972 Automatic examination of welded seams. Part 3:1972 Manual examination of nozzle welds.
3971:1966	Image quality indicators for radiography and recommendations for their use.
4069:1966	Magnetic flaw detection inks and powders.
4129:1967	Resistance welding properties of welding primers and weld-through sealers.
4206:1967	Methods of testing fusion welds in copper and copper alloys,
4397:1969	Methods for magnetic particle testing of welds.
4416:1969	Method for penetrant testing of welded or brazed joints in metals.
4778:1979	Glossary of general terms used in quality assurance (including reliability and maintainability terms).
4870: —	Approval testing of welding procedures. Part 1:1974 Fusion welding of steel.
4871: —	Approval testing of welders working to approved welding procedures. Part 1:1974 Fusion welding of steel.

BS

4872: —	Approval testing of welders when welding procedure approval is not required. Part 1: 1972 Fusion welding of steel. Part 2: 1976 TIG or MIG welding of aluminium and its alloys.
5289: 1976	Code of practice for visual inspection of fusion-welded joints.
5430: —	Periodic inspection, testing and maintenance of transportable gas containers (excluding dissolved acetylene containers). Part 2: 1977 Welded steel containers of water capacity 1 l up to 130 l.

PD

6493: 1980	Guidance on some methods for the derivation of acceptance levels for defects in fusion-welded joints.

Materials

1640: —	Steel butt-welding pipe fittings for the petroleum industry. Part 1: 1962 Wrought carbon and ferritic alloy steel fittings. Part 2: 1962 Wrought and cast austenitic Cr–Ni steel fittings. Part 3: 1968 Wrought carbon and ferritic alloy steel fittings. Metric units. Part 4: 1968 Wrought and cast austenitic Cr–Ni steel fittings. Metric units.
1965: —	Butt-welding pipe fittings for pressure purposes. Part 1: 1963 Carbon steel.
3014: 1958	'As welded' and cold drawn welded austenitic stainless steel tubes for mechanical, structural and general engineering purposes.
3600: 1976	Dimensions and masses per unit length of welded and seamless steel pipes and tubes for pressure purposes.
3602: —	Steel pipes and tubes for pressure purposes: carbon and carbon manganese steel with specified elevated temperature properties. Part 1: 1978 Seamless, electric resistance welded and induction welded tubes. Part 2: 1978 Submerged arc-welded tubes.
3605: 1973	Seamless and welded austenitic stainless steel pipes and tubes for pressure purposes.
3799: 1974	Steel pipe fittings, screwed and socket welding for the petroleum industry.
4360: 1979	Specification for weldable structural steels.
4534: 1969	Weldable chromium–nickel cast steel tubes.

Applications

855: 1976	Welded steel boilers for central heating and hot water supply (rated output 44 kW to 3 MW).
1500: —	Fusion-welded pressure vessels for general purposes. Part 1: Superseded by BS 5500: 1976. Part 3: 1965 Aluminium.
2654: 1973	Vertical steel-welded storage tanks with butt-welded shells for the petroleum industry.
2790: —	Shell boilers of welded construction (other than water-tube boilers). Part 1: 1969 Class 1 welded construction. Part 2: 1973 Class 11 and Class 111 welded construction.
3256: 1969	Small fusion-welded air reservoirs for road and railway vehicles.
4741: 1971	Vertical cylindrical welded steel storage tanks for low-temperature service. Single wall tanks for temperatures down to −50 °C.
5169: 1975	Fusion-welded steel air receivers.
5400: —	Steel, concrete and composite bridges. Part 6: 1980 Specification for materials and workmanship, steel. Part 10: 1980 Code of practice for fatigue.
5500: 1976	Unfired fusion-welded pressure vessels.

BIBLIOGRAPHY

Welding processes

Resistance welding

'Welding Handbook', 6th Edn, Section 1, American Welding Society, 1968.
P. T. Houldcroft, 'Welding Processes', Cambridge University Press, 1967.
A. C. Davis, 'Science and Practice of Welding', Cambridge University Press, 1971.

Fusion welding

'Welding Handbook', 6th Edn, Vol. 1, American Welding Society, 1968.
A. C. Davis, 'Science and Practice of Welding', Cambridge University Press, 1971.
P. T. Houldcroft, 'Welding Processes', Cambridge University Press, 1967.
'Electron Beam Welding', Special Feature, *Met. Constr. Br. Weld. J.*, 1970, **2**, 473.

Friction welding

'Friction Welding', Special Feature, *Met. Constr. Br. Weld. J.*, 1970, **2**, 181.

Steels

G. A. Phipps, 'Projection Welding of Low Carbon Mild Steel', *Br. Weld. J.*, 1958, **5**, 549.
K. S. Irvine, 'High Strength Weldable Steels', *Metallurgia*, 1958, **58**, 13.
D. Séférian, 'The Metallurgy of Welding', Chapman and Hall, London, 1962.
G. E. Linnert, 'Welding Metallurgy', *Carbon and Low Alloy Steels*, 3rd edn, American Welding Society, 1965, and 'Volume 2, Technology', 1967.
K. G. Richards, 'The Weldability of Steel', The Welding Institute, 1972.
N. Bailey, 'Welding Carbon: Manganese Steels', *Met. Constr. Br. Weld. J.*, 1970, **2**, 442.
R. G. Baker, F. Watkinson and R. P. Newman, 'The Metallurgical Implications of Welding Practice as Related to Low Alloy Steels', *Proc. Second Commonwealth Welding Conference*, Institute of Welding, London, 1965.
'General Requirements for the Electric Fusion Welding of Structural Steels', BS 2642: 1965.
J. C. M. Farrar and R. E. Dolby, 'Lamellar Tearing in Welded Steel Fabrication', The Welding Institute, 1972.
T. G. Gooch and D. C. Willingham, 'Weld Decay in Austenitic Stainless Steels', The Welding Institute, Abington, Cambridge, 1975.
R. Castro and J. de Cadenet, 'Welding Metallurgy of Stainless and Heat-Resisting Steels' Cambridge University Press, Cambridge, 1974.

Aluminium and aluminium alloys

'The Gas Welding of Aluminium', Information Bulletin No. 5, Aluminium Development Association, 1967.
'Resistance Welding of Wrought Aluminium Alloys', Information Bulletin No. 6, Aluminium Development Association.
'Welding Kaiser Aluminium', Kaiser Aluminium & Chemical Sales Inc., 1967.
'Manual MIG Welding of Aluminium', Alcan Service Bulletin, 1964.
'Mechanised MIG Welding of Aluminium', Alcan Service Bulletin, 1964.

Copper and copper alloys

'Gas Shielded Arc Welding of Copper and Copper Alloys', Technical Note TN2, Copper Development Association.
'The Bronze Welding Process', Technical Note TN5, Copper Development Association.
P. G. F. du Pré, 'The Gas-shielded Arc Welding of Copper and Copper Alloys', *Philips Welding Reporter*, 1972, **8**, 14.

Magnesium alloys

'Joining', Pamphlet, Magnesium Elektron Ltd.
E. F. Emley, 'The Metallurgical Background to Magnesium Alloy Welding', *Br. Weld J.*, 1957, **4**, 321.
P. Klain, 'The Welding of Magnesium Alloys', *Weld. J.*, 1957, **36**, 321.
'Joining Magnesium', Dow Chemical Co., 1956.

Nickel and nickel alloys

'Welding, Brazing and Soldering of Wiggin Nickel Alloys', Henry Wiggin & Co., Ltd., 1971.
'The Joining of Some Nickel Alloys', Leaflet, International Nickel Co. Ltd.
J. Hinde, 'Welding of Nickel and High Nickel Alloys', *Br. Weld. J.*, 1958, **5**, 311.

Refractory metals

C. A. Terry and E. A. Taylor, 'Welding of Titanium', *Weld. Metal Fabric*, 1958, 26 (June).
J. G. Purchas, D. R. Harris and H. Cobb, 'The Welding of Zircalloy-2', *Br. Weld. J.*, 1957, **4**, 412.
G. L. Miller, 'Zirconium', Butterworths, London, 1957.
F. G. Cox, 'Tantalum', *Weld. Metal Fabric.*, 1957, **25**, 416.
F. G. Cox, 'Welding and Brazing Refractory Metals—Molybdenum, Niobium, Tantalum, Zirconium', *Murex Review.*, 1956, **1**, 429.
L. Northcott, 'Molybdenum', Butterworths, London, 1956.
T. R. C. Gough and D. Roberts, 'The Welding of Uranium', *Br., Weld. J.*, 1957, **4**, 393.
'I.M.I. Titanium Fabrication', Imperial Metals Industry (Kynoch) Ltd., 1966.
M. H. Scott, 'The Joining of the Rarer Metals' edited by G. Isserlis, Chap 4, Columbine Press, 1962.
E. G. Thompson, 'Welding of Reactive and Refractory Metals', Welding Research Council Bulletin, No. 85, 1963.

M. H. Scott and P. M. Knowlson, 'The Welding and Brazing of the Refractory Metals, Niobium, Tantalum, Molybdenum and Tungsten—a review', *J. Less. common Met.*, 1963, **5**, 205.

Non-ferrous metals—general

E. A. Taylor, 'Inert Gas Welding of Non-Ferrous Metals', *Metall. Rev.*, No. 116, 1967.

Dissimilar metals

M. C. T. Bystran, 'Welding Dissimilar Alloy Steels', *Br. Weld. J.*, 1958, **5**, 475.
J. G. Young and A. A. Smith, 'Joining Dissimilar Metals', *Weld. Metal Fabric.*, 1959, **27**, 278, 331.
'Dissimilar Metals', *Met. Constr. Br. Weld. J.*, 1969, **1**, 12s.
'Dissimilar—Metal Joint', International Nickel Co. Ltd.

34 Soldering and brazing

34.1 Introduction

Soldering and brazing are heterogeneous joining processes in which metals are joined together by a dissimilar, lower melting point metal. The American Welding Society defines brazing as a process in which the joining metal has a melting point above 427 °C (800 °F). This temperature may be used to distinguish between brazing and soldering although, in practice, most solder metals are completely molten at temperatures below 300 °C. The majority of soldering and brazing metals are alloys based on binary or ternary eutectic systems and are designed—possibly by means of addition elements—to have a good capillarity at the joining temperatures, and characteristics suitable to meet specific environmental conditions. Another way to differentiate between soldering and brazing is to assume that the lower-temperature soldered joints are non-permanent and are of little strength, whereas brazed joints are designed to be both permanent and strong.

In practice, soldered joints cannot be reworked indefinitely because there is always a reaction which creates a thin metallurgical bond between the solder and the metals being joined. Repeated soldering or long soldering times increase the volume of the reaction zone and this invariably has a deleterious effect on the final properties of the interconnection, particularly when intermetallic compounds are formed. Tin and its alloys rapidly form intermetallics with most base materials during solder-dipping and soldering processes. These intermetallics then continue to grow at a greatly reduced rate by time- and temperature-dependent solid-state reactions during storage and operational life of a joint. Under certain conditions, the tin component of the solder may be totally consumed as intermetallic compound. The higher-temperature brazing operations form much broader reaction zones of modified composition and, although properly chosen braze alloys will not form brittle intermetallics, the modified microstructure will elevate the melting temperature to possibly that of the work-piece liquidus and thus preclude the possibility of rework. Brazing is distinguished from welding by the fact that the brazed joint is formed mainly by capillary action between the surfaces being joined, whereas the welded joint starts by melting and fusing the base metal surfaces together, usually with the aid of a welding filler metal.

34.2 Soldering

34.2.1 Design

The information in this section concerns the soldering of metals utilized in structural applications, gas or liquid sealing applications and electrical interconnections. The low melting points of solder alloys may be benificial in that the joining operation does not modify the metallurgical properties of the materials being joined.

In structural applications, the solder is weaker than the parts being fastened together and therefore the mechanical fittings must be designed to bear the entire load without the solder which will only seal and stiffen the assembly. Care should be taken to avoid placing the soldered joint where it will be subjected to a tensile load or peeling action. Where possible, joints should be designed with large overlapping areas which will be subjected only to shear loads. Clearances of between 0.075 and 0.25 mm are permissible between overlapping parts, but the best strength and ease of soldering is obtained with approximately 0.125 mm clearance. During soldering, this gap must remain reasonably constant and—on solidification of the solder—there should be no relative movements between the parts being joined because these would disturb or crack the joint. Self-locking designs, the use of jigs, spot-welds and temporary solder-tagging are used for maintaining joint clearances.

The soldering process should be performed in the shortest possible time. The method of direct or indirect heating must be compatible with the assembly and any jigging equipment, so that the surfaces to be wetted will have similar rates of heating. The overlapped areas will be penetrated by wetting and capillary action of the liquid solder only if the separate surfaces are clean and have a good solderability. Venting holes can be incorporated in some joint designs to facilitate cleaning and avoid gas or flux entrapment within the final joint.

Joint designs should be such that flux residues can be easily washed from all resulting fillets and that access is possible for visual inspection.

34.2.2 Choice of flux

Solderability is the property of a work-piece surface which permits it to form a good bond with a specific liquid solder alloy at a specified temperature and, usually, in the presence of a liquid flux. The extent of wetting will depend on the chemical nature of the surface being soldered and will be facilitated by the presence of a flux in order to:

1. Form the initial heat bridge between the heat source and the part.
2. React with, and either modify or remove, surface oxides and contamination films.
3. Eliminate air from the liquid solder front because, first, it wets, and second, it spreads by surface tension forces.

Fluxes may be applied as an external liquid, a coating or a predetermined volume incorporated in the core of a strand of solder wire. The type of flux must be chosen to suit the form of tarnish existing on the work-piece surface and the joints should be designed so that the flux can be both properly introduced to the joint area and then totally removed after soldering. The various methods of soldering which transmit heat energy to the work-piece (by means of hand-held irons, torches, resistance soldering, oven or furnace soldering, wave-soldering, etc.) may necessitate special fluxes which avoid factors such as the evolution of corrosive fumes during soldering, low flash-point constituents, irritating or dermatitis-producing liquids and fumes.

When torch-soldering is used, a more active flux may be necessary to deal with the extra tarnish formed during heating by the flame. Table 34.1 may be used as a guide in the selection of flux types for the soldering of common metals. The less noble metals listed require either a more active fluxing system or a more solderable finish, such as copper-plating.

When long storage periods are envisaged before the final assembly of parts, surfaces can be plated with copper or nickel, and subsequently coated with a eutectic composition tin–lead, which can be fused in hot oil to form a 4–12 μm thick pore-free protective layer possessing excellent solderability—even after many years of exposure to industrial atmospheres. The dual metallic finish precludes the need for strongly activated fluxes which are often difficult to remove in post-soldering cleaning operations and may later lead to extensive corrosion of the less noble metals in any joint design.

Table 34.1 RECOMMENDED SOLDER FLUX TYPES FOR ENGINEERING METALS

Group no.	*Metallurgical category*	*EMF** V	*Flux type and possible protective finish with precautions*
1	Gold, solid or plated; gold–platinum alloys; wrought platinum	+0.15	Both gold and platinum have excellent solderability. Soldering to gold and its alloys should be avoided because gold–tin intermetallics embrittle joints. Gold platings can be removed by solder-dipping (to dissolve gold), then pre-tin in second solder pot. Rosin, non-activated flux adequate. Alternatively, use indium–lead solder alloys if gold cannot be removed. Bright, hard gold platings may be difficult to wet due to certain alloying elements or organic additives from plating solution (in that case, use activated rosin flux)

* Calomel electrode/sea water: for galvanic compatibility: max. pot. diff. between soldered metals should be less than 0.5 V.

Table 34.1 RECOMMENDED SOLDER FLUX TYPES FOR ENGINEERING METALS—*continued*

Group no.	*Metallurgical category*	*EMF** V	*Flux type and possible protective finish with precautions*
2	Rhodium	+0.05	Not easy — inorganic flux
3	Silver, solid or plated on copper; high silver alloys	0	Easily soldered with mildly activated flux or, when free of surface sulphide, non-activated rosin is preferable. If chloride-contaminated (from plating bath), may need abrasion. Use silver-loaded solder when joining to thin silver plate. The silver-saturated liquid reduces danger of scavenging. Silver-plated parts are not recommended for electrical circuits due to problem of silver migration and subsequent short circuits
4 (a)	Nickel, solid or plated; Monel	−0.15	Difficult to solder. Use inorganic or organic acid for pre-tinning with solder and non-activated rosin after pre-tinning
4 (b)	Titanium	−0.15	Impossible to solder. Can be copper-plated
5 (a)	Copper, solid or plated, high copper bronzes	−0.20	With red oxide tarnish, can be soldered with mildly activated rosin. Black oxide only removable with activated rosin, organic acids or zinc ammonium chloride solutions (e.g. for radiator plates)
5 (b)	Copper–nickel alloys, Nichrome alloys, austenitic, high corrosion-resistant steels; Nilo-K. Kovar, Monel, etc.	−0.20	Very difficult to solder with inorganic acid fluxes, zinc chloride and ammonium chloride solutions. Some proprietary brands are available. Can be nickel-plated, but deposit must by non-porous or substrate will become oxidized by plating salts. All ionic matter, incl. handling contamination, must be thoroughly and immediately removed because this group is susceptible to stress corrosion cracking
6	Commercial bronze	−0.25	If thin tarnish film, quite easy to solder with mildly activated resin. May be copper or silver-plated, but ensure good plating adhesion
7	Commercial brasses (60Cu 40Zn and 70Cu 30Zn)	−0.30	Difficult to solder. Use activated rosin if tarnish is thin. Impossible to solder — even with inorganic flux — if oxidation is visible due to surface film of zinc oxide. Barrier plating of more than 3 μm nickel or copper recommended for preserving solderability during shelf-life (should prevent zinc diffusion to surface). Barrier of 5 μm necessary if highly leaded brass
8	18% Chromium-type corrosion-resisting steel	−0.35	See Group 5(b)
9	Tin-plated metals	−0.45	New coatings are easily soldered with non-activated rosin flux. Activation of rosin depends on extent of tin oxide. Fused tin is preferred as it is less porous. Pure tin coatings not recommended for electronic applications due to risk of whisker growth. Should exceed 1 μm thickness as otherwise will react with copper and completely convert to intermetallic, which is extremely difficult to solder
10	Tin–lead, solid, plated or fused	−0.50	Most suitable finish for easy soldering with non-activated rosin flux. Porous platings need activated flux. Should exceed 1 μm thickness, but lead slows down intermetallic formation. Good shelf-life for fused coatings on copper wire and printed circuits

* Calomel electrode/sea water: for galvanic compatibility, max, pot. diff. between soldered metals should be less than 0.5 V.

Table 34.1 RECOMMENDED SOLDER FLUX TYPES FOR ENGINEERING METALS—*continued*

Group No.	*Metallurgical Category*	*EMF* V*	*Flux Type and possible protective finish with precautions*
11	Lead, solid or plated; high lead alloys	−0.55	Mildly activated flux required to penetrate surface oxides. High dissolution of lead in tin–lead solder produces joints with extremely low shear strength
12	Aluminium–copper alloys (e.g. Duralumin and most of AA2XXX series)	−0.60	Impossible with tin–lead alloys. Generally, welding or dip-brazing are more suitable. Plate with zincate and copper. The low melting solders are preferred to avoid thermal stressing platings
13	Iron, wrought, grey or Armco; plain carbon and low alloy steels	−0.70	Solderability depends on oxide thickness. Clean, pickled surfaces can be easily soldered with mild or activated rosin flux. Passivated or phosphated steels require special fluxes
14	Aluminium and most alloys other than in group 12 (e.g. AA1XXX, 3XXX and 5XXX series)	−0.75	Can be soldered, but with special alloys (e.g. tin–zinc or cadmium–zinc alloys). May be friction-soldered without flux or, when aluminium oxide is removed, by ultrasound. Some proprietary fluxes are available. Corrosion in aluminium-soldered joints is particularly troublesome; a water-proof coating is essential
15(a)	Aluminium cast alloys other than silicon-type	−0.80	As for group 14, but choice of solder and flux requires specialist advice
15(b)	Cadmium platings	−0.80	Easy to solder with mildly or fully activated rosin fluxes. If passivated by chromate film, they are very difficult and require ammonium chloride-type flux. Fumes are toxic
16	Hot dipped zinc plate	−1.05	As for group 15(b)
17	Zinc, wrought; zinc-based casting alloys; zinc plate	−1.10	Generally, very difficult to solder due to oxidation. Use inorganic acid or special proprietary flux. Water-tight protection necessary to avoid corrosion
18	Magnesium and magnesium-based alloys, cast or wrought	−1.60	Not recommended because of poor strength and corrosion

* Calomel electrode/sea water: for galvanic compatibility, max. pot. diff. between soldered metals should be less than 0.5 V.

The chemical composition of strong fluxes may be based on aqueous solutions of halides, such as ammonium and zinc chlorides or orthophosphoric acid; these are widely employed in general engineering. For use on electronic assemblies, rosin-based fluxes—often in the form of alcoholic solutions—are chosen. A range of rosin types are commonly used, including non-activated halide-free, organically activated and halide-activated rosins.

34.2.3 Control of galvanic corrosion

It is desirable to avoid joining together metals which have greatly separated positions on the galvanic scale. Table 34.1 includes data of the relative activity of metals in relation to a standard reference electrode based on a sea water environment. Many authorities recommend that, to avoid corrosion couples, metallic combinations which will be subjected to uncontrolled terrestrial environments do not exceed a potential difference of 0.25 V. When the working environment is controlled, e.g. in the case of electrical units operating under clean room conditions, compatible corrosion-free couples may be separated by potential differences not exceeding 0.5 V.

Fortunately tin–lead, the most common soldering alloy (Group 10 of Table 34.1) is generally compatible with most solderable metals, except gold. The galvanic series is useful as a first approximation in selecting materials for both solderability and corrosion control, but for many applications it may be too simplistic because it does not provide information about corrosion rates or changes in surface chemistry which may pacify or accelerate corrosion at bi-metallic interfaces.

34.2.4 Choice of solder alloy

SOLDER FORMULATIONS

Generally, solder alloys are based on the metals tin, lead, cadmium, zinc and indium. They are available in a variety of physical forms to facilitate different means of application. Solder ingots are used to replenish large baths for dip or wave-soldering. Wire solder, available with or without a core of flux, finds its use in hand-soldering. Solder creams, containing a gel of solder powder, flux and wetting agent, can be painted or screen-printed for microelectronic applications. Preforms of solder are used for furnace and torch-soldering and include formed wires and parts punched from flat sheet.

Solder alloys can be divided into four groups according to their melting ranges. Melting temperatures are not recommended soldering temperatures. Typically, 20–70 °C above the liquidus temperatures ensures good alloy fluidity and wetting characteristics. Compositions, melting ranges, and typical uses for these types are given in Tables 34.2 to 34.5.

Table 34.2 HIGH TEMPERATURE SOLDERS (370–430 °C)*

Nominal composition			wt %		*Melting range* °C	*Name*
Sn	Zn	Al	Cd	Ag		
73(87)	8(15)	5(12)	—	—	193–510, depending on alloy composition	General purpose solder for aluminium alloys (MIL-S-12204, Comp. C)
—	95	5	—	—	380 m.p.	High strength solder for aluminium
—	100	—	—	—	418 m.p.	High strength solder for aluminium
—	—	—	95	5	337–393	General purpose (DIN 1707, LCdAg5)
19	81	—	—	—	195–385	General purpose (DIN 1707, LZnSn20)

* Rarely used because of high dissolution of many base metals into the solder. Rapid tarnishing of adjacent areas to the joint unless special protection by flux or inert gas is provided. Only high temperature stable inorganic fluxes useful.

Table 34.3 INTERMEDIATE TEMPERATURE SOLDERS (270–370 °C)
Elevated temperature applications where some creep strength is required.

The lead-based alloys, which contain additions of up to 8% tin and/or silver, may be used to about 130 °C. Higher service temperatures may be satisfied by selecting tin-free lead alloys, but these have poor wetting properties and require the use of special organic fluxes which will not decompose at the high soldering temperatures. Alloys containing in excess of 90% lead are selected for cryogenic applications because they retain ductility and can take up mismatches in thermal expansion at the low temperatures. The final inspection of cleaned connections made with these alloys may reveal high contact angles and dull surface finishes, but these should not be construed as a sign of cold solder joints.

Nominal composition						*Melting range* °C	*Name*
Sn	Pb	Sb	Ag	Cd	Zn		
2	98	—	—	—	—	320–325	DIN 1707, L PbSn 2
5	93.5	—	1.5	—	—	296–301	BS 219, Grade 5S and BS AU 90–5S
8	91.7	0.3	—	—	—	280–305	DIN 1707, L PbSn 8 (Sb)
—	—	—	2.0	82	16	270–280	DIN 1707, L CdZnAg 2
—	—	—	10	68	22	270–380	DIN 1707, L CdZnAg 10
—	97.5	—	2.5	—	—	304–305	DIN 1707, L PbAg 3

Table 34.3 INTERMEDIATE TEMPERATURE SOLDERS (270–370 °C)—*continued*

Nominal composition							
Sn	Pb	Sb	Ag	Cd	Zn	*Melting range* °C	*Name*
2	96.2	—	1.8	—	—	304–310	DIN 1707, L PbAg2Sn2
70	—	—	—	—	30	200–300	Solder for aluminium (by friction when liquid)
—	—	—	—	40	60	265–350	DIN 8512, L ZnCd40 (frictional soldering of aluminium)
3	97	5	—	—	—	245–284	BS AU 90–3AX (automobile body-filling)

Table 34.4 COMMON SOLDER ALLOYS (183–270°C)

Nominal composition Sn	Pb	*Other*	*Melting range* °C	*Specification* BS 219 (1977) *grade*	*Typical uses (similar alternative standard)*
Tin solders					
63	37	0.6 Sb max	183 m.p.	A	Soldering of electrical connections to copper, brass and zinc. Capillary joints in copper and stainless steel (QQ–S–571d, type Sn 63*
63	37	0.2 Sb max	183 m.p.	AP	As 'A', but for higher reliability on printed circuit boards (DIN 1707 LSn 63 Pb) (ESA QRM–08, Sn 63)*
60	40	0.5 Sb max	183–188	K	As 'A', but also for pre-tinning of electrical components (DIN 1707 LSn 60 PbSb) (QQ–S–571d, type SN–60)
60	40	0.2 Sb max	183–188	KP	Hand and machine-soldering of electronics and pre-tinning (DIN 1707 LSn 60 Pb) (ESA QRM–08, Sn 60) (ASTM–B–32–66T–60A)
50	50	0.5 Sb max	183–212	F	General engineering work on copper, brass and zinc. Can-soldering (DIN 1707 LSn 50 Pb(Sb), etc.)
45	55	0.4 Sb max	183–224	R	
40	60	0.4 Sb max	183–234	G	
35	65	0.3 Sb max	183–244	H	Joining of electrical cable sheaths (DIN 1707, LPb Sn 35 (Sb))
30	70	0.3 Sb max	183–255	J	
20	80	0.2 Sb max	183–276	V	Lamp solder. Dip-soldering. For low temperature service (also BS 441–20)
15	85	0.2 Sb max	227–288	W	
100	—	—	232 m.p.	(BS 2352)	Food-handling equipment and cans
Tin–Lead–Antimony					
50	50	2.5–3.0 Sb	185–204	B	Hot dip coating and soldering of ferrous metals; jointing of copper conductors
45	55	2.2–1.7 Sb	185–215	M	
40	60	2.0–2.4 Sb	185–277	C	General engineering. Heat exchanges. General dip-soldering

* Eutectic type solders with single discrete melting point; they solidify with smooth bright fillets which can be readily inspected—recommended for highly reliable electrical connections. These are used for hand or machine soldering of printed circuit boards but care should be taken to avoid trace impurities, e.g. Zn or Al $<0.005\%$ to prevent oxide skin; Fe or Au or Cu or As or Sb $<0.2\%$ to prevent hard or embrittled joints.

Table 34.4 COMMON SOLDER ALLOYS (183–270 °C)—*continued*

Nominal composition Sn	Pb	Other	*Melting range* °C	*Specification* BS 219 (1977) *grade*	*Typical uses (similar alternative standard)*
32	68	1.6–1.9 Sb	185–243	L	Plumbing, wiping of lead and lead alloy
30	70	1.5–1.8 Sb	185–248	D	cable-sheathing. Dip-soldering
18	82	0.9–1.1 Sb	185–275	N	Dip-soldering (also BS AU 90–19A)
Tin–Antimony					
95	Rest	4.7–5.0 Sb	236–243	95A	High service temperatures (more than 100°C) and refrigeration equipment. Step-soldering
Tin–Silver					
96	Rest	3.5–3.7 Ag	221 m.p.	96S	High service temperatures (more than 100 °C)
Tin–Lead–Silver					
62.5	Rest	1.8–2.2 Ag	178 m.p.	62S	Soldering of silver-coated substrates (ESA QRM–08, 62 Sn) (approx. DIN 1707 LSn 60 PbAg) (QQ–S–571d, Sn62)
Solder For Aluminium					
90	—	10 Zn	200–210	—	Soldering aluminium by ultrasonics (DIN 8512, LSn Zn 10)
60	—	40 Zn	200–300	—	Frictional soldering of aluminium (DIN 8512 LSn Zn 40)
Gold–Tin Alloy					
20	—	80 Au	280 m.p.	—	Micro-electronics manufacturing
Lead–Indium Alloys					
—	50	50 In	215–230	—	Hybrid micro-electronics (thick-thin film
—	75	25 In	173–190	—	substrates). Prevents scavenging of gold

Table 34.5 VERY LOW MELTING POINT SOLDERS (below 183 °C)
Uses include joining heat-sensitive components or adding components to existing circuits (one solder operation on another where second soldering does not remelt initial joint). Contamination with certain other solder compositions may cause alloying and even lower melting temperatures.

Nominal composition							*Melting range*	
Sn	Pb	Sb	Bi	Cd	In	Ag	°C	*Name*
51	31	—	—	18	—	—	145 m.p.	BS 219, Grade 'T' (DIN 1707, L–SnPb Col 18)
50	—	—	—	—	50	—	117–125	Glass-to-metal sealing (L–SnIn 50)
14.5	28.5	9	48	—	—	—	103–227	Matrix alloy
22	28	—	50	—	—	—	96–110	Rose's
13	27	—	50	10	—	—	70–73	Lipowitz's
12.5	25	—	50	12.5	—	—	70–72	Bending (Wood's)
8.3	22.6	—	44.7	5.3	19.1	—	47 m.p.	—
12.8	25.6	—	48	9.6	4	—	62 m.p.	—
—	—		—	27	73	—	123 m.p.	—

Table 34.5 VERY LOW MELTING POINT SOLDERS (below 183 °C)—*continued*

Nominal composition							*Melting range*	
Sn	Pb	Sb	Bi	Cd	In	Ag	°C	*Name*
37.5	37.5	—	—	—	25.0	—	134–174	—
—	15	—	—	—	80	5	149 m.p.	—
13	27	—	50	10	—	—	70 m.p.	Quarternary eutectics
26	—	—	54	20	—	—	103 m.p.	Ternary eutectics
—	—	—	60	40	—	—	144 m.p.	Binary eutectics

34.2.5 Cleaning

Soldered joints should be cleaned to prevent surfaces corrosion and stress corrosion cracking. Cleaning also facilitates inspection. Immersion cleaning is improved by ultrasonic agitation.

The following solvents are acceptable for the cleaning of electronic equipment provided cleaning takes place immediately after the soldering operation (e.g. before the flux and residues have had time to polymerize or age):

1. Ethyl alcohol, 99.5 or 95% pure by volume.
2. Isopropyl alcohol, 99% pure.
3. Trichlorotrifluorethane, clear, 99.8% pure.
4. Any mixture of the above.
5. De-ionized water at 40°C maximum may be used for certain fluxes. Items shall be thoroughly dried directly after the use of de-ionized water.

34.2.6 Product assurance

Acceptable solder connections are generally characterized as:

1. Clean, smooth, undisturbed surfaces.
2. Concave fillets between solder and joined surfaces.
3. Complete wetting as evidenced by a low contact angle between the solder and the joined surfaces.

Unacceptable solder conditions, which may be cause for rejection, are:

1. Damaged, crushed, cracked, melted, corroded, etc. surfaces.
2. Improper tinning.
3. Flux residues or other contamination.
4. Cold joints.
5. Fractured joints.
6. Pits, holes or voids in the joint which are not attributable to liquid-to-solid solder shrinkage.
7. Excessive or insufficient solder.
8. Splattering of flux or solder on adjacent areas.
9. De-wetting, etc.

34.3 Brazing

34.3.1 General design consideration

Brazing can be considered when one or more of the following design requirements must be met:

1. Joints will be loaded in shear.
2. Joining of thin sections to heavy sections.
3. Joining of sections too thin for welding.

4. Hermetically sealed assemblies.
5. Low distortion with a minimum residual stress distribution.
6. High volume production.
7. Joining dissimilar metal combinations which cannot be welded.
8. Producing complex, permanent assemblies with multiple joints which would be inaccessible to welding.

34.3.2 Joint design

Normally only small or medium sized assemblies can be brazed as the heat required to melt the braze alloy must be applied to a broad, or usually the entire area of the component parts being joined.

Both lap and butt joints can be made with flat or tubular parts. *The strength of the joint* depends on the amount of bonding surface so that the simple butt joint will have a small bonding area that

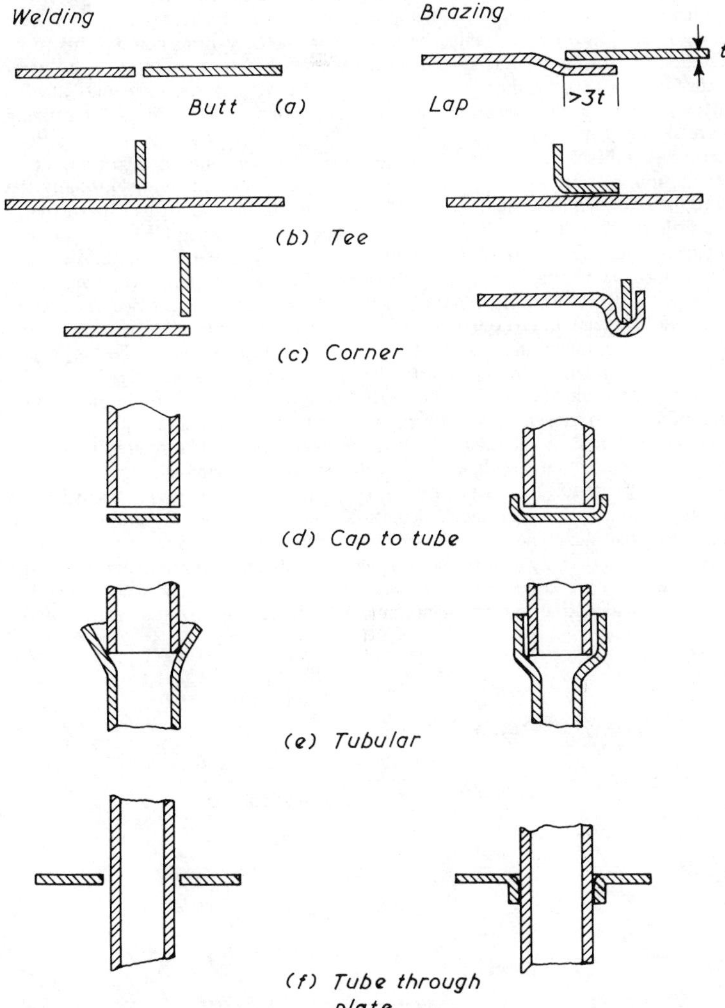

Figure 34.1a *Comparison of joint designs used for welding and brazing.* (From Brooker and Beatson)

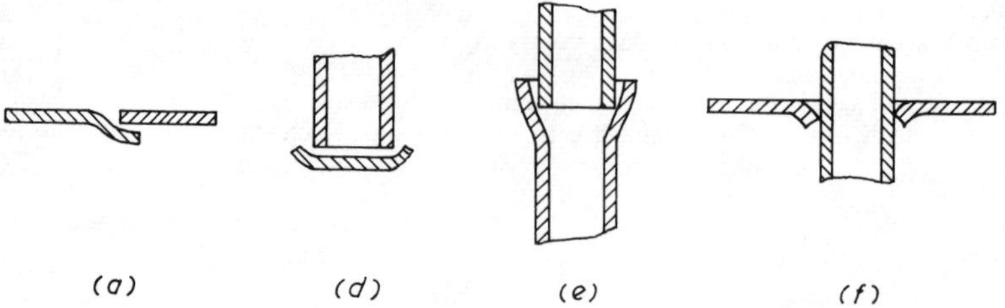

Figure 34.1b *Some of the joint designs from Figure 34.1a amended for aluminium brazing. These half fillet–half lap type joints facilitate flux displacement and removal.* (From Brooker and Beatson)

will not exceed the cross-sectional area of the thinner member. The preferred type of joint loading is in shear and lap or lap–butt joints are the most reliable (shown in Figure 34.1). The lap joint should be designed to be three times as long as the thickness of the thinnest component part. The parts should be designed to prevent stress from being concentrated at a single point.

During the brazing operation provisions must be made to hold the various components to be joined. This is best done by using an *interlocking design* where alignment is obtained by parts fitting into each other under gravity rather than by the use of jigs or fixtures. Parts may also be positioned by means of spot or tack welding, crimping and pinning. Techniques for holding component parts are sketched in Figure 34.2.

The *joint clearance* has a direct bearing on the mechanical strength of any brazed joint. Joint clearance is the dimension between the interfaces of the completed brazed joint. Normally the highest strengths are obtained with the smallest possible thickness of filler and this is particularly true where there is little or no solubility of the parent material in the braze filler (e.g. stainless steels brazed with silver- or copper-based alloys). More clearance is required if material solubility exists, particularly in furnace brazing when long heating cycles are required (e.g. austenitic stainless steels brazed with nickel fillers). It is extremely important to plan a joint clearance based on the brazing temperatures and not room temperature. The coefficients of expansion of the metals being joined must be taken into account, especially in tubular assemblies in which dissimilar metals are being joined. If the metal with the greater expansion coefficient is on the inside there may be no clearance at the brazing temperature and under reverse conditions a room temperature gap can become too large for adequate capillary flow at the brazing temperature. As a design rule the filler metal should be placed in slight compression once the assembly returns to room temperature—thus the metal having the greater coefficient of expansion is normally used as the female member of the joint. Ideally, parts to be brazed should have a similar coefficient of expansion in order to facilitate joint clearance calculation and prevent the generation of large residual stresses caused by badly matched materials. The optimum joint clearance will depend on the nature and sizes of the parts being joined, the heating methods and the configuration of the joint itself. As there are many variables involved in assessing the actual clearance at brazing temperatures it is recommended that simple calculations are made utilizing the coefficients of

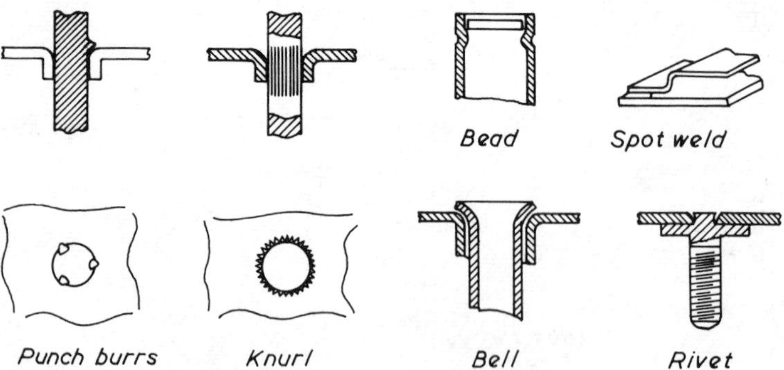

Figure 34.2 *Method of holding component parts together without using external jigs.* (From Brooker and Beatson)

expansion presented in Table 34.6 and these are followed up by trial operations using representative test samples. For most brazing alloys effective capillary action occurs when the braze path gap at the brazing temperature is maintained within the range 0.025–0.2000 mm. Table 34.7 may also be used as a guide.

Table 34.6 COMPARISONS OF MATERIALS: COEFFICIENT OF THERMAL EXPANSION[a]

	10^{-6} °F^{-1}		10^{-5} °C^{-1}	
Material	*High*	*Low*	*High*	*Low*
Zinc and its alloys[c]	19.3	10.8	3.5	1.9
Lead and its alloys[c]	16.3	14.4	2.9	2.6
Magnesium alloys[b]	16	14	2.8	2.5
Aluminium and its alloys[c]	13.7	11.7	2.5	2.1
Tin and its alloys[b]	13	—	2.3	—
Tin and aluminium brasses[c]	11.8	10.3	2.1	1.8
Plain and leaded brasses[c]	11.6	10	2.1	1.8
Silver[b]	10.9	—	2.0	—
Cr-Ni-Fe superalloys[d]	10.5	9.2	1.9	1.7
Heat resistant alloys (cast)[d]	10.5	6.4	1.9	1.1
Nodular or ductile irons (cast)[d]	10.4	6.6	1.9	1.2
Stainless steels (cast)[d]	10.4	6.4	1.9	1.1
Tin bronzes (cast)[c]	10.3	10	1.8	1.8
Austenitic stainless steels[c]	10.2	9	1.8	1.6
Phosphor silicon bronzes[c]	10.2	9.6	1.8	1.7
Coppers[c]	9.8	—	1.8	—
Nickel-base superalloys[d]	9.8	7.7	1.8	1.4
Aluminium bronzes (cast)[c]	9.5	9	1.7	1.6
Cobalt-base superalloys[d]	9.4	6.8	1.7	1.2
Beryllium copper[c]	9.3	—	1.7	—
Cupro-nickels and nickel silvers[c]	9.5	9	1.7	1.6
Nickel and its alloys[d]	9.2	6.8	1.7	1.2
Cr-Ni-Co-Fe superalloys[d]	9.1	8	1.6	1.4
Alloy steels[d]	8.6	6.3	1.5	1.1
Carbon free-cutting steels[d]	8.4	8.1	1.5	1.5
Alloy steels (cast)[d]	8.3	8	1.5	1.4
Age hardenable stainless steels[c]	8.2	5.5	1.5	1.0
Gold[c]	7.9	—	1.4	—
High temperature steels[d]	7.9	6.3	1.4	1.1
Ultra high strength steels[d]	7.6	5.7	1.4	1.0
Malleable irons[c]	7.5	5.9	1.3	1.1
Titanium carbide cermet[d]	7.5	4.3	1.3	0.8
Wrought irons[c]	7.4	—	1.3	—
Titanium and its alloys[d]	7.1	4.9	1.3	0.9
Cobalt[d]	6.8	—	1.2	—
Martensitic stainless steels[c]	6.5	5.5	1.2	1.0
Nitriding steels[d]	6.5	—	1.2	—
Palladium[c]	6.5	—	1.2	—
Beryllium[b]	6.4	—	1.1	—
Chromium carbide cermet[c]	6.3	5.8	1.1	1.0
Thorium[b]	6.2	—	1.1	—
Ferritic stainless steels[c]	6	5.8	1.1	1.0
Gray irons (cast)[c]	6	—	1.1	—
Beryllium carbide[d]	5.8	—	1.0	—
Low expansion nickel alloys[c]	5.5	1.5	1.0	0.3
Beryllia and thoria[e]	5.3	—	0.9	—
Alumina cermets[d]	5.2	4.7	0.9	0.8
Molybdenum disilicide[e]	5.1	—	0.9	—
Ruthenium[b]	5.1	—	0.9	—
Platinum[c]	4.9	—	0.9	—
Vanadium[b]	4.8	—	0.9	—
Rhodium[b]	4.6	—	0.8	—
Tantalum carbide[d]	4.6	—	0.8	—
Boron nitride[d]	4.3	—	0.8	—
Niobium and its alloys	4.1	3.8	0.7	0.68
Titanium carbide[d]	4.1	—	0.7	—
Steatite[c]	4	3.3	0.7	0.6
Tungsten carbide cermet[e]	3.9	2.5	0.7	0.4
Iridium[b]	3.8	—	0.7	—
Alumina ceramics[e]	3.7	3.1	0.7	0.6
Zirconium carbide[d]	3.7	—	0.7	—
Osmium and tantalum[b]	3.6	—	0.6	—
Zirconium and its alloys[b]	3.6	3.1	0.6	0.55
Hafnium[b]	3.4	—	0.6	—
Zirconia[e]	3.1	—	0.6	—
Molybdenum and its alloys	3.1	2.7	0.6	0.5
Silicon carbide[e]	2.4	2.2	0.4	0.39
Tungsten[b]	2.2	—	0.4	—
Electrical ceramics[c]	2	—	0.4	—
Zircon[c]	1.8	1.3	0.3	0.2
Boron carbide[e]	1.7	—	0.3	—
Carbon and graphite[c]	1.5	1.3	0.3	0.2

[a] Values represent high and low sides of a range of *typical* values.
[b] Value at room temperature only.
[c] Value for a temperature range between room temperature and 212–750 °F, 100–390 °C.
[d] Value for a temperature range between room temperature and 1 000–1 800 °F/540–980 °C.
[e] Value for a temperature range between room temperature and 2 200–2 875 °F/1 205–1 580 °C.
Reprinted from *Materials Selector*, Reinhold Publishing Co., Penton/IPC.
See also chapter 14—General Physical Properties.

Table 34.7 RECOMMENDED JOINT CLEARANCES (mm)

	Parent metal			
Brazing material	*Copper*	*Copper-base alloys*	*Ferrous metals*	*Aluminium and its alloys*
Noble metal alloys[a]	0.025–0.10	0.025–0.10	0.025–0.1	—
Copper	—	—	Nil–0.075	—
Brasses	0.075–0.375	0.075–0.375	0.05–0.25	—
Copper–phosphorus	0.075–0.375	0.075–0.375	—	—
Silver–copper–phosphorus	0.05–0.30	0.05–0.30	—	—
Silver brazing alloys	0.03–0.25	0.03–0.25	0.025–0.15	—
Nickel-base alloys	—	—	0.075–0.375[b]	—
Aluminium brazing alloys	—	—	—	0.125–0.6

[a] Ensure that the solidus value of the parent metal is at least 100 °C above the liquidus of the selected brazing material.
[b] Special formulations are available which will permit wider gaps than 0.375 mm to be brazed in a satisfactory manner.
Reprinted from Brooker and Beatson, *Industrial Brazing*, 2nd edn. Newnes-Butterworth, 1975.

34.3.3 Precleaning and surface preparation

Uniform capillary action may only occur when all grease, oil, oxides and dirt have been removed from both the base metals and braze alloy. Fluxes cannot be relied upon to clean the area to be brazed—they are mainly intended to prevent the formation of oxides during brazing and reduce the surface tension of the filler metal. Cleaning may be performed by chemical, mechanical or vacuum methods. Chemical cleaning is most frequently carried out utilizing commercial solvents, vapour degreasing, alkaline mixtures (detergents, soaps, carbonates, hydroxides), electrolytic methods (anodic and cathodic), acidic solutions and salt bath pickling. Mechanical cleaning methods include grinding, filing, machining, shot blasting and ultrasonics — in all cases care must be taken to avoid the embedment of oxides or ferrous materials into the surfaces to be brazed as these will interfere with the proper wetting of the base metal by the liquid braze alloy. Very smooth, polished surfaces are less satisfactory for brazing than slightly roughened surfaces.

34.3.4 Positioning of filler metal

The braze filler metal may be used in wire, ring, shim, clad sheet, powder, paste or slurry form, depending upon the specific application and process. These forms of filler metals can be positioned (Figure 34.3) by grooves, shoulders and/or recesses which serve as reservoirs for filling the braze

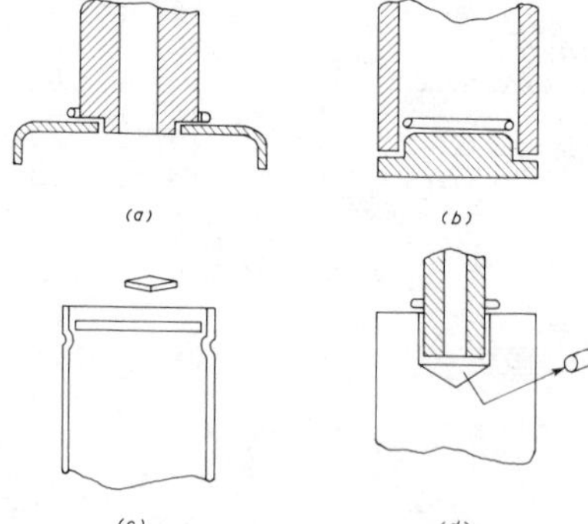

Figure 34.3 *Some methods of pre-placing the alloy (a) Wire ring sited for H.F. induction heating. (b) Wire ring sited for torch, induction or furnace heating. (c) A case where a ring should be used for a large diameter or a slug for up to about* 8 mm *diameter. (d) Another case for either ring or slug; if the latter, the rod must fall freely into the hole and it should have a groove or a flat for venting the closed space.* (From Brooker and Beatson)

paths. Designs for hand feeding and pre-form brazing may be different, as shown in Figure 34.4. All braze assemblies should be designed so that the parent metals and the filler metals are uniformly heated when component parts have different thermal capacities. Two-stage heating with long dwell times may be necessary to ensure that liquation and flow of the filler metal occur simultaneously in all locations. Gravity flow of the filler metal through the braze path should be provided when possible. Machined grooves should be provided when large. ($>50\,mm^2$) flat surfaces are united by brazing. Unless full penetration by the brazing alloy is achieved, an internal crevice will remain as a potential site for flux and residue. In certain configurations the presence of visible filler metal fillets will aid inspection. Common positioning faults are depicted in Figure 34.5.

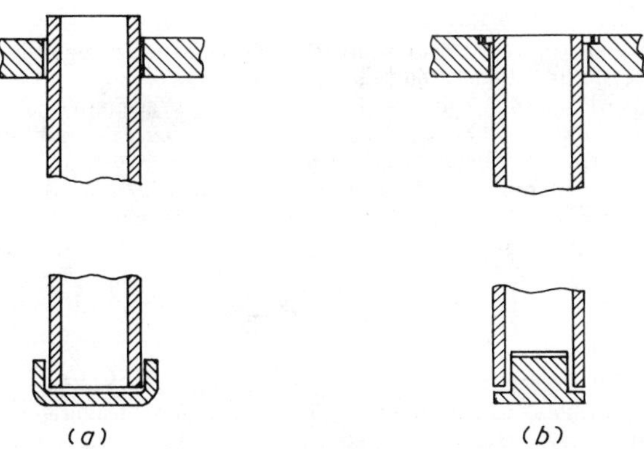

Figure 34.4 *Recommended designs for (a) hand feeding and (b) pre-placed braze alloy rings.* (From Brooker and Beatson)

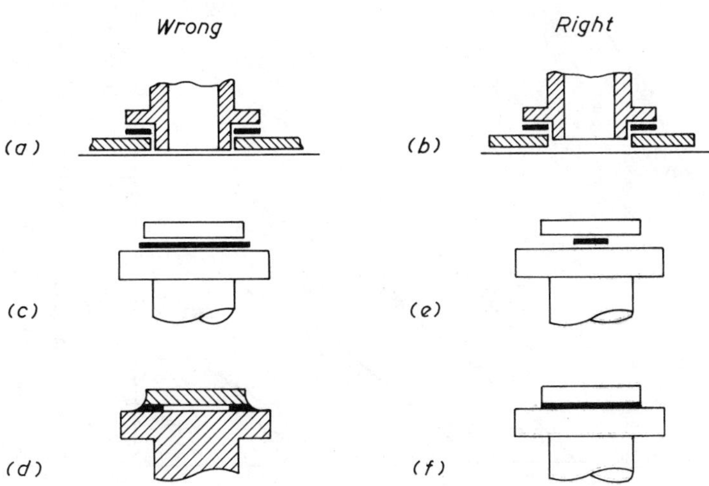

Figure 34.5 *Some 'wrong' and 'right' ways to pre-place braze alloy pre-forms. Obstacles must not prevent the free-falling of component parts (a). Also, if light-weight parts are not free to fall when the alloy melts (c) then brazing atmospheres can become entrapped (d).* (From Brooker and Beatson).

34.3.5 Heating methods

TORCH BRAZING

Parts should be preheated with a neutral or slightly reducing flame so that uniform heating of the surfaces to be joined can be achieved. Localized overheating must be avoided. Flux is usually applied (some proprietary fluxes indicate the parts' surface temperature by a colour change) and the braze alloy is introduced at one of the mating surfaces. Torch brazing is either performed manually or by semi- or fully mechanized equipment. The former method is frequently used for the attachment of small parts on to large surfaces, one-off jobs and repairs. Automation is best suited to high production runs.

FURNACE BRAZING

The furnace atmosphere should be strictly controlled (particularly the Dew point and composition should be continually assessed to prevent oxidation, carburization, etc.). Braze alloy pre-forms are located between pre-assembled parts which are usually fluxed and self-jigging. Vacuum furnaces employ pressures of 1.0 to 1.0×10^{-9} torr and are heated by radiation from resistance elements, or induction from special coils. Metal–metal oxide equilibrium diagrams are useful to show the relationship which exists between the furnace operating temperature and the reducing or oxidizing behaviour of the atmosphere. Figure 34.6 indicates the effect of Dew point of hydrogen and partial pressure of water on the brazeability of pure metals.

INDUCTION BRAZING

Heating is localized, fast, and generally accomplished within one or two minutes. Coils may be shaped to follow the contours of the assembly and adjustment to the coil separation distances can cause heavier sections to heat up at the same rate as thin sections. The mating surfaces can be

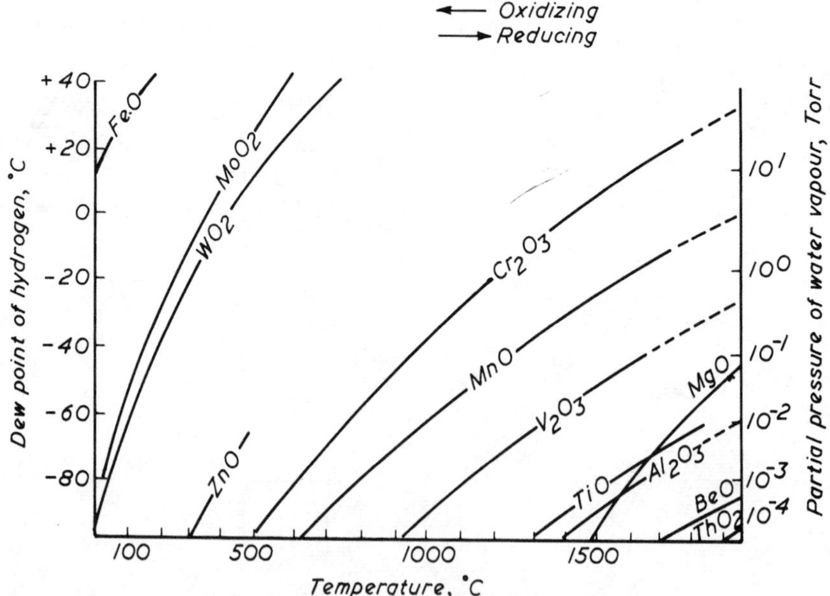

Figure 34.6 *Metal–metal oxide equilibria in hydrogen atmospheres. Oxidizing conditions exist at the left of each curve; reducing conditions which facilitate brazing exist on the right of each curve. Note Au, Pt, Ag, Pd, Ir, Cu, Co, Ni and Os oxides are easier to reduce than those plotted.*

coated with flux, or pre-assembled units with braze alloy pre-forms may be heated in a suitable atmosphere. It must be rembered that high-resistance metals such as steel heat up rapidly whereas low-resistance metals take longer to reach the brazing temperature.

RESISTANCE BRAZING

Suitable only for relatively small joint areas because it is difficult to maintain uniform current distributions. Localized heating does not usually modify the mechanical properties of the parent metals. Success will depend upon the thermal capacities and thermal conductivities of the metals being joined. The presence of an intermediate layer of flux can be detrimental if direct heating is employed (i.e. when heating current passes through two work pieces sandwiching the brazing alloy). Once the optimum brazing schedule has been established this method can become semi-automated and requires minimal operator skills.

DIP BRAZING

The assembled joint, with the filler metal(s) preplaced, is dipped into a bath of molten salt which acts as flux. Occasionally the parts may be dipped into a bath of molten brazing alloy covered with a layer of flux. Dip brazing is widely used for joining aluminium components. Preheating of parts is recommended as it prevents excessive freezing of the salt on a cold surface, expels water entrapped within the assembly thereby minimizing the risk of explosions from immersed parts, and markedly increasing the speed of production. The specific gravity of molten salt is close to that of aluminium, reducing by half the weight of submerged pieces—enabling the dip-brazing of complex aluminium assemblies that are not strong enough to be self-supporting when brazed by other techniques. Personnel must wear protective clothing and face masks.

34.3.6 Brazeability of materials and braze alloy compositions

It is not possible to detail the optimum heating methods and filler metals for the wide range of base metals and combinations of base metals. Literature searches and experimentation must be performed prior to the brazing of uncommon materials. It is strongly recommended that the referenced Standards are consulted for additional information. The brazing alloy systems in common use are listed in Table 34.8. A preliminary guide to the suitability of parent material and filler metal for torch brazing, furnace brazing, induction brazing, electrical resistance brazing, and vacuum brazing is presented in Tables 34.9 to 34.13.

Table 34.8 COMMON BRAZING-ALLOY SYSTEMS (BASED INITIALLY ON BS 1723)

Alloy system	*Melting range* °C	*Refer to* BS 1845 *table no.*	*AWS-ASTM classification (non-exhaustive)*	*DIN designations and Werkstoff numbers (in brackets) (non-exhaustive)*
Copper	1 083	4	BCu–1	L–Cu (2.0081), L–SFCU (2.0091)
Silver	961			
Copper–zinc	860–890	5	RBCuZn–A	L–CuZn46 (2.0413)
Bronze welding type	920–980	5		
Copper–phosphorus	705–800	3	BCuP–1, BCuP–2	L–CuP8 (2.1465)
Silver–copper	779	2	BAg–8	L–Ag72 (2.5151)
Copper–silver–zinc	690–775	2	BAg–5	L–Ag44 (2.5147), L–Ag60
Copper–silver–phosphorus	644–780	3	BCuP–3, BCuP–5	L–Ag15P (2.1210), L–Ag5P (2.1466)
Copper–silver–tin	602–718	2		
Copper–silver–zinc–cadmium–nickel	630–660		BAg–3	L–Ag50CdNi (2.5160)
Copper–silver–zinc–cadmium	620–640	2	BAg–1, BAg–2	L–Ag50Cd (2.5143)
Cobalt–chromium–boron	984–1 240			
Nickel–chromium–boron	1 030–1 175			
Nickel–chromium–silicon–boron	950–1 100		BNi–2	
Nickel–silicon–boron	950–1 070	6	BNi–3, BNi–4	
Nickel–phosphorus	890–1 023		BNi–6	
Silver–palladium–manganese	1 000–1 200	7		(2.5154, 2.5164)
Nickel–palladium–manganese	1 120	7		(2.4304)
Copper–palladium–nickel–manganese	1 060–1 105	7		
Palladium–nickel	1 237			
Copper–nickel	1 100–1 150			
Copper–palladium	1 080–1 090	7		
Silver–palladium	970–1 010	7		
Silver–copper–palladium	807–950	7		
Gold–nickel	950	8	BAu–4	
Gold–copper	910	8	BAu–1, BAu–2	
Aluminium–silver–nickel (or manganese)	558–680			
Aluminium–silicon	565–625	1	BAlSi–2, BAlSi–4	L–AlSil2 (3.2285)
Aluminium–silicon–copper	550–570	1	BAlSi–3	

Note:
Some alloys are available in only one specific form and this should be ascertained before joint-design is considered. The available forms are wire, foil, powder, clad sheet, deposited coating and, in many cases, paste.

Table 34.9 COMBINATIONS OF PARENT MATERIAL AND FILLER METAL FOR FLAME (TORCH) BRAZING*

Filler metal

Parent material	Copper	Silver	Copper–zinc	Bronze welding type	Copper–phosphorus	Silver–copper	Copper–silver–zinc	Copper–silver–phosphorus	Copper–silver–tin	Copper–silver–zinc–cadmium–nickel	Copper–silver–zinc–cadmium	Cobalt–chromium–boron	Nickel–chromium–boron	Nickel–chromium–silicon–boron	Nickel–silicon–boron	Nickel–phosphorus	Silver–palladium–manganese	Nickel–palladium–manganese	Copper–palladium–nickel–manganese	Palladium–nickel	Copper–nickel	Copper–palladium	Silver–palladium	Silver–copper–palladium	Gold–nickel	Gold–copper	Aluminium–silver–nickel (or manganese)	Aluminium–silicon	Aluminium–silicon–copper
Copper	N	R	R	R	R	R	R	R	P	R	R	N	N	N	N	N	N	N	N	N	N	N	N	R	P	R	N	N	N
Copper-base alloys	N	R	P	P	R	R	R	R	P	R	R	N	N	N	N	N	N	N	N	N	N	N	N	R	P	R	N	N	N
Mild steel	N	P	R	R	N	P	R	N	P	R	R	P	R	R	R	N	R	R	P	P	N	R	R	R	P	P	N	N	N
Carbon steels	N	P	R	R	N	P	R	N	N	R	R	P	R	R	R	N	R	R	P	P	N	R	R	R	P	P	N	N	N
Alloy steels	N	P	R	R	N	P	R	N	N	R	R	P	R	R	R	N	R	R	P	P	N	R	P	R	P	P	N	N	N
Stainless steels and irons	N	N	R	R	N	N	R	N	P	R	R	P	R	R	R	N	R	R	R	P	N	R	P	R	P	P	N	N	N
Malleable iron	N	N	R	R	N	N	P	N	N	R	R	N	N	N	N	N	P	P	P	P	N	R	P	P	P	P	N	N	N
Wrought iron	N	N	R	R	N	N	P	N	N	R	R	N	N	N	N	N	P	P	P	P	N	R	P	P	P	P	N	N	N
Cemented carbides	N	N	R	R	N	N	P	N	N	R	R	N	P	P	P	N	P	P	P	P	N	P	N	P	P	P	N	N	N
Nickel-base alloys	N	N	R	R	N	N	R	N	N	R	R	P	R	R	R	N	R	R	R	P	N	R	P	R	P	P	N	N	N
Cobalt-base alloys	N	N	P	R	N	N	P	N	N	P	P	P	P	P	P	N	P	P	P	P	P	P	P	N	P	P	N	N	N
Aluminium and certain aluminium alloys	N	N	N	N	N	N	N	N	N	N	N	N	N	N	N	N	N	N	N	N	N	N	N	N	N	N	N	R	R
Tungsten	N	N	P	P	P	P	R	P	P	R	R	N	P	P	P	N	P	P	P	N	N	N	P	P	P	P	N	N	N
Molybdenum	N	N	P	P	P	P	R	P	P	R	R	N	P	P	P	N	P	P	P	N	N	N	P	P	P	P	N	N	N
Titanium	N	N	N	N	N	N	N	N	N	N	N	N	N	N	N	N	N	N	N	N	N	N	N	N	N	N	N	P	P
Zirconium	N	N	N	N	N	N	N	N	N	N	N	N	N	N	N	N	N	N	N	N	N	N	N	N	N	N	N	N	N
Tantalum	N	N	N	N	N	N	N	N	N	N	N	N	N	N	N	N	N	N	N	N	N	N	N	N	N	N	N	N	N
Beryllium	N	N	N	N	N	N	N	N	N	N	N	N	N	N	N	N	N	N	N	N	N	N	N	N	N	N	N	N	N
Niobium	N	N	N	N	N	N	N	N	N	N	N	N	N	N	N	N	N	N	N	N	N	N	N	N	N	N	N	N	N

R—Recommended and known to be in general use. It should be recognized, however, that in many instances jointing by brazing results in a weakening of the parent material due to intercrystalline penetration, grain growth and other causes.

P—Possible but not known to be in general use.

N—Not recommended.

* From BS 1723.

Table 34.10 COMBINATIONS OF PARENT MATERIAL AND FILLER METAL FOR FURNACE BRAZING*

Parent material	*Filler metal* Copper	Silver	Copper–zinc	Bronze welding type	Copper–phosphorus	Silver–copper	Copper–silver–zinc	Copper–silver–phosphorus	Copper–silver–tin	Copper–silver–zinc–cadmium–nickel	Copper–silver–zinc–cadmium	Cobalt–chromium–boron	Nickel–chromium–boron	Nickel–chromium–silicon–boron	Nickel–silicon–boron	Nickel–phosphorus	Silver–palladium–manganese	Nickel–palladium–manganese	Copper–palladium–nickel–manganese	Palladium–nickel	Copper–nickel	Copper–palladium	Silver–palladium	Silver–copper–palladium	Gold–nickel	Gold–copper	Aluminium–silver–nickel (or manganese)	Aluminium–silicon	Aluminium–silicon–copper
Copper	N	R	R	N	R	R	R	R	R	R	R	N	N	N	N	N	N	N	N	N	N	N	N	R	R	R	N	N	N
Copper-base alloys	N	R	P	N	R	R	R	R	R	R	R	N	N	N	N	N	N	N	N	N	N	N	N	R	R	R	N	N	N
Mild steel	R	P	R	R	N	R	R	N	R	R	R	P	R	R	R	N	R	R	R	R	R	R	R	R	R	R	N	N	N
Carbon steels	R	P	R	R	N	R	R	N	R	R	R	R	R	R	R	N	R	R	R	R	R	R	R	R	R	R	N	N	N
Alloy steels	R	P	R	R	N	R	R	N	P	R	R	P	R	R	R	N	R	R	P	P	R	R	P	R	P	P	N	N	N
Stainless steels and irons	R	P	P	N	N	R	P	N	R	P	P	R	R	R	R	R	R	R	R	R	R	R	R	R	R	R	N	N	N
Malleable iron	R	P	R	N	N	P	R	N	P	P	P	P	P	P	P	N	R	P	P	P	R	R	P	R	P	P	N	N	N
Wrought iron	R	P	R	N	N	P	R	N	P	P	P	P	P	P	P	N	R	P	P	P	R	R	P	R	P	P	N	N	N
Cemented carbides	R	P	P	R	N	P	P	N	P	P	P	N	P	P	P	N	P	P	R	P	P	R	P	R	P	P	N	N	N
Nickel-base alloys	R	P	P	N	N	R	P	N	P	P	P	R	R	R	R	R	R	R	R	R	P	R	R	R	R	R	N	N	N
Cobalt-base alloys	P	P	N	P	N	P	P	N	P	P	P	R	P	P	P	N	P	P	P	P	P	P	P	P	P	P	N	N	N
Aluminium and certain aluminium alloys	N	N	N	N	N	N	N	N	N	N	N	N	N	N	N	N	N	N	N	N	N	N	N	N	N	N	N	R	R
Tungsten	N	N	N	N	N	N	P	P	P	P	P	N	R	R	R	N	R	R	R	R	P	P	P	R	R	P	N	N	N
Molybdenum	N	N	N	N	N	N	P	P	P	N	N	N	R	R	R	N	R	R	R	R	P	R	P	R	R	P	N	N	N
Titanium	N	P	N	N	N	P	N	N	P	N	N	N	N	N	N	N	N	N	N	N	P	P	N	R	P	P	R	P	P
Zirconium	N	P	N	N	N	P	N	N	P	N	N	N	N	P	P	N	P	P	P	P	P	P	P	P	P	P	N	P	P
Tantalum	P	N	N	N	N	P	N	N	P	N	N	N	N	P	P	N	P	P	P	R	P	P	P	P	R	P	N	N	N
Beryllium	N	P	N	N	N	P	N	N	N	N	N	N	N	N	N	N	P	N	N	N	N	P	P	R	P	P	N	N	N
Niobium	N	N	N	N	N	N	N	N	N	N	N	N	N	P	P	N	P	P	P	P	P	P	P	P	P	P	N	N	N

R—Recommended and known to be in general use. It should be recognized, however, that in many instances jointing by brazing results in a weakening of the parent material due to intercrystalline penetration, grain growth and other causes.
P—Possible but not known to be in general use.
N—Not recommended.
Note: Alloys containing zinc or cadmium require a flux.
* From BS 1723.

Table 34.11 COMBINATIONS OF PARENT MATERIAL AND FILLER METAL FOR ELECTRIC INDUCTION BRAZING*

Parent material	*Filler metal*																												
	Copper	Silver	Copper–zinc	Bronze welding type	Copper–phosphorus	Silver–copper	Copper–silver–zinc	Copper–silver–phosphorus	Copper–silver–tin	Copper–silver–zinc–cadmium–nickel	Copper–silver–zinc–cadmium	Cobalt–chromium–boron	Nickel–chromium–boron	Nickel–chromium–silicon–boron	Nickel–silicon–boron	Nickel–phosphorus	Silver–palladium–manganese	Nickel–palladium–manganese	Copper–palladium–nickel–manganese	Palladium–nickel	Copper–nickel	Copper–palladium	Silver–palladium	Silver–copper–palladium	Gold–nickel	Gold–copper	Aluminium–silver–nickel (or manganese)	Aluminium–silicon	Aluminium–silicon–copper
Copper	N	R	P	P	P	R	R	P	P	R	R	N	N	N	N	N	N	N	N	N	N	N	N	R	R	R	N	N	N
Copper-base alloys	N	R	P	P	P	R	R	P	P	R	R	N	N	N	N	N	N	N	N	N	N	N	N	R	R	R	N	N	N
Mild steel	P	N	R	R	N	R	R	N	P	R	R	P	R	R	R	N	R	R	P	P	P	R	R	R	P	P	N	N	N
Carbon steels	P	N	R	R	N	R	R	N	P	R	R	P	R	R	R	N	R	R	P	P	P	R	R	R	P	P	N	N	N
Alloy steels	P	N	R	R	N	R	R	N	P	R	R	P	R	R	R	N	R	R	P	P	P	R	P	R	P	P	N	N	N
Stainless steels and irons	P	N	P	N	N	R	P	N	P	P	P	P	R	R	R	P	R	R	R	R	P	R	P	R	R	R	N	N	N
Malleable iron	P	N	R	R	N	P	R	N	P	P	P	P	P	P	P	N	R	P	P	N	P	R	P	R	P	P	N	N	N
Wrought iron	P	N	R	R	N	P	R	N	P	P	P	P	P	P	P	N	R	P	P	N	P	R	P	R	P	P	N	N	N
Cemented carbides	R	P	R	R	N	P	R	N	P	R	P	N	P	P	P	N	P	P	R	P	P	R	P	R	P	P	N	N	N
Nickel-base alloys	R	P	P	N	N	R	P	N	P	P	P	P	R	R	R	P	R	R	R	R	P	R	R	R	R	R	N	N	N
Cobalt-base alloys	P	N	N	N	N	P	P	P	P	P	P	P	P	P	P	N	P	P	P	P	P	P	P	P	P	P	N	N	N
Aluminium and certain aluminium alloys	N	N	N	N	N	N	N	N	N	N	N	N	N	N	N	N	N	N	N	N	N	N	N	N	N	N	N	P	P
Tungsten	N	N	P	P	P	N	R	P	P	R	R	N	P	R	R	N	P	P	P	R	N	P	P	R	R	P	N	N	N
Molybdenum	N	N	P	P	P	N	R	P	P	R	R	N	P	R	R	N	R	P	P	R	N	R	P	R	R	P	N	N	N
Titanium	N	P	N	N	N	P	P	N	N	P	P	N	N	P	N	N	N	N	N	N	N	N	N	R	P	P	P	P	P
Zirconium	N	N	N	N	N	P	P	N	N	N	N	N	N	N	N	P	P	P	P	P	P	P	P	P	P	P	N	N	N
Tantalum	N	N	N	N	N	N	N	N	P	N	N	N	N	N	N	N	P	P	P	R	P	P	P	P	P	P	N	N	N
Beryllium	N	P	N	N	N	P	N	N	N	N	N	N	N	N	N	N	P	N	N	N	N	P	P	P	P	P	P	N	N
Niobium	N	N	N	N	N	N	N	N	N	N	N	N	N	N	N	N	P	P	P	P	N	P	P	P	P	P	N	N	N

R—Recommended and known to be in general use. It should be recognized, however, that in many instances jointing by brazing results in a weakening of the parent material due to intercrystalline penetration, grain growth and other causes.
P—Possible but not known to be in general use.
N—Not recommended.
* From BS 1723.

Table 34.12 COMBINATIONS OF PARENT MATERIAL AND FILLER METAL FOR ELECTRIC RESISTANCE BRAZING*

Parent material	*Filler metal*																												
	Copper	Silver	Copper–zinc	Bronze welding type	Copper–phosphorus	Silver–copper	Copper–silver–zinc	Copper–silver–phosphorus	Copper–silver–tin	Copper–silver–zinc–cadmium–nickel	Copper–silver–zinc–cadmium	Cobalt–chromium–boron	Nickel–chromium–boron	Nickel–chromium–silicon–boron	Nickel–silicon–boron	Nickel–phosphorus	Silver–palladium–manganese	Nickel–palladium–manganese	Copper–palladium–nickel–manganese	Palladium–nickel	Copper–nickel	Copper–palladium	Silver–palladium	Silver–copper–palladium	Gold–nickel	Gold–copper	Aluminium–silver–nickel (or manganese)	Aluminium–silicon	Aluminium–silicon–copper
Copper	N	R	R	N	P	R	R	R	P	R	R	N	N	N	N	N	N	N	N	N	N	N	N	R	R	R	N	N	N
Copper-base alloys	N	R	R	N	P	R	R	R	P	R	R	N	N	N	N	N	N	N	N	N	N	N	N	R	R	R	N	N	N
Mild steel	P	P	P	N	N	P	R	N	P	R	R	P	P	P	P	N	R	R	P	P	P	R	R	R	P	P	N	N	N
Carbon steels	P	P	P	N	N	P	R	N	P	R	R	P	P	P	P	N	R	R	P	P	P	R	R	R	P	P	N	N	N
Alloy steels	P	P	P	N	N	P	R	N	P	R	R	P	P	P	P	N	R	R	P	P	P	R	P	R	P	P	N	N	N
Stainless steels and irons	P	P	P	N	N	P	R	N	P	R	R	P	P	P	P	P	R	R	R	R	P	R	P	R	R	R	N	N	N
Malleable iron	P	N	P	N	N	P	R	N	P	R	R	P	P	P	P	N	R	P	P	P	P	P	P	P	P	P	N	N	N
Wrought iron	P	N	P	N	N	P	R	N	P	R	R	P	P	P	P	N	R	P	P	P	P	P	P	P	P	P	N	N	N
Cemented carbides	P	N	P	N	N	P	R	N	P	R	R	N	P	P	P	N	P	P	P	P	P	P	P	P	P	P	N	N	N
Nickel-base alloys	P	P	P	N	N	P	R	N	P	R	R	P	P	P	P	P	R	R	R	R	P	R	R	R	R	R	N	N	N
Cobalt-base alloys	P	N	P	N	N	P	P	N	P	P	P	P	P	P	P	N	P	P	P	P	P	P	P	P	P	P	N	N	N
Aluminium and certain aluminium alloys	N	N	N	N	N	N	N	N	N	N	N	N	N	N	N	N	N	N	N	N	N	N	N	N	N	N	N	P	P
Tungsten	N	N	P	P	P	N	R	P	P	R	R	N	N	N	N	N	P	P	P	R	P	P	P	R	R	P	N	N	N
Molybdenum	N	N	P	P	P	N	R	P	P	R	R	N	N	N	N	N	R	P	P	R	P	R	P	R	R	P	N	N	N
Titanium	N	P	N	N	N	N	P	P	P	P	P	N	N	N	N	N	N	N	N	P	N	P	P	P	P	P	P	P	P
Zirconium	N	N	N	N	N	N	P	P	N	P	N	N	N	N	N	P	P	P	P	P	P	P	P	P	P	P	N	P	P
Tantalum	N	N	N	N	N	N	N	N	N	N	N	N	N	N	N	N	P	P	P	R	P	P	P	P	P	P	N	N	N
Beryllium	N	N	N	N	N	N	N	N	N	N	N	N	N	N	N	N	P	R	R	P	N	P	P	R	P	P	N	N	N
Niobium	N	N	N	N	N	N	N	N	N	N	N	N	N	N	N	N	P	P	P	P	N	P	P	P	P	P	N	N	N

R—Recommended and known to be in general use. It should be recognized, however, that in many instances jointing by brazing results in a weakening of the parent material due to intercrystalline penetration, grain growth and other causes.
P—Possible but not known to be in general use.
N—Not recommended.
* From BS 1723.

Table 34.13 COMBINATIONS OF PARENT MATERIAL AND FILLER METAL FOR VACUUM BRAZING*

Parent material	*Filler metal*: Copper	Silver	Copper–zinc	Bronze welding type	Copper–phosphorus	Silver[a]–copper	Copper–silver[a]–zinc	Copper–silver[a]–phosphorus	Copper–silver[a]–tin	Copper–silver[a]–zinc–cadmium–nickel	Copper–silver[a]–zinc–cadmium	Cobalt–chromium–boron	Nickel–chromium–boron	Nickel–chromium–silicon–boron	Nickel–silicon–boron	Nickel–phosphorus	Silver[a]–palladium–manganese	Nickel–palladium–manganese	Copper–palladium–nickel–manganese	Palladium–nickel	Copper–nickel	Copper–palladium	Silver[a]–palladium	Silver[a]–copper–palladium	Gold–nickel	Gold–copper	Aluminium–silver[a]–nickel (or manganese)	Aluminium–silicon	Aluminium–silicon–copper
Copper	N	P	N	N	N	R	N	N	R	N	N	N	N	N	N	N	N	N	N	N	N	N	N	R	P	R	N	N	N
Copper-base alloys	N	P	N	N	N	R	N	N	R	N	N	N	N	N	N	N	N	N	N	N	N	N	N	R	P	P	N	N	N
Mild steel	R	P	N	N	N	R	N	N	R	N	N	P	R	R	R	N	R	R	R	P	P	P	P	R	P	P	N	N	N
Carbon steels	R	P	N	N	N	P	N	N	P	N	N	P	R	R	R	N	R	R	R	P	P	P	P	R	P	P	N	N	N
Alloy steels	R	P	N	N	N	P	N	N	P	N	N	P	R	R	R	N	P	P	P	P	P	P	P	P	P	P	N	N	N
Stainless steels and irons	R	P	N	N	N	P	N	N	R	N	N	R	R	R	R	R	R	R	R	R	R	R	R	R	R	R	N	N	N
Malleable iron	R	P	N	N	N	P	N	N	P	N	N	P	P	P	P	N	P	P	P	P	P	P	P	P	P	P	N	N	N
Wrought iron	R	P	N	N	N	P	N	N	P	N	N	P	P	P	P	N	P	P	P	P	P	P	P	P	P	P	N	N	N
Cemented carbides	R	P	N	N	N	P	N	N	P	N	N	N	P	P	P	N	P	P	P	P	P	P	P	P	P	P	N	N	N
Nickel-base alloys	R	P	N	N	N	R	N	N	P	N	N	R	R	R	R	P	R	R	R	R	R	R	R	R	R	R	N	N	N
Cobalt-base alloys	P	P	N	N	N	P	N	N	P	N	N	R	P	P	P	N	P	P	P	P	P	P	P	P	P	P	N	N	N
Aluminium and certain aluminium alloys	N	N	N	N	N	N	N	N	N	N	N	N	N	N	N	N	N	N	N	N	N	N	N	N	N	N	N	P	P
Tungsten	N	N	N	N	N	N	N	N	P	N	N	N	P	R	R	N	P	P	P	R	P	R	R	R	R	P	N	N	N
Molybdenum	N	N	N	N	N	N	N	N	P	N	N	N	P	R	R	N	P	P	P	R	P	R	R	R	R	P	N	N	N
Titanium	N	P	N	N	N	P	N	N	P	N	N	N	N	N	N	N	N	N	N	N	N	N	N	R	N	P	R	P	P
Zirconium	N	P	N	N	N	P	N	N	P	N	N	N	N	P	P	P	P	P	P	P	P	P	P	P	P	P	N	P	P
Tantalum	R	P	N	N	N	P	N	N	P	N	N	N	N	P	P	N	P	P	P	R	P	P	P	P	R	R	N	N	N
Beryllium	N	P	N	N	N	P	N	N	N	N	N	N	N	N	N	N	P	N	N	N	P	P	P	R	N	P	N	N	N
Niobium	P	P	N	N	N	P	N	N	P	N	N	N	N	P	P	N	P	P	P	P	P	P	P	P	P	P	N	N	N

[a] Some alloys contain elements which are volatile at the temperatures and pressures involved in the brazing process and caution is required under prolonged heating.

R—Recommended and known to be in general use. It should be recognized, however, that in many instances jointing by brazing results in a weakening of the parent material due to intercrystalline penetration, grain growth and other causes.

P—Possible but not known to be in general use.

N—Not recommended.

Note:

This table is intended only as a guide and any further information should be obtained from the manufacturer of the filler metal.

* From BS 1723.

34.4 Bibliography

34.4.1 Bibliography (soldering)

B. M. Allen, 'Soldering Handbook', Iliffe, London, 1969.

D. R. Andrews, 'Soldering, Brazing, Welding and Adhesives', Inst. of Production Engineers, 1978.

B. D. Dunn and C. Chandler, 'The Corrosive Effect of Soldering Fluxes and Handling on some Electronic Materials', *Welding J.*, Oct. 1980.

European Space Agency, 'The Manual Soldering of High Reliability Electrical Connections'. Specification PSS-14/QRM-O8, Issue 1, 1973.

K. T. Hang, 'Creep Strength of Solders', Schweisser und Schneider, Section 17.5, pp. 200–207, 1965.

G. Leonida, 'Handbook of Printed Circuit Design, Manufacture, Components and Assembly', Electrochemical Publications, Ayr, 1981.

C. A. Mackay, 'Some Science in the Art of Soldering', *Welding J.* June 1979, 37–47.

H. H. Manko, 'Solders and Soldering', McGraw-Hill, New York, 1979.

NASA, 'Requirements for Soldered Electrical Connections', NHB 5300–4 (3A–1), 1976.

NASA, 'Soldering Electrical Connections', NASA SP–5002.

C. J. Thwaites and B. T. K. Barry, 'Soldering', Engineering Design Guides, Oxford University Press, 1975.

Westinghouse Defense and Space Centre, 'Development of Highly Reliable Soldered Joints', NASA CR–98433, 1968.

R. N. Wild, 'Properties of Some Low Melt Fusible Solder Alloys', Proc. Tech. Progr. Internepcon, 1971.

34.4.2 Bibliography (brazing)

American Society for Metals, 'Source Book on Selection and Fabrication of Aluminium Alloys', ASM, Ohio, 1978.

American Welding Society, 'Brazing Manual', AWS, New York, 1963.

D. R. Andrews, 'Soldering, Brazing, Welding and Adhesives', Institute of Production Engineers, 1978.

H. R. Brooker and E. V. Beatson, revised by P. M. Roberts, 'Industrial Brazing', 2nd edn, Newnes–Butterworth 1975.

W. Hegmann, 'Aluminium Workshop Practice', Technicopy Ltd, Stonehouse (UK), 1978.

NASA, 'Welding, Brazing and Soldering Technical Brief', NASA TSP 69–10264, 1969.

P. M. Roberts, 'Brazing', Engineering Design Guides, Oxford University Press, 1975.

M. M. Schwartz, 'Modern Metal Joining Techniques', Wiley–Interscience, 1969.

The Technical Information Sheets B1–B15 issued by the British Association for Brazing and Soldering (c/o BNF, Wantage) and the trade literature of both Handy and Harman (The Brazing Book) and Johnson Matthey Metals Ltd (Brazing Materials and Applications) provide valuable information.

34.5 Solder and braze alloy standards

34.5.1 Solder alloy standards

ASTM-B32-1966T	Soldering alloys.
BS 219:1977	Soft solders.
BS 441:1954	Rosin-cored solder wire.
BS AU 90:1965	Soft solders for automobile use.
BS 3252:1960	Ingot tin.
ESA PSS-14/QRM-O8, 1973	Solder composition (chapter 4).
DIN 1707:1976	Alloys for soldering.
DIN 8512:1975	Brazing alloys and soft solders for aluminium.
Federal Specification QQ-S-571-E	'Solder, Tin Alloy; Lead-Tin Alloy and Lead Alloy', with second ammendment of 2 July 1975.
MIL Specification S-12204C, 4 January 1974	'Solder, Lead-Tin'.

34.5.2 Braze alloy standards

BS 1723:1969	Specification for brazing.
BS 1845:1977	Specification for filler metals for brazing.
DIN 8505:1969	Brazing and soldering of metallic pieces.
DIN 8511:1961	Fluxes for brazing and soldering.

DIN 8513:1976	Brazing alloys; T1 ferrous metals; T2 heavy metals with less than 20% silver; T3 heavy metals with greater than 20% silver.
DIN 8514:1976	Brazeability and solderability, definitions.
Federal Specification QQ-B-654, May 1970	Brazing alloys, Silver.
MIL-Specification B-7883B February 1968	Brazing of Steels, Copper, Copper Alloys, Nickel Alloys, Aluminium and Aluminium Alloys'.

35 Vapour deposited coatings

Vapour deposition processes are of two kinds: physical vapour deposition (PVD) and chemical vapour deposition (CVD). Process details and references for elements are given in Tables 35.1 and 35.2, for oxides in Tables 35.3 and 35.4, for nitrides in Tables 35.5 and 35.6, and for carbides in Tables 35.7 and 35.8.

35.1 Physical vapour deposition

Physical vapour deposition processes use a physical effect such as evaporation or sputtering to transport material, usually a metal, from a source to the substrate to be coated. If the material is transformed (e.g. into a carbide) during transport, then the process is described as reactive. All PVD processes are carried out in a relatively high vacuum (i.e. pressure $<10^{-4}$ Torr).

35.1.1 Evaporation (E)

The substrate to be coated is placed in a vacuum chamber with a line-of-sight to the source which is a pool of molten material. The pool is heated either by an electron beam or by resistance heating. The electron beam method is best for source materials with a high melting point.

REACTIVE EVAPORATION (RE)

The same process as evaporation except that the zone through which the evaporated species is passing contains a very small concentration of a reactive gas, usually a hydrocarbon, N_2, NH_3, or O_2. This results in the deposition of the carbide, nitride or oxide of the source material.

ACTIVATED REACTIVE EVAPORATION (ARE)

The same process as reactive evaporation except that an electrically excited plasma is established in the region where the evaporated species encounters the reactive gas.

ION PLATING (IP)

The same process as evaporation except that substrate is biassed negatively with respect to the source. This usually results in a plasma region around the substrate which can be enhanced by r.f. coupling. The ionized species actually represents only a small fraction of the total material flux.

REACTIVE ION PLATING (RIP)

The same process as ion plating except that a reactive gas (usually a hydrocarbon, N_2, NH_3 or O_2) is introduced into the flux travelling towards the substrate so that the carbide, oxide or nitride of the source material is formed.

35.1.2 Sputter plating (S)

A process in which material is transferred from a target and deposited on a substrate by means of ionic bombardment of the target. Usually the ion bombardment is achieved by establishing a d.c. or r.f. plasma at the surface of the target, although ion guns can also be used.

SPUTTER ION PLATING (SIP)

The same process as sputter plating except that the substrate has an electrical negative bias so that it is bombarded with positive ions as well as the neutral flux created by the sputtering.

REACTIVE SPUTTER PLATING (RSP)

The same process as sputter plating except that a reactive gas is introduced into the plasma region so that the transported target material reacts before it arrives at the substrate. For example, the conversion of Ti into a deposit of TiN by the introduction of nitrogen.

35.1.3 Ion cleaning

An argon plasma may be used to bombard a surface with excited argon atoms. These transfer their momentum to atoms in the surface and some of these are ejected from the surface. This has the effect of gradually removing material from a surface, hence cleaning the surface. Ion cleaning is often used as a pretreatment in PVD coating.

35.2 Chemical vapour deposition

Chemical vapour deposition processes use the vapour phase to transport reactive material to the surface of a substrate where a chemical reaction occurs to form a coating, e.g. tantalum by the reaction: $TaCl_5 + H_2 \rightarrow Ta + 5HCl$. Normally the substrate is heated to activate the reaction, but a plasma zone near the substrate is also used in *plasma-activated CVD* for low temperature processes. In many CVD processes the substrate takes no part in the formation of the coating, but in *diffusion coatings* the substrate takes part in the formation of the coating (e.g. aluminizing).
CVD processes can be divided into two groups, hot wall and cold wall, as follows:

A hot-wall reactor is a chamber which is heated from the outside so that both the chamber of the work pieces reach the reaction temperature and so become coated. Consequently there is a certain amount of redundant plating which is undesirable with expensive materials.
A cold-wall reactor is a chamber in which only the work pieces are heated, usually by induction. Only the work pieces become coated.
CVD processes are a rather diffuse group because the physical conditions for each process differ widely so that there are no standard items of equipment as there are for PVD. Many CVD processes use a reduced pressure to improve throwing power (cf. PVD).
In the references *CVD* stands for the International Chemical Vapour Deposition Conference Proceedings published by The Electrochemical Society, Vapor Deposition in the refs. refers to that title by Powell, Oxley and Blocher, J. Wiley, N.Y. 1965. Process details Tables 35.2, 35.4, 35.6, 35.8.

Table 35.1 PHYSICAL VAPOUR DEPOSITION OF ELEMENTS

Element	*Process**	*Coating thickness* μm	*Typical application*	*Ref*
Ag	S	0·02	Window glass	2
	RIP	—	Photographic materials	6
Al	E	~1	IC metallization	1
	S			3
	MS	~1	Mirror coating	8

Table 35.1 PHYSICAL VAPOUR DEPOSITION OF ELEMENTS —*continued*

Element	*Process**	*Coating thickness* µm	*Typical application*	*Ref*
Au	IP	0.2		4
	E	0.02–0.1	Window glass, electron microscopy	7
	S	1.4	Coating micro-spheres for laser fusion	5
Bi	S		Undercoat for photographic	6
	E	~1	materials	
Co (Cr/Al)	E		Turbine blades	14.15
Cr	3	0.1–0.6	Ceramics	11
			Plastics	12
			Ni/Cr for window glass	16
			Switch contacts	
Cu	IP	0.2	Window glass coating	4
	S	0.02	Metallization of plastics prior	11
	MS	0.01–1.0	to electroplating	13
Mo	IBS		Multi-level metallization	17
	S	0.15	for LSI	18
Nb	S	0.4	Superconductors	22
Ni	E, S	—	—	19, 25
Pb	E(EG)	0.4	Pb/oxide/Pb. Josephson junctions	21
				—
Pd	E	0.06	Thermoelectic power	21
	E(EG)	0.1–0.15	Contact layer	23, 24
Pt	MS	1.0	Schottky barrier solar cells	10
	E			—
Rh	MS	0.8	—	—
Si	IBS	0.4	Integrated circuits	—
	E			
Ta	S, E	6–20	Corrosion resistant layer	26, 27
Ti	S	—	—	12
Ti–W	MS	500–5 000 Å	Resistance films	—
V	E	0.086	—	30
W	S	500–5 000 Å	Barrier layer for ICs	28, 29

*ARE—Activated reactive evaporation.
E—Evaporation.
EG—Electron gun.
IBS—Ion beam sputtering.
IP—Ion plating.
MS—Magnetron sputtering.
RIS—Reactive ion sputtering.
RIP—Reactive ion plating.
RE—Reactive evaporation.
RSP—Reactive sputtering.
S—Sputtering.
SIP—Sputter ion plating.
TS—Triode sputtering.

REFERENCES TO TABLE 35.1

1. R. J. Hill, 'Physical Vap. Dep.', Airco Temescal, Berkeley, Cal., 1976, p. 114.
2. T. Abe and T. Yamashina, *Thin Solid Films*, 1975, **30**, 19.
3. C. R. Fuller and P. B. Ghate, *Thin Solid Films*, 1979, **64**, 25.

4. T. Spalvins, *Thin Solid Films*, 1979, **64**, 143.
5. L. Buene *et al.*, *Thin Solid Films*, 1980, **65**, 247.
6. S. K. Sharma and J. Spitz, *Thin Solid Films*, 1980, **65**, 339.
7. A. T. Lowe and C. D. Hosford, *J. Vac. Sci. Tech.*, 1979, **16**, 197.
8. A. J. Learn, *J. Electrochem. Soc.*, 1976, **123**, 894.
9. A. Aronson and S. Weinig, *Vacuum*, 1977, **27**, 151.
10. S. Schiller *et al.*, *J. Vac. Sci. Tech.*, 1977, **14**, 813; 1975, **12**, 858.
11. S. Hurwitt, Trans. Conf. Sputter Mats. Res. Corp. Orangeberg, N.Y., 1974, p. 31.
12. D. W. Hoffmann and J. A. Thornton, *Thin Solid Films*, 1977, **40**, 355.
13. T. Tsutada and N. Hosokawa, *J. Vac. Sci. Tech.*, 1979, **16**, 348.
14. D. H. Boone *et al.*, *Thin Solid Films*, 1979, **64**, 299.
15. C. J. Spengler and S. Y. Lee, *Thin Solid Films*, 1979, **64**, 263.
16. Matsushita Electric Works, Jap. Pat. 5476972.
17. P. H. Schmidt *et al.*, *J. Appl. Phys.*, 1973, **44**, 1833.
18. J. Nagano, *Thin Solid Films*, 1980, **67**, 1.
19. D. McKeown *et al.*, An. Rep. Low Energy Sput, Space Sci. Lab. Gen. Dynamics, 1962.
20. M.Murahami, *Thin Solid Films*, 1980, **69**, 253.
21. G. Wedler and R. Chander, *Thin Solid Films*, 1980, **65**, 53.
22. C. T. Wu, *Thin Solid Films*, 1979, **64**, 103.
23. D. J. Sharp, *J. Vac. Sci. Tech.*, 1979, **16**, 204.
24. V. Köster *et al.*, *Thin Solid Films*, 180, **67**, 35.
25. S. Schiller *et al.*, *J. Vac. Sci. Tech.*, 1975, **12**, 858.
26. S. Kashu *et al.*, *J. Vac. Sci. Tech.* 1972, **9**, 1399.
27. J. Spitzel and J. Chevallier, *CVD*, 1975, V, 204.
28. Kossowsky, *Electrochem. Soc. Ext. Abst.*, 1977, **77–2**, 429.
29. J. B. Bindell and T. C. Tisone, *Thin Solid Films*, 1974, **23**, 31.
30. A. Borodziuk-kulpa *et al.*, *Thin Solid Films*, 1980, **67**, 21.

Table 35.2 CHEMICAL VAPOUR DEPOSITION OF ELEMENTS

Elements	*Process*	*Temp.* °C	*Thickness* μm	*Typical application*	*Ref.*
Al	Al R or Al HR_2 thermal decomp. (R usually isobutyl)	200–600	50	Impervious layer on plastics	6, 9
	Al + Al X_3 (X = halide)	700–1 000	20–100	Aluminizing steel and super alloys	10
	Al Cl_3 + H_2	900–1 000			15
As	$AsCl_3$ thermal decomp.	300–500	—	—	8
	AsH_3 thermal decomp.	230–300			7
	AsH_3 plasma assisted				1
B	B_2H_6 thermal decomp.	400–700		B fibres	13
	BBr_3 thermal decomp.	1 000–1 300			
	BBr_3 + H_2	600–1 100		High purity boron	5
	BCl_3 + H_2	1 000–1 500	5–1 000	Jewelled bearings, fibres	2, 5, 12
C as graphite or carbon	CH_4, C_2 H_6, C_2 H_4 thermal decomp.	1 000–2 500	1–1 000	Barrier layers, conducting layers, free-standing deposits	3, 4, 11 14
Co	Co(acac)$_2^*$ thermal decomp.	300–500	1.5	—	17
Cr	Cr I_2 thermal decomp.	900–1 400		High purity chromium	19
	$Cr(CO)_5$	300–650			20
	Cr dicumene	300–400	25–50	Coating nuclear fuel elements	14
	Cr + NH_4 Cl Pack	950–1 050	20–50	Coating turbine blades	16
Cu	Cu(acac)$_2^*$ thermal decomp.	260–450	—	—	18
Ge	$GeCl_4$ + H_2	600			22
	GeH_4 thermal decomp.	400–900	—	Epitaxial semiconductor layer	21
Hf	HfX_4 thermal decomp. (X = I or Br)	—	—	—	25

*acac = acetylacetonate

Table 35.2 CHEMICAL VAPOUR DEPOSITION OF ELEMENTS—*continued*

Elements	*Process*	*Temp.* °C	*Thickness* μm	*Typical application*	*Ref.*
Ir	$Ir(CO)_2Cl_2$ thermal decomp.	600	—	—	28
Mo	MoX_5+H_2 (X=Cl, F)	650–1 400	8	Thin film resistors	26, 30
	$Mo(CO)_6+H_2$	300–600	0.2	Coating nuclear fuels	29
	$Mo(CO)_6$ plasma			Anti-reflection on surfaces for photo-chem. conversion	27
					24, 23
Nb	$NbCl_5+H_2$	1 000–1 400	5–50	Nuclear fuel cladding	32
Ni	$Ni(CO)_4$ thermal decomp.	180–200	1–1 000	Vapour forming	33
	$Ni(acac)_2^*$ thermal decomp.	250–450		Coating of uranium	34, 38
	$Ni(CO)_4$ plasma assisted				24
	$Ni(CO)_4$ laser assisted		0.055	Localized deposit	31
Pb	Pb Et_4 thermal decomp.	300–500	—	—	35
Pd	$Pd(acac)_2^*$ thermal decomp.	350–450	—	—	39
Pt	$Pt(CO)_2Cl_2$ thermal decomp.	600			46
	$Pt(acac)_2^*$ thermal decomp.	350–450	—	—	40
Re	ReF_5+H_2	850–1 100			
	$ReCl_5+H_2$	600–1 200			42
	Re O Cl_4 thermal decomp.	1 250–1 500	—		37
	Re Cl_5 thermal decomp.	1 100–1 300		Thermionic emitters	41
Rh	$Rh(tfa)_3^\dagger+H_2$	250	—	—	38
	$Rh(CO)_2Cl_2$ thermal decomp.	600			46
Ru	$Ru(CO)_5$ thermal decomp.	200			47
	$Ru(CO)_2Cl_2$ thermal decomp.	600	—	—	46
Sb	$SbCl_3$ thermal decomp.	500–600			49
	SbH_3 thermal decomp.	~150	—	Prod. of high purity Sb	48
Si	SiH_4 thermal decomp.	300–1 200	—	—	51
	$SiCl_4+H_2$	900–1 800		Prod. of polycryst. Si	
	$SiHCl_3+H_2$	950–1 250	30	Solar cells	50
	SiH_4 plasma decomp.	300–1 150		Semiconductor layers	44
	Siliconizing Ni alloys			Si/B corrosion-resistant layers on gas turbine parts	45
					43
Sn	Sn+NH_4 Cl/MgO pack diffusion	450–750	10	Corrosion and wear resistant coating	53
Ta	Ta Cl_5+H_2	900–1 100	5–20	Corrosion-resistant layers	54
				Thin film resistors	58
				Impregnation of C fibres	59
Ti	Ti Br_4 thermal decomp.	1 100–1 400		Metal production	55
	Ti I_4 thermal decomp.	1 200–1 500	—		56
U	UI_6 thermal decomp.	~1 500	—	—	57
V	V Cl_4+H_2	800–1 000			60
	VI_2	1 000–1 200	10–50	—	69, 61
W	W F_6+H_2	400–700	1–1 000	Thin wall components	62
	W Cl_6+H_2	600–700		Erosion-resistant	66
	$W(CO)_6$ thermal decomp.	350–600		coatings on rocket motors	63, 64
				Electron emission surfaces	67, 68
Zr	Zr I_4 thermal decomp.	1 000–1 500	5–50	Nuclear fuel cladding	65

*acac = acetylacetonate †tfa = trifluoro-acetylacetonate

REFERENCES TO TABLE 35.2

1. J. C. Knights and J. E. Matan, *Solid State Corrosion*, 1977, **21**, 983..
2. R. M. Mehalso and R. J. Diefendorf, *CVD*, 1975, V, 84.
3. R. P. Gower and J. Hill, *CVD*, 1975 V, 114.
4. W. F. Knippenberg *et al.*, *Philips Tech. Rev.*, 1977, **37**, 189.
5. D. R. Stern and L. Lynds, *J. Electrochem. Soc.*, 1958, **105**, 676.
6. H. O. Pierson, *Thin Solid Films*, 1977, **45**, 257.
7. K. Tamaru, *J. Phys. Chem.*, 1955, **59**, 777.
8. 'Vapor Deposition', p. 283.
9. E. R. Breining *et al.*, Ger. 1 235 106, 1967.
10. 'Vapor Deposition', p. 277.
11. J. Chin *et al.*, *CVD*, 1975, VI, 364.
12. H. E. Hintermann *et al.*, *Ext. Abs. Electrochem. Soc.*, 1973, **73/2**, 218.
13. R. B. Reeves and J. J. Gelhardt, *SAMPE*, 1979 **10**, (1) 13.
14. J. H. Oxley *et al.*, *Ind. and Eng. Chem: Prod. Res. and Dev.*, 1962, **1**, 102.
15. L. H. Marshall, US Pat. 1 893 782, 1933.
16. H. M. J. Mazille, *Thin Solid Films*, 1980, **65**, 67.
17. E. J. Jablonwski, *Cobalt*, 1962, **14**, 28.
18. P. Palovlyk, US Pat. 2 704 728, 1955.
19. Powell *et al.*, 'Vapor Deposition', p. 290.
20. B. B. Owen and R. T. Webber, Am. Inst. Min. Met. Eng. Tech., 1948, Pub 2306.
21. D. J. Dumin *et al.*, *RCA Rev.*, 1970, **31**, 620.
22. E. C. Cave and B. B. Czorny, *RCA Rev.*, 1963, **24**, 523.
23. B. O. Seraphin, *J. Vac. Sci. Tech.*, 1979., **16**, 193.
24. H. F. Sterling *et al.*, *Vide*, 1966, **21**, 80.
25. F. B. Litton, *J. Electrochem. Soc.*, 1951, **98**, 488.
26. J. E. Cline and J. Wulff, *J. Electrochem. Soc.*, 1951, **98**, 385.
27. J. J. Lander and L. H. Germer, Am. Inst. Min. Met. Eng.: Inst. Met. Div. Met. Tech., **14**, 6. Tech. Prod. 2259, 1947.
28. J. A. M. van Liempt, *Metallwerkschaft*, 1932, **11**, 357.
29. K. Hieber and M. Stolz, *CVD*, 1975, V, 436.
30. R. R. Jaeger and S. T. Cohen, *CVD*, 1972, III, 500.
31. S. D. Allen and M. Bass, *J. Vac. Sci. Tech.*, 1979, **16**, 431.
32. W. A. Jenkins and H. W. Jacobson, US Pat. 3 020 148, 1962.
33. L. W. Owen, *J. less-common Met.*, 1962, **4**, 35.
34. E. C. Marboe, US Pat. 2 430 520, 1947.
35. R. N. Meinert, *J. Am. Chem. Soc.*, 1933, **55**, 979.
36. E. H. Reerin, *Z. f. Anorg. Chem.*, 1928, **173**, 45.
37. A. N. Zelikman *et al.*, *Russian Metall.*, 1963, **4**, 120.
38. R. L. Van Hemert *et al.*, *J. Electrochem. Soc.*, 1965, **112**, 1123.
39. C. F. Powell, J. H. Oxley and J. M. Blocher, in 'Vapor Deposition', Wiley, N. Y., 1968.
40. 'Vapor Deposition', p. 314.
41. L. Yang *et al.*, *CVD*, 1972, III, 253.
42. P. J. Sherwood, *CVD*, 1072, **III**, 728.
43. A. R. Nicoll *et al.*, *Thin Solid Films*, 1974, **64**, 321.
44. H. F. Sterling and R. C. G. Swan, *Solid State Electronics*, 1965, **8**, 653.
45. W. G. Towsend and M. E. Uddin, *Solid State Electronics*, 1973, **16**, 39.
46. E. H. Reerin, US Pat. 1 818 909, 1931.
47. Smelin, 'Handbuch der Anor. Chem.', 8th edn, Verlag Chemie Berlin, p. 63, 1938.
48. K. Tamaru, *J. Phys. Chem.*, 1955, **59**, 1084.
49. 'Vapor Deposition', p. 283.
50. T. L. Chu *et al.*, *CVD*, 1975, V, 653.
51. CVD of Si, Semiconductor Silicon 1977, Electrochem. Soc. Princetown, NJ., 1977.
52. A. K. Praturi, *CVD*, 1977, VI, 20.
53. S. Andisio, *J. Electrochem. Soc.*, 1980, **127**, 2299.
54. C. F. Powell *et al.*, *J. Electrochem, Soc.*, 1948, **93**, 258.
55. I. E. Campbell *et al.*, *J. Electrochem. Soc.*, 1948, **93**, 271.
56. B. W. Gosner, Titanium Rep. Symp. ONR, Washington DC, 1948.
57. G. Derge and G. P. Monet, US Pat. 2 743 173, 1956.
58. J. Spitz and J. Chevallier, *CVD*, 1975, V, 204.
59. C. M. Hollabaugh *et al.*, *CVD*, 1977, VI, 559.
60. A. E. van Arkel, 'Reine Metalle', Springer, Berlin, 1939.
61. A. E. van Arkel *et al.*, US Pat. 1 891 124, 1932.
62. A. Bremer and W. E. Reid, US Pat. 3072983, 1963.
63. C. F. Powell *et al.*, *J. Electrochem. Soc.*, 1948, **93**, 258.
64. J. J. Lander, US Pat. 2516058, 1930.
65. Z. M. Shapiro, 'Metallurgy of Zirconium', McGraw-Hill, N.Y., 1955.
66. P. J. Sherwood *et al.*, *CVD*, 1975, V, 801.
67. A. M. Shroff *et al.*, *CVD*, 1975, V, 351.
68. J. A. Papke and R. D. Stevenson, *CVD*, 1969 I, 193.
69. K. J. Miller *et al.*, *J. Electrochem. Soc.*, 1966, **113**, 902.

Table 35.3 PHYSICAL VAPOUR DEPOSITION OF OXIDES

Compound	*Process**	*Coating thickness μm*	*Typical applications*	*Ref.*
Al_2O_3	S RSP IP	1–6	Passive sealing layers. Wear-resistant layers	1, 2
CuO/Cu_2O	r.f. S $Cu+O_2$			3
Cr_2O_3	r.f. S RE	0.1 0.01	Wear resistant layer anti-reflectance coating	4, 5
NiO	S	1		6
SiO_2	MS r.f. S	0.006 13	Mask layer on IC Multilayer laser fusion test.	7, 8
SiO–Cr	r.f. S		Resistor film	9
SnO_2	RSP		Transparent electrodes for electro-optics	10
Ta_2O_5	r.f. S $Ta+O_2$		Optical wave guides	11
TiO_2	RSP $Ti+O_2$	1–2	Anti-reflectance coatings on Al mirror	12
ZnO	r.f. S		Surface acoustic wave (SAW) devices	13
ZrO_2	d.c./r.f. S	25–48	Thermal barrier coatings for gas-turbine parts	14

* *See* Table 35.1 for the key.

REFERENCES TO TABLE 35.3

1. R. S. Nowicki, *J. Vac. Sci. Tech.*, 1977, **14**, 127.
2. R. F. Bunshah and R. J. Schramm, *Thin Solid Films*, 1977, **40**, 211.
3. M. Samirant *et al.*, *Thin Solid Films*, 1971, **8**, 293.
4. B. Bhushan, *Thin Solid Films*, 1979, **64**, 231.
5. E. Ritter *et al.*, US Pat. 4 172 156, 1976.
6. P. V. Plunkett *et al.*, *Thin Solid Films*, 1979, **64**, 121.
7. K. Urbanek, *Solid State Tech.*, 1977, **20**, 87.
8. S. F. Meyer and E. J. Hsieh, *Thin Solid Films*, 1979, **64**, 383.
9. V. Fronz *et al.*, *Thin Solid Films*, 1980, **65**, 33.
10. E. Leja *et al.*, *Thin Solid Films*, 1980, **67**, 45.
11. W. M. Paulson *et al.*, *J. Vac. Sci. Tech.*, 1979, **16**, 307.
12. L. D. Hartsough and P. S. McLeod, *J. Vac. Sci. Tech.*, 1977, **14**, 123.
13. K. Ohji *et al.*, *J. Vac. Sci. Tech.*, 1978, **15**, 1601.
14. J. W. Patten, *Thin Solid Films*, 1979, **64**, 337.

Table 35.4 CHEMICAL VAPOUR DEPOSITION OF OXIDES

Material	*Process*	*Temp. °C*	*Coating thickness μm*	*Typical applications*	*Ref.*
Al_2O_3	$AlCl_3+CO_2+H_2$	850–1 800	2–8	Hard coating for tool tips Coating for nuclear fuels	1–5
	$AlCl_3+O_2$ plasma assisted	230–350		Passive sealing layer for integrated circuits. Single crystal	
	$Al(CH_3)_3$ or $Al(OC_3H_7)_3+O_2$	275–500		growth	

Table 35.4 CHEMICAL VAPOUR DEPOSITION OF OXIDES—*continued*

Material	*Process*	*Temp.* °C	*Coating thickness* μm	*Typical applications*	*Ref.*
B_2O_3–SiO_2	$B_2H_6+SiH_4+O_2$	350	0.4–0.6	Diffusion source for doping electronic materials. Cladding for optical fibres	23
Cr_2O_3	$Cr(CO)_6+O_2$	400–600	<25	Oxidation-resistant coating	6
SiO_2	SiH_4+O_2	300–430	5–25	Passive layer-encapsulant for electronics	7–12
	$Si(OEt)_4$ thermal decomp.	800–1 000			
	$Si(OEt)_4+H_2O$	500–1 000		Corrosion-resistant coating for	
	SiH_4+N_2O	300–1 500		chemical plant. AGR plant.	
	$SiCl_4+O_2$ plasma assisted	>1 000	20–50	Prep. of optical fibres	
SnO_2	$SnCl_4+O_2$	600–1 000	0.1	Semiconductor assemblies. Optoelectronic applications.	13–15
	$SnCl_4+H_2O$	>430		Transparent electrically conducting coatings	
	$Sn(CH_3)_4+O_2$	400–600	0.1		
TiO_2	$TiCl_4+O_2+$hydrocarbon (in flame)	450–800		Pigment production	16–19
	$TiCl_4+CO$ plasma assisted or $TiCl_4+O_2$ plasma assisted			Dielectric films in MOS structure	
	$Ti(OC_3H_7)_4+O_2$	450			
ZrO_2	$Zr(acac)_4+O_2$	450–700		MOS devices	20–22
	$ZrCl_4+CO_2+CO+H_2$	1 000	5–10	Hard coatings on sintered carbides. Oxidation-resistant coatings	
	$Zr(OC_3H_7)_4$	500–600	12–75		

REFERENCES TO TABLE 35.4

1. R. Funk *et al.*, *J. Electrochem. Soc.*, 1976, **123**, 285.
2. J. N. Lindström and R. T. Johannesson, *J. Electrochem. Soc.*, 1976, **123**, 555.
3. P. S. Schaffer, *J. Am. Ceram. Soc.*, 1965, **48**, 508.
4. H. Katto and Y. Koga, *J. Electrochem. Soc.*, 1971, **118**, 1619.
5. M. T. Duffy and W. Kern, *RCA Rev.*, 1970, **31**, 754.
6. J. J. Lander, US Pat. 2671 739, 1954; 'Vapor Deposition', p. 390.
7. J. Graham, *High Temperatures—High Pressures*, 1974, **6**, 577.
8. J. Middelhoek and A. J. Klinkhamer, *CVD*, 1975, V, 19.
9. 'Vapor Deposition', p. 392.
10. J. Irven and A. Robinson, *Phys. Chem. of Glass*, 1980, **21**, 47.
11. D. Küppers and H. Lydtin, *CVD*, 1977, VI, 461.
12. J. Irven and A. Robinson, *Electronics Letters*, 1979, **15**, 252.
13. O. Tabata, *CVD*, 1975, V, 681.
14. 'Vapor Deposition', p. 398, Wiley, N.Y., 1966.
15. R. N. Ghostagore, *J. Electrochem. Soc.* 1971, **118**, 1619.
16. A. Pechukas and G. Atkinson, US Pat., 2394633; 1946.
17. H. F. Sterling *et al.*, *Vide*, 1966, **21**, 80.
18. J. H. Alexander *et al.*, p. 186, *Thin Film Dielectries*, F. Vratny, Electrochem Soc., N.Y., 1969.
19. C. C. Wang *et al.*, *RCA Rev.*, 1970, **31**, 728.
20. M. Balog *et al.*, *J. Cryst. Growth.*, 1972, **17**, 298.
21. J. N. Lindström *et al.*, US Pat. 3837896; 1974.
22. K. S. Kagdiyasin and C. T. Lynch, USAF Rept. ASD-DR-322, 1963.
23. J. Wong, *J. Electrochem. Soc.*, 1980, **127**, 62.

Table 35.5 PHYSICAL VAPOUR DEPOSITION OF NITRIDES

Compound	*Process**	*Coating thickness* μm	*Typical applications*	*Ref.*
AlN	RIS $Al+N_2$		Electronic material.	1–3
	RIP $Al+N_2$ or NH_3— substrate 1 000 °C	2	Refractory dielectric. Accoustic transducers.	
	RE $Al+NH_3$	0.6–1.5	SAW diodes	
HfN	RSP		Thin film dielectrics	4
	RIP $Hf+N_2$	0.2–0.3		
NbN	RSP		Superconducting layers	5
TaN	RSP	115 Å min^{-1}	Resistor film	6
Si_3N_4	RIS $Si+N_2$	0.25–0.08	Encapsulant for GaAs	1,
	r.f. RSP $Si+N_2$		Optical wave guides	7–9
	RSP $Si_3N_4+N_2$			
TiN	RSP $Ti+N_2$		Hard coatings	10–11
	ARE $Ti+N_2$	15		
W_2N	RSP	100 Å min^{-1}	Superconductivity	6

**See* Table 35.1 for key.

REFERENCES TO TABLE 35.5

1. H. J. Erler *et al.*, *Thin Solid Films*, 1980, **65**, 233.
2. Y. Murayama *et al.*, *J. Vac. Sci. Tech.*, 1980, **17**, 796.
3. S. Yoshida *et al.*, *Appl. Phys. Lett.*, 1975, **26**, 461.
4. F. T. J. Smith, *J. Appl. Phys.*, 1970, **41**, 4227.
5. J. Spitz and A. Aubert, *CVD*, 1975, V, 258.
6. F. M. Kilbane and P. S. Habig, *J. Vac. Sci. Tech.*, 1975, **12**, 107.
7. L. E. Bradley and J. S. Sites, *J. Vac. Sci. Tech.*, 1979, **16**, 189.
8. F. H. Eisen *et al.*, *Solid State Electronics*, 1977, **20**, 219.
9. W. M. Paulson *et al.*, *J. Vac. Sci. Tech.*, 1979, **16**, 307.
10. T. Abe and T. Yamashiua, *Thin Solid Films*, 1975, **30**, 19.
11. K. Nakamura *et al.*, *Thin Solid Films*, 1977, **40**, 155.

Table 35.6 CHEMICAL VAPOUR DEPOSITION OF NITRIDES

Material	*Process*	*Temp.* °C	*Coating thickness* μm	*Typical applications*	*Ref.*
AlN	$AlCl_3+NH_3+H_2$	800–1 000	0.5	Electronic applications Refractory dielectrics SAW devices	1–5
	$Al+N_2$	1 800–2 000	Whiskers	High strength refractories	
	$AlCl_3+N_2+H_2$ plasma assisted	800–1 200		III/V Semi conductors	
	$(CH_3)_3Al+NH_3+H_2$	1 200	Epitaxial		
BN	BCl_3+NH_3	1 000–2 000	1–2 000	Formation of free-standing crucibles, tubes and plates	6–12
	BF_3+NH_3	1 000–2 000		Densification of BN felt. Travelling wave tube isolators	
	$B_3N_3H_3Cl_3$ thermal decomp.	1 000–2 000		Insulating layers in semi-cond. manufacture	
	$B_2H_6+NH_3$ plasma assisted	400–700	0.2–0.6	Diffusion doping of Si	
HfN	$HfCl_n+N_2+H_2$ (n = 2, 3, 4)	900–1300	5–10	Wear-resistant coatings. Diffusion barriers.	13

Table 35.6 CHEMICAL VAPOUR DEPOSITION OF NITRIDES—*continued*

Material	*Process*	*Temp.* °C	*Coating thickness* μm	*Typical applications*	*Ref.*
TaN	$TaCl_5/TaBr_5+N_2$	800–1 500	—	Resistor film	14–15
Si_3N_4	SiH_4+N_2/NH_3 plasma assisted	350	0.15	Encapsulation of III/V semiconductor devices	
	SiF_4+NH_3	1 000–1 500		Anti-reflection coatings on solar cells	12, 16–20
	$SiCl_4+NH_3$	1 000–1 500	10–2 000		
	$SiHCl_3+NH_3$	1 100–1 400		Gas turbine parts	
	$SiH_4+NH_3+N_2H_4$ photo-chem. decomp.				
TiN	$TiCl_4+N_2+H_2$	650–1 700	2–8	Hard coatings for hard-metal cutting tools. Scratch-resistant decorative surfaces. Diffusion barrier layers	21–23
	$Ti(NMe_2)_4$ thermal decomp.	300–500			
ZrN	$ZrCl_4+N_2+H_2$	900–1 200	2–8	Hard coatings for cutting tools	13

REFERENCES TO TABLE 35.6

1. A. J. Noreika and D. W. Ing. *J. Appl. Physics*, 1968, **39**, 5578.
2. T. L. Chu and R. W. Kelm Jr., *J. Electrochem. Soc.*, 1975, **122**, 995.
3. K. M. Taylor and C. Lenie, *J. Electrochem. Soc.*, 1960, **107**, 308.
4. J. Bauer *et al.*, *Physica Status Solid*(a), 1977, **39**, 173.
5. H. M. Manasevit *et al.*, *J. Electrochem. Soc.*, 1971, **118**, 1864.
6. G. Clerc and P. Gerlach, *CVD*, 1975, V, 777.
7. D. Morin and M. Le Clercq, French Pat. 2232613; 1975.
8. N. J. Archer, High Temp. Chem. of Inorg. and Ceram. Mats., Chem. Soc. Pub 30, 1977.
9. H. O. Pierson, *J. Compos. Mat.*, 1975, **9**, 228.
10. R. Francis and E. P. Flint, US Army Report, WAL-766, 41/1, 1961.
11. M. Hirayama and K. Shohno, *J. Electrochem. Soc.*, 1975, **122**, 1671.
12. K. Shohno *et al.*, *J. Electrochem. Soc.*, 1980, **127**, 1546.
13. M. J. Hakim, *CVD*, 1975, V, 634.
14. R. Kieffer *et al.*, *Powder Metall Inst.*, 1973, **5**, 188.
15. K. Hieber, *Thin Solid Films*, 1974, **24**, 157.
16. Dietrich and Reid, *Ext. Abst.*, *Electrochem. Soc.*, 1977 **77–2**, 510.
17. M. Ružička, *Czech J. Phys.*, 1974, **B24**, 465.
18. J. Gebhart, *et al.*, *CVD*, 1975, V, 786.
19. J. Bühler *et al.*, *CVD*, 1977, VI, 493.
20. Hughes Aircraft U.S. Pub. 4181751, 1980.
21. W. Schintlmeister *et al.*, *CVD*, 1975, V, 523.
22. K. Sugiyama and S. Motojima, *CVD*, 1975, V, 147.
23. R. Warren and M. Carlsson, *CVD*, 1975, V, 611.

Table 35.7 PHYSICAL VAPOUR DEPOSITION OF CARBIDES

Compound	*Process**	*Coating thickness* μm	*Typical applications*	*Ref.*
Cr_3C_2	RSP	0.3–2.5	Solar energy collectors	1–2
	IP		Wear-resistant coatings	
Cr_7C_3	ARE			
HfC	ARE/RE $Ar+C_2H_2+Hf$	2.5 μm min^{-1}	—	2
Mo_2C	RSP IP $Ar+CH_4$	0.18	Solar energy absorption	1

**See* Table 35.1 for key.

Table 35.7 PHYSICAL VAPOUR DEPOSITION OF CARBIDES —*continued*

Compound	*Process**	*Coating thickness* μm	*Typical applications*	*Ref.*
NbC	ARE/RE Ar + CH_4 or C_2H_4 + Nb	2.5 μm min^{-1}	Optical/semiconductor	2
SiC	RSP Ar + C_2H_2	1	Electronic barrier coating	3–4
Ta_2C	RSP	0.8	Solar energy absorption	1, 2
TaC	Ar + CH_4, C_2H_2 ARE/RE Ta + C_2H_2	1.5 μm min^{-1} at 590 °C		5
TiC	RSP A + CH_4 ARE ARE/RE Ti + C_2H_4	5.0 μm at 700/900°C 4.0 μm min^{-1} at 450°C	Cutting tools Wear resistant surfaces	2 6, 7, 8
VC	ARE V + C_2H_2	3.0 μm min^{-1} at 555°C	—	2
W_2C	RSP	0.8	Solar energy absorption	1
WC	W + Ar + CH_4			
ZrC	ARE/RE Zr + C_2H_2	5.0 μm min^{-1} at 540 °C	Diffusion barrier	2, 9

* *See* Table 35.1 for key.

REFERENCES TO TABLE 35.7

1. G. L. Harding, *J. Vac. Sci. Tech.*, 1976, **13**, 1070.
2. R. F. Bunshah and A. C. Raghuram, *J. Vac. Sci. Tech.*, 1972, **9**, 1385.
3. K. E. Haq, *Appl. Phys. Lett.*, 1975, **26**, 255.
4. Y. Murayama and T. Takao, *Thin Solid Films*, 1977, **40**, 309.
5. W. Grossklaus and R. F. Bunshah, *J. Vac. Sci. Tech.*, 1975, **12**, 811.
6. F. Shinoki and A. Itoh, *Jap. J. Appl. Phys. Suppl.*, 1974, **2** (1), 505.
7. K. Nakamura *et al.*, *Thin Solid Films*, 1977, **40**, 155.
8. W. R. Stowell, *Thin Solid Films*, 1974, **22**, 111.
9. K. D. Kennedy and Scheuermann, US Pat. 3 900 592; 1975.

Table 35.8 CHEMICAL VAPOUR DEPOSITION OF CARBIDES

Material	*Process*	*Temp.* °C	*Coating thickness* μm	*Typical application*	*Ref.*
B_4C	$BCl_3 + CO + H_2$	1 200–1 800		Nuclear Industry	1
	$B_{10}C_2H_{12}$ thermal decomp.	1 000–1 300		Armament	2
	$B_2H_6 + CH_4$			Abrasive blasting nozzles	3
	BMe_3 thermal decomp.	~550		Rocket nozzles	4
	$BCl_3 + CCl_4$ thermal decomp.	1 200–2 000		Fibres	5
	$BCl_3 + CH_4 + H_2$	1 300–1 900	10–1 000		6
Cr_3C_2	$Cr(CO)_6 + H_2$	300–650			7
Cr_7C_3	$CrCl_2 + CH_4 + H_2$	~1 000	10–40	Wear-resistant coatings HV 2250	8
	$CrCl_2 + H_2$	1 000		Chromizing of steel	9
	Cr dicumene—therm. decomp.	450–650			10
HfC	$HfCl_4 + H_2 + CH_4$	1 000–1 500	5	Diffusion barrier Wear resistant coating	11 12

Table 35.8 CHEMICAL VAPOUR DEPOSITION OF CARBIDES—*continued*

Material	*Process*	*Temp.* °C	*Coating thickness* μm	*Typical application*	*Ref.*
Mo_2C	$Mo(CO)_6$ thermal decomp.	350–475		Wear resistant layer	13
	$MoF_6+C_6H_6+H_2$	400–1 000			14
NbC	$NbCl_5+CCl_4+H_2$	1 500–1 900		Wear-resistant coating; Diffusion barrier	15
SiC	$SiCl_4+C_6H_5CH_3$	1 000–2 000	10–1 000	Oxidation-resistant coatings. Single fibres of high strength	16
	$CH_3SiCl_3+H_2$	1 000–1 600		Heating elements. Electrical heat stylus	17
	$SiH_4+C_2H_4$ or CH_4 plasma assisted		~1	Electrochemical machining tool	18
Ta_2C TaC	$TaCl_5+CH_4$	1100		Coating W filaments for incorporation into superalloy matrices	19
TiC	$TiCl_4+H_2+CH_4$ or C_6H_6	800–1 400	2–12	Wear-resistant coatings on steel and carbides—particularly cutting tools	20–24
	$TiCl_4+CCl_4+H_2$	>1 000			
VC	$VCl_2+CH_4+H_2$	1 050–1 130		Wear-resistant coating	25–26
	Diffusion coating VC_x $x=0.84–0.89$				
WC	$WF_6+C_6H_6+H_2$	400–900		Wear-resistant coatings	27–31
	$W(CO)_6$ thermal decomp.	300–500	5–50		
	WF_6+CO+H_2	600–1 000			
	$WCl_6+CH_4+H_2$	900–1 150			
ZrC	$ZrCl_4+CH_4+H_2$	1 050–1 500	50–100	Wear-resistant coating. Diffusion barrier layer	32–33
	$ZrCl_4+C_3H_6+H_2$	1 150–1 300	50	High temp. insulation. Nuclear fuel coating	34

REFERENCES TO TABLE 35.8

1. M. Formstecher and E. Ryskevic, *Compt. rend.*, 1945, **221**, 558.
2. R. L. Hough, *SAMPE*, 1979, **10**, (1)25.
3. G. R. Martin, US Pub. 2,484,519; 1949.
4. As 3.
5. S. Mierzejewska and T. Niemyski, *J. less-common. Met.*, 1965, **8**, 368.
6. R. G. Bourdean, US Pub. 3,334,967; 1967.
7. 'Vapor Deposition' p. 360.
8. B. B. Owen and R. T. Weber, Am. Inst. Min Met. Tech. Pub. 2306, 1948.
9. S. Csch, Pro. 3rd Int. Symp. Met and Heat Treatment Met., Warsaw, 1967.
10. J. E. Gates *et al.*, US Pat. 3951612; 1980.
11. M. J. Hakim, *CVD* 1975, V, 634.
12. D. Hertz *et al.*, *High Temp.–High Pressure*, 1974, **6**, 423.
13. 'Vapor Deposition', p. 362.
14. R. H. Lewin and C. Hayman, BP 1326769.
15. T. A. Lyndvinskaya *et al.*, 'Refractory Carbides', N.Y. Consultants Bur., 1974.
16. E. Fitzer *et al.*, *CVD*, 1975, V, 523.
17. H. Beutler *et al.*, *CVD*, 1975, V, 749.
18. G. Verspui, *CVD*, 1977, VI, 366.
19. W. J. Heffernan *et al.*, *CVD*, 1973, IV, 498.
20. W. Schintlemeister *et al.*, *CVD*, 1979, V, 523.
21. M. Maillat *et al.*, *Thin Solid Films*, 1979, **64**, 243.
22. Many papers, 1972, *CVD*, III.
23. V. K. Sarin and J. N. Lindstrom, *CVD*, 1977, VI, 389.
24. J. J. Nickl *et al.*, *J. less-common Met.*, 1972, **26**, 335.
25. G. Ebersbach *et al.*, *Die Technik*, 1974, **29**, 273.

26. E. Horvath and A. J. Perry, *Thin Solid Films*, 1980, **65**, 309.
27. N. J. Archer, *CVD*, 1975, V, 556.
28. N. J. Archer and K. K. Yee, *Wear*, 1978, **48**, 237.
29. 'Vapor Deposition' p. 372.
30. D. A. Tarver, US Pat., 3 574 672; 1971.
31. H. Mantle *et al.*, *CVD*, 1975, V, 540.
32. T. C. Wallace, *CVD*, 1973, IV, 91.
33. C. Hollabaugh *et al.*, *CVD*, 1977, VI, 419.
34. A. R. Driesner *et al.*, *CVD*, 1973, IV, 473.

36 Superplasticity

Superplasticity is the ability of a material to sustain extremely large tensile deformations at a temperature around half the melting point expressed in Kelvin. It is only found in metals and alloys, which have, and can maintain during forming, a very fine grain structure. A parameter which indicates the degree of superplasticity is the strain rate sensitivity *m*, given by the high temperature flow equation: $\sigma = K\dot{e}^m$, σ is the stress for plastic flow, $\dot{e}$ the applied strain rate and K is a constant. Superplastic materials have *m* values normally between 0.4 and 0.6, while most other metals and alloys at elevated temperatures have *m* values of 0.2; viscous materials (e.g. glass) behave like a Newtonian fluid and have *m* values of 1.

A full discussion of the mechanism of superplasticity, including methods for determining *m*, can be found in K. A. Padmanabhan and G. J. Davies, 'Superplasticity', Berlin, Springer-Verlag, 1980.

The tables in this chapter give alloy systems with the temperature range over which they show superplasticity, the maximum possible percentage elongation, and the *m* value. Table 36.1 gives mainly non-ferrous systems, Table 36.2 iron and steel systems, and Table 36.3 powdered materials with superplastic properties.

Table 36.1 NON-FERROUS SYSTEMS SHOWING SUPERPLASTICITY

Alloy system	*Temperature range* °C	*Maximum elongation* %	*m*	*References*	*Remarks*
Ag–28.1Cu	675	500	0.53	1	
Al (commercial)	380–580	6 000	0.2	2	
Al–7.6Ca	400–600	850	0.78	3	Euratom alloy
Al–5Ca–5Zn	450	—	—	4	Alcan 08050
Al–17Cu	400–520	600	0.35	5	
Al–33Cu	380–520	1 150	0.9	6	
Al–6Cu–0.5Zr	400–500	2 000	0.5	7	TI Supral
Al–25–33Cu–7–11Mg	420–480	> 600	0.72	8	
Al–Ga–Ti	RT	—	0.3–0.5	9	
Al–4Ge	400–500	230	—	10	
Al–1.56Mg–5.6Zn	530	500	0.7	11	
Al–3Mg–6Zn	340–360	400	0.35	12	TI BA 480
Al–0.93Mg–10.72Zn–0.42Zr	550	1550	0.9	11	
Al–5.8Mg–0.37Zr + others	520	> 800	0.6	13	
Be	600–700	130	0.9	14	
Bi–44.5Pb	20	600	0.42	15	
Bi–31Pb–17Sn	20	600	0.45	15	
Bi–43Sn	20	1 950	—	15	
Cd–17.5Zn	20	400	0.5	16	
Cd–27Zn	20–30	350	0.5	16	
Co–10Al	1 200	450	0.47	17	

Table 36.1 NON-FERROUS SYSTEMS SHOWING SUPERPLASTICITY—*continued.*

Alloy system	*Temperature range* °C	*Maximum elongation* %	*m*	*References*	*Remarks*
Cr–30Co	1 200	160	—	18	
Cu–9.8Al	540–700	700	0.7	19	
Cu–9.5Al–14Fe	800	>800	0.7	19	CDA619
Cu–10Al–3Fe	800	7 200.6	0.6	20	
Cu–10Al–4Fe	750	1 000	0.6	21	
Cu–2.8Al–1.8Si–0.4Co	500–600	320	0.5	22	CDA 638
Cu–10–20Mg	700	250	—	23	
Cu–7P	410–600	>600	0.5	24	Solder, temporary superplasticity
Cu	450	>300	0.45	25	IN836
Cu–9.8Zn	163	570	0.4–0.5	26	
Cu–40Zn	627	>525	0.75	27	
Cu–48Zn	500–800	450	0.9	28	
Cu–38.5Zn–3Fe	500–800	330	0.53	25	
Cu–38Zn–15Ni–0.2Mn	450–565	200	—	29	
In–34Bi	20	450	0.76	30	Eutectic
Mg–33Al	350–400	2 100	0.8	31	Eutectic
Mg–4.3Al–3Zn–0.5Mn				32	Russian MA 15
Mg–30.7Cd	450	250	—	23	
Mg–5.5Zn–0.5Zr	270–310	1 000	0.6	33	ZK 60
Mg–0.5Zr	500	150	0.3	34	
Ni	800	180	0.5	35	
Nichrome*	1 000	190	0.5	35	
Ni–20Cr	800–900	200	—	36	
Ni–15Co–9.5Cr–5.5Al–5Ti–3Mo	810–1 070	1 000	0.5	37	IN 100
Ni–34.9Cr–26.2Fe–0.58Ti	795–855	>1 000	0.5	38	
Ni–38Cr–14Fe–1.75Ti–1Al	810–980	1 000	0.5	38	
Ni–39Cr–10Fe–1.75Ti–1Al	810–980	1 000	0.5	38	
Ni–16Cr–8.3Co–3.4Ti–3.4Al–2.6W–1.78Ta–1.75Mo 0.9Nb–0.1Zr–0.17C–0.01B	800–1 000	500	0.4	39	IN 738
Pb–5Cd	20	—	0.35	40	
Pb–17.4Cd	25–100	>350	0.6	41	Eutectic
Pb–30Cd	25	>900	0.35	41	
Pb–40In	55	215	—	42	
Pb–11Sb	246	1 200	—	43	Eutectic
Pb–19Sn	20–80	—	0.5	44	
Pb–7.9Tl	20	400	0.5	45	
Sn	20	—	0.5	44	
Sn–1Bi	22	500	0.48	46	
Sn–5Bi	20	1 000	0.68	47	
Sn–5Cd	25	350	0.32	48	
Sn–2Pb	20–80	600	0.5	44	
Sn–38Pb	20–170	>4 850	0.7	44	Eutectic
Sn–32Pb–18Cd	25–70	—	0.55	49	
Sn–31.1Pb–3.4Zn	25–70	500	0.55	49	
Sn–2–6Sb	140–210	—	0.4	50	
Sn–9.8Zn	20–180	570	0.5	26	Eutectic
Ti (commercial)	900	—	0.8	37	RC 70
Ti–4Al–0.25$0_2$	950–1 050	—	0.6	37	
Ti–5Al–2.5Sn	900–1 100	450	0.72	37	
Ti–6Al–4V	750–1 000	1 000	0.85	37	Commercial alloy used throughout world
Ti–6Al–5Zr–4Mo–1Cu–0.25Si	800	300	—	23	IMI 700

* Registered Trade Mark.

Table 36.1 NON-FERROUS SYSTEMS SHOWING SUPERPLASTICITY—*continued*

Alloy system	*Temperature range* °C	*Maximum elongation* %	*m*	*References*	*Remarks*
Ti–8Mn	580–900	140	0.95	28	
Ti–15Mo	580–900	450	0.6	28	
Ti–11Sn–5Zr–2.25Al–1Mo–0.25Si	800	500	—	23	IMI 679
Zn (Commercial)	20–70	409	0.2	51	
Zn–0.2Al	23	465	0.8	52	
Zn–0.4Al	20	650	0.5	51	
Zn–4.9Al	200–360	300	0.68	53	Eutectic
Zn–18Al	22–350	—	0.6	54	
Zn–22Al	20–300	2 900	0.7	54	Eutectoid; main commercial alloy, i.e. SPZ
Zn–36Al	22–350	—	0.5	54	
Zn–40Al	250–300	1 300	0.65	55	
Zn–50Al	250–300	1 000	0.3	55	
Zn–22Al–0.1Cu	250	—	0.65	56	
Zn–22Al–4Cu	20–250	1000	0.5	57	
Zn–22Al–0.2Mn	20–250	1000	0.5	57	
Zn–0.1Ni–0.04Mg	100–250	>980	0.51	58	
Zr–2.5Nb	627–827	430	0.6	59	
Zr–1.2–1.7Sn–0.07–0.2Fe–0.05–0.15Cr–0.03–0.08Ni	800–1 050	2 500	0.57	60	Zircaloy 2+0_2
Zr–1.2–1.7Sn–0.12–0.18Fe0.05–0.15Cr–0.007Ni	800–1 050	2 500	0.57	60	Zircaloy 4+0_2
W–15–30Re	2000	260	0.8	17	

REFERENCES TO TABLE 36.1

1. H. E. Cline and D. Lee, *Acta Met.*, 1970, **18**, 315.
2. V. A. Likhachec, M. M. Myshlyaev, S. S. Olevskii and T. N. Chuchman, *Acta Met.*, 1974, **22**, 829.
3. G. Piatti, G. Pellegrini and R. Trippodo, *J. Mat. Sci.*, 1976, **11**, 186.
4. D. M. Moore and L. R. Morris, *Mat. Sci. Eng.*, 1980, **43**, 85.
5. J. R. Cahhoon, *Met. Sci.*, 1975, **9**, 346.
6. G. Rai and N. J. Grant, *Metall. Trans.*, 1975, **6A**, 385.
7. R. Grimes, M. J. Stowell and B. M. Watts, *Met. Technol.*, 1976, **3**, 154.
8. R. Horiuchi, A. B. El-Sebai and M. Otsuka, *Scripta Met.*, 1973, **7**, 1101.
9. S. K. Marya and G. Wyon, *J. Phys. (Paris)*, 1975, **36**, (10 Suppl.), c4.309–c4.313.
10. R. I. Kuznetsova and N. N. Zhukov, *Phys. Met. Metallography*, 1977, **44** (6), 134.
11. K. Matsuki and M. Yamada, *J. Jap. Inst. Met.*, 1973, **37**, 448.
12. R. H. Bricknell and J. W. Edington, *Metall. Trans.*, 1976, **7A**, 153.
13. K. Matsuki, Y. Uetani, M. Yamada and Y. Murakami, *Met. Sci.*, 1976, **10**, 235.
14. C. R. Heiple, *Metall. Trans.*, 1973, **4**, 585.
15. A. M. S. Guthrie, D. E. Newbury and P. M. Hazzledine, *Scripta Met.*, 1972, **6**, 841.
16. C. M. H. Jenkins, *J. Inst. Met.*, 1928, **40**, 41.
17. H. E. Cline, *Trans. Met. Soc. AIME*, 1967, **239**, 1906.
18. J. R. Stephens, and W. D. Klopp, *Trans. Met. Soc. AIME*, 1966, **236**, 1637.
19. D. M. R. Taplin and S. Sagat, *Mat. Sci. Eng.*, 1972, **9**, 53.
20. D. Oelschlaegel and V. Weiss, *Trans. Am. Soc Met.*, 1966, **59**, 143.
21. M. W. A. Bright and D. M. R. Taplin, *Copper Dev. Assn Pubn*, 061/2, 1972.
22. R. G. Fleck and C. J. Beevers and D. M. R. Taplin, *J. Mat. Sci.*, 1974, **9**, 1737.
23. R. Pearce and C. J. Swanson, *Sheet Metal Ind.*, 1970, **47**, 599.
24. G. Herriot, M. Suery and B. Baudelet, *Scripta Met.*, 1972, **6**, 657.
25. J. W. Edington, K. N. Melton and C. P. Cutler, *Prog. Mat Sci.*, 1976, **21**, 61.
26. R. J. Prematta, P. S. Venkatesan and A. Pense, *Met. Trans.*, 1976, **7A**, 1235.
27. M. Suery and B. Baudelet, *J. Mat. Sci.*, 1973, **8**, 363.
28. P. Griffiths and C. Hammond, *Acta Met.*, 1972, **20**, 935.
29. R. D. Schelleng and G. H. Reynolds, *Met. Trans.*, 1973, **4**, 2199.
30. C. Dasarathy, *Z. Metallk.*, 1971, **62**, 612.
31. D. Lee and E. W. Hart, *Met. Trans.*, 1971, **2**, 1245.
32. Yu, V. Gusev, *Tekhnol. Legk. Splavov*, 1978 (1), 9.

33. A. Karim and W. A. Backofen, *Mat. Sci. Eng.*, 1968–69, **3**, 306.
34. D. A. Woodford, *J. Inst. Metals*, 1968, **96**, 371.
35. O. A. Kaibyshev and A. A. Markelov, *Phys. Met. Metallog.*, 1976, **41** (1), 165.
36. I. V. Doronin, *Fiz. Khim. Obrab. Mat.*, 1976 (6), 147.
37. D. Lee and W. A. Backofen, *Trans. Met. Soc. AIME*, 1967, **239**, 1034.
38. H. W. Hayden and J. H. Brophy, *Trans. Am. Soc. Met.*, 1968, **61**, 542.
39. D. A. Woodford, *Met. Trans.*, 1976, **7A**, 1244.
40. T. H. Alden, *Trans. Am. Soc. Met.*, 1968, **61**, 559.
41. S. Srinivasa Rao, O. Sivakesavan, S. H. Ghude and R. V. Tamhankar, *Trans. Ind. Inst. Met.*, 1970, **23** (4), 44.
42. J. C. Wei and W. D. Nix, *Scripta Met.*, 1979, **13**, 1017.
43. J. Gryziecki and J. Jarominek, *Rudy Met. Niezelaz*, 1975, **20** (6), 316.
44. H. E. Cline and T. H. Alden, *Trans. Met. Soc. AIME*, 1967, **239**, 710.
45. R. C. Gifkins, *J. Inst. Met.*, 1967, **95**, 373.
46. M. A. Clark and T. H. Alden, *Acta Met.*, 1973, **21**, 1195.
47. T. H. Alden, *Acta Met.*, 1967, **15**, 469.
48. V. S. Darekar and R. D. Chaudhari, *Trans. Ind. Inst. Met.*, 1970, **23** (3), 56.
49. M. D. C. Moles and G. J. Davies, *Met. Sci.*, 1976, **10**, 314.
50. S. B. Agarwal and M. L. Vaidya, *Scripta Met.*, 1975, **9**, 447.
51. G. R. Edwards, J. C. Shyne and O. D. Sherby, *Met. Trans.*, 1971, **2**, 2955.
52. R. C. Cook and N. R. Risebrough, *Scripta Met.*, 1968, **2**, 487.
53. J. C. Marshall, T. J. Stewart and T. C. Babcock, *Met. Eng. Quart.*, 1973, **13** (4), 12.
54. T. H. Alden and H. W. Schadler, *Trans. Met. Soc. AIME*, 1968, **242**, 825.
55. K. N. Melton and J. W. Edington, *Scripta Met.*, 1975, **9**, 559.
56. H. Naziri and R. Pearce, *Int. J. Mech. Sci.*, 1970, **12**, 513.
57. K. Nuttall, *J. Inst. Met.*, 1973, **101**, 329.
58. J. D. Lee and P. Niessen, *J. Mat. Sci.*, 1974, **9**, 1467.
59. K. Nuttall, *Scripta Met.*, 1976, **10**, 835.
60. A. M. Garde, H.-M. Chung and T. F. Kaussner, *Acta Met.*, 1978, **26**, 153.

Table 36.2 IRON AND STEEL SYSTEMS SHOWING SUPERPLASTICITY

Alloy system	*Temperature range* °C	*Maximum elongation* %	*m*	*References*	*Remarks*
Cast iron					
Fe–2.13C–1.41Mn	650	526	0.5	1	White
Fe–2.36C–1.48Mn	650	291	0.5	1	White
C. steels					
Fe–0.1C	827	—	0.9	2	
Fe–0.4C	827	—	0.6	2	
Fe–0.8C	750–860	100	0.35	3	
Fe–1.3C	540–650	500	0.4	4	
Fe–1.6C	540–650	500	0.45	4	
Fe–1.9C	540–650	500	0.5	4	
Alloy steels					
Fe–0.3C–2Al	1 040	—	0.37	5	
Fe–O.34C–2Al–0.47Mn	900–950	372	0.48	6	
Fe–1.6C–1.5Cr	650	1 220	—	7	AISI 52160
Fe–0.14C–1.93Mn	727–800	—	0.6	2	
Fe–0.44C–2Mn	700–800	460	0.8	2	AISI 340
Fe–2C–30Mn	770–950	250	—	8	
Fe–C–Mn steels + V, Nb, Al and Ti	800–1 000	184	0.7	9	
Fe–0.07C–0.91Mn–0.5P–0.1V	800–950	169	0.31	6	
Fe–0.14C–1.16Mn–0.5P–0.11V	800–950	270	0.57	6	
Fe–0.16C–1.54Mn–1.98P–0.13V	900	376	0.55	6	
Fe–0.18C–1.54Mn–0.9P–0.11V	900	320	0.55	6	
Fe–0.42C–1.87Mn–0.24Si	727	460	0.65	2	AISI 1340
Fe–0.91C–0.45Mn–0.12Si	710–915	142	0.42	10	
Fe–0.13C–1.11Mn–0.11V	700–900	310	0.45	6	
Fe–0.03–0.1C–3.9Ni–3Mo–1.58Ti	850–1 000	820	0.67	11	
Fe–0.01C–6.4Ni–0.35Nb	700–800	—	0.4	12	

Table 36.2 IRON AND STEEL SYSTEMS SHOWING SUPERPLASTICITY—*continued*

Alloy system	*Temperature range* °C	*Maximum elongation* %	*m*	*References*	*Remarks*
Fe–0.2C–3.1Ni–0.29Nb	700–800	—	0.56	12	
Fe–0.12C–1.97Si	800–950	150	0.26	6	
Stainless steels					
Fe–C–12Cr–10Ni+Ti	900–950	—	—	13	Martensitic Russian Kh12Ni 10T
Fe–C–12Cr–10Ni+Ti+Al	900–950	—	—	13	Martensitic Russian Kh12Ni10TYu
Fe–C–13Cr–7Ni	900–950	—	—	13	Martensitic Russian OKh13N7
Fe–0.08C–20Cr–9Ni–7Mn+Ti	990–1 020	195	0.3–0.5	14	Russian O8Kh20N9G7T
Fe–C–26Cr–6.5Ni	700–1 020	>1 000	0.62	15	IN 744

Table 36.3 POWDERED MATERIAL SYSTEMS SHOWING SUPERPLASTICITY

Alloy system	*Temperature range* °C	*Maximum elongation* %	*m*	*References*	*Remarks*
Ni–0.125C–6Al–6Cr–4Mo–2.5Nb–8Ta–4W	980	>600	—	16	NASA–TAZ–8a
Ni–0.15C–18.5Co–15Cr–4.2Al–5.2Mo–3.5Ti	1 020	1 000	0.42	16	U700
Ni–5.4Al–0.015B–0.14C–7.5Co–6Cr–2Mo–0.5Nb–9Ta–1Ti–6W–Hf–Re–Zr	1 090	>300	—	16	NASA-TRW-VIA
Ni–5.5–6.5Al–1.8–2.8Nb–12–14Cr–3.8–5.2Mo up to 2.5Fe–0.5–1Ti	1 090	230	—	16	IN 713

REFERENCES TO TABLES 36.2 AND 36.3

IRON AND STEEL

1. J. Wadsworth, L. E. Eiselstein and O. D. Sherby, *Materials Eng. Appl.*, 1979, **1** (3), 143.
2. H. W. Schadler, *Trans. Met. Soc. AIME*, 1968, **242**, 1281.
3. A. R. Marder, *Trans. Met. Soc. AIME*, 1969, **245**, 1337.
4. O. D. Sherby, B. Walser, C. M. Young and E. M. Cady, *Scripta Met.*, 1975, **9**, 569.
5. E. Snape and N. L. Church, *J. Metals*, 1972, **24** (1), 23.
6. W. B. Morrison, *Trans. Am. Soc. Met.*, 1968, **61**, 423.
7. J. Wadsworth and O. D. Sherby, *J. Mat. Sci.*, 1978, **13**, 2645.
8. S. I. Bulat and A. L. Molochnikova, *Met. Sci. Heat Treat.*, 1979, **21** (6), 420.
9. M. J. Stewart, *Met. Trans.*, 1976, **7A**, 399.
10. G. R. Yoder and V. Weiss, *Met. Trans.*, 1972, **3**, 675.
11. C. W. Humphries and N. Ridley, *J. Mat. Sci.*, 1974, **9**, 1429.
12. T. Hirano, M. Yamaguchi, and T. Yamane, *Met. Trans.*, 1974, **5**, 1245.
13. I. N. Bogachev, L. I. Lepekhina and A. A. Kovaleva, *Phys. Met. Metallog.*, 1977, **44** (6), 138.
14. Ya. M. Okrimenko, *Nauchn. Trud. Moskov. Inst. Stali Splavov*, 1976 (94), 75.
15. H. Hildebrand, G. Michalzik and B. Simmon, *Met. Tech.*, 1977, **4**, 32.

COMPACTED METAL POWDERS

16. J. W. Edington, K. N. Melton and C. P. Cutler, *Prog. Mat. Sci.*, 1976, **21**, 61.

Index